JPC	J. Phys. Chem.
JPR	J. Prakt. Chem.
JPS	J. Pharm. Sci.
JSP	J. Mol. Spectrosc.
JST	J. Mol. Struct.
K	Kristallografiya
KGS	Khim. Geterotsikl. Soedin.
LA	Liebigs Ann. Chem.
M	Monatsh. Chem.
MI	Miscellaneous [book or journal]
MIP	Miscellaneous Pat.
MS	Q. N. Porter and J. Baldas, 'Mass Spectrometry of Heterocyclic Compounds', Wiley, New York, 1971
N	Naturwissenschaften
NEP	Neth. Pat.
NJC	Nouv. J. Chim.
NKK	Nippon Kagaku Kaishi
NMR	T. J. Batterham, 'NMR Spectra of Simple Heterocycles', Wiley, New York, 1973
OMR	Org. Magn. Reson.
OMS	Org. Mass Spectrom.
OPP	Org. Prep. Proced. Int.
OR	Org. React.
OS	Org. Synth.
OSC	Org. Synth., Coll. Vol.
P	Phytochemistry
PAC	Pure Appl. Chem.
PC	Personal Communication
PH	'Photochemistry of Heterocyclic Compounds', ed. O. Buchardt, Wiley, New York, 1976
PIA	Proc. Indian Acad. Sci.
PIA(A)	Proc. Indian Acad. Sci., Sect. A
PMH	Phys. Methods Heterocycl. Chem.
PNA	Proc. Natl. Acad. Sci. USA
PS	Phosphorus Sulfur
QR	Q. Rev., Chem. Soc.
RCR	Russ. Chem. Rev. (Engl. Transl.)
RRC	Rev. Roum. Chim.
RTC	Recl. Trav. Chim. Pays-Bas
S	Synthesis
SA	Spectrochim. Acta
SA(A)	Spectrochim. Acta, Part A
SAP	S. Afr. Pat.
SC	Synth. Commun.
SH	W. L. F. Armarego, 'Stereochemistry of Heterocyclic Compounds', Wiley, New York, 1977, parts 1 and 2
SST	Org. Compd. Sulphur, Selenium, Tellurium [R. Soc. Chem. series]
T	Tetrahedron
TH	Thesis
TL	Tetrahedron Lett.
UKZ	Ukr. Khim. Zh. (Russ. Ed.)
UP	Unpublished Results
USP	U.S. Pat.
YZ	Yakugaku Zasshi
ZC	Z. Chem.
ZN	Z. Naturforsch.
ZN(B)	Z. Naturforsch., Teil B
ZOB	Zh. Obshch. Khim.
ZOR	Zh. Org. Khim.
ZPC	Hoppe-Seyler's Z. Physiol. Chem.

COMPREHENSIVE HETEROCYCLIC CHEMISTRY

IN 8 VOLUMES

COMPREHENSIVE HETEROCYCLIC CHEMISTRY

The Structure, Reactions, Synthesis and Uses of Heterocyclic Compounds

Volume 7

Chairman of the Editorial Board
ALAN R. KATRITZKY, FRS
University of Florida

Co-Chairman of the Editorial Board
CHARLES W. REES, FRS
*Imperial College of Science and Technology
University of London*

Part 5

Small and Large Rings

EDITOR
WALTER LWOWSKI
New Mexico State University

PERGAMON PRESS
OXFORD · NEW YORK · TORONTO · SYDNEY · PARIS · FRANKFURT

U.K.	Pergamon Press Ltd., Headington Hill Hall, Oxford OX3 0BW, England
U.S.A.	Pergamon Press Inc., Maxwell House, Fairview Park, Elmsford, New York 10523, U.S.A.
CANADA	Pergamon Press Canada Ltd., Suite 104, 150 Consumers Road, Willowdale, Ontario M2J 1P9, Canada
AUSTRALIA	Pergamon Press (Aust.) Pty. Ltd., P.O. Box 544, Potts Point, N.S.W. 2011, Australia
FRANCE	Pergamon Press SARL, 24 rue des Ecoles, 75240 Paris, Cedex 05, France
FEDERAL REPUBLIC OF GERMANY	Pergamon Press GmbH, Hammerweg 6, D-6242 Kronberg-Taunus, Federal Republic of Germany

---

First edition 1984

Library of Congress Cataloging in Publication Data

Main entry under title:

Comprehensive heterocyclic chemistry.

Includes indexes.
Contents: v. 1. Introduction, nomenclature, literature, biological aspects, industrial uses, less-common heteroatoms –
v. 2. Six-membered rings with one nitrogen atom – [etc.] –
v. 8. Indexes.
1. Heterocyclic compounds. I. Katritzky, Alan R. (Alan Roy)
II. Rees, Charles W. (Charles Wayne)
QD400.C65 1984 547′.59 83-4264

British Library Cataloguing in Publication Data

Comprehensive heterocyclic chemistry
1. Heterocyclic compounds.
I. Katritzky, Alan R. II. Rees, Charles W.
547′.59 QD400

ISBN 0-08-030707-8 (vol. 7)
ISBN 0-08-026200-7 (set)

Typeset by J. W. Arrowsmith Ltd., Winterstoke Road, Bristol
Printed in Great Britain by A. Wheaton & Co. Ltd., Exeter

Contents

Foreword

Scope

Heterocyclic compounds are those which have a cyclic structure with two, or more, different kinds of atom in the ring. This work is devoted to organic heterocyclic compounds in which at least one of the ring atoms is carbon, the others being considered the heteroatoms; carbon is still by far the most common ring atom in heterocyclic compounds. As the number and variety of heteroatoms in the ring increase there is a steady transition to the expanding domain of inorganic heterocyclic systems. Since the ring can be of any size, from three-membered upwards, and since the heteroatoms can be drawn in almost any combination from a large number of the elements (though nitrogen, oxygen and sulfur are the most common), the number of possible heterocyclic systems is almost limitless. An enormous number of heterocyclic compounds is known and this number is increasing very rapidly. The literature of the subject is correspondingly vast and of the three major divisions of organic chemistry, aliphatic, carbocyclic and heterocyclic, the last is much the biggest. Over six million compounds are recorded in *Chemical Abstracts* and approximately half of these are heterocyclic.

Significance

Heterocyclic compounds are very widely distributed in Nature and are essential to life; they play a vital role in the metabolism of all living cells. Thus, for example, the following are heterocyclic compounds: the pyrimidine and purine bases of the genetic material DNA; the essential amino acids proline, histidine and tryptophan; the vitamins and coenzyme precursors thiamine, riboflavine, pyridoxine, folic acid and biotin; the B_{12} and E families of vitamin; the photosynthesizing pigment chlorophyll; the oxygen transporting pigment hemoglobin, and its breakdown products the bile pigments; the hormones kinetin, heteroauxin, serotonin and histamine; together with most of the sugars. There are a vast number of pharmacologically active heterocyclic compounds, many of which are in regular clinical use. Some of these are natural products, for example antibiotics such as penicillin and cephalosporin, alkaloids such as vinblastine, ellipticine, morphine and reserpine, and cardiac glycosides such as those of digitalis. However, the large majority are synthetic heterocyclics which have found widespread use, for example as anticancer agents, analeptics, analgesics, hypnotics and vasopressor modifiers, and as pesticides, insecticides, weedkillers and rodenticides.

There is also a large number of synthetic heterocyclic compounds with other important practical applications, as dyestuffs, copolymers, solvents, photographic sensitizers and developers, as antioxidants and vulcanization accelerators in the rubber industry, and many are valuable intermediates in synthesis.

The successful application of heterocyclic compounds in these and many other ways, and their appeal as materials in applied chemistry and in more fundamental and theoretical studies, stems from their very complexity; this ensures a virtually limitless series of structurally novel compounds with a wide range of physical, chemical and biological properties, spanning a broad spectrum of reactivity and stability. Another consequence of their varied chemical reactivity, including the possible destruction of the heterocyclic ring, is their increasing use in the synthesis of specifically functionalized non-heterocyclic structures.

Aims of the Present Work

All of the above aspects of heterocyclic chemistry are mirrored in the contents of the present work. The scale, scope and complexity of the subject, already referred to, with its correspondingly complex system of nomenclature, can make it somewhat daunting initially. One of the main aims of the present work is to minimize this problem by presenting a comprehensive account of fundamental heterocyclic chemistry, with the emphasis on basic principles and, as far as possible, on unifying correlations in the properties, chemistry and synthesis of different heterocyclic systems and the analogous carbocyclic structures. The motivation for this effort was the outstanding biological, practical and theoretical importance of heterocyclic chemistry, and the absence of an appropriate major modern treatise.

At the introductory level there are several good textbooks on heterocyclic chemistry, though the subject is scantily treated in most general textbooks of organic chemistry. At the specialist, research level there are two established ongoing series, 'Advances in Heterocyclic Chemistry' edited by Katritzky and 'The Chemistry of Heterocyclic Compounds' edited by Weissberger and Taylor, devoted to a very detailed consideration of all aspects of heterocyclic compounds, which together comprise some 100 volumes. The present work is designed to fill the gap between these two levels, *i.e.* to give an up-to-date overview of the subject as a whole (particularly in the General Chapters) appropriate to the needs of teachers and students and others with a general interest in the subject and its applications, and to provide enough detailed information (particularly in the Monograph Chapters) to answer specific questions, to demonstrate exactly what is known or not known on a given topic, and to direct attention to more detailed reviews and to the original literature. Mainly because of the extensive practical uses of heterocyclic compounds, a large and valuable review literature on all aspects of the subject has grown up over the last few decades. References to all of these reviews are now immediately available: reviews dealing with a specific ring system are reported in the appropriate monograph chapters; reviews dealing with any aspect of heterocyclic chemistry which spans more than one ring system are collected together in a logical, readily accessible manner in Chapter 1.03.

The approach and treatment throughout this work is as ordered and uniform as possible, based on a carefully prearranged plan. This plan, which contains several novel features, is described in detail in the Introduction (Chapter 1.01).

ALAN R. KATRITZKY
Florida

CHARLES W. REES
London

Contributors to Volume 7

Professor A. G. Anastassiou
Department of Chemistry, University of Petroleum & Minerals, Dhahran, Saudi Arabia

Professor F. A. L. Anet
Department of Chemistry, University of California, Los Angeles, CA 90024, USA

Dr G. R. Baker
Department of Chemistry, Louisiana State University, Baton Rouge, LA 70803, USA

Professor E. Block
Department of Chemistry, State University of New York at Albany, 1400 Washington Avenue, Albany, NY 12222, USA

Dr D. R. Boyd
Department of Chemistry, Queen's University of Belfast, Belfast BT9 5AG, UK

Dr D. E. Davies
Department of Organic Chemistry, University of Liverpool, PO Box 147, Liverpool L69 3BX, UK

Professor D. C. Dittmer
Department of Chemistry, Syracuse University, 108 Bowne Hall, Syracuse, NY 13210, USA

Dr G. L. Dunn
Assistant Director, Research Chemistry, Smith Kline & French Laboratories, 1500 Spring Garden Street, PO Box 7929, Philadelphia, PA 19101, USA

Professor A. D. Hamilton
Department of Chemistry, Princeton University, Princeton, NJ 08544, USA

Professor H. Hart
Department of Chemistry, Michigan State University, East Lansing, MI 48824, USA

Dr K. G. Holden
Deputy Director, Research Chemistry, Smith Kline & French Laboratories, PO Box 7929, Philadelphia, PA 19101, USA

Professor E. G. Lewars
Department of Chemistry, Trent University, Peterborough, Ontario K9J 7B8, Canada

Professor W. Lwowski
Department of Chemistry, New Mexico State University, PO Box 3C, Las Cruces, NM 88003, USA

Dr D. N. McGregor
Bristol Laboratories, PO Box 657, Syracuse, NY 13201, USA

CHC 7-A*

Professor J. A. Moore
Department of Chemistry, University of Delaware, Newark, DE 19711, USA

Professor G. R. Newkome
Department of Chemistry, Louisiana State University, Baton Rouge, LA 70803, USA

Professor A. Padwa
Department of Chemistry, Emory University, Atlanta, GA 30322, USA

Professor Dr E. Schmitz
Zentralinstitut fur Organische Chemie, Akademie der Wissenschaften der DDR, Rudowere Chaussee 5, 1199 Berlin, German Democratic Republic

Professor S. Searles
Department of Chemistry, University of Missouri-Columbia, Columbia, MO 65211, USA

Dr J. T. Sharp
Department of Chemistry, University of Edinburgh, West Mains Road, Edinburgh EH9 3JJ, UK

Dr R. K. Smalley
Department of Chemistry and Applied Chemistry, University of Salford, Salford M5 4WT, UK

Dr R. C. Storr
Department of Organic Chemistry, University of Liverpool, PO Box 147, Liverpool L69 3BX, UK

Professor J. W. Timberlake
Department of Chemistry, University of New Orleans, Lake Front, New Orleans, LA 70122, USA

Dr J. G. Traynham
Department of Chemistry, Louisiana State University, Baton Rouge, LA 70803, USA

Dr A. D. Woolhouse
Chemistry Division, Department of Scientific & Industrial Research, Private Bag, Petone, New Zealand

Contents of All Volumes

Volume 8 (Part 6: Indexes)

5.01

Structure of Small and Large Rings

W. LWOWSKI
New Mexico State University

5.01.1 INTRODUCTION

Part 5 of Comprehensive Heterocyclic Chemistry contains two groups of systems, with little of their structural properties in common. This chapter is, therefore, divided into two parts, dealing separately with small and with large heterocyclic rings.

5.01.2 STRUCTURE OF SMALL HETEROCYCLIC RINGS

5.01.2.1 Known Small Heterocycles

A very large number of different arrangements of the same heteroatoms are possible in large rings. For three- and four-membered ones, however, it is still possible to write down all combinations of C, N, O and S, from plausible to outrageous ones. With the help of new techniques, the border between the plausible and the outrageous ones is shifting towards more and more strained and unstable systems. It seems that chemists have often underestimated the activation barriers separating thermodynamically unfavorable species from their more stable isomers or fragmentation products. Tables 1 and 2 list the three- and four-membered rings, respectively, which have been made or for which there is strong evidence (such as from isotope or stereochemical studies). A spectacular addition to the list is a triaziridine, a long sought for molecule, which turns out to be relatively stable after all.

5.01.2.2 Bond Lengths and Bond Angles

The geometry of three- and four-membered rings has been determined mainly by X-ray diffraction on crystalline materials ⟨72PMH(5)1, p. 12⟩, and also by electron diffraction,

Table 1 Structures of Known Three-membered Heterocyclic Compounds

Skeleton	*Name*	*Evidence*[a]	*Ref.*	*Chapter or section number*
	Aziridine	IP, XR, MW		5.04
	1-Azirine	IP, XR		5.04
	Aziridinone	IP, XR	80AG(E)276	5.04
	Alkylideneaziridine	IP	80AG(E)276	5.04
	Aziridineimine		80AG(E)276	5.04
	Aziridinedione	MI	80JA6902	5.04
	Azirinimine	TH	73CC435	5.04
	2-Azirine	CE, TH	80PAC1623	5.04.2.1
	Oxirane	IP, XR, MW		5.05
	Oxiranone	MI, GP, IP	82CC362	5.05
	Alkylideneoxirane	IP	80AG(E)276	5.05
	Oxiranimine	CE	80AG(E)276	5.05
	Oxirene	CE		5.05.6
	Thiirane	IP, XR		5.06
	Thiiranone	IP, XR	80AG(E)276	5.06
	Thiiranimine	IP, XR	80AG(E)276	5.06
	Alkylidenethiirane	IP, XR	80AG(E)276	5.06
	Thiiranediimine	CE	79CC160	5.06
	Thiirene	IP, MI, CE	80PAC1623	5.06
	Diaziridine	IP		5.08
	1-Diazirine	IP, MW, TH		5.08
	Diaziridinone	IP, XR	80AG(E)276	5.08
	Diaziridinimine	IP, XR	80AG(E)276	5.08
	Oxaziridine	IP, XR		5.08

Table 1 *(continued)*

Skeleton	Name	Evidence[a]	Ref.	Chapter or section number
	Dioxirane	MI, MW CE	 80JOC4758, 80MI50100	5.08
	Dioxiranone	MI	80AG(E)276	—
	Thiaziridine 1,1-dioxide	IP		5.08.1
	Oxathiirane	UV		5.08.1
	Thiaziridineimine	MI		5.08.1
	Triaziridine	IP	80CC1197	5.08
	Triazirine	CE	81CC299	—
	Oxadiaziridine	IP	70JOC2482	—
	Thiadiaziridine 1,1-dioxide	IP	77CB1780, 80H(14)1677	—

[a] IP, isolated pure; MI, matrix isolated; GP, data from pure gas phase material; CE, chemical evidence for existence; TH, theoretical calculation; XR, X-ray structure; MW, microwave structure; UV ultraviolet spectrum.

microwave spectroscopy and PE spectroscopy ⟨74PMH(6)53⟩. Three-membered heterocycles generally have shorter C—C bonds than does cyclopropane, thiirane dioxides being an exception. The C—X bonds are longer than in CH_3—X—CH_3. The CXC angles in oxirane and aziridine are close to 60°, and the peripheral HCH bond angles are near 118°. Table 3 gives representative data.

The geometry of four-membered rings is far more complex; the rings are usually not planar, and substituted rings have ring inversion conformers of different energies. Moreover, inversion of substituents on heteroatoms may multiply the number of conformers of different energies. The ring inversion barriers of saturated four-membered systems are often very low. From IR and microwave data the barriers are 5.27 kJ mol^{-1} for azetidine, nearly zero for oxetane, and 3.14 kJ mol^{-1} for thietane ⟨74MI50100, p. 273⟩. Table 4 gives bond lengths and angles for some four-membered heterocycles.

The strain in three- and four-membered rings is mostly due to bond angle deformation. Some conventional ring strain energies (CRSEs) ⟨74PMH(6)199, p. 228⟩ are given in Tables 3 and 4. The ring strain in three- and four-membered rings is of the same magnitude, depending more on the nature of the heteroatom(s) than on the ring size. For comparison, the CRSE is 115 kJ mol^{-1} for cyclopropane and 111 kJ mol^{-1} for cyclobutane. As long as nonbonding interactions are avoided, alkyl substituents stabilize small rings by a few kJ mol^{-1}. For example, 2-methyloxirane is more stable than oxirane by 4 kJ mol^{-1} ⟨74PMH(6)199, p. 229⟩.

Exocyclic unsaturation can stabilize small ring heterocycles. In three-membered rings it is difficult to separate the contributions from increased angle strain and from electronic interactions between the unsaturation and the heteroatom. In four-membered rings such separation has been done ⟨74PMH(6)199, p. 235⟩. The CRSEs change from oxetane

Table 2 Structures of Known Four-membered Heterocyclic Compounds

Skeleton	Name	Evidence[a]	Ref.	Chapter or section number
 N	Azetidine	IP, XR, MW	79CRV331	5.09
 N	1-Azetine	IP, XR		5.09.4.1
 N	2-Azetine	IP		5.09.4.3.3
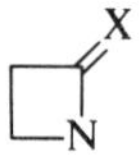 X, N	Azetidin-2-one	IP, XR		5.09
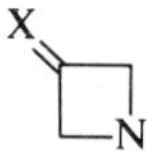 X, N	Azetidin-3-one	IP		5.09
 N	Azete (benz[*b*]-)	IP	78CC9, B-79MI50100	5.09.5
	Azete [tris(dimethylamino)-]	IP, TH	73AG(E)847	
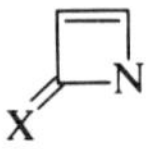 N, X	2-Azetin-4-one	IP		5.09.4.3.5
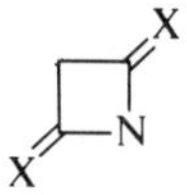 X, N, X	Azetidine-2,4-dione	IP		5.09
	4-Thioxoazetidin-2-one	IP		
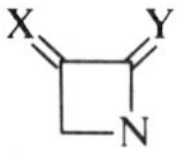 X, Y, N	X = N, Y = O	IP		5.09
	X = CR_2, Y = O	IP		5.09
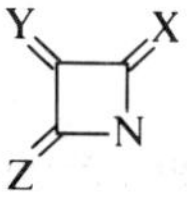 Y, X, N, Z	X = CR_2, Y = Z = O	IP		5.09
	X = CR_2, Y = O, Z = N	IP		5.09
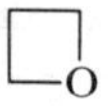 O	Oxetane	IP, XR		5.13
 O	Oxete	IP, MI		5.13
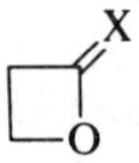 X, O	Oxetan-2-one	IP, XR		5.13
	Oxetan-2-imine	IP		5.13
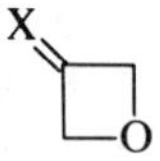 X, O	Oxetan-3-one	IP		5.13
	Oxetan-3-imine	IP		5.13
	3-Alkylidenoxetane	IP		5.13
 S	Thietane	IP, XR		5.14
 S	Thiete	IP		5.14
 S	Thiacyclobutadiene	TH, CE		5.14.2.6

Table 2 (*continued*)

Skeleton	*Name*	*Evidence*[a]	*Ref.*	*Chapter or section number*
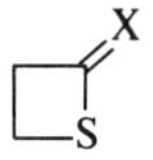	Thietan-2-one (benzo[*b*]-)	MI		5.14.2.5
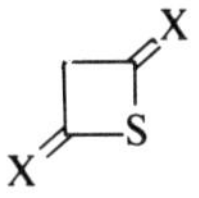	Thietane-2,4-diimine	IP		5.14
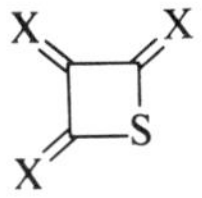	Thietane-2,3,4-triimine	IP, XR		5.14
	1,2-Diazetidine	IP		5.15.1.2.1
	3*H*,4*H*-Diazetine	IP		5.15
	1*H*,2*H*-Diazetine	IP		5.15
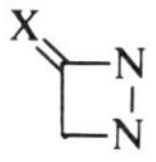	1,2-Diazetidin-3-one	IP		5.15
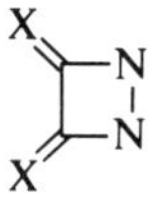	1,2-Diazetidine-3,4-dione	IP		5.15
	1,2-Oxazetidine	IP		5.15.1.2.2
	4*H*-1,2-Oxazetine	IP		5.15.1.2.2
	1,2-Thiazetine (1,1-dioxide, benzo[*b*]-)	IP		5.15.1.2.3
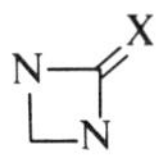	1,3-Diazetidin-2-one 1,3-Diazetidin-2-imine	IP IP		5.15.1.2.4 5.15.1.2.4
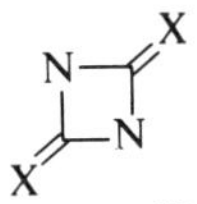	1,3-Diazetidine-2,4-dione	IP, XR		5.15.1.2.4
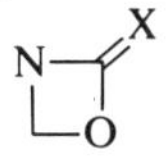	1,3-Oxazetidin-2-one	IP		5.15.1.2.5
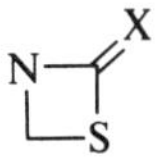	1,3-Thiazetidin-2-imine	IP, XR		5.15.1.2.6
	1,2-Dioxetane	IP, XR		5.15.1.2.2
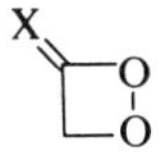	1,2-Dioxetanone	IP		5.15.1.2.2

Table 2 *(continued)*

Skeleton	Name	Evidence[a]	Ref.	Chapter or section number
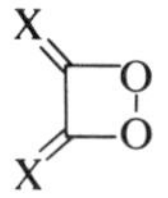	1,2-Dioxetanedione	CE		5.15.1.2.2
	1,2-Oxathietane	IP		5.15.1.2.8
	1,2-Oxathiete (benz[*b*]-)	CE, MI	79JA4755	5.15.1.2.8
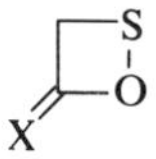	4-Alkylidene-1,2-thietane 1,1-dioxide	IP	68TL6293	5.15.1.2.8
	1,2-Dithietane	IP, TH		5.15.1.2.9
	1,3-Dithietane	IP, XR		5.15.1.2.10

For four-membered heterocycles with more than two heteroatoms see Section 5.15.5

[a] For an explanation of these abbreviations see the footnote to Table 1.

(106 kJ mol^{-1}) by -11 kJ mol^{-1} to oxetan-2-one (95 kJ mol^{-1}) (corrected for electronic effects) and 4-methyleneoxetan-2-one (95 kJ mol^{-1}). In contrast, an increase of 10 kJ mol^{-1} over the value for cyclobutane (111 kJ mol^{-1}) is observed on going to both methylenecyclobutane and 1,3-bismethylenecyclobutane.

For the antiaromatic three-membered heterocycles, experimental data are available only for thiirenes (and there is some doubt about the true antiaromaticity of thiirenes). Bond lengths have been calculated, however, for these antiaromatic 4π-systems ⟨80PAC1623⟩. In comparison with the corresponding saturated heterocycles, the C—X bond lengths are increased by 0.05 to 0.17 Å and the C—C bond length is decreased by 0.2 Å.

Four-membered heterocycles are often puckered rather than planar (Table 4). As expected, *exo*- and *endo*-unsaturation tend to make these systems planar.

5.01.2.3 Stereochemistry and NMR Spectra

5.01.2.3.1 Inversions

Low energy barriers for the inversion of small rings, and for inversion on nitrogen, have made NMR spectroscopy a prime tool for the investigation of the dynamic stereochemistry. Variable temperature studies on aziridines and diaziridines show a remarkable range of nitrogen inversion rates. Electron delocalizing substituents on small-ring nitrogen lower the inversion barrier by lowering the energy of the transitional, 'flat' geometry in which three substituents of the nitrogen are all in the same plane ⟨67JA352⟩. Substituents bearing unshared electron pairs raise the inversion barrier to levels at which enantiomers can be isolated ⟨68JA508⟩. Pure invertomers have been obtained of *N*-haloaziridines ⟨B-73NMR137⟩ and *N*-alkoxyaziridines ⟨70JA1079, 73TL619⟩. Observed coalescence temperatures of aziridines range from below -160 °C for *N*-acetylaziridine in vinyl chloride solution to above 145 °C for *N*-ethylaziridine in D_2O. Substituents on aziridine carbons can decrease the barriers by a few kJ mol^{-1}, but the literature data are difficult to compare, due in part to experimental difficulties, in part to the large solvent effects operating. For example, *N*-ethylaziridine shows coalescence temperatures of 108 °C (neat), 115 °C (in tetrachloroethylene) and

Table 3 Bond Lengths and Bond Angles of Three-membered Heterocyclic Compounds

Structure	Bond lengths (Å) a	b	c	Bond angles (°) ca	ab	bc	CRSE (kJ mol^{-1})	Section number/ref.
 aziridine (N–H)	1.482	1.491	1.482	—	—	—	113	5.04.2.2
 N–CH(OH)CCl_3	1.49	1.49	1.50	59.9	60.4	59.8	—	74PMH(6)1, p. 8
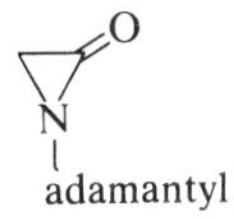 =O, N–adamantyl	1.33	1.45	1.51	60.9	—	53.4	—	5.04.2.2
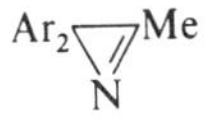 Ar_2, Me, N	1.256	1.463	1.598	60.3	—	48.2	—	5.04.2.2
 O	1.44	1.47	1.44	61.24	59.18	59.18	114	5.05.2.2
 S	1.815	1.484	1.815	48.3	—	—	—	5.06.2.1
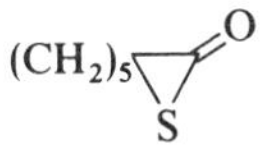 $(CH_2)_5$, =O, S	1.726	1.460	1.916	49.6	73.4	59.7	—	5.06.2.1
 S (unsaturated)	1.79–1.98	1.25–1.29	—	—	—	—	—	5.06.2.1
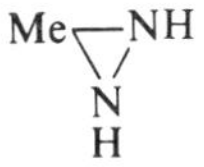 Me, NH, N–H	1.468	1.479	1.479	59.5	59.5	61	—	5.08.2.2
 N, N	1.228	1.428	1.428	64.5	64.5	50.9	—	5.08.2.2
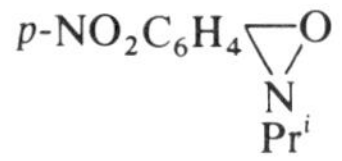 p-$NO_2C_6H_4$, O, N–Pr^i	1.50	1.405	1.434	57.2	59.0	63.8	—	5.08.2.2

>145 °C in D_2O. Extensive tables are found in Lehn's review ⟨70MI50100⟩. Calculations of inversion barriers have met with mixed success. The MNDO SCF method gives results which compare well with experimental values, including the high barriers of *N*-halo- and *N*-amino-aziridines, and the low ones for *N*-trimethysilyl- and *N*-phosphino-aziridines ⟨80JCS(P2)1512⟩.

Some absolute configurations on nitrogen of aziridines are known, such as that of diethyl 1-methoxyaziridine-2,2-dicarboxylate ⟨79DOK(246)1150⟩.

For oxaziridines the N-inversion barrier is considerably higher than that for similar aziridines. *N*-Alkyl-3,3-dialkyloxaziridines are resolvable and absolute configurations have been determined (Section 5.08.2.3.1).

Diaziridines also show slow nitrogen inversion, and carbon-substituted compounds can be resolved into enantiomers, which typically racemize slowly at room temperature (when *N*-substituted with alkyl and/or hydrogen). For example, 1-methyl-3-benzyl-3-methyl-diaziridine in tetrachloroethylene showed a half-life at 70 °C of 431 min ⟨69AG(E)212⟩. Preparative resolution has been done both by classical methods, using chiral partners in salts ⟨77DOK(232)1081⟩, and by chromatography on triacetyl cellulose (Section 5.08.2.3.1).

N-Inversion in azetidine and azetidin-2-one is rapid, even at −77 and −40 °C, respectively ⟨B-73NMR144⟩. Again, halo substituents on nitrogen drastically slow the inversion rate, so that *N*-chloro-2-methylazetidine can be separated into two diastereomers ⟨B-77SH(1)54⟩. Substituent effects on N-inversion are much the same as in the aziridines: *N*-aryl and *N*-acyl

Table 4 Bond Lengths and Bond Angles of Four-membered Heterocyclic Compounds

Structure	Bond lengths (Å) a	b	c	d	Bond angles (°) da	ab	bc	cd	Pucker angle (°)	Ring inversion barrier (kJ mol^{-1})	Section number/ref.
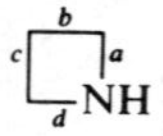 a, b, c, d, NH	1.477	1.560	1.560	1.477	88	—	—	—	33	5.3	5.09.2.1, 80CRV231
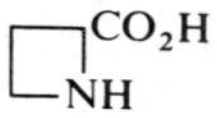 CO_2H, NH	1.51	1.54	1.53	1.52	88	90	88	90	—	—	5.09.2.1
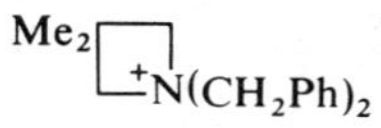 Me_2, $\overset{+}{N}(CH_2Ph)_2$	1.51	1.53	1.51	1.51	84	96	84	97	—	—	B-79MI50100
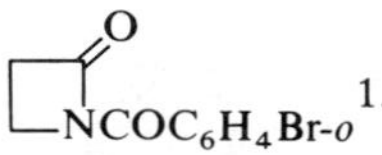 O, $NCOC_6H_4Br$-*o*	1.38	1.51	1.57	1.50	94.0	93.4	86.3	86.3	—	—	5.09.3.1
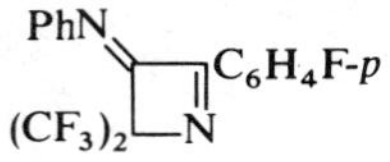 PhN, C_6H_4F-*p*, $(CF_3)_2$, N	1.307	1.467	1.575	1.504	92.4	97.9	83.6	86.1	Flat	—	5.09.4.1
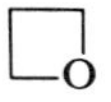 O	1.449	1.091	1.091	1.449	91.98	91.73	88.55	91.98	Flat	0.18	80CRV231, 74MI50100
					CRSE = 106 kJ mol^{-1}						
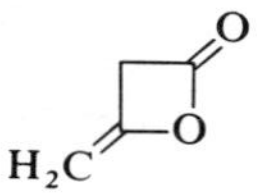 O, O, H_2C	1.39	1.51	1.54	1.47	95.8	83	91.3	90	Flat	—	80CRV231, 72PMH(5)1, p. 12

Compound											
S	1.847	1.549	1.549	1.847	76.8	90.6	95.6	90.6	26	3.3	5.14.2.1, 80CRV231, 74MI50100
O, S	1.826	1.528	1.528	1.826	—	—	100.5	—	Flat	—	5.14.2.1
Me_2, Ph_2, NAr, S	1.777	1.524	1.591	1.868	77	95.8	—	90.1	20	—	5.14.2.1
SO_2	1.79	1.43	1.39	1.77	80.5	—	—	104.5	Flat	—	5.14.2.1
NH, NH	1.427	1.481	1.537	1.471	—	—	—	—	24.3	—	5.15.1.2.1
ade, O, O, ade ade = adamantylidene	1.48	1.549	1.475	1.549	—	—	—	—	21.3	—	5.15.1.2.2
S, S	2.146	1.835	1.564	1.835	99.1	—	80.9	—	—	—	5.15.1.2.3
S, Cl_2, Cl_2, S	1.801	—	1.77	—	83.9	—	96.1	—	Flat	—	5.15.1.2.6

compounds undergo N-inversion faster, whereas *N*-halo, *N*-amino and *N*-nitroso compounds are slower ⟨B-77SH(1)56⟩. By and large, the N-inversion barrier of azetidines is 38 kJ mol^{-1} lower than that of similarly substituted aziridines ⟨B-77SH(1)55⟩. In 1,2-diazetidines one finds the N-inversion rate lowered, and coalescence temperatures and free energies of activation have been reported for a number of 1,2-diaryldiazetidin-3-ones ⟨B-77SH(1)61⟩.

A recent review relates ^{13}C NMR spectra and stereochemistry in heterocyclics ⟨B-79MI50100⟩.

5.01.2.3.2 Chemical shifts and coupling constants

The NMR spectra of three- and four-membered heterocyclics display regularities of great value to structure determination. For protons on adjacent carbons the coupling constants J_{cis} seem to be always greater than J_{trans}. In three-membered rings J_{gem} is almost always smaller than J_{cis} and J_{trans}. Extensive tables are found in ⟨B-73NMR138⟩. The average values for 64 aziridines are $J_{gem} = 1.4$, $J_{trans} = 3.3$ and $J_{cis} = 6.4$ Hz ⟨71PMH(4)121, p. 126⟩. The size of J_{gem} and of the vicinal C—H coupling constants seems to depend more on the number of nonbonding electron pairs at the heteroatom than on its electronegativity. Each electron pair contributes +5.5 Hz to J_{gem}, −2.5 Hz to J_{cis}, and −2.7 Hz to J_{trans} ⟨80OMR(13)45⟩. Table 5 gives some examples, the data being taken from the monograph chapters of the present work and from ⟨B-73NMR, 71PMH(4)121, 80OMR(13)45⟩. Data on some three-membered heterocycles with an *exo* methylene group are found in Table 6 ⟨78RTC214⟩.

Table 5 Ranges of NMR Data of Three-membered Heterocyclic Systems

Skeleton	δ(C—H)	J_{gem}	J_{cis}	J_{trans}	$J(^{13}C—H)$	$\delta(^{13}C)$
Cyclopropane	0.22	−3 to −1	6–12	−4 to 8	164	−2.2
Aziridine	1.48	0.9–4	5–9	2–7	168	18–22
Oxirane	2.54	5–7	2–5	1–3	176	39.7
Thiirane	2.27	−14 to 1	6–7	5–6	170	18
1-Azirine C-2	10	—	—	—	—	160–170
C-3	0.2–2.5					19–45
Diaziridine	1.2	—	—	—	—	56
1-Diazirine	0.4	—	—	—	—	—
Oxaziridine	4.5–5	—	—	—	—	56

Table 6 Ranges of NMR Data for Three-membered Heterocyclic Systems with Exocyclic Unsaturation

Skeleton	$\delta(C_2)$	$\delta(C_3)$	$\delta(C_{ex})$	$J(C_3\text{–}H)$	$J(C_{ex}\text{–}H)$
Cyclopropane	−2.6	−2.6	—	160.5	—
Methylenecyclopropane	131.0	3.0	103.5	161.5	160.8
N-*t*-Butyl-2-methyleneaziridine	134.0	23.8	80.6	170.7	165
2-Methylene-3-*t*-butyloxirane	144.3	68	70.5	—	—
2-Methylenethiirane	130.1	18.5	99.5	174	166

In azetidine derivatives the proton–proton coupling constants J_{gem} on the carbons adjacent to N are 5–7.5 Hz, and J_{cis} is larger than J_{trans}. Long range coupling between ring protons is common. Scheme 1 gives some examples ⟨B-73NMR142, 71PMH(4)121, p. 144⟩.

H_2C—NH (3.7, 3.95); H_2C—CH_2 (3.4, 3.7)

H_2C—NC_6H_{11} (4.19, 3.06); ArCOHC—CHPh (4.39, 4.64)

$J_{2,3} = 9.0$
$J_{3,4}(cis) = 7.4$
$J_{3,4}(trans) = 3.0$
$J_4(gem) = 6.8$

Ar = *p*-PhC_6H_4- (for Ph and ArCO *cis*)

O=C—NH 6.83; Ph—C—C—H 3.23; H 4.39, H 3.66

$J_{4,4} = -5.6$
$J_{3,4}(cis) = 5.9$
$J_{3,4}(trans) = 2.6$
$J_{1,3} = 1.1$

Scheme 1 ^{1}H NMR shifts and coupling constants of azetidine derivatives

5.01.2.3.3 Heteronuclear NMR spectra

Aside from ^{13}C–H coupling constants (often obtained from ^{13}C satellites in ^{1}H NMR spectra) not very much information is available on the ^{13}C NMR spectra of small heterocycles. A recent review ⟨B-79MI50101⟩ devotes only seven pages to the ^{13}C NMR spectra of all three- and four-membered heterocyclics.

Nitrogen-15 NMR spectra of aziridines and azetidines have been measured by Lichter *et al.* ⟨80JOC1277⟩. Relative to anhydrous ammonia the aziridine nitrogen absorbs at −8.5 p.p.m., and *N*-alkylation moves this shift downfield. For N—Me the signal is at 0.7 p.p.m., for N—$CHMe_2$ at 30.2 p.p.m., and for N—Me_3 at 33.5 p.p.m. Substitution on the *β*-carbon shifts the ^{15}N resonance downfield relative to unsubstituted aziridine. This effect decreased with increasing bulk of the substituent, so that 2-methylaziridine has the nitrogen signal at 10.5 p.p.m., and 2-*t*-butylaziridine has it at 3.4 p.p.m. Further substitution on one or both *β*-carbons causes more downfield shift, the effect being only poorly reproduced by assuming group contributions to be additive.

Aziridine ^{15}N shifts parallel the ^{13}C shifts — in a plot of ^{13}C *vs.* ^{15}N shifts of 13 aziridines, the correlation coefficient was 0.953 and the slope 2.1 p.p.m. N/p.p.m. C ⟨80JOC1277⟩.

Azetidine ^{15}N shifts are similar to those of the aziridines. Unsubstituted azetidine has its ^{15}N resonance (relative to anhydrous ammonia) at 25.3 p.p.m., and *N*-*t*-butylazetidine shows the signal at 52 p.p.m. ⟨80JOC1277⟩.

5.01.2.4 Mass Spectra

The mass spectral fragmentation patterns of three- and four-membered heterocycles consist of cleavages typical for substituents, of those due to the formation of particularly stable and accessible fragments (such as N_2), and of more characteristic patterns attributable to fragmentations promoted by ring strain and by stereochemical factors. Thus, small rings usually open after ionization. In aziridines this can be accomplished by loss of the substituent on the nitrogen, *i.e.* of H•, R•, *etc.*, to give ions of the type $R_2C{=}\overset{+}{N}{=}CR_2$ ⟨B-71MS296⟩. More generally, three-membered heterocyclics cleave into a radical and a cation, either of which can contain one or two of the original ring atoms (Scheme 2). (Sections 5.04.2.8, 5.05.2.4 and 5.06.2.3). Especially in thiiranes, this may involve rearrangements, such as path (c) in Scheme 2. *α*-Cleavage, particularly important in oxiranes and thiiranes, may give a substituent radical and a cyclic ion (Scheme 3). *β*-Cleavage, more important in aziridines, gives a radical and an ion (Scheme 3). Longer side chains permit rearrangements, such as that in Scheme 4 ⟨B-71MS11⟩.

$H_2C{-}CH_2$ (X bridging) $\xrightarrow[-e^-]{\text{ioniz.}}$ $H_3C{-}CHX^+$ → $H_3C^+ + {\cdot}CHX$ (a)
→ $H_3C{\cdot} + {}^+CHX$ (b)

$RCH_2HC{-}CH_2$ (X bridging) → $[RCH_2CH(X{\cdot}){-}CH_2]^+$ → $RX{\cdot} + H_2C{=}CHCH_2^+$ (c)

Scheme 2

$RC(H){-}CH(R')$ (X⁺• bridging) → $R'{\cdot} + RC(H){-}CH$ (X⁺ bridging, double bond) *α*-Cleavage

$RCH_2HC{-}CH_2$ (N⁺•H bridging) → $R{\cdot} + H_2C{=}CH{-}\overset{+}{N}(H){=}CH_2$ *β*-Cleavage

Scheme 3

Four-membered heterocycles prefer to cleave, upon ionization, into two fragments, each containing two of the ring atoms. Further cleavages commence from these initial fragments (Scheme 5). Specific details can be found as follows: azetidines ⟨B-71MS296⟩, oxetanes

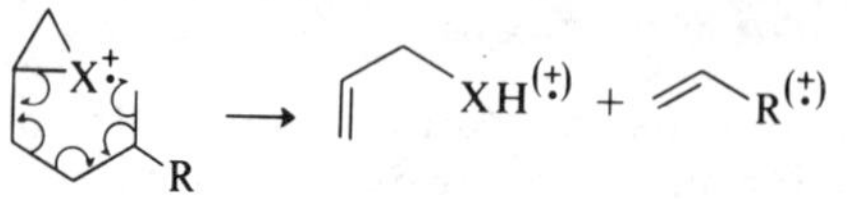

either fragment can be the radical ion

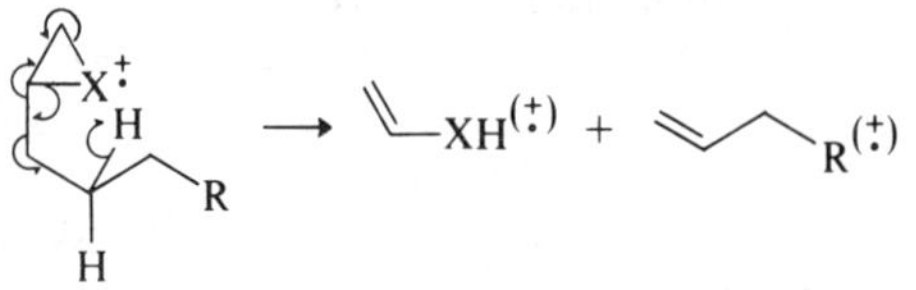

either fragment can be the radical ion

Scheme 4

⟨B-71MS34⟩, thietanes (Section 5.14.2.3) ⟨B-71MS229⟩. The cleavage to two sets of two ring-atom fragments is illustrated by the formation of fragments with the masses of ethylene, methyleneimine and HC≡CNH from azetidine, and that of those with the masses of RNCO, ketenes and imines from azetidin-2-ones ⟨B-71MS300⟩.

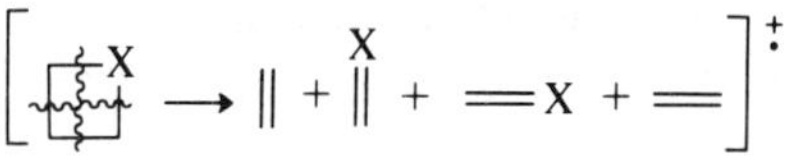

Scheme 5

5.01.2.5 Electronic Spectra

Saturated three- and four-membered heterocyclics absorb but little light in the readily accessible regions of the UV spectrum. Sulfur-containing rings are an exception, as can be seen in Table 7. Despite the lack of absorption of most parent compounds, there is a wealth of photochemistry of small heterocyclics. Light absorption by substituents, and energy transfer from photoexcited molecules present in the photoreactive system make photoconversion of the heterocycles practical. On the other hand, the lack of substantial absorption of their own can be exploited in the preparation of small heterocycles, by designing the system to be unsuitable for destructive energy transfer.

The introduction of a second heteroatom (other than sulfur) does not change drastically the absorption characteristics of small heterocycles. Oxaziridine and diaziridine are still 'transparent' to light of wavelengths above 220 nm (Section 5.08.2.3.2).

5.01.2.6 IR Spectra

IR spectroscopy can give a great deal of information on small ring heterocyclics, because of the effects of ring strain on the frequencies of vibration of substituents attached to the ring, and because the ring vibrations fall into a readily accessible region of the IR spectrum. A wealth of data has been gathered and can be found in the monograph chapters of this work and in the following reviews ⟨71PMH(4)265, 63PMH(2)161, B-75MI50100, B-75MI50101⟩. This section concentrates on vibrations of general diagnostic value. A treatment emphasizing the theoretical foundations is available ⟨63PMH(2)161, 71PMH(4)265⟩.

Small rings show high C—H absorption frequencies for the ring C—H bonds (between 3080 and 3000 cm^{-1}). The asymmetric C—H stretching frequency decreases with increasing ring size, from 3047 cm^{-1} for aziridine to 2966 cm^{-1} for azetidine and 2950 cm^{-1} for pyrrolidine. Analogous changes are found in saturated oxygen heterocycles (3052, 2978, 2958 cm^{-1}) and their sulfur analogs (3047, 2968, 2959 cm^{-1}) ⟨71PMH(4)265, p. 278⟩. The stretching frequencies for exocyclic C=X bonds follow a similar sequence, with the smallest rings having the highest frequencies, as seen in Table 8. Four-membered rings have somewhat lower C=X frequencies; the carbonyl frequency of azetidin-2-one is 1786 cm^{-1} and that of oxetan-2-one is 1832 cm^{-1}.

Ring breathing frequencies are shown in Table 9, together with some C—H IR absorptions. As expected, the ring breathing frequencies are lower for four- than for three-membered rings ⟨63PMH(2)161, 71PMH(4)265⟩.

Table 7 Electronic Absorption Spectra of Small Heterocyclic Systems

Skeleton	λ_{max} (nm)	*Absorption coefficient*	*Section number/ref.*
NH (aziridine)	179 145 118	4200 6100 6300	69JCP(51)52, 72BCJ3026
O (oxirane)	171.3 158 143	5600	5.05.2.5, 76JCP(64)2062
S (thiirane)	260 205	40 4000	5.06.2.4
NH (azetidine)	191.7		76JCP(64)2062
O (oxetane)	187 174 161 153	2000 2750	76JCP(64)2062
S (thietane)	275 218	30 600	5.14.2.4
S–S (1,2-dithietane)	340 238	80 7440	5.15.1.2.9

Table 8 Stretching Frequencies for Exocyclic Double Bonds on Small Rings[a]

Skeleton	Stretching frequencies (cm^{-1}) X=C	X=N	X=O
RR′, X, N, R″	1770 (R=R′=H, R″=Et, X=C)	1805 (R=H, R′=Bu^t, R″=Me, X=NMe)	1837 (R=R′=Me, R″=Bu^t, X=O)
RR′, X, O	1780 (R=R′=Bu^t, X=$CHBu^t$)	—	1890 (R=R′=Bu^t) 1990, 1945[b] (R=R′=CF_3)
RR′, X, S	1738 (R=R′=Me, X=CMe_2)	1700, 1630[c] (R=R′=Ph, X=NTs)	1785 (R, R′=CH_2=)
X, NR, N, R′	1690–1650[d] (R=Ar, R′=SO_2Me, X=CMe_2)	1790 (R=R′=*trans*-Bu^t, X=NBu^t)	1882[e] (R=R′=Me)
O, O, O	—	—	2045

[a] Taken from ⟨80AG(E)276⟩ unless otherwise stated. [b] ⟨82CC362⟩. [c] ⟨78AG(E)195⟩. [d] ⟨77AG(E)475⟩. [e] ⟨75AG(E)428⟩.

Table 9 Ring Breathing and C—H IR Absorptions of Small Heterocycles[a]

Structure	*Symmetry*	*Ring breathing*	*C—H stretch*	*CH_2 scissoring*	*CH_2 wagging*	*Ref.*
 N H	C_s	1268, 1210	3078, 3012	1475, 1455	1128, 1131, 1088, 998	63PMH(2)161, 71PMH(4)265, p. 277
 O	C_{2v}	1266	3079, 3063, 3016, 3005	1490, 1470	1153, 1120	63PMH(2)161, 71PMH(4)265, p. 277
 S	C_{2v}	1112	3080, 3000	1446, 1427	1051, 1025	63PMH(2)161, 71PMH(4)265, p. 277
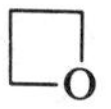 O	—	980–970, 900	—	—	—	63PMH(2)161, 71PMH(4)265, p. 277
 S S	—	738	2980, 2942	1438	1187	79MI50102

[a] cm^{-1}.

5.01.3 STRUCTURE OF HETEROCYCLIC RINGS WITH MORE THAN SIX RING ATOMS

5.01.3.1 Bond Lengths and Bond Angles

Heterocyclics with seven and more ring members display an enormous variety of shapes. Bond lengths are often close to those of open chain counterparts, but bond angles can be greatly different. Aromaticity is possible where the conditions of planarity and Hückel's rule are met, but the majority of fully unsaturated large heterocycles are not aromatic (see below).

Fully unsaturated seven-membered heterocyclics have alternating bond lengths and are normally in boat conformations. Ring inversion barriers are 42.7 kJ mol^{-1} for 3-methyl-3*H*-azepine and 35.6 kJ mol^{-1} for 3*H*-azepin-2-one (Section 5.16.2.3). The barriers for oxepin and thiepin are somewhat lower (27 kJ mol^{-1} for 3-methyl-6-isopropylthiepin 1,1-dioxide; Section 5.17.1.4). The conformational equilibria of many oxepins and thiepins are complicated by their heterapine–heteranorcaradiene equilibria, which are strongly influenced by substituents. Table 10 gives some data on bond lengths, calculated resonance energies and the longest wavelength electronic absorption of seven-membered heterocycles ⟨70T4269, 70JA1453, 81H(15)1569⟩.

Annelation can introduce large conformational barriers, to the extent of making possible the resolution into enantiomers of a tribenzoxepine ⟨71CB2923⟩. Chapters 5.16, 5.17, 5.18 and 5.19 contain much more information on inversion barriers, bond lengths and bond angles.

Fully saturated seven-membered heterocycles with one or two heteroatoms are normally in mobile twist–chair conformations (Section 5.17.1.1, Chapter 5.18) ⟨B-77SH(2)123⟩. Annelation and the introduction of exocyclic double bonds can have profound effects; oxepan-2-one, for example, is in a near chair conformation ⟨67JA5646⟩.

The bond lengths of fully saturated seven-membered rings are the same as those in corresponding open chain compounds, while the bond angles tend to be larger (Section 5.17.1.1). Accordingly, the IR spectra of, for example, oxepanes (C—O—C, C—H stretching, and CH_2 deformation) are the same as in analogous noncyclic compounds. Transannular dimensions can be altered by bonding interactions, such as in some azocin-5-ones, which have a distance between N-1 and 5-C=O, across the ring, of only 2.23 Å, and a lowered stretching frequency of the C—C carbonyl (Section 5.19.2.1).

Table 10 Structure and ^{1}H NMR Data of Seven-membered Heterocyclic Compounds

X	*Y*	*Bond lengths* (Å) a	b	c	d	*Resonance* (kJ mol^{-1})	$\delta(^{1}H)$ (p.p.m.) H-2	H-3	H-4	H-5	H-6	H-7	*Longest wavelength UV absorption* (nm)	log ε
C	C	1.45	1.37	1.48	1.35	—	—	—	—	—	—	—	—	—
O	C	1.39	1.35	1.46	1.35	+0.5	5.7	5.7	6.3	—	—	—	305	2.95
HN	C	1.42	1.35	1.46	1.35	−0.75	—	—	—	—	—	—	—	—
EtOCON	C	—	—	—	—	—	5.95	5.51	6.15	—	—	—	318	2.83
EtOCON	N	—	—	—	—	—	—	6.23	5.75	6.55	6.25	7.40	355	2.38
S	C	1.79	1.35	1.46	1.35	−6.1	—	—	—	—	—	—	—	—

Ring number is 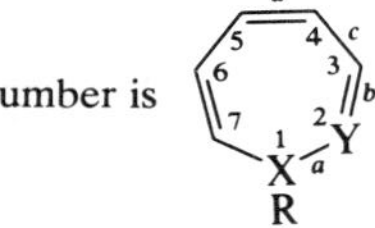

Heterocyclic aromaticity requires a planar or nearly planar, conjugated system containing $4n+2$ π-electrons. The subject has been reviewed ⟨74AHC(17)255, p. 339⟩. Planarity often cannot be achieved because of increased strain in going from puckered to planar geometries. The Hückel condition often can be met, utilizing the system's π-bonds and n electron pairs, and sometimes by forming cations or anions to adjust the number of participating electrons. Aromatic heterocyclic systems are known with 7 to 21 ring members. Still, a great many heterocycles having the right number of electrons are polyenic rather than aromatic, due to the excessive energy required to achieve near planarity. The diagnostic tools for ascertaining the nature of such systems are discussed in Section 5.20.2.2. The aromatic systems are diatropic as seen in their ^{1}H NMR spectra, and show double bond delocalization in their electronic spectra and bond length equalization in their X-ray or electron diffraction structures. Cases intermediate between polyenic and aromatic are found as well (Section 5.20.2.2.1). Aromatic systems known include 1,3-dithiepin anions (Section 5.18.4.2), 1,4-dihydro-1,4-diazocines (Section 5.19.4.4), azonine (Section 5.20.2.2.2), 2,7-methano-aza[10]annulene (Section 5.20.2.3) ⟨78AG(E)853⟩ and *trans,trans,trans*-aza[13]annulene (Section 5.20.2.4).

5.01.3.2 NMR Spectra

The NMR spectra of heterocyclic compounds with seven or more ring members are as diverse as the shape, size and degree of unsaturation of the compounds. NMR is perhaps the most important physical method to ascertain the structure, especially the conformational statics and dynamics, of large heterocycles. Proton–proton coupling constants provide a wealth of data on the shape of the molecules, while chemical shift data, heteroatom–proton coupling constants and heteronuclear spectra give information of the electronic structure. Details are found in Chapters 5.16–5.22. Some data on seven-membered rings are included in Table 10.

5.01.3.3 Electronic Spectra

Electronic spectra, so very important in the characterization of five- and six-membered heterocycles, have played a much lesser role in the study of large heterocyclic rings, and far fewer data are available for comparison. Aromaticity in large heterocycles can be detected by their electronic absorption. Beyond that, much of the information derived from UV and visible spectra is the same as for acyclic compounds, and is most valuable in the detection of the structure of subunits, questions of conjugation, *etc.*, but is not easily discussed in general terms relating to the underlying heterocyclic structures. Bond angle deformation due to the cyclic nature of some systems can be assessed by UV spectroscopy. Flattened angles at bridgehead nitrogens affect the UV spectra (Section 5.19.2.1), and ester groups incorporated in a ring absorb at wavelengths longer than their acyclic counterparts (Section 5.19.3.1).

5.01.3.4 IR Spectra

The use of IR spectra for the structure determination of large heterocyclic rings is not as straightforward as for small ones. The ring breathing vibrations fall into a range well below that commonly used in the laboratory, and the absorptions are often broad and ill defined. Owing to the almost infinite variety of special effects, bond angle deformation and the consequent effect on the absorption frequencies of ring C—H, C=O and C=X bonds are not easily used in the diagnosis of an unknown structure. These frequencies are, of course, quite valuable for the determination of structural details in systems already fairly well understood. Transannular nucleophilic interaction in azocin-5-ones (Section 5.19.2.1) has already been mentioned as an example.

5.02

Reactivity of Small and Large Rings

W. LWOWSKI
New Mexico State University

5.02.1 INTRODUCTION

Chapter 5.02 attempts to give an overview of the reactivity of 'small or large' ring systems treated individually in the monograph chapters of Part 5 of this work. The great diversity of these systems presents a serious problem of organization. No one scheme proved satisfactory, and it was eventually decided not to use ring size, or aromaticity *vs.* nonaromaticity, as primary operators in structuring the chapter. Rather, the nature of the reaction and the distance of the site of attack from the heteroatoms were used, in the hope of arriving at a reasonably adequate organization scheme.

It will be noticed that the various monograph chapters differ in their arrangement of reactions, within their section on reactivity. Many authors, while writing the chapters, came to realize that one uniform scheme of subdividing the material would not do justice to all the various rings systems — small and large, strained and unstrained, flat or puckered, saturated, polyenic or aromatic. This is especially true for the chapters on special systems, many of which have been explored mostly with respect to very special features, be it physiological activity or the complexing of ions, for example.

5.02.2 GENERAL SURVEY

5.02.2.1 Neutral Molecules

The reactivity of small (three- and four-membered) heterocyclic rings is dominated by the effects of ring strain, which facilitates all modes of ring opening. Aromaticity is not observed, antiaromaticity is present in only a few isolated examples, and thus does not play a general role. Many reactions are initiated by unimolecular ring opening, to give diradicals or ylides, whose reaction products are then observed. Extrusion of stable as well as unstable moieties occurs, assisted by the ring strain. Four-membered systems tend to cleave into two two-member moieties (consisting of two former ring atoms and their ligands). Attack on ring carbons concomitant with ring opening is very common, and is usually subject to electrophilic catalysis.

Large heterocyclics, *i.e.* those with more than six ring members, often show little effect of ring strain on the reactivities of the neutral molecules. (It is, of course, impossible to make really general statements on 'large ring reactivities' due to the infinite variety of structures, many of which do have considerable ring strain.) Factors very important for large ring reactivity are unsaturation, especially polyenic and aromatic characteristics, and the steric accessibility of heteroatoms and functional groups, as well as the possibility of transannular reactions. The majority of unsaturated large rings are not aromatic, even where the Hückel rule seems to be formally obeyed. Only the combination of proper electron numbers and proper geometry leads to aromatic characteristics, often ascertained more by spectroscopic (NMR, UV) studies and X-ray data than by chemical properties ⟨74AHC(17)255⟩.

Large systems contain ring atoms enough to allow more than one way of bonding between a given sequence of atoms, and a wealth of valence isomerizations becomes possible. Already for seven ring members, two types of bicyclic isomer are observed, namely the [4.1.0] and the [3.2.0] systems, each with positional ring isomers if more than one heteroatom is present.

5.02.2.2 Cations

Onium ions of small and large heterocyclics are usually produced by electrophilic attack on a heteroatom. In three- and four-membered rings nucleophilic attack on an adjacent carbon follows immediately, in most cases, and ring opening stabilizes the molecule. In large rings the onium ion behaves as would its acyclic analog, except where aromaticity or transannular reactions come into play (each with its electronic and steric pre-conditions). A wide diversity of reactions is observed.

Cations of a different kind may be derived from heterocyclics by removing a leaving group with its bonding electrons, *e.g.* a halide ion from an *N*-halo moiety. Such species have been discussed in reactions of *N*-halo nitrogen heterocycles, but were excluded as intermediates in the ring opening of *N*-haloaziridines, to name just one example (Section 5.04.3.9). Such cations, 'nitrenium ions', were also assumed to be reactive intermediates in reactions of other *N*-halo heterocycles, but it is often very difficult to exclude that observed reactions are not of a radical chain (Hofmann–Loeffler) type ⟨70ACR26, 74CJC2123, 73CC461, 80JOC4069⟩. The stability of azirinyl and diazirinyl cations has been calculated with the result that the former may be stable in nonnucleophilic media ⟨80TL4553⟩.

5.02.2.3 Anions

Anions of small heterocyclics are little known. They seem to be involved in some elimination reactions of oxetan-2-ones ⟨80JA3620⟩. Anions of large heterocycles often resemble their acyclic counterparts. However, anion formation can adjust the number of electrons in suitable systems so as to make a system conform to the Hückel rule, and render it aromatic if flat geometry can be attained. Examples are found in Chapter 5.20. Anion formation in selected large heterocycles can also initiate transannular reactions (see also Section 5.02.7 below).

5.02.2.4 Radicals

Radicals of most small and many large heterocycles are known, but their chemistry has not always been explored in depth. The ESR spectra of small ring radicals have been measured and generally found to be in good agreement with theoretical predictions. Table 1 gives some data for three-membered heterocyclics. Noteworthy is the close similarity of the ESR spectra of 1-aziridinyl, 1-azetidinyl and dimethylaminyl ⟨71TL2247⟩. The radicals in the table are all π-radicals.

Table 1 ESR Spectra of some Small Ring Nitrogen Radicals

Radical	a^N	*g*	*Ref.*
Aziridinyl	12.52	2.0043	71TL2247
Diaziridinyl, 3,3-bis(methoxycarbonyl)-	14.2, 11.7	2.0038	76TL4205
Diazirinyl, 3-methyl-	7.8	2.0042	79JA837
Oxaziridinyl, 3,3-diphenyl-	1.16	2.0041	75CC322
Azetidinyl	13.99	2.0045	71TL2247

Small ring radicals with the unpaired electron at the heteroatom or at a carbon adjacent to a heteroatom undergo ring cleavage as the predominant mode of stabilization, as known for oxaziranes ⟨78CJC2985, 77TL4289, 76HCA880, 76JCS(P2)1044⟩, aziridines (Section 5.04.3.9), diaziridines (Section 5.08.3.2.3), diazirines ⟨79JA837⟩, oxaziridines (Section 5.08.3.1.5) and thietanes (Section 5.14.3.10.1). The heteroatom is usually retained in the product, except in thiirane cleavages (Section 5.06.3.7), where desulfurization occurs. Thietanes, in contrast, are less readily desulfurized (Section 5.14.3.10.1). Oxaziridinyls display a variety of reactions, including N—O and N—C cleavage (Section 5.08.3.1.5); diaziridinyls behave analogously, with C—N and N—N cleavage (Section 5.08.3.2.3).

Large ring heterocyclic radicals are not particularly well known as a class. Their behavior often resembles that of their alicyclic counterparts, except for transannular reactions, such as the intramolecular cyclization of 1-azacyclononan-1-yl (Scheme 1) ⟨72CJC1167⟩. As is the case with alicyclic ethers, oxepane in the reaction with t-butoxy radical suffers abstraction of a hydrogen atom from the 2-position in the first reaction step (Scheme 2) ⟨76TL439⟩.

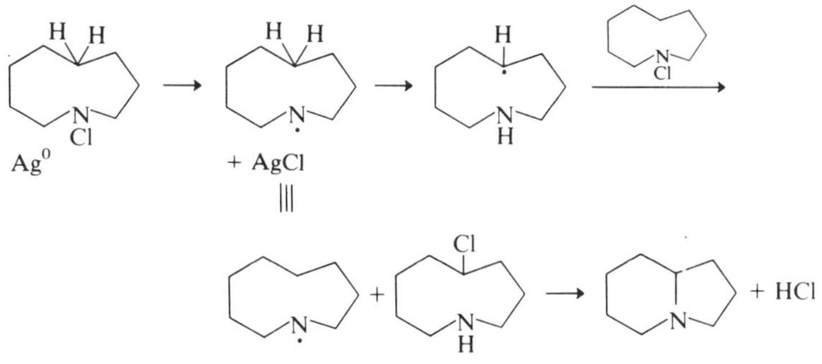

Scheme 1

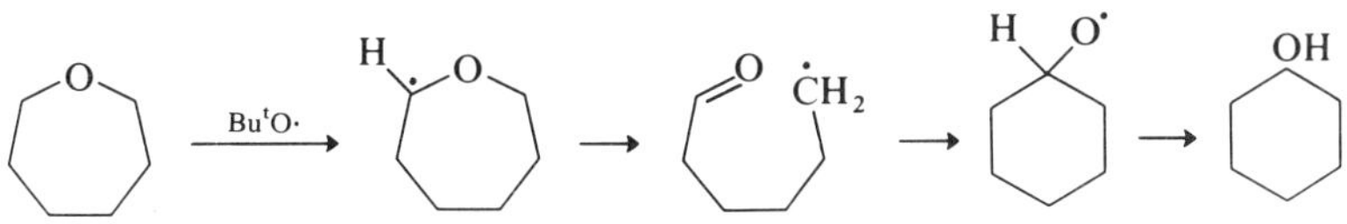

Scheme 2

5.02.3 THERMAL AND PHOTOCHEMICAL REACTIONS, NOT FORMALLY INVOLVING OTHER SPECIES

5.02.3.1 Fragmentation Reactions

Fragmentation reactions are particularly common in small rings. Relief of strain and the gain in stability in forming certain common fragments (such as N_2, CO_2), as felt in the transition state of the rate determining step, are important driving forces. Three-membered rings fragment to give moieties a—b (usually unsaturated) and c. The latter might be a stable molecule, such as CO, but also a carbene or nitrene, atomic sulfur or singlet SO, to name the most common ones. Scheme 3 gives examples.

$\xrightarrow{h\nu}$ X: + C=C X = NR, S, SO_x

$\xrightarrow{h\nu}$ R′N: + R_2C=O

$\xrightarrow{h\nu}$ R_2C: + N_2

→ PhCN + CO (Section 5.04.3.4)

→ RNCO + CO ⟨80JA6902⟩

Scheme 3

Four-membered heterocycles most often give fragments containing two ring atoms with their respective ligands. However, [3 + 1] fragmentation is well known, giving an atom (such as S) or stable species (such as SO_2) and often a three-membered ring. Examples are found in Schemes 4 and 5. The [2 + 2] fragmentations are often stereospecific and the fragmentation can be reversible. The reversibility can lead to interconversions as seen in Scheme 6; the RN= moieties of an isocyanate and a carbodiimide are exchanged *via* a 3-imino-1,3-diazetidin-2-one ⟨69ACR186⟩.

$\xrightarrow{h\nu \text{ or } \Delta}$ a=b + c=d and/or a≡c + b≡d

flash pyrolysis → + RNCO (Section 5.09.3.2.2)

→ ‖ + ‖ S (Section 5.14.3.2.1)

Scheme 4

$\xrightarrow{\Delta \text{ or } h\nu}$ + b

→ ‖ + ‖ SO + + SO (Section 5.14.3.2.1)

SO_2 → + SO_2

Scheme 5

RN=C=O + R′N=C=NR′ ⇄ ⇄ RN=C=NR′ + O=C=NR′

Scheme 6

Nitrogen, CO_2, SO_2, RCN, RNCO and RNCS are particularly common fragments containing two of an original four ring atoms.

The fragmentation of large heterocyclics occurs less readily, since the ring strain is usually less. The most favorable leaving moieties, such as N_2, can of course be extruded easily (often giving 1,*n* diradical species). Thus, 1,2,4-triazepines can lose nitrogen to give pyrroles (Scheme 7; Section 5.18.7.2). Most often, fragmentations of large heterocyclics can be classified either as retrocycloadditions, or they are analogous to what would be expected from acyclic counterparts of the large rings. The fragmentations may be orbital symmetry controlled. For example, 2,7-dihydrothiepin 1,1-dioxide loses SO_2 to give *cis*-hexatriene (Section 5.17.2.3.2). The orbital control of thermal extrusions from thiepins has been studied in some detail (Section 5.17.2.4.1).

N N N Ph R $\xrightarrow{h\nu}$ N_2 + Ph N R

Scheme 7

5.02.3.2 Rearrangements

(a) *Cis–trans* isomerizations are commonly observed upon heating or irradiating three-membered heterocyclics. The formation of 1,3-diradicals leads to rotation about the single bonds, and the isomerization has been used to probe the bond strengths of the 2,3-bond of such heterocyclics (Scheme 8). *Cis–trans* isomerizations in large rings can be due to retrocycloadditions or the temporary conversion of parts of the heterocycle, such as dehydrogenation–hydrogenation reactions. Diradical formations, to give long chain 1,ω-diradicals, usually do not lead to recyclization, due to unfavorable entropy factors.

cis ⇄ ⇄ ⇄ *trans*

Scheme 8

(b) Hydrogen shifts are common in large, unsaturated rings, such as azepines. Series of 1,5-hydrogen shifts, thermally allowed, connect the 1*H*, 2*H*, 3*H* and 4*H* isomers of unsubstituted azepines, oxepins and thiepins. While the sigmatropic mechanism does allow the interconversion of the isomers, base catalysis has been observed in some cases. Ionic mechanism(s) must therefore be considered (Scheme 9).

1*H* ⇄ 4*H* ⇄ 2*H* ⇄ 3*H*

Scheme 9

(c) Ring–chain isomerizations are common with small heterocycles, with the ring strain assisting the opening. The reverse reaction is often found where reactive opening products are obtained (see isomerizations). Scheme 10 gives a few examples of irreversible ring openings, and Scheme 11 shows some which are readily or spontaneously reversed.

$\longrightarrow H_2C{=}N_2$ (Section 5.08.3.3.2)

$\xrightarrow{h\nu}$ (Section 5.09.4.2.2)

$\xrightarrow{h\nu}$ (Section 5.14.3.2.2)

Scheme 10

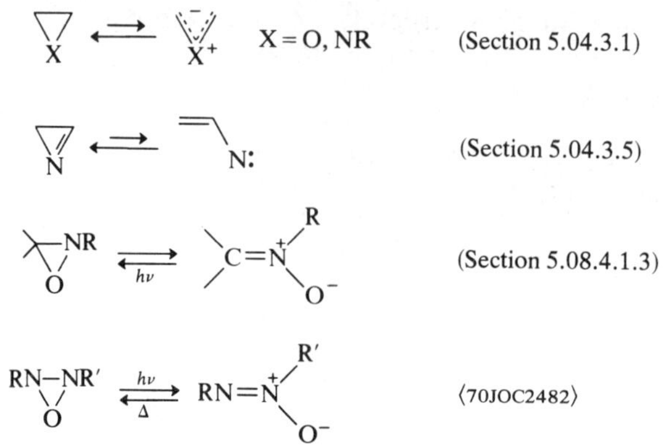

X = O, NR (Section 5.04.3.1)

(Section 5.04.3.5)

(Section 5.08.4.1.3)

⟨70JOC2482⟩

Scheme 11

(d) Ring–ring valence isomerizations occur in small and large rings, with or without changes of ring size. An example of the latter course is the intriguing interconversion of three-membered rings with exocyclic double bonds (Scheme 12) observed with methyleneaziridines ⟨73AG(E)414, 78AG(E)213⟩, aziridinimines ⟨70AG(E)381⟩ and diaziridinimines ⟨69AG(E)449⟩.

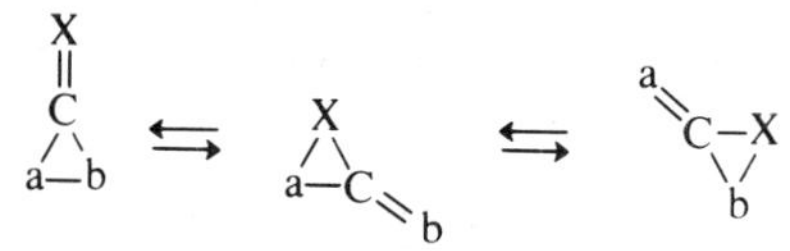

Scheme 12

Ring expansion of small rings is once again favored by ring strain, and many 3→5 conversions are known. Four-membered rings can expand to five- or six-membered ones. Examples are given in Scheme 13.

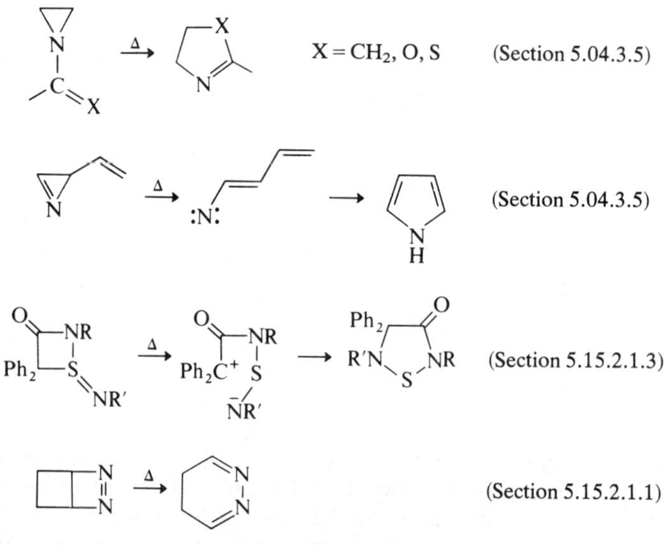

X = CH_2, O, S (Section 5.04.3.5)

(Section 5.04.3.5)

(Section 5.15.2.1.3)

(Section 5.15.2.1.1)

Scheme 13

Large rings isomerize to two condensed smaller ones, by transannular reactions of most any bond-making mechanism, and by electrocyclic reactions. Seven-membered, fully unsaturated systems can convert to [3.2.0] and to [4.1.0] isomers. The former conversion is allowed photochemically, the latter thermally. Consequently, the chemistry of azepines, oxepins and thiepins is often governed by the rate and activation barrier (or the photochemical conditions and parameters) prevailing. Thiepins, for example, extrude sulfur *via* the thianorcaradiene isomer, and sulfur loss is likely to occur when a given system isomerizes to the [4.1.0] isomer more rapidly than competing reactions occur through the monocyclic isomer. Depending on the nature of the heteroatom (and the presence of other heteroatoms in the ring) and the substituents, the heterapine–heteranorcaradiene rearrangement can be fast or slow, and one or the other component can be dominant in the equilibrium. Photoinduced rearrangement leads to [3.2.0] systems, which may revert to the seven-membered monocycle thermally — perhaps by homolytic cleavage of the common bond. Both types of bicyclization are observed in systems containing N, O or S (including SO and SO_2) as heteroatoms. Scheme 14 gives examples.

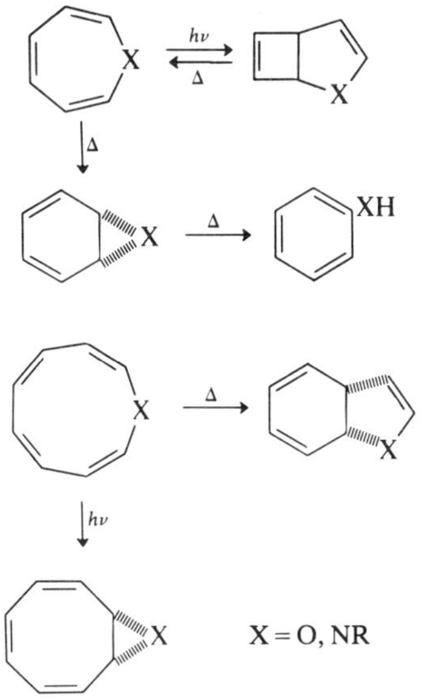

X = NR, O, S (Sections 5.17.2.4.2, 5.17.2.2.2, 5.16.3.2.1)

(Section 5.17.1.3)

X = O, NR (Section 5.20.3.2.1)

Scheme 14

2*H*-Azocinones are in thermal equilibrium with 8-azabicyclo[4.2.0]octa-3,5-dienones, as measured by NMR (Section 5.19.2.3).

5.02.4 ELECTROPHILIC ATTACK ON RING HETEROATOMS

5.02.4.1 Introduction

Electrophilic attack on ring heteroatoms ties up an electron pair (which may have been engaged in resonance in the parent species) and confers positive charge on the system, thereby inviting nucleophilic attack or elimination reactions to follow. In small systems the primary product is usually quite unstable. Nucleophilic attack on protonated or Lewis acid-coordinated species will be treated below (Section 5.2.7), because it is not always clear whether such reactions are preceded by an electrophilic step (*e.g.* protonation) or not.

5.02.4.2 Protonation and Basicity

The basicities of saturated heterocycles are similar to those of analogous open chain systems, with the exception of three-membered heterocycles, in which the basicity is markedly reduced. Table 2 gives pK_a values for the equilibria between free and monoprotonated heterocycles. As the ring size increases, the protonated species become more stable and the pK_a values approach those of the open chain analogs. Data for some small rings are hard to come by, but NMR data suggest the order of increasing basicity as thiirane < oxirane < aziridine (Section 5.06.3.3.2). The same order seems to prevail in gas phase proton affinities, which were calculated ⟨78JA1398⟩ and measured by ion cyclotron resonance. A few data are shown in Table 3 ⟨80JA5151⟩.

Protonation or Lewis acid complexation of a heteroatom invites nucleophilic attack, including nucleophilic attack by a parent molecule. Oligomerization and polymerization are thus often the results of bringing heterocycles into an acid environment without making sure that all of the potentially nucleophilic sites are protonated.

Alkylation, acylation, *etc.* at the heteroatom lead to onium salts. In small systems these are difficult to isolate, and very weakly nucleophilic counterions must be used, such as

Table 2 Basicities for some Heterocycles: pK_a Values for the Equilibria Between Parent and Monoprotonated Species in Water

Parent species	pK_a	*Section number/ref.*	*Parent species*	pK_a	*Section number/ref.*
 aziridine (N–H)	8.04	63PMH(1)1	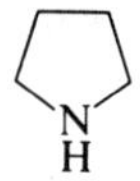 pyrrolidine (N–H)	11.27	63PMH(1)1
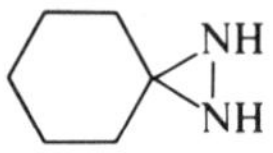 cyclohexane spiro-diaziridine (NH, NH)	4.6	5.08.2.3.1	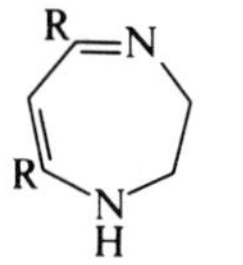 R, R-substituted diazepine (=N, N–H)	13–14	5.18.2.3
 cyclohexane spiro-diaziridine (NH, NBu^n)	6.4	5.08.2.3.1	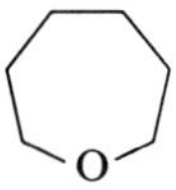 oxepane (O)	−2.02	5.17.2.1.3
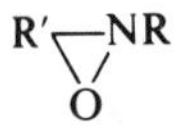 R′, NR oxaziridine (O)	0.13 to −1.81	5.08.3.1.3	 tetrahydrofuran (O)	−2.08	5.17.2.1.3
 azetidine (NH)	11.29	63PMH(1)1			

Table 3 Gas Phase Proton Affinities of Small Heterocycles ⟨80JA5151⟩

Heterocycle	*Proton affinity* ($kJ\ mol^{-1}$)	*Open chain analog*	*Proton affinity* ($kJ\ mol^{-1}$)
 aziridine (N–H)	902.5	Me—NH—Me	922.6
 oxirane (O)	793.3	Me—O—Me	807.9
 thiirane (S)	812.9	Me—S—Me	839.7

BF_4^-. In large rings the fate of the onium ions depends mostly on the structure and degree of unsaturation of the particular compound, and the onium salts range from completely stable to highly unstable.

Electrophiles, such as C—Hal functions, contained in side chains may be well positioned for interaction with ring heteroatoms. Thus, *N*-*t*-butyl-2-tosyloxymethylaziridine in ethanol displaces tosylate ion from the side chain, and nucleophilic opening of the resulting azabicyclobutanonium ion by solvent gives 3-hydroxy- and 3-ethoxy-azetidine (Section 5.09.2.3.2).

5.02.5 NUCLEOPHILIC ATTACK ON RING HETEROATOMS

Nucleophilic attack on ring heteroatoms is found most often in two situations: (1) where the heteroatom in question is sulfur, and (2) in small rings with two heteroatoms. Attack by hydroxide of *S*-methylsulfinyl-1,8-naphthalene gives 1-naphthyl methyl sulfoxide (Section 5.14.3.6.4), and sulfur on sulfur nucleophilic attack is known in thietanones (Section 5.14.3.6.2). Oxaziridines are attacked on oxygen when bulky ring substituents are present, otherwise the nitrogen is attacked, resulting in nitrogen transfer and the formation of a carbonyl compound (Section 5.08.3.1.4), while the ring carbon is altogether inert towards nucleophilic attack.

5.02.6 NUCLEOPHILIC ATTACK ON RING CARBON ATOMS

The ring opening of small heterocycles by nucleophilic attack on a carbon adjacent to a heteroatom is exceedingly common. Only in oxaziridines is the ring carbon inert relative to the two ring heteroatoms (Section 5.08.3.1.4). In the other three-membered rings, nucleophilic ring opening leads to the corresponding heteroanion or to the XH compound, according as the heteroatom is being protonated before or concurrently with the ring opening respectively. The reaction can be stereospecific in any of these cases. Alternatively, ring opening can occur before the nucleophilic attack, either after protonation of the heteroatom (to give a carbocation), or due to ylide formation. In the latter cases the reactions become nonstereospecific or partially stereospecific, depending on the timing of the processes involved. Details are found in the monograph chapters and in ⟨B-77SH⟩.

Nucleophilic attack on ring atoms of large heterocycles is largely confined to saturated systems, saturated parts of partially unsaturated systems, and to carbonyl functions and the like. These reactions are not fundamentally different from those of corresponding acyclic systems, except for transannular reactions.

Transannular nucleophilic attack on ring atoms is best known in systems with seven or more ring members. For example, nucleophilic attack by the ring nitrogen on suitably substituted ring carbons in the 3- or 4-position in azepine derivatives has been studied (Section 5.16.3.5.2). However, transannular nucleophilic attack can be found already in four-membered heterocycles. The nitrogen of *N*-*t*-butyl-3-chloroazetidine is, because of the puckered conformation, close enough to the 3-carbon to displace chloride. This, in turn, opens the azabicyclobutanonium ion to give *N*-*t*-butyl-2-chloromethylaziridine ⟨B-77SH(1)38⟩.

In systems of proper geometry, nucleophiles within a side chain may be well connected for attack on ring atoms. For example, an aminomethyl group at the 5-position of a dibenzazepine-2-one was found to attack the carbonyl group (Section 5.16.3.5.2). Such reactions should be possible in rings of any size.

Apparent nucleophilic attack on large, fully unsaturated rings may occur by way of attack on a valence tautomer, such as the reaction of oxepin with azide ion. Attack on the oxanorcaradiene valence tautomer leads to ring opening of the three-membered ring, and formation of 5-azido-6-hydroxy-1,3-cyclohexadiene (Section 5.17.2.2.4).

5.02.7 NUCLEOPHILIC ATTACK ON PROTONS ATTACHED TO RING ATOMS

The formation of anions by proton abstraction leads to a very diversified set of reactions in small and large heterocycles. However, a few rules do generally apply. In the absence of complicating substituents, nitrogen anions can often be formed from small and large heterocycles, and subsequent alkylation, acylation, *etc.* can be achieved ⟨B-76MI50200, 64HC(19-2)886⟩ (Section 5.15.2.3). The N-anions of large rings are usually unexceptional (barring transannular reaction). N-anions of unsaturated large heterocycles are obtained with difficulty, but they are synthetically useful in the azepine field (Section 5.16.3.6).

Proton abstraction from ring carbons of small and large heterocycles often leads to ring opening. Oxiranes, for example, are usually converted to carbenes under conditions under which a carbanion is formed, with a few exceptions such as the synthetically useful 2-triphenylsilyl-2-lithiooxirane (Section 5.05.3.5). Likewise, oxaziridines can be opened by nucleophilic attack on a hydrogen at carbon-3 ⟨82JOC419⟩ (Section 5.08.3.1.4). Thiirane and thiirene 1,1-dioxides give cyclic carbanions, which are easily ring opened (Section 5.06.3.6). Large heterocyclics offer a greater variety for ring opening, such as the abstraction of protons from not only the 2- but also the 3-position, leading to ring opening by β-elimination and the formation of ω-unsaturated compounds, such as 6-hydroxy-1-hexene from oxepane (Section 5.17.2.1.4).

5.02.8 ATTACK BY RADICALS OR ELECTRON DEFICIENT SPECIES

Surprisingly little is known about the attack of radicals on small and large heterocycles. Hydrogen abstraction from the heteroatom of small rings leads to ring opening, and in the case of thiiranes to removal of the sulfur (*cf.* Section 5.02.2.4 above). Abstraction of H· exocyclic and α to nitrogen in oxiranes leads to N—O cleavage, and the reaction of vanadium(IV) with the oxygen of 1-oxa-2-azaspiro[2.5]octane gives N—O cleavage and ring expansion to caprolactam (Scheme 15) (Section 5.08.3.1.5).

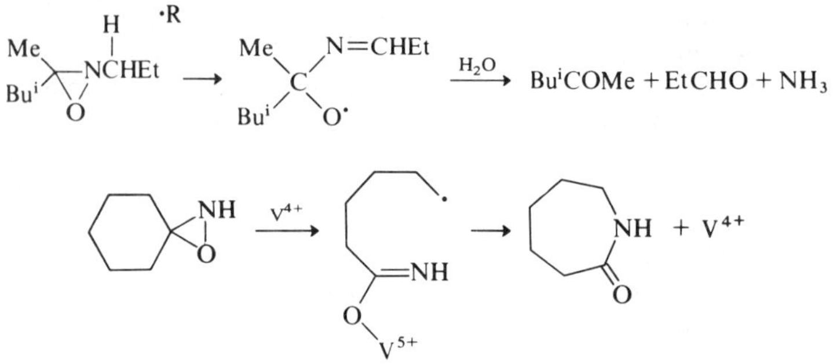

Scheme 15

Thiiranes are desulfurized by radicals (H·, S·), by singlet carbenes and by electrolysis (Section 5.06.3.7.1).

Azetidine derivatives, which are less strained, are less sensitive and removal of a hydrogen in the α-position of a substituent on nitrogen does not necessarily lead to ring opening (Section 5.09.3.2.5). Thietane rings are opened by radicals attacking on S, while the less strained thiolanes are attacked by hydrogen abstraction at a 2-carbon (Section 5.14.3.10.1). In thietane 1,1-dioxides, radicals abstract a hydrogen from the 3-position to give a cyclic radical (Section 5.14.3.10.1). Producing radicals exocyclic in the α-position of *N*-substituents of azetidin-2-ones did not result in ring opening (Section 5.09.3.2.5).

Saturated large rings may form nitrogen radicals by H abstraction from N, or abstraction may occur in the α- or β-positions in nonnitrogen systems. Oxepane gives the radical in the 2-position, with subsequent cleavage and reclosure of the intermediate carbenoid to cyclohexanol (Section 5.17.2.1.5). In unsaturated large systems a variety of reactions, unexceptional in their nature, are found. Some azepines can be brominated by *N*-bromosuccinimide; others decompose under similar conditions (Section 5.16.3.7).

Electron deficient species can attack the unshared electron pairs of heteroatoms, to form ylides, such as in the reaction of thietane with bis(methoxycarbonyl)carbene. The S^+—C^- ylide rearranges to 2,2-bis(methoxycarbonyl)thiolane (Section 5.14.3.10.1). *N*-Ethoxycarbonylazepine, however, is attacked by dichlorocarbene at the C=C double bonds, with formation of the *trans* tris-homo compound (Section 5.16.3.7).

5.02.9 REACTIONS WITH CYCLIC TRANSITION STATES

5.02.9.1 Cycloadditions

Concerted cycloadditions are observed with heterocyclics of all ring sizes. The heterocycles can react directly, or *via* a valence tautomer, and they can utilize all or just a part of unsaturated moieties in their rings. With three-membered rings, ylides are common reactive valence tautomers. Open chain 4π-systems are observed as intermediates with four-membered rings, and bicyclic valence tautomers are commonly reactive species in additions by large rings. Very often these reactive valence tautomers are formed under orbital symmetry control, both by thermal and by photochemical routes.

5.02.9.1.1 [2+4] Cycloadditions

The participation of a single double bond of a heterocycle is found in additions of small and large rings; azirines (Section 5.04.3.3) and thietes (Section 5.14.3.11) furnish examples. Azepines and nonaromatic heteronins react in this mode, especially with electron deficient dienes (Scheme 16; Section 5.16.3.8.1).

(Section 5.04.3.3)

(Section 5.18.9)

A = EtOCO—

(Section 5.16.3.8.1)

Scheme 16

Diene moieties, reactive in [2+4] additions, can be formed from benzazetines by ring opening to azaxylylenes (Section 5.09.4.2.3). 3,4-Bis(trifluoromethyl)-1,2-dithietene is in equilibrium with hexafluorobutane-2,3-dithione, which adds alkenes to form 2,3-bis-(trifluoromethyl)-1,4-dithiins (Scheme 17; Section 5.15.2.4.6). Systems with more than two conjugated double bonds can react by [$6\pi + 2\pi$] processes, which in azepines can compete with the [$4\pi + 2\pi$] reaction (Scheme 18; Section 5.16.3.8.1). Oxepins prefer to react as 4π components, through their oxanorcaradiene isomer, in which the 4π-system is nearly planar (Section 5.17.2.2.5). Thiepins behave similarly (Section 5.17.2.4.4). Nonaromatic heteronins also react in orbital symmetry-controlled [4+2] and [8+2] cycloadditions (Scheme 19; Section 5.20.3.2.2).

(Section 5.09.4.2.3)

(Section 5.15.2.4.6)

Scheme 17

$(NC)_2C{=}C(CN)_2$

A = MeOCO—

(Section 5.16.3.8.1)

Scheme 18

AC≡CA

A = MeOCO—

(Section 5.17.2.2.5)

Scheme 19

5.02.9.1.2 1,3-Dipolar cycloadditions

1,3-Dipolar cycloadditions in which the heterocyclus provides the 1,3-dipole are common with three-membered rings, which can provide ylide intermediates, as has been mentioned above. Ylide formation is usually orbital symmetry controlled, and may be achieved by thermolysis or photolysis, with the expected stereochemical consequences. Rotation about one of the ylide C—X bonds results in loss of the original stereochemistry. These interconversions often are slow compared to the cycloaddition reactions of the ylides, so that partial stereospecificity is observed. Scheme 20 gives a generalized reaction. Extensive work has been done with aziridines (Section 5.04.3.1). Diaziridines behave similarly (Section 5.08.3.2.2) ⟨79AHC(24)63⟩. Azirines produce nitrile ylides upon photolysis (Scheme 21; Section 5.04.3.2). Examples for oxiranes and thiiranes are found in the corresponding monograph chapters (Sections 5.05.3.2.1 and 5.06.3.8).

Scheme 20

Scheme 21

Heterocyclics of all sizes, as long as they are unsaturated, can serve as dipolarophiles and add to external 1,3-dipoles. Examples involving small rings are not numerous. Thiirene oxides add 1,3-dipoles, such as diazomethane, with subsequent loss of the sulfur moiety (Section 5.06.3.8). As one would expect, unsaturated large heterocyclics readily provide the two-atom component for 1,3-dipolar cycloadditions. Examples are found in the monograph chapters, such as those on azepines and thiepines (Sections 5.16.3.8.1 and 5.17.2.4.4).

5.02.10 REACTIVITY OF TRANSITION METAL COMPLEXES

Metal complexes of heterocyclics display reactivities changed greatly from those of the uncomplexed parent systems. All or part of the π-electron system(s) of the parent heterocycle can be tied up in the complex formation, leaving the rest to do 'alkenic' reactions. The system may be greatly stabilized in the complex so that reactions, on a heteroatom for example, can be performed which the parent itself would not survive. Orbital energy levels may be split and symmetries changed, allowing hitherto forbidden reactions to occur. In short, a multitude of new reaction modes may be made possible by using complexes. Since decomplexing is usually practically possible, a great many applications seem to await realization. So far the number of examples is limited and is dispersed in the literature, but some background is available ⟨78JHC1057, 81ACR348, 71JA1123⟩. The chemistry of azepines (Sections 5.16.3.8.1 and 5.16.3.8.2) and diazepines (Section 5.18.2) ⟨78JHC1057, 81ACR348⟩ provides examples.

1*H*-Azepine derivatives form a diene complex with tricarbonyliron, leaving uncomplexed the third of the double bonds. If the 3-position is substituted, two different such complexes are possible, and are in equilibrium, as seen in the ^{1}H NMR spectrum. An ester group in the 1-position of the complex can be removed by hydrolysis, to give an NH compound which, in contrast to the free 1*H*-azepine, is stable. The 1-position can then be derivatized in the manner usual for amines (Scheme 22). The same tricarbonyliron complex can, by virtue of the uncomplexed 2,3-double bond, serve as the dienophile with 1,2,4,5-tetrazines. The uncomplexed *N*-ethoxycarbonylazepine also adds the tetrazine, but to the 5,6-double bond. Thus, two isomeric adducts can be synthesized by using or not using the complex (Scheme 23; Section 5.16.3.8.1).

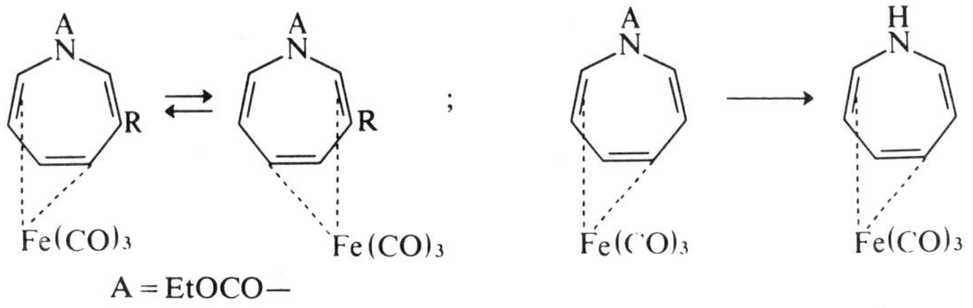

A = EtOCO—

Scheme 22

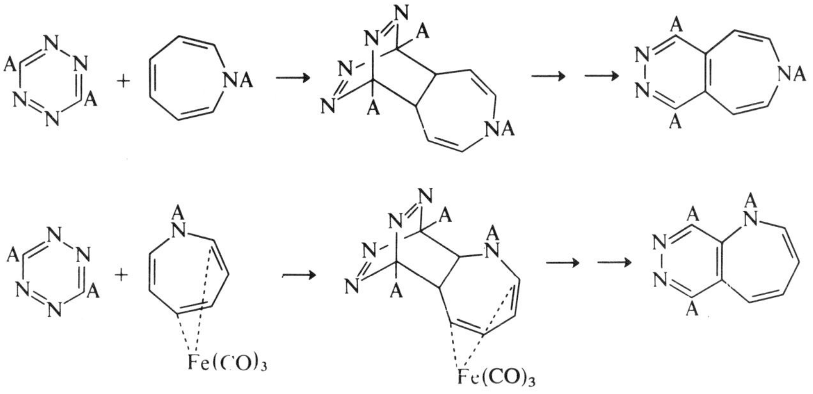

Scheme 23

Tricarbonyliron complexes of 1,2-diazepines do not show the rapid isomerization found in their azepine counterparts (Scheme 22); the iron forms a diene complex with the C=C double bonds in the 4- and 6-positions. The chemistry of the 1,2-diazepine complexes is similar to that of the azepine complexes (Section 5.18.2.1) ⟨81ACR348⟩.

Oxepin also forms a diene complex with tricarbonyliron ⟨78JHC1057⟩.

5.03

Synthesis of Small and Large Rings

W. LWOWSKI
New Mexico State University

5.03.1 INTRODUCTION

A large fraction of the chemical reactions known are used to form heterocyclic compounds. Displacement reactions and cycloadditions are particularly important, and their rates are therefore of great practical interest. The same is true for the rates of reverse reactions — ring opening by displacements or retrocycloadditions. It was realized over the last 40 years that many strained rings are formed rapidly, and that the reverse reactions are also rapid, and that the reaction types and conditions had to be selected accordingly to actually obtain these rings. Triaziridines, α-lactones, 1,2-dioxetanes and many others have now been made, systems a candidate would have proposed only at his peril in an examination 40 years ago.

Some of the factors operating in closing and converting large heterocyclic rings were recognized early. However, only the advent of sophisticated NMR techniques has made it possible to analyze in detail the spatial structure of large rings. These techniques were also instrumental in assessing aromaticity. Chapters 5.19 and 5.20 bear witness to this.

5.03.2 SYNTHESIS FROM NONHETEROCYCLIC COMPOUNDS BY MAKING ONE BOND

5.03.2.1 Formation of One Bond between Two Heteroatoms

5.03.2.1.1 Displacement reactions

The most common reactions for forming bonds in ring closures are displacement reactions (of any mechanistic type). Nucleophilic attack on heteroatoms proceeds readily with nitrogen, halogen or sulfur as the atom being attacked by a nucleophile. Halide is also, of course, a good leaving group in such reactions, and chloride has often been used as a leaving group from nitrogen. However, nucleophilic attack on the chlorine is also facile, leading to reactions of the 'positive chlorine' type, and homolysis of N—Cl bonds can lead to undesired side reactions. In such cases, displacing sulfonate ions from nitrogen is preferred. Examples are given in Scheme 1.

(Section 5.15.3.1.3)

Scheme 1

5.03.2.1.2 Valence isomerization

Two heteroatoms can be joined to form a ring by valence isomerization, such as in the formation of oxadiazirines and triaziridines (Scheme 2) ⟨70JOC2482, 80CC1197⟩.

$$\mathrm{Bu^tN{=}NBu^t(\rightarrow O)} \xrightarrow{h\nu} \mathrm{Bu^tN{-}NBu^t}\ (\text{bridged by O})$$

Scheme 2

5.03.2.1.3 Radical combination

Joining two heteroatoms to a ring by radical combination is not presently a common route to heterocycles. It might become more important if the art of metal-catalyzed redox reactions keeps advancing at the present pace. Current examples are the conversion of 1,5-dithiols to 1,2-dithiepanes by oxidants such as $FeCl_3$, and the oxidation of 1,3-propane-bis-hydrazines to 1,2,3,4-tetrazepines (Sections 5.18.4.1 and 5.18.10.1).

5.03.2.2 Formation of One Bond between Carbon and a Heteroatom

5.03.2.2.1 Displacement reactions

Displacement reactions involving a heteroatom and a carbon atom are very common in the synthesis of heterocyclics of many sizes. Most often, the heteroatom functions as a nucleophile, and halide or sulfonate ions are displaced from carbon. The Gabriel synthesis (Section 5.04.4.1) is widely used 95 years after its publication ⟨1888CB1049⟩. Aziridines and oxiranes are commonly made by such nucleophilic displacements, the detailed mechanisms of which have been studied intensively ⟨64HC(19-1)1, 525, B-50MI50300, B-56MI50300⟩. The formation of three-membered rings by nucleophilic displacement at a carbon in a position β to the heteroatom has become a large family of reactions, in part by the introduction of new leaving groups, some formed *in situ* (Scheme 3) ⟨80CC784⟩.

$$\underset{\text{HO}}{\overset{}{\text{C}}}-\overset{\text{OH}}{\text{C}} \xrightarrow{Ph_3P + RO_2CN{=}NCO_2R} -\underset{\text{O}}{\text{C}-\text{C}}- + Ph_3PO + RO_2CNHNHCO_2R$$

Scheme 3

Bond formation to the γ-position relative to the heteroatom is one of the most common paths to azetidines, oxetanes, thietanes and many of their oxo derivatives. In saturated chains the attack by the heteroatom on the γ-position is disfavored by the entropy factor, since the four-atom chain prefers the zigzag conformation. Reaction occurs only when molecular motion brings the γ-atom close to the attacking nucleophile (the heteroatom), thus the reaction rates for cyclization are about 100 times less than for attack on β-positions, which are always in interactive distance from the heteroatom ⟨B-56MI50300⟩. *exo*-Unsaturation can change this relation drastically. The cyclization rate of bromoacetate ion in dimethyl sulfoxide at 50 °C is 2.41×10^{-3}, while under the same conditions β-bromopropionate ion cyclizes at a rate of 0.6 — about 250 times more rapidly. The entropy of activation for the closure of the four-membered ring is indeed more negative than that for the three-membered lactone (-4.9 *vs.* -2.5 J K^{-1} mol^{-1}), but the enthalpy of activation for the formation of the four-membered lactone is 74 kJ mol^{-1}, while that for the three-membered lactone is 92 kJ mol^{-1} ⟨77JA2591, 78JA550⟩ (*cf.* Section 5.01.1.2).

Cyclization of 3-bromo- or 3-chloro-amines, first employed in 1888, is still a commonly used route to azetidines and azetidinium salts ⟨64HC(19-2)885⟩. Its stereochemistry has been studied and the behavior of the epimeric tropanyl halides provides a classroom example of the stereospecificity of the displacement reaction ⟨B-77SH(1)353⟩. The γ displacement reaction is also used to make azetidin-2-ones (β-lactams), using a variety of β-amino acid derivatives, such as mixed anhydrides, acid halides, *etc.* ⟨64HC(19-2)885⟩. Treating free β-amino esters with two equivalents of Grignard reagent leads to azetedin-2-one *via* the N-anion. Good yields were obtained by using phase transfer catalysis in the reaction of *N*-alkyl-3-bromopropionamides and potassium hydroxide ⟨80H(14)467⟩ (Section 5.09.3.3.1).

The salts of β-halo acids cyclize in ionizing media to oxetan-2-ones, as do β-diazonium carboxylates ⟨64HC(19-2)787⟩. Thietanes are obtained analogously (Section 5.14.4.1.1).

Displacements forming carbon to heteroatom bonds are very important also in the synthesis of large rings. Chain length, chain flexibility and the *cis* or *trans* geometry of double bonds (if any) obviously have a strong effect on the rate of ring formation. Seven-membered rings are obtained readily (Sections 5.16.4.1; 5.17.3.1.1) but larger rings can also be obtained, such as [18]-crown-6 (Section 5.21.4.1.1).

5.03.2.2.2 Valence isomerization

Important synthetic paths to azirines and aziridines involve bond reorganization, or internal addition, of vinylnitrenes. Indeed, the vinylnitrene–azirine equilibrium has been demonstrated in the case of *trans*-2-methyl-3-phenyl-1-azirine, which at 110 °C racemizes 2000 times faster than it rearranges to 2-methylindole ⟨80CC1252⟩. Created in the Neber rearrangement or by decomposition of vinyl azides, the nitrene can cyclize to the β-carbon to give azirines (Scheme 4; Section 5.04.4.1).

Intramolecular ring closure by valence isomerization to a carbon in the 3-position is not common, but known; 1,1-di-*t*-butyl-2-nitrosoethylene, stable at room temperature, cyclizes at 220 °C to the stable 4,4-di-*t*-butyl(4*H*)oxazete ⟨75AG(E)70⟩.

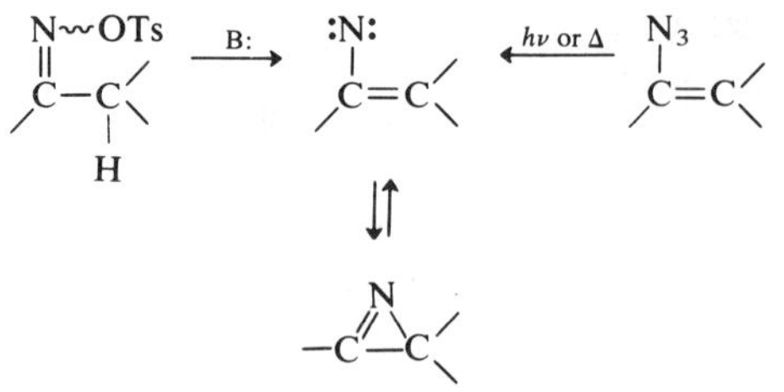

Scheme 4

5.03.2.2.3 Rearrangement of alicyclic compounds

A substantial number of molecular rearrangements lead to the incorporation of heteroatoms into carbon–carbon chains. The Curtius, Beckmann, Hofmann, Lossen and Tiemann rearrangements can accomplish this for nitrogen, the Baeyer–Villiger oxidation does it for oxygen, to name just a few. If the heteroatom in these conversions is attached to a carbocyclic ring, and rearrangement of one ring segment to the heteroatom is effected, a heterocyclus of $n+1$ members is formed (Scheme 5). Especially, rearrangements of the Beckmann type are commonly used to make large nitrogen-containing rings, such as azepinones (Section 5.16.4.1.1). Analogously, the Baeyer–Villiger oxidation converts cyclohexanone to oxetan-1-one (Section 5.17.3.1.1).

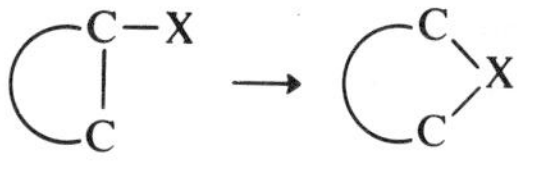

Scheme 5

The advantage of starting with a ring of $n-1$ members lies in the nature of the rearrangements, which proceed through cyclic transition states, so that the system never becomes open-chain — the carbon–carbon bond is broken only while the carbon–heteroatom bond is being made.

While these rearrangements are used most often to prepare large rings, it should be noted that the expansion of cyclopropane derivatives to azetidines is also practical (Scheme 6; Section 5.09.3.3.3.a).

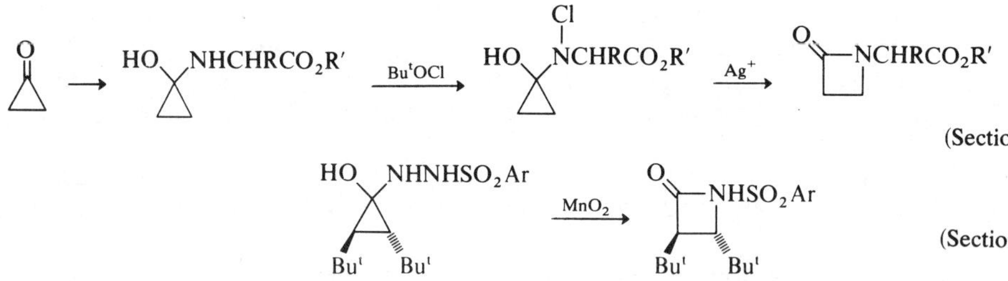

(Section 5.09.3.3.3)

(Section 5.09.3.3.3)

Scheme 6

5.03.2.2.4 Radical reactions

Synthesis of heterocycles by forming C—X bonds by radical reactions is not a generally applicable method, and seems not to be useful for making small rings. However, the attack of thiol radicals on double bonds can be a practical synthetic route, such as in the conversion of 1-hexene-7-thiol to thiepane (Section 5.17.3.3.1).

5.03.2.3 Formation of One Bond between Two Carbon Atoms

5.03.2.3.1 Displacement reactions

Ring closure by making a C—C bond is a common synthetic method for rings of all sizes. Scheme 7 gives a few examples of displacement reactions. Other examples are found in the preparation of benzazepines (Section 5.16.4.1.2) and oxepins (Section 5.17.3.2.1).

PhCHBrSOCHBrPh $\xrightarrow{Et_3N}$ (Section 5.06.4.1.2)

$Ar_2\bar{C}$–S–CAr_2Br → (Section 5.06.4.1.2)

(Section 5.21.4.1.2)

Scheme 7

5.03.2.3.2 Rearrangements

Many rearrangements forming C—C bonds have been applied to the preparation of heterocyclics. The Cope rearrangement is prominent, and an example is shown in Scheme 8. The staring material usually most accessible is an alkene, which is converted to the required cyclopropane at some stage before the rearrangement step.

N=CHAr, N=CHAr $\xrightarrow{125\,°C}$ (Section 5.18.2.3)

Scheme 8

5.03.2.3.3 Radical combination

Making just one C—C bond by radical combination is not a prominent method for preparing heterocyclics, except for those cases in which the radical pair arises from a cleavage or extrusion reaction of a heterocyclic starting material. Such cases are treated below (Section 5.03.5.2.2).

5.03.3 SYNTHESIS FROM NONHETEROCYCLIC COMPOUNDS BY MAKING TWO BONDS

5.03.3.1 From Fragments Providing [1 + *n*] Ring Atoms

5.03.3.1.1 By initial electrophilic attack

Attack on multiple bonds by electrophiles is very common in the synthesis of three-membered heterocycles. Formation of the two new bonds can be simultaneous or in discrete steps, and examples of the same type of net conversion can have large differences in their detailed mechanisms. Alkene epoxidations (Section 5.05.4.2.2) provide good examples. Peracids commonly react stereospecifically — the formation of both bonds overlaps in time, and is followed by loss of carboxylate ion or carboxylic acid. Several proposals have been put forth for the detailed structure of the transition state, but one single such model might not be applicable to all cases. An *ab initio* calculation has been done on the transition state ⟨78JA743⟩. Hydrogen bonding by groups attached to the alkene component steers the peracid to one or the other face of the alkenic double bond, in competition with solvent hydrogen bonding and nonbonding interactions in the activated complex ⟨*e.g.* 77JCS(P1)572⟩.

Transition metal-catalyzed epoxidations, by peracids or peroxides, are complex and diverse in their reaction mechanisms (Section 5.05.4.2.2) ⟨77MI50300⟩. However, most advantageous conversions are possible using metal complexes. The use of *t*-butyl hydroperoxide with titanium tetraisopropoxide in the presence of tartrates gave asymmetric epoxides of 90–95% optical purity ⟨80JA5974⟩.

Peracid oxidation of imines is the most general synthesis of oxaziranes (Section 5.08.4.1.1). Other peroxides and metal catalysis have also been employed.

Electrophilic nitrogen reagents attack alkenes and other carbon π-systems. Nitrenes, R—N, add to alkenes stereospecifically if the nitrenes are in the singlet state, and nonstereospecifically when they are in the triplet state ⟨B-70MI50300, B-78MI50300, B-81MI50300⟩. Aziridines are produced from alkenes (Section 5.04.4.2) and azepines from benzene derivatives (Section 5.16.4.2). Highly complex multicyclic systems have been made by intramolecular nitrene additions ⟨78HCA795, B-81MI50300⟩. Metal catalysis, so important in oxirane synthesis, is little known for the alkene–aziridine conversion, although some examples exist ⟨78CL1403, B-81MI50300⟩.

Electrophilic nitrogen compounds, such as arenesulfonyloxyamines, can convert alkenes to aziridines without the intervention of free nitrenes ⟨80CC560⟩.

Thiiranes can be made from 'nascent sulfur', such as is obtained by the thermolysis of diethyl tetrasulfide ⟨64HC(19-1)591⟩.

5.03.3.1.2 By initial nucleophilic attack

A multitude of reactions commencing with a nucleophilic displacement or addition are used to prepare heterocyclics. Many of these reactions are trivial in the sense that the nucleophilic attack produces a product well suited for ring closure by a second displacement or addition (thus creating a starting material for the kind of process dealt with in Section 5.03.2). For example, 2,3-dibromoketones with amines give 2-bromo-3-alkylaminoketones which immediately cyclize to *C*-acylaziridines ⟨64HC(19-1)535⟩. The epoxidation of alkenes by alkaline hydrogen peroxide proceeds by addition of HOO^-, abstraction of a β-proton and displacement of HO^-, resulting in (nonstereospecific) epoxidation (Section 5.05.4.2.2). Carbonyl groups are attacked by the anion of hydroxylamine-*O*-sulfonic acid, and the resulting oxide ion displaces sulfate from nitrogen to give oxaziridines (Section 5.08.4.1.2). The ylide $Ph_2\overset{+}{S}—\overset{-}{N}H$ adds stereospecifically to *E* and *Z* conjugated alkenes, and chiral sulfimides can transfer chirality to the aziridines formed ⟨80T73⟩. Important methods are based on the attack by the carbon of diazomethane on sp^2 carbons of carbonyls, thiocarbonyls and iminium cations. Loss of nitrogen from these adducts leads to the corresponding three-membered rings. Malonyl dichloride combines with primary amines to give azetidine-2,4-diones ⟨64HC(19-2)953⟩.

The two-step processes mentioned above are very suitable for making small rings, because the ring closure is favored by the proximity of the reactive ends of the primary product. When large rings are to be made that way, the initial product faces the same problems in cyclizing as do other 1,ω-ring closures. Nevertheless, the method is useful. Thiepane is obtained in 59% yield from 1,6-dibromohexane and sodium sulfide (Section 5.17.3.3.1). Carbon as well as heteroatoms can comprise the one-atom fragment, as in the 1,2-diazepine synthesis of Scheme 9 (Section 5.18.2.1).

$$(BrCH_2)(Ar)C=N-N=C(CH_2Br)(Ar) \xrightarrow[Na^+\ OEt^-]{H_2C(CO_2Et)_2} \text{3,7-diaryl-1,2-diazepine-5,5-}(CO_2Et)_2$$

Scheme 9

5.03.3.1.3 By [1+n] addition

Four-membered heterocycles can be formed by the addition of isocyanides to 1,3-dipoles ⟨80AG(E)45⟩ and by the reaction of carbon monoxide with β-haloamines, with the aid of palladium catalysis (Scheme 10) ⟨79CC699⟩.

$BuN^{+}(O^{-})=CHBu$ + CNR $\xrightarrow{\text{(i) } BF_3,\ \text{(ii) } Et_3N,\ -40\,°C}$ [oxazetidine: O, BuN, Bu, =NR]

$H_2C=C(Br)CH_2NHCH_2Ph$ $\xrightarrow{Ph_3P,\ Pd(OAc)_2,\ CO;\ Bu_3N,\ HMPA}$ [H_2C, O, PdL, N, H, CH_2Ph] $\longrightarrow$ [H_2C, O, NCH_2Ph]

L = ligand

Scheme 10

5.03.3.1.4 Radical reactions

Providing a single atom of a heterocyclus by a $1+(n-1)$ radical combination is not particularly common. Some oxirane formations from (triplet) oxygen and alkenes with metal catalysis might be of this class. Thiiranes have been prepared from diaryl thioketones and magnesium–magnesium iodide couple (Section 5.06.4.2), a reaction one might call a $2+(2-1)$ atom fragment combination (Scheme 11). The addition of triplet electron carbenes and nitrenes also belongs to this class.

$$2Ar_2C{=}S + Mg + MgI_2 \longrightarrow MgI_2 + MgS + \text{[thiirane: } Ar_2, Ar_2, S]$$

Scheme 11

5.03.3.2 From Fragments Providing [2+*n*] Ring Atoms

5.03.3.2.1 By two displacement reactions

The reactions of this class are little different conceptually from those of Section 5.03.1.2. A string of individual conversions leads to a net $[2+n]$ synthesis of heterocycles, such as in Scheme 12.

PhCH=NMe + $NCCH_2COCl$ $\longrightarrow$ PhCH(Cl)—NMe($COCH_2CN$) $\xrightarrow{Et_3N}$ [azetidinone: Ph, NMe, NC, O]

Scheme 12

5.03.3.2.2 Cycloaddition

Cycloadditions contributing two ring atoms are very important practically, especially for preparing four-membered rings. Making large rings usually involves cycloaddition to other rings, followed by ring enlargement or extrusion reactions, such as retro-Diels–Alder additions. In many cases the detailed mechanism remains uncertain. Stepwise additions can be difficult to distinguish from concerted ones. Stereochemical criteria are commonly employed, but there are conceivable cases in which stereochemistry is lost even in concerted $[2+2]$ additions ⟨78JA9⟩. Large solvent effects may be observed in stepwise and also in concerted processes (with substantial charge separation in the transition state). Nevertheless, there are many examples in which the presence of discrete intermediates has been ascertained by physical methods or by trapping. Examples are provided by the addition of ketenes to carbodiimides (Scheme 13; Section 5.09.3.3.2) ⟨B-80MI50300⟩, and by the addition of ketenimines to yneamines to give methyleneazetines (Section 5.09.4.3.1) ⟨B-78MI50301⟩.

$Ph_2C{=}C{=}O + RN{=}C{=}NR \qquad R = Pr^i$

$$\downarrow$$

$Ph_2\bar{C}{-}C{=}O$ / $RN{=}\overset{+}{C}{-}NR$ $\xrightarrow{H_2O}$ $Ph_2CHCONRCONHR$

$\downarrow$ (four-membered ring: Ph_2, O, NR, Pr^iN) ; $\xrightarrow{SO_2}$ (five-membered ring: O, NR, Ph_2, S, O_2, NR)

Scheme 13

In stepwise additions, ketenes are usually the nucleophilic component, so that such additions can be catalyzed by Lewis acids, such as the additions of trimethylsilylketenes to aldehydes, catalyzed by BF_3 (Scheme 14) ⟨79JOC733⟩. However, the roles can be reversed, such as in the addition of chlorocyanoketene to benzaldehyde ⟨79JA5435⟩.

$$Me_3SiCH{=}C{=}O + RCH{=}O{\cdot}BF_3 \longrightarrow$$ (oxetanone: Me_3Si, O, R, O) *cis* and *trans*

Scheme 14

A zwitterionic, heterocyclic intermediate was found in the dimerization of propanethial *S*-oxide ⟨80JA2490⟩.

Concerted thermal [2+2] cycloadditions forming heterocycles have been reviewed recently ⟨77AHC(21)245⟩. The Woodward–Hoffmann rules predict high activation energies for the suprafacial–suprafacial addition of two carbon–carbon double bonds, *i.e.* such additions are 'forbidden'. The suprafacial–antarafacial bond formation is allowed on these grounds, but the two components would have to approach each other with a geometry very unfavorable for the required orbital overlap. The applicability of the Woodward–Hoffmann rules to systems containing heteroatoms is limited (if often successful), and it is safer to rely on calculations of a particular system, rather than on the rules. Even in all-carbon systems, however, polar effects may result in acceptable activation energies for the doubly suprafacial [2+2] addition ⟨74AG(E)751⟩. Calculated activation energies sometimes differ from each other ⟨77JA4511⟩, but as computational methods continue to improve, they will play a greater role in predicting the course of our reactions, and their mechanisms. Importantly, Epiotis *et al.* have pointed out that in nonionic $[{}_{\pi}2_s + {}_{\pi}2_a]$ reactions, either cycloaddend may act as the antarafacial component. Stereochemical factors may dictate the antarafacial component; if not, some fraction of each of the addends may act as the antarafacial component and the stereospecificity of the reaction would be lost wholly or partially. Consequently, on the basis of LCFC calculations, Epiotis concludes that 'the stereochemical criterion for distinguishing nonionic $[2\pi + 2\pi]$ cycloaddition pathways is inadequate when applied by itself'. Nonstereospecificity thus does not by itself indicate an open chain intermediate (such as a diradical) ⟨78JA9⟩. The lower the polarity of the transition state, the greater the expected danger of stereorandomization.

The addition of diethyl azodicarboxylate to enol ethers gives diesters of 3-alkoxy-1,2-diazetidine-1,2-dicarboxylic acids. The reaction is stereospecific, the energies of activation are modest (29–46 kJ mol^{-1}) and the entropies of activation are negative (−36 to −46 J K^{-1} mol^{-1}). For *n*-butyl vinyl ether, the volume change of activation is very large and negative (−45.9 cm^3 mol^{-1}). This is explained by electrostriction of the solvent by a highly charge-separated transition state, rather than by the combination of two discrete moieties ⟨75JCS(P2)1555⟩. The volume change of activation of the [2+2] addition of *n*-butyl vinyl ether to diphenylketene is quite similar to that of the above diazetidine formation (−50.7 cm^3 mol^{-1}). In this cyclobutanone formation the rate for the *cis* vinyl ether is 180 times greater than that for the *trans* isomer, pointing to the $[{}_{\pi}2_s + {}_{\pi}2_a]$ mechanism ⟨70CC1061, 75JCS(P2)1555⟩, although the reaction has also been considered to involve a zwitterionic intermediate ⟨78CRV407⟩. This assignment was made because of the lack of stereospecificity in [2+2] carbocyclic ring formations, however.

Lately a third type of transition state has been favored for [2+2] cycloadditions forming carbocyclic and heterocyclic four-membered rings. The experimental data on the addition of diarylketenes to arylethylenes are well accommodated by the $[_{\pi}2_s + _{\pi}2_s + _{\pi}2_s]$ process proposed by Baldwin ⟨70JA4874⟩. The steric effects on the cycloaddition of allenes to ketenes also favor this mechanism ⟨76JA7698⟩.

Another interesting cycloaddition, the detailed mechanism of which is still under investigation, is the addition of singlet oxygen to alkenes producing 1,2-dioxetanes (Section 5.15.3.3.2).

Metal catalysis seems not to have been employed extensively in the preparation of four-membered heterocycles by [2+2] cycloaddition. As has been pointed out for all-carbon systems, processes which are orbital symmetry forbidden can be changed to allowed ones by making certain types of metal complexes of the systems for which the desired process is forbidden. Nonreacting ligands in such a complex may split orbitals involved in the desired reaction, changing the symmetry of the process to an allowed one. Interaction with the ligand fields of nonreacting ligands also may create activation barriers for (perhaps undesired) conversions and dictate a different reaction pattern (or render the system inert) ⟨71JA1123⟩. Exploitation of these possibilities may prove very fruitful. One example is the addition of diethyl azodicarboxylate to the tricarbonyliron complex of cyclobutadiene, which yields 1,2-bis(ethoxycarbonyl)-1,2-diazabicyclo[2.2.0]hex-4-ene (Scheme 15) ⟨78TL2469⟩.

C=C, C=C (cyclobutadiene) · $Fe(CO)_3$ + RO_2C–N=N–CO_2R ⟶ bicyclic C–C–NCO_2R / C–C–NCO_2R

Scheme 15

[2+2] Photocycloadditions are common and usually involve diradical intermediates. Most prominent are reactions of photo-excited ketones with a variety of unsaturated systems (Scheme 16) ⟨B-78MI50302, Chapter 11⟩. Both the singlet and the triplet (n, π^*) excited states of the ketones will form oxetanes with electron rich alkenes. With electron deficient alkenes only the singlet states give oxetanes. Diradicals are the immediate precursors to the oxetanes in all cases, but the diradicals are formed by different mechanisms, depending on the availability of electrons in the two components. A brief description of the process will be given here, because its use as a model for other cases might be fruitful, taking into account that [2+2] photocycloadditions can also be used to prepare large rings, using ring expansion after the addition step.

RR′C=X + abC=Ccd $\xrightarrow{h\nu}$ RR′C–X / abC–Ccd (ring) + RR′C–X / cdC–Cab (ring)

X = O, S, NR

Scheme 16

The initial bond formation between the $n \rightarrow \pi^*$ excited carbonyl compound and an alkene can occur by interaction of the half-filled n-orbital of the [>CO]* with the π-system of the alkene, in a sense transferring a π-electron to the n-orbital and making a bond between an alkene carbon and the carbonyl oxygen. In this process (common for electron rich olefins) the plane formed by the alkene carbons and their four substituents is perpendicular to the plane of the carbonyl groups and its two substituents (Figure 1). In the

O–C· / C·–C· ⟶ O–C / C–C

Figure 1

intermediate 1,4-diradical, rotation about single bonds can destroy the original geometric relation of the substituents on the alkene. This occurs to a much greater extent in triplet diradicals, because their life span is longer (a spin inversion has to precede ring closure) and the interaction between the atoms bearing the radical electrons is weaker, thus the looser arrangement is less subject to steric hindrance interfering with rotation about single bonds. When the alkene is unsymmetrically substituted, the more stable 1,4-diradical is formed preferentially, and ring closure gives an oxetane mixture in which the corresponding isomer is more abundant than that arising from the less stable diradical.

Alkenes predominantly substituted with electron withdrawing groups react with both singlet and triplet $n \rightarrow \pi^*$ excited carbonyl compounds, but only the singlet reactions lead to oxetane, while the triplet is merely quenched. The oxetanes produced are often formed with high stereospecificity, but isomerization of the alkene occurs in the triplet quenching process, so that the steric integrity of the starting material is destroyed. Thus, for practical purposes it is necessary to use the alkene component in large excess if stereospecificity is desired. For an example, see Scheme 17 ⟨B-78MI50303⟩.

S_1 $(n \rightarrow \pi^*)$

stereospecific addition, through singlet diradical

T_1 $(n \rightarrow \pi^*)$

S_0

triplet alkene diradical

+ some *cis* isomerized alkene

Scheme 17

Many [2+2] photocycloadditions have not been assigned a definitive mechanism, but they serve well as synthetic methods. Thiones add vinyl ethers to give thietanes in very good yields (Section 5.14.4.1.2), and interesting wavelength–stereochemistry relations were found in the photoaddition of 2-adamantanone to dicyanoethylene (Section 5.14.4.1.2). Diheterocyclobutanes can also be prepared by [2+2] photocycloadditions (Section 5.13.3.3).

The *electrolysis* of adamantylideneadamantane solutions affords the radical cation, which can add molecular (triplet) oxygen to give the peroxide radical anion, which can react with adamantylideneadamantane to give the 1,4-diradical and another molecule of adamantylideneadamantane radical cation. The latter reacts with oxygen, to continue the chain of the reaction, while the former cyclizes to the corresponding 1,2-dioxetane (Scheme 18) ⟨81JA2098⟩.

ad=ad − e^- → ad$\overset{+}{-}$$\dot{}$ad $\xrightarrow{^3O_2}$ ·O—O / ad$\overset{+}{-}$ad $\xrightarrow{ad=ad}$

ad$\overset{+}{-}$$\dot{}$ad + ·O—O / ·ad—ad → O—O / ad—ad

↓

chain

Scheme 18

5.03.4 SYNTHESIS FROM NONHETEROCYCLIC COMPOUNDS BY MAKING THREE BONDS

5.03.4.1 From Fragments Providing [1+1+*n*] Ring Atoms

Heterocyclics of a ring size $n+2$ can be made by successively transferring two fragments, contributing one ring member each. For example, two methylene moieties can be transferred to ketones to make oxetanes in high yields (Scheme 19) ⟨79JA6135⟩. Arenesulfonylisothiocyanates add to *t*-butylisocyanide to give tris-iminothietanes (Scheme 20) ⟨79CC160⟩.

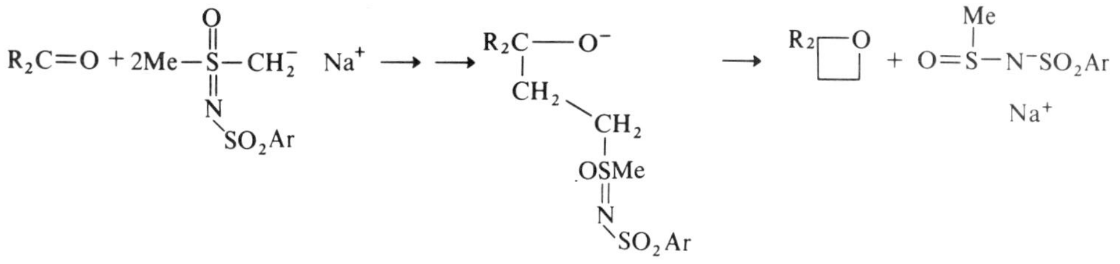

Scheme 19

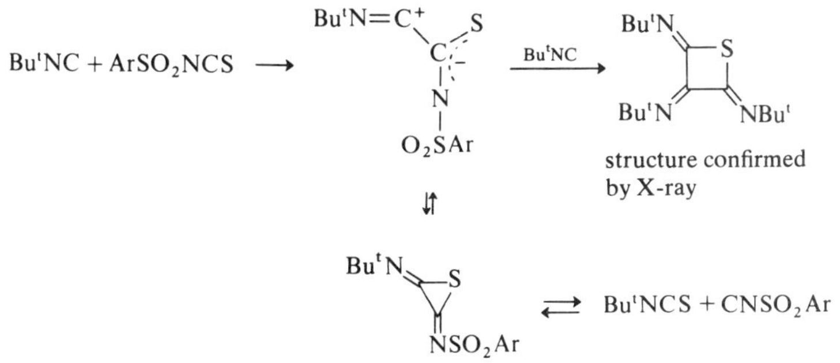

Scheme 20

5.03.5 SYNTHESIS FROM HETEROCYCLIC COMPOUNDS

The preparation of heterocycles from heterocycles is of great practical and theoretical interest. Van der Plas' book provides a wealth of pertinent information ⟨B-73MI50300⟩. A systematic treatment of the topic might well be a mechanistic taxonomer's happy dream, but it is more like a nightmare for an author. Observed conversions involve a large number of reactions, arranged in many different sequences, and virtually all possible combinations of ring sizes for starting materials and products, $n \rightarrow m$. Beyond this, a classification would have to describe the members of a reaction sequence by their ring size or degree of unsaturation, or both. Thus, the conversion of oxiranes to thiiranes, by thiocyanate ion, (Section 5.6.4.3) is a $3 \rightarrow 3$ ring transformation and also a sequence of two subsequent nucleophilic attacks, followed by a rearrangement, and a third nucleophilic attack. The sequence also involves the sequence of a three-membered ring, an open chain molecule, a five-membered ring, an open chain, and finally a three-membered ring. A cheerful taxonomer might thus classify the conversion as a $3r^1$, No^1, N5r, Ro^2, $N3r^2$ sequence. No such symbolism will be used in the following.

5.03.5.1 Isomerization without Change of Ring Size

Changes of ring atom sequence without change of ring size commonly occur in five- and six-membered heterocycles, but are comparatively rare with small and large rings. The latter groups, after ring opening, usually prefer to cyclize to five- or six-membered rings (examples are found in Chapter 5.18). Geometric isomerization of small rings, notably aziridines, occurs by thermal or photoinduced ring opening (to diradicals or ylides), followed by rotation about a single bond and ring closure.

Structural isomerization of three-membered rings with exocyclic unsaturation (Scheme 21) has been found with aziridineimines (Section 5.04.3.13) and diaziridineimines (Section 5.08.3.2.4). This involves an acyclic intermediate, while large ring systems prefer to isomerize through bicyclic ones. The 1,2- to 1,3-diazepine conversion is an example (Scheme 22) (Section 5.18.2.2).

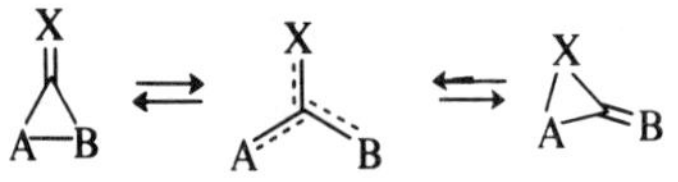

Scheme 21

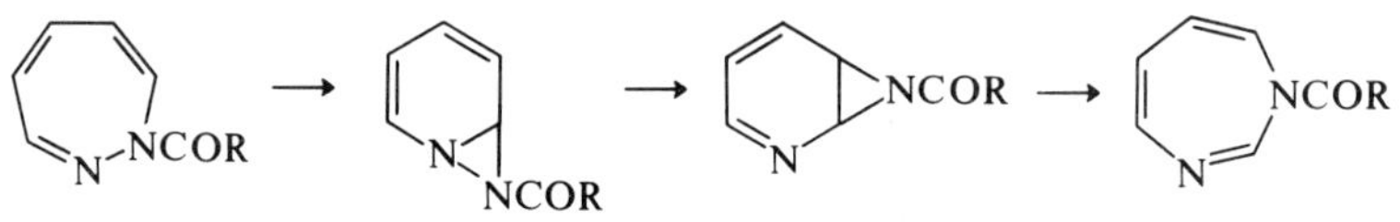

Scheme 22

5.03.5.2 Isomerization with Ring Contraction

The conversion of small rings to smaller ones, without loss, is not common. 3-Chloroazetidine isomerizes reversibly to 2-chloromethylaziridine (Section 5.09.2.2.5). Flash vacuum pyrolysis can convert isoxazoles to azirines (Section 5.04.4.3). More common is the isomerization of medium-sized, *i.e.* five- or six-membered rings, *e.g.* certain succinimides (Scheme 23) ⟨81JOC27⟩ to azetidinediones, or bicyclic 1,2-dioxetanes to bis-oxiranes (Section 5.05.4.3.2).

Large rings isomerize to bicyclic systems in a variety of ways. Examples are found in Sections 5.07.5, 5.16.2.4, 6.16.3.2.1, 5.17.1.2 and 5.17.1.4.

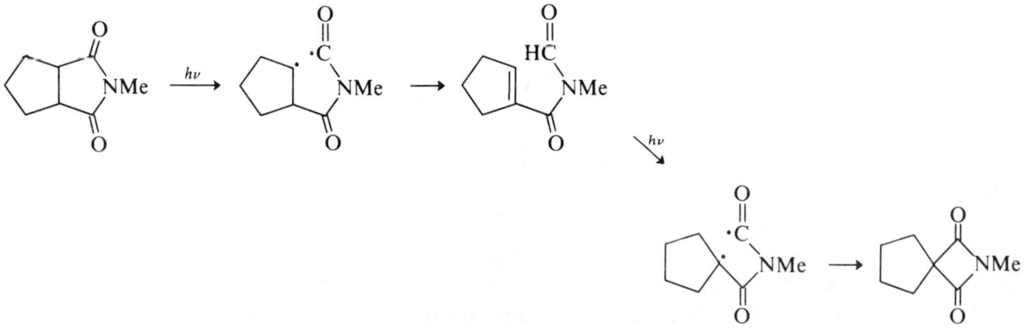

Scheme 23

5.03.5.3 Isomerization with Ring Expansion

2-Chlorocarbonylaziridines can undergo ring expansion by intramolecular nucleophilic displacement of chloride, followed by re-attack and formation of the less strained four-membered system (Scheme 24; Section 5.09.3.3.3). Much more general are ring expansions of cyclic conjugated systems with a partially exocyclic ylide function, such as 1-pyridinium imines. Incorporation of the exocyclic half of the ylide function expands the ring by one member (Scheme 25) (numerous examples in Sections 5.18.2.1, 5.18.2.2, 5.18.5.1, 5.18.5.2 and 5.18.6.2). Other ring expanding isomerizations rely on the Cope rearrangement, *e.g.* the 2,3-divinylaziridine to 2,3-dihydro-1*H*-azepine conversion (Section 5.16.4.2.1). More complex is the isomerization of 2-oxa-3-azabicyclo[3.2.0]hept-3-enes to 1,3-oxazepines (Section 5.18.5.2).

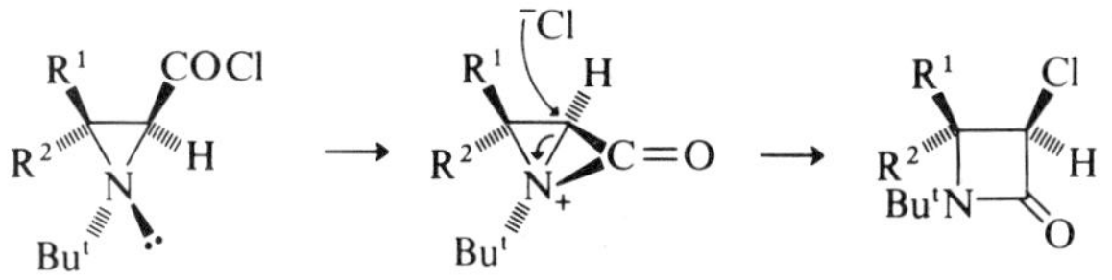

Scheme 24

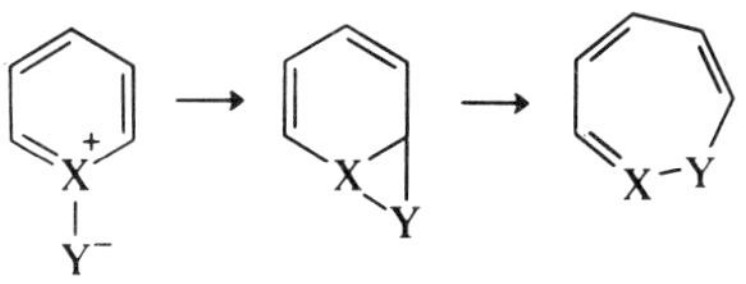

Scheme 25

5.03.5.4 Fragmentation of Heterocyclics

The loss of one or two (or sometimes more) ring members from heterocyclics, concerted with or followed by formation of a new ring, is a highly versatile method for heterocyclic synthesis. Loss of N_2, CO, CO_2, S, SO, SO_2, $H_2C{=}CH_2$, *etc.* is common. Diradical or dipolar intermediates are often encountered, and valence isomerization before the actual fragmentation is characteristic for some systems.

5.03.5.4.1 Extrusion of one ring atom

The loss of CO, S, SO or SO_2 by thermolysis or photolysis has been used to make three- and four-membered rings. Scheme 26 gives examples.

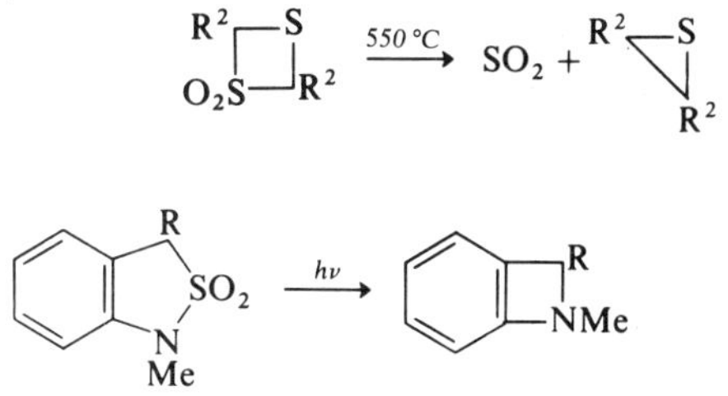

(Section 5.06.4.4)

⟨80CC471⟩

Scheme 26

5.03.5.4.2 Extrusion of two ring atoms

The loss of two ring atoms is the most common class in the reactions considered here, and nitrogen is the most common leaving group. Five-membered nitrogen heterocycles can be converted to a multitude of three-membered rings ⟨for a review, see 77AG(E)835⟩. The conversion of Δ^1-1,2,3-triazolines to aziridines has been studied thoroughly (Section 5.04.4.3 and Chapter 4.11). A whole family of reaction mechanisms operates in extensions of the triazoline decomposition to five-membered rings of higher unsaturation, five-membered rings with exocyclic unsaturation, and larger rings. Examples are shown in Scheme 27. Nitrogen extrusion has been used to make fragile molecules — 2-thiirene has been obtained by matrix photolysis of 1,2,3-thiadiazole (Section 5.06.4.5).

Carbon dioxide is readily extruded from five-membered rings, by thermolysis or photolysis (Scheme 28).

There seem to be few limits to the kind of fragment extruded (Scheme 29), or to the nature of reactions preceding the fragmentation step. An illustrative example is the formation of a zwitterionic isomer before loss of ethylene (Scheme 30; Section 5.14.4.2.4) ⟨70JOC584, B-73MI50301⟩.

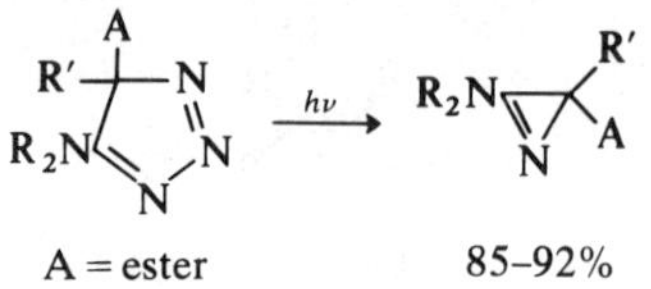

A = ester 85–92% ⟨80CC940⟩

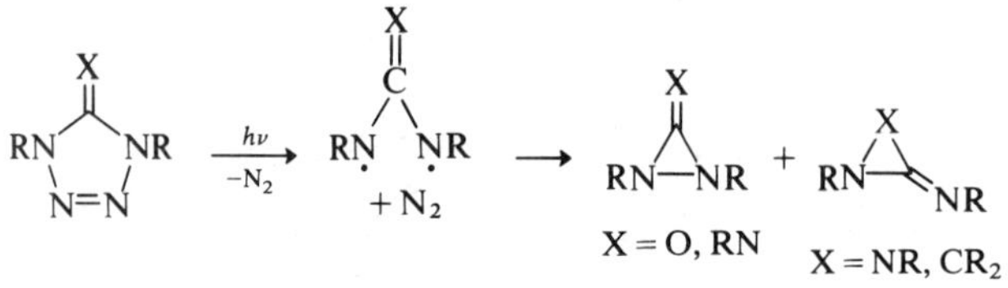

$X = O$, RN $X = NR$, CR_2 ⟨80AG(E)49⟩

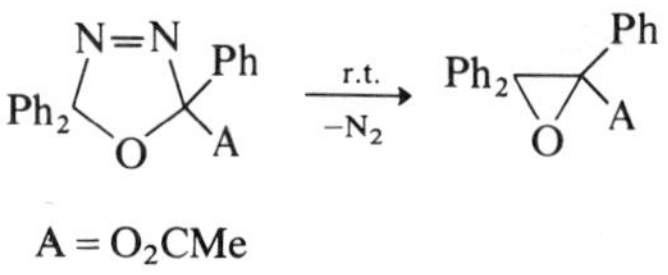

$A = O_2CMe$ (Section 5.04.4.3.1)

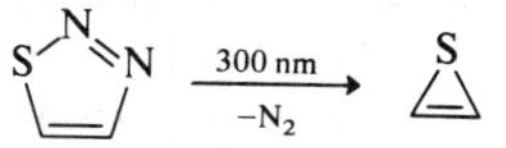

(Section 5.06.4.5)

Scheme 27

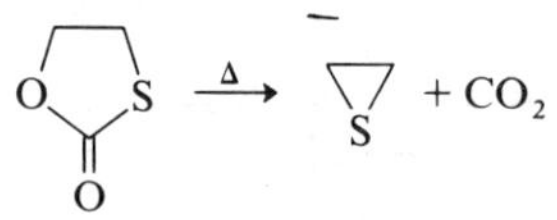

(Section 5.06.4.5)

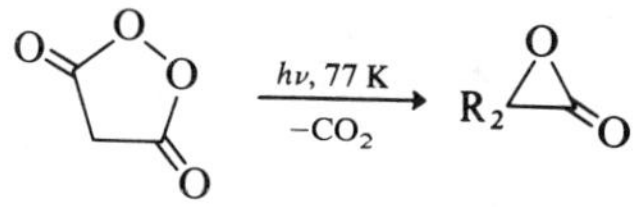

⟨72JA1365⟩

Scheme 28

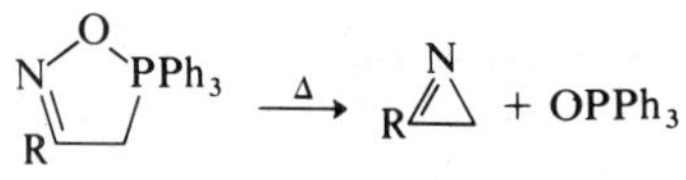

⟨79JOC3861⟩

Scheme 29

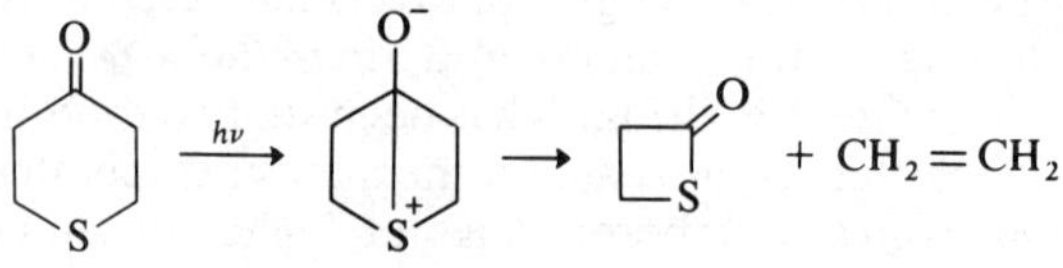

Scheme 30

5.03.5.4.3 Loss of exocyclic groups or parts thereof

Loss of ring substituents or parts thereof can initiate changes in the ring size and skeleton. The Wolff rearrangement can induce ring contraction ⟨B-73MI50302⟩, and an intriguing decomposition of 4-azido-2-pyrrolinones gave azetidin-2-one (Scheme 31) ⟨81JA1769⟩.

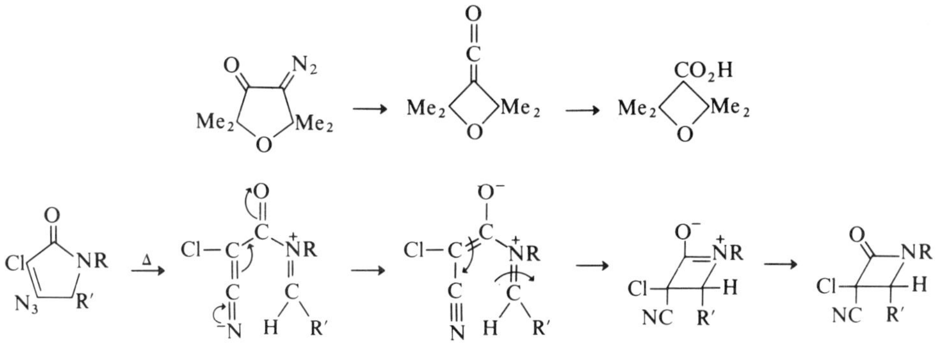

electrocyclic ring closure

Scheme 31

5.03.5.5 Intermolecular Reactions

5.03.5.5.1 Electrophilic attack

Anions of heterocyclics may attack heterocumulenes to set up systems which can incorporate new atom sequences into the ring by nucleophilic attack on a ring site (such as a carbonyl group). Scheme 32 gives an example ⟨80AG(E)466⟩.

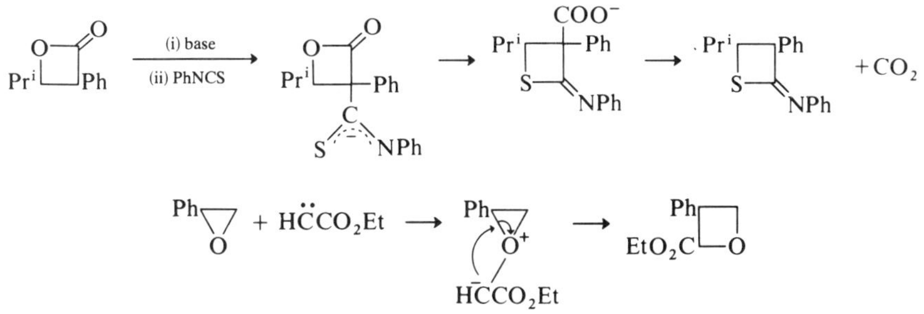

(i) base

(ii) PhNCS

$+CO_2$

Scheme 32

5.03.5.5.2 Nucleophilic attack

Since ring opening of small rings by nucleophiles is very facile, it is possible to generate, by using complex nucleophiles, intermediates which can close a new ring, incorporating parts of the nucleophile. The classic prototype of these reactions is the conversion of oxiranes to thiiranes by thiocyanate ion (Scheme 33; Section 5.06.4.3). Inversion at both ring carbons makes the reaction stereospecific with respect to the E/Z relation of the substituents on the oxirane carbons.

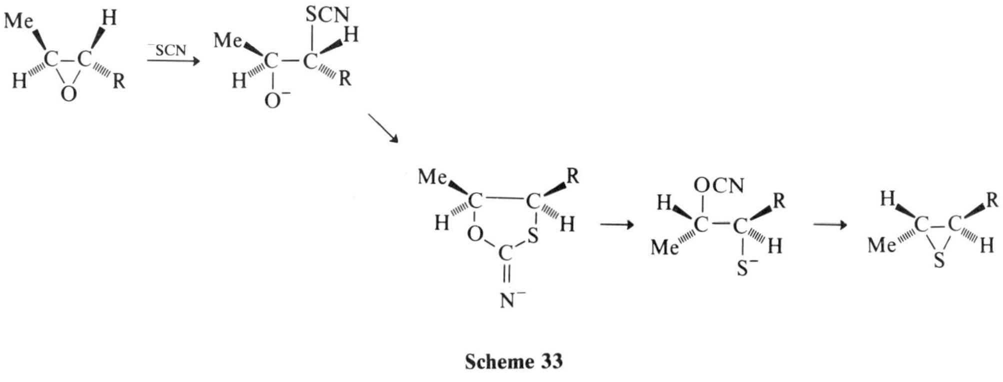

^{-}SCN

Scheme 33

5.03.5.5.3 Oxidation or reduction

Oxidation and reduction can initiate changes leading to heterocycle–heterocycle conversions. The reaction of tetraphenylfuran with singlet oxygen (Scheme 34) ⟨B-73MI50303⟩ and that of isoxazoles with LAH (Scheme 35) are examples.

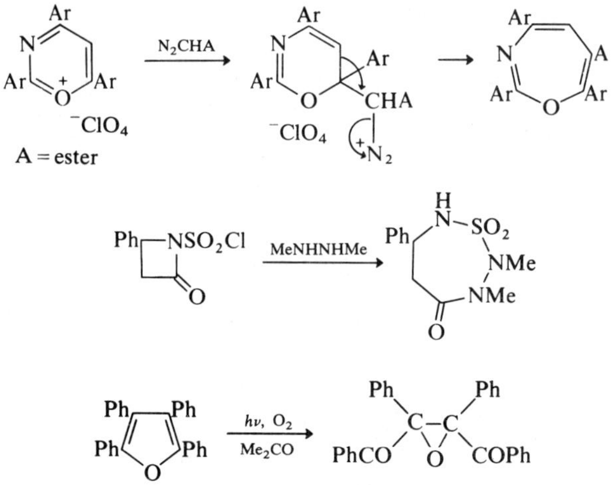

A = ester

N_2CHA

MeNHNHMe

$h\nu$, O_2 / Me_2CO

Scheme 34

$LiAlH_4$ / THF

$LiAlH_4$

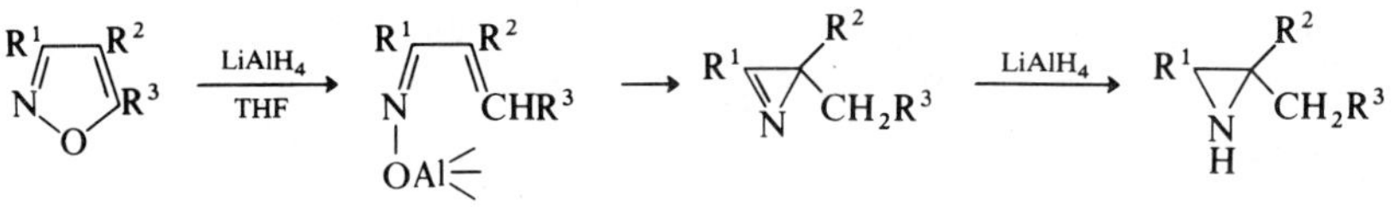

Scheme 35

5.04

Aziridines, Azirines and Fused-ring Derivatives

A. PADWA and A. D. WOOLHOUSE
Emory University, Atlanta

5.04.1 INTRODUCTION

Azirines and their dihydro derivatives, aziridines, can be regarded as representatives of the first and most simple of all heterocyclic systems, one which is characterized by the presence of two carbon atoms and one nitrogen atom in a three-membered ring. Whilst numerous members of the aziridine (**1**) and 1-azirine (**2**) ring systems are known and have been fully characterized, derivatives of the 2-azirine ring system (**3**) are known only as transient intermediates. Interest in these nitrogen-containing small rings is due to the general influence of ring strain upon chemical reactivity, to the degree to which the 2-azirine ring, for example, is destabilized by conjugation of the nitrogen lone pair electrons with the π-bond, and to the potential of derivatives of these compounds to act as precursors to more elaborate heterocyclic molecules. The stabilities and overall profiles of chemical reactivity of these heterocycles are attributable not only to the combined effects of bond shortening and angle compression, but also to the presence of the electron rich nitrogen atom. With the 2-azirines, cyclic delocalization of the lone pair electrons is believed to destabilize the ring to an extent which precludes isolation but not detection of the 4π-electron containing antiaromatic ring system.

(1)  **(2)** 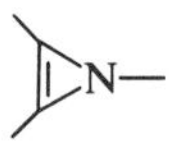 **(3)**

5.04.2 STRUCTURE

5.04.2.1 Theoretical Considerations

The structure and relative stabilities of the parent heterocycles have been the subjects of considerable theoretical attention. Recent *ab initio* MO calculations ⟨73MI50403⟩ have shown that for 1-azirine (**2**), in which an overall C_s symmetry is assumed, a C—C bond shorter and a C—N bond longer than those commonly encountered in acyclic molecules were predicted. Experimentally, these predictions have been substantiated from an X-ray crystallographic structure determination reported for a highly substituted derivative ⟨80CC98⟩. For the 2-azirine, in which both C_s symmetry and non-planarity of the nitrogen substituents were considered, the C—N bond is found to be slightly longer and the C=C bond shorter than that in cyclopropene (1.28 Å). The degree of non-planarity about nitrogen is predicted to be *ca.* 72°, a value which is significantly larger than that calculated for the aziridine ring. It has been suggested that as a planar species, the 2-azirine ring is unstable because of the antiaromatic character which would result from complete delocalization of its 4π-electrons. Simple Hückel calculations with enamines ($DE = 0.30\beta$) and 2-azirine ($DE = 0.00\beta$) also predict destabilization resulting from the change from acyclic to cyclic conjugation ⟨68JA2875⟩. The difference between the calculated barriers to nitrogen inversion in 2-azirine (146.7 kJ mol^{-1}) and aziridine (64.8 kJ mol^{-1}) also supports the concept of an energetically unfavorable planar molecule ⟨B-70MI50400, 69MI50402⟩. In fact, ground state energy calculations predict 1-azirine to be 169.3 kJ mol^{-1} more stable than 2-azirine. The calculated values of the heats of hydrogenation, often taken as a measure of destabilization due to antiaromaticity, decrease from *ca.* 270 kJ mol^{-1} to *ca.* 180 kJ mol^{-1} on going from a planar to a non-planar configuration ⟨73CC688⟩.

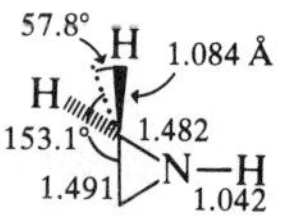

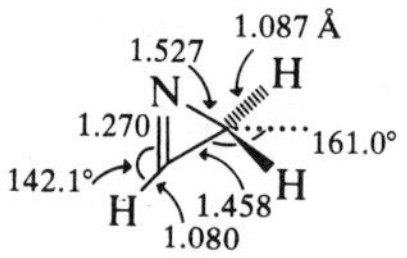

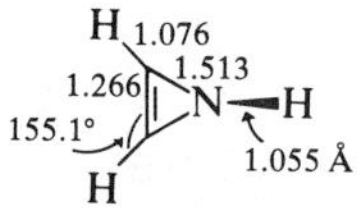

(1) **(2)** **(3)** **(4)**

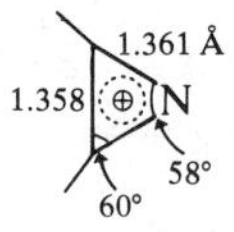

For the aziridine ring, calculated structures ⟨73MI50403⟩ are generally in good agreement with those derived experimentally from microwave spectral data ⟨71JST(10)385⟩. Constraining the nitrogen into a three-membered ring has the effect of increasing the deviation from planarity from 66.4° (NH_3) to 69.7° ⟨53JPC564⟩. The azirinyl cation (**4**), for which C_{2v} symmetry was assumed, is predicted to have extensive π-electron delocalization which suggests that this species might be a stable 2π-electron aromatic ion ⟨74JOC373⟩.

5.04.2.2 Bond Lengths, Bond Angles, Ring Strain

The dimensions of monocyclic aziridine and 1-azirine rings have been determined by single crystal X-ray diffraction ⟨72PMH(5)1⟩ and, for aziridine, also by electron diffraction ⟨61BCJ369⟩, microwave and photoelectron spectroscopy ⟨74PMH(6)1, 53⟩. All of the methods demonstrate a pronounced C—C bond shortening (1.48 Å) and considerable C—N bond lengthening (1.49 Å) when compared to normal open chain C—C (1.54 Å) and C—N (1.46 Å) bond lengths. Endocyclic angles are all close to 60° and the geometry at nitrogen is essentially pyramidal. Bicyclic aziridines with a common C—C bond, *e.g.* (**5**), follow general C—C and C—N bond lengthening trends to 1.54 Å and 1.50 Å ⟨68JHC343, 70JHC755⟩, respectively, as the fused carbocycle increases from cyclopentyl to cyclotridecyl. Aziridines such as (**6**) possessing exocyclic unsaturation are characterized by an even greater strain component than in the parent heterocycle ⟨80AG(E)276⟩. The π-resonance stabilization normally ascribed to amides would be expected to be inhibited somewhat by the non-planar configuration about the ring nitrogen. As yet, no precise structural parameters for fused bicyclic aziridines possessing a common C—N bond are available. Crystallographic data for the 1-azirines (**7**) ⟨80CC98⟩ and (**8**) ⟨78JOC57⟩ reveal a lopsided triangular configuration, the asymmetry of which is apparently more accentuated in (**7**) than in the palladium π-complex (**8**). The C=N bonds suffer no significant shortening when compared to normal acyclic imine bonds (1.237–1.281 Å) ⟨71HCA1255⟩. Accommodation of the double bond in the three-membered ring causes the C—N bond to stretch to unusually long distances (*e.g.* 1.598 Å in **7**), a fact which is reflected in the tendency of 1-azirines to react by preferential scission of the C—N bond in thermal reactions. Fused 1-azirines have been synthesized but no structural data are available.

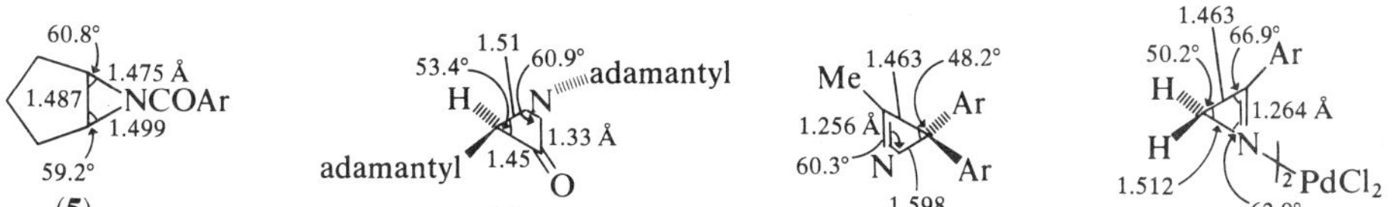

(**5**) (**6**) (**7**) (**8**)

The strain energy associated with both saturated and unsaturated three-membered rings is due principally to deformation of normal bond angles between the atoms of the ring. The total strain energy, of course, may be somewhat higher than that of the unsubstituted ring because of contributions from interactions of substituents with the ring and with other substituents. The replacement of a methylene residue of cyclopropane by a nitrogen atom causes an insignificant change in the conventional ring strain energy (CRSE); values of 115–116 kJ mol^{-1} for cyclopropane and 112–113 kJ mol^{-1} for aziridine are standard ⟨74PMH199⟩. The strain energy associated with the 1-azirine ring has been described simply as being in excess of 170 kJ mol^{-1} ⟨B-76MI50400⟩. However, the ring might be reasonably expected to parallel cyclopropene, for which a ring strain value of 228 kJ mol^{-1} has been reported ⟨70JA2377⟩. Addition across the imine bond readily reduces this value by 109 kJ mol^{-1} ⟨79ACR310⟩. The presence of ring strain within small ring compounds results in the formation of bent bonds which are high in *p*-character. This phenomenon can be determined quantitatively from a knowledge of NMR coupling constants. The *s*-character of the endocyclic aziridine bonds has been determined from $J(^{15}N–^{13}C)$ values to be very low (20%) and indicative of sp^4 hybridization ⟨79JOC2041⟩. Lone pair hybridization has been estimated similarly to be $sp^{2.2}$ (31% *s*) and $sp^{2.3}$ (30% *s*) for the exocyclic bonds in the parent aziridine ⟨75JA4137⟩. For the 1-azirine ring, the exocyclic methylene bonds have been estimated to possess *ca.* 37% *s*-character and the exocyclic methine bond to possess *ca.* 49% *s*-character ⟨74OMR(6)483⟩. These data also infer that the hybridization of the nitrogen atom approaches *sp* in character ⟨77OMR(9)559⟩.

5.04.2.3 Dipole Moments

MO methods have been used to calculate dipole moments of each of the three ring systems ⟨73MI50403, B-70MI50400⟩. Calculated values for aziridine are somewhat higher (2.09–2.40 D) than the known experimental value (1.89 D). Dipole moment studies on a few simple aziridines have led to the determination of the preferred conformation of *N*-arylaziridines in solution and in the vapor state ⟨71JCS(C)2104, 66DOK(169)839⟩. For the 1-azirine system, no values have been determined experimentally, but values of 2.40–2.56 D for 1-azirine and 2.50–2.51 D for 2-azirine have been calculated ⟨73MI50403⟩.

5.04.2.4 Electronic Absorption Spectroscopy

Simple aziridines are optically transparent in the UV region of the electromagnetic spectrum ⟨B-69MI50401⟩. In more highly substituted aziridines, such as the 2-aroyl-3-arylaziridines (**9**), there is an interaction between the carbonyl and aryl ring substituents which is transmitted through the ring and which manifests itself in the appearance of longer wavelength absorption maxima and greater molar absorptivities for the *trans* isomers than for the *cis* isomers ⟨60JA4241⟩. These differences, depicted with *cis*- and *trans*-aziridines (**9**), are of use for distinguishing pairs of stereoisomers and are, in fact, observable in both the $\pi \rightarrow \pi^*$ (3–5 nm) and $n \rightarrow \pi^*$ (*ca.* 25 nm) transitions ⟨66JOC1244, 75JA2822⟩. This interaction has been attributed to the existence of a more favorable orbital overlap between the carbonyl π-bond and the bent *p*-type σ-bonds in the *trans* rather than in the *cis* isomer.

H, Ph, H, COPh, N, C_6H_{11} (*cis*) H, H, Ph, COPh, N, C_6H_{11} (*trans*)

λ_{max}	ε	λ_{max}	ε
247 $\pi \rightarrow \pi^*$	12 000	253 $\pi \rightarrow \pi^*$	14 600
327 $n \rightarrow \pi^*$	150	349 $n \rightarrow \pi^*$	297

(**9**)

1-Azirines bearing alkyl substitution at C-2 display only weak absorption ($\varepsilon \approx 100$) at *ca.* 230 nm ⟨62JOC3557⟩. Electronic absorption spectra of substituted and unsubstituted 2-arylazirines generally exhibit intense ($\varepsilon \approx 20\,000$) absorption maxima at *ca.* 245 nm together with a weaker inflection (*ca.* 285 nm, $\varepsilon \approx 1000$) on the longer wavelength side of the principal absorption band ⟨76ACR371⟩. This latter band is attributable to an $n \rightarrow \pi^*$ transition as a result of a shift in wavelength maxima and a change in extinction coefficient as a function of solvent polarity ⟨B-66MI50400⟩.

5.04.2.5 Infrared Absorption Spectroscopy

Considerable attention has been devoted to the assignment of the fundamental vibrations of simple, low molecular weight aziridines. Absorptions specifically attributable to the various modes of ring deformation have been assigned in the 1230–1250 and 850 cm^{-1} regions ⟨B-69MI50401⟩. However, these are of little real diagnostic value in more highly substituted systems. 1-Azirines, on the other hand, exhibit absorptions very characteristic of the unsaturated ring. Alkyl and aryl substitution at C-2 give rise to intense absorption (C=N stretching) at 1740 and 1770 cm^{-1}, respectively. Azirines bearing no substituent at C-2 exhibit C=N absorption in the region commonly observed (*ca.* 1650 cm^{-1}) for those of normal acyclic imines ⟨68JA2869, 70T453, B-75MI50400⟩.

5.04.2.6 Nuclear Magnetic Resonance Spectroscopy

NMR spectroscopy has found wide application to the study of structure and stereochemistry of aziridines, which unlike 1-azirines possess a conformationally flexible nitrogen atom. Whilst ^1H, ^{13}C, ^{14}N and ^{15}N nuclei and the consequent spin–spin interactions which derive from them have all been applied to the general problem of structure elucidation, the most useful method involves ^1H NMR spectroscopy ⟨69MI50400⟩. The use of internuclear coupling constants between certain pairs of nuclei to determine exo- and endo-cyclic bond orders has been outlined in the preceding sections. Vicinal proton–proton coupling constants of 5–9 Hz for *cis* related protons and of 2–6 Hz for *trans* related protons generally enable the stereochemistry at C-2 and C-3 to be determined in many aziridines. Extensive tables of vicinal and geminal *J*-values show that the average magnitudes follow the order: $J_{cis} = 6.4$ Hz, $J_{trans} = 3.3$ Hz and $J_{gem} = 1.4$ Hz ⟨B-69MI50401, B-73NMR135⟩. Protons attached to the parent aziridine ring carbons resonate at *ca.* δ 1.5 p.p.m. and substituents almost invariably have a deshielding influence upon them. Hydrogen atoms attached to nitrogen tend to be at low δ values (*ca.* 1.0 p.p.m.), even lower than those observed for acyclic secondary amines and larger cyclic amines. It is frequently possible to assign ring stereochemistry on the basis of chemical shift data. Resonances due to the *cis* isomers appear at higher fields (*ca.* 3.2 p.p.m.) than do those of the *trans* isomers (*ca.* 3.6 p.p.m.) ⟨65TL4369, 75JA2822⟩. Compilations of ^1H NMR spectral data for representative bicyclic aziridines having either common C—C or C—N bonds have been published ⟨71JOC31,

B-73NMR135⟩. The magnetic anisotropy associated with the saturated three-membered ring also influences the substituents in the vicinity of the ring ⟨66TL2921⟩. By way of illustration, the bridge protons of the *exo*-fused tricycles (**10**) and (**11**) *anti* to the aziridine ring, H_a, are shielded by 0.4–0.7 p.p.m. whilst those *syn* to the ring, H_s, are only marginally affected relative to those of the corresponding alkenes ⟨64CJC926⟩. A detailed theoretical rationale for the origin and nature of this anisotropy effect (magnetic anisotropy and electric field effects) has been published ⟨67JA6605⟩. An anisotropic effect associated with the 1-azirine ring, which results in a deshielding effect of 2.0–2.5 p.p.m., has been demonstrated from

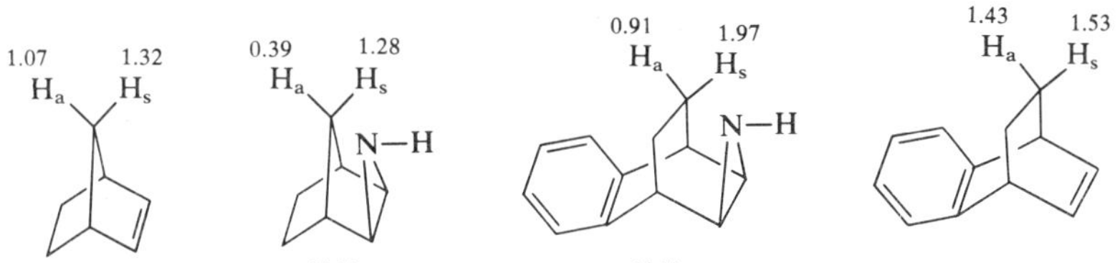

(**10**) (**11**)

^{1}H NMR spectra of azirines which are unsubstituted at C-2 ⟨69TL4073⟩. The imine protons are commonly observed at *ca.* 10 p.p.m. whereas those of normal acyclic aldimines are observed in the vicinity of 8 p.p.m.; imine (**12**), for example, displays a singlet at 7.63 p.p.m. ⟨62CJC882⟩. A similar but less pronounced shielding effect of *ca.* 0.4 p.p.m. has been observed for the protons at C-3; compare, for example, the resonances assigned to the azirine (**13**) and aziridine (**14**) ⟨66JOC3907⟩. Applications of ^{13}C NMR spectroscopy to the study of aziridines and azirines are becoming more frequent. Coupling constants [$J(^{13}C-^1H)$ and $J(^{13}C-^{15}N)$] have been used as probes of the extent of hybridization of endo- and exo-cyclic orbitals ⟨79JOC2041, 75JA4137⟩.

$MeCH{=}NPr^i$
(**12**) (**13**) 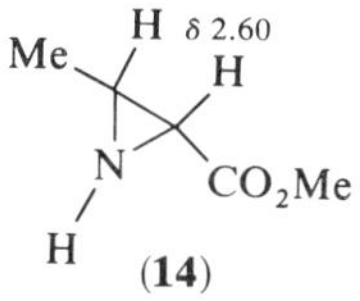 (**14**)

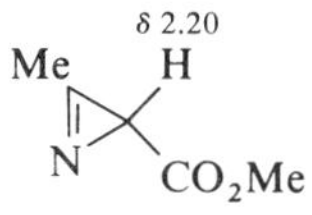

Full chemical shift data recorded for a large number of simple mono- and bi-cyclic aziridines, unsubstituted at nitrogen, have led to the derivation of a general empirical formula for predicting stereochemical details within related groups of aziridines ⟨76OMR(8)79, 90⟩. The parent aziridine ring carbons are observed at 18.2 p.p.m. and typically in the 30–50 p.p.m. region for those bearing alkyl and/or aryl substituents. Resonances due to the imine carbon of the 1-azirine system fall within the range 160–170 p.p.m. customarily observed for acyclic aldimines and ketimines, and those due to the methylene carbon fall within the 19–45 p.p.m. range ⟨74OMR(6)483, 77OMR(9)559⟩. ^{14}N and ^{15}N NMR spectroscopy has only recently been applied to certain aspects of organic structure elucidation ⟨B-73MI50400⟩. In view of the fact that ^{15}N nuclei are much less naturally abundant and sensitive, and because ^{1}H and ^{13}C NMR spectroscopy already provide so much structural information, attention is likely to focus more, in the future, upon phenomena associated with polyaza compounds and biologically important systems ⟨77JA714⟩. Nonetheless, the effects of ring substitution upon chemical shifts (relative to ammonia) of the aziridinyl nitrogen have been determined by ^{15}N NMR. *N*-Alkylation has the effect of shifting the resonances to lower field, *e.g.* NH, −8.5; NEt, 16.4; NPr^i, 30.2; NBu^t, 33.5 p.p.m. *C*-Alkylation has the reverse effect, *e.g.* CMe, 10.5; CEt, 7.9; CPr^i, 7.3; CBu^t, 3.4 p.p.m. ⟨80JOC1277⟩. In general, ^{15}N shifts parallel the ^{13}C shifts of the corresponding carbon atoms in the same compounds.

5.04.2.7 Nitrogen Inversion

In contrast to most cyclic and acyclic amines, the bonding constraints of the aziridine ring have the effect of retarding nitrogen inversion rates to extents where they are measurable by NMR spectroscopy. In fact, dynamic NMR spectroscopy has enabled the relatively easy determination of rates of nitrogen inversion through the measurement of resonance coalescence temperatures ⟨B-73NMR135⟩. In spite of the fact that determinations of thermodynamic parameters by coalescence temperature methods are known ⟨B-73NMR135, 71MI50400⟩ to be inferior to other methods (*e.g.* total line shape analysis ⟨69JA642⟩), it is qualitatively evident that substituents bound to nitrogen have pronounced effects upon the barriers to nitrogen inversion. Those able to delocalize the lone pair electrons, which occupy a *p*-orbital in the transition state, lower the barrier to (or increase the rate of) inversion ⟨67JA352⟩. An increase in substituent bulk has the same effect since destabilizing non-bonded interactions in the ground state configuration are partially relieved in the transition state ⟨70AG(E)400⟩. Halogens or other substituents bearing unshared electron pairs have the effect of dramatically raising the barriers to inversion to an extent which often enables the physical separation and isolation of invertomers, *e.g.* (**15**) ⟨68AG(E)224⟩ and (**16**) ⟨73TL619⟩, at ambient temperature ⟨68JA506, 71JOC230⟩.

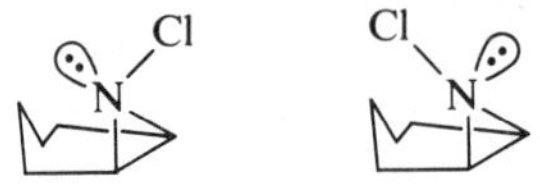

(**15**) 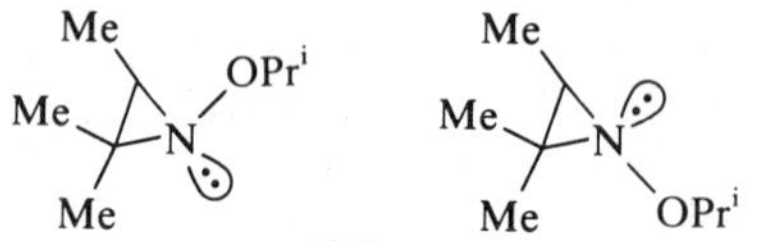 (**16**)

Whilst comparisons of literature data are sometimes difficult to make as a result of experimental differences and marked solvent effects, several energy barriers ($\Delta G^{\neq}$) which reflect the different substituent effects are summarized in Table 1.

Table 1 Barriers to Nitrogen Inversion in Substituted Aziridines[a]

Aziridine	$\Delta G^{\neq}$ (kJ mol^{-1})	*T* (°C)
 NEt	43.1	−65
 NCO_2Me	29.7	−138
 NPh	49.0	−40
 NBu^t	71.2	52
 NEt	81.2	108
 NCl	99.5	29
 NCl	111.8	80

[a] 71MI50400.

5.04.2.8 Mass Spectrometry

A systematic study of the behavior of simple monocyclic aziridines towards electron impact has shown that certain fragmentation pathways are preferred ⟨70OMS1259⟩. Spectra of *C*-alkyl-substituted aziridines, *e.g.* (**17**), show a prominent β-cleavage process which can be rationalized as resulting from concerted fragmentation and ring collapse. α-Cleavage appears to be a significant but less important pathway. *N*-Alkyl-substituted aziridines (**18**) show ions resulting principally from α-cleavage, but sequential β-, γ- *etc.* cleavages within the side chain are also observed. The mass spectrum of *N*-phenylaziridine demonstrates a tendency to 1,2-aryl migration prior to fragmentation. The spectra recorded for a series of *cis*- and *trans*-arylaziridines have been detailed and shown to be only of marginal use in distinguishing isomeric pairs ⟨72JHC587⟩. Although there are apparently no diagnostic fragmentation pathways observed for 1-azirine derivatives, most display prominent parent ions and, in certain instances, nitrile (RCN^+) or nitrilium ($RCNH^+$) ions or ions which correspond to residues resulting from the loss of the RCN moiety from the azirine nucleus ⟨77JA1514⟩. A detailed analysis of the behavior of 2,3-diphenyl-1-azirine under electron impact shows that alternative pathways, resulting in HCN and H_2CN losses, are preferred ⟨70OMS925⟩.

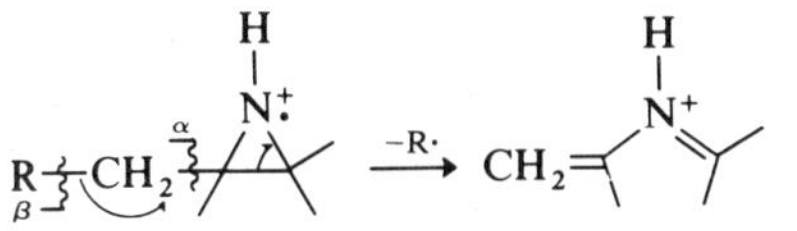

(17) 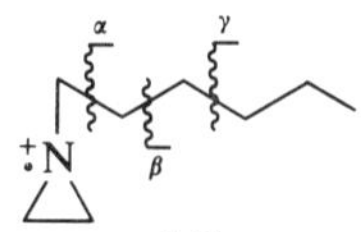 (18)

5.04.3 REACTIONS

A major and characteristic reaction of the aziridine ring is its reactivity towards a wide variety of reagents, an effect undoubtedly resulting from the necessary compression of bond angles in the three-membered ring. Thus, this system is extremely susceptible towards ring cleavage because of the favorable release of strain energy involved. For this reason, aziridines may be converted to a wide variety of functionalized compounds. Cycloaddition reactions of aziridines have also invoked considerable interest in recent years, and the synthetic applications of these reactions have attracted a great deal of attention. Conclusive evidence has been presented that the reactions proceed *via* intermediate azomethine ylides, formed from aziridines by conrotatory ring opening in a thermal process and disrotatory ring opening when photochemically generated.

The 1-azirine ring represents another reactive and versatile substrate as a result of certain inherent features within its structure. These include high ring strain, a reactive π-bond, a lone pair of electrons on the nitrogen, and the ability to undergo ring cleavage on thermal or photochemical excitation to give reactive species such as vinyl nitrenes and nitrile ylides. 1-Azirines are capable of acting in reactions as nucleophiles and electrophiles, as 2π-components in thermal cycloadditions, and as 4π-components in photochemical cycloadditions. These reactions can be regarded in general terms as involving the participation of the C=N, C—C and C—N bonds of the azirine ring.

5.04.3.1 Dipolar Cycloadditions

Cycloaddition reactions of aziridines have invoked considerable interest and the synthetic applications of these reactions have been reviewed ⟨71MI50402⟩. Conclusive evidence has been presented that the reactions proceed *via* intermediate azomethine ylides, formed from aziridines by conrotatory ring opening in a thermal process and disrotatory ring opening when photochemically generated. A number of papers dealing with thermally initiated cycloadditions of aziridines (**19**) to alkenes and alkynes have appeared ⟨71MI50402⟩. The reactions are generally stereospecific, and in the case of alkenes lead to pyrrolidines (**20**). In the absence of a reactive dipolarophile the aziridines equilibrate *via* azomethine ylide intermediates. Kinetic studies show that the rate of cycloaddition is independent of the dipolarophile concentration ⟨71JA1777⟩. A scheme involving a reversible first-order reaction of aziridine to give a small concentration of azomethine ylide was proposed. The difficulty of studying such small concentrations of intermediate azomethine ylides was overcome by generating a high concentration of these species by flash photolysis ⟨71JA1779⟩. Photolytic ring opening takes place by a disrotatory process, so that azomethine ylide (**24**) with the same configuration as starting aziridine is obtained. The thermal recyclization which follows is conrotatory to give the azomethine ylide (**23**) of opposite configuration. Half-lives for the *trans*-ylide (**23**) and the *cis*-ylide (**24**) were 5.4 and 7.8 s, respectively. The data obtained were used to construct an energy profile, and the equilibrium concentration of ylide was determined as 1 in 30 000 at 120 °C and 1 in 50×10^6 at 25 °C. The thermal cycloaddition process is not always stereospecific. Stereochemical mixtures of pyrrolidines are encountered when less reactive dipolarophiles are used since isomerization of azomethine ylides (**23**) and (**24**) can become competitive. It has also been noted that the additions of (**23**) are faster than those of (**24**). Aromatic nitrogen substituents appear more likely to give stereospecific addition than aliphatic nitrogen substituents. The energetics of these interconversions have been deduced from a series of carefully designed experiments ⟨71JA1779⟩. Activation energies of approximately 121.8 and 92.4 kJ mol^{-1} are required for the (**21**) → (**23**) and (**23**) → (**24**) interconversions, respectively.

Cycloaddition reactions of aziridines with a wide assortment of dipolarophiles have been studied. The reaction of dialkyl azodicarboxylates with the *cis*-aziridine (**27**) is stereospecific

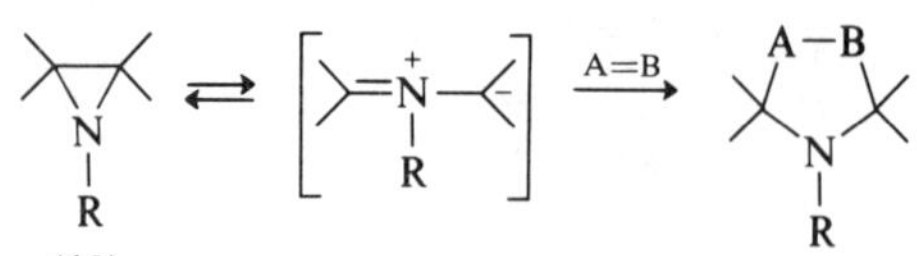

(19) (20)

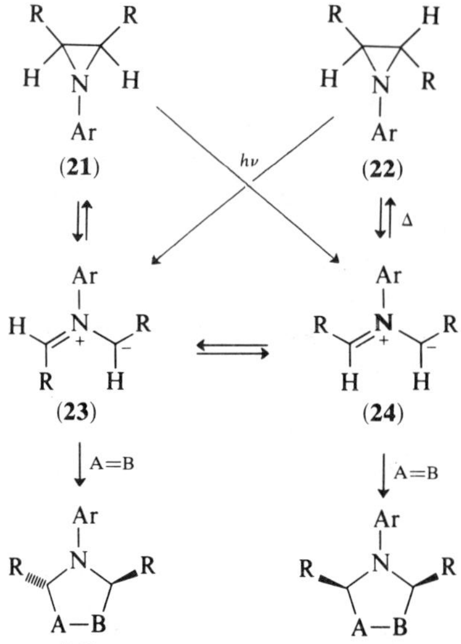

(21) (22) $h\nu$ Δ (23) (24) A=B A=B (25) (26)

to give the *trans* product (**29**). A similar addition of the *trans*-aziridine is less stereospecific, and this is explained by partial isomerization of the intermediate *cis*-azomethine ylide to the *trans* being allowed by the relative unreactivity of the dipolarophile involved ⟨71TL473⟩.

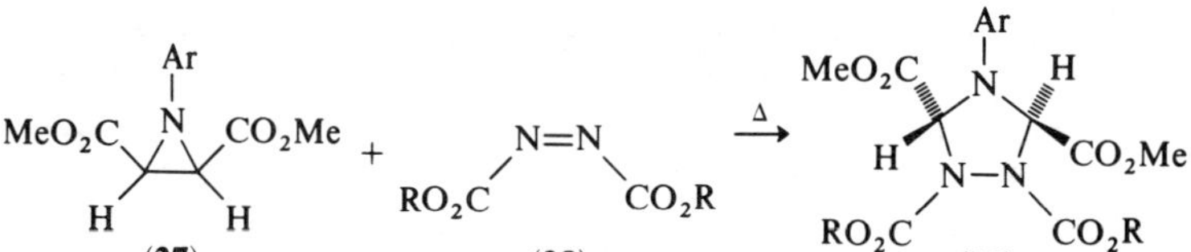

(27) (28) (29)

The high reactivity of azomethine ylides allows addition to aromatic systems ⟨71TL481⟩. For example, *trans*-aziridine (**30**) adds to phenanthrene to give the *trans*-phenanthropyrrolidine (**31**). The reversal of expected stereochemistry is again attributed to azomethine ylide interconversion being allowed by the low reactivity of the aromatic system.

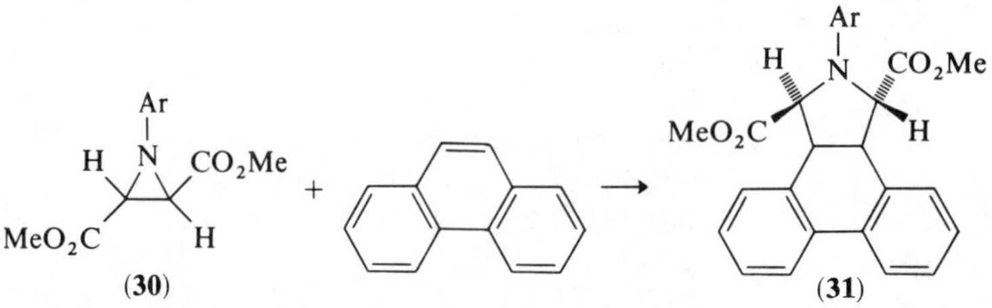

(30) (31)

2-Aroylaziridines (**32**) and aromatic aldehydes react to give oxazolidines (**33**), the stereochemistry of which suggests reaction very largely through the *trans*-azomethine ylide, irrespective of the aziridine configuration ⟨70JCS(C)2383⟩.

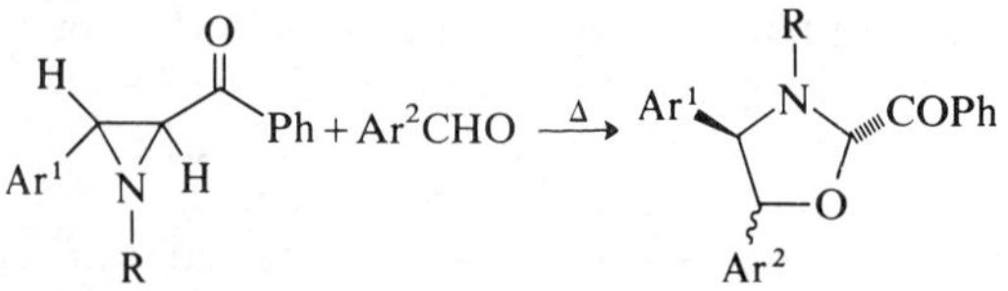

(32) (33)

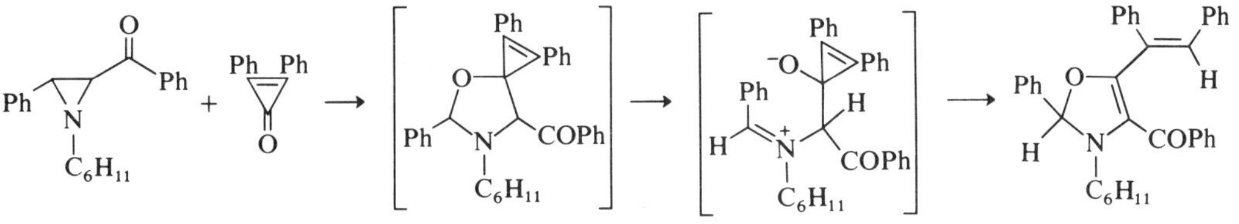

(34) (35) (36)

Cycloadditions of aziridines to diphenylcyclopropenone lead to 4-oxazolines (**36**) ⟨70CJC89⟩. A mechanism involving initial addition to the cyclopropenone carbonyl group followed by ring opening and recyclization was suggested.

The reactions of azomethine ylides with aromatic nitroso groups have also been investigated ⟨71CJC1610⟩. 1-Hydroxy-2-nitrosonaphthalene and aziridines give compound (**38**), whereas reaction with 1-nitroso-2-aminonaphthalene produces a variety of products depending on the aziridine substituents. A mechanism involving initial 1,3-dipolar addition to the aryl nitroso group was postulated to explain all these products. Cycloaddition reactions with carbon disulfide, carbon dioxide and phenyl isothiocyanate have also been studied ⟨71MI50402⟩.

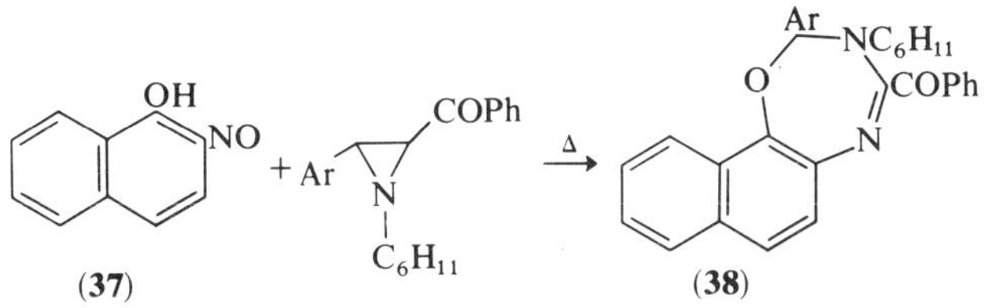

(37) (38)

5.04.3.2 Photoinduced Dipolar Cycloadditions of 1-Azirines

1-Azirines undergo irreversible ring opening on electronic excitation to give nitrile ylides (**40**) as reactive intermediates ⟨76ACR371, 77H(6)143⟩. Nitrile ylides may be classified as nitrilium betaines, a class of 1,3-dipoles containing a central nitrogen atom and a π-bond orthogonal to the 4π-allyl system. They can be intercepted with a wide variety of dipolarophiles to form five-membered heterocyclic rings, *e.g.* (**41**).

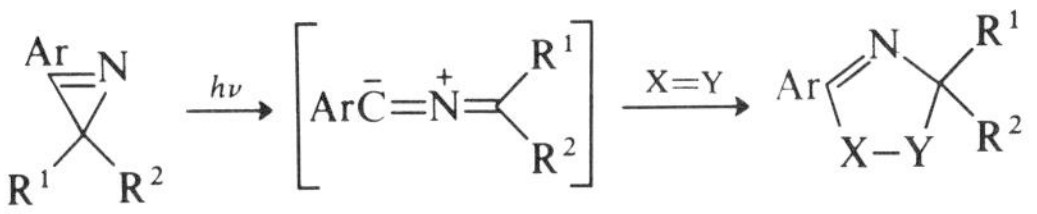

(39) (40) (41)

The photocycloaddition of arylazirines with electron deficient alkenes to produce Δ^1-pyrrolines ⟨73JA1945⟩ exhibits all the characteristics of a concerted reaction, including stereospecificity and regioselectivity. 1,3-Dipolar additions proceed *via* a 'two-plane' orientation complex where the dipole and dipolarophile approach each other in parallel planes ⟨68JOC2291⟩. For the case of diphenylazirine and methyl acrylate, two possible orientation complexes (**42** or **44**) exist. The interaction of substituent groups in the *syn* complex (**42**) can be of an attractive (π-overlap, dipole–dipole interaction) or a repulsive nature (van der Waals' strain). Both effects are negligible in the *anti* complex (**44**). The ratio of the products obtained gives insight into the interplay of steric and electronic substituent effects in the transition state of 1,3-dipolar addition and emphasizes the important role these effects have in controlling the stereochemical distribution of the products obtained.

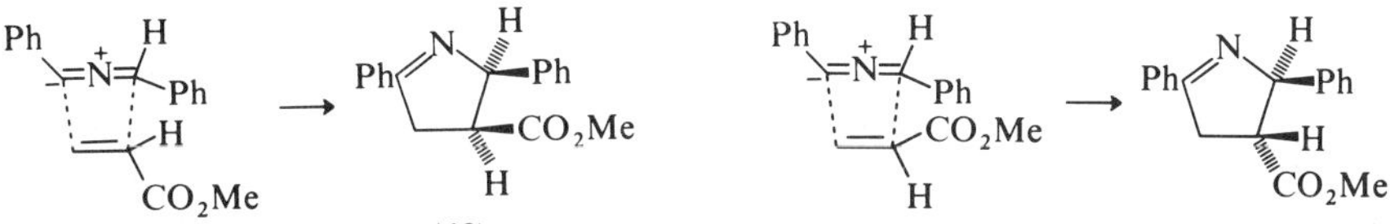

(42) *syn* (43) (44) *anti* (45)

Frontier molecular orbital theory correctly rationalizes the regioselectivity of most 1,3-dipolar cycloadditions ⟨73JA7287⟩. When nitrile ylides are used as 1,3-dipoles, the dipole highest occupied (HOMO) and dipolarophile lowest unoccupied (LUMO) orbital interaction importantly stabilizes the transition state. The favored cycloadduct is that formed by union of the atoms with the largest coefficient in the dipole HOMO and dipolarophile LUMO. An electron deficient alkene has the largest coefficient on the unsubstituted carbon in the LUMO. In order to predict regioselectivity in the photocycloaddition of arylazirines, the relative magnitudes of the coefficients in the HOMO of the nitrile ylide must be known. The photoconversion of arylazirines to alkoxyimines (**46**) indicates that in the HOMO of the nitrile ylide the electron density at the disubstituted carbon is greater than at the trisubstituted carbon atom. With this conclusion, all the regiochemical data found in the photoaddition of arylazirines with dipolarophiles can be explained.

(**39**) $\xrightarrow[\text{MeOH}]{h\nu}$ $Ar\bar{C}{=}\overset{+}{N}{=}CR^1R^2$ (**40**) → ArCH=N–C(OMe)R^1R^2 (**46**)

The photochemical addition of azirines to the carbonyl group of aldehydes, ketones, and esters is also completely regiospecific ⟨77H(6)143⟩. Besides the formation of the isomeric oxazolines (**50**) from (**39**) and ethyl cyanoformate, there is also formed the imidazole (**51**) from addition to C≡N in the expected regioselective manner. Thioesters lead to thiazolines (**52**), while isocyanates and ketenes produce heterocycles (**53**).

(**39**) + PhCHO, $h\nu$ → (**47**); + MeCOMe, $h\nu$ → (**48**); + CF_3CO_2Me, $h\nu$ → (**49**)

(**39**) + $NCCO_2Et$, $h\nu$ → (**50**) + (**51**); + $PhCS_2Me$, $h\nu$ → (**52**); + R^3–Y=C=X → (**53**)

The photocycloaddition of arylazirines with a variety of multiple bonds proceeds in high yield and provides a convenient route for the synthesis of five-membered heterocyclic rings. Some of the dipolarophiles include azodicarboxylates, acid chlorides, vinylphosphonium salts and *p*-quinones.

Irradiation of arylazirines with alkenes of low dipolarophilic activity produced no photoadduct but instead gave dimers ⟨73JA1954⟩. Originally it was reported that photolysis of phenylazirine gave azabicyclo[2.1.0]pentane ⟨68AG(E)130⟩. The dimer actually isolated was subsequently shown to be diazabicyclo[3.1.0]hexane. In the absence of an added dipolarophile the nitrile ylide generated photochemically simply adds to the ground state azirine molecule. A crossed dimerization of azirines has also been realized ⟨72CC409⟩.

Additional work showed that the dimerization of arylazirines to 1,3-diazabicyclo[3.1.0]hex-3-enes is a general reaction which is independent of the nature of the substituent groups attached to the C atom of the azirine ring. Care is required in the choice of solvent, photolysis time and substituents since the 1,3-diazabicyclohexenes are themselves photochemically labile ⟨72JA7788⟩.

Whereas the cycloaddition of arylazirines with simple alkenes produces Δ^1-pyrrolines, a rearranged isomer can be formed when the alkene and the azirine moieties are suitably arranged in the same molecule. This type of intramolecular photocycloaddition was first detected using 2-vinyl-substituted azirines ⟨75JA4682⟩. Irradiation of azirine (**54**) in benzene afforded a 2,3-disubstituted pyrrole (**55**), while thermolysis gave a 2,5-disubstituted pyrrole (**56**). Photolysis of azirine (**57**) proceeded similarly and gave 1,2-diphenylimidazole (**58**) as the exclusive photoproduct. This stands in marked contrast to the thermal reaction of (**57**) which afforded 1,3-diphenylpyrazole (**59**) as the only product.

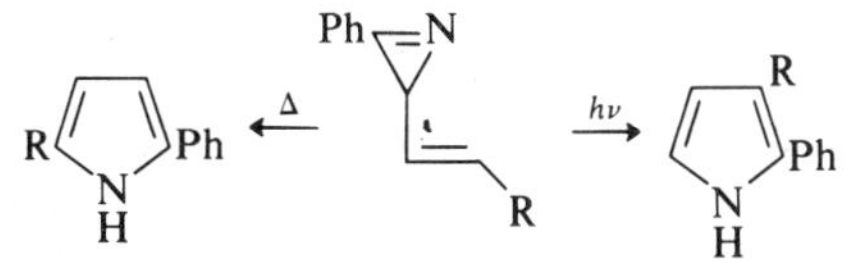

(56) (54) (55)

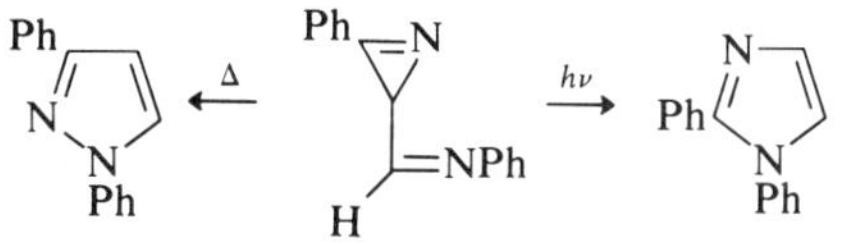

(59) (57) (58)

The evidence obtained clearly indicates that the above photorearrangements proceed by a mechanism involving a nitrile ylide intermediate since cycloadducts could be isolated when the irradiations were carried out in the presence of trapping agents. Intramolecular cycloaddition of the nitrile ylide followed by a 1,3-sigmatropic hydrogen shift of the initially formed five-membered ring readily accounts for the formation of the final product.

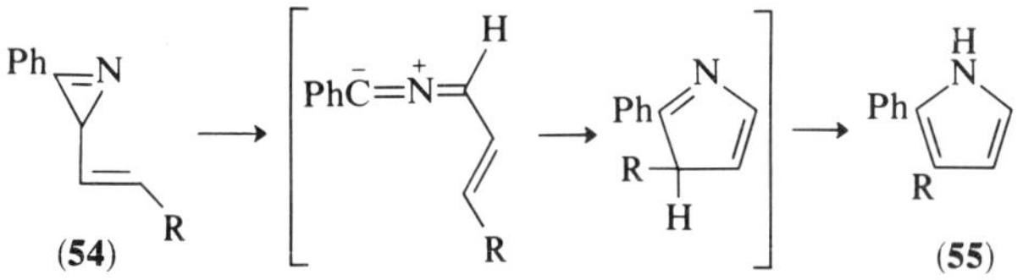

(54) (55)

The thermal transformations observed with these systems can be rationalized in terms of an equilibration of the 1-azirine with a transient vinylnitrene which subsequently rearranges to the 2,5-disubstituted pyrrole (**56**).

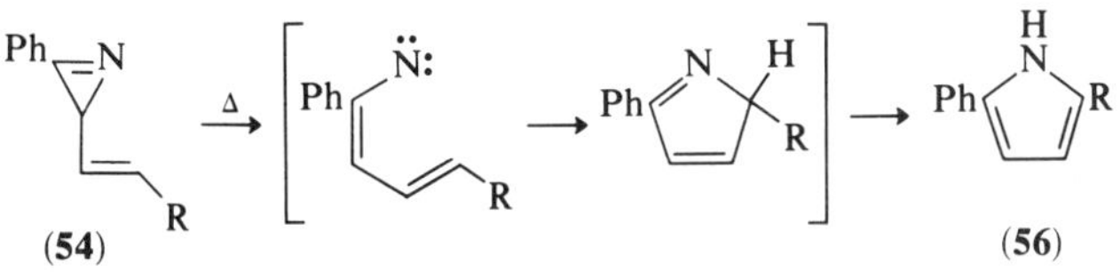

(54) (56)

In contrast to the photochemical results encountered with (**54**), the presence of a *cis*-styryl side chain in the 2-position of the azirine ring (**60**) leads to ring expansion and gives benzazepine (**61**). The 2-[2-(α- and β-naphthyl)vinyl]3-phenyl-1-azirine systems behaved similarly and proceeded with complete regiospecificity. The photolysis of the isomeric *trans*-styrylazirine followed an entirely different course and produced 2,3-diphenylpyrrole as the major product. This observation requires that opening of the azirine ring followed by intramolecular cyclization proceeds faster than isomerization about the C=C bond. The formation of (**61**) indicates that the nitrile ylide obtained from (**59**) cyclizes more easily *via* a seven-membered transition state and leads to the preferential formation of benzazepine (**61**). Cyclization of the nitrile ylide derived from the *trans* isomer to a seven-membered ring is precluded on structural grounds, and formation of 2,3-diphenylpyrrole occurs instead.

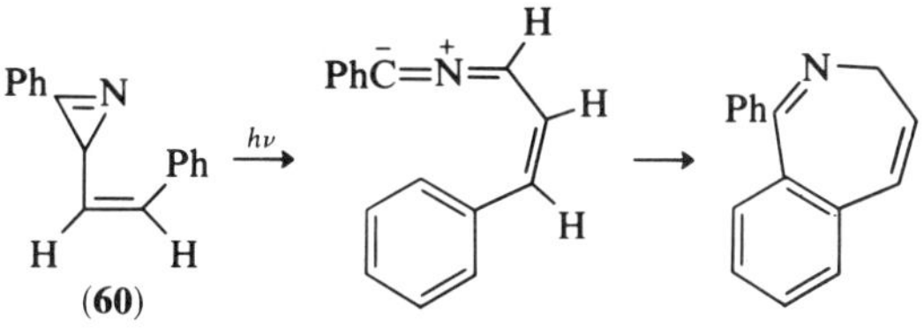

(60) (61)

The photochemical rearrangement of 3,5-diphenylisoxazole (**63**) to 2,5-diphenyloxazole (**64**) was shown to proceed *via* 3-phenyl-2-benzoyl-1-azirine (**62**) as a transient intermediate ⟨67JA6911⟩. The photobehavior of (**62**) was found to be dramatically controlled by the wavelength of the light used. With 3130 Å light, (**62**) rearranges almost quantitatively to oxazole (**64**), whereas 3340 Å light causes rearrangement to isoxazole (**63**). The formation of the isoxazole was suggested to occur *via* the $n\pi^{*3}$ state of the carbonyl chromophore. Oxazole formation was attributed to selective excitation of the $n\pi^*$ state of the azirine ring.

(63) (62) (64)

Nitrile ylides derived from the photolysis of 1-azirines have also been found to undergo a novel intramolecular 1,1-cycloaddition reaction ⟨75JA3862⟩. Irradiation of (**65**) gave a 1:1 mixture of azabicyclohexenes (**67**) and (**68**). On further irradiation (**67**) was quantitatively isomerized to (**68**). Photolysis of (**65**) in the presence of excess dimethyl acetylenedicarboxylate resulted in the 1,3-dipolar trapping of the normal nitrile ylide. Under these conditions, the formation of azabicyclohexenes (**67**) and (**68**) was entirely suppressed. The photoreaction of the closely related methyl-substituted azirine (**65b**) gave azabicyclohexene (**68b**) as the primary photoproduct. The formation of the thermodynamically less favored *endo* isomer, *i.e.* (**68b**), corresponds to a complete inversion of stereochemistry about the π-system in the cycloaddition process.

(65) (66)

a; $R^1 = R^2 = H$
b; $R^1 = H, R^2 = Me$

(68) (67)

The photochemical behavior of the isomeric 3-methyl-2-phenyl-2-allyl-1-azirine (**66**) system was also studied. Irradiation of (**66**) in cyclohexane gave a quantitative yield of azabicyclohexenes (**67**) and (**68**). Control experiments showed that (**65**) and (**66**) were not interconverted by a Cope reaction under the photolytic conditions. Photocycloaddition of (**66**) with an added dipolarophile afforded a different 1,3-dipolar cycloadduct from that obtained from (**65**). The thermodynamically less favored *endo* isomer (**68b**) was also formed as the exclusive product from the irradiation of azirine (**66b**).

(65) (68) (66) (67)

1,3-Dipolar cycloadditions proceed *via* a 'two-plane' orientation complex in which the dipole and dipolarophile approach each other in parallel planes. Inspection of molecular models of the allyl-substituted nitrile ylides indicates that the normal 'two-plane' orientation approach of the nitrile ylide and the allyl π-system is impossible as a result of the geometric restrictions imposed on the system. Consequently, the normal mode of 1,3-dipolar addition does not occur. In these systems attack by the double bond is constrained to occur perpendicular to the plane of the nitrile ylide. The LUMO, which is perpendicular to the ylide plane, is low-lying and presents a large vacancy at C-1 for attack by the more nucleophilic terminus of the neighboring double bond, without the possibility of simultaneous bonding at the C-3. This attack will generate a six-membered ring dipole which contains a secondary carbonium ion as well as an azallyl anion portion. Collapse of this new 1,3-dipole results in the formation of the observed azabicyclohexenes. Several additional examples demonstrate the generality of the intramolecular 1,1-cycloaddition reaction of unsaturated 1-azirine derivatives. Thus, photolysis of azirine (**69**) gave rise to a mixture of *endo*- and *exo*-benzobicyclohexenes (**70**) in quantitative yield ⟨76JA1048⟩. No detectable quantities of the isomeric 1,3-dipolar adduct were observed. Similarly, irradiation of azirine (**71**) resulted in an analogous 1,1-cycloaddition.

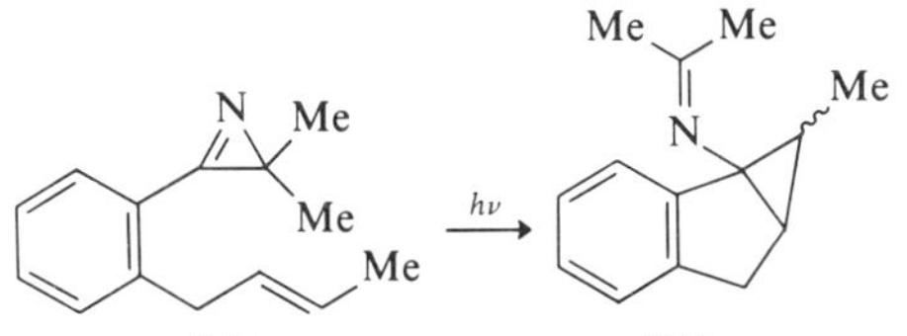

(**69**) → $h\nu$ → (**70**)

(**71**) → $h\nu$ → (**72**)

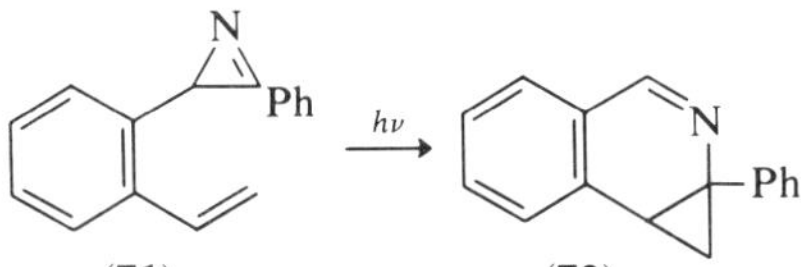

When the chain between the azirine ring and the alkene end is extended to three carbon atoms, the normal mode of 1,3-intramolecular dipolar cycloaddition occurs. For example, irradiation of azirine (**73**) gives Δ^1-pyrroline (**74**) in quantitative yield ⟨77JA1871⟩. In this case the methylene chain is sufficiently long to allow the dipole and alkenic portions to approach each other in parallel planes.

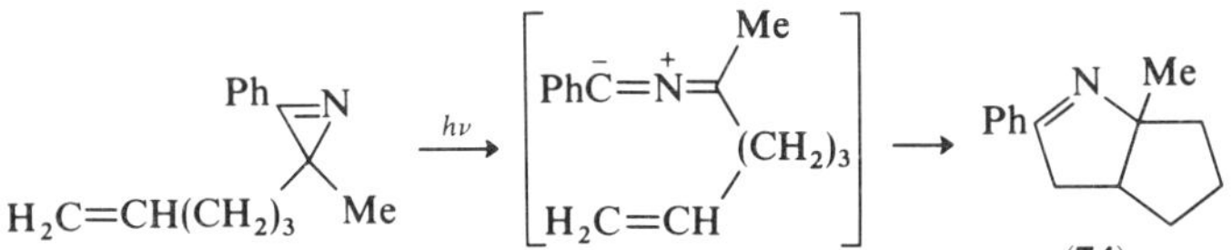

(**73**) → $h\nu$ → [$Ph\bar{C}{=}\overset{+}{N}{=}C(Me)(CH_2)_3CH{=}CH_2$] → (**74**)

5.04.3.3 [4+2] Cycloadditions

The 2π-electrons of the carbon–nitrogen double bond of 1-azirines can participate in thermal symmetry-allowed [4+2] cycloadditions with a variety of substrates such as cyclopentadienones, isobenzofurans, triazines and tetrazines ⟨71AHC(13)45⟩. Cycloadditions also occur with heterocumulenes such as ketenes, ketenimines, isocyanates and carbon disulfide. It is also possible for the 2π-electrons of 1-azirines to participate in 'ene' reactions ⟨73HCA1351⟩.

One of the first examples of cycloaddition of the 1-azirine ring system involved the thermal reaction of (**75**) with cyclopentadienone (**76**) to give 3*H*-azepine (**80**) ⟨72JOC802, 72JA8255⟩. The first step of this reaction involves a [4+2] cycloaddition to give the *endo* adduct (**77**). Cheletropic fragmentation of adduct (**77**) furnishes azanorcaradiene (**78**). This material undergoes a disrotatory electrocyclic ring opening followed by a 1,5-suprafacial hydrogen shift to give the thermodynamically most stable 3*H*-azepine ring.

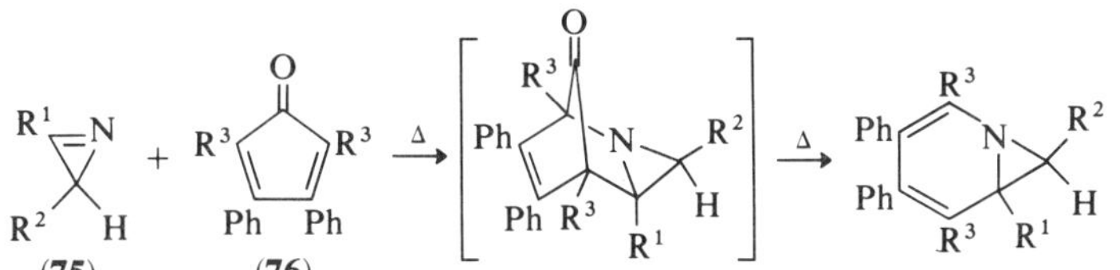

(**75**) + (**76**) → Δ → [(**77**)] → Δ → (**78**) 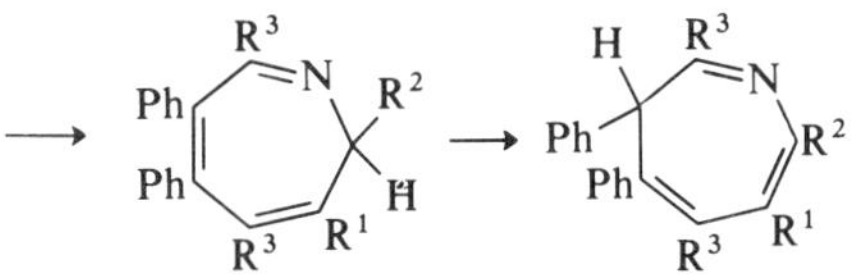 → (**79**) → (**80**)

1,3-Diphenylisobenzofuran has been reported to react with azirine (**75**) in refluxing toluene to give cycloadduct (**81**), the primary product of a [4+2] cycloaddition ⟨72JOC2508, 74JOC2031⟩.

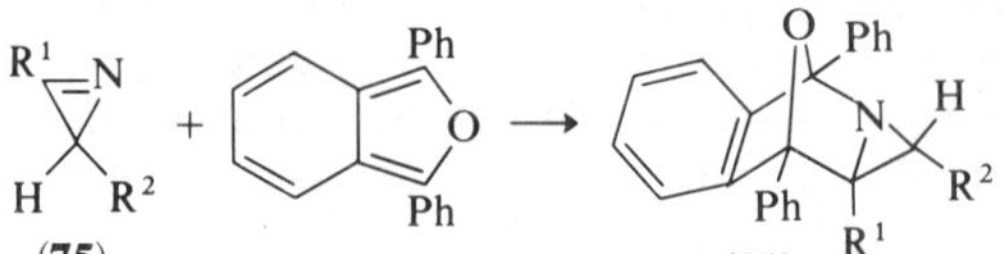

(**75**) (**81**)

A particularly interesting system where nitrogen is lost cheletropically after formation of the initial [4+2] cycloadduct involves the thermal reaction of azirines with tetrazines (**82**) ⟨74CC45, 74TL2303, 74CC782, 75JHC183⟩. A variety of heterocyclic products are produced depending on the structure of the azirine and tetrazine used and the reaction conditions.

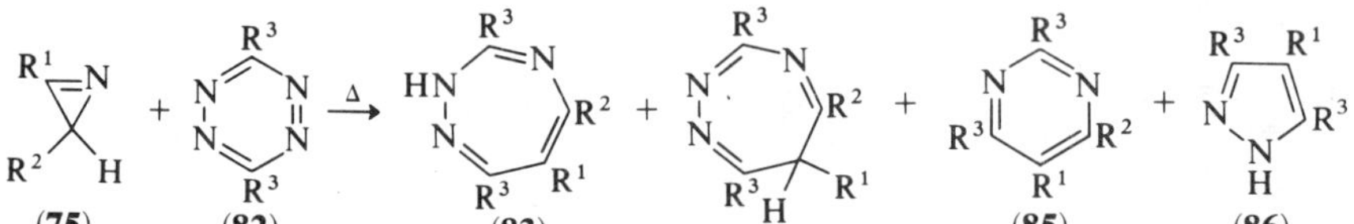

(**75**) (**82**) (**83**) (**84**) (**85**) (**86**)

The interaction of diazomethane with 1-azirines was the first example of a 1,3-dipolar cycloaddition with this ring system ⟨64JOC3049, 68JOC4316⟩. 1,3-Dipolar addition produces the triazoline adduct (**87**). This material can exist in equilibrium with its valence tautomer (**88**), and allylic azides (**89**) and (**90**) can be produced from these triazolines by ring cleavage.

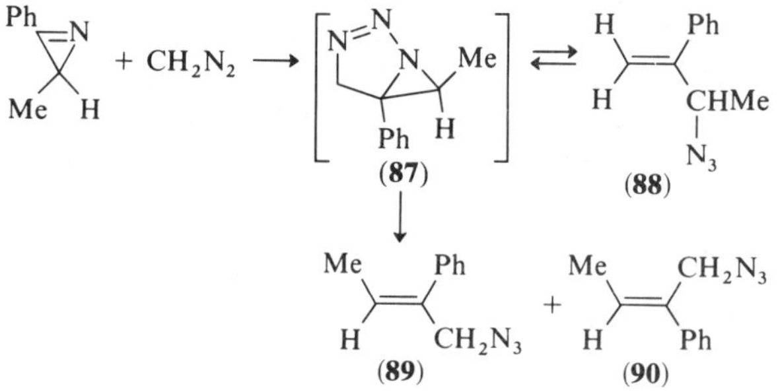

(**87**) (**88**) (**89**) (**90**)

Aziridines, *e.g.* (**91**), undergo thermal ring opening in a conrotatory manner to generate azomethine ylides. These azomethine ylides are 4π-components and can participate in [4+2] cycloadditions with 1-azirines acting as the 2π-component ⟨73HCA1351⟩.

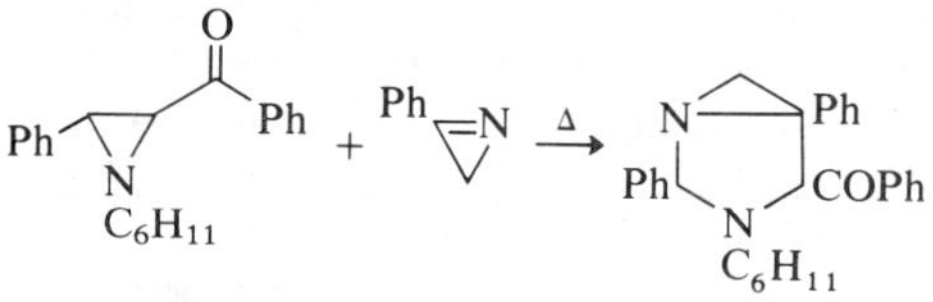

(**91**) (**92**)

The 4π-electron system of nitrile oxides can also participate in 1,3-dipolar cycloaddition with 1-azirines ⟨71TL4831⟩. For example, aromatic nitrile oxides react exothermically with 1-azirines to furnish carbodiimides in high yield. A possible mechanism for the formation of the carbodiimide assumes the initial formation of a cycloadduct from a 1,3-dipolar addition between the nitrile oxide and the azirine. Ring cleavage of the bicyclic adduct or its valence tautomer is followed by a 1,2-migration of the R group of the nitrile oxide in a Beckmann-type rearrangement to give the carbodiimide (**97**).

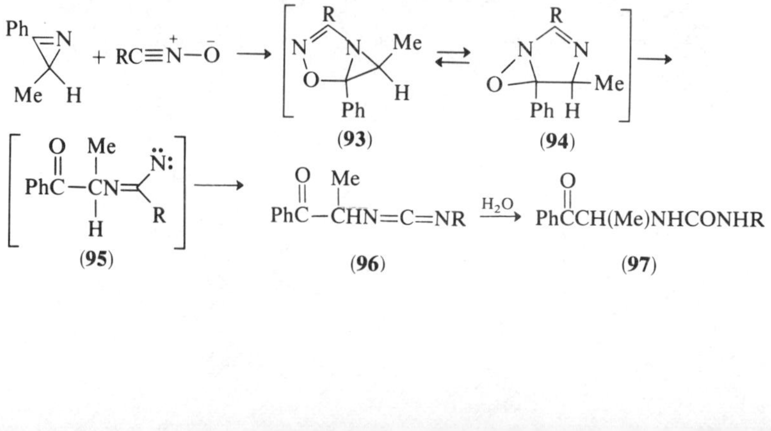

(**93**) (**94**)

(**95**) PhC(O)–CH(Me)N=C=NR (**96**) $\xrightarrow{H_2O}$ PhC(O)CH(Me)NHCONHR (**97**)

The reactions of 1-azirines with ketenes and ketenimines represent non-concerted additions and are formally different from the additions to 4π-systems of dienes and 1,3-dipolar compounds ⟨73JOC3466, 71CB2786⟩.

5.04.3.4 Photochemistry

Investigations on the photochemistry of the aziridine ring have shown that this system is exceptionally reactive under the influence of UV light ⟨B-76PH41⟩. Irradiation may lead to geometrical isomerization, rearrangement, internal hydrogen abstraction, photofragmentation or photochemical valence tautomerization. In those cases that have been fully investigated it has usually been found that the photolysis can best be described by a cleavage of the C—C bond of the aziridine ring followed by a multitude of possible second steps.

Irradiation of 1,2,3-triphenylaziridine (**98**) in various alcohols has been reported to give benzaldehyde acetals and *N*-benzylaniline ⟨68T2193⟩. Competitive fragmentation to give *N*-(benzylidene)aniline and phenylcarbene, which is trapped as the alkyl benzyl ether, also appears to occur.

(**98**) $\xrightarrow[\text{ROH}]{h\nu}$ $PhCH(OR)_2 + PhCH_2NHPh$

(**98**) $\xrightarrow[\text{ROH}]{h\nu}$ $PhCH{=}NPh + PhCH_2OR$

A pair of stereoisomeric 1,3-cycloaddition products having the 1,2,3-triphenyloctahydroisoindole skeletal structure (**100**) is formed upon photolysis of (**99**) in cyclohexene ⟨68T2193⟩.

(**99**) + cyclohexene $\xrightarrow{h\nu}$ (**100**)

(**101**) $\xrightarrow{h\nu}$ (**102**) + PhCHO

In contrast to the behavior observed for (**99**), irradiation of a methanol solution of 1-(2,4,6-trinitrophenyl)-2,3-diphenylaziridine (**101**) affords 1-hydroxy-4,6-dinitrobenzimidazole (**102**) and benzaldehyde ⟨68TL4801⟩.

The photochemical behavior of a number of related *N*-substituted diphenylaziridines has been studied ⟨72JA303⟩. In the case of *N*-cyanodiphenylaziridine, irradiation at 254 nm leads not to fragmentation but rather to clean isomerization. While both stereoisomers rearrange to (**105**) on irradiation, only the *trans* isomer undergoes detectable stereoisomerization to the *cis* isomer. Under similar photolysis conditions the stereoisomeric urethanes (**103b**) and (**104b**) undergo clean geometrical isomerization without any detectable fragmentation (photostationary state, (**103b**) : (**104b**) = 9 : 1). Prolonged irradiation of the analogous *N,N*-dimethyl amido system (**103c**) and (**104c**) in methanol resulted in the formation of (**106**), the product of methanol-assisted C—C bond scission. The data obtained indicate that the chemical fate of the electronically excited aziridine is particularly sensitive to the nature of the substituent groups on nitrogen. The various transformations were rationalized in terms of a preferential C—C cleavage of the aziridine ring.

[(**103**) $\overset{h\nu}{\rightleftharpoons}$ (**104**)] $\longrightarrow$ (**105**) + (**106**)

a; R = CN
b; R = CO_2Et
c; R = $CONMe_2$

Three basic reaction paths have been proposed to occur in the photochemistry of aziridinyl ketones, namely: (a) photodeamination ⟨67JA102⟩, (b) a photochemically induced 1,5-hydrogen shift ⟨71JA1400⟩, and (c) photochromism in rigid glasses at 77 K ⟨72JA4046⟩. To account for the products obtained from the irradiation of *trans*-*N*-benzyl-2-phenyl-3-benzoylaziridine (**107**), it was proposed that the reaction proceeds by intramolecular hydrogen transfer from carbon to the p_y orbital of the $n\pi^*$ excited state. The course of the overall photoreaction was noted to be drastically altered for the corresponding *cis* isomer. Irradiation of the *cis*-aziridine gave acetophenone and *N*-benzylbenzalimine as the major photoproducts. The excited state of the *cis*-aziridine was suggested to undergo intermolecular hydrogen abstraction followed by homolytic ring cleavage to give a radical that can lead to the observed products by a succession of hydrogen abstraction and photoelimination reactions.

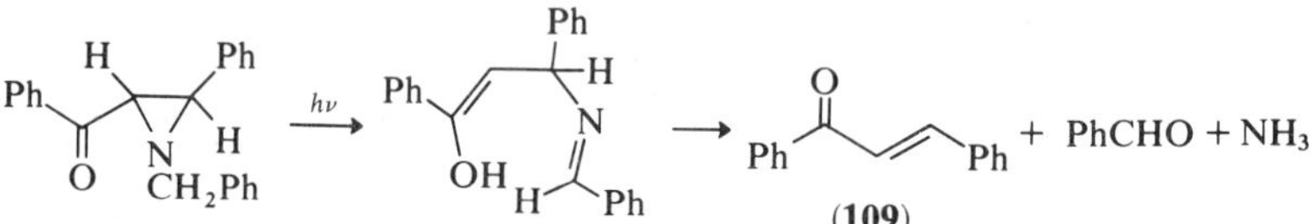

(**107**) $\xrightarrow{h\nu}$ (**108**) → (**109**) + PhCHO + NH_3

The photochemistry of *cis*- and *trans*-1-*t*-butyl-2-phenyl-3-benzoylaziridines has also been investigated ⟨71JA1400⟩. Irradiation of the *trans* isomer (**110**) in pentane afforded (**111**), (**112**), *N*-*t*-butylbenzalimine and benzaldehyde. The last two compounds were found as minor components of the reaction mixture. At least two mechanisms were proposed which are consistent with the experimental results. One path involves a C—N ring opening followed by a 1,2-hydrogen shift. Alternatively, a prior 1,5-hydrogen transfer may occur with subsequent ring opening. In order to distinguish which mechanism is operative, the photoisomerization of *trans*-1-*t*-butyl-2-phenyl-3-benzoylaziridine-2-d_1 (**113**) was examined. The results obtained with (**113**) suggest that deuterium loss occurs before the formation of the final product, which is consistent with the 1,5-hydrogen shift mechanism. Photolysis of the isomeric *cis*-aziridine gave oxazole (**111**) and *t*-butylbenzalimine as the exclusive photoproducts. The deuterium labeling experiments and the total absence of (**112**) from the irradiation of the *cis* isomer provide strong support for the 1,5-hydrogen transfer mechanism. Sensitization and emission studies showed that these reactions are derived from the $n\pi^*$ triplet state. The failure to quench the reactions implies that the rearrangement of the ketone is too rapid for diffusion of the excited state to the quencher molecule.

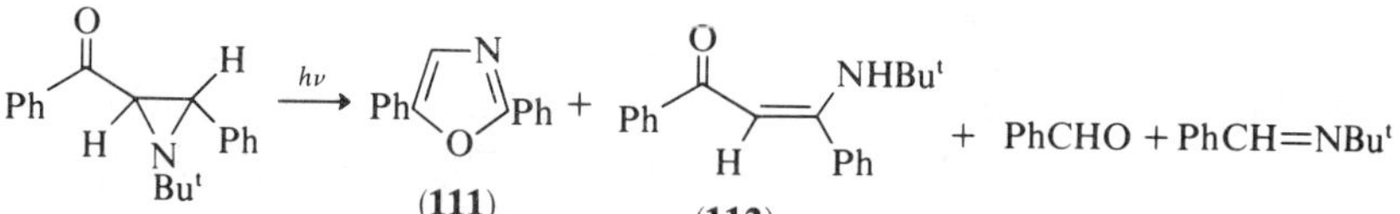

(**110**) $\xrightarrow{h\nu}$ (**111**) + (**112**) + PhCHO + PhCH=NBu^t

(**113**) $\xrightarrow{h\nu}$ $\xrightarrow[\text{transfer}]{\text{1,5-D–O}}$ → $\xrightarrow{H_2O}$ (**112**)

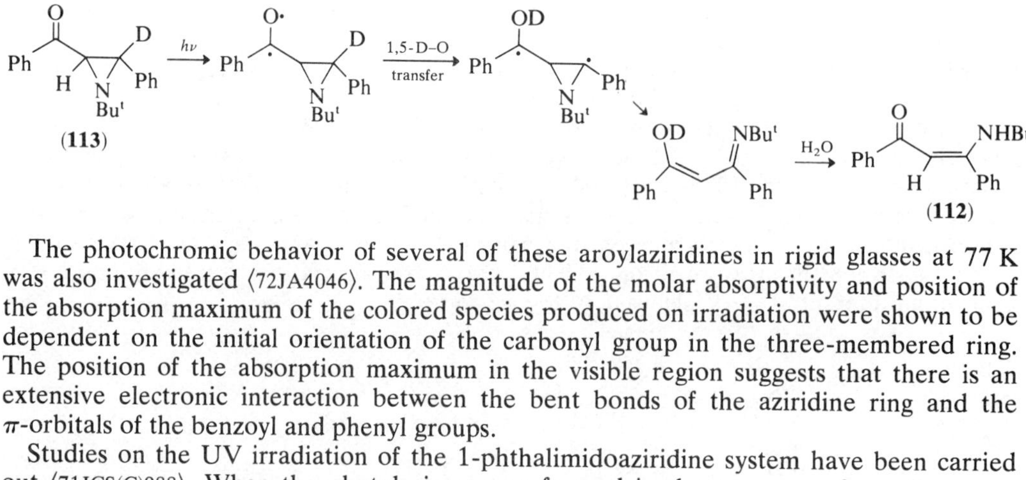

The photochromic behavior of several of these aroylaziridines in rigid glasses at 77 K was also investigated ⟨72JA4046⟩. The magnitude of the molar absorptivity and position of the absorption maximum of the colored species produced on irradiation were shown to be dependent on the initial orientation of the carbonyl group in the three-membered ring. The position of the absorption maximum in the visible region suggests that there is an extensive electronic interaction between the bent bonds of the aziridine ring and the π-orbitals of the benzoyl and phenyl groups.

Studies on the UV irradiation of the 1-phthalimidoaziridine system have been carried out ⟨71JCS(C)988⟩. When the photolysis was performed in the presence of an alkene, an exchange reaction was observed. Evidence was presented which showed that the reaction involved concerted fragmentation of the phthalimidoaziridine (**114**) to an alkene and phthalimidonitrene which was trapped by the added alkene. The reaction is not general for all aziridines with a heteroatom at the N-position of the ring.

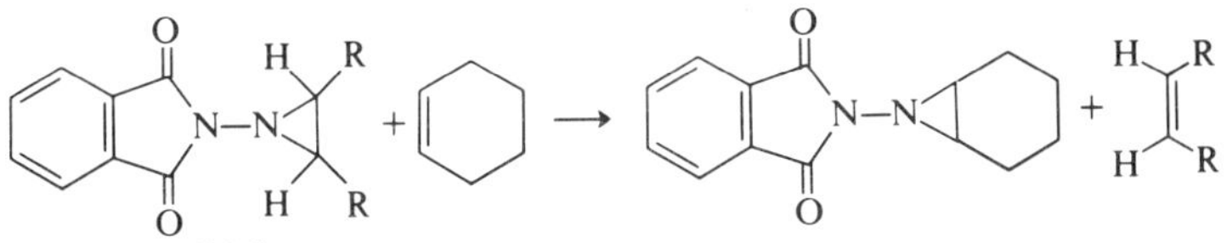

Photolysis of aziridine (**115**) leads to the fused ring system (**116**) *via* an intramolecular 1,3-dipolar cycloaddition reaction ⟨71AG(E)273⟩. Conrotatory ring opening is obviously restricted by the ring system and, in contrast to the photolysis, heating (**115**) at 160 °C gives the rearranged product (**117**).

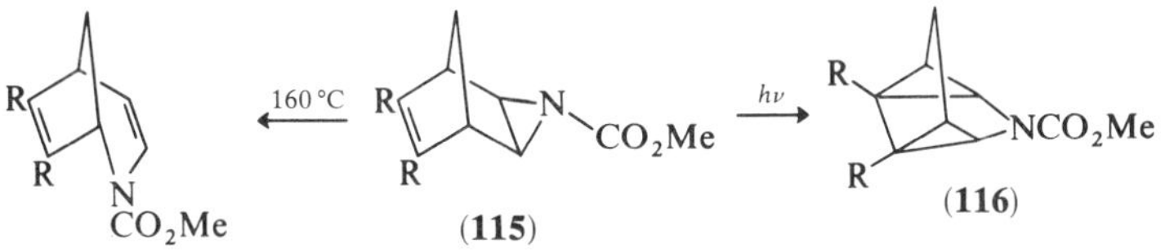

Azomethine ylides have been generated from a wide variety of aziridines by both thermal and photochemical methods. An example of an apparent conrotatory photolytic ring opening of an aziridine has been reported ⟨70JA6997⟩. Irradiation of the bicyclic aziridine (**118**) caused rapid development of an intense color, which faded after 12 h at room temperature but could be regenerated on further photolysis. After irradiation for several hours the solution became yellow and enediimine (**120**) could be isolated. The colored intermediate was confirmed as azomethine ylide (**119**) by trapping with dimethyl acetylenedicarboxylate. The *cis* configuration was assigned to adduct (**121**) thereby indicating a conrotatory ring opening of aziridine (**118**), which is a symmetry-disallowed photochemical process. The authors, by analogy with previous suggestions ⟨70CC692⟩, consider that a 'hot' ground state, rather than the lowest excited state, is involved in this reaction.

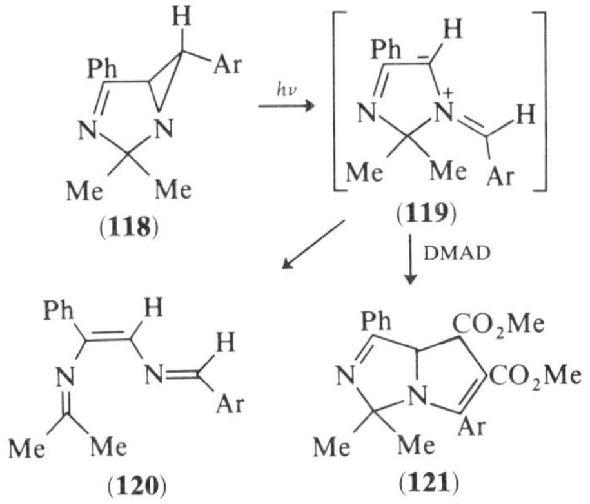

As was discussed in an earlier section, 1-azirines undergo irreversible ring opening on photolysis to give nitrile ylides as reactive intermediates ⟨76ACR371⟩. A wide assortment of studies have rigorously established that these 1,3-dipoles function as useful intermediates in a variety of synthetic applications. Recent work in the literature has shown that the photolysis of certain 1-azirines can also lead to new and interesting photochemistry. One of the most interesting examples involves 2-phenyl-1-azaspiro[2.2]pent-1-ene (**122**) ⟨75JA5912⟩. Irradiation of this spiroazirine in methanol results in a Griffin fragmentation and produces ethylene and 2-phenylazirinylidene (**123**). This novel carbene undergoes subsequent reaction with methanol to produce 2-methoxy-2-phenyl-1-azirine (**124**). When the irradiation of (**122**) is carried out in the presence of oxygen, benzonitrile and carbon monoxide are formed. In the absence of oxygen an extremely small quantity of benzonitrile is formed and is presumably derived by competitive cycloelimination from (**122**). The formation of benzonitrile was attributed to the intermediacy of 2-phenylazirinone (**125**) as a transient species. Azirinones are known to be unstable and readily lose carbon monoxide to form nitriles ⟨69T1637⟩.

(122) (123) (124) (125)

$\xrightarrow{h\nu}$ + $CH_2{=}CH_2$; MeOH; O_2; $\longrightarrow$ PhCN + CO

Irradiation of a series of hydroxymethyl-1-azirine derivatives (**126**) which contain good leaving groups was found to give *N*-vinylimines (**128**) *via* a novel 1,4-substituent shift ⟨78JA4481⟩. The results indicate that the migrating substituent (X) must be a reasonably good leaving group in order for the rearrangement to occur. Involvement of a nitrile ylide intermediate (**127**) in these reactions was demonstrated by trapping experiments. The quantum efficiency and rate of reaction were shown to be directly related to the leaving group ability.

$\xrightarrow{h\nu}$ $[PhC{\equiv}\overset{+}{N}{-}\overset{-}{C}HCH_2{-}X]$ (**127**) $\longrightarrow$ (**128**)

(**126**) X = Cl, Br, OAc, $OCOCF_3$, OBz

5.04.3.5 Thermal Chemistry

Aziridines bearing unsaturation on nitrogen rearrange under a variety of conditions ⟨71MI50401⟩. For example, *N*-acyl derivatives of aziridines are readily converted upon distillation to 2-substituted 2-oxazolines (**129**) ⟨59JA2202⟩. The rearrangement occurs by intramolecular attack of the carbonyl oxygen at the ring carbon to cause rupture of the system. The driving force for this process is found in the relief of strain which the opening of the three-membered ring provides. Similar rearrangements occur under the influence of acid catalysts.

(**129**) (**130**) (**131**)

Iodide and thiocyanate ion are effective catalysts for inducing a related rearrangement ⟨62AG(E)528⟩. This reaction can be envisioned as proceeding by nucleophilic attack on the lesser substituted aziridinyl carbon atom by iodide ion to give an iodoethyl intermediate such as (**132**) which is subsequently converted to the final product.

$\xrightarrow{I^-}$ $ArC(O^-){=}NC(CH_2I)Me_2$ (**132**) $\longrightarrow$ (**133**)

Pyrolysis of *N*-acyl derivatives of 2-alkyl-substituted aziridines (**134**) results in isomerization to *N*-allylamides ⟨60JA4930⟩. Such rearrangements proceed *via* transition states in which intramolecular hydrogen transfer from a side chain carbon to oxygen occurs concomitantly with cleavage of the three-membered hetero ring and involves a stereospecific *cis* elimination. Considerable evidence has been accumulated in support of the concerted six-center transition state ⟨64JHC293⟩. First order kinetics, a high entropy of activation, the lack of a solvent effect along with a Hammett equation indicative of negative charge accumulation on nitrogen are also consistent with the proposed mechanism.

(**134**) $\longrightarrow$ (**135**) $\longrightarrow$ $R^2C(O)NHCH_2C(R^1){=}CH_2$ (**136**)

Aziridines bearing pendant unsaturation on the carbon atom of the three-membered ring also display interesting and diverse thermal chemistry. The actual reaction course depends upon the geometry and nature of the substituents. Thermolysis of compounds with structure (**137**) yields the seven-membered heterocyclic ring (**139**) ⟨78BSF(2)97, 67JOC2628⟩. In contrast, isomers of structure (**140**) give 3-pyrrolines (**141**) upon heating. Although either concerted or diradical mechanisms may be written, the lack of crossover between (**137**) and (**140**) decompositions suggests a concerted type mechanism.

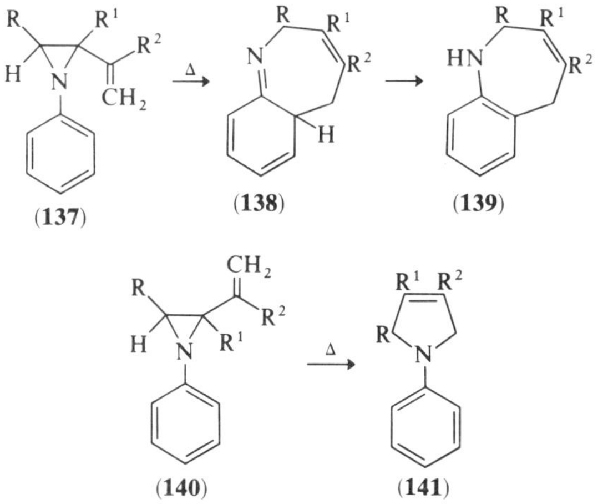

(**137**) (**138**) (**139**)

(**140**) (**141**)

Cis- and *trans-*2,3-divinylaziridines have also been studied in various laboratories ⟨65JOC88, 74TL3897⟩. The *cis* isomers (**142**) have been found to undergo an extremely rapid room temperature Cope rearrangement. By contrast, the *trans* isomers require much higher temperatures (*ca.* 300 °C) to undergo this rearrangement. Although the stereochemistry of the final product was not defined, an azomethine ylide was probably involved in this reorganization ⟨76CJC1571⟩. The alkynic analog (**144**) has also been found to rearrange to (**146**) *via* (**145**) ⟨77T2399⟩.

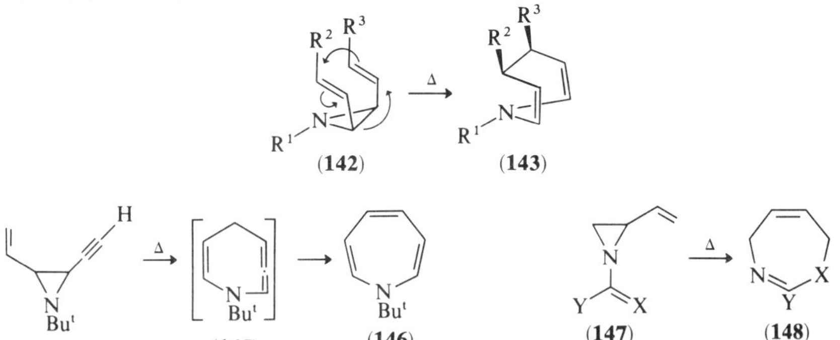

(**142**) (**143**)

(**144**) (**145**) (**146**) (**147**) (**148**)

Thermal Cope-type rearrangements between *C*-vinyl and nitrogen substituents bearing appropriately labeled unsaturation are also known, *e.g.* (**147**) → (**148**) ⟨67JA605, 71JOC3076⟩.

In contrast to the well-defined photochemical behavior of 1-azirines the thermal reactions of these compounds have been studied less thoroughly ⟨68TL3499⟩. The products formed on photolysis of azirines can best be rationalized in terms of an equilibration of the heterocyclic ring with a transient vinylnitrene. Thus, products formed from the thermolysis of azirines are generally consistent with C—N cleavage. For example, the vinylnitrene generated from the thermolysis of azirine (**149**) can be trapped with phosphines ⟨72CC565⟩.

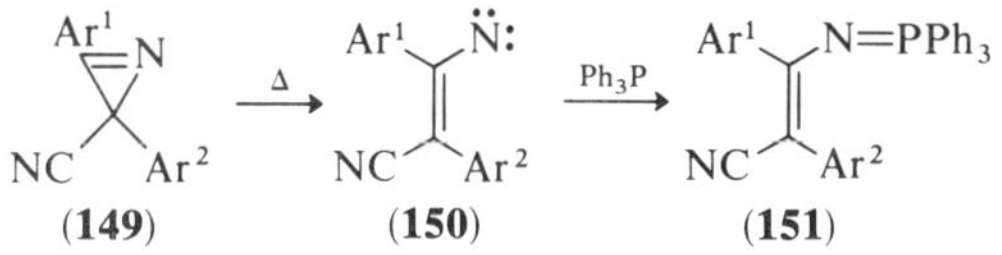

(**149**) (**150**) (**151**)

The thermal chemistry of a number of aryl-substituted azirines often results in the formation of indole derivatives ⟨68TL3499⟩. Thus, heating a sample of azirine (**152**) gave 3-phenylindole (**155**) and dihydropyrazine (**156**). The formation of (**155**) was suggested to

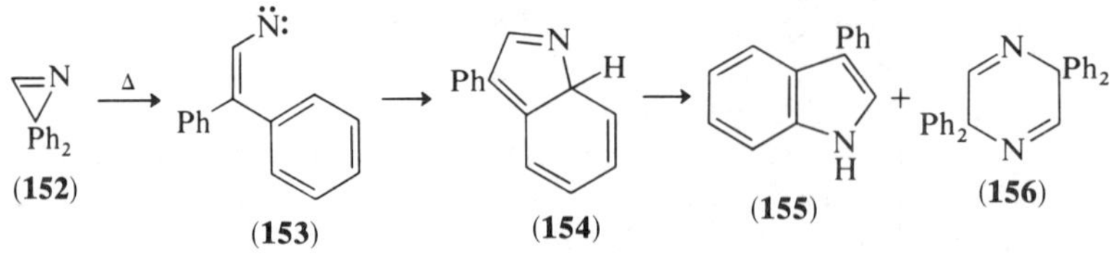

(152) (153) (154) (155) (156)

proceed *via* a vinylnitrene intermediate which cyclizes and then undergoes a 1,5-sigmatropic shift ⟨77H(6)1563⟩.

The 1-azirines obtained from the vapor phase pyrolysis of 4,5-disubstituted 1-phthalimido-1,2,3-triazoles (**157**) have been found to undergo further thermal reactions ⟨71CC1518⟩. Those azirines which contain a methyl group in the 2-position of the ring are cleaved to nitriles and phthalimidocarbenes, whereas those azirines which possess a phenyl substituent in the 2-position rearrange to indoles.

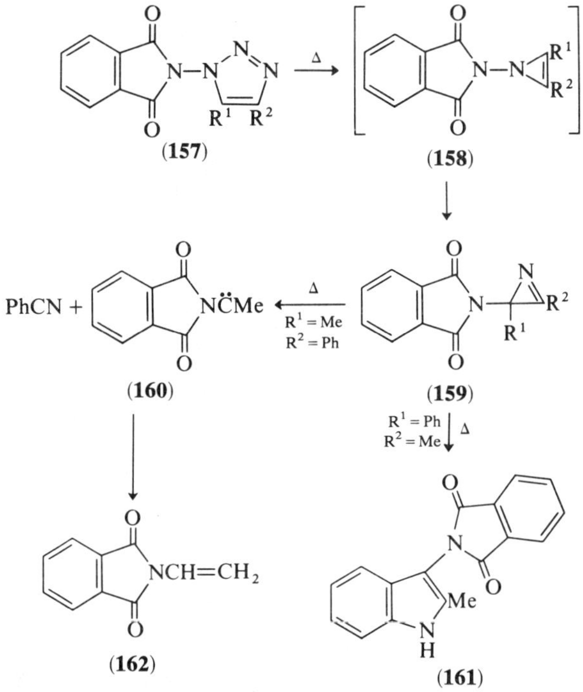

(157) (158) (159) (160) (161) (162)

Products formed during photochemical isomerizations involve C—C bond rupture, while thermal isomerization products usually arise from initial C—N bond cleavage. Unambiguous evidence for a thermal C—C cleavage has been obtained, however, in the vapor phase pyrolysis of several azirines ⟨76JOC831⟩. The diradical species (**163**) generated from C—C cleavage is postulated to undergo a 1,4-hydrogen transfer to produce azadiene (**164**) which ultimately fragments to a nitrile and an alkene *via* an azacyclobutene intermediate (**165**), or undergoes an electrocyclic ring closure followed by a 1,5-hydrogen shift to produce dihydroisoquinoline (**167**).

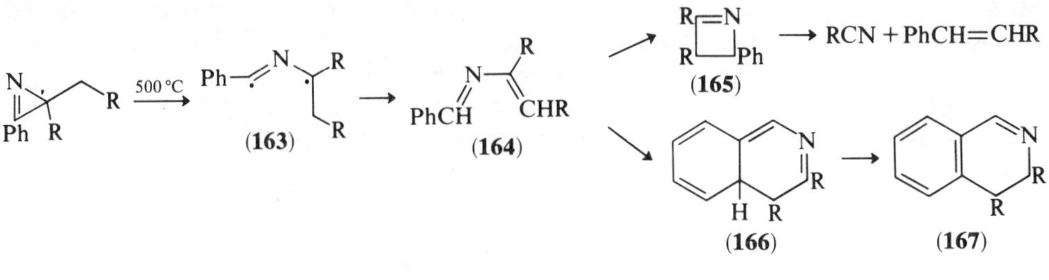

(163) (164) (165) RCN + PhCH=CHR (166) (167)

Vapor phase pyrolysis of 2-dimethylaminoazirine (**168**), on the other hand, proceeds in a similar manner at 340 °C to give substituted azadiene (**169**) in high yield ⟨75JA4409⟩. Azadiene (**169**) has been employed in the construction of heterocyclic rings such as pyridines *via* a [4+2] cycloaddition–elimination sequence.

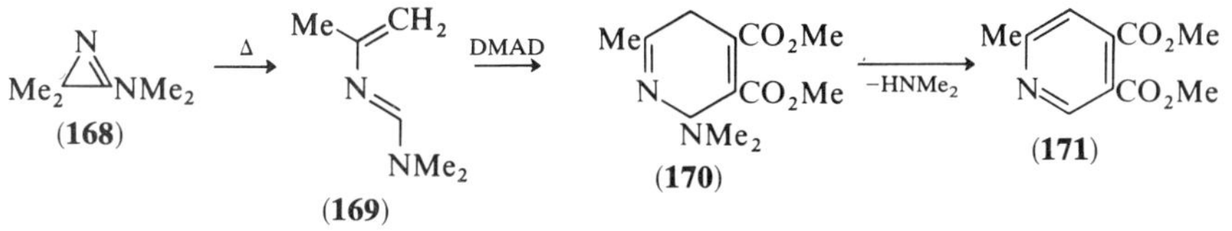

The synthesis of a number of pyrrole derivatives has been effected by the thermal conversion of 3-vinyl-1-azirines to butadienylnitrenes followed by 1,5-electrocyclization ⟨75JA4682⟩. The thermal transformations observed on thermolysis of azirine (**53**) were rationalized in terms of an equilibration of the azirine with a butadienylnitrene, which subsequently rearranged to the final products. The rearrangement of (**54**) to pyrrole (**56**) was envisaged as occurring by an electrocyclic reaction followed by a 1,5-sigmatropic ethoxycarbonyl shift and subsequent tautomerization ⟨76JOC543⟩. Other examples of this rearrangement have been reported to occur with ester, acyl and cyano groups ⟨72TL3479⟩. The formation of pyridine (**175**) was postulated to arise by insertion of the butadienylnitrene into the neighboring allylic methyl group followed by oxidation of the transient dihydropyridine (**174**).

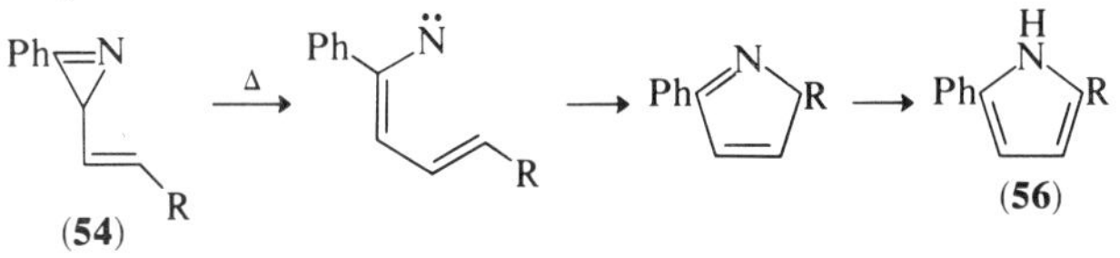

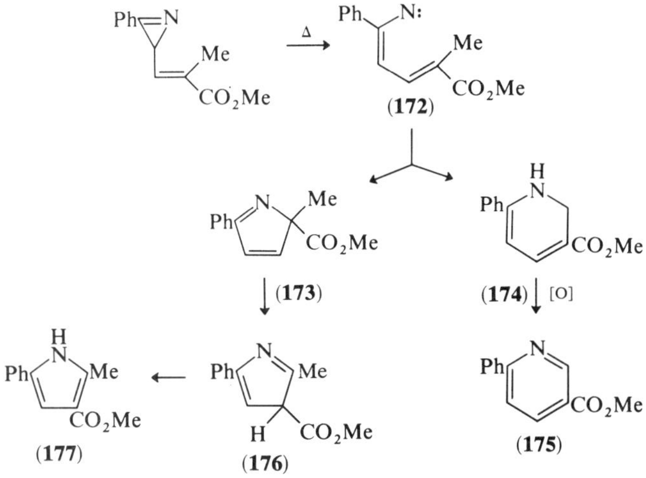

A number of allyl-substituted azirines were found to undergo smooth rearrangement on heating to give 2-azabicyclo[3.1.0]hexenes (**180**) ⟨78JOC2029⟩. The formation of (**180**) can be explained in terms of an equilibration of the 1-azirine with a transient vinylnitrene, which subsequently adds to the adjacent π-bond. The initially formed bicycloaziridine rearranges to the 3-azabicyclohexene ring system by means of a 1,3-sigmatropic shift. Evidence favoring this pathway is provided by the isolation of Δ^1-pyrroline (**181**) from the thermolysis of azirine (**65**; $R^1 = H$, $R^2 = Me$). The formation of (**181**) was rationalized as proceeding *via* a homo[1,5]-hydrogen migration from a 6-*endo*-methyl-substituted bicycloazirine intermediate (**179**).

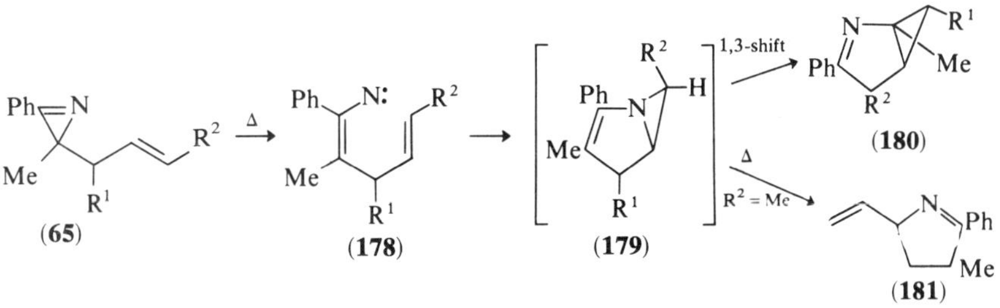

The thermal behavior of the closely related homoallyl-substituted azirine (**182**) has also been studied ⟨77JA1871⟩. Heating a solution of (**182**) in toluene gave 2-methylbiphenyl (**183**)

and 2,5-dimethyl-6-phenylpyridine (**184**). The formation of these products is explained in terms of vinylnitrene (**185**) which undergoes a 1,4-hydrogen transfer from the neighboring methylene group to generate azatriene (**186**). This reactive intermediate undergoes a thermally allowed 1,5-sigmatropic shift to give triene (**187**). Electrocyclic closure to cyclohexadiene (**188**) followed by loss of ammonia readily accounts for the formation of the substituted biphenyl derivatives. The formation of (**184**) involves an insertion of the vinylnitrene into the vinyl group followed by tautomerization and a subsequent oxidation.

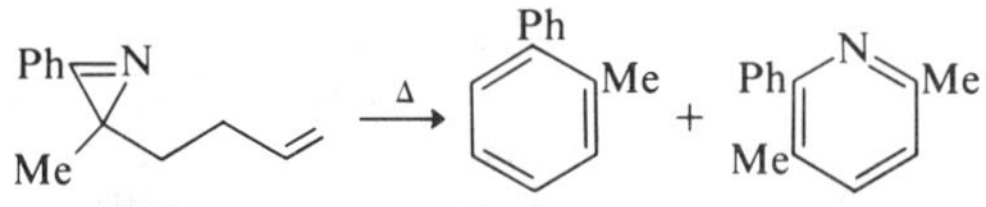

(**182**) (**183**) (**184**)

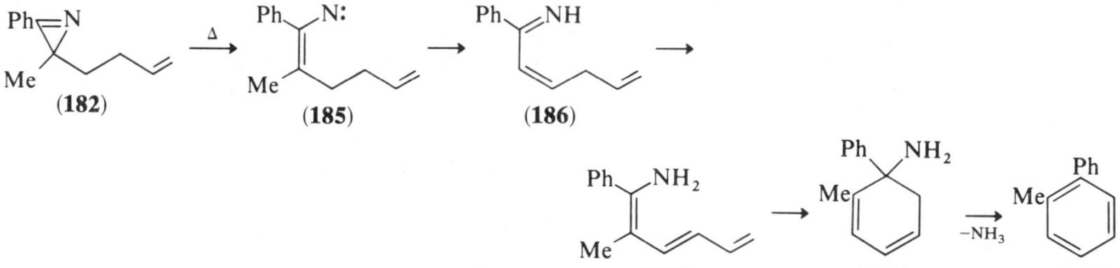

(**182**) (**185**) (**186**)

(**187**) (**188**) (**183**)

5.04.3.6 Reactions with Electrophilic Reagents

Ring opening reactions of aziridines are greatly accelerated in acidic media as exemplified by the observation that ethyleneimine polymerizes explosively in the presence of acids under non-controlled conditions. From the stereochemical point of view, the ring opening generally occurs stereospecifically with inversion of configuration at the point of attack. This reflects the fact that fully developed carbonium ions are customarily not generated in such electrophilic processes, and agrees with the mechanistic interpretation based on the 'push–pull' theory ⟨64HC(19-1)524⟩. Bimolecular ring opening of aziridines by electrophilic reagents can be governed either by steric factors or, where electron releasing groups are attached to carbon, by positive charge development in the transition state. Acid-catalyzed ring opening of rigid steroidal aziridines (**189**) results in *trans*-diaxial ring opening ⟨65JOC1748⟩. Comparison of acid-catalyzed ring opening of epoxides and aziridines led to the conclusion that aziridines are more likely to give *trans* opening ⟨72TL2591⟩.

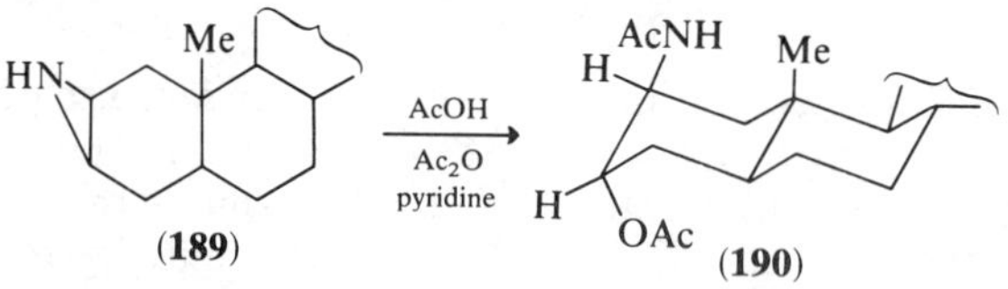

(**189**) (**190**)

Certain bifunctional nucleophiles allow cyclization after ring opening. The formation of 2-thiazolium salts ⟨71JHC405⟩ and the analogous production of 2-amino-2-thiazolines (**191**) from aziridines and thiocyanic acid fall into this category ⟨72JOC4401⟩.

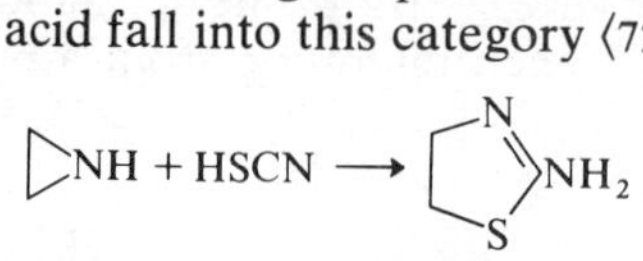

(**191**)

When acid-catalyzed ring opening is not synchronous with nucleophilic attack, the intermediate carbonium ion can undergo rearrangement (**193, 195**) ⟨66JOC3941, 73CJC1448⟩.

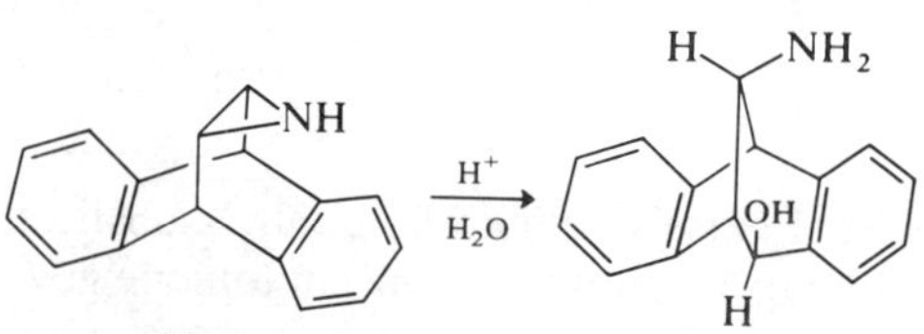

(**192**) (**193**)

(**194**) (**195**)

Protonated aziridines or quaternary aziridinium salts are exceptionally reactive towards nucleophiles and attempts to prepare them generally result in ring cleavage ⟨65JA749⟩. Generally, however, isolation of such compounds has proven feasible when anions of low nucleophilicity are employed. The mechanism of biological action of nitrogen mustards is believed to involve the alkylation of functional groups of metabolic importance by intermediate aziridinium salts ⟨B-62MI50400⟩. The reaction of aziridinium salts with various nucleophilic reagents results in the formation of ring cleavage products. The reactions of aziridinium salts can be generally described according to the nucleophilicity of the attacking reagent. If the attacking species is very nucleophilic, the product will be that in which cleavage of the less substituted C—N bond of the three-membered ring occurs, *e.g.* (**197**). If the attacking species is a relatively poor nucleophile the reaction can be viewed as an ionization with cleavage of the three-membered ring to yield the most stable carbonium ion which then reacts with the poor nucleophile.

(**196**) (**197**) (**198**)

The 1-azirine ring also undergoes a number of reactions in which the heterocycle plays the role of the nucleophile. Although the basicity of the nitrogen atom in the azirine ring is much lower than in simple aliphatic amines, this system can still function as a nucleophilic reagent. One example of this involves the acid-catalyzed hydrolysis of 1-azirines to α-aminoketones (**200**) which represents a well-established reaction. In fact, in many reactions of 1-azirines where acid catalysis is used, formation of α-aminoketones is difficult to avoid ⟨67JA4456⟩.

(**199**) (**200**)

The reaction of 2-phenyl-1-azirine (**201**) with benzoic acid gave *N*-benzoylphenacylamine (**204**) ⟨67BCJ2938⟩. The overall mechanism of the reaction involves initial protonation on nitrogen followed by addition of the nucleophile to the azirinium ion and finally ring opening.

(**201**) (**202**) (**203**) (**204**)

The protonated azirine system has also been utilized for the synthesis of heterocyclic compounds ⟨67JA4456⟩. Thus, treatment of (**199**) with anhydrous perchloric acid and acetone or acetonitrile gave the oxazolinium perchlorate (**207**) and the imidazolinium perchlorate (**209**), respectively. The mechanism of these reactions involves 1,3-bond cleavage of the protonated azirine and reaction with the carbonyl group (or nitrile) to produce a resonance-stabilized carbonium–oxonium ion (or carbonium–nitrilium ion), followed by attack of the nitrogen unshared pair of electrons to complete the cyclization.

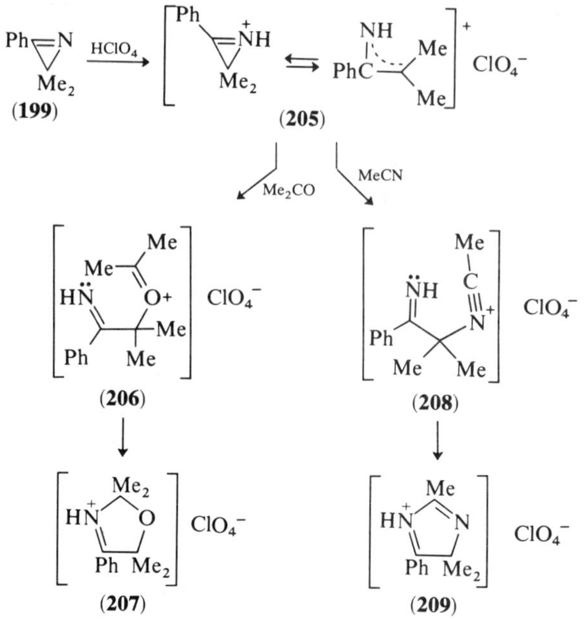

(199) (205) (206) (207) (208) (209)

1-Azirines react with carboxylic acid chlorides in benzene to give aziridines where RCOCl has been added to the C=N double bond ⟨B-69M150401, 74JOC66⟩. For example, azirine (**210**) reacts with benzoyl chloride presumably through the azirinium ion (**211**) to give a stereoisomeric mixture of *N*-benzoyl-2-chloroaziridine (**212**). This system is unstable and is converted in polar solvents or by heating into a mixture of oxazole (**214**) and dichloroamide (**213**). It should be noted that the rearrangement to the oxazole proceeds with opposite regiochemistry to that observed in the formally similar acid-catalyzed reaction of 1-azirines with ketones ⟨67JA4456⟩.

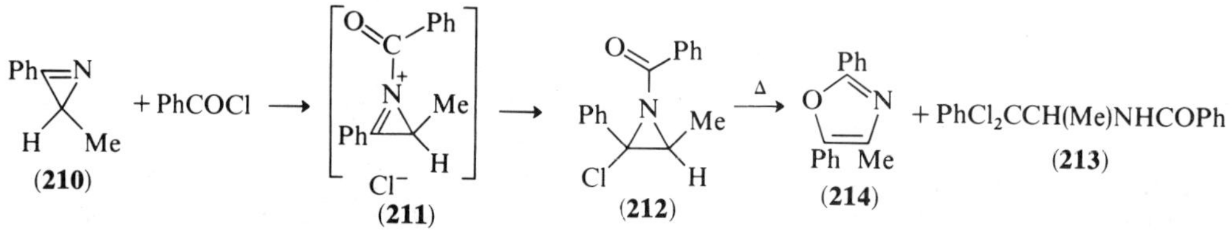

(210) + PhCOCl → (211) → (212) → (214) + $PhCl_2CCH(Me)NHCOPh$ (213)

5.04.3.7 Reactions with Nucleophilic Reagents

Ring opening reactions of aziridines which are initiated by nucleophilic reagents have been shown to proceed with extensive, if not complete, inversion of configuration at the point of attack ⟨57JA734⟩. When unsymmetrical aziridines are involved, ring opening can occur in either of two different directions. Frequently, the nucleophile attacks the less hindered carbon atom preferably with the result that one direction of ring opening is predominant ⟨59CRV737⟩. However, such reactions are generally difficult to predict because the product ratio can easily be affected by changes in the solvent and in the proportion of the reagents. The diverse and often seemingly contradictory facts pertaining to the opening of this strained ring can be correlated in terms of a 'push–pull' mechanism ⟨64HC(19-1)524⟩. According to this concept the major factors involved in such processes are approach of the nucleophilic reagent, the rupture of the C—X bond, and the effect of the electrophilic reagent. As a result, steric factors are less influential than usual, while sensitivity to factors such as solvent, resonance and the presence of electron releasing substituents is substantially increased.

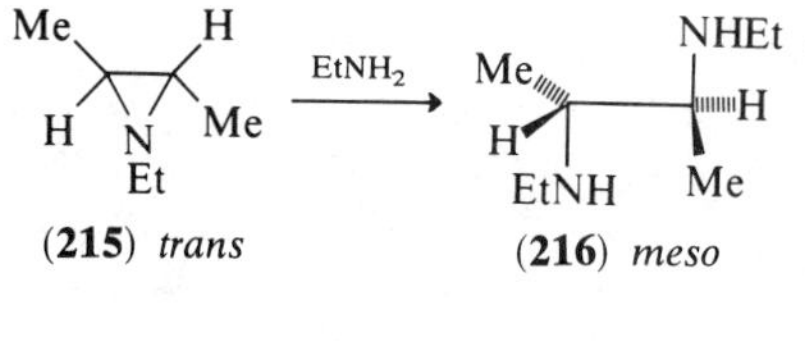

(**215**) *trans* —$EtNH_2$→ (**216**) *meso*

Most of the aziridines which undergo direct nucleophilic ring opening bear strong electron accepting groups (RCO, RSO_2, CN, Ar, *etc.*) on nitrogen. Those aziridines which do not bear such substituents require very strong nucleophiles or vigorous reaction conditions for ring opening. It is possible that some of the latter examples actually involve general acid catalysis. Competition occurs between attack on the ring and the carbonyl group of (**217**) and thus presents a complication in the synthetic utilization of such reactions ⟨70TL4623⟩.

(**217**) $+ ArNH_2 \longrightarrow EtO_2CNHCH_2CH_2NHAr$ (**218**)

The most useful reactions combine carbanion nucleophiles with activated aziridines. For example, the ring expansion which occurs on treatment of aziridines (**219**) with malonate salts typifies the heterocyclic synthesis possible. The conversion is quite general since many analogous transformations have been observed in which different carbanion stabilizing substituents were employed ⟨73S546⟩.

(**219**) $+ \bar{C}H(CO_2R^2)_2 \longrightarrow$ (**220**)

Attack by nucleophilic reagents also occurs on the 1-azirine ring. The electrophilic character of the azirine ring is associated with the polarized nature of the C—N double bond. Lithium aluminum hydride reduces 1-azirines in a highly stereospecific manner to give aziridines ⟨68JA2869, 53JA33⟩. For example, 3-methyl-2-phenyl-1-azirine (**210**) is stereospecifically reduced to the *cis*-aziridine (**222**). Approach of the hydride occurs exclusively from the less hindered side of the azirine molecule. The observed reactivity of 1-azirines towards Grignard reagents represents an anomalous reaction of an imine. Generally, Grignard reagents react by proton removal to give the enamine anion which on work-up generates the starting imine ⟨68JA2875⟩. The failure of 1-azirines to follow this behavior can be explained in terms of the instability of the enamine derived from the 1-azirine. The reaction of azirines with Grignard reagents exhibits similar stereospecificity as observed for hydride reductions ⟨69TL4001⟩.

(**221**) $\xleftarrow{PhMgBr}$ (**210**) $\xrightarrow{LAH}$ (**222**)

Azirines react with enolate anions. Initial nucleophilic attack on phenyl 1-azirine by the enolate anion derived from acetophenone gives intermediate (**223**) which undergoes 1,2-bond cleavage, cyclization and hydroxyl group elimination to give pyrrole (**226**).

$+ MeCOPh \xrightarrow{Me\bar{S}OCH_2}$ [(**223**)] $\longrightarrow$ [(**224**)] $\longrightarrow$ [(**225**)] $\longrightarrow$ (**226**)

Carbanions in the form of ylides also add to azirines. For example, treatment of 1-azirine (**227**) with dimethylsulfonium methylide gives 1-azabicyclobutane (**229**) in good yield ⟨72JA2758⟩. The addition of the methylene group occurs by initial nucleophilic attack by the ylide to give intermediate (**228**) which cyclizes with expulsion of dimethyl sulfide.

(**227**) $\xrightarrow{Me_2\overset{+}{S}-\bar{C}H_2}$ [(**228**)] $\longrightarrow$ (**229**)

Addition of trichloromethide ion to azirine (**210**) generates aziridine (**230**). When this aziridine was treated with base, cyclization and rearrangement occurred and the azetidine (**233**) was isolated ⟨73JA2982⟩.

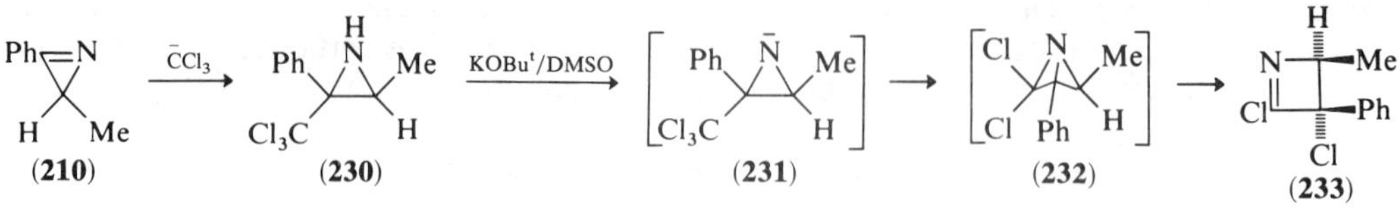

(**210**) (**230**) (**231**) (**232**) (**233**)

1-Azirines react with alcohols in the presence of alkoxides to give alkoxyaziridines ⟨67JA4456⟩. Further treatment with alcohol and alkoxide results in the formation of amino ketone acetals. Alkoxyaziridines are not isolated in general from the acid-catalyzed addition of methanol to azirines. Azirines are also known to react with amines ⟨66JOC1423⟩. Frequently the initially produced adducts undergo subsequent transformations.

5.04.3.8 Ring Isomerization

There are at least two mechanisms available for aziridine *cis–trans* isomerism. The first is base-catalyzed and proceeds *via* an intermediate carbanion (**235**). The second mechanism can be either thermally or photochemically initiated and proceeds by way of an intermediate azomethine ylide. The absence of a catalytic effect and interception of the 1,3-dipole intermediate provide support for this route. A variety of aziridinyl ketones have been found to undergo equilibration when subjected to base-catalyzed conditions ⟨65JA1050⟩. In most of these cases the *cis* isomer is more stable than the *trans*. Base-catalyzed isotope exchange has also been observed in at least one molecule which lacks a stabilizing carbonyl group ⟨72TL3591⟩.

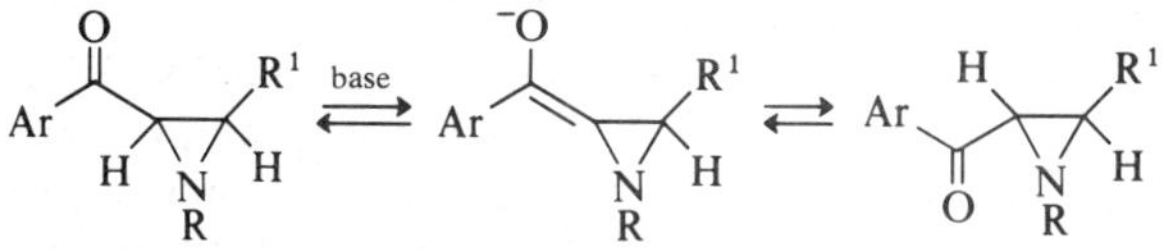

(**234**) (**235**) (**236**)

Equilibration of aziridines *via* azomethine ylides has been reported for a variety of structures ⟨67JA1753⟩. Most aziridines equilibrated by this method show greater *cis* stability. An energy barrier has been detected between the two isomeric azomethine ylides ⟨69AG(E)602⟩.

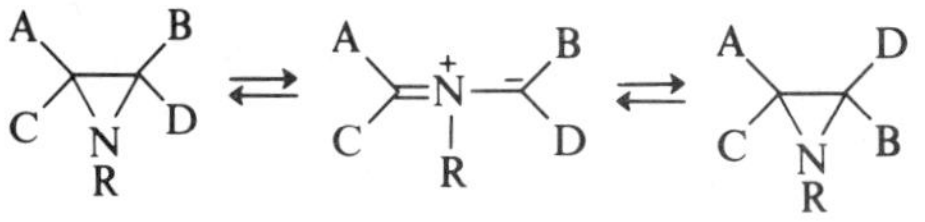

5.04.3.9 Ring Opening Reactions

Interest in the opening of the cyclopropyl ring to an allyl-substituted system sparked similar studies with aziridines containing leaving groups on both the nitrogen and carbon atoms. Investigations dealing with solvolysis of carbon-substituted *N*-chloroaziridines have demonstrated that nitrogen–chlorine and carbon–carbon cleavage occur simultaneously in a disrotatory fashion ⟨70JA2084⟩. For example, aziridine (**237**) is quite labile towards aqueous conditions whereas (**238**) is stable. The relative rates of (**239**), (**240**) and (**241**) also support this conclusion.

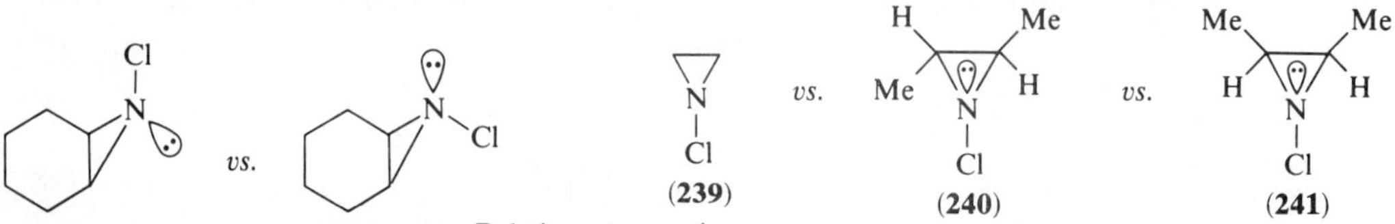

(**237**) *vs.* (**238**) Relative rates (**239**) 1 *vs.* (**240**) 1500 *vs.* (**241**) 150 000

Other approaches to the generation of the azallyl cation have been found. One of the most useful involves the use of lead tetraacetate ⟨73TL2143⟩. The anodic oxidation of aziridines also leads to the azallyl cation intermediate ⟨75JA1600⟩.

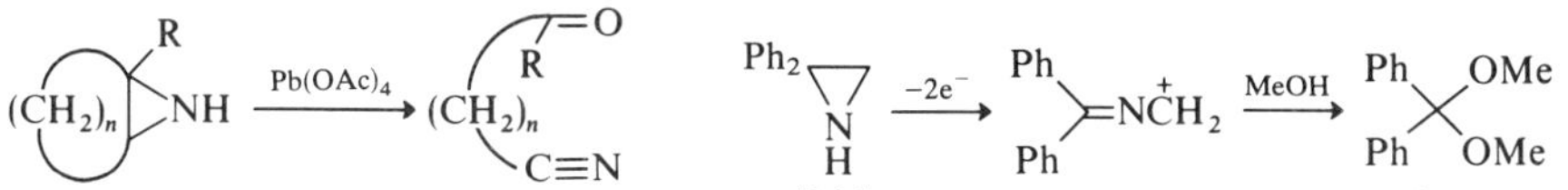

(242) (243) (244) (245) (246)

The anionic version of these ring openings is also known ⟨77CB651⟩. Thus, 1,3-(*cis,trans*)-diphenyl-2-azallyllithium (**248**) has been generated by heating *N*-lithio-*cis*-2,3-diphenylaziridine and trapped by reaction with *trans*-stilbene to give either (**251**) or (**252**). If the reaction is cooled to 0 °C before stilbene is added, the pyrrolidine (**250**) is formed, presumably since the primary ring-opened product (**248**) is converted into *trans,trans*- or *cis,cis*-(**249**). An alternative route, involving reaction of benzylidenebenzylamine with LDA at 0 °C, has been used to investigate further the stereochemistry of the carbanionic cycloaddition ⟨72AG(E)290⟩. The pyrrolidines obtained from the reaction of the azallyllithium are compatible with a symmetry-allowed $[4\pi_s+2\pi_s]$ cycloaddition reaction.

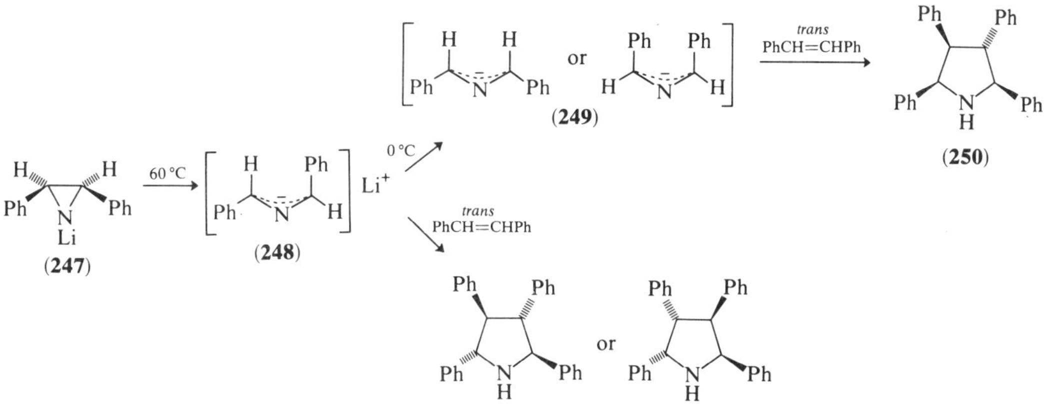

(249) (250) (247) (248) (251) (252)

The corresponding aziridinyl radical has been generated by thermolysis of the appropriate perester and was also found to undergo ring opening to give (**256**) and (**257**) ⟨75CB1527⟩.

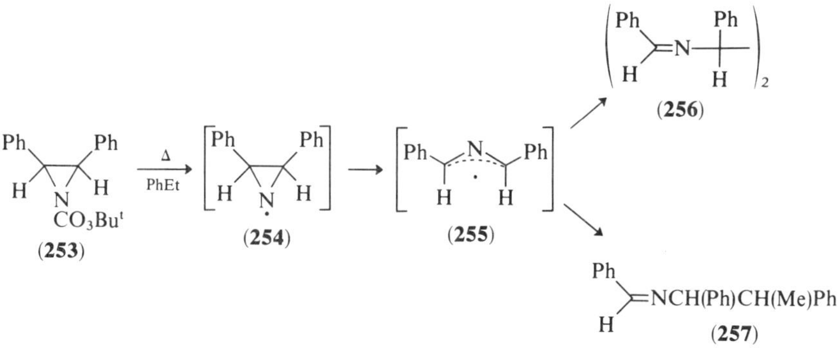

(256) (253) (254) (255) (257)

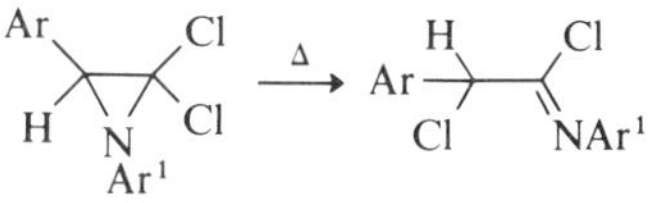

(258) (259)

Thermolysis of dihaloaziridines results in an analogous ring opening reaction and produces α-haloimidoyl halides in good yield ⟨75JOU583⟩. Although it has been claimed that (**259**) is formed directly in a concerted electrocyclic ring opening, the evidence for this assertion may not be adequate.

A detailed mechanistic study of the hydrolysis of 2,2-dichloroazirines (**260**) has been carried out and shown to be compatible with the route shown ⟨77JPS1653⟩.

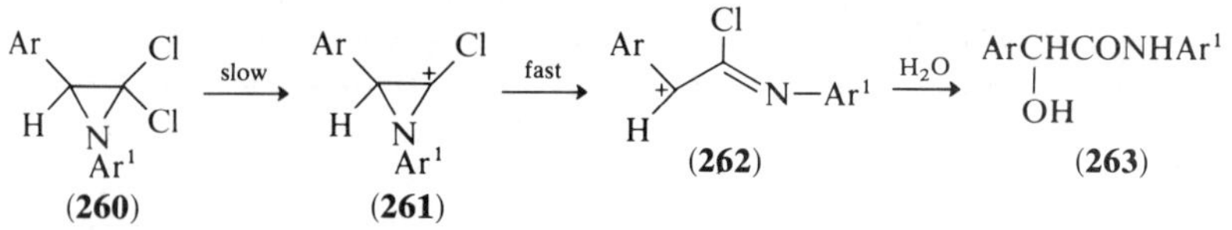

(260) (261) (262) (263)

Substituted 2-haloaziridines are also known to undergo a number of reactions without ring opening. For example, displacement of chlorine in (**264**) with various nucleophilic reagents has been found to occur with overall inversion of stereochemistry about the aziridine ring ⟨65JA4538⟩. The displacements followed first order kinetics and faster rates were noted for (**264**; R = Me) than for (**264**; R = H). The observed inversion was ascribed to either ion pairing and/or stereoselectivity.

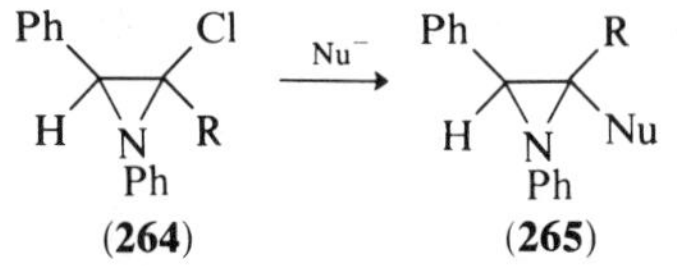

Nu = Me (from MeLi), OMe (from NaOMe), CN, PhS

(264) (265)

The tributyltin hydride reduction of dihaloaziridines, *e.g.* (**266**), represents another example where the ring system has been maintained ⟨79CJC1958⟩. Especially noteworthy is the retained configuration associated with the reaction. This behavior differs from the cyclopropyl analog and was explained on the basis of increased *s*-character in the exocyclic bond caused by the nitrogen atom.

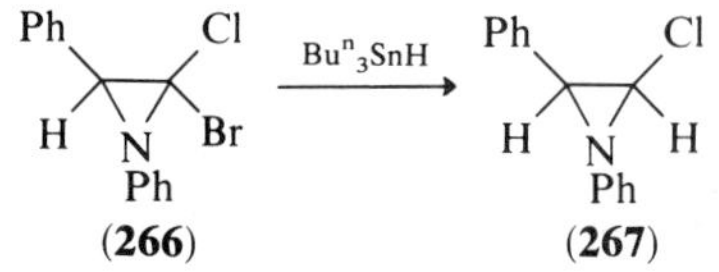

(266) (267)

5.04.3.10 Reactions Involving Extrusion of the Heteroatom

Aziridines can undergo loss of a nitrogen atom in a number of ways and this process is frequently stereospecific in terms of the alkene formed. For example, the reaction of aziridines, *e.g.* (**268**), with nitrosating agents such as nitrosyl chloride or methyl nitrite results in the formation of alkenes with greater than 99% stereoselective deamination ⟨64JOC1316⟩. Such transformations proceed *via* *N*-nitrosoaziridine intermediates which are isolable at temperatures below −20 °C, but which decompose to the observed products at higher temperatures.

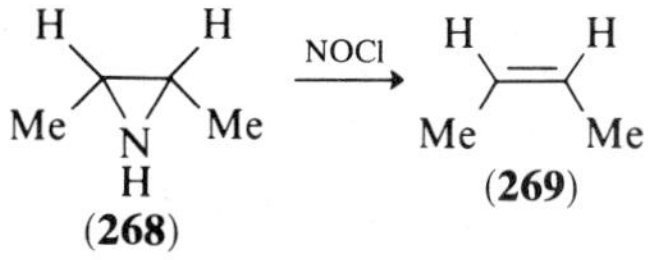

(268) (269)

N-Aminoaziridines have been converted to alkenes by reaction with a variety of oxidizing agents ⟨70JA1784⟩. Usually, the deamination reaction is stereospecific. The oxidation of 1-amino-2,3-diphenylaziridines with manganese dioxide, however, was not stereospecific. The *trans* compound gives entirely *trans*-stilbene, whereas the *cis*-aziridine forms a mixture of 85% *trans*- and 15% *cis*-alkene. *cis*-Stilbene is not isomerized to *trans* under the reaction conditions, and the results are explained in terms of an azamine intermediate which can isomerize through a tautomeric equilibrium.

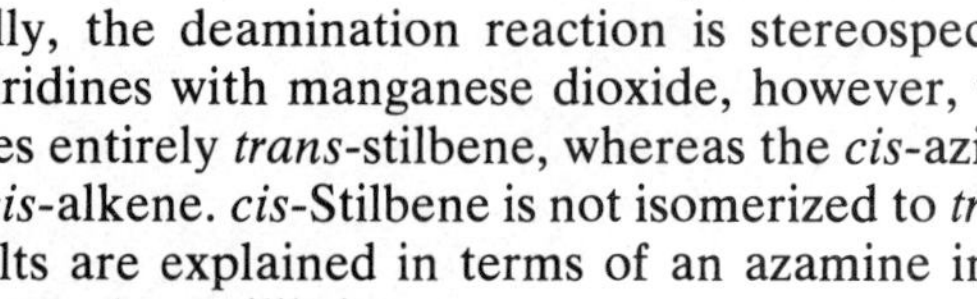

$\xrightarrow{-N_2}$ PhCH=CHPh

N-Alkylaziridines can be stereospecifically deaminated to alkenes by reaction with *m*-chloroperbenzoic acid ⟨70AG(E)374⟩. The reaction and work-up are carried out in the dark to avoid isomerization of the *cis*-alkene, and the mechanism is thought to involve an initial oxidation to an amine oxide followed by a concerted elimination. Aziridine oxides have been generated by treating aziridines with ozone at low temperatures ⟨71JA4082⟩. Two pathways exist for this decomposition, one involving stereospecific formation of alkene and the other a rearrangement to a hydroxylamine derivative (**270**). The latter route is generally preferred when a choice is available, although highly hindered aziridines often give products from both reactions.

$CH_2{=}CHCH_2N(R)OH$ (**270**) $\quad$ $CH_3CH{=}CH_2 + RNO$

Certain *N*-substituted aziridines are particularly labile towards deamination. *N*-Aminoaziridines (**271**) decompose with high stereospecificity to alkenes and diimide between 20 and 60 °C in good yield ⟨70HCA1479⟩.

(**271**) $\xrightarrow{\Delta}$ + HN=NH

Aziridinyl hydrazones function as masked diazo compounds. For example, thermally induced fragmentation of the hydrazones derived from phenylglyoxal and diastereomeric 2,3-disubstituted 1-aminoaziridines takes place with high stereospecificity to give alkenes and diazoketone (**273**) ⟨70HCA1479⟩. The most important application of this procedure has been to cyclic α,β-epoxyketones ⟨72HCA1276⟩. Although temperatures of about 150 °C are required, the product is removed by distillation as formed. Photochemical decomposition has also been employed with success ⟨75HCA1444⟩. This procedure has been particularly useful in a recent attempt to synthesize vitamin B_{12} ⟨76JA6313⟩.

(**272**) $+ PhCOCHO \rightarrow RCH{=}CHR + PhCOCHN_2$ (**273**)

(**274**) $\xrightarrow{\Delta}$ (**275**) $\rightarrow$ (**276**) $\rightarrow$ (**277**)

The closely related *N*-arylazoaziridine system (**278**) decomposes in refluxing benzene to give aryl azides and alkenes, again stereospecifically ⟨70T3245⟩. However, biaryls, arenes and other products typical of homolytic processes are also formed in a competing reaction, although this pathway can be suppressed by the use of a polar solvent and electron withdrawing aryl substituents.

(**278**) $\xrightarrow{80\,°C}$ $RCH{=}CHR + ArN_3$

Other non-oxidative procedures have also been used to deaminate aziridines. For example, aziridines react with carbenes to yield ylides which subsequently decompose to the alkene. Dichlorocarbene and ethoxycarbonylcarbene have served as the divalent carbon source. The former gives dichloroisocyanides, *e.g.* (**281**), as by-products ⟨72TL3827⟩ and the latter yields imines ⟨72TL4659⟩. This procedure has also been applied to aziridines unsubstituted on the nitrogen atom although the decomposition step, in this case, is not totally stereospecific ⟨72TL3827⟩.

The reaction of carbon atoms with *N*-unsubstituted aziridines leads to alkenes and hydrogen cyanide ⟨72JA3455⟩, probably *via* extrusion from the initially formed adduct (**285**). The fragmentation does not appear to be concerted, although this would be a symmetry-allowed process, since only about half the alkene formed retains the aziridine stereochemistry in the case of *cis*-2,3-dimethylaziridine.

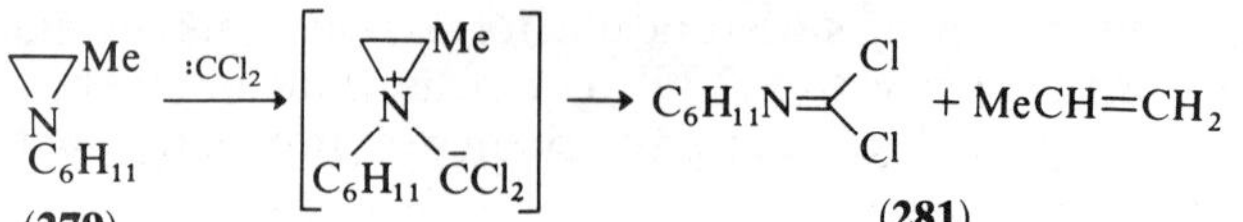

(279) (280) (281)

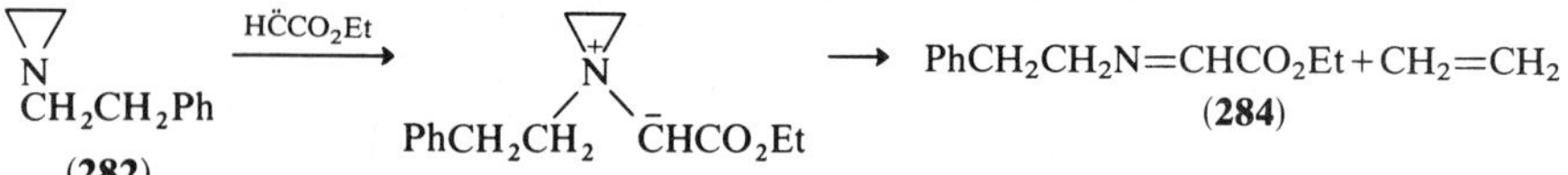

(282) (283) (284)

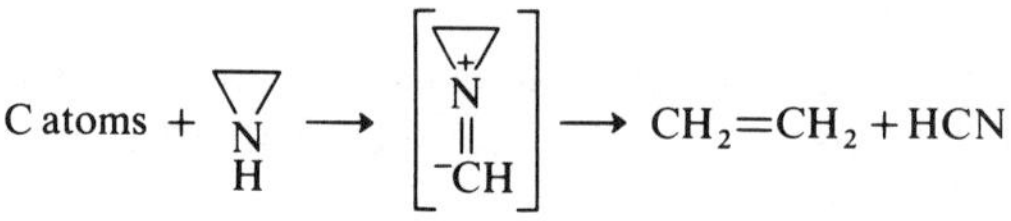

(285)

Finally, one last system worthy of mention involves the thermolysis of aziridine (**286**) in refluxing benzene to generate phthalimidonitrene (**287**) which can be trapped in the usual way ⟨72CC884⟩.

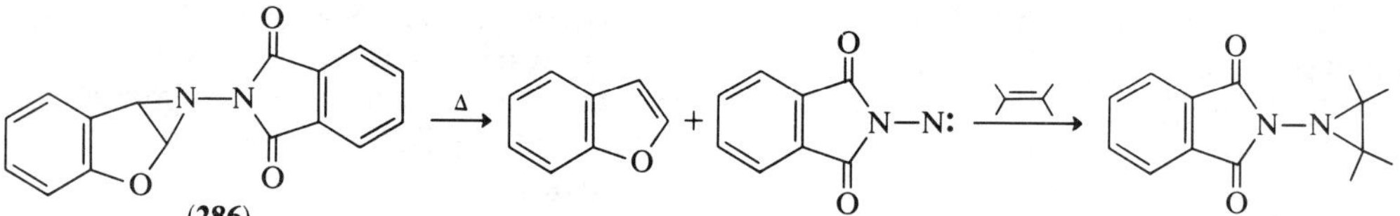

(286) (287) (288)

5.04.3.11 Metal-induced Reactions

The synthesis of metal-coordinated 1-azirines and the reactions of azirines induced by metals have opened a new area in the chemistry of this small ring heterocycle. Many of the reactions encountered bear resemblance to previously discussed thermally and photochemically induced reactions of 1-azirines. The reaction of a series of diiron enneacarbonyls in benzene results in coupling and insertion to give diimine complexes and ureadiiron complexes as well as pyrroles and ketones ⟨76CC191⟩. A mechanism for the formation of these products which involves initial 1,3-bond cleavage and generation of a nitrene–iron carbonyl complex as an intermediate was proposed.

Dimerization reactions of 1-azirines with several transition metal complexes have been studied ⟨76TL2589⟩. Reaction of 2-arylazirines (**289**) with an equimolar amount of a Group VI metal carbonyl gives 2,5-diarylpyrazines (**290**) in good yield. On the other hand, these compounds are converted to 2-styrylindoles (**291**) with rhodium carbonyl compounds or with dicobalt octacarbonyl in benzene.

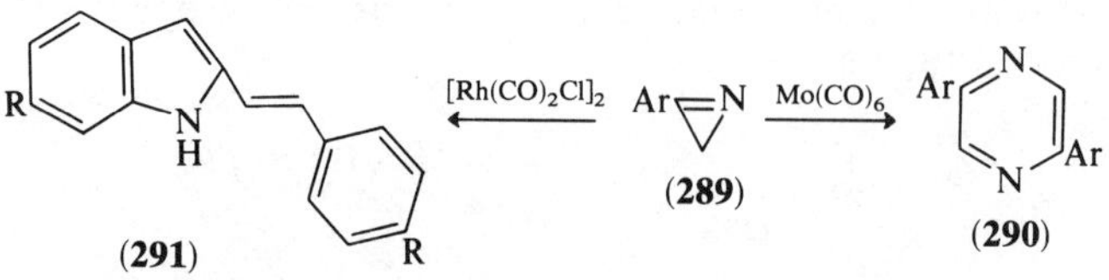

(291) (289) (290)

Treatment of 1-azirine (**292**) with catalytic quantities of dichlorobis(benzonitrile)palladium(II) gave a quantitative yield of the indole (**293**) ⟨77CC664⟩. This transformation proceeds through the intermediacy of a 2 : 1 azirine–palladium chloride complex. Conversion of the 1-azirine ring to indoles under uncatalyzed thermolytic conditions provides a mechanistically interesting comparison with the Pd(II)-catalyzed conversions. The C—N bond cleavage in the latter is apparently accelerated as a result of the coordination of the azirine to palladium.

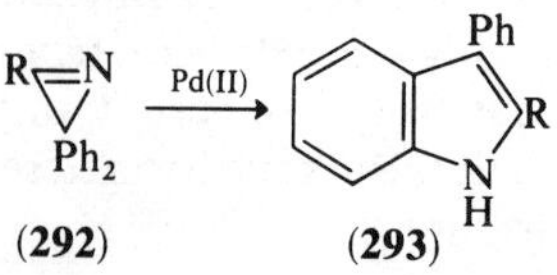

(292) (293)

Finally, the bimolecular cycloaddition of alkynes with 2-phenylazirines in the presence of molybdenum hexacarbonyl has been studied ⟨79TL2983⟩. The pyrrole derivatives (**294**) obtained appear to arise from an initial [2+2] cycloaddition followed by a ring opening reaction.

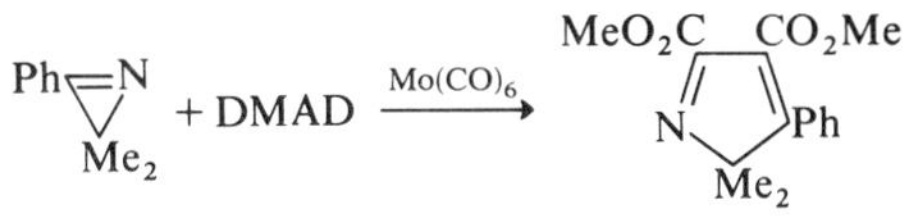

(**294**)

Transition metals have also been inserted into the aziridine ring affording derivatives (**295**). Stereochemical studies suggest that transfer of a proton is followed by bimolecular attack on the ring with subsequent closure on the carbonyl group ⟨76AG(E)495⟩.

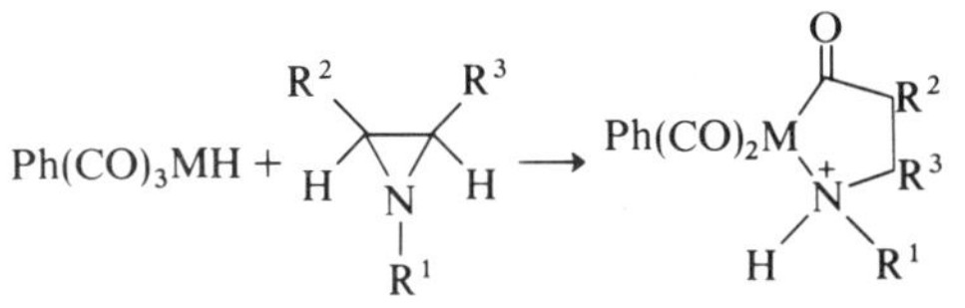

M = Co, Mo, Mn

(**295**)

5.04.3.12 Reactions on Substituents

Side chain modification has been carried out on several aziridines; for example, various aziridine methanols have been prepared from 2-methoxycarbonylaziridines ⟨70JOC3424⟩. Aziridine methanols (**296**) react with thionyl chloride in the presence of base to give both rearranged (**299**) and unrearranged (**298**) chlorides ⟨70JOC3428⟩. The rearrangement product is thought to be formed *via* the azabicyclobutane intermediate (**297**).

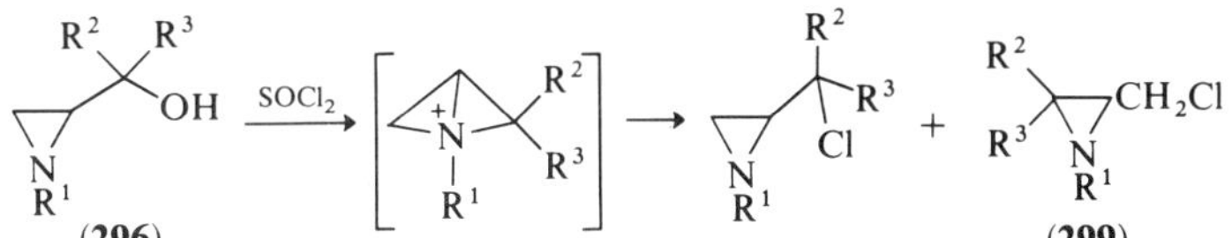

(**296**) (**297**) (**298**) (**299**)

Equilibration of (**300**) and (**302**) in acetic acid has also been observed ⟨68TL5919⟩.

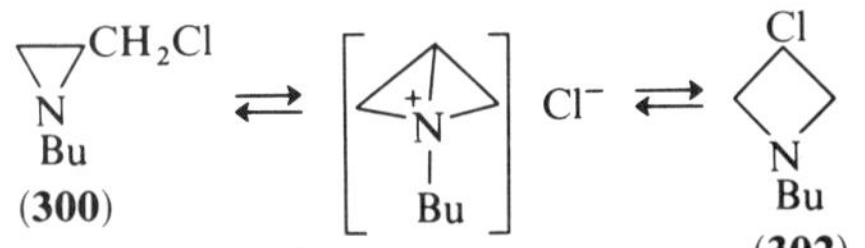

(**300**) (**301**) (**302**)

Solvolysis of tosylate (**303**) yields, in addition to aziridinyl alcohol (**304**), ring expanded products (**305**) and (**306**) ⟨68TL6179⟩. These products and the observed rates of solvolysis were explained in terms of bicyclic intermediate (**301**).

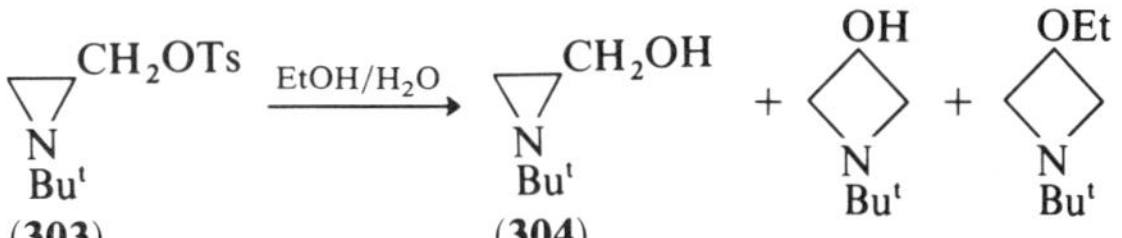

(**303**) (**304**) (**305**) (**306**)

A variety of carboxylate activating groups convert certain aziridine carboxylates (**307**) into 3-halogenoazetidin-2-ones (**309**). The reaction is stereospecific and is believed to proceed *via* a 1-azabicyclo[1.1.0]butan-2-one cation ⟨74JOC902⟩.

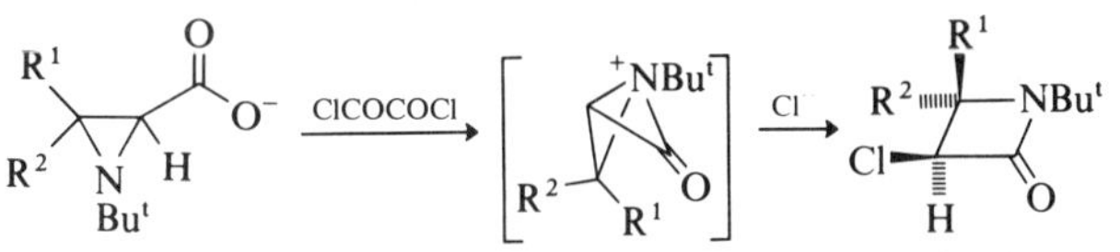

(**307**) (**308**) (**309**)

(310) (311) (312)

Aziridine hydrazides, *e.g.* (**310**), undergo an interesting fragmentation reaction which results in ring opening and diimide formation ⟨68JA3592⟩.

5.04.3.13 Reactions of Aziridines with Exocyclic Unsaturation

Thermal decomposition reactions of alkyl- and aryl-substituted derivatives of the three keto, imino and methylene aziridine ring systems proceed to give isocyanides together with ketones (or aldehydes), imines and alkenes, respectively ⟨80AG(E)276⟩. The thermolyses are thought to proceed *via* an initial 1,3-sigmatropic shift followed by isocyanide formation. Thus, the thermal decomposition of (**313**) proceeds *via* cyclopropanimine (**314**) ⟨73AG(E)414⟩ while the fragmentation of α-lactam (**315**) involves the elusive iminolactone (**316**) as a reactive intermediate ⟨68AG(E)25⟩. A similar rearrangement occurs with aziridinimine (**317**) ⟨77TL1057⟩.

(313) (314) (315) (316)

(317) (318)

(320) (319)

Alternative degradation pathways are operative when the initial aziridines bear β-hydrogen atoms. In these instances α,β-unsaturated amides (**322**) (from aziridinones) and amidines, *e.g.* (**320**), from aziridinimines are frequently isolated as the major products.

(321) (322)

An interesting variation was uncovered on thermolysis of the *N*-ethoxycarbonyl-substituted alkylideneaziridine (**323**) ⟨75JOC224⟩. Heating this material at 130 °C gave oxazoline (**325**) as the exclusive product. In this case cyclization of the diradical occurs *via* a five-membered ring.

(323) (324) (325)

In contrast to the thermolysis experiments, the main products of the photoinduced decomposition of α-lactams (**326**) are carbon monoxide and the corresponding imines ⟨69JA1176⟩. This pathway only occurs to a minor extent in some thermal degradations.

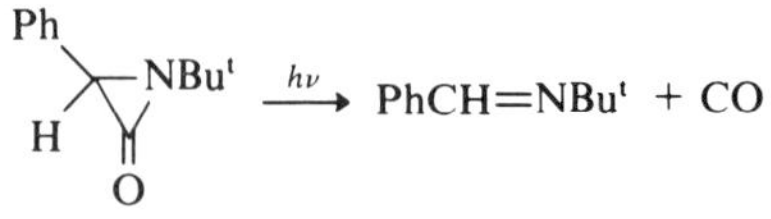

Ph, H, NBu^t, O $\xrightarrow{h\nu}$ PhCH=NBu^t + CO

(**326**)

Several cycloaddition reactions on *N*-alkylideneaziridines have been carried out. Thus, (**327**) reacts with TCNE to give the spiro adduct (**328**), and the reaction with dimethyl acetylenedicarboxylate at room temperature is reported to yield pyrrole (**330**) ⟨67CC928⟩. The course of this reaction has not been completely elucidated. The authors suggest that the primary stages involve successively [2+2] cycloaddition, ring opening of the cyclobutene, valence isomerization and tautomerization to give (**329**). The pathway leading from (**329**) to (**330**) is not understood.

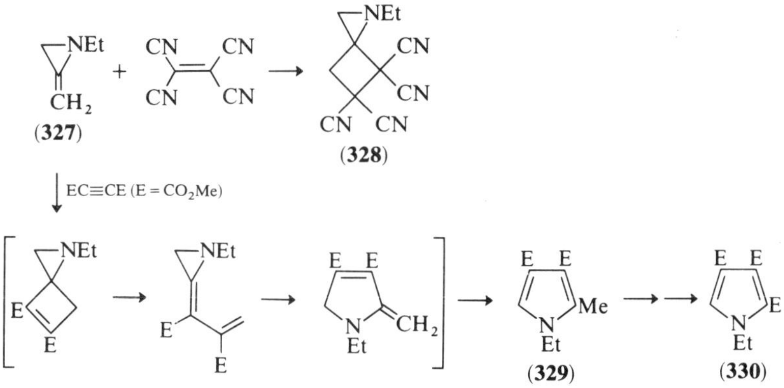

(**327**) (**328**) (**329**) (**330**)

EC≡CE (E = CO_2Me)

Addition on to the exocyclic C—C double bond of an alkylideneaziridine also occurs when this compound is allowed to react with organic azides ⟨75JOC2045⟩. The initially formed spirotriazolines (**332**) are converted into four-membered ring amidines (**334**) with extrusion of molecular nitrogen. In the case of phenyl azide, the amidine (**334**) is obtained alongside the triazoline (**333**).

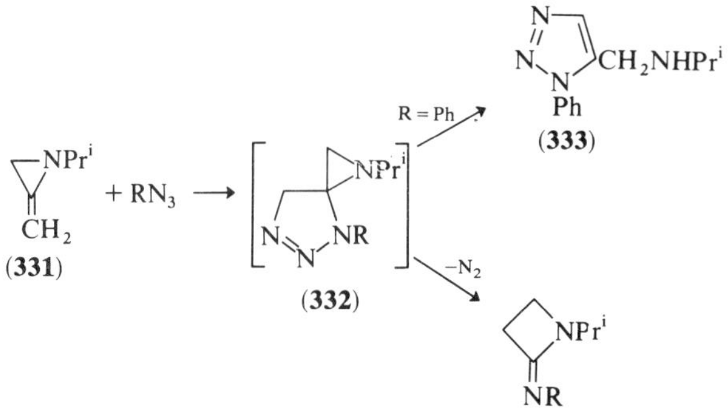

(**331**) + RN_3 → (**332**); R = Ph → (**333**); $-N_2$ → (**334**)

Aziridinones are also known to undergo cycloaddition reactions. Thus, treatment of (**321**) with phenyl isocyanate gave adducts (**335**) and (**336**) ⟨76CL47⟩.

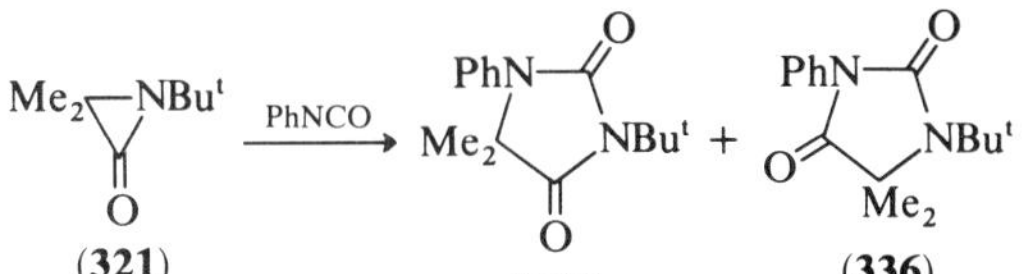

(**321**) $\xrightarrow{PhNCO}$ (**335**) + (**336**)

The sterically stabilized α-lactam (**337**) reacts with adamantyl isocyanide on warming to give the four-membered ring azetidinone (**338**) in good yield ⟨73CC48⟩. The latter is also formed along with the corresponding aldehyde and isocyanide when (**337**) is decomposed at 160 °C. The weakness of the sp^3 C—N bond, which is broken in this reaction, is apparent from an X-ray crystallographic study ⟨72CC43⟩. Furthermore, the substituents attached to the nitrogen atom of (**337**) are not coplanar as they are in amides, thereby inhibiting the π-resonance stabilization which is normally ascribed to amides.

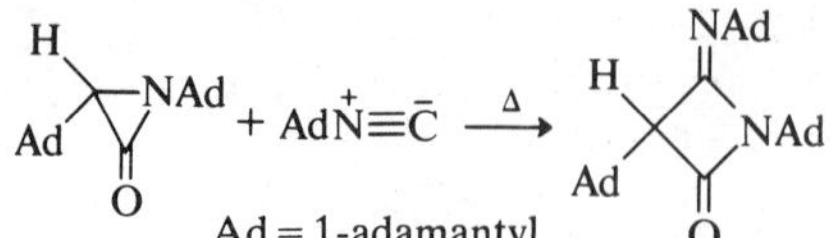

Ad = 1-adamantyl

(**337**) (**338**)

Aziridinones undergo two types of selective ring opening by nucleophiles ⟨68AG(E)25⟩. Reaction with proton-containing nucleophiles, *e.g.* water, alcohols, thiols, amines and mineral acids, leads exclusively to amides (**339**), corresponding to an sp^3 C—N bond rupture. In contrast reaction with aprotic nucleophiles, *e.g.* alkoxides, $LiAlH_4$ and sulfur ylides (Z^-), yields amino acid derivatives (**341**), resulting from sp^3 C—N bond scission. The third possible way of ring opening, namely at the C—C bond, has also been observed in certain cases, *i.e.* (**342**) → (**343**) ⟨67TL5033⟩.

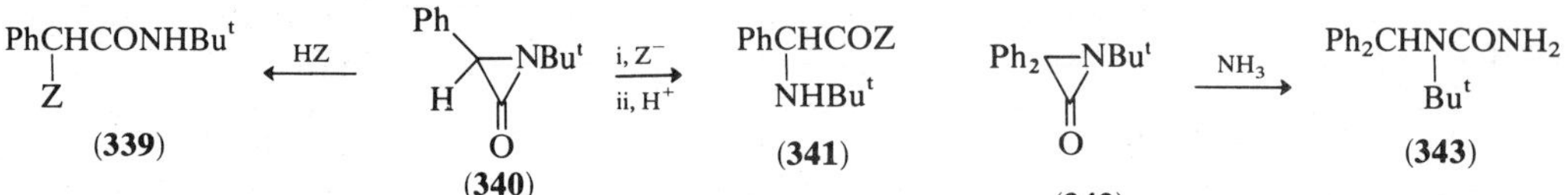

(**339**) (**340**) (**341**) (**342**) (**343**)

The ring opening reactions of *N*-alkylideneaziridines, *e.g.* (**313**), with hydrogen, dilute hydrochloric acid and phenol have also been reported and are outlined below ⟨80AG(E)276⟩.

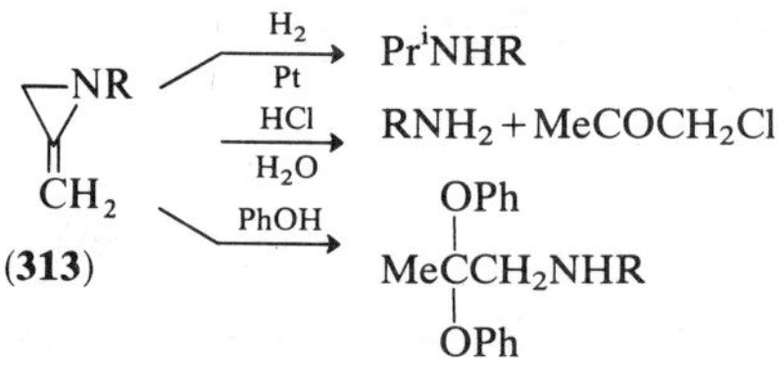

(**313**)

5.04.4 SYNTHESIS OF AZIRIDINES AND AZIRINES

Extensive reviews covering many synthetic and mechanistic aspects of reactions leading to the syntheses of aziridines and azirines have been published in recent years ⟨64HC(19-1)524, B-69MI50401, 71AHC(13)45⟩. Only those reactions to which a measure of generality and/or uniqueness can be assigned will be outlined in the following discussion.

5.04.4.1 Intramolecular Syntheses Forming One Bond

Preparative routes to aziridines and 1-azirines are derived from cycloelimination processes in which one, and sometimes two, bonds are formed directly to the nitrogen atom (Scheme 1). For aziridines these include the two intramolecular cyclization pathways involving either nucleophilic displacement by the amine nitrogen (or nitrenium anion) on the β-carbon (route a) or nucleophilic displacement by a β-carbanionic centre on the amine nitrogen

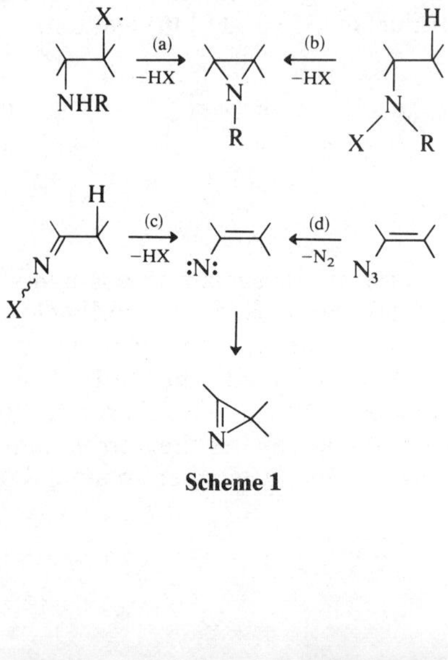

Scheme 1

(route b) with concomitant departure of a suitable leaving group, X. Of the two routes, the latter finds much less application as a general route to aziridines but nonetheless has been used successfully for the preparation of aziridines bearing carbonyl-containing substituents ⟨68JOC887⟩. Intramolecular cyclization initiated by the nitrogen residue is stereospecific, with inversion at the β-carbon, and is prone to failure under circumstances which preclude the formation of the *trans*-diaxial configuration. Most of the routes which are commonly used are derived from the base-induced dehydrohalogenation of β-haloamines (**344**; X= halide), known as the Gabriel synthesis, or from the more recent Wenker modification which utilizes the β-aminohydrosulfate ester (**344**; X=OSO_3H), prepared simply by esterification of the β-aminoalcohol with sulfuric acid or chlorosulfonic acid ⟨62JA605⟩. Numerous *C*-alkyl- and -aryl-substituted aziridines have been synthesized by this method. Both of the above precursors are readily accessible from β-amino alcohols (**344**; X=OH) which are in turn conveniently available from aminolyses of epoxides or from the reduction of α-amino esters. β-Haloamines possessing the halide substituent at tertiary and even at vinyl centers lead to the corresponding aziridines in excellent yields ⟨60JA6068⟩. Sulfonic acid esters (**344**; X=OSO_2R) have also been employed as leaving groups where bulky or electron withdrawing substituents impair the basicity of the amine. Numerous epiminomonosaccharides have been prepared in high yields by treating β-cyanamino mesylates and tosylates with base ⟨65JOC4053⟩. The reductive cyclization of β-azidotosylates might be considered simply as a variation of this route. An intermediate species with the approximate stoichiometry of a β-aminotosylate is not unreasonable. This particular approach has been employed successfully for the preparation of the novel tris-homoaromatic *cis* benzenetriimine (**345**) ⟨73AG(E)989⟩.

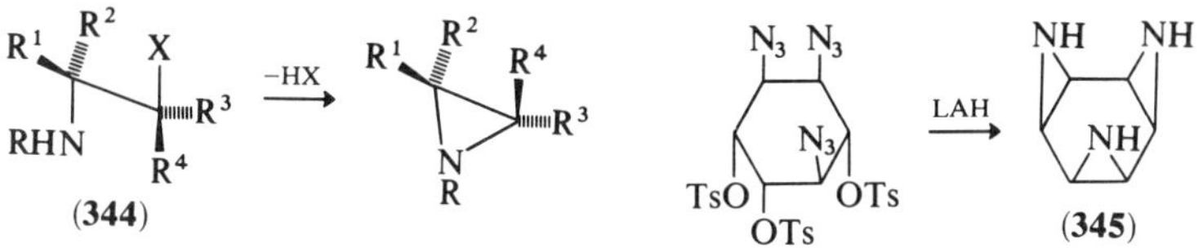

Non-derivatized β-amino alcohols can be converted under very mild conditions directly to aziridines. Typically, treatment of the amino alcohol, either with triphenylphosphine dibromide in the presence of triethylamine ⟨70BCJ1185⟩ or with the reagent prepared by the interaction of triphenylphosphine with diethylazodicarboxylate ⟨78TL5153⟩, affords the corresponding aziridine. The effectiveness of these reagents undoubtedly derives from the driving force associated with the formation of the strong P—O bond in (**346**) and subsequent elimination of a very stable organic fragment.

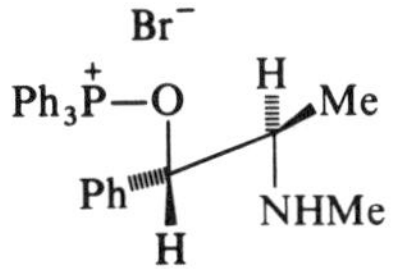

Many variations of the Gabriel synthesis *via* cyclodehydrohalogenation have been described. In many instances the β-haloamines can be regarded as incipient intermediates which are generated *in situ* and which undergo subsequent cyclization under the reaction conditions. For example, a number of steroids bearing pendant aziridine functionality (at C-3 or C-17) have been prepared by the reduction of precursors possessing the chloroacetamide ($NHCOCH_2Cl$) residue ⟨70T1967⟩. Imines and nitriles possessing α-halo substitution serve as suitable precursors to a number of monocyclic aziridines. Thus, reductive cyclization of α-halo- and α,α-dihalo-aldimines, α-haloketimines (**347**) ⟨77RTC242⟩ and α-halonitriles (**348**) ⟨67BCJ432⟩ with excess LAH affords the corresponding aziridine, presumably *via* β-haloamine-type intermediates. β-Haloamines are known to be intermediates in the amine-induced transformations of both α,β-dibromoketones (**349**) and α-bromo α,β-unsaturated aldehydes, ketones and esters, *e.g.* (**350**), into aziridines ⟨74JHC347⟩. This reaction, which lacks any degree of stereospecificity, involves sequential amine-induced dehydrobromination to give (**350**), nucleophilic addition of the amine to give the vicinal bromoamine (**351**) and ultimate cyclization to the aziridine ⟨77JHC459⟩. Precursors such as (**349**) and (**350**) bearing other potently electron withdrawing substituents such as nitro, cyano and phenylsulfonyl can be similarly converted to aziridines. The base-induced ring

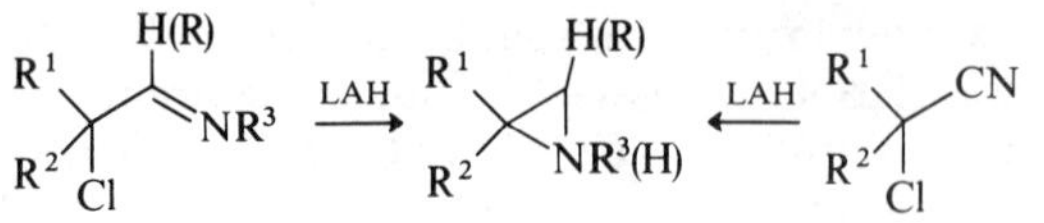

(347) (348)

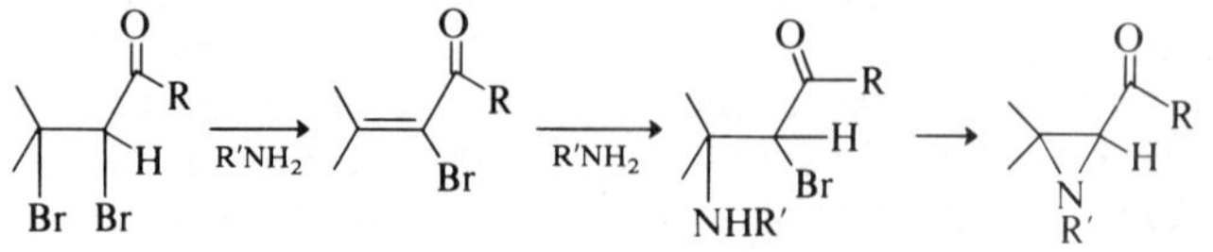

(349) (350) (351)

contraction of 3-haloazetidin-2-ones (**352**) (route a) furnishes aziridine-2-carboxylic acid derivatives ⟨74JOC902⟩. A mechanism involving the intermediacy of a β-haloamine-type species (**353**) is the most plausible. The Darzens condensation between an aldimine and an α-halocarboxylic acid derivative or nitrile (route b) has been extended to provide a non-stereospecific synthesis of *cis*- and *trans*-aziridines bearing pendant amide, ester or nitrile functionality. The stereochemical course of this reaction is evidently influenced by the base–cation system employed ⟨77CC602⟩, and an intermediate (**354**) possessing similar stoichiometry to that (**353**) outlined above has been postulated ⟨69JOC2724⟩.

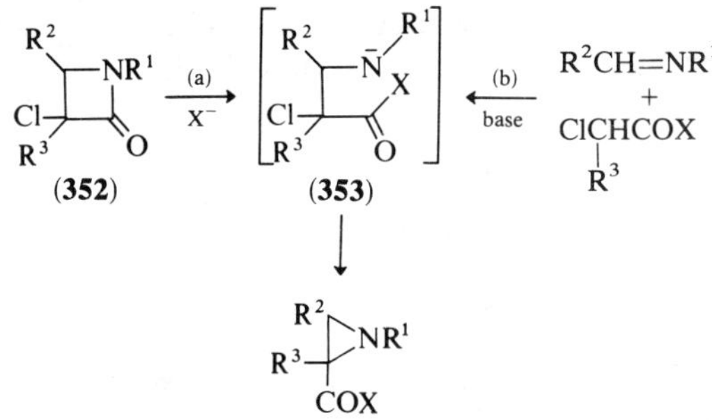

(352) (353)

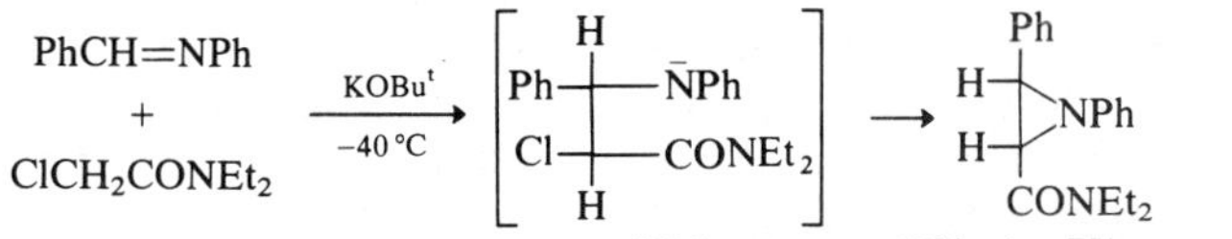

(354) 58% *cis* + 7% *trans*

Cycloelimination reactions initiated by carbanionic displacement at nitrogen (route b, Scheme 1) have been utilized in certain instances for the preparation of aziridines. The proton attached to the β-carbon is rendered suitably acidic by the presence of a contiguous carbonyl function ⟨51JA1044⟩. Leaving groups (X) attached to nitrogen include alkoxy, trimethylammonium (iodide) and halide. For example, quaternization of the hydrazine (**355**) with methyl iodide followed by treatment of the methiodide salt (**356**) with sodium methoxide afforded the aziridine 2-carboxamide in good yields. Synthetic routes to precursors, such as (**355**), derive from the nucleophilic addition of 1,1-dimethylhydrazine, *O*-alkylhydroxylamines or haloamines to the appropriate vinyl ketone derivatives ⟨68JOC887⟩.

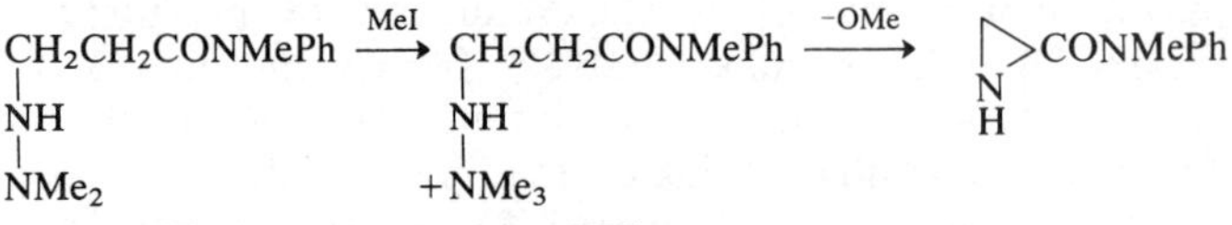

(355) (356)

One of the more important approaches to 1-azirines involves a similar base-induced cycloelimination reaction of a suitably functionalized ketone derivative (route c, Scheme 1). This reaction is analogous to route (b) (Scheme 1) used for the synthesis of aziridines wherein displacement of the leaving group at nitrogen is initiated by a β-carbanionic center. An example of this cycloelimination involves the Neber rearrangement of oxime tosylate esters (**357**; X = OTs) to 1-azirines and subsequently to α-aminoketones (**358**) ⟨71AHC-(13)45⟩. The reaction has been demonstrated to be configurationally indiscriminate; both *syn* and *anti* ketoxime tosylate esters afforded the same product mixture of α-aminoketones

(Scheme 2) ⟨63JOC2271⟩. For this reason vinylnitrenes (**359**) were considered to be the most plausible intermediates in the sequence, following the loss of the tosyloxy residue ⟨B-70MI50402⟩. Concerted cycloelimination processes which would involve both backside and frontside displacement cannot however be excluded. Electrocyclization of the 4π-electron vinylnitrene would then give the azirine which, under the conditions of the Neber reaction, would be hydrolyzed to the amino alcohol. 1-Azirines have in fact been isolated and characterized as intermediates in the Neber rearrangement, although as a preparative route to such compounds this synthesis lacks generality. Ketone dimethylhydrazone methiodides (**360**) also undergo the Neber rearrangement and 1-azirines can be isolated in excellent yields from reactions performed under mildly basic conditions ⟨77JA1514⟩. Evidence which supports the intermediacy of vinylnitrenes in this type of modified Neber rearrangement has been demonstrated ⟨67JA5724⟩. Such reactions are frequently complicated by product rearrangement and/or solvent incorporation; however, the use of non-hydroxylic bases such as NaH in DMSO usually minimizes the formation of deleterious by-products.

$ArCH_2C(=NX)R$ (**357**) $\xrightarrow{base}$ [$Ar\bar{C}HC(=NX)R \xrightarrow{-X^-} ArCH{=}C(\ddot{N}{:})R$ (**359**)] $\rightarrow$ azirine (Ar, H, R) $\rightarrow ArCH(NH_2)C(=O)R$ (**358**)

$X = OTs, \overset{+}{N}Me_3I^-$

$Ar^1CH_2C(=NOTs)CH_2Ar^2 \rightarrow Ar^1\bar{C}HC(=NOTs)CH_2Ar^2 \rightarrow Ar^1CH{=}C(\ddot{N}{:})CH_2Ar^2 \rightarrow$ azirine (H, Ar^1, CH_2Ar^2) $\rightarrow Ar^1CH(NH_2)COCH_2Ar^2$

$Ar^1 = p\text{-}C_6H_4NO_2$
$Ar^2 = p\text{-}C_6H_4OMe$

Scheme 2

$PhC(=NNMe_2)CH(Me)CH(R^1)CH{=}CR^2R^3$ (**360**) $\xrightarrow[R\,=\,alkyl]{i,\ ii}$ 2-methyl-2-[$CH(R^1)CH{=}CR^2R^3$]-3-phenyl-2H-azirine

i, MeI; ii, NaH/DMSO

Alkenes which are susceptible to electrophilic attack often serve as starting materials for stereospecific aziridine syntheses which ultimately proceed through a base-induced cyclodehydrohalogenation of a β-haloamine. The reactions leading to the β-haloamine can be characterized mechanistically by the attack of electrophilic halogen followed by *anti* addition of the nucleophilic nitrogen residue to halonium ion (**361**; Scheme 3). Those systems which have found application to various aziridine syntheses include cyanamide/NBS ⟨70TL1125⟩, *O*-alkylhydroxylamine/NCS ⟨70JA1079⟩, *N,N*-dihaloamines (possessing a strongly electron withdrawing substituent) ⟨75JGU1219⟩ and the more versatile pseudohalogens, iodine isocyanate (prepared *in situ* from silver cyanate and iodine) ⟨67JOC540⟩ and iodine azide (prepared *in situ* from ICl/NaN_3) ⟨71ACR9⟩. The addition of nitrosyl chloride to alkenes which are especially susceptible to electrophilic attack, *e.g.* tetraalkyl-substituted alkenes, proceeds nominally in the reverse manner to give the vicinal chloronitrosoalkane which, after reduction with tin(II) chloride or LAH, can be cyclized to the aziridine, *e.g.* Scheme 4 ⟨64AG(E)642⟩.

In some cases, the β-haloamine undergoes spontaneous cyclization to the aziridine ⟨70TL1125⟩. Most of these routes, however, require either reductive or base-induced cyclization to the aziridine. Access to a vast number of aziridines and 1-azirines has been

(361)

Scheme 3

Scheme 4

gained through the cyclization of vicinal haloisocyanates and haloazides, which are in turn prepared by the addition of iodine isocyanate and iodine azide (and to a lesser extent bromine and chlorine azides) to alkenes. Iodine isocyanate adds stereospecifically *trans* to give the β-iodoisocyanates, the formation of which has been rationalized as resulting from opening of a three-membered ring iodonium ion intermediate by backside approach of the isocyanate anion. Whilst the regiospecificity with which the addition occurs is of no consequence inasmuch as it pertains to aziridine syntheses, the reactions proceed in some instances regiospecifically (*e.g.* styrene, Scheme 5) and in others only regioselectively (*e.g.* 1-hexene) ⟨70JA1326⟩. A dependency on steric factors has been clearly demonstrated by the regiospecific manner in which INCO adds to *t*-butylethylene to give the terminal isocyanate derivative. Conjugating unsaturated acids, esters, ketones and nitriles are unreactive towards this reagent. The β-iodoisocyanates are converted quantitatively to the corresponding carbamic acid and ester derivatives by treatment with aqueous base or with alcohols. Cyclization to the aziridine, hydrolysis and decarboxylation to give the *N*-unsubstituted derivative can be effected in a single step with aqueous base ⟨67JOC540⟩.

Scheme 5

β-Iodoazides are formed quantitatively and stereospecifically *trans* from the *anti* addition of IN_3 to alkenes ⟨71ACR9⟩. Regiochemically, the addition is consistent with the intermediacy of the iodonium ion which opens *via* the lower energy transition state to give the non-terminal iodoazide (Scheme 6). In contrast to INCO, IN_3 adds to α,β-unsaturated ketone

Scheme 6

systems to give β-iodoazides. However, in a few instances regioisomeric mixtures are obtained. Reactions of alkenes with bromine azide and chlorine azide predictably exhibit degrees of free radical character which are markedly dependent upon solvent polarity. Polar media favor the formation (*via* the halonium ion) of non-terminal haloazides whereas non-polar media favor the formation (*via* initial attack by $\cdot N_3$) of the regioisomeric terminal haloazides. With the appropriate choice of reaction conditions it is possible to prepare almost exclusively each of the regioisomeric β-haloazides. Treatment of the resulting haloazides with LAH effects a clean, stereospecific reductive cyclization to the *N*-unsubstituted aziridines. Cyclization and simultaneous derivatization at nitrogen can also be accomplished with phosphines, phosphites ⟨70JA3733⟩ and certain chloroborane reagents ⟨73JA4067⟩. β-Iodoazides possessing a hydrogen β to the iodo substituent can be stereospecifically dehydrohalogenated (in the direction of the azide function) very smoothly with base to give vinyl azides which can be converted, through the elimination of nitrogen, to 1-azirines (route d, Scheme 1) ⟨67JA2077⟩. This haloazide route (Scheme 6) is therefore particularly useful for the synthesis of both the aziridine and 1-azirine rings. Vinyl azides can be decomposed thermally and photochemically to give 1-azirines. Although vinylnitrenes are commonly implicated as discrete intermediates in these fragmentation reactions (route a, Scheme 7) and indeed in the Neber rearrangement, alternative mechanistic pathways are as plausible. That which is believed to be operating, at least in the primary stages of the azide decomposition, is one involving anchimeric assistance to nitrogen elimination by the contiguous π-bond as depicted in (**362**) ⟨75AG(E)775⟩. Consideration has also been given to a similar pathway (route c) involving the loss of nitrogen from an unstable 1,2,3-4*H*-triazole intermediate ⟨62JOC3557⟩. The thermal mode of nitrogen elimination has been used successfully for the preparation of a large number of 1-azirines ⟨62JOC3557⟩. In certain cases, especially when the substituent groups are thermally labile, the preferred mode of elimination is generally photochemical. Reactions can be performed at much lower temperatures and in dilute solution, factors which lead to higher yields and less decomposition of often unstable azirines. The spiro 1-azirine (**363**), for example, can only be prepared by the photochemical decomposition of the terminal vinyl azide (**364**); thermolysis of the same azide leads only to dimer (**365**) formation (25%) ⟨69AG(E)772⟩.

(a)
(b)
(c)
(**362**)

Scheme 7

$h\nu$, −15 °C
Δ, 78 °C

(**363**) (**364**) (**365**)

A variety of 1-azirines, including the first fused derivatives, have been prepared by the selective irradiation at 350 nm of the appropriate vinyl azide precursors ⟨68JA2869⟩.

Azirines have been shown to be discrete intermediates in the Hoch–Campbell synthesis of *N*-unsubstituted aziridines (Scheme 8). Ketoximes possessing an α-hydrogen atom, when treated with an excess of a Grignard reagent, initially give the 1-azirine which undergoes subsequent nucleophilic attack by an additional equivalent of Grignard reagent to give the corresponding aziridine. Vinylnitrenes have been implicated as intermediates in this reaction. An analogous reaction which is mechanistically very similar to the Hoch–Campbell synthesis can also be effected with complex metal hydrides. Unlike the Neber reaction, this cyclization is regiospecific in that proton abstraction occurs from the side *syn* to the oxime hydroxyl ⟨70BSF3003⟩. Subsequent nucleophilic attack by Grignard reagent or by hydride upon the azirine ring occurs stereoselectively from the least hindered side ⟨73BSF2680⟩. Thus, treatment of 2-phenylcyclohexanone oxime (**366**) with an excess (>3 mole equivalents) of methylmagnesium bromide affords both the *cis*- and *trans*-azabicyclo[4.1.0]heptanes (**367**) and (**368**) in the ratio of 3 : 1. This route has also been applied successfully using α-hydroxy, α-keto and α,β-unsaturated oximes. Moreover, the

Scheme 8

(366)

(367) *cis*

(368) *trans*

reaction is not restricted to the use of oximes: *N,N,N*-trimethylhydrazonium salts such as (**357**; X = $\overset{+}{N}Me_3\ I^-$) also serve as precursors to aziridines.

5.04.4.2 Intermolecular Syntheses Forming Two Bonds

All of the conceivable modes of intermolecular 1,1-cycloaddition reaction (Scheme 9) which can lead to aziridines and to 1- and 2-azirines have been utilized with varying degrees of success for the syntheses of these heterocycles. The electrophilic addition of singlet nitrenes, generated from organic azides, from the 1,1-elimination from aryloxysulfonylurethanes or from the oxidation of certain primary amines, to suitably substituted alkenes has been employed as a stereospecific entry to the aziridine ring system ⟨B-70MI50402⟩. The facility with which the initially formed singlet nitrenes undergo intersystem crossing to the triplet species (and therefore to non-stereospecific addition and competing abstraction and insertion processes) is dependent to a large extent upon the nitrene substituent. Nitrenes which are adjacent to a carbonyl group, to a saturated nitrogen atom and, to a lesser extent, to an oxygen atom are species which are preparatively the most useful even though overall yields are sometimes very low. For example, oxidation of a series of *O*-alkylhydroxylamines with LTA in the presence of simple alkenes affords the corresponding alkoxyaziridines in yields of 9–35%, some of which have been isolated at ambient temperature as stable invertomers, *e.g.* (**369**) and (**370**) ⟨73TL619⟩. Nitrenes with aryl or potently electron withdrawing groups do not normally add to alkenes to give aziridines. Singlet alkoxycarbonylnitrenes

Scheme 9

RONH$_2$ ⟶ RO$\ddot{N}$ ⟶ (**369**) + (**370**)

R = Me, Et, Pr^i

add stereospecifically to alkenes while triplet nitrenes add non-stereospecifically. Monoadditions to 1,3-dienes and α,β-unsaturated esters have also been reported to occur in yields of 60–70%. Certain *N*-aminonitrenes, the singlet configurations of which derive stabilization from adjacent functionality, add stereospecifically and in high yields to alkenes ⟨70JCS(C)576⟩. It has been suggested that the singlet configuration can be stabilized by delocalization of the lone pair electrons of the trivalent nitrogen into the vacant *p*-orbital in the monovalent nitrogen, thus conferring more nucleophilic character upon the terminal nitrogen. Oxidation of the readily accessible *N*-amino heterocycles (**371**)–(**373**) with LTA in the presence of a wide range of electrophilic and nucleophilic alkenes provides an excellent entry to the aziridine ring system. Subsequent removal of the larger heterocyclic moiety and rederivatization of the aziridine ring can be accomplished by reaction with hydrazine ⟨78HCA795⟩.

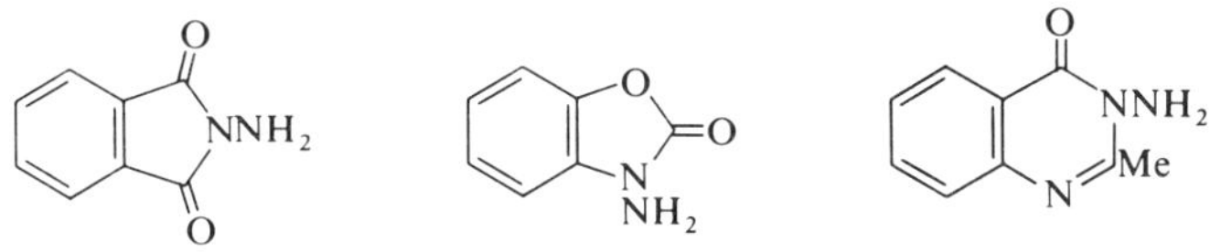

(**371**) (**372**) (**373**)

The addition of phthalimidylnitrene (**374**) to simple alkynes affords 1-azirines in yields of 1–15% (Scheme 10). In this reaction, which is of no real preparative value, the symmetrical 2-azirines (**375**) were suggested as the most plausible intermediates and unequivocal proof of the existence of such species was demonstrated from a series of 1,2,3-triazole pyrolysis reactions ⟨71CC1518⟩. Extrusion of nitrogen from the regioisomeric 4,5-disubstituted 1,2,3-triazoles (**376**) during flash vacuum pyrolysis furnished identical product mixtures which included both regioisomeric 1-azirines (**377**).

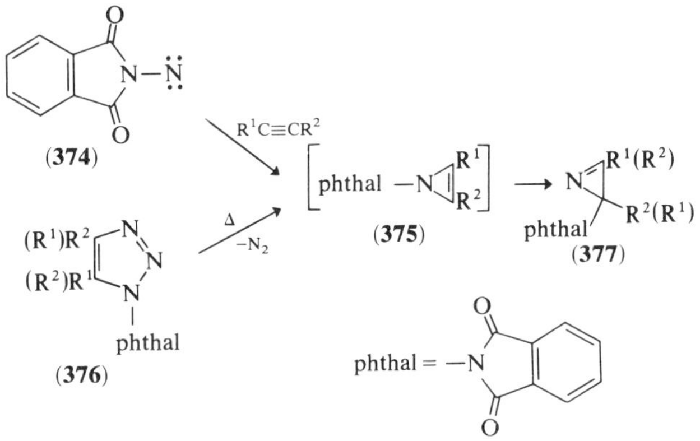

(**374**) $R^1C{\equiv}CR^2$ (**375**) (**377**) (**376**) Δ, $-N_2$

Scheme 10

Intramolecular alkylnitrene addition to an alkenic moiety situated δ,ε to the electron deficient center has been utilized for the preparation of bi- and tri-cyclic aziridines (Scheme 11) ⟨68JA1650⟩. Oxidation of the primary alkylamine can be effected cleanly with NCS, LTA or mercury(II) oxide.

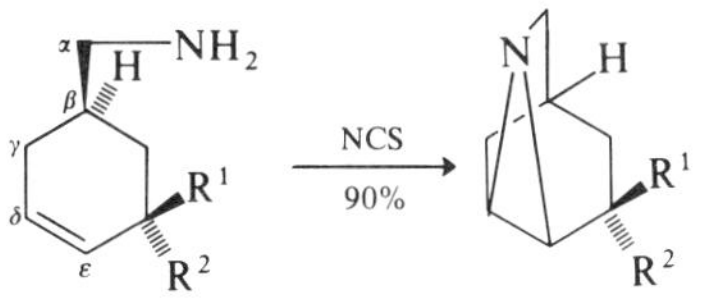

NCS, 90%

Scheme 11

Aziridines have been prepared stereospecifically by the nucleophilic addition of the nitrogen residue to alkenes ⟨80T73⟩. Introduction of the 'nitrene' is accomplished readily *via* a Michael-type addition with free diphenylsulfilimine (Scheme 12), and where a chiral sulfilimine is used the chirality is transferred to the aziridine with optical yields in excess of 25%.

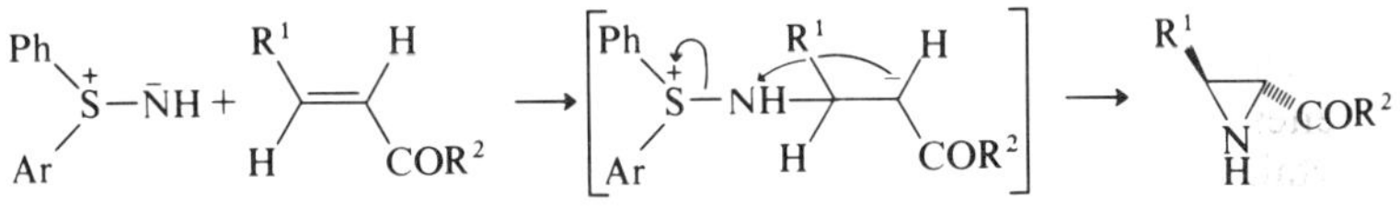

R^1, R^2 = aryl

Scheme 12

Corresponding syntheses of aziridines and 1-azirines have been accomplished by carbene additions to imines and nitriles (Scheme 9). Symmetrical and unsymmetrical dihalocarbenes, prepared by standard methods, undergo facile addition to the C=N bond of arylaldimines and ketimines to give *gem*-dihaloaziridines ⟨78JOC1346⟩. 'Methylenation' of simple *N*-arylaldimines and 1-azirines, *e.g.* (**378**), with dimethyl sulfonium and sulfoxonium methylides leads to the formation of aziridines ⟨65JA1353⟩, and in the presence of benzonitrile leads to 2-phenyl-1-azirine (**378**) in moderate yields ⟨65AF49⟩. Methylene addition, either through the copper-catalyzed decomposition of diazoalkanes or *via* the Simmons–Smith procedure for cyclopropanation, finds little application as a preparative route to aziridines.

$$(378) \xrightarrow{Me_2\overset{+}{S}-\overset{-}{C}H_2} \text{2-phenylazetidine-type product} \qquad PhCN \xrightarrow{Me_2\overset{+}{S}(=O)\overset{-}{C}H_2} (378)$$

5.04.4.3 Syntheses Involving Ring Contraction

The thermally or photochemically induced extrusion of stable, neutral molecules from five-membered heterocycles has also been used as an entry to each of the aziridine, 1-azirine and 2-azirine ring systems. Δ^2-1,2,3-Triazolines (**379**), prepared by the 1,3-dipolar cycloaddition of organic azides to alkenes or, less commonly, by the addition of diazoalkanes to imines, serve as precursors to aziridines. Extrusion of nitrogen can be accomplished by either mode of decomposition ⟨B-73MI50402⟩, but the photochemical route is generally preferred since by-product formation (*e.g.* isomeric imines) is minimized. Diradical, zwitterionic and concerted fragmentation mechanisms have been proposed to account for the formation of aziridines and of isomeric by-products ⟨B-70MI50401⟩. Of the possible mechanisms for the thermal ring contraction, that which is the most consistent with observed kinetic data involves an initial heterolysis of the weak N-1—N-2 bond which leads to the zwitterionic diazonium species (**380**). Cyclization by nucleophilic displacement by the amine anion and expulsion of nitrogen would lead to the aziridine, whereas displacement of nitrogen by a migrating group (from C-5) would lead to the isomeric imine (**381**). The Δ^2-triazoline photodecomposition process, on the other hand, is highly consistent with an initial homolysis of the N-1—N-2 bond which leads to a singlet diradical (**382**) which then loses nitrogen to give diradical (**383**). Ring closure either preceding or succeeding C—C bond rotation then affords the aziridines. The photochemical extrusion of nitrogen from monocyclic Δ^2-triazolines generally displays a degree of stereoselectivity in which the triazoline stereochemistry is preserved in the aziridine ring ⟨B-70MI50401⟩. Monocyclic, fused and spiro aziridines, *e.g.* (**384**) ⟨74JOC63⟩, have been synthesized by these routes. The photolytic extrusion of nitrogen from 4*H*-1,2,3-triazoles (**385**), synthesized quantitatively by treatment of α-chloroenamines (**386**) with sodium azide, affords the corresponding 2-amino-substituted 1-azirines (**387**) in yields of 85–90% ⟨80CC940⟩.

Although by no means a preparative route to either 1- or 2-azirines, the elimination of nitrogen (by flash vacuum pyrolysis at 400 °C) from the regioisomeric 4,5-disubstituted 1*H*-1,2,3-triazoles (**376**) leads to similarly regioisomeric 1-azirines (**377**) ⟨73JCS(P1)550⟩.

(379) —Δ→ [(380)] —$-N_2$→ (381) + aziridine (NR)

(379) —$h\nu$→ [(382)] —$-N_2$→ [(383)] → aziridine (NR)

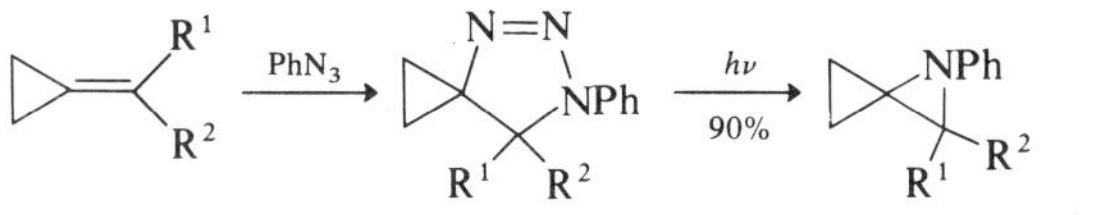

(384)

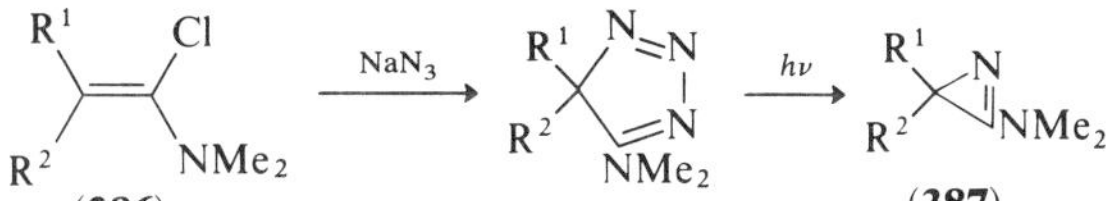

(386) (385) (387)

A variety of 1-azirines are available (40–90%) from the thermally induced extrusion (>100 °C) of triphenylphosphine oxide from oxazaphospholines (**388**) (or their acyclic betaine equivalents), which are accessible through 1,3-dipolar cycloaddition of nitrile oxides (**389**) to alkylidenephosphoranes (**390**) ⟨66AG(E)1039⟩. Frequently, the isomeric ketenimines (**391**) are isolated as by-products. The presence of electron withdrawing functionality in either or both of the addition components can influence the course of the reaction. For example, addition of benzonitrile oxide to the phosphorane ester (**390**; $R^1 = CO_2Et$) at room temperature affords only the ketenimine (**391**; $R^1 = CO_2Et$), whereas addition of ethoxycarbonylfulminate (**389**; $R = CO_2Et$) to the same ylide at 80 °C, affords exclusively the 1-azirine. Oxazaphospholines, prepared by an alternative route from α-bromoketoximes, have been used as precursors to 1-azirines which are not easily accessible through the more general vinyl azide decomposition route ⟨79JOC3861⟩. This particular approach circumvents the necessity of handling potentially hazardous, low molecular weight vinyl azides.

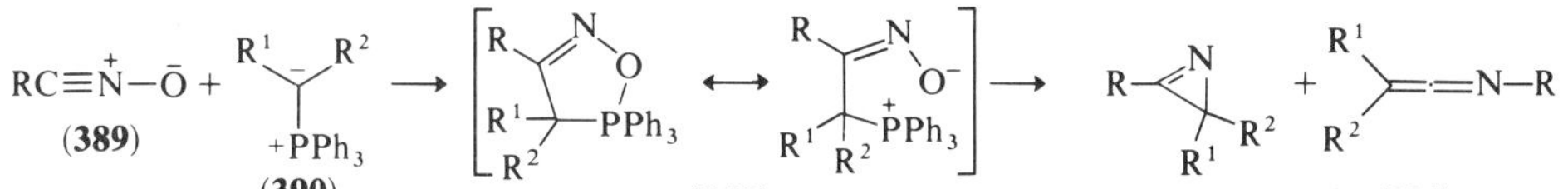

(389) (390) (388) (391)

The facile, thermally induced ring contraction of 3,5-disubstituted isoxazoles (**392**) has been utilized as an entry to the 1-azirine ring system (Scheme 13). Similar photoinduced isomerizations have also been observed with several 3,5-diaryl-substituted isoxazoles, but yields are generally very low ⟨75S20⟩. Azirines possessing pendant ester and thioester functions at C-3 have been prepared by thermolysis (neat at 200 °C) of the corresponding 5-alkoxy- and 5-alkylmercapto-3-phenylisoxazoles ⟨70T453⟩. 3,5-Diamino-substituted isoxazoles can be isomerized in high yields thermally and photochemically to 2-amino-1-azirines. Both the thermal and photochemical rearrangements are believed to proceed *via* an initial homolytic cleavage of the weak N—O bond, followed by ketone formation and 1,3-diradical reunion.

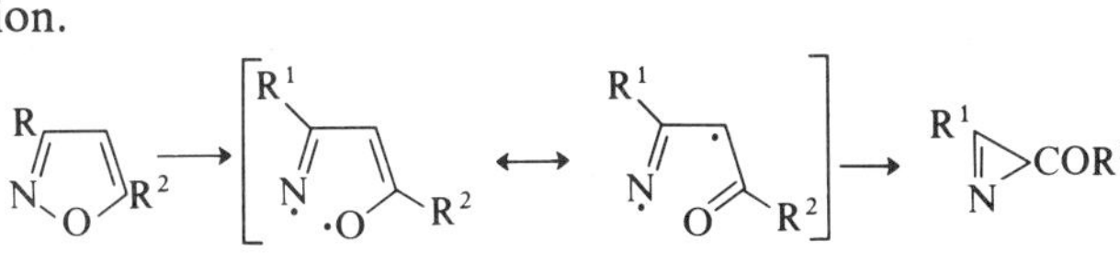

(392)

Scheme 13

In a manner analogous to that observed for the LAH-induced cyclization of oximes and oxime ethers, the LAH reduction of Δ^2-isoxazolines (**393**) results in a ring contraction which leads to the formation of *N*-unsubstituted aziridines ⟨70T539⟩. The yields depend principally upon the substituents at C-4 and C-5 of the isoxazolines, which are in turn easily accessible from the 1,3-dipolar cycloaddition of nitrile oxides to appropriately substituted alkenes. Aziridine formation is favored from those Δ^2-isoxazolines with phenyl groups at both the C-4 and C-5 positions. Considerable attention has been devoted to a study of the mechanism of this reaction (Scheme 14) ⟨B-73MI50401⟩. 1-Azirines are believed to be discrete intermediates which derive from the metallated species (**394**) by hydride attack at C-5 followed by cyclization and simultaneous cleavage of the N—O bond. Subsequent hydride attack from the least hindered side of the ring then furnishes the corresponding aziridine.

Δ^4-Isoxazolines, which are readily accessible by 1,3-dipolar addition of nitrones and nitronic esters to activated alkynes, undergo facile rearrangement upon warming (<110 °C)

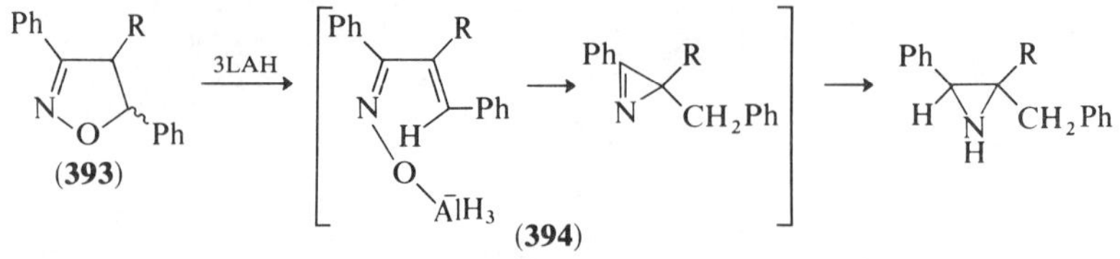

Scheme 14

to give *N*-substituted aziridinyl ketones in moderate yields. Aziridines possessing additional side chain functionality can be prepared by this route ⟨68JA5325⟩. For example, the labile Δ^4-isoxazoline (**395**) isomerizes smoothly to the aziridinyl α-hydroxyketone derivative (**396**) after two hours at 78 °C. In some cases the aziridines are isolated directly from the cycloaddition reaction. The reaction of nitronic esters (**397**) with alkynic ketones and esters allows the stereoselective synthesis of invertomers of *N*-methoxyaziridines (**399**). The rehybridization which accompanies the addition of the nitrones leads to the formation of diastereoisomeric Δ^4-isoxazolines (**398**) under kinetic control. The asymmetric center which is created at the nitrogen atom is maintained during the ring contraction which has been rationalized in terms of thermally allowed 1,3-sigmatropic pathways ⟨77JA6667⟩.

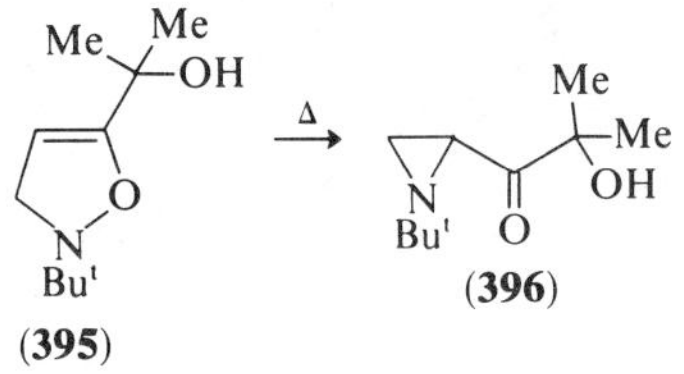

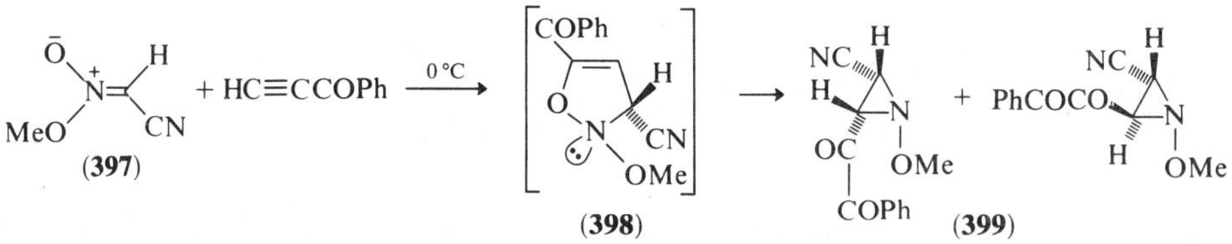

The preparative routes to aziridines so far described have all involved either intra- or inter-molecular cyclization processes. In the latter routes, which derive from oxime and Δ^2-isoxazoline precursors, aziridines are formed as a result of nucleophilic addition of Grignard reagent or hydride to the C=N bond of the intermediate 1-azirines ⟨73BSF2680⟩. Several additional routes to aziridines take advantage of this electrophilic character of azirines and various C, N, O, P and halide nucleophiles have been added across the imine bond. Such addition reactions are generally stereospecific and occur optimally by approach of the nucleophile from the less hindered side of the ring. Reduction of 1-azirines by LAH or by catalytic hydrogenation affords the *cis*-aziridines exclusively, and in high yields. Acyl and aroyl chlorides add to azirines in excellent yields to give stable *N*-acylhaloaziridines (**401**) which can be efficiently refunctionalized at C-2 with azido and acetoxy substituents ⟨75JA4692⟩. A mechanism involving halide addition to the initially formed *N*-acylazirinium ion (**400**) has been demonstrated.

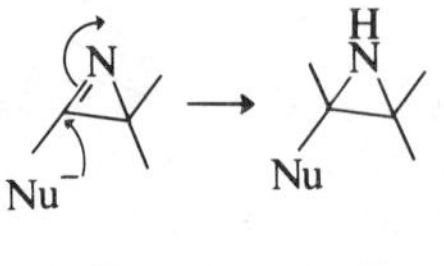

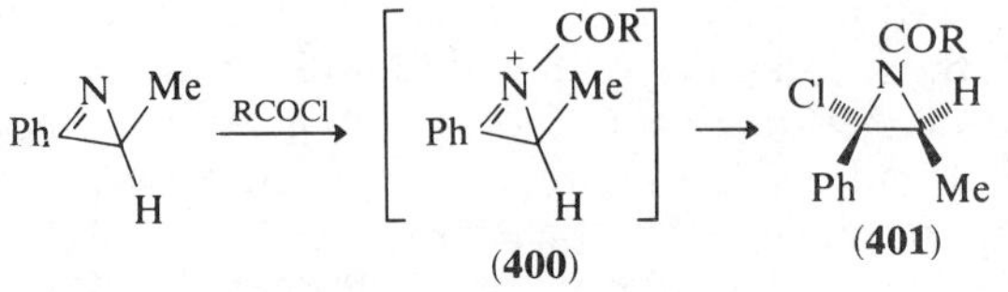

The addition of carbon nucleophiles through organometallic reagents has been shown to be involved in the mechanism by which oximes are converted to aziridines ⟨73BSF2680⟩. In fact, this particular reaction of 1-azirines with Grignard reagents has been utilized as a general synthesis of *N*-unsubstituted aziridines ⟨69TL4001⟩. Organolithium, *e.g.* $LiCCl_3$, and organozinc (Reformatsky) reagents ⟨78T3291⟩ have been used for the preparation of aziridines bearing pendant functionality (Scheme 15). 2-Phenylazirines, on heating (145 °C) in the presence of the azlactone (**402**), afford the aziridinyl-Δ^2-oxazolinones (**403**) in high yields. An ene reaction mechanism involving the enol form of the azlactone has been demonstrated. The adducts derived from similar reactions with dimedone spontaneously isomerize under these conditions ⟨73HCA1351⟩.

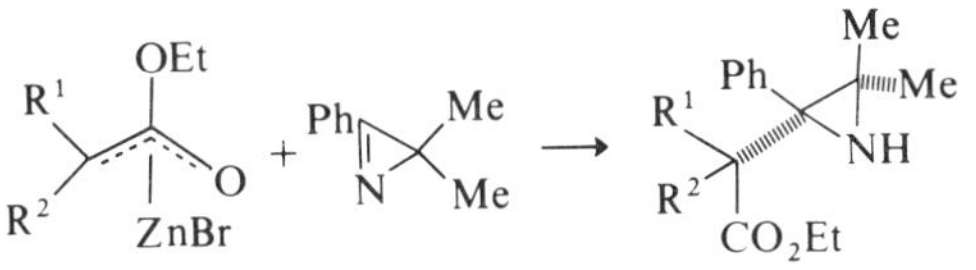

Scheme 15

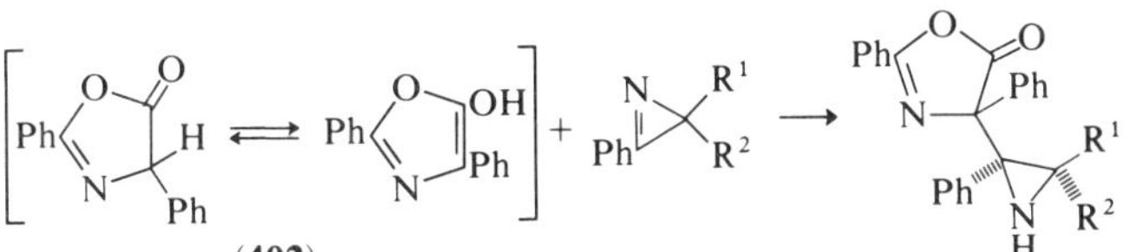

(**402**) (**403**)

Aziridines are readily accessible from epoxides in a single transformation with iminophosphoranes ⟨76CB814⟩ and phosphoramidate esters ⟨76TL4003⟩. The reactions, illustrated here with the latter reagent (**404**), involve nucleophilic attack by nitrogen upon the lesser substituted carbon atom of the epoxide followed by ring closure to give the 1,3,2λ^5-oxazaphospholidine (**405**), which fragments with the extrusion of the stable phosphate ester residue. Reactions are performed typically in the region 140–200 °C and aziridine yields are consistently *ca.* 60% ⟨76TL4003⟩.

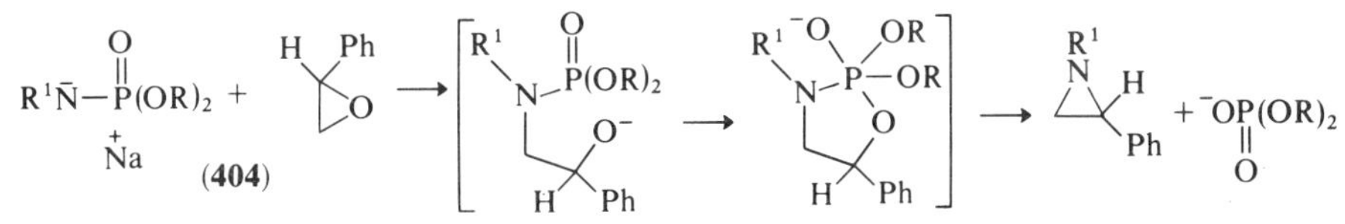

(**404**) (**405**)

Amine nucleophiles such as hydrazine, hydroxylamine, pyrazole and azide add to 1-azirines from the less hindered side of the ring as previously described to give the corresponding aziridines in high yields. For example, the reaction of the 1-azirine (**406**) with the pyrazole (**407**) in refluxing toluene affords (80%) the 1:1 adduct (**408**) ⟨74CC782⟩. Oxygen nucleophiles such as alkoxide add to the imine bond to give mixtures of the *cis*- and *trans*-aziridinyl alkyl ethers ⟨68BCJ1440⟩. Trialkyl phosphites also react with 1-azirines by nucleophilic addition of phosphorus to the C=N bond to give the corresponding aziridin-2-yl phosphonates, *e.g.* (**409**) ⟨71JCS(C)3021⟩.

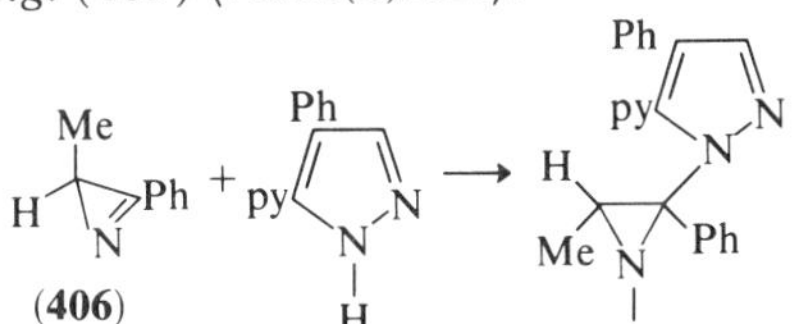

(**406**) (**407**) py = 2-pyridyl (**408**) 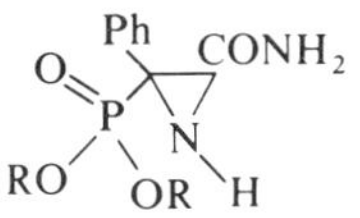 (**409**)

Bicyclic aziridines possessing a common C—N bond are readily accessible from 1-azirines by cycloaddition. As 2π-electron components, azirines are capable of reaction either as dipolarophiles (to give fused five-membered heterocycles) or as dienophiles (to give fused six-membered heterocycles) in the [4+2] cycloaddition process ⟨B-79MI50400⟩. The fused aziridines produced by 1,3-dipolar addition of certain dipoles, *e.g.* nitrile oxides ⟨71TL4831⟩ and diazoalkanes ⟨64JOC3049⟩, are frequently so unstable that only skeletally rearranged isomers are isolated. Nitrile ylides, derived either from the irreversible photochemical ring opening of 1-azirines ⟨72JA1395⟩ or by the standard procedure from the α-haloimine precursors ⟨73HCA1351⟩, readily add to azirines to give the novel diazabicyclo[3.1.0]hexenes (**410**). Azomethine ylides, produced by the thermally induced ring opening of aziridines, behave similarly to afford (>70%) mixed isomers of the diazabicyclohexanes (**411**) ⟨73CL759⟩. Diels–Alder reactions with 1-azirines occur primarily with electron deficient dienes

⟨B-79MI50400⟩; however, in most instances the aziridine ring so formed is destroyed in subsequent rearrangements. 1,3-Diphenyl isobenzofuran reacts with 1-azirines to give the stable *exo*-fused aziridines (**412**) in excellent yields (80–90%). Cycloaddition reactions of 1-azirines with certain heterocumulenes also lead to novel fused heterocyclic aziridines in good yields. In some cases the product aziridines result from nominally [4+2] cycloaddition processes, *e.g.* (**413**) ⟨74TL1487⟩, and in others the net incorporation of two equivalents of heterocumulene to give (**414**) ⟨72TL1353⟩ is suggestive of a stepwise union.

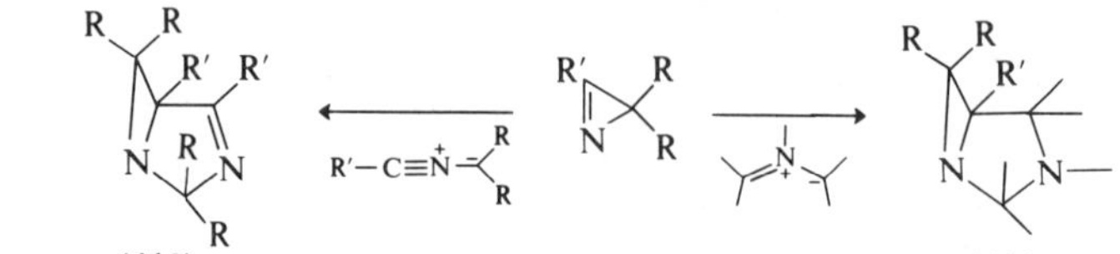

(**410**) (**411**)

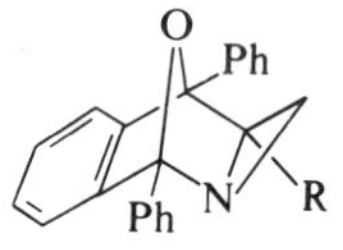

(**412**)

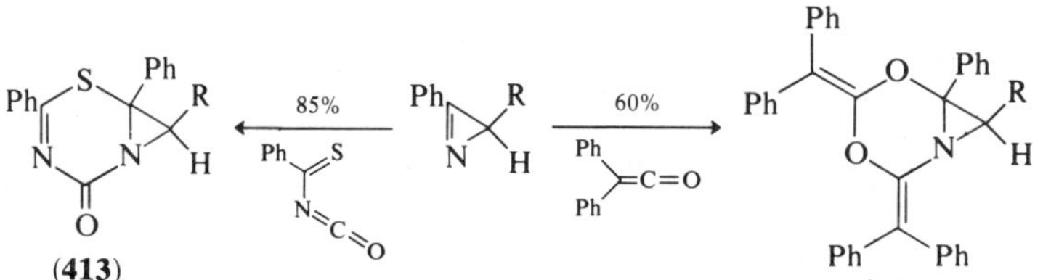

(**413**) 85% 60% (**414**)

5.04.4.4 Synthesis of Aziridines Containing Exocyclic Unsaturation

Principal synthetic routes to the more highly strained aziridines possessing exocyclic unsaturation, namely methyleneaziridine, aziridinone (α-lactam) and aziridinimine, involve an ultimate base-induced cycloelimination reaction. Derivatives of these ring systems generally bear bulky substituents, the interactions between which tend to offset ring strain forces and, at the same time, shield the ring from nucleophilic attack ⟨80AG(E)276⟩. *N*-Alkyl-substituted alkylideneaziridines (**417**) are readily prepared in good yields by dehydrohalogenation of *N*-(2-bromoallyl)alkylamines (**415**) with sodium amide in ammonia. *C*-Alkyl-substituted derivatives are also easily accessible by this route ⟨73AG(E)414⟩. The intermediacy of allenic amines, *e.g.* (**416**), in such reactions has been firmly established ⟨62JA195⟩.

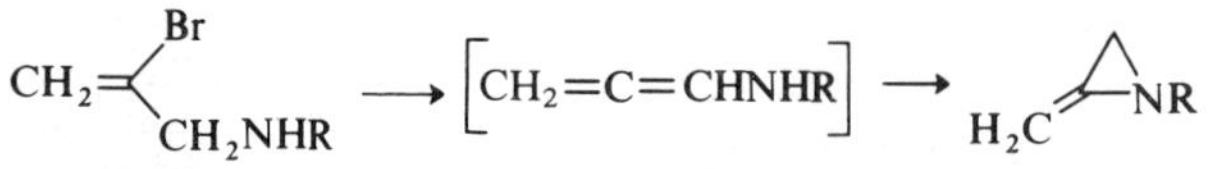

$CH_2{=}C(Br)CH_2NHR$ (**415**) ⟶ [$CH_2{=}C{=}CHNHR$] (**416**) ⟶ (**417**)

Access to the aziridinone ring system can be accomplished in preparative yields by dehydrohalogenation either of *N*-halo- or α-halo-amide precursors of which (**418**) and (**419**) are representative ⟨68AG(E)25⟩. Of the two types of precursor, α-haloamides are generally regarded as being better suited to the preparation of aziridinones for two reasons ⟨61JA4792⟩. First, the greater solubility difference between the product aziridinone and the α-haloamide than between the aziridinone and the *N*-haloamide facilitates an easier separation. Secondly, the formation of an intermediate anionic species is believed to be more favored when the charge resides upon nitrogen than upon the α-carbon. A considerable number of stable aziridinones having bulky nitrogen substituents have been synthesized by these routes ⟨71TL373⟩.

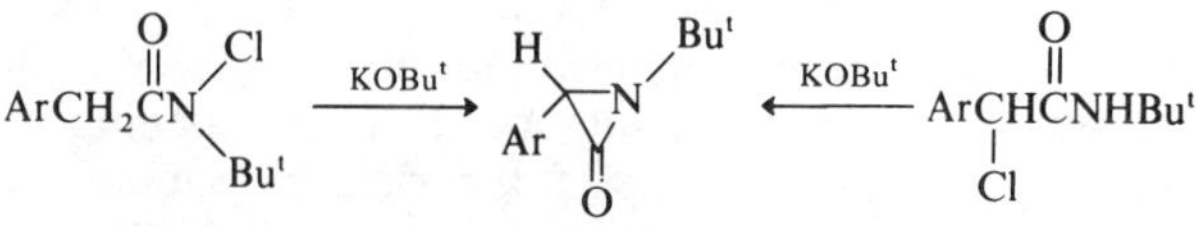

$ArCH_2C(O)N(Cl)Bu^t$ (**418**) $\xrightarrow{KOBu^t}$ (**420**) $\xleftarrow{KOBu^t}$ $ArCH(Cl)C(O)NHBu^t$ (**419**)

Under conditions similar to those already outlined, stable aziridin imine derivatives, *e.g.* (**422**) and (**423**), can be prepared in excellent yields (70–80%) by treating the appropriate α-bromoamidines (easily accessible from the amide precursor) with potassium *t*-butoxide in ether ⟨70AG(E)381⟩. At low temperatures the elimination proceeds with high regio- and stereo-selectivity; at −40 °C (**421**) yields predominantly (**422**).

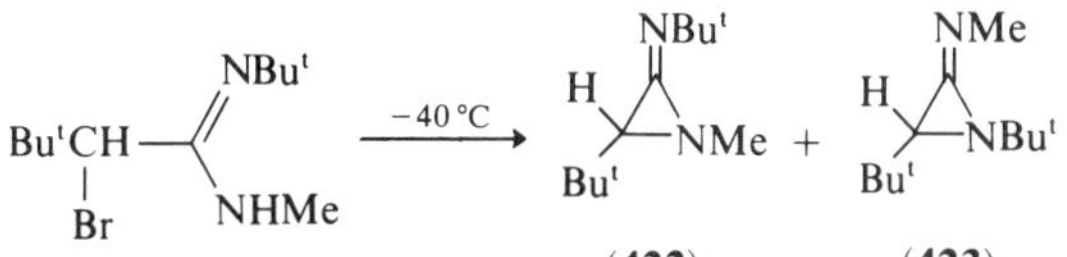

(**421**) (**422**) (**423**)

Aziridinimine derivatives are also accessible by the photochemically induced extrusion of nitrogen and ring contraction of simple alkylidene tetrazolines (**424**) ⟨75AG(E)428⟩. Thus, Pyrex-filtered photolysis of (**424**) at −60 °C quantitatively afforded the aziridinimines (**425**).

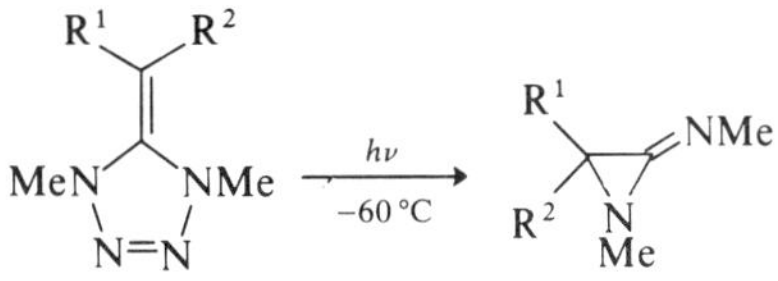

(**424**) $R^1, R^2 = Me$; $R^1 = H, R^2 = Me$ (**425**)

Other potential synthetic routes to these unsaturated aziridine derivatives which involve the addition of nitrenes to allenes ⟨75JOC224⟩, carbenes to imines with subsequent hydrolysis ⟨67JA362⟩, and of carbenoid species to ketenimines ⟨76TL1317, 79TL559⟩ have been investigated but are collectively of little or no preparative value.

5.04.5 NATURAL OCCURRENCE AND BIOLOGICAL PROPERTIES

Only fused derivatives of the saturated aziridine ring system have been found to occur in nature. A group of structurally related substances (**426**) possessing wide-ranging antibiotic and antitumor activity have been isolated as metabolites of several strains of *Streptomyces*. From *S. caespitorsus* and *S. verticillatus*, for example, mitomycins-A, -B and -C (**426a–c**) have been isolated and characterized as pure compounds; from the broth of cultures of *S. ardus*, porfiromycin (**426d**) has been isolated ⟨62JA3185, 3187⟩. A complete X-ray crystallographic analysis of the brosylate derivative of mitomycin-A has been published ⟨62JA3188⟩. It has been demonstrated that the antibiotic activity shown by the mitosanes (**426a–e**) depends upon the presence of the aziridine ring, and accordingly many synthetic *N*-derivatized systems have been synthesized and tested for biological activity.

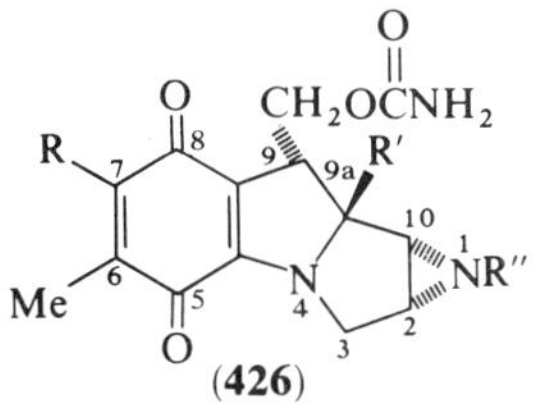

(**426**)

a; R = R′ = OMe, R″ = H (Mitomycin-A)
b; R = OMe, R′ = OH, R″ = Me (Mitomycin-B)
c; R = NH_2, R′ = OMe, R″ = H (Mitomycin-C)
d; R = NH_2, R′ = OMe, R″ = Me
e; R = R′ = OMe, R″ = Me

The antitumor activity displayed by the mitosanes and many synthetic aziridines stems from their ability to act as alkylating agents which chemically modify (crosslink) DNA. For this reason, a large number have been screened for antitumor activity, the mechanism of which has been the subject of considerable research effort ⟨75CJC2891⟩. An excellent account of the broad spectrum of biological properties of a multitude of compounds containing the aziridinyl moiety has been published ⟨B-69MI50401⟩.

5.05

Oxiranes and Oxirenes

E. G. LEWARS
Trent University, Ontario

5.05.1 INTRODUCTION TO OXIRANES AND OXIRENES

Oxiranes are compounds with a three-membered ring composed of an oxygen atom and two formally sp^3-hybridized carbon atoms, while oxirenes possess a three-membered ring made of an oxygen atom and two mutually double-bonded formally sp^2 carbon atoms. Thus oxiranes are ethers with the oxygen in a saturated three-membered ring, while oxirenes are ethers with the oxygen in an unsaturated three-membered ring. The parent compound of the saturated series, oxirane (**1**), is also known as ethylene oxide, epoxyethane or oxacyclopropane. The last appellation is based on a simple and effective system of replacement nomenclature ⟨74MI50500⟩ which is, ironically, as yet little used to name monocyclic heterocycles, although it is widely used in connection with polycyclic systems. The currently favored 'oxirane' nomenclature is based on the cumbersome Hantzsch–Widman system ⟨74MI50500⟩. The parent compound of the unsaturated series, oxirene (**2**), has also been called acetylene oxide; in replacement nomenclature it is oxacyclopropene.

Derivatives of (**1**) are often named as alkene oxides, a term based on the (actual or conceptual) derivation of the compound from an alkene, and occasionally as epoxyalkanes, reflecting the most widely used generic name for this class of compound, *epoxides*. These various nomenclature systems are illustrated in the naming of (**3**) and (**4**). The reader should be warned that *Chemical Abstracts*, in keeping with its practice of naming polycyclic fused-ring systems as derivatives of a fully conjugated parent system, denotes as oxirene derivatives some compounds which are clearly *oxiranes*, derivatives of (**1**) rather than of (**2**). An example is compound (**5**), which is really a derivative of the oxirane phenanthrene 9,10-oxide, but is called in *Chemical Abstracts* ⟨81CA(94)15453⟩ 1a,9a-dihydro-1a,9b-diphenylphenanthro[9,10-*b*]oxirene.

(**1**)

(**2**)

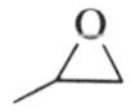

(**3**)
Propylene oxide
1,2-Epoxypropane
2-Methyloxirane
2-Methyloxacyclopropane

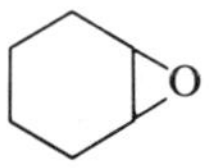

(**4**)
Cyclohexene oxide
1,2-Epoxycyclohexane
7-Oxabicyclo[4.1.0]heptane

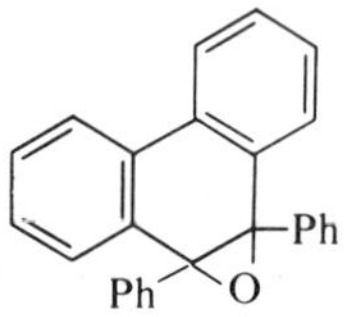

(**5**)

Compounds of type (**6**), (**7**), (**8**) and (**9**), although not strictly derivatives of a *saturated* heterocyclic system, will be discussed in this chapter. Our discussion of (**7**) begins and ends here, since oxiranethiones or α-thiolactones are apparently unknown ⟨80AG(E)276⟩. Little is known of (**8**) and its derivatives, oxiranimines or α-iminolactams. They have been postulated as intermediates in the thermal decomposition of aziridinones (α-lactams) (Scheme 1) but there is no well-established case of the isolation of an oxiranimine ⟨80AG(E)276⟩.

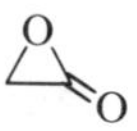

(**6**)  (**7**)  (**8**)  (**9**)

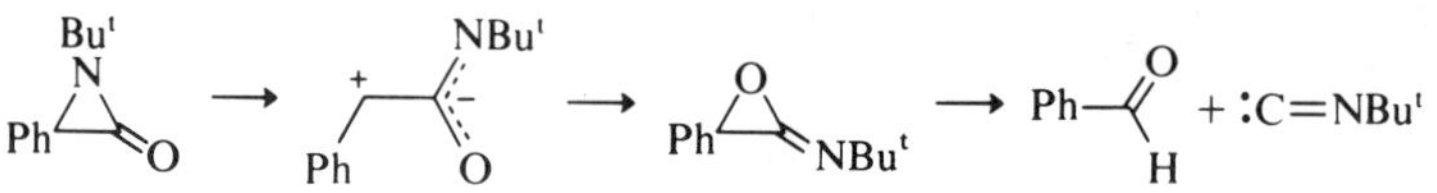

Scheme 1

Including oxiranes and oxirenes under the same rubric is somewhat akin to attempting to present an integrated discussion of the chemistries of cyclohexane and benzene; the structural similarities are superficial and resemblances based on chemical kinship are even more tenuous. Oxiranes are well-known compounds, the parent molecule (**1**) having been first prepared by Wurtz in 1859 ⟨1859CR(48)101⟩. They are , for the most part, reasonably stable compounds, well-represented in the realm of natural products, and of considerable technical utility and industrial importance. The more prosaic members of the class are usually accessible by well-established synthetic reactions, and the outcome of their encounters with heat, photons or reagent molecules can often be predicted with tolerable accuracy. In sharp contrast to this comfortable situation, not only has not a single oxirene ever been observed, it is not even certain if oxirene is a true molecule or merely a transition state or activated complex. In other words, it is not clear if the structure (**2**) represents a relative minimum or maximum on the potential energy–reaction coordinate hypersurface. Although a species with the symmetry of (**2**) and derivatives of this species have been shown to be involved in several reactions, only one mode of reaction, isomerization to oxocarbenes, has been established for these intermediates or transition states. Because of this sharp contrast in the status and behavior of oxiranes and oxirenes, they will be discussed in separate sections of this chapter. A short list of reviews on the oxirane literature follows.

(1) ⟨64HC(19-1)1⟩ The classical, extensive review of the oxirane literature from earliest times up to 1962.
(2) ⟨59CRV737⟩ A critical discussion of the mechanisms of ring-opening of epoxides by nucleophiles.
(3) ⟨B-72MI50500⟩ Comparable in scope to review (2) above.
(4) ⟨73CI(L)109⟩ Similar to reviews (2) and (3) above, but more concerned with steroidal rearrangements.
(5) ⟨B-73MI50500⟩ A thorough review of oxirane synthesis.
(6) ⟨75MI50500⟩ Deals mainly with 1970–1974.
(7) ⟨80MI50500⟩ The manufacture and uses of oxirane.
(8) ⟨80MI50501⟩ A review of epoxidation, industrially oriented.
(9) ⟨81H(15)517⟩ Reagents for the oxidation of alkenes to oxiranes.
(10) ⟨B-81MI50500⟩ A general review of oxiranes; deals mainly with 1965–1977.

For a review on oxirene, thiirene and 1*H*-azirine, see ⟨80PAC(52)1623⟩.

5.05.2 OXIRANES: STRUCTURE AND PROPERTIES, INCLUDING SPECTRA

5.05.2.1 Carcinogenicity and Some Physical Properties

For the experimentalist, a most important property of oxiranes is that several of them appear to be carcinogenic and mutagenic ⟨B-75MI50501⟩. This point is discussed in Chapter 5.07 and briefly dealt with in Section 5.05.5. To provide a feel for the physical reality of this class of compound, the boiling and melting points of a few oxiranes, encompassing all three states of matter under ambient conditions, are given in Table 1. For comparison, the boiling and melting points are also given of alcohols which are similar in structure and molecular mass. The oxiranes are significantly more volatile than their 'corresponding' alcohols, but higher molecular mass oxiranes can have melting points comparable to or higher than those of these alcohols.

Oxirane (**1**) and methyloxirane (**3**) are miscible with water, ethyloxirane is very soluble in water, while compounds such as cyclopentene oxide and higher oxiranes are essentially insoluble ⟨B-73MI50501⟩ (for a discussion of the solubilities of heterocycles, see ⟨63PMH(1)177⟩). Other physical properties of heterocycles, such as dipole moments and electrochemical properties, are discussed in various chapters of PMH. The optical activity of chiral oxiranes has been investigated by *ab initio* molecular orbital methods ⟨81JA1023⟩.

5.05.2.2 Bond Lengths and Angles

The geometries of oxiranes have been determined mainly by X-ray diffraction on crystalline natural products, the oxirane ring being widespread in nature (Section 5.05.5.3). However, the detailed structure of the parent compound (Figure 1) has been secured by microwave spectroscopy and electron diffraction studies ⟨64HC(19-1)1⟩. The strain in this

Table 1 Melting and Boiling Points of some Oxiranes

Compound	*B.p.* (°C) (100 kPa)	*M.p.* (°C)	*Ref.*
O	11	−112	B-67MI50500
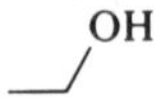 OH	78	−117	B-67MI50500
 O	35	—	B-67MI50500
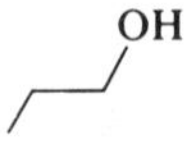 OH	97	—	B-67MI50500
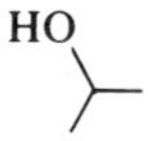 HO	82	−90	B-67MI50500
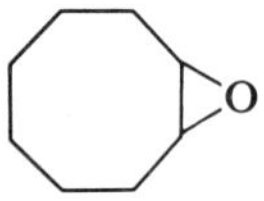 O	190	57	B-74MI50501
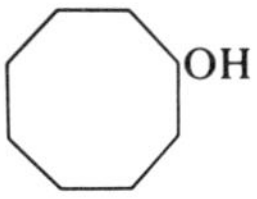 OH	*ca.* 290 (est.)	15	B-44MI50500
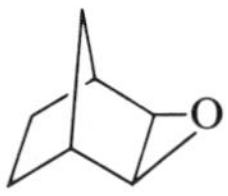 O	158	127	B-74MI50502
 O	*ca.* 120 (est.)	171	B-74MI50502
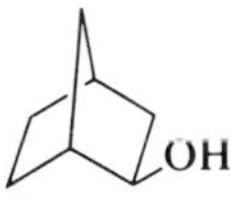 OH	177	129	B-65MI50500
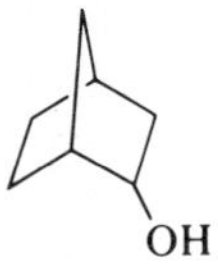 OH	—	150	B-65MI50501

system is mostly angle strain, although torsional interactions probably contribute. The 'conventional ring strain energy' (CRSE) ⟨74PMH(6)199⟩ of oxirane is 114 kJ mol^{-1}, which can be compared with 115 kJ mol^{-1} for cyclopropane.

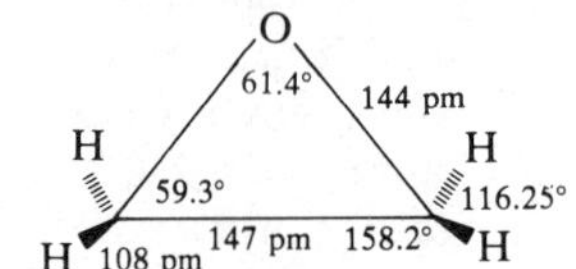

Figure 1 The geometry of oxirane

5.05.2.3 NMR Spectra

Useful compilations of ^{1}H NMR spectra are ⟨B-75MI50502⟩, ⟨B-74MI50503⟩ and ⟨B-63MI50500⟩. The ^{1}H and ^{13}C NMR chemical shifts and the geminal and vicinal proton–proton coupling constants for oxirane and other heterocycles are given in a very readily compared manner, along with many useful spectroscopic references, in ⟨B-72MI50501⟩. Chemical shifts and coupling constants for oxiranes are summarized in Table 2. A useful compilation of ^{1}H and ^{13}C NMR data for several oxiranes is given in ⟨B-67MI50501⟩.

Table 2 Some Oxirane δ and J Values[a]

Compound	$\delta(^{1}H)^{b}$ (p.p.m.)	$\delta(^{13}C)^{b}$ (p.p.m.)	J_{gem} (Hz)	J_{cis} (Hz)	J_{trans} (Hz)	$J(^{13}C\text{–}H)$ (Hz)
Oxirane	2.54	39.7	—	4.45	3.1	176
Oxiranes	—	—	4.5–6.3	2.2–4.5	1.4–3.1	—

[a] ⟨B-72MI50501, B-73NMR135, 71PMH(4)121⟩. [b] Relative to TMS.

Recent papers devoted to the NMR spectra of oxiranes, which provide leading references, are: ⟨79MI50502⟩ on the relation between structure and ^{17}O chemical shifts; ⟨79TL1833⟩ and ⟨80OMR(14)534⟩ on the NMR spectra of alkoxycarbonyloxiranes ('glycidic esters'); ⟨80OMR(14)145⟩ on the ^{13}C spectra of stereoisomeric oxiranes spiro-fused to the cyclohexane ring; and ⟨81H(15)1007⟩ on structure determination with the aid of ^{13}C chemical shift changes induced by converting a carbon–carbon double bond to an oxirane moiety.

5.05.2.4 Mass Spectra

These are discussed in ⟨B-71MS4⟩. Oxirane itself shows a strong molecular ion peak and a slightly stronger base peak at m/e 29 (CHO^{+}) due to isomerization to ethanal and loss of a methyl radical. Substituted oxiranes tend to show only weak molecular ion peaks, because of rearrangement and fragmentation.

5.05.2.5 UV Spectra

Oxirane and its simple alkyl derivatives do not absorb in the region of the UV spectrum routinely used for structure determination (above 200 nm). Oxirane itself shows λ_{max}^{vapor} at 171 nm (ε 5600) ⟨63PMH(2)10⟩, and methyloxirane shows a strong continuous absorption beginning at 175 nm ⟨64HC(19-1)17⟩. The oxirane ring can, however, delocalize electrons in a manner analogous to that of the carbon–carbon double bond or the cyclopropane ring (though its conjugating power is apparently less than that of the latter ⟨64HC(19-1)17⟩), so that ethenyloxiranes and oxiranyl ketones show a bathochromic shift in their UV spectra. The UV spectra of several carbonyl-conjugated oxiranes are given in ⟨64HC(19-1)17⟩.

5.05.2.6 IR Spectra

Oxiranes show three characteristic bands in the IR spectrum at *ca.* 8.0 μm (1250 cm^{-1}) (in-phase stretching and shrinking of all ring bonds, *i.e.* ring breathing), 10.5–12.4 μm (950–810 cm^{-1}) (asymmetrical ring stretch) and 11.9–13.3 μm (840–750 cm^{-1}) (unassigned in this reference) ⟨B-67MI50502⟩; the C—H stretching of these rings gives rise to bands at 3.28–3.34 μm (3050–2995 cm^{-1}). The IR spectra of several oxiranes are presented and interpreted in ⟨B-67MI50503⟩; from these spectra it appears that the three characteristic bands mentioned above are in the intensity ratio of about medium : strong : strong, the medium band being roughly as intense as the C—H stretch for oxiranes like methyloxirane. A complete analysis of vibrational modes of oxirane and a summary of the characteristic IR bands of oxiranes is given in ⟨63PMH(2)176⟩. Oxiranes derived from terminal alkenes show a C—H stretch at 3.28 μm (3050 cm^{-1}) and near-IR bands at *ca.* 1.1, 1.35, 1.65 and 2.2 μm ⟨B-72MI50501⟩. The positions of the characteristic IR bands of many oxiranes are tabulated in ⟨64HC(19-1)1⟩ and this reference also gives the carbonyl stretching frequencies of several $\alpha\beta$-epoxy ketones. Compilations of IR spectra which include those of several oxiranes are ⟨B-75MI50503⟩ and ⟨B-74MI50504⟩.

5.05.3 OXIRANES: REACTIVITY

5.05.3.1 General Survey of Reactivity

The eight modes of reactivity are outlined in Figure 2.

(1, 2)	Ring cleavage (probably homolytic), thermal and photochemical (Section 5.05.3.2)
(3)	Electrophilic attack on ring oxygen (Section 5.05.3.3)
(4)	Nucleophilic attack on ring carbon (Section 5.05.3.4)
(5)	Nucleophilic attack on ring hydrogen (proton abstraction) (Section 5.05.3.5)
(6)	Reactions with electrons and surface reactions (Section 5.05.3.6)
(7)	Cycloadditions (Section 5.05.3.7)
(8)	Reactions of a substituent (Section 5.05.3.8)

Figure 2 The modes of reactivity of oxiranes

5.05.3.2 Reactions Formally Involving No Other Species (Except as Trapping Agents)

5.05.3.2.1 Thermal and photochemical reactions

The gas phase thermal chemistry and photochemistry of oxiranes is reviewed in ⟨77CRV473⟩. References to thermal reactions of oxiranes are given in ⟨B-80MI50502⟩ and ref. 10 in ⟨76TL1449⟩.

Oxirane on thermolysis or photolysis suffers C—O homolysis to give a plethora of products (Scheme 2). Substituted oxiranes behave similarly on thermolysis although some C—C cleavage is observed (Scheme 3). Cyclopentene and cyclohexene oxides undergo only C—O cleavage (Scheme 4).

$$\longrightarrow H_2\dot{C}-CH_2\dot{O} \longrightarrow [CH_3CHO]^* \longrightarrow$$

$$H_2+CO+HCHO+CH_4+C_2H_4+C_2H_6+C_3H_8+C_4H_{10}+C_2H_2+C_3H_6(!)$$

Scheme 2

(a) (b) (c)
Major product

Scheme 3

Scheme 4

Thermal or photochemical C—C cleavage generates a 1,3-dipolar species (**10**) called a carbonyl ylide (Scheme 5), and this mode of cleavage tends to dominate the thermal chemistry and photochemistry of oxiranes substituted with electronegative groups, which stabilize the zwitterion ⟨76T2165, 81JA1066⟩. These ylides can be trapped by 1,3-dipolarophiles to give heterocycles ⟨77AG(E)572⟩ and can suffer photochemical fragmentation to carbenes ⟨71AG(E)537, 76JA5697⟩. Ring-opening to carbonyl ylides appears to occur predominantly in accord with the Woodward–Hoffmann rules ⟨B-70MI50500⟩ (Schemes 6 and 7).

Δ or $h\nu$

(**10**)

Scheme 5

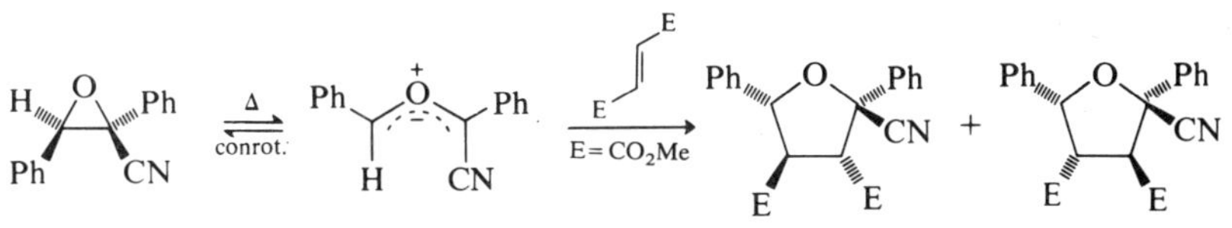

Scheme 6

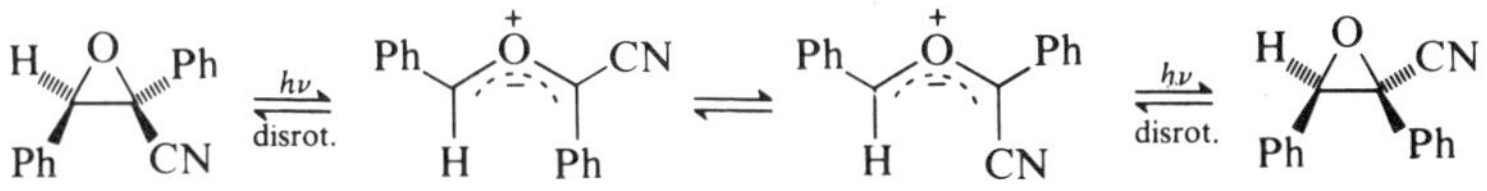

Scheme 7

Eberbach and coworkers have made an extensive study of the thermal chemistry and photochemistry of ethenyloxiranes ⟨81CB1027 and refs. therein⟩. Thermal isomerization can involve mainly either C—O (Scheme 8) or C—C (Scheme 9) cleavage. The reluctance of (**11**) to generate an ylide may be due to the fact that allowed, conrotatory opening would give a strained species (**12**) (*cf.* (*E*)-cycloheptene). The photoisomerization of ethenyloxiranes can involve either C—O (Scheme 10) or C—C (Scheme 11) cleavage. Photolysis of (**13**) led to the synthesis of the ketone tautomer (**14**) of a phenol (Scheme 12) ⟨80TL463⟩.

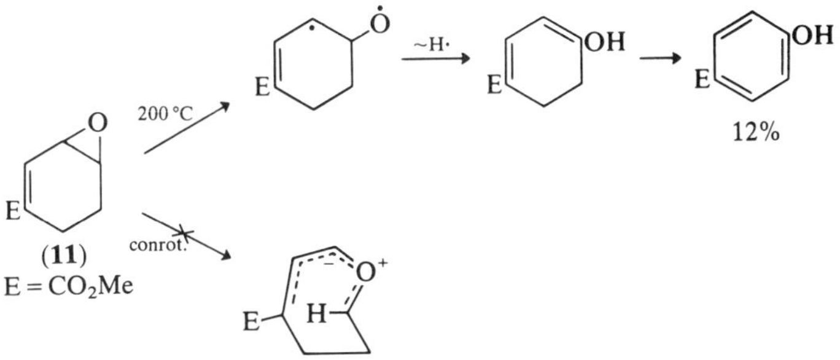

(**11**)
E = CO_2Me

(**12**)

Scheme 8

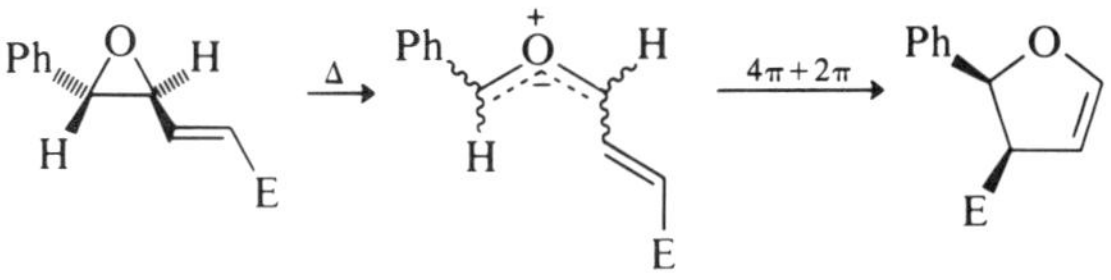

E = CO_2Me Family of interconverting ylides

Scheme 9

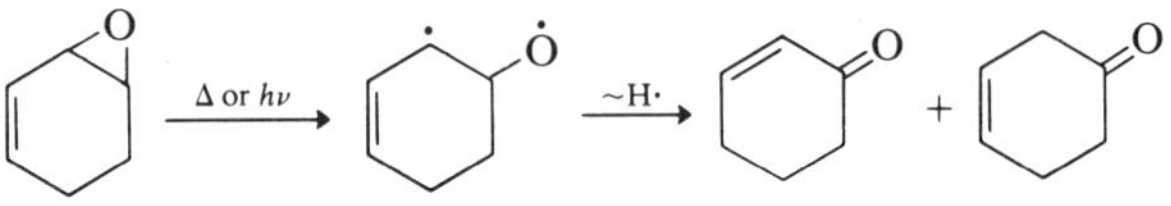

Scheme 10

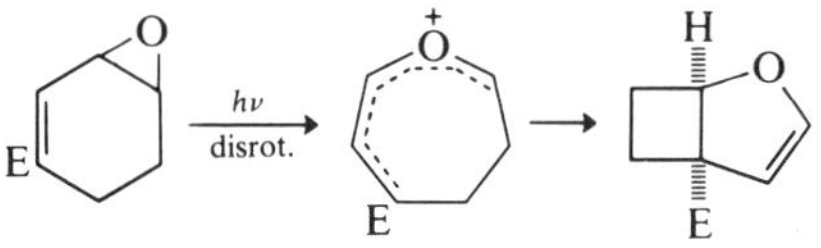

E = CO_2Me

Scheme 11

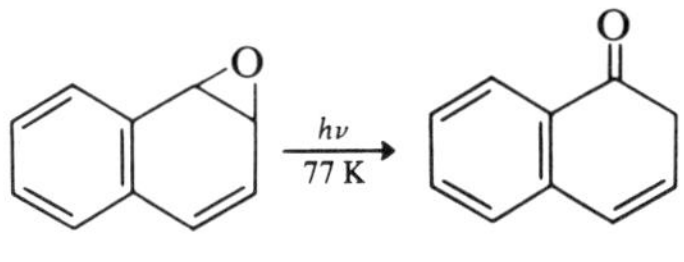

(**13**) (**14**)

Scheme 12

Warrener and coworkers give several referencs to reactions of ethenyloxiranes in their discussion of the isomerization of certain of these to 1,4-oxides ⟨81AJC421⟩.

Jeger, Wolf and coworkers extensively studied the photochemistry of butadienyloxiranes ⟨81HCA198⟩ and oxiranyl enones ⟨81HCA224⟩. Butadienyloxiranes give both C—O and C—C cleavage (Scheme 13); the paths open to oxiranyl enones have been systematically analyzed by Hart and coworkers ⟨77JOC3635⟩, who point out that by far the commonest process is C—O cleavage (*e.g.* Scheme 14).

hν 254 nm $^1(\pi \rightarrow \pi^*)$

hν (?)

hν > 280 nm propanone $^3(\pi \rightarrow \pi^*)$

Scheme 13

hν

Scheme 14

Acyloxiranes can follow a variety of photochemical pathways ⟨81MI50501, 81CL201, 80CC1263, 79TL2723, 78JOC4256⟩. In the oxirane (**15**) a remote double bond participates in ring opening (Scheme 15) ⟨80JOC2091⟩.

(**15**)

Scheme 15

2,3-Diethenyloxiranes undergo thermal Cope-type rearrangements to oxacycloheptadienes (Scheme 16) ⟨67AG(E)385⟩ even when these are very strained intermediates (Scheme 17) ⟨80JOC428⟩. This process is particularly facile in fused-ring diethenyloxiranes like (**16**) (Chapter 5.17) and (**17**) (Chapter 5.07).

Scheme 16

Ph + other products

Scheme 17

(**16**) (**17**)

Oxiranones (α-lactones) ⟨81JA686, 80AG(E)276⟩, *e.g.* (**6**), are highly reactive, readily polymerizing (Scheme 18), possibly *via* a zwitterion (**18**). Such a species (**19**) would also account for the rearranged products (**20**) and (**21**) from (**22**; Scheme 19).

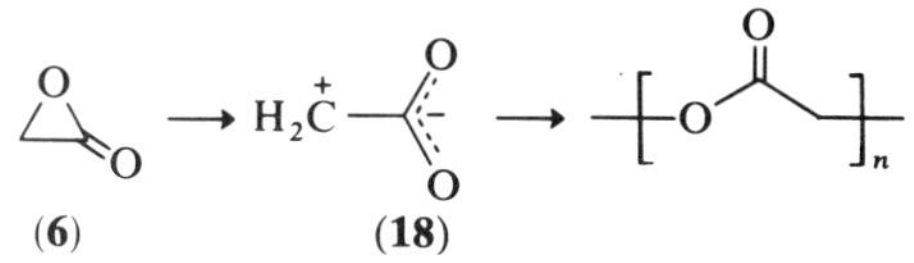

(**6**) (**18**)

Scheme 18

(**22**) (**19**) (**20**) (**21**)

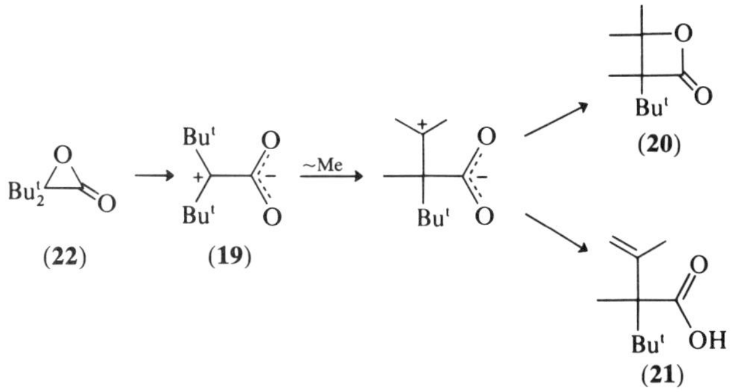

Scheme 19

Methylideneoxiranes (allene oxides) ⟨80T2269⟩, *e.g.* (**9**) and (**23**), are also highly reactive, and undergo facile thermal isomerization to cyclopropanones, possibly *via* an 'oxyallyl' intermediate (**24**; Scheme 20).

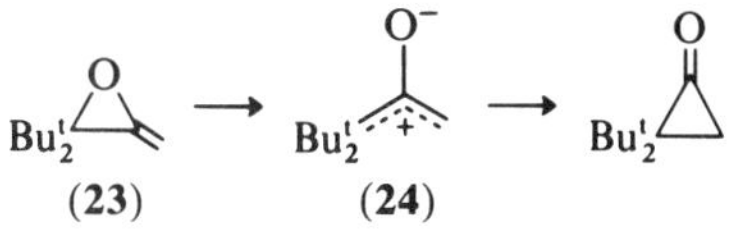

(**23**) (**24**)

Scheme 20

Oxiranes with lone-pair atoms like nitrogen ⟨67JA3084⟩ or halogens ⟨B-71MI50500⟩ on the ring are thermally labile (*e.g.* Scheme 21).

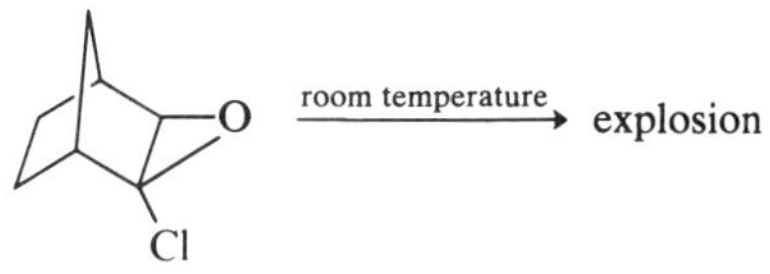

room temperature → explosion

Scheme 21

5.05.3.2.2 Catalyzed reactions

Oxiranes react with protic or Lewis acids to give the kinds of products expected from the carbocation (**25**) (not always a discrete intermediate), namely (**26–29**; Scheme 22). (a) Quenching by external nucleophiles giving (**26**) is discussed in Section 5.05.3.4, except for dimerization and polymerization (Scheme 23) ⟨64HC(19-1)451⟩. A remarkable example of this process is the synthesis of the first 'topologically nonplanar' molecule (**30**; Scheme 24), shown here for heuristic reasons as going through a tris(zwitterion) (**31**) ⟨81JA1581, 81TL287, 81TL291⟩. (b, c) A 1,2-hydride or carbon shift converts (**25**) into a ketone (**27**) or an aldehyde (**28**) ⟨64HC(19-1)230⟩. An important application of carbon migration is the conversion of a spirooxirane (*e.g.* **32**) into a cyclobutanone (**33**), a versatile synthetic intermediate (Scheme 25) ⟨B-75MI50504⟩. In steroids the initial carbocation can induce remarkable 'backbone' rearrangements (Scheme 26) ⟨73CI(L)109⟩. (d) Proton loss from R^1 or R^2 gives an allylic alcohol, but the synthesis of these from oxiranes more commonly involves base (below).

The reaction of oxiranes with base can follow several paths, giving products of type (**34–38**; Scheme 27). (a) Formation of an oxiranyl anion (**34**) is rare (Section 5.05.3.5). (b) Nucleophilic ring opening to give (**35**) is common with unhindered bases (Section 5.05.3.4). (c) α-Elimination to give a carbene or carbenoid (**36**) is favored by alkyllithium bases and

(26) Nu⁻ (a) (28) ~C (c) (25) ~H (b) −H⁺ (d) (27) (29)

Scheme 22

A dimer

A 'polyglycol'

Scheme 23

acid or Δ (31) (30)

Scheme 24

$O + Ph_2S=$ (32) H^+ (33) NaOMe i, MeLi ii, NaOMe

Scheme 25

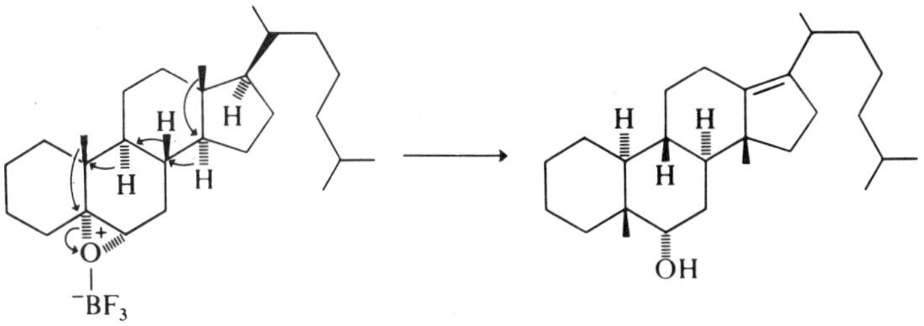

Scheme 26

oxiranes lacking β-hydrogens [see (d)] ⟨67JA4526, 4527, 66QR119⟩. (d) β-Hydrogen abstraction gives an allylic alcohol (**37**) ⟨73JA2697, refs. 12, 13, 14⟩. Recent improvements of this reaction are the use of HMPT as solvent ⟨78T1541⟩ and diethylaluminum 2,2,6,6-tetramethylpiperidide as base ⟨78AG(E)169⟩. Hydroxymethyloxiranes can be isomerized to allylic alcohols with high stereochemical control by titanium isopropoxide ⟨81JA462⟩. Other routes to allylic alcohols are given in Section 5.05.3.4.3. (e) Ketone (**38**) formation is seen with acyloxiranes, aryloxiranes and cyclooctatetraene oxide ⟨64HC(19-1)262⟩

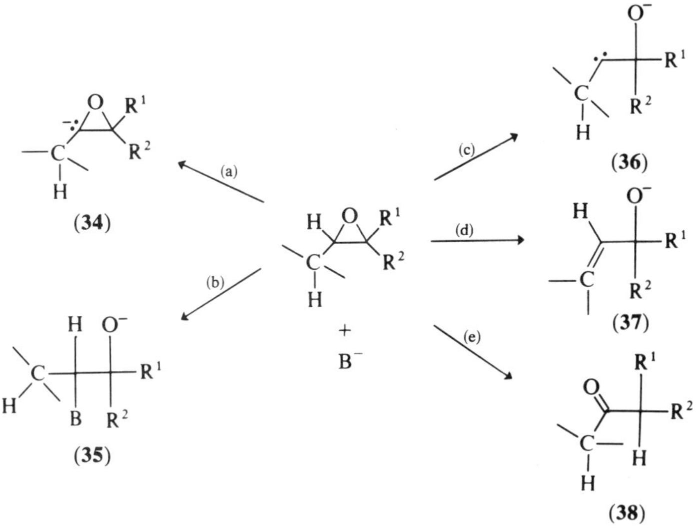

Scheme 27

Oxiranes have been isomerized by palladium compounds to allylic alcohols and enones ⟨79JA1623⟩, and to 1,3-diketones ⟨80JA2095⟩.

5.05.3.3 Electrophilic Attack on Ring Oxygen

5.05.3.3.1 Protonation

Molecular orbital calculations predict that oxirane forms the cyclic conjugate acid (**39**), which is 30 kJ mol^{-1} stabler than the open carbocation (**40**) and must surmount a barrier of 105 kJ mol^{-1} to isomerize to (**40**) ⟨78MI50500⟩. The proton affinity of oxirane was calculated ⟨78JA1398⟩ to be 807 kJ mol^{-1} (*cf.* the experimental values of 773 kJ mol^{-1} for oxirane and 777–823 kJ mol^{-1} for dimethyl ether ⟨80MI50503⟩). The basicity of cyclic ethers is discussed in ⟨B-67MI50504⟩.

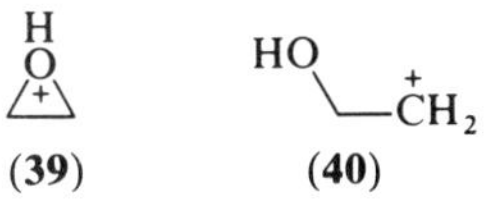

(**39**) (**40**)

5.05.3.3.2 Lewis acids

Typical Lewis acids like BF_3 and $SbCl_5$ coordinate with oxirane oxygen to give (presumably) a cyclic oxonium ion (**41**) which reacts further (Scheme 28) ⟨64HC(19-1)446, B-67MI50505⟩.

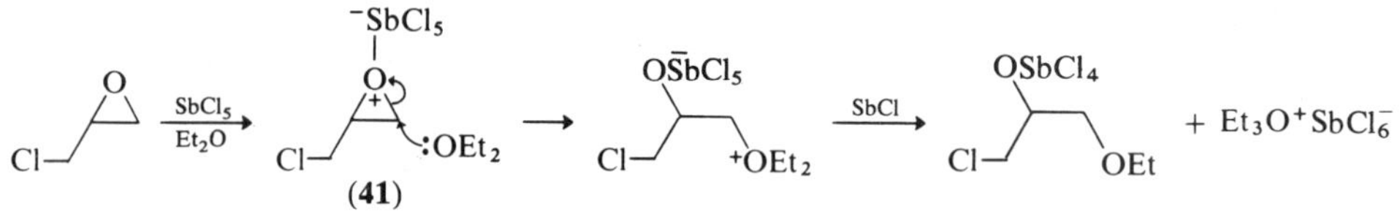

Scheme 28

5.05.3.3.3 Alkyl halides and related compounds

These give the products expected from electrophilic attack on oxygen by the electrophilic reagent atom, followed by nucleophilic opening of the cyclic oxonium ion (*e.g.* **42**; Scheme 29) ⟨64HC(19-1)436⟩.

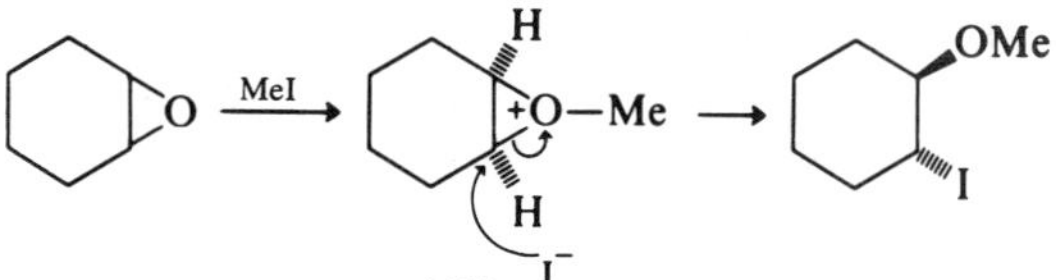

Scheme 29

5.05.3.3.4 Halogens

Oxirane and chlorine allegedly form complexes $C_2H_4O \cdot Cl$ and $C_2H_4O \cdot 3Cl$ ⟨64HC(19-1)445⟩ at −80 °C, but in this and other cases explosions occurred at room temperature, perhaps caused by hypohalites.

5.05.3.3.5 Peroxy acids and other oxidants

Oxygen donors like peroxy acids, ozone, and pyridine *N*-oxides cause carbon–carbon cleavage, perhaps by formation of a 'perepoxide' (**43**; Scheme 30) ⟨81JCS(P1)1871⟩. Other oxidants have also been reported to react with oxiranes ⟨64HC(19-1)228⟩.

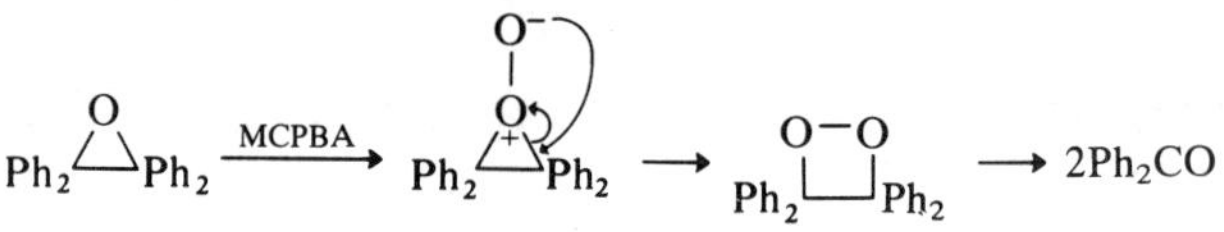

Scheme 30

5.05.3.3.6 Carbonyl and related groups

Aldehydes, ketones and cumulated double bonds react with oxiranes to form five-membered heterocycles (*e.g.* Schemes 31 and 32) ⟨64HC(19-1)453⟩.

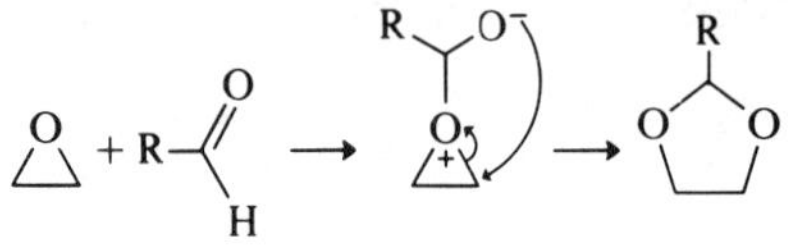

Scheme 31

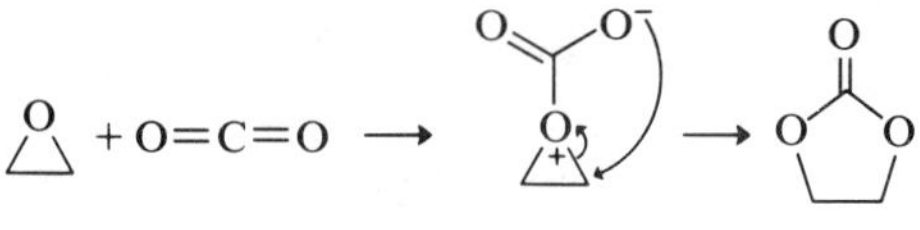

Scheme 32

5.05.3.3.7 Oxiranes

Dimerization and polymerization can be initiated by electrophilic attack on an oxirane by a protonated or Lewis acid coordinated oxirane (Section 5.04.3.2.2).

5.05.3.3.8. Electron deficient carbon and silicon species

Anchimeric assistance by oxirane oxygen may accelerate the solvolysis of (**44**; Scheme 33) (by a factor of *ca.* 10^4–10^5) compared to the *exo* isomer (**45**; Scheme 34) ⟨81JOC3512⟩.

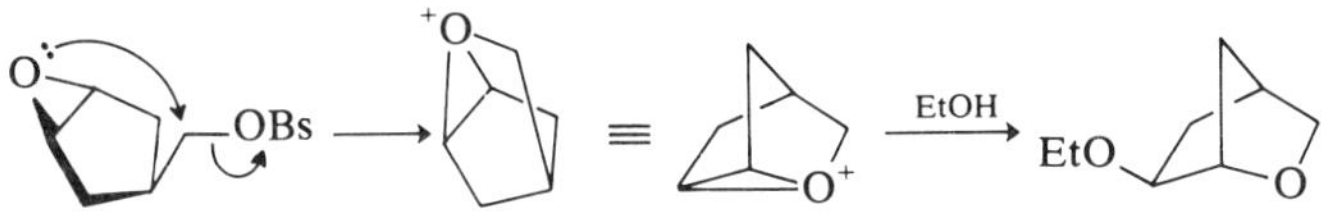

(44)

Scheme 33

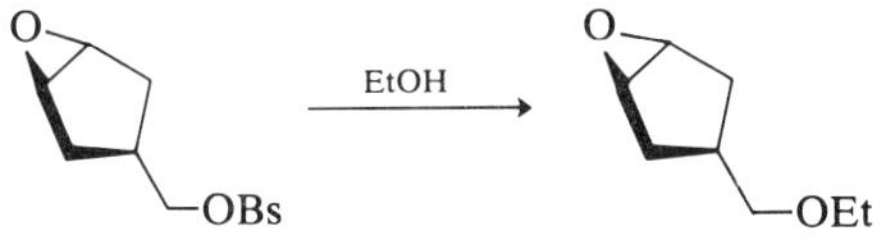

(45) Main product

Scheme 34

The bis(diene) (**46**) adds dienophiles preferentially on the side *syn* to the oxirane moiety (Scheme 35) ⟨80T149⟩. This may be due to formation of a charge-transfer complex by donation of electron density from oxygen into an antibonding orbital on the dienophile.

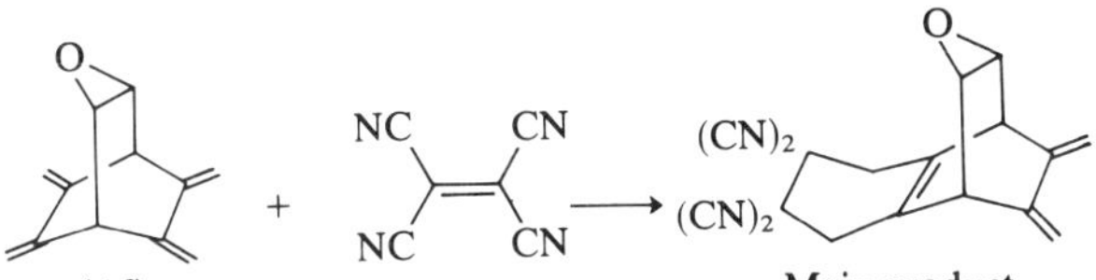

(46) Main product

Scheme 35

Carbenes coordinate with oxirane oxygen to give ylides (*e.g.* **47**) which undergo ring cleavage or expansion (Scheme 36) ⟨B-71MI50501⟩. Intramolecular trapping of a carbenoid by an oxirane has been reported ⟨80T2145⟩.

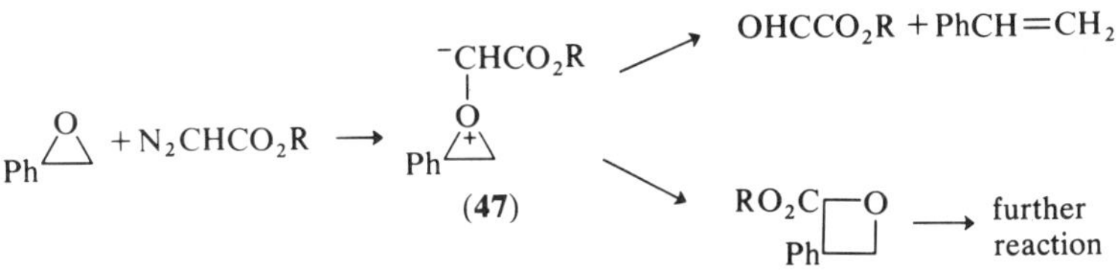

(47)

Scheme 36

Atomic carbon deoxygenates oxiranes (Scheme 37) ⟨B-71MI50501, 76JA3820⟩. Silylenes can also deoxygenate oxiranes (Scheme 38) ⟨80JA1451⟩.

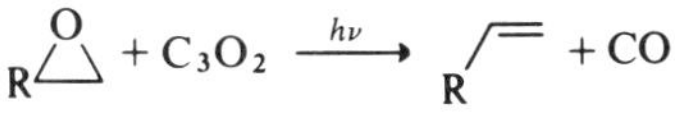

Scheme 37

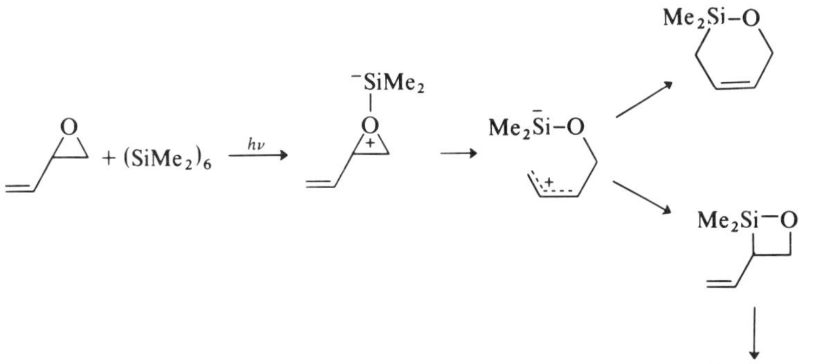

Scheme 38

5.05.3.4 Nucleophilic Attack on Ring Carbon

5.05.3.4.1 Introduction

Nucleophilic attack on ring carbon (Scheme 39) is the most important reaction of these compounds (the electrophile may bond to oxygen either before or after nucleophilic attack). For vinylogous nucleophilic opening by S_N2' attack on ethenyloxiranes see Section 5.5.3.8.

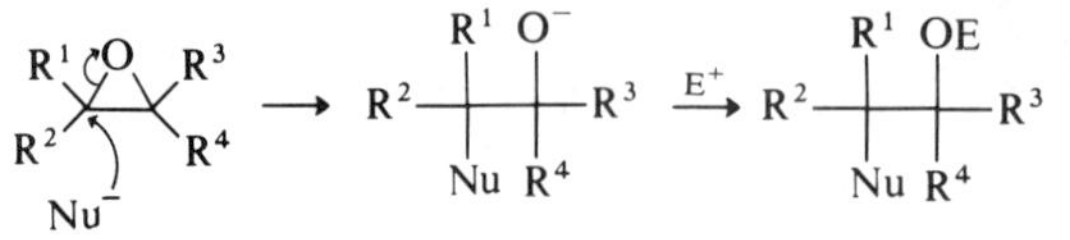

Scheme 39

5.05.3.4.2 Mechanistic aspects

This topic has been thoroughly reviewed ⟨72MI50500, 64HC(19-1)1, 59CRV737⟩.

(i) Regiochemistry

The main features of the effect of structure on the site of attack are summarized in Table 3, and can be understood in terms of a 'borderline S_N2' ⟨59CRV737⟩ transition state (**48**) which somewhat resembles an S_N1 transition state in charge distribution because C—O bond breaking runs ahead of Nu—C bond making.

Table 3 Favored Position of Nucleophilic Attack on Substituted Oxiranes[a]

Substituent group	*Favored position of attack*[b] *Nonacidic conditions*	*Acidic conditions*
Alkyl	C_β	C_α or C_β
Electronegative, nonconjugating	C_β	C_β
π-Conjugating	C_α or C_β	C_α or C_β
n-Conjugating	C_α	C_α

[a] ⟨59CRV737⟩. [b] C_α is the carbon bearing the substituent.

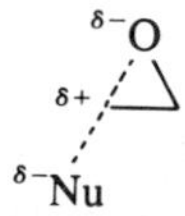

(**48**)

Alkyl groups under nonacidic conditions sterically deflect nucleophiles from C_α, but under acidic conditions this steric effect is to some extent offset by an electronic one; the protonated oxirane opens by transition states (Scheme 40) which are even more S_N1-like than the borderline S_N2 one of the unprotonated oxirane. Thus electronic factors favor cleavage at the more substituted carbon, which can better support a partial positive charge; the steric factor is still operative, however, and even under acidic conditions the major product usually results from C_β attack.

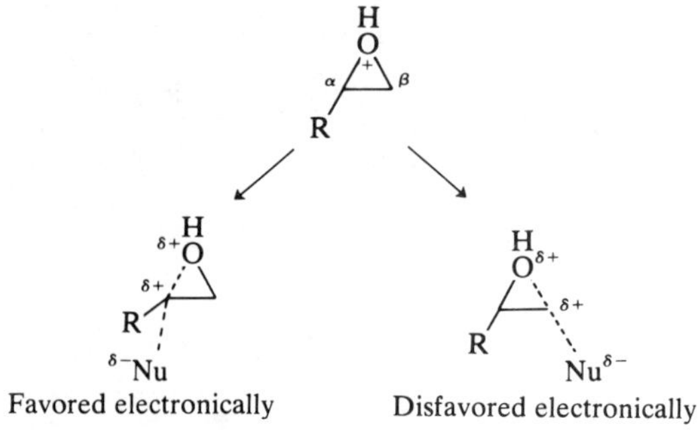

Favored electronically

Disfavored electronically

Scheme 40

Electronegative, nonconjugating groups (which interact with an incipient carbocation only by an inductive or field effect) discourage attack at C_α. This is due to destabilization of transition state (**49**) by the juxtaposition of positive charge.

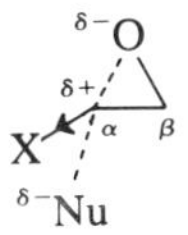

(**49**) 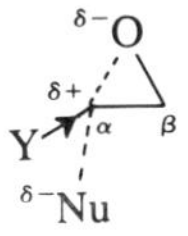 (**50**)

π-Conjugating groups tend to favor attack at C_α, but the ratio of $C_\alpha:C_\beta$ attack depends strongly on a balance of steric and electronic factors arising from both substituent and nucleophile (Table 4). The results can be rationalized, to a first approximation, by assuming that with good π-donors stabilization of the incipient carbocation in (**50**) offsets steric hindrance.

Table 4 Relative Amounts of Products from Nucleophilic Attack on Phenyloxiranes[a]

Oxirane	*Nucleophile*	*Product ratio, $C_\alpha:C_\beta$ attack*
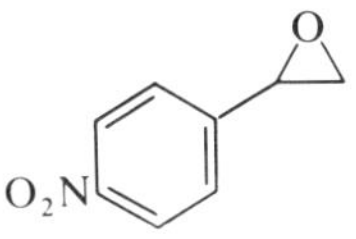 4-O_2N-phenyloxirane	NaOPh	36:64
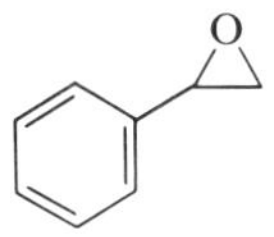 phenyloxirane	NaOPh	76:24
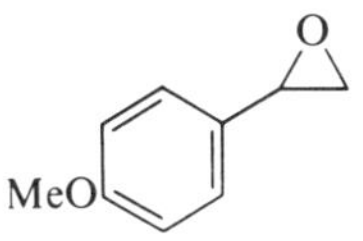 4-MeO-phenyloxirane	NaOPh	100:0
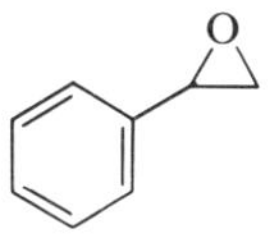 phenyloxirane	NaOMe	30:70

[a] ⟨59CRV737⟩.

n-Conjugating groups consistently favor attack on C_α, as expected from their ability to stabilize carbocations (*cf.* **50**).

Oxiranones (α-lactones) (Section 5.05.3.2.1), *e.g.* (**6**), in accord with their putative facile opening to zwitterions (Scheme 18), usually react readily with nucleophiles by alkyl–oxygen cleavage (Scheme 41) ⟨71JA557⟩, rather than the acyl–oxygen cleavage common for esters.

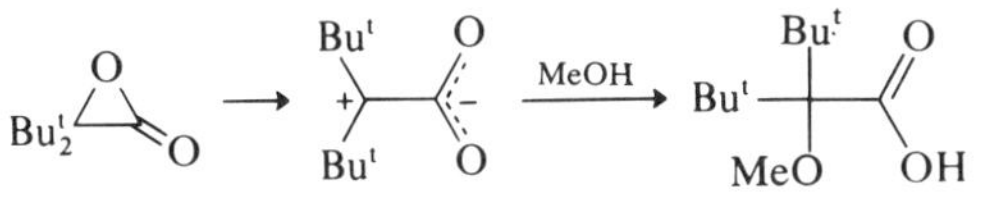

Scheme 41

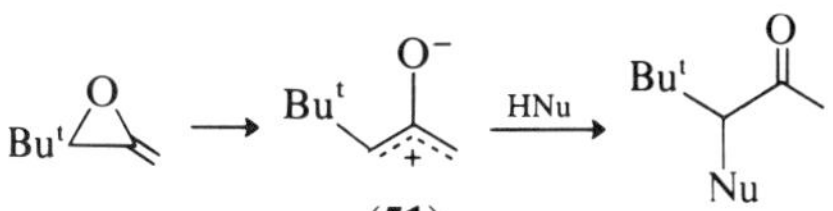

(**51**)

Scheme 42

Methylideneoxiranes (allene oxides; Section 5.05.3.2.1) react with nucleophiles as if ring opening occurs to give a zwitterion (*e.g.* **51** or **52**), which may be captured by the nucleophile before (Scheme 42) or after (Scheme 43) isomerization to a cyclopropanone.

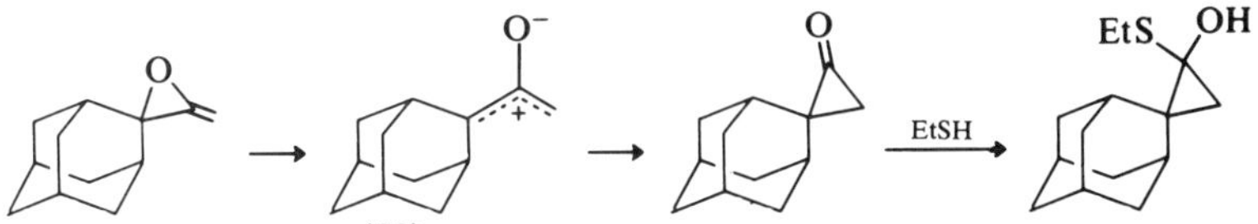

(**52**)

Scheme 43

(*ii*) *Stereochemistry*

Nucleophilic attack on oxirane carbon usually proceeds with inversion of configuration (Scheme 44) as expected for S_N2 reactions, even under acid conditions (Scheme 45). Scheme 45 also illustrates the fact that cyclohexene oxides open in a *trans*-diaxial manner; this is known as the Fürst–Plattner rule ⟨49HCA275⟩ and there are very few exceptions to it.

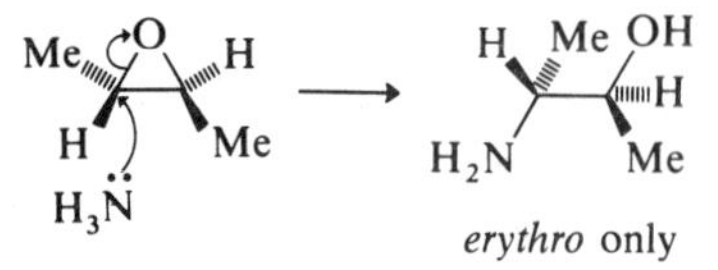

erythro only

Scheme 44

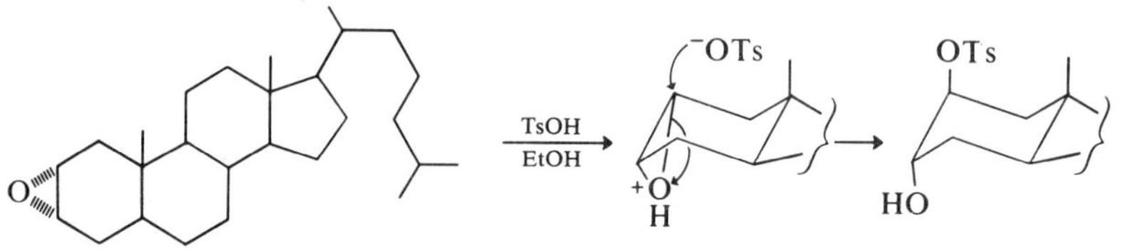

Scheme 45

A relatively few oxirane ring openings have been reported to give retention or a mixture of inversion and retention (see the general reviews on the mechanisms of oxirane ring opening, and the specialized one devoted to retention ⟨68RCR448⟩).

5.05.3.4.3 Synthetic aspects

Only a few reactions, of outstanding utility or special mechanistic interest, can be mentioned here.

(*i*) *Oxygen nucleophiles*

Nucleophilic opening of oxiranes to give ultimately 1,2-diols is usually effected without isolation of the oxirane: oxiranation (epoxidation) of alkenes with unbuffered peroxyethanoic acid or hydrogen peroxide in methanoic acid (Section 5.05.4.2.2(*i*)) tends to give monoesters of 1,2-diols (*e.g.* **53**), which can be hydrolyzed to the diols (Scheme 46).

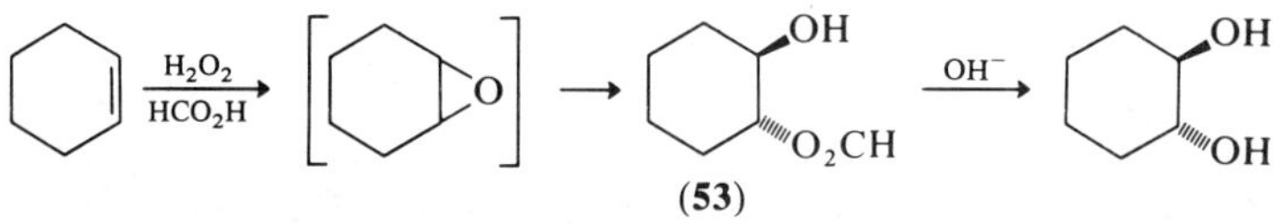

(**53**)

Scheme 46

Opening by trimethylsilyl trifluoromethanesulfonate yields an adduct (**54**) from which trifluoromethanesulfonic acid can be eliminated to give an allylic alcohol (Scheme 47) ⟨79JA2738⟩ [*cf.* base-promoted isomerization to allylic alcohols (Section 5.05.3.2.2)].

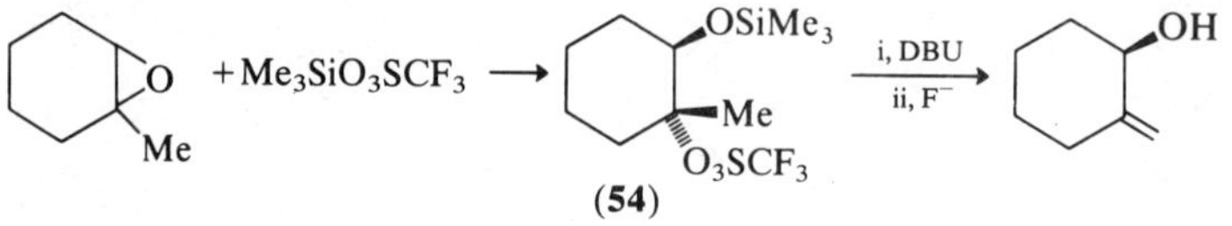

(**54**)

Scheme 47

Cleavage of silyloxiranes (*e.g.* **55**) by methanol, leading to aldehydes or ketones (*e.g.* **56**; Scheme 48), has found use in natural products synthesis ⟨81JA1831, 80JA5004⟩.

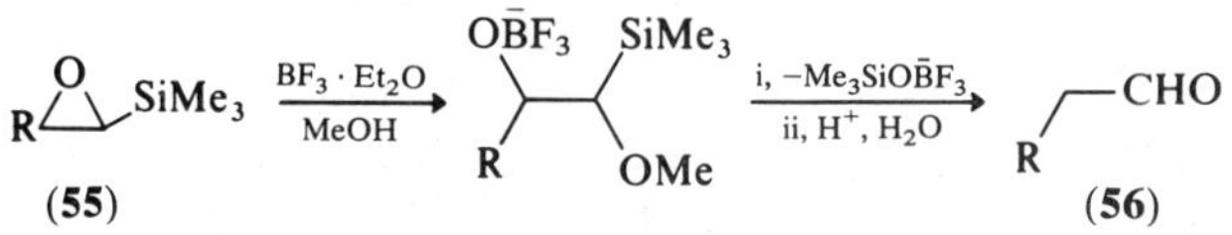

(**55**) (**56**)

Scheme 48

(ii) Nitrogen nucleophiles

The commonest of these for oxirane opening are amines and azide ion [amide ions promote isomerization to allylic alcohols (Section 5.05.3.2.2)]. Reaction with azide can be used in a sequence for converting oxiranes into aziridines (Scheme 49) and this has been employed in the synthesis of the heteroannulenes (**57**) and (**58**) ⟨80CB3127, 79AG(E)962⟩.

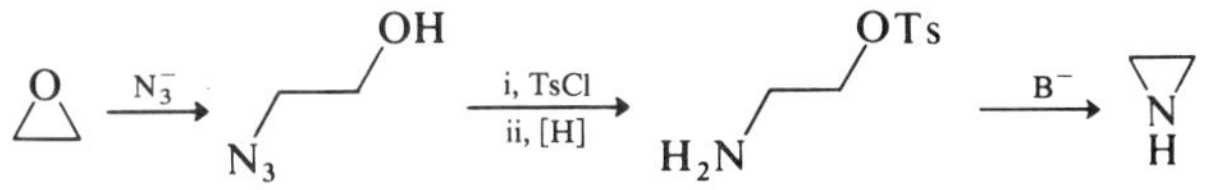

Scheme 49

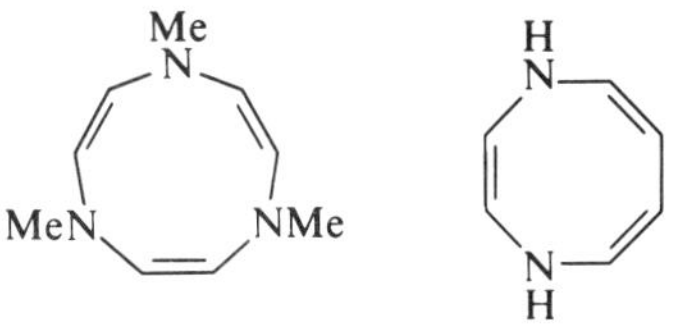

(**57**) (**58**)

(iii) Sulfur, selenium and tellurium nucleophiles

Cleavage of spirooxiranes (*e.g.* **59**) by thiophenol is a key step in a new route to *trans*-fused γ-lactones (Scheme 50) ⟨81TL279⟩.

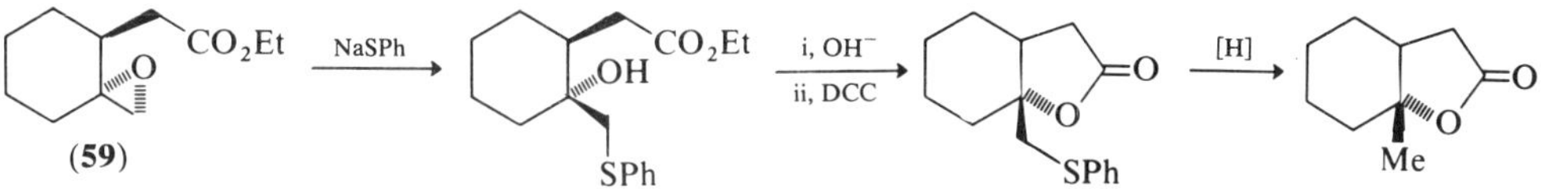

Scheme 50

Reaction with selenophenoxide ion is one step in a method of converting oxiranes to allylic alcohols under mild conditions (Scheme 51) ⟨73JA2697⟩ (*cf.* Sections 5.05.3.2.2 and 5.05.3.4.3(*i*)).

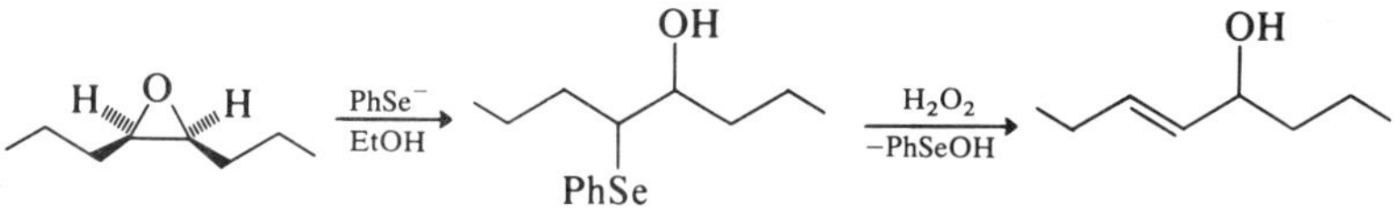

Scheme 51

Deoxygenation of oxiranes to alkenes with selenium ⟨80S636⟩ or tellurium (Scheme 52) ⟨80JOC2347⟩ reagents gives retention of configuration, since two S_N2 displacements intervene (reagents which utilize as their nucleophilic atom an element which has an avidity for oxygen and can expand its octet, *e.g.* phosphorus or silicon (Section 5.05.3.4.3(*viii*)), give inversion).

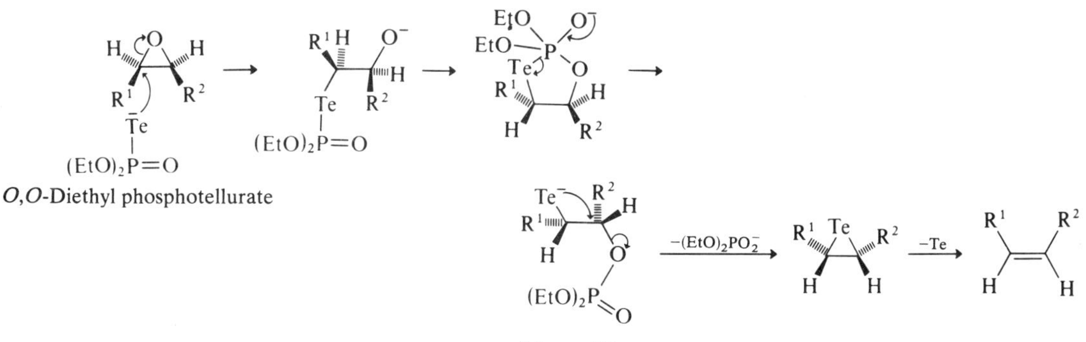

O,O-Diethyl phosphotellurate

Scheme 52

(iv) Halides

Oxiranes react with iodotrimethylsilane to give silylated halo alcohols (*e.g.* **60**) which can be converted to allylic alcohols (Scheme 53) ⟨80JOC2579, 80TL2329⟩; *cf.* other syntheses of allylic alcohols (Sections 5.05.3.2.2, 5.05.3.4.3(*i*) and (*iii*)).

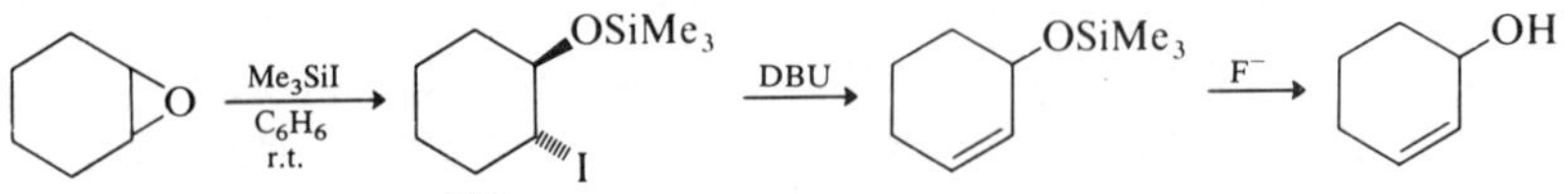

Scheme 53

(v) Carbanions

The reaction of oxirane itself with a wide range of Grignard reagents (Scheme 54) is a classical homologation reaction ⟨64HC(19-1)394⟩, but complications can arise from isomerization or cleavage due to MgX_2 in the Grignard reagent (Scheme 55).

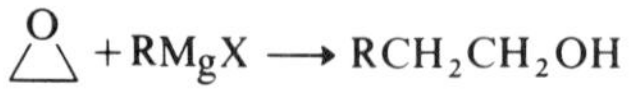

$$\text{oxirane} + RMgX \longrightarrow RCH_2CH_2OH$$

Scheme 54

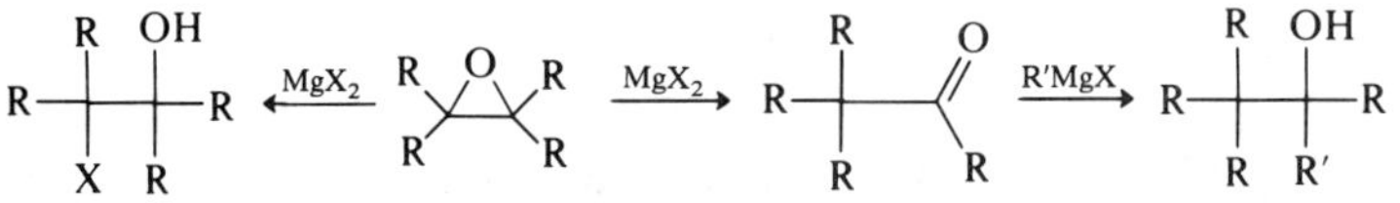

Scheme 55

Oxirane opening by functionalized ethenylaluminum ⟨81JA1851⟩ and ethenylcopper ⟨79JCS(P1)2954⟩ reagents has found applications in natural products synthesis.

Conjugate bases of relatively strong carbon acids like esters and nitriles react with oxiranes. The former give lactones (Scheme 56) ⟨64HC(19-1)418⟩ while the latter provide a route to cyclobutanes (Scheme 57) ⟨74JA5270⟩.

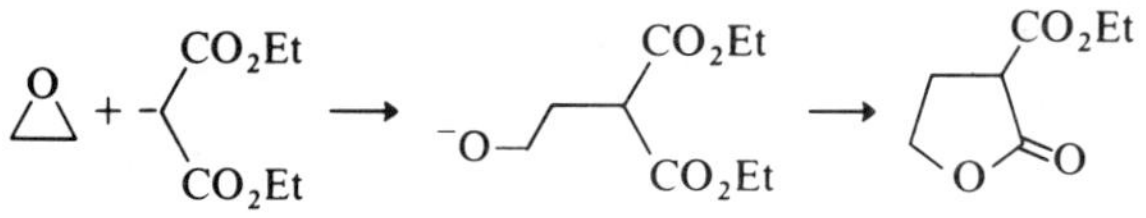

Scheme 56

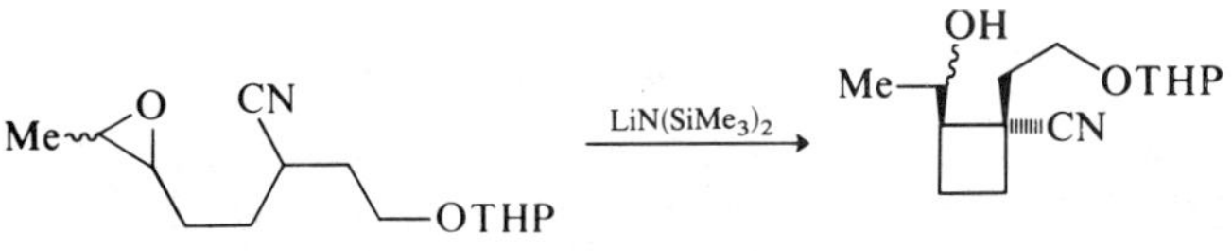

Scheme 57

(vi) π-Systems

Oxiranes undergo Friedel–Crafts reactions ⟨64HC(19-1)432⟩ but these do not appear to be synthetically useful.

Cyclization of polyenes by intramolecular π-attack on an oxirane is an important biochemical reaction (Section 5.05.5.2).

(vii) Hydride ion

Reductive cleavage of oxiranes to alcohols by lithium aluminum hydride is an important reaction ⟨64HC(19-1)199⟩, but the most powerful hydride donor for this purpose is lithium triethylborohydride ⟨73JA8486⟩.

(viii) Phosphorus and silicon nucleophiles

Attack on oxiranes by trivalent phosphorus ⟨64HC(19-1)431⟩ provides a method of deoxygenation to alkenes with inversion (*cf.* Section 5.05.3.4.3(*iii*)) and this makes possible the interconversion of (*Z*)- and (*E*)-alkenes (Scheme 58) ⟨B-74MI50505⟩. Silicon nucleophiles behave analogously ⟨76JA1265, 76S199⟩.

Scheme 58

5.05.3.5 Nucleophilic Attack on Ring Hydrogen (Proton Abstraction)

This usually generates a carbene (Section 5.05.3.2.2), but the oxirane (**61**) is deprotonated by butyllithium to the versatile oxiranyllithium (**62**; Scheme 59) ⟨76JA4646⟩.

Scheme 59

5.05.3.6 Reactions with Electrons and Surface Reactions

Reductive cleavage of oxiranes to alcohols by sodium or lithium ⟨64HC(19-1)181⟩ has been largely displaced by hydride reagents (Section 5.05.3.4.3(*vii*)), but lithium in refluxing THF has been advocated as a general reagent for reduction to alkenes with retention of configuration ⟨80TL1173⟩ (*cf.* Sections 5.05.3.4.3(*iii*) and (*viii*)).

Reductive cleavage to alcohols by catalytic hydrogenation ⟨64HC(19-1)188⟩ has largely given way to reduction with hydride reagents (Section 5.05.3.4.3(*vii*)).

5.05.3.7 Cycloaddition Reactions

These are due to carbonyl ylides (Section 5.05.3.2.1).

5.05.3.8 Reactions Involving Substituents

Isomerization to allylic alcohols was discussed in Sections 5.05.3.2.2, 5.05.3.4.3(*i*), (*iii*) and (*iv*).

S_N2' ring opening by attack of a nucleophile on an ethenyl substituent (*e.g.* Scheme 60) has assumed importance in natural products synthesis ⟨80JCS(P1)2084, *cf.* 81JA5969⟩. The cyclopentene oxide (**63**) is a versatile synthon for alkylated, functionalized cyclopentanes (Scheme 61) ⟨81JA2112⟩.

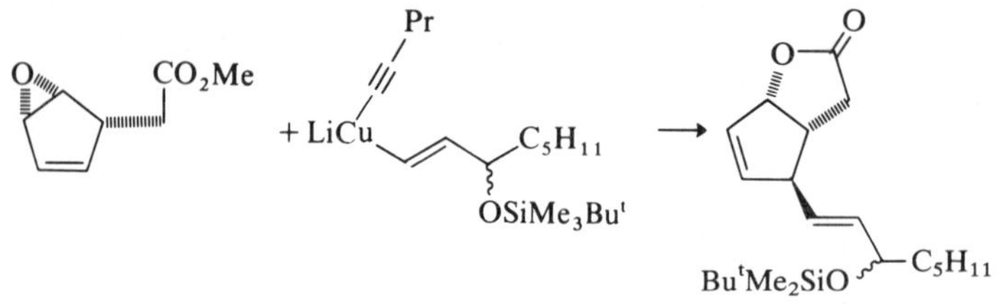

A prostaglandin intermediate

Scheme 60

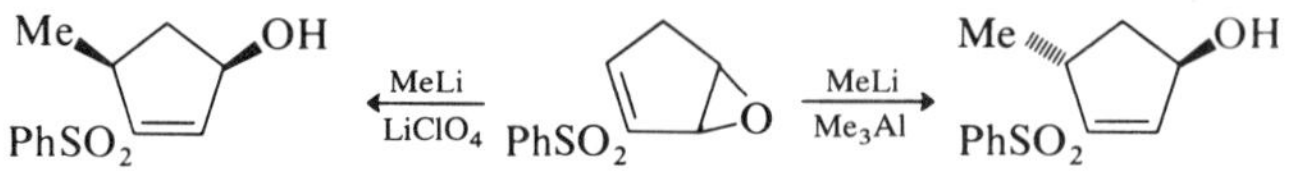

(63)

Scheme 61

Ring opening by intramolecular attack by an electron source located on a β-carbon is represented by the Wharton reaction (Scheme 62) ⟨81JCS(P1)2363⟩ and the Eschenmoser fragmentation (Scheme 63) ⟨81S276⟩.

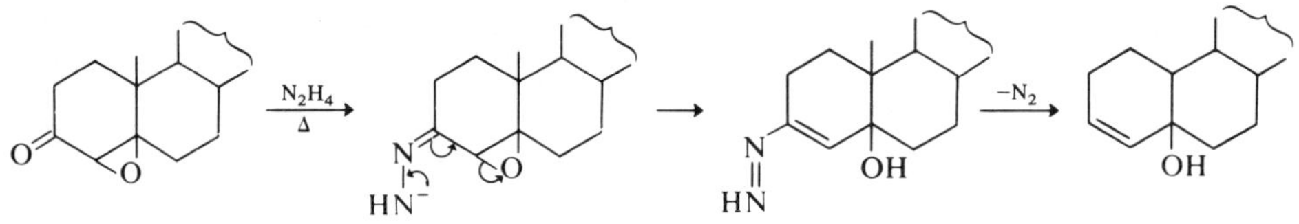

Scheme 62

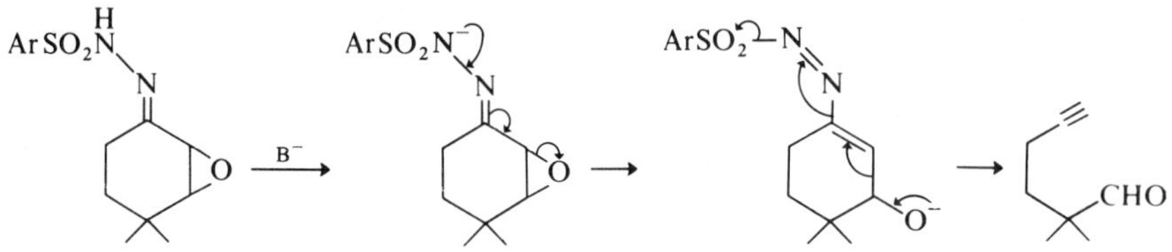

Scheme 63

Carbon–carbon cleavage induced by chemical attack is rare for oxiranes [*cf.* thermal and photolytic cleavage (Section 5.05.3.2.1)], but a few examples are known, *e.g.* Scheme 64 ⟨70TL3025⟩.

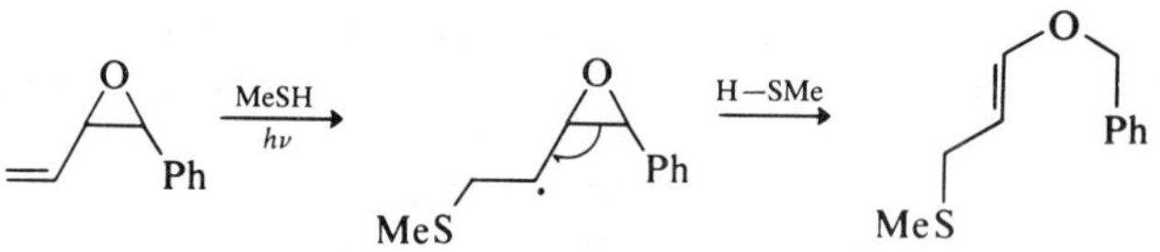

Scheme 64

5.05.4 OXIRANES: SYNTHESIS

5.05.4.1 General Survey of Synthesis

The most important oxirane syntheses are by addition of an oxygen atom to a carbon–carbon double bond, *i.e.* by the epoxidation of alkenes, and these are considered in Section 5.05.4.2.2. The closing, by nucleophilic attack of oxygen on carbon, of an OCCX moiety is dealt with in Section 5.05.4.2.1 (this approach often uses alkenes as starting materials). Finally, oxirane synthesis from heterocycles is considered in Section 5.05.4.3; one of these methods, thermal rearrangement of 1,4-peroxides (Section 5.05.4.3.2), has assumed some importance in recent years. The synthesis of oxiranes is reviewed in ⟨B-73MI50500⟩ and ⟨64HC(19-1)1⟩.

5.05.4.2 Synthesis from Non-heterocycles

5.05.4.2.1 By nucleophilic displacement by oxygen of a β leaving group

The critical intermediate (**64**) for this route (Scheme 65) can be reached in several ways. The commonest is by the reaction of an alkene with hypohalous acid (Scheme 66) which proceeds with overall retention.

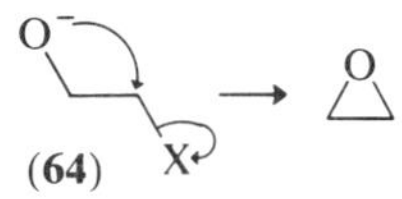

(**64**)

Scheme 65

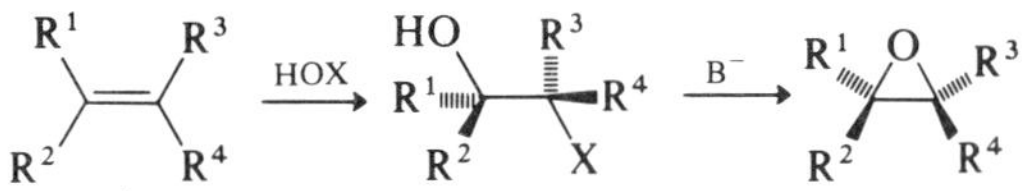

Scheme 66

Attack on the caronyl group of a 2-halo ketone by a nucleophile also generates oxiranes *via* (**64**), but this method has been relatively little used (Scheme 67).

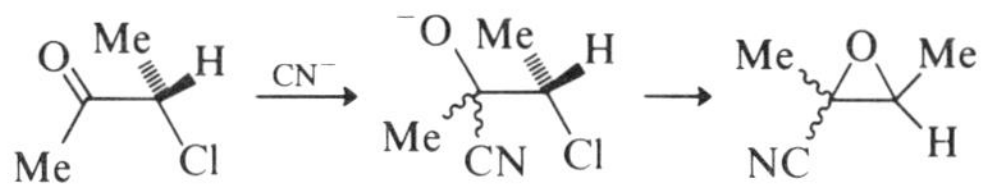

Scheme 67

A very useful route to (**64**) is the Darzens reaction, the nucleophilic attack on an aldehyde or ketone carbonyl group of an α-halo carbanion (**65**; Scheme 68) ⟨75CC95⟩. The 'glycidic ester' (**66**) is useful since it can be eliminatively rearranged to an aldehyde or ketone homologous to the starting ketone (**67**) ⟨79JOC450, 79TL1321, 63JA955⟩.

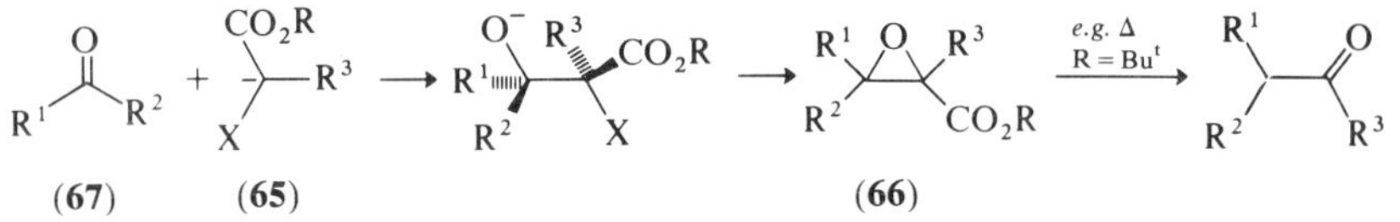

(**67**) (**65**) (**66**)

Scheme 68

Other syntheses of oxiranes which are occasionally useful and involve a species (**64**) are the dehydration of 1,2-diols (Scheme 69) ⟨78TL5153, 78OS(58)12⟩ and the reductive coupling of aldehydes (Scheme 70) ⟨B-73MI50500⟩.

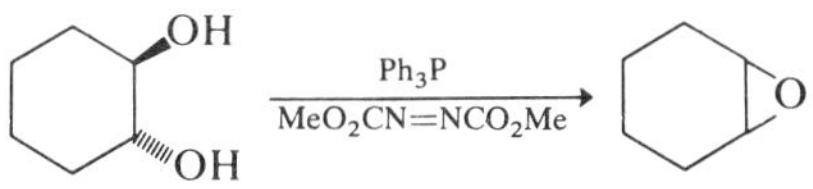

Scheme 69

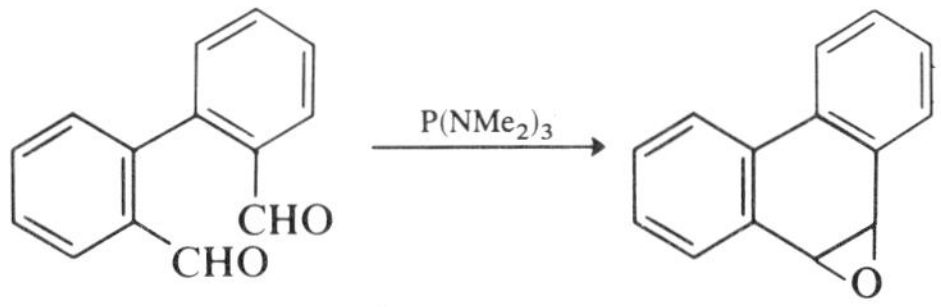

Scheme 70

A versatile oxirane synthesis *via* (**64**) is the sulfur ylide approach ⟨B-75MI50504, *cf.* 76TL457⟩, which in effect inserts a CR_2 group into a carbonyl group (Scheme 71). Older, less generally useful versions of this 'insertion' utilize diazomethane or dibromomethane–butyllithium.

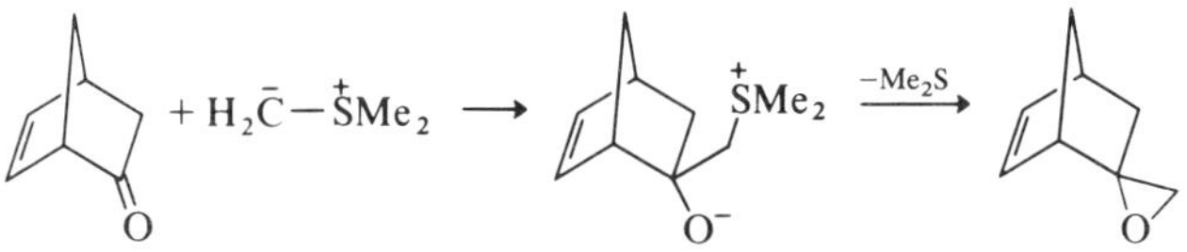

Scheme 71

There are several examples of the synthesis of optically active oxiranes by β-elimination, e.g. ⟨80TL2733, 81HCA687, 81JOC3348⟩.

5.05.4.2.2 From [2+1] fragments

(i) By initial electrophilic attack (alkene epoxidation)

The conversion of an alkene to an oxirane by an electrophilic oxidant (Scheme 72) is the commonest route to oxiranes ⟨81H(15)517⟩. The usual laboratory reagent is a peroxy acid (a peracid), and the most widely used of these nowadays is 3-chloroperoxybenzoic acid (*m*-chloroperbenzoic acid, MCPBA; **68**). The oxirane made in greatest quantity, oxirane itself (ethylene oxide), is manufactured by oxidizing ethylene with molecular oxygen ⟨B-80MI50500, B-80MI50501⟩.

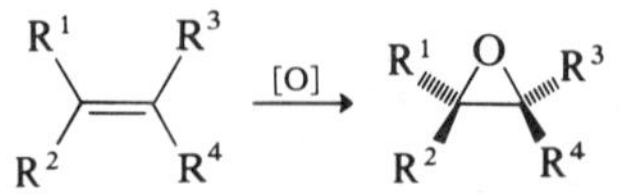

Scheme 72

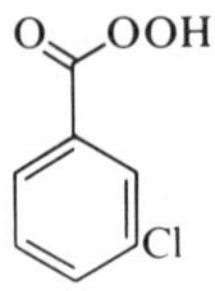

(68)

Recently ⟨79MI50500⟩ Sharpless and coworkers have shown that *t*-butyl hydroperoxide (TBHP) epoxidations, catalyzed by molybdenum or vanadium compounds, offer advantages over peroxy acids with regard to safety, cost and, sometimes, selectivity, *e.g.* Scheme 73, although this is not always the case (Scheme 74). The oxidation of propene by 1-phenylethyl hydroperoxide is an important industrial route to methyloxirane (propylene oxide) ⟨79MI5501⟩.

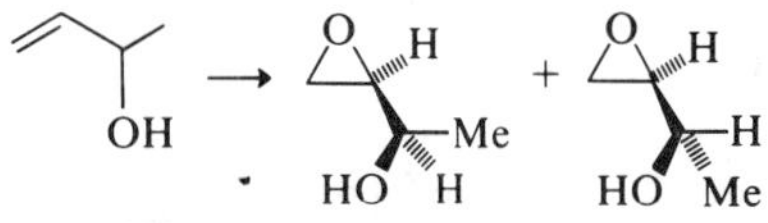

TBHP, V^{5+}	20	80
MCPBA	60	40

Scheme 73

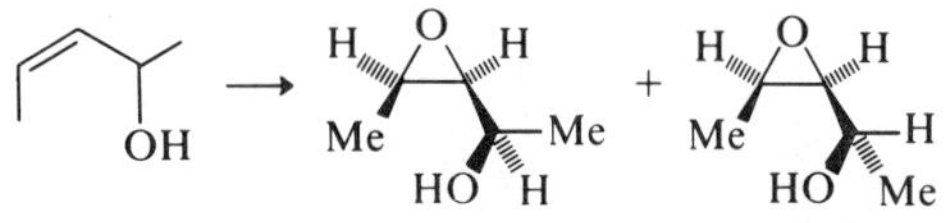

TBHP, V^{5+}	71	29
MCPBA	95	5

Scheme 74

Peroxy acid epoxidation shows the characteristics of attack on alkenes by a weakly ⟨81JA7106⟩ electrophilic reagent and the transition state has been represented ⟨78JA743⟩ as shown in (**69**) for the prototype reaction, although a more symmetrical transition state has been traditionally favored. The stereospecificity shown by some TBHP epoxidations of allylic alcohols may arise from intramolecular electrophilic attack by an oxygen in a metallacycle coordinated to the alkene (Scheme 75). As expected from these mechanisms, these reactions give an oxirane with the same configuration as the starting alkene.

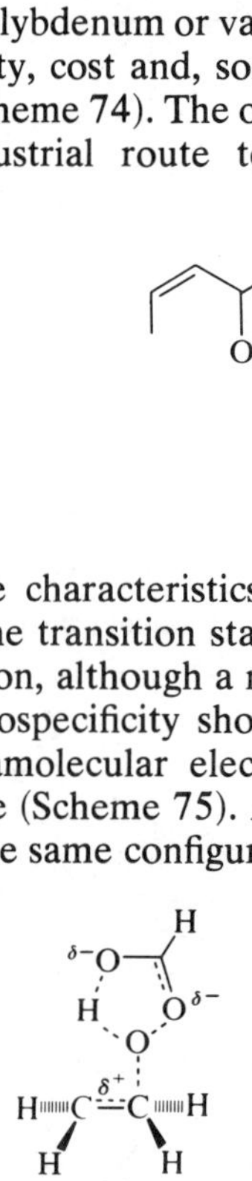

(69)

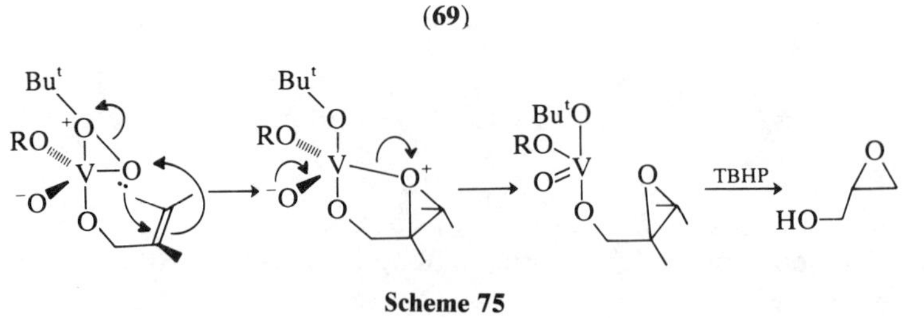

Scheme 75

The remarkable stereospecificity of TBHP-transition metal epoxidations of allylic alcohols has been exploited by Sharpless' group for the synthesis of chiral oxiranes from prochiral allylic alcohols (Scheme 76) ⟨81JA464⟩ and for diastereoselective oxirane synthesis from chiral allylic alcohols (Scheme 77) ⟨81JA6237⟩. It has been suggested that this latter reaction may enable the preparation of chiral compounds of *complete* enantiomeric purity (*cf.* Scheme 78)!

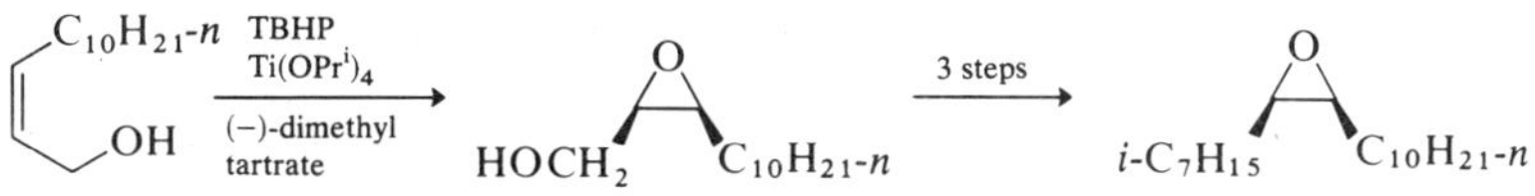

80%

91% enantiomeric excess (+)-dispalure

Scheme 76

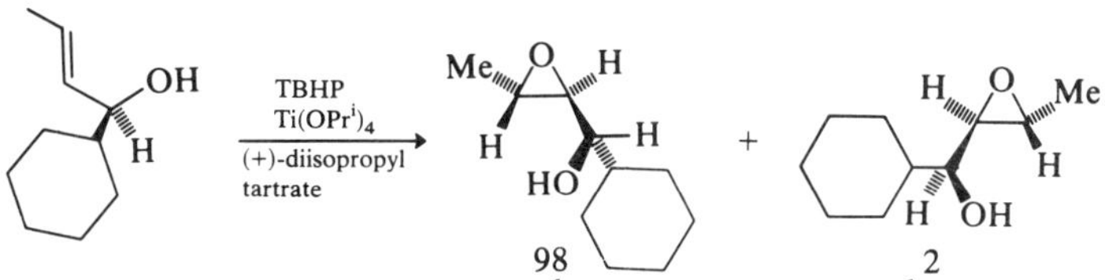

98 *erythro* 2 *threo*

Scheme 77

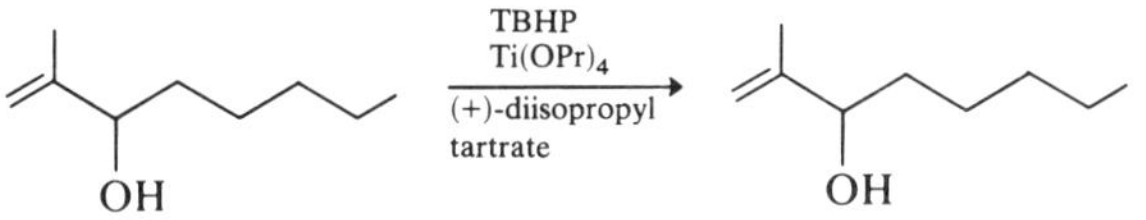

Racemic 60% conversion Remaining alkene 99.999 999% (calc.) optical purity

Scheme 78

(ii) By initial nucleophilic attack

Electron deficient carbon–carbon double bonds are resistant to attack by the electrophilic reagents of Section 5.05.4.2.2(*i*), and are usually converted to oxiranes by nucleophilic oxidants. The most widely used of these is the hydroperoxide ion (Scheme 79). Since epoxidation by hydroperoxide ion proceeds through an intermediate α-carbonyl anion, the reaction of acyclic alkenes is not necessarily stereospecific (Scheme 80) (unlike the case of epoxidation with electrophilic agents (Section 5.05.4.2.2(*i*)); the stereochemical aspects of this and other epoxidations are reviewed at length in ⟨B-73MI50500⟩).

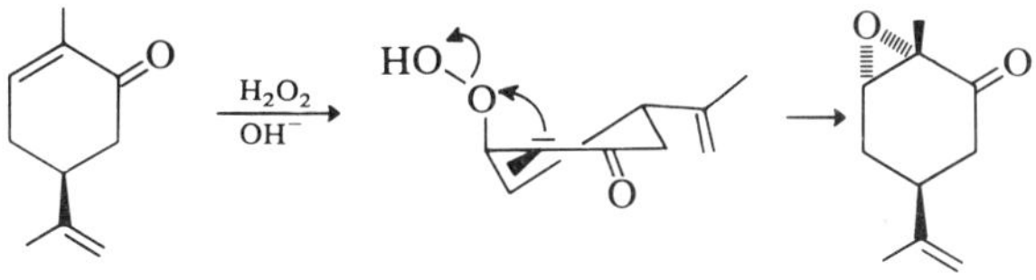

Scheme 79

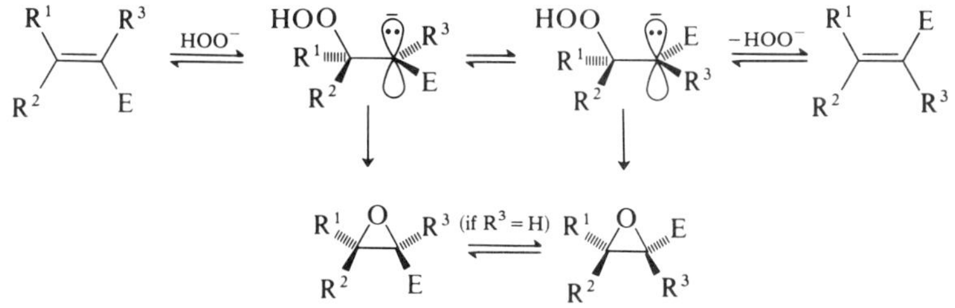

Scheme 80

5.05.4.3 Synthesis from Heterocycles

5.05.4.3.1 From five-membered heterocycles by extrusion

Oxiranes have been made by extrusion of nitrogen from oxadiazolenes (Scheme 81) ⟨B-73MI50502⟩, by flash thermolysis of dioxazolones (Scheme 82) ⟨81JA5414⟩, and (α-lactones) by photolysis of 1,2-dioxole-3,5-diones (Scheme 83) ⟨73JOC2269⟩.

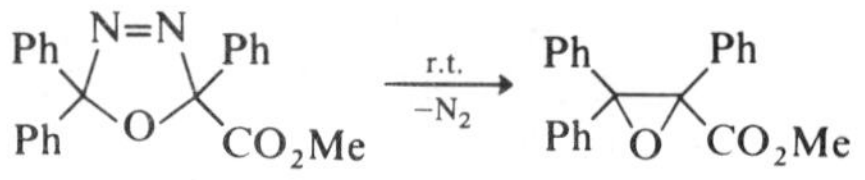

Scheme 81

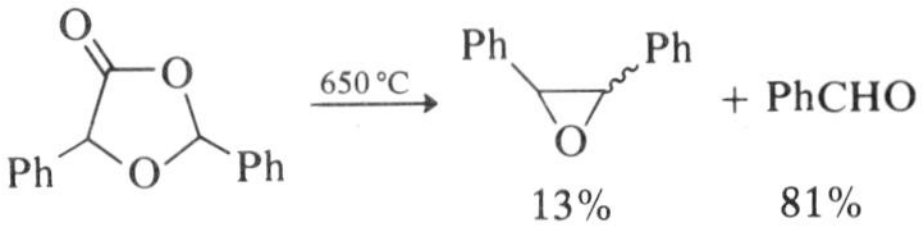

Scheme 82

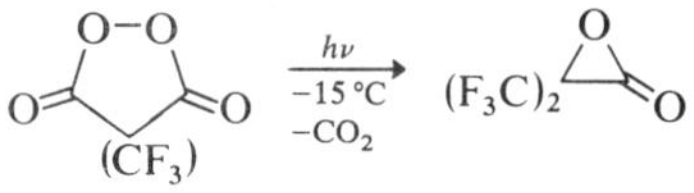

Scheme 83

5.05.4.3.2 From six-membered heterocycles by rearrangement or extrusion

The thermal fragmentation of unsaturated bicyclic 1,4-peroxides, often readily made from 1,4-dienes (Scheme 84), has become an important route to novel bis(oxiranes) ⟨80T833, 81CRV91⟩.

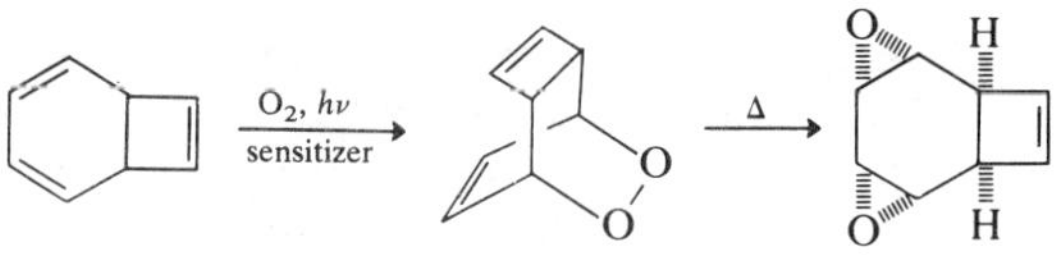

Scheme 84

Oxiranes have been made by photolysis of cyclic oxalates (Scheme 85) ⟨80TL1021⟩.

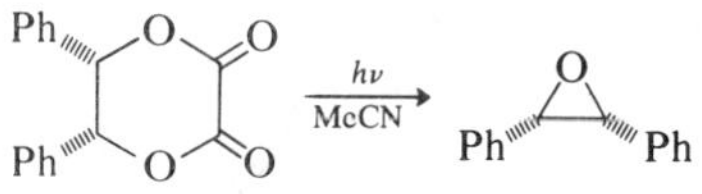

Scheme 85

5.05.5 OXIRANES: APPLICATIONS, BIOLOGICAL ASPECTS, OCCURRENCE

5.05.5.1 Applications

The manufacture and uses of oxiranes are reviewed in ⟨B-80MI50500, B-80MI50501⟩. The industrially most important oxiranes are oxirane itself (ethylene oxide), which is made by catalyzed air-oxidation of ethylene (*cf.* Section 5.05.4.2.2(*i*)), and methyloxirane (propylene oxide), which is made by β-elimination of hydrogen chloride from propene-derived 1-chloro-2-propanol (*cf.* Section 5.05.4.2.1) and by epoxidation of propene with 1-phenylethyl hydroperoxide (*cf.* Section 5.05.4.2.2(*i*)) ⟨79MI50501⟩.

Oxirane is used as a fumigant for grain and a sterilant (*e.g.* for space vehicles bound for putative abodes of life), but mainly for the manufacture of 1,2-ethanediol ('ethylene glycol'), emulsifiers, plastics and resins (below), plasticizers, synthetic rubber and synthetic fibers. Methyloxirane is used mainly to make detergents, hydraulic fluids and lubricants.

The term 'epoxy' is familiar to nonchemists because of the widespread use of epoxy glues and resins. These are crosslinked polyether thermoplastics made from a liquid resin which is typically a mixture of bisphenol A diglycidyl ether (**70**) and a polymer (**71**; Scheme 86). The liquid resin is cured or hardened to the final resin by mixing with a crosslinking reagent, which can be an acid, a di- or poly-alcohol, or a di- or poly-amine (Scheme 86).

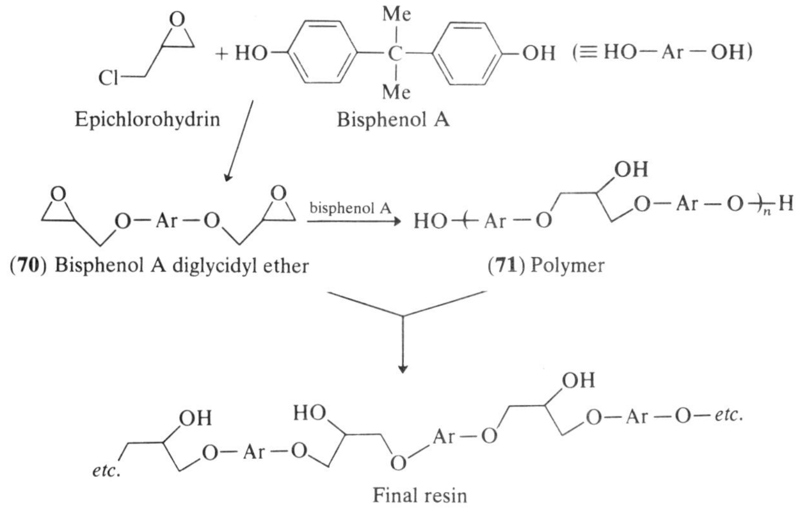

Scheme 86

5.05.5.2 Biological Aspects

The most important oxirane, from an anthropocentric viewpoint, is probably squalene oxide (**72**), a precursor of lanosterol (**73**) and thus of the maligned but essential cholesterol (**74**; Scheme 87) ⟨78MI50501⟩. The cyclization of (**72**) to (**73**) represents nucleophilic π-attack on oxirane carbon (*cf.* Section 5.05.3.4.3(*vi*)), and the process has also been extensively investigated *in vitro* ⟨68ACR1⟩. Oxiranes are even more ubiquitous in steroid biosynthesis than had been thought, for a cholesterol epoxide has been shown to be a product of mammalian steroid biosynthesis ⟨81JA6974⟩.

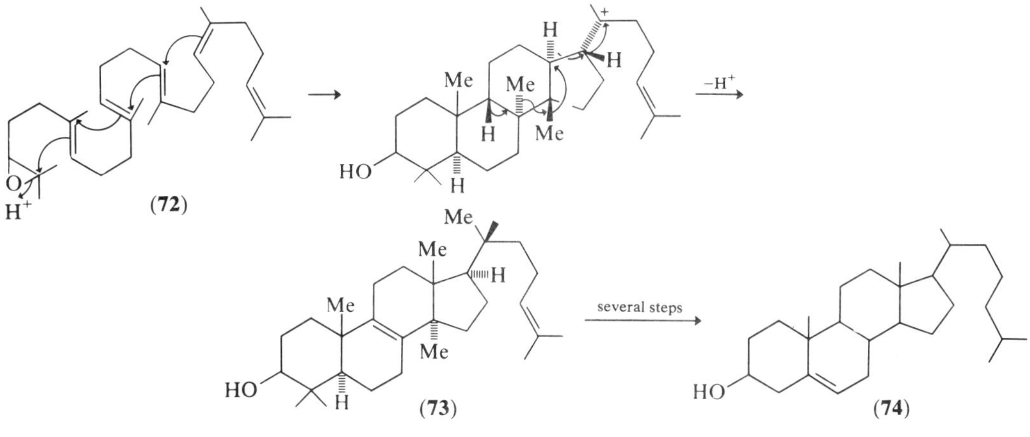

Scheme 87

Several oxiranes are carcinogenic ⟨B-80MI50504, 77MI50500, 75MI50501⟩, those derived from polycyclic aromatic hydrocarbons being of particular interest (Chapter 5.07). This carcinogenicity seems to derive from the alkylation of nucleic acids by nucleophilic nitrogen ⟨82JA544⟩ (*cf.* Section 5.05.3.4.3(*ii*)).

The epoxide of vitamin K is involved in the regeneration of the anticoagulant vitamin (a naphthoquinone) from the active hydroquinone form ⟨81JA5939⟩.

The mechanisms of formation of oxiranes by biological systems (which are relevant to carcinogenesis — see the penultimate paragraph) are the subject of continuing interest ⟨81JA6744, 81JA7371⟩.

5.05.5.3 Occurrence (Natural Products)

The demarcation between the compounds of Section 5.05.5.2 and those of Section 5.05.5.3 is somewhat artificial, but the overwhelming majority of naturally-occurring oxiranes have no known biological function, being among the products of secondary metabolism to which the term 'natural products' is conventionally limited.

In 1956 the oxirane group was described as occurring in natural products 'exceedingly rarely' ⟨57JA1726⟩; by 1963 naturally-occurring oxiranes were recognized as being 'widely distributed if not very numerous' ⟨B-63MI50501⟩. They are indeed very numerous, but appear to be largely confined to terpenoids. Since the appearance of two early reviews on oxirane natural products ⟨B-63MI50501, 60QR317⟩, many more of these compounds have been characterized; a plethora of leading references may be obtained by surveying recent issues of *Annual Reports of the Royal Society of Chemistry* and of the terpenoids section of *Chemical Abstracts*.

Many of these compounds have marked physiological properties, showing toxic ⟨e.g. 81JA6686, 73JA5423⟩, carcinogenic ⟨82JA544⟩, anticancer ⟨80JA7991⟩ and antibiotic ⟨81JA6526, 81JA7380⟩ properties.

5.05.6 OXIRENES

5.05.6.1 Introduction

Oxirene (**2**) is one of a number of heterocycles in which the CH_2 group of cyclopropene has been replaced by a group or element associated with Groups V or VI of the periodic table. Replacement of the CH_2 group of cyclopropene by an NH group gives 1*H*-azirine (**75**), while replacement by a sulfur atom results in thiirene (**76**). Compounds (**2**), (**75**) and (**76**) have been by far the most extensively investigated heterocyclopropenes. While (**2**) and (**75**) remain elusive, (**76**) has recently been observed spectroscopically at low temperatures. The systems (**75**) and (**76**) are reviewed elsewhere in this work (Chapters 5.04, 5.06 and 5.07).

H N (**75**) S (**76**)

The first phase of the study of oxirenes is marked by three mistaken reports of the synthesis of this system. Oxirenes first appeared in the chemical literature in 1870, when Berthelot claimed that oxidation of propyne gave methyloxirene ⟨1870BSF(14)113⟩. They reappeared half a century later, with the report by Madelung and Oberwegner ⟨31LA(490)201⟩ that the reaction of an α-chloro ketone with base gave diphenyloxirene. In 1952, Schubach and Franzen reported that 5-decyne (dibutylacetylene) reacted with peroxyethanoic acid (peracetic acid) to give dibutyloxirene ⟨52LA(577)60⟩.

The second, and current, phase of oxirene chemistry began in 1968 with the demonstration by Strausz and coworkers ⟨68JA7360⟩, as outlined in Scheme 88, that a species with the symmetry of an oxirene is involved in the decomposition of α-diazo ketones (the Wolff rearrangement), and the reality of this species has been confirmed by further work by the research groups of Strausz and of Zeller (see Section 5.05.6.3.4(*i*)).* However, no oxirene has ever been isolated or spectroscopically observed, there is but one problematic example of the trapping of an oxirene (Section 5.05.6.3.4(*i*), and it is not known if the species (**77**) is a real intermediate or merely a transition state. This distinction, although important, has only rarely been explicitly recognized in experimental work on oxirene; see, for example, the perceptive comments by Greenberg and Liebman ⟨B-78MI50502⟩ and by Zeller ⟨79LA2036⟩.

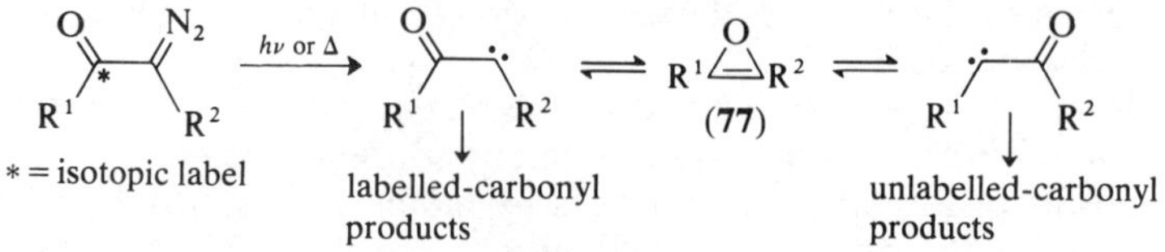

O N_2 R^1 R^2 * = isotopic label — $h\nu$ or Δ → O R^1 R^2 ↓ labelled-carbonyl products ⇌ R^1 O R^2 (**77**) ⇌ O R^1 R^2 ↓ unlabelled-carbonyl products

Scheme 88

We may now be on the verge of the third phase of oxirene chemistry in which modern matrix isolation techniques ⟨80CSR1⟩ will permit the spectroscopic observation of this system, theory will serve as a guide to the synthesis of relatively stable oxirenes (*cf.* a fairly stable

* It has very recently been suggested that this species is not oxirene, but the authors do not make it clear how their intermediate differs from an oxirene canonical form ⟨82JA216⟩.

thiirene ⟨78JOC2490⟩), and these elusive heterocycles may even become recognized as genuine, albeit transient, natural products (Section 5.05.6.3.1(*iii*)).

5.05.6.2 Theoretical Investigations

The simplest, qualitative, theoretical understanding of the nature of oxirene is provided by Breslow's concept of antiaromaticity. Whatever criticisms may be levelled at this notion ⟨78JA6920⟩, it does correctly predict that oxirene should be unusually unstable.

Sophisticated *ab initio* MO calculations with full geometric optimization indicate that oxirene is less stable than was suggested by some of the earlier computations. The results of such work are presented in a recent review on (**2**), (**75**) and (**76**) ⟨80PAC1623⟩ and, along with leading references, in an excellent report on state-of-the-art calculations on (**2**) ⟨80JA7655⟩. The salient conclusions of these more recent computations are summarized in Figure 3. Oxirene is predicted to be a true intermediate, but with an activation barrier of only 8 $kJ\,mol^{-1}$ (31 $kJ\,mol^{-1}$ according to ⟨80PAC1623⟩; even an activation of 31 $kJ\,mol^{-1}$ corresponds to a room temperature lifetime of *ca.* 10^{-8} s). The surmounting of this small barrier converts oxirene into methanolycarbene (**78**) (formylcarbene), which is predicted to be merely a transition state on the way to ketene (**79**) as shown in Figure 3, or at best an unstable intermediate which isomerizes to ketene with an activation energy of only 24 $kJ\,mol^{-1}$ ⟨80PAC1623⟩. The reverse process, the conversion of the oxocarbene to oxirene, requires vibrationally excited carbene. This is accessible photochemically by an electronically excited oxocarbene falling to an upper vibrational level of its electronic ground state (internal conversion), or, less efficiently, by generating the carbene at elevated temperatures (Section 5.05.6.3.4(*i*)). The oxirene–oxocarbene interconversion, and other potential routes to oxirene, are limited to the singlet hypersurface (for a review of theroetical and experimental work on oxirene, see ⟨80PAC1623⟩).

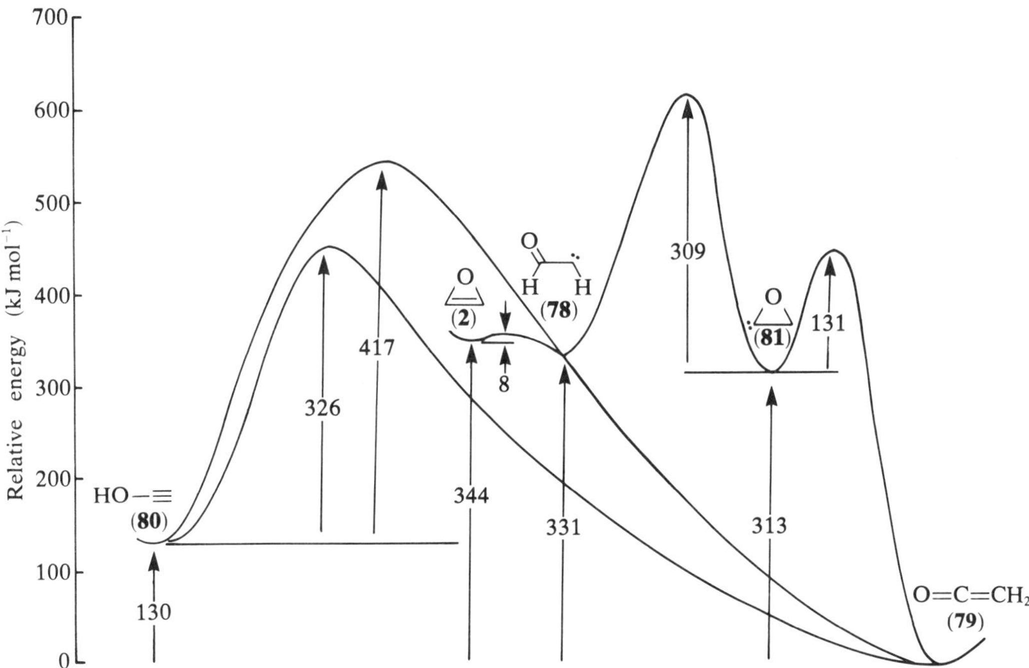

Figure 3 The C_2H_2O hypersurface

Neither ground-state ethynol (hydroxyacetylene) (**80**) nor carbenaoxirane (**81**) appears to be a viable point of ingress to the oxirene–methanoylcarbene system, as both can isomerize to ketene by lower-energy pathways. The limited experimental information available on carbenaoxirane (Section 5.05.6.3.4(*ii*)) indicates that it is indeed largely isolated from the oxirene–methanoylcarbene manifold (but note the photolysis of ketene in Section 5.5.6.3.4(*ii*)); appropriate labelling experiments with (the unknown) ethynol have not been performed.

5.05.6.3 Reactions Possibly Involving an Oxirene, or with the *a priori* Potential of Generating an Oxirene

5.05.6.3.1 Formation of two bonds to oxygen (i.e. involving [2+1] fragments)

(*i*) *By peroxy acids* (*mainly*)

The literature on oxirene began with Berthelot's claim ⟨1870BSF(14)113⟩ to have oxidized propyne to methyloxirene. This compound, however, should certainly not be isolable under normal conditions, and the reaction of alkynes with strong oxidants is known to give other products ⟨B-69MI50500⟩. The next claim of a successful oxirene synthesis ⟨31LA(490)201⟩ (Section 5.05.6.3.4(*i*)) involved a non-oxidative approach, but in 1952 Schlubach and Franzen ⟨52LA(577)60⟩ claimed the oxidation of 5-decyne to dibutyloxirene with peracetic acid. However, Franzen soon showed that the product of the oxidation was in fact 6-decen-5-one, and went on to show that the oxidation of alkynes with peracetic acid can produce, besides α,β-unsaturated ketones, products plausibly derivable from them ⟨55CB717⟩ or from ketenes ⟨54CB1218⟩. These various products can be rationalized as arising from an oxocarbene, possibly formed by rearrangement of an oxirene (*cf.* Section 5.05.6.3.4(*i*)). Further evidence for oxocarbenes in the peroxy acid oxidations of alkynes was obtained by later workers who employed phenylacetylene ⟨64JA4866⟩ and diphenylacetylene ⟨64JA4871⟩.

The demonstration of the involvement of an oxirene species in α-diazo ketone decompositions (Sections 5.05.6.1, 5.05.6.2, 5.05.6.3.4(*i*)) opened up the possibility of probing the involvement of oxirenes in peroxy acid oxidations of alkynes by attempting to generate the same oxirene in these two quite different ways. The obtention of similar product distributions would be consistent with the common formation of an oxirene, and oxygen scrambling (*cf.* Scheme 88) would require the intervention of this symmetrical (or pseudosymmetrical, $R^1 \neq R^2$) species. Such experiments were reported in 1970 by Ciabattoni and coworkers ⟨70JA3826⟩. They showed that the oxidation of 2,2,5,5-dimethyl-3-hexyne (di-*t*-butylacetylene) (**82**) with *m*-chloroperbenzoic acid gave (allowing for further reaction of the product (**83**) with peroxy acid) a product distribution virtually identical with that from thermolysis of the diazo ketone (**84**; Scheme 89). The products of these reactions can be rationalized as arising from a common oxocarbene (**85**) (which may arise from di-*t*-butyloxirene, although the results do not require this) which can undergo a 1,2-methyl migration to give the enone (**83**), or effect insertion of the carbene carbon into a carbon–hydrogen bond on the adjacent *t*-butyl group, giving the cyclopropane (**86**). Ciabattoni and coworkers also reported product comparison studies of the oxidation of cyclic alkynes and the thermal decompsition of cyclic diazo ketones ⟨73JA3284⟩. Although both types of reaction gave the same products, the relative yields in this case (in contrast to the acyclic, di-*t*-butyl compounds, above) were considerably different. The products are of three types: ring-contracted ketenes (or the ketones derived from them by peroxy acid oxidation), bicyclic ketones and enones. The diazo ketones gave mainly ring-contracted products and enones, but the cycloalkynes yielded as their major products enones and, in the cases of cyclooctyne and cyclodecyne, bicyclic ketones as well. All three kinds of products can be explained in terms of a common oxocarbene, as shown for cyclooctyne in Scheme 90, although the fact that the two kinds of reaction gave very significantly different ratios of ring-contracted product, bicyclic ketone and enone shows that this must be an oversimplification.

$Bu^t-\equiv-Bu^t$ (**82**) —MCPBA→ (**85**) ←Δ, $-N_2$— (**84**)

(**85**) → (**83**) + (**86**); (**83**) —MCPBA→ epoxy ketone

Scheme 89

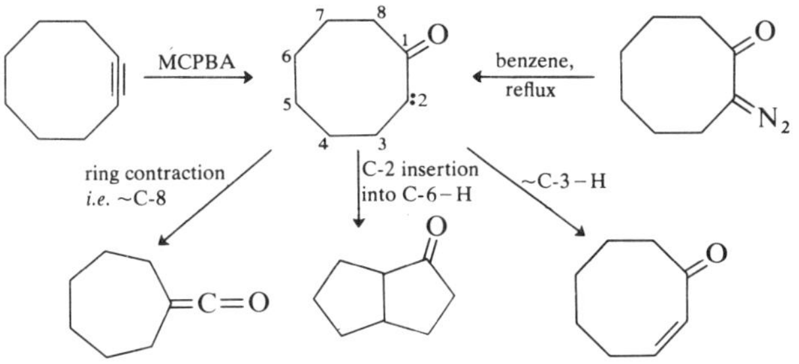

Scheme 90

More definitive evidence for the formation of an oxirene intermediate or transition state was presented recently by Cormier ⟨80TL2021⟩, in an extension of his earlier work on diazo ketones ⟨77TL2231⟩. This approach was based on the realization that, in principle, the oxirene (**87**) could be generated from the diazo ketones (**88**) or (**89**) (*via* the oxocarbenes **90** or **91**) or from the alkyne (**92**; Scheme 91). If the carbenes (**90**) (from **88**) and (**91**) (from **89**) equilibrate through the oxirene (**87**), and if (**87**) is also the initial product of epoxidation of (**92**), then essentially the same mixture of products (hexenones and ketene-derived products) should be formed on decomposition of the diazo ketones and on oxidation of the alkyne; this was the case.

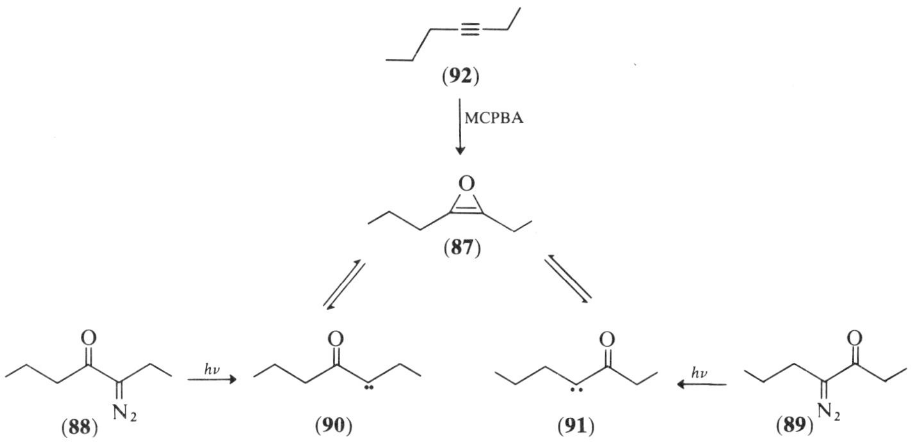

Scheme 91

That an oxirene species is the first-formed entity in the oxidation of alkynes by peroxy acids was strongly indicated by kinetic studies ⟨73JA7894⟩ which showed that the transition state for attack on alkynes is analogous to that for attack on alkenes, which latter process leads to oxiranes (Section 5.05.4.2.2(*i*)). Although the authors concluded that the oxirene-like transition state (**93**) does indeed lead to oxirenes (which quickly decompose), it is not clear that this is necessarily the case; it seems conceivable that the *incipient* oxirene, still partially bonded to the departing carboxylic acid moiety, might begin to suffer the ring-opening reactions which lead to the observed products of alkyne epoxidation. If this is the case then an oxirene-like transition state, but not an intermediate, is involved in alkyne epoxidation. In other kinetic studies, the effect of ring substituents on the rate of reaction of phenylacetylenes with perbenzoic acid was investigated ⟨73JOC1044⟩. The reaction was accelerated by electron-donating groups, showing that the peroxy acid attacks the alkyne as an electrophile, rather than as a nucleophile (the latter being an *a priori* possibility for alkynes).

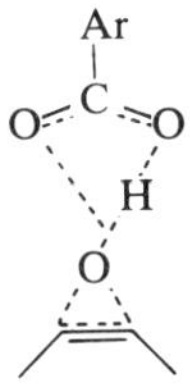

(**93**)

The molybdenum-catalyzed oxidation of alkynes by *t*-butyl hydroperoxide has been investigated ⟨73JCS(P1)2851⟩ (the epoxidation of *alkenes* by this system has become an important reaction; Section 5.05.4.2.2(*i*)) but the formation of oxirenes was excluded.

(ii) By O_2 or O

Acetylene is oxidized to ketene by hot air ⟨24JCS(125)1534⟩. Oxirene is probably not involved since molecular oxygen is not a donor of singlet oxygen atoms (see the remarks on this point in ⟨64JA4866⟩).

Alkynes have been reacted with oxygen atoms from the photodecomposition of nitrous oxide ⟨62JA2855, 72JCS(F1)512⟩ but oxirenes were probably not formed for reasons of spin conservation ⟨*cf.* 73JA133⟩. Other work on the reaction of alkynes with oxygen atoms ⟨34MI50500, 33CB(66B)1815, 31MI50500⟩ may not have involved oxirenes either, for the same reasons.

(iii) By biochemical systems

Probably the most surprising aspect of the putative oxidation of alkynes to oxirenes is the fact that evidence has recently been obtained that oxirenes may be formed, enzymatically, in living cells. The oxidation of labelled 4-biphenylylacetylene (**94**) to the carboxylic acid (**95**; Scheme 92) may involve an oxirene, since the alkyne gave acid with complete deuterium retention (path a); path (b) should have given unlabelled acid ⟨80JA7373⟩.

(a) → Ar, D oxirene $\xrightarrow{\sim D}$ Ar(D)C=C=O → $ArCHDCO_2H$ (**95**)

(**94**) —≡—D

(b) → Ar—≡—OD $\xrightarrow{H_2O}$ Ar(H)C=C=O → $ArCH_2CO_2H$

Scheme 92

The alkynyl steroid (**96**) inactivates aromatase, an enzyme which catalyzes the conversion of androgen to estrogen. It has been suggested ⟨81JA3221⟩ that the inactivation process involves the oxidation of (**96**) to an oxirene which rearranges to an oxocarbene which then binds to the enzyme prosthetic group, thus inactivating it.

(**96**)

5.05.6.3.2 β-Elimination reactions

Little work appears to have been done on the introduction of a double bond into a preformed oxirane ring by β-elimination (Scheme 93). The most obvious candidates for this transformation, halooxiranes, have apparently not been subjected to a systematic study of their reactions with nonnucleophilic bases (for a thermal reaction of a halooxirane, see Scheme 21). The chlorooxirane (**97**) gave on dehydrohalogenation the cyclopropane (**98**), evidently *via* an oxocarbene (**99**; Scheme 94), but it is not known if an oxirene is the precursor of, or equilibrates with, the carbene ⟨75JOC1694⟩. The situation seems to be similar, too, for the β-elimination of other simple groups ⟨77JOC3840⟩, although Padwa and coworkers ⟨67JA4435⟩ have rationalized some of the products from photolysis of a β,γ-epoxy ketone by invoking an oxirene intermediate (Scheme 95). A Norrish Type II photoreaction of (**100**) was suggested to lead to formation of phenyloxirene (**101**), which on aqueous work-up gave rise to phenylacetic acid (**102**), one of the observed products. The validity of this mechanism has, however, been called into question by Cormier in his paper comparing alkyne epoxidation and diazo ketone decomposition (Section 5.05.6.3.1(*i*)) ⟨80TL2021⟩.

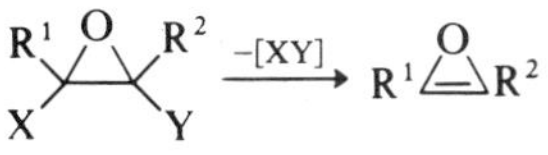

Scheme 93

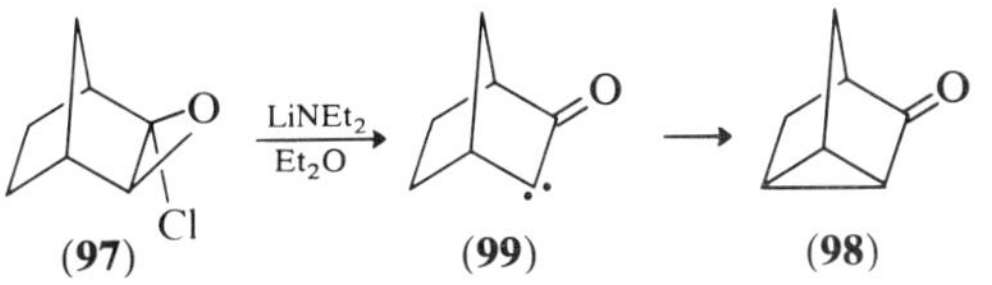

(97) (99) (98)

Scheme 94

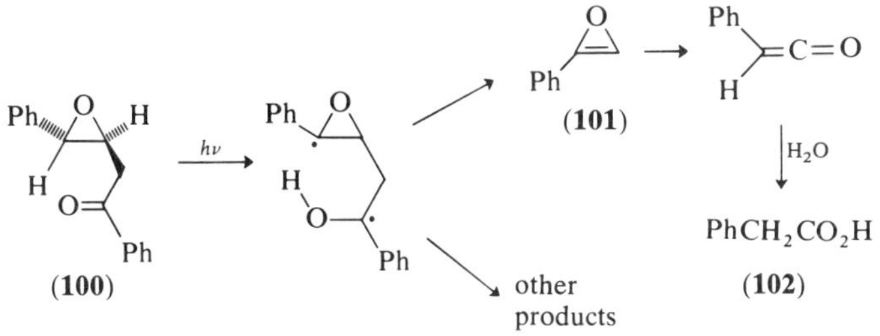

(100) (101) (102)

Scheme 95

5.05.6.3.3 Direct extrusion of an oxirene by retrocycloaddition

The reverse Diels–Alder reaction ⟨68CRV415⟩ offers a possible route to oxirene by extrusion from molecules with the 1,4-cyclohexadiene oxide moiety (Scheme 96). Lewars ⟨77UP50500⟩ found that flash thermolysis ⟨B-80MI50502⟩ of (**103**) gave ketene, and Lewars and Morrison ⟨77CJC966⟩ showed that the *a priori* more favorably constructed barrellene oxide system (*e.g.* **104**) also gave some ketene on flash thermolysis. However, it is not known if the ketene arose from extrusion and isomerization of oxirene. The dibenzobarrellene oxide (**105**) rearranged on heating ⟨77JOC3840⟩, but the bicycloheptene derivative (**106**) gave evidence of possible oxirene extrusion ⟨75TL2363⟩. Attempted extrusion of benzooxirene from (**107**) led to rearrangement ⟨73AG(E)840⟩.

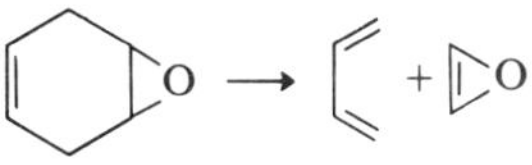

(103)

Scheme 96

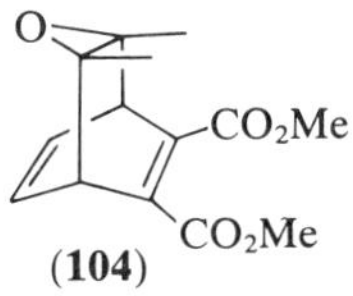

(104) 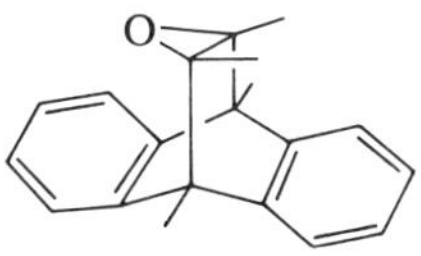 (105) 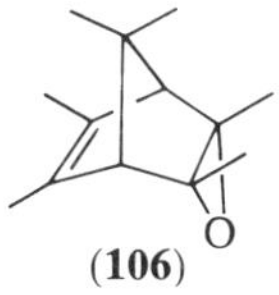 (106) 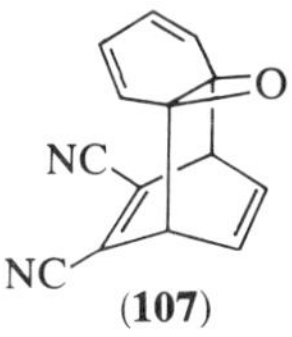 (107)

The reverse [2+2] cycloaddition of the formal benzene adduct (**108**) is a very tempting approach to oxirene, since Warrener and coworkers have shown that this kind of process, the photoaromatization reaction, can be used to generate alkenes from their formal [2+2] adducts with benzene derivatives (references in ⟨81AJC855⟩). The adduct (**108**) might be photolyzed at low temperatures, offering, should the desired reaction occur, a much greater change of actually observing an oxirene than would be expected from the Diels–Alder reactions mentioned above. In a cleverly conceived and skillfully executed synthesis, Warrener and coworkers ⟨81AJC855⟩ prepared the formal [2+2] oxirene–arene adduct (**109**);

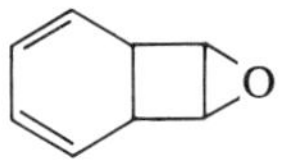

(108) 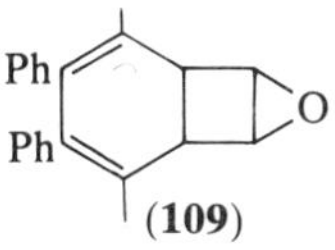 (109)

unfortunately however, photolysis (or thermolysis) of (**109**) caused conversion to a bicyclic isomer with no indication of oxirene extrusion.

5.05.6.3.4. Isomerization reactions

(i) Of oxocarbenes ('ketocarbenes'); the Wolff rearrangement

After the initial claim of the synthesis of an oxirene (by the oxidation of propyne; Section 5.05.6.3.1) this system reappeared with the claim ⟨31LA(490)201⟩ that 2-chloro-1,2-diphenylethanone (**110**) reacted with sodium methoxide to give diphenyloxirene (**111**), but it was later shown ⟨52JA2082⟩ that the product was the prosaic methoxy ketone (**112**; Scheme 97) (the formation of **111** from **110** would be an α-elimination carbene-type reaction). Even with strong, nonnucleophilic bases, (**110**) failed to provide evidence of diphenyloxirene formation ⟨64JA4866⟩.

(**112**) ← (**110**) + NaOMe —✕→ (**111**)

Scheme 97

The decomposition of an α-diazo ketone (**113**) to an oxocarbene (**114**), which by migration of a group on the carbonyl carbon gives a ketene (**115**; Scheme 98), is the *Wolff rearrangement*, a reaction of great synthetic and theoretical interest ⟨75AG(E)32, B-71MI50501⟩. The oxirene intermediate (or transition state) appears in thermal and both sensitized and unsensitized photolytic decompositions of α-diazo ketones ⟨80PAC1623⟩ but apparently not in the metal ion-promoted decomposition of diazo ketones ⟨72CC11⟩, a not surprising finding since these reactions may proceed through organometallic carbenoids.

(**113**) $\xrightarrow{-N_2}$ (**114**) → (**115**)

Scheme 98

Oxocarbenes in which an adjacent carbon has a C—H bond undergo considerable enone formation (from a 1,2-hydrogen shift) in adiition to the Wolff rearrangement, and this forms the basis of a test for oxirene participation based on labelling the oxirene with alkyl groups rather than the isotopic carbon (*cf.* Section 5.05.6.3.1(*i*)). By this means the involvement of a common oxirene species has been demonstrated in the peroxy acid oxidation of 'dialkylalkynes' (Section 5.05.6.3.1(*i*)) and in the thermal and photochemical decomposition of a 'dialkyl α-diazo ketone' ⟨72JCS(P1)2623, 77TL2231⟩. Labelling an oxirene with different substituent groups, rather than isotopically, does have the disadvantage that the extent of oxirene participation is not necessarily reflected in the degree of oxygen migration, since a substituent can bias the oxirene ring towards cleavage of the proximate or the distal C—O bond. Thus, in the extreme, formation of the oxirene (**116**; Scheme 99) would not be detected should R^1 and/or R^2 coerce (**116**) into undergoing cleavage of bond (a) to the exclusion of bond (b). The amount of oxygen scrambling observed in reactions involving unsymmetrical oxirenes thus represents a minimum value for the extent of oxirene participation ⟨80PAC1623, 75CC317⟩.

products ← $\overset{(a)}{\rightleftharpoons}$ (**116**) $\overset{(b)}{—✕→}$ → no products from this carbene

Scheme 99

For the oxygen migration test for oxirene participation to be valid, it must be shown ⟨80ZN(B)1040⟩ that intermolecular oxygen transfer does not occur, and that oxygen scrambling does not happen at the ketene stage. Dilution studies excluded the former for photolysis of simple aliphatic α-diazo ketones ⟨72CC11⟩ and it appears to be taken for granted that this complication does not intrude; oxygen scrambling at the ketene stage was ruled out for a series of alkyl/aryl-labelled diazo ketones ⟨79CB678⟩.

There appear to be no reports of deliberate attempts to trap oxirenes, but one adventitious case of the possible trapping of such an intermediate has appeared. Photolysis of the diazo amide (**117**) in methanol gave the methanol adduct (**118**) ⟨75AG(E)32⟩. This was regarded as strong evidence for a true intermediate (**119**; Scheme 100), but incorporation of methanol at an earlier stage seems to be conceivable, *e.g.* as outlined in Scheme 101. Less ambiguous would be the observation of a cycloaddition reaction such as the trapping of an oxirene by a diene to give a Diels–Alder adduct, or the isolation of the dihydro derivative (an oxirane) in the presence of diazene (diimide).

(**117**) (**119**) (**118**)

Scheme 100

Scheme 101

The most promising technique for the spectroscopic observation of oxirene is matrix isolation ⟨80CSR1⟩, in conjunction with photolysis of a suitable precursor. However, Krantz found ⟨73CC670⟩ that photolysis of diazoethanal or ethyl diazoethanoate in an argon matrix at 8 K gave only ketene or ethoxyketene (plus a carbonyl compound which was probably a lactone from intramolecular trapping of an oxocarbene), respectively. Either oxirenes are photochemically or thermally (!) unstable under these conditions, they are not true intermediates, or a solid matrix prevents oxirene formation, perhaps by rapid deactivation of vibrationally excited oxocarbene (*cf.* Section 5.05.6.2) to a low-lying level which can only slide energetically downhill to a ketene (Figure 3). Labelling experiments would show if an oxirene species is involved at all in low-temperature matrix photolysis.

Hoffmann and Schüttler ⟨75CB844⟩ attempted to generate methanoylcarbene by high-temperature chelotropic extrusion from a bicycloheptadiene derivative, but the only identified product was a very low yield of benzene.

Other routes to oxocarbenes are the α-elimination of bromine from α,α-dibromo ketones ⟨73JA2708, 73JA5416⟩, and the elimination of CO_2 or COS from dioxolenones (vinylene carbonates) or their sulfur analogues. The former reaction has not been investigated with regard to oxirene involvement; the latter is discussed in Section 5.05.6.3.5. The formation of an oxocarbene from a chlorooxirane was mentioned in Section 5.05.6.3.2.

(ii) Of a carbenaoxirane (an oxiranylidene)

The isomerization of carbenaoxirane (**81**) to oxirene seems unlikely, as *ab initio* calculations predict a more facile conversion to ketene (Figure 3). However, oxirene involvement has been demonstrated for ketene photolysis (Scheme 102) ⟨80PAC1623⟩ and this may proceed through vibrationally excited (**81**); 'cold' carbenaoxiranes would appear to be even

(**81**)

Scheme 102

more reluctant than 'cold' oxocarbenes to isomerize to oxirenes, and the oxocarbene–oxirene interconversion appears to be confined to 'hot' (vibrationally excited) oxocarbenes (Section 5.05.6.2). It should be noted that the generation of oxirenes from ketenes involves less oxirene participation than generation from oxocarbenes (*i.e.* diazo ketones).

Flash thermolysis of compounds of the type (**120**), derivatives of 'Meldrum's acid', is a fairly general synthesis of ketenes (Scheme 103). Brown and coworkers ⟨77AJC179⟩ found that the spirooxirane (**121**) gave ketene, possibly *via* the expected carbonyloxirane (**122**) and probably by isomerization of carbenaoxirane (Scheme 104).

(**120**) —400–600 °C→ $R^1R^2C{=}C{=}O$ + $O{=}CMe_2$ + CO_2

Scheme 103

(**121**) —510 °C, 10 Pa→ (**122**) —−CO→ (**81**) → CH_2CO

Scheme 104

Hoffmann and Schüttler showed by ^{14}C labelling ⟨75CB844⟩ that (**81**) generated by chelotropic extrusion from a bicycloheptadiene derivative did not isomerize to oxirene. This is as expected from MO calculations (Section 5.05.6.2).

In a study of the deoxygenation of carbonyl compounds by atomic carbon, Dewar and coworkers ⟨81JA2802⟩ presented experimental and theoretical evidence that the carbonyl group can react with carbon atoms to form a carbenaoxirane.

5.05.6.3.5 Formation from other heterocyclic systems

Any heterocycle containing the OCH=CH moiety can *in principle* extrude the superfluous fragment and form oxirene, as illustrated for a five-membered ring in Scheme 105. Probably the most propitious AB fragment would be nitrogen, but the required 1,2,3-oxadiazole (**123**) is unknown (see Chapter 4.21), probably because of ready valence tautomerization to diazoethanal (Scheme 106) (this approach has been spectacularly successful with the sulfur analogue of (**2**) ⟨81JA486⟩). The use of (**123**) as an oxirene precursor is thus closely linked to the important diazo ketone decompositions discussed in Section 5.05.6.3.4(*i*).

→ oxirene + AB

Scheme 105

(**123**) → $OCH{-}CH{=}\overset{+}{N}{=}N$

Scheme 106

The flash thermolysis of the dioxolene (**124**) gave diphenylketene among the products (Scheme 107) ⟨70CC206⟩, and it is conceivable that diphenyloxirene is the precursor of this product.

(**124**) —530 °C, 30 Pa, silica→ $Ph_2C{=}C{=}O$ + HCOOEt + other products

Of much greater promise than thermolysis for the detection of oxirene is low-temperature photolysis. Strausz and coworkers ⟨80JOC2271⟩ investigated the photolysis, in an argon matrix at low temperatures, of 1,3-dioxolenone (vinylene carbonate; **125**). However, the only C_2H_2O species detected was ketene. The sulfur derivatives of (**125**), compounds (**126**) and

(**127**), behaved similarly, as did the benzo derivatives of (**126**) and (**127**). Flow pyrolysis of these compounds also provided ketenes, indicating that these cyclic carbonates are a new source of oxo carbenes.

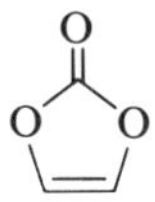

(**125**) 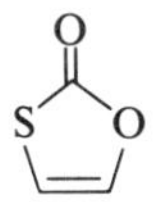 (**126**)

5.05.6.4 Conclusions

Oxirene is probably a true intermediate, but is separated from ketene by only a very low barrier. Since its instability results from unimolecular isomerization rather than from attack of other molecules, the only viable current technique for its direct observation seems to be generation and spectroscopic examination in an inert matrix at temperatures near absolute zero.

Note Added in Proof

Gasteiger and Herzig have investigated the reactions of halooxiranes with nucleophiles; the oxiranes function as halo ketone equivalents ⟨81AG(E)868, 81T2607⟩.

An extensive series of low-temperature matrix isolation experiments has failed to provide any evidence of oxirene formation, either by diazo ketone photolysis ⟨82CB2192⟩ or by attempted photo-retro-cycloaddition ⟨82CB2202⟩.

A new series of high-level *ab initio* calculations on the C_2H_2O hypersurface is in essential agreement with Figure 3 ⟨82JOC1869⟩.

5.06

Thiiranes and Thiirenes

D. C. DITTMER

Syracuse University

5.06.1 INTRODUCTION

Three-membered rings containing one sulfur atom are named as thiiranes (**1**) or thiirenes (**2**), ring numbering starting at the sulfur atom. In more complex systems such as (**3**), the substitution method of nomenclature is used; the position of the sulfur atom which replaces a carbon atom in the parent hydrocarbon is indicated by number and the prefix 'thia'. Thus (**3**) is 7-thiabicyclo[4.1.0]heptane. Other systems of nomenclature which have been used are (1) name of alkene + sulfide; (2) name of alkene + episulfide; (3) episulfide of 'name of alkene'; (4) epithioalkane with position of the functional group being given by numbers. By these systems compound (**3**) may be called cyclohexene sulfide, cyclohexene episulfide, the episulfide of cyclohexene or 1,2-epithiocyclohexane. The designation episulfide or epithio has been used for larger rings and their use may be ambiguous unless the ring size is made clear. 2-Methylthiirane is propylene sulfide or 1,2-epithiopropane. Thiirane 1-oxides and 1,1-dioxides are sometimes called episulfoxides and episulfones, and the simplest examples are often referred to as ethylene sulfoxide and ethylene sulfone. The thiirane, thiirene or substitution system of nomenclature is preferred, although the episulfide terminology is often useful in general discussions of complex systems. Thioethylene oxide is a poor name for thiirane.

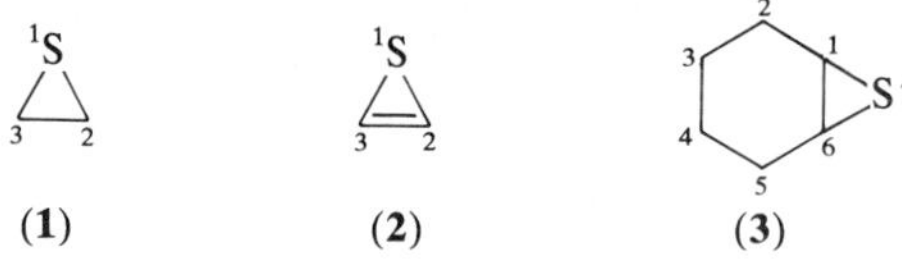

(**1**) (**2**) (**3**)

5.06.2 STRUCTURE

5.06.2.1 Bond Lengths and Bond Angles

Bond lengths and bond angles for some thiiranes and thiirenes are given in Table 1. The data for other thiirane derivatives are similar although the C—C and C—S bond lengths are as high as 1.60 and 1.92 Å respectively in (**4**) ⟨78CC555⟩, presumably because of steric crowding, and as low as 1.375 and 1.729 Å respectively in 2-triphenylsilylthiirane because of thermal motion of the ring during X-ray analysis ⟨79JOM(172)285⟩. The C-3—S bonds in α-thiolactone (**5**) ⟨77AG(E)722⟩, thiiranimine (**6**) ⟨80AG(E)276⟩ and 2-methylenethiirane (1.849 Å) ⟨78JA7436⟩ are also unusually long, no doubt because of the large CCS angle associated with the sp^2-hybridized carbon atoms. The exocyclic C—H bonds in thiiranes are comparable in length (1.08 Å) to those in ethylene.

(4) (5) (6)

The increase in C—C bond length and decrease in C—S bond length noted in Table 1 on going from thiirane to thiirane 1,1-dioxide is explained by a bonding scheme which considers these compounds as complexes of ethylene with sulfur or sulfur dioxide respectively ⟨73JA7644⟩. The dominant interaction is electron donation from the SO_2 fragment to the antibonding π^*-orbital of the ethylene fragment. This weakens the C—C bond and lengthens it. Participation of sulfur $3d$-orbitals, which is more important in the sulfone, depletes bonding electron density from the ethylene fragment, further weakening it. The C—S and, for the sulfoxide–sulfone, S—O bond shortening and strengthening also is explained by invoking $3d$-orbitals which when mixed with available 3s- and $3p$-orbitals provide better overlap with atoms, carbon or oxygen, bound to sulfur. This model for bonding predicts that substitution of π-electron donors (*e.g.* OR, NR_2, hal) on thiirane or thiirane 1,1-dioxide will decrease the C—C bond length and that substitution of π-electron acceptors (*e.g.* CN, COR, NO_2) will increase it, this being due principally to raising or lowering the energy of the antibonding π^*-orbital. *Ab initio* SCF-MO calculations with incorporation of d-orbitals give good agreement for the C—C, C—S and S—O bond lengths with the experimentally observed values ⟨75JA2025⟩.

Table 1 Bond Lengths and Angles for some Thiiranes and Thiirenes

Compound	*Bond length* (Å) C—C	C—S	S—O	*Bond angle* (°) C—S—C	O—S—O	*Ref.*
thiirane	1.484	1.815	—	48.3	—	74MI50600
thiirane 1-oxide	1.504	1.822	1.483	48.8	133.6[a]	76AX(B)2171
thiirane 1,1-dioxide	1.590	1.731	1.439	54.7	121.4	76AX(B)2171
thiirene	1.288[b]	1.790[b]	—	—	—	76AX(B)2171
	1.251[b]	1.978[b]	—	—	—	80JA2507
	1.27[b]	1.81[b]	—	—	—	80PAC1623
diphenylthiirene 1-oxide (Ph Ph)	1.305	1.784	1.467	42.9	126.0	76AX(B)2171
diphenylthiirene 1,1-dioxide (Ph Ph)	1.354	1.716, 1.703	1.444, 1.453	46.7	116.1	76AX(B)2171
1-methyl-di-t-butylthiirenium BF_4^- (Me, S^+, Bu^t Bu^t)	1.277	1.820	1.802[c]	41.1	—	79MI50600

[a] Twice the angle between the S—O bond and the ring.
[b] Calculated value.
[c] S—Me.

5.06.2.2 Stereochemistry and NMR Spectra

5.06.2.2.1 Stereochemistry

The configurations of chiral thiiranes and thiirane 1-oxides can be determined by ^1H NMR experiments in chiral solvents such as (R)-(−)-1-phenyl-2,2,2-trifluoroethanol. The protons α to sulfur usually are shifted more downfield for the (*S*)-configuration than for the (*R*)-configuration, all shifts being relative to the shifts in carbon tetrachloride ⟨77T999⟩. The *cis* and *trans* isomers of 2,3-diphenylthiirane can be distinguished easily by the difference in ^{1}H NMR shifts caused by shielding of the protons in the *trans* isomer by a phenyl group; other *cis–trans* isomers may be differentiated by the fact that the *cis* coupling constant is generally larger than the *trans*.

In thiirane 1-oxides and 1-substituted thiiranium salts the stereochemistry around the sulfur atom is pyramidal. In the sulfoxides the oxygen atom is bent about 67° out of the plane of the ring. The anisotropy of the S—O bond is useful in distinguishing alkyl groups *syn* or *anti* to the oxygen atom by ^{1}H NMR, although the increased thermal (room temperature) instability of thiirane 1-oxides with alkyl groups *cis* to oxygen presents difficulties. Relative to the episulfide, hydrogens on alkyl groups *syn* to the oxygen are deshielded (downfield shifts) and those *anti* are shielded (upfield shifts) ⟨71T4821⟩. Hydrogens directly bonded to the ring of the sulfoxide did not show this behavior to a significant extent although the ring hydrogens in *trans*-2,3-di-*t*-butylthiirane 1-oxide are well differentiated, $\delta = 3.11$ (*syn*), 1.87 p.p.m. (*anti*), $J = 12$ Hz. Assignments were confirmed by benzene solvent assisted shifts (relative to CCl_4) which were more upfield for protons *anti* to oxygen. Both *cis*- and *trans*-2,3-dimethylthiirane 1,1-dioxides are differentiated by the chemical shifts for the ring hydrogens which absorb at δ 3.36 and 2.78 p.p.m. respectively ⟨70S393⟩. The ^{1}H NMR spectrum of 2-methylthiirane in FSO_3H–SbF_5–SO_2 shows two doublets for the methyl group for *syn* and *anti* configurations in the protonated thiirane ⟨71JOC1121⟩. The barrier to inversion at sulfur in *S*-protonated thiirane is calculated to be high (326.8 kJ mol^{-1}) ⟨75JCS(P2)1722⟩. The calculated barrier for pyramidal inversion for 1-methyl-2,3-di-*t*-butylthiiranium tetrafluoroborate is similar (313.8 kJ mol^{-1}) ⟨79MI50600⟩. Barriers calculated for 2-methylthiirene 1-oxide and the 1,2-dimethylthiirenium ion are greater by about 6 kJ mol^{-1} than those for 2-methylenethiirane 1-oxide and the 1-methyl-2-methylenethiiranium ion, the range of barriers being from 355.6 to 272.4 kJ mol^{-1} ⟨71IJS(A)(1)66⟩.

5.06.2.2.2 Chemical shifts and coupling constants

Chemical shifts (^{1}H, ^{13}C) and coupling constants for some thiiranes and thiirenes are given in Table 2. The shifts of substituted derivatives may be calculated by the use of additivity relationships found in textbooks on NMR.

In the series of tetralin 2,3-thiiranes, oxiranes and aziridines, the deshielding effects of the heteroatoms are in the sequence S > O > NH ⟨80JST(63)73⟩ which differs from the order found in the parent three-membered heterocycles (O > S > NH) ⟨B-73NMR138, 366, 425⟩. The difference in shielding in both systems between sulfur and oxygen is not great. Thiirane hydrogens in steroidal α-episulfides are at higher field than the hydrogens of the β-isomers. Vicinal hydrogen–hydrogen coupling constants ($J^{vic}_{H\text{–}H}$) are larger and geminal hydrogen–hydrogen coupling constants ($J^{gem}_{H\text{–}H}$) are smaller for thiiranes than for oxiranes, but care should be taken in comparing $J^{gem}_{H\text{–}H}$ values because of solvent influences ($J^{gem}_{H\text{–}H}$ is more negative in more polar solvents). Three-bond carbon–hydrogen and hydrogen–hydrogen coupling constants ($^3J_{C\text{–}H}$, $^3J_{H\text{–}H}$) are larger in thiiranes than in aziridines or oxiranes, the values paralleling the length of the ring C—C bond ⟨78JOC4696⟩. The $^3J_{C\text{–}H}$ values for *cis* Me and H ring substituents are larger than for *trans* substituents, which is an aid in the assignment of stereochemistry.

The ^{1}H NMR spectrum of thiirane 1-oxide is complex (AA′BB′); at 60 MHz 24 lines are observed consisting of two sets of 12 centered about a midpoint. The ^{1}H NMR chemical shift in thiirane 1,1-dioxide is fairly sensitive to solvent variations partly because of the high dipole moment (4.4 D) of the sulfone. The benzene-induced shift, $\Delta\delta$ (C_6D_6–CCl_4), is large (−1.04 p.p.m.), as expected from the presence of a sulfone group. Oxygen-17 chemical shifts for thiirane 1-oxide and thiirane 1,1-oxide are −71 and +111 p.p.m. respectively, relative to H_2O.

In *S*-protonated 2,3-di-*t*-butylthiiranes (**7**)–(**9**) the more crowded S—H proton in (**9**) appears at higher field (δ 2.68 p.p.m.) than the S—H protons of (**7**; δ 3.01 p.p.m.) and (**8**, δ 3.54 p.p.m.) in the solvent FSO_3H. The methine hydrogens of the ring are coupled principally to the S—H proton ($^3J_{H\text{–}H} = 6$–8 Hz). Proton and carbon chemical shift differences for C—H in *cis* and *trans* derivatives (**7**)–(**9**) are very small, although $^1J_{C\text{–}H}$ for (**7**; 167 Hz)

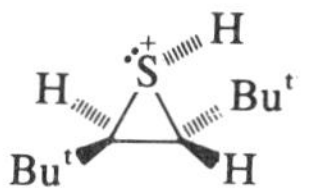

(7) 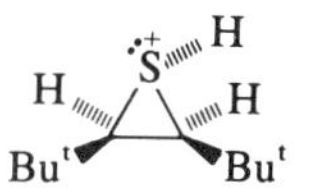 (8) 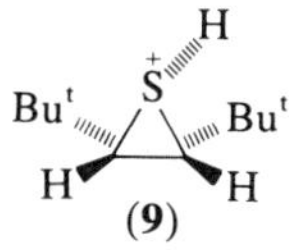 (9)

is significantly larger than for (**8**; 127 Hz) ⟨74JA3146⟩. Similar small differences in ^{1}H NMR chemical shifts are observed for *cis–trans* isomers of *S*-methylthiiranium ions ⟨75TL2603⟩.

Palladium, platinum and iridium complexes of thiirene 1,1-dioxides (**10**) show upfield shifts of about 5 p.p.m. for vinyl protons and 1 p.p.m. for methyl protons, the shifts being attributed to back-donation of electrons by the metal and its associated ligands ⟨73JOM(57)403⟩.

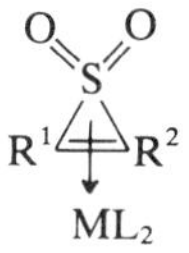

(10)

^{19}F NMR measurements on 2,3-fluorophenylthiirene 1-oxides and 1,1-dioxides show that electron withdrawing conjugative effects are greater for the sulfone than for the sulfoxide and that both are less conjugated than cyclopropenones ⟨79JA390⟩.

5.06.2.3 Mass Spectra and Photoelectron Spectra

The principal ions observed in the mass spectra (electron impact) of thiiranes are due either to the loss of a hydrogen atom or alkyl group or to isomerization to a thioaldehyde or thioketone followed by loss of a neutral alkyl group or hydrogen (Scheme 1). Loss of neutral HS is responsible for the base peak of 2-methylthiirane ⟨B-71MS225⟩. In 2-vinylthiirane the tendency for isomerization is reduced because of stabilization of ion (**11**) by conjugation. Ion cyclotron resonance mass spectrometry of thiirane with ammonia shows that the main reaction of the thiiranium cation radical is the transfer of sulfur ⟨82MI50600⟩. An extensive study of the mass spectra of epithio carbohydrate derivatives shows that most of the fragmentation paths do not involve the thiirane ring; those that do result in the formation of thiophenes or loss of sulfur ⟨78OMS113⟩.

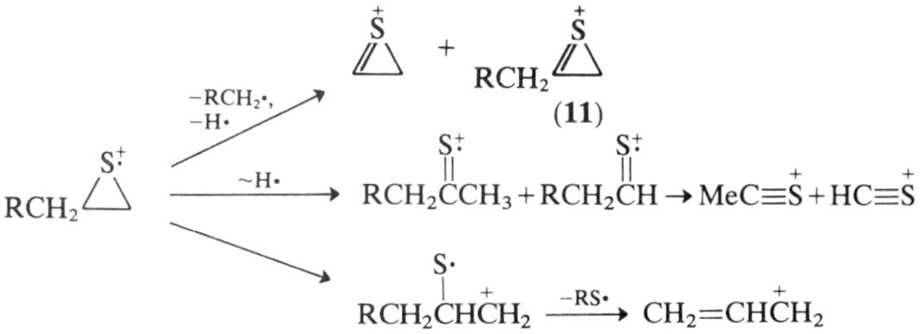

$$\text{RCH}_2\text{-thiirane}^{+\cdot} \xrightarrow{-\text{RCH}_2\cdot,\ -\text{H}\cdot} \text{thiirenium}^{+} + \text{RCH}_2\text{-thiirenium}^{+}\ (\mathbf{11})$$

$$\xrightarrow{-\text{H}\cdot} \text{RCH}_2\text{C}(=\text{S}^{+\cdot})\text{CH}_3 + \text{RCH}_2\text{CH}=\text{S}^{+\cdot} \rightarrow \text{MeC}{\equiv}\overset{+}{\text{S}} + \text{HC}{\equiv}\overset{+}{\text{S}}$$

$$\rightarrow \text{RCH}_2\text{CH(S}\cdot)\overset{+}{\text{C}}\text{H}_2 \xrightarrow{-\text{RS}\cdot} \text{CH}_2{=}\text{CH}\overset{+}{\text{C}}\text{H}_2$$

Scheme 1

Ions corresponding to protonated thiiranes are observed in chemical ionization mass spectrometry. Theoretical calculations confirmed by appearance energy measurements show that the protonated thiirane structure is comparable but somewhat lower in energy to two isomers, $\text{MeCH}{=}\overset{+}{\text{S}}\text{H}$ and $\text{Me}\overset{+}{\text{S}}{=}\text{CH}_2$ ⟨79OMS543⟩. 1-Methyl- and 1-phenyl-thiiranium ions are stable in the gas phase and can be identified from their collisional activation spectra. The corresponding oxygen analogs could not be generated, acyclic ions, *e.g.* $\text{Me}\overset{+}{\text{O}}{=}\text{CHMe}$, being preferred probably because of greater strain in the transition state for cyclic oxonium ion formation ⟨77T1785⟩.

2,3-Diphenylthiirene 1-oxide and several thiirene 1,1-dioxides show very weak molecular ions by electron impact mass spectrometry, but the molecular ions are much more abundant in chemical ionization mass spectrometry ⟨75JHC21⟩. The major fragmentation pathway is loss of sulfur monoxide or sulfur dioxide to give the alkynic ion. High resolution mass measurements identified minor fragment ions from 2,3-diphenylthiirene 1-oxide at *m/e* 105 and 121 as $PhCO^+$ and $PhCS^+$, which are probably derived *via* rearrangement of the thiirene sulfoxide to monothiobenzil (Scheme 2).

Table 2 ^{1}H and ^{13}C NMR Data[a] for some Thiiranes and Thiirenes

Compound	*Solvent*[b]	$\delta(^{1}H)$	$\delta(^{13}C)$	$^{2}J_{H-H}$
S; H^3, H^2, H^4, H^1	Nematic solvent	2.27[c]	18.1[d]	0 (H^1–H^2)
O, S:; H^3, H^2, H^4, H^1	CCl_4	2.42 (H^1, H^4), 1.92 (H^2, H^3)	33.8[d]	−6.4 (H^1–H^2)
O, O, S; H^3, H^2, H^4, H^1	$CDCl_3$	3.15 ($CDCl_3$), 2.00 (C_6D_6)	31.6[d]	−9.2
Me, $\overset{+}{S}$; Bu^t, Bu^t	$CDCl_3$	4.40 (ring CH), 2.80 ($\overset{+}{S}$Me)	76.2 (ring C), 27.6 ($\overset{+}{S}$Me)	—
Me, $\overset{+}{S}$; Me Me	SO_2	2.77 (CMe), 2.50 ($\overset{+}{S}$Me)	107.0 (ring C), 25.6 ($\overset{+}{S}$Me), 9.0 (CMe)	—
O, O, S; Me	$CDCl_3$	9.04 (H), 2.50 (Me)	—	—
S; H^3, H^4, 1, 2, 3, C, H^2, H^1	$CFCl_3$, $CDCl_3$	5.71 (H^1), 5.34 (H^2), 2.68 (H^3, H^4)	130.1 (C^2), 99.5 (C^3), 18.5 (C^1)	±1.55 (H^1–H^2)

[a] δ in p.p.m. relative to TMS, *J* in Hz.
[b] Unless stated otherwise.
[c] Neat ⟨B-73NMR423⟩.
[d] In $CDCl_3$ ⟨80JOC4807⟩.

[O, S, Ph, Ph]‡ → [PhC(=O)—C(=S)Ph]‡ → Ph$\overset{+}{C}$O + Ph$\overset{+}{C}$S

Scheme 2

The vertical ionization potentials from the photoelectron spectra of some thiirane and thiirene derivatives are given in Table 3. A Walsh localized scheme of bonding is generally preferred. There is a strong hyperconjugative interaction in thiirene 1,1-dioxides between the occupied C=C π-MO and the occupied SO_2 σ-MO, and modest mixing between the C=C π-MO and a vacant SO_2 σ^*-MO, which is a nearly pure sulfur *d*-AO. Thiirene oxides are suggested to be less 'aromatic' than cyclopropenones and tropone.

5.06.2.4 Electronic Spectra, Optical Rotatory Dispersion–Circular Dichroism

Absorptions in the UV spectra of thiiranes are observed around 260 nm ($n \rightarrow \sigma^*$; $\varepsilon \approx 40$) and 205 nm ($\varepsilon \approx 4000$) ⟨75MI50600⟩. A number of other transitions are reported in the vacuum UV spectrum, and the calculated lowest singlet transition energies correspond to $n \rightarrow \sigma_1^*$, $n \rightarrow \sigma_2^*$ and $\sigma \rightarrow \sigma^*$ transitions ⟨75BCJ33⟩. Thiirane and oxirane groups behave as electron withdrawing substituents when attached to aromatic rings as indicated by the UV spectra of 2-arylthiiranes.

$^3J_{H\text{-}H}$	$^4J_{H\text{-}H}$	$^1J_{C\text{-}H}$	$^2J_{CCH}$ or $^3J_{CCCH}$	*Ref.*
7.15 (H^1–H^4), 5.65 (H^1–H^3)	—	170.26	−1.50	75JST(24)85
11.7 (H^1–H^4), 11.5 (H^2–H^3), 10.5 (H^1–H^3)	—	171.8 (C–H^2), 171.9 (C–H^1)	—	70TL2877
11.87 (H^1–H^4), 6.89 (H^1–H^3)	—	170.4	—	72OMR441
0	—	165 (ring CH), 128 ($\overset{+}{S}$Me)	—	74JA3146
—	—	—	—	79MI50600
—	1.1	230 (=CH)	—	71JA476
—	±1.6, ±1.3	174 (C–H^4, H^3), 166 (C–H^1, H^2)	10 (*trans* CC=CH), 2 (*cis* CC=CH)	78RTC214

Table 3 Vertical Ionization Energies for some Thiiranes and Thiirenes

Compound	*Vertical ionization energies*[a] (eV)	*Ref.*
S	9.03 (n), 11.37 (σ_a), 11.93 (σ_s), 13.51 (π_{CH_2}), 15.33 (n), 16.58 (π_{CH_2})	80JA5151
O S:	9.66 (n_S), 9.78 (π_{SO}), 12.91 (n_O), 13.30, 15.9, 16.8	74CB2299
O O S	10.20, 11.57, 11.98, 12.03, 13.92, 14.62	75CB897
O S: Ph Ph	8.29 (π_1), 8.89 (π_{SO}), 9.07 (n_S), 9.38 (π_2, π_3), 9.92 (π_4), 10.86 (π_5)	78JA8056
O O[b] S Ph Ph	8.62 (π_1), 8.82 (π_{SO_2}), 10.02 (π_2, π_3), 10.42 (π_4), 11.13 (σ_{SO_2}), 11.38 (σ_{SO_2}), 12.20 (π_{SO_2}), 12.45 (π_5)	78JA8056

[a] Where assignments of orbitals were made they are indicated in parentheses.
[b] Calculated values only.

The 260 nm band of chiral thiiranes is optically active and a Cotton effect is observed; (R)-(+)-methylthiirane shows a negative Cotton effect at *ca.* 250 nm followed by a positive effect below 200 nm. An MO analysis indicates that charge transfer contributions are most important in determining the optical activity of the $n \rightarrow \sigma^*$ transition ⟨81JCS(F2)503⟩. The sign and amplitudes of the Cotton effect in epithiosteroids have been related to the disposition of the atoms in space about the sulfur atom ⟨66T1039⟩. In 16,17α-epithiopregnenolone, X-ray analysis and Cotton effects indicate a *trans* orientation is preferred between the thiirane ring and the acetyl side chain at C-17 ⟨82T165⟩.

The UV spectra of thiirane 1-oxide and (1*S*,2*S*)-(+)-2-methylthiirane 1-oxide show a broad maximum at about 205 nm ($\varepsilon \approx 23\,000$). The latter shows a positive Cotton effect at low energy followed by a negative effect at high energy. The lowest excited states of thiirane 1-oxide involve excitations from the two lone pairs of the oxygen atom ⟨79G19⟩. 2,3-Diphenylthiirene 1-oxide and 1,1-dioxide show absorption due to the 1,2-diphenylethylene chromophore.

5.06.2.5 Infrared, Raman and Microwave Spectra

The IR spectrum of thiirane has been extensively analyzed and satisfactory agreement of calculated vibrational frequencies with observed has been achieved ⟨77JA1685⟩. Ring deformation frequencies are reported at 626 and 685 cm^{-1}. The barrier to rotation of a phenyl group in 2-phenylthiirane from its preferred orthogonal orientation (bisected) to the thiirane ring was calculated as 77 kJ mol^{-1}. The bisected conformation also is preferred in 2-phenyloxirane and phenylcyclopropane, the preference being explained by an electronic interaction between the two rings. Raman and microwave spectra show that the barrier to rotation of the methyl group in 2-methylthiirane is much smaller, being 13.6 kJ mol^{-1}.

The S—O bond in thiirane 1-oxides absorbs near 1060 cm^{-1} in solution but the stretching frequency is sensitive to solvation effects, moving to higher frequency in more polar solvents. The vapor phase IR spectrum of thiirane 1-oxide shows a band at 1050 cm^{-1} and a shoulder at 1065 cm^{-1} which were attributed to associated and free molecules respectively ⟨68TL959⟩. However, an IR and Raman study of thiirane 1-oxide assigns the band at 1062 cm^{-1} to a C—C stretch and a band at 1120 cm^{-1} to the S—O stretch. The sulfone group in thiirane 1,1-dioxide exhibits symmetric and antisymmetric bending frequencies at 1159 and 1308 cm^{-1} respectively and a bending frequency at 446 cm^{-1}. The stretching frequencies are essentially the same as in other aliphatic sulfones ⟨72MI50600⟩. The frequency of the methylene scissoring vibrations decreases in the order thiirane (1450 cm^{-1}) > thiirane 1-oxide (1418 cm^{-1}) > thiirane 1,1-dioxide (1388 cm^{-1}).

The IR spectra of thiirene and several substituted thiirenes have been analyzed ⟨80PAC1623⟩. In thiirene itself the following bands are observed: 3208 (w, ν^{s}_{C-H}), 3169 (m, ν^{as}_{C-H}), 1660 (w, $\nu_{C=C}$), 912 (m, C—H in-plane bend), 563 (s, C—H out-of-plane bend) cm^{-1}. The sulfoxide group of 2,3-diphenylthiirene 1-oxide absorbs at 1061 cm^{-1} ($CHCl_3$), and the corresponding sulfone shows absorption at 1155 and 1258 cm^{-1} ($CHCl_3$). The antisymmetric vibration frequency of the sulfone group in thiirene 1,1-dioxides is reduced by 50–60 cm^{-1} from that of the saturated derivatives. This behavior is not observed in acyclic sulfones. The C=C double bond in 2,3-dimethylthiirene 1,1-dioxide absorbs at 1685 cm^{-1} (Raman, solid) and that in 2-methylthiirene 1,1-dioxide at 1608 cm^{-1} ($CHCl_3$). The IR spectra of palladium, platinum and iridium complexes (**10**) of thiirene 1,1-dioxides lack observable absorption for the C=C double bond and show absorptions for the sulfone group at 1040–1050, 1118–1140 and 1225–1250 cm^{-1}. The C—C stretching frequency in methylenethiirane is tentatively assigned to a weak absorption at 1717 cm^{-1}. The carbonyl group of α-thiolactones (2-ketothiiranes) absorbs in the region 1785–1810 cm^{-1}.

5.06.2.6 Thermodynamic Properties

Thiirane has a lower strain enthalpy (73.7 kJ mol^{-1}) and entropy (89.8 J K^{-1} mol^{-1}) than oxirane (160.8 kJ mol^{-1}, 247.6 J K^{-1} mol^{-1}) ⟨75RCR138⟩. The decomposition of thiirane to acetylene and hydrogen sulfide is calculated to be endothermic, and the decomposition to ethylenethiol is calculated to be exothermic ⟨73RRC1353⟩. The heat of formation of thiirane is 82.4 kJ mol^{-1} and its heat of vaporization is 30.3 kJ mol^{-1}. Calculations (MINDO) of the heat of formation of thiirane and its *S*-protonated ion give reasonably good agreement with experimental quantities ⟨79JA783⟩. The stability of thiiranium ions is calculated to decrease in the order: pyramidal ion (**12**), sp^3-hybridized ion (**13**) and sp^2-hybridized ion (**14**) ⟨74T2197⟩. Ion (**15**) is calculated to be more stable than (**16**) by 276 kJ mol^{-1}.

$\overset{+}{C}H{=}CHSH$

(12)  **(13)**  **(14)**  **(15)** **(16)**

MO calculations for the gas phase indicate that sulfurane intermediate (**17**) is more stable than the ion (**18**) by about 380 kJ mol^{-1}, which suggests that sulfuranes may be important in the reaction of sulfenyl halides with alkenes in non-polar solvents ⟨77JCS(P2)1019⟩.

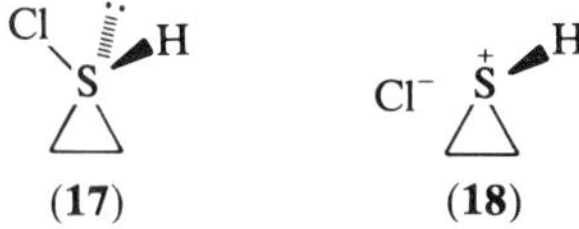

(17) **(18)**

Thermodynamic stabilities of all six C_2H_2S isomers, including thiirene, have been computed and are shown in Scheme 3 in order of ascending energy from left to right with energy separations in kJ mol^{-1} above the arrows ⟨80PAC1623⟩. Charge densities on sulfur and carbon in thiirene are calculated as +0.1122 and −0.1552 respectively, indicating a contribution from resonance structure (**19**). Calculations of resonance energies of thiirene and the thiirenium ion indicate the former is antiaromatic with minimum conjugation between the sulfur atom and the double bond and that the latter, a two-π-electron system, is relatively stabilized; a theoretical study of several thiirenium ions shows them to be more stable by 67.7–99.1 kJ mol^{-1} than isomeric thiovinyl cations ($RSCR{=}CR^+$) ⟨78G543⟩.

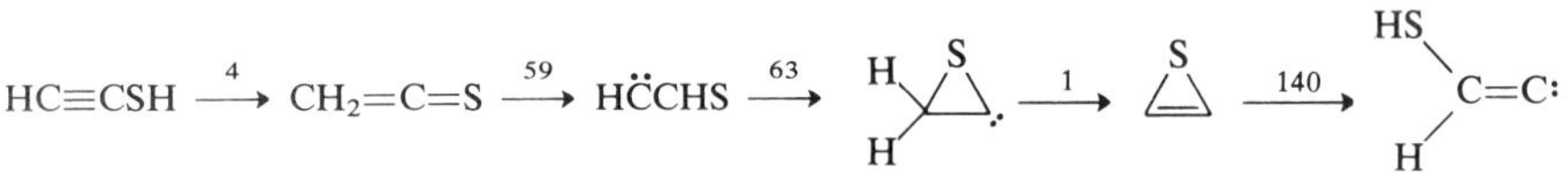

$$HC{\equiv}CSH \xrightarrow{4} CH_2{=}C{=}S \xrightarrow{59} H\ddot{C}CHS \xrightarrow{63} \cdots \xrightarrow{1} \cdots \xrightarrow{140} \cdots$$

Scheme 3

(19)

5.06.2.7 Miscellaneous Properties: Dipole Moments; Magnetic Properties

Dipole moments of thiirane, thiirane 1-oxide and thiirane 1,1-dioxide are 1.66, 3.72 and 4.41 D respectively. The dipole moment of oxirane is 1.88 D. The C—S and S—O bond moments in thiirane 1-oxide are 1.06 and 2.53 D respectively. The dipole moment of 2,3-diphenylthiirene 1,1-dioxide is 5.63 D as compared to 5.12 D for diphenyl sulfone.

A dramatic decrease in the magnitude of the magnetic susceptibility anisotropy is observed on going from thiirane to the open-chain analog, dimethyl sulfide, and has been attributed to non-local or 'ring-current' effects ⟨70JCP(52)5291⟩. The decrease also is observed to a somewhat lesser extent in oxirane relative to dimethyl ether.

5.06.3 REACTIVITY

5.06.3.1 Introduction

The ring strain of thiiranes and thiirenes, although less than their oxygen analogs, accounts for much of the reactivity of these compounds. Thermal and photochemical reactions may result in cleavage of a carbon–sulfur bond and even extrusion of sulfur, sulfur monoxide or sulfur dioxide respectively from the cyclic three-membered sulfides, sulfoxides and sulfones. Several reactions demonstrate the weakness of the C—C bond in the episulfones. Electrophilic reagents attack the sulfur atom of the sulfides. Nucleophiles attack at sulfur or carbon to cleave the rings. Thiirene sulfoxides and sulfones are thermally more stable than their saturated analogs.

5.06.3.2 Thermal and Photochemical Reactions

5.06.3.2.1 Extrusion of sulfur, sulfur monoxide or sulfur dioxide

Thermolysis of thiiranes may cause extrusion of elemental sulfur with the formation of alkenes ⟨76RCR25, 66CRV297, 64HC(19-1)576⟩, but other reactions may occur which involve cleavage of only one C—S bond leading to products of rearrangement, isomerization or polymerization. Thiiranes which are highly substituted or which are substituted with groups which can stabilize free radicals, such as phenyl, vinyl, acetyl, or chloro, are more prone to lose sulfur. *trans*-1,2-Diphenylthiirane is unstable at room temperature; the *cis* isomer is more stable possibly because of steric inhibition of resonance of the benzene ring with a developing radical center in the transition state. 2,3-Divinylthiirane (**20**, *cis* or *trans*) loses sulfur at 90 °C to give a mixture of *cis*- and *trans*-1,3,5-hexatrienes (ratio 1:4) although about 75% of the thermolysis reaction goes by rearrangement to dihydrothiepins (**21**–**23**; Scheme 4) ⟨79JA254⟩. However, thermolysis at 100 °C of *trans*-2,3-diethynylthiirane (**24**) gives a good yield of *trans* alkene (**25**) with only a small amount of the *cis* alkene (Scheme 4). The *cis* diethynylthiirane gives mainly *cis* alkene (**26**). At 395 °C (gas phase) the stereospecificity is lower, a 4:3 ratio of (**25**) to (**26**) being obtained from the *trans* thiirane and a 1:2 ratio from the *cis* thiirane ⟨73JA7538⟩. The retention of stereochemistry may be explained by a simple chelotropic extrusion of a sulfur atom in which both C—S bonds break simultaneously. In cases of non-stereospecificity, a mechanism involving stepwise cleavage of C—S bonds is probable; and the kinetics of desulfurization of (**24**) indicates a bimolecular process may occur under some conditions. The extrusion reaction in other favorable cases in which competing reactions are minimized can give good yields of alkenes (Scheme 5) ⟨74CHE623⟩. 2,2,3,3-Tetrafluorothiirane is extremely stable to heat or light. Below 300 °C very little decomposition occurs; at 430 °C, difluorothioformaldehyde (49%), tetrafluoroethylene (30%) and unreacted thiirane (11%) are formed ⟨65JOC4188⟩.

(**20**) *cis* or *trans* —90–100 °C→ hexatrienes + (**21**) + (**22**) + (**23**)(from **22**)

(**24**) —100 °C, PhMe→ (**25**) + (**26**)

Scheme 4

—heat, 98%→

Scheme 5

Photolysis of thiirane (>220 nm) causes decomposition *via* a singlet excited state to ethylene, hydrogen sulfide, hydrogen and methane. Intersystem crossing occurs from the singlet excited state to a triplet state which undergoes desulfurization by collision with a molecule of ground state thiirane. Both *cis*- and *trans*-2,3-dimethylthiirane are stereospecifically desulfurized to *cis*- and *trans*-2-butene respectively ⟨81JPC1089⟩. Photolysis of *cis*-2,3-diphenylthiirane (with iodine) or 2,2,3,3-tetraphenylthiirane gives good yields of phenanthrene and 9,10-diphenylphenanthrene *via* stilbene intermediates in which dehydrogenation is accomplished by iodine or the extruded sulfur ⟨73MI50600, 73JHC879⟩. The photochemistry of 2,3-dibenzoyl-2,3-diphenylthiirane and its *S*-oxide actually involves the chemistry of 2-benzoyl-2,4,5-triphenyl-1,3-oxathiole, which may rearrange to thiirane intermediates during reaction ⟨74JOC2722⟩. A photochemical extrusion of sulfur which goes in good yield is shown in Scheme 6.

2537 Å, MeOH, 56%

Scheme 6

Heating thiirane 1-oxides results in extrusion of sulfur monoxide ⟨B-81MI50600⟩. Temperatures of 100–150 °C are needed for unsubstituted and alkyl-substituted episulfoxides. Aryl substituents lower the decomposition temperature to 20–40 °C. The stereospecificity of the extrusion of sulfur monoxide decreases as the temperature is increased. A diradical intermediate is suggested. 1,2-Oxathietane is indicated as an intermediate in the flash vacuum thermolysis of thiirane 1-oxide at above 1000 K ⟨82JCS(P2)279⟩. Stereospecific extrusions are observed with 2,3-diaryl- and 2,2,3-triaryl-thiirane 1-oxides which decompose at room temperature or slightly above (Scheme 7). The sulfur monoxide reacts with dienes, diazoalkanes, ylides, azides, pyridine *N*-oxides and (triphenylphosphine)rhodium chloride or bromide to give, respectively, five-membered cyclic sulfoxides (Scheme 8) ⟨77JOC2127⟩, sulfines, sulfines, *N*-sulfinylamines, pyridines and rhodium complexes ⟨81MI50601⟩ of sulfur monoxide. Another path for the thermolysis of thiirane 1-oxides involves intramolecular proton abstraction by the oxygen atom leading to a sulfoxylic acid intermediate ⟨71JA2810⟩. *cis*-2,3-Dimethylthiirane 1-oxide (**27**) is stable at 35 °C but the *trans* isomer (**28**) decomposes at this temperature to give the thiosulfinate (**29**) and thiosulfoxylate (**30**; Scheme 9).

40 °C, CH_2Cl_2 + SO

Scheme 7

PhMe, refl., N_2, $-CH_2{=}CH_2$, 100%

Scheme 8

35 °C, CH_2Cl_2 N.R.

(**27**)

(**28**) 35 °C, CH_2Cl_2 SOH (**29**) (**30**)

Scheme 9

Thiirane 1,1-dioxides extrude sulfur dioxide readily ⟨70S393⟩ at temperatures usually in the range 50–100 °C, although some, such as *cis*-2,3-diphenylthiirane 1,1-dioxide or 2-*p*-nitrophenylthiirane 1,1-dioxide, lose sulfur dioxide at room temperature. The extrusion is usually stereospecific (Scheme 10) and a concerted, non-linear chelotropic expulsion of sulfur dioxide or a singlet diradical mechanism in which loss of sulfur dioxide occurs faster than bond rotation may be involved. The latter mechanism is likely for episulfones with substituents which can stabilize the intermediate diradical. The Ramberg–Bäcklund reaction ⟨B-77MI50600⟩ in which α-halosulfones are converted to alkenes in the presence of base, involves formation of an episulfone from which sulfur dioxide is removed either thermally or by base (Scheme 11). A similar conversion of α,α'-dihalosulfones to alkenes is effected by triphenylphosphine. Thermolysis of α-thiolactone (**5**) results in loss of carbon monoxide rather than sulfur (Scheme 12).

Scheme 10

Scheme 11

(5)

Scheme 12

Some thiirene derivatives also undergo thermal and photochemical extrusion reactions to give alkynes. Most of the relatively few known thiirenes rearrange thermally or photochemically to thioketenes *via* thioketocarbenes, but 2,3-bis(trifluoromethyl)thiirene photochemically extrudes sulfur to give hexafluoro-2-butyne ⟨80PAC1623⟩. Photolysis of 2,3-diphenylthiirene 1-oxide gives diphenylacetylene (87%); thermolysis in air gives benzil *via* a rearrangement. Thiirene 1,1-dioxides lose sulfur dioxide readily (*ca.* 100 °C) to give alkynes ⟨79JA390⟩. Examples of these extrusion reactions of thiirenes are given in Scheme 13. In the thermolysis of 2,3-diarylthiirene 1,1-dioxides the rate of loss of sulfur dioxide correlates best with the sum of σ_p^+ substituents ($\rho = -0.86$) and a stepwise homolytic cleavage of the C—S bonds is inferred ⟨77CC713⟩. 2,3-Diphenylthiirene 1,1-dioxide decomposes *ca.* 10^3 times slower than the saturated *cis*-2,3-diphenyl episulfone. The thermal stability of 2,3-dimethylthiirene 1,1-dioxide is similar to that of the diphenyl derivative. Derivatives of Pd(0), Pt(0) and Ir(I) catalyze the thermal extrusion of SO_2 from thiirene sulfones *via* complexes of the C=C double bond with the transition metals, the latter acting as electron donors to decrease the bond order of the C=C double bonds. As a result, the thermal reactivity of the complex approaches that of the less stable saturated thiirane dioxides (Scheme 14) ⟨73JOM(57)403⟩.

Scheme 13

Scheme 14

5.06.3.2.2 Rearrangement and isomerization

Thermolysis of thiirane at temperatures above 250 °C gives unstable ethylenethiol *via* C—S bond fission. Hydrogen sulfide and acetylene are produced as stable products. At 500 °C thiirane yields thiophene (7%) and benzothiophene in addition to ethylenethiol (3%) ⟨79BAU1936, 79JA3000⟩. In substituted thiiranes C—S bond cleavage can result in rearranged products exemplified in Schemes 15–18 ⟨72CC1298, 277, 71CC979, 79CC881,

160–170 °C, PhCN or $h\nu$, MeCN
84%

Scheme 15

$h\nu$ PhCOMe, Et_2O
56 °C

Scheme 16

$h\nu$ Et_2O
(31)

Scheme 17

C_6H_6 reflux, −HBr

Scheme 18

(32)
C_6H_6 reflux
$Pr^iOCSC(Ph)_2COPr^i$
100%

Scheme 19

72JOC1537⟩. The mechanisms of these transformations may involve homolytic or heterolytic C—S bond fission. A 'sulfur-walk' mechanism has been proposed to account for isomerization or automerization of Dewar thiophenes and their *S*-oxides (*e.g.* **31** in Scheme 17) ⟨76JA4325⟩. Calculations show that a symmetrical pyramidal intermediate with the sulfur atom centered over the plane of the four carbon atoms is unlikely ⟨79JOU1401⟩. Reactions which may be mechanistically similar to that shown in Scheme 18 are the thermal isomerization of thiirane (**32**; Scheme 19) ⟨70CB949⟩ and the rearrangement of (**6**) to a benzothiophene ⟨80JOC4366⟩.

The C—C bond in thiiranes is cleaved thermally in a few cases. The conversion of Dewar thiophene (**31**) to a thiophene (Scheme 17), the formation of dihydrothiepins from 2,3-divinylthiirane (Scheme 4), and the formation of the fused thiophene derivative (**33**) (along with products of sulfur extrusion) from *cis*-2,3-diethynylthiirane (Scheme 20) are examples ⟨73JA7538⟩. The thermal rearrangement of *cis*-2,3-divinylthiirane to 4,5-dihydrothiepin is said to be a ($_{\sigma}2_s + _{\pi}2_s + _{\pi}2_s$) process, but the isomerization of the *cis*-divinylthiirane to *trans* and the formation of other products (Scheme 21) may involve diradicals or thiocarbonyl ylides ⟨79JCR(S)56, 79JA254⟩. The C—C bond of 2,2,3,3-tetraphenylthiirane 1,1-dioxide also is cleaved thermally, a dihydroisothianaphthene sulfone being formed.

HC≡C C≡CH
250 °C 5%
(33)

Scheme 20

Thiirenes are photochemically converted to thioketenes or to alkynylthiols (Scheme 22) ⟨80PAC1623⟩. 2-Acetyl-3-methylthiirene rearranges on irradiation at 310 nm to a mixture of acetylmethylthioketene and thioacetylmethylketene, identified by their IR spectra ⟨80NJC703⟩. 2,3-Diphenylthiirene 1-oxide at 130 °C is believed to be isomerized to monothiobenzil which is air-oxidized to benzil (Scheme 23) ⟨79JA390⟩.

360–460 °C

Scheme 21

$CH_2{=}C{=}S$ ← ← → → HC≡CSH

Scheme 22

130 °C → O_2 → PhC(=O)–C(=O)Ph

Scheme 23

5.06.3.2.3 Polymerization and oligomerization

Both heat and light effect polymerization of thiiranes. Thiirane and 2-phenylthiirane slowly polymerize to a white polymer at room temperature ⟨66CRV297⟩. 2-Methyl- and 2,3-dimethyl-thiirane are more stable, but 2-methylthiirane has been polymerized by light or an electric discharge. Thioacetamide is said to stabilize thiirane to polymerization ⟨71USP3557145⟩. The photopolymerization of 2-methylthiirane is catalyzed by maleic anhydride which may form a charge transfer complex with the thiirane. Cyclohexene episulfide is polymerized in the solid phase by γ-irradiation at −196 °C. The light-catalyzed polymerization of tetrafluorothiirane requires the presence of bis(trifluoromethyl) disulfide, and copolymerizations with ethylene and propylene give polymers in which the monomer units alternate ⟨65JOC4188⟩. Thermal and photochemical polymerizations of thiiranes probably proceed through diradical intermediates which may be intercepted by reaction with various acceptors or which form dimers and oligomers in cases where polymerization to high molecular weight material is unfavorable (Scheme 24). Abstraction of hydrogen atoms by the diradical intermediate may occur as in the thermolysis of cyclohexene episulfide which yields dicyclohexyl sulfide and disulfide, cyclohexanethiol, 2-cyclohexenyl cyclohexyl sulfide, cyclohexyl phenyl sulfide and 2,3,5,6-bis(tetra methylene)-1,4-dithiane. Desulfurization to cyclohexene also occurs under the reaction conditions (210 °C) and the formation of the observed products may in part be derived from reactions of elemental sulfur ⟨75BCJ1665⟩.

$-(SCH_2CH_2)-_n$

$\cdot SCH_2CH_2\cdot$ → $MeCH{=}CH_2$ → [MeĊH, H, $H_2C{-}S$, $CH{-}CH_2\cdot$] → $MeCH_2CH_2SCH{=}CH_2$

Scheme 24

Thiirene intermediates in the photolysis of 1,2,3-thiadiazoles readily undergo ring opening to a diradical which can be trapped by reaction with an alkyne (Scheme 25) ⟨79CB1769⟩. Significant polymerization is not observed.

Me, Ph-1,2,3-thiadiazole $\xrightarrow[(HOCH_2CH_2)_2O]{240\,°C}$ Ph, Me-thiirene ⇌ ·S–C(Ph)=C(Me)· + S·–C(Me)=C(Ph)· $\xrightarrow{PhC\equiv CPh}$ 2-Me-3,4,5-Ph thiophene + 2,3,5-Ph-4-Me thiophene

Scheme 25

5.06.3.3 Electrophilic Attack on Sulfur

5.06.3.3.1 Introduction

Attack on the sulfur atom of thiiranes or thiirenes by electrophiles can yield cyclic sulfonium salts which are expected to be pyramidal about sulfur and which are calculated to be more stable than isomeric ring-opened cations. Ring-opened ions are more likely with oxiiranes. In the gas phase neutral sulfurane structures, *e.g.* (**17**), are favored over ionic structures, *e.g.* (**18**); the latter may be important in polar solvents. A general outline of the course of electrophilic reactions of thiiranes is given in Scheme 26. As shown in the scheme, an open carbenium ion may be an intermediate especially if it is stabilized by the substituent. In this event a single product (**34**) is obtained. An S_N2 mechanism would predict that the nucleophile would attack the thiiranium ion at the least sterically hindered site to give mainly (**35**). Contact or intimate ion-pair intermediates have been suggested in several cases. At present no electrophilic addition to the sulfur atom of thiirenes is known, although a zwitterionic intermediate (see Scheme 22) is proposed in the isomerization of thiirenes to alkynylthiols. Stable thiirenium salts are prepared by other methods.

R-thiirane + E^+ ⟶ [thiiranium (S$^+$–E, R) ⇌ $ESCH_2\overset{+}{C}HR$] $\xrightarrow{Nu^-}$ $ESCH_2CH(Nu)R$ + $NuCH_2CH(SE)R$
(**34**) (**35**)

Scheme 26

5.06.3.3.2 Protonation and basicity

The proton affinities (gas phase) of thiirane and other three-membered heterocycles have been determined: azirane (902.5), thiirane (819.2), phosphirane (815.0), oxirane (793.3 kJ mol^{-1}) ⟨80JA5151⟩. Increasing *s* character in the lone electron pairs decreases proton affinities. Data derived from ^1H NMR chemical shifts in chloroform indicate the order of decreasing basicity is azirane > oxirane > thiirane ⟨73CR(B)(276)335⟩. The base strengths of four-, five- and six-membered cyclic sulfides are greater than that of thiirane.

Fluorosulfonic acid at −60 °C protonates *cis*-1,2-di-*t*-butylthiirane to give an 80:20 mixture of *anti*- and *syn*-*S*-protonated species. The *trans*-di-*t*-butylthiirane also is protonated ⟨74JA3146⟩. Protonation of 2,2,3,3-tetramethylthiirane or of thiirane itself by fluorosulfonic acid gives considerable polymerization in addition to protonation. Polymerization or oligomerization also is common with other protic acids ⟨66CRV297, 76RCR25⟩. According to ^1H NMR thiirane 1-oxide apparently protonates on the sulfur atom rather than oxygen when treated with fluorosulfonic acid ⟨71JOC1121⟩.

The acid-catalyzed addition of nucleophiles, *e.g.* hal^-, $RCOO^-$, MeOH, H_2O, RSH, to give ring-opened products *via* either S_N1 or S_N2 mechanisms is common (Scheme 27). Polymerizations and oligomerizations occur when the sulfhydryl group of the ring-opened product attacks another molecule of thiirane. Interestingly, *cis*-2,3-di-*t*-butylthiirane is inert to HCl. The ring-opening of one of the thiirane rings in (**36**) generates a sulfhydryl group in proximity to the second thiirane ring which results in an intramolecular cyclization (Scheme 28) ⟨78JOU2003⟩. Nitriles may act as nucleophiles as shown in Scheme 29 to give fair to good yields of 1,3-thiazolines ⟨78BSF(2)539⟩.

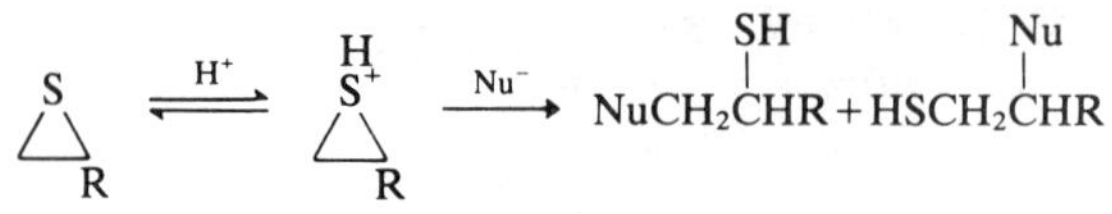

Scheme 27

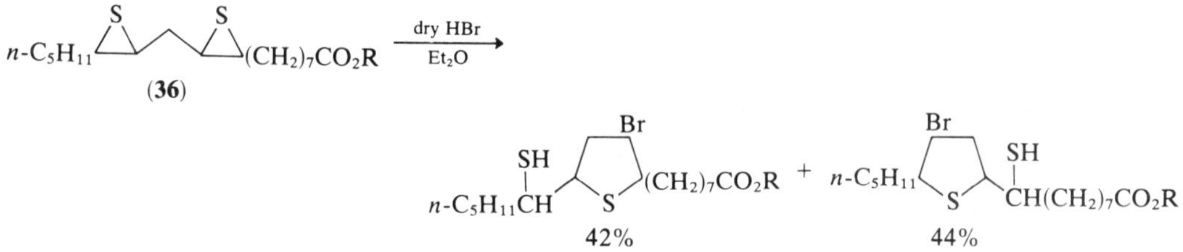

Scheme 28

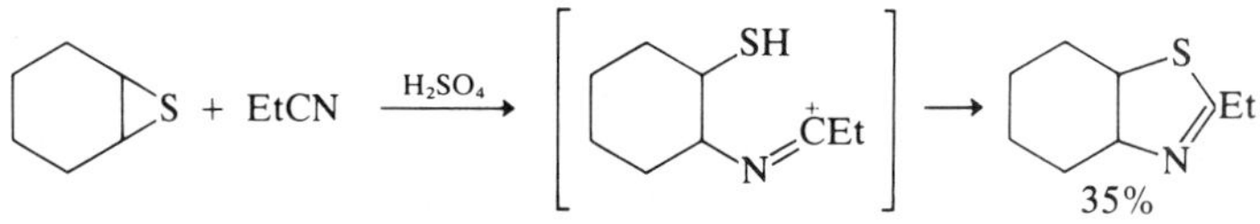

Scheme 29

Acids are poor catalysts for ring cleavage of thiirane 1,1-dioxides but are good catalysts for reactions of thiirane 1-oxides with nucleophiles. These reactions of episulfoxides are believed to proceed by protonation of the oxygen atom (but see the NMR evidence cited above for *S*-protonation in fluorosulfonic acid) and will be treated in the section on nucleophilic reactions.

5.06.3.3.3 Lewis acids

The most important reaction with Lewis acids such as boron trifluoride etherate is polymerization (Scheme 30) ⟨72MI50601⟩. Other Lewis acids have been used: $SnCl_4$, Bu^i_2AlCl, Bu^i_3Al, Et_2Zn, SO_3, PF_5, $TiCl_4$, $AlCl_3$, Pd(II) and Pt(II) salts. Trialkylaluminum, dialkylzinc and other alkyl metal initiators may partially hydrolyze to catalyze the polymerization by an anionic mechanism rather than the cationic one illustrated in Scheme 30. Cyclic dimers and trimers are often products of cationic polymerization reactions, and desulfurization of the monomer may occur. Polymerization of optically active thiiranes yields optically active polymers ⟨75MI50600⟩.

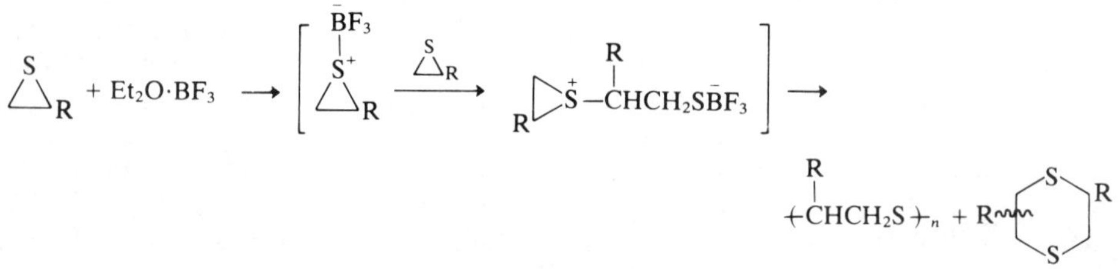

Scheme 30

Treatment of tetrafluorothiirane with aluminum chloride gave a 64% yield of 2,4-bis(pentafluoroethylthio)-2,4-bis(trifluoromethyl)-1,3-dithietane, the dimer of pentafluoroethyl dithiotrifluoroacetate formed by rearrangement of the thiirane *via* trifluorothioacetyl fluoride ⟨65JOC4188⟩. Thiiranes form complexes with manganese, cobalt, nickel, mercury, silver, gold, palladium and platinum salts; with cadmium, zinc and magnesium porphyrins; and with diethylzinc and diethylcadmium. The complex of 2-methylthiirane with diethylzinc is less stable than the corresponding complex with 2-methyloxirane. The metal ions may catalyze nucleophilic attack on the ring carbon atoms.

Tungsten hexachloride and molybdenum pentafluoride desulfurize 2-methylthiirane to propene ⟨72DOK(207)899⟩ and a ruthenium(II) complex desulfurizes thiirane ⟨73JA4758⟩.

5.06.3.3.4 Alkyl halides, oxonium salts and related compounds

Electrophilic attack on the sulfur atom of thiiranes by alkyl halides does not give thiiranium salts but rather products derived from attack of the halide ion on the intermediate cyclic salt ⟨B-81MI50602⟩. Treatment of *cis*-2,3-dimethylthiirane with methyl iodide yields *cis*-2-butene by two possible mechanisms (Scheme 31). A stereoselective isomerization of alkenes is accomplished by conversion to a thiirane of opposite stereochemistry followed by desulfurization by methyl iodide ⟨75TL2709⟩. Treatment of thiiranes with alkyl chlorides and bromides gives 2-chloro- or 2-bromo-ethyl sulfides (Scheme 32). Intramolecular alkylation of the sulfur atom of a thiirane may occur if the geometry is favorable; the intermediate sulfonium ions are unstable to nucleophilic attack and rearrangement may occur (Scheme 33).

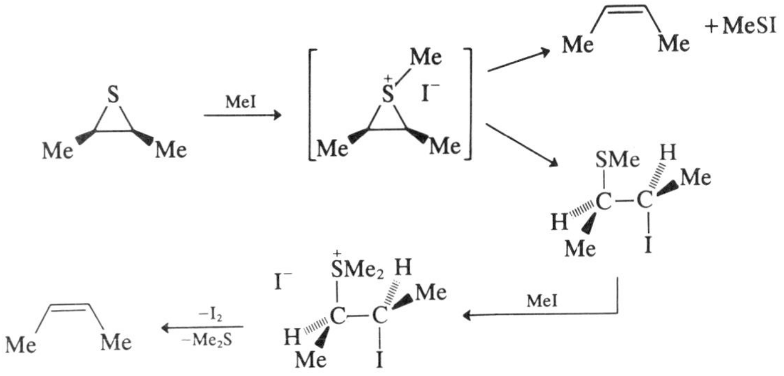

Scheme 31

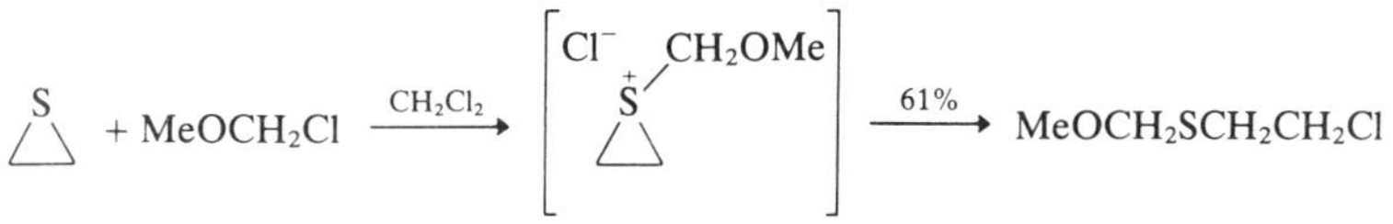

Scheme 32

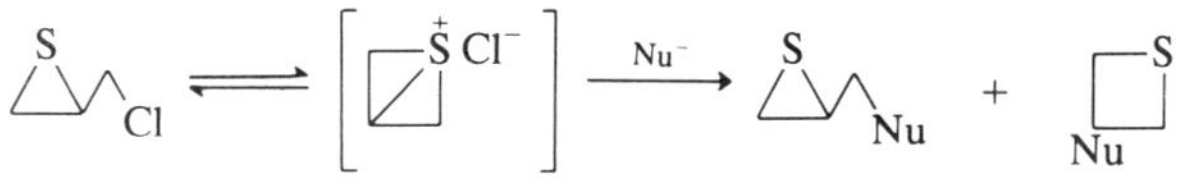

Scheme 33

Alkylating agents which involve non-nucleophilic anions, *e.g.* $Et_3O^+ BF_4^-$, $MeOSO_2F$, Me_3O^+ 2,4,6-$(NO_2)_3C_6H_2SO_3^-$, are required for isolation of *S*-alkylthiiranium salts (Scheme 34). These salts are frequently unstable in the presence of excess thiirane and rapid polymerization occurs when the thiirane is treated with triethyloxonium tetrafluoroborate or dimethyl sulfate ⟨B-77MI50601⟩. Unfortunately the fast, quantitative polymerization is followed by degradation to low molecular weight polymer and oligomers by the action of the alkylating agent ⟨78MI50600⟩. Methylenethiirane polymerizes on treatment with methyl fluorosulfonate. Cationic catalysts for polymerization of thiirane may be generated electrochemically ⟨72USP3645986⟩.

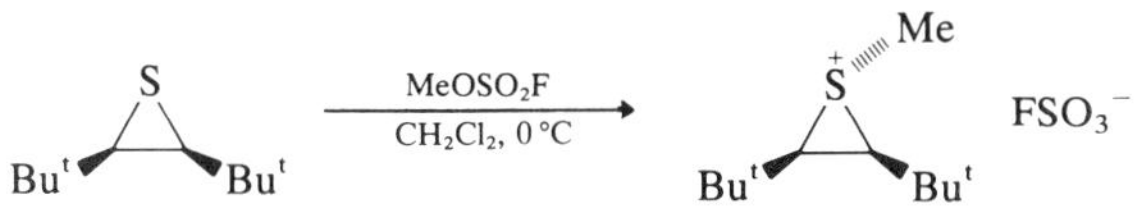

Scheme 34

5.06.3.3.5 Acyl halides and related compounds

A variety of acid halides react with thiiranes to give β-haloethylthiolacyl derivatives (Scheme 35) ⟨76RCR25, 66CRV297⟩. The reaction probably involves electrophilic attack on sulfur by the acyl group, but ring opening of thiirane by traces of hydrogen halide may occur followed by acylation of the thiol group. This latter path may be especially important in the presence of amine hydrohalides. Acid anhydrides may react by similar mechanisms, except that nucleophilic attack on carbon is by the carboxylate anion instead of halide. The reaction with acid bromides and iodides is exothermic and aroyl halides generally are less reactive than aliphatic acid halides. The reaction with acid anhydrides may require a pyridine catalyst, but epithiosteroids are resistant. The presence of other more reactive functional groups may preclude the reaction of acyl halides with a thiirane.

$$\text{thiirane-R} \xrightarrow{MeCOCl} ClCH_2CH(SAc)R + AcSCH_2CH(Cl)R$$

Scheme 35

Phosgene reacts exothermically with thiirane in two steps (Scheme 36) ⟨77MI50602⟩. 3,5-Dinitrobenzoyl chloride and benzoyl fluoride initiate polymerization of thiirane. A novel reaction of benzoyl isocyanate or trichloroacetyl isocyanate, which yields ethylenethiol derivatives from epithiochlorohydrin (2-chloromethylthiirane), 2-methylthiirane or cyclohexene episulfide, has been reported (Scheme 37) ⟨71BAU2432⟩.

$$\text{thiirane} \xrightarrow[C_5H_5N,\ -10\,°C]{COCl_2} ClCH_2CH_2SCOCl \xrightarrow[Et_3N,\ 50\,°C]{\text{thiirane}} (ClCH_2CH_2S)_2CO$$

Scheme 36

Scheme 37

5.06.3.3.6 Halogens

Chlorination (Cl_2, SO_2Cl_2) of thiiranes under anhydrous conditions proceeds smoothly in carbon tetrachloride or other inert solvent (Scheme 38). Either 2-chlorosulfenyl chloride or bis(2-chloroethyl) disulfide may be formed depending on the ratio of chlorine to thiirane. Substituted thiiranes give mixtures (see Scheme 26). Chlorination of cyclohexene episulfide under vigorous conditions gave 1,2-dichlorocyclohexane and polymer. Brominations and iodinations proceed under mild conditions as for chlorination except that only disulfide is obtained with iodine. Tetrafluorothiirane is unreactive under these conditions but readily undergoes free radical halogenation with ring cleavage. The reaction of iodine with thiiranes may be used to effect desulfurization *via* an intermediate 2-iodoethylsulfenyl iodide. Treatment of *cis*-2,3-dimethylthiirane with iodine in refluxing benzene gives 80% of mostly (98%) *cis*-2-butene. Iodine is observed to form charge transfer complexes with thiiranes. If a carbon–carbon double bond is situated near a thiirane ring, rearrangements are observed on halogenation with chlorine, bromine, iodine or sulfuryl chloride (Scheme 39) ⟨80CC100⟩.

$$\text{thiirane} \xrightarrow[CCl_4]{Cl_2} ClCH_2CH_2SCl \text{ or } ClCH_2CH_2SSCH_2CH_2Cl$$

Scheme 38

Scheme 39

Chlorination of thiiranes in hydroxylic solvents gives β-chloroethylsulfonyl chlorides due to further oxidation of the intermediate sulfenyl chloride by chlorine or hypochlorous acid (Scheme 40). Polymer is usually obtained also unless the reaction is done in concentrated hydrochloric acid, which causes rapid ring cleavage to 2-chloroethylthiols which are subsequently oxidized to the sulfonyl chlorides. An 85% yield of (**37**) is obtained in concentrated hydrochloric acid–HCl(g) whereas only a 15% yield is obtained in CCl_4–H_2O.

Cl_2 / CCl_4–H_2O → $ClCH_2CH_2SCl$; HCl → $ClCH_2CH_2SH$; Cl_2 / H_2O → $ClCH_2CH_2SO_2Cl$ (**37**)

Scheme 40

Treatment of a carborane derivative of thiirane with *N*-bromosuccinimide gives a β-bromodisulfide ⟨79MI50601⟩. Chlorination of *cis*-2,3-di-*t*-butylthiirane by *t*-butyl hypochlorite proceeded differently to the reaction with chlorine itself (Scheme 41) ⟨74JA3146⟩.

Bu^tOCl / CH_2Cl_2 → $Bu^tCHCH(SCl)Bu^tS$–CH=CHBu^t

Scheme 41

5.06.3.3.7 Electrophilic sulfur, nitrogen, phosphorus and arsenic

2-Chloroethyldisulfides are obtained by electrophilic attack on the sulfur atom of thiiranes by sulfenyl halides (Scheme 39). Sulfur dichloride and disulfur dichloride react similarly to give more sulfur-rich derivatives: di- and tri-sulfenyl halides, and tri- and tetra-sulfides (Scheme 42). A 1:1 ratio of sulfur halide to thiirane gives the di- or tri-sulfenyl halide; a 2:1 ratio the tri- or tetra-sulfide. Thiirane 1-oxides are cleaved by sulfenyl halides to thiolsulfinates (Scheme 43) ⟨74JAP7440461⟩.

SCl_2 → $RCHClCH_2SSCl$ or $(RCHClCH_2S)_2S$; S_2Cl_2 → $RCHClCH_2SSSCl$ or $(RCHClCH_2S)_2S_2$

Scheme 42

PhSCl / Et_2O → PhSS(O)CH_2CH_2Cl 77%

Scheme 43

The reaction of *N*-chlorobenzenesulfonylformimidoyl chloride with cyclohexene episulfide may involve attack by an electrophilic nitrogen on sulfur (Scheme 44) ⟨74TL837⟩. Attempts to react *cis*-2,3-di-*t*-butylthiirane with chloramine-T or *p*-toluenesulfonyl azide failed ⟨74JA3146⟩. Either *cis*- or *trans*-2-methyl-3-phenyloxaziridine desulfurizes thiiranes *via* attack on sulfur by electrophilic nitrogen to give thionitrosomethane as a reactive intermediate (Scheme 45) ⟨80JOC1691⟩. Treatment of thiiranes with both organic and inorganic phosphorus(III) halides yields 2-haloethylthiophosphines in which one or more of the halogens has been replaced (Scheme 46). *N,N*-Diethylaminophosphorus dichloride gives different products depending on the temperature (Scheme 47). Formation of the phosphine sulfide at higher temperatures suggests the occurrence of desulfurization of the episulfide, which phosphines are known to do (see the section on nucleophilic attack on sulfur). Sulfur is eliminated from thiirane by treatment with $RPCl_3^+ AlCl_4^-$ to give $RP(S)Cl_2$ and 1,2-dichloroethane. Treatment of thiirane, 2-methylthiirane and 2-chloromethylthiirane with dichlorophosphoranes yields 2-chloroethylthiophosphonium salts (Scheme 48) ⟨81ZN(B)447⟩. Polymerization occurs on treatment of thiiranes with phosphorus pentachloride. Arsenic(III) halides and related compounds, *e.g.* $AsCl_3$, $RAsCl_2$, R_2AsCl,

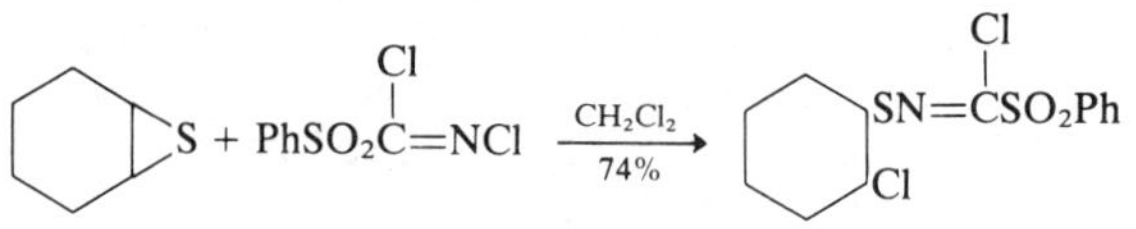

Scheme 44

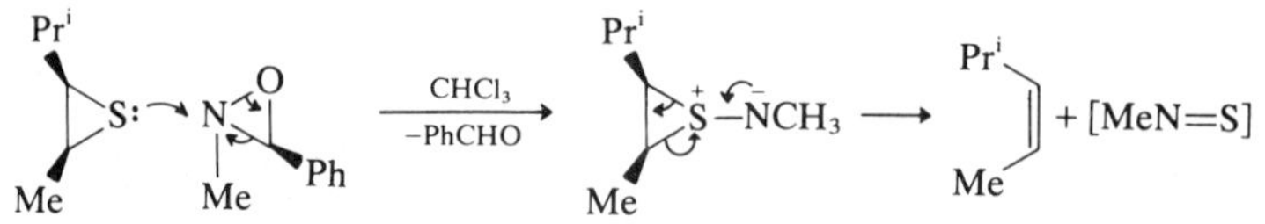

Scheme 45

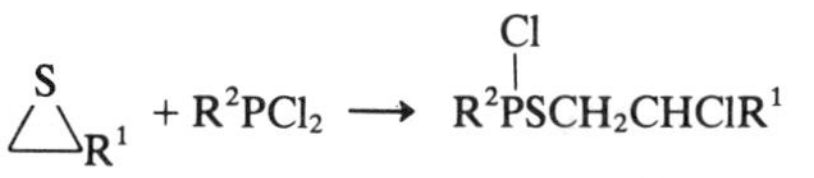

Scheme 46

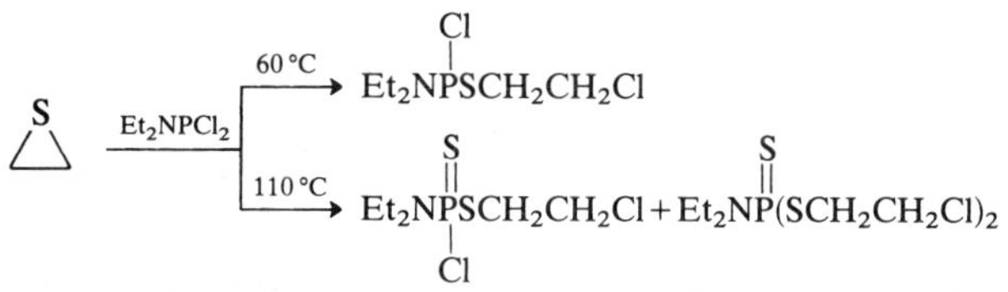

Scheme 47

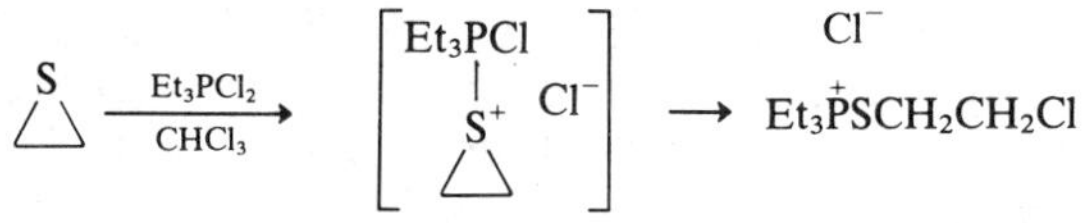

Scheme 48

R_2AsNCS, react similarly to the phosphorus halides ⟨76RCR25⟩. Halosilanes do not react with thiiranes.

5.06.3.3.8 Peracids and other sources of electrophilic oxygen

Oxidation of thiiranes with peracids or sodium metaperiodate affords thiirane 1-oxides in good yield (Scheme 49) ⟨B-81MI50600⟩. Perbenzoic or *m*-chloroperbenzoic acid is more general. The stereochemistry of sulfoxide formation is sensitive to steric factors, the oxygen being delivered preferentially to the least hindered face of the sulfur atom to give the *anti* isomer. The oxidation may be done without affecting carbon–carbon double bonds ⟨81TL4815⟩ although oxidation of thiirane (**38**) gives a thietane 1-oxide by rearrangement (Scheme 50) ⟨69JOC3998⟩. *cis*-2,3-Di-*t*-butylthiirane can be oxidized to the *anti* sulfoxide but it is resistant to hydrogen peroxide and to aqueous potassium permanganate. Oxidation of 2,2-dichloro-3,3-diarylthiirane with MCPBA yields benzothiophene derivatives by rearrangement of the initially formed thiirane 1-oxide (Scheme 51) ⟨72RTC1345⟩. Thiirene 1-oxides are readily converted to thiirene 1,1-dioxides by oxidation with MCPBA.

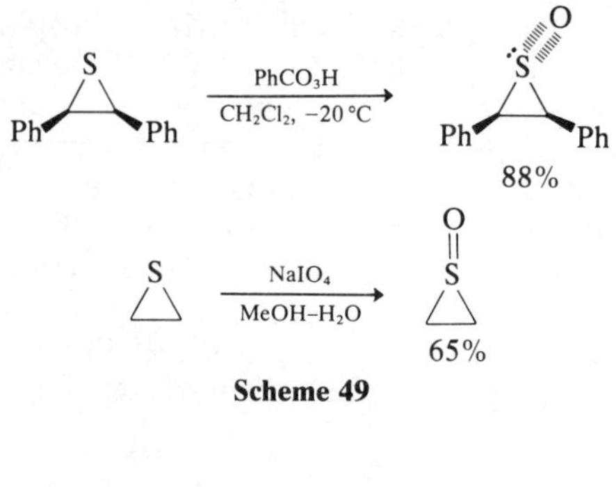

Scheme 49

(38)

KIO_4, MeOH–H_2O

Scheme 50

MCPBA, CH_2Cl_2

$-H^+$

9% 40%

Scheme 51

Hydrogen peroxide is less satisfactory as an oxidizing agent. Oxidation of thiirane with hydrogen peroxide in the presence of a catalytic amount of vanadium pentoxide gives a mixture of sulfone and sulfoxide ⟨69JCS(C)2334⟩. The reported oxidation of 2-hydroxymethylthiirane to the 1,1-dioxide by hydrogen peroxide is in error because the purported thiirane was in fact 3-hydroxythietane ⟨65JOU731, 69JCS(C)2334⟩. Treatment of thiirane or 2-methylthiirane with hydrogen peroxide in the absence of catalyst gives poly(ethylene sulfone) and 2-hydroxypropanesulfonic acid respectively.

Ozone is reported to yield a sulfoxide from norbornene episulfide ⟨74JAP7435375⟩, but its gas phase reaction with thiirane gives sulfur dioxide, ethylene, formaldehyde and carbon dioxide ⟨80MI50600⟩. Nitric acid and potassium permanganate give extensive oxidative ring opening of thiirane, carboxysulfonic acids being isolated in the reaction with nitric acid.

5.06.3.3.9 Carbenes

Treatment of several thiiranes with ethyl diazoacetate in the presence of copper(II) acetoacetate effects desulfurization to give alkenes in good yield. The mechanism is believed to involve electrophilic attack by ethoxycarbonyl carbene on the sulfur atom (Scheme 52) ⟨75JA2553⟩. The desulfurization is stereospecific, *cis*- and *trans*-thiiranes giving *cis*- and *trans*-alkenes respectively. When di(*p*-methoxyphenyl)diazomethane was used the diaryl thioketone could be isolated along with the alkene. Treatment of methylenethiirane with diazomethane results in polymerization ⟨78RTC214⟩. The diazotricyclic thiirane derivative (**39**) decomposes, probably *via* the diazoalkane derivative (**40**) and the thiete (**41**), to thioketone (**42**; Scheme 53) ⟨80JA6634, 80JOC2962⟩.

N_2CHCO_2Et, Cu(acac)$_2$, 110 °C; $\overset{+}{S}-\overset{-}{C}HCO_2Et$; $-HC(=S)CO_2Et$; 70%

Scheme 52

5.06.3.4 Nucleophilic Attack on Sulfur

5.06.3.4.1 Oxygen nucleophiles

Oxygen nucleophiles usually attack a ring carbon atom rather than the sulfur atom of a thiirane, and those cases in which desulfurization is observed on treatment of a thiirane with oxygen bases probably involve the extrusion of sulfur by mechanisms other than a nucleophilic attack on sulfur, *e.g.* thermal. Desulfurization of thiirane intermediate (**43**)

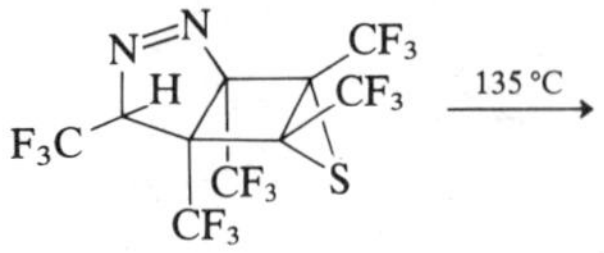

(39)

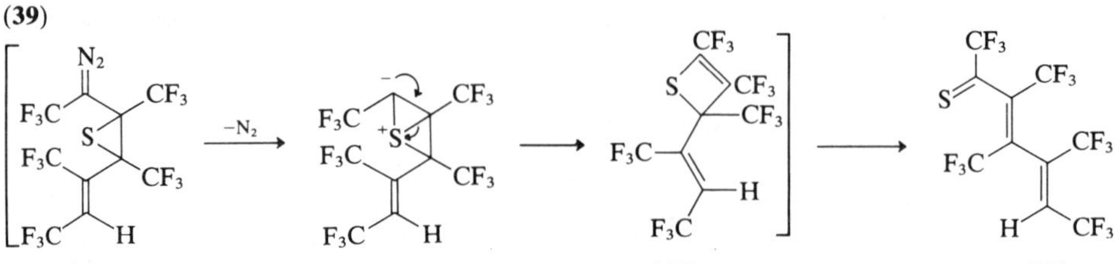

(40) (41) (42)

Scheme 53

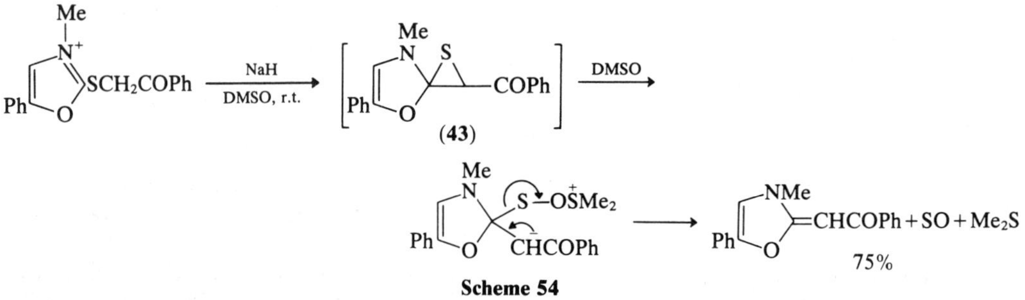

(43)

75%

Scheme 54

may involve nucleophilic attack by oxygen of the solvent, dimethyl sulfoxide, although the formation of sulfur monoxide or dimethyl sulfide was not mentioned (Scheme 54) ⟨72BCJ1797⟩.

Episulfoxides and episulfones are readily attacked by oxygen nucleophiles at the sulfur atom, which is more electron deficient due to attachment to electronegative oxygen atoms. Attack at carbon or a proton may also occur. The desulfurization of episulfoxide (**44**) by dimethyl sulfoxide (Scheme 55) ⟨76JA4325⟩ is analogous to that of the episulfide in Scheme 54. Acid-catalyzed ring openings of thiirane 1-oxide involve attack by nucleophiles on both sulfur and carbon; however, a nucleophile first attacks carbon to give an acyclic sulfenic acid which undergoes nucleophilic attack at the sulfur atom to give the observed products. This reaction will be discussed under nucleophilic attack on a ring carbon atom. Episulfones may be attacked by oxygen bases at the sulfur atom to give sulfonates, *e.g.* (**45**, Scheme 56) ⟨67JA4487⟩, although the reaction is not usually competitive with the extrusion of sulfur dioxide (Ramberg–Bäcklund reaction). Thiirene 1,1-dioxides give α,β-unsaturated sulfonates and alkynes by similar mechanisms. 2,3-Diphenylthiirene 1,1-dioxide reacts about 5000 times faster with alkoxide ions than does 2,3-diphenylcyclopropenone because of conjugative stabilization in the latter ('aromatic' two-π-electron system in the zwitterionic resonance structure for the carbonyl group) ⟨69JA2084⟩. These ring openings generally occur to produce the most stable carbanion intermediate.

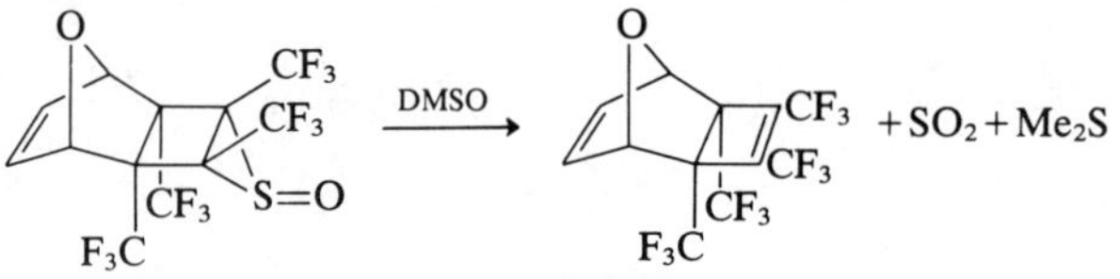

(44)

Scheme 55

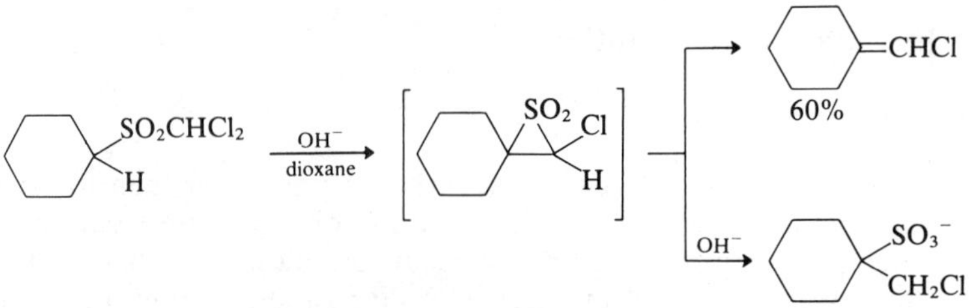

60%

(45) 20%

Scheme 56

5.06.3.4.2 Nitrogen nucleophiles

There is no evidence that nitrogen nucleophiles attack the sulfur atom of thiiranes except for the reported desulfurization of an episulfide of a derivative of the antibiotic kanamycin by hydrazine ⟨77JAP(K)7771445⟩. Certain thiiranium salts do undergo attack on sulfur by various nucleophiles, including azide ion or triethylamine, to give the alkene ⟨78CC630⟩. Thiirane 1-oxides also are attacked at the sulfur atom by lithium *N,N*-diisopropylamide to give the alkene ⟨B-81MI50600⟩.

5.06.3.4.3 Sulfur nucleophiles

Treatment with thiocyanic acid of a steroid 16α,17α-episulfide substituted with an acetyl group results in desulfurization with subsequent addition of the nucleophile to the α,β-unsaturated ketone produced ⟨79BAU168⟩. Although the reaction occurs at 20 °C, no evidence was given to show that the desulfurization involved attack on sulfur by the thiocyanate ion. Methanethiolate ion attacks the sulfur atom of the *S*-methyl salt of the episulfide of adamantylideneadamantane to give the alkene and dimethyl sulfide ⟨78CC630⟩, and thiourea behaves similarly with the *S*-methyl salt of cyclooctene episulfide ⟨80T1361⟩. Dimethyl sulfide attacks the sulfur atom of *S*-alkylthiirenium salts to give the alkyne and dimethyl(alkylthio)sulfonium ions, $Me_2\overset{+}{S}SR$ ⟨79MI50600⟩.

5.06.3.4.4 Phosphorus nucleophiles

One of the most widely used reactions of thiiranes is the stereospecific desulfurization by trivalent phosphorus compounds to give an alkene and a sulfur–phosphorus derivative (Scheme 57). The reaction is not sensitive to solvent polarity and probably proceeds by a concerted cleavage of both sulfur–carbon bonds. Triphenylphosphine is the most common reagent; it and trialkylphosphines often are effective at room temperature while phosphites usually require higher temperatures. Tris(alkylamino)phosphines may react exothermically ⟨75BAU878⟩. Triphenylphosphine causes polymerization of methylenethiirane (allene episulfide) ⟨78RTC214⟩, but tris(trimethylamino)phosphine effects desulfurization of tetramethylallene episulfide at 65–75 °C to tetramethylallene in 45% yield ⟨76JA7081⟩. The desulfurization of thiiranes has been used in the synthesis of alkenes which are not easily prepared by other methods as exemplified in Schemes 58–60 ⟨80JOC2966, 79JCS(P1)2401, 80JOC1481⟩. Optically active benzyl-*n*-butylmethylphosphine sulfide is obtained by treatment of the optically active phosphine with cyclohexene episulfide ⟨64TL359⟩. Treatment of a Dewar thiophene, tetrakis(1,2,3,4-trifluoromethyl)-5-thiabicyclo-2-pentene, with diphenylphosphorus chloride or other trivalent phosphorus derivatives catalyzes the rearrangement to the thiophene by cleavage of the carbon–carbon bond of the thiirane ring *via* a sulfur–phosphorus intermediate. The thiiranium salt shown in Scheme 61 is converted to cyclooctene by tri-*n*-butylphosphine ⟨69JA3606⟩.

R_3P; $+R_3PS$

Scheme 57

Ph_3P, *n*-pentane; $+Ph_3PS$

Scheme 58

$(MeO)_3P$, 110 °C

Scheme 59

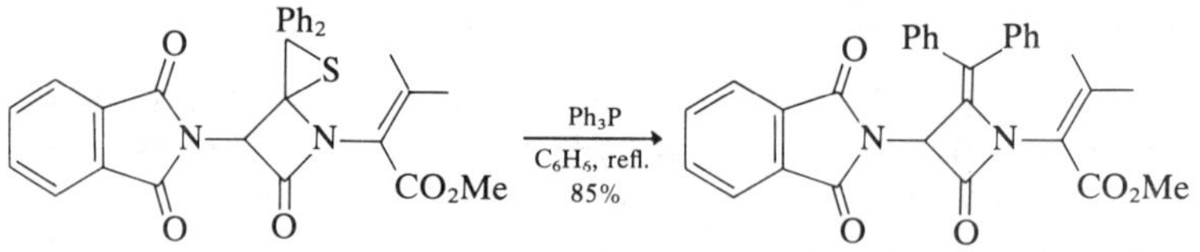

Scheme 60

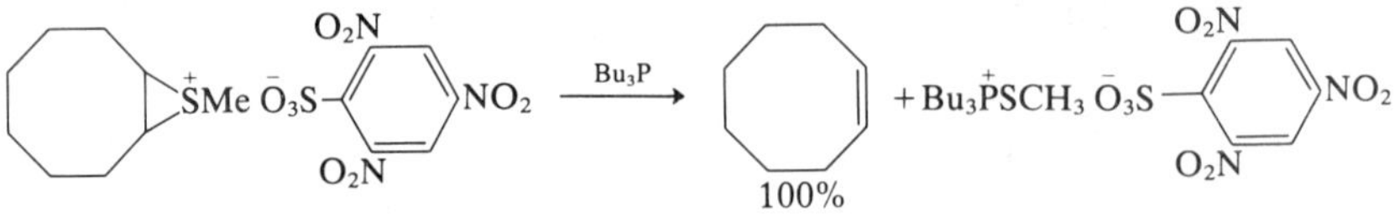

Scheme 61

5.06.3.4.5 Halide ions

S-Alkylthiiranium salts, *e.g.* (**46**), may be desulfurized by fluoride, chloride, bromide or iodide ions (Scheme 62) ⟨78CC630⟩. With chloride and bromide ion considerable dealkylation of (**46**) occurs. In salts less hindered than (**46**) nucleophilic attack on a ring carbon atom is common. When (**46**) is treated with bromide ion, only an 18% yield of alkene is obtained (compared to 100% with iodide ion), but the yield is quantitative if the methanesulfenyl bromide is removed by reaction with cyclohexene. Iodide ion has been used most generally. Sulfuranes may be intermediates, although in only one case was NMR evidence observed. Theoretical calculations favor a sulfurane structure (*e.g.* **17**) in the gas phase, but polar solvents are likely to favor the thiiranium salt structure.

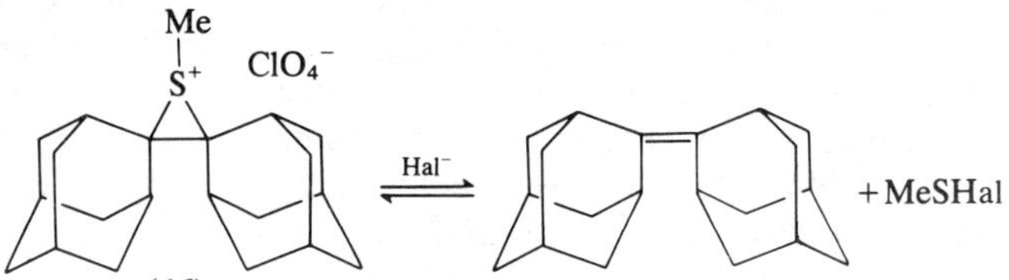

Scheme 62

Fluoride ion attacks the sulfur atom in 2,3-diphenylthiirene 1,1-dioxide to give *cis*-1,2-diphenylethylenesulfonyl fluoride (23%) and diphenylacetylene (35%). Bromide or iodide ion does not react ⟨80JOC2604⟩. Treatment of *S*-alkylthiirenium salts with chloride ion gives products of carbon attack, but the possibility of sulfur attack followed by addition of the sulfenyl chloride so produced to the alkyne has not been excluded ⟨79MI50600⟩. In fact the methanesulfenyl chloride formed from 1-methyl-2,3-di-*t*-butylthiirenium tetrafluoroborate has been trapped by reaction with 2-butyne. A sulfurane intermediate may be indicated by ^{1}H NMR experiments in liquid sulfur dioxide.

5.06.3.4.6 Carbanions

Treatment of thiiranes with alkyllithium reagents yields alkenes and the lithium salt of the alkanethiol. If the sulfur atom is hindered to attack, a proton may be removed from the thiirane carbon atom instead. The desulfurization is stereospecific. *cis*-2,3-Diphenylthiirane gives *cis*-stilbene (47%) and the *trans*-thiirane gives *trans*-stilbene (99%) ⟨79TL3987⟩. A sulfurane intermediate is suggested which extrudes RSLi by a concerted, disrotatory process (Scheme 63). *n*-Butyl- and phenyl-lithium are commonly used and attack the sulfur atom, but the lithium salts of isocyanides attack mainly the carbon atoms of the ring. Attack on the sulfur atom of the iminothiirane (**47**) is observed with the anion of diethyl malonate, a phosphous ylide, *N*-methylindole and enamines or ynamines (Scheme 64) ⟨80JOC4366⟩.

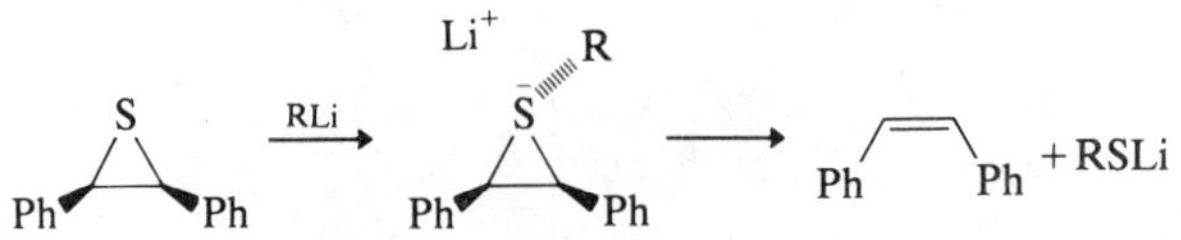

Scheme 63

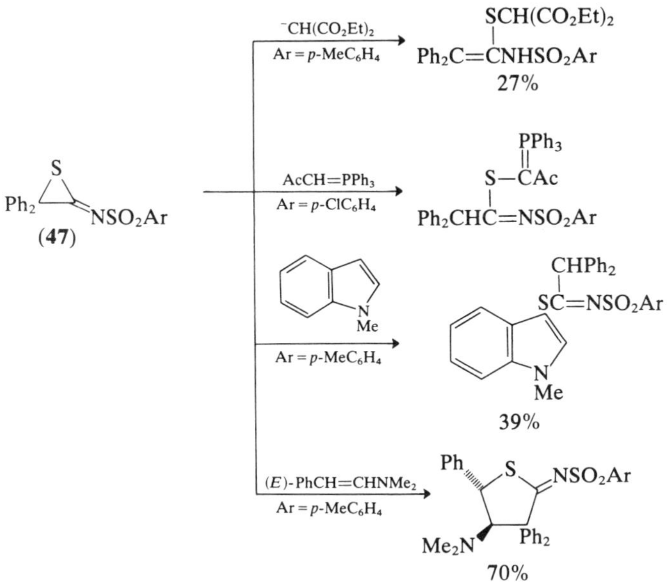

Scheme 64

Thiirane 1-oxides are desulfurized stereospecifically by treatment with *n*-butyl- or phenyllithium (Scheme 65) ⟨B81MI50600⟩. Yields are good except when attack on the sulfur atom is sterically hindered, as in *cis*-2,3-diphenylthiirane 1-oxide, in which case attack on a proton occurs to give salts of vinylsulfenic acids in addition to *cis*-stilbene (Scheme 66). A sulfurane (**48**) is suggested as an intermediate. Thiirane 1,1-dioxides are also desulfurized by treatment with alkyllithium or Grignard reagents (to give an alkene and lithium or magnesium salts of sulfinic acids). The reaction with methyl- and *n*-butyl-lithium is stereospecific with respect to alkene formation. Methyllithium quantitatively desulfurizes *S*-methylthiiranium salt (**46**) to the alkene and dimethyl sulfide, and the anion of dimethyl malonate removes sulfur from the *S*-methyl salt of cyclooctene episulfide ⟨80T1361⟩. Treatment of 2,3-dimethylthiirene 1,1-dioxide with α-metalated nitriles gives products from attack on both sulfur and carbon, *e.g.* (**49**) and (**50**) respectively (Scheme 67) ⟨79JOC830⟩.

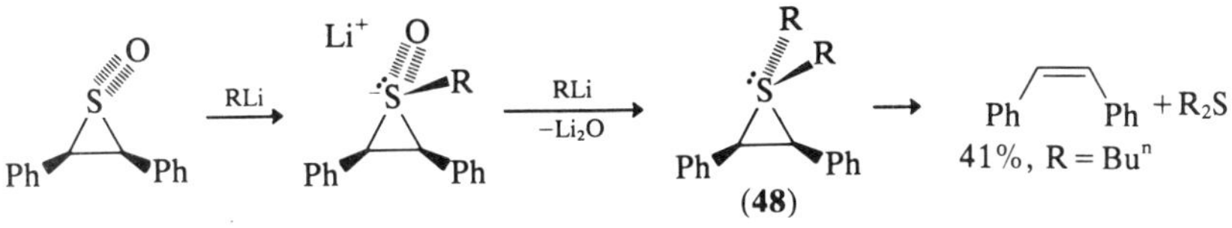

Scheme 65

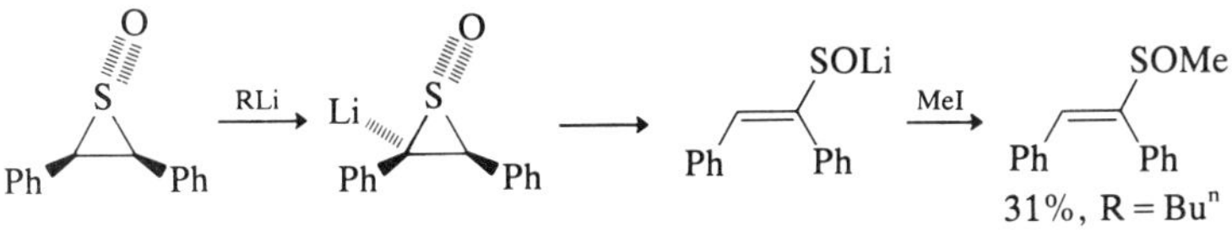

Scheme 66

5.06.3.4.7 π-Systems

One example of nucleophilic attack by a π-electron system on a sulfur atom of a thiirane 1-oxide is shown in Scheme 51. *S*-Alkylthiirenium ions react with tetramethylethylene to transfer the *S*-alkyl group yielding the alkyne and an *S*-alkyl-2,2,3,3-tetramethylthiiranium ion ⟨79MI50600⟩.

5.06.3.4.8 Hydride

Lithium aluminum hydride normally reacts with thiiranes *via* nucleophilic attack on carbon, but where that process is hindered sulfur is attacked to give the alkene, usually in good yield, and lithium sulfide ⟨70JPR421⟩.

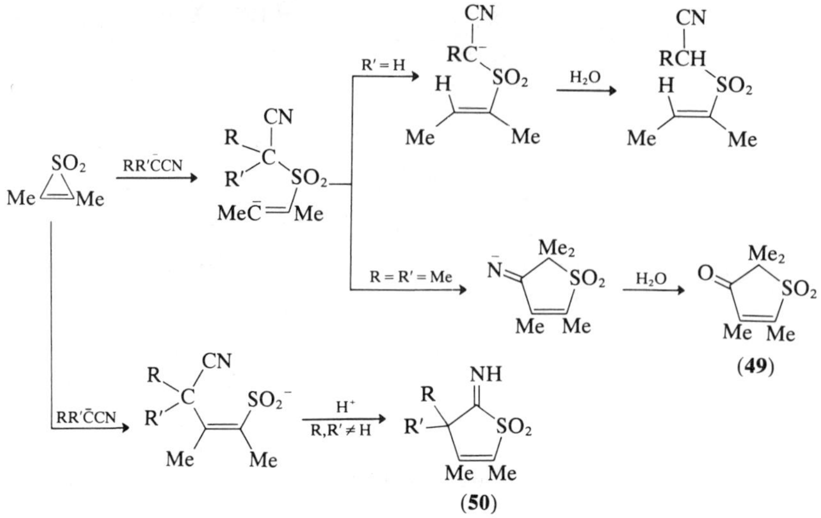

Scheme 67

5.06.3.4.9 Low-valent metals

Treatment of thiiranes with Raney nickel affords the alkene in usually excellent yield, but the desulfurization may be accompanied by reduction of the alkene or by hydrogenolysis in which both the carbon–carbon bond and a carbon–sulfur bond of the thiirane are cleaved. Bis(1,5-cyclooctadiene)nickel desulfurizes cyclohexene episulfide in 67–80% yield ⟨73TL2667⟩. Other reagents which desulfurize thiiranes are copper bronze or copper powder, amino complexes of germanium ⟨79HCA152⟩, and complexes of manganese, molybdenum, tungsten, iron, platinum and palladium. Iron carbonyls give highly but not completely stereospecific desulfurization. A molybdenum nitrido complex yields the thionitrosyl derivative (**51**; Scheme 68) ⟨79JCS(D)1⟩. Treatment of 2,3-diphenylthiirane 1-oxide with tris(triphenylphosphine)rhodium chloride or bromide gives a rhodium complex of sulfur monoxide ⟨81MI50601⟩.

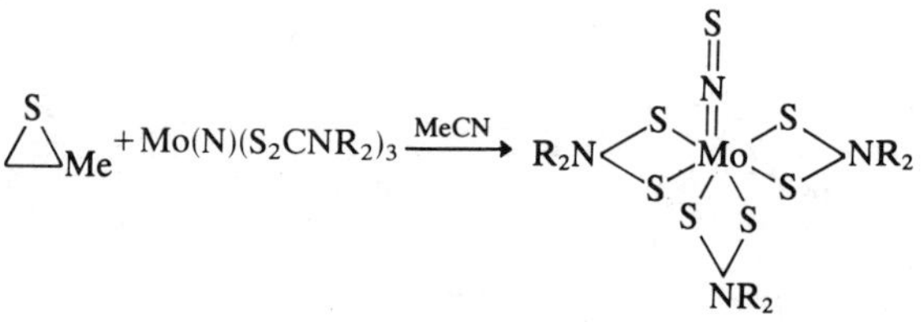

(**51**) 80%

Scheme 68

5.06.3.5 Nucleophilic Attack on Carbon

5.06.3.5.1 Introduction

A wide variety of nucleophiles attack the carbon atoms of thiirane derivatives to give ring-opened products (Scheme 69). Lewis or Brønsted acids may catalyze the reactions of episulfides (Scheme 26) and episulfoxides, and the mechanism can vary from S_N2 to S_N1. Anionic polymerization of thiiranes, commonly initiated by attack on carbon by a nucleophile and propagated by nucleophilic attack by thiolate ion produced in the initiation step, gives high molecular weight material; and chiral polymer can be produced from chiral monomers or chiral catalysts. The acid-catalyzed ring opening of episulfoxides yields unstable sulfenic acids. Carbocationic intermediates are common in reactions of episulfonium ions (especially at elevated temperatures) and both anti-Markownikoff (**52**) and Markownikoff (**53**) products are observed in nucleophilic displacements on thiiranium ions (Scheme 69). Nucleophiles add to the carbon–carbon double bond of thiirene 1,1-dioxides. The few known α-thiolactones react with nucleophiles at the carbonyl group ⟨77AG(E)722⟩.

Scheme 69

5.06.3.5.2 Oxygen nucleophiles

Nucleophilic attacks have been observed with water, alcohols, hydroxide ion, alkoxide and phenoxide ions, carboxylate ions and oximes. Water is a weak nucleophile and its addition to thiirane at 95 °C to give 2-mercaptoethanol is catalyzed by 4A molecular sieves ⟨68USP3369019⟩. Polymerizations are catalyzed by water- and alcohol–diethylzinc (or trialkylaluminum) and by alkoxides or phenoxides frequently complexed to zinc or other metal ions. Crown ethers facilitate these polymerizations. Anionic and cationic polymerization mechanisms occasionally coexist. Stereospecific anionic polymerizations of optically active thiiranes give optically active polymers ⟨75MI50600⟩. Diethylzinc in concert with (*R*)-(−)-3,3-dimethyl-1,2-butanediol is useful in the formation of crystalline, optically active polymers from optically inactive substituted thiiranes, *e.g.* *cis*-2,3-dimethylthiirane ⟨79NJC669⟩. One enantiomer of a racemic thiirane mixture can be polymerized leaving the unreacted monomer enriched in the other enantiomer. Chiral additives, *e.g.* (*R*)-(+)-limonene, increase the optical purity of recovered monomer. In the polymerization of racemic 2-methylthiirane the system selects the enantiomer having the same relative configuration as the chiral glycol ligand in the initiator ⟨76MI50600⟩.

The reaction of thiirane 1-oxides with water or methanol is usually acid-catalyzed and gives *β*-substituted sulfenic acids which dimerize to thiolsulfinates (**54**; Scheme 70) ⟨72JA5786⟩. If acetic acid is used a mixture of disulfide (**55**) and thiolsulfonate (**56**) is obtained. Treatment of thiirane 1,1-dioxides with hydroxide ion may involve attack on carbon as well as on sulfur as exemplified by 2-phenylthiirane 1,1-dioxide (Scheme 71).

$[NuCH_2CH_2SOH] \longrightarrow NuCH_2CH_2S(O)SCH_2CH_2Nu \rightarrow NuCH_2CH_2SSCH_2CH_2Nu + NuCH_2CH_2S(O_2)SCH_2CH_2Nu$

(**54**) $Nu^- = \bar{O}Me$ (**55**) (**56**) $Nu^- = Ac\bar{O}$

Scheme 70

$PhCH(SO_2^-)CH_2OH \xrightarrow{MeI} PhCH(SO_2Me)CH_2OH \xrightarrow[-CH_2O]{OH^-} PhCH_2SO_2Me$

Scheme 71

S-Substituted thiiranium ions react with water and alcohols to give *trans* ring opening (Scheme 72). A report that oxygen nucleophiles attack sulfur as well as carbon has been shown to be incorrect ⟨79ACR282⟩. The intermediate thiiranium ion (**57**) in the presence of lithium perchlorate readily yields the carbenium ion which undergoes a transannular hydride migration (Scheme 73) ⟨80T1361⟩. In the absence of perchlorate ion the intermediates formed by addition of sulfenyl halides to alkenes in solvents of low polarity exist either as sulfuranes or as intimate and solvent-separated ion pairs in which product is formed by attack of chloride ion, even in solvent acetic acid. When perchlorate ion is added to the acetic acid solution, chloride ion diffuses away in the more polar medium and acetates are obtained (Scheme 74). Covalent perchlorates are said to be obtained in a few cases ⟨B-81MI50603⟩.

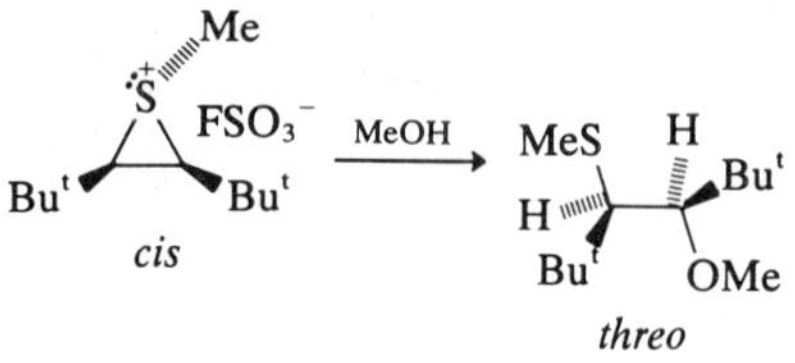

Scheme 72

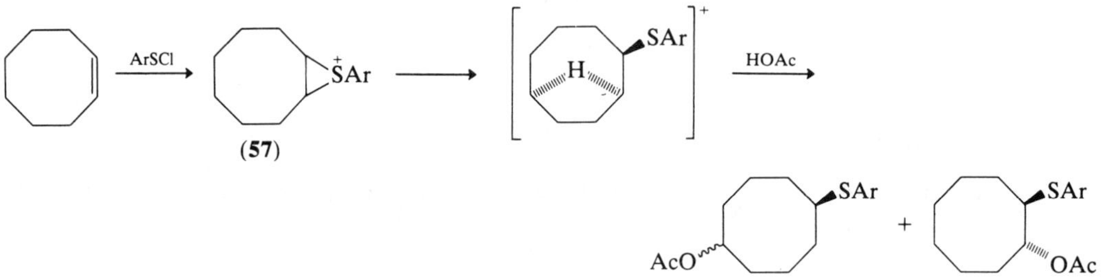

(57)

Scheme 73

Ar = 2,4-$(NO_2)_2C_6H_3$

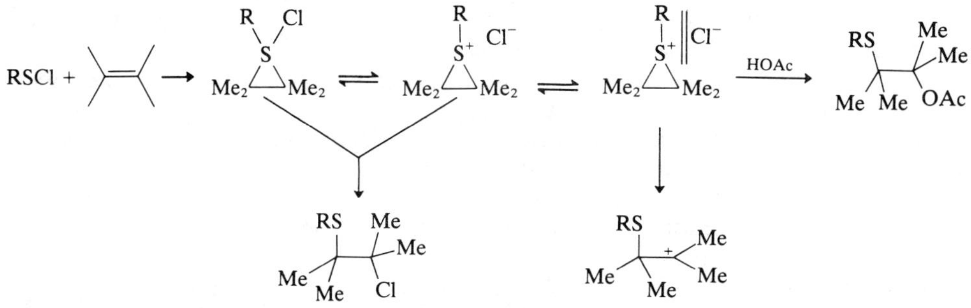

Scheme 74

Thiirenium ions formed as intermediates in the solvolysis of *trans*-β-thiovinyl sulfonates react with methanol to give product with retention of configuration (Scheme 75) ⟨79MI50600⟩.

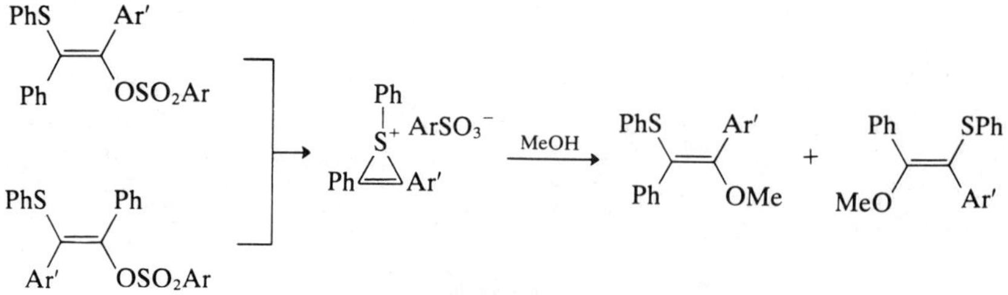

Scheme 75

5.06.3.5.3 Nitrogen nucleophiles

Primary and secondary aliphatic and aromatic amines react readily with thiiranes to give 2-mercaptoethylamine derivatives (Scheme 76) ⟨76RCR25, 66CRV297⟩. The reaction fails or gives poor yields with amines which are sterically hindered (*e.g.* *N,N*-dicyclohexylamine) or whose nitrogen atom is weakly basic (*e.g.* *N,N*-diphenylamine). Aromatic amines are less reactive and higher reaction temperatures are usually required for them. The reaction mechanism is S_N2 and substituted thiiranes are attacked preferentially at the least hindered position. Side reactions include polymerization and oligomerization initiated by the free thiol of the mercaptoethylamine, bis(mercaptoalkylation) of primary amines, and other reactions *e.g.* nucleophilic displacements (Scheme 77), oxidation, additions to double bonds involving the free thiol group. A high concentration of the amine reduces the occurrence of polymerization and bis(mercaptoalkylation). Unhindered, basic tertiary amines (*e.g.* triethylamine) and amide ions (*e.g.* KNH_2) initiate anionic polymerization of thiirane. The reactions of primary and secondary amines are catalyzed by triethylamine and Lewis acids such as silver ion. The reaction of 2,2-disubstituted thiiranes with ammonia is sluggish, but the addition of hydrazine is said to promote the formation of mercaptoalkylated product ⟨63MI50600⟩. Ammonia (aqueous) causes rapid polymerization of thiirane. The primary product of ring opening of tetrafluorothiirane by morpholine undergoes further reactions occasioned by the loss of hydrogen fluoride.

$+ R'NH_2 \rightarrow R'NHCH_2CH(SH)R + R'N(CH_2CH(SH)R)_2$ (major)

Scheme 76

Me_2NH → Me_2NCH_2, Me, Me, $CHOHCO_2Pr^i$; Me Me, CO_2Pr^i

Scheme 77

Other nitrogen nucleophiles (Scheme 78) which attack thiiranes are 2-chloroimidazoles ⟨78FRP2364218⟩, *N*-alkyl- and *N,N'*-dialkyl-hydrazines ⟨80JOU1663, 80JCS(P2)279⟩, azomethines (imines) ⟨80KGS1569, 79SC201⟩, oximes ⟨80KGS1569⟩, hydrazones ⟨79KGS1637⟩ and nitriles ⟨78BSF(2)539⟩ (Scheme 29).

i, ii: $RNCH_2CH_2SH$ (NH_2); iii: R^1, R^2, EtN; iv: $R^1R^2C{=}NNHCH_2CH_2SH \rightleftharpoons$ R^1, R^2, HN, N, H

i, R, N, Cl, NH; ii, $RNHNH_2$; iii, $R^1R^2C{=}NEt$; iv, $R^1R^2C{=}NNH_2$

Scheme 78

S-Substituted thiiranium ions react with secondary amines to give ring-opened products. Nitriles also react with thiiranium ions, probably *via* an open carbenium ion whose formation is favored by increasing the polarity of the medium by the addition of lithium perchlorate (Scheme 79) ⟨79ACR282⟩. An intramolecular displacement by an amide nitrogen atom on an intermediate thiiranium ion has been invoked ⟨80JA1954⟩.

i, ii → SAr, NHC(O)Me

i, ArSCl, MeCN, $LiClO_4$; ii, H_2O

Scheme 79

2,3-Diphenylthiirene 1-oxide reacts with hydroxylamine to give the oxime of benzyl phenyl ketone ⟨79JA390⟩. The reaction probably occurs by addition to the carbon–carbon double bond followed by loss of sulfur monoxide (Scheme 80). Dimethylamine adds to the double bond of 2,3-diphenylthiirene 1,1-dioxide with loss of sulfur dioxide (Scheme 81) ⟨75JOC3189⟩. Azide ion gives seven products, one of which involves cleavage of the carbon–carbon bond of an intermediate cycloadduct (Scheme 81) ⟨80JOC2604⟩.

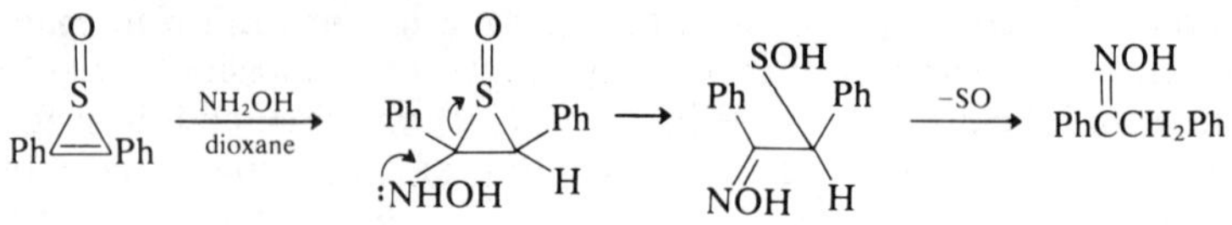

Scheme 80

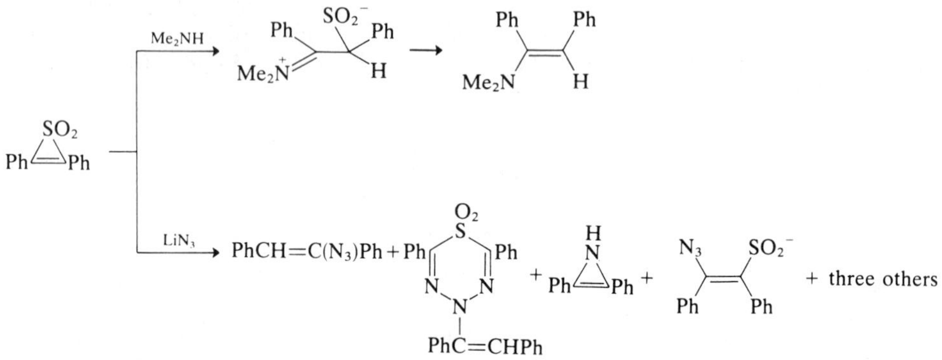

Scheme 81

5.06.3.5.4 Sulfur, selenium and phosphorus nucleophiles

Thiiranes react readily with aliphatic and aromatic thiols even in the absence of an electrophilic catalyst, and the reaction is faster if the thiolate anion is used. Sulfur nucleophiles are generally more reactive toward thiiranes than oxygen or nitrogen nucleophiles. The reaction mechanism is classified as S_N2, the least hindered carbon atom being most commonly attacked. Hydrogen sulfide reacts in the presence of 4A molecular sieves. Anionic polymerizations of thiiranes are propagated by attack of a sulfhydryl anion from a ring-opened thiirane on another molecule of thiirane. Initiation can occur by attack of an external thiolate anion. For example, cadmium thiolates of esters of cysteine initiate the polymerization of racemic 2-methylthiirane to give optically active and highly isotactic polymer ⟨80NJC95⟩. Side reactions, in addition to polymerization and oligomerization, involve attack of the sulfhydryl group liberated from the thiirane on other functional groups. Copolymerization of thiiranes with elemental sulfur has been observed. The thiolate nucleophiles may be generated from isothiocyanates, carbon disulfide and carbon oxygen sulfide in the presence of a base (Scheme 82) ⟨76RCR25⟩. Sulfur–phosphorus and acylthiol derivatives react well, and proteins with sulfhydryl groups (*e.g.* wool, hair) bind thiirane. Iminothiirane (**6**) reacts with ethanethiol possibly *via* an S_N1 process to give a quantitative yield of the thiono derivative (**58**; Scheme 83) ⟨80JOC4366⟩.

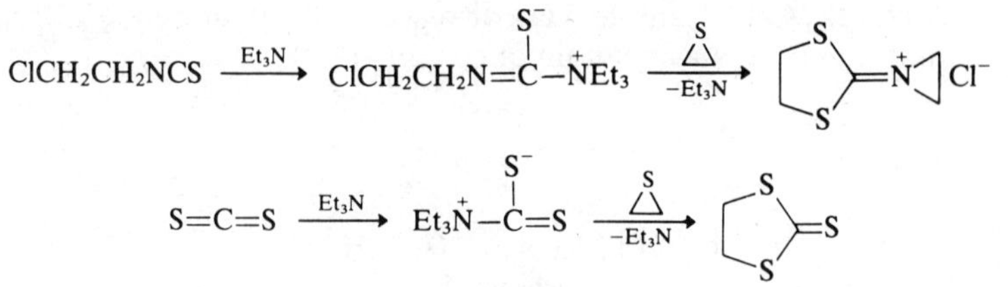

Scheme 82

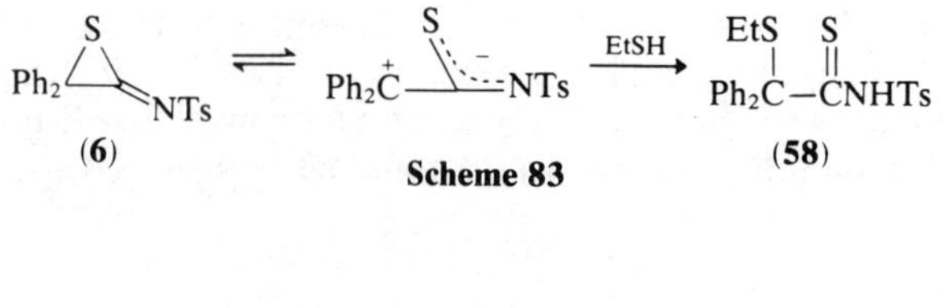

Scheme 83

Sodium or potassium hydrogen sulfite reacts with several thiiranes to give disulfides of β-mercaptosulfonic acid salts ⟨76EGP122086⟩. Potassium thiocyanate in dimethylformamide or aqueous ethanol isomerizes thiiranes (Scheme 84) ⟨72CJC3930⟩. 1,2-Dithiols are obtained by treatment of thiiranes with $NaBH_2S_3$ obtained from sodium borohydride and sulfur ⟨73TL1401⟩.

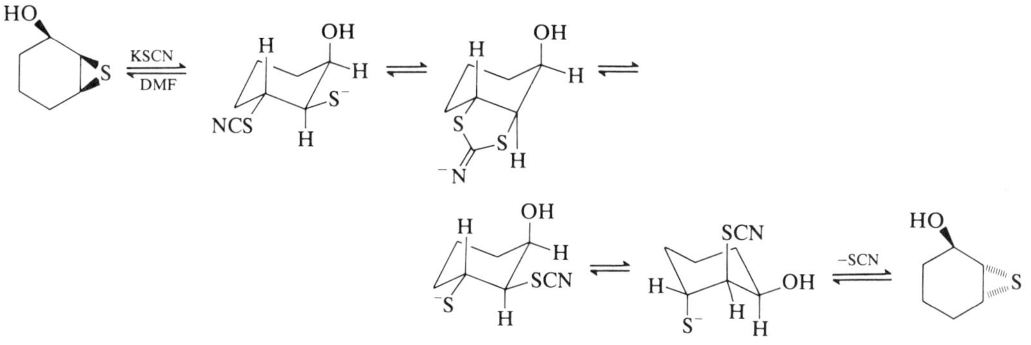

Scheme 84

Selenium derivative (**59**) desulfurizes thiiranes stereospecifically *via* nucleophilic ring opening by selenium (Scheme 85) ⟨76S200⟩. Disubstituted phosphines and their sodium salts, as well as the sodium salt of phosphine itself (R_2PH, R_2PNa, $NaPH_2$, respectively), react at carbon to give ring-opened products. Tertiary phosphines cause desulfurization (Section 5.06.3.4.4). Treatment of thiirane with a germaphosphine derivative, $Me_2Ge{=}PR'$, gives a 20% yield of 2-dimethylgerma-3-thio-1-phenylphospholane ⟨78JOM(157)C35⟩.

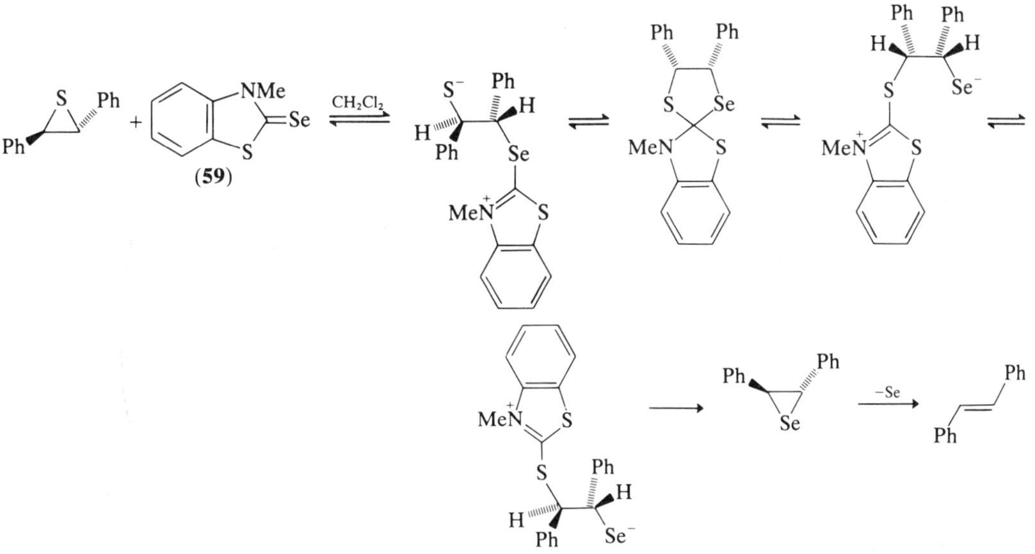

Scheme 85

Thiirane 1-oxide undergoes acid-catalyzed ring opening by ethanethiol to give ethyl 2-ethylthioethyl disulfide. Treatment of thiirane 1,1-dioxide with thiolate anions, sodium sulfide or thiourea gives β-mercaptosulfinic acid derivatives ⟨75S55⟩. Thiiranium ions are attacked at carbon by most sulfur nucleophiles ⟨79ACR282⟩, but see Section 5.06.3.4.3 for exceptions.

Addition–elimination reactions occur on treatment of 2,3-diphenylthiirene 1,1-dioxide with benzenesulfonate ion or with trisubstituted phosphines (Scheme 86) ⟨75JOC3189⟩.

5.06.3.5.5 Halide ions

The lithium chloride-catalyzed addition of 2-phenylthiirane to diphenylketene may involve attack on carbon by chloride ion followed by addition of the anion of 2-chloro-1-phenylethanethiol to the ketene, but no data about the mechanism were given ⟨69TL259⟩.

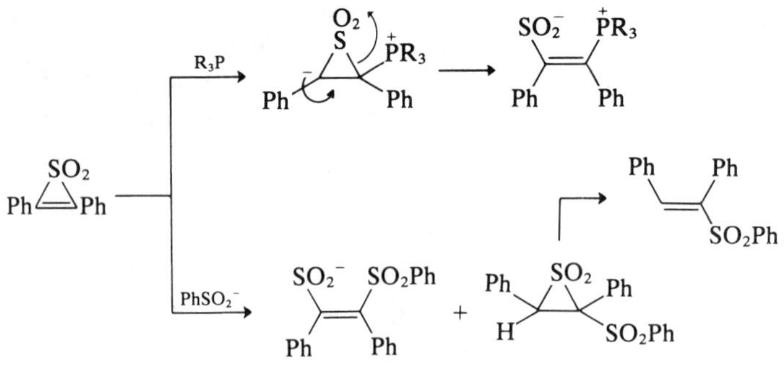

Scheme 86

The acid-catalyzed additions of bromide and chloride ion to thiiranes occurs readily, with halide preferentially but not exclusively attacking the most substituted carbon atom of the thiirane. The reaction of 1-substituted thiiranes with acetyl chloride shows a slight preference for halide attack at the less substituted carbon atom ⟨80MI50601⟩. For further discussion of electrophilic catalysis of halide ion attack see Section 5.06.3.3.2. The reaction of halogens with thiiranes involves electrophilic attack on sulfur (Section 5.06.3.3.6) followed by nucleophilic attack of halide ion on carbon.

The acid-catalyzed addition of halide ions to thiirane 1-oxide gives β-haloalkyl derivatives of sulfenic acids. Copper(II) bromide and chloride are useful electrophilic catalysts for halide addition, but other electrophilic reagents such as sulfenyl chlorides or α-chloroethers also provide β-halosulfenic acid derivatives (Scheme 87) ⟨69TL2743, 71S590⟩. Treatment of thiirane 1,1-dioxides with lithium, magnesium or zinc halides yields β-halosulfinic acid derivatives or sulfones (Scheme 88) ⟨74TL3275, 71S428⟩.

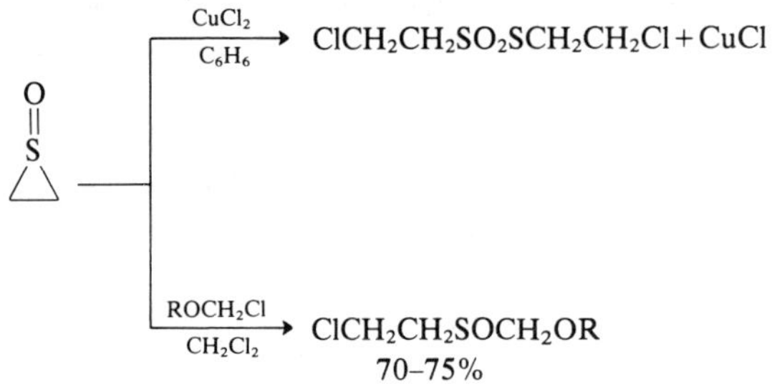

Scheme 87

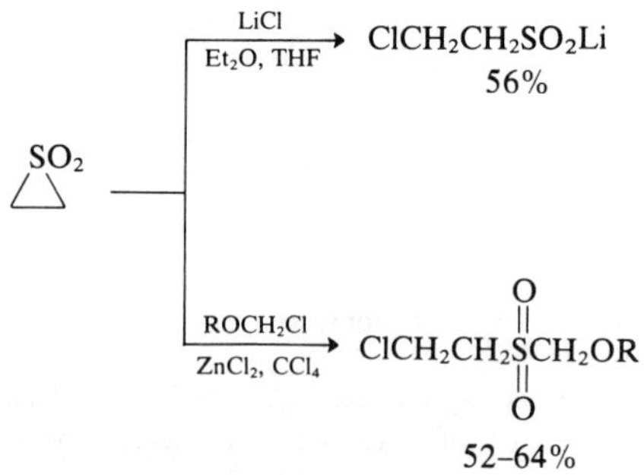

Scheme 88

Halide ions may attack *S*-substituted thiiranium ions at three sites: the sulfur atom (Section 5.06.3.4.5), a ring carbon atom or an *S*-alkyl carbon atom. In the highly sterically hindered salt (**46**) attack occurs only on sulfur (Scheme 62) or the *S*-methyl group (Scheme 89). The demethylation of (**46**) by bromide and chloride ion is the only example of attack on the carbon atom of the sulfur substituent in any thiiranium salt ⟨78CC630⟩. Iodide and fluoride ion (the latter in the presence of a crown ether) prefer to attack the sulfur atom of (**46**). *cis*-1-Methyl-2,3-di-*t*-butylthiiranium fluorosulfonate, despite being somewhat hindered, nevertheless is attacked at a ring carbon atom by chloride and bromide ions. The *trans* isomer could not be prepared; its behavior to nucleophiles is therefore unknown ⟨74JA3146⟩.

(**46**) + $Ph_3\overset{+}{P}Me\ Br^-$ → [structure] + MeBr

82%

Scheme 89

With the exception of iodide ion, halide ions attack other, less hindered thiiranium ions at a ring carbon atom to give generally *trans* or *anti* 2-halothioethers. Addition of alkyl and aryl sulfenyl halides, as well as thiocyanogen bromide and chloride, to carbon–carbon double bonds generates electrophilic intermediates represented as thiiranium halides which are involved in equilibria between sulfurane structures, intimate ion pairs, solvent-separated ion pairs, dissociated ions and open carbenium ions (Scheme 74) ⟨79ACR282, 81ACR227, B-77MI50603, B-81MI50603⟩. Rearrangement of the carbon skeleton of the thiiranium ions *via* open carbenium ions can occur in highly polar media (*e.g.* in the presence of lithium perchlorate) or when structural features are present which can stabilize carbenium ions (*e.g.* bridging groups) (Scheme 90) ⟨B-81MI50603⟩. Fluoride ion may be introduced into the products from tetrafluoroborate ions, which are usually considered to be nucleophilically inert. The regioselectivity of addition of sulfenyl halides to carbon–carbon double bonds depends on the polarity of the system. Under non-polar conditions (no free ions) chloride ion attack occurs predominantly at the least hindered position of the thiiranium ion. Under conditions where free ions may occur (addition of RS^+) attack is predominantly at the more positive carbon atom. Isomerizations of the β-halosulfides or thiiranium ions *via* open carbenium ions may occur to complicate stereochemistry especially if reaction times and temperatures are not carefully controlled, longer times and higher temperatures leading to more isomerization (Scheme 91) ⟨79ACR282⟩.

OMe, OMe →(RSCl) OMe, SR, Cl^-, $\overset{+}{O}Me$ → OMe, SR, Cl, OMe

Scheme 90

Me Me →(ArSCl, $AgSbF_6$, −50 °C) Ar, S^+ SbF_6^-, Me, Me ⇌(20 °C) Ar, S^+, Me, Me

↓HOAc ↓HOAc

ArS Me H Me H OAc — *threo*

ArS H Me Me H OAc — *erythro*

Scheme 91

The reaction of thiirenium ions with chloride ion is sensitive to steric effects, the reaction being immeasurably slow with 1-methyl-2,3-di-*t*-butylthiirenium ion. In some cases, attack on sulfur may occur to give a sulfenyl halide and an alkyne which recombine to yield ultimately the β-chloroalkenylthio ether (Scheme 92; see Section 5.06.3.4.5) ⟨79MI50600⟩. The reaction of 1-methyl-2-*t*-butyl-3-phenylthiirenium hexachloroantimonate with chloride ion gives the *anti*, regiospecific Markownikoff product (Scheme 93) ⟨81JOC4720⟩. These displacement reactions by chloride ion are believed to occur in the plane of the molecule rather than perpendicularly because of the *anti* stereospecificity. The most electrophilic of the two ring carbon atoms is the one bearing the phenyl substituent which stabilizes the charge by resonance.

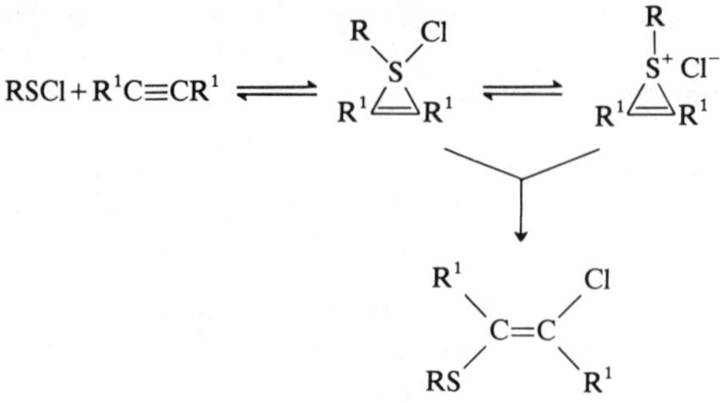

Scheme 92

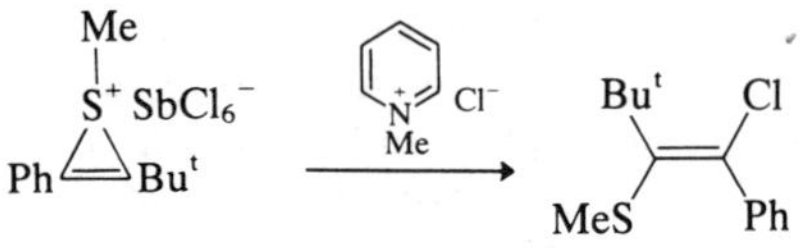

Scheme 93

5.06.3.5.6 Carbon nucleophiles

Grignard reagents attack the least hindered carbon atom of thiiranes. With allylic Grignard reagents the thiol product can be cyclized under free radical conditions (Scheme 94) ⟨78M609⟩. Treatment of thiiranes with other carbanions (*e.g. n*-butyllithium) frequently causes polymerization, but ring opening followed by cyclization has been observed with lithium enolates of esters, anions of cyanoacetic ester and malononitrile, the dianion of ethyl acetoacetate and the anion of the chromium pentacarbonyl complex of methoxy methyl carbene (Scheme 95). The lithium salts of isocyanides give normal ring-opened products, and the dimerization of selenonium ylides is catalyzed by thiiranes *via* ring-opened intermediates. The reaction of octafluoroisobutene with thiirane in the presence of fluoride ion gives products derived from ring cleavage by the tris(trifluoromethyl) carbanion.

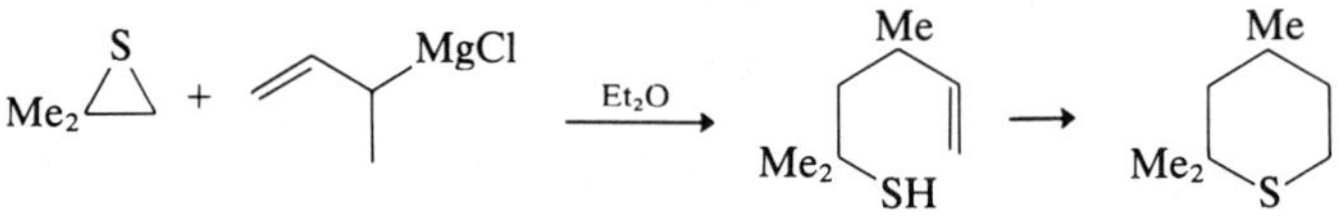

Scheme 94

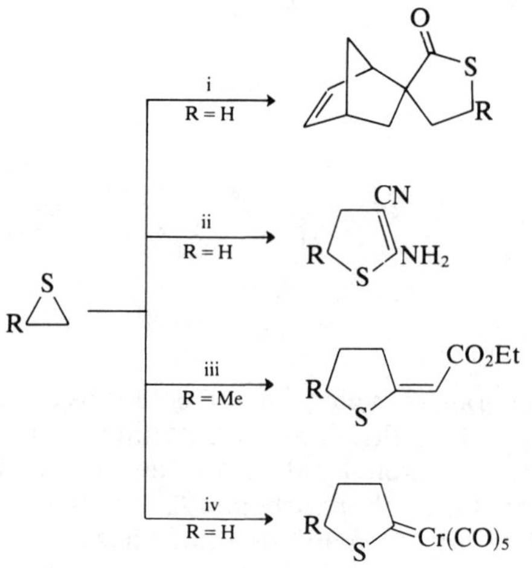

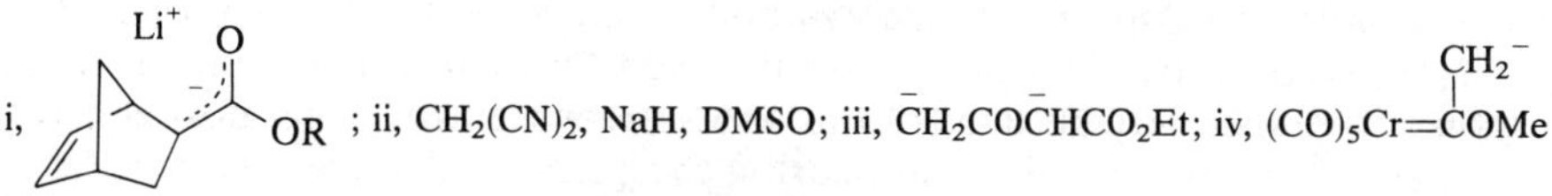

i, ; ii, $CH_2(CN)_2$, NaH, DMSO; iii, $\bar{C}H_2CO\bar{C}HCO_2Et$; iv, $(CO)_5Cr{=}C(CH_2^-)OMe$

Scheme 95

α-Metalated nitriles and enamines attack the carbon atoms of thiirene 1,1-dioxides to give products derived from ring opening followed by recyclization to five-membered cyclic sulfones (Scheme 96). A sulfur ylide of acetophenone reacts with 2,3-diarylthiirene 1,1-dioxides to give four- and six-membered cyclic sulfones by cleavage of the carbon–carbon bond in the thiirene sulfone (Scheme 97). Reaction of a pyridinium ylide of acetophenone is similar, but sulfur dioxide is lost from the intermediate adduct ⟨73BCJ667⟩. Thiiranium ions may be attacked at carbon or sulfur by the anion of diethyl malonate.

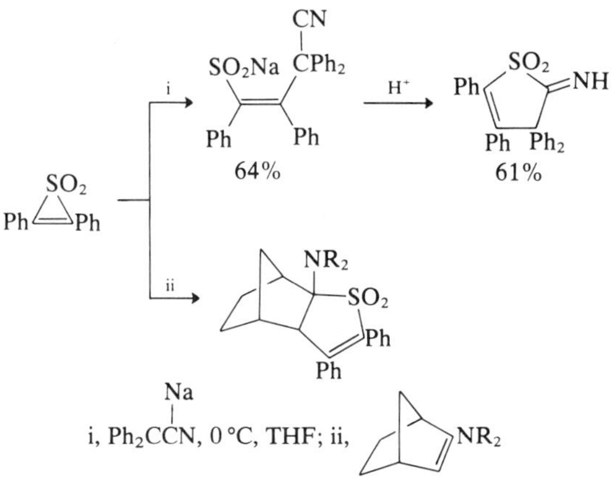

64% 61%

i, $Ph_2C(Na)CN$, 0 °C, THF; ii, NR_2

Scheme 96

$PhCO\bar{C}H\overset{+}{S}Me_2$

$-Me_2S$

13% 15%

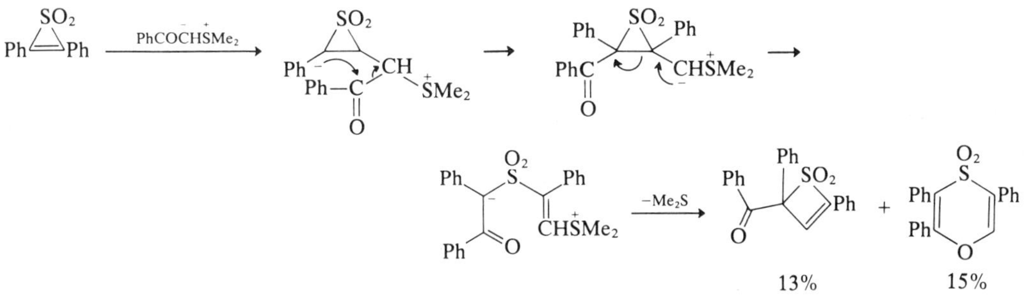

Scheme 97

5.06.3.5.7 π-Systems

The π-electron systems of anisole, mesitylene, thiophene and 2,4,6-trimethylpyridine attack the carbon atoms of S-alkylthiiranium salts (Scheme 98) ⟨81IZV1929⟩.

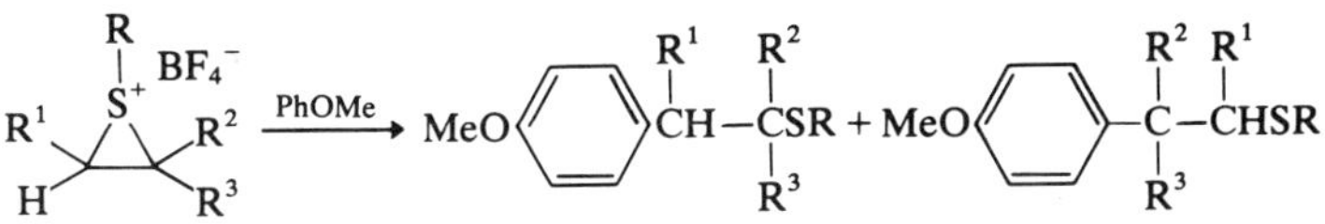

PhOMe

Scheme 98

5.06.3.5.8 Hydride and equivalents

Treatment of thiiranes with lithium aluminum hydride gives a thiolate ion formed by attack of hydride ion on the least hindered carbon atoms ⟨76RCR25⟩. The mechanism is S_N2, inversion occurring at the site of attack. Polymerization initiated by the thiolate ion is a side reaction and may even be the predominant reaction, *e.g.* with 2-phenoxymethylthiirane. Use of THF instead of ether as solvent is said to favor polymerization. Tetrahydroborates do not reduce the thiirane ring under mild conditions and can be used to reduce other functional groups in the presence of the episulfide. Sodium in ammonia reduces norbornene episulfide to the *exo* thiol.

An interesting cleavage of the carbon–carbon bond by lithium or sodium borohydride is observed with *cis*-2,3-diphenyl- and 2,2,3,3-tetraphenyl-thiirane 1,1-dioxide (Scheme 99). Lithium aluminum hydride gives at best only a very small amount of carbon–carbon bond cleavage, presumably because it forms a relatively unreactive complex with solvent (THF, ether, diglyme). At least two phenyl substituents are required for carbon–carbon bond cleavage since 2-phenylthiirane 1,1-dioxide undergoes carbon–sulfur bond cleavage (Scheme 99) ⟨68BCJ635⟩. Lithium hydride and sodium hydride are unreactive.

$R^1R^4CHSO_2CHR^2R^3$ (LiBH$_4$, THF): $R^1 = R^2 = Ph$, $R^3 = R^4 = H$; $R^1 = R^2 = R^3 = R^4 = Ph$

$R^1R^4CHC(R^2)(R^3)SO_2Li$ (LiBH$_4$, THF): $R^1 = Ph$, $R^2 = R^3 = R^4 = H$

Scheme 99

The carbon–carbon double bond of 2,3-diphenylthiirene 1,1-dioxide is reduced by aluminum amalgam in wet ether to *cis*-2,3-diphenylthiirane 1,1-dioxide (16%).

5.06.3.5.9 Others

The triphenyllead anion reacts with thiirane to give 2-mercaptoethyltriphenyllead.

5.06.3.6 Nucleophilic Attack on Hydrogen (Proton Abstraction)

The protons of thiirane 1-oxides, 1,1-dioxides and thiirene 1,1-dioxides are acidic and several examples of nucleophilic attack on hydrogens of these compounds are shown in Schemes 66 and 100. Episulfones undergo hydrogen–deuterium exchange, and the elimination of hydrogen halide from 2-halothiirane 1,1-dioxides is the final step in the synthesis of thiirene 1,1-dioxides. Episulfoxide and episulfone carbanions displace the sulfenate or sulfinate anion by a ring opening elimination. The carbanion formed by proton abstraction from thiirene 1,1-dioxides behaves similarly, forming an alkynic sulfinate ⟨71JA476⟩.

DBN, −40 °C; BuLi; NaOH; $^-O_2SC{\equiv}CMe$

Scheme 100

5.06.3.7 Reactions with Radicals

5.06.3.7.1 Radical attack on sulfur

Hydrogen atoms desulfurize thiiranes non-stereospecifically to alkenes and hydrogen sulfide; alkanes and thiols may be formed also ⟨79MI50602⟩. The reaction of ground state sulfur atoms with thiirane gives ethylene and S_2; and a similar transfer of sulfur is observed for the reaction of the radical cation of thiirane with a neutral molecule of thiirane, which gives $C_2H_4S_2^{+\cdot}$ and ethylene ⟨80MI50602⟩. The trifluoromethylthio radical attacks the sulfur atom of tetrafluorothiirane to initiate polymerization ⟨65JOC4188⟩, and polymerizations of other thiiranes are initiated by radicals, radical anions, radical cations, light, heat, γ-rays, electrolysis and an electric discharge.

Singlet carbon atoms (arc-generated at liquid nitrogen temperature), methyl radicals (from photolysis of azomethane) and phenyl radicals (from thermolysis of phenylazotriphenylmethane) effect desulfurization of thiiranes to alkenes. The reaction of *cis*-2,3-dimethylthiirane with these radicals gives predominantly *cis*-2-butene but considerable *trans*-2-butene is obtained with carbon atoms and phenyl radicals, which suggests an open carbon radical intermediate for reactions with these species. The reaction with methyl radicals may be a concerted process. When aryl radicals are generated from aryldiazonium chlorides and copper(II) chloride in the presence of thiirane, no desulfurization of the latter is observed. Instead, a ring-opened carbon radical (**60**) is apparently formed which yields β-chloroethyl aryl sulfide by reaction with the copper chloride (Scheme 101) ⟨75BAU1090⟩. Chlorine and bromine react with tetrafluorothiirane on irradiation with UV light to give products derived from the β-chloro- or -bromo-sulfenyl halides.

$$\text{thiirane} \xrightarrow[\text{CuCl}_2]{\text{ArN}_2^+\,\text{Cl}^-} \underset{(\mathbf{60})}{\text{ArSCH}_2\text{CH}_2\cdot} \begin{cases} \xrightarrow{\text{CuCl}_3^-} \text{ArSCH}_2\text{CH}_2\text{Cl} + \text{CuCl}_2^- \\ \longrightarrow \text{CH}_2{=}\text{CH}_2 + \text{ArS}\cdot \rightarrow \text{ArSSAr} \end{cases}$$

Scheme 101

Hydroxyl radicals, generated from hydrogen peroxide and titanium trichloride, add to the sulfur atom of 2-methylthiirane 1-oxide leading to the formation of propene and the radical anion of sulfur dioxide (Scheme 102) ⟨75JCS(P2)308⟩.

$$\text{2-methylthiirane 1-oxide} \xrightarrow[\text{TiCl}_3]{\text{H}_2\text{O}_2} \text{S(OH)(O}\cdot\text{) adduct} \longrightarrow \text{CH}_2{=}\text{CHMe} + \text{SO}_2^{\cdot-} + \text{H}^+$$

Scheme 102

5.06.3.7.2 Radical reactions involving ring substituents

Treatment of 2-methylthiirane with *t*-butyl hydroperoxide at 150 °C in a sealed vessel gave very low yields of allyl disulfide, 2-propenethiol and thioacetone. The allyl derivatives may be derived from abstraction of a hydrogen atom from the methyl group followed by ring opening to the allylthio radical. Percarbonate derivatives of 2-hydroxymethylthiirane decompose *via* a free radical pathway to tar. Acrylate esters of 2-hydroxymethylthiirane undergo free radical polymerization through the double bond.

5.06.3.7.3 Electrochemical reactions

Electrochemical reduction of 2,3-diphenylthiirene 1-oxide yields acetylene (80%) and benzil (10%). Electrolysis of 2,3-diphenylthiirene 1,1-dioxide in DMF gives *trans*-stilbene (30%); but in the presence of acetic acid, 1,2-diphenylvinylmethyl sulfone (27%) is obtained in addition to the stilbene (40%) ⟨81CC120⟩.

5.06.3.8 Reactions Involving Cyclic Transition States

5.06.3.8.1 Dipolar additions to thiirene 1-oxides and 1,1-dioxides

Diazoalkanes add to the carbon–carbon double bonds of 2,3-diphenylthiirene 1-oxide and 1,1-dioxide. The adducts lose SO or SO_2 to give pyrazoles and related compounds (Scheme 103) ⟨80CB1632⟩. Mesoionic oxazolones ⟨75CL1153⟩, 4-methyl-5-phenyl-1,2-dithiolene-3-thione ⟨80JOU395⟩ and pyrylium betaines ⟨72JOC3838⟩ react similarly *via* intermediate adducts (Scheme 104). Enamines (Scheme 96) and ynamines add to the double bond of 2,3-diarylthiirene 1,1-dioxides to give acyclic and cyclic sulfones by a thermal, two-step [2+2] cycloaddition mechanism. The reaction is faster than that of diphenylcyclopropenone. The initially formed cyclobutane derivative may cleave to an acyclic β-amino-α,β-unsaturated sulfone or may eliminate the amine and cleave to give a cyclic sulfone (Scheme 105) ⟨74JOC3805, 72USP3706769⟩. The enamine addition shown in Scheme 96 probably goes by a Michael addition of the enamine to the thiirene episulfone, ring opening to give a sulfinate anion, and addition of the sulfinate to the electron deficient α-carbon of the original enamine.

Scheme 103

Scheme 104

Another type of reaction involving cyclic transition states originates in ring opening of the thiirane to give a zwitterion (or possibly a diradical) which may add to carbon–oxygen, carbon–nitrogen and carbon–carbon double or triple bonds (Scheme 106) ⟨80AG(E)276⟩ or which may rearrange to larger rings (Scheme 107) ⟨80TL3579, 79JCR(S)56⟩. Ring opening of thiirane is predicted to proceed by conrotatory motions ⟨74JA5005, 73MI50601⟩. The formation of zwitterions imposes the requirement that substituents stabilize the charges as much as possible as illustrated by the thiiranimine (**47**; Scheme 106) for which several resonance structures are given. Thiiran-2-one 1,1-dioxide is a possible intermediate in the reaction of ketene with sulfur dioxide, and its reactions with the carbon–nitrogen double bond of azines, ketenimines and p-toluenesulfonyl isocyanate can be explained on the basis of the intermediate $\overset{+}{O}{\equiv}CCH_2SO_2^-$ ⟨76JOC3925⟩. However, the structure of the purported thiiran-2-

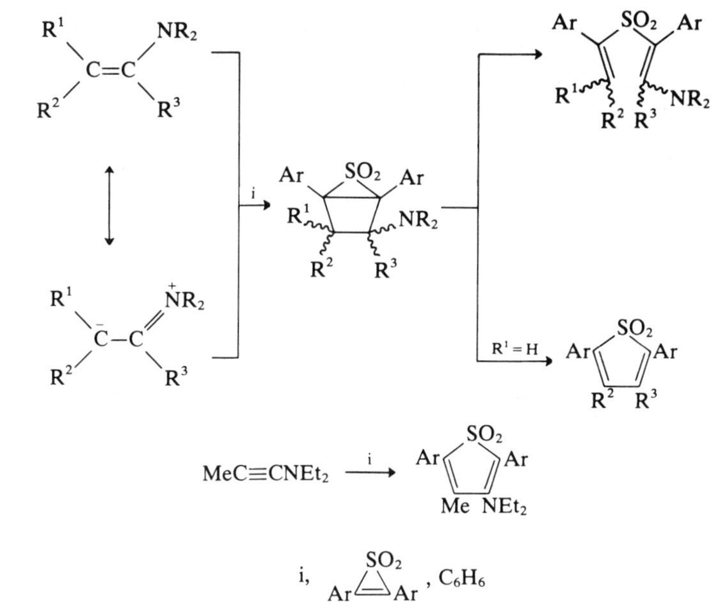

Scheme 105

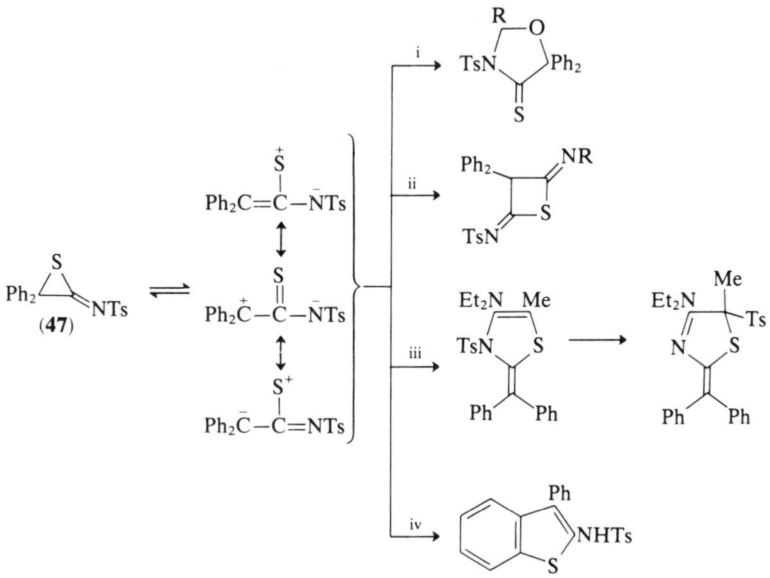

i, RCHO; ii, $R\overset{+}{N}\equiv\overset{-}{C}$; iii, $MeC\equiv CNEt_2$; iv, CCl_4, reflux

Scheme 106

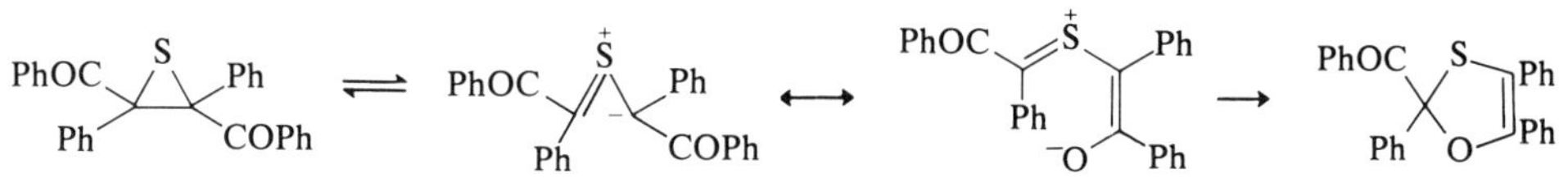

Scheme 107

one 1,1-dioxide has been questioned; the compound, isolated in a matrix of argon or nitrogen, is said to be 1,2-oxathietan-4-one 2-oxide ⟨78CC1020⟩.

The rearrangement (automerization) of Dewar thiophene *S*-oxide (**61**), observed by ^{19}F NMR, occurs so much more rapidly than that of the corresponding episulfide that special mechanisms have been invoked. The one which involves a zwitterionic intermediate (Scheme 108) is favored over a pseudopericyclic 'sulfur-walk' mechanism in which the electrons of the carbon–sulfur σ-bond and the pair of electrons on sulfur exchange places as the sulfur atom migrates around the ring ⟨80JA2861⟩.

(61)

Scheme 108

5.06.3.9 Reactions of Substituents on Thiirane Rings

5.06.3.9.1 Reactions on carbon

2-Chloromethylthiirane undergoes nucleophilic displacement reactions at the chloromethyl group by a variety of nucleophiles (water, alcohols, acetate, phosphate, phenolate, oximes, secondary amines, thiols, thiolphosphate, dithiocarbamates, thiocyanate, selenophenolate) ⟨75RCR138⟩. Ring opening is a side reaction and primary amines yield polymers. Care must be taken in identifying the products because rearrangement to 3-substituted thietanes is common (Scheme 33). The rearrangement occurs more readily in polar solvents and with 2-bromomethylthiirane. Acrylate esters of 2-hydroxymethylthiirane can be copolymerized under free radical conditions with other acrylate esters without affecting the thiirane ring. The hydroxyl group at C-17 in 2,3-epithioandrostanol can be oxidized with chromium trioxide in pyridine or acetic acid without prejudice to the episulfide; the resulting ketone may be treated with carbanionic reagents, *e.g.* methyllithium, also without harm to the episulfide. Methyl esters of thiirane-2-carboxylic acids can be hydrolyzed and the acids converted to acid chlorides and amides without ring opening. Hemiacetals of epithiosugars can be hydrolyzed with catalysis by acid ion exchange resins ⟨77CPB1140⟩. A C-17 alkyne may be hydrogenated over the Lindlar catalyst without hydrogenolysis or desulfurization of the thiirane ring ⟨69BRP1162742⟩. Lithium aluminum hydride or sodium borohydride reduces the carbonyl group of 4,4,6,6-tetramethyl-1-thiaspiro[2.3]hexan-5-one in 82% yield without reacting with the thiirane ⟨70JOC1501⟩, and photochemical reactions involving the carbonyl group also can be performed selectively as long as the wavelength of the light is 280 nm; shorter wavelengths (253.7 nm) cause desulfurization.

Treatment of several Dewar thiophenes with aryl or alkyl azides ⟨80JOC2966⟩, 2,2,2-trifluorodiazoethane and dienes (Diels–Alder) ⟨79CC881, 81CC1289, 82JA847⟩ results in addition to the carbon–carbon double bond. Generally the thiirane ring remains intact, but the reaction of 1,3-cyclohexadiene with (**62**) causes its rupture (Scheme 109) ⟨77JCS(P1)2355⟩. The azide adducts can be converted to Dewar pyrroles (**63**) ⟨80JOC2966⟩.

(62) (63)

i, $CH_2{=}CHCH{=}CH_2$; ii, (1,3-cyclohexadiene); iii, RN_3, CH_2Cl_2; iv, $h\nu$, *n*-pentane; v, Ph_3P

Scheme 109

5.06.3.9.2 Reactions on oxygen

The 17-hydroxy group of 2,3-epithiosteroids and the hydroxy groups of some epithiosugars may be acylated with acid anhydrides or chlorides without affecting the episulfide ⟨77CPB1140⟩.

5.06.4 SYNTHESIS

5.06.4.1 From Non-heterocyclic Sulfur Compounds

5.06.4.1.1 By formation of a C—S bond

Schemes 110–113 outline the most common general methods for accomplishing the synthesis of thiiranes by formation of a C—S bond ⟨75RCR138, 66CRV297, 64HC(19-1)576⟩. The methods in Schemes 111–113 are variations of Scheme 110; they differ in the details of the generation of the thiolate anion which effects the ring closure by a displacement reaction. The methods of converting oxiranes to thiiranes, to be discussed separately (Section 5.06.4.3), involve a displacement like that of Scheme 110 as the final step.

X = Hal, SCN, OSO_2Me, OAc, $OCO(CF_2)_nCF_3$, $\overset{+}{N}_2$, $\overset{+}{N}Me_3$

Scheme 110

Y	*B*	*X*	*Y*	*B*	*X*
Ac	OH^-, HCO_3^-	OAc, OTs, Cl	2,4-$(O_2N)_2C_6H_3$—	OMe^-	Cl
EtOC(=S)	OH^-	OAc			
CN	OBu^{t-}, OMe^-, OH^-	I, OSO_2Me, OTs	succinimido (N—)	H^-	Cl
$H_2NC(=NH)$—	OH^-	Br			
1-methylpyridinium-2-yl	OH^-	Br	R_3Si	Br^-	Br
1-R-tetrahydropyrimidin-4-one-2-yl	MeOH, H^+	OSO_2Me	ArS, RS	S^{2-}, NH_2^-, e^- (Al-Hg), Na_2SO_3, $PhCO_2^-$, glucose–OH^-	Cl

Scheme 111

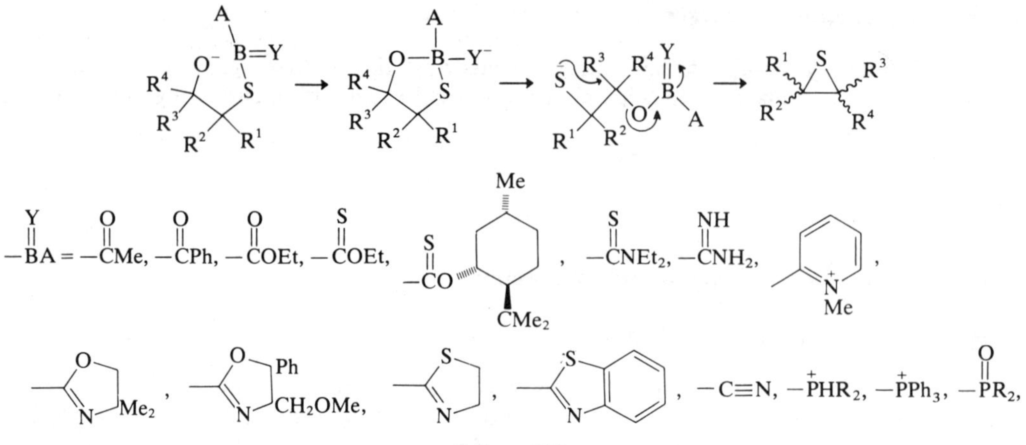

Scheme 112

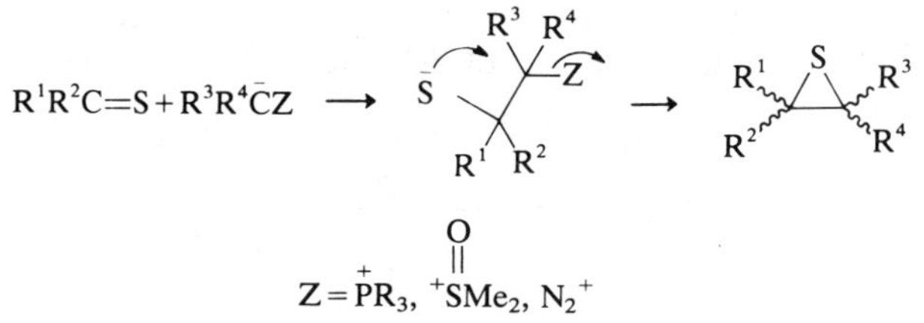

$Z = \overset{+}{P}R_3$, $^{+}\overset{O}{\overset{\|}{S}}Me_2$, N_2^{+}

Scheme 113

Treatment of 2-haloalkanethiols with bases gives thiiranes; yields are good if high concentrations of strong bases, which can cause ring opening, are avoided. 2-Haloalkanethiolate ions are intermediates in the conversion of 1,2-dihaloalkanes to thiiranes by sulfide ion. The formation of 2-vinylthiirane from *trans*-1,4-dibromo-2-butene involves an S_N2' mechanism (Scheme 114) ⟨76RTC153⟩. 2-Chloromethylthiirane can be prepared in good yield from 2,3-dichloro-1-propanethiol by treatment with aqueous sodium bicarbonate. Internal displacements by thiolate ion of acetates and methanesulfonates (mesylates), and the acid- or heat-catalyzed reactions of 2-mercaptoethyl acetates or mesylates also give thiiranes. Esters of perfluorocarboxylic acids and 2-mercaptoethanol are claimed to give better yields (triethylamine catalyst) of thiirane than the acetates ⟨80BCJ2097⟩. The acid-catalyzed (H_2SO_4, $KHSO_4$, HCl, B_2O_3, TiO_2) dehydration–cyclization of 2-mercaptoethanols to thiiranes has been used relatively infrequently. Optically active thiiranes have been obtained from cysteine derivatives (Scheme 115) ⟨79JCS(P1)1852⟩ and from the quaternary methiodide of optically active 2-dimethylamino-1-phenylpropanethiol by the internal displacement of a nitrogen-containing functional group.

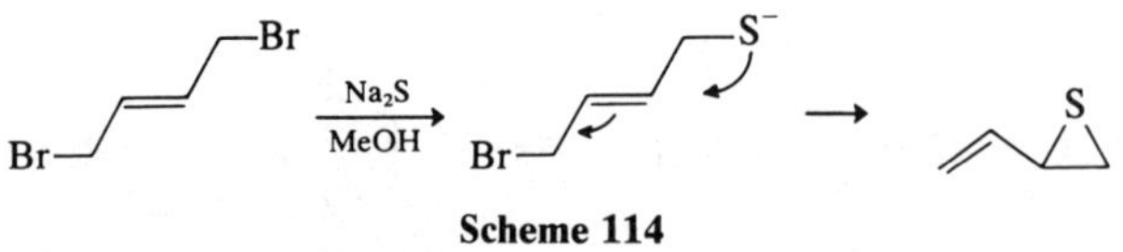

Scheme 114

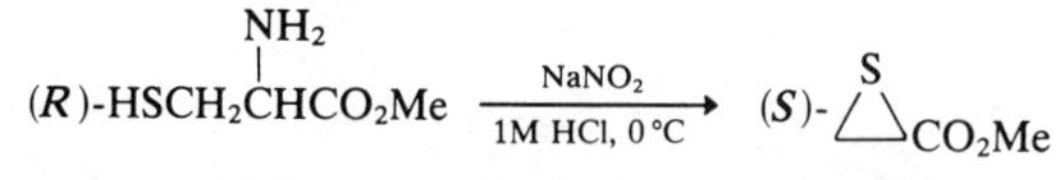

Scheme 115

Generally, good to excellent yields of thiiranes are obtained from 2-mercaptoethyl derivatives involving a masking group on the sulfur atom (Schemes 111 and 112). Examples of these methods are given in Schemes 116–123 ⟨79JCS(P1)765, 81JCS(P1)1934, 77S884, 79JCS(P1)3013, 72CL1065, 76JOC1735, 79BCJ3371, 76CC667⟩. Norbornadiene episulfides are difficult to make except by the method of Scheme 120. Chiral thiiranes of poor to fair optical purity may be obtained by the use of a chiral oxazoline (Scheme 121), a chiral solvating agent derived from (*S*)-proline (Scheme 122), or from lithiated *O*-(−)-menthyl

S-methyl dithiocarbonate ⟨76S413⟩. Stereoselective isomerization of 1,2-disubstituted alkenes may be achieved by a sequence such as the following: *trans*-alkene → bromohydrin → β-hydroxythiocyanate → *cis*-thiirane → *cis*-alkene ⟨75TL2709⟩.

I_2, $(SCN)_2$; KOBu^t, Et_2O; SCN; I; S; 57%

Scheme 116

$SiMe_3$; SCN; OSO_2Me; $LiAlH_4$, Et_2O; S^-; S; 70%

Scheme 117

Br; $+BrCH_2CHCH_2CH_2Ph$; EtOH; N, Me, S; Br; CH_2CH_2Ph; OH^-; Ph; 74%

Scheme 118

NSCl; Cl; S—N; $LiAlH_4$, −78 °C; S; 58%

Scheme 120 precursor: **Scheme 119**

p-MeC_6H_4SSCl; SSAr; Cl; Na_2S; S; 78% (*exo*:*endo* 77:23)

Scheme 120

Li; RCHS; O; N; Me_2; R^1CR^2 (C=O); H; R; S; R^1; R^2; O^-; S^-; R; R^1; R^2

Scheme 121

N; S; SCH_2Li +; CH_2; N; Me; CH_2OLi; PhCHO, −100 °C; S; Ph; 50% (20% opt. pure)

Scheme 122

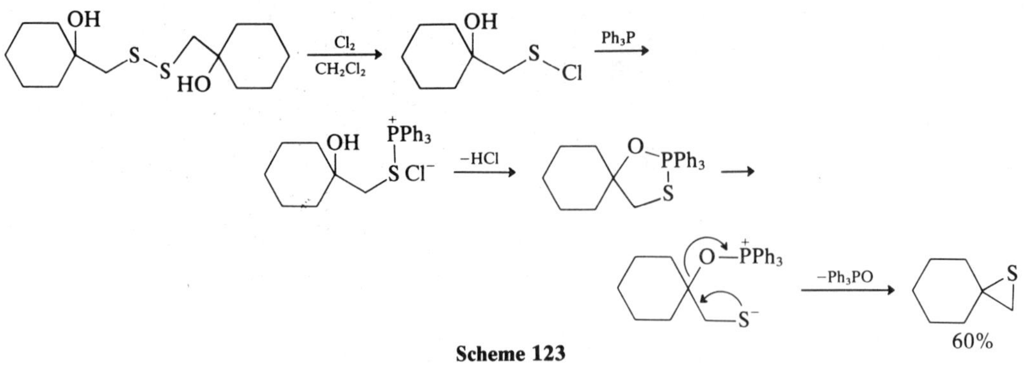

Scheme 123

The addition of anions to the carbon atom of a thiocarbonyl group generates a thiolate anion which can displace a leaving group in the β-position to give a thiirane (Scheme 113). Phosphonium and sulfonium ylides have been used (Scheme 124) ⟨73CR(C)(276)875⟩, but by far the most common reagents are diazoalkanes. The mechanism given in Scheme 113 is probably more complex for diazoalkanes ($Z = \overset{+}{N}_2$). The reaction may involve dihydrothiadiazole intermediates (some are isolated) which decompose by loss of nitrogen to form diradicals which cyclize to thiiranes, or it may proceed by carbene formation from the diazoalkane followed by addition of the carbene across the carbon–sulfur double bond. A thiophilic attack (on sulfur) may occur instead of attack on carbon. These possibilities are discussed in Sections 5.06.4.1.2, 5.06.4.2 and 5.06.4.5. Non-enolizable thiocarbonyl compounds apparently are required for successful thiirane formation with phosphonium and sulfonium ylides; and while it is true that most reactions of diazoalkanes also involve non-enolizable thiocarbonyl compounds, they are not essential ⟨79JCS(P1)1166⟩. Among the miscellaneous syntheses of thiirane derivatives are the intramolecular nucleophilic attack by carbon on the sulfur atom of a sulfenyl chloride ⟨80IZV1692⟩ and the oxygen transfer from nitrones to thioketenes (Scheme 125) ⟨80AG(E)276⟩. Electron spin resonance reveals the presence of an *S*-alkylthiirane radical cation in reactions of an alkyl vinyl sulfide with hydroxyl radicals or the chlorine radical anion ⟨81JCS(P2)1066⟩.

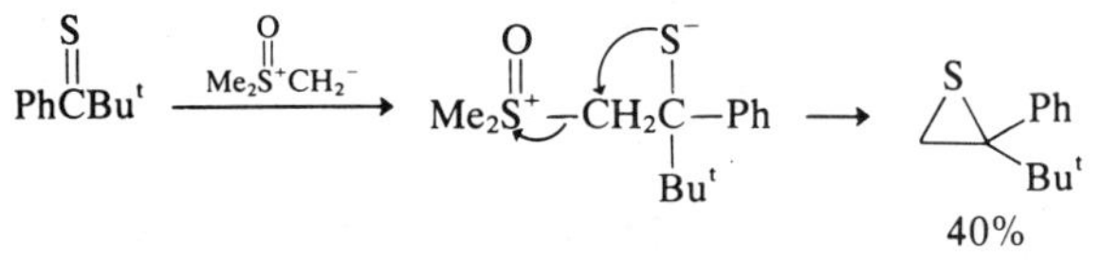

Scheme 124

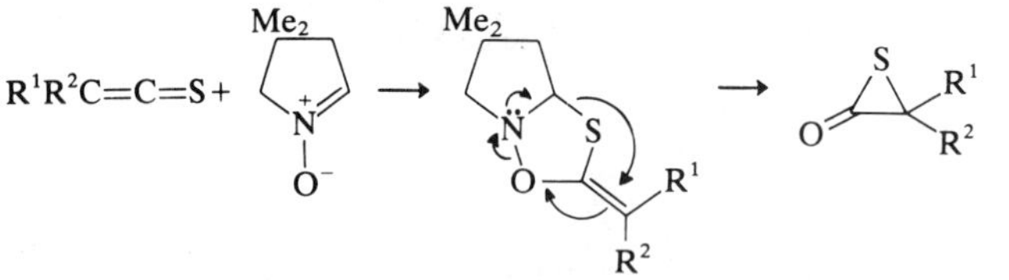

Scheme 125

The synthesis of thiiranium salts by formation of carbon–sulfur bonds can be accomplished by treatment of β-halothioethers with Lewis acids ($AgBF_4$, $AgSbF_6$, AgOTs, 2,4,6-$(NO_2)_3C_6H_2SO_3Ag$, $SbCl_5$, $AlCl_3$; Scheme 126). Both *S*-alkyl and *S*-aryl salts have been obtained ⟨81MI50603, 79ACR282⟩. Since thiiranium salts are powerful electrophiles it is important that non-nucleophilic anions and solvents (SO_2, $ClCH_2CH_2Cl$, CH_2Cl_2, $MeNO_2$, but *not* MeCN) be used with the Lewis acids. The salts generally should be handled at low temperatures and away from moist air. If the salts are hindered to nucleophilic attack they are more stable. The salt 1,2,2,3,3-pentamethylthiiranium chloride is formed from the β-chlorothioether in liquid sulfur dioxide and is observed by NMR to be in equilibrium with its acyclic precursor ⟨82JOC590⟩. Thiiranium ions are frequently invoked as intermediates in reactions of β-thioethyl derivatives ⟨B-77MI50603⟩. If a carbenoid center is generated β to an *S*-alkyl group short-lived thiiranium ylides are formed which decompose by a ring opening elimination (Scheme 127) ⟨73BCJ1539⟩.

Scheme 126

Scheme 127

The formation of 2,3-di-*t*-butyl-1-methylthiirenium chloride from *trans*-3-chloro-4-methylthio-2,2,5,5-tetramethyl-3-hexene is quantitative (by NMR) in liquid sulfur dioxide (Scheme 128) ⟨82JOC590⟩. Similar thiirenium ions are intermediates in the reactions of *β*-thiovinyl derivatives ⟨79MI50600⟩.

Scheme 128

S-Alkylation of thiiranes is usually unsuccessful in the preparation of thiiranium salts, but see Scheme 34 for an exception.

5.06.4.1.2 By formation of a C—C bond

Thiophilic addition of carbanions derived from aryl Grignard reagents or phenyllithium to diarylthioketones yields tetraarylthiiranes *via* carbon–carbon bond formation (Scheme 129) ⟨72JA597⟩. Thiocarbonyl ylides also may cyclize to thiiranes (Scheme 130) ⟨79JOC2244, 81TL3305⟩, but if the anion is conjugated with a carbonyl group an oxathiole may be formed (*cf.* Scheme 107) ⟨80TL3579⟩. The cyclobutene episulfide (**64**) is obtained by a photochemical [2+2] cyclization (Scheme 131) ⟨69JOC896⟩.

$$Ar_2C{=}S + Ar^1M \rightarrow Ar_2\bar{C}{-}S{-}Ar^1 \xrightarrow{Ar_2CS} \ldots \rightarrow$$

Scheme 129

$$Bu^t_2C{=}S{=}O + RCH_2MgX \rightarrow Bu^t_2C{=}\overset{OMgX}{S}CH_2R \xrightarrow{-OMgX^-} Bu^t_2C{=}\overset{+}{S}{-}CH_2R \xrightarrow{RCH_2MgX} Bu^t_2C{=}\overset{+}{S}{-}\bar{C}HR \rightarrow$$

Scheme 130

$$(PhCH{=}CH)_2S \xrightarrow[34\%]{h\nu}$$ (**64**)

Scheme 131

Episulfoxides are formed by the non-stereospecific addition of sulfines to diazoalkanes (Scheme 132) ⟨81MI50600⟩. Episulfones are obtained by the reaction of sulfur dioxide or sulfenes with diazoalkanes (Scheme 133) ⟨75RCR138, 70S393⟩. The reaction is usually not stereospecific, but a preference is shown for a *trans* configuration of bulky groups in the thiirane 1,1-dioxide. Side products are dihydro-1,3,4-thiadiazole 1,1-dioxides. A mechanism involving a zwitterionic intermediate analogous to that in episulfoxide formation is supported by MO calculations, and the formation of the dihydrothiadiazole 1,1-dioxides is predicted to occur by a concerted [3+2] cycloaddition ⟨81CB787⟩. More polar solvents and groups which stabilize carbanions give more episulfone and less thiadiazole sulfone. Acetonitrile was the best solvent for formation of *trans*-2,3-di-*t*-butylthiirane 1,1-dioxide from *t*-butyldiazomethane and sulfur dioxide. Treatment of phosphorus ylides having no α-hydrogens with sulfenes yields episulfones or their decomposition products ⟨67T2137⟩.

$$R^1R^2CN_2 + R^3R^4C{=}S{=}O \rightarrow R^3R^4\bar{C}{-}S(O){-}CR^1R^2(\overset{+}{N}_2) \rightarrow$$

Scheme 132

i, SO_2; ii, $R^1R^2CN_2$; iii, $R^3R^4C{=}SO_2$

Scheme 133

The Ramberg–Bäcklund reaction ⟨B-77MI50600⟩ of α-halosulfones with bases goes *via* an episulfone intermediate (*e.g.* Scheme 11). The reaction with α,α'-dihalosulfoxides or α,α'-dihalosulfones gives good yields of thiirene 1-oxides and thiirene 1,1-dioxides (Scheme 134) ⟨79JA390, 70OS(50)65⟩. A variation is the reaction of an $\alpha,\alpha,\alpha',\alpha'$-tetrabromosulfone with phosphines which also yields thiirene 1,1-dioxides ⟨74JOC2320⟩.

$(PhCHBr)_2SO_n$, $n = 1, 2$ $\xrightarrow[CH_2Cl_2]{Et_3N}$ [episulfone intermediate] $\xrightarrow{Et_3N}$ thiirene SO_n

Scheme 134

Thiirane 1,1-dioxides substituted with highly electron withdrawing substituents are predicted to be unstable ⟨73JA7644⟩.

5.06.4.2 Formation By Making Two Bonds

Sulfur atoms (from photolysis of COS) in both singlet ($^1D_2$) and triplet (^3P_j) states react in the vapor phase with alkenes to give thiiranes ⟨79JA3000⟩, generally not a very practical method of synthesis. Triplet sulfur atoms add stereospecifically unlike triplet carbenes, nitrenes and oxygen atoms ⟨72MI50602⟩. Thiirene was suggested as an intermediate in the reaction of sulfur atoms (^1D) with acetylene ⟨80PAC1623⟩.

Heating or irradiating alkenes in the presence of sulfur gives relatively low yields of thiiranes. For example, a mixture of sulfur and norbornadiene in pyridine–DMF–NH_3 at 110 °C gave a 19% yield of the monoepisulfide of norbornadiene as compared with a 78% yield by the method of Scheme 120 ⟨79JCS(P1)228⟩. Often 1,2,3-trithiolanes are formed instead of thiiranes. The sesquiterpene episulfides in the essential oil of hops were prepared conveniently by irradiation of the terpene and sulfur in cyclohexane (Scheme 135) ⟨80JCS(P1)311⟩. Phenyl, methyl or allyl isothiocyanate may be used as a source of sulfur atoms instead of elemental sulfur.

The reaction of thiocarbonyl compounds with diazoalkanes (alkyl, aryl substituted) frequently gives good to excellent yields of thiiranes. The mechanism may involve addition of a carbene across the thiocarbonyl group, especially in the presence of rhodium(II) acetate

$\xrightarrow[h\nu]{S_8}$ S + S

20% 5%

Scheme 135

⟨80JOC1481⟩, but mechanisms involving a zwitterion (Scheme 113) or a dihydrothiadiazole also are possible ⟨75RCR138⟩. The thiocarbonyl compound may be generated *in situ* by reaction of the diazoalkane with elemental sulfur or by treating ketone hydrazones with mercury(II) oxide, elemental sulfur and a catalytic amount of potassium hydroxide. In both these cases the thiocarbonyl compound produced reacts with the carbene derived from the diazoalkane or hydrazone. Scheme 136 illustrates the general reaction of diazoalkanes with thiocarbonyl compounds, a number of which are indicated in the scheme. Phenyl(trihalomethyl)mercury compounds (halogen = Cl, Br) are source of dihalocarbenes or carbenoids which react with elemental sulfur or thiocarbonyl compounds to give thiiranes (Scheme 137). Azibenzil and diarylthioketones give 1,3-oxathioles instead of thiiranes ⟨78JOC3730⟩. The reaction of diazoalkanes with sulfines, sulfenes and sulfur dioxide has been discussed previously (Section 5.06.4.1.2).

$R^1R^2CN_2 \xrightarrow{S_8} R^1R^2C{=}S \xrightarrow{R^1R^2CN_2}$ thiirane (R^1, R^2, R^1, R^2)

$R^1R^2CN_2 \longrightarrow R^1R^2C\!: \xrightarrow{XC(=S)Y}$ thiirane (R^1, R^2, X, Y)

X	*Y*	*X*	*Y*
H, R, Ar	H, R, Ar		
alkenyl	R, Ar	ArS	SAr
Me_3Si	Ph		
Cl	Cl	RO	SC(=S)OR
R, Ar	OR′	$ArSO_2$	SPh
R, Ar	SR′	RCON(Ar)	CN
		$ArSO_2{=}N$	

Scheme 136

$$PhHgCCl_2Br + Ph_2C{=}S \xrightarrow[C_6H_6,\ N_2]{70\,^{\circ}C} Ph_2\overset{S}{\triangle}Cl_2 \quad 75\%$$

Scheme 137

Thiiranium or thiirenium ions formally can be derived by addition of RS^+ in a non-nucleophilic solvent across a carbon–carbon double bond or triple bond respectively ⟨B-81MI50602⟩. Thiiranium ions have long been postulated as intermediates in the addition of sulfenyl halides to alkenes ⟨81ACR227, B-77MI50603⟩, but recent evidence indicates that a variety of intermediates (sulfuranes, thiiranium ion pairs, dissociated thiiranium salts and open carbocations) must be considered ⟨79ACR282⟩. If the counter-anion of the sulfenyl cation is non-nucleophilic, *i.e.* not halide, the additions to double and triple bonds yield relatively stable ions (Scheme 138) ⟨79ACR282, 79MI50600⟩. Episulfonium ions are powerful electrophiles and are known to remove fluoride ion from tetrafluoroborate. Stable chloride or tribromide salts were obtained only from the hindered adamantylideneadamantane and methanesulfenyl chloride or bromide ⟨78CC630⟩ and from di-*t*-butylacetylene and methanesulfenyl chloride. The latter salt can be converted to its tetrafluoroborate by treatment with silver tetrafluoroborate ⟨77TL911⟩. The additions to carbon–carbon double bonds are stereospecific.

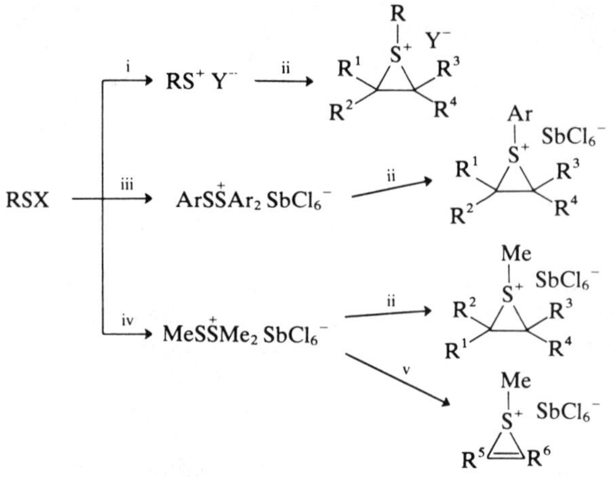

i, AgY (Y = BF_4^-, SbF_6^-, 2,4,6-$(NO_2)_3C_6H_2SO_3^-$, ClO_4^-), R = alkyl, aryl, X = Cl, Br; ii, $R^1R^2C{=}CR^3R^4$; iii, $SbCl_5$; R = aryl, X = Cl; iv, Me_2S, $SbCl_5$ or SbF_5, R = Me; v, $R^5C{\equiv}CR^6$

Scheme 138

5.06.4.3 Formation from Oxiranes

One of the most convenient methods for the preparation of thiiranes is the reaction of oxiranes with thiocyanate ion or thiourea (Schemes 139 and 140) ⟨75RCR138, 66CRV297, B-66MI50600⟩. Reactions of 2,3-disubstituted oxiiranes are often slow (but yields are good); thiocyanic acid (HSCN) is effective in acid catalysis of ring opening, especially of steroid epoxides. However, if thiocyanic acid or acidified thiourea is used the synthesis of thiiranes must be completed by treatment of the *trans*-2-hydroxythiocyanates or thiuronium salts with base (*e.g.* Scheme 141) ⟨75AG(E)252⟩. Tri- and tetra-substituted oxiranes generally are not satisfactory. The reactions of epoxides fused to five-membered rings are also sluggish and a two-step procedure *via* the 2-hydroxythiocyanate is used, particularly in steroid and carbohydrate systems. The synthesis of thiirane carboxylic acid derivatives (thioglycidic acids) from the corresponding oxiranes failed with thiocyanate, thiourea and thiolacetic acid ⟨79JCS(P1)1852⟩. The mechanisms of Schemes 139 and 140 show that the *trans* (or *cis*) stereochemistry of the oxirane is preserved in the thiirane and that optically active (*R,R*)-oxiranes should yield optically active (*S,S*)-thiiranes, which has been demonstrated experimentally ⟨75MI50600⟩.

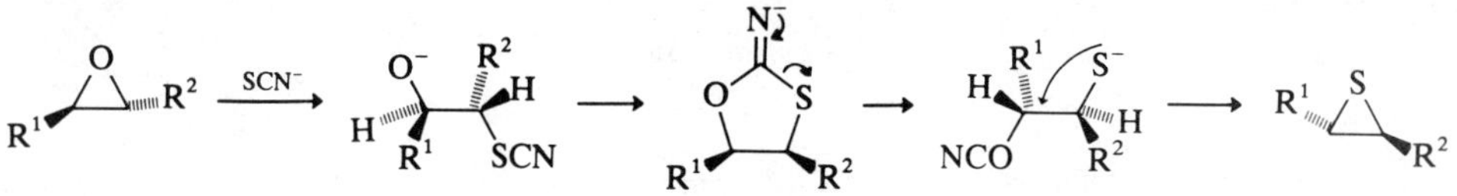

Scheme 139

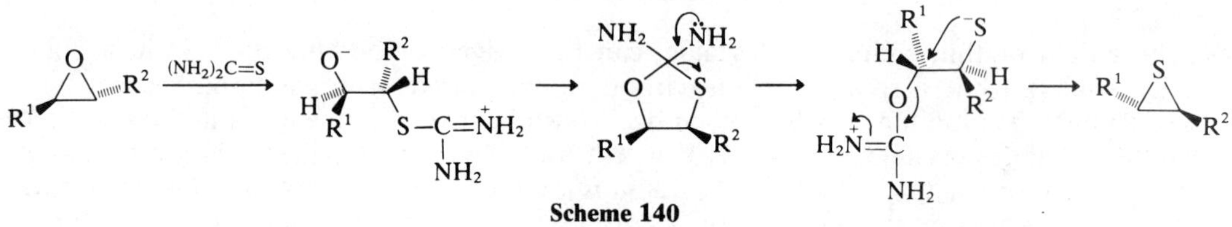

Scheme 140

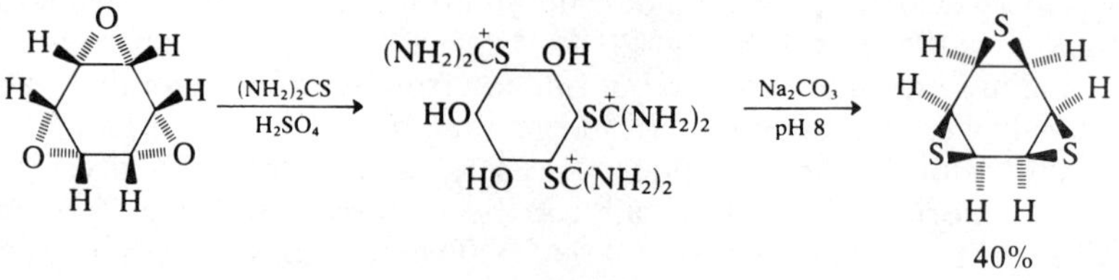

40%

Scheme 141

The reactions of oxiranes with thiocyanate ion or with thiourea are usually done in homogeneous solution in water, alcohols or alcohol–acetic acid. The use of silica gel as a support for potassium thiocyanate in toluene solvent is advantageous for the simple work-up (filtration and evaporation of solvent) ⟨80JOC4254⟩. A crown ether has been used to catalyze reactions of potassium thiocyanate.

The acid-catalyzed reaction of oxiranes with triphenylphosphine sulfide gives thiiranes of opposite configuration like the reactions with thiocyanate and thiourea (Scheme 142) ⟨73IJS(8)45⟩. Derivatives of phosphoro-mono- and -di-thioic acid also convert oxiranes to thiiranes. Treatment of oxiranes with a variety of thiocarbonyl compounds results in exchange of oxygen for sulfur. These compounds include carbon disulfide, carbon oxysulfide, thiocarbonyl difluoride ⟨65JOC4188⟩, pyrrolidine-2-thione, 1-thiazolidine-2-thiones ⟨81JCS(P1)52⟩, thioacetone and potassium dithiocarbonates ($ROCS_2K$). Fluorinated thiiranes are prepared from oxiranes and fluorinated thiocarbonyl compounds such as thiocarbonyl difluoride. 2-Triphenylsilylthiirane is obtained by treatment of 2-triphenylsilyloxirane with 3-methylbenzothiazole-2-thione. Treatment of oxiranes with thiolacetic acid gives 2-hydroxy-*S*-acetyl derivatives which are converted to thiiranes by treatment with base. Oxiranes are also converted to thiiranes by treatment with the sodium salt of 1-phenyl-5-mercaptotetrazole.

Ph_3PS, CF_3CO_2H, C_6H_6; $-Ph_3PO$

Scheme 142

5.06.4.4 Formation from Four-membered Heterocycles

A few special syntheses of thiirane derivatives from four-membered sulfur heterocycles are known. Methylenethiiranes are derived by thermolysis of the tosylhydrazones of thietan-3-ones (Scheme 143) ⟨81TL4815⟩, and 2,2,3,3-tetrakis(trifluoromethyl)thiirane is obtained by thermolysis of 2,2,4,4-tetrakis(trifluoromethyl)-1,3-dithietane 1,1-dioxide with loss of sulfur dioxide. 2,2,3,3-Tetrakis(trifluoromethyl)thiirane 1,1-dioxide is obtained in low yield by heating a mixture of 3,3-bis(trifluoromethyl)-1,2,4-oxadithietane 2,2,4,4-tetroxide and bis(trifluoromethyl)ketene. The mono-episulfide of norbornadiene is obtained from thietane derivative (**65**; Scheme 144) ⟨73JOC649⟩.

130–160 °C, 10^{-4} Torr; 60–70%

Scheme 143

NaCN

(**65**)

Scheme 144

5.06.4.5 Formation from Five-membered Heterocycles

Treatment of cyclic carbonates of 1,2-diols with thiocyanate ion at temperatures of 100 °C or higher yields thiiranes (Scheme 145) ⟨66CRV297, 75RCR138⟩. Thiourea cannot replace thiocyanate satisfactorily, and yields decrease as the carbonate becomes more sterically hindered. The reaction mechanism is similar to the reaction of oxiranes with thiocyanate (Scheme 139). As Scheme 145 shows, chiral thiiranes can be derived from chiral 1,2-diols ⟨77T999, 75MI50600⟩.

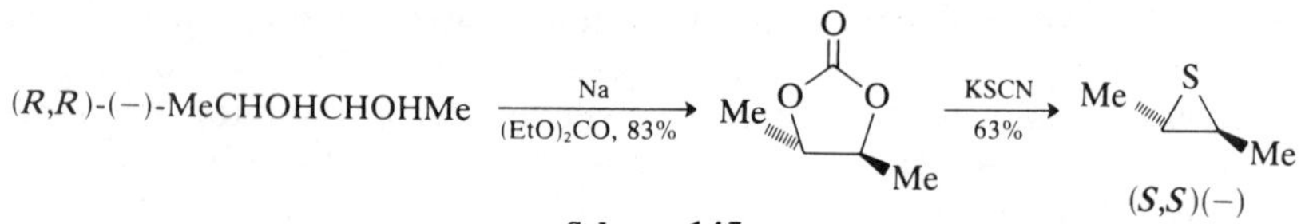

Scheme 145

Heating cyclic mono-, di- or tri-thiocarbonates, usually in the presence of base, gives thiiranes and carbon dioxide, carbon oxysulfide or carbon disulfide respectively. The methiodide salts of 2-methylimino-1,3-oxathiolanes are converted to thiiranes with high stereoselectivity, except for 5-aryl-substituted oxathiolanes (Scheme 146) ⟨80LA1779⟩. Flash vacuum thermolysis of 1,3-oxathiolan-5-ones causes loss of carbon dioxide and nearly quantitative formation of thiiranes of inverted configuration (Scheme 147) ⟨80JA744⟩. For example, thermolysis of *cis*-2,4-diphenyl-1,3-oxathiolan-5-one gives *trans*-2,4-diphenylthiirane.

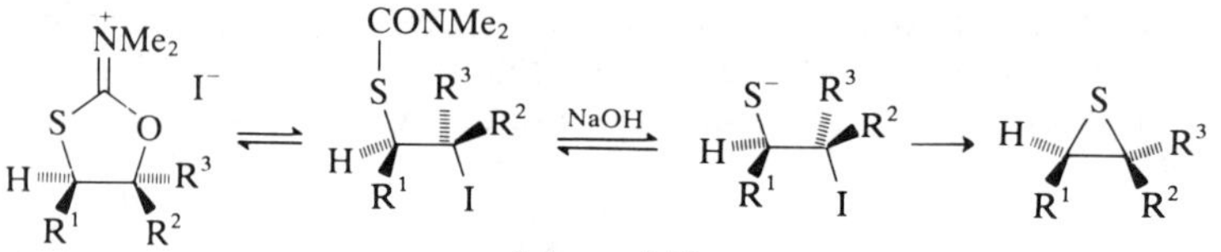

Scheme 146

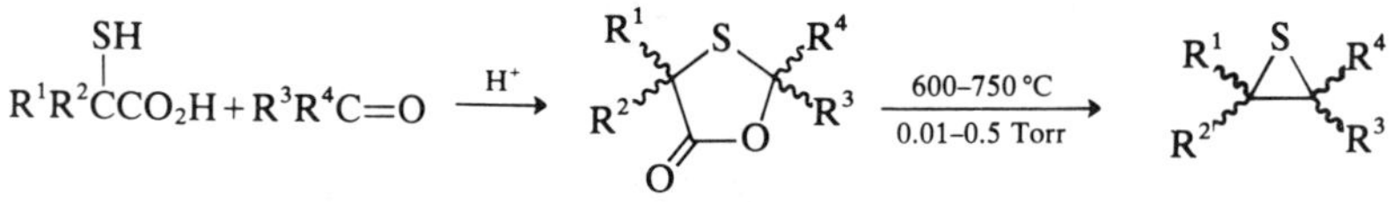

Scheme 147

The reactions of thiocarbonyl compounds with diazoalkanes may proceed *via* 2,5-dihydro-1,3,4- or possibly 4,5-dihydro-1,2,3-thiadiazoles, since these are known to decompose with loss of nitrogen to give thiiranes ⟨75RCR138⟩. Several examples are given in Scheme 148 ⟨80JOC1481, 75JOC2212⟩. Thiocarbonyl ylides are believed to be intermediates in the conversion of 2,5-dihydro-1,3,4-thiadiazoles to thiiranes, a conrotatory ring closure being indicated ⟨76T1641⟩. The 2-phenylimino-2,5-dihydro-1,3,4–thiazole (**66**) gives quantitatively a spirothiirane derivative (**67**) on treatment with diphenylketene (Scheme 149) ⟨81S813⟩. The *S*-oxide of 2,2-di-*t*-butyl-3-(1-methylethylidene)thiirane is obtained directly in 13% yield by irradiation of 3,3-di-*t*-butyl-5,5-dimethyl-1-pyrazoline-4-thione *S*-oxide ⟨81T219⟩. Thermolysis or photolysis of 2,5-dihydro-1,3,4-thiadiazole 1-oxides and 1,1-dioxides, obtained by addition of diazoalkanes to sulfines and sulfenes respectively, is not a satisfactory method of generating thiirane 1-oxides and thiirane 1,1-dioxides. The thiadiazole 1-oxides and 1,1-dioxides either revert to the sulfine or sulfene and diazoalkane, or lose sulfur monoxide or sulfur dioxide to give azines ⟨81CB802⟩. At high temperatures some alkene is obtained from thiadiazole oxides and dioxides suggesting the possibility of thiirane sulfoxide or sulfone formation, probably *via* the dissociated sulfine or sulfene and diazoalkane.

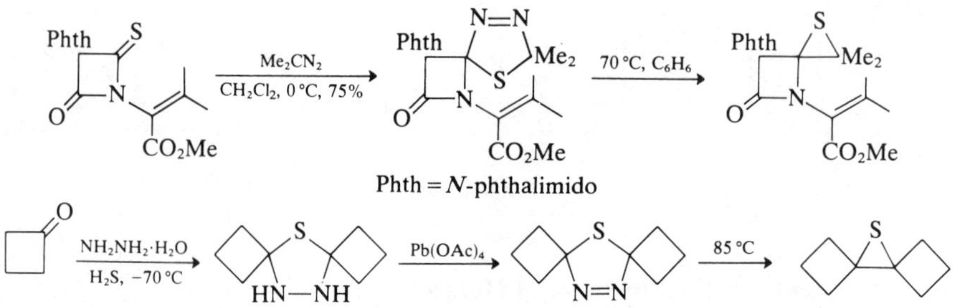

Phth = *N*-phthalimido

Scheme 148

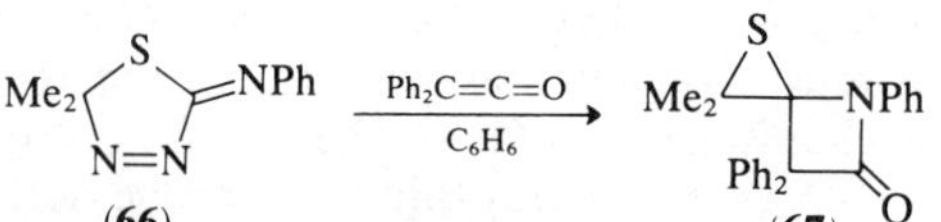

Scheme 149

The stable Dewar thiophene (**68**) is obtained by irradiation of 2,3,4,5-tetrakis(trifluoromethyl)thiophene (Scheme 150) ⟨72CJC2721⟩. Dewar thiophenes are proposed as intermediates in the photochemical isomerizations of substituted thiophenes (Scheme 17).

(**68**)

Scheme 150

Thiirenes have been isolated in argon matrices at 8 K by photolysis of 1,2,3-thiadiazoles or vinylene trithiocarbonates (Scheme 151) ⟨80PAC1623, 81JA486⟩. They are highly reactive and decompose to thioketenes and alkynes (Scheme 22). Electron withdrawing substituents stabilize thiirenes somewhat, but no known thiirene is stable at room temperature unlike the relatively stable thiirene 1-oxides and thiirene 1,1-dioxides.

Scheme 151

5.06.4.6 Formation from Six-membered Heterocycles

Thermolysis of trithiane (**69**) or carbonate (**70**) at reduced pressure yields methylenethiirane which is stable in cold, dilute solution (Scheme 152) ⟨78JA7436, 78RTC214⟩. A novel acenaphthylene episulfide is obtained by treatment of the six-membered sulfoxide (**71**) with acetic anhydride (Scheme 153) ⟨68JA1676⟩, and photolysis of (**72**) gives a low yield of episulfide (**73**; Scheme 154) ⟨72JA521⟩. Low yields may be due to the desulfurization of the thiiranes under the reaction conditions.

(**69**) 600–700 °C; 495 °C, 0.5 Torr, 90% (**70**)

Scheme 152

(**71**) Ac_2O, 100 °C 40%

Scheme 153

(**72**) $h\nu$, pentane 28% (**73**)

Scheme 154

5.06.4.7 Miscellaneous Methods

Optically active thiiranes have been obtained by resolution of racemic mixtures by chiral tri-*o*-thymotide. The dextrorotatory thymotide prefers the (*S*,*S*)-enantiomer of 2,3-dimethylthiirane which forms a 2:1 host:guest complex. A 30% enantiomeric excess of (*S*,*S*)-(−)-2,3-dimethylthiirane is obtained ⟨80JA1157⟩.

Methylene thiirane is obtained by thermolysis of several spirothiirane derivatives which are formally Diels–Alder adducts of methylenethiirane and cyclopentadiene or anthracene ⟨78JA7436⟩. They were prepared *via* lithio-2-(methylthio)-1,3-oxazolines (*cf.* Scheme 121). A novel synthesis of the allene episulfide derivatives, 2-isopropylidene-3,3-dimethylthiirane (good yield) or its *S*-oxide (poor yield), involves irradiation of 2,2,3,3-tetramethylcyclopropanethione or its *S*-oxide ⟨81AG293⟩. Substituents on the thiirane ring may be modified to give new thiiranes (Section 5.06.3.9). The synthesis of thiirane 1-oxides and thiirane 1,1-dioxides by oxidation is discussed in Section 5.06.3.3.8; and the synthesis of *S*-alkylthiiranium salts by alkylation of thiiranes is discussed in Section 5.06.3.3.4. Thiirene 1-oxides and 1,1-dioxides may be obtained by dehydrohalogenation of 2-halothiirane 1-oxides and 1,1-dioxides (Section 5.06.4.1.2).

5.06.5 APPLICATIONS AND IMPORTANT COMPOUNDS

5.06.5.1 Naturally Occurring Thiiranes

Terpene episulfides (derived from caryophyllene and humulene) are found in the essential oil of hops (*cf.* Scheme 135) ⟨80JCS(P1)311⟩. Several thiirane nitriles, *e.g.* 2-cyanomethylthiirane, are among the hydrolysis products of rapeseed glucosinolates and other seeds of the *Cruciferae* family and are found also in cabbage and rutabagas. One of these, 2-(2-cyanoethyl)thiirane, was nephrotoxic and caused embryonal death and decreased fetal weight in laboratory animals. Thiirane and 2-methylthiirane have been detected by GC–MS in the aroma of canned beef. Thiirane is in the aroma of cooking mutton and is formed in the browning of foods. Thiirane carboxylic acid is found in white asparagus. Some of these thiiranes may be formed by degradation of the sulfur-containing amino acids, cysteine, cystine and methionine.

5.06.5.2 Polymers

Poly(thiirane) is a crystalline thermoplastic which can be injection molded. Its trade name is 'Thiolon' and it is suggested for applications which require good solvent resistance and a high modulus of elasticity ⟨B-77MI50601, 72MI50601, 71MI50600⟩. A variety of polymers and copolymers of thiiranes have been investigated. The polymer obtained by treating 2,2-dimethylthiirane with *p*-vinylaniline is a scavenger of silver in photographic film, and copolymers of acrylic esters of 2-hydroxymethylthiirane with other acrylates adsorb mercury(II) ions. A copolymer of cyclohexene episulfide and cyclohexadiene disulfide is a reuseable carrier for diborane which can be stored at ambient conditions under argon ⟨75USP3928293⟩, and a graft copolymer of thiirane and poly(ethyleneimine) is said to be a useful corrosion inhibitor for iron. Thiirane-2,3-dicarboxylic acid esters and bis-thiiranes have been used as rubber plasticizers, and other thiiranes have been used as curing agents and as oxidation inhibitors for epoxy resins. Adducts of 2-methylthiirane and ether derivatives of 2-hydroxyethylamine are said to be useful in treating woolens to render them 'permanent press'.

5.06.5.3 Drugs

Epithioandrostanol derivatives, *e.g.* 'Mepitiostane' (**74a**) and 'Epitiostanol' or 'Thiodrol' (**74b**), are antitumor drugs effective against breast cancer ⟨80MI50603, 78MI50601⟩. Cyclohexene episulfide has been investigated as an inhibitor of L-glutathione transferase and aryl hydrocarbon hydroxylase ⟨78MI50602⟩. An epithiocardenolide (**75**) is said to increase blood pressure and to be useful as a respiratory stimulant ⟨68JAP6809058⟩ and thiiranes (**76**) are claimed to be hypoglycemic agents ⟨80USP4196300⟩.

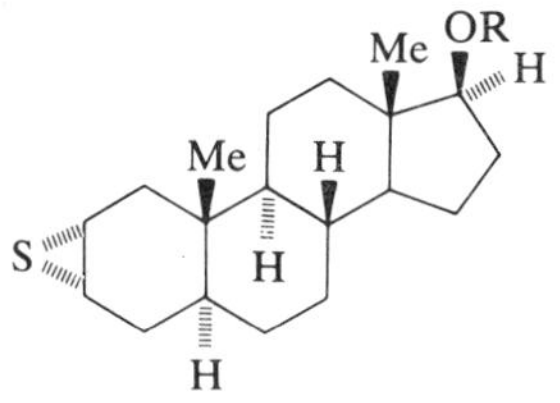

(74) a; R = H

b, R = 1-methoxycyclopentyl

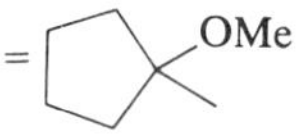

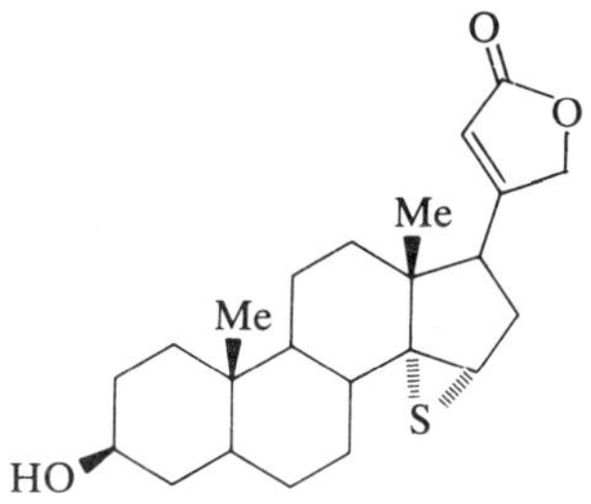

(75)

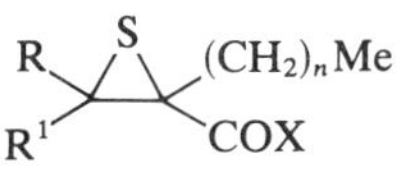

(76)

R = R[1] = H, alkyl
X = OH, OR, NH_2, NHR, NR_2, $NHCH_2OH$
$n = 10–15$

Thiirane is more bactericidal than oxirane, and derivatives of 2-mercaptomethylthiirane inhibit tuberculosis. The following pharmacological uses have been reported for compounds derived from thiirane derivatives: gold complexes of the adducts of diethylphosphine and thiirane (antiarthritic), adducts of thiiranes and malononitrile (antibacterial, blood vessel dilators, muscle relaxants, sedatives), thermolysis products of thiirane 1-oxides and adducts of thiirane 1-oxides with sulfenyl chlorides (antibacterial), adducts of 2,3-diarylthiirene 1,1-dioxides with ynamines (antibacterial, parasiticidal), adducts of 2,3-diarylthiirene 1,1-dioxides with enamines (antifertility), adducts of *p*-aminophenylacetic esters with thiirane (immunosuppressants), adducts of amines and thiiranes (radioprotective drugs).

5.06.5.4 Toxicity

The carcinogenic activity of thiirane is less than that of oxetane; only a few tumors were observed in rats subjected to subcutaneous injections once a week ⟨70MI50600⟩. Inhalation of thiiranes by rats produces respiratory irritation, death occurring from pulmonary edema and depression of the central nervous system. The LD_{50} values for a 30 minute survival time were 4000 p.p.m. for thiirane, 5000 p.p.m. for 2-methylthiirane and 750 p.p.m. for 2-chloromethylthiirane ⟨64MI50600⟩. The oral LD_{50} values for these three thiiranes are 178, 254 and 68 mg kg^{-1} respectively; and the intraperitoneal LD_{50} values are 42, 44 and 41 mg kg^{-1} respectively. Thus, these thiiranes show moderately acute toxicity to the rat by inhalation and ingestion. Skin and eye contact with thiiranes should be avoided and they should be handled in good hoods. A Russian study gives the LD_{50} of thiirane for mice as 35.6 mg kg^{-1} and a mean lethal concentration in the atmosphere of 1400 mg m^{-3}. The danger of poisoning by absorption through the skin is stressed, and a maximum concentration of 0.1 mg m^{-3} is recommended for an industrial atmosphere ⟨69MI50600⟩. A slightly more recent study confirms that 2-chloromethylthiirane is about twice as toxic as thiirane and 2-methylthiirane ⟨71MI50601⟩.

5.06.5.5 Insecticides and Herbicides

Thiiranes (**77**) show some juvenile hormone activity, but the epoxide is often more active. Thiophosphates of 2-mercaptomethylthiirane are strong contact insecticides; 2-chloromethylthiirane and 4-vinyl-1,2-epithiocyclohexane are nematocides. Several thiirane 1-oxides are reported to be insecticides, molluscicides and herbicides ⟨68USP3413306⟩. 1,2-Epithio-1,2,3,4-tetrahydronaphthalene is a mild herbicide.

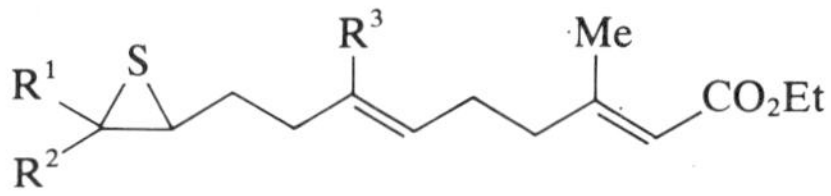

(77)

5.06.5.6 Miscellaneous

Ethers, esters, amides and imidazolidines containing an epithio group are said to be effective in enhancing the antiwear and extreme pressure performance of lubricants. Other uses of thiiranes are as follows: fuel gas odorant (2-methylthiirane), improvement of antistatic and wetting properties of fibers and films [poly(ethyleneglycol) ethers of 2-hydroxymethyl thiirane], inhibition of alkene metathesis (2-methylthiirane), stabilizers for poly(thiirane) (halogen adducts of thiiranes), enhancement of respiration of tobacco leaves (thiirane), tobacco additives to reduce nicotine and to reduce phenol levels in smoke [2-(methoxymethyl)thiirane], stabilizers for trichloroethylene and 1,1,1-trichloroethane (2-methylthiirane, 2-hydroxymethylthiirane) and stabilizers for organic compounds (*O*,*O*-dialkyldithiophosphate esters of 2-mercaptomethylthiirane). The product of the reaction of aniline with thiirane is reported to be useful in the flotation of zinc sulfide.

5.07

Fused-ring Oxiranes, Oxirenes, Thiiranes and Thiirenes

H. HART
Michigan State University

This chapter differs in organization from other chapters. It is short, but covers a variety of rather dissimilar systems, most of them occupying a printed page or less. Structure, reactivity and synthesis are presented intertwined, as one encounters them in the laboratory. Separating structure, reactivity and synthesis as is done in other chapters would give an extremely subdivided presentation, of limited utility. Therefore, organization according to ring size was adopted.

5.07.1 INTRODUCTION

In principle, virtually any double bond in a ring can be epoxidized. Consequently one can envision a large number of fused-ring oxiranes, and indeed a very large number are known. They are derived from mono-, bi- and poly-cyclic natural or synthetic precursors. The oxirane can be fused with a carbocyclic or heterocyclic ring. It is impossible to review here all such possibilities, nor is it necessary to do so since in many examples the chemistry of the oxirane moiety is unexceptional and can be anticipated. Therefore, this review is restricted to selected examples of fused-ring oxiranes which, because of special structural features such as strain, novelty, elusiveness or biological significance, have been the targets of recent study.

Although for every fused-ring oxirane a corresponding thiirane is theoretically possible, the actual number is much smaller. On the other hand, being larger than oxygen, sulfur can often bridge where oxygen cannot. Attention will be given here to strained fused thiiranes whose oxygen counterparts are not yet known.

The literature on fused oxirenes and thiirenes, in contrast to their saturated analogs, is small and will be covered comprehensively.

5.07.2 FUSION TO THREE-MEMBERED RINGS

Oxidation of cyclopropenes by peroxy acids ⟨67HCA1669, 69JA6534, 70JOC450⟩ or photochemically ⟨80JOC2334, 81H(15)1643⟩ gives enones. The products and kinetics are consistent with the intermediacy of an oxabicyclobutane. For example, oxidation of (**1**) gave (**3**) and (**4**), presumably *via* (**2**) ⟨69JA6534⟩. Nevertheless, to date no example of this fused-ring oxirane has been trapped or detected spectroscopically ⟨80JOC3198⟩. 2,4-Dioxabicyclobutanes have been proposed as intermediates in acetylene oxidations ⟨64JA4866, 64JA4871, 82JA216⟩ but the evidence is tenuous.

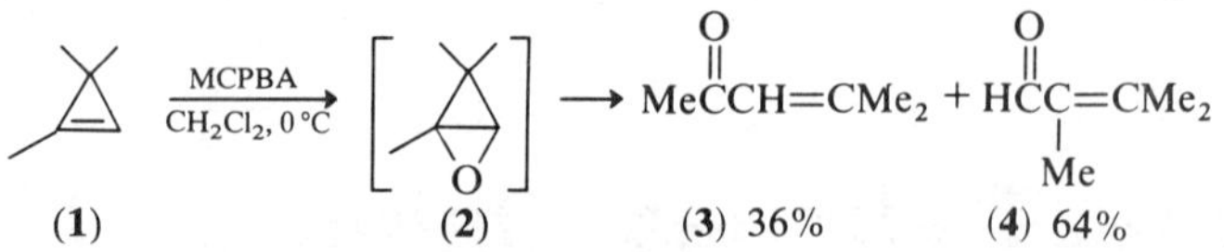

MCPBA, CH_2Cl_2, 0 °C → MeC(=O)CH=CMe₂ + HC(=O)C(Me)=CMe₂

(**1**) (**2**) (**3**) 36% (**4**) 64%

Evidence for the formation of oxabicyclobutane (**6**) in the irradiation (**5**) was negative ⟨72JA1193⟩. The striking difference in photochemical behavior of enones such as (**5**) and the analogous conjugated dienes, which readily form bicyclobutanes, is attributed to the substantially greater endothermicity of the isomerization in the oxa series.

CD_3, O, Me — $h\nu$ ↛ D_3C, O, Me → Me, O, CD_3

(**5**) (**6**) (**7**)

5.07.3 FUSION TO FOUR-MEMBERED RINGS

Cyclobutene oxide (**8**) and several methyl analogs were prepared by conventional epoxidation ⟨65TL979, 65BSF2755⟩. These oxiranes are sensitive to acids. For example, (**8**) rearranges quantitatively to (**9**); with Lewis acids, some cyclobutanone is also formed. Oxirane (**10**) ⟨68TL5855⟩, which has 'inverted tetrahedral' geometry at the bridgehead carbons ⟨72JA7396⟩, is remarkably stable, though it rearranges to (**11**) at 190 °C ($t_{1/2} = 4$ h) or with acid.

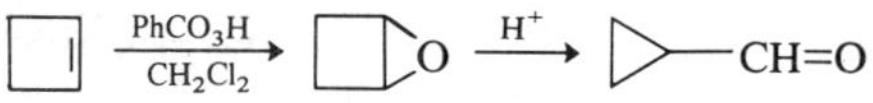

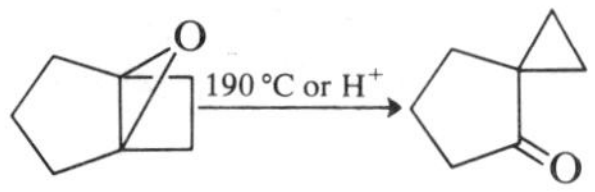

$PhCO_3H$, CH_2Cl_2 → O; H^+ → CH=O; 190 °C or H^+ →

(**8**) (**9**) (**10**) (**11**)

The first simple thiirane fused to a saturated four-membered ring is (**13**), prepared in 34% yield by irradiation of the acyclic sulfide (**12**) ⟨69JOC896⟩. The parent compound is not known.

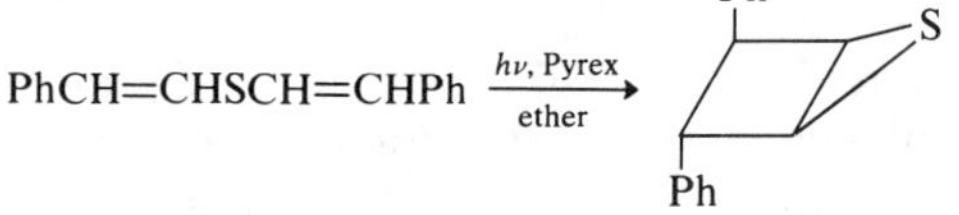

PhCH=CHSCH=CHPh —($h\nu$, Pyrex, ether)→ Ph, S, Ph

(**12**) (**13**)

The unsaturated analog of (**8**) (*i.e.* **14**) can be regarded either as the monoepoxide of cyclobutadiene or as the 'Dewar' isomer of furan. Neither (**14**) nor its derivatives are known, although its thermal disrotatory reorganization to furan is disallowed ⟨80MI50700⟩*. The oxygen 'walk' isomerization of (**14**) to (**14′**) also has a high barrier ⟨80JA2861⟩ which, however, can be lowered appreciably in the hypothetical *O*-oxide (**15**). These calculations are relevant to the sulfur analogs (*vide infra*) where such 'walks' have been observed.

O ⇌ O $\overset{+}{O}$–$\bar{O}$

(**14**) (**14′**) (**15**)

In contrast to (**14**), several examples of 'Dewar' thiophenes are known. The first example was (**17**), prepared by irradiation of (**16**) ⟨72CJC2721, 75CPB2773⟩. The degenerate sulfur 'walk' to (**17′**) occurs at a rate of 55 s^{-1} at 157 °C ⟨77JA629⟩, appreciably faster than thermal

* After this chapter was written, a derivative of (**14**) was reported ⟨82JA847⟩.

rearomatization to (**16**) ($t_{1/2}$ = 5.1 h at 160 °C in benzene). However, aromatization is rapid at room temperature if catalyzed by triphenylphosphine ⟨75TL1639⟩.

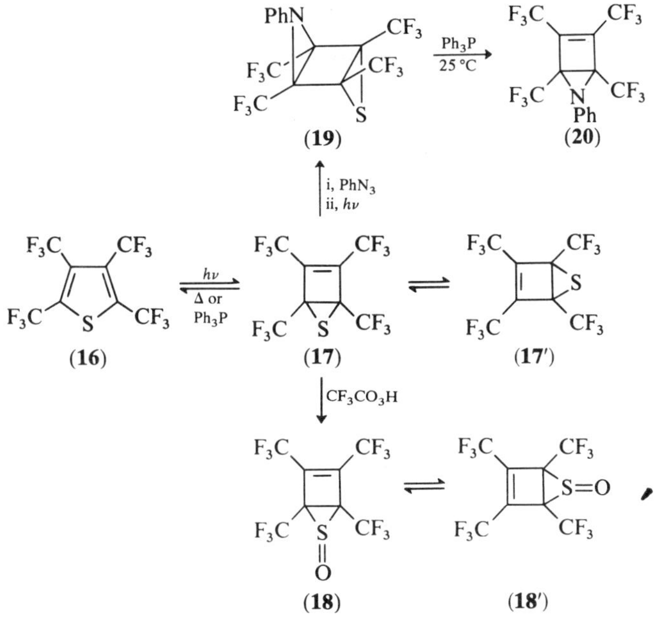

Peroxy acid oxidation of (**17**) gave sulfoxide (**18**) whose ^{19}F NMR spectrum showed equivalent CF_3 groups even at −95 °C ⟨76JA4325⟩. The rate ratio for the sulfur 'walk' in (**18**/**17**) is an astounding 10^{10} at 25 °C; theoretical reasons for the difference have been discussed ⟨80JA2861⟩.

Both (**17**) ⟨74TL2841⟩ and (**18**) ⟨76JA4325⟩ undergo cycloadditions at the double bond without disruption of the thiirane ring ⟨80JOC2962, 80JA6633⟩. This type of reaction has been used to convert (**17**) to the 'Dewar' pyrrole (**20**) *via* the tricyclic thiirane (**19**) ⟨77JA7350⟩.

Dewar thiophenes (*i.e.* **22** and **23**) are intermediates in the photoisomerization of cyanothiophenes. Their presence has been demonstrated by trapping and by direct NMR observation ⟨79CC881, 79CC966⟩. The rapid sulfur 'walk' (*i.e.* **22** → **23**) fully explains the substituent scrambling in the room temperature irradiations (*i.e.* **21** → **24**).

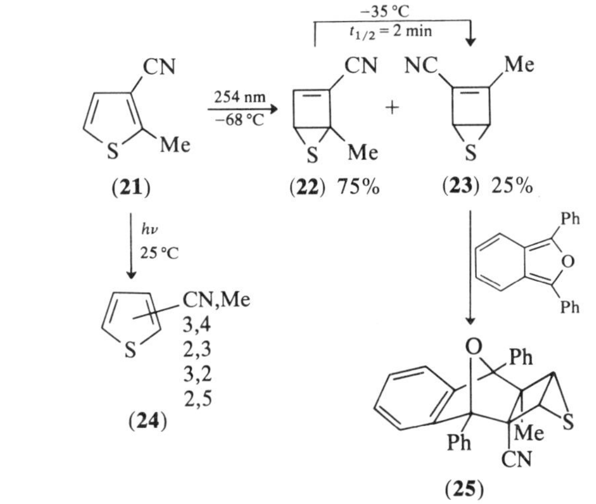

5.07.4 FUSION TO ORDINARY AND LARGE RINGS

In general, the methods of synthesis and reactions of oxiranes and thiiranes fused to ordinary or large rings are not particularly affected by the ring fusion. The reader is referred to Chapters 5.05 and 5.06, and reviews cited there, many of which include fused-ring examples.

The conformations of simple fused systems (**26–28**) have been studied. The five-membered ring in (**26**; X = O, S) assumes a boat conformation with C(3) and X on the same side of the C(1)C(2)C(4)C(5) plane ⟨74JST(22)247, 76JSP(60)179⟩. The angles of 'flap' are about 30° for C(3) and 60–70° for X. The motivation for the boat conformation is to stagger the ring-juncture protons with respect to the adjacent methylene protons. As expected, (**27**; X = O, S) prefers a twisted half-chair in the six-membered ring ⟨72JSP(44)459, 73CJC529⟩. NMR studies suggest that (**28**; X = O) exists in two interconverting chair conformers of comparable energy ⟨74JA4185⟩. Chiral (**29**; X = O) has been prepared; the (−)-enantiomer has the 1*R*,8*R* configuration ⟨72T1955⟩.

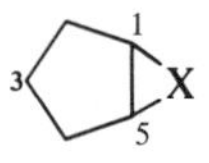

(**26**) 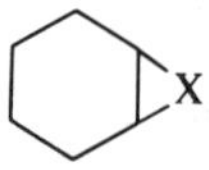 (**27**) 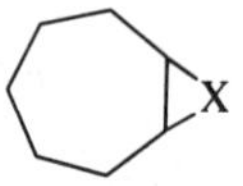 (**28**) 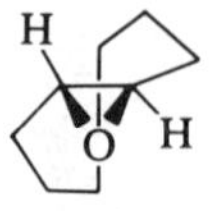 (**29**)

5.07.5 ARENE OXIDES OF POLYNUCLEAR AROMATIC HYDROCARBONS

Benzene monoxide–oxepin and its sulfur analog are treated elsewhere (Chapter 5.1.7) ⟨67AG(E)385⟩. However, we point out here that electron-withdrawing substituents often favor the benzene oxide tautomer. The first study on oxides of the environmentally hazardous polychloro- and polybromo-biphenyls shows that they exist mainly in the benzene oxide form ⟨81JOC3721⟩. Oxides of polynuclear aromatic hydrocarbons (PAH) also exist mainly in the fused-ring oxirane form.

Because of their involvement in carcinogenicity, oxiranes of PAH have been extensively studied and reviewed ⟨73H(1)267, 74MI50701, 76ACR378, 78PAC1033, B-78MI50700, B-78MI50701⟩. In recent years, attention has focused on arenediol epoxides, as a consequence of the identification of (**30**) as the probable active form of the common environmental carcinogen benzo[*a*]pyrene (**31**). Initially, it was found that diol (**32**) bound to DNA to a much greater extent than (**31**) ⟨73JMC502⟩, and that (**32**) could be further activated by epoxidation ⟨74MI50700⟩. Subsequently, (**30**) was found to be the bound form of the carcinogen ⟨76PNA2679, 76MI50700⟩, the 10-position of the (+)-7*R*,8*S*,9*S*,10*S* isomer ⟨77JA258⟩ being linked to the 2-amino group of guanosine ⟨76JA5714, 76JA6720⟩.

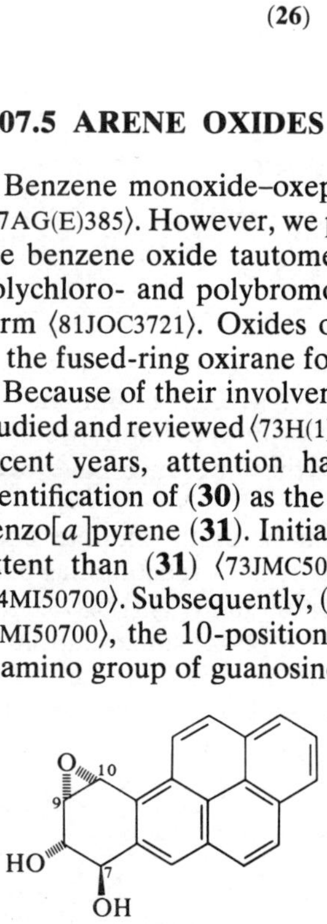

Bay Region

K-Region

(**30**) 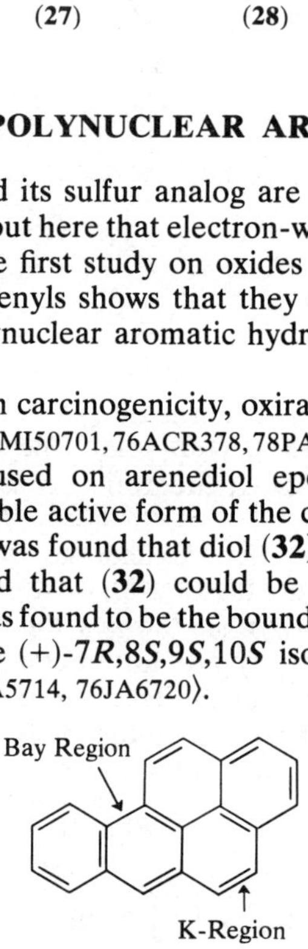 (**31**)

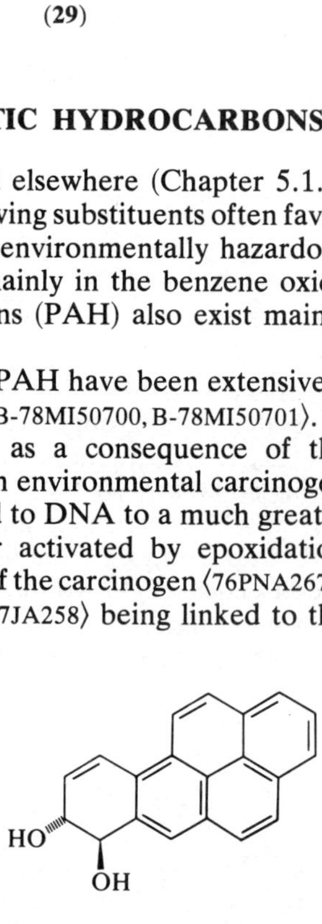

Although details vary for particular cases, a common synthetic route to diol epoxides such as (**30**) frequently begins with the ketone (**33**) ⟨78MI50700⟩. The final epoxidation is often highly stereoselective. A general route to non-K-region arene oxides has been described ⟨75JA3185⟩.

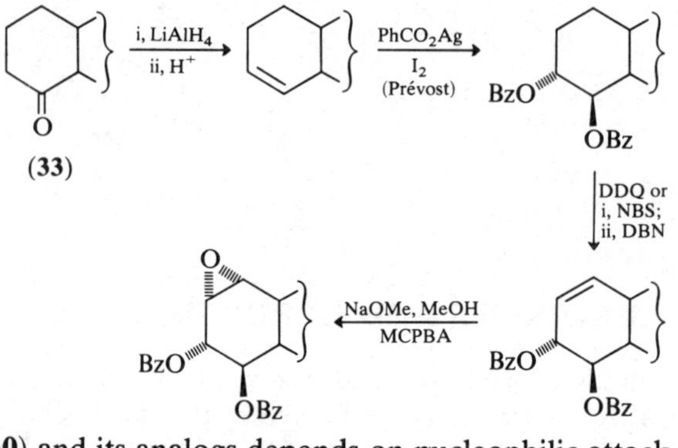

(**33**) —i, $LiAlH_4$; ii, H^+→ —$PhCO_2Ag$, I_2 (Prévost)→ BzO, OBz —DDQ or i, NBS; ii, DBN→ BzO, OBz —NaOMe, MeOH; MCPBA→ BzO, OBz epoxide

DNA binding of (**30**) and its analogs depends on nucleophilic attack on the oxirane ring at C(10), the preferred position because it is benzylic. With strong nucleophiles such as

RS^-, ring-opening is stereospecific and S_N2 to give (**34**), whereas with weaker nucleophiles (H_2O, MeOH) the S_N1 path contributes significantly. It has been suggested on the basis of MO calculations that 'bay region' epoxides are more reactive than other diol epoxides as a consequence of the greater ease of forming benzylic carbocations in the 'bay region', and that this greater reactivity and carcinogenicty are related ⟨78PAC1033⟩. The diol moiety is thought to provide a metabolic route to the arene oxide, and also provide some steric control on the epoxidation and on the ring-opening.

(30) — strong nucleophile S_N2 → (34)

(30) — weak nucleophile S_N1 → (34) + (35)

Arene oxides of heterocycles such as pyridine have often been postulated as photo-isomerization intermediates ⟨81H(15)1569, 72JOC3592⟩ but to date they have not been isolated or extensively studied.

5.07.6 COMPOUNDS WITH TWO OR MORE FUSED OXIRANES AND/OR THIIRANES

Cyclic polyoxiranes have, in recent years, provided a fruitful playground for the synthesis of fascinating structures ⟨80T833⟩. Most significant from the mechanistic, synthetic and biological viewpoints are the di- and tri-epoxides of benzene, all four of which (**36–39**) are known.

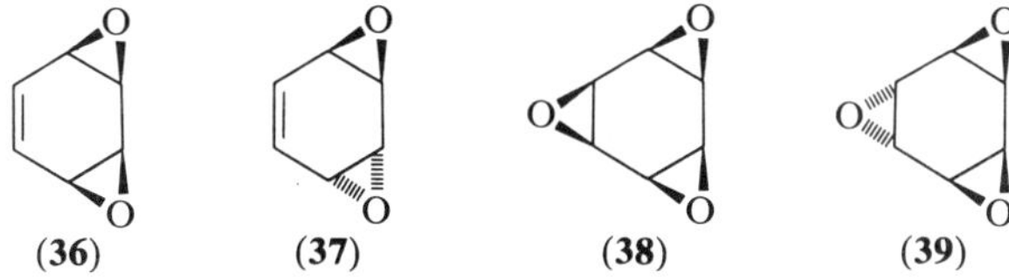

(**36**) (**37**) (**38**) (**39**)

Syn-Diepoxide (**36**) ⟨72AG(E)937, 73AG(E)838⟩ can be obtained from 1,4-cyclohexadiene monoxide (**40**). Although the *anti*-isomer (**37**) can be obtained from (**42**) ⟨73AG(E)838⟩, a better route is available from quinone ⟨78TL3333⟩. *Syn*-Triepoxide (**38**) was first made independently by Vogel ⟨72AG(E)939⟩ and Prinzbach ⟨72AG(E)942⟩ from (**41**). The *anti*-triepoxide (**39**) has been obtained by several routes. Epoxidation of (**36**) gave (**39**) directly ⟨72AG(E)939⟩, but the route *via* (**47**) is better ⟨78TL3333⟩. Alternatively, (**39**) can be obtained by thermal rearrangement of peroxide (**53**). Prinzbach prepared (**39**) from dibromide (**48**) which, though essentially unavailable by direct bromination of (**40**), can be obtained by equilibrating the mixture of (**41**) and (**42**) with quaternary ammonium bromides. Full experimental details for improved syntheses of (**38**) and (**39**) (including the conversion of **38** to **39**) have been published ⟨79CB3318, 79CB3347⟩.

An X-ray structure of *syn*-triepoxide (**38**) ⟨74AG(E)539⟩ shows that the cyclohexane ring is planar, with the three oxygens and six hydrogens lying in parallel planes 1.19 and 0.42 Å respectively from the carbon plane. With all oxygens on one 'face', (**38**) forms a crystalline 2:1 complex with potassium iodide.

Syn-Epoxides (**36**) ⟨72AG(E)937⟩ and (**38**) ⟨72AG(E)939, 72AG(E)942, 79CB3318⟩ readily isomerize to their monocyclic valence-bond tautomers. The kinetics for (**38**) → (**55**) have been measured ⟨76JCS(P2)36⟩. The activation energy for isomerization of (**36**) and (**38**) is 113 and 176 kJ mol^{-1} respectively, compared to only 38 kJ mol^{-1} for benzene oxide itself. The *anti*-epoxides (**37**) and (**39**) do not tautomerize.

Br
OAc
Br
OAc
(43)
KOH / MeOH → (36)
Ac$_2$O, H$^+$
(40) NBS / CCl$_4$ → (41) + (42)
Br O Br
KMnO$_4$
HO HO Br Br O
(44) KOH / MeOH → (38)

(45) Br$_2$ / 0 °C → (46) NaBH$_4$ → (47) KOH / MgSO$_4$ ether → (37)
MCPBA
(36) CF$_3$CO$_3$H → (39) ← KOH / MgSO$_4$ ether

(48) Me$_4$N$^+$OAc$^-$ / acetone → (49) KMnO$_4$ / pH 8 → (50)
i, TsCl, py
ii, NH$_3$, NH$_4$OAc
(51)
(52) ^{1}O$_2$ → (53) heat → (39) ← NaOCH$_2$CH$_2$OH / THF (51)

(**36**) $\xrightarrow[C_6H_6]{60\,°C}$ (**54**) (**38**) $\xrightarrow[\text{vapor}]{400\text{–}550\,°C}$ (**55**)

Perhaps the greatest utility of (**36–39**) (and many of their precursors, such as **41**, **42** and **48**) is as starting points for the synthesis of highly functionalized cyclohexanes. For example, (**38**) has been converted in five steps (overall yield 70–75%) to streptamine (**57**); the key step is selective ring-opening with hydrazine to give intermediate (**56**) ⟨75AG(E)630⟩.

(**38**) $\xrightarrow{N_2H_4}$ (**56**) $\xrightarrow{\text{4 steps}}$ (**57**)

The thiiranes (**58**) ⟨74AG(E)736⟩, (**59**) and (**60**) ⟨75AG(E)253⟩ and (**61**) ⟨75AG(E)252⟩ have been prepared by manipulation of the oxirane rings in (**36**) and (**38**).

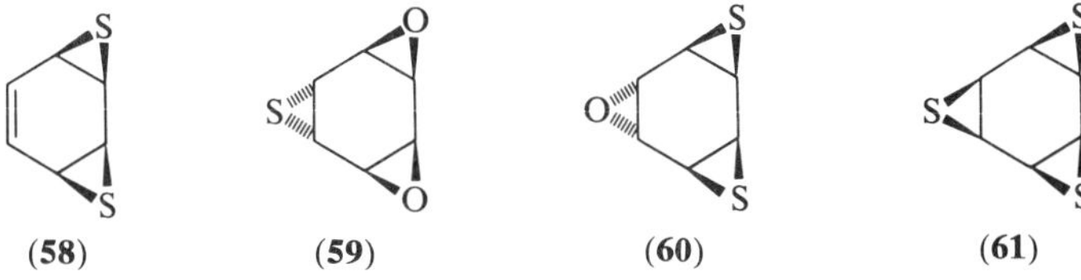

(**58**) (**59**) (**60**) (**61**)

The naphthalene polyepoxides (**62–66**) are known ⟨76AG(E)229, 77AG(E)169, 77AG(E)171⟩. Simple anthracene analogs are relatively unknown, although substituted derivatives have been prepared, mainly by thermal rearrangement of endoxides ⟨68PAC(16)169⟩ but also by direct oxidation ⟨77AG(E)171⟩.

(**62**) (**63**) (**64**) (**65**) (**66**)

Polyoxiranes of many other classic unsaturated carbon frameworks are known. *syn*-Cyclopentadiene dioxides ⟨67MI50700⟩ are better known than the *anti*-isomers, although both *syn*- and *anti*-diepoxides of tetraphenylcyclopentadienone have been characterized ⟨70CC1550⟩. The *syn*- and *anti*-diepoxides of 1,3- ⟨72JOC4246⟩ and 1,4-cyclohexadiene ⟨67JOC3743⟩ are known. Recently the various di- and tri-epoxides of cyclohetatriene were prepared ⟨76AG(E)559, 78BCJ2131, 78CL395, 80JA1961⟩. Although several di- and tri-epoxides of cyclooctatetraene are known ⟨72JOC2421⟩, the four possible tetraepoxides have not yet been described. The pinwheel shaped triepoxide (**68**), obtained directly from 'barrelene' (**67**), undergoes a truly remarkable rearrangement to (**69**), which contains six tetrahydrofuran-like rings ⟨76AG(E)686⟩. Perhaps the most unusual of all fused-ring polyoxiranes

(**67**) $\xrightarrow[CH_2Cl_2,\ KHCO_3]{\text{MCPBA}}$ (**68**) $\xrightarrow[\text{5 min}]{BF_3,\ -20\,°C}$ (**69**)

(**70**) $\xrightarrow{^1O_2}$ [(**71**)] $\longrightarrow$ (**72**)

is (**72**), obtained in 75% yield on singlet oxygen oxidation of furanophane (**70**), presumably *via* the endoperoxide (**71**) ⟨69TL5315⟩.

5.07.7 NATURAL PRODUCTS WITH FUSED OXIRANE OR THIIRANE RINGS

A rather large number of natural products containing one or more fused oxirane rings are known ⟨64HC(19-1)1, p. 24⟩, examples fused to six-membered carbocyclic rings being most common. Some with rather simple structures include pulegone oxide (**73**) ⟨56JA5022⟩, the antibiotic terreic acid (**74**) ⟨58JA5536⟩ and epoxydon (**75**) ⟨66HCA204⟩, which shows anti-tumor activity. Of more recent interest are the monoepoxides senepoxide (**76**) ⟨68T1633, 74TL4235, 78JA6483⟩, pipoxide (**77**) ⟨79TL715⟩ and the related crotepoxide (**78**) (also called futoxide) ⟨68JA2982⟩. All three have been synthesized, crotepoxide by several ingenious routes ⟨75TL3187, 76JA634, 77TL3361⟩. Possible routes for the biosynthesis of these anti-tumor agents have been suggested ⟨78T3353⟩. These compounds, which show some similarity to arene oxides, are thought to be cytotoxic as a consequence of protein binding *via* nucleophilic attack of SH groups on the oxirane rings. A simple antibiotic LL-Z1220 (**79**), containing a *syn*-benzene dioxide moiety, has been isolated ⟨74JOC435⟩. Considerably more complex are the diepoxide stemolide (**80**) ⟨76TL2489, 80JA1202⟩ and the triepoxide triptolide (**81**) ⟨72JA7194, 80JA1200, 80JA5424⟩, both of which have been synthesized.

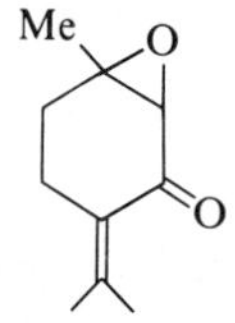

(**73**) 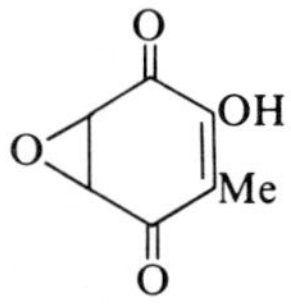 (**74**) 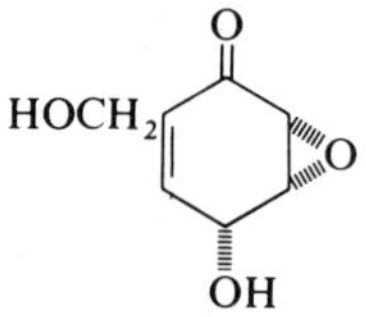 (**75**)

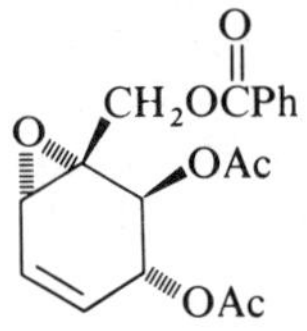

(**76**) 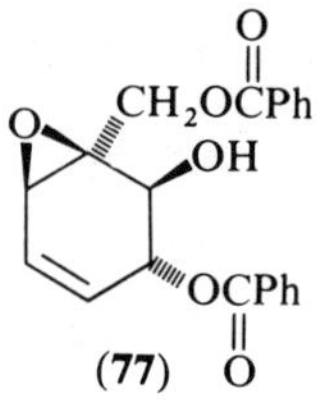 (**77**) 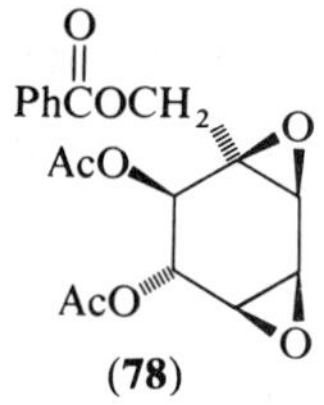 (**78**)

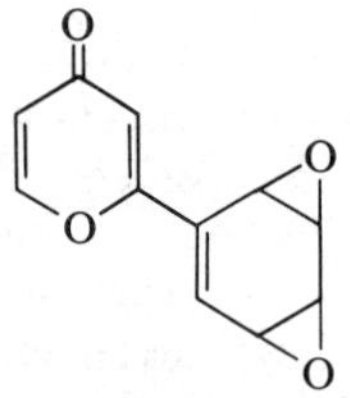

(**79**) 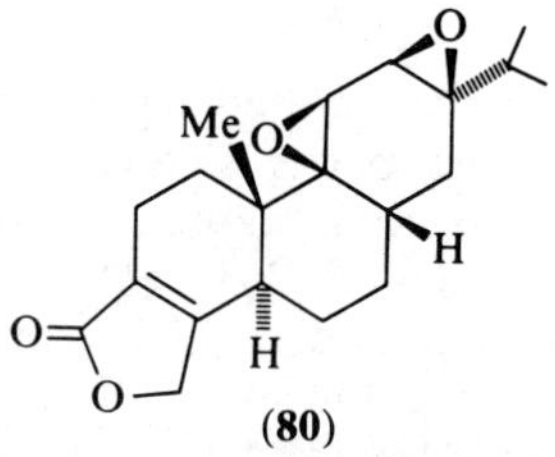 (**80**) 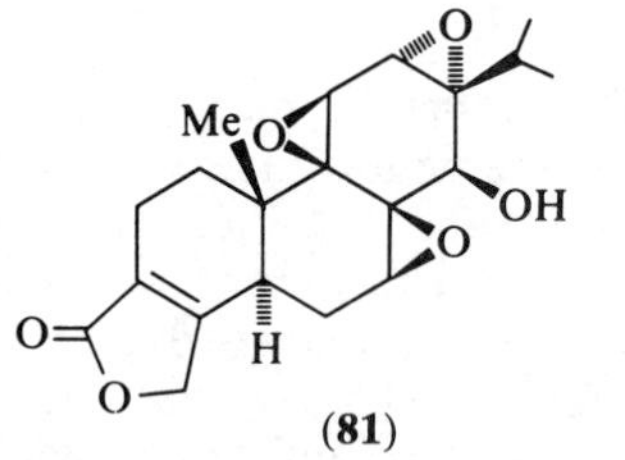 (**81**)

A few natural products with oxirane rings fused to five-membered rings are known. These include the antibiotic methylenomycin-A (**82**) ⟨79H(13)353, 79JOC4210, 80JA3904, 81CC714⟩ and the truly remarkable fungal metabolite trichoviridine (**83**), which appears to be the first example of an isocyanide epoxide ⟨76CPB832⟩.

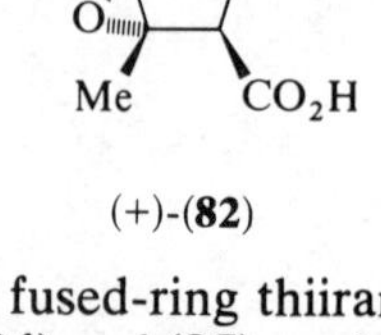

(+)-(**82**) 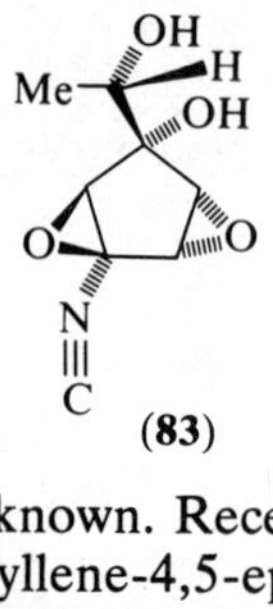 (**83**)

Few naturally occurring fused-ring thiiranes are known. Recently, however, humulene-1,2- and -4,5-episulfides (**84**) and (**85**), and caryophyllene-4,5-episulfide (**86**) were isolated from hops which had been heavily dressed with sulfur during the growing period, to control mildew ⟨80JCS(P1)311⟩.

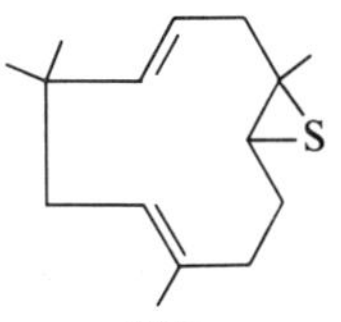

(84) 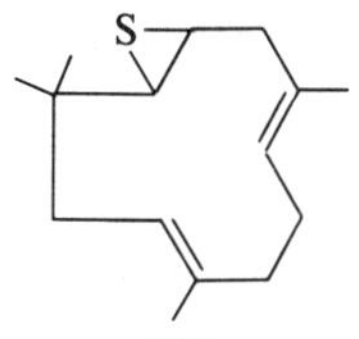 (85) 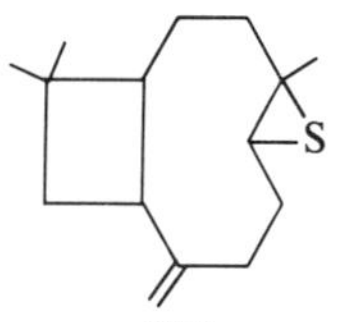 (86)

5.07.8 FUSED-RING OXIRENES AND THIIRENES

Evidence suggests that oxirenes are intermediates in Wolff rearrangements of acyclic α-diazo ketones ⟨75AG(E)32⟩ (see also Chapter 5.05). However, until recently no evidence for their intermediacy has been detected in cyclic diazo ketones such as (**87**) ⟨77T453⟩, (**88**) or (**89**) ⟨75CB3566⟩ or (**90**) ⟨72CC694, 76ZN(B)586⟩. The first evidence for an oxirene intermediate in cyclic systems was encountered in the Wolff rearrangement of (**91**), which gives products derived from both ketocarbenes (**92**) and (**94**); presumably oxirene (**93**) is an intermediate in their interconversion ⟨78CB1549⟩. Despite several attempts ⟨73AG(E)840, 74CC900, 79TL59⟩, no evidence for an aromatic oxirene of the type (**95**) has yet been obtained. The best designed attempt, which involved the retro-Diels–Alder reaction of (**96**), was thwarted by an oxygen 'walk' to give (**97**) ⟨73AG(E)840⟩.

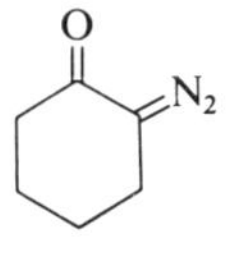

(87) 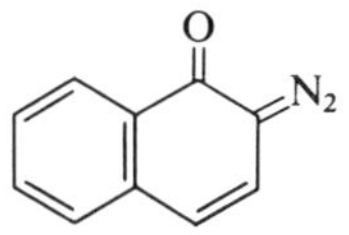 (88) 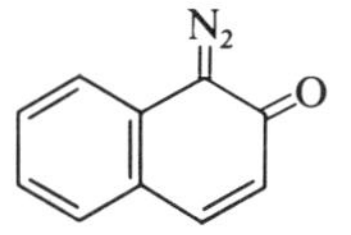 (89) 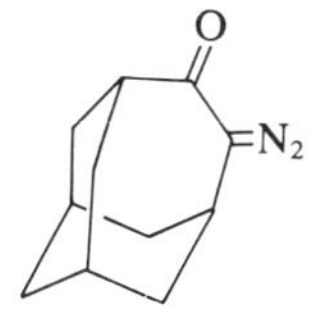 (90)

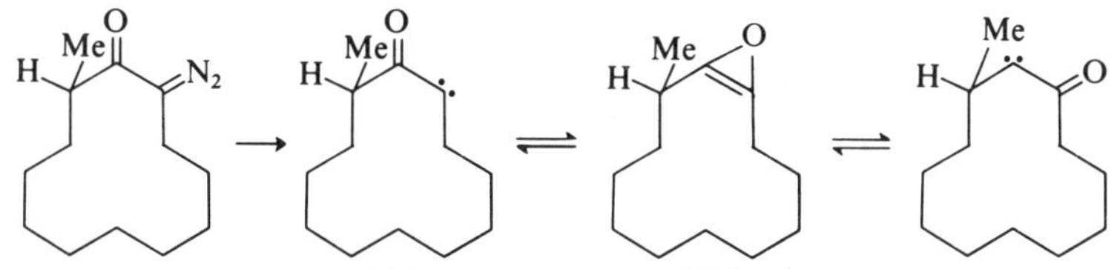

(91) (92) (93) (94)

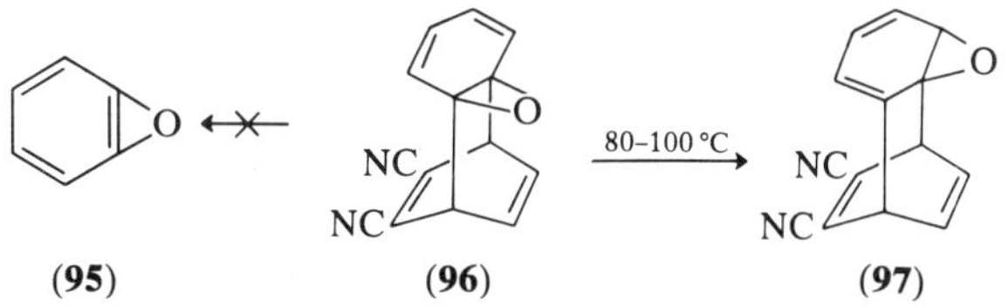

80–100 °C

(95) (96) (97)

The greater size and polarizability of sulfur compared with oxygen might increase the possibility for forming fused-ring thiirenes. They have indeed been proposed to account for scrambling in the decomposition of (**98**) ⟨74CC900⟩ and (**100**) ⟨77TL2643⟩; arguments against these proposals ⟨77JOC575⟩ have been countered ⟨79TL2785⟩. Careful studies on the thermal decomposition of ^{13}C-labelled thiadiazoles (**102**) ⟨78JHC697⟩ and (**103**) ⟨79NJC715⟩ find no evidence for a thiirene intermediate, although it is still possible that the photochemical decomposition of such compounds produces fused-ring thiirenes ⟨79TL2785, 78JOC2490⟩. At this point, however, no fused-ring thiirene has been unequivocally identified. In this connection, an old and widely quoted report on the preparation of (**104**) as an isolable crystalline compound ⟨62JPR(18)255⟩ requires re-examination.

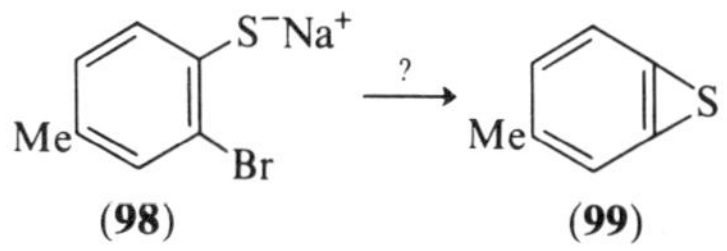

(98) (99) 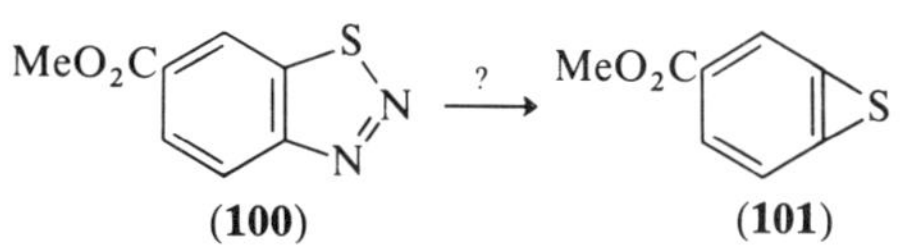 (100) (101)

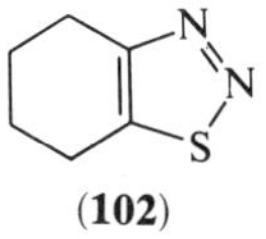

(102) 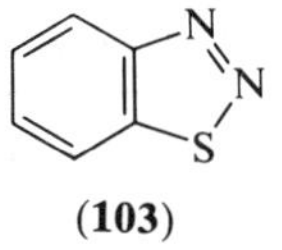 (103) 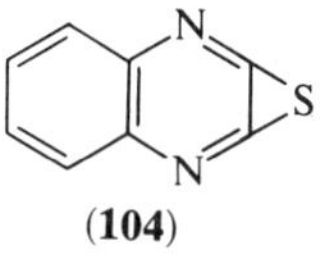 (104)

5.08

Three-membered Rings with Two Heteroatoms and Fused-ring Derivatives

E. SCHMITZ
Akademie der Wissenschaften der DDR, Berlin

5.08.1 INTRODUCTION

The first authentic three-membered rings with two heteroatoms were synthesized only after 1950. Because they became available at once by simple and generally applicable procedures, oxaziridines (**1**) ⟨64HC(19-1)624, 68HOU(10-4)449⟩, diaziridines (**2**) ⟨67HOU(10-2)71⟩ and diazirines (**3**) ⟨68HOU(10-4)895⟩ are now extensive, well investigated classes of compound.

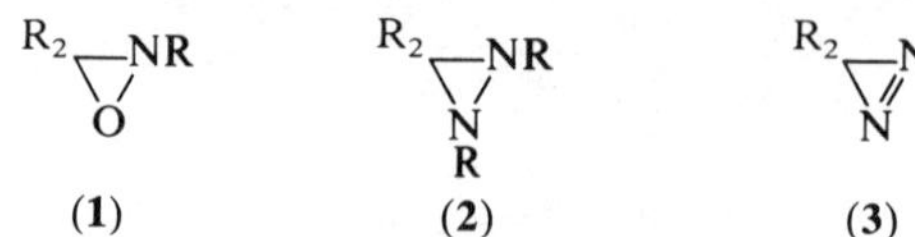

All older structural proposals containing these ring systems have turned out to be erroneous ⟨B-67MI50800⟩. Nitrones had been envisaged as three-membered rings but are now formulated as *N*-oxides of Schiff bases. A three-membered structure was also claimed for hydrazones in the older literature, but we know now that a diaziridine ring is in no case obtained from a carbonyl compound and a hydrazine. The three-membered ring structure played a special role in the discussion on the structure of aliphatic diazo compounds, written as diazirines by the discoverers. Since no satisfactory alternative structure could be given to aliphatic diazo compounds using only bond lines, the cyclic diazomethane structure survived many discussions and was replaced only recently by a linear formula postulated by physical measurements.

It is interesting to note that the existence of an isomer of diazomethane with a diazirine structure was never discussed, so the report on its synthesis ⟨61TL612⟩ simultaneously started and ended the discussion on its existence.

In the oxaziridines (**1**) ring positions 1, 2 and 3 are attributed to oxygen, nitrogen and carbon respectively. The latter almost always is in the oxidation state of a carbonyl compound and only in rare cases that of a carboxylic acid. Oxaziridinones are not known. The nitrogen can be substituted by aryl, alkyl, H or acyl; the substituent causes large differences in chemical behavior. Fused derivatives (**4**), accessible from cyclic starting materials (Section 5.08.4.1), do not differ from monocyclic oxaziridines.

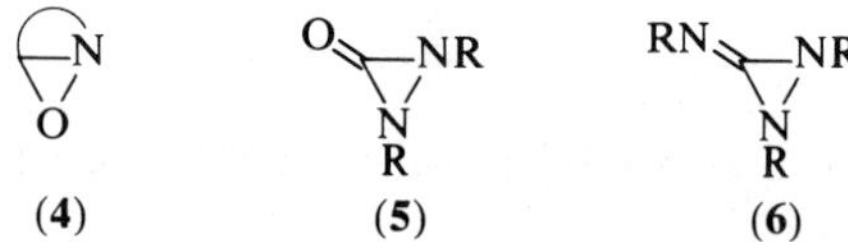

In the diaziridine field many compounds are known bearing *N*-H, *N*-alkyl and *N*-acyl groups, but here no dramatic changes in reactivity are caused by *N*-substituents. *N*-Aryldiaziridines are underrepresented. The ring carbon is in the oxidation state of a carbonyl compound or, in the diaziridinones (**5**) and the diaziridinimines (**6**) that of carbonic acid. In single cases, diaziridine carbon bears chlorine or fluorine.

Fused diaziridines may be bridged from either C to N as in (**7**) or from N to N as in (**8**). The latter type usually contains the *N*-substituents *cis* to each other.

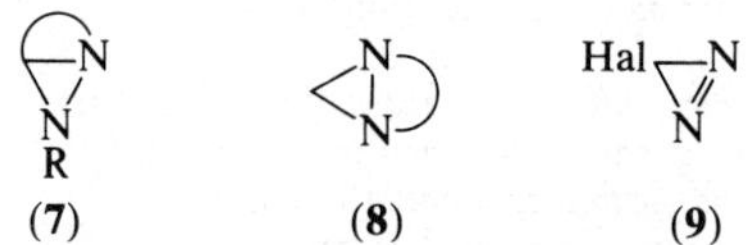

Among unsaturated C—N—N three-membered rings only the 3*H*-diazirines (**3**) are known. 1*H*-Diazirines with a C=N double bond were never obtained. Diazirines with one or two alkyl groups at carbon were prepared in many cases, aryldiazirines only in some cases. An important role is played by difluorodiazirine as well as by diazirines containing chlorine or bromine (**9**).

An intense search for further three-membered rings with two heteroatoms has been successful during recent years. 1,2-Di-*t*-butyldiphosphirane (**10**) was synthesized from the potassium compound of a diphosphane with 2,2-dichloropropane ⟨78ZN(B)1208⟩. Azaphosphiridine (**11**) ⟨81AG(E)385⟩ and oxaphosphirane (**12**) ⟨78CB3105⟩ have been described, as was the thiaziridine dioxide (**13**) which was stable at −30 °C ⟨80H(14)1677⟩. A C—P—S three-membered ring (1,2λ^5-thiaphosphirane) was also obtained recently ⟨81CC72⟩.

Me_2 PBu^t P Bu^t (**10**) — Bu^t NBu^t P NPr$^i_2$ (**11**) — $(CF_3)_2$ O P R_2N NR (**12**) — Bu^t NBu^t S O_2 (**13**)

Two further compounds are on the borderline of stability. The oxathiirane (**14**), obtained by photorearrangement of a thioketone *S*-oxide, could be characterized by its UV spectrum ⟨76JCS(P1)1404⟩. Thiazirine (**15**) was stable only at 15 K ⟨78JCS(P1)746⟩.

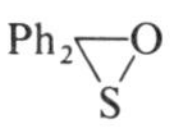

(14) 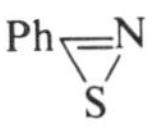 (15) 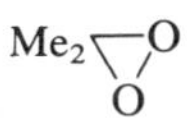 (16) 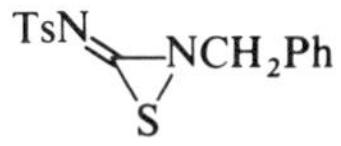 (17)

The peroxidic three-membered dioxirane (**16**) ⟨80JOC4758⟩ and the thiaziridinimine (**17**) ⟨76JOC3403⟩ were made plausible as intermediates, but there was no direct proof of their existence.

Beyond the scope of this chapter, but similar in mode of formation and reactivity to the title compounds, are some three-membered rings with three heteroatoms. Oxadiaziridine (**18**) was obtained by photorearrangement of an azoxy compound ⟨70JOC2482⟩. Triaziridine (**19**) was also made by photo valence isomerization of its open chain dipolar isomer ⟨80CC1197⟩. By 1,3-elimination of hydrogen halide, thiaziridine dioxides (**20**) were produced by several authors; the diazaphosphiridine derivative (**21**) was obtained analogously from a phosphorus-containing precursor ⟨80H(14)1677⟩. An azadiphosphiridine (**22**) was obtained by HF elimination from an open chain compound containing a P—F bond ⟨81AG(E)385⟩, and a P—P—S three-membered ring and its Se analog from diphosphanes ⟨81AG(E)1044⟩.

(18) 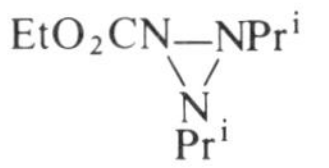 (19) 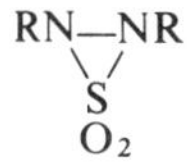 (20) 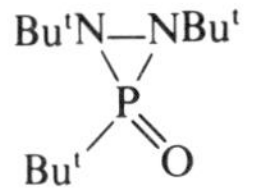 (21) 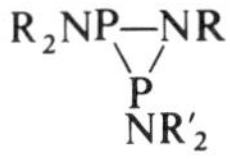 (22)

5.08.2 STRUCTURE OF THREE-MEMBERED RINGS WITH TWO HETEROATOMS

5.08.2.1 Theoretical Methods

Whereas oxaziridine and diaziridine were partial subjects of comprehensive theoretical studies on cyclic compounds ⟨73MI50800⟩, diazirine and some of its simple derivatives were the special target of quantum chemical investigations. Since diazirine, the lowest molecular weight heterocycle, has only five atoms and is of high symmetry, there was a chance for *ab initio* calculations, which followed some semiempirical studies.

In an early investigation ⟨66T539⟩ the two highest occupied and the two lowest unoccupied orbitals were calculated on the basis of an extended Hückel theory to determine the electron transition responsible for the long wavelength UV absorption. An $A_1 \rightarrow B_1$, $\sigma \rightarrow \pi^*$ transition was discussed.

Later there was an attempt by *ab initio* calculation to fit the electron structure of diazirine into the Walsh model of cyclopropane ⟨69MI50800⟩. According to these SCF–LCAO–MO calculations three MOs add to the description of the lone electron pairs, all of which also contribute to some extent to ring bonding. As to strain, π-character and conjugative effect, the term 'pseudo-π-character' was used.

An extensive investigation on structure and stability of three-membered rings by *ab initio* MO theory included discussion of older results ⟨73MI50800⟩. Treatment of 30 three-membered rings, constructed from C, N and O, included oxaziridine, diaziridine and diazirine. These SCF–LCAO–MO calculations used three different basis sets: STO-3G, 4-31G and 6-31G*. The STO-3G set yielded good values for bond lengths: C—N and C—H bond lengths agreed with experimental values within 1%; the calculated N—N bond length (1.266 Å) exceeded the experimental value by 3%. The calculated dipole moment of diazirine, determined by several authors to be in the range 1.29–2.47 D, compares with an experimental value of 1.59 D.

There was less agreement between calculated and experimental energy values. The use of 6-31G*, the best procedure in energy calculations of three-membered rings, yielded a value too low by more than 40 kJ mol^{-1} in the case of diazirine: bond separation energy was calculated as −45 kJ mol^{-1}; the experimental value is +0.4 kJ mol^{-1}. Vibrational correction and extrapolation to 0 K would reduce this difference by several kJ mol^{-1}.

The heat of formation (332 kJ mol^{-1}) was calculated to be 376 kJ mol^{-1}. Since a correction of about 35 kJ mol^{-1} is not unusual in calculations of three-membered rings, a corrected value of 340 kJ mol^{-1} appears reasonable.

Even more recent calculations using STO-3G and 6-31G basis sets could not safely predict diazomethane as the more stable compound in comparison with diazirine, although there is an experimental energy difference of 125 kJ mol^{-1} ⟨79JST(52)275⟩.

Some theoretical investigations have attempted to follow reactions of diazirines, among them reactions of 3-chloro- and 3-fluoro-diazirines.

There were also several calculations of geometrical parameters, dipole moment and energy of oxaziridine ⟨71MI50800⟩. Since the parent compound of oxaziridines is unknown, comparison of calculated and experimental values is still lacking.

Some remarkable results were obtained in theoretical treatment of reactions of oxaziridines. The inversion barrier of the N—H group in oxaziridine was calculated as 134 kJ mol^{-1} ⟨68MI50800⟩, which is in agreement with an experimental value of 136 kJ mol^{-1}, obtained with isolated 3-benzyl-2,3-dimethyloxaziridine ⟨69LA(727)224⟩. The experimental fact that *O*-protonated oxaziridines may react by either O—C or O—N cleavage was correctly matched by an extended Hückel MO calculation ⟨78MI50800⟩. A MINDO/3 calculation claimed that oxaziridine hydrolysis may proceed by either *O*- or *N*-protonation according to the substitution pattern, which is in agreement with experiments ⟨78JOC2380⟩. An *ab initio* SCF calculation regarded the oxaziridine as being subject to nitrone isomerization and correlated it with the *cis–trans* isomerization of nitrones, which is assumed to involve the same intermediate ⟨79JA2560⟩.

Calculations of diaziridines in most cases treated the parent compound, which at present is known only in solution ⟨65TL1857⟩, so that rigorous proof of results is still lacking. The SCF–LCAO–MO study ⟨73MI50800⟩ with STO-3G basis set yielded N—N bond lengths of 1.483 and 1.488 Å for (*E*)- and (*Z*)-diaziridine. The *E* compound was predicted to be more stable by 30 kJ mol^{-1}. Interaction between nitrogen lone pair dipoles is assumed to cause the difference. The difference appears fairly high in view of the easy formation of fused diaziridines which must have the *Z* configuration. The value of 10.5 kJ mol^{-1} is more plausible ⟨72TL2451⟩. The latter investigation predicts an inversion barrier of 70 kJ mol^{-1} when a transition state with one N—H bond in the ring plane is assumed, and one of 140 kJ mol^{-1} when both N—H bonds lie in the plane of the ring.

5.08.2.2 Molecular Dimensions

True three-membered rings with two heteroatoms were discovered and structurally proved by chemical methods only in the late 1950s. However, the older literature contains many erroneously assigned three-membered ring structures. Thus, additional structural proof by physical methods was desired and was given promptly. These investigations confirmed the new three-membered ring structures and yielded data on molecular geometry and absolute configuration.

The results of an X-ray structural investigation of oxaziridine (**23**) are given in detail together with some comments by the authors ⟨72JCS(P2)1137⟩. The measurements were carried out with the pure *Z* compound. There are, however, no significant differences in comparable bond lengths between (**23**) and an (*E*)-oxaziridine investigated earlier. The dihedral angle between the arene ring and the oxaziridine ring is 118.8°; the C atoms attached directly to the ring lie out of the plane bisecting the respective oxaziridine ring angles by 0.14 Å.

H, N, *p*-$O_2NC_6H_4$, O, Pr^i

(**23**)

Bond lengths (Å): O—C 1.405 'Fairly normal'
N—C 1.434 'Fairly short'
N—O 1.50 'Fairly long'

Bond angles (°): N—C—O 63.8
C—N—O 57.2
C—O—N 59.0
H—C—Ar 113
Pr^i—N—O 110.5

X-ray analysis of an optically active oxaziridine substituted at nitrogen with the 1-phenylethyl group of known configuration led to the absolute configuration (+)-(2*R*,3*R*)-2-(*S*-1-phenylethyl)-3-(*p*-bromophenyl)oxaziridine of the dextrorotatory compound; as expected, *C*-aryl and *N*-alkyl groups were *trans* to each other ⟨79MI50800⟩.

Structural data of a diaziridine come from gas phase electron diffraction measurements ⟨74CC397⟩. The N—N bond of 3-methyldiaziridine (**24**) is longer than in hydrazine (1.449 Å); the C—N bond distances in (**24**) and in diazirine are nearly equal (1.479 *versus* 1.482 Å), indicating that the double bond has a negligible effect on the adjacent single bonds. This would be unusual for open chain molecules.

(24)

Bond lengths (Å): C—N 1.479
N—N 1.468
C—C 1.497
Bond angles (°): N—C—N 61
C—N—N 59.5

X-ray analysis of diaziridinone (**25**) gave an N—N bond length of 1.60 Å, and an N—CO bond length of 1.325 Å. The bonds from nitrogen to substituent form angles of 56° with the plane of the ring; the substituents are *trans* to each other ⟨78JOC922⟩.

(**25**) R = $-CMe_2C_6H_4Br$-*p*

(**3**) R = H

Bond lengths (Å): C—H 1.09
C—N 1.428
N—N 1.228
Bond angles (°): C—N—N 64.5
H—C—H 117
N—C—N 50.9

The molecular geometry of diazirine (**3**; R = H) was analyzed by microwave spectroscopy ⟨62JA2651⟩. Rotatory spectra of diazirine, of (^{13}C)diazirine and of (^{14}N–^{15}N)diazirine yielded the bond lengths and bond angles shown. The dipole moment of diazirine is 1.59 D.

Bond lengths and bond angles of 3-methyldiazirine, 3,3-dimethyldiazirine, 3-chloro-3-methyldiazirine and 3-bromo-3-methyldiazirine were also obtained by microwave spectroscopy ⟨70JCP(53)1543⟩.

The bond lengths of difluorodiazirine, obtained by electron diffraction, are 1.315 Å for C—F and 1.426 Å for C—N, and the bond angles are 111.84° (F—C—F) and 53.95° (N—C—N) ⟨67JA5527⟩.

5.08.2.3 Molecular Spectra

5.08.2.3.1 NMR investigations on oxaziridines and diaziridines; configurational stability at nitrogen

NMR investigations of oxaziridines followed by analogous investigations of diaziridines revealed an unprecedented degree of configurational stability of pyramidal nitrogen. Whereas the nitrogen of amines inverts rapidly, and only quaternization or fixation into an aziridine ring brings inversion rates into the NMR time scale, inversion barriers were found in saturated three-membered rings with two heteroatoms, which offered a chance to isolate invertomers. The discovery of an AB spectrum of the methylene group of 2-*t*-butyloxaziridine ⟨57JA5739⟩ was followed by measurements of the inversion barrier of oxaziridines; separation of invertomers by chromatography led to (*E*)- and (*Z*)-(**26**) for which NMR and kinetic data are as shown ⟨69LA(727)224⟩. *E–Z* Geometry was attributed by comparison with structurally similar diaziridines of known geometry.

(*E*)-(**26**) 79%: H_3C 1.34, $PhCH_2$ 2.79, $N-CH_3$ 2.66

$\Delta G^{\neq} = 135.6$ (forward); $\Delta G^{\neq} = 131.5$ (reverse)

(*E*)-(**26**) 21%: H_3C 1.22, $PhCH_2$ 3.04, $N-CH_3$ 2.81

NMR data in p.p.m. (CCl_4), $\Delta G^{\neq}$ values in kJ mol^{-1}

An *E–Z* discrimination between isomeric oxaziridines (**27**) was made by NMR data ⟨69JCS(C)2650⟩. The methyl groups of the isopropyl side chains in the compounds (**27**) are nonequivalent due to the neighboring carbon and nitrogen centres of asymmetry and possibly due to restricted rotation around the exocyclic C—N bond in the case of the *Z* isomer. The chemical shift of a methyl group in (*Z*)-(**27**) appears at extraordinarily high field, an effect probably due to the anisotropic effect of the *p*-nitrophenyl group in the isomer believed to be *Z*.

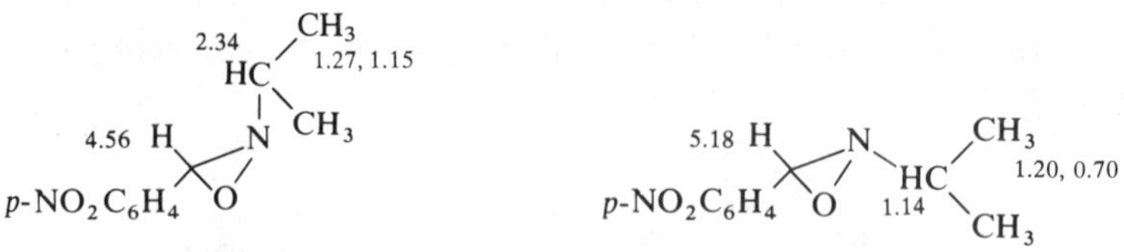

(*E*)-(**27**) (*Z*)-(**27**)

NMR data in p.p.m. (CCl_4)

Z Arrangement was also ascribed to the isomer absorbing at higher field in the case of the ethyl compounds. CH and CH_2 protons near the ring nitrogen are shielded by the aromatic ring in the *Z* compound. The protons at the ring carbon absorb at lower field (near 5.2 p.p.m.) in the *Z* compounds than in the *E* compounds (4.50–4.70 p.p.m.). The chemical shift of this proton may be used for *E–Z* discrimination in further substances.

Also due to the high barrier of inversion, optically active oxaziridines are stable and were prepared repeatedly. To avoid additional centres of asymmetry in the molecule, symmetrical ketones were used as starting materials and converted to oxaziridines by optically active peroxyacids *via* their ketimines ⟨69CC1086, 69JCS(C)2648⟩. In optically active oxaziridines, made from benzophenone, cyclohexanone and adamantanone, the order of magnitude of the inversion barriers was determined by racemization experiments and was found to be identical with former results of NMR study. Inversion barriers of 128–132 kJ mol^{-1} were found in the *N*-isopropyl compounds of the ketones mentioned; inversion barriers of the *N-t*-butyl compounds lie markedly lower (104–110 kJ mol^{-1}). Thus, the *N-t*-butyloxaziridine derived from adamantanone loses half of its chirality within 2.3 days at 20 °C ⟨73JCS(P2)1575⟩.

Nitrogen chirality may also be produced by the action of an achiral peroxyacid on a Schiff base containing a chiral amine ⟨75JOC3878⟩. In this case the oxaziridine contains a configurationally known centre of chirality; relative to this, absolute configurations of the centres of chirality at nitrogen and carbon, and thus the complete absolute configuration of the molecule, can be determined (see Section 5.08.2.2).

NMR spectra of oxaziridine enantiomers may be different from each other in chiral media. In the presence of chiral arylperfluoroalkylcarbinols, shift differences of up to 0.35 p.p.m. are observed, which may be used for discrimination of enantiomers ⟨77JOC3217⟩.

The centre of protonation can be found by NMR in some di- and tri-alkyloxaziridines. These oxaziridines have p*K* values between 0 and −2 and are preferentially protonated at nitrogen. Low field shifts on protonation are roughly proportional to the distance between protonated nitrogen and the hydrogen under discussion ⟨77JHC1393⟩. Coordination of rare earth shift reagents to nitrogen also occurs in sterically unhindered oxaziridines ⟨79JMR(34)251⟩.

In ^{13}C NMR spectra of oxaziridine invertomers, signals of carbon atoms directly bound to ring position 3 and *trans* to the electron lone pair at nitrogen were found at higher field by 3.4–9 p.p.m., compared with analogous compounds with *cis* arrangement of these groups ⟨77OMR(9)322⟩.

NMR investigations in the diaziridine field also led to the problem of inversion stability at nitrogen. Further investigations paralleled those of oxaziridines; NMR investigation in solution ⟨67CB1178⟩ was followed by preparative separation of invertomers and finally preparation of optically active individuals.

Nitrogen inversion in a diaziridine, *e.g.* (**28a**), normally does not lead to a *cis*-diaziridine (**28b**), but its formation is followed by a second inversion step yielding another *trans* compound (**28c**) isomeric with the starting compound. Isomer separation was carried out initially by TLC to give the isomers (**28a**) and (**28c**) ⟨69AG(E)212⟩; cellulose triacetate chromatography was used subsequently ⟨79CB2028⟩. The latter publication gives references on further attempts of invertomer separation of diaziridines.

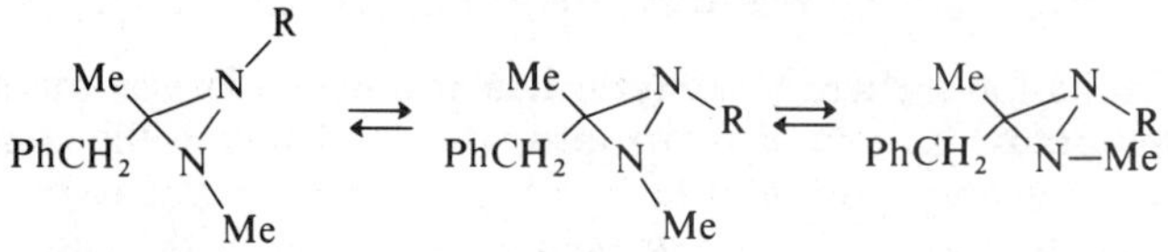

(**28a**) *trans* ⇄ (**28b**) *cis* ⇄ (**28c**) *trans*

The inversion barrier between (**28a**) and (**28c**) is 114 kJ mol^{-1}, the invertomers having the same free energy by chance. NMR data of oxaziridine carbon and substituent groups of the ring for compounds (**29**) and (**30**), taken from the above publications, are as shown.

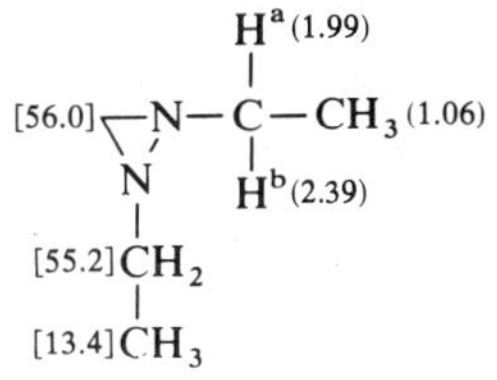

(29)

$^{2}J(H^{a}\text{-}H^{b}) = 11.6$ Hz
$^{1}J(^{13}C\text{-}H^{a}) = 7.4$ Hz
$^{1}J(^{13}C\text{-}H^{b}) = 6.6$ Hz

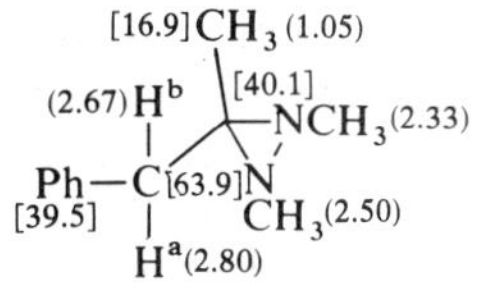

$^{2}J(H^{a}\text{-}H^{b}) = 13.5$ Hz

(30)

^{1}H ( ) and ^{13}C [ ] NMR chemical shift data in p.p.m. (CCl_4)

The *cis*-fused diaziridines (**31a**) and (**31b**) are also an equilibrium system, interchanging *exo* and *endo* positions of methyl and ethyl groups. The ^{1}H NMR spectrum shows two methyl peaks at 0 °C, coalescing to a single sharp peak at 75 °C. The ethyl group shows the sharp characteristic quartet–triplet splitting pattern at 75 °C ⟨74JOC3187⟩.

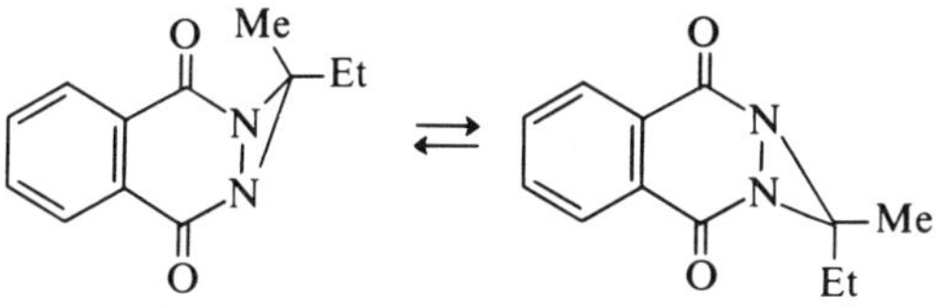

(**31a**) ⇄ (**31b**)

Optically active diaziridines with nitrogen as the centre of chirality were obtained either by reacting a diaziridine with less than one mole of a chiral isocyanate or by stereospecific synthesis using an aminating reagent with a chiral leaving group. By monitoring the rate of racemization an inversion barrier of 117 kJ mol^{-1} was found for compound (**32**) ⟨74DOK(219)873⟩.

(**32**) 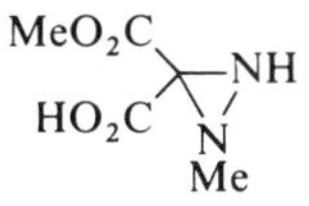 (**33**)

Diaziridine (**33**) was obtained with more than 90% optical purity by separation of the α-phenylethylamine salts ⟨76IZV1898⟩.

5.08.2.3.2 Molecular spectra of oxaziridines and diaziridines

Simple oxaziridines and diaziridines do not absorb in the near UV. Lack of absorption was one argument to distinguish between true three-membered ring structures and unsaturated open chain isomers like nitrones or hydrazones.

To a certain degree, however, the oxaziridine ring can undergo conjugation with unsaturated groups. There is, for example, a remarkable agreement as to position and extinction of UV absorption between 2-ethyl-3-(4-methoxyphenyl)oxaziridine (**34**) and the analogous oxirane (**35**) ⟨B-67MI50800⟩.

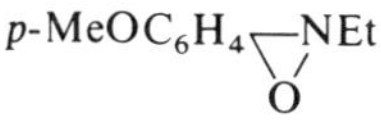

(**34**)	(**35**)
233 (14 000)	230 (12 400)
275 (2000)	275 (1700)
282 (1700)	285 (1500)

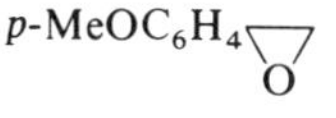

UV spectroscopic data: λ_{max} in nm, ε in l mol^{-1} cm^{-1} (in parentheses)

An investigation of acylaziridines was carried out by comparison of IR, NMR and MS data and included some 1,2-dibenzoylaziridines as well as 2-*p*-nitrobenzoyl-3-phenyloxaziridine ⟨68IZV1530⟩. Amide conjugation in acylated nitrogen-containing three-membered rings is weaker than in open chain acid amides.

The IR spectra of oxaziridines often show a band between 1430 and 1470 cm^{-1}, which is assumed to be due to C—H bending ⟨57JA5739⟩.

IR spectra of diaziridinones show carbonyl absorption at relatively high wave numbers, as found for aziridinones and cyclopropanones, absorbing at 1855–1880 cm^{-1} ⟨69JOC2254⟩.

Vibrational spectra including Raman data of 3,3-dimethyldiaziridine and its hexadeutero compound were recorded in the gas phase and in the crystalline state. Assuming C_2 symmetry and employing isotopic shifts and comparison with azetidine, a classification of bands which regarded 33 normal modes could be given ⟨75SA(A)1509⟩.

On the basis of IR data the basicity of some diaziridines was discussed. The basicity of the spirodiaziridines (**36**) increased on alkylation of one nitrogen from pK = 4.6 (R = H) to pK = 6.4 (R = Bu^n). pK values of diaziridines alkylated on both nitrogens were always below 5.0 ⟨69IZV1847⟩.

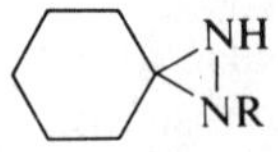

(**36a**) R = H
(**36b**) R = alkyl

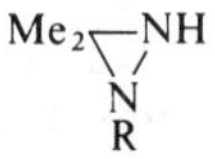

(**37a**) R = H
(**37b**) R = Me

Mass spectra of the same diaziridines were reported later ⟨74JOU1140⟩. Whereas in longer alkyl side chains of diaziridines the typical amine degradation by $n-1$ carbon atoms predominates, successive elimination of NH and methyl was observed in (**37a**). The 1-methyl compound (**37b**) undergoes competitive elimination of NH and MeN.

A series of mass spectra of oxaziridines was reported, including E and Z isomers of (**38**)–(**40**) ⟨69JCS(C)2650⟩. Besides the mole peak M and $M-1$ there are strong peaks at $M-16$ (loss of oxygen) and $M-31$ (loss of oxygen and a methyl group) as well as peaks indicating cleavage to give an aldehyde and the fragment alkyl—N.

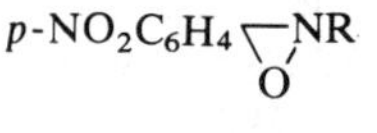

(**38**) R = Et
(**39**) R = Pr^i
(**40**) R = Bu^t

ArCH=N(O)R

(**41**)

In the discussion of some mass spectra of nitrones (**41**), intermediate isomerization to oxaziridines was concluded from the occurrence of aldehyde fragments.

The PE spectra of simple diaziridines exhibit two bands at lower ionization potentials; these are assumed to arise from removal of an electron from one or other of the lone pair MOs. An E arrangement of the two N-bound hydrogens in 3,3-dimethyldiaziridine (**37a**) is assumed, since the difference of the ionization potentials (0.51 eV) is much smaller than in the fused diaziridine (**42**) (1.40 eV) which necessarily has the Z configuration ⟨73HCA1614⟩.

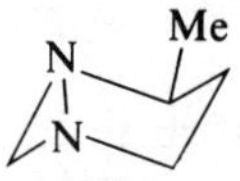

(**42**)

$(EtO_2C)_2$—NH, N·

(**43**)

ESR measurements on irradiation of 3,3-diethoxycarbonyldiaziridine in t-butyl peroxide showed signals of the diaziridinyl radical (**43**). The structure of the radical followed from coupling with two N atoms with $a_N = 14.2$ and $a_N = 11.7$ g and one proton with $a_H = 45.7$ g ⟨76TL4205⟩.

5.08.2.3.3 Molecular spectra of diazirines

^{1}H NMR signals of simple diazirines appear at very high field. Protons directly attached to the ring absorb at 0.40 p.p.m. in diazirine (**44**), at 0.63 p.p.m. in isopropyldiazirine (**45**), and at 0.60 p.p.m. in t-butyldiazirine (**46**) ⟨B-67MI50800⟩. Anisotropic magnetic susceptibility due to the diazirine ring is responsible for high field shift of protons at exocyclic carbon atoms. The methyl group in (**47**) absorbs at 1.1 p.p.m. The values reported for the

(**44**) R = H
(**45**) R = Pr^i
(**46**) R = Bu^t

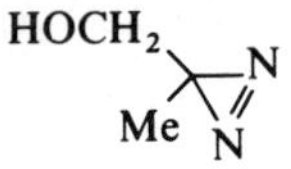

(**47**)

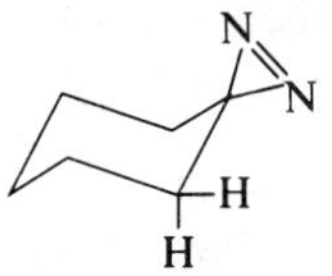

(**48**)

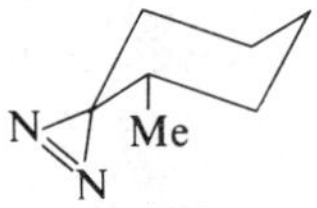

(**49**)

spirodiazirine (**48**) demonstrate appreciably different shielding of axial *versus* equatorial protons in the vicinity of the three-membered ring (0.6 p.p.m. for equatorial, 2.0 p.p.m. for axial protons) ⟨64JA4618⟩.

According to measurements on steroidal diazirines (**49**) there is also a high field shift of equatorial methyl groups, although their protons are more distant by one position from the diazirine ring. Reported values lie between 0.1 and 0.2 p.p.m., about 0.8 p.p.m. higher than in alkanes ⟨65JA2665⟩.

Chemical shifts of fluorine attached to the diazirine ring were reported to be 122.5 (another fluorine atom as the second substituent), 121.8 (OMe), 157.8 (CN) and 159.8 p.p.m. (NF_2) ⟨65JHC371⟩.

Diazirines exhibit very characteristic UV spectra, especially in the gas phase. That of the parent compound (**44**) is shown in Figure 1 ⟨62JA1063⟩. Absorption between 280 and 350 nm consists in a series of intense, relatively sharp bands. The spectra of other diazirines, including fluorodiazirines, bear great resemblance to that of the parent compound. The series of many bands is the result of excitation from different levels of vibration. The connections between vibrational structure of UV spectra of diazirines and energy levels known from IR investigations have been analyzed repeatedly ⟨69JCP(51)33, 72JCP(56)2919⟩. An exact analysis of the UV spectra of diazirine and isotopically labelled derivatives shows the existence of two systems separated by 220 cm^{-1} ⟨66JSP(19)372⟩.

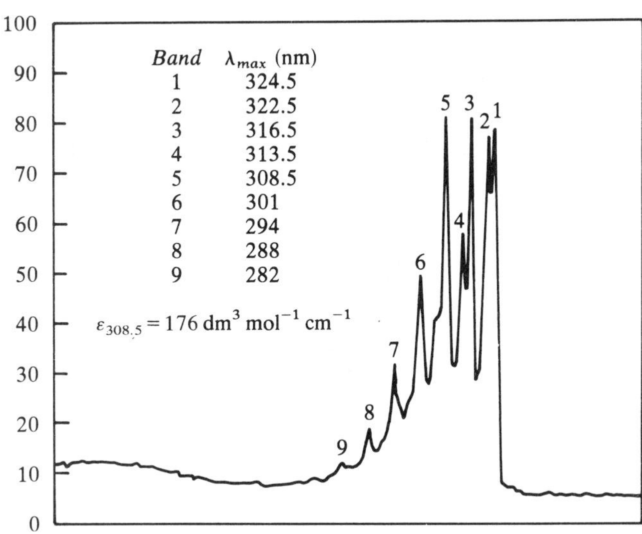

Figure 1 UV spectrum of diazirine in the gas phase

Further investigations were carried out with 3-methyldiazirine ⟨72JCP(57)941⟩ and with 3-chloro-3-methyldiazirine ⟨72JSP(42)403⟩. The high resolution electronic spectra were submitted to vibrational analysis; deuterated derivatives were included. All excitation and ground state fundamentals observed could be assigned.

In the IR spectrum of diazirine (**44**), complete assignment of vibrations could be performed when deuterated derivatives were used (Table 1) ⟨64JCP(40)1693⟩.

Table 1 Assignment of IR Bands of Diazirine (**44**) in the Ground State

No.	*Vibration*	*Frequency* (cm^{-1})
1	C—H stretching	3020
2	N—N stretching	1626
3	CH_2 deformation	1458
4	C—N stretching	991
5	CH_2 twisting	970
6	CH_2 wagging	967
7	C—N stretching	807
8	C—H stretching	3131
9	CH_2 rocking	1125

A compilation of N—N stretching frequencies of 15 halo substituted diazirines shows a range from 1540 cm^{-1} (3-chloro-3-fluorodiazirine) to 1600 cm^{-1} (difluoroaminofluorodiazirine) ⟨B-67MI50800⟩. A complete assignment of ground state vibrations of difluorodiazirine was given on the basis of IR and Raman data ⟨79SA(A)895⟩. Use of wave numbers of the diazirine spectrum allowed band assignments for more complicated diazirines like protonated dimethyldiazirine, 3-chloro-3-methyldiazirine and methyldiazirine ⟨69JSP174⟩.

The following bands of the PE spectrum of dimethyldiazirine were reported : $IP(\pi)$ 12.1, $IP(n_+)$ 13.3, $IP(n_-)$ 9.75 eV. Thus the $\psi(\pi)$-orbital lies between the orbitals of the lone electron pairs. Interpretation of the spectrum suggested that the 'lone' electron pairs are not so lonely, but are localized at the N atoms to only 56% and 43% respectively ⟨73HCA1933⟩. This effect is known, but less markedly, from (*Z*)-azomethane. PE spectra of the parent diazirine and the difluoro compound were also recorded. Comparison with calculated spectra gave good agreement ⟨72JCP(57)1758⟩.

The energy difference between diazirine and diazomethane, interesting from the point of view of their isomerism, came from MS measurements ⟨63JCP(39)3534⟩. The appearance potentials of the CH_2^+ ion, common to both compounds, yielded a difference in heats of formation of 125 kJ mol^{-1}. A strong peak in the mass spectrum of 3-chloro-3-methyldiazirine (**50**) with relative mass 55 was ascribed to the methyldiazirinium ion (**51**).

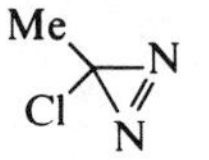

(**50**) 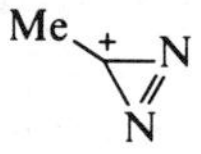 (**51**)

5.08.3 REACTIVITY OF THREE-MEMBERED RINGS WITH TWO HETEROATOMS

5.08.3.1 Reactivity of Oxaziridines

Thermal and photochemical treatment of oxaziridines, action of electrophiles and nucleophiles, as well as homolytic reactions almost always lead to destruction of the three-membered ring. Acid amides, carbonyl compounds, ammonia and amines are the main products of these reactions in the field of 2-alkyloxaziridines, thoroughly investigated and mechanistically classified by one of the discoverers of this class of compound ⟨57JA5739⟩. In the field of 2-aryloxaziridines the most important reactions like isomerization to nitrones and to acid amides were also described by the discoverers ⟨58JOC651, 65JOC3427⟩. As to reactivity, oxaziridines unsubstituted at nitrogen and their *N*-acyl derivatives form a special class of compound; they are characterized by their ability to transfer their nitrogen function to reaction partners ⟨B-67MI50800⟩.

5.08.3.1.1. Reactions of oxaziridines with conservation of the three-membered ring

There are only few reactions known introducing substituents to the H-bearing nitrogen of oxaziridines. *N*-Alkylation of 1-oxa-2-azaspiro[2.5]octane (3,3-pentamethylene-oxaziridine; **52**) with *t*-butyl chloride to give (**53**) was carried out for structure proof of (**52**). This reaction is of no preparative importance, since *N*-alkylated oxaziridines are easily obtained by ring synthesis.

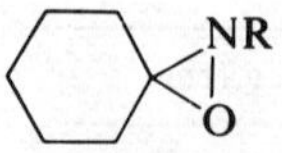

(**52**) R = H
(**53**) R = Bu^t

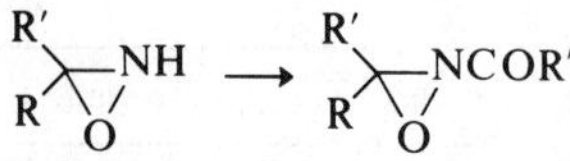

(**54**) (**55**) R″ = Me, Ph, NHPh, NH$_2$

N-Acylation of oxaziridine (**54**) is of more importance, yielding 2-acyloxaziridines which were unaccessible otherwise until recently. Oxaziridines (**54**) derived from cyclohexanone, butanone or benzaldehyde are acylated readily by acetic anhydride, acid chlorides or isocyanates. Oxaziridines from aliphatic aldehydes, too unstable to be isolated, may be trapped *in situ* by benzoylation ⟨67CB2593⟩.

5.08.3.1.2 Reactions of oxaziridines involving no partner

Some oxaziridines easily rearrange to the isomeric nitrones. The half reaction time of 2-*t*-butyl-3-phenyloxaziridine (**56**) is 212 min in diethylcarbitol at 100 °C; the enthalpy of activation is 117 kJ mol^{-1} and the entropy of activation −12 J K^{-1} mol^{-1} ⟨57JOC1263⟩. The scope of the reaction is fairly limited. A smooth reaction is only observed when an aryl substituent at carbon stabilizes the developing positive charge. Oxaziridine (**57**), arylated at carbon, rearranges by warming for only 1 h in isopropyl alcohol ⟨62JOC4671⟩. In the group of the especially unstable 2-aryloxaziridines the effective electron donor 4-dimethylaminophenyl causes rearrangement of (**58**) in the record time of 5 s at room temperature in ethanol ⟨65JOC3427⟩.

Ph—NBu^t (O) ⟶ PhCH=N(O)Bu^t (**56**)

(**57**)

p-$Me_2NC_6H_4$—$NC_6H_4NO_2$-*m* (O) (**58**)

Retardation by a factor of 1000 by use of benzene in place of ethanol in the rearrangement of (**58**) points to appreciable charge separation in the transition state.

Rearrangement involving cleavage of the C—O bond is also observed with the phenyl isocyanate adduct (**59**). In place of an *O*-acylated nitrone its product (**60**) of acyl migration is isolated ⟨67JPR(36)86⟩.

Ph—NCONHPh (O) (**59**) —toluene, 1 h, 100 °C→ PhCH=NOCONHPh (**60**)

Some 2-acyloxaziridines (**61**) undergo thermal ring enlargement to form the 1,3,4-dioxazoles (**62**) ⟨67CB2593⟩. From the conditions necessary for complete reaction a polar cleavage of the C—N bond was deduced, since better stabilization of a negative charge at nitrogen on going from benzoyl to 4-nitrobenzoyl enhances the reaction whereas decreasing cation stabilization at carbon on going from 3,3-dialkyl- to 3-monoalkyl-oxaziridines hinders the reaction.

R_2—NCOR′ (O) (**61**) ⟶ (**62**)

The isomerization of oxaziridines (**1**) to acid amides with migration of a substituent from C to N is a general reaction and is always observed when no other reactions predominate under the relatively harsh conditions (heating to above 150 °C or photolysis). Even then one can make acid amide formation the main reaction by working at 300 °C ⟨57JA5739⟩ and by dilution techniques. For example, caprolactam (**63**) is formed in 88% yield by flash pyrolysis of oxaziridine (**52**) at about 300 °C, whereas decomposition of (**52**) at lower temperatures gives almost no (**63**) ⟨77JPR274⟩.

R_2—NR (O) (**1**) ⟶ $RCONR_2$

(**52**) —300 °C→ (**63**)

In the course of acid amide or lactam formation, migrations of H, alkyl or aryl proceed, in many cases competitively. The migration tendency was poorly understood for a long time but appears now to depend less on the nature of the migrating group but on its geometrical position; both thermal and photochemical experiments have demonstrated that migration of the substituent *trans* to the nitrogen lone pair must be assumed (**64** to **65**) ⟨80JHC1025⟩.

(**64**) ⟶ (**65**)

In the 2-aryloxaziridines, acid amide formation proceeds under mild conditions and is the most often observed stabilization reaction of these very unstable compounds. Heating at 75 °C for 30–45 min is sufficient to convert *N*-arylspirooxaziridines (**64**; R = Ar) to the isomeric lactams (**65**) in 75–90% yield.

2,3-Diaryloxaziridines are even less stable. With oxaziridines (**66**; Ar = Ph or 4-$Me_2NC_6H_4$) acid amide formation at 25 °C proceeded in the course of 66 and 10 h respectively; in the latter case there were equal amounts of H and aryl migration. Ethanol as solvent again accelerated the reaction, in the latter case by a factor of 10^4.

Ar–NPh (O) → ArCONHPh + HCON(Ph)Ar

(**66**)

In some cases acid amide formation was observed on attempted deprotonation at oxaziridine ring carbon. 2-*t*-Butyl-3-(4′-nitrophenyl)oxaziridine (**67**) was converted to the anion of acid amide (**68**) by sodium amide ⟨69TL3887⟩, while 2-(4′-nitrobenzoyl)-3-phenyl-oxaziridine (**69**) afforded the diacylimide (**70**) by addition of cyclohexylamine to its benzene solution at room temperature ⟨67CB2593⟩.

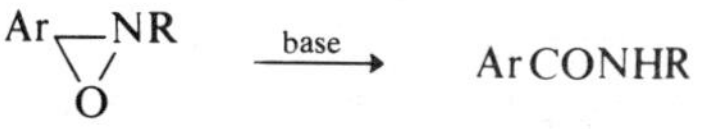

Ar–NR (O) —base→ ArCONHR

(**67**) Ar = p-$NO_2C_6H_4$, R = Bu^t (**68**) R = Bu^t
(**69**) Ar = Ph; R = p-$NO_2C_6H_4CO$ (**70**) R = Ph

Oxaziridines are claimed as intermediates in photolysis of aromatic *N*-oxides in many investigations. Although they are plausible intermediates, direct evidence of their existence was given only recently. Isolated products are mostly formulated to arise from a primarily formed oxaziridine (**71**) either by electrocyclic ring enlargement to (**72**) or by 1,5-shift of a bond issuing from oxaziridine oxygen to give (**73**), followed by stabilization reactions of (**72**) or (**73**) ⟨*e.g.* 70TL2329⟩.

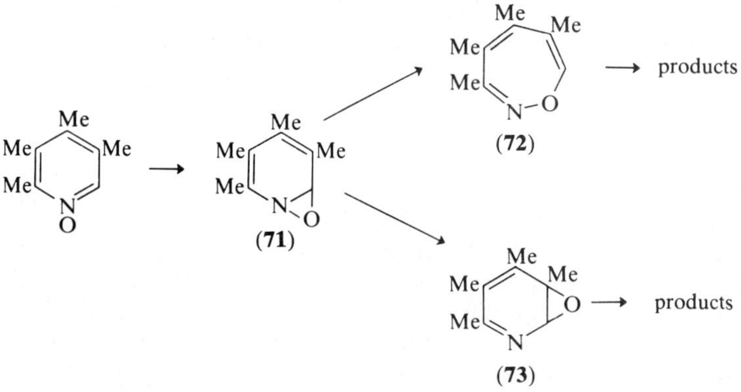

(**71**) → (**72**) → products; (**71**) → (**73**) → products

Direct proof of an oxaziridine intermediate was achieved in photolysis experiments in an organic glass at 77 K ⟨80JA5643⟩. Oxaziridine (**75**), formed by photolysis of *N*-oxide (**74**) and evidenced by UV spectroscopy under the above conditions, decomposed at higher temperature to form the imino ether (**76**) by N—O bond cleavage and C → O migration of an aryl group.

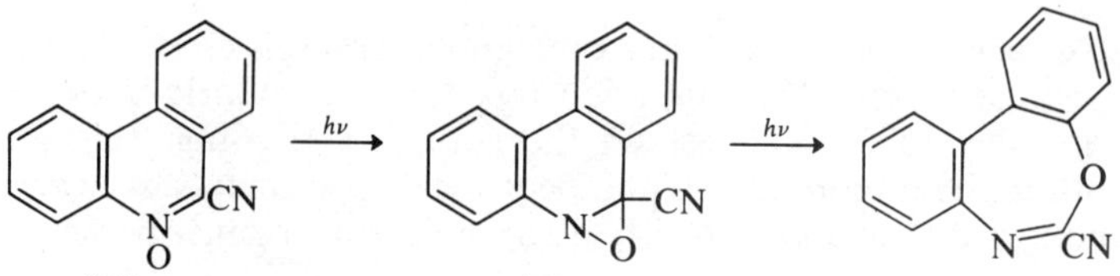

(**74**) —hν→ (**75**) —hν→ (**76**)

5.08.3.1.3 Reactions of oxaziridines with electrophiles

Like isomerization of oxaziridines to nitrones, acid hydrolysis of oxaziridines proceeds with conservation of the N—O bond. Both reactions are related mechanistically and are subject to the same limitation of scope and the same substituent effects. 3-Aryloxaziridines with *t*-alkyl groups at nitrogen are hydrolyzed to the aromatic aldehyde and *t*-alkylhydroxylamine in 70–85% yield, although under unexpectedly severe conditions — action of at least 2M sulfuric acid at 20 °C for 20 h ⟨57JA5739⟩.

According to a kinetic study which included (**56**), (**56a**) and some oxaziridines derived from aliphatic aldehydes, hydrolysis follows exactly first order kinetics in 4M $HClO_4$. Proton catalysis was observed, and there is a linear correlation with Hammett's H_0 function. Since only protonated molecules are hydrolyzed, basicities of oxaziridines ranging from $pK_A = +0.13$ to -1.81 were found from the acidity rate profile. Hydrolysis rates were $1.49 \times 10^{-3}\ min^{-1}$ for (**56**) and $43.4 \times 10^{-3}\ min^{-1}$ for (**56a**) ⟨71JCS(B)778⟩. *O*-Protonation is assumed to occur, followed by polar C—O bond cleavage. The question of the place of protonation is independent of the predominant *N*-protonation observed spectroscopically; under equilibrium conditions all protonated species are thermodynamically equivalent.

Ar—NBu^t (oxaziridine) ⟶ ArCHO + Bu^{t}NHOH

(**56**) Ar = Ph
(**56a**) Ar = *p*-$NO_2C_6H_4$

(**77**)

The reason for the slow hydrolysis compared to that of structurally similar compounds like nitrones or *O,N*-acetals might be the following ⟨B-67MI50800⟩: in the protonated species (**77**) assistance of the lone pair of electrons at nitrogen is sterically hindered due to the large angle of its orbital to the plane of the ring.

The small tendency of oxaziridines to undergo hydrolytic C—O cleavage is underlined by the observation that oxaziridine (**52**) in a mixture of acid and ^{18}O-labeled water yields up to 50% labeled hydroxylamine. NH transfer to water successfully competes with hydrolysis ⟨B-67MI50800⟩.

(**52**) $\xrightarrow{H^+,\ H_2{}^{18}O}$ $H_2N^{18}OH$

Oxaziridine hydrolysis corresponds to acetal hydrolysis mechanistically. Thus, its rate decreases with decreasing substitution by cation stabilizing groups at position 3. In oxaziridines derived from aliphatic aldehydes, N—O cleavage competes successfully with C—O cleavage. N—O cleavage is the main reaction in oxaziridines prepared from formaldehyde. 2-*n*-Butyloxaziridine in ethanolic solution is decomposed by sulfuric acid to formaldehyde, butyraldehyde and ammonia. Polar N—O cleavage with concomitant H-migration according to the arrow in (**78**) is assumed to occur ⟨57JA5739⟩.

(**78**) ⟶ $\xrightarrow{H_2O}$ $H_2CO + PrCHO + NH_3$

Migration to the developing electron sextet at nitrogen is not restricted to hydrogen. In (**79**) there is methyl migration with formation of methylamine and acetone; in the acid-catalyzed decomposition of (**80**), phenyl migration leads to aniline and acetaldehyde.

(**79**) (**80**)

With a peroxyacid, the reagent used in their preparation, oxaziridines further react to yield aliphatic nitroso compounds. An electrophilic attack to ring nitrogen is plausible, leading to an intermediate oxaziridine *N*-oxide (**81**), which immediately decomposes to a nitroso compound and an aldehyde ⟨57JA6522⟩.

Me—NPr^i (oxaziridine) $\xrightarrow{AcOOH}$ (**81**) ⟶ MeCHO + $(Pr^iNO)_2$

Phenyl isocyanate reacts with (**82**) to form (**83**) by ring enlargement ⟨74JOC948⟩. Reaction of oxaziridines with diphenylketene proceeds less obviously. Isolated products (**84**) and (**85**) contain only the R—N group of the starting material ⟨69TL263⟩.

Ph—NR (O) —PhNCO→ Ph—NR / PhN—O / C=O

(**82**) (**83**)

R″, R′ —NR (O) —Ph_2CCO→ Ph_2, O, NR, =C(Ph)Ph + Ph_2, Ph_2, NR, O, O

(**84**) (**85**)

5.08.3.1.4 Reactions of oxaziridines with nucleophiles and reducing agents

The oxaziridine ring itself is stable towards alkali; there is, for instance, no substitutive ring opening by hydroxyl ions as in oxiranes. 2-*t*-Butyl-3-phenyloxaziridine (**56**) is not attacked by methoxide ion in methanol during 12 h at room temperature; 3-isopropyl-2-*t*-octyloxaziridine does not react at room temperature with either solid potassium hydroxide or potassium methoxide solution ⟨57JA5739⟩.

In some cases, especially in the presence of strongly electron attracting substituents, isomerization to acid amides has been observed, probably preceded by deprotonation at ring carbon. Even (**56**), known for its stability towards common alkali, undergoes this rearrangement when a lithium amide is used as the base ⟨80JOC1489⟩.

If, however, hydrogen is present in the α-position of the *N*-alkyl substituent, 2-alkyl-oxaziridines are easily decomposed by alkali. Base attack on this H atom effects 1,2-elimination at the C—N bond. From (**86**) and aldimine (**87**) forms, and a mixture of ammonia and two carbonyl compounds is finally obtained, one of them containing the carbon atom of the original oxaziridine ring, the other originating from the *N*-alkyl group ⟨57JA5739⟩.

$$R_2C(O)N{-}CH(H)R' \xrightarrow{^-OH} R_2C(N{=}CHR')O^- \longrightarrow HN{=}CHR' + R_2CO \rightarrow R_2CO + R'CHO + NH_3$$

(**86**) (**87**)

The decomposition of an oxaziridine by brucine, originally assumed to be a deoxygenation, proved to be a base-catalyzed decomposition ⟨79TL3353⟩.

Oxaziridines are powerful oxidizing agents. Free halogen is formed from hydrobromic acid ⟨B-67MI50800⟩. Reduction by iodide in acidic media generally yields a carbonyl compound, an amine and two equivalents of iodine from an oxaziridine (**1**). With 2-alkyl-, 2-acyl and with *N*-unsubstituted oxaziridines the reaction proceeds practically quantitatively and has been used in characterization. Owing to fast competing reactions, iodide reduction of 2-aryloxaziridines does not proceed quantitatively but may serve as a hint to their presence.

$$R_2C(O)NR' \xrightarrow{2I^-,\ 3H^+} R_2CO + R'\overset{+}{N}H_3 + I_2$$

$$R_2C(O)NR' \xrightarrow{Bu_3P} R_2C{=}NR' + Bu_3PO$$

(**1**)

Strong reducing agents were reacted with oxaziridines for structural proof ⟨57JA5739⟩. Lithium alanate yielded either Schiff bases by deoxygenation or secondary amines by further hydrogenation. Platinum-catalyzed hydrogenation also led to secondary amines. Tertiary phosphines reduced oxaziridines to Schiff bases forming phosphine oxides themselves, thus suggesting direct attack of phosphorus on oxaziridine oxygen.

The competitive attack at either oxygen or nitrogen by nucleophiles was the subject of a recent investigation ⟨81JOC610⟩. Whereas triphenylphosphine reacted with oxaziridine (**88**) preferentially by attack on oxygen (78%), attack on nitrogen predominated in reactions of the same oxaziridine with Ph_3As, Ph_2S and Ph_2Se. Increasing size of the *N*-alkyl group favored *O*-attack in these cases also.

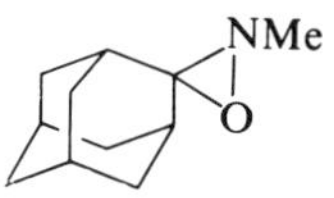

(88) (89) (90) (91)

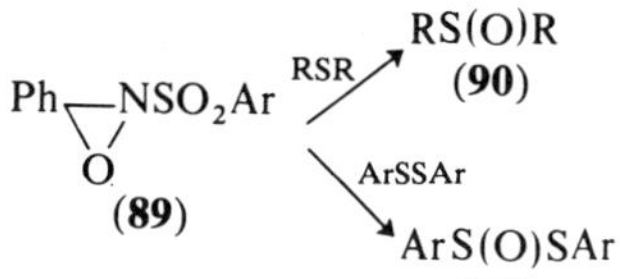

2-Sulfonyloxaziridines were recently proposed as O-transferring reagents. Oxaziridine (**89**) converted thioethers to sulfoxides (**90**) and diaryl disulfides into their *S*-oxides (**91**) ⟨78TL5171⟩. Epoxidations are also possible ⟨81TL917⟩.

Experiments with cyclic thioethers ⟨80JCS(P1)1693⟩, thiourea, thiocyanate and ethyl xanthate always led to destruction of oxaziridines ⟨73AJC2159⟩. Products of complicated consecutive reactions could be isolated but only with some difficulty, *e.g.* (**92**) from a reaction with carbon disulfide ⟨74JOC957⟩, and (**93**), obtained by trapping with butadiene a product of a reaction between an oxaziridine and a thiirane ⟨80JOC1691⟩.

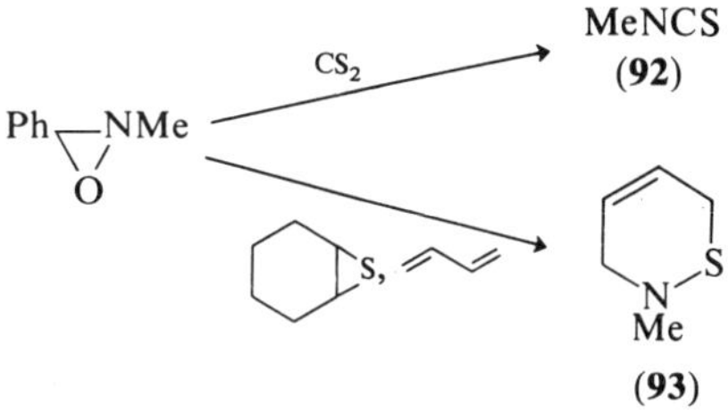

(92) (93)

The marked tendency of 1-oxa-2-azaspiro[2.5]octane (**52**) to transfer its NH group gives rise to smooth reactions with sulfur compounds. Aminothiocyanate (**94**) is formed in 91% yield at room temperature within one minute and 4-methylbenzenesulfonamide (**95**) is formed within five minutes in 84% yield from the sulfinate ⟨68TH50800⟩.

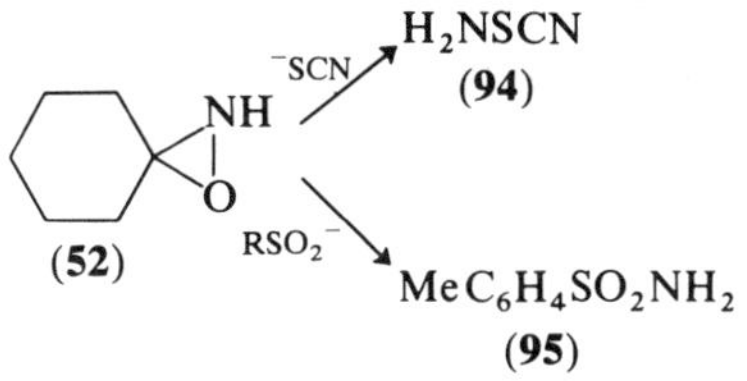

(52) (94) (95)

It was reported only recently that *N*-methyl transfer from an oxaziridine to an amine occurs with formation of an N—N bond ⟨79JA6671⟩. N—N bond forming reactions with *N*-unsubstituted oxaziridines had been found immediately after discovery of this class of compound ⟨64CB2521⟩ and have led to simple hydrazine syntheses ⟨79AHC(24)63⟩. Secondary amines like diethylamine or morpholine are *N*-aminated by (**52**) in the course of some minutes at room temperature with yields exceeding 90% ⟨77JPR195⟩. Further examples are the amination of aniline to phenylhydrazine, and of the Schiff base (**96**) to the diaziridine (**97**).

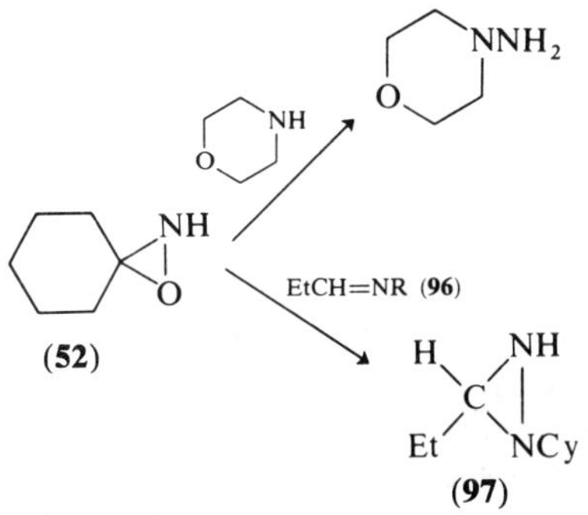

(52) (97)

3-Trichloromethyloxaziridine (**98**) transfers its NH group to primary amines under the conditions of its formation from hydroxylamine-*O*-sulfonic acid. Thus the slow alkylhydrazine synthesis from amines and hydroxylamine-*O*-sulfonic acid is effectively catalyzed by chloral hydrate, which takes up and transfers an NH group. Oxaziridine (**98**) quickly forms hydrazine with ammonia at room temperature.

$$Cl_3CCH(OH)_2 + H_2NOSO_3H + RNH_2 \longrightarrow Cl_3C\text{-oxaziridine-NH}\ (\mathbf{98}) \xrightarrow{RNH_2} RNHNH_2$$

Several alkenes are converted to aziridines by treating with oxaziridine (**52**) at elevated temperatures. Styrene, α-methylstyrene and their derivatives substituted in the benzene ring react smoothly, and so do 1,1-diphenylethylene, indene and acrylonitrile ⟨74KGS1629⟩.

Ph(Me)C=CH₂ —(**52**), toluene, 3 h, 110 °C→ 2-Ph-2-Me-aziridine (N–H), 46%

The mechanism of the reaction is unknown. The stereospecificity observed with (*E*)- and (*Z*)-1-methyl-2-phenylethylene points to a one-step reaction. The very low Hammett constant, −0.43, determined with phenylethylenes substituted in the benzene ring, excludes polar intermediates. Yields of only a few percent are obtained in the reaction of aliphatic alkenes with (**52**). In the reaction of cyclohexene with (**52**), further amination of the aziridine to aminoaziridine (**99**) is observed. Instead of diphenylazirine, diphenylacetonitrile (**100**) is formed from diphenylacetylene by NH uptake from (**52**) and phenyl migration.

(**52**) —cyclohexene→ cyclohexene-fused aziridine with N–NH₂ (**99**)

(**52**) —PhC≡CPh→ Ph_2CHCN (**100**)

N-Acetyl derivatives of 3-phenyloxaziridine can also transfer their nitrogen function to nucleophiles. 2-(4′-Nitrobenzoyl)-3-phenyloxaziridine (**69**) converts piperidine to the acylhydrazine (**101**) in 92% yield within some minutes at room temperature ⟨67JPR(36)86⟩. Since (**69**) is stable in the absence of a nucleophile a nitrene is not involved in the reaction, which is assumed to occur by nucleophilic attack of the amine on the oxaziridine nitrogen.

Ph-oxaziridine-$NCOC_6H_4NO_2$-*p* (**69**) —piperidine→ piperidine-$NNHCOC_6H_4NO_2$-*p* (**101**)

Smooth acylaminations are also given by isocyanate adducts of 3-phenyloxaziridine. Ammonia and primary as well as secondary amines form the corresponding semicarbazides, *e.g.* (**102**). Owing to the fast amination reaction, shortlived substances like triazanes become accessible, *e.g.* the cyclohexyl compound (**103**), which is stable for only a few minutes at room temperature.

oxaziridine-NCONHR (R = H, Ph) —$C_6H_{11}NH_2$→ $C_6H_{11}NHNHCONHPh$ (**102**)

oxaziridine-NCONHR (R = H, Ph) —$C_6H_{11}NHNH_2$→ $H_2N(C_6H_{11})NNHCONHPh$ (**103**)

If there is no phenyl substituent in the 3-position the amination ability decreases. The acyloxaziridine (**104**) yields only 11% of a semicarbazide derivative with piperidine. In the presence of strong bases an intramolecular amination competes. Compound (**104**) reacts with methoxide within a couple of seconds to give phenylhydrazine carboxylic ester (**106**), and with cyclohexylamine to give the substituted semicarbazide (**107**). A diaziridinone (**105**) is assumed to be the common intermediate, formed by an intramolecular reaction from deprotonated (**104**) ⟨67CB2600⟩.

Ph HN CO N O (104) $\xrightarrow{-H^+}$ Ph N CO N O^- (105) $\xrightarrow{MeOH}$ $PhNHNHCO_2Me$ (**106**); $\xrightarrow{C_6H_{11}NH_2}$ $PhNHNHCONHC_6H_{11}$ (**107**)

A comparison of (**102**) with (**107**) shows that two similar starting materials reacting with the same compound give two semicarbazides with interchanged positions of substitution.

Marked reactivity towards nucleophiles like alcohols, amines and thiols was observed with the perfluorodiazirine (**108**) ⟨79MI50802⟩.

F_2—NCF_3 O (**108**) $\xrightarrow{Nu}$ FC(=O)N(Nu)CF_3

5.08.3.1.5 Homolytic reactions of oxaziridines

Oxaziridines substituted in the 2-position with primary or secondary alkyl groups undergo decomposition at room temperature. In the course of some weeks, slow decomposition of undiluted compounds occurs, the pattern of which is analogous to that of acidic or alkaline N—O cleavage (Sections 5.08.3.1.3 and 4). Radical attack on a C—H bond in (**109**) effects N—O cleavage, probably synchronously ⟨57JA5739⟩. In the example presented here, methyl isobutyl ketone and ammonia were isolated after two hour's heating at 150 °C.

Me, Bu^i, N—CHEt (H, R·) O (**109**) ⟶ Me, Bu^i, C(N=CHEt, HR)(O·) $\xrightarrow{(109)}$ $MeCOBu^i + EtCHO + NH_3$

Homolytic oxaziridine decomposition can be easily initiated by iron(II) ion in acidic media. Catalytic amounts are sufficient because chain reactions proceed. The reaction proceeds obviously in the case of 2-*t*-alkyloxaziridines like (**56**), where it yields the isomeric acid amide (**111**) ⟨57JA5739⟩.

Ph—NBu^t O (**56**) $\xrightarrow[H^+]{Fe^{2+}}$ H, $NHBu^t$, C, Ph, O· (**110**) $\xrightarrow[-H^+]{Fe^{3+}}$ $PhCONHBu^t$ (**111**)

When β-scission can occur in the radical, further reactions compete with acid amide formation. Thus oxaziridine (**112**) with iron(II) ion and acid yields stabilization products of the isopropyl radical. If α-hydrogen is present in the *N*-alkyl group, radical attack on this position in (**113**) occurs additionally according to the pattern of liquid phase decomposition.

Pr^i—NBu^t O (**112**) ⟶ propane + propene; Et_2—NEt O (**113**) ⟶ ⟶ $Et_2CO + NH_3$

In spirooxaziridines like (**114**), β-scission proceeds with ring opening. Stoichiometric amounts of iron(II) salt in acidic solution lead to the dicarboxylic acid derivative (**115**). The radical undergoes some interesting reactions with added unsaturated compounds. For example, pyridine yields a mixture of 2- and 4-alkylation products in 80% yield. Catalytic amounts of iron(II) ion are sufficient here since the adduct of the radical with pyridine is oxidized by iron(III) ion to the final product (**116**), thus regenerating iron(II) ion ⟨68TL5609⟩.

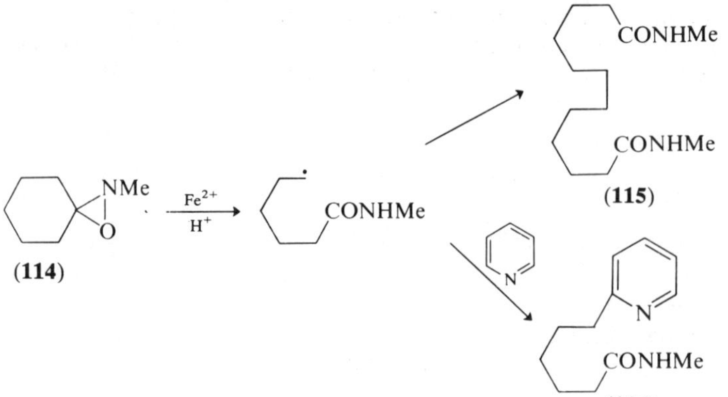

(114) (115) (116)

β-Hydroxyoxaziridines like (**117**) undergo double ring cleavage to form, for example, a nine-membered keto lactam ⟨81AG(E)670⟩.

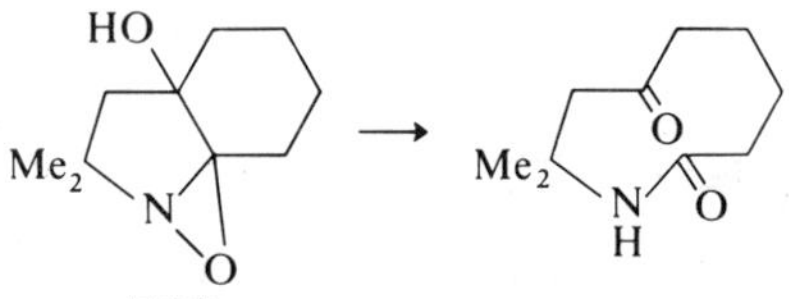

(117)

Action of catalytic amounts of vanadium compounds on oxaziridine (**52**) yields caprolactam almost quantitatively. Reductive opening of the oxaziridine ring and β-scission yield radical (**118**), which recyclizes with elimination of the metal ion to form the lactam (**63**) ⟨77JPR274⟩.

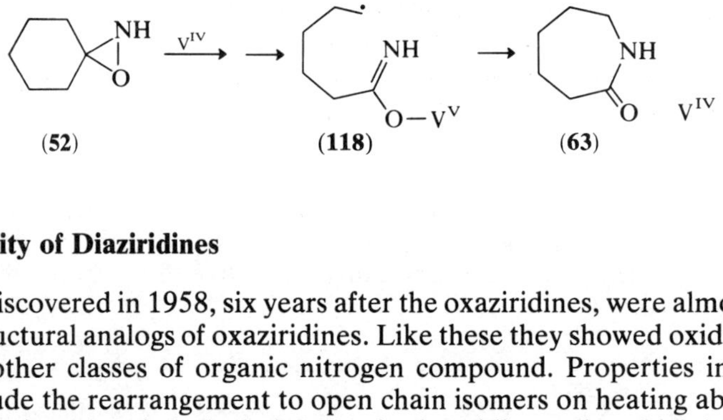

(52) (118) (63)

5.08.3.2 Reactivity of Diaziridines

Diaziridines, discovered in 1958, six years after the oxaziridines, were almost immediately realized to be structural analogs of oxaziridines. Like these they showed oxidizing properties unexpected for other classes of organic nitrogen compound. Properties in common with oxaziridines include the rearrangement to open chain isomers on heating above 100 °C (for several diaziridines), and their hydrolytic behavior in acidic media, which leads to carbonyl compounds with conservation of the hetero–hetero bond.

There are differences in the high temperature behavior. While oxaziridines almost always isomerize to acid amides, a similar reaction of diaziridines, which should lead to amidines, has not been observed. Sensitivity towards bases, often encountered in oxaziridines, is observed only in some special substituted diaziridines. The tendency of some classes of oxaziridines to transfer the nitrogen function also lacks in the diaziridine field. On homolytic reactions of diaziridines there are only a few observations.

Substitution with conservation of the three-membered ring is more versatile in diaziridine chemistry, because the ring is somewhat more stable and because of the presence of two nitrogen atoms.

Diaziridinones and diaziridinimines, discovered about 10 years later, add much to the versatility of the diaziridine field, especially due to valence isomerizations.

5.08.3.2.1 Reactions of diaziridines with conservation of the three-membered ring

Alkylation of diaziridines with simple alkylating agents has never been attempted, because *N*-alkyldiaziridines are as easily obtained by ring synthesis as the unsubstituted ones. There are reports on alkylation with acrylonitrile, butenone, alkenyl ketones, and vinyl sulfones ⟨79AHC(24)63⟩, *e.g.* with formation of compounds (**119**) ⟨74ZOR1510⟩ and (**120**) ⟨68LA(720)98⟩. Two successive alkylations were possible in 3,3-dialkyldiaziridines ⟨67BRP1081292⟩, *e.g.* with formation of (**121**) and (**122**).

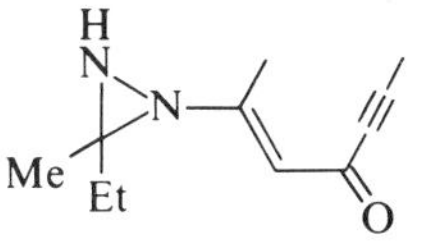

(**119**) 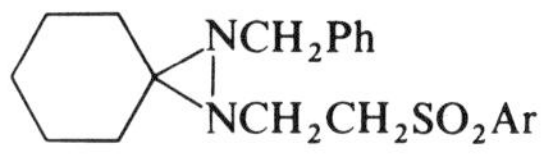 (**120**) 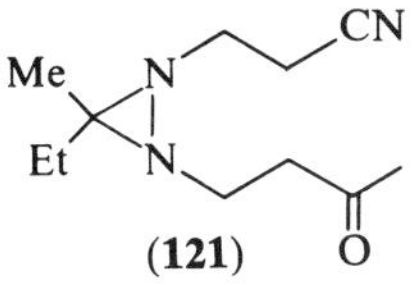 (**121**) 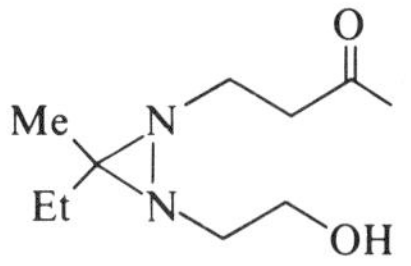 (**122**)

Telomerization of 3,3-dimethyldiaziridine with butadiene catalyzed by Pd complexes yielded 2:1 adducts (**123**) and (**124**) ⟨80IZV220⟩.

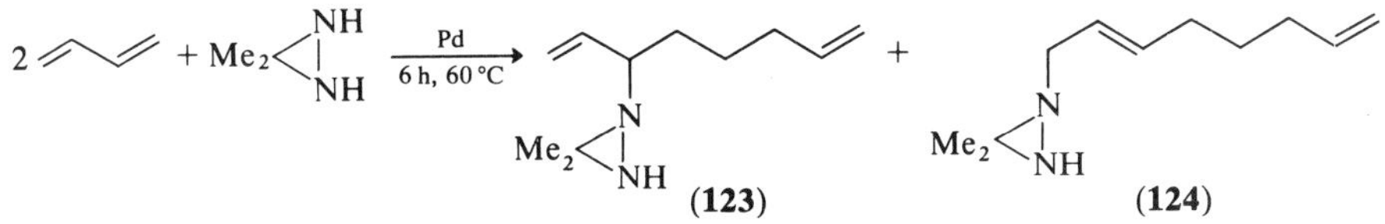

Pd, 6 h, 60 °C

(**123**) (**124**)

1-Aryldiaziridines (**125**), difficult to obtain by standard diaziridine syntheses, were prepared by arylation with 2,4-dinitrofluorobenzene ⟨72JOC2980⟩.

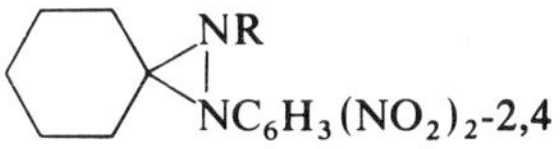

(**125**)  (**126**)

Chloral forms well-crystallized adducts (**126**) with diaziridines containing at least one NH group ⟨B-67MI50800⟩. Carbonyl addition products to formaldehyde or cyclohexanone were also described. Mixtures of aldehydes and ammonia react with unsubstituted diaziridines with formation of a triazolidine ring (**128**). Fused diaziridines like (**128**) are always obtained in ring synthesis of diaziridines (**127**) from aldehyde, ammonia and chloramine. The existence of three stereoisomers of compounds (**128**) was demonstrated ⟨76JOC3221⟩. Diaziridines form Mannich bases with morpholine and formaldehyde ⟨64JMC626⟩, *e.g.* (**129**).

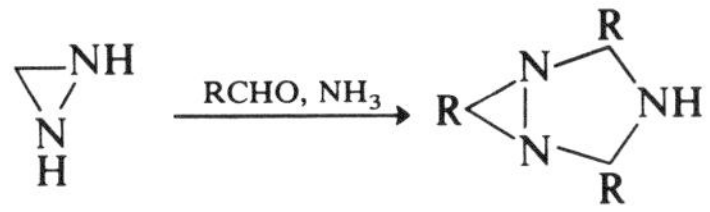

RCHO, NH_3

(**127**) (**128**)

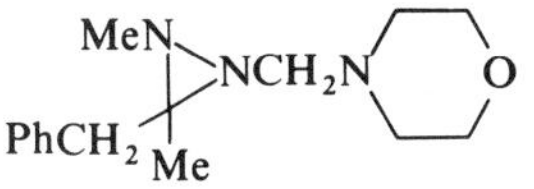

(**129**)  (**130**)  (**131**)

Acylations were often carried out with diaziridines. Twofold acylation is normally observed when two NH groups are present. Most acylations were performed with benzoyl chloride, acetyl chloride or phenyl isocyanate ⟨B-67MI50800⟩. Ring opening reactions during acylation, foreseeable for intermediates of electrophilic attack on nitrogen, were observed only seldom, provided mild conditions were used.

Phosphorylations of diaziridine nitrogen are also possible without ring opening. Compounds (**132**) ⟨67JMC101⟩ and (**133**) ⟨80ZOB1502⟩ were prepared using $POCl_3$ and $(RO)_2P(O)Cl$ in presence of triethylamine.

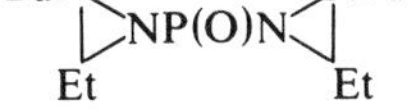

(**132**) 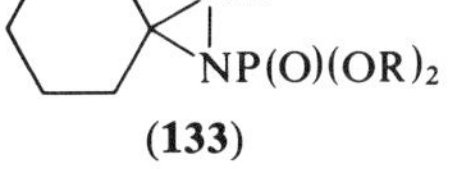 (**133**)

Tosylations, as well as silylations using trialkyl- or dialkyl-chlorosilanes in the presence of triethylamine, proceed with conservation of the three-membered ring ⟨69DOK(187)335⟩.

There are indications that smooth deacylations, known for 1-acylaziridines, are also encountered in diaziridine chemistry. The two benzoyl groups in (**130**) can be removed with 1M NaOH at room temperature during 20 h ⟨B-67MI50800⟩. Acyl cleavage proceeds with special ease in the *cis*-fused acyldiaziridine (**134**), where it is observed even at room temperature in methanol; thus (**134**) must be handled in *t*-butyl alcohol ⟨76JPR946⟩.

MeOH → MeO_2C ... HN ... Ph

(**134**) (**135**)

The diaziridine ring exhibits a surprising stability towards strong oxidizing agents. Diaziridines unsubstituted at both N atoms can be transformed into diazirines by dichromate in acidic solution or by chlorine (Section 5.08.4.3). Radical attack by decomposing peroxide converts (**136**) to the diaziridinyl radical (**137**), as evidenced by ESR spectroscopy ⟨76TL4205⟩.

$(EtO_2C)_2$ NH, NH → (RO·) $(EtO_2C)_2$ NH, N·

(**136**) (**137**)

5.08.3.2.2 Reactions of diaziridines with conservation of the N—N bond

Several reactions lead to opening of the diaziridine ring leaving the N—N bond intact. Besides the generally possible hydrolysis to hydrazines there are some thermal reactions of acylated diaziridines proceeding especially cleanly.

Dibenzoyldiaziridine (**130**) thermally rearranges with ring opening and benzoyl migration to the dibenzoylhydrazone (**138**) ⟨67CB142⟩.

NCOPh, NCOPh → $=NN(COPh)_2$

(**130**) (**138**)

Ring enlargement of *N*-acyl compounds, generally observed with oxaziridines, is observed only occasionally with diaziridines. Under more forced conditions of acylation with acetic anhydride, oxadiazolines like (**139**) were obtained ⟨76MIP50800⟩. A 4-nitrobenzoyl derivative rearranged at room temperature ⟨76JOC3229⟩.

R′, R, NH, N, H → $(R''CO)_2O$, 150 °C → HN—N, R′, R, O, R″

(**139**)

N-Carbamoyldiaziridines can open the three-membered ring and recyclize through the carbamoyl nitrogen, as demonstrated in a benzoyl isocyanate adduct ⟨74JOC3198⟩ and in the formation of (**140**) ⟨79JOC3935⟩.

Me, Ph, NC_6H_{11}, N, CONHPh → 110 °C → C_6H_{11}, N—N, Me, Ph, N, Ph, O

(**140**)

In the investigation of a ring enlargement, discrimination was possible between the reaction paths *via* a 1,3-dipole (**142**) and a quaternary acyl product (**143**), precursors of different triazolidones. Exclusive formation of (**144**) from (**141**) and phenyl isocyanate decided in favor of the dipolar intermediate (**142**) ⟨81JOC320⟩.

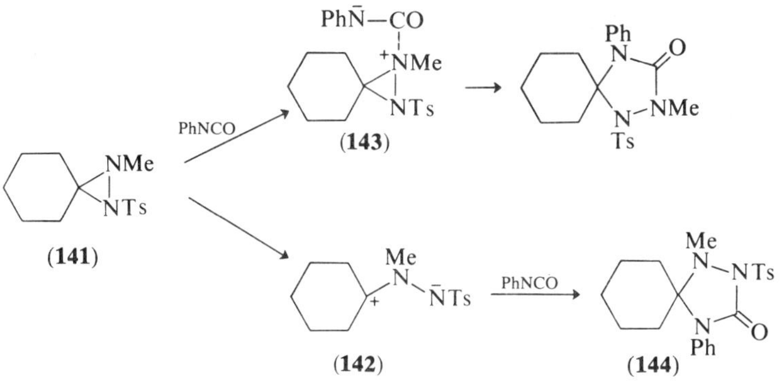

(141) (143) (142) (144)

Clean examples of diaziridine to hydrazone rearrangements are rare. Diaziridine (**119**) mentioned above rearranges to the isomeric enhydrazone in boiling toluene, and 2,4-dinitrophenyldiaziridine (**125**) under the same conditions affords the 2,4-dinitrophenylhydrazone (**145**) within 4 h. On blocking this rearrangement by *N*-methyl, conversion with loss of cyclohexanone occurred to give benzotriazole *N*-oxide (**146**) ⟨72JOC2980⟩.

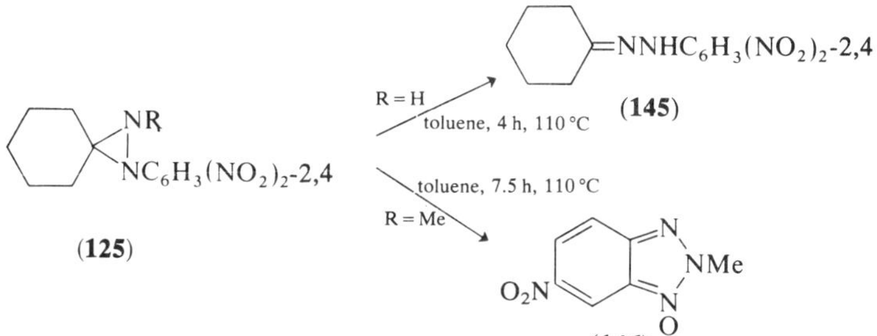

(125) (145) (146)

Another ring opening reaction was observed on heating diacyldiaziridine (**147**) in benzene for 10 h. 1,2-Elimination led to enhydrazide (**148**) ⟨69TH50801⟩.

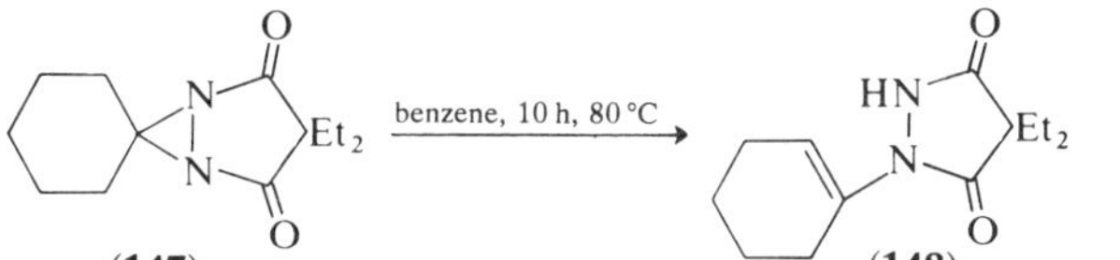

(147) (148)

Explanation of the above isomerizations through 1,3-dipoles is plausible. In some of the following reactions unassisted ring opening of the diaziridines has not been ruled out. Rate comparisons of diaziridine ring opening in the absence and presence of reaction partners would be desirable. Since diaziridine ring opening is in no case reversible, rate increases in the presence of the partner would indicate its participation in ring opening.

Formal addition products of 1,3-dipoles are compiled in a recent review ⟨79AHC(24)63⟩. Examples include addition to a nitrone forming a six-membered ring, to an enamine forming a pyrazolidine ring, and to the C=O bond of diphenylcyclopropenone.

Maleic hydrazide (**149**), produced on a large scale as a herbicide, can be obtained, according to a patent, from a simple diaziridine, which on an industrial scale could be less expensive than hydrazine hydrate. In the proposed manufacture of diamide (**150**) from a diaziridine and urea it must be taken into account that the reaction of urea is preceded by solvolytic hydrazine formation, which probably proceeds under the reported conditions (several hours, 70–80 °C, acidic media) ⟨79AHC(24)63⟩.

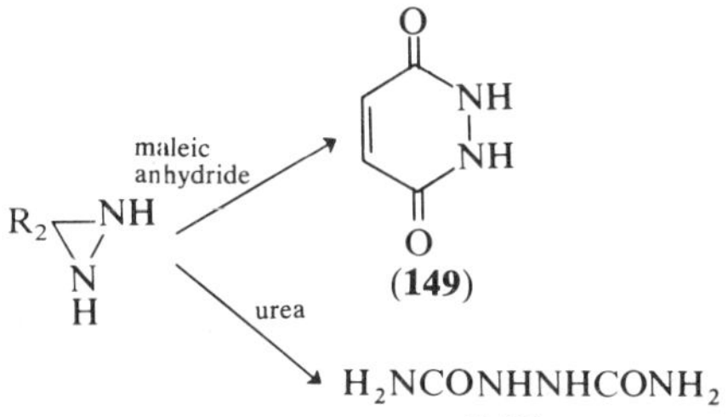

(149)

$H_2NCONHNHCONH_2$
(150)

Practically all diaziridines (**151**) can be hydrolyzed by acids to a carbonyl compound and a hydrazine derivative. The only exceptions are diaziridines derived from formaldehyde, in which acid catalyzed N—N cleavage successfully competes with slow hydrolysis. Monoalkylhydrazines are formed in 80–95% yield, *N,N'*-dialkylhydrazines in 65–85% yield ⟨B-67MI50800⟩.

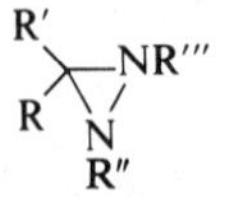

(**151**) ⟶ RCOR' + R''NHNHR'''

The influence of temperature, acidity and substituents on hydrolysis rate was investigated with simple alkyldiaziridines ⟨62CB1759⟩. The reaction follows first order kinetics. Rate constants and activation parameters are included in Table 2.

Table 2 Rate Constants and Activation Parameters of Diaziridine Hydrolysis

Diaziridine[a]		$k_{35} \times 10^5$ (s^{-1})	$\Delta H^{\neq}$ (kJ mol^{-1})	$\Delta S^{\neq}$ (J K^{-1} mol^{-1})
 1-Cy-diaziridine	(**152**)	<0.03	>110	*ca.* 0
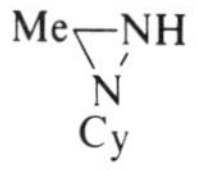 Me, 1-Cy-diaziridine	(**153**)	0.71	109	11
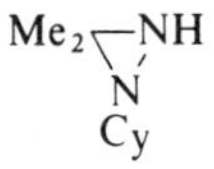 Me_2, 1-Cy-diaziridine	(**154**)	58.4	97	10
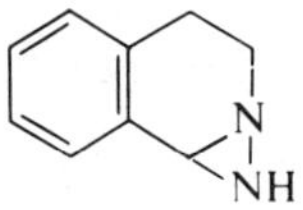 tetrahydroisoquinoline-fused diaziridine	(**155**)	*ca.* 200 (25 °C)	—	—

[a] Cy = cyclohexyl.

Successive introduction of two methyl groups at ring carbon increases the hydrolysis rate by a factor of 10^2 in each step, indicating cation formation in the transition state as in acetal hydrolysis. Equilibrium protonation before hydrolysis becomes evident from an increasing rate of hydrolysis with a decreasing pH value (Table 3). Below pH 3 no further increase of rate is observed, so that protonation is assumed to be complete.

Table 3 Hydrolysis Rate Dependency of Diaziridine (**114**) on pH

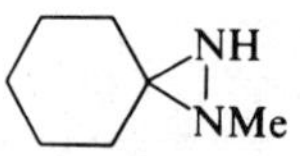

(**114**)

pH	$k_{35} \times 10^5$ (s^{-1})
7.2	0.03
6.0	0.9
4.6	5.38
3.0	17.7

The repeatedly attempted commercialization of hydrazine synthesis *via* dialkyldiaziridines required hydrolysis of the latter without the use of stoichiometric amounts of acid. Catalytic amounts of acid, *e.g.* an acidic ion exchanger, were used in the presence of excess ketone, yielding the azine as an intermediate, which could be hydrolyzed without acid above 100 °C to give hydrazine hydrate and the ketone ⟨62GEP1126395, 71JAP7102008⟩.

5.08.3.2.3 Reactions of diaziridines involving N—N cleavage

Diaziridines are also very strong oxidizing agents, even liberating chlorine from hydrochloric acid. The reaction with iodide in acidic solution proceeds almost quantitatively in most cases. The two equivalents of iodine obtained from a diaziridine (**151**) are of analytical value together with the number of acid equivalents consumed ⟨B-67MI50800⟩.

$$(\mathbf{151}) + 2I^- + 4H^+ \rightarrow RCOR' + R''NH_3^+ + R'''NH_3^+ + I_2$$

The reduction by iodide fails when fast hydrolysis competes, *e.g.* in the case of 1,2-diacyldiaziridines with two alkyl groups at carbon as in (**130**). In such cases elimination of the acyl groups by alkali must precede reduction.

1,2-Dialkyldiaziridines react with acidic iodide solution only on heating; catalysis by traces of copper salts permits quantitative reduction at room temperature in these cases also.

These reductions are not important for preparative purposes. The same can be said for reductions with $LiAlH_4$, and with hydrogen over a catalyst, converting diaziridines to a mixture of amines including products of reductive amination of the former carbon atom of the diaziridine ring.

Diaziridines (**156**) unsubstituted on both nitrogens decompose at 125 °C by a redox reaction yielding one mole of a diazirine (**157**) together with two moles of ammonia and one mole of ketone from two moles of (**156**). The reaction proceeds below 60 °C when copper salts are present ⟨64AG(E)229⟩.

2 (**156**) $\xrightarrow{Cu^{2+},\ H_2O}$ (**157**) + EtCOEt + $2NH_3$

1,2-Dialkyldiaziridines (**158**) undergo thermal decomposition, which may be observed even at room temperature over the course of some months and bears some resemblance to the liquid phase decomposition of oxaziridines (Section 5.08.3.1.5). 1,2-Dibutyl-3-hexyldiaziridine (**158**) underwent 70% decomposition upon storage at 0 °C for 9 months, forming heptanal, butyraldehyde, butylamine and ammonia ⟨B-67MI50800⟩. A radical chain reaction is assumed to give (**159**) by attack of a radical on an *N*-methylene group involving N—N cleavage and dealkylation of one atom.

R· + (**158**) → (**159**) → $C_6H_{13}CHO + PrCHO + BuNH_2 + NH_3$

A radical reaction is sometimes observed to compete with the particularly slow acid hydrolyses of diaziridines derived from formaldehyde. With other diaziridines the radical reaction can be made to predominate by using HCl in carbon tetrachloride solution. Acetaldehyde, butyraldehyde, butylamine and ammonia are obtained from (**160**) ⟨64CB49⟩.

Acid hydrolysis of (**161**) yields acetone and methylamine by N—N cleavage and methyl migration.

(**160**) (**161**) → MeCOMe + $MeNH_2$

Most diaziridines are not sensitive towards alkali. As an exception, diaziridines derived from 2-hydroxyketones are quickly decomposed by heating with aqueous alkali. Acetaldehyde, acetic acid and ammonia are formed from (**162**). This reaction is not a simple N—N cleavage effected intramolecularly by a deprotonated hydroxy group, since highly purified hydroxydiaziridine (**162**) is quite stable towards alkali. Addition of small amounts of hydroxybutanone results in fast decomposition. An assumed reaction path — Grob fragmentation of a hydroxyketone–diaziridine adduct (**163**) — is in accord with these observations ⟨B-67MI50800⟩.

(162) $\xrightarrow{OH^-}$ MeCHO + $MeCO_2H$ + $2NH_3$

(163)

Cleavage of N—N by potassium *t*-butoxide to give amidine (**164**) was observed with diaziridine (**134**). This is the only known analog of the generally observed acid amide formation from oxaziridines ⟨74JPR999⟩.

(134) $\xrightarrow{Bu^tOK}$ (164)

A modified amidine structure can be seen in (**166**), the product of reaction of 1,2-diethyldiaziridine (**165**) with diphenylketene. A second molecule of diphenylketene may have formed the four-membered ring of (**166**) ⟨75JAP75117765⟩.

(165) $\xrightarrow{Ph_2CCO}$ $Ph_2CHOC(Et)N$ (166)

5.08.3.2.4 Reactions of diaziridinones and diaziridinimines

While diaziridinones unsubstituted at nitrogen were proposed but not isolated as intermediates of N—N bond forming reactions ⟨79AHC(24)63⟩, diaziridinones with two tertiary *N*-alkyl groups exhibit marked stability. 1,2-Di-*t*-butyldiaziridinone (**167**), the best studied compound, decomposes only at 180 °C to yield the (*E*)-azo compound (**168**) together with carbon monoxide. The half reaction time is 2 h ⟨69JOC2254⟩. Compound (**167**) is also stable towards nucleophiles. Heating for 16 h with potassium *t*-butoxide in *t*-butyl alcohol effects only 50% ring opening; 0.4M sodium methoxide in boiling methanol converts (**167**) to the methyl ester (**169**; R = Me) within 40 min. There are faster reactions in acidic media. Hydrogen chloride in pentane opens the ring within a few minutes at room temperature to give the acid chloride (**170**); benzoic acid in boiling benzene reacts within 1 h to give anhydride (**171**). In reactions with reducing nucleophiles, reduction to the substituted urea (**172**) predominates over nucleophilic attack. For example, phenylhydrazine reduces (**167**) to (**172**) quantitatively; thiols, even phenol and benzhydrol, effect reductive N—N cleavage.

$Bu^tN{=}NBu^t$ + CO (**168**)

$Bu^tNHCONHBu^t$ (**172**)

(**167**)

$Bu^t(H)N{-}N(Bu^t)CO_2R$ (**169**)

$Bu^t(H)N{-}N(Bu^t)COX$

(**170**) X = Cl
(**171**) X = $PhCO_2$

The *cis*-fused diaziridinone (**173**) undergoes the same reactions at temperatures lower by almost 100 °C. Extrusion of carbon monoxide proceeds at 25 °C with $t_{1/2}$ = 25 h to give the (*Z*)-azo compound. Solvolysis and reduction are also much faster. Ground state destabilization by unfavorable arrangement of the nitrogen lone pairs is assumed to be responsible ⟨76JOC2813⟩. Inversion of the six-membered ring through the plane of the three-membered ring proceeds at a temperature also about 100 °C lower than in the *E* compound (**167**). The inversion barrier of (**173**) was estimated as 20–25 kJ mol^{-1}.

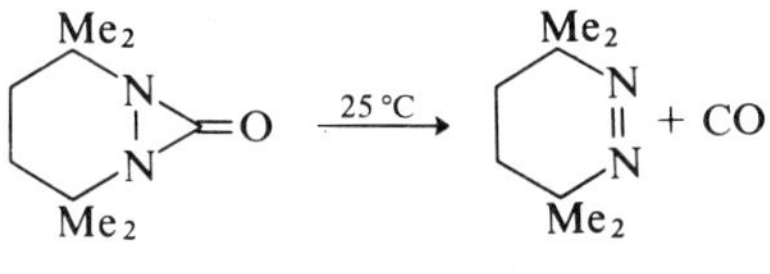

(173)

Five-membered rings are formed from diaziridinones in some cases by ring expansion, which proceeds by insertion of a multiple bond from a reagent. Opening of either the N—N bond or one C—N bond can be involved. With benzoyl isocyanate or benzonitrile ⟨74CL383⟩ a C—N bond reacts with formal cycloaddition to give (**174**); with diphenylketene, additions involving N—N as well as C—N cleavage compete to give (**174**) and (**175**). A P—S bond can also be inserted into a five-membered ring of type (**174**) ⟨79BSB737⟩.

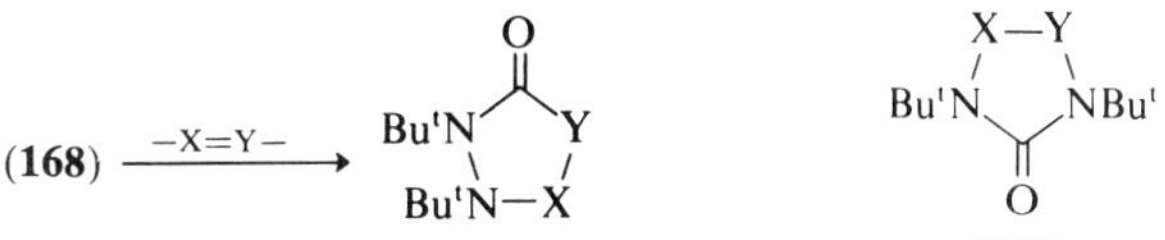

(**174**) (**175**)

In the diaziridinimine field there is no sharp contrast between stable *t*-alkyl and unstable primary alkyl compounds. Both types are known as stable compounds ⟨80H(14)1677⟩.

Below 100 °C tri-*t*-butyldiaziridinimine (**176**) only undergoes inversion at the exocyclic nitrogen, as evidenced by coalescence of ^{1}H NMR signals at about 50 °C. Heating for 1 h to 150 °C, however, results in clean formation of (*E*)-azoisobutane and *t*-butyl isocyanide.

On the other hand, mutual conversion of (**177**) and (**178**), prepared independently, is observed slightly above room temperature. Compound (**178**) predominates in the equilibrium. Activation parameters are $\Delta H^{\ddagger} = 104\ \text{kJ mol}^{-1}$, $\Delta S^{\ddagger} = 0.4\ \text{J K}^{-1}\ \text{mol}^{-1}$ for (**177**), and $\Delta H^{\ddagger} = 123\ \text{kJ mol}^{-1}$, $\Delta S^{\ddagger} = 11\ \text{J K}^{-1}\ \text{mol}^{-1}$ for (**178**). Opening of the N—N bond and recyclization involving the former exocyclic N atom are assumed as the mechanism ⟨70CB1234⟩.

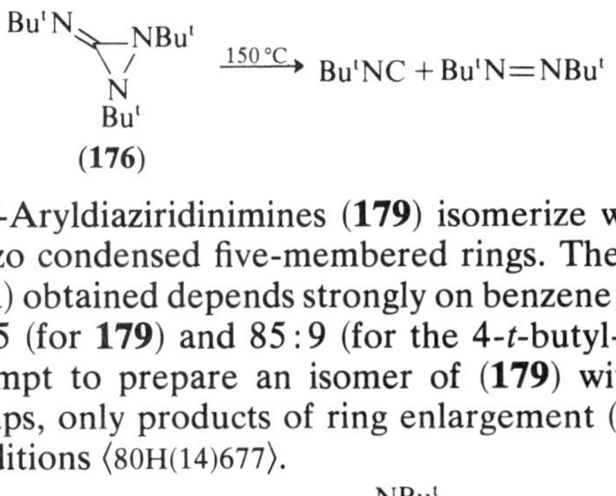

(**176**) 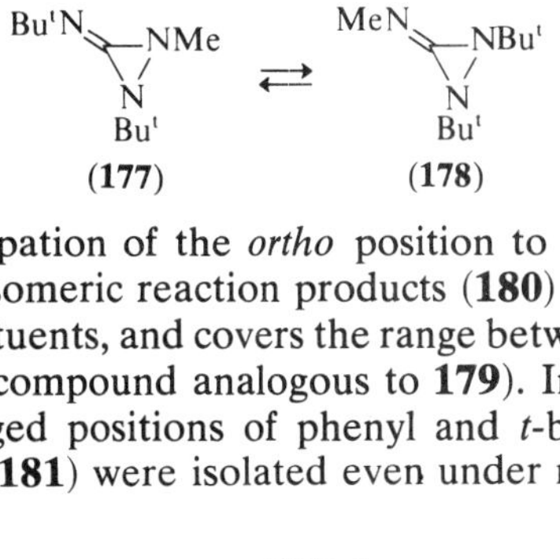 (**177**) (**178**)

N-Aryldiaziridinimines (**179**) isomerize with participation of the *ortho* position to give benzo condensed five-membered rings. The ratio of isomeric reaction products (**180**) and (**181**) obtained depends strongly on benzene ring substituents, and covers the range between 2:85 (for **179**) and 85:9 (for the 4-*t*-butyl-2-methyl compound analogous to **179**). In an attempt to prepare an isomer of (**179**) with exchanged positions of phenyl and *t*-butyl groups, only products of ring enlargement (**180**) and (**181**) were isolated even under mild conditions ⟨80H(14)677⟩.

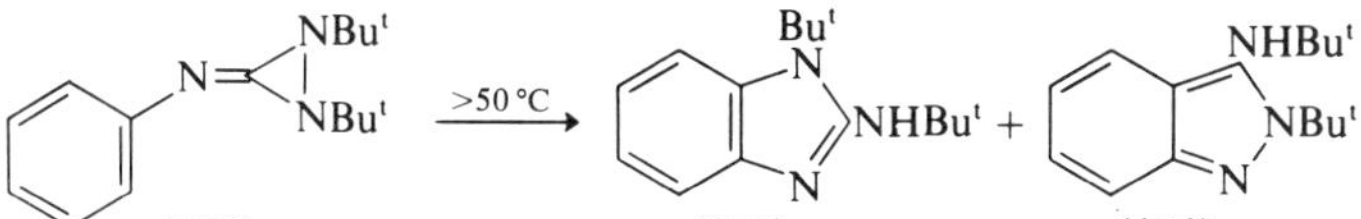

(**179**) (**180**) (**181**)

Intermolecular reactions with typical cycloaddition components are also possible. Phenyl isocyanate in ether converts triisopropyldiaziridinimine (**182**) to the 1,2,4-triazolidine under mild conditions. Labeling with a deuterated isopropyl group revealed that cycloaddition is not preceded by N—N cleavage, which should have resulted in deuterium randomization ⟨77AG(E)109⟩.

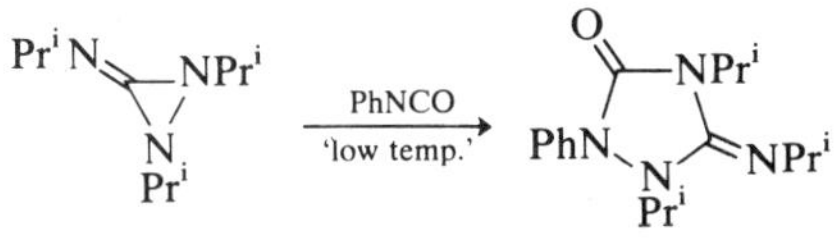

(**182**) (**183**) 82% 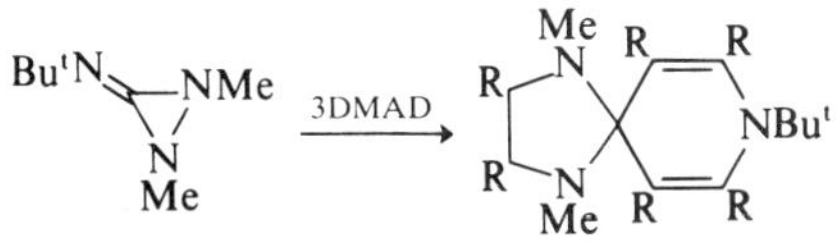 (**184**) R = CO_2Me (**185**) 27%

Acetylenedicarboxylic acid ester reacts with (**184**) in benzene–pentene at 20 °C to give the 3:1 adduct (**185**) ⟨77AG(E)177⟩.

5.08.3.3 Reactivity of Diazirines

5.08.3.3.1 Reactions with conservation of the three-membered ring

It was not their reactivity but their chemical inertness that was the true surprise when diazirines were discovered in 1960. Thus they are in marked contrast to the known linear diazo compounds which are characterized by the multiplicity of their reactions. For example, cycloadditions were never observed with the diazirines. Especially surprising is the inertness of diazirines towards electrophiles. Strong oxidants used in their synthesis like dichromate, bromine, chlorine or hypochlorite are without action on diazirines. Diazirine formation may even proceed by oxidative dealkylation of a diaziridine nitrogen in (**186**) without destruction of the diazirine ring ⟨75ZOR2221⟩. The diazirine ring is inert towards ozone; simple diazirines are decomposed only by more than 80% sulfuric acid ⟨B-67MI50800⟩.

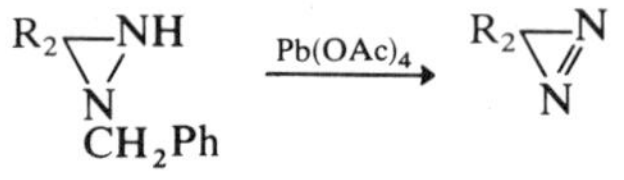

R_2C(NH)(N–CH_2Ph) (**186**) —$Pb(OAc)_4$→ R_2C(N=N)

The chemical inertness of the three-membered ring permitted many conversions of functional groups in diazirines. Esterifications, cleavage of esters and acetals, synthesis of acid chlorides, oxidation of hydroxy groups to carboxyl groups as well as Hofmann alkenation all left the three-membered ring intact ⟨79AHC(24)63⟩.

Reductions of the N—N double bond yield diaziridines and were carried out for proof of structure, using for example sodium amalgam or catalytic hydrogenation. They are unimportant beyond that, because most diazirine syntheses start with diaziridines.

Diazirines (**3**) smoothly add Grignard compounds to the N—N double bond, giving 1-alkyldiaziridines. Reported yields are between 60 and 95% without optimization ⟨B-67MI50800⟩. The reaction is easily carried out on a preparative scale without isolation of the hazardous diazirines and may serve as an easy access to alkylhydrazines. The reaction was also used routinely to detect diazirines in mixtures. The diaziridines formed are easily detected by their reaction with iodide. Phenyllithium or ethylzinc iodide also add to (**3**) with diaziridine formation.

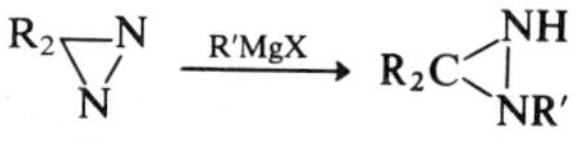

R_2C(N=N) (**3**) —R′MgX→ R_2C(NH)(NR′)

Metal carbonyls react with diazirines with complex formation at one or both N atoms ⟨75JOM(94)75⟩. The 1 : 2 complex (**187**) is converted to (**188**) by N—N cleavage in acidic media.

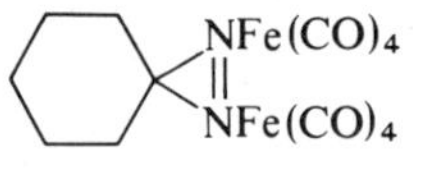

$NFe(CO)_4$ / $NFe(CO)_4$ (**187**) $(CO)_3$Fe / =N N= / Fe$(CO)_3$ (**188**) N=N (**189**)

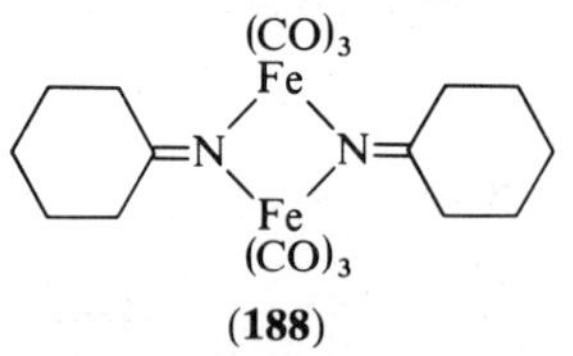

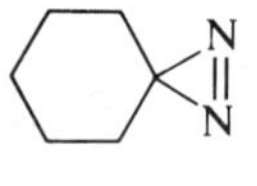

Analogous complexes are formed between 1,2-diazaspiro[2.5]oct-1-ene (**189**) and the $Ru(NH_3)_5$ cation in a 1 : 1 or 1 : 2 ratio. An X-ray analysis of the 1 : 1 complex (**189a**) gives evidence of a nitrogen–metal bond. Oxidation converts the 1 : 1 complex to the known dinitrogen complex (**190**) by liberation of cyclohexanone ⟨80MI50800⟩.

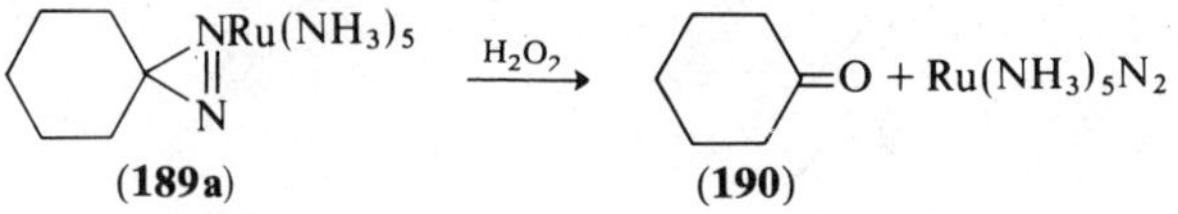

$NRu(NH_3)_5$ / N (**189a**) —H_2O_2→ =O + $Ru(NH_3)_5N_2$ (**190**)

Complexes (**191**) and (**192**) are formed from dimethyldiazirine with carbonyls of chromium, molybdenum and tungsten. They show no tendency towards N—N cleavage ⟨80JOM(193)57⟩. Complex (**193**) is made from a mixed complex by displacement of norbornadiene.

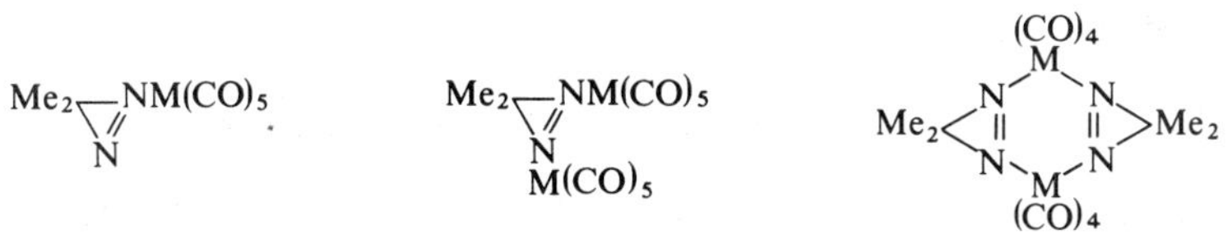

(191) (192) (193)

5.08.3.3.2 Reactions of diazirines leaving nitrogen in the molecule

Photolytic transformation of diazirines to diazoalkanes was observed in some cases. The parent compound (**44**) on irradiation in the gas phase with light ($\lambda = 3200$ Å) yields diazomethane. The quantum yield is 0.2 ⟨64JA292⟩.

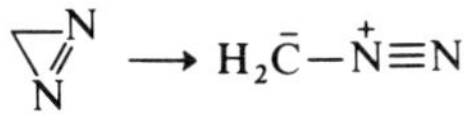

$$\longrightarrow H_2\bar{C}-\overset{+}{N}\equiv N$$

(**44**)

Formation of diazomethane from diazirine was also observed in a solid nitrogen matrix on irradiation ⟨64JCP(41)3504⟩. Labeling experiments demonstrated that elimination and uptake of nitrogen occurred.

With some special diazirines interconversion with diazoalkanes was observed on illumination ⟨79AHC(24)63⟩, *e.g.* with diazirinecarboxylic acid piperidide (**194**), spirodiazirine (**195**) and the tetracyclic ketodiazirine (**196**) ⟨78CC442⟩.

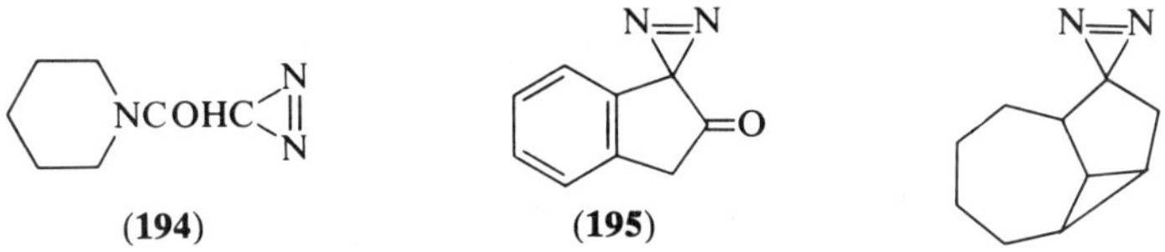

(**194**) (**195**) (**196**)

Thermal conversion of diazirines to linear diazo compounds was postulated occasionally and proved by indirect methods. The existence of a diazo compound isomeric to diazirine (**197**) was proved spectroscopically on short thermolysis in DMSO ⟨76JA6416⟩. An intermediate diazoalkane was trapped by reaction with acetic acid, yielding the ester (**198**) ⟨77JCS(P2)1214⟩.

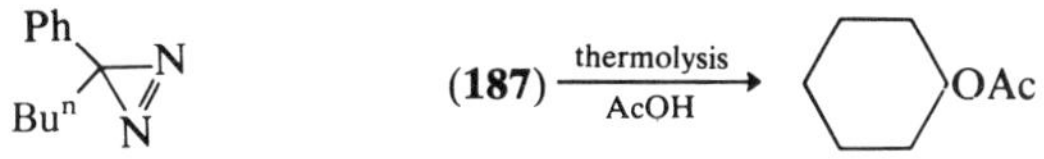

(**197**) (**187**) $\xrightarrow[\text{AcOH}]{\text{thermolysis}}$ (**198**)

Methylvinyldiazirine (**199**) rearranges at room temperature in the course of some days. Formation of the linear isomer is followed by electrocyclic ring closure to give 3-methylpyrazole. The linear diazo compound could be trapped by its reaction with acids to form esters, while the starting diazirine (**199**) is inert towards acids ⟨B-71MI50801⟩.

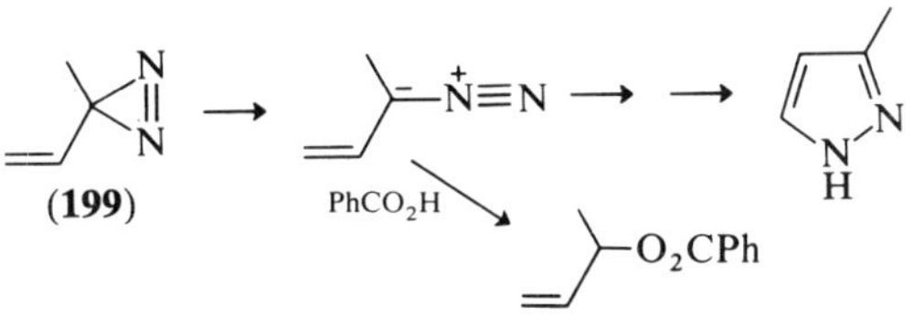

(**199**)

A formal ring enlargement of diazirines to five-membered rings is also observed with some hydrazones of ketodiazirines ⟨65CB2509⟩. On attempted preparation of hydrazones (**201**) from ketodiazirine (**200**) at 0 °C the diazo compounds (**202**) are plausible intermediates since their transformation to aminotriazoles (**203**) is known.

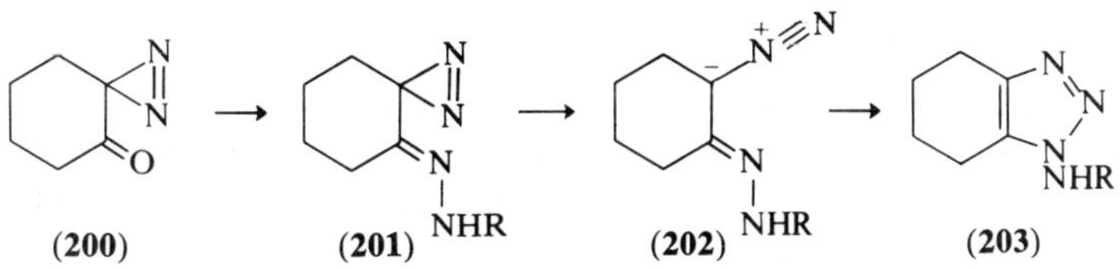

(**200**) (**201**) (**202**) (**203**)

Formation of azines on thermolysis of diazirines was reported repeatedly ⟨B-67MI50800⟩, *e.g.* with perfluorodimethyldiazirine (**204**), with chloromethoxydiazirine and with chlorophenyldiazirine.

$$(F_3C)_2C\text{-diazirine} \longrightarrow (F_3C)_2C{=}N{-}N{=}C(CF_3)_2$$

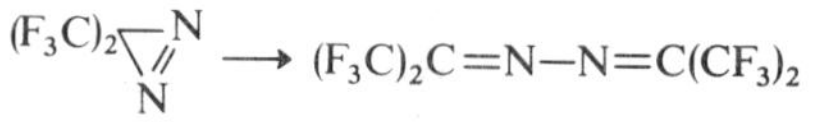

(**204**)

There are some reports on reactions involving complete N—N cleavage in diazirine reactions such as formation of amidine (**205**) from chlorophenyldiazirine, or on formation of products containing only one nitrogen atom. Betaine (**206**) was described as a product from difluorodiazirine and triphenylphosphine. Compound (**207**) is formed from decomposing (**204**) and cyclohexane ⟨79AHC(24)63⟩.

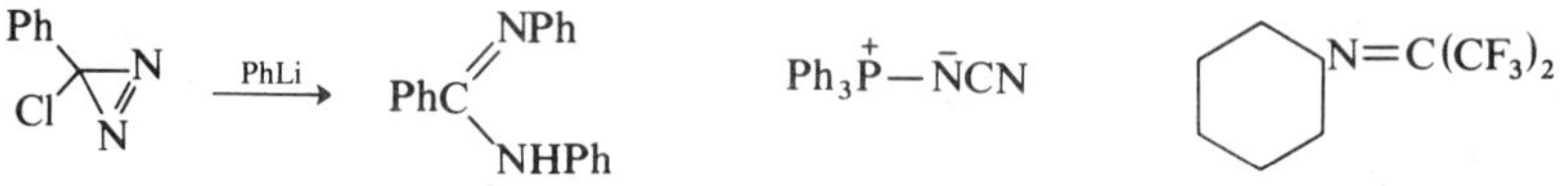

$$\xrightarrow{\text{PhLi}} PhC(=NPh)NHPh \quad (\mathbf{205}) \qquad Ph_3\overset{+}{P}{-}\bar{N}CN \quad (\mathbf{206}) \qquad C_6H_{11}N{=}C(CF_3)_2 \quad (\mathbf{207})$$

5.08.3.3.3 Reactions of diazirines with electrophiles

A direct attack on the diazirine ring was observed only by use of almost concentrated sulfuric acid or strong Lewis acids, *e.g.* aluminum chloride in the case of (**208**). Some hydrazine was found together with cyclohexanone and cyclohexanol after decomposition of the spirodiazirine (**189**) with 80% sulfuric acid ⟨B-67MI50800⟩.

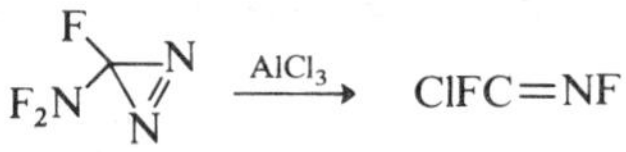

$$(\mathbf{208}) \xrightarrow{AlCl_3} ClFC{=}NF$$

Increased sensitivity towards acid is observed when protonation occurs on a functional group outside the diazirine ring, giving rise to electron dilution at the carbon atom adjacent to the diazirine carbon. The products isolated are in accord with the proposal ⟨79AHC(24)63⟩ that cation formation at this carbon atom leads to nitrogen extrusion, probably with formation of a vinyl cation. Thus protonated hydroxydiazirine (**209**) yields acetone, and methylvinyldiazirine (**199**) on treatment with acids yields butanone ⟨67CB2093⟩.

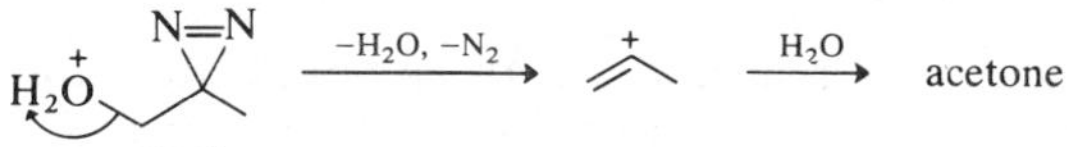

$$(\mathbf{209}) \xrightarrow{-H_2O,\ -N_2} \text{vinyl cation} \xrightarrow{H_2O} \text{acetone}$$

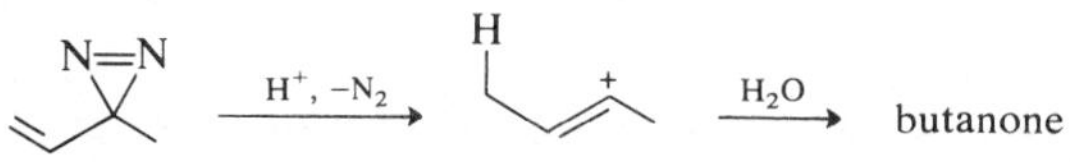

$$(\mathbf{199}) \xrightarrow{H^+,\ -N_2} \text{cation} \xrightarrow{H_2O} \text{butanone}$$

Acidic cleavage of the oxirane ring in (**210**) results in formation of the alkynic aldehyde (**211**) ⟨68TL4905⟩.

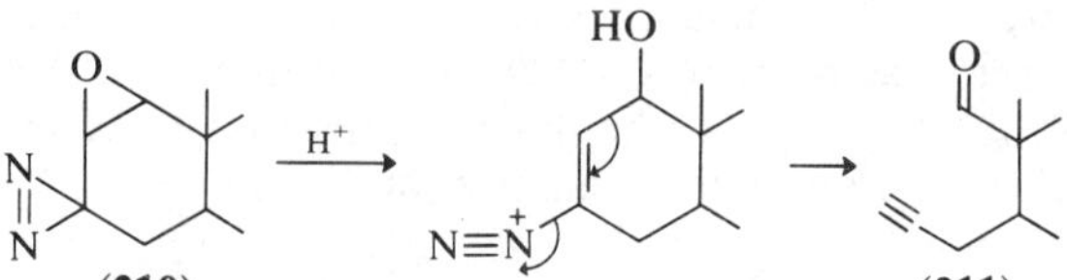

$$(\mathbf{210}) \xrightarrow{H^+} \text{intermediate} \longrightarrow (\mathbf{211})$$

Especially sensitive towards acid is the ketodiazirine (**200**). It decomposes on contact with 0.1N acid, forming methylenecyclopentanone (**213**) and cyclopentanecarboxylic acid (**214**). The products may be formed from a developing vinyl cation (**212**) by C—C bond shift either to the double bond (**213**) or across the double bond (**214**) ⟨B-67MI50800⟩.

(200) (212) (213) (214)

5.08.3.3.4 Thermal extrusion of nitrogen from diazirines

Diazirines tend to decompose, in some cases without obvious cause. Gaseous methyldiazirine exploded at normal temperature and liquid α-hydroxyethylmethyldiazirine exploded on taking it from a dry ice bath ⟨67CB2093⟩. Phenylchlorodiazirine is reported to be more shock sensitive than nitroglycerol ⟨79AHC(24)63⟩.

The thermal cleavage to nitrogen and a carbene, suggested by the structural formula, is observed with almost all diazirines and was investigated thoroughly. An early review was given in ⟨B-67MI50800⟩. Thermal stability of diazirines was found to be remarkable. Although they are more energy rich by about 125 kJ mol^{-1} compared with diazoalkanes, thermal cleavage of simple diazirines proceeds with activation energies of more than 120 kJ mol^{-1}. The carbenes formed undergo the known intramolecular stabilization to alkenes and cyclopropanes.

First order decomposition was established for dimethyldiazirine (**215**) and ethylmethyldiazirine (**216**). The activation energy is 139 kJ mol^{-1} for (**215**); the half life at 100 °C is 97 h. On decomposition of (**216**) the products formed and their respective yields are as indicated. The products correspond qualitatively and quantitatively with the results of thermal decomposition of 2-diazobutane formed *in situ* in aprotic solvents. Analogous comparisons of decomposition products of diethyldiazirine, isopropylmethyldiazirine, *n*-butyl- and *t*-butyl-diazirine agree equally well ⟨66TL1733⟩.

(215) (216) 3.6% 66.5% 29.5% 0.4%

Diazirines are in most cases more easily available than linear diazo compounds. Moreover, their decomposition *via* true carbenes is free of side reactions, whereas linear diazo compounds in presence of H-donors may react by a cationic pathway. Only where reactions of linear diazo compounds are optimized for carbene formation do they give the same products as do decomposing diazirines.

Thermal decomposition of alkyldiazirines proceeds through the lowest singlet state of the carbene ⟨79JA5082⟩.

The stabilization reactions of alkylcarbenes were used preparatively in some cases. The diazirine derived from adamantanone gave the dehydroadamantane (**217**) thermally in 96% yield ⟨73ZOR430⟩. Alkene formation was reported for a steroid with its C-3 atom part of a diazirine ring. At 140 °C a Δ-2-unsaturated steroid was formed ⟨65JA2665⟩.

(217)

Hydroxymethylmethyldiazirine (**209** unprotonated) formed propionaldehyde as the sole product by thermal nitrogen extrusion; 4-hydroxy-1,2-diazaspiro[2.5]oct-1-ene (**218**) formed a mixture of cyclohexanone (73%), cyclohexenol (21%) and cyclohexene oxide (5%). Thermal decomposition of difluorodiazirine (**219**) was investigated intensively. In this case there is no intramolecular stabilization possible. On heating for three hours to 165–180 °C hexafluorocyclopropane and tetrafluoroethylene were formed together with perfluoroformaldazine ⟨64JHC59⟩.

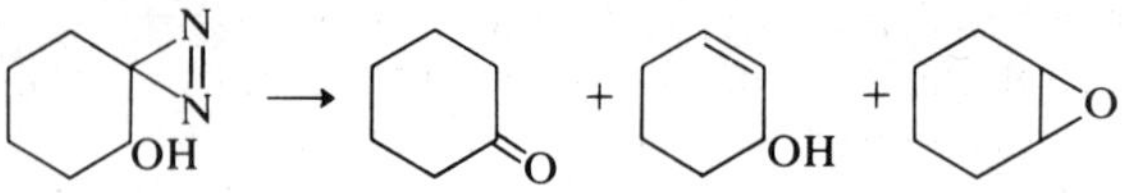

(218)

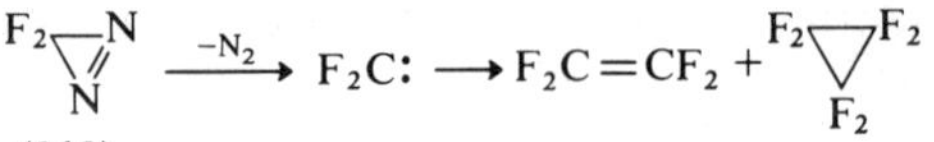

(219)

The thermally formed difluorocarbene adds stereospecifically to either (*Z*)- or (*E*)-butene to form the corresponding dimethyldifluorocyclopropanes, which points to a singlet difluorocarbene. With butadiene mono- (**220**) and bis-cyclopropanated product (**221**) are formed, and with perfluorobutadiene the mono-adduct (**222**) results ⟨65JA758⟩.

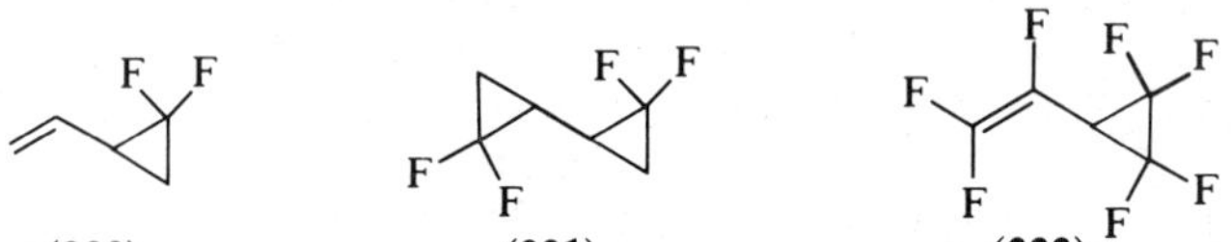

(220) (221) (222)

Two other fluorine-containing diazirines give similar results. Cyanofluorodiazirine (**223**) and fluoromethoxydiazirine (**224**) release nitrogen at 100 °C; the carbenes formed can be trapped by tetrafluoroethylene to give cyclopropanes ⟨65JHC371⟩. Without a carbene trap alkenes are obtained.

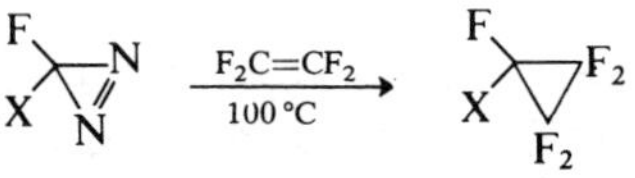

(**223**) X = CN
(**224**) X = OMe

Difluoroaminofluorodiazirine (**225**) extrudes nitrogen at only 75 °C. Intramolecular stabilization gives trifluoromethylenimine (**226**); added tetrafluoroethylene is cyclopropanated. This type of dichotomy is not often found in carbene chemistry; alkylcarbenes undergo intramolecular stabilization as a rule, whereas intermolecular stabilization is observed exclusively with alkoxycarbonylcarbenes and with difluorocarbene. In the latter case CF_2 attacks its precursor when no other reaction partner is present.

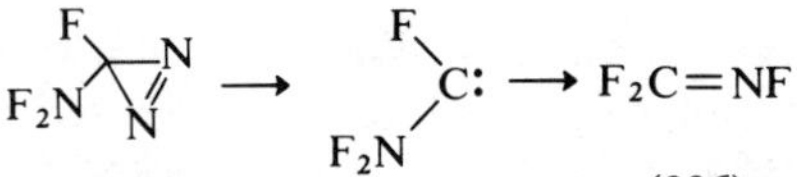

(225) (226)

The same dichotomy was observed with hexafluorodimethylcarbene (**228**), formed by thermolysis of diazirine (**227**) at 150 °C. The carbene (**228**) can stabilize itself either intramolecularly to perfluoropropene (**229**), or intermolecularly by addition to multiple bonds. Oxirane (**230**) is formed with hexafluoroacetone, cyclopropene (**231**) with 2-butyne ⟨66MI50800⟩.

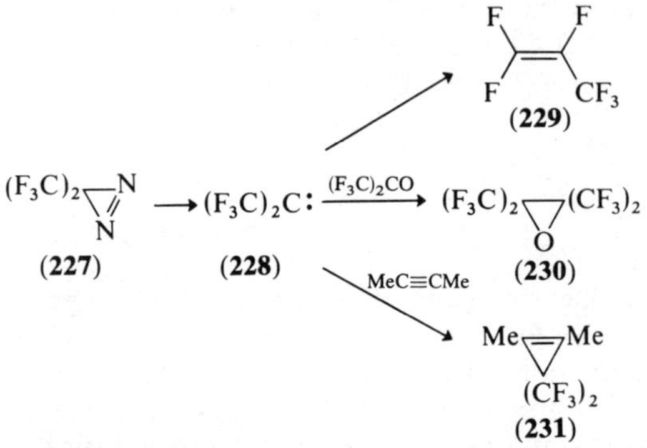

(227) (228) (229) (230) (231)

Thermal decomposition of chlorodiazirines has been investigated intensively during recent years. This was favored by the easy accessibility of chlorodiazirines with almost any additional substituent in the 3-position, including aryl and alkoxyl groups, and also by the identity of chlorine-containing carbenes with intermediates from the solvolysis of halomethanes, thus connecting to a classical line of carbene chemistry.

Kinetic investigations of alkylchlorodiazirine thermolysis were carried out in the gas phase ⟨70JCS(A)1916⟩. Chloromethyldiazirine (**232**) decomposition follows first order kinetics giving nitrogen and vinyl chloride, easily interpretable as isomerization of a carbene.

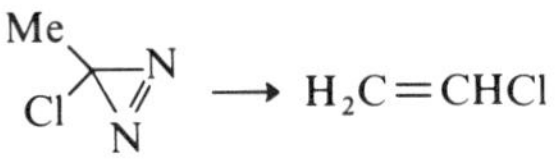

(**232**) ⟶ $H_2C{=}CHCl$

Analogous products are obtained from thermolyses of chlorodiazirines bearing ethyl, propyl, isopropyl or *t*-butyl groups.

Decomposition of more complex diazirines follows first order kinetics also. Chlorophenylcarbene adds to cyclohexene to give a norcarane derivative. Substituent effects of *m*-Cl, *m*-NO_2 or *m*-Me groups, as well as solvent effects, are small. Chlorotrichloromethyldiazirine yields tetrachloroethylene; chlorocyclooctyldiazirine also leads to an alkene ⟨74CJC246⟩.

The stabilization of chloromethoxycarbene (**234**) was intensively studied. It is formed from diazirine (**233**) in a first order reaction with $t_{1/2} = 34$ h at 20 °C. It reacts either as a nucleophile, adding to electron poor alkenes like acrylonitrile with cyclopropanation, or as an electrophile, giving diphenylcyclopropenone with the electron rich diphenylacetylene. In the absence of reaction partners (**234**) decomposes to carbon monoxide and methyl chloride ⟨78TL1931, 1935⟩.

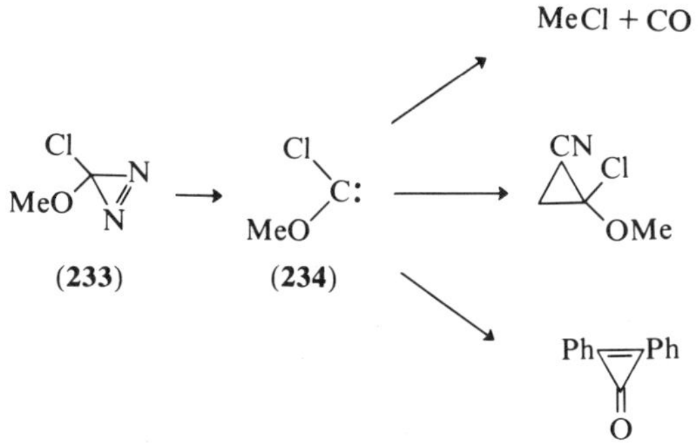

(**233**) ⟶ (**234**) ⟶ MeCl + CO

5.08.3.3.5 Photolysis of diazirines

Photolysis of diazirines to nitrogen and carbenes is a general reaction and plays a greater role in carbene chemistry than photolysis of linear diazo compounds. Whereas the latter are often obtained only under the conditions of their thermal decomposition from suitable precursors, diazirines are obtainable in a pure state in most cases. Photolysis has the further advantage to permit nitrogen extrusion at atmospheric pressure, even with low-boiling materials.

Photolysis of the parent compound (**44**) yields singlet methylene, as evidenced by its stereospecific addition to (*E*)-butene. The cyclopropane is formed together with the characteristic insertion products of methylene ⟨62MI50800, 64PAC(9)527⟩

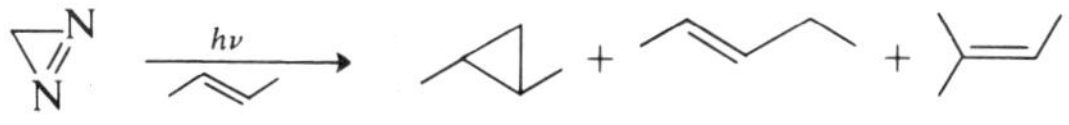

(**44**) $\xrightarrow{h\nu}$

Methylene from diazirine has higher energy of vibration than the product from photolysis of ketene, but it is more discriminating in insertion reactions into primary and secondary C—H bonds.

Photolysis of dialkyl- and alkyl-diazirines has been broadly investigated since the products formed gave evidence of shortlived intermediates possessing excess vibrational energy. As shown in Scheme 1 diazirines (**3**) were photoactivated to (**235**) by irradiation with a medium pressure mercury lamp with quantum yields smaller than one; this means reversibly. Nitrogen extrusion yielded carbenes (**236**). These rearranged to the primary products, alkenes and cyclopropanes with excess vibrational energy, which could be dissipated by collisions or result in secondary conversions. From the pressure dependence of the final products the competition between stabilization of primary products by collision and formation of the secondary products could be reconstructed.

$R_2C(N{=}N)$ (3) ⇄ $R_2C(N{=}N)^*$ (235) $\xrightarrow{-N_2}$ R_2C: (236)

(236) → vibrationally excited primary products → secondary products

vibrationally excited primary products $\xrightarrow{M}$ stable primary products (alkenes, cyclopropanes)

M = collision partner

Scheme 1

The primary and secondary products of photolysis of common diazirines are collected in Table 4. According to the table secondary reactions include not only isomerization of alkenes and hydrogen elimination to alkynes, but also a retro-Diels–Alder reaction of vibrationally excited cyclohexene, as well as obvious radical reactions in the case of excited propene.

Table 4 Primary and Secondary Products of Diazirine Photolysis

Diazirine	*Primary products*	*Secondary products*
Methyl-	Ethylene	Acetylene, H_2
t-Butyl-	1,1-Dimethylcyclopropane, 2-methyl-2-butene	Isomeric methylbutenes
Dimethyl-	Propene	Ethane, ethylene, propane, 1-butene, 2,3-dimethylbutane, 2-methyl-4-pentene, isobutane, 1,5-hexadiene
Ethyl-methyl	1-Butene, (*Z*)- and (*E*)-butene, methylcyclopropane	(*E*)-Butene, isobutene, methylcyclopropane
Cyclohexane-spiro-diazirine (N=N)	Cyclohexene	1-Methylcyclopentene, butadiene, ethylene

An extrapolation of product composition to zero pressure showed considerable but not complete absence of primary products. A second path to the primary products was assumed, leading from excited diazirine directly to the products without involving a carbene.

In the photolysis of difluorodiazirine (**218**) a singlet carbene was also observed ⟨65JA758⟩. Reactions of the difluorocarbene were especially studied with partners which are too reactive to be used in the presence of conventional carbene precursors, such as molecular chlorine and iodine, dinitrogen tetroxide, nitryl chloride, carboxylic acids and sulfonic acids. Thus chlorine, trifluoroacetic acid and trifluoromethanesulfonic acid reacted with difluorodiazirine under the conditions of its photolysis to form compounds (**237**)–(**239**) ⟨64JHC233⟩.

F_2CCl_2 (**237**) $F_3CCO_2CHF_2$ (**238**) $F_3CSO_2OCHF_2$ (**239**)

The photolysis of chlorodiazirine was investigated in several cases. From chloromethyldiazirine (**232**) vinyl chloride was formed as the stable primary product of stabilization of chloromethylcarbene, with acetylene and hydrogen chloride as secondary products. Some 1,1-dichloroethane was assumed to have been formed through a linear diazo compound by reaction with HCl. Added HBr yielded 1-bromo-1-chloroethane ⟨76MI50800⟩.

Photofragmentation of chloromethyldiazirine in the presence of several gases was assumed to involve C—N bond cleavage to form a diradical, which undergoes a second C—N bond cleavage to form the carbene ⟨80MI50801⟩.

In chlorodiazirine photolysis there is also evidence of competition between intramolecular stabilization and reaction with added chemicals. Apparently carbenes are stabilized by chlorine to such a degree that intermolecular reactions have a chance. Chlorocyclopropyldiazirine in pentane yields 1-chlorocyclobutene. The intermediate carbene (**240**) does not insert into C—H bonds of pentane. With added (*Z*)- or (*E*)-butene stereospecific formation of cyclopropylcyclopropanes occurs ⟨78JA6788⟩.

hν, −20 °C

(**240**)

Cyclopropanations by carbenes from chlorodiazirines were observed in several cases, *e.g.* with the *t*-butyl compound. Cyclopropanation and stabilization by ring enlargement and by elimination compete in chlorocyclobutyldiazirine photolysis.

In laser-impulse experiments with chlorophenyldiazirine the carbene could be observed by UV spectroscopy. On addition of defined amounts of alkene the rate of cyclopropanation was measured directly. The rate constants with various alkenes were (10^6 mol^{-1} s^{-1}): 1-hexene, 1.3; (*E*)-2-pentene, 34; 2-methyl-2-butene, 77; 2,3-dimethyl-2-butene, 130 ⟨80JA7576⟩.

Photochemically produced chloromethoxycarbene revealed ambiphilicity. Relative rates of cyclopropanation showed that electron donating as well as electron attracting substituents favor the reaction (Scheme 2) ⟨79JA4736⟩.

CO_2Me CN

12.6 5.43 1.00 29.7 54.6

Scheme 2 Relative rates of cyclopropanation

Rate differences observed between the same bromophenylcarbene (**241**) when prepared by two different routes, diazirine photolysis and the reaction of benzylidene dibromide with potassium *t*-butoxide, vanish when a crown ether is added to the basic solution in the latter experiment. In this case the complexing potassium bromide is taken over by the crown ether, and selectivity towards alkenes reaches the values of the photolytic runs ⟨74JA5632⟩.

Bu^tOK, crown ether

(**241**)

5.08.4 SYNTHESIS OF THREE-MEMBERED RINGS WITH TWO HETEROATOMS

Although the three-membered rings with two heteroatoms were discovered only after 1950, they are accessible by very simple procedures familiar to the chemists decades before. The most common starting materials, peracids and *N*-haloamines, were available to the chemist before 1900; preparation, as well as isolation, follows standard procedures.

Most ground state syntheses of oxaziridines and diaziridines follow the general scheme outlined for (**242**). Syntheses from starting materials containing a hetero–hetero bond, reported in the older literature, have not been confirmed.

(**242**)

Gross rate comparisons indicate that these ring closures proceed considerably faster than analogous substitution reactions not leading to three-membered rings ⟨64AG(E)333⟩.

Ground state syntheses of diazirines involve dehydrogenation of diaziridines, either isolated or generated *in situ*.

The three types of three-membered ring are also accessible by photoisomerization of open chain 1,3-dipoles: nitrones, azomethinimines and linear diazo compounds respectively. All three-membered rings prepared prior to 1967 were included in a comprehensive review ⟨B-67MI50800⟩.

5.08.4.1 Synthesis of Oxaziridines

5.08.4.1.1 Peracid procedure

Oxaziridines are generally formed by the action of a peracid on a combination of a carbonyl compound and an amine, either as a Schiff base (**243**) or a simple mixture. Yields are between 65 and 90%. Although oxygenation of Schiff bases is formally analogous to epoxidation of alkenes, the true mechanism is still under discussion. More favored than an epoxidation-type mechanism is formation of a condensation product (**244**), from which an acyloxy group is displaced with formation of an O—N bond.

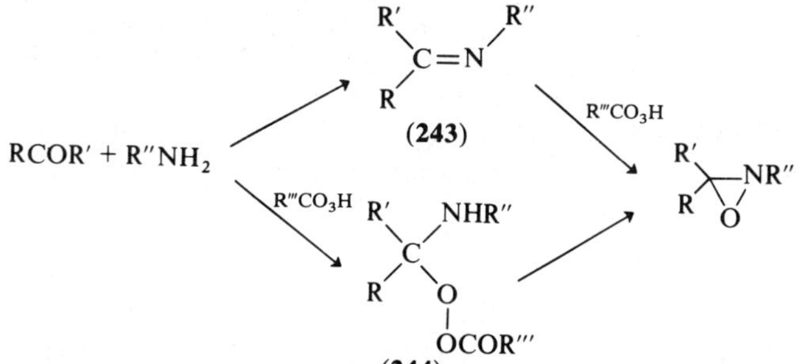

$$RCOR' + R''NH_2 \longrightarrow RR'C{=}NR''\ (\mathbf{243}) \xrightarrow{R'''CO_3H} \text{oxaziridine}$$

$$RCOR' + R''NH_2 \xrightarrow{R'''CO_3H} RR'C(NHR'')(O{-}OCOR''')\ (\mathbf{244}) \longrightarrow \text{oxaziridine}$$

Formation of mixtures of (*E*)- and (*Z*)-oxaziridines from sterically defined Schiff bases fits a two step mechanism through (**244**) ⟨70CC745⟩.

There is almost no limitation of oxaziridine synthesis by the structure of starting materials. Thus, aliphatic as well as aromatic aldehydes (including formaldehyde) and ketones on the one hand, and aliphatic and aromatic amines and even sulfonamides on the other were successfully used as starting materials. The discoverers ⟨58CB1057, 57JA5739, 57CB2184⟩ used peracetic acid as oxidant, which has now been widely replaced by the commercially available MCPBA. Oxaziridine syntheses proceed at room temperature and can compete successfully with common peracid reactions, like epoxidation of C—C double bonds present in the molecule, *N*-oxide formation and lactonization of keto groups ⟨79AHC(24)63⟩.

Preparation of spirooxaziridines from cyclic ketones poses no problems nor does oxaziridine synthesis from cyclic Schiff bases, which was preferably carried out with pyrrolines to give, for example (**245**) ⟨59JCS2102⟩ and, in connection with tranquilizer synthesis, with heterocyclic seven-membered rings to give, for example, (**246**) ⟨63JOC2459⟩.

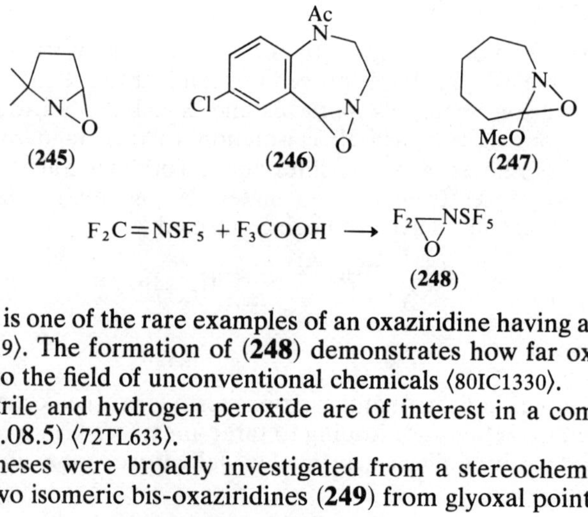

(**245**) (**246**) (**247**)

$$F_2C{=}NSF_5 + F_3COOH \longrightarrow \text{(248)}$$

(**248**)

Compound (**247**) is one of the rare examples of an oxaziridine having an orthoacid carbon in the ring ⟨71TL4519⟩. The formation of (**248**) demonstrates how far oxaziridine synthesis can be extended into the field of unconventional chemicals ⟨80IC1330⟩.

Mixtures of a nitrile and hydrogen peroxide are of interest in a commercial hydrazine synthesis (Section 5.08.5) ⟨72TL633⟩.

Oxaziridine syntheses were broadly investigated from a stereochemical point of view. The formation of two isomeric bis-oxaziridines (**249**) from glyoxal pointed to a tetrahedral carbon in the oxaziridines just discovered ⟨57JA5739⟩. Formation of two isomers (**250a** and **250b**) from simple aldimines gave evidence of the extreme configurational stability at the oxaziridine nitrogen which was already known from NMR experiments ⟨68TL4561⟩. Optically active oxaziridines, *e.g.* (**251**), with only nitrogen as a centre of chirality were obtained from benzophenonimides and optically active peracids ⟨73G681⟩. Partial asymmetric syntheses of oxaziridines from achiral materials were carried out in chiral solvents like menthol or 2-octanol. Optical yields of up to 30% were obtained ⟨77CC731⟩.

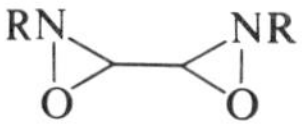

(**249**) 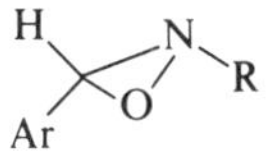 (**250a**) 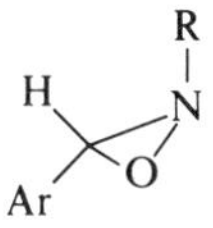 (**250b**) 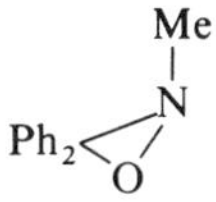 (**251**)

5.08.4.1.2 Amination of carbonyl compounds

Whereas the peracid procedure, involving formal addition of oxygen to a C—N double bond, is a general reaction, formal addition of carbon to an N—O double bond is known in only one case, when diphenyldiazomethane reacts with perfluoronitrosoisobutane ⟨79IZV1788⟩. The third possibility, formal addition of nitrogen to a C—O double bond, has a fair degree of scope and involves action of electrophilic amination reagents on carbonyl compounds. A strong base is required to make the oxygen in the intermediate (**252**) sufficiently nucleophilic to displace the leaving group X.

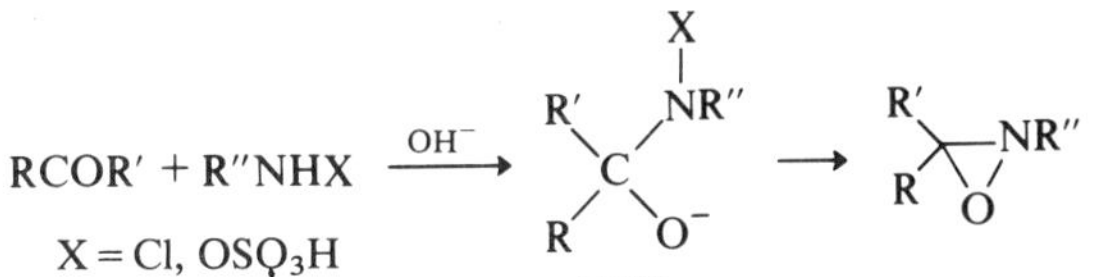

RCOR′ + R″NHX $\xrightarrow{OH^-}$ (**252**) ⟶

X = Cl, OSO_3H

This synthesis works especially well with cyclohexanone giving 80% oxaziridines with either chloramine ⟨77JPR195⟩ or *N*-chloromethylamine. Simple aliphatic ketones and *ortho* substituted aromatic aldehydes yield 30–50% oxaziridines with *N*-chloromethylamine ⟨65CB2516⟩.

Compound (**253**) is formed from benzaldehyde and methylhydroxylamine-*O*-sulfonic acid in 35% yield. With ethyl-substituted chloramine or hydroxylamine-*O*-sulfonic acid yields do not exceed 10%, which is assumed to be due to steric hindrance and is foreseeable for both carbonyl addition and O—N bond formation.

PhCHO + $MeNHOSO_3H$ $\xrightarrow{OH^-}$ (**253**)

5.08.4.1.3 Photoisomerization of nitrones

Oxaziridine formation by photoisomerization of nitrones was discovered almost simultaneously with the peracid procedure ⟨58JOC651⟩. The *t*-butyl nitrones of benzaldehyde and 4-nitrobenzaldehyde yielded about 90% of oxaziridines (**254**) and (**255**) on UV irradiation.

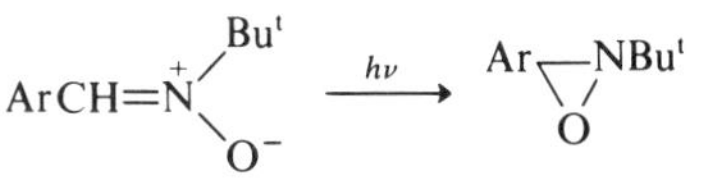

$ArCH=\overset{+}{N}(Bu^t)O^-$ $\xrightarrow{h\nu}$

(**254**) Ar = Ph
(**255**) Ar = $C_6H_4NO_2$-*p*

Compound (**256**) is formed from the isomeric nitrone by visible light. The arrangement of substituents *cis* to the C—N double bond is retained during ring closure to (**256**) ⟨71JA4075⟩. When R′ and R″ are parts of a ring, bicyclic oxaziridines are formed, as reported for (**245**), (**246**), the steroidal compound (**257**) ⟨68TL5087⟩ and the isoindoline derivative (**258**) ⟨68JA3893⟩. The latter was also formed on irradiation of the oxime ether (**259**).

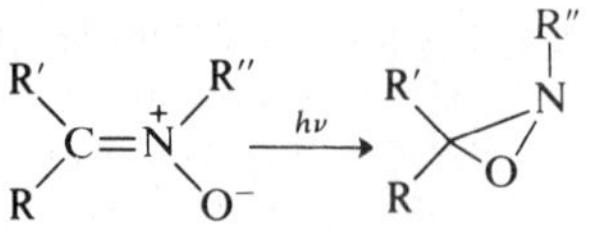

(256)

(257) 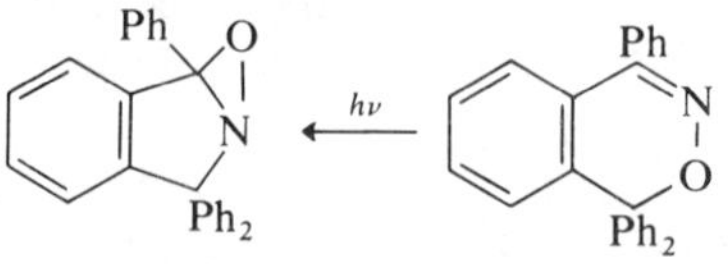 (258) (259)

From *N*-oxides of aromatic bases oxaziridines were obtained only at very low temperatures, but oxaziridines were often postulated as intermediates in the photoconversion of such *N*-oxides (Section 5.08.3.1.2). Isolation of the more stable photoisomers of nitrones also causes some problems due to their thermal and photochemical instability leading to acid amides, *e.g.* ⟨69TL2281⟩, or, by fragmentation, to carbonyl compounds and products of stabilization of nitrenes, *e.g.* from (**260**) ⟨69ZN(B)477⟩.

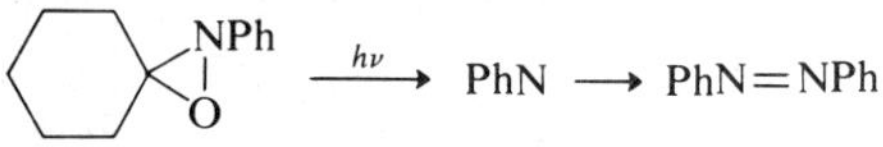

$\xrightarrow{h\nu}$ PhN ⟶ PhN=NPh

(**260**)

5.08.4.2 Synthesis of Diaziridines

5.08.4.2.1 Synthesis of diaziridines by electrophilic amination of carbonyl–amine adducts

Diaziridines (**151**) are formed from a carbonyl compound, an amine and an electrophilic aminating reagent ⟨B-67MI50800, 79AHC(24)63⟩.

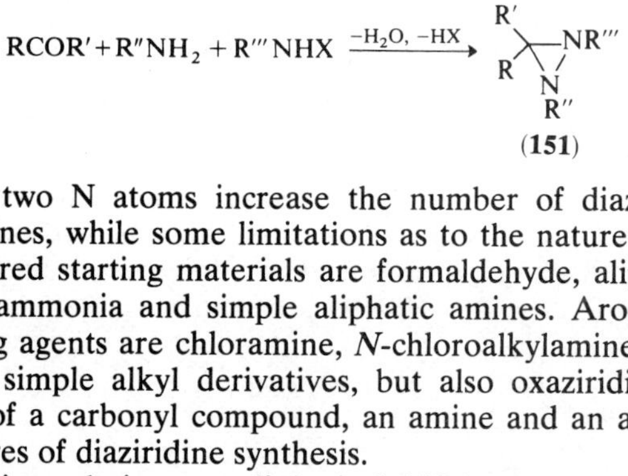

$$\text{RCOR}' + \text{R}''\text{NH}_2 + \text{R}'''\text{NHX} \xrightarrow{-\text{H}_2\text{O},\ -\text{HX}}$$

(**151**)

Two substituents on two N atoms increase the number of diaziridine structures as compared with oxaziridines, while some limitations as to the nature of substituents on N and C decrease it. Favored starting materials are formaldehyde, aliphatic aldehydes and ketones, together with ammonia and simple aliphatic amines. Aromatic amines do not react. Suitable aminating agents are chloramine, *N*-chloroalkylamines, hydroxylamine-*O*-sulfonic acid and their simple alkyl derivatives, but also oxaziridines unsubstituted at nitrogen. Combination of a carbonyl compound, an amine and an aminating agent leads to the standard procedures of diaziridine synthesis.

(a) Action of chloramine solutions on aliphatic Schiff bases leads to 1-alkyldiaziridines (**261**) in 50–70% yield ⟨62CB680⟩. *N*-Chloroalkylamines react with comparable ease to give 1,2-dialkyldiaziridines (**262**) ⟨64CB49⟩. The type (**262**) was prepared mainly from aldimines (R′ = H).

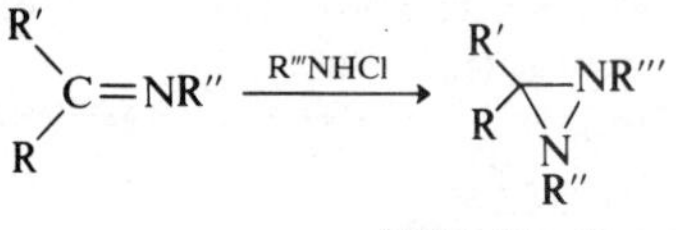

$\xrightarrow{\text{R}'''\text{NHCl}}$

(**261**) R″ = H
(**262**) R‴ = alkyl, R′ = H

(b) Aliphatic ketones, mostly acetone or butanone, are treated with ammonia and chloramine or with ammonia and chlorine in solution or in the gas phase affording (**263**) ⟨59AG283, 61CB968⟩. Diaziridines with both N atoms unsubstituted are still of interest in connection with hydrazine production, *e.g.* ⟨76MIP50801⟩.

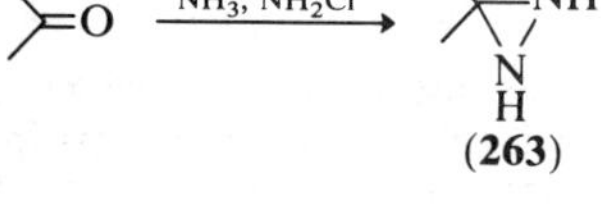

$\xrightarrow{\text{NH}_3,\ \text{NH}_2\text{Cl}}$

(**263**)

(c) Aqueous solutions of aliphatic primary amines and formaldehyde on addition of hypochlorite form 1,2-dialkyldiaziridines unsubstituted at carbon ⟨66CB2104⟩.

(d) Azomethines form diaziridines with aliphatic amines or ammonia; the azomethine (**265**) can be made *in situ* from a dichloramine (**264**) by HCl elimination ⟨69TH50800⟩.

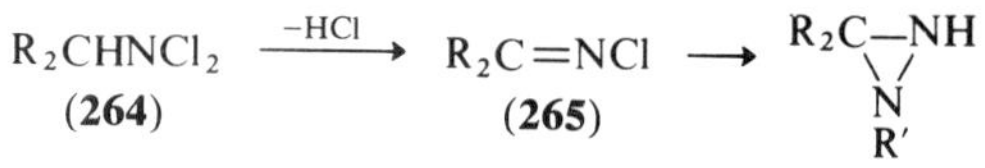

R_2CHNCl_2 (**264**) $\xrightarrow{-HCl}$ $R_2C{=}NCl$ (**265**) $\longrightarrow$ $R_2C{-}NH$ / N / R′

(e) Hydroxylamine-*O*-sulfonic acid in place of chloramine expands the scope of diaziridine synthesis, making them available, for example, from hydroxyketones (**266**), from a Mannich base (**267**) ⟨67CB2093⟩ and from steroidal ketones (**268**) ⟨65JA2665⟩.

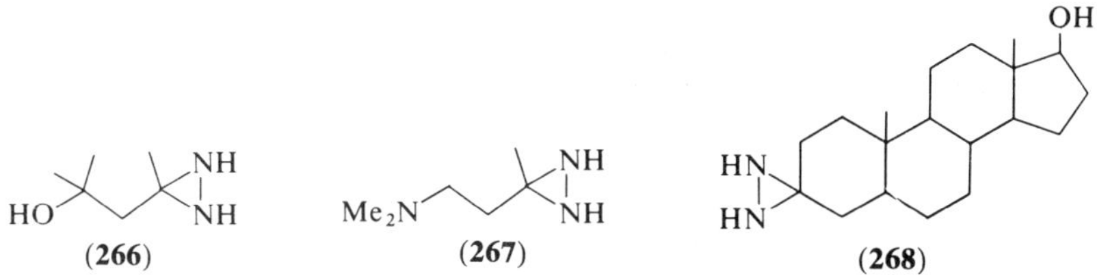

(**266**) (**267**) (**268**)

(f) *O*-Sulfonylated oximes, prepared from ketones and hydroxylamine-*O*-sulfonic acid or from oximes and sulfonyl chlorides, yield diaziridines with aliphatic amines. Diaziridines from electron deficient carbonyl compounds have become accessible by this procedure, *e.g.* (**269**) from mesoxalic ester ⟨74IZV2823⟩.

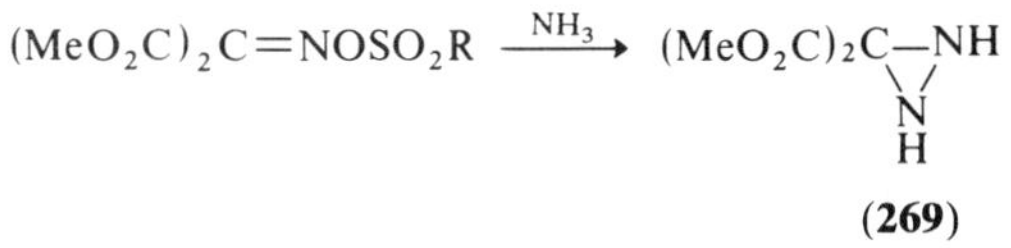

$(MeO_2C)_2C{=}NOSO_2R \xrightarrow{NH_3} (MeO_2C)_2C{-}NH$ / N / H

(**269**)

(g) Ethylmethyloxaziridine transfers its NH group to a Schiff base with diaziridine formation (Section 5.08.3.1.4).

It has been assumed that these diaziridine syntheses proceed through an aminal-like intermediate (**270**), in which the leaving group of one nitrogen is replaced by the other nitrogen forming an N—N bond. Similarity of diaziridine yields and equilibrium position of acetal formation in various carbonyl compounds have been discussed as an argument for this ⟨B-67MI50800⟩.

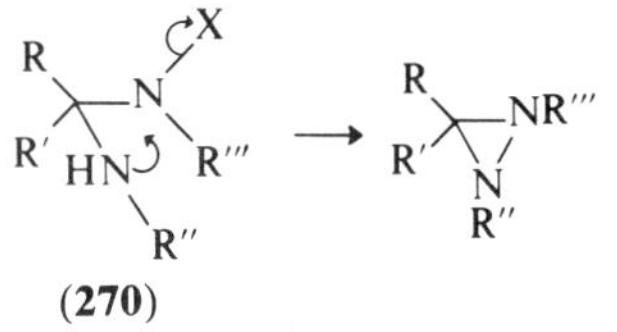

(**270**)

5.08.4.2.2 Photochemical diaziridine syntheses

Photoisomerization of azomethinimines to diaziridines was observed in several classes of compound. Diazepine (**271**) was converted, even by sunlight, to its diaziridine isomer ⟨68JA4738⟩. Further photoisomerizations were observed with azomethinimines (**272**) obtained from pyrazolidones ⟨70JPR161⟩, and with some pyridazinium betaines (**273**).

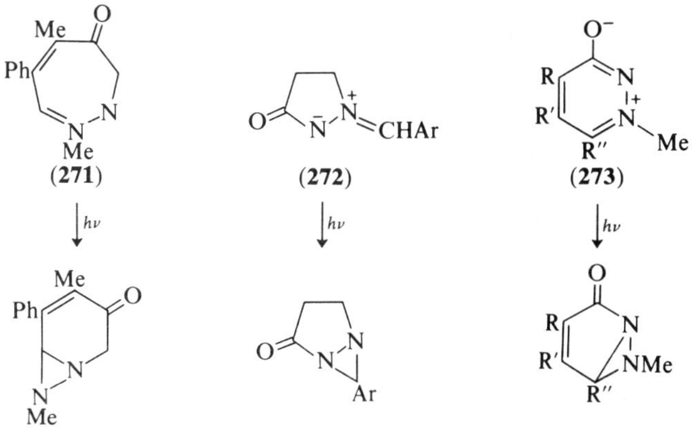

(**271**) (**272**) (**273**)

$\downarrow h\nu$

Even *N*-aryldiaziridines can be obtained. Compound (**274**) is formed on irradiation of its 1,3-dipolar isomer, which is in thermal equilibrium with its head-to-tail dimer ⟨82TH50800⟩.

hexahydrotetrazine ⇌ (65 °C) ... ⇌ ($h\nu$ / Δ) (**274**)

A novel principle of diaziridine synthesis was found in nitrogen extrusion from tetrazolines (**275**). *N*-Aryldiaziridines (**276**) almost unaccessible by older procedures thus become available ⟨74CL185⟩.

(**275**) → ($h\nu$, $-N_2$) (**276**)

A photochemical preparation of fluorodiaziridine (**278**) involves addition of HN_3 to the imine (**277**) followed by UV irradiation ⟨65JOC1398⟩.

$(F_3C)_2C{=}NH$ (**277**) → (HN_3) → ($h\nu$) (**278**)

5.08.4.2.3 Synthesis of diaziridinones and diaziridinimines

Diaziridinones (**167**) are obtained from di-*t*-alkylureas by *N*-chlorination and base-promoted elimination of hydrogen chloride from (**279**) ⟨69JOC2254⟩. The very unstable *cis*-fused diaziridinone (**280**) could be prepared by a similar but milder procedure ⟨76JOC2813⟩.

(**279**) → (Bu^tOK) (**167**) (**280**)

Diaziridinimines were made analogously. Tri-*t*-butylguanidine (**281**) with *t*-butyl hypochlorite gave (**176**) in 80% yield. (**281**) itself is a sufficiently strong base to deprotonate its *N*-chloro derivative ⟨69AG(E)448⟩. This procedure also works with methyl- or phenyl-substituted guanidines.

(**281**) → (Bu^tOCl) (**176**)

Diaziridinone (**167**) was also obtained in attempted syntheses of oxaziridinimine (**282**) from either di-*t*-butylcarbodiimide and peracid, or nitrosoisobutane and *t*-butyl isocyanide, probably by valence isomerization of the transient (**282**) ⟨70JOC2813⟩.

$Bu^tN{=}C{=}NBu^t$ (peracid) → (**167**); $Bu^tNO + Bu^tNC$ → (**167**); (**282**)

Photolytic nitrogen extrusion from tetrazoline derivatives (**283**) and (**284**) permits further syntheses of diaziridin-ones (**285a**) and -imines (**285b**) respectively ⟨75AG(E)428⟩. Thermal carbon dioxide extrusion from a five-membered precursor can also give a diaziridinimine ⟨75JOC3112⟩.

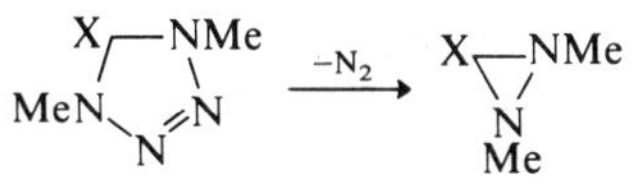

(283) X=O (285) a; X=O
(284) X=NMe b; X=NMe

5.08.4.3 Synthesis of Diazirines

Most diazirines are easily obtained from diaziridines. Dialkyldiazirines are simply formed by dehydrogenation of 3,3-dialkyldiaziridines ⟨60AG781⟩. For example, the spirodiazirine (**187**) can be prepared in 65–75% yield from the diaziridine with silver oxide ⟨65OS(45)83⟩.

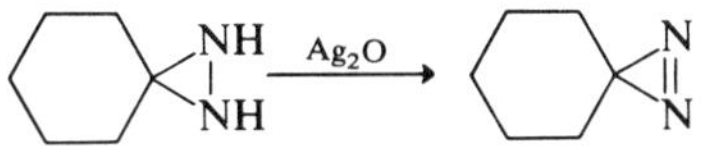

(**187**)

By similar procedures diazirines were prepared not only from simple aliphatic ketones but also from hydroxyketones and β-aminoketones ⟨B-67MI50800⟩, and so were a large number of diazirines from steroidal ketones ⟨65JA2665⟩. Permanganate, bromine, chlorine and hypochlorite were used as oxidants. A one-step preparation of diazirines from ketones like 3-nonanone, ammonia and chlorine has been claimed in a patent ⟨66USP3290289⟩. 3,3-Diazirinedicarboxylic acid derivatives like (**286**) were obtained directly from oxime tosylates by the action of two moles of *O*-ethoxyamine ⟨81AG(E)200⟩.

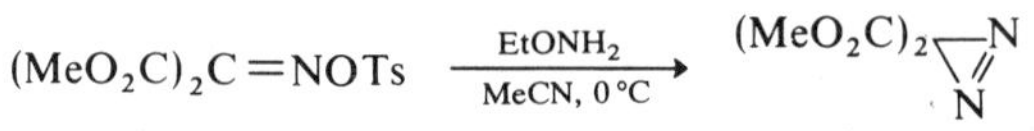

$$(MeO_2C)_2C{=}NOTs \xrightarrow[\text{MeCN, 0 °C}]{\text{EtONH}_2} (MeO_2C)_2CN_2$$

(**286**)

Syntheses of alkyldiazirines (**287**) must start from the diazirinotriazolidines (**128**) formed from an aldehyde, ammonia and *t*-butyl hypochlorite. Since the three-membered ring in (**128**) is much more stable towards acids than the five-membered ring (Section 5.08.3.2.2), the diazirines are obtained by acid hydrolysis of compounds (**128**) in the presence of an oxidant ⟨62CB795⟩.

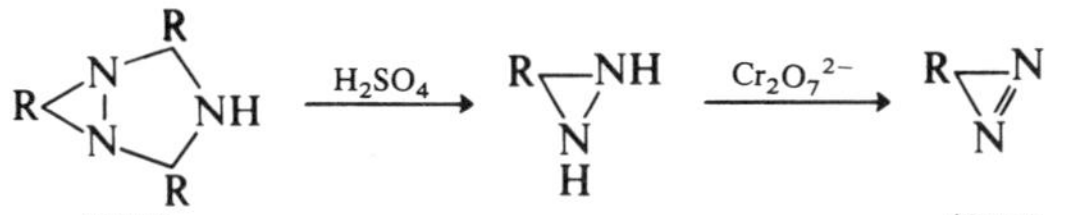

(**128**) (**287**)

The parent compound, 'cyclic diazomethane', was first obtained from formaldehyde, ammonia and chloramine by dichromate oxidation of the initially formed higher molecular diaziridine–formaldehyde condensation product ⟨61TL612⟩. Further syntheses of (**44**) started from Schiff bases of formaldehyde, which were treated with either difluoramine or dichloramine to give (**44**) in a one-pot procedure. Dealkylation of nitrogen in the transient diaziridine was involved ⟨65JOC2108⟩.

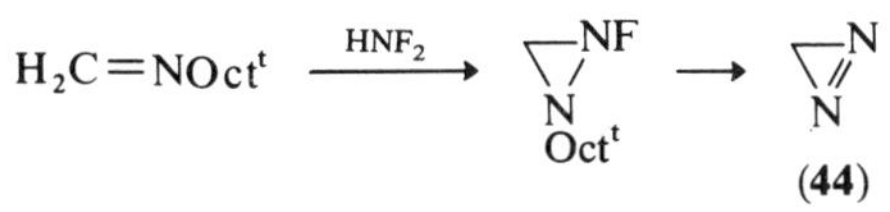

$$H_2C{=}NOct^t \xrightarrow{HNF_2}$$

(**44**)

Another one-pot procedure has given chlorodiazirines in many cases. Amidines are reacted with hypochlorite to form, for example, chlorophenyldiazirine (**289**) from (**288**) ⟨65JA4396⟩.

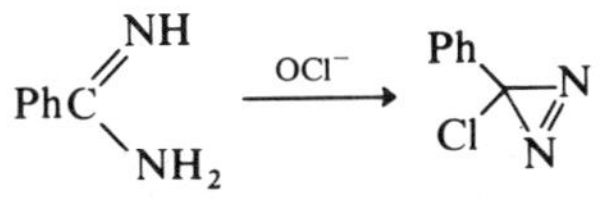

(**288**) (**289**)

Fluorodiazirines are obtained from suitable perfluoro compounds by reduction. Difluorodiazirine (**218**) was prepared from perfluoromethylenediamine (**290**) by reduction with ferrocene ⟨66JHC245⟩, or from perfluoroformamidine (**291**) by iodide reduction ⟨67JOC1944⟩.

(290) (218) (291)

From trifluorotriaminomethane (**292**) and also from perfluoroguanidine (**294**) fluorodifluoroaminodiazirine (**293**) was obtained. The presence of chloride leads to the formation of chlorine-containing diazirines.

$(F_2N)_3CF$ (292) (293) (294)

Since photoconversion of diazirines to diazoalkanes, described in Section 5.08.3.3.2, is a reversible reaction, diazirines may be obtained in certain cases from their linear isomers.

5.08.5 APPLICATIONS OF THREE-MEMBERED RINGS WITH TWO HETEROATOMS

Since syntheses of three-membered rings with two heteroatoms almost always proceed under mild conditions with good yields and from simple starting materials, they are useful intermediates in synthesis. The easy formation of N—O and N—N bonds offers synthetic potentialities. Furthermore, the three-membered ring may serve as a protecting group during additional operations of substitution.

Even the trivial decomposition of oxaziridines may have some importance. In the oxidation of *s*-alkylamines to ketones conversion to the Schiff base of 2-pyridinealdehyde was proposed, followed by peracid oxidation to the oxaziridine (**295**). Decomposition by alkali yields the ketone; added excess acetone suppresses condensation of pyridinealdehyde with dialkyl ketone ⟨75AJC2547⟩.

R_2CHNH_2 → → R_2CHN (295)

$R = C_6H_{13}$

73%

By working in anhydrous solvents even aldimines like (**296**) could be prepared ⟨79TL3201⟩.

DABCO, CCl_4 (296)

One-electron cleavage of the oxaziridine ring is especially useful in the case of the oxaziridine (**114**), easily obtained from cyclohexanone, methylamine and hypochlorite. The radical formed from (**114**) by uptake of one electron from the iron(II) ion can be trapped, for example, by acrylonitrile to form products derived from radical (**297**) ⟨68TL5609⟩.

(114) Fe^{2+}, H^+ CONHMe → CONHMe CN (297)

Because of analogy of radical formation by iron(II) ion from either peroxides or oxaziridines, the latter were proposed repeatedly as initiators of radical chains, *e.g.* in styrene polymerization and in treatment of unsaturated polyesters. Oxaziridines appear to be easier to prepare than peroxides and to be less hazardous in handling ⟨76MI50801⟩.

Preparation of alkylhydroxylamines *via* oxaziridines was proposed early ⟨57JA5739⟩. *N*-*t*-Butylhydroxylamine (82%) and *N*-cyclohexylhydroxylamine (70%) were obtained from the easily hydrolyzing 3-phenyloxaziridines. Syntheses of, for example, *N*-hydroxyamino acids *via* oxaziridines were worked out later. Amino acid esters were condensed with 4-methoxybenzaldehyde, followed by peracid oxidation to give oxaziridines, which were cleaved to the *N*-hydroxyamino acids. A yield of 42% has been reported for *N*-hydroxy-L-phenylalanine (**298**); valine, leucine and glutamic acid react analogously ⟨79BAP459⟩.

PhCH(NH_2)CO_2R ⟶ ⟶ PhCH(CO_2R)–N(oxaziridine, C_6H_4OMe-*p*) ⟶ PhCH(NHOH)CO_2H
(**298**)

Aliphatic nitroso compounds are obtained either by hypobromite oxidation of alkylhydroxylamines made by oxaziridine hydrolysis, or directly from oxaziridines by action of a second mole of peracid (Section 5.08.3.1.3).

Oxaziridines unsubstituted at nitrogen as well as some *N*-acylated oxaziridines offer synthetic potentialities due to their ability to transfer their nitrogen function to nucleophiles (Section 5.08.3.1.4). The simplicity of preparation of some aziridines from alkenes and the spiro oxaziridine (**52**) equals the simplicity of epoxidation. Aziridine (**299**), for example, is obtained by simple heating of indene with (**52**) in toluene ⟨74KGS1629⟩.

(**52**)
NH
O
NH
(**299**)

This type of amination by an oxaziridine is assumed to be the key step of a novel process for hydrazine manufacture, in the course of which butanone in solution with ammonia is reacted with hydrogen peroxide and acetonitrile. The smooth formation of oxaziridines from Schiff bases and hydrogen peroxide–nitrile mixtures is as well known as NH transfer from an oxaziridine like (**300**), suggesting the intermediacy of (**300**) as the N—N forming agent ⟨72TL633⟩.

MeCOEt + NH_3 + H_2O_2 + MeCN ⟶ [Me, Et, NH, O] ⟶(NH_3) ⟶ N=, =N
(**300**)

In the 'modified Raschig process', used by Bayer A.G. and by Mobay Chem. Co. for large scale production of hydrazine, the intermediacy of an oxaziridine could be clearly evidenced ⟨81MI50800⟩. In this process ammonia and hypochlorite are reacted in the presence of acetone to form ketazine (**302**). Nitrogen–nitrogen bond formation is faster by a factor of about 1000 in the presence of acetone than in its absence. Thus acetone does not merely trap hydrazine after formation, but participates in the N—N bond forming reaction. Very fast formation of oxaziridine (**301**), which is isolable, is followed by its likewise fast reaction with ammonia.

MeCOMe + NH_3 + NaOCl ⟶ Me_2C(NH)(O) ⟶(NH_3) N=, =N
(**301**) (**302**)

Use of diaziridines in synthesis is assumed to be especially attractive for synthesis of substituted hydrazines; large scale production of the parent compound, however, was not undertaken after an optimistic development (Section 5.08.3.2.2).

To prepare alkylhydrazines, cyclohexanone is treated with a primary amine and hydroxylamine-*O*-sulfonic acid in a one-pot procedure; hydrolysis to the alkylhydrazine is carried out without isolation of the diaziridine ⟨68JPR(37)257⟩. Yields are between 60 and 70%.

While alkylation of hydrazine brings several alkyl groups to the same nitrogen atom, *N,N'*-dialkylhydrazines can be made advantageously from 1,2-dialkyldiaziridines, easily obtained from aliphatic Schiff bases and *N*-chloroalkylamines (Section 5.08.4.2.1). To suppress side reactions the larger alkyl group is introduced by the Schiff base and the smaller one by the *N*-chloramine. All groups introduced into diaziridines (Section 5.08.3.2.1) remain in the hydrazines obtained by hydrolysis, *e.g.* in (**303**)–(**305**). This enhances the preparative importance of the approach.

$C_6H_{11}NHNHCONHPh$ (**303**) $C_6H_{11}NHNHTs$ (**304**) (**305**)

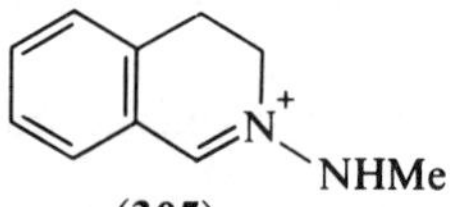

Preparative use of diazirines is also attractive, since they are available from simple starting materials in good yield and can be handled without risk, when their isolation is circumvented. Addition of Grignard reagent with formation of diaziridines (Section 5.08.3.3.1) gives yields of up to 95%, which suggests an alkylhydrazine synthesis from Grignard reagents.

Synthesis of steroidal diazirines has led to some biologically active compounds ⟨65JA2665⟩. The observation that diazirines and their parent ketones are identical in smell ⟨B-67MI50800⟩, points to the possibility that a diazirine group may stand for a keto group *vis-à-vis* a receptor.

A cyclodiazomethane laser has been described in a patent ⟨74USP3842369⟩.

Diazirines have been widely used as carbene precursors (Sections 5.08.3.3.4 and 5). Recently diazirines have found an interesting application in biochemistry. A subunit of a complex structure, *e.g.* a fatty acid, is exchanged by a similar subunit bearing a diazirine group. Photolysis leads to formation of a carbene, which inserts into a bond of an adjacent subunit. Subsequent degradation reveals linkage of the two subunits involved and points to their neighborhood in the original complex structure. Use is made of diazirine groups yielding carbenes with no tendency towards intramolecular alkene formation, such as diazirines derived from adamantanone (**306**), fluorinated diazirines (**307**) or aryldiazirines (**308**).

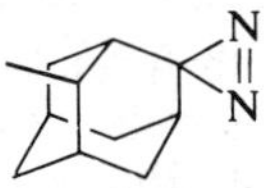

(**306**) 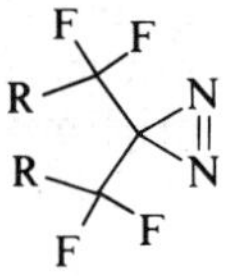 (**307**) 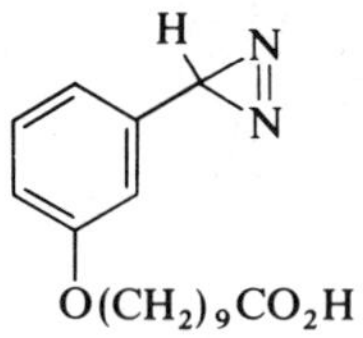 (**308**)

Thus, during structural investigations of lactate dehydrogenase, a diazirine group was substituted for the $CONH_2$ group of the nicotine amide part. After photolysis degradation products were analyzed with special regard to novel covalent bonds between parts of the original structure and information about the active centre was obtained ⟨80B2811⟩.

In another experiment tritiated adamantane diazirine fixed to the hydrocarbon core of a membrane gave rise to carbene insertion into the catalytic subunit of ATP-ase. After protolytic degradation, adjacent areas of the original structure became evident ⟨80JBC(255)860⟩.

Further investigations were carried out at lipid double layers and at phospholipids of membranes. Lipid–lipid and lipid–protein interactions were recognized by diazirine labeling ⟨79PNA2595⟩.

5.09

Azetidines, Azetines and Azetes

D. E. DAVIES and R. C. STORR
University of Liverpool

5.09.1 INTRODUCTION

This chapter deals with four-membered rings containing one nitrogen atom. Azetidine (**1**), 1-azetine (**2**) and 2-azetine (**3**), and azete (**4**) are the nitrogen analogues of cyclobutane, cyclobutene and cyclobutadiene, respectively. Azetidines are well studied, in particular their derivatives the azetidin-2-ones (β-lactams) have received considerable attention mainly because of the antibacterial properties of penicillins and cephalosporins. Both azetidines ⟨64HC(19-2)885, 79CRV331⟩ and azetidin-2-ones ⟨73S327, 78T1731 and refs. therein⟩ have been thoroughly reviewed. In contrast, the chemistry of both 1- and 2-azetines has been developed only since the mid-1960s and these systems have not yet been comprehensively reviewed. With the possible exception of the benzo fused derivatives, azetes remain chemical curiosities.

(**1**) (**2**) (**3**) (**4**)

In view of the uneven attention which azetidines, azetines and azetes have received and because of their lack of chemical similarity, they are treated separately in this chapter. Furthermore, because of the considerable literature on azetidin-2-ones, these have been dealt with in their own right, rather than as derivatives of azetidine.

5.09.2 AZETIDINES

5.09.2.1 Structure

Azetidine (**1**) is a colourless, mobile liquid, b.p. 62.5 °C/747 mmHg ⟨56JA4917⟩, which is completely miscible with water. Its density $d_4^{24} = 0.8412$ and refractive index $n_D^{25} = 1.4278$ ⟨37HCA109⟩. Table 1 gives b.p. and m.p. data for other representative azetidines.

Table 1 Representative Azetidines

Azetidine	*B.p.* (°C/Torr)	*M.p.* (°C)	*Azetidine*	*B.p.* (°C/Torr)	*M.p.* (°C)
1-Methyl	40/735	—	1-Benzyl	78/5.5	—
2-Methyl	74/755	—	1-*p*-Nitrophenyl	—	119
3,3-Dimethyl	90–92	—	1-*t*-Butyl-2-carboxylic acid	—	173–175
2,4,4-Trimethyl	86–88	—	1-*t*-Butyl-3-hydroxy	—	45–46
1,2,4,4-Tetramethyl	93–97	—	3,3-Diphenyl	—	95–96
1-Phenyl	70/1	—			

Azetidine itself has been studied by electron diffraction, which reveals a non-planar structure (Figure 1) ⟨73CC772⟩. The enhanced length of the bonds reflects the strain in the ring and the angle between the CCC and CNC planes of 37° is similar to that found for cyclobutane (35°), but quite different from that for oxetane (4°).

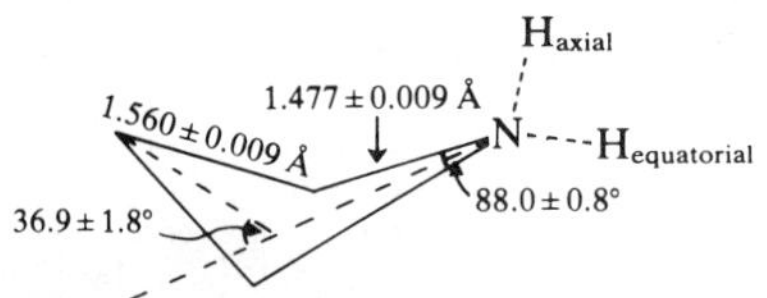

Figure 1 Azetidine

X-Ray crystallographic studies have been reported for a number of azetidines and azetidinium salts ⟨71JA1761 and refs. therein, 75AX(B)49, 77AX(B)824, 74AX(B)534, B-79MI50900⟩. These also generally indicate non-planarity of the ring, although interaction between large ring substituents can enforce planarity ⟨B-77SH(1)54⟩; some representative bond lengths and angles are given in Section 5.01.1.2. A barrier of 5.27 kJ mol^{-1} has been estimated for ring flipping in azetidine compared with 6.02 kJ mol^{-1} for cyclobutane.

The substituent on the nitrogen in azetidine can be either axial or equatorial and rapid inversion at nitrogen normally occurs ⟨B-77SH(1)54⟩. Both axial and equatorial H forms appear to be present in azetidine from far-IR studies, and estimates of the barrier to pyramidal N-inversion have been made from NMR studies for a large number of azetidines (see Section 5.01.1.2) ⟨71MI50900⟩. The extensive NMR studies of azetidines support a conformationally mobile, non-planar structure in which rapid N-inversion occurs. NMR spectral parameters have been summarized in Section 5.01.1.2. The magnitude of the vicinal coupling constant ($J_{3,4}^{cis}$ = *ca.* 7.5 Hz, $J_{3,4}^{trans}$ = *ca.* 3.0 Hz) is widely used in the assignment of stereochemistry.

The absolute configuration of naturally occurring *S*(−)-azetidine-2-carboxylic acid has been established ⟨73CL5⟩, and the DL form has been resolved ⟨69JHC993⟩. ORD and CD curves have been determined for 2-methylazetidine and an octant rule has been proposed for the *N*-chloro- and *N*-cyano-2-methylazetidines ⟨74T39⟩.

Infrared, electronic and mass spectral properties of azetidines are discussed in Sections 5.01.1.4–5.01.1.6.

Photoelectron spectroscopic studies show that the first ionization potential (lone pair electrons) for cyclic amines falls in the order aziridine (9.85 eV) > azetidine (9.04) > pyrrolidine (8.77) > piperidine (8.64), reflecting a decrease in lone pair *s*-character in the series. This correlates well with the relative vapour phase basicities determined by ion cyclotron resonance, but not with basicity in aqueous solution, where azetidine (pK_a 11.29) appears more basic than pyrrolidine (11.27) or piperidine (11.22). Clearly, solvation effects influence basicity ⟨74JA288⟩.

5.09.2.2. Reactivity

5.09.2.2.1 General survey

Azetidine can be considered as a fairly typical cyclic amine. Strain in the four-membered ring is less than that in the three-membered aziridine system, and as a result azetidines show few, if any, of the exceptional properties associated with aziridines. Thus, ring cleavage reactions occur with greater ease than in larger ring cyclic amines, but much less readily than with aziridine; for example, unlike aziridines, azetidines do not function as alkylating agents.

The ring is generally thermally stable and is unreactive towards reducing agents. Although ring cleavage by nucleophiles occurs, the fact that many azetidines are prepared under basic conditions indicates that the ring strain does not make them especially labile. This means that a wide variety of transformations can be performed on attached functional groups without effect on the azetidine ring.

5.09.2.2.2 Thermal and photochemical reactions

There are few reports of the thermal and photochemical decomposition of simple azetidines.

Methyleneimine has been observed directly and trapped at low temperatures in the gas phase pyrolysis of azetidine ⟨78TL577⟩. The azetidine (**5**) undergoes ring cleavage on distillation at 160 °C ⟨64LA(661)187⟩. The formation of pyrrole derivatives (**7**) on irradiation of 3-acylazetidines (**6**) is a general, stereoselective process, which involves an intramolecular H abstraction ⟨70JA100, B-76PH93⟩.

Ph, Et, N, CO_2Et (**5**) ⟶ Ph, Me, $NHCO_2Et$

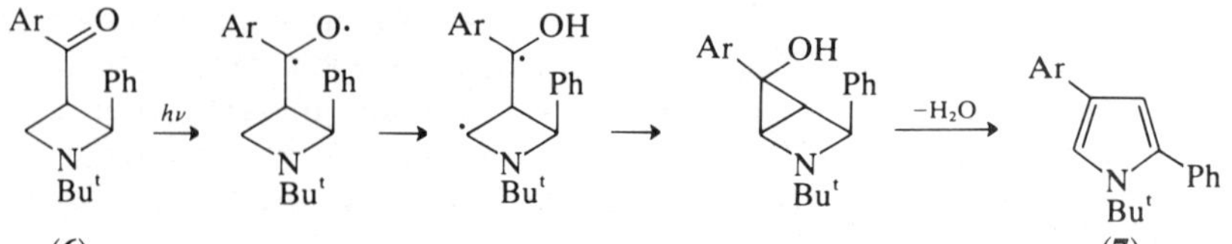

(6) (7)

5.09.2.2.3 Electrophilic attack

In aqueous solution, azetidine (pK_a 11.29) is slightly more basic than pyrrolidine and larger-ring cyclic amines and appreciably more basic than aziridine. It forms an addition compound (m.p. −9 to −6 °C) with trimethylboron which is more stable than that formed by pyrrolidine ⟨50JA2926, 64HC(19-2)885⟩. Azetidinium salts are well known (Section 5.09.2.2.7).

Azetidines are unstable towards mineral acid, ring cleavage occurring by attack of a nucleophile on the protonated azetidine and being favoured by relief of strain. Since ring strain is less in geminally substituted azetidines, some 3,3-disubstituted derivatives are remarkably stable. Ring opening initiated by electrophilic attack on nitrogen can compete with reaction at substituents, *e.g.* 1-cyclohexyl-3-azetidinol (**9**) gives the cleaved product (**10**) rather than the expected tosylate with tosyl chloride in pyridine. The more hindered 1-*t*-butyl-3-azetidinol, however, does give the *O*-tosyl derivative under these conditions. The 1-cyclohexyl-3-*O*-tosyl derivative (**8**) can be made by treatment of azetidinol (**9**) with sodium hydride in ether followed by tosyl chloride ⟨67BCJ2401, 71JHC1063⟩.

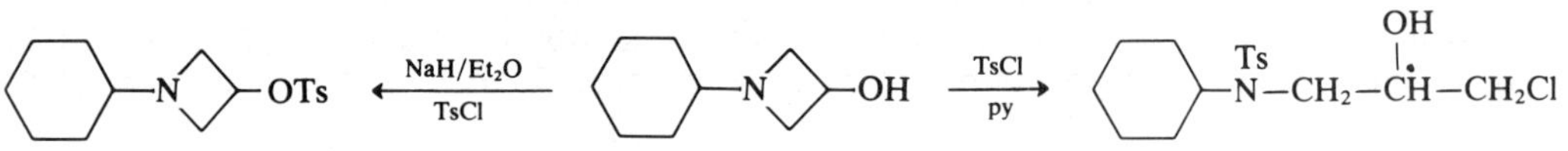

(**8**) (**9**) (**10**)

N-Unsubstituted azetidines generally behave like secondary amines and can be converted by standard methods into amides, carbamates, sulfonamides and *N*-guanyl derivatives. *N*-Alkylation gives tertiary azetidines and reaction with ethylene oxide gives *N*-hydroxyethylazetidines. Conjugate addition to activated acetylenes, *e.g.* dimethyl acetylenedicarboxylate, and activated alkenes is also general. *N*-Arylation with activated aryl halides has also been reported. Condensation occurs with ethyl acetoacetate and acetylacetone to give the enamines (**11**) and (**12**) respectively. Enamines are also produced in 22% and 15% yield with cyclohexanone and cyclopentanone respectively ⟨67BCJ1964⟩.

(**11**) (**12**)

Azetidine and a number of its derivatives have been converted to *N*-nitroso derivatives with nitrous acid ⟨79CRV331 pp. 334, 338⟩. These *N*-nitroso compounds are readily reduced to the *N*-amines which behave as typical hydrazines in condensation reactions ⟨79CRV331 p. 338⟩. *N*-Nitrosoazetidine has been oxidized with trifluoroperacetic acid to *N*-nitroazetidine ⟨61JA4417⟩. *N*-Chloro- and *N*-bromo-azetidines have been reported from reaction of azetidines with *t*-butyl hypochlorite ⟨71IZV893⟩ or *N*-halosuccinimides ⟨74T39⟩. Little information concerning the oxidation of azetidines is available; azetidine is reported to give acrolein and ammonia on treatment with hydrogen peroxide ⟨38JGU1545⟩ and the direct oxidation of an azetidine to a β-lactam has been achieved using ruthenium tetroxide ⟨74JOC2264⟩.

5.09.2.2.4 Nucleophilic attack

Nucleophilic ring opening of azetidines, *e.g.* (**13**), can proceed in good yield depending on the nucleophile and catalyst ⟨69JHC273, 79CRV331⟩. Cleavage of the azetidine ring by nucleophiles under the influence of Lewis acid catalysts has already been discussed (Section 5.09.2.2.3).

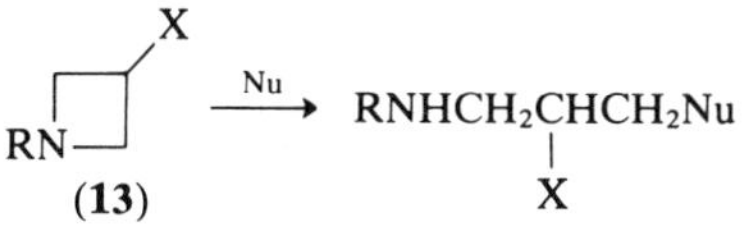

RN (13) —Nu→ $RNHCH_2CHCH_2Nu$ (X)

However, the azetidine ring is relatively stable towards basic conditions and nucleophiles as evidenced by the number of routes to azetidines which employ such conditions. The reduction of azetidin-2-ones to azetidines with LAH and the formation of azetidin-3-ols from azetidin-3-ones and organolithium reagents also testifies to this ⟨73S153⟩. This relative stability of the azetidine ring also allows nucleophilic replacement of groups attached to ring carbons. For example, the azetidin-3-ols (**14**; R = Bu^t, cyclohexyl) can be converted to the chloro compounds (**15**) on treatment with triphenylphosphine in carbon tetrachloride. Replacement of the chlorine in (**15**; R = Bu^t) can then be effected with a variety of nucleophiles (Scheme 1). Similar displacements occur with mesylate and tosylate as leaving groups ⟨79CRV331 pp. 344–346⟩.

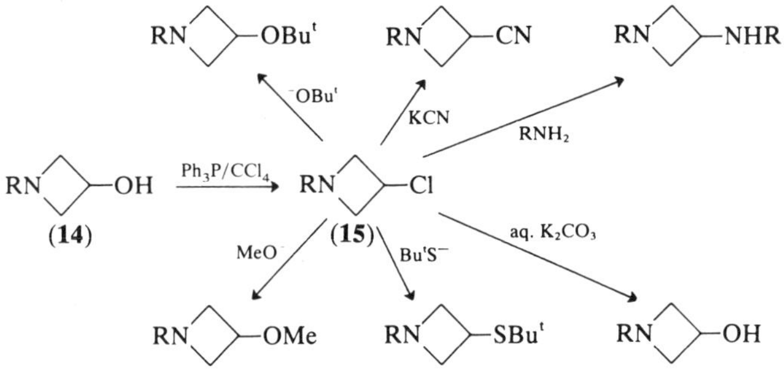

Scheme 1

5.09.2.2.5 Ring transformations

On heating in solution, azetidine (**16**; X = Cl) undergoes reversible ring contraction to the aziridine (**17**), presumably *via* the azabicyclobutyl cation (**18**). The tosylate (**16**; X = Ts) in ethanol gives aziridine (**17**; X = OEt) ⟨B-73MI50901⟩.

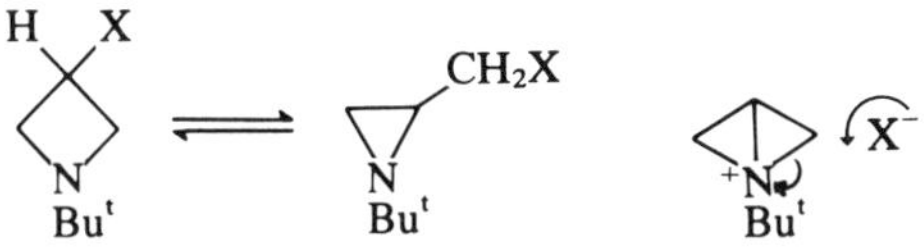

Oxidation of *N*-aminoazetidines (**19**), deoxygenation of *N*-nitrosoazetidines (**20**) and direct deamination of azetidines (**21**) with difluoroamine leads to cyclopropanes (**23**) by extrusion of nitrogen from a diazine intermediate (**22**) ⟨63JA97⟩. A further interesting ring contraction occurs in the Ag^+ catalysed solvolysis of the *N*-chloroazetidine (**24**), which appears to involve the intermediate cation (**25**) ⟨71TL109⟩.

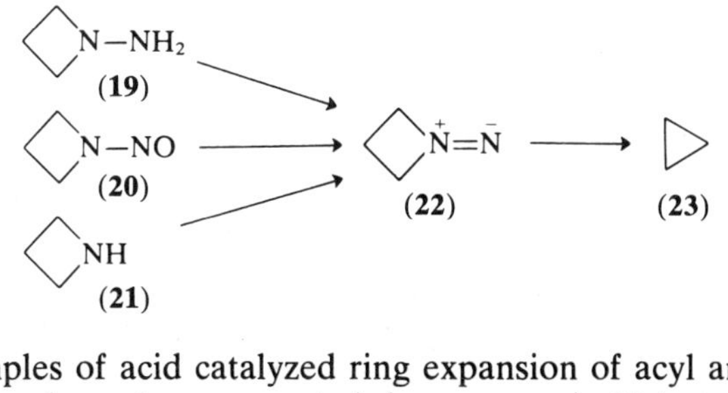

A number of examples of acid catalyzed ring expansion of acyl and thioacyl azetidines to six-membered rings have been reported ⟨B-73MI50903⟩. This type of rearrangement (Scheme 2) is similar to the more general vinylaziridine to pyrroline ring expansion.

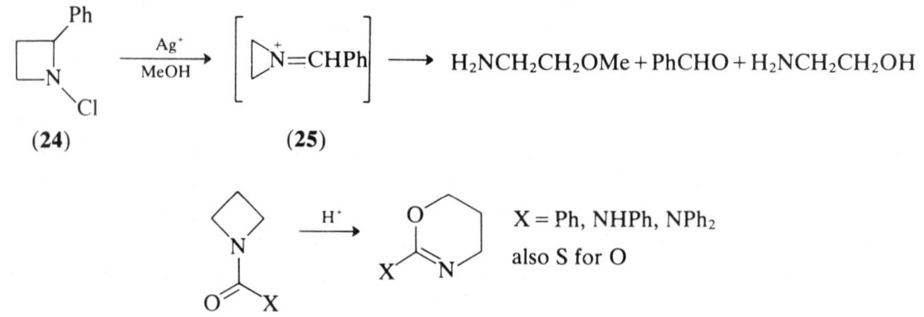

(24) $\xrightarrow[\text{MeOH}]{Ag^+}$ [(25)] ⟶ $H_2NCH_2CH_2OMe + PhCHO + H_2NCH_2CH_2OH$

$\xrightarrow{H^+}$ X = Ph, NHPh, NPh_2; also S for O

Scheme 2

5.09.2.2.6 Modification of substituents

Table 2 summarizes some of the transformations of substituents which have been carried out on azetidines without effect on the ring ⟨79CRV331⟩. Other transformations of interest are the base catalyzed epimerization, H exchange and alkylation of the activated H-3 in azetidines (**26**) ⟨69JHC153⟩ and the nitration, reduction, diazotization and hence further modification of the aromatic ring in 3-phenyl-*N*-acetylazetidine (**27**) ⟨61LA(647)83⟩.

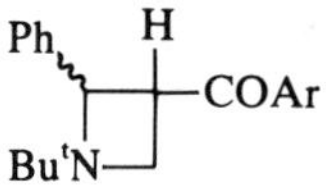

(**26**) 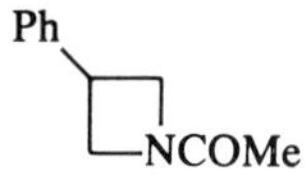 (**27**)

Table 2 Some Transformations of Substituents in Azetidines

Substituent		*Product (reagent)*	*Substituent*		*Product (reagent)*
N—CH_2Ph	→	NH (H_2, Pd/C)	3-CN	→	CH_2NH_2 ($LiAlH_4$)
N—Ts	→	NH (Na, alcohol)	3-CN	→	CO_2H (i, EtOH, NaOH; ii, H_2O)
N—CO_2CH_2Ph	→	NH (H_2, Pd/C)	3-CN	→	COAr (ArMgBr)
N—NO	→	NNH_2 ($LiAlH_4$)	3-OH	→	OCOR (RCOCl, pyridine)
N—NO	→	NNO_2 ($MeCO_3H$)	3-OH	→	OSO_2R (RSO_2Cl)
N—NH_2	→	NN=CHR (RCHO)	3-OTs	→	H ($LiAlH_4$)
N—COR	→	NCH_2R ($LiAlH_4$)	3-OTs	→	CHXY ($Na^+\bar{C}HXY$)
N—$CHRCH_2CH_2NH_2$	→	$NCHRCH_2CH_2OH$ ($NaNO_2$, aq. MeOH)	3-OH	→	C=O ($H_2Cr_2O_7$)
2- or 3-CO_2H	→	CH_2OH ($LiAlH_4$)	3-CH_2OCOR	→	CH_2OH (OH^-)
2-CO_2H	→	$CONH_2$ (i, $ClCO_2Et$; ii, NH_3)	3-CH_2OH	→	CH_2OCOR (RCOCl)
2-$CONH_2$	→	CH_2NH_2 ($LiAlH_4$)	2-COAr	→	CHOHAr ($LiAlH_4$)
2-CO_2Me	→	$CONHNH_2$ (NH_2NH_2, EtOH, room temp.)	2-CH_2CH_2Br	→	CH_2CH_2OH (Ag_2O, H_2O)
2-CO_2Me	→	CO_2H (OH^-, H_2O)	2-CH_2CH_2Br	→	$CH_2CH_2NH_2$ (Gabriel synth.)
2-CO_2Me	→	CH_2OH ($LiAlH_4$)	3-SMe	→	SO_2Me ($KMnO_4$)

5.09.2.2.7 Azetidinium salts

Azetidinium salts can be obtained by quaternization of tertiary azetidines with alkyl halides or by direct cyclizations involving tertiary amines (Section 5.09.2.3). They are crystalline solids, but not surprisingly tend to undergo ring cleavage on heating, as in the case of azetidinium salt (**28**). 1,1-Dimethylazetidinium bromide decomposes rapidly to a linear polymer; the higher homologues polymerize less readily. Hofmann elimination of simple quaternary azetidinium hydroxides, *e.g.* (**29**), leads to ring cleavage ⟨64HC(19-2)904⟩. On the other hand, more labile *N*-alkyl groups can leave the ring intact, *e.g.* alkylation and Hofmann elimination in the ethoxycarbonylethyl derivative (**30**) has been used to give the azetidine (**31**) quantitatively ⟨66JHC230⟩. Ring expansion *via* a Stevens rearrangement has also been observed (Scheme 3) ⟨B-73MI50902⟩.

Scheme 3

5.09.2.3 Synthesis

5.09.2.3.1 From non-heterocyclic precursors

(i) Formation of 1,2-bond

Intramolecular displacement of a leaving group on carbon by a γ-amino function comprises one of the most general routes to azetidines (Scheme 4). The variation in X and Z and in the substituents on the carbon atoms is wide. Yields tend to be better when the amino function carries a large substituent and when the carbon bearing the leaving group is primary. Competing side reactions are dimerization, polymerization, fragmentation, elimination and solvolysis, depending on the conditions ⟨79CRV331, 64HC(19-2)885 and refs. therein⟩.

R^1	R^2	R^3	R^4	R^5	R^6	X	Z	*Yield*
H	H	H	H	H	H	Br	H	v. low
H	Me	H	H	H	H	OTs	Ts	68%
H	H	H	H	H	CO_2Et	Br	Ts	95%
H	H	Me	Me	H	H	OSO_3^-	CH_2Ph	92%
H	H	H	H	H	H	OSO_3^-	CH_2Ph	5–9%

Scheme 4

(ii) Formation of 2,3-bond

Photolysis of phenacylamines (**32**) gives azetidin-3-ols (**33**) as minor products ⟨66CC289⟩. Higher and more consistent yields are obtained from photolyses of α-*N*-alkylamidoacetophenones (**34**), which gives *N*-tosylazetidines (**35**) in 74–95% yield ⟨71JA2793⟩. The key step

$PhCOCH_2N(CHR^1R^2)(Ts)$ $\xrightarrow[Et_2O]{h\nu}$ (35)

(34) (35)

in these reactions would appear to be H-abstraction by the photoexcited carbonyl group to give a diradical which collapses to the azetidine.

(iii) Formation of two bonds

Dialkylation of an amine or sulfonamide with a 1,3-dihalide provides a further route to azetidines ⟨79CRV331, 64HC(19-2)885⟩. Examples of this approach are the formation of *N*-tosylazetidine from tosylamide and 1-bromo-3-chloropropane and the formation of *N*-alkylazetidinyl esters (**36**). The latter reaction works well except for R=Me; the former provides a useful route to azetidine since the tosyl group can be removed by reductive methods.

$BrCH_2CH_2CH(Br)CO_2Me \xrightarrow{RNH_2}$ (36)

(36)

There are few examples of formation of simple azetidines from [2+2] fragments, although this type of approach is important for certain azetidine derivatives, *e.g.* azetidin-2-ones (see Section 5.09.3). Nitrogen analogues of the Paterno–Buchi route to oxetanes are rare; an example involves the photoaddition of 3-ethoxyisoindolone (**37**) to the enol ether (**38**) ⟨75JA7288, 72CC1144⟩.

(37) $\xrightarrow{h\nu}$ (38)

(37) (38)

5.09.2.3.2 From heterocyclic precursors

(i) From three-membered rings

There are several examples of ring opening of azabicyclobutanes (**39**) with electrophiles or nucleophiles to give azetidines (**40**) ⟨69AG(E)70, 69CB3148⟩. Solvolytic ring expansion of the aziridine tosylate (**41**) also gives the alcohol and ether (**42**; R = H, Et) ⟨68TL6179⟩.

(39) $\xrightarrow{R'X}$ (40) R'X = acid halide, anhydride, thiophenol, amine

(39) (40)

(41) $\xrightarrow[Et_3N]{95\%\ EtOH}$ (42) R = H, 38%; R = Et, 4%

(41) (42)

Ring expansion of activated aziridines (**43**) with sulfur ylides also provides a synthesis of azetidines ⟨75JOC2990, 58BSF345, 81CC417⟩. The highly reactive sulfonium methylide (**44**; $R^1 = R^2 = H$) undergoes further reaction with the azetidines (**46**), but the reaction is satisfactory for substituted methylides. The less reactive sulfoxonium methylide (**45**; $R^1 = R^2 = H$)

(43) $\xrightarrow{(44)\ \text{or}\ (45)}$ (46) E = CO_2Me

(43) (46)

gives the azetidine without complication, but substituted sulfoxonium methylides (**45**; $R^1 = H$, $R^2 = COPh$) are unreactive.

The reaction of amines with halogenoalkyloxiranes (**47**) gives 3-hydroxyazetidines (**49**) *via* ring closure of intermediate 1-alkylamino-3-chloro-2-alkanols (**48**) ⟨66TL4691, 67JOC2972, 68CC93, 67TL2155, 71JHC961, 71JHC1059⟩. The *N*-alkyl groups must provide moderate steric hindrance to suppress side reactions and *N*-aryl groups are insufficiently nucleophilic. In some cases, better yields are obtained when the intermediate alkylamino-3-halo-2-alkanols are isolated and purified prior to cyclization.

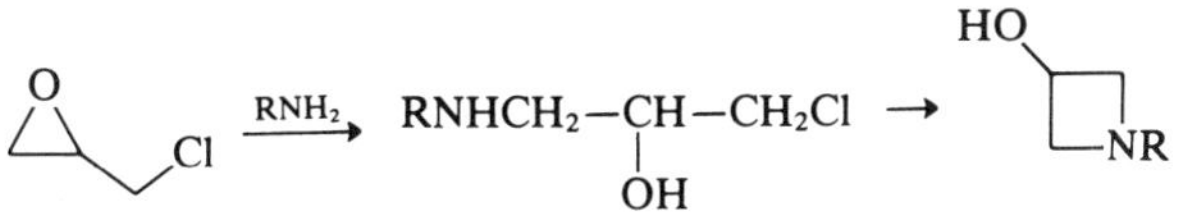

RNH_2; $RNHCH_2-CH(OH)-CH_2Cl$ → HO, NR

(**47**) (**48**) (**49**)

(ii) From four-membered rings

The importance of β-lactams has led to the development of many methods for their formation (Section 5.09.3). It is not surprising, therefore, that reduction of the carbonyl group in β-lactams (**50**) is one of the most general routes to azetidines ⟨79CRV331 pp. 332, 337, 335, 64HC(19-2)885⟩. The reaction is, however, satisfactory only when the nitrogen bears no substituent, otherwise LAH, Raney nickel, $NaBH_4/AlCl_3$ or diborane all give ring cleavage to 3-aminopropanols. The reagent of choice is normally LAH. Reduction of both carbonyl groups of malonimides (**51**) is also useful.

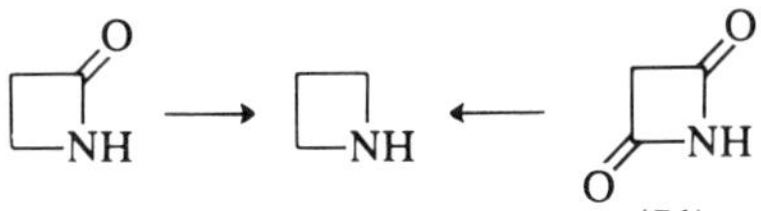

O, NH → NH ← O, NH, O

(**50**) (**51**)

(iii) From five-membered rings

A number of specialized routes to azetidines involving contraction of five-membered rings have been reported. Photolysis of pyrroline *N*-oxides (**52**) gives *N*-acylazetidines (**54**) *via* intermediate oxaziridines (**53**) ⟨63JOC511, 78TL175⟩. Ring contraction of tetramethylpyrrolidinedione to azetidine derivatives has been achieved both by benzilic acid rearrangement and by Wolff rearrangement (Scheme 5) ⟨58BSF345, 67BCJ2398⟩. 4-Oxazolines (**55**) on reaction with sulfur ylides (**44**) and (**45**) also give azetidines (**46**) ⟨75JOC2990, 73TL1923⟩.

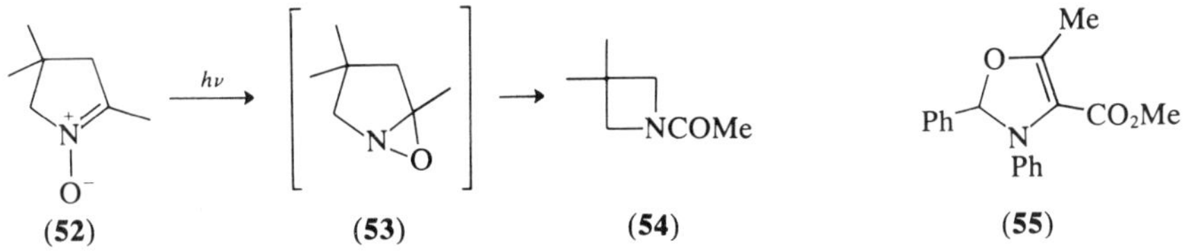

$h\nu$; N⁺, O⁻; N, O; NCOMe; Me, O, Ph, N, Ph, CO_2Me

(**52**) (**53**) (**54**) (**55**)

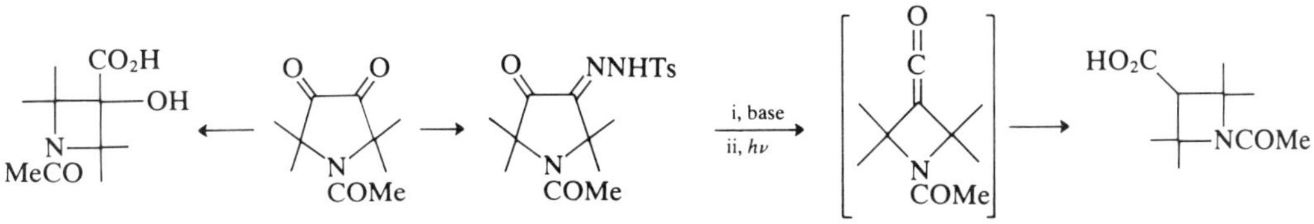

CO_2H, OH, N, MeCO ← O, O, N, COMe → O, NNHTs, N, COMe; i, base; ii, $h\nu$; O=C, N, COMe → HO_2C, NCOMe

Scheme 5

5.09.2.3.3 Derivatives

(i) Azetidine

Azetidine was first prepared in low yield and impure form in 1888 by treatment of 3-bromopropylamine with base ⟨1888CB2669⟩. Various modifications of the standard cyclization procedures have been reported ⟨79CRV331⟩, but by far the most efficient synthesis to date, which utilizes only readily available materials, is that outlined in Scheme 6.

$$H_2N(CH_2)_3OH \xrightarrow[\text{ii, SOCl}_2\text{, DMF, CHCl}_3]{\text{i, CH}_2\text{=CHCO}_2\text{Et, }\Delta} Cl(CH_2)_3N(CH_2CH_2CO_2Et)_2 \xrightarrow[\text{dry distillation 165–205 °C}]{\text{Na}_2\text{CO}_3}$$

$$\text{azetidine-}NCH_2CH_2CO_2Et \xrightarrow[\text{ii, dry distillation 200–270 °C}]{\text{i, KOH, 80 °C}} \text{azetidine-}NH$$

Scheme 6

(ii) Alkyl-, aryl- and acyl-azetidines

Alkyl and aryl substituted azetidines are normally prepared from appropriate precursors bearing the substituents (Sections 5.09.2.3.1(*i*) and (*iii*)) and by reduction of *N*-unsubstituted *β*-lactams and malonimides with LAH (Section 5.09.2.3.2(*ii*)). However, alkyl and aryl groups have been introduced into the azetidine ring by nucleophilic addition of organometallic reagents to azetidin-3-ones (Section 5.09.2.2.4). Alkylation of suitably activated azetidines, *e.g.* (**26**), has also been employed. *N*-Alkyl and *N*-aryl groups are also generally introduced into the precursors before cyclization, but *N*-alkylazetidines can be obtained by alkylation or by reduction of *N*-acyl derivatives. *N*-Arylation using activated aryl halides has also found application.

A number of 2-acylazetidines have been prepared by reaction of 1,3-dihaloacyl compounds with amino derivatives (Section 5.09.2.3.1(*iii*)). This is illustrated for azetidine 2-carboxylic acid (**56**), the only known naturally occurring azetidine. Ring expansion of activated aziridines (**43**) and contraction of 4-oxazolines (**55**) has also found limited use (Section 5.09.2.3.2(*i*) and (*iii*)).

$$BrCH_2CH_2CH(Br)CO_2CH_2Ph \xrightarrow[\text{ii, HCl, Et}_2\text{O; iii, base}]{\text{i, Ph}_2\text{CHNH}_2\text{, MeCN}} \text{2-}CO_2CH_2Ph\text{-azetidine-}NCHPh_2 \xrightarrow{H_2\text{,Pd/C}} \text{2-}CO_2H\text{-azetidine-}NH\ \textbf{(56)}$$

3-Acylazetidines were first synthesized by cyclization of 2-acyl substituted 3-bromopropylamines with base ⟨69JOC310⟩.

(iii) Heteroatom substituted azetidines

Ring expansion of haloalkyloxiranes provides a simple two-step procedure for the preparation of azetidin-3-ols (Section 5.09.2.3.2(*i*)) which can be extended to include 3-substituted ethers and *O*-esters ⟨79CRV331 p. 341⟩. The availability of 3-hydroxyazetidines provides access to a variety of 3-substituted azetidines, including halogeno, amino and alkylthio derivatives, by further substitution reactions (Section 5.09.2.2.4). Photolysis of phenylacylamines has also found application in the formation of azetidin-3-ols (**33**). Not surprisingly, few 2-*O*-substituted azetidines are known. The 2-methoxyazetidine (**57**) has been produced by an internal displacement, where the internal amide ion is generated by nucleophilic addition to an imine.

$$Me_2C(CHO)(CH_2OTs) \xrightarrow{PhNH_2} Me_2C(CH{=}NPh)(CH_2OTs) \xrightarrow{KOMe} \text{3,3-Me}_2\text{-2-OMe-azetidine-}NPh\ \textbf{(57)}$$

N-Halogeno-, *N*-nitroso- and hence *N*-amino-azetidines have been prepared from azetidines by reaction with positive halogen reagents and nitrosating agents, respectively (Section 5.09.2.2.3).

5.09.2.4 Applications

A number of azetidines have been investigated for pharmacological activity, but no important derivatives have emerged ⟨64HC(19-2)908, 68JMC466, 69JMC196, 79JMC183⟩.

5.09.3 AZETIDIN-2-ONES AND RELATED AZETIDINE DERIVATIVES

5.09.3.1 Structure

Azetidin-2-one (**50**) is a hydrolytically sensitive, colourless solid, m.p. 73–74 °C ⟨75LA2195⟩. Other simple azetidin-2-ones are usually low-melting solids or oils ⟨74LA539⟩ (*cf.* Table 3).

X-Ray crystallographic studies of a number of monocyclic azetidin-2-ones indicate that the ring is essentially planar with the nitrogen atom slightly (*ca.* 0.003 Å) out of the mean plane of its substituents except where steric factors enforce greater deviations from planarity ⟨73JCS(P2)2042, 70AX(B)1283, 74MI50900, 80AX(B)1507, 69AG(E)990, 80JA2060⟩. Representative data are given for 1-(2-bromophenyl)azetidin-2-one (**58**) ⟨77JCS(P2)547⟩.

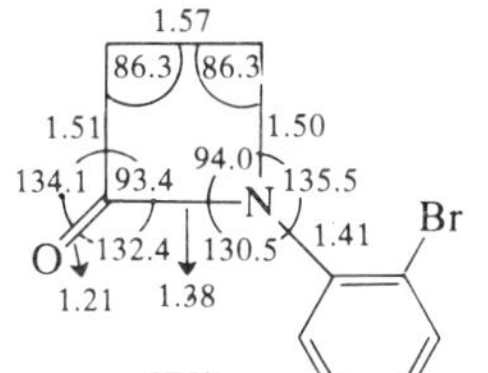

(**58**)

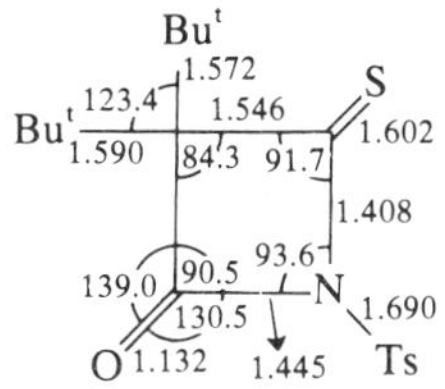

(**59**)

The N—CO distance of 1.38 Å in (**58**) is rather greater than that of a 'normal' amide (*ca.* 1.32 Å); this has been attributed to ring strain and to inhibition of normal amide resonance by interaction with the *N*-aryl substituent. This inhibition of resonance is more pronounced in the *N*-tosyl-4-thioxoazetidin-2-one (**59**), which exhibits very short C=O and C=S distances as well as the unusually long C—N bonds ⟨80TL4247⟩. NMR investigations

Table 3 Physical Properties of Some Azetidin-2-ones

	M.p. (°C)	*B.p.* (°C/Torr)	$\bar{\nu}_{C=O}$ (cm^{-1})	*Ring proton NMR signals* (δ)
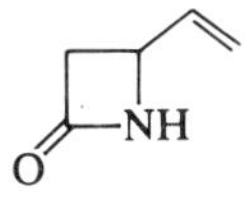	—	76/0.25	1780	2.67 (1H, m), 3.17 (1H, m), 4.1 (1H, m)
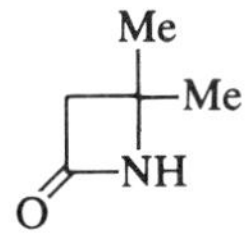	16–17	112/1.5		2.66
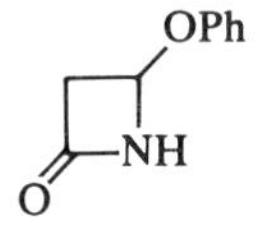	110–111	—	1780	2.8–3.6 (2H, m), 5.6 (1H, dd)
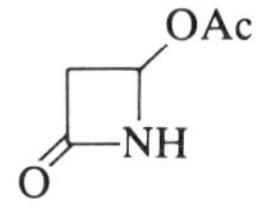	34	80–82/10^{-3}	1785	2.5–3.5 (2H, m), 5.85 (1H, dd)
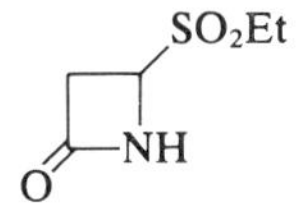	79–81	—	1755	2.8–3.5 (2H, m), 4.85 (1H, dd)
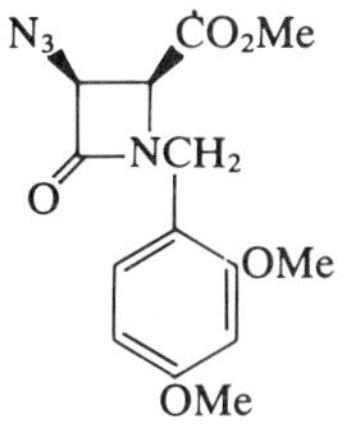	82–84	—	1780	4.08 (1H, d, *J* 5.5 Hz), 4.68 (1H, d, *J* 5.5 Hz)

have shown low barriers to inversion at nitrogen in β-lactams; these have been discussed elsewhere (Section 5.01.1.3.1). Lanthanide induced shift studies on penicillin and cephalosporin derivatives have been used to determine the conformation and stereochemistry of such systems ⟨72T5977, B-74MI50901, B-78MI50900 and refs. therein⟩.

The NMR spectra of a series of 1-benzyl-3,4-disubstituted azetidin-2-ones indicate that the *N*-benzyl substituent lies out of the plane of the β-lactam ring. The marked magnetic non-equivalence of the benzylic CH_2 protons and their stereospecific coupling to the C-3 protons in 1-benzyl-4-substituted azetidin-2-ones suggests highly preferred conformations of the *N*-benzyl moiety ⟨65TL3325⟩. In general, $J_{3,4}(cis)$ is *ca.* 5.0–5.9 Hz, while $J_{3,4}(trans)$ is *ca.* 2.2–2.8 Hz; this provides a simple method for determining the relative stereochemistry of 3,4-disubstituted β-lactams. Typical values for chemical shifts in substituted β-lactams are given in Section 5.01.1.3.1 (for references, see Table 5).

The mass spectra of several azetidin-2-ones exhibit both molecular ions and $[M+1]^+$ ions, although in some cases only one of these is observable. This is in contrast to the behaviour of the azetidines, which generally show M^+ rather than $[M+1]^+$ ions in their mass spectra. Ring cleavages of azetidin-2-ones to give isocyanate plus alkene and ketene plus imine are common processes ⟨68OMS(1)857, 68JOC3448, 69T4421⟩. Related systems such as 4-iminoazetidin-2-ones ⟨77CC302⟩, 2-imino-4-alkylideneazetidines ⟨72JHC1147⟩ and 2-iminoazetidines ⟨79AG(E)320⟩ exhibit analogous ring cleavages.

The most informative feature of the IR spectra of azetidin-2-ones is generally the β-lactam carbonyl absorption, the frequency of which is affected by substitution and by fusion of the ring (*cf.* Table 3). Thus, IR spectra of simple monocyclic β-lactams generally have absorption maxima in the region 1730–1760 cm^{-1}, while the fused 2- and 3-cephem systems (**60**) and (**61**) show IR maxima in the regions 1772–1784 and 1782–1792 cm^{-1}, respectively ⟨B-72MI50900 p. 318⟩.

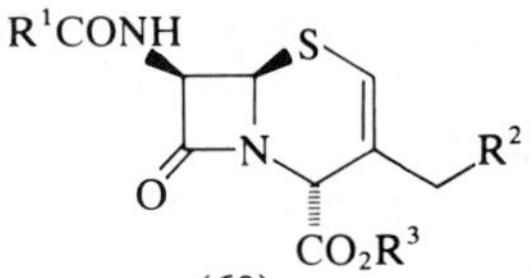

(60) 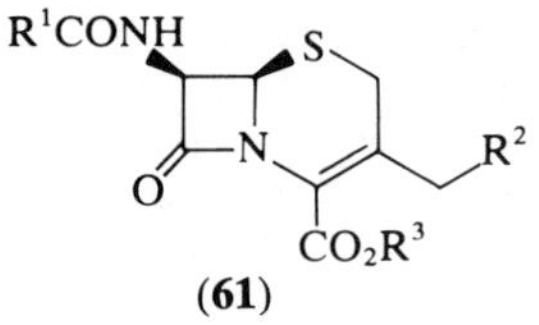 **(61)**

The higher frequencies of the β-lactam carbonyl absorption in fused systems has been attributed to increased inhibition of amide resonance as the β-lactam ring becomes less planar ⟨B-72MI50900 p. 303⟩. For the 3-cephems (**61**) there is also the possibility of enamine resonance which could further reduce the ability of the β-lactam nitrogen to contribute to amide resonance.

IR spectra of systems related to β-lactams show the expected trends in the frequency of the carbonyl absorption, where present. For example, the presence of an exocyclic double bond at C-4 in an azetidin-2-one raises the value of $\bar{\nu}_{C=O}$ considerably. Thus the 4-thioxoazetidin-2-one (**62**; Z=S) and the derived 4-alkylidene systems (**62**; $Z=CR^1R^2$) exhibit β-lactam carbonyl absorptions at 1835 and 1800–1810 cm^{-1} respectively ⟨80JOC1477, 80JOC1481⟩, while the 4-iminoazetidin-2-ones (**63**) have $\bar{\nu}_{C=O}$ at 1800–1825 cm^{-1} ⟨81CC41⟩. Additional spectral data for these and similar systems may be found in the references in Table 5.

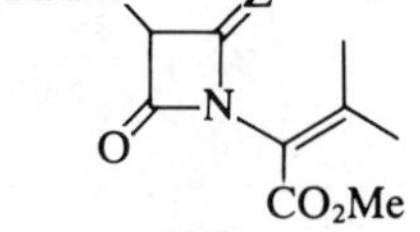

(62) 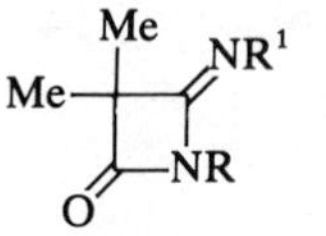 **(63)**

The circular dichroism of monocyclic β-lactams has been studied ⟨78JOC4434⟩.

5.09.3.2 Reactivity

5.09.3.2.1 General survey

Azetidin-2-ones are the most extensively studied derivatives of azetidine, largely as a result of the discovery of the antibacterial properties of penicillins, cephalosporins and cephamycins. The chemistry of these bicyclic β-lactams has been widely investigated and there are numerous reviews on the subject ⟨71MI50903, B-72MI50900, 73S327, 75T2321, 75S547, 76CRV113, 76CSR181, 78T1731⟩. Recently, there has been considerable interest in other fused β-lactams, such as clavulanic acid, thienamycin and the related olivanic acid derivatives, and the penems ⟨B-80MI50900⟩ (*cf.* Chapters 5.10, 5.11, 5.12). Non-fused β-lactam containing natural products include the nocardicins (**64**) ⟨80CC770⟩ and the monobactams (**65**) ⟨82JOC176, 82JOC179⟩, as well as the more complex pachystermines A and B ⟨65TL3473⟩, wild-fire toxin ⟨71MI50901⟩ and the bleomycins ⟨71MI50902⟩.

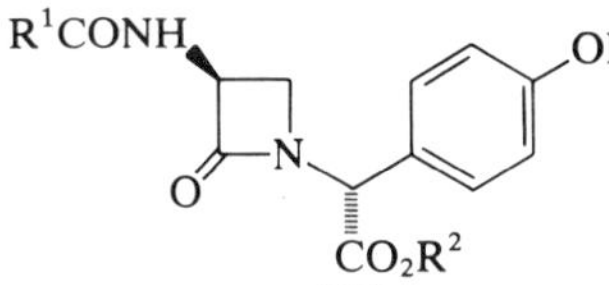

(**64**) 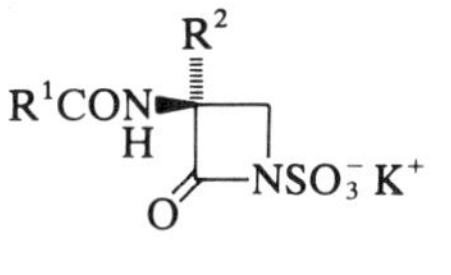 (**65**)

Incorporation of an amide linkage into a four-membered ring results in angle strain and some degree of inhibition of amide resonance, rendering β-lactams more susceptible than normal amides to nucleophilic attack at the carbonyl group. Not surprisingly, β-lactams undergo N(1)—C(2) cleavage on treatment with a variety of nucleophiles ⟨75S547 p. 579⟩ and this ability of a β-lactam to act as an acylating agent is generally considered to be, at least in part, responsible for the antibacterial properties of penicillins and cephalosporins. These strained bicyclic β-lactams inhibit bacterial cell wall biosynthesis, apparently by acylating transpeptidases ⟨72MI50900 p. 472⟩.

Polymerization of β-lactams, involving cleavage of the amide bond, can be induced by treatment with strongly basic catalysts or by acylating agents ⟨75S547 p. 581⟩.

Introduction of a heteroatom substituent at C-4 tends to destabilize the β-lactam, promoting N(1)—C(4) bond cleavage ⟨75S547, p. 583, 78T1731 p. 1760⟩. This lability is illustrated by the facile epimerization of the 4-alkoxyazetidin-2-ones (**66**) and (**67**) ⟨71CB1987⟩ and the ready acid-catalyzed ring opening of (**68**) ⟨63AG(E)152⟩.

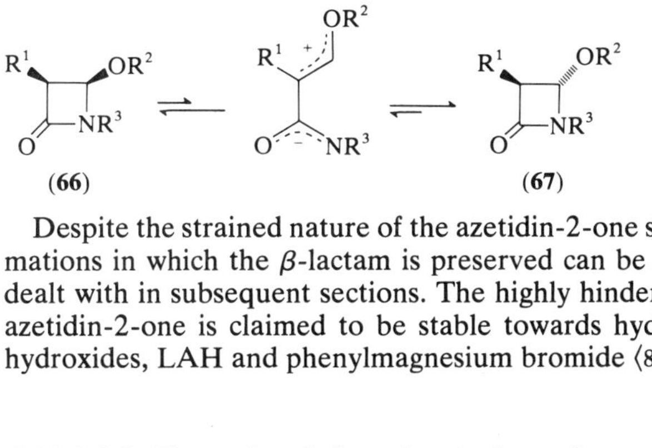

(**66**) (**67**) 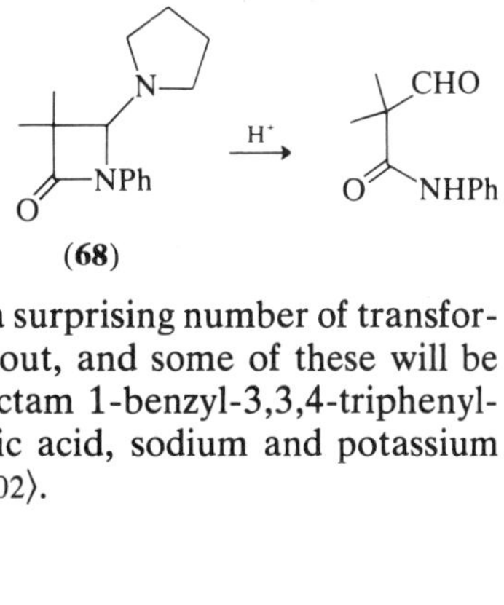 (**68**)

Despite the strained nature of the azetidin-2-one system, a surprising number of transformations in which the β-lactam is preserved can be carried out, and some of these will be dealt with in subsequent sections. The highly hindered β-lactam 1-benzyl-3,3,4-triphenylazetidin-2-one is claimed to be stable towards hydrochloric acid, sodium and potassium hydroxides, LAH and phenylmagnesium bromide ⟨80IJC(B)702⟩.

5.09.3.2.2 Thermal and photochemical reactions

Monocyclic β-lactams undergo thermolysis or photolysis to give alkenes and isocyanates or ketenes and imines depending on the substitution pattern ⟨75S547 p. 586⟩. Apparently, thermolysis favours the former pathway while photolysis favours the latter ⟨68CB2669⟩.

Gas-phase thermolysis of (**69**) and (**70**) at 600 °C/12 Torr occurs to give the corresponding *cis* and *trans* alkenes respectively, suggesting a concerted $[_{\sigma}2_s + _{\sigma}2_a]$ cycloreversion process ⟨70JA1763⟩.

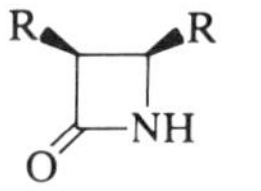

(**69**) R = Me, Et 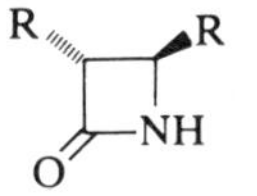 (**70**) 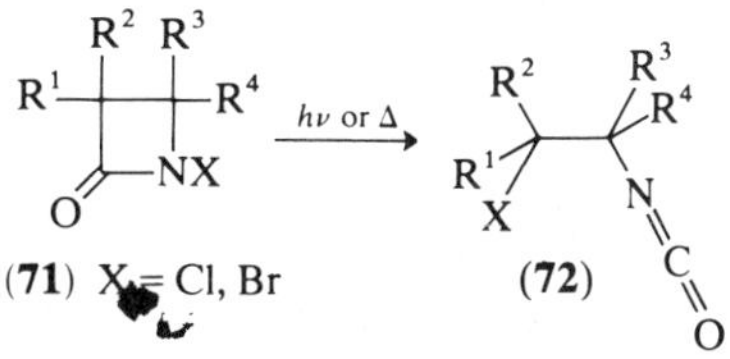 (**71**) X = Cl, Br (**72**)

Irradiation of the *N*-bromo- or *N*-chloro-azetidin-2-ones (**71**) in the presence of alkenes, alkynes or radical donors induces rearrangement to the β-haloalkyl isocyanates (**72**) *via* a radical process in which racemization at C-3 occurs. This reaction gives higher yields of isocyanates with *N*-bromoazetidin-2-ones and is also promoted thermally ⟨69TL117, 71LA(752)142⟩.

5.09.3.2.3 Electrophilic attack

O-Alkylation of *N*-unsubstituted β-lactams to give the corresponding 2-alkoxy-1-azetines can be achieved by reaction of the azetidin-2-ones with 'hard' electrophiles (trialkyloxonium tetrafluoroborates) followed by treatment with base (*cf.* Section 5.09.4.3.1) ⟨67JHC619, 69LA(725)124⟩. In contrast, reaction of the *N*-unsubstituted azetidin-2-ones (**73**) or their derived anions with a variety of softer electrophiles results in *N*-substitution, and some representative reactions are illustrated in Scheme 7.

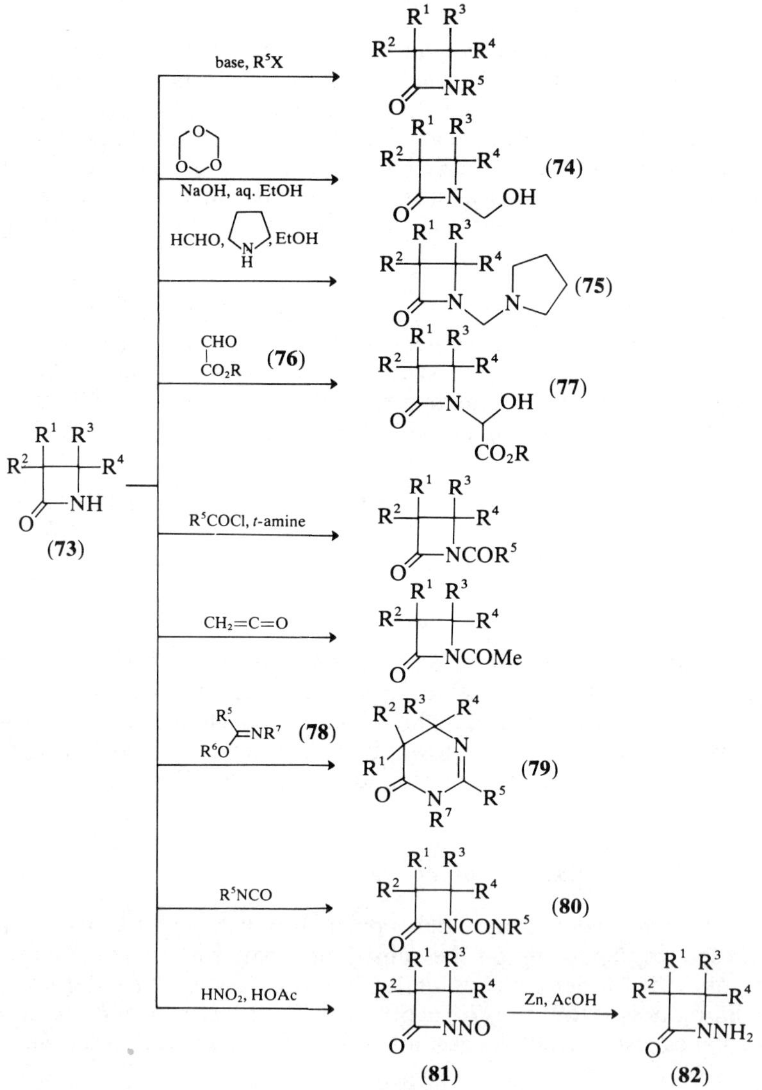

Scheme 7

N-Alkylation of (**73**) occurs in the presence of a strong base such as sodamide ⟨66LA(696)108⟩, sodium hydride ⟨63BRP924589⟩ or potassium *t*-butoxide ⟨81JCS(P1)2539⟩; yields are usually high. With reactive alkylating agents such as bromoacetic esters, anhydrous potassium carbonate in DMF has been used successfully ⟨80MI50900 p. 88⟩. Recently, phase-transfer catalyzed alkylation of β-lactams using aqueous potassium hydroxide and an alkyl halide in the presence of tetra-*n*-butylammonium bromide has been effected ⟨78TL615⟩. Phase-transfer catalyzed alkylation using 18-crown-6 has also been reported ⟨81H(15)1101⟩.

N-Alkylation also occurs with aldehyde derivatives such as *s*-trioxane ⟨66LA(696)108⟩ and *N*-hydroxymethylpyrrolidine ⟨63LA(661)181⟩ to give the corresponding amidol (**74**) and aminomethyl β-lactams (**75**) respectively. Glyoxylate esters (**76**) undergo a similar reaction and the resulting amidols (**77**) ⟨81JCS(P1)3242⟩ have been used as precursors of novel fused β-lactams ⟨B-80MI50900 p. 291⟩.

N-Acylation of (**73**) can be achieved with acyl chlorides and pyridine or triethylamine, or with ketene ⟨75S547 p. 560⟩. *N*-Trimethylsilylazetidin-2-ones have been reported to undergo acyldesilylation on treatment with acetyl chloride. Acylation of (**73**) with imidates (**78**) at 150–160 °C was followed by ring expansion to give the pyrimidin-4-ones (**79**). Pyrimid-4-ones were obtained from a similar reaction of imidates with 4-alkoxyazetidin-2-ones ⟨74CB270⟩. Acylation with subsequent ring expansion also occurs with dithioimidates ⟨73CPB1305⟩.

Reaction of (**73**) with isocyanates gives the corresponding ureas (**80**) in low yields ⟨63LA(661)181; see also 78JCS(P1)817⟩.

N-Unsubstituted β-lactams are nitrosated in acetic acid under nitrogen at 15 °C to give *N*-nitroso-β-lactams (**81**), which can be reduced by zinc dust in acetic acid to afford the corresponding *N*-amino compounds (**82**) ⟨67G1719⟩.

N-Substituted β-lactams are deprotonated at C-3 using a strong base, such as LDA, at low temperatures. The resulting anions have been quenched with various electrophiles including ketones, esters or reactive alkyl halides ⟨72CJC3196⟩, alkyl nitrates ⟨71GEP1960157⟩, acetylimidazole ⟨81JOC2208⟩ and trimethylsilyl chloride ⟨78S746⟩. In the last example, the 3-trimethylsilyl-β-lactam (**83**) undergoes the Petersen alkenation reaction with aldehydes and ketones to yield the 3-alkylidene-β-lactams (**84**). Deprotonation of the corresponding position in penicillins and cephalosporins with subsequent epimerization and/or quenching with electrophiles has been reviewed ⟨75S547 p. 565, 78T1731 p. 1754⟩.

Me$_3$Si … NPh (**83**) —i, LDA; ii, R^{1}R^{2}CO→ R^1, R^2 … NPh (**84**)

CO$_2$H … NPh (**85**) —LTA, pyridine→ OAc … NPh

PhOCH$_2$CONH, SO$_2$H, N, CO$_2$R (**86**) —X^{+}→ PhOCH$_2$CONH, X, N, CO$_2$R (**87**) —*trans* isomer→ OPh, N, O, N, CO$_2$R (**88**)

Other interactions of β-lactams with electrophiles include the oxidative decarboxylation of the azetidin-2-one-4-carboxylic acid (**85**) on treatment with LTA and pyridine ⟨81M867⟩, and the reaction of the azetidin-2-one-4-sulfinic acid (**86**) with 'positive halogen' reagents. This affords a mixture of *cis*- and *trans*-4-halogeno-β-lactams (**87**), the latter undergoing cyclization to give the bicyclic β-lactam (**88**) ⟨81JOC3568⟩.

Reaction of β-lactams with phosphorus pentasulfide affords the corresponding azetidine-2-thiones (*cf.* Section 5.09.4.3.1) ⟨68JHC433⟩.

5.09.3.2.4 Nucleophilic attack

As might be expected, the strained β-lactam ring is more readily cleaved by hydroxide ion than are comparable acyclic amides and γ- and δ-lactams ⟨72JCS(P2)1366⟩. Hydrolysis also occurs under acidic conditions to give β-aminocarboxylic acids, and treatment of β-lactams with anhydrous hydrogen chloride results in ring opening by chloride ion to give β-amino acid chloride hydrochlorides ⟨76TL4053⟩.

Azetidin-2-ones are also cleaved by amines and other nucleophiles ⟨75S547 p. 579⟩. In some cases other heterocycles are produced (*cf.* Section 5.09.3.2.6).

N-Substituted β-lactams undergo reductive N(1)—C(2) fission on treatment with LAH to give the corresponding *N*-substituted 3-aminopropan-1-ols ⟨75S547 p. 581⟩. However, reaction of the azetidin-2-one (**89**) with ethanolic sodium borohydride was reported to result in ethanolysis to give (**90**) rather than in reduction ⟨65ZOR1099⟩. In contrast, reduction of *N*-unsubstituted β-lactams with LAH or diborane readily occurs to give azetidines, and conditions can be chosen so as to minimize competing ring cleavage ⟨75S547 p. 588⟩ (*cf.* Section 5.09.2.4).

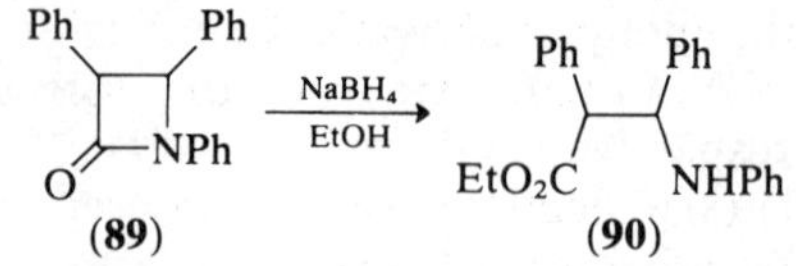

Another mode of reductive ring cleavage is observed for 4-phenylazetidin-2-ones. These undergo N(1)—C(4) bond fission on hydrogenolysis in the presence of Raney nickel to yield the corresponding 3-phenylpropanamides ⟨75S547 p. 583⟩.

In contrast to the nucleophilic ring cleavages discussed above, there are several reactions in which a substituent attached to the β-lactam is displaced without disruption of the ring. For example, the readily available 4-acetoxyazetidin-2-one (**91**) (prepared from vinyl acetate and chlorosulfonyl isocyanate, *cf.* Section 5.09.3.3.2) readily undergoes displacement of the acetoxy group with alkyl- and aryl-thiolates, alkoxides, phenoxides, alkyl- and aryl-sulfinates, azide and phthalimide ⟨74LA539⟩, stabilized tertiary carbanions ⟨81TL1161⟩, α,β-unsaturated carboxylates ⟨81H(16)243⟩ and thiocarboxylates ⟨81HCA1303⟩, α,β-unsaturated thiolates ⟨79JA6306⟩, thiocarboxylates ⟨79JA6296⟩, trithiocarbonates ⟨80CPB3232⟩, β-ketodithioesters (alkylation on sulfur) ⟨79H(12)1315⟩ and aluminum enolates ⟨80CPB3494⟩ (Scheme 8).

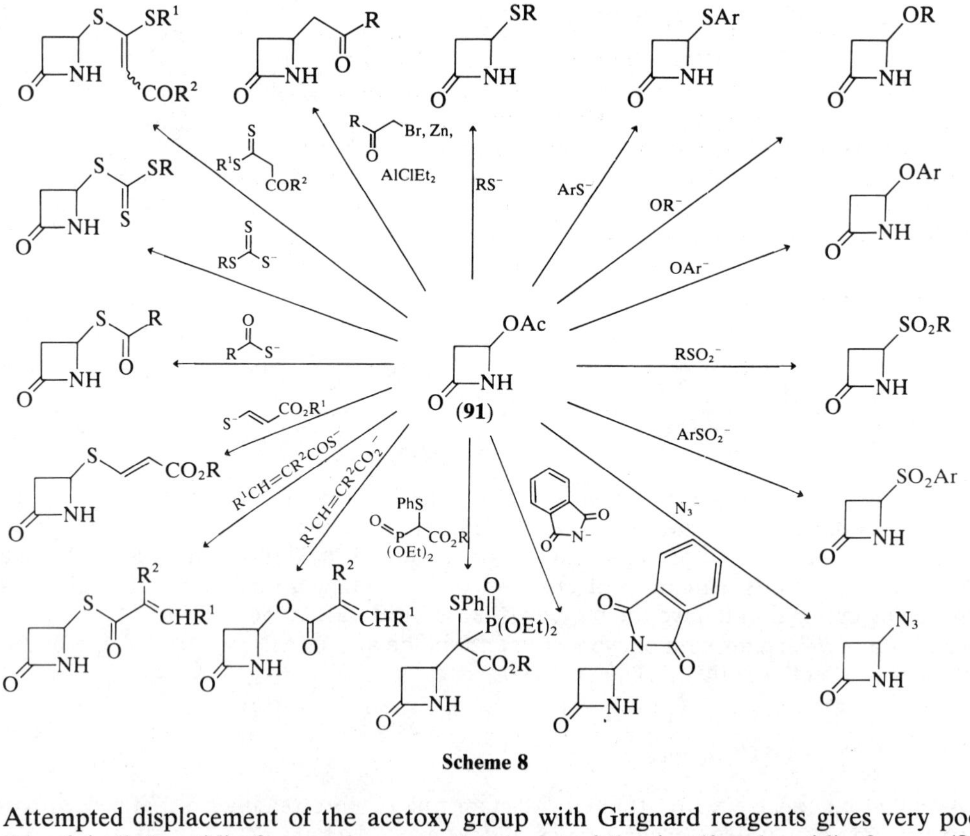

Scheme 8

Attempted displacement of the acetoxy group with Grignard reagents gives very poor yields of 4-alkylazetidin-2-ones; however, treatment of the 4-sulfonylazetidin-2-ones (**92**)

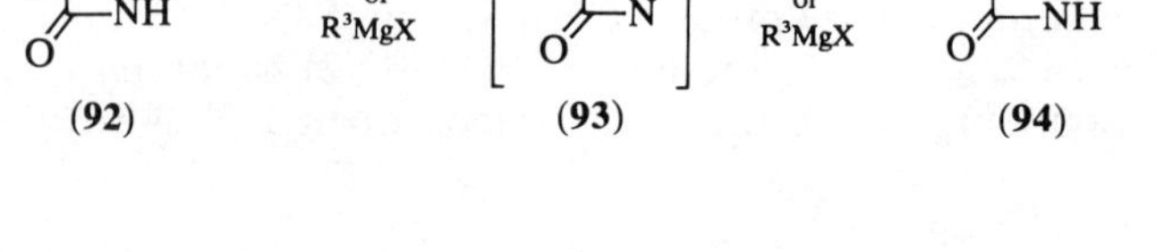

with Grignard reagents or with lithium dialkylcuprates affords high yields of the 4-alkyl-β-lactams (**94**) ⟨80CC736⟩. The reaction apparently occurs *via* an elimination–addition mechanism involving azetinone (**93**).

In a formally similar reaction, the 4-methanesulfonyl-β-lactam (**95**) undergoes displacement of methanesulfinate on treatment with propargylic or allylic alcohols in the presence of a Lewis acid catalyst to give a mixture of the *cis*- and *trans*-4-alkoxy-β-lactams (**96**) and (**97**) ⟨79JCS(P1)2268⟩.

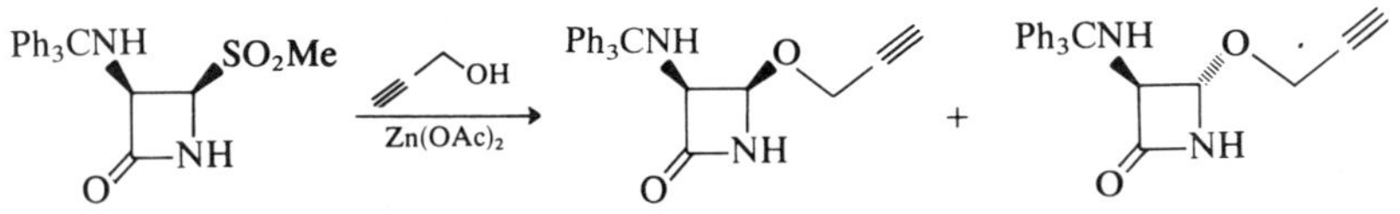

(**95**) (**96**) (**97**)

1-Azetin-4-ones similar to (**93**) (*cf.* Section 5.09.4.3.2) have also been suggested as intermediates in displacement reactions of azetidin-2-one-4-sulfoxides. Thus, treatment of the β-lactam sulfoxide (**98**) with pentan-3-ol at 100 °C gives the *trans*-β-lactam (**100**) by regio- and stereo-selective addition to the azetin-4-one (**99**) ⟨81CC959⟩.

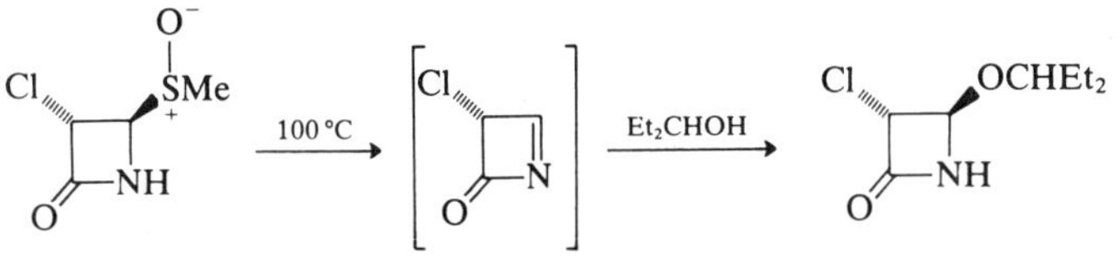

(**98**) (**99**) (**100**)

β-Lactams bearing a 4-chloro substituent (available from 4-alkylthio-β-lactams by chlorinolysis ⟨75S547 p. 573⟩ readily undergo displacement of halide with a variety of nucleophiles ⟨78T1731 p. 1757⟩. The 4-chloro-β-lactam (**101**) reacts with allylcopper reagents by chloride displacement to give (**102**). However, treatment of (**101**) with lithium diallylcuprates gives mainly the products of Michael addition to the α,β-unsaturated ester rather than displacement of halide ⟨79TL3867⟩. There appear to be no examples of nucleophilic displacement of a 3-substituent in a monocyclic β-lactam, although a number of analogous reactions in the penicillin field are known ⟨75S547 p. 563⟩. For example, the penicillin trifluoromethanesulfonate (**103**) undergoes S_N2 displacement of triflate with soft nucleophiles such as iodide ion ⟨B-80MI50900 p. 320⟩.

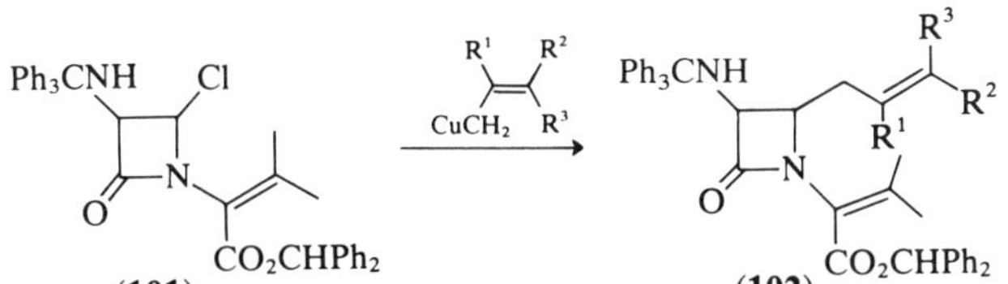

(**101**) (**102**)

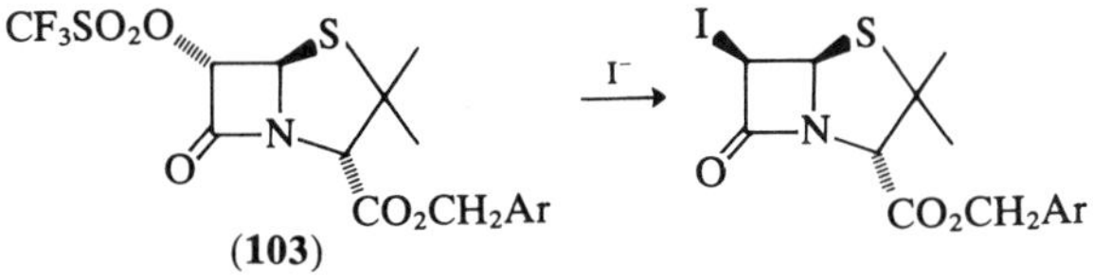

(**103**)

In an unusual application of the Wittig reaction, treatment of clavulanic acid derivatives and esters of penicillin V with methoxycarbonylmethylenetriphenylphosphorane afforded the corresponding *exo*-alkylideneazetidines. Thus penicillin V benzyl ester (**104**) gave (**105**) as a mixture of *E* and *Z* isomers. The β-lactam could be regenerated by low-temperature ozonolysis ⟨81CC929⟩.

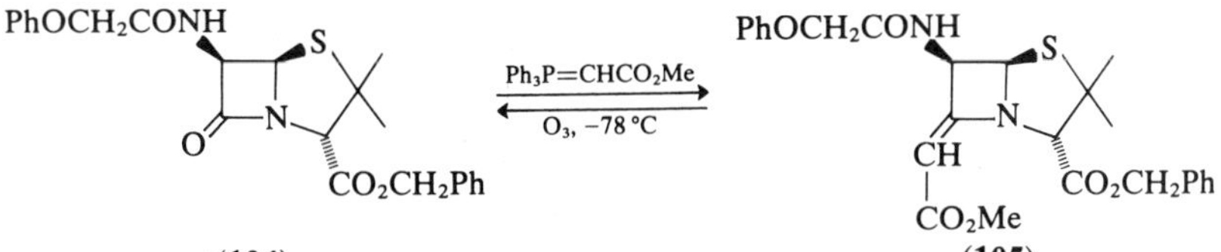

(**104**) (**105**)

5.09.3.2.5 Reactions with carbenoids and radicals

There are several examples of intramolecular reactions of monocyclic β-lactams with carbenes or carbenoids; most of these involve formation of olivanic acid or clavulanic acid derivatives. Thus treatment of the diazo compound (**106**) with rhodium(II) acetate in benzene under reflux gives (**107**), an intermediate in the synthesis of thienamycin ⟨80H(14)1305, 80TL2783⟩.

$ArCH_2O_2CO$ … $Rh_2(OAc)_4$, PhH, Δ … OCO_2CH_2Ar

(**106**) (**107**) $Ar = p\text{-}NO_2C_6H_4$

Cyclization of the diazo compound (**108**) with a copper catalyst affords the clavulanic acid derivatives (**110**) and (**111**), possibly *via* rearrangement of the sulfur ylide (**109**) ⟨80H(14)1999⟩. Similar reactions have been reported in the recent literature ⟨80H(14)1967, 81H(16)1305, 80TL31⟩.

(**108**) (**109**) (**110**) (**111**)

An intermolecular carbenoid reaction followed by intramolecular displacement of acetate gives the clavulanic acid derivative (**112**) in one step from 4-acetoxyazetidin-2-one (**91**) ⟨80CC1257⟩. Carbene-induced reactions of penicillins and cephalosporins have been reviewed ⟨75S547, 78T1731⟩.

(**91**) $Rh_2(OAc)_4$ (**112**)

Intramolecular reactions of β-lactams containing radical centres have been investigated ⟨81TL2689, 81TL2693⟩. Generation of the radical (**116**) from the three precursors (**113**), (**114**) and (**115**) with tributylstannane and azobisisobutyronitrile gives the products (**117**) and (**118**). The ratio of cyclization to hydrogenolysis product is approximately the same for the three precursors, cyclization being increasingly favoured at high dilutions.

Bu^n_3SnH, AIBN, PhH, Δ

(**113**) X = Cl
(**114**) X = PhS
(**115**) X = PhSe

(**116**) (**117**) (**118**)

5.09.3.2.6 Ring transformations

The *N*-chlorosulfonyl-4-alkenyl-β-lactams (**119**), formed by cycloaddition of CSI with the corresponding dienes, undergo a thermal rearrangement to give the formal [4+2] cycloadducts (**120**) and (**121**) as well as the acyclic systems (**122**) ⟨71JOC2841⟩. Thermally induced ring expansion of the 3-azido-β-lactam (**123**) affords the imidazolin-2-one (**124**) ⟨71CC1496⟩. A similar ring expansion to give (**126**) occurs on heating the benzocinnoline-*N*-imide (**125**) in the presence of acid ⟨75CC788⟩.

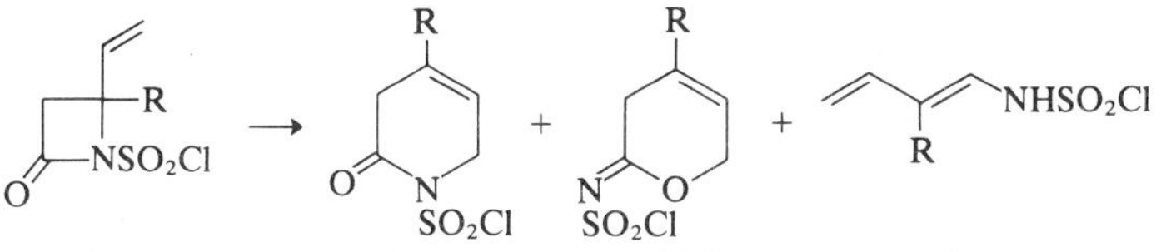

(**119**) (**120**) (**121**) (**122**)

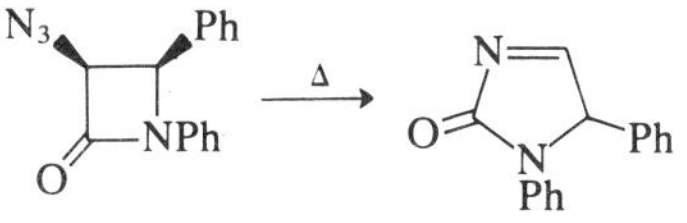

(**123**) (**124**)

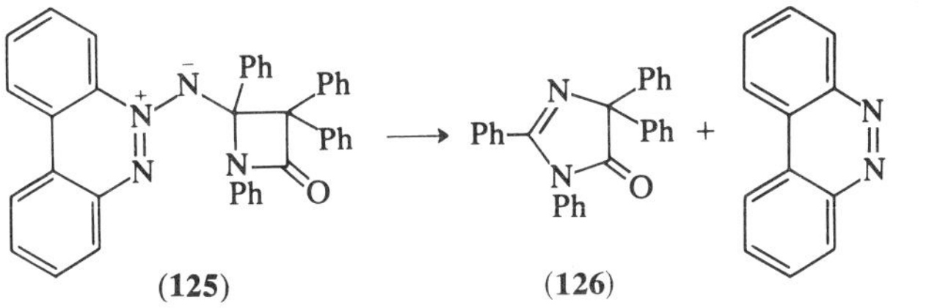

(**125**) (**126**)

Irradiation of 1-phenylazetidin-2-one (**127**) results in a photochemical Fries rearrangement to give (**128**). Similar results are obtained using acid catalysis ⟨78T1731 p. 1758⟩. The more heavily substituted β-lactam (**129**) undergoes boron trifluoride catalyzed ring expansion to yield the indenone (**130**) ⟨73JCS(P1)2664⟩. Treatment of the epoxides (**131**) with methanesulfonic acid gives the butenolides (**132**) ⟨79JOC3946⟩.

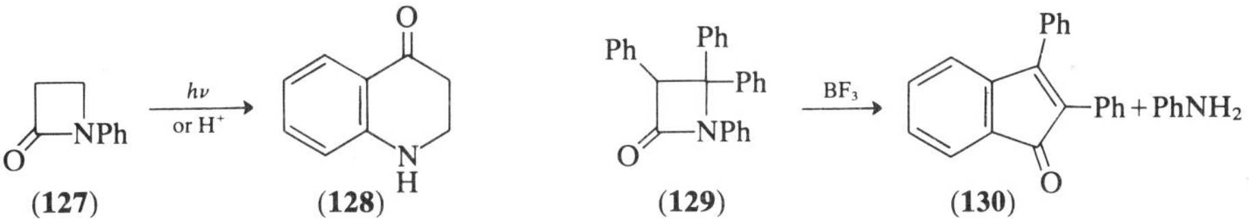

(**127**) (**128**) (**129**) (**130**)

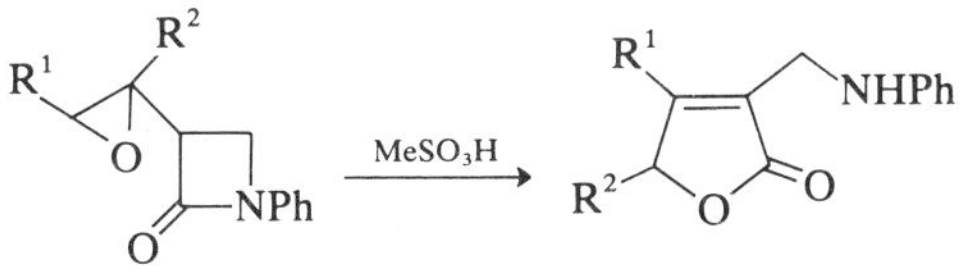

(**131**) (**132**)

Ring expansions of suitable β-lactams can also be achieved on treatment with base; rearrangement of the *N*-substituted azetidin-2-ones (**133**) occurs in the presence of LDA to give (**134**) ⟨72JA9261⟩. Aminolysis of the β-lactam epoxide (**135**) and the aldehyde (**137**) affords (**136**) and (**138**) respectively ⟨81JHC1239⟩.

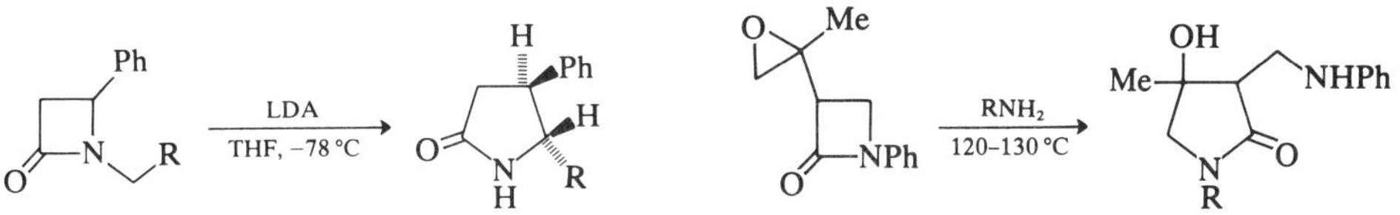

(**133**) (**134**) (**135**) (**136**)

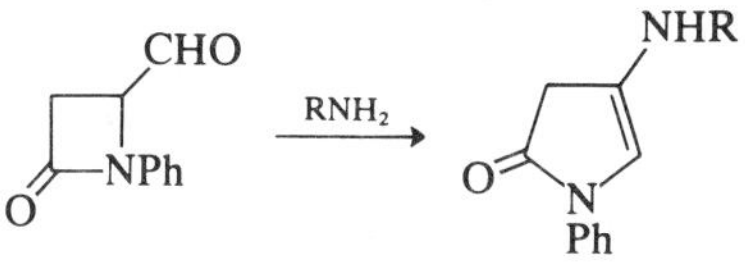

(**137**) (**138**)

5.09.3.2.7 Related azetidine derivatives

A number of azetidine derivatives related to azetidin-2-ones are known (Table 5; see Section 5.09.3.3.5). For many of these, minimal studies of reactivity have appeared. The information available, however, suggests that their reactivity is fairly predictable. This is illustrated for a few specific examples; Table 5 gives leading references for other systems.

As expected, the carbonyl group of azetidin-3-ones undergoes the characteristic reactions of a ketone ⟨73S153⟩. 2-Iminoazetidine (**139**) is stable towards heat, light, acids, bases and reduction. Thus it appears to be a fairly typical trisubstituted amidine ⟨75CC436⟩. The carbon–carbon double bond in the 3-methylene-β-lactam (**140**; R = CH_2Ph) can be selectively reduced catalytically or with borohydride ⟨79CC698, 78CC903⟩. Conjugate additions to the α,β-unsaturated carbonyl system in (**140**; R = CH_2Ph, Ph) also occur ⟨78CC903⟩. Azetidine-2,4-diones are readily cleaved by nucleophiles and their reduction parallels that of azetidin-2-ones, LAH causing ring cleavage of *N*-substituted derivatives but giving azetidines in the case of the *N*-unsubstituted systems ⟨64HC(19-2)900⟩. 4-Thioxoazetidin-2-ones undergo nucleophilic cleavage by attack at the carbonyl group and base catalyzed alkylation with methyl iodide occurs preferentially on sulfur ⟨80TL4247⟩. The thiocarbonyl group can be converted to a carbonyl group by ozone and to an alkylidene group with diazo compounds ⟨80JOC1481⟩.

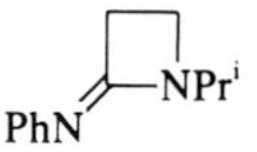

(**139**) 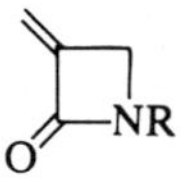 (**140**)

5.09.3.3 Synthesis

5.09.3.3.1 From non-heterocyclic precursors by closure of one bond

(i) Formation of N—C(2) bond

The direct thermal cyclization of β-amino acids (**141**) to β-lactams fails because of β-elimination resulting in deamination, although there is a report of the preparation of the parent compound from heating of 3-aminopropanoic acid in DMSO at 150 °C ⟨74ZOR1556⟩. Cyclization can, however, be effected using a variety of activating agents such as acetic anhydride, acetyl chloride, phosphorus trichloride, thionyl chloride, sulfonyl chlorides, triphenylphosphine/pyridyl disulfide, dicyclohexylcarbodiimide and triphenylphosphine/carbon tetrachloride ⟨80M1117⟩. *N*-Acyl-β-amino acids cyclize directly on heating, possibly *via* a hydroxylactone. β-Amino acid chlorides have been cyclized on treatment with organic bases, for example triethylamine. Treatment of β-amino acid esters with Grignard reagents has been fairly widely applied to β-lactam synthesis (see for example ⟨64CI(L)2097, 73BSF1668, 78T1731, 76CSR181⟩). Modifications include the use of trimethylsilyl esters of *N*-trimethylsilyl-β-amino acids ⟨75LA2195⟩ and the use of triisobutylaluminum as cyclizing agent.

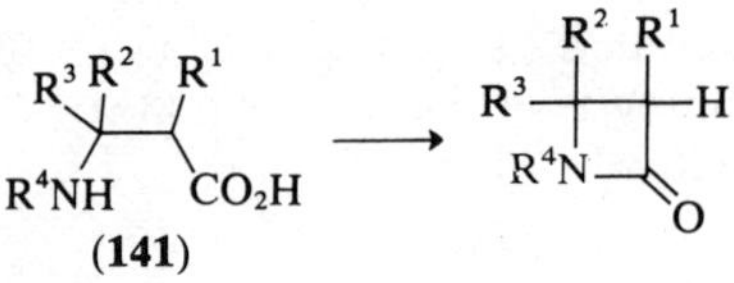

(**141**)

(ii) Formation of N—C(4) bond

A number of examples of cyclization of *N*-substituted 3-halopropanamides (**142**) in the presence of a strong base ($NaNH_2$ or KNH_2) to give β-lactams have been reported ⟨78T1731 p. 1736⟩. Yields are usually high and unaffected by substituents on nitrogen. Particularly for unfused β-lactams, elimination to give α,β-unsaturated acid amides is a side reaction, but this can be minimized ⟨79TL549⟩. Phase transfer catalysis has been applied to this reaction ⟨81CPB1063⟩. An application of this type of cyclization to produce 4-benzoylazetidinones is illustrated in Scheme 9 ⟨80TL997⟩.

A further example employing displacement of a suitably activated hydroxyl group rather than halogen is also shown in Scheme 10. The use of titanium trichloride in the final step

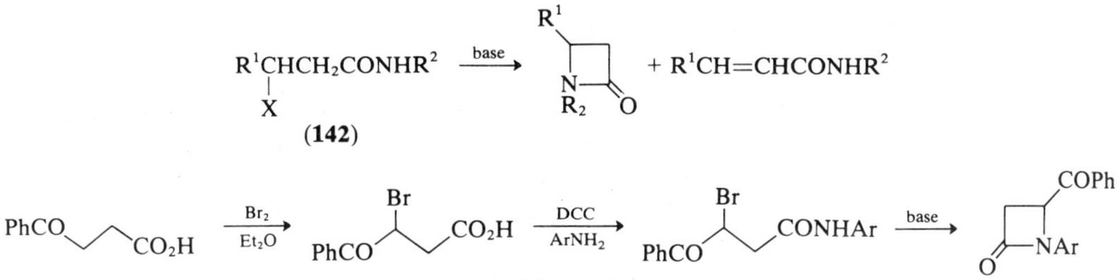

Scheme 9

of this sequence allows reduction of the N—O bond without affecting the *t*-butoxycarbonyl group and the base sensitive chiral centre at C(3) ⟨80JOC410⟩.

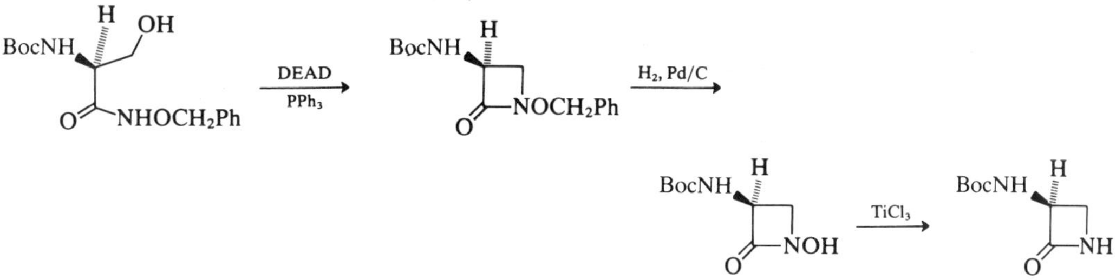

Scheme 10

Azetidin-2-one formation by N—C(4) ring closure has also been observed in the irradiation of *cis*-α-phenylcinnamides (**143**) in degassed benzene. Both *cis* and *trans* lactams are formed ⟨68JA2333⟩.

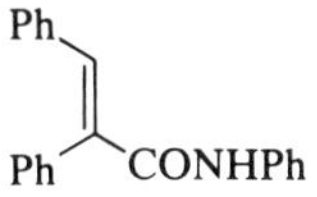

(**143**)

(iii) Formation of C(2)—C(3) bond

There appear to be few examples of the formation of azetidin-2-ones by closure of the C(2)—C(3) bond. One reaction which fits into this category involves reaction of the iron carbonyl lactone complexes (**144**) with an amine to give the allyl complexes (**145**) which on oxidation are converted in high yield to 3-vinyl-β-lactams (**146**) ⟨80CC297⟩.

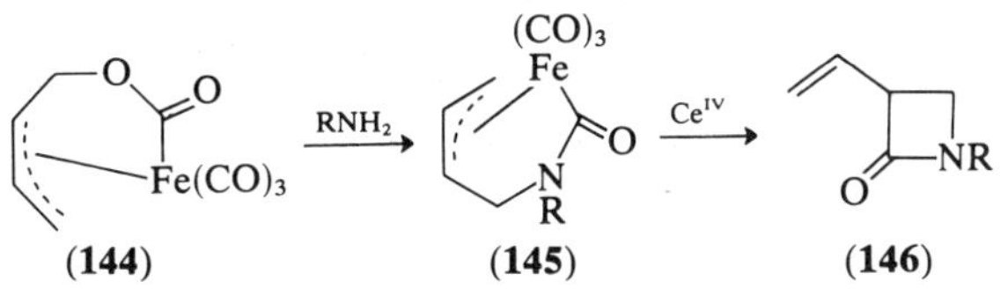

(**144**) (**145**) (**146**)

(iv) Formation of C(3)—C(4) bond

α-Haloacetanilidomalonates (**147**) can be cyclized to β-lactams in the presence of base ⟨73S327 p. 338, 78T1731 p. 1736⟩. Most examples involve activation of the methine H by two electronegative groups (R^1 and R^2), and bases used include triethylamine, diethylamine, benzylamine, alcoholic NH_3, alcoholic KOH, ion exchange resin and DMF. In the case of $R^1 = COAr$, cyclization to give a dihydro-1,4-oxazine (**148**) also occurs depending on the aryl group.

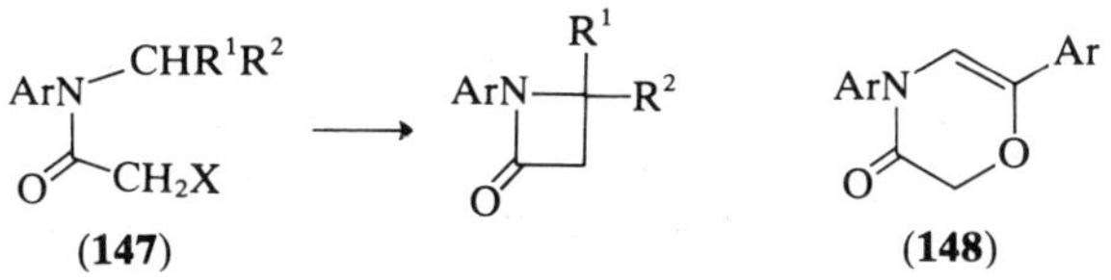

(**147**) (**148**)

Intramolecular Michael addition of acrylamides (**149**) in the presence of piperidine gives β-lactams ⟨65T449⟩. The method requires that R^3 must be a strongly electron withdrawing group (CO_2Et, CO_2Me, *o*- or *p*-$O_2NC_6H_4$) and R^1 and R^2 must be similarly activating ($R^1 = R^2 = CO_2R$; $R^1 = Ph$, $R^2 = CN$).

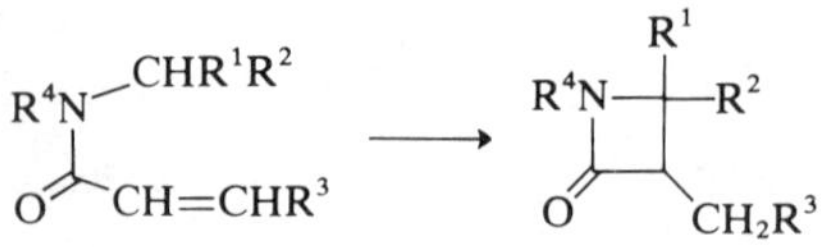

(149)

A number of related photochemical processes resulting in ring closure of the 3,4-bond of β-lactams have been reported. Irradiation of *N,N*-dialkyl-α-oxoamides (**150**), especially in the solid state, gives β-lactams by a Norrish type II process ⟨79JA5343, 79SC179, 78JOC419⟩. There are several examples of bicyclic β-lactams produced in low yield by this route ⟨78T1731 p. 1739⟩. A related reaction occurs with α-thioxamides (**151**) ⟨79CC899⟩.

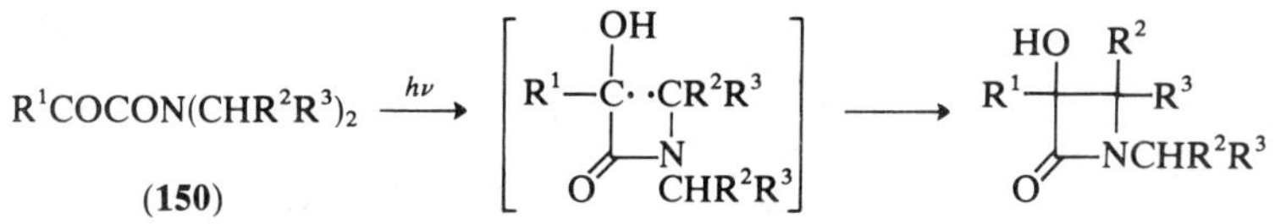

(150)

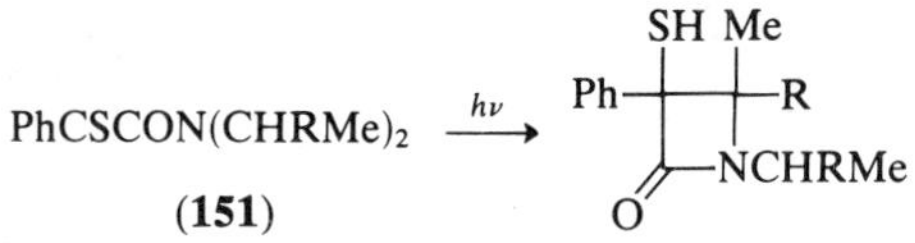

(151)

Photocyclization of α,β-unsaturated amides (**152**) also gives β-lactams. This proceeds by cleavage of the intermediate biradical (**153**) to give ketene and imine which recombine. A small amount of amide (**154**) results from reaction of the ketene with amine that is produced by hydrolysis of the imine. This type of photodealkylation detracts from the generality of the process ⟨79JCS(P1)263, 77T485⟩.

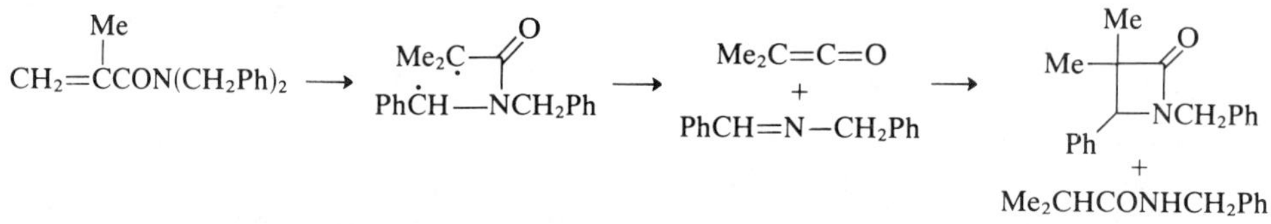

(152) (153) (154)

Bicyclic β-lactams have been produced by photochemical closure of the 3,4-bond on the irradiation of pyridone (**155**) and related derivatives ⟨72CB2780, 72TL3563⟩. In the case of pyrimid-4-ones (**156**) in alcoholic solution, the intermediate bicyclic azetine (**157**) undergoes ring opening ⟨79JOC2083, 80JOC936⟩.

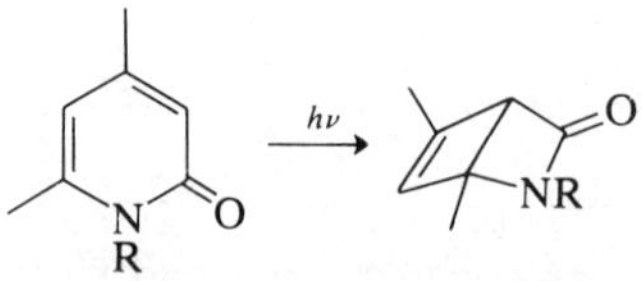

(155)

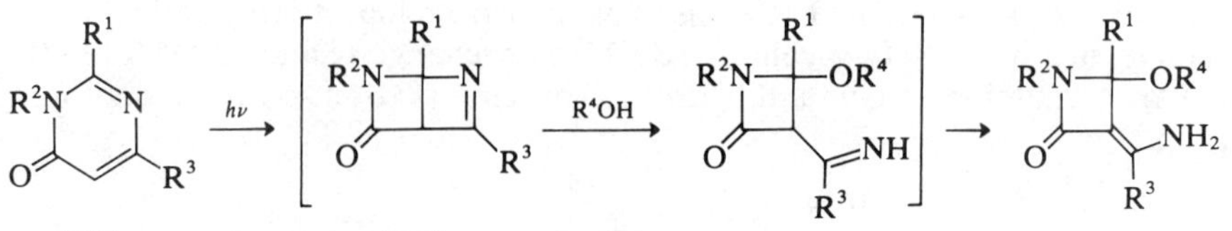

(156) (157)

A final method of β-lactam 3,4-bond formation which has found fairly wide application is based on carbenic insertion ⟨78T1731 p. 1739⟩. The carbenic centre can be generated by photolysis of a diazo compound as in the case of (**158**) ⟨72JA1629, 79CC846⟩ or from organometallic precursors, for example (**159**) ⟨71ACS1927⟩.

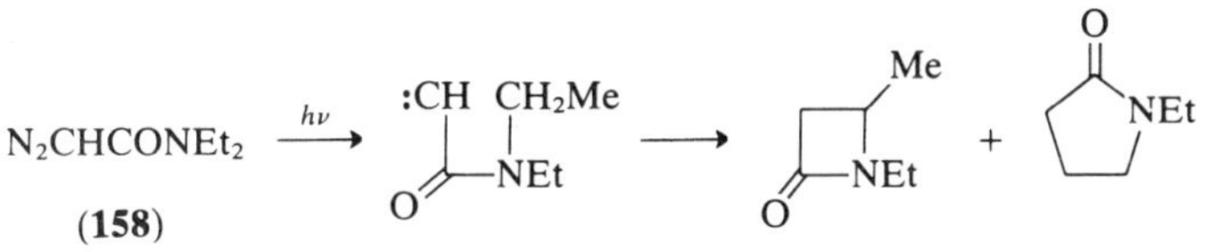

$N_2CHCONEt_2$ (158) —hν→ :CH CH_2Me / NEt / O → Me / NEt / O + O / NEt

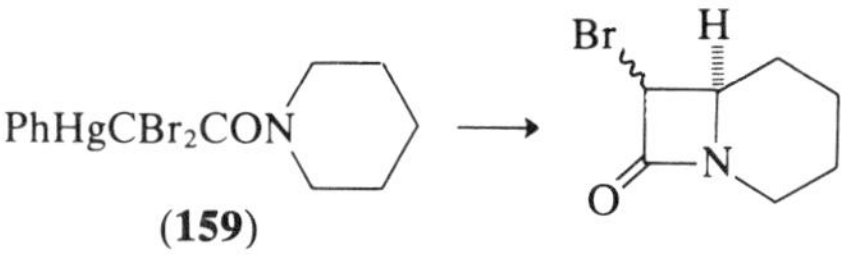

$PhHgCBr_2CON$ (159) → Br H / N / O

5.09.3.3.2 From non-heterocyclic precursors by closure of two bonds

(i) Formation from [3+1] fragments

Formation of the β-lactam (**161**) by reaction of the dianion (**160**) with methylene diiodide provides an example of a [3+1] type of ring closure ⟨79TL2031⟩. The insertion of carbon

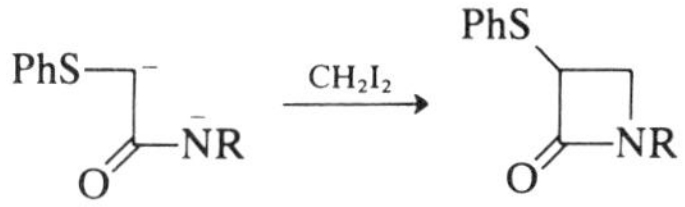

PhS / NR / O (**160**) —CH_2I_2→ PhS / NR / O (**161**)

monoxide in the sequence shown in Scheme 11 could also formally be fitted into this category ⟨77JA2823⟩.

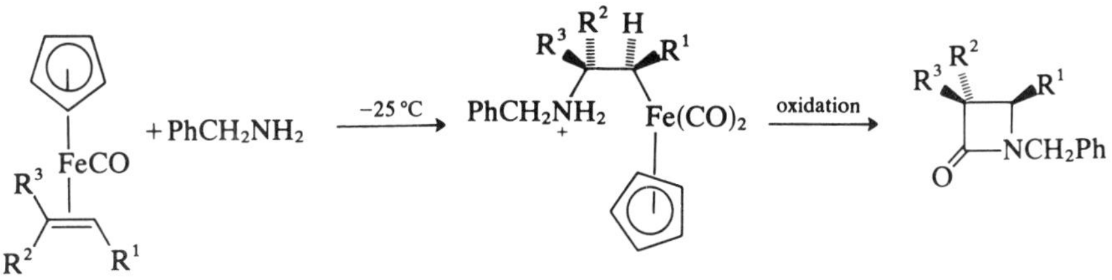

FeCO / R^3 / R^2 / R^1 + $PhCH_2NH_2$ —−25 °C→ R^3 R^2 H R^1 / $PhCH_2\overset{+}{N}H_2$ $Fe(CO)_2$ —oxidation→ R^3 R^2 R^1 / NCH_2Ph / O

Scheme 11

(ii) Formation from [2+2] fragments

(a) Formation of N—C(2) and C(3)—C(4) bonds. Addition of a two-atom component to an imine function provides one of the most important and versatile routes to β-lactams ⟨78T1731, 73S327⟩.

The first β-lactam was produced by addition of a ketene to an imine and there are now many examples of this type of approach. The ketenes are most frequently generated *in situ* from acid chlorides by dehydrohalogenation, but have also been produced from diazo ketones, by heating of alkoxyacetylenes and in the case of certain cyanoketenes by thermolysis of the cyclic precursors (**162**) and (**163**).

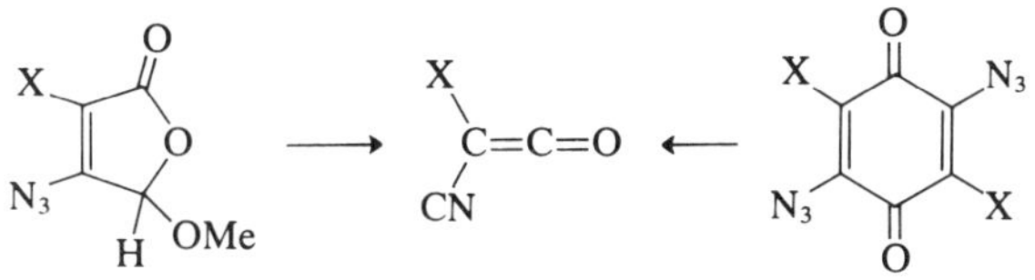

X / O / O / N_3 / H OMe (**162**) → X / C=C=O / CN ← O / X / N_3 / N_3 / X / O (**163**)

Two extreme mechanisms can be envisaged (Scheme 12), concerted [2+2] cycloaddition or the more generally accepted formation of a dipolar intermediate (**164**) which closes to a β-lactam or which can interact with a second molecule of ketene to give 2:1 adducts (**165**) and (**166**) which are sometimes found as side products. In some cases 2:1 adducts result from reaction of the imine with ketene dimer.

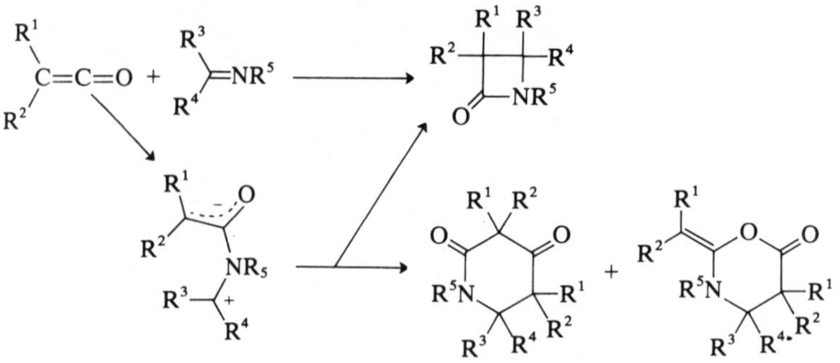

Scheme 12

The interaction of acid chlorides (**167**; X = Cl) with imines in the presence of bases such as triethylamine may involve prior formation of a ketene followed by cycloaddition to the imine, but in many cases it is considered to involve interaction of the imine with the acid chloride to give an immonium ion (**168**). This is then cyclized by deprotonation under the influence of the base. Clearly, the distinction between these routes is a rather fine one and the mechanism involved in a particular case may well depend on the reactants and the timing of mixing. Particularly important acid chlorides are azidoacetyl chloride and phthalimidoacetyl chloride, which provide access to β-lactams with a nitrogen substituent in the 3-position as found in the penicillins and cephalosporins.

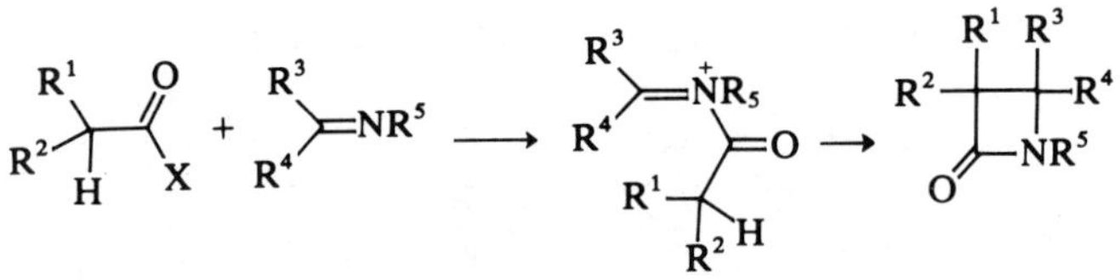

(**167**) (**168**)

There are numerous modifications of this acid chloride route in which the substituted acetic acid itself is used as the ketene equivalent under the influence of a suitable activating agent. Thus, mixed anhydrides formed with chloroformates, trifluoroacetic acid ⟨73TL2319⟩, chlorophosphates ⟨82S63, 78TL4587⟩, chlorophosphamides ⟨82S63⟩, cyanuric chloride ⟨81S209⟩, saccharyl chloride ⟨77S407⟩, triphenylphosphine and carbon tetrabromide ⟨76S689⟩ have all been employed.

A wide variety of β-lactams are available by these routes because of the range of substituents possible in either the ketene or its equivalent substituted acetic acid derivative. Considerable diversity in imine structure is also possible. In addition to simple Schiff bases, imino esters and thioethers, amidines, cyclic imines and conjugated imines such as cinnamylidineaniline have found wide application in the synthesis of functionalized β-lactams. *N*-Acylhydrazones can be used, but phenylhydrazones and *O*-alkyloximes do not give β-lactams. These β-lactam forming reactions give both *cis*- and *trans*-azetidin-2-ones; some control over stereochemistry can, however, be exercised by choice of reactants and conditions.

The reactions of ketenes or ketene equivalents with imines, discussed above, all involve the imine acting as nucleophile. Azetidin-2-ones can also be produced by nucleophilic attack of enolate anions derived from the acetic acid derivative on the electrophilic carbon of the imine followed by cyclization. The reaction of Reformatsky reagents, for example

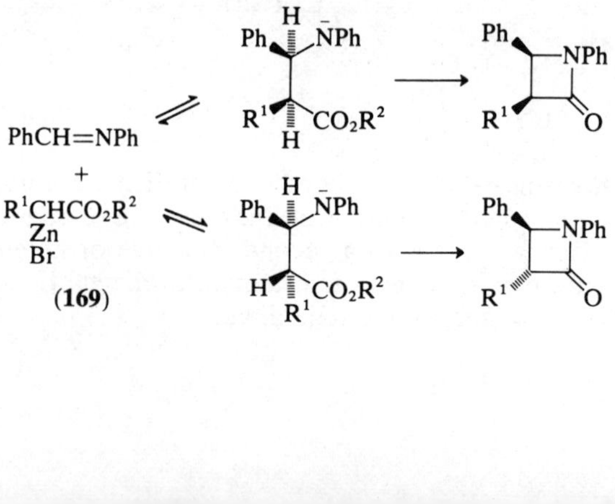

(**169**)

(**169**), with Schiff bases falls into this category. The mechanism is believed to be as shown, the stereochemistry depending on the solvent polarity, increased polarity favouring the *cis* β-lactam ⟨73S327 p. 338, 71BSF2260⟩. Other examples involve the direct addition and cyclization of ester enolate anions ⟨80JOC3413⟩ and the related reaction of ketene silylated acetals (**170**) with imines in the presence of titanium tetrachloride. In the last procedure the cyclization can occur *in situ* ⟨77TL3643, 80TL2081⟩, or the intermediate β-amino ester can be isolated and cyclized with LDA ⟨81S545, 81CPB1747⟩.

OMe, OSiMe$_3$ (**170**) $\xrightarrow[\text{TiCl}_4,\ -78\text{ to }20\ °\text{C}]{\text{R}^1\text{CH=NR}^2}$ R^1, NR^2, O

(*b*) *Formation of N—C(4) and C(2)—C(3) bonds.* The [2+2] cycloaddition of isocyanates (**171**) to alkenes provides one of the best routes to β-lactams ⟨78T1731, 73S327, 74S461, 76CSR181⟩. By far the most important isocyanate is chlorosulfonyl isocyanate, which because of its exceptionally high reactivity will react with a wide variety of alkenes. Also the resulting β-lactams can easily be converted into the *N*-unsubstituted derivatives, *e.g.* with aqueous sodium bisulfite/sodium bicarbonate or with benzenethiol and pyridine in acetone. The vinyl (**172**) ⟨81TL1021, 81TL2025⟩ and acetoxy (**91**) ⟨74LA539, 79CC532⟩ β-lactams, prepared from CSI and butadiene and vinyl acetate respectively, are examples of β-lactams which are important precursors to fused derivatives.

+RN=C=O ⟶ NR, O (**171**)   NH, O (**172**)

The addition of CSI to alkenes is mechanistically related to the addition of ketenes. Formally disallowed, the low lying orthogonal π-bond of the cumulene renders concerted [2+2] addition energetically feasible, although the high electrophilicity of the isocyanate carbon makes a stepwise addition through a dipolar intermediate highly attractive. A pseudo-concerted mechanism in which an intermediate 1,4-dipole is formed in a conformation favourably aligned for ring closure seems to be a reasonable compromise between these two extremes.

Other isocyanates undergo [2+2] cycloaddition, but only with very electron rich alkenes. Thus phenyl isocyanate gives β-lactams with ketene acetals and tetramethoxyethylene. With enamines, unstable β-lactams are formed; if the enamine has a β-H atom, ring opened amides are produced; 2:1 adducts are also found. Photochemical addition of *cis-* and *trans*-stilbene to phenyl isocyanate has also been reported ⟨72CC362⟩.

5.09.3.3.3 Miscellaneous routes to azetidin-2-ones

(i) Ring expansions

There are several mechanistically related ring expansion reactions of cyclopropanones which lead to β-lactams. The conversion of cyclopropanone to β-lactam (**174**) *via* the cyclopropanolamine (**173**) ⟨75JOC1505⟩ is just one modification, but it illustrates the strategy of this type of approach ⟨73TL4855, 69JA2375⟩ which has been applied to the synthesis of 3-amino-nocardicinic acid ⟨81JOC2999⟩.

O $\xrightarrow{\text{RCH(NH}_2\text{)CO}_2\text{Et}}$ HO NHCHRCO$_2$Et (**173**) $\xrightarrow{\text{Bu}^t\text{OCl}}$ H—O NCl(CHRCO$_2$Et) $\xrightarrow{\text{Ag}^+}$ O, NCHRCO$_2$Et (**174**)

HO NHNHSO$_2$Ar, Bu^t, Bu^t (**175**) $\xrightarrow{\text{MnO}_2}$ O, NNHSO$_2$Ar, Bu^t, Bu^t

Oxidative ring expansion occurs when the hydrazine (**175**) is treated with manganese dioxide ⟨73TL4091⟩. Diphenylcyclopropenone (**176**) reacts with ammonia and methylamine to give β-lactams ⟨69BCJ1777⟩. Initial attack could occur at either C-1 or C-2 of the cyclopropenone.

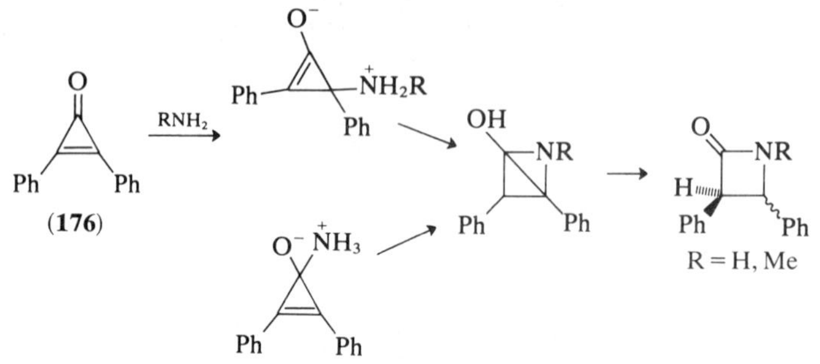

(**176**) R = H, Me

The aziridinecarboxylate (**177**) undergoes ring expansion in good yield, possibly *via* a mixed anhydride, on treatment with thionyl or oxalyl chlorides in benzene ⟨69JA4590⟩.

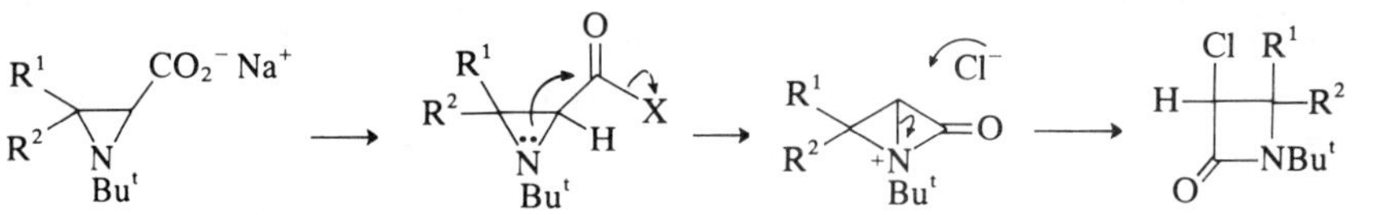

(**177**)

(ii) Ring contractions

Several ring contraction routes to β-lactams have been developed. One of the most important is the photochemical Wolff rearrangement of 3-diazopyrrolidine-2,4-diones (**178**), which provides a general route to 3-carboxyazetidin-2-ones (**179**). Lack of stereoselectivity is a problem, but facile epimerization is possible because of the electron withdrawing 3-substituent ⟨78T1731⟩.

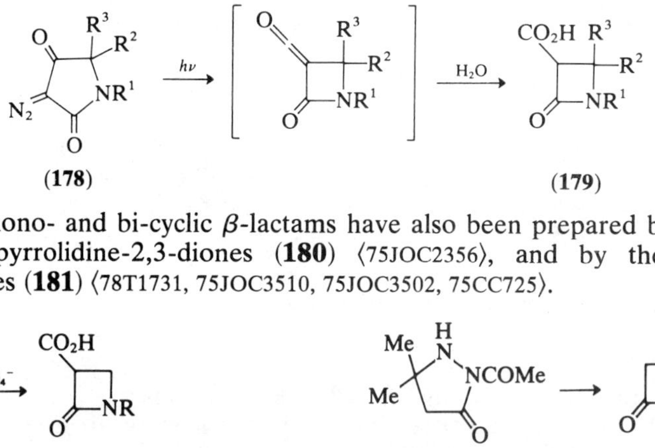

(**178**) (**179**)

A variety of mono- and bi-cyclic β-lactams have also been prepared by oxidative ring contraction of pyrrolidine-2,3-diones (**180**) ⟨75JOC2356⟩, and by the photolysis of pyrazolidin-3-ones (**181**) ⟨78T1731, 75JOC3510, 75JOC3502, 75CC725⟩.

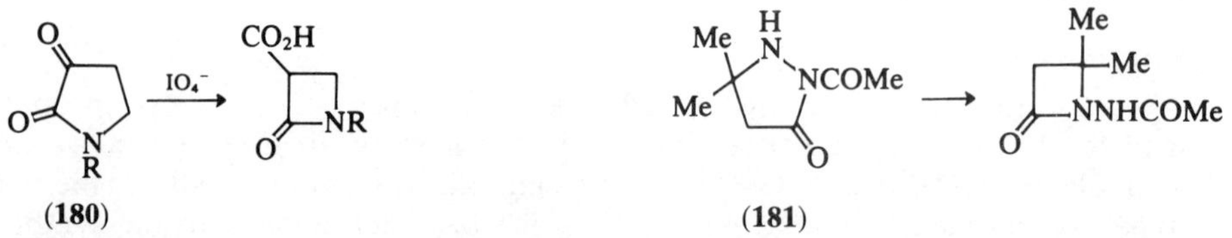

(**180**) (**181**)

Other rather specialized ring contractions which have not found general use include desulfurization of the heterocycles (**182**) ⟨74JOC1210⟩ and (**183**) ⟨78TL2037⟩ with Raney nickel, photolysis of 3-oxo-1-pyrroline *N*-oxides (**184**) ⟨74CC129⟩, oxidation of 2-aminopyrazolidin-3-ones (**185**) ⟨75TL4089⟩, thermolysis of azides (**186**) ⟨76JA3728⟩ and treatment of the oxazolium salts (**187**) with dialkyl acylphosphonates in the presence of triethylamine ⟨74CPB1526⟩.

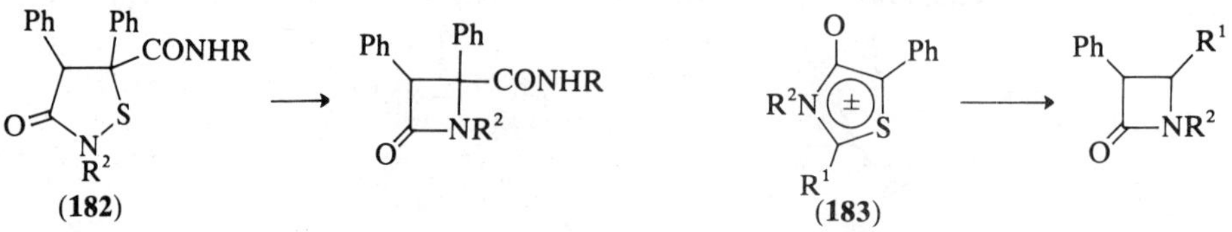

(**182**) (**183**)

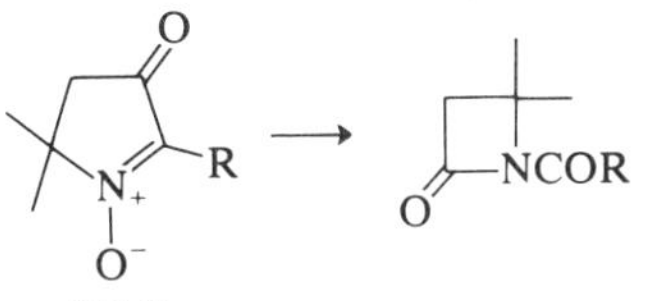

(**184**)

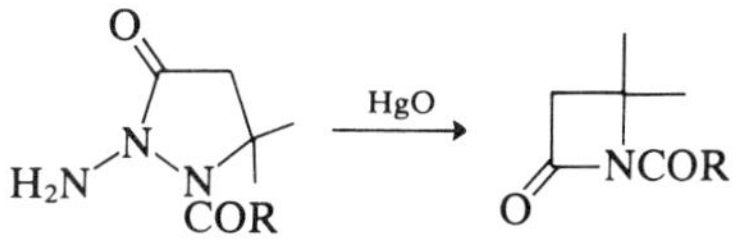

(**185**)

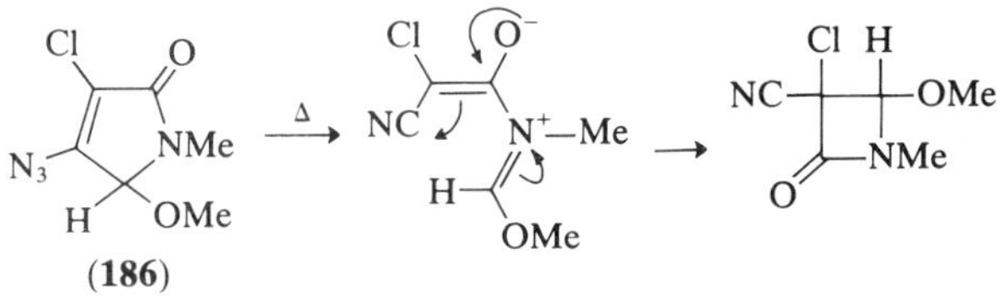

(**186**)

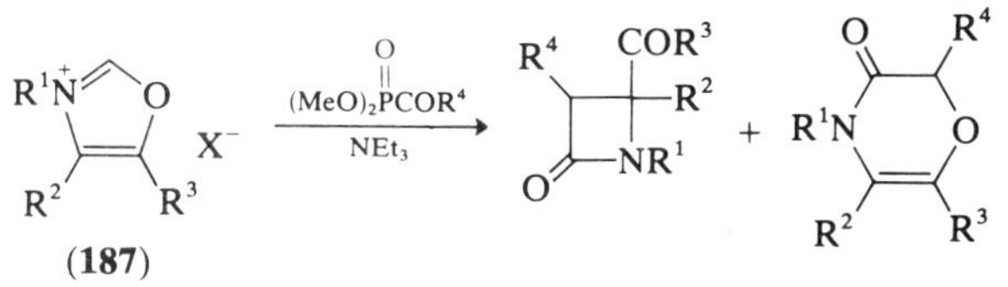

(**187**)

(iii) From azetidines

Azetidine-2-carboxylic acids (**188**) can be converted into azetidin-2-ones *via* (**189**) by conversion to the dianion followed by reaction with oxygen ⟨76TL4613⟩. An improved alternative route involves treatment with oxalyl chloride to give the azetinium salt (**190**), followed by reaction with MCPBA and pyridine ⟨77TL1449⟩. Addition of perfluoroisobutene to benzylidineaniline under forcing conditions gives a [2+2] adduct, which can be hydrolyzed to the β-lactam (**191**) ⟨73S327⟩. The highly reactive chloroenamines give azetidinimmonium salts (**192**) with imines, which give β-lactams on hydrolysis ⟨74AG(E)267⟩. A β-lactam has also been obtained by hydrolysis of the unstable 2-amino-2-azetine resulting from [2+2] addition of 1-diethylaminoprop-1-yne to dihydroisoquinoline ⟨70CB573⟩.

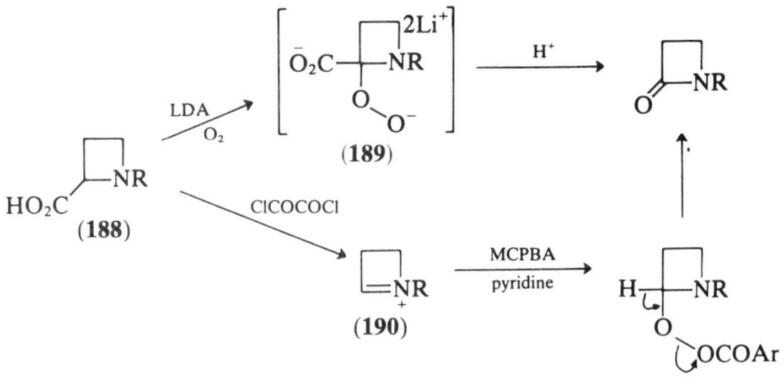

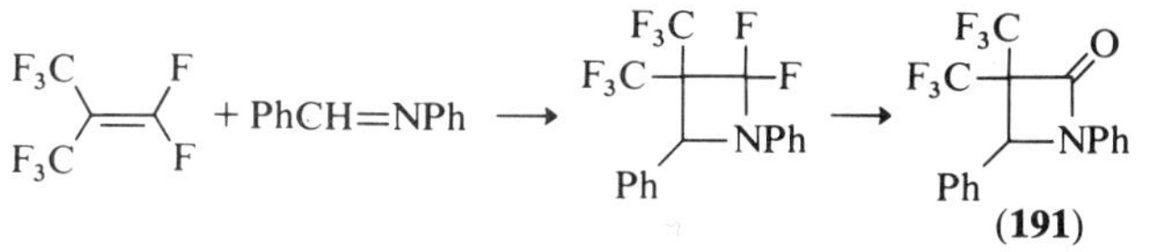

(**188**) (**189**) (**190**)

(**191**)

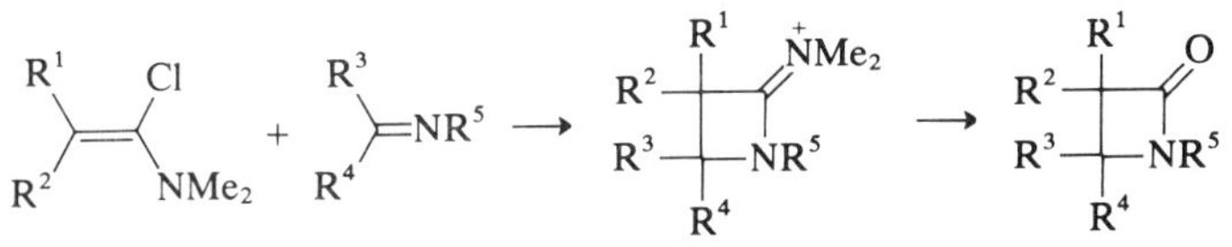

(**192**)

Several routes to β-lactams starting from nitrones have been reported. The reaction of copper(I) phenylacetylide with diaryl nitrones (**193**) gives *cis*-β-lactams ⟨72CC466⟩. The reaction of ketenes with nitrones proceeds by deoxygenation of the nitrone to give an imine which then undergoes [2+2] addition with more ketene ⟨75JCS(P1)498⟩. The isoxazolines (**195**) formed by 1,3-dipolar cycloaddition of cyanonitroethene to *C*-phenyl *N*-*t*-butyl nitrone (**194**) are converted photochemically into β-lactams ⟨81JA4974⟩.

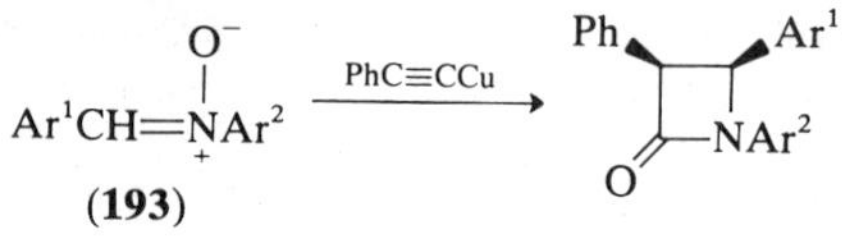

5.09.3.3.4 Modifications of azetidin-2-ones

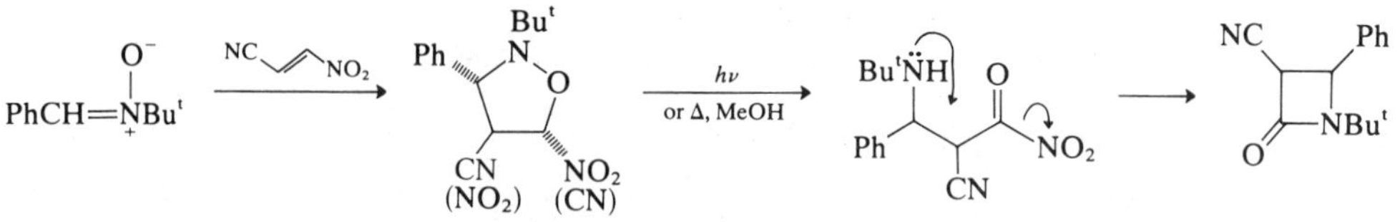

It is often necessary to prepare β-lactams with particular substituents at N-1, C-3 and C-4, *e.g.* in the preparation of fused β-lactams from monocyclic precursors. For reasons of space it is not possible to give an exhaustive list of the variously substituted β-lactams available; however, Table 4 summarizes the most general routes to β-lactams bearing particular substituents at C-3 and C-4. *N*-Substitution of azetidin-2-ones has already been dealt with (*cf.* Section 5.09.3.2.3). *N*-Unsubstituted β-lactams and protected 3-amino-β-lactams are particularly important synthetic intermediates and methods of preparing these are discussed below.

Table 4 Preparation of *C*-Substituted Azetidin-2-ones

Substituent	*C-3*	*C-4*	*Substituent*	*C-3*	*C-4*
H	B, C, D	C, D	acyloxy		B
Alkyl	A, C, D	A, B, C, D	amino	A, J	A, B
Aryl	A, C, D	A, B, C, D	acylamino	A, J	A, G
Vinyl	A	B	azido	A	G
Acyl	I	E	nitro	I	
Alkoxycarbonyl	F	E	halogeno	A, H	G, J
Cyano	A		alkyl/arylthio	A, H	B, G
Alkoxy/aryloxy	A, H	B	alkyl/arylsulfonyl		G

(A) Imine/ketene equivalent cycloaddition (*cf.* Section 5.09.3.3.2). (B) i, CSI/alkene cycloaddition; ii, reductive dechlorosulfonylation (*cf.* Section 5.09.3.3.2). (C) Cyclization of β-amino acid derivatives (*cf.* Section 5.09.3.3.1). (D) Cyclization of β-haloamides (*cf.* Section 5.09.3.3.1). (E) Cyclization of α-haloamides (*cf.* Section 5.09.3.3.1). (F) Wolff rearrangement (*cf.* Section 5.09.3.3.3). (G) Nucleophilic displacement of 4-substituent in β-lactams (*cf.* Section 5.09.3.2.4). (H) Reaction of diazotized 3-amino-β-lactams with nucleophile (*cf.* Section 5.09.3.3.4). (I) i, Deprotonation of C-3 (*e.g.* with LDA); ii, electrophilic attack on C-3 anion (*cf.* Section 5.09.3.2.3). (J) Degradation of readily available penicillin derivatives (*cf.* Section 5.09.3.3.4).

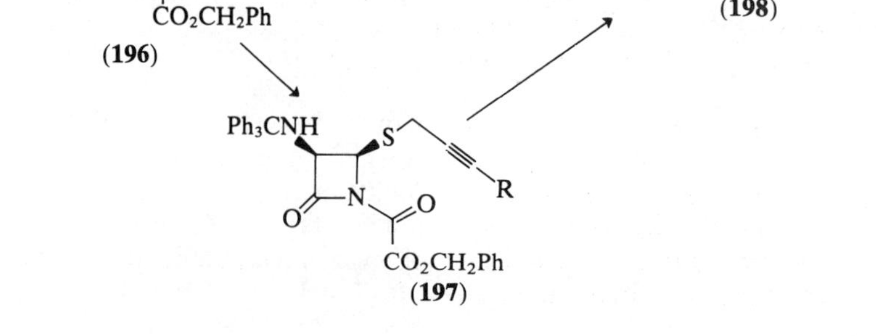

N-Unsubstituted azetidin-2-ones are versatile intermediates in the preparation of a variety of novel β-lactam containing systems. They are usually made either by reductive dechlorosulfonylation of alkene/chlorosulfonyl isocyanate cycloadducts (*cf.* Section 5.09.3.3.2), which can be achieved conveniently with sodium sulfite under mildly basic conditions ⟨75S547 p. 556⟩, or by cleavage of penicillin derivatives. For example, oxidative cleavage of the secopenicillins (**196**) with potassium permanganate in moist pyridine/DMF gives the *N*-unsubstituted β-lactams (**198**), presumably via the hydrolytically sensitive oxamides (**197**) ⟨76JCS(P1)447, 76JCS(P1)1612⟩. The versatile imine/ketene equivalent reaction (Section 5.09.3.3.2) can also be used to prepared *N*-unsubstituted β-lactams if a suitable group is used to protect the imine nitrogen. Thus the *N*-(2,4-dimethoxybenzyl)-β-lactam (**199**) (prepared *via* imine/azido ketene equivalent cycloaddition) undergoes oxidative cleavage on treatment with buffered potassium persulfate in aqueous acetonitrile to give (**200**) ⟨77JA2352⟩. Other methods for oxidative and reductive removal of β-lactam *N*-substituents have been reviewed ⟨75S547, 78T1731⟩.

PhOCH$_2$CONH CO$_2$Me — (**199**) — $K_2S_2O_8$, Na_2HPO_4, aq. MeCN, Δ → PhOCH$_2$CONH CO$_2$Me NH (**200**)

As many natural and synthetic β-lactams bear 3-acylamino substituents, the corresponding free amines or protected forms thereof are versatile synthetic intermediates. They may be prepared in several ways, for example by deacylation of the 7-amido group in naturally occurring penicillins by enzymic or chemical means. Chemical degradation usually involves conversion of the amide to a chloroimidate followed by cleavage with aqueous alcohols ⟨75S547 p. 560, 78T1731 p. 1753⟩.

Alternatively, the 3-phthalimido- and 3-azido-β-lactams are available *via* the imine/ketene equivalent cycloaddition reaction and these may be transformed into the corresponding 3-amino-β-lactams by dephthaloylation and reduction respectively. The former process may be achieved by hydrazinolysis in dichloromethane ⟨68JHC779⟩ or by treatment with sodium sulfide followed by ethyl chloroformate and triethylamine, trifluoroacetic anhydride or carbodiimides to give iminolactones which undergo selective hydrazinolysis without affecting the β-lactam ring ⟨75JA5582, 75JA5583⟩. Reduction of the azido group can be carried out by catalytic hydrogenation ⟨75S547 p. 560, 78T1731 p. 1754⟩ or by treatment with ammonium sulfide ⟨76JOC1112⟩.

A third approach to 3-amino-β-lactams is by Curtius rearrangement of the corresponding acyl azides. These are readily prepared from *t*-butyl carbazides, available *via* photochemical ring contraction of 3-diazopyrrolidine-2,4-diones in the presence of *t*-butyl carbazate (*cf.* Section 5.09.3.3.2). Thus treatment of (**201**) with trifluoroacetic acid followed by diazotization gives the acyl azide (**202**) which, in thermolysis in benzene and subsequent interception of the resulting isocyanate with *t*-butanol, yields the protected 3-amino-β-lactam (**203**) ⟨73JCS(P1)2907⟩.

Bu^{t}O$_2$CNHNHCO NR (**201**) — i, TFA; ii, NaNO$_2$, HCl → N$_3$ NR (**202**) — i, Δ, PhH; ii, Bu^{t}OH, Δ → Bu^{t}O$_2$CNH NR (**203**)

The 3-amino substituted β-lactams undergo the normal reactions of a primary amine. In the case of 6-aminopenicillanic acid derivatives, diazotization can give the corresponding 6-diazopenicillanates ⟨72CC192⟩ and the reactions of these systems with various nucleophiles, electrophiles and dipolarphiles have been studied ⟨75S547, 78TL995, 78JOC4856, 79TL1441, 80TL395, 81CC10⟩.

5.09.3.3.5 Related azetidine derivatives

The methods of preparation of azetidine derivatives with exocyclic unsaturation, other than azetidin-2-ones, are summarized in Table 5.

Table 5 Related Azetidine Derivatives

Basic structure	*Preparation*	*Ref.*
 NR	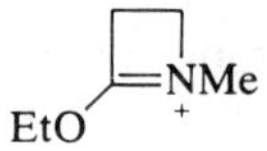 EtO, =NMe^{+} + $H\bar{C}R_2$	68AC(R)1283
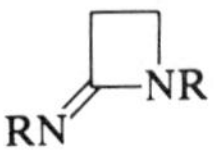 RN, NR	Imine + ketenimine Methylene aziridine + azide	79AG(E)320 75CC436
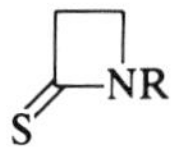 S, NR	β-Lactam + P_4S_{10} Imine + thioketene	67JHC619 79CB1000
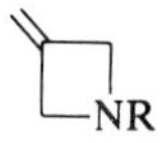 NR	3-Chloroazetidine: i, $R_2\bar{C}CN$; ii, hydrolysis; iii, Hofmann degradation of amine; iv, elimination	81GEP3027169
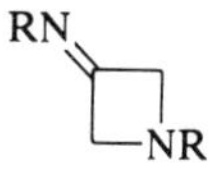 RN, NR	Isocyanide + azomethine ylide	79TL1381, 75S250
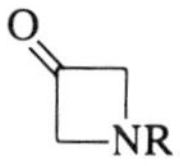 O, NR	Oxidation of azetidin-3-ol $TsNHCHRCOCHN_2 + H^+ \rightarrow$	79JMC183, 73S153, 73CPB228 80ZOR2447
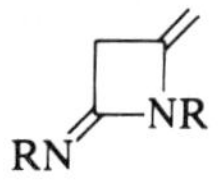 RN, NR	Ketenimine dimerization	72JHC1147, 72ZOR856, 71CJC1792
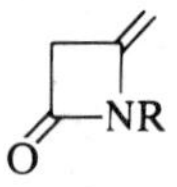 O, NR	4-Thioxoazetidin-2-ones + diazoalkane and desulfurization	80JOC1481
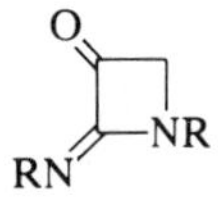 O, RN, NR	α-Lactam + isocyanide	80DOK(253)886
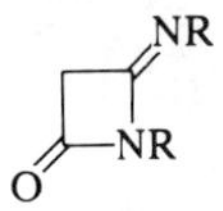 NR, O, NR	Ketene + carbodiimide Isocyanate + ketenimine	81CC41, 79LA1322, 77CC302 73CC271; see also 73CC48, 72CCC4035
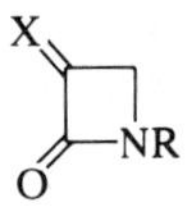 X, O, NR	Intermediates in penicillin and cephalosporin synthesis X = NR, O	78T1731, 75S547
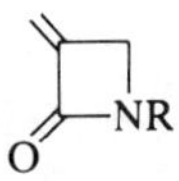 O, NR	Cyclization of 3-bromo-2-bromomethylpropionamides CO insertion into 2-bromo-3-aminopropenes Petersen alkenation of 3-trimethylsilyl-β-lactam	78CC903 79CC698, 80CC770 78S746

Table 5 (cont.)

Basic structure	*Preparation*	*Ref.*
	Malonoyl chloride + amine	64HC(19-2)951
	Ketene + isocyanate	64HC(19-2)951
	Photolysis of 4-hydroxyazetidin-2-one pyruvate ester	79CC344
	Photolysis of trialkylsuccinimides	79JA3670; see also 80JOC1481
	From β-lactam-4-sulfenic acids	76JA7864, 80JOC1477
	Thioketene + isocyanate	80TL4247
	Photolysis of 4-mercaptoazetidin-2-one pyruvate ester	76TL3975
	$Ph_3\overset{+}{P}-\overset{-}{C}=C=NR$; i, CO_2; ii, RCHO	77AG(E)349
	$Ph_3\overset{+}{P}-\overset{-}{C}=C=NR$: i, RNCO; ii, RCHO	77AG(E)349

5.09.3.4 Applications

The most important applications of β-lactams are as antibiotics related to penicillins and cephalosporins, and a vast amount of research effort has gone into increasing the efficacy of such drugs against resistant organisms, such as Gram-negative bacteria. In connection with this, there has been much recent interest in other bicyclic β-lactams such as clavulanic acid and olivanic acid derivatives, which show β-lactamase inhibitory activity ⟨80MI50900⟩ (see also Chapters 5.10, 5.11, 5.12). Certain monocyclic β-lactams, including the nocardicins and the monobactams (*cf.* Section 5.09.3.2.1), also have antibacterial properties. Simple β-lactams may also show other forms of biological activity, including inhibition of aspartate oxidase and other enzymes, and as antidepressants ⟨76CSR181⟩.

Polymerization of β-lactams to yield linear polyamides could have industrial applications if the cost of the starting materials can be reduced sufficiently to make the process economically attractive ⟨75S547 p. 581⟩.

High anti-inflammatory activity has been found for 3,3-dialkyl-1-aminoazetidine-2,4-dione derivatives such as (**204**) ⟨64HC(19-2)885⟩.

(204)

5.9.4 AZETINES

5.9.4.1 Structure

The parent 1-azetine is a colourless liquid, stable at −70 °C, which undergoes polymerization within a few seconds at 20 °C ⟨81JA468⟩. Substituted 1-azetines possess greater stability, and physical data for representative derivatives of 1-azetine are shown in Table 6.

Table 6 Physical Properties of Representative 1-Azetines

	M.p. (°C)	*B.p.* (°C)	$\bar{\nu}_{C=N}$ (cm^{-1})	*Ring proton NMR signals* (δ, p.p.m.)
N	—	—	1570	3.09 (2H, t, J 2.8 Hz), 4.00 (2H, t, J 2.8 Hz), 8.33 (1H, s)
Ph, N	—	30–35/10^{-3} Torr	1568	3.25 (2H, t, J 3.5 Hz), 3.85 (2H, t, J 3.5 Hz)
Cl, Cl, Ph, N	38–40	65/10^{-3} Torr	1572	4.50 (2H, s)
Me, Me, MeO, N	—	113/760 Torr	1615	2.72 (2H, s)
Ph, Ph, EtO, N	—	120/10^{-1} Torr	1620	3.76 (2H, s)
Ph, Ph, EtS, N	70–72	—	1535	4.25 (2H, s)
Cl, Ph, Me, Cl, N	—	—	1580	4.42 (1H, q)

X-Ray diffraction studies on the 3-imino-1-azetine (**205**; Ar = *p*-FC_6H_4), show that the four-membered ring is planar with an unusually long endocyclic C=N bond ⟨74ZN(B)399⟩. The structure of the 1-azetine *N*-oxide (**275**) has also been determined by X-ray crystallographic techniques ⟨79CC993⟩.

N 1.238, CF_3, 1.511, 1.575, CF_3, 83.6, 86.1, 1.517, 1.467, 1.504, 97.9, 92.4, N, 1.453, 1.307, Ar

(**205**)

Further support for the planarity of the 1-azetine ring comes from variable temperature NMR studies of 2-ethylthio-1-azetines ⟨67JHC583⟩. NMR data for representative 1-azetines are given in Table 6. Vicinal coupling constants in 2-alkoxy- and 2-ethylthio-1-azetines are similar to those of analogous β-lactams (*cf.* Section 5.09.3.1), while geminal coupling constants in 3-substituted 1-azetines are of greater magnitude (J_{44} = *ca.* −10 Hz).

The mass spectra of 2-aryl-1-azetines generally show cleavage to give alkene plus nitrile ⟨79CB3914, 71JA2051⟩, although the 2-aryl-3-imino-1-azetine (**205**) undergoes a [3 + 1] cycloreversion to give isocyanide plus nitrile ylide ⟨74ZN(B)399⟩.

IR spectra of 2-substituted 1-azetines show the expected trends in the frequency of the imine absorption; thus 2-aryl-1-azetines generally show $\bar{\nu}_{C=N}$ at *ca.* 1550–1580 cm^{-1}, while 2-ethylthio and 2-alkoxy-1-azetines show absorptions at *ca.* 1535 and 1620 cm^{-1} respectively ⟨79CB3914, 67JHC619, 69LA(725)124⟩. The parent 1-azetine shows $\bar{\nu}_{C=N}$ at 1570 cm^{-1}

⟨81JA468⟩, and the 3-imino-1-azetine (**205**; Ar = p-FC_6H_4) has imine absorptions at 1600 and 1715 cm^{-1}, assigned to the endo- and exo-cyclic imine functions respectively ⟨74ZN(B)399⟩.

Little is known about the structure of 2-azetines. The parent system has not been reported, but INDO molecular orbital calculations suggest that the ring is planar with the N—H bond less than 48° out of the plane of the ring. The barrier to inversion at nitrogen has been calculated to be *ca.* 17 kJ mol^{-1} ⟨77CC806⟩. NMR studies of the 1-substituted 2-azetines (**237**; Z = Ts, CO_2Me) suggest that the molecules are either planar or undergo inversion at nitrogen at a rate which is rapid on the NMR timescale. On the limited data available, chemical shifts for protons at C-2, C-3 and C-4 are *ca.* δ 6.6–6.8, 5.0–5.7 and 4.1–4.5 respectively ⟨77CC806, 68JOC4381, 63JA3524⟩.

The IR spectrum of the 2-azetin-4-one derivative (**301**) shows $\bar{\nu}_{C=O}$ at 1750 cm^{-1} ⟨63JA3524⟩. The corresponding absorptions in the benzazetidinones (**251**; R = Bu^t, adamantyl) are 1810 ⟨71JA1543⟩ and 1790 cm^{-1} ⟨73JCS(P1)868⟩ respectively.

5.09.4.2 Reactivity

5.09.4.2.1 General survey

Prior to 1966 there were no reports of adequately characterized 1- or 2-azetines, and few systematic studies, particularly of 2-azetines, have appeared to date. The 1- and 2-azetines are strained cyclic imines and enamines respectively, and exhibit the types of reactivity expected for such systems. In addition, both 1- and 2-azetines undergo electrocyclic ring opening; this should occur in a conrotatory manner in accordance with the Woodward–Hoffmann rules. However, the ring opening of certain fused 2-azetines is apparently disrotatory, and various theories have been advanced which might explain this (*cf.* Section 5.09.4.2.3) ⟨80JOC1341, 74AG(E)751⟩.

As 1- and 2-azetines exhibit quite different types of reactivity, they are dealt with separately in subsequent sections.

5.09.4.2.2 1-Azetines

(i) Thermal and photochemical

The recently prepared 1-azetine (**2**), which is highly unstable towards polymerization, undergoes electrocyclic ring opening on flash vacuum pyrolysis to give the hitherto unknown 2-azabutadiene ⟨81JA468⟩. The formation of the imidate (**208**) on vapour phase pyrolysis of the 2-alkoxy-1-azetine (**206**) at 200 °C can be rationalized on the basis of conrotatory ring opening followed by a 1,5-hydrogen shift in the resulting 2-azadiene (**207**) ⟨75JOC1349⟩. Ring opening of 2-alkoxy-1-azetines under more vigorous conditions is apparently non-stereospecific; this is possibly due to isomerization of the initially formed azadienes ⟨70JA1763⟩. Whereas the 2-alkoxy-1-azetine (**209**) undergoes electrocyclic ring opening on pyrolysis in the vapour phase, in solution Chapman rearrangement to give the isomeric azetidin-2-one (**210**) is observed ⟨75JOC1349⟩. Predictably, fusion of the 1-azetine ring to a cyclohexadienyl system greatly facilitates ring opening, and the fused 2-methoxy-1-azetines (**211**) exist mainly as the azocines (**212**) at room temperature ⟨71JA152⟩.

MeO, N; 200 °C, 10^{-2} Torr; H, MeO, N; MeO, N

(**206**) (**207**) (**208**)

MeO, N; 130 °C, CN; O, NMe; R, OMe, N, R; disrotatory; R, OMe, N, R

(**209**) (**210**) (**211**) (**212**) R = H, Me

In an attempt to generate (**214**), a derivative of the elusive monocyclic azete system (*cf.* Section 5.09.5.3.1), the fused 1-azetine (**213**) was pyrolyzed, but gave only intractable tars in addition to low yields of dimethyl phthalate ⟨71JOC435⟩.

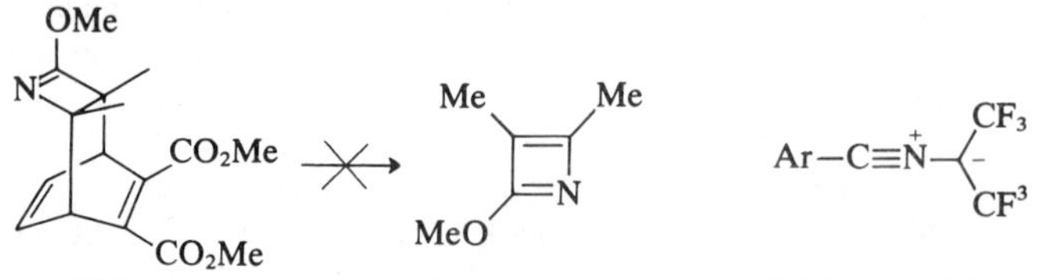

(**213**) (**214**) (**215**) Ar = 4-XC_6H_4, X = H, Me, Cl, F

2-Phenyl-1-azetine undergoes ring opening to give 3-phenyl-2-azabutadiene on prolonged irradiation, although 1- and 2-naphthyl-1-azetines are apparently photostable under similar conditions ⟨76CC729, 72JA5929, 77JOC4238⟩. In contrast, the 3-cyclohexylimino-1-azetines (**205**), prepared by cycloaddition of the nitrile ylides (**215**) with cyclohexyl isocyanide (*cf.* Section 5.09.4.3.1), do not undergo photochemical ring opening to give iminoazadienes, but regenerate the nitrile ylides (**215**) on irradiation ⟨73AG(E)155⟩.

(ii) Electrophilic attack

The parent 1-azetine (**2**) (*cf.* Section 5.09.4.3.1) is extremely sensitive to acid- or oxygen-catalyzed polymerization at room temperature, but undergoes addition of hydrogen cyanide to the imine function at −50 °C to give (**216**) ⟨81JA468⟩. The less heavily substituted 2-alkoxy-1-azetines are also highly susceptible to acid catalyzed polymerization ⟨69LA(725)124⟩. In contrast, the 2-dimethylamino-1-azetines (**217**) are stable towards polymerization, and react readily with electrophiles at the imine nitrogen to give, after hydrolysis, the azetidin-2-ones (**218**) ⟨80CC173⟩. Similarly, *N*-methylation of (**219**) gives the *N*-methyl-2-ethoxyazetinium salt (**220**), which affords the 2-exomethylene azetidines (**221**) on treatment with active methylene compounds ⟨68AC(R)1283⟩. Further examples of reactions initiated by attack of an electrophile on the imine nitrogen of 2-alkoxy-1-azetines are to be found in those between the 1-azetine (**209**) and DMAD ⟨75JOC2360⟩, *t*-butylcyanoketene ⟨75JOC2552⟩ and tosyl isocyanate ⟨75JOC2356⟩. The less electrophilic phenyl isocyanate fails to react with (**209**).

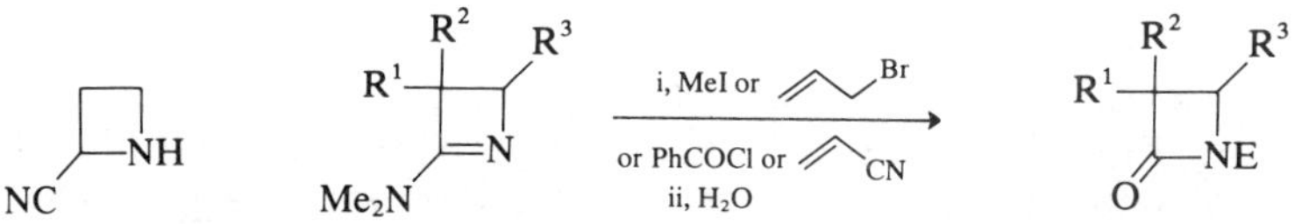

(**216**) (**217**) (**218**) E = Me, allyl, PhCO, CH_2CH_2CN

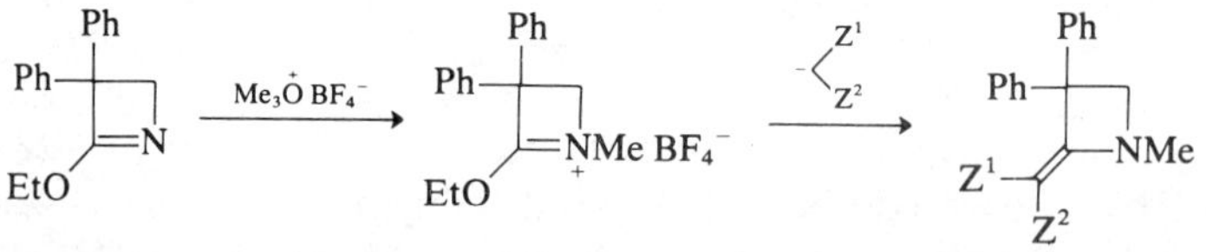

(**219**) (**220**) (**221**) Z^1, Z^2 = CN, CO_2Me, $CONH_2$

Treatment of 2-phenyl-3,3-dichloro-1-azetine (**222**) with MCPBA gives the nitrone (**223**) and the oxaziridine (**224**) in low yield ⟨79CB3914⟩. However, similar treatment of the 2-alkoxy-1-azetines (**209**) and (**227**) gives the blue β-nitroso ester (**226**) and the oxazoline (**228**) respectively. In the former case the presumed oxaziridine intermediate (**225**) can be observed spectroscopically by ^{1}H NMR at −20 °C before it undergoes ring opening and further oxidation to give (**226**) ⟨73TL1807⟩. Reaction of *N*-benzylazetinium perchlorate (**190**)

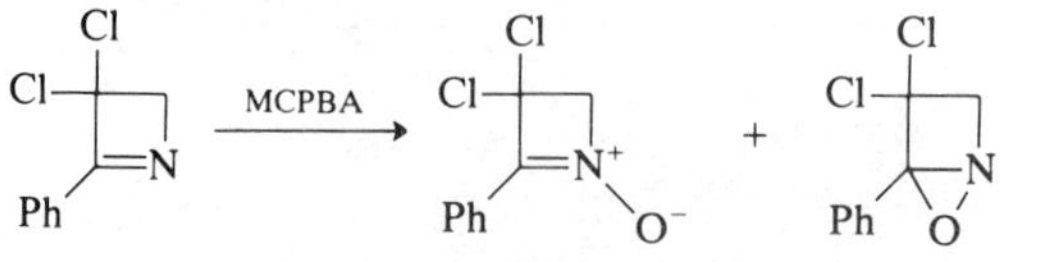

(**222**) (**223**) 19% (**224**) 3%

(**209**) MCPBA → (**225**) [O] −20 °C → MeO_2C N=O (**226**) 50%

(**227**) MCPBA −20 °C → (**228**)

with MCPBA in the presence of pyridine affords the corresponding azetidin-2-one in good yield ⟨77TL1449⟩ (*cf.* Section 5.09.3.3.3(iii)).

2-Phenyl-3,3-dichloro-1-azetine (**222**) forms a 2:1 complex with silver tetrafluoroborate and 1:1 complexes with silver trifluoromethanesulfonate and with boron trifluoride. The dibromo analogue of (**222**) affords a similar 2:1 complex with silver tetrafluoroborate ⟨79CB3914⟩.

(iii) Nucleophilic attack

Not surprisingly, the strained imine function in 1-azetines is readily attacked by nucleophiles. Simple 2-aryl-1-azetines (**229**; X = Ar) give the expected β-amino ketones on mild acid hydrolysis ⟨71JA2051⟩ while the 2-alkoxy-1-azetines (**229**; X = OR) yield the corresponding β-amino esters ⟨69LA(725)124, 67JHC619⟩. In contrast, the 2-chloro-1-azetines (**229**; X = Cl) behave like acid chlorides, mild acid hydrolysis giving azetidin-2-ones and treatment with sodium methoxide yielding the corresponding 2-alkoxy-1-azetines (Scheme 13) ⟨70AG(E)731⟩. Nucleophilic displacement rather than ring cleavage has exceptionally been observed for 2-alkoxy-1-azetines ⟨70CB1797⟩. A synthetically useful application of the nucleophilic ring cleavage of a 2-alkoxy-1-azetine involves treatment of (**209**) with the bidentate nucleophiles (**230**) under conditions of mild acid catalysis to give the benzo fused heterocycles (**231**) ⟨76LA336⟩.

(**229**) X = Ar, OR, H⁺ aq. → ; X = Cl, H⁺ aq. → ; X = Cl, MeO⁻ →

Scheme 13

1-Azetines can be reduced to azetidines with LAH ⟨81JA468, 79CB3914⟩ or sodium borohydride ⟨70JA2178⟩. Selective reduction of the endocyclic imine bond of the 3-imino-1-azetine (**205**; Ar = Ph) can be achieved using one equivalent of LAH ⟨72AG(E)47⟩. Treatment of the 3,3-dichloro-1-azetine (**222**) with an excess of LAH results in reductive dechlorination as well as imine reduction; sodium borohydride leaves the chlorine substituents unaffected ⟨71JA2051⟩. Reduction of the nitrone (**223**) (*cf.* Section 5.09.4.2.2) with one equivalent of LAH gives the *N*-hydroxyazetidine (**232**) ⟨79CB3914⟩. Nucleophilic addition of *n*-butyllithium to 2-phenyl-1-azetine gives the azetidine (**233**) in moderate yield ⟨79CB3914⟩.

(**230**) + (**209**) PTSA MeOH → (**231**) (**232**) (**233**)

5.09.4.2.3 2-Azetines

The most important thermal reaction of 2-azetines is ring opening to give an azadiene. In the case of the azetines (**234**) this even precludes their isolation ⟨66AG(E)585⟩. The photochemical ring closures of the type shown for the diazepinone (**235**) are reversed thermally. Simple orbital symmetry considerations would suggest that ring opening of a 2-azetine should be conrotatory and, therefore, not a favoured process in a fused 2-azetine such as (**236**). However, it has been pointed out that introduction of N at the terminus of the butadiene system can change the nodal characteristics of the orbitals and so render the disrotatory process more favoured ⟨80JOC1341⟩; see also ⟨74AG(E)751⟩.

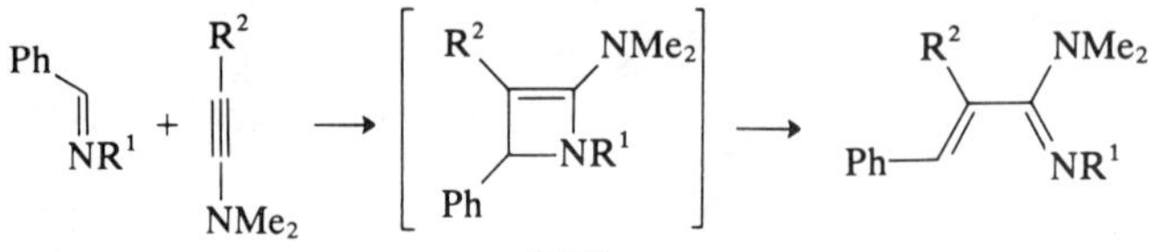

(**234**)

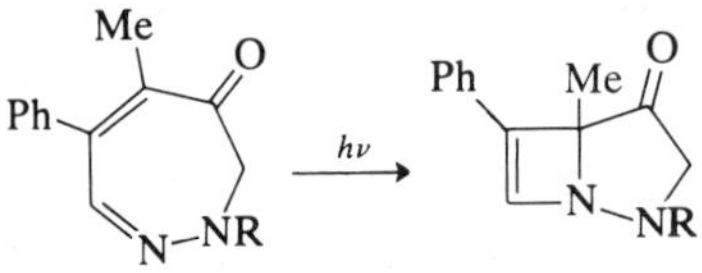

(**235**) (**236**) R = H, Me, Ac, COPh, Ts

The double bond in *N*-methoxycarbonyl-2-azetine (**237**; Z = CO_2Me) undergoes acid or photocatalyzed hydration and subsequent ring opening to give the aldehyde (**238**). In cycloadditions it is inert to TCNE and diphenylisobenzofuran but it does react with dipyridyltetrazine ⟨77CC806⟩.

(**237**)  (**238**)

There are dubious claims for the displacement of the *N*-fluorine in perfluoroazetine (**240**) ⟨60DOK(131)1338⟩ by ethanol and for the formation of an azete by reaction of the bicyclic azetine (**241**) with base. An attempt to produce azete by elimination of *p*-toluenesulfinic acid from azetine (**237**; Z = Ts) failed ⟨71JOC435⟩.

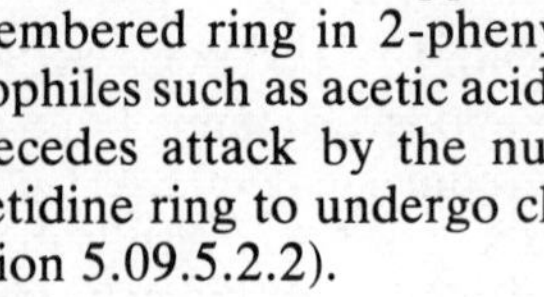

(**239**) (**240**) 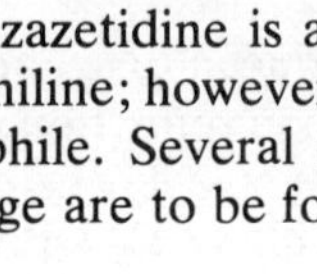 (**241**)

With benzazetidines, thermal ring opening to give azaxylylene intermediates is the predominant reaction. For example, the azaxylylene (**244**) produced by thermal ring opening of 2-phenylbenzazetidine (**243**) can be trapped in a Diels–Alder reaction with *N*-phenylmaleimide. The four-membered ring in 2-phenylbenzazetidine is also cleaved by heating in the presence of nucleophiles such as acetic acid or aniline; however, it is not clear whether or not ring opening precedes attack by the nucleophile. Several more examples of this tendency of the benzazetidine ring to undergo cleavage are to be found in the reactions of benzazete adducts (Section 5.09.5.2.2).

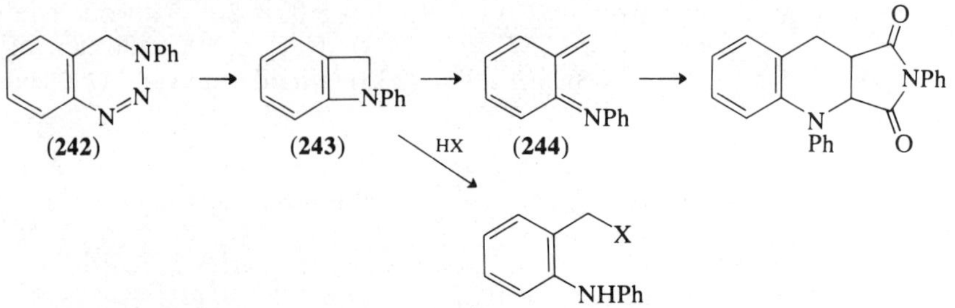

(**242**) (**243**) (**244**)

The 2-azetin-4-one system also appears to be unstable towards ring cleavage; indeed an attempt to isolate the simple azetinone (**246**) failed. The adducts (**248**; X = MeO, MeNH) were obtained in the presence of methanol and methylamine by nucleophilic attack either on the four-membered ring or its iminoketene isomer (**247**) ⟨75TL1335⟩. In contrast, the phenyl substituted analogue (**301**) is isolable. A different mode of ring nucleophilic cleavage involving attack on C-2 has been reported for the methyleneazetine (**249**) ⟨78IZV880⟩.

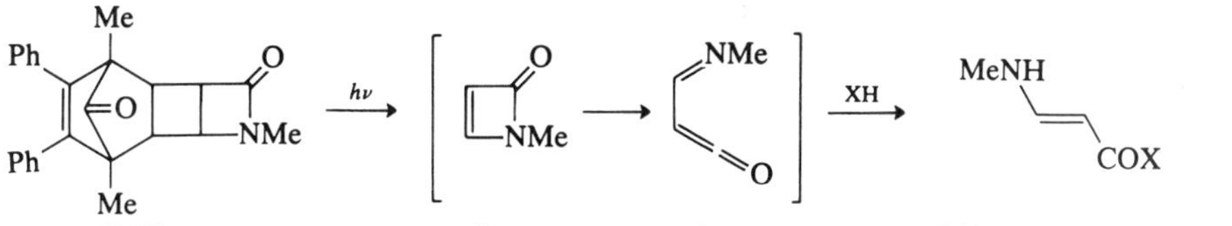

(**245**) —$h\nu$→ [(**246**) → (**247**)] —XH→ (**248**) X = MeO, MeNH

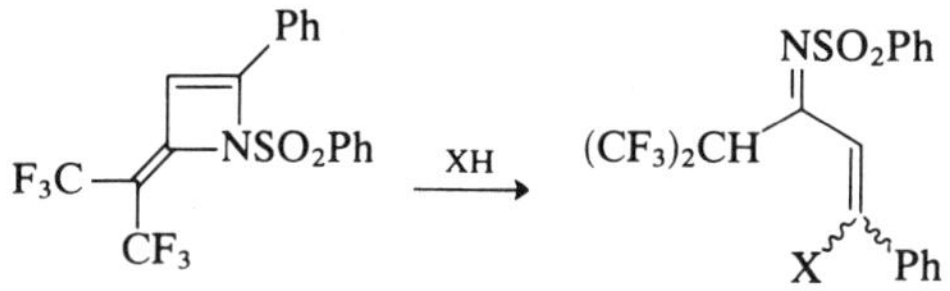

(**249**) —XH→ X = Cl, PhNH, OH

The chemistry of benzazetidin-2-ones (**251**) can also be explained in terms of facile ring opening to the iminoketenes (**252**) which dimerize, rearrange or can be intercepted by nucleophiles or in cycloadditions depending on the conditions. Indeed, this ring opening precludes their isolation in all but exceptional cases (Section 5.09.4.3.5) ⟨76AHC(19)215⟩.

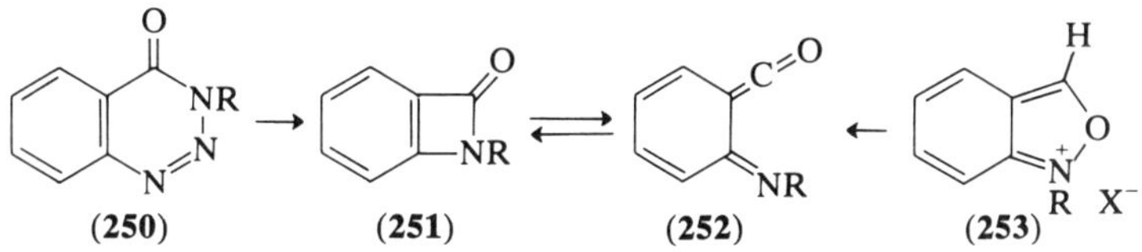

(**250**) → (**251**) ⇌ (**252**) ← (**253**)

5.09.4.3 Synthesis

5.09.4.3.1 1-Azetines

(i) From non-heterocyclic precursors

Thermal ring expansion of cyclopropyl azides (available from diazo transfer to the corresponding metallated cyclopropylamines or from carbenoid addition to vinyl azides) provides a general route to 1-azetines bearing simple alkyl or aryl substituents ⟨73AG(E)161, 79CB3914⟩. In the specific examples of the dichloroazidocyclopropanes (**254**) the more electron-rich methylene group, rather than the dichloromethylene group, migrates towards the incipient nitrene centre to give the 3,3-dichloro-1-azetines (**255**) rather than the isomeric 4,4-dichloro species ⟨71JA2051⟩.

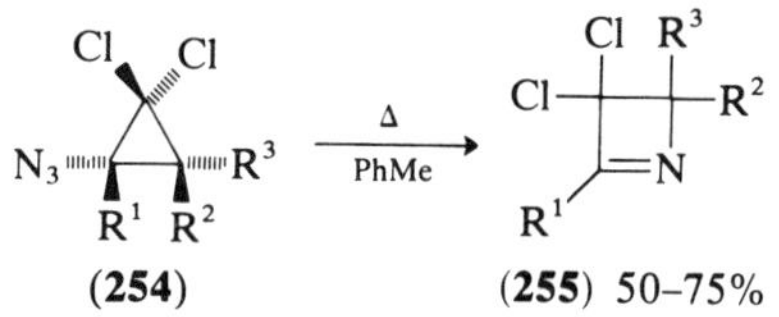

(**254**) —Δ, PhMe→ (**255**) 50–75%

The reaction of imines with α-chloroenamines in the presence of base provides a fairly general preparation of 2-dimethylamino-1-azetines. Thus treatment of the α-chloroenamines (**256**) with the benzhydryl imines (**257**) in the presence of triethylamine, followed by counterion exchange with perchloric acid, gives the amidinium perchlorates (**258**). Hydrogenolytic removal of the benzhydryl group and subsequent basification yields the 2-dimethylamino-1-azetines (**217**) as stable liquids ⟨80CC173⟩. This sequence is analogous to the versatile imine/ketene equivalent route to β-lactams (*cf.* Section 5.09.3.3.2). The α-chloroenamines (**256**) are available from treatment of the corresponding dimethylamides with phosgene.

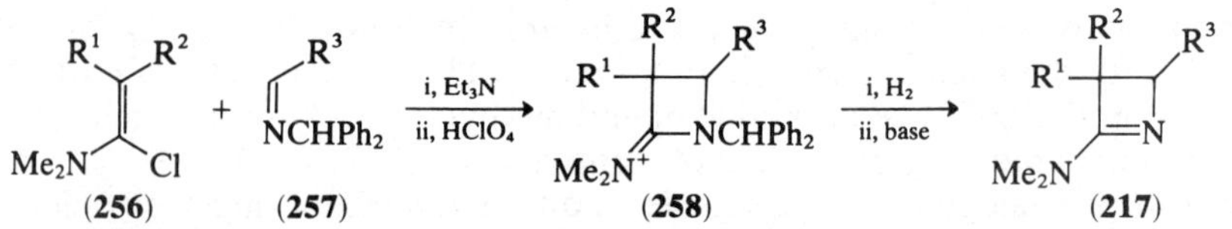

(256) (257) (258) (217)

Photochemical [2+2] cycloaddition of benzonitrile and of 1- and 2-naphthonitriles to electron-rich alkenes such as 2,3-dimethylbut-2-ene gives the corresponding 2-aryl-1-azetines in poor yield ⟨72JA5929, 76CC729, 77JOC4238⟩. This does not appear to be a versatile route to 1-azetines.

(ii) From heterocyclic precursors

Nucleophilic addition of trichloromethyllithium to the azirines (**259**; R = Me, Ph) leads to the aziridines (**260**; R = Me, Ph), which on treatment with potassium *t*-butoxide in DMSO give the 2,3-dichloro-1-azetines (**262**; R = Me, Ph) *via* the azabicyclobutanes (**261**; R = Me, Ph). In the case where R = Ph the azetine (**262**) cannot be isolated but undergoes spontaneous ring opening to the azadiene (**263**). Interestingly, reaction of (**259**; R = Me) with the carbenoid trichloromethylphenylmercury does not lead to the azabicyclobutane (**261**; R = Me) and hence the azetine (**262**; R = Me), but gives the azadiene (**265**) *via* collapse of the ylide (**264**) formed by electrophilic attack of the carbenoid on the azirine nitrogen (Scheme 14) ⟨70AG(E)731, 73JA2982⟩.

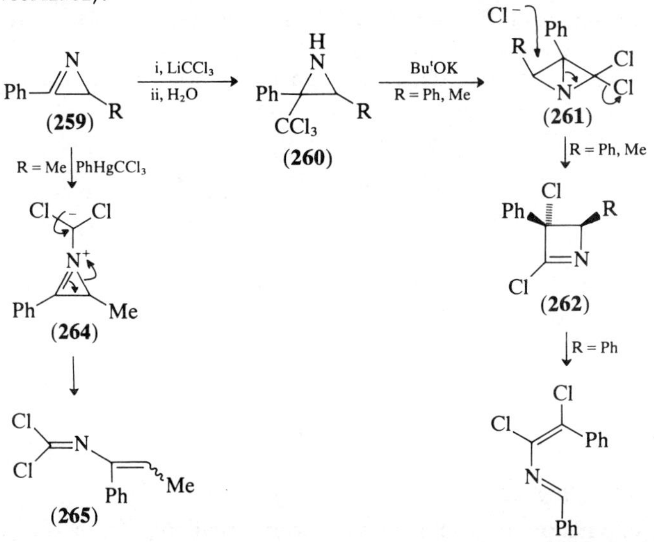

(259) (260) (261) (262) (264) (265) (263)

Scheme 14

O-Alkylation of the readily available *N*-unsubstituted azetidin-2-ones (β-lactams) constitutes a versatile route to 2-alkoxy-1-azetines (*cf.* Section 5.09.3.2.3). Thus treatment of the β-lactams (**266**) with trialkyloxonium tetrafluoroborates followed by basification affords the 2-alkoxy-1-azetines (**267**) in moderate yields ⟨67JHC619, 69LA(725)124⟩. Similar treatment of the azetidine-2-thiones (**268**) (available from thiation of the corresponding β-lactams with phosphorus pentasulfide) affords the analogous 2-ethylthio-1-azetines (**269**) ⟨67JHC619⟩, which are generally more stable than their 2-alkoxy analogues.

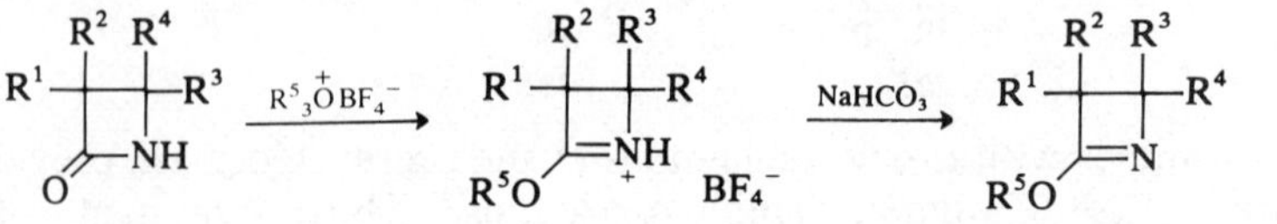

(266) (267) R^{1-4} = H, alkyl, phenyl

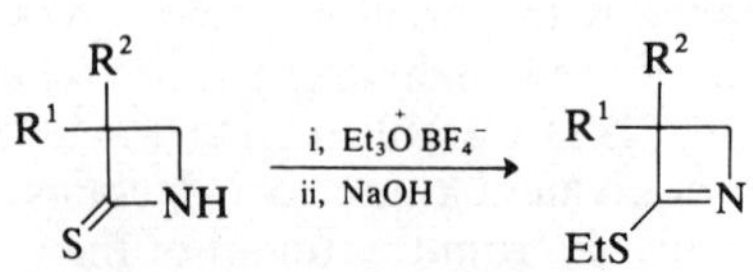

(268) (269)

The low-pressure gas-phase dehydrohalogenation of *N*-chloroazetidine (**270**) using potassium *t*-butoxide supported on silica gives the parent 1-azetine (**2**) in excellent yield ⟨81JA468⟩. This can be trapped at −196 °C, but rapidly undergoes polymerization at room temperature (*cf.* Section 5.09.4.2.2). The 2-phenyl analogue of (**2**) can be prepared *via* a similar route ⟨71IZV893⟩.

(**270**) (**271**)

A number of photochemical ring closures of heterocyclic systems containing a 2-azadiene moiety yield fused 1-azetines. For example, irradiation of pyridine gives 'Dewar pyridine' (**271**) ⟨70JA2178⟩ (*cf.* Section 5.09.3.3.1). Similar photochemical ring closures have been reported for a fluorinated pyridazine ⟨76CC1005⟩ (*cf.* Section 5.09.5.3.1), an oxazinone ⟨74AG(E)425, 77TL1053⟩ (*cf.* Section 5.09.5.3.1) and a dihydroazepine ⟨71JOC1934⟩.

Reaction of the azocine (**212**; R = Me) with DMAD yields the fused 1-azetine (**213**) by Diels–Alder cycloaddition to the bicyclic valence tautomer (**211**; R = Me) ⟨71JOC435, 71JA152⟩.

5.09.4.3.2 Derivatives of 1-azetines

Three preparations of 1-azetine *N*-oxides have been reported. Oxidation of the 2-aryl-1-azetine (**222**) with MCPBA gives the nitrone (**223**) ⟨79CB3914⟩. However, similar treatment of 2-alkoxy-1-azetines fails to give the corresponding *N*-oxides but yields products derived from oxaziridines (*cf.* Section 5.09.4.2.3).

Cyclization of 3,3-dimethyl-2-hydroximino-4-tosyloxybutane (**272**) with 1,6-bis-(dimethylamino)naphthalene is reported to occur *via* *N*-alkylation to give the 1-azetine *N*-oxide (**273**) as an unstable oil which decomposes over a few days at room temperature ⟨74TL4283⟩.

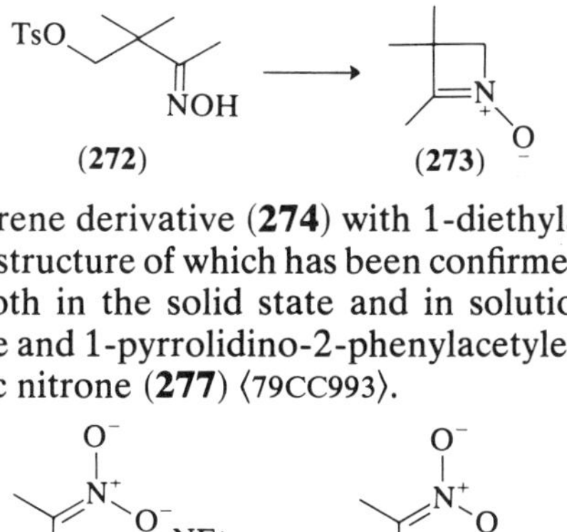

(**272**) (**273**)

Reaction of the β-nitrostyrene derivative (**274**) with 1-diethylaminoprop-1-yne gives the 1-azetine *N*-oxide (**275**), the structure of which has been confirmed by X-ray crystallography. Although (**275**) is stable, both in the solid state and in solution, a similar system (**276**), prepared from β-nitrostyrene and 1-pyrrolidino-2-phenylacetylene, undergoes ring opening in solution to give the acyclic nitrone (**277**) ⟨79CC993⟩.

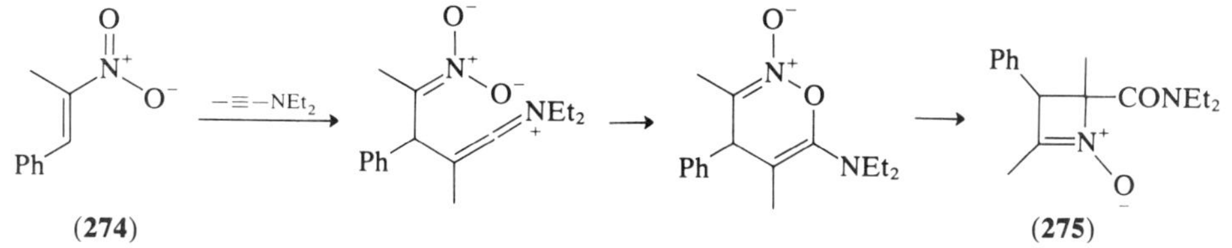

(**274**) (**275**)

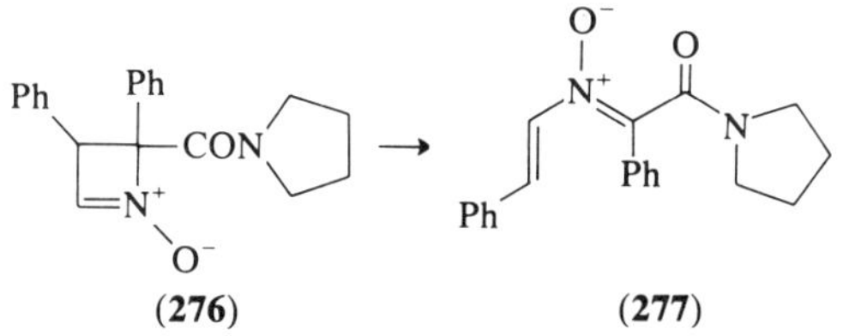

(**276**) (**277**)

1-Azetin-4-ones have been postulated as intermediates in some nucleophilic substitution reactions of *N*-unsubstituted β-lactams (*cf.* Section 5.09.3.2.4) and there are two reports of isolable examples of this system. Treatment of the cyanoacetate derivative (**278**) with an excess of benzoyl chloride in the presence of aluminum chloride is claimed to result in cyclization to give the unstable 2-chloro-4-oxo-1-azetines (**279**) and (**280**) in low yield. Similar results have been claimed for ethyl 2-cyanopropionate ⟨73BCJ1755⟩.

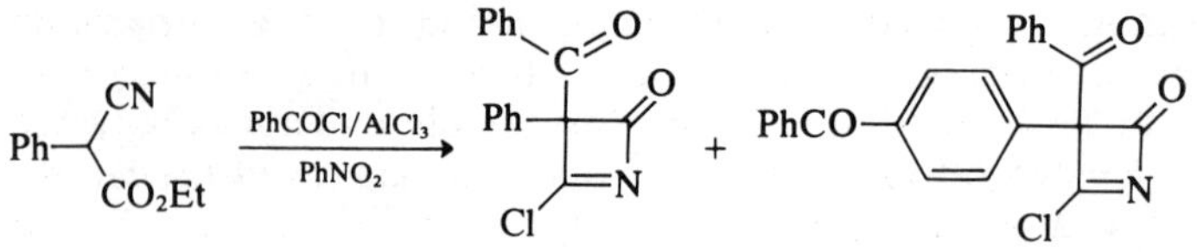

PhCOCl/AlCl$_3$, PhNO$_2$

(278) (279) 6% (280) 28%

The 4-thioxoazetidin-2-one (**281**) (*cf.* Section 5.09.3.3.5) is reported to undergo exclusive *S*-alkylation on treatment with methyl iodide in the presence of base to yield the 2-methylthio-1-azetin-4-one (**282**) ⟨80TL4247⟩.

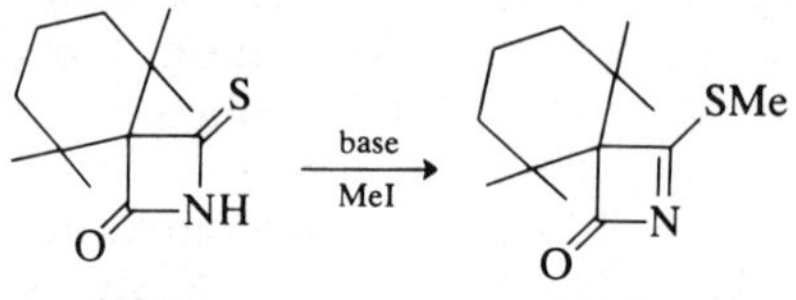

(281) (282)

A series of 3-imino-2-aryl-1-azetines has been prepared by cycloaddition of nitrile ylides to isocyanides. Thus generation of the nitrile ylides (**215**) in the presence of cyclohexyl isocyanide gives the 3-cyclohexylimino-2-aryl-1-azetines (**205**). Similar 1-azetines are formed by interception of nitrile ylides with α-methylbenzyl isocyanide ⟨72AG(E)47⟩ (*cf.* Section 5.09.4.2.2).

5.09.4.3.3 2-Azetines

The first reported 2-azetine (**285**) was obtained by [2+2] cycloaddition of the sulfonylimine (**283**) to the reactive ketene aminal (**284**) followed by elimination of one molecule of amine ⟨66AG(E)416⟩. The very similar 2-dimethylamino-2-azetines (**234**) have been postulated as intermediates in the formation of vinylamidines from ynamines and imines ⟨66AG(E)585⟩. The *N*-tosyl- and *N*-methoxycarbonyl-2-azetines (**237**) were obtained by photofragmentation of the precursors (**286**; Z = Ts, CO$_2$Me), but this route failed for Z = H or Me ⟨77CC806⟩. A number of bicyclic 2-azetines, for example (**236**) and (**241**), have been produced by photochemical ring closure of heterocycles which incorporate a 1-azabutadiene moiety ⟨65CC468, 68JOC4381, 70JOC3615, 78TL1543⟩. There is a claim for the formation of the perfluoro-2-azetine (**240**) from the vinyl azide (**239**) ⟨60DOK(131)1338⟩.

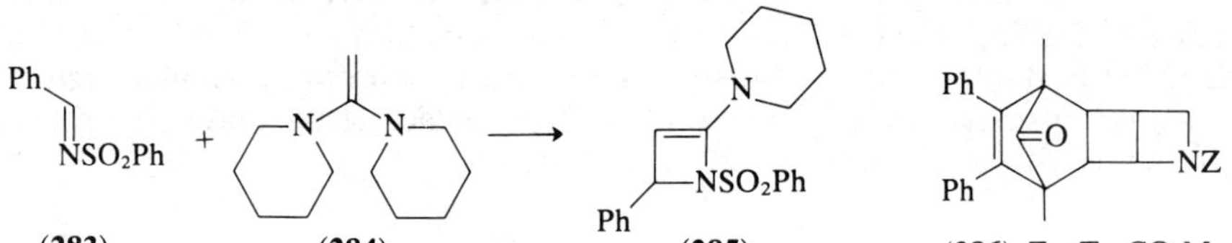

(283) (284) (285) (286) Z = Ts, CO$_2$Me

5.09.4.3.4 Benzazetidines

The first benzazetidine (**243**) was isolated from the photolysis of 3-phenyldihydrobenzotriazine (**242**) ⟨66JA1580⟩. Another route to benzazetidines involving formation of the N to aryl-C bond utilizes intramolecular nucleophilic substitution *via* aryne (**287**). It is not general, however, and is only satisfactory when R^1 and/or R^2 are alkoxy groups. The reaction also fails for *N*-alkylamines ⟨78LA608⟩.

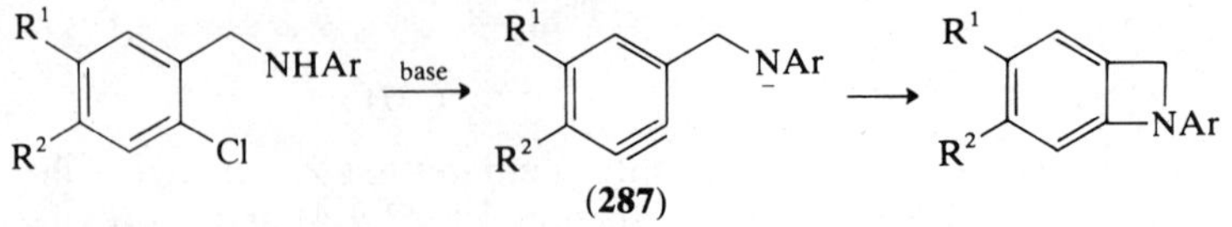

base

(287)

A number of benzazetidines have been produced by nucleophilic and cyclo-additions to the highly reactive benzazetes. In some cases these are isolated, but in others they rearrange by reactions which are initiated by ring opening of the benzazetidine to an azaxylylene (Section 5.09.5.2.2).

By analogy with the ready formation of benzocyclobutenes from *o*-xylylenes, azaxylylenes could be precursors to benzazetidines. Thermal and photochemical extrusion of sulfur dioxide from the cyclic sulfonamides (**288**) does give the benzazetidines (**290**; R = H, Me), possibly *via* an azaxylylene (**289**) ⟨80CC471⟩. Formation of the benzazetidine (**292**) in the short-term irradiation of the dihydroquinoline (**291**) can also be explained by such a cyclization ⟨75CC575⟩. On the other hand, flash pyrolytic 1,4-elimination of water from the amino alcohols (**293**) ⟨80JOC1547, 82TL4501⟩ and of carbon dioxide from the dihydrobenzoxazinones (**294**) gives rearrangement products of the azaxylylenes and no benzazetidine ⟨82TL4501⟩. Photodecarbonylation of amide (**295**) and fluoride ion induced 1,4-elimination in the silylamine (**296**) gives dimers of the azaxylylene and no benzazetidine ⟨69CB3486, 69CB3495, 81JA5250⟩.

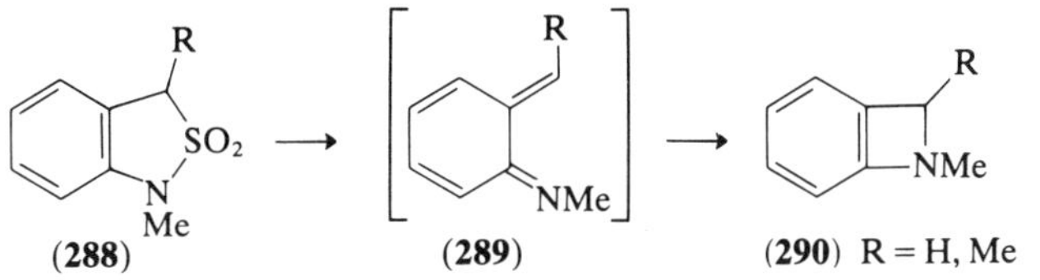

(**288**) (**289**) (**290**) R = H, Me

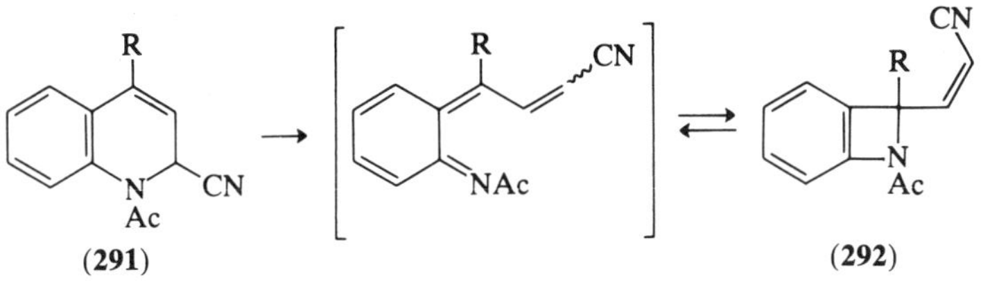

(**291**) (**292**)

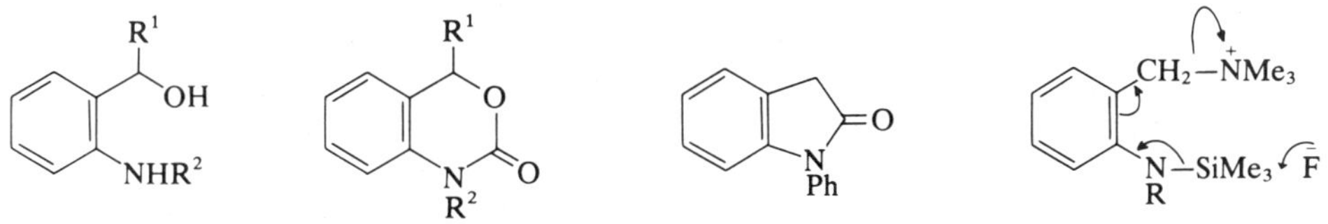

(**293**) (**294**) (**295**) (**296**)

There is some evidence for the formation of unstable benzazetidines from [2+2] cycloaddition of benzyne to imines ⟨75BCJ1063⟩. A novel formation of a benzazetidine is reported in the solvolysis of the *exo N*-chloro compound (**297**). Neighbouring group participation by the benzene ring leads to the cation (**298**), which is intercepted by methanol to give the benzazetidine (**299**) ⟨81CC1028⟩.

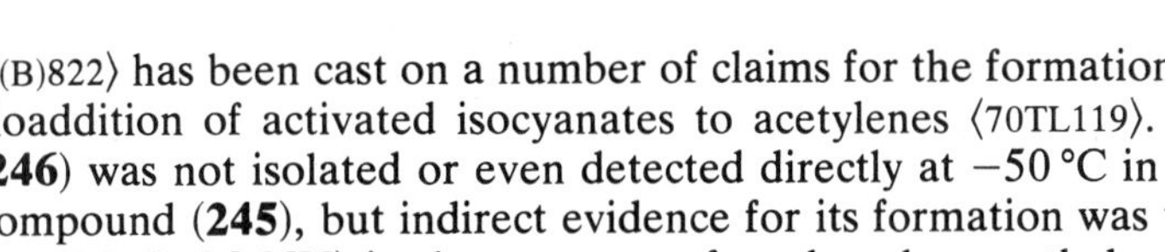

(**297**) (**298**) (**299**)

5.09.4.3.5 2-Azetin-4-ones and benzazetidin-2-ones

Doubt ⟨75ZN(B)822⟩ has been cast on a number of claims for the formation of 2-azetin-4-ones from cycloaddition of activated isocyanates to acetylenes ⟨70TL119⟩. The simple 2-azetin-4-one (**246**) was not isolated or even detected directly at −50 °C in the photofragmentation of compound (**245**), but indirect evidence for its formation was the isolation of adducts (**248**; X = MeO, MeNH) in the presence of methanol or methylamine ⟨75TL1335⟩. The most convincing evidence for an isolable 2-azetin-4-one involves treatment of the

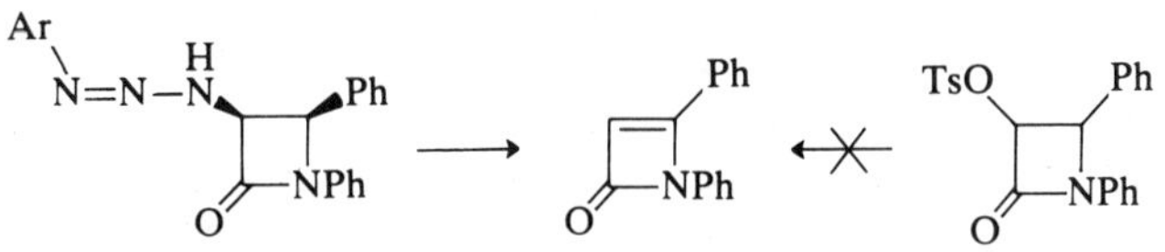

(**300**) (**301**) (**302**)

β-lactam triazene (**300**) with boron trifluoride. The resulting azetinone (**301**) was characterized by hydrogenation to give the known β-lactam ⟨63JA3524⟩. An attempt to prepare the same azetinone by elimination of *p*-toluenesulfonic acid from (**302**) failed. 4-Methylene-2-azetines ⟨71AG(E)184⟩ and 4-imino-2-azetines ⟨74JOC1707⟩ have also been reported.

The formation of transient benzazetidinones (**251**) in the photolysis and thermolysis of benzotriazin-4-ones (**250**) is well established ⟨76AHC215⟩ and the highly hindered adamantyl derivative has actually been isolated from flash pyrolysis of 4-adamantylbenzotriazinone ⟨73JCS(P1)868⟩. A second route to such hindered benzazetidinones involves cyclization of the iminoketene valence tautomer (**252**; R = Bu^t), the latter being generated by deprotonation of the anthranilium salts (**253**) ⟨71JA1543⟩.

5.09.4.4 Applications

No significant applications of azetines or their derivatives have yet been reported.

5.09.5 AZETES

5.09.5.1 Structure

Azete (**4**) is the simplest heterocyclic analogue of cyclobutadiene, the prototypical antiaromatic system. Simple Hückel treatment for square cyclobutadiene indicates that the molecule has zero resonance energy and a triplet ground state. Distortion to a rectangular geometry removes the degeneracy of the non-bonding orbitals and so allows for a singlet ground state, in line with experimental observation. Introduction of a nitrogen atom into the cyclobutadiene ring as in azete removes the degeneracy of the two non-bonding orbitals, making a singlet ground state possible without distortion. More sophisticated treatments suggest that both cyclobutadiene and azete have negative resonance energies, the latter being slightly less antiaromatic ⟨70MI50900, 76JOC3058⟩.

The only report of an isolable, unfused azete is for trisdimethylaminoazete (**304**). The flash pyrolysate from triazine (**303**) contained *ca.* 30% of an unstable red compound which decomposed on warming to room temperature. The spectral data indicated in (**304**) were obtained at low temperature and cited as evidence for the azete structure ⟨73AG(E)847⟩.

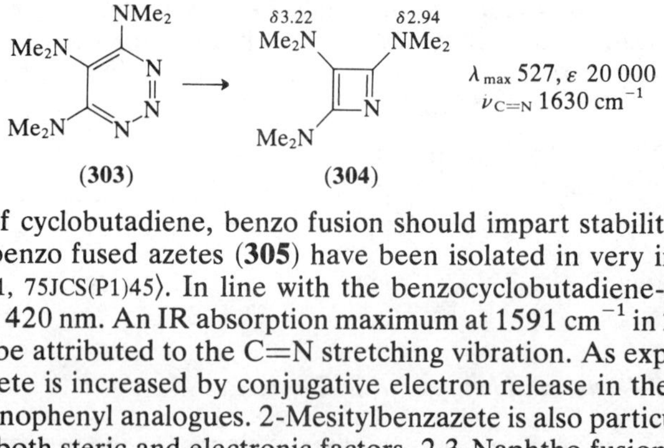

NMe$_2$ Me$_2$N Me$_2$N (**303**) ⟶ δ3.22 Me$_2$N, δ2.94 NMe$_2$, Me$_2$N (**304**)

λ_{max} 527, ε 20 000
$\nu_{C=N}$ 1630 cm^{-1}

As in the case of cyclobutadiene, benzo fusion should impart stability to the azete ring and a number of benzo fused azetes (**305**) have been isolated in very impure form below −40 °C ⟨75JCS(P1)41, 75JCS(P1)45⟩. In line with the benzocyclobutadiene-like structure they are all red, λ_{max} *ca.* 420 nm. An IR absorption maximum at 1591 cm^{-1} in 2-phenylbenzazete (**305**; R = Ph) can be attributed to the C=N stretching vibration. As expected, the stability of 2-phenylbenzazete is increased by conjugative electron release in the *p*-methoxyphenyl and *p*-dimethylaminophenyl analogues. 2-Mesitylbenzazete is also particularly stable, probably as a result of both steric and electronic factors. 2,3-Naphtho fusion in the azete (**306**) renders it sufficiently stable for isolation as an impure and unstable solid at room temperature ⟨75JCS(P1)45⟩.

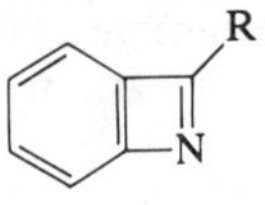

(**305**) R = Ph, *p*-MeOC$_6$H$_4$, *p*-NMe$_2$C$_6$H$_4$, *p*-NO$_2$C$_6$H$_4$, thiazol-2-yl, mesityl, SMe

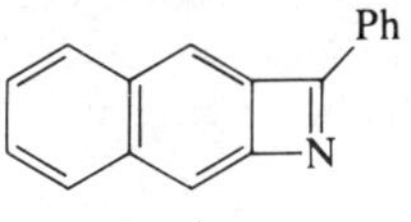

(**306**)

Unlike the situation with unfused azetes, there is considerable support from reactivity studies for the existence of benzazetes which can, therefore, be regarded as firmly established.

5.09.5.2 Reactivity

5.09.5.2.1 Unfused azetes

No unambiguous reactions of unfused azetes have been reported; rather, the existence of transient azetes has been inferred from the observation of products which might reasonably be attributed to their further reaction. Separate discussion of reactivity and preparation is, therefore, inappropriate, and apart from a few general comments is deferred to Section 5.09.5.3.

Extrapolation from the known reactivity of cyclobutadiene would suggest that azetes should be highly reactive towards dimerization and as dienes and dienophiles in cycloaddition reactions and the presence of a polar C=N should impart additional reactivity towards attack by nucleophiles. Isolation of formal dimers of azetes has been claimed as evidence for the intermediacy of such species, but no clear reports of their interception in intermolecular cycloaddition reactions or by nucleophiles have yet appeared.

One, possibly general, reaction of azetes which emerges is fragmentation to nitrile and acetylene. There is evidence that this occurs spontaneously even in an argon matrix, although under these particular conditions it is not clear whether the fragmentation is a thermal or photochemical process. Such a cycloreversion is thermally forbidden and significantly cyclobutadiene shows no comparable thermal fragmentation, and is only converted into acetylene on prolonged irradiation.

5.09.5.2.2 Benzazetes

The evidence for the existence of benzazetes is much more firmly based than that for azetes. 2-Arylbenzazetes can be isolated at −80 °C and typically are stable up to *ca.* −40 °C. Thus they can be easily handled and their reactions have been extensively studied ⟨75JCS(P1)45⟩.

In many ways the chemistry of benzazetes parallels that of benzocyclobutadiene, for example their ready dimerization and reactivity as a 2π but not a 4π component in cycloadditions. The presence of the nitrogen atoms also confers the reactivity expected of a highly strained imine and susceptibility towards addition of nucleophiles is their other dominant feature. Not surprisingly, the reactivity of the imine function is lowered by conjugative electron release. Thus *p*-anisylbenzazete (**305**; R = *p*-$MeOC_6H_4$) survives in solution at room temperature and is markedly less reactive than 2-phenylbenzazete as a dienophile in its reaction with cyclopentadiene. The red color of 2-*p*-dimethylaminobenzazete (**305**; R = *p*-$Me_2NC_6H_4$) and 2-mesitylbenzazete persists at room temperature for several hours and trapping of these molecules is very inefficient compared with that of the 2-phenyl derivative.

(i) Thermal reactions

In contrast to unfused azetes, benzazetes do not show any marked tendency to undergo fragmentation. Such cycloreversion would, however, lead to a very unstable alkyne (benzyne) or to loss of aromatic stability if cleavage were to proceed to give the cyanoacetylene (**307**). It may be significant that in the flash pyrolytic generation of benzazetes, too high an oven temperature gives increasing amounts of benzyne, but there is no evidence to suggest that this involves fragmentation of a benzazete rather than direct fragmentation of the benzotriazine (*cf.* Section 5.09.5.3.2).

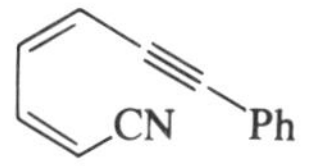

(**307**)

In solution or in the condensed phase, on warming from −80 °C to *ca.* −40 °C, 2-phenylbenzazete undergoes dimerization to give the angular dimer (**308**). This dimerization is analogous to that of benzocyclobutadiene and appears to involve a Diels–Alder reaction, one molecule acting as diene the other as dienophile. The regioselectivity is that expected; 2-anisylbenzazete dimerizes similarly, but more slowly. In the presence of Lewis acid catalysts such as aluminum chloride an alternative mode of dimerization occurs to give the diazocine (**309**).

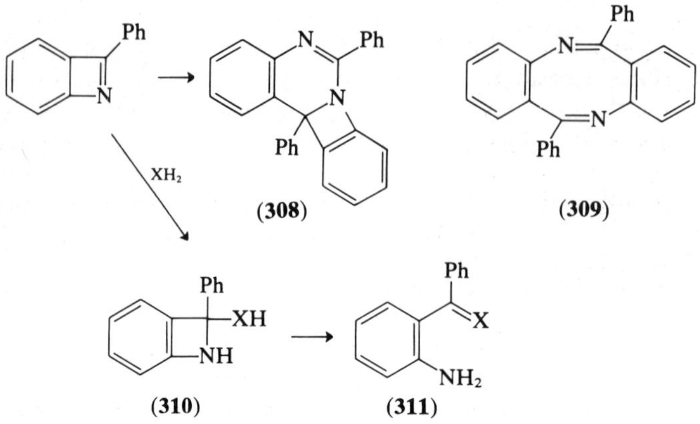

(**308**) (**309**)

(**310**) (**311**)

(ii) Towards nucleophiles

Hydrazine and phenylhydrazine react rapidly with 2-arylbenzazetes to give the o-aminohydrazones (**311**; $X=NNH_2$, NNHPh). Presumably initial addition gives the benzazetidines (**310**), which undergo rapid ring cleavage and hydrogen shift. Hydration also occurs to give the aminoketones (**311**; X = O), but only under the influence of acid catalysis. Addition of other nucleophiles such as Grignard and alkyllithium reagents and attempted reduction with LAH gave no characterizable products.

(iii) Cycloaddition

Like benzocyclobutadiene, benzazetes are reactive dienophiles, but fail to react as dienes except in the special case of dimerization. Thus 2-phenylbenzazete is inert to cyclopentene, but-2-yne, *N*-phenylmaleimide and *N,N*-dimethylaminopropyne.

(*a*) *Dienes.* Cycloaddition of 2-arylbenzazetes occurs with the reactive dienes cyclopentadiene, diphenylisobenzofuran, tetraphenylcyclopentadienone and dipyridyltetrazine

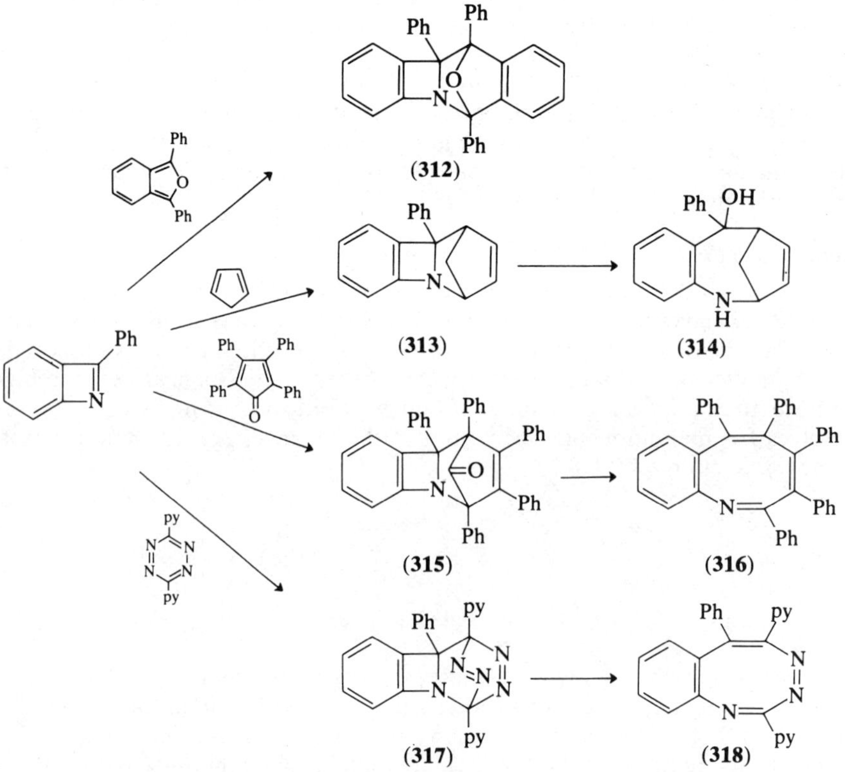

(**312**) (**313**) (**314**) (**315**) (**316**) (**317**) (**318**)

Scheme 15

(Scheme 15). In all cases the primary cycloadducts are benzazetidines, but the only one which has been isolated is the isofuran adduct (**312**). In all other cases the strain in the four-membered ring leads to further reaction. In the case of the cyclopentadiene adduct (**313**) the azetine ring is cleaved by hydration to give the amino alcohol (**314**), and in the cyclopentadienone (**315**) and tetrazine adducts (**317**) opening of the four-membered ring occurs in concert with extrusion of carbon monoxide and nitrogen to give the benzazocine (**316**) or triazocine (**318**) respectively. Even diphenylketene acts formally as a diene in its reaction with 2-phenylbenzazete, although there is evidence that the reaction proceeds through the zwitterion (**319**). Aromatization of the initial adduct (**320**) in ether occurs by a hydrogen shift to carbon to give the benzazetidine (**321**), but in dichloromethane diphenylanthracene (**322**) is formed, most likely *via* hydrogen shift to nitrogen followed by ring closure and extrusion of HNCO ⟨76CC125⟩.

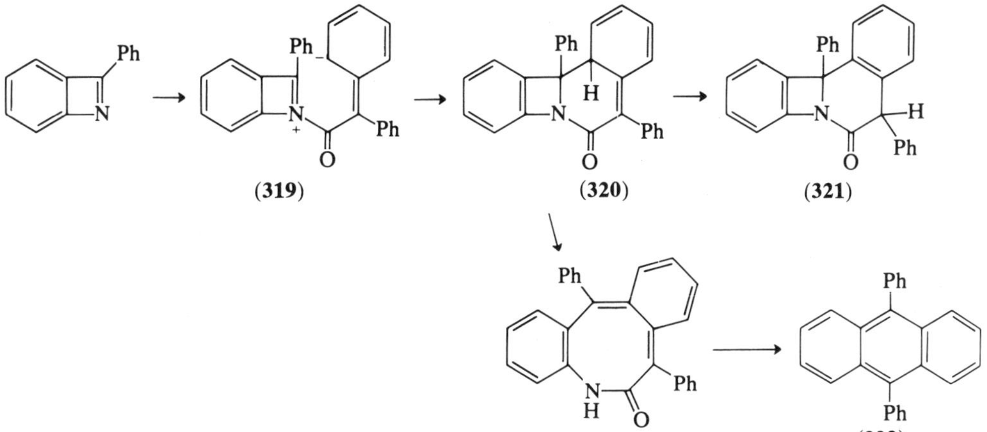

(**319**) (**320**) (**321**)

(**322**)

(*b*) *1,3-Dipoles.* Unstable benzazetidines have also been isolated from the addition of 1,3-dipolar systems to benzazetes. Thus labile adducts (**323**; X = O) are obtained from nitrile oxides ⟨75CC740⟩. On heating or attempted chromatography these rearrange to benzoxadiazepines (**325**; X = O) presumably as shown, the key step being cleavage of the benzazetidine ring to give a transient azaxylylene (**324**) (see Section 5.09.4.2.3). A synthetically useful extension of this reaction is the formation of otherwise inaccessible benzo-1,3,5-triazepines (**325**; X = NAr) using nitrileimines. In this case the benzazetidines (**323**; X = NAr) were not isolable, rearrangement occurring spontaneously. The adduct (**326**) from

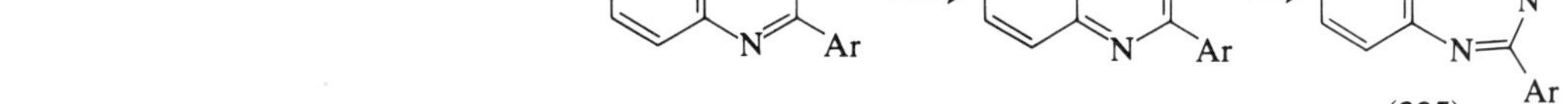

$ArC{\equiv}\overset{+}{N}{-}\overset{-}{X}$

(**323**) (**324**) (**325**)

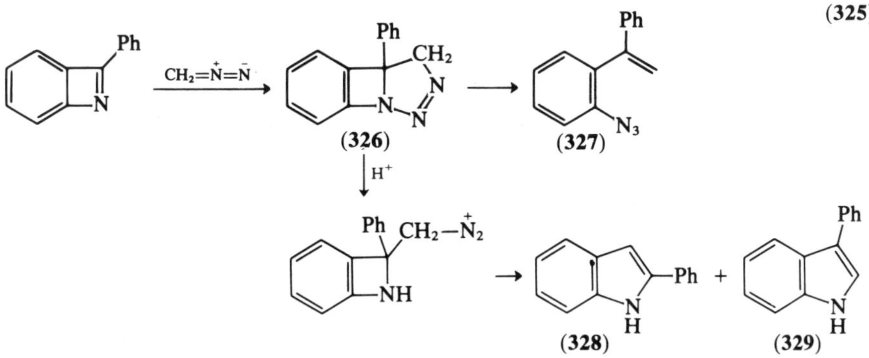

$CH_2{=}\overset{+}{N}{=}\overset{-}{N}$

(**326**) (**327**)

H^+

(**328**) (**329**)

diazomethane undergoes thermal cycloreversion to give the azidostyrene (**327**), but in the presence of acidic catalysts loss of nitrogen and formation of indoles (**328**) and (**329**) occurs ⟨78CC396⟩.

With munchnones, the initial adducts (**330**) spontaneously lose carbon dioxide with ring cleavage to give the benzo-1,3-diazepines (**331**) ⟨81TH50901⟩.

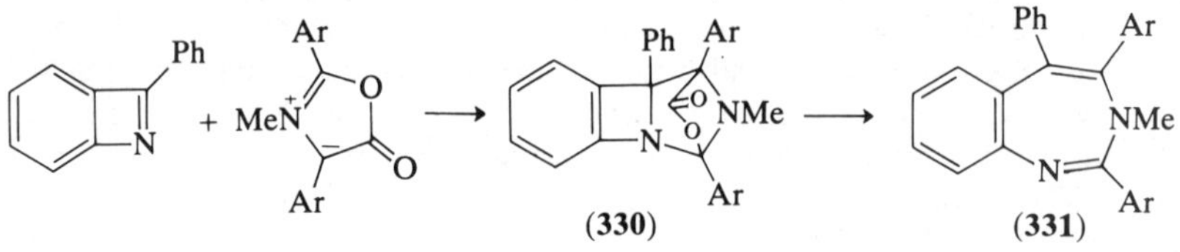

(**330**) (**331**)

5.09.5.3 Synthesis

5.09.5.3.1 Unfused azetes

Several approaches to unfused azetes have appeared, the most widely used being extrusion of small molecules from heterocyclic precursors.

Perhaps the most firmly based report for the formation of an azete involves flash pyrolysis of tris(dimethylamino)triazine (**303**). This gave a red pyrolysate believed to contain the highly stabilized azete (**304**) on the basis of spectroscopic data. The putative azete decomposed only slowly at room temperature, but all attempts to trap it failed ⟨73AG(E)847⟩. Flash pyrolysis of other 1,2,3-triazines gives only acetylenes and nitriles and it is not possible to tell whether these are formed by direct $_{\sigma}2+_{\sigma}2+_{\sigma}2$ fragmentation of the triazine or by prior extrusion of nitrogen and collapse to an azete ⟨81JCR(S)162⟩.

Oxazinones have also been considered as precursors to azetes. Irradiation of (**332**) at −78 °C and 270 nm gives the bicyclic lactone (**333**), which on further irradiation in an argon matrix loses carbon dioxide to give the products (**334**)–(**337**). The formation of all four fragments was interpreted in terms of the formation of a transient azete which undergoes rapid thermal or photochemical cycloreversion even at this extremely low temperature ⟨80LA798⟩. Similar results were also obtained in the flash pyrolysis of oxazinone (**338**); no azete could be isolated but the distribution of fragments indicated the possible intervention of such a species ⟨81JCR(S)162, 81JCS(P1)2878⟩.

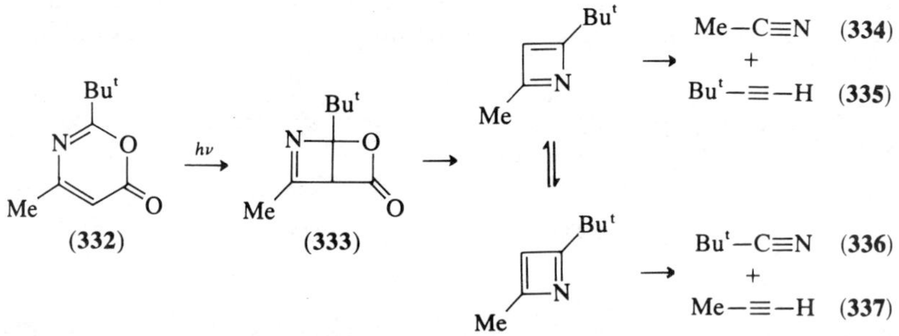

(**332**) (**333**)

Me—C≡N (**334**) + Bu^t—≡—H (**335**)

Bu^t—C≡N (**336**) + Me—≡—H (**337**)

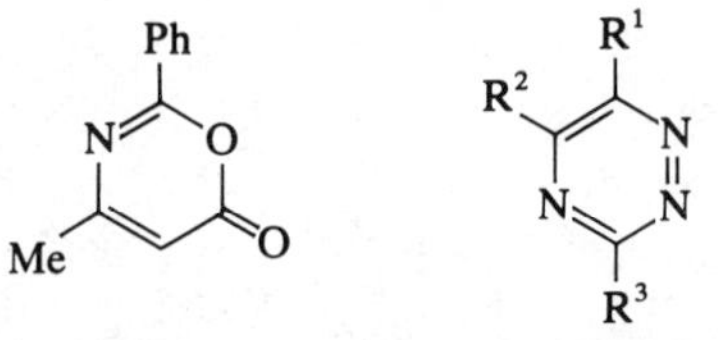

(**338**) (**339**)

Thermal fragmentation of 1,2,4-triazines (**339**) also occurs to give nitriles, nitrogen and acetylenes, but in this case the product distribution does not indicate the intervention of transient azete intermediates ⟨81JCR(S)162⟩. On the other hand, their involvement has been suggested in the photofragmentation of 1,2,4-triazines ⟨80LA798⟩. Trichloroazete has also been invoked as an intermediate to account for the products of flash pyrolysis of trichloro-1,2,4-triazine, which gave perchloroacrylonitrile rather than dichloroacetylene and cyanogen chloride ⟨79JCS(P1)1978⟩.

Photolysis of perfluoro-3,5-diisopropylpyridazine (**340**) in the gas phase gives products (**343**)–(**345**), which are formally dimers of azete (**342**) formed by loss of perfluoroisobutyronitrile from the valence isomer (**341**). However, the azete was not detected directly ⟨76CC1005⟩. Photolysis of the triazine (**346**) and the triazafulvenes (**347**) and (**348**) gave the same distribution of products and triphenylazete was suggested as a common intermediate ⟨73JOC176⟩. Several other potential routes to azetes are worth a brief mention, although in the event they failed. These include attempted cycloreversion in the fused azetine (**213**) ⟨71JOC435⟩ and elimination of sulfinic acid from the sulfonylazetine (**237**; Z = Ts) ⟨77CC806⟩, and of hydrogen chloride from the azetine (**349**) ⟨73JA2982⟩.

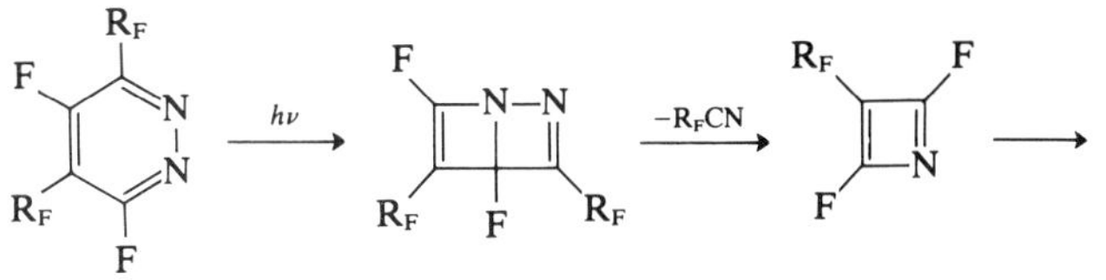

(**340**) (**341**) (**342**)

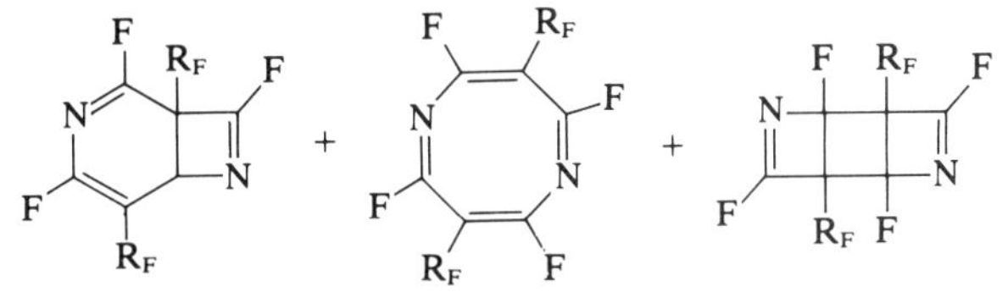

(**343**) (**344**) (**345**)

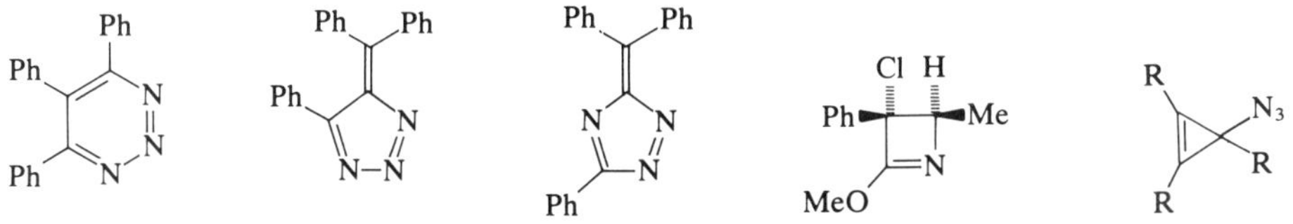

(**346**) (**347**) (**348**) (**349**) (**350**)

Cyclopropenyl azides (**350**), the obvious precursors to cyclopropenylnitrene and hence possibly azetes by ring expansion of the latter, give 1,2,3-triazines on heating or irradiation ⟨78HC(33)5⟩.

A number of other suggestions for formation of azetes seem most unlikely or have been definitely refuted ⟨B-50MI50900, 76TL3931, 78TL1543⟩.

5.09.5.3.2 Benzazetes

The only available route to benzazetes involves extrusion of nitrogen from 1,2,3-benzotriazines. Sublimation of 4-phenylbenzotriazine (**351**) (10^{-3} Torr) through a hot oven (450 °C) and trapping of the pyrolysate at −80 °C gives 2-phenylbenzazete in approximately 60% yield ⟨75JCS(P1)45⟩. A higher oven temperature leads to complete fragmentation of the triazine to give benzonitrile and benzyne, while at lower temperatures the triazine is unchanged.

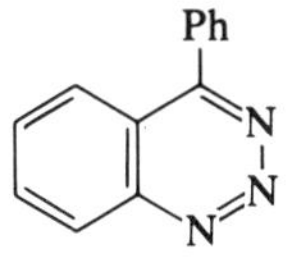

(**351**)

Other 2-arylbenzazetes have been produced by this method. The methylthiobenzazete (**305**; R = SMe) can also be obtained in low yield, but the method fails for 2-phenylthiobenzazete ⟨81TH50900⟩. Although 4-alkylbenzotriazines are available, these cannot be converted to alkylbenzazetes with the possible exception of the *t*-butyl derivative ⟨76TL4647⟩.

The benzazetes produced by this method are of insufficient stability for purification from the other components in the pyrolysate. However, they can be washed from the cold receiver using solvents cooled to below −40 °C and their reactions studied.

2-Arylbenzazetes can also be produced by irradiation of 4-arylbenzotriazines. Irradiation at room temperature leads only to good yields of the benzazete dimer, but at −80 °C solutions containing the benzazete can be obtained. As a practically useful procedure, however, the photochemical route suffers from the disadvantage that benzazete itself, many useful trapping agents, and the benzazete derived products are photolabile ⟨76CC411⟩.

5.09.5.4 Applications

The benzodiazepine fused azetes (**353**; R = Et, Pr) have been claimed as useful tranquilizers and sedatives ⟨76BRP1448895⟩. However, their stability and formation from benzodiazepines (**352**) by reaction with an aldehyde and base is hard to reconcile with the proposed structure.

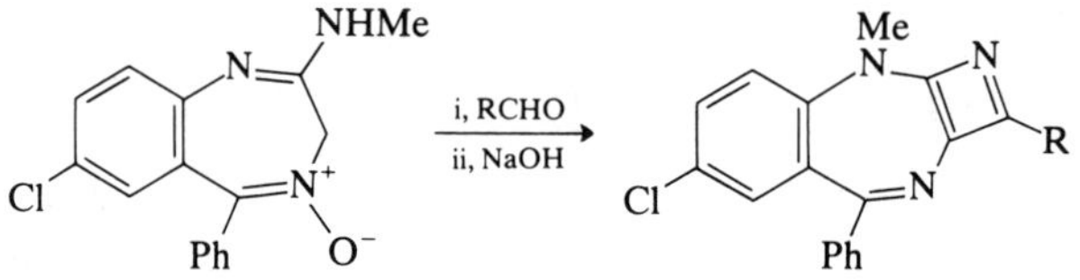

(**352**) (**353**)

5.10

Cephalosporins

K. G. HOLDEN

Smith Kline & French Laboratories, Philadelphia

5.10.1 INTRODUCTION

Cephalosporins belong to a growing class of β-lactam antibiotics that, starting with penicillins in the 1940s, has proved to be the dominant class of drugs used for the treatment of bacterial diseases in man and animals. The discovery of cephalosporins began with the observation of antibiotic activity produced by culture filtrates of a *Cephalosporium sp.* isolated from seawater near a sewage outlet on Sardinia. One of the antimicrobial substances isolated from the filtrate, cephalosporin C (**1a**), proved to be a new β-lactam antibiotic structurally related to penicillins. Following the discovery of cephalosporin C, methods for its production and chemical modification have led to the partial synthesis of a large number of useful agents exemplified by cefazolin (**1b**). A closely related family of antibiotics, the cephamycins which bear a methoxy group on the β-lactam ring, is exemplified by cefoxitin (**1c**).

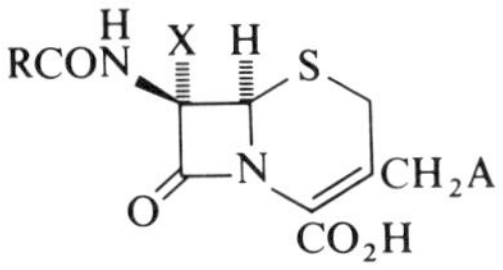

RCON(H)–[X, H]–S … N, CH_2A, CO_2H

	R	*X*	*A*
(**1a**)	$HO_2CCH(NH_2)CH_2CH_2CH_2-$	H	OAc
(**1b**)	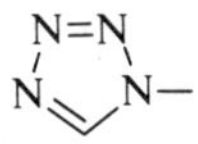 tetrazol-1-yl (N=N, N, N–)	H	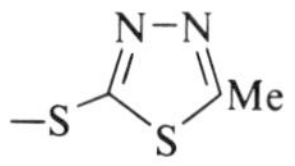 –S–(5-methyl-1,3,4-thiadiazol-2-yl) (N–N, S, Me)
(**1c**)	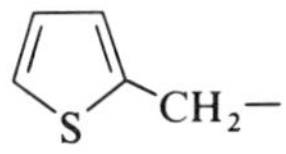 2-thienyl–CH_2-	OMe	$OCONH_2$

The nomenclature and numbering system for cephalosporins is based on the cepham ring system (**2**). Although this does not correspond to the systematic nomenclature adopted by *Chemical Abstracts* (8-oxo-5-thia-1-azabicyclo[4.2.0]octane), it is far more convenient and is used exclusively in the cephalosporin literature. For this chapter the carbon attached to C-3 has been designated C-3′ although this is not a uniform literature practice; it has also been assigned C-10 or has been left unnumbered. The acylamine group at C-7 and the C-3′ substitutent have been the major sites of cephalosporin modification. For convenience and consistency these will be abbreviated RCONH and A respectively, and, when not defined in the text or figures, can be assumed to be uninvolved in the chemistry under discussion. A variety of esters have been used to temporarily block the C-4 carboxyl, and although their use is discussed in the text (Section 5.10.3.4), they will be generalized as CO_2R' and will not be explicitly defined unless critical to the discussion. Many related bicyclic β-lactams have now been produced, largely through total synthesis, but, except for the very closely related isocephams (**3**), these will be covered separately (see Section 5.12).

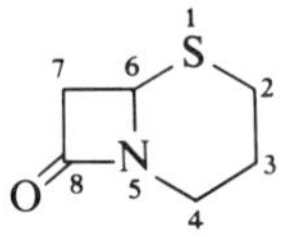

(**2**) 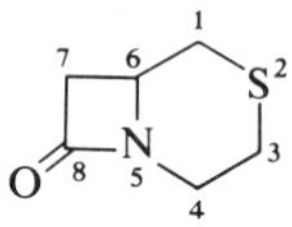 (**3**)

5.10.2 STRUCTURE

The structure and stereochemistry of cephalosporin C were determined largely by chemical degradation experiments carried out by Abraham and coworkers at Oxford ⟨B-72MI51000⟩ and were confirmed by X-ray crystallography ⟨61BJ(79)393⟩. The X-ray crystal structure of a number of β-lactams has now been determined, and the degree of reactivity and antibacterial activity has been correlated with a decrease in amide resonance due to strain imposed by the bicyclic system ⟨74JMC523⟩. The unusual reactivity of penicillins was originally explained in this way by Woodward ⟨B-49MI51000⟩. For normal amide resonance to occur the nitrogen atom and its three substituents must lie in a plane. While this is possible for monocyclic β-lactams, bicyclic β-lactams such as penicillins and cephalosporins (**4**) are seen to have a somewhat folded conformation. A convenient way to express the amount of deformation of the lactam nitrogen from planarity is by the sum of the bond angles about nitrogen. Thus, for a completely planar system the sum of the angles is 360°, while for strained bicyclic systems the sum is somewhat smaller. For example, in penicillins the sum of the angles is in the range of 337–343°, while in cephalosporins the corresponding value is slightly higher (345–351°) ⟨B-72MI51001⟩.

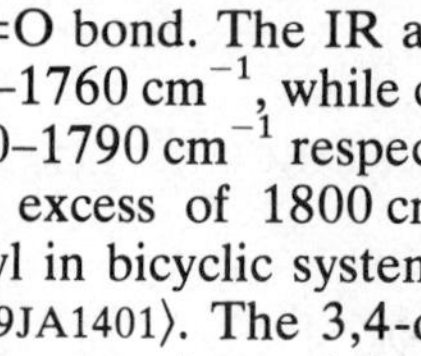

(**4**)

The deformation of the β-lactam nitrogen bonds from planarity and the concomitant loss of resonance are also reflected in an increased N—CO bond length and a slight diminution in the length of the C=O bond. The IR absorption frequencies of monocyclic β-lactams are in the region of 1730–1760 cm^{-1}, while cephalosporins and penicillins exhibit bands at 1782–1792 cm^{-1} and 1770–1790 cm^{-1} respectively. The corresponding sulfoxides and sulfones have absorptions in excess of 1800 cm^{-1}. The increase in IR stretching frequency of the β-lactam carbonyl in bicyclic systems has been roughly correlated with reactivity and biological activity ⟨69JA1401⟩. The 3,4-double bond in cephalosporins is an important factor in distinguishing the chemistry of this ring system from that of penicillins. Conjugation with the carboxyl group at C-4 and the enamide nature of the system produce a chromophore at approximately 260 nm ($\varepsilon \approx 7000$) in the UV spectrum of cephalosporins. The sulfur of the dihydrothiazine ring, although insulated by a methylene group, and suppression of amide resonance in the fused β-lactam are thought to contribute to the observed bathochromic shift for cephalosporins compared to simple *N*-acyl-α,β-unsaturated amino acids. The presence of the 3,4-double bond is required for biological activity, and reduction or isomerization to the 2,3-position leads to compounds which are ineffective as antibacterial agents. This can be explained by decreased reactivity of the β-lactam carbonyl resulting from a reduction in both strain and electronic activation. The presence of the 3,4-double bond also allows relatively facile displacements to occur at the C-3 methylene group. Cephalosporin C (**1a**) has served as the source of a multitude of semisynthetic antibacterial agents because the acetoxy group is readily displaced by a variety of nucleophiles ⟨70MI51000⟩.

Other structural features of cephalosporins that appear to be required for good antibacterial activity are a free carboxyl group at C-4, *cis* fusion of the β-lactam ring (i.e. a β-orientation of the C-7 amide function) and the presence of a hydrogen on the C-7 amide nitrogen. The free carboxyl at C-4 which normally has a pK_a in the range of 4.7–6.7 appears to be required for recognition by the enzymes involved in bacterial cell wall synthesis that are the target of cephalosporins. The *cis* nature of the β-lactam protons at C-6 and C-7 is most easily distinguished by NMR spectroscopy because of their relatively large coupling constant (4–5 Hz) compared to the *trans* isomer (1.5–2 Hz). Another useful NMR resonance is the two-proton AB quartet (δ3.23–3.70 p.p.m.) for the C-2 protons of 3-cephems. This distinguishes them from their 2-cephem isomers which display a one-proton vinyl multiplet (δ 5.90–6.71 p.p.m.) for the C-2 proton and a broadened singlet (δ 4.57–4.95 p.p.m.) originating from the C-4 proton. Mass spectrometry of cephalosporins is usually performed on derivatives in which the polar carboxy group has been esterified. Under these conditions the major fragmentations of diagnostic value are those of the β-lactam ring to yield dihydrothiazine ions and cleavage of the C-3 substituent in either of two ways (**5**) ⟨B-72MI51002⟩.

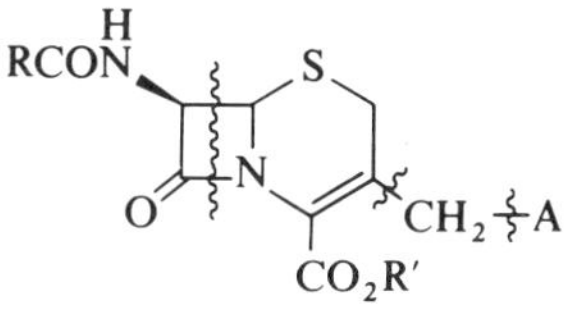

H
RCON S
N
O CH$_2$–A
CO$_2$R′

(**5**)

5.10.3 REACTIVITY

5.10.3.1 β-Lactam Cleavage

Cleavage of the β-lactam carbonyl–nitrogen bond is of cardinal significance, because the antibacterial activity of β-lactam antibiotics is dependent on this reaction. The ability of β-lactam antibiotics to act as acylating agents within the active site of bacterial enzymes critical for cell wall biosynthesis is, in turn, dependent on the reactivity of the β-lactam carbonyl. At the same time the carbonyl must not be so reactive that it is hydrolyzed by adventitious nucleophiles prior to its arrival at the target enzymes. Thus, the proper balance between reactivity and stability of the β-lactam ring must be struck to achieve maximum antibacterial potency. Furthermore, since the more reactive β-lactams are cleaved under relatively mild conditions, this feature must be carefully considered in any strategy for the total synthesis or chemical modification of penicillins and cephalosporins. Compared to penicillins, cephalosporins are relatively more stable to β-lactam cleavage. Thus, while penicillins are normally stable only near neutrality (pH 5–8) and react readily with alcohols, cephalosporins are hydrolyzed less readily, particularly under acid conditions, and have sufficient stability to permit their recrystallization from methanol. However, the β-lactam of cephalosporin does react readily with good nucleophiles. For example, cleavage by hydroxylamine to form a hydroxamic acid is the basis of a rapid assay for β-lactam antibiotics. Reaction is essentially complete within 3 min at pH 7 ⟨B-72MI51003⟩.

One of the major differences between penicillins and cephalosporins is the possibility for a concerted elimination of the C-3′ substituent in the case of cephalosporins (**6**→**7**). There is now considerable evidence to support the idea that an increase in the ability of the C-3 substituent to act as a leaving group results in an increased reactivity of the β-lactam carbonyl ⟨75JMC408⟩. Thus, both the hydrolysis rate of the β-lactam and antibacterial activity increase as the ability of the C-3′ substituent to act as a leaving group increases. However, because increased reactivity contributes to instability, a parabolic relationship between antibacterial activity of cephalosporins and β-lactam reactivity is observed ⟨80JA1812⟩. When the C-3′ substituent cannot function as a leaving group (**6**; A = H), antibacterial activity is usually somewhat reduced since β-lactam reactivity is still adequate. However, for cephamycins (**6**; X=OMe) this trend is much more pronounced, and in the case where A = H the derivatives are essentially inactive as antibacterial agents. When the fragment released as A is itself an antimicrobial agent, an antibiotic with a dual mechanism of action results ⟨76MI51000⟩.

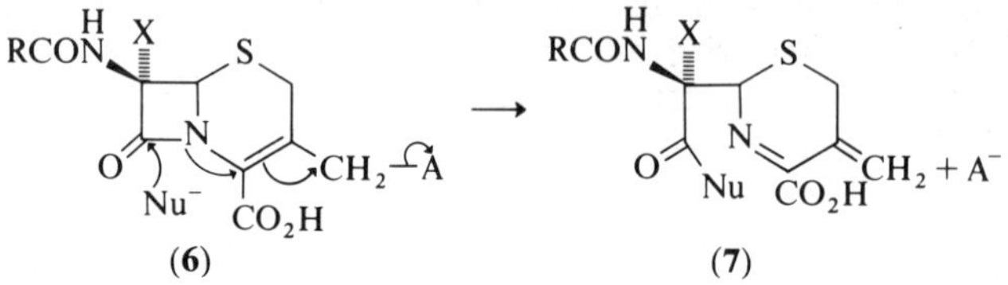

(**6**) → (**7**)

5.10.3.2 Modifications at C-3′ and C-3

Although the concerted displacement of the C-3′ substituent concomitant with β-lactam opening is an important feature of cephalosporin reactivity under physiological conditions, direct displacement at C-3′ has proved to be particularly useful for the synthesis of cephalosporin analogs. Of the thousands of cephalosporins produced by partial synthesis, all but a small fraction differ only in the nature of the 7-acyl and C-3′ substituents. The starting material for semisynthetic cephalosporins is cephalosporin C (**8a**) which bears an acetoxy substituent at C-3; therefore, it is the displacement of acetate that has received the most attention. The facility with which the acetoxy group is displaced by nucleophiles was first observed when a solution of cephalosporin C in aqueous pyridine acetate buffer was found to produce a new component with increased antimicrobial activity. This proved to be the corresponding 3′-pyridinium derivative (**8b**). This eventually led to the development of cephaloridine (**8c**) as a broad spectrum antibiotic. Displacements with other nitrogen nucleophiles such as azide and anilines have been reported but have not led to useful antibacterial agents. Another early observation was the displacement of acetate by sodium thiosulfate to give (**8d**). Since then a large number of derivatives have been prepared by displacement with alkyl, aryl and heterocyclic thiols as well as thioureas, xanthates, dithiocarbamates and sulfinates. In particular, those analogs derived from heterocyclic thiols have proved to be most important in terms of providing useful antibacterial agents.

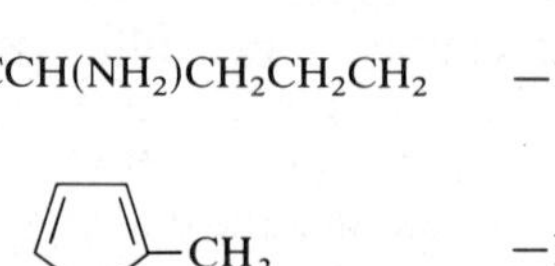

(**8**)

	R	*A*
(**a**)	$HO_2CCH(NH_2)CH_2CH_2CH_2$	OAc
(**b**)	$HO_2CCH(NH_2)CH_2CH_2CH_2$	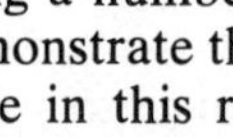 $-\overset{+}{N}$
(**c**)	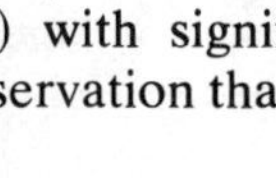 S $-CH_2$	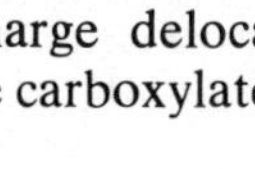 $-\overset{+}{N}$
(**d**)	$HO_2CCH(NH_2)CH_2CH_2CH_2$	$-S_2O_3Na$

Kinetic data on acetate displacement from C-3′ using a number of sulfur and nitrogen nucleophiles in aqueous solution at near neutral pH demonstrate that the reaction proceeds by an S_N1 mechanism ⟨B-72MI51004⟩. The intermediate in this reaction is depicted as a dipolar allylic carbonium ion (**9**) with significant charge delocalization. Of particular significance in this regard is the observation that the free carboxylate at C-4 is required since the corresponding esters do not undergo C-3′ displacement reactions. The carbonium ion character of the intermediate is demonstrated by its ability to participate in aromatic electrophilic substitution reactions with, for example, resorcinol to produce C-3′ arylated derivatives (**10**). While displacement with sulfur and nitrogen nucleophiles often proceeds readily, oxygen nucleophiles do not react, and direct formation of lactone (**12**) is not observed. However, since displacements are normally carried out in water or mixtures of water with polar organic solvents, substantial hydrolysis of the acetate ester to the corresponding alcohol (**11**) is often observed. The alcohol (**11**) readily lactonizes, and (**12**) often precipitates as an undesired byproduct during displacement reactions. Acid-catalyzed displacements in anhydrous organic solvents (*e.g.* acetonitrile, 1,2-dichloroethane) show promise as an alternative that minimizes this side reaction ⟨80MI51000⟩. Also, the 3′-halo derivatives (**8**; A = Cl, Br, I) are usually more readily displaced than the corresponding 3′-acetate ⟨B-72MI51004⟩.

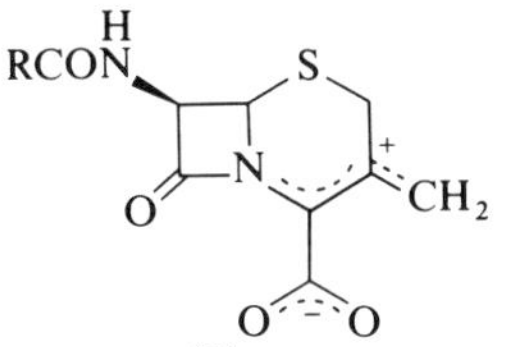

(**9**) 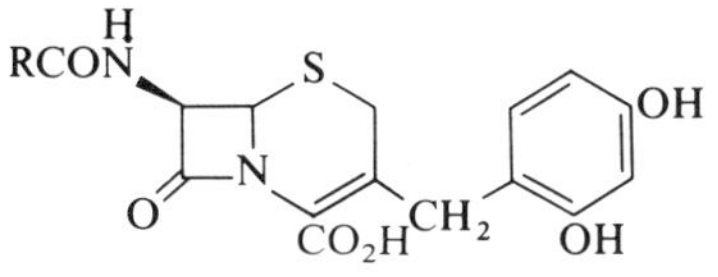 (**10**)

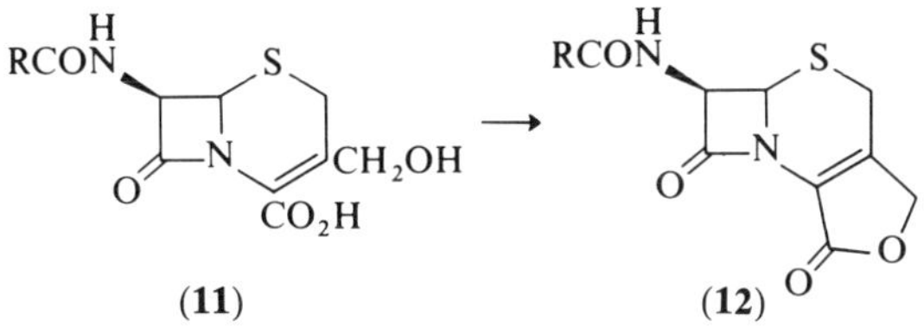

(**11**) → (**12**)

Only a few cephalosporins are efficiently absorbed when administered orally. These are usually derivatives of 3′-deacetoxycephalosporanic acid in which the 3-substituent is a methyl group (**13**). 3′-Deacetoxy analogs may be prepared by catalytic hydrogenolysis of the 3′-acetoxy group but are more economically derived from penicillins by way of a ring expansion reaction which will be described in Section 5.10.4.2. Since orally active cephalosporins are of considerable importance, new methods for producing 3′-deacetoxy derivatives were investigated by several research groups. From this work emerged 3-methylenecephams (**14**) which could be readily isomerized to the desired 3-deacetoxy analog (**13**). Perhaps more importantly, these intermediates provide access to cephalosporin analogs in which the C-3 methyl group is replaced by a heteroatom. 3-Methylenecephams (**14**) can be prepared by several methods: (1) reductive elimination of acetate from 3′-acetoxycephalosporins by chromium salts or electrolysis, (2) desulfurization of 3′-thiocephalosporins with Raney nickel, (3) dehydrohalogenation of 3-methyl-3-bromocepham 1-oxides and (4) rearrangement of penicillin sulfoxides ⟨B-82MI51000⟩. Careful ozonolysis of the 3-methylenecepham (**14**) as its

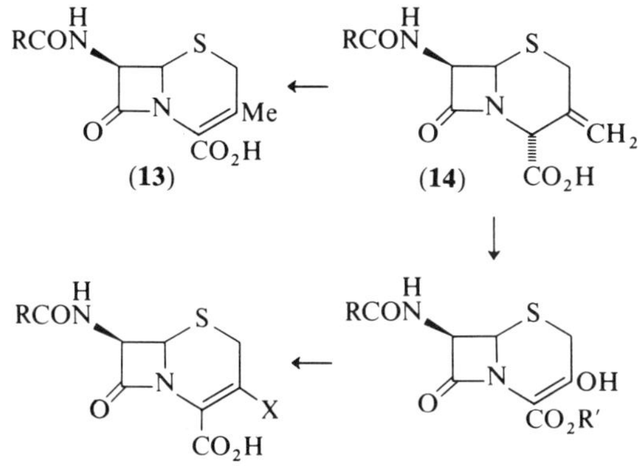

(**13**) ← (**14**)

(**16**) ← (**15**)

(**16**) **a**; X = OMe
b; X = Cl
c; X = H
d; X = CHO
e; X = CN, CO_2R, CH_2F, CHF_2, CF_3, $CH{=}CH_2$, SR, NH_2, NHCOR

p-nitrobenzyl or benzhydryl ester gives the 3-ketone which exists largely in the enol form (**15**). Treatment of (**15**) with, for example, diazomethane followed by liberation of the free acid gives the corresponding 3-methoxycepham (**16a**), while triphenyl phosphite dichloride gives the 3-chlorocepham (**16b**). These derivatives are of substantial interest as potent, orally effective cephalosporin antibiotics (**16a** and **b**; R = PhCH(NH_2) [CGP 3940 and cefaclor respectively]). Other modifications of the C-3 substituent include replacement by hydrogen (**16c**), which can be accomplished by reductive dehalogenation of (**16b**) or decarbonylation of (**16d**). A variety of other functionalities have also been introduced (**16e**).

5.10.3.3 Modifications at C-7

The 7-acylamino substituent of cephalosporins obtained by fermentation has exclusively the β configuration. However, some routes to the total synthesis of cephalosporins and their nuclear analogs have led to intermediates with the 7α configuration. Since the 7α epimers are inactive as antibacterial agents, methods were developed to invert the stereochemistry at this position. Under equilibrating conditions only approximately 15–30% of the β-epimer is obtained; however, kinetically controlled protonation of an anion at C-7 has given a β/α ratio as high as 2:1 ⟨74JOC437⟩. This method was used to invert the C-7 configuration of the synthetic intermediate (**17a**) used in the total synthesis of (±)-cephalothin ⟨73TL4653⟩.

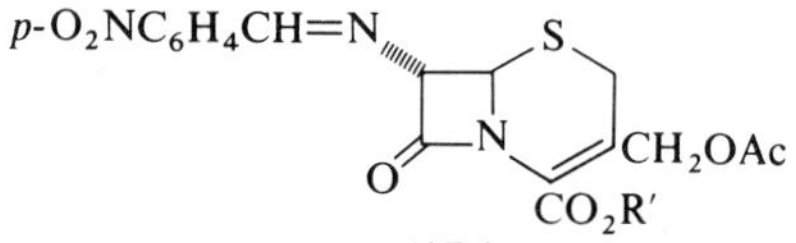

p-$O_2NC_6H_4CH{=}N$ ... S ... N ... CH_2OAc ... CO_2R'

(**17a**)

The increased resistance to degradation by bacterial β-lactamases and concomitant increase in antibacterial spectrum of cephamycins (*e.g.* **1c**) led to a search for chemical methods of introducing a 7α-methoxy substituent into the cephalosporin nucleus. At least six different methods have been developed, but many share the general feature of the addition of methanol from the less hindered α-face of an intermediate of the type (**17b**) to give the desired functionality and stereochemistry (**18**) ⟨B-80MI51001⟩. Using intermediates such as (**17a**) or the anion of Schiff's bases related to (**17a**), a variety of 7α-substituted cephalosporin derivatives including 7,7-spiro structures have been prepared, but their antibacterial activity has been disappointing.

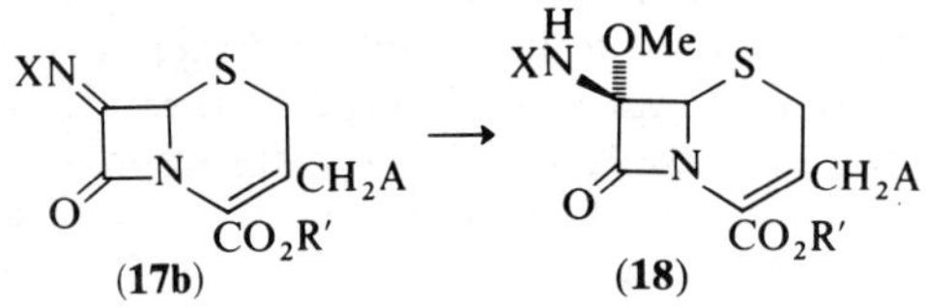

(**17b**) → (**18**)

X = *e.g.* RCO, RS, O=⟨2,6-R₂-cyclohexadienylidene⟩=CH–

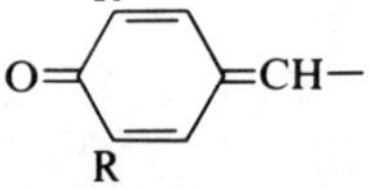

Modifications in which the nitrogen at C-7 has been replaced have also been investigated. One of the more useful intermediates to compounds of this type is the 7-diazo derivative (**19**) which provides access to a variety of analogs lacking a nitrogen at C-7. For example, (**19**), generated from the *N*-nitrosoamide, gives 7α-hydroxycephalosporins (**20**) on treatment with aqueous acid. Oxidation to the ketone (**21**) followed by reduction (borohydride) produces the β-epimer, which on esterification and deblocking of the C-4 carboxyl gives cephalosporin analogs (**22**) in which the usual C-7 amide function is replaced by an ester. The ester is very sensitive to hydrolysis, which may explain the low antibacterial potency

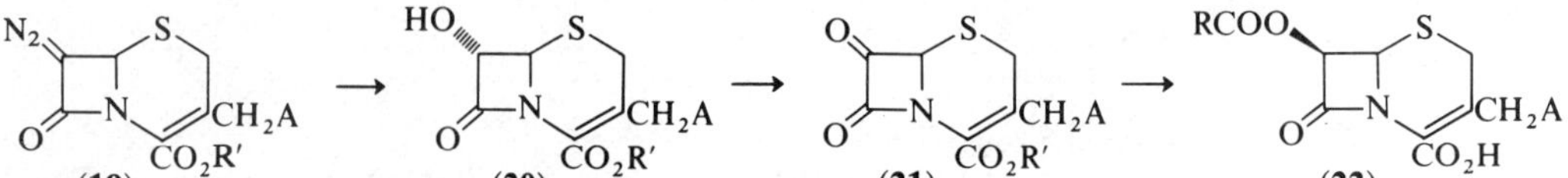

(**19**) → (**20**) → (**21**) → (**22**)

of such derivatives. Other substitutions of this type in which the C-7 nitrogen has been replaced by sulfur, halogen and carbon have also been investigated, but the products have not displayed useful antibacterial activity.

5.10.3.4 Dihydrothiazine Ring Modifications

Because the integrity of the dihydrothiazine ring and its C-4 carboxyl substituent is crucial to useful antimicrobial activity, reactions involving this part of the cephalosporin molecule are usually undesirable. The possibilities for sulfur oxidation or alkylation, substitution at C-2 which is adjacent to both sulfur and a double bond, double bond isomerization and addition reactions, and the influence of a free carboxylic acid must all be considered in designing reactions to selectively modify other cephalosporin functionalities.

Cephalosporin sulfoxides can be obtained by oxidants such as peracids, hydrogen peroxide and sodium periodate. The corresponding sulfones result from further oxidation of the sulfoxides. Peracid oxidation normally gives a mixture of (*S*)- and (*R*)-sulfoxides (**23** and **24**) with the *S* isomer predominating (9:1). The *S* isomers have considerably reduced antimicrobial activity. However, when (*R*)-sulfoxides (**24**) are selectively formed by preventing hydrogen bonding between the 7-NH and oxidant which normally directs it to the β face of the molecule, these isomers were found to have antimicrobial activity comparable to the parent sulfides ⟨B-77MI51000⟩. When 2-cephemes are oxidized, migration of the double bond occurs to give 3-cepheme sulfoxides (**23** and **24**). Reduction of sulfoxides to the corresponding sulfides can be achieved with a variety of reducing agents (iodide, thiosulfate, tin(II) and iron(II) ions) if the sulfoxides are first activated by reaction with an acid halide to form a sulfoxonium intermediate. Reagents such as imino chlorides, silanes and phosphorus trihalides serve as both activating and reducing agents.

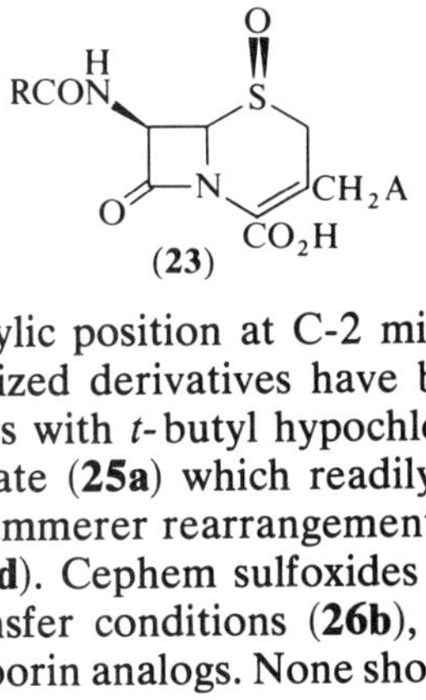

(**23**) 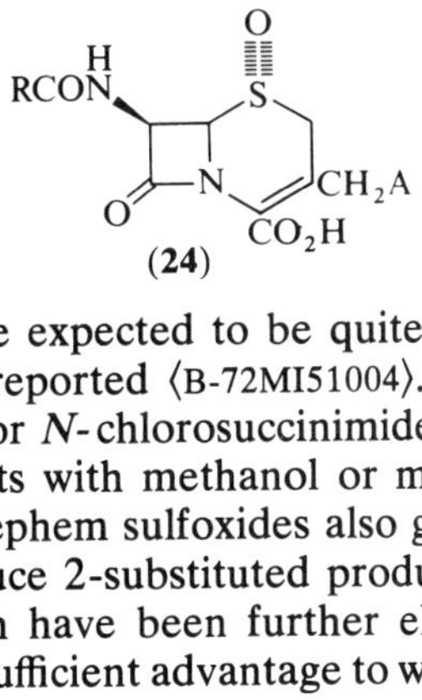 (**24**)

Although the allylic position at C-2 might be expected to be quite reactive, relatively few C-2 functionalized derivatives have been reported ⟨B-72MI51004⟩. Direct reaction of cephalosporin esters with *t*-butyl hypochlorite or *N*-chlorosuccinimide gives the unstable 2-chloro intermediate (**25a**) which readily reacts with methanol or methanethiol to give (**25b**) and (**25c**). Pummerer rearrangement of cephem sulfoxides also gives 2α-substituted derivatives (*e.g.* **25d**). Cephem sulfoxides produce 2-substituted products under Mannich (**26a**) or diazo-transfer conditions (**26b**), which have been further elaborated to give a variety of cephalosporin analogs. None showed sufficient advantage to warrant development as therapeutic agents.

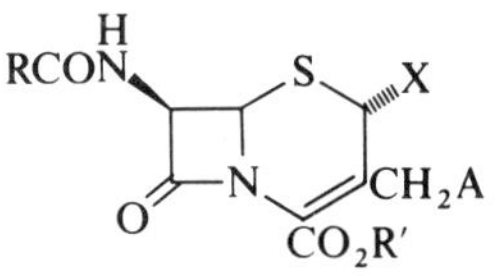

(**25**) **a**; X = Cl
b; X = OMe
c; X = SMe
d; X = OAc

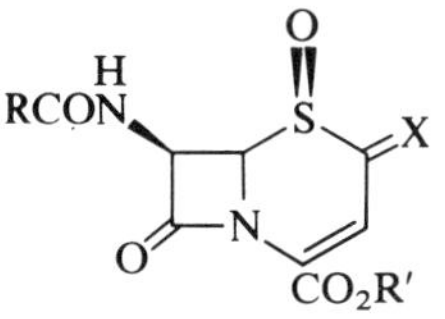

(**26**) **a**; X = CH_2
b; X = N_2

Isomerization of 3-cephems (**27**) to 2-cephems (**28**) takes place in the presence of organic bases (*e.g.* pyridine) and is most facile when the carboxyl is esterified. Normally an equilibrium mixture of 3:7 (3-cephem/2-cephem) is reached. Since the 2-cephem isomers are not active as antibacterial agents, the rearrangement proved to be an undesirable side reaction that complicated acylation of the C-7 amine under certain conditions. A method for converting such mixtures to the desired 3-cephem isomer involves oxidation with concomitant rearrangement to the 3-cephem sulfoxide followed by reduction. Additions to the 3-cephem double bond are not common. Diazomethane gives a pyrazoline, but electrophilic reagents such as bromine fail to add. Hydrogenation produces 3,4-dihydrocephalosporins which have no antibacterial activity.

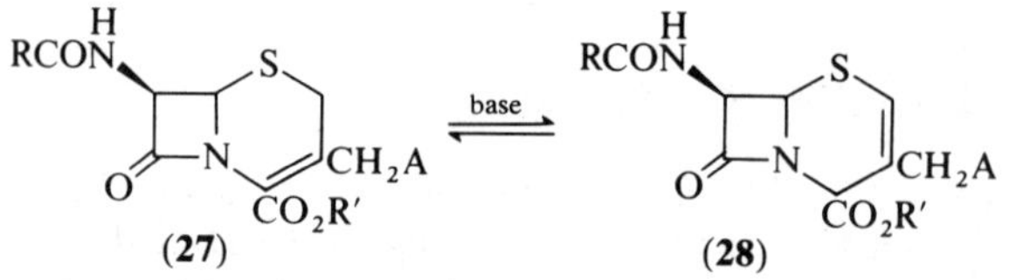

(27) ⇌ base (28)

For the chemical manipulation of other functionalities it is often desirable to temporarily block the C-4 carboxylic acid. Various esters have been used for this purpose, some of which have been adapted from peptide chemistry. The primary concerns in choosing the correct protecting group are (1) formation under conditions that do not isomerize the double bond, (2) stability to planned reaction conditions and (3) removal by reagents compatible with maintaining the integrity of the β-lactam ring. Some of the more useful esters have proved to be: *t*-butyl, trichloroethyl, benzhydryl, *p*-nitrobenzyl and trimethylsilyl. Other esters which are cleaved *in vivo* have been used in the design of prodrugs. More drastic modification of the C-4 carboxyl generally leads to compounds with little or no antimicrobial activity.

5.10.4 SYNTHESIS

5.10.4.1 Fermentation

All β-lactam antibiotics in commercial production are obtained by semisynthesis from fermentation products. The microorganisms used industrially are mutants of *Penicillium chrysogenum* and *Cephalosporium acremonium* for penicillins and cephalosporins respectively ⟨B-78MI51000⟩. The details of the biosynthetic pathways by which these fungi elaborate β-lactams are still being worked out, but the overall scheme for cephalosporins appears to involve a penicillin intermediate ⟨77T1545⟩. Accumulated evidence suggests that α-aminoadipic acid, cysteine and valine are first assembled into a tripeptide (**29**). Cyclization to (**30**) appears to proceed *via* formation of the β-lactam ring followed by the thiazolidine, although no intermediates in this sequence have been isolated.

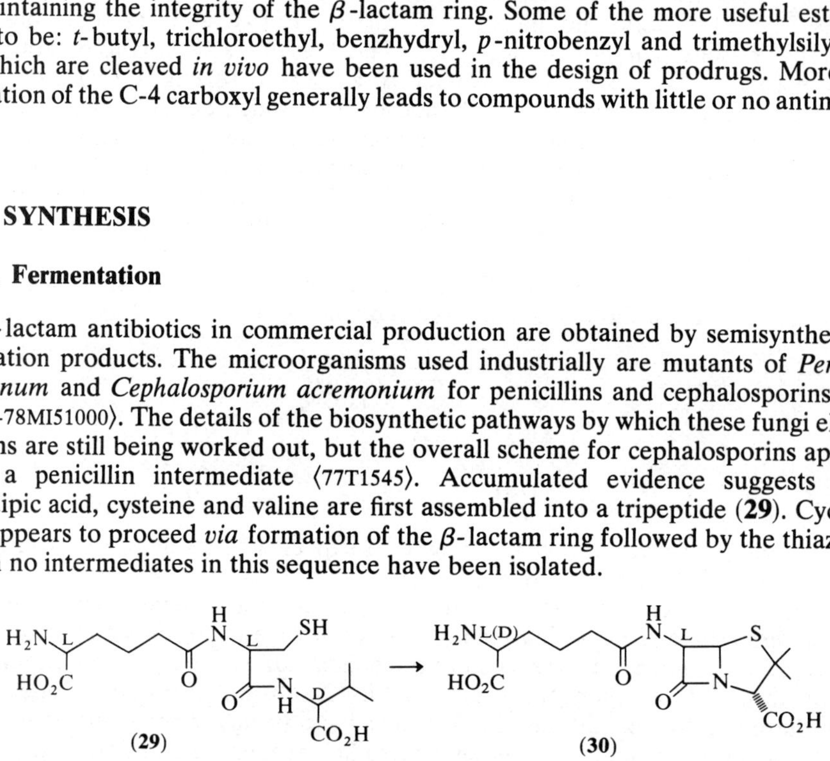

(**29**) → (**30**)

In organisms which produce cephalosporin and cephamycins, the configuration of the α-aminoadipyl side chain of (**30**) is D, while penicillin producers yield the L isomer. The exact point at which the configuration is inverted is unknown. Subsequent steps in cephalosporin biosynthesis are believed to involve ring expansion to deacetoxycephalosporin C (**31**), which may proceed by a mechanism analogous to the chemical pathway (see Section 5.10.4.2), followed by hydroxylation and acetylation at C-3′ to produce cephalosporin C (**32**).

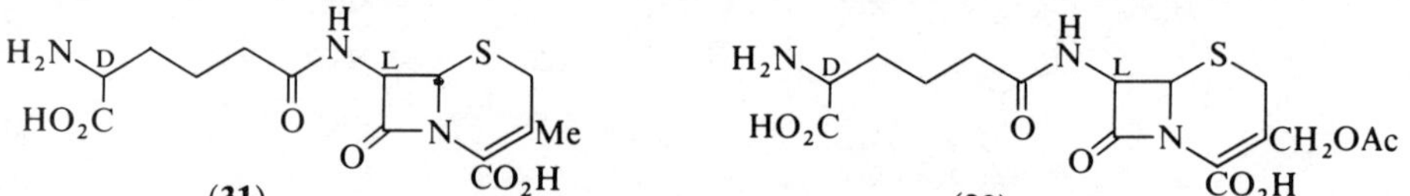

(**31**) (**32**)

Early attempts to produce cephalosporin analogs by varying the 7-acylamino substituent were frustrated because, in contrast to previous experience with penicillins, a good method for producing the necessary 7-amino compound (**33a**) could not be found. This problem was finally solved when it was discovered that diazotization of the α-aminoadipyl residue produces an iminolactone (**33b**) which can be hydrolyzed to the free amine in good yield. Subsequently an improved procedure was developed which involves silylation of the carboxyl groups followed by reaction with phosphorus pentachloride to yield iminochloride (**33c**) which can be hydrolyzed to (**33a**) in excellent yield ⟨B-72MI51005⟩. These methods do not work well for cephamycins, so the introduction of a 7α-methoxy substituent into the cephalosporin nucleus has been developed as an alternative (see Section 5.10.3.3).

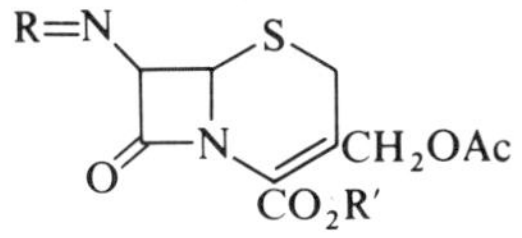

R=N ... S ... N ... CH_2OAc ... O ... CO_2R'

(**33**) **a**; $R = H_2$, $R' = H$

b; R = [HO_2C–tetrahydropyranylidene], $R' = H$

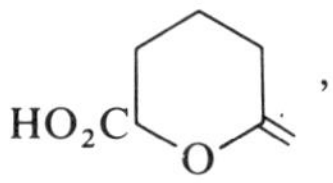

c; $R = Me_3SiO_2CCH(NH_2)(CH_2)_3C(Cl){=}$, $R' = SiMe_3$

Acylation of the 7-amino group of (**33a**) and related cephems having various C-3′ substituents has produced a vast array of cephalosporin analogs. General procedures for acylation utilize acid chlorides, mixed anhydrides, activated esters or the free acid with a coupling agent such as *N,N*-dicyclohexylcarbodiimide. Reactions are often run in aqueous solution with an organic cosolvent and an acid acceptor such as sodium bicarbonate or an organic base. When nonaqueous conditions are required the C-4 carboxyl is usually blocked as an ester which subsequently can be removed under mild conditions. Many of the more useful 7-acylamino groups contain reactive functionalities (amino, hydroxyl, carboxyl), and these must also be temporarily blocked during the acylation procedure. When the acyl group bears a chiral center α to the carboxyl, care must be taken to prevent racemization. Also, rearrangement to a 2-cephem must be considered as a side reaction, especially when the C-4 carboxyl is esterified (see Section 5.10.3.4).

5.10.4.2 Penicillin Rearrangement

Only a few cephalosporins are efficiently absorbed when administered orally. Most of these are derivatives of 3′-deacetoxycephalosporins (**39**). Since reductive removal of the acetoxy group from cephalosporins is not economically attractive (see Section 5.10.3.2), another route to (**39**) was sought. Morin and coworkers ⟨69JA1401⟩ found that penicillin (*S*)-sulfoxides (**34**) rearrange to a mixture of (**37**) and (**38**) on heating at reflux in acetic anhydride. The latter eliminates the elements of acetic acid on treatment with triethylamine to give the desired 3′-deacetoxycephalosporin (**39**). Trapping experiments have confirmed that sulfenic anhydride (**35**) is an intermediate. This then forms episulfonium intermediate (**36**) which is attacked by acetate to give the observed products (**37**) and (**38**) with the expected regio- and stereo-chemistry. Because of the commercial importance of this reaction, it has been studied in detail ⟨B-72MI51006⟩ and improved conditions have been developed that allow the industrial production of 3′-deacetoxycephalosporins by this route.

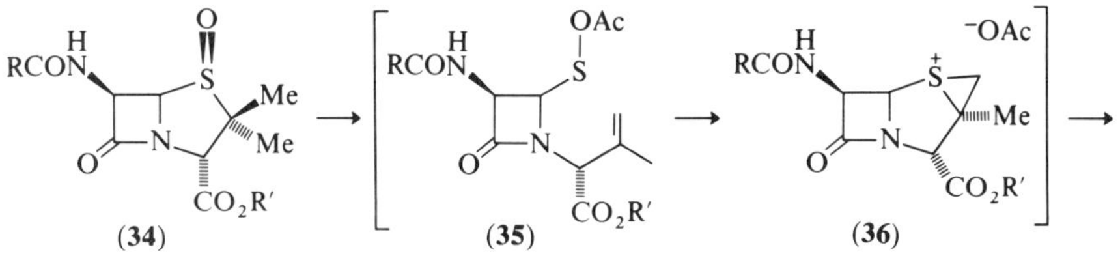

RCONH ... S=O ... Me ... Me ... CO_2R' (**34**) → [RCONH ... S–OAc ... CO_2R' (**35**) → RCONH ... $\overset{+}{S}$ ... Me ... ^-OAc ... CO_2R' (**36**)] →

RCONH ... S ... CH_2OAc ... Me ... CO_2R' (**37**) + RCONH ... S ... OAc ... Me ... CO_2R' (**38**) → RCONH ... S ... Me ... CO_2R' (**39**)

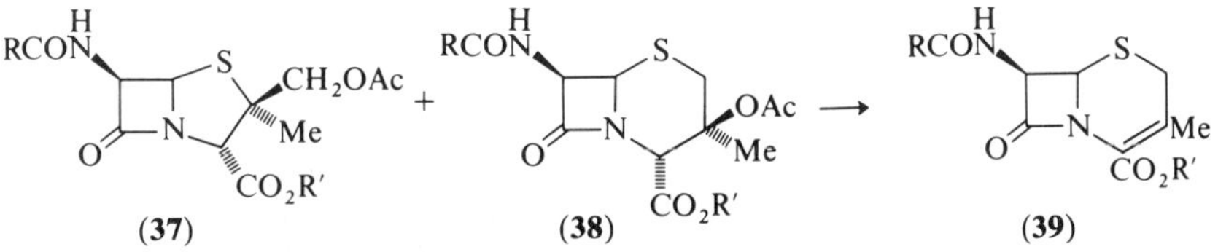

A related rearrangement developed by Kukolja and coworkers ⟨B-77MI51001⟩ gives 3-methylenecepham (**43**) which is a key intermediate for the production of cefaclor (see

Section 5.10.3.2). Treatment of methyl 6-phthalimido penicillinate (R)-sulfoxide (**40**) with N-chlorosuccinimide in refluxing carbon tetrachloride gives an epimeric mixture of sulfinyl chlorides (**41**) which are ring closed to epimeric 3-methylenecepham sulfoxides (**42a**) using tin(IV) chloride. Reduction with phosphorus tribromide gives the desired methyl 7-phthalimido-3-methylenecepham 4-carboxylate (**42b**).

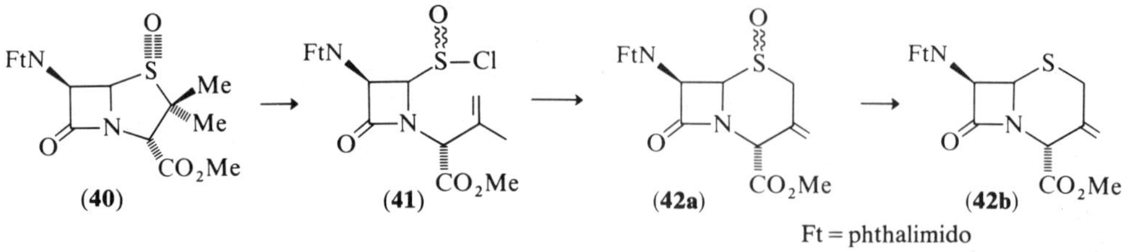

(**40**) → (**41**) → (**42a**) → (**42b**)

Ft = phthalimido

A very efficient 'one-pot' procedure for the production of 3-hydroxy-3-cephems (**45**) has been developed which gives the desired product in almost 80% overall yield from (**43a**) which is readily available from penicillin. The sequence of reactions is: (1) mesylation to give (**43b**), (2) formation of enamine (**43c**), (3) bromination to afford (**44**) and (4) hydrolysis/cyclization with hydrochloric acid in methanol to afford (**45**) which, in some cases, crystallizes directly from the reaction mixture ⟨B-82MI51000⟩.

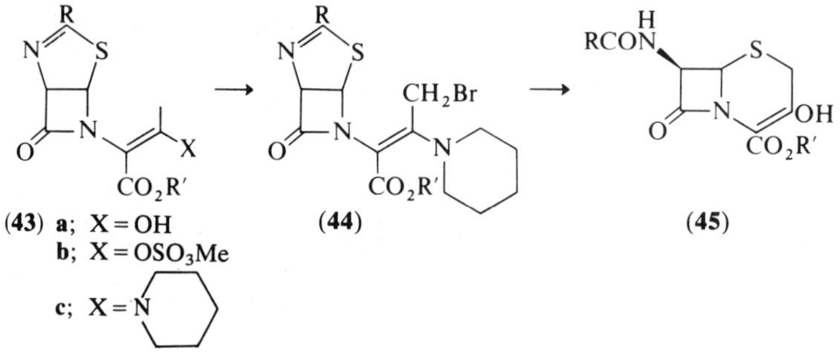

(**43**) **a**; X = OH
b; X = OSO_3Me
c; X = N(piperidino)

(**44**) → (**45**)

5.10.4.3 Total Synthesis

Sheehan and coworkers' total synthesis of penicillin in 1962 was followed by the total synthesis of cephalosporin C in 1966 by Woodward and coworkers ⟨B-72MI51007⟩. These syntheses were based on two different general approaches: (1) formation of the β-lactam ring late in the synthetic scheme (penicillin) and (2) formation of the β-lactam early in the sequence (cephalosporin). Since it is the bicyclic nature of the β-lactam that imparts reactivity, both approaches have the advantage of delaying introduction of this source of instability until much of the necessary manipulation has been completed. However, early formation of the β-lactam ring has two added advantages: (1) the necessary *cis* configuration of the incipient ring juncture and acylamine function can be established early in the synthesis and (2) the monocyclic β-lactam can serve as a common intermediate for the synthesis of nuclear analogs of penicillins and cephalosporins ⟨B-82MI51001⟩.

Early attempts to apply the Sheehan penicillin synthetic strategy to the total synthesis of cephalosporins were not particularly successful. Although the key step, formation of the β-lactam CO—N bond, could be carried out efficiently (**46**→**47**), subsequent conversion of the lactone to a free C-4 carboxyl could only be accomplished in poor yield ⟨B-72MI51007⟩.

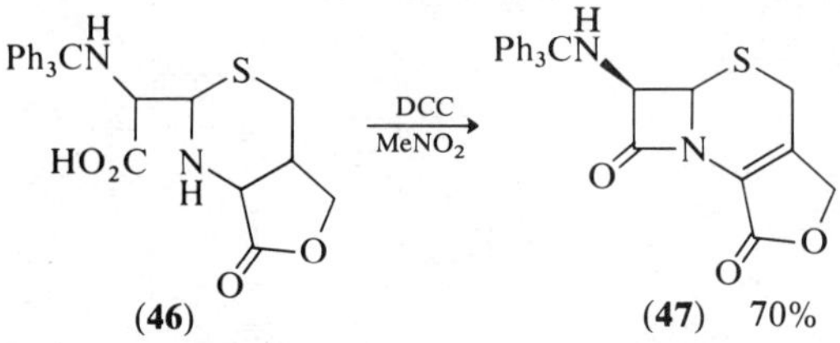

(**46**) —DCC, $MeNO_2$→ (**47**) 70%

Woodward's total synthesis of cephalosporin C begins with L-cysteine (**48**) which establishes the chiral center at C-7. The *cis* geometry at C-6,7 is achieved in intermediate (**49**) which is cyclized to (**50**) by treatment with triethylaluminum. The dihydrothiazine ring is constructed by Michael addition to the condensation product of trichloroethyl glyoxylate with malondialdehyde to give (**51**). Removal of the N- and S-protecting group (TFA) effects ring closure to (**52**) which is further elaborated to cephalosporin C ⟨66JA852⟩. In a related approach Kishi and coworkers devised a biomimetic synthesis which proceeds *via* (**53**) to (**54**) which cyclizes to (**55**) ⟨75JA5008, 5010⟩.

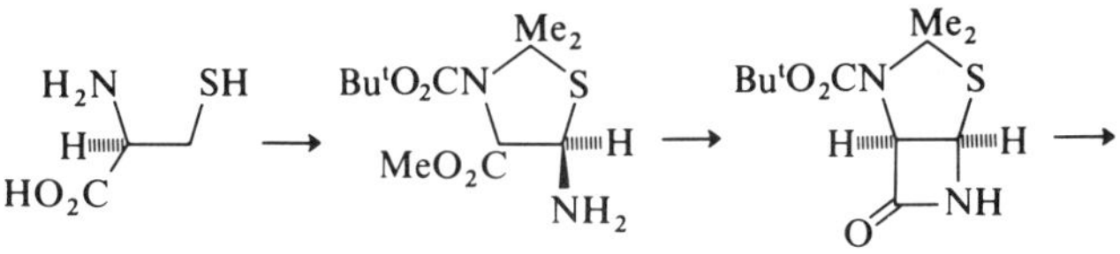

(**48**) (**49**) (**50**)

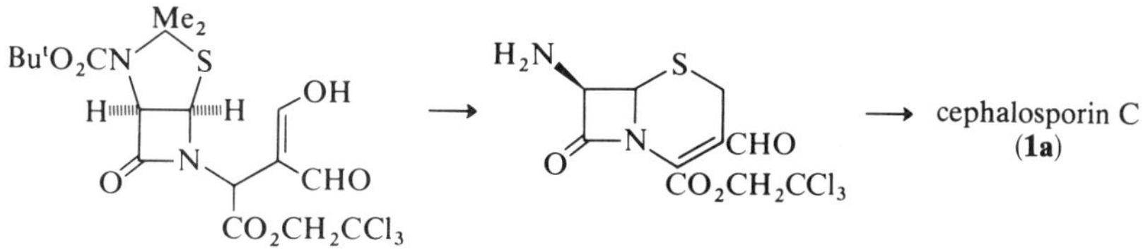

→ cephalosporin C (**1a**)

(**51**) (**52**)

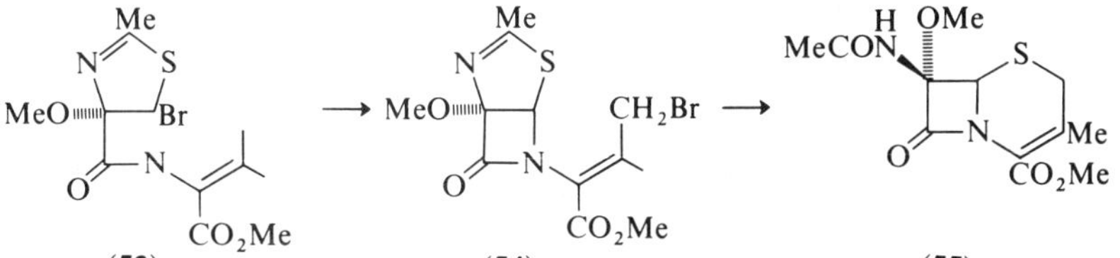

(**53**) (**54**) (**55**)

Scientists at Merck developed a cephalosporin synthesis based on the addition of azidoacetyl chloride to 1,3-thiazines (**56**). Although this gives the incorrect 7α-epimer (**57**), it could be equilibrated to a mixture of 7-amino epimers (see Section 5.10.3.3) from which the desired 7β-isomer could be separated and further elaborated to cephalosporins ⟨B-82MI51001⟩.

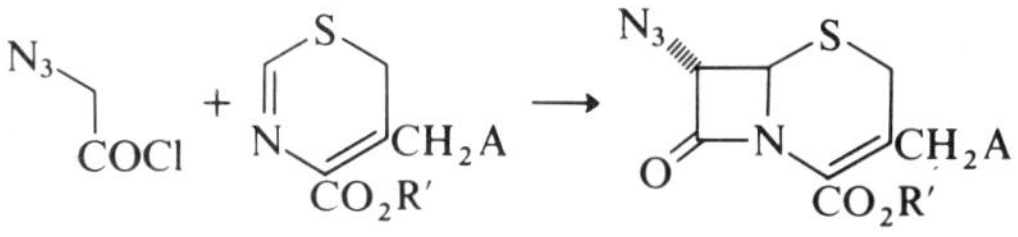

(**56**) (**57**)

At Smith Kline & French a general approach to cephalosporin and penicillin nuclear analogs was developed that utilizes a monocyclic β-lactam (**59**) with the correct *cis* stereochemistry as a key intermediate. This is prepared by reaction of the mixed anhydride of azidoacetic acid and trifluoroacetic acid with imine (**58**) followed by oxidative removal of the dimethoxybenzyl group. This product could be further elaborated to intermediate (**60**) which, on reaction with α-bromoketones, provides isocephalosporins (**61**). These nuclear analogs displayed antibacterial properties similar to cephalosporins ⟨B-79MI51000⟩.

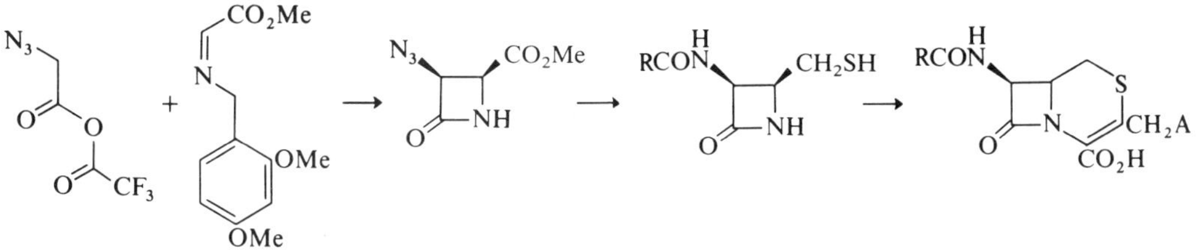

(**58**) (**59**) (**60**) (**61**)

In spite of the considerable progress in developing methods for total synthesis, this route to cephalosporins cannot compete with fermentation or penicillin rearrangement (see Sections 5.10.4.1 and 2) for the industrial production of cephalosporin antibiotics. While total synthesis does provide access to nuclear analogs not readily obtainable from fermentation products, none of the totally synthetic materials have displayed sufficient advantages to warrant their development as new drug products ⟨B-81MI51000⟩.

5.10.5 APPLICATIONS

5.10.5.1 Mechanism of Action

A major difference between mammalian and bacterial cells is that the latter are protected by an outer semirigid covering called a cell wall. The cell wall is a polymeric, cross-linked network composed of sugars and amino acids. Glycan strands are constructed of alternating units of *N*-acetylglucosamine and *N*-acetylmuramic acid linked $\beta 1 \rightarrow 4$. The carboxyl function of the muramic acid residues serves as a branch point on the glycan strands to permit cross-linking to parallel strands by short peptide chains. It is the final cross-linking step that is believed to be the major site of action of β-lactam antibiotics. This step is carried out by an enzyme called a transpeptidase which first cleaves D-alanine from a peptide chain with a D-alanyl-D-alanine terminus (**62**). The resulting acyl–enzyme intermediate (**63**) then reacts with an amino function of a neighboring peptidoglycan to form a cross-link through a new amide bond (**64**). The D-alanyl-D-alanine terminus (**62**) has been drawn to resemble the structure of β-lactam antibiotics because it is believed that they are recognized as the normal substrate by the transpeptidase enzyme. Note that the amide bond cleaved by the enzyme corresponds to the reactive one in β-lactam antibiotics. D-Alanine carboxypeptidase (see below) is functionally very similar to the transpeptidase except that the acyl–enzyme intermediate (**63**) reacts with water to produce a carboxylic acid instead of an amide. This enzyme, which is also present in bacterial membranes and is inactivated by β-lactam antibiotics, has been studied in more detail than the transpeptidase. For D-alanine carboxypeptidase the acyl–enzyme intermediate has been characterized as a serine ester, and it has been demonstrated that the same serine residue is acylated by β-lactam antibiotics ⟨80JBC(255)3964⟩. Trapping of these peptidases as stable acyl–enzyme intermediates inactivates both enzymes and, in the case of the transpeptidase, halts cross-linking of the peptidoglycan chains. The incomplete cell wall results in exposure of the fragile underlying cytoplasmic membrane to environmental insult with eventual lysis of the cell and cell death. The relatively low toxicity of β-lactam antibiotics presumably derives from the fact that mammalian cells have neither cell walls nor the enzymes required for their synthesis ⟨74MI51000⟩.

RCONH–D–C(=O)–NH–D–CO_2H → RCONH–D–C(=O)–Enz → RCONH–D–C(=O)–NHR′

(**62**) (**63**) (**64**)

It has now become clear that the mechanism of action of β-lactam antibiotics is more complex than originally proposed ⟨65PNA(54)1133⟩. Multiple binding sites for penicillins and cephalosporins have been detected in the membranes of all bacteria that have been examined, and several β-lactam-sensitive enzymes that apparently correspond to the binding sites have been identified ⟨78MI51001⟩. The observation that the concentration of a given β-lactam antibiotic required to compete with radiolabeled penicillin for irreversible binding sites parallels the concentration required for growth inhibition implies that the binding sites are synonymous with killing sites ⟨79MI51001⟩. Six distinct penicillin binding proteins have been identified in the inner membrane of *E. coli.* On electrophoresis in gels containing sodium dodecylsulfate they separate according to molecular weight and have been numbered in order of decreasing size (91 000→40 000). The two smallest proteins, 5 and 6, have been shown to be equivalent to D-alanine carboxypeptidase which appears as a doublet under these conditions of electrophoresis. Although sensitive to penicillin this enzyme is not believed to be a normal killing site for most β-lactam antibiotics. Protein 4 is also an unlikely candidate because mutants have been reported that lack this protein but grow normally. Protein 3 is believed to be involved in septum formation and cell division. Selective inhibition of this protein causes the normally rod-shaped cells to become filamentous. Protein 2 has been shown to be necessary for maintenance of cell shape. Mutants which lack this protein, or normal bacteria treated with an agent that selectively inhibits it, grow as osmotically stable spherical cells. Finally, protein 1, which appears as a doublet (1A and 1B), has been identified as the peptidoglycan transpeptidase, a major killing site for penicillin. Selective inhibition of this enzyme preferentially inhibits cell elongation and causes cell lysis.

Although most β-lactam antibiotics bind covalently to some or all of the same six proteins, there are decided differences among them in terms of their relative affinities. For example, cefoxitin (see Table 1 for structures) fails to bind to protein 2 while cephacetrile binds very slowly to proteins 5 and 6. Cephaloridine binds most avidly to protein 1, the transpeptidase, and inhibits cell elongation and causes lysis at its minimum inhibitory concentration. On the other hand, cephalexin binds preferentially to protein 3 and causes inhibition of cell division and filament formation ⟨75PNA2999, 77MI51002⟩.

Studies on the mechanism of action of β-lactam antibiotics have shed considerable light on how these agents kill bacteria. They also help explain qualitative differences between various agents and why there is a correlation between the reactivity of the β-lactam and antibacterial activity. However, it is also clear that reactivity is only one factor in determining how effectively a given β-lactam antibiotic will inactivate bacterial enzymes ⟨82BJ(203)223⟩.

5.10.5.2 Clinically Useful Cephalosporins

Of the thousands of cephalosporins synthesized as potential antibacterial agents, most only differ in the nature of the 3′- and 7-substituents. This is reflected in the structures of the clinically useful cephalosporins of which examples are illustrated in Table 1.

Table 1 Examples of Clinically Useful Cephalosporin Antibiotics

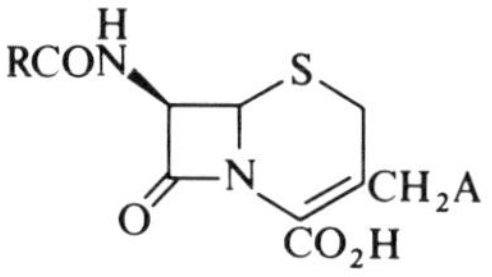

RCONH, S, N, O, CH_2A, CO_2H

Generic name	*R*	*A*
Cephalothin	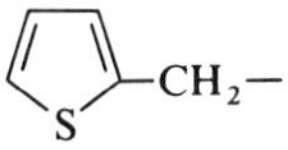 2-thienyl–CH_2–	OAc
Cephacetrile	$NCCH_2$–	OAc
Cephaloridine	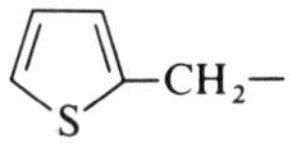 2-thienyl–CH_2–	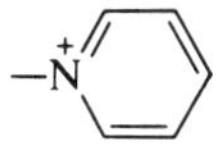 –$\overset{+}{N}$ (pyridinium)
Cefazolin	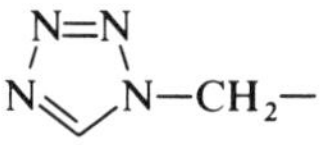 N=N, N–CH_2– (tetrazolyl)	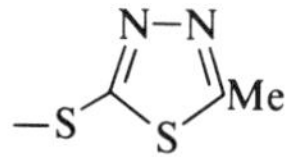 –S– (N–N, S, Me thiadiazolyl)
Cephamandole	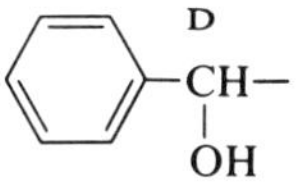 phenyl–CH(OH)– (D)	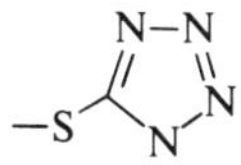 –S– (N–N, N, N tetrazolyl)
Cephalexin	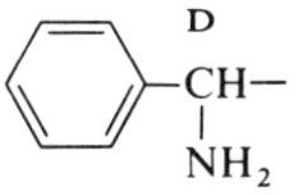 phenyl–CH(NH_2)– (D)	H
Cefoxitin	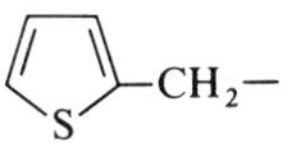 2-thienyl–CH_2– (7α-OMe)	–$OCONH_2$
Cefotaxime	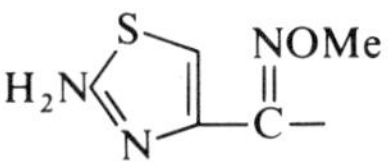 H_2N (S, N thiazolyl)–C(=NOMe)–	OAc

Compared to penicillins, cephalosporins are generally effective against a broader range of organisms and are more resistant to β-lactamases. β-Lactamases are bacterial enzymes that efficiently hydrolyze β-lactam antibiotics to inactive species in which the β-lactam bond has been cleaved. Cephalothin was the first cephalosporin to be marketed and continues to account for a significant portion of cephalosporin use. Cephacetrile and cephalothin have similar antimicrobial activities and pharmacokinetic behavior. Cephaloridine has a somewhat broader spectrum and causes less pain on injection. Cephazolin and cephamandole have somewhat increased spectrum and potency compared to earlier cephalosporins and display increased blood levels when given by injection. Cephalexin is one of the few cephalosporins that is effective when given orally. Cefoxitin and Cefotaxime are among the newer cephalosporins that show increased potency, broader spectrum and better resistance to β-lactamases ⟨B-79MI51002⟩. There are many new cephalosporins presently undergoing clinical trial that show promise against a variety of resistant organisms. Some of these are characterized by a wider range of structural diversity, *e.g.* removal of the C-3′ carbon, replacement of the sulfur of the dihydrothiazine ring with oxygen ⟨81MI51001⟩. Because of their relative safety and broad spectrum of activity, β-lactam antibiotics have become the therapy of choice for most bacterial infections.

5.11

Penicillins

D. N. McGREGOR
Bristol Laboratories, Syracuse

5.11.1 INTRODUCTION

The penicillins are a class of compound having the general structure (**1**). Because of their unique effectiveness in the treatment of bacterial infections in humans, these compounds have been investigated intensively from the chemical, microbiological and clinical points of view since about 1940. The early history of these developments ⟨see especially B-80MI51100, B-49MI51100⟩ contains the following pivotal studies:

1929 Fleming's serendipitous observation of the lysis of staphylococcus colonies in the vicinity of *Penicillium notatum* ⟨29MI51100⟩.
1940 Recognition of the therapeutic potentialities of penicillin ⟨40MI51100⟩.
1945 Confirmation of the structure of benzylpenicillin by X-ray diffraction studies ⟨B-49MI51101⟩.
1948 Production of penicillin V by biosynthesis ⟨48JBC(175)793⟩.
1959 Discovery and isolation of 6-aminopenicillanic acid ⟨59MI51100⟩.
1960 Production of methicillin by semisynthesis ⟨60MI51100⟩.
1962 Total synthesis of 6-aminopenicillanic acid ⟨62JA2983⟩.
1963 Conversion of the penicillin ring system into the cephalosporin ring system ⟨63JA1896⟩.

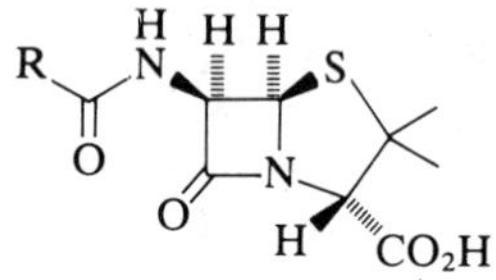

(**1**)

These studies trace the evolution of penicillin investigations from microbiological curiosity through the development of increasing therapeutic utility to increasingly sophisticated chemical manipulations. This chapter will focus primarily on the chemical aspects of this area. Because of the vast amount of relevant literature, it has been possible to discuss only selected studies which hopefully are representative of the different investigative directions. The reader is urged to consult the cited references for more detailed discussion and for references to related studies.

The nomenclature of penicillins requires special comment. Compound (**2**) can be named as follows: (a) penicillin G; (b) benzylpenicillin (note that the term 'penicillin' may refer to the compound class (**1**), to the structural fragment (**3**) or, especially in the medical literature, to compound (**2**) itself); (c) 6β-phenylacetamidopenicillanic acid; (d) 2,2-dimethyl-6β-phenylacetamidopenam-3α-carboxylic acid; (e) (2*S*,5*R*,6*R*)-3,3-dimethyl-7-oxo-6-(2-phenylacetamido)-4-thia-1-azabicyclo[3.2.0]heptane-2-carboxylic acid; and (f) [2*S*-(2α,5α,6β)]-3,3-dimethyl-7-oxo-6-[(phenylacetyl)amino]-4-thia-1-azabicyclo-[3.2.0]heptane-2-carboxylic acid. The numbered system shown in (**2**) is the one most commonly used in the penicillin literature and will be used in this chapter; note that different number is used when (**2**) is named according to (e) and (f) above.

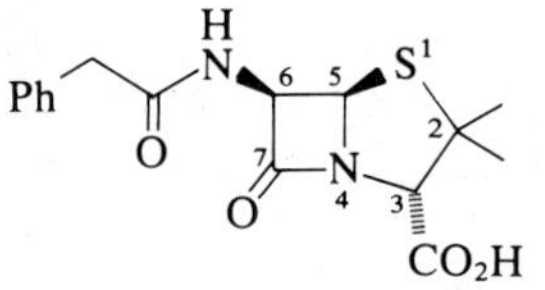

(**2**) 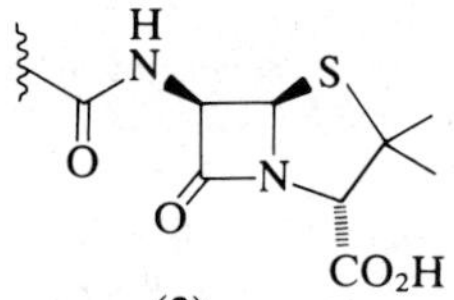 (**3**)

5.11.2 STRUCTURE

5.11.2.1 X-ray Crystallographic Studies

During the early 1940s the structures of benzylpenicillin and other members of the penicillin class were studied primarily through chemical investigations (both degradative and synthetic) and the new (at that time) technique of IR spectroscopy ⟨B-49MI51102⟩. These studies led to three proposed structures: (**4**), (**5**) and (**1**). It remained for X-ray crystallographic analysis to show that (**1**) was the correct structure ⟨B-49MI51101⟩. Since this pioneering work, benzylpenicillin has been reexamined ⟨78JCS(P1)185⟩, and various other penicillins have been subjected to X-ray analysis: phenoxymethylpenicillin (penicillin V) ⟨63BJ(86)514⟩, D-α-aminobenzylpenicillin (ampicillin) ⟨68MI51100, 76AX(B)2279⟩ and 2,6-dimethoxyphenylpenicillin (methicillin) methyl ester ⟨77AX(B)580⟩ and D-α-amino-*p*-hydroxybenzylpenicillin (amoxicillin) ⟨78AX(B)461⟩. Some of the bond distances and bond angles found in these studies are summarized in Table 1.

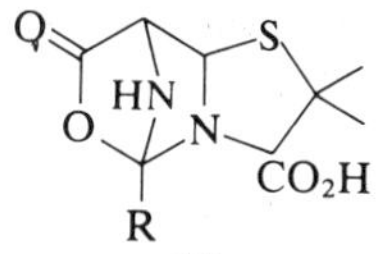

(4) 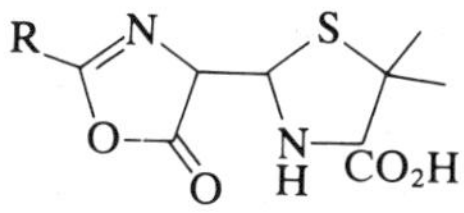 (5)

Table 1 Interatomic Distances and Bond Angles in Penicillins

Interatomic distances (Å)		*Bond angles* (°)	
S(1)—C(2)	1.84–1.87	C(5)—S(1)—C(2)	89.8–96
C(2)—C(3)	1.55–1.57	S(1)—C(2)—C(3)	102.6–105
C(3)—N(4)	1.45–1.47	C(2)—C(3)—N(4)	104–106.4
N(4)—C(5)	1.45–1.52	C(3)—N(4)—C(5)	117.1–120
C(5)—S(1)	1.77–1.82	N(4)—C(5)—S(1)	103–103.7
C(5)—C(6)	1.53–1.58	N(4)—C(5)—C(6)	86.8–92
C(6)—C(7)	1.51–1.55	C(5)—C(6)—C(7)	83–85.6
C(7)—N(4)	1.36–1.46	C(6)—C(7)—N(4)	91.8–96
C(6)—N(9)	1.42–1.44	C(7)—N(4)—C(5)	88–93.6
C(7)—O(8)	1.18–1.21		

Several interesting factors emerge from these studies:

(1) The configuration of the thiazolidine rings of different penicillins is different ⟨76AX(B)2279⟩, probably due to varying steric requirements of the 6β side chains. Four of the ring atoms are approximately coplanar, with the fifth atom being significantly out of the plane. Which atom is out-of-plane, however, depends on the compound: in anhydrous ampicillin, C(2) is out-of-plane; in ampicillin trihydrate, S(1) is out-of-plane; in benzylpenicillin and phenoxymethylpenicillin, C(3) is out-of-plane. Alternatively, the thiazolidine ring can be considered to have two principal conformations; (a) the conformation in which the α-methyl is equatorial and the β-methyl and α-carboxyl are axial (*e.g.* benzylpenicillin), and (b) the conformation in which the α-methyl is axial and the β-methyl and α-carboxyl are equatorial (*e.g.* ampicillin trihydrate) ⟨B-72MI51100⟩. Note, however, that NMR studies ⟨75JCS(F1)1145⟩ indicate that, in solution, both benzylpenicillin and ampicillin adopt conformation (b).

(2) The structure of the β-lactam amide group found in the penicillins is much different from that of a normal amide ⟨B-72MI51100⟩. The carbonyl bond length of the β-lactam is shorter than that of a normal amide (1.18–1.21 Å *versus* 1.20–1.23 Å) while the C—N bond length is longer (1.36–1.46 Å *versus* 1.333–1.346 Å). Even more striking, however, are the geometry differences about the amide N. In a normal amide the N atom and the three atoms attached to it are coplanar, presumably because of orbital overlap of the N p-electrons and the carbonyl π-electrons. Because of the fused ring structure of the penicillins, however, such overlap is inhibited, leading to an amide in which the N and its three substituents, C(7), C(5) and C(3), are not coplanar. The resulting pyramidal geometry is usually expressed in terms of the distance between the N atom and the plane formed by its substituents, and for penicillins this is 0.35–0.40 Å. Taken together, these data indicate that in the penicillin β-lactam the carbonyl has greater double bond character and the C(7)—N(4) bond has greater single bond character than does a normal amide. This is felt to explain the much greater susceptibility of the penicillin β-lactam to nucleophilic attack at the carbonyl, resulting in its behavior as a potent acylating agent. This acylating reactivity is directly associated with much of the chemical and biological behavior of the molecule.

5.11.2.2 Nuclear Magnetic Resonance Spectroscopic Studies

^{1}H NMR spectroscopy has become a very important tool in penicillin research, and virtually every chemical paper in this field since about 1965 has utilized this technique for structural assignments and confirmations. The amount of ^{1}H NMR data on penicillins and their derivatives, therefore, is quite large. Some of these data have been collected, along with a discussion of the use of nuclear Overhauser effects, aromatic solvent induced shifts and hydrogen bonding effects ⟨B-72MI51101⟩. In general, however, original papers should be consulted for ^{1}H NMR data on a particular class of penicillin derivative. As a reference point the chemical shifts (δ, in p.p.m. relative to TMS) and coupling constants (J, in Hz) for phenoxymethylpenicillin methyl ester in $CDCl_3$ are: 2α-CH_3, 1.49; 2β-CH_3, 1.60; H(3), 4.47; H(5), 5.58 (d, J = 4.5); H(6), 5.74 (q, J = 4.5, 9). Note particularly the coupling constant of H(5) and H(6). For normal penicillins in which H(5) and H(6) are *cis*, J = 4–5 Hz. For 6-epipenicillins in which H(5) and H(6) are *trans*, J = 1.5–2 Hz.

^{13}C and ^{15}N NMR spectroscopy have been used much less frequently than ^1H NMR for the characterization of penicillins and their derivatives. Nevertheless, some studies have made use of these techniques, and some of the ^{13}C NMR (δ, in p.p.m. relative to TMS) ⟨see 76JCS(P1)1772, 77JCS(P2)1749, B-72MI51101⟩ and ^{15}N NMR (δ, in p.p.m. relative to NH_4Cl) ⟨78JOC2013⟩ data for benzylpenicillin are: C(2), 65.2; 2α-CH_3, 27.3; 2β-CH_3, 31.7; C(3), 73.9; CO_2^-, 174.7; C(5), 67.4; C(6), 58.9; C(7), 175.3; N(4), 143.7; 6β-C=O, 174.1; 6β-NH, 90.7.

5.11.2.3 Infrared Spectroscopic Studies

The role of IR spectroscopy in the early penicillin structure studies has been described ⟨B-49MI51103⟩ and the results of more recent work have been summarized ⟨B-72MI51101⟩. The most noteworthy aspect of a penicillin IR spectrum is the stretching frequency of the β-lactam carbonyl, which comes at approximately 1780 cm^{-1}. This is in contrast to a linear tertiary amide which absorbs at approximately 1650 cm^{-1}, and a β-lactam which is not fused to another ring (*e.g.* benzyldethiopenicillin), which absorbs at approximately 1740 cm^{-1} (the exact absorption frequency will, of course, depend upon the specific compound and technique of spectrum determination). The β-lactam carbonyl absorptions of penicillin sulfoxides and sulfones occur at approximately 1805 and 1810 cm^{-1} respectively. The high absorption frequency of the penicillin β-lactam carbonyl is interpreted in terms of the increased double bond character of that bond as a consequence of decreased amide resonance, as discussed in the X-ray crystallographic section. Other aspects of the penicillin IR spectrum, *e.g.* the side chain amide absorptions at approximately 1680 and 1510 cm^{-1}, and the carboxylate absorption at approximately 1610 cm^{-1}, are as expected.

5.11.2.4 Mass Spectrometric Studies

The fragmentation patterns of relatively volatile derivatives of penicillins (*e.g.* benzylpenicillin methyl ester) under electron impact ⟨B-72MI51101⟩ and chemical ionization ⟨75MI51100⟩ conditions have been described. For both techniques the primary fragmentation is that shown in Scheme 1.

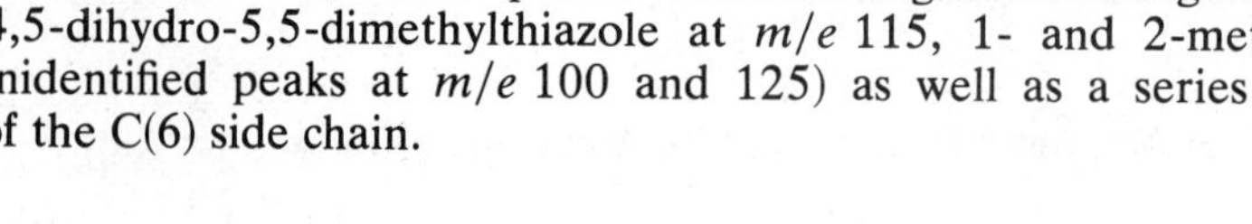

Scheme 1

Pyrolysis mass spectrometry, which does not require a volatile derivative, has been applied to various penicillins ⟨78MI51100⟩. These spectra contained fragments arising from the bicyclic ring system (4,5-dihydro-5,5-dimethylthiazole at m/e 115, 1- and 2-methylpyrrole at m/e 81 and unidentified peaks at m/e 100 and 125) as well as a series of fragments characteristic of the C(6) side chain.

5.11.2.5 Circular Dichroism Studies

The circular dichroism curves for a variety of penicillanic acid derivatives have been published and discussed ⟨B-77MI51100⟩ and have been used to support extended Hückel MO calculations ⟨77T711⟩.

5.11.2.6 Molecular Orbital Calculations

The β-lactam antibiotics, including various structures related to penicillins, have been the subject of an ongoing series of MO calculations, especially through the extended Hückel method ⟨see 79T1499, B-82MI51104 and refs. therein⟩. Of particular interest are calculations which show a structural similarity between penicillins and the tetrahedral intermediate obtained during hydrolysis of the amide bond in glycylycine ⟨79JMC533⟩. The antibacterial activity of penicillins involves the inhibition of a transpeptidase enzyme, and these calculations lend support to the concept that the penicillins are acting as transition state analogs. In another study, CNDO/2 calculations have been used to support the idea that differences in the antibacterial effectiveness of different β-lactam antibiotics are due to variations in binding energies with the inhibited enzyme and variations in the activation energies of acylation ⟨74JMC342⟩.

5.11.2.7 Other Physical Properties

As a class the penicillins do not have a characteristic UV absorption spectrum. UV spectra did, however, play an important role in the structure elucidation of some of the penicillin degradation products during the early structure studies ⟨B-49MI51104⟩.

The acid dissociation constants ($pK \approx 2.7$), equilibrium solubilities in several solvents ⟨71JPS503⟩ and lipophilicity properties ⟨77JPS1675⟩ for various penicillins have been summarized.

5.11.3 REACTIVITY

5.11.3.1 Introduction

During the early and middle 1960s the investigative emphasis in penicillin chemistry was on the modification of the 6β side chain in an effort to obtain penicillin derivatives with improved therapeutic properties. Since that time it has become increasingly apparent that the fermentable penicillins, such as benzylpenicillin and phenoxymethylpenicillin, are relatively inexpensive and readily available starting materials with defined chirality which can be chemically manipulated to afford a surprising variety of transformation products. In particular, these penicillins can serve as efficient precursors to other β-lactam-containing antibiotics such as the cephalosporins (see Chapter 5.10) and the carbapenems (see Chapter 5.12), and the reader should consult those chapters for the full reaction sequences leading to those compound classes.

There have been a number of reviews which have covered recent developments in penicillin chemistry. The reader is especially referred to ⟨B-80MI51102⟩ for a thorough review through 1978. Other reviews are also of value: ⟨B-72MI51104, B-72MI51106, B-77MI51103, B-81MI51101, 74FOR(31)1, 76CRV113, 78T1731, 75S547, B-82MI51102⟩.

5.11.3.2 Early Studies

The early investigations of the reactions of the penicillin class of compounds were largely of a degradative nature, and were primarily associated with structure elucidation. These have been discussed in detail ⟨B-49MI51102⟩ and some of the principal transformations are outlined in Schemes 2, 3 and 4 using benzylpenicillin as an example. Some of these reactions will be discussed in greater detail later in this section.

The following trivial names have been assigned to the penicillin transformation products shown: (**6**), benzylpenicilloate; (**7**), benzylpenaldic acid; (**8**), D-penicillamine; (**9**), α-methyl D-α-benzylpenicilloate; (**10**), benzylpenillic acid; (**11**), benzylpenillamine; (**12**), benzylisopenillic acid; (**13**), benzylpenilloic acid; (**14**), benzylpenilloaldehyde; (**15**), benzylpenicillenic acid; (**16**) benzylpenillonic acid methyl ester; and (**17**), dethiobenzylpenicillin methyl ester.

(6) (7) (8) (9)

i, OH^-; ii, $HgCl_2$; iii, MeOH; iv, $PhCH_2NH_2$

Scheme 2

(13) (12) (2) (10) (11) (15) (14) (8)

i, H_3O^+, heat; ii, $HgCl_2$; iii, $Ba(OH)_2$; iv, H_2O, heat

Scheme 3

(17) (16)

i, MeMgI; ii, $KMnO_4$; iii, $PhCH_2NH_2$; iv, $NaIO_4$; v, 140 °C; vi, HSCN; vii, Raney nickel

Scheme 4

5.11.3.3 Modification at the S(1) Position

5.11.3.3.1 Synthesis and reactions of penicillin sulfoxides

The preparation and reactions of penicillin sulfoxides have been extensively studied because of their utility as intermediates for the preparation of other β-lactam-containing antibiotics such as the cephalosporins ⟨for reviews, see B-80MI51102, 74FOR(31)1, 73ACR32, 76CRV113, B-72MI51106, B-82MI51103⟩. Sulfoxide formation can afford either the α- or β-sulfoxide (*R* or *S* respectively), and the course of the reaction depends on the C(6) side chain and the oxidant. Table 2 illustrates some of these findings ⟨74JOC441, 76JCS(P1)1772, 69JA1528⟩. The observed results have been rationalized in terms of: (a) hydrogen bonding between a 6β amide group and the oxidant leading to β-oxidation; (b) stabilization by hydrogen bonding between a β-sulfoxide and a 6β amide group; and (c) steric destabilization of β-sulfoxides by bulky 6β groups.

Table 2 Sulfoxide Formation of Various Penams

Substrate	*Oxidant*	*β-Sulfoxide*	*α-sulfoxide*
PhO, H, N, S, O, O, N, CO_2CH_2Ph	MCPBA	100	0
	$PhICl_2$	20	80
	O_3	50	50
PhO, H, N, S, O, O, N, CO_2CH_2Ph	MCPBA	100	0
	$PhICl_2$	0	100
	O_3	50	50
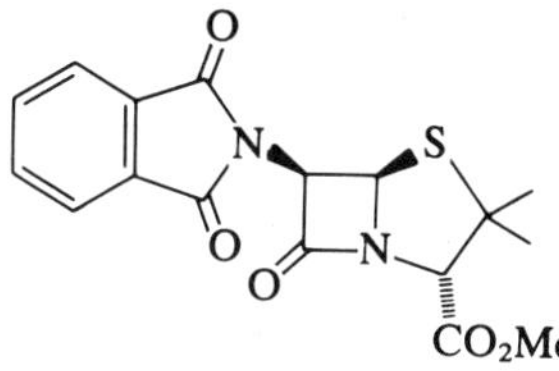 O, N, S, O, O, N, CO_2Me	MCPBA	0	100
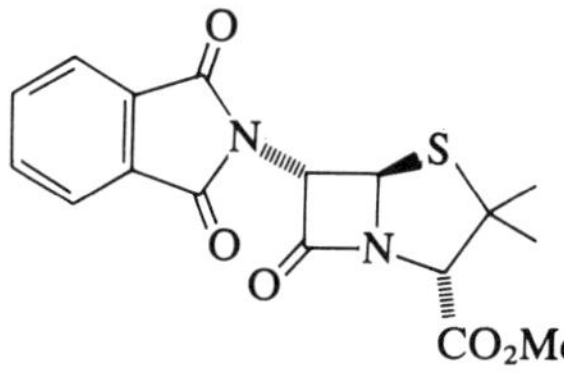 O, N, S, O, O, N, CO_2Me	MCPBA	80	20
Br, Br, S, O, N, CO_2H	$NaIO_4$	27	73
	MCPBA	9	91
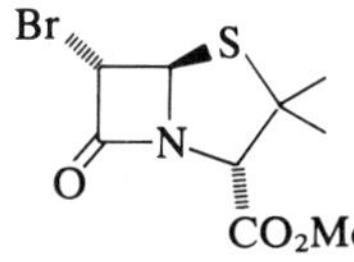 Br, S, O, N, CO_2Me	MCPBA	92	8
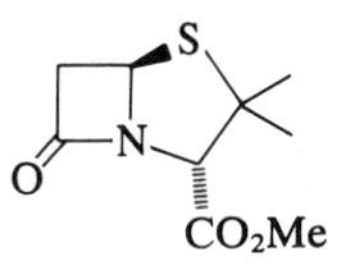 S, O, N, CO_2Me	MCPBA	79	21

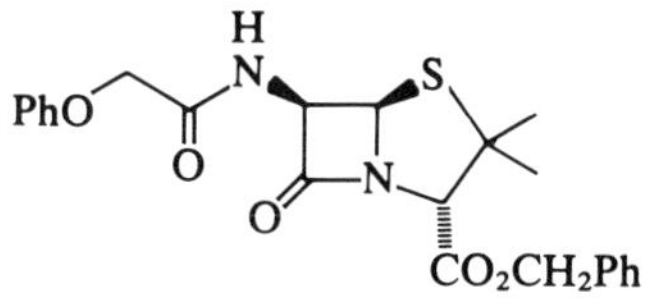

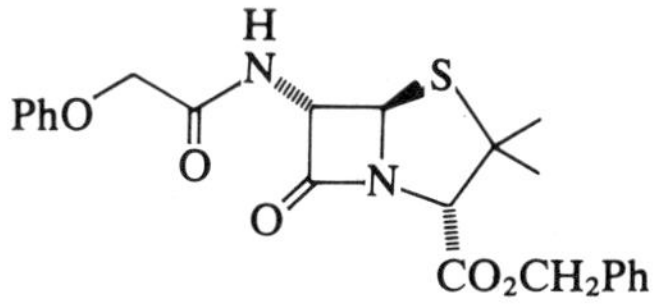

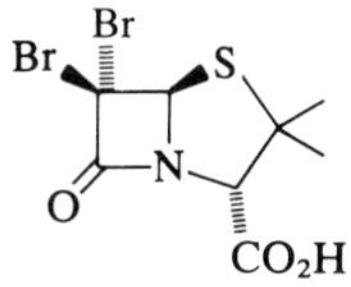

Penicillin sulfoxides can be epimerized by heat to afford thermal equilibrium mixtures of α- and β-sulfoxides, the position of the equilibrium depending on the C(6) side chain (Scheme 5). Deuterium incorporation studies support a sulfenic acid, *e.g.* (**18**), as the intermediate in these transformations. This mechanism is also supported by the finding that when an α-sulfoxide epimerizes to a β-sulfoxide there is a simultaneous epimerization at C(2) ⟨71JCS(C)3540⟩. With irradiation by UV light it is possible to convert a more thermodynamically stable β-sulfoxide to the α-sulfoxide ⟨69JA1530⟩.

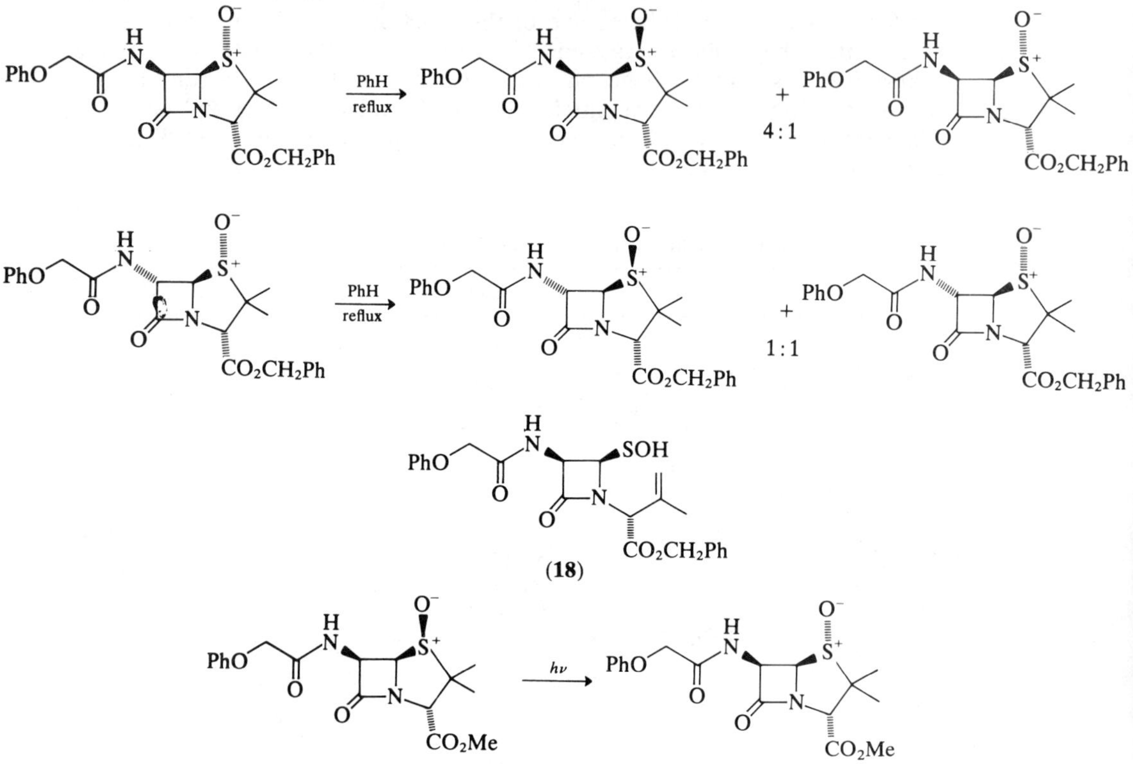

(**18**)

Scheme 5

Penicillin sulfoxides may be reduced back to the sulfide with P_2S_5 ⟨76TL971⟩, PCl_3 or PBr_3 ⟨*e.g.* 71JCS(C)3540⟩.

Scheme 6 depicts a typical penicillin sulfoxide rearrangement ⟨69JA1401⟩. The mechanism probably involves an initial thermal formation of a sulfenic acid which is trapped by the acetic anhydride as the mixed sulfenic–acetic anhydride. Nucleophilic attack by the double bond on the sulfur leads to an episulfonium ion which, depending on the site of acetate attack, can afford either the penam (**19**) or the cepham (**20**). Product ratios are dependent on reaction conditions. For example, in another related study acetic anhydride gave predominantly the penam product, while chloroacetic anhydride gave the cepham product ⟨71JCS(C)3540⟩. The rearrangement can also be effected by acid; in this case the principal products are the cepham (**21**) and the cephem (**22**; Scheme 7). Since these early studies a wide variety of reagents have been found to catalyze the conversion of a penicillin sulfoxide to the cepham/cephem ring system ⟨*e.g.* 77JOC2887⟩.

The intermediate sulfenic acid derived from a penicillin sulfoxide has been trapped by a large assortment of reagents and, in one case, the sulfenic acid itself has been isolated ⟨74JA1609⟩. Only some of these products will be discussed here, and the reader is referred to the cited reviews ⟨especially B-80MI51102⟩ for additional examples.

The fused thiazoline–azetidinone (**23**) has been studied fairly extensively ⟨B-77MI51104⟩. It can be formed either by an intermolecular reduction of the sulfenic acid intermediate with $P(OMe)_3$ followed by cyclization ⟨70JA2575⟩, or by an intramolecular reductive trapping of either the sulfenic acid ⟨76TL975⟩ or a disulfide ⟨75BCJ3429⟩ (Scheme 8). This intermediate can be converted to 3-modified cephem derivatives ⟨*e.g.* 77H(8)283, 80JCS(P1)233, 80TL781⟩ and to 2-modified penam derivatives ⟨*e.g.* 72CC229, 77JCS(P1)2479, 78TL1755, 80JCS(P1)150 and the preceding paper⟩ (Scheme 9). (*N.B.* In this chapter a wavy bond indicates a mixture of isomers.)

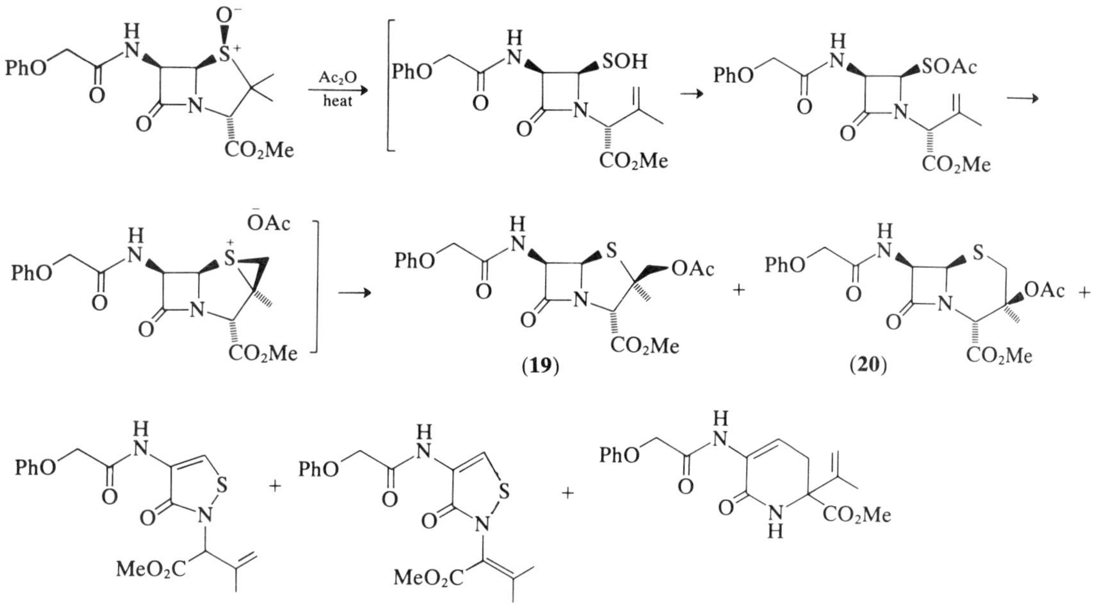

Scheme 6

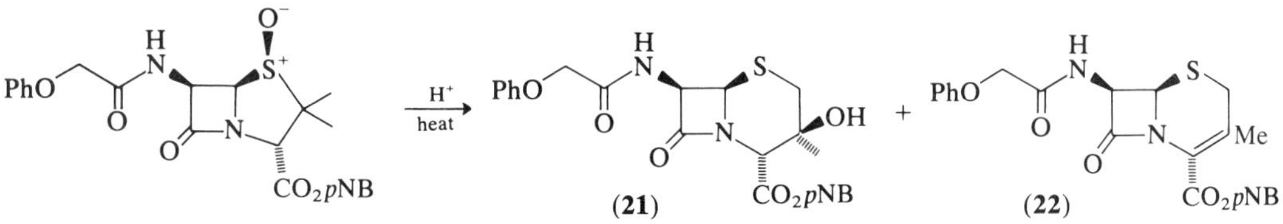

Scheme 7

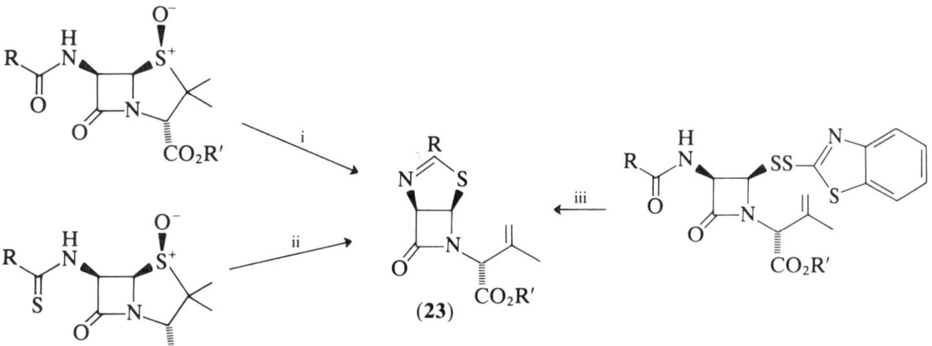

i, $P(OMe)_3$; ii, heat; iii, P_2S_5

Scheme 8

Another synthetically useful transformation is the conversion of a penicillin sulfoxide to the disulfide intermediate (**24**). This compound has been a starting material for several reaction sequences, leading to various modified cephams, cephems, penams and penems (*via* **27**), and some of these are partially shown in Scheme 10 ⟨76JA2342 and refs. cited therein, 79JOC3084, 77JA5091, 80HCA201, 81JCS(P1)2087⟩. It is interesting that the 2β-hydroxymethylpenam (**25**) is relatively unstable, undergoing intramolecular attack of the hydroxy group on the β-lactam carbonyl. Scheme 11 illustrates the use of a similar disulfide intermediate to form a cephalosporin with a substituted methyl at the 3-position ⟨78H(10)99⟩.

Another means of trapping the sulfoxide-derived sulfenic acid is by conversion to a sulfenyl halide ⟨76JA5040⟩. As shown in Scheme 12, these intermediate sulfenyl halides are useful for the preparation of the cepham ring system ⟨75JA5583, 81JOC3568⟩. Scheme 13 illustrates the rather facile interconversions of the penam and cepham ring systems, with the episulfonium ion (**31**) presumably being the common intermediate ⟨75JA3192, 80JOC3205⟩.

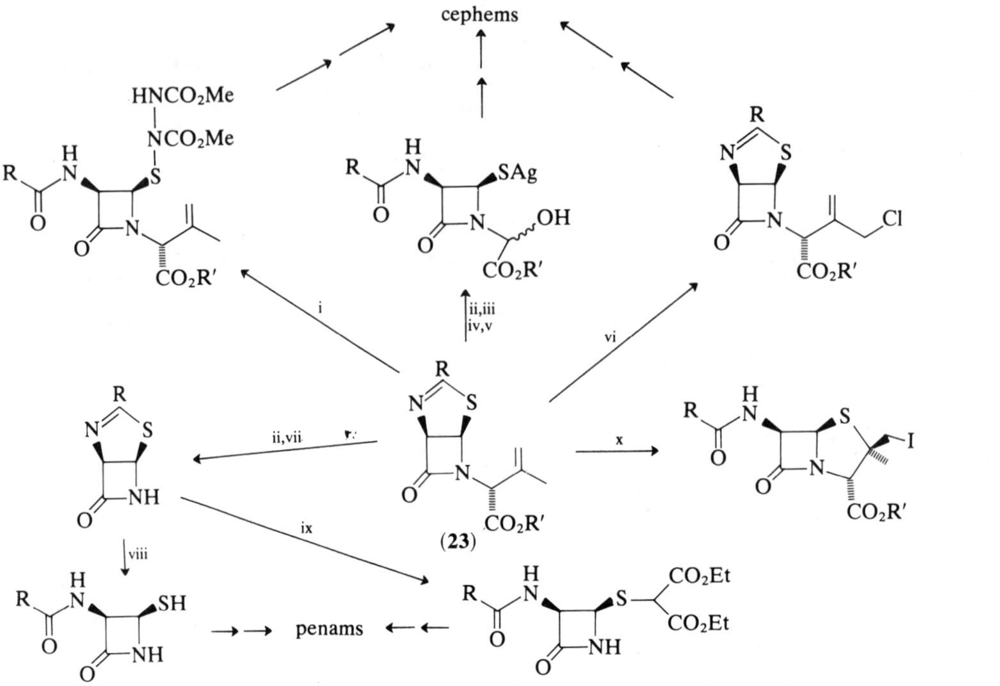

i, $MeO_2CN{=}NCO_2Me$, TsOH; ii, Et_3N; iii, O_3; iv, Zn, H^+; v, $AgClO_4$; vi, Bu^tOCl; vii, $KMnO_4$; viii, H_3O^+; ix, $BrCH(CO_2Et)_2$; x, I_2

Scheme 9

Various other reagents have been used to trap the intermediate sulfenic acid, and Scheme 14 shows some of these ⟨74JCS(P1)1459, 74TL725, 74JCS(P1)1456, 77JCS(P1)1477, 70TL4897, 78TL4167⟩.

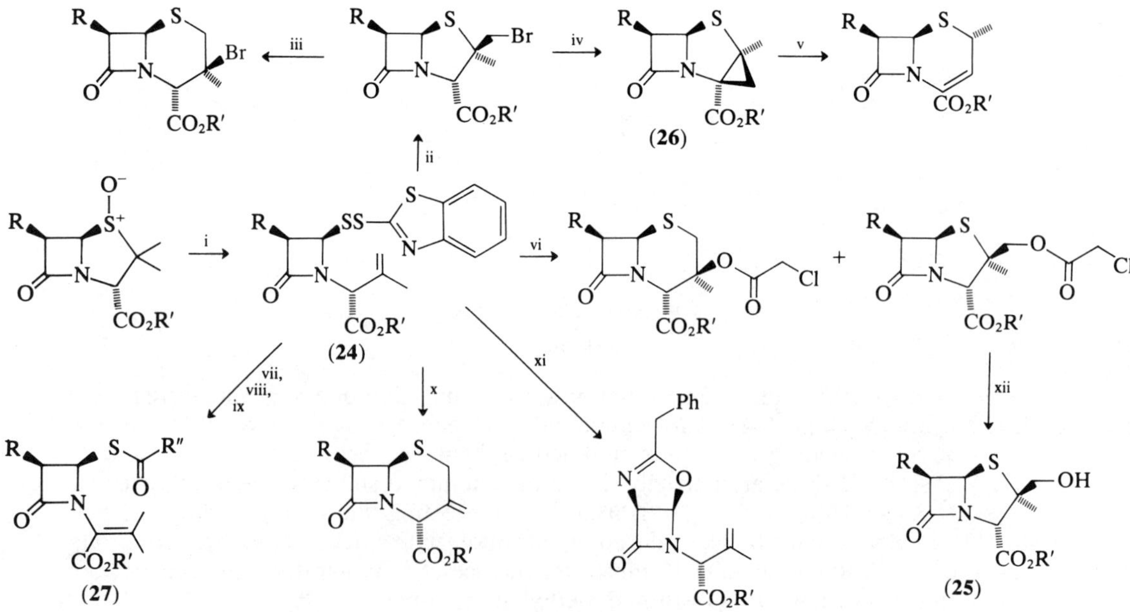

i, 2-mercaptobenzothiazole, heat; ii, Br_2; iii, DMF; iv, DBU; v, $AlBr_3$; vi, $ClCH_2CO_2H$, AgOAc; vii, Et_3N; viii, Ph_3P; ix, R″COCl; x, *hν*; xi, Ph_3P (R = $PhCH_2CONH$); xii, H_2NCSNH_2

Scheme 10

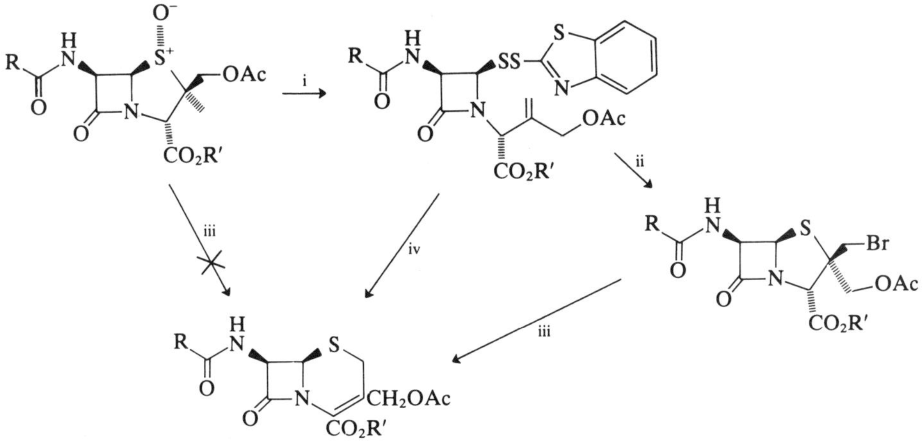

i, 2-mercaptobenzothiazole, heat; ii, Br_2; iii, heat; iv, AgF

Scheme 11

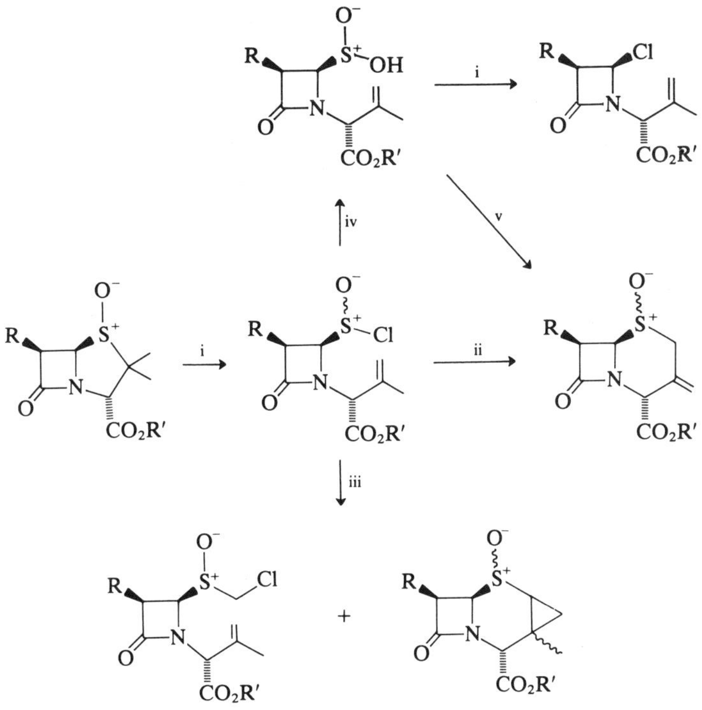

i, *N*-chlorosuccinimide; ii, $SnCl_4$; iii, CH_2N_2; iv, H_3O^+; v, $MeSO_3H$

Scheme 12

5.11.3.3.2 Other electrophilic reactions at the S(1) position

Upon treatment with relatively reactive alkylating agents in the presence of strong base, the thiazolidine ring of the penam ring system may be opened to form '1,2-secopenicillins' ⟨76JCS(P1)1615 and refs. therein⟩. These products have been useful as intermediates in the preparation of cephalosporin derivatives, 1-oxa-1-dethiacephalosporin derivatives ⟨80JMC1108, 79JCS(P1)2268⟩ and the 1-oxa-1-dethiapenem ring system ⟨77CC720⟩. Scheme 15 illustrates the ring cleavage reaction and a few of the subsequent reactions of the initial ring-opened products ⟨76TL3975, 76JA7864⟩. The oxazolidine–azetidinone (**32**) ⟨80JOC3682⟩ can also be prepared by the treatment of a penicillin ester with iodobenzene dichloride ⟨75JCS(P1)1932⟩, and a related compound was previously encountered in Scheme 10.

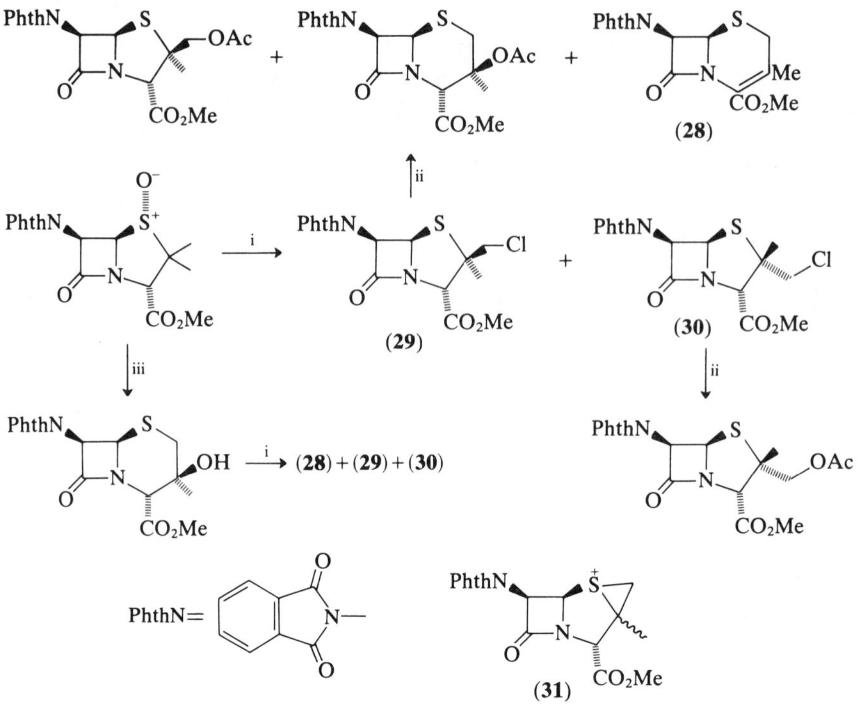

i, $SOCl_2$, Et_3N; ii, AgOAc; iii, H_2SO_4

Scheme 13

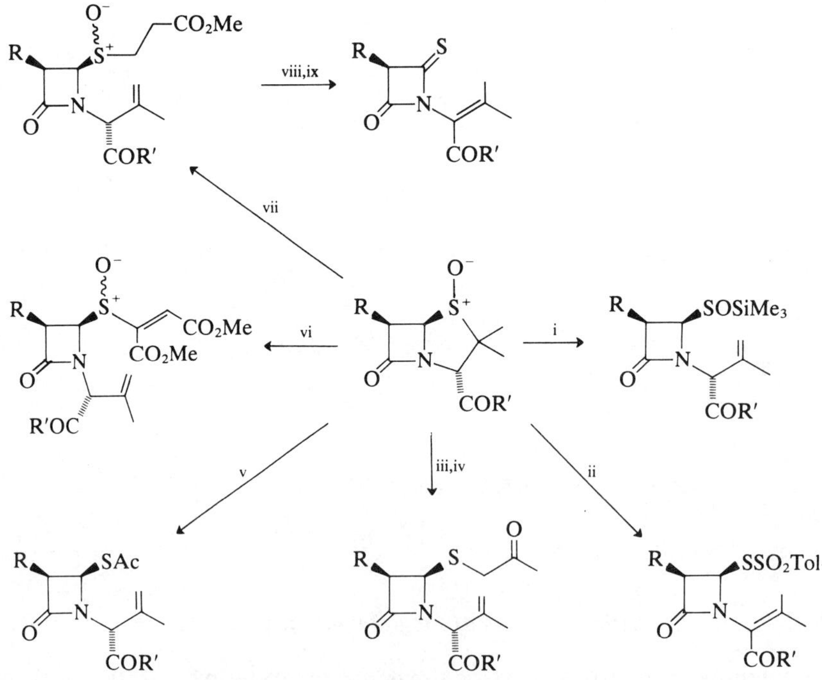

i, Me_3SiCl; ii, p-$TolSO_2H$; iii, $CH_2{=}C(Me)OBu^n$, $AlCl_3$; iv, H_3O^+; v, $P(OMe)_3$, Ac_2O; vi, $MeO_2CC{\equiv}CCO_2Me$; vii, $CH_2{=}CHCO_2Me$; viii, Et_3N; ix, heat

Scheme 14

With certain 6-substituents on the penam nucleus, it has been possible to isolate sulfonium derivatives (Scheme 16) ⟨79JCS(P1)3175⟩. As expected, these are highly reactive compounds.

Scheme 15 shows an example of the cleavage of a carbon–sulfur bond by chlorinolysis. This reaction can also be carried out on the intact penam ring system as shown in Scheme 17 ⟨71JA6269 and the previous paper⟩. In Scheme 18 this reaction is used to form the epimer of

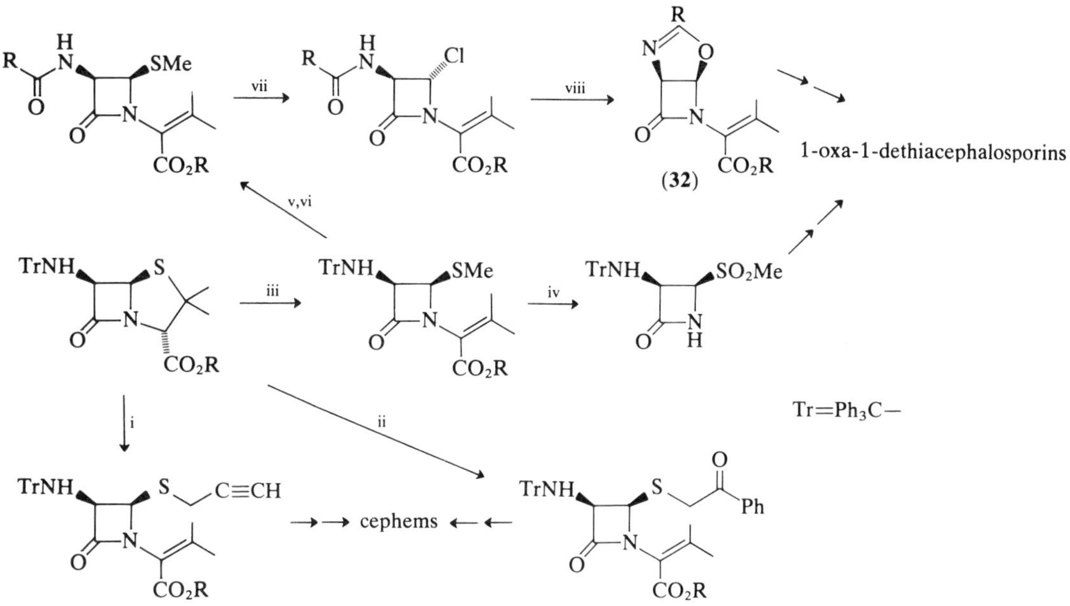

TrNH S C≡CH → → cephems ← ← TrNH S Ph

i, NaH, $BrCH_2C{\equiv}CH$; ii, $KOBu^t$, $BrCH_2COPh$; iii, NaH, MeI; iv, $KMnO_4$, HOAc; v, TsOH; vi, RCOCl; vii, Cl_2; viii, $AgBF_4$, Ag_2O

Scheme 15

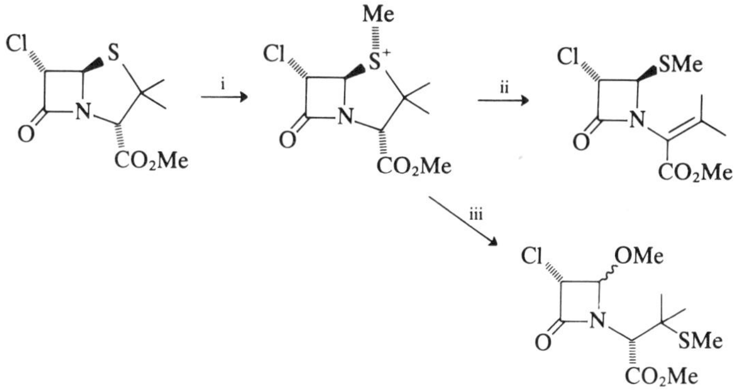

i, Me_3OBF_4; ii, Na_2CO_3; iii, MeOH

Scheme 16

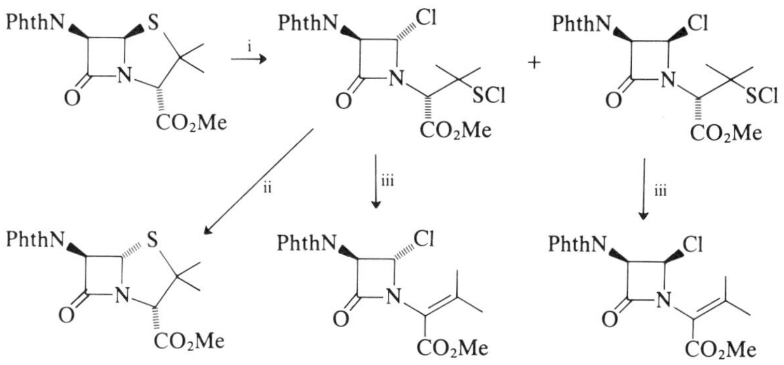

i, 1 eq. Cl_2; ii, $SnCl_2$; iii, Cl_2 or Et_3N

Scheme 17

the fused oxazolidine–axetidinone (**32**) shown in Scheme 15, which has also been useful for the synthesis of 1-oxa-1-dethiacephalosporins ⟨79JA4403, 80TL351⟩. This fused oxazolidine–azetidinone can also be formed by the action of chloramine-T or iodobenzene dichloride on the 6-epipenicillin ester ⟨75JCS(P1)1932⟩.

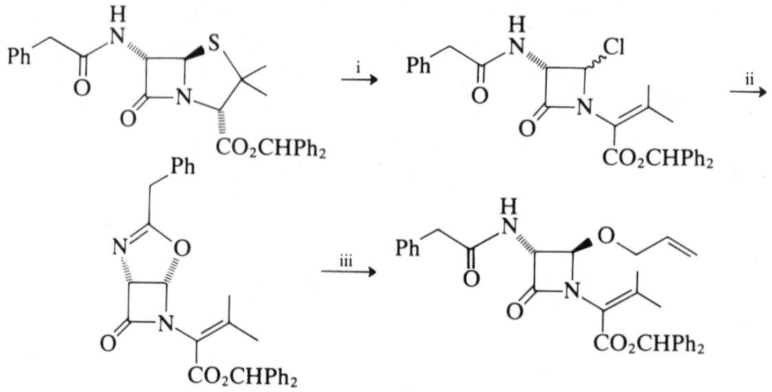

i, Cl_2; ii, NaOH; iii, $CH_2{=}CHCH_2OH$, TfOH

Scheme 18

The chlorinolysis reactions described above led to a cleavage of the S(1)—C(5) bond, while previous examples of thiazolidine ring cleavage described in this section occurred at the S(1)—C(2) bond. Cleavage of S(1)—C(5) can also be achieved by the treatment of the penam ring system with mercury(II) salts as shown in Scheme 19 ⟨79JCS(P1)2268, 74JCS(P1)185, 80CC34⟩. Note especially the use of *t*-butyl as a blocking group for a sulfenic acid in step (viii).

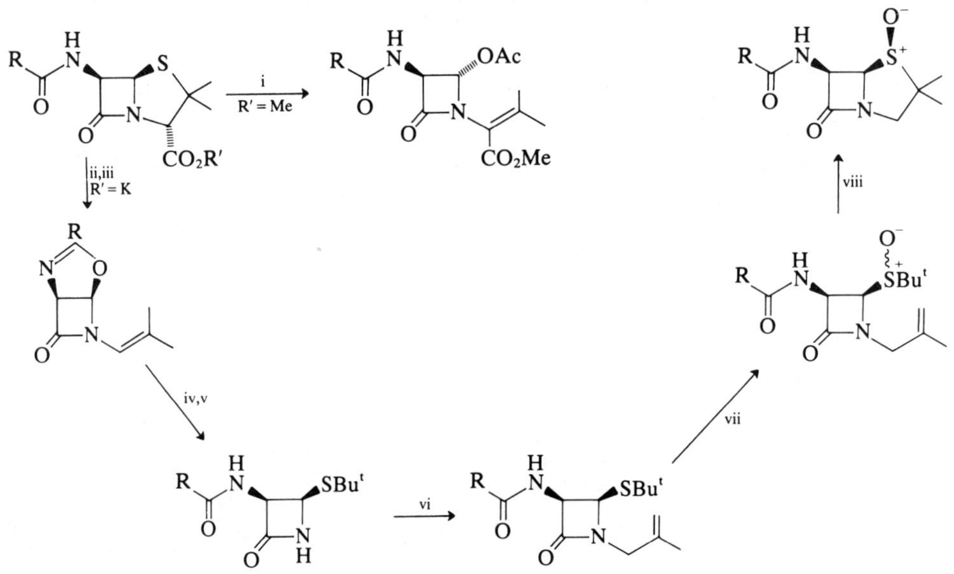

i, $Hg(OAc)_2$; ii, $Hg(OAc)_2$, HOAc; iii, DMSO; iv, NBS, H_2O; v, Bu^tSH, Et_3N; vi, $KOBu^t$, $CH_2{=}C(Me)CH_2Br$; vii, $NaIO_4$; viii, heat

Scheme 19

5.11.3.4 Modification at the C(2) Position

Modifications at this position are described in various other sections. See, for example, the preparation of 2-substituted methylpenams in Section 5.11.3.3.1, the total synthesis of 2-spirocycloalkylpenams in Section 5.11.4.4, and the synthesis of penems in Section 5.11.4.6.

5.11.3.5 Modification at the C(3) Position

Some C(3) modifications are described in other sections: the preparation of penems in Section 5.11.4.6, and the synthesis of the tricyclic compound (**26**) in Scheme 10.

In common with other carboxylic acids, the 3-carboxyl of penicillins can be subjected to a variety of transformations, provided that they can be effected under conditions compatible with β-lactam survival. Many of these have been described in two review articles ⟨B-80MI51102, B-75MI51102⟩, and the reader should consult these for a more complete treatment.

One aspect of carboxyl modification of particular interest is its replacement with other acidic functional groups. One of these, the replacement of the carboxyl with a 5-tetrazolyl group, is of particular interest because of the resultant improved antibacterial properties. This transformation is shown in Scheme 20 ⟨78USP4115385⟩.

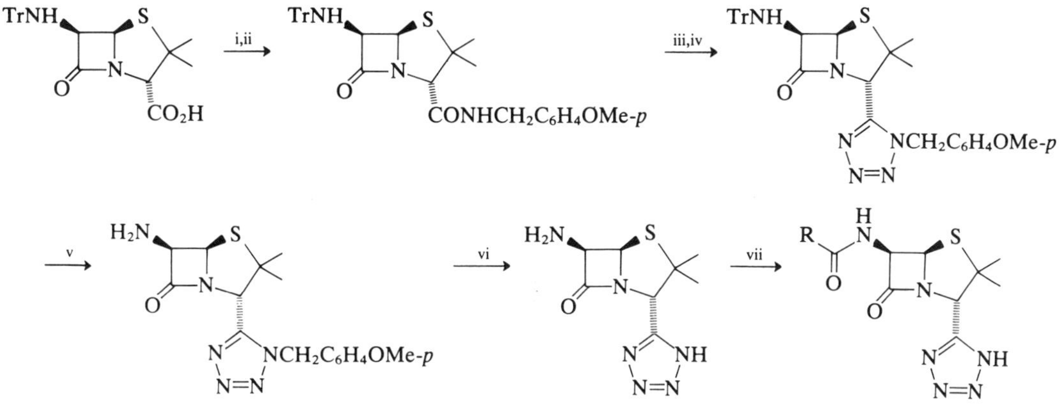

i, $ClCO_2Et$; ii, $H_2NCH_2C_6H_4OMe$-*p*; iii, $SOCl_2$; iv, Me_3SiN_3; v, TsOH; vi, TFA, PhOMe; vii, RCOCl

Scheme 20

Some further transformations involving reduction of the 3-carboxylic acid group are shown in Scheme 21 ⟨66JOC1922, 64JMC483, 70JMC389⟩.

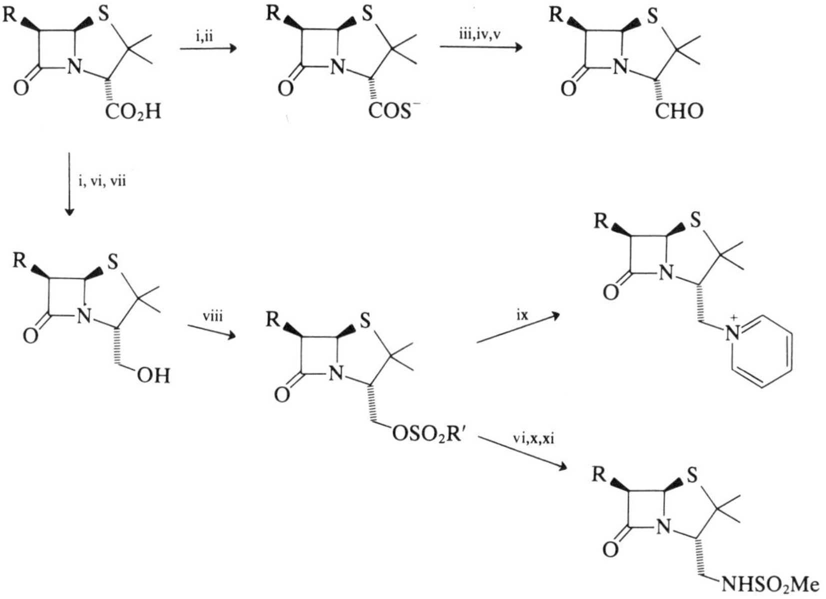

i, $ClCO_2Et$; ii, NaSH; iii, Raney nickel; iv, $(PhNHCH_2)_2$; v, TsOH; vi, NaN_3; vii, $NaBH_4$; viii, $R'SO_2Cl$; ix, pyridine; x, Pd/C, H_2; xi, $MeSO_2Cl$

Scheme 21

Some reactions of penicillanoyl diazomethane derivatives are summarized in Scheme 22 ⟨71JCS(C)3864, 80TL2451 and refs. cited therein⟩.

Application of the Curtius reaction to the 3-carboxyl of a penicillin has provided intermediates which have been used for the construction of cephem derivatives. As can be seen in Scheme 23, this route allows the selective cleavage of the C(3)—N(4) bond of the thiazolidine ring, thereby allowing a reconstruction of that ring in a different form ⟨72HCA388 and the following three papers⟩. The preparation of a related intermediate is shown in Scheme 24 ⟨76HCA2298⟩.

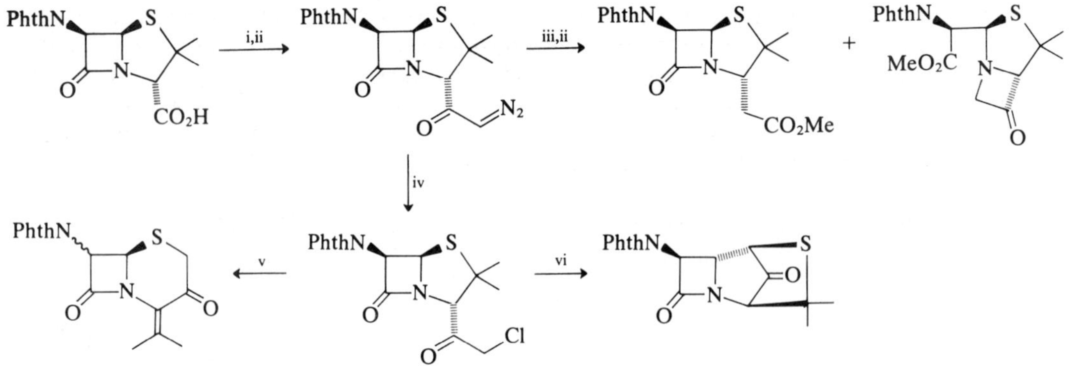

i, $ClCO_2Et$; ii, CH_2N_2; iii, $h\nu$; iv, HCl; v, Et_3N; vi, Cu acetylacetonate

Scheme 22

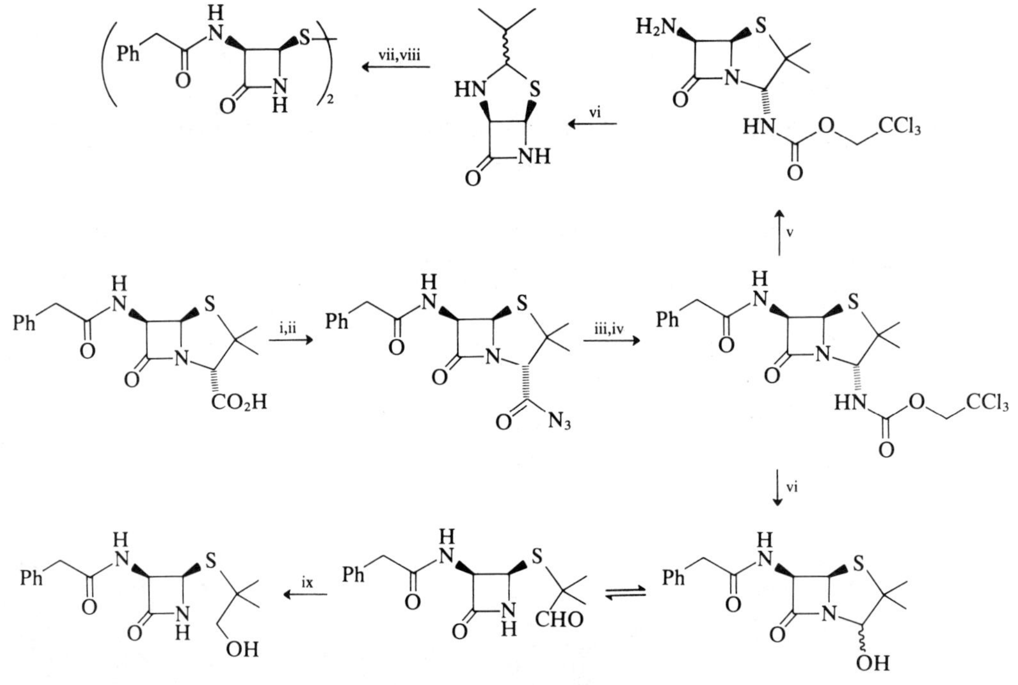

i, $ClCO_2Et$; ii, NaN_3; iii, heat; iv, $HOCH_2CCl_3$; v, PCl_5, MeOH; vi, Zn, HOAc; vii, I_2; viii, $PhCH_2COCl$; ix, $NaBH_4$

Scheme 23

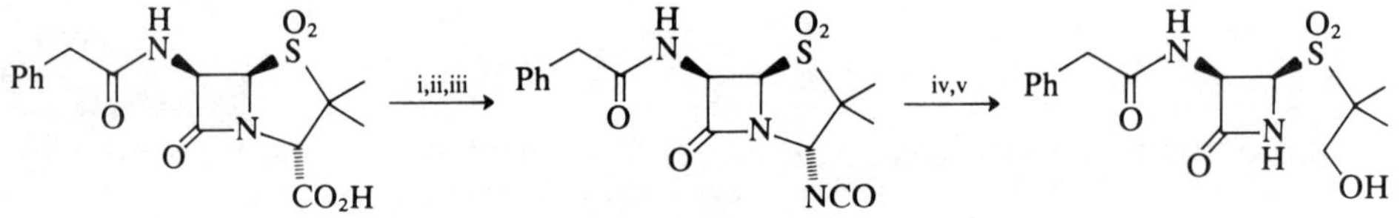

i, $ClCO_2Bu^i$; ii, NaN_3; iii, heat; iv, H_3O^+; v, $NaBH_4$

Scheme 24

When the carboxylic acid of a penicillin is converted to an acid chloride, mixed anhydride or activated ester and then treated with base, a rearrangement occurs, presumably by way of intermediate (**33**), producing a so-called anhydropenicillin (**34**) along with some rearrangement by-products (Scheme 25) ⟨68CJC2549, 69TL3381⟩. The anhydropenicillins have been used to construct analogs of the cephem ring system ⟨72CJC2902 and the preceding two papers, 74CJC3996⟩.

i, $SOCl_2$; ii, Et_3N; iii, Cl_2; iv, MeOH; v, NBS

Scheme 25

5.11.3.6 Modification at the N(4) Position

There are two principal transformations that involve the N(4) position: the cleavage of the C(3)—N(4) bond described in Section 5.11.3.5, and the cleavage of the C(7)—N(4) bond described in Section 5.11.3.9.1.

5.11.3.7 Modification at the C(5) Position

Substitution at C(5) of the penicillanic ring system has been relatively little explored. The 5α-phenyl analog has been prepared by total synthesis as shown in Scheme 26 ⟨75JMC486⟩. It was interesting that epimerization (step iv) occurs without side chain silylation (*cf.* Section 5.11.3.8.4).

i, N_3CH_2COCl, Et_3N; ii, Pt, H_2; iii, $PhOCH_2COCl$; iv, DBN

Scheme 26

A direct substitution at the 5-position of the intact penicillanic acid ring system has been reported as shown in Scheme 27 ⟨79CC485⟩.

Scheme 27

Epimerization at C(5) has not been observed under the conditions discussed for the preparation of C(6) epimers (see Section 5.11.3.8.4). It is possible to prepare 5-epipenicillins, however, as shown in Scheme 28. Note particularly the successful removal of the phthaloyl group (step iii) in this sequence, a procedure which leads to β-lactam cleavage when C(5) is the *R*, or natural, configuration. Silylation of (**35**) followed by DBN treatment afforded (**36**), which corresponds to epimerization at C(3), and (**37**), which corresponds to epimerization at C(3) and C(6). No product corresponding to only C(6) epimerization was observed ⟨76JOC2561⟩.

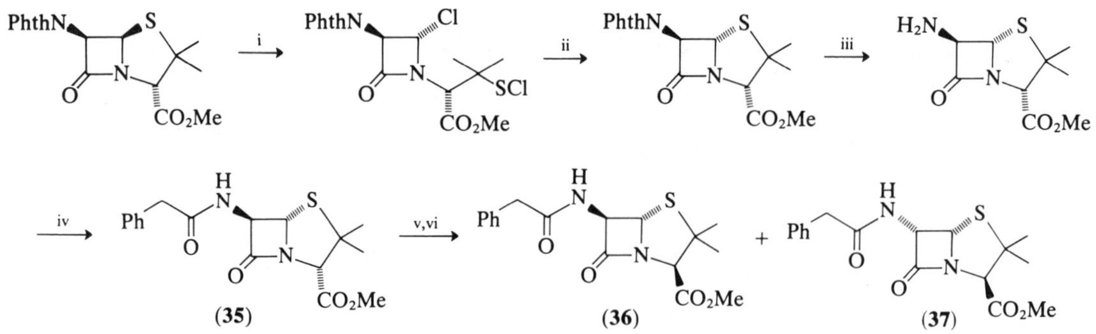

i, Cl_2; ii, $SnCl_2$; iii, N_2H_4; iv, $PhCH_2COCl$; v, *N,O*-bis($SiMe_3$) acetamide; vi, DBN

Scheme 28

5.11.3.8 Modification at the C(6) Position

5.11.3.8.1 Synthesis and reactions of 6-diazopenicillanates

When 6β-aminopenicillanic acid (6-APA) is diazotized in the presence of chloride ion, the principal product obtained is 6α-chloropenicillanic acid (**38**) ⟨62JOC2668⟩, presumably by way of the diazo intermediate (**39**; Scheme 29) ⟨72JCS(P1)895⟩. If the diazotization is carried out in the presence of excess bromide instead of chloride, significant amounts of the 6,6-dibromo derivative are obtained, and in the case of excess iodide the 6,6-diiodopenicillanic acid becomes the predominant product ⟨69JCS(C)2123⟩. The 6,6-dihalo products presumably arise from nitrous acid oxidation of halide to halogen, which then reacts with (**39**).

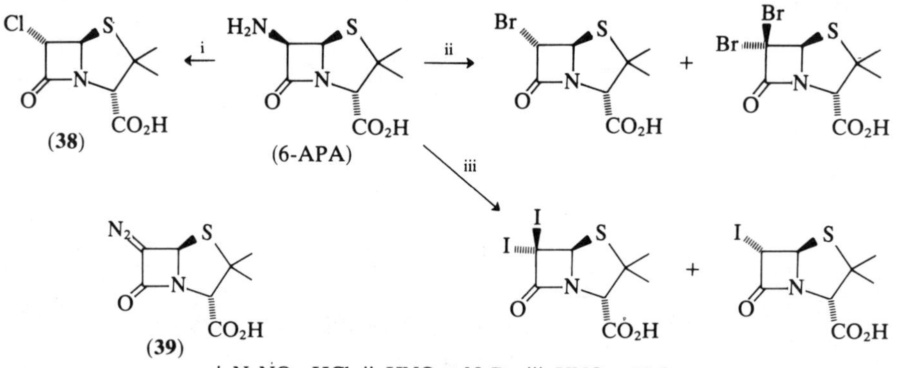

i, $NaNO_2$, HCl; ii, HNO_2+NaBr; iii, HNO_2+NaI

Scheme 29

It is possible to prepare and isolate esters of the diazo intermediate (**39**) by two different routes as shown in Scheme 30. The resulting diazo ester (**40**) can then be reacted with either halide, halogen or pseudohalogen to form the corresponding 6-substituted penicillanic acid derivatives ⟨69JCS(C)2123, 67HCA1327, 74JOC1444, 72JA1408, 77JOC2960⟩.

Some aspects of the reactions of 6-halopenicillanic acid derivatives and related compounds have been briefly reviewed ⟨B-81MI51102⟩. In general, the 6-halo- and 6,6-dihalo-penicillanic acids are resistant to nucleophilic displacement; instead, nucleophilic attack occurs at the β-lactam carbonyl, followed by rearrangement to form, for example, (**45**) from (**38**; Scheme 31 ⟨68JCS(C)2533⟩. The behavior of the pseudohalogen addition products is different, however. Thus, reaction of the isomeric mixture (**44**) with methanol and $AgBF_4$ afforded a single displacement product (**46**), which could then be converted to 6α-methoxybenzylpenicillin. Similarly, not only could benzyl 6β-bromo-6α-methoxypenicillanate (**43**) be equilibrated with LiBr to a mixture of (**43**) and (**47**), but the bromo group in each of these could be displaced by azide ion with inversion (Scheme 32) ⟨72JA1408⟩.

Halogen atoms in the 6-position of penicillanic acid are readily removed under reducing conditions (Scheme 33). Methyl 6,6-dibromopenicillanate is converted first to the 6α-bromo

i, $PhCHN_2$; ii, $NaNO_2$, TFA; iii, N_2O_4; iv, pyridine; v, HI, NaI; vi, Br_2; vii, I_2; viii, AcNHBr, MeOH; ix, BrN_3

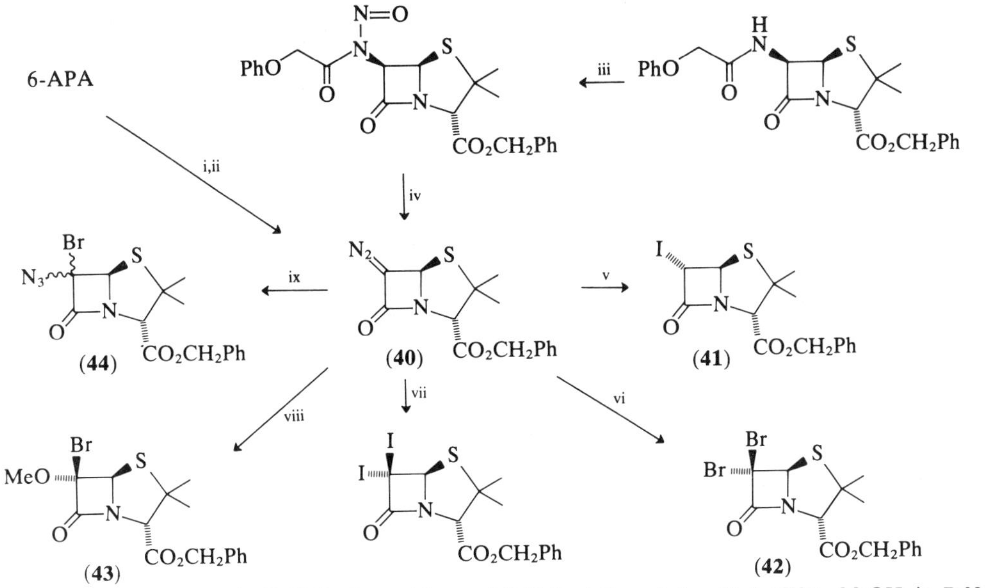

Scheme 30

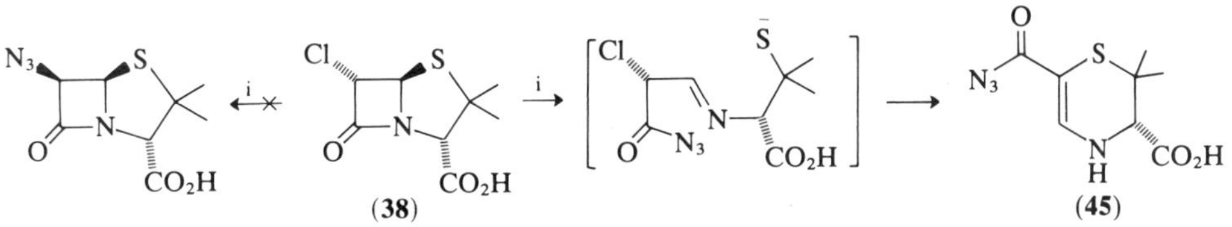

i, NaN_3, DMF

Scheme 31

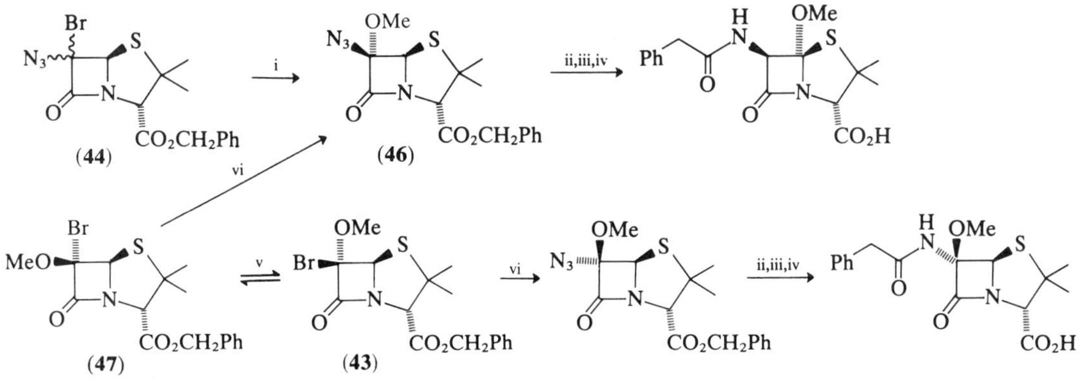

i, MeOH, $AgBF_4$; ii, Pd/C, H_2; iii, $PhCH_2COCl$; iv, Pd/C, H_2; v, LiBr, DMF; vi, LiN_3, DMF

Scheme 32

derivative, then to methyl penicillanate by catalytic hydrogenolysis ⟨69JCS(C)2123⟩. In contrast, tri-*n*-butyltin hydride reduction of trimethylsilyl 6,6-dibromopenicillanate affords first the 6β-bromo ester, then further reduction provides the penicillanate ester ⟨79TL4631⟩. 6β-Halopenicillanic acids can also be obtained by basic equilibration of the 6α-isomer ⟨80MI51103⟩, but the reductive procedure is preferred for obtaining this important (as β-lactamase inhibitors) class of penicillanic acid derivatives.

The 6α-halo- and 6,6-dihalo-penicillanates have been shown to undergo halogen–metal exchange to form enolates which can then react with acetaldehyde to form 6-(1-hydroxyethyl)penicillanates (Scheme 34) ⟨77JOC2960, 79TL3805⟩. From (**41**) the isomeric products were obtained in the ratio (**48**):(**49**):(**50**) = 24:49:27. From (**42**) the isomeric products were obtained in the ratio (**51**):(**52**):(**53**) = 39:1.5:60. Product ratios were very

i, $Pd/CaCO_3$, H_2; ii, $Bu^n{}_3SnH$

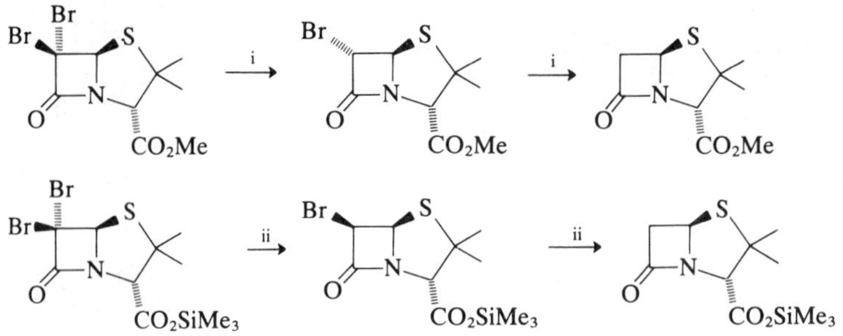

Scheme 33

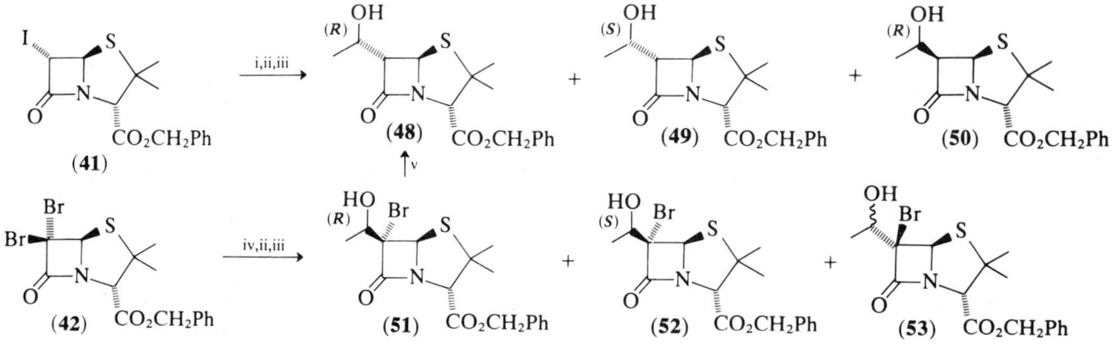

i, MeMgBr; ii, AcH; iii, H^+; iv, Bu^tLi; v, Zn–Ag, MeOH

Scheme 34

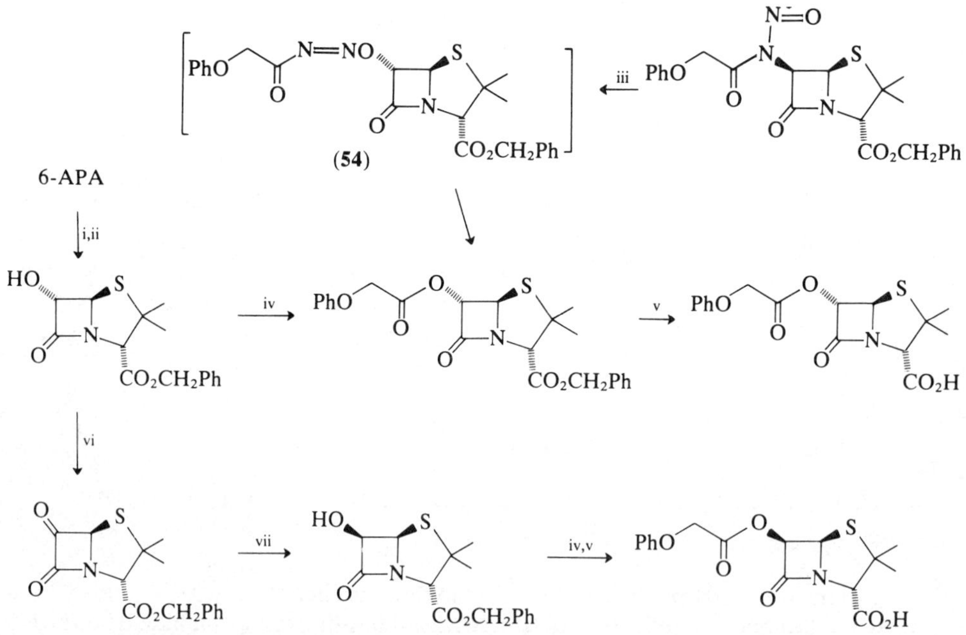

i, $NaNO_2$, TsOH, H_2O; ii, $PhCHN_2$; iii, heat; iv, $PhOCH_2COCl$; v, Pd, H_2; vi, DMSO, Ac_2O; vii, $NaBH_4$

Scheme 35

much dependent on reaction conditions, however. Compound (**48**), which contains the 6α-(*R*-1-hydroxyethyl) substituent, is important as a precursor to penems and carbapenems.

When 6-APA is diazotized in the absence of an anion more nucleophilic than water, or when a 6-diazopenicillanate is treated with aqueous nonnucleophilic acid, 6α-hydroxypenicillanates are produced ⟨67HCA1327, 74JOC1444⟩. Esters of these compounds can be formed either by acylation, or by thermal rearrangement of the corresponding *N*-

nitrosoamides (presumably *via* **54**). The corresponding 6β-hydroxypenicillanates and their esters can be obtained by oxidation of the 6α-hydroxy to 6-oxo, followed by hydride reduction which proceeds stereospecifically (Scheme 35) ⟨72JA8253, 76JCS(P1)704⟩. Alternatively, 6-oxopenicillanates can be prepared directly from 6-diazopenicillanates ⟨77JOC4045⟩ or from 6-APA esters by way of a sulfenimine intermediate ⟨80JA1690⟩. These transformations are shown in Scheme 36.

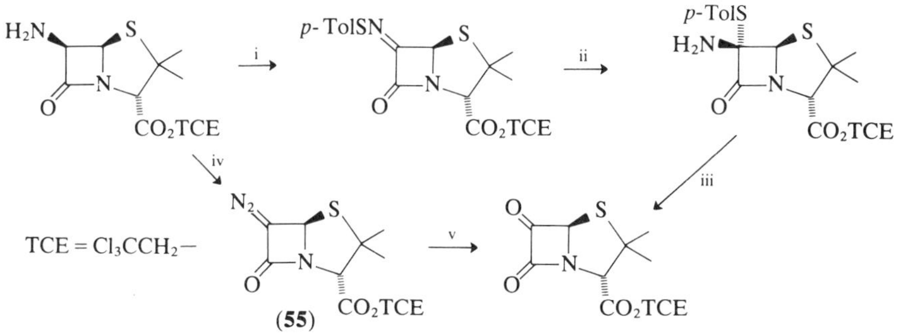

TCE = Cl_3CCH_2—

(**55**)

i, 3 eq. *p*-TolSCl; ii, Ph_3P; iii, $HgCl_2$; iv, HNO_2; v, NBS, H_2O or Ph_3P, HNO_2

Scheme 36

The 6-oxopenicillanates have found use as starting materials for several other penicillanic acid derivatives. As shown in Scheme 37, reaction with an acyliminophosphorane followed by, for example, methanol provides another route to the 6α-methoxypenicillins (*cf.* Scheme 32) ⟨75JOC191⟩. The 6-oxopenicillanates or their cyanohydrins also react with Wittig reagents, thereby providing another route to carbon–carbon bonding at the C(6) of the penicillanic acid nucleus ⟨77JOC4045⟩. An example of this as part of a preparation of a penicillin homolog is shown in Scheme 37.

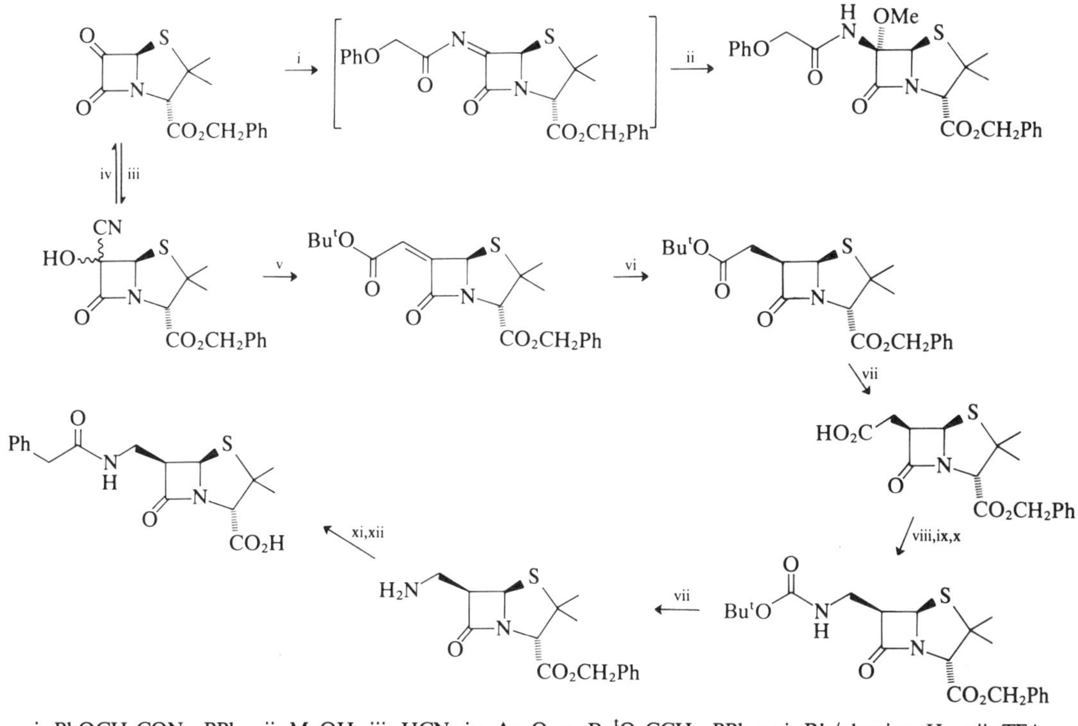

i, $PhOCH_2CON{=}PPh_3$; ii, MeOH; iii, HCN; iv, Ag_2O; v, $Bu^tO_2CCH{=}PPh_3$; vi, Rh/alumina, H_2; vii, TFA; viii, $ClCO_2Et$; ix, NaN_3; x, Bu^tOH; xi, $PhCH_2COCl$; xii, Pd/C, H_2

Scheme 37

It was mentioned earlier that 6-halopenicillanic acids are resistant to nucleophilic displacement. Displacement at the 6-position with soft nucleophiles (*e.g.* halide, RS^-) but not hard nucleophiles (*e.g.* MeO^-, amines) can be carried out, however, on 6-trifloxy- and 6-nonafloxy-penicillanate esters ⟨80TL2991⟩. Some examples are shown in Scheme 38.

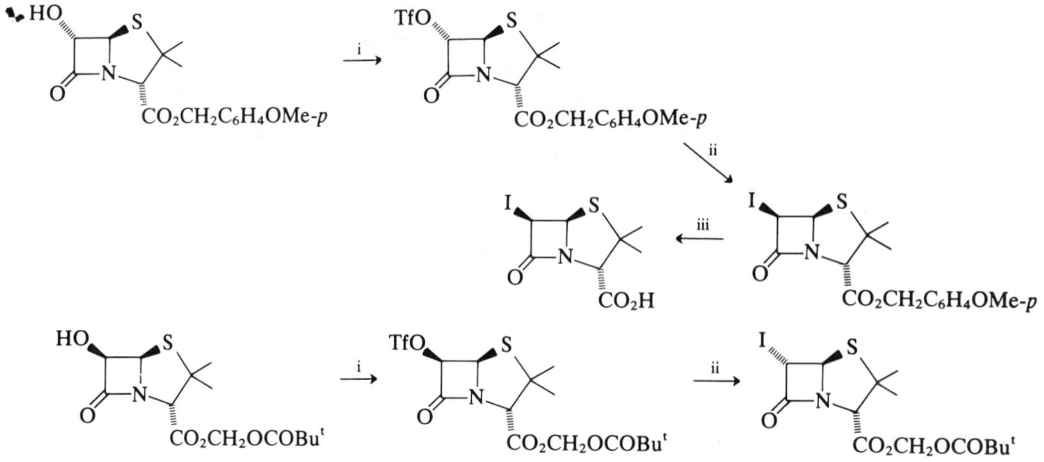

i, TfCl, Et_3N; ii, NaI; iii, TFA

Scheme 38

When 6-diazopenicillanates are irradiated in the presence of sulfur nucleophiles, predominantly 6β-substitution products are obtained ⟨77JOC2224⟩. When $BF_3 \cdot Et_2O$ is used to catalyze the reaction with nucleophiles, however, the products are primarily the 6α-isomers ⟨78TL995⟩. The use of rhodium or copper catalysis led primarily to ring-opened thiazepine products, presumably by way of the intermediate (**56**; Scheme 39) ⟨80CC798⟩.

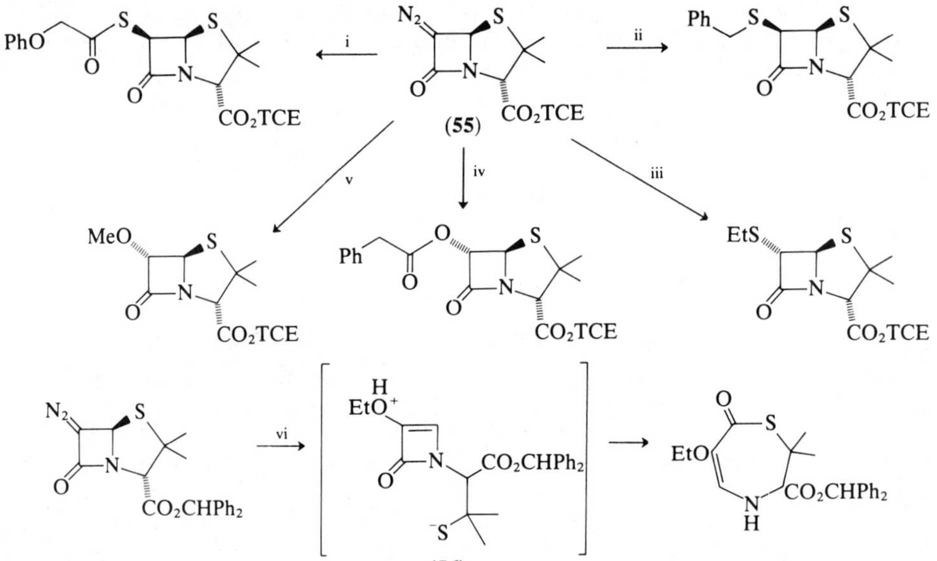

i, $PhOCH_2COSH$, $h\nu$; ii, $PhCH_2SH$, $h\nu$; iii, EtSH, $BF_3 \cdot Et_2O$; iv, $PhCH_2CO_2H$, $BF_3 \cdot Et_2O$; v, MeOH, $BF_3 \cdot Et_2O$; vi, EtOH, $Rh_2(OAc)_4$ or Cu acetylacetonate

Scheme 39

When the reaction of acetaldehyde with a 6-diazopenicillanate is catalyzed by $BF_3 \cdot Et_2O$, an epoxide of unknown stereochemistry is obtained ⟨79H(13)227⟩. With $ZnCl_2$ catalysis, however, the 6α-acetyl derivative is obtained, which can be stereospecifically reduced as the Mg chelate to the 6α-(R-1-hydroxyethyl) derivative as part of an elegant synthesis of the carbapenem thienamycin (Scheme 40) ⟨81JA6765⟩.

5.11.3.8.2 C(6) substitution via C(6) carbanions

As will be pointed out in Section 5.11.3.8.4, epimerization at C(6) under basic conditions involves the formation of a carbanion at C(6). Using Schiff bases to activate the C(6) proton,

(55)

(40)

(48)

i, AcH, $BF_3{\cdot}Et_2O$; ii, AcH, $ZnCl_2$; iii, $Mg(O_2CCF_3)_2$, $BH_3{\cdot}HNPr^i_2$

Scheme 40

advantage has been taken of this anion formation to introduce various alkyl and acyl groups at the 6-position of the penicillanic acid nucleus. An example of such a sequence is shown in Scheme 41 ⟨73JOC230⟩. Using similar procedures it has been possible to introduce a variety of C(6) substituents, and some of these are listed in Scheme 41 ⟨72TL4917, 73TL145⟩. The new substituents have been assigned the α stereochemistry, based on X-ray and NMR evidence, which reflects electrophilic attack on the anion from the less hindered α face. A similar approach can be taken for the introduction of a 6α-methoxy group as shown in Scheme 42 ⟨73JOC2857⟩.

R' = CH_2OH, CH_2F, CH_2NH_2, Ac, CO_2Me, CO_2Na

i, $KOBu^t$, MeI; ii, TsOH; iii, $PhOCH_2COCl$; iv, $Pd/CaCO_3$, H_2

Scheme 41

In a like manner the 6β-isocyano group can be utilized to activate the C(6) proton ⟨79JCS(P1)2455⟩, an example of which is shown in Scheme 43. It was also possible to introduce the 6α-methylthio substituent by reacting the intermediate anion with methyl methoxycarbonyl disulfide. The 6α-methylthio group could subsequently be converted to the 6α-methoxy group by treatment with Cl_2 and MeOH.

i, NaH; ii, $MeSSO_2Me$; iii, TsOH; iv, MeOH, $HgCl_2$; v, $PhCH_2COCl$; vi, Pd/C, H_2

Scheme 42

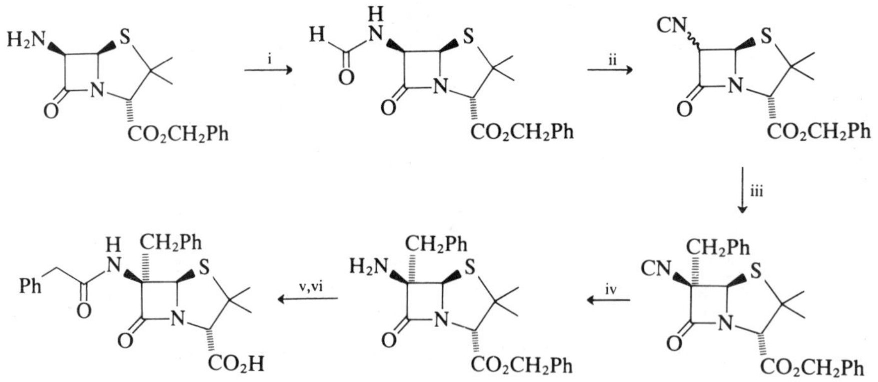

i, HCO_2H, DCC; ii, $COCl_2$; iii, $PhCH_2Br$, K_2CO_3; iv, TsOH; v, $PhCH_2COCl$; vi, Pd/C, H_2

Scheme 43

5.11.3.8.3 C(6) substitution via 6-imino derivatives

6-Iminopenicillanic acid derivatives have been particularly useful in the preparation of 6α-methoxypenicillanates (note that the 6-diazo group is a special case of the 6-imino function, and that a 6-acyliminopenicillanate was obtained from the 6-keto derivative in Scheme 37).

In Scheme 36 the preparation of a 6-sulfeniminopenicillanate was shown. This could be converted directly to the 6α-methoxypenicillanate as shown in Scheme 44. Alternately, the 6-sulfenimino derivative may be generated from the 6β-sulfenamide by MnO_2 oxidation, followed by introduction of the 6α-methoxy group with LiOMe ⟨79CPB2718⟩.

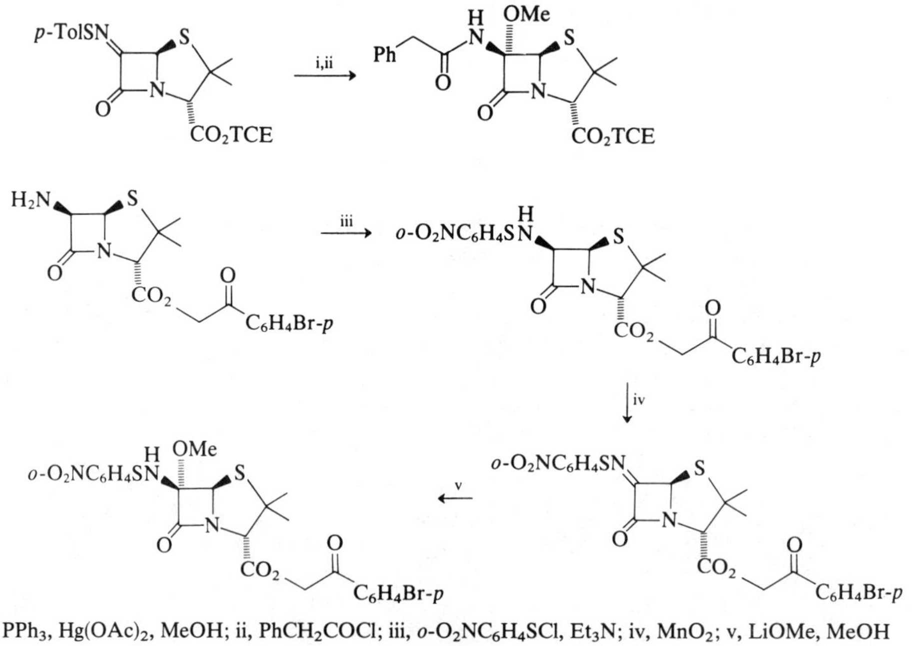

i, PPh_3, $Hg(OAc)_2$, MeOH; ii, $PhCH_2COCl$; iii, o-$O_2NC_6H_4SCl$, Et_3N; iv, MnO_2; v, LiOMe, MeOH

Scheme 44

The Schiff base intermediate (**57**) permits the oxidative formation of an imino intermediate which can then be converted to the 6α-methoxy derivative (Scheme 45) ⟨76MI51100⟩.

Another route to a certain class of 6α-methoxypenicillanates ⟨77TL3831⟩ also probably involves the intermediacy of a 6-imino species. As shown in Scheme 46, 6β-ketenimino intermediates are readily formed with penicillins having this kind of side chain. Chlorination followed by treatment with methoxide affords the 6α-methoxy-6β-ketimine, probably involving the intermediates shown.

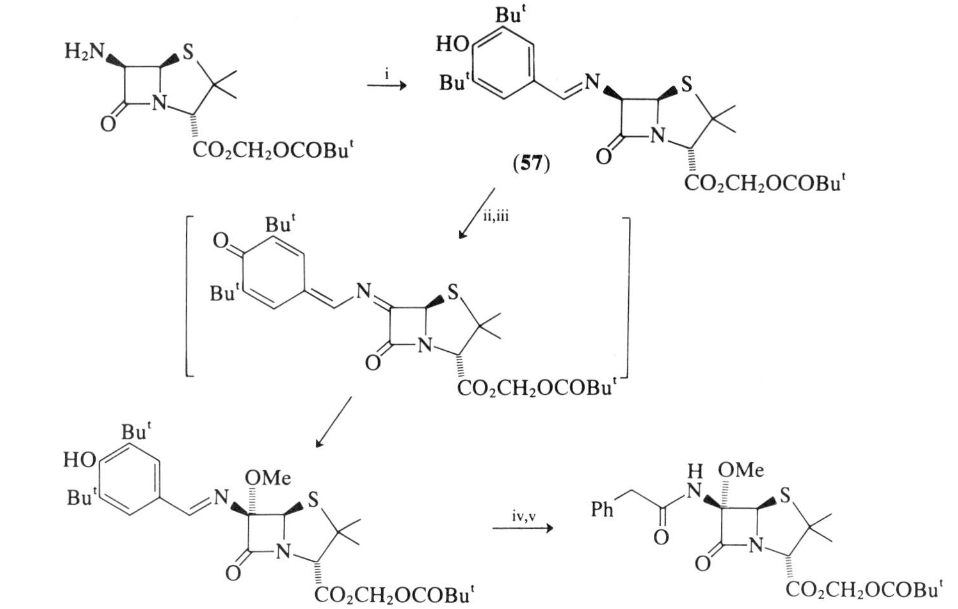

i, 3,5-Bu^t_2-4-HOC_6H_2CHO; ii, LiOMe, MeOH; iii, Bu^tOCl; iv, Girard reagent T; v, $PhCH_2COCl$

Scheme 45

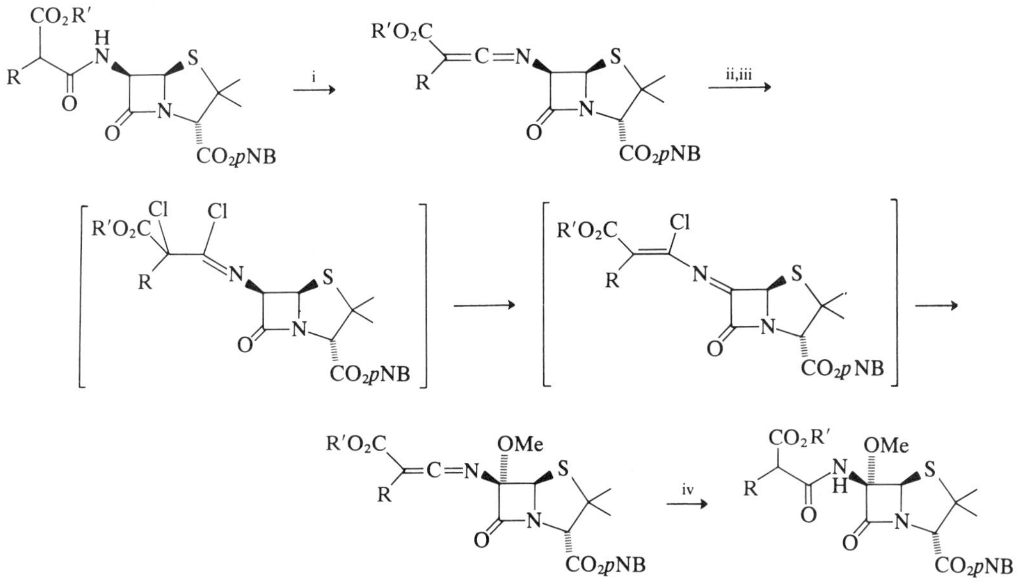

i, PCl_5, pyridine; ii, Cl_2; iii, LiOMe, MeOH; iv, H_3O^+

Scheme 46

A very direct, one-step procedure for the preparation of a particular 6α-methoxypenicillanate, which also presumably goes through a 6-acylimino intermediate, has been described (Scheme 47) ⟨81AG(E)808⟩. Note that this route does not even require blocking of the 3-carboxyl.

The introduction of N substitution at the 6α-position has also been achieved by a sequence possibly incorporating a 6-imino intermediate (Scheme 48) ⟨76JCS(P1)1918⟩.

5.11.3.8.4 Epimerization at C(6)

By virtue of the stereochemistry at C(5) and C(6), the penicillin molecule exists in a configuration that is thermodynamically relatively unstable. Epimerization at either C(5)

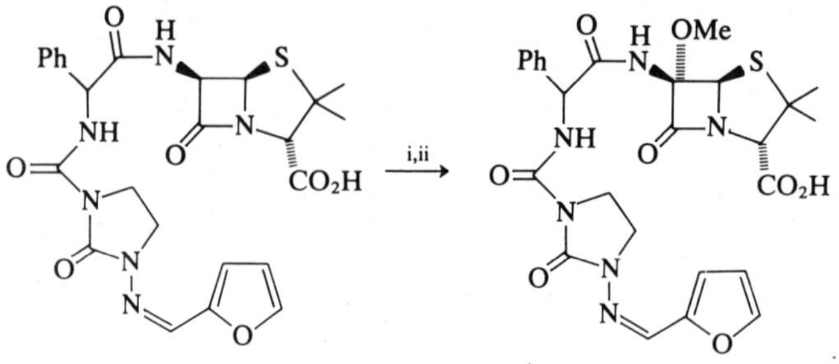

i, LiOMe, MeOH; ii, Bu^tOCl

Scheme 47

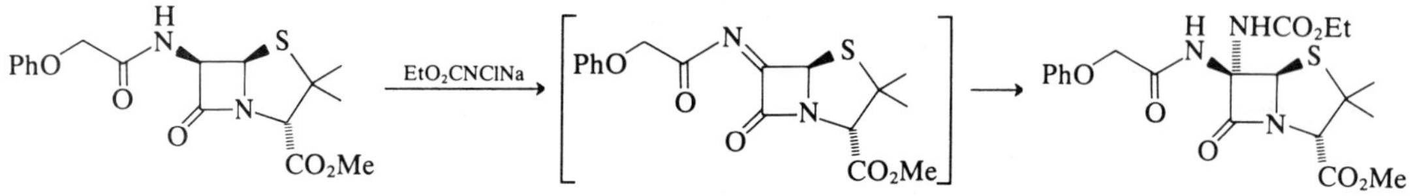

Scheme 48

or C(6) would lead to a species of greater stability because of decreased steric interaction between, especially, the 6β-substituent, the 3β-hydrogen and the 2β-methyl group. In order for epimerization to occur at C(6) under basic conditions it is necessary that the C(6) proton be sufficiently acidic to form the assumed carbanion intermediate, and that the relatively acidic amide hydrogen on the C(6) side chain be masked ⟨69CC129⟩. Thus, epimerization of (**58**) with base, *e.g.* Et_3N, is unsuccessful. When, however, the side chain amide is converted to the imino ether by silylation, both epimerization to (**59**) (presumably *via* **61**) and rearrangement to (**60**) (presumably *via* **62**) are observed (Scheme 49) ⟨72TL285⟩. The amount of rearrangement product can be reduced by using DBN as the base. Similarly, compounds (**63**)–(**66**) epimerize under basic conditions ⟨68TL1903, 71CC647⟩ but 6-APA, (**67**) and (**68**) do not. Sulfoxide formation enhances the ease of 6-epimerization, and either the α- or the β-sulfoxide of (**58**) will epimerize either with DBN treatment alone ⟨74JOC441⟩, or by silylation without added base. The role of stable ion pairs in base-catalyzed sulfoxide epimerization has been discussed ⟨81JOC2046⟩. Sulfone formation also enhances 6-epimerization, and (**69**) can be epimerized at C(6) with DBN without prior silylation ⟨78JCS(P1)1366⟩. It should be pointed out that, at least under some conditions of base treatment, the kinetically favored (as opposed to equilibrium favored) product of anion protonation, the 6β derivative, is obtained ⟨76CC288⟩.

5.11.3.8.5 Other modifications of the 6-amino group

The 6β-amino group of 6-APA may be alkylated either with diazoalkanes ⟨67LA(702)163⟩ or by the reduction of Schiff bases (Scheme 50) ⟨65JCS3616⟩. Two special cases of *N*-alkylation are also shown in Scheme 50: the formation of an imidazolidinone ring upon treating ampicillin with acetone ⟨66JOC897⟩, and the formation of a 6β-amidinopenicillanic acid from 6-APA ⟨77MI51105⟩.

5.11.3.9 Modification at the C(7) Position

5.11.3.9.1 β-Lactam cleavage by Nucleophiles

By virtue of their fused β-lactam–thiazolidine ring structure, the penicillins behave as acylating agents of a reactivity comparable to carboxylic acid anhydrides (see Section 5.11.2.1). This reactivity is responsible for many of the properties of the penicillins, *e.g.* difficult isolation due to hydrolytic instability ⟨B-49MI51102⟩, antibacterial activity due to irreversible transpeptidase inhibition (Section 5.11.5.1), and antigen formation *via* reaction with protein molecules.

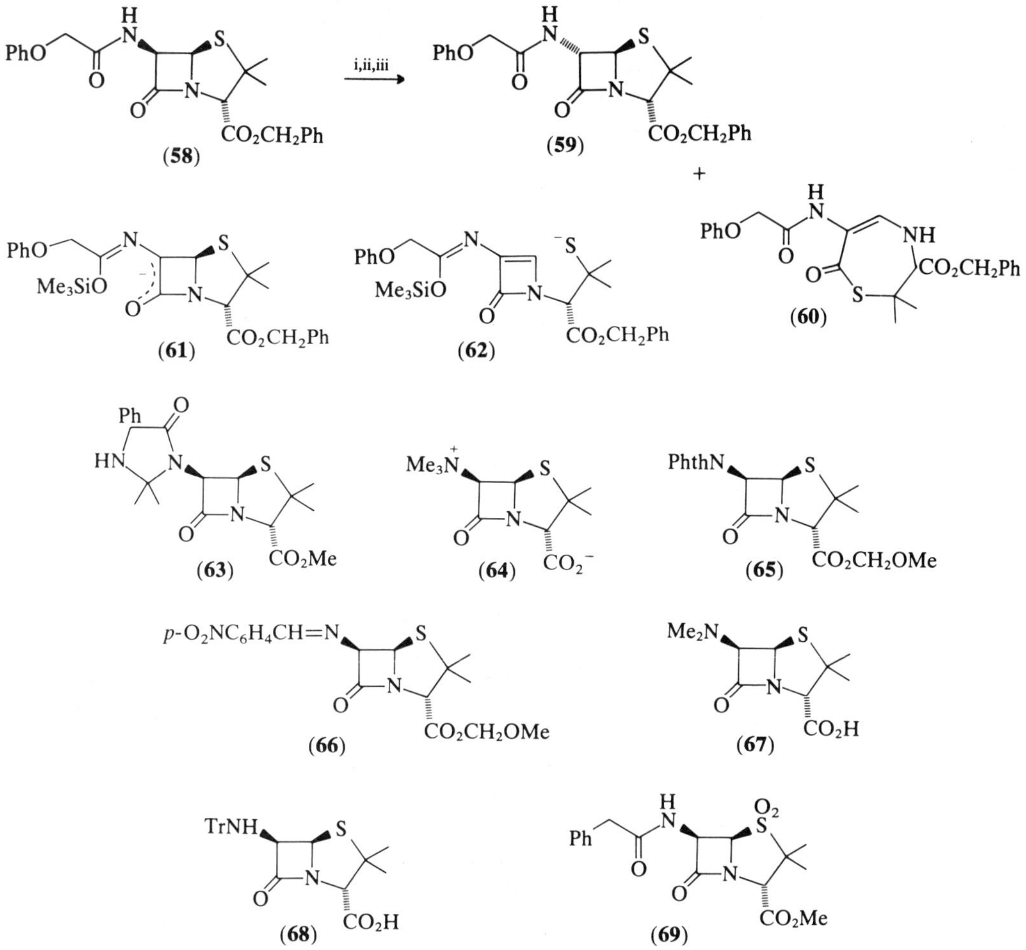

(58) (59) (60) (61) (62) (63) (64) (65) (66) (67) (68) (69)

i, *N,O*-bis(Me_3Si)acetamide; ii, Et_3N; iii, H_2O

Scheme 49

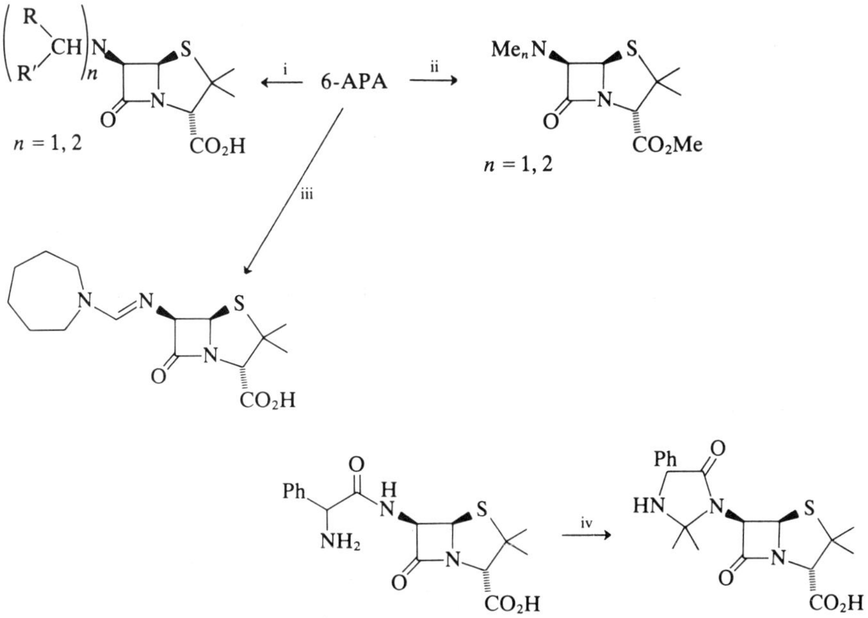

i, RCOR′, Pt, H_2; ii, CH_2N_2; iii, $(CH_2)_6NCH(OMe)_2$; iv, AcMe, pH 7.5

Scheme 50

The behavior of penicillins under hydrolysis conditions has been extensively studied ⟨*e.g.* B-49MI51102, 71JPS503⟩, and some of the early work has been summarized in Section 5.11.3.2. Generally, the rate of β-lactam hydrolysis for various penicillins reaches a minimum between pH 6 and 7, and increases sharply above and below that minimum ⟨74CPB1186⟩. Under basic conditions, hydrolysis rates are relatively insensitive to the 6-substituent, and the product is the relatively stable penicilloate and, at least in some cases, its 5-epimer (Scheme 51) ⟨77JHC503⟩.

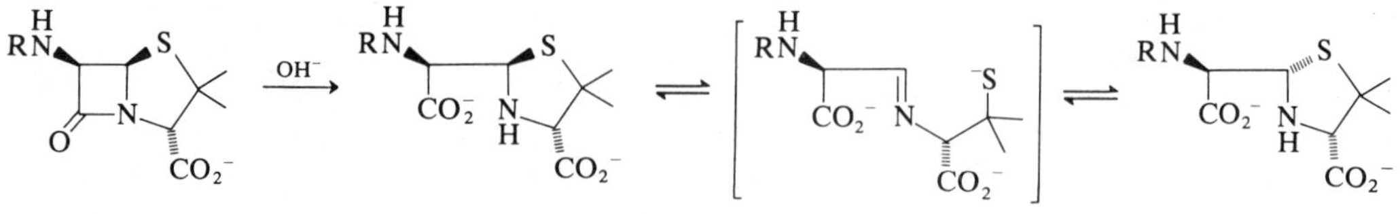

Scheme 51

Under approximately neutral conditions, where the degradation rate is independent of pH for a range of pH values characteristic of the individual penicillin, hydrolysis appears to occur through a general base-catalyzed attack of a water molecule ⟨77JPS861⟩ on the β-lactam carbonyl.

Under acidic conditions the degradation process becomes more complex. The substituent on the 6-position has an important effect on the rate of degradation. Thus, acid stability has been found to increase as follows: 6-APA > ampicillin, cyclacillin > phenoxymethylpenicillin, cloxacillin > benzylpenicillin, methicillin. These effects have been explained by postulating an initial intramolecular attack of the side chain amide on the β-lactam to form a penicillenic acid intermediate (**15**; Scheme 52), with increasing nucleophilicity of the side chain amide correlating with decreasing acid stability. The course of the reaction is not simple, however. Scheme 52 represents the route of benzylpenicillin degradation at pH 2.5 and 37 °C as determined by an NMR and deuterium incorporation study ⟨79JCS(P2)86⟩. After 100 min the benzylpenicillin has essentially disappeared, and the reaction mixture consists of 57% benzylpenillic acid (**10**), 24% benzylpenamaldic acid (**70**) and 19% benzylpenicilloic acid as an epimer mixture (**71**). After 10 days the mixture is 25% (**10**) and 75% benzylpenilloic acid as an epimer mixture (**72**). The intermediate benzylpenicillenic

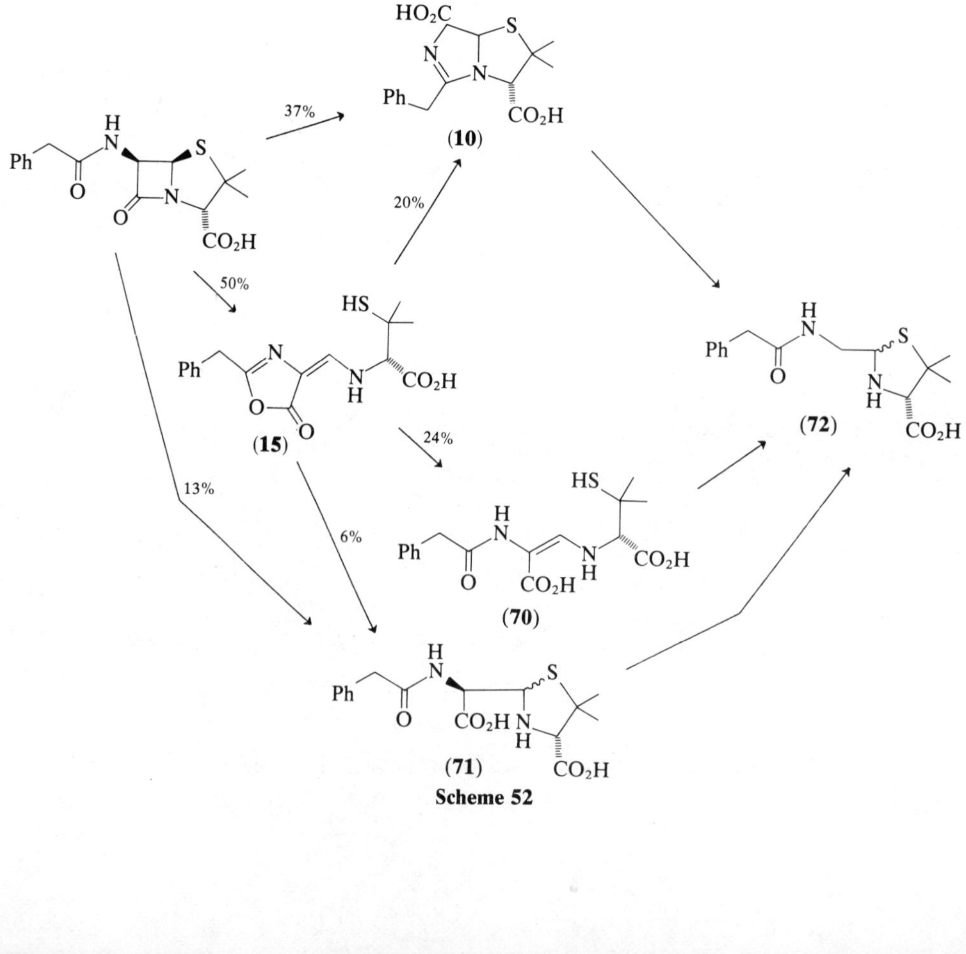

Scheme 52

acid (**15**) can be detected by its characteristic UV spectrum, but is present only in small concentrations. Furthermore, in order to explain the deuterium incorporation pattern, three degradation pathways as shown in Scheme 52 were postulated. Note, however, that mechanistic uncertainty still remains ⟨81JCS(P2)1247⟩.

Various chemical species influence the rates of hydrolysis of penicillins, *e.g.* metal ions ($Cu^{2+} > Zn^{2+} > Ni^{2+} \sim Co^{2+}$) ⟨80JCS(P2)1725⟩, carbohydrates ⟨78MI51101⟩, certain amine-containing catechol derivatives ⟨69JPS1102⟩ and β-cyclodextrin ⟨71JA767⟩. Some of these even show some of the characteristics of enzyme-catalyzed hydrolyses.

The most thoroughly studied mode of β-lactam hydrolysis is that catalyzed by the enzyme β-lactamase (EC 3.5.2.6). The elaboration of this enzyme is one of the three principal ways in which a bacterium can obtain resistance to a β-lactam antibiotic (see Section 5.11.5.1), and much of the chemical work carried out on the penicillin molecule has been related to attempts to deal with this problem. A discussion of the β-lactamases is beyond the scope of this work. The reader is referred to ⟨B-79MI51101⟩ for a recent review.

Another example of nucleophilic cleavage of the β-lactam ring involves the reaction of amines to form penicilloic amides. This subject has been of particular interest because it has been felt that the principal antigenic determinant in penicillin allergy is the penicilloyl group formed by the reaction of the penicillin molecule with protein amines. The reaction of amines with penicillins has been found to involve attack of the amine on the α face of the molecule to form a tetrahedral intermediate (Scheme 53), to be subject to intermolecular general base catalysis and, when conditions and reactants are appropriate, to be subject to intramolecular general base and general acid catalysis ⟨80JCS(P2)212⟩.

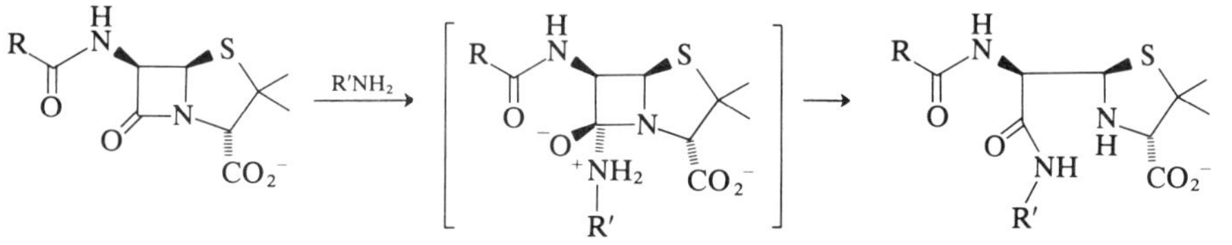

Scheme 53

A special case of β-lactam aminolysis is involved in the polymerization of penicillins, and particularly those penicillins containing amino groups in their side chains. The study of this reaction has also been of some importance because of the possible role of penicillin polymers in the allergic reactions of penicillin-sensitive patients. This subject has been recently reviewed ⟨80MI51104⟩. The structures which have been assigned to the polymers of 6-APA, benzylpenicillin and ampicillin are shown in Scheme 54. Of these, the ampicillin polymerization has been the most thoroughly studied, and it is interesting that the major polymer products are those in which $n = 0$, 2 and 4. It should be noted, however, that it is quite possible that the polymerization products are more complex and less uniform than is indicated in the scheme ⟨78MI51102⟩.

5.11.3.9.2 Other modifications at C(7)

Modification at the C(7) position of the penam ring system (other than ring opening reactions) has not been extensively studied. It was possible, however, to convert the β-lactam to a β-thionolactam in 1% yield as shown in Scheme 55 ⟨75JA5628⟩. The deblocking product (**73**) had greatly reduced antibacterial activity compared to the parent β-lactam.

5.11.4 SYNTHESIS

5.11.4.1 Fermentation Processes

The penicillins began as natural products, and the most common methods for preparing penicillins and their derivatives are still by direct fermentation or by synthetic modification of fermentation-derived starting materials. A number of different microbial species, mostly fungi, have been found to produce penicillins ⟨see B-72MI51102, B-82MI51105 for reviews⟩, but most development work has been carried out on the fungus *Penicillium chrysogenum*. It

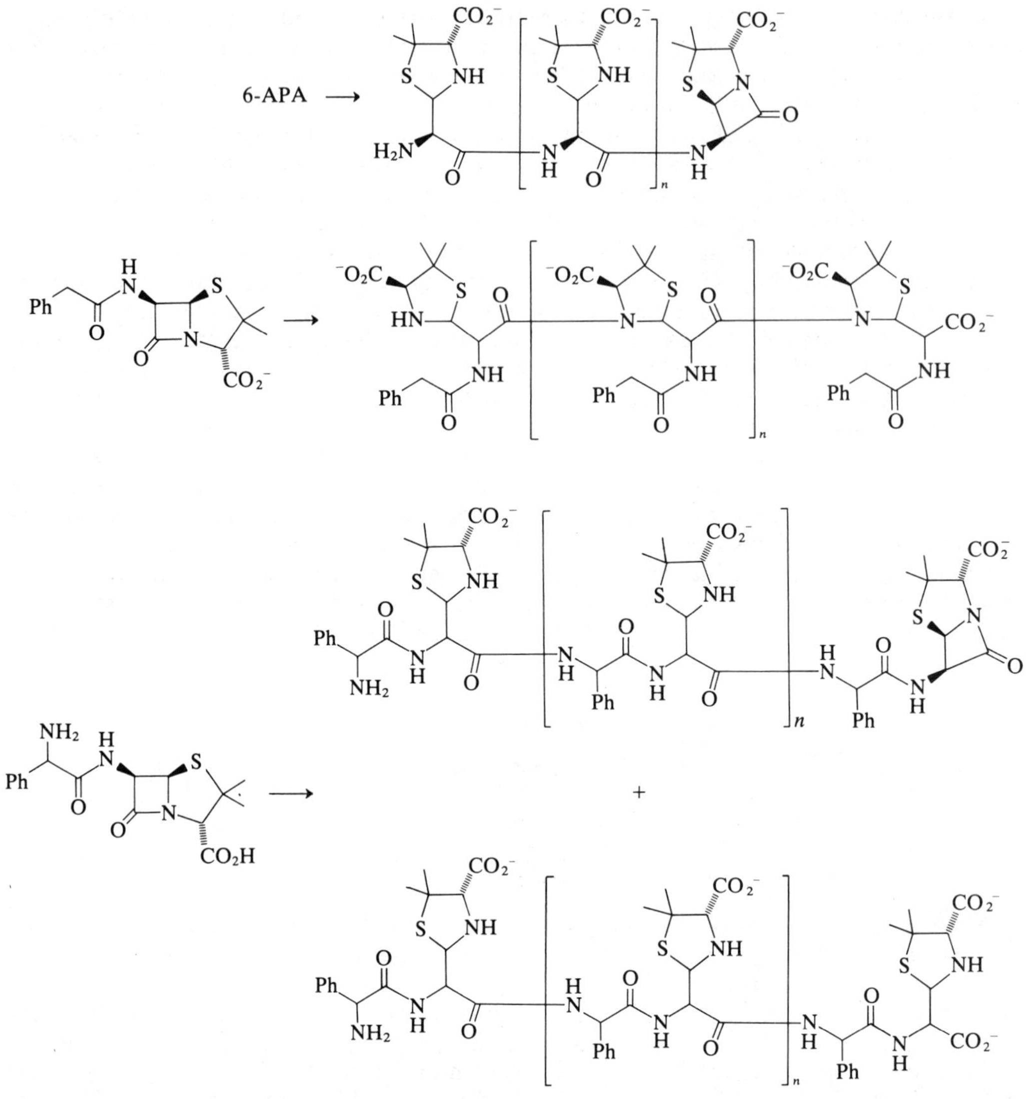

6-APA →

Scheme 54

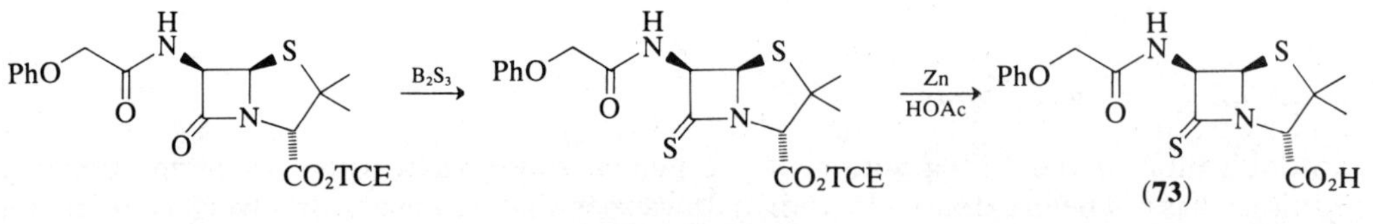

Scheme 55

was recognized during the early penicillin studies that, during fermentations, a series of penicillins was produced which differed in the C(6) side chain, and that the mixture of products was dependent upon the composition of the medium. Furthermore, it was found that, within limits, the penicillin production could be controlled by the addition of side chain precursor acids to the fermentation medium ⟨B-49MI51105⟩. Today benzylpenicillin and phenoxymethylpenicillin are produced commercially with controlled fermentation media and *P. chrysogenum* mutants selected for their high productivity, and recent genetic technology is expected to make this production even more efficient ⟨81MI51103⟩. Almost all the penicillin derivatives described in this chapter are derived from one of these starting materials.

The mechanism by which penicillins are synthesized during *P. chrysogenum* and *Cephalosporium acremonium* fermentation is being extensively investigated ⟨see B-81MI51100 for recent review⟩. Many of the findings can be summarized as shown in Scheme 56. L-α-Aminoadipic acid, L-cysteine and L-valine combine to form (L-α-amino-δ-adipyl)-L-cysteinyl-D-valine (**74**) (the Arnstein tripeptide), which is the only isolated and identified intermediate between the constituent amino acids and isopenicillin N (L-4-amino-4-carboxybutylpenicillin; **75**). During this step L-valine is converted to D-valine, and this is accomplished with exchange of H(2), but without exchange of H(3) ⟨81T1589⟩ or N ⟨76CC110⟩. Then, in an oxidative step which is not understood in detail, (**74**) is converted into (**75**). This is a highly stereospecific step, with there being complete retention of stereochemistry at C(3) of cysteine and at C(2) and C(3) of valine (in Scheme 56 the circled protons are incorporated without exchange). In *P. chrysogenum* isopenicillin N is then presumably converted to other penicillins (perhaps by way of 6-APA), depending on the composition of the fermentation medium. With *C. acremonium* isopenicillin N is converted to penicillin N and then to cephalosporin derivatives.

(**74**)

(**75**)

Scheme 56

The number and kind of different penicillins which can be produced by medium supplementation are quite limited. In general, only monosubstituted acetic acids, such as phenylacetic acid (penicillin G), *p*-hydroxyphenylacetic acid (penicillin X), phenoxyacetic acid (penicillin V), 3-hexenoic acid (penicillin F) and octanoic acid (penicillin K) are incorporated into the penicillin side chain. Under fermentation conditions in which the medium is depleted with respect to side chain precursor acids, however, it is possible to isolate the nucleus, 6-APA. A more efficient method for the production of 6-APA was developed following the discovery that a large number of microorganisms, including many species of bacteria and fungi, produce enzymes, called acylases, which can hydrolyze the side chains of fermentation-produced penicillins, such as benzylpenicillin and phenoxypenicillin ⟨B-72MI51103⟩. In this way, large amounts of 6-APA became available from which, as will be described later (Section 5.11.4.3), a large variety of penicillins could be prepared.

5.11.4.2 Chemical Cleavage of Penicillin Side Chain Amides

At first glance it would seem very difficult to cleave the side chain amide of a penicillin, which is relatively stable to hydrolysis conditions, without at the same time cleaving the β-lactam, which is quite susceptible to hydrolysis. By taking advantage of the fact that the side chain amide is secondary and the β-lactam is tertiary, however, such a selective hydrolysis is indeed possible ⟨70RTC1081, B-72MI51103, 74FOR(31)1⟩. As shown in Scheme 57, a benzylpenicillin ester can be converted by way of an imino chloride to an imino ether which can be hydrolyzed under very mild conditions. Although other carboxyl blocking groups have been used in this sequence, the silyl ester is particularly effective because it does not have to be removed in a separate step. This method has been modified and developed into a highly efficient, large-scale process, which has displaced enzymatic cleavage for the commercial production of 6-APA.

Several approaches have been investigated which would permit the exchange of one acyl group for another on the 6β-amino of 6-APA without actually going through 6-APA as an intermediate. Scheme 58 illustrates one of these ⟨72YZ454, 72YZ459⟩. In principle, this sequence, which presumably goes through the intermediate (**76**), is applicable to the exchange of any two acyl groups.

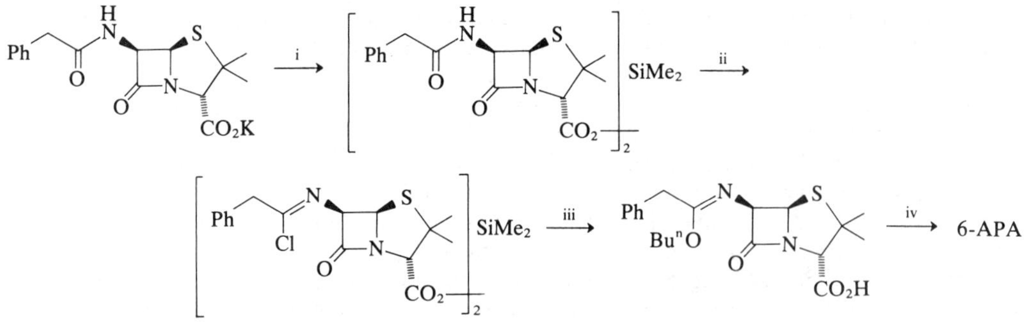

i, Me_2SiCl_2, $PhNMe_2$; ii, PCl_5; iii, Bu^nOH; iv, H_3O^+

Scheme 57

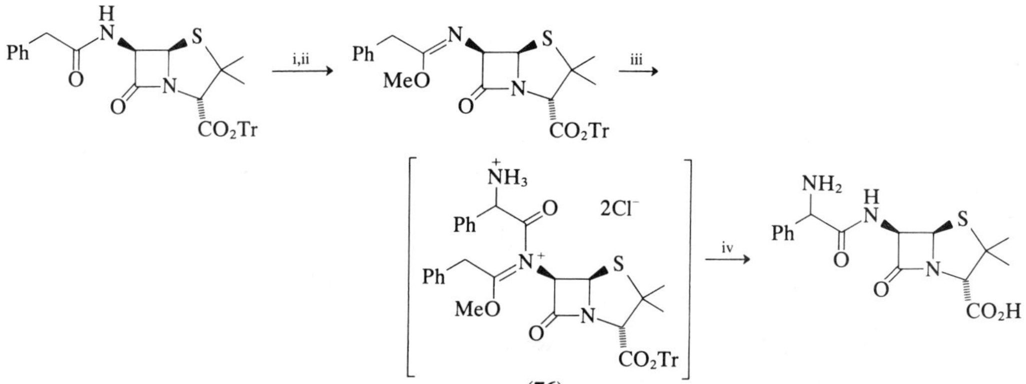

(76)

i, PCl_5, $PhNMe_2$; ii, MeOH; iii, $PhCH(NH_2 \cdot HCl)COCl$; iv, H_3O^+

Scheme 58

5.11.4.3 Acylation of 6β-Aminopenicillanic Acid

The acylation of 6-APA (Scheme 59) has been a very versatile way in which to generate new penicillin derivatives which differ from fermentation-produced penicillins in the 6-side chain. As will be discussed in Section 5.11.5.1, this approach has led to significant improvements in the therapeutic properties of penicillins, and, in fact, of the penicillins in medical use today, only benzylpenicillin and phenoxymethylpenicillin are produced directly by fermentation.

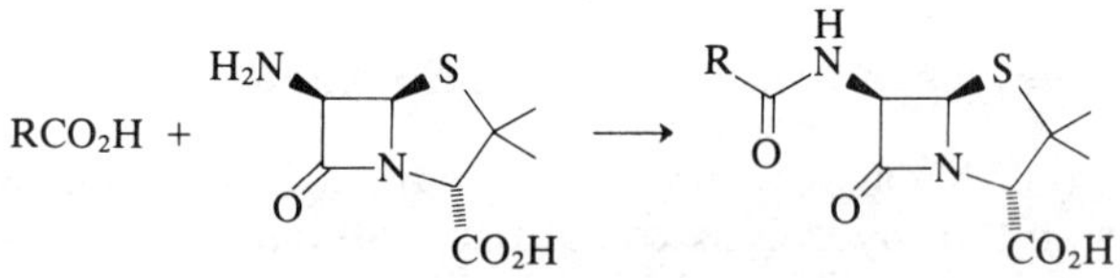

Scheme 59

It is probably safe to say that virtually every method of forming an amide bond that does not involve extremes of temperature or pH has been applied to the acylation of 6-APA ⟨see 71MI51100, B-72MI51104, 74FOR(31)1 for reviews⟩. The most commonly used methods are probably the acid chloride route ⟨*e.g.* 60JA3934⟩ and the mixed carboxylic–carbonic acid anhydride route ⟨*e.g.* 62JMC1016, 79MI51100⟩. Many other methods, however, have been described, *e.g.* mixed anhydrides with pivalic acid ⟨65CB789⟩, diethyl phosphate ⟨76T2211⟩ and dibenzyl phosphate ⟨80MI51101⟩, *N,N′*-dicyclohexylcarbodiimide ⟨68ACS367⟩, 1,1′-carbonyldiimidazole ⟨72JMC333⟩, *N,N*-dimethylchloroformiminium chloride ⟨65E360⟩, '*N,N*-bis(2-oxo-3-oxazolidinyl)phosphorodiamidic chloride' ⟨80TL4179⟩, esters of *p*-nitrophenol and 2,4-dinitrophenol ⟨68JMC70⟩ and various intramolecular anhydrides ⟨64JA3870, 77BSB991⟩. Various amine blocking groups have been found to be useful when acylating 6-APA with an acid which also contains an amino group. The most commonly used blocking groups have probably been the proton ⟨66JOC897⟩, the benzyloxycarbonyl and *p*-nitrobenzyloxycarbonyl groups ⟨71JCS(C)1920, 71MI51101⟩, and the enamines derived from β-dicarbonyl compounds ⟨65CB789, 71JCS(C)1920⟩. Other blocking systems include the *o*-nitrophenylsulfenyl group ⟨65ACS1245⟩, the trityl group ⟨62MI51100⟩, the acetoacetyl group ⟨80YZ29⟩, the 2-methyl-2-(*o*-nitrophenoxy)propionyl group ⟨66JOC2560⟩ and the azide as an amine precursor ⟨65ACS281⟩. Amine blocking groups which have been used to protect the 6β-amine of 6-APA include the phthaloyl group ⟨75JA5582, 77MI51101⟩, the trityl group ⟨70JMC389⟩, the *p*-nitrobenzylidene group ⟨72TL4917⟩ and the 2,2,2-trichloroethoxycarbonyl group ⟨72HCA417⟩.

Various groups have been employed to block the 3-carboxylic acid group of penicillanic acid derivatives while chemical reactions are carried out elsewhere in the molecule. Most commonly used have been the silyl esters referred to previously ⟨see also 64LA(673)166⟩ and the benzyl and *p*-nitrobenzyl groups ⟨*e.g.* 65MI51100, 72CC192, 78JOC1243⟩. Other blocking groups have included phenacyl esters ⟨67ACS2210⟩, tin(II) esters ⟨68ACS367⟩, trityl esters ⟨72YZ454⟩, 3,5-di-*t*-butyl-4-hydroxybenzyl esters ⟨72YZ454⟩, esters of (*E*)-oximes of benzaldehyde and 2-furaldehyde ⟨71JCS(C)1917⟩ and amides of *N*,*N*′-diisopropylhydrazine ⟨72JCS(P1)929⟩ and other hydrazine derivatives ⟨77JCS(P1)1477⟩.

5.11.4.4 Total Synthesis

While total synthesis has not been able to compete with direct fermentation and semisynthesis (*i.e.* synthetic modification of fermentation-produced starting materials) for the production of commercially important penicillins, it has provided an interesting challenge to synthetic organic chemists, and has allowed the preparation of penicillin analogs which would have been very difficult to obtain otherwise. There are a number of reviews which deal with this aspect of penicillin chemistry ⟨B-72MI51105, 76CRV113, 78T1731, B-80MI51102, B-82MI51107⟩.

The early attempts to prepare natural penicillins by total synthesis have been summarized ⟨B-49MI51106, B-49MI51107⟩. The first synthesis to afford sufficient amounts of a natural penicillin to be isolated and characterized is shown in Scheme 60 ⟨59JA3089, 76JOC2556⟩. Using the same basic synthetic strategy and employing the trityl function to block the amine and protecting the carboxyl group as a benzyl ester, 6-APA has been prepared ⟨62JA2983⟩. Similarly, 'bisnorpenicillin V' (**77**) ⟨74JMC389⟩, '2(*S*)-' and '2(*R*)-norpenicillin V' (**78, 79**) ⟨75MI51101⟩ and a series of penicillin analogs in which the 2,2-dimethyl portion of the molecule has been replaced by 2-spirocycloalkanes (**80**; $n = 1$–3) ⟨78CC46⟩ have been prepared.

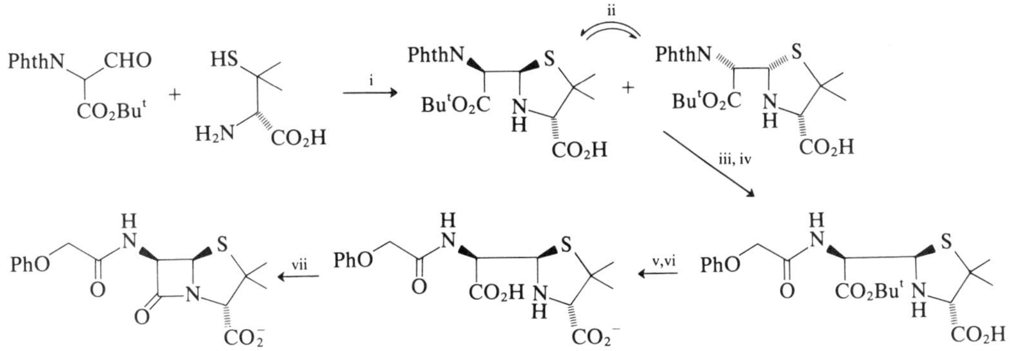

i, NaOAc; ii, pyridine; iii, N_2H_4; iv, $PhOCH_2COCl$; v, HCl; vi, 2 eq. OH^-; vii, DCC

Scheme 60

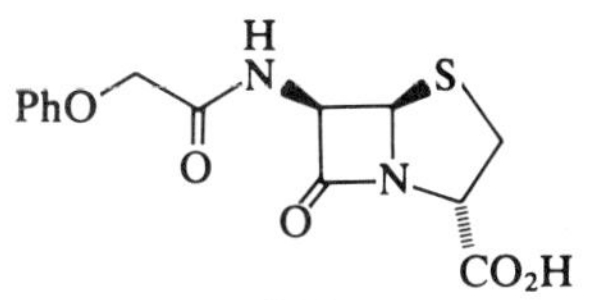

(**77**)

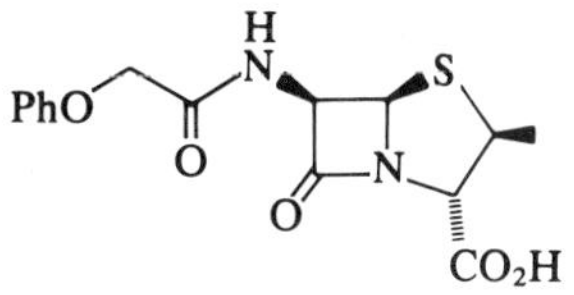

(**78**)

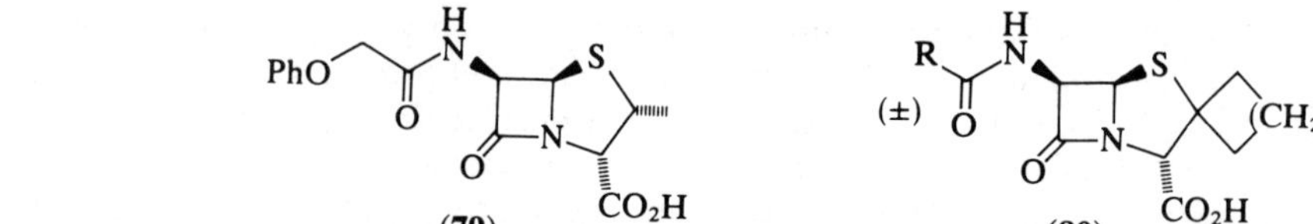

(79) (80)

Another synthetic sequence leading to penicillin derivatives is illustrated in Scheme 61 ⟨77MI51102⟩. Note that the cycloaddition of azidoacetyl chloride to the thiazoline affords an azidopenam with the 6-epi configuration. Equilibration leads to a mixture of (**81**) and (**82**) in a ratio of 4:1, but repeated recycling allowed the isolation of (**82**) in 40% yield. A related synthesis was shown in Scheme 26.

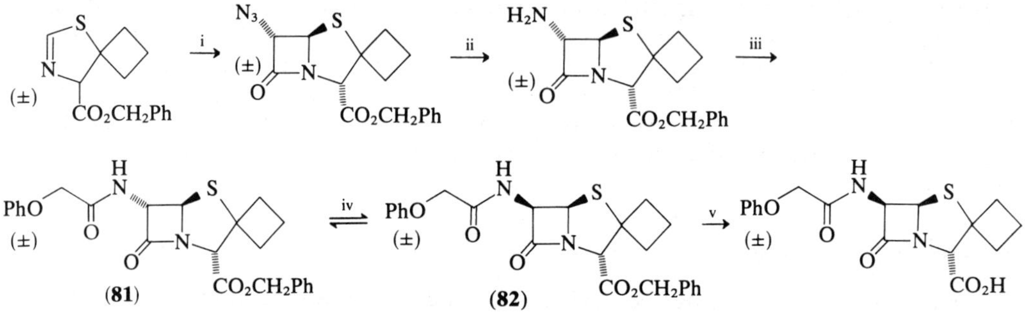

i, N_3CH_2COCl, Et_3N; ii, H_2S, Et_3N; iii, $PhOCH_2COCl$; iv, *N,O*-bis($SiMe_3$)acetamide, DBN; v, Pd/C, H_2

Scheme 61

In contrast to the preceding synthetic routes, the sequence shown in Scheme 62 forms the β-lactam ring before closing the thiazolidine ring ⟨80JCS(P1)2228 and refs. therein⟩. The product (**83**) had been previously reported and converted to the corresponding penicillin derivative (see Scheme 61) ⟨77MI51102⟩.

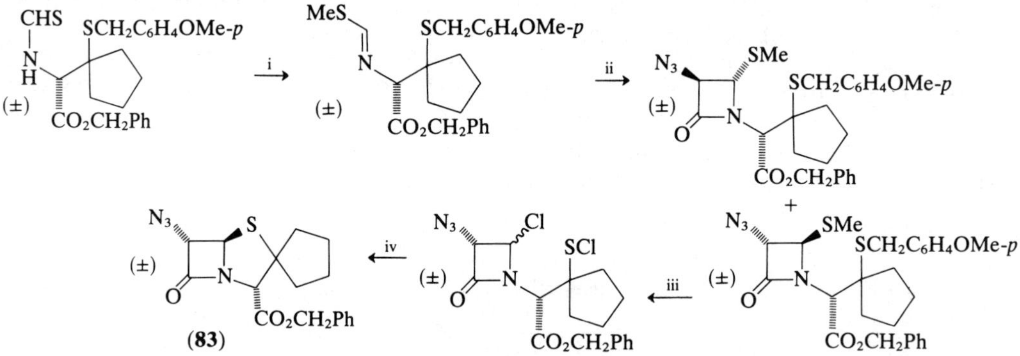

i, MeI, K_2CO_3; ii, N_3CH_2COCl, Et_3N; iii, Cl_2; iv, $SnCl_2$

Scheme 62

Scheme 63 illustrates another synthetic route leading to the penam ring system ⟨78JA4597⟩. This sequence contains some interesting points of synthetic strategy which are discussed in the reference.

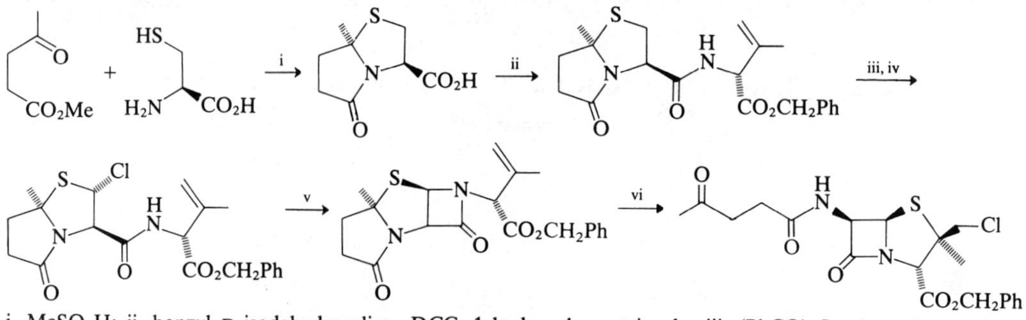

i, $MeSO_3H$; ii, benzyl D-isodehydrovaline, DCC, 1-hydroxybenzotriazole; iii, $(PhCO)_2O_2$; iv, HCl; v, NaH; vi, Bu^tOCl

Scheme 63

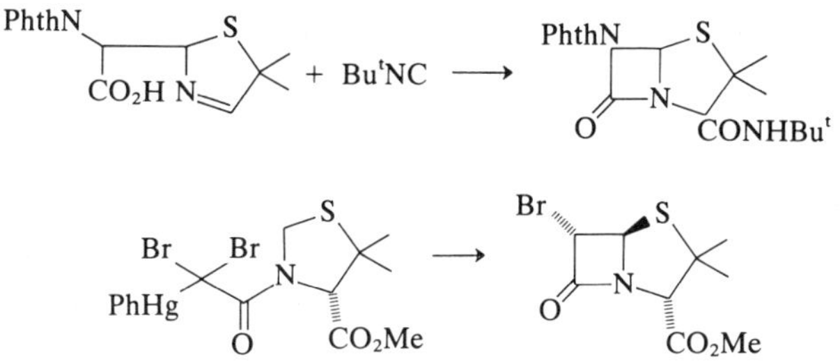

Scheme 64

Scheme 64 shows the key steps in two additional routes leading to the penam ring system ⟨79JCR(S)157, 76ACS(B)377⟩.

Although not qualifying as a total synthesis because the starting material was obtained semisynthetically from 6-APA, the reaction sequence shown in Scheme 65 illustrates an alternative completion of the bicyclic ring system by formation of the C(2)—C(3) bond ⟨77JCS(P1)2479⟩. It is interesting that in intermediate (**84**) both the ester and the sulfoxide functions were needed to provide sufficient activation to the methylene group for anionic ring closure. Also of interest is the apparent isolation of a single isomer from step (vii).

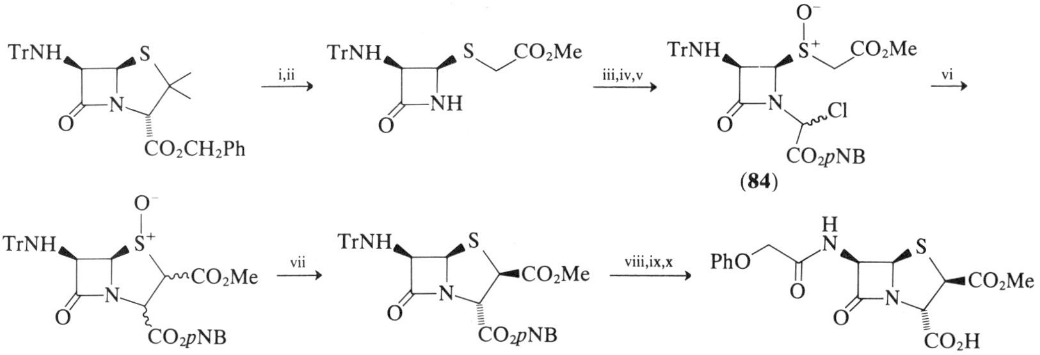

i, $BrCH_2CO_2Me$, 2,6-Bu^t_2PhOK; ii, $KMnO_4$, pyridine; iii, *p*NB glyoxylate; iv, $SOCl_2$; v, MCPBA; vi, $KOBu^t$; vii, PBr_3; viii, TsOH; ix, $PhOCH_2COCl$; x, Pd/C, H_2

Scheme 65

5.11.4.5 Synthesis of Isopenicillins

The trivial name 'bisnorisopenicillin' has been coined for the 6-acylamino-3-thia-1-azabicyclo[3.2.0]heptane-2-carboxylic acid (**85**) structural class. While not extensively investigated, an example of this class has been prepared as shown in Scheme 66 ⟨78JMC413⟩. The following features of this sequence are noteworthy: (a) the exclusively *cis* cycloaddition to form intermediate (**86**) and (b) the use of the 2,4-dimethoxybenzyl amide blocking group.

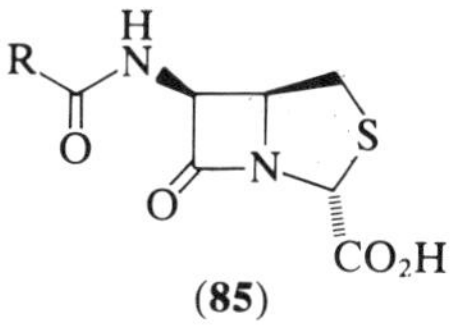

(**85**)

6-Unsubstituted isopenams have been prepared as shown in Scheme 67 ⟨80CC928⟩. It is interesting but not unexpected that the thermodynamically preferred configuration has the 3-carboxylate and the H(5) *cis*.

5.11.4.6 Synthesis of Penems

The trivial name 'penem' has been applied to compounds possessing the ring structure (**88**). Because of their interesting antibacterial properties, derivatives of this ring system have been extensively investigated since about 1976. The following describes some of the

i, $K_2S_2O_8$; ii, $NaBH_4$; iii, TsCl; iv, Zn, HOAc; v, 2-thienylacetic acid; vi, NaI; vii, benzyl glyoxylate, BF_3; viii, $SOCl_2$; ix, KSAc; x, base; xi, cyclohexylamine

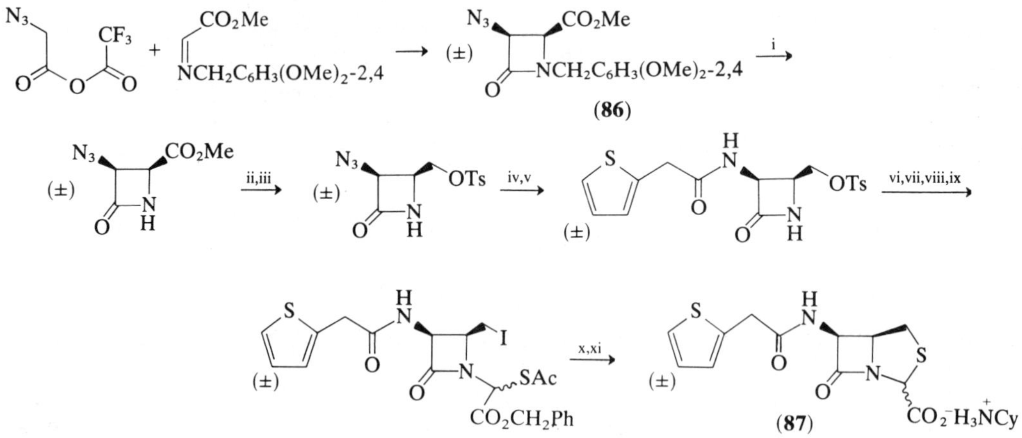

Scheme 66

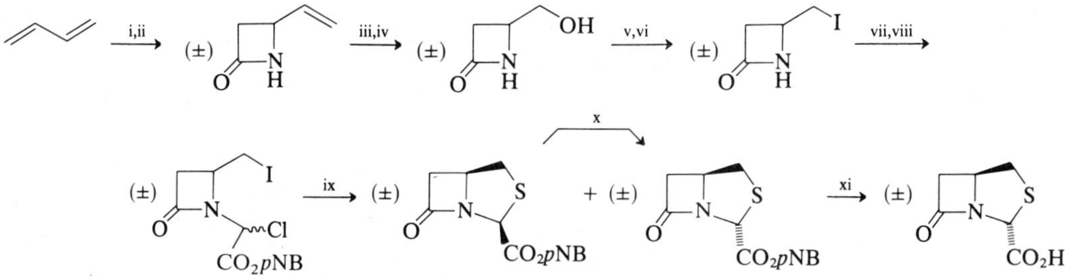

i, CSI; ii, PhSH, pyridine; iii, O_3; iv, $NaBH_4$; v, TsCl; vi, NaI; vii, *p*NB glyoxylate; viii, $SOCl_2$; ix, H_2S; x, DBN; xi, Pd/C, H_2

Scheme 67

approaches, both totally synthetic and semisynthetic, to compounds of this class ⟨for reviews, see B-77MI51107, B-82MI51100⟩.

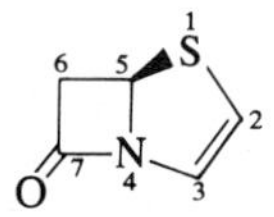

(**88**)

Scheme 68 shows the conversion of the phenoxymethylpenicillin-derived disulfide (see Scheme 10) to penem derivative (**91**) ⟨78JA8214⟩. Of particular interest in this sequence is the reductive acylation step to afford (**89**) and the Wittig ring closure to give (**90**). The rate of the latter reaction was found to be greatly influenced by the steric and electronic character of both the thiol ester and the carboxyl blocking group.

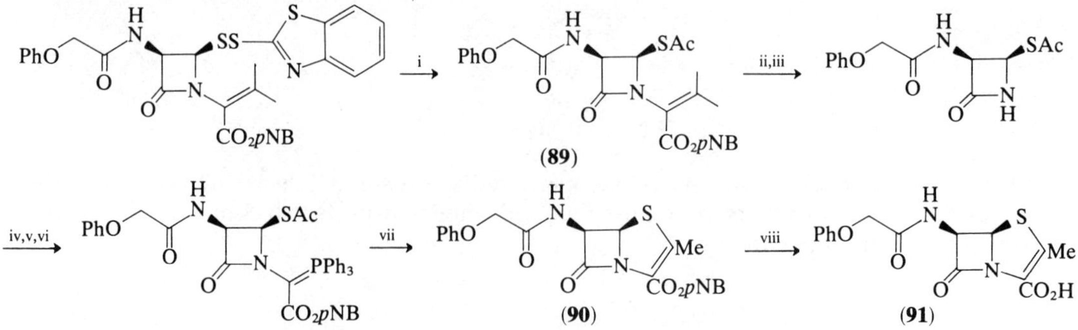

i, PPh_3, Ac_2O; ii, O_3; iii, MeOH; iv, *p*NB glyoxylate ethyl hemiacetal; v, $SOCl_2$; vi, PPh_3, polymeric Hünig base; vii, heat; viii, Pd/C, H_2

Scheme 68

Using reaction sequences which also employed the Wittig ring closure, a series of 6-unsubstituted penems, (**92**; R = *e.g.* Me, Ph, *n*-pentyl), (**93**) (from 6-APA), (**94**) and (**95**), has been prepared ⟨79JA6306 and refs. therein⟩. An X-ray diffraction analysis indicated that the degree of pyramidal character at N(4) is 0.43 Å, which is slightly higher than that of penicillins generally (*cf.* Section 5.11.2.1).

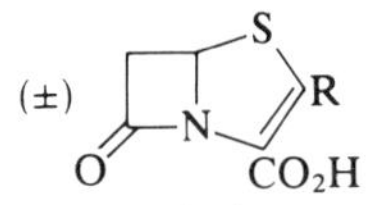

(±) **(92)** 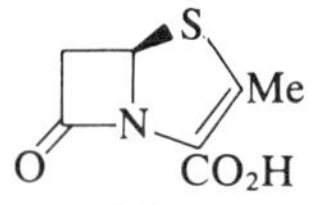 **(93)** 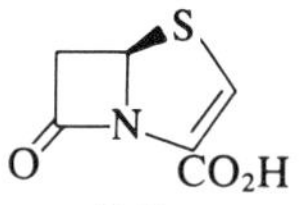 **(94)** 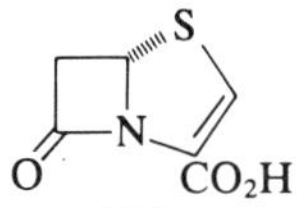 **(95)**

Scheme 69 illustrates another approach which utilizes a 6-APA-derived starting material to provide the (5*R*) configuration necessary for antibacterial activity ⟨80CC70⟩. Note particularly the selective ozonolysis (step ii) and the use of TFA to protonate, and thereby protect, the phosphorane during ozonolysis (step vi). Scheme 70 illustrates a related sequence ⟨79TL3777⟩. It is interesting that, in this case, the double bond migrates to the more stable *exo* configuration. Other derivatives of this type have been reported ⟨80TL561⟩.

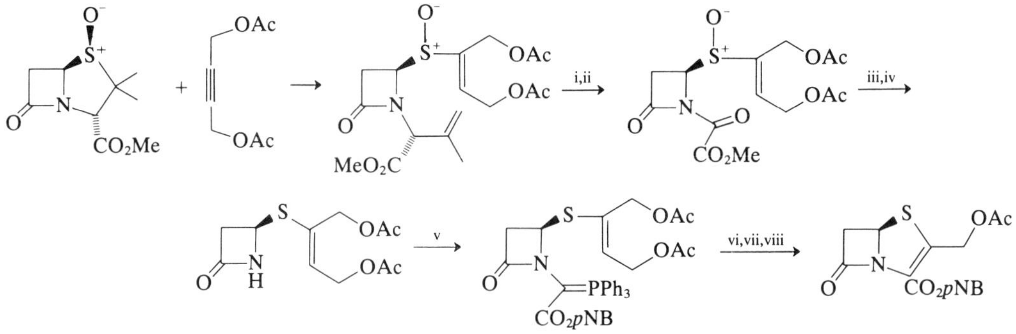

i, base; ii, O_3; iii, PBr_3; iv, MeOH; v, as in Scheme 68; vi, TFA; vii, O_3; viii, base, heat

Scheme 69

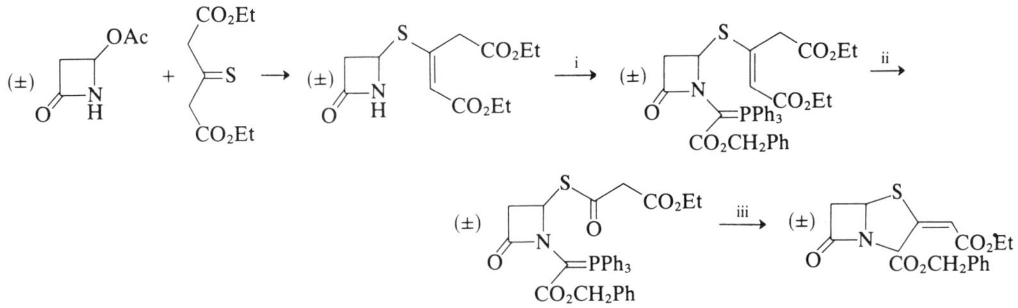

i, as in Scheme 68; ii, TFA, O_3; iii, heat

Scheme 70

The previously described penem syntheses from 6-APA-derived starting materials have been inefficient in the sense that the C(2) and C(3) atoms of the penam are lost during the sequence. Scheme 71 shows a route in which C(2) and C(3) of the penam become C(2) and C(3) of the penem ⟨79CC665⟩. The major product of this sequence is the (5*S*) enantiomer. A related synthetic approach, starting with the natural product clavulanic acid, has been described ⟨79CC663⟩.

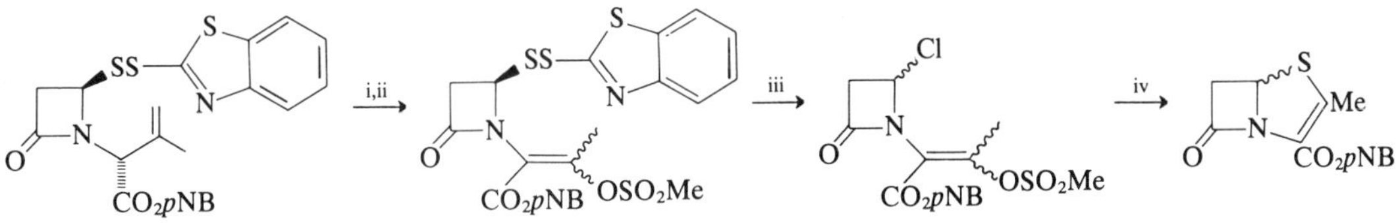

i, O_3; ii, $MeSO_2Cl$; iii, Cl_2; iv, H_2S

Scheme 71

Several routes to 2-alkylthiopenem derivatives have been described, and one of these is shown in Scheme 72 ⟨81TL3485⟩. The key to building this system was the observation that a trithiocarbonate will undergo a Wittig ring closure (step xi). The sequence shown here is complicated by the incorporation of a 6α-[(*R*)-1-hydroxyethyl] group, which poses some interesting problems of stereochemical control. Also in this sequence, note the use of the blocked hydroxyethyl group to control the stereochemistry of the chlorinolysis (step v), and the use of an allyl ester to block the carboxylic acid group.

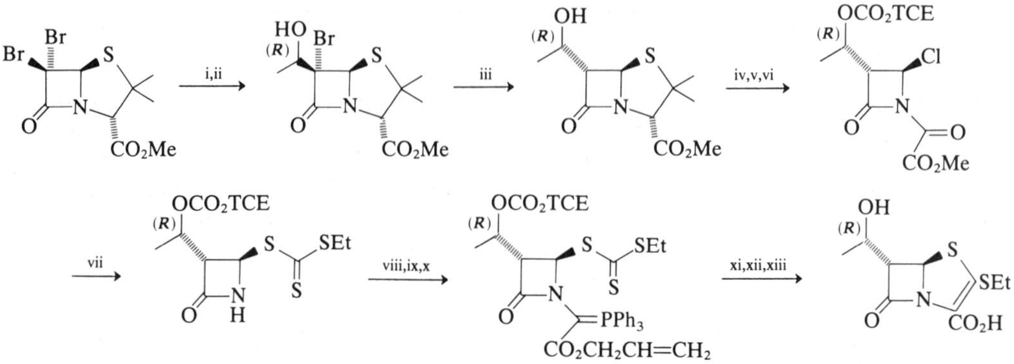

i, MeMgBr; ii, AcH; iii, Pd/$CaCO_3$, H_2; iv, Cl_3CCH_2OCOCl; v, Cl_2; vi, O_3; vii, $EtSCS_2K$; viii, allyl glyoxylate; ix, $MeSO_2Br$; x, PPh_3; xi, heat; xii, Zn; xiii, Pd/PPh_3

Scheme 72

5.11.5 APPLICATION

5.11.5.1 Antibacterial Agents

The importance of the penicillins as a class of heterocyclic compounds derives primarily from their effectiveness in the treatment of bacterial infections in mammals (especially humans). It has been estimated that, in 1980, the worldwide production of antibiotics was 25 000 tons and, of this, approximately 17 000 tons were penicillins ⟨81MI51103⟩. The Food and Drug Administration has estimated that, in 1979 in the U.S.A., 30.1×10^6 prescriptions of penicillin V and 44.3×10^6 prescriptions of ampicillin/amoxicillin were dispensed. This level of usage indicates that, compared to other methods of dealing with bacterial infection, the cost–benefit properties of penicillin therapy are particularly favorable. Stated differently, penicillin treatment leads to the elimination of the pathogen in a relatively high percentage of cases of bacterial infection at a relatively low cost to the patient in terms of toxic reactions and financial resources.

The scientific literature devoted to the medical, microbiological, pharmacological, toxicological and biochemical properties of penicillins is, of course, very large. Most general textbooks of infectious disease, microbiology and pharmacology contain discussions of the penicillins from the perspective of that particular discipline. Only a very brief outline of the antibacterial properties of penicillins will be given here.

Table 3 lists the structures and generic names of the principal penicillins in current medical usage, and Table 4 indicates in summary form the relative antibacterial properties of these compounds. The reader is urged to consult reviews such as ⟨B-77MI51106⟩ for a more complete treatment, since summaries such as Table 4 inevitably omit a large amount of important detail.

The biochemical basis of penicillin action continues to be an area of active investigation. Penicillins are highly specific inhibitors of enzyme(s) involved in the synthesis of the bacterial cell wall, a structure not present in mammalian cells. Three principal factors are thought to be important for effective antibacterial action by a penicillin:

(a) The antibacterial agent must penetrate the outer cell wall of the bacterium and reach the cell membrane where the cell wall synthesizing enzymes are located. Barriers to penetration are a resistance mechanism of certain Gram-negative bacteria.

(b) The penicillin must survive the action of extracellular and intracellular β-lactamases. This is an important source of resistance in both Gram-positive and Gram-negative bacteria. In Table 4 the main difference between group 1 and group 2, and, to a certain extent,

Table 3 Penicillins in Medical Practice

R	*Generic name*	*R*	*Generic name*
	Penicillin G	(*R*)	Ampicillin
	Penicillin V	(*R*)	Hetacillin
	Methicillin	(*R*)	Amoxicillin
	Nafcillin		Carbenicillin
	Oxacillin		Ticarcillin
	Cloxacillin		Mecillinam
	Dicloxacillin	(*R*)	Azlocillin
	Flucloxacillin	(*R*)	Mezlocillin
	Cyclacillin	(*R*)	Piperacillin

Table 4 Antibacterial Properties of Penicillins

Penicillin	*Bacterial group* I[a]	II[b]	III[c]	IV[d]	V[e]
Penicillin G	++[f]	−	±	−	−
Penicillin V	+	−	−	−	−
Methicillin	+	+	−	−	−
Nafcillin	+	+	−	−	−
Oxacillin	+	+	−	−	−
Cloxacillin	+	+	−	−	−
Dicloxacillin	+	+	−	−	−
Flucloxacillin	+	+	−	−	−
Cyclacillin	++	−	±	−	−
Ampicillin	++	−	+	−	−
Hetacillin	++	−	+	−	−
Amoxicillin	++	−	+	−	−
Carbenicillin	+	−	+	±	+
Ticarcillin	+	−	+	±	+
Mecillinam	−	−	+	±	−
Azlocillin	+	−	+	±	+
Mezlocillin	+	−	+	+	±
Piperacillin	+	−	+	+	++

[a] Susceptible Gram-positive organisms such as *Streptococcus pneumoniae* and β-lactamase-negative *Staphylococcus aureus*.
[b] β-Lactamase-positive *Staphylococcus aureus*.
[c] Susceptible Gram-negative organisms such as some *Escherichia coli* and *Proteus mirabilis*.
[d] Resistant Gram-negative organisms such as *Klebsiella pneumoniae* and *Enterobacter aerogenes*.
[e] *Pseudomonas aeruginosa*.
[f] −, not active; +, active; ++, quite active.

between group 3 and group 4, is that groups 2 and, to some extent, 4 are β-lactamase producers.

(c) The antibiotic must 'irreversibly' inhibit one or more of several enzymes associated with cell wall synthesis, presumably *via* acylation. The penicillin binding proteins (PBPs; cell membrane-associated proteins which bind a penicillin) for various bacterial species have been identified. Different β-lactam antibiotics will bind preferentially to different PBPs, resulting in different morphological effects on the bacterium.

The mechanism of action of β-lactam antibiotics has been the subject of several recent reviews ⟨79MI51102, 79MI51103, 81MI51104, B-81MI51105, B-82MI51101⟩.

As can be seen from Table 3, only modifications at the 6β-amino groups have been successful in producing penicillins of medical significance up to this time. Several reviews have dealt with the structure–activity relationship in this area in considerable detail ⟨B-80MI51102, B-77MI51106, B-75MI51102⟩ and should be consulted for the actual effects of structural modification on antibacterial activity.

In brief summary, the following transformations have led to decreased antibacterial activity: epimerization at any point in the penam ring, β-lactam cleavage, thiazolidine ring cleavage, oxidation at S(1), loss of an acidic function at C(3), substitution at C(5), most 6α-substitution (an exception is the 6α-methoxy derivatives of ticarcillin), and most 6β-substitution that is not a carboxamido group (an exception is mecillinam and related compounds). Modifications at C(2) have led to retained antibacterial activity and, in fact, the penems (see Section 5.11.4.6) are a particularly active group of compounds which may acquire medical importance in the future (see ⟨81JA4526⟩ for a discussion of the relationship between β-lactam reactivity and antibacterial activity for penams, penems and other β-lactam antibiotics).

As is well known, the principal toxicity associated with penicillin therapy is the occurrence of hypersensitivity reactions. This and other aspects of pencillin toxicity have been recently reviewed ⟨81MI51106⟩.

5.11.5.2 β-Lactamase Inhibitors

Certain penicillins and penam derivatives have the ability to inhibit the β-lactamase enzyme, and can provide a variable degree of protection to susceptible β-lactam antibiotics from the action of this enzyme. There are two principal compound classes which have this property. The first is the group of penicillins which are effective against β-lactamase-producing *S. aureus*, *i.e.* methicillin, cloxacillin, nafcillin and related compounds. The second is a group of penams which include the 6β-halopenicillanic acids, penicillanic acid sulfone (generic name, sulbactam) and 6α-chloropenicillanic acid sulfone. Many of these compounds have been studied in considerable detail, and the reader is referred to chapters in ⟨B-79MI51101⟩, ⟨B-81MI51105⟩ and ⟨B-82MI51106⟩ for reviews of this topic.

5.11.6 CONCLUSION

In conclusion, the penicillin class of compounds continues to be actively studied by the scientific community. While the penicillins are an important part of the physician's armamentarium against infectious disease and will remain so for a considerable time, there is increasing emphasis on the study and use of fermentable penicillins as starting materials for the production of other β-lactam antibiotics, leading to a wealth of heterocyclic chemistry.

5.12

Other Fused-ring Azetidines, Azetines and Azetes

G. L. DUNN

Smith Kline & French Laboratories, Philadelphia

5.12.1 INTRODUCTION

The literature on fused-ring azetidines and azetidinones has virtually exploded in the last 15 years. A major factor in the expansion of research in this area has been the discovery and exploitation of new classes of β-lactam antibiotics such as the thienamycins, olivanic acids and clavulanic acid. The classic β-lactam antibiotics, cephalosporins, penicillins and their closely related analogs, are discussed separately in Chapters 5.10 and 5.11, respectively. In this chapter fused-ring azetidinones, which encompass these new β-lactams, are described separately from non-carbonyl-containing azetidines. Though the field is too broad to cite all the relevant papers, an attempt has been made to cite leading references for those systems which have been studied extensively.

For the most part only bicyclic fused-ring systems are described here. The nomenclature used in the literature to describe these ring systems can lead to some confusion. For fused-ring azetidines and azetes systematic naming rules are followed. Thus, numbering is begun at a bridgehead position with preference given to a heteroatom and is continued sequentially first around the larger ring. In the absence of a bridgehead heteroatom numbering still is begun at a bridgehead atom and continues as described so that the numbers assigned to heteroatoms are minimized. Thus compounds (**1**) and (**2**) are named 1-azabicyclo[5.2.0]nonane and 3,8-diazabicyclo[5.2.0]nonane, respectively. However, a dual nomenclature exists for fused-ring azetidinone ring systems related to the naturally occurring β-lactam antibiotics. For example, *Chemical Abstracts*, which follows systematic rules of nomenclature, lists compound (**3a**) as a 1-azabicyclo[3.2.0]heptane, whereas workers in the field have adopted a simplified system based on the original numbering sequence assigned to penicillin (see Chapter 5.11). Under this system compound (**3b**) is named a 1-carbapenam indicating the replacement of the penicillin sulfur atom by carbon. This simplified system is commonly used only for azetidinones fused to five- and six-membered rings. Fusions to rings of other size are described using systematic rules.

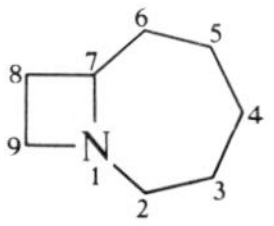

(**1**) 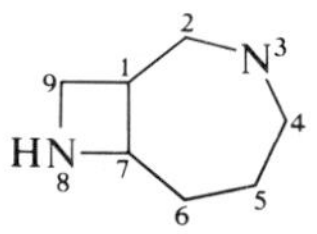 (**2**) 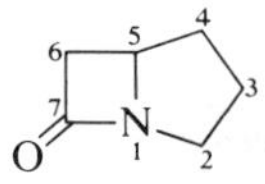 (**3a**) 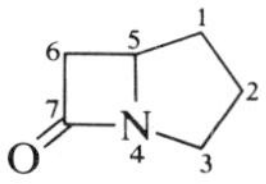 (**3b**)

5.12.2 FUSED-RING AZETIDINES

5.12.2.1 Structure

For the most part the parent unsubstituted bicyclic azetidines have not been reported. Those unsubstituted ring systems which have been reported are 1-azabicyclo[2.2.0]hexane (**4**) ⟨64HCA2145⟩, 1-azabicyclo[4.2.0]octane (**5**) ⟨60JA2609⟩ and the *cis*- and *trans*-6-azabicyclo[4.2.0]octanes (**6**) ⟨66JOC1372⟩.

(**4**) (**5**) (**6**) (**7**)

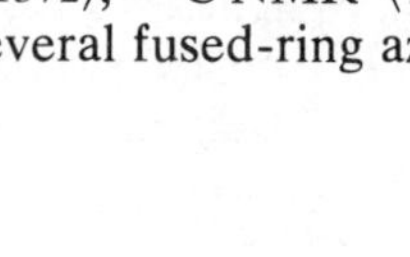

An X-ray crystallographic analysis of the naturally occurring antibiotic (**7**) indicates that the azetidine ring is essentially planar ⟨80ABC73⟩. Selected bond distances (Å) and angles (°) are shown. ^{1}H NMR ⟨74JA4879, 72JOC516, 73TL3727, 66JOC1372⟩, ^{13}C NMR ⟨79JA6677, 81JA3136⟩ and mass spectra ⟨71JOC1934, 72JOC516, 74JA4879⟩ of several fused-ring azetidines have been reported.

5.12.2.2 Reactivity

5.12.2.2.1 Thermal and photochemical reactions

The thermal reactivity of fused-ring azetidines is related to the degree of ring strain. The most highly strained bicyclic azetidine which has been described is 1-azabicyclo[2.2.0]hexane (**4**) though it has eluded isolation due to its high reactivity ⟨64HCA2145⟩. Thus, on heating in solution at 80 °C (**4**) undergoes fragmentation of the N(1)—C(4) bond to give alkene

(**8**). The structurally isomeric 2-azabicyclo[2.2.0]hex-5-ene (**9**), though stable enough to be isolated and characterized, also undergoes ring opening *via* cleavage of the bridging bond when heated either in the gas phase or in an inert solvent to give 1,2-dihydropyridines (**10**; Scheme 1) ⟨79JA6677, 76JA2344⟩. The rate of thermal ring opening is sensitive to the nature of the substituent attached to nitrogen. Thus, (**9**; R = Me) has a half-life of 0.77 h at 125 °C whereas (**9**; R = CO_2Me) has a half-life of 1.13 h at 157 °C. Numerous examples illustrating the generality and usefulness of this transformation have been reported ⟨78JA6696, 78JOC944, 72JOC1321⟩. For example, azetidine (**11**) has been used to synthesize compound (**12**), a potential intermediate in the synthesis of *Strychnos* alkaloids ⟨79JA6677⟩.

Scheme 1

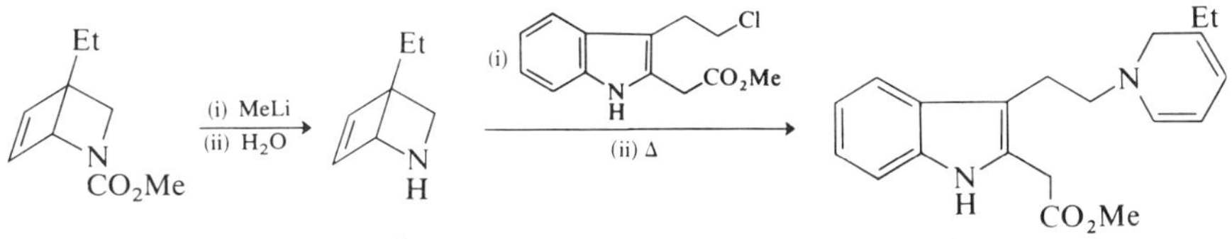

Thermal lability with cleavage of the bridging bond is not limited to azetidines fused only to carbon-containing rings. The 1,2-diazabicyclo[3.2.0]heptanes when heated undergo a number of interesting transformations. Compound (**13**; Scheme 2) upon heating in benzene produces in 77% yield the unusual 1,6-diazabicyclic compound (**14**) ⟨67JOC1346⟩. Subsequent studies ⟨76JOC3156⟩ showed that dihydrodiazepinone (**15**) was a minor by-product of this reaction. In addition, when (**14**) was heated in alcohol, tetrahydropyridazinones (**16**) were obtained suggesting the intermediacy of (**18**), which in the presence of a primary or secondary alcohol undergoes nucleophilic attack at the three-membered ring. When alcohol is not present, (**18**) is free to rearrange to (**14**) as observed.

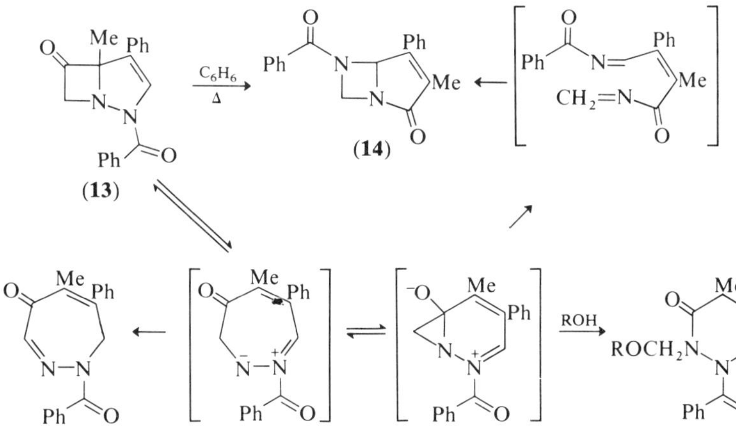

Scheme 2

One report of the photochemical decomposition of a fused-ring azetidine (**19**) utilized photoaromatization to produce 2-azetines (**20**) ⟨77CC806⟩ (see Section 5.09.4.3.3).

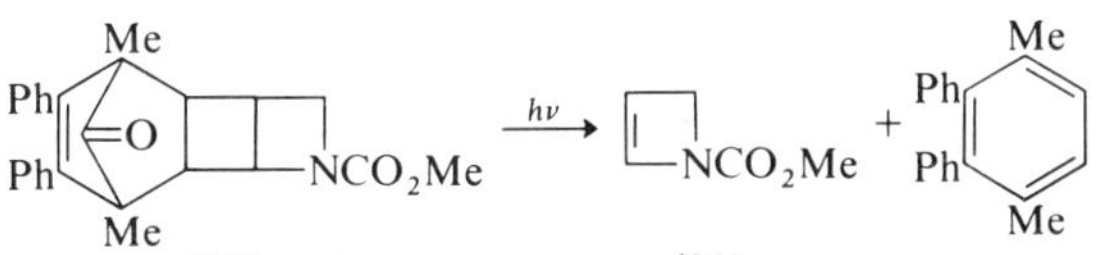

5.12.2.2.2 Electrophilic reactions

The primary mode of electrophilic attack on the azetidine of fused-ring derivatives is acid-catalyzed opening of the bridging bond. For example, 1-azabicyclo[2.2.0]hexane (**4**) gives 4-chloropiperidine (**21**) on treatment with hydrogen chloride ⟨64HCA2145⟩.

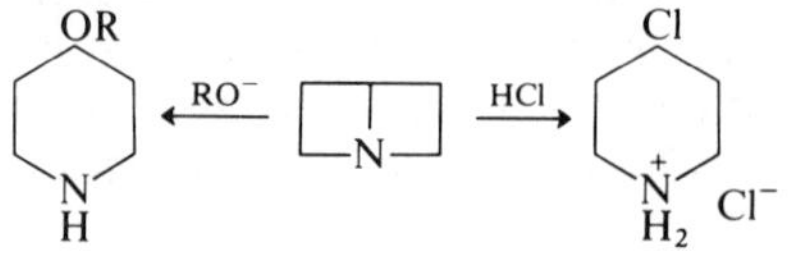

(**4**) (**21**)

5.12.2.2.3 Nucleophilic reactions

Fused-ring azetidines also are susceptible to nucleophilic attack, primarily at the bridging bond. Attack on bicyclohexane (**4**) by hydroxide or ethoxide gives the corresponding 4-substituted piperidine ⟨64HCA2145⟩.

The fused-ring azetidines are stable to the action of some reducing agents. Bicyclooctane (**22**) can be detosylated with sodium in alcohol to give the corresponding free base. Other examples include ring opening of (**13**) ⟨67JOC1353⟩ and (**24**) ⟨75JA7288⟩.

The free bases of the 2-azabicyclo[2.2.0]hexanes are relatively stable to HCl but their *N*-acyl derivatives (**26**) react rapidly to give ring-opened cyclobutenes ⟨79JA6677⟩.

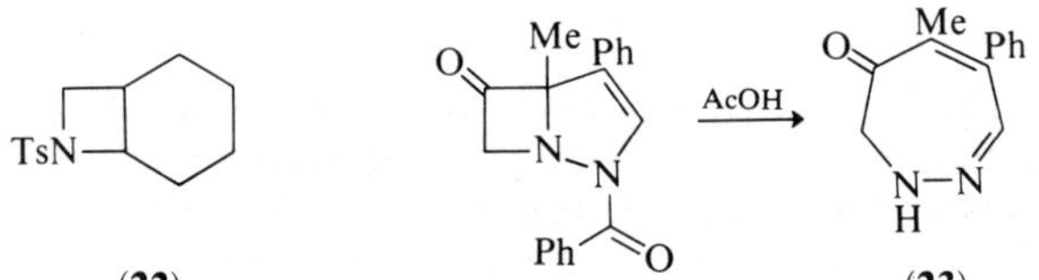

(**22**) (**13**) (**23**)

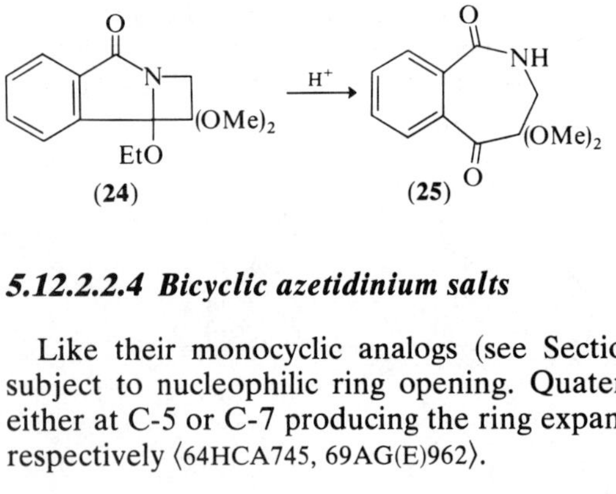

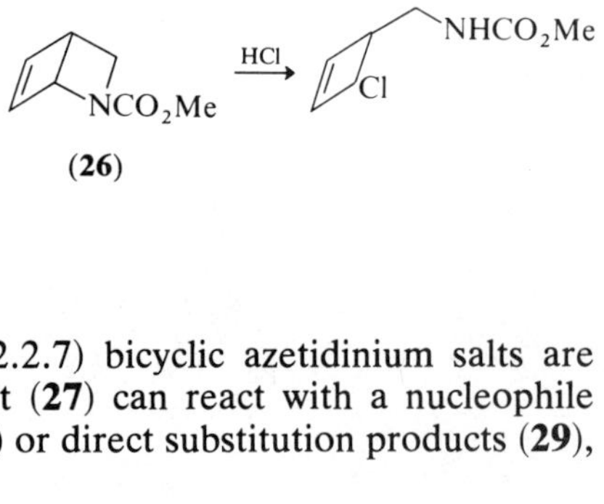

(**24**) (**25**) (**26**)

5.12.2.2.4 Bicyclic azetidinium salts

Like their monocyclic analogs (see Section 5.09.2.2.7) bicyclic azetidinium salts are subject to nucleophilic ring opening. Quaternary salt (**27**) can react with a nucleophile either at C-5 or C-7 producing the ring expansion (**28**) or direct substitution products (**29**), respectively ⟨64HCA745, 69AG(E)962⟩.

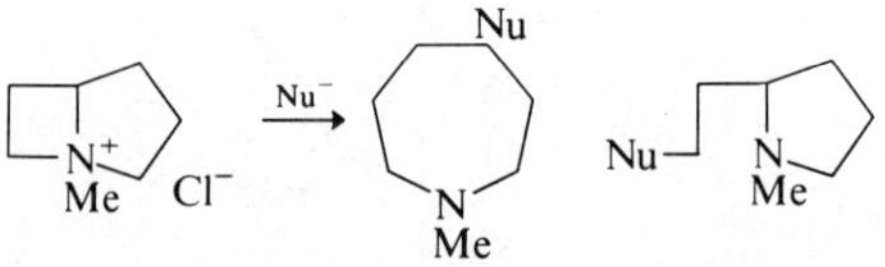

(**27**) (**28**) (**29**)

It has been postulated ⟨60JA2609⟩ that the formation of a quaternary salt of 1-azabicyclo[4.2.0]octane is the first step in the mechanism of its methyl iodide-catalyzed polymerization.

5.12.2.3 Synthesis

5.12.2.3.1 From non-azetidine precursors

Several bicyclic azetidine systems have been synthesized *via* photochemically induced valence bond isomerization of a heterocycle. Photolysis of 2-amino-5-chloropyridine in dilute HCl produces the 2-azabicyclo[2.2.0]hexane (**30**) ⟨61JA2967⟩. Similarly, bicyclohexanes (**31**) and (**26**) are produced in moderate yields upon irradiation of methyl 2-pyridylacetate and 1-methoxycarbonyl-1,2-dihydropyridine, respectively. The bicyclo[3.2.0]heptane (**33**) was isolated in 45% yield upon irradiation of dihydroazepine (**32**) in methanol in the presence of a sensitizer ⟨71JOC1934⟩. No reaction took place in the absence of sensitizer. Bicyclic azetine (**34**) has been suggested as an intermediate in this reaction and is trapped as (**33**) by reaction with solvent. The favored mechanism is an electrocyclic disrotatory ring closure. The 1,2-diazabicyclo[3.2.0]heptane (**36**) is produced analogously from dihydrodiazepine (**35**).

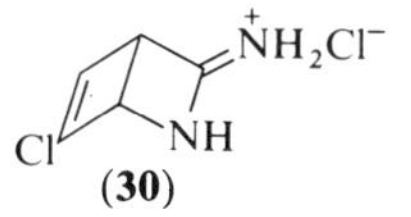

(**30**) 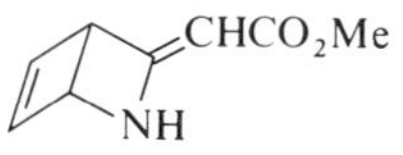 (**31**)

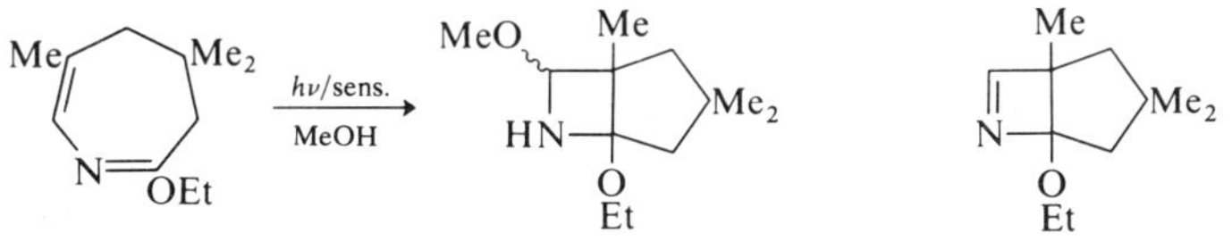

(**32**) (**33**) (**34**)

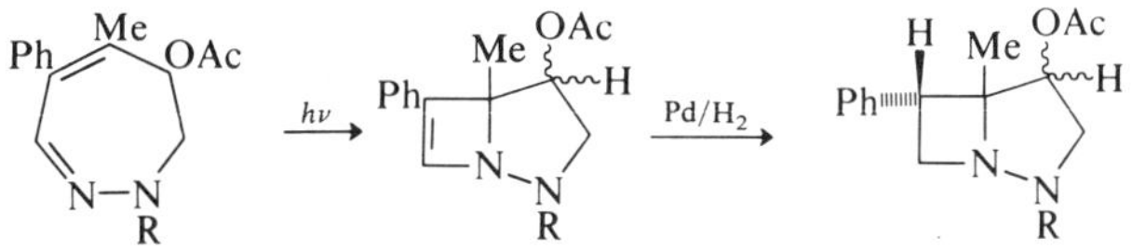

(**35**) (**36**)

The most widely used method for synthesizing bicyclic azetidines is by photoaddition reactions, primarily of the [2+2] type. Scheme 3 lists several examples of this method: (**37**) ⟨81JA3136⟩, (**38**) ⟨75JA7288⟩, (**39**) ⟨77TL431⟩, (**40**) ⟨74JHC453⟩. Cycloadditions to carbon–nitrogen double bonds have been reviewed ⟨78MI51200⟩.

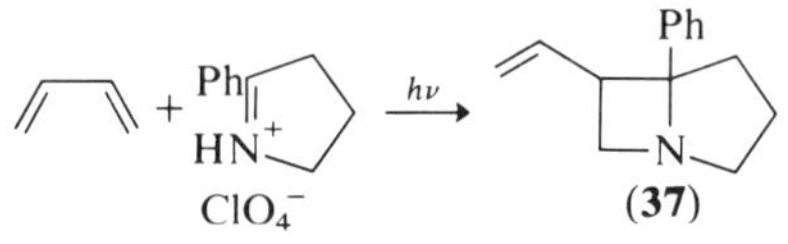

(**37**)

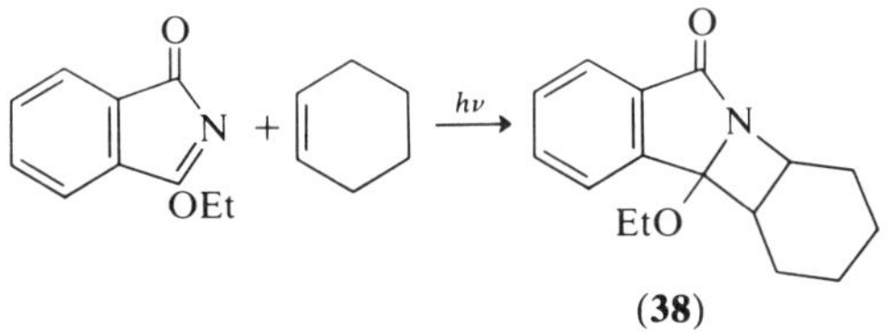

(**38**)

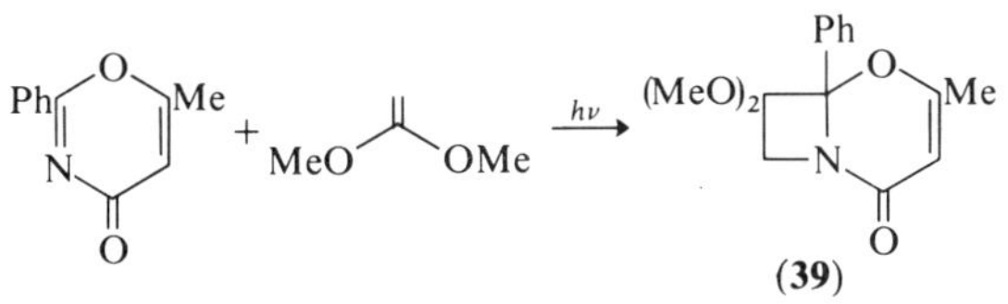

(**39**)

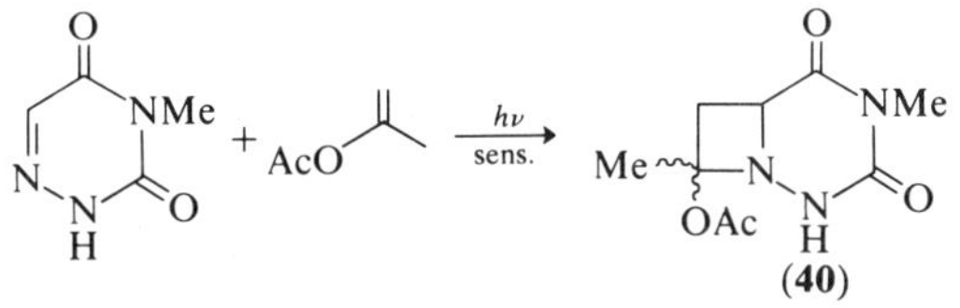

(**40**)

Scheme 3

Photochemical excitation of cyclic α-amino ketones (**41** and **43**) leads to the formation of bicyclic azetidines and azetidinols by abstraction of a hydrogen γ to the ketone followed by cyclization. Production of (**42**) ⟨72CC1108⟩ and (**44**) ⟨75TL2463⟩ is believed to occur as a singlet state reaction.

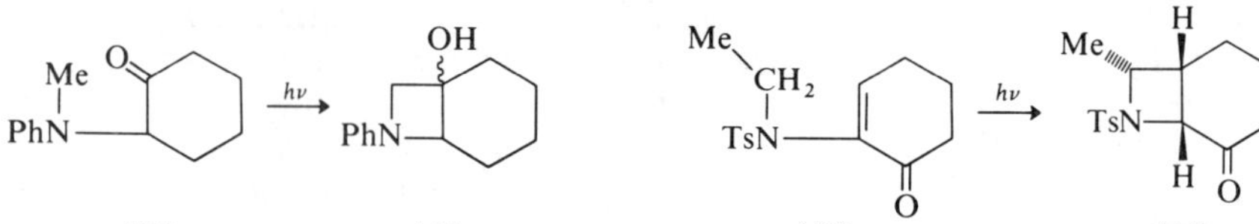

(41) (42) (43) (44)

A general reaction (Scheme 4) for bicyclic azetidine formation involves base catalyzed intramolecular cyclization *via* nucleophilic displacement. Bicyclic azetidines (**4**) ⟨64HCA2145⟩, (**45**) ⟨65T2725⟩, (**46**) ⟨66T(S8)113⟩ and (**47**) ⟨66JOC1372⟩ all were synthesized by this approach. A series of highly substituted 2,7-diazabicyclo[4.2.0]octane derivatives (Scheme 5) were obtained in good yields as cycloaddition–rearrangement products ⟨75JOC3363⟩.

Scheme 4

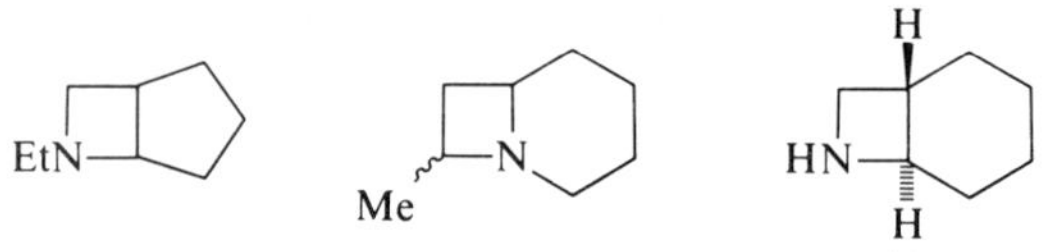

(45) (46) (47)

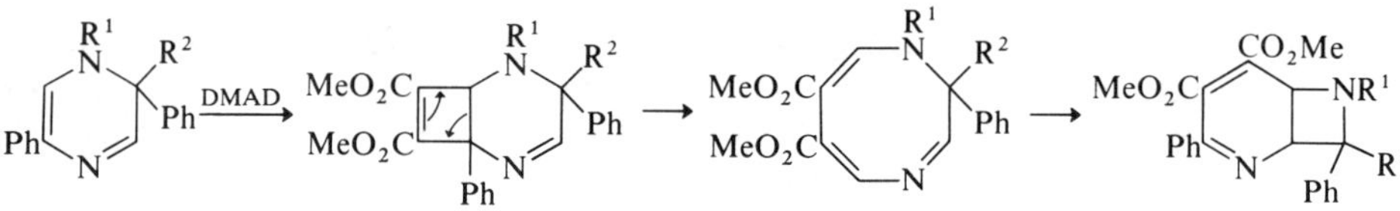

Scheme 5

5.12.2.3.2 From azetidine precursors

Examples of the synthesis of fused-ring azetidines by ring closure onto a preformed azetidine are uncommon, in contrast to approaches used for the preparation of fused ring azetidinones (see Sections 5.12.3, 5.11.4.4 and 5.10.4.3). The 1,3-diazabicyclo[3.2.0]heptane derivative (**48**) was prepared ⟨72JOC516⟩ by cyclization with DCC of the carbamoylazetidine followed by thermal rearrangement of the intermediate thiazolidinone (Scheme 6).

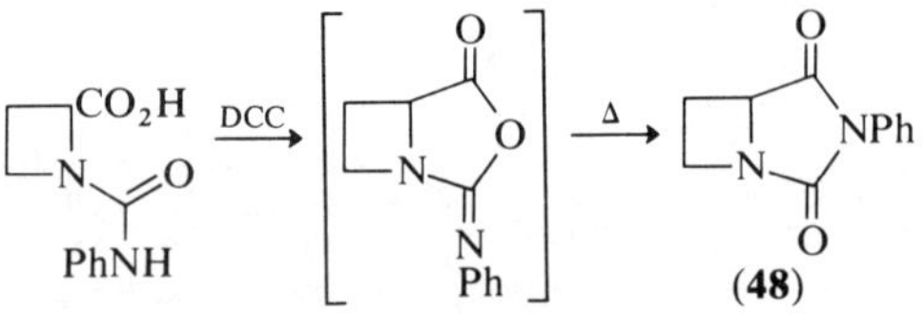

(48)

Scheme 6

Examples of preparations for a number of bicyclic azetidines are summarized in Tables 1–3.

5.12.2.4 Application

The 2-azabicyclo[2.1.0]pentane-3-carboxylic acid (**7**) was isolated from a fermentation broth of *Streptomyces zaomyceticus* SF-1836 ⟨79ABC2279⟩ and was shown to display antimicrobial activity *in vitro* against *Xanthomonas* species. Antagonism of its bioactivity by L-proline suggests that (**7**) is acting as an antimetabolite. Bis-quaternary salts of 1-azabicyclo[4.2.0]octane (conidine) were evaluated as pharmacological agents but were found to be highly toxic to mice ⟨60JA2609⟩.

Table 1 Bicyclic Azetidines having a Carbon-containing Fused Ring

Ring system	Systematic name	Substituents	Yield (%)	Ref.
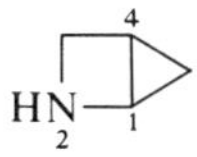	2-Azabicyclo[2.1.0]pentane	3-CO_2H	Natural product	80ABC73
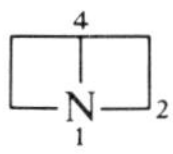	1-Azabicyclo[2.2.0]hexane	—	Not isolated	64HCA2145
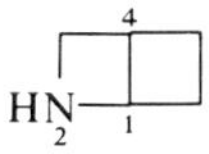	2-Azabicyclo[2.2.0]hexane	2-Methoxycarbonyl-5-ene	50	79JA6677
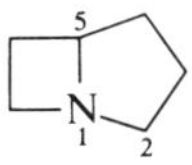	1-Azabicyclo[3.2.0]heptane	5-Phenyl-6-vinyl	9	81JA3136
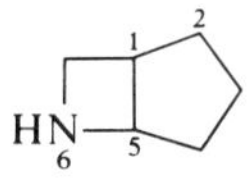	6-Azabicyclo[3.2.0]heptane	1,3,3-Trimethyl-5-ethoxy-7-methoxy	45	71JOC1934
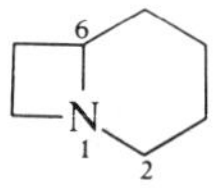	1-Azabicyclo[4.2.0]octane	—	37	66T(S8)113, 60JA2609
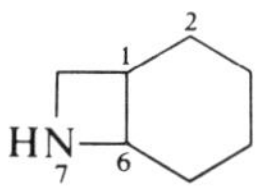	7-Azabicyclo[4.2.0]octane	7-Tosyl	55	66JOC1372
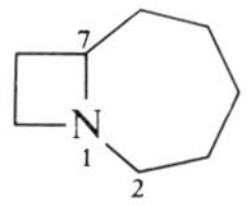	1-Azabicyclo[5.2.0]nonane	2,4,6,8-Tetraene	—	71DIS(B)(31)7191
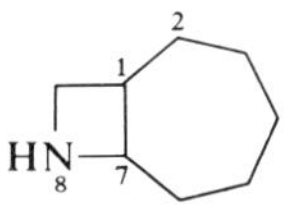	8-Azabicyclo[5.2.0]nonane	—	87	74T2561, 80MI51204

Table 2 Bicyclic Azetidines containing One Heteroatom in the Fused Ring

Ring system	Systematic name	Substituents	Yield (%)	Ref.
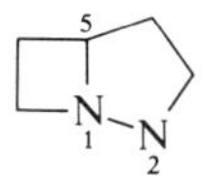	1,2-Diazabicyclo[3.2.0]heptane	4-Phenyl-5-methyl-6-one	34	66JOC34
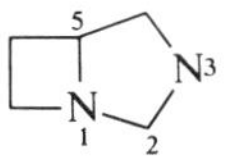	1,3-Diazabicyclo[3.2.0]heptane	3-Phenyl-2,4-dione	66	72JOC516
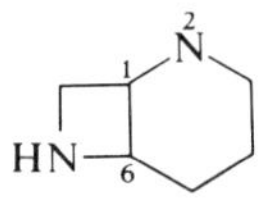	2,7-Diazabicyclo[4.2.0]octane	3,7-Diphenyl-4,5-dimethoxycarbonyl-6,7-di-*t*-butyl	69	75JOC3363
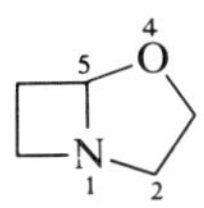	4-Oxa-1-azabicyclo[3.2.0]heptane	5-Phenyl-6,6-dimethoxy-2-one	65	75JA7298

Table 3 Bicyclic Azetidines containing Two Heteroatoms in the Fused Ring

Ring system	Systematic name	Substituents	Yield (%)	Ref.
	1,2,4-Triazabicyclo[4.2.0]octane	2,4-Dimethyl-3,5-dione	91	74JA4879
	1,3,5-Triazabicyclo[4.2.0]octane	3,5-Ditosyl-6-methoxy-8,8-dimethyl-2,4-dione	82	75JOC2356
	3,5-Dioxa-1-aza-bicyclo[4.2.0]octane	2,4-Dicyanomethylene-6-methoxy-7,8-tetramethyl	42	75JOC2552

5.12.3 BICYCLIC AZETIDINONES RELATED TO THE PENICILLINS AND CEPHALOSPORINS

5.12.3.1 Introduction

Comprehensive review of the chemistry and biology of bicyclic β-lactams related to the penicillins and cephalosporins has been the subject of several recent monographs ⟨B-82MI51200, B-82MI51201, B-82MI51202, B-80MI51200⟩. Hence, this chapter will be limited to discussions of only selected examples of this type of fused-ring azetidinone with emphasis on the newer carbapenem and clavulanic acid ring systems. For a more detailed description of the chemistry the reader is referred to the monographs which have served as the source for much of the content of this chapter.

The nomenclature used in this section differs from the strictly systematic rules used by *Chemical Abstracts* to index fused-ring azetidinones, but conforms more to that commonly found in the literature describing the β-lactam antibiotics. A description of this naming system is given in the introduction (Section 5.12.1). Virtually all of the compounds discussed in this section contain a carboxyl group adjacent to the β-lactam ring and conform to general structure (**49**).

(**49**)

5.12.3.2 Structure

An enormous amount of X-ray crystallographic and spectroscopic data has been generated on fused-ring azetidinones, especially those fused to five- and six-membered rings. Two excellent monographs ⟨B-82MI51200, B-82MI51201⟩ provide comprehensive discussion of these results. Only selected data are included here. The bond lengths and the distance of the nitrogen above the plane of the three carbons attached to it (a measure of bond angle strain) are listed in Table 4 for a series of fused ring azetidinones. These data illustrate that ring fusion of a second ring to an azetidine at the N-1/C-4 positions causes an increase in bond angle strain and a lengthening of the lactam bond. As expected, there is greater strain with a five-membered ring than with a six-membered one. Fusion of the second ring to the C-3/C-4 carbon atoms of the β-lactam ring (**55**) produces significantly less strain on the β-lactam bond. Many other bicyclic azetidinones have been reported for which X-ray crystallographic data have been determined ⟨B-82MI51204⟩.

Extensive ^{1}H and ^{13}C NMR studies on bicyclic azetidinones have been reported ⟨B-82MI51205, B-82MI51206, B-72MI51200⟩. Data on acetylthienamycin (**56**), epithienamycin A

Table 4 Structural Properties of some Bicyclic Azetidinones

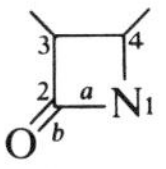

3
4
2
a
N1
O
b

Compound		*Substituents*	*Distance of N above plane* (Å)	*Bond Lengths* (Å) a	b	*Ref.*
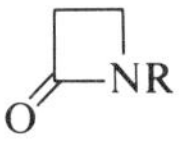	(**50**)	R = C_6H_4Br-*p*	0	1.363	1.207	73JCS(P2)2042
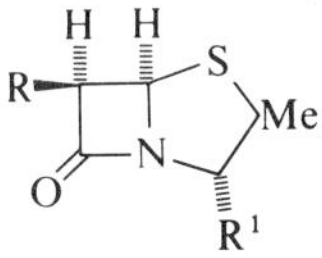	(**51**)	R = $PhCH_2CONH$, $R^1 = CO_2^-$ [procaine]$^+$	0.40	1.38	1.21	78JCS(P1)185
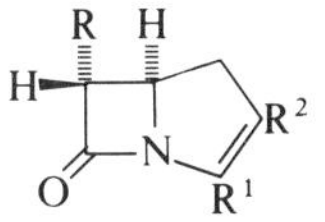	(**52**)	R = CHOHMe, $R^1 = CO_2Me$, $R^2 = SCH_2CH_2NHAc$	0.49	1.416	1.217	78JA6491
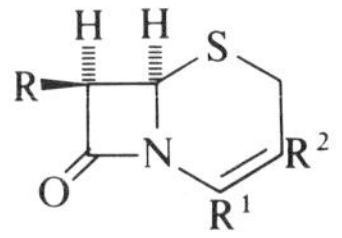	(**53**)	R = 4-py-CH_2CONH, $R^1 = CO_2H$, $R^2 = CH_2OAc$	0.22	1.377	1.211	B-82MI51206
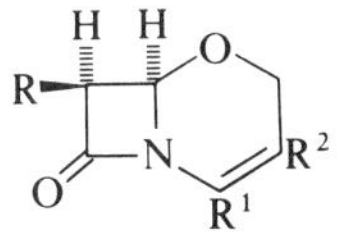	(**54**)	R = $PhCH_2CONH$, $R^1 = CO_2CHPh_2$, $R^2 = CH_2SMTZ$[a]	0.30	1.391	1.195	B-82MI51206
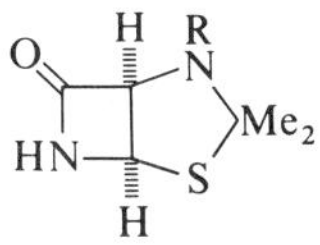	(**55**)	R = CO_2Bu^t	—	1.337	1.201	B-72MI51201
Free amides		—	—	1.32	—	B-72MI51201

[a] SMTZ = 1-methyltetrazol-5-ylthio.

(**57**) and clavulanic acid (**58**) illustrate resonances typically found for fused-ring azetidinones (see also Section 5.11.2.2).

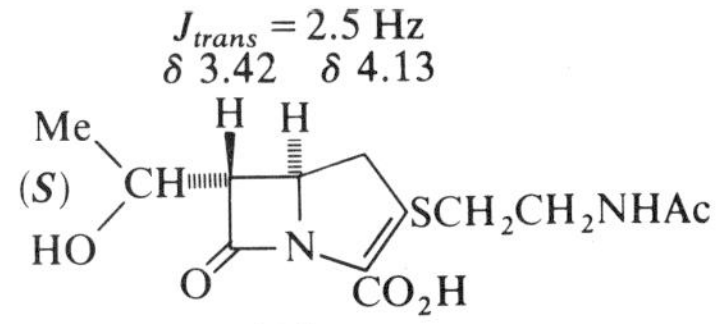

J_{trans} = 2.5 Hz
δ 3.42 δ 4.13
Me
(S)
HO
SCH₂CH₂NHAc
CO₂H

(**56**)

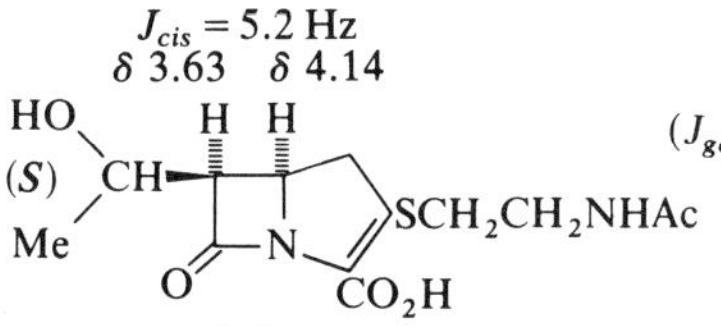

J_{cis} = 5.2 Hz
δ 3.63 δ 4.14
HO
(S)
Me
SCH₂CH₂NHAc
CO₂H

(**57**)

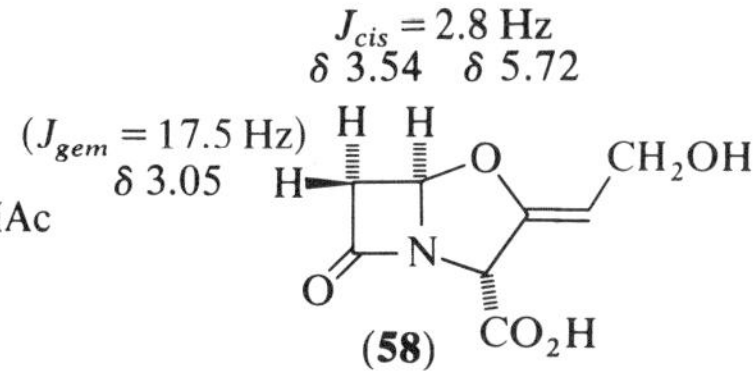

J_{cis} = 2.8 Hz
δ 3.54 δ 5.72
(J_{gem} = 17.5 Hz)
δ 3.05
CH₂OH
CO₂H

(**58**)

IR, UV and mass spectral properties of bicyclic azetidinones, other than penicillins and cephalosporins, have received thorough treatment ⟨B-82MI51201⟩.

Reports of CD studies ⟨78JOC4438⟩, MO ⟨79TL1499⟩ and conformational energy ⟨81MI51206⟩ calculations have appeared.

5.12.3.3 Reactivity

5.12.3.3.1 Modification at ring atoms

The carbapenems (**59**) are a relatively new class of β-lactam antibiotics first reported in 1976 ⟨76MI51200⟩. Three reactions which have been described for this class of compound are reductive removal of the sulfur-containing side chain ⟨78JA8004⟩, isomerization of the double bond by treatment with strong base in aprotic solvents ⟨B-82MI51207⟩ and ozonolysis ⟨80MI51202⟩. Few examples of modification at a ring atom in the natural product clavulanic acid (**60**; R = H) have been reported. One interesting example is aldol condensation in aprotic solvents at the 6-position of esters (**60**) with aldehydes and ketones ⟨78GEP2822001⟩.

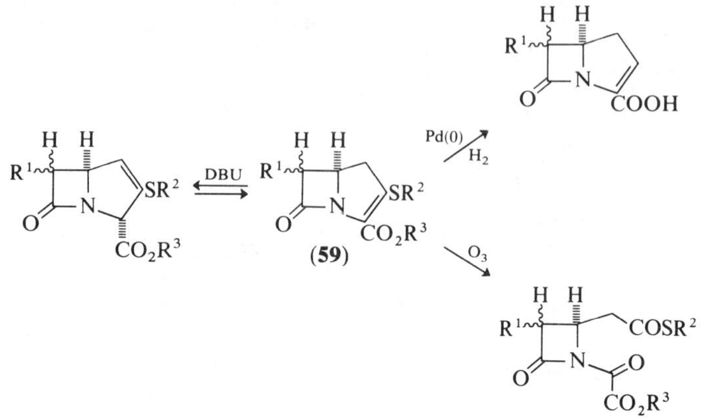

The reactions at the ring atoms of other nuclear analogs of the natural β-lactam antibiotics parallel closely those already described for penicillins (Chapter 5.11) and cephalosporins (Chapter 5.10). The reader also is referred to the monographs cited in Section 5.12.3.1.

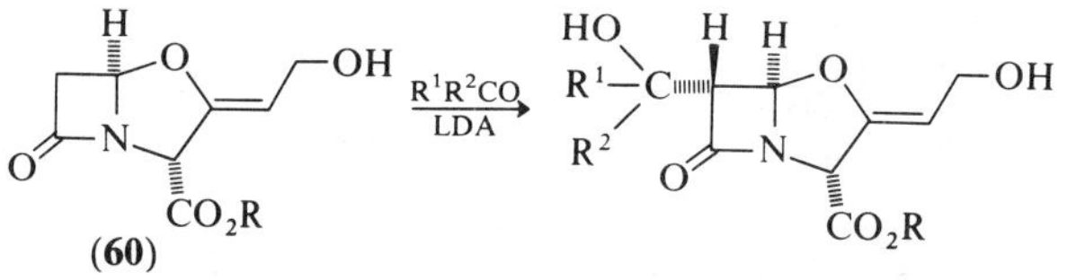

5.12.3.3.2 Modification of ring substituents

Extensive modifications of the substituents attached to these bicyclic ring systems have been accomplished. Using the natural product thienamycin (**61**; R^1–R^4 = H) as an example, the cysteamine side chain has been subject to wide variation which includes *N*-acylation, *N*-alkylation and *N*-imidoylation, and oxidation to the sulfoxide and sulfone. The carboxyl group commonly can be esterified by a variety of standard reagents, usually in aprotic solvents, after first acylating the side chain amino group. The hydroxyethyl side chain has been *O*-acylated and *O*-alkylated. The hydroxyl group has been removed to give (**62**) by initial mesylation of a protected intermediate followed by stereospecific *trans* elimination in base.

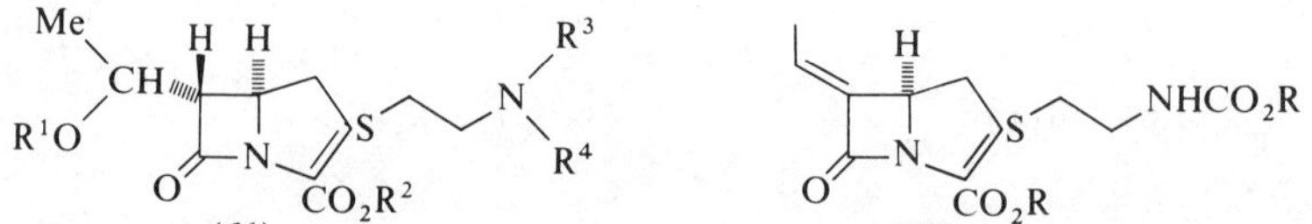

Substituent modifications of another natural, fused-ring β-lactam, clavulanic acid (**58**), also have received considerable attention. For example, the exocyclic double bond (*Z* isomer) can be isomerized to the *E* isomer (isoclavulanates) in about 10% yield under mildly basic conditions ⟨76TL4203⟩ or in 40% yield by photolysis of an appropriate ester ⟨B-77MI51201⟩. Under hydrogenation conditions hydrogenolysis of the allylic alcohol is preferred over saturation of the double bond ⟨B-77MI51201⟩. The double bond can be reduced under more vigorous conditions. Extensive modifications have been carried out at the exocyclic allylic position, usually on intermediates in which the carboxyl is blocked with a removable ester group. These modifications of (**58**) include *O*-alkylation, *O*-acylation, replacement of the hydroxyl by halogen ⟨78CC467⟩ and subsequent conversion to a variety of nitrogen- and sulfur-containing substituents. Attempted oxidation of the allylic alcohol to the corresponding aldehyde with DMSO–DCC–orthophosphoric acid gave instead diene (**63**), most likely *via* a 1,4-elimination ⟨77CC808⟩.

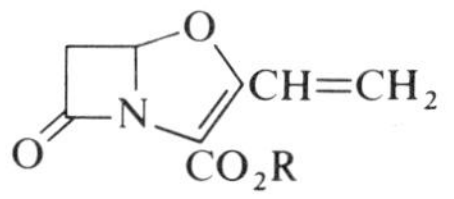

(**63**)

5.12.3.3.3 Cleavage of the β-lactam ring

Like the penicillins (Chapter 5.11) and cephalosporins (Chapter 5.10) the strained β-lactam ring of the bicyclic azetidinones discussed in this chapter is cleaved readily under a variety of conditions. However, though cleavage is facile in protic solvents, under appropriate aprotic conditions the β-lactam ring is sufficiently stable to allow extensive modifications to be made at other sites in the molecule. The β-lactam ring of thienamycin (**64**), which contains a primary amino group in its 2-position side chain, is cleaved more readily at pH 7 than is benzylpenicillin. Acylation of the amino group increases stability but these compounds still are readily susceptible to cleavage by standard reagents like hydroxylamine. Treatment of (**64**) with acetic anhydride at 0 °C in DMF followed by esterification with diazomethane affords ring-opened product (**65**).

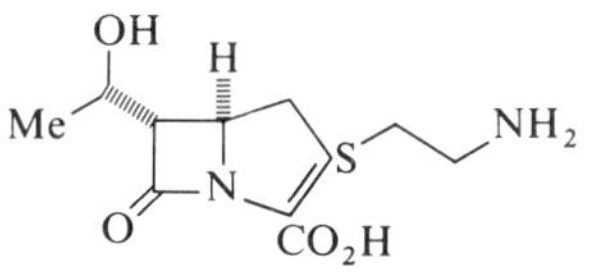

(**64**) 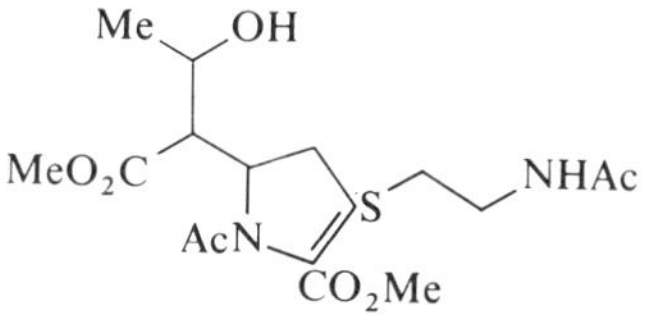 (**65**) 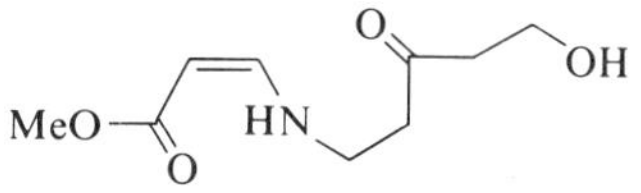 (**66**)

Methanolysis of clavulanic acid (**58**) salts at room temperature results in complete destruction of both rings to give enamine (**66**).

The susceptibility of these newer ring systems as well as other nuclear analogs to cleavage by a variety of nucleophiles and by acids (Section 5.12.3.4) parallels that described in earlier chapters on the penicillins and cephalosporins.

5.12.3.4 Synthesis

5.12.3.4.1 Fermentation processes

Controlled fermentation of microorganisms has provided a rich source of β-lactam antibiotics beginning with penicillin in 1940, followed by cephalosporins in 1953 and the cephamycins in 1971. In 1976 two new bicyclic β-lactams were discovered from the fermentation of *Streptomyces* species. These compounds were given the trivial names thienamycin (**64**; a carbapenem) and clavulanic acid (**58**; an oxapenam). Since their discovery 17 additional carbapenems (Table 5) and four additional clavulanic acid derivatives (**67** and **68**) have been isolated. Olivanic acids MM22380–22383 are identical to epithienamycins A–D which were isolated independently ⟨81MI51201, 81MI51202⟩. Olivanic acid MM4550 is believed to be identical to another carbapenem coded MC-696-SY2-A ⟨77MI51202⟩. Two 8-position sulfated derivatives of olivanic acids MM22380 and MM22382 have also been reported ⟨77CC953, 77CC523⟩.

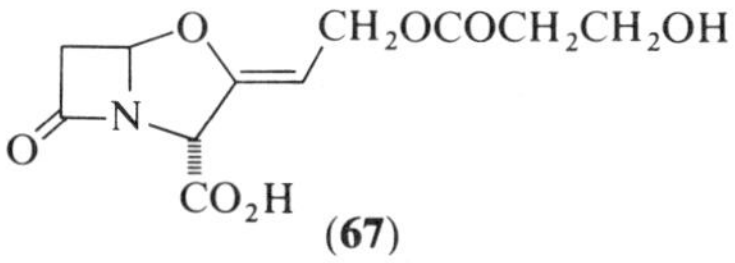

(**67**) 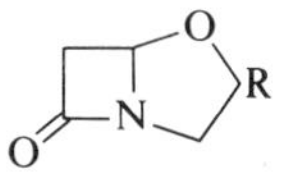 (**68**)

The four oxapenams [(**67**), (**68**; R = CH_2OH, CH_2OCHO, CO_2H)] structurally related to clavulanic acid (**58**) also have been isolated from fermentation broths of *Streptomyces* species.

Table 5 Naturally Occurring Carbapenems

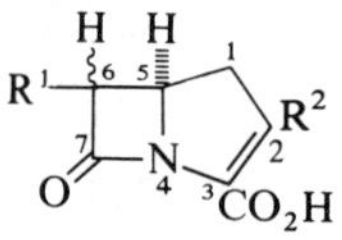

Compound	*H^5/H^6*	*R^1*	*R^2*	*Ref.*
Thienamycin	*trans* (5*R*,6*S*)	Me(OH)CH— (8*R*)	$—SCH_2CH_2NH_2$	79JA1
N-Acetylthienamycin	*trans* (5*R*,6*S*)	Me(OH)CH— (8*R*)	$—SCH_2CH_2NHAc$	B-82MI51205
N-Acetyl-11,12-dehydro-thienamycin	*trans* (5*R*,6*S*)	Me(OH)CH— (8*R*)	—SCH=CHNHAc	B-82MI51205
9-Northienamycin	*trans* (5*R*,6*S*)	$HOCH_2—$	$—SCH_2CH_2NH_2$	B-82MI51205
Olivanic acid MM22381	*trans* (5*R*,6*S*)	Me(OH)CH— (8*S*)	$—SCH_2CH_2NHAc$	79MI51201
Olivanic acid MM22383	*trans* (5*R*,6*S*)	Me(OH)CH— (8*S*)	—SCH=CHNHAc	79MI51201
PS-5	*trans* (5*R*,6*R*)	Et	$—SCH_2CH_2NHAc$	80MI51202
PS-6	*trans* (5*R*,6*R*)	Pr^i	$—SCH_2CH_2NHAc$	80MI51203
PS-7	*trans* (5*R*,6*R*)	Et	—SCH=CHNHAc	80MI51203
PS-8	?	Pr^i	—SCH=CHNHAc	82MI51203
Olivanic acid MM22380	*cis* (5*R*,6*R*)	Me(OH)CH— (8*S*)	$—SCH_2CH_2NHAc$	79MI51201
Olivanic acid MM22382	*cis* (5*R*,6*R*)	Me(OH)CH— (8*S*)	—SCH=CHNHAc	79MI51201
Olivanic acid MM4550	*cis* (5*R*, 6*R*)	$Me(OSO_3H)CH—$ (8*S*)	—S(O)CH=CHNHAc	77CC523
Carpetimycin A	*cis* (5*R*,6*R*)	$Me_2(OH)C—$	—S(O)CH=CHNHAc	81MI51203

However, the low yields obtained for the carbapenems by fermentation processes have engendered intense interest in synthetic approaches to these biologically important compounds.

5.12.3.4.2 Synthesis of carbapenems

Three main synthetic routes have been used to prepare variously substituted carbapenems. Two involve an intramolecular Wittig reaction to form either the C(2)—C(3) bond (equation 1) or the C(1)—C(2) bond (equation 2). The third approach uses a carbene insertion step to form the C(3)—N(4) bond (equation 3).

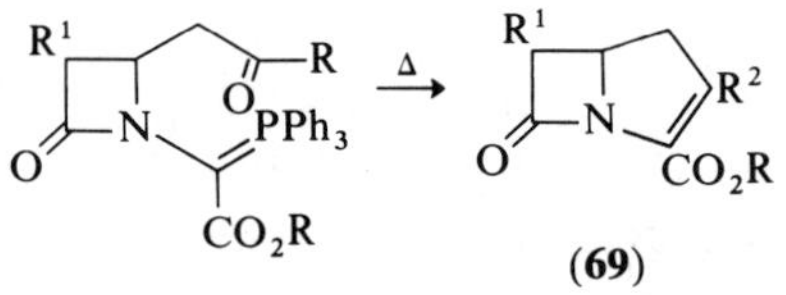

(69) (1)

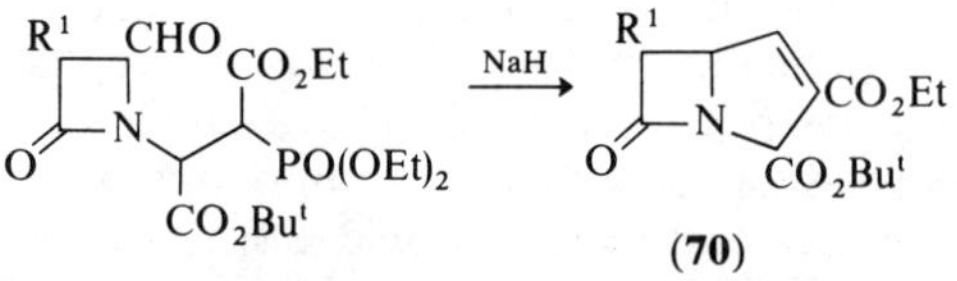

(70) (2)

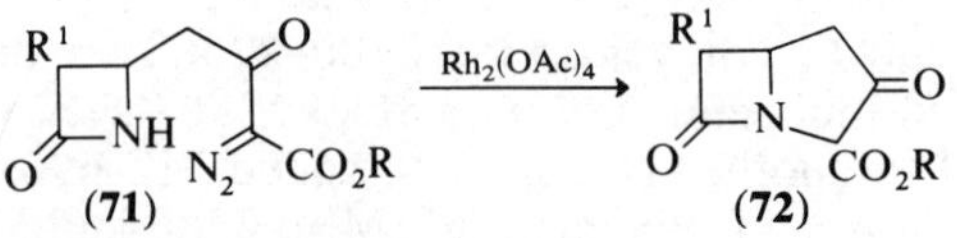

(71) (72) (3)

Several groups have reported the synthesis of carbapen-2-ems *via* equation (1) ⟨78JA8006, 79CC847, 80JA2060, 79CC236, 80TL2013⟩. This internal Wittig reaction also has been applied to the synthesis of cephems (Section 5.10.4.3) and penems (Section 5.11.4.6). The synthesis of carbapen-1-ems by the general reaction of equation (2) has been reported ⟨81TL191, 81TL2025⟩ but has found only limited use. Isomerization of the double bond to the desired 2,3-position has proved to be difficult. The third route *via* α-diazo-β-keto esters (**71**) by a rhodium(II) acetate-catalyzed insertion reaction has found general use ⟨80TL31, 81TL3557, 80H(14)1077⟩. A stereocontrolled enantiospecific synthesis of thienamycin (**64**) using this approach has been described ⟨80JA6161⟩. Further improvements in this process also have been reported ⟨80TL2783, B-81MI51200⟩.

Other synthetic routes to the carbapenem ring system include photolytic Wolff rearrangement ⟨73JCS(P1)2024⟩, aldol condensation ⟨78JA313⟩, addition–cyclization ⟨80JOC1135⟩ and β-lactam formation ⟨78JOC4438, 79TL4359⟩. An excellent review of carbapenem synthesis has recently appeared ⟨82H(17)463⟩.

5.12.3.4.3 Synthesis of clavulanic acid and related oxapenams

The isolation of clavulanic acid (**58**) rekindled interest in developing synthetic approaches to the oxapenams. The most widely used general method involves O(1)—C(5) bond formation illustrated for the synthesis of (**73**).

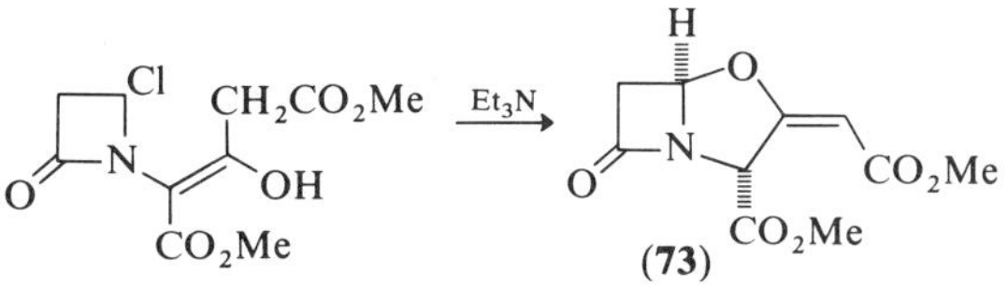

Cl CH$_2$CO$_2$Me OH CO$_2$Me —Et$_3$N→ (**73**) CO$_2$Me CO$_2$Me

Compounds (**74**) ⟨78CPB1761⟩ and (**75**) ⟨77CC905⟩ have been synthesized by this method, as has methyl clavulanate ⟨79TL1889⟩.

Photolytic decomposition of diazo intermediates has been used successfully to form oxapenams. For example, compound (**77**) was prepared from (**76**) by formation of the C(5)—C(6) bond ⟨78CPB448⟩.

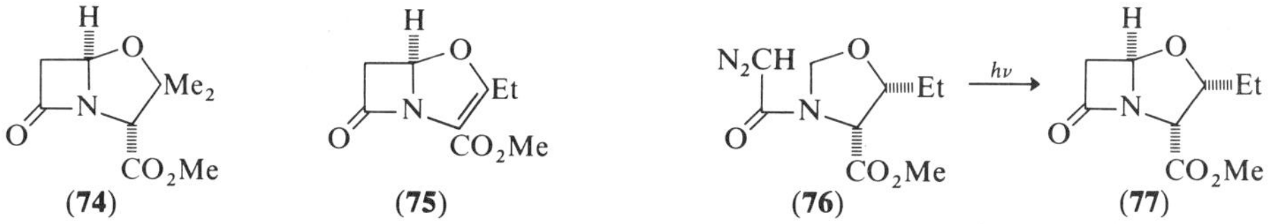

(**74**) (**75**) (**76**) (**77**)

Photochemical cyclization of (**78**) with formation of the C(3)—N(4) bond leads to oxapenam (**79**) ⟨78TL4233⟩ (see also Section 5.12.3.4.2). Another method of oxapenam formation involves closure of the C(3)—N(4) bond by base-induced cyclization of α-bromoesters ⟨77CC359, 77CC405⟩.

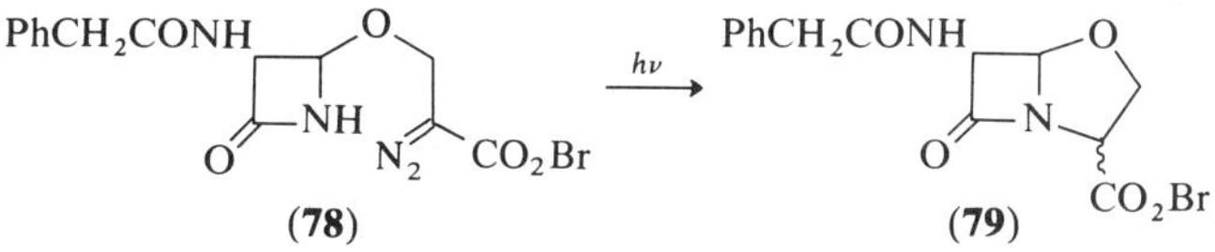

PhCH$_2$CONH … NH N$_2$ CO$_2$Br (**78**) —$h\nu$→ PhCH$_2$CONH … CO$_2$Br (**79**)

5.12.3.4.4 Synthesis of nuclear analogs of naturally occurring β-lactam antibiotics

As a result of the intense interest in exploitation of the biological properties of the β-lactam antibiotics a large number of nuclear analogs of the various naturally occurring structures have been synthesized. In addition to the carbapenems and oxapenams already described (Sections 5.12.3.4.2 and 5.12.3.4.3) a number of other novel bicyclic azetidinones have been reported. Examples of many of these are listed in Table 6. The methods used to synthesize these compounds are too varied to list in any systematic way.

5.12.3.5 Application

The clinical utility of the penicillin and cephalosporin classes of β-lactam antibiotics for treating myriad infectious diseases of man and animals has been thoroughly discussed in the two preceding chapters of this volume. Three newer non-penicillin/non-cephalosporin β-lactam compounds have undergone extensive clinical evaluation. Compound (**104**) is a potent broad-spectrum antibiotic discovered at Shionogi Co. Ltd. and given the generic names moxalactam (USAN) and latamoxef (IWN). It was approved for marketing in the United States in 1982 and is sold commercially by the Eli Lilly Co. The greatly improved spectrum and potency of moxalactam toward Gram-negative bacteria has been attributed

Table 6 Nuclear Analogs of Naturally Occurring β-Lactam Antibiotics

Compound	Structure	Substituents	Ref.
(**80**)	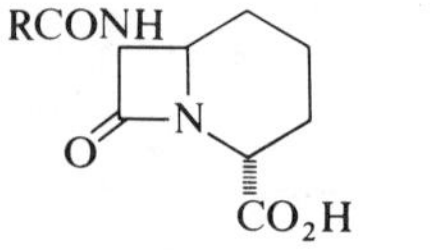	R = $PhCH_2$	71JCS(C)3756
(**81**)	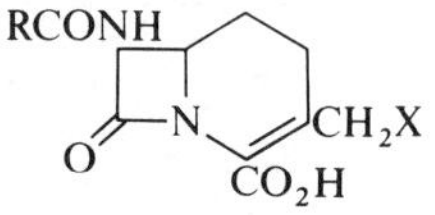	R = $PhCH_2$, X = OAc	74JA7584
(**82**)	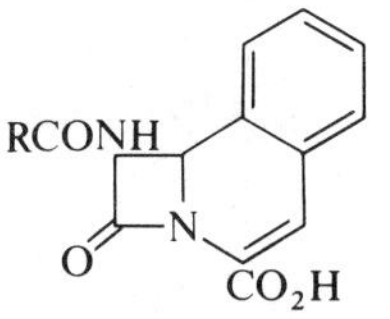	R = $PhOCH_2$	78TL1629
(**83**)	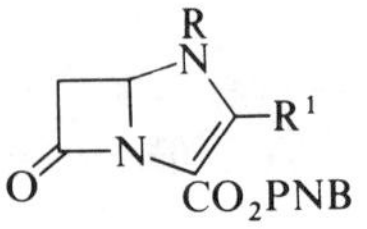	R = Ph, Et, R^1 = Me	81CC1269
(**84**)	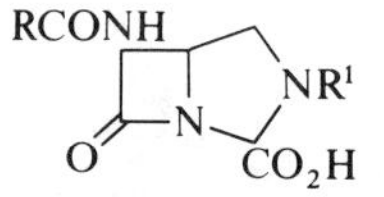	R = $PhOCH_2$, R^1 = Ac	77JA2352
(**85**)	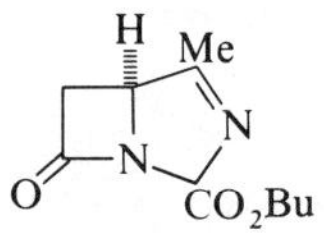		81CC946
(**86**)	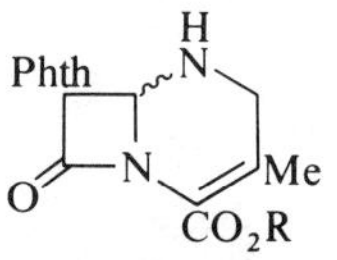		72CJC2902
(**87**)	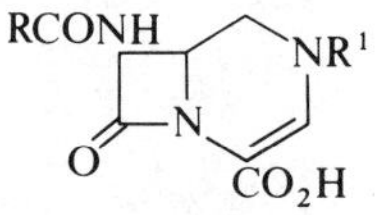	R = $PhOCH_2$, R^1 = CO_2Et, Me	77CJC2719
(**88**)	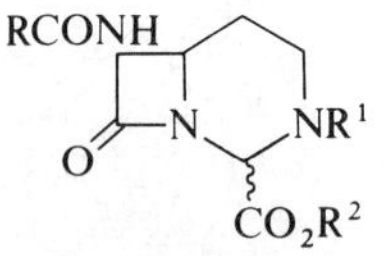	R = $PhOCH_2$, $PhCH_2$, R^1 = CO_2Et, C_6H_4Me, R^2 = H, Me	B-79MI51202 80TL3947
(**89**)	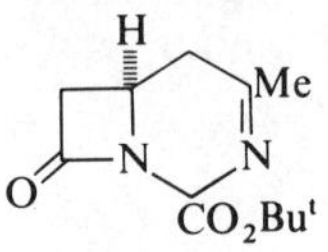		81CC946
(**90**)	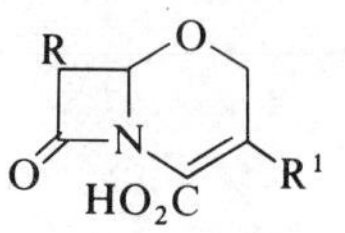	R = $PhCH_2CONH$, R^1 = CH_2OAc R = H, R^1 = Me	74JA7582 78JCS(P1)1450
(**91**)	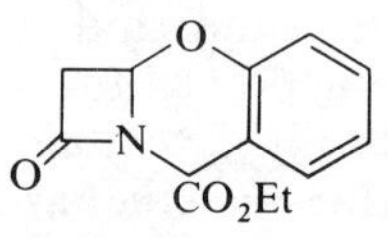		79H(12)947

Table 6 *(continued)*

Compound	*Structure*	*Substituents*	*Ref.*
(**92**)		R = H, CO_2Bu^t	81TL2689
(**93**)		R = N_3	79SC113
(**94**)		R = $ArCH_2CONH$, R^1 = Me, CH_2CO_2Me, $PhCH_2$	77CJC484, 78CJC1335, 80CJC2508
(**95**)		R = $PhCH_2CONH$, $PhOCH_2CONH$, R^1 = H, OMe	79JA4730
(**96**)			79H(12)795
(**97**)		R = PhCH(OCHO)CONH, R^1 = OMe	80TL3947
(**98**)		R = $PhCH_2CONH$	72HCA2567
(**99**)		R = PhOCONH, R^1, R^2 = $-(CH_2)_3-$, $-(CH_2)_4-$	77TL1855
(**100**)		R = $PhCH_2CONH$	78JCS(P1)668
(**101**)		R = Ph, Et	81CC1269
(**102**)		R = $PhOCH_2CONH$	B-81MI51205
(**103**)		R = $PhOCH_2CONH$	B-81MI51205

to its improved stability toward the β-lactamases, bacterial enzymes which can degrade the β-lactam antibiotics. The 1-oxacephem nucleus of moxalactam is responsible for the greater *in vitro* potency in comparison to its classical cephalosporin (1-thia) counterpart. The α-carboxyl group in the 7-position side chain and the 7-α-methoxy group impart stability to the β-lactamases. A thorough review of the chemistry and biological properties of (**104**) has appeared ⟨81MI51204⟩.

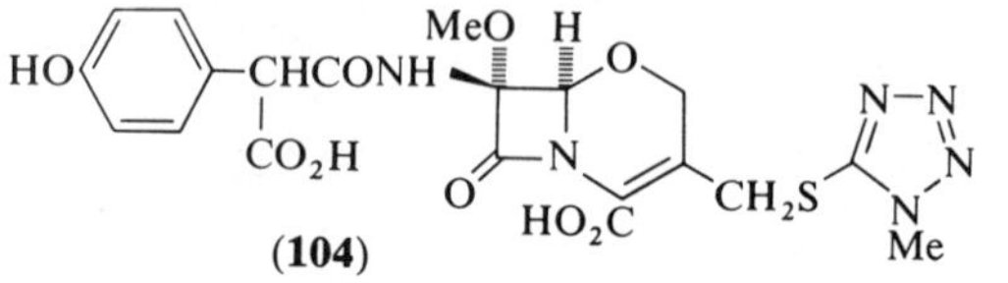

(**104**) 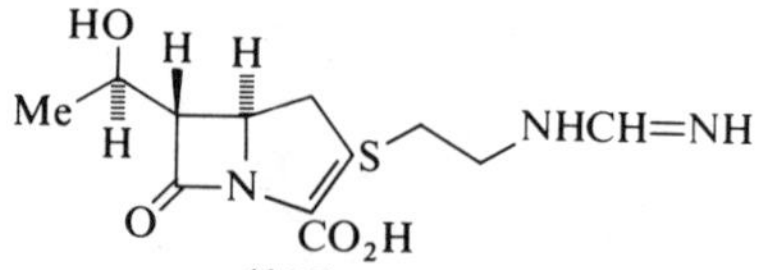 (**105**)

The second compound receiving intensive clinical evaluation is carbapenem (**105**), *N*-formimidoylthienamycin. This semisynthetic derivative of the parent antibiotic thienamycin (**64**) has been shown to possess very potent antibacterial activity against a range of Gram-positive and Gram-negative bacteria, including the normally resistant organism *Pseudomonas aeruginosa* ⟨80MI51201⟩. It also overcomes much of the instability in solution observed with thienamycin itself ⟨79JMC1435⟩. The compound was discovered and is being developed by Merck and Co. It has been given the designation MK-0787.

The third compound, clavulanic acid (**58**), in contrast to (**104**) and (**105**) which are potent antibiotics, is being developed by Beecham Laboratories for its inhibitory properties against a wide range of β-lactamases ⟨77MI51200⟩ (see Section 5.11.5.2). It acts synergistically with many penicillins and cephalosporins, in many cases by lowering significantly the minimum inhibitory concentrations of these antibiotics against a number of organisms. Clavulanic acid has only modest activity as an antibiotic. It is being studied extensively in combination with the oral penicillin, amoxicillin ⟨81MI51207⟩.

5.12.4 OTHER FUSED-RING AZETIDINONES

5.12.4.1 Structure

X-ray crystallographic examination of bicyclic azetidinone (**106**) indicates that the azetidinone ring is essentially planar ⟨73JA4647⟩. Selected bond angles and distances are listed in Table 7. ^{1}H NMR and IR properties for several bicyclic azetidinones are given in Table 8. Theoretical calculation ⟨79T1499⟩ and CD studies ⟨78JOC4438⟩ for the 1-azabicyclo[3.2.0]heptane system have been reported.

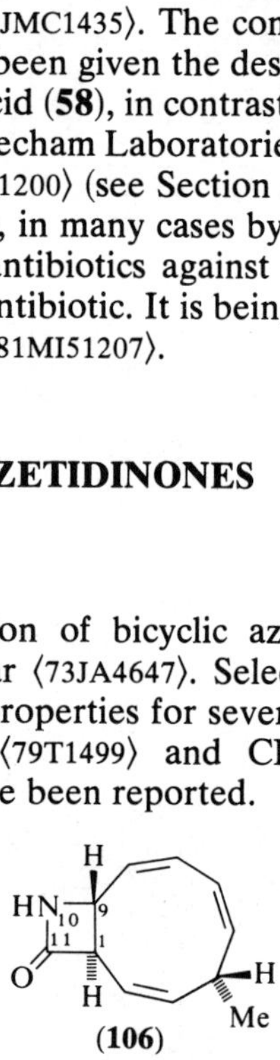

(**106**)

Table 7 Selected Bond Angles and Distances for Compound (**106**)

Bond distance (Å)		*Bond angle* (°)	
C(1)—C(9)	1.570	∠C(11)—C(1)—C(9)	84.6
C(9)—N(10)	1.479	∠C(1)—C(9)—N(10)	86.1
N(10)—C(11) (lactam)	1.341	∠C(9)—N(10)—C(11)	94.6
C(11)—C(1)	1.511	∠N(10)—C(11)—C(1)	93.7

5.12.4.2 Reactivity

5.12.4.2.1 Thermal and photochemical reactions

Under appropriate thermal conditions bicyclic azetidinones can undergo valence bond tautomerism in a behavior similar to that described for bicyclic azetidines in Section 5.12.2.2.1. Thus, the position of equilibrium illustrated in equation (4) strongly favors the

Table 8 Spectroscopic Properties of Selected Bicyclic Azetidinones

Structure		$\delta(^1H)$ (p.p.m.)	$J(^1H-^1H)$ (Hz)	$\nu(C{=}O)$ (cm^{-1})	*Ref.*
		—	—	1770	78JOC4438
Bu^tO_2C	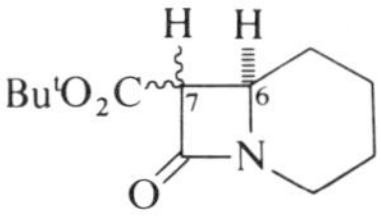 *cis*	H-7 3.90	$J_{6,7}$ 5	1765	71CC577
	trans	H-7 3.43	$J_{6,7}$ 2	1765	71CC577
		H-7 3.61(m), H-8α 2.85, H-8β 2.30	$J_{7,8}$ (α)5, $J_{7,8}$ (β) 3	1745	71CJC1648
MeN		H-1 4.18, H-4 4.38		1742	64JA950
HN		H-1 3.37, H-5 4.46	$J_{1,5}$ 4.1	1755	77JCS(P1)874
HN		H-1 3.25, H-6 3.85		1750	71TL3721
HN		H-1 2.95–3.35, H-9 3.55–3.95		1750	73JA4647
HN		H-1 3.23–2.83, H-12 3.65–3.36		1751	76JOC3583

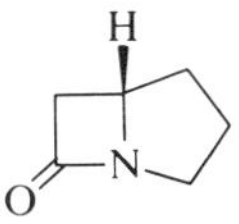

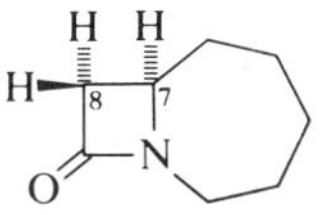

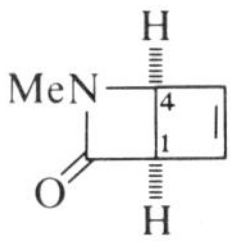

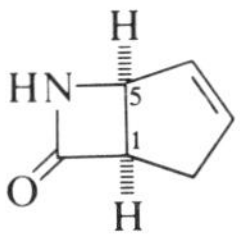

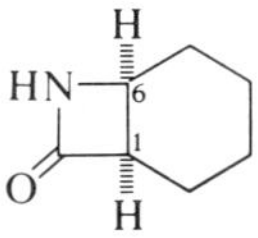

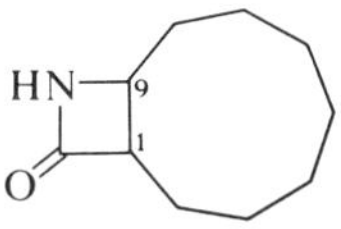

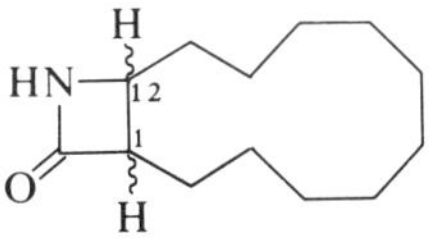

bicyclic compound, but when R = H increasing amounts of the dihydroazocine are observed (15% at 115 °C) ⟨69TL1455⟩. However, when R = Me only the bicyclic structure was present up to 120 °C. Above that the compound decomposed to *o*-xylene and methyl isocyanate.

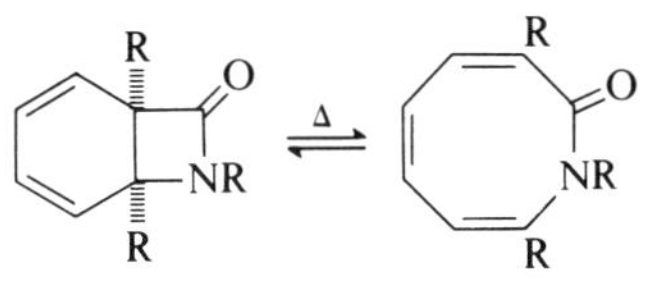

$$\rightleftharpoons^{\Delta} \qquad (4)$$

In concert with the thermal reactivity of bicyclic azetidines the ease of ring opening of bicyclic azetidinones decreases with increasing ring size. For example, chlorosulfonylazetidinone (**107**) is transformed with increasing temperature through the intermediate iminolactone (**108**) to the thermodynamically stable product (**109**) ⟨77JCS(P1)874⟩. The dipolar species (**110**) has been proposed as an intermediate in each of these conversions. In contrast, azetidinone (**111**) having a seven-membered ring gave on heating only the ring-opened product (**112**), plus some of the corresponding nitrile.

Irradiation of azetidinone (**113**) in alcohol affords ether (**114**) ⟨76JOC2808⟩.

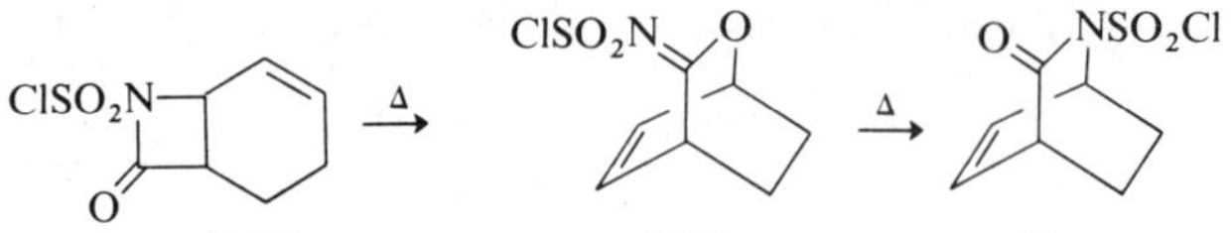

(107) (108) (109)

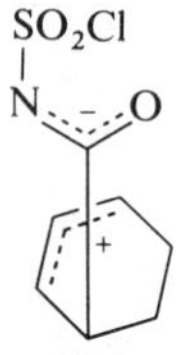

(110) 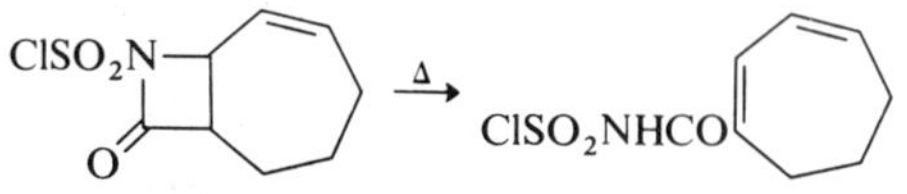 (111) (112) 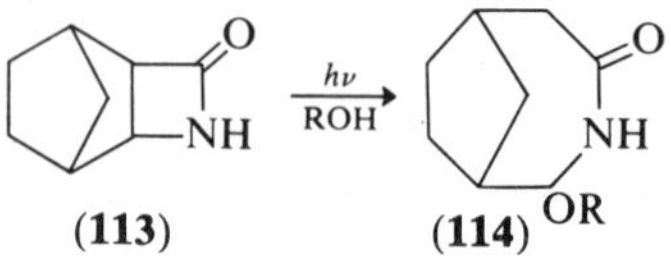 (113) (114)

5.12.4.2.2 Modification of ring substituents

The majority of transformations carried out at ring substituents of bicyclic azetidinones were described in Section 5.12.3.3.2 for azetidinones related to cephalosporins and penicillins. Nucleophilic displacement reactions of 7-halo-1-azabicyclo[4.2.0]octan-8-ones with thallium phthalimide and hydrazine have been reported ⟨76ACS(B)318⟩.

5.12.4.2.3 Cleavage of the β-lactam ring

The β-lactam bond of bicyclic azetidinones is susceptible to ring opening under a variety of conditions. The ease of hydrolysis under acidic conditions also is strongly influenced by the size of the fused ring; the smaller the ring the more facile the hydrolysis ⟨69JCS(C)2093⟩. Treatment with boiling methanolic HCl affords the ring-opened methyl esters (equation 5).

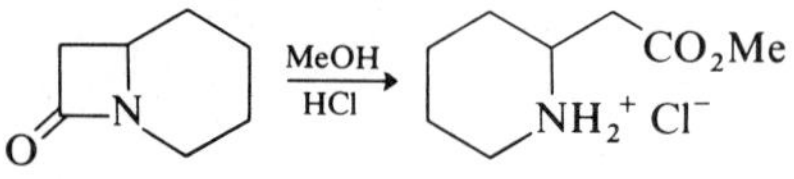

(5)

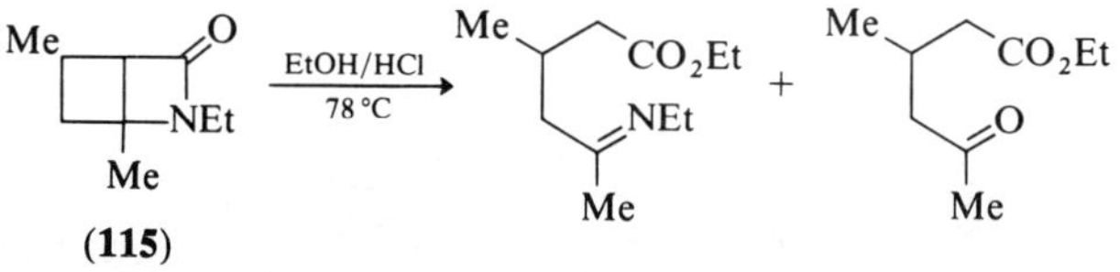

(115)

(6)

Azetidinones having a ring fused at the N-1/C-4 atoms usually are more susceptible to hydrolysis than those compounds in which the ring fusion is at C-3/C-4. The 2-azabicyclo[2.2.0]hexan-3-one (**115**) is cleaved with ethanolic HCl to give ring-opened products (equation 6) ⟨72CB2780⟩.

5.12.4.3 Synthesis

5.12.4.3.1 From azetidinone precursors

In contrast to the synthesis of bicyclic azetidinones containing a carboxyl group on the atom adjacent to the lactam nitrogen (see Section 5.12.3.4), few examples are reported in which bicyclic azetidinones without this carboxyl group are prepared from azetidinone precursors. One unique approach, however, utilized a free radical ring closure to obtain 6-oxa-1-azabicyclo[5.2.0]nonan-9-one (**116**) in modest yield ⟨81TL2689⟩.

(116)

5.12.4.3.2 From non-azetidinone precursors

A number of different methods have been used to prepare bicyclic azetidinones from non-azetidinone precursors. The most widely used and generally applicable method is [2+2] cycloaddition of ketenes and isocyanates to unsaturated systems. Chlorosulfonylisocyanate has found extensive use and adds readily under relatively mild conditions to a variety of unsaturated compounds as illustrated in equations (7) and (8) ⟨68JA3897, 66JA3657, 73JA4647, 76JOC3583, 77JCS(P1)874, 76JOC2808, 68LA(718)94⟩. Addition of ketenes (equation 9) to dihydro-1,3-oxazines ⟨68JHC779⟩, dihydrothiazines ⟨73JHC791, 81JA1769⟩, tetrahydropyrimidines ⟨72JOC3437⟩ and diazepines ⟨73TL4163⟩ has also been employed. Another approach used by several workers involves photochemical cyclization of α-diazoamides (equation 10) ⟨69JCS(C)2093, 75JCS(P1)1517, 65JA2518, 71CC577⟩. Other methods include photochemical cyclization of 2-pyridones (equation 11) ⟨64JA950, 72CB2780⟩, ring contractions of acylpyrazolidinones (equation 12) ⟨75TL4089, 75CC725⟩ and α-ketolactams (equation 13) ⟨75JOC1264⟩, azide reaction with cyclopropanones (equation 14) ⟨71TL3721⟩, acid-catalyzed valence tautomerization ⟨71JOC435⟩ and internal acylation (equation 15) ⟨68AP230, 78JOC4438, 71CJC1648⟩.

$(CH_2)_n$ —CSI→ $(CH_2)_n$ —NSO_2Cl —Na_2SO_3→ $(CH_2)_n$ —NH (7)

—i, CSI; ii, PhSH→ NH (8)

—RCH=C=O→ (9)

—$h\nu$→ (10)

NR —$h\nu$→ NR (11)

RCO, H, HN, O, R^1 —HgO, RSO_2ONH_2→ RCON, H, O, R^1 (12)

O —KIO_4→ HO_2C, H H, O, N (13)

OH, OMe —NaN_3, Δ→ H, NH, H, O (14)

HO_2C NH —Δ or DCC→ O N (15)

5.12.4.4 Application

There appear to be no applications of commercial importance for the types of bicyclic azetidinone described in this section.

5.12.5 FUSED-RING AZETINES AND AZETES

5.12.5.1 Structure

Detailed structural studies of bicyclic azetines have been difficult to carry out because of the instability inherent in this type of ring system. The presence of fluorine-containing substituents though can stabilize some bicyclic azetines sufficiently to obtain physical data in support of these structures. X-ray crystallographic examination of palladium complex (**117**) showed that each 1-azabicyclo[2.2.0]hexadiene ring is planar. Selected bond angles and distances for (**117**) are listed in Table 9 ⟨76CPB2219⟩.

Table 9 Selected Bond Angles and Distances for Compound (**117**)

Bond angle (°)		*Bond distance* (Å)	
∠N(1)—C(2)—C(3)	96	N(1)—C(2)	1.50
∠C(2)—C(3)—C(4)	94	C(2)—C(3)	1.30
∠C(3)—C(4)—N(1)	85	C(3)—C(4)	1.53
∠C(4)—N(1)—C(2)	85	C(4)—N(1)	1.56

Working at low temperatures several groups have been able to observe NMR spectra of bicyclic azetines. ^{1}H NMR chemical shifts (δ downfield from TMS, p.p.m.) are illustrated for azetines (**118**) ⟨70JA2178⟩, (**119**) ⟨80LA798⟩, (**120**) ⟨77CJC56⟩, (**121**) ⟨80JCS(P1)607⟩ and (**122**) ⟨75JOC1683⟩. ^{13}C NMR values for (**119**) also were reported. IR spectra recorded for a series of perfluoro-2,5-diazabicyclo[2.2.0]hexa-2,5-dienes (**123**) ⟨75JCS(P1)1130⟩ show strong C=N stretching absorption at 1660–1690 cm^{-1}, while bicyclic azetine (**124**) shows C=C absorption at 1725 cm^{-1}.

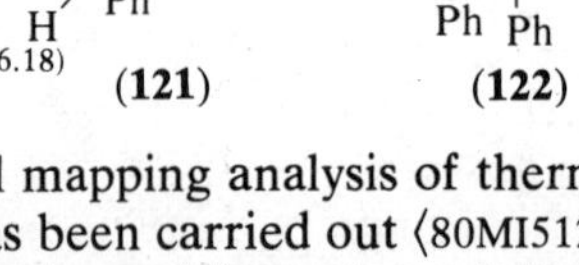

(**117**) 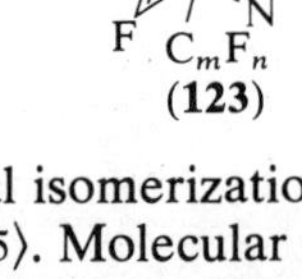 (**118**) 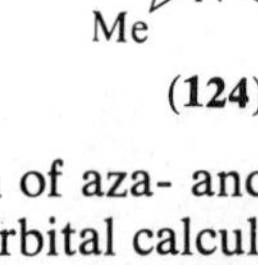 (**119**) 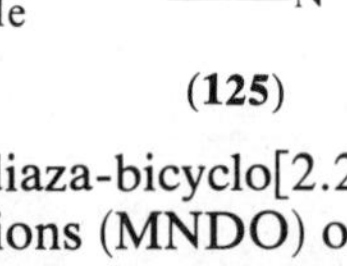 (**120**)

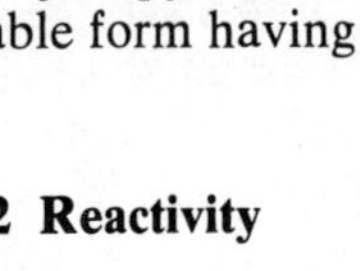

(**121**) 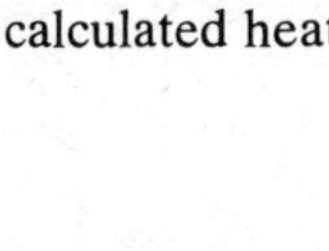 (**122**) 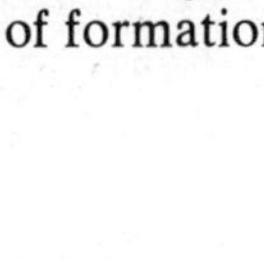 (**123**) (**124**) (**125**)

Orbital mapping analysis of thermal isomerization of aza- and diaza-bicyclo[2.2.0]hexadienes has been carried out ⟨80MI51205⟩. Molecular orbital calculations (MNDO) of a series of bisdehydropyridines showed that the 2-azabicyclo[2.2.0]hexa-1,3,5-triene (**125**) is the least stable form having a calculated heat of formation of 728.4 kJ mol^{-1} ⟨77CC539⟩.

5.12.5.2 Reactivity

Bicyclic azetines tend to be highly unstable species. The more strained the bicyclic system the greater the reactivity. Azetines of the 2-azabicyclo[2.2.0]hexadiene structure have been suggested as transient intermediates in photolytic conversion of pyridines ⟨77JCS(P2)1148, 78JOC944⟩ and of pyrimidines ⟨81JCS(P1)943⟩.

Under thermal conditions bicyclic azetines undergo facile valence bond tautomerization. Typically, 1-azabicyclo[2.2.0]hexadienes give pyridines (equation 16) ⟨76CPB2219⟩, 1,2-diazabicyclo[3.2.0]heptadiene (**126**) gives diazepine (**127**) ⟨77CJC56⟩ and 2-azabicyclo[2.2.0]hexadiene (**118**) produces pyridine ⟨70JA2178⟩.

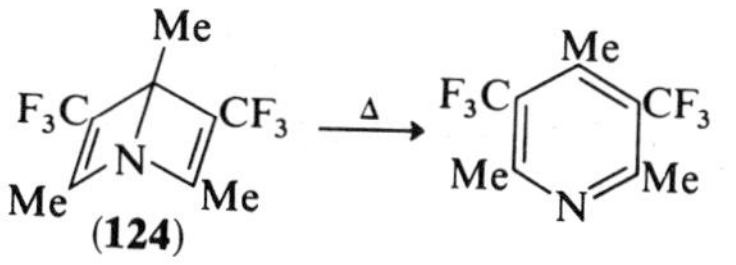

(16)

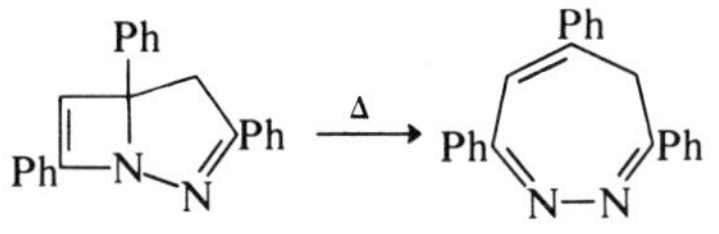

(124) (126) (127) (118) (128)

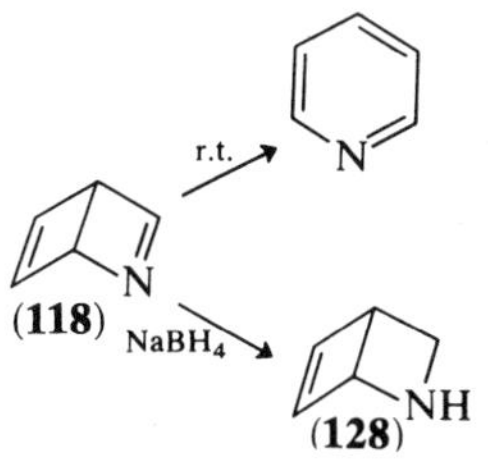

Treatment of (**126**) with ethanolic hydrochloric acid causes fragmentation to give acetophenone and 3,5-diphenyl-2-pyrazoline (equation 17).

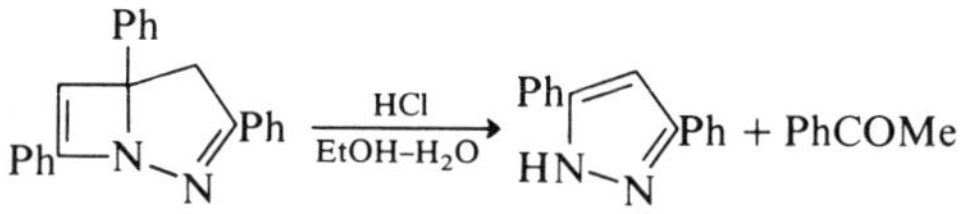

(17)

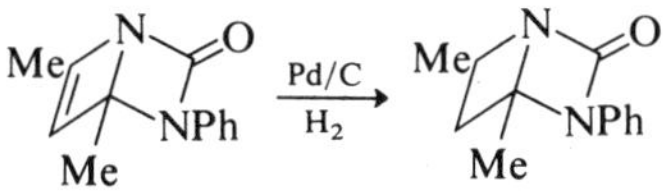

(18)

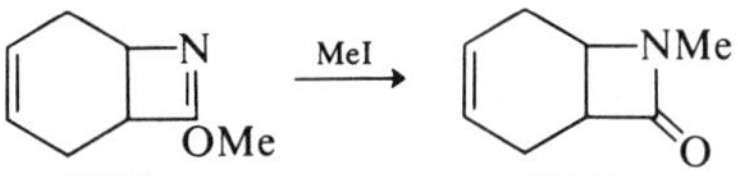

(129) (130)

The double bond of the azetine ring (**118**) can be reduced to the corresponding azetidine (**128**) with borohydride ⟨70JA2178⟩ or by catalytic hydrogenation (equation 18) ⟨80JCS(P1)607⟩.

Azetine (**129**) reacts with methyl iodide at reflux without rupture of the ring to give the *N*-methylazetidinone (**130**) ⟨71JOC435⟩.

5.12.5.3 Synthesis

The most common method of preparing bicyclic azetines is photoinduced valence bond isomerization. Thus, irradiation of pyrimidine (**131**) in benzene gives a 72% yield of bicyclic azetine (**132**) ⟨80JCS(P1)607⟩. Azepine (**133**) can be converted to the thermally labile 1-azabicyclo[3.2.0]hexadiene (**134**) on irradiation at 280 nm ⟨80TL595⟩. Other examples of similar transformations with other ring systems have been reported ⟨77TL4389, 80LA798, 77CJC56, 75JCS(P1)1130, 76CPB2219⟩. Other reported methods include thermal decomposition of cyclopropyl azides (equation 19) ⟨72TL3933⟩, photoinduced addition of nitriles to alkenes (equation 20) ⟨77JOC4238⟩ and *O*-alkylation of β-lactams (equation 21) ⟨71JA152⟩.

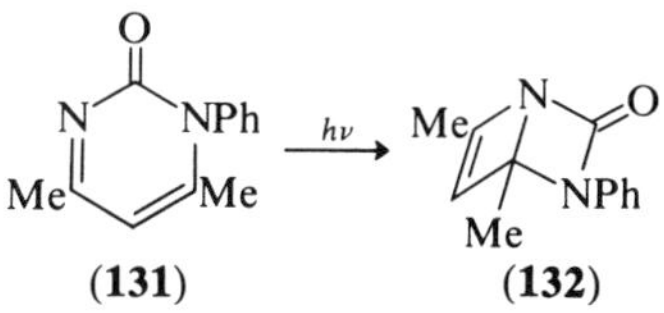

(131) (132) 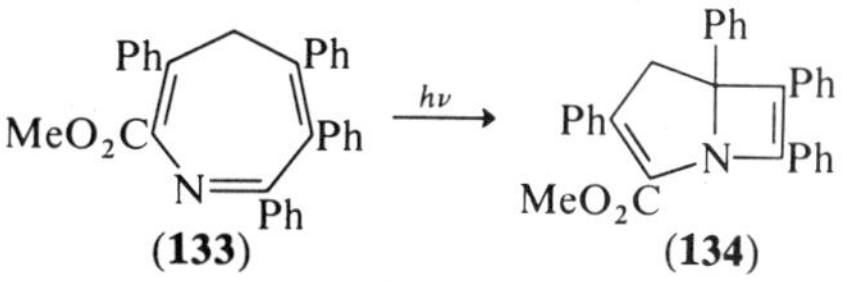 (133) (134)

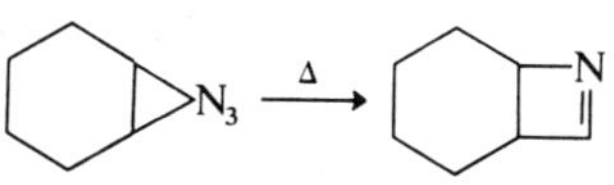

(19)

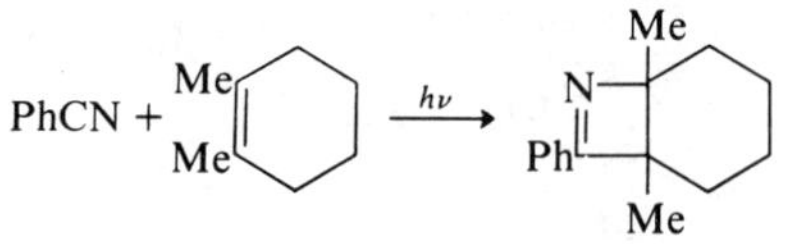

(20)

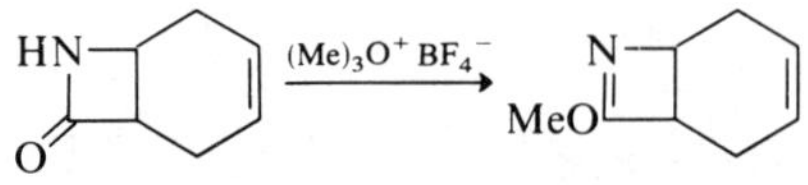

(21)

Finally, a rather unusual cycloaddition–rearrangement reaction produced the 1,2,5-triazabicyclo[4.2.0]octane derivative (**135**) in 64% yield ⟨80JOC4587⟩.

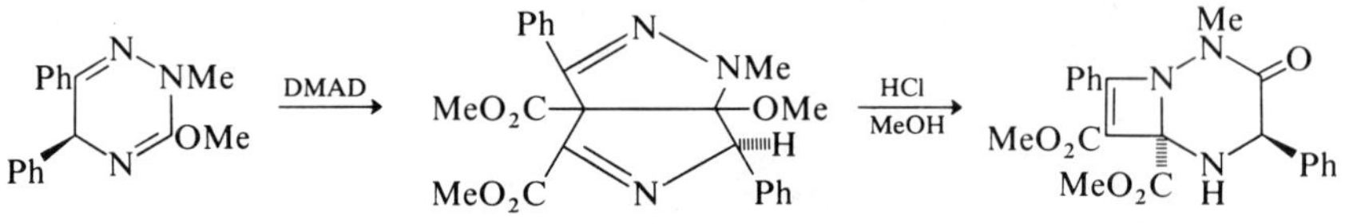

(**135**)

No examples of the synthesis of fused-ring azetes were found (see Section 5.09.5.3.2).

5.12.5.4 Application

At the present time there do not appear to be any commercial/industrial applications of fused-ring azetines.

5.13

Oxetanes and Oxetenes

S. SEARLES
University of Missouri-Columbia

5.13.1 INTRODUCTION

Four-membered cyclic ethers are known as oxetanes. Although they have been known since 1858, when the parent compound was first isolated, they were studied very little until the 1950s. Subsequently they have been investigated in a seemingly ever-increasing rate, due to discovery of their interesting properties and to development of new synthetic methods.

For similar reasons, activity on 2-oxetanones, also known as β-lactones, has been very intense in recent years.

Previous general reviews of oxetanes and of 2-oxetanones were published in 1964 ⟨64HC(19-2)983, 64HC(19-2)729⟩, covering the literature to about 1962. Additional surveys of methods of preparation and reactions of these classes of compounds were published in 1963 and 1965 ⟨63HOU(6/2)511, 65HOU(6/3)489⟩.

In the present work, emphasis is placed on more recent work and on the present state of the field.

5.13.2 STRUCTURE

5.13.2.1 Theoretical Methods

Since theoretical calculations can be carried out best with small molecules, it is not surprising that oxetane and some of its derivatives have been the subject of most types of theoretical methods.

Several semi-empirical methods, including CNDO, INDO and MINDO/3, have been used to calculate the molecular dimensions and the ring-puckering barrier in oxetane, 2-oxetanone, 3-oxetanone and 3-methyleneoxetane ⟨76JST(32)1, 72JCP(57)2318, 80MI51300⟩. The INDO method gave quite good results for oxetane, whereas the predictions made by the CNDO and MINDO/3 methods tended to agree only crudely with experimental data.

The INDO method has also been used with moderate success to predict the directly bonded ^{13}CH spin–spin coupling constants in oxetane and other small ring compounds ⟨70JA5829⟩. Both CNDO and INDO methods have been applied to calculation of the dipole moments of a series of 2-aryloxetanes, resulting in calculated dipole moments which were generally about 10% higher than the measured values ⟨74ZN(A)1907⟩. CNDO/2 calculations on oxetane, with parameters derived for three-membered rings, gave good agreement between predicted and observed energies for electronic transitions of oxetane ⟨76JCP(64)2062⟩.

Ab initio MO calculations with the minimal STO-3G basis set have been carried out for oxetane and other small ring compounds. These have been used to make interpretations of PE spectra by calculating the energies of the different MOs present. For oxetane, such calculations gave moderately good agreement between observed and predicted peaks in the PE spectra ⟨77JA3226⟩, but for 2-oxetanone, 3-oxetanone and 3-methyleneoxetane the agreement between predicted and observed peaks was relatively crude. The semi-empirical MINDO/3 method, however, gave relatively good agreements for the first two peaks in the PE spectra of these compounds ⟨79MI51301⟩.

Extended Hückel MO calculations have been used to determine the variation of the energy of 2-oxetanyl cations, anions and radicals with the torsional angle of the α-hydrogen atom ⟨79DOK(248)654⟩. It was predicted that the α-carbon atom of both the cation and radical would be planar, but not that of the anion. *Ab initio* MO calculations with both STO-3G and 4-31G basis sets have been carried out on the 2-oxetanyl cation to provide detailed molecular dimensions and to indicate considerable stabilization by the π-overlap involving the oxygen atom ⟨76TL4025⟩.

Ab initio MO calculations with the same two types of basis sets have been performed for oxetene (**1**) and for the oxetenyl anion (**2**) and cation (**3**) with excellent agreement observed between the STO-3G and 4-31G results. Complete molecular dimensions were derived for all three species ⟨80TL1807⟩.

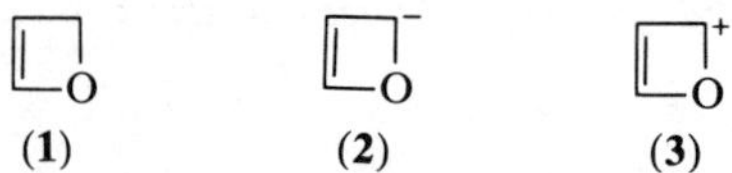

(**1**) (**2**) (**3**)

The force-field or molecular mechanics method has been applied to oxetane, as well as to a number of ethers ⟨76JA6798⟩. The force field method developed for hydrocarbons was modified by treating the lone pairs on the oxygen atoms like atoms themselves, having size, hardness (polarizability) and position relative to the carbon substituents on the oxygen atom. The resulting parameters were determined from known data on a series of ethers, including oxetane. Subsequent application to the structure of oxetane gave a prediction of an essentially planar conformation (which it has) and a heat of formation which differed by only 2% from the experimental value. It will be interesting to see this new and potentially useful method applied to 2-oxetanones and other related oxetanes.

5.13.2.2 Molecular Dimensions

The bond lengths and angles in several oxetanes have been precisely determined by microwave, electron diffraction and X-ray diffraction methods. In Table 1 are listed the bond lengths and angles in the four-membered ring for five representative oxetanes. A notable feature is the long C—O bond length, which is greater in each case than in other types of compound. In fact, an oxetane C—O bond length as long as 1.48 Å has been determined in the X-ray study of 2,2-bis(*p*-ethoxyphenyl)-3,3-dimethyloxetane (**4**) ⟨73CC34⟩.

Table 1 Ring Skeletal Dimensions in Oxetane ⟨79ZN(A)1130, 64HC(19-2)983⟩

Dimension	*Oxetane*	*2-Methyloxetane*	*2-Oxetanone*	*3-Oxetanone*	*3-Methyleneoxetane*
C(2)—O(1) (Å)	1.449	1.448	1.45	1.441	1.449
C(2)—C(3) (Å)	1.549	1.534	1.53	1.524	1.523
C(2)OC(4) (°)	91.8	92.0	89	94.8	—
C(2)C(3)C(4) (°)	84.55	85.5	83	88.2	87.05

Depending on the substituents present, the oxetane ring may be either coplanar or puckered. In oxetane itself the average configuration is planar, but there is an important puckering vibration due to the repulsive interactions of the eclipsed methylene hydrogen atoms. The rocking motion allows the hydrogen atoms to attain a more staggered configuration but is opposed by the resulting increase in bond angle deformation. There is a double minimum in the energy–dihedral angle relationship, but the energy barrier between the two minima is below the lowest vibrational energy level. Thus, the puckered configuration is intrinsically unstable for oxetane, unlike cyclobutane, and it is a state of dynamic equilibrium ⟨74MI51300⟩.

Substitution of the hydrogen atoms at C-3 by fluorine atoms, as in 3,3-difluorooxetane, or introduction of a carbonyl group in place of a methylene group, as in 2-oxetanone and 3-oxetanone, decreases the eclipsing repulsions and results in a single minimum in the energy–dihedral angle relationship. Such compounds are strictly planar ⟨76JSP(59)28⟩.

Substitution of alkyl and aryl groups on the oxetane ring, on the other hand, increases the barrier of the puckering vibration. 2-Methyloxetane and 3-methyloxetane are non-planar, existing as mixtures of stable equatorial and axial conformers at room temperature ⟨79JST(56)157⟩.

Determination of vicinal NMR coupling constants has been a widely used tool for assignment of configuration of substituted oxetanes. In 2-phenyloxetane the ring is puckered with the phenyl group equatorial. *trans*-2-Phenyl-3-methyloxetane exists in a puckered form with both substituents equatorial, while the *cis* isomer is an equimolar equilibrium mixture of the two puckered forms, each with one equatorial and one axial substituent ⟨75JOC473⟩. Recently, formulae have been developed for calculating the dihedral and torsional angles from the NMR coupling constants; values of 7.8 and 8.2° were determined for the dihedral angles in 2-methyl- and 2-phenyl-oxetane, respectively ⟨81ZOR1154⟩.

X-ray crystal structure studies have been reported for three oxetanes: (**4**), 3,3-bis(trimethylaminomethyl)oxetane methanesulfonate and the 3,5-dinitrobenzoate ester of *threo*-α,3,3,4,4-pentamethyloxetanemethanol (**5**). These compounds were all non-planar, with dihedral angles of 16, 29.4 and 8°, respectively ⟨73CC34, 71DIS(B)(32)1445, 78CSC309⟩.

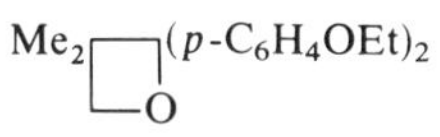

Me_2 / $(p\text{-}C_6H_4OEt)_2$ / O

(**4**)

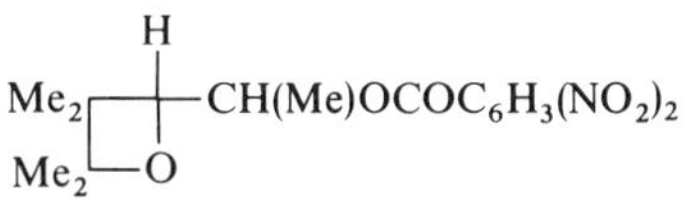

H / Me_2 / $CH(Me)OCOC_6H_3(NO_2)_2$ / Me_2 / O

(**5**)

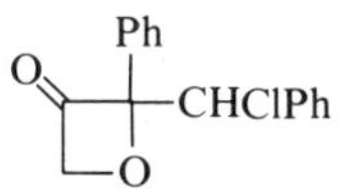

O / Ph / CHClPh / O

(**6**)

Me_2 / O / O / O / O / Me_2 / O / Me / O

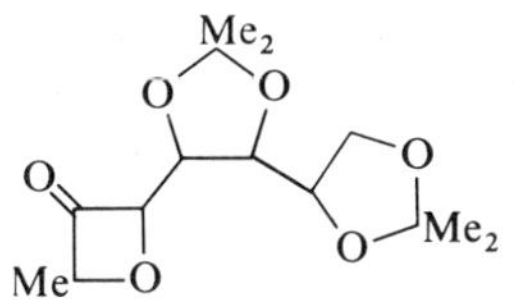

(**7**)

As to be expected, substitution in 2- and 3-oxetanones is less effective in causing puckering of the four-membered ring. The molecular structures of two 3-oxetanones have been determined by X-ray analysis: 2-(α-chlorobenzyl)-2-phenyloxetane (**6**) and the carbohydrate derivative (**7**), which is a *cis*-2,4-dialkyl-3-oxetanone. The oxetane ring of the former appeared to be planar, while that of the latter had a dihedral angle of only 5.6° ⟨81CSC345, 76JCS(P2)1678⟩. The structure of 4-methyl-2-oxetanone has been investigated by microwave spectroscopy and that of 4-(trichloromethyl)-2-oxetanone by dipole moment and Kerr constant studies. Both compounds were found to have planar structures ⟨76JSP(59)28, 73ZOR178⟩.

The molecular dimensions of oxetene have been calculated by *ab initio* MO methods. A tetramethyloxetene has been isolated but is thermally unstable. Oxetene is predicted to have the following skeletal bond lengths and angles: C—O, 1.424; C—C, 1.526; C=C, 1.312 Å; OC(2)C(3), 88.3; C(2)C(3)C(4), 98.1° ⟨80TL1807⟩.

5.13.2.3 Molecular Spectroscopy

5.13.2.3.1 NMR spectroscopy

This technique has been a great value in assigning oxetane structures, particularly because the α-protons in oxetanes have much higher chemical shift values than in any other class of ethers. For the parent compound the chemical shifts are at 4.73 and 2.72 p.p.m. in deuteriochloroform. Proton NMR spectroscopy is very useful for 2-oxetanones also, but has not been given as much study. Table 2 gives data for several representative oxetanes.

(**8**)

Table 2 ^{1}H NMR Chemical Shifts for Oxetanes[a] ⟨75OMR(7)554⟩

Substituent	H_A	H_B	H_C	H_D	H_E	H_F
None	4.73	4.73	2.72	2.72	4.73	4.73
2-Me	1.35 (Me)	4.85	2.24	2.63	4.37	4.49
2,2-Me_2	1.36 (Me)	1.36 (Me)	2.34	2.34	4.32	4.32
3,3-Me_2	4.23	4.23	1.27 (Me)	1.27	4.23	4.23
trans-2,4-Me_2	1.36 (Me)	4.72	2.28	2.28	4.72	1.36 (Me)
cis-2,3-Me_2	1.22 (Me)	4.92	1.13 (Me)	2.94	3.98	4.64
trans-2,3-Me_2	1.37 (Me)	4.52	2.62	1.18	4.59	4.21
4-Me-2-one	1.53 (Me)	4.57	2.98	3.50	—	—

[a] In CCl_4 or $CDCl_3$; proton labels shown in (**8**).

Introduction of substituents often causes special shielding effects, due to proximity effects in the small ring, changes in the puckering of the ring and perhaps other factors. The methyl group in 2-methyloxetane shields the β-*cis* proton by 0.48 p.p.m., the opposite α-*cis* proton by 0.36 p.p.m. and the opposite α-*trans* proton by 0.23 p.p.m. Similar effects are seen in 2,2-, 3,3- and *cis*- and *trans*-2,3-dimethyloxetanes ⟨75OMR(7)554⟩. The phenyl group in 2-phenyloxetane shows a different effect, deshielding the opposite α-*trans* proton by 0.17 p.p.m. with negligible effect on the *cis* protons ⟨75JOC473⟩. The different effects for *cis* and *trans* protons has been used for making stereochemical assignments in various disubstituted oxetanes.

Spin–spin coupling in the proton NMR spectra of oxetanes can be very complex because of cross-ring interactions and the different signs of the coupling constants. In the A_2B_4 system of oxetane the coupling constants have been determined as follows: ^2J_x −6.02, ^2J_B −11.0, $^3J_{cis}$ 8.65, $^3J_{trans}$ 6.60, $^4J_{cis}$ 0.20 and $^4J_{trans}$ 0.14 ⟨B-73NMR369⟩.

The values of the coupling constants have been extensively studied, because they provide a readily available measure of the predominate direction and degree of ring puckering. The vicinal coupling constants for the protons on C-3 and C-4 in 2-substituted oxetanes are modified by the ring puckering, due to changes in the angles of staggering, in accord with the well-known Karplus relationship. Thus a high value of J_{CF}/J_{CE} in a *trans*-AD-disubstituted oxetane structure indicates that it exists largely in the diequatorial puckered conformation ⟨75JOC473⟩. It has been reported that the dihedral angle of puckering, θ, can be calculated fairly exactly from the vicinal coupling constant for the *trans* diaxial protons, B and C (structure **8** substituent at A), by the relationship,

$$^3J_{BC} = k \cos^2 \theta - C$$

where k and C are adjustable constants used to give agreement with microwave values for θ ⟨81ZOR1154⟩.

The magnitude of the long-range coupling constant between α and *meta* protons in 2-aryloxetanes has been used to determine that the plane of the benzene ring is perpendicular to the C(3)—O axis of the oxetane ring. Furthermore, in 2(*o*-chlorophenyl)oxetane the chlorophenyl group is oriented with the chlorine close to the methine proton ⟨75OMR(7)286⟩.

Several studies have been reported on the proton NMR of oxetane and 2-oxetanone in nematic liquid crystal solvents. The direct experiments lend themselves well to analysis of molecular geometry, and the results agreed well with those from microwave data ⟨77JMR(28)243, 72MI51301⟩.

The ^{13}C NMR spectra of oxetanes and oxetanones are also very useful in structure determination. The chemical shift values for the α-carbon atoms tend to be higher than for other cyclic ethers. Some representative ^{13}C chemical shift values are given in Table 3. The ^{13}C isotope effect on ^{13}C chemical shifts has been determined as +0.27 and +0.23 Hz for C-2 and C-3 in oxetane ⟨78OMR(11)157⟩.

Table 3 ^{13}C Chemical Shifts[a] for Oxetanes ⟨75OMR(7)554⟩

Substituent	*C-1*	*C-2*	*C-3*	*Me*
None	72.8	23.1	72.8	—
2-Me	78.3	29.2	66.6	24.1
cis-2,3-Me_2	79.6	31.8	74.5	16.9, 13.0
trans-2,3-Me_2	85.0	37.4	73.5	22.8, 17.5

[a] In p.p.m. relative to TMS for $CDCl_3$ solutions.

Recent ^{17}O NMR studies have given us two values for the ^{17}O chemical shifts in oxetane, −12.0 and −20.5 p.p.m., referenced to water ⟨79BCJ3391, 80T3431⟩. The rather wide band width may explain part of this apparent discrepancy. The ^{17}O chemical shift of oxetane and other ethers was found to correlate with the reciprocal of their ionization potentials.

5.13.2.3.2 *Infrared spectroscopy*

IR has long been used routinely in structure determination of oxetanes and 2-oxetanones. In oxetanes a strong absorption band at about 980 cm^{-1} is a useful characteristic; it is due to a ring-stretching mode with considerable C—O stretch character and is at about 50 cm^{-1} lower frequency than the C—O stretch band of the other ethers. 2-Oxetanones are characterized by an intense absorption, due to carbonyl stretching, at 1840–1820 cm^{-1} which is at a higher frequency than observed with acylic esters or larger-ring lactones.

Recent studies in vibrational spectroscopy have centered on the far-IR and Raman region, in order to study the ring puckering vibration. The results of this work are taken up in Section 5.03.1.4.

5.13.2.3.3 *Ultraviolet spectroscopy*

This has received very little attention. Oxetane absorbs only in the vacuum UV ⟨76JCP(64)2062⟩. 2-Oxetanones apparently absorb slightly at about 300 nm, but the extinction coefficient is too small for this absorption to be useful ⟨64HC(19-2)729⟩ (see also Section 5.01.2.5).

5.13.2.3.4 Photoelectron spectroscopy

The PE spectra of several oxetanes and related small ring compounds have received careful study. The first ionization potential for oxetane occurs at 7.65 eV as a sharp, adiabatic transition. This is at significantly lower potential than for oxirane (10.57 eV) or for acyclic ethers (dimethyl ether, 10.04 eV), showing the potential-lowering effect of the four-membered ring. This is seen also in azetidines and thietanes.

This potential-lowering effect has been explained as due to the antibonding π_{CH} interactions in the HOMO for the four-membered ring. The resulting decrease in π_{CH_2} bonding between the methylene groups allows more mixing of π_{CH_2} orbitals with a lone pair orbital, raising the energy of the HOMO. In addition, the presence of three methylene groups in oxetane, compared to two in oxirane, causes increased antibonding interaction between the methylene groups in the HOMO and further raises the orbital energy by hyperconjugation.

The orbital assignments of the first ionization potential, as well as of the higher bands which are broader and less intense, have been confirmed by *ab initio* MO calculations and by comparison with PE spectra of other small-ring cyclic ethers, amines, sulfides, silanes and cycloalkanes ⟨77JA3226⟩.

The PE spectra for 2- and 3-oxetanone and 3-methyleneoxetane have also been measured and analyzed by means of comparison of the potentials with those calculated for the orbitals by both *ab initio* and semi-empirical methods. The first four ionization potentials and their assignments, as well as those for oxetane, are listed in Table 4.

Table 4 Lowest Ionization Potentials (eV) and Assignments in PE Spectra of Oxetanes

Compound	*IP_1*	*IP_2*	*IP_3*	*IP_4*	*Ref.*
Oxetane	9.63 (b_1–π_O)	11.32 (a_1)	11.98 (b_2)	1.34 (–)	77JA3226
2-Oxetanone	10.78 (a′–n_{CO})	11.30 (a′–n_O)	12.82 (a″–σ)	14.02 (π_{CO})	79MI31501
3-Oxetanone	10.31 (b_2–n_{CO})	10.73 (b_1–n_O)	13.77 (a_1)	14.16 (π_{CO})	79MI31501
3-Methyleneoxetane	9.46 (b_1–π_{CC})	10.60 (b_1–n_O)	11.57 (b_2–σ)	12.20 (σ)	79MI31501

The first ionization band in the PE spectrum of 2-oxetanone can be assigned to an orbital which is mainly a lone pair orbital on the carbonyl carbon atom. The mixing of this orbital with the endocyclic oxygen lone pair raises the energy of the former and lowers that of the latter, with the result that the ionization potential for 2-oxetanone is 0.8 eV lower than that for cyclobutanone. The first ionization bands in the PE spectra of 3-oxetanone and 3-methyleneoxetane are due to a lone pair carbonyl orbital and an exocyclic π-orbital, respectively. The levels of both are much raised by a homoconjugative interaction with a lone pair orbital on the cyclic oxygen atoms, transmitted through the π_{CH_2} orbitals. Thus it appears that these compounds are resonance stabilized in a rather similar manner to 2-oxetanone, due to the unique possibilities for orbital interactions in an oxetane ring ⟨79MI51301⟩.

5.13.2.3.5 Mass spectrometry

The relatively small amount of work published prior to 1968 has been reviewed elsewhere ⟨B-71MS34, B-71MS179⟩. Although the molecular ion of the parent compound is prominent, the molecular ions of substituted oxetanes are generally of low intensity, while those of 2-oxetanones are not seen, due to the great ease of fragmentation.

Cross-ring cleavage to form alkene and carbonyl fragments is the predominate type of fragmentation for oxetanes. The charge may reside on either fragment, as shown for oxetane (equation 1).

$$\text{oxetane} \xrightarrow{e^-} \begin{cases} [CH_2{=}CH_2]^+ + CH_2{=}O \\ [CH_2{=}O]^+ + CH_2{=}CH_2 \end{cases} \qquad (1)$$

In the MS of oxetane the relative abundance of the ethylene radical ion is about eight times greater than that of the formaldehyde radical ion at the usual ionizing potential of 70 eV. However, 2,2-dimethyloxetane fragments much more to the radical ion of acetone than to isobutene cation. The charge site distribution has been found to depend on a number of factors, including the structure of the oxetane, the ionizing voltage, and the nature and temperature of the sample inlet. This complexity has somewhat frustrated rational approaches to explaining fragmentation products simply on the basis of stabilities and electron affinities of the fragments ⟨78JOC2184⟩.

MS fragmentation of the symmetrical pentasubstituted oxetane (**9**) gave principally the diene radical ion (**10**) with loss of acetone. The unsymmetrical hexadeutero analog (**11**) fragmented to a mixture of (**10**) and the hexadeuterodiene radical cation (**12**) in a ratio of 39:61 at 10 eV ionizing potential. From these data an isotope effect, k_H/k_D, of 1.08 per deuterium atom can be calculated. This is comparable in magnitude to the β-deuterium isotope effect observed in limiting solvolyses of secondary and tertiary substrates. While this result might appear to suggest a two-step process for cross-ring cleavage *via* a ring-opened molecular ion like (**13**), the authors favored a single-step mechanism for the cleavage ⟨76CC36⟩.

(**9**) $\xrightarrow{e^-}$ (**10**) + $Me_2C{=}O$

(**11**) $\xrightarrow{e^-}$ { (**10**) + $(CD_3)_2C{=}O$; (**12**) + $Me_2C{=}O$ }   (**13**)

In unsymmetrically substituted oxetanes there are two chemically distinguishable modes for cross-ring cleavage, each with two different charge distributions, resulting in four possible sets of fragments. Generally it will be found that one or two of such potentially competing processes will be dominant. For example, *cis*- and *trans*-2-methyl-3-phenyloxetane (**14**) both give mainly the styrene molecular ion with loss of acetaldehyde ⟨78JOC2184⟩.

(**14**) $\xrightarrow{e^-}$ $[PhCH{=}CH_2]^{+} + MeCH{=}O$

(**15**) $\xrightarrow{e^-}$ $Ph\overset{+}{C}{=}O + MeCH_2^{+}$; $[PhCH{=}O]^{+} + CH_2{=}CH_2$; $[PhCH{=}CH_2]^{+} + PhCH{=}O$

In addition, hydrogen rearrangement processes are commonly seen in the fragmentation of oxetanes, and sometimes they predominate. An example may be seen in the case of 2-phenyloxetane (**15**), which gave the benzoyl cation as its principal cationic fragment. This was identified by high-resolution MS, as well as by deuterium-labeling experiments. It appears to have been produced to at least some extent by a direct rearrangement of the molecular ion, since a metastable ion was seen for this fragmentation; some of it, however, may be derived by loss of an hydrogen atom from the benzaldehyde radical ion which was also seen. In addition, a considerable amount of the styrene cation was observed ⟨77OMS348⟩.

While geometrical isomers usually give very similar MS, this is not always the case. An interesting example is that of *cis*- and *trans*-2,4-dimethyloxetane ⟨71ACS763⟩. The principal fragment ion from the *trans* isomer (**16**) was the propene radical cation (m/e 42), while that from the *cis* isomer (**17**) was protonated acetaldehyde (m/e 45), apparently a hydrogen-rearrangement product.

(**16**) $\xrightarrow{e^-}$ $[MeCH{=}CH_2]^{+} + MeCHO$

(**17**) $\xrightarrow{e^-}$ $[MeCH{=}OH]^{+} + MeCH{=}CH_2$

The MS fragmentation of 2-oxetanones has received relatively little attention. 2-Oxetanone is reported to give both modes of cross-ring cleavage, as well as a hydrogen-rearrangement process (equation 2). These three processes seem to be very competitive, since fragment ions from each are observed in roughly comparable abundances ⟨B-71MS179⟩.

$$\xrightarrow{a} [CH_2{=}C{=}O]^{+} + CH_2{=}O$$
$$\xrightarrow{b} [CH_2{=}CH_2]^{+} + CO_2 \qquad (2)$$
$$\longrightarrow Me\overset{+}{C}{=}O \rightarrow Me + CO$$

MS fragmentation of a series of 13 4-aryl-3,3-dichloro-2-oxetanes (**18**) has been reported to give the dichloroketene cation as the base peak in every case, except for the 4-(*m*-methoxyphenyl) compound. There the base peak was for the *m*-methoxybenzaldehyde cation. The results were in marked contrast to the cleavage seen on thermolysis, which gave carbon dioxide and the corresponding β,β-dichlorostyrene ⟨78JHC1165⟩.

$$\longrightarrow [Cl_2C{=}C{=}O]^{+} + XC_6H_4CH{=}O$$
$$\xrightarrow[X = m\text{-OMe}]{} [m\text{-MeOC}_6H_4CH{=}O]^{+} + Cl_2C{=}C{=}O$$

(**18**)

5.13.2.4 Thermodynamic Aspects

5.13.2.4.1 Thermal stability

Oxetane and alkyl-substituted oxetanes are relatively stable to heat and light, but can be thermally decomposed at 400–500 °C. At that temperature they undergo cleavage to form alkene and carbonyl fragments ⟨67JPC2326⟩.

Phenyl- and vinyl-substituted oxetanes are more labile and tend to decompose at 250–300 °C. 2-Oxetanones are very sensitive to thermolysis, often decomposing with loss of carbon dioxide at temperatures of about 100 °C. Oxetenes are thermally unstable at room temperature.

5.13.2.4.2 Ring strain

The ring strain energy of oxetane, calculated as the difference between the observed and estimated heats of formation, has been determined as 106.3 kJ mol^{-1}, which is in fair agreement with the 'strain correction' of 110.5 kJ mol^{-1} for this ring in Benson's group-additivity method of calculating heats of formation ⟨71ACS141, 69CRV279⟩. This is about the same as for cyclobutanone, about 1 kJ mol^{-1} less than for oxirane and about 20 kJ mol^{-1} more than for tetrahydrofuran.

The ring strain energies for 3,3-dimethyloxetane, 3,3-bis(chloromethyl)oxetane and 3-ethyl-3-chloromethyloxetane have been determined from the heats of formation and all have the same value of 93 kJ mol^{-1}, showing a decrease in ring strain energy of 13.3 kJ mol^{-1}

Table 5 Potential Barrier for Ring Puckering Vibration in Oxetanes

Substitution	*Barrier* (kJ mol^{-1})	*Method*	*Ref.*
None	0.181	MW	79JST(56)157
Perdeutero	0.134	IR	72JSP(43)393
2-Me	1.29	IR	79JST(56)157
2,2-Me_2	0.61	IR	79JST(56)157
3-Me	1.18	IR	78JSP(69)159
3,3-Me_2	0.55	IR	78JSP(69)159
3-CH_2	0.0	MW	70JCP(52)5234
3-one	0.0	MW	72JCP(57)2318
3,3-F_2	0.0	IR	74JCP(61)2071
2-one	0.0	IR	65JCP(43)1190
4-Me-2-one	0.0	MW	76JSP(59)28

for geminal disubstitution ⟨71ACS141⟩. Perhaps this is due to the relief of repulsive steric interaction, caused by the puckering of the substituted oxetane rings.

Table 5 lists values of the potential barrier for the ring puckering vibration in various oxetanes (see also Section 5.13.2.3.2).

5.13.2.4.3 Aromaticity of the oxetenyl anion and cation

These compounds, (**2**) and (**3**), are of theoretical interest, since they have cyclic π-orbital systems, and *ab initio* MO calculations have been performed to estimate what degree of stabilization may result. This was done by calculating the ΔE for their formation by an isodesmic reaction with a hypothetical cyclobutanone enol ion (**19a**,**b**), which also has a four-membered ring with an allylic system conjugated to an oxygen atom.

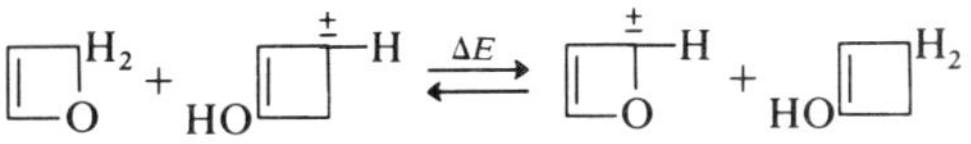

(**1**) (**19a, b**) (**2, 3**)

It was found that anion (**2**) has about 80 kJ mol^{-1} more stabilization than the hypothetical model (**19a**). This shows that there is a substantial amount of stabilizing resonance energy, which is in accord with its having a cyclic 6π-electron system. The stabilization is not great enough, however, for oxetene to be expected to show appreciable acidic properties. Calculations show oxetene to be less acidic than propene.

Classically, cation (**3**) might be anticipated to be considerably stabilized by the neighboring oxygen atom and double bond; on the other hand, its 4π-electron system would suggest anti-aromaticity. The isodesmic calculations showed it to be about 200 kJ mol^{-1} less stable than the model cation (**19b**) and about 40 kJ mol^{-1} less stable than the allyl cation. Thus it appears that it would have an anti-aromatic electronic structure, if it was formed ⟨80TL1807⟩.

5.13.3 REACTIVITY OF OXETANES AND 2-OXETANONES

5.13.3.1 General Survey

The ring strain and polar properties of a four-membered oxygen-containing ring cause significant susceptibility to thermal cleavage, as well as to reaction with a wide variety of electrophilic and nucleophilic reagents. The relatively high basicity of oxetanes makes them quite susceptible to electrophilic reagents and to combinations of nucleophilic and electrophilic reagents. The chemistry of 2-oxetanones is in many respects in contrast to that of oxetanes with regard to reactions with electrophiles and nucleophiles, since 2-oxetanones have relatively low basicity, but the presence of a very reactive carbonyl group in a strained lactone causes high susceptibility to reaction with nucleophilic reagents, both with and without electrophilic catalysis.

5.13.3.2 Thermolysis and Photolysis

5.13.2.1 Oxetanes

Thermal decomposition of oxetanes proceeds at elevated temperatures, usually in the range of 300–450 °C, to give an alkene and a carbonyl compound in practically quantitative yield. Gas phase thermolyses of several alkyl-substituted oxetanes have been studied in detail, because such compounds occur as intermediates in the oxidation of alkanes (Section 5.13.3.3). When the oxetane is unsymmetrically substituted, as in the case of 2,4-dimethyloxetane (**20**), two modes of cleavage, to give two different sets of products, are possible and are generally observed. In fact, when only hydrogen and alkyl substituents are present,

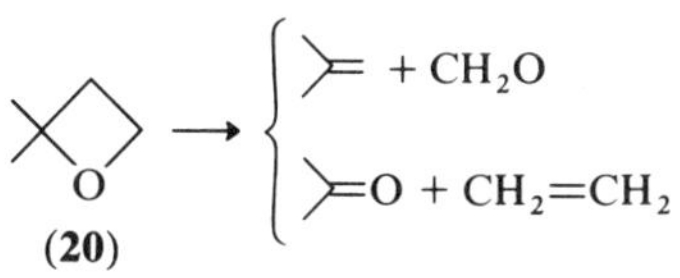

$+ CH_2O$

$=O + CH_2{=}CH_2$

(**20**)

both modes of cleavage occur to similar extents, if the surface of the reactor vessel has been seasoned. With clean glass surfaces and with metal surfaces, however, a high degree of regioselectivity is often found, probably due to surface catalysis effects (Section 5.13.2.5) ⟨77JCS(F1)890⟩.

First-order kinetics were observed for the gas phase pyrolysis of each of nine oxetanes in seasoned reactors. Tests for occurrence of any free radical chain reaction were negative. Substituent effects of alkyl groups on rate were relatively small, but that of the 3-vinyl group was strongly rate-enhancing. It is of particular interest that the 2-butene formed in the thermolysis of *cis*-(**20**) and of *trans*-(**20**) retained the stereochemistry of the starting oxetane to the extent of 68–76%, and there was very little geometrical isomerization of the starting oxetanes during the thermolysis.

All these data are consistent with a two-step mechanism involving a biradical intermediate, which has little tendency to cyclize and has a lifetime of only a few rotations or less before cleavage (equation 3). This mechanism is quite analogous to that which is accepted for the pyrolysis of cyclobutanes. The lack of regioselectivity in oxetane pyrolysis seems attributable to the similar bond dissociation energies for C—C and C—O bonds in oxetanes. The rate-enchancing effect of a 3-vinyl group is explainable on the basis of stabilization of one of the radical sites by resonance, thus lowering the energy of the biradical intermediate and increasing the rate. In fact, *cis*-2,4-dimethyl-*trans*-3-vinyloxetane (**21**) pyrolyses readily at about 100 °C lower than *trans*-2,3-dimethyloxetane ⟨74JCS(F1)43, 80JCS(F1)1849⟩.

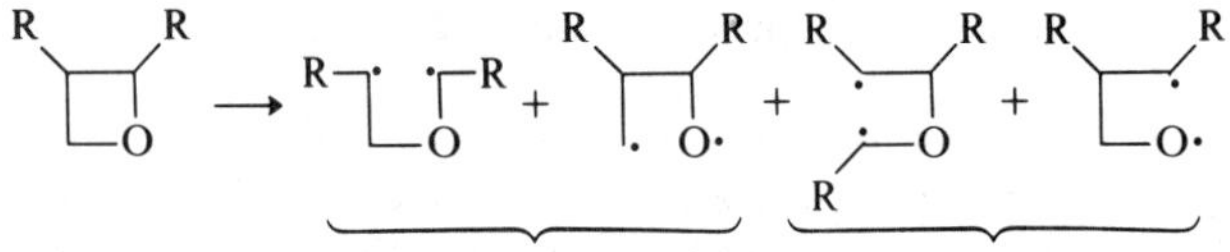

$RCH{=}CH_2 + RCHO$ $\quad$ $RCH{=}CHR + CH_2O$ (3)

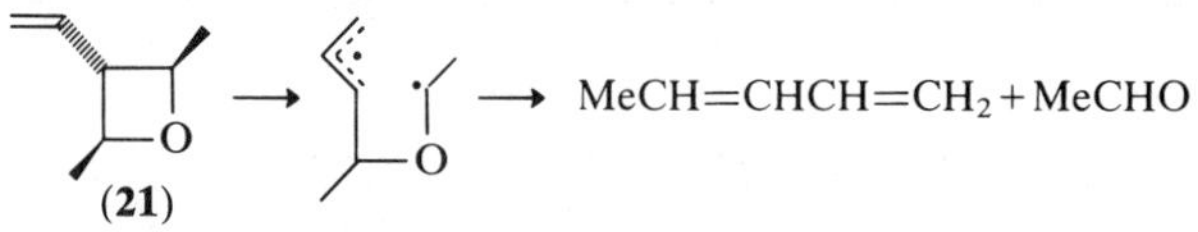

$\rightarrow$ $MeCH{=}CHCH{=}CH_2 + MeCHO$

(**21**)

A similar rate-enhancing effect is probably observed with cyano substitution. 2-Cyano-3,3-dimethyl-2-phenyloxetane (**22**) underwent thermal decomposition at 350 °C to give benzoyl cyanide ⟨68ZOR2055⟩. The regioselectivity in this case would be in accord with cleavage to the intermediate biradical (**23**), which would have extensive spin stabilization at the cyano-substituted carbon atom and would probably be the most stable of the four possible biradical intermediates.

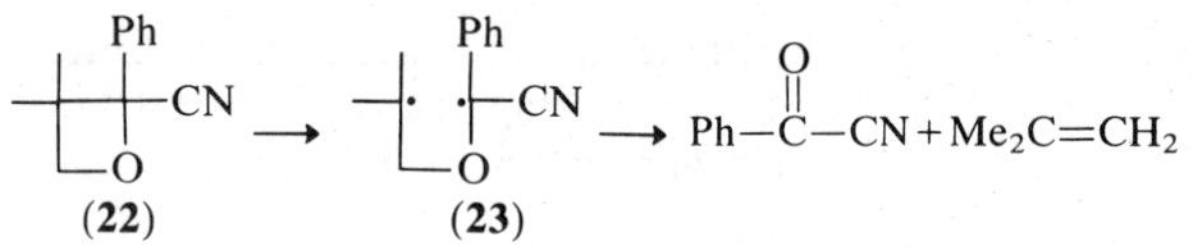

$\rightarrow$ $Ph{-}C({=}O){-}CN + Me_2C{=}CH_2$

(**22**) (**23**)

Even greater rate enhancement was seen with 2-ethoxy substitution, since 2-ethoxy-3,3-dimethyl-2-phenyloxetane (**24**) rapidly cleaved at only 225 °C, forming the enol ether (**26**). This can be rationalized in terms of resonance stabilization of the presumed biradical intermediate (**25**) ⟨67LA(707)1⟩.

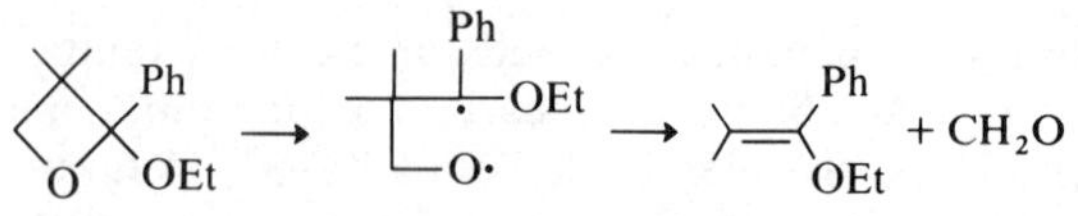

$+ CH_2O$

(**24**) (**25**) (**26**)

An investigation of several phenyl-substituted oxetanes and of *trans*-diethyl oxetane-2,3-dicarboxylate has been carried out in hot solutions rather than in the gas phase. The solvents must, of course, be high boiling and completely free of any acidic impurities, which would cause an acid-catalyzed decomposition. 2,2-Diphenyl-3,3,4,4-tetramethyloxetane (**27**) pyrolyzed at 310 °C at practically the same rate in each of three solvents of quite different polarity — diphenyl ether, DMF and TMEDA — showing that the intermediate is relatively non-polar. High regioselectivity is shown in this reaction, which gave mainly benzophenone and 2,3-dimethyl-2-butene. The lack of solvent effect and the observed regioselectivity seem to agree with a biradical mechanism in this case also ⟨74TL2103⟩.

Ph$_2$ → + Ph_2CO

O

(27)

While the predominant mode of cleavage of 2-methyl-3-phenyloxetane (equation 4) is in accord with the predicted most stable biradical structure and the 3-phenyl substituent seems to have about the same rate-enhancing effect as a 3-vinyl group, the situation with regard to the mode of cleavage and mechanism for pyrolysis of various 3-alkyl-2-phenyloxetanes seems unclear. The *cis* isomers gave predominately one mode of cleavage (that predictable on the basis of the most stable biradical), but the *trans* isomers gave the opposite mode of cleavage. Furthermore, 3,3-dimethyl-2-phenyloxetane showed cleavage in the expected direction, but its *p*-tolyl and *p*-anisyl analogs showed the opposite mode of cleavage. It has been proposed that the cases that do not fit the biradical mechanism fragment by a concerted $[2_s + 2_a]$ retrograde cycloaddition, but the details have not been clarified yet ⟨80CL41⟩.

Ph Me → Ph Me → $MeCHO + PhCH{=}CH_2$ (4)

O O

Photolysis of oxetane, 2,2-dimethyloxetane and 2-phenyloxetane has been found to occur slowly with mercury arc irradiation. The products are similar to those observed from pyrolysis of these compounds. Oxo-substituted oxetanes are probably much more susceptible to photolysis. Irradiation of oxo-oxetane (**28**) at 350 nm isomerized it to a furan (**29**) and a dioxalene (**30**), each of which can be explained by cleavage of the oxetane ring to a 1,4-biradical ⟨69CC572⟩.

O O R H R → R R O + O O CH_2R R

(28) **(29)** **(30)**

It has been predicted from an RRKM calculation that 3-oxetanone would photolyze readily, forming formaldehyde and ketene ⟨75JPC1990⟩. This is in accord with the observation that 2,2,4,4-tetramethyl-3-oxetanone is transformed in high yield to dimethylketene and acetone on irradiation with 3130 Å light in polar solvents, such as acetone and ethanol. In nonpolar solvents, however, photo-decarbonylation to tetramethyloxirane becomes a competing process (equation 5) ⟨66JA1242⟩.

O

Me_2 Me_2 → $Me_2C{=}C{=}O + Me_2CO$

O → Me_2 Me_2 + CO (5)

O

5.13.3.2.2 2-Oxetanones

On mild heating, most 2-oxetanones lose carbon dioxide to form alkenes in high yields (equation 6). This reaction is of synthetic value for conversion of 3-hydroxycarboxylic acids into alkenes, since a good route to the precursor 3-hydroxycarboxylic acids is available by the reaction of ketones with the α-lithiocarboxylate synthon (**31**). The method thus provides an attractive alternative to the now classic Wittig reaction for converting ketones to alkenes. The yields are generally high in each step, with that in the final decarboxylation step being usually 90–100%. The method has been applied to synthesis of several types of enol ethers, as well as a great variety of alkyl- and phenyl-substituted alkenes ⟨72JA2000, 73CJC981, 79S388⟩.

$$R_2C=O + R'CH_2O-\bar{C}H-CO_2^-\,2Li^+ \rightarrow R'CH_2OC(R_2COH)HCO_2H \quad (31)$$

$$\text{(R}_2\text{-substituted 2-oxetanone)} \rightarrow R_2C{=}CR_2 + CO_2 \qquad (6)$$

A clever application of this reaction has recently been carried out to achieve a high yield synthesis of arene oxides and other dihydroaromatic, as well as aromatic, compounds. Fused-ring β-lactones, such as 1-substituted 5-bromo-7-oxabicyclo[4.2.0]oct-2-en-8-ones (**32**) can be readily prepared by bromolactonization of 1,4-dihydrobenzoic acids (obtainable by Birch reduction of benzoic acids) ⟨75JOC2843⟩. After suitable transformation of substituents, mild heating of the lactone results in decarboxylation and formation of aromatic derivatives which would often be difficult to make otherwise. An example is the synthesis of the arene oxide (**33**) shown ⟨78JA352, 78JA353⟩.

HO₂C H → (CH_2O) → HO₂C CH₂OH → (i, Br_2/OH^-; ii, PhCOCl) → (**32**) → (i, CF_3CO_3H; ii, DBU) → → (Δ) → (**33**) $R = CH_2OCOPh$

The mechanism of decarboxylation of β-lactones has attracted much attention. The gas-phase decomposition of 2-oxetanone is a unimolecular first-order process. It has a considerably lower energy of activation than the pyrolysis of oxetane and a much higher entropy of activation, indicating a loose activated complex ⟨69JA7743⟩. The ease of the reaction is greatly affected by the electronic effect of substituents at position-4, but not at position-3. The Hammett treatment of a series of *trans*-4-aryl-3-methyl-2-oxetanones gave a good correlation with σ^+, with a ρ value of -15.2.

The practical meaning of these substituent effects is that compounds with electron-donating substituents at position-4 pyrolyze with extreme ease, such as at 24 °C for the *p*-anisyl derivative (**34**), while those with electron-withdrawing substituents require relatively high temperatures, such as 175 °C for the *p*-nitrophenyl derivative (**35**). In fact, 4-trichloromethyl-2-oxetanes, such as (**36**), do not decarboxylate even at 250 °C; decarboxylation of these was achieved by passing the vapor over an electrically heated wire, with dehydrochlorination also occurring, to give perchloroallene or trichloromethylallenes in 40–60% yields ⟨72JOC3536, 78JOC1305, 79JOC3574⟩.

(**34**) $R = p\text{-}MeOC_6H_4$
(**35**) $R = p\text{-}O_2NC_6H_4$
(**36**)

The observations indicate that the transition state in the decarboxylation of 2-oxetanones has considerable positive charge at position 4. Whether its exact structure is coplanar with only the C(1)—C(4) bond fully broken, or of perpendicular geometry with bond breaking further advanced, is a question which is under debate at present ⟨80JOC2354, 80TL751, 80AG(E)465⟩.

4-Methylene-2-oxetanone, or diketene (**37**), does not decarboxylate on heating but cleaves to form ketene at 250 °C. A thermal rearrangement of 3-aryl-4-benzal-2-oxetanones (**38**) to form naphthalene derivatives appears to involve this type of pyrolysis, followed by a $[_\pi 2 + _\pi 4]$ cycloaddition ⟨78JOC1146, 72LA(765)15⟩.

(**37**) → $2CH_2{=}C{=}O$

(**38**) →

Photolysis of 2-oxetanones gives decarboxylative cleavage to alkenes, similar to pyrolysis, but subsequent photoaddition reactions of the alkene product may lead to complex reaction mixtures. A very useful example of 2-oxetanone photolysis is that of 5-oxabicyclo[2.2.0]oct-2-en-6-one, the photoisomer of α-pyrone; when it was irradiated in a argon matrix at 80 K, carbon dioxide and cyclobutadiene were formed (equation 7) ⟨73JA1337⟩.

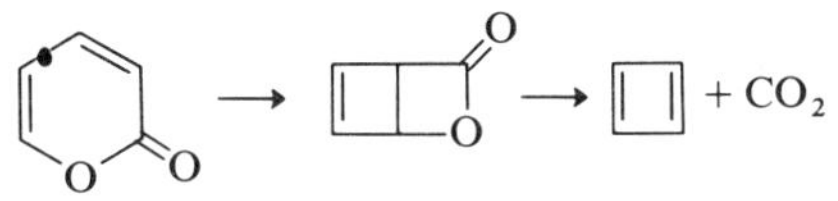

$$\longrightarrow \quad \longrightarrow \quad + CO_2 \qquad (7)$$

5.13.3.2.3 Oxetenes

Decomposition of unfluorinated oxetenes to α,β-unsaturated carbonyl compounds occurs so readily that these oxetenes are not stable at room temperature. This type of thermolysis has not received any detailed experimental study, but it appears to be the oxygen analog of the well-known electrocyclic ring-opening of cyclobutene. The energy and structural changes in the isomerization of oxetene to acrolein have been calculated by the *ab initio* 41-3G method ⟨75MI51300⟩, showing it to be exothermic by 180 kJ mol^{-1}.

5.13.3.3 Reaction with Electrophiles, including Oxidizing Agents

5.13.3.3.1 Electrophilic interactions at the ring oxygen atom

The oxygen atom of oxetanes possesses exceptionally high electron donor ability for an ether and appears to coordinate with all electron acceptors that have been investigated. With relatively weak electron acceptors, equilibria can be observed, but with strong electron acceptors the interaction is often so strong that irreversible ring cleavage occurs.

(i) Basicity

The instability of protonated oxetane has prevented the determination of pK_a or gas-phase basicity values for oxetane itself, whereas these have been obtained for tetrahydrofuran and tetrahydropyran. 3,3-Dialkyl substitution, however, sufficiently reduces the reactivity that pK_a values can be obtained. The pK_a for 3,3-dimethyloxetane has been determined to be −2.56, which is less than that for tetrahydrofuran (−2.08) but more than that for tetrahydropyran (−2.79). Since the 3,3-dialkyl substitution markedly reduces the basicity of oxetane towards iodine, the pK_a of the parent oxetane must be greater than these values (see equation 8) ⟨71BSF4576, 80JOC1166⟩.

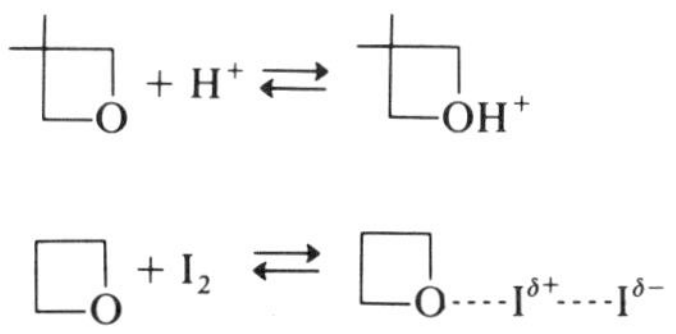

$$O + H^+ \rightleftarrows OH^+$$

$$O + I_2 \rightleftarrows O\cdots\cdot I^{\delta+}\cdots\cdot I^{\delta-} \qquad (8)$$

In 2-oxetanone, as in lactones and esters generally, the ring oxygen atom is a significantly poorer electron-donor site than the carbonyl oxygen atom, and thus is not directly attacked by electrophilic agents. Because of the resonance interaction between these two oxygen atoms, however, it is indirectly involved. This resonance interaction is sterically inhibited in β-lactones, such as 2-oxetanone, with the result that they are much weaker bases than larger-ring lactones (see equation 9) ⟨76JOC49⟩.

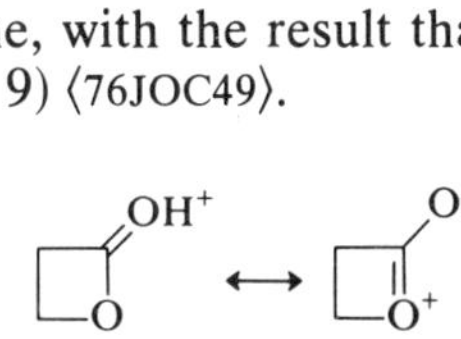

$$OH^+ \quad O \longleftrightarrow OH \quad O^+ \qquad (9)$$

β-Propiolactone and β-butyrolactone are completely protonated in 1:1 FSO_3H–SbF_5 diluted with SO_2 at −80 °C, and their conjugate acids have been studied by NMR spectroscopy. The acid protons appear in the C=OH$^+$ region of δ 13–14 and give two absorptions of approximately equal intensity. Thus two isomeric species are present, which are undoubtedly the *cis* and *trans* isomers (equation 10) ⟨70JOC3916⟩.

$$\text{2-oxetanone (R)} \xrightarrow[\text{SbF}_6]{\text{FSO}_3\text{H}} \text{O-protonated (}cis\text{)} + \text{O-protonated (}trans\text{)} \quad (\text{R} = \text{H or Me}) \tag{10}$$

(ii) Weak electrophiles

As it involves only a partial transfer of a proton, hydrogen bonding can be considered a weak electrophilic interaction. The hydrogen bonding of oxetanes and other cyclic and acyclic ethers to chloroform, alcohols (particularly methanol-d_1) and various phenols (equation 11) is easily measured by various spectrometric techniques and has been extensively studied. Charge-transfer complexation with iodine in inert solvents has also been carefully investigated. The results from both types of interactions agree that the ring size effect with respect to electron-donor ability is 4- > 5- > 6- ≫ 3-membered ring (with acyclic ethers being comparable to or slightly stronger than epoxides) ⟨B-67MI51300, 71BSF4576, 73CR(B)(276)335⟩.

$$\text{oxetane} + \text{H—Z} \rightleftarrows \text{oxetane-O}\cdots\text{H}\cdots\text{Z} \tag{11}$$

It has recently been proposed that the relatively high electron-donor ability in hydrogen bonding associated with the oxetane ring is due to steric factors, rather than to an electronic effect. This is based on the fact that the gas phase basicity of tetrahydropyran is greater than that of tetrahydrofuran and on a report that the shift of the O—H stretching frequency of phenol hydrogen bonded to cyclic ethers in carbon tetrachloride solvent is in the order THP > THF > oxetane ⟨80JOC1166⟩. This latter result is the opposite to that previously reported for the same system (except for being in tetrachloroethylene solution ⟨62LA(659)81⟩), which agrees with all the other hydrogen bonding data reported. The relatively low first IP for oxetane, which has been explained on a molecular orbital basis ⟨77JA3226⟩, would indicate the validity of an electronic explanation, and it is hoped that future studies will clarify the relative importance of steric and electronic factors.

Studies of hydrogen bonding and iodine complexation on alkyl-substituted oxetanes have shown that 2-alkyl and 2,2-dialkyl substitution enhances the electron donor ability, while 3,3-dialkyl substitution substantially diminishes it. This has been attributed to a decrease in the bond angle strain at the oxygen atom, due to a Thorpe–Ingold effect at C-3, since the effect increases with the size of the alkyl substituent at that position ⟨71BSF4576⟩.

Another type of hydrogen bonding of oxetane is clathrate formation with water. Clathrate compositions of 1:6.5 and 1:17 moles of water have been observed ⟨72CR(C)(274)1108⟩. Oxetane has been found to be more effective than other cyclic ethers in solvating uranyl hexafluoroacetonylacetate ⟨81IC1415⟩.

Hydrogen bonding of 2-oxetanone with phenol and methanol-d_1 show an interaction of about half the magnitude for oxetane, judging from the shift of the O—H stretching frequency in each case (which is considered a measure of the strength of interaction). By the same criterion, 2-oxetanone is about 20% weaker as an electron donor in hydrogen bonding than γ-butyrolactone; in both cases, of course, the hydrogen bond is formed at the exocyclic oxygen atom ⟨76JOC49, 68BCJ767⟩.

The only electrophile towards which oxetane has been reported to be a relatively poor electron donor is dimethylzinc. Measurements of association in benzene solution showed oxetane to be a poorer donor than THF, which was poorer than THP ⟨66MI51300⟩.

5.13.3.3.2 Acid-induced reactions producing mostly rearrangement

Coordination of the ring oxygen atom with a strong electrophile, which withdraws electrons from the oxygen quite effectively and weakens the C—O ring bonds, results either in the scission of one of them or in nucleophilic attack on an adjacent carbon atom. When bond scission occurs, fragmentation or molecular rearrangement may occur.

Interesting examples of both of these processes can be seen in the reactions of a series of 2-aryl-3,3-dimethyl-4-*t*-butyloxetanes when treated with boron trifluoride etherate in ether or with aluminum chloride in benzene. When the aryl group was phenyl or a negatively substituted phenyl group, ring expansion to the corresponding 2-aryl-3,3,4,5,5-pentamethyl-THF occurred in good yield, but when the aryl group had an electron-donating *para* substituent, fragmentation to alkene and carbonyl compounds took place, also in good yield. These products suggest a reaction mechanism involving bond scission to carbonium ion intermediates, the outcome depending on which site has the greater cation stabilization. Scission of the carbon(aryl)–oxygen bond can lead to ring cleavage *via* a carbonium ion (equation 12), whereas breaking the carbon(t-butyl)–oxygen bond allows for a 1,2-methyl shift in the resulting carbonium ion (**39**). This shift must be rapid compared to ring fragmentation, since no aromatic aldehyde was detected ⟨79CC382⟩.

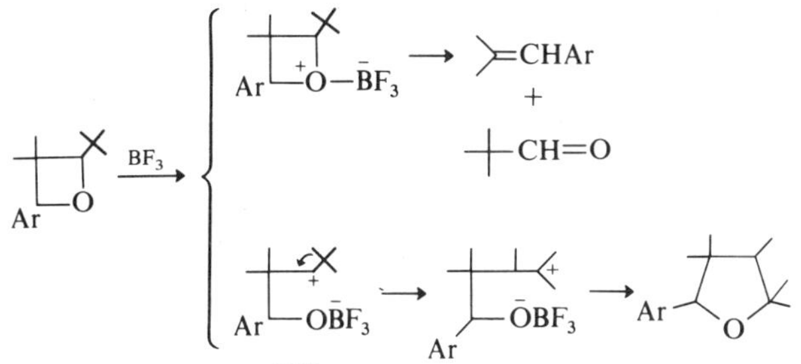

(12)

(**39**)

The rearrangement of 3,3-dimethyl-4,4-diphenyl-2-isopropylideneoxetane on treatment with neutral alumina is somewhat different. Here the products consisted of two isomeric cyclobutanones, and so the initial cleavage that might be expected to give the diarylmethyl cation must be followed by a recyclization to an intermediate which rearranges in two ways (equation 13). The proposed oxycyclopropylcarbinyl cation seems to serve this purpose. When the spiro-xanthene structure (**40**) is treated with alumina, ring expansion of the oxetane ring occurred to give a six-membered ring. In this case the initial carbonium ion is greatly stabilized by the interaction of the xanthyl oxygen atom, and attack of the alkoxide oxygen atom on an *ortho* position of the aromatic ring system becomes favored.

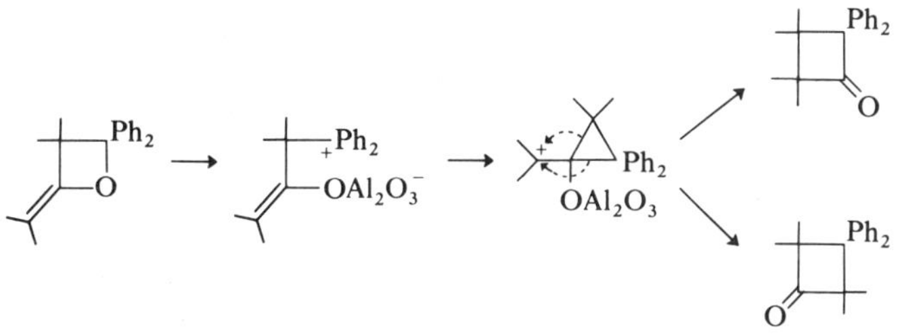

(13)

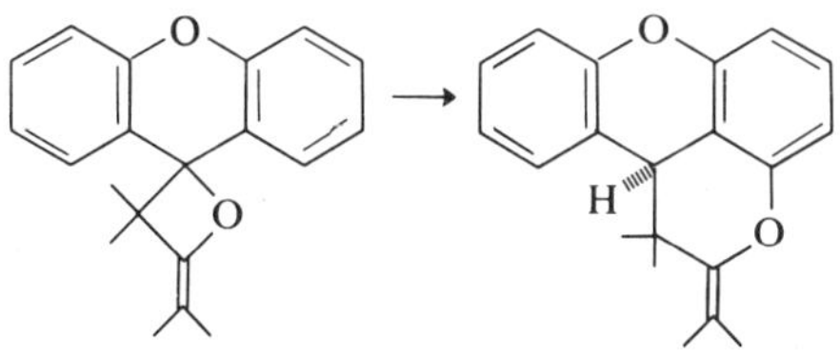

(**40**)

A third type of rearrangement of polysubstituted 2-isopropylideneoxetanes on treatment with neutral alumina was observed when there was at least one methyl substituent on C-4. The initial carbonium ion formed then underwent deprotonation, giving the unsaturated ketone in excellent yield (equation 14) ⟨74CB3922⟩. With saturated oxetanes this type of reaction gives allylic alcohols, as is often observed with strong acids in the absence of effective nucleophiles. Thus 2,2-dialkyloxetanes are isomerized by sulfuric acid in ether mainly to the corresponding primary allylic alcohols (equation 15), with minor amounts of the tertiary allylic alcohol being detectable ⟨73ZOR890⟩. The interesting rearrangement of 2,2,3-trimethyl-3-oxetanol in dilute sulfuric acid to 1-hydroxy-2,2-dimethyl-3-butanone may be of this type, with protonation at the ring oxygen atom followed by cleavage to a carbonium ion and a subsequent rapid hydroxymethyl migration favored stereoelectronically (equation 16). Other mechanisms, however, are also compatible with the evidence ⟨73T1163⟩.

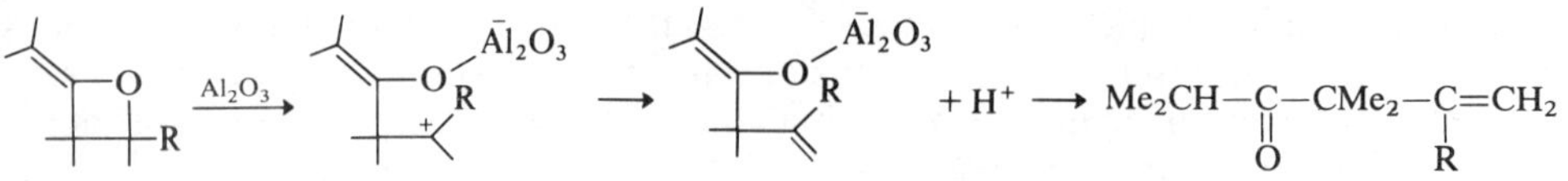

Al_2O_3 → $+ H^+$ → $Me_2CH{-}C({=}O){-}CMe_2{-}C(R){=}CH_2$ (14)

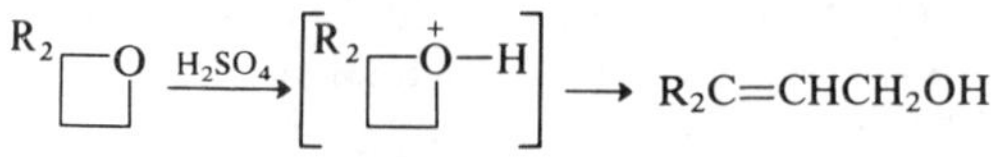

R_2 oxetane $\xrightarrow{H_2SO_4}$ $[R_2 \text{ oxetane}{-}\overset{+}{O}{-}H]$ → $R_2C{=}CHCH_2OH$ (15)

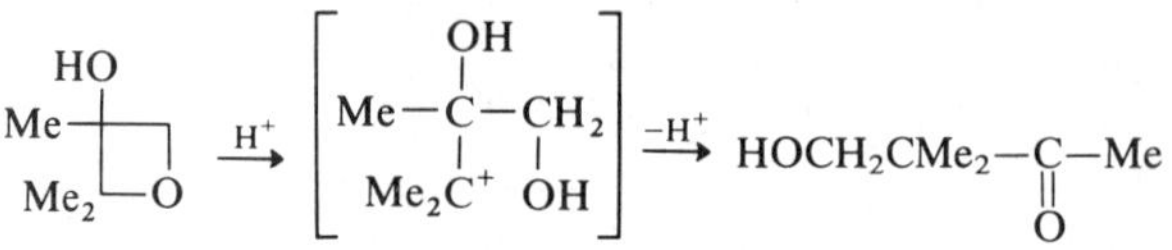

HO, Me, Me_2 oxetane $\xrightarrow{H^+}$ $[Me{-}C(OH){-}CH_2(OH){-}Me_2C^+]$ $\xrightarrow{-H^+}$ $HOCH_2CMe_2{-}C({=}O){-}Me$ (16)

Protonated β-butyrolactone in FSO_3H–SbF_5–SO_2 undergoes a cleavage and rearrangement at temperatures above −50 °C. Analysis of NMR spectra indicated that alkyl–oxygen cleavage occurs to give the secondary carbonium ion, which further underwent deprotonation and dehydration to give the 2-butenoyl cation (equation 17) ⟨70JOC3916⟩.

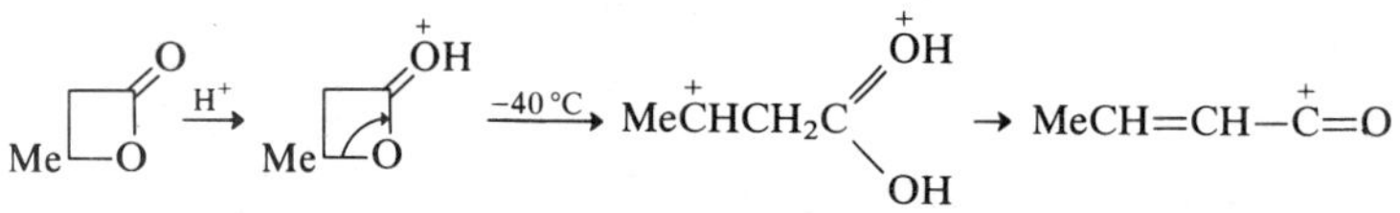

$\xrightarrow{H^+}$ $\xrightarrow{-40\,°C}$ $Me\overset{+}{C}HCH_2C(\overset{+}{O}H)OH$ → $MeCH{=}CH{-}\overset{+}{C}{=}O$ (17)

4-Vinylic substituted 2-oxetanones undergo rearrangement in the presence of Pd(II) salts, preferably in the presence of trimethyl phosphite, to give butadienecarboxylic acids in generally excellent yields. It has been proposed that the metal atom chelates with the ether oxygen and the double bond, leading to oxygen–alkyl scission to a five-membered cyclic organopalladium compound. A 1,3-hydride shift from carbon to palladium with simultaneous C—Pd breaking then follows (equation 18) ⟨76JOC2527⟩.

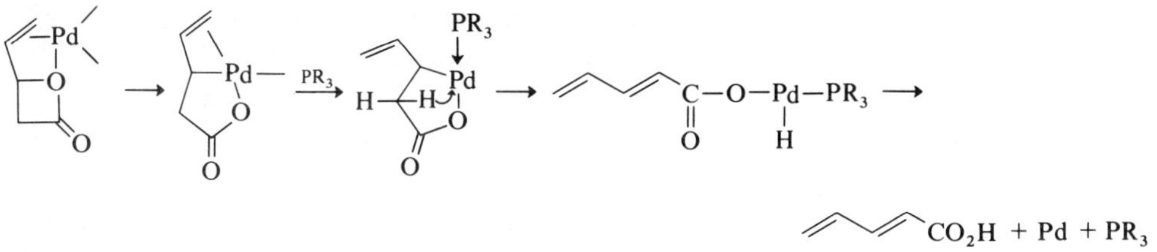

→ $\xrightarrow{PR_3}$ → $C({=}O){-}O{-}Pd(H){-}PR_3$ →

$CO_2H + Pd + PR_3$ (18)

5.13.3.3.3 Acid-induced ring cleavage followed by nucleophilic attack

This is a widely observed type of reaction for oxetanes and 2-oxetanones and has many variations. The typical pattern involves electrophilic attack on the ring oxygen atom of oxetanes to form an unstable oxetanium ion. The latter either undergoes C—O ring scission to a carbonium ion which then combines with a nucleophile, or nucleophilic attack by even very weak nucleophiles at the α-carbon atom. For clarity, these reactions will be arranged according to the nature of the overall process.

(i) Cleavage by hydrogen halides

The reaction of oxetanes with hydrogen halides, giving 3-halogeno-1-propanols, is well known and very general. It has been reviewed previously ⟨64HC(19-2)983⟩; a more recent application reported is the cleavage of 2-oxaspiro[2.3]hexane with aqueous hydrochloric or hydrobromic acids to give the expected 3-halogenomethylcyclopropylcarbinols in good yields ⟨68IZV670⟩.

The mechanism of the reaction of hydrochloric acid and 2-methyloxetane has been investigated by a kinetic study in different solvents. The principal product is 4-chloro-2-butanol (equation 19), with a relatively small amount of 3-chloro-1-butanol formed as the by-product. This product distribution and the relatively small effect of the solvating power of the solvent indicates the mechanism involves an S_N2 attack of the chloride ion on the α-carbon atom of the protonated oxetane ⟨67MI51301⟩.

$$\text{2-methyloxetane} + HCl \longrightarrow \text{protonated 2-methyloxetane} \xrightarrow{Cl^-} MeCH(OH)CH_2CH_2Cl \quad (19)$$

$$\text{diketene} \xrightarrow{HF} MeCOCH_2COF \xrightarrow[AlCl_3]{ArH} MeCOCH_2COAr \quad (20)$$

2-Oxetanone is also cleaved readily by hydrogen halides, giving 3-halogenopropionic acids by alkyl–oxygen scission, while 4-methylene-2-oxetanone (diketene) is cleaved by acyl–oxygen scission to give acetoacetyl halides. Acetoacetyl fluoride, obtained by reaction of anhydrous hydrogen fluoride with diketene, has been useful in Friedel–Crafts acylations ⟨64HC(19-2)729, 73GEP2209692⟩.

(ii) Acid-catalyzed solvolysis

The perchloric acid-catalyzed methanolysis of 2-methyloxetane gives a mixture of 4-methoxy-2-butanol and 3-methoxy-1-butanol, with the former somewhat predominating (equation 21). The effect of solvent on the product distribution and the reaction rates indicated that protonated 2-methyloxetane was reacting by a borderline S_N1–S_N2 mechanism ⟨67MI51302⟩. Similar studies with the acid-catalyzed methanolysis of oxetane itself indicated that methanol reacted with protonated oxetanium ion by the S_N2 process. The same type of studies with a series of 2-aryloxetanes indicated that methanolysis of these compounds involved the borderline mechanism for the protonated oxetanium ions ⟨69MI5101, 72MI5102, 73MI5100⟩.

$$\text{2-methyloxetane} \xrightarrow{H^+} \text{Me}\overset{\delta+}{\text{C}}\cdots\overset{\delta+}{\text{O}}\text{—H} \xrightarrow{MeOH} MeCH(OH)(CH_2)_2OMe + MeCH(OMe)(CH_2)_2OH \quad (21)$$

The solvolyses of several 2,2-disubstituted oxetanes, including 6-substituted 7-oxabicyclo[4.2.0]octanes (**41**), with several alcohols has been studied in some detail, with general agreement that reaction of the protonated oxetanes by the S_N1 mechanism is involved. A range of intimate structures for the carbonium ion is suggested, depending on the structure and the solvents used. The loss of formaldehyde by retrograde Prins reaction to give (**42**) was observed mainly in non-polar solvents ⟨78G619, 81ZOR408⟩.

(**41**) $\xrightarrow{H^+}$ [protonated oxetane X^- → $\delta+$/$\delta-$ ion pair X^- + $X^-\cdots H$–O–R ion pair] → cis-diol (OH, CH_2OH) + trans-diol (OH, CH_2OH) + cyclohexene with R and CH_2OH + 1-R-cyclohexene (**42**)

The mechanism for the acid-catalyzed hydrolysis of 2-methoxy-3,3-dimethyloxetane (**43**), which is a cyclic acetal, has received detailed study ⟨74JA819⟩. Kinetic studies in buffered media and solvent deuterium isotope effects demonstrated that this reaction showed general acid catalysis, which had previously been observed only when C—O bond breaking is facilitated by a very good leaving group or by formation of a very stable oxocarbonium ion. In the case of (**43**) the relief of ring strain in the opening of the protonated oxetane ring must be the cause of the general acid catalysis, as well as the extremely high rate of hydrolysis observed. In fact it is proposed that ring cleavage occurs simultaneously with protonation in the rate-determining step, as shown in equation (22). This mechanism resembles that for the hydrolysis of oxetane in that both are of the general class A_1, but the hydrolysis of oxetane is considerably less sensitive to acid catalysis and is *specific* proton-catalyzed ⟨58JA4162, 67MI51302⟩. Both compounds show virtually no reaction with neutral water, in contrast to the hydrolysis of 2-oxetanone for which the reaction with neutral water is dominant between pH 1 and 9 (Section 5.13.2.4).

$$\text{Me}_2\text{-oxetane-OMe}\ (\mathbf{43}) + \text{HA} \longrightarrow [\text{oxetane} \cdots \text{H}\cdots\text{A}^{\delta-}] \xrightarrow[-\text{HA}]{\text{H}_2\text{O}} \text{Me}_2\text{C}(\text{CH(OH)OMe})(\text{CH}_2\text{OH}) \longrightarrow \text{HOCH}_2\text{CMe}_2\text{CHO} \quad (22)$$

$$\text{Me}_2\text{Me}_2\text{-oxetan-2-one} + \text{H}_2\text{O}_2 \xrightarrow{\text{H}^+} \text{Me}_2\text{Me}_2\text{-peroxylactone} \quad (23)$$

3,3,4,4-Tetramethyloxetan-2-one is solvolyzed, with recyclization, by means of hydrogen peroxide in ether containing sulfuric acid, to give a 60% yield of the corresponding peroxylactone in high purity (equation 23) ⟨72S616⟩.

The reaction of oxetanes with aldehydes in the presence of sulfuric acid has been extensively studied, and procedures giving good yields of 1,3-dioxanes have been developed. The corresponding reaction occurs with ketones, but higher yields of 1,3-dioxanes are obtained with these when boron trifluoride etherate is used as the catalyst ⟨81MI51300⟩. Since these reactions occur readily with the very available 3,3-bis(chloromethyl)oxetane (**44**), and the products can be converted back to aldehydes by hydrolysis, this method has potential synthetic value for masking carbonyl groups. On the basis of a kinetic study of the reaction of (**44**) with aromatic aldehydes, it has been proposed that the mechanism involves rate-determining electrophilic attack on the oxetane oxygen atom by the oxycarbonium ion, $[\text{RCHOH}]^+$, derived from the aldehyde ⟨79MI51302⟩.

$$(\text{ClCH}_2)_2\text{-oxetane}\ (\mathbf{44}) + \text{R}_2\text{C=O} \longrightarrow (\text{ClCH}_2)_2\text{-1,3-dioxane-R}_2$$

A similar type of acid-catalyzed condensation of aldehydes with 4-methylene-2-oxetanone (diketene), giving 4-oxo-6-methyl-1,3-dioxins, has been patented ⟨73GEP2149650⟩. However, other work has established that δ-hydroxy-β-keto acids or unsaturated keto acids are formed as the principal products (equation 24) ⟨78CPB3877, 78CL409⟩. The latter reaction probably involves electrophilic attack of the protonated aldehyde on the nucleophilic exocyclic methylene carbon atom of the diketone. A closely related reaction of acetals with diketene, catalyzed by titanium tetrachloride, gives the corresponding δ-alkoxy-β-keto esters ⟨74CL1189⟩.

$$\text{RCHO} + \text{diketene} \xrightarrow{\text{H}^+} \text{RCH(OH)CH}_2\text{CH=O}^+\text{(ring)} \longrightarrow \text{RCH(OH)CH}_2\text{C(O)CH}_2\text{CO}_2\text{H} \rightarrow \text{RCH=CHC(O)CH}_2\text{CO}_2\text{H} \quad (24)$$

With formaldehyde, diketene reacts quite vigorously in water solution without acid catalysis, to give carbon dioxide and 2,6-heptanedione, isolated in 40% yield ⟨75JOC675⟩. This reaction can be explained analogously, with an initial electrophilic attack of the formaldehyde carbon atom on the exocyclic methylene carbon atom of the diketene, followed by ring opening and then reaction with a second molecule of diketene (equation 25).

$$\text{CH}_2\text{=O} + \text{diketene} \longrightarrow \text{CH}_2(\text{OH})\text{CH}_2\text{C(O)CH}_2\text{CO}_2\text{H} \rightarrow \text{CH}_2\text{=CHC(O)CH}_2\text{CO}_2\text{H} \xrightarrow[\text{CO}_2]{\text{diketene}}$$

$$\text{HO}_2\text{CCH}_2\text{C(O)CH}_2\text{CH}_2\text{CH=C(OH)CH}_2\text{CO}_2\text{H} \longrightarrow (\text{MeC(O)CH}_2)_2\text{CH}_2 + 2\text{CO}_2 \quad (25)$$

The sulfuric acid-catalyzed condensation of oxetanes with nitriles gives dihydro-1,3-oxazines (**45**). This reaction has been reported for a wide variety of 2,2-disubstituted oxetanes and nitriles, generally giving fair yields ⟨74ZOR95, 82ZOR181⟩. The nitrogen atom of the nitrile always bonds to the more substituted α-carbon of the oxetane.

$$R_2\text{-oxetane} + R'CN \xrightarrow{H_2SO_4} \text{(45)}$$

(45)

The acid-catalyzed reaction of oxetanes with *t*-butyl isocyanide gives 2-iminotetrahydrofurans, boron trifluoride etherate being used as catalyst (equation 26). The authors proposed that the isocyanide reacts with the protonated oxetanes by an S_N2 mechanism ⟨70S475⟩.

$$R_2\text{-oxetane} + RNC \xrightarrow{BF_3} R_2\text{-2-(NR)-tetrahydrofuran} \qquad (26)$$

The Friedel–Crafts alkylation of aromatic compounds by oxetanes in the presence of aluminum chloride is mechanistically similar to the solvolyses above, since the first step is electrophilic attack on the ring oxygen by aluminum chloride, followed by a nucleophilic attack on an α-carbon atom by the aromatic compound present. The reaction of 2-methyloxetane and 2-phenyloxetane with benzene, toluene and mesitylene gave 3-aryl-3-methyl-1-propanols and 3-aryl-3-phenyl-1-propanols as the main products and in good yields (equation 27). Minor amounts of 3-chloro-1-butanol and 4-chloro-2-butanol are formed as by-products from 2-methyloxetane, and of 3-phenyl-1-propanol from 2-phenyloxetane ⟨73ACS3944⟩.

$$R\text{-oxetane} \xrightarrow{AlCl_3} R\text{-oxetane-}\overset{+}{O}\text{-}\overset{-}{Al}Cl_3 \xrightarrow[\text{ii, } H_2O]{\text{i, ArH}} ArCHR(CH_2)_2OH \qquad (27)$$

(*iii*) *Alkylative cleavage*

Several methods for cleaving oxetanes by reaction with alkyl carbonium ions have been discovered. It is reported that 2-methyloxetane reacts with α-chloroethers in ether solution containing zinc chloride to form a mixture of two isomeric chloro-substituted acetals (equation 28) ⟨72IZV125⟩.

$$Me\text{-oxetane} + RCH(Cl)OR' \xrightarrow{ZnCl_2} MeCH(Cl)(CH_2)_2OCH(R)OR' + Cl(CH_2)_2CH(Me)OCH(R)OR' \qquad (28)$$

Another approach has been to treat oxetane or a 3,3-dialkyloxetane with an equimolar mixture of a reactive alkyl halide and a mercury(II) salt. The products are aliphatic ethers containing the alkyl group derived from the alkyl halide and the anion of the mercury salt (equation 29). The yields were fair to good when ethyl iodide, allyl bromide and benzyl bromide served as the alkyl halides. When mercury(II) thiocyanate was used as the salt, the products were very predominately the isothiocyanate compounds, with the thiocyanate isomers present as minor coproducts. This gives support for the cleavage of the *O*-alkyloxetanium ion intermediate to a carbonium ion. The corresponding reaction with THF also takes place but gives equal amounts of thiocyanate and isothiocyanate isomers, showing how the ring strain in the oxetanes promotes carbonium ion formation from the ring scission ⟨79BCJ3611⟩.

$$R_2\text{-oxetane} + R'X + HgY_2 \longrightarrow \left[R_2\text{-oxetane-}\overset{+}{O}\text{-}R\right] \longrightarrow R'OCH_2CR_2CH_2Y \qquad (29)$$

Chlorination of mixtures of alkenes and oxetanes also gives rise to alkylative cleavage of the oxetane ring, forming ethers in fair yields (equation 30). The reaction is carried out by introducing chlorine gas into a solution of the alkene and oxetane in carbon tetrachloride. Here the *O*-alkyloxetanium intermediate probably forms by a nucleophilic attack of the oxetane oxygen atom on the alkene–chlorine π-complex ⟨76ZOR2331, 78ZOR1193⟩.

$$R_2\text{-oxetane} + RCH{=}CH_2 + Cl_2 \longrightarrow RCH(Cl)CH_2OCH_2CR_2CH_2Cl \qquad (30)$$

(iv) Polymerization of oxetanes

Polymerization is frequently observed as a side reaction in acid-catalyzed reactions, and under appropriate conditions many oxetanes can be quantitatively transformed into high molecular weight linear polyethers with useful properties. Polymerization is very general for oxetanes and is closely related to Lewis acid-catalyzed polymerization of oxiranes, THF and oxepane, but oxetanes generally polymerize much more rapidly than THF and oxepane and at a rate similar to oxiranes ⟨72MI51300⟩.

The degree and type of substitution has much effect on the properties of the polymer, as well as the rate of polymerization. Increasing methyl substitution at position-3 gives a regular increase in the rate of polymerization, and the most useful properties are observed with polymers of 3,3-dialkyloxetanes, with melting points ranging from 50 to nearly 300 °C. The polymers of 3,3-bis(chloromethyl)oxetane ('BCMO') have been commercial products for many years, known as 'Penton' and 'Pentaplast' (equation 31).

$$(ClCH_2)_2\text{C}_3\text{H}_4\text{O (oxetane)} \xrightarrow{BF_3} (-O-CH_2-C(CH_2Cl)_2CH_2-)_n \quad (31)$$

High polymers are generally obtained on treatment with Lewis acids at low temperatures in an inert solvent. Boron trifluoride and boron trifluoride etherate are the most widely used catalysts, but a small amount of water must be present, which is termed a 'promoter'; triethylaluminum and triisobutylaluminum are also useful initiators and are generally used with addition of water. Alkylating agents, such as ethyl triflate, triethyloxonium tetrafluoroborate and hexafluorophosphite, and 2-methyl-1,3-dioxolenium perchlorate, are also effective initiators ⟨76MI51301, 72MI51304⟩.

The principal side reaction is telomerization, which is mainly formation of a cyclic tetramer, a 16-membered cyclic ether. A great deal of study has been given to suppression of this undesired by-product by variations in catalysts, promoters and other conditions.

The mechanism of the polymerization of oxetanes has been extensively investigated by means of kinetics, trapping and coupling experiments and NMR studies. It is now generally accepted that the reaction is initiated by electrophilic attack of the initiator on the ring oxygen atom and propagated by an S_N2 type chain mechanism, as shown below in equation (32) ⟨74BCJ2699⟩.

Initiation: X or $X-A^- +$ O◇R_2 ⟶ $X-\overset{+}{O}$◇R_2 (with A^- associated) **(44)**

Propagation: **(44)** + O◇R_2 ⟶ $X-\overset{\delta+}{O}\cdots\cdots\overset{\delta+}{O}$◇$R_2$ (transition state, ring $CH_2CR_2CH_2$) ⟶ $X-O-(CH_2CR_2CH_2)_n-\overset{+}{O}$◇$R_2$ (32)

Poly(trimethylene oxide) is a very regular polyether chain in a very densely packed random coil. The two C—C skeletal bonds show preference for the *gauche* configuration while the two C—O bonds show preference for a *trans* configuration. Polymers of BCMO and 3,3-bis(azidomethyl)oxetane have analogous structures, with the substituents on the central carbons oriented away from each other ⟨77JCP(66)1901, 78MI51302⟩.

Oxetanes can be copolymerized very successfully with a wide range of other monomers and even carbon dioxide, to give plastic compositions with a wide variety of properties. Naturally, this area has been the subject of a great deal of research, and many patents have appeared which are beyond the scope of this survey.

(v) Cationic polymerization of 2-oxetanones

β-Lactones are easily polymerized by both cationic and anionic initiators, and in fact they (particularly β-propiolactone) are known to polymerize explosively and unpredictably without addition of any initiators. Inhibitors, such as trialkylborates, have been used to prevent undesired polymerization ⟨73GEP2255194⟩.

At low temperatures in inert solvents (such as methylene dichloride) a controlled polymerization can be effected using various acids and alkylating agents. These initiators include boron trifluoride etherate, triethylaluminum, trityl hexachloroantimonate, triethylammonium hexachloroantimonate, diethyloxonium hexafluoroantimonate, *p*-toluenesulfonic acid and diethylzinc or cadmium–1,2-diol complexes. Crystalline, high molecular weight polyesters with useful properties can be obtained (equation 33). Stereospecific polymerization of 3-alkyl-2-oxetanones with triethylaluminum initiator has been reported ⟨74MI51301, 79MI51300⟩.

$$\text{2-oxetanone} + XY \longrightarrow [\text{O-complexed oxetanone}]^{+}\,Y^{-} \longrightarrow [\ldots] \longrightarrow \left[CH_2CH_2\overset{O}{\overset{\|}{C}}-O-\right]_n \quad (33)$$

The mechanism for the polymerization probably fits the general pattern of electrophilic attack by the initiator on the carbonyl oxygen atom for the initiation step, with propagation taking place by attack of the electron deficient C-4 of the last unit of the growing chain on the carbonyl group of a monomer molecule. Certain modifications of this general mechanism are apparently observed with specific catalysts, and the subject appears to need further study.

(vi) Cleavage with acid halides

The easy cleavage of the oxetane ring with acyl chlorides is well known and was reviewed previously ⟨64HC(19-2)983⟩. More recently it has been observed that 2-oxaspiro[3.2]hexane was cleaved by acyl bromides to give 1-bromomethylcyclopropylcarbinol esters in high yield ⟨68IZV670⟩.

Much work has been done on the cleavage of oxetanes with phosphorus halides. For example, phosphorus trichloride reacts with two moles of 2-methyloxetane to give bis(4-chloro-2-butyl) chlorophosphite and with three moles to give the corresponding trialkyl phosphite. Dichlorophosphines ($RPCl_2$; R = ethyl, phenyl, ethoxy, *etc.*) react in an analogous manner ⟨73IZV627, 73IZV2345⟩.

(vii) Cleavage by Group IV compounds

Both oxetanes and 2-oxetanone are cleaved by bromotrimethylsilane, the former very vigorously and exothermically but the latter requiring heat (equation 34) ⟨81S383, 79AG(E)689⟩. 2-Oxetanone is cleaved in the same manner by diethylaminotrimethylsilane, diethylaminotrimethylgermane, and chloro-, bromo-, methoxy- and methylthio-trimethylstannane. The opposite regioselectivity, however, is observed with diethylaminotrimethylstannane (**45**), which reacts with exceptional vigor (equation 35) ⟨69JOC459⟩.

$$\text{3,3-}R_2\text{-oxetane} + Me_3SiBr \rightarrow Me_3SiCH_2CR_2CH_2Br$$

$$\text{2-oxetanone} + Me_3SiBr \rightarrow Me_3SiCH_2CH_2CO_2SiMe_3 \quad (34)$$

$$\text{2-oxetanone} + \underset{(\mathbf{45})}{Me_3SnNEt_2} \rightarrow Me_3SnCH_2CH_2CONEt_2 \quad (35)$$

It seems reasonable that bromotrimethylsilane and the other Group IV compounds except (**45**) are reacting by initial complexation of the metal atom with both oxygen atoms, with simultaneous release of the halogen atom or other attached group as an anion, as shown in structure (**46**). The released anion may then attack the adjacent methylene group in an S_N2-like process. The amino group of (**45**) is quite nucleophilic, since $(p\text{–}d)\pi$-bonding between the metal and the nitrogen atom is relatively low; thus it may proceed by nucleophilic attack on the carbonyl carbon atom of 2-oxetanone, which would be followed by transfer of the tin atom from the now-positive nitrogen atom to the just-released alkoxide ion (structure **47**).

(**46**) (**47**)

Trialkylsilanes and carbon monoxide, in the presence of octacarbonyldicobalt, cleave oxetane upon heating in benzene solution at 140 °C (equation 36). Tetrahydrofuran and oxiranes show the same type of reaction ⟨77AG818⟩.

$$\text{oxetane} + HSiEt_2Me + CO \xrightarrow{Co_2(CO)_8} Et_2MeSiO(CH_2)_3CHO \quad (36)$$

5.13.3.3.4 Oxidizing agents

Oxetanes and 2-oxetanones seem to be generally not very reactive with oxidizing agents, and such reactions have received relatively little study. Photooxidation of oxetane has been found to occur readily, presumably to form 2-oxetanone (equation 37), although this was not positively identified. The reaction was similar to, but more rapid than, the photooxidation of THF, which was shown to give the corresponding lactone ⟨69JOC1345⟩.

$$\text{oxetane} + O_2 \xrightarrow{h\nu} \text{2-oxetanone} \quad (37)$$

$$\text{oxetane} + Cl_2O_7 \rightarrow O_3ClO(CH_2)_3OClO_3 \quad (38)$$

Dichlorine heptoxide, which can be a powerful oxidizing agent, reacts with oxetane in carbon tetrachloride solution at 0 °C to give a fair yield of propane-1,3-diperchlorate (equation 38). It was suggested that the mechanism involved formation of a perchlorooxetanium ion, which subsequently reacted with perchlorate ion ⟨75JOC81⟩.

5.13.3.4 Reaction with Nucleophiles, including Reducing Agents

5.13.3.4.1 Oxetanes

Oxetanes are much less susceptible to cleavage by nucleophiles than oxiranes, except in the presence of acids. Several types of acid-catalyzed nucleophilic reactions are described in the previous section, such as reaction of hydrogen halides to give 3-halogeno-1-propanols and various acid-catalyzed solvolysis reactions. Another example of this type is the reaction of thiourea with 2-alkyloxetanes in the presence of hydrochloric or perchloric acid to give excellent yields of 3-alkyl-3-hydroxybutylisothiouronium salts (equation 39) ⟨67CR(C)(264)1309⟩.

$$\text{2-R-oxetane} + (H_2N)_2C{=}S \xrightarrow{HX} [RCH(OH)(CH_2)_2SC(NH_2)_2]^+X^- \quad (39)$$

With reactive nucleophiles, catalysis by relatively weak acids often suffices, especially with heating. Thus primary and secondary amines have been found to react with a series of 3-alkyloxetane-3-methanols by heating at 120–125 °C in the presence of silica gel, boric acid or cadmium hydroxide (equation 40) ⟨73LA365, 74JAP7420071⟩.

$$\text{3-R-3-}HOCH_2\text{-oxetane} + RNHR' \rightarrow (HOCH_2)_2CRCH_2NRR' \quad (40)$$

$$\text{2-R-3-H-3-OH-oxetane} + PhCH_2S^-Na^+ \rightarrow RCH(OH)CH(OH)CH_2SCH_2Ph \quad (41)$$

Reaction of very reactive nucleophiles with oxetanes has been achieved without any acid catalysis through use of heat and a favorable solvent system. Thus the oxetane ring in certain carbohydrate derivatives has been cleaved in good yield by toluene-α-thiolate in DMF at 150 °C (equation 41), providing a route to sulfur-containing carbohydrates. Thiosulfate ion also gave a similar cleavage reaction ⟨68JOC396⟩.

Enol alkylation of cyclohexanone by oxetane has been achieved by the reaction of oxetane with an enamine salt, bromomagnesium *N*-cyclohexyliminocyclohexane, in THF. An 80%

$$\text{1-[N(MgBr)}C_6H_{11}\text{]cyclohexene} + \text{oxetane} \rightarrow \text{hexahydrochroman-8a-ol (OH)} \quad (42)$$

yield of the hemiacetal of 2-(3-hydroxypropanoyl)cyclohexanone was obtained (equation 42) ⟨75JOC2963⟩.

The reaction of oxetanes with Grignard reagents and organolithium compounds is well known and has been useful, particularly with oxetane itself. A recent application involves use of the reaction of isopentylmagnesium bromide with oxetane as the first step of a synthesis of an insect pheromone (equation 43) ⟨76CB3375⟩. The Grignard reaction with 2-methyloxetane and 2-phenyloxetane is less satisfactory, due to the extensive formation of by-products.

$$\mathrm{Me_2CHCH_2CH_2MgBr} + \text{oxetane} \rightarrow \xrightarrow{H^+} \mathrm{Me_2CH(CH_2)_5OH} \qquad (43)$$

Recent work has been reported on the copper-catalyzed Grignard reaction with oxetane. This reaction, which probably involves the intermediacy of organocuprates, proceeds under milder conditions and gives yields of 60–75% of the expected alcohols, which are somewhat better than those obtained with the usual Grignard reagents ⟨79TL1503⟩. 2-Alkoxyoxetanes are especially reactive towards Grignard reagents, reaction occurring at the alkoxy-substituted carbon atom to give the expected hydroxyalkyl ethers (equation 44) ⟨73USP3715401⟩.

$$\text{2-OR-oxetane} + \mathrm{R'MgX} \rightarrow \mathrm{HOCH_2CH_2CHR'OR} \qquad (44)$$

It has recently been found that oxetane reacts with phosphorus ylides in a manner similar to ethylene oxide. Trimethylmethylenephosphorane and oxetane give the six-membered heterocyclic compound 2,2,2-trimethyl-1-oxa-2-phospholane in 49% yield (equation 45). The cyclic ylides (**46a,b**) give 39–47% yields of the corresponding spirobicyclic oxaphosphoranes. An NMR study showed these compounds to be fluxional as a result of turnstile pseudorotation ⟨79CB501⟩.

$$\mathrm{Me_3P{=}CH_2} + \text{oxetane} \longrightarrow \mathrm{Me_3P}\text{(1-oxa-2-phosphacyclohexane)} \qquad (45)$$

$$\mathrm{(CH_2)_n P(Me){=}CH_2} + \text{oxetane} \longrightarrow \mathrm{(CH_2)_n P(Me)}\text{(spiro, O-ring)}$$

(**46**) **a**; $n = 4$
b; $n = 5$

Reduction of oxetanes can usually be achieved with lithium aluminum hydride in THF. The reaction is almost general, though polyalkyl substitution may decrease the rate to an impractical level. As is to be expected with a strong nucleophile, ring cleavage occurs generally between the oxygen atom and the least substituted α-carbon atom (equation 46) ⟨72MI51303⟩.

$$\text{2-R-oxetane} + \mathrm{LAH} \xrightarrow{\mathrm{THF}} \xrightarrow{\mathrm{H_2O}} \mathrm{RCH(OH)CH_2Me} \qquad (46)$$

Reductive ring cleavage of 3-aryloxetanes by aluminum hydride in THF or ether may occur between the oxygen atom and the aryl-substituted α-carbon atom (equation 47), if the aryl group is *para* substituted by an electron donating group (such as methyl or methoxy). This less nucleophilic reagent is therefore prone to react by a borderline S_N1 mechanism. Dichloroaluminum hydride in ether cleaves 2-aryloxetanes, predominately by the S_N1 mechanism, thus providing synthetic versatility ⟨69BSF3653⟩.

$$\text{2-Ar-oxetane} + \mathrm{AlCl_2H} \xrightarrow{\mathrm{Et_2O}} \xrightarrow{\mathrm{H_2O}} \mathrm{ArCH_2CH_2CH_2OH} \qquad (47)$$

A convenient new reagent, lithium tri-butoxyaluminum hydride–triethylborane, has been found to reduce oxetane rapidly to 1-propanol in 98% yield, but has not yet been studied with substituted oxetanes ⟨79JOC3678⟩.

5.13.3.4.2 2-Oxetanones

There are an extremely large number of reactions of 2-oxetanones with nucleophilic reagents, and space will allow inclusion of only representative examples. β-Lactones show the interesting $B_{AL}2$ mechanism for base-catalyzed hydrolysis and the $A_{AL}2$ mechanism for acid-catalyzed hydrolysis, according to data on kinetics and optical rotation studies of optically active lactones. The mechanistic interpretations are complicated, however, by the possibilities for subsequent elimination and addition reactions to occur, so that both of the two sites for nucleophilic attack on the β-lactone skeleton, C-2 and C-4, may become involved. In fact β-lactones are unusually insensitive to base, as well as acid, catalysis, the slow reaction with neutral water predominating between pH 1 and 9 ⟨74JCS(P2)377⟩.

Methanolysis of β-propiolactone with a trace of hydroxide catalyst gives a good yield of methyl 3-hydroxypropionic acid, but with an equimolecular quantity of sodium methoxide the product is sodium 3-methoxypropionate (equation 48) ⟨64HC(19-2)729⟩.

$$\text{(2-oxetanone)} + \text{MeOH} \xrightarrow{OH^-} HOCH_2CH_2CO_2Me \quad ; \quad \xrightarrow{\text{1 eq. NaOMe}} MeOCH_2CH_2CO_2Na \tag{48}$$

$$\text{(2-oxetanone)} + R_2NH \rightarrow HOCH_2CH_2CONR_2 \quad ; \quad \rightarrow R_2NCH_2CH_2CO_2H \tag{49}$$

Reaction with ammonia and primary and secondary amines can also give two types of products, β-hydroxyamides or amino acids (equation 49). The amide is obtained from the reaction of β-propiolactone with ammonia in water, while the amino acid is obtained from the reaction in acetonitrile, both in good yield. β-Lactones react very generally with both aliphatic and aromatic amines, but the type of product does not correlate with polarity of solvent or basicity of amine. Fortunately, conditions can usually be found for the formation of the desired product.

The reaction of several α-amino acid esters with β-lactones in chloroform solution gave hydroxyamide esters (equation 50), said to be useful fungicides ⟨74JAP(K)74127918⟩. Enamines derived from cyclohexanone react smoothly with β-propiolactone to give 3-(2-cyclohexanonyl)propionamides in reasonable yields (equation 51). No acylation of the enamine is observed. This reaction has been used as a key step in a new synthesis of 8-aza steroids ⟨75JOC50⟩. Cyclohexanone imines react in the same manner, except that the expected initial product cyclizes to give bicyclic lactams and enaminones (equation 52) ⟨80T3047⟩.

$$\text{(4-}C_{11}H_{23}\text{-2-oxetanone)} + RCH(NH_2)CO_2Et \rightarrow C_{11}H_{23}CH(OH)CH_2CONHCHRCO_2Et \tag{50}$$

$$\text{(1-}NR_2\text{-cyclohexene)} + \text{(2-oxetanone)} \rightarrow \left[\text{(2-}(CH_2)_2CO_2^-\text{-cyclohexylidene-}\overset{+}{N}R_2)\right] \rightarrow \text{(2-}(CH_2)_2CONR_2\text{-cyclohexanone)} \tag{51}$$

$$\text{(cyclohexylidene-NR)} + \text{(2-oxetanone)} \rightarrow \left[\text{(2-}(CH_2)_2CO_2^-\text{-cyclohexylidene-}\overset{+}{N}HR)\right] \rightarrow \text{(bicyclic lactam, N-R)} + \text{(bicyclic enaminone, N-R)} \tag{52}$$

The enolate ions of acetoacetic ester and other active methylene compounds react with β-propiolactone to give the ethoxycarbonyl derivative, but the yields are generally not high. Application of this reaction to 2-ethoxycarbonyldodecanone (equation 53) has been recently patented, with the product reported to be a useful perfume intermediate ⟨77JAP(K)77133952⟩. The reaction is used quite widely with diketene, which gives *C*-acylation rather than alkylation of the enolate ion, followed by cyclization ⟨72CPB1574⟩.

Thiols and thiophenols generally react readily with β-lactones at C-4 to give excellent yields of 3-alkylmercapto- or 3-arylmercapto-propionic acid (equation 54). For example,

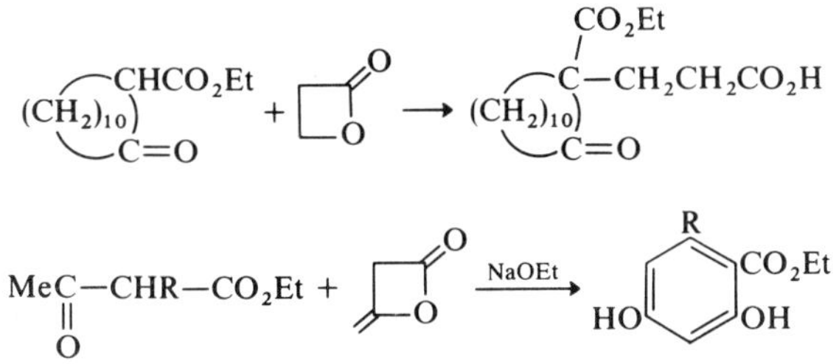

$$(CH_2)_{10}\langle CHCO_2Et, C{=}O\rangle + \beta\text{-propiolactone} \longrightarrow (CH_2)_{10}\langle C(CO_2Et){-}CH_2CH_2CO_2H, C{=}O\rangle$$

$$MeC(=O){-}CHR{-}CO_2Et + \text{diketene} \xrightarrow{NaOEt} \text{(R)(}CO_2Et\text{)(OH)(HO)-benzene} \qquad (53)$$

addition of 3,3-dimethyl-2-oxetanone (pivalolactone) to an alcoholic solution of potassium thiophenoxide gave a 90% yield of 3-phenylmercaptopivalic acid ⟨73JOC1361⟩. Similar yields of these products can also be obtained by aluminum chloride catalyzed addition of thiols to β-propiolactone ⟨81JOC5163⟩. Analogous reactions are observed with selenium compounds, *e.g.* potassium selenocyanate and sodium toluene-α-selenate (equation 55) ⟨67ACS837⟩. Monoalkylthioureas also react analogously, except that cyclization occurs to give 1,3-thiazine derivatives (equation 56) ⟨77AP273⟩.

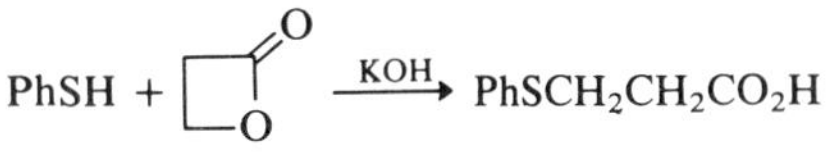

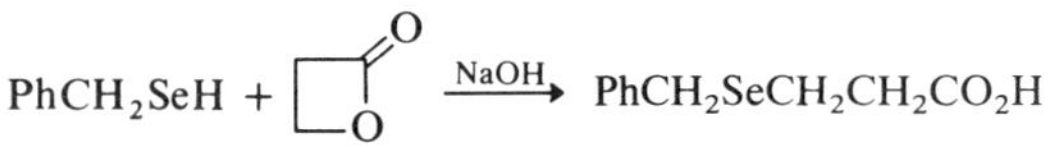

$$PhSH + \beta\text{-propiolactone} \xrightarrow{KOH} PhSCH_2CH_2CO_2H \qquad (54)$$

$$PhCH_2SeH + \beta\text{-propiolactone} \xrightarrow{NaOH} PhCH_2SeCH_2CH_2CO_2H \qquad (55)$$

$$RNHCSNH_2 + \beta\text{-propiolactone} \longrightarrow \text{3-R-2-imino-1,3-thiazin-4-one} + \text{2-(RN=)-1,3-thiazin-4-one} \qquad (56)$$

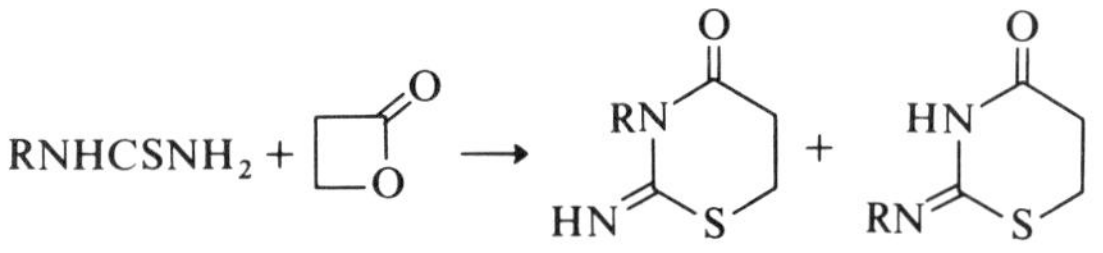

β-Propiolactone reacts with the thiol glutathione in aqueous solution to give both thioester and thioether in about equal quantities, whereas β-butyrolactone reacts with this thiol to give only thioether. The greater tendency of β-butyrolactone to react at the carbonyl carbon may be related to its being highly carcinogenic, whereas β-propiolactone is not ⟨75MI51301⟩. The acute toxicity of β-propiolactone to most, if not all, living organisms, including microorganisms, has led to its wide use as a chemical sterilant ⟨70MI51302⟩.

As can be seen in some of the above examples, the reactions of nucleophiles with β-lactones are subject to modifications depending on the structure of the lactone. A good example can be seen in reactions of 3,3-dichloro-2-oxetanones, which react with various nucleophiles only by acylation (acyl–oxygen cleavage) (equation 57) ⟨73JHC239⟩.

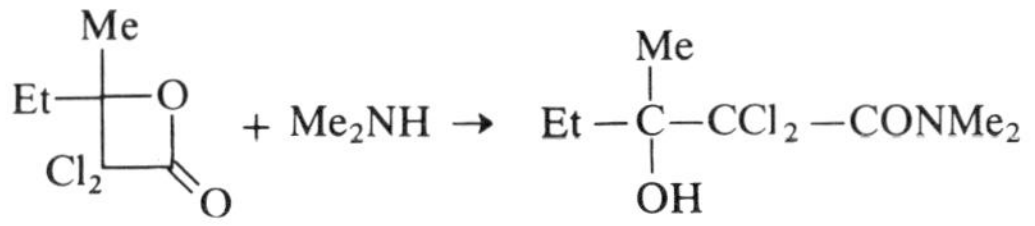

$$\text{4-Et-4-Me-3,3-}Cl_2\text{-2-oxetanone} + Me_2NH \rightarrow Et{-}C(Me)(OH){-}CCl_2{-}CONMe_2 \qquad (57)$$

An outstanding example of the effect of substitution is to be found in the reactions of diketene, where alkyl–oxygen cleavage is repressed by the 4-methylene group, but a number of new reactions become possible because of carbanionoid character of this methylene group, as well as other changes. The reader is referred to an excellent recent review of the subject ⟨74ACR265⟩.

The reaction of a very strong amide to give carbanion formation, rather than nucleophilic attack, was recently reported. Lithium diisopropylamide in THF at −78 °C causes deprotonation of 3,4-disubstituted 2-oxetanones to give a surprisingly stable carbanion derivative (**47**). These are useful synthetic intermediates, since they react readily with alkyl and acyl halides to give good yields of the corresponding alkyl or acyl derivatives. The observed stability of (**47**) has been attributed to the unshared electron pair on C-3 being in a *p*-orbital which overlaps with the carbonyl π-orbital and is orthogonal to the orbitals comprising the bonds in the ring skeleton. Thus elimination involving cleavage of a ring bond does not occur as readily as might otherwise have been expected, although this does occur on warming to room temperature, giving the acrylic acid anion (equation 58) ⟨80JA3620⟩.

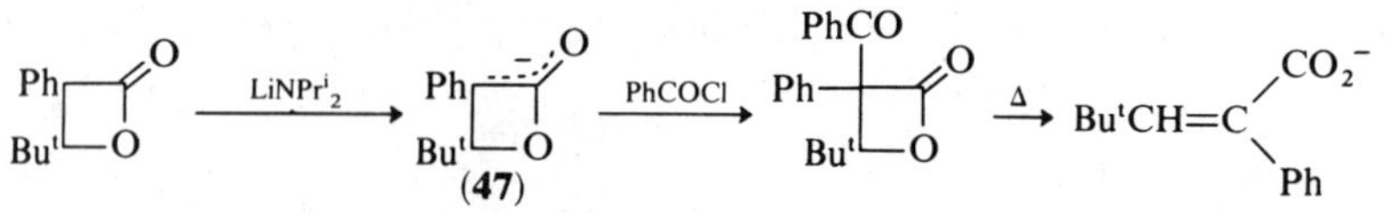

$$\text{(58)}$$

The reaction of Grignard reagents and organolithium compounds with β-propiolactone is rather complex, yielding a mixture of diols, ketones, acids and polymer. Recently several groups have found that conversion of the Grignard reagent to a cuprate, by addition of copper(I) bromide or iodide, makes this reaction a useful, high-yield synthesis of acids. This gives three-carbon homologation from the organic halide used to make the Grignard reagent. The reaction can be carried out with either catalytic amounts of copper(I) bromide in THF–dimethyl sulfide solvent, or with full conversion to the cuprate by use of a full equivalent of copper(I) iodide. The reaction works equally well with a variety of methyl-substituted 2-oxetanones (equation 59) ⟨80TL935, 80TL2181⟩. Extension of this method to diketene is also very successful; here, however, $NiCl_2$ and $PdCl_2$ catalysis is more satisfactory than CuI catalysis. 3-(Trimethylsilylmethyl)-3-butenoic acid prepared in this way (equation 60) has been found to be a valuable synthon for synthesis of unsaturated acids and δ-lactones ⟨77CC500, 77CL103⟩.

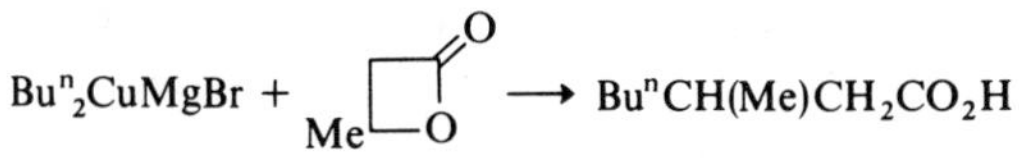

$$Bu^n_2CuMgBr + \text{[4-methyl-2-oxetanone]} \longrightarrow Bu^nCH(Me)CH_2CO_2H \quad (59)$$

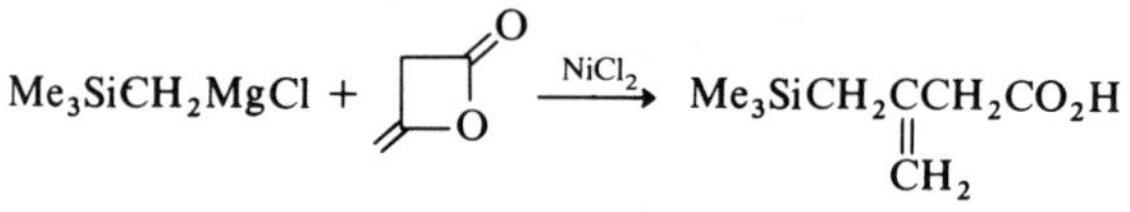

$$Me_3SiCH_2MgCl + \text{[diketene]} \xrightarrow{NiCl_2} Me_3SiCH_2C(=CH_2)CH_2CO_2H \quad (60)$$

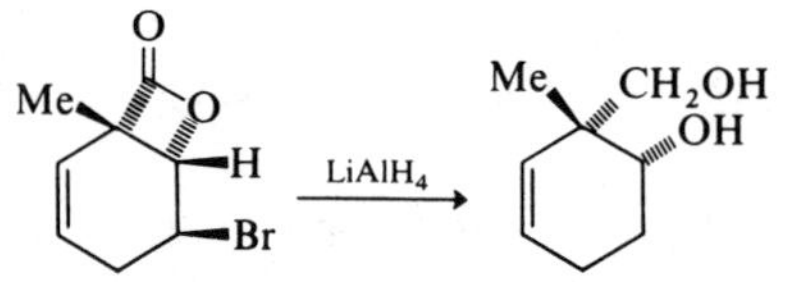

$$\text{(61)}$$

Lithium aluminum hydride in ether readily reduces β-propiolactones to 1,3-diols in fair to good yield. A recent application of this reaction is with the γ-bromo-β-lactone in equation (61), in which reductive cleavage of the bromine atom occurred simultaneously ⟨78CC961⟩.

5.13.3.4.3 Anionic polymerization of β-lactones

Salts of carboxylic acids react with β-lactones to give β-(alkylcarboxy)propionate anion, which can react subsequently with an additional molecule of lactone (equation 62). This is the basis for the anionic polymerization of β-lactones, which has been developed to give highly crystalline high polymers of high molecular weight and high melting points.

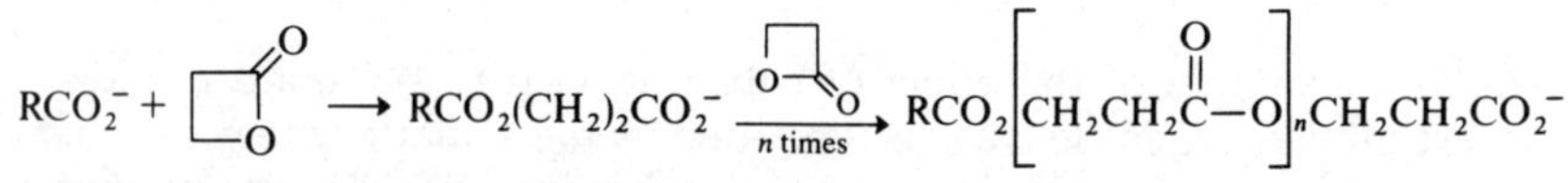

$$RCO_2^- + \text{[2-oxetanone]} \longrightarrow RCO_2(CH_2)_2CO_2^- \xrightarrow[n\ \text{times}]{\text{[2-oxetanone]}} RCO_2\left[CH_2CH_2\overset{O}{\overset{\|}{C}}{-}O\right]_nCH_2CH_2CO_2^- \quad (62)$$

While high polymers of β-lactones can also be formed by cationic polymerization, most of the commercial production seems to be by the anionic route. Carboxylate salts such as sodium acetate or benzoate are commonly the initiators, but other nucleophiles, such as triethylamine, betaine, potassium *t*-butoxide, aluminum and zinc alkoxides, various metal oxides and tris(dimethylamino)benzylphosphonium chloride (the anion of which is the initiator), are of value. Addition of crown ethers to complex the counter cation increases the rate of reaction. When the reaction is carried out in inert but somewhat polar organic solvents, such as THF or ethyl acetate, or without solvent, chain propagation is very fast and proceeds without transfer reactions.

Anionically formed polymers of β-lactones have helical macrostructures, in which the ester groups are approximately in the planar configuration. Copolymers of different β-lactones have the same basic structure but show variations of other properties ⟨73AG(E)432, 77MI51301, 78MI51302, 74CJC3742⟩.

5.13.3.5 Reactions with Free Radicals, Carbenes and Nitrenes, and at Surfaces

Oxetane is readily photochlorinated by *t*-butyl hypochlorite, forming 2-chloroxetane (equation 63). This product, an α-chloroether and very labile in moist air, was not isolated but was hydrolyzed to 3-hydroxypropanol. The rate of photochlorination was significantly greater than for other unsubstituted cyclic and acyclic ethers, probably due to better stabilization of the intermediate free radical by the adjacent oxygen atom ⟨67JA1515⟩. The photoaddition of dimethyl acetylenedicarboxylate and dimethyl azodicarboxylate to cyclic ethers is also a free radical chain reaction and oxetane again shows a high degree of reactivity. Both geometric isomers of the acetylenedicarboxylate adduct with oxetane were obtained in a total yield of 80% (equation 64) ⟨74TL2779⟩.

$$\text{oxetane} + Me_3COCl \xrightarrow{h\nu} \text{2-chloroxetane} + Me_3COH \qquad (63)$$

$$\text{oxetane} + MeO_2C{-}C{\equiv}C{-}CO_2Me \longrightarrow (E)\text{-adduct} + (Z)\text{-adduct} \quad (R = CO_2Me) \qquad (64)$$

The reaction of ethoxycarbonylcarbene with phenyloxetane gives an excellent yield of 2-ethoxycarbonyl-3-phenyltetrahydrofuran, showing insertion between the oxygen atom and C-2 had occurred. The first step is believed to be attack of the carbene on the oxygen atom ⟨66T3393⟩. In contrast, ethoxycarbonylnitrene inserts into an α-CH bond of oxetane, as with other cyclic ethers ⟨76BCJ2572⟩.

$$\text{Ph-oxetane} + :CHCO_2Et \longrightarrow [\text{Ph-oxetane-}O^+{-}\bar{C}HCO_2Et] \longrightarrow \text{EtO}_2\text{C-(Ph)tetrahydrofuran} \qquad (65)$$

$$\text{oxetane} + :C{=}C{=}O \longrightarrow [O{=}C{=}\bar{C}{-}\overset{+}{O}(\text{oxetane})] \longrightarrow 2CO + \text{cyclopropane} + \text{propene}$$
$$\downarrow$$
$$O{=}C{=}(\text{tetrahydrofuran-2-ylidene}) \xrightarrow{h\nu} \text{:(tetrahydrofuran-2-ylidene)} \longrightarrow CH_2{=}C{=}O + CH_2{=}CH_2 \qquad (66)$$

Carbonylcarbene reacts with oxetane in a complex manner, giving cyclopropane and propene by a deoxygenation process and ketene plus ethylene as apparent insertion–fragmentation products (equation 66). The reaction is carried out by irradiation of a solution of carbon suboxide in oxetane; under these conditions, photodecomposition of intermediates is possible. The initial attack is believed to be at the ring oxygen atom to give a zwitterionic species ⟨78JA6425⟩.

Methylene insertion in the ring of the perfluorodiketene (**48**) by means of diazomethane has been observed between the ring oxygen atom and C-4 ⟨73IZV2562⟩. Benzoylnitrene insertion in diketene, on the other hand, has been found to take place between the carbonyl group and C-3 (equation 68) ⟨78CL697⟩.

$$(\mathbf{48}) + CH_2N_2 \longrightarrow \text{(F}_3\text{C)}_2\text{C=, (CF}_3)_2\text{-}\gamma\text{-lactone} \qquad (67)$$

$$\text{diketene} + PhC(=O){-}N: \longrightarrow \text{HO-pyrrolinone-}N\text{-COPh} \qquad (68)$$

$$\text{2,2,4,4-}Me_4\text{-oxetane} \xrightarrow[\Delta]{Pt} MeC(=O)CH_2CMe_3 \qquad (69)$$

Isomerization of 2,2,4,4-tetramethyloxetane on platinum, palladium and rhodium catalysts at 100 °C to 4,4-dimethyl-2-pentanone has been observed (equation 69). A mechanism involving noble metal cleavage of C—O and a 1,3-methyl shift has been proposed ⟨79CC139⟩.

5.13.3.6 Reactivity of Substituents

As seen in the preceding sections, 2-chloro and 2-alkoxy substituents are very labile, due to the effect of the adjacent ether oxygen atom and the small ring. 2-Oxetanol has no significant existence, the tautomeric 3-hydroxypropanal being much more stable.

Substituents at C-3 are much more normal, though the general susceptibility of oxetanes for ring opening under acid conditions precludes study of their reactivity in acidic media. 3-Oxetanol displays normal reactivity in base-catalyzed ester interchange and in reaction with acid halides under basic conditions, but under acidic conditions the ring is too labile for reactions with hydrogen halides and for acid-catalyzed reactions of the hydroxyl group to be observed.

3-Oxetanol can be oxidized to 3-oxetanone in fair yield by means of the chromic oxide–pyridine complex. This ketone reacts with base in an unknown manner and polymerizes on standing.

3-Bromo-, 3-chloro-, 3-fluoro- and 3-iodo-oxetanes have all been prepared in good yield by the reaction of 3-oxetanyl tosylate with alkali metal halides in hot triethylene glycol (equation 70). Substitution reactions of the halogen atom have not been reported, except for the reaction of 3-iodooxetane with diethylamine. A low yield of 2-diethylaminooxetane was obtained from this reaction at 200 °C, but its chemical properties are not known ⟨73JOC2061⟩.

H O—CH=CHMe; $\xrightarrow[\text{MeOH}]{H^+}$ H OH; $\xrightarrow{CrO_3}$ O; TsCl, py; H OTs $\xrightarrow[\Delta]{\text{MHal}}$ H Hal (70)

(Hal = F, Cl, Br, I)

A halogen substituent on a 3-methyl group has the properties of a normal, primary halogen atom. 3,3-Bis(chloromethyl)oxetane gives spiro substitution products when treated with diamines, diols and dithiols in the presence of base (equation 71). Reactions with long-chain diols and diamines gave macrocyclic polyethers and azapolyethers which could have interesting applications for forming cryptate complexes ⟨74JOC2351, 75JOC1205⟩.

$+ HOCH_2CH_2O(CH_2CH_2)_nOH \xrightarrow{OH^-}$ (71)

5.13.4 SYNTHESIS OF OXETANES AND 2-OXETANONES

5.13.4.1 Ring Synthesis by Formation of One Bond

5.13.4.1.1 Oxetanes

The intramolecular Williamson reaction has been widely applied to the synthesis of oxetanes and is one of the two most commonly used methods. Starting materials are alcohols having a good leaving group at the γ-position, such as bromine, chlorine or tosylate, and cyclization is achieved by treatment with base, usually with heat. An S_N2 displacement of the leaving group by the alkoxide occurs with inversion of configuration at the carbon bearing the leaving group. In some cases, esters of the alcohol, from which the same alkoxide

$$(ClCH_2)_3CCH_2OH \xrightarrow[\Delta]{OH^-} \left[(ClCH_2)_2C\begin{matrix}CH_2O^-\\CH_2Cl\end{matrix}\right] \rightarrow (ClCH_2)_2\langle\rangle O \quad (72)$$

is generated *in situ* by hydrolysis, give better yields of oxetanes. A commercial synthesis of BCMO is illustrative (equation 72) ⟨64HC(19-2)983⟩.

A common side reaction is a fragmentation process of the intermediate alkoxide ion, which results in C—C bond cleavage as well as loss of the leaving group with simultaneous formation of alkene and carbonyl double bonds (equation 73). When the leaving group is on a tertiary carbon atom, its enhanced tendency for S_N1 ionization ensures that this fragmentation process will be dominant with little or no oxetane formed. For oxetane synthesis, therefore, the leaving group generally should be on a primary or secondary carbon atom, but the carbinol group can be primary, secondary or tertiary ⟨78HCA2336⟩.

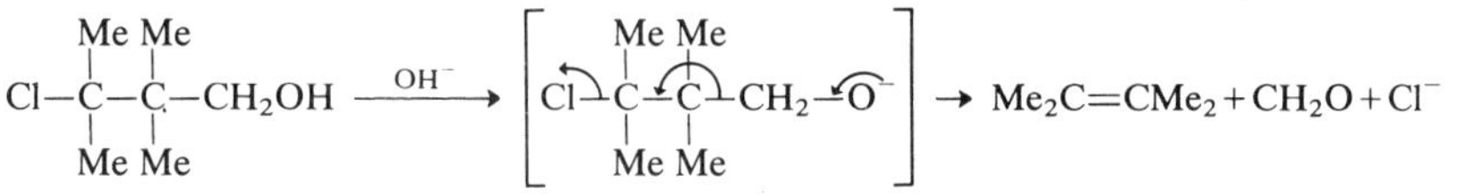

$$Cl-CMe_2-CMe_2-CH_2OH \xrightarrow{OH^-} [Cl-CMe_2-CMe_2-CH_2-O^-] \rightarrow Me_2C{=}CMe_2 + CH_2O + Cl^- \quad (73)$$

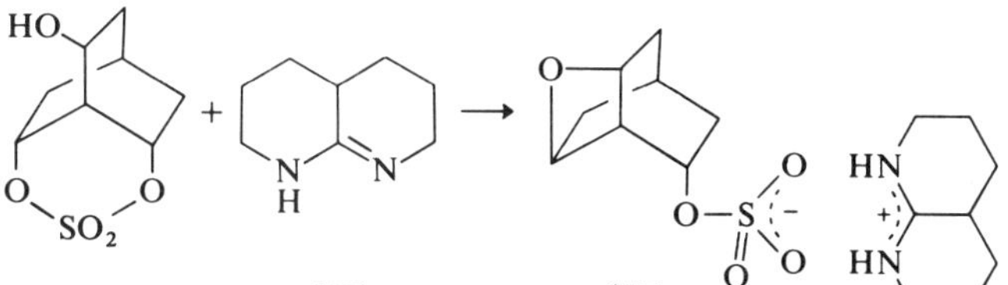

(74)

(49) (50)

There has been considerable variation in the choice of base used. In most cases, hot, concentrated potassium hydroxide gives good results, but sometimes potassium hydroxide in aqueous DMSO, potassium *t*-butoxide in DMSO or NaH in DMSO is more satisfactory ⟨80TL585⟩. The tercyclic oxetane (**50**), however, was not formed from the hydroxy-cyclic sulfate precursor under any of these conditions, but use of the amidine base (**49**) at room temperature produced it in 46% yield ⟨81JOC3144⟩.

DBU in methylene chloride has been used to give a good yield of 3-fluoro-3-nitrooxetane from the monotriflate of the corresponding diol (equation 75) ⟨80JOC4853⟩. On the other hand, no base (other than possibly water) was needed for oxetane formation from the carbohydrate tosylate (**51**) during processing. The rigidity of the ring system and the favorable *trans* configuration of the tosyloxy and reacting hydroxyl groups is probably responsible for the unusual case of cyclization in this case ⟨78TL49⟩.

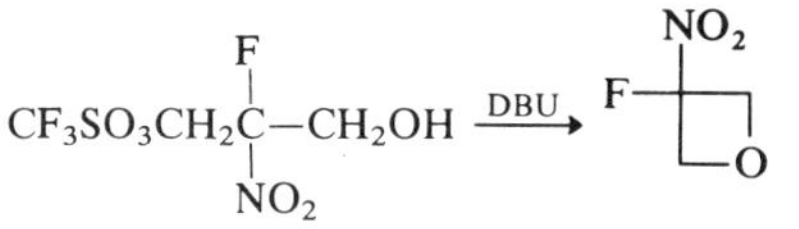

$$CF_3SO_3CH_2C(F)(NO_2)-CH_2OH \xrightarrow{DBU} \text{3-fluoro-3-nitrooxetane} \quad (75)$$

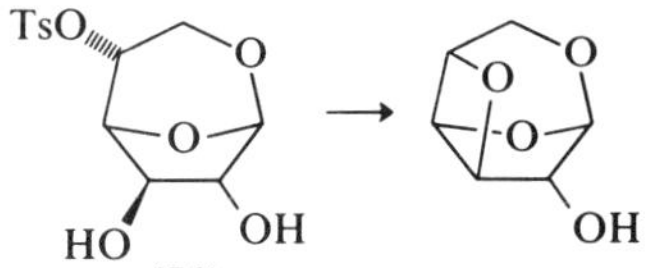

(76)

(51)

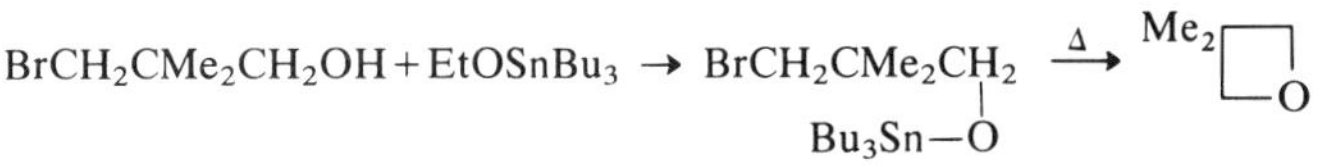

$$BrCH_2CMe_2CH_2OH + EtOSnBu_3 \rightarrow BrCH_2CMe_2CH_2-O-SnBu_3 \xrightarrow{\Delta} \text{3,3-dimethyloxetane} \quad (77)$$

An alternative to the use of base for effecting ring closure is reaction with ethoxytributyltin, followed by thermal decomposition of the intermediate at 200 °C (equation 77) ⟨73JOM(47)337⟩.

Precursor γ-halogeno alcohols are frequently prepared by the classic sequence of addition of hydrogen halide to α,β-unsaturated aldehydes, ketones, acids or esters, followed by Grignard reaction or hydride reduction. Recently a novel and general synthesis of 3-methoxyoxetanes from 3-phenylseleno-2-propenal was reported. This method comprises a sequence of Grignard addition to the aldehyde function, treatment with two equivalents of MCPBA, and then reaction with methanolic sodium hydroxide (equation 78) ⟨80JOC4063⟩.

A number of other useful modifications of the intramolecular Williamson synthesis have been developed. Reaction of α,α-dialkyl-β-tosyloxy aldehydes and ketones with potassium cyanide or with sodium alkoxide gives the corresponding 2-cyano- or 2-alkoxy-oxetanes (equation 79) ⟨71CB3607⟩. If the carbonyl group is enolizable in the direction away from the displaceable group, treatment with potassium hydride in THF gives the 2-alkylidene-oxetane as the dominant product (equation 80) ⟨75JOC2692⟩.

$$\text{PhSeCH=CHCHO} + \text{RMgX} \rightarrow \text{PhSeCH=CHCHROH} \xrightarrow{\text{2MCPBA}} \xrightarrow[\text{MeOH}]{\text{NaOH}} \left[\text{PhSeO}_2\overset{-}{\text{C}}\text{H}-\underset{}{\overset{\text{OMe}}{\text{CH}}}-\text{CHROH}\right] \rightarrow \text{MeO/R-oxetane} \quad (78)$$

$$\text{TsOCH}_2\text{CR}_2\text{CHO} \xrightarrow{\text{KCN}} \text{R}_2\text{-oxetane-CN}; \quad \xrightarrow{\text{NaOR}'} \text{R}_2\text{-oxetane-OR}' \quad (79)$$

$$\text{TsOCH}_2\text{CMe}_2\underset{\text{O}}{\underset{\|}{\text{C}}}-\text{CH}_2\text{Ph} \xrightarrow{\text{KH}} \text{Me}_2\text{-oxetane=CHPh} \quad (80)$$

In a new method of ring enlargement of oxiranes to oxetanes, the intramolecular Williamson reaction is the essential last step. Reaction of an oxirane with selenomethyllithium reagents, followed by bromination, gives γ-bromo alcohols, which can be cyclized by base to oxetanes. Since the oxiranes can be prepared readily from ketones and the selenomethyllithium reagent, this is also a synthesis of oxetanes from ketones ⟨80TL585⟩.

$$\text{PhSeCH}_2\text{Li} + \text{CH}_2\text{—CR}_2\text{(O)} \xrightarrow[\text{HMPT}]{\text{THF}} \text{PhSeCH}_2\text{CH}_2\underset{\text{OH}}{\text{CR}_2} \xrightarrow{\text{Br}_2} \text{BrCH}_2\text{CH}_2\underset{\text{OH}}{\text{CR}_2} \xrightarrow{\text{Bu}^t\text{OK}} \text{R}_2\text{-oxetane} \quad (81)$$

Since it is well known that an epoxy oxygen atom can serve as a good leaving group, the conversion of 3,4-epoxy alcohols has received much investigation. The very promising synthetic usefulness of this approach is illustrated by the examples in equation (82), the last of which is a crucial step in a synthesis of thromboxane A_2 analogs ⟨80BCJ2895, 80TL3051, 80TL3061⟩.

$$\text{1-(oxiranylmethyl)cyclohexanol (OH)} \xrightarrow[\text{DMSO}]{\text{KOH}} \text{spiro-oxetane-CH}_2\text{OH} \; (49\%)$$

$$\text{Me-epoxide-CH}_2\text{CMe}_2\text{OH} \xrightarrow{\text{Bu}_3\text{SnOMe}} \xrightarrow{\text{RCO}_2\text{H}} \text{Me}_2\text{-oxetane-}\underset{\text{OH}}{\text{MeCH}} \; (100\%) \quad (82)$$

$$\text{epoxy bicyclic alcohol (OCH}_2\text{Ph, OH)} \rightarrow \text{HO, OCH}_2\text{Ph oxetane-bridged bicycle} \; (85\%)$$

A rapid one-pot method for converting 1,3-diols into oxetanes by the intramolecular Williamson reaction has recently been described. The monolithium salt is generated by treatment of the diol with one equivalent of butyllithium in cold THF, followed by addition of one equivalent of tosyl chloride to give a monotosylate, which is cyclized by addition of a second equivalent of butyllithium (equation 83). Yields of 70–90% are reported for a variety of alkyl- and aryl-substituted oxetanes ⟨81S550⟩. Another simple method for converting 1,3-diols into oxetanes consists of converting them to cyclic carbonate esters by ester interchange with ethyl carbonate, followed by thermal decomposition of the resulting cyclic carbonate esters (equation 84). The latter gives 60–80% yields of various 3,3-dialkyl-oxetanes, when KCN or K_2CO_3 is used as a catalyst for the thermolysis. There is a yield-limiting tendency for side reactions (particularly fragmentation to alkene and aldehyde or ketone), which have received recent study ⟨72BSF4655⟩. Direct dehydration of 1,3-diols to oxetanes has been achieved by passing the diols over pumice at 350 °C, but complex mixtures of products were obtained with relatively low yields of oxetanes ⟨74BSF595⟩.

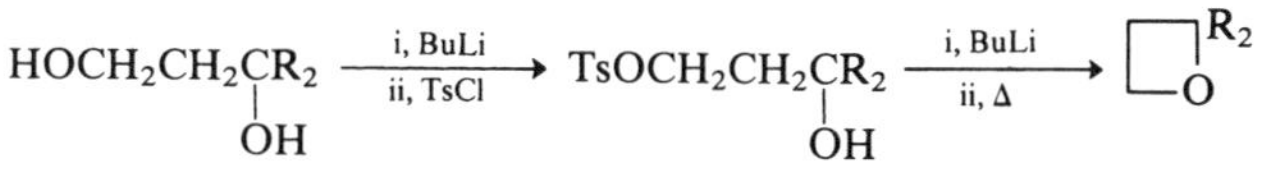

$$HOCH_2CH_2CR_2(OH) \xrightarrow[\text{ii, TsCl}]{\text{i, BuLi}} TsOCH_2CH_2CR_2(OH) \xrightarrow[\text{ii, }\Delta]{\text{i, BuLi}} \text{2,2-}R_2\text{-oxetane} \qquad (83)$$

$$R_2C(CH_2OH)_2 \xrightarrow{(EtO)_2CO} R_2C(CH_2O)_2C{=}O \xrightarrow[K_2CO_3]{\Delta} \text{3,3-}R_2\text{-oxetane} \qquad (84)$$

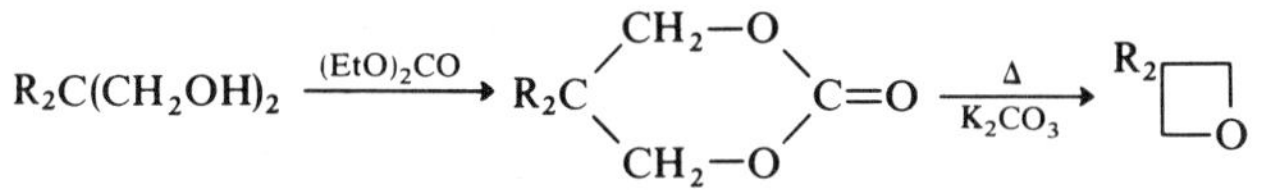

A very promising new method for converting oxiranes, as well as ketones, into oxetanes has recently been reported. This method uses the carbanion of dimethyl(*N*-tosyl)sulfoximine and gave good yields in the several cases reported. When this reagent is employed with ketones, oxirane formation is presumably an intermediate stage, but the oxirane is not isolated. The method thus provides an excellent synthesis of spiro-oxetanes from ketones, as the example with camphor in equation (85) shows ⟨79JA6135⟩.

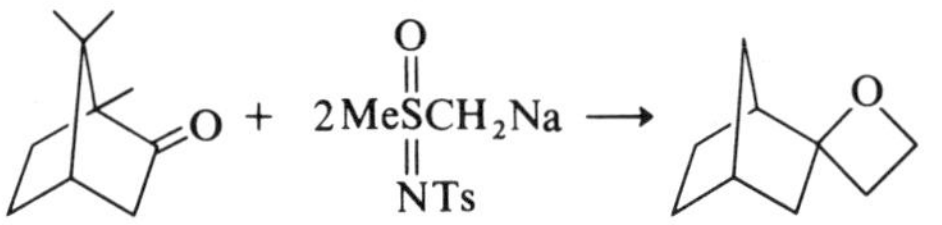

$$\text{camphor} + 2MeS(O)(NTs)CH_2Na \longrightarrow \text{spiro-oxetane} \qquad (85)$$

The above synthetic methods for oxetane all involve formation of a new C—O bond. Cyclization by formation of a new C—C bond has been applied with compounds having benzylic or alkylic CH groups. Recent examples of this type of ring closure are the rearrangement of *trans*-2,3-epoxycyclohexyl allyl ether by means of *s*-butyllithium and the dehydrochlorination of α-cyanobenzyl 2-chloroethyl ether with aqueous base and phase transfer catalyst (equation 86). Both reactions probably involve carbanion intermediates ⟨76TL2115, 75MIP51300⟩.

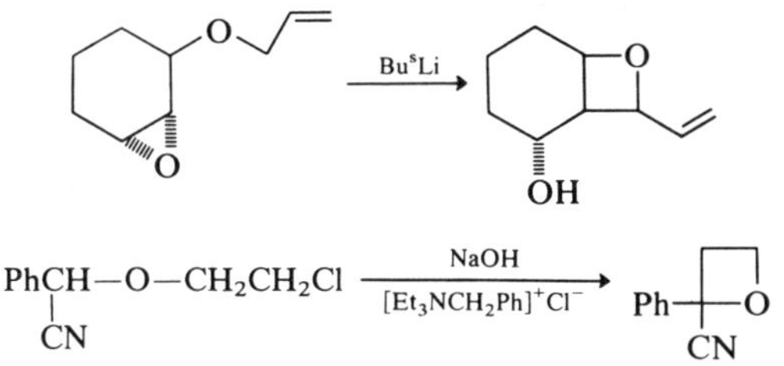

$$\xrightarrow{Bu^sLi} \qquad (86)$$

$$PhCH(CN)-O-CH_2CH_2Cl \xrightarrow[[Et_3NCH_2Ph]^+Cl^-]{NaOH} \text{2-Ph-2-CN-oxetane}$$

There are other reactions which have been found to yield oxetanes but have not been developed as synthetic methods. The thermolysis and photolysis of γ-methyl-γ-peroxy-valerolactone gives 2,2-dimethyloxetane, presumably by way of a diradical intermediate (equation 87). This is the same type of intermediate involved in pyrolysis of the oxetane at a much higher temperature ⟨71CC1299⟩. Somewhat related are photochemical syntheses

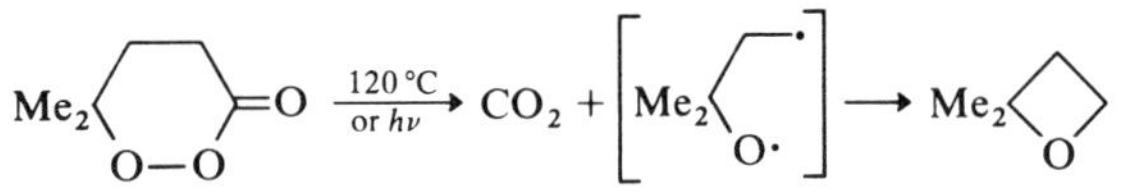

$$\xrightarrow[\text{or } h\nu]{120\,°C} CO_2 + [\text{diradical}] \longrightarrow \text{2,2-}Me_2\text{-oxetane} \qquad (87)$$

$$PhC(O)-CH(OCH_3)CH_2Ph \xrightarrow{h\nu} [Ph-\dot{C}(OH)-CH(CH_2Ph)-O-\dot{C}H_2] \longrightarrow \text{3-OH-3-Ph-2-}PhCH_2\text{-oxetane } (80\%) \qquad (88)$$

$$PhC(O)CH_2O\text{-cyclopropyl} \xrightarrow{h\nu} [Ph-\dot{C}(OH)-CH_2O\text{-}\dot{c}\text{-propyl}] \longrightarrow \text{spiro oxetane, HO, Ph } (78\%)$$

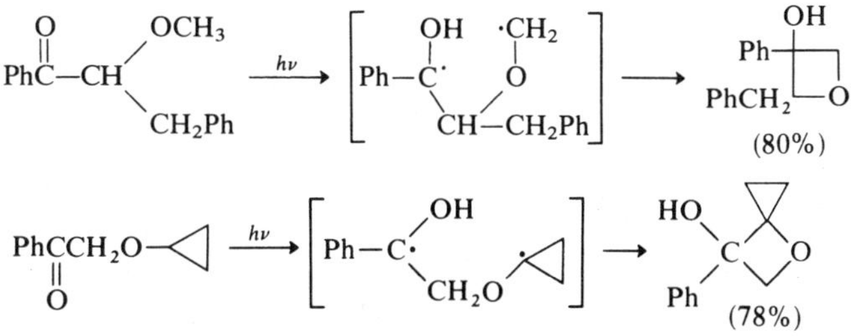

from α-alkoxy ketones which give Norrish type II photocycloadditions. Owing to the simultaneous elimination process, the yields of oxetanes are highly variable and usually are only fair. Two interesting recent cases are shown in equation (88) ⟨79T2501, 70MI51301⟩.

Appropriately substituted diazo ketones have been converted into oxetanes in two instances by Wolff rearrangement processes. The structure of compound (**52**) was established by X-ray crystallography ⟨69MI51300, 81CSC345⟩. Reaction of 4,4-dibromo-2,2,5,5-tetramethyltetrahydro-3-furanone with aqueous base is a good method of preparation for 3-hydroxy-2,2,4,4-tetramethyloxetane-3-carboxylic acid (equation 90) ⟨66JA1242⟩.

N_2, R_2, R_2, O, O + H_2O $\xrightarrow{h\nu}$ H, CO_2H, R_2, R_2, O

Ph, PhCH—C—C—CHN_2, Cl HO O $\xrightarrow{BF_3 \cdot OEt_2}$ O, Ph, PhCHCl, O (**52**) (89)

Br_2, O, Me_2, Me_2, O $\xrightarrow{OH^-}$ HO, CO_2H, Me_2, Me_2, O (90)

5.13.4.1.2 2-Oxetanones

2-Oxetanones can be generally prepared by displacement processes on various β-substituted carboxylic acids or by halolactonization of β,γ-unsaturated acids. A very general and reliable method consists of treatment of a β-hydroxy acid with benzenesulfonyl chloride and pyridine at 0 °C (equation 91). The yields of β-lactones are usually in excess of 80% ⟨79JOC356, 74JOC1322⟩. An alternative method involves cyclization of the benzenethiol ester of a β-hydroxy carboxylic acid by means of mercury(II) methanesulfonate in acetonitrile (equation 92). The yields were excellent in the two cases reported ⟨76JA7874⟩.

$R_2C—CR_2CO_2H$, OH $\xrightarrow[py]{PhSO_2Cl}$ R_2, R_2, O, O (91)

O, CH_2C—SPh, OH $\xrightarrow{Hg(O_3SMe)_2}$ =O, O (92)

3,3-Dimethyl-2-oxetanone (β-pivalolactone), which is an important monomer for polymerization, can be prepared in good yield by reaction of 3-chloro-2,2-dimethylpropionic acid with various bases, including aqueous sodium hydroxide in a biphasic medium (equation 93) ⟨73JAP7306455⟩. A method of cyclizing 3-acetoxy-2,2-dimethylpropionic acid by passing its vapor in a stream of nitrogen gas over lithium fluoride at 300 °C has been reported ⟨72JAP7219791⟩.

$ClCH_2CMe_2CO_2H$ $\xrightarrow{OH^-}$ Me_2, O, O (93)

$H_2C{=}CHCMe_2CO_2H + I_2 \longrightarrow$ Me_2, ICH_2, O, O (94)

Treatment of β,γ-unsaturated acids with iodine in sodium bicarbonate solution may give γ-iodo-β-lactones in good yield (equation 94), but if potassium iodide is added then β-iodo-γ-lactones are produced. One compound, 4-phenyl-3-butenoic acid, gave the γ-lactone even when KI was not present, presumably due to the phenyl group favoring development of carbonium ion character at C-4 in the intermediate iodine complex. With

1,4-dihydrobenzoic acids, only low yields of the iodolactones were obtained and the method cannot be considered general ⟨72TL1777, 75JOC2843⟩.

Iodolactonization of β,γ-unsaturated acids can also be achieved by the reaction of iodine with the thallium(I) carboxylates in ether solution. This reaction also has the possibility of giving β- or γ-lactones, with the substitution on the double bond having the major influence. Thus thallium 3-butenoate gave the β-lactone, while thallium 4-methyl-3-pentenoate gave the γ-lactone, both in high yields ⟨74JCS(P1)1864⟩.

Bromolactonization of β,γ-unsaturated acids has proven to be a much more satisfactory method of synthesis of β-lactones, giving good yields of stable crystalline γ-bromo-β-lactones, except when the substitution at the γ-carbon atom can favor development of carbonium character there. Thus 1,4-dihydrobenzoic acid and 2-methyl-1,4-dihydrobenzoic acid form β-lactones (equation 95), while 3-methyl-1,4-dihydrobenzoic acid forms the γ-lactone ⟨75JOC2843⟩. The reaction of the sodium salt of α-methylcinnamic acid with bromine in water or methanol also gives β-lactone, but the yield is low ⟨78JOC3131⟩.

H CO$_2$H $\xrightarrow[\text{OH}]{\text{Br}_2}$ (bicyclic β-lactone with H, Br) (95)

$$\text{N}_2\text{C(SiMe}_3\text{)–C(=O)OMe} \rightarrow \text{N}_2 + \text{:C(SiMe}_3\text{)–C(=O)–OMe} \longrightarrow \text{Me}_3\text{Si-oxetan-2-one} \quad (96)$$

3-Trimethylsilyl-2-oxetanone has been isolated in unstated yield following pyrolysis of methyl trimethylsilyldiazoacetate (equation 96). This appears to be an interesting example of carbene insertion in the proximate carbon–hydrogen bond of the methoxyl group ⟨75JA3818⟩.

The most recent method for β-lactone synthesis involves a very novel approach. Phenylethynolate anion (which may be in equilibrium with phenylketene anion) can be generated by the reaction of n-butyllithium or t-butyllithium in THF at −78 °C. This highly nucleophilic species reacts with cyclohexanone to give the spiro-β-lactone (**54**), probably through intermediacy of the lithium alkoxide of the β-hydroxyketene (**53**). The β-lactone (**54**) was isolated in about 50% yield ⟨79LA219, 82JA321⟩.

$$\text{PhC(=O)CH}_2\text{Cl} \xrightarrow{\text{RLi}} \text{PhC(OLi)=CHCl} \xrightarrow{\text{RLi}} \text{PhC}\equiv\text{COLi}$$

PhC≡COLi + cyclohexanone → [LiO–C$_6$H$_{10}$–C(Ph)=C=O] (**53**) → spiro-β-lactone, Ph (**54**) (97)

5.13.4.2 Ring Synthesis by Formation of Two Bonds

5.13.4.2.1 From fragments each contributing two ring atoms

(i) Thermal [2+2] cycloaddition reactions

These occur readily between electron-rich alkenes and electron-poor carbonyl compounds. The first example, reported in 1959 ⟨64HC(19-2)729⟩, was the formation of 4,4-diaryloxetane-2,2-dicarbonitriles by the room temperature reaction of 1,1-diarylethylenes and carbonyl cyanide. Continued investigation of this reaction shows that a telomerization product is also formed, the tetraphenylpentadienedinitrile (**55**) from 1,1-diphenylethylene and carbonyl cyanide. This may be interpreted to indicate that carbon–carbon bond formation may commence somewhat ahead of carbon–oxygen bond formation ⟨75MI51302⟩. This type of reaction also occurs readily when tetraalkoxyethylenes and acyl cyanides are heated. 3,3,4,4-Tetralkoxyoxetane-2-carbonitriles are formed in excellent yields (equation 99) ⟨76JCS(P1)1048⟩.

$$Ph_2\overset{\delta^+}{C}=\overset{\delta^-}{C}H_2 + (NC)_2\overset{\delta^+}{C}=\overset{\delta^-}{O} \rightarrow Ph_2\overset{\delta^+}{C}\cdots C\cdots C(CN)_2(O^{\delta^-}) \rightarrow Ph_2\text{(oxetane)}(CN)_2$$

$$\downarrow Ph_2C=CH_2$$

$$Ph_2\overset{\delta^+}{C}\cdots CH_2\cdots C(Ph)_2-CH_2\cdots C(CN)_2(O^{\delta^-}) \rightarrow Ph_2C=CH-CPh_2-CH=C(CN)_2 \quad \mathbf{(55)} \qquad (98)$$

$$(RO)_2C=C(OR)_2 + R'COCN \longrightarrow \text{3,3,4,4-}(RO)_2(OR)_2\text{-2-}R'\text{-2-CN-oxetane} \qquad (99)$$

It has now been found that further extension of this reaction to ketone acetals and a wide range of aldehydes and ketones is possible when catalytic amounts of zinc chloride are added to the reaction mixture. The catalyzed reaction generally proceeds at room temperature and usually gives good yields of 2,2-alkoxyoxetanes (equation 100). This appears to be the method of choice for such compounds, since it is highly regiospecific, whereas the photochemical addition of aldehydes or ketones to ketene acetals is not ⟨77JOC3128⟩.

$$MeCH=C(OMe)_2 + PhCHO \xrightarrow{ZnCl_2} \text{oxetane: Me, }(OMe)_2\text{, Ph, O} \ (75\%) \qquad (100)$$

$$Me_3C-CH=CH_2 + Cl_3CCHO \xrightarrow{AlCl_3} \text{oxetane: H, Bu}^t\text{, Cl}_3C\text{, O} + \text{oxetane: H, Bu}^t\text{, Cl}_3C\text{, O} \qquad (101)$$

The cycloaddition of chloral with 3,3-dimethyl-1-butene to form a good yield of oxetanes can be effected with aluminum chloride catalysis (equation 101). This reaction, however, is not regioselective ⟨72GEP2133564⟩.

(ii) Photocycloaddition reactions

The photochemical reaction of carbonyl compounds and alkenes to give oxetanes is very well known. It is one of the two most used methods for oxetane synthesis, and due to its synthetic value and theoretical interest it is definitely the most thoroughly studied synthesis of oxetanes. The subject was reviewed by Arnold in 1968, and modern concepts of the mechanism were presented by Turro in 1978 ⟨68MI51300, B-78MI51301⟩. Although this method is unfortunately not completely general, it is capable of giving high yields and unusual structures that would be difficult to synthesize by other methods. The reaction of benzophenone with 2-methyl-2-butene under UV irradiation is representative of the most common type of this reaction, which involves electron-rich alkenes. The oxetane product, obtained in 90% yield, consists of isomers (**56**) and (**57**) in a 9:1 ratio. This illustrates the regioselectivity, which can be predicted for this type of cycloaddition on the basis of the relative stability of the diradical intermediates resulting from carbon–oxygen bond formation, (**58**) and (**59**). The more stable diradical (**58**) corresponds to the major product formed, (**56**).

$$Ph_2CO + CH_2=CMe_2 \xrightarrow{h\nu} \text{oxetane: Me}_2\text{, Ph}_2\text{, O} + \text{oxetane: Me}_2\text{, Ph}_2\text{, O} \qquad (102)$$

(**56**) 90% (**57**) 10%

(**58**) (**59**)

Another characteristic of photocycloaddition to electron-rich alkenes is the loss of any stereochemistry of the starting alkene in the oxetane structure. An example is the formation of practically the same mixture of geometric isomers of 2,2-diphenyl-3,4-dimethyloxetane from benzophenone and either *cis*- or *trans*-2-butene (equation 103). This is understandable on the basis of the diradical intermediate having a sufficiently long lifetime for bond rotations to occur.

$Ph_2CO + MeCH{=}CHMe$ (*cis* or *trans*) $\xrightarrow{h\nu}$ (68%) + (11%) (103)

$Me_2CO + CH_2{=}C(Me){-}CH{=}CH_2 \longrightarrow$ **(60)** + **(61)** (104)

PhCHO + $\xrightarrow{h\nu}$ + (105)

With conjugated dienes, photocycloaddition of carbonyl compounds occurs at one of the double bonds to give vinyloxetanes. An interesting example is the reaction of acetone with 2-methyl-1,3-butadiene, which gave the two oxetanes (**60**) and (**61**) in a ratio of 3 : 1 and a total yield of about 20% ⟨72JA8761⟩. Other alkenes which have been used for photosynthesis of oxetanes include enol ethers, ketene acetals, enamines, allenes and diketene, with the reaction of the last compound with benzaldehyde illustrated in equation (105) ⟨75CPB365⟩.

The double bonds in certain heterocyclic compounds, such as furans, *N*-acylpyrroles and *N*-acylindoles are also susceptible to photoaddition of carbonyl compounds to form oxetanes (equation 106) ⟨77JHC1777⟩. A wide range of carbonyl compounds can be used, including quinones, α-diketones, acyl cyanides, perfluorinated aldehydes and ketones and esters. A remarkable case of asymmetric induction in oxetane formation has been reported from optically active menthyl phenylglyoxylate and 2,3-dimethyl-2-butene; the oxetane product obtained after hydrolysis of the ester group had an optical purity of 53% ⟨79AG(E)868⟩.

Ph_2CO + (N-COMe imidazole) $\xrightarrow{h\nu}$ (50%) (106)

Irradiation of β- and γ-ketoalkenes may produce some highly strained and unusual polycyclic oxetanes. For example, from 7,8-dihydro-β-ionone (**62**) at 25 °C in pentane solution there was obtained a 70% yield of (**63**), while in glyceryl triacetate solution at 60 °C a 10% yield of (**64**) was obtained (as well as other products, of course). This is an example of the possibilities for rather complex cycloadditions to occur, in the course of generating an oxetane structure ⟨80HCA154⟩.

(62) $\longrightarrow$ **(63)** + **(64)** (107)

$Me_2CO + CH_2{=}C(Me)(CN) \longrightarrow$ (108)

Photocycloaddition of carbonyl compounds with electron-poor alkenes, such as halogeno- and cyano-alkenes, is generally highly regiospecific and highly stereospecific. The regioselectivity is generally predictable on the basis of consideration of the relative stability of possible diradical intermediates formed by initial carbon–carbon bond formation, as can be seen in the case of the reaction of acetone with 2-methylacrylonitrile (equation 108) ⟨72JA1951⟩. There is an attractive alternative explanation that the stereochemistry is determined by dipole orientation of the reactants in an exciplex intermediate ⟨77JA3023⟩.

The photocycloaddition reaction does not always give good yields. Other photochemical processes compete, such as dimerization of the alkene reactant, photoenolization of the ketone, and hydrogen abstraction processes. Cyclohexene is prone to give low yields of oxetanes, because of hydrogen abstraction, resulting in radical coupling. In addition, steric blocking of the allowed approach of the reactants to each other can prevent reaction.

(*iii*) *2-Oxetanones*

Carbonyl cyanide reacts readily with ketene and dialkylketenes to give the corresponding dicyano-β-lactones (equation 109). This reaction seems entirely analogous to the addition of carbonyl cyanide with alkenes to give oxetanes, described in the preceding section ⟨75MI51302⟩.

$$(NC)_2CO + R_2C{=}C{=}O \longrightarrow \text{3,3-}R_2\text{-4,4-}(NC)_2\text{-oxetan-2-one} \qquad (109)$$

The most widely used synthesis of β-lactones is the acid-catalyzed addition of aldehydes with ketenes. This reaction was reviewed in 1963 and 1964 ⟨63HOU(6/2)511, 64HC(19-2)729⟩. Patents have continued to be issued on refinements of catalysts and conditions for this commercially important reaction. It is very general with regard to the ketene reactant, with ketene itself, dialkyl ketenes, diphenylketene, dichloroketene and silylketenes all reacting successfully with formaldehyde and aliphatic aldehydes. The usual catalyst for laboratory synthesis has been zinc chloride, but boron trifluoride finds increasing use (equation 110). With this catalyst, excellent yields of 4-aryl-2-oxetanones have been obtained from ketene and aromatic aldehydes, which had previously been found not to react satisfactorily. The reaction usually does not proceed well with ketones, but hexafluoroacetone reacted with diphenylketene in the presence of potassium fluoride as catalyst ⟨79JOC733, 71ZOR2295, 76MI51300⟩.

$$RCHO + R'_2C{=}C{=}O \xrightarrow[\text{or } BF_3]{ZnCl_2} \text{3,3-}R'_2\text{-4-}R\text{-oxetan-2-one} \qquad (110)$$

Ketene dimerization is the principal synthetic route to 4-methylene-2-oxetanones. This reaction proceeds very satisfactorily for ketene and methylketene, but disubstituted ketenes dimerize only to cyclobutane-1,3-diones. The cycloaddition reaction of *t*-butylcyanoketene to ketene and to methylketene gives α-cyanoalkylidene-β-lactones in about 40% yield in addition to the cyclobutane-1,3-dione dimer of *t*-butylcyanoketene. A mechanism has been proposed for the formation of both types of dimers from a common zwitterionic intermediate (equation 111), with the relative amount of each product determined by the configurational equilibrium of the intermediate ⟨80JOC4483, 75JOC3417⟩.

$$Bu^t(NC)C{=}C{=}O + Me(H)C{=}C{=}O \longrightarrow \left[Bu^t(NC)C^-{-}C(=O){-}C(H)(Me){-}C^+{=}O \right] \longrightarrow \text{2-}Bu^t\text{-2-CN-4-Me-4-H-cyclobutane-1,3-dione} \;\; + \;\; \text{3-Me-4-}[{=}C(CN)Bu^t]\text{-oxetan-2-one} \qquad (111)$$

5.13.4.2.2 From fragments contributing one and three ring atoms

(*i*) *Oxetanes*

A novel synthesis of an oxetane from a 1,4-diene by means of the new reagent 'PhSeOH' has been reported. From 2,4-dimethyl-1,4-pentadiene there was obtained 2,4-dimethyl-2,4-bis(phenylselenomethyl)oxetane (**65**) in 34% yield. This compound can be deselenated with Raney nickel to 2,2,4,4-tetramethyloxetane ⟨79JOC1742⟩.

Me Me + 2PhSeOH ⟶ Me Me PhSeCH$_2$ O CH$_2$SePh (**65**) (112)

Me Me Bu^t $\xrightarrow{MeCO_3H}$ Me$_2$ O O Bu^t $\xrightarrow{H^+}$ Me$_2$ O O Bu^t (113)

3-Oxetanones have been obtained from peroxy acid oxidation of *t*-butyl substituted allenes to form the allene diepoxide, which then either rearranged spontaneously, or upon addition of acid, to the 3-oxetanone in about 30% yield (equation 113). Thus the insertion of the ring oxygen between the ends of the allene system comes about indirectly.

Mention should be made of studies of slow, controlled combustion of alkanes, where formation of oxetanes can be detected. For example, oxetane is observed during combustion of propane, while 2-*t*-butyl-3-methyloxetane and 2-isopropyl-3,3-dimethyloxetane are observed from combustion of isooctane. While the yields are extremely low, the presence of these compounds, along with the other products found, have provided evidence for the mechanism of combustion. The oxetanes are believed to result from rearrangement of peroxy radicals in the radical chain process (equation 114) ⟨70MI51300, 73MI51301⟩.

RCH$_2$CH$_2$CHR(O—O·) ⟶ RĊHCH$_2$CHR(O—OH) ⟶ RCH(C H$_2$)(O)CHR (114)

(*ii*) *2-Oxetanones*

A very promising synthesis of β-lactones has been recently reported, involving the palladium-catalyzed carbonylation reaction of halogeno alcohols. For example, 3-phenyl-2-oxetanone was obtained in 63% yield from 2-phenyl-2-bromoethanol in DMF solution at room temperature under 1 atmosphere pressure of carbon monoxide (equation 115). A proposed mechanism, in which palladium metal inserts into the carbon–halogen bond, followed by insertion of a molecule of carbon monoxide into the carbon–palladium bond and then ring closure, fits kinetics data ⟨80JA4193⟩.

Ph(Br)CHCH$_2$OH + CO $\xrightarrow{Pd}$ Ph O O (115)

O + Fe(CO)$_5$ ⟶ O C (CO)$_3$Fe O (**66**) $\xrightarrow{Ce(IV)}$ O=C O (116)

A recently described method for insertion of a carbon monoxide molecule into the monoepoxide of a conjugated diene gives β-lactones in high yield. This is achieved by reaction of iron pentacarbonyl with the starting vinyloxirane to give the π-allyl iron complex (**66**), which on oxidation with cerium(IV) ammonium nitrate gives the β-lactone. In some cases, γ-lactone products can also be obtained from this reaction ⟨81JCS(P1)270⟩.

Benz-2-oxetanone, a very unstable molecule, has been prepared in an argon matrix at 8 K by irradiation of phthaloyl peroxide (equation 117) ⟨73JA4061⟩.

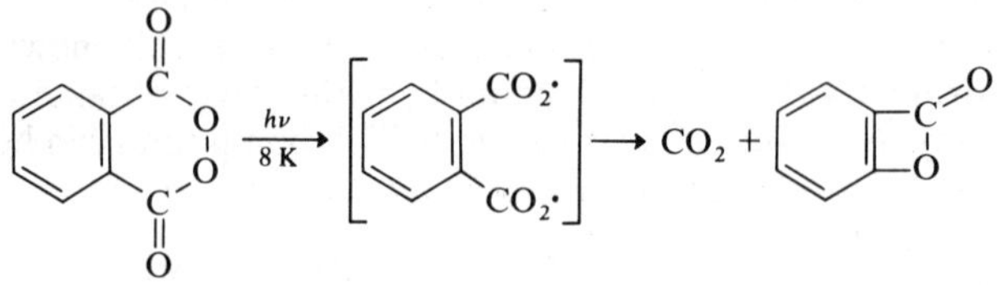

(117)

5.13.5 NATURAL PRODUCTS

The four-membered oxetane ring is not common in nature, but it occurs in some biologically interesting compounds. An outstanding example is the bicyclic oxetane thromboxane A_2, TXA_2 (**67**). This is a naturally occurring prostaglandin and a potent aggregatory agent for platelets in human blood, as well as a constrictor of vascular and bronchial smooth muscle. A synthetic analog, 11a-carbathromboxane A_2 (**68**), is also an oxetane and has related physiological activity ⟨80TL445⟩.

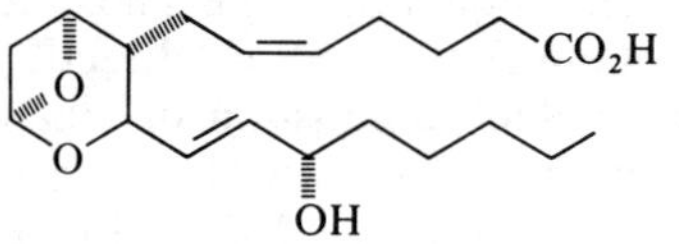

(**67**) TXA_2

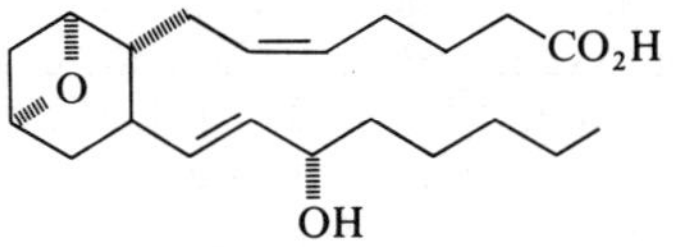

(**68**) 11a-Carba-TXA_2

The taxane alkaloids have a tetracyclic skeleton containing an oxetane ring. The general structure is given by (**69**), in which R^3 is one of three types of an α-hydroxy-β-amino acid structure and the other R groups are hydrogen and hydroxyl. Six of these taxane alkaloids have been identified to date. They are obtained by extraction of a Southeast Asian plant and certain of them shown antileukemic properties ⟨81JOC1469, 81MI51301⟩.

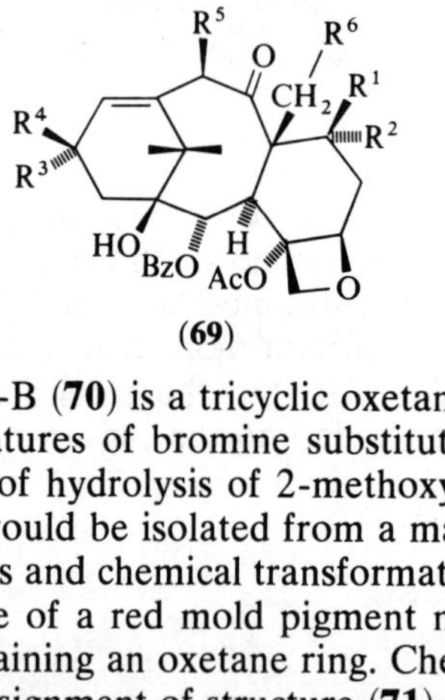

(**69**)

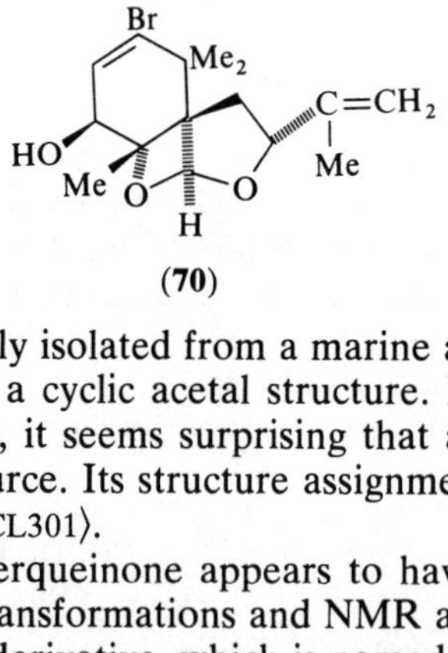

(**70**)

Laureacetal-B (**70**) is a tricyclic oxetane recently isolated from a marine alga. It has the interesting features of bromine substitution and a cyclic acetal structure. In view of the extreme ease of hydrolysis of 2-methoxyoxetane, it seems surprising that a strained ring cyclic acetal would be isolated from a marine source. Its structure assignment is based on spectral studies and chemical transformations ⟨79CL301⟩.

A derivative of a red mold pigment named herqueinone appears to have a polycyclic structure containing an oxetane ring. Chemical transformations and NMR and MS studies support the assignment of structure (**71**) for this derivative, which is named 'trimethylherqueinone B' ⟨70JOC179⟩. The oxetane structure that had been assigned to the trichothecane family of fungal metabolites, however, has been shown to be incorrect. An X-ray crystallographic study showed that the cyclic ether portion of the structure was an oxirane, rather than an oxetane ⟨66ACS1044⟩.

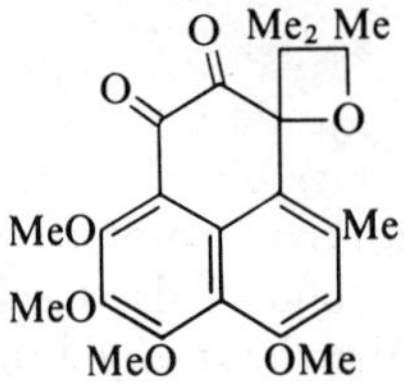

(**71**)

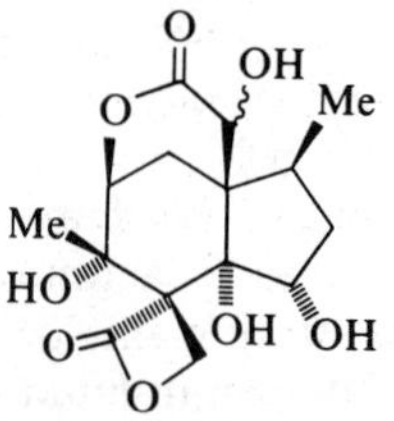

(**72**)

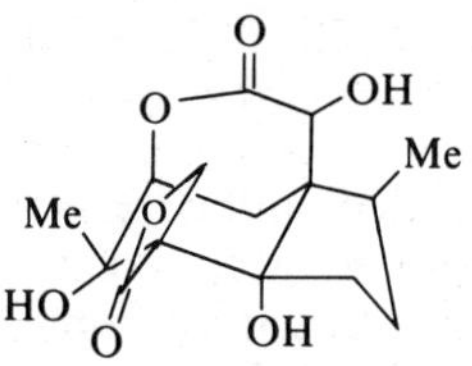

(**73**)

There are also a few β-lactone structures in natural products. Two toxic compounds, anisatin and neoanisatin, were isolated from a toxic plant, Japanese star anise, and have been assigned to the β-lactone-containing structures (**72**) and (**73**), respectively. These structures are based on extensive study of spectral data and chemical transformations ⟨68T199⟩.

Metabolism in the rat has been found to transform the steroid drug furazol (**74**) into a β-lactone derivative (**75**) ⟨72CPB1243⟩. A spiro-oxetane analog of estrone (**76**) has been synthesized for possible antiestrogenic properties ⟨67JMC546⟩.

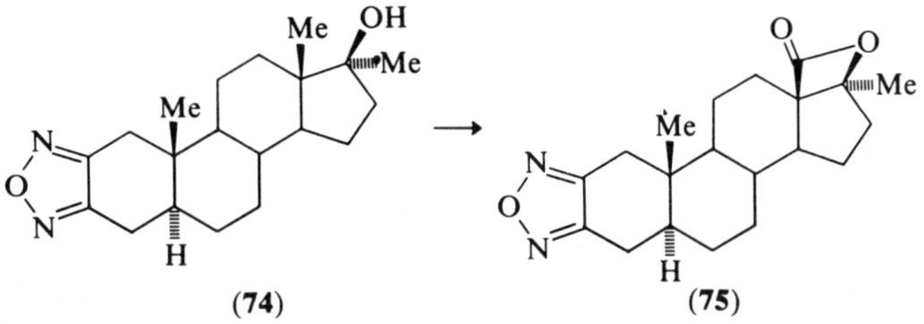

(**74**) → (**75**)

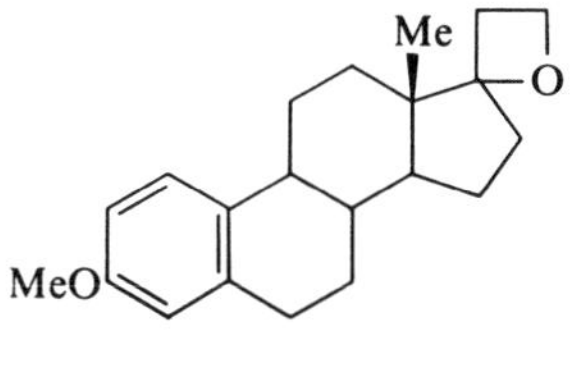

(**76**)

β-Lactone structures have recently been determined for two microbial metabolites which have antibiotic properties. Antibiotic 1233A (**77**) was the first β-lactone to be isolated from a fungus. Esterastin (**78**), an inhibitor of esterase, is produced by a strain of *Streptomyces* ⟨71JCS(C)3888, 78MI51300⟩.

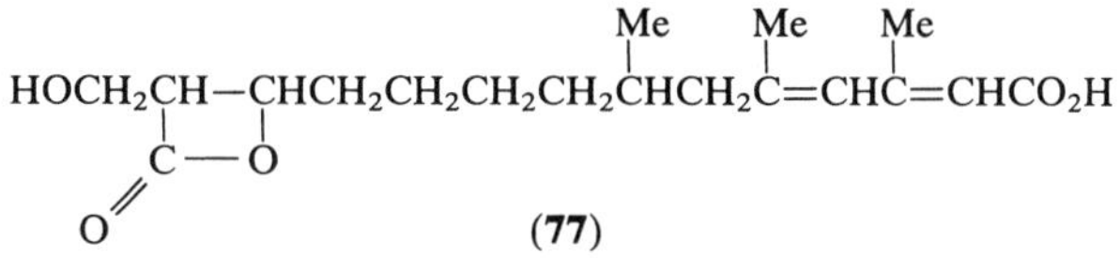

$HOCH_2CH{-}CHCH_2CH_2CH_2CH_2CH(Me)CH_2C(Me){=}CHC(Me){=}CHCO_2H$ (β-lactone ring: C(=O)—O)

(**77**)

$Me(CH_2)_4CH{=}CHCH_2CH{=}CHCH_2CHCH_2$–[β-lactone ring, H]–$(CH_2)_5Me$, with O–C(=O)–CHNHCOMe–$CH_2CONH_2$

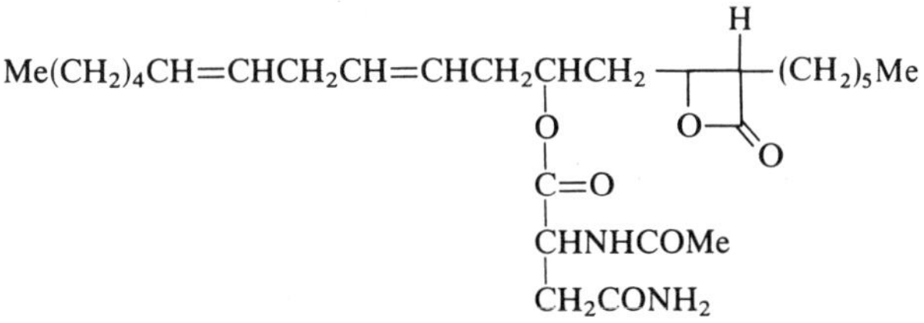

(**78**)

5.13.6 APPLICATIONS

Practical applications of oxetanes and 2-oxetanones are numerous, as evidenced by the thousands of patents that have been issued in this area. Space permits mention of only a few of what seem major applications.

Polymers of these compounds are in widespread use. Poly-BCMO is a very tough, durable plastic, which is estimated to have an average lifetime of 70 years in water. As most oxetanes and β-lactones can be polymerized and copolymerized to form a variety of useful solid compositions, this application will probably grow. One unusual application mentioned is the preparation and polymerization of 3,3-bis(azidomethyl)oxetane, which could be an explosive polymer ⟨81MI51302⟩.

β-Propiolactone has excited a great deal of attention for biological applications, as it seems to be highly toxic to all living organisms, besides being a hazardous carcinogen. Its application as a general, cold sterilant for instruments, cell cultures, milk and many other items have been repeatedly and enthusiastically described. It also has found considerable application in the treatment of cotton and wool fibers, because of its ability to react with the functional groups in those substances, giving cross-linking. The resulting modification in chemical structure provides protection against bacteria, molds and insects. For the same reason, it is very effective as a wood preservative.

Diketene also is widely employed as a natural and synthetic fiber cross-linking agent, wood preservative and paper-sizing agent. Both it and β-propiolactone have wide application as chemical intermediates. The synthetic applications of diketene for the synthesis of aromatic, heterocyclic and aliphatic compounds is exceptionally extensive ⟨74ACR265⟩.

The biological properties of oxetanes are slowly attracting attention, although many have been known for a long time. For example, 2,2-bis(*p*-ethoxyphenyl)-3,3-dimethyloxetane is an excellent insecticide, 25 times as potent as DDT ⟨71MI51300⟩. Many oxetanes have been patented as fungicides. The recent discovery of thromboxane A_2 and the antileukemic toxanes will undoubtedly stimulate further study of oxetane derivatives for possible medical and biological applications.

5.14

Thietanes, Thietes and Fused-ring Derivatives

E. BLOCK
State University of New York at Albany

5.14.1 INTRODUCTION

Thietanes, like other heterocyclobutanes, have received somewhat less attention than their synthetically somewhat more accessible counterparts with three-, five- or six-membered heterocyclic rings. The ring strain of thietane (80.3 kJ mol^{-1}) is virtually the same as that of the three-membered ring thiirane due to bond eclipsing effects in the former, which cause ring puckering with a concomitant increase in bond angle strain. Strain effects, together with the entropic difficulty of bringing together the ends of a four atom chain (less of a problem with the three atom chain in thiirane formation), make thietanes more difficult to prepare than thiiranes, thiolanes and thianes. Thus, while there are a variety of syntheses of thietanes with particular substitution patterns, there is no simple and completely general method for preparing thietanes of a wide range of ring substitution patterns. Interest in the synthesis of thietanes should grow with the discovery of the substantial biological activity of certain bicyclic thietane analogues of the prostaglandin-related compound thromboxane A2, and with the discovery of various thietanes as natural products (scent markers?) in several mammals of the mustelid family (mink, ferret, weasel).

The geometry of thietanes (puckered or planar) has attracted considerable attention from spectroscopists. The strong, geometry-dependent interaction of magnetic nuclei in a small ring makes thietanes excellent candidates for detailed NMR studies. Such NMR studies have led to the discovery of unusual deshielding effects which seem to be characteristic of four-membered rings in general.

As might be imagined, the reactions of thietanes are often dominated by the effects of ring strain. Thietanes undergo facile polymerization with either Lewis acids or bases as well as ready ring opening with a variety of reagents. Occasionally the ease of ring opening of thietanes has been exploited in synthesis, although in general, limited use has been made of these interesting and versatile heterocycles as reagents for organic synthesis.

There are no recent, general reviews of thietanes. Surveys have appeared covering work on thietanes through the mid-1960s ⟨64HC(19-2)647, 66CRV341⟩, and brief general reviews ⟨B-73MI51400⟩ or reports of the chemistry of thietes ⟨B-79MI51402⟩ have been published. In addition each volume of the Royal Society of Chemistry's Specialist Periodical Report on Organic Compounds of Sulphur, Selenium and Tellurium contains a summary of research on thietanes and related compounds for the years covered in each volume.

5.14.2 STRUCTURE

5.14.2.1 Bond Lengths and Bond Angles

Representative bond lengths and bond angles for various thietanes and thietes and their derivatives as determined by X-ray crystallography, microwave spectroscopy and electron diffraction methods are collected in Table 1. Thietane itself is nonplanar (pucker angle 26°), although the barrier to planarity is only 274 cm^{-1} (3.28 kJ mol^{-1}); the 1.847 Å long C—S bond is slightly (0.008 Å) longer than comparable C—S bonds in unstrained rings, while the C—S—C angle of 76.8° is considerably smaller than the corresponding angle in thiolane (93.4°). The severely strained bicyclic thietane 5-thiabicyclo[2.1.1]hexane possessing a 51.3° pucker angle has even longer C—S bonds (1.865 Å) and an even smaller C—S—C angle (69.5°).

Thietane 1-oxide is nonplanar with a pucker angle of 34.9° and with oxygen occupying an equatorial position. The other structural parameters of thietane 1-oxide and several of

Table 1 Bond Lengths and Bond Angles for Thietane and Thiete Derivatives

Compound	*Bond lengths* (Å)	*Bond angles* (°)	*Pucker angle* (°)	*Method*[a]	*Ref.*
	C—S 1.847, C—C 1.549, C—H (α) 1.09, C—H (β) 1.12	CSC 76.8, CCC 95.6, CCS 90.6, HCH (α) 112, HCH (β) 114	26	ED	75BCJ1691
	C—S 1.865	CSC 69.5	51.3	ED	71JA2799
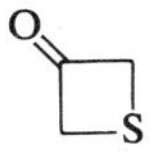	C—S 1.826, C=O 1.224, C—C 1.528	CCC 100.5	0	MW	75JSP(55)464
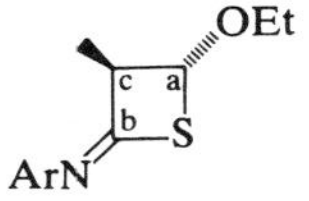	C^a—S 1.871, C^b—S 1.739	CSC 76.4, SC^aC^c 91.1, SC^bC^c 97.6	0	X	81AG(E)613
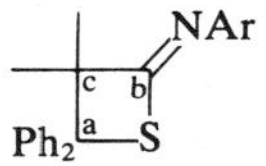	C^a—S 1.868, C^b—S 1.777, C^a—C^c 1.591, C^b—C^c 1.524	CSC 77.0, SC^aC^c 90.1, SC^bC^c 95.8	20	X	78AX(B)3403
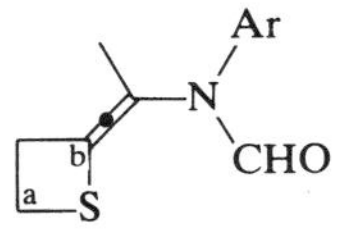	C^a—S 1.843, C^b—S 1.749	CSC 76.5	5	X	72JCS(P2)248
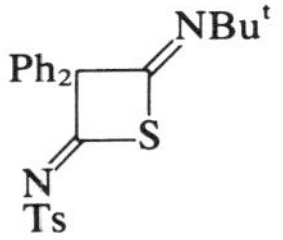	C^a—S 1.74, C^b—S 1.83, C^c—C 1.54, C^d—C 1.53	—	—	X	79TL3213
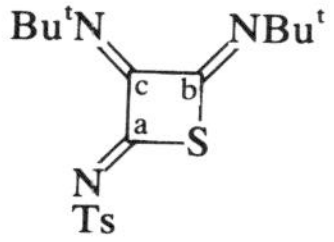	C^a—S 1.52, C^b—S 1.53, C^c—C 1.52	—	—	X	79CC160

Table 1 *(continued)*

Compound	*Bond lengths* (Å)	*Bond angles* (°)	*Pucker angle* (°)	*Method*[a]	*Ref.*
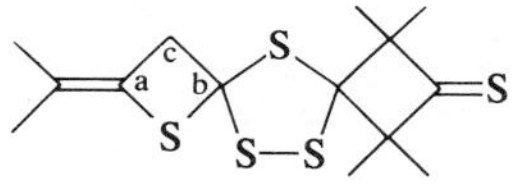	C^a—S 1.761, C^b—S 1.834	CSC 78.0, $C^aC^cC^b$ 93.7	9.7 (thietane ring)	X	73AX(B)2128
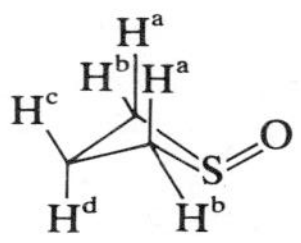	C—S 1.836, C—C 1.542, S—O 1.475, C—H^a 1.099, C—H^b 1.099, C—H^c 1.087, C—H^d 1.098	CSC 75.7, CCC 93.9, CSO 113.3, CCS 89.6, HCH (α) 110.7, HCH (β) 108.7	34.9	MW	77MI51400
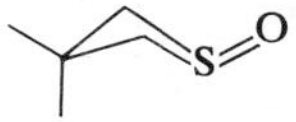	C—S (av) 1.845, S—O 1.48, C—C (av, ring) 1.555	CSO (av) 111.4, CSC 75.8, CCS (av) 89.5, CCC 93.6	35.4	X	73JA6046
	C—S (av) 1.84, S—O 1.526, C—C (av, ring) 1.53	CSC 76.6, CSO (av) 112, SCC (av) 90.3, CCC 96.6 ring	27	X	72ACS494
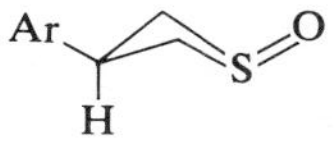 Ar = *p*-BrC_6H_4	C—S (av) 1.84, S—O 1.482, C—C (av) 1.54	—	33	X	75CC133
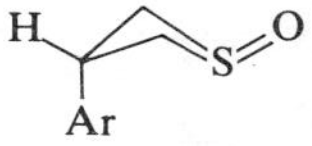 Ar = *p*-BrC_6H_4	C—S (av) 1.83, S—O 1.492, C—C (av) 1.52	—	32	X	75CC133
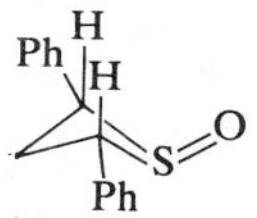	C—S 1.847, S—O 1.466, C—C 1.565	CSC 76.5, OSC 114.1, SCC 86.9, CCC 93.9	41.9	X	74JOC246

	C^a—S 1.777, C^b—S 1.807, S—O (av) 1.456	CSC 80.2, OSO 117.6, CCC (ring) 97.3, C^cC^aS 89.8, C^cC^bS 88.3	22.5	X	73LA1702
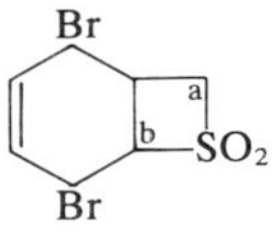	C^a—S 1.80, C^b—S 1.76, S—O (av) 1.42	CSC 79, OSO 118	—	X	75ACS(B)1071
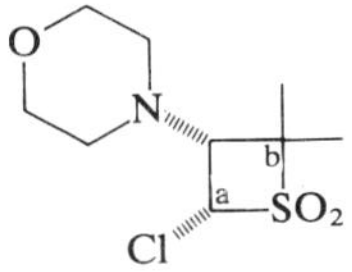	S—O (*cis* to Cl) 1.462, S—O (*trans* to Cl) 1.439, S—C^a 1.802, S—C^b 1.839	CSC 81.6, OSO 119.6	26.6	X	74G1207
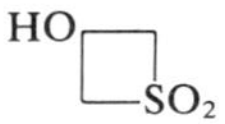	S—O (av) 1.436, C—S (av) 1.785	CSC 82.4, OSO 117.5, CCC 99.1, CCS 89.1	6.8	X	74G1207
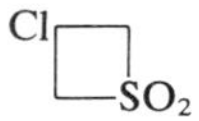	S—O (av) 1.423, C—S 1.791	CSC 81.5, OSO 117.9, CCC 99.1, CCS 89.3	9.4	X	74G1207
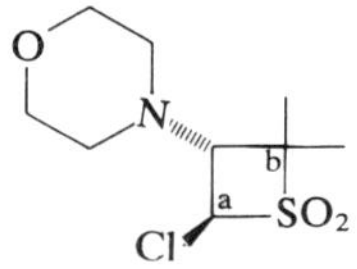	S—O (av) 1.422, S—C^a 1.833, S—C^b 1.835	CSC 80.9, OSO 118.1	25.1	X	74G1207
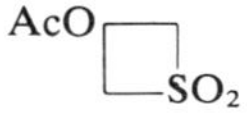	S—O (av) 1.423, C—S 1.789	CSC 80.6, OSO 117.1, CCC 97.8, CCS 89.4	17.9	X	74G1207
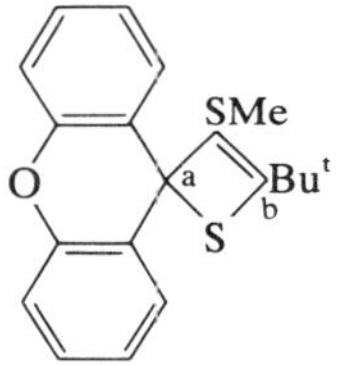	C^a—S 1.926, C^b—S 1.788, C^b=C^c 1.360	CSC (ring) 74.2	0	X	80AX(B)484

Table 1 *(continued)*

Compound	*Bond lengths* (Å)	*Bond angles* (°)	*Pucker angle* (°)	*Method*[a]	*Ref.*
 c, a, b, SO_2	C^a—S 1.79, C^b—S 1.77, S—O 1.43, C=C 1.39	CSC 80.5, OSO 120, CCC 104.5	0	X, MW	73ACS3128
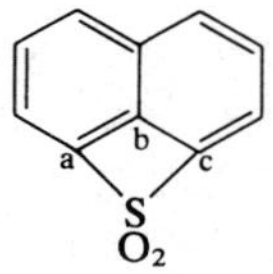 a, b, c, S, O_2	C—S (av) 1.82, S—O (av) 1.435, C^a—C^b (av) 1.396	CSC 75.4, $C^aC^bC^c$ 106	0	X	76JA6643

[a] ED = electron diffraction, MW = microwave spectroscopy, X = X-ray crystallography.

its derivatives which have been examined are rather similar to those of thietane itself. 2,2-Dimethylthietane 1,1-dioxide is nonplanar with a pucker angle of 22.5°, a C—S—C angle of 80.2° (slightly larger than that of thietane itself) and C—S bonds of 1.777–1.807 Å (somewhat shorter than in thietane). Other thietane 1,1-dioxides show pucker angles ranging from 7 to 27°.

5.14.2.2 Stereochemistry and NMR Spectra

5.14.2.2.1 ¹H NMR chemical shifts and coupling constants

Selected ^{1}H NMR chemical shifts and coupling constants for a variety of thietane and thiete derivatives are collected in Table 2. In general the α-protons of thietanes and their derivatives are deshielded relative to analogous derivatives with larger rings ⟨80JOC4807⟩. Thus, the α-protons of thietane and thietane 1,1-dioxide appear at δ 3.21 and 4.09 p.p.m. respectively, while the corresponding signals for the α-protons in thiolane and thiolane 1,1-dioxide appear at δ 2.82 and 3.01 p.p.m. respectively. The α-proton chemical shifts of a series of cycloalkanes, oxacycloalkanes, azacycloalkanes and cycloalkanones also show a maximum in deshielding at the four-membered ring. This effect was noted as early as 1954 ⟨54JA4242⟩ and apparently has never been adequately explained. The four-membered ring deshielding effect is larger in the case of the sulfones than the sulfides, an effect which is particularly dramatic with ^{13}C NMR spectra. While the β-protons of thietane (δ 2.94 p.p.m.) are also deshielded compared to those of thiolane (δ 1.93 p.p.m.), the β-protons of thietane 1,1-dioxide (δ 2.14 p.p.m.) are slightly shielded by comparison with the corresponding β-protons of thiolane 1,1-dioxide (δ 2.21 p.p.m.), an effect which is also magnified in the ^{13}C NMR spectra (see below).

A number of detailed ^{1}H NMR studies of thietane and thiete derivatives have appeared in which all of the coupling constants have been determined and some of these data are summarized in Table 2. Liquid crystal NMR studies of thietane ⟨75MI51400⟩ and thietane 1,1-dioxide ⟨79JMR(36)343⟩ suggest that the former is puckered by 26.5° while the latter is planar or slightly distorted from planarity with a low barrier to planarity. A study of lanthanide-induced chemical shifts of a series of substituted thietane 1-oxides indicates that oxygen prefers the equatorial position ⟨79JOC4757⟩. A related study of these types of compound utilizing aromatic solvent-induced chemical shifts indicates that the signal assigned to the α-hydrogen exhibits the larger shift to high field in benzene ⟨71T341⟩.

5.14.2.2.2 Heteronuclear NMR spectra

A summary of ^{13}C NMR chemical shift data for thietanes, thietes and related compounds is given in Table 3. An analysis of these data reveals that thietane 1,1-dioxides, thiete 1,1-dioxides and thietane sulfoximines show unusual α-carbon deshielding and β-carbon shielding when compared to other thiacycloalkane 1,1-dioxides and acyclic counterparts. This has been referred to as the 'four-membered-ring sulfone effect' ⟨80JOC4807⟩. Thus, the α-carbon shifts of thiolane and its 1-oxide and 1,1-dioxide appear at 31.7, 54.3 and 51.1 p.p.m. while the corresponding α-carbon shifts for thietane and its *S*-oxides appear at 26.1, 52.7 and 65.6 p.p.m.

Decreasing the C—S—C angle in thiolane 1,1-dioxides does not bring about an effect similar to that seen with the thietane 1,1-dioxides [*e.g.* the bridgehead carbon in 7-thiabicyclo[2.2.1]heptane 7,7-dioxide (**1**), which should have a C—S—C angle very similar to that in thietane 1,1-dioxide, appears at 51.3 p.p.m. compared to a shift of 53.2 p.p.m. for the bridgehead carbons in the less strained 9-thiabicyclo[3.3.1]nonane 9,9-dioxide (**2**)]. The α-deshielding effect in thietane 1,1-dioxides does not seem to be directly attributable to interactions involving the α-hydrogens, since replacing these hydrogens with methyl groups does not eliminate the effect (the quaternary carbons of 2,2,4,4-tetramethylthietane and its 1,1-dioxide appear at 37.1 and 75.5 p.p.m. respectively, a difference that is virtually identical to that seen with the parent thietanes). Nonplanar thietane 1,1-dioxides such as (**3**) ⟨79JOC4757⟩ (α-carbon shift of 76.8 p.p.m.) show an α-deshielding effect that is comparable to that seen in the near-planar parent thietane 1,1-dioxide [thus, the α-carbon shifts for sulfoxides (**4**) and (**5**) are 68.0 and 64.6 p.p.m. respectively, differing from the value

Table 2 ^{1}H NMR Spectral Data for Thietane and Thiete Derivatives

Compound	*Chemical shifts* (p.p.m.)	*Coupling constants* (Hz)	*Solvent*	*Ref.*
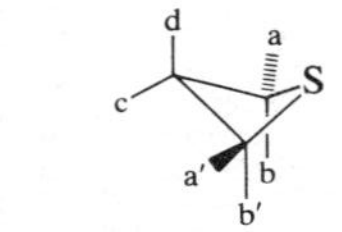 d, c, a, a′, b, b′, S	3.21 (a), 2.94 (c)	J_{ab} −8.7, J_{cd} −11.7, $J_{aa'}$ 0.9, $J_{ab'}$ −0.2, J_{ad} 9, J_{bd} 6.3	—	74JMR(15)367
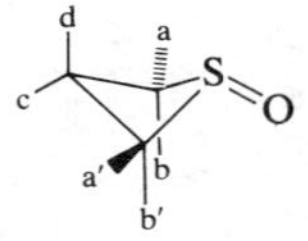 d, c, a, a′, b, b′, S=O	3.46 (a), 3.09 (b), 2.27 (c), 1.95 (d)	$J_{aa'}$ 6.10, $J_{bb'}$ −0.57, $J_{ab'}$ 0.95, J_{ab} −10.29, J_{cd} −12.69, J_{bd} 12.53, J_{ac} 1.83, J_{bc} 10.63, J_{ad} 7.47	Deuteroacetone	74JMR(15)367
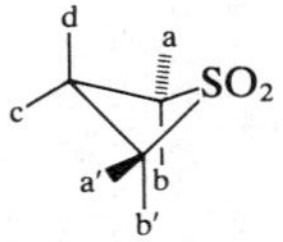 d, c, a, a′, b, b′, SO_2	4.09 (a), 2.14 (c)	J_{ab} 14.0, J_{cd} −12.63, J_{ad} 10.34, J_{bd} 6.33, $J_{aa'}$ 2.24, $J_{ab'}$ −1.24	—	74JMR(15)367
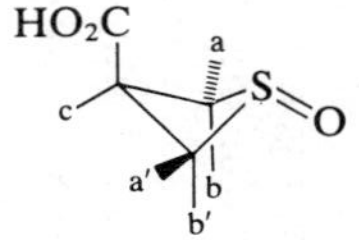 HO_2C, c, a, a′, b, b′, S=O	3.83 (a), 3.44 (b), 3.79 (c)	J_{ab} −11.95, J_{ac} 10.84, J_{bc} 3.46, $J_{aa'}$ 5.27, $J_{bb'}$ 0.59, $J_{ab'}$ −1.01	MeOH	74JMR(15)367
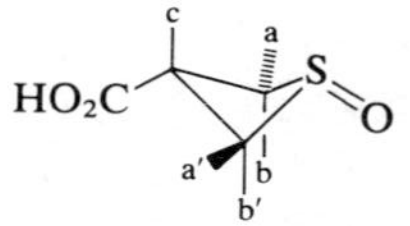 c, HO_2C, a, a′, b, b′, S=O	3.24 (a), 3.55 (b), 4.08 (c)	J_{ab} −9.14, J_{ac} 9.38, J_{bc} 8.04, $J_{aa'}$ 0.99, $J_{bb'}$ 0.01, $J_{ab'}$ −0.26	MeOH	74JMR(15)367
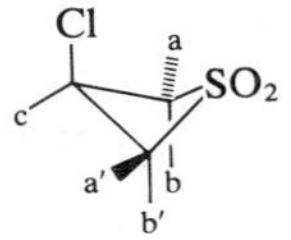 Cl, c, a, a′, b, b′, SO_2	4.31 (a), 4.85 (b), 4.92 (c)	$J_{aa'}$ 3.88, $J_{bb'}$ 2.48, $J_{ab'}$ −1.32, J_{ac} 3.89, J_{bc} 8.34, J_{ab} −14.56	Deuteroacetone	75JMR(17)219
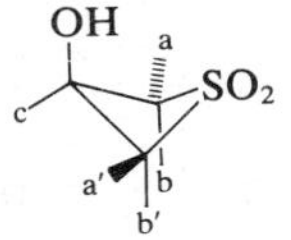 OH, c, a, a′, b, b′, SO_2	3.99 (a), 4.43 (b), 4.73 (c)	$J_{aa'}$ 3.94, $J_{bb'}$ 2.86, $J_{ab'}$ −1.18, J_{ac} 3.50, J_{bc} 7.53, J_{ab} −13.95	Deuteroacetone	75JMR(17)219

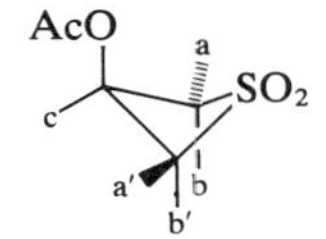	4.16 (a), 4.62 (b), 5.33 (c)	$J_{aa'}$ 4.29, $J_{bb'}$ 2.59, $J_{ab'}$ −1.25, J_{ac} 3.21, J_{bc} 7.79, J_{ab} −14.62	Deuteroacetone	75JMR(17)219
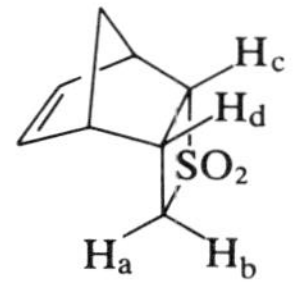	3.29 (a), 3.84 (b), 4.83 (c), 2.92 (d)	J_{ab} −13.6, J_{cd} 9.0	Chloroform-*d*	80OMR(14)192
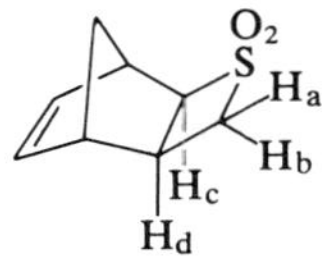	3.92 (a), 3.37 (b), 4.08 (c), 2.35 (d)	J_{ab} −14.0, J_{cd} 7.5	Chloroform-*d*	80OMR(14)192
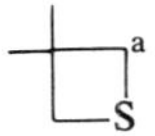	2.92 (a) 2.72 (a)	—	Chloroform-*d* Benzene	71T341
	3.32	—	Chloroform-*d*	80TH51400
	3.20 (a)	—	Chloroform-*d*	74LA1328
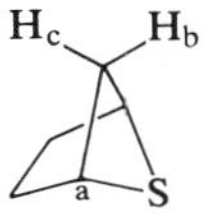	3.84 (a), 2.89 (b), 1.72 (c)	J_{bc} 6.8	—	70TL2931
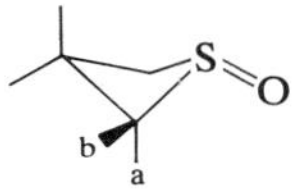	3.02 (a), 3.52 (b) 2.63 (a), 2.87 (b)	J_{ab} −11.8, $J_{bb'}$ 5.5	Chloroform-*d* Benzene	71T341 73JA6046

Table 2 (*continued*)

Compound	*Chemical shifts* (p.p.m.)	*Coupling constants* (Hz)	*Solvent*	*Ref.*
b, a, SO_2	3.95 (a), 1.95 (b)	—	Chloroform-*d*	75S333
a, SO_2	3.80 (a) 3.38 (a)	—	Chloroform-*d* Benzene	71T341
O_2S, SO_2	4.72		Trifluoroacetic acid	74LA1328
S, NTs, a, b	3.92 (a), 3.52 (b) 3.67 (a), 2.68 (b)	—	Chloroform-*d* Benzene	71T341
b, a, SF_2	3.85 (a), 3.58 (b)	—	—	73JA4064
b, a, SF_4	4.4–5.0 (a), 2.0–2.35 (b)	—	—	73JA8191
O, S	3.05 and 4.02	Two triplets, *J* 6.5	Carbon tetra-chloride	70JOC584
b, a, c, S	3.80 (a), 6.50 (b), 5.60 (c)	—	Chloroform-*d*	79MI51402
c, a, S	3.70 (a), 6.10 (c)	—	Chloroform-*d*	79MI51402
Ph, a, c, S	4.10 (a), 6.76 (c)	—	Chloroform-*d*	79MI51402

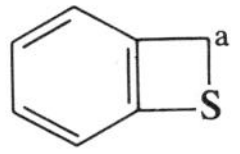 a S	4.28 (a), 6.8–7.8 (ring)	—	—	77RTC282
 b a c SO_2	4.57 (a), 7.23 (b), 6.80 (c)	J_{ab} 1.60, J_{ac} −0.52, J_{bc} 4.10	Deuteroacetone	72OMR(4)107
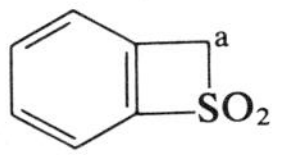 a SO_2	5.10 (a), 7.56 (ring)	—	—	77RTC282

Table 3 ^{13}C NMR Chemical Shifts for Thietanes, Thietes and Related Compounds (p.p.m. from TMS)

Compound	$n = 0$	$n = 1$	$n = 2$	*Ref.*
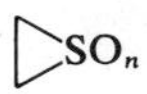 SO_n	18.1	33.8	31.6	80JOC4807
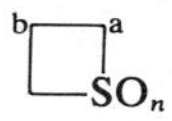 b, a, SO_n	26.1 (a), 28.0 (b)	52.7 (a), 10.4 (b)	65.6 (a), 5.8 (b)	80JOC4807
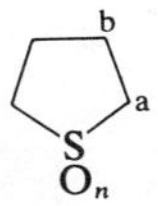 b, a, S, O_n	31.7 (a), 31.2 (b)	54.3 (a), 25.4 (b)	51.1 (a), 22.7 (b)	80JOC4807
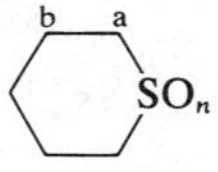 b, a, SO_n	29.1 (a), 27.9 (b)	49.0 (a), 19.3 (b)	52.6 (a), 25.1 (b)	80JOC4807
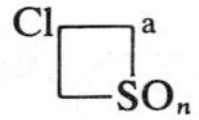 Cl, a, SO_n	38.9 (a)	63.3 (a) (*cis*); 60.4 (a) (*trans*)	75.4 (a)	80JOC4807, 82OMR(18)82
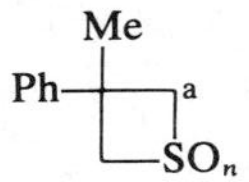 Me, Ph, a, SO_n	38.2 (a)	63.2 (a) (O/Ph *cis*); 62.9 (a) (O/Ph *trans*)	75.6 (a)	80JOC4807
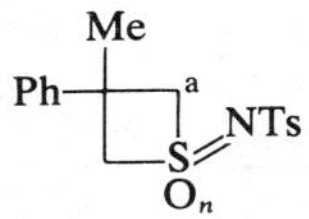 Me, Ph, a, S, NTs, O_n	61.8 (a, N/Ph *cis*)	75.6 (a) (N/Ph *cis*)	—	80JOC4807
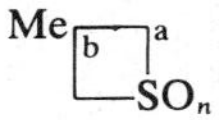 Me, b, a, SO_n	33.1 (a), 36.1 (b)	59.7 (a), 21.8 (b) (*cis*); 57.8 (a), 22.9 (b) (*trans*)	71.1 (a), 20.5 (b)	82OMR(18)82, 80UP51401
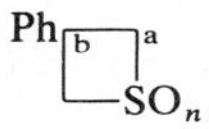 Ph, b, a, SO_n	32.0 (a), 44.1 (b)	59.5 (a), 28.0 (b) (*cis*); 57.0 (a), 34.2 (b) (*trans*)	71.8 (a), 24.4 (b)	82OMR(18)82, 80UP51401
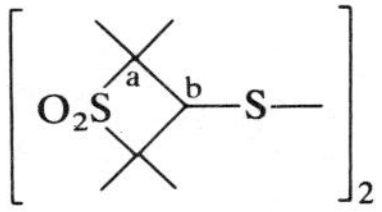 [O_2S, a, b, S—]$_2$	—	—	79.8 (a), 59.6 (b)	75JCS(P1)2513

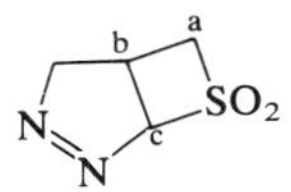	—	—	72.9 (a), 120.8 (c), 17.9 (b)	80JOC3646
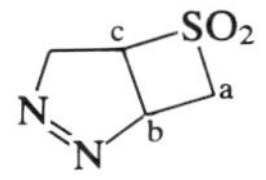	—	—	69.6 (a), 72.9 (c), 73.7 (b)	80JOC3646
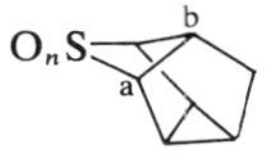	—	68.0 (a), 27.4 (b) (*exo*); 64.6 (a), 45.1 (b) (*endo*)	76.8 (a)	79JOC4757
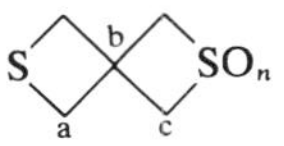	39.7 (a), 52.1 (b)	38.6 (a), 64.6 (c), 38.4 (b)	37.8 (a), 75.88 (c), 33.7 (b)	80UP51400
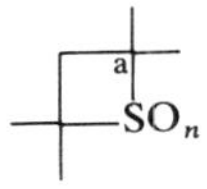	37.1 (a)	—	75.5 (a)	80JOC4807
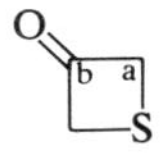	55.2 (a), 194.6 (b)	—	—	80UP51400
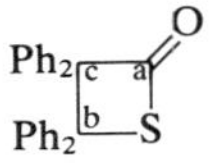	194.0 (a), 65.0 (b), 92.7 (c)	—	—	78JOC4961
	46.4 (a), 123.9 (b), 134.3 (c)	—	72.8 (a), 138.6 (b), 148.5 (c)	72OMR(4)107, 79MI51402
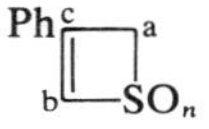	35.8 (a), 122.3 (b), 135.3 (c)	—	69.9 (a), 127.3 (b), 147.1 (c)	82OMR(18)82, 79PS(7)5
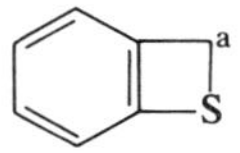	36.4 (a)	—	—	77RTC282
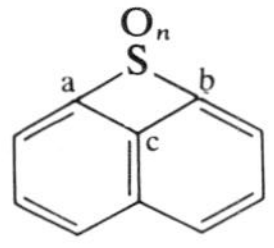	140.5 (a, b), 138.6 (c)	164.8 (a, b), 151.3 (c)	162.5 (a, b), 138.3 (c)	76JA6643

for sulfone (**3**) by an amount (8.8–12.2 p.p.m.) comparable to the α-carbon shift difference between thietane 1-oxide and its 1,1-dioxide (12.9 p.p.m.)].

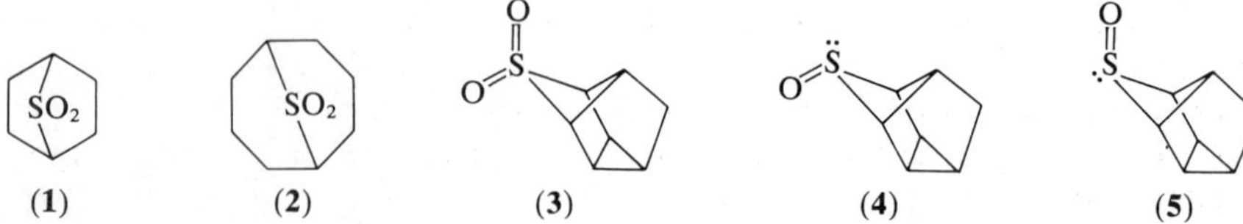

(**1**) (**2**) (**3**) (**4**) (**5**)

A limited amount of ^{17}O NMR chemical shift data is available for thietane 1-oxides and 1,1-dioxides and thiete 1,1-dioxides. These data, summarized along with selected data for rings of other sizes in Table 4, indicate that there is once again a maximum in deshielding at the four-membered ring. As can also be seen, discrete ^{17}O NMR chemical shifts are observed for the nonequivalent sulfone oxygens in several of the thietane 1,1-dioxides.

Table 4 ^{17}O NMR Chemical Shifts for Thietane and Thiete *S*-Oxides and Related Compounds

Compound	*Chemical shift* (p.p.m.)	*Solvent/^{17}O reference*	*Ref.*
PhS SO_2	182, 198	Acetone	80JOC4807
Cl_2 SO_2	186.6	Chloroform/water	81UP51400
SO_2	170.7	Chloroform/water	81UP51400
HO SO_2	184, 187 (SO_2)	Acetone	80JOC4807
Me_3SiO SO_2	183, 187 (SO_2)	Acetone	81UP51400
SO_2	180	Acetone	80JOC4807
SO_2	182	Acetone	80JOC4807
SO	68	Acetone	80JOC4807
SO_2	164	Acetone	80JOC4807
SO	12	Acetone	80JOC4807
SO_2	149	Acetone	80JOC4807
SO	−2	Acetone	80JOC4807
SO_2	111	Acetone	80JOC4807
SO	−70	Acetone	80JOC4807
SO_2	154	Acetone	80JOC4807
SO	11	Chloroform/water	80JOC4807

5.14.2.3 Mass Spectra

Scheme 1 summarizes some of the common electron-impact induced fragmentation pathways available to thietanes, thietes and their *S*-oxides. Major processes include retro-2 + 2 fragmentation for thietanes, thiolactones and iminothietanes, loss of sulfur oxides for thietane *S*-oxides, and loss of an α-hydrogen atom from thietes and benzothietes. Certain of these processes, considered to involve electron-impact induced pathways, may actually involve thermolytic reactions occurring in the source region of the mass spectrometer ⟨78JHC421, 71JA676⟩.

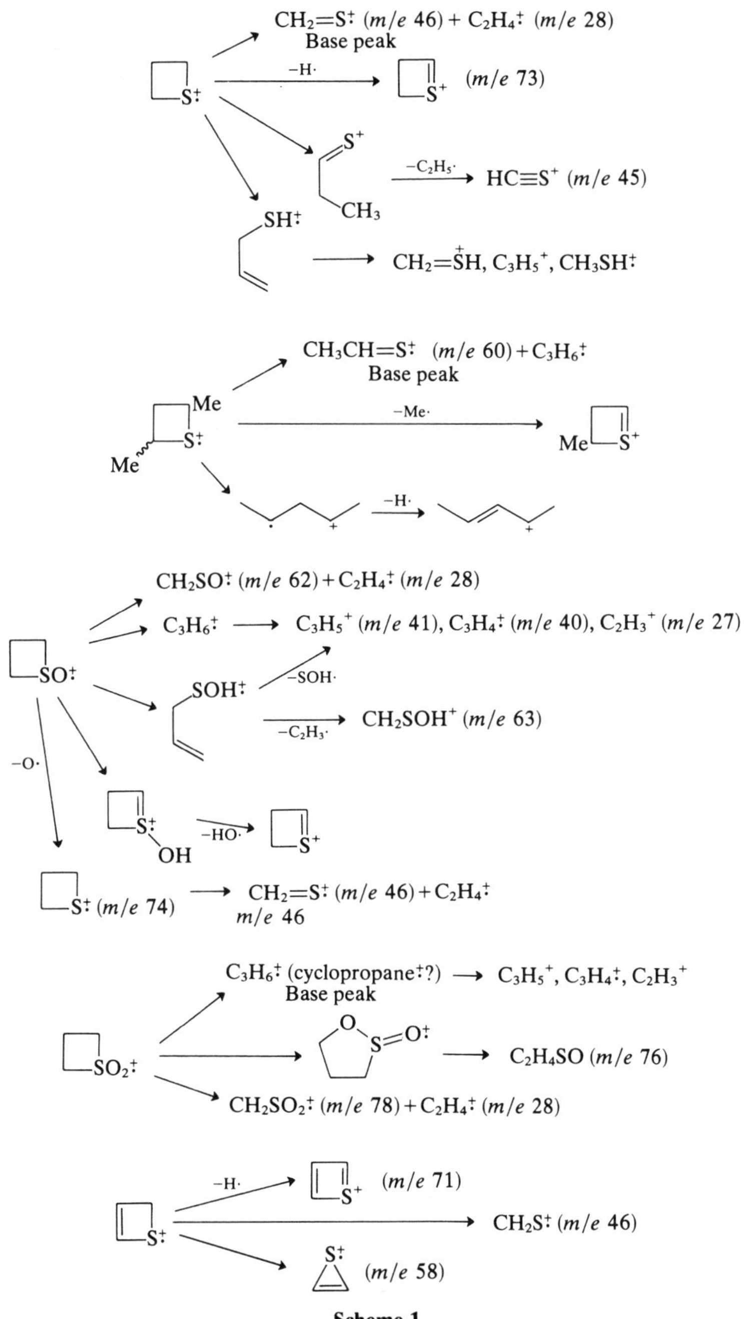

Scheme 1

5.14.2.4 Electronic Spectra: UV, PE

Selected UV data for thietanes, thietes and their derivatives are summarized in Table 5. Thietanes absorb at somewhat longer wavelengths than larger ring sulfides such as thiolanes

Table 5 UV Spectral Data for Thietane and Thiete Derivatives

Compound	*UV absorption*[a] *(extinction coefficient)*	*Ref.*
Thietane	273	67MI51401
3,3-Dimethylthietane	265	67MI51401
	278	67MI51401
	268 (123)	74LA1328
	268 (123)	74LA1328
	231 (2050), 289 (440)	81TL4421
	245 (92), 330 (166)	67JA2761
	221 (4200), 320 (5960), 331 (6560)	80JOC4804
	302.5 (27)	74JA2964
Thiete	215 (1920), 236 (3000), 285 (50)	72JOC1111
	228sh (1100), 248 (2000), 252 (2000), 256 (1715), 294 (567)	72JOC1111
3-Phenylthiete	301 (12 800)	79MI51402
	220 (32 000), 260 (5000), 280 (2800), 290sh (1700)	76JA6643
	225 (38 000), 274 (4300), 284 (5000), 316 (410)	67LA(703)96
	253 (13 800)	72TL4781
	262 (16 600)	72TL4781

[a] nm.

(λ_{max} at 270 and 239 nm respectively); alkyl-substituted thietes show an even greater bathochromic shift having long-wavelength maxima at 285–294 nm.

PE spectroscopy indicates that thietane has a first ionization potential (8.67 eV) only slightly larger than that of dimethyl sulfide (8.65 eV) and significantly lower than that of thiirane (9.05 eV) ⟨77JA3226⟩. The PE spectrum of benzothiete has also been determined ⟨80TL343⟩.

5.14.2.5 IR and Raman Spectra

IR data on a number of thietanes are contained in Table 6. There have been a number of far-IR and Raman studies of thietanes and their derivatives ⟨66JCP(45)61, B-73MI51400⟩.

5.14.2.6 Thermodynamic Aspects: Stability, Ring Strain, Aromaticity

The strain energies of thiirane, thietane and thiolane are reported to be 81.1, 80.3 and 5.0 kJ mol^{-1} respectively ⟨79MI51400⟩. Molecular mechanics calculations indicate strain energies for thietane, 5-thiabicyclo[2.1.1]hexane, thiolane and thiane to be 82.4, 135.6, 17.2 and 2.5 kJ mol^{-1} respectively ⟨75JA5167⟩.

Hückel MO calculations predict that both the thietide anion (**6**) and the benzothietide anion (**7**) will be highly reactive and rather unstable ⟨65CCC3016⟩. Electron repulsions would be severe with six π-electrons crowded into the small ring. Theoretical calculations suggest that thiacyclobutadiene (**8**) will be nonaromatic rather than antiaromatic, with a partial negative charge on the carbon adjacent to sulfur and a partial positive charge on sulfur (*e.g.* ylide-like) and pyramidal sulfur ⟨77T3061⟩. Evidence has been presented for the formation of thiacyclobutadiene (**9**); further spectroscopic support for this structure would be desirable ⟨74TL3911⟩.

(**6**) (**7**) (**8**) (**9**)

5.14.3 REACTIVITY

5.14.3.1 Thermal and Photochemical Reactions not Formally Involving Other Species

5.14.3.1.1 Fragmentations and eliminations

Two major thermal and photochemical decomposition pathways available to thietane derivatives (**10**) are fragmentation into thiocarbonyl compounds and alkenes (retro-2+2 reaction) or expulsion of a sulfur oxide and a three-carbon fragment, typically a cyclopropane (retro-3+1 reaction). The pathway preferred depends on the relative stabilities of the possible fragments, *e.g.* thietane 1,1-dioxides generally lose sulfur dioxide rather than sulfenes while thietanes lose thiocarbonyl compounds rather than atomic sulfur. The reactions most likely involve diradical intermediates. Representative examples of thietanes whose fragmentation has been studied include thietane (**11**) itself ⟨75MI51400, 78CJC114⟩, dialkyl (**12**) ⟨75CJC1744⟩ and aryl (**13**) derivatives ⟨70JOC1781⟩ and β-thiolactones (**14**) ⟨80JOC4359⟩.

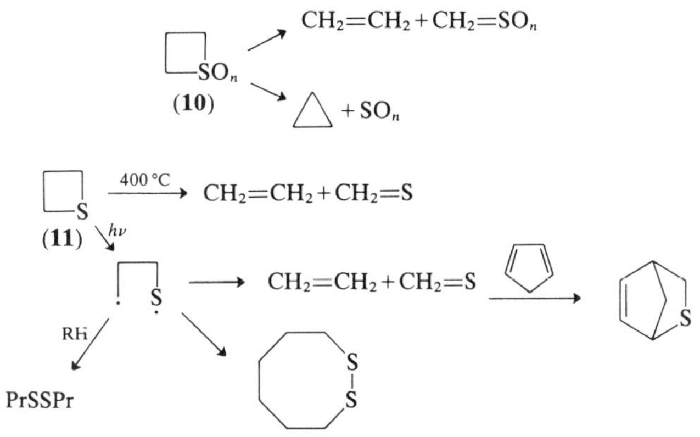

(**10**) → $CH_2{=}CH_2 + CH_2{=}SO_n$; → △ + SO_n

(**11**) —400 °C→ $CH_2{=}CH_2 + CH_2{=}S$

(**11**) —$h\nu$→ diradical → $CH_2{=}CH_2 + CH_2{=}S$ —(cyclopentadiene)→ 2-thiabicyclo adduct; diradical → cyclic disulfide; diradical —RH→ PrSSPr

Table 6 IR Spectral Data for Thietane and Thiete Derivatives

Compound	*IR bands*[a] *(assignment)*	*Notes*	*Ref.*
Thietane	698, 675s (C—S stretch)	Vapor	64HC(19-2)647
3,3-Dimethylthietane 1-oxide	1087, 1017s (S=O) 1060, 1010s (S=O)	CCl_4 CCl_4	73JA6046 71T341
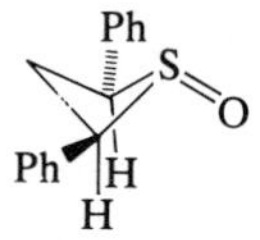 Ph S O Ph H H	1069 (S=O)	Nujol	70JOC2520
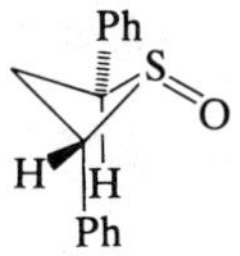 Ph S O H H Ph	1064 (S=O)	Nujol	70JOC2520
 (exo) O S	1049 (S=O)	CH_2Cl_2	79JOC4757
 (endo) S O	1097 (S=O)	CH_2Cl_2	79JOC4757
3-Methylthietane 1,1-dioxide	1395, 1310, 1220, 1195sh, 1160 (SO_2)	Film	70JOC3657
3-*t*-Butylthietane 1,1-dioxide	1320, 1140 (SO_2)	CCl_4	70JOC3657
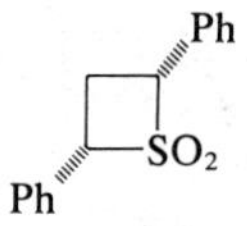 Ph SO_2 Ph	1309, 1178, 1139 (SO_2)	Nujol	70JOC2520
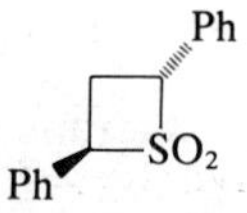 Ph SO_2 Ph	1312, 1183, 1159, 1117s (SO_2)	Nujol	70JOC2520
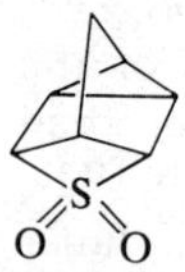 S O O	1300, 1175, 1097 (SO_2)	CH_2Cl_2	79JOC4757
3-Phenylthietane 1,1-dioxide	1324, 1130 (SO_2)	$CHCl_3$	70JOC3657
3-Phenylthiete 1,1-dioxide	1563w (C=C), 1310, 1125 (SO_2)	$CHCl_3$	70JOC3657
3-*t*-Butylthiete 1,1-dioxide	1590 (C=C), 1305, 1150 (SO_2)	$CHCl_3$	70JOC3657
3-Bromothiete 1,1-dioxide	1540s (C=C), 1310, 1210, 1120s (SO_2)	KBr	76JOC3044
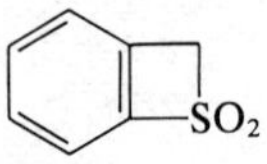 SO_2	1300, 1195, 1120s (SO_2), 720, 710s	—	76JOC3044
2-Thietanone	1776s (C=O)	—	70JOC584
3,3,4,4-Tetramethyl-2-thietanone	1757, 1721 (C=O)		73TL1221
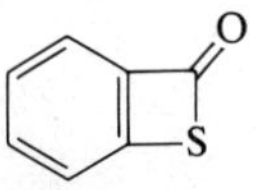 O S	1803 (C=O)	At 77 K	70JA7001

Table 6 *(continued)*

Compound	*IR bands*[a] *(assignment)*	*Notes*	*Ref.*
	1820, 1785 (C=O), 1680 (C=C)	—	73TL1221
	1850, 1760 (C=O)	—	72JOC2514
2,2-Dimethyl-3-thietanone	1762 (C=O)	Film	79LA1768
	1740 (C=O), 1642 (C=C)	—	80JOC4804
3-Thietanone 1,1-dioxide	1845, 1779 (C=O), 1314, 1196, 1129s (SO_2)	—	63JA3236

[a] cm^{-1}.

(12) → [Et/Pr diradical intermediates] → Et/Pr thietane + Et–CH=CH–Pr (7) + Et–CH=CH–Pr (1) + EtCH=CH_2

(13) $\xrightarrow[\text{or } 230\,°C]{h\nu}$ PhCOCH=CHPh

(14) $\xrightarrow{\Delta}$ Ph_2C=S + [Ph_2C=C=O]

The decomposition of thietane 1-oxide (**15**) is somewhat more complex, involving both retro-2+2 and -3+1 pathways as well as deoxygenation and ring opening following rearrangement to a sultene ⟨81JCS(P2)1166⟩. The retro-2+2 process, which may be only a minor pathway, leads to the formation of ethylene and sulfine (thioformaldehyde *S*-oxide; detected by microwave spectroscopy) ⟨76JA1264⟩. Thietane (**16**) is reported to lose sulphur monoxide on UV irradiation ⟨78BCJ3403⟩.

(15) → CH_2=CH_2 + CH_2=S=O

(15) → C_3H_6 + SO

(15) → thietane

(15) → [sultene] → [$HSCH_2CH_2CHO$] → CH_2=CHCHO

(16) $\xrightarrow{h\nu}$ 100%

Thietane 1,1-dioxide decomposes cleanly to cyclopropane and sulfur dioxide at 400 °C with an activation energy of 117.5 kJ mol^{-1} ⟨75MI51401⟩; at higher temperatures some propylene is produced. Cyclopropanes and sulfur dioxide are also the major products on irradiation of thietane 1,1-dioxide and its 2-phenyl derivatives with light of wavelength 147 nm for the parent compound ⟨79JPC2025⟩ or 254 nm for 2-phenyl derivatives (*e.g.* **17**) ⟨78S579, 70JOC1781⟩. *cis*-2,4-Dimethylthietane 1,1-dioxide at 350 °C gives a mixture of *cis*- and *trans*-1,2-dimethylcyclopropane (50%) and the isomeric 2-pentenes (25%) ⟨71JA676⟩. The 3-alkylidene thietane 1,1-dioxide (**18**) gives a mixture of an alkylidene cyclopropane and a 1,3-diene on pyrolysis above 600 °C ⟨78TL3855⟩. UV irradiation of 3-thietanone 1,1-dioxide affords a mixture of ketene and sulfene, as demonstrated by low temperature IR analysis and trapping studies ⟨74JA2964⟩. Tetracyclic sulfone (**19**) has been found to decompose at 150 °C to yield norbornadiene and sulfur dioxide; this reaction is presumably an example of a retro-2+2+2 process ⟨80UP51400, 82CC1105⟩.

Ph, R, SO_2; $h\nu$, CH_2Cl_2 → SO_2 + Ph, R
(**17**) R = Me, Et, $PhCH_2$ 83–95%

SO_2; 600 °C → +
(**18**)

S O_2; 150 °C ⇄ 25 °C; + SO_2
(**19**)

5.14.3.1.2 Rearrangements

2-Imino- and 2-thiono-thietanes such as (**20**) ⟨80AG(E)466⟩, (**21**) ⟨77TL4343⟩ and (**22**) ⟨80CC243, 80JOC4532⟩ are known to undergo light- and heat-induced rearrangements and ring opening. 3-Carbenathietanes such as (**23**) ⟨76JA7081, 78JA7436⟩ and (**24**) ⟨81TL4815⟩ rearrange with ring contraction giving alkylidene thiiranes (allene episulfides) and dialkylidene thiiranes (thiiranoradialenes) respectively.

Ph, Pr^i, S, PhN; 120 °C → $Pr^iCH{=}C(Ph)CSNHPh$
(**20**)

O, S, =NMe, Ph_2; $h\nu$ → O, Me N, =S, Ph_2 + O =CPh_2 + MeN=C=S
(**21**)

S, S; $h\nu$ >300 nm → S·, S; C_6H_6 → S, S 60%
(**22**)

↓ MeOH

S, S → S, S, OMe 80%

TsN(Na)N R₂ R₂ S R = H or Me Δ (23) R₂ S⁺ R R

(24) 70%

Alkyl-substituted thietane 1-oxides may isomerize *via* reversible ring opening as in the case of (**25**) ⟨77JCS(P1)1574⟩ or *via* pyramidal inversion as with (**26**) ⟨69JA2796⟩. 2,2,4,4-Tetramethyl-3-thietanone 1-oxide (**27**) rearranges to thiolane derivatives on heating; an unsaturated sulfenic acid similar to that formed from sulfoxide (**25**) is believed to be an intermediate in this interesting reaction ⟨76JCS(P1)2590⟩.

(25)

175 °C

(26)

C_6H_6 heat

67%

Me_3SiCl, $(Me_3Si)_2NH$

(27)

Thietes, benzothietes and their *S,S*-dioxides may undergo thermal and photochemical electrocyclic ring opening to the corresponding *α,β*-unsaturated thiocarbonyl compounds. This ring opening reaction may be reversible as in the case of spirothiete (**28**) ⟨78TL3617, 4839⟩; it may lead to dimer formation as in the case of benzothiete (**29**) ⟨77RTC282, 70JA7001⟩ or to ring expansion in the case of thiete (**30**) ⟨80JOC2962⟩, thiete 1,1-dioxide (**31**) ⟨69CC32, 70CJC3704, 72TL4781, 77JOC1910⟩ and various benzo- and naphtho-thiete 1,1-dioxides ⟨76JOC3044, 69JOC1310⟩. Naphtho[1,8-*bc*]thiete 1,1-dioxide (**32**) also undergoes ring expansion, among other interesting heat- and light-induced reactions ⟨67LA(703)96⟩.

=C(SR)CS(SR)

(28)

(29)

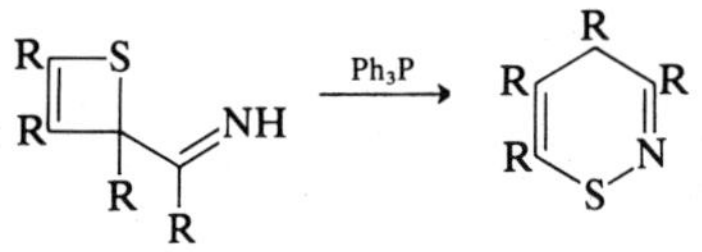

(30) R = CF_3

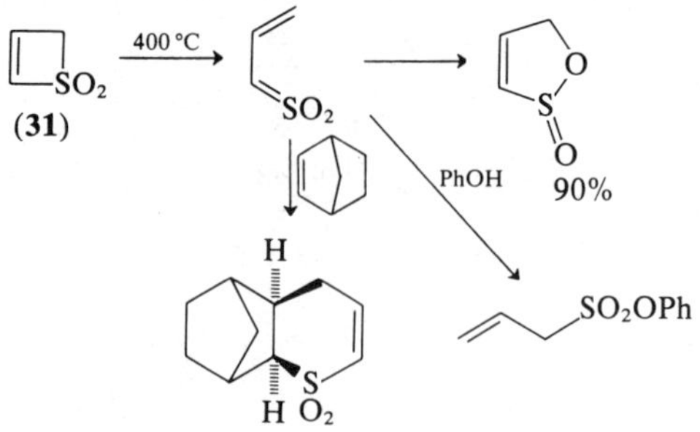

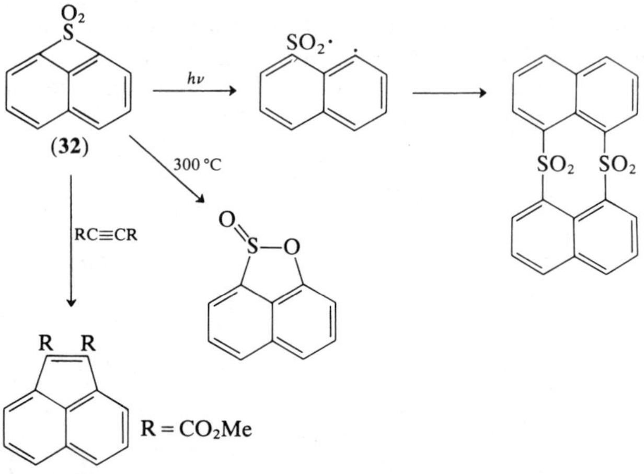

(31)

(32)

R = CO_2Me

5.14.3.2 Electrophilic Attack on Ring Heteroatom

5.14.3.2.1 Protonation and basicity

The order of decreasing Lewis basicity toward trimethylaluminum is thiane, thiolane, diethyl (or dimethyl) sulfide, thietane and 2-methylthiirane ⟨67IC1461⟩. Thietane can be protonated at −60 °C with fluorosulfonic acid in antimony pentafluoride–liquid sulfur dioxide; the ^{1}H NMR spectrum shows bands at δ 7.40 (S—H) and 3.20–4.40 p.p.m. (C—H) ⟨71JOC1121⟩.

5.14.3.2.2 Alkyl halides and related compounds

Thietane undergoes ring opening upon heating with methyl iodide affording dimethyl(3-iodopropyl)sulfonium iodide ⟨B-81MI51400⟩. Examples of useful applications of this reaction are preparations of sulfides (**33**) ⟨78JOC1185⟩ and (**34**) ⟨71S429⟩. Alkylation of thietanes and thietes without ring opening can be accomplished using trimethyloxonium fluoroborate ⟨71HCA2297, 71JA676, 76JA6643⟩.

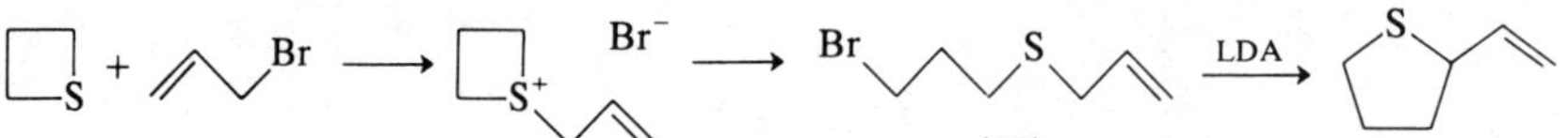

(**33**)

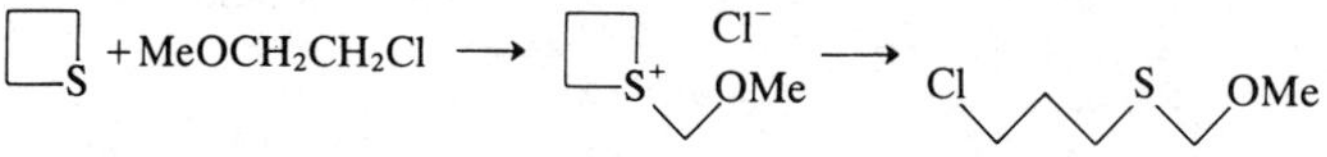

(**34**) 90%

5.14.3.2.3 Acyl halides

Benzoyl chloride reacts with thietane giving *S*-(3-chloropropyl) thiobenzoate in 86% yield ⟨71CHE557⟩.

5.14.3.2.4 Halogen

Thietanes and β-thiolactones generally undergo ring opening with chlorine and various sources of Cl^+ ⟨66CI(L)1555⟩. These ring opening reactions can be of synthetic utility, *e.g.* as in the preparation of ketodithiolane (**35**) ⟨68BAU1796⟩, sulfonyl chloride (**36**) ⟨69BSB191⟩ or disulfide (**37**) ⟨71LA(749)62⟩. An exception to this tendency for ring opening is seen in the reaction of thietane with 1-chlorobenzotriazole in methanol giving thietane 1-oxide in 90% yield ⟨69CC365⟩.

Cl_2 ⟶ ClS COCl ⟶ (H_2S) (**35**)

Cl_2, HOAc, H_2O ⟶ Cl … SO_2Cl (**36**) 51%

+ pyridine·ICl_2 ⟶ $(ClCH_2CH_2CH_2S)_2$ (**37**) 64%

5.14.3.2.5 Nitrogen

Electrophilic nitrogen compounds can be used to prepare *S*-aminothietanonium salts, as illustrated by the synthesis of (**38**) ⟨72TL4137⟩.

+ SO_2ONH_2 ⟶ (CH_2Cl_2, −20 °C) S^+ NH_2 $ArSO_3^-$ (**38**) 77%

5.14.3.2.6 Oxygen

The reaction of thietanes with electrophilic oxygen will be covered in Section 5.14.4.3 (Synthesis of Derivatives of the Ring Systems).

5.14.3.2.7 Carbocations

Intramolecular interaction of thietane sulfur atoms with carbocations (or developing carbocations) can lead to either ring expansion, as in the case of (**39**) ⟨66CCC1831⟩, or to ring contraction, as with (**40**) ⟨73JOC649, 69JOC3998⟩.

S OH ⟶ (HCl, Δ) S_+ ⟶ Cl S

(**39**)

X X ⟶ (CN^-) S

(**40**) X = Br or Cl

5.14.3.2.8 Sulfur

β-Thiolactones such as (**41**) undergo ring opening on treatment with the electrophilic sulfur compound methanesulfenyl chloride ⟨72CHE432⟩.

O S ⟶ (MeSCl) O S^+ SMe ⟶ MeSS … COCl

(**41**) 67%

5.14.3.3 Electrophilic Attack on Ring Atoms Adjacent to Heteroatom or Conjugated with Heteroatom

5.14.3.3.1 Protonation

Examples of reactions of the carbon–carbon double bond of thietes with protic acids are illustrated by the formation of mercaptoaldehyde (**42**) ⟨72JOC1111⟩, hydrogen cyanide adduct (**43**) ⟨77JOC3506⟩ and sulfide (**44**) ⟨62JA399⟩.

$\xrightarrow{H_3O^+}$ $HSCH_2CH_2CHO$ (**42**)

$\xrightarrow[KCN]{HCN}$ (**43**)

$\xrightarrow{H_2S}$ (**44**)

5.14.3.4 Insertion of Sulfur

A limited number of examples have been published of the insertion of sulfur into thietane rings giving 1,2-dithiolanes such as (**45**) ⟨76BCJ2491⟩ and (**46**) ⟨70TL441, 71CPB1022⟩.

$\xrightarrow[Et_2NH,\ DMF]{S_8}$ (**45**) 57%

$\xrightarrow[DMF]{H_2S}$ (**46**) 91%

5.14.3.5 Nucleophilic Attack on Ring Heteroatom

5.14.3.5.1 Carbanions

Ethyllithium catalyzes the polymerization of 2-methylthietane (**47**), presumably through initial ring opening of the thietane by attack of the organolithium compound at sulfur ⟨70JA3217⟩. Both butyllithium and phenylmagnesium bromide cause the ring opening of thietane (**48**), most likely by initial nucleophilic attack at sulfur ⟨76BCJ2491⟩. Thietane 1-oxide is reported to undergo ring expansion upon treatment with benzylmagnesium chloride ⟨75DOK(223)1379⟩. Ylide (**49**) is a probable intermediate in this reaction. Butyllithium effects the stereospecific desulfurization of thietanonium salts (**50**) and (**51**) ⟨71JA676⟩. The stereospecific collapse of tetravalent sulfur(IV) species has been invoked for this reaction. Methyllithium brings about the ring opening of naphthothiete (**52**) ⟨76JA6643⟩.

(**47**) $\xrightarrow[-78\,°C]{EtLi}$ $\longrightarrow$ $Et[SCH(Me)CH_2CH_2]_nLi$

(**48**) $\xrightarrow[-60\,°C]{BuLi}$ $\xrightarrow{-LiF}$ 53%

(**48**) $\xrightarrow{PhMgBr}$ 67%

(49)
33%

(50)
9 : 1
25%

(51)
13 : 87
25%

(52)
$n = 0$ or 2

5.14.3.5.2 Sulfur

Thietanone (**53**) ⟨79LA1768⟩ and thietane (**54**) ⟨72BAU505⟩ react with sulfur-containing nucleophiles giving ring-opened products. It is thought that these reactions are initiated by nucleophilic attack by the sulfur nucleophile on the ring sulfur.

(53)
$(MeCOCH_2S)_2$

$RSSCH_2CH{=}CH_2$
R = Bu or Ph

(54)

5.14.3.5.3 Phosphorus

Certain thietane derivatives such as (**54**) ⟨67JOC2009⟩, (**55**) ⟨81JCR(M)1132⟩ and (**56**) ⟨80JOC2962⟩ are desulfurized on treatment with triarylphosphines. These reactions probably involve initial attack by phosphorus on sulfur.

(55)
43%

(**56**) $R = CF_3$

5.14.3.5.4 Oxygen

Sodium hydroxide brings about the conversion of *S*-methylnaphthothiete derivative (**57**) to 1-methylsulfinylnaphthalene by a process undoubtedly involving nucleophilic attack of oxygen on sulfur ⟨76JA6643⟩.

(**57**)

5.14.3.6 Nucleophilic Attack on Ring Atoms Adjacent to Heteroatom or Conjugated with Heteroatom

5.14.3.6.1 Oxygen

The carbonyl group of β-thiolactones is quite susceptible to nucleophilic attack by oxygen and other heteroatoms leading to ring opening, *e.g.* as shown for thiolactones (**58**) ⟨80JOC4359⟩, (**59**) ⟨67BAU1365⟩ and (**60**) ⟨70JA7001⟩. Ring opening of (**59**) has been utilized in a synthesis of esters and amides of thiiranecarboxylic acids, which are hard to prepare by other methods. Addition of hydroxide or alkoxide ions to the carbon–carbon double bond of thiete 1,1-dioxides such as (**61**) ⟨62JA399⟩ and (**62**) ⟨67JOC3872⟩ also results in ring opening.

(**58**) $\xrightarrow{\text{MeONa}}$ $\longrightarrow$ $COS + Ph_2C{=}CPh_2 + NaOMe$

(**59**) $\xrightarrow{X^-}$ $\xrightarrow{\text{base}}$

$X = MeO$, 65%
$X = Et_2NH$, 76%

(**60**) $\xrightarrow{\text{MeOH}}$

(**61**) $\xrightarrow{OH^-}$ $\longrightarrow$ $^-CH_2SO_2CH_2CHO \longrightarrow MeSO_2Me$; $\xrightarrow{Me_2NH}$; $\xrightarrow{H_2S}$

(**62**) $\xrightarrow{\text{KOBu}^t}$ $\longrightarrow$ $+ C_4H_8$ 66%

5.14.3.6.2 Nitrogen

Nucleophilic attack by an amino group on the carbonyl function of a β-thiolactone forms the basis for a novel peptide synthesis, as illustrated by the preparation of (**63**) ⟨64RCR493⟩. Amines undergo Michael addition to the carbon–carbon double bond of thiete 1,1-dioxide (**61**) ⟨62JA399⟩.

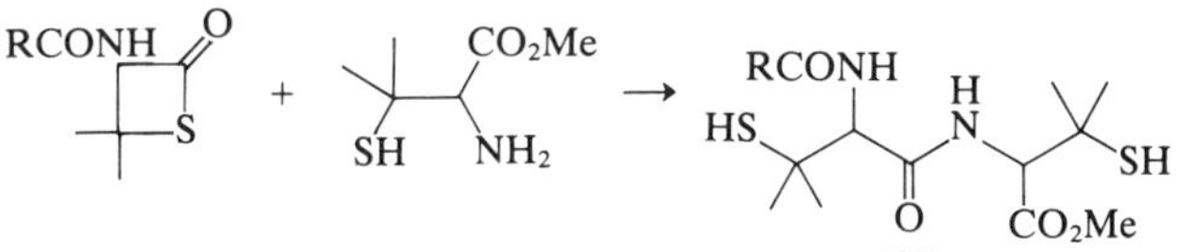

(**63**)

5.14.3.6.3 Hydride

Lithium aluminum hydride has been found to effect reductive ring opening of a number of thietes and thiete 1,1-dioxides (*e.g.* (**64**) ⟨70JOC3657⟩, (**65**) ⟨76JOC3044⟩ and (**66**) 76JA6643⟩). On the other hand, sodium borohydride in isopropyl alcohol at 60 °C reduces the carbon–carbon double, *e.g.* in (**64**) ⟨70JOC3657⟩, leaving the ring intact.

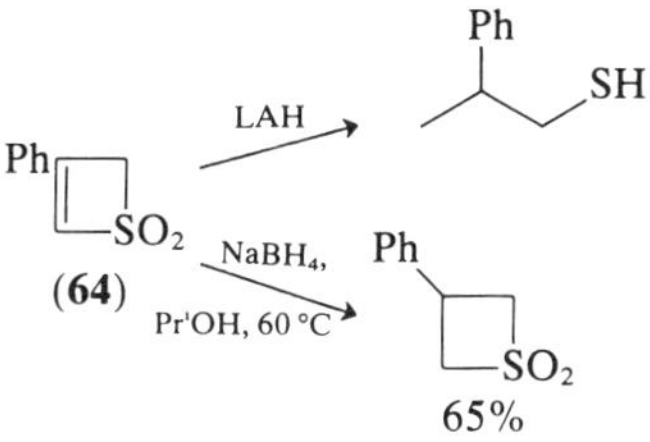

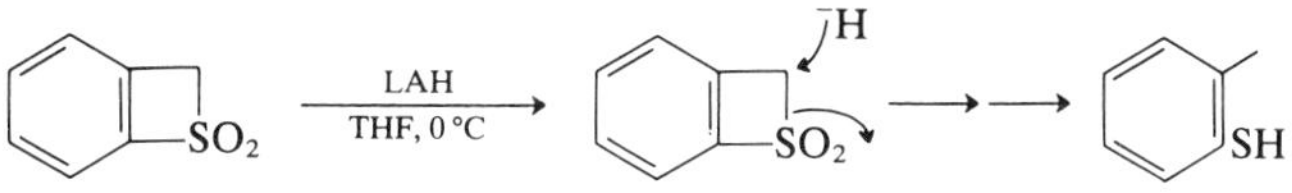

(**64**) 65%

(**65**)

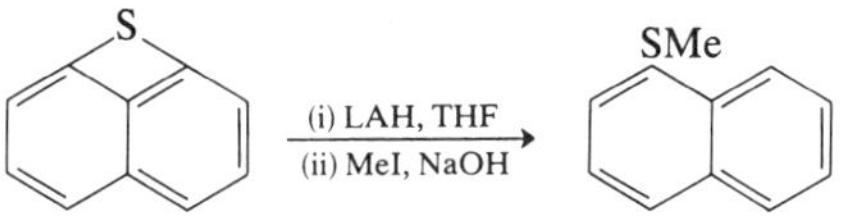

(**66**)

5.14.3.6.4 Sulfur

Hydrogen sulfide adds to thiete 1,1-dioxide (**61**) producing a sulfide ⟨62JA399⟩.

5.14.3.7 Nucleophilic Attack on Other Ring Sites

Examples of nucleophilic attack by oxygen, sulfur and hydride on the 3-position of thietane derivatives are illustrated for thietanes (**67**) ⟨72BAU505⟩, (**68**) ⟨75JCS(P1)2513⟩ and (**69**) ⟨63JA3236⟩ respectively.

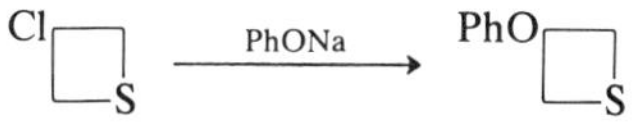

(**67**)

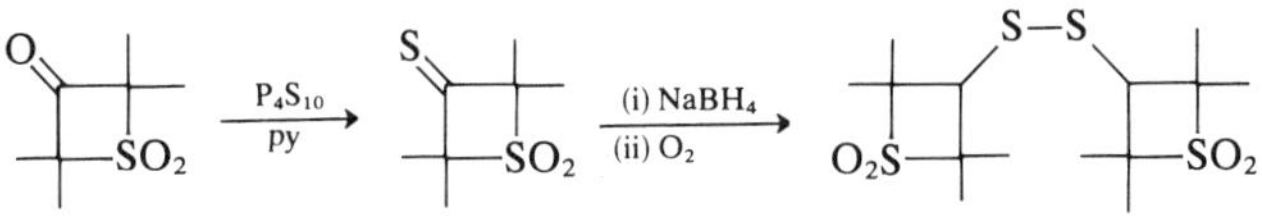

(**68**)

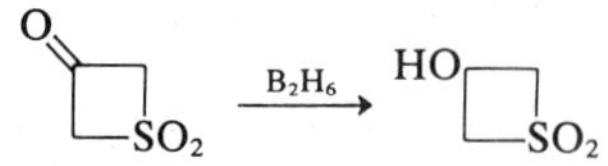

(69) $\xrightarrow{B_2H_6}$ 56%

5.14.3.8 Nucleophilic Attack on Hydrogen (Proton Abstraction)

2-Iminothietanes undergo deprotonation with LDA and stereospecific alkylation as illustrated by the synthesis of (**70**) ⟨80AG(E)466⟩. *S*-Ethylthietanonium tetrafluoroborate (**71**) has been reported to exchange its α-ring hydrogens 10^5 times faster than analogous open chain sulfonium salts. Thietane 1-oxides such as (**72a**) do not incorporate deuterium on exposure to deuterium oxide–sodium deuteroxide mixtures at 50 °C in contrast to the corresponding 1,1-dioxides (**72b**) ⟨70JOC3657⟩. 2,4-Diphenylthietane 1-oxide (**73**) undergoes an unusual base-induced ring contraction which presumably involves an α-sulfinyl carbanion ⟨71JOC2703⟩. Alkylthietane 1-oxides undergo a similar reaction ⟨82CC589⟩.

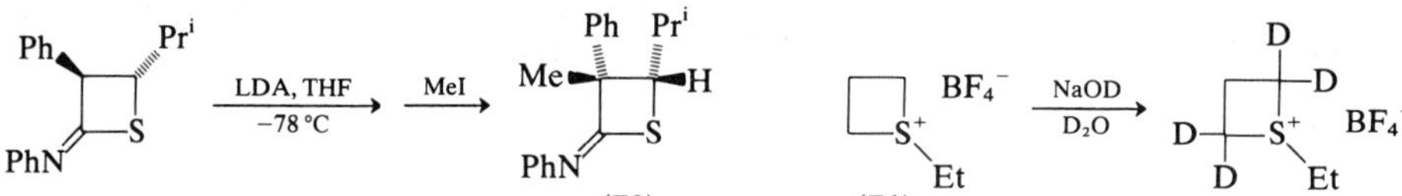

LDA, THF, −78 °C; MeI → (70)

(71) BF_4^- $\xrightarrow{NaOD,\ D_2O}$ BF_4^-

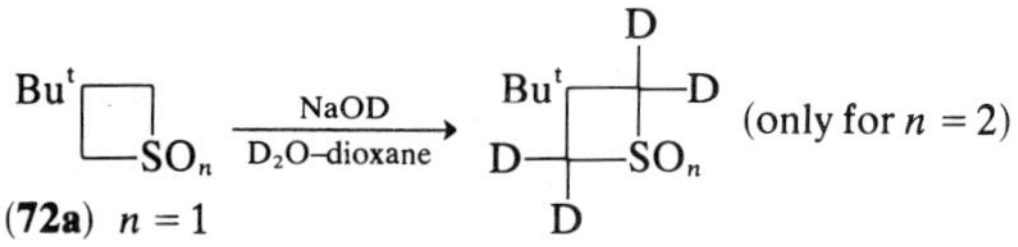

(72a) $n = 1$
(72b) $n = 2$

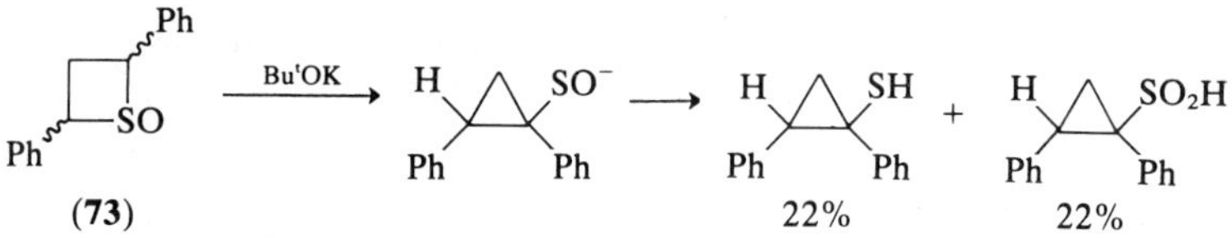

NaOD, D_2O–dioxane (only for $n = 2$)

(73) $\xrightarrow{Bu^tOK}$ → 22% + 22%

Thietane 1,1-dioxide and its derivatives can be deprotonated with *n*-butyllithium, and the mono- or di-anion (the latter generated with two equivalents of base) can be captured with a variety of alkylating agents as illustrated by preparation of (**74**) and (**75**) ⟨81TH51400, 81UP51400⟩. Specific applications of this reaction will be given in Section 5.14.4.2.5 below. 3-Thietanone 1,1-dioxide is particularly acidic, undergoing vigorous effervescence with aqueous sodium bicarbonate ⟨63JA3236⟩. Proton abstraction in 3-substituted thietane 1,1-dioxides is employed in several syntheses of thiete 1,1-dioxides, *e.g.* (**76**) ⟨76JOC3044⟩. Deprotonation of thietane 1,1-dioxides can also result in rearrangement, as in the ring contraction of (**77**) ⟨71JOC2693⟩. Deprotonation of bicyclic thiete 1,1-dioxide (**78**) is undoubtedly the key step in the base-catalyzed isomerization of this compound ⟨67JOC3872⟩.

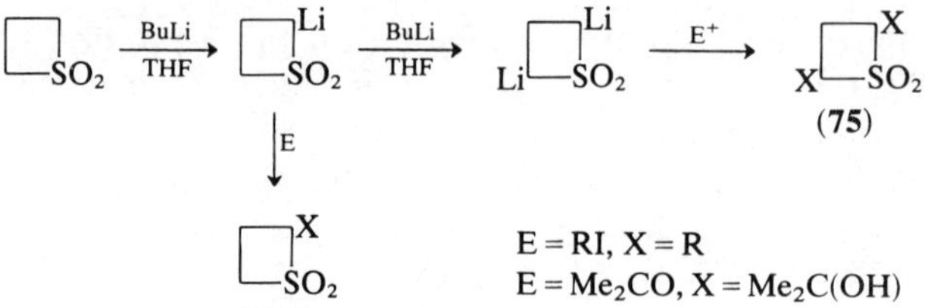

BuLi, THF; BuLi, THF; E^+ → (75)

E → (74)

E = RI, X = R
E = Me_2CO, X = $Me_2C(OH)$

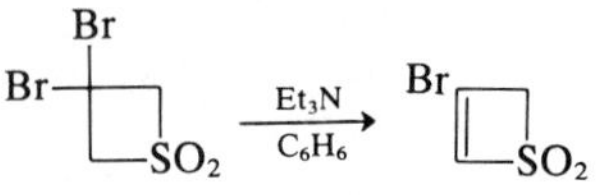

$\xrightarrow{Et_3N,\ C_6H_6}$ (76) 88%

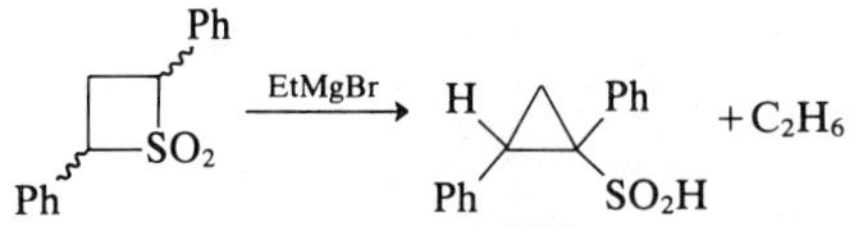

(77) $\xrightarrow{EtMgBr}$ 75% + C_2H_6

(78) $\xrightarrow{KOH,\ THF}$ 90%

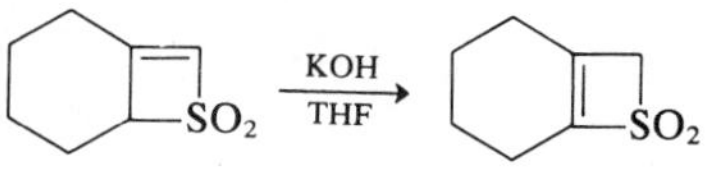

5.14.3.9 Reaction with Radicals or Electron Deficient Species

5.14.3.9.1 Radical attack on ring heteroatom

ESR evidence suggests that free radicals attack thietane sulfur leading to ring opening, possibly *via* a tricoordinate intermediate or transition state such as (**79**) ⟨76CC407, 77JCS(P2)1708⟩. The less strained thiolane undergoes abstraction of an α-hydrogen atom under these same conditions.

$(CH_2)_n$—S; RO·, $n = 1$ → [(**79**), OR] → $ROSCH_2CH_2CH_2\cdot$; RO·, $n = 2$ → thiolanyl radical (H)

5.14.3.9.2 Radical attack on other ring atoms

Free radical bromination of thietane 1,1-dioxide takes place at the 3- rather than the 2-position, giving ultimately 3,3-dibromothietane 1,1-dioxide (**80**) ⟨76JOC3044⟩. This preference for abstraction of a hydrogen atom β to a sulfonyl group reflects the instability of a free radical α to a sulfone function.

SO_2 —(Br_2–Cl_2, CCl_4, $h\nu$)→ Br, SO_2 —(Br_2–Cl_2, CCl_4, $h\nu$)→ Br_2, SO_2 (**80**)

5.14.3.9.3 Attack of carbenes and related species on ring heteroatom

Reaction of a carbenoid reagent with thietane has been found to lead to ring expansion to thiolane (**81**) ⟨72JOC1721, 75BCJ1490⟩. Atomic carbon desulfurizes thietanes in a stereoselective manner, perhaps *via* ylidic species (**82**) ⟨73JA1547⟩.

S + $N_2C(CO_2Me)_2$ —($CuSO_4$, 110 °C)→ S^+–$\bar{C}(CO_2Me)_2$ → $(CO_2Me)_2$ S (**81**) 26%

S —(C_1)→ [S^+–C^- (**82**)] → [$\overset{+}{S}$=C^-] —(–CS)→ [ ] → 26% + 7% + C_5H_{10} + C_5H_8 26% 41%

5.14.3.9.4 Reaction on surfaces and with transition metal complexes

Examples of the desulfurization of thietane and thiete derivatives upon treatment with Raney nickel are given in Table 7. In addition to replacement of the C—S bonds with C—H bonds other types of reactions occur including cyclization and retro-2+2 decomposition. Other reactions which presumably involve heterogeneous processes include ring expansion of thietane to give 1,2-dithiolane (**83**) and of 2-methylthietane to yield thiolane (**84**) ⟨B-73MI51401⟩ and catalytic hydrogenation of thiete derivatives (**85**) ⟨68JOC3027⟩ and (**86**) ⟨75S333⟩.

R, S —(Al_2O_3, H_2S, 350 °C, R = H)→ S, S (**83**)

R, S —(Al_2O_3, H_2S, 350 °C, R = Me)→ S (**84**) 16%

Table 7 Desulfurization of Thietane Derivatives with Raney Nickel

Compound	*Product (yield)*	*Ref.*
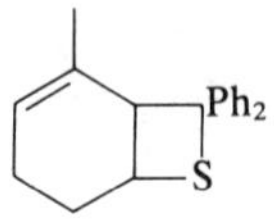 Ph2, S	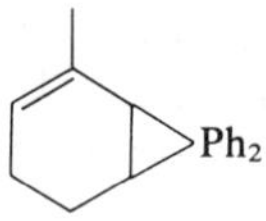 Ph2	67JOC3676
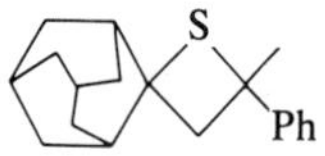 S, Ph	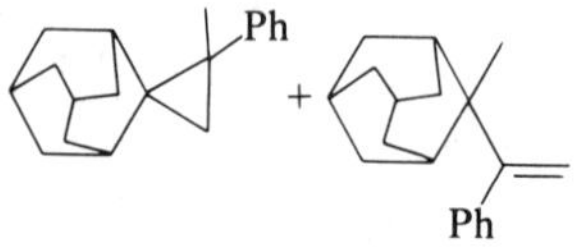 Ph + Ph	76JA2219
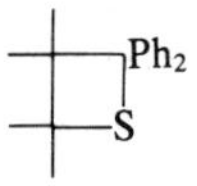 Ph2, S	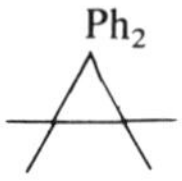 Ph2 (79%)	72MI51400
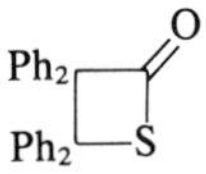 Ph2, O, Ph2, S	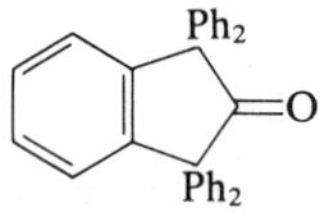 Ph2, O, Ph2 (49%) + $Ph_2C{=}CPh_2$ (47%)	80JOC4359
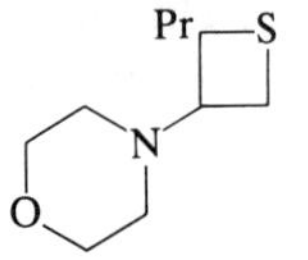 Pr, S, N, O	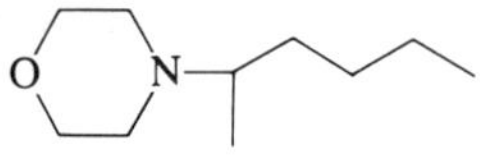 O, N	65LA(684)92
 S, S	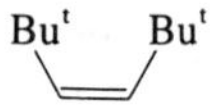 Bu^t Bu^t (66–72%) + *trans* isomer (37%)	67JA2059
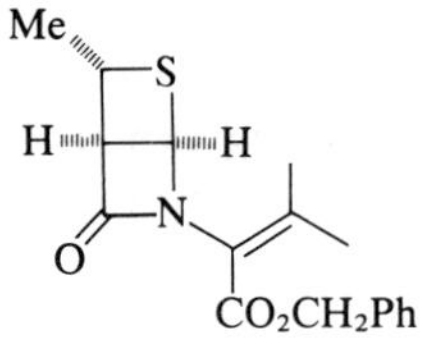 Me, S, H, H, O, N, CO_2CH_2Ph	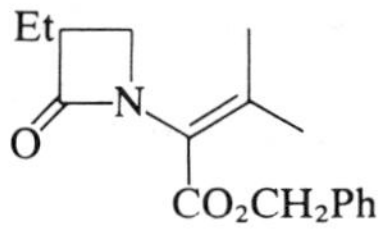 Et, O, N, CO_2CH_2Ph	78JA3251
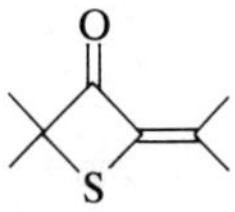 O, S	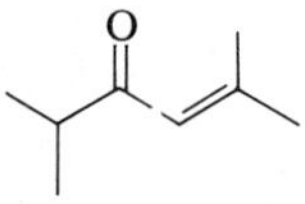 O	80JOC4804

Ph, SO_2 —H_2→ Ph, SO_2
(85)

SO_2 —H_2, 10% Pd/C→ SO_2
(86) 100%

The spirodithietanes 2,6-dithiaspiro[3.3]heptane, 2,8-dithiadispiro[3.1.3.1]decane and 2,10-dithiatrispiro[3.1.1.3.1.1]tridecane form diruthenium complexes which serve as models for the study of electron transport in biological systems ⟨82JA2596⟩. 2,2,4,4-Tetramethyl-3-thietanone has been reported to form the diiron complex (**87**), which in turn undergoes cycloaddition with 1,3-dienes ⟨76JA6696⟩. Thiete 1,1-dioxides form complexes with nonacarbonyldiiron (*e.g.* **88**) ⟨73JOC3963⟩ and with platinum compounds ⟨81JOM(205)417⟩ while thietes themselves form ring-opened thioacrolein complexes on treatment with iron and cobalt carbonyls (*e.g.* **89**) ⟨76JA2795, B-79MI51400⟩.

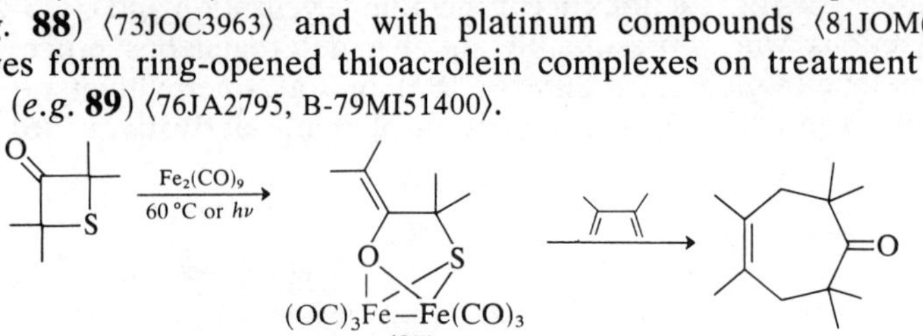

O, S —$Fe_2(CO)_9$, 60 °C or $h\nu$→ O, S, $(OC)_3Fe{-}Fe(CO)_3$ **(87)** —→ O

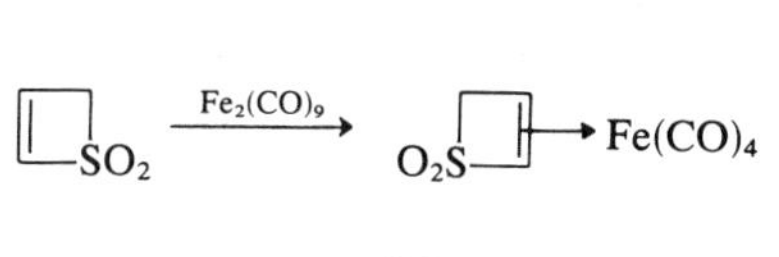

(88) 50%

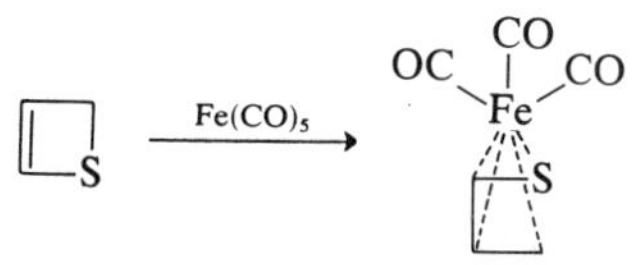

(89) 30%

5.14.3.10 Reactions with Cyclic Transition States, Formally Involving a Second Species

Thiete 1,1-dioxides readily form Diels–Alder adducts with 1,3-butadiene (**90**) ⟨76JOC3044⟩, cyclopentadiene (**91**), furan (**92**) ⟨78TL3337⟩, anthracene (**93**) ⟨62JA399⟩ and isonaphthofuran derivatives (**94**) ⟨65JOC629⟩ and form 1,3-dipolar cycloadducts with diazoalkanes, *e.g.* (**95**) and (**96**) ⟨80JOC3646, 70JOC999⟩. Thiete 1,1-dioxide also forms 2+2 cycloadducts such as (**97**) and (**98**) with enamines and ynamines ⟨70JOC905⟩. A 2+2 photodimer (**99**) is formed on irradiation of a methylenethiete 1,1-dioxide ⟨68JOC3027⟩. Cycloadducts are formed from the reaction of tris(imino)thietane (**100**) with enamines and ynamines ⟨78BSB893, 79BSB297⟩.

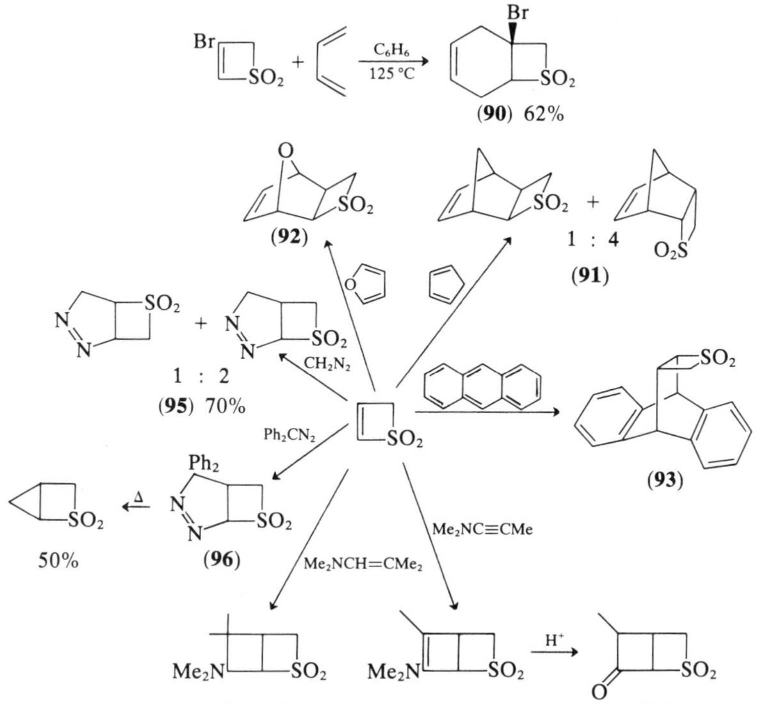

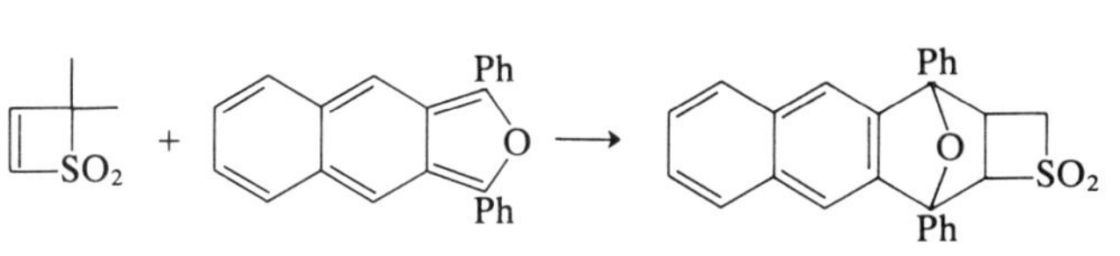

(90) 62%

(92)

1 : 4

(91)

1 : 2

(95) 70%

50%

(96)

(93)

(97) 60%

(98)

45%

(94) 91%

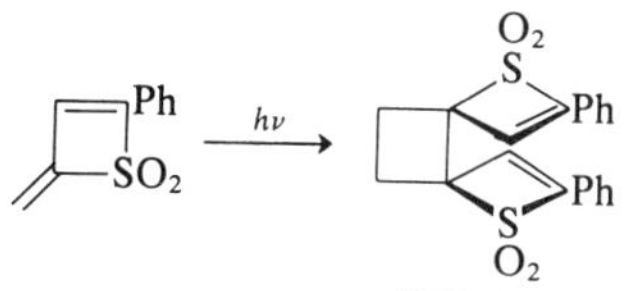

(99) 12%

71%

(100)

74%

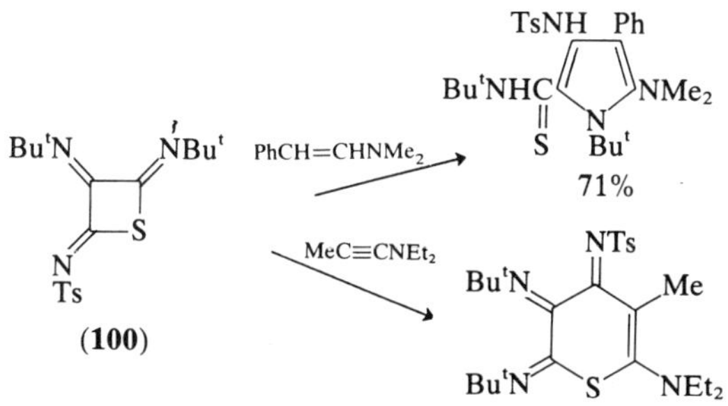

5.14.4 SYNTHESIS

5.14.4.1 Ring Synthesis from Nonheterocyclic Compounds

5.14.4.1.1 Synthesis via formation of one bond

This section considers syntheses of thietanes and thietes in which acyclic precursors undergo cyclization with formation of one ring bond, typically the C—S bond. Simple examples of this approach are illustrated by the preparation of 2,2-dimethylthietane from chlorothiol precursor (**101**) ⟨74HCA2514⟩ and the highly strained 5-thiabicyclo[2.1.1]hexane and 6-thiabicyclo[3.1.1]heptane from halothioacetates (**102a**) ⟨70TL2931⟩ and (**102b**) ⟨58JOC1026⟩ respectively. The requisite thioacetates can be prepared through base-catalyzed addition of thioacetate ion to conjugated enones, as illustrated by the syntheses of the biologically very potent sulfur analogues of thromboxane A2, (**103**) ⟨81TL5301⟩ and (**104**) ⟨81JA4597⟩, or from free radical addition of thioacetic acid to unsaturated compounds, such as allyl chloride (**105**) in a synthesis of thietane ⟨64HC(19-2)647⟩.

H_2S, $AlCl_3$; OH^-

(**101**) 90%

SCOMe; $(CH_2)_n$; X; KOH; S

(**102a**) $n = 1, X = Cl$
(**102b**) $n = 2, X = Br$

O; CO_2Me; OAc; MeCOSH, Bu^{t}OK, −78 °C; AcS; (i) $Zn(BH_4)_2$ (ii) MesCl

OMes; AcS; NaOMe, MeOH, 55 °C; S; CO_2Me; OH

(**103**)

O; CO_2Me; S; $CH(OMe)_2$; $MeO_2CCH_2CH_2SH$, R_3N; S; MeO_2C; KOBu^t

OMes; S; MeO_2C; H←:B; CO_2Na; OH

(**104**)

Cl + MeCOSH; $h\nu$; AcS; Cl; KOH; S

(**105**)

3-Hydroxythiocyanates can be converted into thietanes (*e.g.* (**106**)) with base ⟨71JA676⟩. A mechanistically related reaction is presumably involved in the conversion of (**107**) into 2,4-diphenylthietane ⟨81S547⟩. While free radical cyclization of 3-butenethiol (**108**) gives thiolane with only a trace of the four-membered ring 2-methylthietane ⟨67CR(C)(264)610⟩, 3-butenesulfenic acid (**109**) cyclizes to a mixture of *cis*- and *trans*-2-methylthietane 1-oxide. This cyclization is reversible since the neat, pure sulfoxides interconvert (even at room temperature!), ultimately giving a 2:1 equilibrium mixture of the *cis* and *trans* isomers ⟨77JCS(P1)1574⟩.

OH SCN —NaH, triglyme, Δ→ [O, S, N⁻] → [NCO S⁻] → (**106**)

PhCH=CHCOPh —$(EtO)_2P(S)SH$→ $PhCH_2+CH_2COPh$ ($SP(S)(OEt)_2$) —(i) $NaBH_4$, 60 °C; (ii) NaH→

Ph Ph S O⁻ S=P(OEt)₂ (**107**) → [Ph Ph S O P ⁻S OEt OEt] → Ph S Ph

(**108**) SH —Δ, ROOR→ S 43% + S trace

Bu^t S O —140 °C→ [S O H] (**109**) → S O 31%, 3:1 *cis*:*trans*

Thietanes are formed by intramolecular cyclization of thiocarbenes (**110**) ⟨75BCJ1490⟩, on pyrolysis of the copolymer of sulfur, tetrafluoroethylene and thiocarbonyl fluoride (**111**) ⟨69CC1274⟩, and upon electrolysis or base treatment of dithione (**112**) ⟨67JOC1562, 78JOC1980, 79JCR(S)320⟩.

S Ph NN(Na)Ts —hν, DME→ [S Ph] (**110**) →

S⁺ Ph → S Ph 25%

$[-CF_2CF_2SCF_2CF_2CF_2S-]_n$ (**111**) —500 °C, vac.→ F_2 F_2 F_2 S 20%

S S (**112**) —electrolysis or base→ S S

β-Thiolactones and thietane-2,4-diones (malonic thioanhydride) are formed from the corresponding β-mercaptocarboxylic acids or thiomalonic acids through the use of dehydrating conditions, as illustrated by the preparation of (**113**) ⟨64RCR493⟩, (**114**) ⟨80JCS(P1)388⟩ and (**115**) ⟨72JOC2514, 74JOC2946⟩. Thietes can be formed by electrocyclic ring closure of certain enethiones such as (**116**) ⟨78TL3617, 4839⟩ and acid treatment of a γ-ketosulfide (**117**) ⟨74TL3911⟩.

$HSCMe_2CH(NHR)CO_2H$ $\xrightarrow{DCC}$ (**113**) 84%

ClCO₂Et, py → (**114**)

Δ or $(CF_3CO)_2O$/py or $MeC(OAc){=}CH_2$; H^+ → (**115**)

(**116**) R = Bu^t

(**117**) $\xrightarrow[POCl_3]{HClO_4}$ ClO_4^-

5.14.4.1.2 Synthesis via formation of two bonds

(i) From 3+1 atom fragments

The oldest method for the preparation of thietanes involves the reaction of 1,3-dihalides and sodium (or potassium) sulfide ⟨64HC(19-2)647⟩. While this reaction involves the intermediacy of 3-halopropanethiolates, insofar as two carbon–sulfur bonds are formed without isolation of intermediates, it is convenient to classify this and related reactions as examples of '3+1' bond forming sequences. A variety of solvents and conditions can be used for this reaction, as illustrated by the preparation of thietane itself (**118**) in liquid ammonia ⟨63RTC63⟩, of 2,5-dithiaspiro[3.3]heptane (**119**) in water using a phase transfer catalyst ⟨80TH51400, 80UP51400⟩, of alkynic dithietane (**120**) using dry sodium sulfide in ethyl cellosolve ⟨67JA2059⟩, of dithietane (**121**) using dimethoxyethane ⟨74LA1328⟩, and of 3,3-disubstituted thietanes using DMSO ⟨78JOC3827⟩. It might be noted here that the formation of 3,3-disubstituted thietanes is sterically favored due to the '*gem*-effect' of the 3-substituents. Thietane specifically deuterated at either the 2- or 3-position has been prepared by treatment of appropriately labelled 1,3-dibromopropane with sodium sulfide ⟨77MI51400⟩. Thiourea or 2-pyridinethiones and base can sometimes be used in place of the alkali sulfide in reactions with 1,3-dihalides ⟨71BSF3064, 77S884⟩.

$\xrightarrow[NH_3]{Na_2S}$ (**118**) 32%

$C(CH_2Br)_4$ $\xrightarrow[C_{16}H_{33}PBu_3^+\ Br^-]{Na_2S{\cdot}9H_2O,\ H_2O,\ 80\,^\circ C}$ (**119**) 74%

$(ICH_2)_2CMeCH_2C{\equiv}CCH_2C(Me)(CH_2I)_2$ $\xrightarrow[\text{ethyl cellosolve}]{Na_2S,\ 110\text{–}125\,^\circ C}$ (**120**) 85%

$\xrightarrow[DME]{Na_2S{\cdot}9H_2O}$ (**121**) 52%

Other processes which can be classified as '3+1' syntheses of thietanes include reaction of sulfur dichloride (**122**) with norbornadiene ⟨79JOC4757, 69JOC3998⟩ and methylenenorbornene ⟨68JOC2620⟩, reaction of thionyl chloride with ketone (**123**) ⟨70JA5258, 75JOC3179, 80PS(8)1⟩, photochemical insertion of sulfur dioxide into cyclopropane (**124**) ⟨77JOC1251⟩, and addition of hydrogen sulfide to acrolein giving 2-hydroxythietane ⟨73GEP2314103⟩.

SCl$_2$ (122)

53%

PhCH$_2$CH$_2$COMe (123) $\xrightarrow{SOCl_2}$ =CHPh 38%

(124) $\xrightarrow[h\nu,\ -10\ °C]{SO_2}$

(ii) From 2+2 atom fragments

Thermal and photochemical cycloaddition reactions of 2π-electron species represent an important synthetic approach to four-membered rings. The reactions summarized in this section include 2+2 cycloaddition reactions of thioketones, thioketenes, isothiocyanates, sulfenes and iminosulfenes with alkenes, allenes, ketenes, ketenimines and alkynes.

Hexafluorothioacetone, produced by the action of fluoride ion on the dimer of the thioketone, adds to ethyl vinyl ether giving a good yield of the thietane (**125**) ⟨76BCJ2491⟩. Less reactive thiones will undergo 2+2 cycloaddition to alkenes upon irradiation ⟨76ACR52⟩. The yield, stereospecificity and regiospecificity of thietane formation are quite dependent on the wavelength of irradiation and the nature of the alkene. Thietanes are themselves photolabile, so irradiation at short wavelengths (*e.g.* 254 nm in the case of 2,2-diphenylthietanes) is likely to result in substantial loss of product. Irradiation of aromatic thiones at somewhat longer wavelength (*e.g.* 360 nm for thiobenzophenone) can lead to population of a reactive upper (S_2) singlet state which adds stereospecifically to electron deficient alkenes such as the 1,2-dichloroethylenes (**126a** and **b**) ⟨69JA5038⟩ and, in the case of xanthione, dimethyl maleate and fumarate (**127a** and **b**) ⟨76TL3999⟩. With electron rich alkenes such as 1,2-dimethoxyethylene (**128**) xanthione undergoes nonstereospecific photocycloaddition. The latter reaction is thought to involve the triplet (T_1) state of the thione rather than the S_2 state, since there is a better match-up of energies of the thione and alkene orbitals with the former state. In the case of the aliphatic thione adamantanethione (**129**), irradiation at short wavelengths (250 nm) leads to excitation into S_2 which adds stereospecifically but nonregiospecifically to alkenes. Irradiation of (**129**) at longer wavelengths (500 nm) leads to nonstereospecific but regiospecific thietane formation with these same alkenes. It is suggested that the reactive state in these latter reactions is the triplet T_1, formed by intersystem crossing of the singlet S_1 state ⟨78CRV125⟩.

$\xrightarrow[DMF]{KF}$ [(CF$_3$)$_2$C=S] $\xrightarrow{CH_2=CHOEt}$ (**125**) 72%

Ph$_2$C=S $\xrightarrow{h\nu,\ 360\ nm}$ (**126a**) 83%; (**126b**) 90%

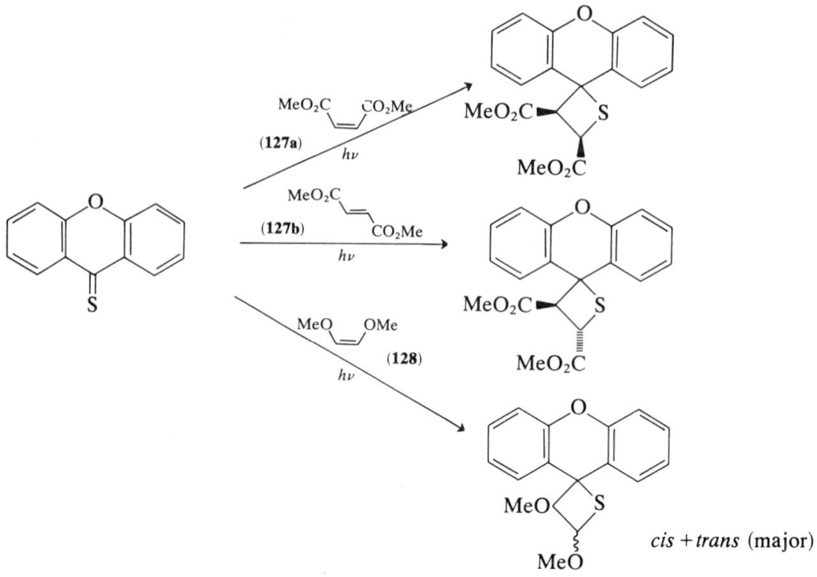

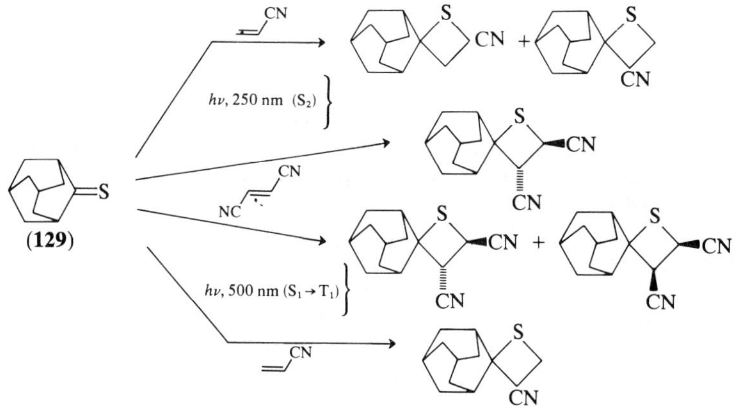

Thiophosgene undergoes photocycloaddition to alkenes and allenes affording reactive dichlorothietanes (**130**) and (**131**) ⟨74CB2544⟩. Other examples of photochemical and thermal 2+2 cycloadditions between thiocarbonyl compounds and cumulenes and heterocumulenes are illustrated by the formation of cumulene adduct (**132**) ⟨79TL4857⟩, β-thiolactone (**133**) ⟨78JOC4961, 79ZOR2602⟩, 2-iminothietanes (**134**) ⟨80JOC3766⟩ and (**136**) ⟨81AG(E)613⟩, 2,4-bis(imino)thietane (**137**) ⟨79TL3213⟩ and 2-alkylidenethietane (**138**) ⟨70JOC3470⟩.

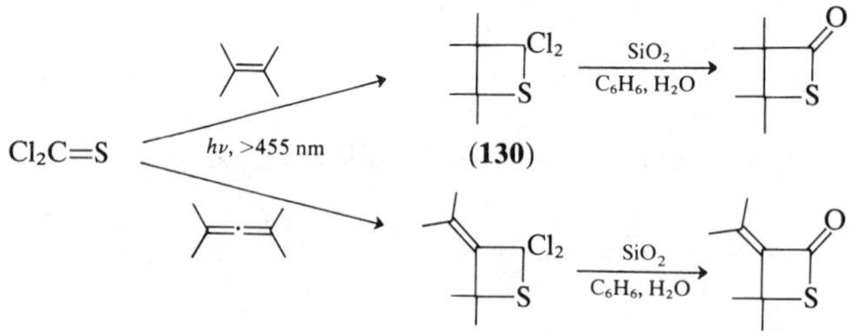

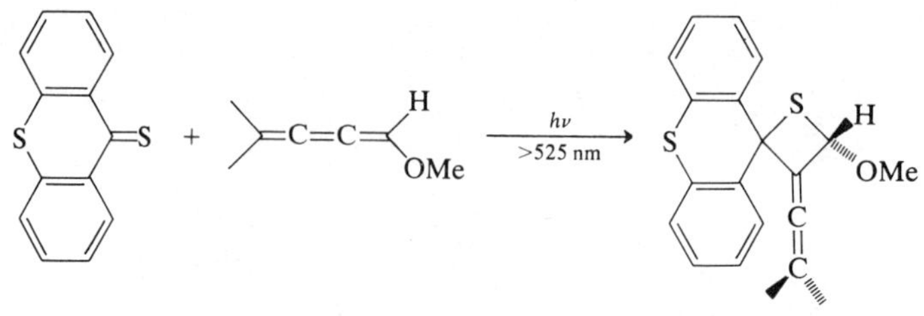

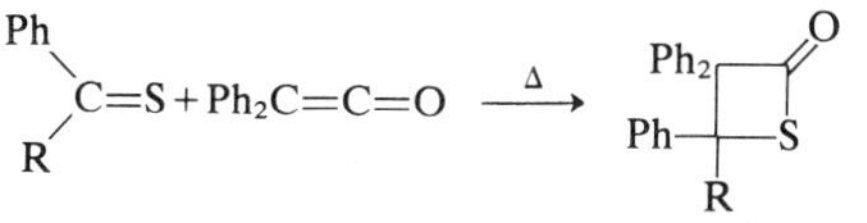

(133) R = Ph, 74%
R = MeS, 61%

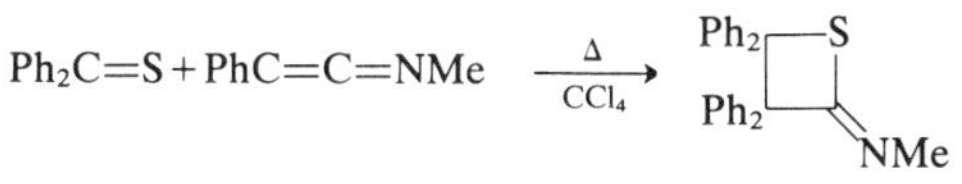

(134) 50%

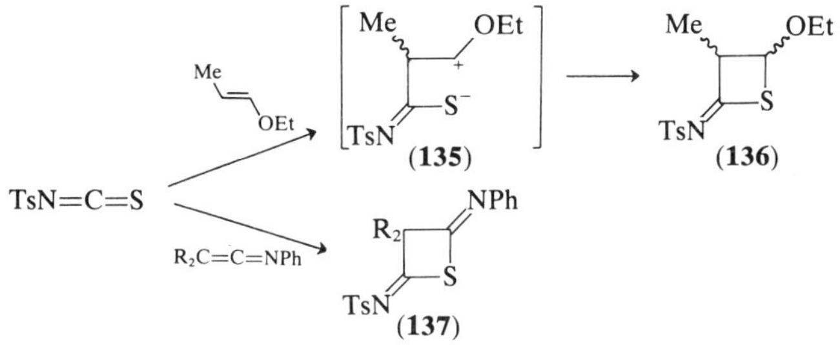

(135)

(136)

(137)

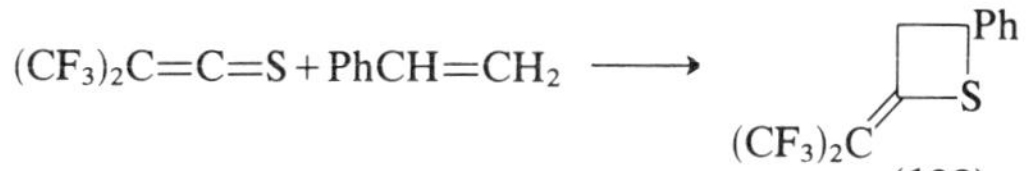

(138)

Thietane 1,1-dioxides are available in good yield through addition of sulfenes to electron rich alkenes, as illustrated by the preparation of (**139**) ⟨64HC(19-2)647⟩, (**140**) ⟨77BCJ1179⟩ and (**141**) ⟨71T2599⟩. The sulfenes may be generated *in situ* by base-induced dehydrochlorination of alkanesulfonyl chlorides (*e.g.* **139**, **141**), interaction of diazoalkanes with sulfur dioxide (*e.g.* **140**), or more recently by fluoride ion-induced desilylation of trimethylsilylmethanesulfonyl chloride (*e.g.* **139**) ⟨82TL4203⟩. Iminosulfenes, generated in a manner analogous to sulfenes, can be trapped with electron rich alkenes yielding thietane sulfoximines, *e.g.* (**143**) ⟨79JOC2055⟩.

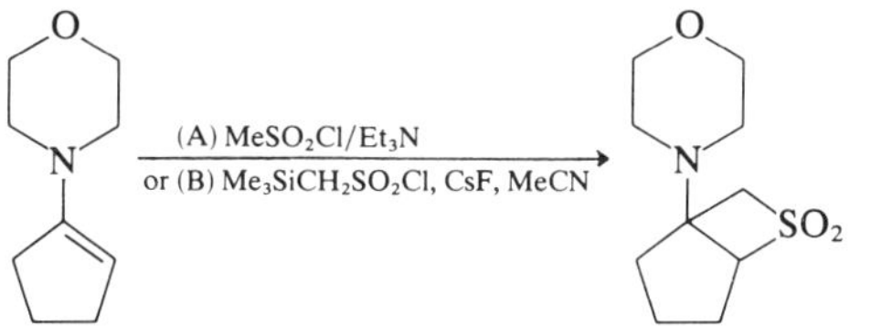

(A) $MeSO_2Cl/Et_3N$ or (B) $Me_3SiCH_2SO_2Cl$, CsF, MeCN

(139) A, 77%; B, 81%

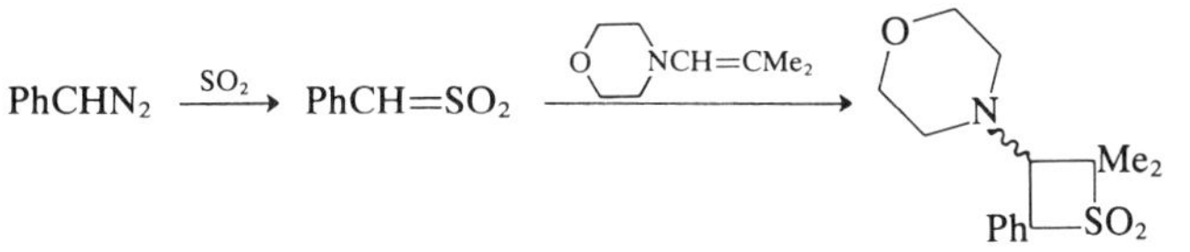

$PhCHN_2 \xrightarrow{SO_2} PhCH{=}SO_2$

(140) 62%

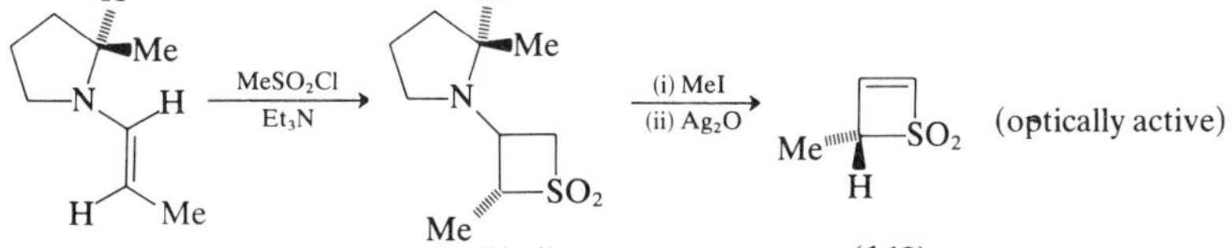

(141) (142) (optically active)

$MeS(O)(NTs)Cl \xrightarrow[CH_2=C(OEt)_2]{DABCO}$

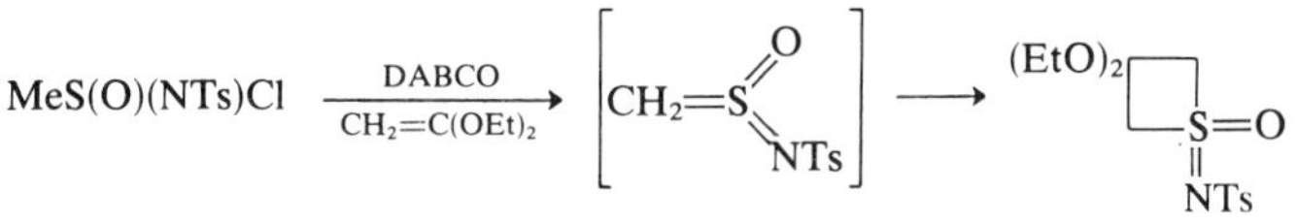

(143) 46%

When enamines are employed as sulfene traps, thiete 1,1-dioxides (*e.g.* 142) can be prepared by subsequent Hofmann elimination of amine. An alternative route to thiete 1,1-dioxides involves trapping sulfenes with ynamines, as illustrated by preparation of (**144**) ⟨73S534⟩. Certain thietes such as (**145**) ⟨80LA873, 78TL3617, 78TL4839⟩ are available directly from thioketones by 2+2 photocycloaddition with alkynes.

$PhCH_2SO_2OAr + MeC{\equiv}CNEt_2 \xrightarrow[THF]{Bu^tOK}$ (**144**) 79%

(**145**)

5.14.4.2 Ring Synthesis by Transformation of Other Heterocycles

5.14.4.2.1 Formation from three-membered heterocycles

Thietanes may be prepared from epichlorohydrin and its derivatives, *e.g.* (**146**) ⟨74ACS(B)701, 78PS(4)167, 78CJC71⟩, and by ring expansion of thiiranes such as (**147**) ⟨74BCJ1455⟩, (**148**) ⟨69JOC3998⟩, (**149**) ⟨79TL3213⟩ and (**150**) ⟨79CC160⟩. Three-membered-ring sulfur compounds (**151**) ⟨80JOC2962⟩ and (**152**) ⟨73BCJ667⟩ are starting materials for unusual syntheses of thiete derivatives.

(**146**) $\xrightarrow[OH^-]{H_2S}$ R = R′ = H, 65%; R = H, R′ = Ph, 36%; R = Ph, R′ = H, 63%

(**147**) $\xrightarrow[NaOAc]{HOAc}$ 82%

(**148**) $\xrightarrow{KIO_4}$

(**149**) $\xrightarrow{ArNC}$ 33%

TsN=C=S + RNC, R = Bu^t ⟶ (**150**) ⟶ 60%

(**151**) R = CF_3, R′ = H, Ph, Bu^t $\xrightarrow{\Delta}$ $\xrightarrow{-N_2}$ ⟶ 83–93%

(**152**) + $Me_2\overset{+}{S}$–$\overset{-}{C}$HCOAr ⟶ 13% + · · ·

5.14.4.2.2 Formation from four-membered heterocycles

While there are few examples of syntheses of thietanes from nonsulfur-containing four-membered heterocycles, *e.g.* (**153**) ⟨80AG(E)466⟩, there are numerous examples of the interconversion of thietanes and thietes, *e.g.* as exemplified by the preparation of (**154**) ⟨72JOC1111, 79PS(7)5⟩, (**155**) ⟨76JOC3044⟩, (**156**) ⟨68JOC3027⟩, (**157**) ⟨70JOC3657⟩, (**158**) ⟨69JOC2791⟩ and (**159**) ⟨76AG(E)242⟩.

PhN=C=S, −78°C; 20 °C; H_2O — (**153**) → 41%

(i) MeI (ii) $KOBu^t$, DMF — (**154**) R = H, 30%; R = Ph, 85%

(i) Et_3N (ii) butadiene; (i) Br_2, CCl_4 (ii) DBN, C_6H_6 — (**155**)

$PhCH_2SO_2Cl + Me_2NCH{=}CHCH_2NMe_2$ (Et_3N) → (H_2O_2, HOAc–Ac_2O) → (**156**) 37%

H_2O_2, HOAc — (**157**)

$Me_2NCH{=}CH_2$ ($MeSO_2Cl$, Et_3N) → (i) MeI (ii) Ag_2O → (**158**); H_2/Pd, 100 atm; LAH → (**159**)

5.14.4.2.3 Formation from five-membered heterocycles

Examples of the formation of thietanes and thietes from five-membered heterocycles include the desulfurization of 1,2-dithiolanes (**160**) ⟨70JOC3259, 78TL5233, 70JOC1781⟩ and (**161**) ⟨72JOC2514, 81IZV1633⟩ with trivalent organophosphorus compounds, the pyrolysis or photolysis of 1,2-dithiolane 1,1-dioxides (**162**) ⟨70JOC1781⟩ and (**163**) ⟨76JA6643⟩, photochemical or thermal Wolff rearrangement of diazoketones (**164**) ⟨80JOC4804⟩ and (**165**) ⟨80TL343, 77CB2242⟩, pyrolytic decarbonylation of (**166**) ⟨80TL343, 79TL59⟩, photochemical rearrangement of bicyclic sulfide (**167**) ⟨81TL4421, 74CL887, 70JOC584⟩, flash vacuum pyrolysis at 1000 °C of benzo[*b*]thiophene 1,1-dioxide (**168**) ⟨77RTC282⟩, solvolysis of 3-chloro-2-methylthiolane (**169**) ⟨B-73MI51401⟩, base-induced contraction of (**170**) ⟨80JOC2151⟩, acid-induced contraction of naphtho[1,8-*cd*][1,2]oxathiole *S*-oxide (**171**) ⟨67LA(703)96⟩ and the remarkable rearrangement of bromopenam (**172**) ⟨78JA3251⟩.

$(R_2^nN)_3P$ — (**160**)

R = H, R′ = $(CH_2)_4CONHPr$, 68%
R = H, R′ = H, 82%
R = Ph, R′ = H, 87%
R = H, R′ = Pr
R = PhCO, R′ = Ph, 60%

(161) $\xrightarrow{R''_3P}$ R = R′ = Me or Et, X = O, 83–86%
R = Me, R′ = H, X = H_2, 59%

(162) $\xrightarrow[C_6H_6]{230\,°C}$ 55%

(163) $\xrightarrow[C_6H_6]{h\nu}$

(164) $\xrightarrow[MeOH]{h\nu}$; $\xrightarrow{100\,°C}$

(165) $\xrightarrow[MeOH]{h\nu}$; $\xrightarrow{350\,°C}$

(166) $\xrightarrow[-CO]{700\,°C}$ X = H, H, 80%
X = O

(167) $\xrightarrow[350\ nm]{h\nu}$ 55%

(168) $\xrightarrow[1000\,°C,\ 0.05\ mmHg]{-CO_2}$ 45%

(169) $\xrightarrow{H_2O/Me_2CO}$ CHOHMe

(170) $\xrightarrow[(ii)\ HCl]{(i)\ NaOH}$ 67%

(171) $\xrightarrow{H^+}$

(172) $\xrightarrow[Ph_3P,\ THF,\ 25\,°C]{EtO_2CN{=}NCO_2Et}$

5.14.4.2.4 Formation from six-membered heterocycles

Among the examples of formation of thietanes and thietes from six-membered heterocycles are ring opening of cyclic carbonate (**173**) with thiocyanate ⟨71JA676, 70JOC2249⟩, photochemical or thermal loss of nitrogen from (**174**) ⟨79JA7684, 67LA(703)96⟩, photochemical fragmentation of carbonyl compounds (**175**) ⟨69CC13⟩, (**176**) ⟨69JOC1566⟩, (**177**) ⟨70JOC584,

73T2135⟩ and (**178**) ⟨70JA7001⟩ and thermal decomposition of 1,3-dithiane (**179**) under mild conditions ⟨78CRV125⟩.

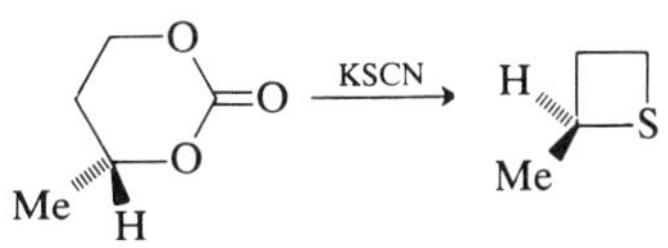

KSCN
(**173**) 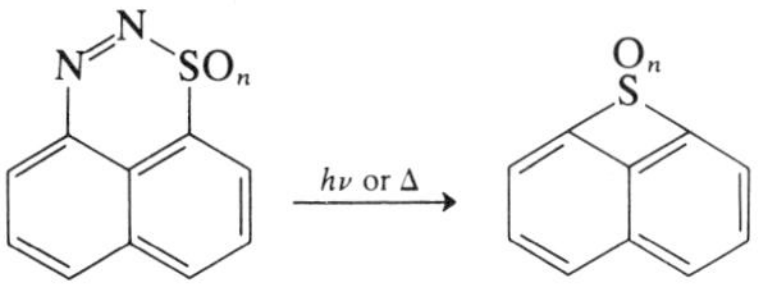
$h\nu$ or Δ
(**174**)
$n = 0$, 1 or 2

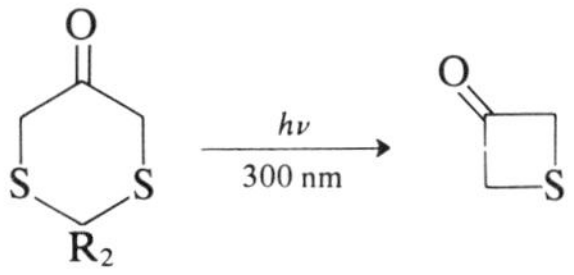

$h\nu$ 300 nm
R = H or R_2 = —$(CH_2)_5$— 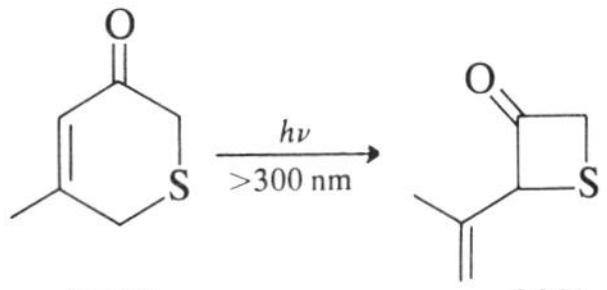
(**175**)
$h\nu$ >300 nm
(**176**) 30%

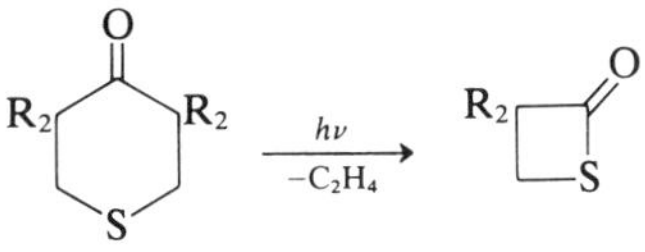

$h\nu$ $-C_2H_4$
R = H, 27%
R = Me, 57% 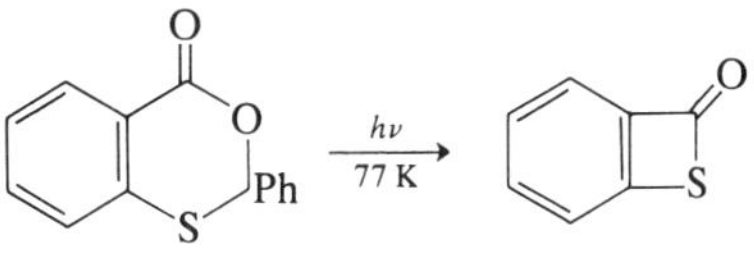
(**177**)
$h\nu$ 77 K
(**178**)

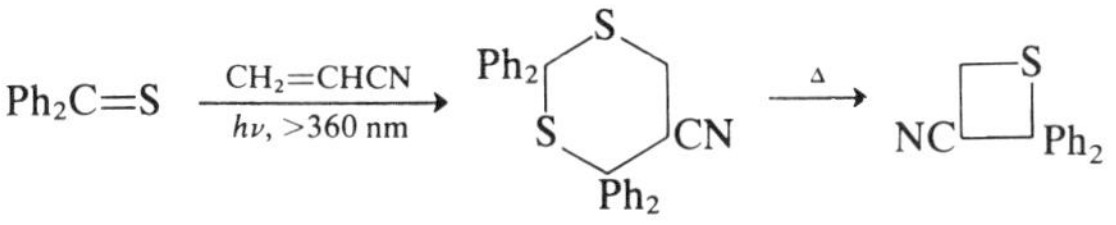

$Ph_2C{=}S$ $\xrightarrow[h\nu,\ >360\ nm]{CH_2{=}CHCN}$ (**179**) $\xrightarrow{\Delta}$

5.14.4.2.5 Formation from seven-membered heterocycles

One instance of the formation of a thietane derivative from a seven-membered heterocyclic precursor is seen in the photochemical fragmentation of dione (**180**) ⟨73T2135⟩.

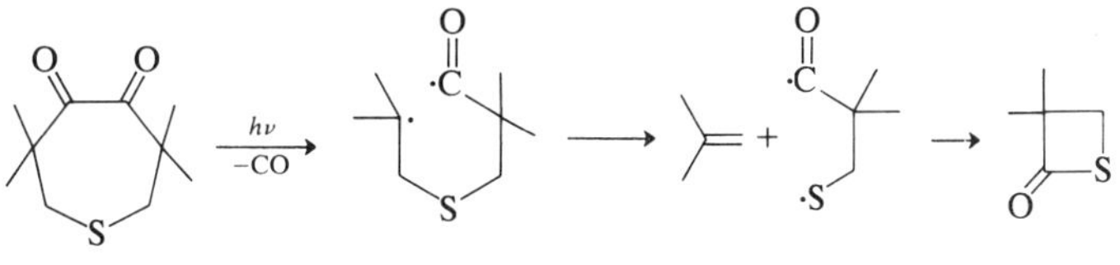

$h\nu$ $-CO$
(**180**)

5.14.4.2.6 Formation from eight-membered heterocycles

Eight-membered ring disulfone (**181**) ⟨67LA(703)96⟩ is reported to undergo conversion to naphtho[1,8-*bc*]thiete 1,1-dioxide upon UV irradiation.

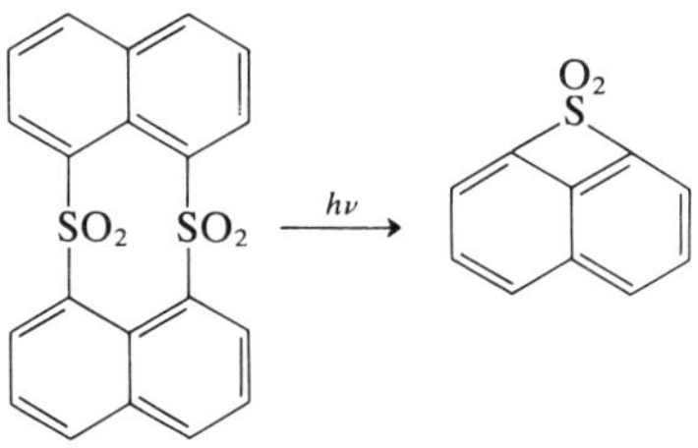

$h\nu$
(**181**)

5.14.4.3 Synthesis of Derivatives of the Ring System

Thietanes can be readily oxidized or otherwise substituted at sulfur. Oxidation of thietanes to their 1-oxide can be accomplished in good yields with such oxidants as MCPBA, hydrogen peroxide/acetic acid, dinitrogen tetroxide, sodium metaperiodate, *t*-butyl hypochlorite, chromium trioxide/pyridine ⟨70JOC3657⟩, photochemically with pyridine *N*-oxide and related *N*-oxides, the monooxygenase enzyme from *Aspergillus niger* ⟨80JCS(P1)1693⟩, oxaziridines ⟨78TL5171, 80JCS(P1)1693⟩ and 1-chlorotriazole ⟨69CC365⟩, as illustrated in the case of thietane (**182**) ⟨70JOC3657⟩ and thiete (**183**) ⟨76JA6643⟩. Stereochemical studies done on the oxidation of 3-substituted thietanes have shown that in general these systems are less sensitive to the nature of the oxidant than the 4-alkylthiane system ⟨70JOC3657⟩. Thus, while the *cis*:*trans* equilibrium ratio for 3-*t*-butylthietane 1-oxide is 82:18, dinitrogen tetroxide gives an 82:18 sulfoxide ratio on oxidation of the corresponding thietane (**182**), chromium trioxide a 70:30 ratio, sodium metaperiodate a 51:49 ratio, MCPBA a 45:55 ratio, and hydrogen peroxide/acetic acid a 43:57 ratio.

N_2O_4 (**182**) *cis* + *trans*

82:18 *c*:*t* for R = Bu^t

(**183**) MCPBA, C_6H_6, 15 °C → 70% → RCO_3H; excess MCPBA, 15 °C

The *cis–trans* sulfoxide mixtures can be separated by column chromatography (the less polar *cis*-sulfoxide elutes first). Inversion of sulfoxides can be achieved by *O*-alkylation (typically with a trialkyloxonium salt) followed by displacement at sulfur with aqueous base, as illustrated with sulfoxide (**184**) ⟨69JA5644, 78JOC3827⟩. Oxygen-18-substituted thietane 1-oxide can be conveniently prepared through the use of sodium metaperiodate which has been equilibrated with oxygen-18 water ⟨77MI51400⟩. Thietane 1,1-dioxides can be prepared by further oxidation of thietane 1-oxides, or directly from thietanes by treatment with an excess of an oxidant such as a peracid or hydrogen peroxide in acetic acid. Precautions are of course necessary in the work-up of products when peroxides are used as oxidants in these reactions, as explosions have been reported ⟨71JOC1324⟩.

(**184**) $Et_3O^+ BF_4^-$ → BF_4^-, OEt →

S-Alkyl derivatives of thietane can be prepared through treatment of thietanes with trialkyloxonium fluoroborates ⟨71JA676, 71HCA2297⟩. *S*-Amino derivatives of thietanes and sulfilimines can be prepared by treatment of the thietane with *O*-mesitylsulfonylhydroxylamine (see the preparation of (**38**) above) and chloramine T respectively, as in the synthesis of (**185**) ⟨78JOC3827⟩. Oxidation of the sulfilimines with potassium permanganate affords sulfoximines such as (**186**). Other *S*-substituted derivatives which can be prepared from thietanes include *S*,*S*-difluoro compound (**187**) ⟨73JA4064⟩, which undergoes ready hydrolysis to thietane 1-oxide, and *S*,*S*,*S*,*S*-tetrafluoro compound (**188**) ⟨73JA8191⟩.

chloramine T → (**185**) NTs → $KMnO_4$ → (**186**) S=NTs

CF_3OF, CaH_2, CH_2Cl_2 → SF_2 (**187**) → $NaHCO_3$, H_2O

excess CF_3OF → SF_4 (**188**)

Conversion of thietane 1-oxides and 1,1-dioxides back to thietanes can be accomplished with aqueous sodium bisulfite (limited to thietane 1-oxides ⟨72JOC919⟩) or more generally with LAH, as illustrated for sulfoxide (**189**) ⟨79JOC4757⟩ and sulfone (**190**) ⟨65LA(684)92, 69JA2796⟩. Ring-substituted thietanes are often conveniently prepared by ring functionalization of the thietane 1,1-dioxide followed by LAH reduction, as will be illustrated below.

Cl Cl $\xrightarrow[Na_2CO_3]{H_2O}$ $\xrightarrow{LAH}$
S O S
(**189**)

$\xrightarrow{LAH}$
N SO$_2$ N S
(**190**) 70%

Deuterium can be readily incorporated into the 2- and 4-positions of thietane 1,1-dioxides as already seen with (**71**) and (**72**) ⟨69JA2796, 70JOC3657, 74JCS(P2)1676⟩. Alkylation, acylation and other substitution by carbon and other Group IV elements in the α-position of thietane 1,1-dioxides can be achieved once or twice by treatment of thietane 1,1-dioxide itself (see formation of (**74**) and (**75**) above) or of derivatives such as (**191**) ⟨78S579⟩, (**192**) ⟨75S333⟩ or (**194**) ⟨75JOU1301, 77JOU353⟩ with an alkyllithium followed by the appropriate electrophile. Naphthothiete (**195**) ⟨68JOC1364⟩ and 3-thietanone condense with aldehydes in the presence of base affording 2-alkylidene thietane derivatives (**196**) ⟨68JOC1364⟩ and (**197**) ⟨74CB2115⟩ respectively. Carbon–carbon bond formation at the position remote from sulfur can be achieved by alkylation of anions of β-thiolactones such as (**198**) ⟨80JCS(P1)388⟩ or of 2-iminothietanes (see formation of (**70**) above) ⟨80AG(E)466⟩, or by transformations of the carbonyl group of 3-thietanones affording (**199**) ⟨77S201⟩, (**200**) ⟨76CZ440⟩ and (**201**) ⟨78TL3855, 75JCS(P1)2513⟩. Wittig condensations of β-thiolactones are known ⟨82CC995⟩.

Ph SO$_2$ $\xrightarrow[\text{(ii) RBr}]{\text{(i) BuLi, }-78\,^{\circ}\text{C}}$ Ph SO$_2$ R R = Me, Et, PhCH$_2$, CH$_2$=CHCH$_2$
(**191**)

SO$_2$ $\xrightarrow[\text{(ii) MeCO}_2\text{Et}]{\text{(i) BuLi, }-70\,^{\circ}\text{C}}$ O CMe SO$_2$ $\xrightarrow[\text{NaOH, dioxane}]{Br_2}$ Br SO$_2$
(**192**) (**193**)

O N R SO$_2$ $\xrightarrow[\text{(ii) E}]{\text{(i) BuLi}}$ O N R X SO$_2$
(**194**)

R = Me, E = Cl$_3$CCCl$_3$, X = Cl, 38%
R = H, E = MeI, X = Me, 64%
R = H, E = Me$_3$SiCl, X = Me$_3$Si, 43%
R = H, E = Me$_3$SnCl, X = Me$_3$Sn, 48%

Ph SO$_2$ Ph $\xrightarrow[\text{NaOEt, EtOH}]{\text{PhCHO}}$ Ph CHPh SO$_2$ Ph
(**195**) (**196**) 96%

O S + HC≡C–(thiophene, S)–C≡CCHO $\xrightarrow[\text{MeOH}]{\text{KOH}}$ HC≡C–(thiophene, S)–C≡CCH= O S
(**197**)

(i) NaH, DMF
(ii) EX

E = Me, $CH_2{=}CHCH_2$, $Bu^tO_2CCH_2$

(198)

(i) NH_2NH_2
(ii) H_2S, H^+
(iii) RN=NR, Δ

Ph_3P

(199)

=PPh_3

(200)

(i) P_4S_{10}, py
(ii) Me_2CHN_2
(iii) Ph_3P

(201)

Halogenation in the α-position of thietane 1,1-dioxides can be achieved by trapping the corresponding α-sulfonyl carbanions with positive halogen sources [see preparation of (**202**) ⟨73CC861⟩ and reaction of (**194**)], by modification of a side chain [see preparation of (**193**)] or rarely by direct halogenation [see (**203**) ⟨B-76MI51400⟩]. Thietane 1,1-dioxide undergoes light-induced halogenation in the β-position [see (**80**) above].

(i) BuLi
(ii)

(202)

(i) Br_2
(ii) H_2O

(203)

Oxygen can be introduced into the α-positions of thietanes by treatment of the thietane with lead tetraacetate as in the formation of (**204**) ⟨75JOC3046⟩. Thietanones can be prepared by hydrolysis of thiete enamines such as (**205**) ⟨65LA(684)103⟩ or by oxidation of 3-thietanol with DMSO/benzoic anhydride ⟨78PS(4)167⟩. Thietanones can be converted into various ring-substituted sulfur derivatives as already illustrated in reactions of (**68**) and preparation of (**199**).

$Pb(OAc)_4$
py, 25 °C

(204) 30%

H_2O

(205)

5.14.5 APPLICATIONS AND IMPORTANT COMPOUNDS

A number of thietanes occur in nature. For example, 2,2-dimethylthietane is found in the anal glands of the mink (*Mustela vison*) and ferret (*Mustela putoris*) ⟨81CZ273, 80MI51400, 80MI51401⟩, 2-ethylthietane occurs in the mink and stoat (*Mustela erminea*) ⟨80MI51402, 78TL5233⟩, 2-propylthietane occurs in the mink, ferret and stoat, 2-pentylthietane occurs in the ferret and stoat, and *cis*- and *trans*-2,3-dimethylthietane occur in the ferret. These odoriferous compounds probably function as scent markers for these animals. It might be noted here that because of their odors, which can be detected even at very low concentrations, thietanes have been considered as odorants for natural gas. The plants *Berkheya angustifolia* and *Cullumia squarrossa* contain the remarkable thiophene–thietanone system (**197**) and related compounds ⟨75CB515⟩.

Bicyclic sulfides (**103**) and (**104**) have biological activities very similar to natural thromboxane A2, *e.g.* in the case of (**104**) causing contraction of rat aorta strips at concentrations as low as 7×10^{-10} M and causing marked irreversible aggregation of human platelets at a concentration of 4.3×10^{-6} M. Other thietane derivatives with physiological activity are (**206**) and (**207**), which are said to be useful as antiinflammatory agents ⟨73USP3729487⟩, (**208**), which is claimed to be useful as a sedative ⟨72USP3632579⟩, and (**209**), which is patented as an insecticide ⟨63USP3071594⟩. Thietanes may be involved as intermediates in the photochemical crosslinking of 4-thiouridine and cytidine units in bacterial transfer-RNA ⟨72B1, 74TL3005⟩.

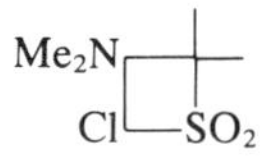

Me_2N Cl SO_2 (**206**) Et_2N Cl SO_2 (**207**) O N SO_2 (**208**) $(EtO)_2P(S)O$ S (**209**)

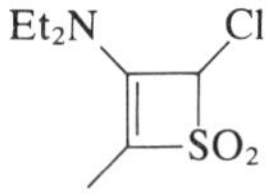

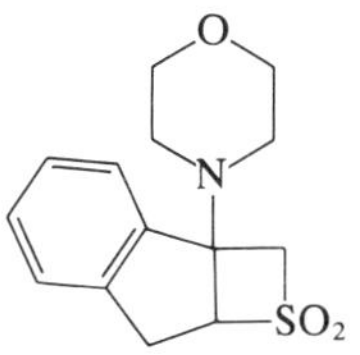

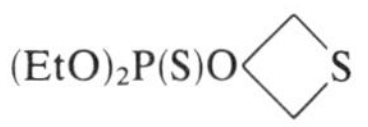

The block copolymer of ethylene oxide and 3,3-dimethylthietane shows useful properties of complexing halogen and heavy metal salts ⟨79MI51402⟩. Thietanes can be polymerized with methylmagnesium iodide as well as with a variety of electrophiles such as methyl sulfate, trimethyloxonium tetrafluoroborate, triethylaluminum, boron trifluoride and phosphorus trifluoride ⟨67IC1461, 67MI51400⟩. Thietane (**210**) has been patented as a stabilizer for poly(vinyl chloride) ⟨73USP3767615⟩.

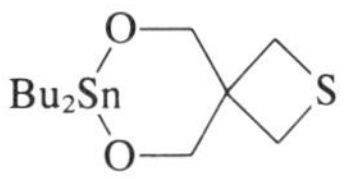

O Bu_2Sn O S

(**210**)

Only a limited number of examples are known of applications of thietanes in organic synthesis. Prominent among these examples would be electrophilic ring opening reactions leading to polyfunctional sulfur compounds (**33**)–(**37**), utilization of 3-thietanones (**55**) and metal complexes (**87**) derived therefrom as oxyallyl zwitterion equivalents in cycloaddition reactions, synthesis of dipeptide (**63**) with a β-thiolactone, Raney nickel desulfurization of thietanes (*e.g.* **120**; *cf.* Table 7) as a route to *gem*-dimethyl compounds, and desulfurization of thietanes (*e.g.* **17**) in the synthesis of cyclopropanes (also see Table 7).

Acknowledgment: Support during the preparation of this chapter by the National Science Foundation and the Petroleum Research Fund, administered by the American Chemical Society, is gratefully acknowledged.

5.15

Four-membered Rings with Two or More Heteroatoms and Fused-ring Derivatives

J. W. TIMBERLAKE and E. S. ELDER
University of New Orleans

5.15.1 STRUCTURE OF FOUR-MEMBERED HETEROCYCLES WITH TWO HETEROATOMS

5.15.1.1 Structure Determination

Examples of virtually all four-membered rings with two heteroatoms of nitrogen, sulfur and oxygen have been prepared. In some cases simple derivatives of the parent systems are unknown and in others the literature is too extensive to cover in a single review. Therefore, selections have been made to represent the various classes without being encyclopedic.

Table 1 contains structures with selected bond lengths and angles for compounds determined either by X-ray, electron diffraction or theoretical treatment. Some parameters have been purposely omitted, and complete data can be found in the references.

The most surprising aspect is the number of rings that are planar or nearly so. The notable exceptions are the 1,2-diazetidine, the 1,2-dioxetane and the 1,2-oxathietane oxide rings. However, the reason for lack of planarity is not clear.

5.15.1.2 Spectroscopic Data for Four-membered Rings

5.15.1.2.1 1,2-Diazetidine derivatives

The complete NMR line shape analysis of 1,2-dialkyl-1,2-diazetidines shows little difference in the rate of inversion for methyl, ethyl and isopropyl but a significantly slower rate for *t*-butyl. For example, at 66 °C the relative rates are $1:0.78:0.51:0.74\times10^{-3}$ respectively ⟨78JOC2785, 78JA2806⟩. One explanation is that nitrogen inversion is slow relative to ring inversion and at lower temperatures concentrations of (**2**) and (**3**) are small because of diaxial interactions. However, when $R = Bu^t$, (**2**) and (**3**) predominate because of alkyl group interaction in (**1**) and (**4**; Scheme 1). The barrier to inversion for 1,2-bis(trifluoromethyl)-3,3,4,4-tetrafluoro-1,2-diazetidine is some 17 kJ mol^{-1} less than for 1,2-dimethyl-1,2-diazetidine ⟨69CC1084⟩, although no explanation is obvious. For 3,3,4,4-tetramethoxy-1,2-diethoxycarbonyl-1,2-diazetidine the coalescence of methoxy group signals was attributed to an amide rotational barrier rather than an inversion process ⟨73JOC1605⟩. The inversion of one aryl group on nitrogen in 1,2-diaryl-1,2-diazetidin-3-one gives a barrier of $\Delta G^{\ddagger} = 53.6$ kJ mol^{-1}.

(1) —slow→ (2)

(1) ⇅ fast (3); (2) ⇅ fast (4)

(3) ←slow— (4)

Scheme 1

The most characteristic spectral detail for Δ^1-1,2-diazetines is the fine structure in the UV spectrum around 350 nm. This is not typical, as other cyclic azo compounds usually show unstructured broad absorptions at 360 nm ⟨62JCS3276, 78TL2469, 75JOC1409⟩. The PE spectrum of the tetramethyl derivative is also unique and shows an inversion in the normal n^+, π, n^- energy level for cyclic azo compounds with the π and n^+ levels inverted ⟨77JOC609⟩.

The carbonyl absorption for 1,2-diazetidin-3-one is higher than that of β-lactams, 1780–1790 cm^{-1} *versus* 1730–1760 cm^{-1}, and indicates less resonance interaction with the

Table 1 Bond Lengths and Angles for Four-membered Heterocycles

Compound	*Bond length* (Å)	*Bond angle* (°)	*Method*	*Ref.*
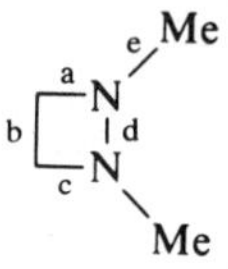	a 1.471 b 1.537 c 1.481 d 1.427 Ring puckered 24.3°	de 113.0	ED	80JCS(F2)1293
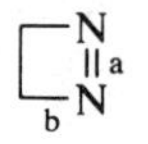	a 1.285 b 1.521 Planar ring	ab 95	TH (STO-3G)	79JA5546
	a 1.262(FF) a 1.267(MO) b 1.491(FF) b 1.495(MO)	ab 96.5(FF) 94.5(MO)	TH (force field, MO)	80CB1201
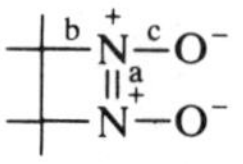	a 1.292 b 1.527 c 1.250	ac 129.2 cb 135.4 ab 95.4	XR	78AX(B)2602
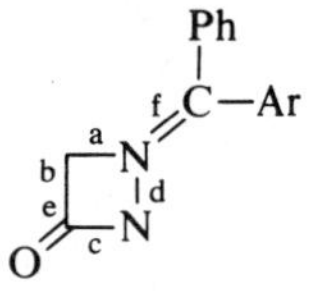	a 1.47 b 1.54 c 1.38 d 1.40 e 1.20 Planar ring	bc 94 dc 90 cb 94 ab 81	XR	68CC917
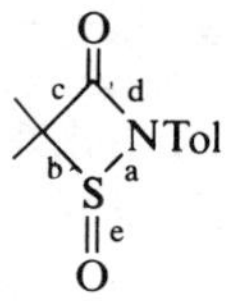	a 1.741 b 1.863 c 1.519 d 1.353 Planar ring	ab 75.7 ad 97.1 dc 100.9 bc 86.4 eb 110.0 ae 110.9	XR	81CC350
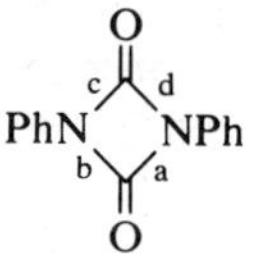	a 1.434 b 1.425 c 1.160 Planar ring	ab 89.6 bc 90.5 ae 135.9	XR	74AX(B)691 55JCS2931
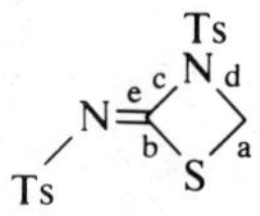	a 1.86 b 1.39 d 1.49 e 1.26 Planar ring	ab 75.2 bc 95.5 cd 101 ad 88.6 ce 127	XR	77RTC263
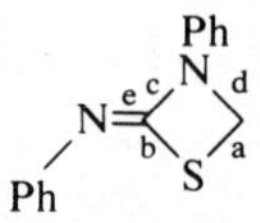	a 1.826 b 1.780 c 1.375 d 1.460 e 1.260 Planar ring	ab 74.3 bc 95.0 cd 100.4 ce 130.0	XR	78AX(B)988
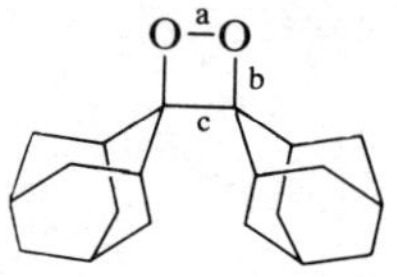	a 1.48 b 1.549 c 1.475 Puckered ring 21.3°		XR	77CC591
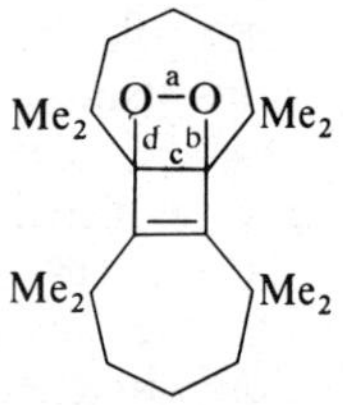	a 1.448 b 1.488 c 1.552 d 1.492	ab 91.2 bc 89.0 cd 88.5 ad 91.3	XR	80TL3171

Table 1 *(continued)*

Compound	*Bond length* (Å)	*Bond angle* (°)	*Method*	*Ref.*
	a 1.502 b 1.392 c 1.503 d 1.473 e 1.180 Planar ring	ab 90.5 bc 92.6 cd 87.4 ad 89.5	TH (4-31)	81JA1292
	a 1.669 b 1.81 c 1.581 d 1.43 Planar ring	ab 100.6 bc 89.3 cd 92.7	TH (CNDO)	77JA2931
	a 1.487 b 1.667 c 1.856 d 1.588 e 1.466	ab 97.1 ad 93.8 bc 109.6 cb 79.0 ce 109.5	XR	81JCS(P1)1826
	a 2.146 b 1.835 c 1.564	ab 99.1 bc 80.9	TH (CNDO)	77JA2931
	a 2.05 b 1.73 c 1.40 Planar	ab 89.5 bc 100.8	ED	76JA899
	a 1.785 Planar with D_{2h} symmetry	bc 83.2	ED	77JPC1682 79JST(54)287 76ACS(A)759 78SA(A)489 63SA769
	a 1.801 c 1.77 Planar with D_{2h} symmetry	ad 83.9 ab 96.1	XR	68ZN(B)739 80LA1981 64CJC2107
	a 1.90 b 1.85 c 1.80 d 1.72 e 1.36 f 2.64 Ring is coplanar with α,β unsaturation	da 83.7 ab 92.3 cd 106.6 de 130.8	XR TH	67TL5307 69JA1120
	a 1.766 b 1.764 e 1.324 f 2.640 Ring is planar and *trans*, 2.64 Å non-bonded interaction is less than van der Waals radii	ad 82 de 128.5 ce 133.4	XR	67TL373
	a 1.815 b 1.816 c 1.815 d 1.819 Planar ring	ad 85.4 ab 93.6	XR	77JOC2345

Table 1 *(continued)*

Compound	*Bond length* (Å)	*Bond angle* (°)	*Method*	*Ref.*
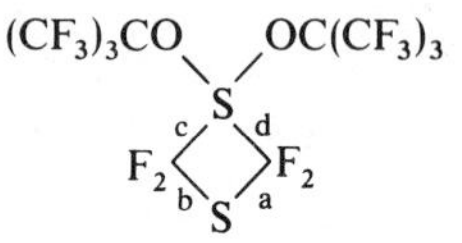 $(CF_3)_3CO$, $OC(CF_3)_3$ on S; ring S(c,d), F_2C, F_2C, S(b,a)	a 1.736 b 1.752 c 1.932 d 1.887 e 1.816 Planar ring with distorted bipyramid at S	ad 99.2 ab 86.5 bc 96.9 cd 77.4	XR	81JA406
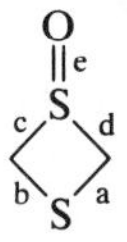 O=S (e) ring with S (a, b, c, d)	a 1.81 b 1.82 c 1.473 Puckered ring 39.3°	ab 91.2 ad 81.7 ac 112.4	XR	76JA5715
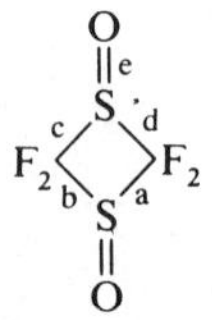 O=S(e), F_2, F_2, S=O ring (a, b, c, d)	a 1.788 b 1.784 e 1.420 Planar ring	ab 94.1 ae 109.1 ef 124.7	XR	80LA1981
 O_2S (e, f) ring with SO_2 (a, b, c, d)	a 1.813 b 1.804 e 1.43 Planar ring	ab 91.5 ad 88.5 ae 111.4 ef 119.4	XR	76JA5715
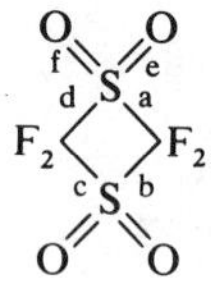 O_2S (e, f), F_2, F_2, SO_2 (a, b, c, d)	a 1.788 b 1.784 e 1.420 Planar ring	ab 94.1 ae 109.1 ef 124.7	XR	80LA1981

nitrogen lone pair ⟨66TL5245, 76AG(E)432⟩. A similar oservation has been recorded for 1,2-di-*t*-butyl-1,2-diazetidine-3,4-dione where the carbonyl at 1813 cm^{-1} implies a considerably more strained amide bond with essentially no nitrogen lone pair delocalization ⟨67JOC2360⟩.

5.15.1.2.2 1,2-Oxazetidine derivatives

The rate of ring inversion of pentafluoro-1,2-oxazetidine as determined by ^{19}F NMR is very fast ($E_a \simeq 4.2$ kJ mol^{-1}), with the nitrogen inversion barrier much higher ⟨70JOC927, 1607⟩.

Several 4*H*-1,2-oxazete structures have been assigned from spectral and chemical data. The IR band at 1585–1595 cm^{-1} is most characteristic, although ^{13}C, ^{1}H NMR and MS have been carefully analyzed ⟨80TL2025, 75AG(E)69⟩.

The carbonyl band at 1785 cm^{-1} for *N*-substituted 4,4-diphenyl-1,2-oxazetidin-3-ones is higher than that of β-lactams. Here, as with 1,2-diazetidin-3-ones, the additional heteroatom reduces amide conjugation ⟨68JOC3619⟩.

5.15.1.2.3 1,2-Thiazetidine derivatives

The characteristic SO_2 stretch of thiadiazetidine dioxides occurs at 1160 and 1330 cm^{-1} ⟨70BCJ3543, 62HCA717⟩. Carbonyls in the 3-position absorb at 1750 cm^{-1}, which is approximately the same place as β-lactams ⟨62HCA717, 79CB1012⟩. Sulfur, unlike oxygen or nitrogen, does not raise the frequency of absorption when substituted for carbon in the β-lactam ring.

5.15.1.2.4 1,3-Diazetidine derivatives

Uretidinediones have carbonyl absorptions similar to thoe of β-lactams (1730–1760 cm^{-1}) ⟨71JOC3056, 78JOC4530⟩ which implies nomal ring conjugation. *exo*-Imino groups absorb at 1680 cm^{-1} ⟨68AG(E)72, 78LA1111⟩.

5.15.1.2.5 1,3-Oxazetidine derivatives

The carbonyl of the 1,3-oxazetidin-2-one ring absorbs between 1840 and 1950 cm^{-1} depending upon the substituent. For example, perfluorinated alkyls at the 4-position lead to very high carbonyl absorption. Mass spectral data show the expected modes of ring cleavage ⟨67JOC2960, 80S571, 80DOK(253)886⟩.

Carbonyl isocyanates and carbodiimides add [2+2] and give 2,4-diimino-1,3-oxazetidines as IR (1630–1680 cm^{-1}, C=N), NMR and MS data indicate ⟨72JA3484⟩.

5.15.1.2.6 1,3-Thiazetidine derivatives

It was originally reported that isothiocyanates added to carbodiimides giving thione derivatives. The products are now recognized to be 2,4-diimino-1,3-thiazetidines which display the expected C=N absorptions between 1620 and 1630 cm^{-1} ⟨70CC1629, 78CCC3258⟩. The UV absorptions, λ_{max} 223 and 260 nm, are reported to be too short for expected C=S $\pi \rightarrow \pi^*$ transitions ⟨73BCJ2559⟩.

5.15.1.2.7 1,2-Dioxetane derivatives

1,2-Dioxetanes are usually yellow and since the λ_{max} occurs at 280 nm the actual origin of the color is not clear ⟨75CJC1103⟩. The only characteristic band in the IR is at 845–895 cm^{-1} ⟨75JA7110⟩. The quaternary carbons usually absorb (^{13}C NMR) between 88 and 90 p.p.m. from TMS, this being perhaps the most unique spectral characteristic ⟨77AHC(21)437⟩.

For α-peroxylactones (1,2-dioxetan-3-ones) which are usually quite unstable, the IR band at 1875 cm^{-1} is the best indicator of the ring's presence ⟨72JA2894, 77JA5836, 5768⟩. PES and MS studies on dioxetanes have also been reported ⟨77AHC(21)437⟩.

5.15.1.2.8 1,2-Oxathietane derivatives

The IR of 1,2-oxathietane dioxide shows an asymmetric SO_2 stretch at 1394 and a symmetric one at 1176 cm^{-1} ⟨67DOK(172)1327⟩. The corresponding monoxides appear at 1150 cm^{-1} ⟨75CC724⟩. The chemical shifts and coupling constants have been used to assign the *cis* and *trans* structures to 3,4-dimethyl-1,2-oxathietane dioxide ⟨73CC841⟩.

5.15.1.2.9 1,2-Dithietane derivatives

The only truly stable derivative of this system, 3,4-bis(trifluoromethyl)-1,2-dithietene, has a characteristic IR absorption at 1629 cm^{-1} and UV λ_{max} of 238 ($\varepsilon = 7440$) and 340 nm ($\varepsilon = 80$) ⟨60JA1515⟩. An extended Hückel calculation shows the favored isomer to be the ring-opened tautomer, dithioglyoxal, while the CNDO/2 method favors the ring-closed 1,2-dithietene tautomer ⟨75JCS(P2)559⟩.

5.15.1.2.10 1,3-Dithietane derivatives

Spectroscopically, tetrafluoro-1,3-dithietane is one of the most extensively studied molecules known. Electron diffraction ⟨76ACS(A)759⟩, ^{19}F NMR ⟨71JCP(54)1563⟩ and IR and Raman studies ⟨63SA769, 73MI51501, 78SA(A)489, 79JST(54)287⟩ all are consistent with a planar ring with D_{2h} symmetry. Similarly, tetrachloro-1,3-dithietane is of the same point group

⟨64CJC2107⟩, and CNDO calculations have matched the three observed λ_{max} positions (207, 230 and 246 nm) to the corresponding molecular orbitals ⟨79SA(A)509⟩. Other simple 1,3-dithietanes have been characterized by ^{1}H ⟨73BCJ2253⟩, ^{19}F ⟨77JA4194, 67BAU2467⟩ and ^{13}C NMR ⟨77JOC2345⟩, mass spectroscopy ⟨73BCJ2253⟩ and UV spectroscopy (with accompanying CNDO/2 calculations) ⟨71JCS(B)312⟩.

The parent 1,3-dithietane tetraoxide ⟨76JA5715⟩ as well as the tetrafluoro, tetrachloro and tetrabromo derivatives ⟨80LA1981⟩ have all had X-ray structures determined. IR, UV and NMR spectral data are also reported for these planar, almost square, rings ⟨80AG(E)203⟩.

Desaurins (2,4-dialkylidene-1,3-dithietanes; Scheme 2) have been shown by X-ray to have a *trans* structure with the unsaturated carbonyl portion and the two sulfur atoms being coplanar ⟨67TL373⟩. They have longer wavelength carbonyl absorptions (1650 cm^{-1}) than normal α,β-unsaturated ketones, leading to the suggestion that sulfur interacts with the carbonyl through space or through the π-system ⟨71CJC1467⟩. Other UV and IR studies ⟨58JA5577, 71CJC1467, 66T3001, 76BAU1913⟩ of desaurins report data that are consistent with this conclusion. They are very stable compounds, and often the MS base peak is the molecular ion ⟨68CJC365, 76BAU1547⟩.

Scheme 2

5.15.2 REACTIVITY OF FOUR-MEMBERED HETEROCYCLES WITH TWO HETEROATOMS

Four-membered rings display a variety of reactivity modes ranging from thermolysis to attack by nucleophiles and electrophiles. Almost all reactions are predictable, simply from consideration of orbital symmetry or simple acid–base concepts.

5.15.2.1 Thermal and Photochemical Reactions

5.15.2.1.1 1,2-Diazetidine derivatives

The thermolysis of the fused-ring diazetidine shown in Scheme 3 has a number of allowed paths that could be followed. However, the exclusive product from flash vacuum pyrolysis of both unsaturated and saturated analogs is the diimine, which probably arises through a diradical intermediate. The symmetry allowed $[_{\sigma}2_s + _{\sigma}2_a]$ is unlikely because of steric requirements ⟨80JA837⟩.

Scheme 3

The addition of *cis*-azobenzenes to ketenimines gives 3-imino-1,2-diazetidines which cleave to give carbodiimides and imines upon thermolysis (Scheme 4) ⟨67JHC155, 72JHC555, 74JHC409⟩. 1,2-Diphenyl-1,2-diazetidin-3-one has similarly been reported to fragment to isocyanate and imine ⟨56AG71⟩.

$Ph_2C{=}C{=}NAr$ + PhN=NPh → [3-(arylimino)-4,4-diphenyl-1,2-diphenyl-1,2-diazetidine] → $Ph_2C{=}NPh$ + PhN=C=NAr

Scheme 4

The thermal decomposition of several Δ^1-1,2-diazetines has been studied kinetically. The activation parameters for tetramethyl and tetrafluoro are $\Delta H^{\ddagger} = 132.7$ and 159.4 kJ mol^{-1} and $\Delta S^{\ddagger} = 1.3$ and 51.8 J K^{-1} mol^{-1} respectively ⟨78JA1876, 73JCS(F1)1817⟩. *cis*-3,4-Diethyl-3,4-dimethyl-Δ^1-1,2-diazetine was prepared by Greene and coworkers in an attempt to elucidate the mechanism of decomposition. However, the stereospecific alkene prouction could imply either a formally disallowed retro $[2_a+2_s]$ or production of a diazenyl diradical with sufficiently short lifetime to allow rotation (Scheme 5). Not all diazetines decompose by extrusion of nitrogen. An interesting bicyclic analog undergoes a valence tautomerization to a dihydropyridazine which was characterized as its trimeric form (Scheme 6) ⟨80JA7994⟩.

Scheme 5

Scheme 6

5.15.2.1.2 1,2-Oxazetidine derivatives

Simple 1,2-oxazetidines show different degrees of thermal stability, but all consistently fragment to give carbonyl derivatives and imines (Scheme 7) ⟨61JCS1351, 64JCS4066, 65JGU855, 72BCJ3187⟩. 3-*exo*-Unsaturated derivatives also cleave across the N—O bond: for example, 3-methylene ⟨65JCS5602, 73JCS(P1)1561⟩, 3-keto ⟨62DOK(142)596⟩, 3-imino ⟨70JHC1203, 74JHC633, 79BAU1648⟩.

$$\xrightarrow{500\,^{\circ}\mathrm{C}} CF_3N{=}CH_2 + F_2CO$$

Scheme 7

$$\xrightarrow{\Delta} Bu^tCN + (MeS)_2CO$$

Scheme 8

One example of decomposition of an oxazete shows identical behavior (Scheme 8) ⟨80TL2025⟩.

5.15.2.1.3 1,2-Thiazetidine derivatives

N,N-Disubstituted sulfur diimides, diaryl ⟨69TL447⟩, di-*t*-butyl ⟨72JOC3810⟩ and bis-*p*-tolylsulfonyl ⟨70TL1427⟩, are reported to add [2+2] to diphenylketene. The initially formed 1-imino-4,4-diphenyl-1,2-thiazetidin-3-one derivatives are relatively unstable and rearrange thermally to five-membered rings at 70 °C (Scheme 9).

Scheme 9

The reaction of *N*-sulfinylsulfonamides with ketenimines was originally reported to add N—S across C—N to give an alkylidene-1,2,4-thiadiazetidine 1-oxide ⟨73BCJ2156⟩. More recently, L'abbé has reassigned the structure as a 3-imino-1,2-thiazetidine 1-oxide derivative resulting from N=S addition to the C=C of the ketenimine. The adduct is quite stable although flash thermolysis at 400 °C gives a 64% isolated yield of amidine *via* loss of SO and hydrogen shift (Scheme 10) ⟨81TL583⟩.

$$\xrightarrow{400\,^{\circ}\mathrm{C}} CH_2{=}C(Me){-}C({=}NTs){-}NHPh$$

Scheme 10

Simple saltams, 1,2-thiazetidine 1,1-dioxides, are known to polymerize thermally ⟨73MI51500⟩, although in one case the alkene product was reported to be formed with retention of stereochemistry (Scheme 11) ⟨75BCJ480⟩.

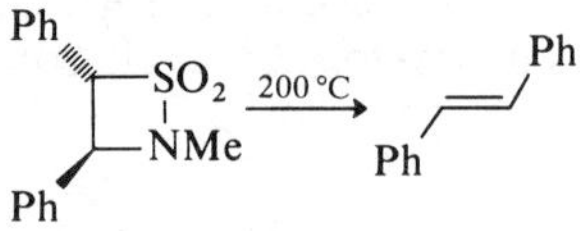

Scheme 11

5.15.2.1.4 1,3-Diazetidine derivatives

Uretidinediones or 1,3-diazetidine-2,4-diones are formed by phosphine-catalyzed dimerization of isocyanates. Both the dimerization and 'dedimerization' are characterized by low $\Delta H^{\ddagger}$ and highly negative $\Delta S^{\ddagger}$ values ⟨66JA3582, 76JGU799⟩. An analogous study on dimerization of phenyl isocyanate and *N,N*-diphenylcarbodiimide displays similar kinetic parameters ⟨68JOC1913⟩. The only reported stable dialkyluretidinedione also fragments and serves as a shelf stable precursor of *t*-butyl isocyanate (Scheme 12) ⟨71JOC3056⟩.

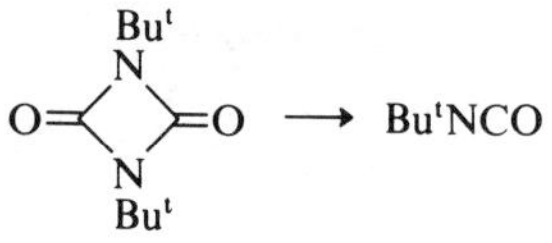

Scheme 12

Carbonyl-substituted isocyanates have been reported to add [2+2] to carbodiimides, although in one case the reported 4-imino-1,3-diazetidin-2-one ⟨64AP623⟩ was reassigned an oxadiazine structure ⟨72BCJ1534⟩. It is clear that the initially formed [2+2] kinetically favored adducts are thermally unstable and ultimately form more stable [4+2] products, oxadiazines (Scheme 13) ⟨79BSF(2)499⟩.

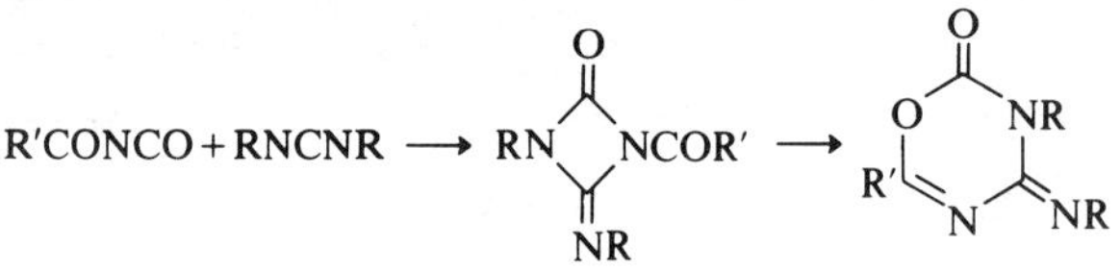

Scheme 13

5.15.2.1.5 1,3-Oxazetidine derivatives

Thermal studies on 1,3-oxazetidin-2-ones, as expected, show the major pathway to be extrusion of carbon dioxide to give imines ⟨70BAU1479, 80S571⟩. These systems are surprisingly stable; for example, for *N*-phenyl-1,3-oxazetidin-2-one $\Delta H^{\ddagger} = 128.9\ \mathrm{kJ\ mol^{-1}}$ and $\Delta S^{\ddagger} = 54.4\ \mathrm{J\ K^{-1}\ mol^{-1}}$. There is one report of fragmentation to give ketone and isocyanate ⟨80DOK(253)886⟩.

5.15.2.1.6 1,3-Thiazetidine derivatives

The equilibrium between isothiocyanates and carbodiimides has been determined. The cycloaddition is favored by 59 to 79 kJ mol^{-1} with decomposition of the adduct being facilitated by electron donating groups on the aryl isothiocyanates ⟨75JCS(P2)1475⟩. Steric factors have also been shown to control the mode of addition and decomposition of isothiocyanates and carbodiimides ⟨72JA3484⟩.

One example of an iminoalkylidene-1,3-thiazetidine thermolysis is reported to fragment in the opposite direction to give isothiocyanate and ketenimine (Scheme 14) ⟨70JOC3470⟩.

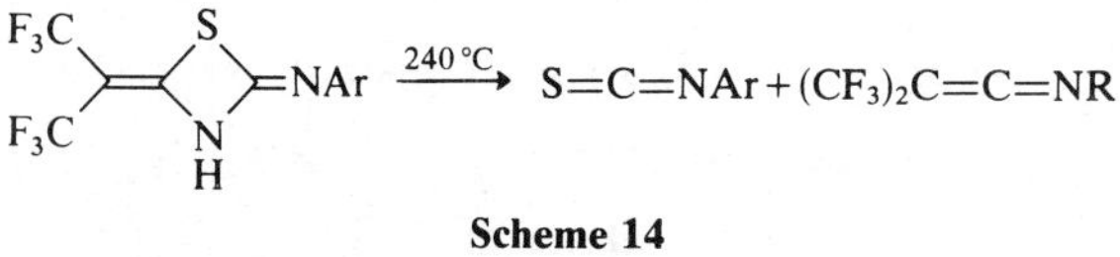

Scheme 14

An interesting example of a $2H$-1,3-thiazete in equilibrium with its very reactive ring-opened valence tautomer has been reported ⟨75AG(E)766, 77CB2114⟩.

5.15.2.1.7 1,2-Dioxetane derivatives

The chemistry of these fascinating molecules has been extensively reviewed ⟨79MI51503, 76CSR149, 77AHC(21)437⟩.

The most remarkable property of both dioxetanes and dioxetanones is their ability to chemiluminesce when heated to their decomposition point. In general this behavior can be summarized as in Scheme 15 ⟨79MI51502⟩. Both 1,2-dioxetanes and 1,2-dioxetan-3-ones (α-peroxylactones) give more triplet than singlet ketones ($\Phi_T:\Phi_S = 2:2500$ ⟨77AHC(21)437⟩) and evidence suggests that both decompose by two or more competitive pathways with discrete transition states ⟨79MI51502, 77MI51501⟩. The stepwise decomposition involves initial scission of the O—O bond followed by C—C cleavage ⟨77JA5403, 78JOC2236, 78CB1413, 81JA1292, 80MI51500⟩. Electron donor activators show a chemically initiated electron exchange luminescence (CIEEL) via a mechanism originally proposed by Schuster and coworkers (Scheme 16) ⟨80JA7100, 81JA1292, 79JA6511⟩. Dioxetanes have even been described as capable of effecting photochemistry without light ⟨73JA7050⟩.

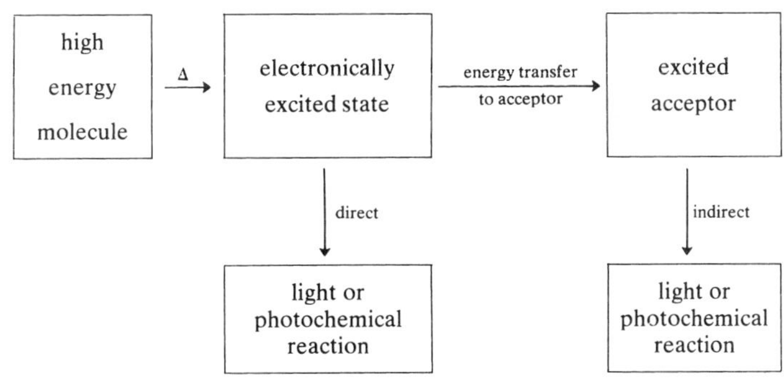

Scheme 15

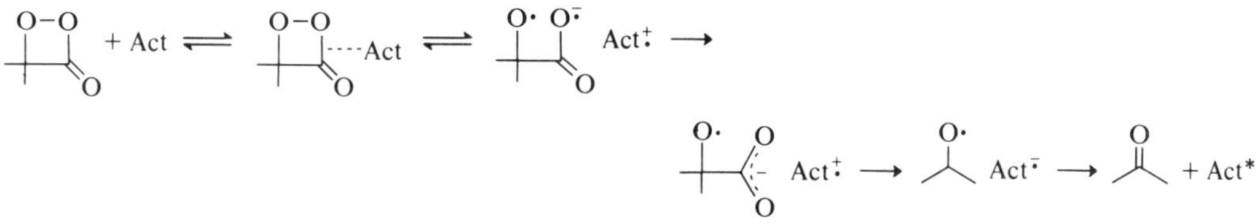

Scheme 16

5.15.2.1.8 1,2-Oxathietane derivatives

The thermal decomposition of 1,2-oxathietane 2-oxides invariably leads to loss of sulfur dioxide and accompanying alkene formation (Scheme 17) ⟨73JA3420, 78BAU142⟩. The lack of any solvent polarity effect on rate ⟨75CC724⟩ as well as supporting theoretical calculations favor a concerted loss of sulfur dioxide without requiring a strained $[_{\sigma}2_s + _{\sigma}2_a]$ pathway ⟨77TL2045⟩.

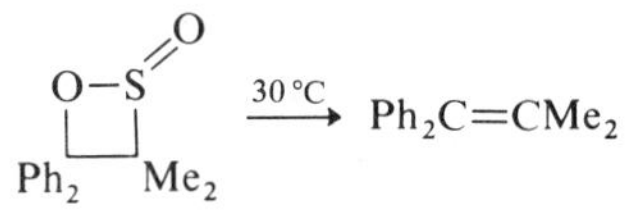

Scheme 17

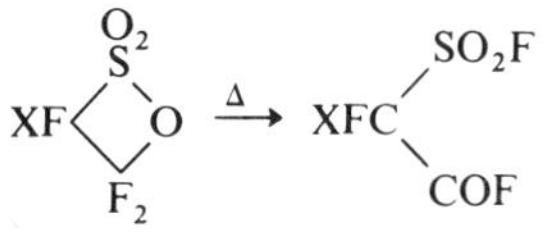

Scheme 18

The thermolysis of fluorinated β-sultones, the most stable derivatives of oxathietane dioxides, generally gives isomerization rather than elimination (Scheme 18) ⟨72AG(E)583⟩.

5.15.2.1.9 1,3-Dithietane derivatives

Bis-alkylidene and perfluoromethylated 1,3-dithietanes are generally quite stable and only decompose at temperatures exceeding 500 °C, resulting in the formation of monomers, thioketenes and thioketones (Scheme 19) ⟨61JA2589, 70JOC3470, 77CB1225⟩.

$R_2C{=}$(1,3-dithietane)${=}CR_2 \xrightarrow{\Delta} R_2C{=}C{=}S$

Scheme 19

The flash vacuum pyrolysis of 1,3-dithietane 1-oxide has been observed to give sulfine ⟨76JA1264⟩ which has been isolated and characterized in a matrix ⟨79JPC1890⟩.

5.15.2.2 Electrophilic Attack on Ring Heteroatoms

5.15.2.2.1 1,2-Diazetidine derivatives

An interesting *ortho* semidine-type rearrangement of a 1,2-diazetidin-3-one has been reported. However, it is probably not general for other derivatives that are unable to stabilize positive charge (Scheme 20) ⟨63JCS674⟩, and a different course has been reported for 1,2,4-triphenyl-1,2-diazetidin-3-one ⟨67AG(E)630⟩.

Ph_2 / PhN–NPh (diazetidinone) $\xrightarrow[\text{MeOH}]{\text{HCl}}$ $Ph_2\overset{+}{C}C(O)NPhNHCO_2Et$ ⟶ (indolinone: Ph_2, =O, N–$NHCO_2Et$)

Scheme 20

O, O / $Bu^tN{-}NBu^t$ $\xrightarrow[\text{MeOH}]{\text{HCl}}$ $MeO_2CCONBu^tNHBu^t$

Scheme 21

Similarly, 1,2-di-*t*-butyl-1,2-diazetidine-3,4-dione opens at the carbon–nitrogen bond in acidic methanol (Scheme 21) ⟨67JOC2360⟩.

5.15.2.2.2 1,2-Oxazetidine derivatives

4-Imino-1,2-oxazetidines are quite sensitive to hydrolysis even though the paths to product formation are mechanistically complex (Scheme 22) ⟨77CC831, 80AG(E)45⟩.

NC_6H_{11} / O / N–Bu^t / Bu^t $\xrightarrow{H_3O^+}$ $Bu^tNHCHCONHC_6H_{11}$ (Bu^t)

Scheme 22

N–O / Bu^t / $(SMe)_2$ $\xrightarrow[\text{MeOH}]{H_2SO_4}$ Bu^t–C(=NOH)CO_2Me

Scheme 23

The one case reported for hydrolysis of an oxazete gives an oxime ester; however, it does not appear to be a synthetically useful transformation (Scheme 23) ⟨80TL2025⟩.

5.15.2.2.3 1,2-Thiazetidine derivatives

The hydrolyses of 1-imino-1,2-thiazetidin-3-ones, which are themselves the result of addition of ketenes to sulfur diimides, are usually characterized by loss of the imino sulfur linkage (Scheme 24) ⟨69TL447, 70TL1427⟩. However, one account where the iminothiazetidine

NAr / ArN–S / Ph_2 / O $\xrightarrow[\text{HBr}]{\text{EtOH}}$ $ArNHCOC(OEt)Ph_2 + ArNH_3^+Br^-$ 67%

Scheme 24

O=S–NPh / Me_2 / NTs $\xrightarrow{\text{MCPBA}}$ O_2S–NPh / Me_2 / NTs

Scheme 25

was not isolated reports cleavage of the N—S bond and further isomerization to give a sulfinamine ⟨81H(16)1187⟩.

The relatively rare 1,2-thiazetidine 1-oxides can be oxidized to sulfoxides with MCPBA (Scheme 25) ⟨81TL583⟩.

5.15.2.2.4 1,3-Diazetidine derivatives

Aqueous acid treatment of uretidinediones leads to formation of ureas ⟨B-67MI51500⟩; labile groups on nitrogen can be removed with anhydrous HCl with subsequent ring opening (Scheme 26) ⟨66CB3103⟩.

$CF_3SN(CO)_2NSCF_3 \xrightarrow{HCl} CF_3SCl + NH_2CONHCOCl$

Scheme 26

5.15.2.2.5 1,3-Thiazetidine derivatives

Peroxide oxidation of 2-imino-1,3-thiazetidine does not lead to simple sulfur oxidation. Oxidative cleavage and ring closure result in 3-oxo-1,2,4-thiadiazolidine 1-oxide formation (Scheme 27) ⟨76AG(E)103⟩.

Scheme 27

5.15.2.2.6 1,2-Dioxetane derivatives

The reaction of dioxetanes with numerous Lewis acids has been studied. With divalent tin, tetramethyldioxetane gives varying amounts of acetone and pinacol. Other metals also catalyze ring opening at rates which correlate well with their Lewis acidity ⟨76CSR149⟩. Treatment with boron trifluoride leads to an interesting set of rearrangement products. Pinacolone and dimeric peroxide imply intermediacy of a carbonyl oxide of pinacolone (Scheme 28) ⟨79MI51503⟩. Divalent metals give only acetone ⟨74JA5557⟩.

70% 5% 25%

Scheme 28

5.15.2.2.7 1,3-Dithietane derivatives

1,3-Dithietanes, dimers of thioketones, are best stabilized by electron withdrawing groups on the ring. They are oxidized to bis-sulfoxide derivatives quite readily, and in at least one case the intermediate dioxide was isolated (Scheme 29) ⟨80AG(E)203⟩.

$\xrightarrow[HOAc]{KMnO_4}$ 60% $\xrightarrow[HNO_3]{CrO_3}$ 27%

Scheme 29

Further evidence of facile substitution on sulfur is shown by the reaction with trifluoromethyl hypochlorite. The sulfurane is stable to 100 °C where it reverts to 1,3-dithietane and peroxide (Scheme 30) ⟨77JA4194⟩.

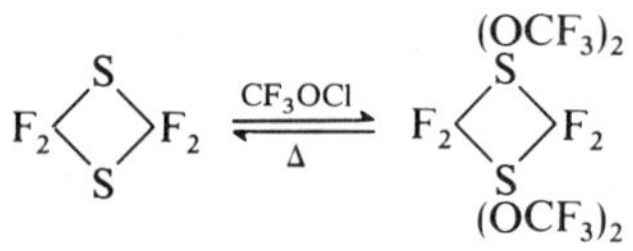

Scheme 30

5.15.2.3 Nucleophilic Attack on Ring Atoms

5.15.2.3.1 1,2-Diazetidine derivatives

Carbonyl derivatives of diazetidinones are readily attacked at the carbonyl by nucleophiles with subsequent ring cleavage (Scheme 31) ⟨64JOC3188⟩.

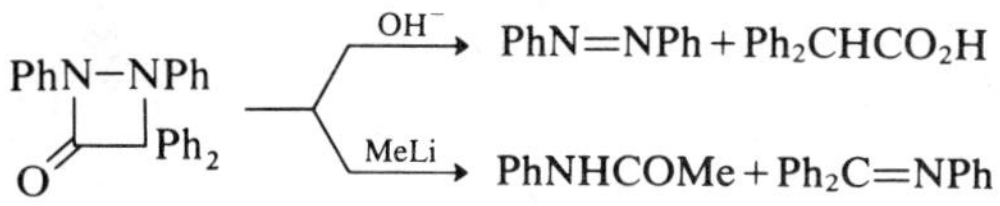

$\xrightarrow{OH^-}$ PhN=NPh + Ph_2CHCO_2H

$\xrightarrow{MeLi}$ PhNHCOMe + Ph_2C=NPh

Scheme 31

5.15.2.3.2 1,2-Oxazetidine derivatives

As part of the structure proof *N-t*-butyl-1,2-oxazetidine was treated with lithium aluminum hydride with cleavage at the O—N bond (Scheme 32) ⟨71JA4082⟩.

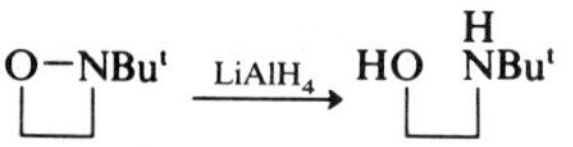

Scheme 32

5.15.2.3.3 1,2-Thiazetidine derivatives

N-Sulfonyl-3-alkoxy-1,2-thiazetidine 1-oxides illustrate the sensitivity of the N—S bond towards nucleophiles such as Grignards (Scheme 33) ⟨66CB3903⟩.

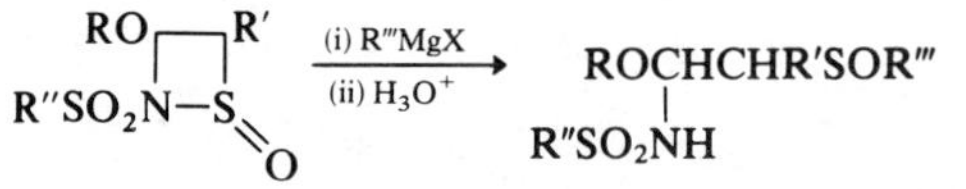

Scheme 33

Treatment of 1,2-thiazetidine dioxides with nucleophiles, base ⟨75BCJ480⟩ or methoxide ⟨72LA(763)46⟩ also cleaves the N—S bond (Scheme 34), although at least one account reports carbon–nitrogen scission ⟨70BCJ3543⟩.

Scheme 34

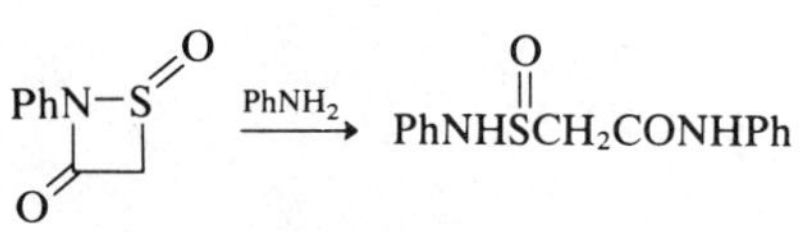

Scheme 35

1,2-Thiazetidin-3-one 1-oxides are nucleophilically attacked at the carbonyl with subsequent formation of amide (Scheme 35) ⟨78JOC3066⟩.

The corresponding dioxides show the characteristic cleavage of the weaker amide bond by both hydrolysis and reduction as well as attack at carbonyl by amines (Scheme 36) ⟨62HCA717, 63HCA450⟩.

$R_2C(CO)NHSO_2$ (ring) $\xrightarrow{H_3O^+}$ $R_2C(CO_2H)(SO_2NH_2)$; $\xrightarrow{LiAlH_4}$ $R_2C(CH_2OH)(SO_2NH_2)$; $\xrightarrow{NH_3}$ $R_2C(CONH_2)(SO_2NH_2)$

Scheme 36

5.15.2.3.4 1,3-Diazetidine derivatives

Simple uretidinediones, as expected, are attacked at the carbonyl with ultimate formation of urea derivatives (Scheme 37) ⟨71JOC3056⟩. More complex derivatives react similarly, although subsequent reaction on labile substituents also occurs ⟨69M1860, 66LA(692)151, 62CJC935, 53JA5439⟩.

Bu^{t}N(CO)$_2$NBu^t $\xrightarrow{MeONa}$ Bu^{t}N(CO$_2$Me)(CONHBu^t)

$CF_3SN(CO)_2NSCF_3$ $\xrightarrow{MeOH}$ $CF_3SOMe + CF_3SNHCONHCO_2Me$

Scheme 37

1,3-Diphenyl-4-(phenylimino)-1,3-diazetidin-2-one reacts with alcohols to give two products, either by attack on monomer or by reaction with isocyanate and carbodiimide, which are in equilibrium with monomer (Scheme 38) ⟨68JOC1913⟩.

PhN(CO)(C=NPh)NPh $\xrightarrow{ROH}$ $RO_2CNHPh + PhNHC(=NPh)OR$

Scheme 38

5.15.2.3.5 1,2-Dioxetane derivatives

At low temperatures dioxetane adducts with trivalent phosphorus have been isolated by Bartlett and coworkers. These cyclic phosphoranes gradually decompose to give the corresponding epoxide and phosphine oxide (Scheme 39) ⟨73JA6486⟩. Aryl-substituted phosphines

Tetramethyl-1,2-dioxetane $\xrightarrow[C_6H_6]{PPh_3}$ dioxaphospholane–PPh_3 $\xrightarrow{55\,°C}$ tetramethyloxirane $+ OPPh_3$

Scheme 39

show a ρ-value of -0.82, which has led to the speculation that insertion is concerted, not nucleophilic in nature ⟨80JOC3593⟩.

Similar reactivity is known for divalent sulfur nucleophiles ⟨75JA3850⟩. For 3,3-dimethyl-1,2-dioxetane, initial attack at carbon by azide ion has been postulated to explain the formation of acetone and nitrogen and imine *N*-oxide (Scheme 40) ⟨71TL749, 72JA1747⟩.

N_3^- + N_2 + $CH_2{=}NO^-$

N_3

Scheme 40

5.15.2.3.6 1,2-Oxathietane derivatives

Basic ring opening of 1,2-oxathietane dioxides is characterized by breaking the oxygen–sulfur bond by base ⟨69CI(L)457⟩, amines ⟨78AP698, 67DOK(172)1327⟩ or water ⟨67JOU2089⟩. Perfluorinated derivatives, in addition to ring opening, have been shown to undergo halogen displacement (Scheme 41) ⟨67BAU1685, 1471, 61JOC4002⟩.

$O{-}SO_2$ Cl_3C $\xrightarrow{NH_2R}$ $Cl_3CCH(OH)CH_2SO_2NHR$ $O{-}SO_2$ F_2 F C_5F_{11} $\xrightarrow{OH^-}$ $C_5F_{11}CF(SO_3^-)CO_2^-$

Scheme 41

The intermediacy of a 4-imino-1,2-oxathietane 2-oxide from the addition of sulfur dioxide to ketenimines was inferred from the trapping with methanol. The final isolated product in absence of methanol, 1,2-thiazetidin-3-one 1-oxide, was established by X-ray (Scheme 42) ⟨81CC350⟩.

$MeOS(O)C(Me)_2CONHR$

MeOH

$R_2C{=}C{=}NR$ $\xrightarrow[-78\,°C]{SO_2}$ O–S(O) Me₂ RN

RN–S(O) Me₂ O

Scheme 42

5.15.2.3.7 1,2-Dithietane derivatives

The structure proof of the stereochemistry of the lachrymatory factor of onions was determined by reduction of the 1,2-dithietane dioxide with LAH. The (±)-hexane-3,4-dithiol product establishes the ring stereochemistry as *trans* (Scheme 43) ⟨80JA2490⟩.

H(Et)C=S⁺–O⁻ → → S–SO_2 Et Et $\xrightarrow{LiAlH_4}$ EtCH(SH)–CH(SH)Et (±)

Scheme 43

5.15.2.3.8 1,3-Dithietane derivatives

Tetrakis(trifluoromethyl)-1,3-dithietanes react with nucleophiles to give products expected from monomeric bis(trifluoromethyl)thione. In fact, fluoride ion in DMF produces monomer ⟨76BCJ2491⟩, anilines give imines, and phenylhydrazine gives hydrazone ⟨73BCJ3260⟩. Triphenylphosphine gives a Wittig ylide that reacts with aldehydes to produce alkenes (Scheme 44) ⟨79TL3997⟩. Tetrahalo-1,3-dithietanes replace two halogens with water (Scheme 45) ⟨57JCS614, 77CB916⟩.

$(CF_3)_2C(S)_2C(CF_3)_2 \xrightarrow[\text{ether, } -78\,^\circ C]{Ph_3P} Ph_3\overset{+}{P}-\overset{-}{C}(CF_3)_2 \xrightarrow{RCHO} RCH{=}C(CF_3)_2$ 26–64%

Scheme 44

$Cl_2C(S)_2CCl_2 \xrightarrow{H_2O} Cl_2C(S)_2C{=}O$

Scheme 45

Dicyanomethylene-1,3-dithietane rings open with nucleophiles with retention of both sulfur atoms and, while NaH produces hydrogen, the anion does not alkylate (Scheme 46) ⟨64JOC497⟩.

$(CN)_2C{=}C(S)_2 \xrightarrow{Ph_3P} (NC)_2C{=}C(S^-)SCH_2\overset{+}{P}Ph_3$

$(CN)_2C{=}C(S)_2 \xrightarrow{\text{pyridine}} (NC)_2C{=}C(S^-)SCH_2\overset{+}{N}C_5H_5$

Scheme 46

A number of interesting reactions of desaurins with nucleophiles have been investigated by Peseke *et al.* (Scheme 47) ⟨76T483, 76ZC16, 75ZC19⟩.

$RO_2C(NC)C{=}C(S)_2C{=}C(CO_2R)CN$

with o-phenylenediamine → benzodiazepine-thione (RO_2C, NH_2, NH) 89%

$\xrightarrow{NH_2NH_2}$ $(RO_2C(NC)C{=}C(NH-)(S-))_2 + \overset{+}{N}H_3NH_2$ 94%

$\xrightarrow{4RNH_2}$ $RO_2C(NC)C{=}C(NHR)S^-\ \overset{+}{N}H_3R$ 30–78%

Scheme 47

5.15.2.4 Miscellaneous Reactions

5.15.2.4.1 1,2-Diazetidine derivatives

Several examples of cleavage of the N—N bond in diazetidine derivatives are known; one by hydrogenation ⟨67TL161⟩, and the other by electroreductive cleavage. This reductive potential is comparable to that of other acyclic hydrazines ⟨78LA1505⟩. However, many cases are known where carbon–carbon double bond reductions elsewhere in the molecule leave the N—N bond intact. The latter has been used to prepare 1,2-diazetidines derived from cycloadditions with diazodicarboxylate derivatives ⟨68CC686, 72CC818, 78TL2469, 79AJC2659, 80TL3251⟩. One of the earliest representative examples is illustrated in Scheme 48 ⟨69JA5668⟩.

$\xrightarrow{RO_2CN{=}NCO_2R}$ (NCO₂R)₂ adduct $\xrightarrow{H_2/Pd}$ (NCO₂R)₂ saturated $\xrightarrow{\text{(i) KOH; (ii) } Cu^{2+}\text{; (iii) } OH^-}$ N=N

Scheme 48

Reduction of diazetidine *N,N*-dioxides in a two-step process is the other common method used to generate cyclic nitrogen–nitrogen double bonds (Scheme 49) ⟨75JOC1409⟩.

(i) $Cl_3SiSiCl_3$ (ii) $LiAlH_4$

Scheme 49

5.15.2.4.2 1,2-Oxazetidine derivatives

Hydrogenation of *N*-substituted 4,4-diphenyl-1,2-oxazetidin-3-ones gives ring opening in a manner dependent upon nitrogen substitution. Alkyl groups direct cleavage between oxygen and carbon, and aryl groups direct nitrogen–oxygen cleavage (Scheme 50) ⟨68JOC3619⟩. It has been reported that treatment of this same substrate with trialkyl fluoroborate followed by base gives ketone, carbamate and urea, although the mechanism by which these products arise is unclear (Scheme 50) ⟨74ZN(B)231⟩.

H_2: $Ph_2CHCON(OH)R$ (R = alkyl) or $Ph_2C(OH)CONHR$ (R = aryl)

(i) $R_3O^+ BF_4^-$ (ii) OH^-: $Ph_2CO + PhNHCO_2R + PhNHC(O)NHPh$ (57%, 46%, 11%)

Scheme 50

5.15.2.4.3 1,2-Thiazetidine derivatives

1,2-Thiazetidin-3-one 1,1-dioxides are conveniently alkylated either on nitrogen ⟨62HCA717⟩ or on oxygen ⟨63HCA450⟩ depending upon reagents (Scheme 51). Hydrolysis of the alkylated derivatives gives the expected sulfonamide ester.

$R'X, K_2CO_3$; Ag_2O, $R'X$; Al_2O_3, H_2O → $R_2C(SO_2NH_2)CO_2R'$

Scheme 51

5.15.2.4.4 1,3-Diazetidine derivatives

Dialkylcarbodiimides undergo [2+2] additions with chlorosulfonylisocyanate at least to some extent. The other product is a triazinedione (2 eq. of CSI with 1 eq. of carbodiimide) and is the exclusive product with diphenylcarbodiimide ⟨77JCS(P1)47⟩. The SO_2Cl group can be removed with KI, and if careful hydrolysis of the imine could be effected, it might be an entry to monoalkylated uretidinediones (Scheme 52).

KI

Scheme 52

5.15.2.4.5 1,2-Oxathietane derivatives

Perfluorinated 1,2-oxathietane dioxides, β-sultones, are the most stable derivatives and, as such, the most widely studied. They undergo some of the more unique reactions, especially those with acid halides which are regarded mechanistically as a concerted displacement (Scheme 53) ⟨67BAU1935⟩.

$F_2C{=}C(F)OSO_2Cl$

ClCOR

$CH_2{=}CH_2$

$F_2C{=}CF_2$ + ⟶ ⟶ ⟶ $(CH_2{=}C)_2SO_3H$

Scheme 53

The reaction with ethylene is visualized as an alkene exchange to give the unstable parent β-sultone, which undergoes ring opening and concomitant proton shift to give ethylenesulfonic acid. The latter reacts with the fluorinated β-sultone with ultimate formation of ethylenebis-sulfonic acid (Scheme 53) ⟨70BAU574⟩.

5.15.2.4.6 1,2-Dithietane derivatives

1,2-Dithietenes are usually quite unstable and can be regarded as being in equilibrium with their valence tautomers, dithiones. The 3,4-bis(trifluoromethyl) derivative, the most stable analog, forms stable complexes with indium ⟨73JOM(51)113⟩, iron ⟨75JCS(D)701⟩, phosphorus ⟨74JA317⟩ and molybdenum ⟨74JA3502⟩. It also adds [4+2] to alkenes (Scheme 54) ⟨60JA1515⟩. Similar addition to vinyl ethers has been observed for 3,4-dicyano-1,2-dithietene ⟨62JA4782⟩.

Scheme 54

5.15.2.4.7 1,3-Dithietane derivatives

1,3-Dithietane tetraoxide, called disulfene, is completely stable to acids even though it is very reactive towards base. It exchanges readily with D_20 and forms a tetrabromide with bromine (Scheme 55) ⟨69AG(E)73⟩.

Scheme 55

5.15.3 SYNTHESIS OF FOUR-MEMBERED RINGS WITH TWO HETEROATOMS

The literature contains examples of virtually all heterocyclic systems with two heteroatoms of nitrogen, sulfur and oxygen. However, in some cases simple derivatives of the parent rings are unknown and for others the literature is very extensive so that selections have been made to represent the various classes. Under each synthetic procedure the same order of coverage will be followed: 1,2-diazetidines (1,2-N,N), 1,2-oxazetidines (1,2-O,N), 1,2-thiazetidines (1,2-S,N), 1,3-diazetidines (1,3-N,N), 1,3-oxazetidines (1,3-O,N), 1,3-thiazetidines (1,3-S,N), 1,2-dioxetanes (1,2-O,O), 1,2-oxathietanes (1,2-O,S), 1,2-dithietanes (1,2-S,S) and 1,3-dithietanes (1,3-S,S).

5.15.3.1 Formation of One Bond between Two Heteroatoms

Only a limited number of systems lend themselves to closure *via* this method, and in some cases the mechanism is unclear. For this reason, unless known, no speculation is made regarding the mechanism.

5.15.3.1.1 1,2-Diazetidine derivatives

Oxidative closure of dioximes leads to diazetine dioxides which can be further reduced in two steps to Δ^1-1,2-diazetines (Scheme 56) ⟨75JOC1409, 75KGS748⟩.

One example of direct conversion to diazetines is known (Scheme 57) ⟨62JCS3276, 4840⟩.

R^1,R^2—NO_2 / R^1,R^2—NO_2 —Zn→ R^1,R^2—NHOH / R^1,R^2—NHOH —Br_2, H_2O_2→ R^1,R^2—$\overset{+}{N}$—O^- / R^1,R^2—$\overset{+}{N}$—O^- —(i) $Cl_3SiSiCl_3$, (ii) $LiAlH_4$→ R^1,R^2—N / R^1,R^2—N

$R^1 = R^2 = Me$
$R^1 = Me, R^2 = Et$

Scheme 56

NC—CN —AgF_2→ F_2—N / F_2—N

Scheme 57

5.15.3.1.2 1,2-Oxazetidine derivatives

A modified Pummerer reaction has been used to prepare a dithio-4*H*-oxazetine, presumably by a valence tautomerization (Scheme 58) ⟨80TL2025⟩.

$Bu^tC(=NOH)CH(SMe)_2$ —MCPBA→ $Bu^tC(NO)=C(SMe)_2$ → N—O ring, Bu^t, $(SMe)_2$

Scheme 58

5.15.3.1.3 1,2-Thiazetidine derivatives

Closures between sulfur and nitrogen are somewhat rare, although it would appear to be a relatively facile approach to the 1,2-thiazetidines (Scheme 59) ⟨75BSF807, 72TL213, 73MI51500, 63JA560⟩.

$R_2C(SO_2Cl)$—C(=O)—Cl —NH_2X, X = Ph, C_4H_9, C_6H_{11}→ R_2—SO_2 / O=—NX (20–70%)

$\overset{+}{N}H_3CH_2CH_2SO_2Cl$ —Na_2CO_3→ —SO_2 / —NH (90%)

Scheme 59

5.15.3.1.4 1,3-Diazetidine derivatives

While 1,3-diazetidine-2,4-diones are well known and readily prepared by dimerization (*vide infra*), very few closures from ureas are documented. One simple dialkyl derivative stable enough to be isolated is illustrated in Scheme 60. Smaller alkyl groups like isopropyl are much less stable ⟨71JOC3056⟩.

$Bu^tNHC(O)NHBu^t \xrightarrow[\text{pyridine}]{COCl_2}$ 1,3-di-Bu^t-1,3-diazetidine-2,4-dione 60%

Scheme 60

5.15.3.1.5 1,2-Dioxetane derivatives

Perhaps no field of organic research has expanded in recent years like this one dealing with the chemistry of dioxetanes ⟨76CSR149, 77AHC(21)437, 79MI51503, 78MI51501⟩. They are of interest for their chemiluminescent behavior and have great significance in connection with living species that exhibit bioluminescence. A myriad of these dioxetanes and dioxetanones have been prepared by a clever method conceived by Kopecky ⟨69CJC709, 73CJC468, 75CJC1103, 80CJC2089⟩. The first stable cyclic peroxide was obtained from cyclization of the β-bromohydroperoxide of 2-methyl-2-butene, which was in turn obtained from the alkene, concentrated hydrogen peroxide and 1,3-dibromo-5,5-dimethylhydantoin (Scheme 61).

$Me_2C{=}CHMe \xrightarrow[H_2O_2]{\text{1,3-dibromo-5,5-dimethylhydantoin}} Me_2C(OOH)CH(Br)Me \xrightarrow[Ag^+]{OH^- \text{ or}}$ Me_2, Me-1,2-dioxetane (O–O)

Scheme 61

α-Peroxylactones or 1,2-dioxetan-3-ones were first prepared by Adam and Liu ⟨72JA2894⟩ and have also been implicated in bioluminescence. For the most part they are exceedingly unstable and must be generated at low temperatures. Several precursors have been used in preparing the α-hydroperoxy acid which is then closed with dicyclohexylcarbodiimide (Scheme 62) ⟨77JA5768⟩.

$Me_2C{=}C(OSiMe_3)_2$

$Me_2C{=}C{=}O \rightarrow Me_2C(OOH)CO_2H \xrightarrow{DCCD}$ Me_2-1,2-dioxetan-3-one (O–O) 68%

$Me_2\bar{C}CO_2^-\ Li^+$

Li^+

Scheme 62

5.15.3.2 Formation of One Bond to a Heteroatom

5.15.3.2.1 1,2-Diazetidine derivatives

Taylor *et al.* have been successful in preparing diazetidinum salts, presumably by displacement. Further derivatization provides an entry into aza analogs of β-lactams (Scheme 63)

$Ph_2C{=}NNHC(O)CH_2Cl \xrightarrow[THF]{NaH}$ $Ph_2C{=}N{-}N^-{-}C(O)CH_2Cl \rightarrow$ 1-($^+{=}CPh_2$)-2-(N^-)-1,2-diazetidin-3-one

$\xrightarrow{NaBH_4}$ 1-$NCHPh_2$-2-NH-1,2-diazetidin-3-one

$\xrightarrow{TsOH \cdot H_2O}$ 1,2-diazetidin-3-one (NH, NH) $\cdot$TsOH

Scheme 63

⟨81JA7743⟩. The benzhydryl imine function can be replaced by treatment with other aldehydes and ketones ⟨81JA7660⟩. One other hydrazinyl cyclization to a diazetidinum salt has been reported (Scheme 64) ⟨64JA2224⟩.

$Me_2\overset{+}{N}HNHCH_2CH(Br)Ph \xrightarrow[NaClO_4]{Na_2CO_3}$ Ph-substituted 1,1-dimethyldiazetidinium (NH, $\overset{+}{N}Me_2$) 65%

Scheme 64

It is difficult to determine how general the reaction of α-halo ketones with hydrazines is but one such example has been reported (Scheme 65) ⟨31RTC669⟩.

$PhCOCH_2Cl \xrightarrow{NH_2NHPh}$ Ph, =N, NPh (diazetine)

Scheme 65

5.15.3.2.2 1,2-Oxazetidine derivatives

The little known 4*H*-1,2-oxazete *N*-oxide ring has also been prepared by cyclization ⟨75AG(E)69⟩, as has the parent 1,2-di-*t*-butyl-1,2-oxazete and several derivatives (Scheme 66) ⟨75AG(E)70⟩.

$Bu^t{}_2C{=}CRNO_2 \longrightarrow$ R, =$\overset{+}{N}$–O^-, $Bu^t{}_2$, O (oxazete *N*-oxide) 35–78%

Scheme 66

β-Halo hydroxylamines can be converted into oxazetidines ⟨71JA4082⟩ and similar cyclizations give *N*-substituted 4,4-diaryl-1,2-oxazetidin-3-ones (Scheme 67) ⟨68JOC3619⟩.

$Bu^tN(OH)CH_2CH_2Br \xrightarrow[0\,°C,\ THF]{Bu^tOK}$ NBu^t, O (oxazetidine)

Scheme 67

5.15.3.2.3 1,3-Diazetidine derivatives

The conversion of carbodiimides to allophanyl chlorides as illustrated in Scheme 68 appears to be a general synthesis of uretidinediones and the method of choice for dialkyl derivatives ⟨78JOC4530⟩.

$RN{=}C{=}NR \xrightarrow{COCl_2} RN{=}C(Cl){-}N(R){-}C(O)Cl \xrightarrow{H_2O} RNHC(O){-}N(R){-}C(O)Cl \xrightarrow{DABCO}$ RN, NR, O, O (uretidinedione) 55–84%

Scheme 68

Another procedure for preparation of allophanyl chlorides from isocyanates and subsequent cyclization in 35–95% yield to uretidinediones has also been reported ⟨70AG(E)372⟩.

5.15.3.2.4 1,3-Oxazetidine derivatives

Imines from α-halo carbonyls treated with oxalyl chloride are reported to give α-halo isocyanates, which upon careful hydrolysis cyclize to 1,3-oxazetidin-2-ones. However, it is difficult to tell how general such a procedure might be since decarboxylation could compete with cyclization (Scheme 69) ⟨80S571⟩.

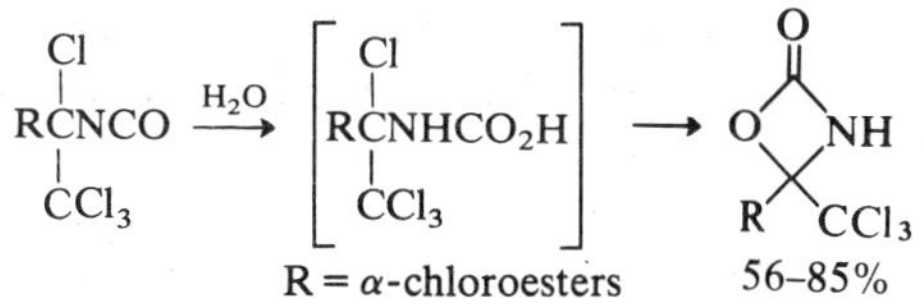

R = α-chloroesters 56–85%

Scheme 69

5.15.3.2.5 1,3-Thiazetidine derivatives

Two examples of thioureas have been reported to give imino-1,3-thiazetidin-2-ones, one ⟨66AG(E)704⟩ in 97%, the other ⟨64CB1232⟩ in 35% yield, presumably by γ-halo displacement (Scheme 70).

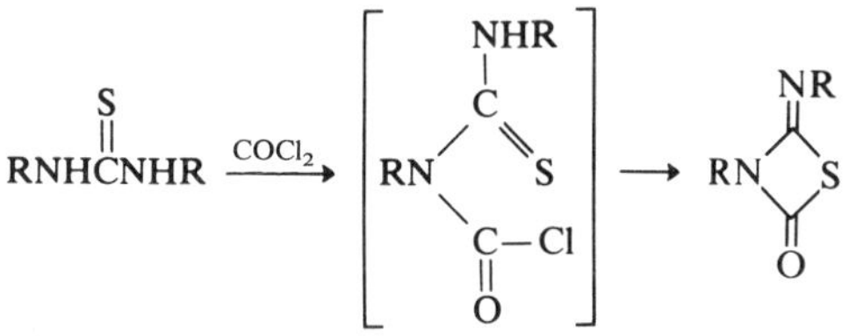

Scheme 70

A similar cyclization utilizing MeSOH as a leaving group in strong acid has been reported to give iminothiazetidinium salts (Scheme 71) ⟨74TL2577⟩.

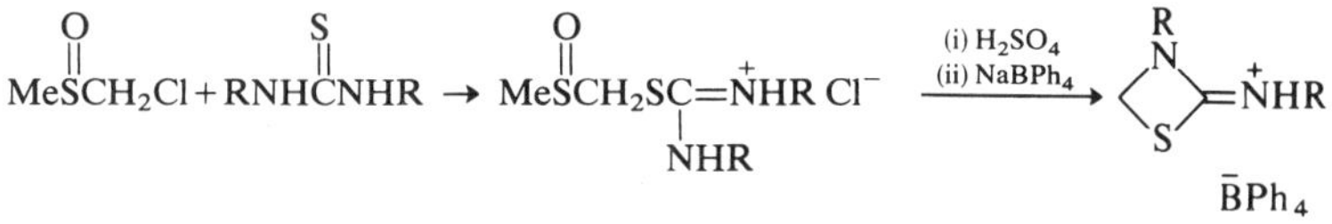

Scheme 71

An example of the first isolated β-sultine makes use of this γ-cyclization where *t*-butyl chloride formally acts as the leaving group ⟨73JA3420, 75CC724⟩. Of the 13 examples reported, all eliminate SO_2 readily to give the corresponding alkenes in 50–90% yields (Scheme 72).

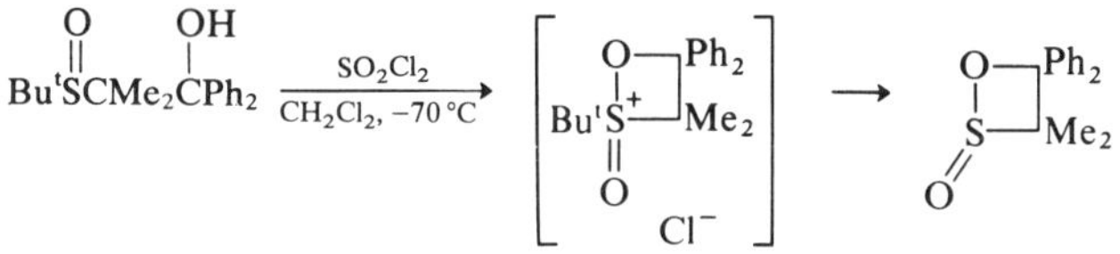

Scheme 72

5.15.3.3 Formation of Four-membered Rings by Making Two Bonds from [2+2] Atom Fragments

This method is by far the most successful and most widely used in the synthesis of four-membered heterocycles with two heteroatoms. In principle these formal additions can be envisioned to occur by (a) two displacements, (b) thermal concerted [2+2] cycloadditions, (c) photoinduced [2+2] additions, (d) stepwise zwitterionic intermediates or dipolar reactants, and (e) radical intermediates. In some cases the mechanisms have been elucidated but in many others the actual mode of addition is not known. For this reason, no attempt has been made to formally divide the latter four (b–e) into categories and only the first [(a), two displacement reactions] is covered separately.

5.15.3.3.1 By two displacement reactions

(i) 1,2-Diazetidine derivatives

Dihaloethanes have been reported to give *N,N*-dialkyl-1,2-diazetidines in reasonable yields (28–60%), but for di-*t*-butylhydrazine the yield falls off drastically (2%; Scheme 73) ⟨71DIS(B)(31)5869, 78JOC2785⟩.

$BrCH_2CH_2Br + RNHNHR \xrightarrow[\text{xylene, 100 °C}]{Na_2CO_3}$ 1,2-diazetidine (NR, NR)

Scheme 73

One example of an α-chloro acid chloride conversion to a diazetidinone has been reported (Scheme 74) ⟨63JCS674⟩. Similarly, oxalyl chloride and di-*t*-butylhydrazine give 1,2-di-*t*-butyl-1,2-diazetidine-3,4-dione which is moderately stable in absence of air. The corresponding diisopropyl is much less stable ⟨67JOC2360⟩.

$Ph_2C(Cl)COCl + PhNHNHCO_2Et \longrightarrow$ O, NPh, Ph_2, NCO_2Et

Scheme 74

5.15.3.3.2 [2+2] cycloadditions, concerted, diradical, dipolar and photochemical

(i) 1,2-Diazetidine derivatives

Azodicarboxylates undergo [2+2] additions to alkenes activated by alkoxy ⟨69TL1133, 71CB873, 71JOC2838, 70TL1925⟩, thioalkyl ⟨72TL4713⟩ or dialkylamino ⟨66AG416⟩ substituents to give substituted 1,2-diazetidines in reasonable yields (Scheme 75). The reaction shows some solvent sensitivity ⟨70TL1929⟩, and in one case a dipolar intermediate has been trapped ⟨72TL4713⟩. Frequently, 1,4-addition to the carbonyl of the azo moiety occurs giving dihydrooxadiazines ⟨70JOC1155⟩.

$(MeO)_2C{=}C(OMe)_2 + EtO_2CN{=}NCO_2Et \longrightarrow$ $(MeO)_2$, NCO_2Et, $(MeO)_2$, NCO_2Et 95%

Scheme 75

Tetraalkoxylallenes give similar additions ⟨72CB2437⟩ and under forcing conditions tetrafluoroethylene ⟨60JA852, 71JOU2361⟩ and tetrafluorobutadiene ⟨61JA386⟩ add [2+2] to azodicarboxylates.

Several workers have made use of the ability of cyclobutadiene to add [2+2] to azodicarboxylates ⟨72CC818, 75JA918, 78TL2469, 79AJC2659⟩ to make fused diazetine compounds (Scheme 76).

$Fe(CO)_3$ + $RO_2CN{=}NCO_2R \xrightarrow{Ce^{IV}}$ NCO_2R, NCO_2R $\longrightarrow \longrightarrow$ N, N

Scheme 76

Nortricyclane also adds to azodicarboxylates in a symmetry allowed fashion ($_{\pi}2_{\sigma}2_{\sigma}2$) ⟨69JA5668, 79JOC2288, 81JOC501⟩, and indene adds [2+2] to *N*-phenyltriazolinedione (Scheme 77) ⟨80TL3251⟩. *cis*-Azo compounds react with ketenes giving diazetidinones (Scheme 78) ⟨56AG71, 64JCS5284, 66TL5245, 67TL161⟩ by what has been described as (1) nearly concerted but not synchronous [$_{\pi}2_s + _{\pi}2_a$] ⟨74JOC1215⟩, (2) concerted ⟨70TL703⟩ and (3) diradical ⟨66JOC1079⟩. Vinylazo compounds give both [2+2] and [2+4] adducts ⟨72AG(E)583⟩.

$\xrightarrow{RO_2CN{=}NCO_2R}$ N, N $\longrightarrow \longrightarrow$ N, N

\+ N, N, NPh, O, O $\xrightarrow[\text{(ii) CuCl; (iii) KOH}]{\text{(i) KOH}}$ N, N

Scheme 77

$$\text{ArN=NAr} + \text{Ph}_2\text{C=C=O} \xrightarrow{h\nu} \text{1,2-diaryl-1,2-diazetidin-3-one (4,4-Ph}_2\text{)}$$

Scheme 78

Ketenimines with *cis* arylazo compounds give iminodiazetidines by addition across the carbon–carbon double bond ⟨67JHC155, 72JHC555, 74JHC409⟩.

(ii) 1,2-Oxazetidine derivatives

A number of oxazetidines have been obtained by [2+2] addition of nitroso compounds to appropriate alkenes, although it may not be a general reaction for all alkenes or nitroso compounds. For example, the trifluoromethylnitroso compound reacts with polyhalogenated ethylenes ⟨61JCS1351, 63DOK(152)1104, 65MI51500, 65JCS5602, 69JCS2119⟩, with styrene ⟨65JGU855⟩ or with allenes ⟨65JCS5602, 73JCS(P1)1561⟩ to give polyhalogenated oxazetidines, probably by a radical mechanism (Scheme 79) ⟨74JOU1427⟩.

$$CF_3NO + F_2C{=}CF_2 \longrightarrow \text{F}_2\text{C–O–NCF}_3\text{–CF}_2 \text{ (ring)}$$

Scheme 79

Only two other reports of non-halogenated nitroso compounds adding to alkenes have been made ⟨64AG(E)380, 72BCJ3187⟩.

The addition of nitrosobenzene to diphenylketene was reported as early as 1911 ⟨11CB365⟩. A re-examination using substituted nitrosobenzenes showed two products with the relative amounts greatly dependent upon substituent but not dependent on solvent polarity. The 4-one predominates with *p*-dimethylamino and the 3-one with *p*-methoxycarbonyl. Again, several mechanisms have been suggested (Scheme 80) ⟨63T133, 74JOC2552⟩.

Ketenimines also add [2+2] to nitrosobenzene (Scheme 81) ⟨70JHC1203, 72JHC77, 74JHC633⟩.

$$Ph_2C{=}C{=}O + ArNO \longrightarrow \text{4,4-Ph}_2\text{-2-Ar-1,2-oxazetidin-3-one} + \text{3,3-Ph}_2\text{-2-Ar-1,2-oxazetidin-4-one}$$

Scheme 80

$$Ph_2C{=}C{=}NAr + ArNO \longrightarrow \text{4,4-Ph}_2\text{-2-Ar-3-(NAr)-1,2-oxazetidine}$$

Scheme 81

(iii) 1,2-Thiazetidine derivatives

Sulfur diimides have been reported to undergo [2+2] additions with diphenylketene to give 1-imino-4,4-diphenyl-1,2-thiazetidin-3-ones (Scheme 82) ⟨69TL447, 69TL4227, 70TL1427, 72JOC3810⟩. However, the products are unstable, and in one case ⟨70BCJ1905⟩ X-ray analysis showed the structure to be a rearranged five-membered isomer.

$$RN{=}S{=}NR + Ph_2C{=}C{=}O \longrightarrow \text{1-(NR)-2-R-4,4-Ph}_2\text{-1,2-thiazetidin-3-one}$$

$R = Bu^t$, Ar, Ts

Scheme 82

Several *N*-sulfinyl derivatives have been shown to add to enol ethers to give thiazetidine 1-oxides in 60–76% yield ⟨66CB3903, 67AG(E)951⟩. Other examples have appeared showing that the sulfinyl group will add [2+2] to ketenes ⟨62T1527, 73JOC2652, 78JOC3066⟩ and ketenimines ⟨81TL583, 81CC350⟩, although usually the adducts are thermally unstable (Scheme 83).

$$Ph_2C{=}C{=}O + RN{=}S{=}O \longrightarrow \text{2-R-4,4-Ph}_2\text{-1,2-thiazetidin-3-one 1-oxide}$$

Scheme 83

Burgess ⟨67JA2502, 72JA4386, 72JA6135⟩ and others ⟨75BCJ480, 79BCJ1102⟩ found that sulfonylamines generated *in situ* display a high degree of electrophilicity and add to nucleophilic alkenes to give β-sultams or sulfur analogs of lactams (Scheme 84). For carbonyl derivatives of sulfonylamines, [2+4] addition also occurs to give a mixture of products in amounts sensitive to both solvent and temperature. The specificity seems to require $[_{\pi}2_s + _{\pi}2_a]$ and $[_{\pi}2_s + _{\pi}4_a]$ reactions, although the solvent effect indicates a 'rotationally rigid dipole' ⟨72JA4386, 75BCJ480⟩.

$EtNHSO_2Cl \xrightarrow[-78\,°C]{Et_3N} [EtN{=}SO_2]$ + (2-dichloromethylene-1,3-dioxolane) → (spiro product: Cl_2, SO_2, NEt, O, O)

CCl_2

Scheme 84

(iv) 1,3-Diazetidine derivatives

Simple derivatives of 1,3-diazetidines are relatively rare and one of the earliest examples prompted Ingold to state that '... it is recognized that the closure of a four-membered ring from a single open-chain molecule is, in general, one of the most difficult operations in ring formation ...' (Scheme 85) ⟨22JCS2793, 23JCS2745⟩.

CH_2O + (NH_2, Me-substituted benzene) → (1,3-di-Tol-1,3-diazetidine: Tol, N, N, Tol)

80%

Scheme 85

Cyclohexylamine is reported to react with formaldehyde giving a 60% yield of 1,3-diazetidine. One additional example reports a photochemical closure of *N*-cyclohexylbenzaldehydeimine ⟨68JA1666⟩. An excellent review covers much of the earlier work on dimerizations of isocyanates ⟨69ACR186⟩. Aromatic isocyanates readily dimerize to uretidinediones (1,3-diazetidine-2,4-diones) when catalyzed by trialkyl phosphites or pyridine presumably by a dipolar intermediate; $\Delta H^{\ddagger} = 2.1$ to $4.2\ \text{kJ mol}^{-1}$, $\Delta S^{\ddagger} = -251$ to $-293\ \text{J K}^{-1}\ \text{mol}^{-1}$ (Scheme 86) ⟨66JA3582, 76JGU799⟩.

$ArNCO \xrightarrow[toluene]{R_3P}$ ArN⟨(C=O)₂⟩NAr

Scheme 86

Triaryl phosphites are not effective as catalysts and alkyl isocyanates usually are not cyclized by phosphites at all, although benzyl isocyanates are reported to yield the corresponding uretidinediones with 1,2-dimethylimidazole as a catalyst ⟨75S463⟩.

Arylcarbodiimides also dimerize with alkyl phosphites ⟨68CB174⟩ and in special cases with triethylamine ⟨76CC567⟩. HBF_4 is effective in giving the corresponding fluoroborate salts which can be converted to the bis(imino)-1,3-diazetidines with base (Scheme 87) ⟨68AG(E)72, 72LA(762)167⟩.

$RN{=}C{=}NR \xrightarrow{HBF_4}$ [RN=(ring: NR, +, NHR, NR) BF_4^-] → RN=(ring: NR, NR)=NR

70–90%

Scheme 87

Imines have also been reported to add to isocyanates ⟨69CB938, 78T101, 80CC313⟩, although here again carbonyl isocyanates frequently give [4+2] additions to form six-membered rings (Scheme 88) ⟨73BAU2542, 74S461, 77BAU878⟩. Diphenylcarbodiimide and phenyl isocyanate have been reported to give the 1,3-diphenyl-4-(phenylimino)-2-uretidinone ⟨68JOC1913⟩ while carbonyl isocyanates give mostly [2+4] addition ⟨72BCJ1534, 79BSF(2)499⟩. Chlorosulfonyl isocyanate also adds [2+2] to carbodiimides ⟨77JCS(P1)47⟩. If isothiocyanates are activated by electron withdrawing groups, they give 2-imino-1,3-diazetidine-2-thiones (Scheme 89) ⟨65AG(E)520, 66T1565⟩.

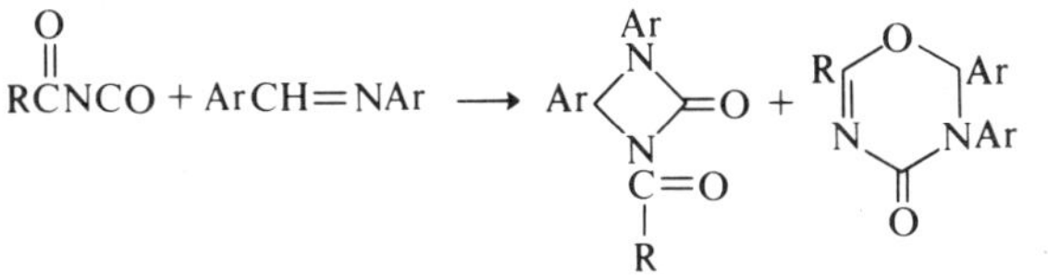

Scheme 88

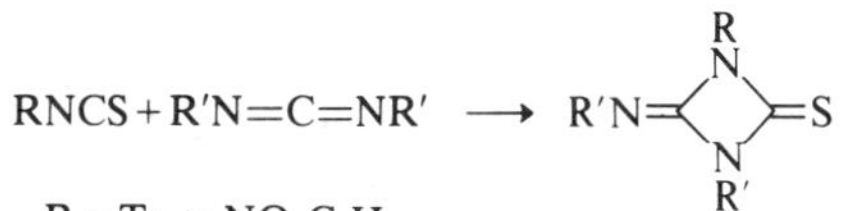

$R' = Pr^i$, C_6H_{11}, Ph
R = Ts, p-$NO_2C_6H_4$

Scheme 89

Several fluorinated ketenimines have been reported to dimerize with amines or trialkyl phosphites (Scheme 90) ⟨72JOU864, 73BAU1740, 76BAU337⟩.

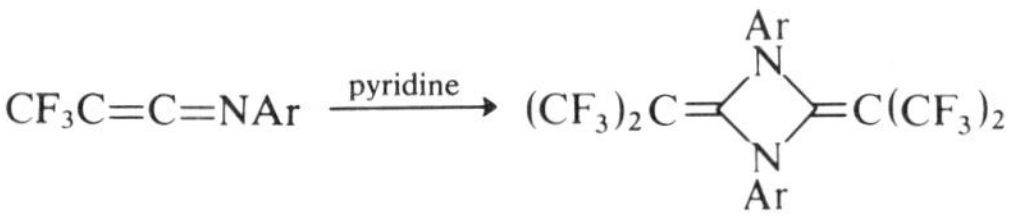

40–70%

Scheme 90

(v) 1,3-Oxazetidine derivatives

Examples of [2+2] conversions to 1,3-oxazetidine derivatives are rare. The one case reported by Ulrich goes contrary to additions previously reported (see previous section) (Scheme 91) ⟨72JA3484⟩.

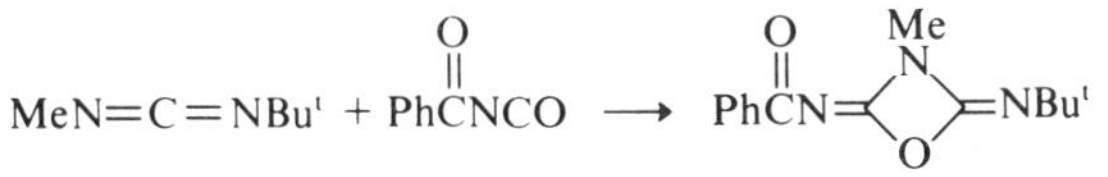

81%

Scheme 91

The second example is more reasonable but 3-phenyl-1,3-oxazetidin-3-one is obtained in low yield from formaldehyde tetramer and phenyl isocyanate (Scheme 92) ⟨67TL3637⟩. Perhalogenated (F or Cl) ketones also add to methylisocyanate catalyzed by CsF, BF_3 or in liquid SO_2 solvent in 5–70% yield ⟨67JOC2960⟩. Dichloroethylene is reported to add to benzoyl isocyanate in 23% yield, also by addition across the carbon–oxygen bond of the isocyanate ⟨70BAU1479⟩.

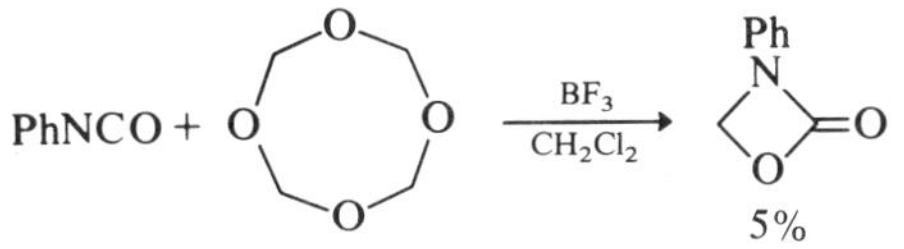

5%

Scheme 92

(vi) 1,3-Thiazetidine derivatives

Isothiocyanates, unlike isocyanates, add across the carbon–sulfur bond to carbodiimides with the stereochemistry shown (Scheme 93), as determined by comparing experimental and calculated dipole moments. The reaction has been described as a thermally allowed [${}_{\pi}2_a$ (isothiocyanate) + ${}_{\pi}2_s$ (carbodiimide)] with steric factors controlling the product formation ⟨76CCC562⟩. The reaction works best for aryl substituents with electron withdrawing groups ⟨73BCJ2559, 75JCS(P2)1475⟩ and does not give thione derivatives as was originally reported ⟨70CC1629⟩. Carbonyl derivatives of isothiocyanates give a mixture of [2+2] and [2+4] adducts with the latter predominating in more polar solvents ⟨78CCC3258⟩.

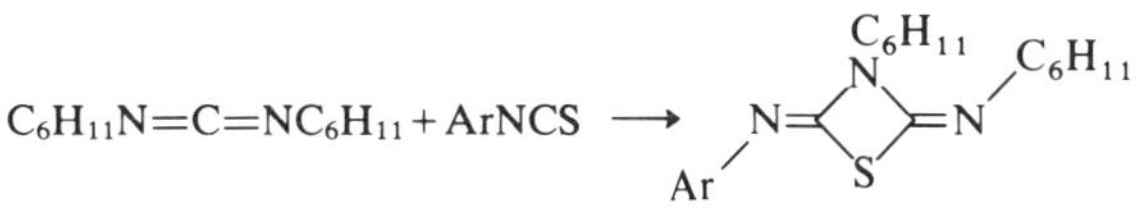

Scheme 93

Isothiocyanates ⟨70LA(731)120⟩ and bis(trifluoromethyl)thioketene ⟨78JOC2500⟩ have also been reported to add to benzalimines but these are undoubtedly special cases (Scheme 94).

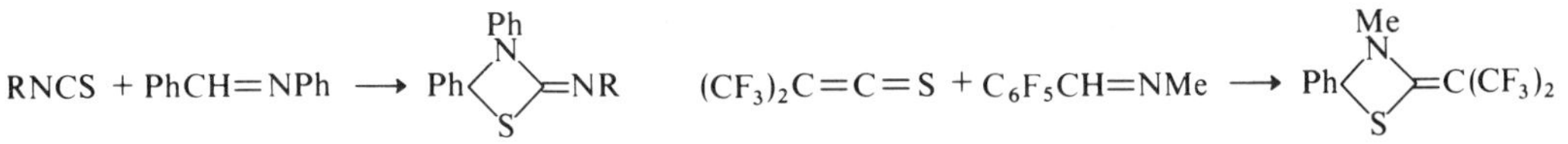

Scheme 94

(vii) 1,2-Dioxetane derivatives

The addition of singlet oxygen to alkenes also gives dioxetanes. A number of mechanisms have been proposed and the literature abounds with theoretical and experimental results supporting one or more possible intermediates: (a) 1,4-diradicals, (b) 1,4-dipolar, (c) perepoxides, or (d) concerted (Scheme 95). Both *ab initio* and semi-empirical calculations have been done and to date the controversy is still not resolved. These mechanisms have been reviewed extensively ⟨77AHC(21)437, 80JA439, 81MI51500 and references therein⟩ and will not be discussed here, except to point out that any one mechanism does not satisfactorily account for the stereospecificity, solvent effects, isotope effects and trapped intermediates observed. The reaction is undoubtedly substrate-dependent and what holds for one system does not always hold for another.

(a) (b) (c) (d) $_{\pi}2_s + _{\pi}2_a$

Scheme 95

The general procedure involves generation of singlet oxygen with an appropriate sensitizer like methylene blue or more commonly polymer-bound rose bengal at low temperature in methylene chloride. Numerous examples are reported ⟨79MI51503⟩, but only two examples are shown here: one using biadamantylidene as an example of an unusually stable dioxetane ($E_a = 144.8$, compared to 107.9 kJ mol^{-1} for tetramethyldioxetane ⟨72TL169, 75JA7110⟩), and one giving a bis-dioxetane (Scheme 96). As illustrated by the above reaction and others ⟨77JA3870, 78JA7127, 6437⟩, enol ethers form peroxides which fragment to ketones and esters accompanied by emission of light. Several chiral dioxetanes have been generated ⟨78JA7127, 80TL785⟩, and in one exceptional case luminescence gave circularly polarized light.

Ⓟ—RB; O_2, $h\nu$, CH_2Cl_2

Ⓟ—RB; O_2, $h\nu$, CH_2CL_2

Scheme 96

Several ketenes have been converted into 1,2-dioxetan-3-ones by reaction with triphenyl phosphite ozonide ⟨77JA5836⟩, and ketenimines form cyclic iminoperoxides with photochemically generated singlet oxygen (Scheme 97) ⟨79AG(E)788, 80CC898⟩.

$Bu^tCH{=}C{=}NBu^t$ → (Ⓟ—RB; O_2, $h\nu$) $t_{1/2}$ at 0 °C = seconds

$R_2C{=}C{=}O$ → (Ph_3PO_3); R = Me, Ph, CF_3

Scheme 97

An interesting case has been reported by Turro where triplet oxygen oxidation of dimethylketene is suggested to involve a dual pathway: a diradical intermediate on the path to polyester formation and a zwitterion leading to dimethyldioxetanone (Scheme 98) ⟨78JA5580⟩.

$Me_2C{=}C{=}O$ → ($^3O_2$) → $+CMe_2C(=O)-O-O+$ >90%; or → 1–5%

Scheme 98

(viii) 1,2-Oxathietane derivatives

The parent 1,2-oxathietane ring has not been made by formal [2+2] additions; however, both SO_2 and SO_3 are known to add alkenes under certain conditions.

Several examples of polyhalogenated alkene addition to sulfur dioxide ⟨60JA6181, 66JCS(C)1171⟩ are known to give moderately stable 1,2-oxathietane 2-oxide adducts (Scheme 99) ⟨78BAU142, 79BAU106⟩.

$CF_2{=}CFCF{=}CF_2 \xrightarrow{SO_2}$ (1,2-oxathietane 2-oxide with F, F_2 and $CF_2{=}CF$ substituents) 15–20%

Scheme 99

Ketene is reported to add photochemically in an argon matrix to sulfur dioxide giving 1,2-oxathietan-4-one 2-oxide ⟨75CC724⟩.

The addition of sulfur trioxide to alkenes is reported to occur stereospecifically although no yields are reported ⟨73CC841⟩. Acid-catalyzed rearrangement of 1,1-diphenyl-3,3-dimethylpropanol in presence of sulfur trioxide gives a stable 1,2-oxathietane 2,2-dioxide in 96% yield (Scheme 100) ⟨77JCS(P1)247⟩. Fluorinated alkenes, because of their electrophilic character, add at high temperature to sulfur trioxide, although the β-sultones (oxathietane dioxides) are usually unstable ⟨72BAU2441, 75BAU2416, 77JA1214⟩.

cis-MeCH=CHMe $\xrightarrow[\text{dioxane}]{SO_3}$ (O–SO_2 ring, Me, Me)

$Me_3CC(OH)Ph_2 \xrightarrow{H^+} CH_2{=}C(Me){-}CPh_2 \xrightarrow{SO_3}$ (O–SO_2 ring, Me, $CMePh_2$)

Scheme 100

Another approach to β-sultones involves the *in situ* generation of sulfene, or its zwitterionic equivalent, and its subsequent addition to perhalogenated ketones or aldehydes in yields of 60–80% (Scheme 101) ⟨67DOK(172)1327, 69CI(L)457, 76CJC2652⟩. The chemistry of fluorinated β-sultones has been reviewed ⟨72AG(E)583⟩.

$MeSO_2Cl \xrightarrow[-15\,^\circ C]{Et_3N} [CH_2{=}SO_2 \text{ or } \bar{C}H_2SO_2\overset{+}{N}Et_3] \xrightarrow{CCl_3CHO}$ (O–SO_2 ring, Cl_3C)

Scheme 101

(ix) 1,2-Dithietane derivatives

One of the very few examples of a four-membered heterocycle with sulfur atoms substituted 1,2 was reported by Dupont workers ⟨60JA1515, 61JA3434⟩. Its synthesis involves the high temperature reaction of fluorinated alkynes and elemental sulfur, probably *via* a diradical intermediate (Scheme 102). The chemistry of these 1,2-dithietenes has been reviewed ⟨76RCR639⟩. As discussed later, these dithietenes form complexes with a number of transition metals.

$CF_3C{\equiv}CCF_3 \xrightarrow[\text{sulfur vapor}]{450\,^\circ C}$ (S–S ring, CF_3, CF_3) 80%

Scheme 102

(x) 1,3-Dithietane derivatives

A number of cases of dimerization of thiones have been reported to give 1,3-dithietane derivatives (Scheme 103) ⟨75BSF1670, 72CJC3923, 61CB368⟩. One particularly intriguing approach generates sulfenyl chlorides from carbonyl derivatives and thionyl chloride with subsequent dimerization through thioketones giving dithietanes. The exact mechanism is not clear and the reaction is limited to carbonyl-substituted dithietanes (Scheme 104) ⟨79JOC1736, see also 74JOC1944⟩. Thio derivatives of phosgene photochemically dimerize and can be hydrolyzed to dithietanones (Scheme 105) ⟨77CB916⟩.

$Et_2C{=}S \xrightarrow{\text{pyridine}} Et_2\langle S_2 \rangle Et_2$

$(PhCH_2)_2C(OEt)_2 \xrightarrow{H_2S} (PhCH_2)_2CS \xrightarrow{h\nu} (PhCH_2)_2\langle S_2 \rangle (CH_2Ph)_2$

$Ph_2CO \xrightarrow{H_2S} Ph_2CS \xrightarrow{Na^+\bar{C}{\equiv}CH} Ph_2\langle S_2 \rangle Ph_2$

Scheme 103

$R_2C(Cl)SCl \xrightarrow{Ph_3P} R_2C{=}S \longrightarrow R',R\langle S_2 \rangle R',R$

R = R′ = CO_2Et; 72–93%
R = PhCO, R′ = Ph;
R = Ph, R′ = CO_2Et

Scheme 104

$X,Y{>}C{=}S \xrightarrow{h\nu} X,Y\langle S_2 \rangle X,Y \xrightarrow{H_2O} X,Y\langle S_2 \rangle{=}O$

10–78%

X = Y = SCF_3; X = Y = Br;
X = F, Y = Cl; X = Cl, Y = Br

Scheme 105

Several different groups have devised methods of generation of hexafluoropropane-2-thione which readily dimerizes to 2,2,4,4-tetrakis(trifluoromethyl)-1,3-dithietane photochemically or thermally. The common precursor to all is perfluoropropene, as originally illustrated by Middleton *et al.* ⟨61JA2589, 65JOC1375⟩ and subsequently improved by Knunyants and coworkers ⟨67BAU2467, 73T2759⟩ and others ⟨64JCS2944, 73BCJ3260⟩ (Scheme 106). Some thioketenes have also been shown to dimerize, particularly bis(trifluoromethyl) ⟨68JOC1850, 70JOC3470, 71BAU2386⟩, although alkyl- ⟨67JOC1562⟩, aryl- ⟨77CB1225⟩ and ester-substituted derivatives ⟨66CC577⟩ also dimerize (Scheme 107). Sulfonyl-substituted isothiocyanates, like many 1,3-dithietanes, dissociate to monomers at higher temperatures (Scheme 108) ⟨65AG(E)430⟩. Bunte salts generated from α-chloroamide can be converted into thiones which dimerize quantitatively (Scheme 109) ⟨77JCR(S)244, 77JOC2821⟩.

$CF_3CF{=}CF_2 \xrightarrow[F^-,\ 130\,^\circ C]{PhNO_2} [(CF_3)_2CS] \longrightarrow (CF_3)_2\langle S_2 \rangle (CF_3)_2$

70%

Scheme 106

$(CF_3)_2C{=}CF_2 \xrightarrow[DMF,\ 0\,^\circ C]{K_2S} (CF_3)_2C{=}\langle S_2 \rangle{=}C(CF_3)_2$

62%

Scheme 107

$MeSO_2N{=}C(S^-K^+)_2 \xrightarrow{COCl_2} MeSO_2NCS \underset{\Delta}{\rightleftharpoons} MeSO_2N{=}\langle S_2 \rangle{=}NO_2SMe$

Scheme 108

$PhCH(Cl)CONHR \xrightarrow[DMF]{Na_2S_2O_3} PhCH(SSO_3^-Na^+)CONHR \xrightarrow{OH^-} RNHOC(Ph)\langle S_2 \rangle(Ph)CONHR$

Scheme 109

One particular class of α,β-unsaturated keto-substituted 1,3-dithietanes called ‘desaurins’ has been known since 1888 ⟨1888CB337⟩. While several specific routes to their synthesis will be outlined later, the one illustrated here (Scheme 110) presumes the intermediacy of an α,β-unsaturated keto-substituted thioketone which then dimerizes. The methods used to obtain the thioketone vary, but all lead to the same final result ⟨58JA5577, 71CJC1456, 76BAU1547, 76BAU1913⟩. In many cases the stereochemistry has not been established, although there is some apparent sulfur–carbonyl interaction that gives preference to the structure shown ⟨67TL5307⟩.

$$\mathrm{MeCOCH{=}CCl_2} \xrightarrow{\mathrm{NaSC(=S)NEt_2}} [-\mathrm{CH{=}C{=}S}] \longrightarrow \text{Me(CO)CH=C}\langle\mathrm{S_2C}\rangle\text{C=CH(CO)Me} \quad 66\%$$

Scheme 110

The *in situ* generation of sulfene gives rise to an interesting dimer, 1,3-dithietane tetraoxide ⟨69AG(E)73⟩, which readily brominates and exchanges with D_2O (Scheme 111).

$$\mathrm{MeSO_2Cl} \xrightarrow[\mathrm{THF,\ -20\,°C}]{\mathrm{Et_3N}} \text{1,3-dithietane 1,1,3,3-tetraoxide} \quad 18\%$$

Scheme 111

5.15.3.4 Formation of Small Heterocycles by Forming Two Bonds from [3+1] Atom Fragments

Not many examples of [3+1] combinations to form four-membered heterocycles are known, and it is apparently not a general method.

5.15.3.4.1 1,2-Oxazetidine derivatives

N-Neopentylidene-*t*-butylamine *N*-oxide adds to isocyanides to give 4-imino-1,2-oxazetidines in reasonable yields provided the reaction is catalyzed by Lewis acids (Scheme 112) ⟨80AG(E)45⟩.

$$\mathrm{Bu^tN^+(O^-){=}CHBu^t + CNR} \xrightarrow[\mathrm{(ii)\ Et_3N}]{\mathrm{(i)\ BF_3}} \text{Bu}^t\text{N–O ring, Bu}^t\text{, =NR}$$

Scheme 112

5.15.3.4.2 1,2-Thiazetidine derivatives

One example of a β-lactam derivative isolated in 1% yield has been reported from dianion addition and cyclization with carbon dioxide (Scheme 113) ⟨73JOC20⟩.

$$\mathrm{ArS(O)(=NH)Me} \xrightarrow[\mathrm{(ii)\ CO_2\ (iii)\ H_3O^+}]{\mathrm{(i)\ 2BuLi}} \text{N=S(=O)(Ar) ring}$$

Scheme 113

Two studies of amination of mixed diacid chlorides to give 1,2-thiazetidin-3-ones are known (Scheme 114) ⟨62HCA717, 76BSF277⟩.

$$\mathrm{(R_2CHCO)_2O} \xrightarrow[\mathrm{(ii)\ NaOH}]{\mathrm{(i)\ H_2SO_4}} \mathrm{R_2CH(CO_2^-\,Na^+)(SO_3^-\,Na^+)} \xrightarrow[\mathrm{(ii)\ NH_3}]{\mathrm{(i)\ SOCl_2}} \text{HN–SO}_2\text{ ring, R}_2\text{, C=O} \quad 13\text{–}55\%$$

Scheme 114

5.15.3.4.3 1,3-Thiazetidine derivatives

Thioureas, in special cases, react [3+1] with dihalomethanes but usually give a mixture of isomers (Scheme 115) ⟨78CB143⟩.

$$R^1NHC(=S)NHR^2 \xrightarrow[Et_3N]{CH_2I_2} \text{1,3-thiazetidine}(=NR^1, N\text{-}R^2) + \text{1,3-thiazetidine}(=NR^2, N\text{-}R^1)$$

Scheme 115

5.15.3.4.4 1,3-Dithietane derivatives

Active methylenes react with carbon disulfide and base to form reactive salts which undergo [3+1] additions to a variety of alkylating agents, even *gem*-dihaloethylenes ⟨77CC207, 80S907, 64JOC497⟩. They can also be oxidized to the symmetrical desaurin derivatives (Scheme 116) ⟨62CB2861⟩.

$$X_2CH_2 \xrightarrow[CS_2]{base} X_2C{=}C(S^-)_2 \xrightarrow{[O]} X_2C{=}\text{(1,3-dithietane)}{=}CX_2$$

$X = CO_2R$ or CN

$$(NC)_2C{=}C(S^-)_2 \xrightarrow{R_2CCl_2} R_2\text{(1,3-dithietane)}{=}C(CN)_2$$

R = Ph, 40%
R = H, 65%

Scheme 116

Phosgene acts as an acylating agent to give methylene-substituted 1,3-dithietan-2-ones ⟨66AG(E)970, 79LA1715⟩.

5.15.3.5 Formation of Small Heterocycles by Forming Three Bonds from [2+1+1] Atom Fragments

It has been reported that amines ⟨32JCS1142⟩ or imines ⟨53JCS143⟩, when reacted with formaldehyde and hydrogen sulfide, give rise to *N*-substituted 1,3-thiazetidines. However, in at least one case the structure was established to be an eight-membered ring dimer, which calls into question the structure of these compounds.

5.15.3.6 Formation of Four-membered Heterocycles from Three- and Four-membered Heterocycles

5.15.3.6.1 1,2-Oxazetidine derivatives

Oxaziridines are known thermally to revert to imine *N*-oxides, and it is not surprising that when heated in presence of isocyanides they undergo the equivalent of a [3+1] addition ⟨79BAU1648⟩. Although the imino 1,2-oxazetidine was not isolated, its intermediacy was inferred from the isolated decomposition products (Scheme 117).

Similarly, the products isolated from the peracid oxidation of *cis*-1,2,3-triphenylaziridine imply the presence of a 1,2-oxazetidine (Scheme 117) ⟨66JOC1995⟩.

$$(CF_3)_3CN\text{–}Ph_2\text{(oxaziridine)} \xrightarrow[60\,°C]{C_6H_{11}NC} (CF_3)_3CN\text{(1,2-oxazetidine, }{=}NC_6H_{11}, Ph_2) \longrightarrow (CF_3)_3CNCNC_6H_{11} + Ph_2CO$$

56%

$$\text{Ph,Ph-aziridine(N-Ph)} \xrightarrow{RCO_3H} [\text{Ph,Ph-aziridine N(Ph)–O}] \longrightarrow [\text{Ph,Ph-1,2-oxazetidine (N–O), Ph}] \longrightarrow PhCHO + PhCH{=}NPh$$

Scheme 117

5.15.3.6.2 1,3-Oxazetidine derivatives

Nitrones act as oxygen transfer agents to α-lactams, giving 1,3-oxazetidin-2-ones in high yields although only two cases are known (Scheme 118) ⟨80DOK(253)886⟩.

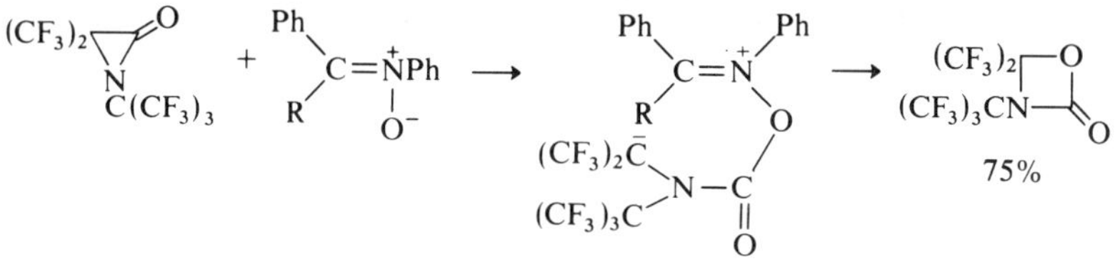

Scheme 118

5.15.3.6.3 1,3-Thiazetidine derivatives

Many thioketene dimers crack at high temperature to give monomers, which then give normal [2+2] additions (Scheme 119) ⟨70JOC3470⟩.

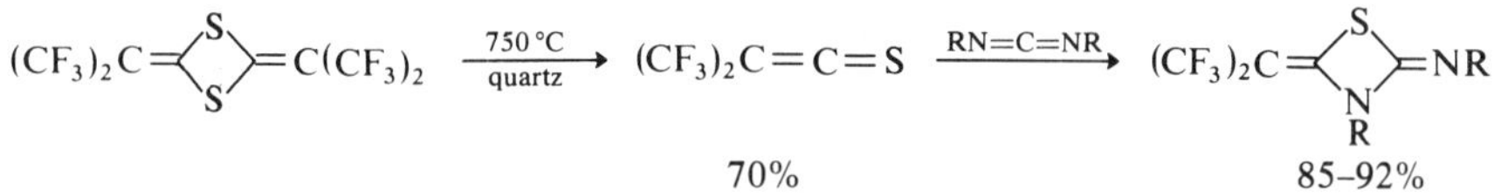

Scheme 119

5.15.3.6.4 1,3-Dithietane derivatives

Dithietanes are themselves readily functionalized to give interesting higher valence state derivatives (Scheme 120) ⟨78CC154, 80AG(E)203, 77JA4194⟩.

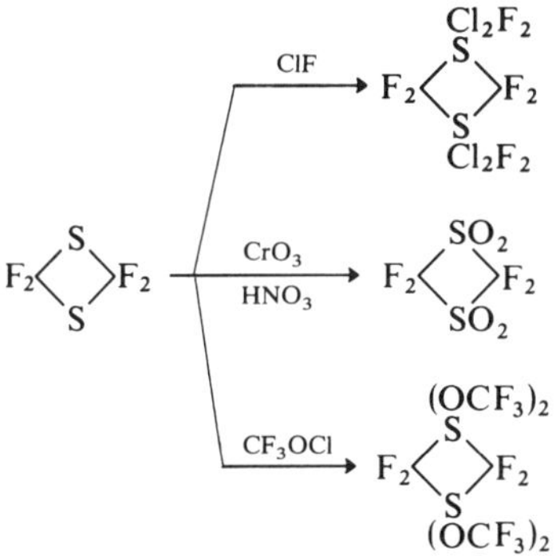

Scheme 120

5.15.3.7 Formation of Four-membered from Five-membered Heterocycles

5.15.3.7.1 1,2-Dithietane derivatives

Thioaldehyde oxides are reported to be one of the lachrymatory agents in onions and are speculated to dimerize *via* an initial [4+2] cycloaddition to a five-membered ring that rearranges to a 1,2-dithietane dioxide. The latter has also been isolated from extracts of onions ⟨80JA2490⟩ and is reported to be the only characterized derivative of a 1,2-dithietane (Scheme 121).

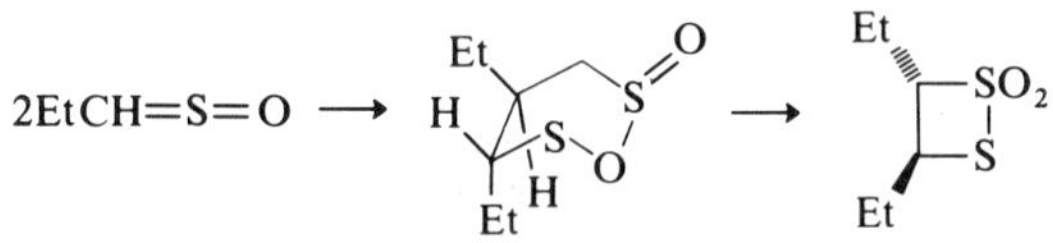

Scheme 121

Photochemically induced extrusion of carbon monoxide from five-membered rings has been used to generate 1,2-dithietanes which are in equilibrium with their valence tautomers, dithiones. They are labile ligands and form complexes with transition metals. The relatively small activation energy for ring closure (93.7 kJ mol^{-1}) has been cited as evidence for delocalization of π-electrons in the ring system (Scheme 122) ⟨73JA2383, 74JA3502⟩.

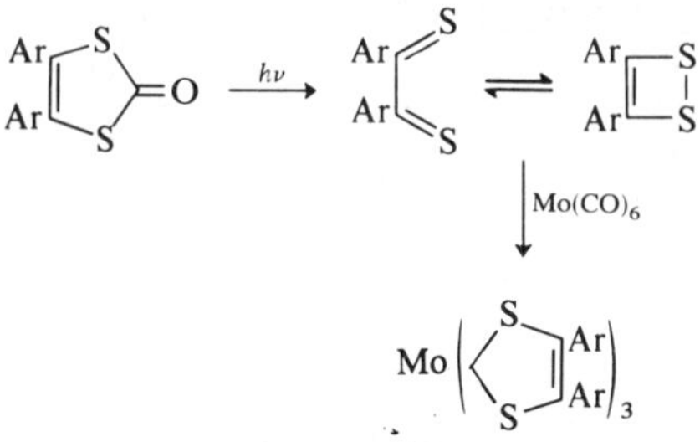

Scheme 122

An earlier study on 3,4-dicyano-1,2-dithietene showed similar results and LCAO-MO calculations established that electron withdrawing groups favored ring closure ⟨62JA4782⟩, this being consistent with the fact that the only stable isolated 1,2-dithietene is the bis(trifluoromethyl) derivative.

5.15.3.8 Formation of Four-membered from Six-membered Heterocycles

5.15.3.8.1 1,2-Diazetidine derivatives

Photoextrusion of nitrogen ⟨70JA6218⟩ and carbon monoxide ⟨71JCS(C)2920⟩ has been used to prepare 1,2-diazetidines (Scheme 123). Photoisomerization has alternately been reported to give fused derivatives of 1,2-diazetidines (Scheme 124) ⟨68CC686, 69JA2818⟩.

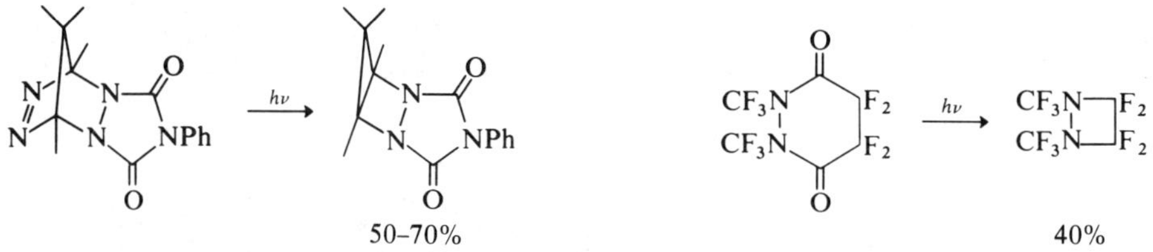

Scheme 123

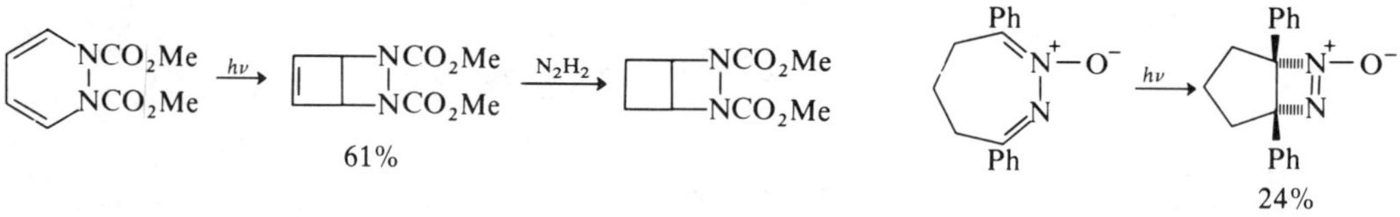

Scheme 124

5.15.3.8.2 1,2-Thiazetidine derivatives

Two groups have prepared 2-mesityl-2*H*-benzo[*c*][1,2]thiazetidine 1,1-dioxide from the corresponding benzothiatriazine by photochemical deazatization. It was found that the mesityl group was necessary for isolation purposes (Scheme 125) ⟨71JA5299, 72LA(763)46⟩.

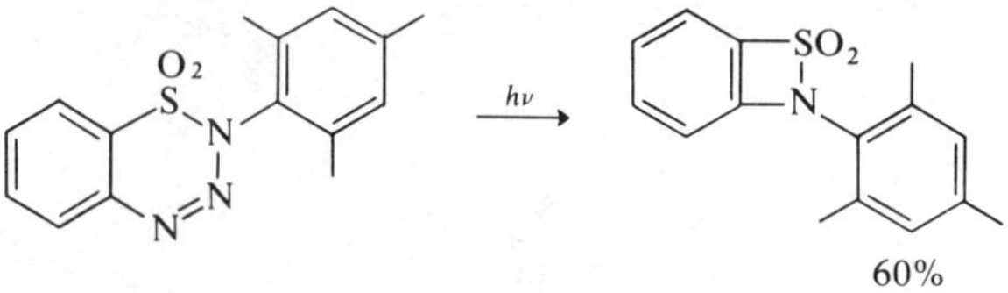

Scheme 125

5.15.3.8.3 1,3-Thiazetidine derivatives

One example of a 2*H*-1,3-thiazete stabilized by trifluoromethyl groups has been reported to arise from pyrolytic expulsion of hexafluoroacetone. The thiazete is in equilibrium with the open chain tautomer as established by trapping experiments with norbornene ⟨77CB2114⟩, isocyanides ⟨75AG(E)766⟩ and TCNE ⟨80AG(E)810⟩ (Scheme 126).

Scheme 126

5.15.4 APPLICATIONS OF FOUR-MEMBERED HETEROCYCLES WITH TWO HETEROATOMS

Four-membered rings with two heteroatoms are a unique and important class of compounds if for no other reason than for their medicinal, agricultural and industrial use.

5.15.4.1 1,2-Diazetidine Derivatives

3,3,4,4-Tetramethyl-Δ^1-1,2-diazetine, first prepared by Greene and Gilbert ⟨75JOC1409⟩, in principle bears structural similarity to tetramethyldioxetane. As described in previous sections, this dioxetane decomposes with the emission of light (it chemiluminesces). Like dioxetane, tetramethyl-1,2-diazetine is also a strained molecule ($SE = 102.5$ kJ mol^{-1}) and has a reasonably high activation energy for loss of nitrogen ($\Delta H^{\ddagger} = 132.6$ kJ mol^{-1}). Thus, from the heat of reaction it can be calculated that the thermolysis is exothermic by 209 kJ mol^{-1} and sufficient energy is available to produce triplet but not singlet excited tetramethylethylene ⟨78JA1876⟩. However, no chemiluminescence is observed, although a more rigid model would undoubtedly have a better chance ⟨69JA5668, 78TL2469⟩ (Scheme 127).

Scheme 127

The mode of decomposition of diazetines, diradical or concerted, has been probed by using *cis*-3,4-diethyl-3,4-dimethyl-Δ^1-1,2-diazetine as a model. The stereospecificity of product formation would imply a formally disallowed decomposition, but it is also conceivable that a short lived diradical intermediate is involved (Scheme 127) ⟨78JA6760⟩.

The corresponding tetramethyl *N*,*N*-dioxide, actually a precursor to the diazetine, has an interesting property in that it is an efficient triplet state quencher. For example, it inhibits the photopinacolization of benzophenone and can be used instead of dienes where wavelengths above 300 nm are employed ⟨72JA5077⟩.

Azomethine ylides studied by Taylor *et al.* have the potential of allowing an entry into highly strained aza analogs of *β*-lactam antibiotics (Scheme 128) ⟨81JA7660⟩.

Scheme 128

5.15.4.2 1,2-Thiazetidine Derivatives

The most characteristic property of derivatives of 1,2-thiazetidine 1,1-dioxides (ethanesultams) is their tendency to polymerize upon treatment with a variety of nucleophiles (Scheme 129) ⟨78MI51500⟩. These high molecular weight polymers have good solubility in polar aprotic solvents (DMSO, HMPT) as well as acids and bases ⟨79MI51501⟩. Bis-sultams also form crystalline poly(amidesulfonamides) of high viscosity and high molecular weight with aliphatic diamines (Scheme 130) ⟨79MI51500⟩.

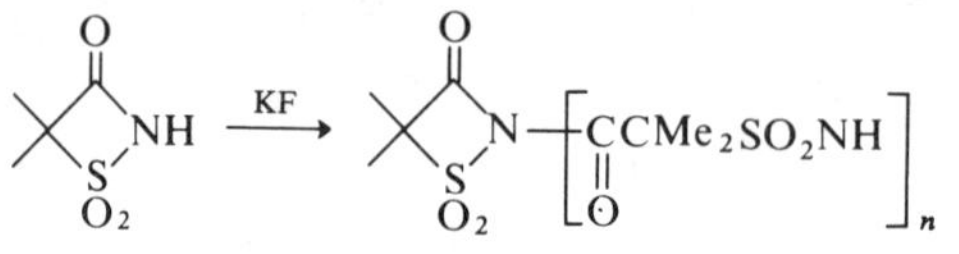

Scheme 129

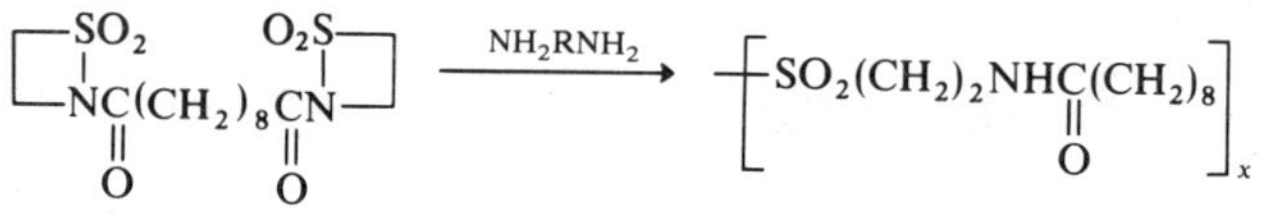

Scheme 130

5.15.4.3 1,3-Diazetidine Derivatives

One patent describes the inhibition of growth of six strains of *Mycobacterium tuberculosis* by 1,3-dicyclohexyl-1,3-diazetidine ⟨78MIP51500⟩ and the corresponding diphenyl derivative shows similar activity ⟨80MIP51500⟩.

Perhaps the four-membered heterocycle with the most industrial interest is the 1,3-diazetidine-2,4-dione or uretidinedione. In particular, 2,4-toluene diisocyanate has been used in films and coatings ⟨61GEP1105155⟩, to vulcanize rubber ⟨75GEP2349726, 75MIP51501, 75MIP51500⟩, to cure polyurethanes imparting additional thermal stability ⟨66FRP1430508, 71GEP1934763, 73GEP2312960⟩, and it forms crosslinked polymers with diamines and diols ⟨72GEP2044838⟩. Other crosslinked TDI copolymers have been used as leather substitutes ⟨80MIP51501⟩ and because of their thermal stability as rocket casing liners ⟨80BRP2033410, 80USP4232608⟩. Aryl-substituted uretidinediones are known to increase heat resistance in polyester fibers ⟨66NEP6509836⟩, have been used in wire insulation ⟨69BRP1147383⟩, give increased moisture stability to coatings ⟨77GEP2710252⟩, are fungicides ⟨66USP3223584⟩, have herbicidal activity ⟨77USP4045473⟩, and are insecticides ⟨78JAP(K)7872826⟩. *N*-Methyl-*N*-(3,4-dichlorophenyl)uretidinedione shows pre- and post-emergent control of weeds and grasses ⟨71USP3579500⟩, particularly for crops like rice, corn and soybeans ⟨71GEP2027345⟩.

5.15.4.4 1,3-Oxazetidine Derivatives

N-Phenyl-1,3-oxazetidin-2-one has been described in a Japanese patent as being 'useful as a drug' ⟨70JAP6927025⟩. This ring system has even been suggested as the fused ring intermediate in flavins responsible for the light producing intermediate in bacterial luciferase chemiluminescence ⟨76BBR(73)465⟩.

5.15.4.5 1,3-Thiazetidine Derivatives

Bis(trifluoromethyl)thioketene reacts with some imines to form 1,3-thiazetidines ⟨78JOC2500⟩. One derivative has been used by Dupont in surface treatment of water-repellent textiles ⟨71USP3592811⟩.

5.15.4.6 1,2-Dioxetane Derivatives

The major mechanistic interest in dioxetanes and dioxetanones comes from their chemiluminescent behavior. This process has been reviewed in the reactions section.

Studying these basic systems has led to a greater understanding of bioluminescence in biological species ⟨80JA3199, 78PNA(75)30, 77MI51502, 77AHC(21)437, 79MI51503⟩. The most extensively studied substrate, luciferin, and the proposed intermediacy of a dioxetanone are illustrated in Scheme 131 ⟨80JA3199⟩.

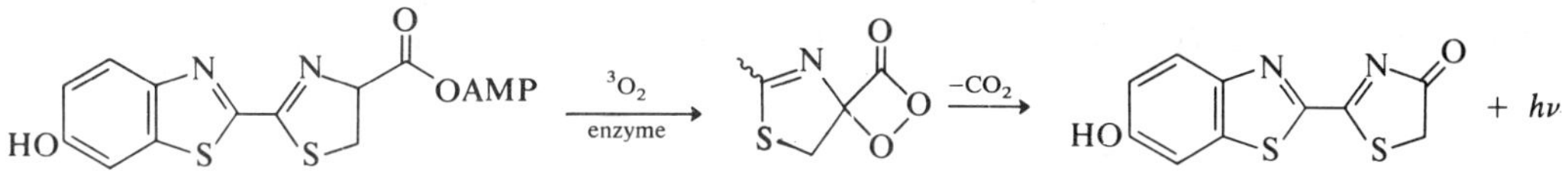

Scheme 131

5.15.4.7 1,2-Oxathietane Derivatives

Fluorinated β-sulfones, and particularly 3,3,4,4-tetrafluoro-1,2-oxathietane 2,2-dioxide, stabilize sulfur trioxide against polymerization ⟨68BAU807⟩ and have been tested as insecticides against the black carpet beetle ⟨64USP3163656⟩. Other derivatives of these fluorinated sulfones inhibit the growth of microorganisms, form nonflammable polymers, are used to produce ion exchange resins, and form polyethers of high thermal stability ⟨72AG(E)583⟩.

5.15.4.8 1,2-Dithietane Derivatives

The parent 1,2-dithietane system is not known, apparently because of the repulsive p,π electron interactions ⟨62JOC3995⟩. To date the most stable derivative is the unsaturated bis(trifluoromethyl)-1,2-dithietene. Its practical use is that of a ligand with a number of metals such as molybdenum ⟨63IC641, 74JA3502⟩, iron ⟨72JA8375⟩, indium ⟨73JOM(51)113⟩ and phsophorus ⟨76JA2924, 77JCS(P1)80, 78JA7300⟩. The latter forms both trivalent ⟨74JA317⟩ and pentavalent ⟨72CC395⟩ derivatives (Scheme 132).

Scheme 132

5.15.4.9 1,3-Dithietane Derivatives

The dimer of bis(trifluoromethyl)thioketene is most useful as a precursor of its monomeric form, which is used to make a number of useful products (Scheme 133). The one illustrated is an antistatic agent for wool ⟨67USP3337586⟩. Others are also useful as textile treating agents ⟨67USP3336334⟩.

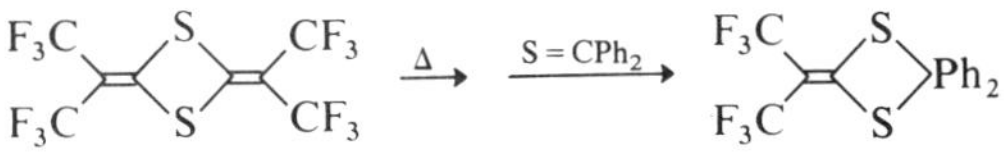

Scheme 133

Other *exo* methylene derivatives have been tested as fungicides ⟨74EGP104170, 75GEP2365511⟩ and have been incorporated on β-lactam derivatives as penicillin derivatives ⟨79GEP2824575, 79GEP2824559⟩.

A large number of imino derivatives display a variety of applications, including use as growth promoting agents for cows, sheep and poultry ⟨75GEP2448207, 75USP3914428⟩; as insecticides, particularly against the southern armyworm, mexican bean beetle and southern corn rootworm ⟨75GEP2448207, 75USP3915962⟩; as a very effective control of southern cattle ticks ⟨73GEP2305517, 75USP3928382, 76USP3954801⟩; as fungicides ⟨75MIP51500⟩; as herbicides ⟨73USP3772331⟩; as nematocides ⟨79FRP2413395⟩; and in controlling other plant disorders like root knots and galling of tomato plants ⟨78USP4070372, 79JAP(K)7988255, 79NEP7712731⟩.

5.15.5 FOUR-MEMBERED RINGS WITH THREE AND FOUR HETEROATOMS

There are relatively few literature reports of four-membered heterocycles containing three or four heteroatoms. Furthermore, even though many of the earlier accounts supply only minimal spectral or chemical evidence, little follow-up work has resulted. This section will list those reports and leave structural substantiation to future researchers.

5.15.5.1 1,2,3-Triazetidine Derivatives

Three early reports of triazetidine derivatives are supported only by analyses (Scheme 134) ⟨22JA1538, 24CB653, 33RTC378⟩.

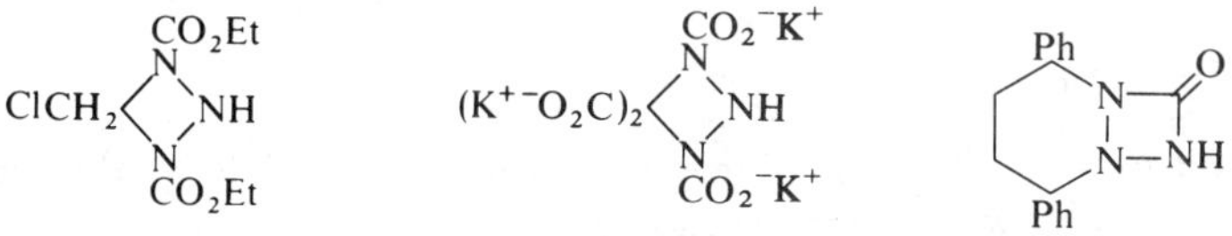

Scheme 134

5.15.5.2 1,2,3-Oxadiazetidine Derivatives, 1,2,3-Thiadiazetidine Derivatives

These two rings are virtually unknown. One German patent report of a 1,2,3-oxadiazetidine was later shown to be a methoxymethyl methylnitrosamine ⟨73TL2671⟩. An *ab initio* calculation (STO-3G) of the decomposition of nitrosyl cyanide to give nitrogen and carbon monoxide discusses the possibility of a 1,2,3-oxadiazetidine carbene as an intermediate ⟨78AJC2349⟩.

5.15.5.3 1,2,4-Oxadiazetidine Derivatives

A French patent claims 1,2,4-oxadiazetidin-4-one is effective in the treatment of psoriasis ⟨76FRP2341309⟩. The reaction of nitrosobenzenes and arylmethyleneimines was reported to give 1,3-diaryl-1,2,4-oxadiazetidines with only C,H,N analysis to support their formation. They also supposedly decompose to diarylcarbodiimides with loss of water, which mechanistically is hard to envision ⟨24JCS87⟩.

5.15.5.4 1,2,4-Thiadiazetidine Derivatives

Both carbodiimides and ketenimines are reported to add to *N*-sulfinylsulfamides to give 3-imino- and 3-alkylidene-1,2,4-thiadiazetidine 1-oxides, (Scheme 135) ⟨73BCJ2156, 75BCJ3259⟩. The adducts were identified from both IR and NMR spectra and from hydrolysis products.

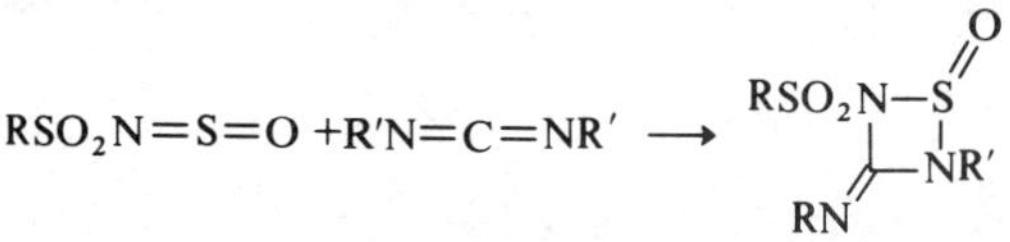

$RSO_2N{=}S{=}O + R'N{=}C{=}NR' \longrightarrow$

Scheme 135

Several examples of 1,2,4-thiadiazetidin-3-one derivatives have been prepared from the addition of 'Y triene' iminosulfur compounds to sulfonyl isocyanates. The reaction sequence is complex and apparently involves addition, loss of *t*-butyl isocyanate and subsequent addition to a second molecule of iminosulfur compound. One intermediate has been trapped, and the final product structure was firmly established by X-ray determination to be the 3-one derivative shown in Scheme 136 ⟨78AG(E)677, 80CB2434⟩.

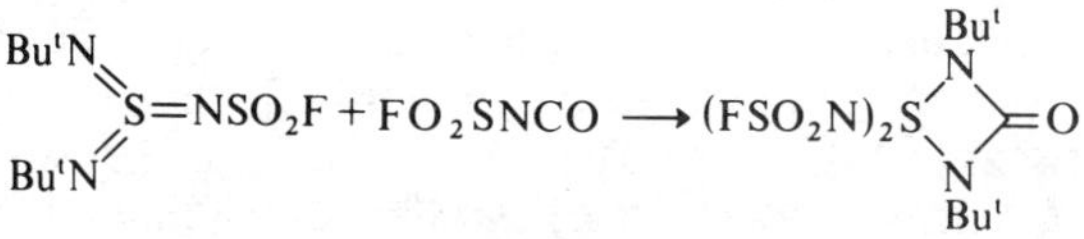

$(Bu^tN)_2S{=}NSO_2F + FO_2SNCO \longrightarrow (FSO_2N)_2S$

Scheme 136

5.15.5.5 1,2,3-Dioxazetidine Derivatives

One reference reports that the addition of oxygen to phenylhydrazones gives a 1,2,3-dioxazetidine which is explosive. However, it could just as readily be a hydroperoxide ⟨14CB3277⟩.

5.15.5.6 1,2,3-Oxathiazetidine Derivatives

The addition of *N*-sulfinylanilines to aldehydes supposedly gives 3-aryl-1,2,3-oxathiazetidines identified from nitrogen analysis, lack of an IR carbonyl stretch, the presence of an oxygen–sulfur band at 1160–1260 cm^{-1}, and by the fact they eliminate sulfur dioxide upon heating ⟨71JOU966⟩.

5.15.5.7 1,2,3-Dithiazetidine Derivatives

The reaction of thiobenzamide and sulfur dichloride is reported to give 3-methyl-4-phenyl-1,2-dithiazetinium chloride (Scheme 137). This salt forms a trityl cation with triphenylmethane, establishing its highly electrophilic nature. However, the analysis is poor (>1.3% off for Cl, N and S) and no other structural proof is provided ⟨70JOU394⟩.

$$PhC(=S)NHMe + SCl_2 \longrightarrow \text{3-methyl-4-phenyl-1,2-dithiazetinium } (Me\text{-}N^+)\ Cl^-$$

Scheme 137

5.15.5.8 1,2,3-Dioxathietane Derivatives

No isolated derivatives of 1,2,3-dioxathietanes have been reported. It is reasonable to assume their intermediacy in the reaction of singlet oxygen with thioketones and sulfines to give ketones ⟨71CC118, 70TL4683⟩, and energy profiles have been calculated (CNDO/B) on the supposed 1,2,3-dioxathietane intermediate ⟨80JCS(P2)188⟩.

5.15.5.9 1,3,2-Dioxathietane Derivatives

The only stable 1,3,2-dioxathietanes known are fluorinated sulfate derivatives formed by addition of sulfur trioxide to bis(trifluoromethyl)ketene. These structures are fairly well characterized from spectral data and from reactions with nucleophiles. Hexafluoroisopropylidene-1,3,2-dioxathietane 2,2-dioxide acts as a sulfur trioxide transfer agent to alkenes and is in equilibrium with a dimeric form as indicated by ^{19}F NMR (Scheme 138) ⟨71KGS1645, 72KGS306, 73KGS178, 1321⟩.

$$(CF_3)_2C{=}C{=}O \xrightarrow[0\,°C]{SO_3} (CF_3)_2C{=}\langle O_2SO_2\rangle \xrightarrow{H_2O} (CF_3)_2CHCO_2H$$

$$\downarrow CF_2{=}CF_2$$

$$\text{O–SO}_2 \text{ ring with } F_2, F_2$$

Scheme 138

5.15.5.10 1,2,4-Oxadithietane Derivatives

Heating hexafluoroisopropylidene-1,3,2-dioxathietane 2,2-dioxide (Scheme 138) to 120 °C forms a six-membered anhydride which upon further heating to 160 °C gives a good yield of 3,3-bis(trifluoromethyl)-1,2,4-oxadithietane 2,2,4,4-tetraoxide. This is the only reported derivative of this ring system ⟨73KGS1321⟩.

5.15.5.11 1,2,3-Trithietane Derivatives

Dithiobenzoic acid is reported to react with sulfur dichloride to give a compound $C_7H_5S_3Cl$ which was suggested to be 4-phenyltrithietanylium chloride (Scheme 139). The reaction of the latter with aniline gives sulfur and thiobenzanilide ⟨68JHC141⟩.

$$PhC(=S)SH + SCl_2 \longrightarrow Ph\text{-}[C S_3]^{+}\ Cl^{-}$$

Scheme 139

5.15.5.12 1,2,3,4-Tetrazetidine Derivatives

As unlikely as this ring system might seem, there are two reports of its existence. The methanolysis of 2,3,5,6-tetrabenzyl-1,4,2,3,5,6-dithiatetrazine is proposed to give tetrabenzyltetrazetidine ⟨75JOU2660⟩, and the other is described as a pink salt from reaction of 3-hydroxy-1,3-diphenyltriazene and α-naphthylamine. No spectral data exist for the latter ⟨73MI51502⟩. Calculations on N_4 or dimeric nitrogen have also been carried out ⟨70CC1233, 74JA4753⟩.

5.15.5.13 1,2,3,4-Dioxadiazetidine Derivatives

No accounts have reported any derivatives of this system. However, *ab initio* calculations of this supposed nitrous oxide dimer have been performed ⟨72JA8250⟩.

5.15.5.14 1,3,2,4-Dioxadiazetidine Derivatives

This ring probably does not exist even though one patent reports it ⟨67USP3340314⟩.

5.15.5.15 1,3,2,4-Dithiadiazetidine Derivatives

Of all the four-membered rings with more than two heteroatoms, the dithiadiazetidines are the most numerous and best characterized.

It is known that ammonia and sulfur dichloride react to give S_4N_4 which thermally cracks to form S_2N_2. This compound is unstable but appears to be planar with D_{2h} symmetry as determined by IR ⟨66SA1371⟩. CNDO/2 calculations have also been carried out on the monomer ⟨81JA7458⟩. It has also been reported to undergo a number of interesting reactions with alkynes, but it is difficult to determine if the reaction occurs from monomer (S_2N_2) or dimer (S_4N_4) ⟨79JHC1009⟩. The monomer forms a stable complex with antimony pentachloride, and an X-ray structure shows the ring to be planar with bond lengths of 1.619 Å, an NSN angle of 84.9°, and an SNS angle of 95.1° ⟨69IC2426⟩.

One other report of 2,4-di-*t*-butyl-1,3,2,4-dithiadiazetidine being useful as a fungicide ⟨68USP3365495⟩ probably has been misassigned ⟨76AG(E)614⟩.

$$FSO_2N{=}S{=}O \xrightarrow{SO_3} FSO_2N\langle S(O_2), S(O_2)\rangle NSO_2F \xrightarrow{\text{pyridine}} FSO_2NSO_2\overset{+}{N}C_5H_5$$

Scheme 140

More complex derivatives with sulfur in a higher oxidation state are known. These compounds have been referred to as cyclodiaza-λ^6-thianes and are quite stable if substituted with fluorosulfonyl groups (Scheme 140) ⟨76AG(E)614, 81ZN(B)1465⟩. They are reactive towards pyridine ⟨78AG(E)129⟩ and form six-membered adducts with benzonitrile ⟨78ZN(B)1072⟩.

Iminosulfur compounds described as 'Y triene' derivatives react with perfluorinated isocyanates to form cyclodiaza-λ^6-thianes (Scheme 141) ⟨79AG(E)235⟩. Several of these have been analyzed by X-ray analysis, and the nature of the substituent determines whether the ring is planar or not.

$(Bu^tN{=})_3S \xrightarrow{C_2F_5NCO}$ [cyclic product: $C_2F_5N{=}S$ / NBu^t / $S{=}NBu^t$ bridged ring; labels: Bu^t, N, C_2F_5N, 1.67, 95.7, 1.73, NBu^t, S 84.3, 80.9 S, C_2F_5N, 1.66, 1.72, NC_2F_5, N, Bu^t]

Scheme 141

5.15.5.16 1,2,4,3-Oxadithiazetidine Derivatives

There are no literature accounts of this ring system.

5.16

Azepines

R. K. SMALLEY
University of Salford

5.16.1 INTRODUCTION

Four tautomeric forms, designated as 1*H*-, 2*H*-, 3*H*- and 4*H*-azepine may be drawn for azacycloheptatriene. 1*H*-Azepine (**1**; R = H) is known as a very unstable (even at −78 °C in $CDCl_3$ solution) red oil which in the presence of acid or base rearranges to the marginally more stable colorless 3*H*-azepine (**2**; $R^1 = R^2 = H$) ⟨80AG(E)1016⟩.

The stability of 1*H*-azepines is enhanced by electron-withdrawing substituents, especially at the 1-position as these decrease the electron density in this 8π antiaromatic ring system ⟨73ACR393, 74AHC(17)255⟩.

N-Unsubstituted 1*H*-azepines are rare since, like the parent system, they tautomerize readily to the 3*H*-isomers in whose preparation they are often considered as transient intermediates (see Section 5.16.4.1.2(ii)). This rearrangement is particularly apparent with 2-amino- and 2-alkoxy derivatives since stabilization of the 3*H*-azepine is then possible by amidine and imidate type resonance. For the CH_2-containing tautomers the order of stability appears to be 3*H* > 4*H* > 2*H*, a fact attested to by the facile thermal and base-catalyzed rearrangements of 4*H*-azepines to the 3*H*-tautomers ⟨72CB982⟩ and the rarity and inherent instability of 2*H*-azepines. The latter are well established as intermediates in the formation of 3*H*-azepines ⟨74JOC3070⟩ but have been characterized only as their benzologues.

Reduced and partially reduced azepines are more common. Perhydroazepine (hexamethylenimine) was first prepared in 1905 and chemically it behaves as a typical secondary amine ⟨B-67MI51600⟩. Its 2-oxo derivative (caprolactam) is a bulk chemical and is of great industrial importance as an intermediate in the manufacture of nylon. Many oxo and dioxo derivatives of azepines and benzazepines have been prepared, often with difficulty, in a mainly fruitless search for aromatic properties in these 'azatropone' and 'benzazatropone' systems (see Section 5.16.3.1.2).

Annulation of a benzene ring to the azepine nucleus can be achieved in three different ways and examples of 1- (**3**), 2- (**4**) and 3-benzazepines (**5**) are well documented ⟨74AHC(17)45⟩. Dibenzazepines, of which there are four isomers (**6**)–(**9**) are equally well known, especially those derivatives (see Section 5.16.5) that possess pharmacological activity. In contrast, there is only a solitary reference ⟨72CB880⟩ to the tribenzazepine system (**10**). The recommended system of nomenclature and ring numbering for azepines and their benzologues accompanies formulae (**1**)–(**10**) in Table 1, the position of saturation (*i.e.* the floating hydrogen) being denoted by a locant.

As expected, annulation to one or two benzene rings increases the overall stability of the azepine system (Section 5.16.1.2). Even so, some systems, *e.g.* (**4**) and (**9**), are too unstable as the NH tautomers to permit isolation, and as yet are known only as the oxo or hydro derivatives or, as in the case of the dibenzazepine (**9**), as the more stable fully benzenoid 5*H*-tautomer.

5.16.2 STRUCTURE OF AZEPINES

5.16.2.1 Theoretical Studies

The conformation of 1*H*-azepines has provoked much theoretical interest since a planar 1*H*-azepine would involve a destabilizing overlap of the nitrogen lone pair with the triene system and hence formation of an antiaromatic 8π electron arrangement. In fact, early calculations by the Hückel LCAO-MO method predicted extensive localization of electrons at the nitrogen and the C—C double bonds, from which it was concluded that 1*H*-azepines should be non-planar and exhibit marked polyene character ⟨B-69MI51600⟩. More recent ⟨70T4269⟩ data on 1*H*-azepine and its benzo and dibenzo derivatives by more refined methods (SCF-MO) indicate clearly that azepines are non-aromatic and that 3*H*-azepine (**2**; $R^1 = R^2 = H$), in which the offending nitrogen lone pair is no longer in conjugation with the π-system, is more stable than the 1*H*-tautomer (**1**; R = H) (see Table 1). Not surprisingly, the most stable fused ring systems are those in which the carbocyclic rings retain their fully benzenoid character (Table 1).

Theoretically derived resonance energies (RE) ⟨70T4269⟩, resonance energies per π-electron (REPE) ⟨72T3657⟩ and topological resonance energies (TRE) ⟨77JA1692⟩ are listed

Table 1 Nomenclature System, Theoretically Derived Resonance Energies and X-ray Data for Azepines, Benzazepines and Dibenzazepines

Structure	*Name (comments)*	*RE* (kJ mol^{-1})	*REPE* [*TRE*] (units)	*Ref.*
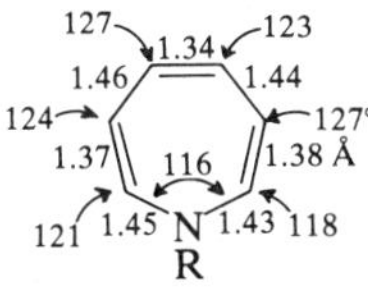 (**1**)[a]	1*H*-Azepine (R=H; known but unstable)	−7.5[b]	−0.036[b] [−0.029][b]	80AG(E)1016 B-69MI51600[c]
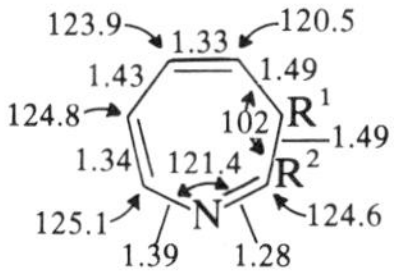 (**2**)[d]	3*H*-Azepine (R=H; known but unstable).	—	—	80AG(E)1016 73CB1033[c]
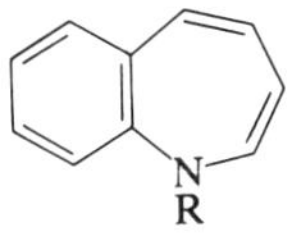 (**3**)	1*H*-1-Benzazepine (known as *N*-acyl derivatives)	76.7[b]	0.014[b] [0.006]	80TL3403 80AG(E)1015
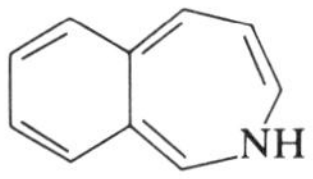 (**4**)	2*H*-2-Benzazepine (known only as 1*H*, 3*H* and 5*H* derivatives)	−3.2	−0.024 [−0.023]	74JOC3070 75JA4682 77HCA1644
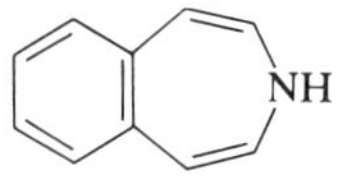 (**5**)	3*H*-3-Benzazepine (known as *N*-alkoxycarbonyl derivatives and 1*H*-tautomers)	73.2	0.015 [0.005]	78H(11)401 81HCA373
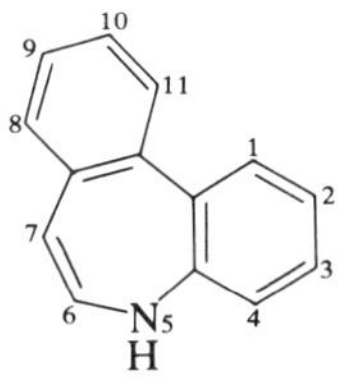 (**6**)	5*H*-Dibenz[*b*,*d*]azepine	158.4	0.034	
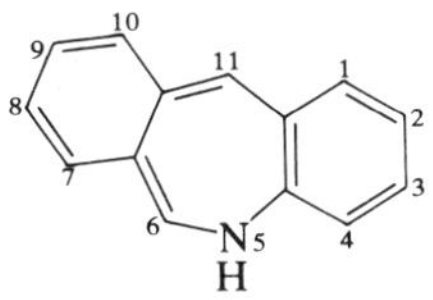 (**7**)	5*H*-Dibenz[*b*,*e*]azepine	81.5	0.013	
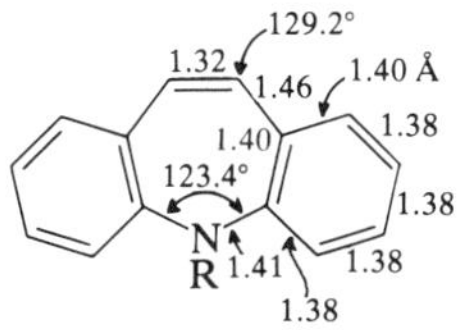 (**8**)	5*H*-Dibenz[*b*,*f*]azepine (known)	160.5[b]	0.034[b]	74CRV101

[a] Bond lengths and angles refer to (**1**; R=*p*-$BrC_6H_4SO_2$).
[b] R=H compound.
[c] X-ray structure.
[d] Bond lengths and angles refer to (**2**; $R^1 = CPh_2OH$, $R^2 = H$) with the substituent in the equatorial position.

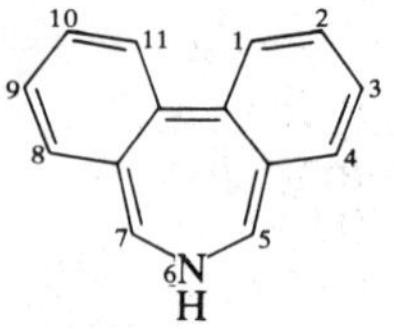 (9)	6*H*-Dibenz[*c*,*e*]azepine (known only as the 5*H*-tautomer)	4.9	0.002	81LA240
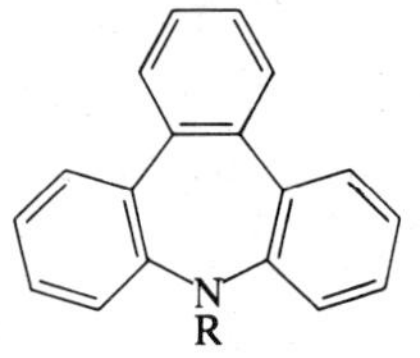 (10)	9*H*-Tribenz[*b*,*d*,*f*]azepine (known only as the 9-Ph derivative)	—	—	72CB880

in Table 1. REPE values for cyclopentadienoazepines, *i.e.* azaazulenes, have also been estimated ⟨75T295⟩.

5.16.2.2 X-ray Data

Bond lengths and bond angles for some azepine systems, as derived by X-ray analysis, are appended to structures (**1**)–(**10**) in Table 1. X-ray crystallographic data on a variety of 1*H*-azepines ⟨B-69MI51600, 72CB434, 81AG(E)699⟩ and 3*H*-azepines ⟨72CB982, 73CB1033⟩ have confirmed the non-planar character of the azepine ring, and have established unequivocally their boat conformations. Carbons 2,3 and 5,6 form the planar 'hull' as indicated for the 1*H*-azepine (**11**; Ar = *p*-BrC_6H_4CO) ⟨81AG(E)699⟩ and the 3*H*-azepines (**12**; R^1 = Ph, R^2 = CO_2Me and R^1 = CO_2Me, R^2 = Ph) ⟨72CB982⟩.

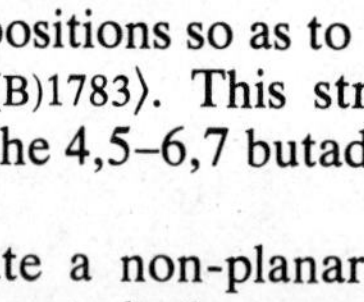

(**11**) 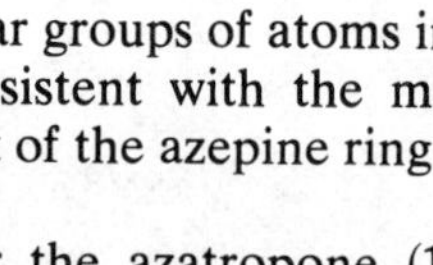 (**12**)

In contrast, in the tricarbonyliron–1-methoxycarbonyl-1*H*-azepine complex (see Section 5.16.3.8.2) the boat conformation is lost and the seven-membered ring becomes hinged about the C-4, C-7 positions so as to give two planar groups of atoms inclined at a dihedral angle of 59° ⟨70JCS(B)1783⟩. This structure is consistent with the metal carbonyl being complexed through the 4,5–6,7 butadiene fragment of the azepine ring rather than through the ring nitrogen.

X-ray data indicate a non-planar structure for the azatropone (**13**; Ar = *p*-BrC_6H_4) ⟨79H(12)1423⟩. In contrast, the benzazatropolone (**14**; R^1 = Me, R^2 = OH) is planar (to within ±0.04 Å) but does not display any significant π-electron delocalization ⟨72T1795⟩. It is not clear whether the non-planarity of azepin-2-one (**13**) is inherent to the system or whether it is brought about by steric interactions between the adjacent phenyl rings. Interestingly, the azepinone appears to crystallize as a conglomerate of the (+) and (−) forms, all the crystals examined having the *S*-configuration.

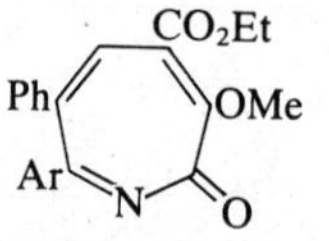

(**13**) 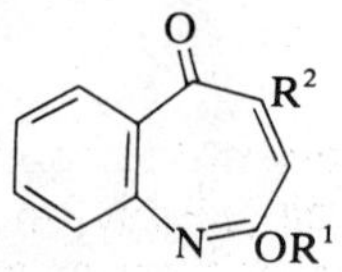 (**14**) 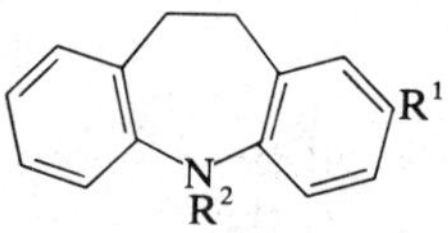 (**15**)

X-ray analysis of 5H-dibenz[b,f]azepine reveals a dihedral angle of 144° 4′ between the planes of the benzene rings ⟨80AX(B)2683⟩; other structural parameters are appended to structure (**8**). Crystal and molecular structure data are also available for the physiologically active 10,11-dihydro-5H-dibenz[b,f]azepines (**15**; $R^1 = CH_2NC_5H_{10}$, $R^2 = H$) ⟨79AX(B)996⟩ and (**15**; $R^1 = H$, $R^2 = (CH_2)_3NMe_2HCl$) ⟨75AX(B)1008⟩ and for more esoteric hydroazepine derivatives ⟨74PMH(5)1⟩.

5.16.2.3 NMR Spectra and Stereochemistry

Historically, ^{1}H NMR spectroscopy was vital in determining the structure of the first azepines ⟨66T81⟩ and has since proved to be invaluable in the conformational analysis of azepines and their benzologues, and in determining ring inversion activation energies of these conformationally mobile ring systems.

Chemical shifts for the azepine ring protons, as might be anticipated from their non-planar, polyene character, lie in the vinyl proton region (δ 4.5–6.8 p.p.m.). Chemical shifts and coupling constants of representative examples of azepines and their hydro, oxo, benzo and dibenzo derivatives are listed in Table 2.

For further extensive and detailed information on the ^{1}H NMR spectra of 1H-azepines ⟨69JOC2866⟩, 3H-azepines ⟨66T81, 72JA513, 78JCS(P1)191⟩, 2,5-bridged azepines (**16**; X = S, Se or NMe) ⟨B-73NMR229⟩, benzazepines ⟨74AHC(17)45⟩ and dibenz[b,f]azepines ⟨74CRV101, 74JCS(P2)1648⟩ the references cited should be consulted. Detailed ^{1}H NMR data on the iron tricarbonyl complex of 1-ethoxycarbonyl-2,3-homo-1H-azepine are also available ⟨78CB3927⟩.

Long-range couplings are evident in methyl-substituted 3H-azepines. For example, there is allylic coupling ($J \simeq 1$ Hz) between the 7-Me and H-5, and between the 5-Me and H-4 ($J \simeq 1.5$ Hz) in azepine (**17**; R = Me, X = OEt) ⟨B-69MI51600⟩, and in some systems between the 2-Me and the 7-H (J 0.8 Hz) ⟨74JOC3070⟩. Homoallylic coupling is also seen in (**17**; R = Me, X = OEt) between H-3 and the 5-Me. As such coupling is seen only if the protons involved are axial it is concluded that conformers, *e.g.* (**17**), are the more densely populated. In contrast however, the abnormally high field resonance (δ 0.7 ± 0.05 p.p.m.) of the 3-methyl group in some 2-diethylamino-3H-azepines (**17**; R = H, X = NEt_2) suggests that the methyl groups lie in the shielding zone of the C-6=C-7 double bond, *i.e.* are axial ⟨72JA513⟩.

Also noteworthy is that the CH_2 units of a 2-ethyl or 2-ethoxy group in azepines of type (**17**) appear as AB systems, due to the close proximity of the asymmetric centre at C-3 and also to restricted rotation about the C-2—CH_2 or C-2—O bonds ⟨74JOC3070⟩. Significantly, in the 2-ethylthio analogue (**17**; R = Me, X = SEt), in which the larger size of sulfur allows bond rotation, the CH_2 unit appears as a normal quartet ⟨B-69MI51600⟩.

The 100 MHz ^{1}H NMR and the 25.16 MHz ^{13}C NMR spectra of 5-azaazulene (**19**) have been analyzed in detail; H–H and C–H couplings, and long-range C–H couplings are cited ⟨80OMR(14)466⟩. A theoretical study of the ring current and excess charge effects on the chemical shifts of the conjugate bases of dibenz[b,f]azepines has been carried out ⟨78MI51600⟩.

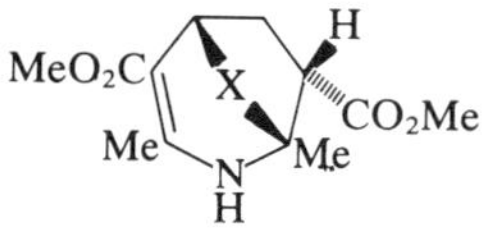

(**16**)

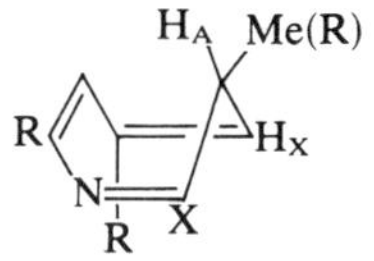

(**17**)

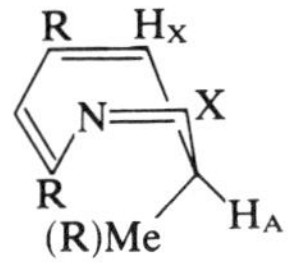

(**18**)

Major references to ^{13}C NMR data on azepines are cited in Table 3, along with chemical shift values for a selection of systems.

Ring inversion of the azepine ring between the two stable boat forms has been studied extensively by ^{1}H NMR spectra ⟨B-77SH(2)305⟩. Early work on 2-anilino-3H-azepine demonstrated that at 25 °C the methylene group appears as an AX system ($J(H_AH_B\text{–}H_X)$ 7 Hz), whereas at −75 °C in acetone-d^6 an ABX system (J_{AB} 11 Hz, J_{AX} 6 Hz, J_{BX} 8 Hz) is evident

Table 2 ^{1}H NMR Chemical Shifts and Coupling Constants for Typical Azepines

Azepine (solvent)[a]		H-2	H-3	H-4	H-5	H-6	H-7	*Other protons* (p.p.m.) [*J values*] (Hz)	*Ref.*
		Ring protons[b] (p.p.m.)							
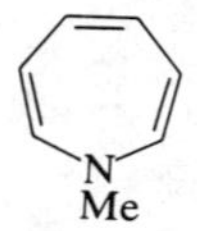 N-Me azepine	(CCl_4)	5.62	4.47	2.32	—	—	—	Me 3.0 [$J_{2,3}$ 9.5, $J_{3,4}$ 6]	76CJC1571
(**1**; R = H)		5.22	4.69	5.57	—	—	—	[$J_{2,3}$ 7.97, $J_{3,4}$ 5.57, $J_{4,5}$ 10.89]	80AG(E)1016
(**1**; R = Me)	(CCl_4)			4.5–5.0					B-73NMR229
(**2**; $R^1 = R^2 = H$)			2.42	5.25			7.55	H-2, -5 and -7 in the range 6.2–6.7	80AG(E)1016
(**2**; $R^1 = H$, $R^2 = NEt_2$)	(CCl_4)	—	2.53	4.93	6.18	5.57	6.92		69JCS(C)2819
(**2**; $R^1 = H$, $R^2 = OMe$)	(CCl_4)	—	2.59	5.19	5.86	6.15	6.86		77BCJ2013
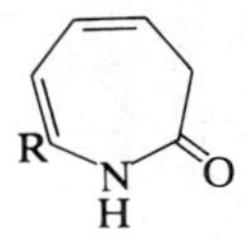 Azepinone (R, NH, O)	R = H	—	2.8		5.7–6.2				B-73NMR229
	R = Ac	—	2.97	5.98	6.36	6.82	—	NH 8.4 [$J_{3,4}$ 7, $J_{4,5}$ 10, $J_{5,6}$ 6]	73HCA1852
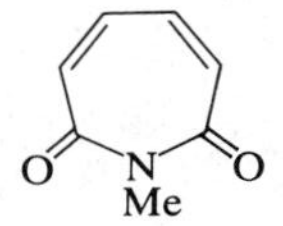 N-Me azepinedione	(C_6D_6)	—	6.35	5.79	—	—	—	Me 3.42	B-73NMR229
(**35**; R = H)		—	—	—	6.90	8.03	—	[$J_{5,6}$ 10]	81H(16)363

		—	6.61	a 2.15 b 2.95	6.39	—	—	[$J_{3,4a}$ 6.01, $J_{3,4b}$ 8.0, $J_{5,4a}$ 6.96, $J_{5,4b}$ 7.74, $J_{4a,4b}$ 13.0]	72JA2770
		7.51	5.80	—	6.83	—	—	[$J_{2,3}$ 8, $J_{7,5}$ 3]	72JA513
	(CCl_4)	1.16	3.60	—	3.28	—	—	ArH 6.52–7.0	78BSF(2)97
(**3**; R = Ac)		6.49	5.91	6.24	6.84	—	—	[$J_{2,3}$ 7.5, $J_{3,4}$ 6.0, $J_{4,5}$ 11.0]	80TL3403
(**14**; R^1 = Et, R^2 = H)		—	6.65	6.81	—	—	—	[$J_{3,4}$ 12]	72JOC208
(**15**; R^1 = R^2 = H)		—	—	—	—	—	—	H-10,11 2.83, NH 5.6, ArH 6.4–7.0	74CRV101
(**8**; R = H)		—	—	—	—	—	—	H-10,11 6.25, ArH 7–7.2	74JCS(P1)52
(**8**; R = Ac)		—	—	—	—	—	—	H-10,11 6.88, ArH 7.2–7.5	74JCS(P1)52
		—	—	—	4.34	—	8.38	ArH 7.19–7.27	81LA240

[a] $CDCl_3$ unless stated otherwise.
[b] TMS internal reference.

Table 3 ^{13}C NMR Chemical Shifts of Typical Azepines

Azepine	C-2	C-3	C-4	C-5	$\delta(^{13}C)$ (p.p.m.) C-6	C-7	Others	*Ref.*
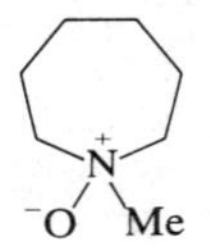	71.3	21.8	26.7	—	—	—	CH_3 60.3	77OMR(9)333
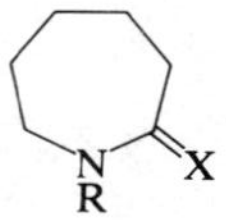								(76JA5082)
R = H, X = O	179.8	36.9	23.2	29.8	30.6	42.65		79AJC567
R = Me, X = O	176.0	36.9	23.4	29.9	27.7	51.4	CH_3 35.8	
R = $SiMe_3$, X = O	182.6	37.4	23.1	29.4	30.0	44.1		79JCS(P1)2478
R = H, X = S	210.5	45.1	24.5	28.1	30.4	47.1		79AJC567
R = Me, X = S	205.7	46.6	24.2	28.9	26.3	55.9	CH_3 45.1	79AJC567
R = Me, X = Se	208.3	50.3	23.2	28.6	25.3	56.4	CH_3 49.35	79AJC567
(**1**; R = H)	138.0	113.0	132.3	—	—	—		80AG(E)1016
(**1**; R = CO_2H)		120.2			120.2		C-2, -4, -5, -7 129.6, 130.6	80AG(E)1016
(**1**; R = Bu^t)	114.4	132.1	135.8				Me_3C 52.4	77JA1272
(**2**; $R^1 = R^2$ = H)	113.3	34.3	117.5	127.3	136.4	141.0		80AG(E)1016
(**2**; $R^1 = CO_2Me$, R^2 = OMe)	145.7	49.3	114.6	126.2	116.4	137.1	OCH_3 51.74	79TH51600
(**15**; $R^1 = R^2$ = H)	119.0	126.2	117.5				C-1 130.1, C-10 34.7, C-9a 128.1, C-4a 141.9	74JCS(P2)1648
(**15**; R^1 = H, R^2 = Me)	118.3	125.9	121.3				C-1 129.2, C-10 32.8, C-9a 132.6, C-4a 148.2	74JCS(P2)1648
(**15**; R^1 = H, $R^2 = CO_2Et$)	126.9	126.1	127.9				C-1 129.3, C-10 30.6, C-9a 135.5, C-4a 140.3	74JCS(P2)1648
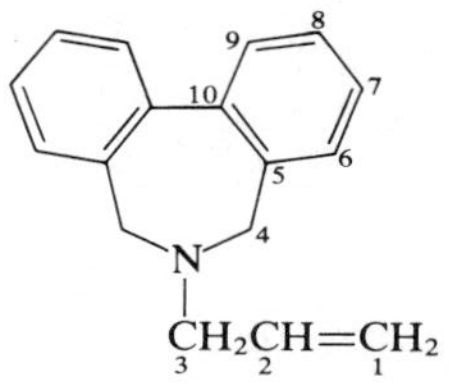	136.1	58.2	54.5	134.3	129.5	127.7	C-8,9 127.4, C-10 140.9	79JPS890

A coalescence temperature of -55 ± 7 °C was observed for the two conformers ($\mathbf{17} \rightleftharpoons \mathbf{18}$; R = H, X = NHPh) and ΔG_c^* for the ring inversion was calculated to be 42.7 kJ mol^{-1}. This figure is some 7.1 kJ mol^{-1} higher than for 3*H*-azepin-2-one ⟨67CB335⟩. The relative stabilities of the two boat conformations depend very much on the nature of the substituents and some substantial energy barriers to inversion have been determined. For example, ΔG^* for the 3*H*-azepine (**20**) is found to be 57.3 kJ mol^{-1} ⟨71JOC978⟩.

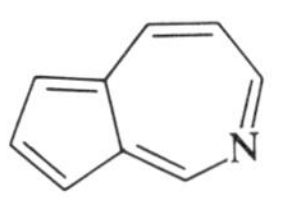

(**19**) 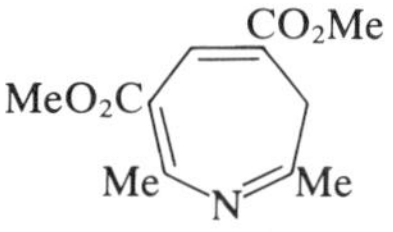 (**20**)

4*H*-Azepines exhibit ring inversion activation energies of similar magnitude, *e.g.* 2,6,7-trimethoxycarbonyl-4*H*-azepine has a ΔG^* (at $T_c - 40 \pm 5$ °C) of 46.8 ± 2 kJ mol^{-1} ⟨72JA2770⟩.

Conformation studies have also been carried out on perhydroazepines ⟨B-77SH(2)305⟩, benzazepines ⟨74AHC(17)45, B-77SH(2)305⟩, dibenz[*b,e*]azepines ⟨74CS(5)52⟩, dibenz[*b,f*]-azepines ⟨74CRV101, 74JCS(P2)1648, 76T1081⟩ and their 10,11-dihydro derivatives ⟨78JA4982⟩, and dibenz[*c,e*]azepines ⟨71TL2515⟩.

N-Acyltetrahydro-1-benzazepines (**21**) are of interest in that at room temperature the equatorial proton (H_e) at position 2 shows a noticeable downfield shift (*ca.* 1.9 p.p.m.) relative to the other CH_2 groups ⟨77TL3023⟩. This result is compatible with the amide displaying a large degree of N=C character, which, coupled with a chair conformation of the azepine ring, positions H_e in the deshielding zone of the $\overset{+}{N}$=C—O$^-$ system. Similar restricted rotation about the amide C—N bond has been noted with other *N*-acylazepines, in particular with *N*-acyl-10,11-dihydrodibenz[*b,f*]azepines, which, in addition, also display conformational equilibria due to inversion of the azepine ring, and rotation about the ethane bridge ⟨74CRV101⟩.

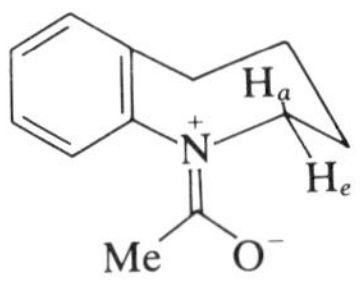

(**21**) 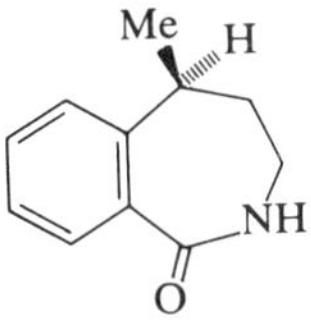 (**22**)

Conformational analysis of hexahydro-1*H*-azepin-2-one (caprolactam) confirms that it, like other five- to eight-membered lactams, has the *cis* conformation. The ^{15}N NMR shift for the seven-membered lactam in $CHCl_3$ appears at 259.7 p.p.m. (upfield from HNO_3) ⟨76JA5082⟩. Solvent shifts for the ^{15}N resonance of caprolactam have been determined ⟨78MI51601⟩. In TFA $J(^{15}N-^{1}H)$ is 92.5 ± 0.5 Hz.

Considerable interest has been shown in dibenz[*c,e*]azepines as examples of optically active 2,2-bridged biphenyls ⟨68MI51600⟩. The absolute configurations of some dibenz[*c,e*]azepines have been determined ⟨B-77SH(2)305⟩, and the CD spectra of the hydro-2-benzazepin-1-one (**22**) recorded ⟨78CHE538⟩. In contrast, to analogous five- and six-membered lactams, azepinone (**22**) has a (+) Cotton effect.

A study on the conjugate acids of dihydro-6*H*-dibenz[*c,e*]azepines has demonstrated the validity of the Karplus equation for determining H—N—C—H coupling constants in secondary and tertiary amines of known rigid conformation ⟨73CJC2433⟩.

5.16.2.4 Valence Isomerism

A feature of 1*H*-azepine chemistry that intrigued early researchers ⟨B-69MI51600, 71AG(E)11⟩ was the possibility of azepine–benzeneimine (or azanorcaradiene) valence isomerism ⟨67AG(E)402⟩ akin to that in the oxepin–benzene oxide system.

X-ray studies on a number of azepines (Section 5.16.2.2) preclude the existence of the bicyclic tautomer in the crystal state, and for simple azepines variable temperature ^{1}H NMR evidence (−90 to 130 °C) is overwhelmingly in favour of the azepine structure. For example, from recent low temperature (−120 °C) ^{1}H NMR studies on 1*H*-azepine (**23**; $R^1=R^2=R^3=$ H) it is concluded that the maximum amount of benzeneimine (**24**; $R^1=R^2=R^3=$ H) present at equilibrium must be <1% ⟨80AG(E)1016⟩.

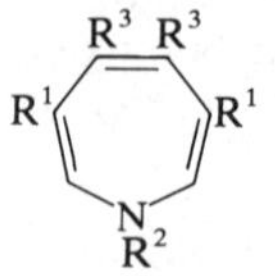

(**23**) 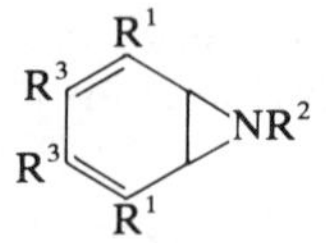 (**24**)

Prinzbach and coworkers ⟨76CB3505⟩ were more successful in providing evidence for the existence of the azepine–benzeneimine equilibrium (**23**⇌**24**) in the more highly substituted azepines (**23**; R^1 = H, R^2 = *p*-Ts, Ac or CO_2Me, $R^3=CO_2Me$) as they were able to trap the benzeneimine tautomers (**24**; R^1 = H, R^2 = *p*-Ts or mesyl, $R^3=CO_2Me$) as their bis-1,3-dipolar adducts with diazomethane. It was found also that the ^{1}H NMR spectrum ($CDCl_3$) of the *p*-Ts derivative (**23**; R^1 = H, R^2 = *p*-Ts, $R^3=CO_2Me$) shows temperature dependence between −68 °C and +40 °C. Subsequently, the diphenyl derivative (**23**; R^1 = Ph, R^2 = *p*-Ts, $R^3=CO_2Me$) was found to be a 97:3 azepine (**23**):benzeneimine (**24**) mixture at room temperature. Additional evidence was obtained from low temperature (−70 °C) ^{1}H and ^{13}C NMR studies on the dichloro compound (**23**; R^1 = Cl, R^2 = *p*-Ts, $R^3=CO_2Me$) ⟨77TL1355⟩. The bicyclic isomer constitutes about 10% of the equilibrium mixture at −70 °C ($\Delta G^*=$ 50.16 kJ mol^{-1}).

Paquette and coworkers ⟨69JOC2866⟩ have demonstrated that 1*H*-azepines can be forced into adopting the benzeneimine structure by bridging the 2,7-positions with a trimethylene chain (**25**). The length of the methylene chain is, however, critical and the tetramethylene derivative (**26**) exists solely as the bridged azepine. A suggestion that benzeneimines may be involved as intermediates in the formation of 3*H*-azepines from 2*H*-azirines and cyclopentadienones (Section 5.16.4.2.1(i)) has been discounted ⟨74JOC3070⟩.

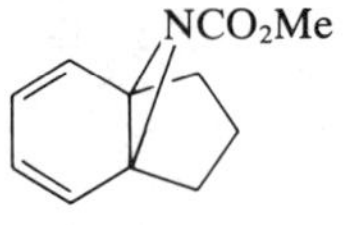

(**25**) 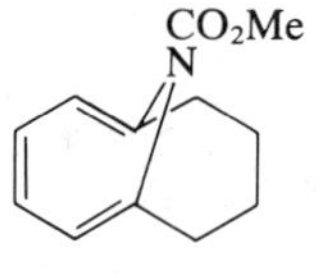 (**26**)

The equilibrium position of the azepine–benzeneimine valence isomerism together with the effect of electron-withdrawing substituents on the stability of 1*H*-azepines and their dimerizations (Section 5.16.3.2.3) have been considered theoretically using extended Hückel calculations ⟨73CB970⟩. More recent MINDO/3 calculations on benzeneimine have been performed and indicate that it is a highly non-planar bicycle with the plane of the three-membered ring inclined at an angle of 72.6° to the plane of the six-membered ring. Calculated heats of formation (ΔH_f°) predict that 1*H*-azepine (175.98 kJ mol^{-1}) is more stable than the benzeneimine by −35.5 kJ mol^{-1} ⟨80JA1255⟩.

The 4*H*-azepine–3-azanorcaradiene (**27**⇌**28**) valence isomerism has also come under close scrutiny, the conclusion being that in general the 4*H*-azepine is energetically more favourable than its bicyclic isomer ⟨72JA2770⟩. However, the 3-azanorcaradiene tautomer (**28**; R^1 = H, $R^2=CO_2Me$, R^3 = Ph) has now been trapped as the [4+2] adduct with cyclopropane ⟨80TL599⟩. Moreover, in the 3,3-dimethyl derivative (**27**; R^1 = Me, $R^2=CO_2Me$, R^3 = Ph) the equilibrium is displaced towards the 3-azanorcaradiene form (**28**; R^1 = Me, $R^2=CO_2Me$, R^3 = Ph) and only at 178 °C does the 4*H*-azepine become visible in the ^{1}H NMR spectrum. Absence of strain between the *gem*-dimethyl groups and the bridgehead protons is thought to be responsible for the unprecedented stability of the bicyclic isomer.

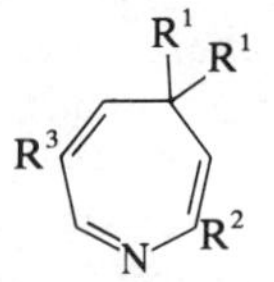

(**27**) 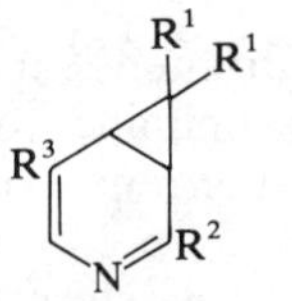 (**28**) 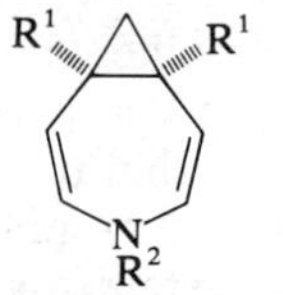 (**29**) 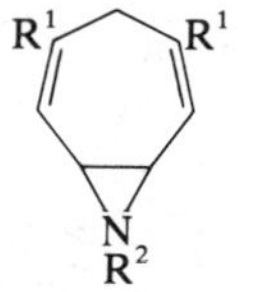 (**30**)

Homo-1*H*-azepines (**29**), unlike their carbocyclic counterparts, show little tendency to isomerize to the thermodynamically less favoured bicyclic aziridine tautomers (**30**). Presumably, the homoazepine gains its stability from delocalization of the nitrogen lone pair through the dienamine system ⟨71AG(E)11⟩. However, the aziridine tautomers (**30**; $R^1 = CO_2Me$, $R^2 = Ac$, CO_2Me or *p*-Ts) have been trapped as their all-*cis* bis-1,3-dipolar cycloadducts with diazomethane ⟨76CB3505⟩.

5.16.2.5 Mass Spectra

Apart from the detailed studies referenced below, mass spectral data on azepines are fragmented and confined to isolated reports on specific compounds ⟨B-71MS432⟩.

Caprolactam (hexahydro-1*H*-azepin-2-one) fragments by cleavage of the C-6—C-7 and C-2—N bonds with simultaneous transfer of a hydrogen atom to give $[\dot{C}H{=}\overset{+}{N}H_2]$ (*m/e* 30) as the base peak. Similar fragmentations, confirmed by deuterium labelling studies, characterize its alkyl derivatives ⟨71OMS(5)651⟩. Not surprisingly, the dominant feature in the mass spectral breakdown of *N*-substituted 1*H*-azepines is loss of the *N*-substituent and formation of the azatropylium cation (*m/e* 92) as the base peak ⟨69JOC2866⟩.

A careful study of the fragmentation of a series of *N*-*p*-toluenesulfonamides, that includes several *N*-(*p*-tosyl)-benz- and -dibenz-azepinones, reveals that loss of $(M-SO_2)$ occurs only in certain systems, and is diagnostic for those structures in which the carbonyl and tosylamino groups are both attached directly to the benzene ring at positions *ortho* to each other, *e.g.* the 1-benzazepin-5-one (**31**; R = *p*-Ts) ⟨69OMS(2)337⟩.

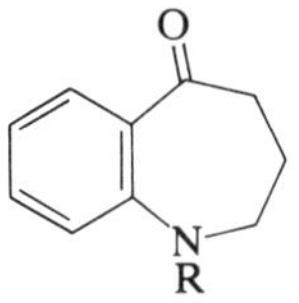

(**31**)

Loss of ethylene and subsequent formation of the stable 3*H*-pyrrolinine cation dominates the mass spectrum of 2,3,4,5-tetrahydro-1*H*-1-benzazepine ⟨75OMS(10)992⟩. Biphenylene (*m/e* 152), acridinium cation (*m/e* 179) and fluorene cation (*m/e* 165) are the major fragments in the mass spectra of dibenz[*b,f*]azepines ⟨74CRV101⟩.

3*H*-Azepine (*m/e* 93) was first detected and characterized by the application of MS–GLC techniques to the thermolysis of the carbinol (**2**; $R^1 = CPh_2OH$, $R^2 = H$) ⟨73CB2084⟩.

5.16.2.6 UV and IR Spectra

As with mass spectral data, detailed information of the UV spectra of azepines is confined to a few sources and for information on particular systems recourse has to be made to the primary literature. Available are compilations of UV data on *N*-acyl- and *N*-sulfonyl-1*H*-azepines ⟨B-69MI51600, 69JOC2866⟩, benzazepines ⟨74AHC(17)45⟩, dibenz[*b,f*]azepines and their 10,11-dihydro derivatives ⟨74CRV101⟩, and the 'azatropolones' (**13**) and their 1,5-dihydro derivatives ⟨79H(12)1427⟩.

1*H*-Azepines and 5*H*-dibenz[*b,f*]azepines display three major absorptions in their UV spectra at 210–215, 240–247 and 285–330 nm, and 240–262, 280–315 and 343–403 nm, respectively. The medium intensity 240 and 280 nm absorptions are seen generally only as shoulders on the strong end absorption band. The weak long wavelength band often extends into the visible region and is responsible for the orange-yellow colour of many of these derivatives. This long wavelength absorption is due to interaction of the nitrogen lone pair with the triene system and is both solvent and substituent dependent ⟨69JOC2866, 74CRV101⟩. *N*-Acylation or *N*-sulfonation causes a lightening in color, particularly with dibenz[*b,f*]azepines whose *N*-acyl derivatives are colorless, as do substituents at the 2- and 7-positions, their effect being to constrain the 1*H*-azepine in the boat conformation.

Marked differences in the UV spectra of 1-alkyl- and 1-acyl-dibenz[*b,f*]azepines support the view that the *N*-acyl derivatives are best described as the polar canonical dibenzazepinium mesomers (**32**), in which the nitrogen lone pair is delocalized onto the oxygen, thus transforming the destabilized 8π-azepine ring into an aromatic 6π-tropylium-like cation ⟨74JCS(P1)52⟩.

The PE spectra of 10,11-dihydro-5*H*-dibenz[*b,f*]azepine ⟨74MI51600⟩ and some 5*H*-dibenz[*b,f*]azepines ⟨74JCS(P1)52⟩ have been recorded.

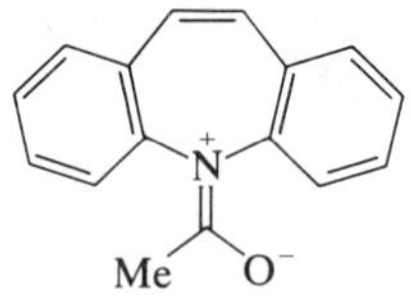

(32)

Apart from two review articles ⟨74AHC(17)45, 74CRV101⟩ IR details of azepines and their benzologues are widely scattered. However, IR data are of no great importance in the structural elucidation of azepines although it appears that ν(C=O) absorptions are useful in distinguishing between 1-benzazepin-2-ones and 2-benzazepin-1-ones, the former absorbing at 15–20 cm^{-1} higher than the 1-ones ⟨74AHC(17)45⟩.

5.16.2.7 Miscellaneous

The ESR spectra of dibenz[*b,f*]azepine nitroxide and its 10,11-dihydro derivative have been measured ⟨74CB2362⟩.

5.16.3 REACTIVITY OF AZEPINES

5.16.3.1 General Survey

5.16.3.1.1 Neutral rings, cations, anions and radicals

The chemistry of unsaturated azepines is dominated by their polyene character. The absence of π-delocalization confers instability on the ring system as witnessed by the many and various ring transformations undergone in acid and base solution, and under thermal and photolytic conditions. Most of the major reactions of azepines involve the neutral molecule: *e.g.* cycloadditions (Section 5.16.3.8.1), metal carbonyl complexation (Section 5.16.3.8.2), dimerizations (Section 5.16.3.2.3) and photo- and thermo-induced valence isomerizations (Section 5.16.3.2.1).

Stable cations, anions and radicals derived from azepines are rare species, although many ring contractions and some valence isomerizations appear to involve prior formation of a cationic species, especially in those partially saturated systems in which an enamine moiety forms an integral part of the azepine ring. It is only recently that azepine anions have been generated and their synthetic potential exploited (Section 5.16.3.6).

5.16.3.1.2 Azepines bearing exocyclic double bonds to O, N and S

One of the facets of azepine chemistry that has attracted organic chemists over the past two decades has been the synthesis of, and search for aromatic character in, the so-called azatropones and azatropolones. To this end all the monocyclic azepin-2-, -3- and -4-ones, several diones, and their benzo and dibenzo derivatives have been prepared.

The early, sometimes erroneous, often unsuccessful work on these 6π-heterocyclic congeners of tropones has been summarized and commented on by Moriconi and Maniscalco ⟨72JOC208⟩. They synthesized (generally by alkylating the dione with Meerwein's reagent) several examples of azatropones including the 4*H*-azepin-4-one (**33**) and the 2*H*-azepin-2-one (**34**) along with various benz- and dibenz-ologues.

The diagnostic tests for aromaticity in these systems would be their enhanced stability, *e.g.* towards hydrogenation, and more particularly, the presence of a substantial diatropic ring current, the latter being reflected in the deshielding of ring hydrogens and methyl groups. However, no such evidence was found. The azepinones (**33**) and (**34**) and their

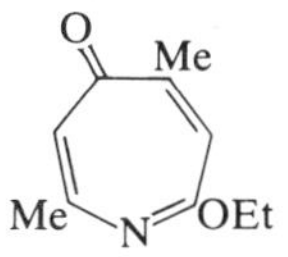

(33)

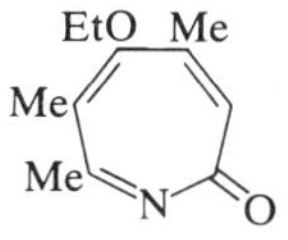

(34)

benzologues had azepine ring proton resonances in the vinyl region (δ 6.13–6.81 p.p.m.), and CH_3 resonances at 1.97–2.30 p.p.m., values not significantly different from those seen in the parent azepinediones. In addition, they all undergo facile hydrogenation to the perhydro derivatives, and acid hydrolysis back to the dione, behavior not in keeping with 'aromatic character'.

True 'azatropolone character' is claimed for the benzazepinone (**14**; R^1 = Me, R^2 = OH) and its 2-ethoxy analogue ⟨72AJC2421⟩. The marked differences in their ^{1}H NMR and UV spectra from the hydroxyazepinedione are credited to their enhanced planarity brought about by intramolecular hydrogen bonding. However, neither of these compounds shows any significant aromatic properties, and X-ray analysis in the case of (**14**; R^1 = Me, R^2 = OH), although confirming its planar nature, reveals no bond delocalization (Section 5.16.2.2).

More convincing evidence of aromatic character is forthcoming from the ^{1}H NMR and UV spectra of the ethoxyazepines (**35**) which represent the first examples of an azepin-3-one system. The long wavelength absorption of (**35**) shows a red shift (λ 370–400 nm) relative to the corresponding 2-ones and to the 3-hydroxy-2-alkoxy-5*H* derivatives. In addition, chemical shifts for (**35**; R = H, Ph, OEt or OAc) are in the region 7.8–8.2 with an *ortho* coupling for (**35**; R = H) of $J_{5,6}$ = 10 Hz, results which suggest the presence of a ring current ⟨81H(16)363⟩. Azepin-2-ones, *e.g.* (**13**; Ar = *p*-BrC_6H_4), have also been prepared ⟨79H(12)1427⟩, but X-ray analysis reveals their non-planar character (Section 5.16.2.2). Chemically, however, both systems are reactive and in protic solvents (*e.g.* MeOH) they ring-contract rapidly and quantitatively to pyridine derivatives (Section 5.16.3.5.1).

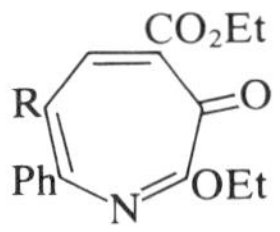

(35)

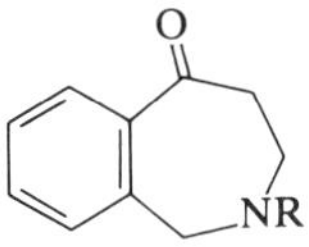

(36)

The hydrobenzazepinones are also of interest as their chemistry is dependent on the relative positions of the nitrogen and the carbonyl functions. The 2-ones are of course cyclic amides and behave as such; the most important member of this series is caprolactam. The 3- and 5-ones are typical aminoketones, although the latter have a lesser degree of reactivity than expected, possibly due to transannular amine–carbonyl interactions ⟨74JCS(P1)1828, 75JCS(P1)622⟩. However, transannular interactions (see Section 5.16.3.5.2) are not as pronounced as they are in the larger ring systems, and in general these azepinones resemble tetrahydro-4-quinolones in their chemistry ⟨B-67MI51600⟩. The 2-benzazepin-5-one (**36**; R = H) can be viewed as a cyclic Mannich base and is stable only as the *N*-acyl or *N*-Ts derivative ⟨73JCS(P1)1084⟩.

In common with many other nitrogen heterocycles azepin-2-ones and their benzo and dibenzo derivatives exist solely as the amide tautomers ⟨76AHC(S1)554⟩.

5.16.3.2 Thermal and Photochemical Reactions (not Formally Involving Other Species)

The non-planar polyene nature of azepines renders them susceptible to a variety of intra- and inter-molecular pericyclic processes. The azepine–benzeneimine valence isomerization has been discussed in Section 5.16.2.4, and the ring contractions of azepines to benzenoid compounds in the presence of electrophiles is covered in Section 5.16.3.3. In this section the thermal and photochemical ring contractions of azepines to bicyclic systems, their dimerizations and their isomerizations *via* sigmatropic hydrogen shifts are discussed. Noteworthy is a recent comprehensive review which compares and contrasts the many and varied valence isomerizations, dimerizations and cycloadditions of heteroepins (conjugated seven-membered heterocycles) containing one, two and three heteroatoms ⟨81H(15)1569⟩.

A recent concise review on the photochemistry of azepines and larger ring heterocycles is also available ⟨B-76PH288⟩.

5.16.3.2.1 Rearrangements

Early work ⟨B-69MI51600⟩ on *N*-substituted-1*H*-azepines revealed that they undergo photoinduced ring contraction to bicyclic valence tautomers as indicated in Scheme 1. Subsequently, it has been found that 3*H*- and 4*H*-azepines enter into analogous ring contractions, as do some of their oxo and benzo derivatives. These transformations, which parallel those undergone by cycloheptatriene, are often thermally reversible and occur by an orbital symmetry-controlled disrotatory electrocyclic process.

Unsubstituted *N*-methoxycarbonyl-1*H*-azepine (**37**; R = H) yields only one product, namely 2-azabicyclo[3.2.0]hepta-3,6-diene (**38**; R = H), which was isolated in 45% yield (as a pale yellow oil) after careful vacuum distillation. However, heating the bicycle above 125 °C brings about the retro-reaction. Latterly, this facile reversible process has been reinvestigated as a possible system for the storage of radiant energy ⟨76JOC2362⟩. The photoisomerism, for which quenching experiments implicate a singlet excited state, with radiation of λ 325–385 nm proceeds cleanly in 97% yield. The exothermic reverse reaction is first order in the temperature range 113–143 °C (E_a 120 kJ mol^{-1}) and is essentially quantitative. Unfortunately, however, the system proved inefficient as an energy store, as after 10 passes the thermal reaction gave a 50% build-up of polymeric material.

If the 1*H*-azepine bears a ring substituent then a mixture of two isomeric 2-azabicycloheptadienes, (**38**) and (**39**), is possible, corresponding to the two allowed electrocyclic processes indicated in Scheme 1 (paths a and b, respectively). In fact, the ratio of the two isomers varies with the position of the substituent. For example, the 3-methylazepine (**37**; R = 3-Me) affords a 1:1 mixture of the 4- and 7-methyl-2-azabicycloheptadienes, whereas the 4-methylazepine (**37**; R = 4-Me) yields the 5- and 6-methyl derivatives in the ratio of 1.5:1 ⟨B-69MI5600⟩.

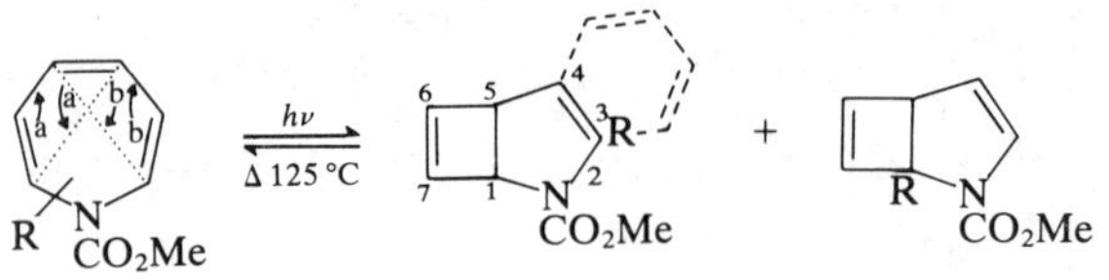

(**37**) ⇌ (hν; Δ 125 °C) (**38**) **a**; benzo + (**39**)

Scheme 1

Recently, a new reactivity index has been proposed ⟨80H(14)1717⟩ which predicts accurately the site selectivity of photocyclization of substituted cycloheptatrienes to their bicyclic valence tautomers. Unfortunately, application of the method to substituted 1*H*-azepines is far less successful. For example, for 2-methyl-1-methoxycarbonyl-1*H*-azepine (**37**; R = 2-Me) ΔG_{RS} values for C-2—C-5 and C-4—C-7 cyclization are calculated as 0.093 and 0.040 kJ mol^{-1}, respectively, *i.e.* predicting the 1-methyl isomer (**39**) as the major product. Experimentally, however, the reverse is true, the yields being 93.5% for 3-methyl (**38**; R = Me) and 6.5% for 1-methyl (**39**; R = Me). The corresponding photoinduced valence isomerizations of 1-benzazepines to 3,4-benz-2-azabicyclo[3.2.0]hepta-3,6-dienes (**38a**) have been recorded ⟨80JOC462⟩. These isomerizations have also been achieved thermally in the presence of silver ion ⟨80TL3403⟩.

Ring contractions of 3*H*-azepines analogous to those outlined in Scheme 1 can give rise to either a 2-azabicyclo[3.2.0]hepta-2,6-diene (**40**) or the fused azetine (**41**). Odum and Schmall found exclusive formation of the 2-azabicycloheptadienes (**40**) for the 3*H*-azepines (**2**; R^1 = H, R^2 = OEt, NH_2 or NMe_2), and argued that the alternative pathway to the bicyclic azetines (**41**) would involve loss of amidine or imidate resonance in the product ⟨69CC1299⟩.

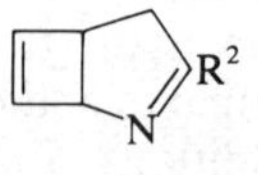

(**40**) 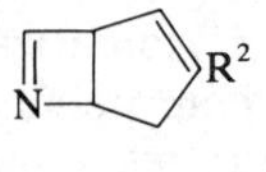 (**41**)

Similar results have been observed with 1-alkyl-3*H*-azepin-2-ones (**42**) which yield only 2-azabicyclo[3.2.0]hept-6-en-3-ones (**43**) ⟨73JCS(P1)1079⟩.

These photoisomerizations can be solvent sensitive. For example, photolysis of the azepinone (**42**; R^1 = aryl, R^2 = Bu^t, R^3 = H) in cyclohexane affords the azabicycloheptadiene (**43**; R^1 = Ar, R^2 = Bu^t, R^3 = H) in 80% yield. Irradiation in methanol solution, however, affords a mixture of the two isomeric enols of the cyclopentenone (**44**; 60%), and only 5% of the bicycle (**43**) ⟨81JOC4077⟩. Apparently, in aprotic solvents the azepine behaves photolytically as a heterocyclic conjugated triene, whereas in methanol its β,γ-unsaturated ketone character predominates.

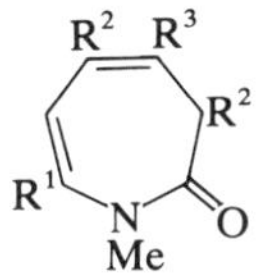

(**42**) 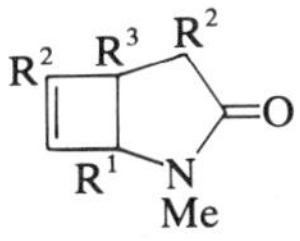 (**43**) 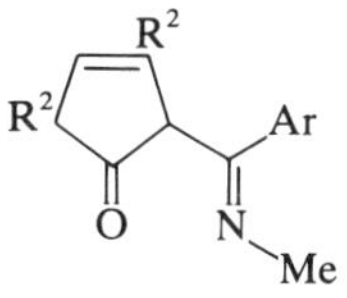 (**44**)

Attempts to isomerize the 2-methylene-3*H*-azepine (**45**; R^1 = H, R^2 = PhCO) to the *N*-benzoyl-1*H*-azepine under acid- or base-catalyzed conditions failed. Irradiation of the 3*H*-azepine resulted only in a photo-Fries type rearrangement of the benzoyl group to give aminoketone (**45**; R^1 = PhCO, R^2 = H) ⟨74JOC3076⟩.

Valence tautomerization to the bicyclic azetine (**47**; R^1 = OEt, R^2 = Me) has been observed in the photosensitized (Ph_2CO or PhCOMe) ring contraction of the otherwise photostable 2-ethoxy-4,5-dihydro-3*H*-azepine (**46**; R^1 = OEt, R^2 = Me) ⟨71JOC1934⟩. In contrast, sensitized photolysis of the dimethylamino derivative (**46**; R^1 = NMe_2, R^2 = H) gave only polymers. Unsensitized photolysis in pyrex yields several cyclic products, all of which are attributable to the intermediacy of an unstable 7-azabicyclo[3.2.0]hepta-3,6-diene (**47**; R^1 = NMe_2, R^2 = H) ⟨73CC327⟩.

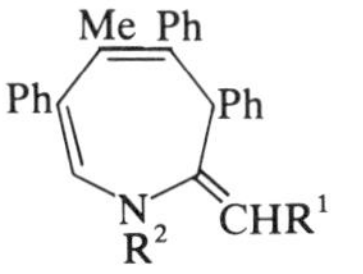

(**45**) 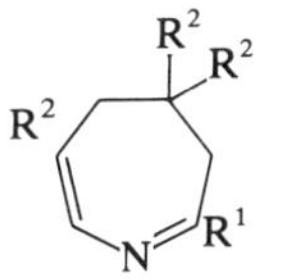 (**46**) 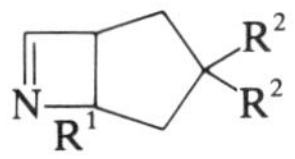 (**47**)

Photorearrangement of the dihydro-2-methoxyazepine (**48**) in heptane is complex, yielding 19 products all in <5% yield! In contrast, irradiation in benzene in the presence of a sensitizer (PhCOMe) yields, as sole product, not the expected bicyclic valence tautomer, but a photodimer of (**48**) *via* a symmetry allowed $[_{\pi}2_s + _{\pi}2_s + _{\sigma}2_s]$ ene reaction ⟨73JOC1090⟩. Sensitized photolysis in methanol affords the 3,4-methanol adduct.

Anomalous isomerizations have been noted during the photolytic and thermal rearrangements of 3-acyl-2-methoxy-3*H*-azepines (**2**; R^1 = acyl, R^2 = OMe) and 3-acyl-3*H*-azepin-2-ones ⟨69T5217⟩. Irradiation in methanol solution produces mixtures of 3-azabicyclo[4.1.0]hepta-2,4-dienes (**28**; R^1 = acyl and H, R^2 = OMe, R^3 = H) (or -4-ene-2-ones) and 3-phenacylpyridines (or pyridones), albeit in poor yields. Detailed, but tentative, arguments involving azanorcaradiene and/or diradical intermediates are presented to explain the formation of these unusual products.

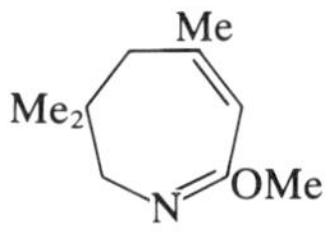

(**48**) 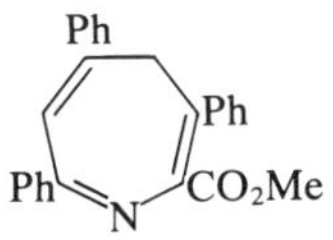 (**49**) 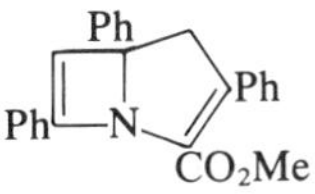 (**50**)

A few examples of the photoisomerization of 4*H*-azepines to 1-azabicyclo[3.2.0]hepta-2,6-dienes have been noted, *e.g.* (**49**) → (**50**) ⟨80TL595⟩.

The thermolysis of monocyclic azepines generally results in dimerization (see Section 5.16.3.2.3). However, in those derivatives where dimerization is sterically inhibited, *e.g.* (**37**; R = 4,5-Me_2), the azepine ring collapses, probably *via* a benzeneimine valence tautomer, to an aromatic urethane in a manner analogous to that outlined in Scheme 8 (Section 5.16.3.3.3) ⟨69JOC2879⟩. The 2,7-dimethyl isomer (**37**; R = 2,7-Me_2) is thermally stable, and resists dimerization and isomerization even on heating at 200 °C for 24 hours.

Thermolysis (180 °C) of the bridged azepine (**26**) and the structurally related benzeneimine (**25**) results in isomerization in high yields (87–93%) to the new bicyclic azepines (**53**; $n = 2$ and 1 respectively). A mechanistic rationale involving either a concerted suprafacial sigmatropic, or diradical, 1,5-nitrogen shift of the benzeneimine (**51**) (*i.e.* a 'nitrogen-walk') ⟨81H(15)1569⟩, followed by electrocyclic ring opening of the new isomeric benzeneimine (**52**; Scheme 2) is proposed ⟨69JOC2879⟩.

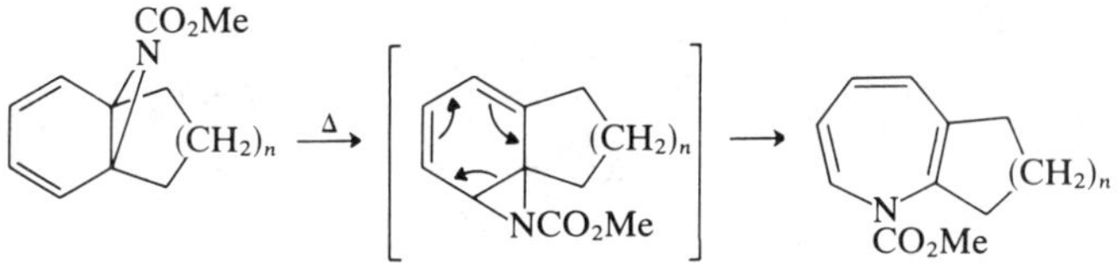

(51) **(52)** **(53)**

Scheme 2

Ring contractions of yet a different kind are observed with 1*H*-azepines of type (**54**), which on heating yield 6-substituted fulvenes as illustrated in Scheme 3 for the dicyano-1*H*-azepine ⟨71JCS(C)1237⟩. Analogous reactions occur with 3*H*-3-benzazepines.

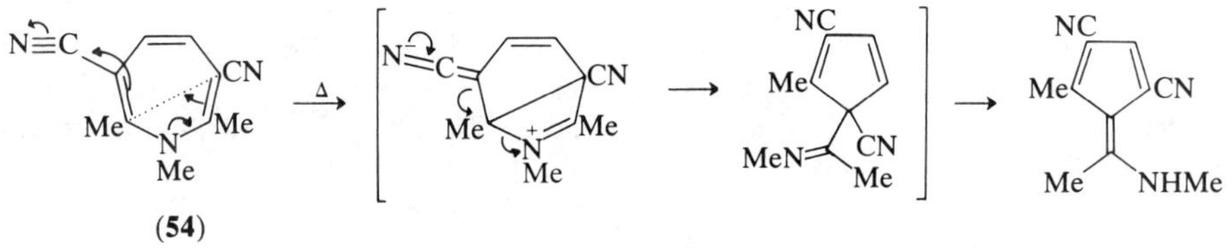

(54)

Scheme 3

A unique thermally induced valence tautomerism (Scheme 4) exists between the vinylcyclopropyl imidate (**55**) and the air and moisture sensitive dihydro-1*H*-azepine (**56**). The small energy difference (4.18 kJ mol^{-1}) between the two isomers is attributed to the unfavourable strain energy of the cyclopropane ring being counterbalanced by the loss of resonance energy in going to the deconjugated imidate of the azepine (**56**) ⟨78JA2908⟩.

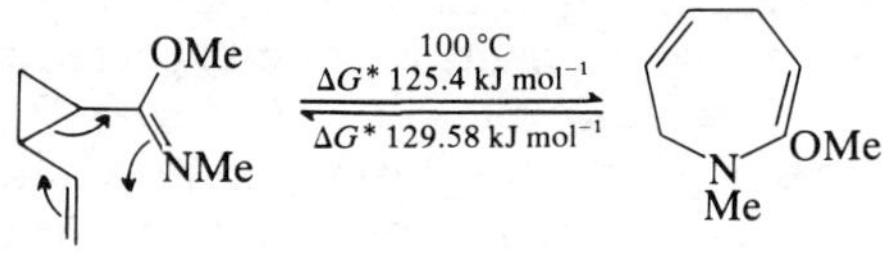

(55) **(56)**

Scheme 4

This behavior is in marked contrast to that of vinylcyclopropyl isocyanate, which in boiling benzene undergoes an irreversible Cope-type rearrangement to 3*H*-azepin-2-one ⟨65LA(682)1⟩.

5.16.3.2.2 Sigmatropic rearrangements

The four tautomeric monocyclic azepines are formally interchangeable by a series of 1,5-H shifts. Often, *e.g.* the parent 1*H*-azepine ⟨80AG(E)1016⟩, this isomerization is acid- and base-catalyzed. Many examples however, are known to occur under thermal non-catalytic conditions, and the accepted order of stability of azepines (*i.e.* 3*H* > 4*H* > 2*H* > 1*H*) is based on such observations. For example, the 4*H*-azepine (**57**) on heating for a few minutes at 190 °C undergoes consecutive 1,5-*H* shifts to give ultimately the 3*H*-azepine (**58**) ⟨72CB982⟩. The facile interconversion of 2*H*-azepines to 3*H*-azepines is similarly explained ⟨76JOC543⟩.

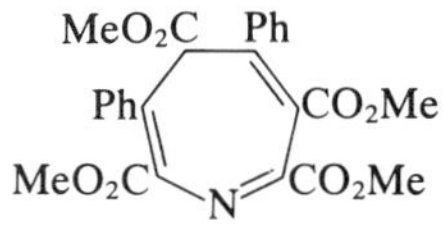

(57) 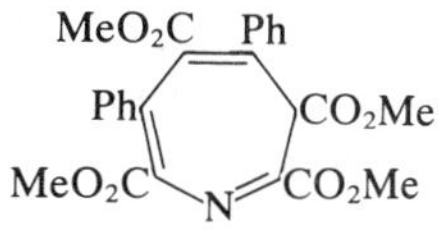 **(58)**

Of particular interest is the *syn*-1,6-imino-8,13-methano[14]annulene (**59**) which represents the first authenticated example of a stable 1*H*-azepine with a free NH group ⟨80AG(E)1015⟩. The annulene with aluminum oxide undergoes a remarkable isomerization to the *anti* isomer (**61**). Investigation shows that the isomerization is not a thermal reaction but involves alumina-catalyzed prototropic shifts *via* the 3*H*-azepine tautomer (**60**). This system is unique in that it is the first example of a 3*H* → 1*H* azepine tautomerism, and is a consequence of the high degree of strain in the 'anti-Bredt' 3*H*-azepine (**60**).

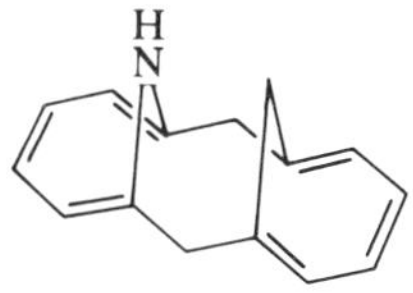

(59) 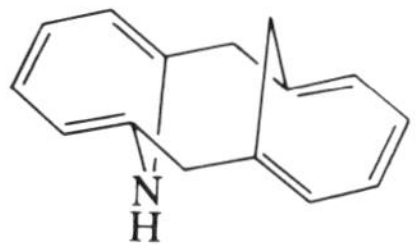 **(60)** 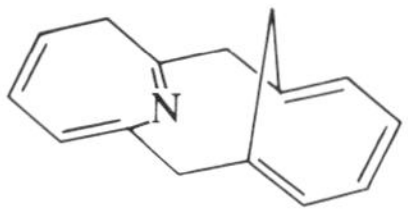 **(61)**

On flash vacuum pyrolysis the 1,6-imino[10]annulenes (**62**; R = Me, Me_3Si, CO_2Me) isomerize to the 1*H*-1-benzazepines.

On the basis of a study of the thermally induced isomerizations of several 2*H*- and 3*H*-azepines a generalization concerning the stability of azepines has been proposed which states that 'sigmatropic 1,5-H shifts in the isomerizations of azepines always proceed in the direction of the C=N rather than in the direction of the N=C bond' ⟨74JOC3070⟩. Occasionally mixtures of 3*H*-azepines are produced.

The generalization also holds for benzazepines. In fact, the only stable 2*H*-2-benzazepines synthesized so far are those in which the 5,6-carbon bond forms part of a fused ring system (*e.g.* **63**). 1,5-H shift to produce the more stable 3*H*-2-benzazepine is suppressed as it involves disruption of the benzenoid system (see Section 5.16.3.6).

(62) 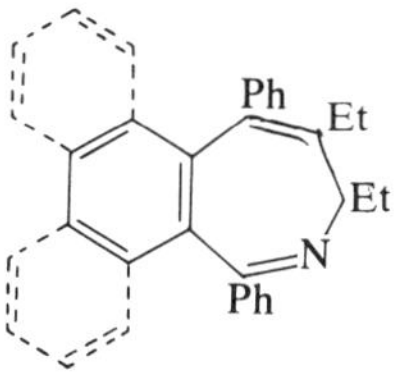 **(63)** **(64)**

a; benzo
b; phenanthro

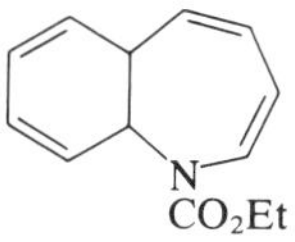

Not unexpectedly the 5a,9a-dihydro-1-benzazepine (**64**) on heating isomerizes to the conjugated 2,3-dihydro isomer ⟨78JOC315⟩. Likewise, ^{19}F NMR spectra reveal that after 24 hours at 25 °C, 30% of the unconjugated bis-(trifluoromethyl)dihydroazepine (**65**) (see Section 5.16.4.2.1(ii)) is converted *via* prototropic shift to the conjugated 2,3-dihydro-3*H*-azepine ⟨67JA605⟩.

The $[_{\pi}2_s + _{\pi}4_s]$ electrocyclic rearrangement of the deuterated *anti*-2-azatricycloheptane (**66**) at 350 °C yields 1-methoxycarbonyl-6,7-dihydro-1*H*-azepine in which the deuterium is statistically distributed between the C-2, C-3, C-6 and C-7 positions. Presumably at 350 °C rapid 1,5-H(D) shifts are taking place involving the thermodynamically less stable 2,7-dihydro-1*H*-azepine (**67**) ⟨73JA7320⟩.

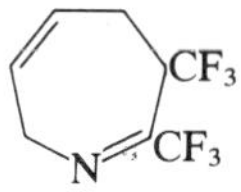

(65) 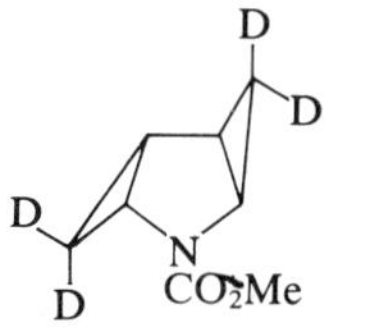 **(66)** 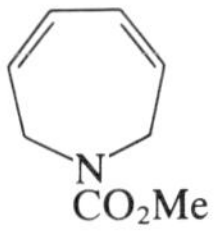 **(67)**

The formation of 2,4-dimethoxy-7-phthalimido-3*H*-azepine (**69**) by the *N*-nitrenophthalimide-induced ring expansion of 1,3-dimethoxybenzene has been explained by successive 1,7-phthalimido and 1,5-H shifts in the first formed *N*-phthalimido-1*H*-azepine (**68**; R^1R^2 = phthalimido) as illustrated in Scheme 5 ⟨73CC67⟩.

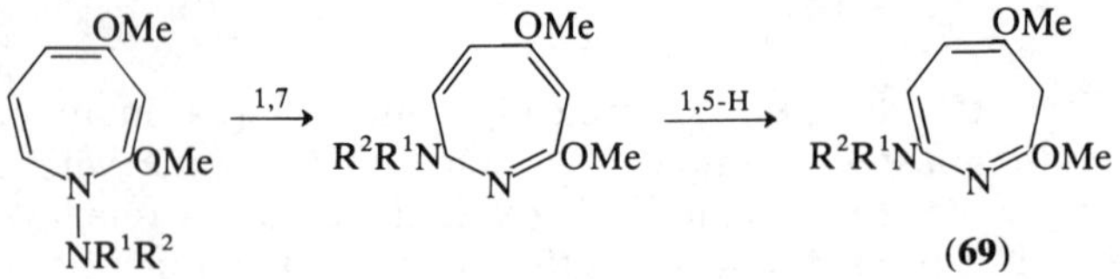

Scheme 5

2-Allyloxytetrahydro-3*H*-azepines (**70**) undergo thermal Claisen rearrangements to the 3-allylazepinones (**72**) *via* the tautomeric 1*H*-azepine (**71**) as outlined in Scheme 6. 4-Allyloxytetrahydro-1*H*-azepin-2-ones behave similarly ⟨80AJC2717⟩.

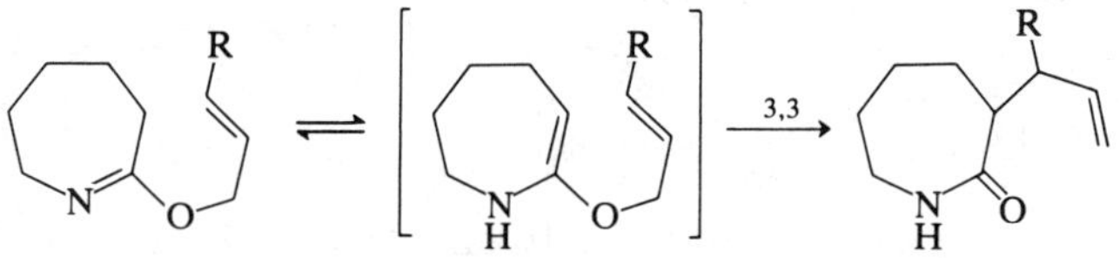

Scheme 6

5.16.3.2.3 Dimerizations

Dimerization of 1*H*-azepines is an extensively studied phenomenon and involves a temperature dependent cycloaddition process. At low (0 °C for **1**; R = Me) or moderate (130 °C for **1**; R = CO_2R or CN) temperatures a kinetically controlled, thermally allowed [6+4] dimerization to the *exo*-adduct (**73**) takes place, accompanied by a small amount (<10%) of symmetrical dimer (**74**). The latter are thermodynamically favored and become the major products (83%) when the 1*H*-azepines are heated briefly at 200 °C. The symmetrical dimers probably arise by a non-concerted diradical pathway since their formation from the parent azepines by a concerted $[6+6]\pi$ cycloaddition, or from dimer (**73**) by a 1,3-sigmatropic C-2, C-10 shift are forbidden on orbital symmetry grounds. Dimerization is subject to steric restraint and is inhibited by 2-, 4- and 7-substituents. In such cases thermolysis of the 1*H*-azepine brings about aromatization to the correspondingly substituted *N*-arylurethane ⟨69JA3616⟩.

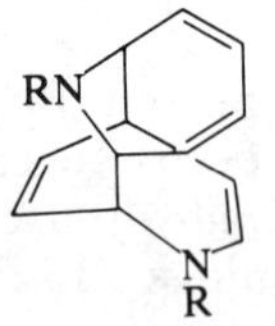

The oxazolo[3,4-*a*]azepin-2-one (**75**) is unusual in that, because of 5,6-ring fusion, it has considerable planarity unlike most other monocyclic *N*-acylazepines. This enhanced antiaromatic character is reflected in its spontaneous, highly peri-, site and regio-specific [6+4] dimerization to the *exo–anti* cycloadduct (**76**) ⟨80CC1161⟩.

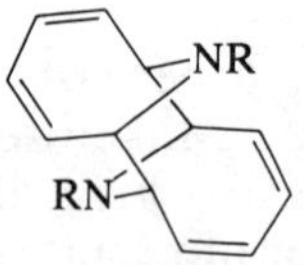

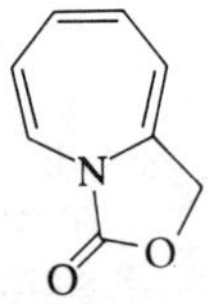

In contrast to 1*H*-azepines the photolysis of *N*-substituted 5*H*-dibenz[*b,f*]azepine under sensitized (Ph_2CO) conditions yields [2+2] cyclobutane dimers (**77**; R = Ac, CO_2Et, $CONH_2$, COPh). The *N-p*-tosyl derivative, however, undergoes photo-Fries reaction to the 2-*p*-tosylamino derivative ⟨74CRV101⟩.

Bischler–Napieralski cyclization of *N*-formyl-3-phenylpropylamine yields not the desired 4,5-dihydro-3*H*-2-benzazepine, but its dimer, 3,4,10,11-dibenz-1,8-diazacyclotetradeca-1,3,8,10-tetraene (**78**) ⟨73AP271⟩.

(76)

(77)

(78)

5.16.3.3 Electrophilic Attack at Nitrogen

5.16.3.3.1 Protonation and basicity

Hexamethyleneimine and its benzo derivatives are typical secondary amines and undergo electrophilic substitution at nitrogen, *e.g.* alkylation and acylation, under standard conditions. Most form stable hydrochlorides and simple derivatives such as picrates ⟨B-67MI51600⟩. Table 4 contains pK_a values of some azepines.

Table 4 pK_a Values of Azepines and their Benzo and Dibenzo Derivatives

Azepine	*Solvent*	*Temp.* (°C)	pK_a	*Ref.*
Hexamethyleneimine	H_2O	20.2	11.29	63BSF1805
2,3,4,5-Tetrahydro-1*H*-1-benzazepine	H_2O	18.6	5.60	63BSF1805
1*H*-1-Benzazepine (**3**; R=H)	H_2O	18.6	5.50	63BSF1805
10,11-Dihydro-5*H*-dibenz[*b,f*]azepine	H_2O–EtOH (4:1)	18.6	−0.78	63BSF1805
5*H*-Dibenz[*b,f*]azepine (**8**; R=H)	H_2O–EtOH (4:1)	18.6	0.87	63BSF1805
11*H*-Dibenz[*b,e*]azepine	H_2O	21±2	4.46	64HCA1163
6,11-Dihydro-5*H*-dibenz[*b,e*]azepine	H_2O	21±2	4.13	64HCA1163

Partially unsaturated azepines, like the fully conjugated systems, tend to be unstable in acid solution and undergo either rearrangement or ring contraction, generally to an aromatic system. Protonation takes place at nitrogen or, as in those hydroazepines in which delocalization of the nitrogen lone pair is possible, at the β-carbon of the enamine system.

1*H*-Azepines bearing *N*-alkyl substituents are unstable in acid solution and yield resinous products. In contrast, *N*-acyl, *N*-*p*-tosyl and *N*-ethoxycarbonyl derivatives undergo rapid aromatization to *N*-arylurethanes ⟨69JOC2879, 73CB3824⟩. The ring contraction is much slower with 2-substituted *N*-acyl-1*H*-azepines, a feature made use of in the separation of 1*H*-azepine mixtures ⟨71JHC729⟩.

Azepines (**26**) and (**53**) are very acid sensitive and with a trace of sulfuric acid rearrange quantitatively to 1,2,3,4-tetrahydronaphthyl carbamate. In contrast, the benzeneimine tautomer (**25**) yields 5-indanol (**81**) as the major product. A mechanistic rationale involving conjugate attack by water on the carbocation intermediate (**80**) formed as shown (**79**)→(**80**) has been proposed. A similar reaction sequence but involving carbocation rearrangements explains phenol formation from some monocyclic 1*H*-azepines ⟨B-69MI51500, 69JOC2879⟩.

5*H*-Dibenz[*b,f*]azepine undergoes acid-catalyzed rearrangement to 9-methylacridine under mild conditions. The initial stage of this interesting ring contraction involves proton

(79) (80) (81)

attack at nitrogen and then rearrangement to the quinone imonium ion (**82**). Subsequently, contraction of the heterocyclic ring takes place *via* the carbocation (**83**) and its valence tautomer (**84**) as illustrated in the generalized Scheme 7 ⟨74CRV101⟩.

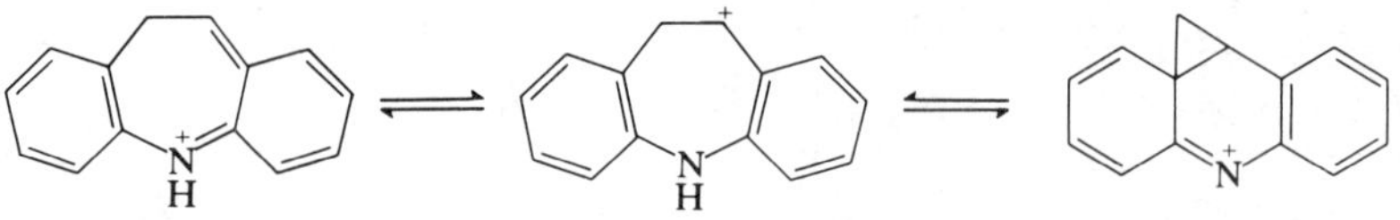

(**82**) (**83**) (**84**)

Scheme 7

This rearrangement, however, is complex and further studies show it to be dependent on temperature, solvent, pH and, most importantly, the absence or presence of oxygen. A similar overall mechanistic scheme operates for the acid-catalyzed rearrangement of *N*-nitroso-5*H*-dibenz[*b*,*f*]azepine to acridine derivatives (*e.g.* acridine-9-aldehyde) ⟨74CRV101⟩.

2-Amino-3*H*-azepines are readily soluble in dilute acid from which they are precipitated unchanged by base. Prolonged boiling with acid, however, effects hydrolysis to the 3*H*-azepin-2-one ⟨75JCS(P2)554⟩. 2-Dialkylamino-3-benzazepinones, *e.g.* (**85**), despite the amidine unit, are acid sensitive and hydrolyze readily to the stable 2,4-dioxo derivatives ⟨80JOC2951, 81HCA373⟩. In the absence of the 2-amino function, 3*H*-azepines aromatize to pyridylacetic acid derivatives, probably *via* proton attack, not at the ring nitrogen, but at the acyl group of the homopyridine valence tautomer, *e.g.* (**86**) ⟨66JCS(C)1075⟩. Alternatively, aniline derivatives are formed in a process analogous to that outlined in Scheme 8 ⟨74JOC3076⟩.

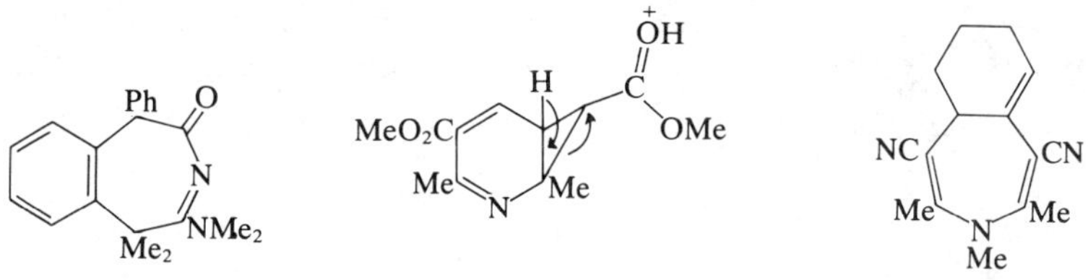

(**85**) (**86**) (**87**)

2-Alkoxy-3*H*-azepines in acid solution hydrolyze to 3*H*-azepin-2-ones ⟨69T5205⟩.

4*H*-Azepines, prepared by base-promoted ring expansion of dihydropyridines, undergo a retro-reaction on treatment with HCl or HBr (see Scheme 37; Section 5.16.4.2.3). In contrast, 4*H*-azepine (**87**) isomerizes in $CHCl_3$–dilute hydrochloric acid or $CHCl_3$–TFA to the 1*H*-azepine ⟨71JCS(C)1237⟩.

5.16.3.3.2 Lewis acids

N-Ethoxycarbonyl-1*H*-azepine is converted rapidly (5 min) and quantitatively into ethyl *N*-phenylcarbamate by boron trifluoride ⟨81JCS(P1)447⟩. The dichloroborane (**15**; $R^1 = H$, $R^2 = BCl_2$) results from the action of boron trichloride on 10,11-dihydro-5*H*-dibenz[*b*,*f*]azepine ⟨74IC2783⟩.

5.16.3.3.3 Alkylation and acylation

Fully reduced azepines and their benzo derivatives alkylate and acylate at nitrogen under the usual conditions employed for secondary amines ⟨B-67MI51600, 76CB2743, 80SC213⟩. Caprolactam undergoes *N*- and *O*-alkylation with dialkyl sulfates in benzene solution, but only *N*-alkylation in the presence of sodamide. Other examples of substitution at the lactam nitrogen have been summarized ⟨76AJC2651⟩. 1*H*-Azepin-2-one acetylates (Ac_2O), phenylates (PhI–Cu powder), cyanoethylates and hydroxymethylates at nitrogen, whereas alkylation with dialkyl sulfates yields the *O*-alkyl derivatives. Quaternary salts of hexamethyleneimine are stable at elevated temperatures, whereas ring fission can be achieved by traditional procedures, *e.g.* von Braun cyanogen bromide method. Apparently, however, tetrahydro-1*H*-1-benzazepinium methiodide suffers demethylation rather than ring opening ⟨B-67MI51600⟩.

Quaternization of 2-dialkylamino-3*H*-azepines with methyl iodide occurs at the ring nitrogen. Hydrolysis of the isolable, crystalline azepinium methiodides provides a useful route to *N*-alkyl-3*H*-azepin-2-ones ⟨73JCS(P1)1079⟩. The 6-chloro and 6-phenyl derivatives, however, yield *o*-bis(dialkylamino) benzenes as by-products, by a ring contraction process reminiscent of that occurring with acylating agents (Scheme 8). Alkylation (MeI or EtI) of the dimer (**78**) results in dedimerization and formation of the quaternary iodide of the 4,5-dihydro-3*H*-2-benzazepine in 90% yield ⟨73AP271⟩.

Some difficulty has been encountered with the *N*-alkylation (*e.g.* with ethyl bromoacetate) of 2,3,4,5-tetrahydro-1*H*-1-benzazepin-5-one (**31**; R = H) ⟨74JCS(P1)1828⟩. The pharmacological activity (Section 5.16.5) of *N*-substituted 5*H*-dibenz[*b,f*]azepine and its 10,11-dihydro derivatives has resulted in an intensive study of the *N*-alkylation of these ring systems ⟨74CRV101⟩. Generally, alkylation is effected with an alkyl halide or tosylate in the presence of base. Phase transfer *N*-alkylations of 5*H*-dibenz[*b,f*]azepine have been reported. The method, however, is less successful with the 10,11-dihydro derivatives ⟨79MI51600⟩.

Huisgen found that 2-amino-3*H*-azepines are unstable under normal acylating conditions. Schotten–Baumann benzoylation, for example, brings about aromatization and, ultimately, formation of 2-(benzamido)diphenylamine, whereas *p*-nitrobenzoyl chloride in cold pyridine yields 2-(*p*-nitrophenyl)-1-phenylbenzimidazole. Benzenesulfonylation, however, proceeds normally at the exocyclic nitrogen. The mechanistic rationale for these ring contractions is outlined in Scheme 8 ⟨B-69MI51600⟩.

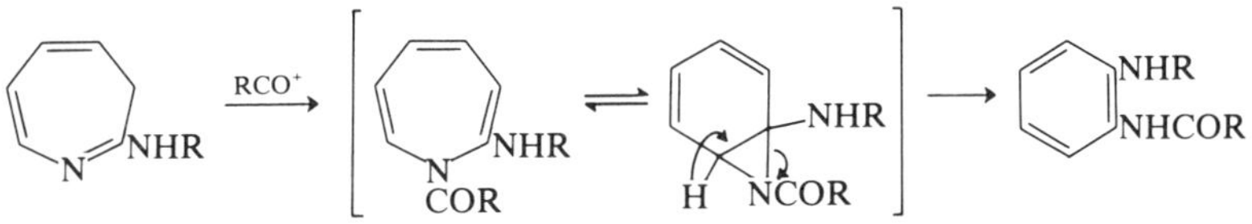

Scheme 8

Atypically, 2,5-dimethyl-3,4,6-triphenyl-3*H*-azepine, on treatment with benzoyl chloride in the presence of DABCO, benzoylates at nitrogen, probably *via* the 2-methylene-1*H*,3*H*-tautomer (**45**; $R^1 = R^2 = H$). Attempts to isomerize the benzoyl derivative (**45**; $R^1 = H$, $R^2 = PhCO$) to the conjugated *N*-benzoyl-2-methyl-1*H*-azepine failed. Weak bases provoked no reaction whereas strong bases regenerated the 3*H*-azepine. In marked contrast, boiling acetic acid caused rapid (2 h) aromatization to 2,5-dimethyl-3,4,6-triphenylaniline ⟨74JOC3076⟩. Ring contraction is much slower with the pentaphenyl-substituted 3*H*-azepine, and is complex if the azepine bears a 7-substituent. Once again isomerization of the 3*H*-azepine to its 1*H* tautomer followed by azanorcaradiene formation (as in Scheme 8) is a plausible explanation for these rearrangements.

5*H*-Dibenz[*b,f*]azepine and its 10,11-dihydro congener acylate under standard conditions ⟨74CRV101⟩.

5.16.3.3.4 Peracids

There are only a few reports concerning the peracid oxidation of azepines. 30% Hydrogen peroxide oxidizes *N*-alkyl-6,7-dihydro-5*H*-dibenz[*c,e*]azepines to their *N*-oxides ⟨75ZN(B)926⟩, whereas MCPBA has been used to prepare the 3*H*-2-benzazepine *N*-oxide (**88**) ⟨74JOC2031⟩. *N*-Oxidation of 2,3,4,5-tetrahydro-1*H*-1-benzazepine with MCPBA is accompanied by dehydrogenation to the 4,5-dihydro *N*-oxide (**89**) ⟨79JOC4213⟩.

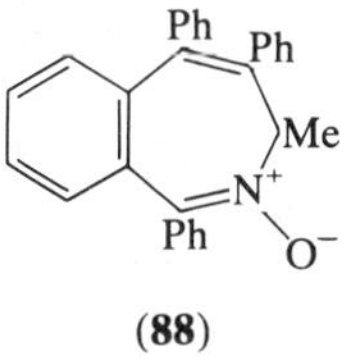

(**88**) 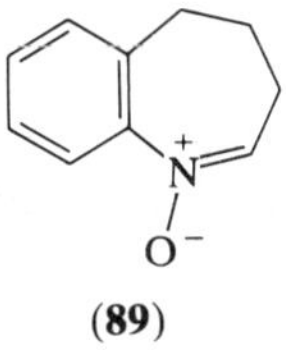 (**89**)

The oxidation of *N*-substituted 5*H*-dibenz[*b,f*]azepines with MCPBA is complex and depends upon the nature of the *N*-substituent. *N*-Acyl derivatives do not form the *N*-oxide but suffer epoxidation of the 10,11-bond. *N*-Aryl derivatives undergo hydroxylation of the phenyl ring, whereas *N*-alkyl congeners, with the exception of the *N*-methyl compound, yield mixtures of diphenylamines and acridones. The *N*-oxide is obtained from the *N*-methyl derivative along with ring-opened and ring-contracted products ⟨81CPB1221⟩.

5.16.3.3.5 Others

N-Nitrosations of 5*H*-dibenz[*b,f*]azepine and its 10,11-dihydro derivative are possible using sodium nitrite in acidified ethanol and ethereal amyl nitrite respectively ⟨74CRV101⟩. N-Nitroso-5,7-dihydro-6*H*-dibenz[*c,e*]azepine (**90**; R = NO) results by careful nitrosation ($NaNO_2$/HCl) of the parent amine ⟨73CJC1109⟩. This amine also couples successfully with aryldiazonium salts to give diazoamino compounds (**90**; R = N=NAr) ⟨70T3245⟩. The nitramine derivative (**90**; R = NO_2) is available in 95% yield by treating the *N*-mesyl compound (**90**; R = $MeSO_2$) successively with acetic acid–acetic anhydride and sodium nitrate in acetic acid ⟨81LA240⟩. Direct *N*-amination of the conjugate base of 10,11-dihydro-5*H*-dibenz[*b,f*]azepine has been achieved, albeit in low yield (20%), using chloramine in ether solution ⟨73JCS(P1)1041⟩.

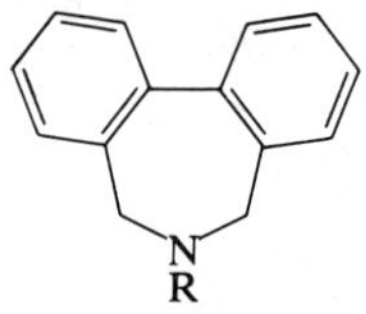

N
R

(90)

^{13}C NMR spectroscopy confirms that caprolactam trimethylsilylates at nitrogen rather than at oxygen ⟨79JCS(P1)2478⟩.

5.16.3.4 Electrophilic Attack at Other Ring Atoms, or at Heteroatoms Conjugated with the Nitrogen

Reports on the electrophilic substitution of the azepine nucleus at positions other than at nitrogen are uncommon, and in the main involve protonation of the hydroazepine, particularly those containing an enamine or dienamine moiety. In such cases protonation occurs predictably at the electron-rich β- (and possibly δ-) positions. For example, Paquette demonstrated spectroscopically that the dienamine (**91**) in perchloric acid exists as the 6,7-dihydro-4*H*-azepinium perchlorate (**92**) ⟨B-69MI51600⟩.

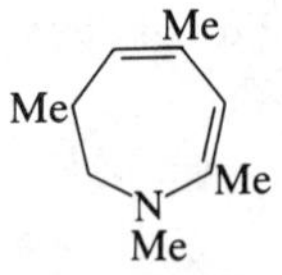

Me Me Me N Me

(91)

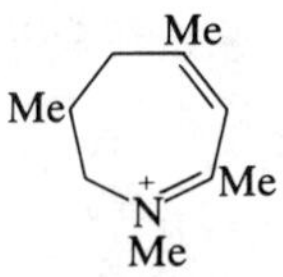

Me Me Me $\overset{+}{N}$ Me

(92)

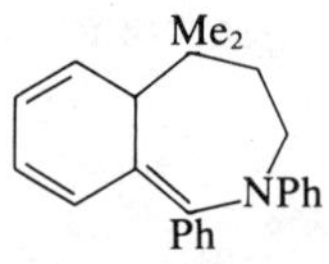

Me_2 NPh Ph

(96)

The rearrangement of the 5-cyano-4,5-dihydro-1*H*-azepine (**93**) to furo[2,3-*b*]pyridine (**95**; Scheme 9) with sodium nitrate in glacial acetic acid or with silver nitrate in aqueous ethanol proceeds by initial protonation at either C-3 or C-6 followed by hydrolysis to the cyanooctanedione ester (**94**). By carrying out the rearrangement with an acid ion exchange resin it is now possible to isolate the dione ester ⟨77CJC4061⟩. Likewise, the hydrolysis of the tetrahydro-2-benzazepine (**96**) to an *o*-(anilinoalkyl)benzophenone is an example of proton attack at the β-carbon of the enamine system ⟨77JA5045⟩.

There is continuing interest in the coordination compounds of cyclic polyenes, such as 1*H*-azepines, as it has become apparent that complexation with tricarbonyliron can lead to marked changes in the reactivity of the polyene system towards electrophilic substitution. In fact, 1-methoxycarbonyl-1*H*-azepinetricarbonyliron at 0 °C with acetic anhydride and

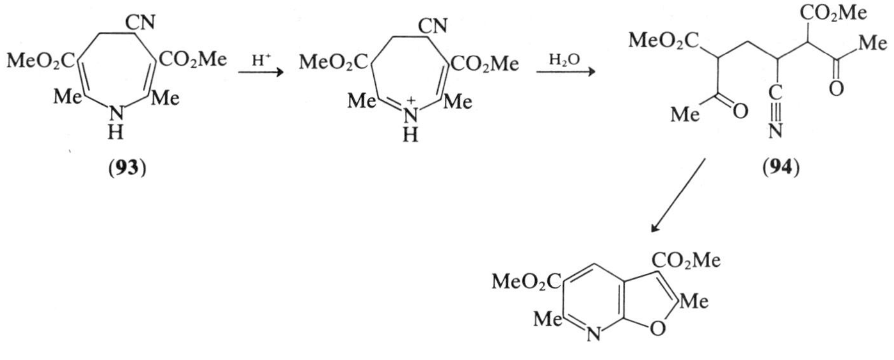

(93) (94) (95)

Scheme 9

borofluoric acid gives the 3-acetyl derivative (**97**; R = Me) in 42% yield. Similarly, propionic anhydride furnishes the ethyl ketone (**97**; R = Et), while with Vilsmeier's reagent (DMF–$POCl_3$) aldehyde (**97**; R = H) is formed, albeit in poor yield (19%) ⟨69CC631⟩. These substitutions represent the first examples of a direct electrophilic attack at the azepine nucleus.

(97) (100)

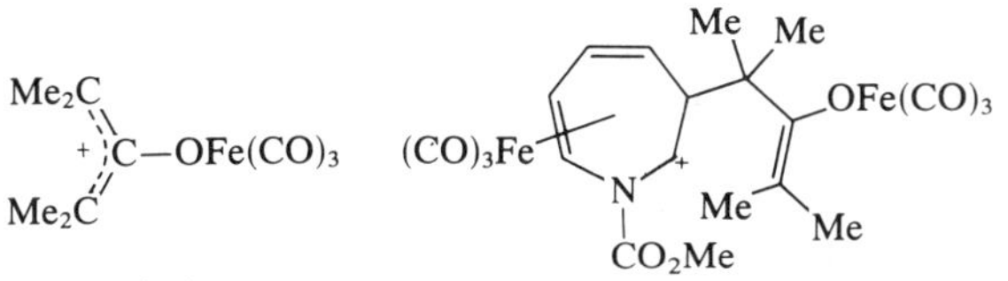

(98)

The azepine–carbonyliron complex is also attacked at C-3 by the 2-oxyallyl cation (**98**) to give initially the charged intermediate (**99**), and finally either the furanoazepine (**100**) or the ketone (**97**; $Me_2CHCOCH_2$ in place of COR), as their tricarbonyliron complexes ⟨81JOC3630⟩.

The action of DMAD on the dihydroazepine (**101**) is highly solvent dependent. In non-polar solvents [4+2] cycloadducts are formed (see Section 5.16.3.8.1), whereas in polar solvents regioselective electrophilic attack by the alkyne at the C-3 position dominates. The resulting zwitterion (**102**) can, depending on the solvent used, yield either the azacyclononatriene (**104**) by ring expansion of the cyclobutenoazepine (**103**), or the 6-vinyl-2,3-dihydroazepine (**105**) directly, as outlined in Scheme 10 ⟨80TL1145⟩.

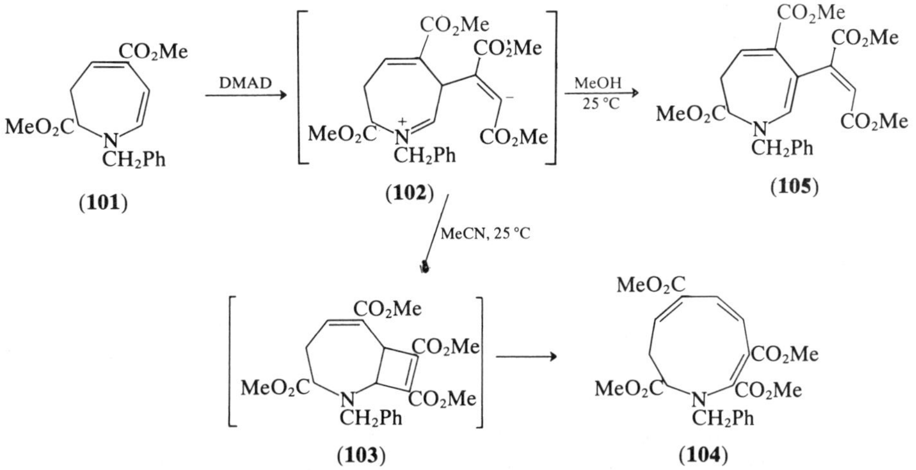

(101) (102) (105) (103) (104)

Scheme 10

Examples of electrophilic substitution (other than protonation) at the heterocyclic ring of benz- and dibenz-azepines appear to be confined to a few Vilsmeier reactions. 8-Chloro-1*H*-1-benzazepin-2-one with a mixture of DMF and $POCl_3$ yields the 2,8-dichloro aldehyde (**106**) ⟨72CPB1325⟩. Under similar conditions *N*-mesyl-4,5-dihydro-3*H*-3-benzazepine formylates at the 1-position (**107**; R^1 = CHO, R^2 = H) ⟨71BSF3985⟩. In contrast, *N*-mesyl-1,2,4,5-tetrahydro-3*H*-3-benzazepin-1-one yields a mixture of the 1-chloro dihydro compound (**107**; R^1 = Cl, R^2 = H) and the chloro aldehyde (**107**; R^1 = Cl, R^2 = CHO).

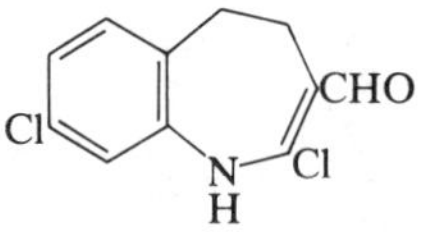

(**106**) 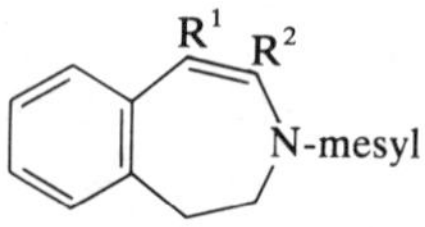 (**107**) 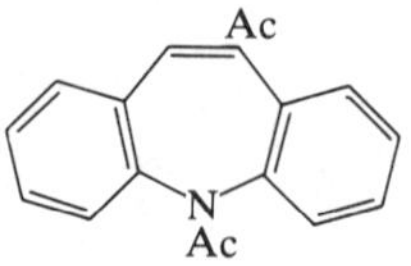 (**108**)

In agreement with MO calculations *N*-acylation of 5*H*-dibenz[*b,f*]azepine alters considerably the pattern of electrophilic substitution. In the *N*-unsubstituted heterocycle the sites of electrophilic substitution are at C-2 and C-8 (*i.e. ortho* and *para* to the free NH; see Section 5.16.3.9.1). However, as predicted theoretically, *N*-acylation deactivates the carbocyclic nuclei towards substitution *via* mesomers of structure (**32**). As a result Friedel–Crafts acetylation furnishes the 5,10-diacetyl derivative (**108**). Electrophilic bromination ($Br_2/CHCl_3$), unlike the free radical process (see Section 5.16.3.7), yields the 10,11-dibromo compound. In contrast, nitration of the *N*-acetyl derivative at low temperature affords only the 3-nitro isomer ⟨74CRV101⟩.

The Grignard reagent derived from 10-bromo-1-methyl-10,11-dihydrodibenzo-[*b,f*]azepine has been used to synthesize the 10-carboxylic acid ⟨78MI51602⟩.

O-Alkylations of azepin-2-ones, and benzazepin-2-ones, are most efficiently brought about by trialkyloxonium tetrafluoroborates (Meerwein's reagents) ⟨B-69MI51600, 73JOC1090, 81HCA373⟩. *S*-Alkylation of the thiones is effected similarly. These oxonium salts have also proved useful for the alkylation of azepinedione precursors of azatropones and azatropolones ⟨72JOC208⟩ (see also Section 5.16.3.1.2). 5*H*-Dibenz[*b,e*]azepine-6,11-dione with triethyloxonium tetrafluoroborate *O*-ethylates at the amide carbonyl and not at the ketonic carbonyl as was first proposed ⟨72AJC2421⟩.

5.16.3.5 Nucleophilic Attack at Ring Atoms

5.16.3.5.1 Oxygen, nitrogen and sulfur nucleophiles

Nucleophilic substitutions at the azepine nucleus are confined mainly to derivatives of hydroazepines such as lactim ethers, imidoyl chlorides and amidines, and to hydride ion reductions of carbonyl and imine groups (see Section 5.16.3.5.2). In addition some transannular nucleophilic displacements have been described, but such reactions are not as common with azepines as with larger ring heterocycles.

Lactim ethers (**109**; X = *O*-alkyl) of caprolactam are readily hydrolyzed to the lactam, and with amines yield the corresponding amidines (**109**; X = NHR). The ethers are themselves best prepared from the lactam with Meerwein's reagents ⟨70AHC(12)185⟩, but can also be prepared by alcoholysis of the imidoyl chloride (**109**; X = Cl) ⟨B-68MI51601⟩. Analogous reactions are known with the benzo ⟨81HCA373⟩ and dibenzo ⟨66HCA1433, 66JHC206⟩ congeners of (**109**). 1,2-Annulation of the lactim ethers by bifunctional nucleophiles has been exploited for the synthesis of polycyclic azepines ⟨79IJC(B)107⟩.

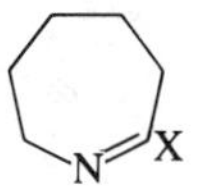

(**109**) 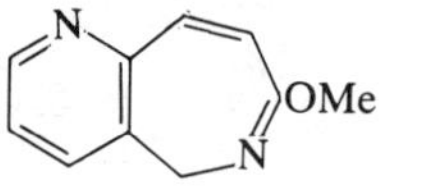 (**110**)

Some nucleophilic displacement reactions of 2-alkoxy- and 2-amino-3*H*-azepines are outlined in Scheme 11. Similar substitutions are undergone by 7-methoxy-5*H*-pyrido-[3,2-*c*]azepine (**110**) ⟨82JCS(P1)431⟩.

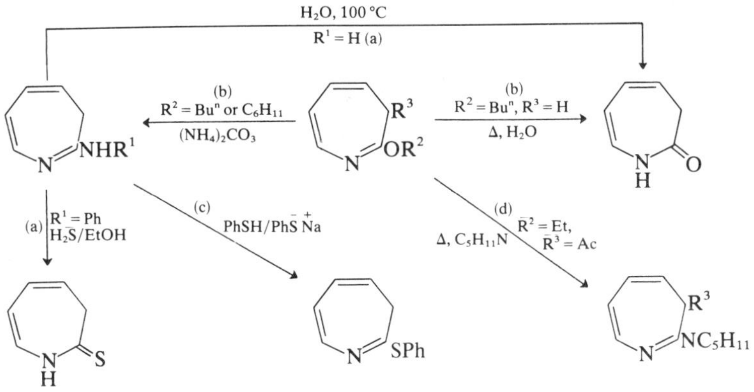

Ref. (a) 66T81; (b) 77BCJ2013; (c) 59CB2961; (d) 69T5205, 71JA5780

Scheme 11

The amino and bromo substituents in the 3*H*-azepine (**111**; X = Br) are susceptible to nucleophilic displacement and are hydrolyzed surprisingly easily by aqueous methanol or DMF to the azepine-2,7-dione. Sodium thiocyanate in DMF yields the aminothiocyanate (**111**; R = SCN) ⟨67JOC2367⟩. It has been noted that amidine formation with the lactim thioether (**112**) is much slower than with isomer (**113**) ⟨77JHC933⟩.

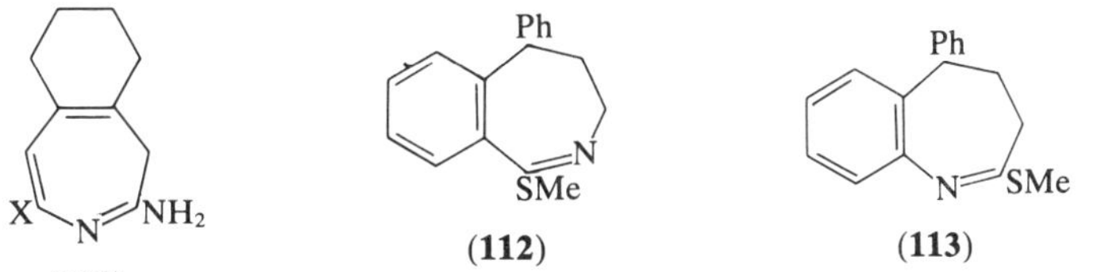

(**111**) (**112**) (**113**)

Alkaline hydrolysis of 5*H*-dibenz[*b*,*e*]azepine-6,11-diones provides a useful route to *o*-aminobenzoylbenzoic acids ⟨72JOC3755⟩.

Hydroazepinones in which the carbonyl group is not conjugated with the nitrogen undergo the usual ketone reactions with nitrogen (*e.g.* hydrazine) and carbon (*e.g.* Grignard reagents) nucleophiles ⟨B-67MI51600⟩. An exception is the tetrahydro-3*H*-3-benzazepin-1-one (**114**; X = H_2) which condenses with hydroxylamine only under forcing conditions to yield the 2-(**114**; X = NOH) rather than the 1-oxime. The *N*-tosyl derivative of (**114**; X = H_2) behaves normally and yields the 1-ketoxime ⟨75JCS(P1)622⟩. 1*H*-1-Benzazepin-5-ones functionalized at the 4-position, *e.g.* (**115**), are useful intermediates for the synthesis of fused ring azepines, *e.g.* pyrazolo[4,3-*d*][1]benzazepine ⟨78JCS(P1)862⟩. Schmidt reaction on 1*H*-1-benzazepin-5-one yields predominantly (65%) the aryl migration product, pyrrolo[1,2-*a*]benzimidazole, and only a little of the [1,6]benzdiazocine (the alkyl migration product) ⟨73JHC689⟩.

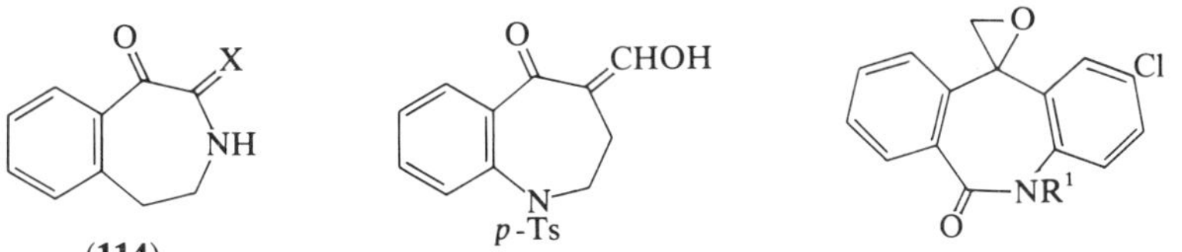

(**114**) (**115**) (**116**)

Dimethylsulfoxonium methylide reacts normally with the ketone carbonyl of dibenz[*b*,*e*]azepine-5,11-diones to yield epoxides (**116**) ⟨75JOC3602⟩.

5.16.3.5.2 Transannular nucleophilic substitutions

Azatropolones (Section 5.16.3.1.2) are surprisingly unstable in methanol solution and rapidly rearrange to pyridines initially by an intramolecular transannular nucleophilic attack by the ring nitrogen of the dioxotautomer (**117**) at the 3-keto group as outlined in Scheme 12 ⟨81H(16)363⟩.

The intriguing ring transformation of the tetrabromotetrahydro-1-benzazepin-5-one (**118**) to 6,8-dibromoquinoline-4-aldehyde in the presence of lithium chloride in DMF involves successive transannular nucleophilic substitutions as outlined in Scheme 13 ⟨72JCS(P1)2012⟩.

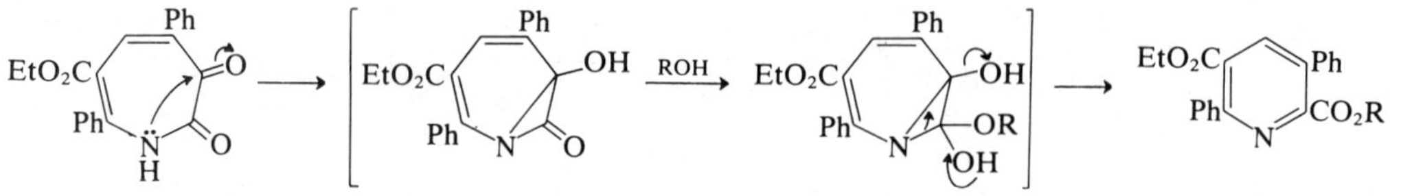

(**117**)

Scheme 12

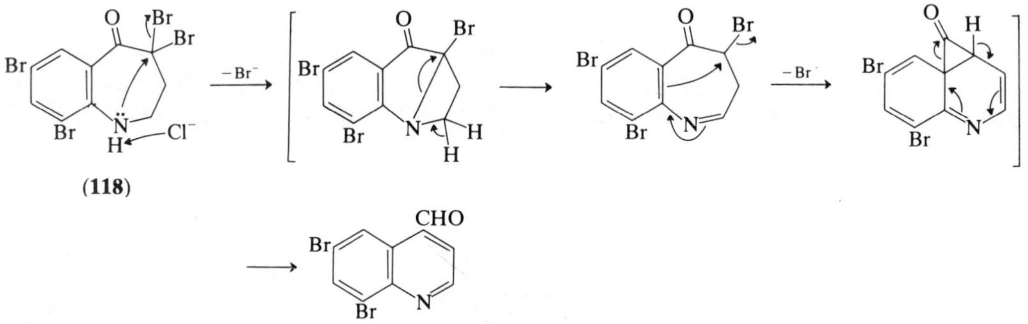

(**118**)

Scheme 13

Intramolecular nucleophilic substitutions are evident in the rearrangement of modified dibenz[*b,e*]azepinediones. The spirooxiranes (**116**) with ammonia ring open to the alkylamino derivatives (**119**) which on heating rearrange to the isoquinolones (**120**; $R^1 = H$) *via* nucleophilic attack of the amine at the amide carbonyl ⟨75JOC3602⟩. The hydrazine (**119**; $R = NH_2$) behaves similarly and yields the *N*-aminoisoquinolone (**120**; $R = NH_2$), whereas the parent dione with ammonium chloride affords isoindolin-1-one *via* intermediate (**119**; NH_2 in place of CH_2NHR). At 180 °C the hydroxy ester (**119**; CO_2Et in place of NHR) undergoes analogous ring contraction to give ultimately the spiroquinolone (**121**). 11-Hydroxy-5*H*-dibenz[*b,e*]azepin-6-ones behave similarly ⟨79AP662⟩. Also related is spiroisoquinolone, *e.g.* (**123**), formation from the hexahydrodibenz[*b,e*]azepine (**122**) ⟨79HCA1103⟩.

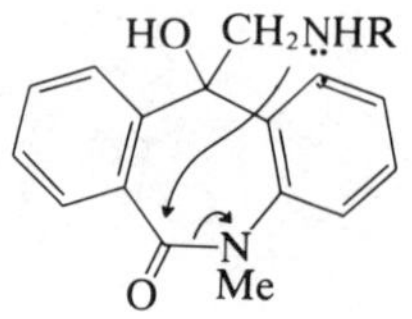

(**119**) 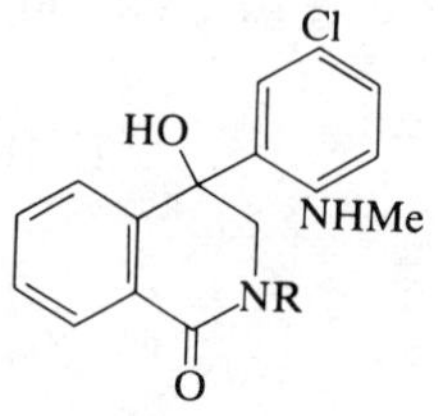 (**120**) 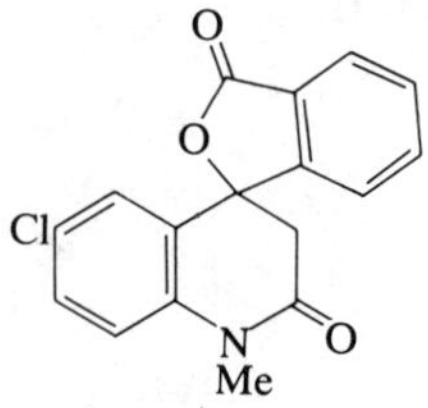 (**121**)

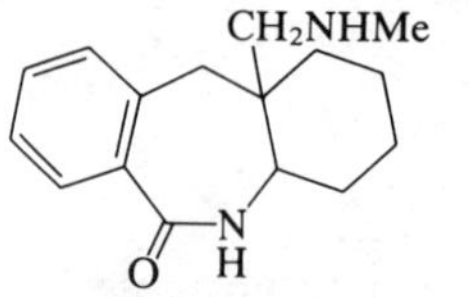

(**122**) 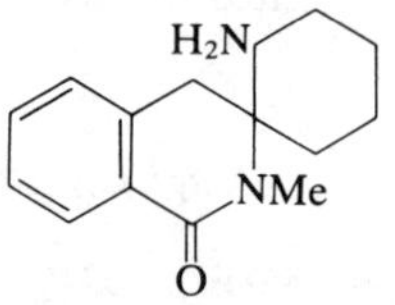 (**123**)

5.16.3.5.3 Hydride ion

Lithium aluminum hydride reduction of azepinones is a standard route to hydroazepines ⟨B-67MI51600, B-69MI51600⟩ and their benzo ⟨75HCA2228⟩ and dibenzo derivatives ⟨65CCC445, 73JPS1204, 76JCS(P1)1279⟩. Sodium borohydride is useful for the selective (ketone carbonyl) reduction of 5*H*-dibenz[*b,e*]azepine-6,11-diones ⟨62JMC1199⟩, tetrahydro-2*H*-2-benzazepine-1,5-diones ⟨78CL769⟩ and tetrahydro-3*H*-3-benzazepine-1,4-diones ⟨75CPB114⟩ to their respective secondary alcohols. The stereoselectivity of the hydride ion reduction of 2-benzylidene-3-*p*-tosyl-1,2,4,5-tetrahydro-3*H*-3-benzazepin-1-ones has been investigated in detail ⟨76BSF455⟩.

Sodium borohydride reduction of the azomethine bond has been noted for 3*H*-2-benzazepines ⟨75JA4682, 76HCA623⟩ and for 11*H*-dibenz[*b,e*]azepines ⟨65CCC445⟩.

Reduction of 5*H*-dibenz[*b,f*]azepine with sodium borohydride in acetic acid yields the *N*-ethyl-10,11-dihydro derivative (**124**; $R^1 = Et$, $R^2 = H$), reduction of the double bond being accompanied by reduction of the *N*-acetyl group formed *in situ* ⟨74JA7812⟩. In contrast, sodium borohydride in methanol reduces 5-acetyl-10-cyanodibenz[*b,f*]azepine to the 5-acetyl-10,11-dihydro derivative (**124**; $R^1 = Ac$, $R^2 = CN$) ⟨76T1345⟩.

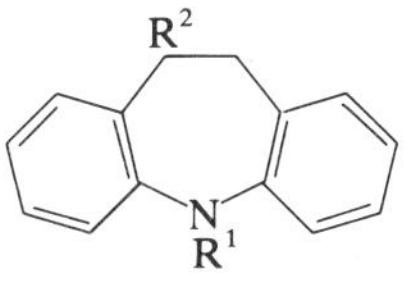

(124)

Diborane (or BH_3 in THF) has been used to reduce tetrahydro-3*H*-3-benzazepin-2-ones ⟨79JHC1525⟩ and tetrahydro-3*H*-2-benzazepin-3-ones ⟨79JHC1317⟩.

5.16.3.5.4 Others

Included in this section are the conversions of azepin-2-ones to imidoyl chlorides and to thiones, as these processes may be viewed as nucleophilic attack by halide or sulfur anions on the modified 2-hydroxyazepine.

Hexahydroazepin-2-one is an atypical lactam in its behavior towards chlorinating agents. With phosgene the initially formed 2-chloroazepine (**109**; X = Cl) is not isolated as it reacts further to give the 1-chlorocarbonyl derivative (**125**). With a mixture of phosphorus penta- and oxy-chlorides the azepinone yields the *N*-(2-azepinyl)azepine (**126**) by condensation of the imidoyl chloride (**109**; X = Cl) with its enamine tautomer ⟨B-68MI51601⟩. 3-Hydroxy-1*H*-1-benzazepine-2,5-dione (**127**; R = OH) and phosphorus pentachloride yield an unstable trichloro compound which readily loses HCl to give the dichloroketone (**128**). The action of thionyl chloride is more simple and yields solely the 3-chloro derivative (**127**; R = Cl), a product also obtained by alkaline hydrolysis of the dichloroketone (**128**) ⟨74CJC610⟩.

5*H*-Dibenz[*b,e*]azepine-6,11-dione and phosphorus pentachloride yield the 6-chloro derivative, ethanolysis of which furnishes the non-planar 'azatropone' (**129**), identical to the product formed by ethylation of the dione with triethyloxonium tetrafluoroborate ⟨72AJC2421⟩.

(125) 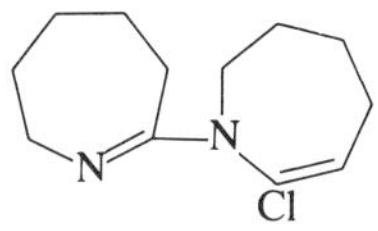 **(126)**

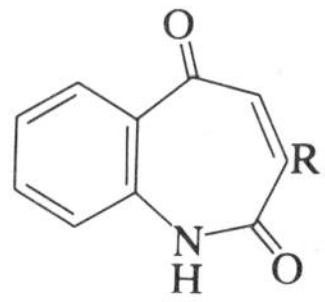

(127) 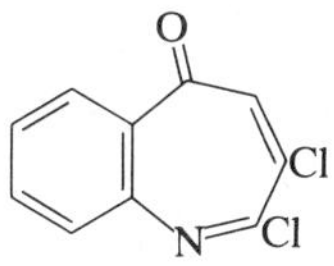 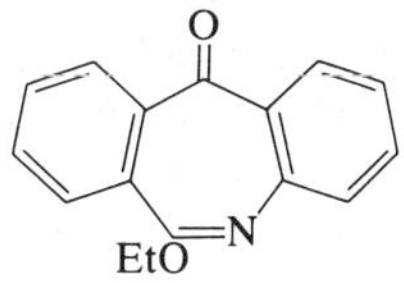 **(129)**

The 1-benzyl-tetrahydro-3*H*-3-benzazepin-2-one (**130**) with phosphorus oxychloride in boiling toluene cyclizes directly to the indenoazepine (**131**), a valuable intermediate for the synthesis of protopine alkaloids ⟨80T317⟩ (see Section 5.16.5).

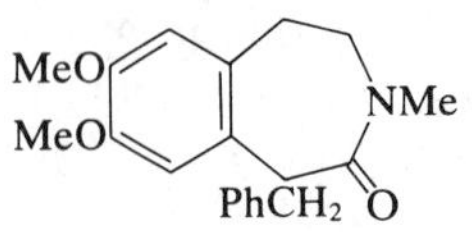

MeO MeO NMe PhCH$_2$ O

(**130**)

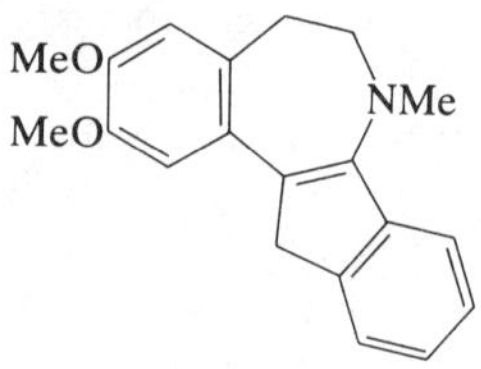

MeO MeO NMe

(**131**)

Azepinethiones are conveniently prepared from the azepinones using phosphorus pentasulfide in hot pyridine ⟨B-69MI51600, 73JCS(P1)1079, 77JHC933, 79JPR1034⟩.

5.16.3.6 Nucleophilic Attack at Hydrogen

N-Alkylation of azepinones has been discussed in Section 5.16.3.3.3. In many cases alkylation involves prior formation of the amide anion, which is readily generated by treating the lactam with strong base, *e.g.* NaH or Bu^{t}OK.

Surprisingly, instances of base-catalyzed alkylations of hydroazepinones at carbon adjacent to the carbonyl group are infrequent. Apparently, alkylations of tetrahydro-1*H*-1-benzazepin-5-one (**132**; $R^1 = R^2 = R^3 = H$) at C-4 are unpredictable, whereas its *N*-tosyl derivative is alkylated and acylated by a variety of reagents, *e.g.* MeI and $ClCO_2Et$, in the presence of base ⟨74JCS(P1)1828⟩. With LDA, however, the aminoketone (**132**) forms the 1,4-dianion which with methyl iodide affords the 1,4-dimethyl compound (**132**; $R^1 = R^2 =$ Me, $R^3 = H$) in 70% yield. However, efforts to prepare a 1,4-dimethylene-bridged azepine using 1,2-diiodoethane failed, as did attempts to form a bridged amide by intramolecular cyclizations of the acetic acid derivative (**132**; $R^1 = p$-Ts, $R^2 = CH_2CO_2H$, $R^3 = CO_2H$) ⟨79JCS(P1)2009⟩.

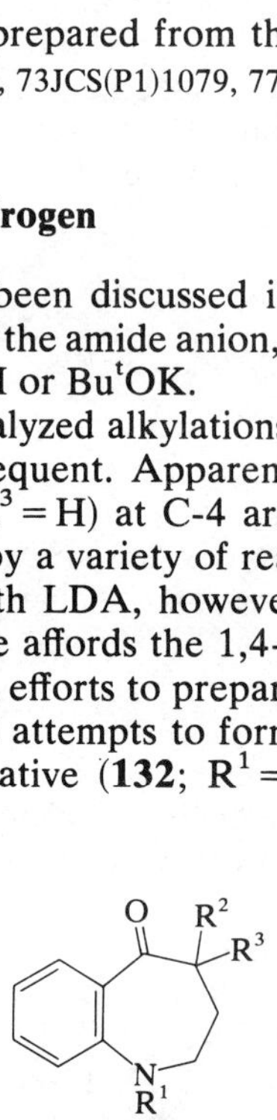

O R^2 R^3 N R^1

(**132**)

Alkylation of 3-methyl-1,2,4,5-tetrahydro-3*H*-3-benzazepin-2-one in THF–DMF solution containing sodium hydride, with primary and secondary alkyl halides and with α-bromoesters, results predominantly in 1-monoalkyl derivatives, whereas with ω,ω-dibromoalkanes, 1,1-spiro derivatives are formed ⟨80T1017⟩. Apparently, 6,7-dihydro-5*H*-dibenz[*b*,*d*]azepin-6-one does not condense with benzaldehyde or with nitrosobenzene at the active methylene group ⟨55JA3393⟩.

3-Hydroxy-2-azepinones are readily methylated at the OH by diazomethane ⟨79H(12)1427⟩.

α-Metallation of *N*-nitroso secondary amines to produce 'α-amino-carbanion' equivalents ⟨75AG(E)15⟩ is a valuable synthetic technique and has been applied to *N*-nitrosohexamethyleneimine and the *N*-nitrosodibenz[*c*,*e*]azepine (**90**; R = NO). In THF at −78 °C, with LDA and with methyllithium respectively, α-proton exchange is rapid, and the resulting carbanions react in high yield with various electrophiles (MeI, Ph_2CO, CO_2, D_2O) ⟨78OS(58)113, 73CJC1109⟩.

As expected, removal of a proton from fully unsaturated 1*H*-, 3*H*- or 4*H*-azepines is difficult as it results in formation of an antiaromatic 8π anionic species. In many cases these systems with strong base undergo anionotropic shifts and/or ring transformations to five- or six-membered aromatic systems ⟨B-73MI51601⟩. The instability of the azepine anion is tempered by annulation, and 5*H*-dibenz[*b*,*f*]azepines are alkylated at nitrogen by a variety of electrophiles including methyl iodide and dimsylsodium, and dialkyl sulfates in alkaline solution ⟨74CRV101, 76CPB2751⟩.

Recently, however, it has been shown that 2-diethylamino-5-phenyl-3*H*-azepine with LDA in 1,2-dimethoxyethane at room temperature yields a black-violet solution of anion (**133**), which on quenching into deuterium oxide yields the 3*D*-azepine (**134**; R = D) ⟨79TL2287⟩. Other electrophiles (MeI, Me_2S and $PhCH_2Br$) yield the 3-methyl, 3-methylthio and 3-benzyl derivatives respectively. Claisen condensations at the 3-position with methyl pivaloate or ethyl trifluoroacetate are also successful. However, the ketones (**134**; R = Bu^{t}CO or CF_3CO) so formed ring-contract under the basic conditions to the phenacylpyridines (**135**; R = Me or F) in accord with the known behavior of 3-acyl-3*H*-azepines in strongly basic media ⟨66JCS(C)1075⟩.

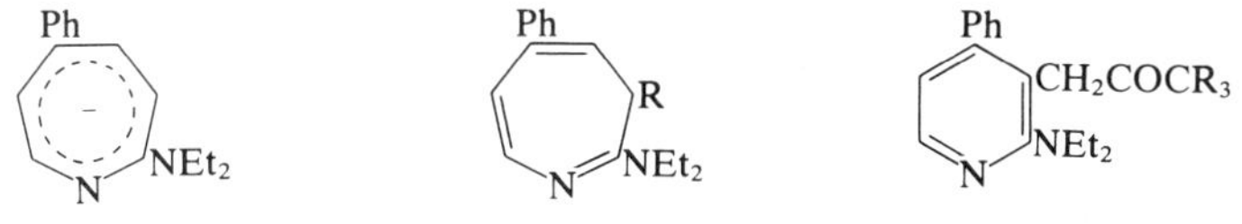

Ph, NEt₂, N — (**133**) Ph, R, NEt₂, N — (**134**) Ph, CH_2COCR_3, NEt₂, N — (**135**)

Base-catalyzed (NaOMe) isomerizations of 1*H*- and 4*H*-azepines to 3*H*-azepines are well known and most likely proceed *via* allylic anion intermediates ⟨80TL595, 72CB982⟩.

As mentioned in Section 5.16.3.2.2 phenanthroazepine (**63b**) with strong base (Bu^{t}OK) isomerizes to the 4*H*-tautomer (a 3*H*-azepine) whereas the benzazepine (**63a**) in which a similar proton shift would involve loss of aromatic character in the solitary benzene ring, resists all isomerization attempts ⟨74JOC3070⟩.

5.16.3.7 Reaction with Radicals or Electron Deficient Species

Radical substitutions at the azepine ring are few and are confined to the bromination of 5-substituted 10,11-dihydro-5*H*-dibenz[*b*,*f*]azepines, with NBS, at the benzylic C-10(11) position ⟨74CRV101, 78MI51602⟩. Other systems, *e.g.* 4,5-dihydro-1*H*-azepines, with NBS yield only bromine-free and often uncharacterized products ⟨66JCS(C)1075⟩.

Surprisingly, there are only a few reports concerning the action of carbenes on azepines. *N*-ethoxycarbonyl-1*H*-azepine and dichlorocarbene, generated by the action of 50% potassium hydroxide on chloroform, furnish the *trans*-trishomoazepine (**136**; $R^1 = R^2 = Cl$) in 35% yield. Under similar conditions the 2,3- and 4,5-homoazepines yield the *trans* homoazepines (**136**; $R^1 = Cl$, $R^2 = H$ and $R^1 = H$, $R^2 = Cl$ respectively). From a careful study of the addition of dichlorocarbene, generated by thermal decomposition of (dichloromethyl)phenylmercury, it is concluded that carbene addition to the 1*H*-azepine takes place sequentially in the order C-4—C-5, C-2—C-3 and C-6—C-7 ⟨74JOC455⟩.

The addition of dichlorocarbene, produced from $CHCl_3$–NaOH in the presence of a phase transfer catalyst, to the 10,11-bond of 5*H*-dibenz[*b*,*f*]azepine is accompanied by formylation at nitrogen ⟨78CPB942⟩, whereas the *N*-methyl derivative with diiodo- or dichloro-methane and a zinc–copper catalyst forms the 10,11-cyclopropanoderivative (**137**; X = CH_2, R = Me) in 55% yield ⟨76CPB2751⟩.

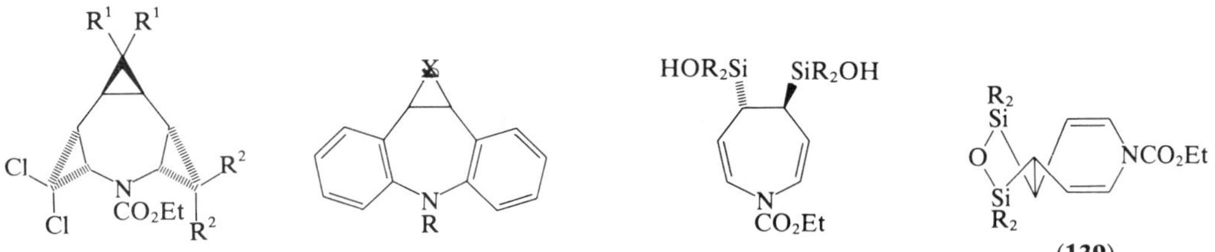

R¹ R¹, Cl, Cl, R², R², N, CO_2Et — (**136**) X, N, R — (**137**) HOR_2Si, SiR_2OH, N, CO_2Et — (**138**) R_2Si, O, SiR_2, NCO_2Et — (**139**)

The additions of alkyl or aryl silyl chlorides to 1-ethoxycarbonyl-1*H*-azepine in HMPA solution in the presence of magnesium proceed *via* anion radicals to yield *trans* adducts (**138**). On heating, these bis-silanols undergo quantitative dehydration to the siloxanes (**139**) ⟨79H(12)263⟩.

The electrochemical oxidation of 5*H*-dibenz[*b*,*f*]azepine and related compounds has been investigated ⟨75MI51600⟩.

5.16.3.8 Reactions with Cyclic Transition States Formally Involving a Second Species

5.16.3.8.1 Cycloadditions

The polyene character of azepines, particularly 1*H*-azepines, is apparent not only in the ease with which they dimerize (Section 5.16.3.3) but also in their propensity towards cycloaddition with a variety of reagents. The behavior of 1*H*-azepines in these mainly concerted electrocyclic cycloadditions is somewhat unique in that they display both diene and dienophilic character, and that they can, and do, function as the 2π, 4π or 6π component.

Diels-Alder $[4+2]\pi$ cycloadditions of 1-alkoxycarbonyl-1*H*-azepines are successful with all but the weakest dienophiles (*e.g.* maleic anhydride). Early work showed that with tetracyanoethylene (TCNE) cycloaddition at C-2—C-5 takes place readily in benzene solution at room temperature to yield adducts (**140**) ⟨69JOC2888, 70JHC1249⟩. The structure of the TCNE adduct with 5-bromo-1-ethoxycarbonyl-1*H*-azepine has been confirmed by X-ray studies ⟨67JCS(B)112⟩.

Unsymmetrically substituted 1*H*-azepines, *e.g.* 3-methyl-1-methoxycarbonyl-1*H*-azepine, afford a mixture of isomeric adducts, *e.g.* (**140**; R^1=H, R^2=Me and R^1=Me, R^2=H) resulting from addition at C-2—C-5 and C-4—C-7 respectively. However, substituents at C-3 and C-6 sterically hinder the cycloaddition so much so that the azepine functions as a triene and enters into a competing $[6+2]\pi$ cycloaddition at the C-2 and C-7 positions to furnish adducts (**141**). As expected, the yield of $[6+2]\pi$ adduct increases with increasing size of R, and is the sole product with R = *t*-butyl ⟨71JHC729⟩. In boiling xylene the $[6+2]\pi$ adducts (the kinetically favoured products) rearrange to the thermodynamically favoured $[4+2]\pi$ adducts (**140**). Rearrangement *via* a concerted $[_{\pi}6_s+_{\pi}2_s]$ process is thermally forbidden on orbital symmetry grounds, and a stepwise mechanism *via* an ionic intermediate of type (**142**) is proposed.

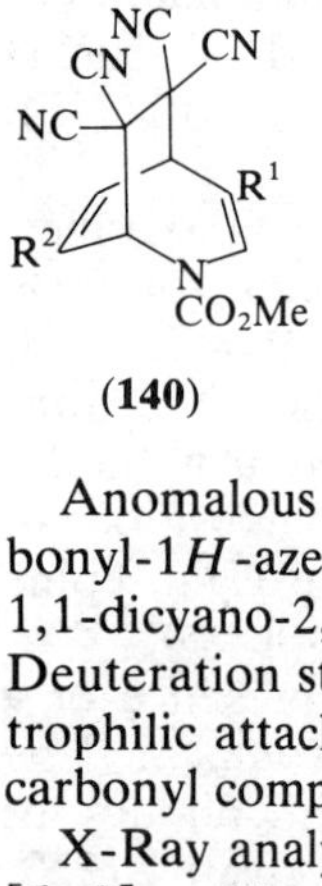

(**140**) 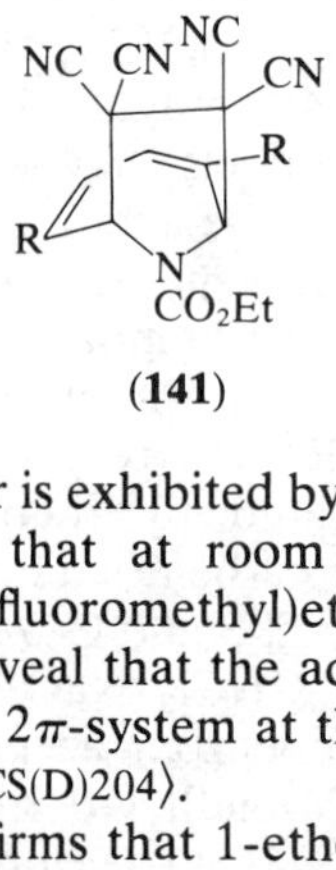 (**141**) 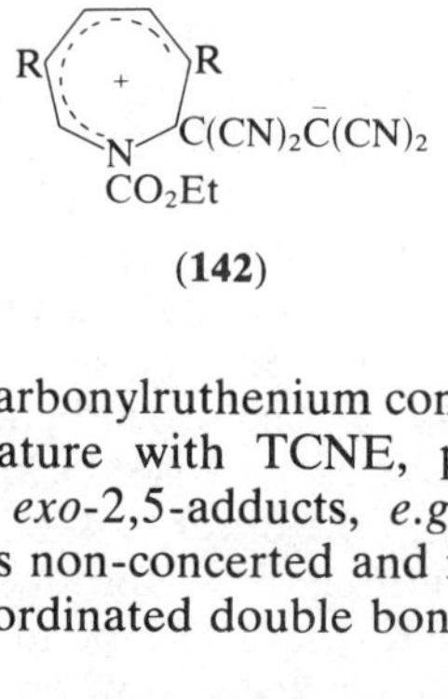 (**142**) 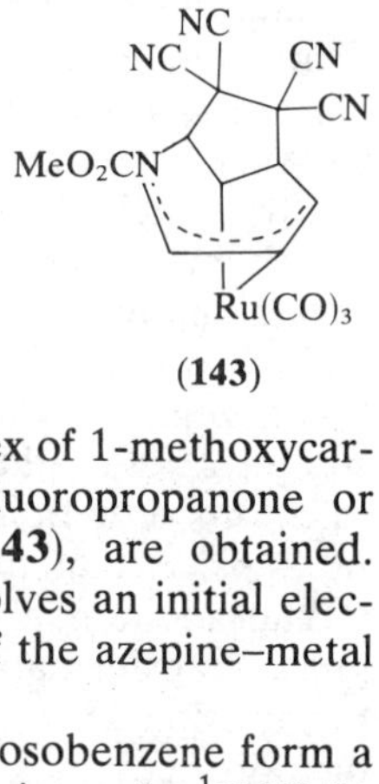 (**143**)

Anomalous behavior is exhibited by the tricarbonylruthenium complex of 1-methoxycarbonyl-1*H*-azepine in that at room temperature with TCNE, perfluoropropanone or 1,1-dicyano-2,2-bis(trifluoromethyl)ethylene, *exo*-2,5-adducts, *e.g.* (**143**), are obtained. Deuteration studies reveal that the addition is non-concerted and involves an initial electrophilic attack by the 2π-system at the uncoordinated double bond of the azepine–metal carbonyl complex ⟨77JCS(D)204⟩.

X-Ray analysis confirms that 1-ethoxycarbonyl-1*H*-azepine and nitrosobenzene form a $[6+2]\pi$ adduct. Subsequently, however, careful analysis of the reaction mixture by ^{1}H NMR has shown that the $[4+2]\pi$ adduct is also formed albeit in low yield (17%). No interconversion of the two adducts could be detected even after eight weeks, from which fact the authors conclude that both adducts derive directly from a zwitterionic intermediate of type (**142**) rather than in a concerted manner ⟨80TL319⟩.

1-Ethoxycarbonyl-1*H*-azepine forms normal $[4+2]\pi$ adducts with 4-phenyl-1,2,4-triazoline-3,5-dione and with diethyl azodicarboxylate ⟨81H(15)1569, 77JCS(P1)1824⟩, but unlike cycloheptatriene, which yields a $[6+2]\pi$ adduct, it gives no cycloadduct with *N*-ethoxycarbonylmethyleneimine ($CH_2{=}NCO_2Et$) ⟨81JCS(P1)447⟩. The critical solvent dependence of the addition of DMAD to 1*H*-azepines has been referred to in Section 5.16.3.4; only in hot carbon tetrachloride is the $[4+2]\pi$ cycloadduct obtained ⟨80TL1145⟩.

DMAD does not form cycloadducts with 1-phenyl-3*H*-2-benzazepine nor with the anion of 1-tosyl-2,3,4,5-tetrahydro-1*H*-1-benzazepin-5-one. Instead, the former yields 5*H*-2-benzazepine (**144**) *via* a concerted ene reaction ⟨75JA4682⟩, whereas the latter affords the benzazonine (**145**) by ring expansion of the initially formed C-4—C-5 [2+2] adduct ⟨80JCS(P1)1251⟩.

The dienophilic character of 1*H*-azepines is demonstrated by the ease with which they enter into inverse electron-demand Diels–Alder cycloadditions with electron-poor dienes.

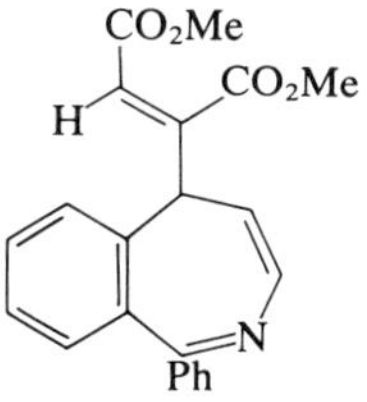

(144)

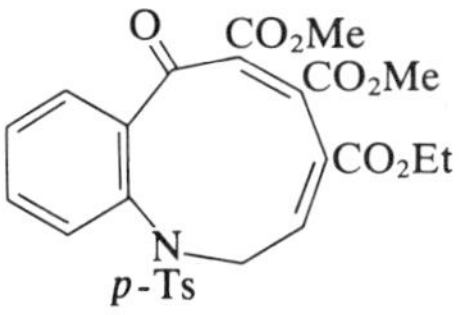

(145)

Such cycloadditions are dependent on the interactions of the azepine HOMO and the diene LUMO. Theoretical consideration of these orbitals reveals that bonding overlap is favourable for C-6—C-7 and C-4—C-5 additions and that, on the basis of secondary orbital interactions, the *endo* product is favored. Experimentally, however, it is found that additions are periselective and C-4—C-5 addition predominates in the cycloaddition of 1*H*-azepines with cyclopentadienones, isobenzofurans, tetra- and hexa-chlorocyclopentadienes, 1,2,4,5-tetrazines, α-pyrones and 3,4-diazacyclopentadienones ⟨81H(15)1569⟩.

1-Ethoxycarbonyl-1*H*-azepine with phenylcyclone undergoes rapid peri- and regioselective cycloaddition to the [2+4]π *endo* adduct (**149**) and not a [6+2]π adduct as proposed originally. However, a reinvestigation of this cycloaddition reveals that *endo* adduct (**146**) at room temperature undergoes a Cope rearrangement to the new *anti–endo* [4+2]π adduct (**147**) ⟨80JOC2368⟩. Adducts with more simple cyclopentadienones behave similarly ⟨81H(15)1569⟩.

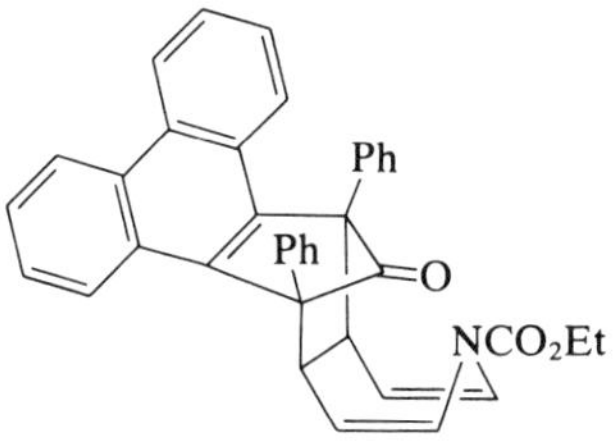

(146)

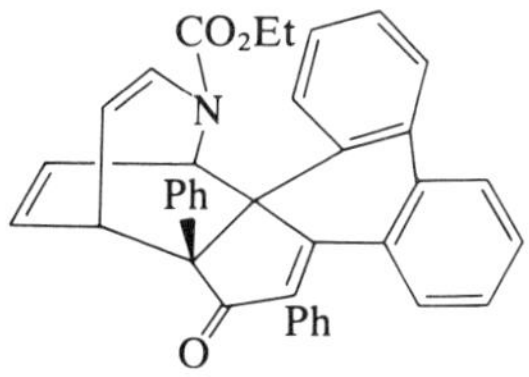

(147)

The reactivity of 1-ethoxycarbonyl-1*H*-azepine towards cyclopentadienones has been studied in terms of frontier MO theory. FMO energy levels for the reactants have been calculated using CNDO/2 MO methods, and from the results it is predicted that the diethyl 3,4-diphenylcyclopentadiene-2,5-dicarboxylate should be more reactive towards 1*H*-azepines than other more commonly used cyclopentadienones, as its LUMO lies at a significantly lower energy level. In fact, with 1-ethoxycarbonyl-1*H*-azepine in boiling benzene two cycloadducts are obtained, the *anti–endo* [4+2]π adduct (**148**) (82%), by Cope rearrangement of the initially formed 'normal' [2+4]π adduct (**149**), and the novel *exo* [6+4]π adduct (**150**) (17%). At room temperature only the adducts (**148**) and (**149**) are obtained, the latter rearranging to the [4+2]π adduct on warming ⟨80JOC4455⟩. A detailed examination of frontier and secondary orbital interactions provides a sound rationale for the products, the structures of which were established unequivocally by NMR and X-ray analysis.

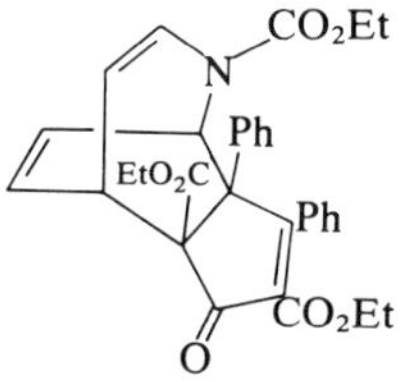

(148)

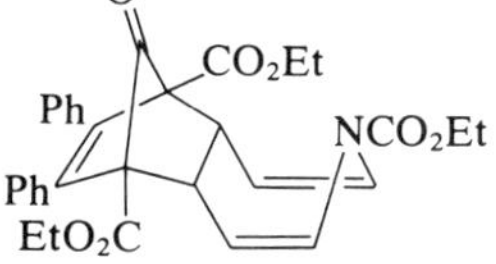

(149)

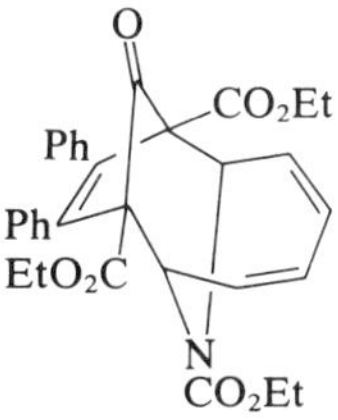

(150)

In contrast, the structurally similar and equally powerful diene, 2,5-diphenyl-3,4-diazacyclopentadienone affords only the [4+2] adduct, corresponding to (**148**), in moderate yield.

$[2+4]\pi$ cycloaddition of 1*H*-azepines and 1,2,4,5-tetrazine-3,5-dicarboxylates, followed by oxidation (DDQ) of the C-4—C-5 adducts (**151**) (formed by loss of N_2 from the original cycloadduct), provides useful access to pyridazine[4,5-*d*]azepines (**152**) ⟨78AP786⟩.

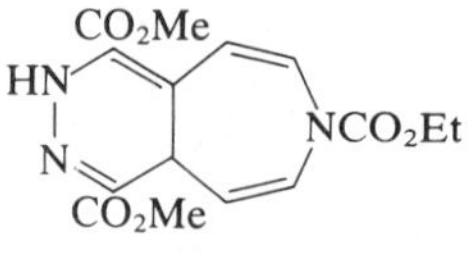

(151) 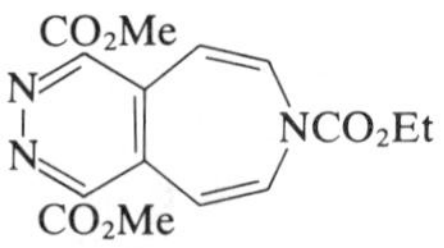 **(152)**

In contrast, 1-methoxy- (or ethoxy-) carbonyl-1*H*-azepine–tricarbonyliron complex with the tetrazine diester undergoes unprecedented $[2+4]\pi$ cycloaddition at the C-2—C-3 azepine double bond to give, after loss of nitrogen and decomplexation and dehydrogenation using *o*-chloranil or trimethylamine *N*-oxide, the isomeric pyridazino[4,5-*b*]azepine system ⟨82JOC110⟩. Similar adducts are also obtained with tetrachlorothiophene 1,1-dioxide.

Attempts to effect cycloadditions with α-pyrone as the diene system failed ⟨78JOC315⟩. However, the more electron-deficient α-pyrone-5- (or 6-) carboxylates add slowly (5 days in boiling toluene) to the 1*H*-azepine to furnish a mixture of the $[2+4]\pi$ (25%) and the $[6+4]\pi$ (20%) adducts (**153**) and (**154**) respectively. On prolonged heating (7 days) in toluene solution adduct (**153**) extrudes CO_2 to form the benzazepine (**155**; $R^1 = CO_2Me$; $R^2 = Et$) ⟨78H(11)401⟩.

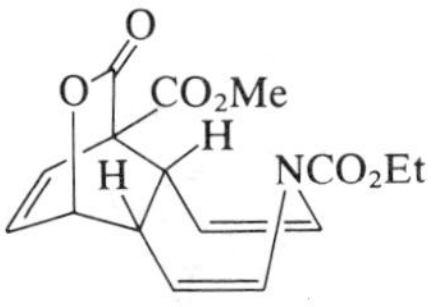

(153) 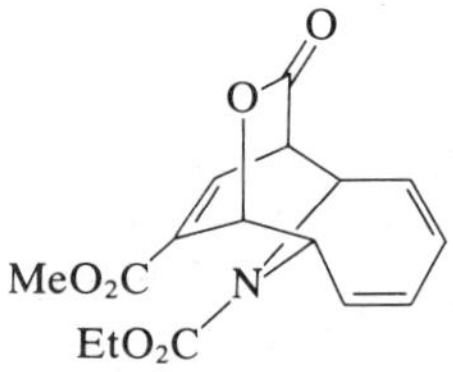 **(154)** 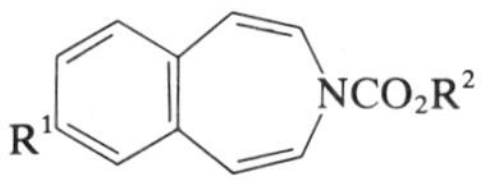 **(155)**

Although formal $[6+6]\pi$ dimers of 1*H*-azepines are common (Section 5.16.3.2.3) the adduct (**156**) formed from 1-methoxycarbonyl-1*H*-azepine and tropone is the only example so far of a $[6+6]\pi$ adduct between an azepine and a different addend ⟨81H(15)1569⟩.

Reports of pericyclic cyloadditions to other azepine systems are rare. Addition to the diene system of 6,7-dihydro-1*H*-azepines occurs readily with DMAD ⟨72CPB1740⟩ and with *N*-phenylmaleimide ⟨73JA7320⟩. The 5,5a-dihydro-3-benzazepin-2-one (**157**), a suspected but non-isolable intermediate in the formation of 1,2,4,5-tetrahydro-3*H*-3-benzazepine-2,4-diones by photoaddition of diphenylketen to amino-2*H*-azirines, has been trapped in the photolysate by *N*-phenyl-1,3,4-triazoline-2,4-dione as the $[4+2]\pi$ adduct (**158**). Its structure was confirmed by X-ray analysis ⟨80JOC2951⟩.

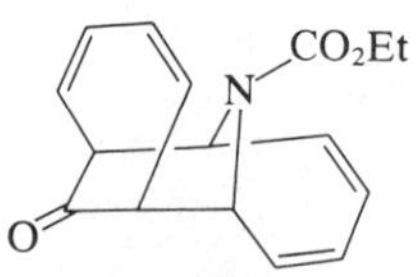

(156) 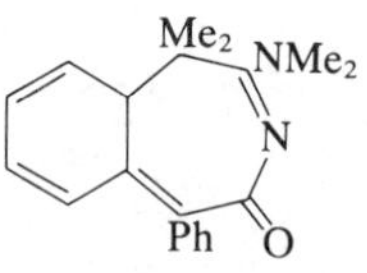 **(157)** 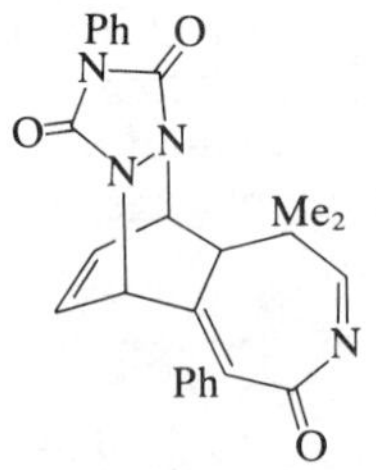 **(158)**

Cycloadditions to 3*H*-azepines appear to be limited to the two examples illustrated in Scheme 14 in which the azepine behaves as the electron-rich 2π-component ⟨78AP786⟩, and to the $[4+2]\pi$ adduct (**159**; R = *N*-phthalimido) formed by the electron-rich dimethoxy-3*H*-azepine (**69**) and 1-phenyl-1,3,5-triazoline-2,4-dione ⟨73CC67⟩.

1,3-Dipolar cycloadditions to azepines are confined to diazomethane and diphenylnitrilimine. The former reagent, depending on the nature of the substituents on the 1*H*-azepine, either adds at the 4,5-bond to yield pyrazolines (**160**) or traps the benzeneimine tautomer of the azepine as the bis-pyrazoline (Section 5.16.2.4) ⟨76CB3505⟩. A pyrazoline is also the product from the addition of diphenylnitrilimine to 5*H*-dibenz[*b,f*]azepine ⟨B-67MI51600⟩.

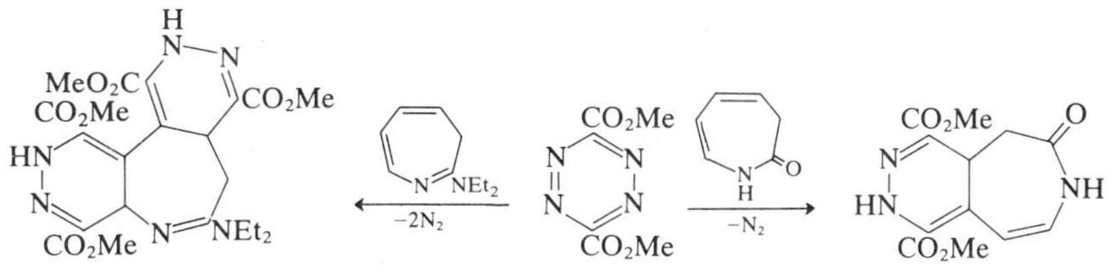

Scheme 14

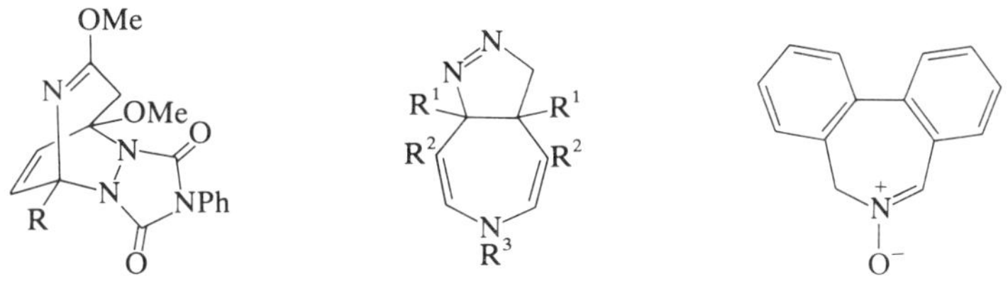

(159) **(160)** **(161)**

The 1,3-dipolar nitrone character of dibenz[*c*,*e*]azepine *N*-oxide (**161**) is apparent in its cycloadditions with phenyl isocyanate, *N*-phenylmaleimide ⟨74ZN(B)425⟩ and *N*-sulfinylarylsulfonamides ($ArSO_2NSO$) ⟨81TL2141⟩.

The methylene blue sensitized photoaddition of singlet oxygen to 1*H*-azepines was reported originally to yield the 2,5-epidioxide (**162**) as the major product (75%). However, a reinvestigation reveals that this [4+2] adduct is accompanied by the hitherto unknown [6+2] adduct (**163**) (30%) ⟨82TL873⟩. On heating with potassium hydroxide, epidioxide (**162**) aromatizes to ethyl *N*-phenylurethane. In hot ethanol adduct (**163**) ring contracts to a pyrrolinone, whereas in boiling benzene it fragments to the diene-aldehyde [$OHC(CH{=}CH)_2N(CHO)CO_2Me$] *via* the dioxetane (**164**).

Sensitized photoaddition of singlet oxygen to 3,5-di-*t*-butyl-1-ethoxycarbonyl-1*H*-azepine, which is known to give a [6+2]π adduct with TCNE, was also investigated. However, in this instance only the [4+2]π adduct (**162**; R = Bu^t) is isolated, which on heating at 170 °C undergoes an unusual rearrangement to the epoxy-epioxide (**165**) in 95% yield.

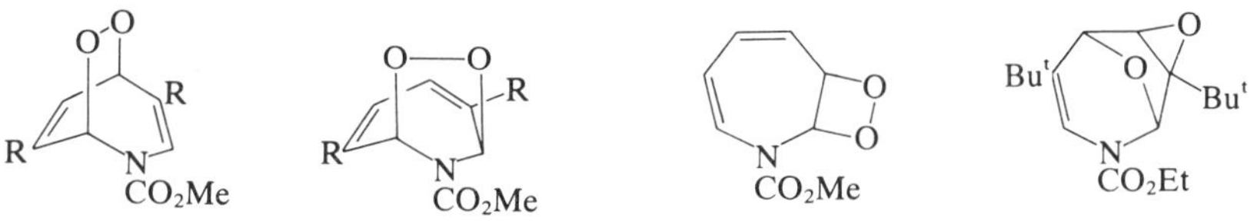

(162) **(163)** **(164)** **(165)**

5.16.3.8.2 Metal carbonyl complexes

The facility of 1*H*-azepines to form transition metal carbonyl complexes was realized soon after they were first synthesized. Variable temperature ^{1}H NMR studies on the tricarbonyliron complex formed either by photolysis of 1-ethoxycarbonyl-1*H*-azepine with tricarbonyliron in THF, or by heating the azepine with nonacarbonyldiiron in hexane, demonstrated that it undergoes rapid reversible valence tautomerism and that there is considerable restricted rotation about the N—CO bond ⟨B-69MI51600⟩. The molecular geometry of the complex has been determined by X-ray analysis (see Section 5.16.2.2).

Initially, attempts to prepare the corresponding chromium, molybdenum and tungsten complexes from the appropriate metal carbonyls were unsuccessful. In each case only the metal carbonyl complexes of the ring-contracted product, *i.e.* ethyl *N*-phenylcarbamate, were obtained. However, these complexes have since been prepared by treating the azepine-1-carboxylate with the metal tricarbonyl tris-acetonitrile complexes in THF solution ⟨77ZN(B)408⟩. The tricarbonylruthenium complex has also been prepared and undergoes some interesting cycloadditions (see Section 5.16.3.8.1) with electron-deficient alkenes, and ketones ⟨77JCS(D)204⟩.

Irradiation of 1-ethoxycarbonyl-1*H*-azepine with tricarbonyl(cyclobutadiene)iron produces the novel metallocene (**166**) as a stable orange crystalline solid, the structure of which was confirmed by X-ray crystallography ⟨71JA262⟩.

Carbonylation of 2,3-homo-1*H*-azepines has been effected by means of their metal and carbonyl complexes and provides a useful route to a variety of isomeric azabicyclononadienones. For example, the tricarbonyliron complex with carbon monoxide at 80 °C and 160 atm yields the 9-oxo-2-azabicyclo[3.3.1]nona-3,7-diene (**167**) (57%) or the 9-oxo-2-azabicyclo[3.2.2]nona-3,6-diene (**168**) (60%) depending on the *exo* or *endo* configuration of the tricarbonyliron complex. A third isomer, namely ethyl 7-oxo-9-azabicyclo[4.2.1]nona-2,4-diene-9-carboxylate (**169**), is formed on heating (125 °C) the azepine with carbon monoxide under pressure in the presence of the rhodium carbonyl complex [$Rh(CO)Cl_2$] ⟨78CB3927⟩.

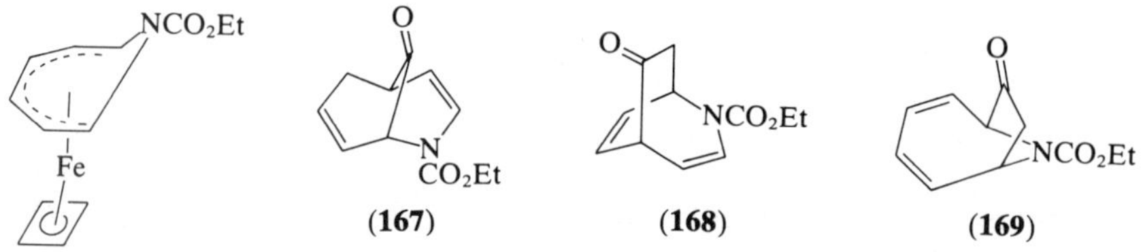

NCO_2Et Fe (**166**) O N CO_2Et (**167**) O NCO_2Et (**168**) O NCO_2Et (**169**)

5.16.3.8.3 Electrophilic and nucleophilic additions

Only a few polar additions to azepines have been reported of which the most common are the electrophilic additions to the 10,11-bond of 5*H*-dibenz[*b,f*]azepine; these have been reviewed ⟨74CRV101⟩. The *N*-acetyl derivative adds fluoroxypentafluoroethane and the adduct on treatment with sodium hydroxide eliminates HF to yield *N*-acetyl-10-perfluoroethoxy-5*H*-dibenz[*b,f*]azepine ⟨80JOC4122⟩.

Addition of iodine azide to 1-ethoxycarbonyl-1*H*-azepine is regioselective and in the presence of sodium azide yields a 10:1 mixture of the *cis*-4,5- and *cis*-2,3-diazides. The *cis* stereochemistry of the adducts is adduced to have arisen by initial *trans* addition of IN_3, substitution of iodide by azide, and then conversion to the *cis* isomers *via* the azanorcaradiene valence tautomer ⟨72JOC890⟩.

The action of anions on 4-(chloromethyl)-1,4-dihydropyridines provides a useful route to 4*H*-azepines (Section 5.16.4.2.3). However, in several instances the 4,5-dihydro-4*H*-azepines (**170**) suffer further cyclization to the 2,5-bridged azepines (**171**; X=NH, NMe, NCH_2Ph, S, Se) *via* an intramolecular Michael type addition at the 2,3-double bond ⟨68JCS(C)2311, 79CJC44⟩. Occasionally, *e.g.* (**170**; X = *p*-MeC_6H_4S), the 4,5-dihydroazepine can be isolated.

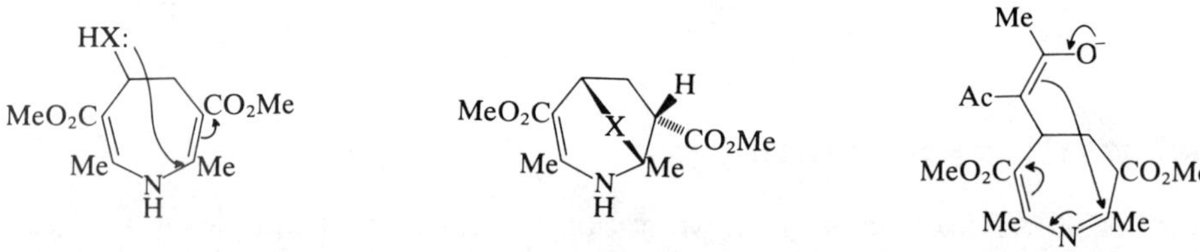

HX: MeO_2C CO_2Me Me N H Me (**170**) MeO_2C H X CO_2Me Me N H Me (**171**) Me O⁻ Ac MeO_2C CO_2Me Me N Me (**172**)

Subsequently, it has been noted that these intramolecular nucleophilic additions involve a disfavored 6-*exo*-trig (with respect to the six-membered ring) process. Consequently, an alternative mode of cyclization has been proposed in which the initially formed 4,5-dihydroazepine, *e.g.* (**170**; X = CAc_2), rearranges to the 3,4-dihydro tautomer (**172**). Cyclization can now take place *via* a favored 5-*exo*-trig (for the five-membered ring) or 6-*endo*-trig (for the six-membered ring) process. At low temperatures (0–5 °C) the 4,5-dihydroazepine (**170**; X = CAc_2) is isolable ⟨79CC1070⟩.

5.16.3.8.4 Eliminations

The synthesis of fully unsaturated azepines by dehydrohalogenation or dehydrogenation of hydroazepines is an obvious ploy. However, such methods have met with mixed success, dehydrogenation methods being particularly disappointing ⟨B-67MI51600⟩ (see also Section 5.16.3.8.5).

The lack of reliable non-destructive methods for the halogenation of azepines means that dehydrohalogenations are relatively unexplored, although where the bromo compounds are available, dehydrobromination generally proceeds well ⟨72JCS(P1)2012, 73JCS(P1)1084⟩ and is the best synthetic route to dibenz[*b,f*]azepines ⟨74CRV101⟩. Dehydrations and dehydroalkoxylations are also known mainly with benzazepinols ⟨*e.g.* 72CJC3886⟩ or their ethers ⟨68JCS(C)2481⟩. The remarkable tendency of the dibromobenzazepine (**118**) to undergo debromination rather than dehydrobromination has been noted ⟨72JCS(P1)2012⟩.

The elimination of HX from 4,5-dihydroazepines (**170**; CN, OEt or halogen in place of XH) is a useful route to 4*H*-azepines ⟨71JCS(C)1237, 79CJC44⟩.

5.16.3.8.5 Oxidations and reductions

Oxidation of hydroazepines to azepines has been effected mainly on the benzo and dibenzo derivatives, DDQ ⟨81HCA373, 81H(16)363⟩, chloranil ⟨78JOC315, 76JCS(P1)1279⟩ and manganese dioxide ⟨72JCS(P1)2012, 68JCS(C)2481, 74JCS(P1)1828⟩ being popular oxidants. Hydroxyazepines are oxidized to azepinones by chromium oxides and their pyridine complexes ⟨75CPB114, 75JCS(P1)622⟩.

Oxidation of the sulfur- or selenium-bridged azepines (**171**; X=S or Se) with mercury(II) oxide in methanol yields ultimately the 4*H*-azepine ⟨68JCS(C)2311⟩; with hydrogen peroxide as oxidant, the sulfur compound furnishes the sulfoxide (**171**; X=SO). Selenium dioxide oxidation of 7,8-dimethyl-1*H*-1-benzazepin-2-one affords the 2,3-dioxo derivative (**173**) that displays no evidence of enol tautomers or heteroaromaticity ⟨71CI(L)1439⟩.

Oxidation of 5*H*-dibenz[*b,f*]azepines by Fremy's salt has been examined closely. At pH 7.5 the quinonoid dibenz[*b,f*]azepin-2-one is produced together with acridine-9-aldehyde ⟨74CRV101⟩. 3-Chloro-10,11-dihydro-5*H*-dibenz[*b,f*]azepine yields a mixture of the quinoneimines (**174**; R=Cl) and (**175**) which with sodium dithionite reduce to the corresponding chlorohydroxydihydrodibenzazepines ⟨76JHC269⟩.

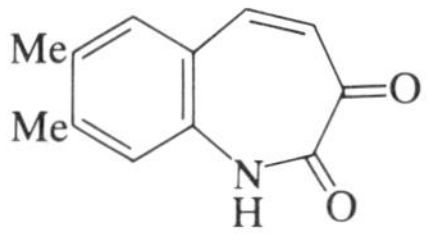

(**173**) 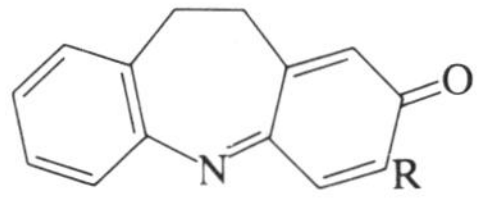 (**174**) 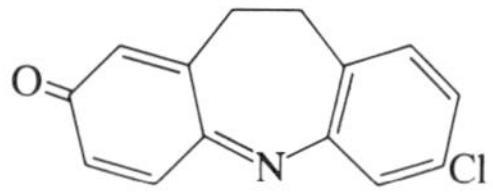 (**175**)

Ozonolysis of 3*H*-azepin-2-one affords malonic acid in quantitative yield ⟨65LA(682)1⟩.

Oxidation of 11*H*-dibenz[*b,e*]azepine with peracetic acid yields the dibenzazepin-6-one and *N*-formylacridan as the major products. Both products most probably result from an oxaziridine intermediate ⟨76JCS(P1)1291⟩.

Complete reduction of the azepine ring in the monocycles and in their benzologues is easily achieved with hydrogen and the usual metal or metal oxide catalysts. ⟨B-67MI51600, B-69MI51600⟩. Extensive use has been made of Raney nickel ⟨78JAP(K)78132586⟩, palladium–charcoal ⟨77BCJ2013, 78JOC315, 77JA5045⟩ and platinum oxide ⟨80JOC1950, 69JA658⟩. The marked resistance of dibenz[*b,d*]azepin-7-one towards reduction (Pt–H_2 or Pd/C–H_2) has been noted ⟨62JCS3468⟩.

Hydrodebromination accompanies the partial reduction of the aminobromoazepine (**111**; X=Br) to its amidine hydrobromide ⟨67JOC2367⟩. Partial reduction of 2-dimethylamino-3*H*-azepine to the 4,5-dihydro derivative is possible with hydrogen and 5% Pd–C ⟨73CC327⟩. Wolff–Kishner reduction of 2,3,4,5-tetrahydro-2-phenyl-1*H*-1-benzazepin-4-one proceeds normally to give the tetrahydroazepine ⟨79JOC4213⟩.

The reduction of dibenz[*b,e*]azepine-5,11-dione has been studied intensively ⟨65CCC445⟩. Clemmensen reduction, or reduction over palladium in acetic acid, yields the -11*H*-5-one, whereas reduction over Adams catalyst furnishes the 11-hydroxy-5-one. Sodium in butanol effects reduction to a mixture of the -11*H*-5-one and its 7,8,9,10-tetrahydro derivative.

5*H*-Dibenz[*b,f*]azepine is reduced quantitatively to its 10,11-dihydro derivative by magnesium in methanol ⟨79JOC3972⟩. In contrast, reduction of its *N*-acetyl-10-dibromo derivative to the *N*-ethyl compound without reduction of the C-10—C-11 double bond or loss of bromine is possible with aluminium hydride ⟨76JMC1345⟩. Platinum oxide in ethanol is an effective catalyst for the reduction of *exo* methylene groups to the corresponding methylazepines ⟨79JCS(P1)771, 79TL1133⟩.

5.16.3.9. Reactions of Substituents on the Azepine Ring

5.16.3.9.1 Reactions at carbon

Reactions at carbon substituents attached directly to the azepine ring are confined mainly to the modifications of carbonyl groups.

Reduction of 1-ethoxycarbonyl-1*H*-azepine with lithium aluminum hydride in ether at −15 °C yields the thermally unstable 1-(hydroxymethyl) derivative: in boiling ether further reduction occurs to *N*-methyl-1*H*-azepine, which subsequently dimerizes to the $[6+6]\pi$ adduct (see Section 5.16.3.2.3) ⟨B-69MI51600⟩. Surprisingly, the action of phenyllithium on 1-ethoxycarbonyl-1*H*-azepine produced the carbinol (**2**; R^1=CPh_2OH, R^2 = H), thermal decomposition of which led to the first characterization of 3*H*-azepine ⟨73CB1033⟩.

Hydrolysis of the trimethylsilyl urethane (**1**; R=CO_2SiMe_3), prepared by the action of trimethylsilyl iodide on the methyl carbamate (**1**; R = CO_2Me), with methanol at −78 °C yields the carbamic acid (**1**; R = CO_2H). On allowing a $CDCl_3$ solution of the carbamic acid to warm to room temperature, decarboxylation takes place to yield the deep red, highly unstable 1*H*-azepine (**1**; R=H) ⟨80AG(E)1016⟩.

In common with π-deficient heterocycles, *e.g.* 2-methylpyridine, it is found that the 2-methyl protons in 3*H*-azepines, *e.g.* (**176**), exchange rapidly (20 min at 80 °C) in D_2O. 7-Methyl protons also exchange but at a much slower rate (24 h at 80 °C) ⟨74JOC3076⟩. The acidity of 2-methyl protons is also demonstrated by the ease with which they undergo base-catalyzed condensation with benzaldehyde to give the benzylidene derivative.

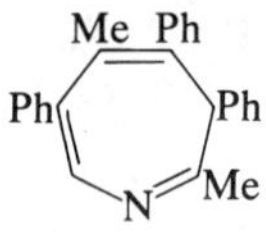

Me Ph
Ph Ph
N Me

(**176**)

The reactivity of substituents on the heterocyclic and the carbocyclic nuclei of dibenz[*b,f*]azepines has been reviewed ⟨74CRV101⟩. Polymers of *N*-acryloyl-10,11-dihydrodibenz[*b,f*]azepines have been prepared ⟨78MI51603⟩.

Nucleophilic ring opening of 10,11-epoxydibenz[*b,f*]azepine (**137**; X=O, R = Ac) (Section 5.16.3.3.4) is a practicable route to 10-substituted (Cl, Br, SCN) 11-hydroxy derivatives ⟨78CPB951⟩.

5.16.3.9.2 Reactions at heteroatoms

From the few examples reported, it is apparent, and to be expected, that functional groups on the carbocyclic ring(s) of benz- and dibenz-azepines behave normally. For example, nitro compounds can be reduced, and the amines diazotized and then modified by classical methods ⟨71JHC779⟩; ethers may be dealkylated ⟨80SC213, 80JMC975⟩ and bromo compounds converted into trifluoromethyl derivatives *via* an Ullmann reaction with CF_3I/Cu in DMF at 150 °C ⟨80SC213⟩. 10-Aminodibenz[*c,e*]azepine-6,11-dione has been used as a precursor for the synthesis of azepines annulated to other heterocycles, *e.g.* the benzazepinoquinoline (**177**) ⟨79AP708⟩.

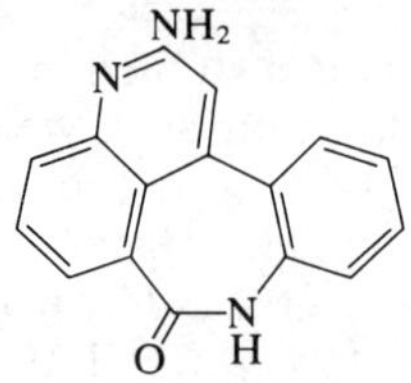

NH_2
N
O N H

(**177**)

HO Ac
N_2
N O
H

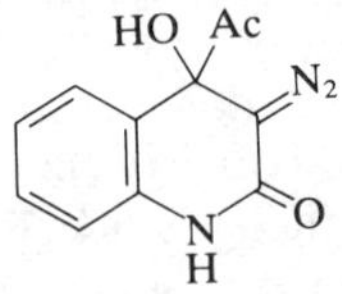

(**178**)

N-substituted azepines include the *N*-aminodibenz[*c,e*]azepine (**90**; R = NH_2), obtained by hydrolysis and decarboxylation of the *t*-butylurethane (**90**; R = $NHCO_2Bu^t$), and *N*-nitrosodibenz[*b,f*]azepine and its 10,11-dihydro congener ⟨74CRV101⟩. The amino compound reacts normally with phenyl isocyanate to yield the thiourea (**90**; R = NHCSNHPh) ⟨77MI51600⟩, and when subjected to the Bamford–Stevens reaction yields 9,10-dihydrophenanthrene ⟨57JA4427⟩. Interestingly, 10,11-dihydro-*N*-nitrosodibenz[*b,f*]azepine, unlike the fully conjugated azepine, undergoes a Fischer–Hepp rearrangement to the 2-nitroso derivative ⟨72AJC2451, 74CRV101⟩.

Primary amino groups attached to the azepine ring are uncommon and behave atypically. For example, attempts to diazotize the 2-amino-3*H*-azepine (**111**; X=Br) with sodium nitrite in 6M sulfuric acid failed ⟨67JOC2367⟩. Also, 3-amino-4-methyl-1*H*-1-benzazepine-2,5-dione with sodium nitrite in acetic acid suffers ring contraction to the diazoquinolone (**178**) and not formation of the diazoazepinone as reported originally ⟨75JOC3874⟩.

5.16.3.9.3 Reactions at fused benzene and other aromatic carbon rings

Direct electrophilic substitution of benz- and dibenz-azepines remains relatively unexplored. Most substituted benzazepines have been prepared from benzene precursors bearing the desired substituents ⟨74AHC(17)45⟩. The bulk of the reported electrophilic substitutions have been carried out on 5*H*-dibenz[*b,f*]azepine ⟨74CRV101⟩, MO calculations on which predict that substitution should occur at the 2- and 4-positions, *i.e. para* and *ortho* to the azepine ring nitrogen. These predictions are borne out by Friedel–Crafts alkylation and acylation studies, although it is apparent that a second alkyl group enters at the 8- rather than at the 4-position. Formylation under Vilsmeier conditions yields the 2-aldehyde. As noted earlier (Section 5.16.3.4), however, the 10,11-dihydro system exhibits different behavior and acylates at the benzylic 10,11-positions. Nitration with mixed acids of the 10,11-dihydro compound at low temperature affords the 3-nitro derivative ⟨74CRV101⟩. No direct halogenations of 5*H*-dibenz[*b,f*]azepine have been attempted. However, the 10,11-dihydro system yields initially the 2,8-dibromo, and subsequently, with an excess of bromine, the 2,4,6,8-tetrabromo compound.

Photobromination of the *N*-acetyl derivative (**124**; R^1=Ac; R^2 = H) with NBS as the brominating agent produces the 10-bromo derivative (**124**; R^1 = Ac, R^2 = Br) whereas with an acyl group at the 4-position, a mixture of the 2-bromo, 8-bromo and 2,8-dibromo derivatives is formed ⟨75JMC976⟩.

Calculations by the Pariser–Parr–Pople method on the quinoneimine obtained by Fremy's salt oxidation of dibenz[*b,f*]azepine indicate that the carbon adjacent to the carbonyl group is the most nucleophilic center. This is confirmed by nitration [$Cu(NO_3)_2$–AcOH] and bromination [NBS–$(PhCO)_2O_2$] studies, in which the 1-substituted derivatives are obtained. However, acylation by Vilsmeier or Friedel–Crafts reaction fails, extensive decomposition of the ring system taking place.

It is difficult to generalize on the behavior of benzazepines towards electrophilic and nucleophilic substitution as few examples have been reported ⟨74AHC(17)45⟩. However, it seems reasonable to assume that electrophilic substitution of hydrobenzazepines will be governed by the electronic influence of the nitrogen, in a manner similar to tetrahydroquinolines, and by the nature of any substituents already present on the carbocyclic ring. For example, Vilsmeier formylation of *N*-benzyl-2,3,4,5-tetrahydro-1*H*-1-benzazepine produces the 7-aldehyde (**179**; R^1=CHO, R^2 = H, R^3 = $PhCH_2$) ⟨76CB2743⟩, whereas chloromethylation of the *N*-acetyl-3-benzazepine (**180**; R^1=R^2 = H) produces a mixture of the mono- (**180**; R^1 = H, R^2 = CH_2Cl) and bis-chloromethyl (**180**; R^1 = R^2 = CH_2Cl) derivatives ⟨72JHC617⟩. 1,2,3,4-Tetrahydro-3*H*-3-benzazepine and its *N*-methyl and *N*-acetyl derivatives nitrate with ammonium nitrate and concentrated sulfuric acid at the 8-position ⟨71JHC779⟩.

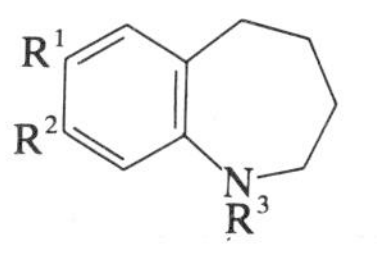

R^1, R^2, N, R^3

(179)

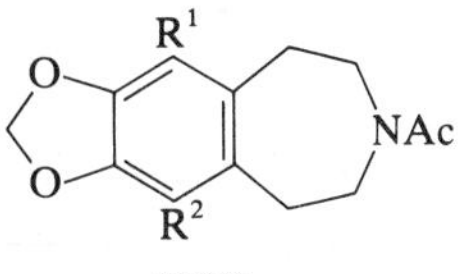

R^1, O, O, NAc, R^2

(180)

An exception is *N*-acetyl-6-hydroxy-1-benzazepine which brominates quantitatively at the 8-position, unlike the corresponding *N*-acetyl-5-hydroxytetrahydroquinoline which furnishes the 5-bromo derivative ⟨73T1115⟩.

Metallation of benz- and dibenz-azepines has been little investigated. 10,11-Dihydro-5*H*-dibenz[*b,f*]azepine lithiates at the 4-position, this being a convenient route to the 4-carboxylic acid ⟨74CRV101⟩.

Of current interest are the ring expansions of (2-halophenethyl)-3*H*-2-benzazepines, *e.g.* (**181**), in strong base (NaH–DMSO) to dibenz[*b,f*]benzazecines ⟨77CPB2875⟩ and indolo[1,2-*a*][3]benzazocines ⟨78TL1209⟩. The reaction proceeds *via* intramolecular ring closure of the azepine ring onto a benzyne intermediate.

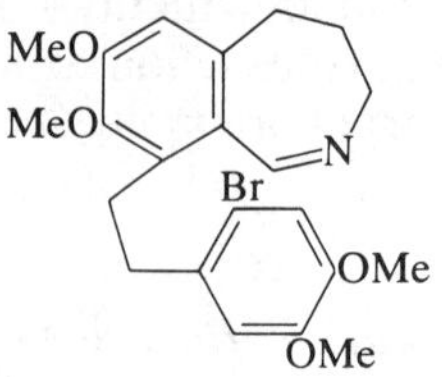

MeO MeO N Br OMe OMe

(**181**)

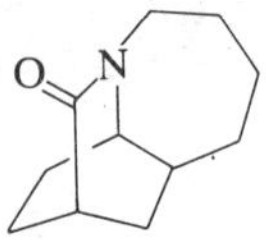

O N

(**182**)

The amino acid resulting from the reduction (Ru/C–H_2) of the benzene ring in the tetrahydrobenzazepinecarboxylic acid (**179**; $R^1 = CO_2H$, $R^2 = R^3 = H$) undergoes spontaneous lactamization to the azatricyclododecanone (**182**) ⟨76CB2743⟩.

5.16.4 SYNTHESIS OF AZEPINES

5.16.4.1 From Non-heterocyclic Compounds

5.16.4.1.1 By formation of one bond adjacent to nitrogen

(i) By displacement reactions

Intramolecular nucleophilic displacements of halogen from ω-halohexylamines is an obvious route to saturated azepines and was the route by which hexamethyleneimine was first prepared. The method has since been superseded as a route to simple hexamethyleneimines but is still used for highly substituted derivatives ⟨67MI51600⟩. However, high dilution techniques are necessary in order to avoid polymerization; intermolecular *N*-alkylations are minimized by using amine protecting groups, *e.g.* *N*-tosyl.

The procedure has been extended to the preparation of 1*H*-1-benzazepines from ω-(*o*-aminophenyl)butyl chlorides, 2*H*-2-benzazepines from benzylamines (**183**; R^1=Cl, $n = 3$, $m = 1$) ⟨74AHC(17)45⟩ and 5-hydroxy-3*H*-benzazepin-2-ones (**185**) from the keto acetamides (**184**; R^1=Me, R^2 = Br, $n = 1$, $m = 0$) ⟨75CPB114⟩.

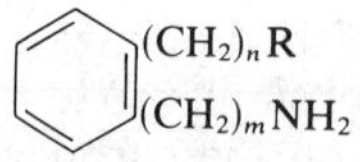

$(CH_2)_nR$ $(CH_2)_mNH_2$

(**183**)

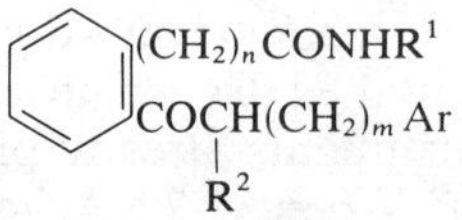

$(CH_2)_nCONHR^1$ $COCH(CH_2)_mAr$ R^2

(**184**)

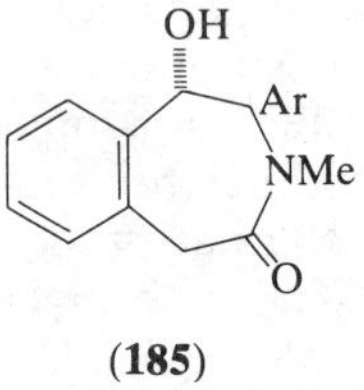

OH Ar NMe O

(**185**)

A novel route to 1*H*-1-benzazepin-9-carboxylic acid, which capitalizes on the reactivity towards metal exchange of a methoxy group *ortho* to an oxazoline function, is outlined in Scheme 15 ⟨81JOC783⟩.

Good yields of tetrahydro-5*H*-pyrido[2,3-*b*]azepines (**186**) are obtained by intramolecular S_N2 Chichibabin type reaction of δ-(3-pyridyl)butylamines in the presence of various basic catalysts ⟨73JHC39⟩.

A general route to azepinones is by intramolecular cyclization of ε-aminohexanoic acids or their derivatives ⟨80TL2443⟩. For caprolactam, however, yields are low and superior preparative methods are available ⟨70MI51600, B-75MI51601⟩. Surprisingly few methods are known for the synthesis of *C*-substituted caprolactams. A useful summary of existing

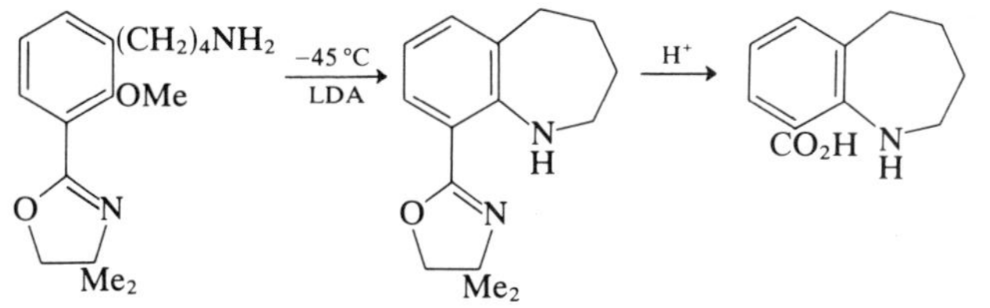

(CH2)4NH2 OMe O N Me2 —(−45 °C, LDA)→ N H O N Me2 —(H+)→ CO2H N H

Scheme 15

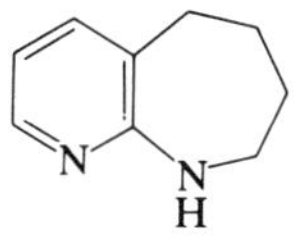

(186) 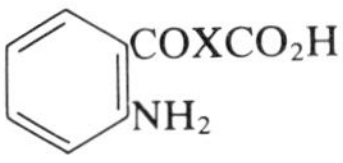 (187; COXCO2H, NH2)

methods and details of new and improved procedures is available ⟨76AJC2667⟩. For benzazepinones it has been found that the ease of lactam formation depends on the relative positions of the amino and carbonyl groups ⟨74AHC(17)45⟩. Cyclization is easy if either function is attached directly to the benzene ring, but becomes more difficult with increasing distance of the participating groups from the aromatic nucleus. γ-(*o*-Aminophenyl)butanoic acids cyclize spontaneously and in high yield to tetrahydro-1*H*-1-benzazepin-2-ones ⟨80CL487, 77JHC933⟩, whereas the isomeric *o*-(γ-aminopropyl)benzoic acids require heating at 190 °C to effect cyclization to tetrahydro-2*H*-2-benzazepin-1-ones. Similarly, more forcing conditions are needed to bring about cyclization of the *o*-(aminoethyl)phenylacetic acids (**183**; R=CO_2H, $n = 1$, $m = 2$) to 3*H*-benzazepin-2-ones ⟨72JHC609, 80T617, 80JHC417⟩. Likewise, β-(*o*-aminobenzoyl)propanoic acids (**187**; X = $(CH_2)_2$) yield 1*H*-1-benzazepine-2,5-diones ⟨78AJC439⟩, whereas the β-(*o*-aminobenzoyl)acrylic acid (**187**; X = CH=CH) fails to cyclize ⟨74AHC(17)45⟩.

Preparative routes to 5*H*-dibenz[*b,e*]azepine-6,11-diones (morphanthridinediones) are based mainly on the cyclization of 2-aminobenzophenone-2′-carboxylic acids and their derivatives ⟨55LA(594)89⟩. Studies on α-aminodiphenylmethane-2′-carboxylic acid reveal that cyclization to 5,11-dibenz[*b,e*]azepinone (**188**) is much slower at room temperature than the cyclizations of the analogous 2-aminobiphenyl-2′-carboxylic acid and the 2′-aminobiphenylacetic acid (**189**), which at room temperature cyclize spontaneously to phenanthridone and dibenz[*b,d*]azepin-6-one (**190**) respectively ⟨61JOC1329⟩. The hydrogen bromide-induced cyclization of dinitriles (Scheme 16) is adaptable to the synthesis of 2-amino-7-bromo-3*H*-azepines and 5*H*-dibenz[*c,e*]azepin-7-ones ⟨67JOC3325⟩. Apparently, for unsymmetrical dinitriles cyclization is such as always to give the azepine with the bromo substituent attached to the carbon of the α,β-unsaturated nitrile as exemplified in Scheme 16.

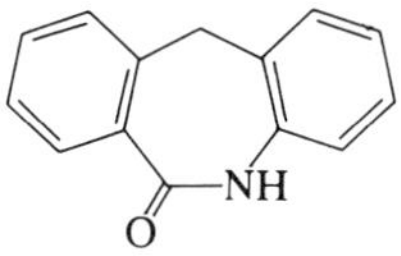

(188) 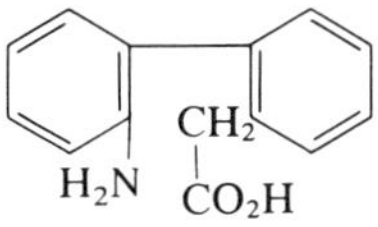 (189) 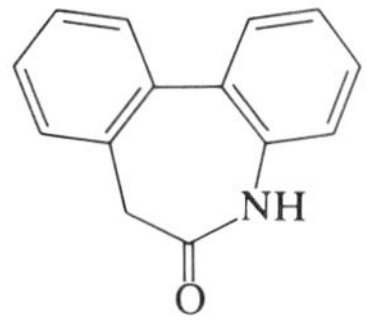 (190)

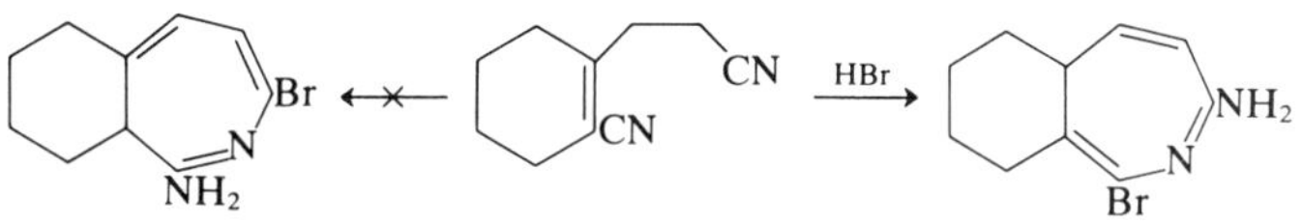

Br NH2 ←×— CN CN —(HBr)→ NH2 N Br

Scheme 16

Analogous ring closures are to be found in the cyclizations of cinnamonitriles (**191**; R^1 = Cl, R^2 = H and R^1 = OH, R^2 = Ar) to 2-benzazepin-1-ones (**192**) and (**193**; R = aryl) with hydrogen chloride and polyphosphoric acid respectively ⟨74AHC(17)45⟩. Polyphosphoric acid has also been used to effect cyclization of the mono- or di-nitriles (**191**; R^1 = OH,

R^2 = H or CN) to 2-benzazepine-1,3-diones (**193**; R = H or $CONH_2$) ⟨72JOC3955⟩, and γ-(*o*-aroylphenyl)propionamides (**184**; R^1=R^2 = H, *n* = 2) to the 1-methylene-2-benzazepin-3-ones (**194**) ⟨75HCA2228⟩.

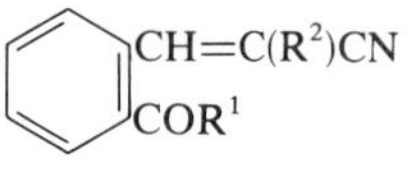

(191) 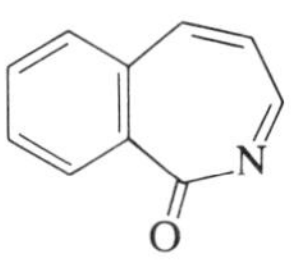 **(192)** 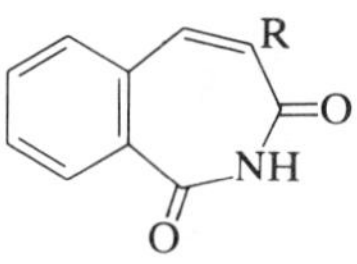 **(193)** 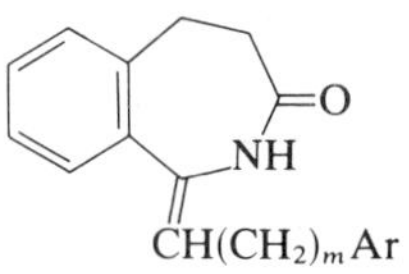 **(194)**

Beckmann rearrangement of cyclohexanone oximes is a useful route to azepin-2-ones and many examples are known ⟨60OR(11)1⟩, of which the most important is the industrial preparation of hexahydro-1*H*-azepin-2-one (caprolactam) ⟨B-75MI51601, 70MI51600⟩. Many catalysts have been employed but polyphosphoric acid is the reagent of choice for laboratory scale reactions. Problems arise with the oximes of unsaturated ketones as in many cases competing Semmler–Wolff aromatization dominates. Rearrangements of the oximes of α-tetralones and 1,4-naphthoquinones pose interesting problems as alkyl (path a) and aryl (path b) migrations are possible with formation of either 2-benzazepin-1-ones (**197**) or 1-benzazepin-2-ones (**196**; Scheme 17).

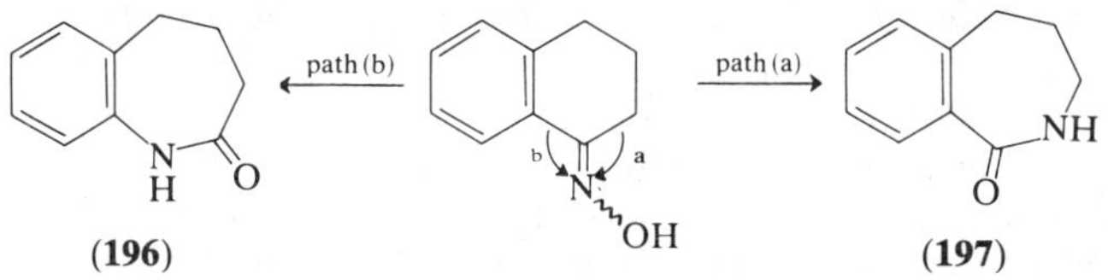

(196) **(197)**

Scheme 17

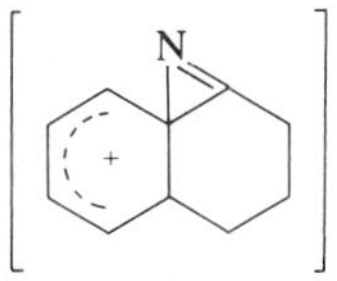

(195) 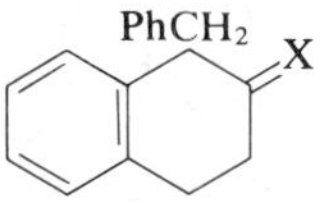 **(198)**

Generally, aryl migration is preferred ⟨*e.g.* 77JHC933⟩, and mechanistic studies by Huisgen and coworkers implicate the resonance stabilized cation (**195**) as a reaction intermediate ⟨74AHC(17)45⟩. However, many cases are known in which substantial alkyl migration also occurs, with the result that mixtures of benzazepinones are formed ⟨75CPB1917, B-71MI51600⟩. Oximes of β-tetralones also offer the possibility of two Beckmann rearrangement products, namely 2-benzazepin-3-ones and 3-benzazepin-2-ones. Normally mixtures are obtained but there is evidence ⟨79JHC1317⟩ to show that bulky substitutents (*e.g.* benzyl) at the α-position exert steric control on the rearrangement, and, by forcing the oxime (**198**; X = NOH) into the least hindered configuration, *i.e.* *anti* to the benzyl group, promote C-2—C-3 bond migration and hence selective formation of 2-benzazepin-3-ones. Beckmann rearrangements on phenanthraquinone and anthraquinone monoximes are practicable routes to dibenz[*b,d*]- ⟨70ACS2252⟩ and dibenz[*b,e*]- ⟨79MI51601, 73JOC809⟩ azepinones respectively.

A novel variation on the Beckmann rearrangement is the ring expansion of the benzenium ion (**199**; X = *p*-Ts), prepared as outlined in Scheme 18, to the 1*H*-azepine (**200**; X=*p*-Ts) ⟨81AG(E)699⟩.

Of equal importance to the Beckmann rearrangement for the preparation of azepines and azepinones is the Schmidt reaction of ketones. Again, however, with unsymmetrically substituted cyclohexanones or with bicyclic ketones such as α-tetralone, two azepines are possible. α-Substituted cyclohexanones invariably yield the product of attack by azide ion at the least hindered side of the protonated ketone as exemplified in Scheme 19 ⟨B-67MI51600⟩.

Three of the four possible azepinediones have been isolated from the Schmidt reaction on 2-methoxy-5-methyl-*p*-benzoquinone; 3-methyl-6-methoxy-1*H*-azepine-2,5-dione is the missing isomer ⟨74CJC3327⟩.

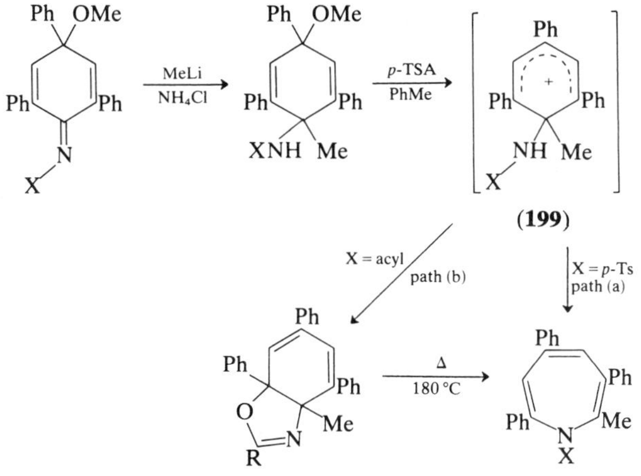

Scheme 18

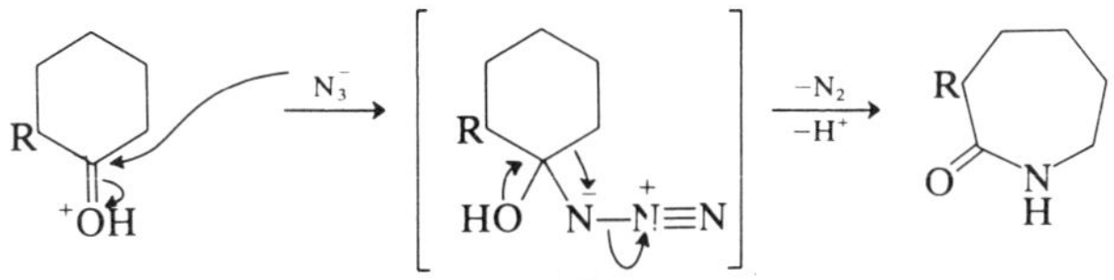

Scheme 19

Investigations by Uyeo and coworkers ⟨74AHC(17)45⟩ on the effect of substituents *para* to the carbonyl group on the Schmidt ring expansion of α-tetralones were inconclusive, but suggested that strong electron donors (*e.g.* MeO) favor alkyl migration ⟨*e.g.* 73JMC516⟩. *meta*-Substituents appear to favor aryl migration.

Schmidt reaction on β-tetralone yields a 2:1 mixture of the 2-benzazepin-3-one and 3-benzazepin-2-one, whereas the 1-benzyl-2-tetralone (**198**; X = O) yields only 1-benzyl-2-benzazepin-3-one ⟨76HCA2059⟩.

Anomalous ring expansions occur with 2-alkoxynaphtho-1,4-quinones, *e.g.* (**202**; R^1 = Me, R^2 = H), in that attack by azide ion takes place at the more hindered carbonyl function (*i.e.* C-1) to give the 1*H*-1-benzazepine-2,5-dione (**203**; R^1=OH, R^2 = R^3 = H). In contrast, the 6-methoxy derivatives (**202**; R^1 = alkyl; R^2 = MeO) suffer attack at the 4-carbonyl group to yield benzazepinedione (**203**; R^1 = H, R^2 = OMe, R^3 = OH) ⟨74CJC610⟩. These results have led the authors to conclude that electronic as well as steric factors are important in the Schmidt reaction in determining the direction of ring expansion.

1-Benzazepine-2,5-diones and not as was thought originally the isomeric 2-benzazepine-1,5-diones are the major products from the Schmidt reaction on naphthoquinones ⟨74AHC(17)45, 72JOC208⟩. Unsymmetrically substituted anthraquinones produce mixtures of isomeric dibenz[*b*,*e*]azepinediones ⟨65JMC74, 79AP662⟩.

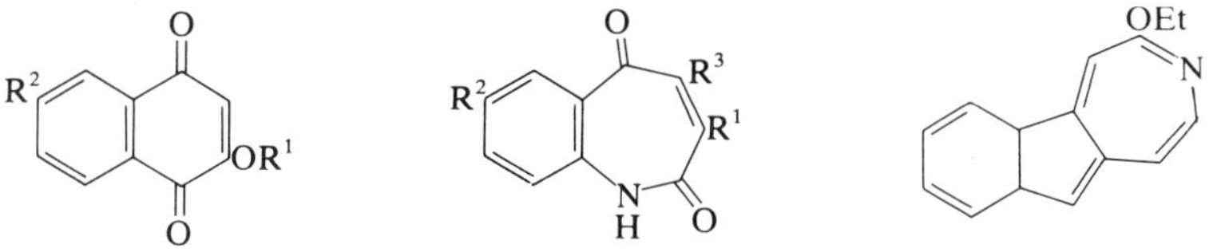

(**202**) (**203**) (**204**)

The Beckmann and Schmidt reactions are applicable to the synthesis of azepines fused to other heterocyclic and carbocyclic systems, such as pyridoazepinones ⟨70JA203, 76AP550, 78JHC249⟩, indoloazepinones ⟨78CB1780⟩ and the 14π aromatic azabenzazulennones (**204**) ⟨80BCJ3232⟩. The general chemistry of azaazulenes has been reviewed ⟨81H(15)547⟩.

The acid-catalyzed ring expansions of tertiary cyclohexylazides to tetrahydroazepines may be viewed as an intramolecular version of the Schmidt reaction and in general proceed in high yield. Unfortunately, the reaction is of little synthetic value as the imines formed are readily hydrolyzed to ω-aminoketones in the strong acid media ⟨B-67MI51600⟩. However, the reaction is successful with 9-azido-9-phenylacridone, which on thermolysis or photolysis ring-expands to a mixture of 9-phenyliminoacridone (15–35%) and 6-phenyl-11*H*-dibenz[*b*,*e*]azepin-11-one (65–85%) ⟨76TL3141⟩.

5.16.4.1.2 By formation of one bond β, γ or δ to nitrogen

(i) By displacement reactions

Under this heading are to be found examples of benzazepine formation by the classical named syntheses normally associated with the formation of six-membered heterocycles, such as intramolecular Friedel–Crafts alkylation and acylation, Dieckmann, acyloin, Bischler–Napieralski and Pictet–Spengler cyclizations. However, only selected and recent examples of these reactions are given as the preparations of benzazepines by these and other routes have been reviewed comprehensively ⟨74AHC(17)45⟩. In fact, these methods have met with only limited success in forming seven-membered rings since, if the opportunity arises, five- and six-membered rings are formed preferentially. For example, cyclization of *N*-tosylphenylaminobutanoic acid (**205**; R=*p*-Ts) or its acid chloride yields pyrrolidones rather than the desired 1-benzazepinones, even with electron-donating substituents in the phenyl ring. More successful is the cyclization of *N*-(benzenesulfonyl)-*N*-(γ-phenylpropyl)-glycine chloride (**206**) in nitrobenzene with an aluminum chloride catalyst. Cyclization is accompanied by loss of CO and HCl to give the tetrahydro-2-benzazepine (**207**; $R^1=SO_2Ph$, $R^2=R^3=H$) in 65% yield. Alternatively, the carboxylic acid can be cyclized directly by heating with phosphorus pentoxide in xylene. In contrast, cyclization of the analogous 3-(benzylamino)propanoic acids or their acid chlorides to the 2-benzazepin-5-ones proved to be much more difficult. After many failures it was found that cyclization could be achieved with aluminum chloride catalyst only at temperatures between −70 and +20 °C ⟨74HCA(17)45⟩.

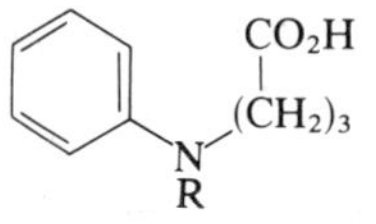

CO_2H $(CH_2)_3$ N R

(**205**)

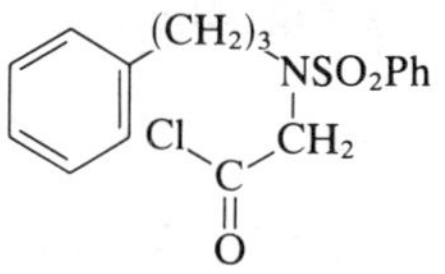

$(CH_2)_3$ NSO_2Ph Cl CH_2 C O

(**206**)

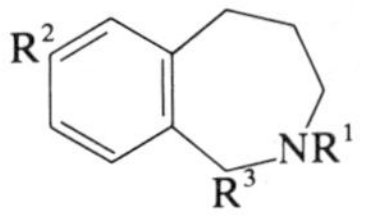

R^2 NR^1 R^3

(**207**)

The formation of 3-benzazepinones from β-(phenethyl)glycine chlorides is also temperature sensitive. Early efforts using aluminum chloride or PPA as catalyst produced mainly tetrahydroquinolones as the major products along with only trace amounts of benzazepinones. Eventually, it was found that the amount of catalyst is critical in determining the size of the heterocyclic ring produced, and that for benzazepine formation an optimum temperature of −10 °C is necessary.

A report that *N*-(3-halopropyl)benzylamines cyclize to tetrahydro-2-benzazepines in hot decalin in the presence of an aluminum chloride catalyst is incorrect; 4-methyltetrahydroisoquinolines are the major products (>60%) along with only minor amounts (<20%) of the seven-membered heterocycle ⟨80JOC2000⟩.

Related to these intramolecular Friedel–Crafts cyclizations are the aluminum chloride and PPA catalyzed ring closures of 3-phenylpropyl iso- and thioiso-cyanates to 2*H*-2-benzazepin-1-ones and their -1-thiones respectively ⟨79H(12)1311, 77JCS(P1)2357⟩. The method has also been applied to the cyclization of *o*-isocyanatodiphenylmethanes to 5,6-dihydro[*b*,*e*]azepin-6-ones ⟨65HCA336⟩.

Bischler–Napieralski cyclizations of *N*-acyl-γ-phenylpropylamines, with P_2O_5 in xylene, with PPA (or its ester), or with $POCl_3$ and $AlCl_3$ mixtures, are valuable routes to 4,5-dihydro-3*H*-2-benzazepines ⟨76HCA623, 77HCA1644⟩, and to 11*H*-dibenz[*b*,*e*]azepines ⟨76JCS(P1)1279⟩.

Condensation of β-phenylpropionamides with arylaldehydes, and then cyclization to 1-aryl-2-benzazepin-3-ones (Pictet–Spengler reaction) is successful in PPA. The reaction fails, however, with aliphatic aldehydes. An analogous condensation between β-(*m*-hydroxyphenyl)propylamine and then cyclization in hot butanol yields 1-aryltetrahydro-2*H*-

2-benzazepines (**207**; R^1=H, R^2 = OH, R^3 = Ar). Yields are claimed to be superior to those obtained by Bischler–Napieralski reaction, but the success of the cyclization is dependent on the presence of the hydroxy group; in fact, the reaction fails with the phenol ether ⟨74JCS(P1)2602⟩.

Cyclizations of secondary and tertiary alcohols in strong acid media (H_2SO_4, PPA, $MeSO_2OH$) proceed *via* electrophilic attack of the carbocation intermediate so-formed, at an activated aromatic ring, and are useful synthetic routes to 3-benzazepines, as typified in the cyclization of (**208**) to (**209**) ⟨78CR(C)(286)83, 80JMC975⟩, 3-benzazepin-2-ones ⟨81HCA373, 79JHC1525⟩ and 11-phenyl-6,11-dihydro-5*H*-dibenz[*b*,*e*]azepines ⟨81H(15)421⟩.

Dieckmann cyclizations of diesters of type (**210**) are catalyzed by potassium *t*-butoxide in toluene, sodium in xylene, or sodium hydride in DMF, and produce 1-benzazepinones in good yields. The method is also applicable to the synthesis of azepines fused to other heterocycles, *e.g.* pyrrolo[2,3-*b*]azepin-4-ones (**211**) ⟨81H(16)399⟩.

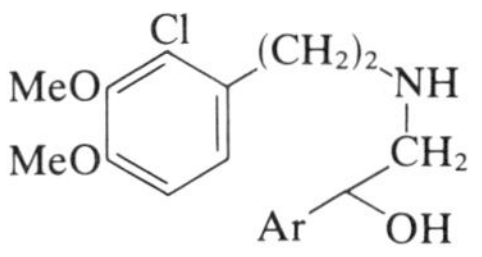

(**208**) 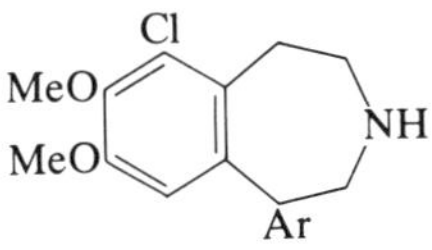 (**209**) 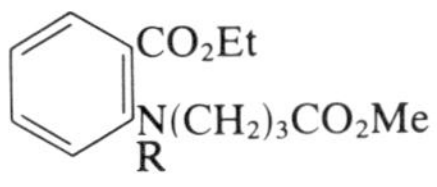 (**210**) 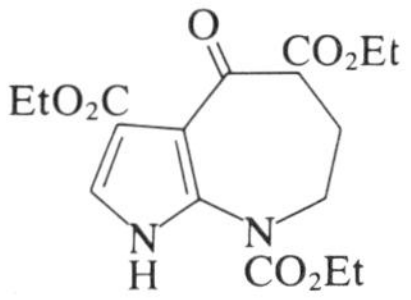 (**211**)

So far only moderate yields of benzazepines have been obtained by metal complex-catalyzed cyclizations of *N*-substituted *o*-haloanilines, *e.g.* Scheme 20 ⟨79TL1133⟩. However, the indications are that such techniques could become highly practicable synthetic routes to benzazepines ⟨79JCS(P1)771⟩.

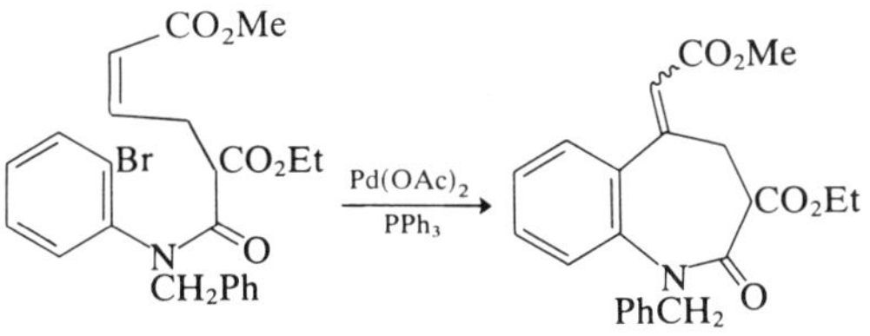

Scheme 20

The ring closure of 2-amino-2′-halobiphenyls with copper or copper–bronze in boiling DMF is a standard route to 10,11-dihydrodibenz[*b*,*f*]azepines ⟨74CRV101⟩.

Pschorr cyclization of diazotized *N*-mesyl-*o*-aminodiphenylamine yields *N*-mesyl-5,7-dihydro-6*H*-dibenz[*c*,*e*]azepine ⟨73AJC1307⟩. Diazonium compounds also feature in the preparation of 11*H*-dibenz[*b*,*e*]azepines *via* nitrilium ions (Scheme 21) ⟨72CB1634⟩, and in the only synthesis reported so far of a tribenz[*b*,*d*,*f*]azepine (**10**; R=Ph) ⟨72CB880⟩.

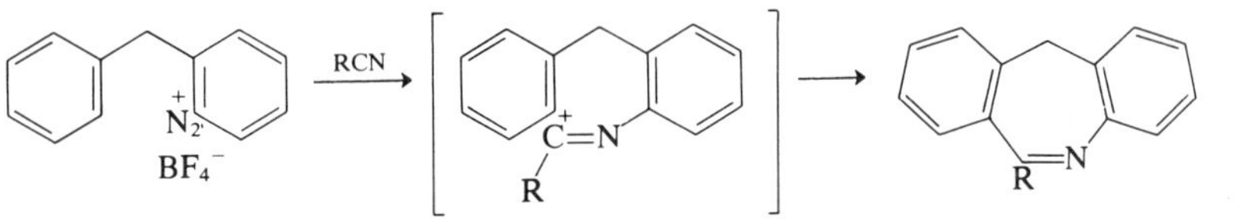

Scheme 21

(ii) By valence isomerization

Historically the first fully unsaturated azepine was obtained by Wolff in 1912 by the decomposition of phenyl azide in aniline. However, the actual structure of this product, named 'dibenzamil', remained unknown until 1955, when Huisgen and coworkers showed it to be an anilinoazepine, originally formulated as the 7-anilino-2*H*-azepine, but later corrected, mainly on the basis of ^{1}H NMR evidence, to the 2-anilino 3*H*-tautomer (**221**; R^1=PhNH; R^2 = H). Subsequently, the thermolysis and photolysis of aryl azides in primary, secondary and, in certain instances, tertiary amine solution has become a standard method for the preparation of 2-amino-3*H*-azepines ⟨79AG(E)900, 81AHC(28)231⟩.

The reaction pathway for the ring expansion as proposed by Huisgen and coworkers is outlined in Scheme 22, and involves generation of the electron deficient singlet nitrene (**213**), which is held to be in dynamic equilibrium with the benzazirine (**214**). Nucleophilic addition of amine to the reactive imine bond followed by electrocyclic ring opening of the azanorcaradiene (**215**) is thought to give initially a 1*H*-azepine, which subsequently rearranges to the more stable 3*H*-tautomer.

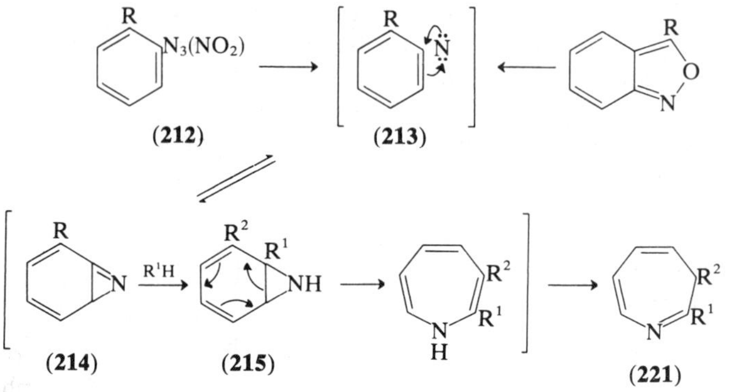

(**212**) (**213**) (**214**) (**215**) (**221**)

Scheme 22

Despite patient and exhaustive effort by many researchers, all attempts to isolate or trap a benzazirine intermediate (**214**) have so far failed, and unequivocal evidence for their participation in either the photolytic or thermal decomposition of aryl azides is still awaited. Evidence in favor of the proposed reaction pathway (Scheme 22) comes from the work of Sundberg and coworkers, who succeeded in identifying 3-alkyl-2-diethylamino-1*H*-azepines as oxygen-sensitive, metastable intermediates in the photolysis of *o*-alkylphenyl azides in diethylamine ⟨72JA513⟩. Later studies on the flash photolysis of aryl azides in dialkylamine solution provided kinetic data which not only confirmed the 1*H*- to 3*H*-azepine tautomerism, but also strongly supported the involvement of a benzazirine intermediate ⟨74JA7491⟩.

However, from elegant labelling studies by Wentrup and coworkers on the gas-phase thermal interconversion of phenylnitrene and 2-pyridylcarbene, it became evident that an alternative and equally viable intermediate in these nitrene ring expansions is the azacycloheptatrienylidene (**216**) or its tautomer, the cumulated system (**217**) ⟨81AHC(28)231⟩.

(**216**) (**217**) (**218**)

Chapman and coworkers ⟨79RTC334⟩ from studies on the low temperature photolysis of phenyl azide in an argon matrix at 8 K produced convincing IR spectroscopic evidence for the formation of 1-aza-1,2,4,6-cycloheptatetraene (**217**), rather than a benzazirine intermediate. In fact, these workers have reinterpreted the formation of 2-alkylamino-3*H*-azepines on the basis of amine addition to the cumulated system (**217**) rather than the benzazirine.

The problem, however, is not yet settled as contemporary studies on the photolysis of bicyclic azides, *e.g.* 1-naphthylazide, in an argon matrix at 12 K demonstrate clearly the existence of two transient species, the IR spectra of which correspond in one case to the cumulated system analogous to (**217**) and in the other, which absorbs strongly at 1708–1736 cm^{-1}, to the azirine structure (**218**) ⟨80CC499⟩.

For a detailed and authoritative review of these, and related, azide decompositions reference ⟨81AHC(28)231⟩ should be consulted.

It is of interest to note that a cumulated system rather than a benzazirine was suggested by Sato ⟨68BCJ2524⟩ to explain the photoinduced ring expansion of the azidocyclohexenone (**219**) in aqueous THF to the azepinedione (**220**; Scheme 23). Better yields of the dione have since been obtained by irradiating the azide in wet benzene ⟨72CPB871⟩.

Scheme 23

The nature of the intermediate involved in the thermolysis and photolysis of aryl azides in solution under the much less extreme conditions employed for the synthesis of 3*H*-azepines is still an open question. Notwithstanding, the method has been used extensively for the synthesis of 2-amino-3*H*-azepines. In addition, and despite early disappointments, the method is now applicable to the synthesis of benzazepines from naphthyl azides and that of pyridoazepines from quinolyl and isoquinolyl azides ⟨82JCS(P1)431, 79AG(E)900⟩.

The decomposition (photolytic or thermal) of aryl azides in the presence of nucleophiles other than amines, as a preparative route to 3*H*-azepines, has met with only limited success. For example, photolysis of phenyl azide in liquid ammonia gives 2-amino-3*H*-azepine (**221**; $R^1 = NH_2$, $R^2 = H$) in poor yield (25%), whereas in ethereal hydrogen sulfide only a 5% yield of 3*H*-azepine-2-thione is produced ⟨66T(22)81⟩. However, by photolyzing aryl azides bearing electron-withdrawing groups (**212**; R=CONHR, CO_2R, SO_2NH_2, CN, CF_3), particularly at the *ortho* position, in alcohol solution practicable yields (50–80%) of 3-substituted 2-alkoxy-3*H*-azepines are obtained ⟨78JCS(P1)191, 71JCS(C)2317, 79MI51602, 79TH51600⟩. Improved yields of 2-alkoxy-3*H*-azepines and 3*H*-azepin-2-ones are also noted if photolyses are carried out in strongly basic media or in the presence of crown ethers ⟨79AG(E)900⟩.

An alternative and major method of preparing 2-amino-3*H*-azepines utilizes the deoxygenation of nitroso- and nitro-arenes in the presence of a secondary amine. Deoxygenation is accomplished using a trivalent phosphorus compound, *e.g.* triethyl phosphite, with or without an inert solvent, and with nitroso compounds proceeds under mild conditions. The more readily available nitro compounds, however, require much harsher conditions, a factor which originally hampered their synthetic potential. However, the discovery of more powerful deoxygenating agents, *e.g.* see Scheme 24, now enables the deoxygenative conversion of nitroarenes to 3*H*-azepines to be carried out smoothly and efficiently under relatively mild conditions ⟨B-79MI51603⟩.

i, $(EtO)_2MeP$ or $Ph_2(OEt)P$; ii, R_2NH

Scheme 24

The intermediacy of a discrete nitrene species in these deoxygenations has been the subject of some debate. Supporting evidence comes from studies on the deoxygenation of a series of mainly *m*-nitroarenes in the presence of diethylamine, which reveal that the percentage composition of the resulting mixture of 4- and 6-substituted 3*H*-azepines (Scheme 24) is, within experimental error, identical to that of those obtained by thermolysis of the corresponding aryl azides under similar conditions. Results suggest that when substituent X is electron withdrawing, the 6-substituted 3*H*-azepine (**223**) is preferred, whereas electron donors favor formation of the 4-substituted isomer (**222**). Directing effects of substituents in arylnitrene ring expansions have been discussed critically ⟨81AHC(28)231⟩.

The deoxygenative ring expansion of nitrobenzene with tri-*n*-butylphosphine in butanol ⟨77BCJ2013⟩, or with phosphorus trichloride and di-*n*-butylamine in hexane, followed by catalytic reduction and hydrolysis of the resulting 2-butoxy- or 2-butylamino-3*H*-azepines have been patented as methods for the production of caprolactam ⟨78JAP(K)78132586, 77GEP2647936 respectively⟩.

Arylnitrenes and subsequently 2-dialkylamino-3*H*-azepines have also been generated by the decomposition of oxaziranes ⟨B-70MI51601⟩ and (in almost quantitative yield) by thermolysis (100 °C) of *N,O*-bis(trimethylsilyl)phenyl hydroxylamine, in the presence of dialkylamines ⟨76JOC3381⟩.

A report that 3-cyclopropyl-2-diethylamino-1*H*-azepine is formed either by thermolysis of *o*-azidocyclopropylbenzene, or by deoxygenation of *o*-nitrosocyclopropylbenzene with trimethyl phosphite in diethylamine, is unique, since, if correct, it is the first example of a stable *N*-unsubstituted 1*H*-azepine ⟨79JOU1272⟩.

(iii) By radical combination

A recent development in the synthesis of 3*H*-3-benzazepin-2-ones has been the photocyclization of *N*-(chloroacetyl)phenethylamines (Scheme 25). Ring closure is by homolysis of the alkyl halide followed by intramolecular coupling of the alkyl radical with an aromatic radical cation. Yields are good, especially with a stabilizing electron-donating group (MeO, NMe_2) at the position *meta* to the ethylamino function (*i.e. ortho* or *para* to the site of cyclization). Isomeric benzazepinones are normally obtained (Scheme 25) with *meta*-substituted phenethylamines ⟨80H(14)11⟩.

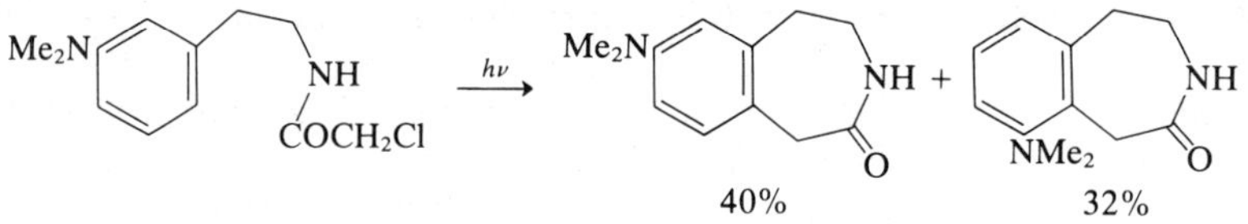

Me_2N NH $COCH_2Cl$ $\xrightarrow{h\nu}$ Me_2N NH O (40%) + NH NMe_2 O (32%)

Scheme 25

Other light-induced ring closures to benzazepines include the intramolecular radical arylation of *N*-substituted enaminones (**224**) to hydrodibenz[*b,d*]azepinones (**225**) ⟨82JOC482⟩ and the formation of dibenz[*c,e*]azepines from *o*-iodo-bis-benzylamines (**226**) ⟨75JOC2883⟩.

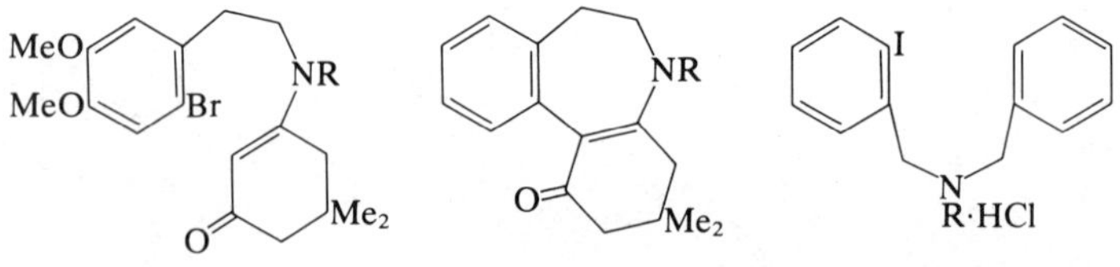

MeO MeO Br NR O Me_2 **(224)**   NR O Me_2 **(225)**   I N R·HCl **(226)**

5.16.4.1.3 By formation of two bonds from [6+1] fragments

(i) By initial electrophilic attack

The ring expansion of arenes by electron-deficient singlet nitrenes is by far the most versatile synthetic route to 1*H*-azepines. The first 1*H*-azepines were prepared independently in 1963 by Hafner, and by Lwowski, and their coworkers. They found that ethoxycarbonylnitrene (Scheme 26, path a; $R=CO_2Et$), generated by photolysis of ethyl azidoformate, adds to benzene to give initially the unstable azanorcaradiene (**227**), electrocyclic ring opening of which yields the stable 1*H*-azepine (**228**; R=CO_2Et). The reaction can be regarded as the intermolecular equivalent of the rearrangement of arylnitrenes to 3*H*-azepines (Section 5.16.4.2(ii)), the main difference being that prototropic shift to the 3*H*-azepine is prevented by the *N*-substituent, a feature that also stabilizes the 1*H*-azepine.

RN̈ path (a) path (b) ii, i NR NCO₂R path (b) i I NHCO₂R

(**227**) (**247**) (**248**)

N R

(**228**)

i, powdered NaOMe, THF; ii, Br_2, CH_2Cl_2 at −70 °C

Scheme 26

The nitrene (RN) is usually generated by thermal or photolytic decomposition of an azidoformate, although other useful methods such as base-catalyzed α-elimination of arylsulfonate ion from *N*-(arylsulfonyloxy)urethanes ⟨B-70MI51601, 74CL1209⟩ and thermal deoxysilylation of *N,O*-bis(trimethylsilyl)-*N*-ethoxycarbonylhydroxylamine[EtO_2CN-($SiMe_3$)$OSiMe_3$] ⟨81JOC342⟩ have also been employed.

In general, the reaction is limited to the preparation of 1*H*-azepines bearing an electron-withdrawing group, and nitrenes (R=CO_2R, CO_2Ph, CN, SO_2Me, SO_2Ar) and, more recently, imidoylnitrenes [R = C(=NCN)OEt] ⟨80TL727⟩ have been used with varying degrees of success. Usually arylsulfonylnitrenes yield sulfonamides rather than azepines, although moderate yields of *N*-sulfonylazepines have been obtained with alkylsulfonyl azides ⟨81JA1525, B-73MI51600⟩. A twenty-fold increase in the yield of *N*-(*p*-tosyl)-1*H*-azepine (**228**; R=*p*-Ts) has been achieved, however, by carrying out the thermolysis of *p*-tosyl azide in benzene under pressure (80–90 atm) in a nitrogen atmosphere ⟨81CC790⟩. Significant increases in the yield of 1-ethoxycarbonyl-1*H*-azepines are possible by photolyzing the azidoformate in the presence of a singlet nitrene stabilizer such as dichloromethane ⟨78JOC2725⟩.

Formation of 1-aryl-1*H*-azepines is rare and occurs only with those arylnitrenes made sufficiently electrophilic by an electron-withdrawing, *e.g.* CN, NO_2 or CF_3, *ortho* or *para* substituent. Even so, these docile nitrenes attack only electron-enriched arenes (*e.g.* mesitylene or *N,N*-dimethylaniline) and are of minor synthetic importance ⟨B-73MI51600⟩. More reactive are π-deficient heteroarylnitrenes, and moderate yields (15–40%) of 1-heteroaryl-1*H*-azepines, *e.g.* (**228**; R=4,6-dimethoxy-1,3,5-triazin-2-yl), may be obtained by the photodecomposition of 2-azido-4,6-dimethoxy-1,3,5-triazine in a variety of aromatic substrates ⟨81BCJ301⟩. Interestingly, intramolecular insertion of arylnitrenes into arenes is more common and has been used for the synthesis of fused azepines, *e.g.* the azepinoindoles (**229**) from *o*-azidodiphenylmethanes ⟨81JCS(P1)1132⟩.

R N N–N O O O

(**229**) (**230**)

There are two distinct disadvantages associated with these ring expansions. Firstly, nitrene attack on a mono- or di-substituted benzene is non-regiospecific and mixtures of difficultly separable, isomeric 1*H*-azepines are formed ⟨70JHC1249⟩, although some measure of regiospecificity is observed if bulky groups, *e.g.* *t*-butyl, are present ⟨71JHC729⟩. Secondly, 1*H*-azepines are acid sensitive, and traces of acid in the reaction mixture cause aromatization (see Section 5.16.3.3.1). This is clearly demonstrated in the ring expansion of 1,3-dimethoxybenzene with *N*-phthalimidonitrene. Generation of the *N*-nitrene by the usual method, *i.e.* lead tetraacetate oxidation of *N*-aminophthalimide, gives mainly the insertion product,

N-(2,4-dimethoxylanilino)phthalimide, whereas formation of the *N*-nitrene under acid-free conditions, *i.e.* by thermolysis of the preformed benzofuran adduct (**230**), produces the 7-(*N*-phthalimido)-3*H*-azepine (**69**) as discussed in Section 5.16.3.2.2 ⟨73CC67⟩.

An interesting development of this azepine synthesis is the intramolecular addition of acylnitrenes to arenes, as exemplified in Scheme 27 for the nitrene (**231**) generated from benzyl azidoformate by spray pyrolysis ⟨80CC1161⟩.

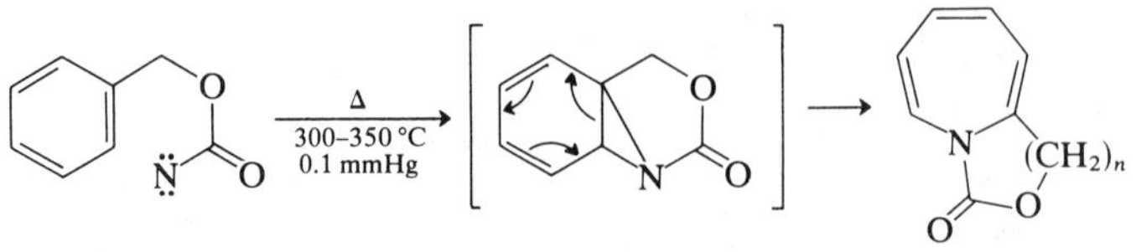

Δ, 300–350 °C, 0.1 mmHg

(**231**) (**232**) $(CH_2)_n$

Scheme 27

Molecular models reveal that the five-membered ring constrains the azepine nucleus in a near planar conformation with the result that the oxazoloazepine (**232**; $n=1$), unlike most other *N*-acyl-1*H*-azepines, is unstable and dimerizes spontaneously in a highly selective [6+4] manner (see Section 5.16.3.2.3). In contrast, the oxazinoazepine (**232**; $n=2$) formed in 26% yield by FVP of phenethyl azidoformate is less strained and is stable as the monomer ⟨81CC1087⟩. Likewise, the benzoxazinoazepine (**233**), derived from biphenyl-2-yl azidoformate, is thermally quite stable, but in acid rearranges to 2-(*o*-hydroxyphenyl)-3*H*-azepine ⟨81CC241⟩.

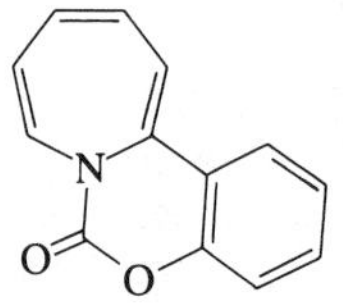

(**233**) 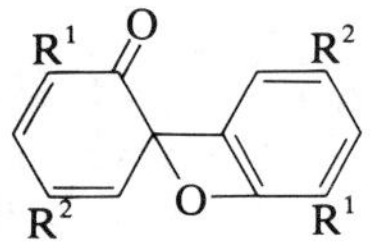 (**234**)

3*H*-Azepin-2-ones are prepared by the ring expansion of 2,6-dialkylphenolates with ethereal chloramine at −70 °C (Scheme 28) ⟨B-69MI51600⟩, and by the mechanistically similar base-promoted ring opening and rearrangement of spiroquinol ethers (**234**) ⟨81JOC4077⟩.

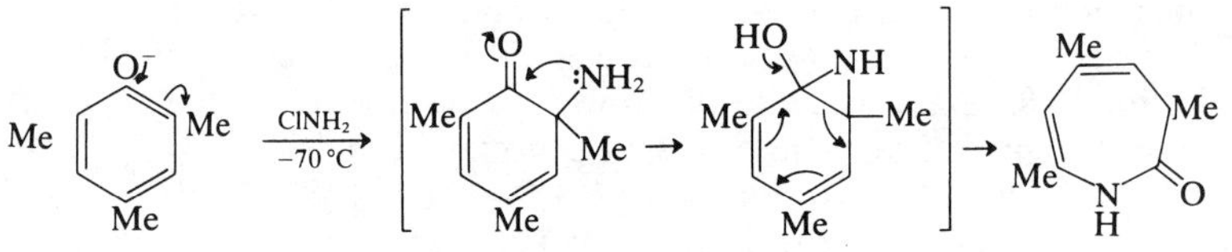

$ClNH_2$, −70 °C

Scheme 28

(*ii*) *By initial nucleophilic attack*

The standard synthetic route to *N*-substituted dibenz[*c,e*]azepines is by condensing bis(2,2′-bromomethyl)biphenyls with primary amines or acid hydrazides ⟨51JOC1475, 57JA4427, 77MI51600, 66LA(697)188⟩. The 5,7-diones are equally accessible by condensing 2,2′-diphenic anhydride with amines in a mixture of glacial acetic acid and sodium acetate ⟨79JPR1034⟩. Related are the cyclizations of 2,2′-aminobibenzyls to dibenz[*b,f*]azepines with polyphosphoric acid or methanesulfonic acid ⟨74CRV101⟩. The action of primary amines on dihalides (**235**; $n=3$, $m=1$ and $n=m=2$) provides a versatile and practicable synthesis of 2- and 3-benzazepines respectively ⟨74AHC(17)45⟩.

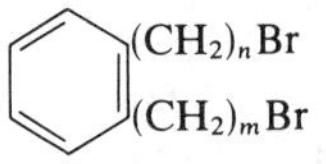

(**235**) 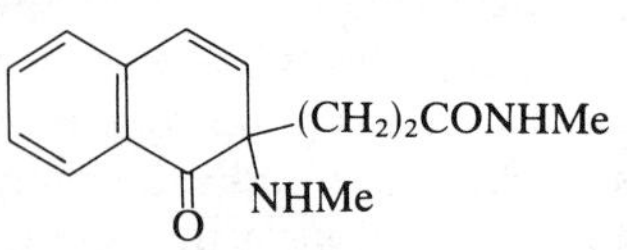 (**236**) 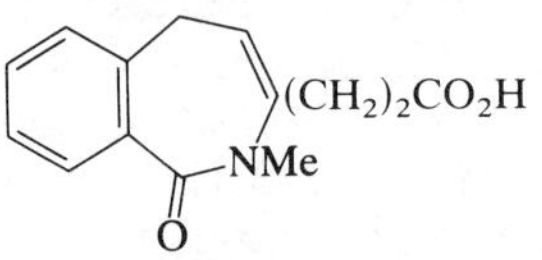 (**237**)

The carbonylation of *N*-benzyl-γ-(*o*-bromophenyl)propylamine to give *N*-benzyltetrahydro-2-benzazepin-1-one is catalyzed by palladium acetate–triphenylphosphine complex ⟨78JOC1684⟩. On treatment with 2M sodium hydroxide the aminoketone (**236**) undergoes ring expansion to the 2-benzazepin-1-one (**237**) in a manner analogous to that outlined in Scheme 28 ⟨80HCA924⟩.

5.16.4.2 From Heterocyclic Compounds

5.16.4.2.1 From three-membered rings

(i) By addition reactions

Extrusion of carbon monoxide from the $[4+2]\pi$ cycloadducts of 2*H*-azirines and cyclopentadienones (Scheme 29) constitutes another useful synthesis of 3*H*-azepines. Careful experimentation reveals that loss of CO from the *endo*-adduct is concerted with the disrotatory cleavage of the C—N bond, and that no azanorcaradiene intermediate is formed ⟨74JOC3070⟩.

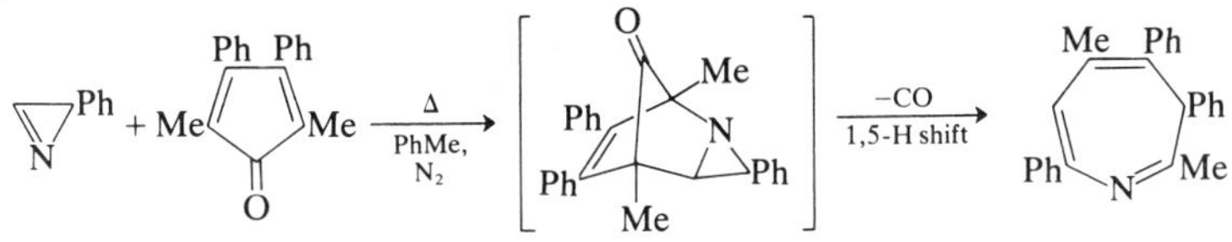

(**238**)

Scheme 29

By using a modification of this method Hassner and Anderson have been able to prepare derivatives of the unstable and elusive 3*H*-2-benzazepine system. They argued that the 2*H*-azepine formed initially by cleavage of the aziridine ring in adduct (**238**) might be prevented from rearranging to the more stable 3*H*-isomer if the C-5—C-6 double bond were incorporated as an integral part of a benzenoid system. Accordingly, the 3*H*-2-benzazepines (**63a** and **b**) were prepared by cycloaddition of 2*H*-azirines and 1,3-diphenylindan-2-one and phenylcyclone respectively. The latter could be isomerized to the 4*H*-tautomer (a 3*H*-azepine) whereas the benzazepine (**63a**), in which a similar proton shift would involve loss of aromatic character in the solitary benzene ring, resisted all isomerization attempts and remains as the only example of a stable 2*H*-azepine system ⟨74JOC3070⟩.

Isobenzofuran and 2*H*-azirines yield the stable adducts, *e.g.* (**239**). However, on treatment with neutral alumina these adducts undergo rapid and efficient isomerization to the novel 4,5-epoxy-4,5-dihydro-3*H*-2-benzazepines (**240**) ⟨74JOC2031⟩. 3-Benzazepin-2-ones are one of the products from the addition of diphenylketen to 3-dimethylamino-2*H*-azirines ⟨80JOC2951⟩.

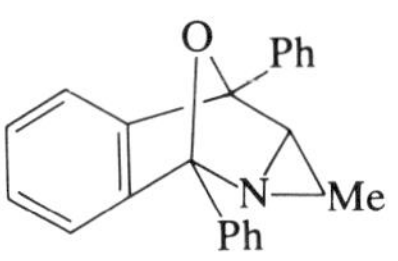

(**239**) 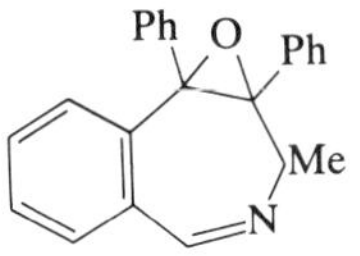 (**240**)

(ii) By rearrangements

The formation of dihydroazepines and 3*H*-azepin-2-one by novel Cope-type rearrangement of vinylcyclopropyl isocyanates and imidates has been noted in Section 5.16.3.2.1. Related are the thermal rearrangements of 1,2- or 2,3-divinylaziridines. *cis*-2,3-Divinyl-*N*-ethylaziridine rearranges spontaneously to 4,5-dihydro-1-ethyl-1*H*-azepine. Likewise the 1,2-divinylaziridine (**241**; R^1=H, $R^2=CF_3$), formed by the addition of perfluorobut-2-yne to 2-vinylaziridine at −70 °C, on warming to room temperature rearranges to the relatively unstable non-conjugated bis(trifluoromethyl)dihydro-3*H*-azepine (**65**) (see also Section 5.16.3.2.2) ⟨67JA605⟩.

Addition of DMAD to 2-phenyl-2-vinylaziridine at −20 °C results in an analogous rearrangement of the unstable *cis*-1,2-divinyl adduct (**241**; R^1 = Ph, $R^2 = CO_2Me$) to diethyl

6-phenyl-4,7-dihydro-1*H*-azepine-2,3-dicarboxylate. Similar additions and subsequent rearrangements to tetrahydro-1*H*-azepines are also observed with electron-deficient alkenes ⟨81TL3691⟩. Attempts to isolate aziridinylallyltriphenylphosphonium salts from the addition of vinylaziridines to propenyl Wittig reagents have failed, as the initially formed aziridinylphosphonium salts (**242**) rearrange spontaneously to the methylenotetrahydro-1*H*-azepines (**243**) ⟨78JOC4207⟩.

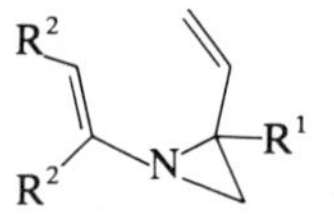

(**241**) 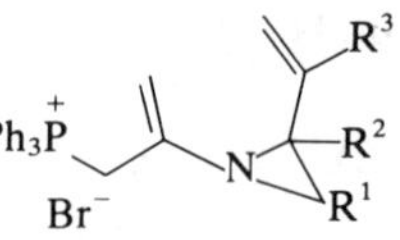 (**242**) 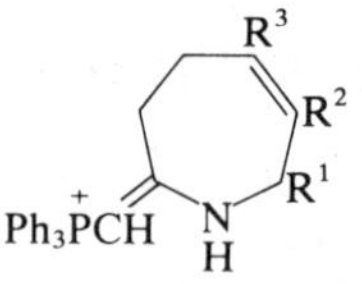 (**243**)

It is now clear that *N*-alkyl-*cis*-2,3-divinylaziridines rearrange rapidly to dihydroazepines at low to moderate temperatures (25–100 °C), whereas the *trans* isomers require vigorous heating (300 °C) in order to effect cyclization ⟨77JA1272⟩. The sterically favorable *cis*-*N*-*t*-butyl-2-vinyl-3-ethynylaziridine (**244**) likewise rearranges rapidly (20 min) at 90 °C to 1-*t*-butyl-1*H*-azepine *via* a [3,3] sigmatropic rearrangement, and then a [1,3] hydrogen shift in the highly strained allene intermediate (**245**). Again the *trans* isomer rearranges only at much higher temperatures (320–340 °C) ⟨77T2399⟩.

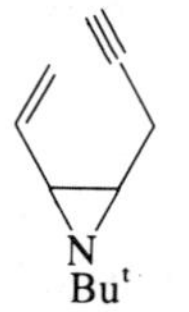

(**244**) 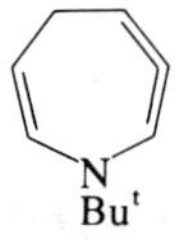 (**245**)

If the vinylaziridine bears an *N*-phenyl substituent then thermal or room temperature acid-catalyzed ($HClO_4$) rearrangement of the ***R***,***R***-isomers (**246**) yields 2,5-dihydro-1*H*-1-benzazepines quantitatively by an intramolecular 1,7-dipolar cyclization as shown in Scheme 30. Rearrangement of the *R*,*S*-conformers yields only pyrrolinines ⟨78BSF(2)97⟩. The preparative value of these rearrangements has been increased considerably by the discovery that *N*-heteroarylvinylaziridines of correct stereochemistry also isomerize on heating to yield isothiazolo- pyrido- and thieno-azepines. As expected, and in accord with the mechanism outlined in Scheme 30, systems (**246**) with Ar=a π-excessive heterocycle (*e.g.* 3-thienyl) rearrange more rapidly than those in which Ar represents a π-deficient system, *e.g.* 3- or 4-pyridyl ⟨81TL637⟩.

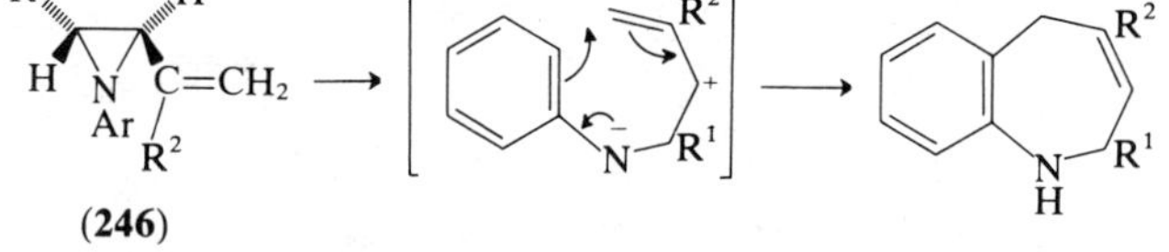

(**246**)

Scheme 30

On irradiation, (*Z*)-2-styryl-2*H*-azirines, *e.g.* (**249**), undergo analogous ring fission, followed by intramolecular 1,7-dipolar cycloadditions of the resulting nitrile ylide to give 1-phenyl-3*H*-2-benzazepines in high yield (80%) ⟨75JA4682⟩. Naphthoazepines have been prepared similarly.

(**249**) 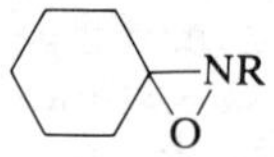 (**250**)

The propensity of azanorcaradienes (**227**; Scheme 26) to rearrange to 1*H*-azepines has been utilized by Paquette and coworkers as the basis for a versatile route to 1*H*-azepines that allows the preparation of specifically substituted 1-alkoxycarbonyl- and 1-arylsulfonyl-

1*H*-azepines without concurrent formation of troublesome isomers (see Section 5.16.4.1.3(i)). The main strategy of the synthesis is to prepare the hydroazanorcaradiene (**247**) by monoaddition of iodine isocyanate to a 1,4-dihydrobenzene (available by Birch reduction), followed by addition of methanol to the isocyanate, and then low temperature, base-catalyzed cyclization of the resulting iodocarbamate (**248**) as illustrated in Scheme 26 (path b). In the final step careful dehydrobromination of the dibromo adduct of (**247**) produces the azanorcaradiene (**227**) which as before (Scheme 26) undergoes spontaneous valence tautomerization to the 1*H*-azepine ⟨69JOC2866⟩.

N-Alkylcaprolactams are formed in high yield by the photorearrangement of spirooxaziridines, *e.g.* (**250**; R=CH_2Ph) ⟨72CR(C)(275)581⟩. In contrast, the *N*-aryl congener (**250**; R=Ph) photofragments to phenylnitrene, which in the presence of dimethylamine undergoes ring expansion to 2-dimethylamino-3*H*-azepine (**221**; R^1=NMe_2, R^2 = H) as outlined in Scheme 22 ⟨B-69MI51600⟩.

5.16.4.2.2 From five-membered rings

(i) By rearrangement

Valence tautomerism of azepines and azabicyclo[3.2.0]heptadienes has been discussed in Section 5.16.3.2.1. In most cases the process is reversible and the azepine can be regenerated by gentle thermolysis of the bicycle. Sano and coworkers have capitalized on this thermal lability of azabicycloheptadienes to synthesize several azatropone and azatropolone derivatives (see Section 5.16.3.1.2), by ring expanding the 2,3-dioxopyrrole [2+2] cycloadducts, *e.g.* (**251**), as illustrated in Scheme 31 ⟨81H(16)363⟩.

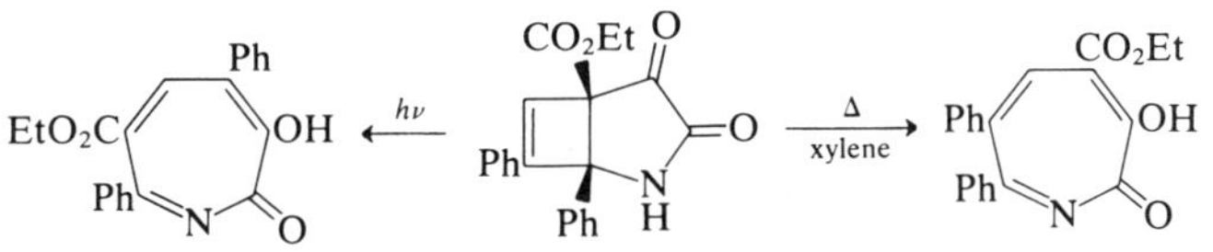

(**251**)

Scheme 31

In a similar manner the modified [2+2] photoadducts (**252**; R = acyl) of *N*-substituted indoles and ethyl acrylate on heating (230–310 °C) rearrange, probably *via* diradical intermediates, to *N*-acyl derivatives of the little known 1*H*-1-benzazepine system (**3**) ⟨80TL3403⟩. Previously, the photorearrangement of the 3a,7a-dihydrobenzazabicycloheptadiene (**252**; R = CO_2Et) had been shown to yield a stereoisomeric mixture of the *cis*- and *trans*-5a,9a-dihydro-1-benzazepines (**64**). On heating, the *trans* isomer isomerizes to the conjugated 2,3-dihydro-1*H*-1-benzazepine (**253**) ⟨78JOC315⟩.

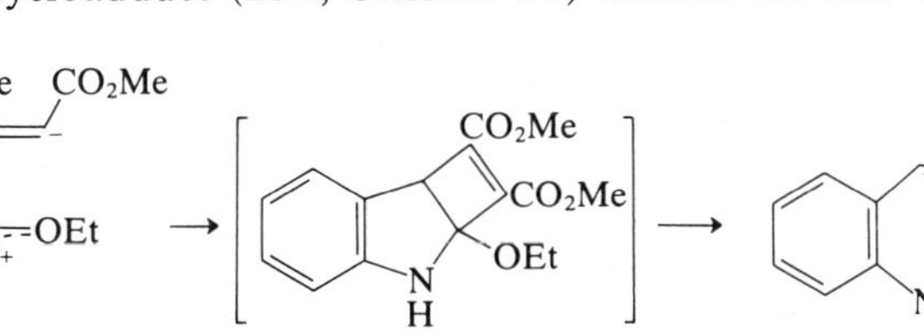

(**252**) (**253**)

The action of DMAD on indoles is often complex but can, in some cases, provide practicable yields of benzazepines. For example, ring expansion of the 2-ethoxyindole–DMAD 1,4-dipolar cycloadduct (**254**; Scheme 32) affords the 3*H*-1-benzazepine (**256**).

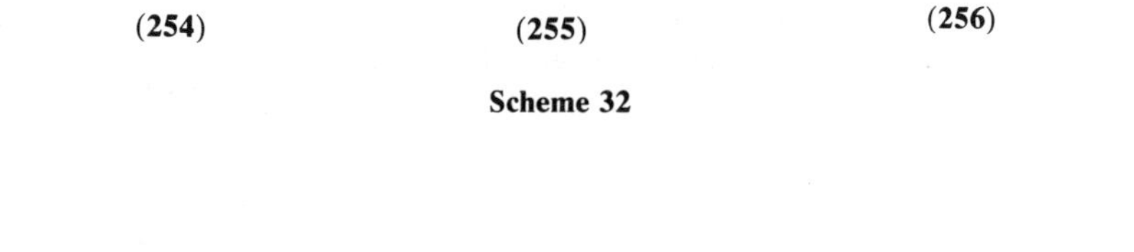

(**254**) (**255**) (**256**)

Scheme 32

When the reaction is carried out on 1-acetyl-3-piperidinoindole, the tricyclic intermediate corresponding to (**255**) can be isolated ⟨78AHC(23)263, 80JOC462⟩. Enhanced yields of benzazepines are reported by carrying out the cycloadditions under high pressures ⟨80H(14)1959, 81H(16)1367⟩.

Acid-catalyzed ring expansions of related 6–5–4 tricyclic systems are also known. For example, the acetal (**257**), formed by cycloaddition of keten acetal and 3-ethoxyisoindolone, in hydrochloric acid rearranges to the 2-benzazepine-1,5-dione (**258**) ⟨75JA7288⟩.

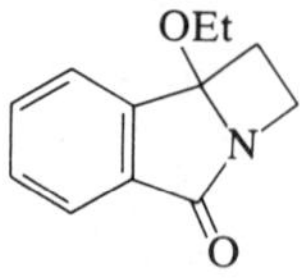

OEt N O (**257**)

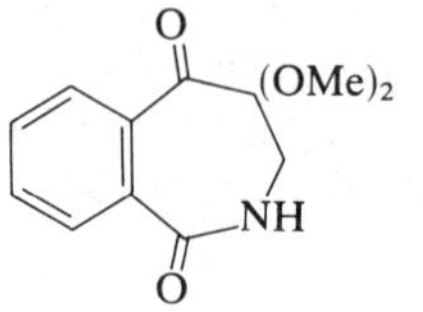

O $(OMe)_2$ NH O (**258**)

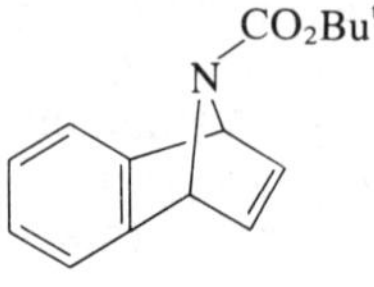

CO_2Bu^t N (**260**)

Cycloadditions of DMAD and pyrroles bearing an electron-withdrawing group on the nitrogen, although not giving azepines directly, yield the thermally labile valence tautomeric 3-azaquadricyclanes (**259**), which are useful intermediates for the synthesis of 4,5-disubstituted 1*H*-azepine-3,6-dicarboxylates as illustrated in Scheme 33 ⟨76CB3505⟩. Related is the photolytic ring expansion of the 2,3-benz-7-azabicyclo[2.2.1]hepta-2,5-diene (**260**) in cyclohexane to the 3*H*-benzazepine (**155**; R^1=H, $R^2 = Bu^t$) ⟨71TL3947⟩.

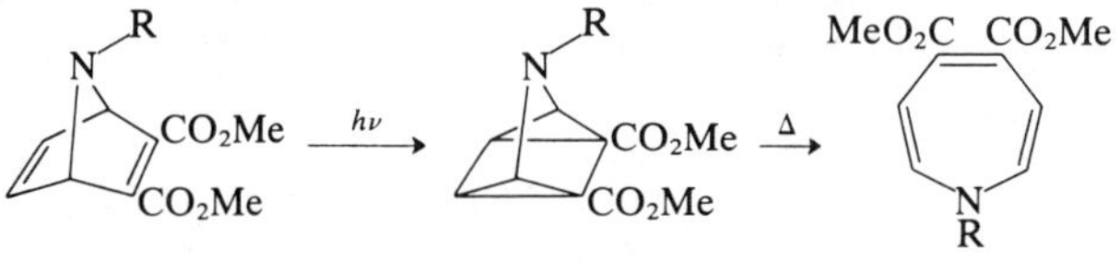

N–R CO_2Me CO_2Me $\xrightarrow{h\nu}$ N–R CO_2Me CO_2Me (**259**) $\xrightarrow{\Delta}$ MeO_2C CO_2Me N R

R = *p*-Ts, Ac, CO_2Me, $CONH_2$ or *p*-$NO_2C_6H_4$

Scheme 33

Two other azepine syntheses involving the interconversion of valence isomers are worthy of note, namely the isomerization of the novel 7-azatetracycloheptane (**261**), prepared from benzvalene as indicated in Scheme 34, to 1-phenyl-1*H*-azepine ⟨79TL1553⟩ and the rearrangement of the *syn* isomer of the 2-azatricycloheptane (**70**; H in place of D), which at 120 °C yields 1-ethoxycarbonyl-2,3-dihydro-1*H*-azepine by a $[_{\pi}2_s + _{\pi}4_s]$ concerted process. As expected, the *anti* isomer, for which a concerted process would require a geometrically unfavoured $[_{\pi}2_a + _{\pi}4_a]$ process, rearranges only at 350 °C, and then by a non-concerted polar reaction ⟨73JA7320⟩.

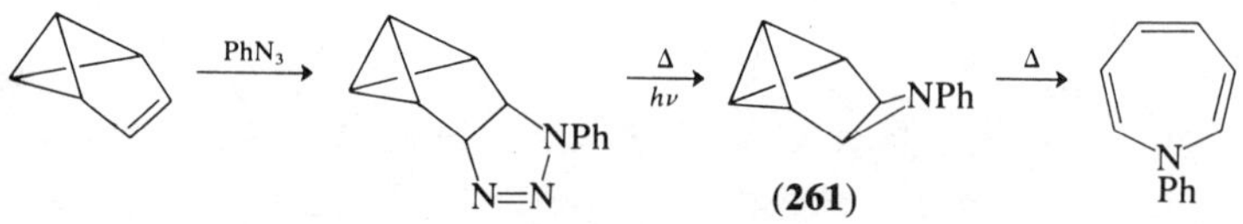

$\xrightarrow{PhN_3}$ NPh N=N $\xrightarrow[h\nu]{\Delta}$ NPh (**261**) $\xrightarrow{\Delta}$ N Ph

Scheme 34

Benzenium ions (**199**; X = acyl) on treatment with *p*-TSA in toluene undergo intramolecular *O*-alkylation to give the oxazolines (**201**; X=acyl; Scheme 18, path b), rather than ring expansion to the 1*H*-azepine as mentioned earlier for the *N*-tosyl derivative (Scheme 18, path a). However, these oxazolines are thermally labile and on heating at 180 °C rearrange in high yields (84–93%) to 1-acyl-1*H*-azepines (**200**; X = ArCO or Ac) ⟨81AG(E)699⟩.

Related to the azide decompositions reported in Section 5.16.4.1.2(ii) are the formation of 3-acyl-2-alkoxy-3*H*-azepines and -3*H*-azepin-2-ones by the photoinduced ring expansions of 3-substituted 2,1-benzisoxazoles (Scheme 22), in alcohol and aqueous alcohol solution, respectively ⟨81AHC(29)1⟩, and the preparation of 3- and 7-acetyl-3*H*-azepin-2-ones by photolyses of 2,3-dimethylindazoles in dilute sulfuric acid ⟨73HCA1852⟩.

In both these reactions it is assumed that ring opening to an *o*-nitrenoaryl ketone, *e.g.* (**213**; R=Ac), followed by benzazirine formation and then nucleophilic addition as in Scheme 22, is responsible for azepine formation.

A recent development in the synthesis of 2-benzazepine-1,5-diones is the photoaddition of alkenes to *N*-alkylphthalimides. *N*-Alkylphthalimides capable of undergoing a Norrish Type II process follow the route outlined in Scheme 35 (path a) to yield hydrobenzazepinediones (**264**) by ring opening of the azacyclobutanol (**262**) ⟨78ACR407⟩.

More interesting are the photoadditions of selected alkenes, dienes, vinyl ethers and vinyl esters to the C(O)—N bond of *N*-methylphthalimide with subsequent ring expansion to benzazepinediones (Scheme 35, path b) ⟨B-81MI51600⟩. Originally, the cycloadditions were thought to involve a diradical intermediate, but the finding that *cis*- and *trans*-2-butene add stereospecifically to the phthalimide has led Mazzochi and coworkers to conclude that the addition is formally a concerted $[2+2]\pi$ cycloaddition to the C—N bond of the zwitterionic mesomer (**263**) ⟨81JOC4846⟩. The regiochemistry of the cycloaddition has been determined; with propene only the 4-methylbenzazepine-2,5-dione (**264**; $R^1=R^3=Me$, $R^2=H$) is formed ⟨81TL4189⟩.

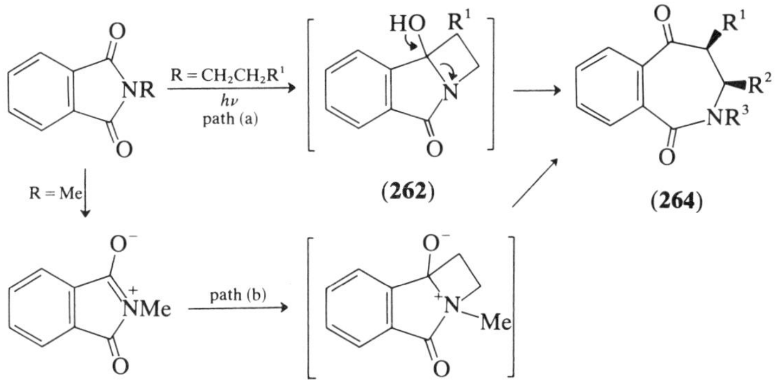

Scheme 35

(ii) By extrusion reactions

The 1,3-dipolar cycloadducts (**266**) from cyclobutenes with oxazolium 5-oxides (**265**) (münchnones) ⟨80JHC1593⟩ extrude carbon dioxide to yield 4,5-dihydro-1*H*-azepines as illustrated in Scheme 36.

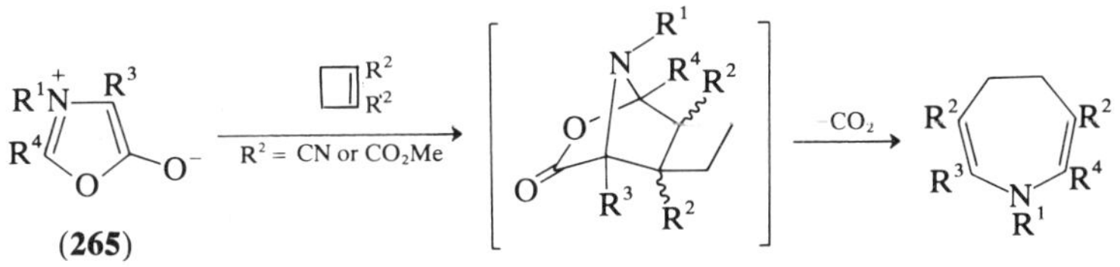

Scheme 36

5.16.4.2.3 From six-membered rings

4*H*-Azepines are more rare than 1*H*- or 3*H*-azepines and only a few synthetic approaches have been developed. Of these the two main methods involve the ring expansion of six-membered heterocycles. Early studies revealed that highly substituted 4*H*-azepines (**269**) result from the base-catalyzed ring expansion of 4-(chloromethyl)-1,4-dihydropyridines (**267**; Scheme 37). The reaction was found to be temperature and solvent sensitive, and azepines (**268**)–(**270**) have been isolated and characterized. However, later studies ⟨68JCS(C)1675⟩ on cyano derivatives (**267**; E=CN) show the reaction to be even more complex than was first thought and that the 3-azabicyclo[4.1.0]heptenes (**271**) are intermediates in the primary process rather than carbonium or carbenoid species. The azabicycloheptene (**271**) has been isolated, whereas the methylene-1*H*-azepine (**272**) has been detected only by ^{1}H NMR, all attempts at isolation resulting in rearrangement to the 1*H*-azepine (**273**). There is also UV evidence for participation of the 4-methylene isomer of (**272**).

The ring expansion to 4*H*-azepines is also induced by other nucleophiles, *e.g.* CN^-, NH_3, MeOH–NEt_3, thiols, potassium phthalimide, Bu^{t}OK and enolates ⟨79CJC44, 71JCS(C)1237, 79CC1070⟩; with NaSH or H_2Se, however, 2,5-bridged azepines (**171**; X = S and Se) are formed ⟨68JCS(C)2311, 79CJC44⟩.

Related to the aforementioned rearrangements are the ring expansions of 1,2-dihydro-2-(*p*-tosyloxymethyl)pyridines in pyridine to 3*H*-azepines ⟨71JOC978⟩, and of the *p*-tosyloxy derivative (**274**) to the unstable 4*H*-azepine (**275**) by solvolysis in acetonitrile ⟨73T391⟩.

[4+2]π cycloaddition of cyclopropenes to 1,2,4-triazines provides alternative access to 4*H*-azepines (Scheme 38) ⟨80TL595, 72CB982⟩. Base or heat treatment (190 °C) of the metastable 4*H*-azepines brings about rearrangement to the more stable 3*H*-isomers.

(**270**) (**267**) (**268**) (**269**)

(**271**) (**272**) (**273**)

a; $R^1 = R^2 = H$, $E = CO_2Me$
b; $R^1 = Me$, $R^2 = Ph$, $E = CN$

i, NaOMe, cold MeOH; ii, warm EtOH; iii, NaOMe, Et_2O; iv, NaOMe, hot EtOH; v, H^+; vi, C_5H_5N; vii, Bu^{t}OK at 0 °C

Scheme 37

The ring expansions of quinolines ⟨74JCS(P1)1828⟩, 1,2-dihydroisoquinolines ⟨80JOC1950⟩ and 1,2-dihydroisoquinolones ⟨80RTC271⟩ by use of dihalocarbenes have some merit for the synthesis of 1- and 2-benzazepines and benzazepinones. However, the report that 1-phenyl-2,3,4-trioxotetrahydroquinoline on treatment with diazomethane ring-expands to 1-phenyl-1*H*-1-benzazepine-2,4,5-trione is incorrect; the product is in fact the spirooxirane (**276**) ⟨76ZN(B)1716⟩.

(**274**) (**275**)

Other synthetic routes to benzazepines involving ring expansion of six-membered heterocycles include the action of diazomethane ⟨77CPB321⟩, sulfonium ylides ⟨77H(7)37⟩ or acyl halides ⟨75T1991⟩ on quaternary 3,4-dihydroisoquinolines; that of sulfoxonium ylides on quaternary quinolines ⟨74IJC(B)1238⟩; and the zinc–acetic acid reduction of quaternary 1-acyltetrahydroisoquinolines ⟨77BSF893⟩. Photoaddition of acyl- or aryl-nitrenes to the exocyclic alkene bond of 2-methylene-1,2-dihydroquinolines results in ring expansion to

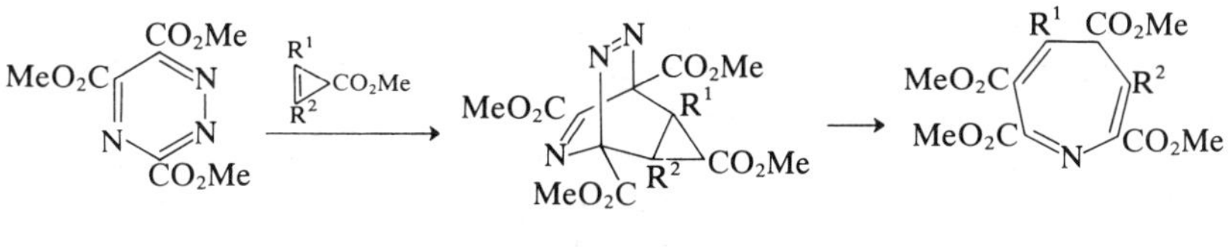

Scheme 38

(276) (277)

2-imino-2,3-dihydro-1*H*-1-benzazepines (**277**; R=CO_2Et, Ph or Bz) in practicable yields (68–94%) ⟨76JOC3325⟩.

Mechanistically the ring expansion of the 2-pyridone (**278**) to the novel 2,2-disubstituted azepin-3-one (**281**) is reminiscent of the preparation of 3*H*-azepin-2-ones outlined in Scheme 28. Steric crowding in the carbanion (**279**) is held to be responsible for cyclization to the bicycle (**280**) and hence ring expansion (Scheme 39) ⟨80JCS(P1)2851⟩.

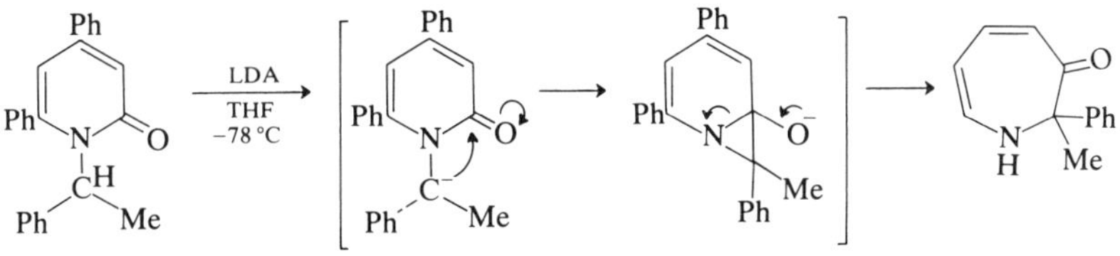

(278) (279) (280) (281)

Scheme 39

5.16.5 USEFUL AND NATURALLY OCCURRING AZEPINES

In this section, in order to avoid a multiplicity of references, reference to the naturally occurring and biologically active azepines is in most cases made by means of a code, *e.g.* (M399), which refers to the reference number of that compound in the Merck index ⟨B-76MI51600⟩.

Caprolactam or (hexahydroazepin-2-one) is, without doubt, the most important azepine derivative. This seven-membered lactam is produced in vast quantities as an intermediate for the manufacture of nylon 6 ⟨B-75MI51601, B-70MI51601⟩. Polymerization, which is carried out at high temperatures with water as the initiator or at low temperatures with a strong base (*e.g.* NaH), proceeds by attack at the caprolactam carbonyl by the amino function of the open-chain monomer, ε-aminohexanoic acid.

C-Substituted caprolactams are known to have significant CNS activity ⟨76AJC2667⟩, and hexahydrozepin-4-ol derivatives possess useful analgesic properties. For example, proheptazine (**282**) (M7571) has twice the analgesic effect of morphine and is only marginally addictive ⟨64JMC57⟩. *N*-Substituted hexahydroazepines have found use as antitussives (amicibone—M399), mydriatics (M3334), antispasmodics (prozapine—M7692) and oral hypoglycemics (tolazamide—M9206). 6,7,8,9-Tetrahydrotetrazolo[1,5-*a*]azepine (**283**) (M6935) is a CNS stimulant and a narcotic antagonist.

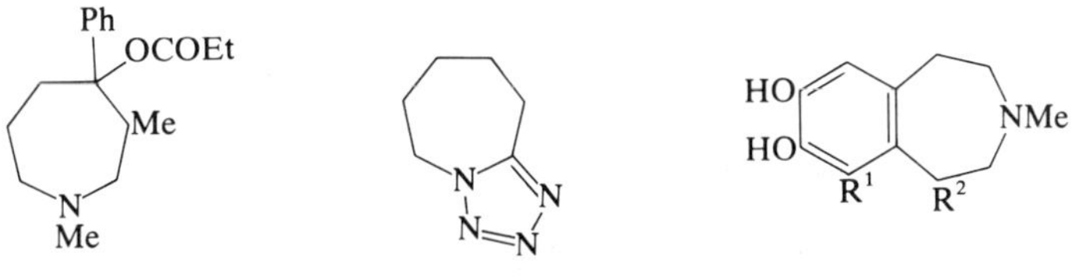

(282) (283) (284)

The pharmacological value of benzazepines is not as great as their more celebrated analogues, the benzdiazepines, but anticonvulsant, antiarrhythmic, antiinflammatory and analgesic activity has been noted for several 1-, 2- and 3-benzazepines ⟨81CCC148⟩. For a synopsis of the relevant patent literature reference ⟨74AHC(17)45⟩ should be consulted. 7,8-Dihydroxy-1,2,4,5-tetrahydro-3*H*-3-benzazepines, *e.g.* (**284**; R^1=SPh, R^2 = H), represent a novel class of dopamine receptor antagonists and neuroleptics. In contrast, the 1-phenyl derivative (**284**; R^1 = H, R^2 = Ph) is a potent antagonist of both central and peripheral dopamine receptors ⟨80JMC975⟩.

The most useful pharmacological agents based on the azepine nucleus are the 10,11-dibenz[*b,f*]azepines such as desipramine (M2881), clocapramine (M2335), trimipramine (M9391), imipramine (**285**; R=H) (M4813) and clomipramine (**285**; R = Cl) (M2350), which are widely used for the treatment of depressive illness ⟨74CRV101⟩. Recently, thienoisosteres have been prepared and promise some advantages, such as fewer toxic side effects, over the dibenzo derivatives ⟨81H(15)1227⟩. 5*H*-Dibenz[*b,f*]azepine-5-carboxyamide (**8**; R=$CONH_2$) (carbamazepine—M1781) is a well-known analgesic and anticonvulsant.

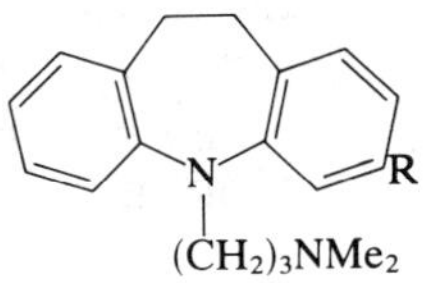

(**285**)

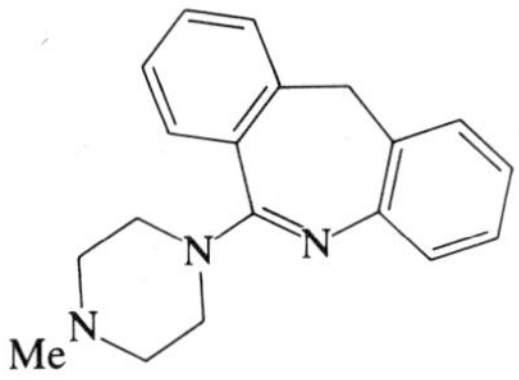

(**286**)

Other noteworthy pharmacologically active dibenzazepines are the antiadrenergic dibenz[*c,e*]azepine (**90**; R=$CH_2CH{=}CH_2$) (azapetine—M913) and the dibenz[*b,e*]azepine (**286**) (perlapine—M6964), a hypnotic.

A group of alkaloids originally thought of as spirobenzylisoquinolines are in fact indenobenzazepines, *e.g.* lahorine (**287**) ⟨81TL3143⟩. Other alkaloids incorporating an azepine nucleus are: astrocasine, a pyrido[2,1-*a*][2]benzazepin-6-one (M878); ibogaine (M4792), an indoloazepine with marked psychological effects; chilenine, an isoindolobenzazepine ⟨82TL39⟩; and rhoeadine (M7988), a system related to (**287**) but with a pyran unit in place of the indene ring.

Muscaflavin, a yellow pigment from the poisonous mushroom *Amanita muscaria* has the unusual dihydroazepine–amino acid structure (**288**) ⟨81LA2164⟩.

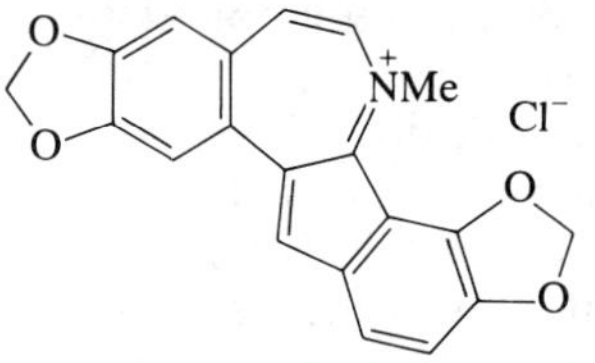

(**287**)

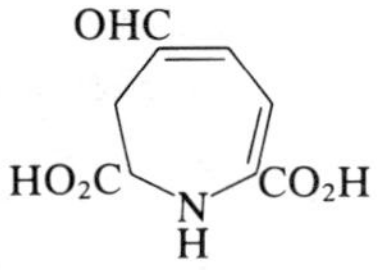

(**288**)

5.17

Oxepanes, Oxepins, Thiepanes and Thiepins

D. R. BOYD
Queen's University of Belfast

The subject matter of the present chapter has been given comprehensive coverage in an excellent volume dealing with the literature up to 1970 ⟨72HC(26)1⟩. While frequent reference will be made to this and other reviews (and references therein), in the current article a greater emphasis will be placed upon the considerable amount of work reported during the past decade on oxepins and thiepins. Among the reviews which have appeared during the latter period three on oxepins ⟨73H(1)267, 74MI51700, 76ACR378⟩ and two on thiepins ⟨71IJS(B)(6)267, 81MI51700⟩ were found to be particularly useful.

The vast proliferation of literature reports on oxygen- and sulfur-containing seven-membered heterocycles which has occurred since 1970 is, in part, the result of improved experimental methods for the synthesis and detection of new oxepins and thiepins (which are generally unstable). Much of the interest in this area of research has been based upon: (a) attempts to verify theoretical predictions concerning the antiaromatic character of oxepins and thiepins, (b) the establishment of an obligatory role for oxepin–arene oxide intermediates in the metabolism of aromatic substrates in plants, animals and microorganisms, and (c) the synthesis of seven-membered oxygen and sulfur heterocycles in the quest for useful chemotherapeutic properties of the type found in nitrogen-containing analogues.

I wish to gratefully acknowledge the valuable assistance given by Mr R. T. Kimber and Miss R. Stewart of the Library staff at The Queen's University of Belfast, who conducted a computer search of *Chemical Abstracts* for the period 1967–1981. From the vast number of references thus obtained it has obviously been necessary to omit a large proportion of material and the present chapter is of necessity selective.

5.17.1 STRUCTURE OF OXEPANES, OXEPINS, THIEPANES AND THIEPINS

5.17.1.1 Structure of Oxepanes

The molecular geometry of oxepane (**1**) has been determined experimentally by the gas-phase electron diffraction method. Average lengths for the C—O (1.419 Å), C—C (1.531 Å) and C—H (1.105 Å) bonds were identical for both of the preferred twist-chair conformations of oxepane which had mean ring valency angles of 112.0° ($^3TC_4$ conformation) and 116.2° ($^2TC_3$ conformation). Good agreement was obtained between these measured values and those resulting from molecular mechanics calculations ⟨80JST(64)239⟩. The bond angles obtained for the two preferred conformations of oxepane are indicated in Table 1. All of the bond lengths in oxepane [C(sp^3)—C(sp^3), C(sp^3)—O, C(sp^3)—H] had the normal values associated with acyclic molecules although the average ring valency angles were larger than the normal tetrahedral values.

Table 1 Measured Bond Angles in Oxepane (**1**)

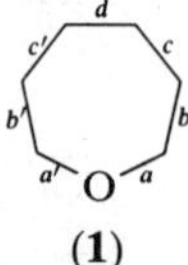

(**1**)

Angle	$^3TC_4$ *conformation* (°)	$^2TC_3$ *conformation* (°)
aa′	112.1	115.9
ab	109.0	116.4
bc	112.2	117.2
cd	112.6	115.4
c′d	111.9	116.5
c′b′	112.7	116.7
b′a′	113.9	115.4

Among the possible conformations of oxepane the boat, twist-boat and chair forms were calculated to be of higher energy than the predominant twist-chair conformations (which were found to exist in approximately equal proportions).

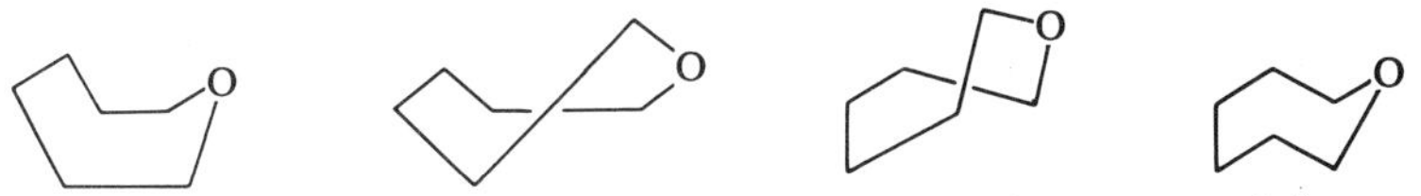

boat twist-boat twist-chair chair

Further experimental confirmation of the preferred geometry of oxepane (**1**) has been obtained from its IR and Raman spectra. These vibrational and rotational spectra have been rationalized in terms of the lowest energy twist-chair conformations pseudorotating *via* a chair form ⟨77JA2866, 2876⟩.

The similarity between the conformations of cycloheptane and oxepane (**1**) has been noted ⟨74T1683⟩ although the smaller steric requirements of an oxygen atom (relative to a methylene group) may result in a conformational selectivity of the oxygen atom for a more hindered position. In general, the conformations of heterocyclic seven-membered rings have been studied and rationalized to a lesser degree than those of other simple monoheterocyclic systems.

The ^{1}H NMR spectrum of oxepane (**1**) was not particularly informative and consisted of two broad bands at δ 1.7 (8H, $H_3H_4H_5H_6$) and 3.7 (4H, H_2H_7) p.p.m. due to coupling of the ring protons and to the electronegativity effect of the oxygen atom.

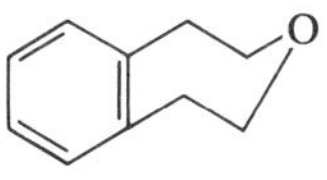

(**2**)

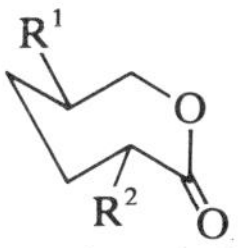

(**3**) R^1=H, R^2 = Me
(**4**) R^1 = Me, R^2 = H

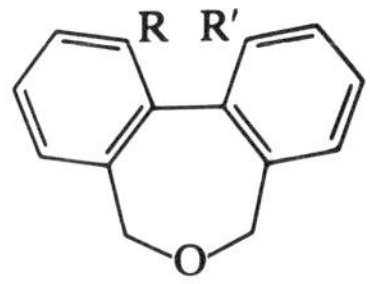

(**5**)

The ^{13}C NMR chemical shifts have been studied over a range of temperatures but showed no evidence of splitting due to individual conformers even at −150 °C. The latter study ⟨77JA2876⟩ suggested a low barrier to pseudorotation of *ca.* 17 kJ mol^{-1}.

Benzoannelation of the oxepane ring system leads to an increase in the barrier to ring inversion. Thus variable temperature NMR spectra on 1,2,4,5-tetrahydro-3-benzoxepin (**2**) suggested that a chair conformation was preferred with a barrier to ring inversion of *ca.* 39.7 kJ mol^{-1}.

Slightly distorted chair conformations have also been assigned to the ε-caprolactones (**3**) and (**4**) on the basis of their chiroptical properties ⟨67JA5649⟩. Both variable temperature NMR and racemization methods have been used to determine the rate of conformational interchange in dihydrodibenz[*c*, *e*]oxepins (**5**). Barriers were found to be in the range 38–71 kJ mol^{-1} for a single *ortho* R substituent (**5**; R′ = H) increasing to 117–146 kJ mol^{-1} for two *ortho* substituents (**5**, R, R′ ≠ H).

While strain energies for a range of cyclic ethers from three- up to six-membered rings were obtained from measurements of the heats of combustion and heats of formation, the corresponding data for oxepane appear not to be available. However, by analogy with cycloheptane and thiepane (which are reported to have very similar strain energies) the oxepane ring is expected to have a strain energy of comparable magnitude. Preferred conformations and transannular interactions are similar in oxepane, thiepane and cycloheptane.

The fragmentation pathways responsible for the mass spectrum of oxepane have been studied using specifically deuterated oxepane and are comparable to those found in a non-branched acyclic ether (Scheme 1) ⟨68OMS(1)403⟩.

The predominant initial bond fission in 2,3,4,5-tetrahydro-1-benzoxepin (**6**) was again at the C—O bond with ring contraction steps occurring to yield five- or six-membered ring products ⟨68OMS(1)403, 73OMS(7)479⟩.

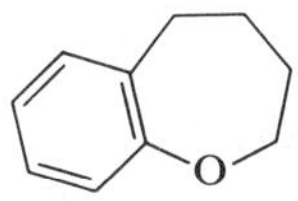

(**6**)

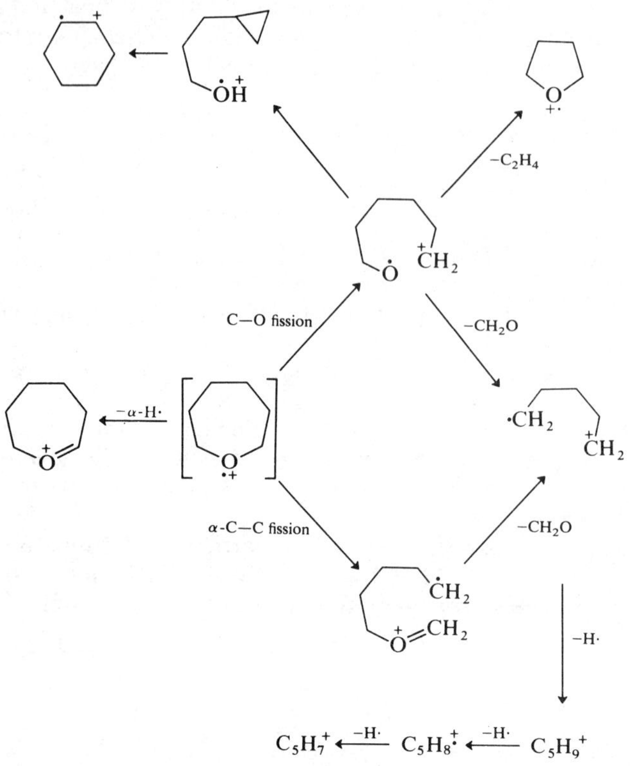

Scheme 1

The UV spectrum of oxepane indicated only end absorption and photochemical reactions were thus carried out at 185 nm ⟨80MI51701⟩. Absorption bands in the IR spectrum of oxepane are characteristic of simple acyclic ethers ⟨52M1502; 57BSF789⟩: C—O—C stretching 1130, 1090, 1070, C—H stretching 2920, CH_2 deformation 1450, and ring vibration 998, 940 cm^{-1}. The IR vibrational frequencies have been measured in the vapour phase and obtained by calculation as part of a conformational study of oxepane ⟨77JA2866⟩.

5.17.1.2 Structure of Oxepins

The parent structure, oxepin (**7**), exists in a state of spontaneous equilibration with the valence bond isomer benzene oxide at ambient temperature.

Accurate molecular dimensions on oxepin have not been obtained experimentally but are available from a range of calculations (Table 2).

Table 2 Calculated Bond Lengths in Oxepin (**7**)

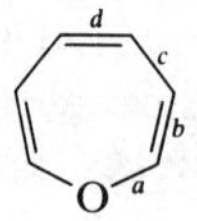

(**7**)

Calculated bond length (Å)					
a	*b*	*c*	*d*	*Method*	*Ref.*
1.389	1.345	1.463	1.351	SCFMO	70T4269
1.333	1.345	1.471	1.349	MINDO/3	80JA1255
1.434	1.284	1.444	1.332	*ab initio*	80JA1255

Molecular orbital calculations have indicated that oxepin, which is isoconjugate with the tropylium anion, will have a very small resonance energy in the range +0.50 to −6.52 kJ mol^{-1} according to the method used ⟨70T4269, 75BCJ1501, 76JA2750⟩. The degree of

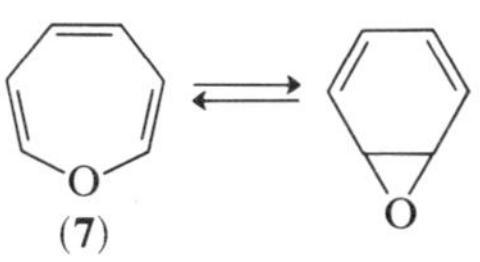

(7)

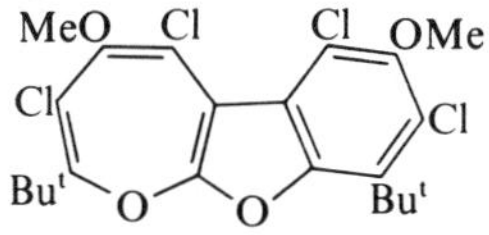

(8)

antiaromatic character anticipated for oxepin based upon these calculated resonance energies is partly reflected in its relative instability.

The oxepin ring structure (**8**) appeared to be formed in preference to the arene oxide tautomer according to the NMR spectrum, and the oxepin form was unequivocally established by X-ray structure analysis ⟨78AG(E)121⟩. Unfortunately the molecular dimensions of the oxepin ring in structure (**8**) were not included in the original report. In contrast, the valence isomeric arene oxide form of oxepin (**9**) was found to predominate and detailed information about the arene oxide molecular geometry was provided by the X-ray diffraction method (Table 3) ⟨80LA1889⟩.

Table 3 Experimentally Determined Bond Lengths and Bond Angles in Oxepin–Arene Oxide (**9**)

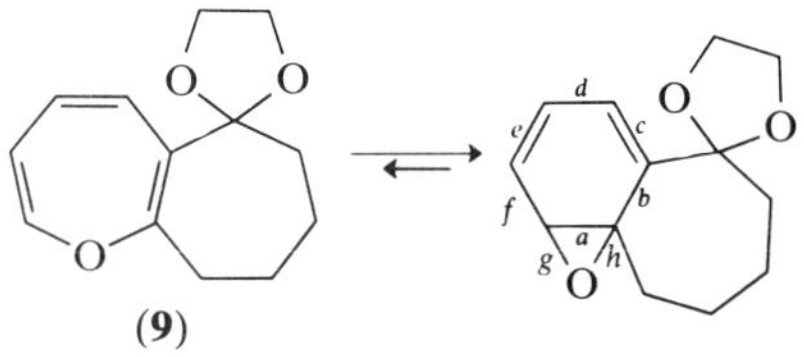

(**9**)

Bond lengths (Å)				*Bond angles* (°)			
a	1.494	e	1.330	ab	115.6	ef	122.4
b	1.518	f	1.464	bc	120.9	fa	118.8
c	1.335	g	1.445	cd	122.2	ag	59.4
d	1.478	h	1.456	de	120.1	gh	62.0
						ha	58.6

NMR spectral studies have been used to investigate the oxepin–benzene oxide equilibrium ⟨67AG(E)385⟩. The ^{1}H NMR spectrum of (**7**) at ambient temperature consisted of three multiplets of equal intensity centered at δ 6.1 (H_α), 5.65 (H_β) and 5.2 (H_γ) p.p.m. due to the fast exchange process. At lower temperatures (−127 °C, CF_3Br/pentane) the rate of isomerization decreased until individual signals for the oxepin tautomer (**7**) were observed. The α, β and γ proton signals in oxepin are in typical chemical shift positions for a range of monocyclic substituted oxepins (Table 4). The oxepin (**7**) and benzene oxide valence isomer were also distinguishable in the low temperature (−134 °C, $EtCl/(CD_3)_2CO$) ^{13}C NMR spectrum ⟨74CB3149⟩.

^{1}*H NMR* (δ values in p.p.m.) ^{13}*C NMR* (δ values in p.p.m.) γ (130.8)

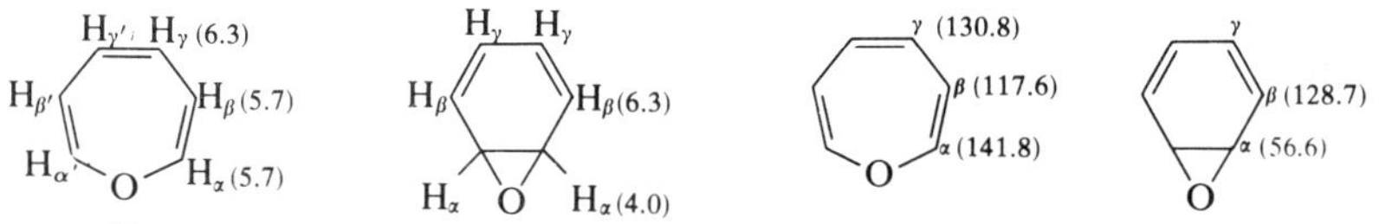

$H_{\gamma'}$ H_γ (6.3) $H_{\beta'}$ H_β (5.7) $H_{\alpha'}$ H_α (5.7) (**7**)

H_γ H_γ H_β H_β (6.3) H_α H_α (4.0)

γ (130.8) β (117.6) α (141.8)

γ β (128.7) α (56.6)

UV spectroscopy has been particularly valuable in the study of the oxepin–benzene oxide equilibration. The oxepin isomer showed absorption at 305 nm (ε 900) while benzene oxide absorbed at a shorter wavelength (271 nm, ε 1430) (Table 5).

The spontaneous oxepin–benzene oxide isomerization proceeds in accordance with the Woodward–Hoffmann rules of orbital symmetry control and may thus be classified as an allowed thermal disrotatory electrocyclic reaction. A considerable amount of structural information about both oxepin and benzene oxide has been obtained from theoretical calculations using *ab initio* SCF and semiempirical (MINDO/3) MO calculations ⟨80JA1255⟩. Thus the oxepin ring was predicted to be either a flattened boat structure (MINDO/3) or a planar ring (SCF), indicative of a very low barrier to interconversion between boat conformations. Both methods of calculation indicated that the benzene oxide tautomer

Table 4 NMR Chemical Shift Values for Oxepins (**10**)–(**20**)

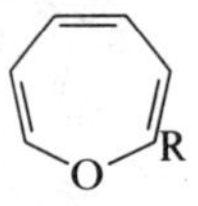

(**10**)–(**15**) 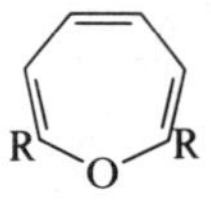 (**16**)–(**20**)

Compound	*R*	H_β	H_γ	$\delta(^1H)$ (p.p.m.) $H_{\gamma'}$	$H_{\beta'}$	$H_{\alpha'}$	*Ref.*
(**10**)	Me		5.3–6.4			5.8	67AG(E)385
(**11**)	COMe		5.5–6.7			5.9	67AG(E)385
(**12**)	CN	6.0	5.8	6.5	6.3	6.4	81JOC813
(**13**)	CHO	6.0	5.7	6.5	6.4	6.6	81JA898
(**14**)	CO_2H	6.9	6.3	6.5	5.8	6.0	79JA2470
(**15**)	CO_2Me	6.9	6.4	6.5	5.8	5.9	79JA2470
(**16**)	Me	5.2	5.8	5.8	5.2	—	67AG(E)385
(**17**)	CHO		6.6			—	76TL1167
(**18**)	CO_2H		6.1			—	76TL1167
(**19**)	CO_2Me		6.7			—	76TL1167
(**20**)	CN		6.7			—	76TL1167

Table 5 UV Absorption Characteristics of Oxepins (**7**), (**10**)–(**13**) and (**15**)–(**20**)

Oxepin	λ_{max}	ε_{max}	*Ref.*
(**7**)	305 (sh)	900	67AG(E)385
(**10**)	297	1800	67AG(E)385
(**11**)	222	10 900	67AG(E)385
	315	2500	
(**12**)	204	10 950	81JOC813
	307	1680	
(**13**)	218	12 700	81JA898
	320	2690	
(**15**)	318	2390	79JA2470
(**16**)	297	1800	67AG(E)385
(**17**)	225	14 370	76TL1167
	330	2200	
(**18**)	224	21 560	76TL1167
	330	2880	
(**19**)	226	27 200	76TL1167
	335	2800	
(**20**)	323	2560	76TL1167

should be non-planar with the oxirane ring being at an angle of 73° to the cyclohexadiene ring. Direct experimental evidence for the boat form of the oxepin is obtained from the magnitude of coupling constants observed, for example in the NMR spectrum of 2,7-dimethyloxepin (**16**) ⟨67AG(E)385⟩. The rate of inversion of the boat conformation of the oxepin ring was found to be markedly decreased by aromatic ring annelation ⟨69AG(E)68⟩. 1-Methyltribenzoxepin-3-carboxylic acid (**21**) has thus been resolved into enantiomers and was found to racemize spontaneously with a barrier ($\Delta G^{\ddagger}_{20.5\,°C}$) of 86.94 kJ mol^{-1}.

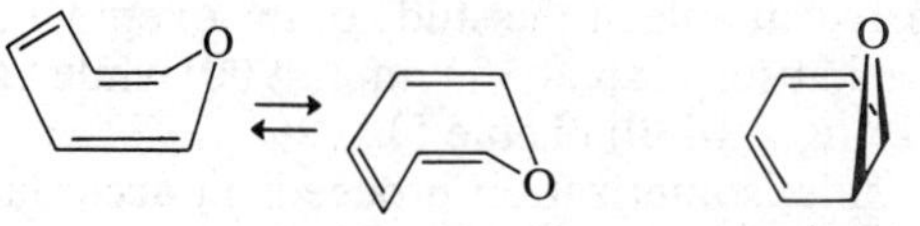

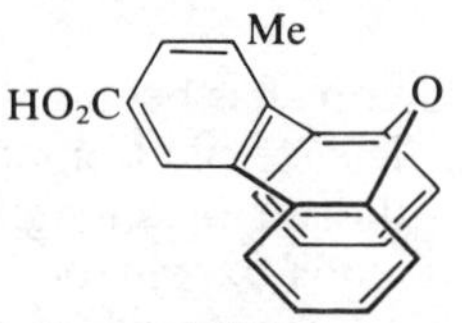

(**21**)

NMR analysis does not provide an accurate estimate of the proportions of oxepin and benzene oxide present at ambient temperature due to a fast rate of isomerization (on the NMR timescale). The oxepin–benzene oxide ratio was found to be dependent upon solvent, temperature, substituents (electronic and steric effects) and the resonance energy of the molecule.

The dipole moments of oxepin and benzene oxide have been calculated to be in the range 0.76–1.36 D and >1.5 D respectively using the *ab initio* SCF and MINDO/3 methods ⟨80JA1255⟩. The lower calculated dipole moment would be in accord with experimental observations where the equilibrium was found to favor oxepin (**7**) in less polar solvents. Coordination between the oxirane oxygen atom and polar solvent molecules would also strengthen the C—C bond of the epoxide and thus lead to a preference for the benzene oxide isomer ⟨72AG(E)825⟩. Thus the proportion of oxepin (**7**) was found by UV spectral analysis to be higher in isooctane solvent (70%) than in water–methanol (10%).

The effect of temperature upon the oxepin–benzene oxide equilibrium has been investigated by NMR and UV methods ⟨67AG(E)385⟩. At lower temperatures the benzene oxide tautomer is preferred. At higher temperatures it is not possible to estimate the proportion of oxepin tautomer by NMR due to the rapid exchange rate. UV spectroscopy however indicates that the proportion of oxepin increases with increasing temperature.

The nature and position of substitution has a profound influence upon the oxepin–arene oxide equilibrium position. The effect of substituents on the relative energies of each valence tautomer has been calculated ⟨80JA1255⟩ and these theoretical results are in accord with the limited experimental data which are available. In general terms, oxepins substituted at the 3-position are less favored than the corresponding arene oxides, while the reverse obtained for 2- and 4-substituted oxepins. This substituent effect has been rationalized in terms of a preference for the maximum number of alternative resonance contributors. The influence of both π donating and π withdrawing substituents on the oxepin contribution is summarized in Scheme 2. This latter effect may be considered as an 'electronic' substituent effect.

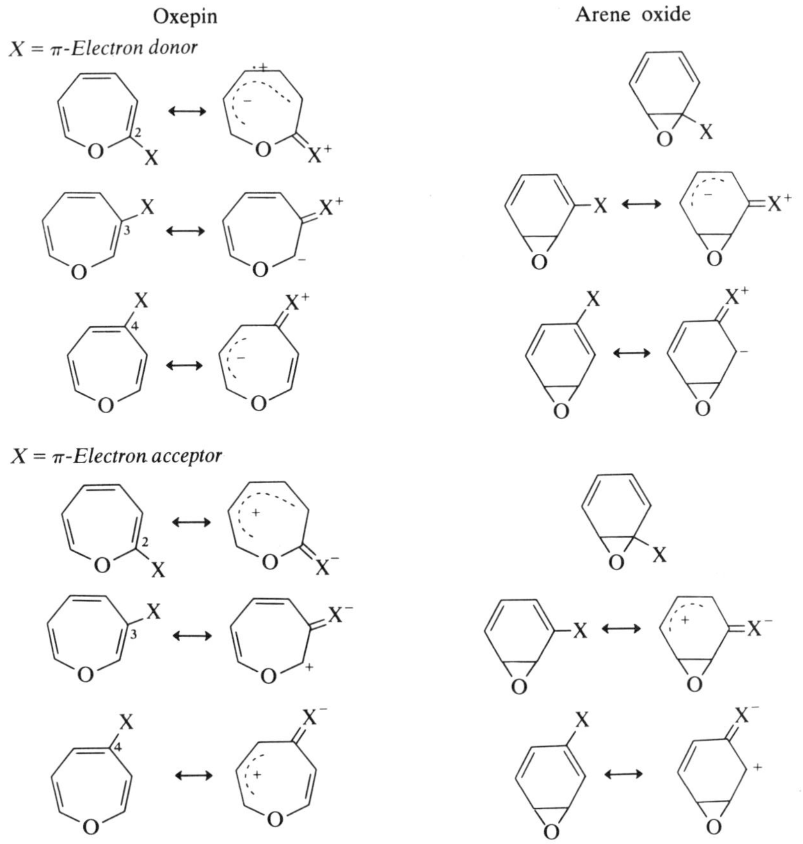

Oxepin | Arene oxide

X = π-*Electron donor*

X = π-*Electron acceptor*

Scheme 2 Resonance structures for monosubstituted oxepins–arene oxides

The preference of 2,7-disubstituted oxepins for this tautomeric form at equilibrium may be rationalized in terms of a 'steric' substituent effect. The eclipsing interactions of the 2,7-substituents in the arene oxide form will be diminished by isomerization to the oxepin. When the 2,7-substituents form part of an annelated ring system, *e.g.* (**22**)–(**24**), the tautomeric preference will be determined by the size of the methylene bridge ⟨67AG(E)385⟩. Thus when $n = 5$ the annelated oxepin (**24**) was present in approximately equal proportions with the arene oxide form. However with $n = 4$ (**23**) tetralin 9,10-oxide was dominant. The

(22) $n = 3$
(23) $n = 4$
(24) $n = 5$

(25)

oxepin tautomer (**22**; $n = 3$) could not be detected in the presence of indane 8,9-oxide due to the severe steric constraint imposed by the small bridge.

Oxepins having annelated aromatic rings may in principle equilibrate with the valence tautomeric arene oxide forms. In practice, the equilibrium distribution will be dominated by a resonance effect, *i.e.* the preference for the isomer having the maximum degree of aromaticity.

Oxepin (**25**) was not observed since this structure would incorporate the antiaromatic cyclobutadiene ring system, *i.e.* the arene oxide tautomer appeared to be formed exclusively ⟨74AG(E)277⟩.

Similarly, the fusion of an aromatic ring to the oxepin–benzene oxide system was found to drive the equilibrium toward extremes in either direction. The calculated resonance energies for oxepins (**26**), (**27**) and (**28**) were 4.81, 78.46 and 81.72 kJ mol^{-1} respectively ⟨70T4269⟩. These calculated values concur with experimental observations since oxepins (**27**) and (**28**) have been synthesized and are relatively stable compounds. The formation of 2-benzoxepin (**26**) from naphthalene 1,2-oxide would involve a considerable loss in resonance energy to the system and has not been detected spectroscopically ⟨67AG(E)385⟩.

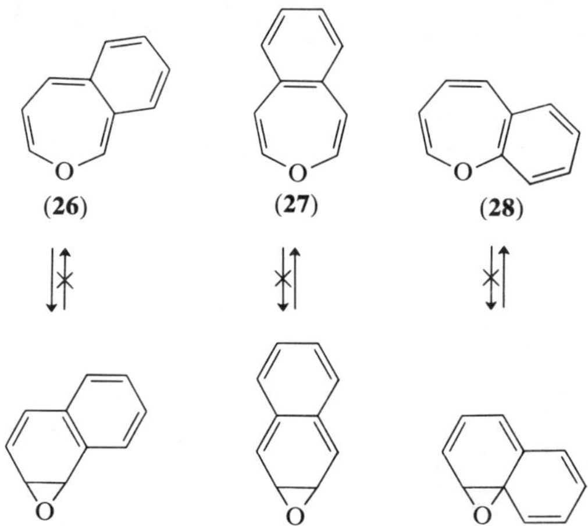

(26) (27) (28)

In the aromatic-ring-annelated oxepin series the resonance effect is clearly the major influence dominating other factors (*e.g.* temperature, solvent, *etc.*) which affect the oxepin–arene oxide equilibrium. It is however very difficult to exclude the presence of a minor (spectroscopically undetectable) contribution from either tautomer at equilibrium. This problem has been investigated by the synthesis of chiral arene oxides from polycyclic aromatic hydrocarbons (PAHs). The presence of oxepin (**26**) in equilibrium with naphthalene 1,2-oxide has been excluded by the synthesis of the optically active arene oxide which showed no evidence of racemization in solution at ambient temperature *via* the achiral oxepin (**26**) ⟨79JCS(P1)2437⟩.

Perturbation molecular orbital (PMO) calculations have been used to predict which oxepins in the PAH series may be present in a rapid state of equilibration with the corresponding arene oxide form at ambient temperature ⟨77CC873⟩. These prognostications have been verified experimentally by the observation of spontaneous racemization of some arene oxides in the phenanthrene ⟨77CC873⟩, chrysene ⟨81CC838⟩ and benz[*a*]anthracene ⟨81JCS(P1)94, 2233, 81UP51700⟩ series. Thus oxepins (**29**)–(**34**) appear to be present as very minor components in equilibrium with the corresponding arene oxide tautomers, despite the resonance effect.

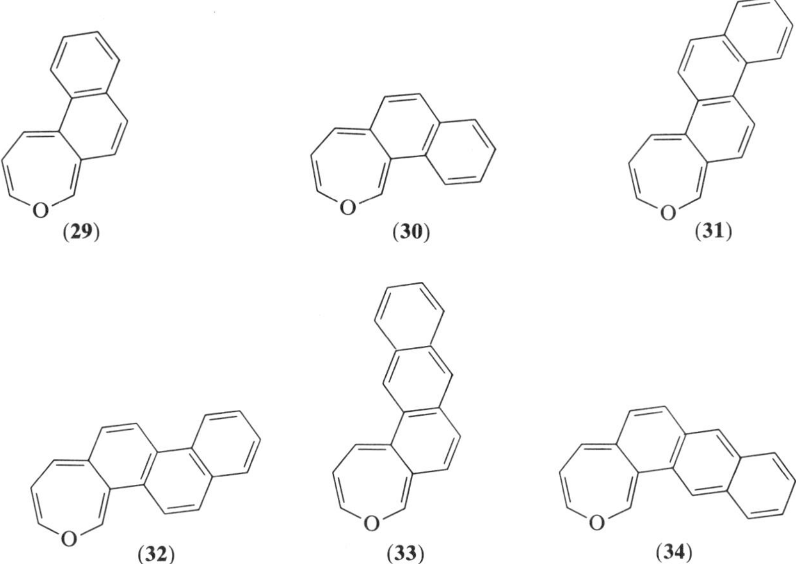

Kinetic data on the oxepin-benzene oxide equilibration have been obtained from the temperature-dependent NMR studies. Low values were observed for the enthalpy of isomerization of oxepin (7.1 kJ mol^{-1}) and 2-methyloxepin (1.7 kJ mol^{-1}) to the corresponding benzene oxides ⟨67AG(E)385⟩. The relatively small increase in entropy associated with oxepin formation (5–11 J K^{-1} mol^{-1}) is as anticipated for a boat conformation in a rapid state of ring inversion. Thermal racemization studies of chrysene 1,2- and 3,4-oxides have allowed accurate thermodynamic parameters for the oxepin–arene oxide equilibration process in the PAH series to be obtained ⟨81CC838⟩. The results obtained from racemization of the 1,2- (E_a 103.7 kJ mol^{-1}, $\Delta S^{\ddagger}$ 3.7 J K^{-1} mol^{-1} and 3,4- (E_a 105.3 kJ mol^{-1}, $\Delta S^{\ddagger}$ 0.7 J K^{-1} mol^{-1}) arene oxides of chrysene are as anticipated for the intermediacy of the oxepins (**31**) and (**32**) respectively.

Thermochemical measurements on the oxepin–benzene oxide system are unavailable. However, based upon experimental observations it would appear that the oxepin tautomer is generally more thermally stable than the benzene oxide as a result of the additional ring strain present in the oxirane ring.

The elevated temperatures normally necessary for obtaining mass spectral data generally result in aromatization of the oxepin–benzene oxide system and this accounts for the general lack of MS information on this system.

UV spectroscopy has been particularly informative about the oxepin–arene oxide ratio at room temperature ⟨67AG(E)385⟩ where NMR spectroscopy has been of little value (due to fast exchange processes). In order to use the UV method of analysis indane 8,9-oxide (the valence tautomer of oxepin **22**) was assumed to be totally in the oxirane form while 2,7-dimethyloxepin (**16**) was assumed to exist exclusively in the oxepin form. The characteristic UV absorption peaks for a series of monocyclic oxepins substituted in the 2- or 2,7-positions are summarized in Table 5. With the exception of oxepin (**7**) the mono- and di-substituted oxepins (**10**)–(**20**) all seem to exist at ambient temperature without any detectable contribution from the arene oxide tautomer.

5.17.1.3 Structure of Thiepanes

While the molecular dimensions of the parent molecule thiepane (**35**) have not been reported, the tetrahydrothiepin ring system in (**36**) has been examined extensively by X-ray crystallography ⟨74AG(E)740, 76CB2558, 77AG(E)187⟩. The dimeric tetrahydrothiepin structure (**36**) was obtained from 3,3,6,6-tetramethyl-1-thia-4-cycloheptyne (**37**), which was the first seven-membered cycloalkyne to be isolated ⟨70TL761⟩. Dimer (**36**) was of particular interest since the bulky *gem*-dimethyl substituents stabilized the cyclobutadiene ring system. X-ray crystallographic analysis indicated that cyclobutadiene structure (**36**) rather than (**38**) was present and that the tetrahydrothiepin ring preferred a chair conformation in the crystalline state. Both PE spectroscopy and LCMO calculations on (**36**) were indicative of a rectangular

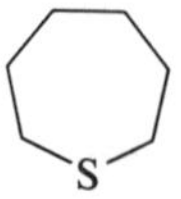

(35) 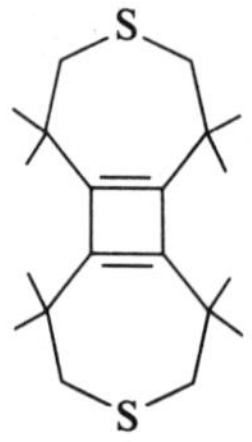 (36) 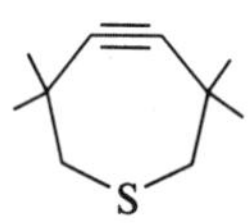 (37) 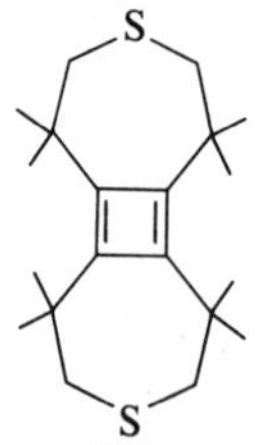 (38)

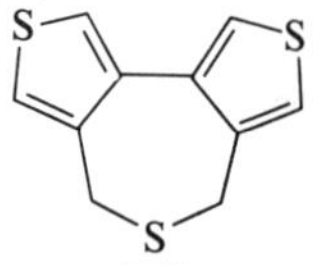

(39) 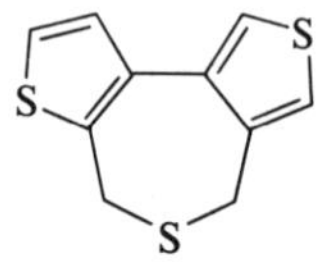 (40) 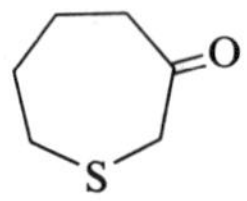 (41) 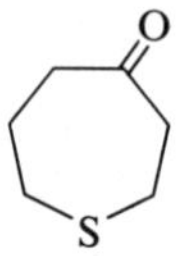 (42) 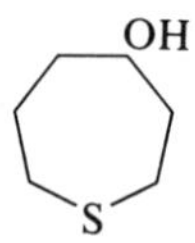 (43)

[4]annulene system existing in a singlet ground state ⟨74AG(E)544⟩. PE spectroscopy of the dihydrothiepin ring system in (**39**) and (**40**) indicated a preference for the twist conformation in the gas phase ⟨77CJC2867⟩.

The ^{1}H NMR spectrum of thiepane was similar to that of oxepane and showed only two broad signals at δ 2.5–2.8 (4H; H-2, H-7) and 1.5–2.1 (8H; H-3, H-4, H-5, H-6) p.p.m. ⟨67JCS(C)1130⟩.

IR spectroscopy has been of value in estimating the transannular interactions in the monosubstituted thiepane ring system. Thus, intramolecular interactions between the sulfur atom in thiepanes (*e.g.* **41, 42** and **43**) and a carbonyl group at the 3- ⟨75JCS(P2)1294⟩ or 4- ⟨60JA4075⟩ position or a 4-hydroxy group ⟨69RTC995, 67AG(E)872⟩ have been detected by IR.

5.17.1.4 Structure of Thiepins

Thiepin (**44**), like oxepin (Section 5.17.1.2), is a member of the 8π-electron series. While oxepin (**7**) has been synthesized, isolated and characterized at ambient temperature, the sulfur analogue, thiepin, is a highly unstable structure which has not been detected to date. The thiepin ring system may be stabilized by formation of the 1,1-oxide (**45**), or by fusion to an aromatic ring (*e.g.* **46**).

HMO ⟨72AG(E)423⟩ and semiempirical SCF MO ⟨70JA1453⟩ calculations have been used to obtain the molecular geometry of the thiepin ring system for the parent thiepin structure (**44**) and 1-benzothiepin (**46**). These theoretically derived bond lengths compare well with the accurate dimensions obtained experimentally by X-ray methods for thiepin 1,1-dioxide (**45**) ⟨68JA4501⟩ and 1-benzothiepin (**46**) ⟨76AG(E)297⟩ (Table 6). X-ray structure analysis of the 1-benzothiepin mono- (**47**) ⟨78CB309⟩ and di- (**48**) ⟨75AX(B)2729⟩ oxides indicates that the molecular geometry is in each case remarkably similar to that found for the benzothiepin (**46**). Thus by analogy it is anticipated that the molecular dimensions for thiepin (**44**) will be almost identical to those of thiepin 1,1-dioxide (**45**). Bond localization in the thiepin ring is evident from the X-ray structural data where both carbon–carbon single and double bonds and carbon–sulfur single bonds showed little change from normal values.

The preferred conformation for both thiepin and thiepin oxides was deduced (by both calculation methods and X-ray crystallography) to be a relatively flat boat form.

NMR spectral analysis of the monocyclic thiepins (**49**) ⟨79JA5059⟩, (**50**) ⟨72CC1233⟩ and (**51**) ⟨70JA5263⟩ indicates that the thiepin ring system in solution generally prefers a non-planar (boat) conformation. The boat conformation would be anticipated for a thiepin

^{1}*H NMR* (δ values in p.p.m.)

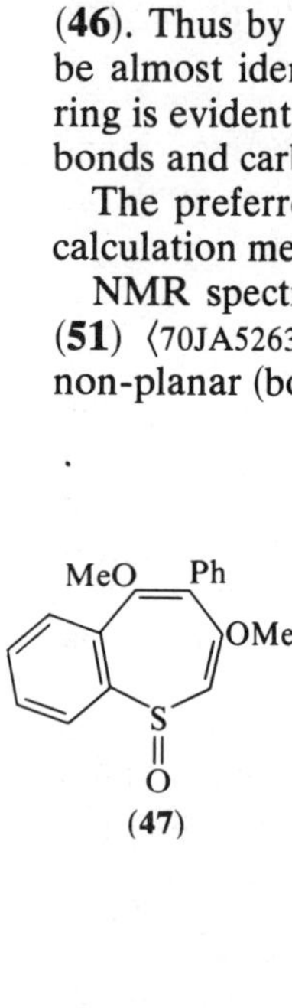

(47)  (48) 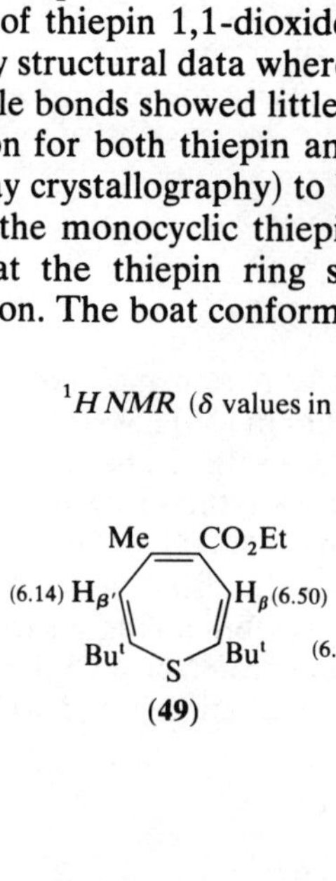 (49) 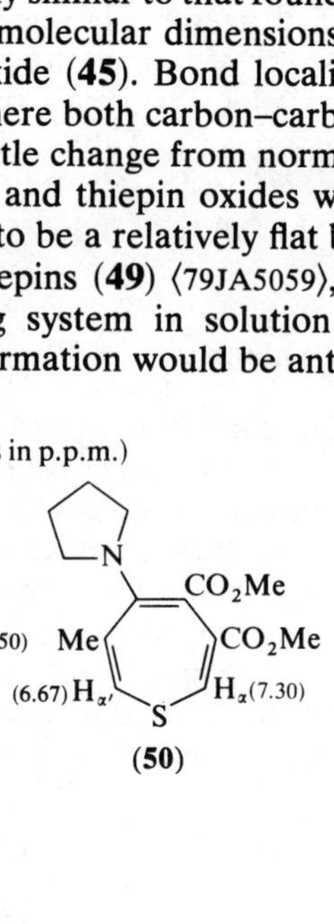 (50) 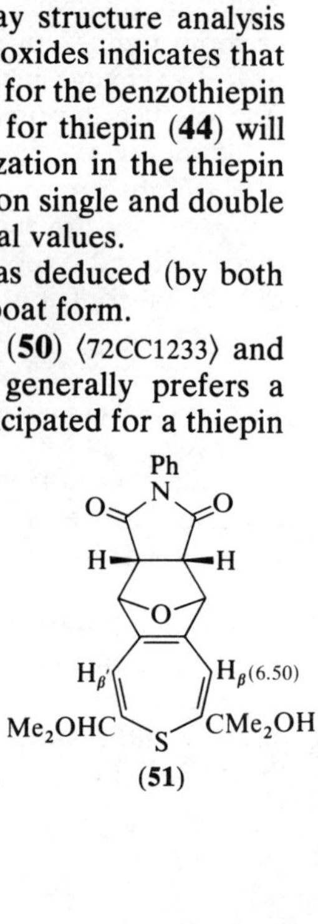 (51)

(49): $H_{\beta'}$ (6.14), H_β (6.50); (50): $H_{\alpha'}$ (6.67), H_α (7.30); (51): H_β (6.50)

Table 6 Theoretically and Experimentally Derived Bond Lengths and Bond Angles for Thiepins (**44**)–(**46**)

Compound	*Bond*	*Calcalulated bond length* (Å)	*Observed bond length* (Å)	*Angle*	*Observed bond angle* (°)
(**44**)	a, a′	1.791			
	b, b′	1.347			
	c, c′	1.466			
	d	1.350			
(**45**)	a, a′		1.723, 1.716	aa′	103.3
	b, b′		1.344, 1.340	ab, a′b′	122.8, 124.4
	c, c′		1.429, 1.438	bc, b′c′	129.4, 128.1
	d		1.333	cd, c′d′	127.8, 128.2
(**46**)	a	1.742	1.781	aa′	101.1
	b		1.321	ab	122.9
	c		1.448	bc	127.1
	d		1.308	cd	127.8
	c′		1.493	dc′	129.1
	b′		1.404	c′b′	123.6
	a′	1.763	1.773		

ring having a minimal degree of conjugation. Clear evidence for bond alternation is obtained from the NMR spectrum of (**49**) where the chemical shifts of oxepin ring protons correlate well with those of appropriate substituted ethylenes. The ring inversion process in thiepins has been studied by low temperature NMR spectroscopy and a barrier of *ca.* 26.8 kJ mol^{-1} has been determined for 3-methyl-6-isopropylthiepin 1,1-dioxide (**52**) ⟨69JA7782⟩.

(**44**) (**52**)

NMR spectral evidence for the presence of the valence tautomeric thianorcaradiene (benzene episulfide) form (analogous to the arene oxide tautomer of an oxepin) has yet to be observed. The latter negative evidence, however, does not exclude the possibility of a rapid thiepin–thianorcaradiene equilibration since the number of monocyclic thiepins which have to date been synthesized is very small, and since the examples examined by NMR probably owe their stability to the very low proportion of benzene episulfide tautomer present.

Theoretical studies on thiepin (**44**) have provided estimates of the resonance energy which are consistently negative as expected for an antiaromatic compound: −29.7 ⟨73JA3907⟩, −26.8 ⟨75BCJ1501⟩ and −6.1 kJ mol^{-1} ⟨70JA1453⟩. Resonance energy calculations ⟨73JA3907⟩ also predict that 1-benzothiepin (**46**; E_{res} + 24.7 kJ mol^{-1}) and 3-benzothiepin (**53**; E_{res} + 26.8 kJ mol^{-1}) should have little antiaromatic character (and should thus be isolable) whereas 2-benzothiepin (**54**; E_{res} − 29.8 kJ mol^{-1}) is anticipated to be highly unstable.

Factors which affect the oxepin–benzene oxide equilibrium positions are similarly expected to influence the thiepin–benzene episulfide distribution at equilibrium. Since however the thianorcaradiene tautomer has not to date been detected, the main evidence for this form is based upon the thermal instability and reactions of the thiepin system. Thus it is assumed that where the thianorcaradiene isomer is present, a spontaneous thermal decomposition involving extrusion of a sulfur atom will occur. Substitution at the 2,7-positions in the oxepin–arene oxide system leads to a preference for the seven-membered ring form and this effect was further enhanced by bulky substituents (*e.g.* Bu^t). A similar effect was observed in thiepins and thus the remarkable thermal stability of (**49**) (2,7-*t*-butyl groups) and (**51**) (2,7-hydroxyisopropyl groups) contrasts with the behavior of thiepin (**55**)(2,7-isopropyl groups), which was thermally unstable even at −70 °C ⟨78CL723⟩. The stability of thiepin (**49**) results from the 2,7-steric (eclipsed) interactions which obtain in the thianorcaradiene form but which are diminished in the thiepin tautomeric form (relative to the episulfide tautomer).

S (46) S

S (53) S

S (54) S

Me CO_2Et Me_2CH S $CHMe_2$ (55)

R S R S (56)

R S R S (57)

As in the oxepin–arene oxide system, the resonance effect will also influence the position of equilibrium in the analogous organosulfur series. Compounds (**46**) and (**53**) thus appear to exist exclusively in the thiepin form. Since the resonance factor would favor the 1,2-episulfide of naphthalene over the thiepin tautomer (**54**) it is highly improbable that this thiepin will be detectable at ambient temperature. Both thiepins (**46**) and (**55**) have been isolated as thermolabile compounds ⟨78JOC3379, 81MI51700⟩.

A further example of the resonance effect and its influence on the thiepin–thianorcaradiene equilibrium is found in compounds (**56**) and (**57**). Compound (**56**; $R = CO_2Et$) is much more stable ($t_{1/2}$ 72 min/94 °C) than thiepin (**57**; $R = -CO_2Et$, $t_{1/2}$ 33 min/−27 °C) since formation of the thianorcaradiene tautomers would be much more difficult in (**56**) where a total loss of aromaticity would be involved ⟨81MI51700⟩.

Thieno[3,4-*d*]thiepin (**58**) superficially resembles the benzothiepins although X-ray structure analysis reveals that the thiepin ring is in fact planar ⟨69JA3995⟩. The bond length for the central C—C bond (1.46 Å) is also consistent with the valence bond structure (**59a**). The stability of this system has been attributed disparately to contributions from the oppositely charged separated resonance structures (**59a** and **59b**) ⟨70JA1453, 73JA2860⟩.

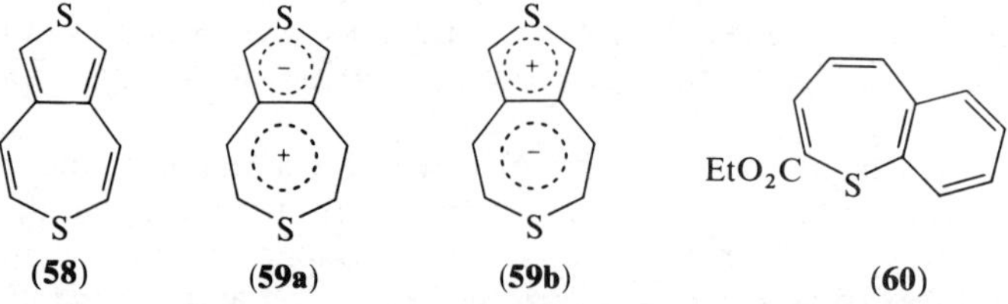

(**58**) (**59a**) (**59b**) (**60**)

The electron donating or withdrawing effect of substituents upon the thermal stability of oxepins has not been systematically studied. However, it would appear from available evidence that electron withdrawing substituents (*e.g.* alkoxycarbonyl) in the 2-position lead to very stable oxepins ⟨79JA2470⟩. A similar effect of an electron withdrawing group at the 2-position in thiepins has been observed and thus 2-ethoxycarbonylbenzo[*b*]thiepin (**60**) was more stable than the unsubstituted benzo[*b*]thiepin (**46**) ⟨81MI51700⟩. This effect has been studied theoretically ⟨76AG(E)297⟩ in terms of orbital interactions and appears to be consistent with a weakening of episulfide C—C bond when an electron withdrawing group is present at position 2, *i.e.* a shift of the tautomeric equilibrium toward the thiepin form in (**60**).

Unfortunately, however, some of the observed effects of substitution upon the thermal stability of thiepins would appear to be in conflict. Thus the presence of electron withdrawing groups such as the alkoxycarbonyl groups in thiepins (**50**) and (**60**) was considered to have a stabilizing influence by decreasing the electron density (and thus antiaromatic character) of the 8π-electron ring system in accordance with the predictions from MO calculations ⟨77T2683⟩ and with experimental observations ⟨74T2093⟩. By contrast, it has also been observed that electron donating groups stabilize the 1-benzothiepin systems over a range of examples ⟨75TL2697, 78TL3567, 78JOC3379⟩. No satisfactory explanation for these apparently conflicting results is currently available.

Thermodynamic parameters have been obtained from kinetic studies of the thermal decomposition of thiepins ⟨74T2431⟩. The sulfur extrusion reaction of (**53**) was examined kinetically and showed a large negative $\Delta S^{\neq}$ value ($-24\ \mathrm{J\,K^{-1}\,mol^{-1}}$) consistent with a decrease in entropy in the transition state which would be anticipated to be close to a thianorcaradiene structure. Thus while the latter valence tautomer of any thiepin has yet to be detected directly, the kinetic data, allied to the other evidence discussed, support the view that thiepins are analogous to oxepins in undergoing spontaneous tautomerization to their episulfide or epoxide forms.

UV spectroscopy has been used to confirm the presence of a non-planar triene structure for the thiepin (**49**) [234 nm (12 880), 356 nm (891)] and thiepin 1,1-dioxide (**45**) [215 nm (13 000), 232 nm (2070), 262 nm (4610)]. The IR spectra of (**49**) and (**45**) have been used to identify the α,β-unsaturated ester ($1715\ \mathrm{cm^{-1}}$) and sulfone (1120, $1300\ \mathrm{cm^{-1}}$) components.

5.17.2 REACTIVITY OF OXEPANES, OXEPINS, THIEPANES AND THIEPINS

5.17.2.1 Reactivity of Oxepanes

5.17.2.1.1 Introduction and general survey

Oxepane (**1**), in common with other ethers including the solvents diethyl ether and tetrahydrofuran, may be considered as a compound of low chemical reactivity. Oxepanes, being relatively strain free cyclic ethers, are reluctant to undergo C—O bond cleavage reactions. The oxygen atom in (**1**) is however slightly basic and is thus vulnerable to attack by electrophiles including protons to yield oxepinium ions (**61**). Oxonium ion intermediates (**61**) have been proposed as unstable intermediates in oxepane synthesis ⟨57BSF789⟩ and in polymerization reactions ⟨75MI51700⟩. An oxepinium ylide species (**62**) has also been postulated as a transient intermediate resulting from photolysis ⟨78JCS(P1)1029⟩. The oxepane carbocation (**63**), carbocation radical (**64**), radical (**65**) and carbene (**66**) intermediates have been reported as unstable moieties resulting from photolysis ⟨81JOC2267⟩, electron bombardment ⟨68OMS(1)403⟩, radical abstraction ⟨76TL439⟩ and pyrolysis ⟨72JA5777⟩ reactions respectively. The majority of reactions on oxepane (**1**) involve ring cleavage *via* one of the transient intermediates (**61**)–(**66**).

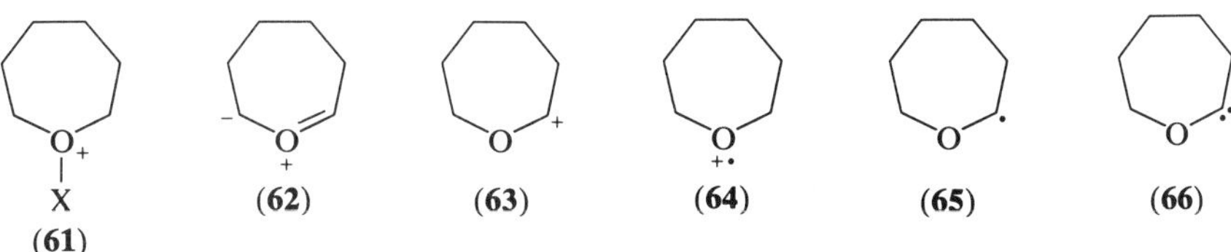

(**61**) (**62**) (**63**) (**64**) (**65**) (**66**)

5.17.2.1.2 Thermal and photochemical reactions

The thermal stability of the parent oxepane (**1**) is evident from the extreme reaction conditions (14 h at 190 °C) reported in one synthetic route ⟨64JOC123⟩. Pyrolysis of the sodium salt of the tosylhydrazone (**67**) derived from oxepan-2-one proceeded *via* oxycarbene (**66**) and diradical intermediates to yield a range of rearrangement products including 2,3,4,5-tetrahydrooxepin (**68**; Scheme 3) ⟨72JA5777⟩. 2-Acetoxy-2,3,4,5-tetrahydrooxepin (**69**) was transformed into 4,5-dihydrooxepin (**70**) by a pyrolytic elimination reaction which occurred *via* a six-membered cyclic transition state ⟨60JA4087⟩ as shown in equation (1). Similar types of thermal elimination reactions (and mechanisms) on the *trans* diacetate (**71**) and the tertiary amine oxide (**72**) yielded 2,3-dihydrooxepin (**73**) as the major product (Scheme 4) ⟨60JA4087⟩.

The photolysis of oxepane (**1**) in the neat liquid state using UV light of wavelength 185 nm gave hex-5-en-1-ol, hexanal and 3′-hydroxypropylcyclopropane as major photoproducts (Scheme 5). Diradicals have been suggested as possible intermediates leading to

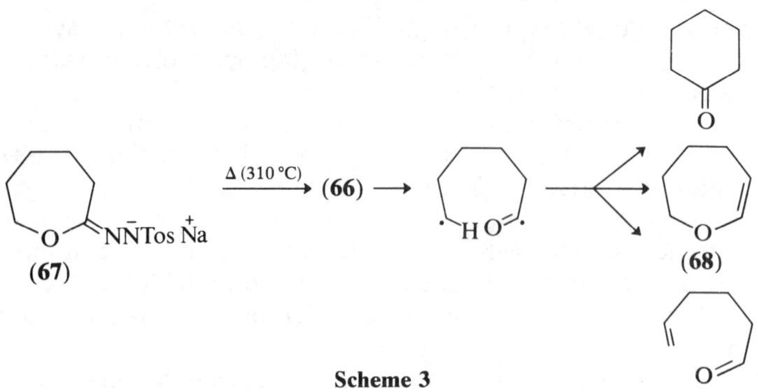

Scheme 3

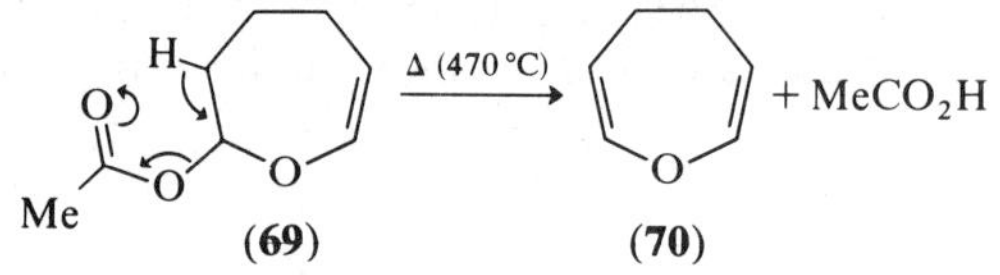

$$\text{(69)} \xrightarrow{\Delta\ (470\ ^\circ\text{C})} \text{(70)} + MeCO_2H \qquad (1)$$

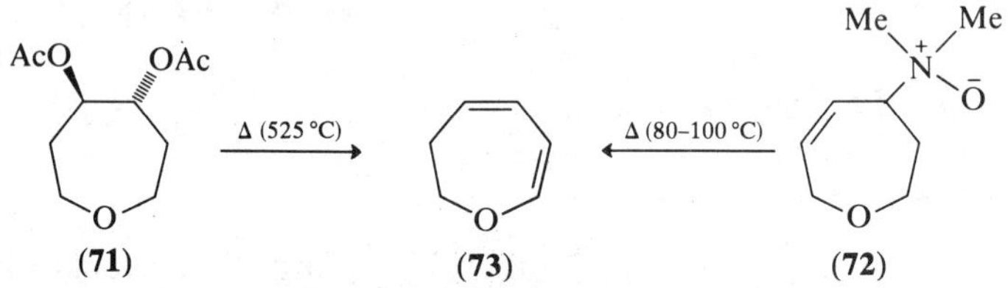

Scheme 4

these isomeric products ⟨80MI51701⟩. Compounds (**68**), (**70**) and (**73**) all contain unsaturated carbon–carbon bonds and thus can absorb light in a more accessible wavelength range (>200 nm) to yield a number of photoisomerization products (Scheme 6). Photochemically induced isomerization of (**68**) to the highly strained *trans* form resulted in the addition of methanol to give 2-methoxyoxepane (**74**) ⟨81JOC2267⟩. The range of isomeric products resulting from photolysis of (**70**) was consistent with homolytic C—O bond cleavage to form diradical intermediates ⟨70TL3539⟩. By contrast, the absorption of UV light by the diene (**73**) resulted in the formation of a bicyclic cyclobutene structure by a concerted disrotatory electrocyclic mechanism ⟨65JA3417⟩.

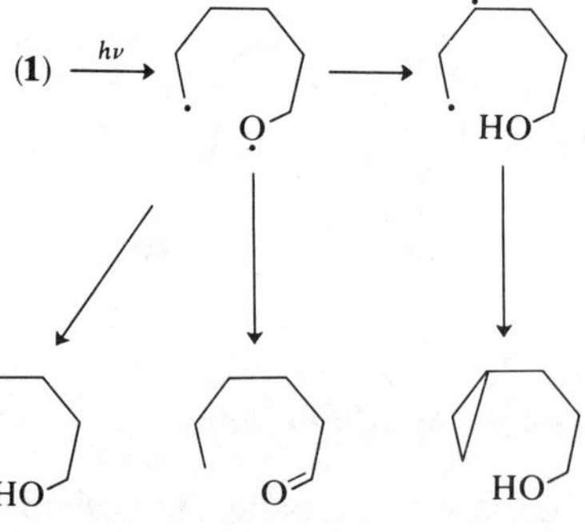

Scheme 5

5.17.2.1.3 Electrophilic attack on the ring oxygen atom

The distribution of cyclic ethers between an inert organic solvent and an aqueous acid solution has been used to determine their pK_a values ⟨62JA1684⟩. According to these studies oxepane (**1**; pK_a −2.02) is slightly more basic than THF (pK_a −2.08) but considerably more basic than THP (pK_a −2.79) or diethyl ether (pK_a −3.59). The basic strength of oxepane (**1**) appears to result from (i) the non-bonding electron pairs in sp^3 hybrids behaving like normal bonds and being repelled by electrons in neighboring C—H bonds, (ii) a reduction of these repulsive interactions upon protonation, and (iii) the relief of these repulsive

Scheme 6

interactions gained upon protonation being greater for oxepane (or THF) since it does not have the almost perfect staggered conformation of THP.

In common with other ethers oxepane (**1**) undergoes heterolytic C—O bond cleavage reactions in the presence of strong acids. Electrophilic attack on (**1**) yielded both 1,6-dihalohexanes and polymers according to the experimental conditions (Scheme 7). When oxepane (**1**) was treated with hydrobromic acid (R—X≡H—Br) both 1,6-dibromohexane and bis(6-bromo-1-hexyl) ether were formed ⟨44CB669⟩. Similarly reaction of (**1**) with methyl iodide at elevated temperatures gave 1,6-diiodohexane which was isolated as the 1,6-bis(2-naphthoxy)hexane derivative ⟨53M1206⟩. Polymerization reactions of (**1**) initiated by mineral acids (*e.g.* HBr), Lewis acids (*e.g.* BF_3), trialkyloxonium salts (*e.g.* $Et_3\overset{+}{O}\overset{-}{B}F_4$, $Et_3\overset{+}{O}Sb\overset{-}{C}l_6$) or methylating agents (*e.g.* $MeOSO_2CF_3$) have been investigated extensively ⟨72MI51700, 75MI51700⟩. The mechanism postulated is similar to that shown in Scheme 7, *i.e.* an S_N2 reaction between the cyclic oxonium ion of the electrophilic propagating species and the nucleophilic monomer.

Scheme 7

2-Hydroxyoxepane (**75**), a cyclic hemiacetal, is particularly susceptible to acid-catalyzed ring opening and has been found to exist as the minor contributor (25%) in equilibrium with 6-hydroxyhexanal (75%) ⟨52JA5324⟩. The mechanism of electrochemical fluorination of oxepane (**1**) involved the initial formation of an oxonium ion upon dissolution in HF (equation 2). Perfluorooxepane (**76**) was obtained as a major product among a range of ring isomerization products ⟨79MI51700⟩.

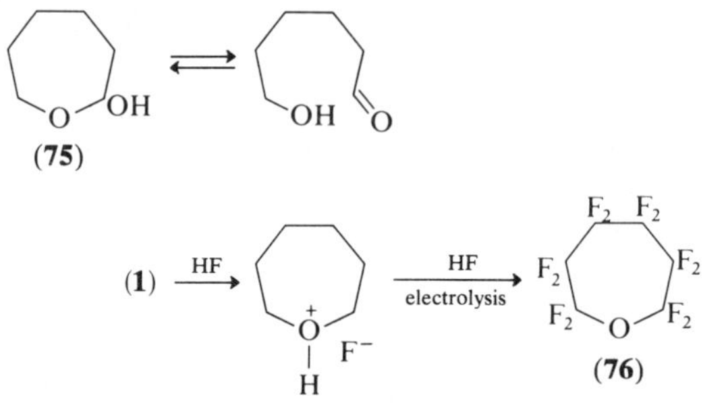

(2)

5.17.2.1.4 Nucleophilic attack on ring atoms adjacent to a heteroatom or conjugated with a heteroatom

Oxepane (**1**) appears to be relatively stable in the presence of nucleophiles and reducing agents. However, a strong nucleophilic reagent, *e.g.* *n*-butyllithium, can abstract a ring proton leading to β-cleavage and isomerization to yield hex-5-en-1-ol (equation 3) ⟨72JOC560⟩. Similarly, a carbanionic intermediate was formed by treatment of 2,3-dihydrooxepin (**73**) with potassium amide in liquid ammonia (equation 4). The resulting oxaheptatrienyl anion was found to have a remarkable degree of configurational stability ⟨70RTC667⟩.

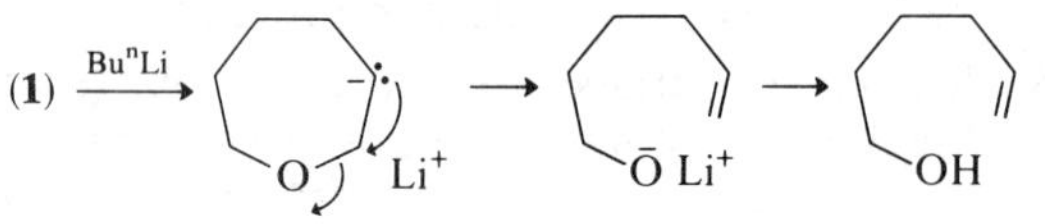

(3)

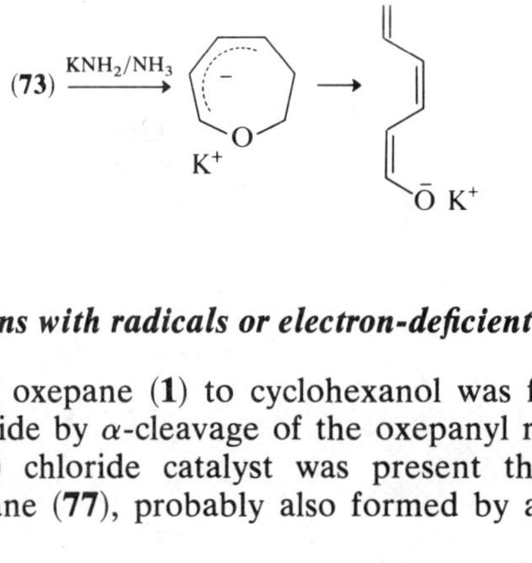

(4)

5.17.2.1.5 Reactions with radicals or electron-deficient species

Isomerization of oxepane (**1**) to cyclohexanol was found to occur in the presence of *t*-butyl hydroperoxide by α-cleavage of the oxepanyl radical intermediate (**65**) ⟨76TL439⟩. When a copper(I) chloride catalyst was present the major product was 2-(*t*-butylhydroperoxy)oxepane (**77**), probably also formed by a free radical pathway (Scheme 8) ⟨80CR(291)223⟩.

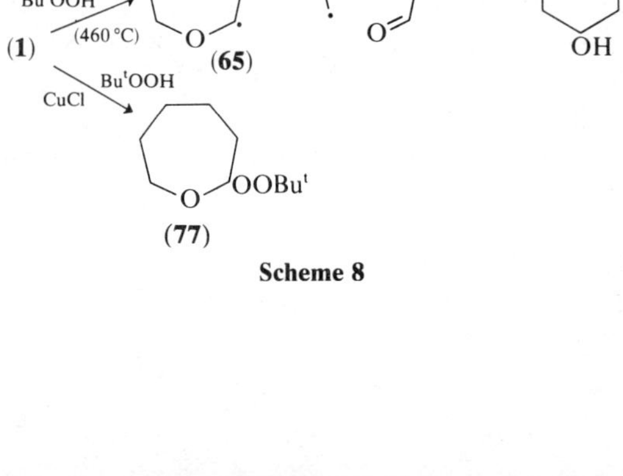

Scheme 8

5.17.2.1.6 Oxidations, reductions and reactions involving cyclic transition states

Oxepane (**1**), as a typical ether, is susceptible to oxidation and yields oxepan-2-one (**78**) as the initial product. Adipic acid was the product finally isolated after oxidation with RuO_4 and $NaIO_4$ in a two-phase system ⟨80SC205⟩ or oxygen in the presence of a Pt catalyst ⟨76CB3707⟩ (Scheme 9). Oxidation of 2,3,6,7-tetrahydrooxepin (**79**) has been reported with peroxybenzoic acid or osmium tetroxide to yield the epoxide (**80**) or the *cis* diol (**81**) respectively (Scheme 10) ⟨58JA3132⟩.

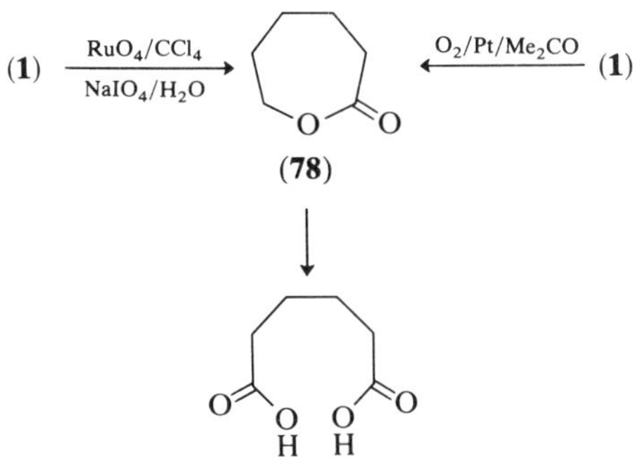

Scheme 9

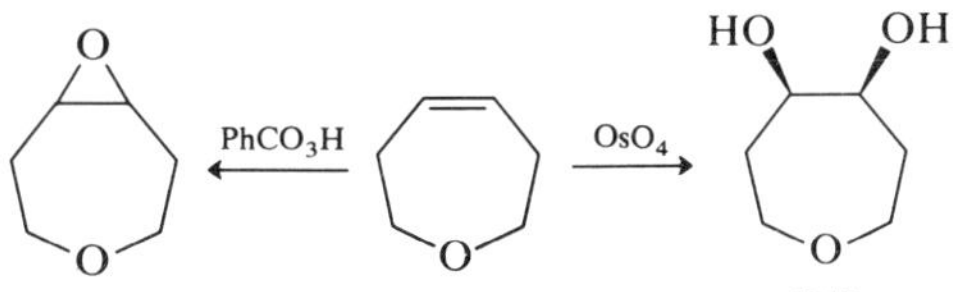

Scheme 10

Oxepane (**1**) has been obtained by catalytic reduction (H_2/Pt) of 2,3,4,5-tetrahydrooxepin (**68**), 2,3,6,7-tetrahydrooxepin (**79**), 2,3-dihydrooxepin (**73**) and 4,5-dihydrooxepin (**70**) ⟨60JA4087, 63JOC1383⟩. 2-Hydroxyoxepane (**83**) has been synthesized by DIBAL reduction of an oxepan-2-one (**82**; Scheme 11). The product, being a hemiacetal similar to compound (**75**), was also found to exist in equilibrium with the acyclic 6-hydroxyhexanal form ⟨76SC465⟩. A key step in the synthesis of the naturally occurring oxepane zoapatanol was the conversion of enol ether (**84**) to the 3-hydroxyoxepane (**85**) by a hydroboration–oxidation sequence as shown in equation (5) ⟨81TL3027⟩. 2,3-Dihydrooxepin (**73**) behaved as a typical conjugated diene in forming a [4+2] cycloadduct with TCNE (equation 6) ⟨60JA4087⟩.

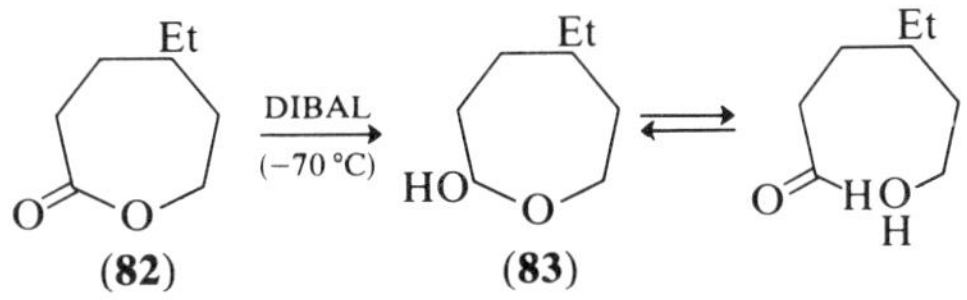

Scheme 11

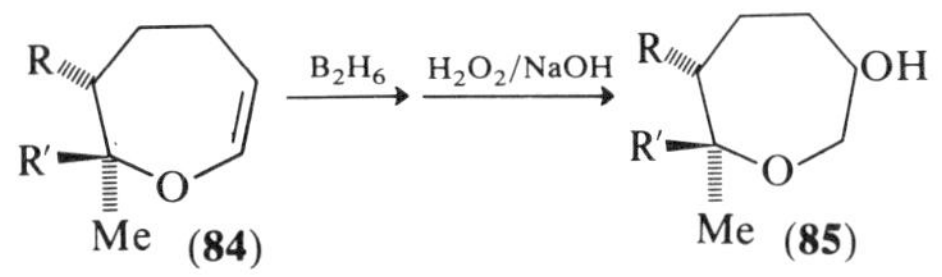 (5)

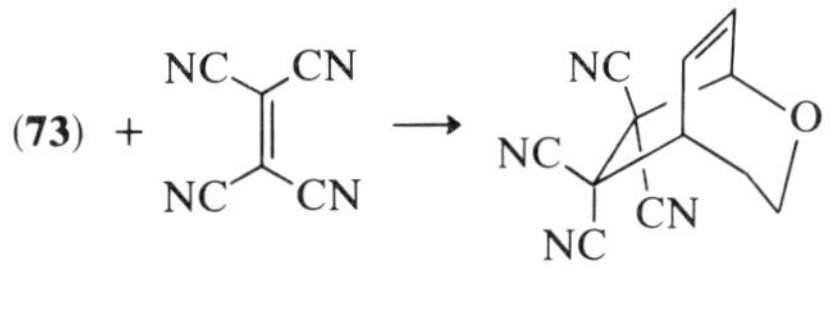 (6)

5.17.2.2 Reactivity of Oxepins

5.17.2.2.1 Introduction and general survey

The existence of a rapid oxepin–arene oxide equilibration process (see Section 5.17.1.2) does not permit the oxepin valence tautomer to be treated as a separate entity. Thus the chemical reactions of oxepin (**7**) include those expected for a conjugated cyclic triene or enol ether (*e.g.* addition, oxidation and reduction). The reactivity anticipated for a conjugated diene or a *cis*-1,2-divinyl epoxide (*e.g.* isomerization, nucleophilic and electrophilic attack, cycloaddition, deoxygenation) is also found with oxepin (**7**). While the reactions discussed will generally use only the term oxepin, it is assumed that the arene oxide tautomer will also be present in most examples. Oxepin (**7**) and substituted derivatives in general tend to be highly reactive molecules due to a number of factors including the ring strain present in the arene oxide tautomer and the gain in resonance stabilization of the products produced by aromatization. The involvement of oxonium (**86, 87**) and carbanion (**88, 89**) intermediates has been postulated in acid-catalyzed atomatization ⟨76ACR378⟩ and substitution reactions respectively ⟨79JOC468⟩ of oxepins.

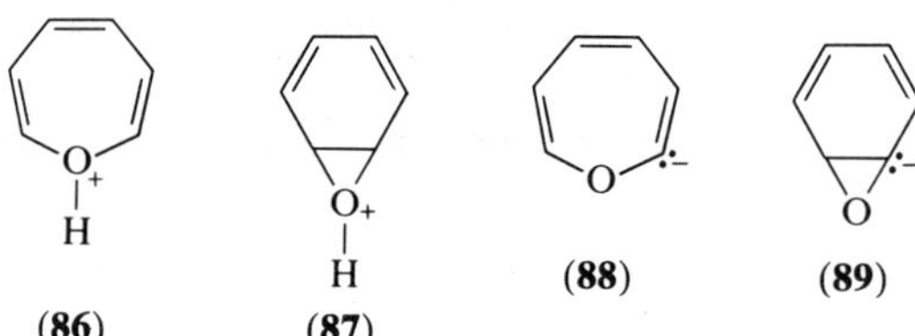

5.17.2.2.2 Thermal and photochemical reactions

The thermal ring closure reaction of a 1,3,5-triene to a 1,3-cyclohexadiene occurs by a concerted disrotatory electrocyclic mechanism. An example of the latter is the oxepin–benzene oxide equilibrium (**7**) which favors the oxepin tautomer at higher temperatures (Section 5.17.1.2). Oxepin (**7**) was found to rearrange to phenol during attempted distillation at normal pressure ⟨67AG(E)385⟩. This aromatization reaction may be considered as a spontaneous rearrangement of the oxirane ring to the dienone isomer followed by enolization (equation 7).

(7)

Photochemical rearrangement of oxepin (**7**) using UV light (>310 nm) gave only the cyclobutene (**90**) *via* an excited singlet state. Using shorter wavelength UV light (254 nm, −80 °C), both phenol (triplet state) and benzene (singlet state) were obtained as major products (Scheme 12) ⟨67JA6390⟩.

Scheme 12

A further type of photoisomerization mechanism for oxepin (**7**) was found by using material specifically labelled with ^{2}H at the 3- and 6-positions ⟨74JA5578⟩. The location of the ^{2}H atoms in the finally isolated sample of (**7**) was consistent with the 'migration' of the oxirane ring to a neighboring position. The latter process (Scheme 13) has been described as an oxygen-walk ⟨74JA5578⟩. Other related examples of photochemically induced oxygen-walks include the irradiation of the K-region arene oxides of phenanthrene (**91**) and pyrene ⟨79T1059⟩. While the initial oxepin–arene oxide equilibrium appeared to favor the arene oxide tautomer (**91**) of the reactant exclusively, the reverse preference was shown by the photoisomerization product (**92**; Scheme 14).

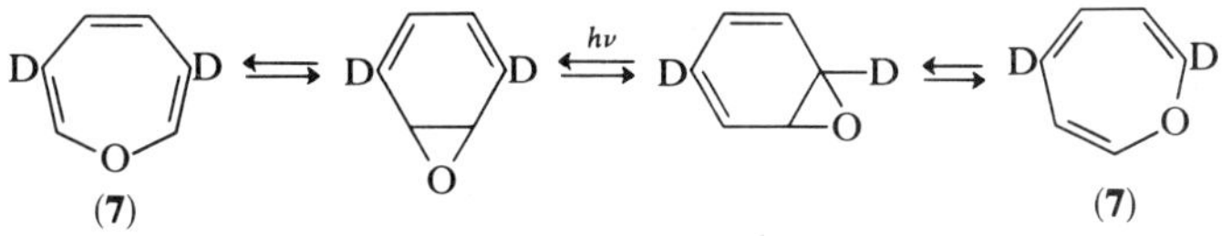

(**7**) (**7**)

Scheme 13

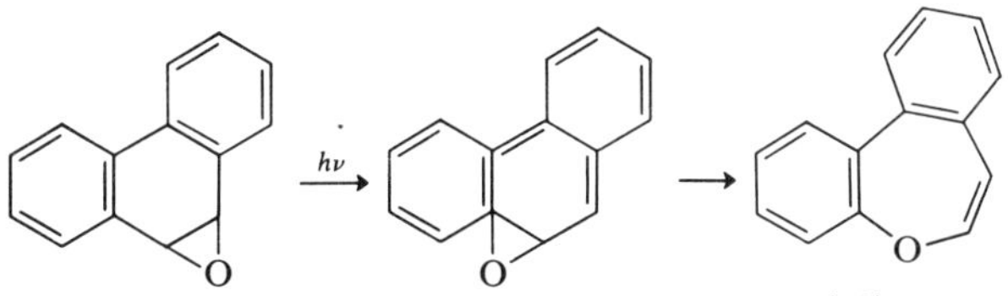

(**91**) (**92**)

Scheme 14

5.17.2.2.3 Electrophilic attack on the ring oxygen atom

The protonation of oxepins–arene oxides at the ring oxygen atom and subsequent acid-catalyzed isomerization are very dependent upon the nature of substituents on the ring. This is exemplified by a comparison of the relative stabilities of oxepin (**7**) and oxepin-2,7-dicarboxylic acid (**18**) under acid conditions. Thus oxepin (**7**) spontaneously decomposed on clean glassware at ambient temperature (unless prewashed with base) while (**18**) showed no change over a period of 10 days in trifluoroacetic acid ⟨79JA2470⟩.

Acid-catalyzed isomerization is among the most extensively studied aspects of oxepin reactivity. This topic has been covered in depth in several review articles ⟨73H(1)267, 76ACR378⟩ and thus the present discussion will be limited to a survey of the major aspects. Protonation on the oxirane ring oxygen atom generally results in C—O ring cleavage and the formation of several carbocationic intermediates before the product phenol is isolated (Scheme 15). Evidence for the involvement of carbocation intermediates in the acid-catalyzed isomerization of oxepins has been found in the substitution patterns of phenols formed and in the effect of substituents and solvents upon the rates of aromatization. The reaction sequence shown in Scheme 15 includes a 1,2-shift of the hydrogen isotope X (^{2}H, ^{3}H) and an enolization step involving loss of either ^{1}H or X depending upon the magnitude of the primary kinetic isotope effect ($k(^1\text{H})/k(^2\text{H})$, $k(^1\text{H})/k(^3\text{H})$]. The migration and retention of substituent X was first encountered during studies of the *para*-hydroxylation of phenylalanine by enzymes present in mammalian liver and bacterial systems. Since the latter observations were made at the National Institutes of Health (Bethesda, Md, USA), the process has been described as the 'NIH Shift' ⟨74MI51700⟩.

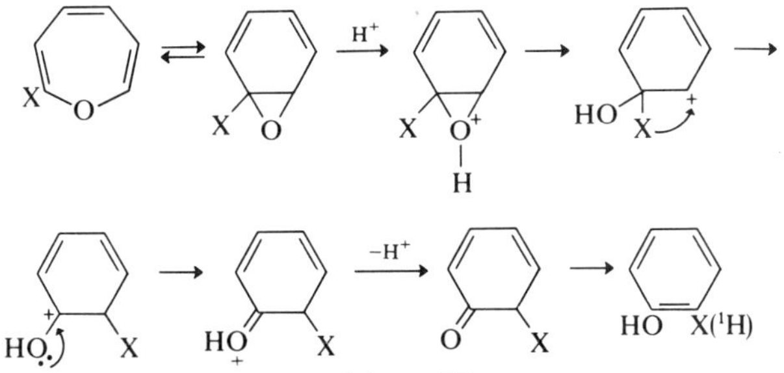

Scheme 15

The proportion of hydrogen isotope X which migrates and which is retained in the phenolic product depends not only upon the choice of X (^{1}H, ^{2}H, ^{3}H) but also upon the nature of ring substituents, pH, temperature, *etc*. The arene oxide tautomeric forms of ^{2}H-labelled oxepins (**93**) and (**26**) were found to show ^{2}H migration and retention values of 40–85 %D ⟨68JA6523⟩ and 80 % ^{2}H ⟨72B1961⟩ respectively according to the experimental conditions used (Scheme 16).

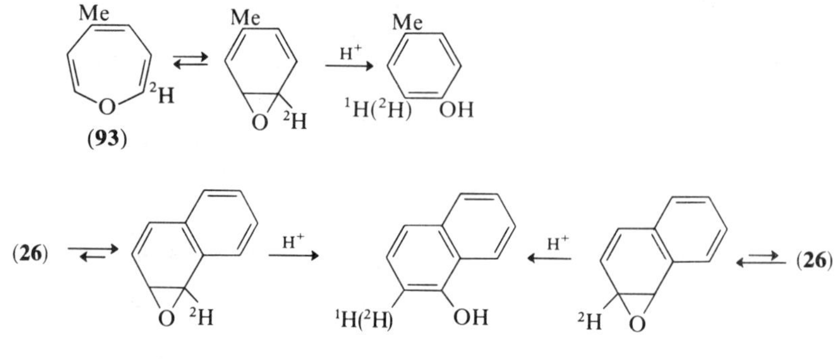

Scheme 16

The NIH shift has been found to occur during aromatic hydroxylations catalyzed by enzymes present in plants, animals, fungi and bacteria. It is thus evident that the acid catalyzed (or spontaneous) isomerization of oxepins–arene oxides is a very important type of *in vivo* reaction. It should be emphasized that the NIH shift may occur under either acid-catalyzed or neutral (spontaneous) conditions ⟨76ACR378⟩. The direct chemical oxidation of aromatic rings has also yielded both phenols (obtained *via* the NIH shift) and arene oxides ⟨80JCS(P1)1693⟩.

The nature and position of substituents relative to the ring oxygen atom have an important effect on the acid-catalyzed isomerization of oxepins. It has been observed that, in addition to the hydrogen isotopes (^{1}H, ^{2}H, ^{3}H), chloro, bromo, methyl ⟨72E1129⟩ and alkoxycarbonyl ⟨79JA2470⟩ substituents also show the migration–retention sequence found in the NIH shift.

The fate of the substituent at the 2-position in the oxepin ring during acid-catalyzed aromatization reactions (Scheme 17) may include (a) loss of substituent R, (b) loss of substituent X, (c) 1,6-migration and retention of substituent X, and (d) 1,3-migration and retention of substituent X. Pathway (a), *i.e.* loss of substituent R (where R = ^{1}H or ^{2}H) is followed by a wide range of arene oxides [R′ = ^{1}H or ^{2}H, X = Me (**10**), CN (**12**), CH_2CO_2Me (**94**), CH=$CHCO_2Me$ (**95**), CO_2Me (**15**), CO_2H (**14**), CHO (**13**), CH_2OH (**96**), CMe_2OH (**97**)] (see Section 5.17.3.2). The X substituents CO_2H (**14**), CHO (**13**), CH_2OH (**96**) and CMe_2OH (**97**) appeared to be 'lost' (route b) by the evolution of carbon dioxide, carbon monoxide, formaldehyde and acetone respectively during acid-catalyzed aromatization of the appropriate 2-substituted oxepins (R, R′ = ^{1}H or ^{2}H) ⟨81JA898⟩. Oxepins bearing Me

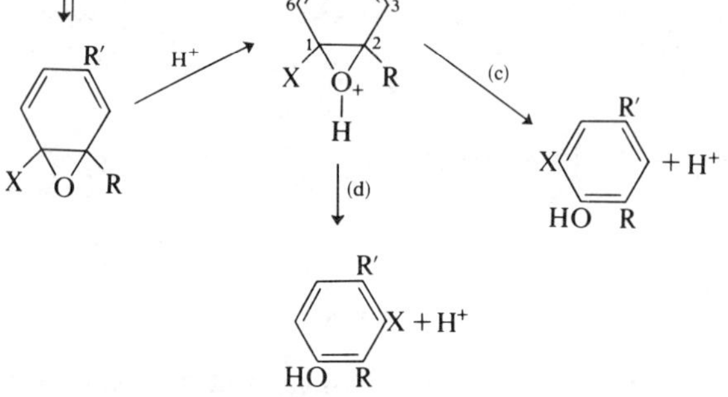

Scheme 17

(**10**), CO_2Me (**15**) or CH_2CO_2Me (**94**) as X groups also aromatized by a 1,6-shift mechanism (route c; R, R′ = ^1H or ^2H). The only example of a 1,3-migration sequence (pathway d) involved oxepin (**98**) and it occurred by two single migration steps (R = Me, R′ = ^1H) ⟨79JA2470⟩.

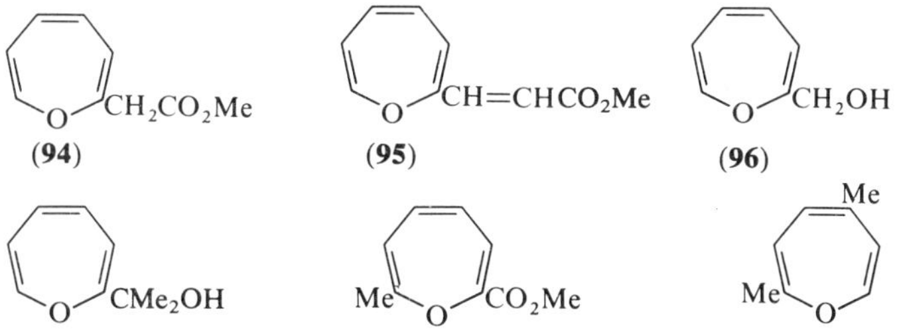

CH_2CO_2Me (**94**) CH=$CHCO_2Me$ (**95**) CH_2OH (**96**)

CMe_2OH (**97**) Me CO_2Me (**98**) Me Me (**99**)

An alternative mechanism for the acid-catalyzed aromatization of oxepins involves solvent capture of the intermediate carbocation. This reaction pathway has been proposed for aromatization of oxepin (**99**) and is shown in Scheme 18 ⟨72JA7876⟩.

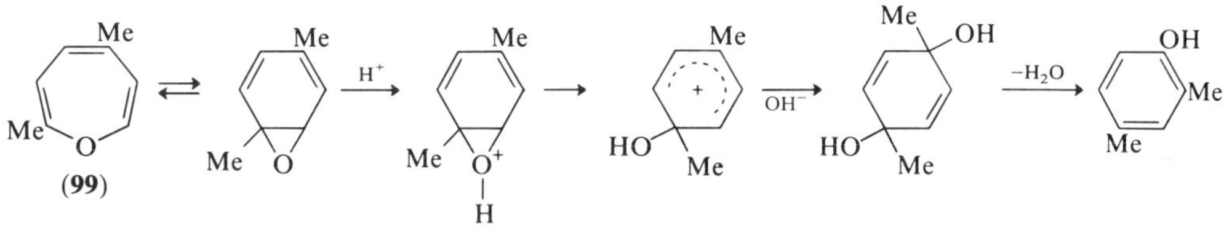

Me Me (**99**) ⇄ Me Me O —H^+→ Me Me O^+ H → Me HO Me —OH^-→ Me OH HO Me —$-H_2O$→ OH Me Me

Scheme 18

A similar solvent trapping mechanism has been proposed to account for formation of 5-indanol from acid-catalyzed isomerization of the arene oxide tautomer of oxepin (**22**) ⟨73JA60641⟩.

5.17.2.2.4 Nucleophilic attack on ring atoms adjacent to a heteroatom or conjugated with a heteroatom

The arene oxide valence tautomer of oxepins in principle should undergo nucleophilic substitution reactions (S_N2) which are characteristic of simple epoxides. In reality oxepin–benzene oxide (**7**) is resistant to attack by hard nucleophiles such as OH^-, H_2O, NH_2^- and RNH_2. Attempts to obtain quantitative data on the relative rates of attack of nucleophiles on (**7**) in aqueous solution have been thwarted by competition from the dominant aromatization reaction.

The attack of oxygen nucleophiles on oxepins is of particular interest (Scheme 19) since the hydration reaction to yield *trans*-dihydrodiols is a favored pathway in the metabolism of aromatic compounds. This nucleophilic addition of water to oxepins occurs in the presence of epoxide hydrolase enzyme by a mechanism involving general base catalysis ⟨80JBC(255)4698⟩. Despite strenuous efforts, to date the non-enzymatic hydration of (**7**) has not been observed. Similarly, ethanol and methanol adducts of oxepin (**7**) have been observed to form only at a very slow rate. Hydroperoxide anion, the strongest of the oxygen nucleophiles to be examined, did however react more rapidly with (**7**) to give the expected *trans* adduct (Scheme 19) ⟨74JA6929⟩.

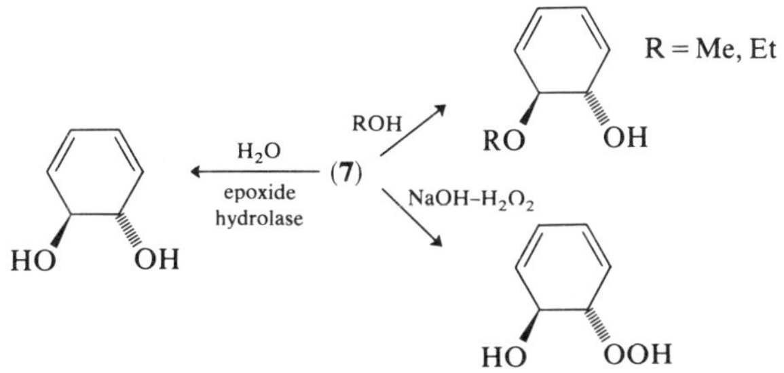

R = Me, Et

HO OH ←H_2O, epoxide hydrolase— (**7**) —ROH→ RO OH; (**7**) —$NaOH$–H_2O_2→ HO OOH

Scheme 19

While nitrogen nucleophiles such as NH_2^- did not react with oxepin (**7**), the larger and more polarizable (softer) N_3^- nucleophile rapidly attacked (**7**; equation 8) to yield a hydroxy azide product ⟨74JA6929⟩. Attempts to induce nucleophilic attack of an amine on a substituted form of oxepin–benzene oxide (**7**) by an intramolecular reaction have not to date been successful, although this type of nucleophilic attack has been postulated to occur under enzyme control (equation 9) during the biosynthesis of gliotoxin (**100**) and a range of epidithiadiketopiperazine metabolites in fungi ⟨80JOC3149⟩. While oxepins are generally very resistant to nucleophilic attack by amines and alcohols, this reaction occurs readily with some polycyclic arene oxides, *e.g.* the arene oxide tautomer of oxepin (**26**), in the presence of a basic alumina catalyst ⟨77JA8214⟩.

(7) $\xrightarrow{NaN_3}$ HO N3 (8)

:N H (100) HO NMe OH (9)

HO S OH ← Na_2S (7) $\xrightarrow{RSNa}$ HO SR
R = Et, Ac

Scheme 20

Sulfur nucleophiles such as thiolate anion generally appear near the top of the nucleophilicity scale and thus readily attack oxepins (Scheme 20). While adduct formation between oxepin (**7**) and thiols occurs spontaneously upon mixing ⟨74JA6929⟩, a similar reaction between the naturally occurring tripeptide thiol, glutathione, and oxepin–arene oxide intermediates formed during mammalian metabolism is catalyzed by a glutathione epoxide transferase.

Dimethylmagnesium or methyllithium provide carbon nucleophiles which react readily with oxepin (**7**) to produce alcohol adducts ⟨74JA6929⟩. While methyllithium (X = Li) gave only the *cis* product (by 1,6-addition) as shown in equation (10), dimethylmagnesium (X = MgMe) gave the latter alcohol (by 1,6-addition) accompanied by the anticipated *trans* isomer (by *trans* 1,2-addition). The 1,6-addition mechanism of nucleophilic attack to yield a *cis* adduct was found to occur when MeLi, Me_2Mg and NaN_3 were used as sources of nucleophiles in conjunction with a doubly labelled (^2H) sample of (**7**).

(7) ⇄ $Me^{\delta-}$—$X^{\delta+}$ → $\xrightarrow{H_2O}$ Me OH (10)

(7) $\xrightarrow{LAH}$ H OH $\xrightarrow{-H_2O}$ (11)

The reduction of (**7**) with LAH yielded the interesting hydrate adduct of benzene, which rapidly aromatized with elimination of a water molecule (equation 11) ⟨67AG(E)385⟩.

5.17.2.2.5 Reactions involving cyclic transition states

For classification purposes the addition of hydrogen to oxepin (**7**) on the surface of a platinum catalyst has been included in the present section. The isolation of oxepane (**1**) from this catalytic hydrogenation provides good evidence for the presence of the oxepin isomer ⟨67AG(E)385⟩.

Attempts to epoxidize oxepin (**7**) gave only (Z,Z)-muconaldehyde ⟨77JCS(P1)1346⟩ although the unsymmetrical oxepin oxide (**101**) was probably present as an unstable intermediate (equation 12). Further evidence for the latter assumption was provided by the isolation of (**103**), a substituted form of monoepoxide (**101**), and the corresponding diepoxide (**104**; equation 13). The successful epoxidation of oxepin (**102**) and isolation of products (**103**) and (**104**) were facilitated by the presence of both a $(CH_2)_6$ bridge and two electron withdrawing alkoxycarbonyl groups. These structural features moved the equilibrium toward a total preference for the oxepin tautomer and led to the stabilization of both reactants and products ⟨80TL4905⟩.

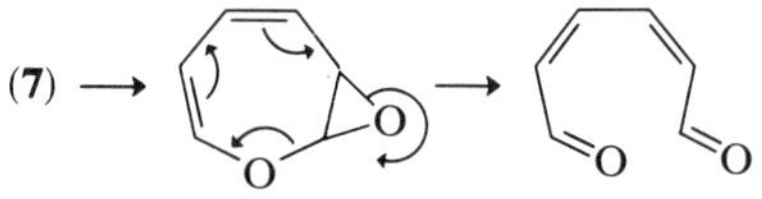

(**7**) → (**101**) → (12)

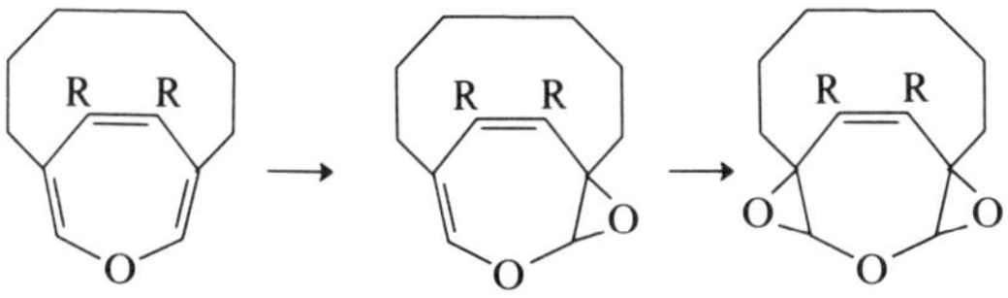

(**102**) R = CO_2Et → (**103**) → (**104**) (13)

The symmetrical oxepin oxide (**105**) has not to date been prepared by direct epoxidation of (**7**) but has been synthesized indirectly by the thermal extrusion of either carbon monoxide ⟨74AG(E)672⟩ or nitrogen ⟨76JA6350⟩ from appropriate precursors (Scheme 21).

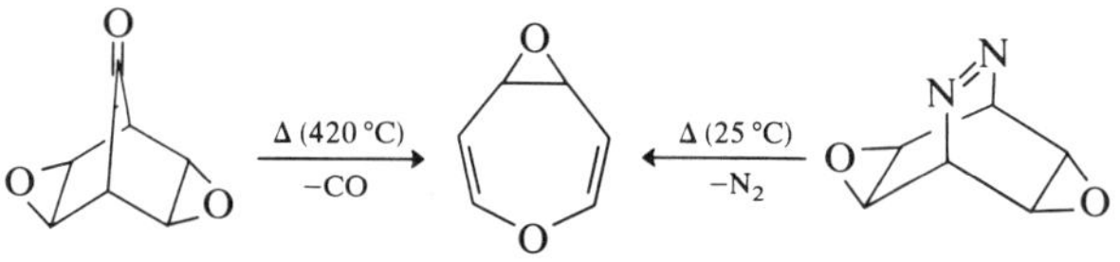

Δ (420 °C), −CO → (**105**) ← Δ (25 °C), $-N_2$

Scheme 21

While both hydrogenation and epoxidation reactions of (**7**) (and substituted forms) occur on the oxepin valence tautomer, cycloaddition reactions proceed more readily on the arene oxide form (where the diene is closer to planarity). Thus the dienophiles DMAD and maleic anhydride (MA) readily yielded [4+2] cycloadducts with (**7**) as shown in Scheme 22 ⟨67AG(E)385⟩. A similar type of singlet oxygen cycloaddition reaction gave an unstable endoperoxide (**106**) which upon heating yielded *trans*-benzene trioxide quantitatively (equation 14). ⟨75JOC3743⟩.

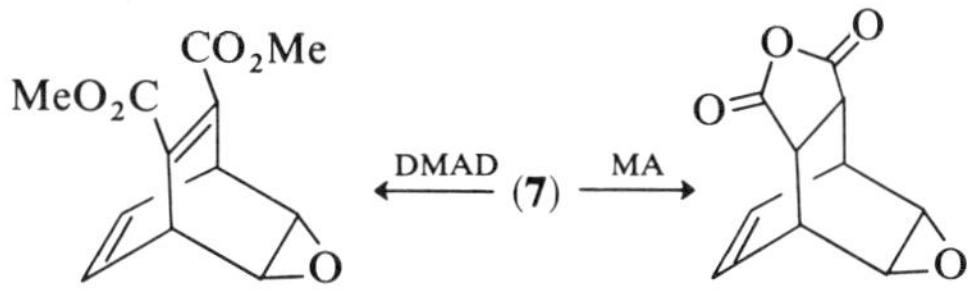

MeO_2C, CO_2Me ← DMAD (**7**) MA →

Scheme 22

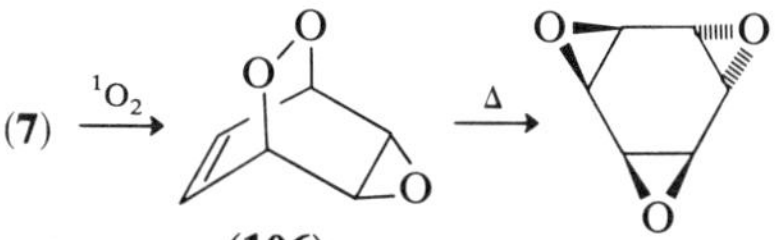

(**7**) $\xrightarrow{^1O_2}$ (**106**) $\xrightarrow{\Delta}$ (14)

The formation of the highly unstable cycloadduct (**107**) from reaction of (**7**) with a nitroso compound has been postulated. This intermediate, (**107**), was proposed to account for the isolation of the *trans*-dioxazahomobenzene product ⟨78AG(E)936⟩. Intermediate (**107**) and the final diepoxide product are similar in structure to the endoperoxide (**106**) and the derived *trans*-benzene trioxide respectively (equation 15).

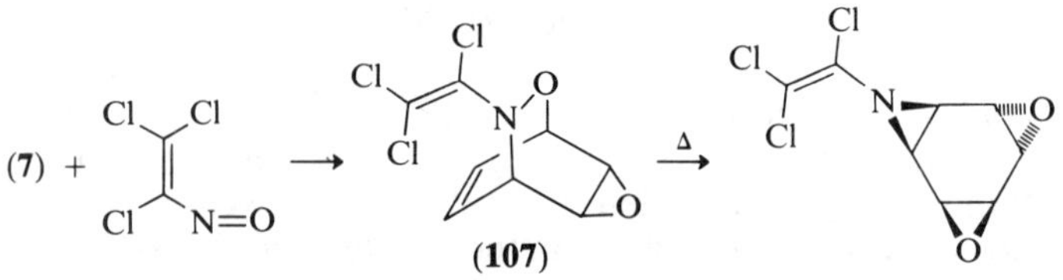

(7) + → (107) → (15)

The presence of substituents at the 2,7-positions of (**7**) results in an almost total preference for the oxepin form and this may explain why the arene oxide forms of oxepins (**98**) ⟨78TL2999⟩ and (**108**) ⟨79JOC468⟩ apparently did not participate in cycloaddition reactions. Oxepins (**16**) and (**108**) thus adopted the role of dienes in [4+2] cycloaddition reactions with azo compounds (equation 16).

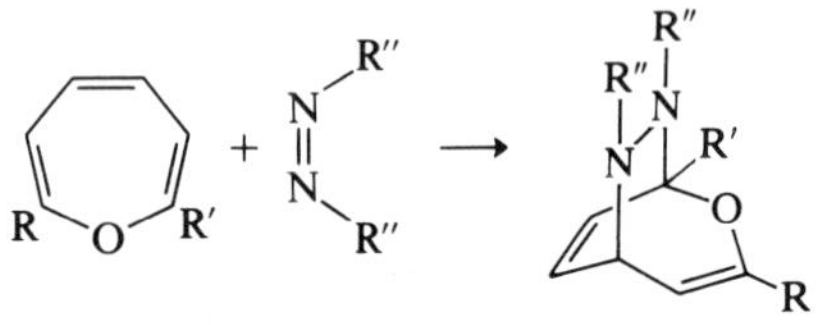

(16)

(**16**) R = R′ = Me, R″ = CO_2Me
(**108**) R = $SiMe_3$, R′ = Me, R″,R″ = —CONMeCO—

In the reaction of (**7**) with 3,4-diphenylcyclopentadienones the oxepin tautomer acted as a dienophile and the cycloaddition reaction occurred either exclusively (R = Me) ⟨78JOC315⟩ or mainly (R = CO_2Me) ⟨80JA5415⟩ by a concerted [4+2] cycloaddition mechanism to yield *endo* adducts (**109** and **110**; equation 17). The minor cycloadduct (**111**) was formed by a [6+4] cycloaddition pathway.

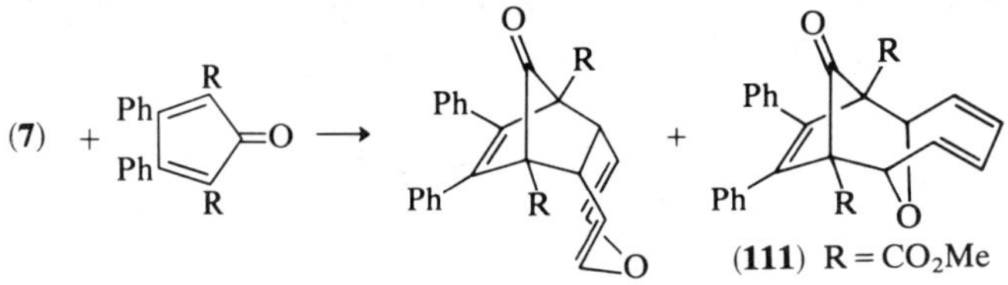

(17)

(**109**) R = Me
(**110**) R = CO_2Me
(**111**) R = CO_2Me

A mechanism involving a metallaoxycyclobutane intermediate has been proposed to account for the deoxygenation reaction of (**7**) with a rhodium carbonyl chloride catalyst (equation 18) ⟨77TL343⟩. Similar deoxygenation reactions of oxepins have been reported using iron ⟨75CB3336⟩ and chromium ⟨67AG(E)385⟩ complexes as catalysts or in the presence of the enzyme epoxide reductase which is found in animal liver ⟨80MI51700⟩.

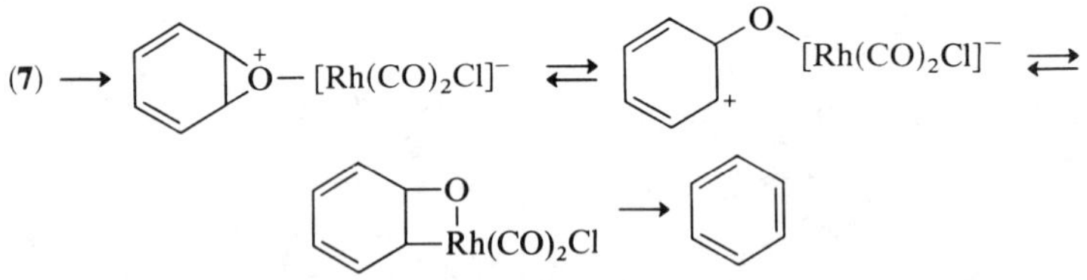

(7) → $\overset{+}{O}$—$[Rh(CO)_2Cl]^-$ ⇄ $[Rh(CO)_2Cl]^-$ ⇄ $Rh(CO)_2Cl$ → (18)

5.17.2.3 Reactivity of Thiepanes

5.17.2.3.1 Introduction and general survey

The reactivity of thiepane (**35**) is generally similar to other dialkyl thioethers where the sulfur atom provides the major site for chemical transformation. The thiepane seven-membered ring system is of particular interest however since it can undergo transannular reactions and can form cycloalkyne derivatives.

The sulfoxide, sulfone and sulfonium salts of thiepane (**35**) are relatively stable derivatives and their chemical reactivity will be considered together with the parent thioether (**35**). The transannular types of sulfonium salt (*e.g.* **112**) have been proposed as short lived reaction intermediates derived from substituted thiepanes. Similarly, the thiepanyl radical (**113**) and thiepane 1-oxide carbanion (**114**) (although stabilized by the neighboring sulfoxide group) are again considered to exist only as transient intermediates during chemical reactions of thiepanes.

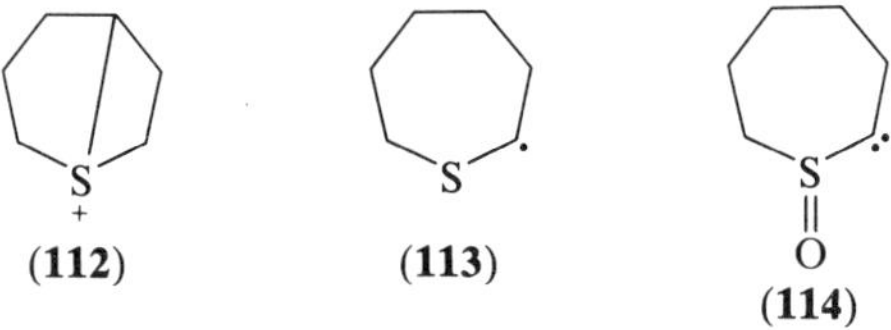

(**112**) (**113**) (**114**)

5.17.2.3.2 Thermal and photochemical reactions

Thiepane (**35**) has been purified by distillation at normal pressure (b.p. 170–174 °C) and may therefore be considered a thermally stable compound. At higher temperatures (400 °C) and in the presence of an aluminum silicate catalyst, however, thermal decomposition does occur to give hydrogen sulfide as one product ⟨72HC(26)573⟩.

As a consequence of the conformational mobility of the thiepane 1-oxide ring (**115**) it was possible to form the necessary planar five-membered cyclic transition state for a thermal E_i elimination reaction (equation 19). The acyclic sulfenic acid intermediate was not isolated but rearranged to *cis*-2-methylthiane 1-oxide by an intramolecular cyclization mechanism ⟨75TL2235⟩.

(**115**) —Δ (140 °C)→ $(CH_2)_4$ CH=CH$_2$:S̈—OH —Δ→ $(CH_2)_4$ CH—Me S=O (19)

A further example of a concerted thermal elimination reaction of a thiepane derivative was the formation of *cis*-hexatriene and the extrusion of sulfur dioxide from heating 2,7-dihydrothiepin 1,1-dioxide (**116**) ⟨67JA1281⟩. That this reaction was under orbital symmetry control was deduced from the results obtained by heating *cis*- (**117**) and *trans*- (**118**) 2,7-methyl analogues of (**116**; equation 20). The reaction was stereospecific as expected for a cheleotropic 1,6-elimination proceeding in an antarafacial manner ⟨75JA3666⟩. Thermal elimination of sulfur dioxide from a 2,7-bridged 2,7-dihydrothiepin 1,1-dioxide (**119**) also yielded a triene product (equation 21) ⟨70JA3807⟩.

—Δ→ + SO_2 (20)

(**116**) R, R′, R″ = H
(**117**) R, R′ = Me, R″ = H
(**118**) R, R″ = Me, R′ = H

(**119**) —Δ→ + SO_2 (21)

Photolysis of the bicyclic 2,7-dihydrothiepin 1-oxide (**120**; equation 22) gave cyclooctatetraene and sulfur monoxide from which products (**120**) had originally been synthesized ⟨71CC979⟩. The photochemical transformation of 4-thiepanones has been utilized as a general synthesis of γ-thiolactones. Thus, early studies on the unsubstituted (R = R′ = H) 4-thiacycloheptanone (**121**) gave a low yield (13%) of the γ-thiolactone among other products ⟨70JOC584⟩. The method has however been improved to provide an acceptable (62% yield) synthetic route to the substituted γ-thiolactone, pantothiolactone, using (**122**) as reactant (Scheme 23) ⟨75JOC3046⟩.

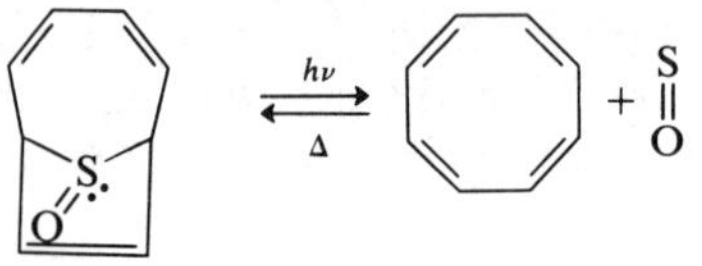

(22)

(120)

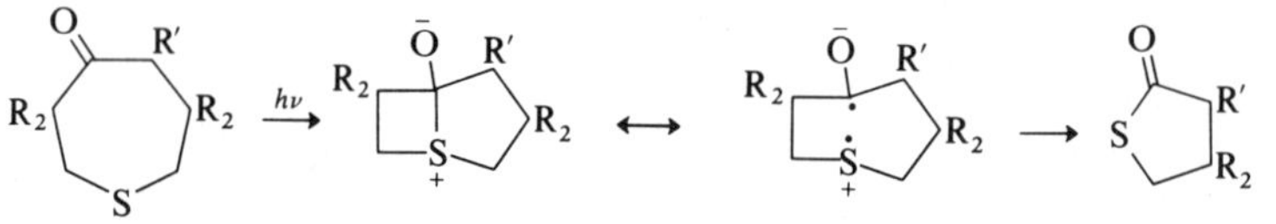

(121) R, R′ = H
(122) R = Me, R′ = OAc

Scheme 23

5.17.2.3.3 Electrophilic attack on the ring sulfur atom

Sulfonium salts of thiepanes are readily formed by electrophilic attack of alkyl halides on the cyclic thioether. Thus, thiepane (**35**) was found to yield a sulfonium iodide (**123**), which at elevated temperatures and in the presence of excess methyl iodide underwent ring cleavage to yield 1,6-diiodohexane (isolated as the 1,6-diphenoxy derivative; Scheme 24) ⟨53M1206⟩. The alkoxysulfonium salt (**124**) formed by reaction of (**35**) with *t*-butyl hypochlorite (equation 23) was characterized as a stable hexachloroantimonate ⟨67JOC2014⟩. Reduction of thiepane 1-oxide (**115**) to thiepane has been achieved using an aqueous solution of $NaHSO_3$ ⟨72JOC919⟩. A hydroxysulfonium salt intermediate (**125**) has been proposed in the latter reduction reaction which provides a general method for sulfoxide reductions under mild conditions (equation 24).

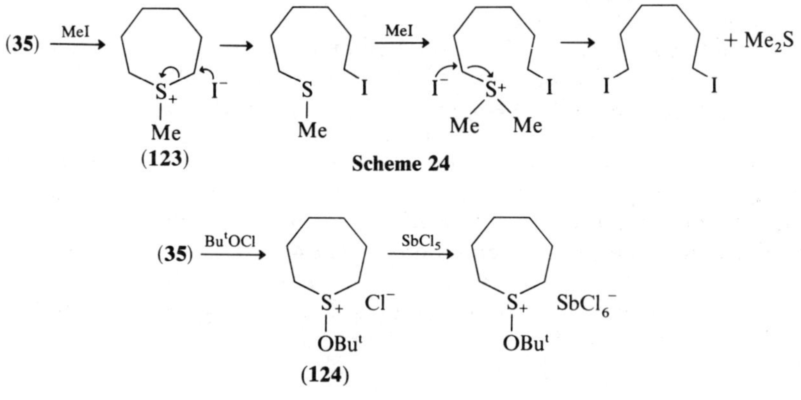

(123)

Scheme 24

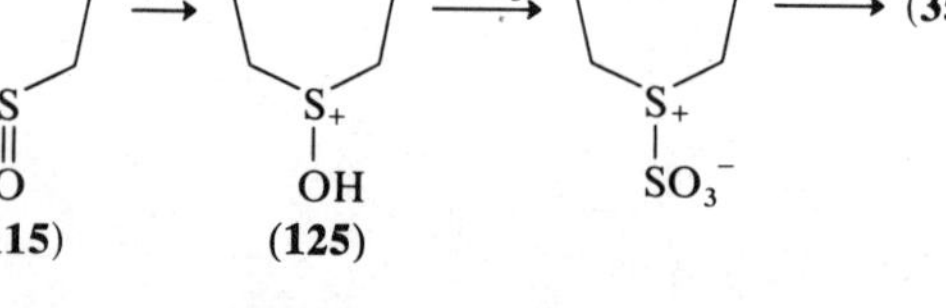

(124)

(23)

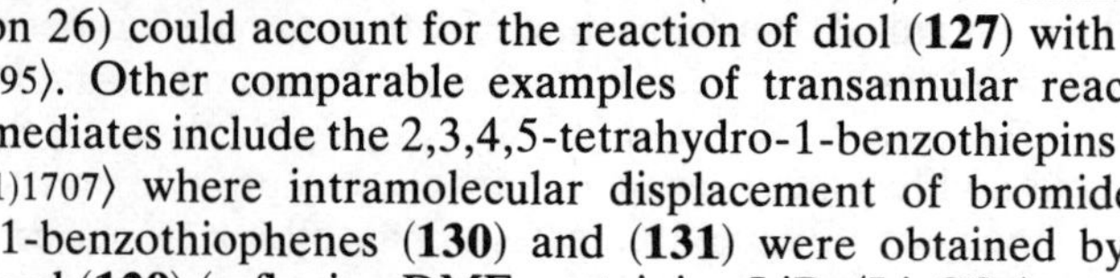

(115) (125)

(24)

The availability of 5-hydroxy-3,3,6,6-tetramethylthiepan-4-one (**126**) by a single step cyclization reaction ⟨72HC(26)573⟩ has resulted in a large number of reports on the reactivity of 3,3,6,6-tetramethylthiepane derivatives. Reaction of (**126**) with triphenylphosphine dibromide gave a thiacyclopentane derivative (equation 25) by a transannular reaction involving an unstable sulfonium salt intermediate ⟨69RTC995⟩. A similar transannular mechanism (equation 26) could account for the reaction of diol (**127**) with sulfuric acid in acetic acid ⟨69RTC995⟩. Other comparable examples of transannular reactions involving sulfonium salt intermediates include the 2,3,4,5-tetrahydro-1-benzothiepins (**128**) ⟨75CC784⟩ and (**129**) ⟨81JCS(P1)1707⟩ where intramolecular displacement of bromide ion occurred (Scheme 25). The 1-benzothiophenes (**130**) and (**131**) were obtained by heating (**128**) (refluxing dioxane) and (**129**) (refluxing DMF containing $LiBr/Li_2CO_3$) respectively.

(126)

Ph_3PBr_2

(25)

(127)

H_2SO_4 AcOH

(26)

(128) R = H, R′ = OH
(129) R,R′ = O

(130) (131)

Scheme 25

The conversion of thiepanes to the corresponding 1-oxides ⟨78TL5239⟩ and 1,1-dioxides ⟨67JA1281⟩ has been achieved using peroxyacids as oxidants.

5.17.2.3.4 Nucleophilic attack on ring carbon atoms

Nucleophilic substitution of the chlorine atom present in 2-chlorothiepane (**132**) using Grignard reagents has provided a synthetic route to 2-methyl- (**133**) or 2-phenyl- (**134**) thiepanes (equation 27) ⟨69JHC115⟩. An α-sulfinyl carbanion (**114**), generated by alkyllithium attack on thiepane 1-oxide (**115**), was found to act as nucleophile in the synthesis of *cis* and *trans* sulfoxides of thiepane (**133**; equation 28) ⟨78TL5239⟩. Polymerization of 2-thiepanone (**135**) has been initiated by attack of the *t*-butoxide anion and concomitant liberation of a thiolate anion (Scheme 26) ⟨64MI51700⟩.

(132) RMgBr (133) R = Me
(134) R = Ph

(27)

(115) RLi (114) (133)

(28)

(135) → → $Bu^tO[CO(CH_2)_5S]_n$—

Scheme 26

The reduction of 3- and 4-thiepanones (**41** and **42**) was reported using either hydride (LAH) ⟨67AG(E)872, 70JOC584⟩ or horse liver alcohol dehydrogenase enzymes which gave the 3-hydroxy- (**136**) and 4-hydroxy- (**43**) thiepane in optically active form ⟨81CJC1574⟩.

(**136**)

5.17.2.3.5 Reactions with radicals

Thiepane (**35**) has been converted to 2-acetoxythiepane (**137**) by a homolytic mechanism using *t*-butyl peracetate in the presence of a copper(I) ion catalyst ⟨67JCS(C)1130⟩. Similarly, α-chlorination of thiepane (**35**) by *N*-chlorosuccinimide (NCS) to yield 2-chlorothiepane (**132**) probably occurred by a free radical pathway (Scheme 27) ⟨69JHC115⟩.

(**137**) $\xleftarrow[Cu^+]{MeCO_3Bu^t}$ (**35**) $\xrightarrow{NCS}$ (**132**)

Scheme 27

5.17.2.3.6 Reactions involving cyclic transition states

The cycloaddition reactions of 3,3,6,6-tetramethylthiepane derivatives have received much attention. As the first example of a relatively stable seven-membered cycloalkyne, 3,3,6,6-tetramethyl-1-thiacycloheptyne (**37**; Section 5.17.1.3) and its synthesis from (**138**; equation 29) were of particular interest ⟨70TL761⟩.

(**138**) $\xrightarrow{N_2H_4}$ [bis-hydrazone] $\xrightarrow{Ag_2O}$ (**37**) (29)

Thiacycloheptyne (**37**) showed typical alkyne reactivity in the formation of a [4+2] cycloadduct with diphenylbenzo[*c*]furan (DPBF) ⟨70TL761⟩. The unusual strain, and thus reactivity, of (**37**) was evident from its ability to form a [2+2] dioxetene adduct (**139**) with singlet oxygen at low temperature (−90 °C). As expected the dioxetene (**139**) was found to decompose spontaneously on warming to yield the 4,5-dione (**138**) (from which (**37**) had been synthesized) with an intense chemiluminescence (Scheme 28) ⟨76JA6758⟩. Ring strain in cycloalkyne (**37**) may again be a major factor in the dimerization reaction which occurred by heating in the presence of $(PhCN)_2PdCl_2$ (equation 30). The resulting cyclobutadiene complex (**140**; $X = PdCl_2$) on treatment with ethylenebis(diphenylphosphane) was found to yield the cyclobutadiene (**36**) ⟨72AG(E)932⟩. The latter compound, in common with cycloalkyne (**37**) and dioxetene (**139**), owes its remarkable stability to the presence of bulky *gem*-dimethyl substituents at the 3- and 6-positions. The nickel complex (**141**) was formed by treatment of (**37**) with nickel dibromide ⟨76CB2558⟩ whereas a similar reaction with pentacarbonyliron yielded the alkyne complex (**142**) ⟨78LA431⟩.

[DPBF adduct] $\xleftarrow{DPBF}$ (**37**) $\xrightarrow{^1O_2}$ (**139**) $\xrightarrow{\Delta}$ (**138**) + $h\nu$

Scheme 28

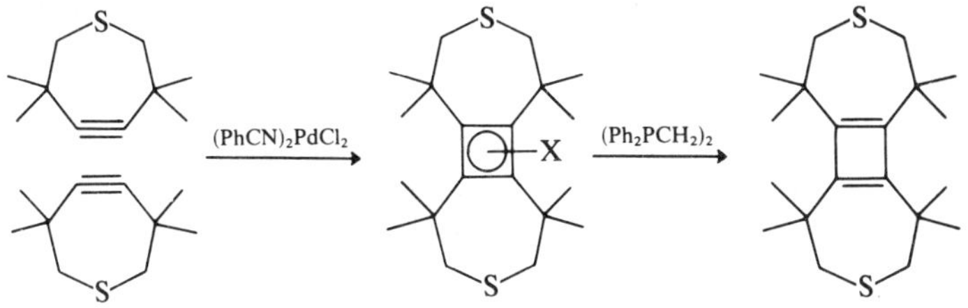

(37) $\xrightarrow{(PhCN)_2PdCl_2}$ (140) X = $PdCl_2$ / (141) X = $NiBr_2$ $\xrightarrow{(Ph_2PCH_2)_2}$ (36) (30)

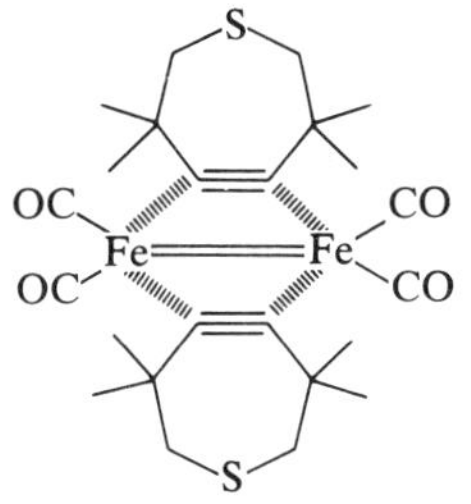

(142)

The cyclobutadiene structure (**36**), like cycloalkyne (**37**), also formed an adduct with singlet oxygen (dioxetane **143**), which rearranged to a dione product (**144**; equation 31) ⟨79LA473⟩.

(36) 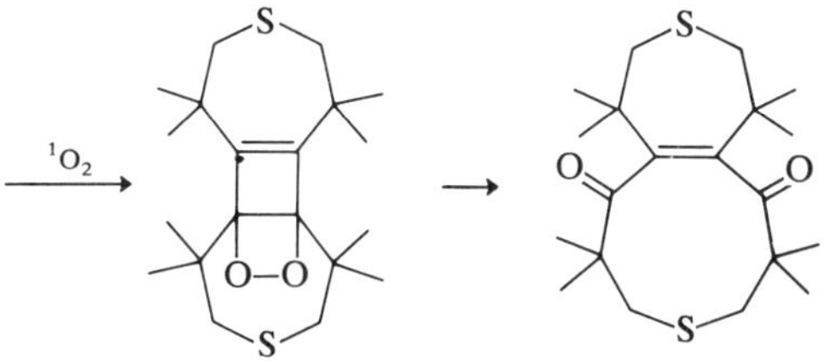 $\xrightarrow{^1O_2}$ (143) → (144) (31)

5.17.2.4 Reactivity of Thiepins

5.17.2.4.1 Introduction

As stated in Section 5.17.1.4, simple thiepins (*e.g.* **44**) are generally too reactive to be isolable under ambient conditions. Thiepins (**49**), (**50**) and (**51**) are among the relatively few stable monocyclic thiepins to have been reported and the majority of reactivity studies on thiepins have been carried out on polycyclic thiepins. The chemical reactivity of thiepins can be considered separately from the reactivity of the valence tautomeric thianorcaradienes more readily than was the case for oxepins–arene oxides. A spontaneous thermal extrusion of sulfur appears to occur from the episulfide tautomer of thiepins and the stable thiepins (**49**)–(**51**) would thus appear to exist exclusively in this valence isomeric form.

Factors known to stabilize thiepins include the presence of bulky substituents at the 2,7-position, and electron donating or withdrawing groups (Section 5.17.1.4). A further mechanism for increasing the thermal stability of thiepins involves chemical transformation at the sulfur atom to form *S,S*-dioxides (sulfones) and thiepinium salts.

5.17.2.4.2 Thermal and photochemical reactions

Thermal extrusion of a sulfur atom is the most common thermal reaction of a thiepin. The mechanism of this thermal process involves two orbital symmetry controlled reactions ⟨69CC1167⟩. The initial concerted step involving a reversible disrotatory electrocyclic rearrangement is followed by a concerted cheleotropic elimination of sulfur (Scheme 29). Similar aromatization reactions occur with thiepin 1-oxides and thiepin 1,1-dioxides, accompanied by the extrusion of sulfur monoxide and sulfur dioxide respectively. Since only a summary of the major factors influencing the thermal stability of thiepins was given in Section

5.17.1.4, several further examples are provided in the present section. Oxidation of the sulfur atom in thiepins may have a marked effect on thermal stability. Thus, while the thiepin (**44**) was too unstable to be detected, the corresponding 1,1-dioxide (**45**) was quite stable ($t_{1/2}$ 180 min at 100 °C) under normal conditions ⟨67JA1281⟩. By contrast, thiepin 1-oxides are less stable than the parent thiepin, as exemplified by thiepin (**47**; $t_{1/2}$ 45 min at 90 °C) and the corresponding thiepin 1-oxide ($t_{1/2}$ 45 min at 50 °C) ⟨75AG(E)812⟩. The stabilizing influence of fused benzene rings on thiepins is evident from a comparison of temperatures at which sulfur extrusion occurred for compounds (**46**; 47 °C), (**145**; 250 °C) and (**146**; 380 °C) ⟨72HC(26)667⟩. It is noteworthy that the most common thermal reaction of oxepins involves rearrangement to the isomeric phenols whereas thiepins generally appear to aromatize with sulfur extrusion on heating. The thermal reactivity of thiepins (**147**) and (**148**) indicates that either reaction pathway may be followed by thiepins if the correct substituents are present (Scheme 30) ⟨74T2431⟩.

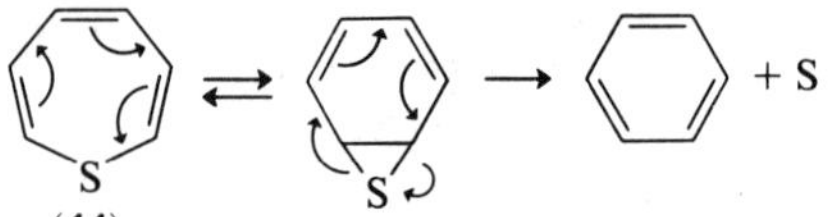

S ⇄ S → + S

(**44**)

Scheme 29

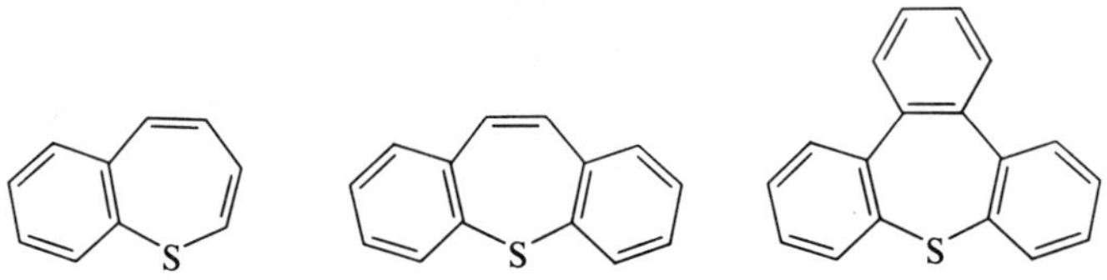

(**46**) (**145**) (**146**)

N CO_2Me CO_2Me ← R CO_2Me CO_2Me S → OH CO_2Me CO_2Me SH

(**147**) R = OH
(**148**) R = —N

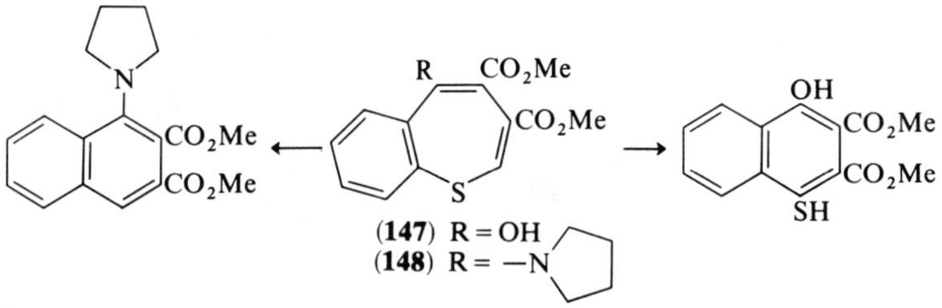

Scheme 30

The orbital symmetry controlled photoisomerization of 1-benzothiepin (**46**) and derivatives (equation 32) is one of the few reported photochemical reactions on thiepins ⟨72TL4597, 76BCJ825⟩.

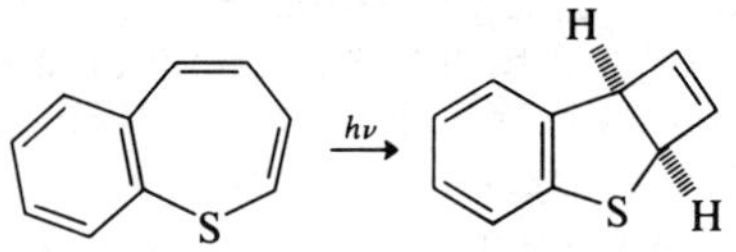

S $\xrightarrow{h\nu}$ H S H (32)

(**46**)

This reaction is directly analogous to that found with 1-benzoxepin and the valence tautomeric cyclobutene product obtained from (**46**) and was formed by a concerted disrotatory ring closure mechanism.

5.17.2.4.3 Electrophilic attack on the ring sulfur atom

Methyl fluorosulfate has been used as a methylating agent in the formation of an *S*-methylthiepinium salt which was found to be more stable than the parent thiepin (**149**; equation 33) ⟨77TL1985⟩.

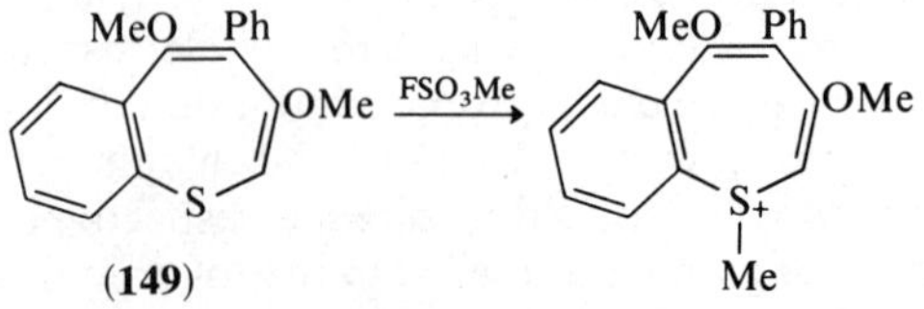

MeO Ph OMe S $\xrightarrow{FSO_3Me}$ MeO Ph OMe S^+ Me (33)

(**149**)

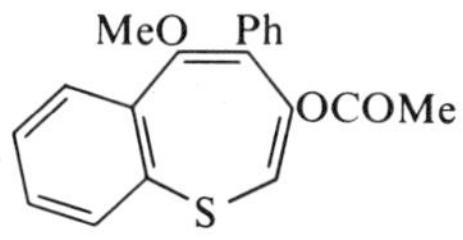

MeO Ph
OCOMe
S
(150)

MCPBA oxidation of thiepins (**149**) and (**150**) to form the 1-oxide and 1,1-dioxide derivatives ⟨75AG(E)812⟩ occurs in high yield. The peroxyacid oxidant assumes the role of electrophile in the initial oxidation step.

5.17.2.4.4 Reactions involving cyclic transition states

4-Substituted 1,2,4-triazoline-3,5-diones are excellent dienophiles which react rapidly at room temperature with oxepins, but particularly with the arene oxide valence tautomer. A similar [4+2] cycloaddition reaction between the episulfide tautomer of thiepin (**44**) and 4-phenyl-1,2,4-triazoline-3,5-dione has been reported ⟨74AG(E)736⟩. Benzene episulfide (the valence tautomer of thiepin **44**) was generated *in situ* by thermal decomposition of the diepisulfide (**151**) at 20 °C and trapped as a cycloadduct at the same temperature (equation 34). A 1,3-dipolar cycloaddition reaction between thiepin (**152**) and diazomethane has been reported ⟨56CB2608⟩. Two possible cycloadduct products are shown since the final structure has not been unequivocally established (equation 35).

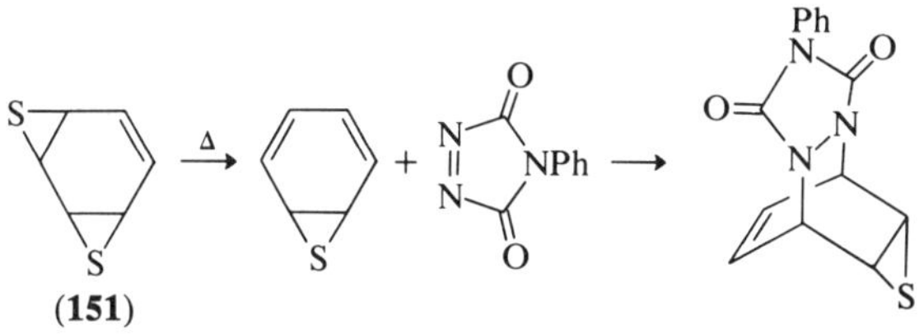

(34)

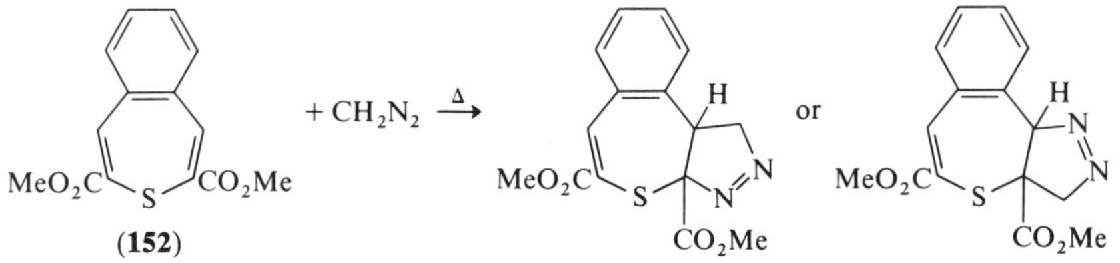

(151)

$+ CH_2N_2 \xrightarrow{\Delta}$ or (35)

MeO_2C S CO_2Me
(152)

5.17.3 SYNTHESIS OF OXEPANES, OXEPINS, THIEPANES AND THIEPINS

5.17.3.1 Synthesis of Oxepanes

5.17.3.1.1 From non-heterocyclic compounds

(*i*) *By Formation of One Bond*

(*a*) *By displacement reactions.* Intramolecular nucleophilic displacement reactions on 1,6-disubstituted hexanes (equation 36) have been widely used in the synthesis of oxepanes ⟨72HC(26)1⟩. The most common synthesis of (**1**) involves the elimination of water from hexane-1,6-diol (R = H, X = OH) using strong acids, *e.g.* H_2SO_4, HBr ⟨72HC(26)1⟩, or alumina ⟨80BCJ3031⟩, dimethyl sulfoxide ⟨64JOC123⟩ and *N*-methyl-*N,N*-di-*t*-butylcarbodiimidium ion ⟨75JA464⟩ as dehydrating agents. Similar cyclization reactions involving the bromine atom (X) as leaving group in conjunction with an iron(III) chloride catalyst (R = Me) or potassium hydroxide (R = H) also yield oxepane (**1**) ⟨72HC(26)1⟩. The tetrahydrobenzoxepin (**6**) has also been synthesized by this route (equation 37) ⟨61JCS2516⟩.

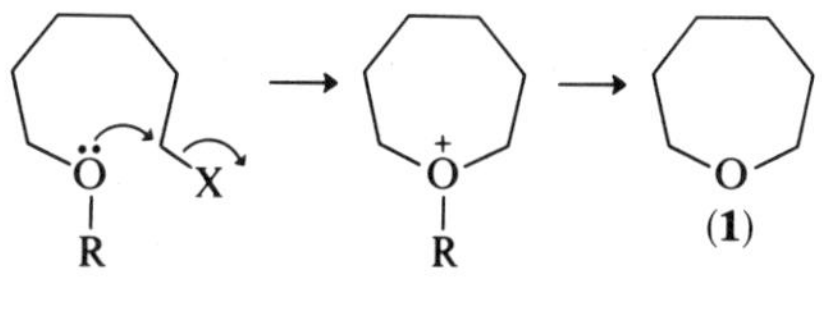

(36)

(1)

Na_2CO_3 / MeCOEt (37)

(6)

Hexane-1,6-diol was found to undergo an oxidation–cyclization process at elevated temperatures (250 °C) in the presence of a Cu–Cr catalyst supported on kieselguhr to yield 2,3,4,5-tetrahydrooxepin (**68**) ⟨65JOC335⟩. The final stage of the latter reaction involves a dehydration of the hemiacetal 2-hydroxyoxepane (**75**) as indicated in equation (38). An alternative type of base-induced cyclization (equation 39) involving intramolecular nucleophilic attack has been used in the synthesis of 4-ethoxycarbonyl-2,3,6,7-tetrahydrooxepin (**153**) ⟨73JOC1767⟩.

Δ / Cu-Cr (38)

(**75**) (**68**)

EtO_2C / EtO_2C–CH / NaOEt / CO_2Et (39)

(**153**)

$PhCO_3H$ (40)

(**78**)

The Baeyer–Villiger oxidation of cyclohexanone to form ε-caprolactone (**78**) using peroxybenzoic acid, as indicated in equation (40), may be considered as a further type of intramolecular displacement reaction of value in seven-membered ring synthesis ⟨49JA2571⟩.

(*b*) *By valence isomerization.* The thermal rearrangement of *cis*-2-vinylcyclopropanecarbaldehyde to form 2,5-dihydrooxepin (**154**) and the reverse reaction (Scheme 31), which occurs spontaneously under ambient conditions ($t_{1/2} \simeq 24$ h), proceed by a concerted pericyclic [3,3] sigmatropic mechanism ⟨69JA2815⟩.

Δ (50 °C)

(**154**)

Scheme 31

(*c*) *By radical combination.* The photochemical generation of both cyclopropyl and other diradical intermediates (equation 41) has been proposed to account for the formation of several cycloaddition products including the 2,3,4,7-tetrahydrooxepin (**155**) ⟨74JA6456⟩.

hν (41)

(**155**)

(ii) By formation of two bonds

Several synthetic routes to oxepanes involving the formation of two C—C bonds from bifunctional components (equation 42) have been reported ⟨72HC(26)1⟩. Thus oxepanes (**156**; X = CH_2Br, Y = $CH(CO_2Et)_2$) and (**157**; X = $CH(CO_2Et)_2$, Y = CH_2Br) have been synthesized by a double nucleophilic displacement of bromide ion using appropriately substituted ether components ⟨72HC(26)1⟩. An alternative method of cyclization of bifunctional components involving C—O bond formation (equation 43) has been used in the synthesis of the 4,5-dihydrooxepin (**158**) ⟨72HC(26)1⟩.

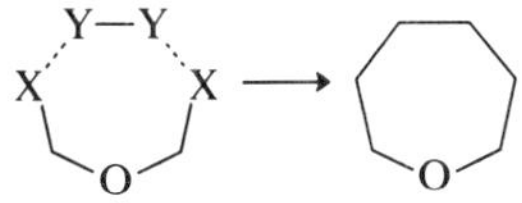

(42)

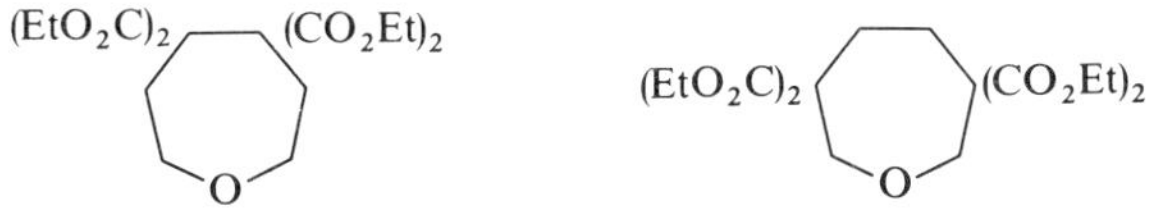

(EtO$_2$C)$_2$ (CO$_2$Et)$_2$

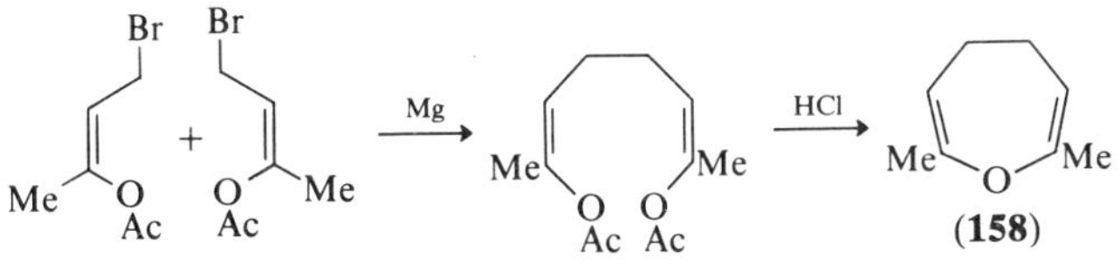

(**156**)

(EtO$_2$C)$_2$ (CO$_2$Et)$_2$

(**157**)

(43)

(**158**)

5.17.3.1.2 From heterocyclic compounds

(i) By formation of seven- from three-membered rings

In a typical concerted thermal [3,3] sigmatropic (Cope) rearrangement (equation 44), 4,5-dihydrooxepin (**70**) was formed from *cis*-1,2-divinylethylene oxide which was generated *in situ* from *sym*-divinylethylene carbonate ⟨63JOC1383⟩.

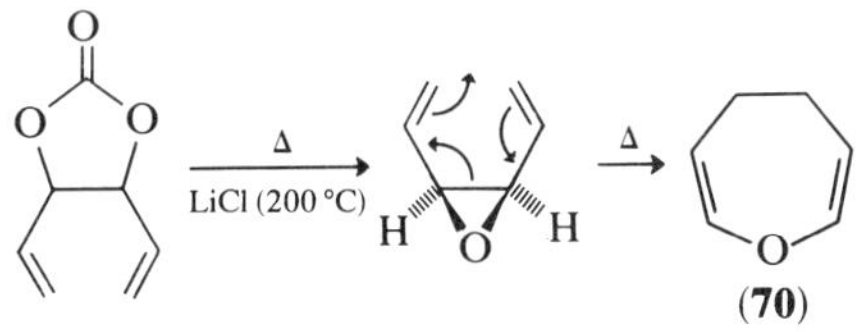

(44)

(**70**)

Oxepane (**159**) was obtained in high yield as an unexpected product of rearrangement from the peroxyacid oxidation of the substituted furan shown in equation (45) ⟨81JOC2589⟩. The isomerization of a bis-epoxide intermediate to the ring expansion product (**159**) occurs in the final step of the latter reaction.

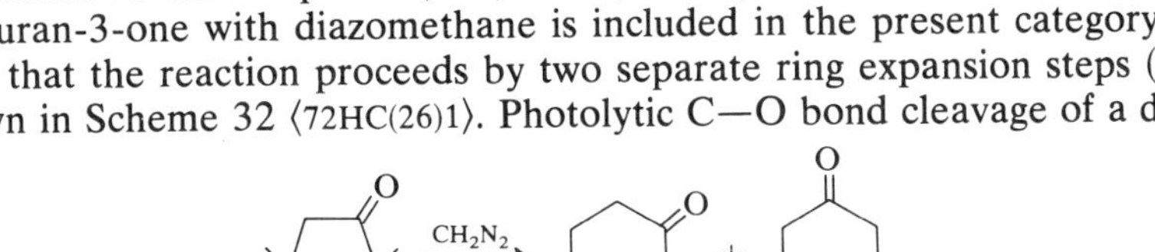

(45)

(**159**)

(ii) By formation of seven- from five-membered rings

The formation of the oxepanes (**160**) and (**161**) by treatment of 2,2,5,5-tetramethyltetrahydrofuran-3-one with diazomethane is included in the present category although it is probable that the reaction proceeds by two separate ring expansion steps (*cf.* equation 47) as shown in Scheme 32 ⟨72HC(26)1⟩. Photolytic C—O bond cleavage of a dihydrofuran

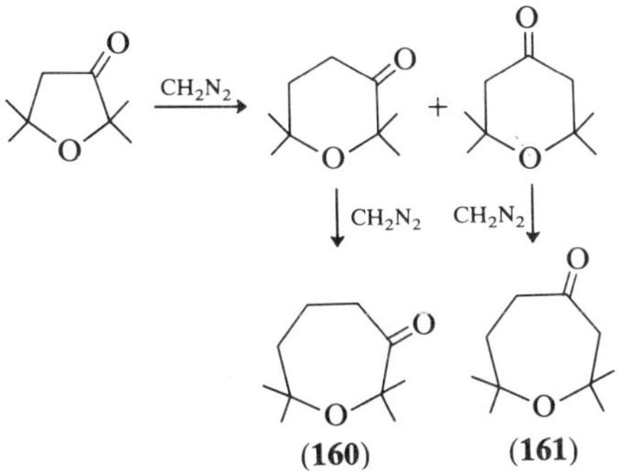

(**160**) (**161**)

Scheme 32

followed by cyclization of a diradical intermediate (equation 46) to form the 2,5-dihydrooxepin (**162**) provides a better example of a direct expansion of a five- to a seven-membered heterocyclic ring ⟨70TL4375⟩.

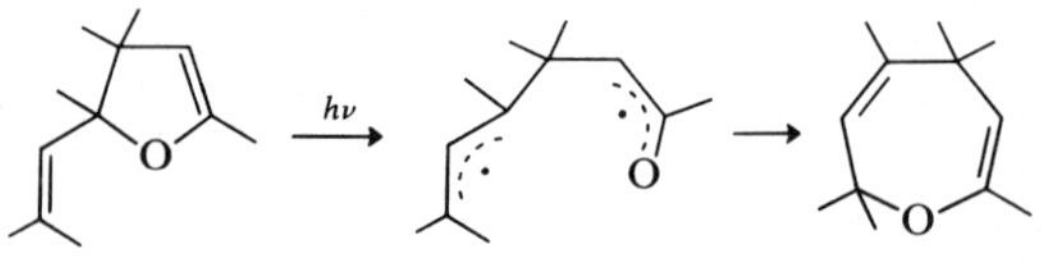

(46)

(162)

(*iii*) *By formation of seven- from six-membered rings*

The ring expansion reaction of tetrahydropyran-4-one with diazomethane to yield oxepan-4-one (**163**) occurs *via* the betaine intermediate shown in equation (47) ⟨58CB1589⟩.

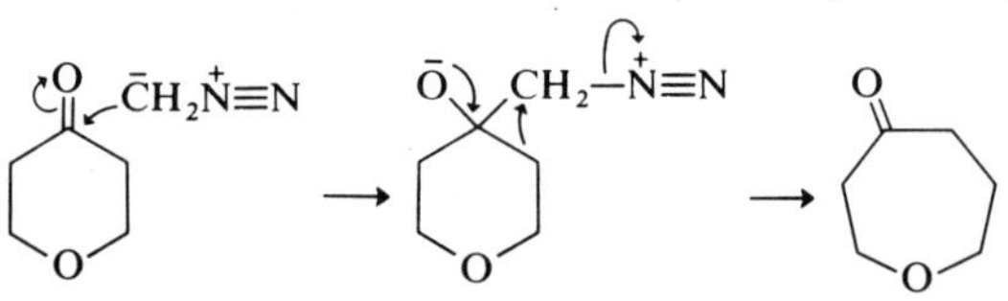

(47)

(163)

5.17.3.1.3 Summary of synthetic routes to benzo-, mono- and di-substituted oxepanes

References to the preferred synthetic routes to the majority of the oxepanes shown in Table 7 have already been mentioned in the synthesis (5.17.3) and reactivity (5.17.2) sections. Nevertheless, these have again been included along with references to the synthesis of oxepanes (**163**)–(**173**) in Table 7.

Table 7 Synthetic Routes to Benzo-, Mono- and Di-substituted Oxepanes

Compound number	*Name*	*Ref.*
(**1**)	Oxepane	64JOC123
(**6**)	2,3,4,5-Tetrahydro-1-benzoxepin	61JCS2516
(**164**)	1,3,4,5-Tetrahydro-2-benzoxepin	67CR(C)(264)1665
(**2**)	1,2,4,5-Tetrahydro-3-benzoxepin	66CB634
(**165**)	10,11-Dihydrodrodibenz[*b*,*f*]oxepin	69TL2007
(**166**)	6,7-Dihydrodibenz[*b*,*d*]oxepin	25CB78
(**5**)	5,7-Dihydrodibenz[*c*,*e*]oxepin	53JOC801
(**68**)	2,3,4,5-Tetrahydrooxepin	66TL5385
(**167**)	2,3,4,7-Tetrahydrooxepin	60JA4087
(**79**)	2,3,6,7-Tetrahydrooxepin	58JA3132
(**73**)	2,3-Dihydrooxepin	60JA4087
(**154**)	2,5-Dihydrooxepin	69JA2815
(**70**)	4,5-Dihydrooxepin	64JOC1275
(**78**)	Oxepan-2-one	49JA2571
(**162**)	Oxepan-4-one	58CB1589
(**75**)	2-Hydroxyoxepane	65JOC335
(**168**)	4-Hydroxyoxepane	58CB1589
(**74**)	2-Methoxyoxepane	81JOC2267
(**169**)	2-Ethoxyoxepane	53JOC1356
(**77**)	2-*t*-Butylhydroperoxyoxepane	80CR(C)(291)223
(**170**)	2-Acetoxyoxepane	58JA3064
(**171**)	4-Acetoxyoxepane	58CB1589
(**172**)	Oxepane-2,7-dione	30JA4110
(**80**)	4,5-Epoxyoxepane	60JA4087
(**81**)	*cis*-4,5-Dihydroxyoxepane	60JA4087
(**173**)	*cis*-4,5-Diacetoxyoxepane	60JA4087
(**71**)	*trans*-4,5-Diacetoxyoxepane	60JA4087
(**174**)	*trans*-4,5-Dibromooxepane	60JA4087

5.17.3.2 Synthesis of Oxepins

5.17.3.2.1 From non-heterocyclic compounds

(i) By formation of one bond

Benzoxepins are frequently synthesized by cyclization of alkyl aryl or diaryl ether precursors. An intramolecular Wittig reaction (equation 48) is used to provide the ring closure step in the synthesis of 1-benzoxepin (**28**) ⟨68JOC2591⟩. An internuclear cyclization reaction of an aromatic sulfonyl chloride (equation 49) occurred upon heating (250 °C) in the presence of a copper(I) chloride catalyst to yield tribenz[*b,d,f*]oxepin (**175**). The analogous thiepin (see equation 71) may also be synthesized by this route ⟨65T1299⟩.

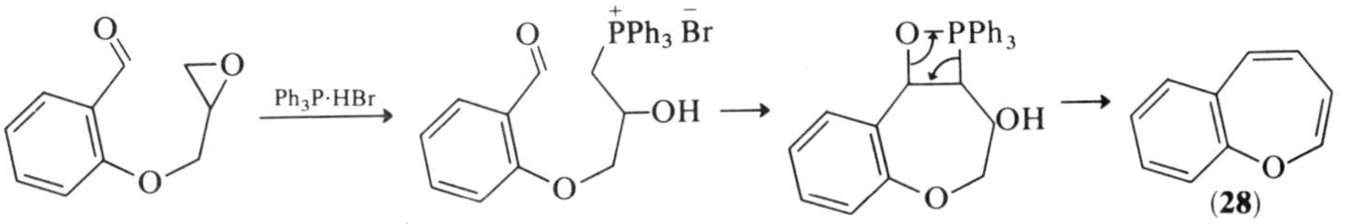

$Ph_3P{\cdot}HBr$ → $\overset{+}{P}Ph_3\,\overset{-}{Br}$ ... OH → O–PPh$_3$ ... OH → (**28**) (48)

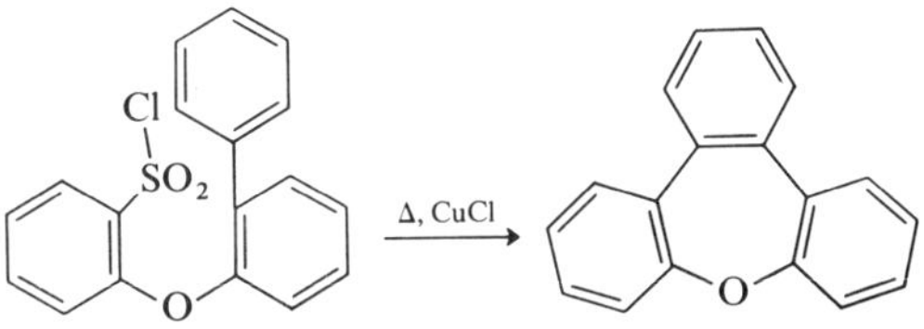

Cl–SO_2 ... Δ, CuCl → (**175**) (49)

(ii) By formation of two bonds

3-Benzoxepin (**27**) has been synthesized from phthalaldehyde and a bis-phosphonium salt by a double Wittig condensation reaction as shown in equation (50) ⟨66CB634⟩.

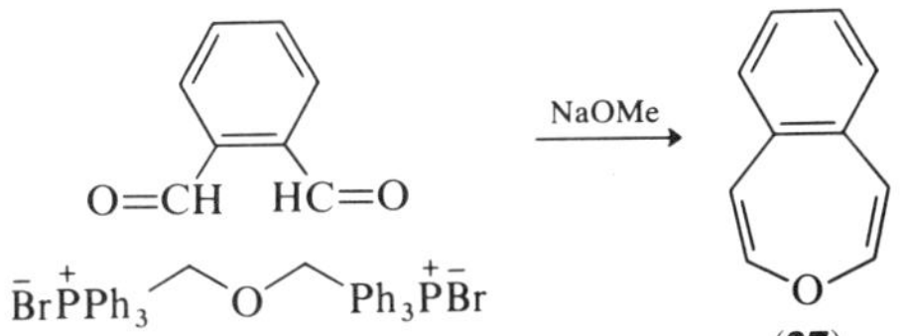

O=CH HC=O; $\overset{-}{Br}\overset{+}{P}Ph_3$ O $Ph_3\overset{+}{P}\overset{-}{Br}$ —NaOMe→ (**27**) (50)

5.17.3.2.2 From heterocyclic compounds

(i) By formation of seven- from three-membered rings

This is the most widely used synthetic route to monocyclic oxepins. The key step in the synthesis of oxepin–benzene oxide (**7**) is the dehydrohalogenation of a dibromoepoxide precursor ⟨64AG(E)510⟩. Since the benzene oxide valence tautomer is formed initially the valence tautomerization of the latter to oxepin (equation 51) may be considered as a ring expansion reaction.

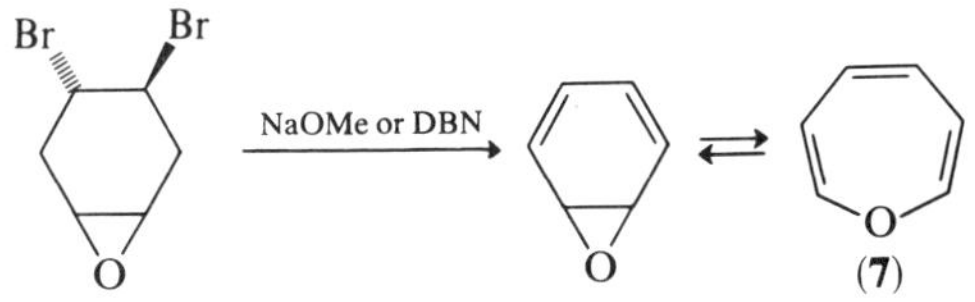

Br Br O —NaOMe or DBN→ O ⇄ O (**7**) (51)

With minor variations on the haloepoxide route a remarkable number of mono- and di-substituted oxepins have been prepared and are included in Table 8. The electron donating or withdrawing capacity of substituent X present in the diene precursor will determine which bond will be more readily epoxidized or brominated (Scheme 33).

The final step in the synthesis of oxepins from a dibromoepoxide precursor (Scheme 33) involved the base-catalyzed elimination of two molecules of HBr. The bicyclic oxepins (**28**) and (**176**) have similarly been synthesized by base treatment of the tetrabromoepoxide

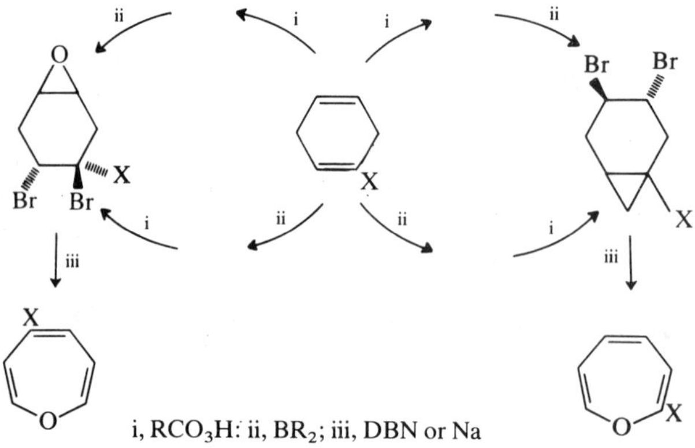

i, RCO_3H: ii, BR_2; iii, DBN or Na

Scheme 33

precursor shown in equation (52) ⟨64JA3168, 64AG(E)642⟩. The elimination of one molecule each of HBr and CO_2 (under mild conditions) from a bromo-β-lactone epoxide (equation 53) has been used as an alternative final step in the synthesis of oxepins (**177**)–(**179**) ⟨78JA6483⟩.

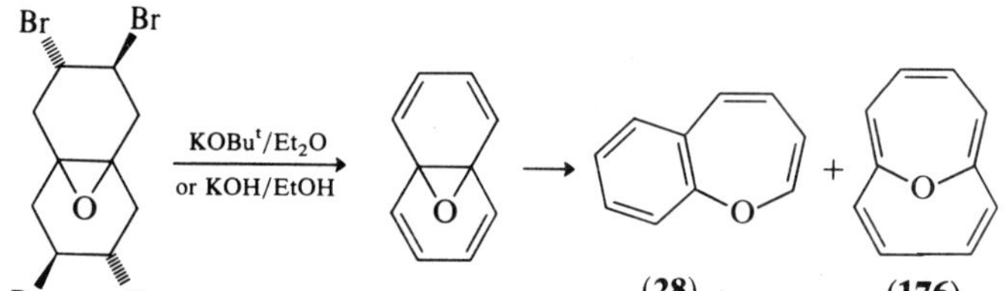

KOBu^t/Et_2O or KOH/EtOH (52)

(**28**) (**176**)

DBU (53)

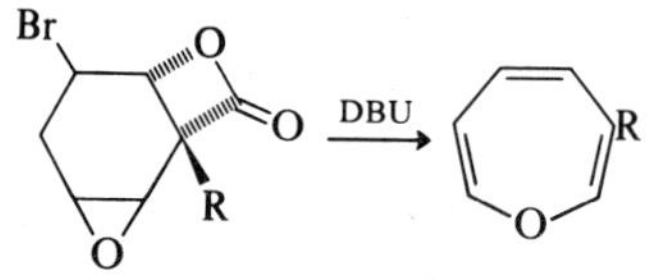

(**177**) R = Me
(**178**) R = CH_2OCOEt
(**179**) R = CH_2OCOPh

The oxepins (**7**; equation 54) and (**92**; equation 55) resulted from the spontaneous isomerization of their valence tautomeric arene oxide forms which were produced by photorearrangement of 2,3-epoxybicyclo[2.2.0]hex-5-ene ⟨67JA3922⟩ and phenanthrene 9,10-oxide (**91**) ⟨73CC37⟩ respectively. A rather specific synthetic route to the relatively stable oxepins (**180**)–(**182**) was based upon the acid-catalyzed dehydration and ring expansion of 2,6-di-*t*-butylcyclohexadiene-1,4-diols (Scheme 34) ⟨71AG(E)425, 71TL1257⟩.

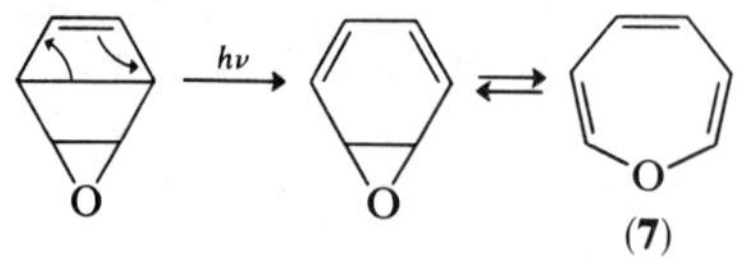

$h\nu$ (54)

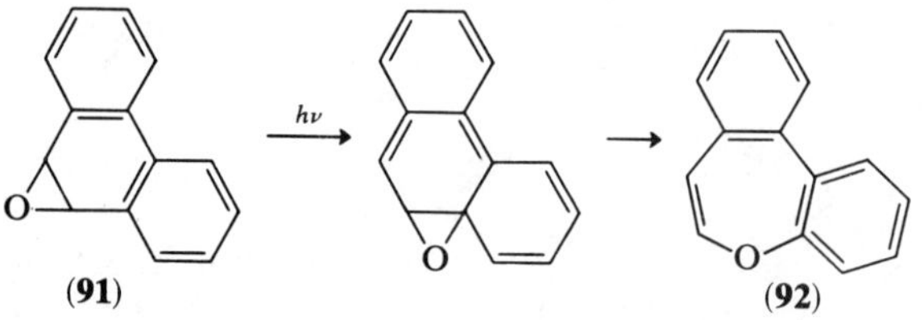

(**7**)

$h\nu$ (55)

(**91**) (**92**)

(ii) By formation of seven- from five-membered rings

The photochemically induced isomerization of a series of 7-oxanorbornadienes to the corresponding 3-oxaquadricyclanes, followed by a thermal rearrangement (equation 56) provides a relatively short and convenient route to the parent oxepin (**7**) ⟨75AG(E)753⟩ and a range of substituted oxepins (**183**)–(**192**) ⟨75CJC138, 76CB2823⟩. While oxepin (**188**) was

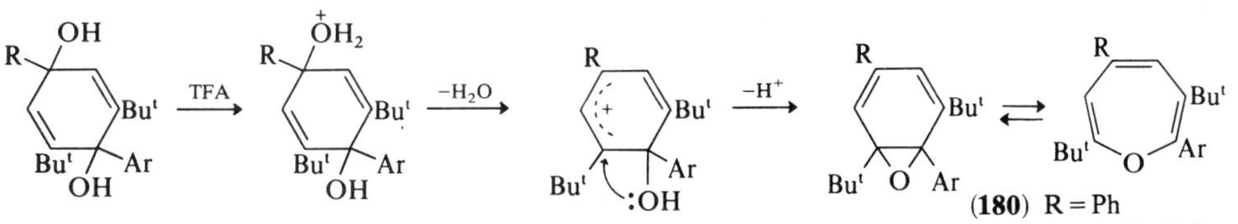

(180) R = Ph
(181) R = C_6H_4OMe-*p*
(182) R = $C_6H_4NMe_2$-*p*
Ar = C_6H_4Me-*p*

Scheme 34

formed by the normal thermal isomerization of the corresponding oxaquadricyclane, the unusual substitution pattern found in oxepin (**192**) resulted from a transition metal-catalyzed $[Rh(CO_2)_2Cl_2]$ isomerization ⟨74TL1479⟩.

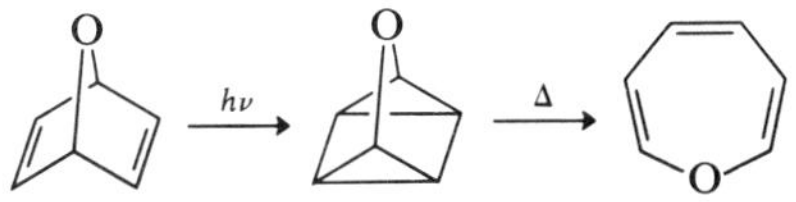

(7) (56)

(183) 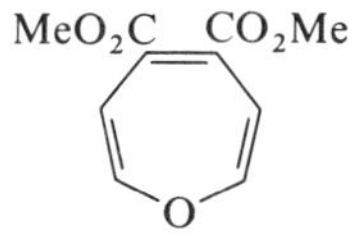 (184) 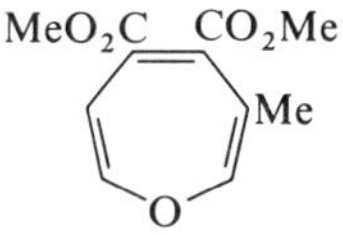 (185) 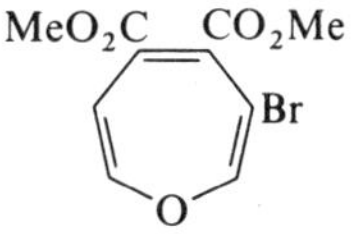 (186) 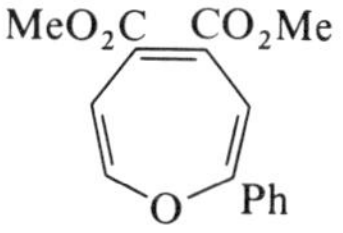 (187)

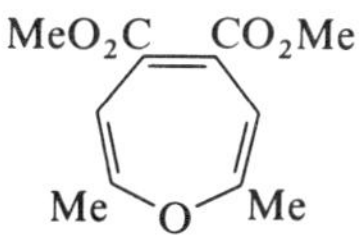

(188) 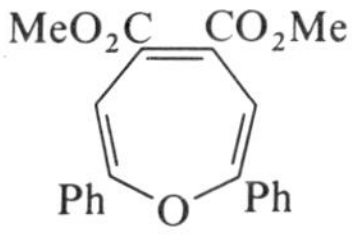 (189) 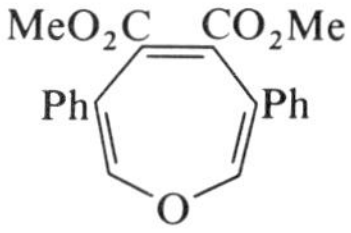 (190) 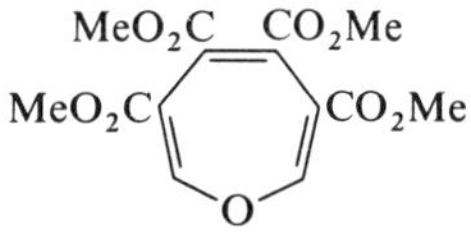 (191) 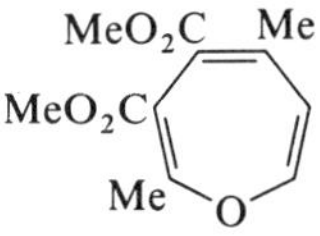 (192)

(iii) By formation of seven- from six-membered rings

The expansion of a di- or a tetra-hydropyran ring fused to a three-membered ring has been used as a synthetic approach to oxepins. Thus the synthesis of oxepin (**193**) was attempted by thermal dehydrochlorination of a tetrahydropyran (equation 57) obtained from a dichlorocarbene addition reaction ⟨65CI(L)184⟩. Unfortunately the equilibrium appeared to favor the keto tautomer to the apparent exclusion of the enolic oxepin form (**193**).

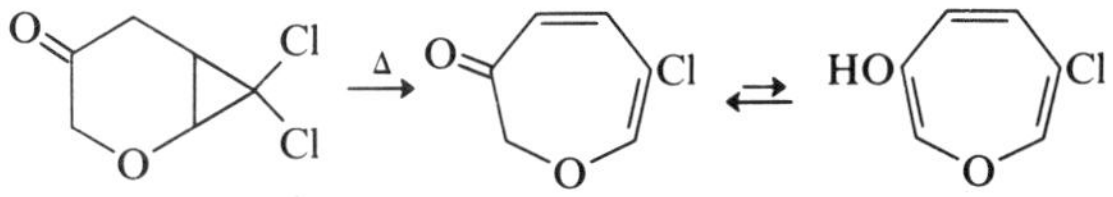

(193) (57)

The acid-catalyzed ring expansion of a spiroepoxide (equation 58) also yielded only the keto tautomer of a 1-benzoxepin. The enolic form of the oxepin in this example however was stabilized by conversion to the diacetate (**194**) ⟨69CB205⟩.

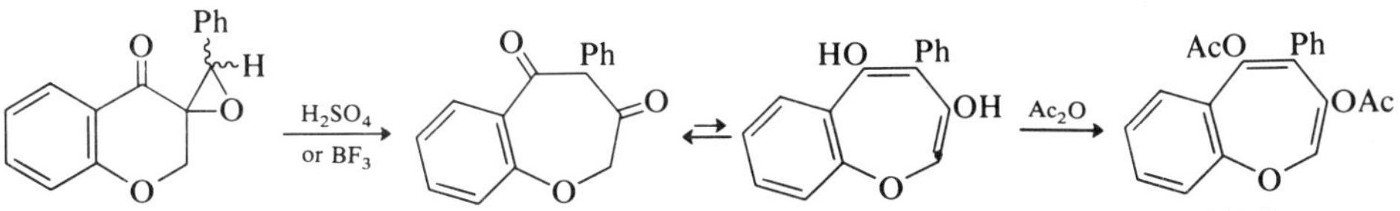

(194) (58)

5.17.3.2.3 Summary of synthetic routes to benzo-, mono- and di-substituted oxepins

While the benzoxepins in Table 8 have been synthesized by different routes, the vast majority of the monocyclic oxepins have been prepared by the general methods shown in Schemes 33, 34 and equation (56).

Table 8 Synthetic Routes to Benzo-, Mono- and Di-substituted Oxepins

Benzoxepins

Compound number	Name	Ref.
(**7**)	Oxepin	67AG(E)385
(**28**)	1-Benzoxepin	68JOC2591
(**27**)	3-Benzoxepin	66CB634
(**195**)	Dibenz[*b,f*]oxepin	65CJ3433
(**92**)	Dibenz[*b,d*]oxepin	73CC37
(**174**)	Tribenz[*b,d,f*]oxepin	65T1299

Monosubstituted oxepins

Compound number	Substituent (position)	Ref.	Compound number	Substituent (position)	Ref.
(**10**)	Me (2)	67AG(E)385	(**177**)	Me (3)	72B3080, 78JA6483
(**196**)	$(CH_2)_2NH_2$ (2)	80JOC3149	(**199**)	Ph (3)	78JOC2711
(**94**)	CH_2CO_2Me (2)	81JOC1191	(**200**)	$(CH_2)_2NH_2$ (3)	80JOC3149
(**96**)	CH_2OH (2)	81JA898	(**178**)	CH_2OCOEt (3)	78JA6483
(**97**)	CMe_2OH (2)	81JA898	(**179**)	CH_2OCOPh (3)	78JA6483
(**95**)	$CH{=}CHCO_2Me$ (2)	81JOC1191	(**201**)	CO_2Bu^t (3)	77JOC2008
(**12**)	CN (2)	81JOC813	(**202**)	Cl (3)	75JA4428
(**13**)	CHO (2)	81JA898	(**93**)	Me (4)	72B3080
(**11**)	COMe (2)	67AG(E)385	(**203**)	$(CH_2)_2NH_2$ (4)	80JOC3149
(**14**)	CO_2H (2)	79JA2470	(**204**)	CN (4)	76ACR378
(**15**)	CO_2Me (2)	79JA2470	(**205**)	CO_2H (4)	74JOC2088
(**197**)	CO_2Bu^t (2)	79JA2470	(**206**)	CO_2Bu^t (4)	74JA1193
(**198**)	$SiMe_3$ (2)	81JOC1817	(**207**)	Cl (4)	75JA4428
			(**208**)	Br (4)	75JA4428

Monocyclic disubstituted oxepins

Compound number	Substituent (position)	Ref.	Compound number	Substituent (position)	Ref.
(**16**)	Me (2), Me (7)	67AG(E)385	(**17**)	CHO (2), CHO (7)	76TL1167
(**209**)	Me (2), Me (6)	72B3080	(**18**)	CO_2H (2), CO_2H (7)	76TL1167
(**99**)	Me (2), Me (5)	72B3080	(**19**)	CO_2Me (2), CO_2Me (7)	76TL1167
(**210**)	Me (2), Me (4)	72B3080	(**20**)	CN (2), CN (7)	76TL1167
(**211**)	Me (2), Me (3)	72B3080	(**182**)	CF_3 (4), CF_3 (5)	76CB2823
(**212**)	Me (3), Me (4)	72B3080	(**183**)	CO_2Me (4), CO_2Me (5)	67AG(E)385
(**213**)	Me (3), Me (5)	72B3080	(**98**)	CO_2Me (2), Me (7)	79JA2470
(**214**)	Me (3), Me (6)	72B3080	(**217**)	CO_2Me (2), Me (5)	79JA2470
(**215**)	Me (4), Me (5)	67AG(E)385	(**108**)	$SiMe_3$ (2), Me (7)	81JOC1817
(**216**)	CH_2OCOMe (2), CH_2OCOMe (7)	76TL1167	(**218**)	$SiMe_3$ (2), Me (5)	81JOC1817

5.17.3.3 Synthesis of Thiepanes

5.17.3.3.1 From non-heterocyclic compounds

(i) By formation of one bond

Thiepane (**35**) has been synthesized by an intramolecular radical addition of the thiyl radical (equation 59) which was generated by photolysis of a thiol ⟨71TL2025⟩. Similarly, C—S bond formation has been achieved (equation 60) by an intramolecular condensation of 6-mercaptohexanoic acid to give the thiolactone, thiepan-2-one (**135**) ⟨64MI51700⟩. A Dieckmann-type base-catalyzed cyclization of a diester precursor followed by acid-catalyzed hydrolysis and decarboxylation has been used in the synthesis of thiepan-3-one (**41**) as indicated in equation (61) ⟨52JA917⟩.

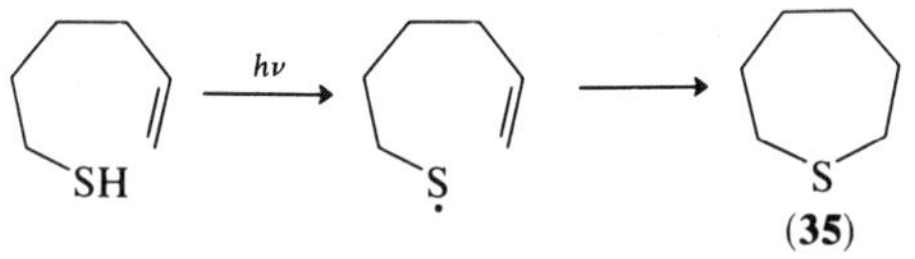

(59)

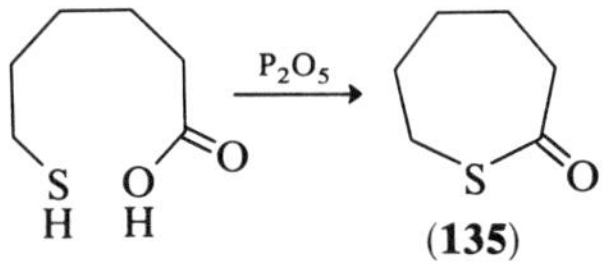

(60)

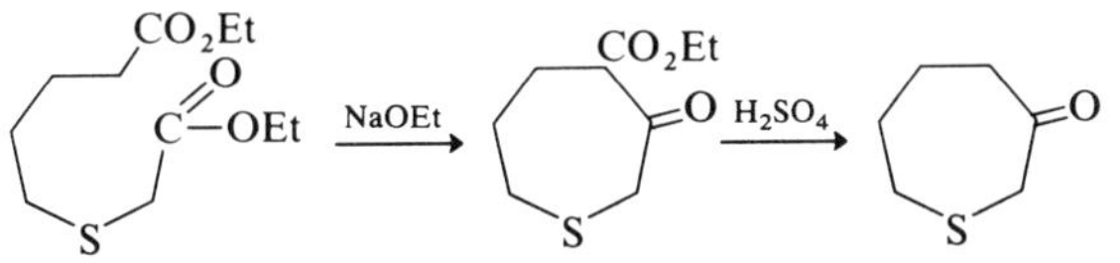

(61)

(ii) By formation of two bonds

A double nucleophilic substitution reaction on 1,6-dibromohexane with sodium sulfide has been found to give an acceptable yield (59%) of the thiepane (**35**; equation 62) ⟨81SC409⟩. The reversible 1,6-addition of sulfur dioxide to *cis*-hexatriene (equation 63) provides a convenient synthetic route to the 2,7-dihydrothiepin 1,1-dioxide (**116**) ⟨67JA1281⟩.

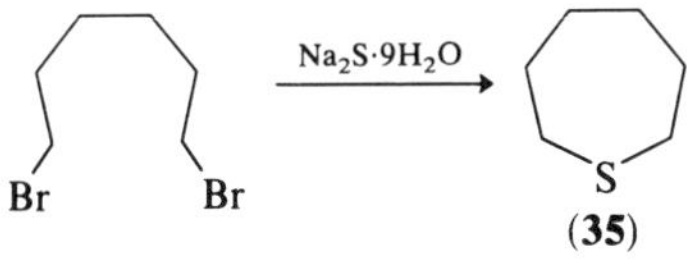

(62)

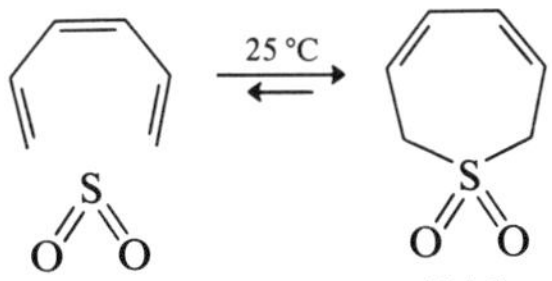

(63)

5.17.3.3.2 From heterocyclic compounds

(i) By formation of seven- from three-membered rings

The concerted thermal rearrangement reactions of *cis*-1,2-divinylthiirane ⟨67JA605⟩ and of *cis*-1,2-divinylthiirane 1,1-dioxide ⟨70CC1254⟩, to yield 4,5-dihydrothiepin (**219**) and 4,5-dihydrothiepin 1,1-dioxide (**220**; equation 64) respectively, are directly analogous to the synthesis of 4,5-dihydrooxepin (equation 44).

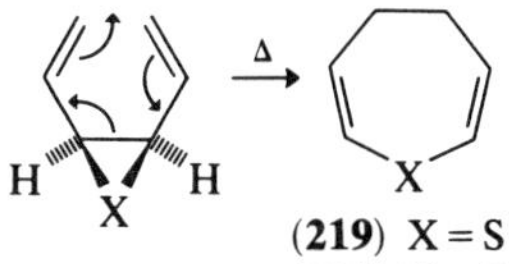

(64)

(ii) By formation of seven- from five-membered rings

The photochemically induced isomerization of cyclic dienes to cyclobutenes (*e.g.* **7**, **46**, **73**) has been discussed in Section 5.17.2. The reverse ring enlargement reaction is thermally allowed and has been used in the synthesis of the 2,3- (**221**) and 2,7- (**222**) dihydrothiepins as indicated in equation (65). The cyclobutene ring in the reactant was formed by a [2+2] cycloaddition reaction between DMAD and the appropriate enamine ⟨73RTC865⟩. The ring expansion of a 1-thiolactone (equation 66) to form 1,3-dihydro-2-benzothiepin-5(4*H*)-one (**223**) (as a result of reaction with vinyllithium and acetic acid) proceeds *via* a thiol intermediate ⟨70JOC3442⟩.

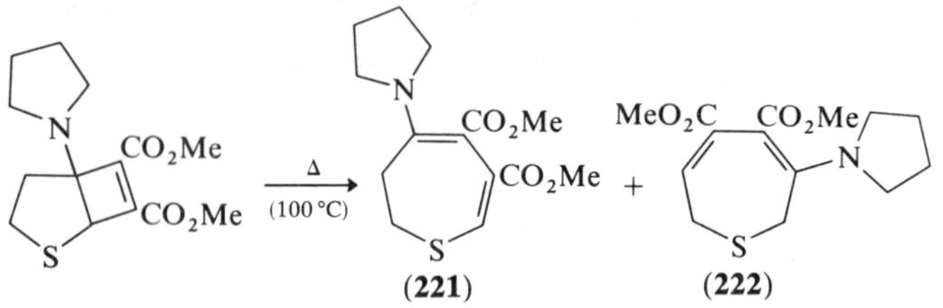

(65)

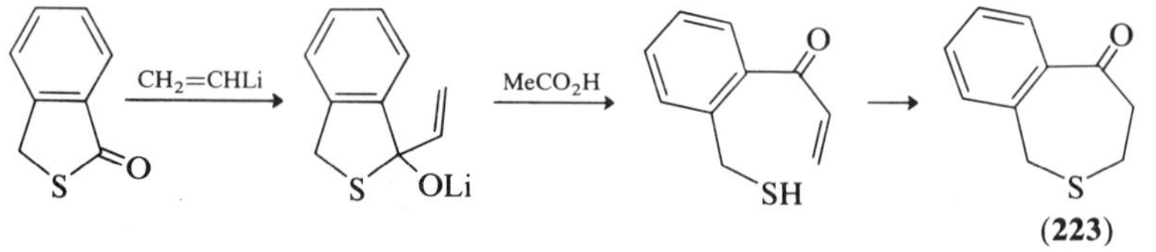

(66)

(iii) By formation of seven- from six-membered rings

The ring expansion reaction of six-membered cyclic ketones using diazomethane, which was previously discussed in the synthesis of oxepan-4-one (**163**; equation 47), has also been applied to the preparation of thiepan-4-one (**42**) ⟨56JA1965⟩. The latter product (**42**) was accompanied by an isomeric spiroepoxide as indicated in equation (67). A similar type of ring expansion has been reported on a benzothiacyclohexanone (equation 68) using ethyl lithiodiazoacetate ⟨77JCS(P1)1822⟩. Acidification of the initially formed adduct resulted in the formation of thiepane (**224**). The regiospecific expansion of a 4-methylenethiochroman ring using thallium(III) nitrate, methanol and trimethyl orthoformate provides a novel alternative synthetic route to (**224**; equation 69) ⟨77TL1827⟩.

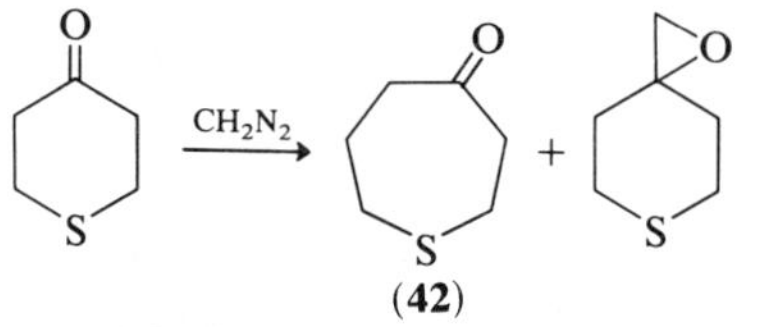

(67)

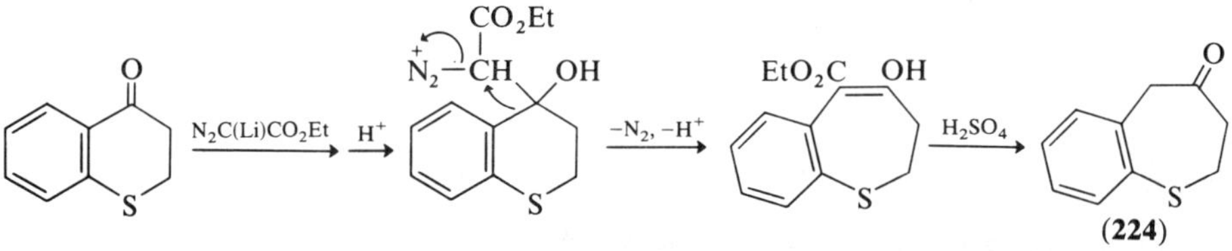

(68)

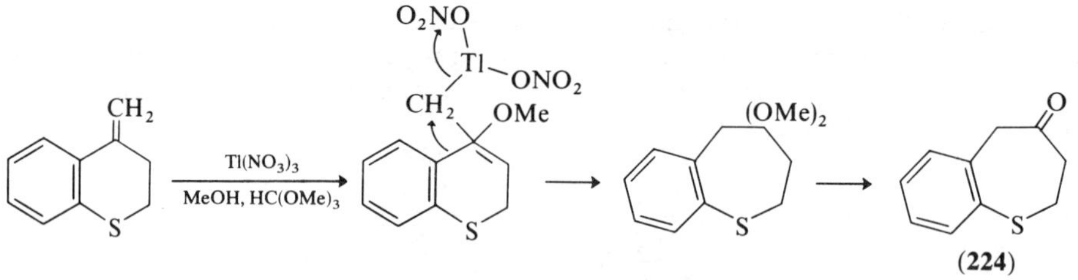

(69)

5.17.3.3.3 Summary of synthetic routes to benzo- and monosubstituted thiepanes

The wide range of synthetic methods used in the synthesis of benzo- and monosubstituted thiepanes is summarized in Table 9.

5.17.3.4 Synthesis of Thiepins

5.17.3.4.1 From non-heterocyclic compounds

(i) By formation of one bond

Ring closure by an intramolecular acylation reaction followed by a halogenation–dehydrohalogenation sequence (equation 70) may account for the isolation of the

Table 9 Synthetic Routes to Benzo- and Monosubstituted Thiepanes

Compound number	Name	Ref.
(**35**)	Thiepane	67JCS(C)1130
(**115**)	Thiepane 1-oxide	60G382
(**225**)	Thiepane 1,1-dioxide	67JA1281
(**226**)	2,3,4,5-Tetrahydro-1-benzothiepin	77JCS(P1)1822
(**227**)	1,3,4,5-Tetrahydro-2-benzothiepin	59BSF(2)1998
(**228**)	1,2,4,5-Tetrahydro-3-benzothiepin	56JA848
(**229**)	10,11-Dihydrodibenzo[*b,f*]thiepin	65M182
(**230**)	5,7-Dihydrodibenzo[*c,e*]thiepin	56JA6130
(**218**)	4,5-Dihydrothiepin	67JA605
(**116**)	2,7-Dihydrothiepin 1,1-dioxide	67JA1281
(**133**)	2-Methylthiepane	69JHC115
(**134**)	2-Phenylthiepane	69JHC115
(**135**)	Thiepan-2-one	64MI51700
(**41**)	Thiepan-3-one	52JA917
(**42**)	Thiepan-4-one	56JA1965
(**231**)	2-Hydroxythiepane	67JCS(C)1130
(**136**)	3-Hydroxythiepane	67AG(E)872
(**43**)	4-Hydroxythiepane	56JA1965
(**137**)	2-Acetoxythiepane	67JCS(C)1130
(**232**)	4-Acetoxythiepane	56JA1965
(**132**)	2-Chlorothiepane	69JHC115

dibenzo[*b,f*]thiepin (**233**) in good yield ⟨67CCC3186⟩. The catalytic decomposition of a sulfonyl chloride (equation 71) at an elevated temperature (250 °C) results in an internuclear cyclization to yield tribenzo[*b,d,f*]thiepin (**146**) ⟨65T1299⟩ (*cf.* equation 49).

CO_2H —($POCl_3$, $ZnCl_2$, $PhNO_2$)→ O → Cl (**233**) (70)

SO_2Cl —(Δ, CuCl)→ (**146**) (71)

(*ii*) *By formation of two bonds*

The thiepin (**234**) has been synthesized from phthalaldehyde by the double Knoevenagel condensation reaction shown in equation (72) ⟨53JA6332⟩.

O=CH HC=O + EtO_2C S CO_2Et —NaOH→ —H^+→ HO_2C S CO_2H (**234**) (72)

5.17.3.4.2 From heterocyclic compounds

(*i*) *By formation of seven- from three-membered rings*

Despite careful control of reaction conditions the base-catalyzed dehydrobromination of the dibromoepisulfide indicated in equation (73) was consistently found to yield only benzene and sulfur ⟨72JOC552⟩. This observation can be explained by the initial formation of benzene episulfide followed by ring expansion to thiepin (**44**) and a rapid spontaneous decomposition by aromatization and sulfur extrusion.

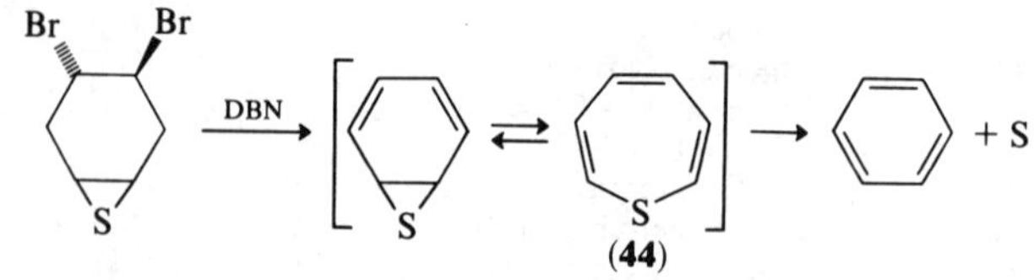

(73)

(ii) By formation of seven- from five-membered rings

The mechanism of thermal isomerization of the cyclobutene compound in equation (74) to thiepin (**50**) is directly analogous to that found in the synthesis of dihydrothiepins (**221**) and (**222**; equation 65). Compound (**50**) was the first monocyclic thiepin to have been identified ⟨74T2093⟩. This synthetic route has also been applied to thiepin (**148**) ⟨74T2431⟩.

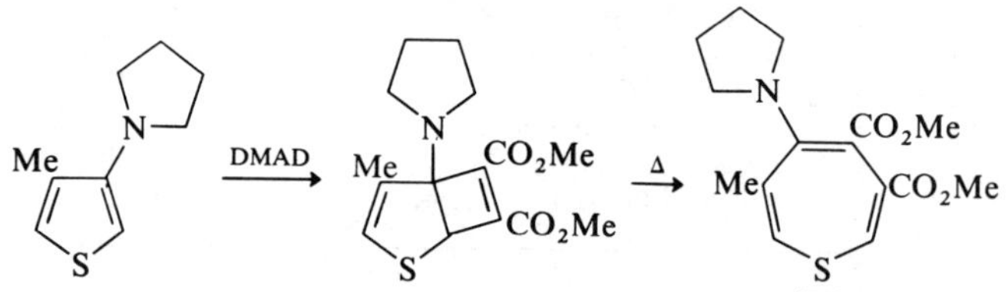

(74)

(iii) By formation of seven- from six-membered rings

The ring expansion reacton of a six-membered thiopyrylium salt using ethyl lithiodiazoacetate as reagent proceeds *via* a carbene intermediate and has been used in the synthesis of a range of thiepins. The remarkably stable thiepin (**49**) has been prepared by this method ⟨79JA5059⟩ from a 2,6-di-*t*-butyl thiopyrylium salt (equation 75). Other thiepins which have been synthesized by this route include the benzothiepins (**60**), (**235**), (**236**), (**237**) ⟨78TL3567⟩ and (**238**) ⟨81MI51700⟩ and the dibenzothiepins (**239**) ⟨76AG(E)611⟩ and (**145**) ⟨61TL593⟩.

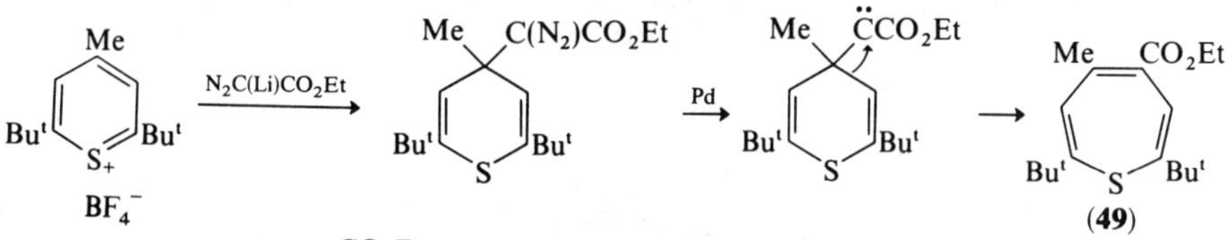

(75)

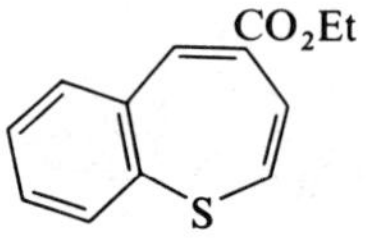

(**235**) 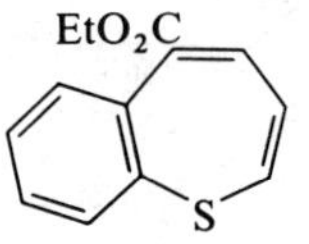 (**236**) 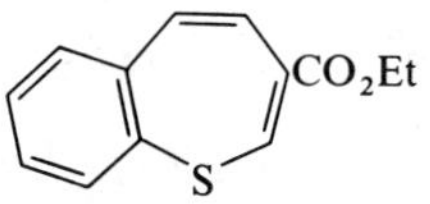 (**237**)

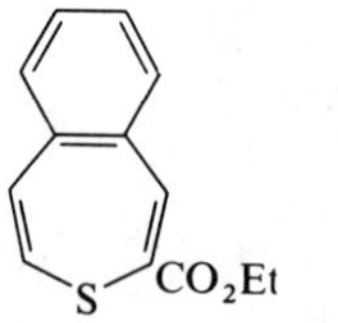

(**238**) 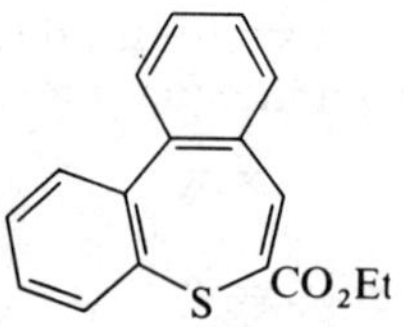 (**239**)

Thiepin 1,1-dioxide (**240**) has been obtained by the reaction of a six-membered cyclic sulfone with diphenylacetylene in the presence of UV light (equation 76). No mechanism has been proposed for this rather unusual ring expansion ⟨74JOC103⟩.

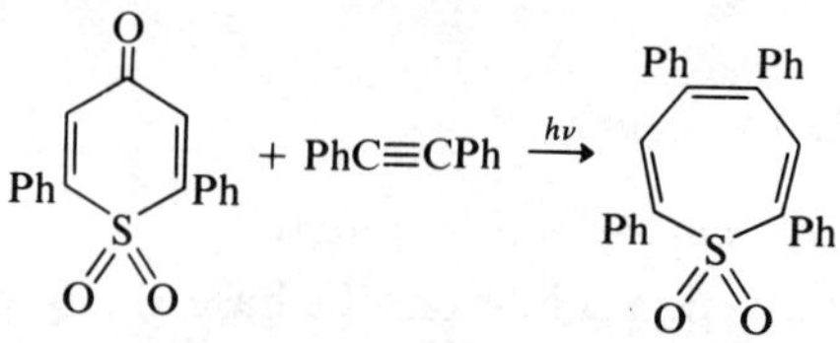

(76)

A valuable synthetic route to 1-benzothiepin (**46**) ⟨74AG(E)142⟩ and a range of substituted 1-benzothiepins (**241**)–(**244**) ⟨75TL2697⟩ is based upon the availability of the corresponding valence isomers containing a bicyclo[1.1.0]butane moiety. Thus, isomerization to 1-benzothiepin (**46**) occurs under mild conditions (~0 °C) in the presence of a dicarbonyl-2,4-pentanedionatorhodium(I) catalyst (equation 77).

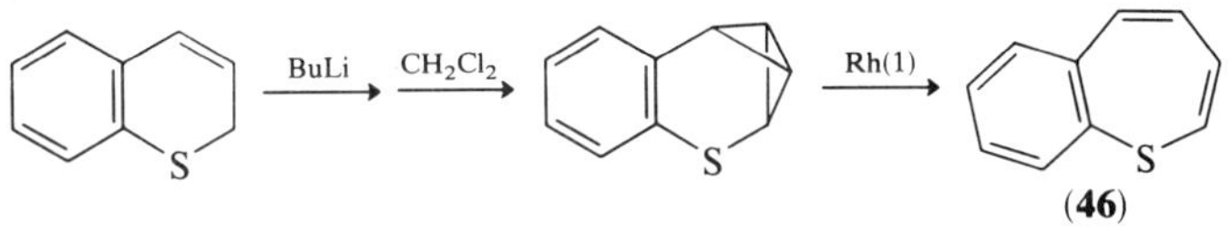

BuLi, CH_2Cl_2 → ; Rh(I) → (**46**) (77)

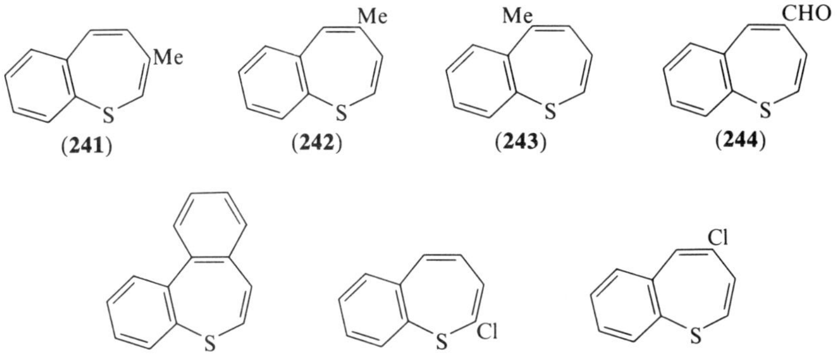

(**241**) (**242**) (**243**) (**244**)

(**245**) (**246**) (**247**)

5.17.3.4.3 *Summary of synthetic routes to benzothiepins*

In view of their instability, relatively few monocyclic thiepins have been synthesized and thus Table 10 provides a summary of synthetic routes to the more stable benzothiepin series. The majority of the thiepins in Table 10 have been prepared by the general methods exemplified by equations (75) and (77).

Table 10 Synthetic Routes to Benzothiepins

Compound number	*Name*	*Ref.*
(**46**)	1-Benzothiepin	74AG(E)142
(**48**)	1-Benzothiepin 1,1-dioxide	61JOC2728
(**53**)	3-Benzothiepin	81MI51700
(**145**)	Dibenzo[*b,f*]thiepin	60JOC828
(**245**)	Dibenzo[*b,d*]thiepin	76AG(E)611
(**146**)	Tribenzo[*b,d,f*]thiepin	65T1299
(**241**)	3-Methyl-1-benzothiepin	75TL2697
(**242**)	4-Methyl-1-benzothiepin	75TL2697
(**243**)	5-Methyl-1-benzothiepin	75TL2697
(**60**)	2-Ethoxycarbonyl-1-benzothiepin	78TL3567
(**237**)	3-Ethoxycarbonyl-1-benzothiepin	78TL3567
(**235**)	4-Ethoxycarbonyl-1-benzothiepin	78TL3567
(**236**)	5-Ethoxycarbonyl-1-benzothiepin	78TL3567
(**244**)	4-Formyl-1-benzothiepin	75TL2697
(**246**)	2-Chloro-1-benzothiepin	73JOC3987
(**247**)	4-Chloro-1-benzothiepin	73JOC3987

5.17.4 APPLICATIONS AND IMPORTANT COMPOUNDS OF OXEPANES, OXEPINS, THIEPANES AND THIEPINS

5.17.4.1 Applications and Important Compounds of Oxepanes

The ease of synthesis of oxepan-2-one (ε-caprolactone; **78**) on a commercial scale has led to its use as a monomer in the production of poly-ε-caprolactone. The latter polymer has been widely used as an additive and plasticizer. Over the last decade a large number of patents have used ε-caprolactone both in monomeric and polymeric form for the production of protective coatings, leather substitutes, elastomers, pigments, adhesives, shrink-resistant agents in wool and dispersing agents.

Among the naturally occurring oxepanes which have important biological effects are the alkaloid strychnine (**248**) which has been used in rodent control and to a limited extent as a tonic and stimulant in humans! The oxepane zoapatanol (**249**) has been used in the induction of labor and thus termination of pregnancy ⟨79JA3404⟩. The fungal metabolite aranotin (**250**) contains two dihydrooxepin rings and has shown antiviral activity ⟨68BBR(33)219⟩.

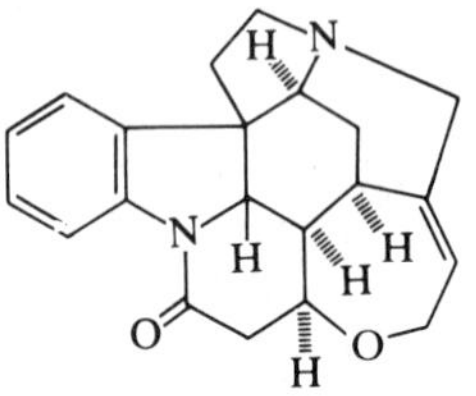

(**248**) 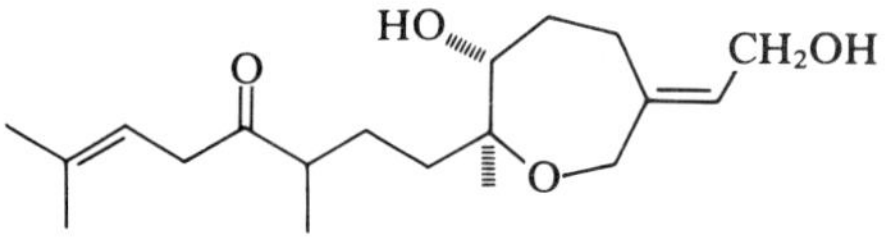 (**249**) 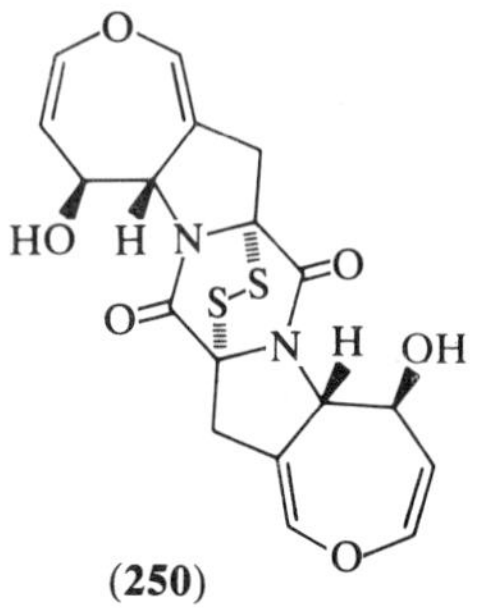 (**250**)

The chemistry of those oxepanes which may be classified as terpenes, steroids, sugars and alkaloids has been covered in depth elsewhere ⟨72HC(26)1⟩ and thus will not be included in the present chapter.

A large number of substituted derivatives of tetrahydro-1-benzoxepin (**6**) and 10,11-dihydrodibenz[*b*,*f*]oxepin (**165**) have been reported in the patent literature to have significant and diverse pharmacological activity. Thus, they have been found to act as muscle relaxants, stimulants of the central nervous system, sedatives, antihistamines, antiepileptics and anticonvulsants. The range of oxepane structures having such activities is too large to permit discussion of individual compounds. It is however noteworthy that the general 10,11-dihydrodibenz[*b*,*f*]oxepinyl piperazine structure (**251**) (and the corresponding thiepin compounds; see Section 5.17.4.3) shows marked neuroleptic activity ⟨75AF1436⟩.

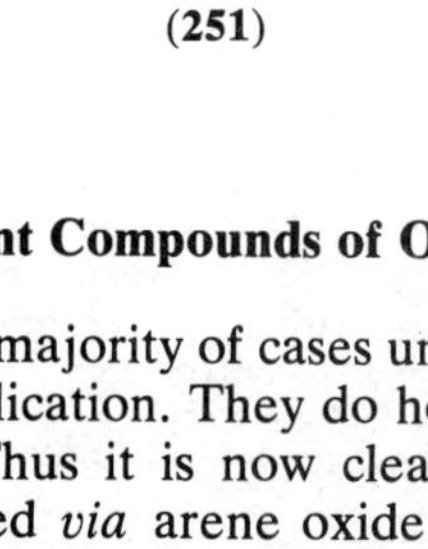

(**251**)

5.17.4.2 Applications and Important Compounds of Oxepins

Monocyclic oxepins, being in the majority of cases unstable compounds, have not to date had any significant commercial application. They do however play a vitally important role in biosynthesis and metabolism. Thus it is now clearly established that the process of aromatic hydroxylation may proceed *via* arene oxide intermediates and this is indicated by the presence of the NIH shift (see Section 5.17.2.2.3). Among the more important oxepins considered to be involved in enzyme-catalyzed hydroxylations are (**252**) and (**253**). The formation of these intermediates during the biosynthesis of tyrosine ⟨73JCS(P1)218⟩ and gliotoxin (**100**) ⟨80JOC3149⟩ has been postulated. It has been assumed that oxepins (**254**) and (**255**) are involved in the conversion of cinnamic acid to *ortho*- ⟨71P3069⟩ and *para*- ⟨67MI51700⟩ coumaric acid respectively. Methyl-substituted oxepins such as (**10**), (**93**), (**177**) and (**209**)–(**215**) have been considered as intermediates in the metabolism of the corresponding arenes by liver microsomal systems ⟨72B3080⟩. Monooxygenase enzymes in the latter system have also been found to catalyze the formation of oxepins (**202**) and (**207**) from chlorobenzene substrate ⟨75JA4428⟩.

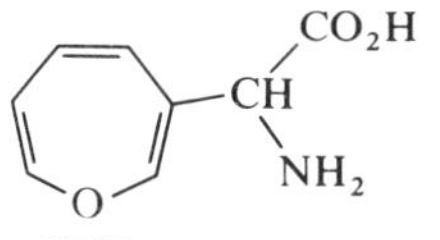

(**252**) 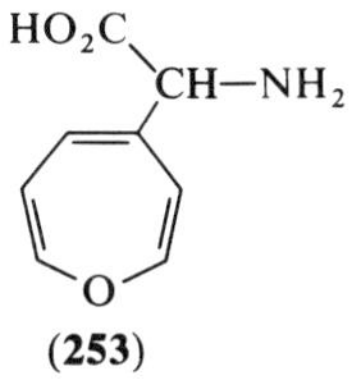 (**253**)

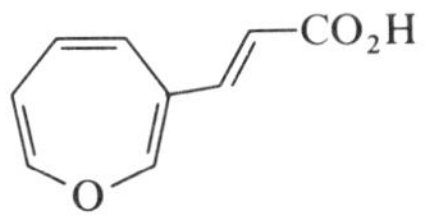

(**254**) 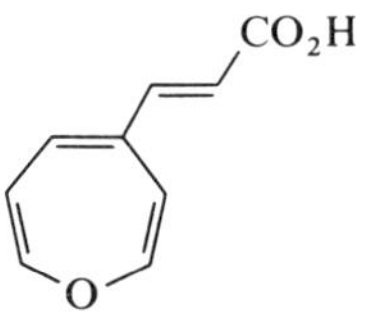 (**255**)

The metabolism of polycyclic aromatic hydrocarbons by enzymes present in animal livers involves epoxidation as the initial step. As indicated in Section 5.17.1.2, evidence is available to suggest that oxepins (**29**)–(**34**) are present as minor contributors to the arene oxide–oxepin equilibrium and thus may legitimately be considered as metabolic intermediates.

The oxepin–arene oxide intermediates formed during metabolism undergo further non-enzymatic and enzyme-catalyzed reactions including spontaneous rearrangement to phenol, hydrolysis catalyzed by epoxide hydrolase, attack by glutathione in the presence of glutathione *S*-epoxide transferase, and deoxygenation in the presence of epoxide reductase enzymes. The latter transformations are summarized for a general oxepin–arene oxide structure as shown on Scheme 35 and are discussed in several review articles ⟨74MI51700, 81ACR218⟩.

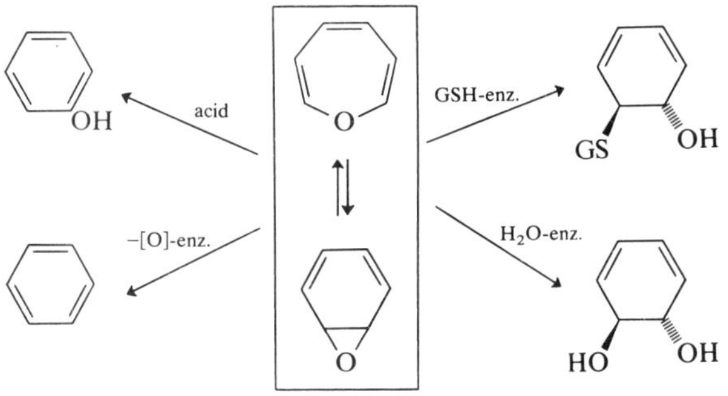

Scheme 35

5.17.4.3 Applications and Important Compounds of Thiepanes

The neuroleptic activity of 10,11-dihydrodibenzo[*b,f*]thiepins has led to the synthesis of a very large number of derivatives. Protiva and coworkers have been responsible for a large proportion of work in this area and have published over one hundred papers in this series ⟨81CCC1199⟩.

Thiepanes of the general structure (**256**) include the drugs octoclothepin (antischizophrenic), chlorotepin (antiinflammatory), trifluthepin (antidepressant), methiothepin (antihistamine) and oxyprothepin (neuroleptic). The slightly modified 10,11-dihydrodibenzo[*b,f*]thiepins (**257**; dithiadene) and (**258**) have a similar range of pharmacological activities.

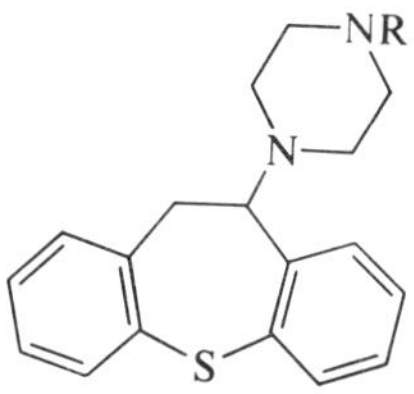

(**256**) 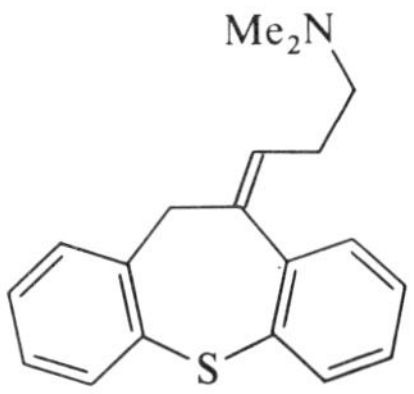 (**257**) 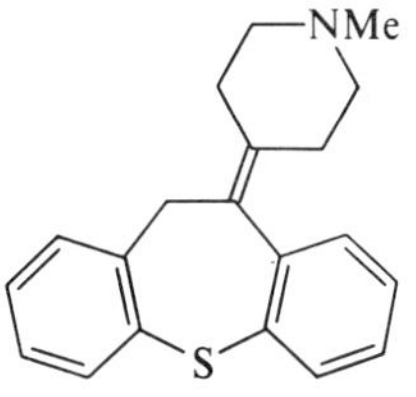 (**258**)

5.17.4.4 Applications and Important Compounds of Thiepins

The monocyclic thiepins are less stable than the corresponding oxepins and the few known examples have to date no significant commercial applications. Several members of the much more stable dibenzo[*b*,*f*]thiepin series have however been synthesized and tested for biological activity. Thus thiepins (**259**) and (**260**) were found to have antiinflammatory and antihistamine properties respectively ⟨74JHC17⟩.

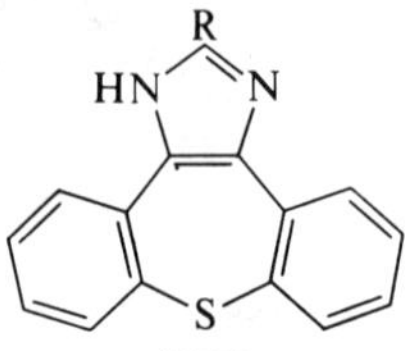

R HN N S

(**259**)

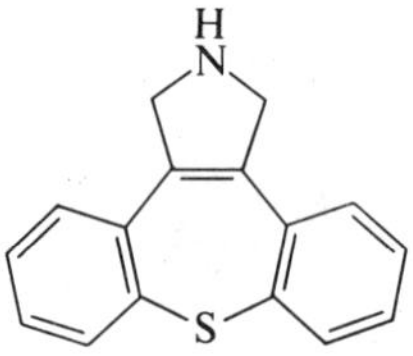

H N S

(**260**)

5.18

Seven-membered Rings with Two or More Heteroatoms

J. T. SHARP
University of Edinburgh

Chapter 5.18 presents over 50 heterocyclic systems whose chemistry usually depends more on the nature of the heteroatoms than on the systems being seven-membered heterocycles. At the present state of knowledge, organizing the chapter in separate sections on structure, reactivity and synthesis would not produce a happy result, especially for the reactivity section. Little is presently known about the reactivities of the majority of the systems, many of which have been prepared to explore their pharmacological activities. The chapter, therefore, has been organized according to the number and nature of the heteroatoms.

5.18.1 INTRODUCTION

This chapter is divided into 10 sections on the major groups of seven-membered heterocyclic systems containing two, three or four of the heteroatoms N, O and S. Each system has its own characteristic chemistry, and although there are some common threads linking both the preparative methods and reactivity of some of these systems, they are neither long nor strong enough to make a combined comparative treatment either practicable or instructive. Emphasis throughout has been laid on preparative methods and on reactivity, and although some mention has been made of spectroscopic study and of applications and uses, space limitation has prevented detailed coverage of these topics.

The preparative routes from non-heterocyclic precursors have been classified by an indication of the bonds formed, using the usual convention of labelling the ring sides by lower case italic letters: *a* for the 1,2-bond, *etc.* Thus in the preparation of 1,2-diazepines (**1**) a reaction involving formation of the N—N bond is a *type a* synthesis while one involving the combination of a C—C—C—C—C fragment with an N—N fragment is a *type bg* process. For each system all the methods involving formation of the *a* bond are given first in the order one-bond, two-bond, *etc.* types, followed by a similar treatment for the *b* bond and so on. This is followed by a section covering preparations from other heterocyclic systems. In each case some indication is given of the range of applications of the method to monocyclic, benzo- or hetero-fused systems.

5.18.2 DIAZEPINES

5.18.2.1 1,2-Diazepines

The synthesis and reactivity of 1,2-diazepines has received increasingly intensive study over the last 15 years and good preparative routes have been developed to the fully unsaturated 1*H*-, 3*H*- and 4*H*-isomers. The 5*H*-isomer (**2**) has not been isolated because its valence tautomeric equilibrium with the diazanorcaradiene (**3**) lies far to the right for all examples so far studied, although (**2**) is thought to exist as a transitory intermediate in the thermal diastereotopomerization of (**3**) ⟨70JA3787, 72JA2770⟩. There are also many synthetic routes to the partially and fully saturated systems. The unsaturated systems adopt flexible, boat-shaped conformations like cycloheptatriene and NMR has been much used to study the energetics of ring inversion of the diazepines themselves, their salts and metal complexes ⟨69ACS3125, 81ACR348, 73JCS(P1)2543, 81TL353, and references cited in the preparative section⟩. The most characteristic feature of the chemistry of 1,2-diazepines is their facility for ring transformation reactions under thermal, photochemical, acidic and basic conditions.

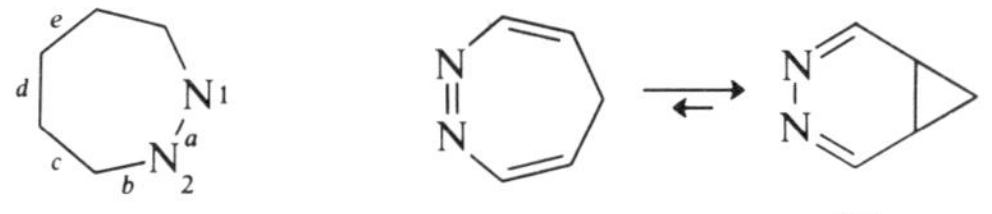

(**1**) (**2**) (**3**)

Reviews: ⟨67AHC(8)21, 68CRV747, 76H(4)1509, 77H(6)2021, 81ACR348, 81H(15)1569⟩.

5.18.2.1.1 Synthesis

(i) Type a *(N—C—C—C—C—C—N)*

Formation of the *a* bond by the reductive cyclization of dinitro compounds has been used in the preparation of dibenzo[*c*,*f*]diazepines ⟨67AHC(8)21⟩.

(ii) Type b *(C—C—C—C—C—N—N)*

The 1,7-electrocyclization of $\alpha\beta,\gamma\delta$-unsaturated diazoalkanes, *e.g.* (**4**), provides a good general route to 3*H*-1,2-diazepines, *e.g.* (**5**), which is applicable to monocyclic systems and to those fused to alicyclic or aromatic rings. In the example shown the 8π-electron cyclization is followed by a 1,5-sigmatropic hydrogen shift to give only (**6**) as the isolated product ⟨82UP51800⟩. However since such hydrogen shifts are rapid at room temperature ⟨79JCS(P1)2209⟩ the isomer ratio will in general reflect the relative thermodynamic stability of (**5**) and (**6**). Similar diazo compounds in which the $\alpha\beta$-double bond is part of an aromatic ring, *e.g.* (**7**), give 1*H*-2,3-benzodiazepines (**8**) ⟨73JCS(P1)2543⟩ and analogous thienodiazepines ⟨80JCS(P1)1718⟩ in high yield. In all cases the ring closure takes place only

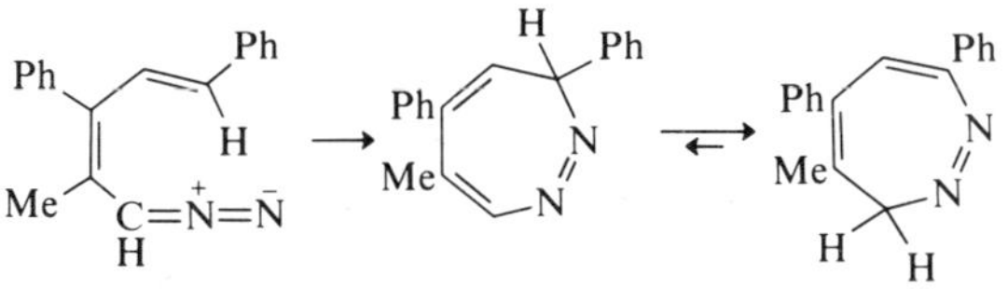

(**4**) (**5**) (**6**)

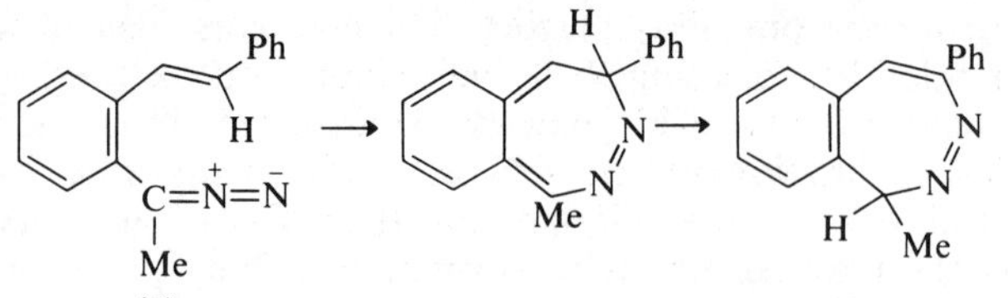

(7) (8)

if the substrate has a *cis* H at the cyclization site; other groups, *e.g.* Me and Ph, obstruct the cyclization transition state, and the diazo compound then reacts either *via* 1,5-cyclization to give pyrazoles or *via* carbenic pathways ⟨80TL4109⟩.

When the $\gamma\delta$-double bond is part of an aromatic ring, *e.g.* (**10**), the balance between 1,5- and 1,7-cyclization is finer and generally favours pyrazole (**9**) formation. However the incorporation of steric constraints which disfavour the 1,5-process, *e.g.* $R^1R^2 = (CH_2)_3$, results in the formation of 3*H*-1,2-benzodiazepines, *e.g.* (**12**), in high yield ⟨75JCS(P1)102, 79JCS(P1)1433⟩. [^{2}H]-Labelling studies have shown that this reaction involves a fast pre-equilibrium between (**10**) and (**11**) followed by an irreversible hydrogen shift ⟨81TL1537⟩. In cyclization on to the unsymmetrically substituted benzene ring in (**10**; $R^1R^2 = (CH_2)_3$, $R \neq H$) the substituent R influences the cyclization site *via* kinetic control when the reaction is done at 80 °C. Thus alkyl, alkoxy and chloro substituents favour cyclization at the more hindered 2′ position while trifluoromethyl favours 6′ ⟨80TL1379⟩. At higher temperatures the kinetically favoured product isomerizes to give the more thermodynamically stable isomer *via* a retro-cyclization process.

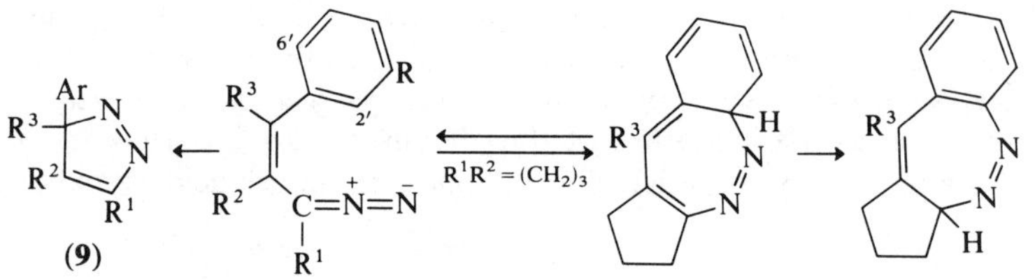

$R^1R^2 = (CH_2)_3$

(9) (10) (11) (12)

(*iii*) *Type* bg (*C—C—C—C—C+N—N*)

A wide range of monocyclic and fused 1,2-diazepines have been prepared by the reactions of 1,5-difunctional compounds with hydrazines or their anions.

(*a*) *1,5-Dihalides and Ditosylates.* These have been used mainly for the preparation of fully saturated monocyclic systems, *e.g.* (**13**), which can be hydrolyzed to give the unsubstituted system, and for 1,2,3,4-tetrahydro-5*H*-2,3-benzodiazepines (**14**) ⟨67AHC(8)21⟩.

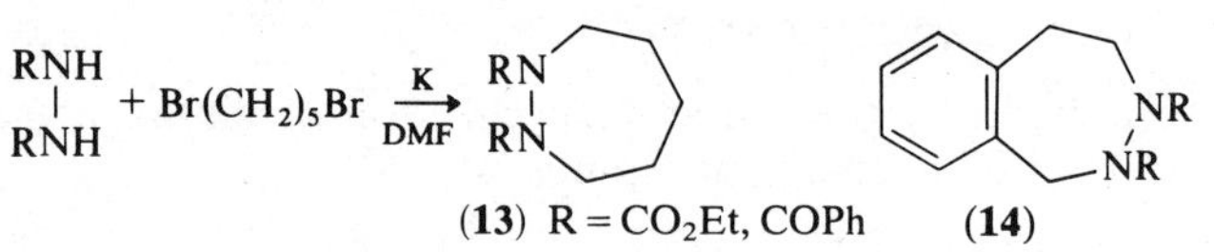

RNH–NHR + $Br(CH_2)_5Br$ $\xrightarrow[\text{DMF}]{\text{K}}$

(**13**) $R = CO_2Et$, COPh (**14**)

(*b*) *1,5-Halogenocarbonyl compounds.* 5-Halogeno-aldehydes and -ketones react with a wide range of substituted hydrazines to give 4,5,6,7-tetrahydro-1,2-diazepines (**15**) ⟨76H(4)1509⟩. Similar reactions with 2-(2-bromoethyl)benzophenone give the 3,4-dihydro-5*H*-2,3-benzodiazepine system (**16**) ⟨67AHC(8)21⟩, and the 5-haloester (**17**) has been used in the synthesis of 2,3-benzodiazepin-4-ones (**18**) ⟨77JHC465⟩.

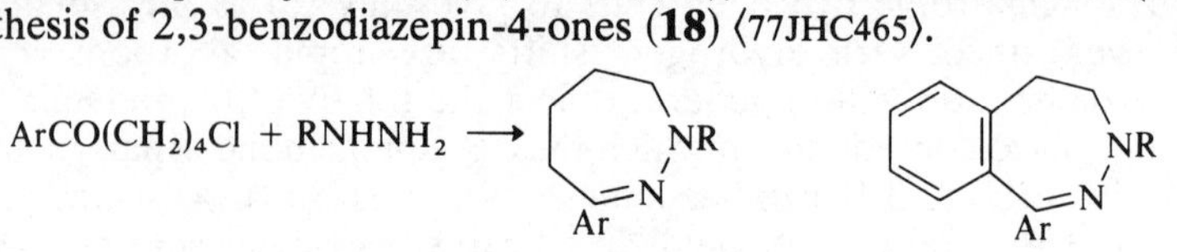

$ArCO(CH_2)_4Cl + RNHNH_2 \longrightarrow$

(**15**) (**16**)

CH_2CO_2Me / $CHBrCO_2Me$ $\xrightarrow{RNHNH_2}$

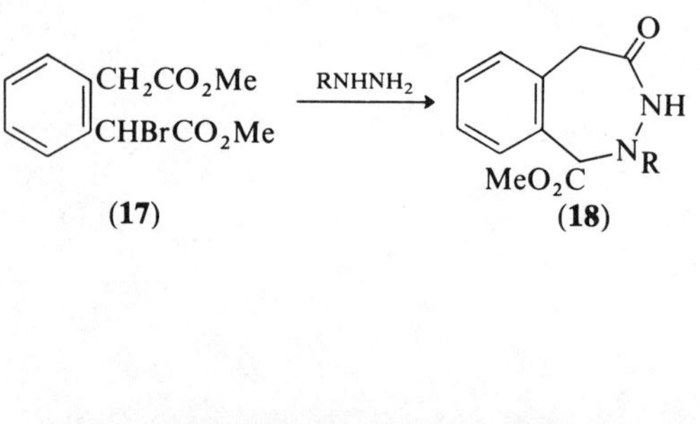

(**17**) (**18**)

(*c*) *1,5-Dicarbonyl compounds.* The reaction of 1,5-diketones with hydrazine has been much used as a source of 5,6-dihydro-4*H*-1,2-diazepines (**19**) ⟨67AHC(8)21, 76H(4)1509⟩ and 5-ketoacids have been similarly used in the preparation of 2,4,5,6-tetrahydro-3*H*-1,2-diazepin-3-ones (**20**), 1,2-benzodiazepin-3-ones, 3,5-dihydro-4*H*-2,3-benzodiazepin-4-ones (**21**) ⟨76H(4)1509, 77MI51804, 76MI51803, 74JHC401⟩, 2,5-dihydro-1*H*-2,3-benzodiazepin-1-ones (**22**) ⟨67AHC(8)21⟩, diazepinoindoles ⟨81JHC889⟩ and pyrazolo[3,4-*d*][1,2]diazepines ⟨81JHC271⟩.

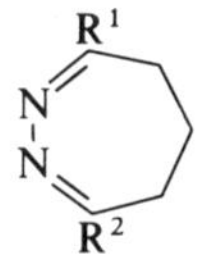

(**19**) 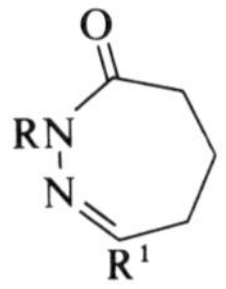 (**20**) 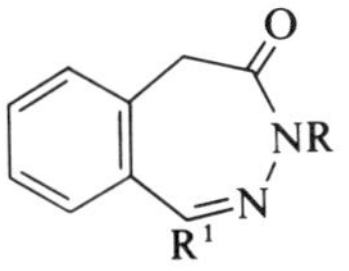 (**21**) 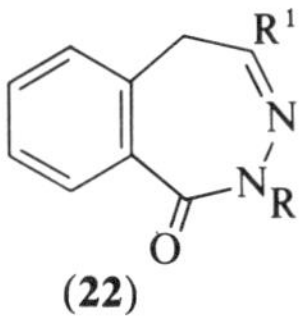 (**22**)

(*d*) *αβ,γδ-Unsaturated ketones.* The acid-catalyzed reactions of various substituted hydrazines with some *αβ,γδ*-unsaturated ketones give 2-substituted 3,4-dihydro-1,2-diazepines (**24**) ⟨79JCS(P1)1640, 81JCS(P1)2761⟩. The ring closure step is however substituent dependent and in some cases, notably where R^2 = aryl, the hydrazone (**23**) could not be cyclized. The products (**24**; R = Ts) are useful as precursors for 3*H*-1,2-diazepines (**25**) ⟨79JCS(P1)2209⟩ (see p. 604).

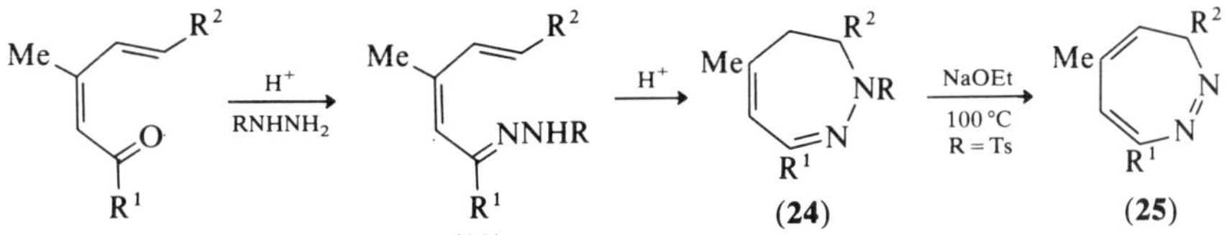

R = $-SO_2R$, COR, CO_2Et, Ph (**23**) (**24**) (**25**)

(*iv*) *Type* c (*C—C—C—C—N—N—C*)

In another example of the use of an intramolecular reaction of a 1,3-dipolar intermediate (see type *b*) the nitrilimines (**27**), generated by the reaction of the hydrazoyl chlorides (**26**) with triethylamine at 80 °C, cyclize to give 1*H*-1,2-benzodiazepines (**30**) ⟨77JCS(P1)2092, 78S603, 79S380⟩. However, when the reaction is carried out at room temperature using silver carbonate, the cyclopropa[*c*]cinnolines (**28**) are obtained; these on heating rearrange *via* ring expansion and hydrogen migration to give the diazepines (**30**) ⟨81JOC1402⟩. The formation of (**28**) is stereospecific and it seems likely that it occurs *via* a 1,1-cycloaddition of the nitrilimine, but an initial 1,7-electrocyclization followed by ring contraction cannot be excluded. The high temperature reaction is a good route to 1*H*-1,2-benzodiazepines provided that, in (**26**), either R or R^1 = H. In cases where the alkenyl moiety is doubly substituted only cyclopropa[*c*]cinnolines are obtained ⟨77JCS(P1)2092⟩, but these can in some cases be rearranged to give 4,5-dihydro-1*H*-1,2-benzodiazepines *via* acid-catalyzed opening of the cyclopropane ring ⟨79JCS(P1)1195⟩. In the above (**26**) to (**30**) conversion the intermediate (**29**)↔(**31**) may be trapped with added nucleophiles to give 5-azido-, -cyano- or -hydroxy-4,5-dihydro-1*H*-1,2-benzodiazepines (**32**) ⟨80JCS(P1)116⟩.

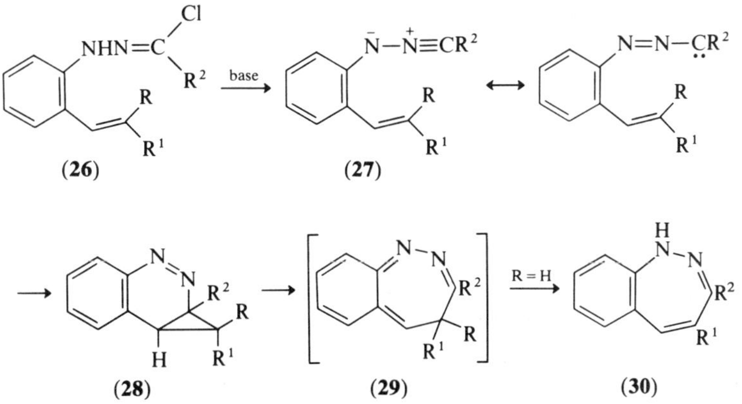

(**26**) (**27**)

(**28**) (**29**) (**30**)

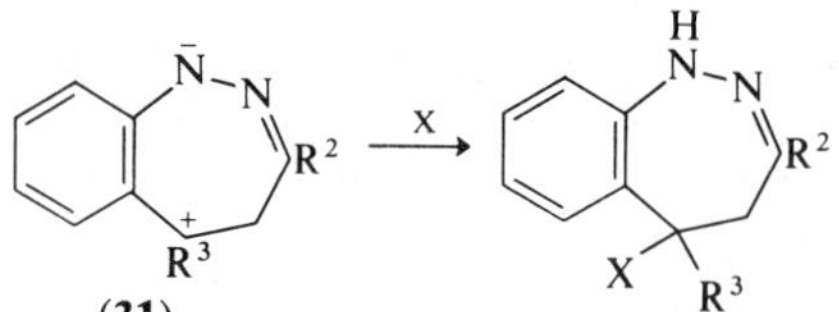

(31) (32)

(*v*) *Type* de (*C—C—N—N—C—C+C*)

5,6-Dihydro-4*H*-1,2-diazepines (**34**) have been prepared in good yields by the reactions of α-bromoazines (**33**) with malonic esters in the presence of base ⟨77BCJ2153⟩.

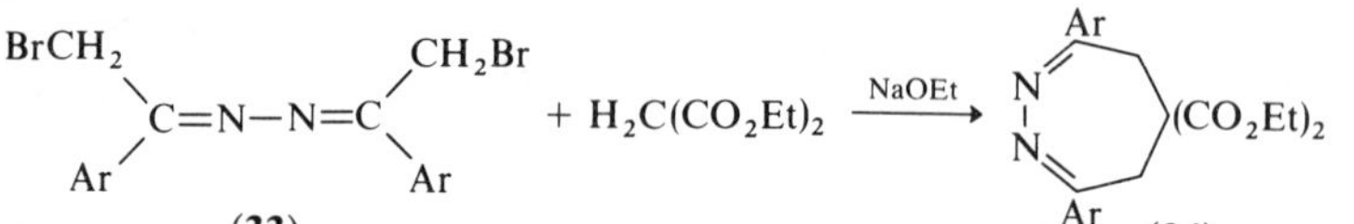

(33) (34)

(*vi*) *Synthesis from other heterocyclic systems*

(*a*) *From 3,4-diazanorcaradienes.* Several examples of the thermal ⟨77TL4393, 70JA7436⟩ and photochemical ⟨73JA3970, 76TL4321⟩ rearrangement of 3,4-diazanorcaradienes (**35**) to 4*H*-diazepines (**37**) are known; both proceed *via* walk-rearrangements to (**36**) followed by valence isomerization. A formally similar rearrangement to give a 1,2-diazepin-4-one has also been observed ⟨73JOC2954⟩.

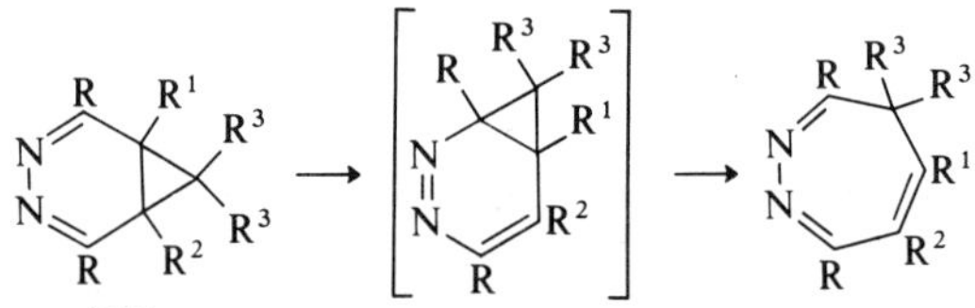

(35) (36) (37)

(*b*) *From pyridine* N-*imides and related species.* Much work has been done over the last 15 years on the photochemical conversion of pyridine *N*-imides, *e.g.* (**38**), to 1*H*-1,2-diazepines (**40**). The convenient preparation of the reactants from pyridines *via* amination with Tamuras' reagent (MSH) makes this a valuable synthetic entry into the 1,2-diazepine system. Yields in many cases are good although side reactions such as N—N fragmentation and isomerization to 2-aminopyridines intervene to varying degrees depending on the nature and position of substituents in the pyridine ring. A recent overview provides convenient access to the considerable literature on this reaction ⟨81ACR348⟩. The mechanism involves electrocyclization of an excited singlet to give the intermediate 1,7-diazanorcaradiene (**39**) followed by a thermal valence isomerization to give (**40**). Evidence for the intermediacy of (**39**) is indirect as it has not yet been trapped or detected spectroscopically. The method has recently been extended to the synthesis of 1,2-diazepin-3-ones ⟨80JOC5095⟩.

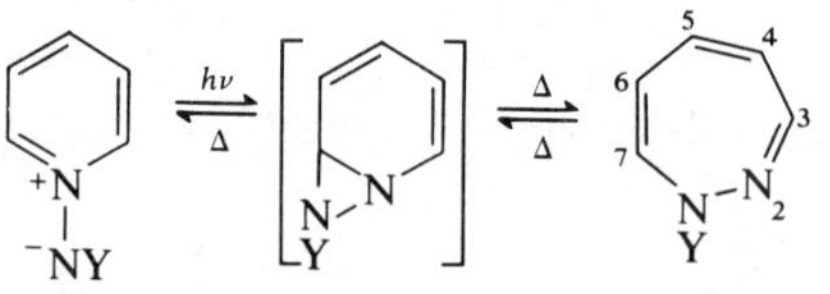

(38) (39) (40)

Studies on the extension of this principle to produce systems with the 1,2-diazepine fused to other rings have had varied and often interesting results (see also the section on 1,3-diazepines). The photolysis of quinoline *N*-imide dimers, which are in equilibrium with the *N*-imides (**41**), has been the most fruitful ⟨77JOC1856⟩ giving 1*H*-1,2-benzodiazepines (**42**) in moderate to good yields. A similar reaction of the *N*-ethoxycarbonylquinolinium betaine in the presence of ethanol gave (**43**) ⟨73T2359⟩. Reactions of this type have also provided routes to pyrido-, thieno- and furo-1,2-diazepines ⟨79CPB2183, 79H(12)1471⟩. Interestingly, in neutral conditions isoquinoline *N*-imides do not give diazepines but in the presence of potassium hydroxide 5*H*-2,3-benzodiazepines are obtained in 30–50% yields ⟨81H(16)1923⟩. 1,2-Diazepines are also formed in the pyrolysis of the *N*-imides of various other six-membered cyclic amines ⟨67AHC(8)21, 76H(4)1509, 81CPB1887⟩.

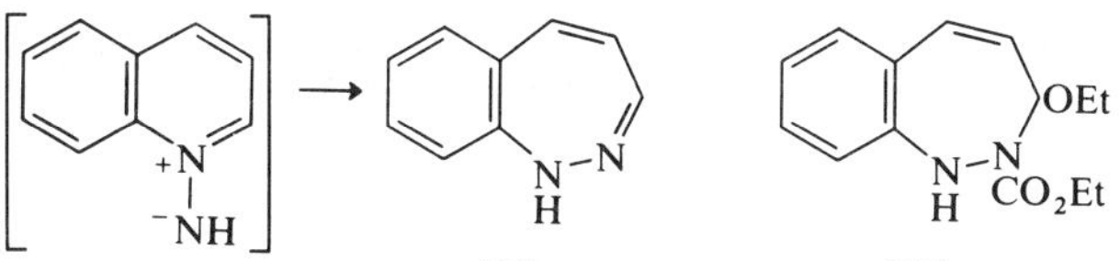

(41) (42) (43)

(*c*) *From isoquinolines.* Reaction of the pseudo-base (**44**) with *O*-sulfonylhydroxylamine derivatives gives 3,4-dihydro-5*H*-2,3-benzodiazepines (**45**) ⟨73S159, 77TL3297⟩.

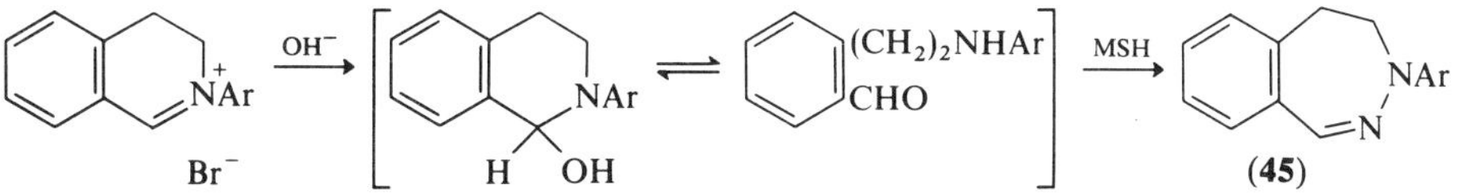

(44) (45)

(*d*) *From pyrylium and thiapyrylium salts.* The much studied reactions of hydrazine and methylhydrazine with pyrylium or thiapyrylium salts, *e.g.* (**46**), provide major routes to 4*H*-1,2-diazepines (**47**) and 1*H*-1,2-diazepines (**48**) ⟨76H(4)1509, 80CJC494⟩. An extension to benzopyrylium salts has produced a commercial synthesis of the CNS active compound Grandaxin (**49**; Ar = 3,4-dimethoxyphenyl) ⟨74CB3883⟩.

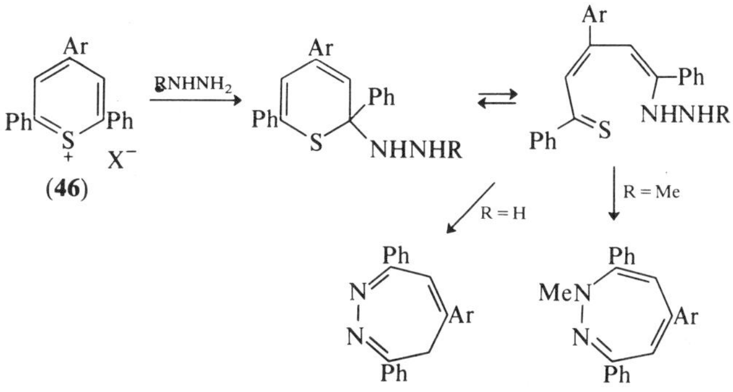

(46) (47) (48)

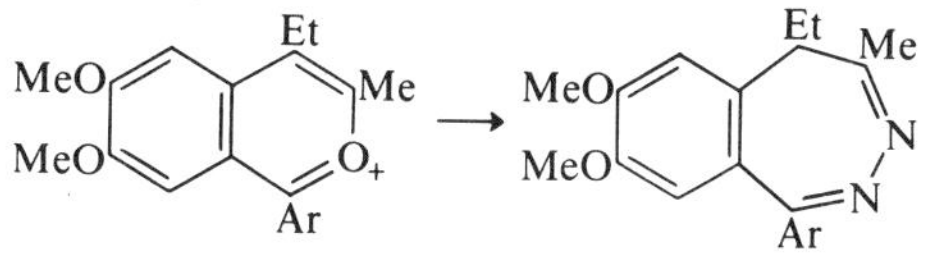

(49)

(*e*) *From pyrazoles, pyrazolines and related species.* The acid-catalyzed conversion of diazoacetylpyrazolines, *e.g.* (**50**), to the diazepinones (**51**) has generated much interesting chemistry ⟨67AHC(8)21, 76H(4)1509, B-67MI51800⟩.

The 1-pyrazolines, *e.g.* (**52**), formed by the cycloaddition of diazomethane to 1,2-disubstituted cyclobutenes are converted quantitatively to 4,5-dihydro-1*H*-1,2-diazepines (**53**) on treatment with hydrogen chloride ⟨76H(4)1509⟩.

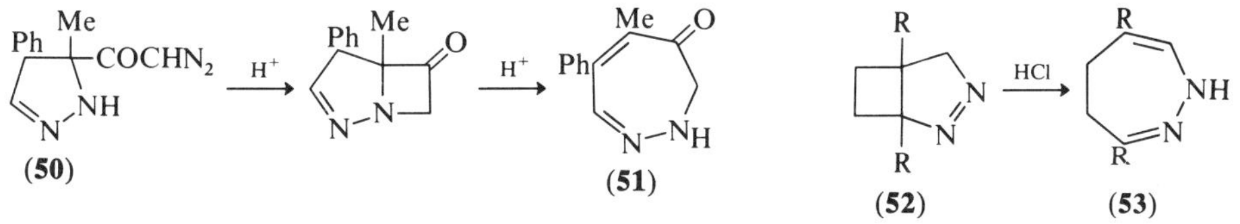

(50) (51) (52) (53)

5.18.2.1.2 Reactivity

(*i*) *Photochemical reactions*

[$_{\pi}2_s + _{\pi}2_s$] Electrocylizations have been effected for all the fully unsaturated isomers as shown overleaf and for some dihydro derivatives, diazepinones and fused systems ⟨76H(4)1509, 80JCS(P1)1230, 80JCS(P1)1718, 76JCS(P1)362⟩. In some cases such reactions are followed by spontaneous extrusion of the two-atom bridge; thus 5*H*-1,2-benzodiazepines (**88**) are converted to indazoles *via* extrusion of a nitrile from a primary product of type (**54**) ⟨80CPB1842⟩, and 4-phenyl-5H-2,3-benzodiazepine (**83**) gives 1-phenylindene *via* the extrusion of nitrogen ⟨72CC827⟩. The further irradiation of (**55**) promoted an aza-di-π-methane rearrangement to give the 1,4-diazepine (**56**) ⟨77TL4389⟩.

(1H-) NR N $\xrightarrow{h\nu}$ N N R (**54**)

(4H-) N N $\xrightarrow{h\nu}$ 1 7 6 2 5 N 4 N 3 (**55**) [$\xrightarrow{h\nu}$ 4 N 1 6 7 5 N 2 3] (**56**)

(3H-) N N $\xrightarrow{h\nu}$ N—N

(ii) Thermally induced reactions

(a) 1H-1,2-Diazepines. The products obtained from the thermal decomposition of 1-acyl-1,2-diazepines (**57**) depend much on the nature and position of the ring substituents ⟨76H(4)1509⟩. In most cases the primary reaction path at >150 °C involves electrocyclization to give the diazanorcaradiene (**58**) followed either by ring opening to give pyridinium N-imides (**59**) (obtained when no, 3-, 5- or 7-, 3,5,7-, and 4- or 6-electron withdrawing groups are present) ⟨81ACR348⟩ or by a walk-rearrangement and ring opening to give the 1,3-diazepine (**61**; 30–70%) (obtained when 4- and/or 6-electron donating groups are present) ⟨79JOC2683, 80CC444⟩. Apparently electron donation by R^2 favours N—N cleavage in (**58**) while electron withdrawal favours C—N cleavage. Both processes are accompanied to varying degrees by the formation of pyridine derivatives and 2-acylaminopyridines.

R^1 R^2 N—N COR (**57**) $\xrightarrow{\Delta}$ R^1 R^2 N NCOR (**58**) → R^1 R^2 $\overset{+}{N}$ $^{-}$NCOR (**59**) ; → R^1 R^2 NCOR N (**60**) → R^1 R^2 NCOR N (**61**)

(b) 3H-1,2-Diazepines. The monocyclic systems (**5**/**6**) are subject to surprisingly rapid 1,5-sigmatropic hydrogen shifts which interconvert (**5**) and (**6**) even at room temperature and below ⟨79JCS(P1)2209⟩. At higher temperatures (~120 °C) these are converted into 3H-pyrazoles, *e.g.* (**62**) to (**63**) ⟨79JCS(P1)2730⟩, which rearrange further *via* 1,5-vinyl and then hydrogen migration to give the 1H-pyrazole (**64**). A similar process in 3H-1,2-benzodiazepines (**65**) ⟨74CC532, 78CPB1890, 79CPB2528⟩ and other analogues with fused aromatic rings ⟨79H(12)1471⟩ is induced either by heat or UV irradiation; if R = H this is followed by hydrogen migration to give the aromatic indazole (**67**). Interestingly, the thermal reaction gives both (E)- and (Z)-alkenylindazoles (**66**) while the photochemical process gives only the former. It has been suggested that the first step in the thermal reaction may be an electrocyclic ring opening ⟨79JCS(P1)2730⟩ or homolytic cleavage ⟨79CPB2528⟩ while the photochemical reaction is a $[{}_{\pi}2_s + {}_{\sigma}2_s]$ process.

(c) 4H-1,2-Diazepines. As with the 1H-isomers the products are dependent on the nature of the substituents on the diazepine ring (**68**). The primary process is similarly an electrocyclization, giving (**69**), followed by either C→C ⟨73JA3970⟩ or C→N walk-rearrangements and in the latter case subsequent ring expansion and hydrogen migration ⟨77CJC56⟩.

(62) → Δ → [(63)] → i, 1,5-vinyl shift; ii, H shift → (64)

(65) ⇌ Δ or hν (66) → R = H → (67)

(68) → Δ → (69) → (a) / (b)

(**a**) R = Me, R′ = H
(**b**) R = H, R′ = Ph

(iii) Reactions with electrophilic reagents

(*a*) *Protonation.* ^{1}H NMR and UV studies on the protonation of the 4*H*- and 1*H*-1,2-diazepines (**70**) and (**72**) have shown that they are converted to the similar species (**71a**) and (**71b**) respectively. Under more vigorous acid treatment the 1*H* isomers give pyridine *N*-imides or *N*-aminopyridinium salts *via* diazanorcaradiene intermediates ⟨81ACR348⟩. Benzodiazepinones (**21**) and (**22**) similarly undergo ring contraction to *N*-aminoisoquinolones ⟨76H(4)1509⟩. The 4*H*-isomers (**70**) give pyridines and pyrazoles *via* (**71a**). The latter, *e.g.* (**75**), are probably formed *via* hydrolytic opening to give (**73**) and the subsequent retro-aldol reaction of (**74**).

(70) → H^+ → (71) ← H^+ ← (72)

(**71**) **a**; R = H
b; R = Me

(**71a**) → H_3O^+ → (73) → (74) → −PhCOMe → (75)

3*H*-1,2-Benzodiazepines (**12**) and 1*H*-2,3-benzodiazepines (**8**), both containing an azo group, react quite differently. The former rearrange readily in the presence of acid (or base) to give the more stable 1*H*-isomers, type (**42**) ⟨75JCS(P1)102, 78CPB1890⟩. The latter (**77**) react *via* two pathways to give the more stable 5*H*-isomers (**76**) and *via* the extrusion of nitrogen to give products derived from the carbonium ion (**78**) ⟨82UP51801⟩.

(*b*) *Acylation.* Some of the problems inherent in the extension of the *N*-imide route to the preparation of 1*H*-1,2-diazepines with a variety of *N*-substituents (**80**) can be avoided by using a procedure involving the alkylation or acylation of the tricarbonyliron complex (**79**) of the unstable parent 1*H*-diazepine ⟨76CC844⟩.

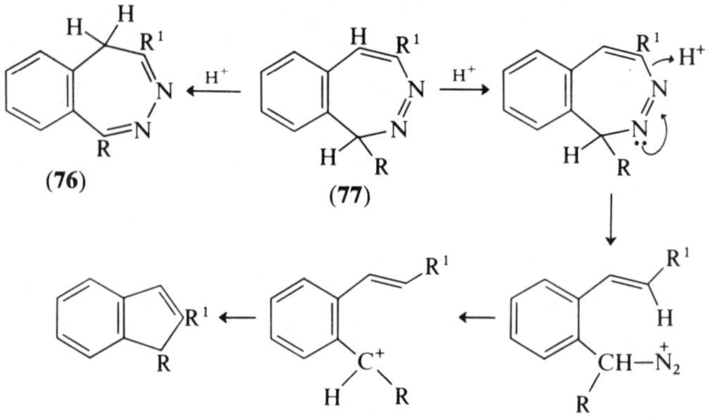

(76) (77) (78)

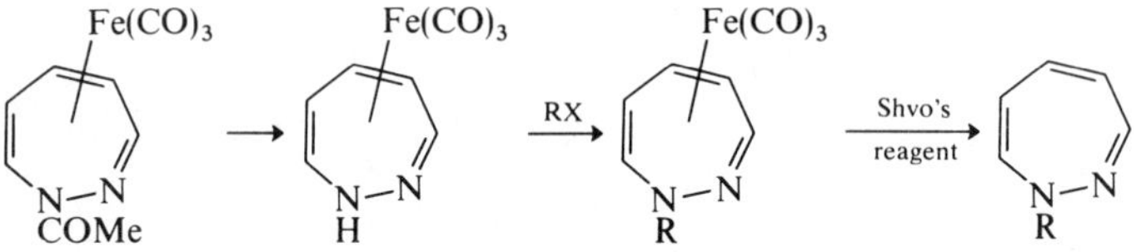

(79) (80)

The acylation of 4*H*-1,2-diazepines (**81**) provides a straightforward route to 1-acyl-3,5,7-triaryl-1*H*-1,2-diazepines (**82**) ⟨75S603⟩. The analogous benzodiazepines might be expected to follow a similar reaction path and in some cases they do so ⟨81H(16)1923⟩, but with Grandaxin (**49**) acylation at N-3 is followed by proton loss from the 4-methyl group to give a 5*H*-4-methylene derivative ⟨74ACH115⟩. 4-Phenyl-5*H*-2,3-benzodiazepine (**83**) behaves quite differently as N-2 is more nucleophilic than N-3; thus with acid anhydrides products of type (**84**) can be obtained, and with acid chlorides the reaction can take two paths depending on conditions: (i) ring contraction of the dehydrochlorinated species (**85**) to isoquinoline *N*-imides (**87**) or (ii) its $[6+4]\pi$ dimerization to give (**86**) ⟨82TL345⟩.

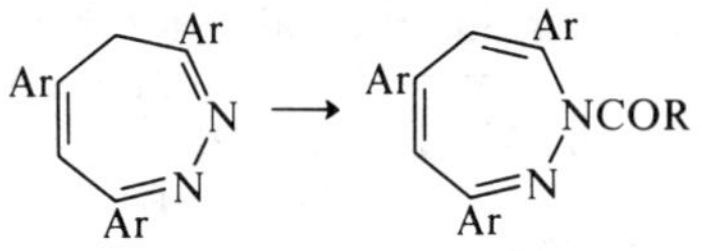

(81) (82)

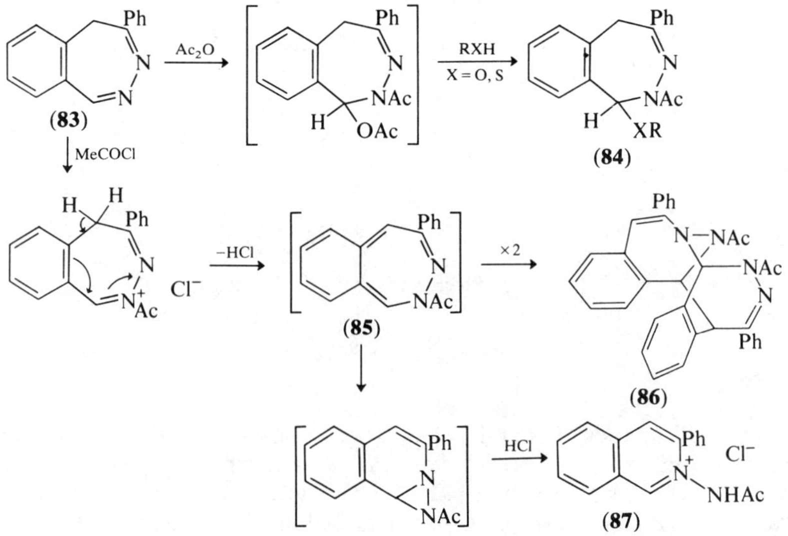

(83) (84) (85) (86) (87)

Typical imine reactivity is also displayed in 1*H*-1,2-diazepines which react with phthaloyl acetyl chloride in the presence of triethylamine to give a β-lactam fused system ⟨76H(5)471⟩.

1*H*-2,3-Dihydro-1,2-benzodiazepines, obtained by the lithium aluminum hydride reduction of (**42**), are acylated preferentially at N-2 by acetic anhydride at room temperature and also at N-1 on heating ⟨77JOC1856⟩. Similar reduction/acylation studies have also been carried out on 1-acyl-1*H*-diazepines and their tricarbonyliron complexes ⟨75CJC519⟩.

(*c*) *Oxidation.* The first example of a 5*H*-1,2-benzodiazepine (**88**) was prepared recently by reaction of (**42**) with LTA; it reacted with acetic acid or methanol *via* 1,4-addition to give (**89**; R^1 = OAc or OMe) ⟨80CPB1842⟩.

The oxidation of 3*H*-1,2-benzodiazepines (**90**; R = H or Me) with MCPBA gave both (**91**) and (**92**). The former have proved useful as precursors for 3-substituted 1,2-benzodiazepines *via* treatment with bases or acids. Thus (**91**) on reaction with alkoxides (R^1O^-) and $^-CH(CO_2Et)_2$ gave respectively (**93**; X = OR^1) and (**93**; X = $CH(CO_2Et)_2$) in *ca.* 70% yields, and treatment with hydrogen chloride and with acetic acid gave (**93**; X = Cl) and (**94**) respectively ⟨78CPB1896⟩.

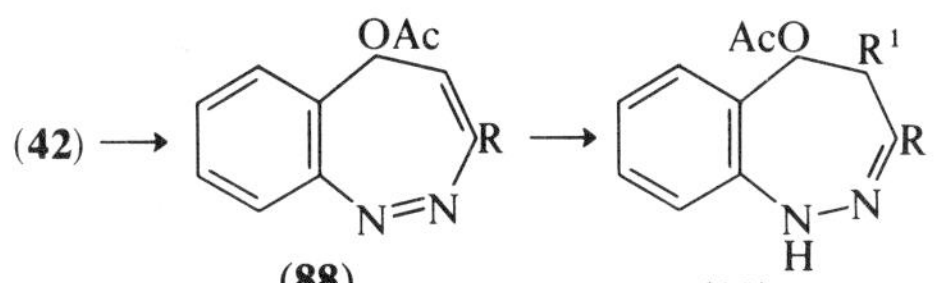

(**88**) (**89**)

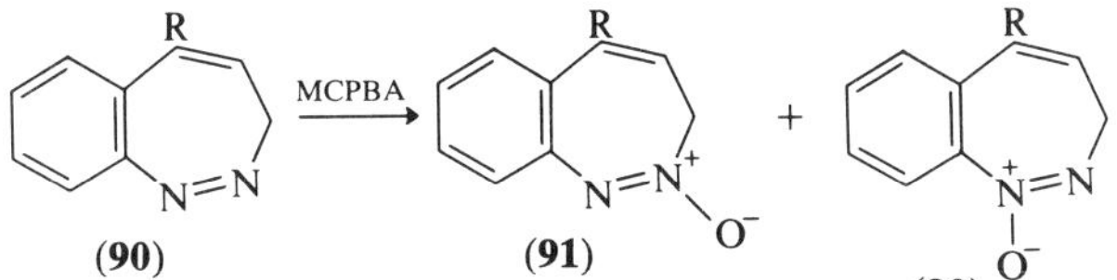

(**90**) (**91**) (**92**)

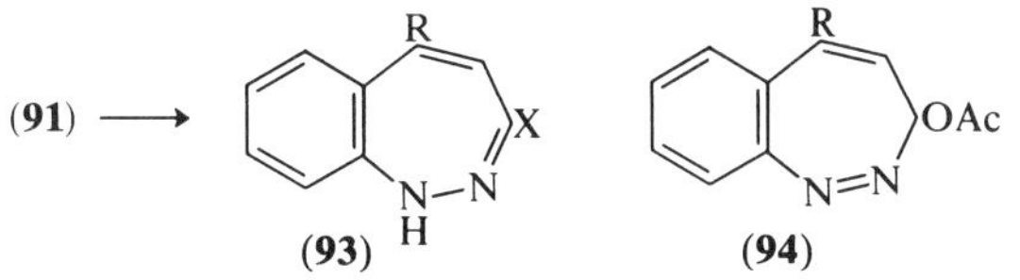

(**93**) (**94**)

The UV irradiation of *N*-oxides of diazepines of type (**19**; $R^1 = R^2$ = Ar) induced $[_\pi 2_s + _\pi 2_s]$ cyclization and subsequent ring opening to give a diazo ketone ⟨69JA2818, 72JA3955⟩.

(*d*) *Halogenation.* A variety of reagents have been used to introduce halogen substituents into diazepines and diazepinones ⟨76H(4)1509, 78CR(C)(286)671⟩. Dehydrohalogenation is often accompanied by rearrangement to diazanorcaradienes.

(*iv*) *Reactions with nucleophilic reagents*

Treatment of 1*H*-1,2-diazepines, *e.g.* (**95**), with alkoxide bases results in ring opening to give (**96**) *via* primary abstraction of the proton at C-3; the product may then undergo cyclization to a 2-aminopyridine (**97**) ⟨81ACR348⟩.

The diazepine (**98**) also reacted with base to give a 2-aminopyridine derivative but *via* a quite different process ⟨71JOC2676⟩.

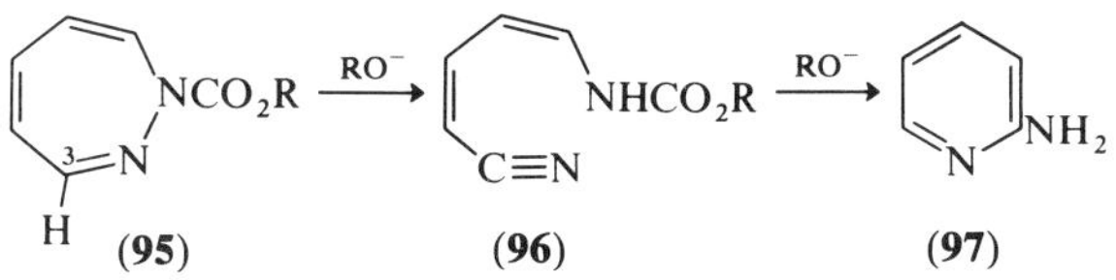

(**95**) (**96**) (**97**)

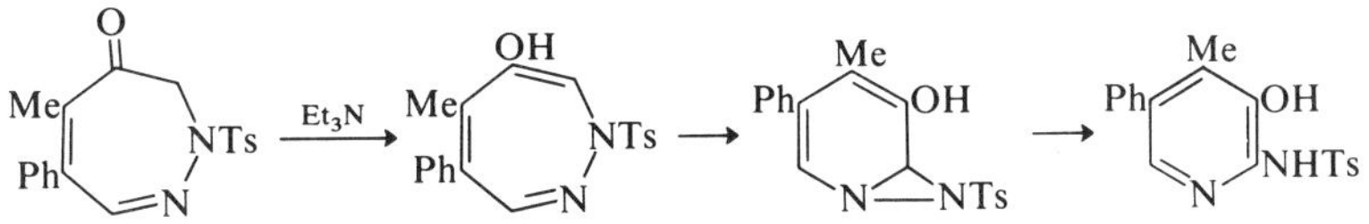

(**98**)

The 4*H*-1,2-diazepine (**99**) reacted with butyllithium at −78 °C *via* addition to the 2,3-imine bond, but LDA gave the anion (**100**) which dimerized in the presence of oxygen. Reaction of (**99**) with Na/K at −20 °C in THF however led to the formation of (**101**) *via* ring contraction of (**100**) ⟨72TL4891⟩. Interestingly, however, when (**100**) was generated with LDA at 0 °C in TMEDA as solvent, no ring contraction was observed and the anion could be reacted with a variety of electrophiles to provide a convenient synthesis of 4-substituted 4*H*-1,2-diazepines, *e.g.* (**102**) ⟨79S130⟩. 3*H*-2,3-Benzodiazepin-4-ones have been substituted at N-3 *via* reaction with sodium followed by an electrophilic reagent.

(**99**) (**100**) (**101**)

MeI, TMEDA, 0 °C

(**102**) 88%

3*H*-1,2-Benzodiazepines (**12**) ⟨75JCS(P1)102, 78CPB1890⟩ and 1*H*-2,3-benzodiazepines (**77**) ⟨73JCS(P1)2543, 80JCS(P1)1718⟩ on treatment with bases in protic solvents undergo proton shifts to give their more stable isomers 1*H*-1,2- (**42**) and 5*H*-2,3-benzodiazepines (**76**) respectively.

2-Toluenesulfonyl-3,4-dihydro-1,2-diazepines (**24**; R = Ts) react with alkoxides under aprotic conditions at *ca.* 100 °C to provide a synthesis of 3*H*-1,2-diazepines (**25**).

The anion (**103**) underwent an interesting 6π-electron electrocyclization to give (**104**) ⟨80TL1223⟩.

i, Bu^{t}OK/benzene; ii, Ac_2O (**103**) (**104**)

(*v*) *Reduction*

Several studies have been made into the selective reduction of C=C and C=N bonds in 1,2-diazepines; in general the former are preferentially reduced by catalytic hydrogenation and the latter by hydride reduction ⟨76H(4)1509, 77JOC1856, 75CJC519⟩. The synthetically useful conversion of 1*H*-1,2-benzodiazepines (**42**) (accessible by rearrangement of quinoline *N*-imides) to the less stable 3*H*-isomers, type (**12**), has been achieved *via* LAH reduction of the C=N bond and subsequent dehydrogenation of the —NHNH— moiety using 4-phenyl-1,2,4-triazoline-3,5-dione ⟨78CPB1890, 1896, 79CPB2188⟩.

(*vi*) *Cycloaddition*

(*a*) *As a 4π component.* The butadiene unit of 1*H*-1,2-diazepines (**40**) is not very reactive to dienophiles but adducts have been obtained with TCNE, 4-phenyl-1,2,4-triazoline-3,5-dione and singlet oxygen. 2,3-Dihydro-1*H*-1,2-diazepines are more reactive. The fully unsaturated systems also undergo an interesting acid-catalyzed dimerization; in this, protonation at N-2 is thought to lower the LUMO energy of the 6,7-double bond making it a reactive enough dienophile to combine with the butadiene unit of a non-protonated molecule ⟨76H(4)1509, 77H(6)2021, 77CPB2749⟩.

(*b*) *As a 2π component.* Diphenylisobenzofuran, cyclopentadienones ⟨75CL423⟩, diazoalkanes, tetrazines ⟨78AP786⟩ and the TOSMIC anion all add selectively to the 4,5-double bond of 1*H*-1,2-diazepines but nitrile oxides and ketenes add first to the imine bond ⟨76H(4)1509, 81ACR348⟩. Cycloaddition of diazepinones and diazepinium betaines has also been studied ⟨76H(4)1509⟩.

As with many other unsaturated seven-membered heterocyclic systems 1,2-diazepines form organometallic complexes of various types. The stable tricarbonyliron complexes of 1*H*-1,2-diazepines, *e.g.* (**79**), have been much studied and found useful in synthesis ⟨81ACR348⟩, and it has been recently shown that 3*H*-1,2-diazepines (**5**) and analogous benzodiazepines form complexes of a quite different type (**105**) involving the azo group only ⟨81TL353⟩.

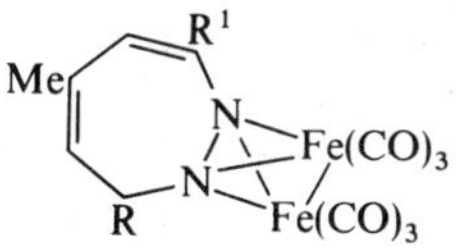

(105)

5.18.2.2 1,3-Diazepines

1,3-Diazepines (**106**) have received rather less attention than their 1,2-isomers, but there are several preparative routes available to the partially and fully saturated systems and recently the first routes to the fully unsaturated monocyclic system (**61**) and to 1*H*-1,3-benzodiazepines (**135**) have been reported.

Reviews: ⟨67AHC(8)21, 68CRV747, 62MI51800⟩.

5.18.2.2.1 Synthesis

(*i*) *Type* a (*C—N—C—C—C—C—N*)

The cyclic urea (**107**) has been prepared by the treatment of tetramethylene diisocyanate with water ⟨67AHC(8)21, p. 38⟩ and an extension of this reaction has recently been used to prepare the 2,5-dione ⟨80JOC485⟩. The ring closure of systems of type (**108**) has been used to prepare 1,3-benzodiazepines. The fully unsaturated system (**109a**) was obtained in moderate yield (25%) by cyclization with molecular sieve but the dihydro analogue (**109b**) was produced in good yield using phosphorus trichloride or oxychloride ⟨73CJC881⟩. 5-Hydroxymethyl-2-methyl-1*H*-1,3-benzodiazepine-4-carboxylic acid γ-lactone has been synthesized by a similar route ⟨68CRV747, p. 750⟩, and dibenzo[*d,f*][1,3]diazepines can be prepared by the treatment of monoacyl derivatives of 2,2′-diaminodiphenyl with phosphorus trichloride ⟨67AHC(8)21, p. 47⟩.

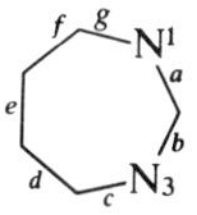

(106) 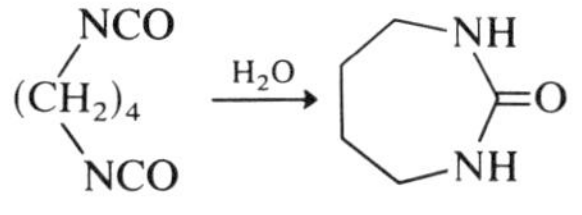 (107)

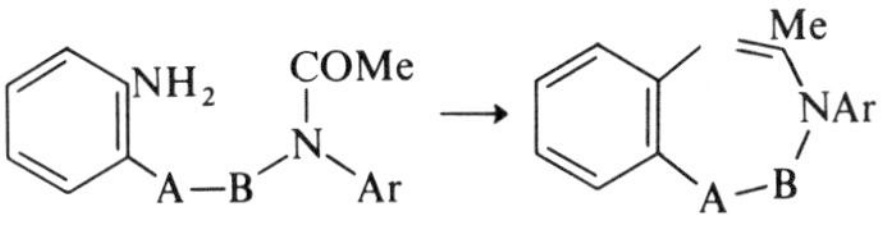

(108) (109)

a; AB = CH=CH (*Z*)
b; AB = CH_2—CH_2

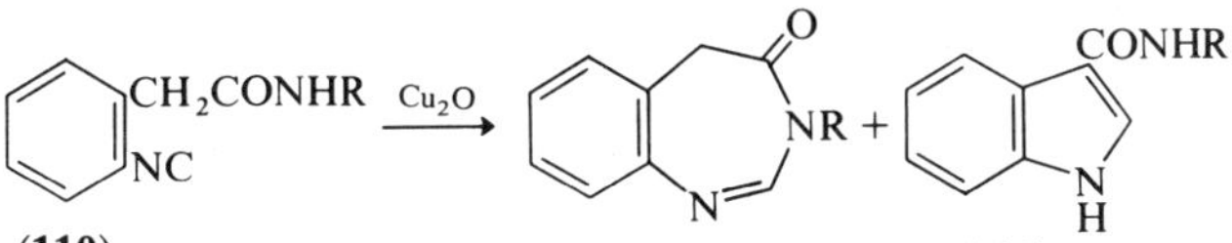

(110) (111) (112)

1,3-Benzodiazepin-4-ones (**111**) have recently been synthesized by the Cu_2O-catalyzed cyclization of (**110**); the competing route to indoles (**112**) is disfavoured by bulky R groups ⟨79TL1039⟩.

(*ii*) *Type* ab (*N—C—C—C—C—N+C*)

These are the most widely used routes to 1,3-diazepines and provide easy access to the monocyclic system, 1,3- and 2,4-benzodiazepines, and dibenzodiazepines by the reactions of (**113**)–(**116**) (and their derivatives) respectively with 1,1-bis-electrophilic reagents. The use of aldehydes produces the rather unstable fully saturated —$NHCH_2NH$— linkage and has been rarely applied to the synthesis of monocyclic systems ⟨67AHC(8)21, p. 40⟩ but provides useful syntheses of 1,3-benzodiazepines, *e.g.* (**117**) ⟨67AHC(8)21, p. 43⟩, and 2,4-benzodiazepines, *e.g.* (**118**) ⟨68CRV747, p.783⟩. A 5,7-dihydroxy-6*H*-dibenzo[*d,f*]

[1,3]diazepine has been synthesized by the reduction of 2,2′-dinitrodiphenyl in the presence of an aldehyde ⟨72ACS1659⟩. The condensation of 1,4-diamines with a variety of carboxylic acid derivatives, *e.g.* imidate esters, orthoformic esters, *N*-ethoxycarbonylthioamides ⟨77JOC2530⟩, nitriles and ethoxyacetylene, produces the cyclic amidine linkage —N=C(R)NH—. This has been used extensively for monocyclic systems ⟨67AHC(8)21 p. 40⟩, 1,3-benzodiazepines, *e.g.* (**119**) ⟨68CRV747, p. 750⟩, and dibenzo[1,3]diazepines ⟨67AHC(8)21, p. 46⟩.

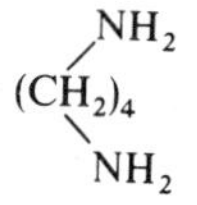

(**113**) 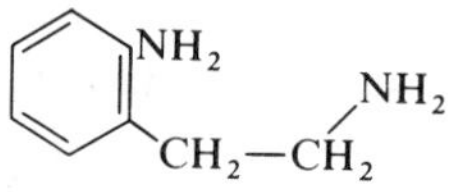 (**114**) 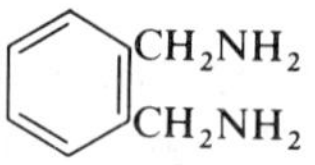 (**115**) 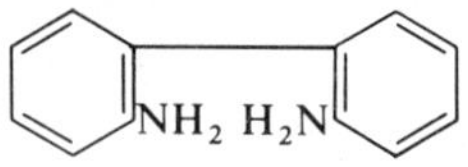 (**116**)

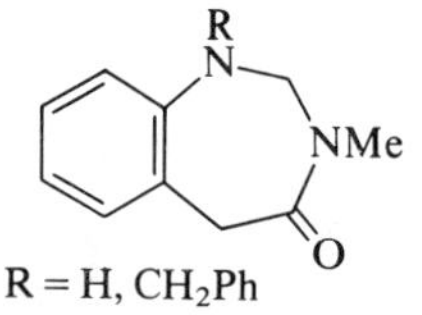

(**117**) R = H, CH_2Ph 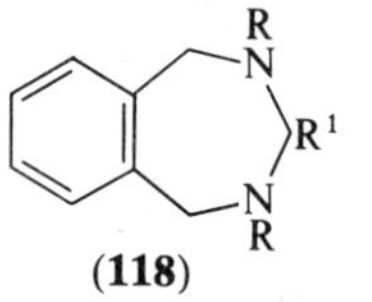 (**118**) 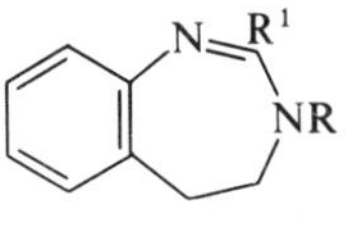 (**119**)

Cyclic ureas, —NHC(O)NH—, have been similarly produced using carbonyl chloride, N,N′-carbonyldiimidazole, carbon monoxide, thiocarbonyl chloride or carbon disulfide ⟨68CRV747, pp. 750, 783, 76JCS(P1)1331, 67AHC(8)21, p. 38⟩.

(*iii*) *Type* ag (*C—N—C—C—C—C+N*)

The reaction of (**120**) with aqueous ammonia or the alternative azide route shown gives the 2,4-benzodiazepin-1-one system (**121**); the longer route gives the better yield ⟨75JHC903⟩.

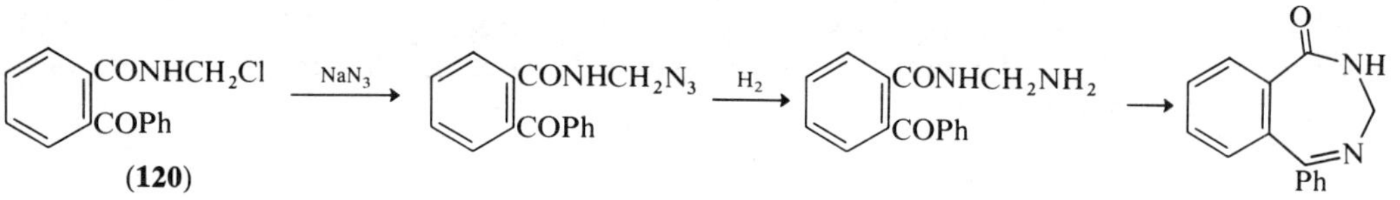

(**120**) (**121**)

(*iv*) *Type* cg (*N—C—N+C—C—C—C*)

The reactions of *N,N*-diarylacetamidines, *e.g.* PhN=C(Me)NHPh, with diacyl halides provide a versatile general synthesis of either monocyclic 1,3-diazepines or those fused to aromatic or heteroaromatic rings ⟨79JOC3843⟩. The reagent (**123**), which is synthetically equivalent to the inaccessible 1,1-diaminomethane, has recently been developed and its reaction with (**122**) gives the 2,4-benzodiazepinone (**124**) ⟨79JHC339⟩. Reactions between ureas and 1,4-dicarbonyl compounds do not generally give diazepines but monocyclic 1,3-diazepine-2,4,7-triones can be prepared by the reaction of succinoyl chloride with *N,N′*-disubstituted ureas ⟨67AHC(8)21, p. 39⟩. There are several examples of the synthesis of pyrido[1,2-*a*][1,3]diazepine derivatives by the reactions of 2-aminopyridines with substituted succinic anhydrides, 4-bromobutanoyl chloride or the analogous 4-bromo imidate ester ⟨67AHC(8)21, p. 45, 72CPB725⟩.

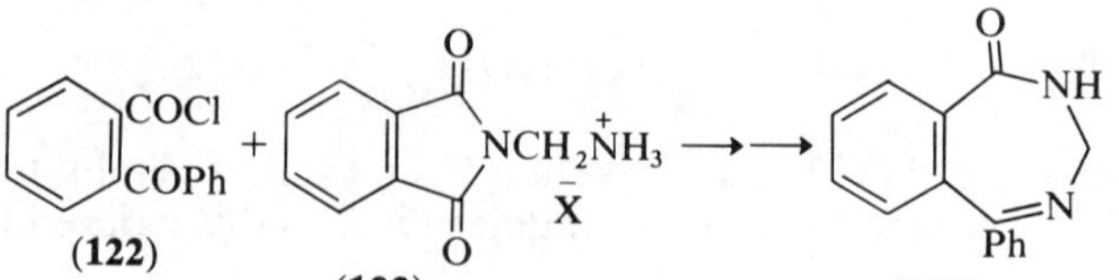

(**122**) (**123**) (**124**)

(*v*) *Synthesis from other heterocyclic systems*

Treatment of 4-chloromethylpyrimidine derivatives, *e.g.* (**125**), with basic nucleophilic reagents causes ring expansion to give 1,3-diazepin-2-ones (**128**; X = OR, CN, $CH(CO_2Et)_2$) in high yield ⟨77CJC895, 75JCS(P1)657⟩. The reaction probably goes *via* (**126**) and (**127**). The halocarbene adducts of 1,3-disubstituted uracil, *e.g.* (**129**; X = Cl, Br), undergo ring expansion on heating in the presence of alcohols to give (**130**). In the analogous fluorochlorocarbene adducts the reaction only takes place when the chlorine atom is in the *exo* configuration ⟨77T2603⟩. The 1,3-dimethyl analogue of (**129**; X = H) undergoes photo-

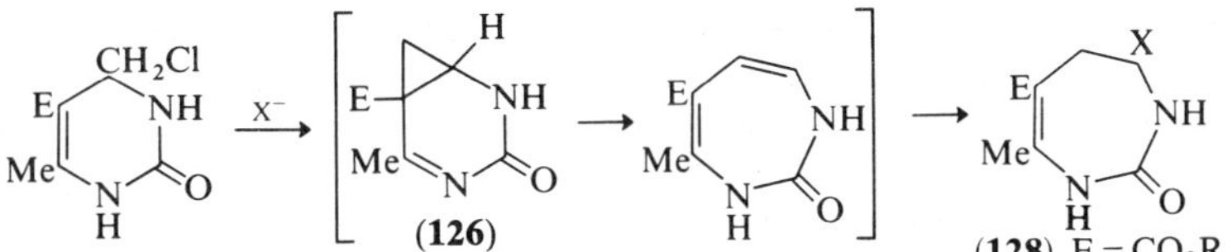

(125) (126) (127) (128) $E = CO_2R$

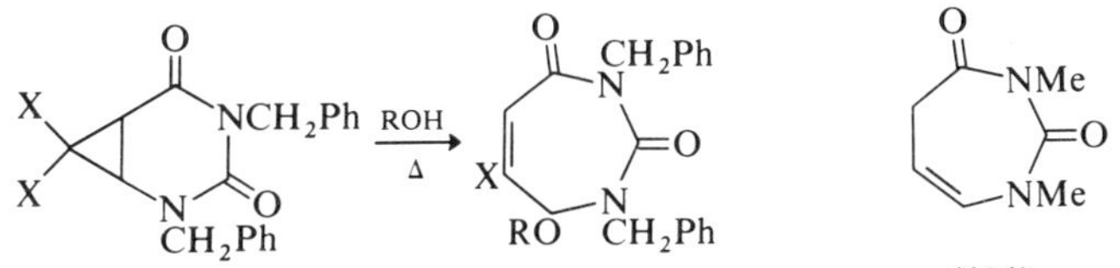

(129) (130) (131)

chemical ring expansion to give (**131**) ⟨71JA3478⟩. The diazepinone (**107**) has been prepared by the Beckmann rearrangement of piperidone oxime ⟨67AHC(8)21, p. 38⟩.

The first fully unsaturated monocyclic 1,3-diazepines (**61**) have been prepared only recently. In the thermolysis of 1*H*-1,2-diazepines it has been found that cleavage of the N—N bond in (**58**) is favoured by electron release from groups in the 4- and 6-positions so that 1,3-diazepines (**61**) are then formed rather than pyridinium *N*-imides (**59**) ⟨79JOC2683, 80CC444⟩. It has also been shown ⟨81H(16)137⟩ that 1,3-diazepines are produced by the thermolysis of 1,2-diazepine photoproducts of type (**54**).

The fully unsaturated 1,3-benzodiazepine (**135**) is formed in a formally similar process by a photoreaction of the 1-substituted isoquinoline *N*-imide (**132**) ⟨80CPB2602⟩. The same principle has been applied to the preparation of thieno-, furo- and pyrrolo-fused 1,3-diazepines ⟨80CC454, 81CPB1539⟩.

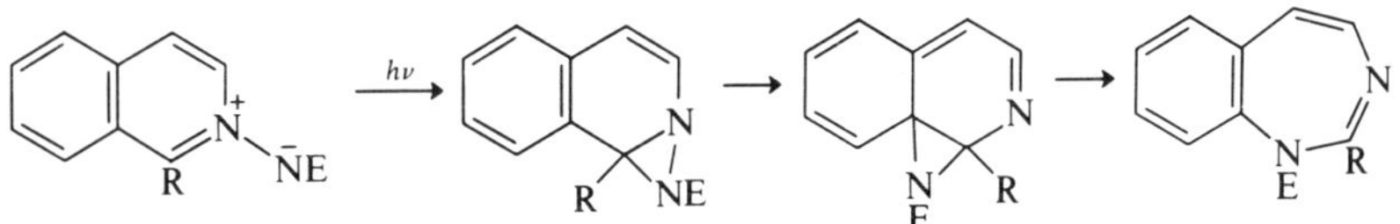

(132) $E = CO_2R$ (133) (134) (135)

5.18.2.2.2 Reactivity

The cyclic urea (**107**) is hydrolyzed by acid to give 1,4-diaminobutane. The $>NCH_2N<$ group shows an acetal-like sensitivity to acid hydrolysis, *e.g.* (**117**) is rapidly decomposed to give an oxindole polymer. The diazepine (**128**; X = OMe) when treated with acid at room temperature was converted rapidly to the 1-carbamoylpyrrole (**136**) probably *via* protonation at N-1 ⟨77CJC895⟩. Compound (**144**; R = Ph) was also unstable to acid and underwent ring contraction to give 3-phenylindolin-2-one ⟨76JCS(P1)1331⟩. The 4-oxo group of (**117**) was reduced with LAH to give the unstable 2,3,4,5-tetrahydro-1*H*-1,3-benzodiazepine system ⟨68CRV747, p. 750⟩.

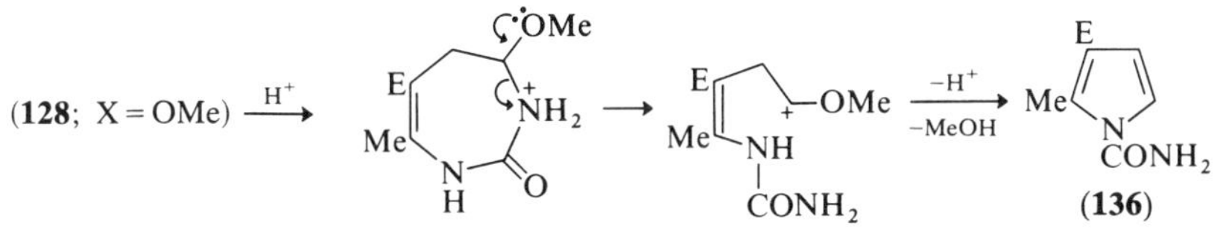

(**128**; X = OMe) $\xrightarrow{H^+}$ ... $\xrightarrow[-MeOH]{-H^+}$ (136)

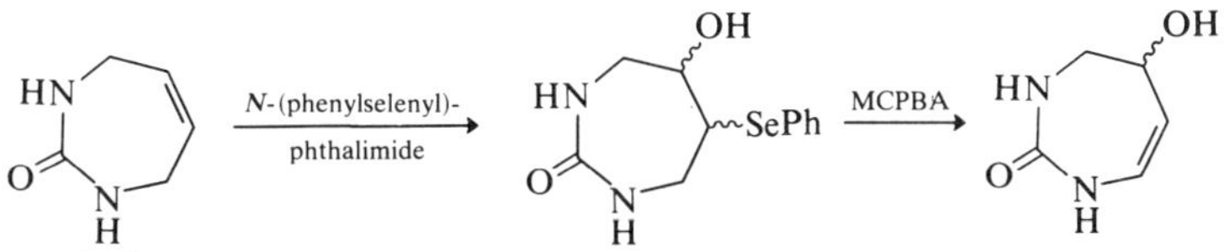

(137) $\xrightarrow{N\text{-(phenylselenyl)-phthalimide}}$ (138) $\xrightarrow{MCPBA}$ (139)

The hydroxy-1,3-diazepinone (**139**) was prepared in high yield by the route shown ⟨80JOC5225, 80JMC713⟩ and its saturated analogue was prepared both by the sodium borohydride reduction of 1,3-diazepine-2,5-dione and by the sequential reactions of (**137**) with diborane and hydrogen peroxide ⟨80JOC485⟩. Synthetic studies have recently shown that the natural product squamalone, thought to be (**140**), is in fact (**141**) ⟨80JOC5308⟩.

Not much is yet known about the reactivity of the recently synthesized fully unsaturated 1,3-diazepines but it has been shown that (**61**) undergoes a thermally reversible photo-isomerization to give (**142**) ⟨81H(16)137⟩.

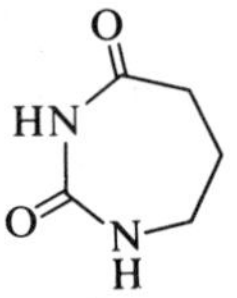

(**140**) 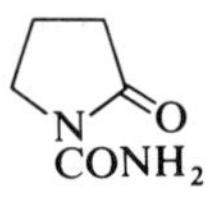 (**141**) 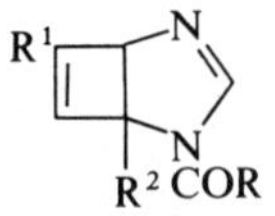 (**142**)

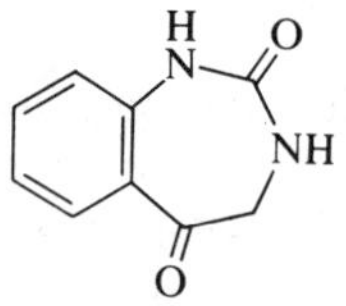

(**143**) 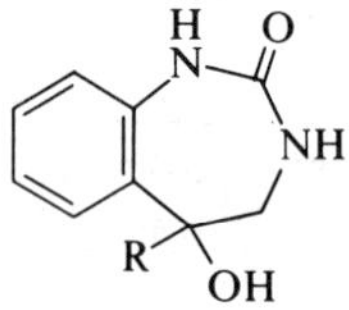 (**144**) 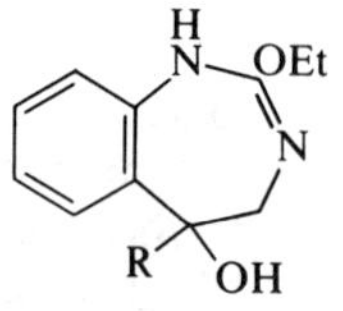 (**145**) 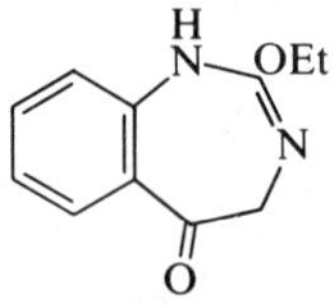 (**146**)

The chemistry of (**143**) has been extensively studied ⟨76JCS(P1)1331⟩. Electrophilic substitution gave the 7-nitro and 7-halogeno derivatives. The 5-oxo group of both (**143**) and (**146**) showed typical ketonic reactivity and borohydride reduction and reactions with Grignard reagents gave the expected products, *e.g.* (**144**) and (**145**). Attempted dehydration of these alcohols was not fruitful but the latter (R = Ph) on treatment with acid did give a small yield of the product with a 4,5-double bond. The diazepine (**143**) could not be alkylated at N using typical conditions but reaction with triethyloxonium fluoroborate gave (**146**) in good yield.

5.18.2.3 1,4-Diazepines

1,4-Benzodiazepines (**148**) have received intensive study since the early 1960s ⟨79JMC1, 78MI51801⟩ because of their value in psychotherapy, and an impressive armoury of synthetic routes have been described. Recently attention has concentrated on the synthesis of analogues having heterocycles in place of the benzene ring and on compounds having additional fused heterocyclic rings (p. 613). In contrast monocyclic 1,4-diazepines, with the exception of the interesting 2,3-dihydro-1,4-diazepine system, have received scant attention in recent years and the fully unsaturated system has proved difficult to prepare. NMR has been used to study the conformation and energetics of ring inversion of 2,3-dihydro-1,4-diazepines ⟨78H(11)549⟩ and 1,4-benzodiazepines ⟨70RCR1064, 79OMR(12)593, 79JHC757⟩ and the relationship between conformation and biological activity has been discussed.

Reviews: general ⟨67AHC(8)21⟩, benzodiazepines ⟨68CRV747⟩, 2,3-dihydro-1,4-diazepines ⟨74AHC(17)1, 78H(11)549⟩, 1,5-benzodiazepines ⟨74AHC(17)27⟩, 1,4-benzodiazepines ⟨64JPS577, 70RCR1064, 71AG(E)34, 72JHC747, 76MI51800⟩, 1,4-benzodiazepines with fused heterocyclic rings ⟨75MI51801, 77MI51800, 80MI51800⟩.

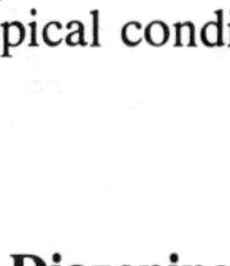

(**147**) 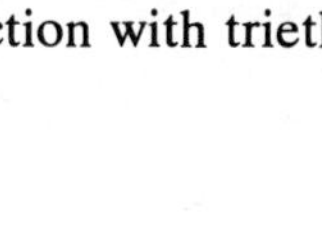 1,4-Benzodiazepine (**148**) 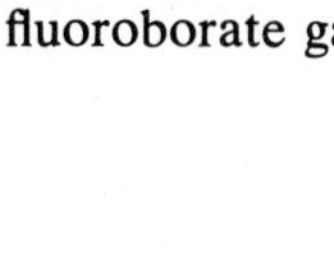 1,5-Benzodiazepine (**149**)

5.18.2.3.1 Synthesis

(i) Type a *(N—C—C—C—N—C—C)*

Formation of this bond as an amide linkage has been much used but most often as the first or last step in a type *ac* synthesis of 1,4-benzodiazepin-2-ones, *e.g.* (**151**) or (**154**). Monocyclic 1,4-diazepines have been prepared by the cyclization of α,ω-aminoesters

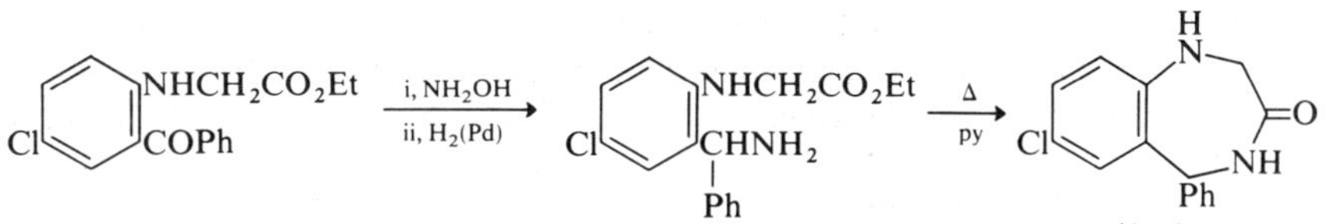

⟨78MI51800, 79AP514⟩ and -cyanoesters ⟨75MI51800⟩. The 1,4-benzodiazepin-3-one (**150**) has been prepared by the sequence shown ⟨68CRV747, p. 763, 77JHC1139⟩.

(ii) Type ac *(N—C—C—C—N+C—C)*

Reactions between 1,5-bis nitrogen nucleophiles and 1,2-bis electrophile equivalents such as 1,2-dihalides, α-halogenoesters and acetylenedicarboxylic acid diesters have been used in the synthesis of the monocyclic system ⟨67AHC(8)21, p. 53⟩, 1,4-benzodiazepines ⟨67AHC(8)21, p. 62, 70RCR1064, p. 1070⟩ and pyridodiazepines. Chloroacetyl chloride has been much used as the two-carbon fragment in the synthesis of 2-oxo-1,4-benzodiazepine systems, *e.g.* (**151**) and (**152**) ⟨68CRV747, pp. 755, 765⟩, and the reaction of ethyl bromoacetate with (**153**) similarly leads to (**154**) ⟨68CRV747, p. 758⟩.

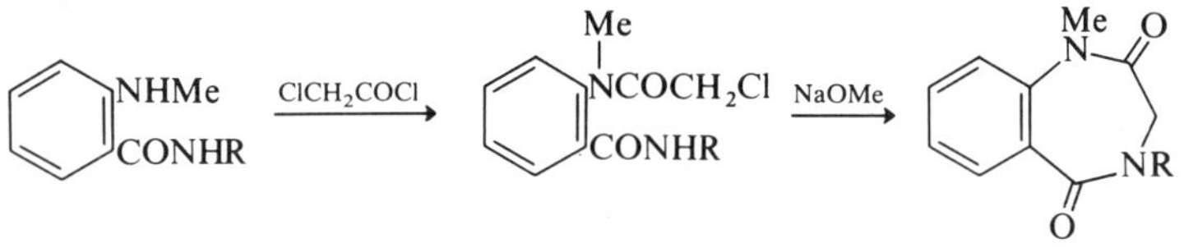

(**151**)

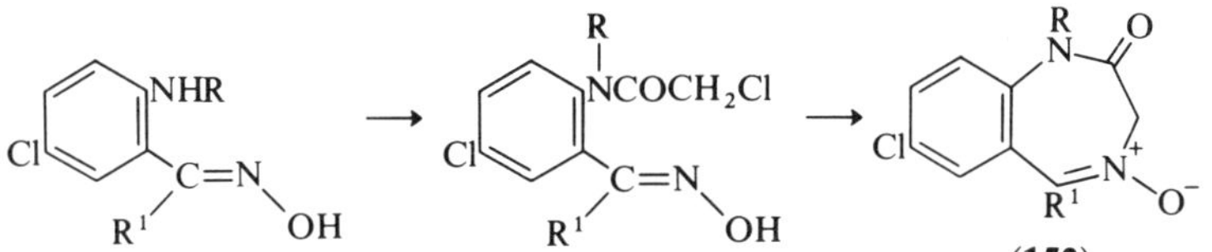

(**152**)

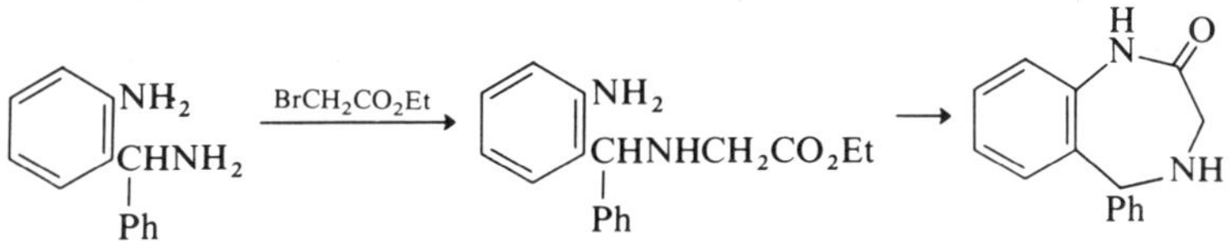

(**153**) (**154**)

(iii) Type ad *(N—C—C—C+C—C—N)*

Reactions of this type provide one of the major routes to the commercially important 1,4-benzodiazepin-2-one system (**156**) *via* the reaction of substituted *o*-aminobenzophenones (**155**) with amino acid esters ⟨68CRV747, p. 756⟩. This synthesis allows the easy preparation of 3-substituted products by the use of readily available α-substituted amino acids, *e.g.* the CNS active 'Tranxene' (**156**; Y = Cl, Ar = Ph, $R^1 = CO_2H$) by using aminomalonic ester. Similar reactions have been carried out using other three-atom components, *e.g.* 2-bromoethylamine ⟨77MIP51800⟩ and ketenimines ⟨77JHC99⟩.

(iv) Type ag *(C—C—N—C—C—C+N)*

One of the commercially important routes to 1,4-benzodiazepin-2-ones (**156**) from 2-aminobenzophenones involves the reaction of (**157**) with a source of nitrogen. Ammonia, ammonium carbonate, hexamethylenetetramine and dinitrosopentamethylenetetramine have all been used and the diazepinones have been obtained in yields of up to 95%. Intermediates, *e.g.* (**158**) or (**159d**) can be isolated under some reaction conditions ⟨68CRV747, p. 756, 80JCR(S)398, 399, 400, 81JHC59, 79MI51800⟩. The same principle has also been applied to the preparation of 1,2-dihydro-3*H*-1,4-benzodiazepines ⟨71JHC845⟩ and 1,4-benzodiazepine-2,5-diones ⟨70RCR1064, p. 1070⟩. It is appropriate to mention here other intermediates related to (**157**) which can serve as precursors to (**158**). Reaction of (**155**) with azidoacetyl chloride, or (**157**) with sodium azide, gives (**159a**) which can be converted to (**158**) by reduction. Reaction of (**155**) with protected glycyl chlorides gives, for example,

(**159b**) and (**159c**) whose protecting groups can be removed by standard techniques, *i.e.* hydrobromic acid hydrolysis for (**159b**) and hydrazinolysis for (**159c**) ⟨79JMC1, 78MI51801, 68CRV747, p. 756, 79JHC445⟩. The hydroxylamine (**157**; X = NHOH) similarly cyclizes to give the 4-oxide of (**156**) ⟨70RCR1104, p. 1066⟩.

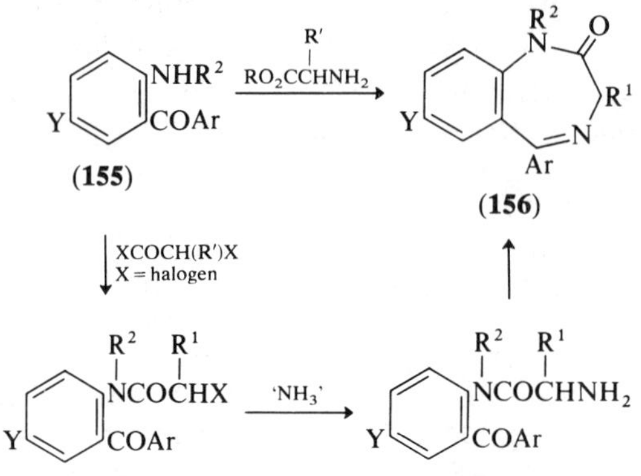

(**155**) (**156**) (**157**) (**158**) 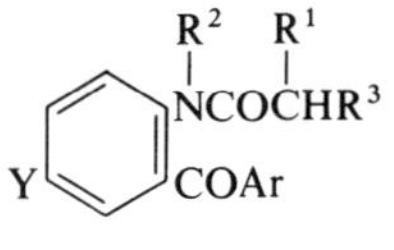 (**159**)

a; $R^3 = N_3$
b; $R^3 = NHCO_2CH_2Ph$
c; R^3 = phthalimido
d; $R^3 = N_4^+(CH_2)_6hal^-$

(*v*) *Type* b (*C—N—C—C—C—N—C*)

Bis-anils of 1,2-diaminocyclopropanes (**160**) undergo a thermal Cope rearrangement followed by hydrogen migration to give the 2,3-diaryldihydrodiazepines (**161**) ⟨78H(11)552⟩.

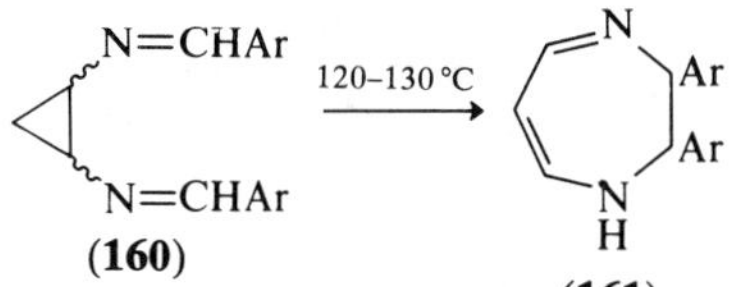

(**160**) (**161**)

(*vi*) *Type* d (*N—C—C—N—C—C—C*)

Formation of the *d* bond by the reductive cyclization of α,ω-nitroketones ⟨68CRV747, p. 778⟩, -cyanoketones and -cyanoesters ⟨74MIP51800⟩ has been used in the preparation of both 1,4- and 1,5-benzodiazepines.

Monocyclic diazepines have also been prepared by the Raney nickel-catalyzed ring closure of compounds of type $H_2NCH_2N(R)(CH_2)_3NH_2$ or dehydrative cyclization of $HOCH_2N(R)(CH_2)_3NH_2$ ⟨67AHC(8)21, p. 62⟩.

(*vii*) *Type* dg (*N—C—C—N+C—C—C*)

Reactions of this type have been much used, particularly in the preparation of monocyclic diazepines and 1,5-benzodiazepines by the reactions of ethylenediamine and *o*-phenylenediamine respectively with a variety of 1,3-bis electrophilic reagents.

(*a*) *1,3,-Dihalides.* 1,2-Diamines and 1,2-bis-sulfonamides or their anions react with 1,3-dihalides to provide a good general route to fully saturated 1,4-diazepines, *e.g.*(**162**) ⟨67AHC(8)21, p. 52⟩, analogous benzo fused systems ⟨67AHC(8)21, p. 66⟩ and 1,4-diazepine-5,7-diones ⟨75MI51802⟩.

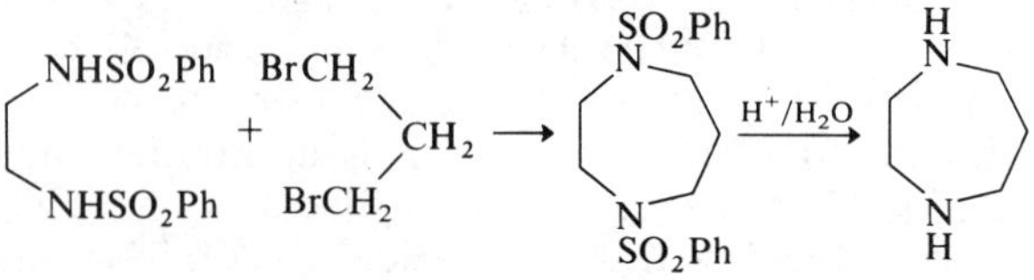

(**162**)

(*b*) *1,3-Dicarbonyl compounds.* The reaction of ethylenediamine (**163**) and its *N*-alkyl, *N,N'*-dialkyl and *N,N'*-diaryl analogues with 1,3-dialdehydes or -diketones gives 2,3-dihydro-1,4-diazepines, *e.g.* (**164**) ⟨74AHC(17)1, p. 2, 78H(11)550⟩. The reaction is best carried out by heating the reactants in acetic acid followed by the addition of either perchloric acid to precipitate the diazepinium perchlorate or of potassium hydroxide to precipitate the dihydrodiazepine itself. An unsaturated analogue, diaminomaleonitrile, gives a similar reaction with pentane-2,4-dione ⟨74JOC2341⟩; this is apparently the only route to a fully unsaturated monocyclic 1,4-diazepine. *o*-Phenylenediamine (**165**) similarly gives 1,5-benzodiazepines (**166**) ⟨74AHC(17)27, p. 28, 77JCS(P1)1901⟩; in cases where this ring closure is unfavourable for steric reasons, benzimidazoles (**213**) are also formed by an alternative cyclization.

NH_2 NH_2 + $MeCOCH_2COMe$ → (Me, Me, H N N)

(**163**) (**164**)

NH_2 NH_2 + $RCOCHR^1COR$ ⇌ (N=CCH(R)(R¹) COR NH_2) ⇌ (N R R¹ N R)

(**165**) (**166**)

The use of derivatives of dicarbonyl compounds such as diazapentadienium or vinamidinium salts, *e.g.* (**167**) and (**168**), is a recent development which offers advantages in providing access to dihydrodiazepines (**164**) not easily prepared in other ways ⟨78H(11)550, 78JCS(P1)1453⟩.

Monocyclic 1,4-diazepin-5-ones and fused 1,5-benzodiazepin-2-ones, *e.g.* (**169**) and (**170**), can be readily prepared by the reactions of 1,2-diamines with *β*-ketoesters ⟨67AHC(8)21, p. 57, 68CRV747, p. 776, 71JHC797⟩. Similarly reactions with malonic acids and esters give 1,4-diazepine-5,7-diones (**171**) ⟨67AHC(8)21, pp. 55, 69, 68CRV747, p. 781⟩.

X^- PhRN NRPh (**167**) X^- Me_2N NMe_2 (**168**)

(**169**) (**170**) (**171**)

(*c*) *α,β-Unsaturated and β-halogeno carbonyl compounds.* 1,2-Diamines also react with *α,β*-unsaturated carboxylic acids; for example, *o*-phenylenediamine reacts with (**172**) to give the 1,5-benzodiazepin-2-one system (**174**) *via* the intermediate (**173**) ⟨68CRV747, p. 778⟩. The same type of product is obtained from *β*-bromocarboxylic acids.

α,β-Unsaturated and *β*-halogeno-aldehydes and -ketones, *e.g.* (**175**) and (**177**), on reaction with *o*-phenylenediamine similarly give the dihydrobenzodiazepine system (**176**) ⟨67AHC(8)21, p. 57, 68CRV747, p. 773, 75JHC825, 72CC962⟩. Alkynic analogues of (**175**) give fully unsaturated 3*H*-1,5-benzodiazepines ⟨76OPP306⟩.

1,4-Benzodiazepines, *e.g.* (**178**; X = NO_2), may be prepared *via* the displacement of the activated halogen in reactions of 2-halogeno-5-nitroacetophenones with 1,2-diamines ⟨68CRV747, p. 767⟩.

(**165**) + HO_2C–CH=CHR → (NHCOCH=CHR, NH_2) → (H N O, N R H)

(**172**) (**173**) (**174**)

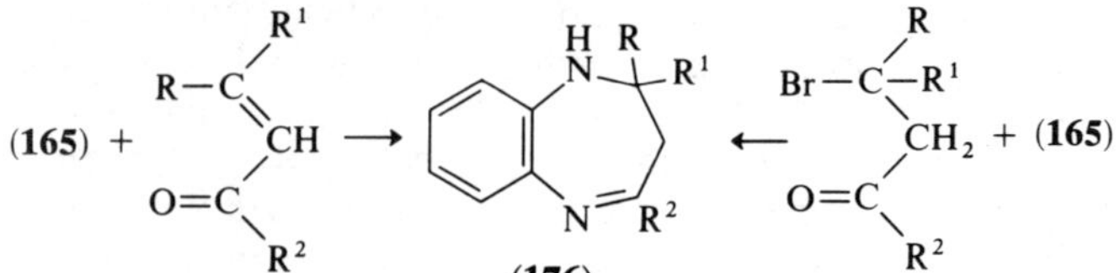

(165) + (175) → (176) ← (177) + (165)

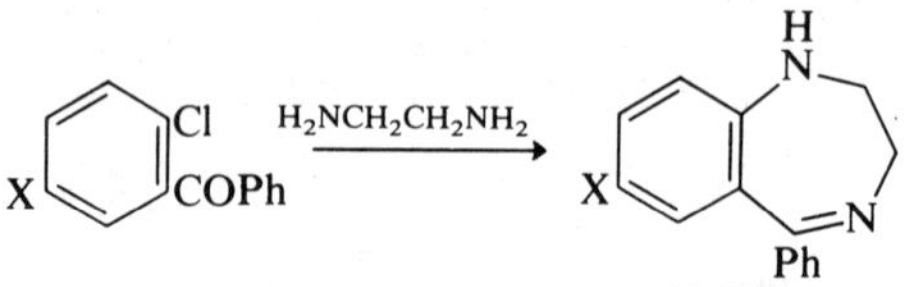

(178)

The 1,4-diazepine system has also been synthesized by the reactions of 1,2-diamines with various other reagents, *e.g.* ketene dimers ⟨68CRV747, p. 778⟩ 1,3-diynes ⟨74AHC(17)1, p. 5⟩, bis(cyanomethylene)-1,3-dithietanes ⟨76T483⟩.

(*viii*) *Type* e (*C—C—N—C—C—N—C*)

Formation of this bond has been achieved in Bischler–Napieralsky cyclizations leading for example from (**179**) to (**180**) ⟨68CRV747, p. 768⟩ and from (**181**) to the dibenzo[*b*,*e*][1,4]diazepine system ⟨67AHC(8)21, p. 74⟩.

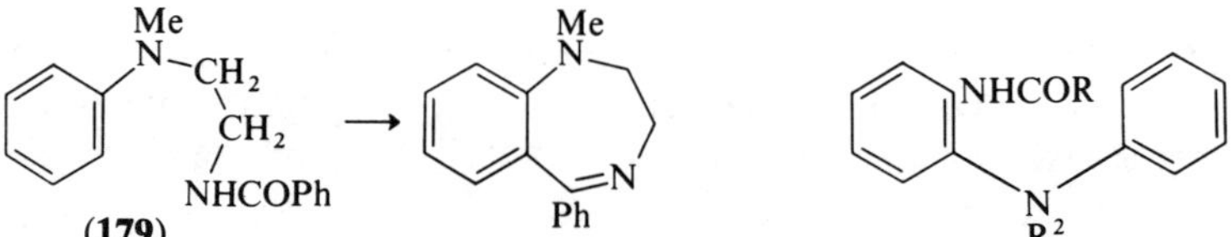

(179) → (180)   (181)

The enediimine (**182**) underwent a high yielding (~90%) thermal transformation to (**184**) apparently *via* an intramolecular enamine addition in the intermediate (**183**) ⟨72CC1116⟩.

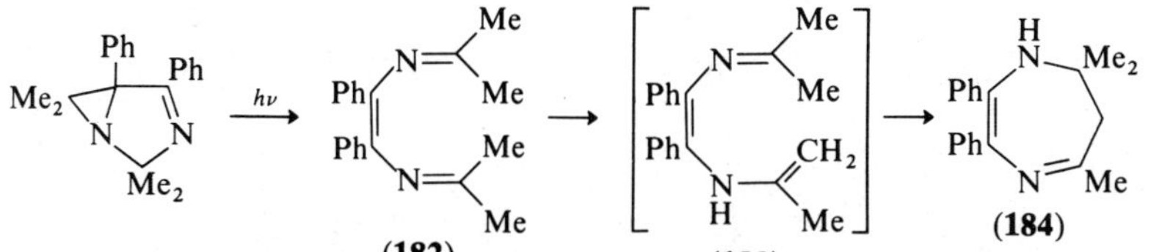

(182) → [(183)] → (184)

(*ix*) *Synthesis from other heterocyclic systems*

(*a*) *From quinazolines.* Quinazoline 3-oxides, *e.g.* (**185**), react with ammonia and primary amines to give 2-amino-1,4-benzodiazepine 4-oxides (**186**). This reaction is valuable as a preparative route as the products are capable of further chemical elaboration, and of historical interest as it stands at the foundation of the benzodiazepine tranquillizer 'industry' ⟨79JMC1, 78MI51801, 68CRV747, p. 751⟩. In general, secondary amines give predominantly 2-methylaminoquinazolines and little of the ring expanded product.

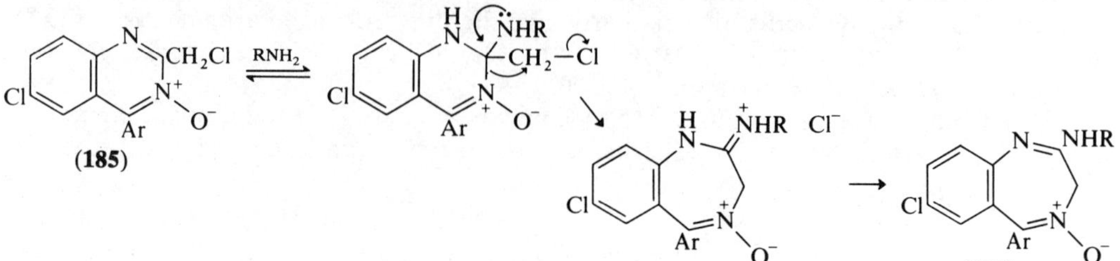

(185) → (186)

Treatment of (**185**) with other nucleophiles, *e.g.* sodium methoxide or the anion of nitromethane ⟨68CRV747, p. 752, 76JHC433⟩, also induces ring expansion and treatment with hydroxide gives the 2-oxo derivative (**188**) ⟨68CRV747, p. 755, 67JOC2417⟩ *via* the oxime (**187**)

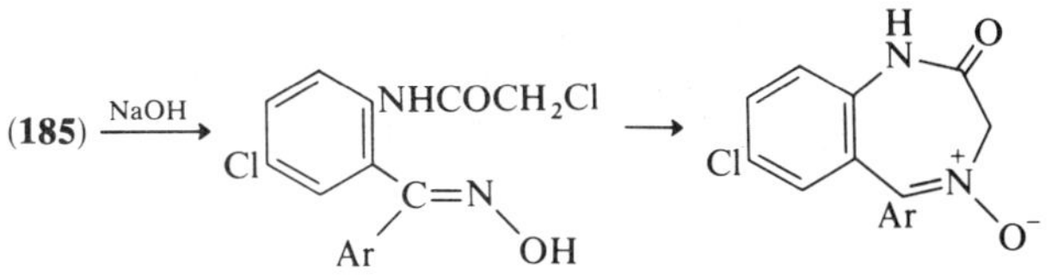

(185) $\xrightarrow{NaOH}$ (187) → (188)

(see also p. 609). An alternative entry to the rearrangement process is *via* the treatment of 1,2-dihydroquinazoline 3-oxides with base ⟨68CRV747, p. 768⟩.

Quinazolinium salts, *e.g.* (**189**) or its 4-oxo analogue, on treatment with diazoalkanes also undergo ring expansion to give 1,4-benzodiazepines ⟨74BCJ343⟩.

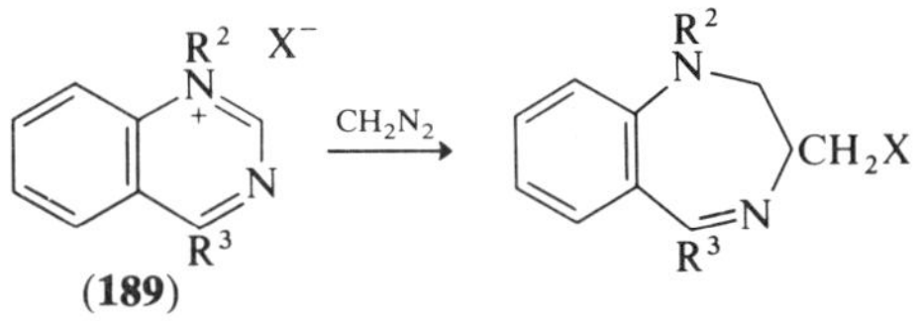

(189) $\xrightarrow{CH_2N_2}$

(*b*) *From indoles.* 1,4-Benzodiazepin-2-ones may be prepared in high yield by the oxidation of 2-aminomethylindoles (**190**). This versatile synthesis has been used for the commercial production of several CNS active compounds ⟨77MI51800⟩. The reaction involves rapid formation of an intermediate of unknown structure which is converted to the diazepine in a first-order process. Other, less well used routes involve the ring expansion of indoles carrying a C—C—N substituent on nitrogen ⟨76CI(L)1067, 77FES33⟩.

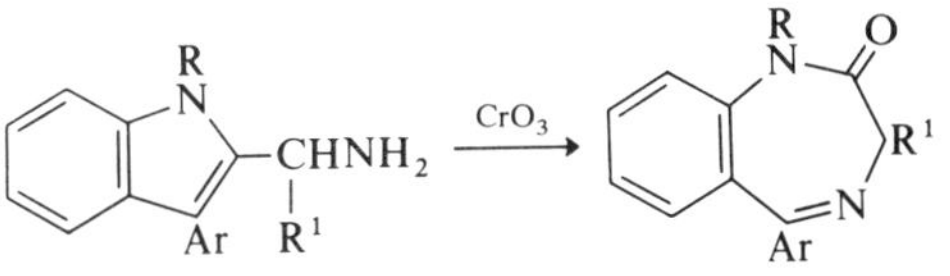

(190) $\xrightarrow{CrO_3}$

(*c*) via *Schmidt and Beckmann rearrangements.* 4-Piperidones and their oximes undergo Schmidt and Beckmann rearrangements respectively to give the fully saturated 1,4-diazepin-5-ones ⟨67AHC(8)21, p. 54⟩. In the analogous benzo fused system (**191**) the product ratio depends on the nature of R ⟨71JHC231⟩, but 2,3-dihydro-1*H*-4-isoquinolone oxime gives only a 1,4-benzodiazepin-2-one ⟨70JOC2455⟩.

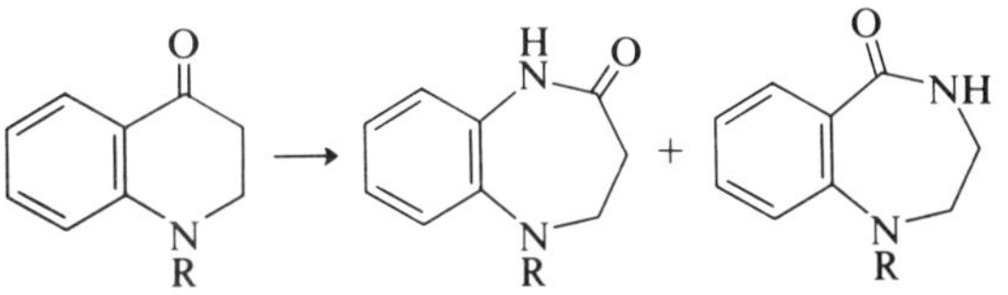

(191)

(*d*) *From isatoic anhydrides.* These, *e.g.* (**197**), are useful substrates for the preparation of 1,4-benzodiazepines; they react with amines having a carbonyl or carboxyl group in the 2-position to give the diazepines, *e.g.* (**198**), directly.

5.18.2.3.2 1,4-Diazepines with Fused Heterocyclic Rings

The clinical importance and commercial success of 1,4-benzodiazepines has led to extensive synthetic studies on related compounds in the hope of finding agents that would be more specific for the various kinds of CNS disturbances. Many analogues have been synthesized in which the benzene ring has been replaced by a heterocyclic moiety and for these the synthetic methods used often closely resemble those used for benzodiazepines themselves ⟨79JMC1, 78MI51801⟩. Another approach, and one which has generated an enormous volume of patent literature over the last 15 years, has been to synthesize 1,4-benzodiazepines having an additional fused heterocyclic ring. Various aspects of this work have been reviewed ⟨76MI51800, 77MI51801, 75MI51801, 80MI51800, 77MI51802⟩.

Synthetic work in this area falls into three types. (i) There are a few methods in which the diazepine ring and the extra heterocyclic ring are constructed in one operation, for example (**192**) can be converted directly into oxazolo[3,2-*d*]- (**193**; X = O) and imidazo[1,2-*d*]- (**193**; X = NH) [1,4]benzodiazepines. Oxazolam (**193**; X = O, Y = H, R = Me) and cloxazolam (**193**; X = O, Y = Cl, R = H) are commercially important examples.

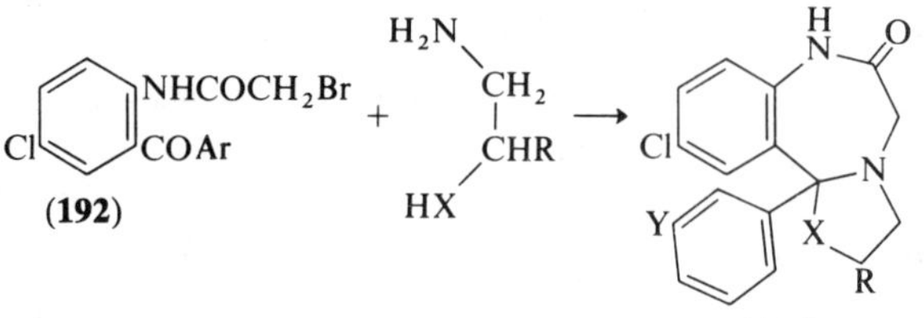

(**192**) (**193**)

(ii) There are a large number of methods in which the diazepine ring is synthesized last by one of the standard routes but from a precursor containing the preformed extra heterocyclic ring. The example shows a route to the *s*-triazolo[4,3-*a*] system (**195**). This method parallels various type *ag* cyclizations and thus a variety of routes from (**194**) to (**195**) are possible. Similarly the *v*-triazolo[1,5-*a*] system (**196**) can be prepared as shown. The pyrrolo[2,1-*c*] skeleton, *e.g.* in (**198**), present in the antitumour antibiotics anthramycin and sibiromycin can be prepared from isatoic anhydride as shown.

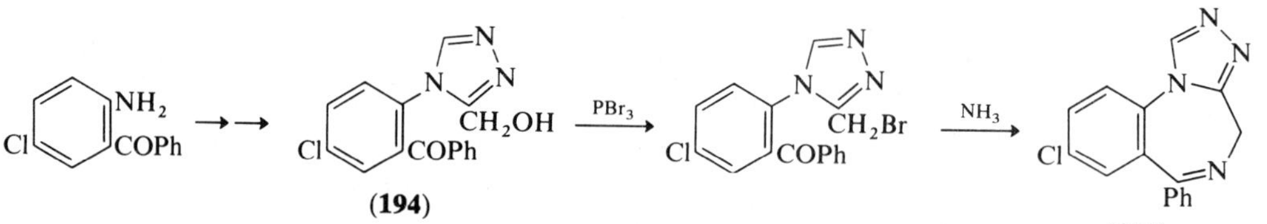

(**194**) (**195**)

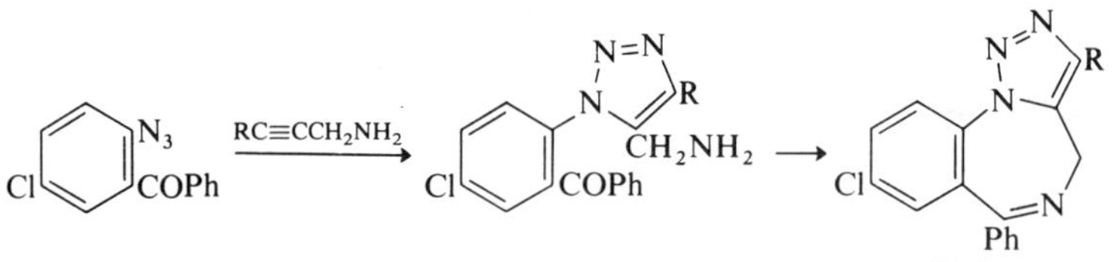

(**196**)

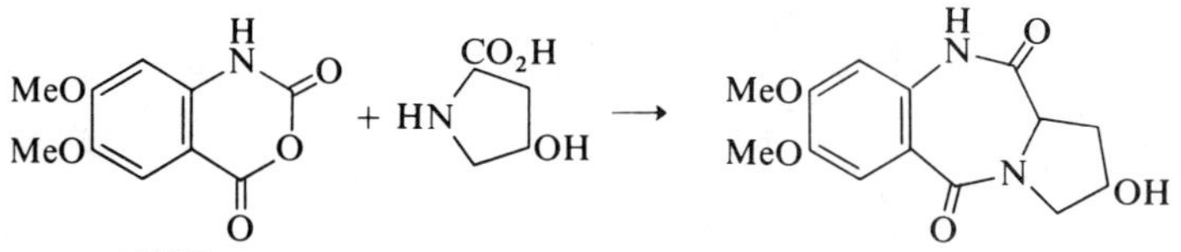

(**197**) (**198**)

(iii) The largest number of synthetic methods involve the construction of the new ring on the appropriate side of a suitably functionalized 1,4-benzodiazepine substrate. This is relatively straightforward on the *a* side because the 1,4-benzodiazepine system has a nucleophilic N at position 1 and can be readily functionalized to create an electrophilic centre at position 2 (see p. 617). Thus a range of ring systems can be built by modifications of standard synthetic routes. Examples are given below for the synthesis of the imidazolo[1,2-*a*] (**199**), *s*-triazolo[4,3,-*a*] (**200**) and tetrazolo[1,5-*a*] (**201**) systems.

Much less work has been done on the fusion of rings on to the *b* side but pyrimidino ⟨75BCJ302⟩ and azetino and imidazo systems ⟨74JOC568⟩ have been made *via* interesting reactions.

c-Fused systems have generally been prepared by method (ii) but a few cycloadditions on to a rather unstable $\Delta^{3,4}$-benzodiazepine have been carried out ⟨76MI51800⟩.

The fusion of rings on to side *d* has usually been achieved *via* adaptations of known reactions of the imine, nitrone or imino chloride functions. Some examples are given for the 1,4-benzodiazepin-2-one (**202**).

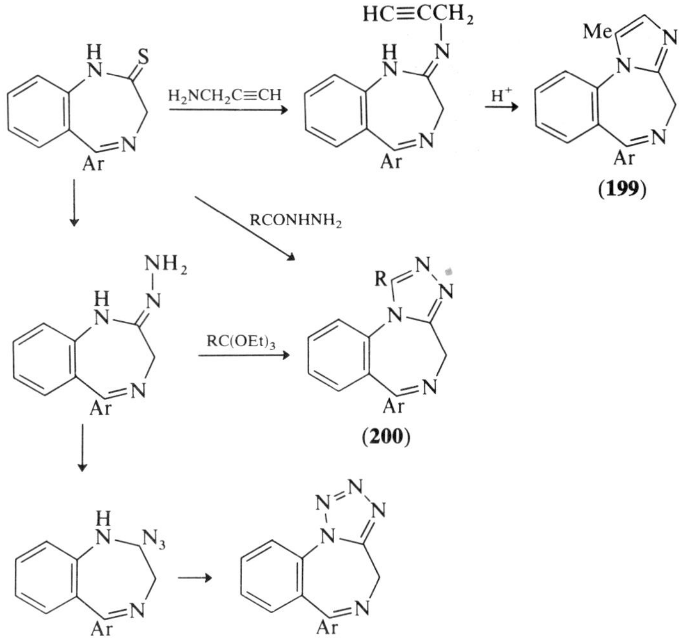

(199)

(200)

(201)

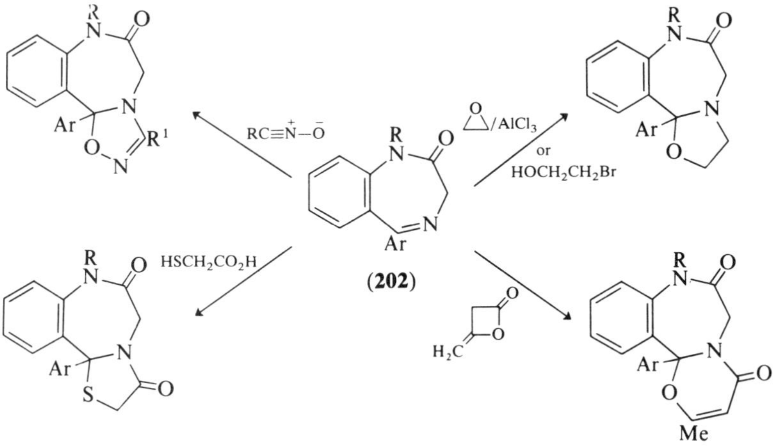

(202)

5.18.2.3.3 Reactivity

(i) Reactions with electrophilic reagents

(a) 2,3-Dihydro-1,4-diazepines (**203**). These are strongly basic ($pK_a \sim 13$–14) and are readily protonated to give the cation (**204**) whose estimated resonance stabilization energy is high at ~80 kJ mol^{-1}. Because of this high degree of resonance stabilization the cations (**204**) have quasi-aromatic properties and their most characteristic reaction is electrophilic substitution at position 6, *e.g.* to give (**206**). They can thus be deuterated, halogenated, nitrated and coupled with diazonium salts under conditions similar to those used for benzene ⟨74AHC(17)1, p. 17, 78H(11)549, p. 555⟩. Dications, *e.g.* (**208**), can be formed in strong acids and under these conditions 6-bromodihydrodiazepinium bromides (**207**) eliminate bromine *via* the protodebromination process shown. An interesting example of a degenerate Cope rearrangement involving one of these dications has recently been reported ⟨78H(11)549, p. 558⟩.

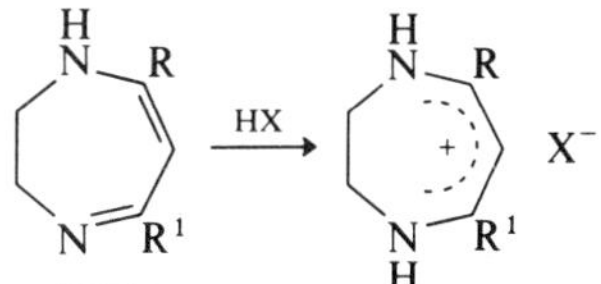

(203) (204)

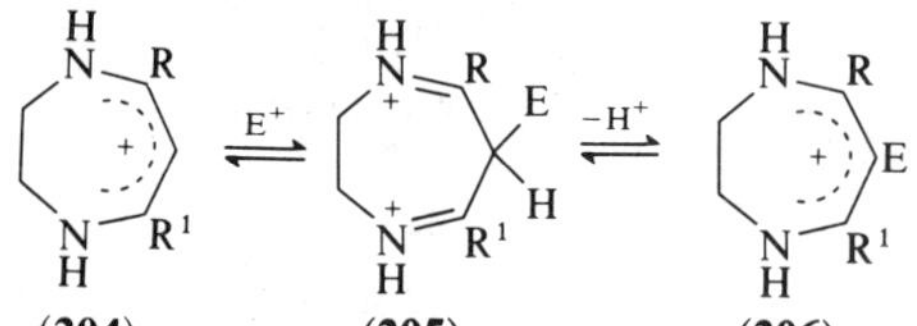

(204) (205) (206)

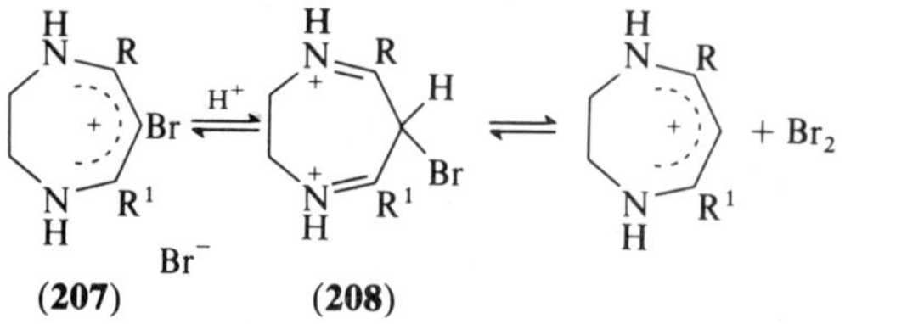

(207) (208)

(*b*) *1,5-Benzodiazepines (**209**).* These are less basic (pK_a ~ 4.5) but protonation produces a cation (**210**) whose purple color suggests extensive delocalization. In stronger acids the colorless dication (**211**) is formed. These compounds and the diazepinone (**170**) rearrange readily under acid conditions to give benzimidazoles. Aqueous solutions undergo the same reaction even in the absence of acid, *e.g.* (**212**) gives 2-methylbenzimidazole (**213**) and acetone *via* the scheme shown. A consequence of the equilibrium existing in the first stage is that aqueous solutions of 1,5-benzodiazepines frequently show the reactions of the dicarbonyl compound and the diamine from which they are formally derived (see p. 611). Thus reaction with phenylhydrazine leads to the formation of pyrazoles derived from the dicarbonyl component and reaction with diacetyl gives quinoxalines derived from the diamine ⟨74AHC(17)27, p. 38⟩.

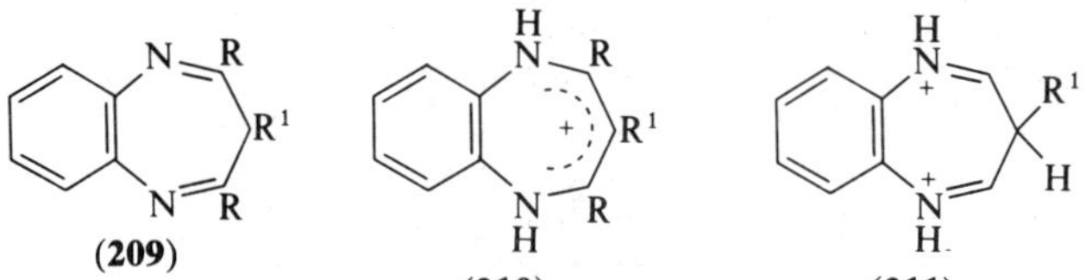

(209) (210) (211)

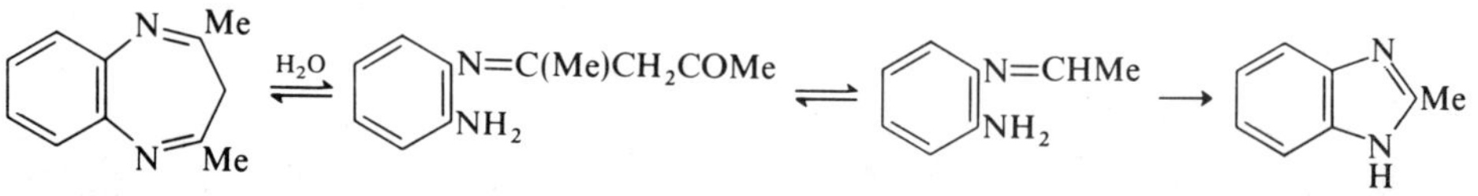

(212) (213)

1,5-Benzodiazepines do not show the high reactivity to electrophilic substitution exhibted by (**204**) but they can be brominated and coupled with diazonium salts at the 3-position. Treatment with sodamide and methyl iodide gives the 3-methyl derivative and by similar means the diazepinone (**170**) can be alkylated at the 5-position.

The photochemical oxidation of (**209**; R′ = H) leads to quinoxalines possibly *via* a diazanorcaradiene intermediate ⟨74AHC(17)27⟩.

(*c*) *1,4-Benzodiazepines.* These undergo several interesting acid-catalyzed ring cleavage and rearrangement reactions. Thus in the acid hydrolysis of (**214**), mild conditions give the 1,4-benzodiazepin-2-one (**215**) while vigorous conditions result in complete hydrolysis of the amide and imine functions. When appropriate substituents are present ring cleavage can be followed by recyclization to give a new heterocyclic system. Thus the 3-amido derivative (**216**) reacts *via* hydrolysis of the imine bond and recyclization to give the indole (**217**); 5-chloro-3-phenylindole is also obtained as a secondary product.

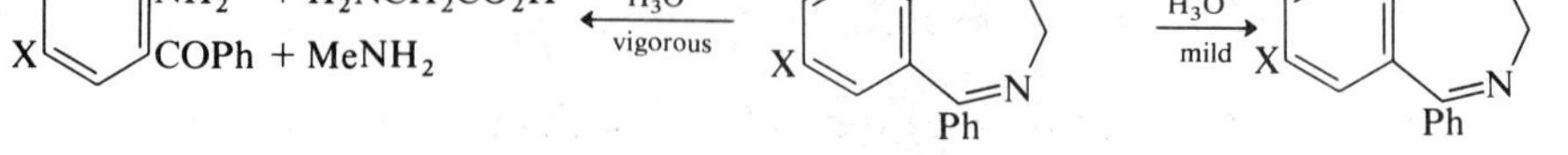

(214) (215)

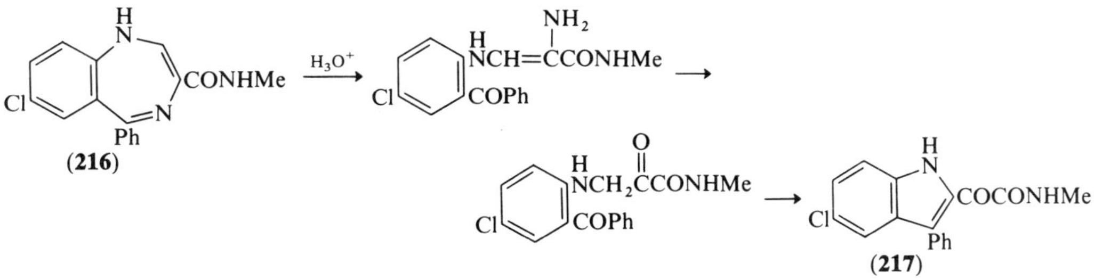

(216) (217) 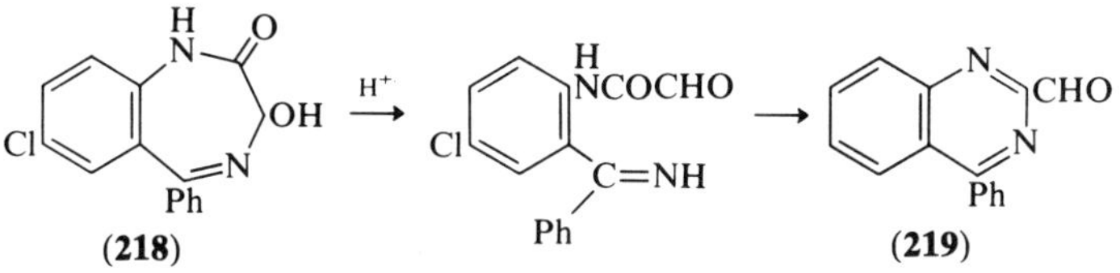 (218) (219) 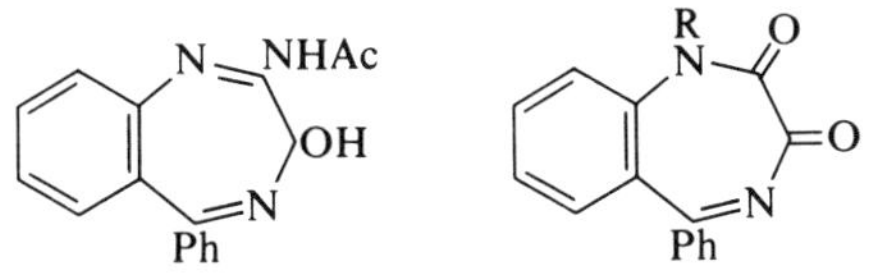 (220) (221)

Another group of acid-catalyzed rearrangements involves cleavage of the 3,4-bond followed by cyclization between positions 2 and 4, for example the conversion of (**218**) to the quinazoline carboxaldehyde (**219**). Similar reactions are given by (**220**) and (**221**).

The nitrogen atoms of various 1,4-diazepines can be alkylated using standard methods; for example, 1,4-benzodiazepines such as (**223**) can be readily alkylated at N-1 by treatment with sodium methoxide followed by an alkyl halide or sulfate. The 1-substituted product (**222**) can be further alkylated or acylated at the 3-position *via* deprotonation with strong base. However, when treated with sodium hydride in DMF at 60 °C 1-alkylbenzodiazepin-2-ones rearrange to give an isoindole carboxamide in high yield by a mechanism similar to that shown for the formation of (**238**) (p. 618). In the alkylation of (**223**) in the absence of base the imine nitrogen is more reactive and (**224**) is obtained. This product can be reduced at the 5-position by sodium borohydride to give (**225**) and is susceptible to nucleophilic attack by amines at the same position.

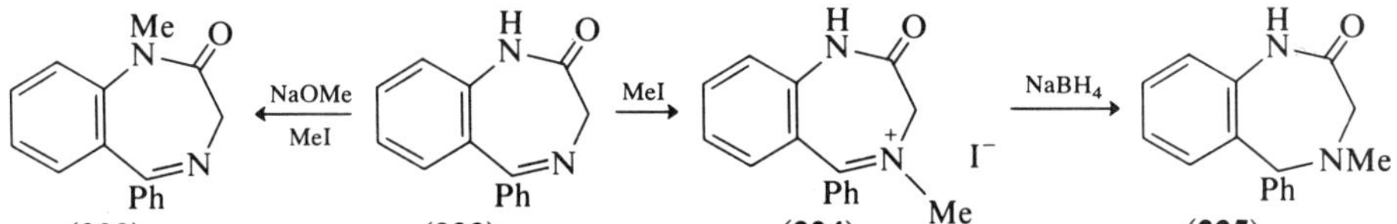

(222) (223) (224) (225)

In the 2-amino compound (**214**) the exocyclic nitrogen can be alkylated directly using aldehydes or *via* reaction with sodium methoxide and an alkyl iodide. The thione (**226**; R = H) is readily alkylated with dimethyl sulfate to give (**229**). The high reactivity of both (**226**) and (**229**) to nucleophilic attack at C-2 makes them valuable substrates for substitution at that position, *e.g.* to give (**227**) or for annelation reactions on to the 1,2-bond (see p. 614). The 1-methyl analogue (**226**; R = Me) reacts with base and ethyl iodide to give (**228**). The phosphonate (**156**; R^2 = Me, R^1 = —P(O)(OEt)$_2$) undergoes Wadsworth-Emmons alkenation at the 3-position ⟨75JOC1508⟩.

The acylation of 1,4-benzodiazepine 4-oxides, *e.g.* (**230**), leads to a Polonovski-type rearrangement giving (**231**) and provides a useful way of functionalizing the 3-position. Treatment of (**230**) with *p*-toluenesulfonyl chloride gives the quinoxalone (**233**) *via* a Beckmann type of rearrangement ⟨72JHC747, p. 755⟩ and similar rearrangements have been observed in the reactions of 4-hydroxy derivatives with thionyl chloride or phosphorus oxychloride.

In compounds of type (**234**), as expected the 4-position is the more reactive and can be acylated with good selectivity, and compounds of type (**178**) can be acylated at N-1.

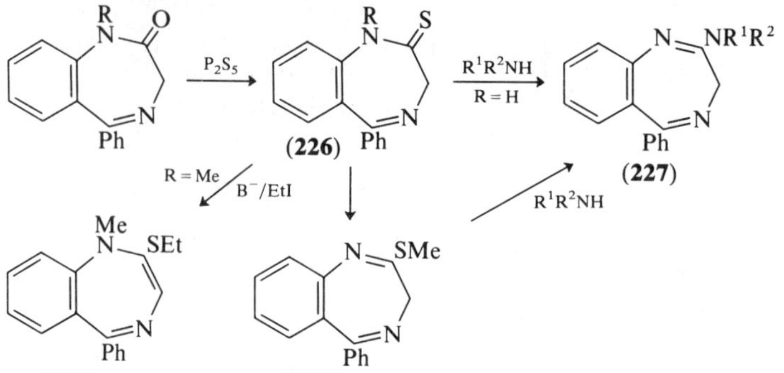

(226) (227) (228) (229)

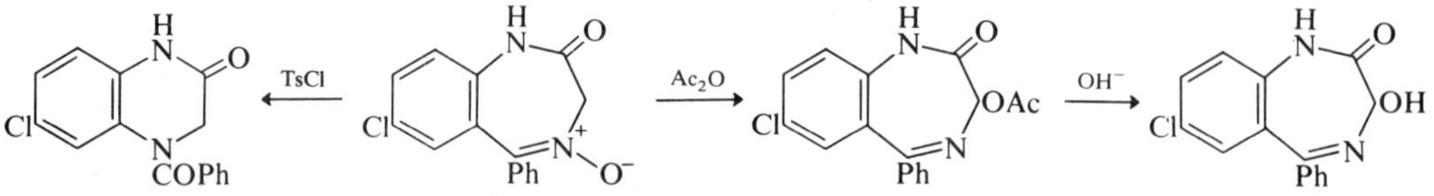

(233) (230) (231) (232)

1-Substituted 1,4-benzodiazepin-2-ones, *e.g.* (**222**), can be formylated at the 3-position by treatment with strong base followed by ethyl formate, or fluorinated by treatment of the anion with ClO_3F. The ambident anion generated from (**236**) is selectively phosphorylated at oxygen to give (**235**) ⟨76JOC2720⟩. This product was found to react quite readily with nucleophiles to give a variety of 2-substituted products; thus this process provides a relatively mild method of chemically activating the secondary amide moiety (see also thiation).

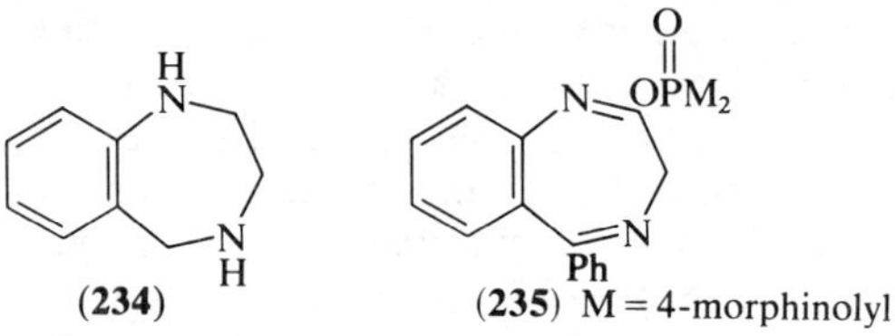

(**234**) (**235**) M = 4-morphinolyl

The direct reaction of (**236**) with isocyanates gives 2-carbamoyl derivatives ⟨76MI51800⟩ but reaction with acetic anhydride leads to two types of rearrangement. In the presence of sodium acetate the oxazoloquinoline (**237**) is formed but in pyridine a different ring contraction gives the isoindole (**238**), which is subsequently acylated under the reaction conditions. The treatment of 1,4-benzodiazepine-2,4-diones with acetic anhydride also induces rearrangement, in this case to 3,1-benzoxazin-4-ones which are probably formed *via* ring opening of the 4-acetyl derivative ⟨76JHC1165⟩.

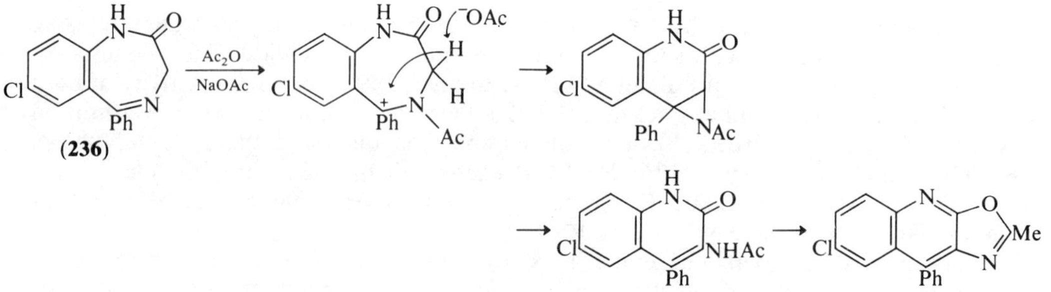

(**236**) (**237**)

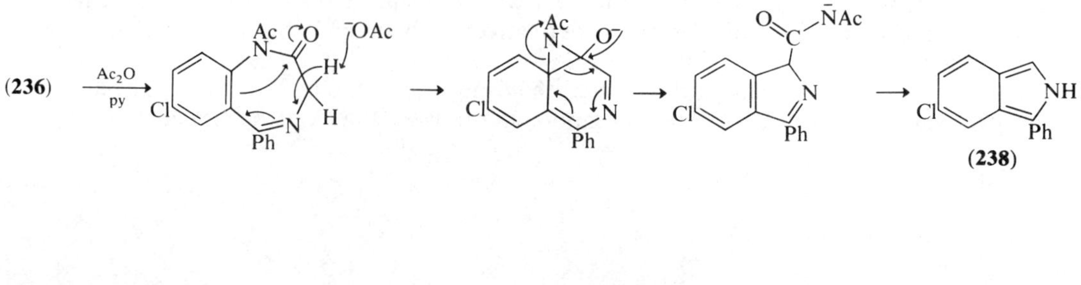

(**236**) (**238**)

Nitration of the aromatic ring of (**223**) occurs at the 7-position. However, its reaction with fuming nitric acid in acetic anhydride followed by treatment with methanol gave the 5-methoxy-4-nitro derivative. This product, on treatment with triethylamine, eliminated nitrous acid from the 3,4-positions.

Peroxyacids oxidize the imine group in (**236**) and related compounds to the nitrone. The 1,4-benzodiazepine (**178**) and its 4,5-dihydro analogue can be oxidized to the 2-oxo derivatives with chromium(VI) oxide. The benzodiazepinone (**236**) can be hydroxylated at the 3-position *via* a two-step process: firstly reaction with LTA gives the 3-acetoxy derivative and this can then be hydrolyzed in high yield using sodium methoxide in methanol/methylene chloride. The 3-acetoxy derivative may also be prepared *via* bromination with NBS and reaction with sodium acetate *in situ* ⟨79JHC1449⟩.

(ii) Reactions with nucleophilic reagents

(*a*) 2,3-*Dihydro*-1,4-*Diazepines*. 6-Bromo derivatives (**239**; X = Br) react readily with alkoxide ions to give 6-alkoxy derivatives (**239**; $X = OR^2$) but in many cases the reaction with nucleophiles leads to the replacement of the bromine with a hydrogen atom *via* nucleophilic attack on the halogen. Both processes are thought to involve reactions of the tautomer (**240**) which is not normally present in a concentration large enough to be detected. The nucleophilic protodehalogenation becomes more favoured as the halogen atom varies from chlorine through to iodine. Interestingly, nucleophilic attack at the 6-position becomes less favoured and attack at the 5- and 7-positions becomes increasingly favoured as the size of the substituents R and R^1 in (**240**) is reduced ⟨78H(11)549, p. 556⟩.

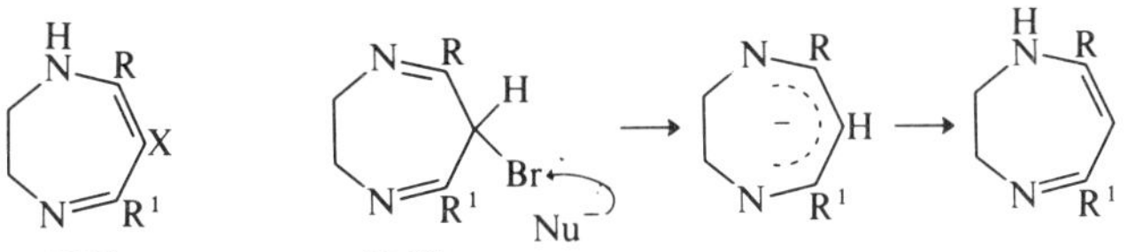

(**239**) (**240**)

(*b*) *1,4-Benzodiazepines*. Various nucleophiles (*e.g.* piperidine, alcohols, hydrogen cyanide, thiols) can be added selectively across the 1,2-imine bond in (**241**), *e.g.* to give (**242**). The functional groups at the 2-position so introduced are capable of useful further transformations and the construction of heterocyclic rings, e.g. in (**243**), fused to the 1,2-bond ⟨74JOC167⟩. The analogous 4-oxide gave a similar reaction when treated with sodium methoxide in methanol. The labile 2-methoxy group in this product could be readily replaced with primary or secondary amines.

Reaction of the 1-tosyl derivative (**244**) with sodium hydride gave (**245**).

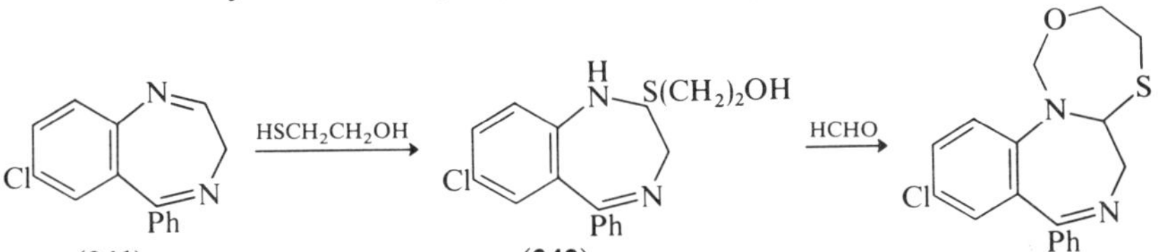

(**241**) (**242**) (**243**)

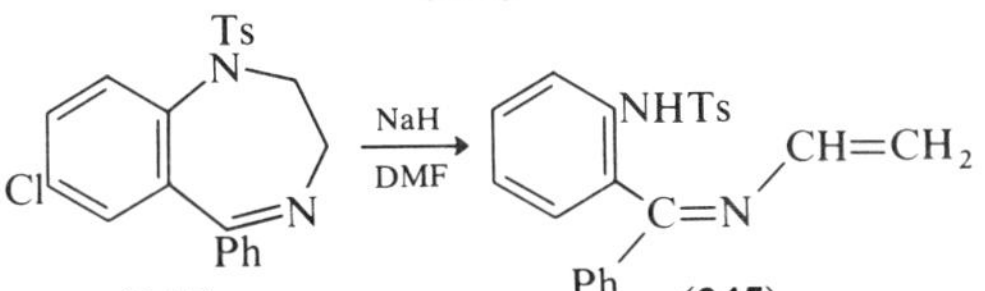

(**244**) (**245**)

The alkaline hydrolysis of the 1,4-benzodiazepin-2-one (**236**) and its 4-oxide resulted in ring opening *via* amide cleavage. Similarly treatment with methylamine gave (**246**) which proved to be a useful intermediate in the synthesis of the 3-carboxamide (**247**).

Mild alkaline hydrolysis of (**186**; R = Me) gave the 1,4-benzodiazepin-2-one 4-oxide. Grignard reagents react with the nitrone (**152**; $R = R^1 = H$) and similar compounds to give the 4-hydroxy-5-phenyl derivative ⟨77ACS(B)701⟩.

The carbonyl group at the 2-position in (**236**) can be aminated directly using methylamine but the reactivity at this position can be easily enhanced by thiation (p. 617), phosphorylation (p. 618) or conversion to the imidoyl chloride. Similar activation of a 5-oxo group by thiation has been reported ⟨77JHC985⟩.

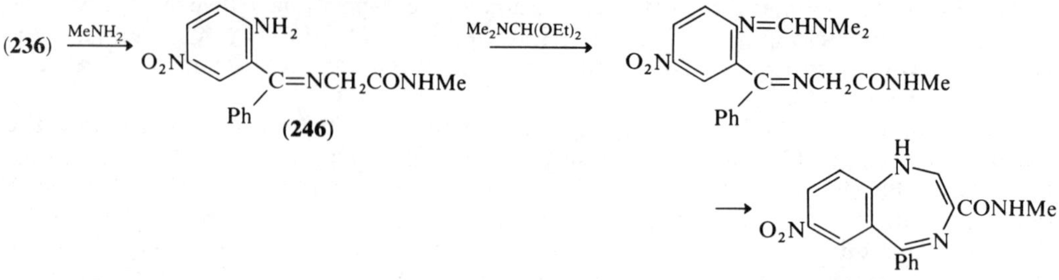

(236) $\xrightarrow{MeNH_2}$ (246) $\xrightarrow{Me_2NCH(OEt)_2}$ → (247)

The benzodiazepine (**236**) can be chlorinated at the 1-position using sodium hypochloride or *t*-butyl hypochlorite, or at the 3-position using NCS. The 3-chloro derivative can also be prepared by treatment of the 3-hydroxy compound with thionyl chloride. Halogens at the 3-position can be displaced to give 3-amino derivatives.

There have been many studies on selective reduction in these systems.

(iii) Thermal and photochemical reactions

Many nitrones in this series are converted to oxaziridines by UV irradiation; for example, a solution of (**248**) in 2-propanol gave (**249**) on exposure to daylight. This isomerization was reversed on heating. UV irradiation of (**248**) in ethanol/benzene induced transformations to (**250**) and (**251**), probably *via* (**249**).

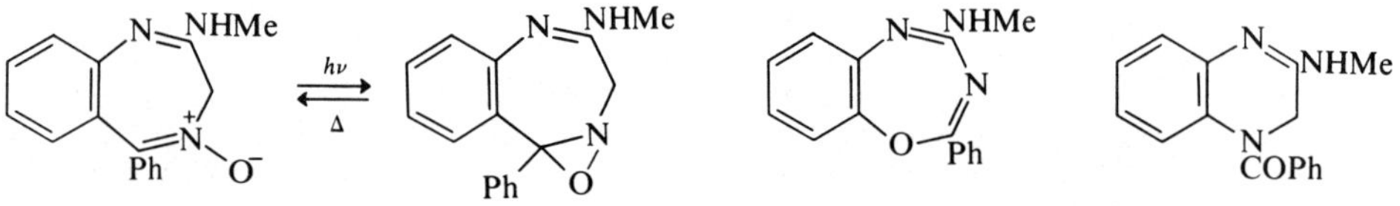

(248) (249) (250) (251)

5.18.3 DIOXEPINS

Dioxepins in general have not attracted extensive study but the 1,3-system has received the most attention. The comprehensive review ⟨72HC(26)319⟩ covers the literature up to *ca.* 1970.

5.18.3.1 1,2-Dioxepins

5.18.3.1.1 Synthesis and reactivity

The monocyclic fully saturated peroxides (**253**; R = H, Me) have recently been prepared by the treatment of the hydroperoxides (**252**) with LTA ⟨81S633⟩, and a 3,7-methano-bridged system (**254**) has been produced in good yield by the reaction of the analogous cyclic azo compound with hydrogen peroxide ⟨81USP4291051⟩. The 1*H*-2,3-benzoxepin (**255**; R = H) can be prepared in high yield by the reaction of indene with ozone at 0–5 °C in ethanol. It is a waxy non-explosive solid which reacts further with ethanol in the presence of acid to give (**255**; R = Et) as a crystalline derivative. Reduction with LAH and zinc/acetic acid gave homophthalyl alcohol and homophthaldehyde respectively, oxidation with hydrogen peroxide gave homophthalic acid, and alkaline hydrolysis gave 2-carboxyphenylacetaldehyde. Various substituted indenes react with ozone to give 1,4-epoxy analogues of (**255**).

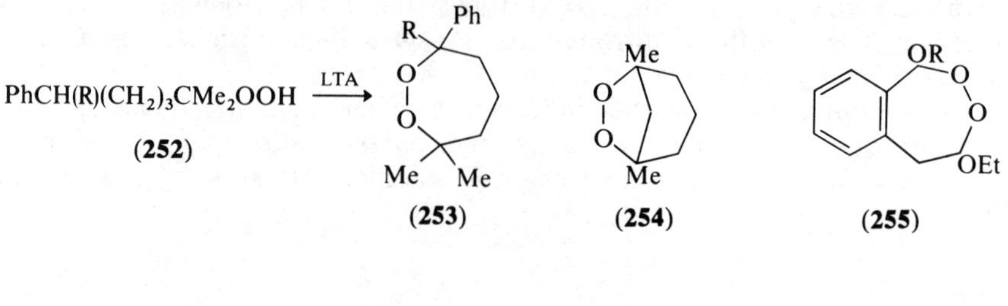

$PhCH(R)(CH_2)_3CMe_2OOH$ $\xrightarrow{LTA}$ (253)

(252) (253) (254) (255)

5.18.3.2 1,3-Dioxepins

5.18.3.2.1 Synthesis

The major route to 4,7-dihydro-1,3-dioxepins (**256**) is by the reaction of *cis*-but-2-ene-1,4-diols with aldehydes or related compounds. Aldehydes themselves give good yields of (**256**; R or $R^1 = H$) but the reaction with ketones is less satisfactory and the preparation of 2,2-disubstituted derivatives is best achieved by an exchange reaction between the diol and a ketal ⟨72HC(26)319, p. 320, 79S975⟩. Orthoformic esters similarly give 2-alkoxy derivatives in good yield. 3*H*-2,4-Benzodioxepins (**257**) can be prepared in a similar way using 1,2-benzenedimethanol.

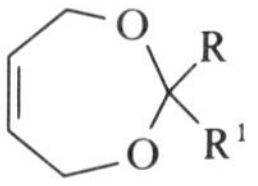

(**256**) 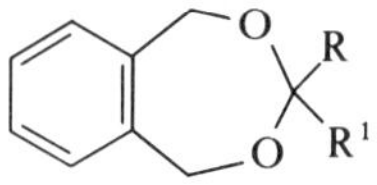 (**257**)

5.18.3.2.2 Reactivity

4,7-Dihydro-1,3-dioxepins are fairly stable to heat and alkaline reagents but treatment with strong bases can lead to isomerization to give the 4,5-dihydro isomers (**258**), or ring contraction to 2-aryl-3-formyl tetrahydrofurans ⟨71CR(C)(273)1655⟩. The isomerization of (**256**) to (**258**) can also be effected using ruthenium hydride complexes ⟨80TL4927⟩. The 1,3-dioxepan-4-ylium ions (**259**), prepared by protonation of (**258**), undergo thermal fragmentation and recyclization to give tetrahydrofuran-3-aldehydes (**260**) on a preparative scale ⟨80CB1472, 80TL4927⟩. Treatment of the fully saturated 1,3-dioxepane (prepared by the hydrogenation of **256**) with 33% sulfuric acid leads to fragmentation to tetrahydrofuran and formaldehyde.

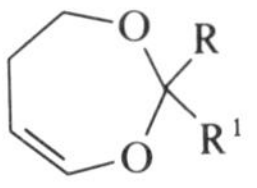

(**258**) 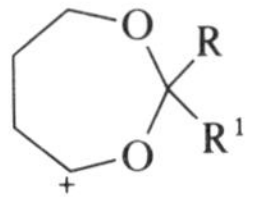 (**259**) 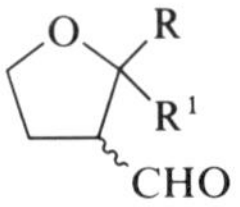 (**260**)

Chlorine and bromine add normally to the double bond in (**256**) and the dibromo compound can be mono-dehydrobrominated using sodium methoxide in methanol to give a mixture of 5-bromo-4,7-dihydro-1,3-dioxepin and 5-bromo-4,5-dihydro-1,3-dioxepin (which is converted *in situ* to the 5-methoxy derivative). More interestingly it can be fully dehydrobrominated using HMPA at 140 °C to give the fully unsaturated 2*H*-1,3-dioxepin (**261**) ⟨76TL2113⟩. This is the only known route to this compound. It gives a Diels–Alder adduct with 4-phenyl-1,2,4-triazoline-2,5-dione in 95% yield and on UV irradiation it isomerizes to (**262**).

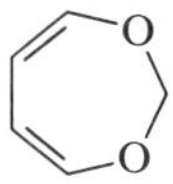

(**261**) 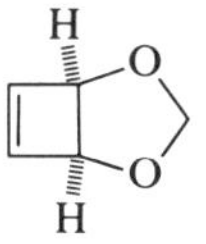 (**262**)

5.18.3.3 1,4-Dioxepins

5.18.3.3.1 Synthesis and reactivity

The 6,7-dihydro-5*H*-1,4-dioxepin (**266**) has been prepared ⟨54CR(38)982⟩. and more recently it has been shown that the 2,3-dihydro-5*H*-1,4-dioxepins (**263**) and (**265**) can be produced from 1,4-dioxine–halocarbene adducts (**264**), either by heating under reflux in xylene or by treatment with bases. The allylic chlorine atom in (**263**) is readily substituted by alkoxide or cyanide ions ⟨77ZC331, 76UKZ968⟩. Saturated rings of type (**267**) have been prepared by the treatment of cyclic acetals of ethane-1,2-diol with vinyl ethers in the presence of boron trifluoride, and 1,4-dioxepan-5-one (**268**) has been prepared by the reaction of bromoform and silver nitrate with aqueous dioxane ⟨60AG415⟩.

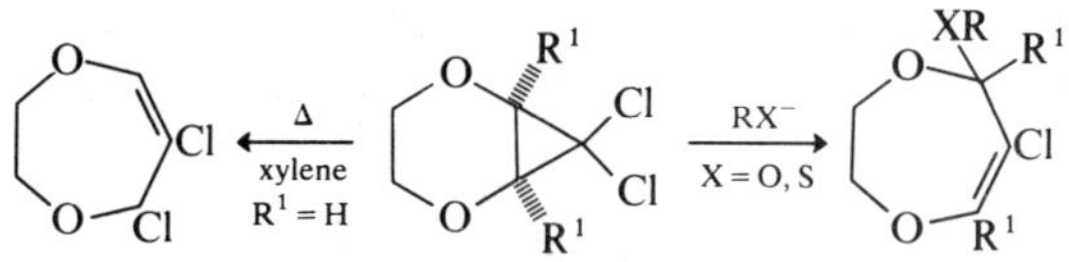

(263) (264) (265)

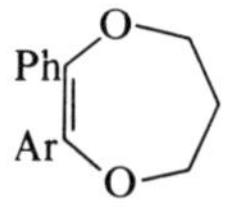

(266) 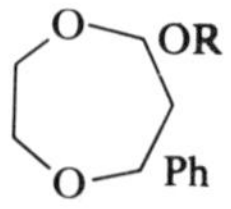 (267) 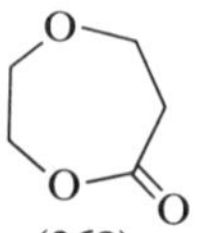 (268)

A number of 5*H*-1,4-benzodioxepin-ones and -diones have been prepared ⟨72HC(26)319, p. 339⟩. The reaction of the sodium salt of salicylic acid with 2-chloroethanol gave (**269**) ⟨75BSF(2)277⟩ and treatment of the methyl ester of 2-acetyl-6-chloro-3,5-dimethoxyphenoxyacetic acid with 3M hydrochloric acid gave (**270**). The dione (**271**) was prepared by heating 2-carboxy-5,6-dimethoxyphenoxyacetic acid in acetic anhydride and (**272**) was prepared from chloroacetylsalicylic acid.

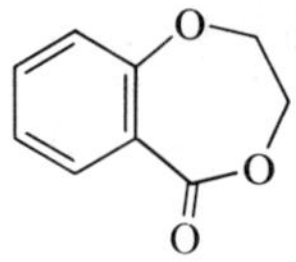

(269) 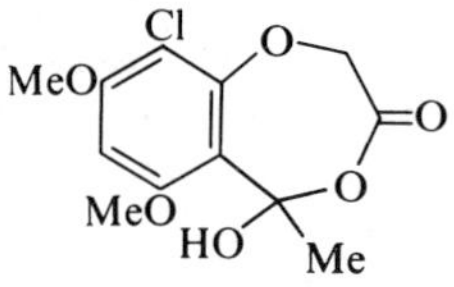 (270) 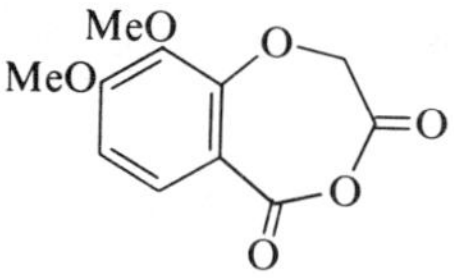 (271) 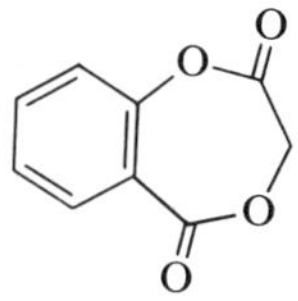 (272)

The two major routes to 3,4-dihydro-2*H*-1,5-benzodioxepins (**274**) from (**273**) and (**275**) are applicable to a wide range of substituted derivatives. The 3-oxo derivative, important as a perfume odorant, can be prepared *via* the reaction of 1,2-dihydroxybenzene with chloroacetonitrile ⟨75CJC2279⟩ or *via* a Dieckmann cyclization ⟨74USP3799892⟩.

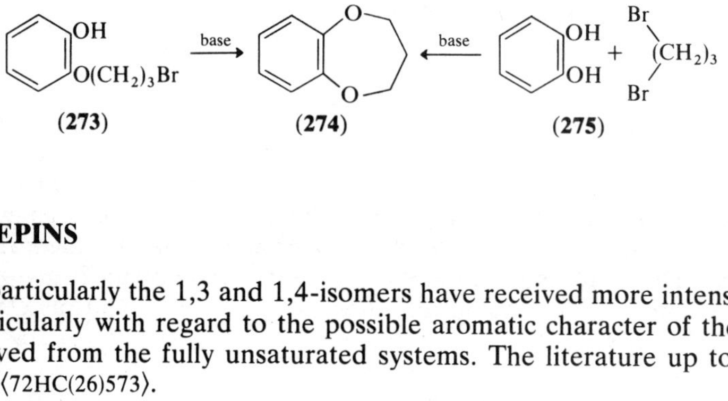

(273) (274) (275)

5.18.4 DITHIEPINS

Dithiepins, particularly the 1,3 and 1,4-isomers have received more intensive study than dioxepins, particularly with regard to the possible aromatic character of the anions (**279**) and (**289**) derived from the fully unsaturated systems. The literature up to *ca.* 1970 has been reviewed ⟨72HC(26)573⟩.

5.18.4.1 1,2-Dithiepins

5.18.4.1.1 Synthesis and reactivity

Only the fully saturated system is known. Simple 1,2-dithiepanes are best prepared by the oxidation of α,ω-alkanedithiols using iron(III) chloride, hydrogen peroxide, iodine or oxygen. The unsubstituted dithiepane has also been prepared by treating the 'Bunte' salt ($NaO_3S \cdot S(CH_2)_5S \cdot SO_3Na$) with hot aqueous copper(II) chloride, and more recently by reaction of the lead dithiolate [$Pb(S(CH_2)_5S)$], prepared from the dithiol and LTA, with sulfur in benzene ⟨73TL655⟩. Another useful general route involves the reaction of 1,5-dibromo compounds with sodium disulfide. For example, 2,6-dibromopimelic acid thus gives 1,2-dithiepane-3,7-dicarboxylic acid and similarly ethyl 2,6-dibromohexanoate reacts with potassium hydrogen sulfide to give 3-ethoxycarbonyl-1,2-dithiepane ⟨72HC(26)573⟩.

1,2-Dithiepane can be oxidized to its 1,1-dioxide by hydrogen peroxide and this is further oxidized to the 1,1,2,2-tetraoxide by hydrogen peroxide in the presence of tungstic acid.

As the oxidation level increases so does the ease of hydrolytic cleavage of the S—S bond. The S—S bond can also be cleaved by UV irradiation ⟨72HC(26)573, p. 607⟩.

5.18.4.2 1,3-Dithiepins

5.18.4.2.1 Synthesis and reactivity

The fully unsaturated 2*H*-1,3-dithiepin (**276**) has recently been prepared ⟨76TL1251⟩. Its deep-red anion (**277**) reacted with various electrophiles to give the 3-substituted derivatives (**278**; R = D, Me_3Si, $C(OH)Ph_2$) in good yields. There has been some theoretical speculation about the possible aromatic character of anions of this type (and analogous ions generated from 1,2- and 1,4-dithiepins) ⟨72HC(26)573, 75PAC(44)767⟩ which can be envisaged as planar 10π-electron systems delocalized through sulfur (**279**). In this case no clear decision concerning aromatic character could be made on NMR evidence, but deuterium exchange studies showed that (**276**) is at least 150 times more acidic at the 2-position than its fully saturated analogue and it was concluded that this corresponds to a small (*ca.* 12 kJ mol^{-1}) stabilization of the anion due to electron delocalization ⟨76JA2005⟩. This conclusion is supported by recent ^1H NMR studies on (**280**) ⟨80CL1427⟩ but no such stabilization was detected in a dibenzo analogue of (**277**) ⟨72HC(26)573, p. 598⟩.

NCS; KOBu^t, ether

(**276**)

(**277**) (**278**) (**279**) (**280**)

The 4,7-dihydro-2*H*-1,3-dithiepin system (**282**) has been prepared by the reactions of either *cis*-but-2-ene-1,4-dithiol ⟨76TL1251, 64TL1929⟩ or the bis-isothiouronium salt (**281**) ⟨78OPP133⟩ with methylene bromide in the presence of base. The analogous 1,5-dihydro-3*H*-2,4-benzothiepin system (**284**) can be prepared by the reaction of (**283**) with either methylene iodide in the presence of base or more generally with aldehydes or ketones in the presence of acid. Compounds of this type (with R and/or R^1 = H) can be readily alkylated ⟨75S720⟩ at position 3 *via* generation of the anion with butyllithium and thus have found use as perhaps the least unpleasant-smelling thioketal carbonyl synthons. Chlorination of (**284**; R = R^1 = H) gave the 3-chloro derivative which reacted with Grignard reagents to give the 3-aryl- or 3-alkyl-substituted derivatives and with triethyl phosphite to give the 3-phosphonate ⟨78JPR255⟩. The latter has been used in Wadsworth–Emmons-type alkenation reactions ⟨78T3081⟩. NMR studies have been carried out on the conformation and energetics of ring inversion in (**282**) and (**284**) ⟨72HC(26)573, p. 614, 79CJC3221⟩.

$CH_2SC(=NH)NH_2$ / $CH_2SC(=NH)NH_2$ (**281**) + 2HCl —(CH_2Br_2, KOH/MeOH)→ (**282**)

(**283**) + $R^1R^2C{=}O$ —(H^+)→ (**284**) —($CuO/CuCl_2$)→ $R^1RC{=}O$

Fully saturated 1,3-dithiepins can be prepared by the Lewis acid-catalyzed exchange reactions of 1,4-butanedithiol with the appropriate diethyl acetal, and 1,3-dithiepan-2-one is obtained as an unstable oil *via* the reaction of 1,4-dichlorobutane with alkali trithiocarbonates ⟨72HC(26)573, p. 616⟩.

5.18.4.3 1,4-Dithiepins

5.18.4.3.1 Synthesis and reactivity

Two examples (**288**) and (**292**) of the fully unsaturated 5*H*-1,4-dithiepin have now been prepared. After attempts to prepare the 6-chloro analogue of (**286**) had resulted in rearrangement, the synthesis of (**288**) was achieved in 60% yield by an elimination reaction of the methyl ether (**287**) using lithium dicyclohexylamide. The product was only moderately stable and decomposed slowly at room temperature. Studies on the anion of this system (**289**) are of interest to compare its capacity for stabilization by electron delocalization with that of the 1,3-system (**279**). It is therefore notable that treatment of (**288**) with butyllithium followed by methyl iodide gave not the expected 5-methyl derivative but (**290**), and that further methylation gave (**291**). The authors concluded that these results provide conclusive evidence for the instability of (**289**) and its lack of aromaticity associated with electron delocalization through sulfur ⟨75TL1895⟩. The enol (**292**) was prepared by the acylation of 5,7-dihydro-1,4-dithiepin-6-one ⟨76CB2928⟩.

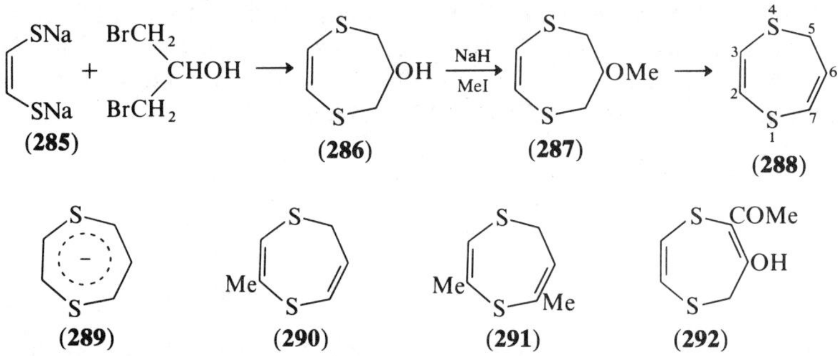

SNa SNa (**285**) + $BrCH_2$ CHOH $BrCH_2$ → (**286**) OH $\xrightarrow[MeI]{NaH}$ (**287**) OMe → (**288**)

(**289**) (**290**) Me (**291**) Me Me (**292**) COMe OH

2,3-Dihydro-5*H*-1,4-dithiepins are rare but one example (**293**) has been prepared by the dehydration of 6-hydroxy-1,4-dithiepane 1,1,4,4-tetraoxide ⟨72HC(26)573, p. 617⟩. The 6,7-dihydro-5*H* system is slightly better known and several examples, *e.g.* (**286**), have been prepared by the reaction of *cis*-ethylene-1,2-dithiol disodium salt (**285**) with 1,3-dihalides ⟨75TL1895, 76CB2928⟩ and also by the reaction of *cis*-1,2-dichloroethylene with the sodium salt of propane-1,3-dithiol. The 3,4-dihydro-2*H*-1,5-benzodithiepin system can be similarly prepared using benzene-1,2-dithiol ⟨72HC(26)667⟩. 6,7-Dihydro-5*H*-1,4-dithiepins may also be prepared by the migration of a thio group in 1,3-dithians. Thus treatment of (**294**; R = H, CO_2Me) with methanesulfonyl chloride gave (**295**), presumably by ring expansion of the sulfonate ester ⟨79JHC1081⟩ and (**296**) on treatment with phosphorus pentoxide gave (**295**; R = H) ⟨70JOC823⟩.

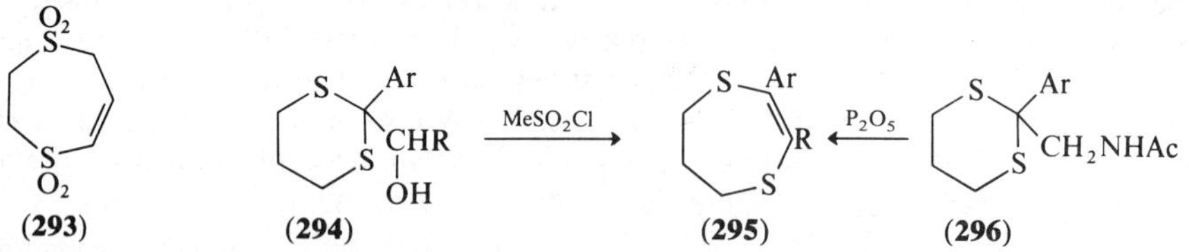

(**293**) O_2 O_2 (**294**) Ar CHR OH $\xrightarrow{MeSO_2Cl}$ (**295**) Ar R $\xleftarrow{P_2O_5}$ (**296**) Ar CH_2NHAc

The fully saturated 1,4-dithiepane system can be prepared by several routes. The reactions of propane-1,3-dithiol with 1,2-dibromoethane, and of ethane-1,2-dithiol with 1,3-dihalopropanes under basic conditions have both been used and have been applied to the synthesis of various substituted derivatives ⟨72HC(26)573, p. 619⟩. The 6-benzoyl derivative has been prepared in good yield by the reaction of ethane-1,2-dithiol with acetophenone and formaldehyde but similar reactions with other ketones gave mostly polymeric products ⟨79ZOR1709⟩. The 6-oxo derivative has been prepared by a Dieckmann cyclization ⟨72HC(26)573, p. 621⟩ and 2-substituted derivatives (**297**) have been prepared as shown ⟨73RTC890⟩. 1,4-Dithian-2-ones, *e.g.* (**299**), have also been prepared recently by the ring expansion of 1,3-dithian derivatives, *e.g.* (**298**) and (**300**), on treatment with LTA ⟨76JA4313, 79CL923⟩.

1,4-Benzodithiepins have not been much studied but 7-chloro-2,3-dihydro-5*H*-1,4-dithiepin has been prepared by the action of base on 2-(2-chloroethylthio)-5-chlorobenzyl isothiouronium chloride ⟨72HC(26)667, p. 787⟩. 1,5-Benzothiepin-2-one has recently been

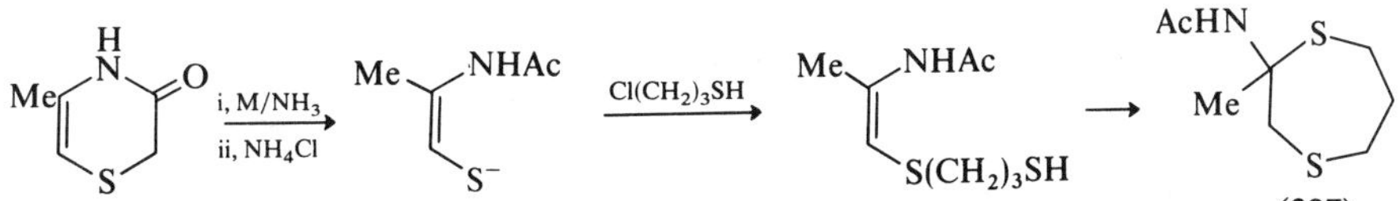

(297)

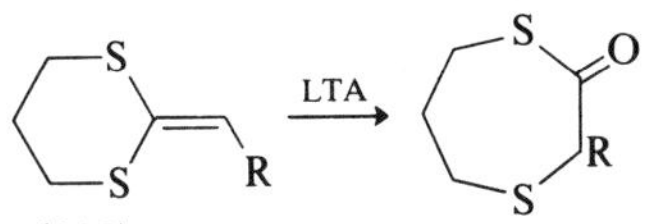

(298) (299) 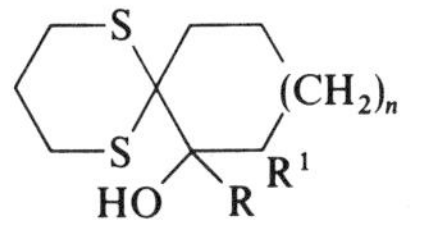 (300)

prepared by the cyclodehydration of 3-(2-mercaptophenylthio)propionic acid with phosphorus pentoxide ⟨77CR(C)(284)795⟩.

5.18.5 OXAZEPINES

Oxazepine chemistry includes some interesting rearrangement reactions both in synthesis and reactivity, many of which have parallels in the chemistry of the analogous diazepine systems. The preparative routes are classified as discussed earlier (p. 595).

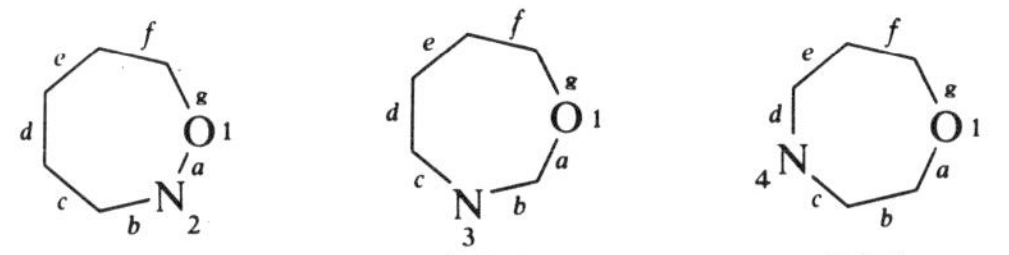

(301) (302) (303)

5.18.5.1 1,2-Oxazepines

5.18.5.1.1 Synthesis

(i) Type b *(N—O—C—C—C—C—C)*

Ethyl 5-bromopentanoxycarbamate (**306**), prepared by the reaction of (**304**) and (**305**), has been cyclized as a dilute solution of its potassium salt in DMF to give (**307**), which on subsequent hydrolysis gave (**308**). The direct one-step conversion of (**304**) and (**305**) to (**308**) was not successful ⟨69JHC111⟩. The same principle has been applied to the synthesis of the 2,3- and 3,2-benzoxazepines ⟨71JHC911⟩.

$Br(CH_2)_5Br$ (**304**) + $KONHCO_2Et$ (**305**) → $Br(CH_2)_5ONHCO_2Et$ (**306**) → (**307**) → (**308**)

(ii) Type bg *(N—O + C—C—C—C—C)*

3,3,7,7-Tetramethyl-1,2-oxazepin-5-one and its *N*-methyl analogue have been prepared by the addition of hydroxylamine and *N*-methylhydroxylamine to phorone ⟨73TL1615⟩.

(iii) Synthesis from other heterocyclic systems

The Meisenheimer rearrangement of tertiary amine *N*-oxides has been applied to the synthesis of both monocyclic 1,2-oxazepines, *e.g.* (**309**) ⟨65JOC3135, 65JCS1653, 82H(19)173⟩, and those fused to benzene ⟨80AJC833⟩ and a variety of heterocyclic rings ⟨80AJC1335⟩.

The unstable dibenz[*c,f*][1,2]oxazepines (**312**; R = CN, Cl) have been isolated as the major products of the UV irradiation of 9-cyano- and 9-chloro-acridine 10-oxides (**310**) in benzene (*cf.* the analogous *N*-imide to 1,2-diazepine conversion on p. 598). Although none of the oxaziridine tautomer (**311**) was detectable by UV spectroscopy, the subsequent deoxygenation of (**312**) to acridine suggests the existence of a thermal equilibrium between (**311**) and (**312**) ⟨79T1273⟩. These dibenzo compounds (**312**) are the only fully unsaturated oxazepines yet isolated but the 2,3-benzoxazepin-1-one system (**314**) has recently been prepared by the reaction of benzonitrile oxide with the benzopyranone (**313**) ⟨80JCS(P1)846⟩.

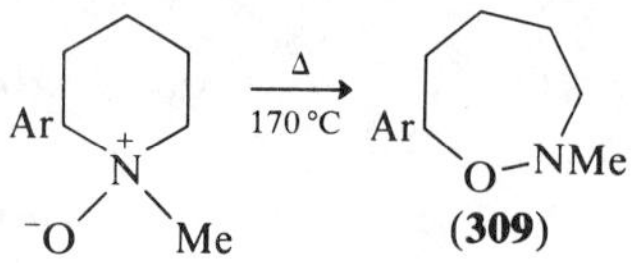

(309)

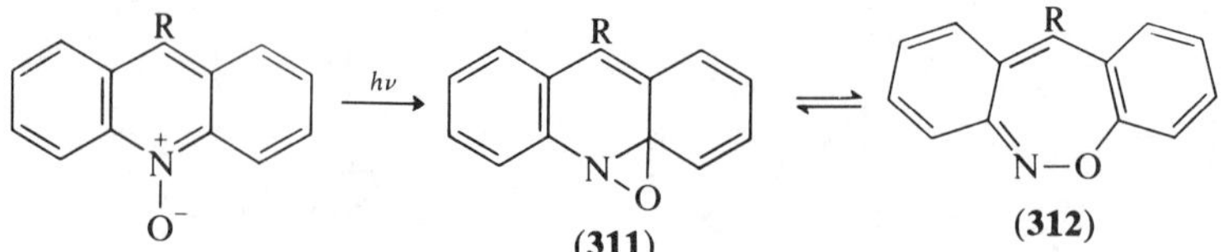

(310) (311) (312)

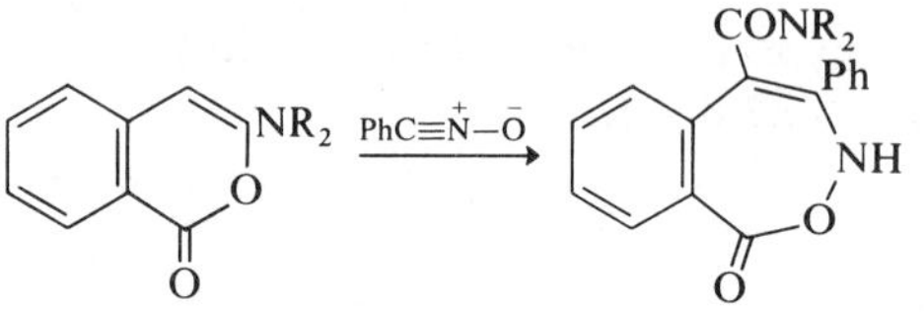

(313) (314)

3,5,7-Triphenyl-1,2-oxazepinium perchlorates, a new class of seven-membered cations with eight π-electrons, have been prepared by the reaction of 2,4,6-triphenylpyrylium perchlorate with nitrones ⟨81ZOR881⟩.

5.18.5.2 1,3-Oxazepines

5.18.5.2.1 Synthesis

(i) Type a *(C—N—C—C—C—C—O)*

Formation of this bond usually involves ring closure *via* nucleophilic attack by oxygen on an electrophilic carbon, *e.g.* a carbonyl group or carbon–halogen bond. Thus 2-methyl-4,5,6,7-tetrahydro-1,3-oxazepine can be prepared by the distillation of 4-acetylaminobutanol. The copper-catalyzed insertion of isocyanides (**315**) into the O—H bond of alcohols gives a high yielding route to 4,5-dihydro-3,1-benzoxazepines (**316**) ⟨78TL2087⟩.

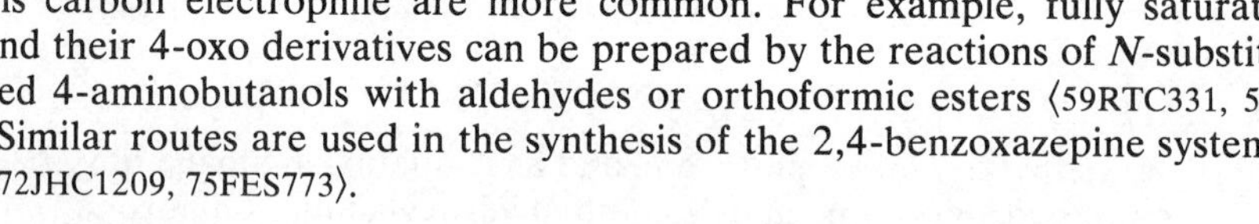

(315) (316)

(ii) Type ab *(N—C—C—C—C—O+C)*

Single stage reactions involving successive nucleophilic attack by oxygen and nitrogen on a 1,1-bis carbon electrophile are more common. For example, fully saturated 1,3-oxepanes and their 4-oxo derivatives can be prepared by the reactions of *N*-substituted or unsubstituted 4-aminobutanols with aldehydes or orthoformic esters ⟨59RTC331, 50JCS978, 77H(7)919⟩. Similar routes are used in the synthesis of the 2,4-benzoxazepine systems (**317**) and (**318**) ⟨72JHC1209, 75FES773⟩.

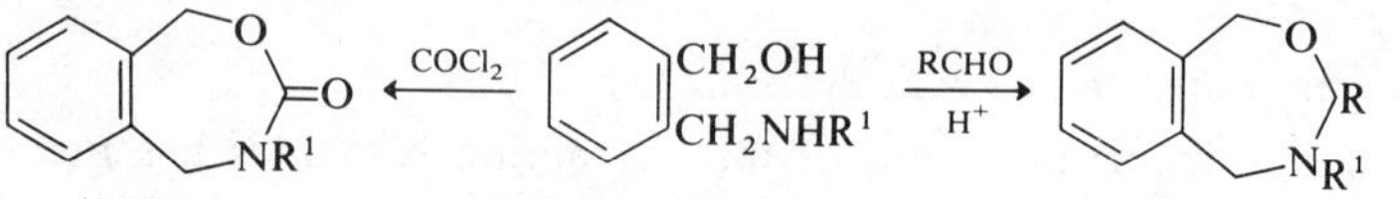

(317) (318)

(iii) Type d *(C—C—C—O—C—N—C)*

The β-lactam fused system (**319**) has recently been prepared in moderate yield by the free radical cyclization shown ⟨81TL2689, 2693⟩.

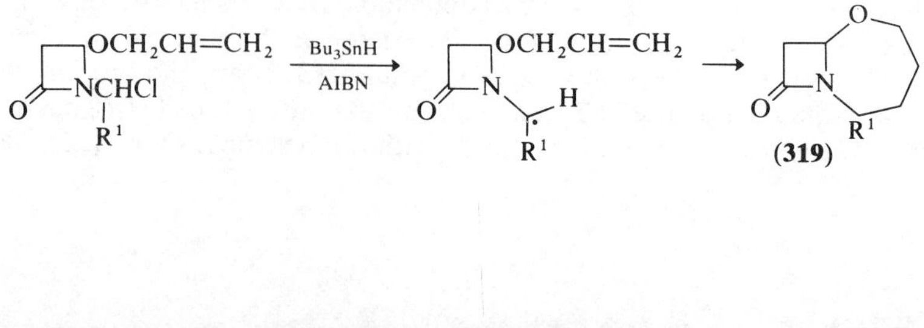

(319)

(iv) Synthesis from other heterocyclic systems

The photochemical rearrangement of aromatic *N*-oxides, *e.g.* (**320**), provides one of the major routes to the fully unsaturated 1,3-oxazepine system, *e.g.* (**323**). The reaction is thought to proceed by the primary formation of an oxaziridine intermediate (**321**) which rearranges by a 1,5-sigmatropic shift to give (**322**) which is then converted to the product by a disrotatory ring opening (*cf.* the route to 1,3-diazepines p. 600). This principle has also been applied to the preparation of 3,1- and 1,3-benzoxazepines and dibenzo[*d,f*]-[1,3]oxazepines from quinoline, isoquinoline and phenanthridine *N*-oxides respectively. This reaction has been well reviewed ⟨76H(4)1391⟩ and more recently has been used to prepare the unsubstituted benzoxazepines ⟨79TL3761⟩.

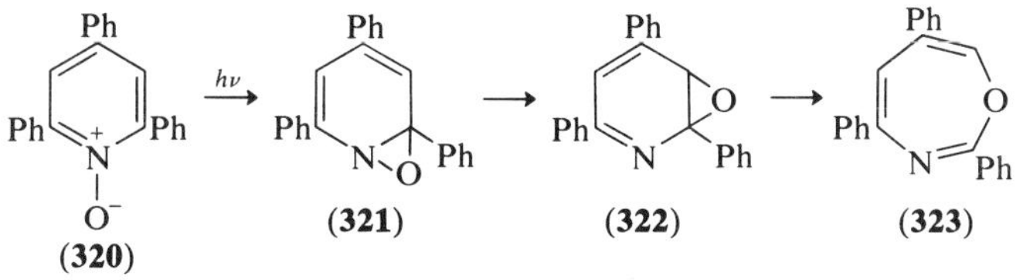

(**320**) (**321**) (**322**) (**323**)

Monocyclic 1,3-oxazepines (**325**) with aryl substituents at the 2-, 4- and 7-positions can be prepared in moderate yield (20–40%) by the reaction of aliphatic diazo compounds with 1,3-oxazinium perchlorates (**324**) ⟨74S187⟩. Tetra- and penta-phenyl-1,3-oxazepines (**328**; R = H or Ph) have been obtained *via* the reaction of azide with pyrylium salts (**326**) ⟨78H(11)331⟩. This principle had earlier been applied to the preparation of 1,3-benzoxazepines ⟨74CR(C)(278)1389⟩ and more recently to 3,1-benzoxazepines ⟨81JHC847⟩. The preparation of 2-phenyl-1,3-oxazepine (**331**) by the UV irradiation of (**329**) is mechanistically interesting in that it apparently involves an intermediate (**330**) of the same type as (**327**) ⟨73TL1835⟩, but the method has only been used in this one case. One of the few examples of a dihydro-1,3-oxazepine (**333**) has been prepared by the thermolysis of the aziridine (**332**) ⟨68JOC4547⟩.

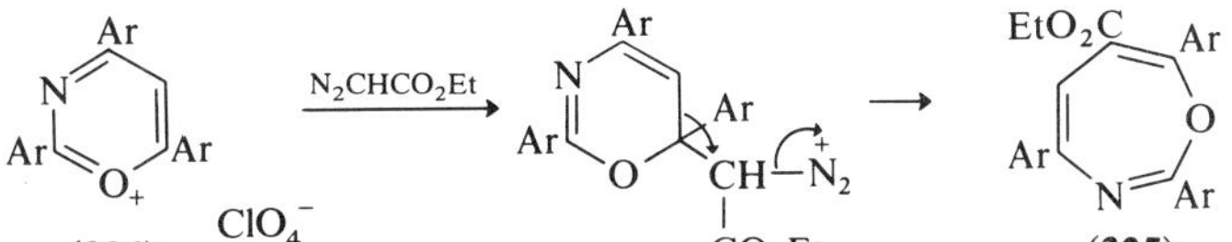

(**324**) (**325**)

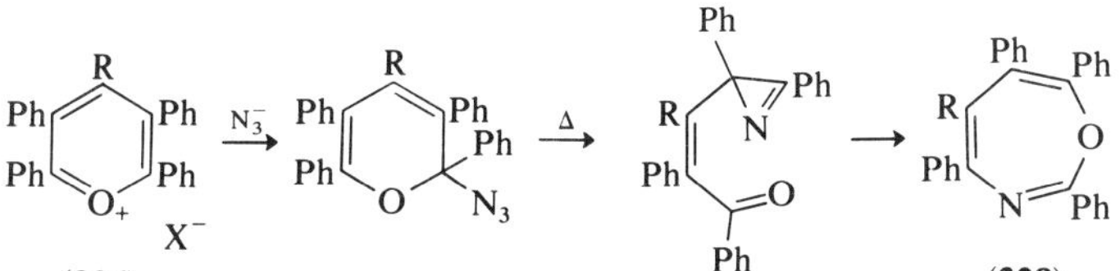

(**326**) (**327**) (**328**)

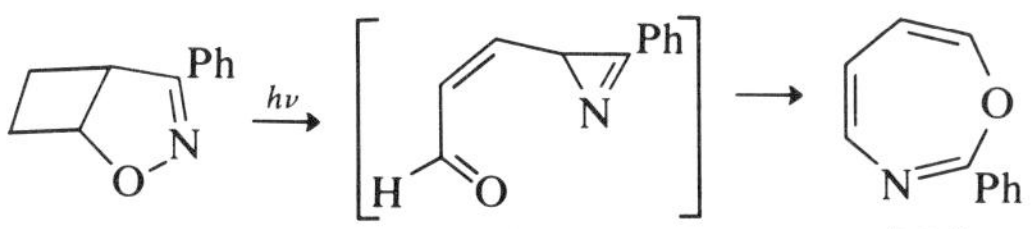

(**329**) (**330**) (**331**)

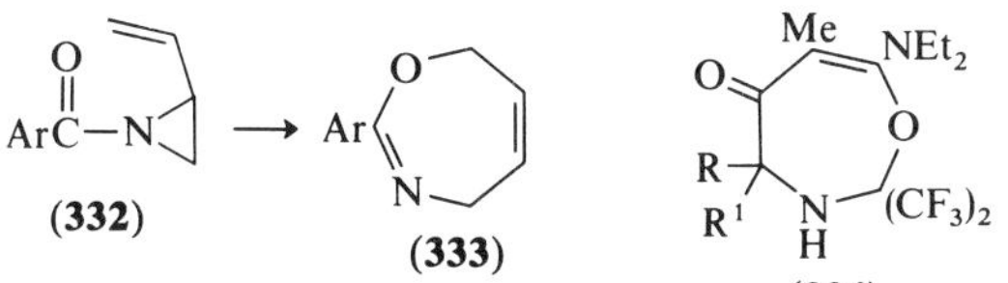

(**332**) (**333**) (**334**)

A range of 2,2-bis(trifluoromethyl)-1,3-oxazepin-5-ones (**334**) has been prepared by the reaction of oxazolidin-5-ones with 1-diethylamino-1-propyne ⟨75TL3223⟩. 1,3-Oxazepan-2-one has been prepared by the Beckmann rearrangement of tetrahydro-2-pyranone oxime.

5.18.5.2.2 Reactivity

1,3-Oxazepines are not very stable compounds and are easily decomposed by heat, light, acids and bases to a variety of products. In some of their reaction paths it seems likely that decomposition takes place *via* hydrolytic opening of the oxazepine ring but in others *via* conversion to the oxazanorcaradiene valence tautomer, *e.g.* (**335**). These conversions and their possible mechanisms have been discussed in depth in a recent review ⟨76H(4)1391⟩ and some examples are given in the reactions shown below. It has recently been confirmed that the photolysis of (**337**) in aprotic solvents gives the labile 3*H*-indole intermediate (**336**) as the primary photoproduct ⟨80CPB1157⟩, but in methanol, for the example shown, the addition reaction which ultimately gives (**338**) is faster ⟨77CC899⟩.

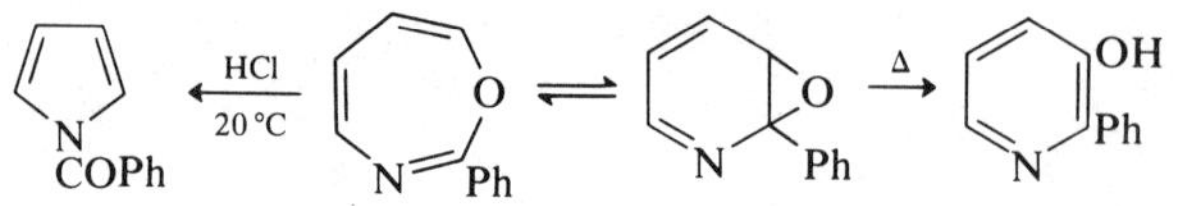

(**335**)

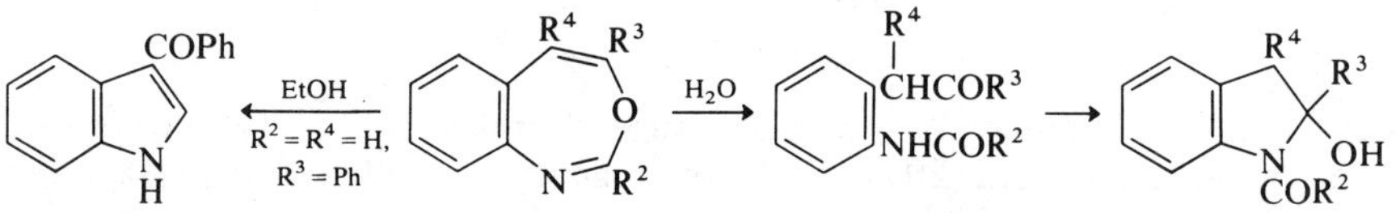

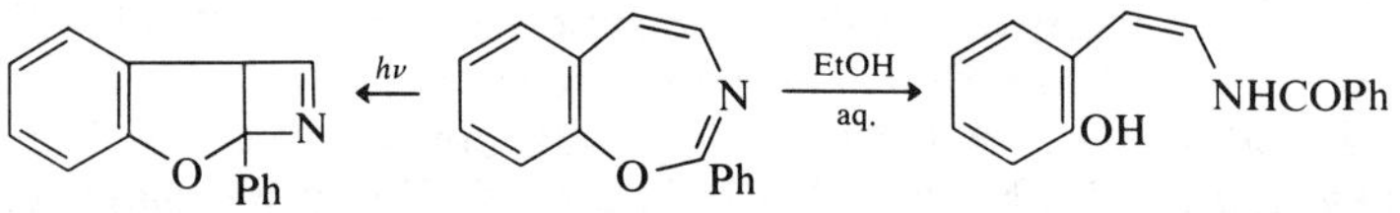

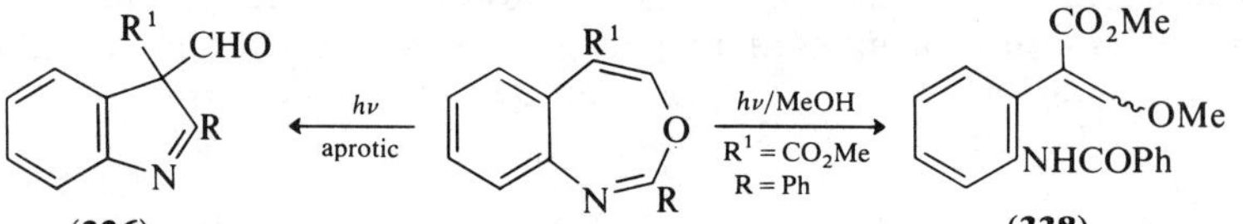

(**336**) (**337**) (**338**)

A systematic study has been made of the reactions of the 4,7-dihydro-1,3-oxazepine (**333**; Ar = 4-$NO_2C_6H_4$) with electrophiles, nucleophiles and basic reagents ⟨71JOC3078⟩. Treatment with electrophilic reagents caused ring contraction to oxazolines, *e.g.* (**339**), and similarly reaction with bromine gave the 5-(1,2-dibromoethyl) analogue of (**339**). Reactions with some nucleophiles also induced rearrangement to (**339**) but more basic reagents gave butadiene derivatives.

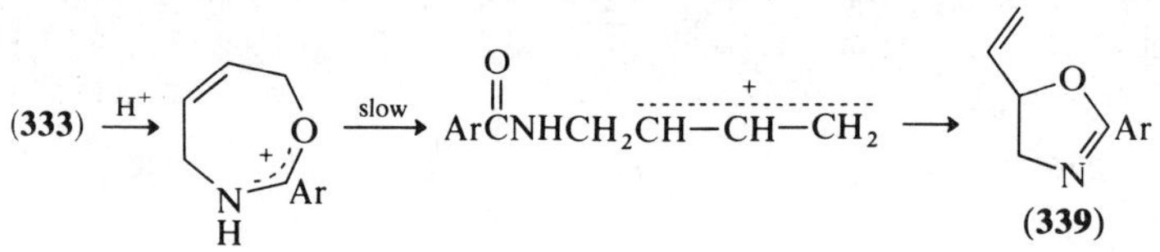

(**339**)

5.18.5.3 1,4-Oxazepines

5.18.5.3.1 Synthesis

(*i*) *Type* ac (*N—C—C—C—O + C—C*)

Many of the most useful syntheses of 1,4-oxazepines are of this type. Either the N—C or the O—C bond may be formed first depending on the nature of the substituents on N and O and in most cases the intermediate of type C—C—N—C—C—C—O or N—C—C—C—O—C—C can be isolated. In this treatment the examples are presented in that order.

(*a*) Via (*C—C—N—C—C—C—O*) *intermediates* (*type* a). The final ring closure can be achieved either by cyclodehydration or by displacement of halogen. The monocyclic systems (**341**) and (**343**) have been obtained by the dehydrative cyclizations of (**340**) and (**342**) respectively.

The use of α-halogenoacyl halides as the C—C fragment leads to 1,4-oxazepin-3-ones, *e.g.* (**345**) from (**344**) ⟨77USP4010166⟩. Similarly *o*-aminobenzyl alcohols give 4,1-benzoxazepin-2-ones (**346**) ⟨65FES323⟩, *o*-hydroxybenzylamines give 1,4-benzoxazepin-3-ones (**347**) ⟨66JHC237⟩, and anthranilic acids give 4,1-benzoxazepine-2,5-diones (**348**). Compounds of type (**346**) are also obtained by the reaction of *o*-aminobenzyl alcohols with chloroacetic esters ⟨71JOC305⟩. The Mannich reaction of phenols, formaldehyde and α-amino acids gives (**349**), which can be cyclized by treatment with thionyl chloride to give the 1,4-benzoxazepin-2-one system (**350**).

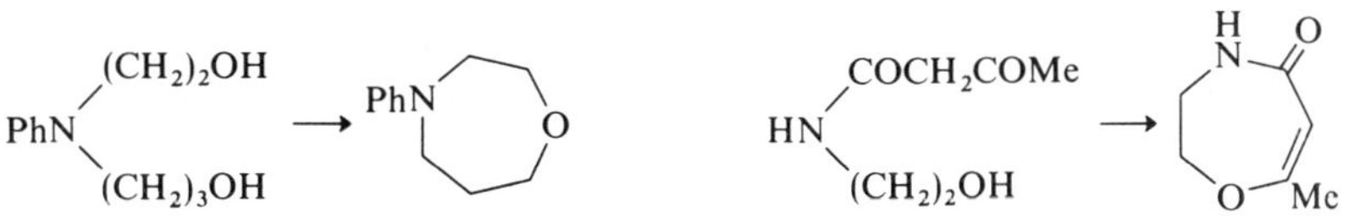

PhN[$(CH_2)_2OH$][$(CH_2)_3OH$] (**340**) → (**341**) HN[$COCH_2COMe$][$(CH_2)_2OH$] (**342**) → (**343**)

HN[CH_2Ph][$(CH_2)_2C(OH)Ph_2$] (**344**) —$ClCH_2COCl$→ $PhCH_2N$[$COCH_2Cl$][$(CH_2)_2C(OH)Ph_2$] → (**345**)

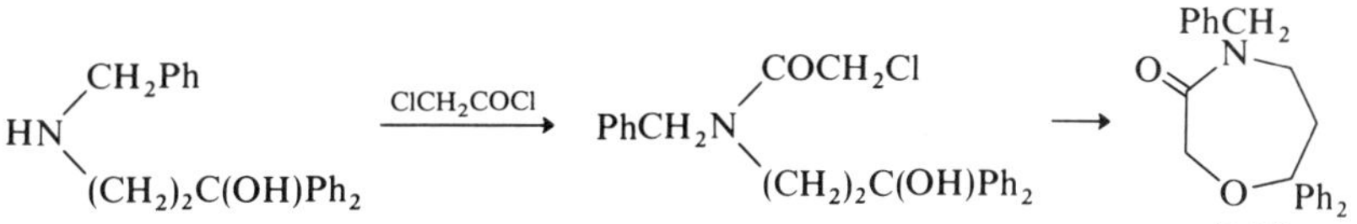

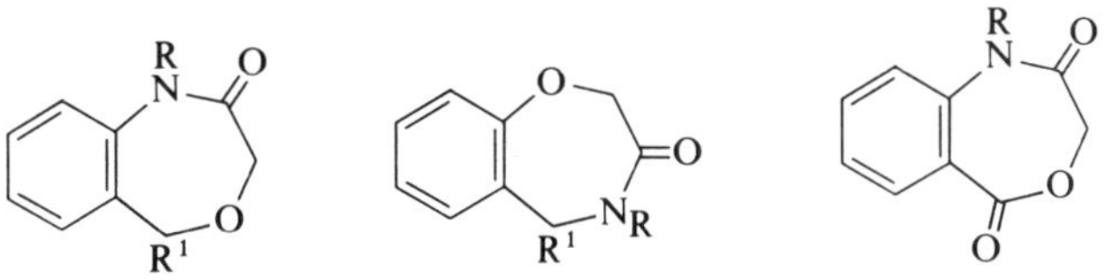

(**346**) (**347**) (**348**)

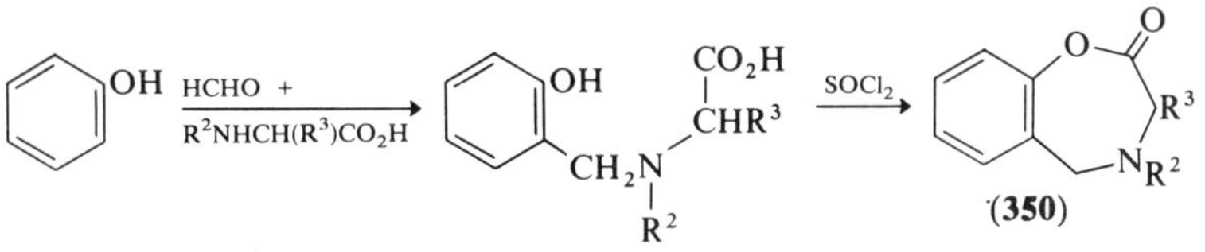

PhOH —HCHO + $R^2NHCH(R^3)CO_2H$→ (**349**) —$SOCl_2$→ (**350**)

(*b*) Via (*N—C—C—C—O—C—C*) *intermediates* (*type* c). Salicylamide (**351**; R = H) can be converted to 1,4-benzoxazepine derivatives, *e.g.* its reaction with 2-chloropheny-lacetic acid gives (**351**; R = $CH(Ph)CO_2H$) which on treatment with acetyl chloride cyclizes to the 3,5-dione (**352**). Similarly reaction of (**351**; R = H) with phenacyl bromide gives (**351**; R = CH_2COPh) which undergoes acid-catalyzed cyclization to give (**353**).

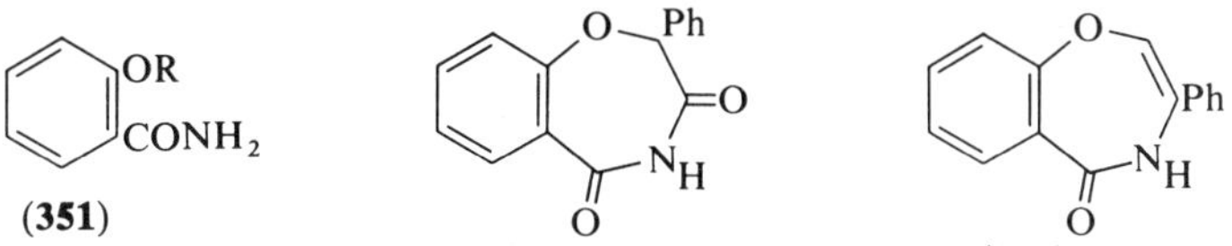

(**351**) (**352**) (**353**)

Dibenz[*b,f*][1,4]oxazepines, which have intense lachrymatory and skin irritant properties, can be prepared in high yield by the Bischler–Napieralsky-type cyclization of *N*-(2-phenoxyphenyl)formamides ⟨76JCS(P1)1279⟩.

(*ii*) *Type* cd (*C—C—O—C—C—C+N*)

These cyclizations can be carried out by the reaction of amines or their derivatives with α,ω-dihalogeno compounds or with halogenocarbonyl compounds. The reaction of 2-chloroethyl 3-chloropropyl ether with acylhydrazines gives (**354**) and the reaction of methylamine with *o*-chloromethylphenyl 2-chloroethyl ether gives (**355**). *o*-(2-Chloroethoxy)acetophenone reacts with amines to give (**356**) ⟨63HCA1696⟩ and the analogous benzophenones react with formamide and ammonium formate to give analogues of (**355**) with a 5-phenyl substituent ⟨69JPS1460⟩.

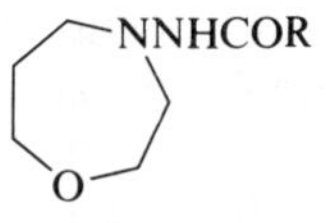

(354) 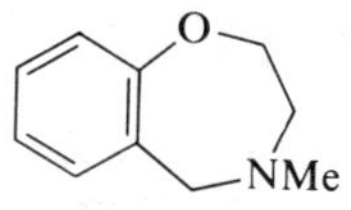 (355) 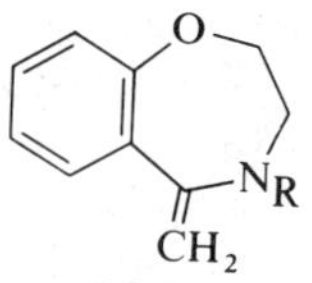 (356)

(iii) *Type* cg (*O—C—C+N—C—C—C*)

The reactions of anthranilic acids and esters ⟨65CB983, 75BSF(2)283⟩ with halohydrins give the 4,1-benzoxazepin-5-one system (**357**). *o*-Methylaminobenzamide reacts with ethylene oxide to give (**358**) which can be cleaved to (**359**) by treatment with aqueous ammonia ⟨66JOC4268⟩.

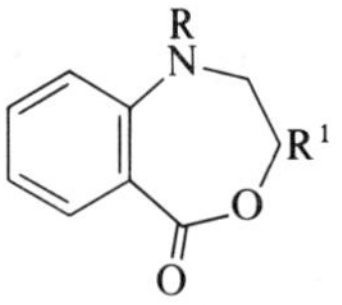

(357) 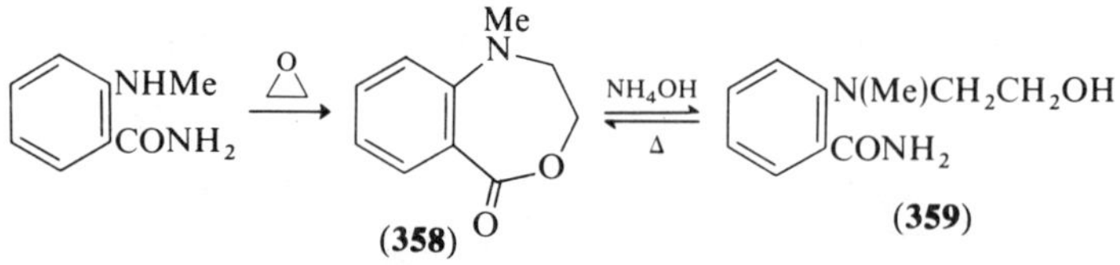 (358) (359)

(iv) *Type* dg (*O—C—C—N+C—C—C*)

Reactions of this type have been widely used for the synthesis of 1,5-benzoxazepines by the reactions of *o*-aminophenols and their derivatives with a variety of functionalized three-carbon chains. Thus reaction with 3-bromo-1-chloropropane gives (**360**) and reaction with 3-chloropropionyl chloride gives the analogous 4-oxo derivative. Similarly α,β-unsaturated ketones give (**361**), β-ketoesters give (**362**), 1,3-oxazolid-5-ones give (**363**), and the reaction of the sodium salt of *N*-methanesulfonyl-*o*-aminophenol with epichlorohydrin gives (**364**). 1,4-Oxazepines fused to heterocyclic rings have been prepared by the reaction of ethanolamine for example with 4-chloro-3-ethoxycarbonylpyridine ⟨71CPB2354⟩ to give (**365**).

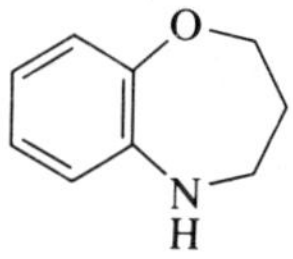

(360) 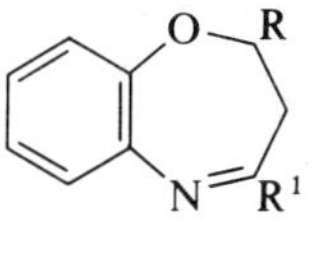 (361) 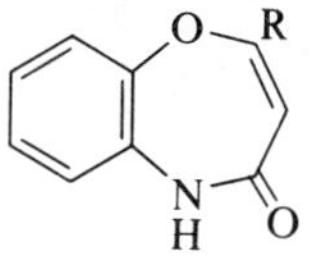 (362) 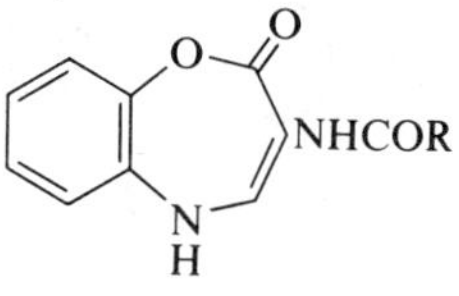 (363)

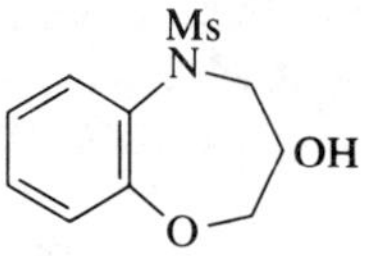

(364) Ms = mesyl 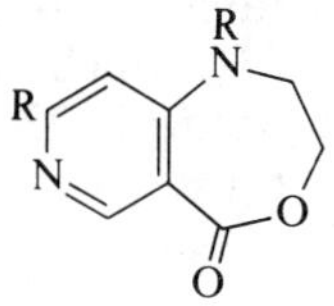 (365) 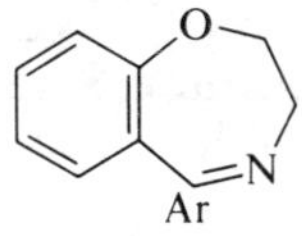 (366)

(v) *Type* e (*C—N—C—C—O—C—C*)

5-Aryl-2,3-dihydro-1,4-benzoxazepines (**366**) have been prepared by a Bischler–Napieralsky-type cyclization ⟨76BSB787⟩.

(vi) *Type* g (*O—C—C—N—C—C—C*)

Almost all examples of this type involve the preparation of 1,4- or 1,5-benzoxazepines, or dibenzoxazepines by the nucleophilic displacement of a halogen from an aromatic ring. For example, (**367**) on treatment with sodium hydride gives (**368**) ⟨77USP4053599⟩.

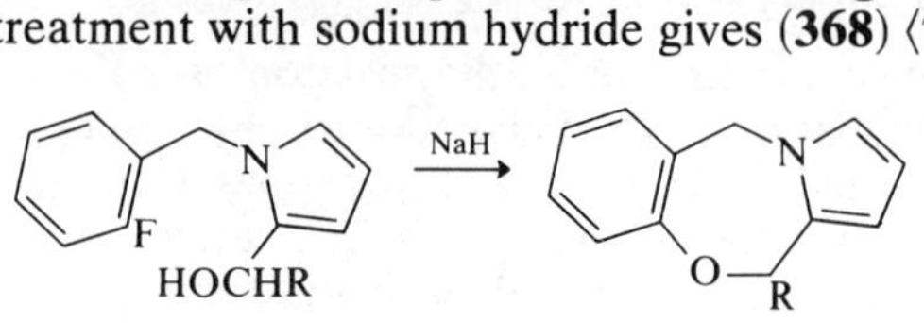

(367) (368)

(*vii*) *Synthesis from other heterocyclic systems*

(*a*) Via *Beckmann or Schmidt reactions.* 1,4-Oxazepan-5-ones can be prepared by the Beckmann rearrangement of tetrahydro-4-pyranone oximes. Both Beckmann and Schmidt reactions on chromanone give mainly (**369**). The Schmidt reaction of chromanones, flavanones and isoflavanones also produces the 1,5-benzoxazepinone (**370**); for example, with isoflavanone the product ratio (**369**) : (**370**) is *ca.* 2 ⟨71G167⟩. The Beckmann rearrangement of xanthone oxime gives dibenz[*b,f*][1,4]oxazepines. Reduction of chromanone or isoflavanone oximes with lithium aluminum hydride leads to 2,3,4,5-tetrahydro-1,5-benzoxazepines ⟨70BCJ1824⟩. The benzofuranone (**371**) on treatment with base rearranged to give (**372**) in high yield ⟨74JHC1087⟩ and in a somewhat similar process *N*-(3-chloropropyl)benzoxazolin-2-one (**373**) was converted to a 1,5-benzoxazepine.

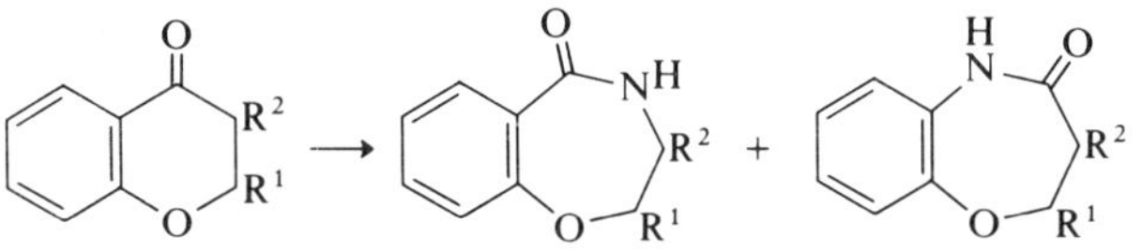

(**369**) (**370**)

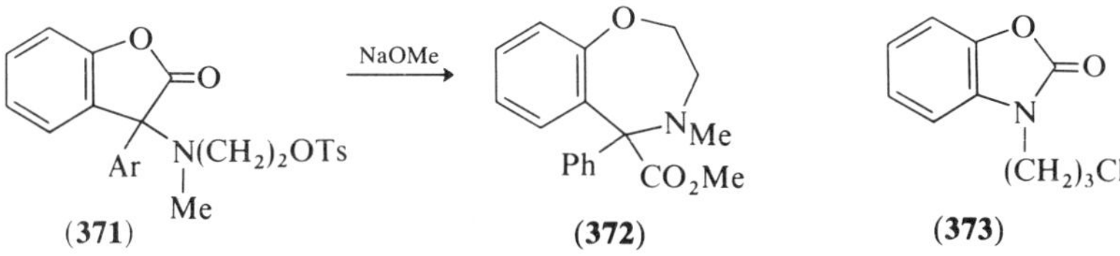

NaOMe

(**371**) (**372**) (**373**)

Enol ethers react with 1,3-oxazolidines in DMSO in the presence of Lewis acids to give 7-alkoxyhexahydro-1,4-oxazepines ⟨76LA1792, 1783⟩.

5.18.5.3.2 Reactivity

There has been little systematic study of the chemistry of 1,4-oxazepines. Vigorous acid hydrolysis cleaves the amide linkage in (**369**; R^1=Ph, R^2=H) and recyclization gives 2-*o*-hydroxyphenyl-5-phenyloxazoline and 1,2,3,4-tetrahydro-1,8-dihydroxy-3-phenylisoquinoline. The 1,4-benzoxazepin-5-one (**353**) can be alkylated at N but on treatment with triethyloxonium fluoroborate it is converted to 5-ethoxy-3-phenyl-1,4-benzoxazepine — one of the very few examples of a fully unsaturated 1,4-oxazepine ring. This product is isomerized to 1-ethoxy-4-hydroxy-3-phenylisoquinoline when boiled in methanol. The 4,1-benzoxazepine-2,5-diones (**348**) are converted to quinazolines by reaction with ammonia. The dihydro-1,4-oxazepin-5-one (**343**) can be acetylated at nitrogen and brominated at the 6-position.

5.18.6 THIAZEPINES

A number of monocyclic and benzo-annelated examples of 1,2- and 1,3-thiazepines have been prepared but there has been little systematic study of these systems. The interesting photochemical interconversions of pyridine *N*-imides into 1,2- and 1,3-diazepines and of pyridine *N*-oxides into 1,2- and 1,3-oxazepines regrettably lack parallels in thiazepine chemistry. There has been more interest in 1,4-thiazepines, as both rearrangement products and possible biogenetic precursors for penicillins and because of the pharmacological value of the benzo- and dibenzo-[1,4]thiazepines as antidepressants and coronary vasodilators. The only review ⟨70ZC361⟩ is excellent but not very recent.

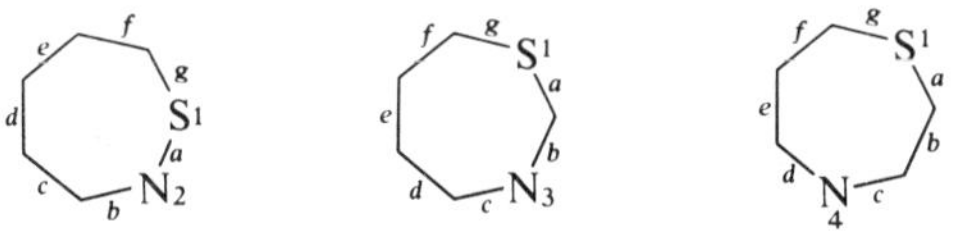

(**374**) (**375**) (**376**)

5.18.6.1 1,2-Thiazepines

5.18.6.1.1 Synthesis

(*i*) *Type* a (*N—C—C—C—C—C—S*)

The cyclic sulfonamide (**377**) can be prepared by heating 5-aminopentanesulfonyl chloride. 4-Chloro-2-benzoylbenzenesulfonyl chloride reacts with acetamidine to give the *o*-benzoyl-*N*-phenylsulfonylamidine which can be cyclized with base to give the fully unsaturated 1,2-benzothiazepine 1,1-dioxide (**378**; $R = R^1 = H$). Alkyl-substituted amidines give the 3-imino derivatives ⟨66USP3377357⟩.

(*ii*) *Type* ag (*S—C—C—C—C—C+N*)

The ketone HC≡CC(O)CH=CHSCN reacts with amines to give the 1,2-thiazepin-5-one system (**379**) ⟨61CB1606⟩.

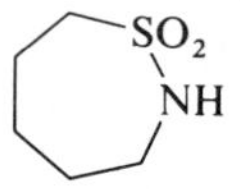

(**377**) 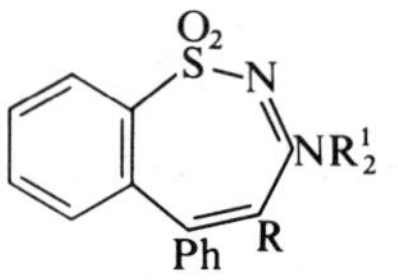 (**378**) 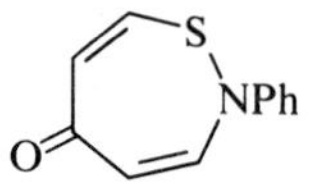 (**379**) 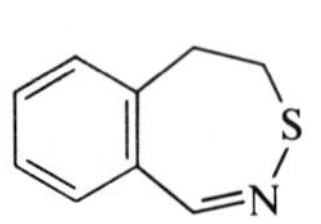 (**380**)

(*iii*) *Type* b (*N—S—C—C—C—C—C*)

Compound (**381**), prepared by direct chlorosulfonation, was converted to (**382**) and thence to the 1,2-benzothiazepin-3-one system (**383**) ⟨76JHC1309⟩.

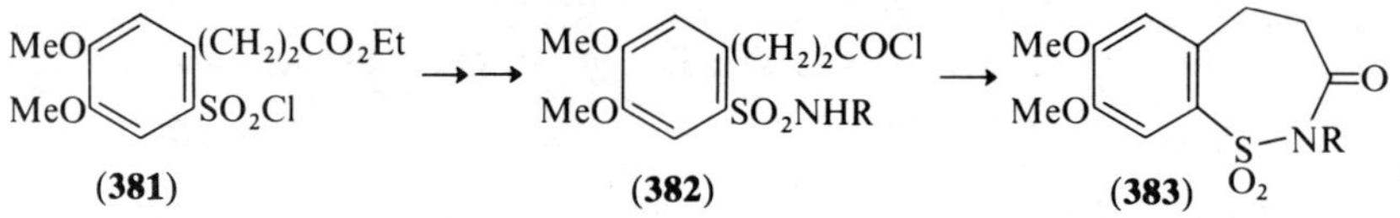

(**381**) (**382**) (**383**)

(*iv*) *Type* bc (*N—S—C—C—C—C+C*)

The 3,2-benzothiazepine 3,3-dioxide (**385**) has been prepared in high yield from (**384**) by intramolecular sulfonyl-amidomethylation ⟨76CC470⟩.

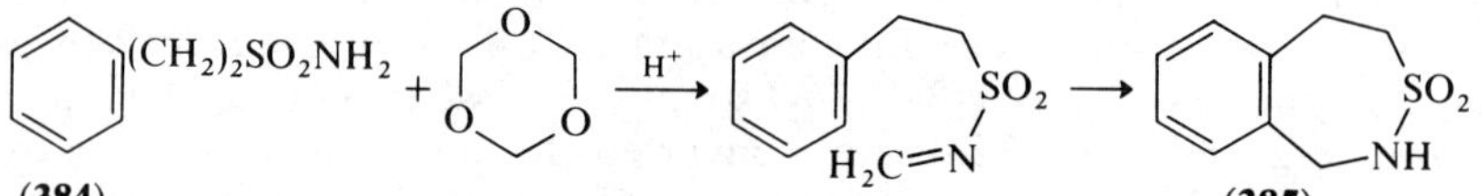

(**384**) (**385**)

(*v*) *Preparation from other heterocyclic systems*

The 4,5-dihydro-3,2-benzothiazepine (**380**) has been prepared by the thermally induced ring expansion of 1-azidoisothiochroman ⟨74LA734⟩. The product reacted with acetyl chloride and then with alcohols to give the 1-alkoxy-2-acyl derivatives, and was oxidized to the 3,3-dioxide.

Benzoisothiazole dioxides undergo [2+2] cycloaddition with 1-diethylaminopropyne followed by ring expansion to give 3-diethylamino-1,2-benzothiazepine 1,1-dioxides (**378**; $R^1 = Me$, $R = Et$) ⟨76H(5)95⟩. *N*-Tosylsulfimines, *e.g.* (**386**), are converted into 1,2-benzothiazepines (**388**) in high yield when treated with triethylamine in benzene under reflux. In the absence of base the intermediate (**387**) can be isolated ⟨80JCS(P1)2830, 81JCS(P1)1037⟩.

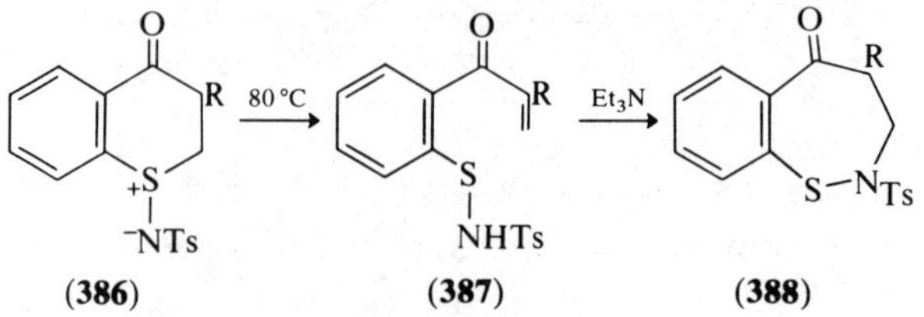

(**386**) (**387**) (**388**)

5.18.6.2 1,3-Thiazepines

5.18.6.2.1 Synthesis

(i) Type cg *(C—C—C—C+S—C—N)*

Several groups have worked on the synthesis of 1,3-thiazepines by the reactions of 1,4-dihalogeno compounds with thioureas or thioamides ⟨77HCA2872, 75CPB1764, 73RTC20⟩. Most recent applications have been to the synthesis of 2,4-benzothiazepines, *e.g.* (**389**) and (**391**), from *o*-xylyl dibromides (**390**). *o*-Chloromethylbenzoyl halides similarly give 2,4-benzothiazepin-5-ones and 4-bromobutyryl chloride gives the 1,3-benzothiazepin-4-one system. In some cases these reactions have been carried out as single stage processes but in others a C—C—C—C—S—C—N-type intermediate has been isolated.

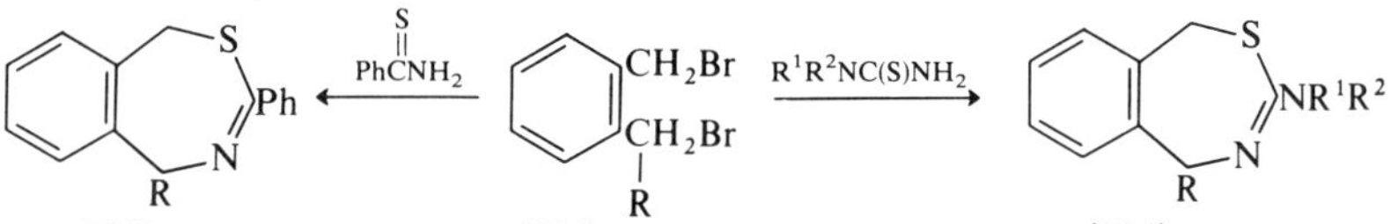

(**389**) (**390**) (**391**)

(ii) Type g *(S—C—N—C—C—C—C)*

ω-Bromo- or -hydroxy-isothiocyanates react with amines to give 2-amino- or 2-imino-1,3-thiazepines depending on the nature of the amine ⟨71OPP27, 71CJC971⟩.

(iii) Preparation from other heterocyclic systems

2-Vinylaziridine (**392**) reacted with phenyl isothiocyanate or 4-chlorothiobenzoylthioglycollate to give (**394**), presumably *via* the intermediate (**393**) (*cf.* the analogous preparation of 1,3-oxazepines, p. 627) ⟨71JOC3076⟩.

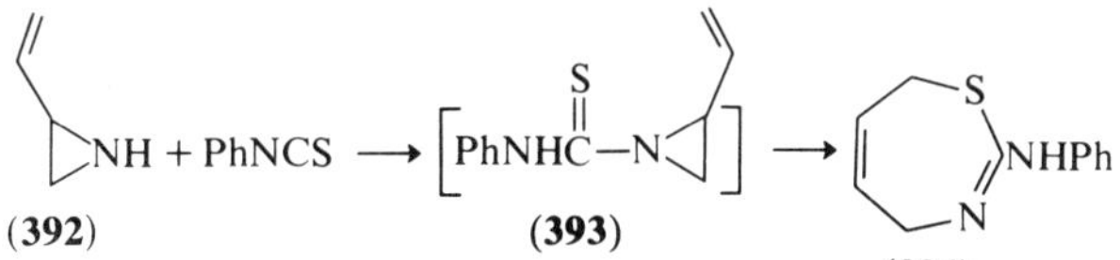

(**392**) (**393**) (**394**)

5.18.6.3 1,4-Thiazepines

5.18.6.3.1 Synthesis

(i) Type ac *(S—C—C—C—N+C—C)*

2-Mercaptobenzamide (**395**) reacts with chloroacetic acid or its ester to give (**396**) which can be cyclized to give (**397**) by treatment with thionyl chloride or acetic anhydride.

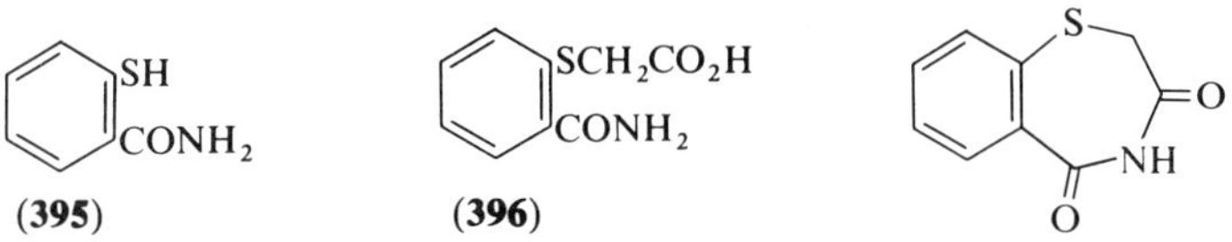

(**395**) (**396**) (**397**)

(ii) Type ad *(S—C—C—C+C—C—N)*

The major application of reactions of this type is in the construction of 1,4-benzothiazepines by the reactions of substrates of type (**398**) with an appropriate source of the C—C—N fragment. Thus the 2,3-dihydro-1,4-benzothiazepin-5-one (**399**) can be prepared by the single stage reaction of (**398**; R = OH) with aziridine. The reaction of the 2-mercaptoaryl ketone (**398**; R = Ph) with 2-bromoethylamine hydrochloride is however a two-stage process giving firstly (**400**), which cyclizes in the presence of pyridine to give (**401**). The reaction of (**398**; R = OMe) with chloroacetonitrile in the presence of alcohols (R^1OH) gives (**402**). Formation of the *d* bond as an amide linkage is of value in the synthesis of dibenzo[*b,f*]thiazepines ⟨74OPP287⟩.

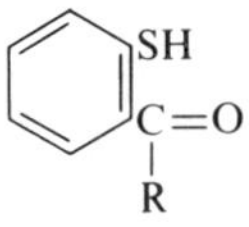

(398) 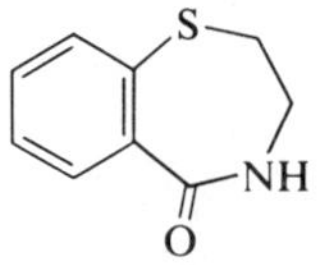 (399) 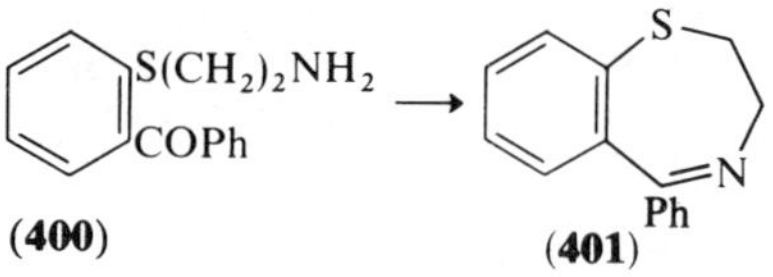 (400) (401) 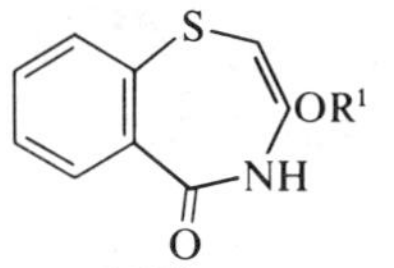 (402)

(iii) Type cg (*S—C—C+N—C—C—C*)

Thioglycollic acid ($HSCH_2CO_2H$) reacts with 3-bromopropylamine to give (**403**) and with 2-aminobenzhydrol under acid conditions to give (**404**). An analogue of (**404**) lacking the 5-phenyl group can be prepared by the reaction of thioglycollic acid with 2-nitrobenzyl bromide, reduction of the nitro group and cyclization.

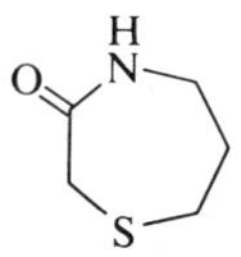

(403) 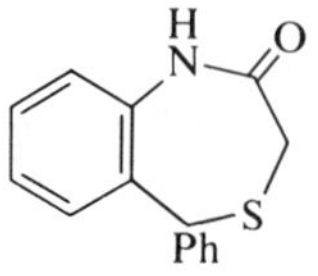 (404)

(iv) Type de (*N—C—C—S—C—C+C*)

The dianion (**405**) reacts with benzophenone to give (**406**) which can be cyclized to give (**407**) by treatment with zinc chloride ⟨75TL1499, see also 79S589⟩.

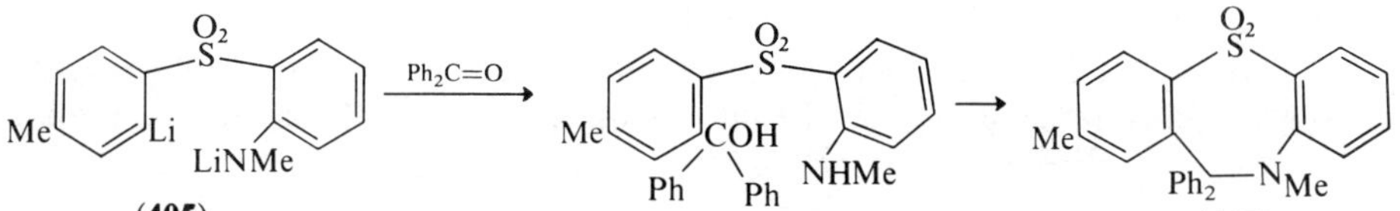

(405) (406) (407)

(v) Type dg (*S—C—C—N+C—C—C*)

Reactions of this type provide major routes to both the monocyclic system and to 1,5-benzothiazepines. In some cases the reactions are single-stage processes but in many cases the intermediate produced by the primary formation of the S—C bond can be isolated. Thus 2-aminoethanethiol reacts with α,β-unsaturated or β-halogeno ketones to give (**408**). Similarly reaction with α,β-unsaturated acids, esters or acid chlorides and with 3-halogenopropionyl halides gives the 5-oxo derivative (**409**). 2-Aminoethanethiol also reacts with activated 2-chlorobenzophenones to give 1,4-benzothiazepines.

In a similar fashion 2-aminothiophenol can be reacted with 1,3-bis carbon electrophiles to give various types of 1,5-benzothiazepine. Thus the benzo analogues of (**408**) and (**409**), *e.g.* (**410**), are prepared by parallel routes, reaction with 1,3-diphenylpropynone gives (**411**), reaction with β-ketoesters gives products of type (**412**), reaction with diketene gives (**413**), and the reaction with methyl 3-arylglycidates gave (**414**) which could not be dehydrated.

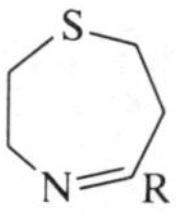

(408) 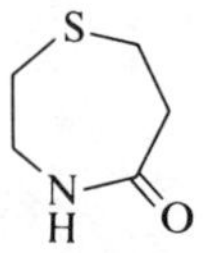 (409) 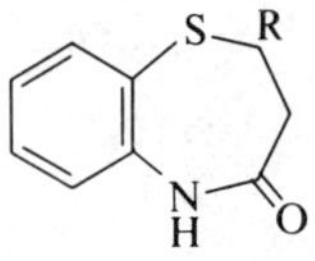 (410) 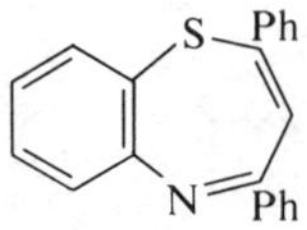 (411)

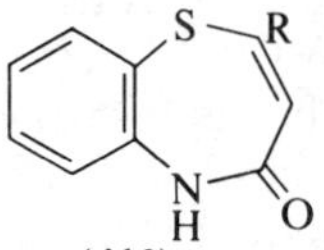

(412) 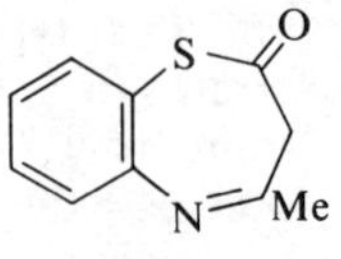 (413) 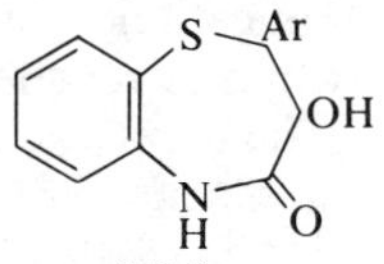 (414)

The acetamido alkenethiolate (**416**), generated by the reductive cleavage of 2,3-dihydro-1,4-thiazin-2-one (**415**) in liquid ammonia (*cf.* p. 625), reacted with 1-chloro-3-bromopropane in a two-stage process to give (**417**).

Dibenzo[*b*,*e*][1,4]thiazepines have been prepared by the thermolysis of 2-azido-2′,6′-dimethyldiphenyl sulfides; the reaction involves formation of the *d* and *g* bonds in a rearrangement reaction *via* a spirodiene intermediate ⟨70CC233⟩.

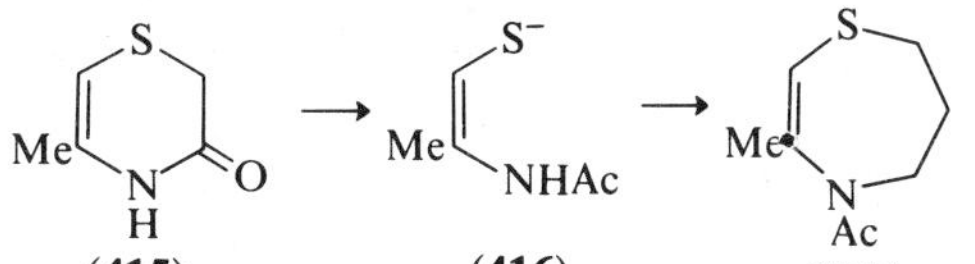

(415) **(416)** **(417)**

(vi) Type e *(C—N—C—C—S—C—C)*

1,4-Benzothiazepines of type (**401**) can be prepared by the Bischler–Napieralsky cyclization of *N*-(arylthioethyl)benzamides and reactions of the same type have been used in the synthesis of dibenzo[*b*,*f*][1,4]thiazepines ⟨70JHC409⟩.

(vii) Synthesis from other heterocyclic systems

As with 1,4-oxazepines the Schmidt reaction of cyclic ketones and the Beckmann rearrangement of their oximes can be applied to the synthesis of monocyclic 1,4-thiazepines, 1,4- and 1,5-benzothiazepines and their 1-oxides and 1,1-dioxides ⟨75CJC276⟩.

Thioxanthen-9-one, 10,10-dioxides, on treatment with sodamide in liquid ammonia, undergo a novel ring expansion to give dibenzo[*b*,*f*][1,4]thiazepin-11-one 5,5-dioxides ⟨75JHC1211⟩. The thiazolium salt (**418**) undergoes base-induced ring expansion to give (**419**) ⟨80TL2429⟩, a direct parallel of the conversion of analogous oxazolium salts to 1,4-benzoxazepines.

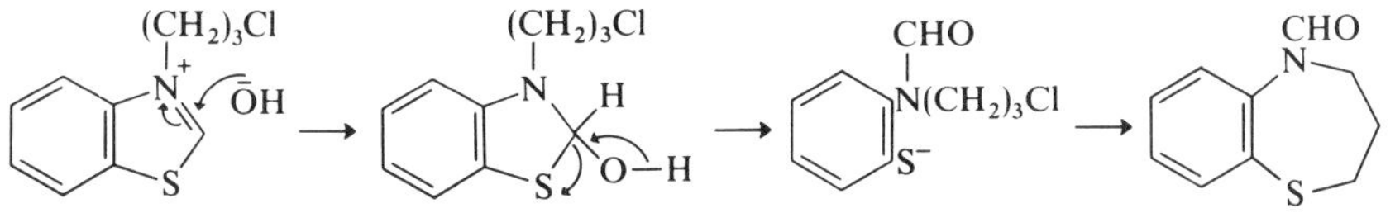

(418) **(419)**

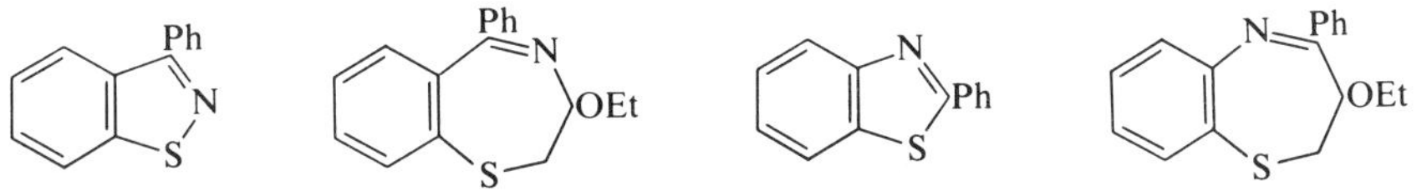

(420) **(421)** **(422)** **(423)**

The 1,4- and 1,5-benzothiazepines (**421**) and (**423**) have been prepared by the photochemical reactions of the benzoisothiazole (**420**) and the benzothiazole (**422**) respectively with ethyl vinyl ether ⟨81TL529, 2081⟩. The mechanisms of these reactions are not fully established but it is interesting to note that the reactions of (**422**) with *cis*- and *trans*-but-2-ene are stereospecific.

It is of interest to note that 1,4-thiazepinones can be produced by the opening of the β-lactam ring in various penicillin derivatives and analogues ⟨73ACS677, 73JHC791, 80CC798, 79H(13)227⟩ but these routes are of little preparative value.

5.18.6.3.2 Reactivity

Many functional groups incorporated into 1,4-thiazepine ring systems show unexceptional reactivity. Lactam linkages, *e.g.* in (**397**), (**399**), (**403**), (**409**) and (**410**; R = H), can be opened by acid hydrolysis, but alkaline hydrolysis of (**410**; R = alkyl) induces cleavage of the 1,2-bond to give an anilide. Alkylation at nitrogen can be carried out under normal conditions; for example, (**399**) can be alkylated in the presence of sodium hydride or sodamide. *N*-Aminoalkylation of compounds of type (**410**) gives products with valuable pharmacological activity; for example, the *N*-dimethylaminoethyl derivative of (**410**; R = Ph) is the antidepressant thiazesim and the analogous 3-acetoxy compound where R = 4-methoxyphenyl is the coronary vasodilator dilitazem. Chlorination of thioethers generally takes place at the α-position and such reactions, *e.g.* of (**424**), followed by subsequent dehydrochlorination provide a good method for introducing unsaturation. Similarly the reactions of (**401**) and of (**410**; R = H) with sulfuryl chloride give the expected 2-chloro derivatives, but reaction of (**410**; R = Ph) with NCS leads to chlorination at the 3-position. Halogenation of these systems has in some cases been undertaken to produce substrates hopefully capable of transannular cyclization to give bicyclo[3.2.0] systems closely related to penicillins. Thus chlorination of (**424**) with sulfuryl chloride at low temperature, followed by warming, gave (**425**) which was converted to (**426**) by reaction with triethyloxonium fluoroborate. However the UV irradiation of (**426**) did not give a $[_{\pi}2_s + _{\pi}2_s]$ electrocyclization but instead gave a triplet-derived dimer in aprotic solvents or the singlet-derived rearrangement product (**427**) in methanol ⟨71CC698⟩. The irradiation of (**429**) also failed to give (**428**) and proceeded by a 1,3-hydrogen shift to give (**430**) ⟨71CC988⟩. The reaction of (**431**) with pyridine in methanol also gave no penam-type product — only the dimethoxy derivative in high yield, and the reaction of (**432**) with silver oxide in methanol resulted in ring opening and desulfurisation to give an enamide ⟨71CJC3866⟩.

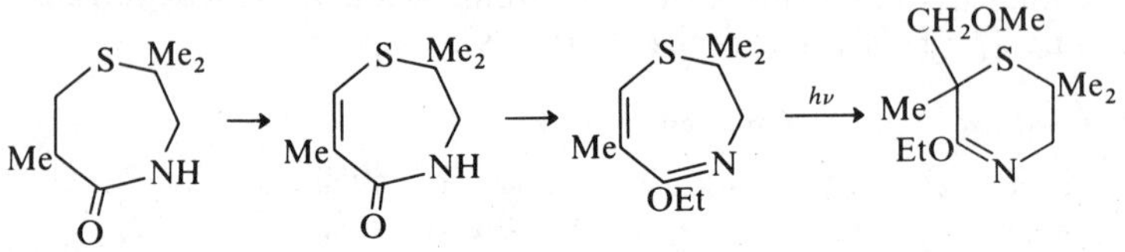

(**424**) (**425**) (**426**) (**427**)

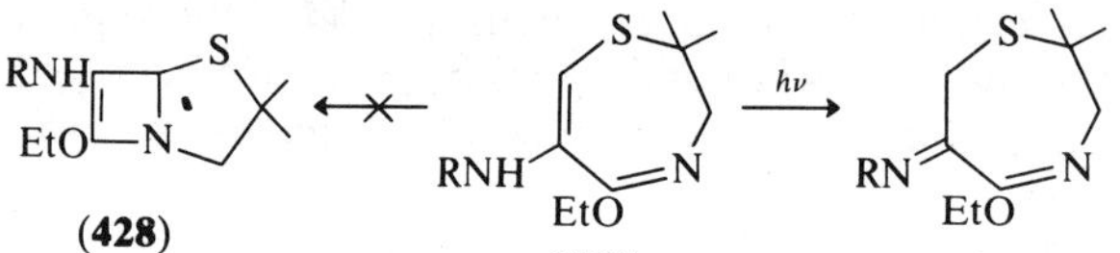

(**428**) (**429**) (**430**)

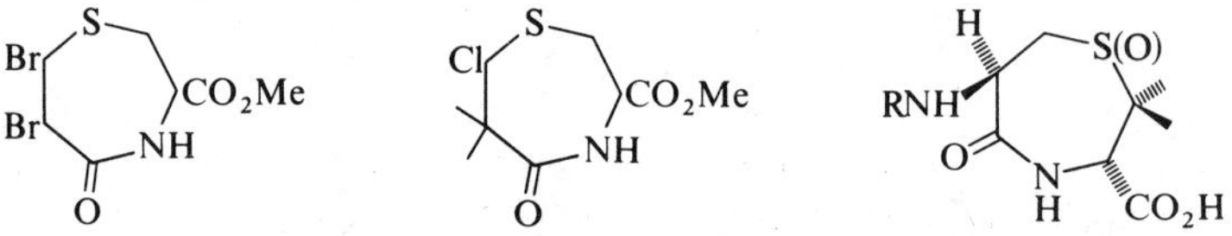

(**431**) (**432**) (**433**)

There has been some speculation on the possibility of the occurrence of 1,4-thiazepine to penicillin conversions in biological systems but it has recently been conclusively shown that neither the thiazepinone (**433**; R = 5-amino-5-carboxypentanoyl) or its 1-oxide are intermediates in the biosynthesis of penicillin N ⟨81CC1146⟩. Another group has also synthesized (**433**) ⟨81CJC406⟩.

5.18.7 TRIAZEPINES

There are four triazepine ring systems (**434**)–(**437**). The 1,2,4-system (**435**) has received the most attention but much less is known about the others. In this section the preparative methods are classified in the usual way and important aspects of the reactivity of the products are discussed in the same section.

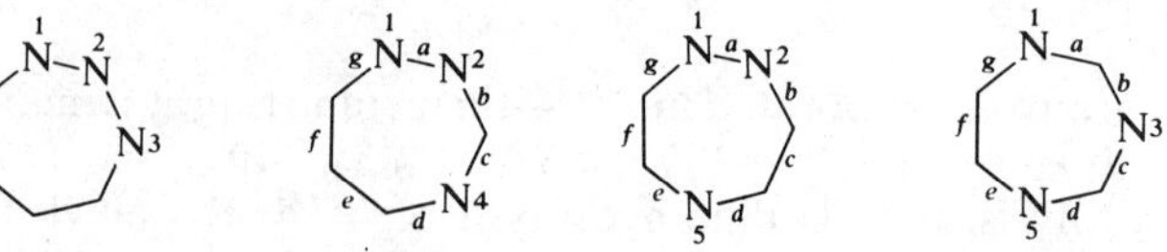

(**434**) (**435**) (**436**) (**437**)

5.18.7.1 1,2,3-Triazepines

The only known compound in this class (**439**) was prepared by treating tetrazotized 2,2′-diaminobiphenyl (**438**) with aqueous ammonia ⟨74JCS(P1)1248⟩. It is a reactive cyclic triazene with a high degree of masked diazonium character. On heating in benzene it rearranged to the stable *N*-imide (**440**), but at higher temperatures gave carbazole (**441**).

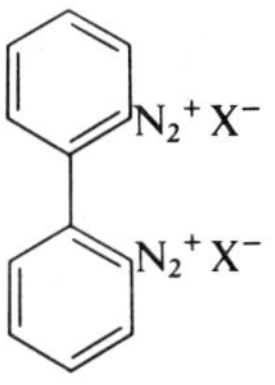

(438) 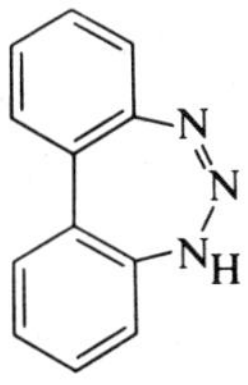 (439) 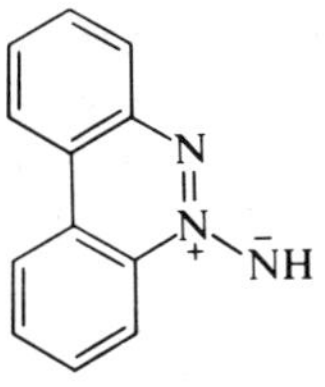 (440) 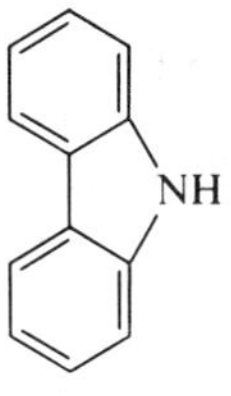 (441)

5.18.7.2 1,2,4-Triazepines

5.18.7.2.1 Synthesis and reactivity

(*i*) *Type* b (*N—N—C—C—C—N—C*)

Cyclization of the diazonium salt (**442**) followed by treatment with base and then with acid gave (**443**; $R^1 = CO_2H$) which was decarboxylated to give (**443**; $R^1 = H$) ⟨77MI51803⟩. These products show typical phenylhydrazone unreactivity to alkylation at N-1 but treatment of (**443**; $R^1 = H$, R = Me) with methyl fluorosulfonate gave a relatively stable quaternary salt *via* methylation at N-2; this isomerized in acid to give 1-imino-3-methylquinazolin-4-one (**444**) ⟨79JHC1411⟩.

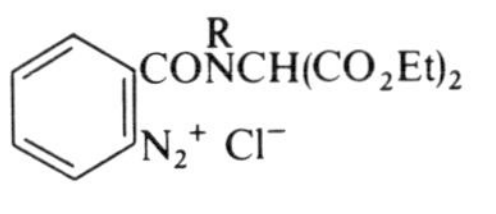

(442) 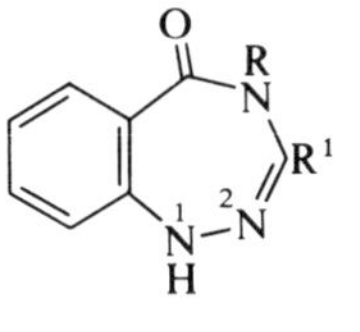 (443) 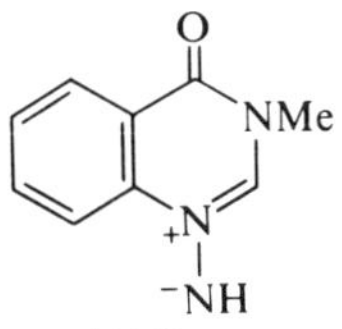 (444)

(*ii*) *Type* bc (*N—C—C—C—N—N+C*)

Reactions of this type provide the major route to the 1,3,4-benzotriazepine system. Thus the hydrazones, phenylhydrazones or semicarbazones of 2-aminoaryl ketones (**445**) react for example with paraformaldehyde to give the 2,3-dihydro-1*H*-1,3,4-benzotriazepine system (**446**) ⟨70BCJ135⟩ or with phosgene or ethyl chloroformate to give the 2-oxo analogue (**447**) ⟨74JPS838⟩. Compounds of type (**447**) can also be prepared by the reaction of 2-aminoaryl ketones with ethoxycarbonylhydrazine. 2-Aminobenzoyl hydrazides, *e.g.* (**449**; $R^3 = Me$), have similarly been used in the preparation of 1,3,4-benzotriazepin-5-ones (**448**) and (**450**) by reaction with ortho esters ⟨76JOC2732, 2736, 76JHC967⟩. However it should be noted that unsubstituted 2-aminobenzoylhydrazide (**449**; $R = R^3 = H$) gives quinazoline derivatives in analogous reactions rather than benzotriazepines ⟨75JOC1909, 76IJC(B)701⟩.

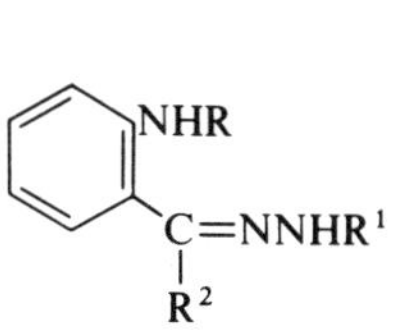

(445) 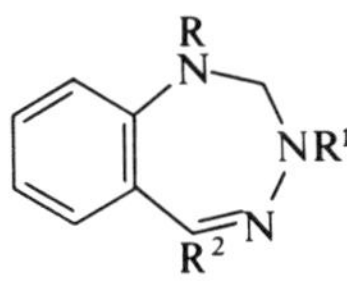 (446) 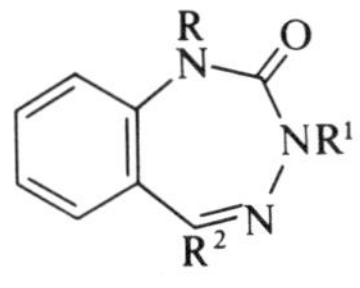 (447)

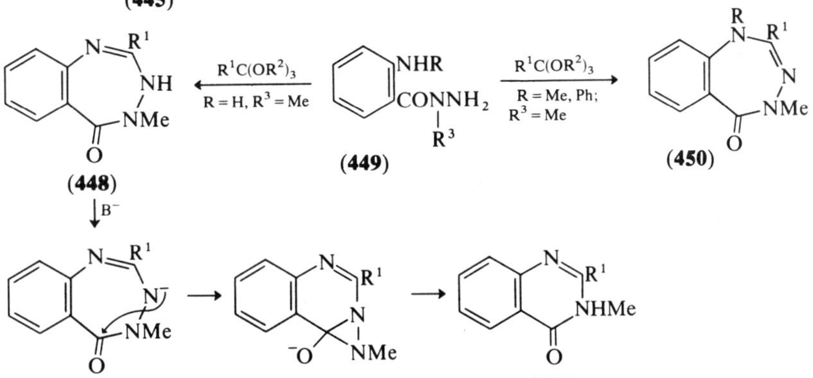

$R^1C(OR^2)_3$, $R = H, R^3 = Me$; $R^1C(OR^2)_3$, $R = Me, Ph; R^3 = Me$

(448) (449) (450)

(451) (452) (453)

o-Aminobenzoylhydrazide (**449**; $R = R^3 = H$) does however react with isothiocyanates in the presence of DCC to give 2-amino-3,4-dihydro-5*H*-1,3,4-benzotriazepin-5-ones ⟨79JHC1435⟩. The reaction of (**449**) with cyanogen bromide gives (**450**; $R^1 = NH_2$) ⟨77JPS605⟩.

Compound (**448**) on treatment with base undergoes ring contraction to give (**453**) *via* (**452**).

(iii) Type bg *(C—N—C—C—C+N—N)*

A major group of reactions in this class involve the reaction of β-functionalized isocyanates or isothiocyanates with hydrazines. Thus for example (**455**; X = O) reacts with methylhydrazine to give (**456**) ⟨77S756⟩, and (**455**; X = S) reacts with both alkylhydrazines and hydrazine itself to give (**454**; R = alkyl or H) ⟨76M1251⟩. 2-Benzoylphenyl isothiocyanate undergoes a similar reaction with hydrazines to give (**458**) ⟨79MI51801⟩. The reaction of 2-alkoxycarbonylphenyl isocyanate (**457**; X = OR) with methylhydrazine gave (**459**) in a two-stage reaction *via* a semicarbazide ester, but in a similar reaction using the acyl halide (**457**; X = Cl) the regioselectivity was reversed giving (**460**) ⟨75JOC1909⟩.

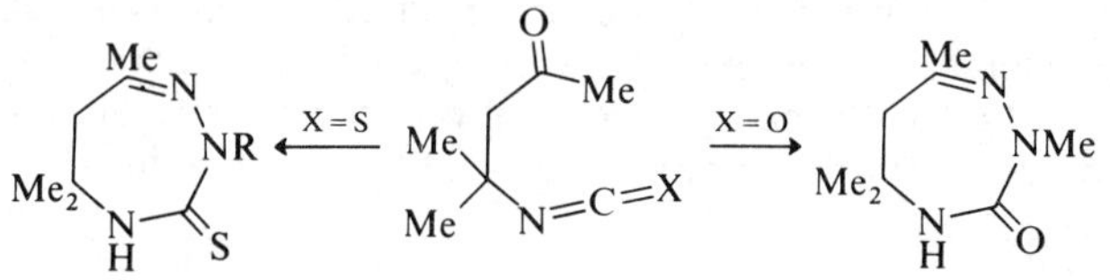

(**454**) (**455**) (**456**)

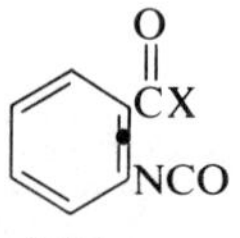

(**457**) 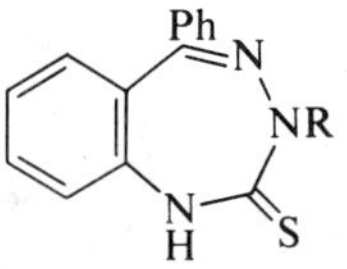 (**458**) 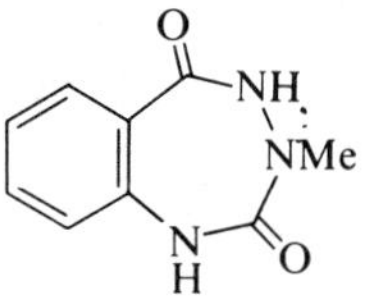 (**459**) 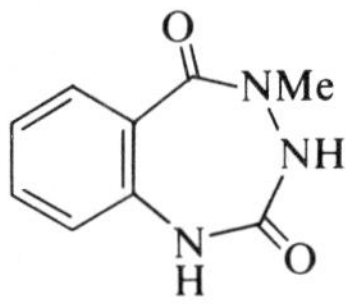 (**460**)

Methods have been developed for selective alkylation at nitrogen in (**459**) and (**460**) ⟨77JOC2551⟩. Compound (**460**) also underwent a base-induced ring contraction analogous to the (**448**) to (**453**) conversion discussed earlier.

(iv) Type ce *(C—C—N—N—C+C—N)*

1*H*-1,2,4-Benzotriazepines can be prepared in good yield by the reaction of some hydrazidoyl chlorides of type (**461**) with cyano compounds. The reaction is thought to proceed *via* the nitrilium salt (**462**). The products (**463**) are stable to alkali but are hydrolyzed in acid to give 1-phenyl-3R-indazoles ⟨74T195⟩.

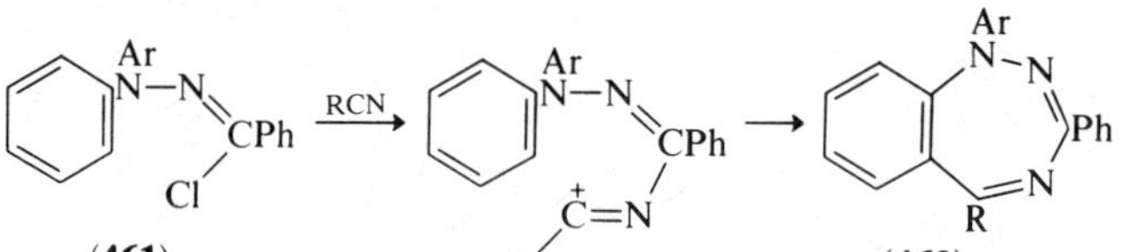

(**461**) (**462**) (**463**)

(v) Type dg *(N—N—C—N+C—C—C)*

Reactions of this type are of value in the synthesis of 1,2,4-triazepines fused to various heterocyclic ring systems. Reactants of type (**464**) combine with β-keto esters to give products such as (**465**) ⟨75JHC661, 1095, 79LA639⟩. Similarly compounds of type (**466**) react with β-halogenoacyl halides, β-diketones and β-keto esters to give a variety of triazepines and triazepinones, *e.g.* (**467**) ⟨77AJC2053⟩.

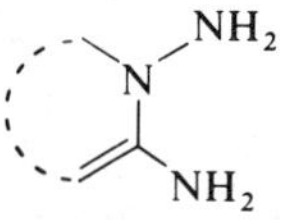

(**464**) 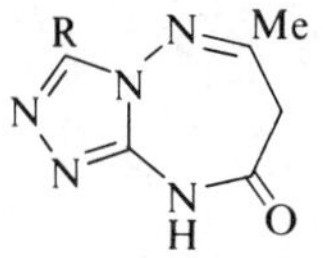 (**465**) 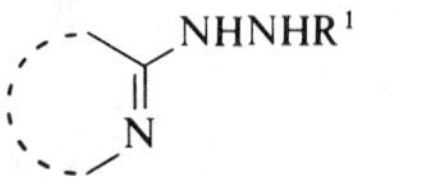 (**466**) 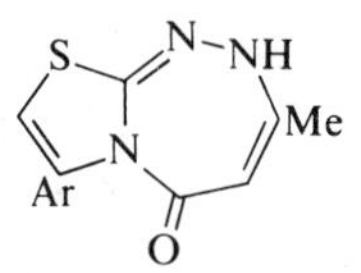 (**467**)

The N—C—N—N fragment can also be supplied by thiosemicarbazides ($NH_2N(R)CSNH_2$) which react with malonic acid derivatives or β-keto esters to give 1,2,4-triazepine-3-thiones ⟨78JHC71, 80RTC301⟩.

(vi) Preparation from other heterocyclic systems

A route to the fully unsaturated 2*H*-1,2,4-triazepine system (**471**) is provided by the cycloaddition of 1-azirines (**468**) to 1,2,4,5-tetrazines (**469**). The initial product (**470**) is not isolable but rearranges by a 1,5-hydrogen shift to give (**471**) ⟨74CC45, 74TL2303⟩. These 2*H*-1,2,4-triazepines are rather unstable and undergo thermal decomposition at temperatures as low as 80 °C in some cases. A variety of products can be obtained depending on the nature of the substituents. Thus for example (**471**; $R^1 = R^2 = Ph$, $R^3 = CO_2Me$) decomposes cleanly in *ca.* 8 h at 80 °C (or rapidly in the presence of a trace of TFA) to give the pyrazole (**472**; $R^1 = Ph$, $R^3 = CO_2Me$) and benzonitrile. Similarly (**471**; $R^1 = Ph$, $R^2 = Me$, R^3 = 2-py) gives (**472**; $R^1 = Ph$, R^3 = 2-py) in high yield. In other cases pyrimidines and the 6*H*-isomer of (**471**) have been obtained. Interestingly, in another study ⟨75JHC183⟩ the triazolylstilbene (**475**) was also isolated as a major product. This is thought to be formed either *via* (**473**) and (**474**) or by a 1,3-carbon to nitrogen migration in (**473**).

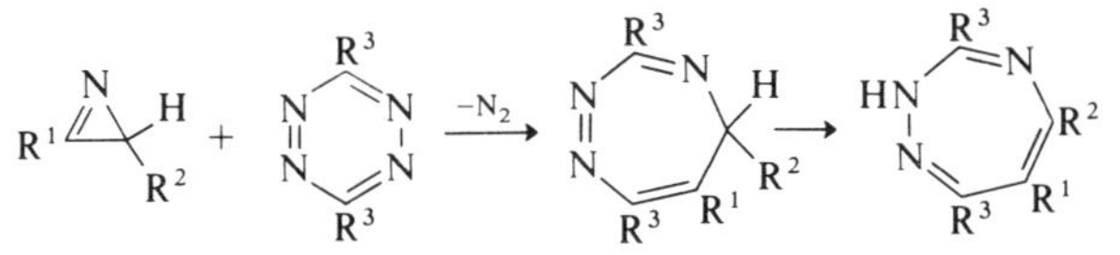

(**468**) (**469**) (**470**) (**471**)

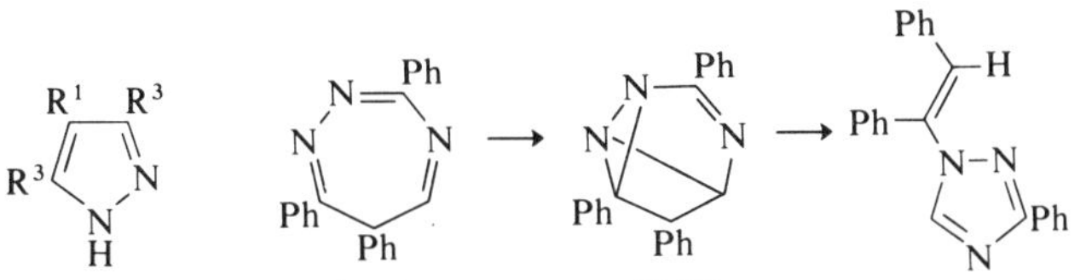

(**472**) (**473**) (**474**) (**475**)

The 4*H*-1,2,4-triazepine system (**477**) can be prepared *via* a photochemical walk-rearrangement of 3,4,7-triaza-2,4-norcaradienes (**476**) ⟨76TL2459⟩. The photochemical and thermal decomposition of these compounds has also been studied. UV irradiation of (**477**; $R = C_6H_{11}$) resulted in the extrusion of nitrogen and quantitative formation of 1-cyclohexyl-2,3-diphenylpyrrole. The thermal decomposition of (**477**; $R = CH_2Ph$) however gave (**481**) as the major product, most likely formed by a 1,5-benzyl migration to give (**479**) followed by electrocyclization to give (**480**) and loss of $PhCH_2N$:. The thermolysis of (**477**; $R = C_6H_{11}$) however took a different course, presumably due to the lower mobility of the cyclohexyl moiety, and gave mainly 1-cyclohexyl-4-phenylimidazole *via* a complex series of interconversions. The presence of traces of water in these decompositions leads to hydrolytic ring opening and recyclization to give (**478**) ⟨76TL4753⟩.

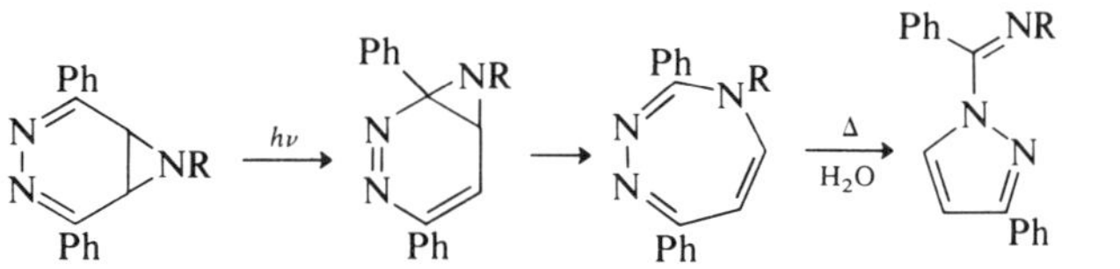

(**476**) (**477**) (**478**)

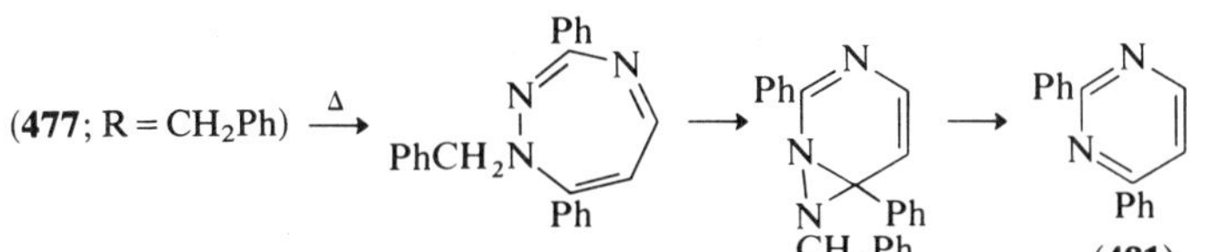

(**477**; $R = CH_2Ph$) $\xrightarrow{\Delta}$ (**479**) (**480**) (**481**)

5,6,7-Trichloro-1,3-diphenyl-1*H*-1,2,4-triazepine has recently been synthesized in high yield by the reaction of tetrachlorocyclopropene with two moles of diphenylnitrilimine ⟨81S322⟩.

5.18.7.3 1,2,5-Triazepines

There has been relatively little study of this system and synthetic routes are few. The fully unsaturated structure (**482**) has not been detected, its valence bond isomer (**483**) being the more stable form (*cf.* 5*H*-1,2-diazepines, p. 595).

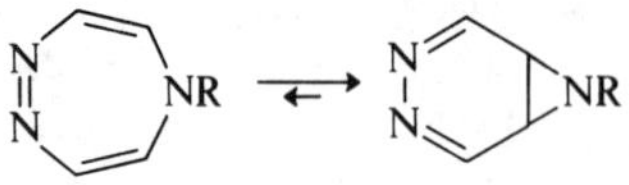

(**482**) (**483**)

5.18.7.3.1 Synthesis and reactivity

(*i*) *Type* b (*C—C—N—C—C—N—N*)

1*H*-4,5-Dihydro-1,2,5-benzotriazepin-4-ones (**485**; $R^1 = CO_2Et$ or CN) can be prepared in good yield by the cyclization of the diazonium salt (**484**). The ester can readily be hydrolyzed and decarboxylated to give (**485**; $R^1 = H$). This compound (R = Me) on treatment with acid undergoes a ring contraction to give (**486**) which is reminiscent of the analogous transformation of the 1,2-diazepine system ⟨68T6395⟩. On treatment with sodium ethoxide in ethanol the ring opens and recyclization occurs to give 3-amino-1-methyl-1,2-dihydroquinoxalin-2-one (**487**) as a major product ⟨74T2765⟩.

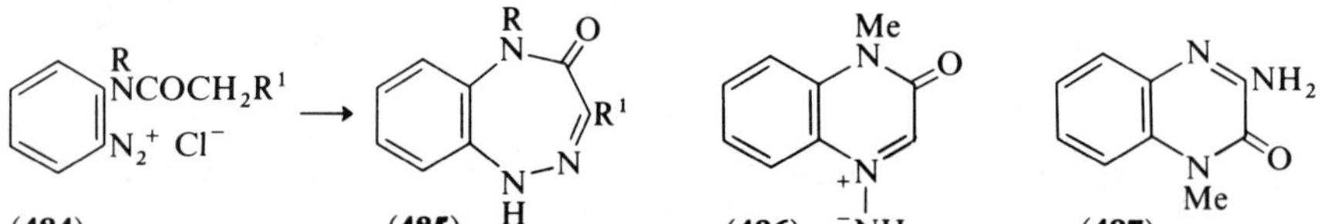

(**484**) (**485**) (**486**) (**487**)

(*ii*) *Type* bd (*N—N—C—C—N+C—C*)

The reaction of *o*-aminophenylhydrazine with α-dicarbonyl compounds to give 1,2,5-benzotriazepines has been known for a long time but its attraction is lessened by the difficulty of synthesizing the hydrazine ⟨55G1098⟩.

The fused system (**489**) has been prepared by the cycloaddition of the extended azomethine imine (**488**) to alkynes ⟨75JCS(P1)556⟩.

(*iii*) *Type* bg (*C—C—N—C—C+N—N*)

1,2-Dialkylhydrazines condense with di(2-chloromethyl)alkylamines to give (**490**) ⟨72MI51800⟩.

(*iv*) *Type* de (*C—C—N—N—C—C+N*)

Bromination of acetophenone ketazine followed by reaction with benzylamine gives (**491**) ⟨71OPP289⟩. Halogenation of the product leads to ring contraction probably *via* an intermediate of type (**482**) ⟨75H(7)547⟩.

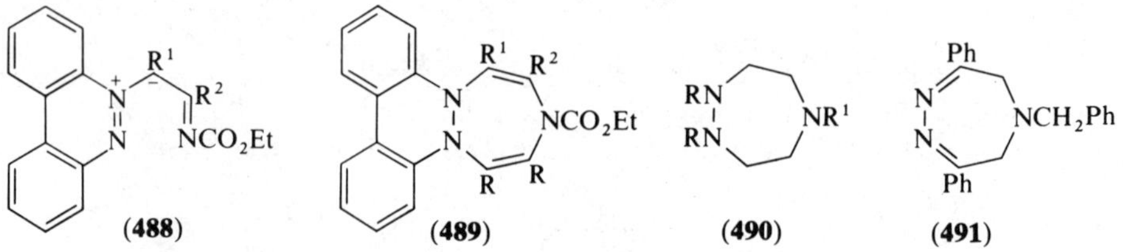

(**488**) (**489**) (**490**) (**491**)

5.18.7.4 1,3,5-Triazepines

5.18.7.4.1 Synthesis

(*i*) *Type* ag (*N—C—N—C—C—N+C*)

Various fused systems, *e.g.* (**494**), have been prepared *via* reactions of compounds of type (**492**) or (**493**) with C_1 reagents such as phosgene, thiocarbonyl chloride or cyanogen bromide ⟨76T57, 839⟩.

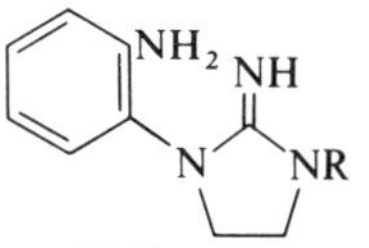

(492) 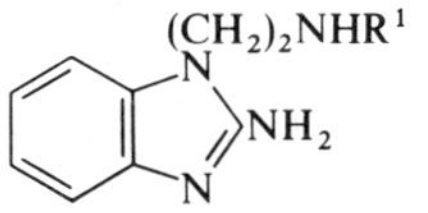 (493) 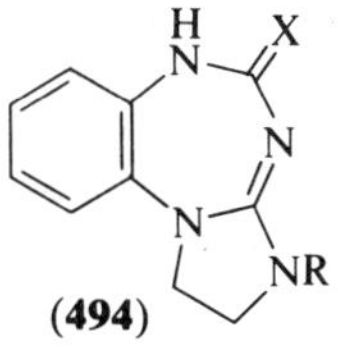 (494)

(*ii*) *Type* bf (*N—C—C—N+C—N—C*)

1,2-Diamines condense with alkoxycarbonyl isocyanates ($RO_2C—N=C=O$), dialkyl iminodicarboxylates [$HN(CO_2Me)_2$] or urea to give triazepinediones (**495**).

(*iii*) *Type* abfg (*N—C—C—N+C+N+C*)

Ethylene dinitramines [$(NO_2)NHCH_2CH_2NH(NO_2)$] and ethylene bis-sulfonamides condense with formaldehyde and primary amines or hydrazines to give triazepanes, *e.g.* (**496**) ⟨7. JOC3846⟩.

(*iv*) *Preparation from other heterocyclic systems*

Diarylnitrilimines ($Ar\overset{+}{C}{\equiv}N—\overset{-}{N}Ph$) react with the benzazete (**497**) to give (**498**) ⟨78CC396⟩. The reaction path appears to parallel that described on p. 643 for the analogous reaction of nitrile oxides except that the initial adduct is not isolable in this case.

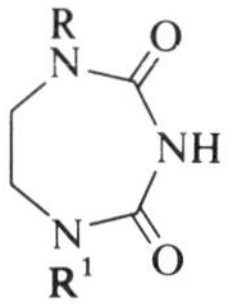

(495) 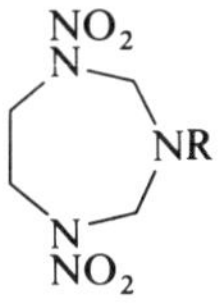 (496) 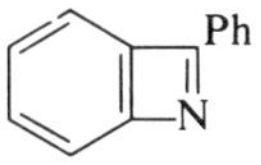 (497) 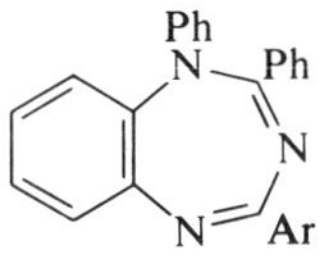 (498)

5.18.8 OXADIAZEPINES

There are nine possible oxadiazepine isomers and examples of all except the two isomers with three adjacent heteroatoms are known. The preparative routes to these systems which were known up to 1972 have been clearly illustrated in a review ⟨72MI51801⟩ and this information will not be given here because of space limitations. In this brief treatment attention has been concentrated on some general aspects of their chemistry and on new preparative methods published in the last 10 years.

Synthesis by the photoisomerization of aromatic N*-oxides.* It was stated earlier that the *N*-oxides of six-membered aromatic rings containing one nitrogen atom, *e.g.* (**310**) and (**320**), could be photoisomerized to give systems containing 1,2- and 1,3-oxazepine rings. Many such reactions have also been observed for analogous systems containing two nitrogen atoms; in some cases the resulting oxadiazepines can be isolated but in others they participate as transient intermediates which react further to give the isolated products. Thus the quinazoline 3-oxide (**499**; R = H or Me) gives the isolable 1,3,5-benzoxadiazepine system (**500**) (see also the conversion of **525** to **500** which also apparently proceeds *via* oxaziridine intermediates of similar types, **526** and **527**). These compounds are converted to *o*-aminophenol on acid hydrolysis and largely to 2-phenylbenzoxazole on solvolysis in boiling methanol ⟨67TL5233⟩. Irradiation of the 1-oxide (**501**; R = Me) however gives the indazole (**503**; R = Me) but it seems likely that this is formed by photochemical ring contraction of the 3,1,4-benzoxadiazepine (**502**; R = Me) particularly since the analogue (**502**; R = NHMe), prepared by the mercury(II) oxide oxidation of (**504**), undergoes a similar photochemical rearrangement ⟨68JOC4438⟩. Similarly, quinoxaline 1-oxides (**505**) are converted to 3,1,5-benzoxadiazepines (**506**) ⟨67TL1873, 68ACS877⟩. The latter (R = H, R[1] = Ph) is sensitive to hydrolysis giving *N*-benzoyl-*N′*-formyl-*o*-phenylenediamine.

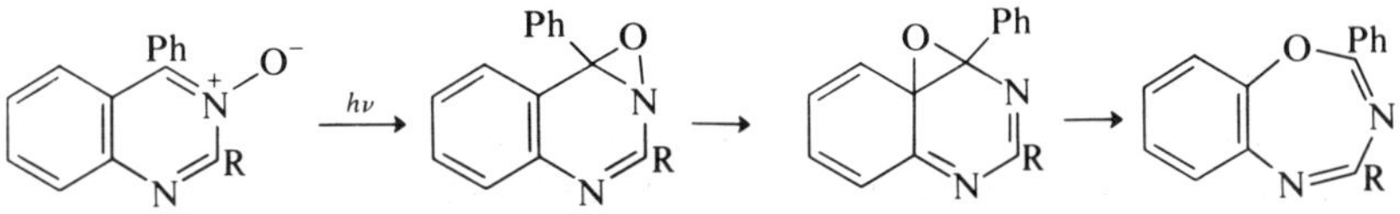

(499) $\xrightarrow{h\nu}$ → → (500)

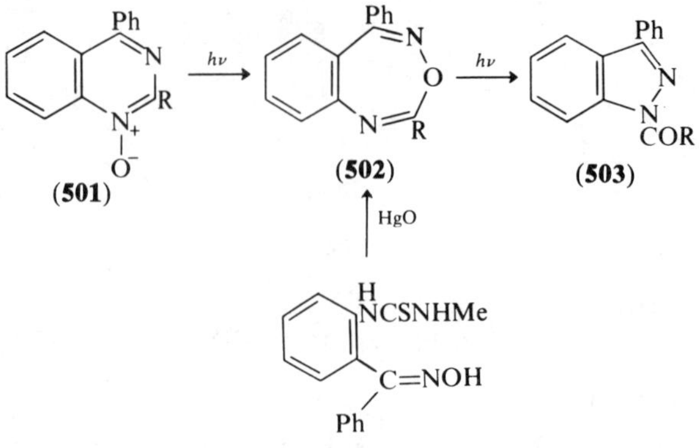

(501) (502) (503)

HgO

(504) 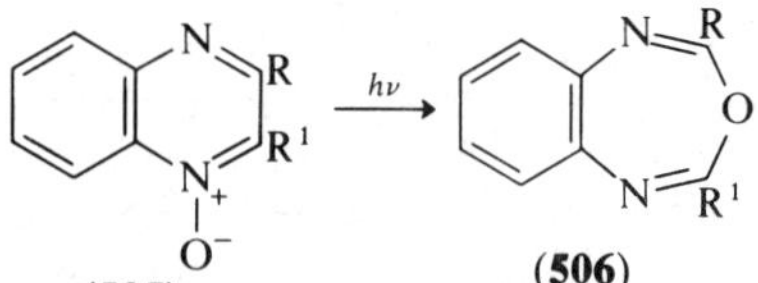 (505) (506)

Monocyclic oxadiazepines have not been isolated in reactions of this type but it has been proposed that the 1,2,6-system (**508**) is an intermediate in the formation of (**509**) from the pyrimidine *N*-oxide (**507**) ⟨69TL4899⟩, and that the 1,2,5-systems (**511**) and (**512**) are intermediates in the formation of the imidazoles (**513**) and (**514**) from the pyrazine *N*-oxide (**510**) ⟨67TL1197⟩.

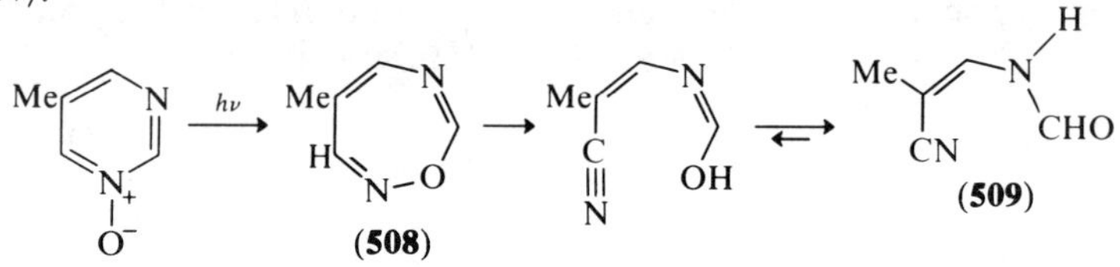

(507) (508) (509)

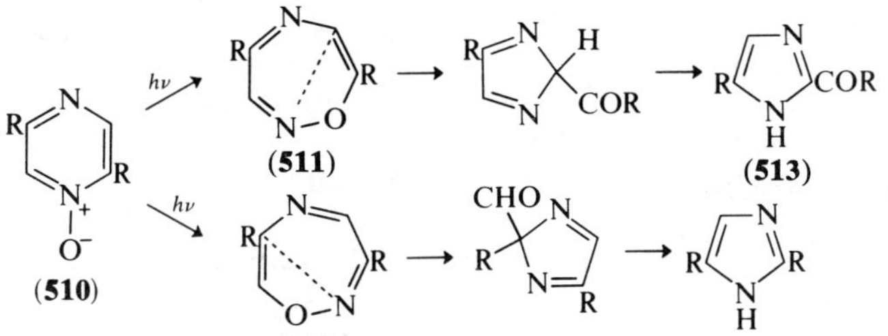

(510) (511) (512) (513) (514)

5.18.8.1 1,2,4-Oxadiazepines

The 3-oxo derivatives (**515**) have been prepared by the reaction of hydroxyurea with some α,β-unsaturated ketones ⟨75TL2979⟩.

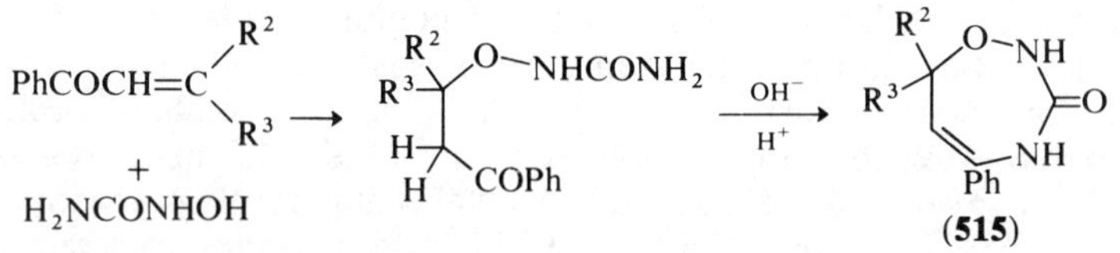

$PhCOCH=C(R^2)R^3$ + $H_2NCONHOH$ → (515)

5.18.8.2 1,2,6-Oxadiazepines

syn-ω-(Benzylamino)propiophenone oximes (**517**) react with phosgene to give the 7-oxo-1,2,6-oxadiazepines (**518**) as a major product ⟨75CB3387⟩, and with formaldehyde to give (**516**) ⟨80CB3373⟩.

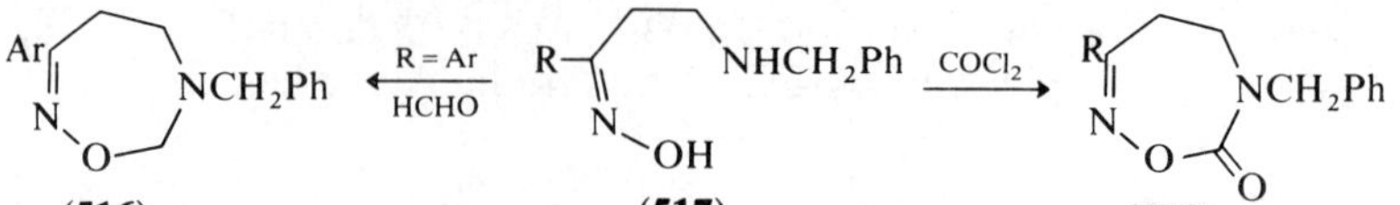

(516) ← R = Ar, HCHO — (517) — $COCl_2$ → (518)

There was confusion for many years regarding the identity of the product obtained by reacting *o*-aminobenzaldehyde with acetic anhydride. It was originally thought to be

1-acetylindazole, and then a 3,1,4-benzoxadiazepine, but more recently it has been shown by an X-ray study to be 2-methylquinazoline 3-oxide ⟨75TL4301⟩. It had been shown earlier that the supposed 3,1,4-benzoxadiazepines prepared from *N*-substituted 2-aminobenzaldoximes were quinazoline 3-oxides.

5.18.8.3 1,3,4-Oxadiazepines

The indeno fused compounds (**521**) have been prepared in moderate yield by the photochemical reaction of 2,5-diaryl-1,3,4-oxadiazoles (**519**) with indene ⟨77JOC1496⟩; the reaction is thought to proceed *via* a diradical or betaine (**520**). These products react with isocyanates (RNCO) to give the triazepine system, *e.g.* (**522**) ⟨76H(4)989⟩.

The compounds (**523**; R = Pr, Bu) were prepared in ~60% yield by the condensation of alkylhydrazones of diacetone alcohol ($MeC(=NNHR)CH_2C(OH)Me_2$) with formaldehyde ⟨77ZOR1563⟩.

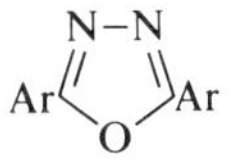

(**519**) 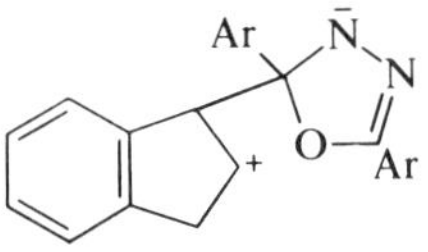 (**520**) 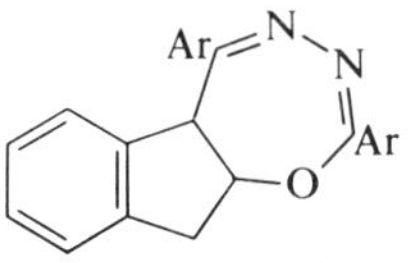 (**521**) 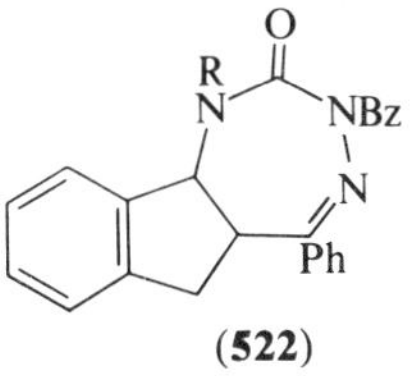 (**522**) 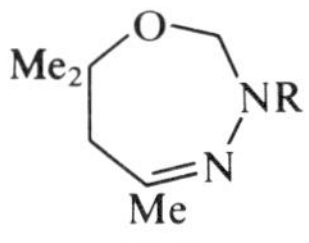 (**523**)

5.18.8.4 1,3,5-Oxadiazepines

2-Phenylbenzazete (**497**) reacts with nitrile oxides ($ArC{\equiv}\overset{+}{N}{-}\overset{-}{O}$) to give the labile adducts (**524**) which rearrange on warming or on exposure to acid to give quantitative conversions to the 1,3,5-oxadiazepines (**500**; R = Ar) ⟨75CC740⟩. It seems likely that the rearrangement goes *via* (**525**), (**526**) and (**527**).

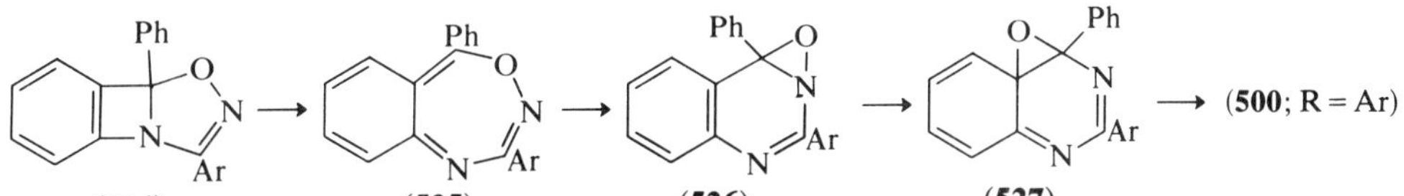

(**524**) → (**525**) → (**526**) → (**527**) → (**500**; R = Ar)

5.18.8.5 1,4,5-Oxadiazepines

In an extension of earlier work with 1,2-disubstituted hydrazines the hexahydro-1,4,5-oxadiazepines (**529**) have been prepared by the reaction of various monosubstituted hydrazines with (**528**) ⟨75MI51803⟩.

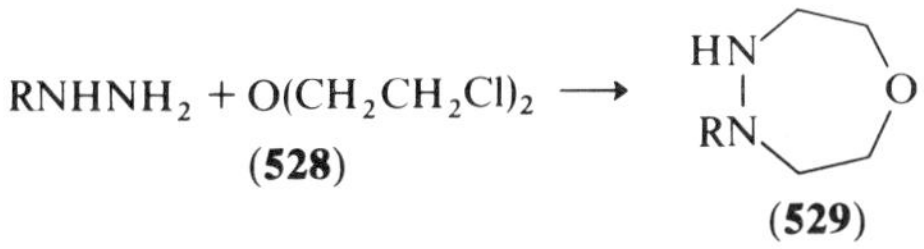

$RNHNH_2 + O(CH_2CH_2Cl)_2 \longrightarrow$ (**529**)

(**528**)

5.18.9 THIADIAZEPINES

5.18.9.1 1,2,3-Thiadiazepines

The only example of this system (**531**) was prepared by the reaction of (**530**) with hydrazine ⟨71CJC943⟩. It is a white solid which decomposed at its m.p. (169–170 °C) with the evolution of nitrogen to give (**532**).

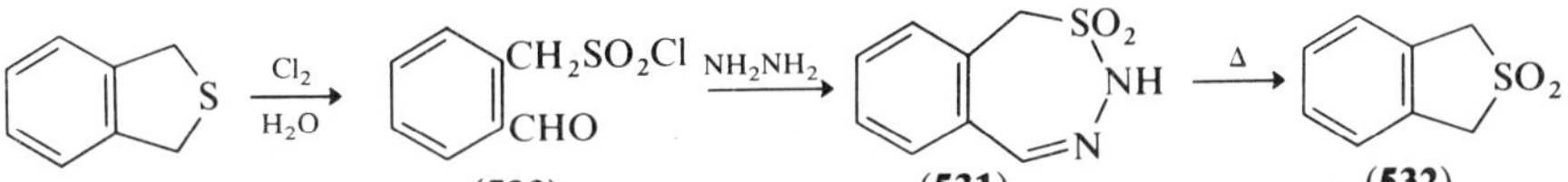

$\xrightarrow[H_2O]{Cl_2}$ (**530**) $\xrightarrow{NH_2NH_2}$ (**531**) $\xrightarrow{\Delta}$ (**532**)

5.18.9.2 1,2,4-Thiadiazepines

Several routes are available to both 1,2,4- and 4,1,2-benzothiadiazepines.

5.18.9.2.1 Synthesis

(*i*) *Type* ad (*S—C—C—C+N—C—N*)

The reactions of 2-benzoylbenzenesulfonyl chlorides (**533**) with amidines, guanidine or *S*-alkylthioureas give benzenesulfenylamidines (**534**; R = Ph, NH_2 or SMe) which cyclize on heating in xylene in the presence of acid to give 1,2,4-benzothiadiazepine 1,1-dioxides (**535**) ⟨68JHC719⟩.

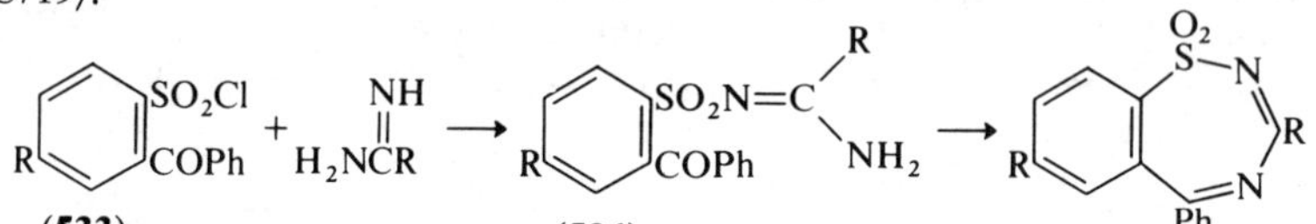

(**533**) (**534**) (**535**)

(*ii*) *Type* bc (*N—S—C—C—C—N+C*)

ω-Aminosulfonamides, *e.g.* (**536**) and (**537**), react with ethyl orthoformate to give the 1,2,4- and 4,1,3-benzothiadiazepines (**538**) and (**539**) respectively ⟨60JA1594⟩. The 2-oxo analogue of (**539**) was similarly prepared using phosgene, and (**541**) was prepared by reaction of the sulfoximide (**540**) with *N,N′*-carbonyldiimidazole ⟨72CB2575⟩. A range of the cyclic sulfoximides (**541**) was also prepared by treatment of (**542**) with hydrazoic acid ⟨72CB2575⟩. Compounds (**540**) were prepared by the hydrolysis of (**541**) and also by an independent route.

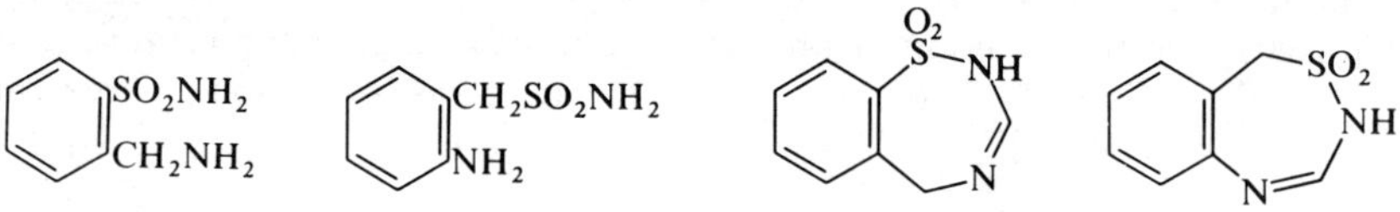

(**536**) (**537**) (**538**) (**539**)

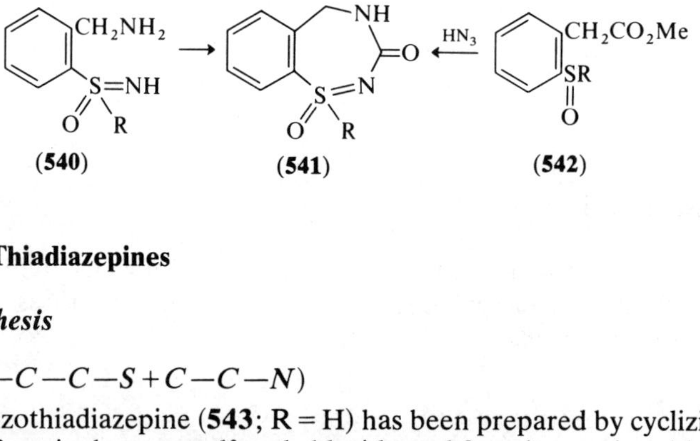

(**540**) (**541**) (**542**)

5.18.9.3 1,2,5-Thiadiazepines

5.18.9.3.1 Synthesis

(*i*) *Type* ad (*N—C—C—S+C—C—N*)

The 1,2,5-benzothiadiazepine (**543**; R = H) has been prepared by cyclizing the condensation product of 2-aminobenzenesulfonyl chloride and 2-aminoacetic acid.

The 1,2,5-benzothiadiazepine 1,1-dioxide (**545**) and its 3,4-dihydro analogue have been prepared by the reaction of 2-nitrobenzenesulfonyl chloride with *ω*-aminoacetophenone to give (**544**) followed by reductive cyclization ⟨79JHC835⟩.

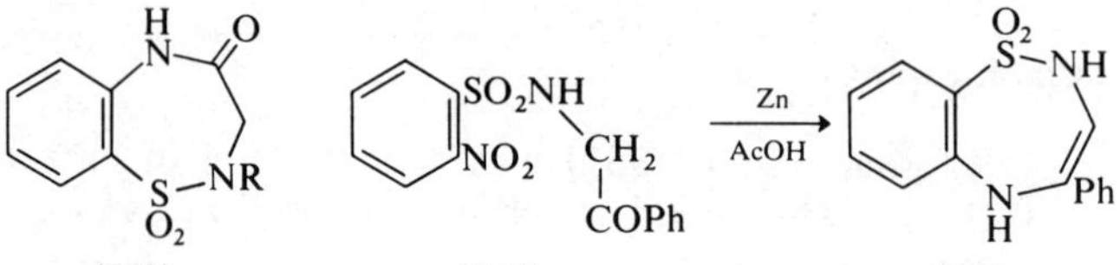

(**543**) (**544**) (**545**)

(*ii*) *Type* ae (*N—C—C—N+S—C—C*)

2,1,5-Benzothiadiazepines, *e.g.* (**546**), can be prepared by reacting *o*-phenylenediamine with appropriate sources of the S—C—C fragment, *e.g.* chlorosulfonyl acetyl chlorides

($RR^1C(COCl)SO_2Cl$) or the adduct of keten and sulfur dioxide ($\bar{S}O_2-CH_2\overset{+}{C}O$) ⟨73JOC2652⟩. The same type of system (**546**), substituted at the 1-position, can also be prepared from (**614**) by cyclization using DCC ⟨76MI51802⟩.

(*iii*) *Type* bd (*N—C—C—S—N+C—C*)

Compounds of type (**543**; R = Ar) have also been synthesized by reacting the ω-aminosulfonamide (**547**) with chloroacetyl chloride and subsequent cyclization of (**548**).

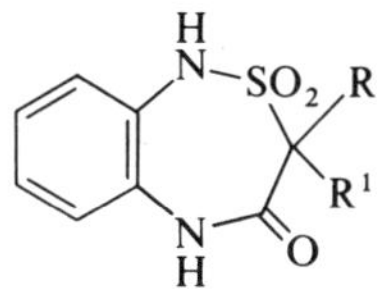

(**546**) 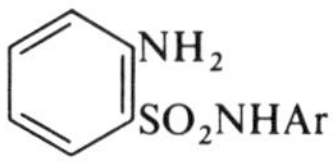 (**547**)  (**548**)

(*iv*) *Preparation from other heterocyclic systems*

The benzothiadiazepine 1-oxides (**550**), formed by the reaction of the diaryl sulfoxides (**549**) with hydrazoic acid, rearrange on treatment with sodium hydroxide to give (**551**) ⟨72CB757⟩. The benzothiadiazine 1,1-dioxide (**552**) also undergoes base-induced ring expansion to give (**553**) ⟨71JOC2968⟩.

(**549**) 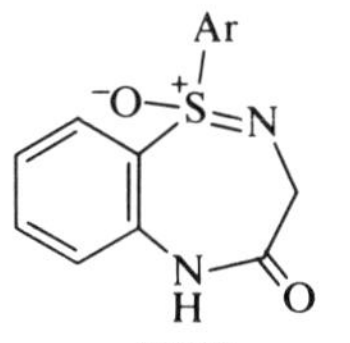 (**550**) 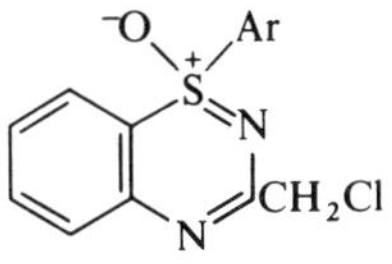 (**551**)

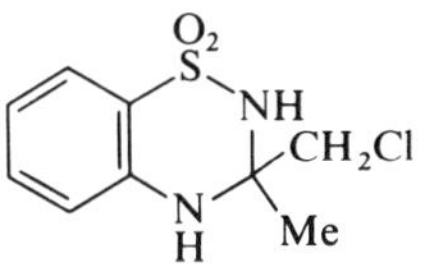

(**552**) 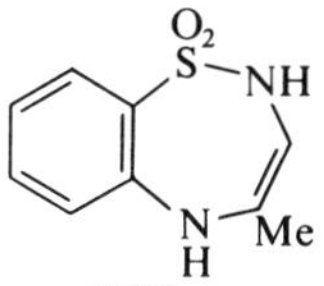 (**553**)

5.18.9.4 1,2,6-Thiadiazepines

5.18.9.4.1 Synthesis

(*i*) *Type* f (*C—S—N—C—C—C—N*)

The 2,1,4-benzothiadiazepine (**554**) can be prepared by the cyclization of 5-chloro-2-(chloromethylsulfonylamino)benzophenone oxime in the presence of sodium methoxide.

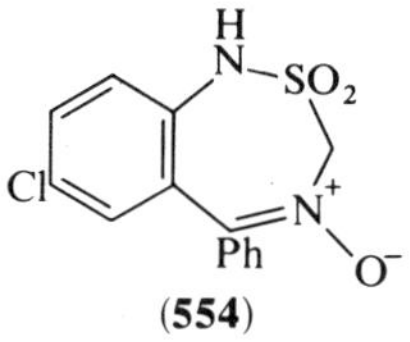

(**554**)

5.18.9.5 1,3,4-Thiadiazepines

5.18.9.5.1 Synthesis

(*i*) *Type* ab (*S—C—C—C—N—N+C*)

2-Mercaptobenzoylhydrazide reacts with formaldehyde to give (**555**) ⟨53JOC1380⟩.

(*ii*) *Type* bd (*C—S—C—C—C+N—N*)

o-Thiocyanatoacetophenone reacts with semicarbazide to give (**556**) ⟨79JCR(S)395⟩.

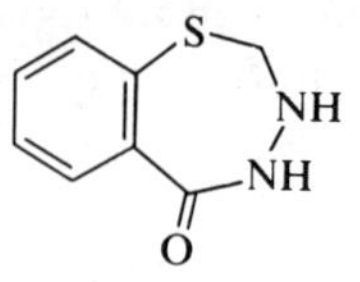

(555) 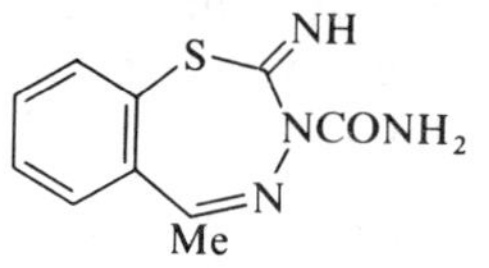 (556)

5.18.9.6 1,3,5-Thiadiazepines

5.18.9.6.1 Synthesis from other heterocyclic systems

The 1,3,5-thiadiazepine (**559**) has been prepared by the thermal rearrangement of the [4+2] cycloadduct (**558**) of the azirine (**557**) and thiobenzoyl isocyanate ⟨74JOC3763⟩. The benzoyl isocyanate adduct does not rearrange to give an oxadiazepine.

The 1,3,5-benzothiadiazepine system (**561**) has recently been prepared for the first time *via* methylation of the sulfimide (**560**; $R = CH_2Ph$) with trimethyloxonium fluoroborate. The rearrangement did not take place however when R = Me or when alkyl halides were used as alkylating agents. The product (**561**) rearranged readily in acid to give a benzothiazoline ⟨79JCR(S)214⟩.

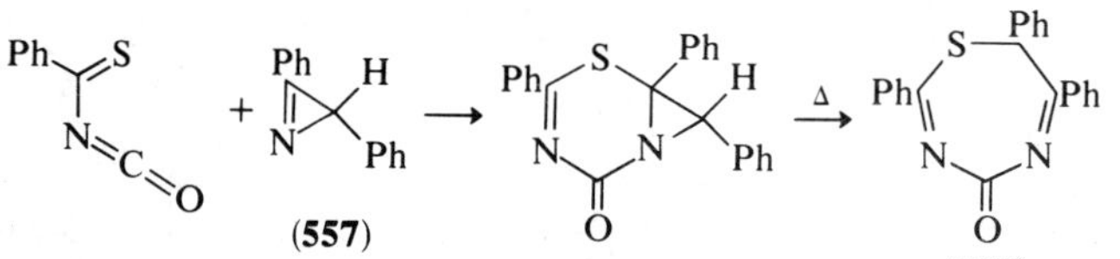

(557) (558) (559)

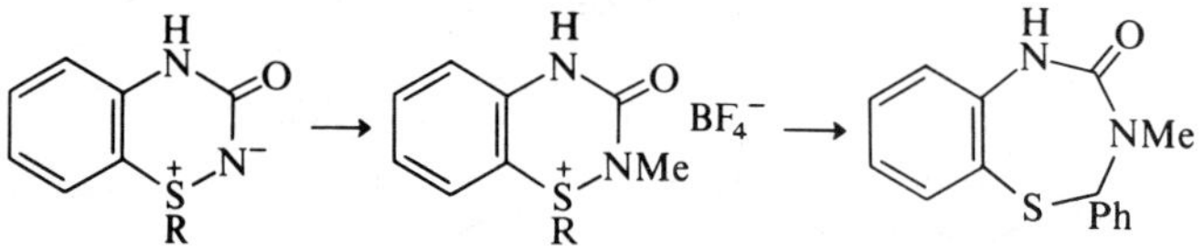

(560) (561)

5.18.9.7 1,3,6-Thiadiazepines

The compound (**562**), which is important as an agricultural fungicide, is prepared by the aerial oxidation of ethylene bis(dithiocarbamate), and (**563**) results from the reaction of 1,3-dialkylimidazolidines with carbon disulfide.

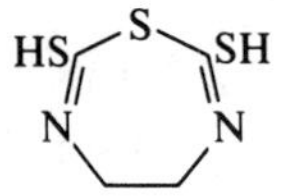

(562) 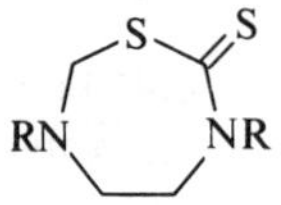 (563)

5.18.9.8 1,4,5-Thiadiazepines

The sulfides or sulfones (**566**) react with hydrazine to give (**565**) or (**567**) depending on the nature of R^1 ⟨70JHC431, 72T2307⟩. When $R^2 = R^3 = Me$ and $R = R^1 = Ph$ both stereoisomers of (**567**) were obtained. When treated with NBS compound (**568**) was converted into 3,6-diphenylpyridazine probably *via* a bromination–dehydrobromination sequence leading to (**569**) and (**570**) ⟨63JCS5496⟩. The 3-oxo derivative (**565**; $R^2 = H$) on treatment with acetic anhydride underwent ring contraction to give (**564**) ⟨72T2307⟩.

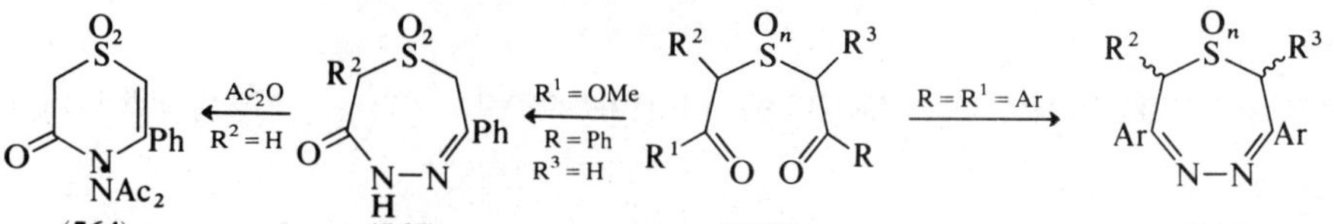

(564) (565) (566) (567)

(568) (569) (570) (571)

5.18.10 SYSTEMS WITH FOUR HETEROATOMS

There has been little systematic study of systems in which the number of heteroatoms exceeds the number of carbon atoms but a number have been prepared over the years. Their chemistry has not been extensively explored but several examples show the expected instability and high susceptibility to acid-catalyzed ring cleavage and contraction reactions.

5.18.10.1 Tetrazepines

5.18.10.1.1 1,2,3,4-Tetrazepines

(*i*) *Type* a (*N—N—C—C—C—N—N*)

The only example of this system (**573**) was prepared by the oxidation of the dihydrazine (**572**) ⟨78JA5978⟩. The thermal stability of (**573**) was markedly lower than that of its six-membered analogue, due apparently to its less rigid structure, and surprisingly its decomposition to give (**574**) had a negative entropy of activation. This indicates either a concerted process or one in which the optimum geometry for decomposition differs from the equilibrium conformation.

MeNNH$_2$—$(CH_2)_3$—MeNNH$_2$ (572) $\xrightarrow{K_3Fe(CN)_6}$ (573) $\xrightarrow[-N_2]{\Delta}$ (574)

5.18.10.1.2 1,2,3,5-Tetrazepines

(*i*) *Type* e (*N—C—N—N—N—C—C*)

The only example known, (**576**), was prepared during an examination of compounds of type (**575**) as models for biologically active alkylating agents ⟨75MI51804⟩.

(575) EtO$_2$C, N=N—N(CH$_2$CH$_2$Cl)$_2$ → (576) EtO$_2$C, N(CH$_2$)$_2$Cl

5.18.10.1.3 1,2,4,5-Tetrazepines

(*i*) *Type* a (*N—C—N—N—C—C—N*)

The benzo fused system (**579**) is produced in good yield by the oxidative ring closure of (**578**) or by the action of zinc and aqueous sodium hydroxide on (**577**) ⟨55CB1284⟩. It rearranged in dilute acetic acid to give 3-phenyl-1,2,4-benzotriazine.

(*ii*) *Type* bc (*N—N—C—C—N—N+C*)

1,2-Dihydrazines react with 1,1-bis electrophile equivalents such as formaldehyde or orthoformic esters to give for example (**581**) from (**580**) ⟨74MI51801⟩. The interesting fused systems (**582**) and (**583**) were prepared by a similar path ⟨73S355⟩.

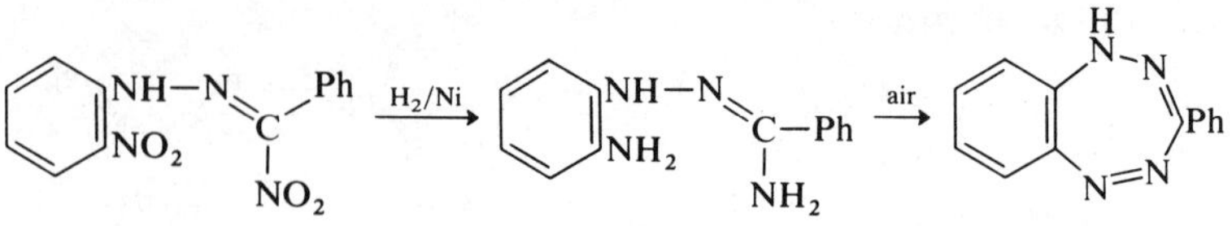

(577) (578) (579)

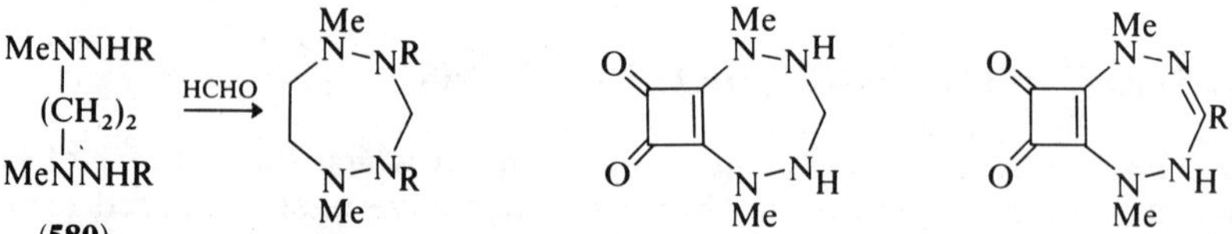

(580) (581) (582) (583)

(*iii*) *Type* eg (*N—N—C—N—N+C—C*)

1,1-Dihydrazines react with 1,2-bis electrophiles to give monocyclic, *e.g.* (**585**) ⟨76MI51801⟩, or fused systems, *e.g.* (**586**) ⟨81H(16)621⟩. It has also been reported that sugar hydrazones react with thiocarbohydrazide to give products of type (**587**) ⟨56CB956⟩.

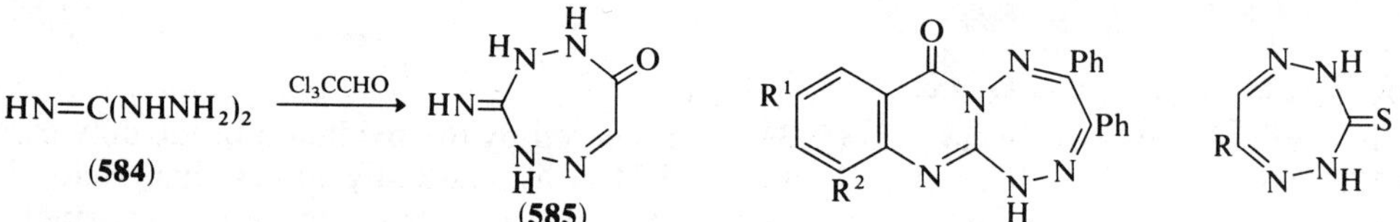

(584) (585) (586) (587)

5.18.10.1.4 1,2,4,6-Tetrazepines

(*i*) *Type* a (*N—C—N—C—N—C—N*)

The 3,7-dione (**589**) results from the elimination of hydrogen chloride from the chlorourea derivative (**588**) ⟨29BSF1013⟩.

(*ii*) *Type* bg (*C—N—C—N—C+N—N*)

The bis-formamide $ArCH(NHCHO)_2$ reacts with hydrazine hydrate to give (**590**) in high yield ⟨63TL1637⟩.

(*iii*) *Type* cf (*C—N—N—C+N—C—N*)

Aryl guanazoles are reported to react with 1,2-diformylhydrazine to give products of type (**591**) ⟨52G735⟩.

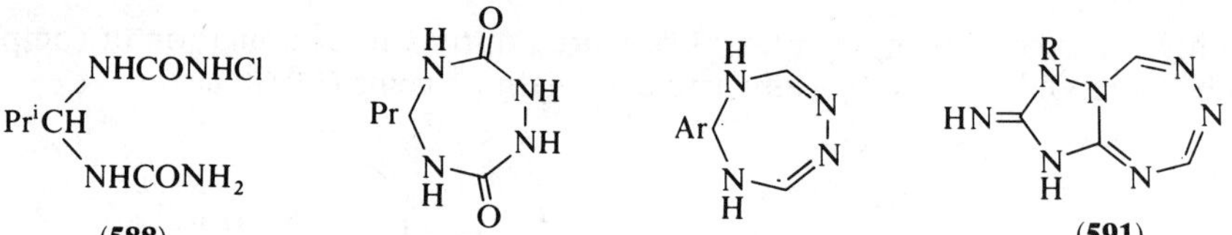

(588) (589) (590) (591)

5.18.10.2 Tetrathiepins

5.18.10.2.1 1,2,4,5-Tetrathiepins

(*i*) *Type* ad (*S—C—C—S+S—C—S*)

A general route to the benzo fused system (**592**) is provided by the reaction of 1,2-bis(chlorosulfanyl)benzene with 1,1-dithiols ⟨71AG(E)331⟩. X-ray structure determinations on (**592**; $R^1=R^2=H$) and (**592**; $R^1=Me$, $R^2=H$) have shown that the seven-membered ring has a chair conformation ⟨79ZN(B)426⟩.

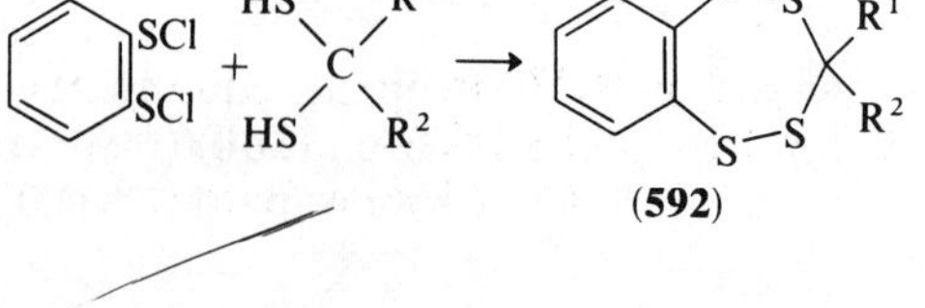

(592)

5.18.10.2.2 1,2,4,6-Tetrathiepins

Several synthetic routes are available to the fully saturated system (**594**) which is also found naturally in some mushrooms and red algae. It can be prepared in low yield by the dilution of a yellow solution of polymeric CH_2S in concentrated sulfuric acid with water ⟨80PS(8)157⟩, or manufactured together with 1,2,3,5-tetrathiane by the reaction of sulfur, sodium sulfide, formaldehyde and chloroform ⟨69JAP6927724; see also 67CPB988⟩. The most recent and best yielding synthesis ⟨81TL1939⟩ involves an interesting reaction between bis-chloromethyl disulfide (**593**) and sodium sulfide in which one molecule of the former apparently functions as a methylene transfer agent. Dynamic NMR studies have shown that the energy barrier to ring inversion in (**594**) is very low ⟨69TL4883⟩. The thione (**595**) is formed in low yield in a complex reaction between methylmagnesium iodide and methyl chlorodithioformate [MeSC(S)Cl] ⟨75RTC1⟩.

$ClCH_2SSCH_2Cl$ (**593**) $\xrightarrow{Na_2S}$ (**594**) (**595**) (**596**) (**597**)

A number of seven-membered rings containing more than four sulfur atoms are known; for example, the unsaturated system (**596**) and its benzo analogue can be prepared in good yields by the reaction of dichlorotrisulfane (S_3Cl_2) with *cis*-mercaptoethylene or 1,2-benzenedithiol respectively ⟨71TL2125⟩. Lenthionine (**597**) and hexathiepane are both found in species of mushroom and can be prepared by reactions of dichloromethane, sodium sulfide and methylene chloride ⟨66TL573⟩.

5.18.10.3 Oxatriazepines

Only two classes appear to be known. The 1,2,5,6-system (**599**) is said to result from the treatment of the semicarbazide (**598**) with iron(III) chloride ⟨28JIC439⟩, and more recently it has been shown that benzoyl isothiocyanate adds to the phenylhydrazones of benzaldehyde and several ketones to give the 1,3,4,6-oxatriazepine-5-thione system (**600**) ⟨67JCS(C)952, 76JCS(P1)2243⟩. The products rearrange to give (**601**) quantitatively on treatment with acid.

(**598**) $\xrightarrow{FeCl_3}$ (**599**)

PhNHN=C(R¹)(R²) + S=C=N–C(=O)Ph → (**600**) $\xrightarrow{H^+}$ (**601**)

5.18.10.4 Thiatriazepines

5.18.10.4.1 1,2,4,5-Thiatriazepines

(*i*) *Type* bc (*N—N—C—C—S—N+C*)

The sulfamoyl-substituted systems (**603**) and (**604**) have been prepared in high yield by, respectively, the reactions of (**602**; R = Me, R¹ = Cl) with formaldehyde in the presence of base ⟨63MI51800⟩ and (**602**; R = H, R¹ = Cl or Me) with urea ⟨63JOC743⟩.

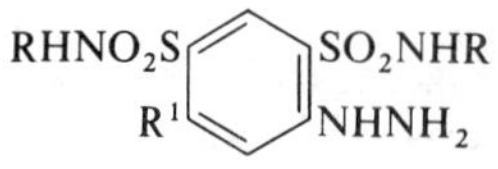

(602) 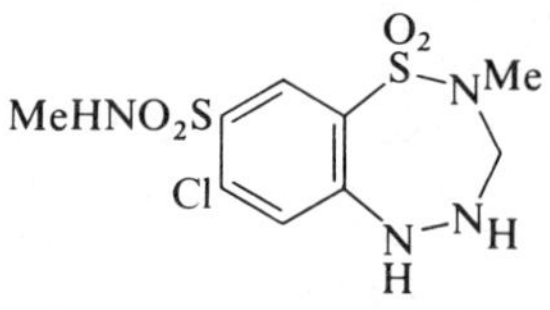 (603) 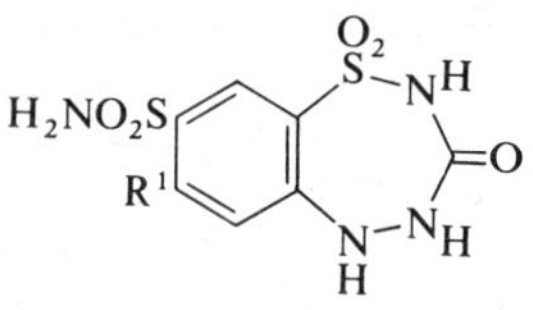 (604)

5.18.10.4.2 1,2,3,7-Thiatriazepines

The monocyclic system (**606**) has been prepared by the reaction of the *N*-chlorosulfonyl β-lactam (**605**) with 1,2-dimethylhydrazine ⟨75CB2137⟩.

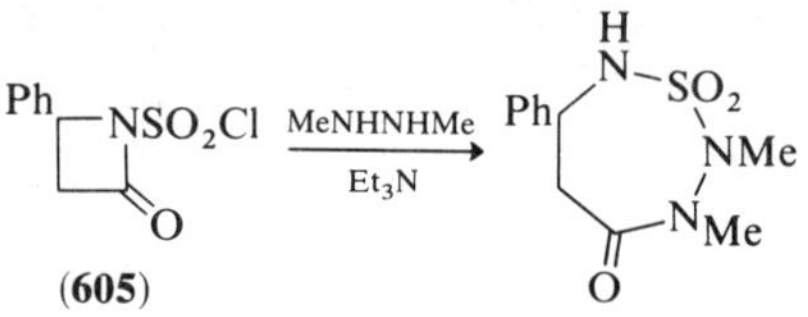

(605) → MeNHNHMe, Et_3N → (606)

5.18.10.4.3 1,2,4,6-Thiatriazepines

(*i*) *Type* af ($N-C-N-C-N+C-S$)

In an examination of the reactions of α-ureido nitrogen heterocycles with chlorothioformyl chloride it was found that (**607**) reacted to give (**608**) in yields of 12–63% depending on the nature of R. The products were rather unstable and decomposed with loss of methyl isocyanate and sulfur both on heating and on long standing at room temperature ⟨73JOC1578⟩.

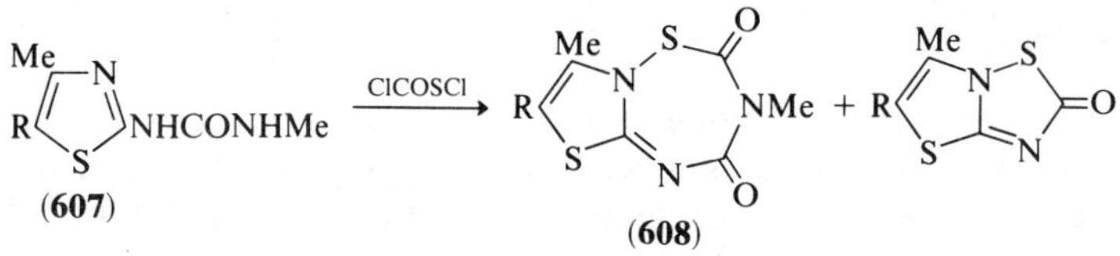

(607) → ClCOSCl → (608) + byproduct

5.18.10.4.4 1,2,4,7-Thiatriazepines

(*i*) *Type* ad ($N-C-C-N+C-N-S$)

Both monocyclic (**609**) and benzo fused systems (**610**) have been prepared by the reactions of 1,2-diamines with chlorosulfonyl isocyanate (ClO_2SNCO) ⟨75GEP2409355⟩.

The system (**611**) was produced in low yield by the reaction of diphenylketene-*N*-phenylimine ($Ph_2C{=}C{=}NPh$) and diphenylsulfurdiimide ($PhN{=}S{=}NPh$) ⟨76CL1379⟩.

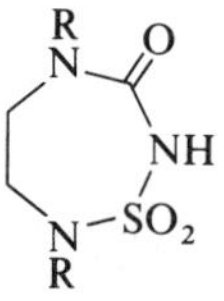

(609) 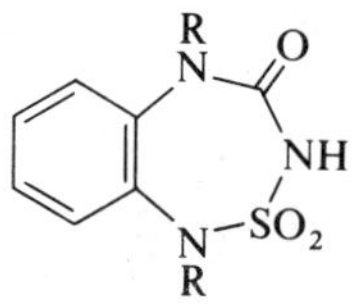 (610) 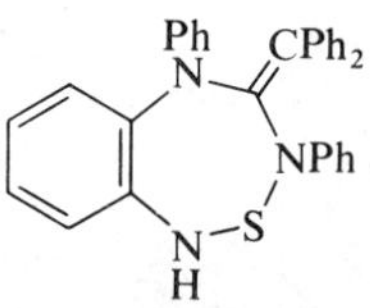 (611)

5.18.10.4.5 1,3,4,6-Thiatriazepines

(*i*) *Type* ae ($C-N-N-C+S-C-N$)

A number of annelated 1,3,4,6-thiatriazepines (**613**) have been prepared by the reactions of the sodium salts of cyclic thioureas with dichlorodiazabutadienes (**612**). Acyclic thioureas take an alternative reaction path *via* the elimination of a carbodiimide to give only 2,5-diphenyl-1,3,4-thiadiazole ⟨80CC156⟩. An X-ray structure determination on (**613**; $n = 2$) shows that the seven-membered ring has the expected boat conformation and that ring fusion does not appear to strain the five-membered ring ⟨81AX(B)486⟩.

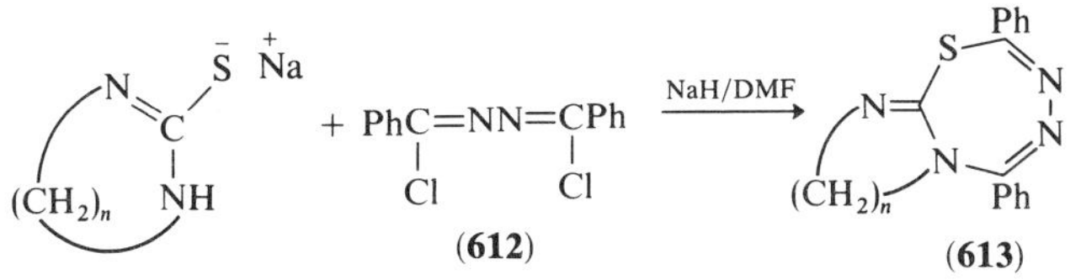

(612)

(613)

5.18.10.4.6 1,2,5,6-Thiatriazepines

(*i*) *Type* f (*C—S—N—C—C—N—N*)

The benzo fused system (**615**) has been prepared by the diazotization of (**614**), a route which parallels that used in the syntheses of the benzotriazepines (**443**) and (**485**) ⟨76MI51802⟩.

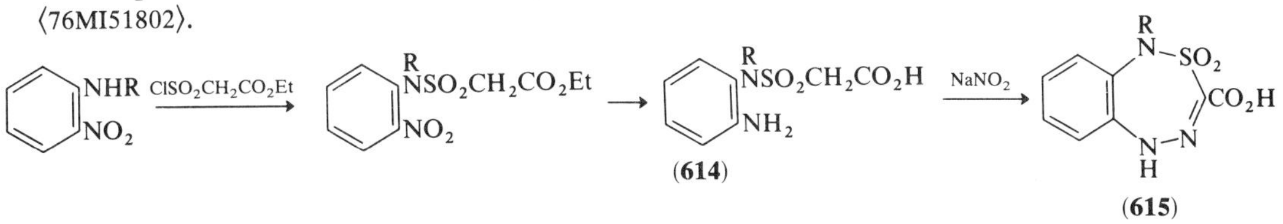

(614)

(615)

5.18.10.5 Oxathiadiazepines

N-Sulfinylbenzamide, produced by the reaction of benzamide and thionyl chloride, reacts with styrene oxide in the presence of tetraethylammonium bromide to give the 1,4,3,5-oxathiadiazepine (**616**) as the major product ⟨71BCJ2836⟩.

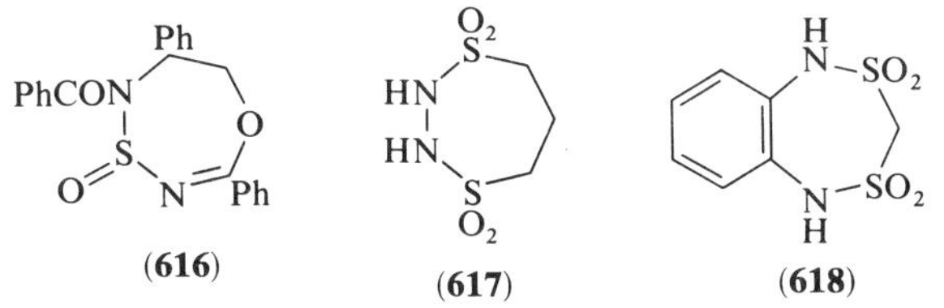

(616) (617) (618)

5.18.10.6 Dithiadiazepines

The 1,4,2,3-dithiadiazepine tetroxide (**617**) is simply prepared by the reaction of the bis-sulfonyl chloride $ClO_2S(CH_2)_3SO_2Cl$ with hydrazine hydrate ⟨73CB713⟩ and the benzo fused 1,3,4,7 system (**618**) is produced by reacting *o*-phenylenediamine with bis(chlorosulfonyl)methane ($ClO_2SCH_2SO_2Cl$) ⟨77JHC493⟩.

5.18.10.7 Dioxathiazepines

The 1,5,2,3-dioxathiazepines (**619**) have been prepared by the reactions of benzene-1,2-diols with chlorosulfonyl isocyanate (ClO_2SNCO) ⟨81PS(11)255⟩ (*cf.* the preparation of **609** and **610**).

5.18.10.8 Dioxadithiepins

The benzo fused 1,5,2,4-system (**620**) is formed by the reaction of benzene-1,2-diol with bis(chlorosulfonyl)methane ($ClO_2SCH_2SO_2Cl$) ⟨44HCA1790⟩ (*cf.* the preparation of **618**).

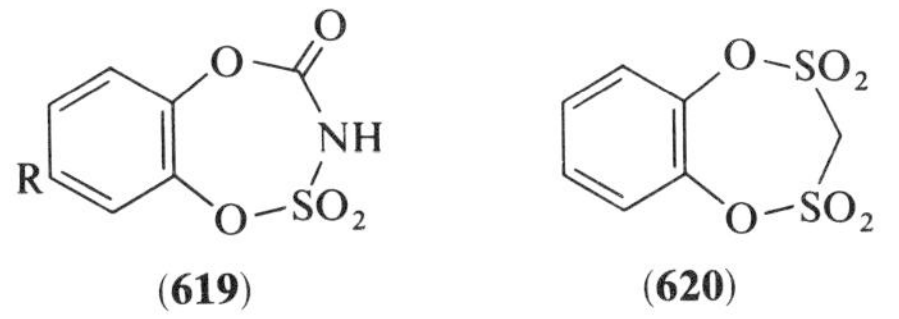

(619) (620)

5.19

Eight-membered Rings

J. A. MOORE
University of Delaware

and

F. A. L. ANET
University of California, Los Angeles

5.19.1 INTRODUCTION

The organization of this chapter is based on the number and type of heteroatoms; only N, O and S are included. Within each heterocyclic system, compounds with reduced rings are discussed before those with higher degrees of unsaturation.

Eight-membered heterocyclic systems embrace a very broad variety of compounds ranging from cyclic ethers and imines to highly labile analogs of cyclooctatetraene, and the properties and chemical behavior of these compounds of course depend entirely on the extent and location of unsaturation. For this reason, preparative methods and reactions are treated together for each ring system and oxidation level.

The presently known eight-membered ring heterocycles fall mainly into saturated and highly unsaturated classes. The former (including compounds with exo unsaturation such as a carbonyl group) resemble cyclooctane and its derivatives in showing 'medium-ring' effects. These include transannular interactions, difficulty of preparation by most cyclization reactions, and a tendency in certain cases for oligomerization. Conformational aspects of the saturated rings provide some points of interest which are treated in a separate section of the chapter. Several eight-membered oxygen-containing rings are found in sesquiterpenoids of marine origin and a few other natural sources; eight-membered cyclic polysulfides are distinctive components of certain fungi.

Most of the highly unsaturated monocyclic eight-membered heterocycles contain one or two nitrogen atoms and have been obtained by bond reorganization processes from strained bicyclic or polycyclic precursors. Although several of the less substituted compounds without stabilizing substituents are highly labile substances, 1,4-dihydro-1,4-diazocines qualify as 'dihetera[8]annulenes' and display distinct aromatic properties.

The systematic names for eight-membered heterocyclic rings provided by the standard Hantzsch–Widman system are based on the stem -oc- with suitable affixes for heteroatoms and degree of unsaturation. Azocine (diazocine, *etc.*), oxocin and thiocin are the names for rings with complete non-cumulative unsaturation, *i.e.* four double bonds for azocines and three for rings containing oxygen or sulfur. The latter rings require a locant for 'indicated hydrogen', *e.g.* 4*H*-oxocin.

Nomenclature for the saturated rings has varied considerably. Both azacyclooctane and azocane have been used in the literature for the fully reduced nitrogen ring, although the traditional Hantzsch–Widman system restricts the ending -ane to saturated oxygen and sulfur rings. *Chemical Abstracts* has followed this practice, and uses octahydroazocine, oxocane, thiocane, *etc.* for the saturated nitrogen, oxygen and sulfur rings. IUPAC recommendations of 1982 now sanction azocane, diazocane, *etc.*

5.19.2 RINGS WITH ONE NITROGEN ATOM

5.19.2.1 Hexahydro- and Octahydro-azocines

The fully saturated compound (**1**) has been named azocane, azacyclooctane or heptamethyleneimine. The most convenient synthesis of (**1**) is by the reduction of the corresponding lactam (enantholactam), hexahydro-2(1*H*)-azocinone (**2**) with LAH ⟨49HCA544, 54JA2317⟩. Hydrogenolysis catalyzed by copper chromite can also be used and is reported to give (**1**) in 93% yield by the use of 100 atm hydrogen at 200 °C ⟨74GEP2347905⟩. Cyclization reactions have been used to make (**1**) and related compounds such as (**3**) and (**4**). Thus 7-bromoheptylsulfonamide under high dilution conditions in amyl alcohol in the presence of K_2CO_3 gives a 60% yield of (**4**), which can be hydrolyzed to the free base (**1**) ⟨52M386⟩. Similarly, the methobromide of the 1-methyl derivative (**3**) of (**1**) has been obtained by cyclization of 7-bromoheptyldimethylamine ⟨30JA287⟩.

(1) R = H
(3) R = Me
(4) R = SO_2Ar

(2) R = H
(11) R = NO

Chloro compounds may require such high temperatures for cyclization that dealkylation to give a tertiary amine occurs, as in the reaction of (**5**) to give (**6**) in 22% yield at 225 °C ⟨60JOC65⟩.

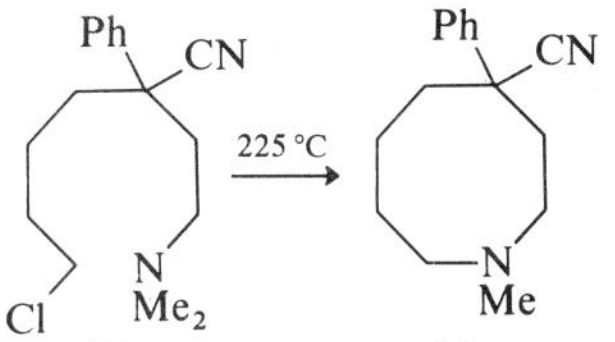

(5) (6)

Methylation of (**1**) to give (**3**) can be carried out either with formaldehyde in formic acid ⟨52M386⟩ or by LAH reduction of the formyl derivative of (**1**) ⟨54JA2317⟩. The latter is conveniently prepared from chloral and (**1**). The reduction of (**2**) in the presence of ethyl acetate provides a convenient one-pot synthesis of the 1-ethyl derivative of (**1**) ⟨79TL3395⟩.

The lactam (**2**) is obtained in 87% yield by treating cycloheptanone with sodium azide and concentrated hydrochloric acid ⟨54JA2317⟩. Beckmann rearrangement of cycloheptanone oxime with concentrated sulfuric acid gives (**2**) in 88% yield ⟨58JA1510⟩. Alternatively, the ketone gives (**2**) directly by reaction with hydroxylamine sulfonate in 95% formic acid ⟨79S537⟩.

The benzenesulfonate of cycloheptanone oxime rearranges readily at room temperature to give the very electrophilic *O*-benzenesulfonate (**7**), which reacts with pyridine to give the salt (**8**) ⟨67CB3024⟩. Treatment of (**8**) with water gives (**2**), and with H_2S the thione (**9**) is produced ⟨58CB972⟩. The reaction of aluminum alkyls with cycloheptanone oxime mesylate provides a direct route to 2-alkyl-substituted octahydroazocines ⟨81JA7368⟩.

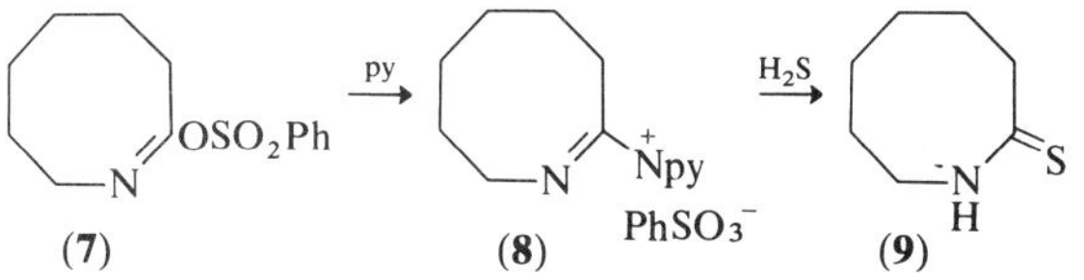

(7) (8) (9)

The lactam (**2**) can be polymerized (Nylon-7), but pyrolysis *in vacuo* of the polymeric amide in the presence of NaOH or KOH at 300–350 °C slowly regenerates the monomeric lactam ⟨59MI51900⟩. The IR carbonyl stretching band (amide I) in (**1**) is at 1658 cm^{-1} ⟨57CB1437⟩ and the amide group has the *cis* configuration. The conjugate acid of (**2**) is a crystalline compound whose structure has been determined ⟨75AX(B)251⟩ and which has a pK_a of 0.5 ⟨57CB1437⟩.

Octahydroazocine (**1**) behaves as a typical secondary amine and forms an azeotrope with water, b.p. 96 °C ⟨52M386⟩. It can be transformed into 1-nitroso and 1-amino derivatives in the usual way ⟨74JMC948⟩. The imine (**1**) can also be cyanoethylated, and adds ethylene oxide ⟨59MI51900⟩ to give the *N*-hydroxyethyl derivative. Attempts to convert (**1**) to an enamine by oxidation with silver acetate gave only a low yield of pyridine ⟨52M386⟩.

Sodium hypochlorite reacts with (**1**) to give the 1-chloro derivative, which gives rise to a variety of products, including the bicyclic amine, pyrrolizidine, when treated with silver salts ⟨72CJC1167⟩. Quaternary salts derived from (**3**) react with sodium methoxide to give mostly the acyclic Hofmann elimination product (83%); minor amounts of (**3**) (14%) are also formed, together with a trace (2%) of the product of S_N2 attack on the α-methylene carbon ⟨80JOC3952⟩.

A number of drugs and drug analogs have an octahydroazocine ring replacing the more common piperidine unit. Probably the most studied of these compounds is [2-(octahydro-1-azocinyl)ethyl]guanidine (**10**) and its salts, especially the sulfate ⟨54JA2317, 59E267⟩. This compound is an antihypertensive agent and is sold under a variety of trade names, *e.g.*

Table 1 Some Physical Properties of Octahydroazocine and Related Compounds ⟨58JA1510, 52M386⟩

Compound	*M.p.* (°C)	*B.p.* (°C)	n_D
(**1**)	−33	162	1.4721
(**2**)	35	155/11 mmHg	—
(**3**)	—	161	1.4598

Guanethidine and Ismelin. The sulfate salt (m.p. 276–281 °C) can be prepared by the reaction of $ClCH_2CN$ with (**1**), followed by reduction of the nitrile group with LAH and treatment with *S*-methylisothiourea sulfate. An extensive literature exists on the pharmacology of this compound.

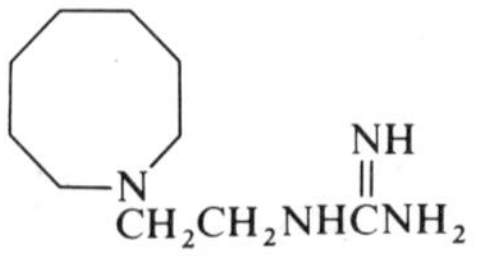

(**10**)

The 1-nitroso derivatives of (**2**) and of other lactams with various ring sizes have very different rates of decomposition, a process which takes place *via* unimolecular rearrangements to give ring expanded cyclic diazotates, which then lose nitrogen. The eight-membered ring compound (**11**) is nearly 40 000 times more reactive than its six-membered ring analog, although it is only half as reactive as the nine-membered ring compound, which is the most reactive in this series ⟨57AG341⟩. The high reactivity of the eight- and nine-membered rings probably reflects medium-ring strain. It is significant that the 10-membered ring compound is much less reactive, probably because the lactam group in this compound can adopt a *trans* configuration. The lactam (**2**) is known to exist in a conformation with a *cis* amide group ⟨79AX(B)1920⟩ (see also Section 5.19.3.1).

Although 3-chloro- and 3-bromo-hexahydroazocin-2(1*H*)-one (**12**) and (**13**) undergo substitution reactions when treated with thiolate ions to give (**14**), only elimination products are formed from other anions; subsequently the anions sometimes add to the α,β-unsaturated intermediate (**15**) to give 4-substituted derivatives such as (**16**) ⟨81AJC569⟩.

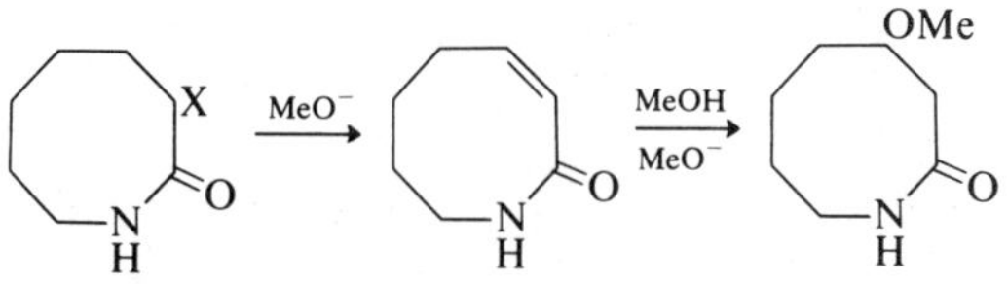

(**12**) X = Cl
(**13**) X = Br
(**14**) X = SPh
(**15**)
(**16**)

Methylation of the lactam (**1**) on oxygen gives an iminoether which can be converted to (**17**) and thence to the nine-membered carbodiimide (**19**) ⟨80JOC1315⟩.

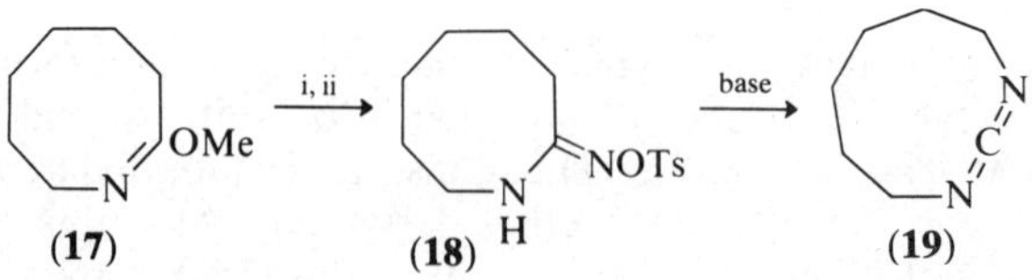

(**17**) (**18**) (**19**)

i, $NH_2OH{\cdot}HCl$, py; ii, TsCl, py

In ethanol, 1-phenylhexahydro-2(1*H*)-azocinone (**20**) undergoes an unusual photochemical rearrangement to give a 10-membered ring (**21**), presumably by α-cleavage of the N—CO bond followed by recombination of the diradical to form a bond between the carbonyl and an *ortho*-position on the benzene ring. A simple proton shift then gives (**21**) ⟨69CB342⟩.

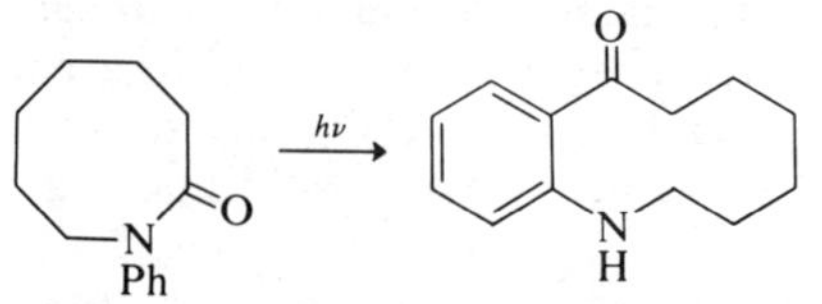

(**20**) (**21**)

In an interesting fragmentation reaction, the hexahydroazocine (**23**) is formed by solvolysis of (**22**) in the presence of $NaBH_4$ in 94% yield ⟨75TL2613⟩. A related compound (**24**) can be prepared from 4-cycloheptenone oxime tosylate *via* the unsaturated lactam (**25**) ⟨79JOC287⟩. Whereas (**25**) adds bromine to the double bond, (**24**) undergoes a transannular reaction to give a 1-substituted pyrrolizidine (**26**). The latter type of reaction also occurs

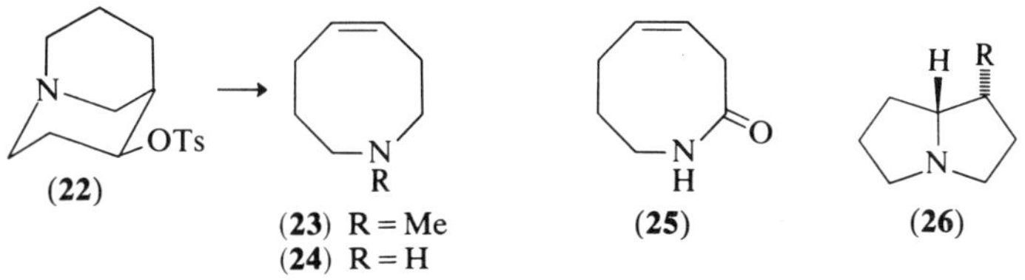

(**22**) (**23**) R = Me (**24**) R = H (**25**) (**26**)

for other electrophiles, such as $HgCl_2$ and PhSBr. The relative configuration of (**26**) has been established by an X-ray diffraction study of the hydrobromide salt of (**26**).

Dieckmann cyclization of the diesters (**27**) under high dilution conditions in the presence of potassium *t*-butoxide followed by hydrolysis and decarboxylation affords in fair yield the interesting aminoketones (**28**) and (**31**)–(**34**) ⟨52JA1704⟩. The *N*-methyl compound (**28**) has an unusually low carbonyl stretching frequency (1683 cm^{-1}, CCl_4), even for an eight-membered ring, which can be ascribed to a transannular interaction between the carbonyl group and the nitrogen lone pair, as depicted in (**29**). This is supported by the finding that (**28**) protonates exclusively on oxygen (although the positive charge is on nitrogen) to give (**30**). The change in pK_a of (**28**) with solvent polarity also shows that protonation takes place on oxygen in solution ⟨54JA630, 55JA6234⟩.

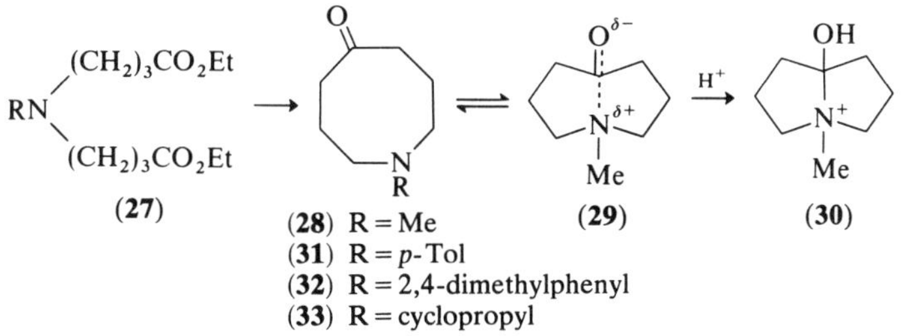

(**27**) (**28**) R = Me (**29**) (**30**)
(**31**) R = *p*-Tol
(**32**) R = 2,4-dimethylphenyl
(**33**) R = cyclopropyl

Related compounds, where the substituent on nitrogen is aryl (**31**), (**32**) ⟨55JA6241⟩, cyclopropyl (**33**) ⟨55JA6245⟩ and tosyl ⟨79JCS(P1)2009⟩ have been synthesized in the same way. These compounds have significantly less transannular interaction than does (**28**), in part because of delocalization of the electron pair on nitrogen and in part because of steric effects, especially when an *ortho*-substituted aryl group is present on the nitrogen. However, the cyclohexyl compound related to (**33**) shows a strong transannular interaction, so that the effect seen in the cyclopropyl compound must be electronic rather than steric in nature. Thus, the salt of the 2,4-dimethylphenyl compound (**32**) shows a carbonyl group in the IR spectrum, whereas the *p*-tolyl (**31**) or the methyl (**28**) compounds (as salts) are transparent in this region of the spectrum. Also, the carbonyl stretching frequencies of these compounds in their base forms are in the range 1694–1698 cm^{-1}, significantly higher than that in (**28**).

The transannular interaction in the eight-membered (and also in nine-membered) ring amino ketones decreases the oxygen (^{17}O) exchange rate with water, as shown by ^{17}O NMR, which also reveals an unusually shielded ^{17}O chemical shift in such compounds ⟨72HCA907⟩.

An interesting new method for the preparation of the methoxycarbonyl (and related) compound (**35**) involves the 'hydroboration–carbon monoxide insertion' in bis-alkenic amines (**34**). The best reagent for this process is thexylborane, followed by cyanidation ⟨82JOC1494⟩. The yields are not very good and seven-membered ring compounds (**36**) can also be formed, but the procedure is short and simple.

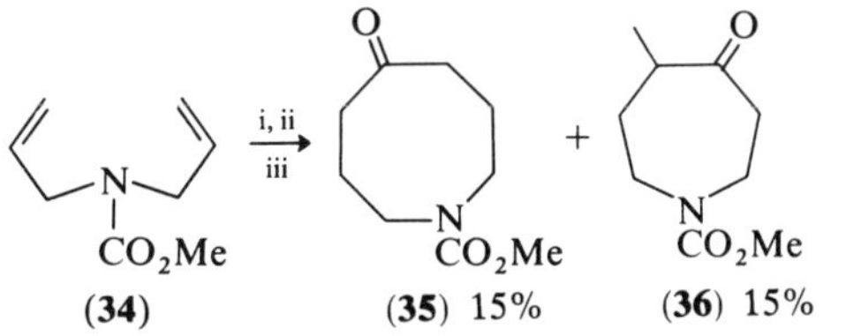

(**34**) (**35**) 15% (**36**) 15%

i, thexylborane; ii, KCN–$(CF_3CO)_2O$; iii, KOH–H_2O_2

Some *Senecio* alkaloids have 'opened pyrrolizidine' ring systems which show transannular interactions and are closely related in structure to the compounds discussed above. The alkaloid senkirkine is a cyclic ester of (**37**) ⟨74JA6165⟩ and its X-ray structure shows a rather short transannular contact of 2.23 Å.

Pyrrolizidine itself can be oxidized by H_2O–Fe^{3+} to an enamine (**38**) which gives (**28**) with methyl iodide ⟨79CPB1981⟩. The alkaloid derivative, (±)-isoretrorecanol, has been synthesized *via* (**39**) ⟨69JOC1066⟩ by the use of transannular reactions of the type described above.

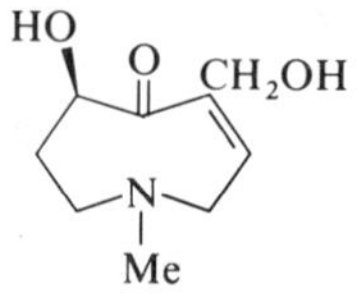

HO O CH_2OH N Me (**37**)   N (**38**)   O CO_2Et N CH_2Ph (**39**)

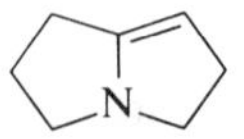

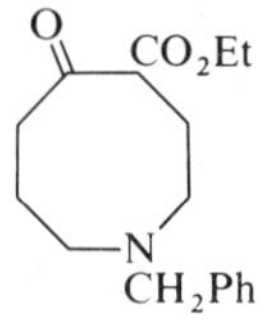

Microbiological oxidation of the 1-benzoyl derivative of (**1**) affords mostly the 5-hydroxy together with some of the 4-hydroxy derivative ⟨68JOC3187⟩. These products are best separated after oxidation to the respective ketones with CrO_3. The 5-isomer (**40**) gives an ethylene ketal (**41**) which may then be reduced to the benzyl derivative (**42**). Hydrolysis of the ketal group then gives (**43**). Hydrogenolysis of (**42**) gives (**44**), which on hydrolysis cyclizes to the bicyclic immonium ion (**45**). Like (**28**), the *N*-benzyl ketone (**43**) shows a strong transannular amine–carbonyl interaction and protonates and acetylates on oxygen to produce bicyclic quaternary ammonium structures. Compound (**43**) can also be synthesized by the Dieckmann procedure in 62% yield, which is unusually high even under the high dilution (55 h addition time) and high speed stirring conditions used ⟨69JOC1066⟩.

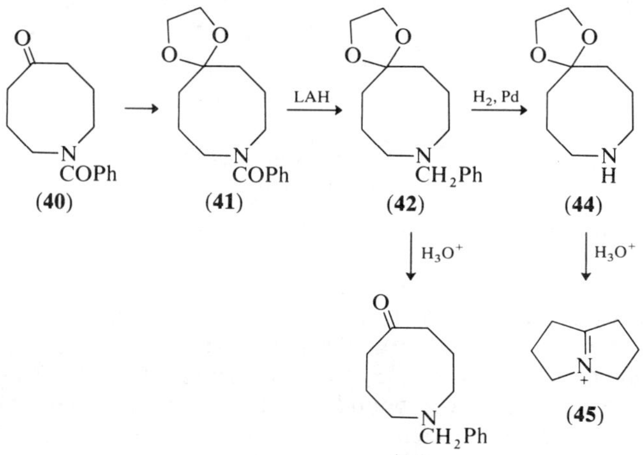

O N COPh (**40**) → O O N COPh (**41**) —LAH→ O O N CH_2Ph (**42**) —H_2, Pd→ O O N H (**44**)

(**42**) —H_3O^+→ O N CH_2Ph (**43**)   (**44**) —H_3O^+→ N+ (**45**)

5.19.2.1.1 Benzo derivatives

The benzo fused derivatives of hexahydroazocine comprise a miscellaneous group of compounds; the eight-membered ring often occurs only as an intermediate stage in a synthetic sequence, as in the case of mitomycin. Some simple benzo fused derivatives can be obtained by methods analogous to those presented for the monocyclic compounds, such as the Schmidt reaction (NaN_3 and acid) on benzocyclohepten-1-one to give (**46**), m.p. 154–155 °C ⟨61JOC27⟩ or by a Beckmann rearrangement of the oxime ⟨62JOC3844⟩; about 20% of the isomeric lactam resulting from alkyl migration is also formed. Reduction of (**46**) with LAH gives the hexahydrobenzazocine (**47**).

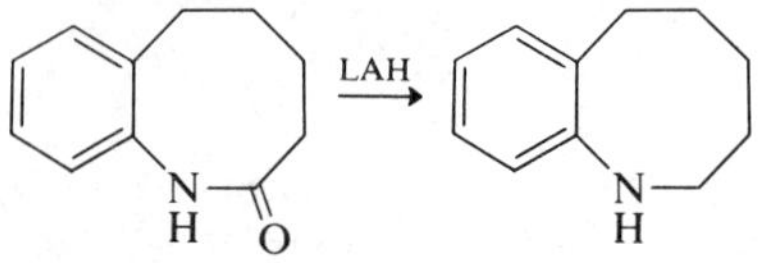

N H O (**46**) —LAH→ N H (**47**)

The ketone (**48**) can be obtained by Dieckmann cyclization, but this reaction failed for other isomers of (**48**) ⟨72JCS(P1)885⟩. Friedel–Crafts cyclization has been used to obtain the *N*-tosyl ketone (**49**) ⟨72JCS(P1)213⟩. Ring expansion reactions based on dibromocarbene additions have also been used to prepare benzo derivatives ⟨72JCS(P1)889⟩.

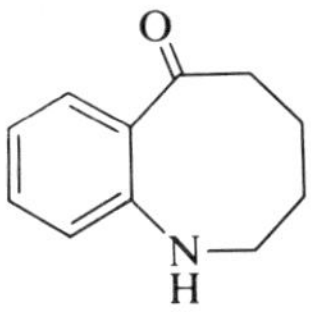

(**48**) 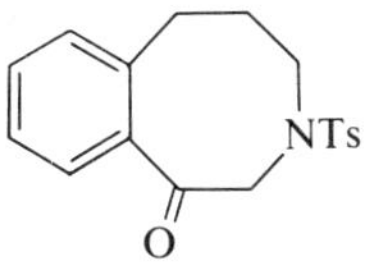 (**49**)

The hydroperoxide (**50**) rearranges in the presence of acid to give keto lactam (**51**) ⟨51JA2196, 51JA2641⟩.

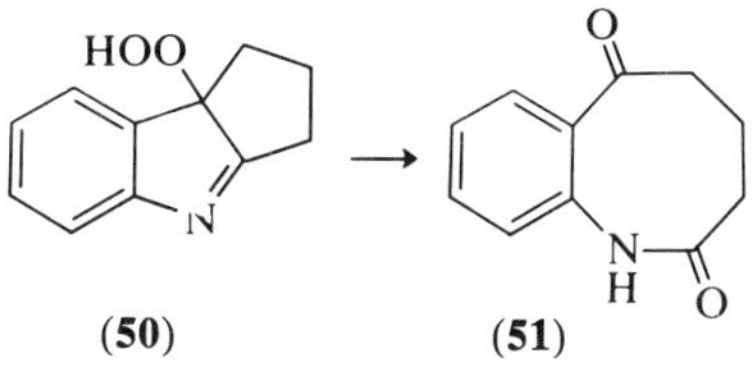

(**50**) (**51**)

Benzazocine intermediates are involved in several of the synthetic approaches to mitomycin, an antibiotic containing a benzo fused pyrrolizidine ring. Conversions between these ring systems can proceed in either direction. Thus cyanogen bromide opens the ring in (**52**) to give (**53**) in 49% yield ⟨78JCS(P1)662⟩. The ketone (**54**), which can be prepared in 25% yield by a Dieckmann cyclization, gives (**55**) on treatment with sodium in ammonia ⟨75CJC960⟩.

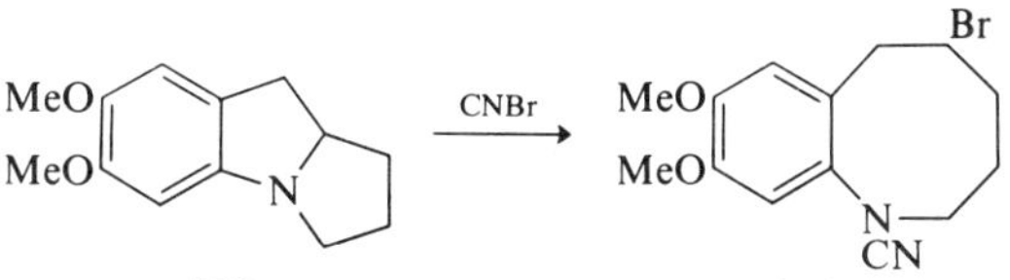

(**52**) (**53**)

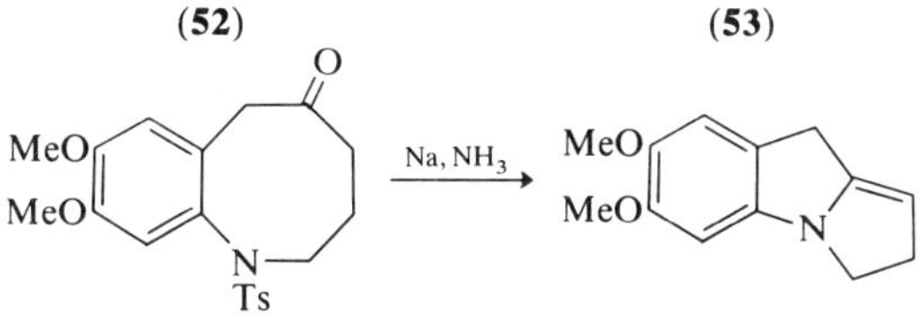

(**54**) (**55**)

The hydroquinone (**56**) cyclizes in 50% yield in the presence of oxygen to give (**57**), a key intermediate in the synthesis of mitomycin and its congeners ⟨77JA4835⟩. Protection of the hydroxyl group in (**57**) and replacement of one of the two geminal methoxy groups by a thiomethyl group give deiminomitomycin A (**58**). The procedure described above has been extended to a synthesis of mitomycin itself ⟨77JA8115⟩. Other workers have also employed eight-membered rings as intermediates in mitomycin syntheses ⟨79JCS(P1)1847⟩.

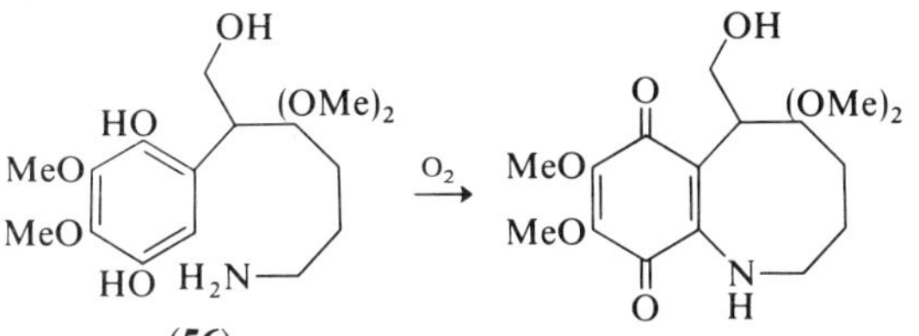

(**56**) (**57**) 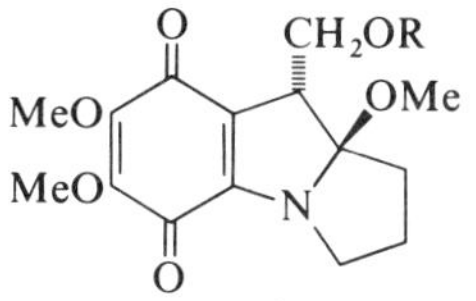 (**58**)

5.19.2.1.2 Nitrogen-bridged derivatives

Manxine (1-azabicyclo[3.3.3]undecane; **60**) is an interesting compound which can be prepared in 60% yield by reduction of the quaternary ammonium salt (**59**) with sodium in liquid ammonia. It has a melting point of 150–152 °C and a remarkably long wavelength absorption [λ_{max} 240 nm, ε 2935 (ether)] for a saturated amine. Its pK_a (9.9) is one unit less than that of quinuclidine, and there is strong evidence that these effects are due to a conformational flattening of the nitrogen atom ⟨72JA7092⟩. An X-ray structure is available only for the HCl salt of manxine ⟨72JA7100⟩ and not for the base itself. PE spectroscopy shows that the nitrogen atom in manxine does not change its geometry very much on ionization to the radical cation, unlike the situation in ammonia or tertiary amines (including quinuclidine); this is consistent with a nearly planar nitrogen atom in manxine ⟨75JA4136⟩.

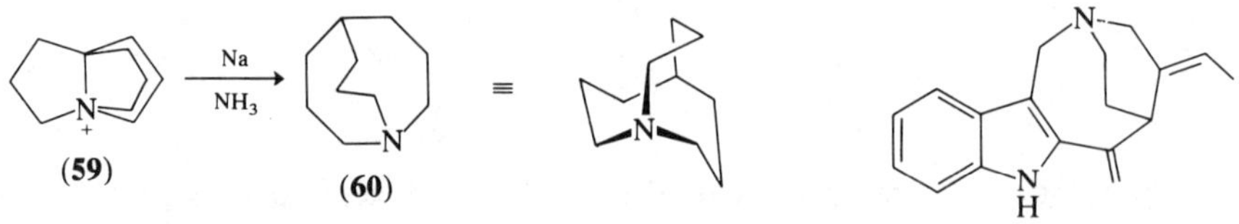

(**59**) (**60**) (**61**)

The indole alkaloid apparicine and its derivatives have an eight-membered heterocyclic ring with the nitrogen at a bridgehead (**61**) but in a [4.2.2] bicyclic arrangement, rather than in a [3.3.3] arrangement as in manxine ⟨80JCS(P2)919⟩. The presence of unsaturation and the sizes of the bridges in (**61**) result in a much lower strain compared to that of manxine.

Conformational aspects of some of the compounds in this section are discussed in Section 5.19.9.

5.19.2.2 Dihydroazocines

A fairly general route to 1,2-dihydroazocines (**64**) is provided by the [2+2] cycloaddition of DMAD to 1,2-dihydropyridines (**62**) ⟨74JCS(P1)2496, 77JOC2903⟩. The reaction gives good yields of the dihydroazocine-6,7-dicarboxylates with *N*-alkyl, -aryl or functionally substituted removable protective groups. Other substituents can be present at C-3 or -4 in the pyridine, but carboxyl substituents at N-1 or C-5 of the pyridine reduce the enamine character, and Diels–Alder addition of the acetylene occurs at the diene system. The [4.2.0] bicyclic intermediates (**63**) can be detected at −10 to 0 °C by NMR; warming to 20 °C causes complete conversion to (**64**).

Cycloaddition of 1-alkyl-1,4-dihydropyridines and DMAD gives cyclobutapyridines which are stable and do not undergo ring opening. However the cycloadducts (**65**) formed from thiocarbamoylmethylenepyridines lead to azocines (**66**) in 40–70% yields ⟨75CPB2749⟩. Similarly, 1-methyl-1,4-dihydroquinoline gives 1,6-dihydrobenzazocine-3,4-dicarboxylate (**67**).

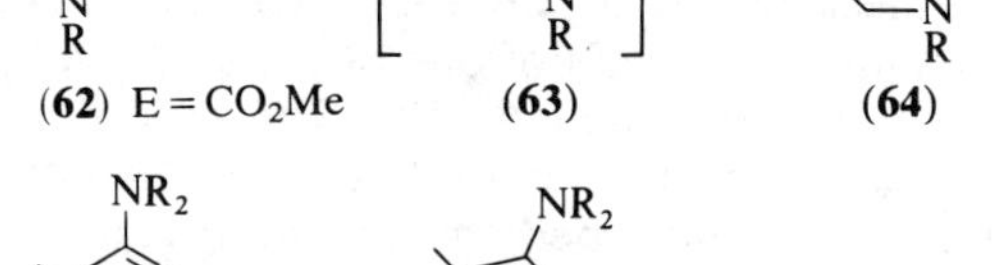

(**62**) E = CO_2Me (**63**) (**64**)

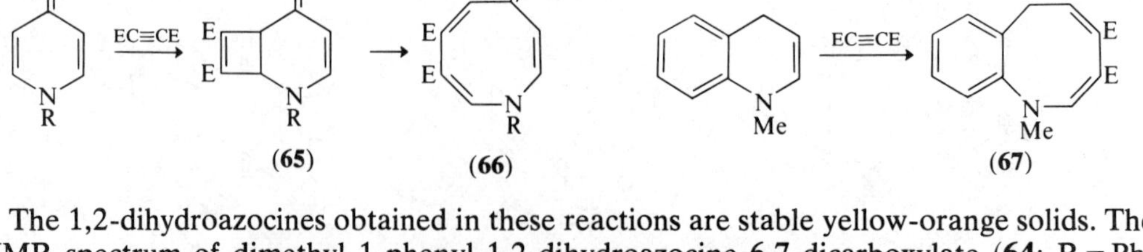

(**65**) (**66**) (**67**)

The 1,2-dihydroazocines obtained in these reactions are stable yellow-orange solids. The NMR spectrum of dimethyl 1-phenyl-1,2-dihydroazocine-6,7-dicarboxylate (**64**; R = Ph) indicates rapid ring flipping at room temperature and freezing of the conformation at −40 °C; bicyclic valence isomers are not detectable ⟨74JCS(P1)2496⟩. The ring is stable to treatment with base at 30 °C, permitting removal of an *N*-β-formylethyl or *N*-formyl group to give the *N*-unsubstituted azocine (**68**), but exposure to mild acid causes rapid decomposition ⟨77JOC2903⟩. Hydrogenation gives 1,2,3,4-tetrahydroazocines (**69**). $NaBH_4$ reduction in one case gave the 1,4,5,8-tetrahydro isomer. Photolysis of the 1-phenylazocine caused cyclization to the [4.2.0] bicyclic diene (**70**).

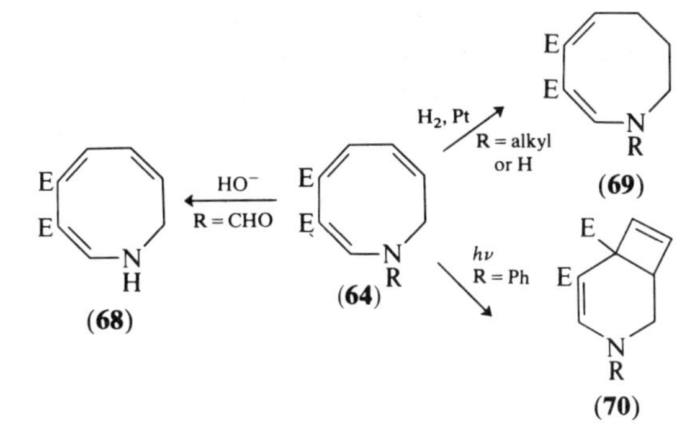

(68) (64) (69) (70)

Derivatives of 1,2-dihydroazocine are potentially available by valence isomerization of the 7-azabicyclo[4.2.0]octadiene system. Sulfonyl derivatives (**71**) of the latter have been prepared, but the equilibria with dihydroazocines lie completely on the side of the bicyclic isomers ⟨71JOC435⟩. On heating, the bicyclooctadienes rearrange to the acyclic tetraenes (**73**), presumably *via* the unstable dihydroazocines (**72**).

The contrasting behavior of the sulfonamides (**71**), which exist in bicyclic form, and the monocyclic dihydroazocines (**64**) must be attributed to the stabilizing conjugation provided by the 7-carboxylate group in (**64**) which would be absent in the bicyclic isomer.

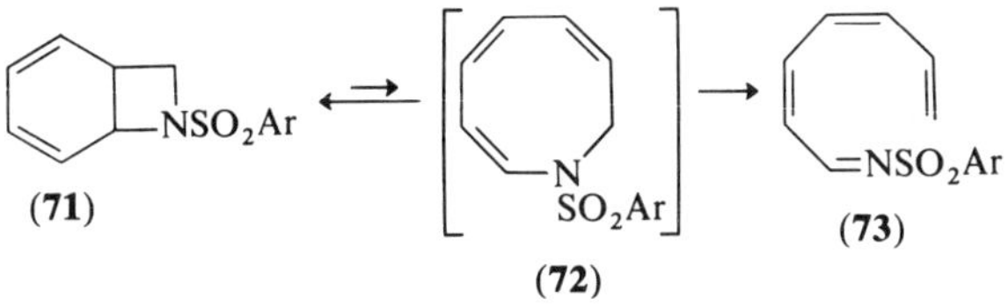

(71) (72) (73)

1-Sulfonyl-1,6-dihydroazocinedicarboxylates (**77**) are formed by valence isomerization of the *cis*-azabishomobenzenes (**76**) at 125 °C, but the compounds are unstable at this temperature and were characterized only by spectra ⟨76CB3505⟩. These azocines are completely monocyclic, and exhibit no valence isomerization, in contrast to the corresponding 6*H*-oxocins.

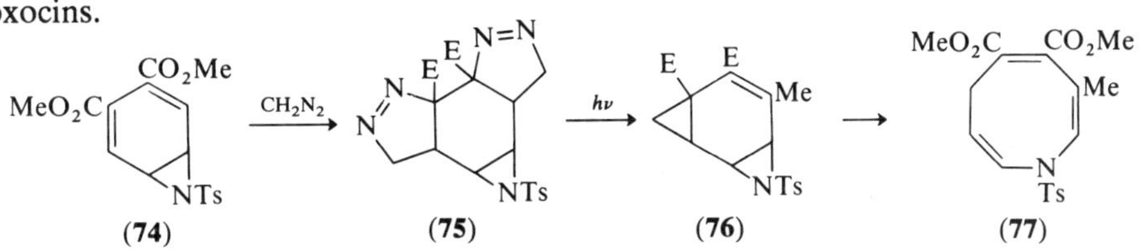

(74) (75) (76) (77)

The dimethylmethoxydihydroazocines, *e.g.* (**78**), obtained by protonation of the dianions, also exist entirely in the monocyclic form. In the presence of base, however, valence isomerization is manifested and elimination of the methanol leads to the cyclobutapyridine (**79**). The dianion (**80**) reacts with benzophenone or benzaldehyde to give a mixture of dihydroazocinemethanol (**81**) and the pyridine (**82**) ⟨71JA168⟩.

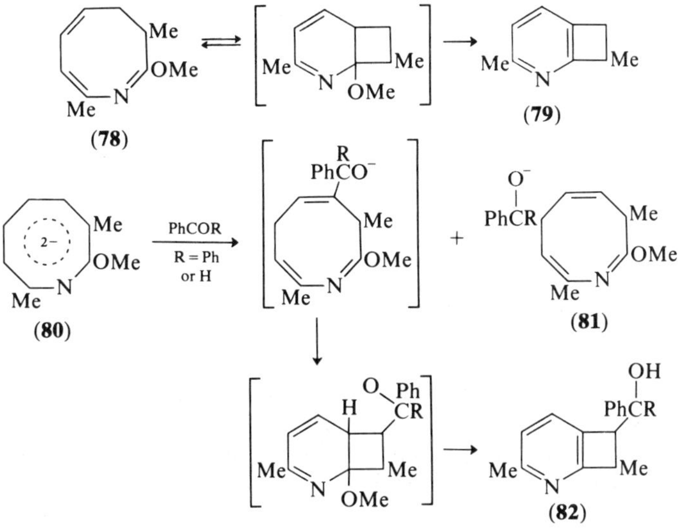

(78) (79)

(80) (81)

(82)

N-Methylbenzodihydroazocines (**83**), (**84**), and (**85**) have been studied from the standpoint of possible stabilization of the 10π-electron anions available by deprotonation ⟨71JA4016⟩. The kinetic acidities, determined by deuterium exchange rates, indicate that (**83**) and (**84**) are converted to the anions about 80 and 30 times faster, respectively, than the corresponding 1,2- and 1,4-dihydroquinoline models; the enhancement in (**85**) was slight. By this criterion the benzazocinyl anion (**86**) was judged to have some degree of aromatic character.

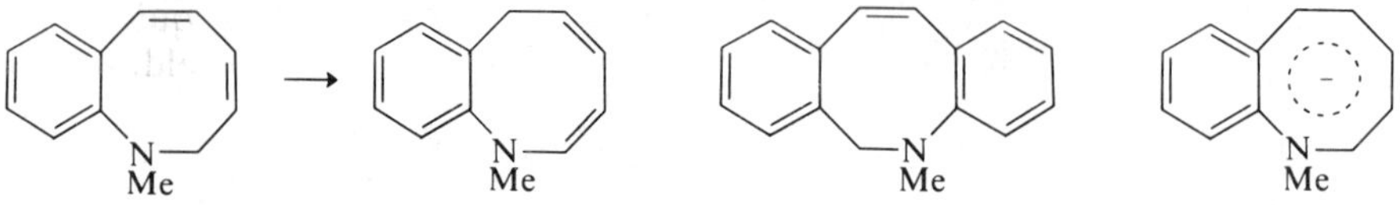

(**83**) (**84**) (**85**) (**86**)

5.19.2.2.1 Dienones

The products formed in the reaction of diphenylcyclopropen-one or -thione and 1-methyl-2-methylthiopyrroline have been shown to be the 1,2,7,8-tetrahydroazocin-2-one (**87**) and the thione, respectively, evidently arising from a 'C,N-insertion' pathway involving an ylide rearrangement ⟨72HCA1759⟩.

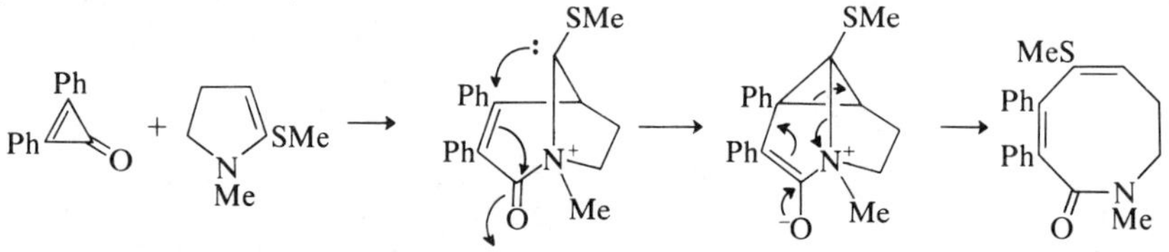

(**87**)

Several *N*-alkyl-2-oxo-1,2,3,4-tetrahydroazocine-3-carbonitriles and -carboxylates (**88**) have been obtained in very low yields by an unusual photocycloaddition of 2-pyridinones and acrylonitrile or methyl acrylate ⟨80T81⟩. The major products are the bicyclo[4.2.0] adducts (**89**).

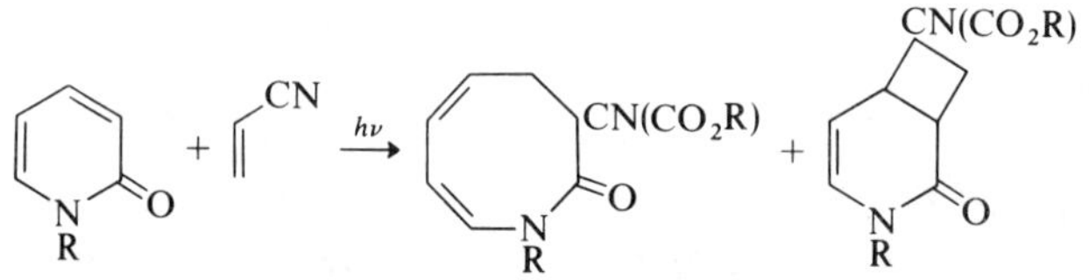

(**88**) (**89**)

5.19.2.3 Azocines

Azocines are a heterogeneous group of compounds. Most of the highly unsaturated azocines have been obtained from bi- or tri-cyclic precursors by bond reorganization processes which often consist of a single example. The properties and reactions of the azocines obtained by various approaches are in large measure characteristic of the substituents associated with a particular method, and for this reason, preparations and reactions have been discussed together.

There has been little systematic or comparative study of azocines as a class, but questions of general interest include the relative stability of the eight-membered rings and bicyclic valence isomers, and the potential aromaticity of 10π-electron systems.

Fully unsaturated azocines are π-equivalent heterocyclic analogs of cyclooctatetraene. Paralleling the latter, the few azocines that have been prepared exist in the monocyclic form, but reactions of azabicyclo[4.2.0] isomers are observed. In the 1,2-dihydroazocine series the equilibrium between mono- and bi-cyclic valence isomers can be controlled by alkoxycarbonyl substituents.

Addition of two electrons to the completely unsaturated azocine or removal of a proton from a dihydroazocine can lead to a dianion or monoanion, respectively, each of which is a 10π-electron system, corresponding to 'π-equivalent' and 'π-excess' analogs of cyclooctatetraenide. Both of these processes have been demonstrated. The dianions are well-characterized aromatic systems, and give rise to dihydroazocines on protonation.

The parent azocine (**91**) was isolated at −190 °C from flash vacuum pyrolysis of 'diazabasketene' (**90**) ⟨71JA3817⟩. The compound, which must be handled in KOH-coated glassware, decomposes at −50 °C to colored tarry material. Characterization was by mass spectrum (*m*/*e* 107), NMR spectrum (see **91**) and conversion with potassium in liquid ammonia to a dianion, which on quenching and hydrogenation gave azocane in low yield.

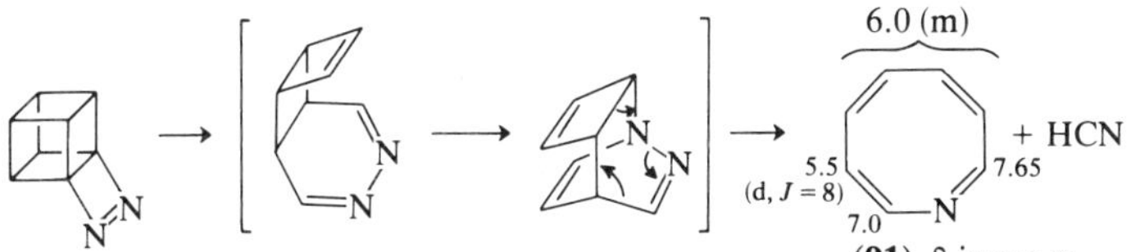

(**90**) (**91**) δ in p.p.m.

The stable diphenylazocinecarboxylates (**94**) were obtained by applying the Diels–Alder reaction of triazine (**93**) and the cyclobutene (**92**) derived from cyclooctatetraene, with two subsequent retro additions leading to the azocine and methyl phthalate ⟨72AJC865⟩. No evidence for double bond alternation or bicyclic valence isomers in (**94**) was seen by NMR. Coalescence of the diastereotopic methylene protons in the ethyl ester (R = Et) from ABX_3 to A_2X_2 at 150 °C indicated ring flipping with a high activation barrier.

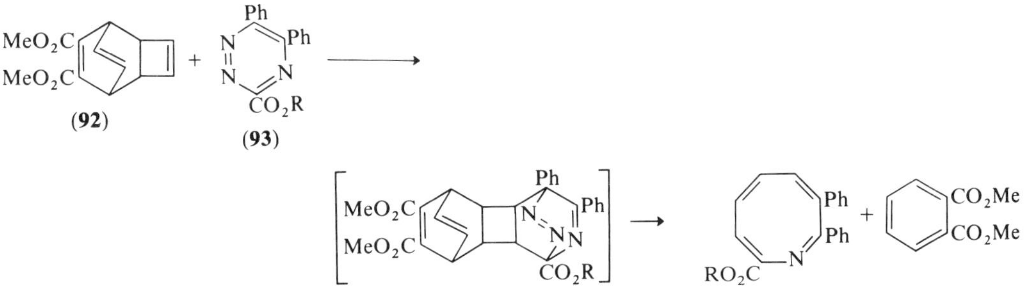

(**92**) (**93**) (**94**)

A similar process with Diels–Alder addition of the benzazete (**95**) and tetraphenylcyclone followed by extrusion of CO provides a route to pentaphenyl-1-benzazocine (**96**) ⟨75JCS(P1)45⟩.

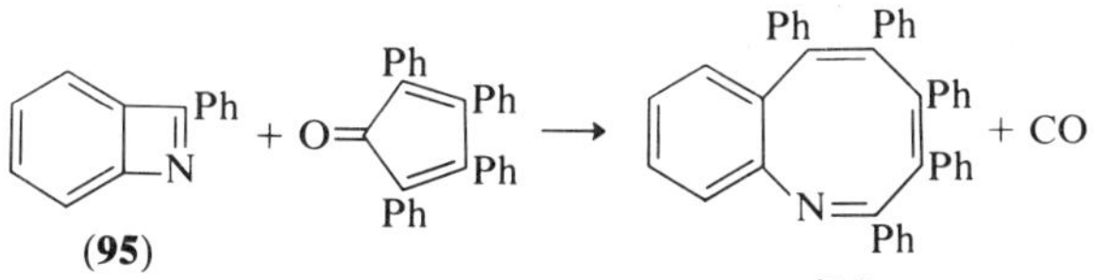

(**95**) (**96**)

By far the most thoroughly investigated azocines are the 2-methoxy derivatives prepared and studied by Paquette and coworkers ⟨71AG(E)11⟩. The synthesis involves addition of chlorosulfonyl isocyanate to a cyclohexadiene, conversion to the imidate and introduction of another double bond by allylic bromination and dehydrohalogenation. Valence isomerization then ensues and 2-methoxyazocine (**98**) or alkyl homologs are isolated from this sequence in multigram quantities as stable yellow oils. NMR data (see **98**) clearly indicate the monocyclic azocine structure; the spectra are invariant from −70 to 180 °C and indicate less than 2% of the bicyclic valence isomers.

Attachment of a polymethylene chain at the bridgehead positions markedly destabilizes the azocine structure, and the 3- and 4-atom bridged compounds exist entirely in the azapropellane form (**99**). With $n = 5$ the NMR spectrum is temperature dependent and

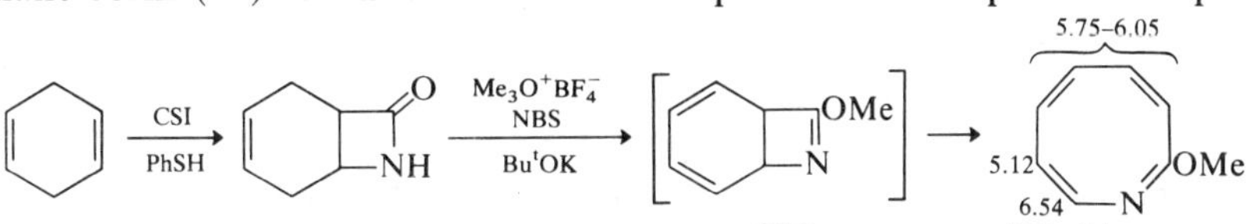

(**97**) (**98**) δ in p.p.m.

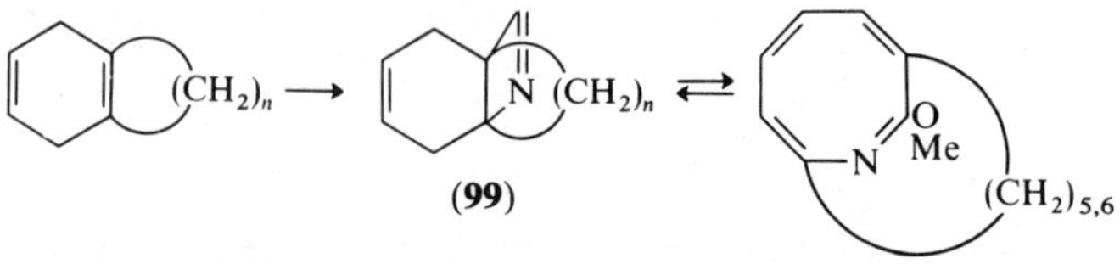

(99)

(100)

indicates a substantial contribution of the bridged azocine (**100**) at elevated temperatures. The equilibrium is displaced completely to the azocine structure when $n = 6$.

Although the monocyclic methoxyazocines are the main contributors to the valence isomeric equilibria between (**98**) and (**101**), most of the reactions lead to products derived from the bicyclic forms ⟨71JOC435⟩. Diels–Alder reaction with *N*-phenyltriazolinedione gives the adduct (**102**). Treatment of the parent methoxyazocine (**98**; R = H) with strong base, or acid hydrolysis of the imidate, leads to benzonitrile and methylbenzoate, respectively, by opening of the four-membered ring in (**101**). The dimethyl compound (**101**; R = Me) undergoes ring opening with rearrangement or elimination to give *o*-xylene or methyl 2,3-dimethylbenzoate, respectively. Demethylation with anhydrous HBr produces the lactams (**103**; R = Me, R′ = H).

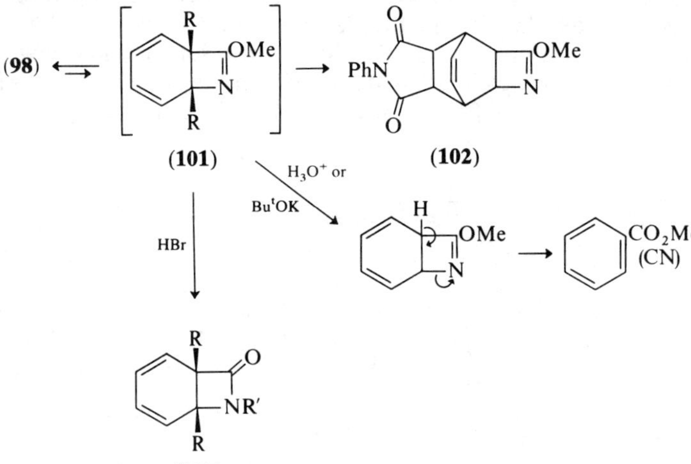

(101) (102) (103)

2-Methoxyazocines exhibit a two-electron polarographic wave corresponding to conversion to the 10π-electron aromatic dianion ⟨71JA161⟩. The HMO delocalization energy for the planar azocinyl dianion has been calculated to be $-5.1\ \beta$, compared to $-3.7\ \beta$ for cyclotetraenide dianion. Voltammetric data for 2-methoxy-3,8-dimethylazocine in acetonitrile–tetramethylammonium fluoroborate have been interpreted in terms of the direct formation of the dianion at the electrode followed by rapid protonation of the strongly basic dianion ⟨76ACS(B)773⟩. In contrast, the reduction of cyclooctatetraene under these conditions takes place in two steps.

The red azocinyl dianions can be generated by means of dissolving potassium in THF or ammonia, and give NMR spectra showing downfield shifts of 0.8 p.p.m. for the protons on methyl substituents, evidence for a considerable diamagnetic ring current ⟨71JA168⟩.

Protonation of the dianions gives mainly the 3,4- (**105**) and 3,6- (**104**) dihydroazocines with small amounts of the 7,8-dihydro isomer (**106**). The product ratio depends on the number and position of methyl substituents, with (**105**) strongly favored in the tri- and tetra-methyl derivatives. Isomerization of the 3,6- (**104**) to the 3,4- (**105**) dihydroazocine occurs with base.

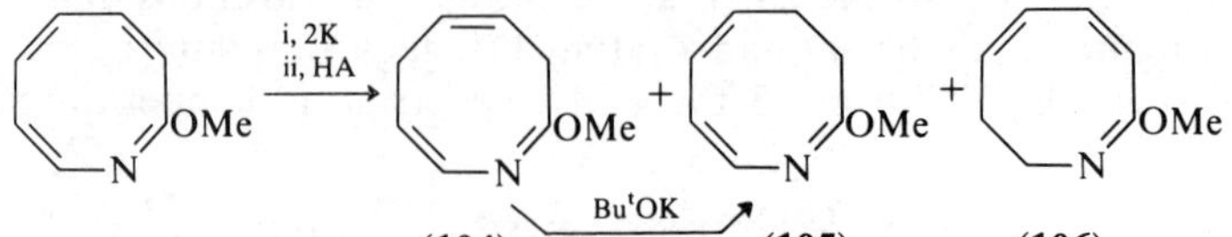

(104) (105) (106)

Several benzo fused methoxyazocines, prepared by conversion of the lactams (benzazazocinones) to imidates, have also been studied ⟨72JA4907⟩. Polarography revealed two-electron reductions, to presumably planar dianions, in the case of 2-methoxy-1-benzazocine (**107**) and the dibenz[*b,f*]azocine (**111**). 4-Methoxy-3-benzazocine (**109**) and the methoxytribenzazocine showed one-electron reduction to radical anions. Chemical reduction of the benzazocines (**107**) and (**109**) gave the 3,6- (**108**) and 1,2- (**110**) dihydro compounds, respectively. The less stable dianion of dibenzazocine (**111**) undergoes further reduction with potassium at −70 °C to give tetrahydrodibenz[*b,f*]azocine; when generated at 25 °C transannular elimination of methanol leads to the indenoindole (**112**). The tribenzazocine, which cannot form a planar dianion because of excessive strain, is reduced by potassium to polymeric products.

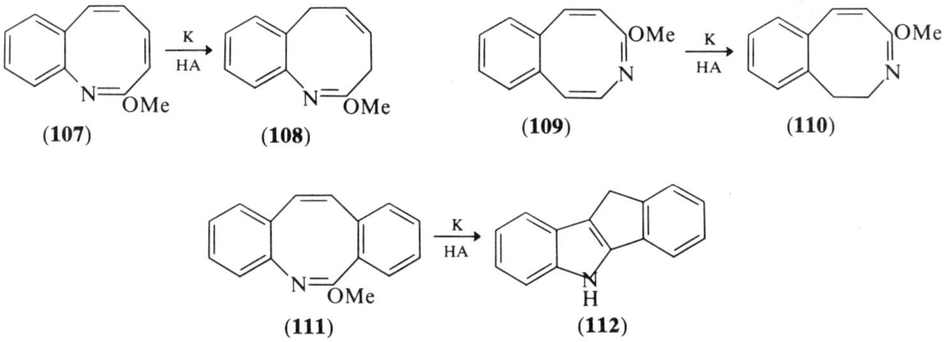

OMe K HA OMe (107) (108) OMe K HA OMe (109) (110)

OMe K HA (111) (112)

5.19.2.3.1 Trienones

The parent 2*H*-azocinone and methyl homologs, available *via* the methoxyazocines, are predominately bicyclic (**113**); however, significant concentrations of the 2*H*-azocinones (**114**) are seen in the NMR spectra. With R = H the azocinone increases from 2% at 60 °C to 15% at 115 °C; the dimethylazocinone (**114**; R = Me, R′ = H) represents 20% of the mixture at 38 °C. Introduction of an *N*-methyl group shifts the equilibrium far to the left, and the azocinone (**114**; R = R′ = Me) is undetectable. The greater stability of the bicyclic valence isomers (**113**) compared to the monocyclic imidates (**98**) and cyclooctatrienone (which is >93% monocyclic) presumably lies largely in the amide resonance of the lactams, which is more fully developed in the bicyclic form than in the twisted tub conformation of the azocinone.

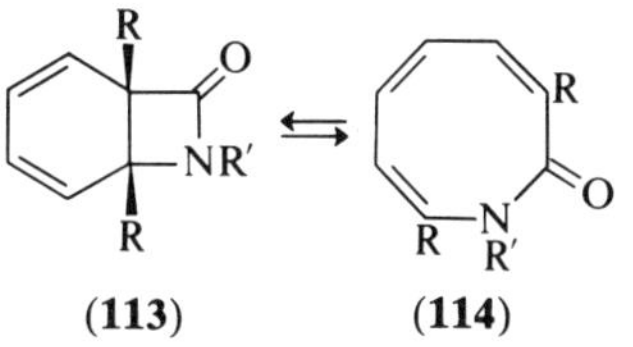

(113) (114)

Benz-, dibenz- and tribenz-azocinones are readily obtained by Beckmann rearrangement of the corresponding benzotropone oximes. The highly substituted 1*H*-2-benzazocine (**115a**) is produced in an electrocyclic reaction when the ketone (**115**) is heated in toluene ⟨72CC680⟩.

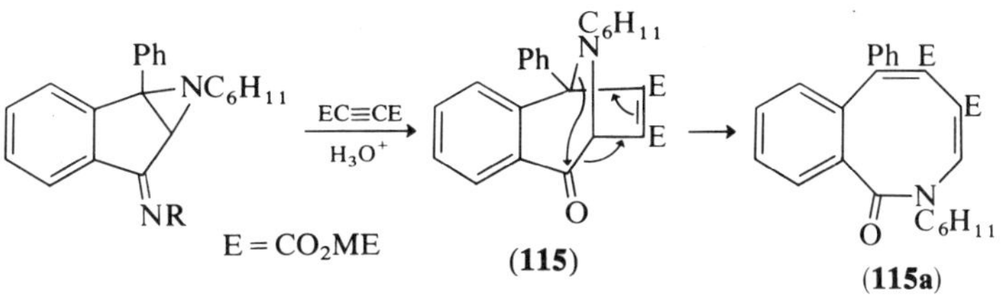

$E = CO_2ME$ (115) (115a)

5.19.3 RINGS WITH ONE OXYGEN ATOM

5.19.3.1 Oxocanes and Tetrahydrooxocins

Oxocane (**116**) is a somewhat inaccessible compound. Cyclization of 7-bromoheptanol under basic conditions gives only a 5% yield of a poorly characterized product, b.p.

131–142 °C, which may contain (**116**) ⟨44CB669⟩. The same authors report that heptane-1,7-diol with hydrogen bromide at 200 °C gives only 1% of an impure product, b.p. 142–150 °C, but treatment of this diol with Nafion-H (a superacidic perfluorinated resin sulfonic acid) at 130 °C has recently been reported to give (**116**), b.p. 140–142 °C, in 51% yield ⟨81S474⟩. This and other cyclic ethers are stated to be pure by GLC and NMR, but no experimental details of the characterization of (**116**) are given. Attempts to prepare (**116**) by a similar dehydration with an ordinary resin sulfonic acid gave a complex mixture of products which contained alkenic and methyl protons, according to the NMR spectrum ⟨76UP51900⟩. Authentic (**116**) can be prepared by the catalytic reduction of 2,3,4,5-tetrahydro-6*H*-oxocin (**117**) ⟨67JOC2723⟩. It has n_D^{26} 1.4486 and its NMR spectrum (CCl_4) shows bands at δ 3.56 p.p.m. for the α-CH_2 groups and a broad band at δ 1.62 p.p.m. for the remaining protons. The reduction of 2-oxocanone (**118**) by LAH or aluminum hydride does not give (**116**) ⟨76UP51900⟩, but the reduction of (**118**) by trichlorosilane under free radical conditions does give (**116**) in about 30% yield ⟨79UP51900⟩.

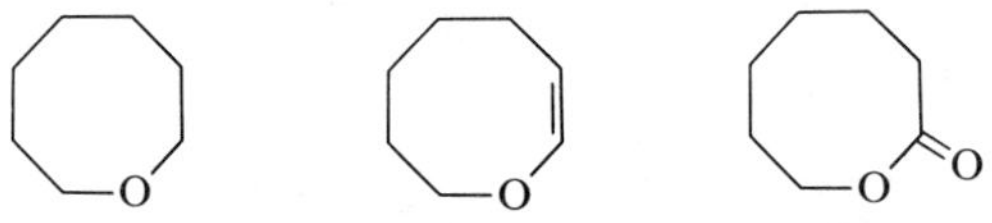

(**116**) (**117**) (**118**)

Tetrahydrooxocin (**117**) has been prepared by two separate and rather long routes. One procedure ⟨67JOC2723⟩ involves the Hofmann degradation of the quaternary methiodide of (**119**) to give the aminodihydrooxocin (**120**) which can be converted into (**117**) by catalytic reduction followed by a Hofmann degradation. Alternatively, (**120**) can be quaternized and catalytically reduced directly to (**116**). Another synthesis of (**117**) starts from dihydropyran and proceeds by a stepwise ring expansion by means of dichlorocarbene. The last intermediate is (**121**) which is converted into (**117**; n_D^{20} 1.4649, IR 1662 and 1644 cm^{-1} (C=C)) ⟨66TL5385⟩. The NMR spectrum of (**117**) shows characteristic chemical shifts and coupling constants for a $CH_2OCH{=}CHCH_2$ moiety.

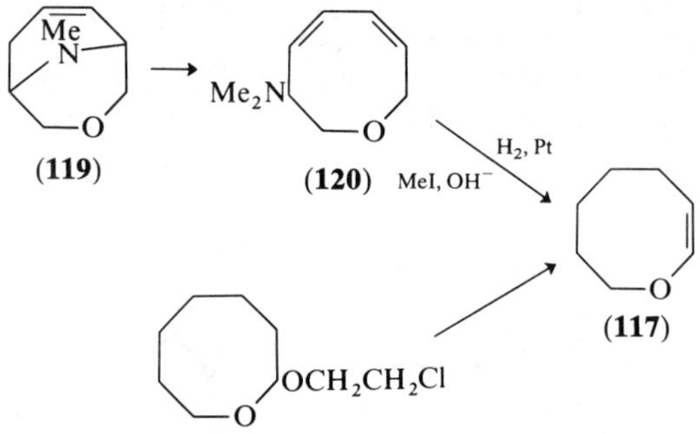

Me N O (**119**) → Me_2N O (**120**) MeI, OH^- ; H_2, Pt → O (**117**) ← OCH_2CH_2Cl O (**121**)

Lead tetraacetate reacts with Δ^7-octen-1-ol to give some five-membered ether (15%) product and almost no six-membered ether, but interestingly, 10–11% each of the two 2-substituted oxocanes (**122**) and (**123**) are formed ⟨73HCA3056⟩. However, this reaction does not appear to be of preparative value.

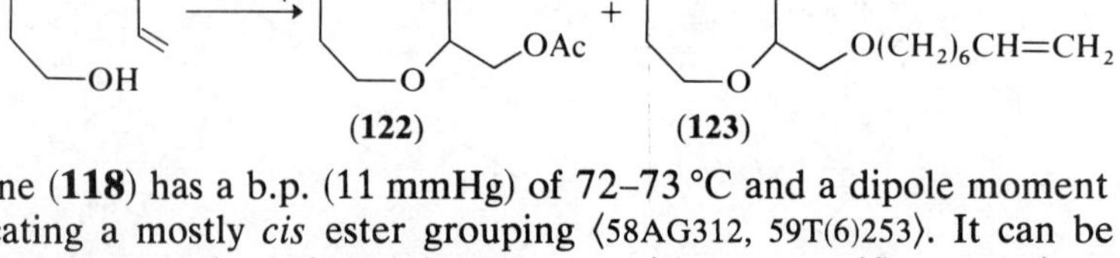

OH $\xrightarrow{Pb(OAc)_4}$ OAc O (**122**) + $O(CH_2)_6CH{=}CH_2$ O (**123**)

The lactone (**118**) has a b.p. (11 mmHg) of 72–73 °C and a dipole moment (benzene) of 3.7 D, indicating a mostly *cis* ester grouping ⟨58AG312, 59T(6)253⟩. It can be prepared in 72% yield by the reaction of cycloheptanone with peroxytrifluoroacetic acid. The UV absorption (λ_{max} 220.4 nm, ε 69, in isooctane) is at longer wavelength than that of aliphatic esters (λ_{max} 212 nm, ε 63) and also indicates a *cis* ester group ⟨67JOC3160⟩.

From the data given in a recent paper ⟨82CC251⟩ dealing with macrolide synthesis, it is expected that cyclization of 7-bromoheptanoate salts to give (**118**) should be very unfavorable. However since (**118**) is readily made by the Baeyer–Villiger method mentioned previously ⟨see also 67JOC3160, 58JA4079⟩, it has been used as a starting material for α,ω-

substituted aliphatic compounds. Thus, the lactone can be converted to methyl 7-bromoheptanoate ⟨75JOC1640⟩; reaction with phenyl selenide and oxidation of the resulting selenoether, gives a terminally unsaturated carboxylic acid ⟨80JA3904⟩.

2,3,4,5-Tetrahydro-8*H*-oxocin (**124**; b.p. 52–53 °C/30 mmHg) is formed, together with ring contracted products, by the acetolysis of 3-oxocanol brosylate ⟨70JA1971⟩. The remaining isomer, 3,4,7,8-tetrahydro-2*H*-oxocin (**125**), is formed by the reaction of 5-oxocanone tosylhydrazone with butyllithium ⟨72JA6751, 6760⟩. The vinylamine (**120**) is hydrolyzed to the unconjugated ketone (**126**) by aqueous acetic acid (HCl and HBr in water give other products). Isomerization of (**126**) with toluenesulfonic acid in benzene gives the conjugated ketone, 2,3,7,8-tetrahydro-6*H*-oxocin-3-one (**127**) ⟨65JA3784, 68JA6148⟩. The saturated ketone 3-oxocanone can be obtained by the reduction of (**126**), and a mixture containing 5-oxacanone (**128**), a little 4-oxocanone and much spiroepoxide product is obtained from the reaction of diazomethane with tetrahydro-4-pyrone ⟨72JA6751⟩. Ketone (**128**) can also be obtained by a Dieckmann cyclization under high dilution conditions in fair yields, together with dimeric products ⟨60JA4075⟩.

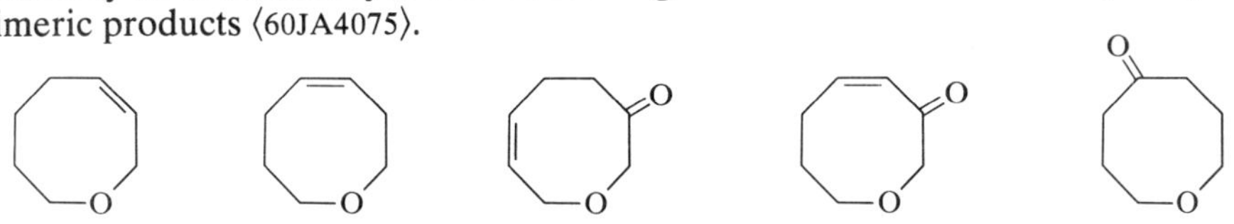

(**124**) (**125**) (**126**) (**127**) (**128**)

5.19.3.1.1 Natural products derived from oxocane

Marine natural products, especially from certain algae, have antimicrobial activity and possess unsaturated chains and often halogens (Br and Cl) attached to an oxygen-containing heterocyclic ring. Various ring sizes occur, but eight-membered rings are quite common. A typical example is laurencin (**129**; m.p. 73–74 °C, $[\alpha_D]+70°$), which has been isolated from *Laurencia glandulifera* Kützing. This compound has been extensively investigated by X-ray, NMR and chemical methods ⟨68T4193⟩. Amongst the more interesting oxocane marine compounds are the isomeric compounds laureatin (**130**) and isolaureatin (**131**) ⟨72JOC680⟩. Treatment of (**130**) with an acidic reagent (the best is $ZnCl_2/CF_3CO_2H/MeCO_2H$) gives (**131**) in 33% yield. An intimate ion-pair has been proposed to explain the stereoselectivity of this rearrangement. Reduction of (**130**) with Raney nickel gives rise to dialkyloxocanols, which can be oxidized to the corresponding oxocanones.

Most of the eight-membered ring oxygen heterocycles derived from marine sources possess a *cis* double bond, although occasionally this is replaced by an epoxide function or its equivalent ⟨some typical recent references are: 81AJC2393, 81TL4081, B-80MI51900, 80CL1177, 80TL2299, 80TL4951, 80T171⟩. A very complex marine product, brevetoxin B (**132**), isolated from organisms responsible for the so called 'red tide' toxin contains an eight-membered oxygen ring in an 11-ring array ⟨81JA6773⟩.

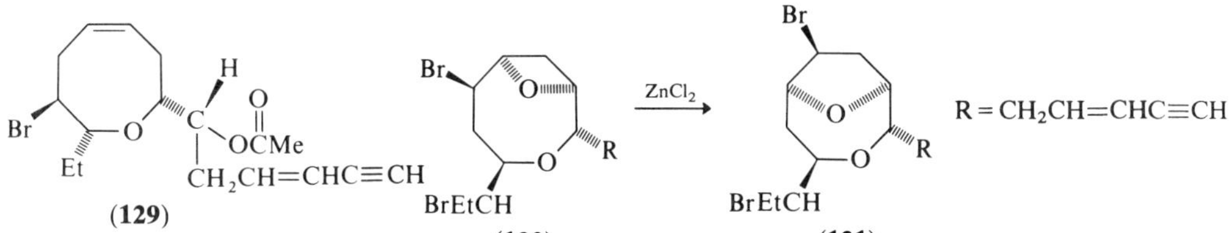

$ZnCl_2$ $R = CH_2CH{=}CHC{\equiv}CH$

(**129**) (**130**) (**131**)

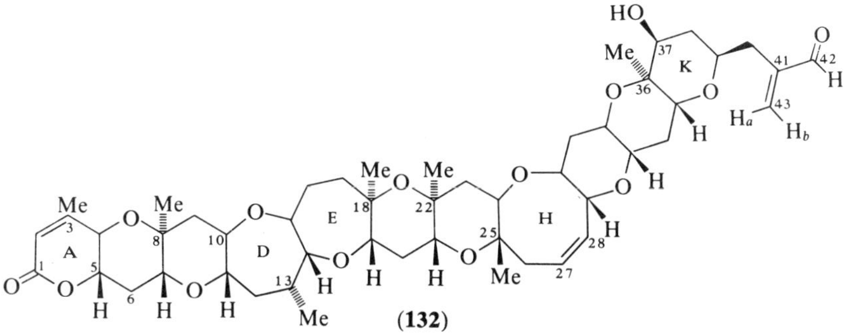

(**132**)

A rather simple bicyclic tetrahydrooxocin derivative (**134**) has been made by slowly distilling the hydroxyketone (**133**). Protonation of (**133**) or (**134**) in concentrated H_2SO_4 gives the oxonium ion (**135**), and heating (**134**) to 280 °C gives a mixture of (**137**) and (**138**), with the latter being the exclusive product at 370 °C. Dehydrogenation of (**134**) with sulfur gives the benzo derivative (**136**) ⟨73BSF2451⟩, which is discussed in the next section.

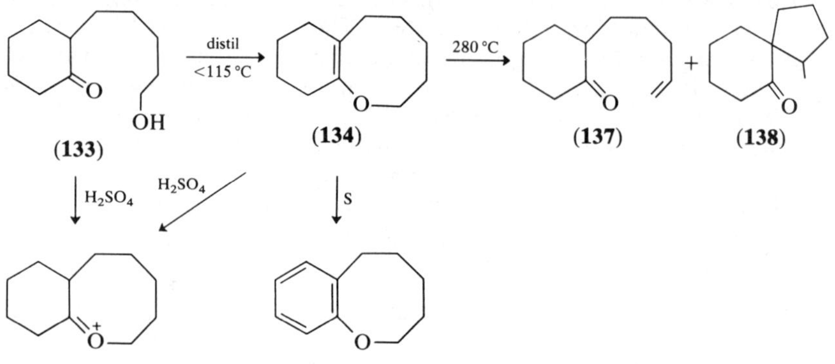

distil <115 °C; 280 °C; H_2SO_4; H_2SO_4; S

(**133**) (**134**) (**137**) (**138**) (**135**) (**136**)

5.19.3.1.2 Benzo derivatives

Only a few benzo derivatives are known, and most of them have the oxygen attached to the aromatic rings, *i.e.* they are cyclic phenol ethers or esters. 3,4,5,6-Tetrahydro-2*H*-1-benzoxocin (**136**), mentioned above, can also be obtained by cyclization of *o*-(bromopentyl)phenol (**139**) with base. The cyclization is rapid in DMSO and does not require high dilution conditions. The yield of (**136**) is 30%; the alkenyl phenol (**140**) is obtained in 57% yield ⟨74JOC2598⟩. The latter product is formed by an intramolecular *E*2 reaction and thus is independent of the concentration of (**139**). In 75% ethanol solution, more of (**136**) is formed (64%) and less of (**140**) (35%), even though the rate of cyclization is much lower than in DMSO ⟨75JA4960⟩.

Cyclization of the ylide derived from phosphonium salt (**141**) gives the 3,4-dihydro-2*H*-benzoxocin (**142**) in low yield as a mixture with the pyran (**143**) ⟨74JOC3038⟩. Hydrogenation of (**142**) leads to the benzoxocin (**136**). The bromobenzoxocin (**145**) is formed on Baeyer–Villager oxidation of the seven-membered ring ketone (**144**) ⟨58JCS2797⟩.

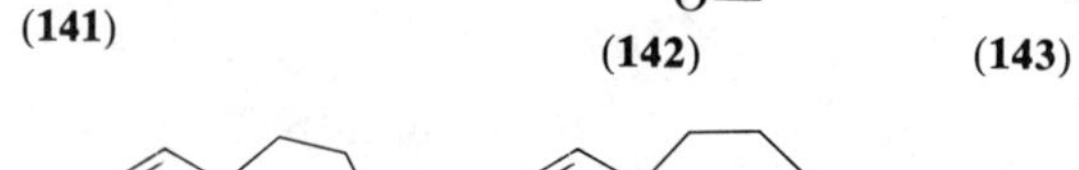

$(CH_2)_5Br$, OH (**139**) $\xrightarrow[\text{DMSO}]{\text{base}}$ (**136**) + $(CH_2)_3CH{=}CH_2$, OH (**140**)

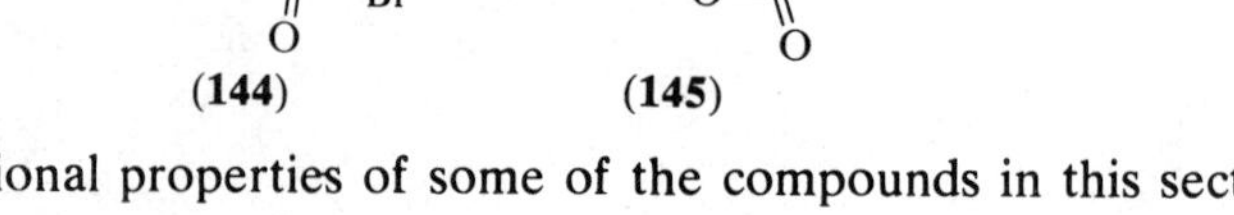

CHO, $O(CH_2)_4\overset{+}{P}Ph_3$ (**141**) → (**142**) + Et (**143**)

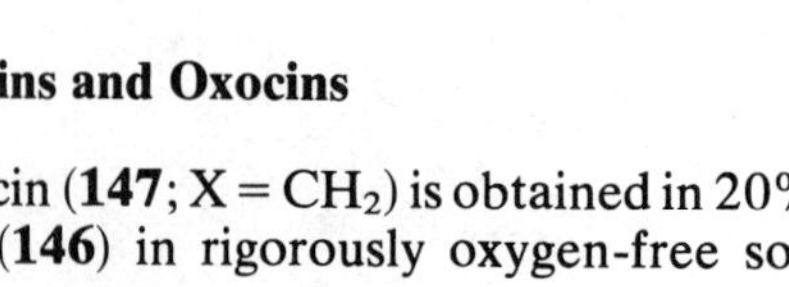

(**144**) Br → (**145**) Br

The conformational properties of some of the compounds in this section are discussed in Section 5.19.9.

5.19.3.2 Dihydrooxocins and Oxocins

The 3,4-dihydrooxocin (**147**; X = CH_2) is obtained in 20% yield by photolytic ring opening of the bicyclic pyran (**146**) in rigorously oxygen-free solution ⟨78CB3608⟩. The reaction, which probably involves a dihydrophenanthrene intermediate, is much more efficient with the dioxane (**146**; X = O), which gives rise to the dioxocin (**147**; X = O) in 88% yield.

Dihydrobenzoxocinone (**149**) is produced by decomposition of the pyrazoline (**148**), from the reaction of 3-acetylcoumarin with excess diazoethane ⟨74JCS(P1)66⟩. The boat-like conformation (**150**) is assigned to the eight-membered ring.

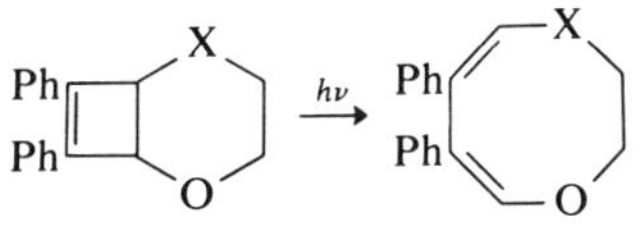

(**146**) (**147**)

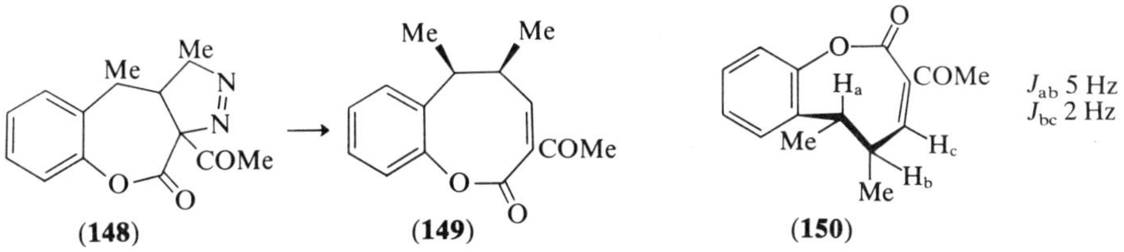

(**148**) (**149**) (**150**)

J_{ab} 5 Hz
J_{bc} 2 Hz

Monocyclic 4*H*-oxocins (**153**) are available by valence isomerization of the *cis*-oxahomobenzenes (**152**), obtained from benzene oxides *via* the monopyrazolines (**151**) ⟨76CB2823⟩. The ring opening is reversible, and an equilibrium is readily established at 100 °C which lies slightly to the side of the oxocins. The activation barriers for the ring opening are about 100 kJ mol^{-1}. Since the interconversion is slow at room temperature the bicyclic oxide and oxocin can be readily separated. The NMR spectrum of (**153**) shows no temperature dependence down to −60 °C; at this temperature the CH_2 protons remain equivalent, indicating a high conformational mobility.

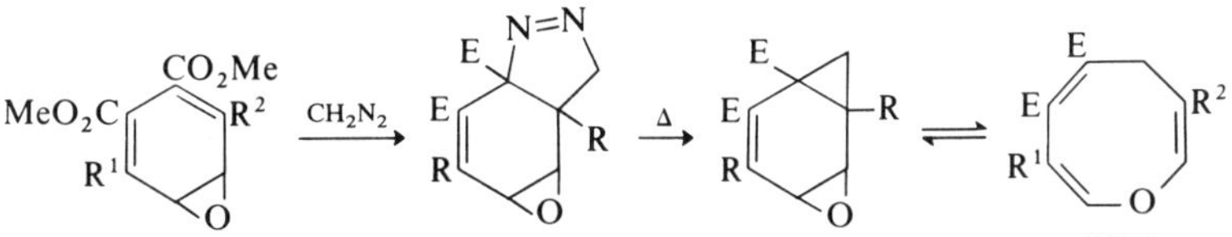

(**151**) (**152**) (**153**)

$R^1 = R^2 = H$
$R^1 = R^2 = Ph$
$R^1 = Br, R^2 = H$
$R^1 = Me, R^2 = H$

Ring expansion of 4-pyrrolidinylcoumarin with acetylenedicarboxylic ester leads to benzoxocinone (**154**) in 43% yield ⟨77JCS(P1)2457⟩.

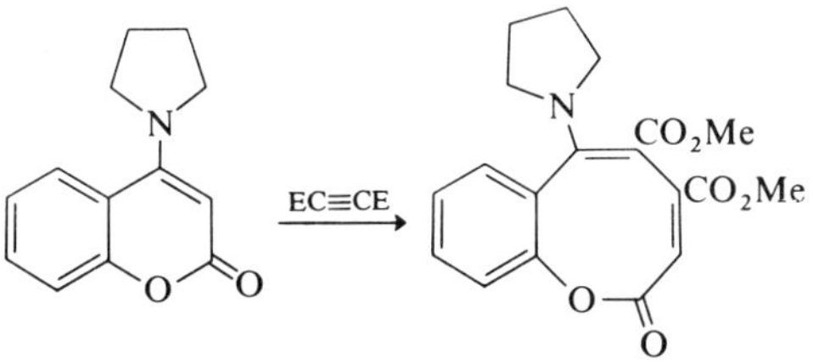

(**154**)

To examine the possibility of aromatic stabilization in an oxocinyl anion, dibenz[*b*,*g*]oxocin (**157**) was prepared from dibenzoxepinmethanol (**155**) by solvolytic ring expansion and dehydration of the oxocinol (**156**) *via* the mesylate ⟨72JOC2161⟩. The relative acidity was determined by pair-wise equilibration of the dibenzoxocin with the monoanions of several reference hydrocarbons. A pK_a value of 27 was estimated for (**157**) placing it between fluorene and xanthene in acid strength. Although the enhanced acidity relative to xanthene suggests 10π-delocalization and some degree of aromaticity in the anion (**158**), NMR data on (**158**) indicate upfield shifts of the benzenoid protons and no aromatic delocalization ⟨80AG(E)393⟩.

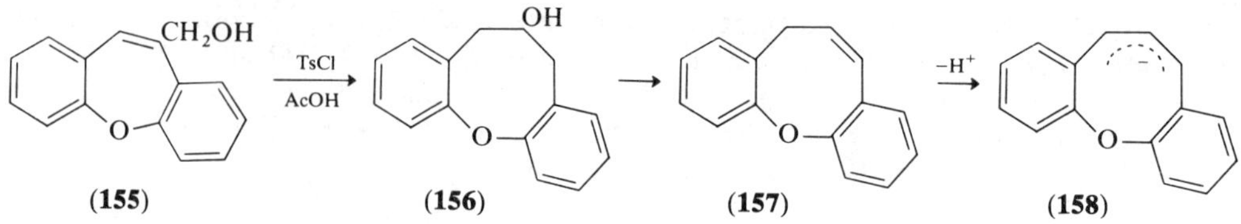

(155) (156) (157) (158)

5.19.4 RINGS WITH TWO HETEROATOMS

5.19.4.1 1,2-Diazocines

The most extensively investigated 1,2-diazocines are 3,4,5,6,7,8-hexahydro derivatives, of interest in connection with studies on the properties of cyclic azo compounds. These compounds are obtained from the hydrazines (**159**) usually not isolated, by oxidation with yellow mercury(II) oxide. 3,8-Diaryloctahydrodiazocines are prepared by reduction of the azines; dialkyl and unsubstituted derivatives are obtained by hydrolysis of the *N,N*-bis(ethoxycarbonyl) compounds ⟨69JA3226, 70JA4922⟩. Cyclization of 2,7-diaminooctane with IF_5 gave the 3,3,8,8-tetramethyl compound ⟨78CB596⟩.

In earlier reports, the two isomers formed in the oxidation of 3,8-disubstituted octahydrodiazocines were assigned *Z*-azo structures with *cis* and *trans* 3,8-groups. It was subsequently recognized that the oxidation products are the *cis-Z* (**161**) and *trans-E* (**162**) structures, and that these give rise to *cis-E* (**160**) and *trans-Z* (**163**) isomers, respectively, on irradiation ⟨81JOC303, 75JA3130, 76CB518⟩. The *Z* and *E* azo groups can be distinguished by their characteristic absorption maxima; *Z*: $\lambda_{max} \sim 380–390$ nm (ε 60–120); *E*: $\lambda_{max} \sim 360–370$ nm (ε 40–60). From molecular mechanics calculations the preferred conformation of the *E*-isomer is the unsymmetrical half-chair (**164**) very similar to that of *Z*-cyclooctene; for the *E* isomer the C_2 twist form (**165**) is 24 kJ mol^{-1} lower in energy than the symmetrical C_{2v} chair ⟨79JA5546⟩.

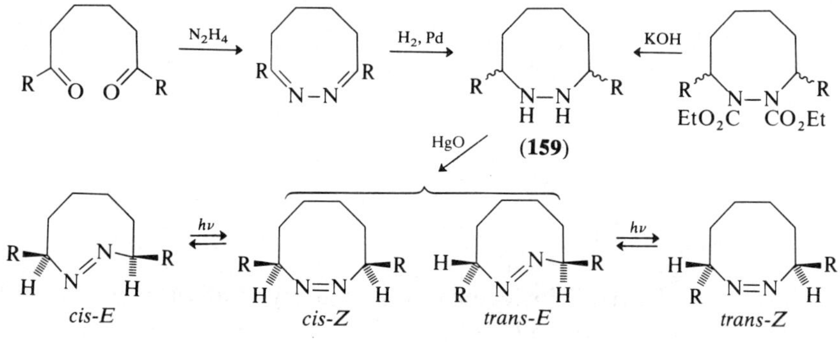

(159)

cis-E (160) *cis-Z* (161) *trans-E* (162) *trans-Z* (163)

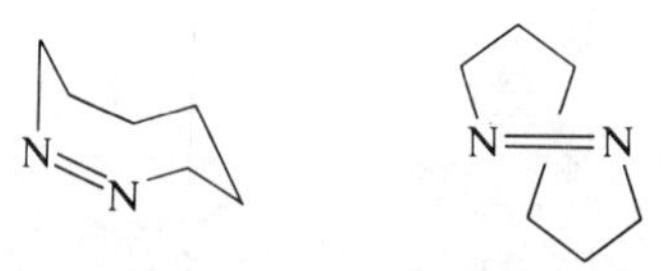

(164) (165)

The stereoisomeric hexahydrodiazocines, as expected, differ considerably in reactivity. The *cis*-3,8-diphenyl *Z* isomer undergoes base-catalyzed rearrangement to the hydrazone (1,4,5,6,7,8-hexahydroazocine) 1000 times faster than the *trans-Z*-diphenyl, and the latter is also very much more stable thermally. The least stable isomer is the *trans-Z* compound, which decomposes explosively at the melting point (62 °C).

Loss of nitrogen occurs on heating or photolysis of (**161**) or (**162**) to give a mixture of stereoisomeric 1,6-disubstituted 1-hexenes and somewhat smaller amounts of 1,2-disubstituted cyclohexanes. Kinetic data suggest that whereas six- and seven-membered cyclic *Z*-azo compounds lose nitrogen in a concerted process, the corresponding *Z*-hexahydrodiazocine undergoes stepwise denitrogenation *via* the intermediate diazenyl diradical ⟨75JA3130⟩.

Tetrahydro-1,2-diazocinones (**166**) are formed in good yield by cyclization of ε-keto acids with hydrazine in dilute solution. Standard transformations lead to the hexahydro derivatives (**167**) and (**168**) and the octahydrodiazocine (**169**) ⟨73BSF2029⟩.

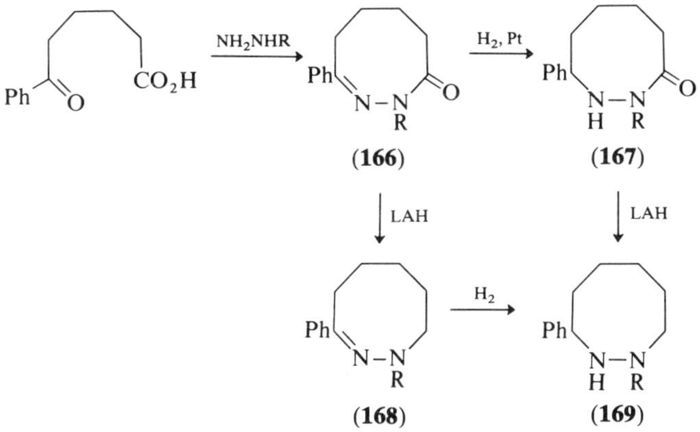

The parent compound 1,2-diazocine has been obtained from the (explosive!) tetracyclic azo compound (**172**) ⟨71JA5573⟩. This precursor can be prepared in 20% overall yield from the DEAD adduct (**170**) ⟨71JA5572⟩ or from the triazolinedione adduct (**171**) of benzvalene ⟨73JA2738⟩. Thermolysis of (**172**) gave quantitatively benzene and nitrogen, but direct photolysis (366 nm) at −78 °C leads to diazocine as a major product along with benzene and minor amounts of Dewar benzene, benzvalene and prismane. Irradiation at 490 nm, with acetophenone sensitizer or at high O_2 pressure, established that the diazocine arises from the T_1 state of (**172**). The yields of diazocine depend on the solvent; the highest diazocine/benzene ratio was 60 in liquid NH_3 ⟨79JOC1264⟩.

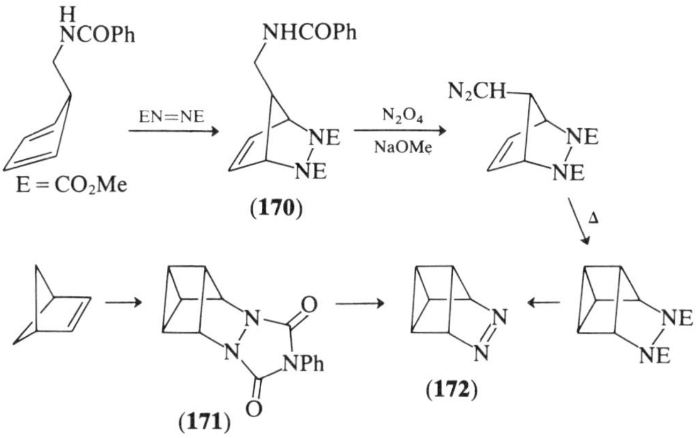

1,2-Diazocine is an unstable yellow oil. The 2,4,6,8-tetraene structure (**173**) is established by the NMR spectra, which show the presence of a single species and are consistent with a tub conformation of (**173**) and not the double bond isomer (**174**). Information on the chemistry of (**173**) is limited. Thermolysis gives mixtures of pyridine and benzene, with approximately equal rates and activation parameters; lower temperatures favor pyridine. Photolysis, however, either direct or sensitized, gives exclusively benzene. No pyridazine is detected in either thermal or photochemical reactions.

The formation of benzene and pyridine presumably occurs *via* the bicyclo[4.2.0] valence isomers (**175**) and (**176**), although adducts of these structures with dienophiles were not obtained on heating or during irradiation. The difference in products in thermal and photochemical reactions could be explained either by interconversion of (**173**) and (**174**) and rate-limiting isomerizations of (**173**) to (**176**) and (**174**) to (**175**), or alternatively, rate-limiting isomerization of (**173**) to (**174**) and symmetry-allowed 4π photocyclization of (**173**) to (**175**) ⟨79JOC1264⟩.

The highly substituted 3,4-diazabicyclo[4.2.0]octatrienes (**177**) have been obtained by cyclization of *cis*-3,4-dibenzoylcyclobutenes with hydrazine ⟨69CB1928⟩. The triene (R = H) undergoes thermal ring opening to give a vinylpyridazine, but with bridgehead substituents (*e.g.* R = Me) conversion to the substituted benzene (**181**) occurs in pyridine at 70–80 °C

(172) $\xrightarrow{h\nu}$ (173) ⇌ [(174)]; (173) $\xrightarrow{4\pi}$ [(175)] → benzene + N_2; (174) $\xrightarrow{4\pi}$ [(176)] → pyridine + HCN; (173) $\xrightarrow{6\pi}$ (176); (174) $\xrightarrow{6\pi}$ (175)

(E_A 117–134 kJ mol^{-1}). Although this reaction must proceed *via* the diazocine (**178**) the latter was not detected spectroscopically. It appears that in this series the barriers for isomerization to the 7,8-diazabicyclo[4.2.0]octatriene (**179**) and its loss of nitrogen are too low to permit isomerization of (**178**) to the presumably more stable and perhaps isolable 2,4,6,8-tetraene (**180**).

(177) → [(178)] → [(179)] → (181) + N_2; (178) ↛ (180)

Dibenzo[*c*,*g*][1,2]diazocine (**184**) is a stable yellow crystalline solid obtained from dinitrobibenzyl by the steps shown ⟨69JOC3237⟩. Reduction of (**184**) gave the dihydro compound (**183**).

1,2-Dihydro-1,2-diazocines such as (**183**) have been referred to as 1,2-diaza[8]annulenes ⟨78AHC(23)55⟩, and are of interest in connection with the potential aromaticity of the 10π-electron system provided by these rings (see Section 5.19.4.4). On the basis of UV and NMR data the dibenzodiazocine (**183**) was suggested to have a 'partially planar' structure; there is no direct evidence of aromatic character.

dinitrobibenzyl $\xrightarrow{\text{Zn, OH}^-,\ \text{HgO}}$ (182) $\xrightarrow{\text{NBS},\ \text{Bu}^t\text{OK}}$ (184) $\xrightarrow{\text{Zn, OH}^-}$ (183)

The 1,6-dihydro-3,4-benzodiazocine (**185**) was prepared from the diketone ⟨60JOC1509⟩. Attempted isomerization of (**185**) to the potentially aromatic 3,4-dihydro compound (**186**) was unsuccessful.

diketone → (185) ↛ (186)

In a valence isomerization approach to the 5,6-dihydrodiazocine series, the all-*cis*-diazabicyclooctadiene (**187**) was isomerized at 120 °C to the thermodynamically more stable *trans*-diene (**188**) ⟨69CB1928⟩. As in the bicyclotriene–tetraene case, ring opening to the diazocine occurs, but is followed by ring flip and recyclization. The monocyclic valence isomer is not detected by NMR.

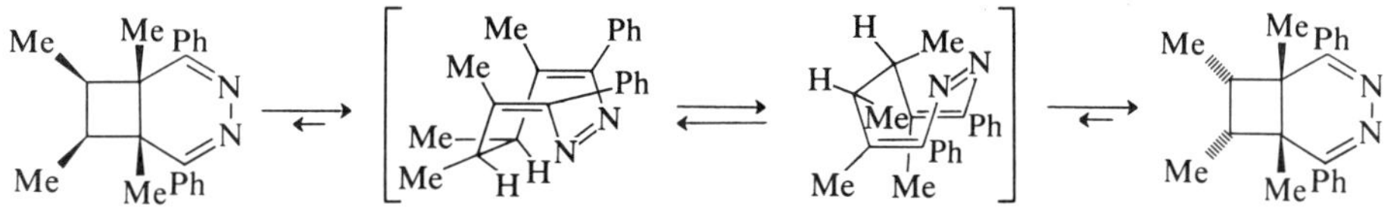

(**187**) (**188**)

In the course of extensive studies on the 1,3-dipolar cycloadditions of heterocyclic betaines, Katritzky and coworkers isolated the benzodiazocinones (**192**) from the reaction of 1-oxido-3-phenylphthalazinium betaine and alkynes ⟨76JCS(P1)2281⟩. The normal cycloaddition products (**189**) were also observed, and the diazocines quite probably arise by electrocyclic rearrangement of (**189**), analogous to that proposed for the similar formation of an azocine (Section 5.19.2.3). The distribution of products depends strongly on the reaction solvent, *e.g.* (**189b**) is obtained in xylene (140 °C) and (**192b**) in refluxing chloroform. Both types of adduct lead to the rearranged bicyclic structures (**190**) on heating in the absence of solvent. The diazocine ring is retained on bromination or catalytic hydrogenation.

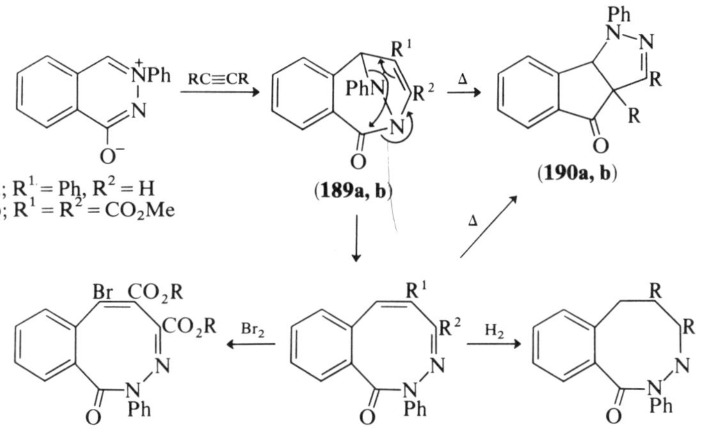

a; $R^1 = Ph$, $R^2 = H$
b; $R^1 = R^2 = CO_2Me$

(**189a, b**) (**190a, b**)

(**191b**) (**192a, b**) (**193a**)

5.19.4.2 1,3-Diazocines

A few scattered examples of hexahydro- and octahydro-1,3-diazocines have been reported, mainly in the context of studies on the effect of chain length on the cyclization of diamines. Reaction of the *N*-aryl-*N′*-aroyldiamines (**194**) with polyphosphoric ester gives the cyclic amidines (**195**) ⟨77JCS(P2)2068⟩. Yields of the diazocines (**195**; $n = 5$) were about 40%, compared to 90% for the 1,3-diazepine (**195**; $n = 4$). The basicities of the products decrease in the order of ring size: $6 \gg 7 > 8 \sim 5$, lower basicity reflecting non-planarity of the amidine system.

Cyclization of pentamethylene bis-urethane (**196**) with formaldehyde gives a very low yield of the 1,3-diazocane (**197**), in contrast to the quite efficient preparation of the tetrazocane by condensation of the tetraaza chain ⟨74JHC937⟩.

Yields of 80–100% are reported for the cyclization of a series of diamines, including (**198**; $n = 5$), with the imidazolium salt (**199**) ⟨79CC117⟩.

$Ar = p\text{-}NO_2C_6H_4$
$Ar' = Ph$ or $p\text{-}NO_2C_6H_4$

(**194**) (**195**)

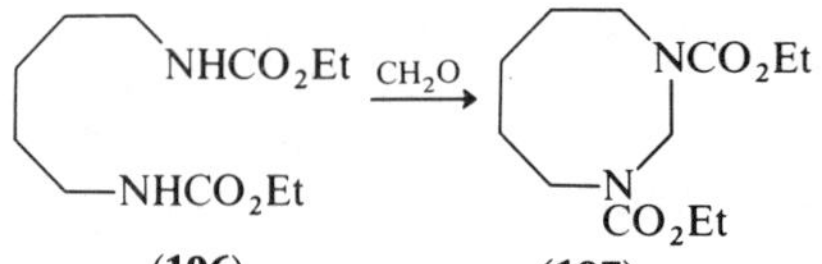

(196) (197) 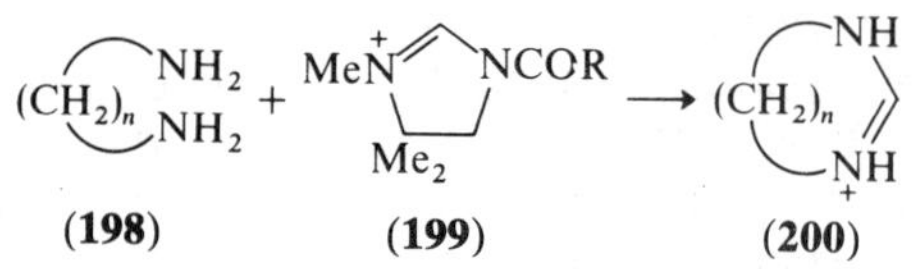 (198) (199) (200)

5.19.4.3 1,4-Diazocines

Reduced 1,4-diazocines and benzologs are available by both direct cyclization and ring-opening routes. The parent octahydro-1,4-diazocine (1,4-diazocane) (**201**) has been prepared in 14% yield by alkylation of the ditosylated diamine with 1,4-dibromobutane and detosylation ⟨74MI51901⟩; it behaves as a typical diamine. In the benzo series, dialkylation was more efficient for the eight-membered ring than either the lower or higher homologs, and led to the ditosylated 1,6-benzodiazocine (**202**; R = Tos) in 90% yield ⟨53CB197⟩. The electrochemical behavior of *N,N*-dimethylhexahydro-1,6-benzodiazocine (**202**; R = Me) (E_1° 0.38; E_2° 0.82 V) resembles much more closely that of the seven-membered lower homolog (E_1° 0.35 V) than that of the nine-membered homolog (E_1° 0.56 V), indicating significantly greater lone pair–lone pair interaction in the eight-membered diamine compared to the nine-membered ring ⟨78JOC2621⟩.

Cyclization to the hexahydro-1,4-benzodiazocinone (**204**) was carried out in 60% yield by hydrogenation of the nitro acid (**203**) and carbodiimide dehydration. The diamine (**205**) was obtained by hydride reduction ⟨71CJC2023⟩.

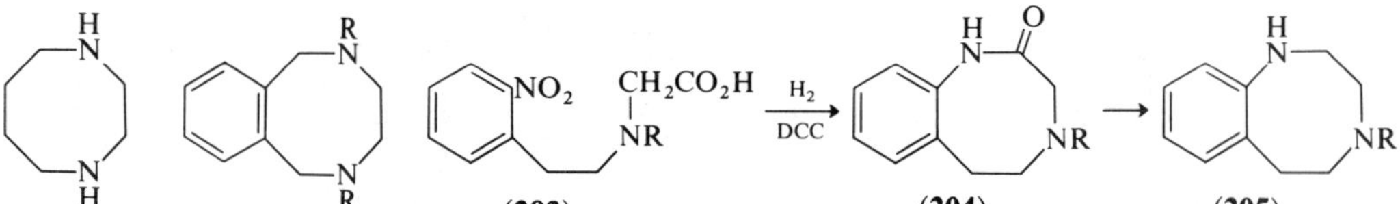

(201) (202) (203) (204) (205)

1-Aryltetrahydro-2,5-benzodiazocines (**207**) are obtained by ring closure of the diaminobenzophenones (**206**) ⟨67JOC3720⟩. These were hydrogenated to hexahydro derivatives (**208**); the latter can be oxidized back to (**207**) in low yield. A more convenient and general route to the hexahydro series is hydride reduction of the imidazoisoindolones (**209**) obtained by condensation of the *o*-benzoyl benzoic acid and diamine ⟨cf. 67JOC2185, 69JOC249⟩. Exposure of the diazocines (**210**) to potassium amide causes ring contraction to *N*-aminoethylisoindolines (**211**).

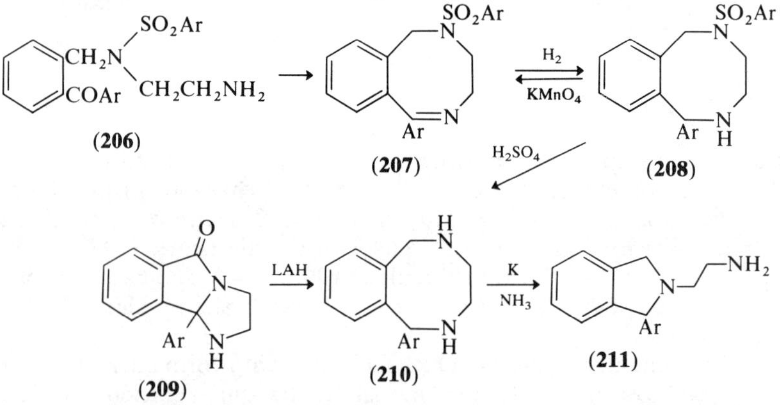

(206) (207) (208) (209) (210) (211)

Aside from the 1,4-dihydro-1,4-diazocines (1,4-diaza[8]annulenes) discussed in Section 5.19.4.4.1, completely unsaturated 1,4-diazocines are represented in the literature only by benzo and dibenzo fused systems.

One of the earliest authentic eight-membered heterocyclic compounds to be described was 6,7-diphenyldibenzo[*e*,*g*][1,4]diazocine (**212**) prepared by Täuber in 1892 by condensation of 2,2′-diaminobiphenyl and benzil. The diazocine ring is opened hydrolytically under vigorous conditions and is reduced to the 5,6,7,8-tetrahydro compound with sodium amalgam. With LAH, ring contraction accompanies reduction to give the carbazole (**215**) ⟨59JOC306⟩.

The main interest in (**212**) and related dibenzo and dinaphtho compounds is in the conformational barrier to racemization of optically active derivatives, which requires deformation of the tub to the planar form. The compounds have proven to be optically stable at very high temperatures ⟨64JCS2326⟩. A minimum $\Delta H^{\ddagger}$ value of 71 kJ mol^{-1} has been calculated for racemization of the resolved 3,10-dicarboxylic esters of (**212**). Thermal decomposition sets in at 240 °C and leads to the phenanthridine (**214**) and benzonitrile, presumably *via* a diradical (**213**) ⟨63JOC3007⟩.

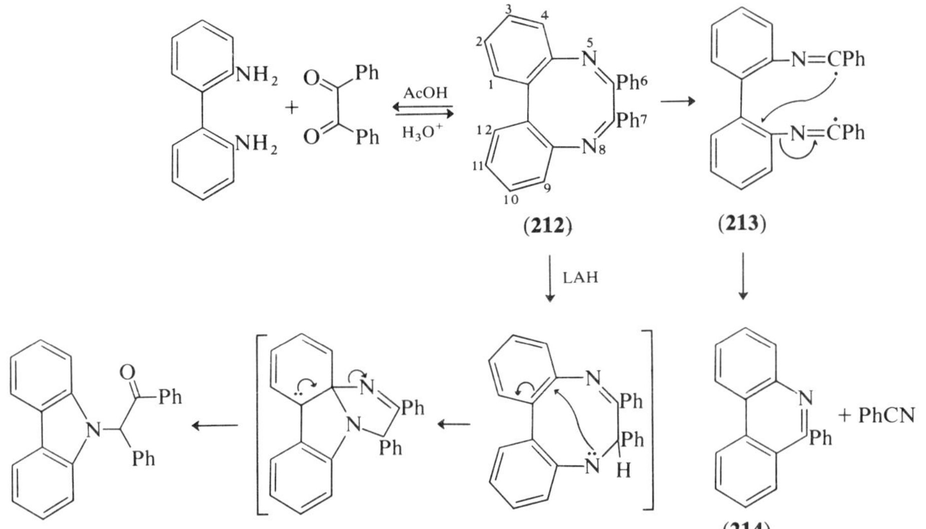

(**212**) (**213**) (**215**) (**214**)

The dibenzo[*b*,*f*][1,4]diazocine system (**216**) has been obtained by condensation of *o*-phenylenediamines with *o*-diaroylbenzenes in the presence of acid catalysts under very specific conditions ⟨68CC1202, 75S246⟩. In the presence of $AlCl_3$ or TsOH the dibenzodiazocines are formed in 70% yields; different conditions lead to formation of *o*-aminoarylisoindoles. Racemization of an optically active aldehyde (**216**; R = CHO) has $\Delta G^{\ddagger}$ 155 kJ mol^{-1} ⟨76CC276⟩.

1,4-Diazocine structures at various oxidation levels have been attributed on numerous occasions to condensation products of *o*-dialdehydes, *o*-acylbenzoic acids or *β*-aroylpropionic acids with 1,2-diamines, but the compounds have later been shown to be imidazoisoindole derivatives (**217**) ⟨74JHC89, 72BSF2868, 67JOC2180, 67JOC2185⟩.

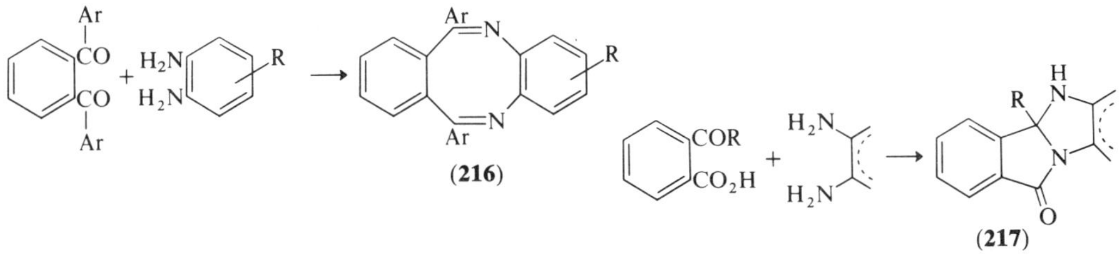

(**216**) (**217**)

1,6-Benzodiazocine and dibenzo[*b*,*f*][1,4]diazocine-diones (**220**) and (**218**) are obtained by condensation of diethyl succinate and phthalate, respectively, with *o*-phenylenediamine ⟨69JOC2138⟩. Contrary to an early report, the use of phthaloyl chloride leads to other products and not the tricyclic diazocine. The dibenzo derivative is also formed in the oxidation of diazabiphenylene (**219**) ⟨78T495⟩. Ring contraction of (**218**) occurs readily on heating or on treatment with PCl_5 or with base.

Several benzodiazocinones have been encountered in processes involving the formation and ring opening of divers benzo fused pyrrolo[1,2-*a*]imidazole intermediates, 1,4-Benzodiazocine-2,5-dione (**221**) has been prepared from 1-azidoacetylisatin by catalytic hydrogenolysis and spontaneous ring opening of the intermediate cyclol ⟨70CJC1946⟩. Addition of 1-alkylbenzimidazoles to the activated triple bond of 4-phenylbutynone gives in low yields a series of 1-alkyl-3-methyl-5-phenyl-1,6-benzodiazocinones (**222**)

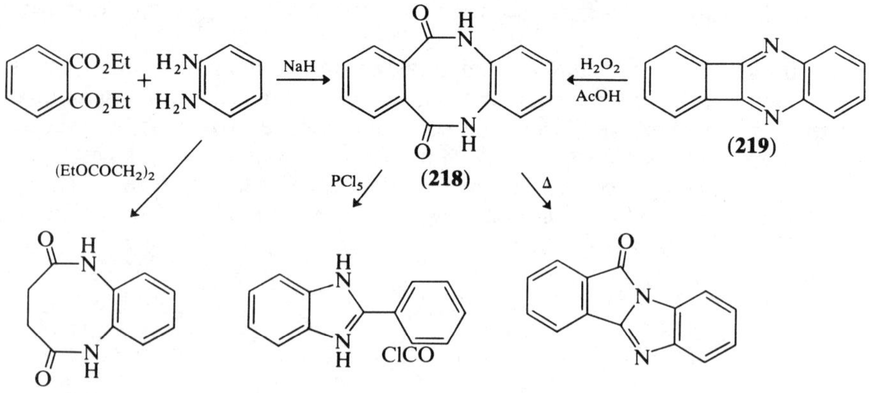

(218)

(219)

(220)

⟨79JCS(P1)584⟩. The 3-amino-2,5-benzodiazocinedione (**223**) is obtained by addition of phthalimide to 3,3-dimethyl-2-dimethylaminoazirine. This diazocine is highly labile and undergoes ring opening on warming in alcohols to the 4*H*-pyrroles (**224**) ⟨77HCA2476⟩.

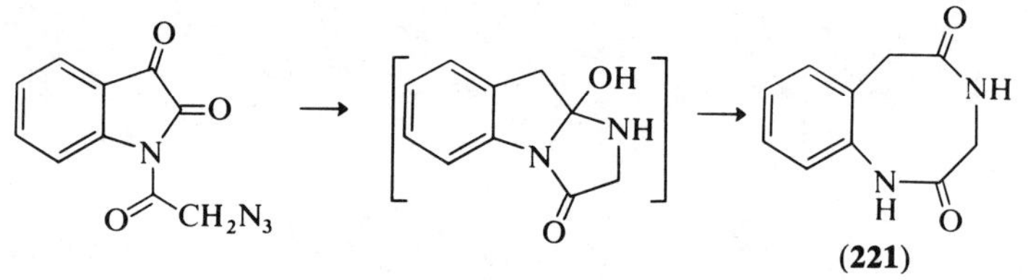

(221)

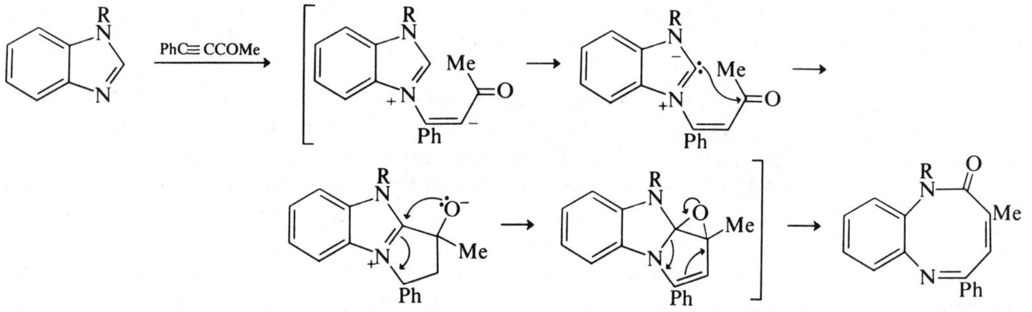

(222)

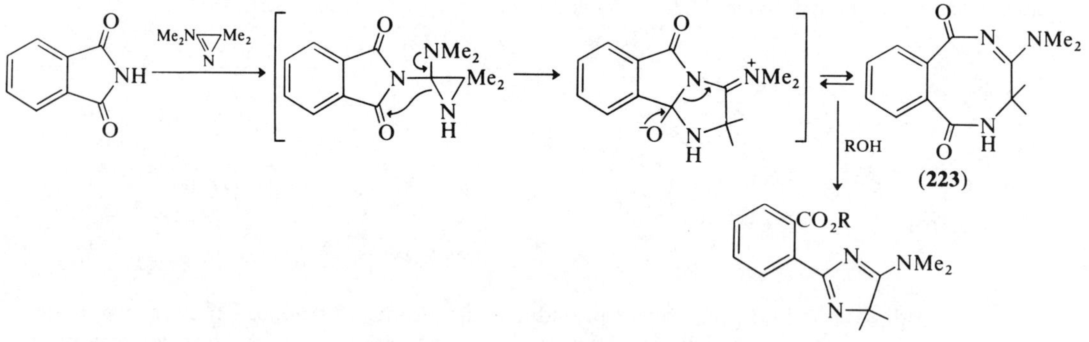

(223)

(224)

5.19.4.4 1,4-Diheterocins

Eight-membered rings with two O, S or N atoms or combinations of these heteroatoms in a 1,2- or 1,4-relationship and three double bonds possess conjugated π-electron frameworks and can be designated as dihetera[8]annulenes ⟨78AHC(23)55⟩. These 1,2- or 1,4-diheterocins are isoelectronic with the cyclooctatetraene dianion, and if planar represent potentially aromatic 10π-electron systems. Considerable interest has attached to the degree of aromaticity of these compounds from both theoretical and experimental standpoints. Most theoretical treatments have led to the conclusion that 1,2-diheterocins, because of the adjacent pairs of *p*-electrons, should be non-aromatic. 1,4-Dioxocin and 1,4-dithiocin have been predicted on other grounds to have no aromatic properties, and experiment bears this out. Certain 1,4-dihydrodiazocines, however, have been found to be distinctly aromatic in character. In this section, 1,4-diheterocins with nitrogen, oxygen and sulfur atoms are treated as a group.

A general approach to 1,4-diheterocins is provided by the valence isomerization of benzene diimines and dioxides. These reactions form part of a general pattern of retro Diels–Alder conversions of *cis*-heteroannelated benzenes to give seven-, eight- and nine-membered rings. The facility with which these reactions occur decreases in the order 7>8>9; the equilibria lie further to the right in the carbocyclic and nitrogen systems than in the oxygen rings.

X, Y, Z = O, NR, S

5.19.4.4.1 1,4-Dihydro-1,4-diazocines

The parent 1,4-dihydro-1,4-diazocine (**232**) and a number of *N,N'*-derivatives have been prepared and thoroughly characterized by Prinzbach and Vogel and their colleagues ⟨79AG(E)962, 964, 80CB3161⟩. The benzene diimine precursors have been obtained by two routes. In one case *cis* benzene dioxide was converted in several steps to the mesyl derivative (**225**) and thence to the diimine dimesylate. The second proceeds from *cis,cis*-benzene trioxide to the triimine (**227**). Two equivalents of acid chloride give the diamides (**228**) in good yield, and nitrosation with cheletropic elimination of N_2O leads to the *N,N'*-disubstituted diimines (**229**).

The diimines (**229**) differ considerably in thermal stability. The sulfonyl compounds (**226**) and (**229a**) are stable crystalline compounds which rearrange at elevated temperatures. The acyl derivatives (**229b** and **c**) and the dianion formed by demesylation of (**226**) undergo retrocyclization below −15 °C at a rate equal to or faster than loss of N_2O from *N*-nitroso intermediates.

The parent dihydrodiazocine (**232**) is a sublimable crystalline solid, m.p. >135 °C, which decomposes rapidly in solution in the presence of air. The UV spectrum [λ_{max} (nm) 284 (sh), 263, 254, 241 (MeCN)] resembles that of dilithium cyclooctatetraenide. NMR data for several diazocines are given in Table 2.

Table 2 NMR Data for 1,4-Dihydro-1,4-diazocines ⟨80CB3161⟩

Compound	*R*	$\delta(^1H)$ (p.p.m.) H-2(3)	H-5(8)	H-6(7)	$^3J(H\text{–}H)$ (Hz) 5, 6	6, 7	$\delta(^{13}C)$ (p.p.m.) C-2	C-5	C-6
(**232**)	H	6.25	6.80	5.72	10.4	11.7	97.9	112.5	92.1
(**230d**)	Me	5.85	6.49	5.64	10.5	10.5	102.5	117.1	91.8
(**230b**)	COPh	6.32	6.58	5.97	8.9	3.9	115.5	128.9	120.1
(**230c**)	CO_2Me	6.15	6.44	5.76	8.5	3.8	112.4	126.5	118.6
(**230f**)	$CONMe_2$	6.23	6.58	5.43	10.5	8.9	105.0	120.2	100.4
(**231**)			6.72	5.33					

Detailed analysis of the NMR data for compounds (**232**), (**230a–f**) and (**231**) permits several conclusions about the structure and aromatic character of 1,4-dihydro-1,4-diazocines. The ^{1}H NMR chemical shifts for the *N,N*-dimethyl compound (**230d**), taken together with those of the benzo[*b*] analog (**236**) and the 1,4-dioxocin counterparts, indicate a downfield shift of 1.1–1.2 p.p.m. for the ring protons of (**230d**) compared to those predicted for a non-planar non-aromatic structure. This shift suggests a diamagnetic ring current similar in magnitude to that observed in furan or thiophene. The nearly equal ^3J(H–H) coupling constants for the diene segment of (**232**) and (**230d**) compared to the unequal values for the *N*-acyl diazocines (**230b** and **c**), as well as the ^{13}C NMR chemical shifts, reveal extensive flattening and equalization of bond lengths in the former and a twist-boat-chair structure in the latter. Significant differences in the data for the bis-carbamate (**230c**) and bis-urea (**230f**) point up the delicate balance in the electron distribution, which leads in (**230c**) to a twisted ring and in (**230f**) to a nearly planar diatropic structure.

Further evidence for aromatic character in this system is found in the crystallographic structures of (**232**) and (**230f**), which have mean bond angles of 135° and mean deviations from planarity of 0.014 and 0.05°, respectively. By contrast, the mean bond angle of (**230c**) is 125.4°.

Chemical evidence for the aromatic character of (**232**) is provided by its stability to acid and base. No tautomerization to imine forms such as (**223**) is detectable by NMR, indicating a lower limit of perhaps 84 kJ mol^{-1} for the stabilization energy. Finally, the 1,4-dimethyl derivative (**230d**) forms a charge transfer complex with trinitrobenzene (λ_{max} 488).

Thus by several criteria 1,4-dihydro-1,4-diazocines without a strongly electron-withdrawing group at the 1,4-positions are aromatic compounds, whereas those with such groups are non-planar systems with localized bonding. Diene character in the latter compounds is seen in the cycloaddition of *N*-phenyltriazolinedione to give the adducts (**234b** and **c**).

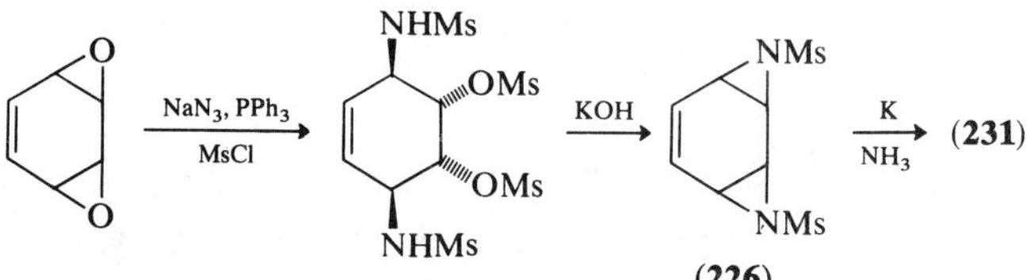

(**225**) Ms = mesyl (**226**) 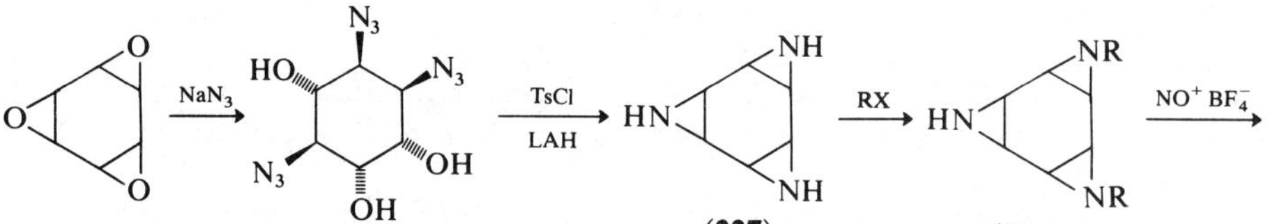 (**227**) (**228**)

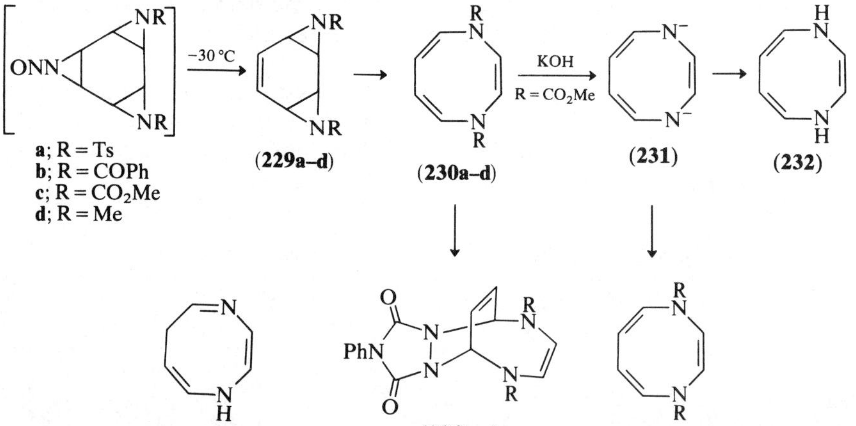

a; R = Ts
b; R = COPh
c; R = CO_2Me
d; R = Me

(**229a–d**) (**230a–d**) (**231**) (**232**)

(**233**) (**234b, c**) (**230**) e; R = $SiMe_3$
f; R = $CONMe_2$

1,6-Dihydro-1,6-benzodiazocine has been prepared as the *N,N*-dimethyl compound (**236**) *via* the bicyclic precursor (**235**) ⟨71TL2437⟩. The diazocine (**236**) is devoid of aromatic properties. The NMR values δ 4.4 (H-3,4) and 5.6 (H-2,5) p.p.m. are those expected for a dienamine; cycloaddition leads to the adduct (**237**).

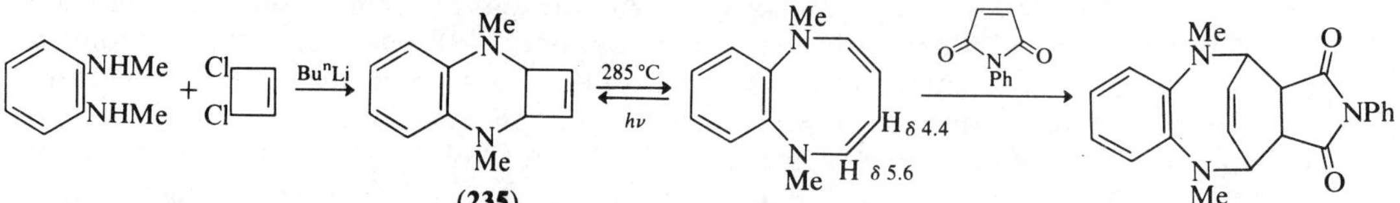

(**235**) (**236**) (**237**)

5.19.4.4.2 1,4-Dioxocins

1,4-Dioxocin (**239**) and 6-bromo derivative (**240**) are obtained by thermal valence isomerization of the respective *cis*-benzene dioxides; the former was also obtained by reduction of the bromodioxocin ⟨72AG(E)935, 937⟩. The precursor of both dioxides is 4,5-epoxycyclohexene, which was transformed *via* the *cis*- and *trans*-3,6-dibromides to dioxides (**238**) and (**241**).

Isomerization of the bromobenzene dioxide (**241**) occurred in refluxing CCl_4 and led to a 1:2 equilibrium mixture of (**241**) and the bromodioxocin (**240**). The unsubstituted dioxide (**238**) in benzene at 60 °C gives a mixture containing 95% of dioxocin (**239**), with E_a for the reaction 115 kJ mol^{-1}. In both cases, the equilibrium can also be reached from the dioxocin side.

Curiously, a benzene dioxide (**242**) has also been found in a fungal antibiotic ⟨74JOC435⟩. Isomerization in warm acetic anhydride gave a 9:1 equilibrium mixture of the pyronyl dioxocin (**243**).

NBS
CF_3CO_3H
H_2SO_4
$NaNH_2$
KOH
Bu_3SnH

(**238**)

(**239**) (**240**) (**241**)

(**242**) (**243**)

1,4-Dioxocin is a colorless liquid with distinctly alkenic properties and no indication of aromatic stabilization. The UV spectrum [λ_{max} (nm) 238 (7400), 285 (320)] is similar in profile to that of 1,3,6-cyclooctatriene; the ^{1}H NMR spectrum [δ 5.12 (H-6,7), 6.00 (H-2,3), 6.59 (H-5,8) p.p.m., $J_{5,6}$ 8.0, $J_{6,7}$ 9.2 Hz] is entirely in the alkenic region. The compound polymerizes rapidly.

1,6-Benzodioxocin (m.p. 5 °C) (**245**), prepared from the dihydro compound (**244**) *via* either the mono- or di-bromide, shows no aromatic character ⟨67AG(E)697⟩. The ^{1}H NMR spectrum contains signals in the alkenic region at δ 5.35 and 6.95 p.p.m. for the dioxocin protons. 1,6-Benzodioxocin gives a Diels–Alder adduct (**247**) and on irradiation undergoes cyclization to the cyclobutene (**246**).

5.19.4.4.3 4H-1,4-Oxazocines

The mixed heteroatom system (**250**), obtained in several steps from (**248**), undergoes ring opening at 60 °C ($t_{1/2}$ 33 min) to the *N*-sulfonyloxazocine (**251**) ⟨80TL3475⟩. The ring

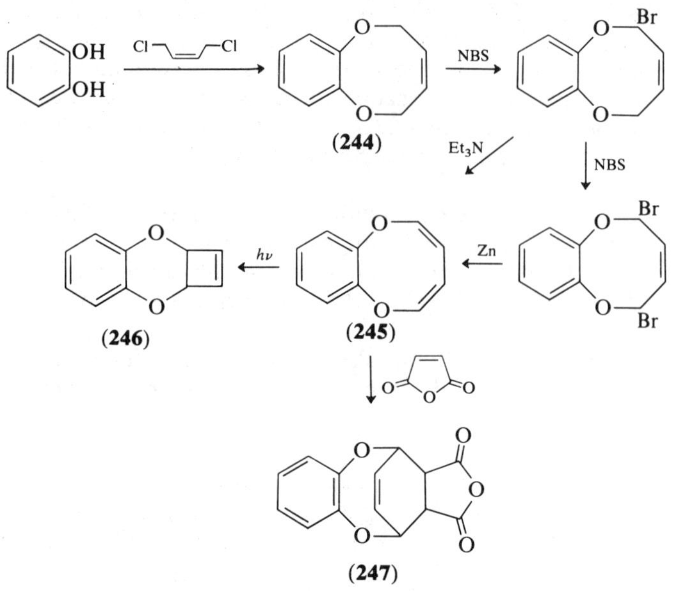

(244)

(245)

(246)

(247)

opening is irreversible as in the case of the dihydro-1,4-diazocines, but the NMR data suggest a half-chair structure resembling that of 1,4-dioxocin rather than the twist-boat-chair adduced for the non-planar diazocines.

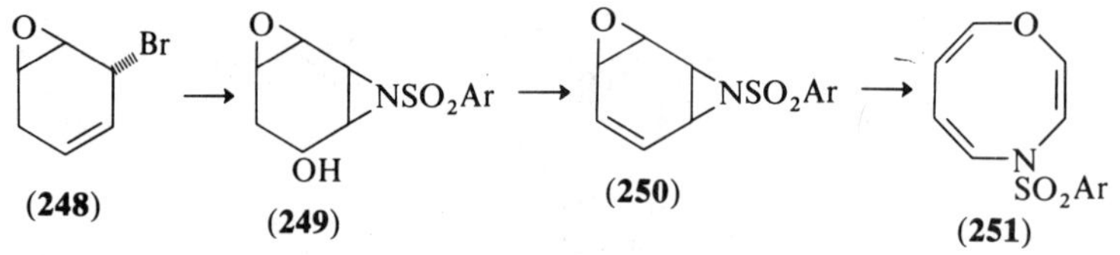

(248) (249) (250) (251)

5.19.4.4.4 1,4-Dithiocins

Several unsuccessful attempts have been made to prepare the parent 1,4-dithiocin by valence isomerization approaches. The diepisulfide (**252**) was obtained from benzene dioxide, but decomposition at 20 °C ($t_{1/2}$ 30 min) gave only benzene by stepwise sulfur extrusion ⟨74AG(E)737⟩. 2,5-Dithiabicyclo[4.2.0]octadiene (**253**) on pyrolysis at 200 °C gave benzene (50%) and a trace of thiophene; a non-concerted ring opening to dithiocin (**254**) followed by immediate destruction of the latter was suggested ⟨71JA4627⟩.

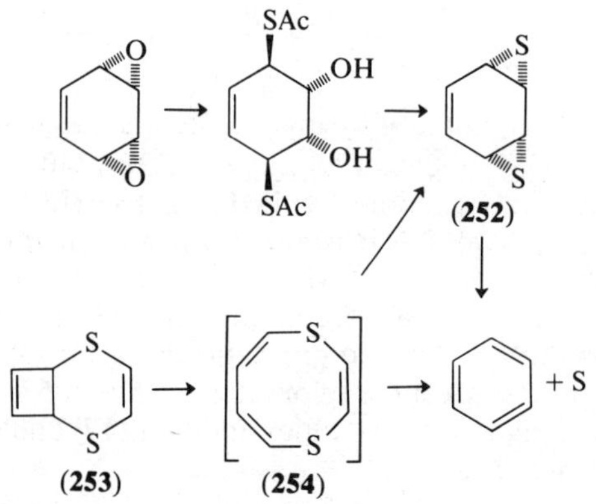

(252) (253) (254)

6-Acetoxy-1,4-dithiocin has been obtained by stepwise introduction of double bonds starting with dithiothreitol ⟨78T3631⟩. The acetal (**255**) is available by cyclization, and elimination of one hydroxyl group was effected in low yield by treatment of the ditosylate with base. Oxidation gave the ketone (**256**). Trapping of the enolate provided the very unstable trimethylsilyl ether or the crystalline acetate (**257**). Slight heating converts the dithiocin (**257**) to phenyl acetate; the dithiabicyclooctadiene (**258**) is produced on irradiation. NMR and crystallographic data show conclusively that 6-acetoxy-1,4-dithiocin is non-aromatic. The ring is non-planar, with the C-7—C-8 bond twisted at an angle of 11°.

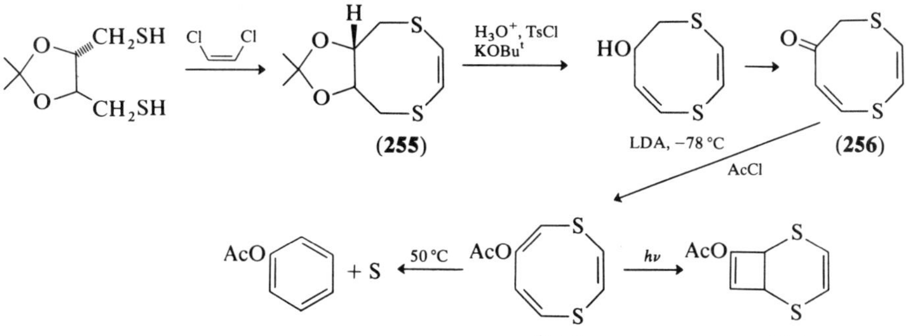

(**255**) (**256**) (**257**) (**258**)

1,6-Benzodithiocin (**260**) was obtained as a yellow oil in 11% yield by thermolysis of (**259**) ⟨71JA4627⟩. Dibenzo [*e*,*g*][1,4]dithiocin (**261**) was prepared by cyclization ⟨69ZC184⟩. Spectra indicate that both of these systems are non-planar with non-aromatic dithiocin moieties.

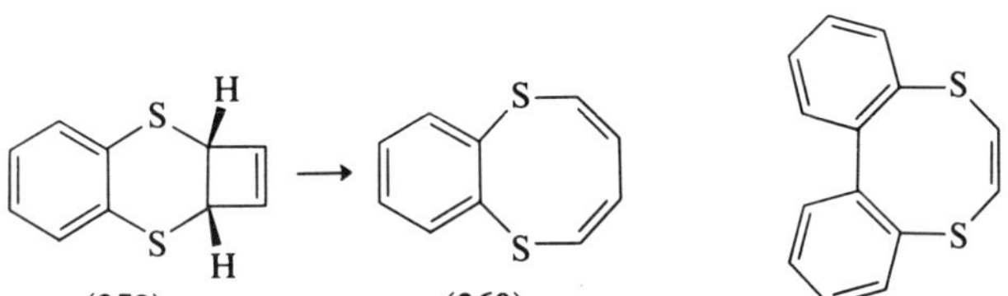

(**259**) (**260**) (**261**)

The perfluorinated tetracyclic 1,4-dithiocin structure (**262**) was assigned to a product obtained in low yield from the reaction of 1,2-dichloro-3,3,4,4-tetrafluorocyclobutene with butyllithium and sulfur chloride ⟨71TL2871⟩. The product is a stable high melting solid. The ^{19}F NMR chemical shifts very closely resemble those of dodecafluorotricyclobutabenzene, suggesting a ring current in (**262**).

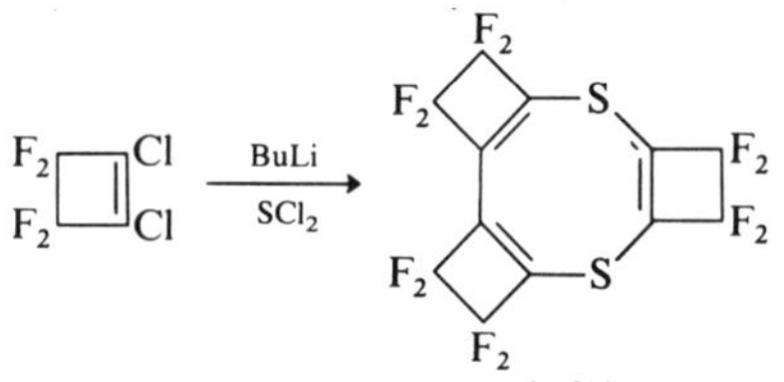

(**262**)

5.19.4:5 1,5-Diazocines

Approaches to fully reduced rings (1,5-diazocanes) include the standard methods of cycloalkylation and ring expansion, and most importantly, ring opening of 1,5-diazabicyclo[3.3.0]octanes. The reaction of equimolar amounts of toluenesulfonamide and 1,3-dibromopropane gave mainly *N*-tosylazetidine and a small amount of ditosyl-1,5-diazocane ⟨1899CB2039⟩. Alkylation of ditosyltrimethylenediamine gave a better yield of the diazocane, accompanied by a considerable amount of oligomer.

The reaction of amines or tosylamide with epichlorohydrins gives mixtures of epimeric 1,5-disubstituted-1,5-diazocane-3,7-diols (**263**) in 20–30% yields. The *t*-alkylamino-epoxides (**264**) can be isolated and dimerized in methanol solution; a sterically demanding R group is essential ⟨67T2123⟩. The diols (**263**; R = Tos, R′ = Me) were converted to dichlorides and thence to the dienes (**265**) and (**266**) ⟨67JOC2425⟩.

The Beckmann rearrangement of 1,4-cyclohexanedionedioxime ditosylate goes smoothly to 1,5-diazocane-2,6-dione (**267**) with none of the isomeric 1,4-diazocanediones ⟨58JA6404⟩. The dione (**267**) is readily polymerizable and is quite susceptible to hydrolytic ring opening. Cyclization of phenylthio *β*-alanyl-*β*-alanine in 10^{-4} M solution gives a low yield of (**267**) accompanied by the 16-membered cyclic dimer ⟨62CB783⟩.

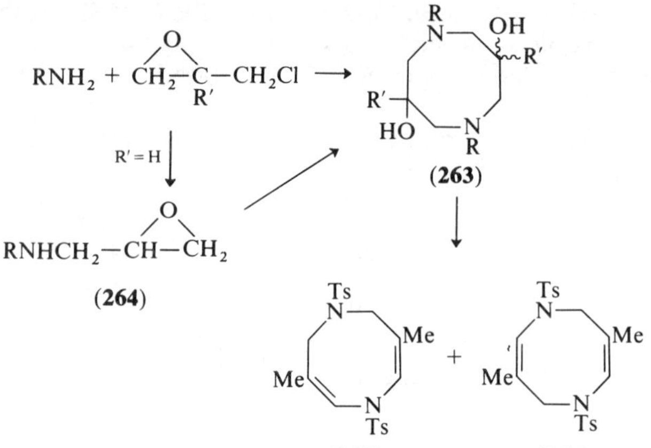

The most flexible and useful routes to 1,5-diazocanes avoid the difficulties inherent in cyclization to an eight-membered ring, and involve reduction and N—N cleavage of oxo derivatives of 1,5-diazabicyclo[3.3.0]octane. In the first examples of this process mono-, di- or tri-alkylated diazabicyclooctanones or 2,6-diones were reduced with LAH and then by catalytic hydrogenation to diazocanes (**268**) ⟨65CB3228⟩. The reduction can be carried out in one step with diborane ⟨79JOC4473⟩. 3,3,7,7-Tetrasubstituted diazocanes (**269**) are conveniently obtained from the tetraones which are prepared with a variety of substituents from malonate precursors and reduced with diborane ⟨78TL543⟩.

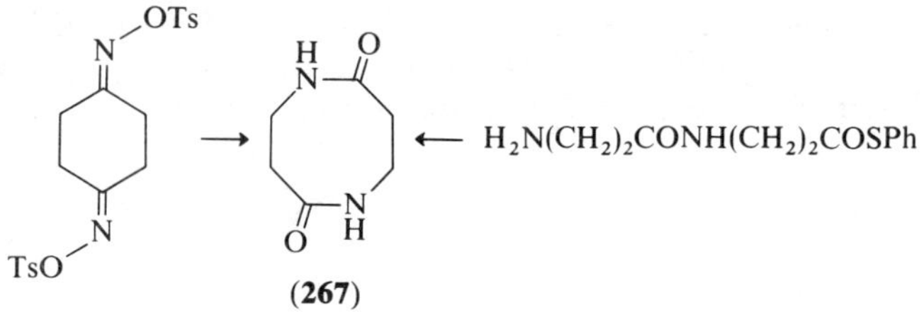

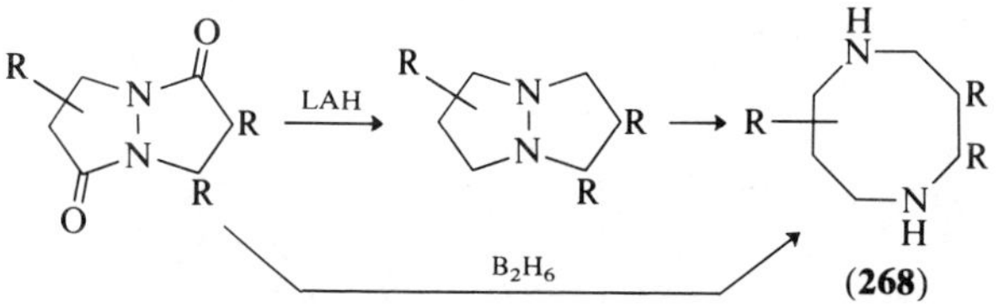

Fully unsaturated monocyclic 1,5-diazocines are represented by the heavily substituted derivatives (**272**) and (**274**), both arising by dimerization of presumed azete intermediates or their equivalents. Irradiation of a triazafulvene, *e.g.* (**270**), or the triazene (**271**) at −78 °C gives the yellow crystalline diazocine (**272**) in 20–25% yields ⟨74JOC940⟩. Thermal cracking of (**272**) gives pentaarylpyridine and nitrile. In a similar process the pyridazine (**273**) is converted on flow photolysis to a complex mixture of dimers of the postulated azete. One of these, also formed by heating two of the others, is assigned the perfluorinated 1,5-diazocine structure ⟨76CC1005⟩.

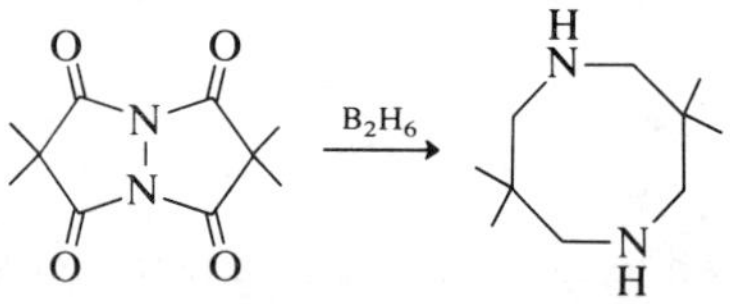

Dimerization of the isolated benzazete (**275**) to 6,12-diphenyldibenzo[*b*,*f*][1,5]diazocine (**276**) occurs in the presence of Lewis acids ⟨76CC411⟩.

Self-condensation of suitable aminocarbonyl compounds in various forms is a potential route to symmetrically substituted 1,5-diazocines of many kinds, but the successful applications are largely limited to benzo fused derivatives. *o*-Aminobenzaldehyde, despite earlier

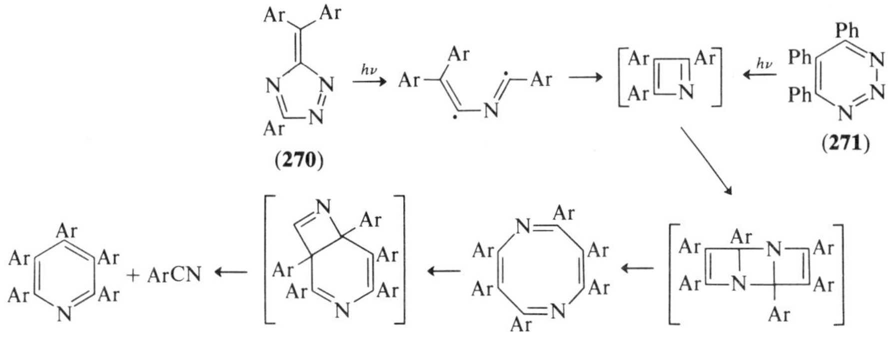

(270)

(271)

(272)

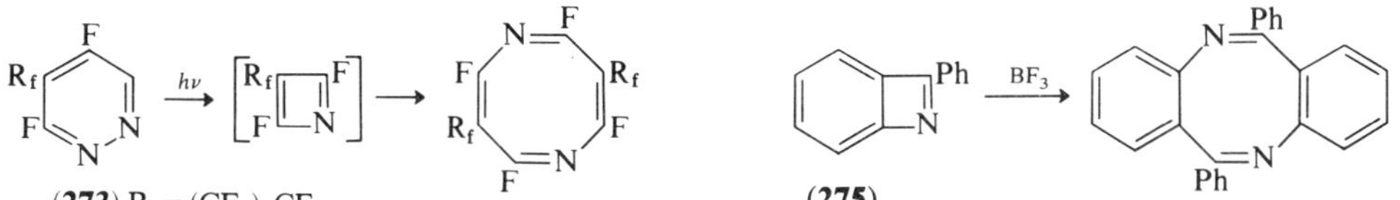

(273) $R_f = (CF_3)_2CF$

(274)

(275)

(276)

claims, does not lead to the simple dibenzo[1,5]diazocine, but rather anhydro-trimers and -tetramers, some of which, *e.g.* (**277**), contain polycyclic diazocine systems ⟨80T2359⟩. Dimerization of aminobenzophenones does occur smoothly and dibenzo[*b,f*][1,5]diazocines (**278**) have been isolated on several occasions as by-products in other reactions. A large number of 6,12-diaryl derivatives have been prepared by refluxing the aminobenzophenone with Lewis acids ⟨66JMC633⟩; these compounds are stable high melting yellow solids.

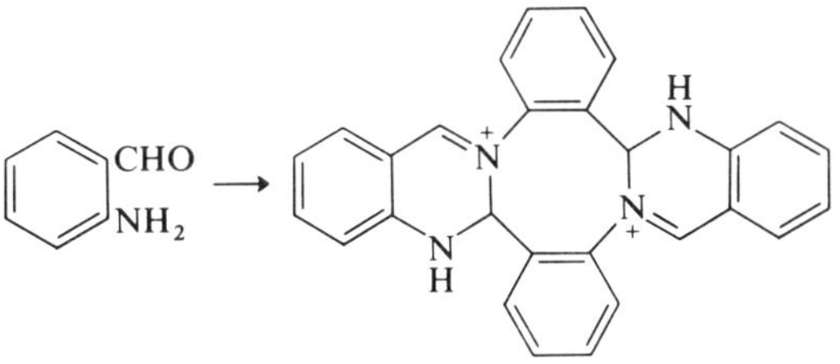

(277)

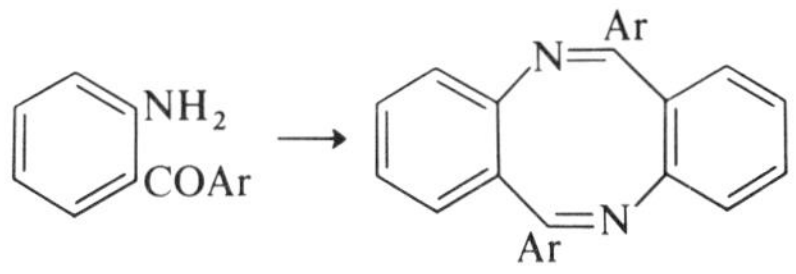

(278)

Condensations of anthranilic acid derivatives lead to the 6,12-diamino- and 6,12-dioxo-dibenzodiazocines, and the dichloro compound (**279**) is available from the latter ⟨54JCS3429⟩. Reduction of (**279**) *via* the dihydrazino compound was used to prepare the parent dibenzo[*b,f*][1,5]diazocine (**280**) ⟨67CC1077⟩. The NMR value (δ 8.53 p.p.m.) of the 6(12) proton is indicative of the diazocine structure, presumably in a tub form, rather than the valence isomeric dibenzodiazapentalene (**281**).

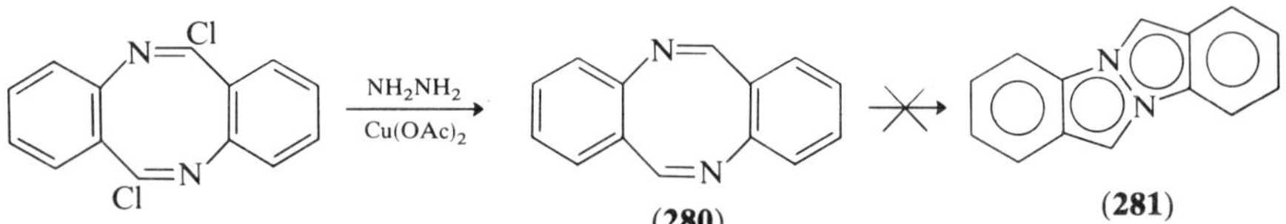

(279)

(280)

(281)

A number of reduction products of 2,8-dichloro-6,12-diphenyldibenzo[*b,f*][1,5]diazocine (**283**) have been obtained ⟨66JOC3356⟩. LAH in ether gives mainly the *trans*-diphenyl-tetrahydro compound (**282**), whereas reduction in pyridine stops at the dihydro stage (**284**) (78%). Catalytic hydrogenation or zinc–acid reduction gives the tetracyclic indolo[3,2-*b*]indole system (**286**) which is readily oxidized back to the diazocine. Transannular cyclization to (**286**) occurs on treatment of the dihydrodiazocine with NaH. Troeger's base derivatives, *e.g.* (**287**), were formed from the tetrahydro derivatives with formaldehyde.

1,5-Benzodiazocines have attracted much interest as homologs of 1,4-benzodiazepine drugs, and several efficient cyclization methods have been developed. The diamine (**289**; R = H), available by several routes from the aminobenzophenone, cyclizes slowly in hot pyridine, but treatment of the halide (**288**) with ammonia leads directly to (**292**), perhaps by prior condensation of the carbonyl group of (**288**) with ammonia. Cyclization of the *N*-methyl derivative (**289**; R = Me) was much more facile. The lower reactivity of (**289**; R = H) was attributed to the unfavorable conformation (**290**) imposed by hydrogen bonding ⟨69JOC179⟩. Dehydration of the amide (**291**) gave only a dimeric diamide and none of the benzodiazocinone, again presumably because of the constraint due to hydrogen bonding. The diazocinone (**293**) was obtained by a multistep approach involving preparation of the methano-bridged lactam (**295**) from the 6-chloroquinazoline, debridging to (**294**) by mild hydrolysis and dehydrogenation with permanganate ⟨69JOC183⟩.

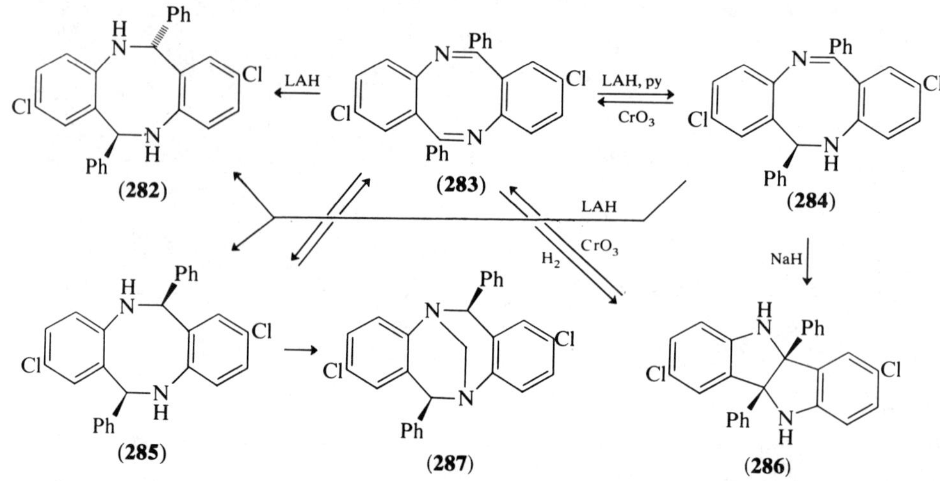

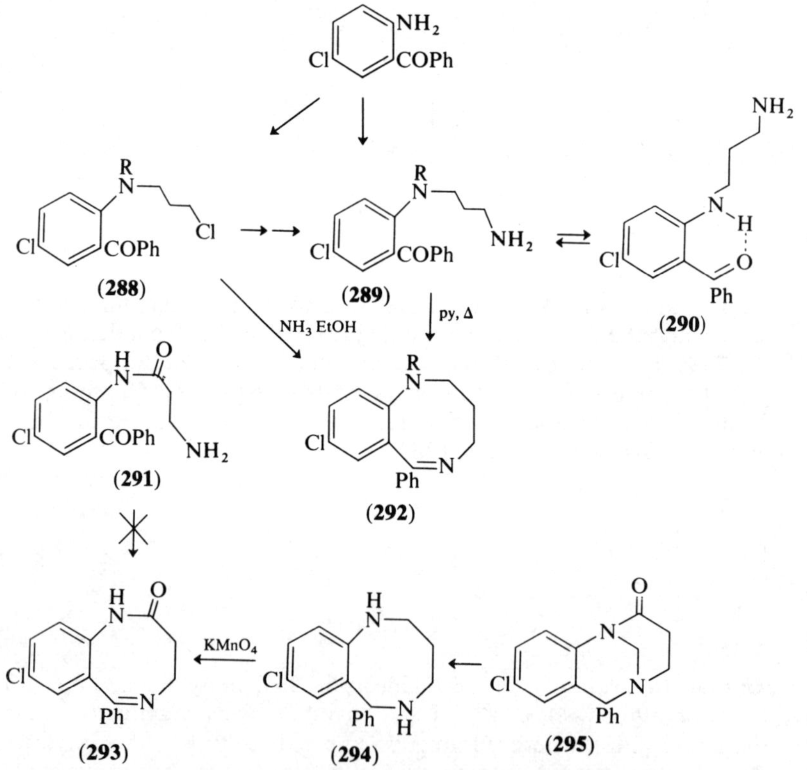

In another cyclization procedure for the 1,5-benzodiazocine system, the nitriles (**296**) are converted to the aminodihydrodiazocines (**297**) ⟨79CPB2589⟩. Attack of nucleophiles on (**297**) occurs at the N-5—C-6 bond to give the 3,4,5,6-tetrahydrodiazocines (**298**) with $NaBH_4$ and the β-aminoethylquinazolines (**301**) on hydrolysis. The diazocines (**297**) behave as typical amidines. Oxidation leads to the amidoximes (**300**) which on hydrolysis are converted to 2,1-benzisoxazoles (**302**), and reaction with diketene leads to the fused pyrimidinones (**299** and 1-methyl-3-one isomer) ⟨79CPB2927⟩.

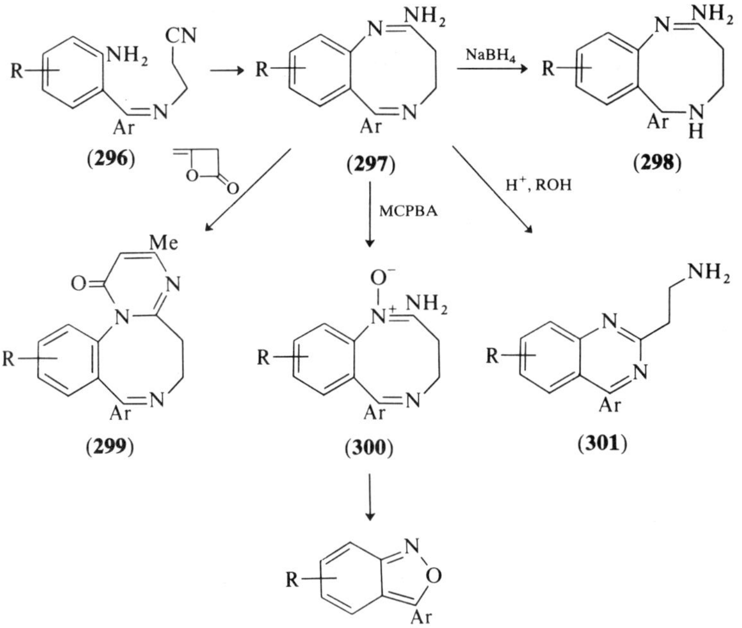

(296) (297) (298) (299) (300) (301) (302)

A quite different approach to the 1,5-benzodiazocine system begins with 1,3-dipolar cycloaddition of a nitrile oxide and an indolenine ⟨78MI51900⟩. The spiroisoxazoline (**303**) is converted by irradiation in good yield to the 1,5-benzodiazocinone (**304**). By using *N,N'*-alkylenediindoles and bifunctional nitrile oxides, the diazocine ring is incorporated as a repeating unit into a photoconducting polymer.

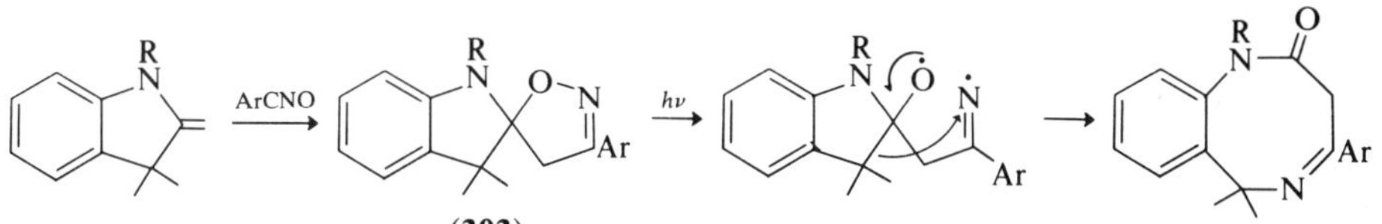

(303) (304)

5.19.5 OXAZOCINES

Four oxazocine rings and 15 benzoxazocine systems are possible; a compilation of the literature on these compounds has been prepared ⟨82H(19)709⟩. Many of these ring systems have been described in patents and rather obscure journals, and in a number of cases the evidence for structural assignments is inadequate. Methods of preparation include cyclization of polyfunctional chains, and ring expansions or rearrangement of several types. Some representative examples of oxazocines and benzoxazocines that have been well characterized are compounds (**305**)–(**310**).

Potentially aromatic 4*H*-1,4-oxazocines are dealt with under Section 5.19.4.4, above.

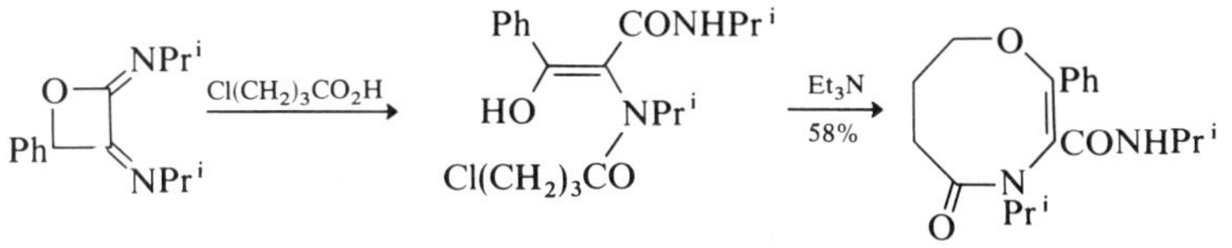

(305) ⟨69LA(730)151⟩

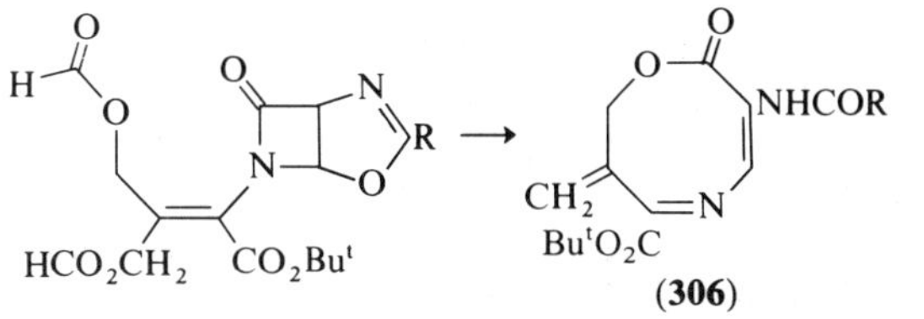

(306) ⟨79TL1257⟩

⟨78TL2087⟩

(307)

⟨65JCS1653⟩

(308)

⟨82UP51900⟩

(309)

⟨70JMC403⟩

(310)

5.19.6 RINGS WITH THREE AND FOUR NITROGEN OR OXYGEN ATOMS

5.19.6.1 Triazocines and Oxadiazocines

The 1,3,4-benzotriazocine system (**312**) arises by condensation of the versatile benzazete (**311**) with activated tetrazines ⟨75JCS(P1)45⟩.

(311)

(312) R = 2-py

Several 1,4,5-benzotriazocines have been described in connection with medicinal chemical studies. The reaction of 2-chloromethylquinazoline oxide (**313**) with hydrazine gives the hydroxylaminotriazocine (**314**) and not a diazepine as originally assigned ⟨73CPB2375⟩. Vigorous acetylation of (**314**) causes a Wolff–Semmler type rearrangement to give the oxazolotriazocine (**315**).

(313) (314) (315)

The related benzotriazocinones (**317**) are obtained by cyclization of dialkylhydrazones (**316**) and mild thermolysis of the quaternary hydrazones ⟨73CPB2375⟩. The latter undergo ring contraction to diazepines by Stevens rearrangement.

(316) (317)

Analogous 4,1,5-benzoxadiazocinones have been prepared by several complementary ring closure processes ⟨67JOC2417⟩. Treatment of the *syn*-oximes (**318**) of haloacetylamidobenzophenones with base gives the oxadiazocinones (**319**) together with dimeric by-products. The eight-membered ring can also be formed by cyclization of the *o*-alkylhydroxylamines (**320**) or (**321**). The oxadiazocine ring undergoes rapid ring opening and reclosure on treatment with excess base to give the hydroxybenzodiazepine (**322**).

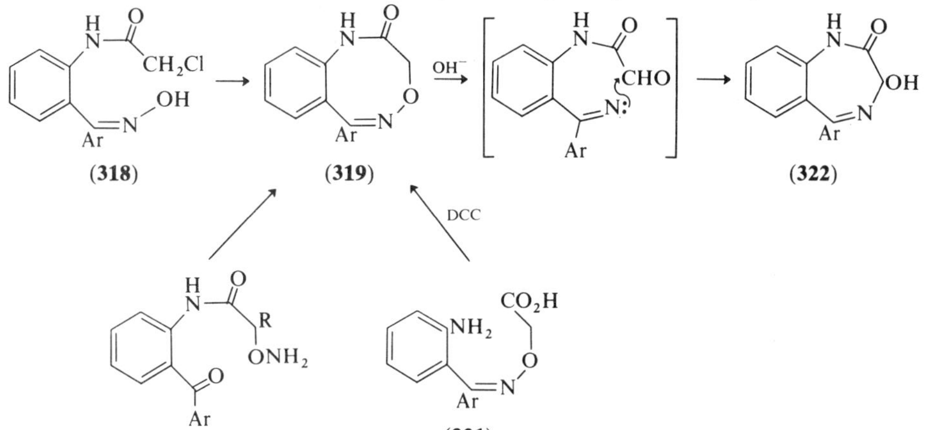

(**318**) (**319**) (**322**) (**320**) (**321**)

5.19.6.2 1,2,5,6-Tetrazocines

A number of compounds obtained by self-condensation of benzil monohydrazones and other α-keto hydrazones have been represented in earlier literature as 1,2,5,6-tetrazocines (**323**). Later work has shown, however, that these substances were in fact 1,3a,4,6a-tetraazapentalenes (**324**) ⟨63CB1827, 67JA2618⟩. HMO calculations indicate substantial resonance stabilization for the aromatic 10π-electron tetraazapentalene system, and significantly greater stability compared to the tetrazocine ⟨67JA2638⟩. The valence tautomeric relationships of polyazocines and polyazapentalenes have been summarized ⟨78AHC(22)183⟩.

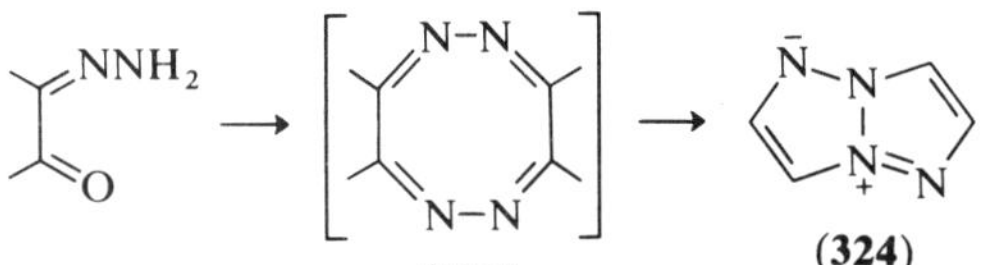

(**323**) (**324**)

5.19.6.3 1,3,5,6-Tetrazocanes

Interest in tetrazocanes developed in the 1940s with isolation of a high melting side-product in the nitrolysis of hexamine (**325**) to produce the explosive 1,3,5-trinitrohexahydrotriazine (**326**), better known as RDX. This impurity was shown by Bachmann and coworkers to be the tetranitrotetrazocane (**327**) and was designated as HMX (high-melting explosive). HMX was found to be superior smokeless oxidizer in gun and rocket propellants, and it is widely used together with ammonium perchlorate in solid-propellant boosters and engines for missiles and space vehicles. A voluminous technical literature has emerged on the solid state properties and burning characteristics of HMX and combinations with various propellant components.

The nitrolysis of hexamine can be controlled by changes in conditions to follow either of two paths. Strongly acidic conditions lead to triazine and linear triamine products. With nitric acid and acetic anhydride the dinitrotetraazabicyclo[3.3.1]nonane (**328**) is produced, and opening of the methylene bridge then gives the tetrazocane (**329**) and finally, in the presence of ammonium nitrate, HMX or the tetramine (**330**) ⟨51JA2773⟩.

Acylation or sulfonation of hexamine leads to the diacyl bicyclic tetramines (**331**) which undergo ring opening in N_2O_4 to give the dinitrosotetrazocanes (**332**) in 50–80% yields. The latter can be further oxidized to nitronitroso or dinitro compounds ⟨73JHC279⟩.

The direct formation of tetrazocane tetracarbamates has been reported to occur in fairly good yields by refluxing methylenebis(carbamates) (**333**) with paraformaldehyde in the presence of toluenesulfonic acid. Under these conditions the tetrazocanes (**335**) appear to

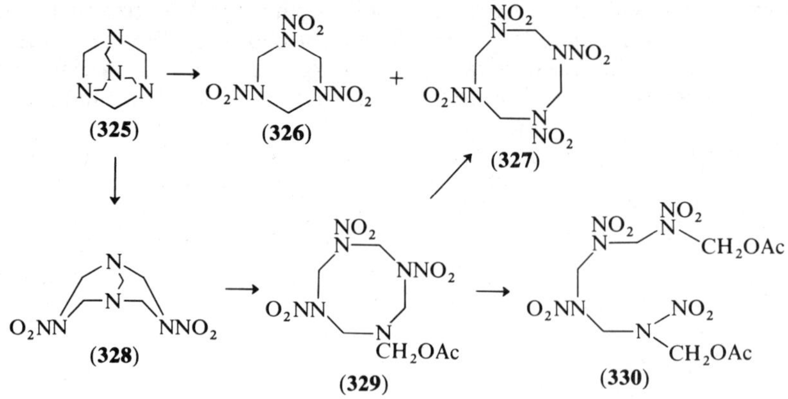

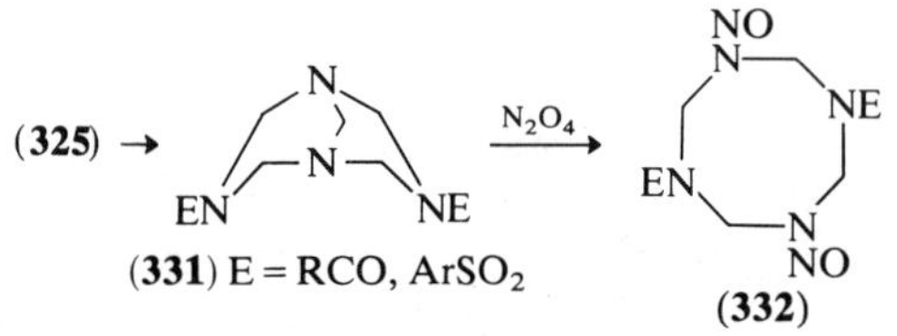

(325) (326) (327) (328) (329) (330)

(325) → (331) E = RCO, $ArSO_2$ —N_2O_4→ (332)

be the kinetically favored products; longer reaction times lead to increasing amounts of the hexahydrotriazine derivatives (**334**). Remarkably, the tetrazocanes are also produced in very high yields by brief treatment of the triazines with cold concentrated H_2SO_4 ⟨74JHC937⟩.

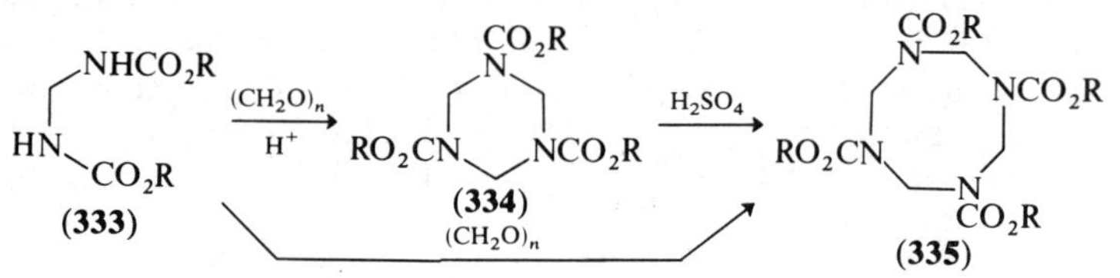

(333) —$(CH_2O)_n$, H^+→ (334) —H_2SO_4→ (335); (333) —$(CH_2O)_n$→ (335)

5.19.7 RINGS WITH TWO OR MORE OXYGEN ATOMS

5.19.7.1 1,2-Dioxocanes

The simplest and apparently only known 1,2-dioxocane is the bridged endoperoxide (**337**; m.p. 50–55 °C) of 1,3-cyclooctadiene (**336**) ⟨74CL345⟩. This peroxide serves as a useful intermediate for the introduction of oxygen functions in a 1,4 relationship in cyclooctane, to give, for example, (**338**).

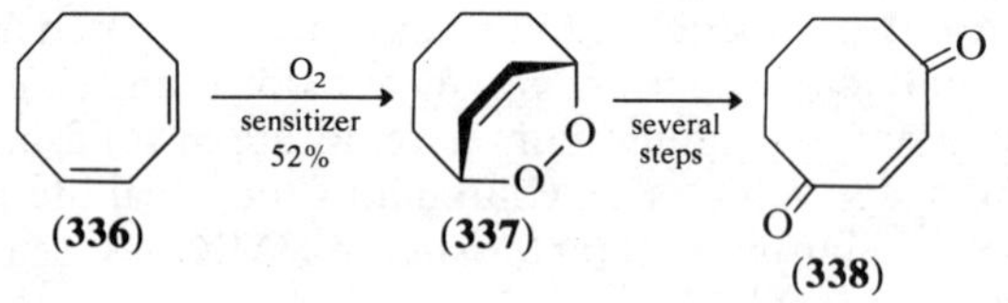

(336) —O_2, sensitizer, 52%→ (337) —several steps→ (338)

5.19.7.2 1,3-Dioxocanes

The 1,3-dioxocanes are relatively common compounds, since they are cyclic acetals or ketals of 1,5-pentanediols, and are easily prepared, at least when they are volatile. Actually, as might be expected from the general difficulty of cyclizations which give eight-membered rings (*cf.* Section 5.19.3), the initial acetalization or ketalization products are not 1,3-dioxocanes but are mostly polymers which, fortunately in this series, can be smoothly cracked at 150–250 °C to give the volatile monomers (**339**) ⟨35JA925, 54MI51900⟩. Some dimeric (16-membered ring) products have also been isolated in some cases ⟨73ACS1519⟩. Instead of a carbonyl component, it is possible to use an equivalent molecule, such as an aliphatic acetal ⟨35JA925⟩ or ketal ⟨73ACS1519⟩. Also, acetylene or vinyl acetate in the presence of mercury(II) salts can be used instead of acetaldehyde ⟨68JCS(B)543, 65JCS1560, *cf.* 23JA3124 and 30JA1018⟩. The acid catalyst can be a sulfonic acid ion-exchange resin ⟨54MI51900⟩, and the water produced in the reaction can be removed by azeotropic distillation with benzene or toluene, and this is a generally useful procedure ⟨73ACS1519⟩. Carbonyl compounds which have been used include acetaldehyde (or its equivalent), dichloroacetaldehyde and isobutyraldehyde. 2,2-Dimethyl-1,3-dioxocanes have been prepared from 2,2-dimethoxypropane rather than from acetone. 3,3-Dimethyl-1,5-pentanediol and diethylene glycol (see Section 5.19.7.2) have been used as the other component in the reaction ⟨73ACS1519⟩. 1,3-Dioxocanes are the products of dehydrohalogenation of the bis(1-chloroalkyl) ethers of 1,5-pentanediol, rather than the expected divinyl ethers ⟨65JCS1560⟩.

R₂ OH OH —(R'R"CO or equivalent, acid)→ polymeric acetal or ketal ⇌(neat / heat) (339) R₂ O O R'R" —(solution)→ (340) $(CH_2)_2$—O—CR'R"—O—$(CH_2)_2$ / CR_2 CR_2 / $(CH_2)_2$—O—CR'R"—O—$(CH_2)_2$

(339) **(340)**

1,3-Dioxocane (b.p. 134 °C) has a heat of combustion as a gas at 20 °C of 3756 kJ mol^{-1} (CO_2 as a gas and H_2O as a liquid) ⟨57MI51900⟩. From group additivity enthalpy values, it can be calculated that the enthalpy of ring formation in this compound is 54 kJ mol^{-1}, as compared to 40 kJ mol^{-1} for cyclooctane or zero for 1,3-dioxane or cyclohexane. The dipole moments of 1,3-dioxocanes are quite low (<1 D) unlike the dipole moment of 1,3-dioxane (1.9 D) and this is a reflection of the different arrangements of the acetal unit in the two-ring system ⟨73ACS1519⟩. Further discussion on the conformational properties of the 1,3-dioxocanes is given in Section 5.19.9.

Most 1,3-dioxocanes are thermodynamically unstable with respect to both dimerization and to oligomerization or polymerization; these reactions are catalyzed by acids and can be inhibited with base. The neat 1,3-dioxocanes, which are generally liquids, give rise to low melting polymers ⟨35JA925⟩, whereas in dilute solutions crystalline dimers are formed ⟨74T1683⟩. The acid-catalyzed hydrolysis of 2-methyl-1,3-dioxocane is much faster than that of its corresponding seven-membered ring analog, which in turn is hydrolyzed much more rapidly than are the five- or six-membered cyclic acetals ⟨68JCS(B)543⟩.

5.19.7.3 1,4-Dioxocanes

1,4-Dioxocane (**341**; b.p. 149 °C) is a reduction product of 1,4-dioxocin ⟨72AG(E)935⟩ (see Section 5.19.4.4). Derivatives of this compound appear to be quite rare and have been the accidental products of reactions rather than the end results of deliberate syntheses. For example, dibutylcadmium reacts with the acid chloride (**342**) to give the lactone (**343**) rather than the expected ketone ⟨67JOC4149⟩. The NMR spectrum of (**343**) is consistent with the assigned structure.

O O —(H_2/Pd)→ O O (341)   O O Me CH_2CH_2COCl (342) —(Bu_2Cd)→ O O O Me Bu (343)

(341) **(342)** **(343)**

5.19.7.4 1,5-Dioxocanes

Several 1,5-dioxocanes, including the parent compound, have been prepared *via* the unsaturated intermediate (**345**), which unfortunately is only formed in very low yield from (**344**) ⟨70LA(736)75⟩. As well as (**345**), there is also formed the 16-membered cyclic dimer of this compound in 14% yield. Catalytic hydrogenation of (**345**) gives 3-methyl-1,5-dioxocane (**346**), whilst ozonolysis provides the ketone (**347**), $\nu = 1725$ cm^{-1}, which is converted to 1,5-dioxocane (**348**) by a three-step reduction procedure *via* the alcohol and the tosylate.

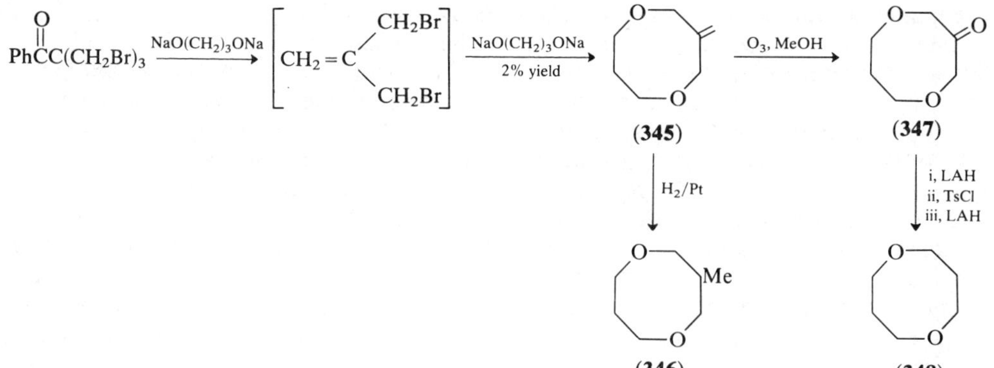

(345) (346) (347) (348)

5.19.7.5 Tetrahydrodioxocins and Benzo Derivatives

Although a tetrahydrodioxocin structure has been suggested for the product of the reaction of bis(2-bromoethyl) ether with diethyl β-oxoglutarate under basic conditions ⟨43JGU352⟩, the reported data are insufficient to establish this structure.

Benzo derivatives of tetrahydrodioxocin, *e.g.* (**349**), can easily be made by the reaction of catechol derivatives with 1,4-dibromobutane under alkaline conditions, although the yields are low ⟨37LA(528)162, 71JCS(B)1231⟩. Much bettter yields are obtained by starting with *o*-(4-bromobutyl)phenol (**350**), which gives mostly monomeric products. The kinetics of this and related cyclizations have been thoroughly studied and it has been shown that an internal *E*2 elimination reaction competes to a slight extent with the internal S_N2 cyclization ⟨80JCS(P2)1774, 74JA1422⟩.

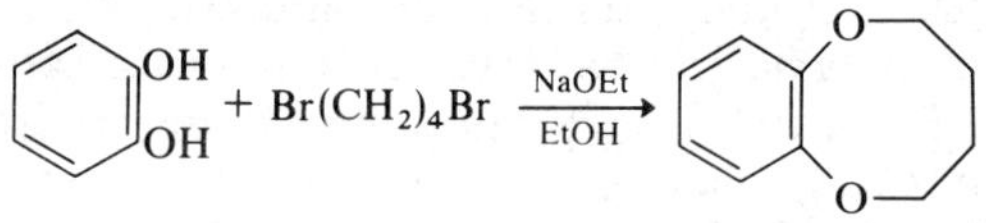

(**349**) 6%

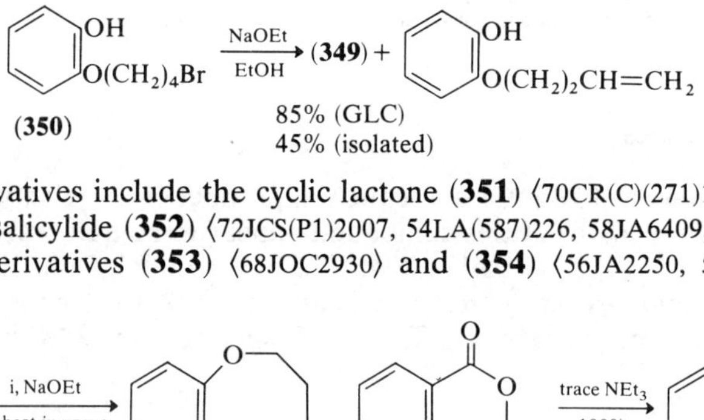

(**350**) 85% (GLC) 45% (isolated)

Other benzo derivatives include the cyclic lactone (**351**) ⟨70CR(C)(271)1262⟩ as well as the cyclic dilactones, disalicylide (**352**) ⟨72JCS(P1)2007, 54LA(587)226, 58JA6409, 51JCS200, 201⟩ and the phthalic acid derivatives (**353**) ⟨68JOC2930⟩ and (**354**) ⟨56JA2250, 58JA3432, 59JA1503, 60JA890⟩.

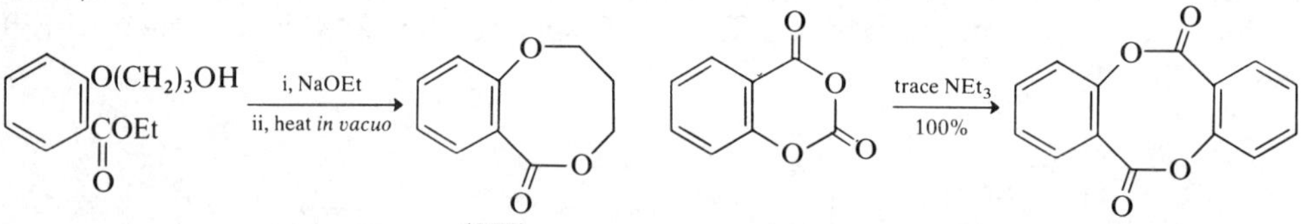

(**351**) (**352**)

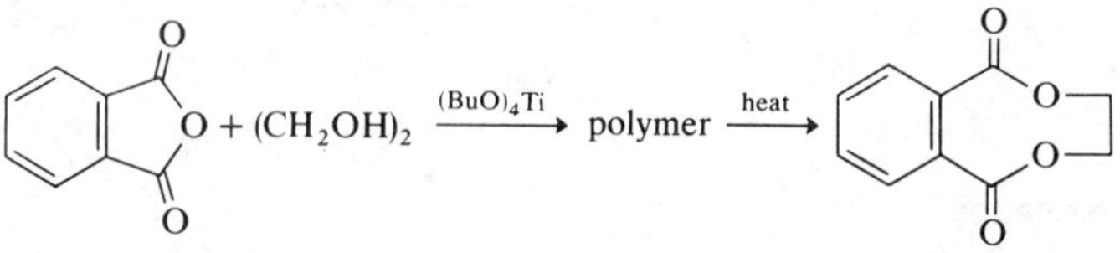

(**353**)

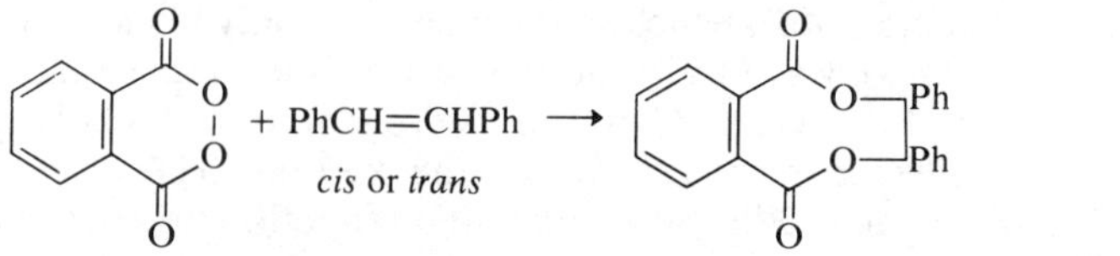

(**354**) *meso* or (±) respectively

5.19.7.6 Trioxocanes

1,3,6-Trioxocane (**355**) can be prepared from diethylene glycol by the methods presented for 1,3-dioxocanes ⟨60BSF718, 54MI51900, 73ACS1519⟩ or by the dehydrohalogenation with quinoline of the bis(chloromethyl) ether of diethylene glycol ⟨46JA1200⟩. Its conformation is discussed in Section 5.19.9.

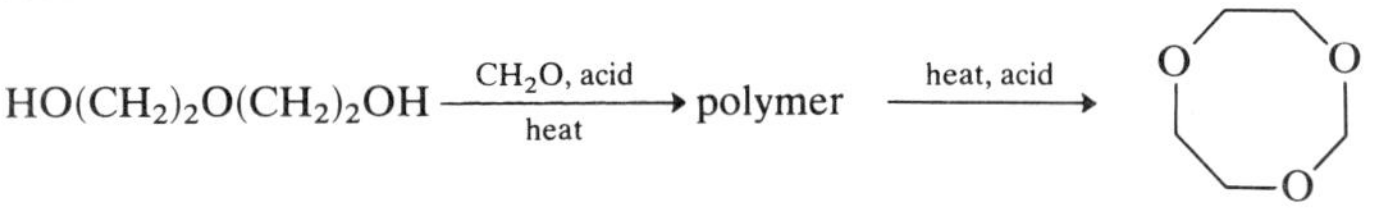

$$HO(CH_2)_2O(CH_2)_2OH \xrightarrow[\text{heat}]{CH_2O,\ \text{acid}} \text{polymer} \xrightarrow{\text{heat, acid}} \text{(355) } 85\%$$

The polymerization of (**355**) in the presence of triethyloxonium fluoroborate in 1,2-dichloroethane proceeds to give a product with a degree of polymerization of about 30 at all conversion percentages. This arises because of the slow initiation step and the presence of chain breaking reactions with this catalyst ⟨60MI51900⟩.

5.19.7.7 Tetroxocanes

The tetramer of formaldehyde, 1,3,5,7-tetroxocane (**356**), is obtained in low yield together with 1,3,5-trioxane (**357**) by the pyrolysis of 'insoluble' poly(oxymethylene)diacetate (**358**) under acidic conditions ⟨25HCA65⟩. It can also be obtained from paraformaldehyde directly with acidic catalysts ⟨62MI51900⟩ and has been observed in very small amounts in reaction of paraformaldehyde carried out under acid conditions ⟨59CB1072⟩. Efforts have been made to optimize the yield of tetroxocane, particularly by varying the nature of the acidic catalyst, and the best reported procedure uses $K_2S_2O_7$ on (**358**) to give a yield of about 30% ⟨69USP3426041, *cf.* 68JAP6816751, 69FRP1551921⟩. The cyclic oligomers of formaldehyde can be conveniently analyzed by GLC, and this has revealed that traces of the pentamer and hexamer are also formed together with formaldehyde, trioxane and tetroxocane in the depolymerization reactions, especially when these are carried out at very high temperatures ⟨68MI51900⟩. Tetroxocane yields trioxane and formaldehyde in a 1:1 ratio on heating with phosphoric acid ⟨62MI51900⟩.

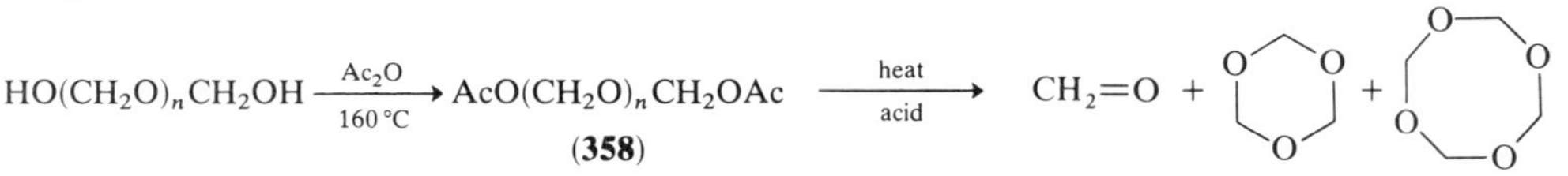

$$HO(CH_2O)_nCH_2OH \xrightarrow[160\,°C]{Ac_2O} AcO(CH_2O)_nCH_2OAc\ \text{(358)} \xrightarrow[\text{acid}]{\text{heat}} CH_2{=}O + \text{(357)} + \text{(356) } 1\text{–}30\%$$

The enthalpies of formation of tetroxocane in the crystalline and gas phases are −700 and −620 kJ mol^{-1}, respectively ⟨69ACS56⟩.

Radiation-induced polymerization of crystalline tetroxocane gives a highly crystalline linear polymer in essentially 100% yield ⟨63MI51900, 68MI51901⟩, and this has attracted some industrial attention. Tetroxocane also gives a crystalline polymer very rapidly under cationic conditions (boron trifluoride etherate catalyst) and the kinetics of the process have been studied ⟨76MI51900⟩.

Other aldehydes also give cyclic tetramers. Indeed, the commercially important 'metaldehyde' (**359**; m.p. 47 °C) is the cyclic tetramer of acetaldehyde, and its structure was determined by X-ray diffraction as early as 1936 ⟨36JA1274⟩. Similarly, chloroacetaldehyde and chloral give the cyclic tetramers (**360**; m.p. 65–67 °C) ⟨66LA(693)117⟩ and (**361**; m.p. 219–220 °C) ⟨61MI51900, 63CJC1893⟩, respectively, although in low yields. These substituted 1,3,5,7-tetroxocanes have the all-*cis* configuration. Actually, minor amounts of other cyclic tetramers are also formed; for example, acetaldehyde gives metaldehyde II (m.p. 45.5 °C) and metaldehyde III (m.p. 52.5 °C). These isomers can be separated by GLC, but their configurations have not been established ⟨62MI51901⟩.

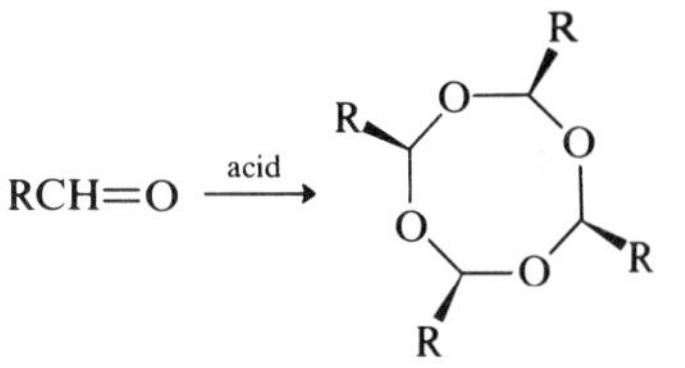

$$RCH{=}O \xrightarrow{\text{acid}} \text{tetramer} + \text{other products}$$

(**359**) R = Me
(**360**) R = CH_2Cl
(**361**) R = CCl_3

The structure, NMR and conformational properties of various tetroxocanes are discussed in Section 5.19.9.

5.19.8 RINGS WITH SULFUR ATOMS

5.19.8.1 Thiocanes

Thiocane (**362**) can be obtained from 1,7-dibromoheptane and sodium sulfide. It has recently been found that neither high dilution in protic solvents, which is reported to give a 34% yield ⟨53M1206⟩, nor moderate dilution in aprotic solvents, which gives a somewhat higher yield (48%) ⟨79SC857⟩, is necessary ⟨81SC409⟩. Concentrated (0.2M) solutions of the dihalide in ethanol with excess sodium sulfide give (**362**) in a 47% yield. The success of this experiment apparently stems from the low solubility of the sulfide salt. In addition, the high nucleophilicity of the RS^- ion promotes the internal S_N2 reaction, and the low basicity of this anion hinders the competing *E*2 elimination ⟨*cf.* 74JOC2598, 74JA1422⟩, and thus cyclization should be much more favorable in the sulfur case than it is for the oxygen analog.

$Br(CH_2)_7Br$ —(Na_2S–EtOH, 47%)→ (**362**) → (**363**)

Thiocane (**362**; m.p. 33–34 °C, b.p., 71 °C/10 mmHg, n_D 1.5152, $HgCl_2$ complex, m.p. 128–129 °C) behaves as a typical sulfide and gives rise to a tertiary sulfonium salt with methyl iodide, and a sulfone (**363**) with H_2O_2 ⟨53M1206⟩. This sulfone is described as having a characteristic, but not unpleasant odor (presumably, unlike **362**).

Thiocane can also be made by the Wolff–Kishner reduction of 5-thiocanone (**364**), a ketone which can be obtained, like its oxygen and nitrogen analogs, by a Dieckmann cyclization under high dilution conditions (*ca.* 20% yield) ⟨60JA4075, 62JA2814⟩. The dihalide route to (**364**) is reported to give a yield of 54% ⟨64JGU1242⟩.

$O{=}C[(CH_2)_3Br]_2$ —(NaSH, 54%)→ (**364**) ←(i, NaOEt; ii, hydrolysis; *ca.* 20%)— $S[(CH_2)_3CO_2Et]_2$

The ketosulfone [**365**; m.p. 124–127 °C, ν 1705 cm^{-1} ($CHCl_3$)], is formed by the oxidation of (**364**) with H_2O_2; no sulfoxide could be isolated when less than one equivalent of the oxidizing agent was used ⟨60JA4075⟩. However, sodium metaperiodate gives an excellent yield of the extremely hygroscopic sulfoxide (**366**) ⟨62JA3701, 62JOC282⟩. The sulfoxide of (**362**) can be prepared in a similar way.

Like its nitrogen analog (see Section 5.19.2.1) the ketosulfide (**364**; m.p. 53-54 °C, ν 1703 cm^{-1} (CCl_4), μ 3.8 D) shows transannular reactions. Although it methylates on sulfur to give (**367**), it protonates on oxygen to give the bicyclic sulfonium ion (**368**) ⟨60JA4075⟩. Reduction of (**364**) with LAH gives an alcohol (**369**) which cyclizes to the sulfonium ion (**370**) with P_2O_5.

Ring expansion of the six-membered ketosulfide (**371**) produces both 5-thiocanone and 4-thiocanone [**372**; ν 1700 cm^{-1} (neat)], but in low yields ⟨70JOC584⟩. The photochemistry of these ketones has been studied and leads to a variety of products in low yields ⟨70JOC584⟩. 3-Thiocanone (**373**; b.p. 66–68 °C/0.35 mmHg, ν 1708 cm^{-1}) is formed in 44% yield by the dihalide method, but in only 3% yield by the Dieckmann reaction under high dilution conditions ⟨70JPR1058⟩.

5.19.8.2 Tetrahydrothiocins and Benzo Derivatives

No simple tetrahydrothiocins appear to be known, but some benzo derivatives have been synthesized. 3,4,7,8-Tetrahydro-2*H*-benzo[*e*]thiocin-4-one (**374**; m.p. 60 °C, ν 1708,

(362)

Wolff–Kishner reduction

$NaIO_4$ H_2O_2

(366) (364) (365)

$HClO_4$ ether LAH MeI

(368) (369) (367)

P_2O_5

(370)

CH_2N_2 (364) +

(371) (372) 2.5%

$Br(CH_2)_5COCH_2Br$ NaSH 44%

$EtO_2C(CH_2)_5SCH_2CO_2Et$ $KOBu^t$ 3%

(373)

1675 cm^{-1}) has been prepared both by the Friedel–Crafts route ⟨29CB2416, 59BSF1998⟩ and by an unusual two-carbon ring expansion from the lower homolog (**375**) ⟨70JOC3442⟩. Wolff–Kishner reduction of (**374**) gives (**376**) accompanied by much ketazine, and Clemmensen reduction gives rearranged products in addition to (**376**) ⟨59BSF1998⟩.

SCH_2COCl $AlCl_3$ 25%

(374) (376)

Li 20%

(375) → (374)

5.19.8.3 Dithiocanes

1,2-Dithiocane (**377**; b.p. 65 °C) is formed by the iron(III) chloride oxidation under high dilution conditions of 1,6-hexanedithiol ⟨57AG713, 58LA(614)66⟩, or, in a low yield, by the steam distillation of the corresponding polymer which can readily be made from 1,6-dibromohexane ⟨63USP3284466, 50JOC865, *cf.* 48JA2611⟩.

Unlike their oxygen analogs (Section 5.19.7.2), which have been much studied, nothing seems to be known about 1,3-dithiocanes.

$HS(CH_2)_6SH$ $\xrightarrow[30\%]{FeCl_3}$ (377) $\underset{\text{steam distillation}}{\overset{AlCl_3 \text{ or peroxides}}{\rightleftharpoons}}$ polymer $\xleftarrow[\text{ii, } Na_2S_2 \text{ or } CuCl_2]{\text{i, } Na_2S_2O_3}$ $Br(CH_2)_6Br$

(377)

1,4-Dithiocanes are represented by several compounds. The simultaneous slow addition of ethanedithiol and 1,4-dibromobutane to sodium ethoxide in ethanol gives 1,4-dithiocane (**378**; b.p. 47–48 °C/3 mmHg) in low yield ⟨78JA6416⟩. Oxidation with $NOBF_4$ gives, presumably *via* the dication (**379**), the monosulfoxide (**380**) in good yield.

$HS(CH_2)_2SH + Br(CH_2)_4Br$ $\xrightarrow[\text{EtOH, } 7\%]{NaOEt}$ (378) $\xrightarrow{2NOBF_4}$ (379) $\xrightarrow{H_2O, NaHCO_3}$ (380)

(378) **(379)** **(380)**

The methylsulfonium salt [**381**; m.p. 115–121 °C (d)] of 1,4-dithiocane is the product of an interesting ring expansion reaction involving ethylene and the sulfonium salt (**382**), the latter being prepared in 87% yield from 1,2-dithiane ⟨71IJS(A)(1)65, *cf.* 69JOC36, 64JOC2702⟩.

1,2-dithiane $\xrightarrow[87\%]{Me_3O^+ X^-}$ (382) $\xrightarrow[78\%]{CH_2{=}CH_2}$ (381); X^- = 2,4,6-trinitrobenzenesulfonate (SO_3^-, NO_2, NO_2, NO_2)

(382) **(381)**

A great deal of work has been carried out on 1,5-dithiocanes, particularly with respect to transannular reactions. This area has been reviewed recently by Musker ⟨80ACR200⟩.

1,5-Dithiocane (**383**; b.p. 45 °C/0.1 mmHg) can be prepared in low yield by the high dilution condensation of propane-1,3-dithiol with 1,3-dibromopropane ⟨34JA2177, 62JA4075⟩. It gives a bis-sulfone (**384**; m.p. 259 °C) with hydrogen peroxide.

$Br(CH_2)_3Br + HS(CH_2)_3SH$ $\xrightarrow[EtOH]{NaOEt}$ (383) $\xrightarrow{H_2O}$ (384)

(383) **(384)**

With $NO^+ BF_4^-$, 1,5-dithiocane undergoes an electron transfer reaction to give NO and a persistent yellow radical cation (**385**) which contains an S—S bond and can be observed by ESR ⟨76JA3055⟩. Further oxidation by the same reagent affords the dication (**386**), which can be isolated as its colorless BF_4^- salt ⟨76JA6745, 78JA6416⟩. Treatment of (**386**) with water yields 1,5-dithiocane monosulfoxide (**387**). Cyclic voltammetry shows that 1,5-dithiocane is very easily oxidized and that the radical cation dimerizes reversibly in solution ⟨80ACR200, 79JA1040⟩. The ESR spectrum of the radical cation (**385**) in MeCN exhibits a triplet of triplets with hyperfine coupling constants of 10.4 and 15.2 G. However, a solution of 1,5-dithiocane in concentrated sulfuric acid, which should contain the same cation radical, gives an ESR spectrum showing equal couplings of 10 G to four protons. It was originally suggested that (**385**) might have the unpaired electron localized on one sulfur ⟨76JA3055⟩, but the data are also compatible with a symmetrical structure in which the free electron is in a σ^*-orbital located on both sulfurs ⟨78JA6416⟩. In the latter case one of the hyperfine coupling constants to the α protons would have to be unobservably small. The difference between the acetonitrile and sulfuric acid spectra has been ascribed to ion-pair and dimerization effects, and this is supported by data on the tetradeuterated 1,4-dithiocane (**388**), which can be obtained from (**387**) ⟨80ACR200⟩.

1,5-Dithiocane forms complexes with Ru(III) and Ru(II). The dimeric mixed-valence complex $[(Ru(NH_3)_5)_2{\cdot}C_6H_{12}S_2]^{5+}$ (**389**) is particularly interesting because its electronic spectrum shows evidence of a transannular sulfur–sulfur interaction ⟨78JA1635⟩. Although the crystal structure of that complex is not known, the nickel complex of 1,5-dithiocane, where this ligand bridges two metal atoms, has a known structure having a cross-ring sulfur–sulfur distance of 3.33 Å ⟨74CS(5)52⟩. This is slightly smaller than the sulfur–sulfur

(383) —NOBF₄/MeCN→ (385) —NOBF₄/MeCN→ (386) —H₂O/NaHCO₃→ (387) —i, D₂O/base; ii, reduction→ (388)

(389)

(390) 3.58 Å

distance in the bis-ketal (**390**) (3.58 Å) or in the complex $C_6H_{12}S_2{\cdot}2I_2$ (3.44 Å) ⟨80ACR200⟩.

The monosulfoxide (**387**) reacts with acetone under acidic conditions to give the acetonyl sulfonium salt (**391**), a type of reaction which is not observed in simple sulfoxides, or indeed in monosulfoxides not having a close proximity of the two sulfur atoms ⟨63MI51900, 69TL177, *cf.* 80ACR200⟩.

(387) —H^+→ → —Me_2CO→ (391)

1,5-Dithiocan-3,7-dione (**392**; m.p. 145–146 °C, ν 1725 cm^{-1}) is formed by the reaction of 1,3-dichloroacetone with H_2S in low yield ⟨67ACS705⟩. Malonyl chloride reacts with trithiocarbonic acid to give colorless crystals of (**393**; m.p. 124 °C), a compound which cannot be recrystallized without decomposition. On heating (**393**) breaks up into carbon suboxide, carbon oxysulfide and thioacetic acid ⟨66AG(E)250⟩.

$O{=}C(CH_2Cl)_2$ —H_2S, HCl / H_2O→ (392)

$CH_2(COCl)_2 + SC(SH)_2$ —$CHCl_3$→ (393)

5.19.8.4 Tetrahydrodithiocins and Benzo Derivatives

The dihalide cyclization method has been used to make the two isomeric tetrahydrodithiocins (**394**) and (**395**), as well as the benzo derivative (**396**). Oxidation of (**394**) and (**395**) gives rise to crystalline disulfones (m.p. 227 and 130 °C respectively), while (**395**) behaves as a dienophile towards cyclopentadiene ⟨64ZC302⟩ (*cf.* **255**, Section 5.19.4.4).

5.19.8.5 Tetrathiocanes

1,2,5,6-Tetrathiocane (**397**; m.p. 63–65 °C) is formed by oxidation of ethanedithiol ⟨82TL3231⟩. A similar reaction on *trans*-1,2-cyclohexanedithiol gives [**398**; m.p. 192 °C, λ_{max} 242 nm (log ε 2.95)] ⟨67AG(E)703⟩. This dithiol also reacts with S_xCl_2 ($x = 2$ or greater)

$(CH_2SH)_2$ + CH₂Cl=CH₂Cl →(base, 20%) (394); $(CH_2Br)_2$ + CH₂SH=CH₂SH →(base, 35%) (394)

$Br(CH_2)_4Br$ + NaS=SNa → (395); $HS(CH_2)_4SH$ + Cl=Cl →(base) (395)

o-$C_6H_4(CH_2Br)_2$ + $(CH_2SH)_2$ → (396)

$(CH_2SH)_2$ →(I_2, NEt_3, $CHCl_3$) (397)

trans-cyclohexane-1,2-dithiol →(I_2) (398)

(399)

to give an eight-membered ring containing a chain of six contiguous sulfur atoms. The 1,2,5,6-tetrathiocane ring system occurs in the alkaloid cassipourine (**399**) ⟨69AX(B)2114⟩.

1,3,5,7-Tetrathiocane (**400**; m.p. 42 °C), unlike its oxygen analog (Section 5.19.7.7), has only been obtained by the halide route and in low yield ⟨59AG407, 66ZN(B)622⟩. It reacts with butyllithium to give mono-, di-, tri- and tetra-anions. The mono-anion has been alkylated with methyl iodide and it reacts with silicon, germanium and tin halides. Two isomeric tetrakis(trimethylsilyl) derivatives have been prepared ⟨76CB1239⟩. The mono-anion also reacts with CO_2, SO_2 and CS_2 to give the expected acids ⟨75CJC3720⟩. Structural and conformational data on 1,3,5,7-tetrathiocane are given in Section 5.19.9.

$CH_2Cl_2 + Na_2S$ →(6%) (400) →(LiR) [anion] →(RX) [R-substituted]

(400)

R = Me, $SiMe_3$, $GeMe_3$, $SnMe_3$

The unsaturated derivative of 1,2,3,4-tetrathiocane (**401**) has been prepared by the halide route with sodium polysulfide ⟨72JOC2367⟩.

Ph(CH₂Br)C=C(CH₂Br)Ph →(NaS_n, DMF) (401)

(401)

5.19.8.6 Saturated Rings with S and O or S and N

A few miscellaneous cyclic esters are known in this class. The cyclic sulfinic ester (**402**) (a sultine) is formed in 40% yield, but no proof has been offered that the compound is not a 16-membered dimer ⟨76CJC3012⟩. The cyclic sulfites (**403**) and (**404**) are obtained, together with their 16-membered dimers, by the action of thionyl chloride on the appropriate glycol ⟨80TL1039⟩.

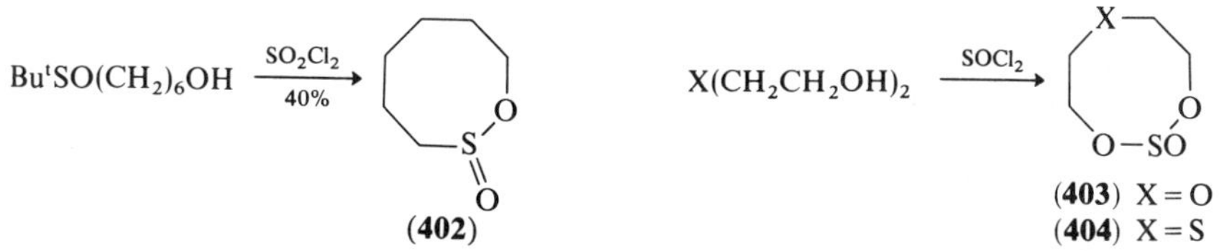

$Bu^tSO(CH_2)_6OH \xrightarrow[40\%]{SO_2Cl_2}$ (402)

$X(CH_2CH_2OH)_2 \xrightarrow{SOCl_2}$ (403) X = O; (404) X = S

The thiazocane (**405**; b.p. 92 °C/25 mmHg) has been made in 13% yield by the halide cyclization method followed by LAH reduction of the quaternary ammonium salt (**406**) ⟨65TL4259⟩. Oxidation with sodium metaperiodate gives the sulfoxide [**407**; m.p. 70–72 °C, ν 1015 cm^{-1} (CCl_4)]. The NMR spectrum shows that the perchlorate of (**407**) is protonated on nitrogen and that the NH group is hydrogen bonded to the sulfoxide oxygen, as shown in (**408**). The X-ray structure of this salt confirms these findings and shows that the ring adopts a boat-chair conformation ⟨65TL4265, 69JCS(B)33, 76JCS(P2)836⟩. Reduction of (**408**) with iodide ion gives back (**405**) and the kinetics of this process has been investigated and seems to be consistent with an intermediate dication (**409**).

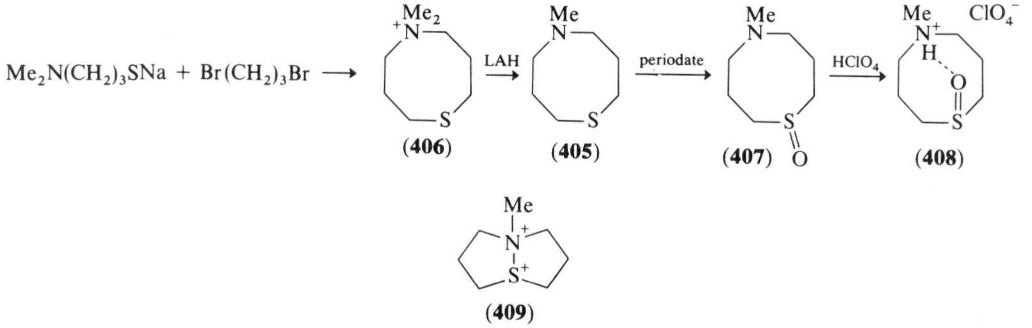

$Me_2N(CH_2)_3SNa + Br(CH_2)_3Br \longrightarrow$ (406) $\xrightarrow{LAH}$ (405) $\xrightarrow{periodate}$ (407) $\xrightarrow{HClO_4}$ (408) ClO_4^-

(409)

In work related to that on 1,5-dithiocane (Section 5.19.8.3), Musker and coworkers ⟨78JA7754⟩ found that (**405**) undergoes a smooth oxidation with $NOBF_4$ to give the dication (**409**). Although no cation radical is observed during the oxidation, a mixture of (**405**) and (**409**) does give the expected ESR spectrum, but the signal disappears on standing. Reaction of (**409**) with water gives the known sulfoxide (**407**).

The conformations of rings with sulfur atoms are discussed in Section 5.19.9.

5.19.9 CONFORMATIONS OF HETEROCYCLIC EIGHT-MEMBERED RINGS

5.19.9.1 General Considerations

This section presents X-ray diffraction, dynamic NMR and theoretical data on the conformations of heterocyclic eight-membered rings. Ideally, it would be desirable to know the whole conformational potential energy surface of the ground electronic state of a molecule. In practice, the lowest (and perhaps the second lowest) energy conformation can be more or less characterized, and some conformational energy barriers can be determined and calculated.

X-Ray diffraction data give the atomic coordinates and thereby the conformations of molecules in the crystalline phase. If many structures of a given type are known, which unfortunately is rarely the case for medium rings, it is likely that an excellent picture of the global conformational energy minimum will be obtained as lattice effects should be more or less random.

Dynamic NMR gives information on the number and symmetries of conformations present in solution and on the energy barriers separating these conformations. This is particularly true for systems with barriers between about 25 and 90 kJ mol^{-1}, a situation which often occurs in the medium ring. The interpretation of the NMR data can be carried out by the examination of molecular models, but this is a relatively crude and sometimes misleading method. Empirical force field (or molecular mechanics) calculations are much superior, even though the parametrization of heteroatoms may be open to question. Quantum mechanical calculations are not very suitable; the semiempirical type, *e.g.* MINDO, do not reproduce conformational properties of even cyclohexane satisfactorily, and the *ab initio* type are currently prohibitively demanding in computation time for molecules of interest in this chapter, especially if the calculations are carried out with a sufficiently large basis set to provide some confidence in the results ⟨B-75MI51900⟩.

It is useful to classify ring systems on the basis of the number of strong torsional constraints present *within* the ring. The dividing line between weak and strong is of course arbitrary, but it might be set at about 80 kJ mol^{-1}. Thus, double bonds and small fused rings are strong constraints, whereas amides lie in the borderline region. Fortunately, only *cis* amide groups occur in eight-membered rings. All other formal single bonds are weak constraints ⟨74MI51900⟩.

Although a good deal is known about the structures and conformational properties of heterocyclic eight-membered rings, it is clear that knowledge in this area is quite spotty compared to the existing situation in heterocyclic six-membered rings ⟨B-80MI51900⟩.

5.19.9.2 Molecules without Strong Ring Torsional Constraints

The conformational properties of such eight-membered ring molecules have been reviewed fairly recently ⟨74MI51900⟩. Cyclooctane is the archetypical molecule in this class, and the heterocyclic analogs, such as the azocanes, oxocanes and thiocanes, as well as carbocyclic derivatives, such as cyclooctanone, all have closely related conformational features and a brief overview of their conformations will now be given.

The medium rings are significantly strained, and different conformations have different distributions of strain. Thus, the crown conformation (**410**), which has a ring torsional angle of 88° (force field calculation, ⟨73TL5029⟩) has mostly torsional strain. It is thus likely to be favored by molecules preferring a torsional angle of about this value (*e.g.* octasulfur), or by molecules having relatively low torsional barriers (*e.g.* tetroxocanes). The symmetrical crown can be distorted by stretching to give the chair-chair (**411**) and the twist-chair-chair (**412**). Force field calculations show that the latter is of slightly lower energy in cyclooctane itself than the other two conformations, but since the conformational interconversion barriers are extremely low, all three conformations will show the same kind of NMR spectra. Experimentally, about 5% of crown family conformations (probably mostly **412**) exist in cyclooctane at room temperature. The remaining 95% consists of the boat-chair conformation (**413**) and therefore the free energy difference between (**412**) and (**413**) is about 7 kJ mol^{-1}. Despite the apparent flexibility of molecular models of this class of eight-membered rings, there is quite a high barrier for ring inversion in the crown family, which includes (**410**), (**411**) and (**412**). Calculations show that ring inversion in this family occurs by a conversion to the boat-chair family with a calculated barrier of about 40 kJ mol^{-1}. The transition state for this crown to boat-chair process has a geometry similar to (**412**), but with the 1,2 bond eclipsed.

(**410**) Crown (**411**) Chair-chair (**412**) Twist-chair-chair

The boat-chair (BC) (**413**) is the commonest conformation in this class. It has less torsional strain than the crown, but more non-bonded repulsions, the latter being largely transferred into internal angle strain. The boat-chair only has C_s symmetry, but it can pseudorotate *via* the twist-boat-chair (**414**) (C_2 symmetry), a process which has a low barrier (calculated for cyclooctane to be 14 kJ mol^{-1}). The transition state for the pseudorotation is unsymmetrical and involves the elipsing of either the 2,3 or the 7,8 bond of the boat-chair. In various analogs of cyclooctane the boat-chair pseudorotation barrier may be much higher than that in the parent compound. When pseudorotation is fast, the boat-chair in cyclooctane has a C_{4v} time-averaged symmetry, *i.e.* the same as that of the crown family. Ring inversion in the boat-chair probably takes place *via* a chair (**415**) or twist-chair (**416**) conformation, and the barrier height is of the order of 30 kJ mol^{-1}.

Boat-chair family *Possible transition states for ring inversion in the boat-chair*

(**413**) Boat-chair (**414**) Twist-boat-chair (**415**) Chair (**416**) Twist-chair

The boat-boat (**417**) and the twist-boat-boat (**418**) have low torsional strains but severe non-bonded repulsions, which, as usual, are transferred to internal angle strains. However, heteroatoms can modify these repulsions and certain transannular interactions can drastically reduce them. Even so, the boat-boat family is relatively unimportant as its energy is calculated to be quite high (12 kJ mol^{-1}) in cyclooctane. It probably serves as an intermediate for certain conformational interconversions of the boat-chair, especially when the twist-boat-chair pseudorotation itinerary is of high energy. In cyclooctane the boat-boat and its twisted partner have nearly the same energies and are not separated by a significant barrier.

Boat-boat family

(**417**) Boat-boat (**418**) Twist-boat-boat

The above discussion provides a background for a discussion of the conformations of various azocanes, oxocanes, thiocanes and related compounds which lack strong ring torsional constraints.

X-Ray diffraction results are available for salts of azocane, but not for the free base, which is a liquid. Unlike the chloride salt, which is disordered and gives little conformational information ⟨B-68MI51902⟩, the tosylate gives excellent data and shows that the cation has the boat-chair conformation (**419**) ⟨77CC939⟩.

Ring torsional angles (°)

$\omega_{1,2}$	55	$\omega_{5,6}$	−68
$\omega_{2,3}$	51	$\omega_{6,7}$	96
$\omega_{3,4}$	−106	$\omega_{7,8}$	−35
$\omega_{4,5}$	70	$\omega_{8,1}$	−65

2.21 H···H 2.26 H N4 1 2 3 6 5 7 8 H···H 2.13

(**419**)

Dynamic ^{13}C NMR measurements show that azocane exists at −112 °C as a 97 : 3 mixture of a boat-chair and a crown family conformation ($\Delta G° = 5$ kJ mol^{-1}, $\Delta G^{\ddagger}$ 44 kJ mol^{-1}) ⟨78JOC3021⟩. From the ^{1}H NMR spectrum the barrier to ring inversion in the boat-chair is 31 kJ mol^{-1}, but the barrier to pseudorotation is apparently too low for measurement. Likely candidates for the crown family and boat-chair conformations of azocane are the chair-chair (**420**), and the BC-1 (**421**), BC-3 (**422**) and BC-7 (**423**) conformations, respectively, but the precise position of the nitrogen atom is not known experimentally. *N*-Methyl- and *N*-chloro-azocane also have predominantly boat-chair conformations ⟨75JOC369⟩.

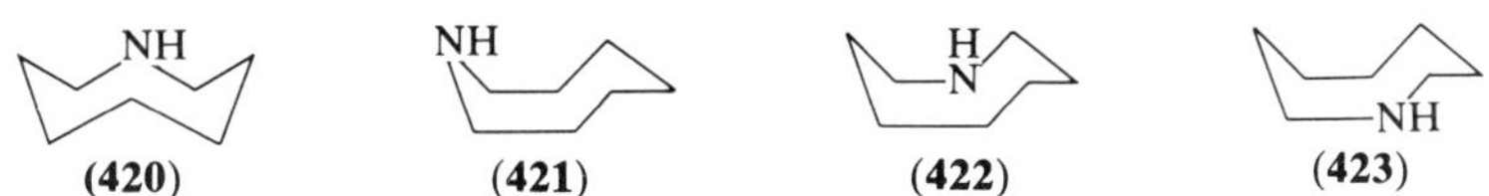

(**420**) (**421**) (**422**) (**423**)

A boat-chair conformation (**424**) having a relatively short nitrogen–carbonyl distance (2.76 Å) has been found by X-ray analysis of *N*-*p*-tolylazocan-5-one ⟨75AX(B)2912⟩. The eight-membered ring in (**425**) has a boat-boat conformation, but this compound may be better regarded as a nitrogen-bridged 10-membered ring ⟨75AX(B)2916⟩. It has a nitrogen–carbonyl distance of 2.46 Å. The related monocyclic azocanone compound exists in the boat-chair conformation (**426**) in aprotic solvents, but there is evidence that it may exist as a boat-boat (**427**) in strong hydrogen-bonding solvents ⟨79UP51900⟩.

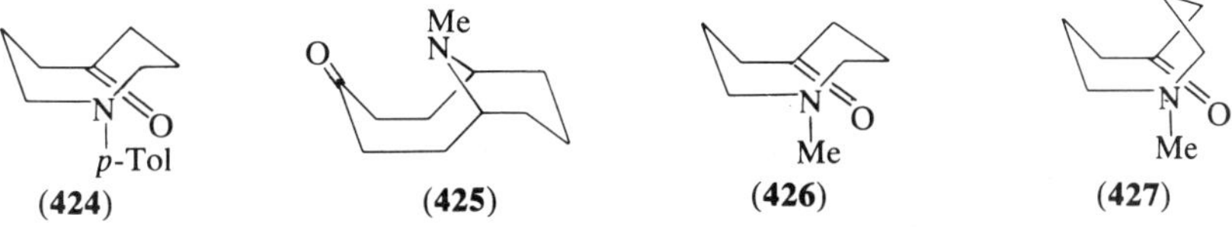

(**424**) (**425**) (**426**) (**427**)

Manxine (**60**) and its derivatives are bridged azocanes (Section 5.19.2.1) in which, at least in the salt of (**60**), three boat-chairs are fused together in a conformation having C_3 symmetry. The barrier for ring inversion in manxine, a process which gives the mirror-image conformation, is about 45 kJ mol^{-1} ⟨72JA7092⟩.

Oxocane shows a dynamic NMR behavior very similar to that of azocane ⟨72JA1390⟩ except that there is some evidence that the boat-chair pseudorotation does become slow at low temperatures ⟨79UP51901⟩. However, the NMR spectra are not well enough resolved to determine the position or positions of the oxygen in the boat-chair. The crown family conformation is somewhat more populated ($\Delta G° = 4$ kJ mol^{-1}) than in the case of azocane or cyclooctane, and the barrier to ring inversion in the predominant boat-chair is 31 kJ mol^{-1}. Empirical force field calculations ⟨80ZN(B)1479, 78UP51900⟩ on oxocane reveal that the BC-3 conformation (**428**) is of lowest energy, followed closely by BC-1 ($\Delta G° = 1.7$ kJ mol^{-1}). The lowest energy crown family conformation is calculated to be the crown itself ($\Delta G° =$ 9 kJ mol^{-1}).

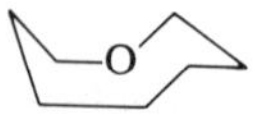

(**428**) BC-3

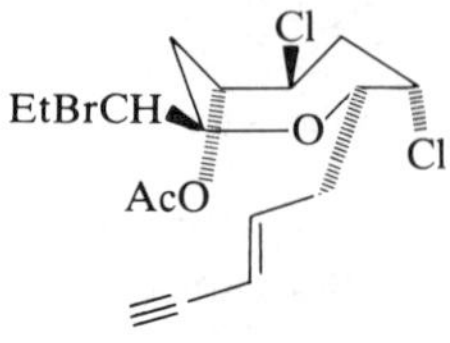

(**429**)

The X-ray structure of the substituted oxocane laurencienyne, (**429**), shows that the ring has the BC-3 conformation (actually, the oxygen is at the mirror image 7 position of the boat-chair) and one chlorine atom is in an axial-like position. The other substituents are in unhindered positions on the ring ⟨80TL2299, 80CSC777⟩.

5-Oxocanone has an unsymmetrical boat-chair conformation (**430**) with a geometry which is suitable for a transannular oxygen–carbonyl interaction. The presence of relatively high barriers for pseudorotation ($\Delta G^{\ddagger} = 33$ kJ mol^{-1}) and for ring inversion ($\Delta G^{\ddagger} =$ 38 kJ mol^{-1}), as compared to those in oxocane or cyclooctanone, is consistent with a transannular stabilization of (**430**). This also shows that the pseudorotation is taking place *via* a twist-boat-chair rather than a boat-boat. Neither the crown family nor the boat-boat is observably populated ⟨72TL3613⟩.

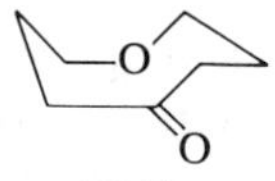

(**430**)

Two distinct dynamic NMR effects occur in 1,3-dioxocane and its derivatives having a *gem*-dimethyl group at the 2- and/or 6-positions (chemical, not conformational numbering) ⟨76JA2059⟩. These molecules have unsymmetrical boat-chair conformations (*e.g.* **432**) which minimize non-bonded and dipole–dipole repulsions ⟨80ZN(B)1479, 81JPC2469, 81JPC2476, 72JA1389, 1390⟩. No crown family conformations have been observed even in minor amounts, unlike the situation in oxocane. The lowest energy conformational process in 1,3-dioxocane is the pseudorotation of (**431**) to its mirror-image form (**432**), *via* the plane symmetrical BC-2,8. This is also true for the 6,6-dimethyl derivative, but the presence of a 2,2-dimethyl group greatly raises the barrier for the pseudorotation process, since there is now a very sterically crowded methyl at the 1-position in the BC-2,8.

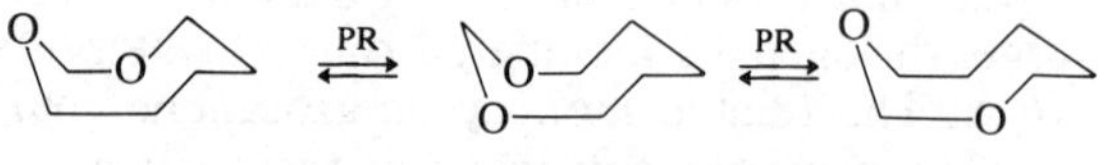

(**431**) BC-1, 3   BC-2, 8   (**432**) BC-1, 7

PR = pseudorotation *via* a twist-boat-chair

The lowest energy conformational process in 2,2-dimethyl-1,3-dioxocane is the conversion of (**433**) to (**434**) *via* the boat-boat intermediate (BB-1, 3) which has C_2 symmetry. The methyl groups (A and B) in (**433**) and (**434**) have exchanged sites. This process is actually similar to the high energy conformational process in 1,3-dioxocane itself. This crossover of mechanisms is shown diagrammatically in Figure 1.

1,3,6-Trioxocane exists as a 1 : 1 mixture (*i.e.* $\Delta G° = 0$) of a boat-chair and a crown family conformation with an interconversion barrier of 36 kJ mol^{-1} ⟨72JA1390, 1389⟩. The ring inversion barrier in the boat-chair is 28 kJ mol^{-1} but the barrier to pseudorotation has not been determined. One set of force-field calculations shows that the BC-1, 3, 6 (**435**) is of

(**433**) BC-1, 3 BB – 1, 3 (C_2) (**434**) BC-1, 3

Figure 1 Relationships of the conformational free energy barriers in 1,3-dioxocane and some of its methylated derivatives. The full lines link barriers for boat-chair to twist-boat-chair processes and the dotted lines link barriers for boat-chair to boat-boat processes. Energies are in kJ mol^{-1}

lowest energy and that the crown (**436**) is the next best conformation ($\Delta G° = 4.2$ kJ mol^{-1}) ⟨80ZN(B)1479⟩ while another set, although also showing the BC-1, 3, 6 as being the lowest in energy, gives the twist-chair-chair ($\Delta G° = 13$ kJ mol^{-1}) as the best crown family conformation ⟨81JPC2469⟩. However, the latter calculation gives a $\Delta G°$ value in poor agreement with experiment.

(**435**) BC-1, 3, 6 (**436**)

The structures and conformations of 1,3,5,7-tetroxocanes and its derivatives have been extensively studied by X-ray diffraction, dynamic NMR, IR and Raman spectroscopy, dipole moment measurements and force-field calculations. 1,3,5,7-Tetroxocane gives monoclinic crystals with four molecules in the unit cell. The conformation is a crown (Figure 2) with approximate C_{4v} and exact C_{2v} symmetry ⟨74BCJ583⟩.

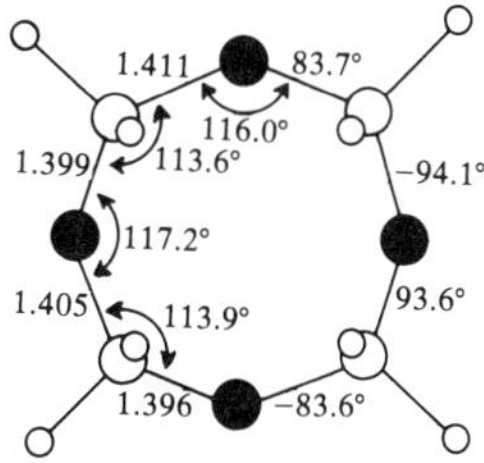

Figure 2 Conformation of 1,3,5,7-tetroxocane. The filled-in circles are the oxygen atoms. Bond lengths (Å), internal and ring torsional angles for the crown (C_{2v}) are shown

Dynamic NMR shows the presence of two conformations in 1,3,5,7-tetroxocane: a crown and a boat-chair. The equilibrium constant is strongly solvent and temperature dependent, with the highly symmetrical crown, which has a high dipole moment, having a lower entropy than the boat chair by about 6 J K^{-1} mol^{-1} ⟨72JA1390, 1389⟩. The interconversion barrier is 50 kJ mol^{-1}, and no ring inversion nor pseudorotation process has been detected in the boat-chair.

IR and Raman spectra also reveal the presence of two conformations in 1,3,5,7-tetroxocane, with the ratio of crown to boat-chair ranging from 3.0 (MeCN) to 0.55 (CCl_4) as the solvent polarity decreases ⟨77SA(A)549, 77JCP(66)32⟩. Force-field calculations show that the crown is 11 kJ mol^{-1} higher in energy than the BC-1, 3, 5, 7 ⟨80ZN(B)1479⟩. These calculations, as in the case of 1,3,6-trioxocane, seem to overestimate the strain energy of the crown.

The value of $^1J(^{13}C–H)$ for the axial proton (174 Hz) in 1,3,5,7-tetroxocane is much larger than that for the equatorial proton (154 Hz) ⟨77JST(36)329⟩. Crown conformations in general have large chemical shift differences between axial and equatorial OCH_2O protons; for 1,3,6-trioxocane and 1,3,5,7-tetroxocane these differences are 0.88 and 0.64 p.p.m., respectively. In contrast, the corresponding shift in the boat-chair of the former compound is only 0.07 p.p.m. ⟨72JA1390⟩.

Metaldehyde (all-*cis*-2,4,6,8-tetramethyl-1,3,5,7-tetroxocane) exists in the crown form in the crystalline phase with all the methyl groups equatorial ⟨36JA1274⟩ and the analogous metachloral is also a crown (in solution) ⟨63CJC1893⟩, as is 2,6-bis(trichloromethyl)-1,3,5,7-tetroxocane ⟨77JST(36)329⟩.

Crown family conformations occur in the crystalline state of the following 1,3,5,7-tetraheteroatom derivatives: (**437**) ⟨71AX(B)185⟩, (**438**) ⟨63AX617⟩ and (**439**) ⟨67CR(C)(265)745⟩. A large axial–equatorial chemical shift difference (0.3 p.p.m.) and a high ring inversion barrier (61 kJ mol^{-1}), which are typical for this kind of conformation, are found in (**439**) ⟨66CC803⟩. In (**437**), which is a crown, the *N*-methyl groups are axial. The conformation of (**438**) is a flattened crown (*i.e.* a chair-chair) with NO_2 groups alternately axial and equatorial around the ring. The crown is also the favored conformation of S_8 and $S_4(NH)_4$ ⟨74MI51900⟩. However, 1,3,5,7-tetrathiocane is entirely in the boat-chair (**440**) in both the crystalline state and in solution.

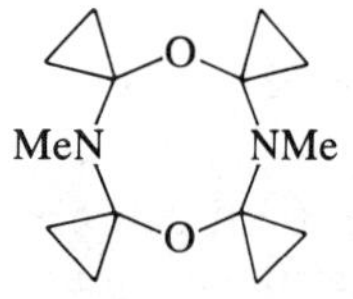

MeN NMe O O
(**437**)

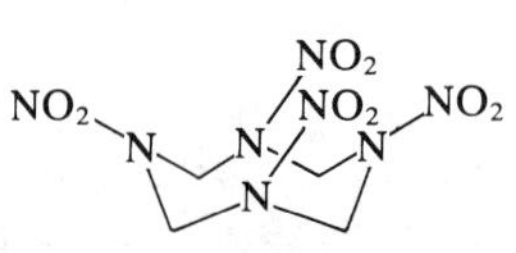

NO_2 NO_2 NO_2 NO_2 N N N N
(**438**)

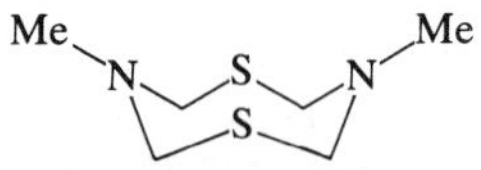

Me Me N S N S
(**439**)

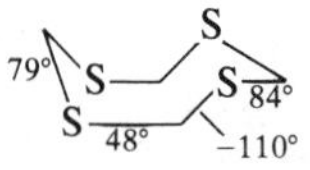

79° S S S S 84° 48° −110°
(**440**)

Several other sulfur eight-membered heterocycles also exist in boat-chair conformations. 5-Thiacanone (**441**) has a conformation similar to its oxygen analog (**430**) and shows barriers to pseudorotation and ring inversion of 28 and 34 kJ mol^{-1}, respectively ⟨80JOC1224⟩. The 1,5-dithiocane derivative (**442**) ⟨70JCS(B)1603⟩ is a boat-chair in the crystalline state, but 1,5-dithiocane, as its nickel(II) chloride complex, exists both as a boat-chair and as a crown in a disordered crystal arrangement ⟨74CS(5)52⟩. The boat-chair structure of (**408**; Section 5.19.8.6) has already been mentioned.

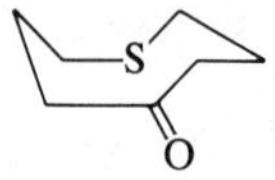

S O
(**441**)

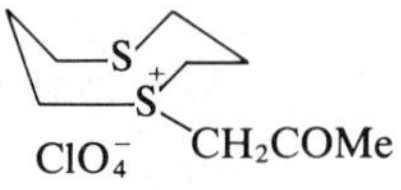

S S+ ClO_4^- CH_2COMe
(**442**)

Finally, 1,2,5,6-tetrathiocanes seem to prefer the twist-boat-chair conformation ⟨82TL3231, 69AX(B)2114⟩. The ring torsional angles in the present compound are given in (**443**).

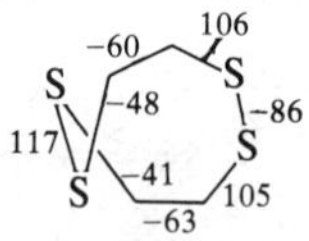

106 −60 S S −48 −86 117 S −41 S 105 −63
(**443**) Torsion angles (°)

5.19.9.3 Molecules with Strong Torsional Constraints

A strong torsional constraint for the purpose of this section is generally an endocyclic *cis* double bond or a related structural feature, such as a *cis* epoxide group, a benzene ring or an amide group. Torsional effects caused by large substituents, although important, are not inherent in a bond and are not considered to fall in the above category. A low energy conformation for a molecule having only one torsional constraint in an eight-membered ring is a distorted boat-chair, where the geometry is close to that of the transition state for the boat-chair to twist-boat-chair pseudorotation in cyclooctane (Section 5.19.9.2). Indeed, enantholactam (**2**; Section 5.19.2.1) and its hydrochloride have this type of conformation, as shown in (**444**) ⟨79AX(B)1920⟩ and (**445**) ⟨75AX(B)251⟩, respectively.

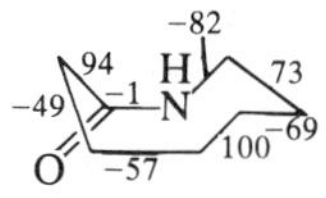

(444) Torsion angles (°) (445)

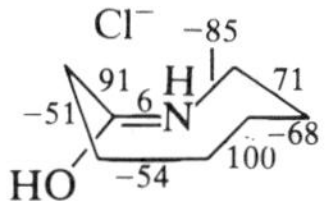

The ring torsional angles of the amide and *O*-protonated amide moieties in these structures are close to zero and the C(O)—N bond length is shorter in the protonated lactam (**445**) (1.298 Å) than it is in the lactam itself (**444**) (1.337 Å), as expected from the dominant resonance structures in these two compounds. The room temperature ^{15}N and ^{13}C NMR spectra of the lactam (**444**) and its *O*-trimethylsilyl derivative have been determined but do not give much information about the solution conformation of these compounds ⟨76JA5082, 82JMR(46)163⟩.

The alkaloid senkirkine has already been mentioned in Section 5.19.2.1 (*cf.* **37**). The eight-membered ring has both an endocyclic double bond and a short 1,5 carbonyl–nitrogen distance, and can best be described as a rather distorted boat-chair (**446**) with the double bond at the 1,2-position and the transannular interaction at the usual 3- and 7-positions (conformational numbering) (*cf.* **426**). A related alkaloid, clivorine, has a very similar structure to senkirkine ⟨72AX(B)2825⟩.

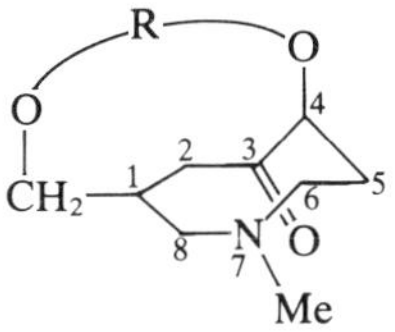

(**446**)

X-Ray structures have been obtained on a number of marine oxygen heterocycles, where the strong torsional constraint is a double bond, as in laurenyne (**447**) ⟨80TL4951⟩ and laurencin (**129**) (Section 5.19.3.1), an epoxide group, as in poiteol (**448**) ⟨80T1747⟩ and epoxyrhodophytin (**449**) ⟨80T171⟩, or a 1,3 bridge, as in chlorofucin (**450**) ⟨80T1747⟩. All these compounds have distorted boat-chair conformations, as expected from the torsional constraints present.

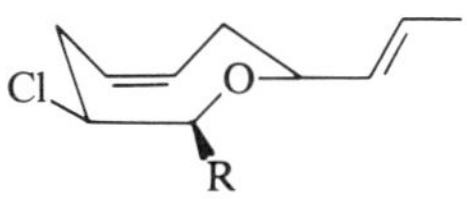

(**447**)  (**448**) 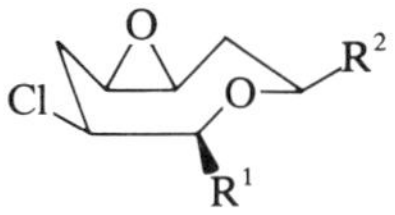 (**449**) 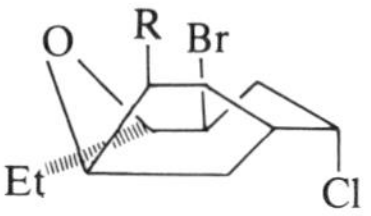 (**450**)

Eight-membered rings with two strong endocyclic torsional constraints exist in a variety of conformations, which fall into two classes, *viz.* rigid and flexible ⟨72JA5645⟩. In this respect, there is an analogy to cyclohexane and related saturated heterocycles which can exist in a rigid chair or a flexible twist-boat. However, the two classes of conformation of an eight-membered ring molecule are often close together in energy, which is a rare situation in six-membered rings. The terms 'rigid' and 'flexible' should be interpreted with some caution; they strictly apply to molecular models rather than to real molecules, because of the neglect of weak torsional constraints and the assumption of stiff internal bond angles in this kind of treatment.

Nearly all the available conformational data on eight-membered heterocycles with two torsional constraints refer to fused dibenzo derivatives. An exception is the lactam di-β-alanyl (**451**), which, like its carbocyclic analog 1,5-cyclooctadiene ⟨74MI51900⟩, exists as a twist-boat ⟨72MI51900⟩. Force-field calculations reproduce well this geometry and show that this conformation is not as flexible as its molecular model might indicate ⟨75JCS(P2)43⟩. The rigid conformation in this system is the chair, and this occurs in some of the compounds described below, but it has not been detected in di-β-alanyl.

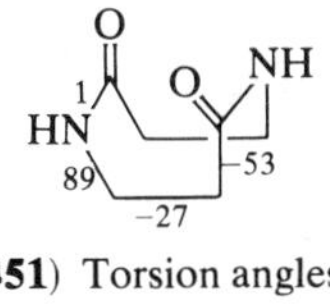

(**451**) Torsion angles (°)

The conformational properties of eight-membered rings linearly fused with two benzene rings have been investigated in some detail by dynamic NMR. The parent carbocyclic compound is (**452**) which exists as a 1:1 mixture of the chair (**453**) and the boat (probably slightly twisted) (**454**) ⟨73JCS(P1)205, 73JA2023⟩. The chair to boat barrier is 43 kJ mol^{-1} and the ring inversion barrier in the boat is 31 kJ mol^{-1}. A third conceivable conformation is the skew form (**455**), which has also been called a twist-boat ⟨73JCS(P1)205⟩.

(**452**) (**453**) (**454**) (**455**)

The introduction of a nitrogen heteroatom into the eight-membered ring of (**452**) to give (**456a**) and (**456b**) results in compounds which no longer give rise to any observable dynamic NMR effects ⟨73JCS(P1)205, 74CS(5)52⟩. However, if the *N* substituent is an acyl group (**456c–e**), two conformational processes are observed and have been assigned to the processes shown in Table 3. These amides exist in only one rotameric form, and thus N—CO rotation cannot be observed.

(**456**)

R
a; H
b; CH_2Ph
c; NAc
d; NBz
e; NTs

If the nitrogen atom is part of a lactam moiety as in (**457a–b**), the need for planarity causes the boat to be the only conformation present. The barrier for ring inversion increases substantially upon alkylation of the nitrogen because of a steric interaction of the *N*-alkyl group and the *ortho* benzene proton (Table 3) ⟨73JCS(P1)205, 74CS(5)52⟩. The ring inversion barriers in di-*o*-thymotide and di-*o*-thiothymotide are 72 and 103 kJ mol^{-1}, respectively ⟨80TL4203, 73CC571⟩, and these compounds, which are closely related to disalicylide (**352**; Section 5.19.7.5), also exist exclusively as boats because of the presence of ester or thioester groups. Actually, (**475a, b**) and the thymotides may be considered to have three and four rather than two strong torsional constraints, respectively, but these compounds are considered here for convenience.

(**457**) **a**; R = H
b; R = CH_2Ph

The tetrahydrodibenzo[*b,f*][1,4]diazocines (**458**) and the tetrahydrodibenzo[*b,f*]-[1,5]diazocines (**459**), which have two nitrogens in the heterocyclic ring, as well as their sulfur analogs (**460**) and (**461**) show conformational properties similar to those of the monoaza compound (**456**). Typical data are included in Table 3. The assignments of conformations and transition states for these dibenzo derivatives are supported by force-field calculations and by detailed dynamic NMR measurements ⟨73JCS(P1)205⟩.

(**458**) **a**; R = NAc
b; R = NBz

(**459**) **a**; R^1 = NAc, R^2 = Me
b; R^1 = NBz, R^2 = Me

(**460**) **a**; X = S, R = Me
b; X = SO_2, R = Me

(**461**) **a**; X = S
b; X = SO_2

Table 3 Conformational Properties of Some Linear Dibenzo Heterocycles ⟨73JCS(P1)205⟩

Compound	*Process*	ΔG° (kJ mol^{-1})	$\Delta G^\ddagger$ (kJ mol^{-1})
(**456c**)	Chair ⇄ boat	2.0	46
	Boat ⇄ boat*	0	84
(**456d**)	Chair ⇄ boat	−0.5	47
	Boat ⇄ boat*	0	74
(**456e**)	Chair ⇄ boat	2.8	44
	Boat ⇄ boat*	0	60
(**457a**)	Boat ⇄ boat*	0	59
(**457b**)	Boat ⇄ boat*	0	90
(**458a**)	Chair ⇄ boat	−0.5	54
	Boat ⇄ boat	0	71
(**458b**)	Chair ⇄ boat	0.3	58
	Boat ⇄ boat	0	60
(**459a**)	Chair ⇄ boat	1	94
	'Inversion'	0	≈92
(**459b**)	Chair ⇄ boat	0	87
	'Inversion'	0	88
(**460a**)	Chair ⇄ boat	3	51
(**460b**)	Chair ⇄ chair*	0	50
(**461a**)	Boat ⇄ boat*	0	51
	Chair ⇄ boat	2	60
(**461b**)	Chair ⇄ chair*	0	78

The interconversion scheme for (**456c–d**) is shown in Figure 3. The boat to chair and boat* to chair* processes involve transition states of the type (**452**) and may be called local inversions ⟨74MI51900⟩. (The asterisk indicates a mirror image conformation.) Only the aromatic rings lacking large benzylic substituents undergo this process in (**456c–d**) or in the analogous (**458a–b**), with a relatively low barrier (44–58 kJ mol^{-1}). Where such a ring is lacking, as in (**459a–b**), the barrier is very high (87–94 kJ mol^{-1}). The boat to boat* processes are ring inversions with transition states of the type (**463**). The transannular repulsions are smaller in (**458**) than they are in (**463**) because they involve a CH_2 and a nitrogen in the former case and two CH_2 groups in the latter case, and this is reflected in the boat inversion barriers, *e.g.* 71 kJ mol^{-1} in (**458a**) and 84 kJ mol^{-1} in (**456c**). In the sulfur heterocycles (**460**) and (**461**) the conformational barriers are relatively low, except for the disulfone (**461b**). Further details on these conformational processes can be found in the excellent paper of Ollis and coworkers ⟨73JCS(P1)205⟩.

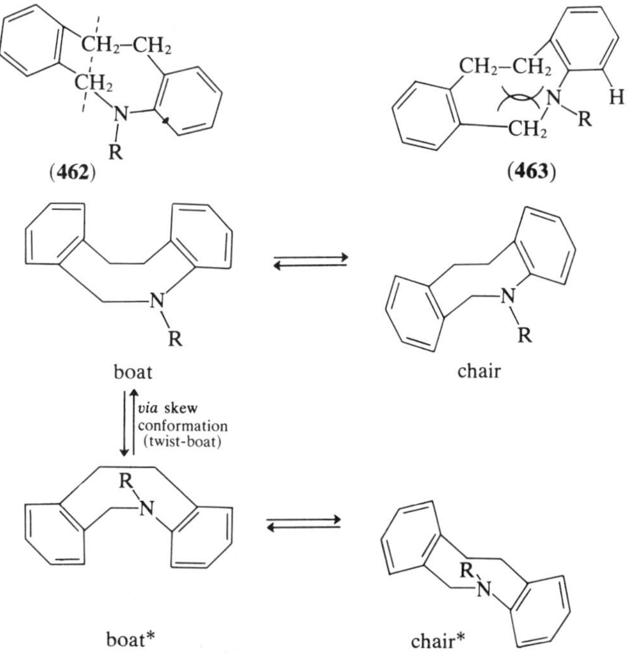

Figure 3 Conformational processes in 5-substituted 5,6,11,12-tetrahydrodibenz[*b,f*]azocines (**456c–e**). The boat to chair process takes place *via* the transition state (**462**), while the boat–ring inversion has the transition state (**463**)

In one instance boat (**464**) and chair (**465**) forms have such a high interconversion barrier that they can be separated, and they only equilibrate on heating above room temperature ⟨77AG(E)185⟩.

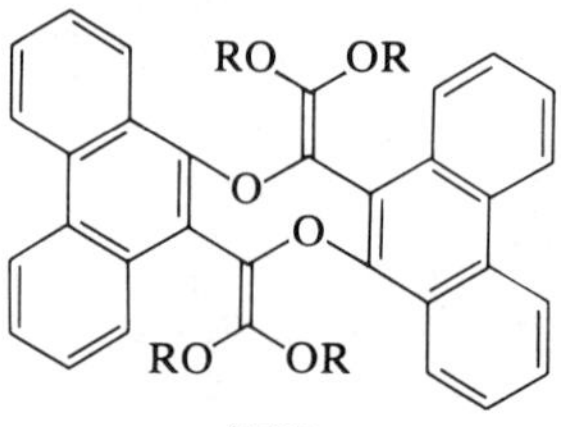

(**464**) 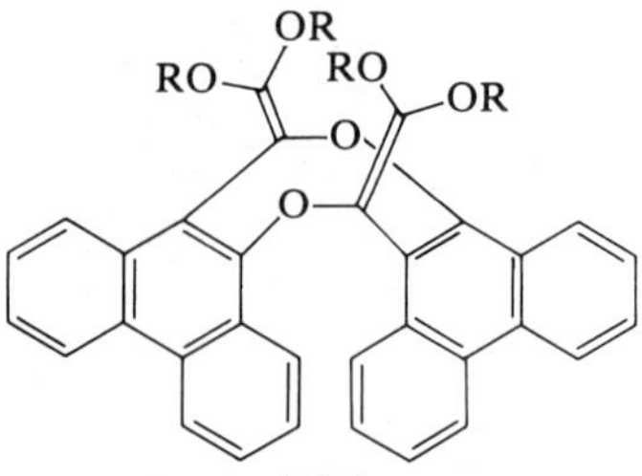 (**465**)

The angularly fused dibenzo derivatives of eight-membered heterocycles have been of as much conformational interest as the linear systems. The parent carbocyclic compound is 5,6,7,12-tetrahydrodibenzo[*a*,*d*]cyclooctene (dibenzo-1,4-cyclooctadiene) and this exists mainly in a rigid boat-chair (**466**) with only about 2% of the flexible twist-boat (**467**), in agreement with force-field calculations ⟨76AG(E)224, 76JCS(P1)913, 77CJC3456⟩. The conformational nomenclature in this system has varied and Ollis and coworkers ⟨76JCS(P1)913⟩ call the boat-chair a chair. We use the nomenclature of Fraser and coworkers ⟨77CJC3456, *cf.* 77JA6986⟩. Actually, because the twist-boat is flexible, this conformation can pseudorotate more or less easily, depending on the precise system, *via* the plane symmetrical boat-boat. All these flexible conformations can be grouped together and called boats.

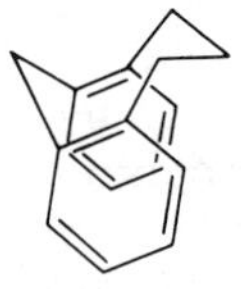

(**466**) 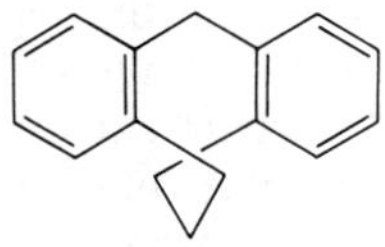 (**467**)

Direct X-ray evidence shows that *N*-methyl-5,6-dihydro-7*H*,12*H*-dibenz[*c*,*f*]azocine exists in the crystalline phase as the rigid plane-symmetrical boat-chair (**468**) ⟨74AX(B)1670⟩. In contrast, the *N*-*t*-butyl analog exists as enantiomeric pairs or twist-boats (**469**) ⟨74AX(B)1674⟩. The first observations of dynamic NMR effects in the angularly fused dibenzo system were made on compounds closely related to (**468**) and (**469**), namely (**470a–c**) ⟨70T1275⟩.

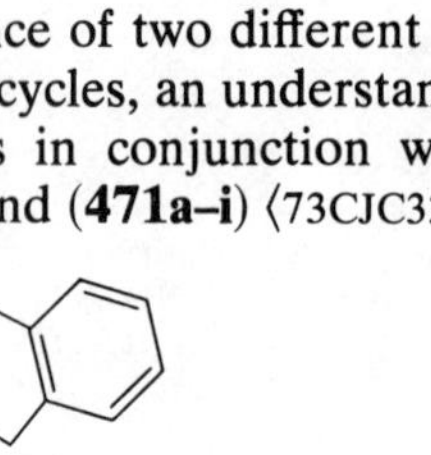

(**468**) Torsion angles (°) (**469**)

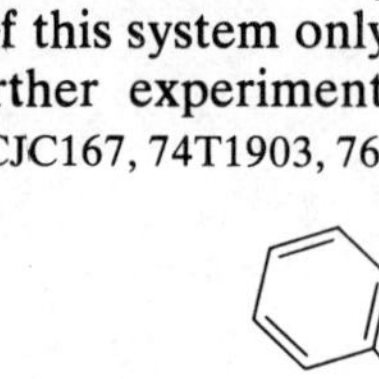

Although the presence of two different conformations is clearly evident from the NMR spectra of these heterocycles, an understanding of this system only came as a result of later force-field calculations in conjunction with further experimental work on the related compounds (**470d–h**) and (**471a–i**) ⟨73CJC3380, 75CJC167, 74T1903, 76JCS(P1)913, 77CJC3456⟩.

(**470**) 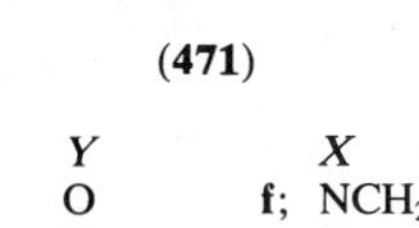 (**471**)

	X		*X*
a;	NH	**e**;	$\overset{+}{N}Me_2$
b;	CN	**f**;	NBu^t
c;	CH_2Ph	**g**;	S
d;	NMe	**h**;	SO_2

	X	*Y*		*X*	*Y*
a;	NCH_2Ph	O	**f**;	NCH_2Ph	SO_2
b;	S	O	**g**;	S	SO_2
c;	SO_2	O	**h**;	SO_2	SO_2
d;	NCH_2Ph	S	**i**;	NCH_2Ph	CO
e;	S	S			

Except for the *N*-*t*-butyl derivative (**470f**) and (**474i**), all the compounds with one heteroatom (**470a–h**) or two heteroatoms (**471a–i**) have the boat-chair as the most stable conformation (Table 4). The minor conformation is a more or less distorted twist-boat, which in some cases can only be observed by the asymmetry which it induces in what would be otherwise a symmetrical NMR line shape. In two compounds, (**471c**) and (**471e**), the pseudorotation barrier is high enough that the twist-boat gives rise to an AB quartet for its CH_2—X—CH_2 protons. In all other compounds these protons give rise to a single line in the twist-boat. The boat-chair, which is a 'rigid' conformation, always gives an AB pattern for the CH_2 groups if it is not interconverting rapidly with the twist-boat. Thus, the twist-boat is an intermediate in the ring inversion of the boat-chair, just as twist-boat cyclohexane is an intermediate in the ring inversion of chair cyclohexane.

Table 4 Conformational Properties of Some Angularly Fused Dibenzo Heterocycles

Compound	*Process*[a]	$\Delta G°$ (kJ mol^{-1})	$\Delta G^{\ddagger}$ (kJ mol^{-1})	*Ref.*
(**470a**)	BC$\rightleftarrows$BC	0	67	75CJC167
(**470c**)	BC$\rightleftarrows$TB	5.6	71	76JCS(P1)913
(**470d**)	BC$\rightleftarrows$TB	6.7	63–69[b]	75CJC167
(**470e**)	BC$\rightleftarrows$TB	3.4	>75	75CJC167
(**470f**)	BC$\rightleftarrows$TB	−4.6		75CJC167
	TB$\rightleftarrows$TB*	0	<46	75CJC167
(**470g**)	BC$\rightleftarrows$TB	5.6	66.1	76JCS(P1)913
(**470h**)	BC$\rightleftarrows$BC*	0	66.5	76JCS(P1)913
(**471a**)	BC$\rightleftarrows$TB	1.5	61.5	76JCS(P1)913
(**471b**)	BC$\rightleftarrows$TB	0.8	56.5	76JCS(P1)913
(**471c**)	BC$\rightleftarrows$TB	1.8	49.8	76JCS(P1)913
	TB$\rightleftarrows$TB*	0	41.8	76JCS(P1)913
(**471d**)	BC$\rightleftarrows$TB	2.4	58	76JCS(P1)913
(**471e**)	BC$\rightleftarrows$TB	3.9	72.4	76JCS(P1)913
	TB$\rightleftarrows$TB*	0	45.6	76JCS(P1)913
(**471f**)	BC$\rightleftarrows$TB	7.5	74.9	76JCS(P1)913
(**471g**)	BC$\rightleftarrows$TB	8.8	80.7	76JCS(P1)913
(**471h**)	BC$\rightleftarrows$BC	0	75.3	76JCS(P1)913
(**471i**)	Boat$\rightleftarrows$boat	0	<33	76JCS(P1)913

[a] BC and TB are boat-chair and twist-boat, respectively, and an asterisk indicates a ring inverted form.
[b] Temperature dependent.

In the series with one heteroatom (**470**), the $ArCH_2Ar$ shows a more positive ^2J(H–H) (−12 to −14 Hz) in the boat-chair than it does in the twist-boat (−19 Hz). Also, the entropy of the twist-boat is higher than that of the boat-chair, and this has been ascribed to the flexibility of the former conformation ⟨75CJC167⟩.

Because of a transannular carbonyl–nitrogen interaction, (**471i**) is postulated to exist in the plane symmetrical boat-boat conformation, rather than in a twist-boat.

The bridged biphenyl (**472**) ⟨67ZN(B)1296⟩ has a 'rigid' twist-boat-chair conformation in the crystal ⟨72CC985⟩ and is an eight-membered heterocycle with two torsional constraints in a 1,3 relationship. It has been resolved by chromatography on cellulose acetate, and $\Delta G^{\ddagger}$ for racemization is 120 kJ mol^{-1} ⟨67M1323⟩.

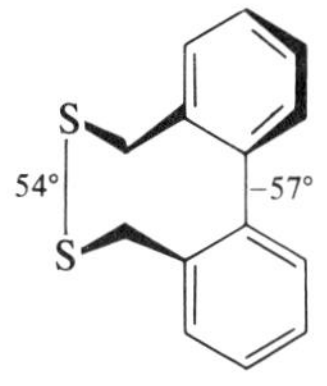

S S 54° −57°

(**472**)

The planarity or non-planarity of aromatic or potentially aromatic eight-membered heterocycles has already been discussed in Section 5.19.4, and will therefore not be further elaborated upon in this section.

5.20

Larger Rings except Crown Ethers and Heterophanes

A. G. ANASTASSIOU
University of Petroleum and Minerals, Dhahran
and *Syracuse University, New York*

5.20.1 INTRODUCTION

This chapter concerns a rather special, relatively rare species of heterocycle, namely those molecules incorporating more than eight peripheral atoms but which are neither heterocyclophanes nor crown ethers. Also excluded from coverage in this review is the large body of naturally occurring substances, or synthetic simulations thereof, incorporating large heterocyclic frames such as macrolides, alkaloids and, possibly, others.

What *is* actually covered in this section is a group of substances belonging to a relatively young (not quite 15 years of age) family of heterocycles whose theoretical conception and subsequent experimental birth resulted chiefly from interest in questions relating to the concept of heteroaromaticity, first formalized by Hückel ⟨32MI52000⟩ half a century ago, with particular emphasis placed on the structural and environmental factors responsible for heteroatomic lone pair participation in cyclic π-electron delocalization. Consequently, one finds the great majority of known specimens to incorporate a full set of π-units, to have been studied primarily with regard to their ability to sustain ring currents, and to have been analyzed and discussed comparatively in terms of overall thermodynamic preferences for buckled or planar geometries ⟨78AHC(23)55⟩. Not unexpectedly then one also finds that many of the partially or fully saturated derivatives of these heteroannulenes have become available for the most part in the course of structural characterization.

Moreover, because of their large size, high unsaturation content and π-excessive character these molecules are naturally associated, respectively, with adverse entropic factors, pronounced affinity to intramolecular pericyclization and high air sensitivity, so that their successful construction necessitated the development and use of rather unorthodox synthetic procedures. Similarly, the study of reactivity has been restricted, for the most part, to (i) internally induced reorganization and (ii) exposure to such mild reagent chemistry as was deemed necessary to induce structurally informative alterations while maintaining undesirable decomposition to a minimum.

5.20.2 STRUCTURE

5.20.2.1 General Survey of Structure

Because of their relatively large size, the heterocycles covered in this section display a natural tendency to possess heavily buckled, frequently flexible frames except in cases where bridging and/or strategically located unsaturation, such as the presence of triple bonds and/or *trans* double bonds, impart a certain degree of skeletal rigidity which quite often results in substantial molecular flattening.

5.20.2.2 Nine-membered Rings

Besides being the smallest among the ring sizes covered in this chapter, the nine-membered family of π-excessive heterocycles, commonly referred to as *heteronins*, is also the oldest known, the description of the first well-characterized such species, the symmetrically dibenzannulated derivatives (**13**), having appeared in the literature as early as 15 years ago ⟨68JA7372⟩. Studies on the heteronin family have been quite extensive and have centered chiefly on the frame's potential toward π-excessive mobilization of its lone pair and the attendant chemical and/or spectroscopic manifestations. The discussion and analysis of the results have been the subject of several reviews ⟨72ACR281,75PAC(44)691,76ACR453⟩.

5.20.2.2.1 Parent (skeletally unrestricted) heteronins

Oxonin (**1**) and a wide variety of azonines, depicted in (**2**), (**3**) and (**4**), are known ⟨72ACR281⟩. These, based on extensive chemical and spectroscopic studies, were classified as polyenic or nonaromatic (**1** and **2**), aromatic (**3**) and nondescript (**4**). The polyenic members of the family, consisting of oxonin and a variety of azonines incorporating electron-withdrawing (−M) *N*-substitution, are uniformly characterized by the presence of molecular frames which are heavily buckled, thermally labile (see Section 5.20.3), π-localized (UV) and decidedly atropic (^{1}H NMR), while those heteronins classified as aromatic display high thermal stability and pronounced air sensitivity, and possess reasonably flat molecular frames which are both well delocalized (UV) and diatropic (^{1}H NMR). The nondescript members of the family display chemical behaviour and spectroscopic characteristics which are intermediate between those of the two extremes; representative ^{1}H NMR spectra have been properly reproduced ⟨75PAC(44)691⟩. Table 1 lists spectroscopic data for four representative 'unrestricted' ⟨78AHC(23)55⟩ heteronins, the tabulated information revealing (i) a bathochromic UV effect, (ii) downfield ^{1}H NMR shifts (onset of diatropicity) and (iii) reduced magnitude of ^{13}C–H coupling constants, indicative of planarization ⟨76JA8266⟩ on passing from (**1**) to (**4a**) to (**3a**) to (**3d**). Besides the information listed in (iii) concerning molecular geometry there exists *direct* evidence bearing on the heavily twisted frame of the polyenic heteronins from an X-ray crystallographic study on (**2d**), the details of which are reproduced in Figure 1 ⟨74JA1636⟩.

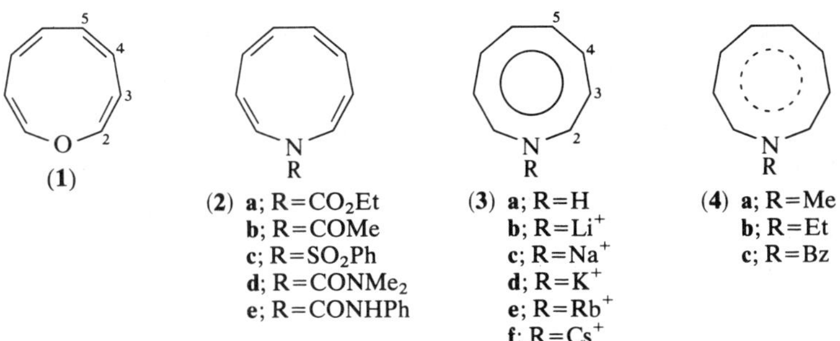

Table 1 Spectroscopic Information on Representative Heteronins ⟨78AHC(23)55⟩

Compound	$\delta(^1H)$ (p.p.m.) H-2	H-3	H-4, H-5	$J(H\text{-}H)$ (Hz) H-2–H-3	$\delta(^{13}C)$ (p.p.m.) (C-4 + C-5)/2	$J(C\text{-}H)$ (Hz) (C-4–H + C-5–H)/2	*Low energy UV band* (nm)
(**1**)	6.25	5.10	5.8–5.9	7.0	127.02	158	253
(**2a**)	6.37	5.32	5.87	9.6	—	—	265
(**4a**)	5.85	4.88	6.0–6.1	10.5	121.75	153	303
(**3a**)	7.07	6.00	6.7–7.0	11.0	117.61	152	335
(**3d**)	8.64	6.63	6.63	—	113.28	140	339

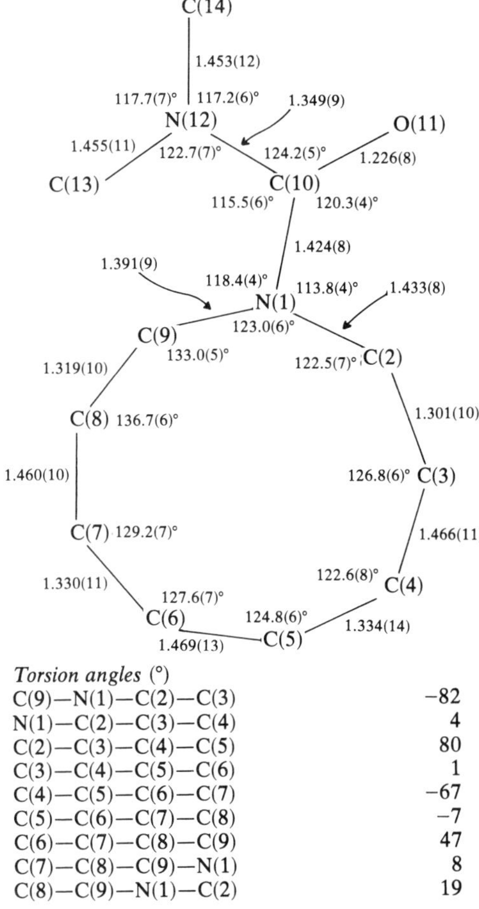

Torsion angles (°)

C(9)—N(1)—C(2)—C(3)	−82
N(1)—C(2)—C(3)—C(4)	4
C(2)—C(3)—C(4)—C(5)	80
C(3)—C(4)—C(5)—C(6)	1
C(4)—C(5)—C(6)—C(7)	−67
C(5)—C(6)—C(7)—C(8)	−7
C(6)—C(7)—C(8)—C(9)	47
C(7)—C(8)—C(9)—N(1)	8
C(8)—C(9)—N(1)—C(2)	19

Figure 1 X-ray crystallographic bond lengths (Å), bond angles and torsion angles for *N*-(*N′*,*N′*-dimethyl)carbamoylazonine (**2d**)

In terms of simple Hückel theory the change from polyenic to aromatic the system experiences on passing from (**1**) and (**2**) to (**3**) was attributed primarily to the lower LVMO energy of the latter ⟨B-73MI52000⟩. A more detailed Hückel calculation ⟨72T3657⟩ utilizing properly selected models in conjunction with thermochemical data also correctly accounts for the differences in π-electron makeup between 1*H*-azonine (**3a**) and oxonin (**1**), which were respectively calculated to possess resonance energies of 0.139 and 0.001 β. The same method of computation yielded a resonance energy of 0.118 β for the as yet unknown thionin ⟨73JA3907⟩, *i.e.* the sulfur analog of (**1**) and (**2**).

Besides the various unrestricted heteronins described in this section there also exist among the known species a number of fully saturated (perhydro) analogs. Since, however, in the majority of cases these paraffinic heterocycles were produced during the structural elucidation of the heteronins, their description will be deferred to the section dealing with reactivity where they more appropriately belong.

5.20.2.2.2 Substituted (skeletally restricted) heteronins

Substituted heteronins generally are more stable than their unrestricted relatives. Substitution in these molecules is incorporated mainly in the form of benzannulation which in all cases but one (benzazonide (**7**)) ⟨75CC149⟩ effectively suppresses the development of aromaticity in the nine-membered moiety.

The few currently available monoannulated heteronins are depicted in (**5**)–(**9**). All but the two anionic members (**5d**) and (**7**) were characterized securely by a variety of spectrosopic methods, but owing to interest chiefly in the presence or absence of ring current effects much of the discussion of spectroscopic data has centered on the NMR information. Pertinent ^{1}H NMR constants which served to classify azulenyl anion (**7**) as the sole aromatic member of the group are given below.

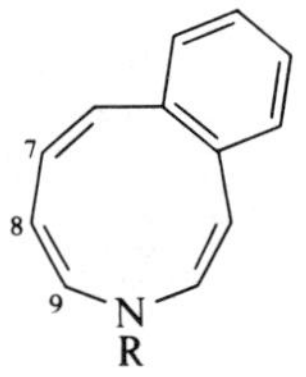

(**5**)

	$\delta(^1H)$ (p.p.m.) H-8	H-9	$J_{8,9}$	$J_{7,8}$ (Hz)
a; R=CO_2Et	4.68	5.73	10.0	5.5
b; R=COMe	—	—	—	—
c; R=H	4.29	5.63	10.0	<2
d; R=K^+	4.77	6.28	8.5	4.5

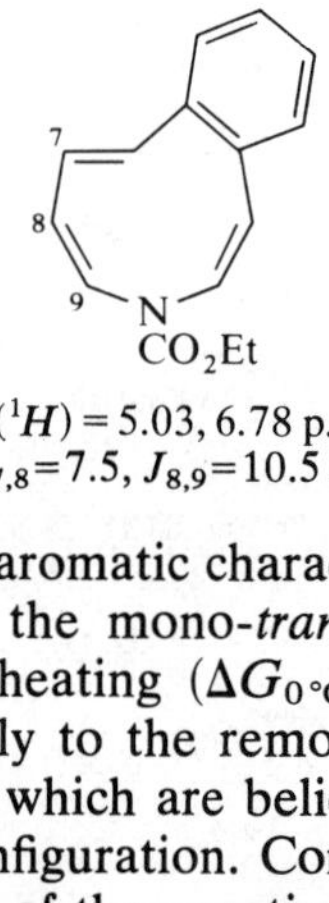

(**6**) $\delta(^1H)=5.03, 6.78$ p.p.m., $J_{7,8}=7.5$, $J_{8,9}=10.5$ Hz

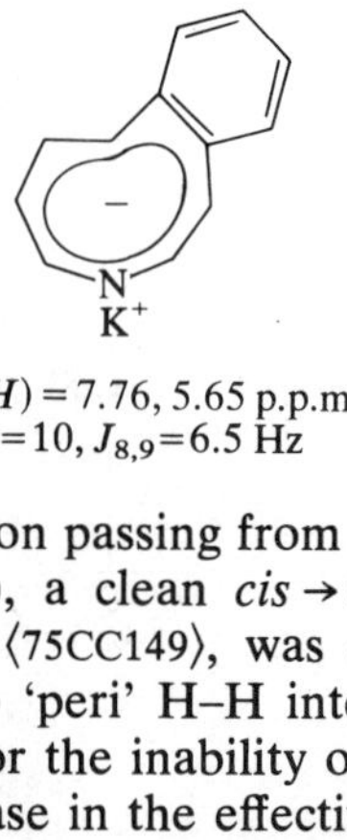

(**7**) $\delta(^1H)=7.76, 5.65$ p.p.m., $J_{7,8}=10$, $J_{8,9}=6.5$ Hz

The acquisition of aromatic character one observes on passing from all-*cis* benzazonide (**5d**) ⟨74AG(E)404⟩ to the mono-*trans* counterpart (**7**), a clean *cis*→*trans* isomerization requiring only mild heating ($\Delta G_{0\,°C}=83.7$ kJ mol^{-1}) ⟨75CC149⟩, was attributed to steric factors and specifically to the removal of one of two 'peri' H–H interactions about the molecule's cross-link which are believed to account for the inability of the all-*cis* variant to adopt a planar configuration. Conversely, an increase in the effective electronegativity of nitrogen from that of the negatively charged amide group in (**5d**) and (**7**) to that of the urethane center of (**5a**) and (**6**) leads to a clean reversal of the geometric isomerization from (**5d**)→(**7**) to (**6**)→(**5a**) ($\Delta G_{140\,°C}=136.0$ kJ mol^{-1}) and (**8**)→(**9b**) ($\Delta G_{61.3\,°C}=106.7$ kJ mol^{-1}) ⟨78TL4755⟩. The reversal in the direction of the process observed with the urethanes was attributed to the release of skeletal strain, which in the absence of

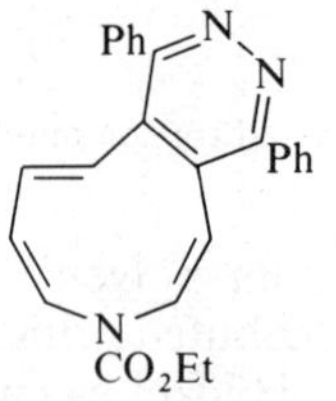

(**8**)

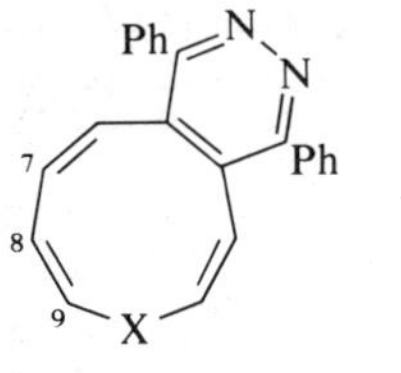

(**9**) **a**; X=NCOMe
b; X=NCO_2Et
c; X=O

the aromatically stabilizing delocalization supplied by the anionic amide group assumes the dominant role in directing the isomerization, in this case from *trans* to *cis*.

A further group of monoannulated heteronins is represented by the three oxygen-bridged azonines (**10**), (**11**) and (**12**) ⟨77JOC2651⟩. They are all characterized by UV absorption at *ca.* 270 nm and NMR (^{1}H, ^{13}C) data consistent with the presence, nature and location of the oxygen bridge and double bonds indicated in each case.

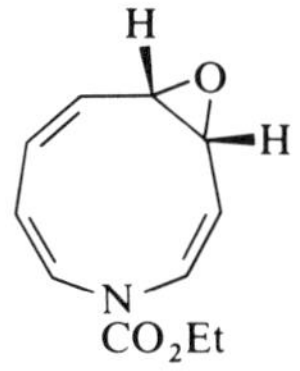

(**10**) 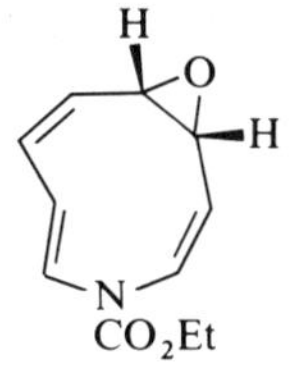 (**11**)  (**12**) **a**; R=Et **b**; R=Me **c**; R=Bu^t

The large variety of known di- and tri-annulated heteronins (**13**)–(**16**) ⟨78AHC(23)55⟩ have been characterized and analyzed on the basis of extensive spectroscopic information (NMR, UV) and are all believed to be heavily buckled, a molecular property which in the case of the three phosphonins (**16**) was elegantly established, along with the presence of the *trans* double bond, from a study of P–H NMR coupling constants ⟨80JA4838⟩. Because of the high, sterically-induced 'peripheral' strain plaguing (**13**) and (**16**) and the electronically induced adversity ⟨B-73MI52000⟩ affecting thionins (**14**) and (**15**), which are believed to prevent these molecules from adopting a planar and π-delocalized frame, it is generally accepted that the question of whether a second-row heteronin is capable of sustaining aromatically induced 10π mobility is, at this writing, largely an open one. The available information on the multiply substituted *P*-phenyl phosphonins (**17**) ⟨70JCS(C)386, 78JHC1319⟩ is not sufficiently specific as to be uniquely implicative of a heteronin frame.

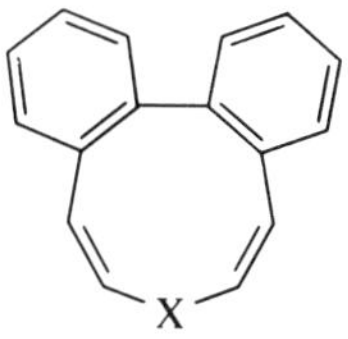

(**13**) **a**; X=O **b**; X=S (**14**) 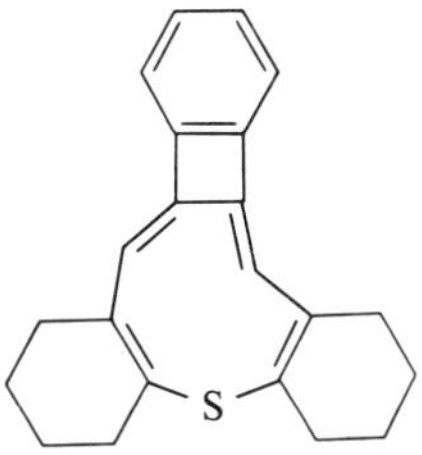 (**15**)

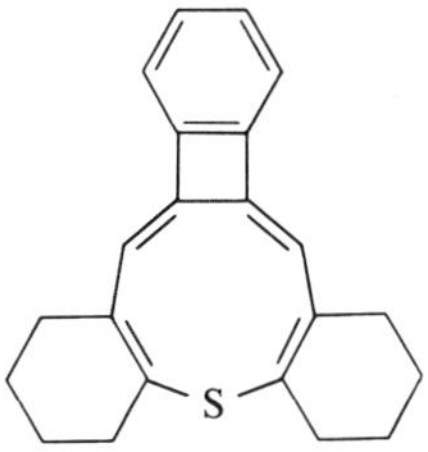

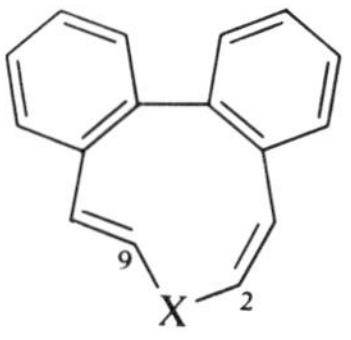

(**16**)

		^3J(P-H^2)	^3J(P-H^9)
a;	X=P(=O)Ph	14	24
b;	X=PPh	11	35
c;	X=$^+$PPhMe	—	25

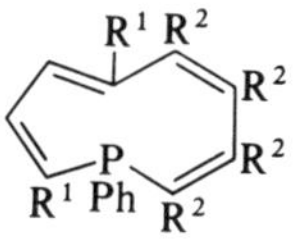

(**17**)

(**17**) **a**; R^1=Ph, R^2=H
b; R^1=H, R^2=CO_2Me
c; R^1=Ph, R^2=CO_2Me

Interestingly and somewhat ironically the first group of molecules to have seriously been assigned a heteronin structure are the heavily annulated, high melting phosphorus ⟨64CB747, 65CB576⟩ and arsenic ⟨64CB769⟩ derivatives shown in (**18**) and (**19**) which were not synthesized tactically but were obtained rather inadvertently as thermal by-products in studies centered on spiro-structured Group VA elements. While the proposed heteronin structures, based chiefly on IR and combustion data, are not necessarily incorrect, their heavily annulated frame unfortunately renders them unsuitable as possible models in studies concerning the nature of the π-frame in second- and third-row heteronins.

(**18**) R^3=H

	X	R^1	R^2
a	$\overset{+}{P}$	Ph	—
b	P	Ph	Me
c	$\overset{+}{P}$	Me	—
d	P	Me	Me
e	P	2-biphenylyl	—
f	$\overset{+}{P}$	$C_6H_4NMe_2$-*p*	—
g	P	$C_6H_4NMe_2$-*p*	Me
h	$\overset{+}{As}$	$C_6H_4NMe_2$-*p*	—
i	As	$C_6H_4\overset{+}{N}Me_3$-*p*	Me
j	As	Ph	—

(**19**)

5.20.2.2.3 Polyheteronins

Despite the wide variety of possible polyheteronins, *i.e.* heteronins incorporating more than one heteroatom, only two types of triheteronin, (**20**) ⟨78AHC(23)55⟩ and (**21**) ⟨72JA534⟩, have been reported thus far. Structurally, these substances basically differ in the number of π-excessive heteroatoms present, with (**20**) incorporating three such units within what may formally be viewed as a 12π periphery and (**21**) containing only one such center within a normal heteronin-like 10π periphery. The multimembered triheteronin family (**20**) ⟨72AG(E)939, 942, 75AG(E)347⟩ has been studied extensively by NMR (^1H, ^{13}C) to reveal the presence of strictly atropic frames adopting crown conformations which are substantially more mobile than that of their hydrocarbon model, trishomobenzene (**20**; X = CH_2), a structural characteristic believed to reflect their resistance toward adopting planar anti-aromatically (12π) delocalized frames ⟨78AHC(23)55⟩. The other known triheteronin (**21**) was found to be so kinetically inert (stable to oxygen and catalytic hydrogenation) as to be pronounced aromatic ⟨72JA534⟩, although it is doubtful that so heavily substituted a molecule would be sterically capable of adopting the type of planar or near planar geometry necessary for the development of π-delocalization ⟨B-73MI52000⟩.

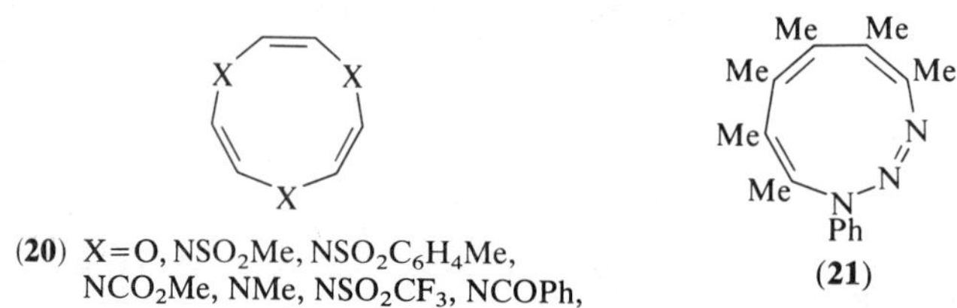

(**20**) X = O, NSO_2Me, $NSO_2C_6H_4Me$, NCO_2Me, NMe, NSO_2CF_3, NCOPh,

(**21**)

5.20.2.3 10-, 11- and 12-membered Rings

There are but few systems known within this range, among which (**22**) ⟨71CC947⟩ apparently represents the only 10-membered π-excessive. It is a colorless substance whose unexceptionally atropic (^1H NMR) and, no doubt, heavily buckled frame may be rationalized in terms of the same stereoelectronic factors advanced to explain the behaviour of its iso-π-electronic triheteronins (**20**). The methano-bridged 4π homologs of pyridine, *i.e.* the 10-membered rings (**23**), were prepared recently ⟨78AG(E)853, 855, 78JOC3813⟩ and were found to be thermally stable and to possess frames which are both distinctly diatropic (^1H NMR) and relatively flat (UV). In fact, parent molecule (**23a**) was determined to be *less* basic ($pK_a = 3.20$) ⟨78AG(E)853⟩ than either pyridine (5.23) or quinoline (4.94); pertinent spectroscopic information relating to this family of compounds is listed in Table 2.

(**22**)

(**23**)

	X	*Y*	*R*
a	N	CH	H
b	N	COT_s	H
c	N	COMe	H
d	CH	N	OMe
e	CH	N	OEt

(**24**) R = CH_2CO_2Me

Table 2 Selected Spectroscopic Data on the Bridged Aza[10]annulenes (**23**) ⟨78AG(E)853, 855⟩

	Bridge proton		*Bridge carbon*		*Low energy UV band*
Compound	$\delta(^1H)$ (p.p.m.) (H_a, H_b)	$J(H_a\text{–}H_b)$ (Hz)	$\delta(^{13}C)$ (p.p.m.)	$J(C\text{–}H)$ (Hz)	(nm)
(**23a**)	−0.40, 0.65	8.4	—	145	364
(**23c**)	0.17, 1.85	9.6	—	—	372
(**23d**)	−0.14, 0.37	9.5	—	—	—
(**23e**)	0.00, 0.54	10	33.84	—	315

The interesting, heavily substituted, diatropically delocalized triaza[10]annulenone formulated as shown in (**24**) is a further known member of the general aza[10]annulene family ⟨71JCS(C)2142⟩.

Hetero[11]annulenes are known either in sterically rigid bridged forms (**25**) ⟨72AG(E)217⟩ and (**26**) ⟨72TL4129⟩ or as the more flexible triannulated varieties (**27**) ⟨70JA5284⟩, (**28**) and (**29**) ⟨71CC1434⟩. Both types of system are strictly atropic by ^{1}H NMR which is not unexpected since the development of a ring current along their $4n$ perimeter would be antiaromatically destabilizing.

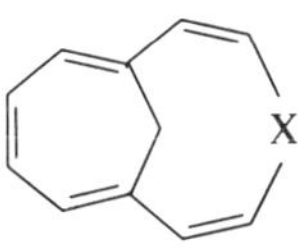

(**25**) **a**; X=O
b; X=S

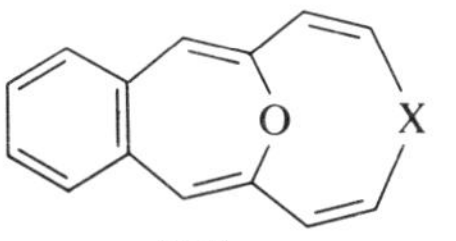

(**26**)

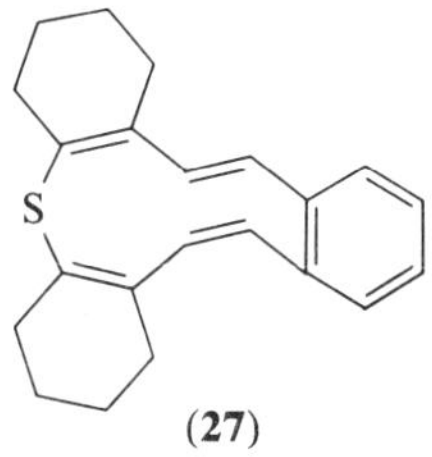

(**27**)

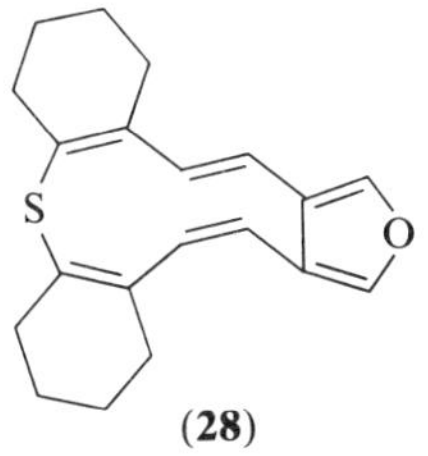

(**28**)

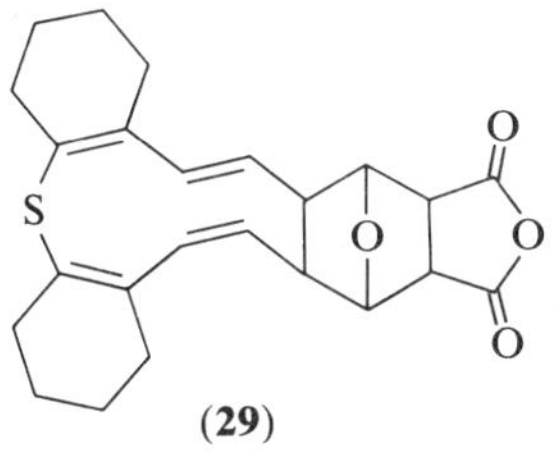

(**29**)

The only known π-excessive 12-membered ring is the multiply annulated thiaannulene (**30**) ⟨70JA5284⟩. It is an unconventional molecule insofar as it incorporates an odd number of electrons (13) within an even-membered frame. As a result, its central π-ribbon cannot be mobilized into π-delocalization without becoming exposed to the adversity of an electronic open shell ⟨78AHC(23)55⟩. The molecule's natural resistance to do so is borne out by its UV and ^{1}H NMR characteristics which are clearly implicative of a nonplanar, strictly atropic central skeleton.

A number of pyridine-like (non-π-excessive) 12-membered systems are also available. Historically, the first member of this family to be isolated in metal-free state apparently is the tetraannulated derivative (**31**) ⟨70T3749⟩. It is believed to be heavily buckled and to possess a strictly localized π frame. Interestingly, its triaza analog is available in metal-bridged form (**32**) ⟨65JA1706, 70JA1929⟩ but apparently has defied all attempts at being isolated metal-free ⟨73MI52001⟩ although its conformationally mobile carbinolamine derivative (**33**) reportedly is isolable and stable. A pyridine-like hetero[12]annulene skeleton is also available in methano-bridged form (**34**) although attempts at preparing the position isomer in which the nitrogen within the eight-membered moiety occupies a position β to the bridghead have thus far been successful ⟨75JOC902⟩. As expected, the $4n$ π-electron perimeter of (**34**) resists the development of antiaromatic delocalization, all available NMR information pointing to the presence of an atropic π-system incorporated within a 12-atom ribbon whose eight-membered segment is, quite likely, heavily buckled. The potentially aromatic $(4n+2)\pi$ dianon of (**34**) (prepared chemically as well as polarographically) has been generated but is apparently not sufficiently stable for purposes of direct spectroscopic observation.

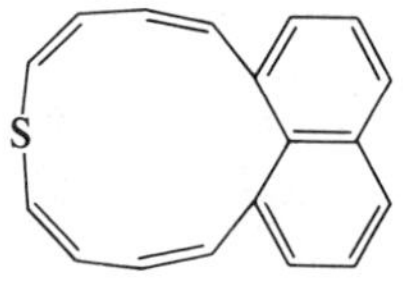

S

(30)

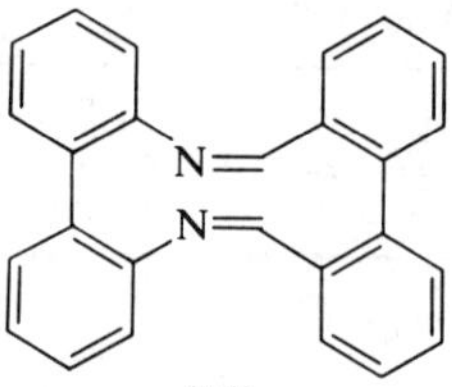

N=
N=

(31)

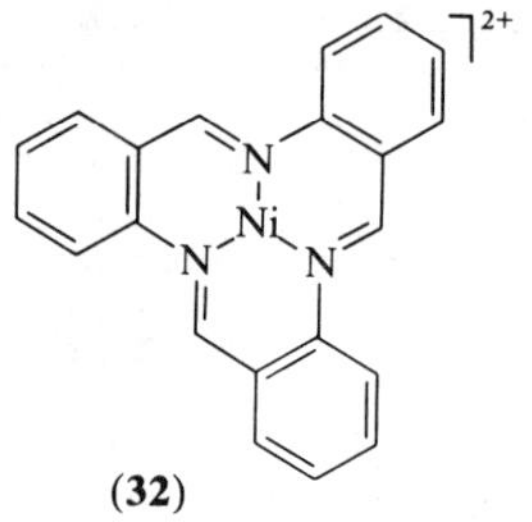

N
Ni
N N

$]^{2+}$

(32)

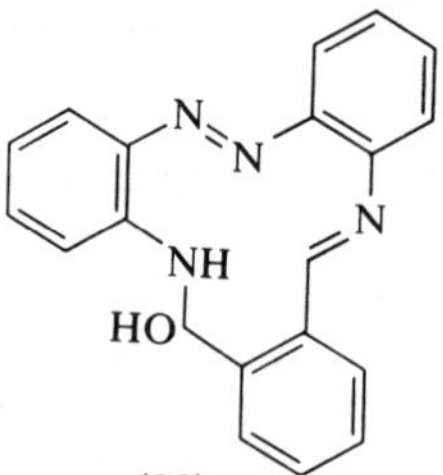

N
N
N
NH
HO

(33)

N
OMe

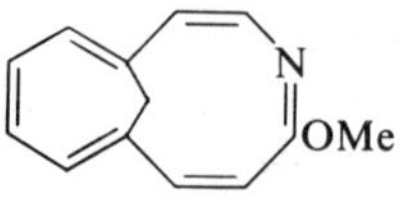

(34)

5.20.2.4 13-membered Rings

The 13-membered ring is available in a wide variety of unrestricted and restricted forms and undoubtedly represents one of the most thoroughly studied large-membered heterocycles. The size of the system's perimeter is sufficiently large to accommodate the presence of as many as *three trans* double bonds which, when strategically implanted, allow the molecular skeleton to adopt a rigidly flat or near-flat geometry in which to aromatically delocalize its 14 π-electrons.

The structures of the various unrestricted hetero[13]annulenes currently available are reproduced in (**35**)–(**40**) ⟨78AHC(23)55⟩. Because of the presence, invariably, of *trans* double bonds in these large π-excessives there always exists the possibility of rotational isomerism about the single links attached to each double bond, and in the case of (**36**), (**37**) ⟨75PAC(44)691⟩ and (**40**) ⟨75CB3753⟩ the available spectroscopic information was deemed insufficient for an unequivocal choice between the rotameric forms shown, although in the cases of the *N*-substituted diatropically delocalized aza[13]annulenes, such as the *N*-alkyl derivatives (**37c**) and (**37d**), a slight preference for form (B) was expressed in view of the reduction of steric congestion an appropriately flattened molecule is believed to experience on passing from (A) to (B). The development of aromaticity as deduced by the onset of molecular flattening (UV) coupled with a distinct rise in diatropicity (^{1}H NMR) is controlled by the same key factor, *i.e.* heteroatom electronegativity, as was already noted with the nine-membered analogs, the heteronins. One thus discovers that whereas ethyl carbamate (**35**) ⟨73AG(E)328⟩, *N*-substituted aza[13]annulenes (**36a**)–(**36d**) and oxa[13]annulene (**40**) are atropic (or at best mildly diatropic in the case of **36c** and **36d**) by ^{1}H NMR and believed to be heavily buckled (UV), their counterparts formulated in (**37**), (**38**) and (**39**) are associated with the types of spectroscopic characteristics which normally are deemed indicative of the presence of aromatic character. Table 3 lists key observables pertaining to the unrestricted system. Simple inspection reveals rather striking, diatropically induced chemical shift differences between the inner and outer α C—H and N—H protons of the aromatic specimens (**37**), (**38**) and (**39**). The interesting thermal isomerization of (**37b**) to (**38b**) which occurs quite readily even at a temperature of −40 °C was attributed to the release of inner proton strain (note that the parent amines **37a** and **38a** are not thermally interconvertible), and it was concluded that anion (**38b**) may well represent the true thermodynamic sink of the family ⟨75PAC(44)691⟩.

Among the restricted members of the family, the dimethylated bisdehydrohetero[13]annulenes (**41**) and (**42**) ⟨75JA640⟩ are the most interesting. Extensive spectroscopic (^{1}H NMR) analysis led to the classification of the di-*trans* isomers (**41**) as mildly diatropic and their thermally labile mono-*trans* variants (**42**) as strictly atropic, the difference between these two close structural relatives apparently stemming from the steric inability of the mono-*trans* analog to accommodate the necessary flat frame! As per expectation, the heavily annulated 13-membered π-excessive shown in (**43**) was found not to sustain a ring current and is thought to be heavily buckled ⟨70JA5284⟩.

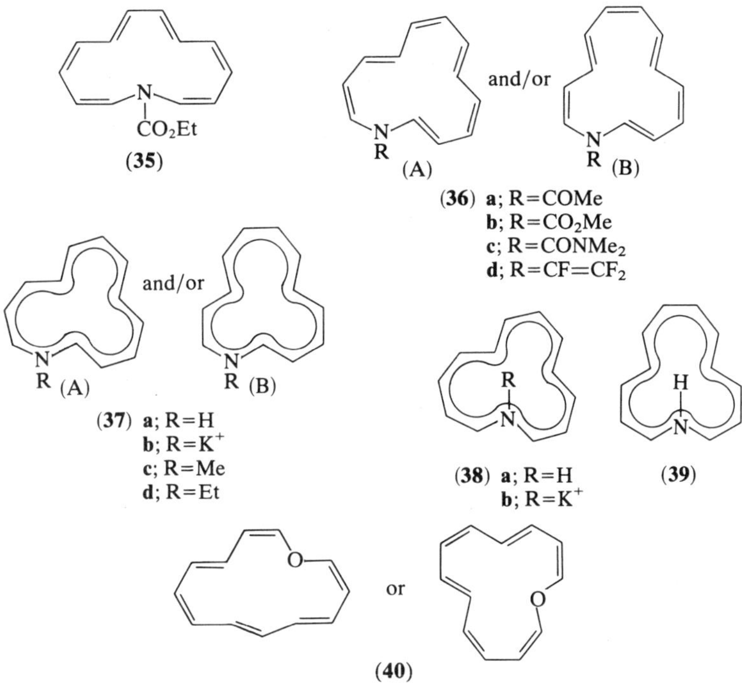

Table 3 Selected Spectroscopic Properties of Certain Unrestricted Aza[13]annulenes ⟨78AHC(23)55⟩

	NMR parameters of inner α-hydrogen (H-13)		*UV characteristics* (nm)	
Compound	*Chemical shift* (p.p.m.)	*Coupling constant* (Hz)	*Weak band*	*Strong band*
(**36a**)	>5.8	—	350	260
(**36b**)	>5.8	—	335	263
(**36c**)	5.50	14.0	350	280
(**36d**)	4.70	14.5	334	278
(**37a**)	2.78	14.5	360	297
(**37c**)	2.63	14.0	360	300
(**37b**)	−2.5	14		

Me Me X 3 4 2

(**41**)

	X	*H*-2	$\delta(^1H)$ (p.p.m.) *H*-3	*H*-4	*Me*
a	O	6.13	4.91	6.86	2.10
b	S	5.51	4.60	7.20	2.29
c	SO_2	4.95	7.10	6.93	2.13
d	SO	4.72	6.68	6.90	2.10

Me Me X

(**42**) **a**; X=O
b; X=S
c; X=SO_2

S

(**43**)

5.20.2.5 14-, 15- and 16-membered Rings

The tetraaza and octaaza bis π-excessive [14]annulenes (**44**) ⟨68LA(717)137⟩ and (**45**) ⟨70TL655, 71TL1601⟩ have been known for quite some time, at least by this chapter's operational standards, but the available spectroscopic information is not sufficiently specific as to suggest geometric detail or implicate the presence or absence of a ring current. Nonetheless, judging from the relatively large size and $4n\pi$ nature of these substances it is doubtful (particularly in the case of the octaaza analog **45**) that the molecules would choose to flatten into a skeletally strained geometry so that it may mobilize its 16 π-electrons into what is certain to be destabilizing (antiaromatic) delocalization.

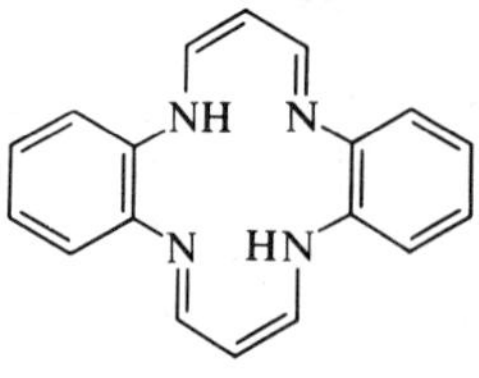

(**44**) 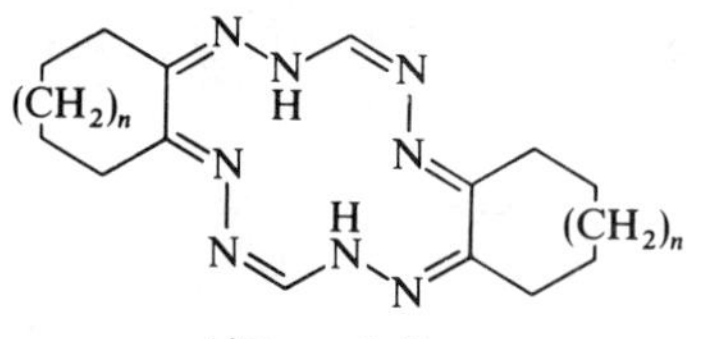 (**45**) $n = 0, 2$ 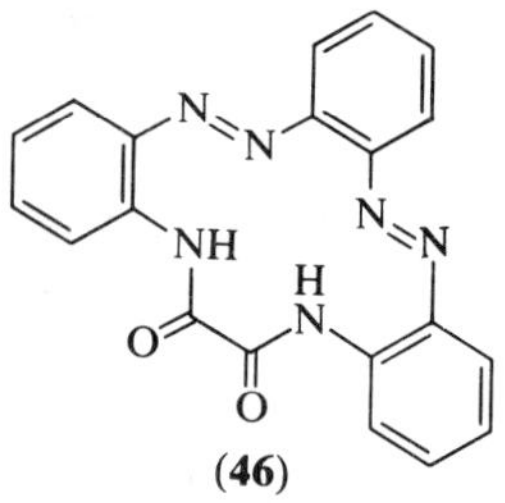 (**46**)

Historically, the first published attempt at constructing a non-π-excessive hetero[14]annulene appears to have involved the elusive hexaaza derivative (**47**) which was found to be unstable and could only be isolated in the form of carbinolamine (**48**) which was characterized as diatropically delocalized and classified as a 14π homoaromatic ⟨71HCA1069⟩! A very interesting and certainly most significant member of this family of compounds undoubtedly is the recently described, twice removed $(4n+2)\pi$ homolog of pyridine, namely unrestricted aza[14]annulene (**49**) ⟨80AG(E)207⟩. It exists as a thermally stable, dark violet crystalline substance whose UV (maxima extending to 620 nm) and ^{1}HNMR ($\Delta\delta$ between resonance manifolds of inner and outer proton signals $= ca.$ 9.5 p.p.m!) spectra clearly implicate the presence of a flat, extensively delocalized and strongly diatropic π-frame. The structural choice in favour of an inner-directed nitrogen lone pair was based on the further observation that the two protons attached to two distinct nitrogen-bound α carbon centers are strictly *cis*-coupled (4.4 and 6.2 Hz) to their immediate neighbours (H_β).

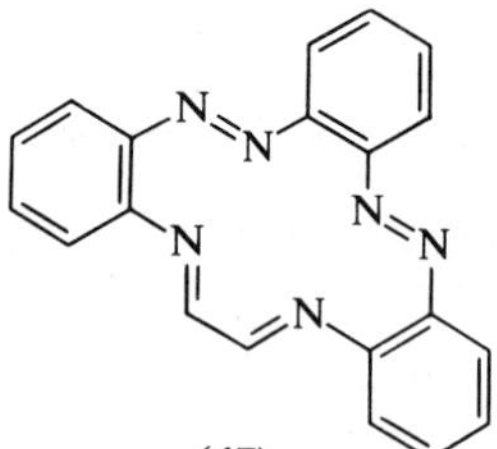

(**47**) 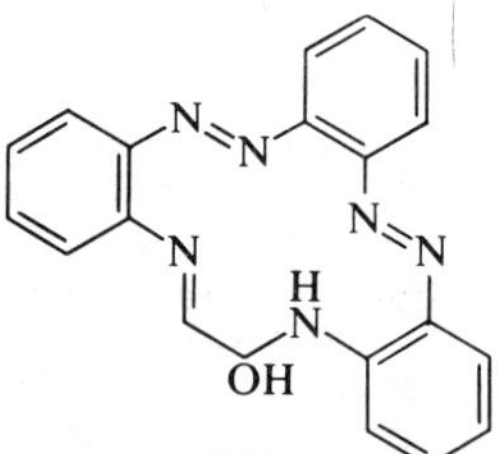 (**48**) 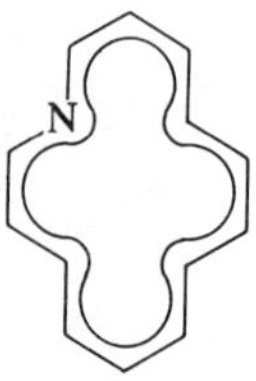 (**49**)

The next higher member of the family, *i.e.* the 15-membered ring, has yet to be prepared in parent form although it is available in the restricted varieties shown in (**50a**) ⟨73T809⟩, (**51**) ⟨75AG591⟩ and (**52**)–(**54**) ⟨75JA641⟩. The bis oxa-bridged oxaannulene (**50a**) is a relatively stable molecule whose ^{1}H NMR resonance manifold experiences the reasonably substantial (for outer protons) upfield shift of 0.8 to 1.2 p.p.m. relative to that of its carbocyclic prototype (**50b**) and may thus be classified as mildly paratropic. The twice-bridged thiaannulene (**51**) is an unstable molecule which readily desulfurizes on heating. Bisdehydrothiaannulenes (**52**) and (**53**) on the other hand are readily isolable paratropically delocalized substances. Interestingly, the oxaannulene analog (**54**) does not sustain a ring current, its inability to do so being attributed not so much to the higher electronegativity of oxygen but rather to the molecule's skeletal inability to adopt the necessary planar geometry possibly because of the presence of only two *trans* double bonds as compared to three such units in (**52**) and (**53**) ⟨78AHC(23)55⟩.

The interesting, highly restricted Ni^{II}-bridged tetraazatetrabenz[16]annulene (**55**) has been available for a number of years ⟨69JA1072⟩ and has been the subject of extensive bridge-centered (Ni^{II}) studies ⟨81JA5107⟩.

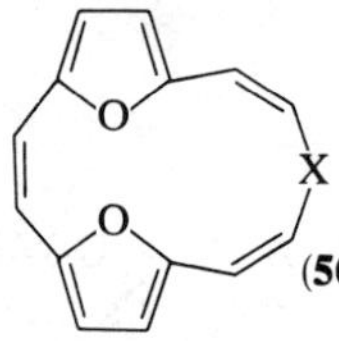

(**50**) **a**; X=O
b; X=CH_2

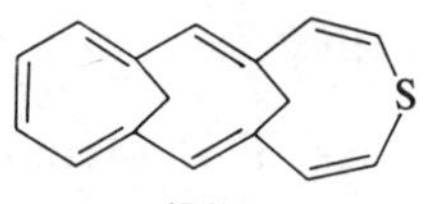

(**51**) 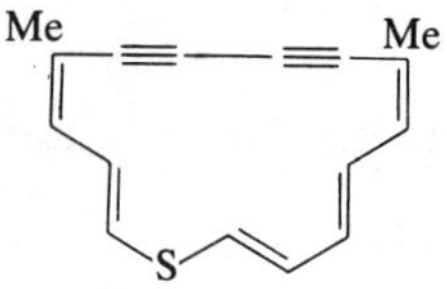 (**52**)

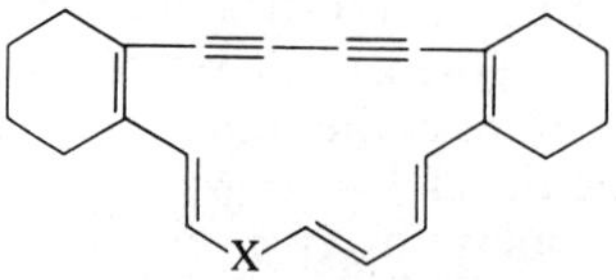

(**53**) X=S, SO_2 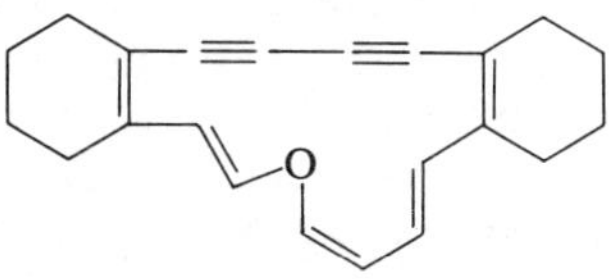 (**54**)

(55)

5.20.2.6 17-membered Rings

The 17-membered ring is available in a wide variety of structural modifications and has the distinction of being the largest odd-membered (π-excessive) heteroannulene to be known in parent form, synthetic access to the unrestricted skeleton being due, as in the case of the lesser $(4n+2)\pi$ homologs, the hetero[9]- and hetero[13]-annulenes, to properly designed pericyclization ⟨76ACR453⟩ involving cyclooctatetraene as the basic synthon.

The oxa and aza prototypes of the family are known in the structurally unspecified or specified isomeric forms (**56**)–(**64**) ⟨75PAC(44)925, 78AHC(23)55⟩ whose ability to sustain a diamagnetic ring current closely parallels that of their smaller $(4n+2)\pi$ homologs insofar as it strictly is a function of heteroatom effective electronegativity, with members (**56**)–(**61**) not sustaining a ring current and those shown in (**62**), (**63**) and (**64**) being decidedly diatropic by ^{1}H NMR; certain key NMR parameters which serve to reveal the stated distinction are listed in Table 4.

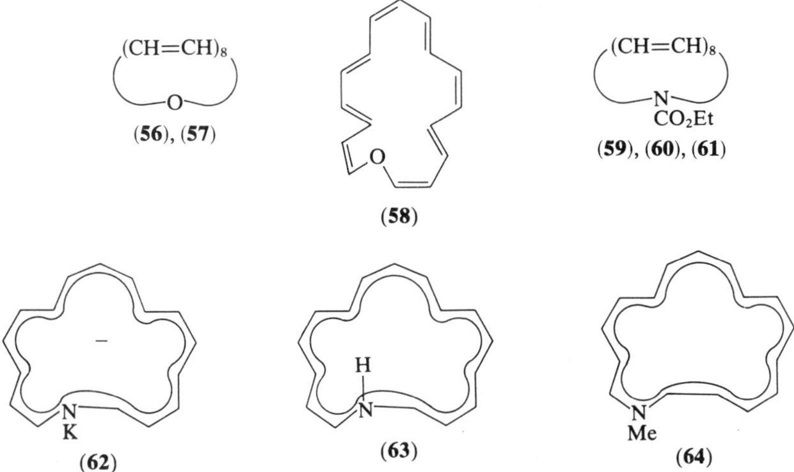

(**56**), (**57**) (**58**) (**59**), (**60**), (**61**)

(**62**) (**63**) (**64**)

Table 4 Inner Proton NMR Shifts of the Unrestricted Hetero[17]annulenes ⟨78AHC(23)55⟩

Compound	*Chemical shift* (p.p.m.)
(**56**)	5.2–5.9
(**58**)	4.5–5.6
(**59**), (**60**)	3.5–5.2
(**61**)	3.6–5.0
(**63**)	1.5–2.4
(**64**)	0.5–2.5
(**62**)	−4.9

Interestingly, the cavity available in a flat or nearly flat frame such as (**63**) does not appear to be sufficiently large to accommodate the bulk of a methyl group insofar as exposure of amide ion (**62**) to methyl iodide readily results in the exclusive formation of (**64**) ⟨75PAC(44)925⟩.

Consistently, the study of heteroaromaticity has also dominated the effort leading to the construction of a rich variety of restricted hetero[17]annulenes ⟨78AHC(23)55⟩. The development of ring current among the restricted thia[17]annulenes (**65**)–(**72**) was found to be a function of both the oxidation state of sulfur and the ability of the ring to adopt a flat or near-flat geometry. Extensive analysis of these substances by ^{1}H NMR led to the classification of skeletally rigid tetra-*trans* sulfides (**66a**) and (**68**) and the rapidly interconverting di-*trans* rotameric pair (**69**) as diatropic, inner-directed di-*trans* sulfones (**65b**) and (**71**) as paratropic, and the remaining specimens, sulfides (**65a**), (**67a**) and (**72**) and sulfones (**66b**), (**67b**) and (**70**) as largely atropic.

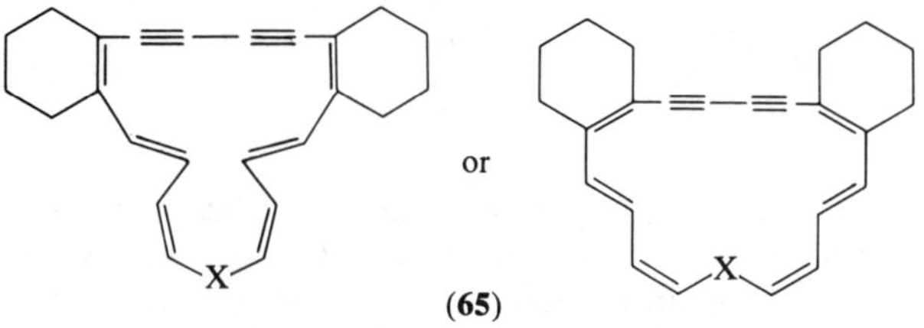

or

(**65**)

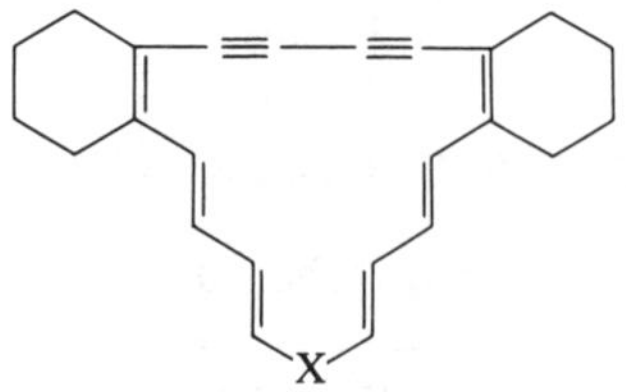

(**66**) **a**; X=S
b; $X=SO_2$

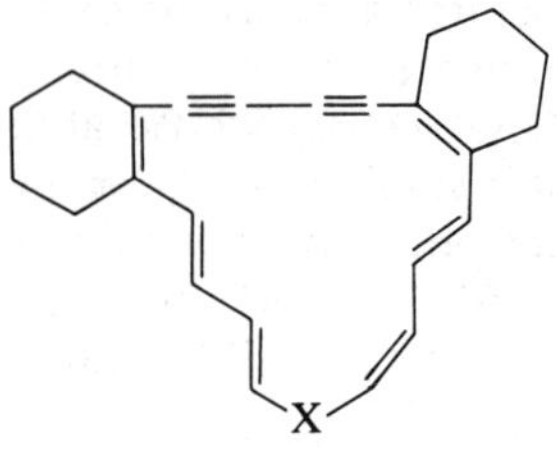

(**67**)

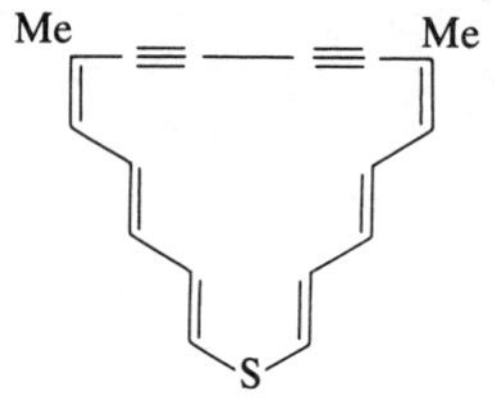

(**68**)

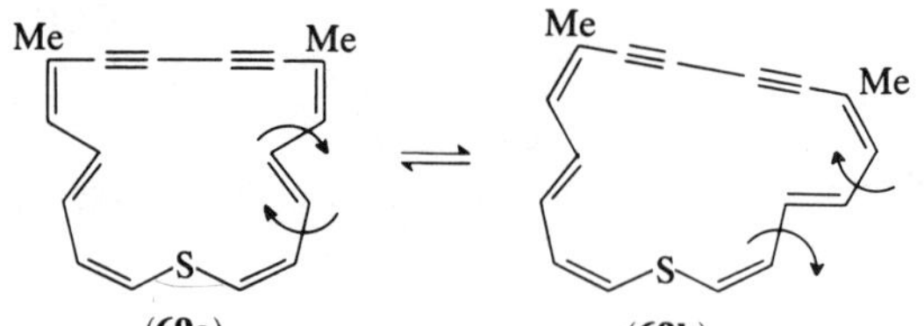

(**69a**) ⇌ (**69b**)

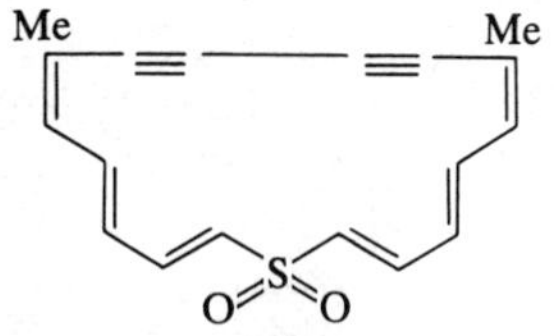

(**70**)

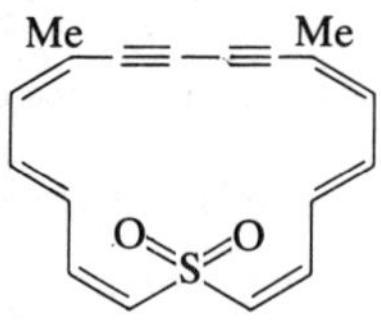

(**71**)

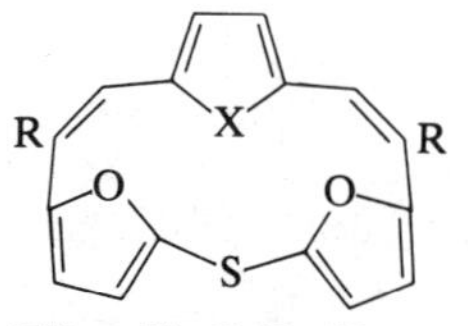

(**72**) **a**; X=O, R=H
b; X=S, R=H
c; X=O, $R=CO_2Me$
d; X=S, $R=CO_2Me$

Because of its relatively large size, the 17-membered system is capable of existing under a new type of multiple restriction, incorporated in the molecules shown in (**73**) and (**74**), which is not structurally available to any of the smaller members of the family. This threefold restriction due to bridging, annulation and the replacement of strategically located double bonds with alkynic functions, combined with the major presence of *trans* double bonds in the system, has the effect of producing a rigidly flat periphery with the interesting result that even strongly electronegative heteroatomic units such as oxygen and the urethane and acetamide functions, *i.e.* groups which normally quench the development of aromatic character, are now capable of mobilizing their lone pairs into aromatic delocalization. Nonetheless, a comparative examination of the ^{1}H NMR shifts listed in Table 5 does reveal the presence of the expected electronegativity effect (the lone pair's effectiveness in promoting delocalization), decreasing in the order $N^- K^+ \gg NMe > NH > S > NCO_2Et > NAc \sim O$, which establishes that members of this triply restricted family not containing a heteroatomic lone pair are either atropic (sulfoxide **74g**) or paratropic (sulfoxide **74f** and sulfone **74d**). Interestingly, the development of paratropicity in the latter two has been attributed to the presence of an 18π-electron 'Möbius'-antiaromatic system ⟨74AG(E)339⟩.

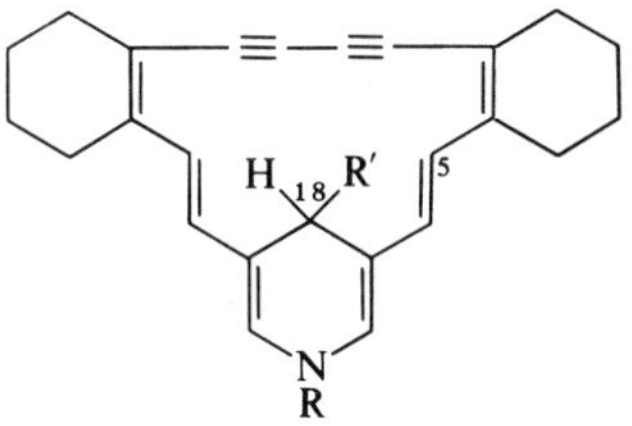

(73)

	R	*R'*
a	K^+	H
b	K^+	alkyl
c	H	H
d	H	alkyl
e	Me	alkyl
f	COMe	H
g	CO_2Et	H

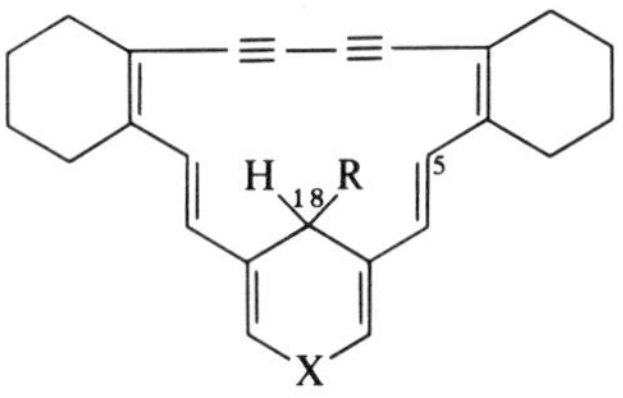

(74)

	X	*R*
a	S	H
b	S	Et
c	O	Et
d	SO_2	H
e	SO_2	Et
f	*syn*-SO	Et
g	*anti*-SO	Et

Table 5 Selected ^{1}H NMR Shifts of the Triply Restricted Hetero[17]annulenes (**73**) and (**74**) ⟨78AHC(23)55⟩

Compound	*Chemical Shift* (p.p.m.) *Inner proton* (H_5)	*Bridge proton* (H_{18})
(**73a**)	−1.51	−4.24
(**73e**)	2.66	−0.83
(**73c**)	3.01	−0.55
(**73h**)	5.22	1.23
(**73g**)	5.49	1.46
(**74a**)	4.78	1.02
(**74c**)	5.73	—
(**74d**)	8.67	4.37
(**74f**)	8.99	5.37
(**74g**)	7.82	4.59

Based on all currently available information on the hetero[17]annulenes it may safely be concluded that irrespective of skeletal restriction, the development of conventional temperature-independent Hückel delocalization among hetero[17]annulenes requires the presence of four *trans* double bonds. Pertinent examples are (**62**) and (**63**) among unrestricted specimens, (**68**) for partially restricted specimens, and (**73**) for the heavily restricted analogs.

5.20.2.7 18-membered and Larger Rings

The only known 18-membered ring is the recently prepared 18π-electron aza[18]annulene (**75**) ⟨79AG(E)312⟩. It is a thermally stable high melting (≥200 °C) dark green crystalline substance endowed with a highly diatropic extensively delocalized frame, its ^{1}H NMR spectrum being characterized by widely separated inner and outer proton absorption manifolds ($\Delta\delta$ between absorption centroids ~11 p.p.m!) and its electronic spectrum consisting of a number of maxima extending from 276 to 682 nm. The molecule's violet-black thermally stable hydrochloride (**76**) exists in two rapidly interconverting forms (**76a**; 20%) and (**76b**; 80%), the striking ^{1}H NMR shift difference (*ca.* 20 p.p.m!) between the inner N—H proton of (**76a**; δ −0.81 p.p.m.) and its outer counterpart in (**76b**; δ + 19 p.p.m.) securely attesting to the system's pronounced diatropicity. The existence of the free amine (**75**) in the form of a single rotameric structure incorporating an inner-directed lone pair and five in-cavity protons is quite likely due to the added nonbonded H-H repulsion the molecule will experience on passing to an outer lone pair, as depicted in (**76b**) for the salt, which must necessarily be associated with six in-cavity protons.

Only a few 19-, 20- and 21-membered ring heterocycles are known at present. They are shown in (**77**)–(**81**). Of these, the 22π-electron tetrafuranosulfide (**79**) was found not to

(75) (76a) (76b)

sustain a ring current, possibly because of the perturbation introduced by the four locally delocalized furanoid units ⟨73JCS(P1)1786⟩, while the 20-membered lactam (**78**) does not appear to have been sufficiently studied yet for possible classification ⟨71CPB97⟩. The heavily complexed tetraaza 20-membered ring (**80**) ⟨77JA8367⟩ represents an ever increasing family of multisized heterocycles incorporating inner-directed nitrogen lone pairs, *e.g.* (**44**), (**55**), (**83**), which are being constructed and studied chiefly because of their relatively wide molecular cavity and high affinity for metal complexation. The rigidly constrained, triply restricted, 19-membered (20π) and 21-membered (22π) urethanes (**77**) ⟨73AG(E)411⟩ and (**81**) ⟨73AG(E)410⟩ were found to possess well developed paramagnetism and diamagnetism respectively. Comparison of key proton shifts listed in Table 6 reveals the existence of rather major ring current differences between these two substances.

(**77**) (**78**) (**79**)

(**80**) (**81**)

Table 6 ¹H NMR Shifts of Aza[19]annulene (**77**) and Aza[21]annulene (**81**) ⟨78AHC(23)55⟩

Compound	*Chemical shifts* (p.p.m.)					
	H-2	*H-4*	*H-5*	*H-6*	*H-7*	*H-14*
(**77**)	6.58	6.00	8.80			9.14
(**81**)	7.42	6.83	5.39	6.90	5.90	
	H-15	*H-16*	*H-17*	*H-19*	*H-20*	*H-22*
(**77**)	5.95	8.21	5.90	6.58	4.63	
(**81**)						2.00

Excepting the very interesting family of bridgehead-nitrogen diazamacrobicyclic amines and their ammonium salts whose fascinating 'in–in', 'in–out' and 'out–out' isomerism was expertly reviewed by its original discoverers ⟨70MI52000⟩, heterocyclic rings incorporating more than 21 peripheral atoms are quite rare and viewed, for the most part, as mere curiosities by the majority of practicing chemists. Two rather interesting such species are the hexaaza[26]annulene (**82**) ⟨73T1399⟩ and a recently synthesized, extensively analyzed (NMR, X-ray) family of bridge-dimeric Ni^{II} complexes exemplified by the bis-complexed hetero[34]annulene (**83**) ⟨81JA5107⟩.

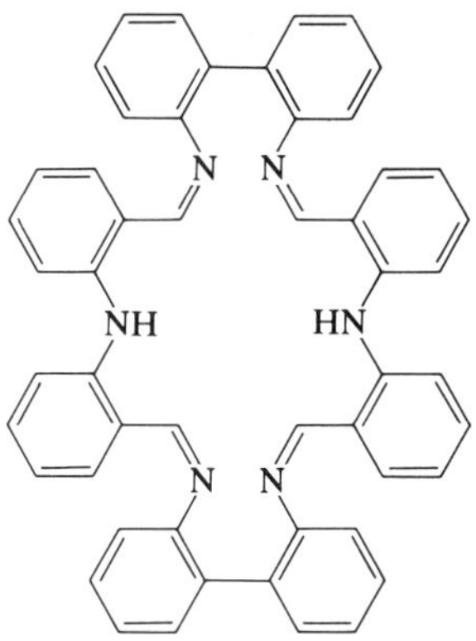

(**82**) 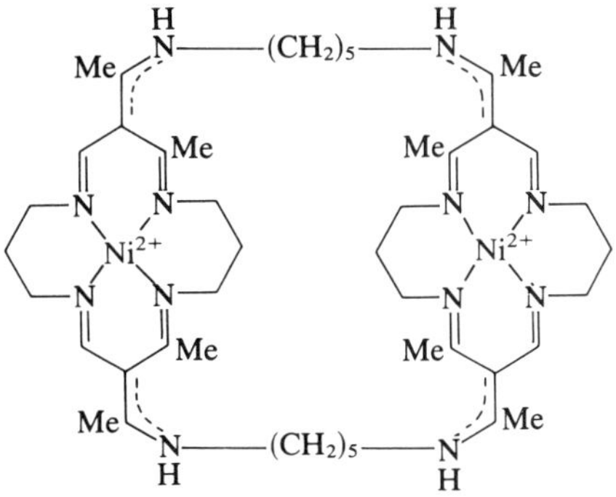 (**83**)

5.20.3 REACTIVITY

5.20.3.1 General Survey of Reactivity

The section of this chapter dealing with reactivity is necessarily short and the coverage unconventional insofar as the substances under discussion possess π-electronic frames which are both rapidly destroyed upon exposure to many of the conventional reagents such as those utilized in aromatic substitution and readily isomerized under conditions of mild thermal and/or photoinduced perturbation.

5.20.3.2 Pericyclic Reactions

5.20.3.2.1 Isomerizations

The presence of a continuous 8π-electron ribbon linked through the heteroatom renders the polyenic heteronins (**1**) and (**2**) with their 'tight' undelocalizable lone pairs amenable to rapid $_{\pi}6_s$ thermal pericyclization to the general *cis*-dihydroindene frame shown in (**84**); experimentally determined activation parameters ⟨72ACR281⟩ for this general transformation are given in Scheme 1. Aromatic and nondescript heteronins, on the other hand, are thermally quite stable (see Scheme 1) but readily darken and polymerize when exposed to air. Photochemically, all three types of heteronin were found to undergo $_{\pi}8_s$ isomerization to the [6.1.0] frames (**85**).

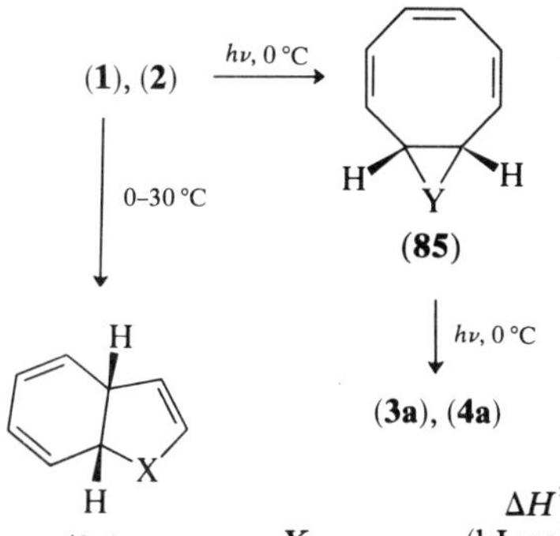

(**1**), (**2**) —$h\nu$, 0 °C→ (**85**) —$h\nu$, 0 °C→ (**3a**), (**4a**)

(**1**), (**2**) —0–30 °C→ (**84**)

(**85**) **a**; Y=O **b**; Y=NCO_2Et **c**; Y=NCOMe **d**; Y=NH **e**; Y=NMe

(**84**)	*X*	$\Delta H^{\ddagger}$ (kJ mol^{-1})	$\Delta S^{\ddagger}$ (J K^{-1} mol^{-1})	$\Delta G^{\ddagger}_{30\,°C}$ (kJ mol^{-1})
a	O	76.1	−54	92.9
b	NCO_2Et	88.7	−29	97.5
c	NCOMe	97.9	~0	98.3
d	$NCONMe_2$	103.8	13	100.8
e	NCONHPh	—	—	—
g	NSO_2Ph	93.3	−13	96.2

Scheme 1

The fleeting *cis*^{3},*trans* azonine rotamers (**86**) and (**87**) were shown through cycloaddition (see following section) to intermediate the thermal isomerization of (**88**) to (**89**) ⟨73JOC1959⟩ and were conjectured to cyclize, respectively, to these [5.2.0] frames *via* symmetry-allowed $_\pi 4_a$ processes. Further, the thermal isomerization of the urethane analog of (**88**) and low-temperature irradiation of cyclooctatetraene epoxide (**85a**) to *trans*-fused [4.3.0] frames (**91a**) and (**91b**) respectively were hypothesized to arise by internal $_\pi 6_s$ pericyclization of a fleeting mono-*trans* (*t,c,c,c*) heteronin frame (**90**) ⟨75PAC(44)691⟩.

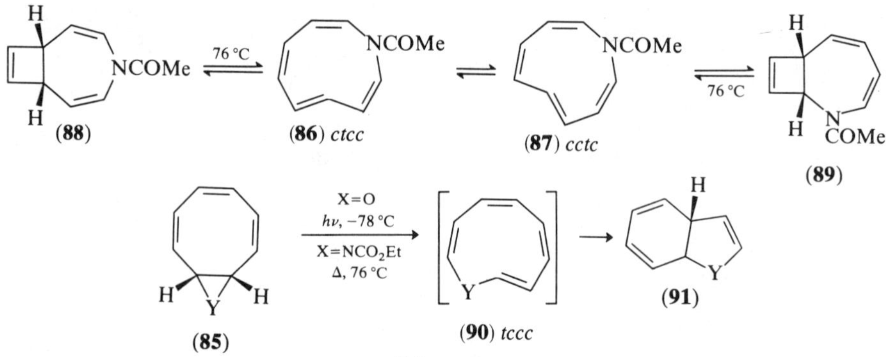

Scheme 2

Annulated heteronins are generally thermally stable since the continuous thermolabile hexatriene segment known to be responsible for the instability of the parent frame toward thermal pericyclization is only present here in benzannulated form and any $_\pi 6_s$ transformation analogous to the (**2**)→(**85**) or (**90**)→(**91**) processes will therefore have to occur at the expense of benzenoid resonance. The only members of the family known thus far to violate this rule are the relatively high energy mono-*trans* specimens (**6**) and (**8**) which when thermally activated undergo *trans*→*cis* isomerization to (**5a**) and (**9b**) along with a $_\pi 6_s$ cyclization–H-migration sequence to (**92**). Moreover, the overall rearrangement of (**6**) and (**8**) was found to proceed at a significantly faster pace in protic media ⟨78TL4755⟩.

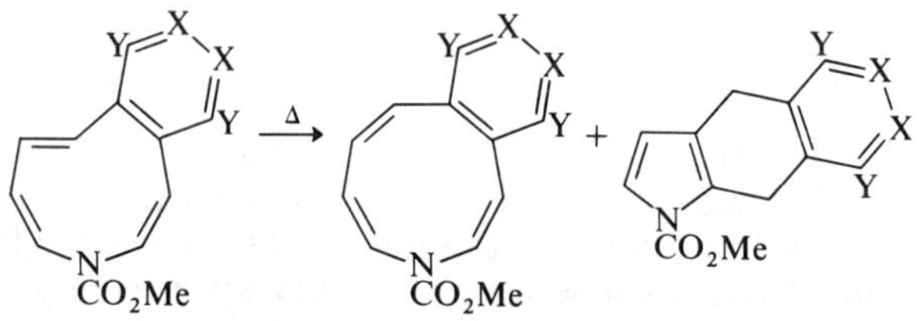

(**92a**) X=CH, Y=H
(**92b**) X=N, Y=Ph

(**6**) $\xrightarrow[106.7 \text{ kJ mol}^{-1}]{\Delta G_{61\,°C}}$ (**5a**)

(**9b**) $\xrightarrow[136.0 \text{ kJ mol}^{-1}]{\Delta G_{140\,°C}}$ (**8**)

Scheme 3

Among other heteronin derivatives, 1,2-azonine epoxide (**10**) undergoes a rich variety of isomerizations, readily transforming (i) to the 1,4 counterpart (**12**) upon exposure to strong base and (ii) to the *trans*-bonded analog (**11**) upon sensitized irradiation ⟨77JOC2651⟩. Furthermore, the photoisomerization is intermediated by a thermally unstable isomer conjectured to be the nitrogen-bridged oxonin (**93**). Also, (**11**) is thermally labile, readily isomerizing ($\Delta G_{56.2\,°C} = 103.8$ kJ mol^{-1}) to (**94**). Upon exposure to Al_2O_3 at −15 °C, (**94**) as well as azonines (**10**) and (**11**) readily fragment to the stable, deep purple pyrindine (**95**) ⟨77JOC2651⟩.

Because of the enhanced rigidity of the higher-membered heteroannulenes, *e.g.* (**36**), imparted chiefly by the presence of properly positioned *trans* double bonds, these frames are generally less thermally labile than their heteronin counterparts although here too aromatic derivatives are substantially more heat resistant than their polyenic relatives. It was noted, for example ⟨75PAC(44)691⟩, that while parent aza[13]annulene (**37a**) is thermally stable at 56 °C, its polyenic acetamide (**36a**) readily rearranges ($t_{1/2} < 1$ h) under these conditions to what is believed to be a tricyclic isomer (**96**).

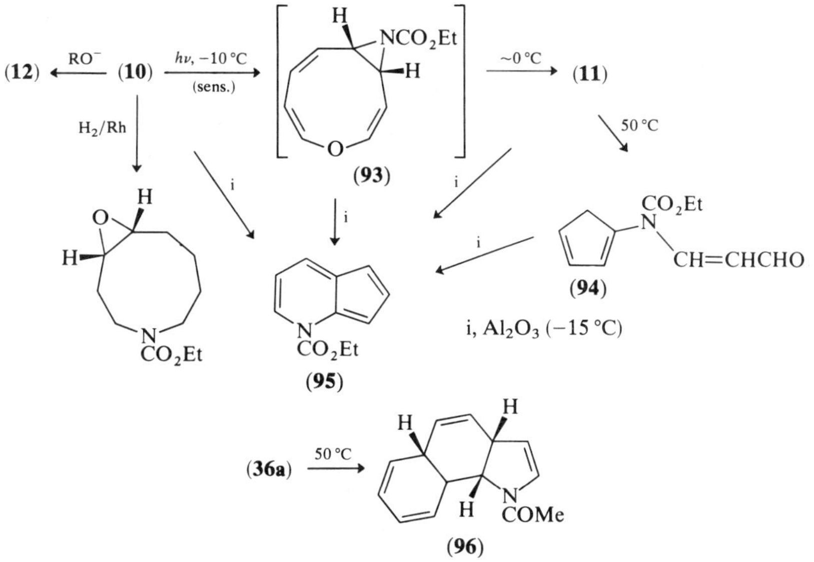

Scheme 4

5.20.3.2.2 Cycloadditions

Nonaromatic heteronins are readily amenable to symmetry-controlled cycloaddition. A number of studies have been reported ⟨78AHC(23)55⟩ on the subject with goals ranging from probing the π-electronic makeup of stable isolable heteronins to establishing the existence of certain fleeting members and, finally, to constructing suitably structured synthons for proper homologation and/or annulation of the basic frame.

The reagent utilized for probing the π-electronic structure of and examining the relative degree of skeletal twist present among the isolable polyenic heteronins invariably has been the powerful dienophile (**97**). The rapid reaction of this dark-red triazolinedione with the $4n$-π system(s) of the polyenic heteronins, to yield [4+2] cycloadducts (**98**) ⟨72CC325⟩ with azonines and the [8+2] counterpart (**99**) with oxonin ⟨70CC484⟩, may be followed visually with titration-like simplicity at subzero temperatures. The formation of the unsymmetrical [4+2] cycloadduct (**98**) instead of the symmetrical alternative (**100**) (not observed) was rationalized in terms of the less pronounced twist experienced by the N-bound butadiene segment compared with that of the more remote unit (see X-ray constants listed in Figure 1). The formation of [8+2] cycloadduct (**99**), on the other hand, was attributed chiefly to the presence of a sterically unencumbered ('bare') heteroatom.

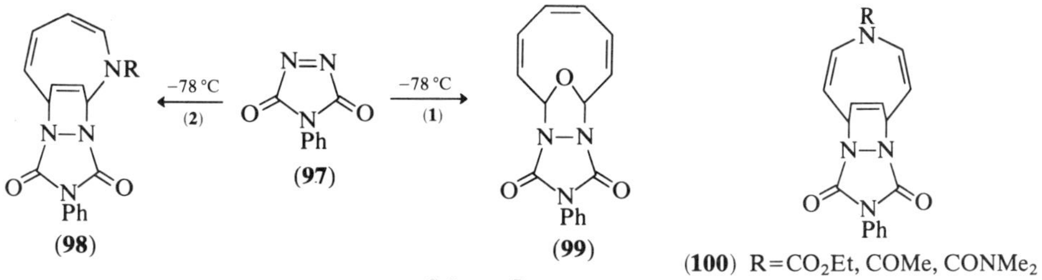

(**100**) R=CO_2Et, COMe, $CONMe_2$

Scheme 5

Utilization of suitably structured dienes (**101**) and (**102**) leads to efficient cycloadditive interception of fleeting mono-*trans* azonine rotamers (**86**) and (**87**) to yield *trans* fused cycloadducts (**103**) ⟨73JOC1959⟩ and (**104**) ⟨73TL3805⟩, the specific diastereomeric arrangement depicted in (**103**) having been deduced from X-ray crystallographic data ⟨73JOC1959⟩.

Related attempts at intercepting the *trans* double bond of the stable mono-*trans* benzazonine (**6**) with use of dienes (**101**) and (**102**) led instead to cycloaddition onto the isolated *cis* double bond with the formation of cycloadducts (**105**) ⟨75CC149⟩ and (**106**) ⟨77TL4465⟩ respectively. Further, attempted thermal extrusion of CO_2 from (**106**) activated deepseated rearrangement to (**107**), the operational details of which were investigated with the use of deuterium-labeled α-pyrone (**102**).

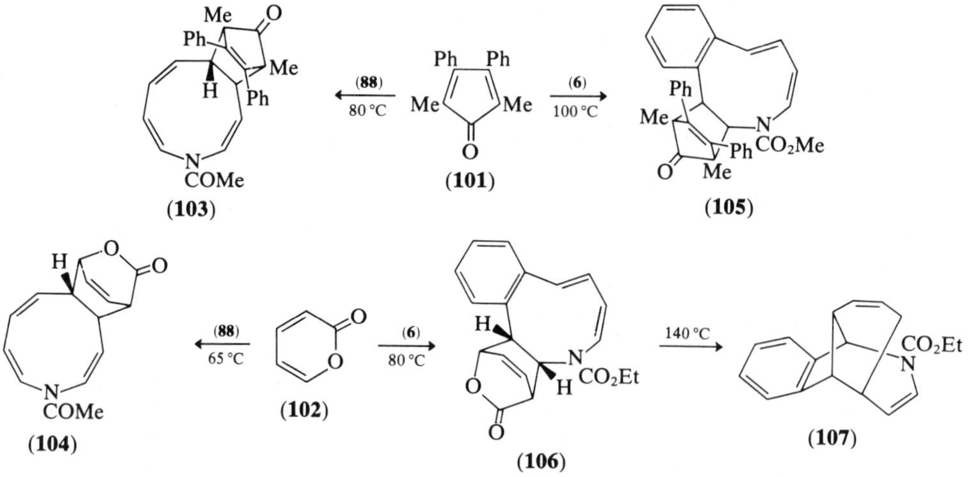

Scheme 6

Maleic anhydride cycloaddition to furanothia[11]annulene (**28**) led to the formation of the triply annulated parent (**29**) (see p. 715).

5.20.3.3 Miscellaneous Reactions

5.20.3.3.1 Reaction with electrophiles

The electrophilic reagents which the various heteroannulenes have been deliberately exposed to, thus far, are substances with strong affinity for the heteroatomic lone pair and have been employed uniformly for purposes of heteroatom functionalization. The reactions may be divided into the three related categories listed in Scheme 7.

Reactive families

(*i*) NY $\xrightarrow{RX}$ NR; Y=H, K^+
(1) Unrestricted aza-[9]-, -[13]-, -[17]- and triaza[9]-annulenes
(2) Benzo[9]- and bisdehydro[17]-annulenes

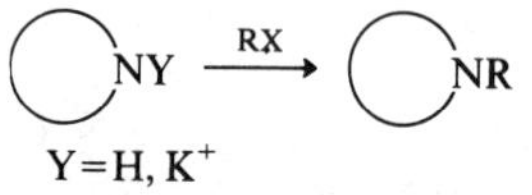

(*ii*) S → SO → SO_2 Bisdehydrothia-[11]-, -[13]- and -[17]-annulenes

(*iii*) XR $\xrightarrow{MeI}$ $^+$X(R)Me; X=P, As; R=Me, aryl
Tetrabenzo-phospha- and -arsa-[9]annulenes

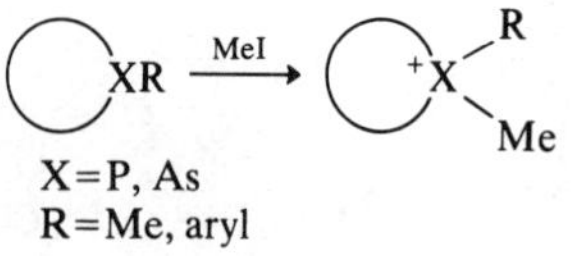

Scheme 7

Protonation of the various known azaannulenyl anions, *i.e.* nine-membered (**3b**)–(**3f**), (**5d**) and (**7**), 13-membered (**37b**) and (**38b**), and 17-membered (**62**), (**73a**) and (**73b**), occurs strictly on nitrogen affording the corresponding, structurally intact free amines ⟨78AHC(23)55⟩. Acid-induced skeletal transformations in the area include the high-yield internal condensation of phosphonane (**108**) to the bicyclic substance (**109**) ⟨77JA8370⟩ and the case of the interesting nine-membered azaketone (**110**) whose transannular effects, clearly established from extensive spectroscopic and dipole measurements, gain chemical significance on exposure to acid which triggers nitrogen-assisted protonation on oxygen with the formation of azabicycle (**111**) ⟨54JA5708, 57JA5476⟩.

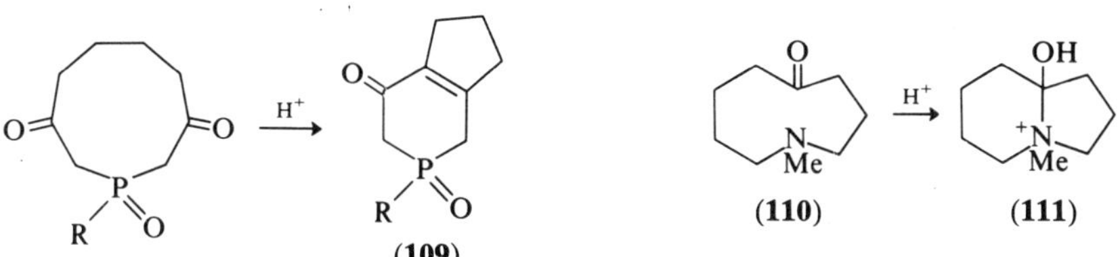

(108) (109) (110) (111)

Scheme 8

5.20.3.3.2 *Reaction with nucleophiles*

Exposure of the heteroannulenes to nucleophiles has been limited largely to the purpose of stripping a suitably *N*-substituted azaannulene of its substituent ⟨78AHC(23)55⟩. Thus, nine-, 13- and 17-membered urethanes (**2a**), (**5a**), (**35**), (**36b**) and (**59**)–(**61**) were readily converted to the corresponding azaannulenides on exposure to potassium *t*-butylate at subzero temperatures while 13-membered acetamide (**36a**) afforded the desired anion (**37b**) on mild subzero treatment with methyllithium.

Proton removal for the purpose of preparing negatively charged conjugate bases in conditions suitable for direct NMR observation may be effected on exposure of the free amine to strongly basic conditions such as potassium amide in liquid ammonia for benzazonine (**5c**) and potassium mirror in THF for aza[13]annulene (**37a**).

5.20.3.3.3 *Catalytic hydrogenation*

All available heteronins ⟨78AHC(23)55⟩, polyenic as well as aromatic, yield the fully saturated counterparts upon exposure to catalytic (Rh/C) hydrogenation at atmospheric pressure ⟨72ACR281⟩, and there is every reason to believe this to be the typical response of other known heteroannulenes as well.

5.20.4 SYNTHESIS

5.20.4.1 General Survey of Synthesis

The synthetic methodology utilized in the construction of the large-membered heterocycles discussed in this chapter has developed along three major tactically conceived, formally unrelated directions dominated by the procedures of (i) pericyclization ⟨76ACR453⟩, (ii) condensation and (iii) oxidative coupling, with the latter two involving primarily heteroatom-containing bis-Wittig reagents ⟨75S765⟩ and, in the case of (iii), terminal alkynes as well.

5.20.4.2 Pericyclization

5.20.4.2.1 *Photoinduced processes*

Given the availability of a suitable chromophore, the use of light offers the mildest possible thermal environment in which to effect a synthetic transformation insofar as photoinduced processes may be activated at temperatures well below 0 °C. Not unexpectedly then, one finds the photosynthetic procedure to be, thus far, the *only* one which has been successfully applied to the construction of the relatively sensitive parent (unrestricted) π-excessive heteroannulenes as shown in Scheme 9a. Similarly, exposure of the tetracyclic azides depicted in Scheme 9b to low-pressure mercury irradiation leads to deazotation and the formation of the pyridine-like 14- and 18-membered parents (**49**) and (**75**).

A notable characteristic of the entire group of unrestricted heteroannulenes displayed in Scheme 9 is that one may trace the synthetic origin of each individual member to the same basic synthon, namely the hydrocarbon cyclooctatetraene.

(a)

(85)

$h\nu$, 0 °C sens.

$h\nu$, −80 °C (n=6)

(n=4)

$(CH{=}CH)_n$

X

(n=6)

$h\nu$, −80 °C

(n=6) ← $h\nu$, 0 °C sens.

COMe N

H

(112)

(n=8)

$h\nu$, −80 °C

$C_{16}H_{16}X$ (tetracyclic)

X=O, NCO_2Et

(b)

$h\nu$ → −70 °C 50% (n=5) $(CH{=}CH)_n$ N (n=7) ← $h\nu$ −80 °C 28%

H N_3

Scheme 9

5.20.4.2.2 Thermally induced transformations

Thermal bond transposition has been utilized successfully in the synthesis of a number of relatively heat-insensitive families of heteroannulenes, such as the triheteronin frame (**20**) constructed thermally from the tetracyclic valence tautomer and the tetrabenzo heteronins (**18**) obtained by thermal bond transposition of their spirostructured valence tautomers (**113**). Similarly, thermally induced bridge extrusion (CO_2) has been utilized as a means of preparing dihydrobenzazonine (**112**) (the direct photoprogenitor of aza[13]annulene **36a**) from cycloadducts (**104**) ⟨73TL3805⟩.

Finally, thermal cycloaddition between benzocyclopropene and triazene (**114**) serves as a conveniently general means of constructing the bridged aza[10]annulene frame depicted in (**115**) (a direct relative of **23d**), the preparation of the last two derivatives (**115c**, **d**) requiring the use of ultra-high pressure (1.5 kbar) ⟨80TL7⟩.

X →Δ (**18**)

(**113**)

+ (**114**) →

(**115**) **a**; $R^1{=}R^2{=}R^3{=}CO_2Et$ (67%)
b; $R^1{=}CN$, $R^2{=}R^3{=}CO_2Et$ (70%)
c; $R^1{=}H$, $R^2{=}R^3{=}CO_2Et$ (41%)
d; $R^1{=}CO_2Et$, $R^2{=}R^3{=}H$ (33%)

Scheme 10

5.20.4.3 Direct Condensation

The usage of the term 'direct' here is in reference to the product which materializes *directly* as a result of the specified condensation process.

5.20.4.3.1 Bis-Wittig coupling

The general process involved in a bis-Wittig coupling reaction (shown in Scheme 11) is the condensation of a bis-Wittig reagent with a suitable dicarbonyl compound (usually a dialdehyde) to produce a ring. The resulting products are normally uniquely accessible by this route although the rather severe entropic adversities characterizing such an intermolecular formation of large rings normally limit the yields to values seldom exceeding 10%. Within the confines of Scheme 11 the most frequently recurring bis-Wittig reagent and dialdehyde are (**116**; Z = O, S) and (**117**) respectively. The bis-Wittig reagent procedure has been successfully applied to the preparation of the following large-membered heteroannulenes ⟨75S765⟩: (**13a**; 6%), (**13b**; 10%), (**14**; 0.7%), (**15**; 1.4%), (**22**; 5%), (**25a**; 9%), (**25b**; 30%!), (**26a**; 2.5%), (**26b**; 12%), (**27**; 2.2%), (**28**; 5.3%), (**30**; 2%), (**43**; 12%), (**50a**; 15%), (**51**; 6%), (**72a**; 15%), (**72b**; 10%), (**79**; 1.3%).

Y(CHO)(CHO) + $Ph_3P{=}CH$–Z–$CH{=}PPh_3$ → [Y/Z ring]

(**116**)

CHO / S / CHO

(**117**)

Scheme 11

5.20.4.3.2 Conventional condensations

The polyaza annulenes depicted in (**31**)–(**33**), (**44**)–(**48**), (**55**), (**80**) and (**82**) have all been prepared by conventional condensations involving suitable aniline and aldehydic components ⟨73MI52001⟩. In some cases, *e.g.* (**32**) and (**55**), the use of a metal template in the condensation largely eliminates the adversity introduced by the entropy factor, thus greatly improving yields, although it appears to be difficult, if at all possible, to generate the polyaza ligand in the free state once it is produced in chelated form.

5.20.4.4 Oxidative Couplings and Aromatizations

A large number of available large size heteroannulenes, namely all those members incorporating a bisdehydro functionality, have been synthesized ⟨74C163⟩ by a combined sequence entailing condensation between a bis-Wittig reagent such as (**117**) and two separate monoaldehydes incorporating terminal alkynic 'tails', followed by cyclization through oxidative coupling of the terminal alkenic functions normally under the influence of a pyridine solution of $Cu(OAc)_2$ at *ca.* 50 °C. An obvious feature of the general process, depicted in Scheme 12, is that a symmetrical final product is to be expected only when Y = Z, *i.e.* when two moles of the *same* monoaldehyde are employed in the initial condensation, a situation exemplified by the 13-, 17- and 21-membered bisdehydro heteroannulene families: (**41**), (**42**); (**65**)–(**71**), (**73**), (**74**); (**81**). The use of different monoaldehydes in the initial step leads to the unsymmetrical 15- and 19-membered families (**52**)–(**54**) and (**77**). Further, in the case of the triply restricted heteroannulenes (**73**), (**77**) and (**81**) the six-membered nitrogen center initially is introduced in the form of a pyridine nucleus which is subsequently reduced to the desired dihydro form upon treatment with lithium aluminum hydride.

(**116**) + CHO–Z′–≡–H + CHO–Z″–≡–H → Z(HC=CH–Z′–≡–H)(CH=CH–Z″–≡–H) $\xrightarrow{Cu(OAc)_2}$ Z(HC=CH–Z′–≡–≡–Z″–CH=CH)

Scheme 12

Oxidative aromatization is normally brought about on exposure of a molecule containing a dihydrobenzene or closely related nucleus to a dehydrogenating agent, normally a quinone such as *o*-chloranil. Thus, the benzazonine skeleton (**5**) and its pyridazinoheteronin counterpart (**9**) were prepared in high yield, 80% and 85% respectively, on exposure of their dihydro derivatives (**112**) ⟨73TL3805⟩, (**118a**) ⟨75AG(E)814⟩ and (**118b**) ⟨76CC313⟩ to *o*-chloranil.

5.20.4.5 Other Transformations

By far the most common synthetic procedures collected under this unclassified section are those involving the construction of heteroannulenes by variously triggered ring enlargements, often by way of cycloaddition onto heterobicycle (**85**) as shown in Scheme 13a. A second, entirely unrelated method of ring enlargement has also been developed in the area. It involves the ozonolysis of properly chosen cross-links and has, thus far, been applied to the synthesis of the general phosphonin skeleton ⟨77JA8370⟩ as shown in Scheme 13b.

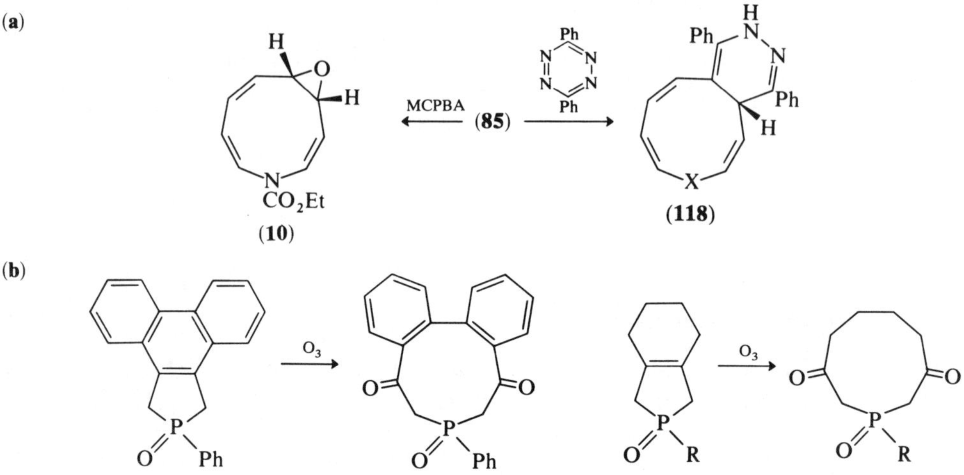

(a) (10) ← MCPBA — (85) → (118)

(b)

Scheme 13

The use of a thermally activated (acid-free) Beckmann rearrangement as a suitable method for preparing an annulene *via* ring expansion dates to the early (possibly the first publicized) unsuccessful attempts to prepare azonine from a suitably functionalized oxime of cyclooctatrienone ⟨67CB3162⟩. On the other hand, good yields of the fully saturated hetero[13]annulene frame (**119**) were recently recorded ⟨81JA7368⟩ by conducting a thermally-induced Beckmann rearrangement on the benzenesulfonate of cyclododecanone oxime.

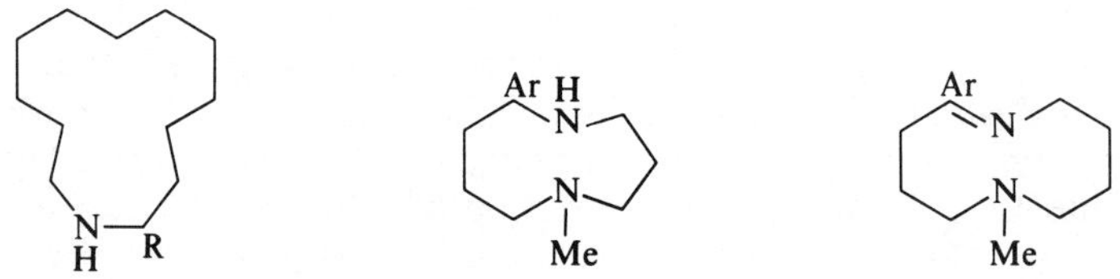

(**119**) R=Me (60%)
R=H (73%)

(**120**) (**122**)

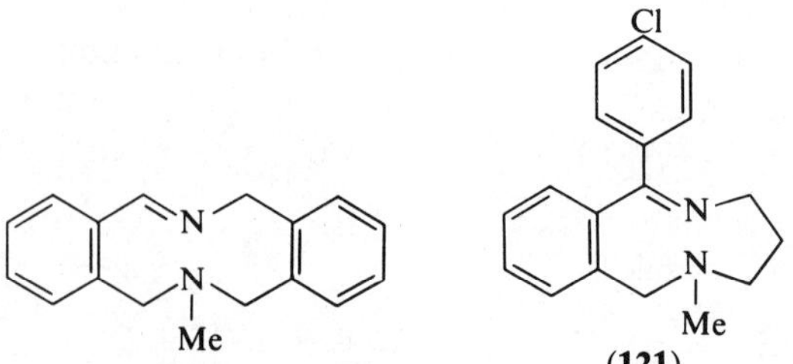

(**123**) (**121**)

Finally it is noted that while the conditions associated with conventional Hoffmann eliminations are obviously much too drastic for preparing heteroannulenes, a rather clever modification of the procedure (one involving the rupture of a cross-link) was successfully applied more than a decade ago to the synthesis of the fully saturated diazonine (**120**) and its 'mostly' saturated benzo counterpart (**121**) as well as the partially saturated diazecine (**122**) and its dibenzo counterpart (**123**) ⟨69JOC2715, 2720⟩. The fully saturated counterparts of the molecules shown in (**121**)–(**123**) were also prepared in that early study.

5.21

Crown Ethers and Cryptands

A. D. HAMILTON
Princeton University

5.21.1 INTRODUCTION

The crown ethers were originally discovered by Pedersen at duPont in 1967 ⟨67JA7017⟩. Although cyclic polyethers had previously been reported ⟨55JOC1147, 56MI52100, 57BRP585229, 59JCS3767⟩, Pedersen was the first person to develop a systematic synthetic approach and also to recognize their remarkable properties of complexing alkaline and alkaline earth cations. The first crown ethers were cyclic hexaethers (**1**) and (**2**), prepared by a variation of the Williamson ether synthesis. They were found to extract alkali metal salts from aqueous solution and to solubilize them in organic solvents. Ethers (**1**) and (**2**) showed a selectivity for potassium ions which was related to the size of the cavity formed by the ether oxygen atoms. This property mirrored that of the known naturally occurring antibiotic ionophores, *e.g.* nonactin (**3**), and caused an explosion of interest in crown ethers and their chemistry ⟨B-79MI52100, B-81MI52107⟩.

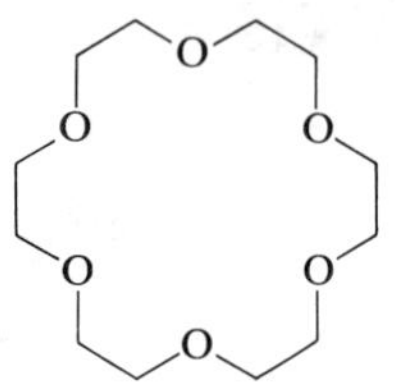

(**1**) 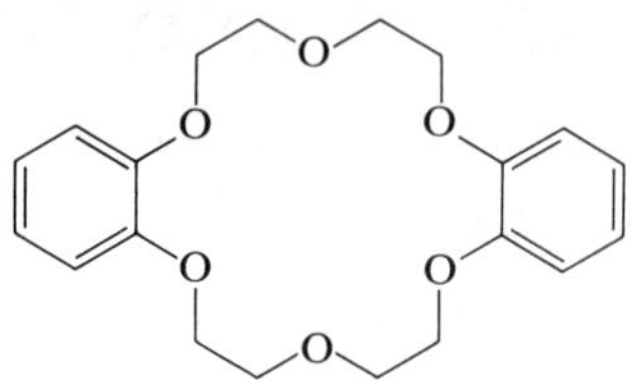 (**2**)

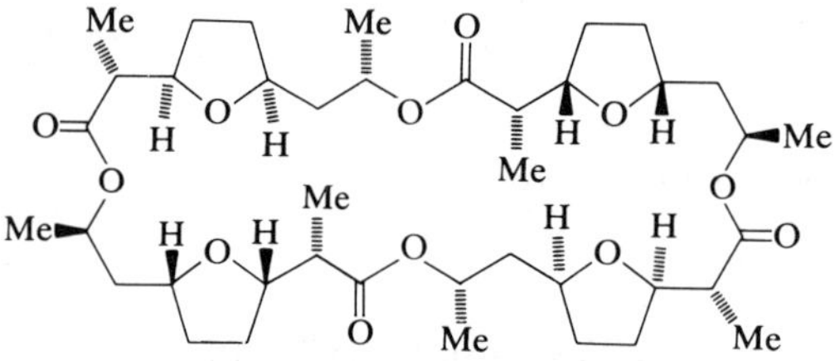

(**3**)

Pedersen prepared a wide range of macrocycles of different sizes and rigidity, *e.g.* (**4**)–(**8**), and further showed that alkylammonium ions could complex into the central cavity ⟨B-78MI52101⟩.

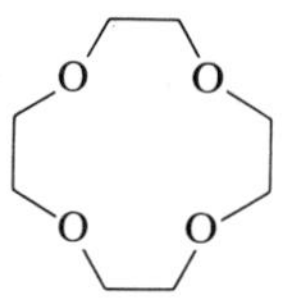

(**4**) 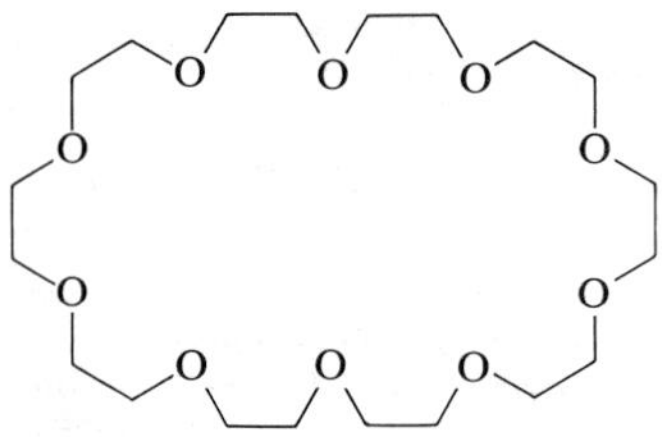 (**5**)

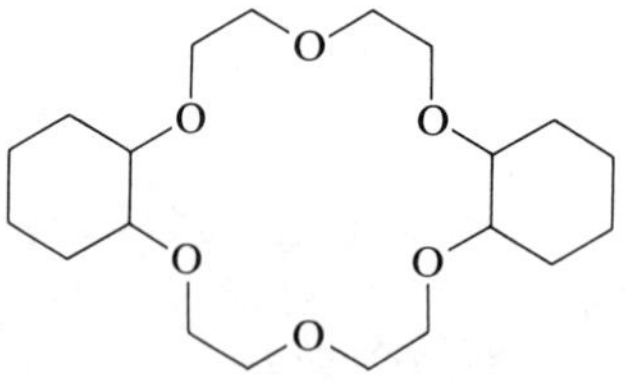

(**6**) 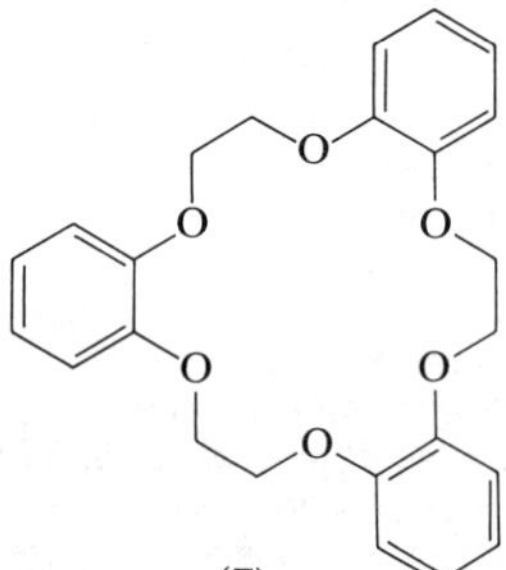 (**7**)

A simple nomenclature was proposed in which, in place of the clumsy IUPAC name 1,4,7,10,13,16-hexaoxacyclooctadecane ⟨72OS(52)66⟩, ether (**1**) becomes [18]crown-6 and (**2**) becomes dibenzo[18]crown-6. The specific classification *crown* is preceded by the total ring size in square brackets and succeeded by the number of heteroatoms in the ring. The dibenzo refers to the two benzene rings fused onto the ring. A new nomenclature system, less ambiguous with regard to the location of the donor atoms, has recently been proposed ⟨B-81MI52102⟩ in which (**1**) is named 18⟨O_6coronand-6⟩ and (**2**) is 18⟨O_6(1, 2)benzeno·2_2-(1, 2)benzeno·2_2coronand-6⟩. Although more specific and illustrative the new nomenclature suffers in more complex cases from name lengths comparable to those of the IUPAC system.

Since the early work a vast array of crown ethers has been synthesized and studied. A representative sample is shown (**8–11**), and includes different donor atoms (**8**) ⟨71JOC254⟩, incorporated heterocyclic rings (**9**), ring esters (**10**) ⟨79LA858⟩ and multicrown ethers (**11**) ⟨79AG(E)219⟩. Two exhaustive lists of synthetic crown compounds have recently been compiled ⟨74CRV351, 80T461⟩.

(**8**) (**9**)

(**10**) (**11**)

An important step in the development of crown ethers came with the recognition by Lehn that crown ether–alkali metal cation complexes could be stabilized by bridging the macrocycle with an additional oligoether chain, *e.g.* (**12**) ⟨78ACR49, 78PAC871⟩. The resulting cage species were called cryptands and the corresponding metal complexes cryptates. A trivial nomenclature was introduced for the diaza cryptands in which the number of oxygen atoms in each chain in square brackets precedes the specific classification, *cryptand*. Thus, cryptand (**12**) is named [2.2.2]cryptand and its derivative with only one oxygen in one of the chains would be [2.2.1]cryptand. The cryptates are formulated by using the mathematical symbol of inclusion $\subset$; for instance, the potassium complex of [2.2.2]cryptand becomes $[K^+ \subset 2.2.2]$ ⟨73MI52100⟩. The original bicyclic cryptands with two bridgehead nitrogen atoms (**12**) ⟨73T1629⟩ were later elaborated to a wide range of ligand systems including spherical tricycle (**13**) ⟨75JA5022⟩ and cylindrical tricycle (**14**) ⟨73AG(E)578⟩.

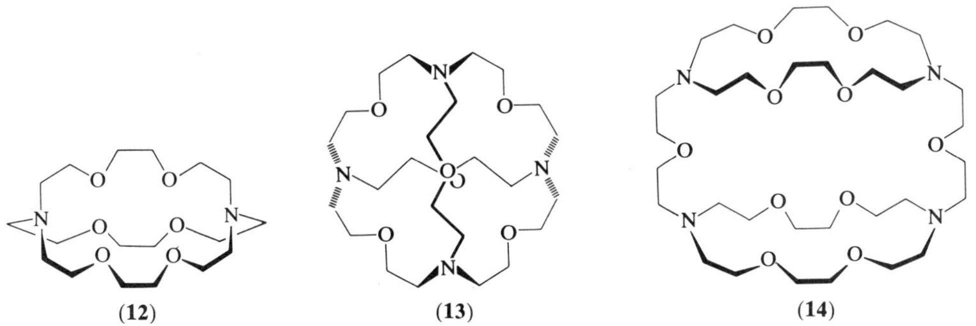

(**12**) (**13**) (**14**)

5.21.2 STRUCTURE OF CROWN ETHERS AND CRYPTANDS

5.21.2.1 Crystal Structure

5.21.2.1.1 Crown ethers

X-Ray crystallography has provided a large amount of data on the structures of crown ethers, cryptands and their complexes in the crystal ⟨B-78MI52102⟩. The uncomplexed

[18]crown-6 ligand does not take up the expected circularly symmetrical structure that gave rise to its name. Instead an elliptical cavity is formed containing three kinds of OCH_2CH_2O unit, four in a *gauche* and two in a *trans* conformation (Figure 1a). The mean carbon–carbon (1.507 Å) and carbon–oxygen (1.410 Å) bond lengths are both short compared to gaseous 1,4-dioxane. However this effect may be due to inadequate treatment of certain thermal motions in the crystal ⟨74AX(B)2733, B-82MI52102⟩.

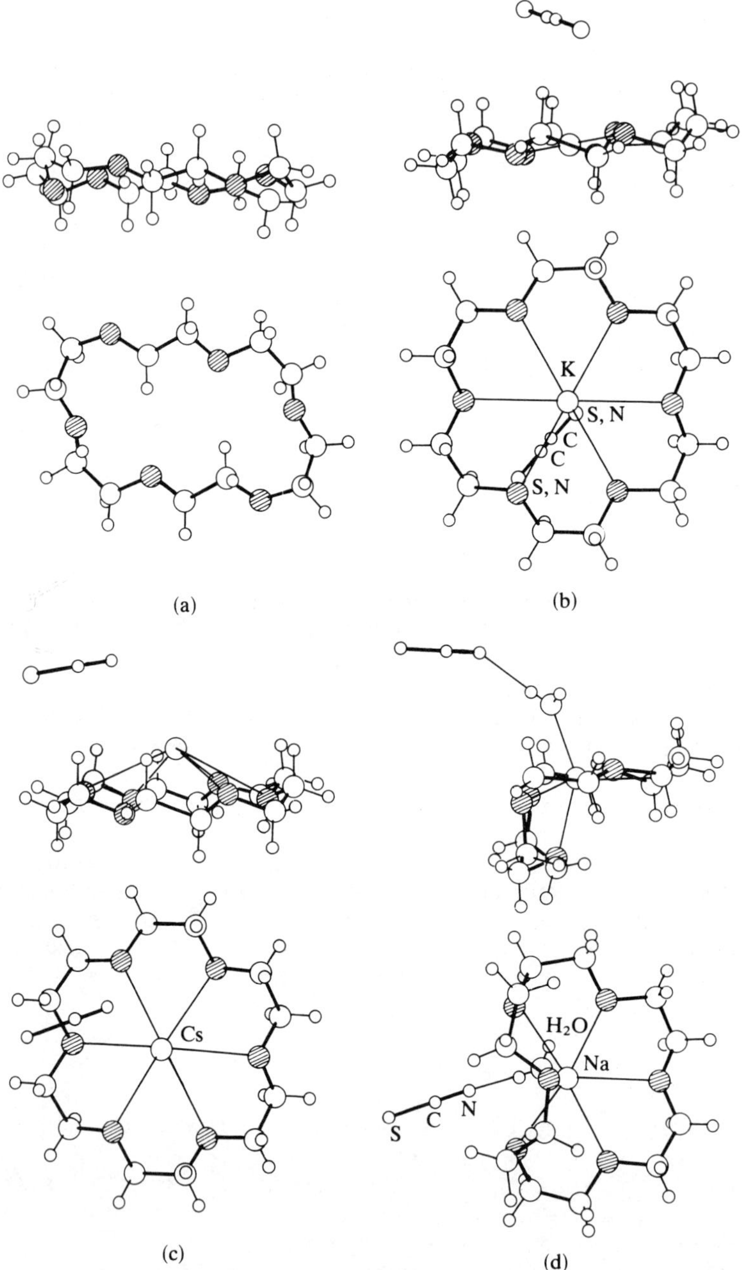

Figure 1 Crystal structures of [18]crown-6 and complexes. Top: view along a direction in the mean plane. Bottom: view in a direction normal to the mean plane. (a) Free ligand; (b) KNCS complex; (c) CsNCS complex; (d) NaNCS·H_2O complex (reproduced with permission from 74AX(B)2733, 2744, 2748, 2741)

On complexation with potassium thiocyanate a conformational change takes place in the macrocycle to expose a cylindrical cavity with D_{3d} symmetry. The potassium ion binds in the center of the ring to all six oxygen atoms and two thiocyanates bind weakly above and below the plane of the ring (Figure 1b) ⟨74AX(B)2744⟩. With the larger rubidium ion the ligand retains D_{3d} symmetry but the cation is displaced 1.20 Å from the mean plane of the hexaether ring. The displacement of the cation is increased to 1.44 Å as the ionic radius of the metal ion enlarges to that of cesium (Figure 1c) ⟨74AX(B)2748⟩. The sodium ion is too small to form a complex of D_{3d} symmetry with [18]crown-6. In fact the Na^+ is coordinated in all six oxygens, but of which only five are in the plane and the sixth lies above the plane. A seventh coordination site is occupied by a water molecule (Figure 1d) ⟨74AX(B)2741⟩. The relationship between cationic radius and complex structure shown by X-ray work confirms the observed selectivity of [18]crown-6 for potassium (see Section 5.21.3).

Similar structural analyses have been carried out on dibenzo[18]crown-6 (**2**) and its alkali cation complexes ⟨73MI52102⟩.

In the case of the sodium ion complex of the smaller benzo[15]crown-5 macrocycle, the cation binds to all five oxygen atoms but lies 0.75 Å above the plane of the macrocycle. With the larger potassium ion, benzo[15]crown-5 forms a 2:1 complex in which the cation sits at the center of a sandwich formed by two macrocycles. The [12]crown-4 ligand with only four coordinating oxygens in each ring forms a 2:1 sandwich complex with sodium ⟨B-78MI52102⟩.

As the size of the ring increases greater conformational flexibility is possible and, consequently, different structural types result. The dibenzo[30]crown-10 ligand forms a 1:1 complex with potassium in which all 10 oxygens wrap around and bind to the cation in a shape resembling the seams of a tennis ball (Figure 2a). In contrast the cavity of dibenzo[24]crown-8 is sufficiently large to complex two sodium ions. Each cation binds to three oxygen atoms in the ring leaving two uncoordinated (Figure 2b).

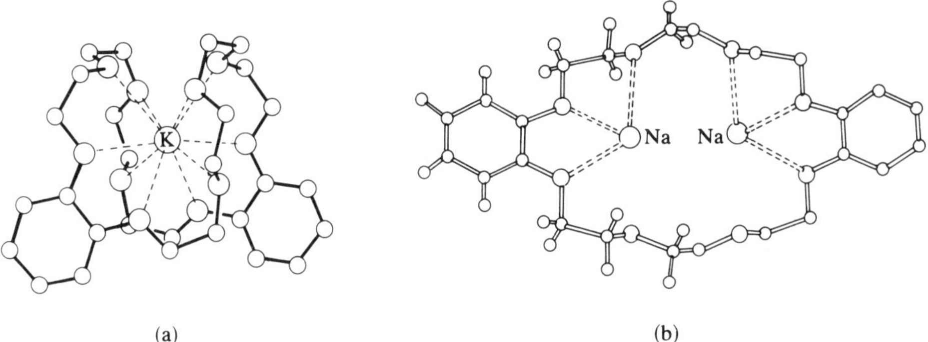

Figure 2 Crystal structures of (a) dibenzo[30]crown-10 complex with KI; (b) dibenzo[24]crown-8 complex with two Na ions (reproduced with permission from 72JCS(P2)345 and 75JCS(D)2374)

Crown ethers also form complexes with alkylammonium ions. The crystal structure of the 1:1 complex between [18]crown-6 and the *t*-butylammonium ion (Figure 3a) confirmed results that had initially been inferred from C.P.K. molecular models ⟨80JCS(P2)1529⟩, namely that the stability of the complex is derived from three hydrogen bonds to three ether oxygen atoms and that the ring OCH_2CH_2O units take up an all-*gauche* conformation with the six oxygen atoms in a single plane ⟨B-81MI52100⟩. The average hydrogen bonded N · · · O distance is 2.88 Å. The remaining three N · · · O distances are short (2.99 Å average) suggesting that favorable electrostatic interactions between the three non-hydrogen bonded oxygens and the cationic nitrogen also play a role in stabilizing the complex. In principle two complexes are possible depending on which of the two sets of three oxygen atoms participates in the tripod arrangement of hydrogen bonds. The importance of hydrogen bonds in these complexes was underlined with the crystal structure of the *t*-butylammonium perchlorate complex of pyrido[18]crown-6 (Figure 3b). As expected, the more favorable $N \cdots HN^+$ hydrogen bond orientates the ammonium ion to give one complex with three hydrogen bonds to two ether oxygen atoms and the pyridine nitrogen atom ⟨79AX(B)2233⟩.

Recently complexes between [18]crown-6 derivatives and the neutral platinum ammine species, *trans*-$[PtCl_2(PMe_3)NH_3]$, have been isolated and their crystal structures determined ⟨81CC847⟩. Dibenzo[18]crown-6 forms a 1:1 complex with mirror symmetry about the plane bisecting the aromatic rings (Figure 4a). Unlike the three-point tripod binding of alkylammonium ions, the NH_3 ligand forms three bifurcated hydrogen bonds in which each hydrogen is equidistant from two ether oxygens (Figure 4b). In contrast [18]crown-6 forms a 2:1 complex with *trans*-$[PtCl_2(PMe_3)NH_3]$ of approximately D_{3d} symmetry (Figure 4c). The crown ether ring adopts its complexing all-*gauche* conformation with a platinum ammine molecule bonded to each face through three NH · · · O hydrogen bonds. All six

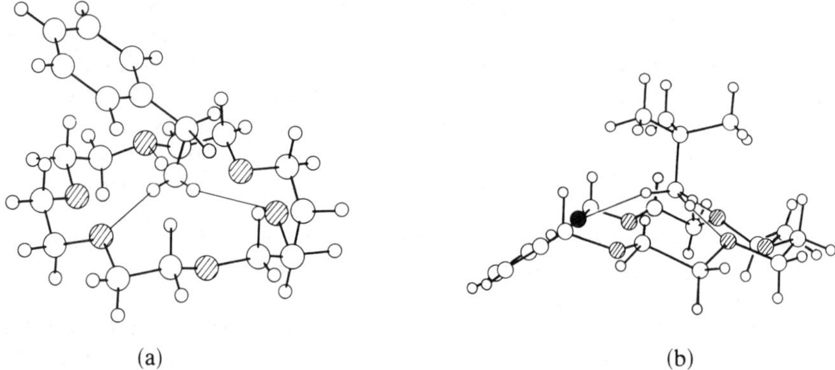

Figure 3 Crystal structures of (a) [18]crown-6 complex of benzylammonium thiocyanate; (b) pyrido[18]crown-6 complex of *t*-butylammonium perchlorate (reproduced with permission from 80JCS(P2)1529 and 79AX(B)2233)

oxygen atoms are, therefore, hydrogen bonded in a trigonal fashion. This six-point binding model with neutral $PtNH_3$ species differs from the three-point model for alkylammonium cations (*e.g.* Figure 3a) where approach of a second RNH_3^+ to the opposite face is presumably inhibited by electrostatic repulsion between the cations. When *trans*-$[Pt(NH_3)_2Cl_2]$ was added to [18]crown-6 a linear hydrogen-bonded polymer with the repeating unit $Pt(NH_3)_2Cl_2$-[18]crown-6$)_n$ was formed. Similar behavior is seen in the polymeric complex of [18]crown-6 with $[Cu(NH_3)_4(H_2O)]^{2+}$ although in this case four additional hydrogen bonds occur between the ring oxygens and the other two *cis*-ammine ligands ⟨81CC849⟩. The bifurcated arrangement of hydrogen bonds has also been reported in a 1:2 complex between [18]crown-6 and 2,4-dinitrophenylhydrazine ⟨81ZN(B)242⟩.

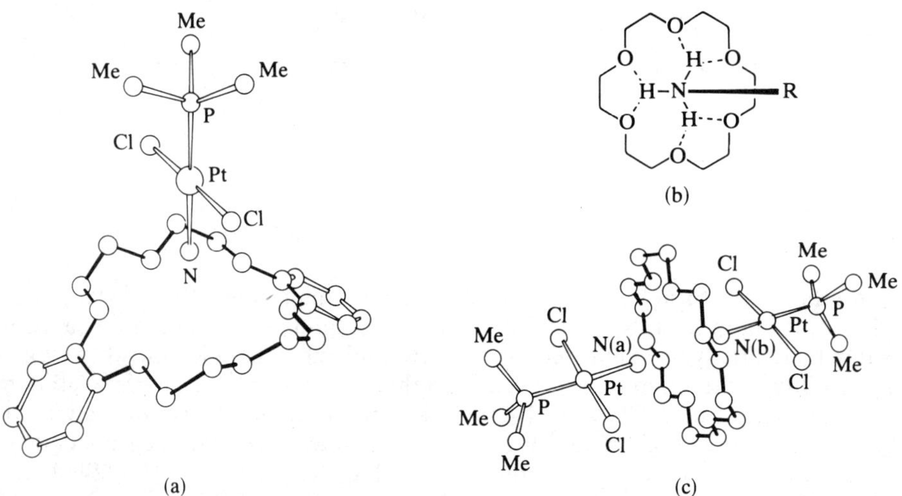

Figure 4 (a) Crystal structure of dibenzo[18]crown-6 complex with *trans*-$PtCl_2(PMe_3)NH_3$; (b) bifurcated arrangement of hydrogen bonds; (c) crystal structure of [18]crown-6 complex with *trans*-$PtCl_2(PMe_3)NH_3$ (a and c reproduced with permission from 81CC847)

Many other neutral molecules form complexes with crown ethers. Like *trans*-$[PtCl_2(PMe_3)NH_3]$, these uncharged guests generally contain acidic NH, OH or CH groups: *e.g.* urea, hydrazines, amides, malonates and nitromethane. The crystal structures of several such complexes have been surveyed in a recent review ⟨B-81MI52105⟩. The crystal structure of the 2:1 complex between malonodinitrile and [18]crown-6, shown in Figure 5a, is typical of this type ⟨77CB2249⟩. Weak CH ··· O hydrogen bonds between malonodinitrile and two ether oxygens stabilize the binding of a guest molecule on each side of the crown ether. An important neutral guest is water and a number of crystal structures of its complexes with crown ethers has been determined. Figure 5b shows the crystal structure of the 1:1 complex formed between water and a binaphthol-containing crown ether. The cavity is only partly filled by the water molecule which appears to form two hydrogen bonds to ether oxygens and a third from the water oxygen to the acidic hydrogen of a naphthol hydroxyl group ⟨78AX(B)3387⟩. A 'neutral' water complex stabilized only by hydrogen bonds to crown ether oxygens has recently been reported ⟨81JA7376⟩.

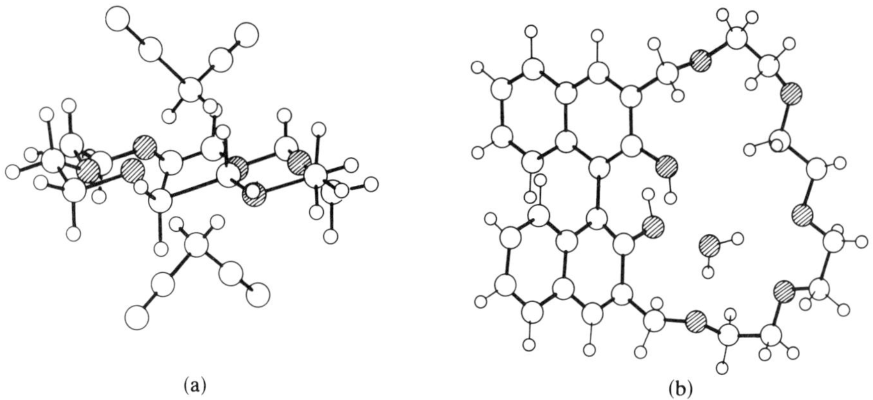

Figure 5 Crystal structures of (a) [18]crown-6 complex with two malonodinitrile molecules; (b) binaphthol crown ether complex with water (reproduced with permission from 77CB2249 and 78AX(B)3387)

5.21.2.1.2 Cryptands

In pioneering NMR work (see Section 5.21.2.2) Simmons and Park showed that uncomplexed diazabicycloalkanes (N[$(CH_2)_{k,l,m}$]N), the parents of cryptands, can exist in three different configurations in solution: the *exo-exo* in which both nitrogen lone pairs point away from the cavity, the *endo-endo* in which both lone pairs point into the cavity and the intermediate *exo-endo* with one lone pair inside and the other outside (Figure 6).

Figure 6 Different conformations of diazabicycloalkanes

The crystal structure of the uncomplexed [2.2.2]cryptand (**12**) shows it to have *endo-endo* conformation (Figure 7a). All three OCH_2CH_2O units are *trans* and the cavity is occupied by two methylene groups. On complexing K^+ the ligand reorganizes (Figure 7b); all three OCH_2CH_2O groups take up a *gauche* conformation which enables the six oxygen atoms to bind to K^+ in the center of the cavity *via* charge–dipole interactions. The *endo-endo* configuration is retained and the nitrogen lone pairs similarly participate in complexation. The K^+ cation is encapsulated by 14 lone pairs of electrons with such an excellent fit that the K^+—N and K^+—O distances are equal to the sum of the van der Waals radii of the atoms and the ionic radius of K^+. As the ionic radius of the cation increases the ligand must distort, within certain limits, to incorporate the larger cations; unlike crown ether complexes, the bicyclic ligand prevents the cation from moving away from the complexing atoms. Table 1 shows that as the ionic radius of the cation becomes larger the N—N and O—O (on different chains) distances and the N—C—C—O dihedral angle increase. This enlargement of the cavity is reflected in the stability of the complexes which shows the trend $K^+ > Rb^+ > Cs^+$. In order to accommodate the smaller Na^+ cation the bicyclic ligand must distort considerably, resulting in a lower stability of the Na^+ complex relative to K^+ ⟨B-79MI52103⟩.

An example of the *exo-exo* configuration is seen with the bis(borane–amine) derivative of [2.2.2]cryptand (Figure 7c) in which both of the large borohydride groups point away from the cavity. In contrast, the mono(borane–amine) derivative of [1.1.1]cryptand takes up an *endo-exo* structure in the crystal ⟨78NJC615⟩.

Table 1 Structural Features of Alkali Metal Complexes of [2.2.2]Cryptand (**12**) ⟨73AX(B)383, 69AX(B)925⟩

Cation	*Ionic radius* (Å)	*N—N distance* (Å)	*O—O (on different chains) distance* (Å)	*N—C—C—O dihedral angle* (°)
K^+	1.38	5.75	4.26	54
Rb^+	1.49	6.00	4.39	57
Cs^+	1.70	6.07	4.50	70

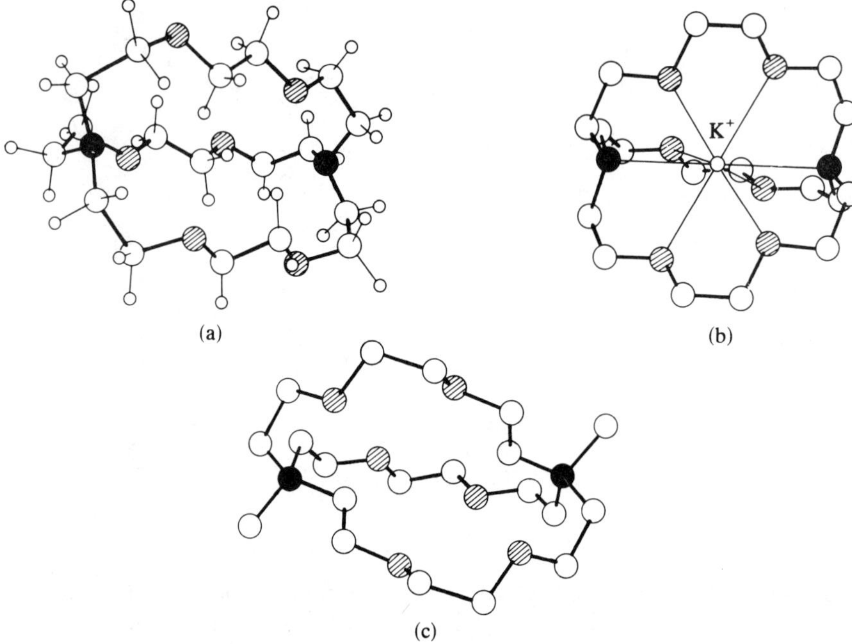

Figure 7 Crystal structures of (a) uncomplexed [2.2.2]cryptand (**12**); (b) potassium complex of [2.2.2]cryptand; (c) bis-borohydride derivative of [2.2.2]cryptand (reproduced with permission from 76JCS(P2)423 and 73AX(B)383)

The propensity for uncomplexed cryptands to fold in on themselves and fill their own cavity is nicely illustrated by the crystal structure of the diaza cryptand (**36**; see Section 5.21.4.2.1) containing a pyridine ring on each bridge. Two of the pyridine rings are located in the expected position on the periphery of the molecule with their planes pointing towards its center whilst the third pyridine is folded in and occupies the inside of the cavity ⟨81TL3035⟩. The nitrogens of this ligand are in the *endo-endo* conformation with a normal C—N—C angle of 114°. Interestingly, a smaller homologue of tris-pyridine cryptand (**36**) possesses bridgehead nitrogens which are planar with crystallographically equivalent C—N—C bond angles of 120° ⟨79JA1047⟩.

The crystal structures of a number of cryptands with bridgehead carbon, as opposed to nitrogen, atoms have also been reported ⟨81CC729⟩.

Ellipsoidal cryptands of type (**12**) generally complex alkylammonium cations relatively weakly. However the spherical cryptand (**13**) contains four nitrogen atoms which define a tetrahedral cavity large enough to complex the tetrahedral NH_4^+ cation. The crystal structure of this complex (Figure 8a) confirms that all four nitrogens are in the *endo* configuration and encapsulate the NH_4^+ cation. The ether oxygens interact electrostatically with NH_4^+, one of them being, in addition, hydrogen bonded to a water molecule ⟨82JA1672⟩. The protonated form of (**13**) can complex anions into the central cavity. Figure 8b shows the crystal structure of the chloride anion complex of tetraprotonated (**13**) with the Cl^- at the center of the cavity hydrogen bonded to all four protonated nitrogens ⟨76CC533⟩.

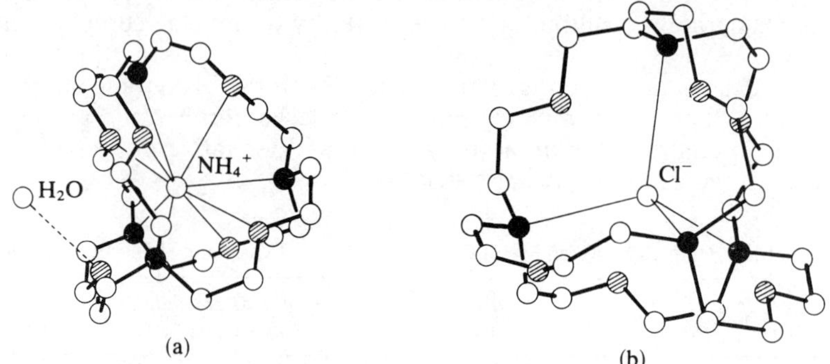

Figure 8 Crystal structures of (a) complex between spherical cryptand (**13**) and NH_4^+; (b) complex between tetraprotonated (**13**) and Cl^- (reproduced with permission from 76CC533)

5.21.2.2 Solution Structure

5.21.2.2.1 Nuclear magnetic resonance spectroscopy

^{1}H and ^{13}C NMR have been widely used in the study of macropolycycles and their complexes. The magnetic resonance frequencies of the ^{13}C or ^{1}H nuclei in both ligand and cation are generally sensitive to conformational changes in uncomplexed and complexed species. More recently, with the advent of multinuclear NMR (*e.g.* ^{15}N, ^{17}O, ^{7}Li, ^{23}Na, *etc.*) it has become possible to study the effect of complexation on the cation itself and also the ligand atoms with which it interacts ⟨B-79MI52103⟩.

The power of ^{1}H and ^{13}C NMR to elucidate solution structures has been demonstrated with a number of benzo substituted crown ethers and their complexes ⟨76JA3769⟩. The ^{1}H NMR spectrum of dibenzo[18]crown-6, shown in Figure 9a, consists of two multiplets in the ether region and a multiplet in the aromatic region. The multiplets for the ether protons are characteristic of a four spin AA'BB' coupled system with coupling constants (in acetone) of $J_{A',B} = 3.5$ and $J_{A,B} = 5.8$ Hz. These values are typical of an XCH_2CH_2Y unit undergoing rapid interconversion between the *syn* and *anti-gauche* rotamers; for example, 1,4-dioxane, which is known to undergo a similar conversion, has $J_{A',B} = 2.7$ and $J_{A,B} = 6.05$ Hz. The solution structure was, thus, concluded to be that shown in Figure 9b with all OCH_2CH_2O units in a *gauche* conformation. This result is contrary to the crystal structure (Figure 9c) which has two OCH_2CH_2O units in a *gauche* arrangement and two in a *trans* arrangement ⟨70JCS(B)1544⟩. On adding K^+ to the free ligand a downfield shift in both ether and aromatic proton resonances occurs in a linear fashion up to a 1:1 K^+:ligand ratio (Figure 9a). The coupling constants of the ether resonances change to $J_{A',B} = 2.2$ and $J_{A,B} = 6.3$ Hz reflecting only a small change in the dihedral angle of the OCH_2CH_2O units. The K^+-dibenzo[18]crown-6 structure is, therefore, the same in both crystal and solution. Similar results were reported for benzo[18]crown-6 suggesting that the rigid benzene ring does not greatly affect solution structure and that the results may be general for [18]crown-6 derivatives. The cesium complex of dibenzo[18]crown-6 gave 1:2 Cs^+:crown ratio in the titration experiment confirming the solution existence of a sandwich complex with the Cs^+ between two crown ethers. The 'wrap-around' structure of K^+ and Cs^+ complexes of [30]crown-10 (see Figure 2a) was similarly demonstrated in this study. All of the above results were supported by ^{13}C NMR measurements.

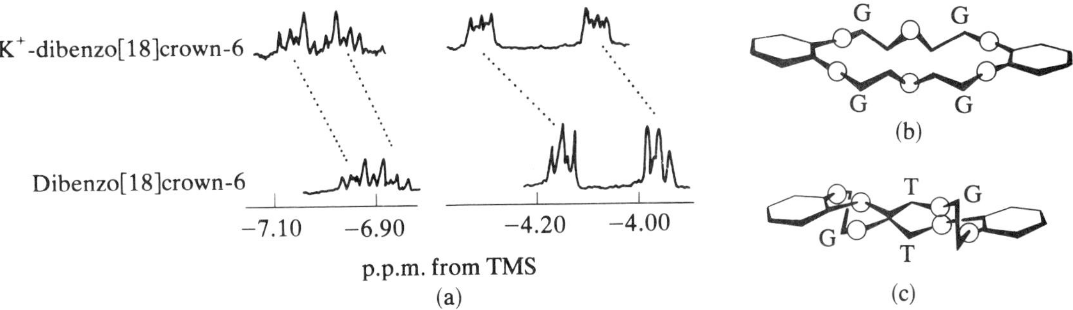

Figure 9 (a) ^{1}H NMR spectrum of dibenzo[18]crown-6 and its potassium complex; (b) proposed solution structure of dibenzo[18]crown-6: (c) crystal structure of dibenzo[18]crown-6 (reproduced with permission from 76JA3769)

^{1}H and ^{13}C NMR have also been widely used to study the binding of alkylammonium cations to crown ethers. The shift in substrate proton resonances can give detailed information of the preferred structure of the complex. This is particularly important in the study of chiral recognition by asymmetric receptors ⟨79CSR85⟩ (Section 5.21.3.2.2) and also the development of bioorganic models and catalysts (Section 5.21.5.1.1).

The conformational interconversion of diazabicyclo [*k.l.m*] alkanes (Figure 6; Section 5.21.3.1.2) has been investigated by variable temperature ^{1}H NMR ⟨68JA2428⟩. The bis-deuterochloride salt of 1,10-diazabicyclo[8.8.8]hexacosane showed a slow isomerization from *exo-exo* to *endo-endo* forms during which the NH proton resonance shifted from 8.0 p.p.m. (*exo-exo*) to 4.6 p.p.m. (*endo-endo*). The equilibrium constant for the process (K_e = *endo-endo*/*exo-exo*) was greater than 100. Similar measurements on differently lengthed diazabicyclo[*k.l.m*]alkanes revealed that even chain lengthed species prefer the *endo-endo* configuration (*e.g.* for [6.6.8]alkane $K_e = 24$) whilst odd chain lengthed species prefer *exo-exo* (*e.g.* for [9.9.9] alkane $K_e = 0.41$). The intermediate *endo-exo* form was only seen in the large [10.10.10] species. Variable temperature measurements on free amine [8.8.8] showed that nitrogen inversion occurred at 25 °C with an activation energy of 32.2 kJ mol^{-1} and a rate constant (k) of $1.4 \times 10^7\ s^{-1}$.

On complexing alkali metals, Lehn's cryptands (*e.g.* **12**) show only small changes in their NMR spectra. Alkaline earth cations however cause a larger downfield shift of the ligand proton resonances of about 0.2 p.p.m. Spherical cryptand (**13**) has a particularly simple ^1H NMR spectrum consisting of two triplets ($J = 5.5$ Hz). This suggests that (**13**) exists in solution as the tetra-*endo*, or tetra-*exo* forms or as a mixture of all forms in rapid equilibrium ⟨78ACR49, B-79MI52103⟩. Cylindrical cryptands of type (**14**) form unsymmetrical complexes with only one metal cation located in one of the crown ether rings. The intramolecular exchange of the cation from one ring to the other has been studied by variable temperature ^{13}C NMR and values for the activation energy and rate constant for exchange have been obtained.

^{13}C relaxation experiments are useful in studying the molecular motions of crown ethers and cryptands. Uncomplexed dibenzo[18]crown-6 shows a segmental mobility which disappears on complexation. Similarly, alkyldiammonium salts have shorter relaxation times when complexed by large cylindrical cryptands (Figure 15). The dynamic coupling of the molecular motions of substrate and receptor increases with their structural complementarity in the complex ⟨81CC833⟩.

Natural abundance ^{17}O and ^{15}N NMR have both been used to study crown ether and cryptand complexation ⟨80HCA668⟩. The magnetic resonances of both nuclei seem relatively insensitive to the electrostatic changes that take place on complexation and the full potential of the techniques remains to be realized. In a recent report, however, the exchange rates, structure and internal mobility of the ammonium complex of spherical cryptand (**13**) have been investigated by ^{14}N and ^{15}N NMR ⟨82JA1672⟩. Alkali metal NMR represents, at present, a more useful approach since all the members possess NMR-active isotopes which are sensitive probes of complexation. Furthermore ^7Li, ^{23}Na, ^{39}K and ^{133}Cs display relatively narrow natural linewidths despite having quadrupole moments. Considerable information on the structure of crown ether and cryptand complexes and also the thermodynamics and kinetics of their formation has been gained from studying these four isotopes ⟨B-79MI52103, 79JA4051⟩.

5.21.2.2.2 Vibrational spectroscopy

The earliest studies on the IR spectra of crown ethers and their complexes were carried out by Pederson ⟨B-78MI52101⟩. Since then IR spectroscopy has been applied a number of times to solve structural problems in crown ether chemistry including ion pair interactions and conformational analysis ⟨B-79MI52103⟩. A low frequency vibration of the cation within the macrocycle has been identified which is dependent on the cation but not on the solvent (214 cm^{-1} for Na^+-dibenzo[18]crown-6 in both DMSO and pyridine). This is in contrast to the frequency of cation vibration within a solvent cage, which for Na^+ is 202 cm^{-1} in DMSO and 181 cm^{-1} in pyridine ⟨72JA5247⟩. Similar results have been obtained with the [2.1.1]cryptand which shows a complexed-Li^+ vibration at 348 cm^{-1} and [2.2.2]cryptand with a complexed-Na^+ vibration at 234 cm^{-1} ⟨75MI52100⟩. Raman spectroscopy has also been used, although not widely, to study crown ether and cryptand complexation and the ion pair interaction of the complexed species ⟨B-79MI52103⟩.

5.21.2.2.3 Electronic spectroscopy

In his early work Pedersen investigated crown ether complexation by UV spectroscopy. He reported that complexation caused a shift in the absorption maximum of dibenzo[18]crown-6 of about 6 nm to a longer wavelength ⟨B-78MI52101⟩. The test was not totally reliable as cadmium caused no change in the spectrum yet gave a crystalline complex. In general, however, UV-visible spectroscopy is of limited use in the study of macrocyclic complexes.

5.21.2.2.4 Theoretical studies

In recent years molecular mechanics (MM) calculations have found increasing use in predicting the preferred conformations of medium and large ring molecules ⟨81T3981⟩. Sutherland and coworkers have adapted the White–Bovill force field (WBFF) for alkanes and alkenes and applied it to crown ethers ⟨80JCS(P2)1529⟩. Their prediction for the lowest energy conformation of uncomplexed [18]crown-6 closely agreed with that in the crystal structure (see Figure 1a) and was 32.80 kJ mol^{-1} lower in energy than the all-*gauche* conformation found in cation complexes. This result appears to disagree with NMR evidence ⟨76JA3769⟩ (Section 5.21.2.2.2) which suggests that uncomplexed [18]crown-6 derivatives take up an all-*gauche* conformation in solution. Interestingly, the MM calculations predict that were it not for the electrostatic repulsion between the oxygen atoms the all-*gauche* conformation would be the lowest in energy by 7.36 kJ mol^{-1}. That most cation complexes of [18]crown-6 have an all-*gauche* structure confirms that the favorable electrostatic interactions between the oxygen atoms and the cation more than compensate for any oxygen–oxygen repulsion.

The molecular orbitals of crown ethers and their complexes have been investigated by the CNDO/2 method ⟨79T1065⟩. The results were used to explain the PE spectra of crown ethers. In addition it was concluded that the charge transfer interaction between ligand and cation molecular orbitals was important for complex formation and that solvation effects played a crucial role in determining the stability of the complex.

5.21.3 REACTIVITY OF CROWN ETHERS AND CRYPTANDS

Crown ethers and cryptands show much of the same functional group chemistry as simple ether- or amine-containing molecules. The remarkable reactivity of these macropolycyclic species is primarily derived not from the composition of functional groups but from their three-dimensional arrangement. The important property of strong cation complexation is determined by the topology of the cavity defined by the ether and amine groups in the molecular superstructure.

5.21.3.1 Alkali and Alkaline Earth Cation Complexation

5.21.3.1.1 Crown ethers

Pedersen was the first person to study in detail the coordination of alkali metal cations by crown ethers and to realize that the stability of the resulting complex depended on the cation diameter and the size of the crown ether cavity ⟨67JA7017⟩. The stability of the complex is measured by the equilibrium constant K_s (sometimes denoted K_f), as defined by the law of mass equilibrium. The relationship between log K_s and the ratio of cation diameter to cavity size for dicyclohexyl[18]crown-6 complexes is seen in Figure 10a. The most stable complex is formed when cation and cavity diameters are the same (*i.e.* with K^+) and any deviation from the perfect fit causes a loss of stability. This difference in stability constants means that dicyclohexyl[18]crown-6 will selectively complex K^+ in the presence of Li^+, Na^+, Rb^+ and Cs^+, the degree of selectivity being determined by the relative sizes of the respective stability constants. As the size of the crown ether increases, the complementarity between cation and cavity diameter changes to give a different selectivity. Figure 10b illustrates how, for a series of crown ethers, the relative stability of Na^+, K^+ and Cs^+ complexes, and thus complexation selectivity, varies with increasing cavity size ⟨73MI52101, 80APO(17)279, B-81MI52102⟩. The simple 1 : 1 binding model cannot be applied to crown ethers larger than [24]crown-8 because their greater flexibility precludes accurate determination of the cavity size. For example, dibenzo[30]crown-10 and dibenzo[60]crown-20, despite their large size, form very stable complexes with K^+ by wrapping around the cation; dinuclear complexes may also form (see Figures 2a and 2b; Section 5.21.2.1.1).

Crown ethers can also form complexes with alkaline earth cations and some divalent transition metal cations. In general, the log K_s values in water for these complexes are greater than those of alkali metal cations. However, when monovalent and divalent cations

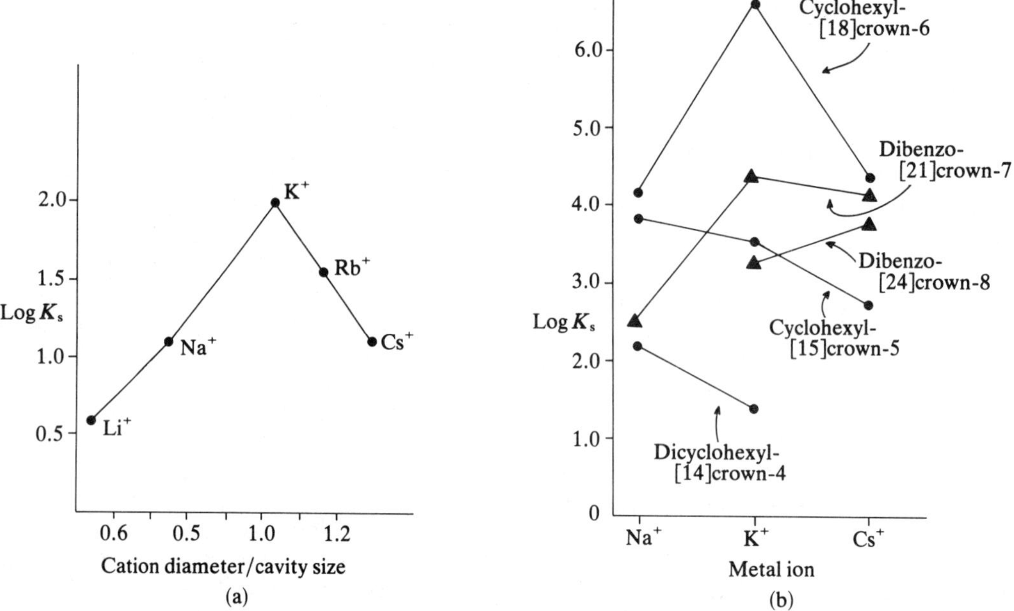

Figure 10 (a) Relation between log K_s and the ratio of cation diameter to cavity size for dicyclohexyl[18]crown-6 in water at 25 °C; (b) log K_s values of several crown ether–alkali cation complexes

of small but similar radii are compared the divalent ion forms less stable complexes due to the increased solvation energies that must be overcome during the process of complexation.

Changes in the electron density of crown ether oxygen atoms greatly affect the stability constants for cation complexation. The electron density on phenol ether oxygens is less than that on alkyl ether oxygens; consequently dibenzo[18]crown-6 (**2**) shows weaker complexing ability than [18]crown-6 (**1**) and hexabenzo[18]crown-6 does not complex cations at all ⟨80APO(17)279⟩.

Complex stability can also vary widely with different substituents on the periphery of the crown ether ring. A tetracarboxylate [18]crown-6 derivative, prepared from tartaric acid, forms exceptionally strong complexes with K^+ ($K_s = 3.0 \times 10^5$; in water). The tetrakis(dimethylcarboxamide) species is a less effective ligand ($K_s = 3.0 \times 10^4$; in methanol/water 95/5) whilst the bis(fused imide) derivative only very weakly complexes K^+ ($K_s < 10$; in methanol/water 95/5) ⟨76CC621⟩. These effects are due to electrostatic and conformational changes in the macrocycle. Particularly, in the last case the fused imide rings cause a distortion of the ether oxygen atoms away from the optimal arrangement for cation binding.

The nature of the donor atom also plays an important role in stabilizing complexation. Substituting one or more sulfur atoms for oxygen in [18]crown-6 (**1**) causes a decrease in K_s for K^+ complexation and an enhanced binding of soft cations such as Ag^+. A similar substitution by nitrogen atoms results in a smaller reduction in K^+ binding but an increased K_s for transition metal complexation.

The importance of solvent in determining the stability of crown ether complexes can be seen in Table 2 in which log K_s values for Na^+ and K^+ complexes of crown ethers in different solvents are compared. The greater values of log K_s in methanol compared to water suggest that the stability of crown ether complexes is mainly enthalpic in origin. Cations are more strongly solvated in water than methanol and the solvation sphere must be removed before complexation can take place.

Table 2 Log K_s Values of Crown Ether Complexes In Methanol and Water at 25 °C ⟨73MI52101⟩

	Na^+		K^+	
Crown ether	*H_2O*	*MeOH*	*H_2O*	*MeOH*
Dicyclohexyl[18]crown-6	1.21	4.08	2.02	6.01
Cyclohexyl[15]crown-5	0.3	3.71	0.6	3.58

5.21.3.1.2 Cryptands

The recognition that a three-dimensional arrangement of donor atoms should enhance complexation led to the development of the macrobicyclic cryptates ⟨77PAC857, 78ACR49, 78PAC871⟩. The early cryptands of Lehn and coworkers ⟨73T1629⟩ showed increased stability constants for alkali metal cation complexation of up to five orders of magnitude. For example, log K_s for the K^+ complex of [2.2.2]cryptand (**15d**) was 9.5 as compared to 4.38 for the monocyclic analogue (**16**). This effect was termed the macrobicyclic cryptate effect, by analogy to the macrocyclic effect.

(**15**)

	j	*k*	*l*
a	1	1	1
b	2	1	1
c	2	2	1
d	2	2	2
e	3	2	2
f	3	3	2
g	3	3	3

(**16**)

As with crown ethers, the complexation selectivity of the cryptands can be altered by varying the size of the cavity defined by the ligand. In the case of the smaller cryptands this behavior is accentuated by the cage-like structure which limits both the expansion and contraction of the cavity, and, as occurs with crown ethers, the movement of large cations away from the donor atoms. Consequently [2.1.1]-, [2.2.1]- and [2.2.2]-cryptands (**15b**, **15c** and **15d**) show a pronounced selectivity for Li^+, Na^+ and K^+, respectively ⟨75JA6700⟩. This sharp difference in stability constants and consequent discrimination against cations which are larger or smaller than the cavity is known as 'peak selectivity' ⟨73MI52100⟩. The larger, more flexible cryptands can distort their cavities to bind ions of different sizes. [3.2.2]-, [3.3.2]- and [3.3.3]-cryptands (**15e**, **15f** and **15g**) have high selectivity for K^+ over Na^+ but show little differentiation between K^+, Rb^+ and Cs^+. This behavior is called plateau selectivity.

Alkaline earth cations are complexed by appropriately sized cryptands even more strongly than alkali cations. The large number of electron pairs involved in binding (14 for **15d**) results in a strong and favorable electrostatic interaction with the divalent cation. Thus the stability constant for Ba^{2+} complexation by [2.2.2]cryptand (**15d**) is 10^4 times greater than that for K^+. The difference drops to 10^{-2} when one OCH_2CH_2O unit in (**15d**) is replaced by $(CH_2)_4$. The reduction in stability is due partly to the loss of two oxygen donor atoms but also to the efficient shielding of the complexed cation from the solvent by the $(CH_2)_4$ unit. This effect, which represents a reversal in selectivity of 10^6, must be cryptate in nature because monocycle (**17**), which has the same binding sites as the $(CH_2)_4$ derivative, has a similar Ba^{2+}/K^+ stability constant ratio to [2.2.2]cryptand (**15d**) ⟨78ACR49⟩. Cryptand (**13**) defines a spheroidal cavity formed by six oxygen and four nitrogen atoms. This is an ideal topology for complexing spherical alkali and alkaline earth cations and complexes of K^+, Rb^+, Cs^+ and Ba^{2+} have been reported. In particular the Cs^+ complex is the most stable known ⟨75JA5022⟩.

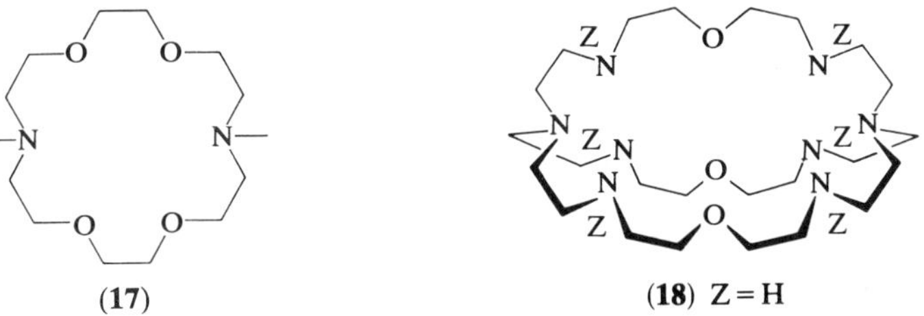

(**17**) (**18**) Z = H

The reduction in electrostatic interaction caused by replacing cryptand oxygen atoms by sulfur or nitrogen also results in a decreased stability of alkali and alkaline earth complexes. However the stability of transition metal cryptates is greatly increased. For instance, replacement of one oxygen in (**15d**) by NMe gives a cryptand which complexes Cd^{2+} more strongly than (**15d**) by a factor of 10^5. Large nitrogen-containing cryptands, such as (**18**), have been synthesized and shown to bind two transition metal ions into the cavity ⟨77JA6766⟩. Dinuclear transition metal complexes of cylindrical macrotricyclic cryptands of type (**14**) have also been studied ⟨77HCA141⟩. Both dinuclear systems of (**18**) and (**14**) present the possibility of either direct or bridging ligand-mediated metal–metal interactions.

The *endo-endo* conformation of cryptands can be internally protonated to form proton cryptates. With the small cryptands, *e.g.* [1.1.1]- and [2.1.1]-cryptand (**15a** and **15b**), the two internal protons are so efficiently shielded from H_2O and OH^- that deprotonation only very slowly occurs even in strong base ⟨81JA6044⟩. Alkali cation cryptates are able to stabilize unusual species as their counterions. Dye and coworkers have isolated several alkali metal anions by this method. The sodium species (Na^+[2.2.2]cryptand Na^-) was obtained as gold metallic crystals and gave a ^{23}Na NMR with a broad Na^+-cryptate resonance and a narrow, upfield Na^- resonance. The other alkali metals show similar behavior and an electride salt (Na^+[2.2.2]cryptand e^-) has even been isolated ⟨B-79MI52105⟩. Crystalline anionic clusters of the heavy post-transition metals (such as Sb_7^{3-}, Pb_5^{2-}, Sn_9^{4-}) were first obtained with alkali metal cryptates as the counterions ⟨75JA6267⟩.

5.21.3.2 Ammonium Ion Complexation

5.21.3.2.1 Complex formation

Crown ethers, and in particular [18]crown-6 derivatives, form complexes with protonated amines (RNH_3^+). The majority of binding energy for this process (~75%) comes from hydrogen bonding between the ammonium hydrogens and oxygen atoms in the ring. Electrostatic interactions, so important in metal ion complexation, between the remaining oxygen atoms and the cationic nitrogen are of minor importance (~25%). For efficient complexation to occur the RNH_3^+ cation should be able to form a tripod arrangement of hydrogen bonds with oxygen atoms on the ring. Strong RNH_3^+ binding to [18]crown-6 thus takes place *via* three N—H···O hydrogen bonds and three N···O electrostatic interactions (Figure 11a; see Section 5.21.2.1.1).

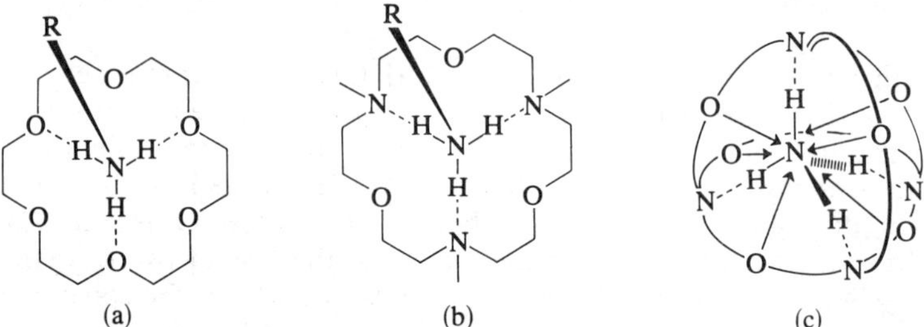

Figure 11 Hydrogen bonding arrangement in various crown ether–alkylammonium ion complexes

In the case of primary ammonium ions a secondary interaction can occur between the macrocycle and substituents on the cation. This interaction is usually steric in nature and destabilizes ammonium ion binding. When the crown ether has a fully accessible cavity (*e.g.* [18]crown-6) changing the nature of the substituents has little effect on the stability constant; for example, the complex between [18]crown-6 and RNH_3^+ has log K_s values of 4.27, 4.25, 3.56 and 2.90 when R = H, Me, Pr^i and Bu^t respectively. The steric interaction between receptor and guest can be increased by incorporating bulky groups into the macrocycle and this concept forms the basis of chiral recognition. Certain receptors, designed as bioorganic models, can undergo favorable secondary interactions with the appropriate primary ammonium substrates ⟨B-82MI52100⟩. These interactions can be electrostatic, hydrophobic or charge transfer in nature.

Substituting nitrogen atoms for oxygen in [18]crown-6 increases the stability of primary ammonium complexes due to the enhanced hydrogen bonding ability of nitrogen over oxygen. When three nitrogen atoms are trigonally arranged in the receptor the RNH_3^+ substrate can form three N–H···N hydrogen bonds and a particularly strong complex results (*e.g.* K_s for Figures 11a and 11b is 2100 and 65 000, respectively; R = Me) ⟨80TL1323⟩.

Spherical cryptand (**13**) contains four nitrogen atoms in a tetrahedral arrangement; this is the topologically ideal receptor for the ammonium ion (NH_4^+) which has a tetrahedral distribution of charge. Consequently a strong complex with NH_4^+ is formed by means of four $NH^+ \cdots N$ hydrogen bonds and six $N^+ \cdots O$ electrostatic interactions (Figure 11c) ⟨82JA1672⟩.

The complexation of primary ammonium ions by [12]crown-4 derivatives has been rationalized in terms of a two-point binding model where only two hydrogen bonds between the small macrocycle and RNH_3^+ are formed; this has been confirmed by X-ray studies ⟨80CC540⟩. Secondary ammonium ions ($R_2NH_2^+$) also bind to [12]crown-4 derivatives by a two-point hydrogen bonding arrangement ⟨77JA8317⟩; the conformational structure of this complex is however in question ⟨80ACS(B)397⟩.

Recently a number of coreceptors containing two potential RNH_3^+ binding sites have been prepared, *e.g.* (**19**). These systems can complex either two alkylammonium cations (**20**) or one alkyldiammonium dication (**21**) inside the cavity ⟨80TL941, 81CC990, 82CC557⟩ and hold considerable potential as models of enzyme binding sites (Section 5.21.5.1.1).

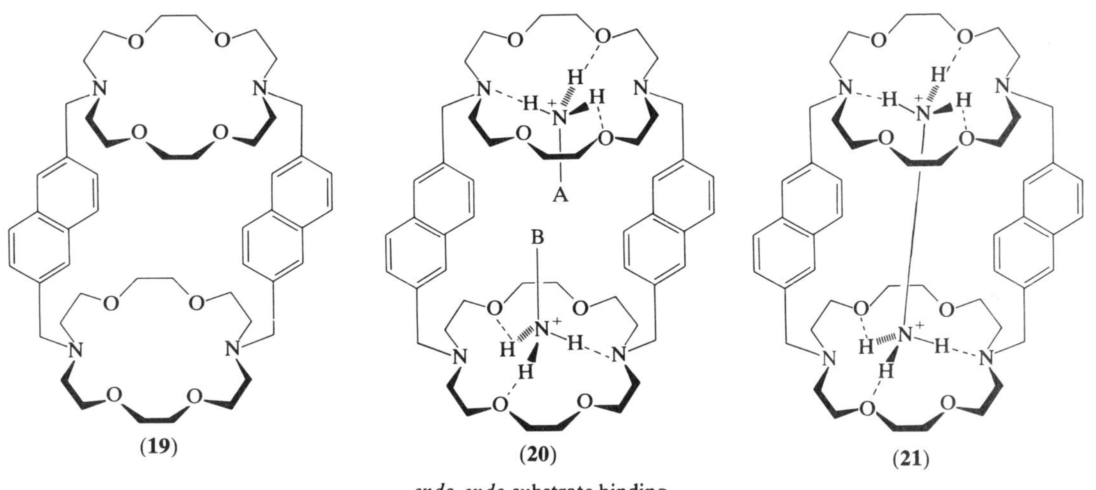

(**19**) (**20**) (**21**)

endo-endo substrate binding
(*endo-exo* and *exo-exo* forms are also possible)

5.21.3.2.2 Chiral recognition

Crown ethers containing chiral groups on the ring can show selective complexation of one enantiomer in a racemic mixture. This chiral recognition is due to unfavorable secondary interactions between chiral host and chiral guest which selectively destabilize the binding of one of the two enantiomers. The degree of destabilization and, thus, selectivity depends on the substrate and, more importantly, the design of the chiral receptor.

Pioneering work in this area has been carried out by Cram and coworkers and has recently been reviewed ⟨78ACR8, B-81MI52104⟩. By incorporating one or more chiral binaphthyl units into crown ethers they were able to discriminate between the (*R*) and (*S*) enantiomers of various protonated amino acid esters. For example, crown ether (**22**), containing two binaphthyl groups, each in an (*S*) configuration, effected a separation of (*S*)-methyl phenylglycinate from its other (*R*) enantiomer ⟨75CC394⟩. This behavior was rationalized in terms of a difference in free energy between the two diastereomeric complexes of about 8.4 kJ mol^{-1}. The (*S*, *S*) receptor (**22**) binds (*S*)-methyl phenylglycinate with the large phenyl group pointing away and the small hydrogen pointing towards the nearby naphthyl group (**23**; the receptor is seen in cross-section). (*R*)-Methyl phenylglycinate cannot take up this low energy arrangement and, thus, its complexation is inhibited. This rationale has been confirmed by a crystal structure of the less stable (*R*)-methyl phenylglycinate complex ⟨77JA6049⟩. The opposite (*R*, *R*) form of (**22**) preferentially complexes (*R*)-methyl phenylglycinate over its (*S*) enantiomer ⟨74JA7100⟩. The chiral recognition properties of (**22**) were further enhanced by placing substituents into the 2,2′-positions of one of the binaphthyl groups ⟨78ACR8⟩.

Many other asymmetric species have been incorporated into crown ethers by different research groups in an attempt to induce chiral recognition ⟨82JHC3⟩. These include amino acids ⟨81JA2091⟩, sugars ⟨79CSR85, B-81MI52106⟩, tartaric acid ⟨80HCA2096⟩ and spiro-bifluorene ⟨78PAC893⟩.

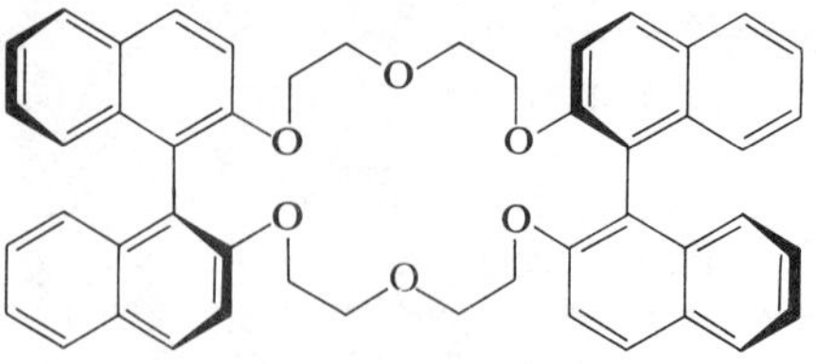

(22)

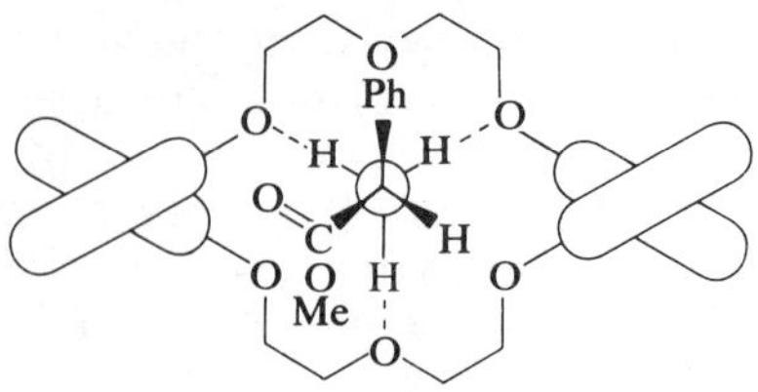

(23)

5.21.3.3 Miscellaneous Cation Complexation

The hydronium ion (H_3O^+), which resembles RNH_3^+ in size and the triangular arrangement of acidic hydrogen atoms, is also complexed by [18]crown-6 ⟨B-81MI52104⟩. Complexation seems to involve a similar trigonal arrangement of hydrogen bonds (**24**).

[18]Crown-6 and other larger crown ethers form complexes with aryldiazonium salts solubilizing them in organic solvents ⟨B-81MI52108⟩. The N_2^+ group projects into the cavity and complexation is stabilized by electrostatic interactions between oxygen electron pairs and the cation (**25**). Interestingly, complexes of arenediazonium salts with dinaphthyl[20]crown-6, unlike those of [18]crown-6, are colored, suggesting a charge transfer interaction between the π-basic naphthyl group and the π-acidic diazonium salt ⟨80APO(17)279⟩.

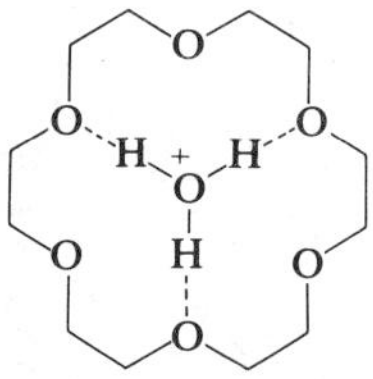

(24) 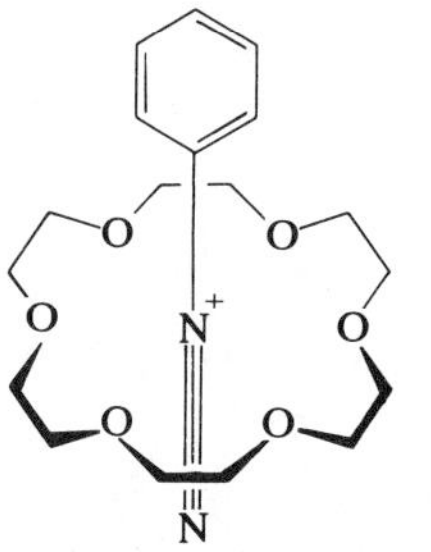 (25) 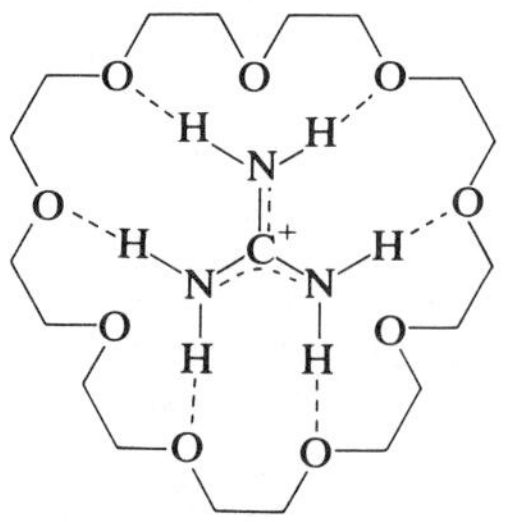 (26)

The flat guanidinium and imidazolium cations also form complexes with large ring crown ethers ⟨75CC427, 79CC296⟩. The [27]crown-9 complex with the guanidinium ion is thought to involve six hydrogen bonds between six ring oxygen atoms and the six acidic hydrogen atoms on the substrate (**26**). The stability constant for this complex is larger than that of the corresponding ammonium complex reflecting a circular discrimination or recognition by the macrocycle. Imidazolium ions can only bind *via* two hydrogen bonds and thus form weaker complexes with [27]crown-9.

A positively charged iodine cryptate has recently been reported ⟨80JA6574⟩. [2.2.2]Cryptand (**12**) appears to complex strongly iodonium ions (I^+) inside the cavity with an equilibrium constant (K_s) of about 10^7.

5.21.3.4 Neutral Molecule Complexation

A wide range of uncharged molecules form complexes with crown ethers and, although stable in the solid state, these species are usually unstable in solution. This subject has been extensively reviewed ⟨B-81MI52105⟩ (Section 5.21.1.1). In general, neutral guest molecules contain acidic hydrogen atoms which can form hydrogen bonds with crown ether oxygen atoms. Thiourea, urea, formamide, hydrazine and other N—H bond-containing species form crystalline complexes with sharp melting points. The crown ether–guest stoichiometry usually ranges from 1:1 to 1:6, although there exists little relationship between these varying ratios and the nature of the crown ether or guest. Similar complexes are formed by acidic C—H-containing compounds such as malonodinitrile, dimethyl sulfoxide and dimethylformamide. The crystalline complex between [18]crown-6 and acetonitrile is commonly used to purify the crown ether ⟨77OS30⟩. Molecules containing acidic C—H groups that readily enolize (*e.g.* acetylacetone) do not appear to form complexes with crown ethers.

A number of crown ether complexes of water have been reported ⟨B-81MI52105⟩ (see Section 5.21.2.1.1).

5.21.3.5 Anion Complexation

Crown ethers and cryptands that contain nitrogen atoms in the ring can be protonated or alkylated to generate a positively charged cavity suitable for the complexation of anionic species ⟨B-81MI52105⟩. In their work on diazabicycloalkanes Simmons and Park ⟨68JA2431⟩ protonated *endo-endo* isomers and showed that a chloride anion could bind inside the cavity (Figure 12a). The complex was stabilized by electrostatic interactions between the anion and the positive cavity and also two $N^+—H \cdots Cl^-$ hydrogen bonds. Similarly, spheroidal cryptand (**13**) in acid solution formed a tetraprotonated cavity which complexed Cl^- ions by a tetrahedral array of $N^+—H \cdots Cl^-$ bonds (Figure 12b; see Section 5.21.1.2 for crystal structure). The stability constant for this macrotricyclic complex (Figure 12b) is considerably larger than that of the macrobicyclic complex (Figure 12a).

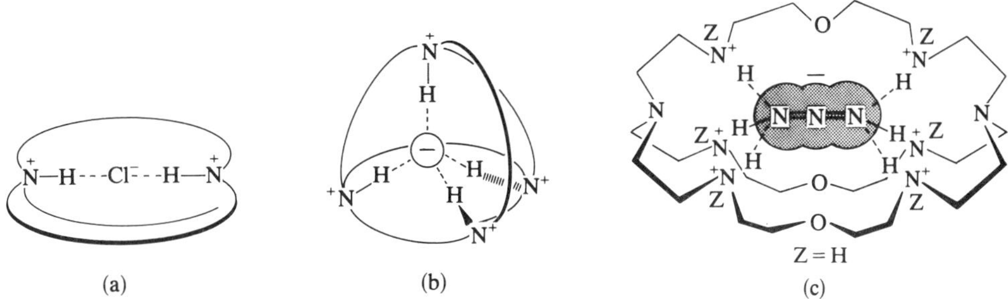

Figure 12 Anion cryptates

Larger anionic species such as sulfate, carboxylate and phosphate necessitate the preparation of larger macropolycycles as receptors.

The elongated 'bis-tren' cryptand (**18**) forms an anion receptor on hexaprotonation. Azide ions bind inside the positively charged cavity to give a 1:1 complex stabilized by two pyramidal arrays of three $N^+—H \cdots N^-$ hydrogen bonds (Figure 12c). Other small anions such as NO_2^-, NO_3^-, HCO_2^- and Cl^- form much weaker complexes demonstrating that the hexaprotonated cryptand is a selective receptor for linear triatomic species ⟨78JA4914⟩.

A number of monocyclic anion receptors have also been prepared by Lehn and coworkers ⟨81JA1282⟩. A series of large ring polyaza macrocycles were synthesized (*e.g.* **27** and **28**) and protonated to generate differently sized cationic cavities. These species formed strong complexes with inorganic and organic polyanions in aqueous solution. Small and highly charged anions are most strongly complexed (*e.g.* oxalate > malonate > succinate > fumarate) confirming that electrostatic interactions are very important for binding. Some complexation selectivity was also shown, large dianions such as fumarate and squarate preferring to bind to the larger macrocycle (**28**-8H$^+$) over the smaller (**27**-6H$^+$).

The protonated polyaza macrocycles (**27**-6H$^+$ and **28**-8H$^+$) also complex metal hexacyanide anions ⟨81CC1067⟩. Both $Fe(CN)_6^{4-}$ and $Ru(CN)_6^{4-}$ form 1:1 complexes which are more difficult to oxidize than the uncomplexed anion. The shifts in anodic oxidation potential are independent of the $M(CN)_6^{4-}$ species involved but vary with the size of the macrocycle (+130 mV for **27**-6H$^+$ and +165 mV for **28**-8H$^+$). This suggests that the redox potential of complexed anions is controlled by electrostatic effects due to the charge of the surrounding macrocycle.

(**27**) (**28**) (**29**)

In a related study ⟨78CC934, 79HCA2763⟩ Lehn and coworkers incorporated guanidinium groups into macrocycles (*e.g.* **29**). The resulting polyguanidium salts form stable complexes in water with phosphate and carboxylate anions. Unlike polyammonium anion receptors, these species remain protonated over a wide pH range and hold considerable potential in the development of anion-selective electrodes.

5.21.4 SYNTHESIS OF CROWN ETHERS AND CRYPTANDS

The various strategies available for the construction of crown ethers and cryptands have been analyzed ⟨73MI52100⟩ and reviewed recently ⟨B-78MI52100, 80T461⟩. In general, the problems encountered are those typical of reactions forming large rings, namely low yields and polymerization. These particular difficulties can be minimized by using, in the cyclization step, a rapid reaction under high dilution conditions. The presence of rigid groups also increases cyclization yields by reducing the number of conformational degrees of freedom in the secomacrocycle. In some cases this effect may be enhanced by using alkali metal cations as templates to keep the chains together during reaction.

5.21.4.1 Crown Ethers

The possible synthetic pathways to monocyclic crown ethers are shown in Figure 13.

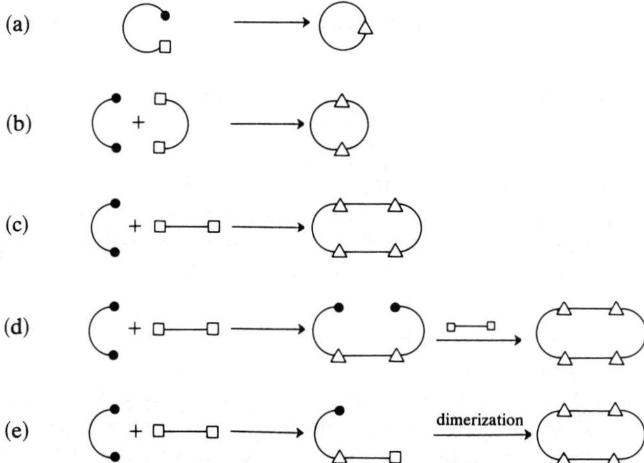

Figure 13 Synthetic strategies to crown ethers

5.21.4.1.1 Cyclization step involving formation of one bond (path a)

An early synthesis of [18]crown-6 (**1**) involved the cyclization of hydroxychloride (**30**) *via* a Williamson ether synthesis. The yield was 2%. In an adaptation of this approach ethylene oxide was oligomerized in the presence of an alkali cation template. By varying the cation template from lithium to sodium to potassium the major macrocyclic product changed from [12]crown-4 to [15]crown-5 to [18]crown-6, albeit only in about 10% yield ⟨76CC295⟩.

(**30**) → [18]crown-6 ← (**31**) + (**32**)

Scheme 1

5.21.4.1.2 Cyclization step involving formation of two bonds (path b)

Considerably higher yields of [18]crown-6 were obtained when dichloride (**32**) was condensed with diol (**31**). Similar procedures for the synthesis of [12]crown-4 and [15]crown-5 have also been developed and the dichloride component can readily be replaced by its ditosylate derivative. Once again the presence of an alkali cation template can considerably increase the yield of the cyclization ⟨77OS30⟩.

Various alkyl-substituted crown ethers have been prepared by cyclic addition across a double bond (*e.g.* bromination ⟨77S854⟩ or oxymercuration ⟨80CC925⟩). Similarly, displacement of oligoethylene glycol dichlorides by cyanamide yields *N*-cyano monoaza crown ethers ⟨81TL4105⟩.

Cyclic dilactones can be prepared in high yields from the corresponding diacid chloride and diol, and have interesting cation binding properties ⟨81JA1821⟩. A useful method for their reduction to crown ethers using lithium aluminum hydride has recently been reported ⟨82CC248⟩.

5.21.4.1.3 Cyclization step involving formation of four bonds (paths c, d and e)

The reaction of a dichloro and dihydroxy component can lead to crown ethers *via* route (b) or (c), and frequently mixtures of products must be separated. However, the presence of rigid benzo groups in the reactants limits the conformational freedom of uncyclized intermediates and an increase in product selectivity results. For example, the condensation of catechol and bis(2-chloroethyl) ether (**33**) in equimolar amounts leads almost exclusively to dibenzo[18]crown-6 (**2**) in 45% yield ⟨B-78MI52101⟩.

The yield of crown ether can be increased to more than 80% by using a longer stepwise route in which the initial catechol is partially protected. Reaction of the phenol (**34**) with bis(2-chloroethyl) ether (**33**) forms intermediate (**35**) which after deprotection can react with a further molecule of bis(2-chloroethyl) ether to give dibenzo[18]crown-6.

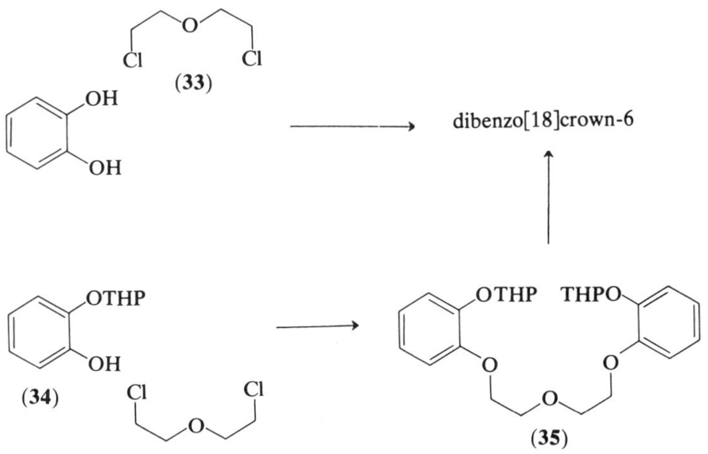

Scheme 2

In synthetic route (e) both components can be mono-protected. This approach has been successfully used in the synthesis of tartrate-containing crown ethers ⟨80HCA2096⟩.

A related synthesis of macrocyclic tetralactones should be mentioned here. By using covalent tin templates, macrocycles containing four lactone groups can be formed in an essentially 'one pot' reaction ⟨80CC176⟩.

5.21.4.2 Cryptands

The various possible strategies for the synthesis of macropolycyclic cryptands are outlined in Figure 14.

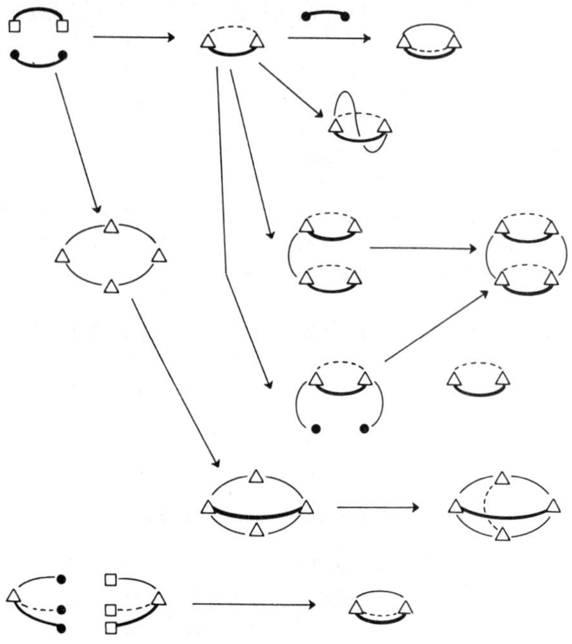

Figure 14 Synthetic strategies to cryptands

5.21.4.2.1 Ellipsoidal cryptands

The classical diaza[2.2.2]cryptand (**12**) of Lehn with bridgehead nitrogen atoms was initially prepared by the stepwise route shown in Scheme 3 ⟨78ACR49⟩. The amide formation and reduction sequence allowed the ready introduction of different heteroatoms and rigidifying groups into the macrobicycle. However, high dilution conditions were necessary and total yields were low. Recently a one step route to (**12**) using simple starting materials has been reported ⟨80TL643⟩. These basic routes have been applied to many diverse cryptands ⟨80T461⟩.

Scheme 3

Ellipsoidal cryptands can also be synthesized by direct alkylation procedures ⟨77AG(E)720, 80CB1487⟩, obviating the need for a diborane or lithium aluminum hydride reduction step. In the case of [1.1.1]cryptand (**15a**) yields of the final amine alkylation step are enhanced by the amine proton itself acting as a template ⟨81CC777⟩.

In a useful extension of the direct alkylation method, ellipsoidal cryptands have been prepared from dimethylated cyclic diamines by the stepwise quaternization–demethylation sequence shown in Scheme 4 ⟨81TL3039⟩.

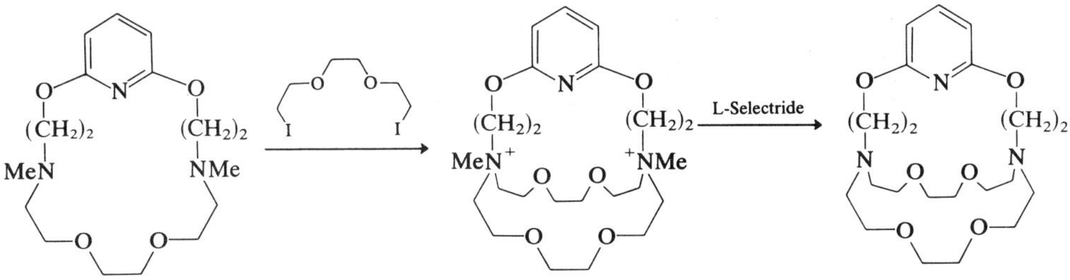

Scheme 4

As seen in Figure 14 the ellipsoidal cryptands may be prepared by the simultaneous formation of three bridges. This route has found limited use due to low yields but has been employed in the synthesis of some cryptands including the tris-pyridino system (**36**) ⟨81TL3035⟩ and bis-tren (**18**) ⟨80UP52100⟩.

The presence of nitrogen bridgeheads in the cryptands discussed above increases both the stability, by participating in complexation, and water solubility of the resulting complexes. However, rapid nitrogen inversion processes decrease the rigidity of the ligand and lead to fast equilibrating thermodynamic mixtures in solution.

A series of cryptands with carbon bridgeheads have been synthesized. Using strategies similar to those above pentaerythritol (**37**) has been incorporated into hydroxymethyl cryptand (**38**) ⟨77JCS(P1)767⟩. Cryptands unsubstituted at the carbon bridgehead pose problems of diastereomeric mixtures in their syntheses and consequently asymmetrically pure starting materials must be used. In a particularly elegant synthesis, 2,3-*O*-isopropylidene-D-glycerol, available in large quantities from D-mannitol, has been incorporated as the bridgehead unit into the *endo-exo* cryptand (**39**) *via* Scheme 5 ⟨79JCS(P1)2577⟩. Analogous routes were used to the benzo substituted *exo-exo* cryptand (**40**) ⟨79CC486⟩.

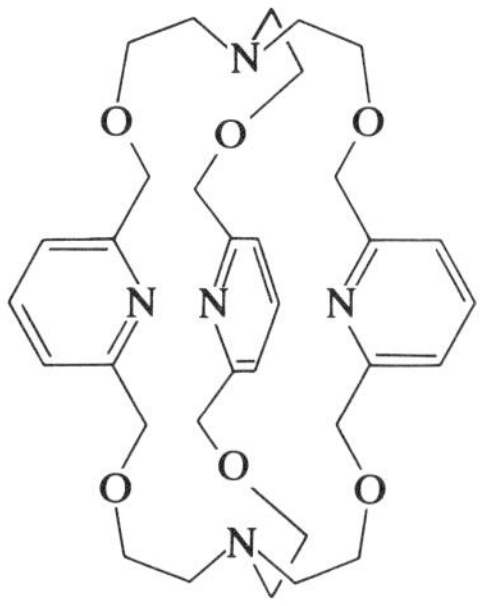

(**36**) 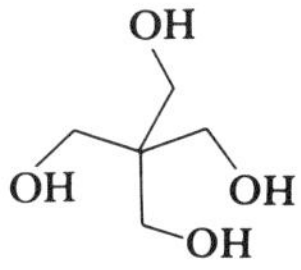 (**37**) 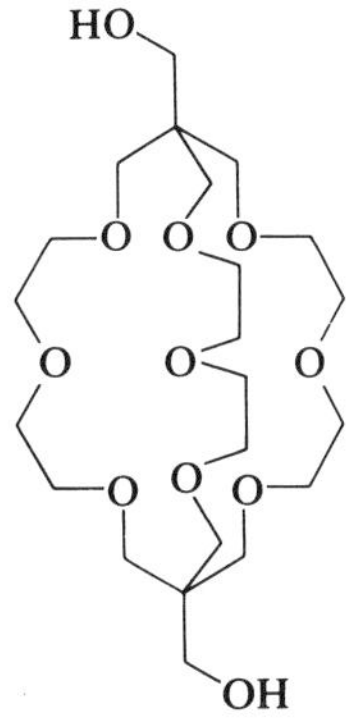 (**38**)

5.21.4.2.2 Cylindrical cryptands

Cylindrical cryptands have been receiving increasing interest recently as multidentate ligands with novel topological properties and also as potential chelates for two metal atoms ⟨80PAC2441⟩.

Most cylindrical cryptands have been based on the readily available diaza crown ether (**41**). By combining either the amide formation and reduction or direct alkylation routes with the strategies outlined in Figure 14 a large number of symmetrical and unsymmetrical bis-diaza crown ethers have been prepared (Figure 15). Although the condensation of diaza crown ether (**41**) with a suitable dihalide will give the cylindrical cryptand in one step, yields are low ⟨81CC990⟩. The stepwise amide formation–reduction route gives good overall yields which in the case of rigid side arms (*e.g.* phenyl, biphenyl) can be as high as 90% ⟨80TL941⟩.

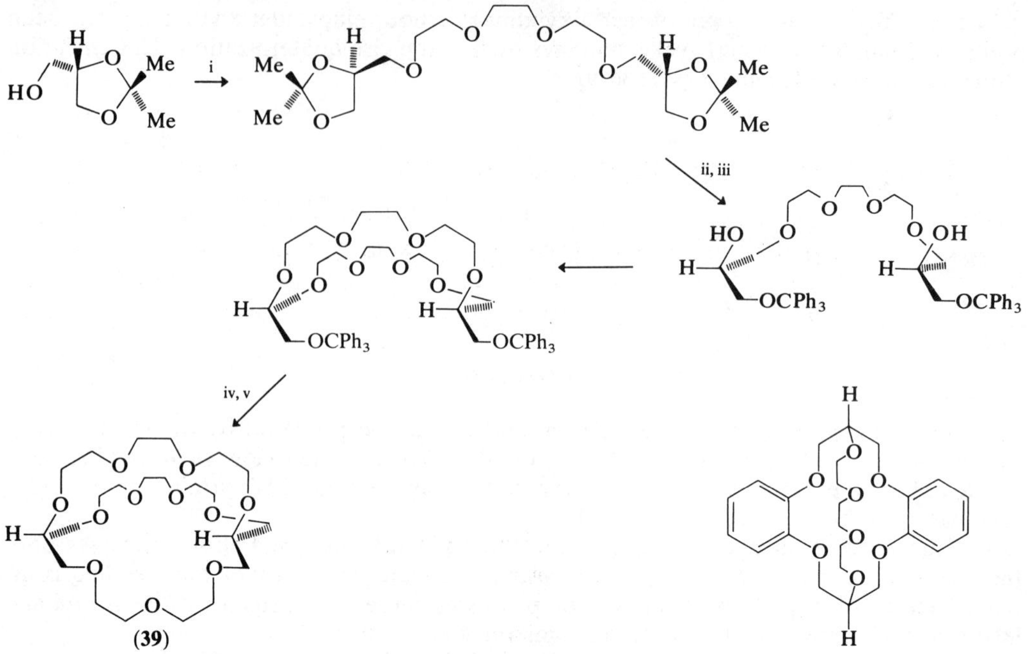

(39)

(40)

i, $TsOCH_2(CH_2OCH_2)_2CH_2OTs$; ii, TFA; iii, Ph_3CCl; iv, $H_2/Pd\text{–}C$; v, $TsOCH_2CH_2OCH_2CH_2OTs$

Scheme 5

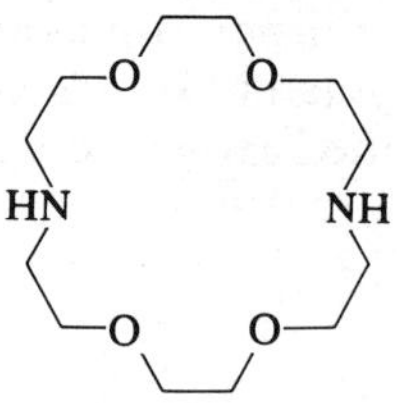

(41)

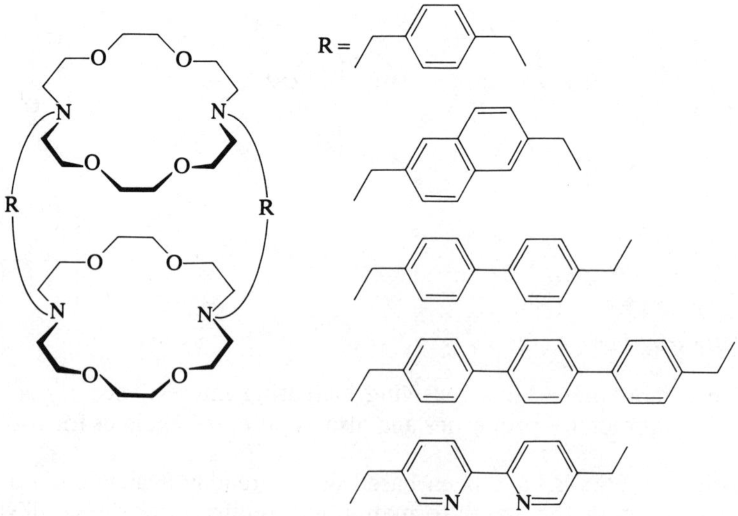

porphyrin

Figure 15 Cylindrical cryptands

The synthesis of unsymmetrical, cylindrical cryptands necessitates a stepwise construction of the tricycle. This is best achieved by initially mono-protecting the diaza crown ether (**41**) as its monobenzyloxycarbonyl derivative ⟨77NJC77⟩.

Several cylindrical cryptands based on other basic crown ether units containing different numbers of oxygen or heteroatoms have been prepared ⟨80T461, 80PAC2441⟩. The synthetic strategies are generally the same.

5.21.5 APPLICATIONS OF CROWN ETHERS AND CRYPTANDS

Interest in crown ethers has exploded since the initial discovery by Pedersen. With the addition of cryptands and other macropolycycles to the armory of coordination chemists there has been feverish activity to seek out new areas in which new and original uses can be devised. Now all branches of chemistry, polymer science, pharmaceutics and industrial processes have been touched by the crown ether revolution.

5.21.5.1 Biological Models

5.21.5.1.1 Binding sites

The property of chiral recognition shown by asymmetric crown ethers towards chiral substrates clearly mirrors the asymmetry of natural interactions and processes. This field is covered in Section 5.21.3.2.2.

The ability of cyclic ethers to complex biologically important alkylammonium cations makes the choice of crown ethers as enzyme binding site models a natural one. In recent years a number of molecules containing both a crown ether-based substrate binding site and a potentially reactive group have been prepared as models for enzyme active sites ⟨79PAC979, B-82MI52100⟩.

Lehn and coworkers have profitably employed tartaric acid-containing crown ethers as enzyme models. The rate of proton transfer to an ammonium-substituted pyridinium substrate from a tetra-1,4-dihydropyridine-substituted crown ether was considerably enhanced compared to that for a simple 1,4-dihydropyridine. The reaction showed first order kinetic data and was inhibited by potassium ions. Intramolecular proton transfer from receptor to substrate was thus inferred *via* the hydrogen bonded receptor–substrate complex shown in Figure 16a ⟨78CC143⟩.

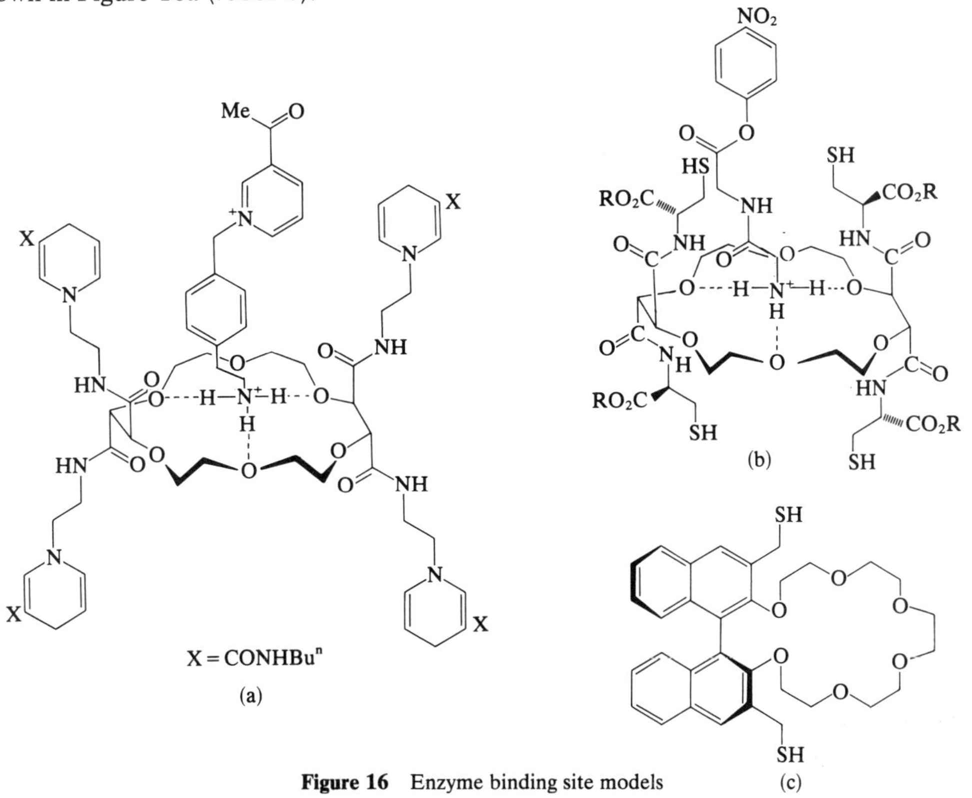

X = CONHBu^n

(a)

(b)

(c)

Figure 16 Enzyme binding site models

A related tetrathiol-substituted crown ether exhibited rate acceleration factors of up to 1.5×10^4, compared with simple thiols, for transacylation reactions with dipeptide *p*-nitrophenyl esters ⟨78CC949⟩. An intramolecular thiolysis was suggested from the kinetic data *via* a receptor–substrate complex (Figure 16b). This interesting model for the protease papain also showed considerable chiral selectivity for amino esters of (L) configuration. A related binaphthyl-based, thiol-substituted crown ether from Cram's group ⟨76JA1015⟩ displayed similar properties of rate acceleration and chiral recognition in amino ester thiolysis reactions (Figure 16c). However, since in both systems the intermediate thiol esters are stable, a subsequent rapid deacylation of the thiol ester must be devised before truly catalytic esterase activity can be claimed. A third thiol-substituted crown ether system has been prepared by a Japanese group ⟨78TL1115⟩.

Recently an unsymmetrical cylindrical cryptand with a porphyrin derivative as one of the side arms has been prepared ⟨81UP52100⟩. Long chain alkyldiammonium picrate salts (*e.g.* C_9) bind inside the cavity and the NMR peaks due to the central methylene protons of the substrate, held above the porphyrin anisotropic ring current, show large upfield shifts of more than 3 p.p.m. (Figure 17a). This model system represents the active site of the important monooxygenase enzyme cytochrome P450 which is known to contain a metalloporphyrin with an adjacent substrate binding site.

1,4-Dihydropyridine units have been incorporated into the macrocyclic framework of crown ethers (*e.g.* Figure 17b) and employed as models for the reactions catalyzed by alcohol dehydrogenases. In the presence of Mg^{2+} ions these molecules reduce carbonyl groups to alcohols presumably *via* a complex in which the substrate carbonyl oxygen binds to the crown-chelated Mg^{2+} ion (Figure 17c). When asymmetric centers are incorporated into the ring impressive enantiomeric excesses of the alcohol product are obtained ⟨81JA2091⟩. Since the initial 1,4-dihydropyridine receptor can be easily regenerated by dithionite reduction of the pyridinium ion product, this system may become a useful catalytic method for the asymmetric reduction of carbonyl groups.

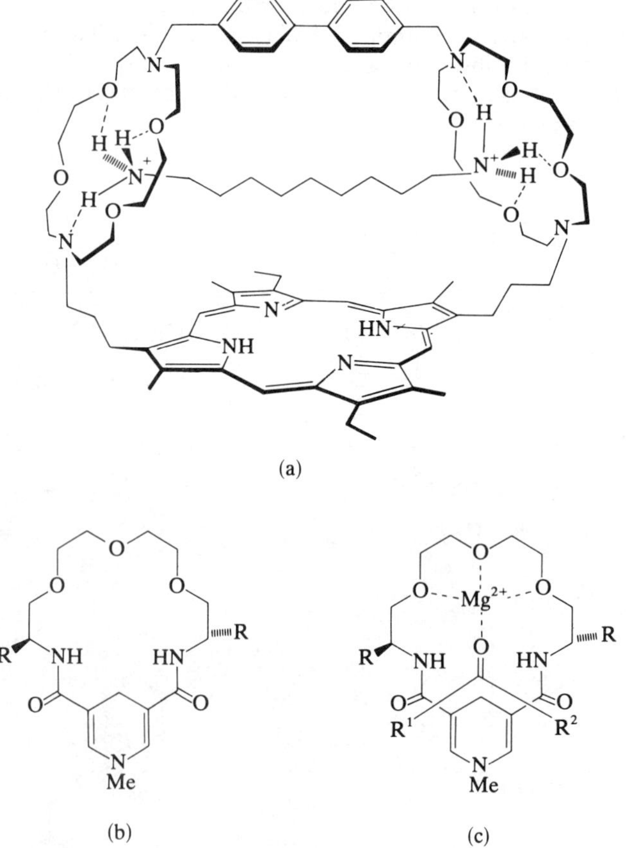

Figure 17 (a) Crown ether based cytochrome P450 model; (b) alcohol dehydrogenase model; (c) proposed transition state in carbonyl group reduction by 1,4-dihydropyridine-containing crown ethers

5.21.5.1.2 Membrane transport

Somewhat before Pedersen's discovery of crown ethers it had been recognized that nature possesses both cyclic and acyclic molecules that bind selectively to the various alkali cations and transport them across membranes. Our knowledge of the structure and chemistry of these biological ion carriers, known as ionophores, has since increased greatly and has been reviewed recently ⟨B-79MI52100, B-81MI52107, B-82MI52101⟩. The structural resemblance between ionophores and crown ethers was soon recognized and research into the membrane transport properties of synthetic macrocycles was commenced.

The mechanism by which cations are transported across a membrane is represented in Figure 18a. A cation–carrier complex is initially formed at the interface. This lipophilic species then diffuses across the membrane as an ion pair and dissociates at the other interface to water soluble ion pair and membrane-soluble carrier. The final step is back diffusion of the free carrier to the initial interface. The factors which influence transport rates and selectivity have been the subject of much research ⟨79PAC979, B-81MI52102⟩.

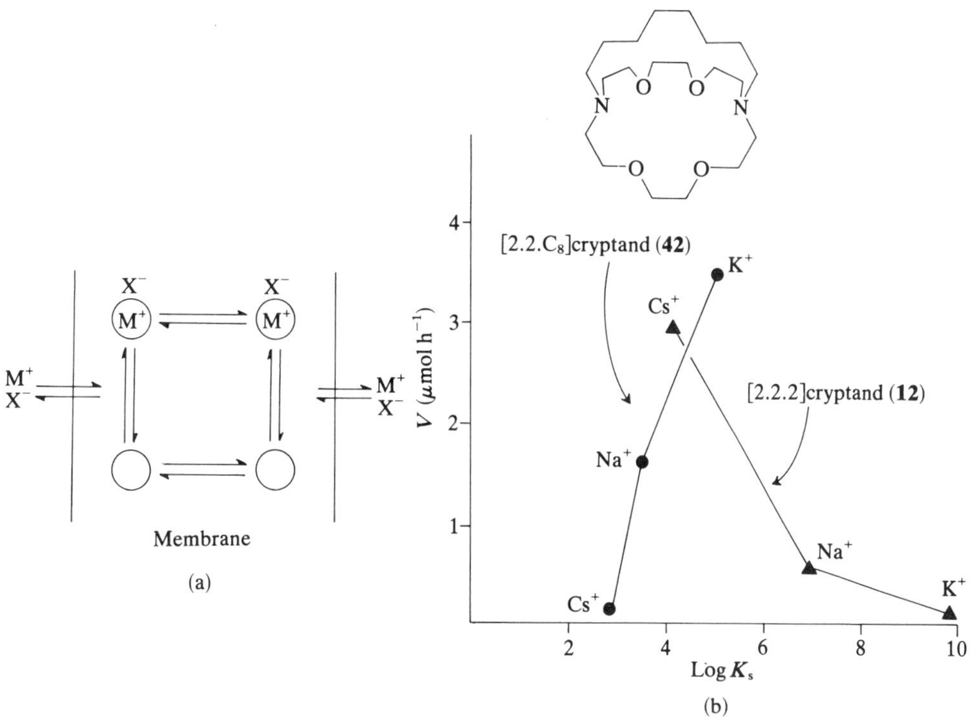

Figure 18 (a) Mechanism of ion pair transport mediated by a cation carrier; (b) plot of initial transport rates (V) of cation picrates as a function of the logarithm of the stability constants

Because of the complex equilibria involved (Figure 18a), transport rates depend on the delicate balance of many factors including complexation properties and lipophilicity of the carrier, the cation and anion being transported, and the nature of the membrane species itself. Importantly, transport rates are not directly proportional to the cation–carrier complex stability but present a maximum as a function of K_s. If the complex stability is too low, insufficient cation will be complexed at the initial interface and, similarly, if the complex stability is too high, insufficient cation will be released at the opposite interface. A compromise between thermodynamics (stability) and kinetics (exchange rates) of complexation is involved.

French ⟨75AG(E)555⟩ and American ⟨80JA6820⟩ groups have studied the transport of cations across a simple chloroform membrane phase using a wide range of synthetic macropolycycles as carriers. They found that, for those particular conditions, the carrier–cation stability constant which gave optimum rates of cation transport was approximately 10^5 (determined in methanol and assumed to be the same in chloroform). The experimental comparison ⟨79PAC979⟩ of two cryptands (**12**) and (**42**) with different cation complex stability constants and their respective transport rates (V) is shown in Figure 18b.

In an elegant series of experiments benzo[15]crown-5 has been connected to a second crown ether or a polar group *via* a photoresponsive diazo linkage ⟨82JA1967⟩. In the *cis* isomer both halves of the molecule participate in cation binding and, consequently, the complex stability constant is higher than in the *trans* isomer (Figure 19). Since the *trans* to *cis* isomerization may be achieved by light irradiation the rate and specificity of cation transport across membranes can be photocontrolled.

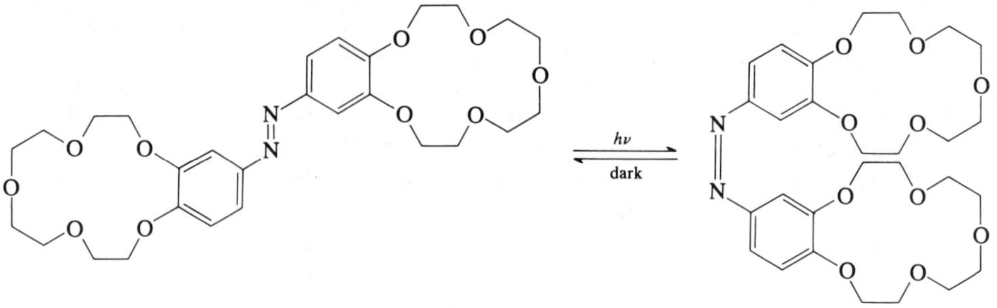

Figure 19 Photochemical control of cation transport across membranes

Apart from alkali metal cations [18]crown-6 derivatives can complex and transport primary alkylammonium cations across chloroform membranes ⟨80JCR(S)136⟩. This process is particularly relevant as many amines (*e.g.* amphetamine) are physiologically important species.

5.21.5.2 Organic Synthesis

5.21.5.2.1 Anion activation

In complexation, crown ethers and cryptands convert hydrophilic alkali or alkaline earth metal ions into large hydrophobic cationic species soluble in solvents of low polarity. Whilst the cation is surrounded by its organic ligand the counteranion which accompanies it is only weakly solvated. The 'naked' anion is highly reactive since it is not strongly solvated and, as a consequence of its smaller size, possesses a higher effective charge. This property of anion activation by crown ethers and cryptands has been widely exploited in organic synthesis ⟨76S168, B-78MI52104, 80PAC2303⟩.

The enhanced nucleophilicity of weakly solvated fluoride ions, solubilized in non-polar solvents as their alkali metal salts by [18]crown-6, has been studied. The wide range of S_N2 reactions possible with this system is illustrated in Table 3. Under equivalent conditions in the absence of crown ether no substitution occurs. Similar effects are seen with many nucleophiles which, even if soluble in the solvent employed, show increased nucleophilic substitution rates in the presence of crown ethers ⟨B-78MI52104⟩. However, the monocyclic crown compound exposes the cation on two sides to approach by the counteranion (see Figures 1b, c and d for illustrations of this effect in the crystalline state). The resultant ion pairs that form in non-polar solvents reduce the reactivity of the anion.

Table 3 [18]Crown-6 Complexed Alkali Metal Fluoride Reactions ⟨74JA2250⟩

Substrate	*Solvent*	*Product* (% *yield*)
Benzyl bromide	MeCN	Benzyl fluoride
1-Bromooctane	MeCN	1-Fluorooctane (92%), 1-octene (8%)
	C_6H_6	1-Fluorooctane (92%), 1-octene (8%)
2,4-Dinitrochlorobenzene	MeCN	2,4-Dinitrofluorobenzene

In macrobicyclic cryptate complexes where the cation is more efficiently encapsulated by the organic ligand these ion pair interactions are diminished and the reactivity of the anion is enhanced. This effect is seen in the higher dissociation constant, by a factor of 10^4, of Bu^tOK in Bu^tOH when K^+ is complexed by [2.2.2]cryptand (**12**) compared to dibenzo[18]crown-6 (**2**). The enhanced anion reactivity is illustrated by the reaction of the hindered ester methyl mesitoate with powdered potassium hydroxide suspended in benzene.

The hydrolysis is much faster (12 h at 25 °C) in the presence of [2.2.2]cryptand (**12**) than in the presence of crown ethers (21 h at 73.8 °C) ⟨80PAC2303⟩.

As well as increasing anion nucleophilicity, crown or cryptand complexation can enhance the basicity of the anion. Table 3 exemplifies this effect with 1-bromooctane where base-promoted elimination to 1-octene competes with nucleophilic substitution. Being small and poorly solvated, 'naked' fluoride is a strong and hard base which causes, in the case of certain substrates (*e.g.* Scheme 6), the elimination product to predominate. As the 'naked' anions increase in size they display less basic characteristics but retain high nucleophilic reactivity ⟨74JA2250⟩.

O Cl Me —K^+F^-, [18]crown-6, MeCN→ O Me (64%) + O Me F (31%)

Scheme 6

5.21.5.2.2 Cation inhibition

Just as anion reaction rates increase, it would be expected that the rates of reactions in which cations are involved would decrease on crown or cryptand complexation ⟨80PAC2303⟩. This effect of cation inhibition is particularly seen in nucleophilic additions to carbonyl groups. The rates of reduction of carbonyl groups by lithium aluminum hydride and sodium borohydride are decreased by the presence of [2.1.1]-and [2.2.2]-cryptands (**15b** and **d**, respectively). The extent to which reaction at a carbonyl group is slowed down by complexation thus becomes a measure of the participation of the cation in the reaction mechanism. For example, α,β-unsaturated carbonyl compounds undergo fast 1,2-addition of hydride when treated with $LiAlH_4$. In the presence of [2.1.1]cryptand the lithium activation effect is diminished, the reaction is slower and gives predominantly 1,4-addition ⟨78TL2571, 80T1937⟩.

The balance between anion activation and cation inhibition of addition reactions to carbonyl groups is exemplified by the reaction of acetonitrile anion with benzaldehyde (Scheme 7). When the cation is Li^+, addition of [2.1.1]cryptand (**15b**) markedly decreases the reaction rate suggesting cation inhibition is dominant. When $M=K^+$, addition of [2.2.2]cryptand (**15d**) increases the rate by, presumably, a predominant anion activation. Finally when $M=Na^+$, the two effects cancel out and the reaction rate is unaffected by [2.2.1]cryptand (**15c**) addition ⟨80PAC2303, 81TL1685⟩.

$$PhCHO + M^+ (CH_2CN)^- \rightarrow PhCHO^- {-} CH_2CNM^+$$

$$M^+ = Li^+, Na^+, K^+$$

Scheme 7

5.21.5.2.3 Reagent solubilization and phase transfer catalysis

Crown ethers and cryptands can solubilize certain normally insoluble reagents in solvents of low polarity. Problems of cosolubility between reagent and reactant can thus be overcome and the reaction conducted under milder conditions.

The now classic example of this type, excluding nucleophilic substitution reactions, is the solubilization of potassium permanganate in benzene by [18]crown-6 derivatives. The resulting solution proved a mild and efficient oxidizing agent for many organic compounds. For example, α-pinene is oxidized to *cis*-pinonic acid in 90% yield (Scheme 8) in contrast to a 40–60% yield in aqueous solution. An additional advantage is that the products of these reactions (*e.g.* potassium salt of *cis*-pinonic acid in Scheme 8) are frequently insoluble in the hydrocarbon solvents and consequently are unable to react further.

The dissolution of potassium permanganate involves a solid–liquid phase transfer with the flat macrocycle, presumably, lining up in the correct complexation geometry on the surface of the crystal lattice. The movement of cation from lattice to ligands is then small.

Scheme 8

Crown ethers and cryptands have also been widely used in liquid–liquid phase transfer catalysis ⟨B-78MI52104, B-82MI52103⟩. They have an advantage over the more usual catalysts (tetraalkylammonium, phosphonium ions, *etc.*) in that, being neutral, no extraneous counter-anions, and potentially side-reacting nucleophiles, are added to the reaction mixture. Furthermore, the lipophilicity of the catalyst can be readily enhanced by the introduction of cyclohexyl rings or alkyl groups onto the exterior of the macropolycycle ⟨76S168⟩. To their disadvantage crown ether and cryptand phase transfer catalysts are expensive to synthesize. In a recent comparative study of all the commercially available phase transfer catalysts, cryptands and crown ethers were pronounced cost ineffective and 'onium' salts and poly(ethylene glycols) the catalysts of choice ⟨80JA4810⟩.

5.21.5.2.4 Control of reaction mechanisms and their regio- and stereo-chemistry

It is not unexpected that reactions involving polar intermediates can be affected by the addition of cationic complexing agents. The effect of cation inhibition on the regioselectivity of $LiAlH_4$ reductions of α,β-unsaturated carbonyls was discussed in Section 5.21.5.2.2. Similarly, the course of alkylation reactions can be influenced by complexing agents. The potassium enolate of ethyl acetoacetate normally takes up the (Z,Z) configuration (Scheme 9) stabilized by bidentate chelation to the cation, and C-alkylation takes place with ethyl iodide. On addition of [2.2.2]cryptand the K^+ is complexed and the (Z,Z) form is destabilized relative to the (E,Z) and (E,E) forms. The oxygen atoms thus become more accessible and reaction with ethyl iodide gives a C- to O-alkylation ratio of 4:1. Furthermore, anion activation accelerates the reaction rate by a factor of 4000 ⟨76TL281⟩.

Scheme 9

In base-promoted β-elimination reactions the size of the approaching base greatly influences the course of the reaction. Of the two possible transition states for β-elimination of 2-butyl tosylates, (**43**) and (**44**), only the former can accommodate a large basic species. In t-butyl alcohol, potassium t-butoxide forms large ion aggregate structures. The elimination of 2-butyl tosylates with Bu^tOK thus favors transition state (**43**) and *cis*-butene is formed. On addition of a crown ether the aggregates are broken up and the smaller 'naked' t-butoxide ion becomes the active base. As a result the proportion of *trans*-butene formed markedly increases ⟨75ACR239; for a differing explanation, see 80CCC3150⟩. Similarly in competing *syn*- and *anti*-β-elimination processes, addition of a suitable crown ether will prevent the formation of the six-membered, ion pair transition state (**45**) necessary for *syn*-elimination and *anti*-elimination will predominate ⟨B-81MI52101⟩.

(43) **(44)** **(45)**

Other commonly used synthetic routes to alkenes are also affected by crown ethers. The Wittig ⟨80TL4831⟩, Wittig–Horner and Wadsworth–Emmons ⟨81S117⟩ reactions all give pure *trans*-alkene products in higher yields in the presence of crown ethers.

The ability of [18]crown-6 derivatives to complex primary alkylammonium ions has been elegantly exploited in the protection of primary amines ⟨80CC300⟩. In the presence of primary amines, secondary amines can be acylated selectively by adding [18]crown-6 and a proton source. This strategy has obvious advantages over normal amine protecting groups which require a deprotection step.

In an important new application of crown ethers Cram and Sogah have recently reported that potassium bases complexed to chiral crown ethers catalyze the stereoselective Michael addition of a β-ketoester to methyl vinyl ketone in high optical yields ⟨81CC625⟩. With chiral crown (**46**), carbanion (**47**) gave alkylated products with an optical yield of about 99% enantiomeric excess. These impressive results were rationalized by complex structure (**48**) in which the crown-complexed K^+ and the carbanion form an ion pair. One face of the associated carbanion is shielded from electrophilic attack by the flanking binaphthyl groups and the approach of methyl vinyl ketone occurs in a stereoselective manner.

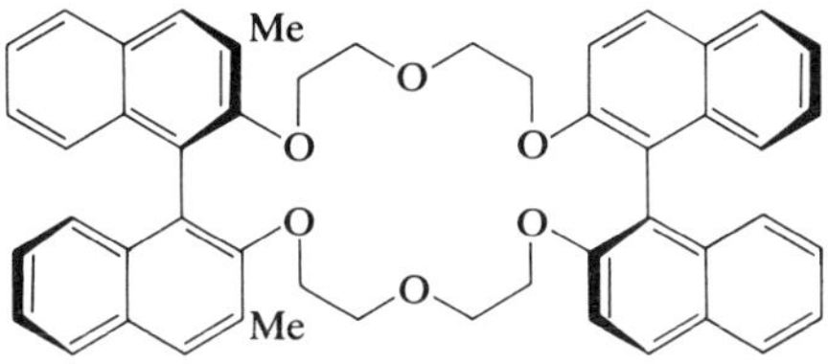

(**46**)

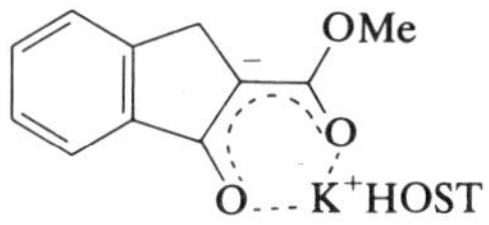

(**47**) 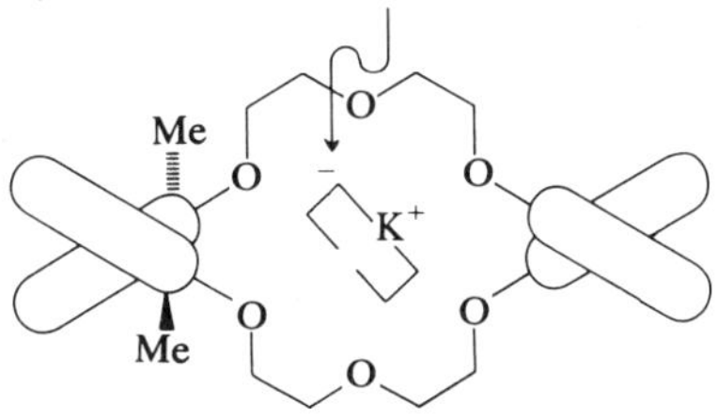 (**48**)

For another example of stereoselective synthesis with chiral crowns see Section 5.2.1.5.1.

5.21.5.3 Analytical Applications

5.21.5.3.1 Separation methods

The selective cation binding properties of crown ethers and cryptands have obvious commercial applications in the separation of metal ions and these have recently been reviewed ⟨B-78MI52103, 79MI52102, B-81MI52103⟩. Many liquid–liquid extraction systems have been developed for alkali and alkaline earth metal separations. Since the hardness of the counterion is inversely proportional to the extraction coefficient, large, soft anions, such as picrate, are usually used.

Interestingly, soluble polymeric crown ethers can have extraction coefficients up to 250 times larger than those of the corresponding monomers. Macromolecules such as polyvinyl [15]crown-5 and polyvinyl [18]crown-6 are readily soluble in organic solvents and many extraction systems involving their use have been developed ⟨B-81MI52103⟩.

Crown ethers have been shown to separate not only different metal ions but also different isotopes of the same metal ion. In a liquid–liquid (organic–aqueous) extraction of a mixture of ^{40}Ca and ^{48}Ca, dicyclohexyl[18]crown-6 was found to enrich the organic phase with ^{40}Ca with a separation factor of 1.008 ± 0.0016 ⟨76MI52100⟩. Similar enrichments have been reported using [2.2.1]- and [2.2.2]-cryptand as the complexing agents ⟨80AG(E)406, 81ZN(B)566⟩. The separation of sodium isotopes, ^{22}Na and ^{24}Na, has been achieved, to varying degrees, by a large number of crown ethers and cryptands in a chloroform–water system ⟨81JA5707⟩. A chloroform solution of [2.2.1]cryptand has also been used preferentially to extract ^{6}Li ions relative to ^{7}Li ions from aqueous solution with a separation factor of 1.036 ⟨79MI52106⟩.

Crown ethers and cryptands can also be used as the stationary phase in column chromatography. A wide range of chromatographic systems with different cation selectivities have been generated by absorbing the complexing agent on a solid support, such as silica gel, by dissolving it in the eluent or, more usually, by preparing the appropriate polymeric crown ether or cryptand ⟨B-81MI52103⟩. For example, the methylene-linked polymer formed by treating dibenzo[18]crown-6 with formaldehyde is a versatile and widely used stationary phase which preferentially binds K^+ among the alkali metals and separates Ra^{2+} from other divalent cations. Small amounts of $^{90}Sr^{2+}$ can also be separated from larger amounts of Ca^{2+} and so the technique is important in determining $^{90}Sr^{2+}$ levels from fall-out in milk. Primary, secondary, tertiary and quaternary ammonium chlorides are also separated from each other by this system.

The enantioselective binding properties of certain chiral crown ethers have been employed in the resolution of amino acid racemates. The racemic amino ester is adsorbed onto silica gel as its ammonium salt and eluted by a chloroform solution of the chiral crown ether. An excellent separation of the two enantiomers is achieved by this method ⟨74JA7100⟩.

5.21.5.3.2 Determination methods

A most important application of crown ethers and cryptands in analytical chemistry has been in the development of ion-selective electrodes. By incorporating different complexing agents into the liquid membrane which separates a reference electrode from the solution under study it is possible to measure selectively the concentration of different cations. The selectivity of the complexing agent determines which cation's concentration is measured ⟨B-79MI52101, B-81MI52103⟩. Earliest electrodes of this type utilized the natural ionophore valinomycin and displayed a selectivity for K^+ over Na^+ that was 200 times greater than simple glass membrane electrodes. Synthetic crown ethers and cryptands have also been incorporated into ion-selective electrodes. Simon and coworkers have prepared a series of acyclic polydentate ligands which bind cations in a similar way to crown ethers. Using these readily available species he has developed commercially important ion-selective electrode systems which can simultaneously and continuously measure Na^+, K^+, Ca^{2+} and β-D-glucose concentrations in the blood ⟨77C404⟩.

5.21.5.4 Polymer Science

The use of crown ethers and cryptands on a large scale for industrial purposes is inhibited by their expense. A potentially useful way around this problem lies in attaching the complexing agent to a polymeric backbone and, thus, facilitating its retrieval. Large numbers of both soluble and insoluble polymers containing crown ethers and cryptands have been developed and have been the subject of a recent review ⟨B-81MI52109⟩. Their utilization in liquid and solid phase separation techniques has been discussed in Section 5.21.5.3.1.

Crown ethers and cryptands have also found widespread use as modifiers in the polymerization process ⟨79MI52104, 81MI52110⟩. Anionic polymerization is strongly affected by the addition of crown ethers and cryptands which cause increases in the polymerization rate and in the yield and molecular weight of the final polymer.

5.21.5.4 Pharmacology

There is currently considerable interest in the potential of crown ethers and cryptands as drugs ⟨B-79MI52107⟩. Their properties of selective metal ion complexation may allow the removal of poisonous heavy metal ions (*e.g.* Hg^{2+} and Pb^{2+}) from an organism without disrupting the delicate balance of physiologically important metal ions. Treatment of lead-poisoned rats with [2.2.2]cryptand (**12**; 10 mg kg^{-1}) has been shown to increase the concentration of lead excreted in the urine by an average factor of 40 ⟨77MI52100⟩. Similarly encouraging results have been seen in the detoxification of ^{85}Sr-poisoned rats by [2.2.2]cryptand (**12**) ⟨70N248⟩.

A major problem in the development of crown ethers and cryptands as drugs is their alleged toxicity. Pedersen has reported that dicyclohexyl[18]crown-6 exhibits significant toxicity to rats with an acute lethal dose, for oral ingestion, of 300 mg kg^{-1} ⟨67JA7017⟩. This problem may, however, be overcome by incorporating appropriate substituents into the complexing agent.

Acknowledgment

The author wishes to thank Professor Jean-Marie Lehn for his help in the preparation of this chapter.

5.22

Heterophanes

G. R. NEWKOME, J. G. TRAYNHAM and G. R. BAKER
Louisiana State University, Baton Rouge

Chapter 5.22 differs in organization from the other chapters in Part 5. Both the nature of the subject and the state of the art made it impractical to deal with the material in the customary order: structure, reactivity, synthesis and applications. The boundless diversity of possible heterophanes leads to a commensurate variety of reactivities, but these reactivities have not yet been broadly explored. Most heterophanes were prepared to study particular effects, and the methods of preparation provide the bulk of our knowledge in this field. The art of making heterophanes has, therefore, been chosen as the first section of the chapter, followed by much briefer sections on reactivity and stereochemistry.

5.22.1 INTRODUCTION

For this review, heterophanes are considered to contain one or more heteroaromatic ring(s) bridged by a non-aromatic chain of atoms. Although a certain arbitrariness identifies (**1**) but not (**2**), as a heterophane, this definition focuses attention on the chemistry appropriate for this review and makes the volume of literature to be reviewed more manageable. Different nomenclature approaches for heterophanes have been proposed ⟨70T5847, 72T5183, 72TL2109⟩ as alternatives for the currently accepted, complex IUPAC or *Chemical Abstracts* names, which actually obscure the phane structural element in these compounds. A proposal

CHC 7-Y*

based on IUPAC replacement nomenclature is perhaps the simplest; the heterocyclic moieties are treated as replacements of single atoms in a monocyclic parent compound ⟨72TL2109⟩. In this chapter, heterophanes will generally be identified by reference to formula representation rather than complex names. Because 'crown ethers' with (sub)heteroaromatic rings are treated in Chapter 5.21 and have been reviewed ⟨77CRV513, 79CRV37, B-79MI52200, B-78MI52201, 83HC(14-S2)⟩, this expansive topic is not included here.

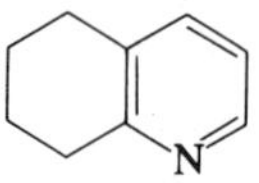

(1) 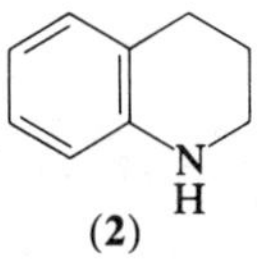 (2)

Pioneering investigations of cyclophanes were apparently prompted by the synthetic challenge of unusual, strained systems and intriguing questions about structure perturbations in them. While these same stimuli have played roles in extensions of phane chemistry to incorporate heteroaromatic systems, they have been joined by others, *e.g.* the identification of naturally occurring heterophanes, some with substantial biological properties, and the potential utility of heterophanes as synthons of macrocyclic compounds or as novel ion complexing reagents.

The structures of a series of antibiotics (virginiamycin family), each possessing a 2,4-oxazole subunit bridged by a 20-membered chain, have been established ⟨77MI52200⟩; Virginiamycin M (**3**) is representative of this family ⟨82TL707⟩. Other natural products include a furanophane (**4**), isolated from an octacoral ⟨75T2513⟩, and several pyrrolophane pigments, illustrated by (**5**) and (**6**), isolated from one or more of the species *Streptomyces*, *Actinomedura* and *Streptovelicillum* ⟨76T1855, 78CJC1115⟩. Samples of (**5b**) and (**6**) multi-labelled with ^{13}C have been biosynthesized, and a possible biogenetic route to these pyrrolophanes from acetate through proline has been proposed ⟨78CJC1155⟩.

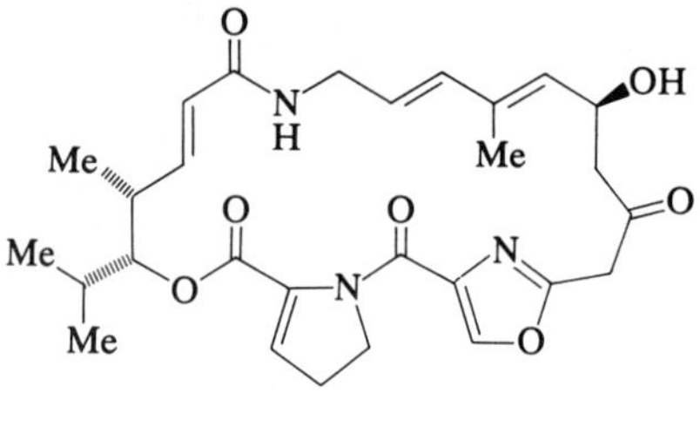

(**3**) 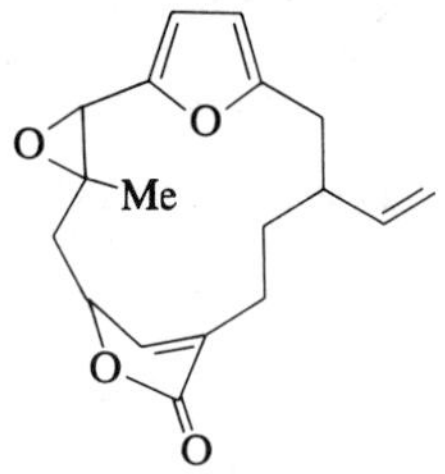 (**4**)

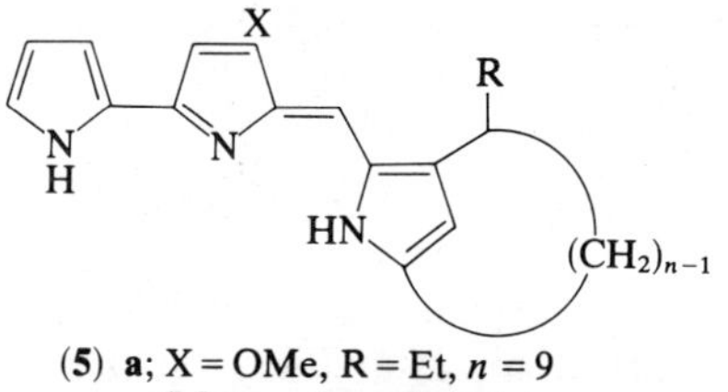

(**5**) **a**; X = OMe, R = Et, n = 9 (Metacycloprodigiosin)
b; X = H, R = Bu, n = 7 (Butylcycloheptylprodiginine)

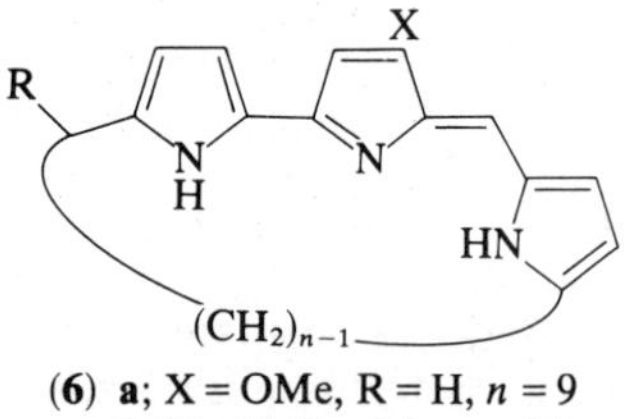

(**6**) **a**; X = OMe, R = H, n = 9
b; X = H, R = Me, n = 10

5.22.2 SYNTHETIC PROCEDURES

5.22.2.1 Synthesis of the Macroring

5.22.2.1.1 Alkyl substituent substitution

Metal–halogen exchange has been commonly used to generate intermediate organometallics, which can subsequently undergo a Wurtz-type cyclocoupling. Thus, phane (**7**) has been prepared from either (**8**) ⟨70CC1558⟩ or (**9**) ⟨58JCS3594⟩ *via* this technique. With electron-deficient heteroaromatics, in which the α-alkyl hydrogens (*e.g.* **10**) are sufficiently acidic, direct hydrogen–metal exchange occurs. A stepwise metallation, followed by an oxidative coupling, usually with copper reagents, gives the desired phane ⟨70AG(E)808⟩. Cyclization of (**10**) with sodium and tetraphenylethylene in THF affords (**7**) along with related oligomers ⟨69C158⟩. Isomers of (**7**) have been similarly prepared ⟨68C139⟩.

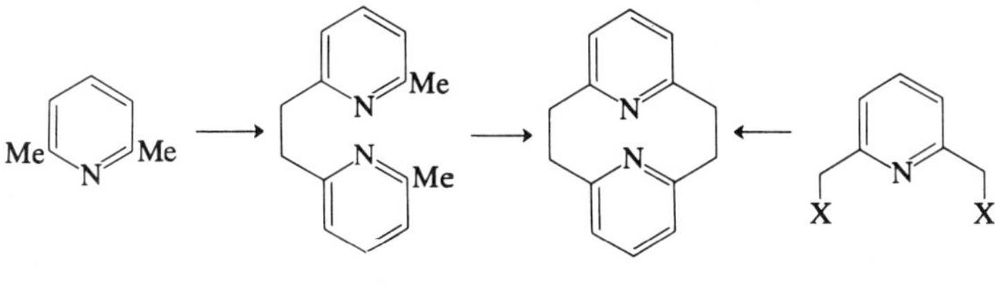

(**10**) (**7**) (**8**)

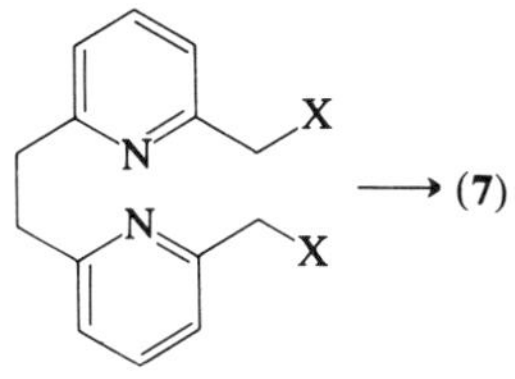

(**9**)

In addition to carbon nucleophiles, oxygen, sulfur ⟨81CL5; 81LA1065⟩, phosphorus and nitrogen anions have been used successfully in this procedure. Carbon–sulfur linkages are the most common bridging combination, since several sulfur expulsion procedures are available (see Section 5.22.3). These reactions are favored by high dilution conditions and exemplified by the formation of (**11**) ⟨70CZ727⟩.

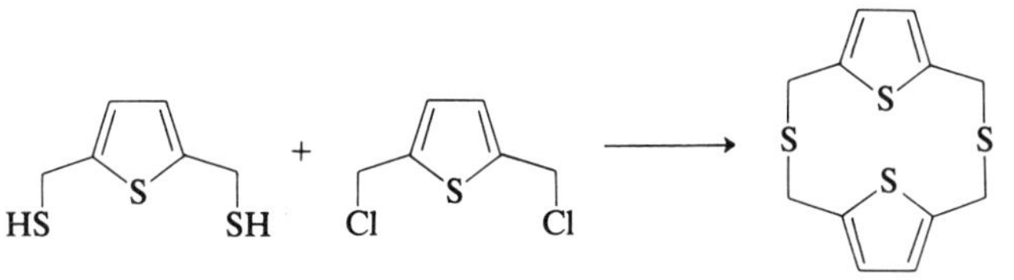

(**11**)

Numerous modifications of this procedure have been used. The synthesis of benzimidazolone monomers (**12**), dimers (**13**) and trimers demonstrate this versatile, simple approach to bridged heterocycles ⟨76TL79⟩. *N*-Alkylated products (*e.g.* **14**) are in many cases the result of side reactions caused by a competitive transmetallation process ⟨78MI52202⟩; polymerization is also a common reaction course.

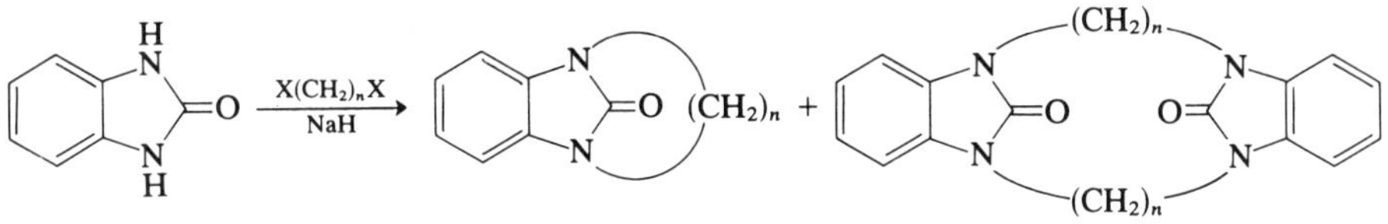

(**12**) (**13**)

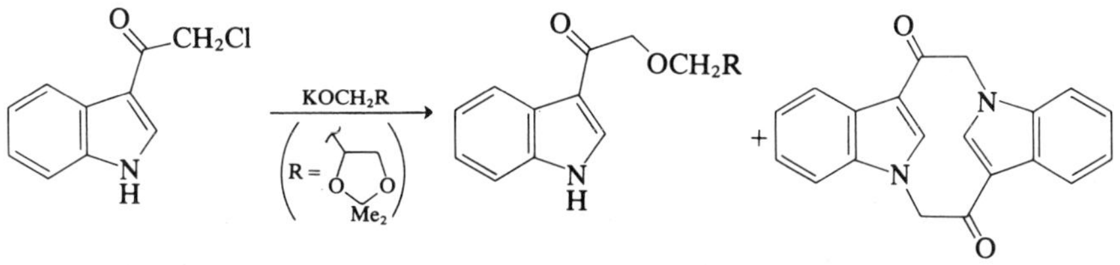

(**14**)

5.22.2.1.2 Heteroaromatic substituent substitution

Although direct nucleophilic ring substitution is unfavored with electron-excessive heteroaromatics, substitution in electron-deficient systems will occur, as illustrated for (**15**), with either carbon nucleophiles and a nickel–phosphine catalyst ⟨75JA4405⟩ or oxygen nucleophiles ⟨77JOC1500, 81JOC833⟩. Inclusion of a 2,4-pyrimidino subunit proceeds smoothly by this direct substitution procedure (**16**) ⟨78JOC3362⟩. Thermal rearrangement of (**16**) gives the corresponding uracil macrocycle, (**17**), which can be synthesized by an alternative procedure ⟨76TL469⟩. Pyrimidinopurinophanes ⟨81TL4101⟩, purinophanes ⟨81TL1123⟩ and quasimetacyclophanes derived from 5-alkyluracils ⟨78MI52200, 76IZV1103⟩ have been prepared similarly.

$BrMg(CH_2)_nMgBr$, Ni(II); HO⁀OH, NaH

(**15**)

NaH, R = H; Δ; R = Me

(**16**) (**17**)

Utilization of oxidative coupling reactions has been reported ⟨74AG(E)291, 78AG(E)755⟩ for the construction of novel heterocyclic arenes such as (**18**).

(**18**)

5.22.2.1.3 Schiff base cyclizations

The application of Schiff base chemistry to the synthesis of heterophanes possessing diverse subunits has been reviewed in part ⟨82MI52202, B-79MI52201⟩. An example of this procedure is the treatment of (**19**) with 2,3-diaminopyridine, with or without metal ion templation, to give (**20**) as an isomeric mixture ⟨75MI52200⟩.

(**19**)

(**20**) X ≠ Y = CH or N

5.22.2.1.4 Electrophilic heteroaromatic substitution

Intramolecular acylations of substituted thiophenes have been used for over 20 years as an entrance into thiophenophanes ⟨63T1851⟩, and the procedure has recently been extended to substituted benzothiophenes ⟨80JOC311⟩. Friedel–Crafts cyclization of long-chain ω-(2-thienyl)alkanoic acids affords the (2,5)thiophenophanes (*e.g.* **21**), in which the free α-position is the preferred site of substitution. If both α-positions are occupied, β-substitution will occur. When the side-chain is short ($n<8$), electrophilic substitution at the α-position of another molecule is favored over intramolecular attack at the adjacent β-position; dimerization thus generates cyclophane (**22**).

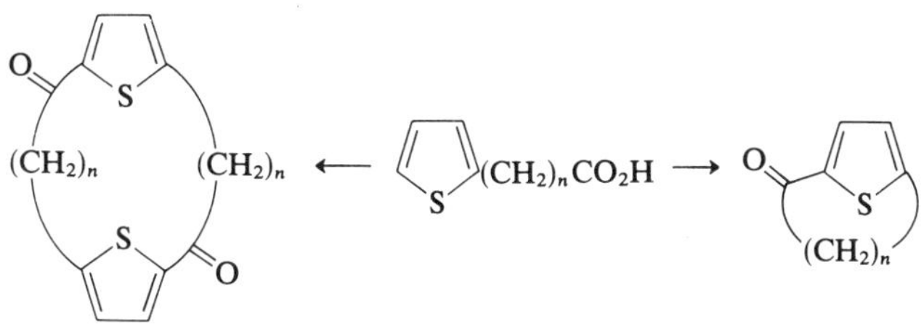

(**22**) (**21**)

The 'one-step', simple acid-catalyzed cyclocondensation of electron-excessive heteroaromatics is a low yield process; however, with metal ions added, enhanced yields have been reported ⟨77CRV513⟩. Pyrrole ⟨1886CB2184⟩, furan ⟨71CJC4017⟩ and thiophene ⟨69TL1493⟩ with acetone in the presence of acid give (**23**) in reasonable yield; if, however, pyrrole is treated with benzaldehyde and propionic acid, a nearly quantitative yield of *meso*-tetraphenylporphin (**24**) results ⟨82MI52200⟩. Incorporation of other heteroaromatics such as tellurophene, selenophene, pyrrole and furan, alone and in combinations, has been accomplished ⟨78TL1885, 167, 79JCS(P1)1066⟩.

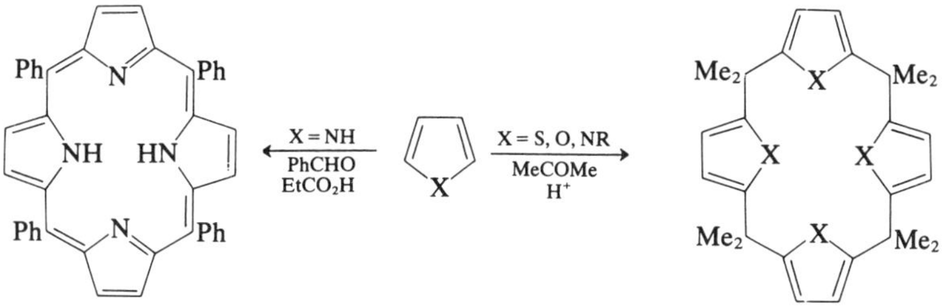

(**24**) (**23**)

Electrophilic cyclodimerization (to **26**) also results when (**25**) is subjected to Bischler–Napieralski conditions ($POCl_3$, reflux), presumably through reaction of pyrazole with the imino chloride intermediate ⟨78JHC1339⟩.

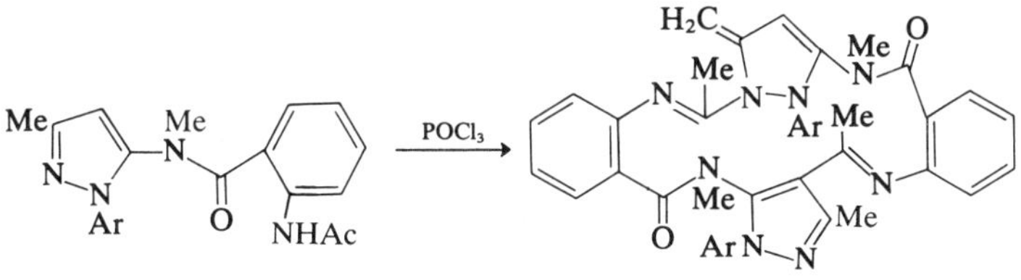

(**25**) (**26**)

5.22.2.1.5 1,6-Elimination–dimerization

The successful synthesis of furanophane (**27**) by pyrolysis of the quaternary ammonium hydroxide (**28**) ⟨60JA1428⟩ has prompted the adaptation of this 1,6-Hofmann elimination procedure to the synthesis of numerous heterophanes ⟨77CRV513, 77H(7)81, 78T1641⟩. Fluoride anion-induced 1,6-elimination of (**29**) or (**30**) gives (**27**) or (**31**) and (**32**), respectively, in good yields ⟨81JOC1043⟩.

Pyridinophanes ⟨72C420⟩ as well as mixed cyclophanes with electron-deficient and -excessive subunits ⟨74JOC2570⟩ have been prepared similarly. Recently, [8+6] and [6+6] cycloadducts (**33**) and (**34**), respectively, have been synthesized by means of a cross-dimerization of vinyl-*p*-xylylene (**35**) with heteroxylylene-type intermediates (**36**) ⟨78TL415⟩.

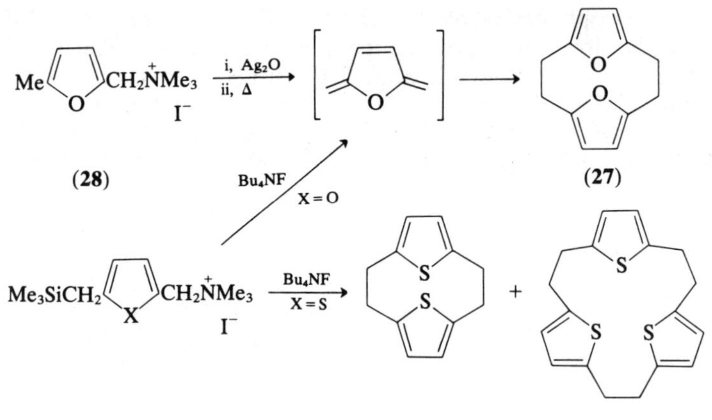

(28) (27)

(29) X=O
(30) X=S
(31) (32)

Thermolysis of (**34**) gives (**33**). Pyrolyses of *o*-(chloromethyl)toluenes give *o*-xylylenes, which thermally dimerize to [2]cyclophanes ⟨80ACR65⟩. This technique has been successfully applied to the preparation of multibridged pyridinophanes ⟨81AG(E)571⟩.

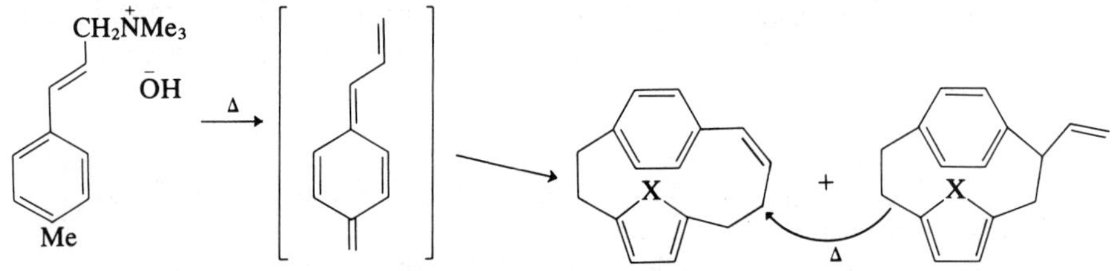

(35) (33) (34)

(36)

5.22.2.1.6 Oxidative cyclization

Thiols under mild oxidative conditions are converted to disulfides, and heating (**37**) with or without an additional oxidizing agent in DMSO gives (**38**) in 82% yield ⟨76JAP(K)08275⟩. Similarly with iodine, 2,4-pyrimidinedithiol affords an isomeric mixture of sulfur-bridged pyrimidinophanes (**39**) and (**40**) ⟨80KGS848⟩.

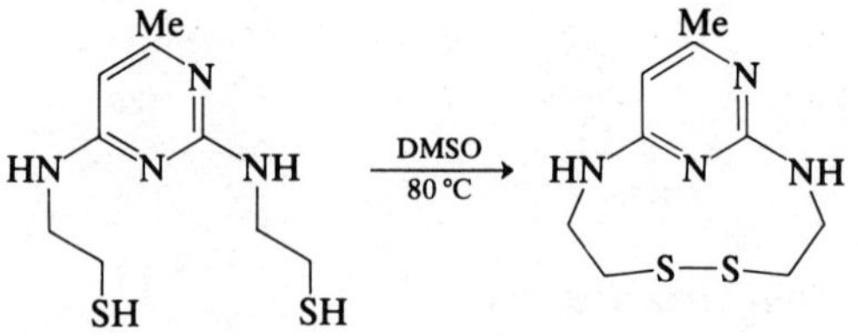

(37) (38)

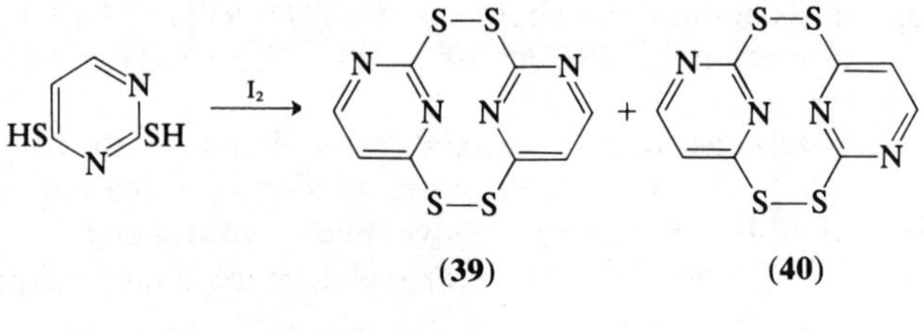

(39) (40)

5.22.2.1.7 Lactone/lactam formation

Under high dilution conditions ⟨81JCR(S)314⟩ bis-acyl chlorides react with diols to generate the corresponding tetralactones (**41**) ⟨77TL2573⟩. This procedure has most commonly been applied to the synthesis of 'crown ethers' with subunits ⟨79CRV37⟩. Lactams have been prepared similarly from suitable diamines ⟨76TL2603⟩.

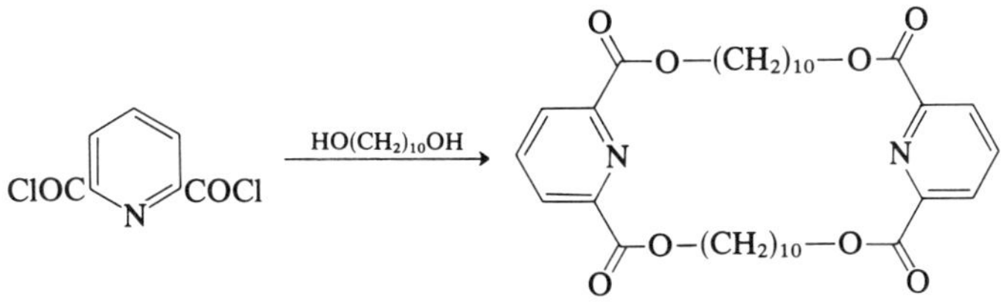

(**41**)

Treatment of (**42**) with sodium hydride in DMSO at room temperature affords the 16-membered ring, bis-amide (**43**) ⟨77JHC561⟩.

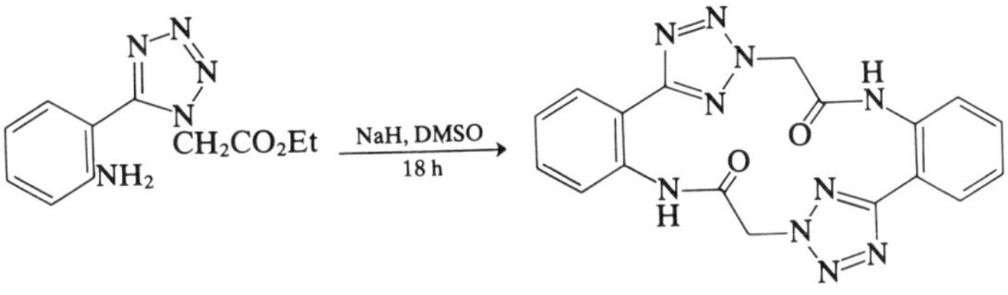

(**42**) (**43**)

5.22.2.1.8 Base-catalyzed condensation

One of the original routes into [18]annulenes (*e.g.* **44**) possessing heterocyclic rings utilized the Perkin condensation (acetic anhydride–triethylamine). This cyclocondensation procedure transformed (**45**) and (**46**) into (**44**) in low yield; the overall results have been reviewed ⟨77CRV513⟩.

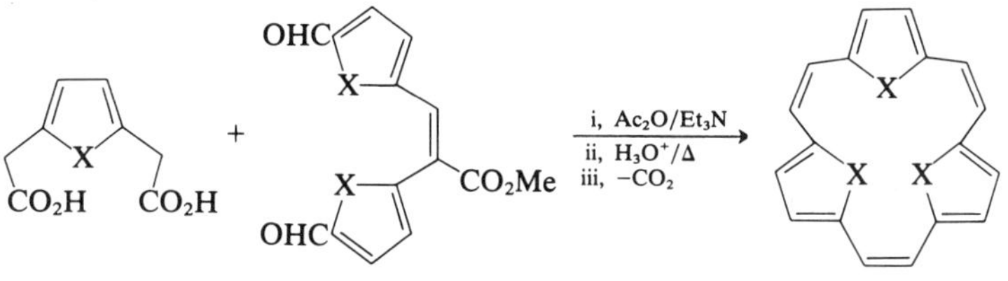

(**45**) (**46**) (**44**)

Nearly all combinations of common electron-excessive heteroaromatics have been achieved by modification of this procedure. However, this procedure has not been successful so far with electron-deficient heteroaromatics ⟨76JOC2536⟩. However, bis-Wittig heterocyclic reagents ⟨75S765⟩ have been condensed with a variety of dialdehydes to generate a series of annulenones (*e.g.* **47**) and annulenes ⟨78HCA2763, for a review see 77CRV513⟩.

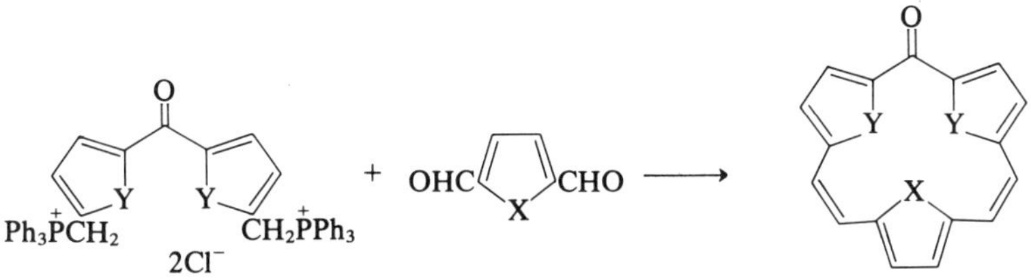

(**47**)

5.22.2.1.9 Miscellaneous procedures

When 2-aminopyrimidine is treated with 1,3-dichloroacetone and sodium cyanide, an unusual condensation dimer (**49**) is formed along with the anticipated product (**48**). The addition of cyanide to (**50**) affords isomeric anions (**51**) and (**52**) that can combine to give (**49**) ⟨76FES731⟩.

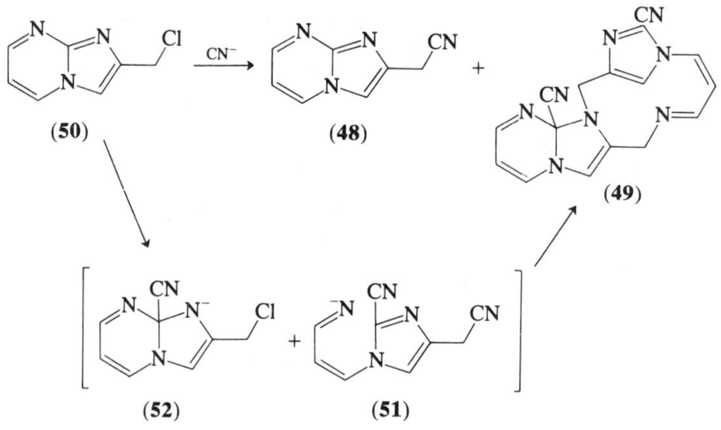

(**50**) (**48**) (**49**) (**52**) (**51**)

5.22.2.2 Construction of Subheterocyclic Ring(s)

Numerous methods are available for the synthesis of each of the common heteroaromatics, many of which have been incorporated into phane structures by the utilization of a functionalized macroring, such as cyclododecanone. The synthetic particulars can be found in most modern heterocyclic texts ⟨B-82MI52201⟩. The syntheses of heterophanes containing pyridine, furan and thiophene units have been reviewed ⟨77CRV513⟩.

5.22.2.2.1 Synthesis from dicarbonyl compounds

The Paal–Knorr synthesis utilizes 1,4-diketones to generate furans, thiophenes and pyrroles, 1,5-diketones to generate pyridines, and 1,3-diketones to generate subunits containing two heteroatoms.

(i) 1,3-Dicarbonyls

1,3-Diketones are synthetically quite versatile, but they have found only modest use in the syntheses of heterophanes. When 1,3-cyclohexadecanedione (**53**) is treated with hydrazine hydrate or phenylhydrazine, pyrazolophane (**54**) or (**55**), respectively, is obtained ⟨78TL2821⟩. Reaction of dione (**56**) with hydroxylamine, followed by acid-catalyzed cyclization of the intermediate monoxime, gives the [3,5]isoxazolophane (**57**) ⟨79TL1875⟩.

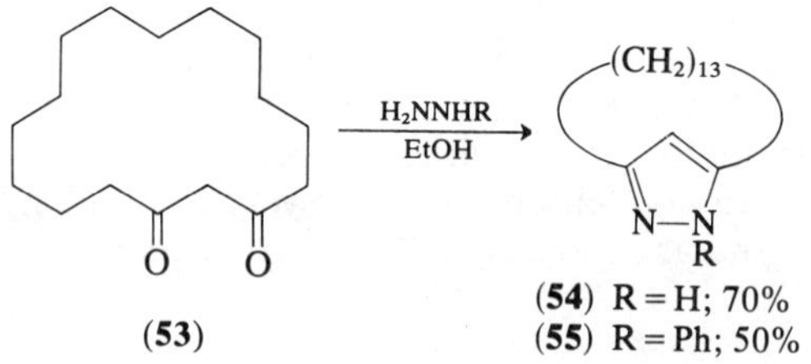

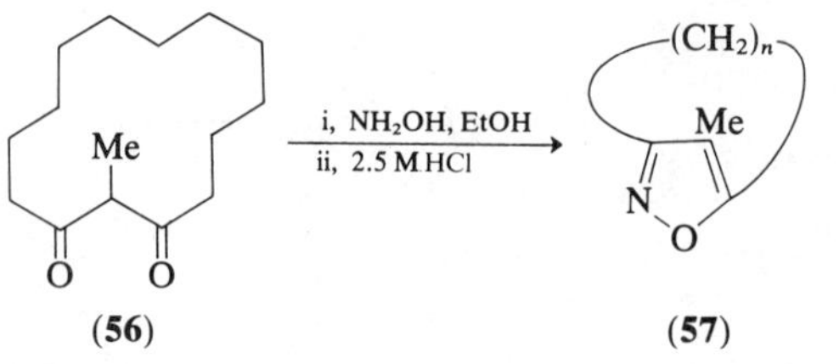

(**53**) $\xrightarrow[\text{EtOH}]{\text{H}_2\text{NNHR}}$ (**54**) R = H; 70%; (**55**) R = Ph; 50%

(**56**) $\xrightarrow[\text{ii, 2.5 M HCl}]{\text{i, NH}_2\text{OH, EtOH}}$ (**57**)

(ii) 1,4-Dicarbonyls

Dehydrations of dione (**61**) *via* the Paal–Knorr synthesis lead to heterophanes (**58**)–(**60**) ⟨68TL2181, 69T5357⟩. Dione (**61**) is readily prepared from cyclododecanone by conventional procedures ⟨64E620⟩.

P_2O_5 (80 °C) or P_2S_5 (80 °C) or $(NH_4)_2CO_3$ (100–120 °C)

(**61**) → (**58**) X = O; 83% (**59**) X = S; 51% (**60**) X = NH; 83%

Pyrrolophanes are of interest because of their natural occurrence, *viz.* porphyrins and prodigiosins. The latter are red compounds with antifungal and antibacterial properties and are present in non-pathogenic bacteria. Following the isolation of metacycloprodigiosin (**5a**), the synthesis of the racemic pyrrolophane verified the structural assignment of the natural product ⟨69JA1264⟩. The cyano ketone (**62**), prepared in several steps from cyclododecanone, is converted to keto aldehyde (**63**) through ketal formation with 1,2-ethanediol, nitrile reduction with DIBAL, and ketal hydrolysis. The aldehyde (**63**) is cyclized with ammonium carbonate to give pyrrolophane (**64**). Condensation of (**64**) with the known bipyrrole fragment (**65**) affords racemic (**5a**). Other synthetic routes to the (2,4)pyrrolophanes ⟨77M233⟩ as well as the synthesis of a prodigiosin derivative (**66**) ⟨78M557⟩ have been reported.

i, 1,2-ethanediol, TsOH; ii, DIBAL; iii, H_3O^+ — $(NH_4)_2CO_3$, DMF

(**62**) (**63**) (**64**)

(**65**) (**5a**) (**66**)

The Paal–Knorr synthesis has also been utilized in the transformation of (**67**) to thiophenophanes (**68**) and pyrrolophanes (**69**) and (**70**) ⟨74T2633⟩. Thiophenophane (**68**; $n = 6$), a labile but distillable liquid, has the shortest 2,4-bridge known in a thiophenophane. And even though the related (2,4)pyrrolophane (**70**; $n = 6$) can be made (>35%), (**69**; $n = 6$) is as yet unknown. Attempts to dehydrate (**67**; $n < 9$) to generate (2,4)furanophanes have failed.

P_2S_5 or $(NH_4)_2CO_3$ or $ArNH_2$

(**67**)

(**68**) X = S; $n = 6, 7, 9$
(**69**) X = NH; $n = 7, 9$
(**70**) X = NAr; $n = 6, 9$

Macrocyclic 1,4-diketones (**71**), generated by hydrolysis of the furan ring in (**27**) ⟨77H(7)81⟩, have been transformed into mixed heterophanes (**72**) by treatment with an amine ⟨73TL4017, 74CC119⟩.

Inclusion of electron-deficient subrings is accomplished by reaction of dione (**61**) with hydrazine, followed by dehydrogenation ⟨74JA5287⟩. Oxidation of (**73**) with peroxybenzoic acid affords the mono-*N*-oxide (**74**), a chiral *ansa* compound.

(27) (71) (72)

(73) (74)

(iii) *1,5-Dicarbonyls and pyrylium salts*

Pyrylium salts are convenient precursors for the synthesis of pyridine rings ⟨B-79MI52202⟩. An example is the bicyclic (heterophane) pyrylium salt (**75**), formed by diacylation of isobutene in the presence of $AlCl_3$; reaction (as the perchlorate) with ammonia gives pyridinophane (**76**) in low yield ⟨71CC32⟩. The major product, first reported to be (**76**), actually is the corresponding dimer; this misidentification illustrates a haunting problem in this field. Similar sequences have been used to prepare (2,6)- ⟨69BCJ1163, 71BCJ2827⟩, (3,5)- ⟨68TL4643, 73RRC1609⟩ and (2,4)-pyridinophanes ⟨71TL671, 73MI52200⟩.

(75) (76)

Alternatively, a pyridine ring can be prepared from precursors to the pyrylium salts, *i.e.* 1,5-diketones. 1,5-Cyclododecanedione (**77**), prepared from boraperhydrophenalene (**78**), with hydroxylamine generates (**79**) ⟨71BCJ2827⟩. Racemic muscopyridine (**80**) has been constructed similarly from dione (**81**), which is generated by a regioselective cyclopentenone annulation sequence involving the conrotatory ring closure of the adduct obtained from propargyl alcohol dianion and 2-cyclododecenone ⟨80TL3897⟩.

(78) (77) (79)

(81) (80)

(iv) *1,2-Dicarbonyls — Hinsberg Thiophene Synthesis*

Macrocyclization of bis-(chloromethyl ketone) (**82**) with sodium sulfide generates (**83**), which by base-catalyzed condensation with ethanedial gives the desired cyclophane (**84**) in 10–91% yields ⟨78CL563⟩. Reduction of the keto groups under Wolff–Kishner conditions, followed by desulfurization with Raney nickel, gives smooth entry into paracyclophanes

(**85**; $n = 3, 10$; 60–95%). Other (2,5)thiophenophanes have been similarly prepared ⟨80CL397⟩.

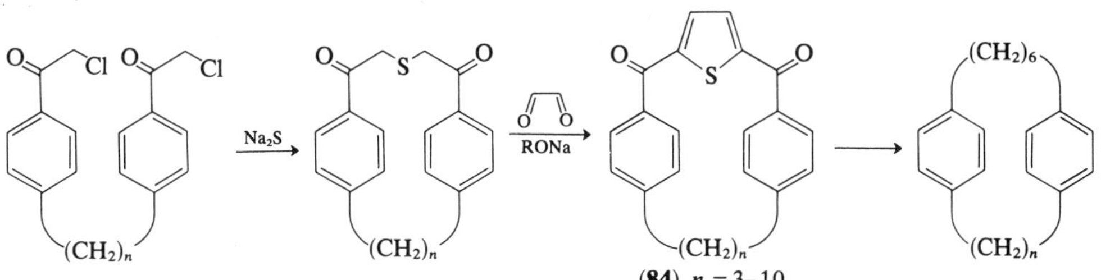

(**82**) (**83**) (**84**) $n = 3$–10 (**85**)

5.22.2.2.2 *α,β-Unsaturated carbonyls and their derivatives*

The use of enones as precursors to heterophanes, *e.g.* (**86**), can be exemplified by the reaction of hydrazine with (**87**). Chlorovinyl ketone (**88**) is transformed into (**86**), (**89**) or (**90**), depending on the nucleophilic reagent used ⟨70TL11, 76T1863⟩.

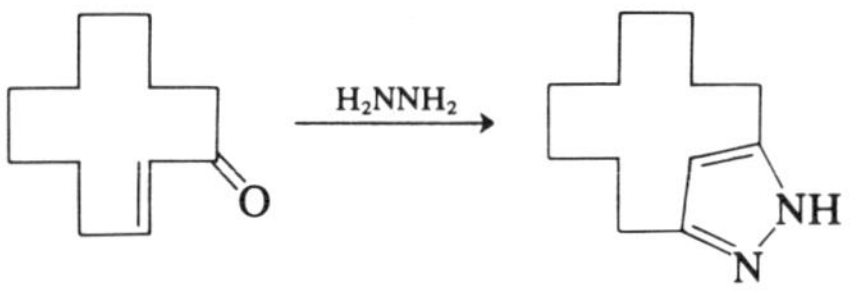

(**87**) (**86**)

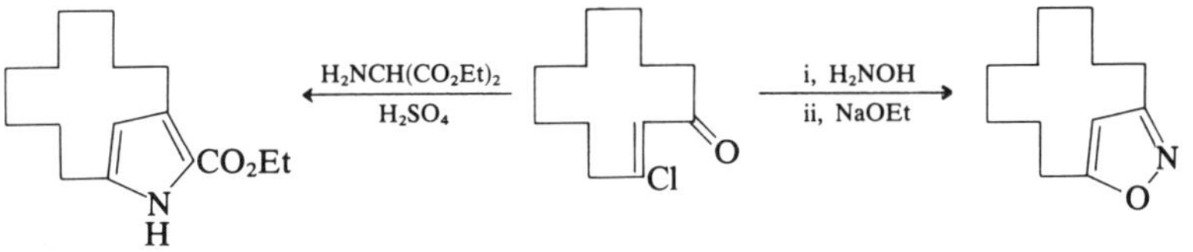

(**89**) (**88**) (**90**)

Reduction of unsaturated γ-lactones (*e.g.* **91**) has been applied to the synthesis of furanophanes; lactonization of (**92**) followed by reduction with DIBAL gives (**93**) ⟨70TL11⟩.

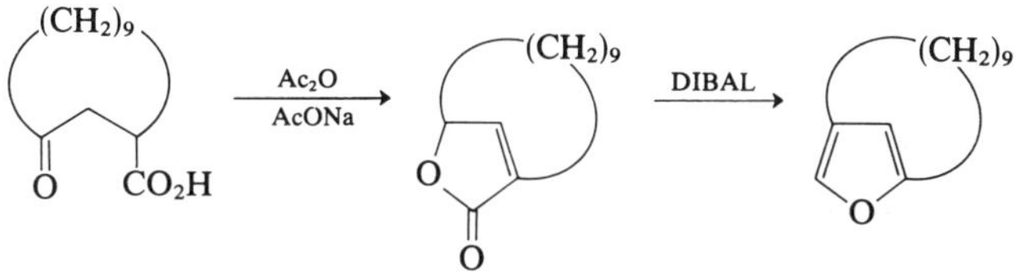

(**92**) (**91**) (**93**)

Cyclization of the α,β-unsaturated *N*-tosylhydrazone (**94**) with carbonate in methanol gives pyrazolophane (**95**) ⟨79JCR(S)246⟩.

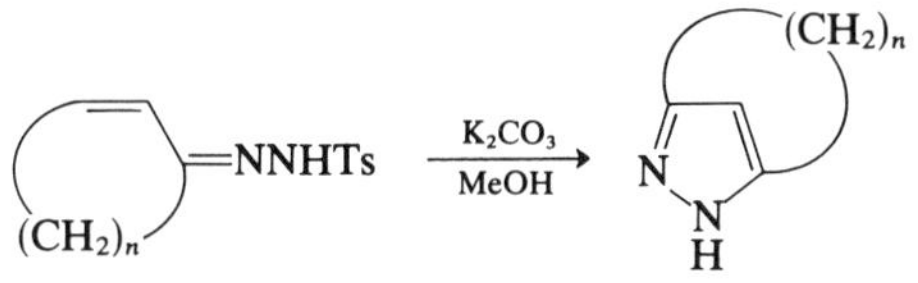

(**94**) (**95**) $n = 10, 12$

Pyridinophanes (**96**) can be constructed from bis-(β-chlorovinyl ketonic) intermediates upon treatment with ammonia. Wolff–Kishner reduction of (**96**) gives the parent (3,5)pyridinophane ⟨68HCA2027⟩. Some β-aminovinyl carbonyls have been condensed with cycloalkanones in the presence of triethylamine and piperidinium acetate to give the (2,3)pyridinophane series (**97**) ⟨70TL3291⟩.

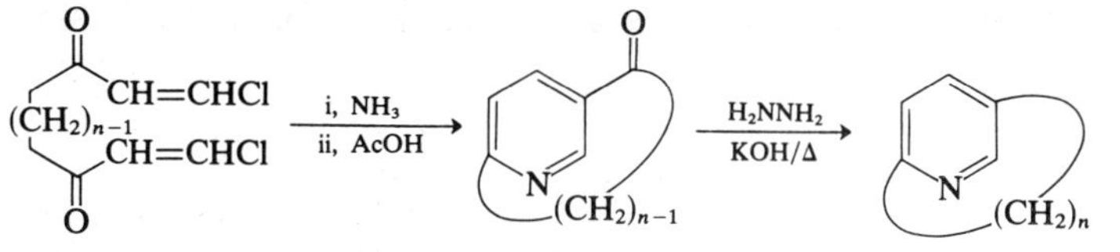

(96)

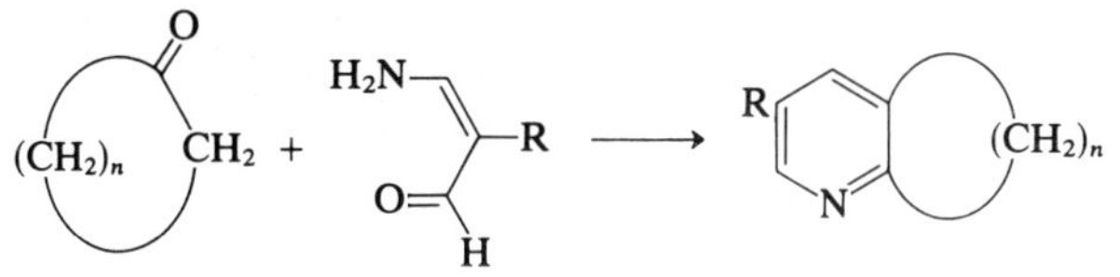

(97)

5.22.2.2.3 gem-*Dihalocyclopropane precursors*

A procedure which uses phenyl(trichloromethyl)mercury (**98**) to generate *gem*-dichlorocyclopropanes from alkenes has been employed to convert (**99**) to the benzopyridinophanes (**100**) in reasonable yields ⟨69TL557, 70JOC3775⟩. The enol acetate (**101**) with (**98**) gives a convenient reagent (**102**), which hydrazine then converts to the pyrazolophane (**95**; $n = 10$) and guanidine to the pyrimidinophane (**103**) ⟨68JOC1476, 67JA985, 69JOC1474⟩.

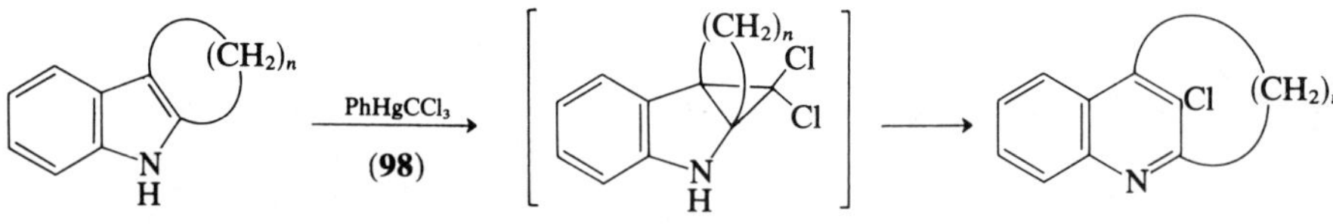

(99) (100)

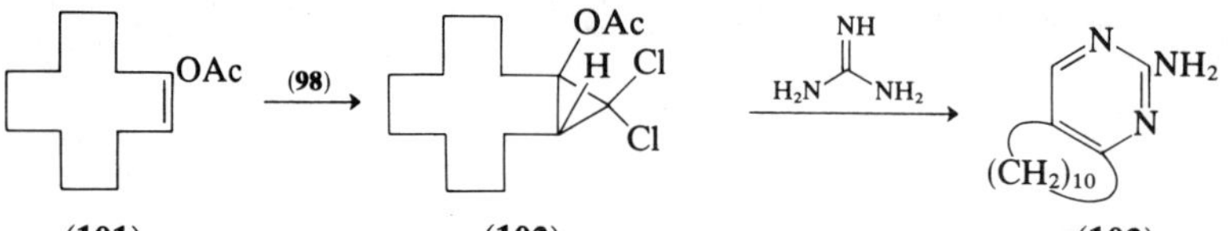

(101) (102) (103)

5.22.2.2.4 *Diyne precursors*

1,3-Cyclohexadecadiyne (**104**) adds arylamines to form N-arylpyrrolophanes such as (**105**), whereas alkylamines produce a mixture of heterophanes such as (**106**) and (**107**) ⟨78TL2821⟩. The reaction is catalyzed by Cu(I).

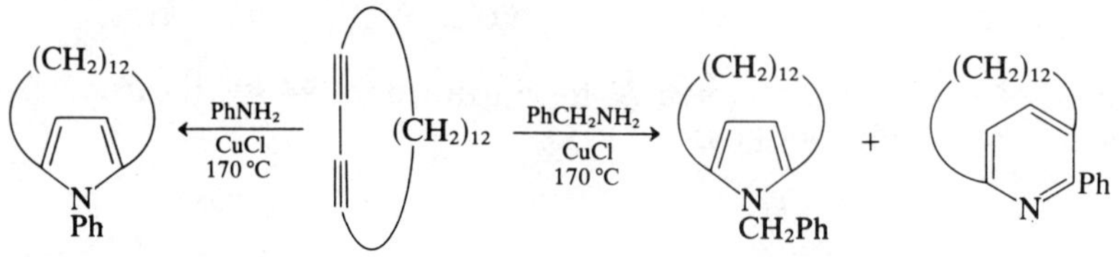

(105) (104) (106) (107)

5.22.2.2.5 *Rearrangement of an existing heterophane*

The Dimroth rearrangement has been utilized for the conversion of (**108**) to the 2,N-polymethylene-bridged 6-aminopyrimidine (**109**) ⟨75AJC119⟩. Photochemical rearrangements of isoxazolophanes (**110**) into oxazolophanes (**111**) have been demonstrated to go through an azirine intermediate (**112**) ⟨79TL1875⟩. Chemical modification of these oxazolophanes by *N*-alkylation, hydrolysis and subsequent cyclization gives the related imidazolophane ⟨80CL659⟩.

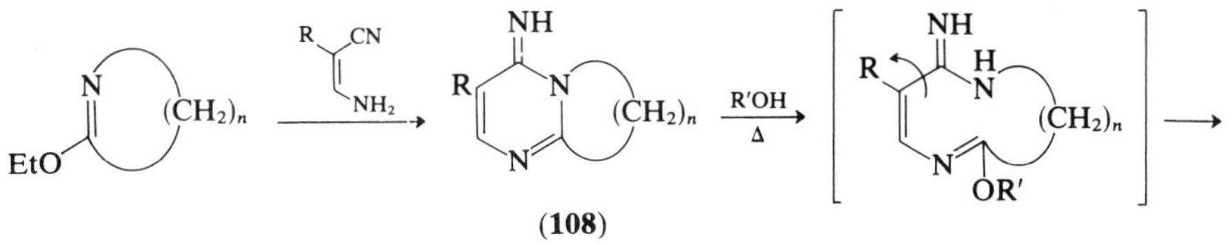

(108)

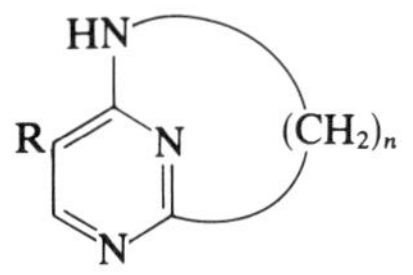

(109)

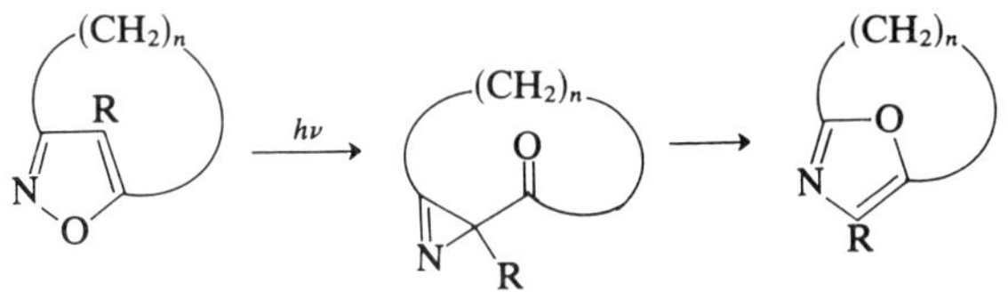

(110) (112) (111)

5.22.2.2.6 Miscellaneous preparations of subring

The Schmidt reaction was used in one of the first routes to pyridinophanes *via* an insertion process. Subsequent macroring modification afforded muscopyridine (**80**) ⟨57JA5558⟩.

Macrocyclic 2-pyrones (**113**) and (**114**) are prepared from enamine (**115**) and ketene ⟨75HCA2409⟩. Reduction of cyano ketone (**116**) with LAH affords an amino ketone which spontaneously cyclizes to generate a tetrahydropyridine (**117**); dehydrogenation by palladium on charcoal produces the aromatized phane (**118**) ⟨71TL671⟩.

Macrocyclic tetrathiafulvalenophanes have recently been prepared ⟨80AG(E)66, 67⟩ by a variation of the synthesis of tetrathiafulvalene ⟨74JOC3608⟩ *via* a high dilution intramolecular cyclization.

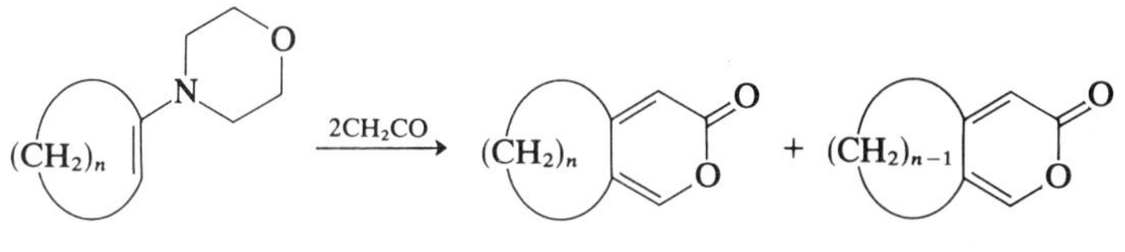

(115) (113) (114)

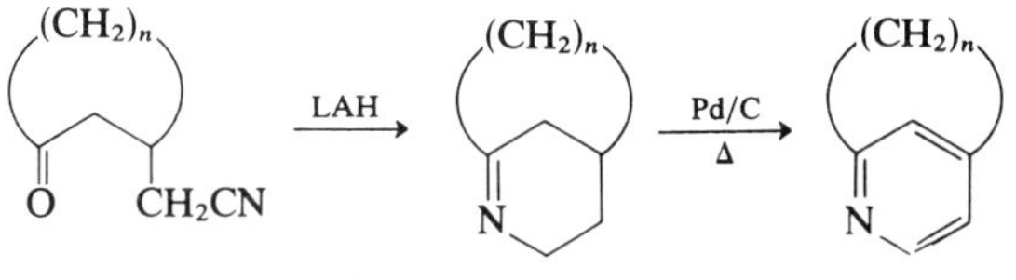

(116) (117) (118)

5.22.3 REACTIONS OF HETEROPHANES

Quite diverse reaction conditions have been used for the final step in the synthesis of different heterophanes, demonstrating the general stability of the multicyclic systems. Heterophanes do undergo reactions, some that are typical of the structural fragments and some that are probably more characteristic of the phane structure itself or the strain associated with it. Some reactions of heterophanes are included in Section 5.22.2.

5.22.3.1 Chemistry of the Subunits

Furanophane (**27**) is reduced by hydrogen and PtO_2 catalyst to (**119**) (2,5-dimethylfuran is reduced under the same conditions to 2-hexanol) and is converted by bromine in methanol to the polyketal (**120**) ⟨60JA1428⟩, which is easily hydrolyzed to (**121**) ⟨77M331⟩. The mixed heterophane (**122**) was hydrolyzed to (**123**) ⟨77H(7)81⟩. As anticipated, electrophilic substitution favors α-substitution ⟨78M137⟩, and (**124**) undergoes formylation ($POCl_3$/DMF) to afford (**125**) ⟨77M233⟩.

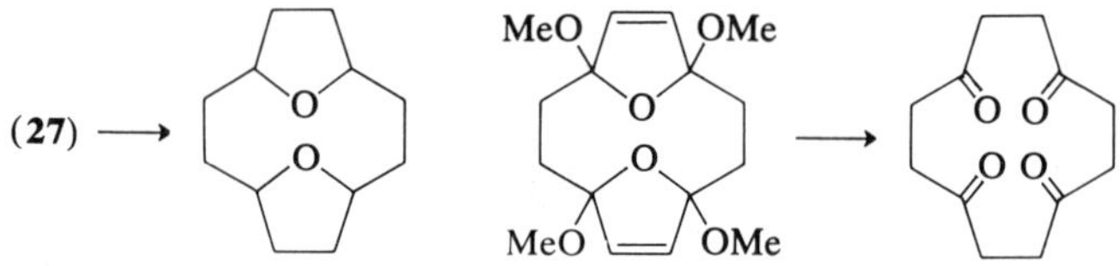

(**27**) → (**119**) (**120**) → (**121**)

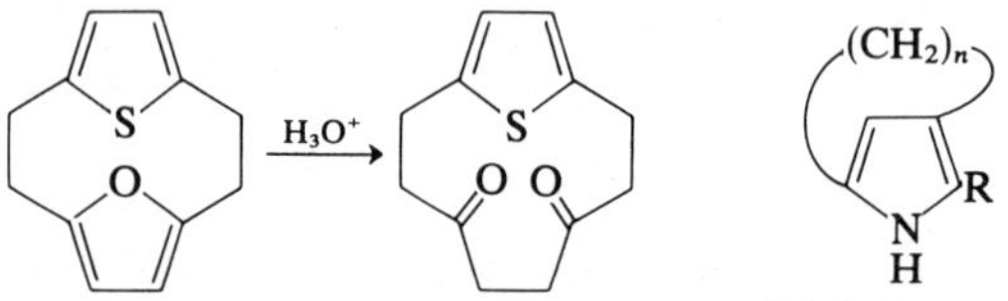

(**122**) $\xrightarrow{H_3O^+}$ (**123**)

(**124**) R = H
(**125**) R = CHO

More reactions have been reported for thiophenophanes than for any other heterophane. Such interest may have been prompted by the ease of desulfurization (Raney nickel) without alteration of the macroring carbonyl functionalities. For example, (**21**; $n = 9$) is converted to cyclotetradecanone. Such compounds are of potential value in perfumery ⟨64BAU1948, 75T2551⟩. Some *C*-alkylations of enolate ions in the bridge occur when potassium metal in HMPA is used, whereas macroring cleavage occurs with usual bases ⟨73CHE13⟩. Wolff–Kishner reduction of (**21**) occurs without ring opening, and Friedel–Crafts acetylation and isopropylation afford (**126**; 70%) and (**127**; 57%), respectively ⟨64BAU1948⟩. Interestingly, nitration of (**21**) with HNO_3/Ac_2O affords mainly the *ipso*-substitution product (**128**; 58%), and (**129**) in only 5% yield ⟨64BAU1948⟩.

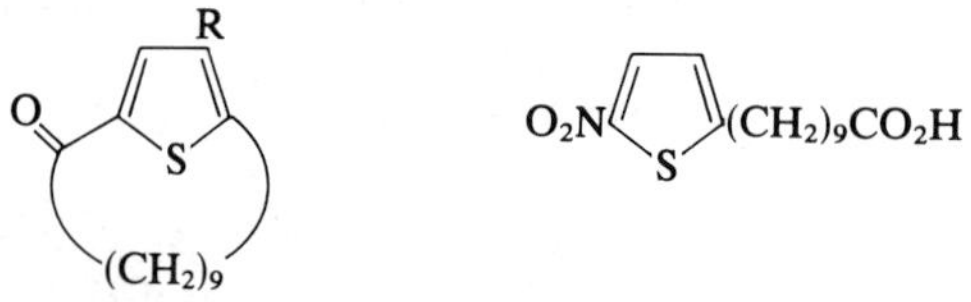

(**71**) R = H
(**126**) R = COMe
(**127**) R = Pr^i
(**129**) R = NO_2

O_2N–(thiophene)–$(CH_2)_9CO_2H$ (**128**)

ipso-Substitution with subsequent rearrangement dominates *t*-butylation of (**130**) ($Bu^tCl/SnCl_4$ in CS_2); the initial product, (**131**), reacts further to afford (**132**) and (**133**) ⟨75T2551⟩. By contrast, *t*-butylation of 2,5-dimethylthiophene occurs at the 3-position in over 90% yield. Raney nickel desulfurization of (**131**) gives (**134**); α-hydrogen–deuterium exchange in (**132**) occurs *via* the lithio intermediate; and (**133**) is dealkylated slowly by treatment with sulfuric acid to give (**132**).

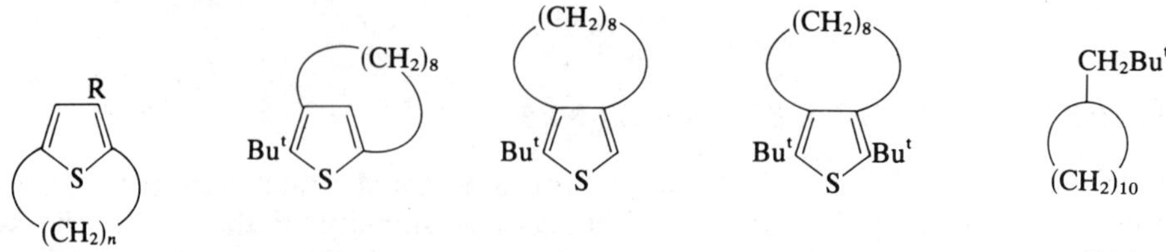

(**130**) R = H; $n = 8$
(**135**) R = H; $n = 11$
(**136**) R = Br; $n = 11$

(**131**) (**132**) (**133**) (**134**)

A variety of reaction conditions have failed to produce the 3-bromo derivative from (**130**), perhaps because ring strain promotes *ipso*-substitution and macroring opening, whereas (**135**; $n = 11$) is brominated (25–30%) with NBS in acetic acid to give (**136**) ⟨76ACS(B)341⟩. Ethyllithium reacts with (**136**) to open the thiophene ring and give (**137**).

Polarographic reductions of (**21**) are easier than those of ketones (**138**) and (**139**). The size of the bridging ring influences both the absorption of the ketone on the dropping mercury electrode (**139**>**138**) and the reduction potential ⟨70MI52200⟩.

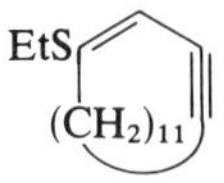

EtS, $(CH_2)_{11}$ (**137**)

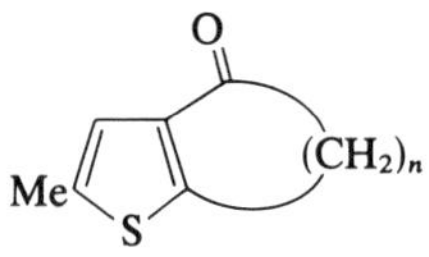

O, Me, S, $(CH_2)_n$ (**138**) $n = 3$ (**139**) $n = 11$

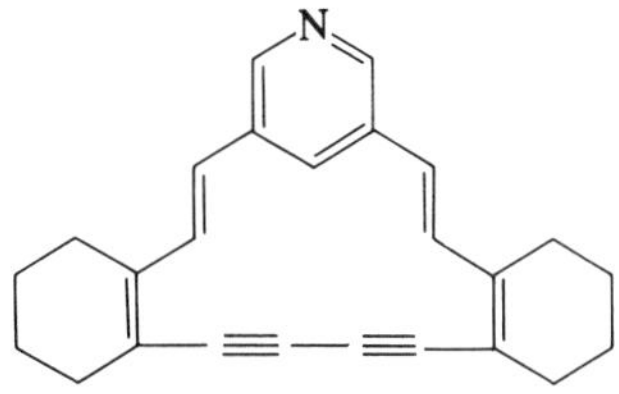

N (**140**)

Annulene (**140**) reacts with LAH and with alkyllithiums by a 1,4-addition mode to give 1,4-dihydropyridines, whose NMR spectra indicate conformational mobility ⟨74TL599⟩.

5.22.3.2 Modification of the Macroring or Connecting Linkage

5.22.3.2.1 α-Functionalization

Typifying α-substitution, quinolinophanes (**141**) and (**142**) ($X = anti$; $Y = syn$) undergo substitution with high stereoselectivity; however, different reagents give different selectivity. With *excess* LAD in ether, (**141a**) and (**141b**) are reduced to (**141c**) *only*; one molar equivalent of LAH did not reduce (**141**) ⟨74JOC2432⟩. *N*-Oxide (**142**) reacts with acetic anhydride or acetyl chloride to give a mixture of (**141d**:**141e**) (with **141f**, the stereoselectivity was 1:1.1); the nearly complete (95–97%) retention of the H/D label excludes a common intermediate. Reactions of (**142**) with acetyl chloride and of (**141f**) with ^{18}O-labelled acetic anhydride indicate that these substitution reactions involve both intra- (74–79%) and inter-molecular (26–21%) acetoxy transfer. With 2-methylpyridine *N*-oxide and ^{18}O-labelled acetic anhydride, only intramolecular acetoxy transfer is detected ⟨74JOC2916⟩. Reaction of (**142**) with *p*-toluenesulfonyl chloride produces *only* (**141g**) ⟨74JOC2916, 73JOC927⟩, although with the model compound 3-chloro-2,4-dimethylquinoline *N*-oxide, only chloro, *not* tosyloxy, substitution on the 2-methyl group occurs ⟨73JOC927⟩.

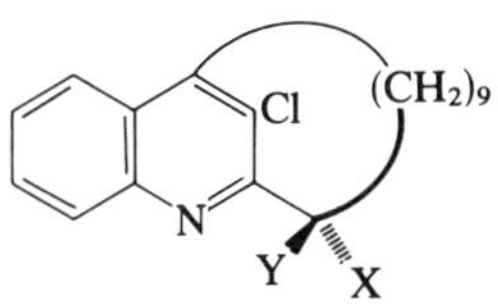

Cl, $(CH_2)_9$, N, Y, X

(**141**) **a**; X = H, Y = Br or OTs
b; X = Br or OTs, Y = H
c; X = H, Y = D
d; X = H, Y = OAc
e; X = OAc, Y = H
f; X = Y = H
g; X = H, Y = OTs

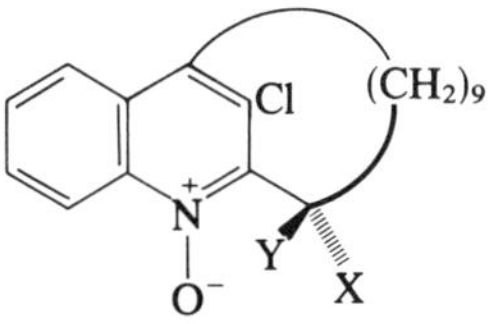

Cl, $(CH_2)_9$, $\overset{+}{N}$, O^-, Y, X

(**142**) X = H, Y = D
X = *anti*, Y = *syn*

5.22.3.2.2 Ring contractions

There are four major routes to contract the macroring portion of heterophanes, all by sulfur expulsion sequences. Since carbon sulfur bridges are easily prepared, their conversion to the carbon bridges is important. Photolysis of pyridinophane (**143**) in the presence of trialkyl phosphite gives (**144**) ⟨73CC406⟩; with polysulfides a stepwise extrusion of sulfur has been demonstrated ⟨75MI52201⟩. Other subheterocyclic rings are retained intact upon photolytic ring contraction ⟨81LA1065⟩. Pyrolytic conditions are most successful when the sulfide is initially oxidized to the sulfone; then thermolysis *in vacuo* (*e.g.* **145**) links the carbons from which sulfur dioxide has been extruded ⟨71TL3843⟩.

hν
$(RO)_3P$

(**143**) (**144**)

i, [O]
ii, Fe/CF_3CO_2H
Δ

680 °C
0.01 mmHg

(**145**)

The creation of an alkenic bridge can be accomplished by two routes. The Stevens rearrangement has been applied to the synthesis of (**146**) *via* an alkylation–elimination sequence. Methylation of (**147**) with trimethyloxonium fluoroborate, followed by treatment with base, affords (**148**); methylation to the onium salt permits a Hofmann elimination to give (**146**) ⟨78JA2457⟩. The Ramberg–Bäcklund rearrangement ⟨77OR(25)1⟩ has been applied to the synthesis of (**149**) ⟨81JOC1747⟩.

i, Me_3OBF_4
ii, Bu^tOK

(**147**) (**148**) (**146**)

(**9**)
X = Br

i, Na_2S
ii, H_2O_2

i, BuLi
ii, Br_2

base

(**149**)

5.22.4 CONFORMATIONAL ASPECTS

The conformational mobility of heterophanes varies widely; some show frozen conformations up to 200 °C ($\Delta G^{\ddagger} > 113$ kJ mol^{-1}), while others are in mobile equilibrium down to −90 °C ($\Delta G^{\ddagger} < 42$ kJ mol^{-1}). The energy barriers are determined by the length of the bridging unit(s), the size of the cavity formed by it, the size of the heteroatom(s) or other fragments that must turn through the cavity, and the bond angles connecting the chain to the heteroaromatic ring. NMR data and X-ray crystallographic data have been used to identify the geometries of the heterophanes, and variable temperature NMR data have been used to assess the energies for conformational change ⟨73TL3989⟩.

The energy barriers for several heterophanes containing one heteroaromatic ring have been calculated from the coalescence temperature of ^{1}H NMR signals. For (**123**), ring flipping is restricted by an energy barrier of 67 kJ mol^{-1} and chain flipping by 47.7 kJ mol^{-1} ⟨77H(7)81⟩. For pyridinophane (**150**), the barrier to ring inversion is 61.5 kJ mol^{-1}, which is higher than that of carbocyclic analog (**151**), 58.2 kJ mol^{-1} ⟨76CB1002⟩. The conformational equilibria of the compounds represented by (**152**) are not frozen even down to −60 °C ⟨74BAU1941⟩.

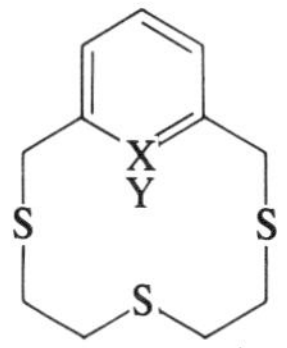

(**150**) XY = $\overset{+}{N}-\overset{-}{O}$
(**151**) XY = CF
(**152**) $n = m = 7$ 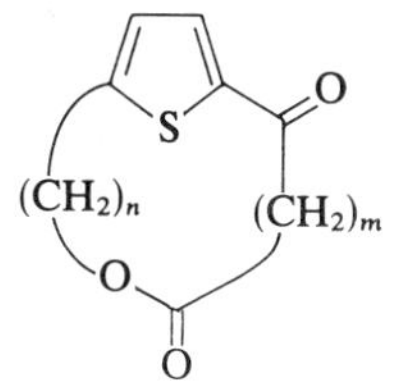
(**153**) 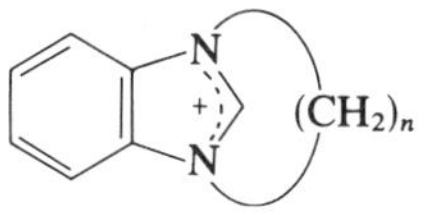

Swinging of the methylene chain in cation (**153**) is restricted when $n = 6$ and 7, but is rapid on the NMR time scale at room temperature for $n > 7$ ⟨73CC427⟩.

A change in heteroatom greatly affects the energy barrier to ring flipping. Furans (**58**) and (**154**) are more mobile than the corresponding thiophenes (**59**) and (**155**) ⟨80CL397, 76TL699⟩. Phane (**155**) exists as a mixture of *syn* and *anti* isomers (ratio 1:20) whose NMR signals coalesce at 105 °C, while (**154**) exhibits no spectral variations down to −90 °C ⟨80CL397⟩. In (**156**) ⟨77JOC1379⟩ and (**144**) ⟨73TL4021⟩, ring rotation is rapid at room temperature ($\Delta G^{\ddagger} = 46$ kJ mol^{-1}), but in (**157**) the rings are frozen up to 110 °C in intersecting planes with the O inclined toward N ⟨74JOC2570, 75JHC433⟩. The barrier to ring rotation in (**158**) is 71 kJ mol^{-1} (higher than in **156**) and in (**159**) and (**160**) it is >113 kJ mol^{-1} ⟨77JOC1379⟩. Apparently transannular interactions between the separate π-systems in the *syn* conformer of (**160**) hinder rotation of the pyrrole nucleus, even though the N—H moiety is able to pass through the similar cavity in (**158**) ⟨77JOC1379⟩.

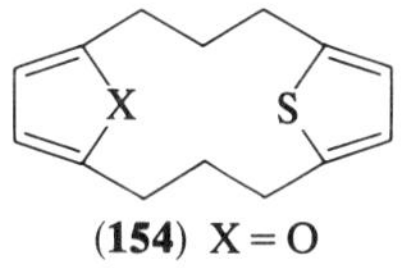

(**154**) X = O
(**155**) X = S
(**156**) X = CH 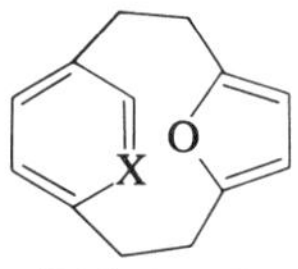
(**157**) X = N
(**158**) R = H 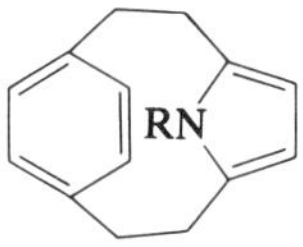
(**159**) R = Me
(**160**) 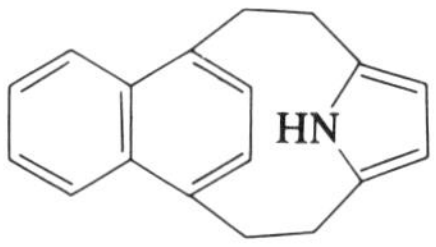

Phanes (**161**)–(**163**) do not exhibit any coalescence of NMR signals up to 190 °C ⟨77JOC1379, 69CC1504⟩, but (**7**) has only a modest barrier to isomerization ($\Delta G^{\ddagger} =$ 61.9 kJ mol^{-1}, $T_c = 13.5$ °C) ⟨69CC1504⟩. The slightly higher barrier of (**164**; $\Delta G^{\ddagger} =$ 51.0 kJ mol^{-1}, *syn* conformation preferred) as compared to (**155**; $\Delta G^{\ddagger} = 46$ kJ mol^{-1}) is attributed to restriction of rotation by overlapping of sulfur-*p*–pyridine orbitals ⟨81CL1781⟩. Thiophenophane (**165**) is conformationally mobile down to −80 °C ⟨74BAU1941⟩.

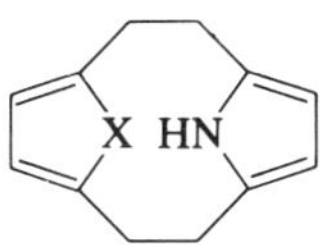

(**161**) X = O
(**162**) X = NH 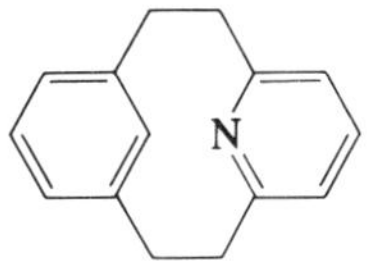
(**163**) 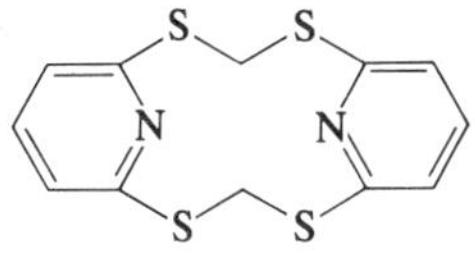
(**164**)

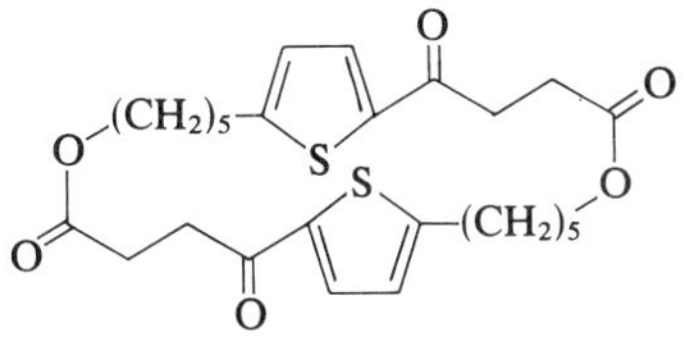

(**165**)

On the basis of NMR data, the conformational equilibrium of (**166a**) ⇄ (**166b**) in solution has been described as 'longitudinal isomerization' rather than a *syn–anti* interconversion of the pyridine rings ⟨79JA7088⟩. With R = H, phane (**166**) has $\Delta G^{\ddagger} = 56.5$ kJ mol^{-1}; with R = $CONMe_2$, $\Delta G^{\ddagger} = 60.7$ kJ mol^{-1}. These barriers are also higher than that for the smaller pyridinophane (**144**) with the hydrocarbon bridges, probably due to the introduction of four rigid imidate linkages.

(**166a**) (**166b**)

NMR data reveal that members of a series of similarly constituted heterophanes (**167**)–(**169**) (core sizes: 18, 20 and 22, respectively) exhibit sharply different conformational preferences ⟨82T665⟩. Phane (**167**) exists preferentially in the *anti* conformation (benzo relationship); (**168**) is saddle-shaped (hetero rings *syn* up, benzo rings *syn* down); and (**169**) is flexible without a conformational preference. For (**168b**), the NMR signals for the methyl hydrogens (R = Me) are not temperature dependent, indicating a locked conformation for the benzo units; however the signals for the methylene hydrogens do change at lower temperatures ($\Delta G^{\neq} = 57.7$ kJ mol^{-1}; $T_c = 5$ °C) ⟨82T665, 81OMR(16)1⟩.

(**167**) (**168**) **a**: R = H
b; R = Me (**169**)

X-Ray data also play an important role in the analysis of conformations in the solid state. The inherent strain in the heterophanes is accommodated by distortion of the bridging bond angles, by puckering of the aromatic subrings, and sometimes by non-parallel orientations of the heterocyclic subunits. The X-ray crystal structure analyses of the multibridged pyridinophanes (**170**) and (**171**), synthesized *via* the dimerization of a xylyene intermediate ⟨81AG(E)571⟩, show that the pyridine subrings are in a boat conformation with the tips pointing outside, and for both compounds the intra-ring distances are *ca.* 2.64 Å ⟨81CSC313, 571⟩. The furan ring in (**172**) is planar (within experimental error) and is inclined towards the best least squares plane of the naphthalene moiety by 22°; the 1,4-carbon atoms lie 0.177 Å out of the plane of the remaining naphthalene carbon atoms ⟨78T1641⟩. The aromatic rings of (**173**) form an angle of 23° with each other, and both the furan and pyridine rings are puckered, by 0.022 and 0.172 Å, respectively ⟨75JHC433⟩. The crystal structures of tetrathiafulvalenoparacyclophanes ⟨80AG(E)66⟩, an alkenyl-bridged pyridinophane ⟨74JA5⟩ and macrocyclic keto-lactones containing a thiophane ring ⟨76IZV1287⟩ have also been reported.

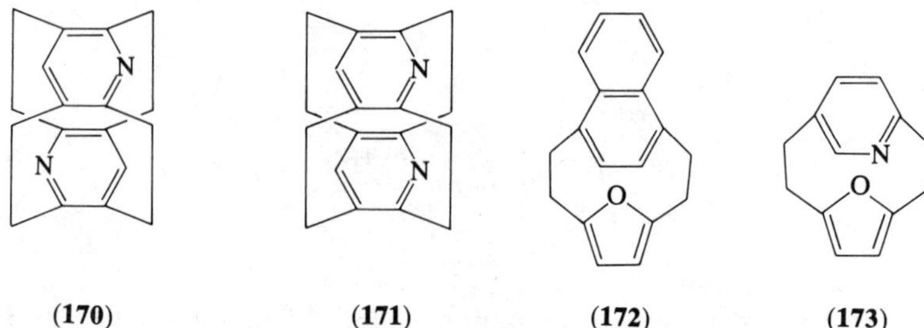

(**170**) (**171**) (**172**) (**173**)

Recently, the X-ray crystal structure of (**164**) has been completed, and it confirmed the *syn* conformational preference of the subheterocyclic rings in the solid state ⟨82UP52200⟩. The related —CH_2SCH_2— bridged analog of (**164**) has been shown to possess a similar orientation ⟨82UP52200⟩.

References

EXPLANATION OF THE REFERENCE SYSTEM

Throughout this work, references are designated by a number–letter coding of which the first two numbers denote tens and units of the year of publication, the next one to three letters denote the journal, and the final numbers denote the page. This code appears in the text each time a reference is quoted; the advantages of this system are outlined in the Introduction (Chapter 1.01). The system is based on that previously used in the following two monographs: (a) A. R. Katritzky and J. M. Lagowski, 'Chemistry of the Heterocyclic N-Oxides', Academic Press, New York, 1971; (b) J. Elguero, C. Marzin, A. R. Katritzky and P. Linda, 'The Tautomerism of Heterocycles', in 'Advances in Heterocyclic Chemistry', Supplement 1, Academic Press, New York, 1976.

The following additional notes apply:

1. A list of journals which have been assigned codes is given (in alphabetical order) together with their codes immediately following these notes. Journal names are abbreviated throughout by the CASSI (Chemical Abstracts Service Source Index) system.
2. A list of journal codes in alphabetical order, together with the journals to which they refer, is given on the end papers of each volume.
3. Each volume contains all the references cited *in that volume*; no separate lists are given for individual chapters.
4. The list of references is arranged in order of (a) year, (b) journal in alphabetical order of journal code, (c) part letter or number if relevant, (d) volume number if relevant, (e) page number.
5. In the reference list the code is followed by (a) the complete literature citation in the conventional manner and (b) the number(s) of the page(s) on which the reference appears, whether in the text or in tables, schemes, *etc.*
6. For non-twentieth century references the year is given in full in the code.
7. For journals which are published in separate parts, the part letter or number is given (when necessary) in parentheses immediately after the journal code letters.
8. Journal volume numbers are *not* included in the code numbers unless more than one volume was published in the year in question, in which case the volume number is included in parentheses immediately after the journal code letters.
9. Patents are assigned appropriate three letter codes.
10. Frequently cited books are assigned codes, but the whole code is now prefixed by the letter 'B-'.
11. Less common journals and books are given the code 'MI' for miscellaneous.
12. Where journals have changed names, the same code is used throughout, *e.g.* CB refers both to *Chem. Ber.* and to *Ber. Dtsch. Chem. Ges.*

Journals

Acc. Chem. Res.	ACR
Acta Chem. Scand., Ser. B	ACS(B)
Acta Chim. Acad. Sci. Hung.	ACH
Acta Crystallogr., Part B	AX(B)
Adv. Phys. Org. Chem.	APO
Agric. Biol. Chem.	ABC

Angew. Chem.	AG
Angew. Chem., Int. Ed. Engl.	AG(E)
Ann. Chim. (Rome)	AC(R)
Ann. N.Y. Acad. Sci.	ANY
Arch. Pharm. (Weinheim, Ger.)	AP
Ark. Kemi	AK
Arzneim.-Forsch.	AF
Aust. J. Chem.	AJC
Biochem. Biophys. Res. Commun.	BBR
Biochemistry	B
Biochem. J.	BJ
Biochim. Biophys. Acta	BBA
Br. J. Pharmacol.	BJP
Bull. Acad. Pol. Sci., Ser. Sci. Chim.	BAP
Bull. Acad. Sci. USSR, Div. Chem. Sci.	BAU
Bull. Chem. Soc. Jpn.	BCJ
Bull. Soc. Chim. Belg.	BSB
Bull. Soc. Chim. Fr., Part 2	BSF(2)
Can. J. Chem.	CJC
Chem. Abstr.	CA
Chem. Ber.	CB
Chem. Heterocycl. Compd. (Engl. Transl.)	CHE
Chem. Ind. (London)	CI(L)
Chem. Lett.	CL
Chem. Pharm. Bull.	CPB
Chem. Rev.	CRV
Chem. Scr.	CS
Chem. Soc. Rev.	CSR
Chem.-Ztg.	CZ
Chimia	C
Collect. Czech. Chem. Commun.	CCC
Coord. Chem. Rev.	CCR
C.R. Hebd. Seances Acad. Sci., Ser. C	CR(C)
Cryst. Struct. Commun.	CSC
Diss. Abstr. Int. B	DIS(B)
Dokl. Akad. Nauk SSSR	DOK
Experientia	E
Farmaco Ed. Sci.	FES
Fortschr. Chem. Org. Naturst.	FOR
Gazz. Chim. Ital.	G
Helv. Chim. Acta	HCA
Heterocycles	H
Hoppe-Seyler's Z. Physiol. Chem.	ZPC
Indian J. Chem., Sect. B	IJC(B)
Inorg. Chem.	IC
Int. J. Sulfur Chem., Part B	IJS(B)
Izv. Akad. Nauk SSSR, Ser. Khim.	IZV
J. Am. Chem. Soc.	JA
J. Biol. Chem.	JBC
J. Chem. Phys.	JCP
J. Chem. Res. (S)	JCR(S)
J. Chem. Soc. (C)	JCS(C)
J. Chem. Soc., Chem. Commun.	CC
J. Chem. Soc., Dalton Trans.	JCS(D)
J. Chem. Soc., Faraday Trans. 1	JCS(F1)
J. Chem. Soc., Perkin Trans. 1	JCS(P1)
J. Gen. Chem. USSR (Engl. Transl.)	JGU
J. Heterocycl. Chem.	JHC
J. Indian Chem. Soc.	JIC
J. Magn. Reson.	JMR

J. Med. Chem.	JMC
J. Mol. Spectrosc.	JSP
J. Mol. Struct.	JST
J. Organomet. Chem.	JOM
J. Org. Chem.	JOC
J. Org. Chem. USSR (Engl. Transl.)	JOU
J. Pharm. Sci.	JPS
J. Phys. Chem.	JPC
J. Prakt. Chem.	JPR
Khim. Geterotsikl. Soedin.	KGS
Kristallografiya	K
Liebigs Ann. Chem.	LA
Monatsh. Chem.	M
Naturwissenschaften	N
Nippon Kagaku Kaishi	NKK
Nouv. J. Chim.	NJC
Org. Magn. Reson.	OMR
Org. Mass Spectrom.	OMS
Org. Prep. Proced. Int.	OPP
Org. React.	OR
Org. Synth.	OS
Org. Synth., Coll. Vol.	OSC
Phosphorus Sulfur	PS
Phytochemistry	P
Proc. Indian Acad. Sci., Sect. A	PIA(A)
Proc. Natl. Acad. Sci. USA	PNA
Pure Appl. Chem.	PAC
Q. Rev., Chem. Soc	QR
Recl. Trav. Chim. Pays-Bas	RTC
Rev. Roum. Chim.	RRC
Russ. Chem. Rev. (Engl. Transl.)	RCR
Spectrochim. Acta, Part A	SA(A)
Synth. Commun.	SC
Synthesis	S
Tetrahedron	T
Tetrahedron Lett.	TL
Ukr. Khim. Zh. (Russ. Ed.)	UKZ
Yakugaku Zasshi	YZ
Z. Chem.	ZC
Zh. Obshch. Khim.	ZOB
Zh. Org. Khim.	ZOR
Z. Naturforsch., Teil B	ZN(B)

Book Series

'Advances in Heterocyclic Chemistry'	AHC
'Chemistry of Heterocyclic Compounds' [Weissberger–Taylor series]	HC
'Methoden der Organischen Chemie (Houben-Weyl)'	HOU
'Organic Compounds of Sulphur, Selenium, and Tellurium' [R. Soc. Chem. series]	SST
'Physical Methods in Heterocyclic Chemistry'	PMH

Specific Books

Q. N. Porter and J. Baldas, 'Mass Spectromety of Heterocyclic Compounds', Wiley, New York, 1971	MS
T. J. Batterham, 'NMR Spectra of Simple Heterocycles', Wiley, New York, 1973	NMR

'Photochemistry of Heterocyclic Compounds', ed. O. Buchardt, Wiley, New York, 1976	PH
W. L. F. Armarego, 'Stereochemistry of Heterocyclic Compounds', Wiley, New York, 1977, parts 1 and 2	SH

Patents

Belg. Pat.	BEP
Br. Pat.	BRP
Eur. Pat.	EUP
Fr. Pat.	FRP
Ger. (East) Pat.	EGP
Ger. Pat.	GEP
Neth. Pat.	NEP
Jpn. Pat.	JAP
Jpn. Kokai	JAP(K)
S. Afr. Pat.	SAP
U.S. Pat.	USP

Other Publications

All Other Books and Journals ('Miscellaneous')	MI
All Other Patents	MIP
Personal Communications	PC
Theses	TH
Unpublished Results	UP

VOLUME 7 REFERENCES

1859CR(48)101 A. Wurtz; *C.R. Hebd. Seances Acad. Sci.*, 1859, **48**, 101. 97

1870BSF(14)113 M. Berthelot; *Bull. Soc. Chim. Fr.*, 1870, **14**, 113. 120, 122

1886CB2184 A. Baeyer; *Ber.*, 1886, **19**, 2184. 767

1888CB337 H. Bergreen; *Ber.*, 1888, **21**, 337. 478
1888CB1049 S. Gabriel; *Ber.*, 1888, **21**, 1049. 33
1888CB2669 S. Gabriel and J. Weiner; *Ber.*, 1888, **21**, 2669. 245

1899CB2039 C. C. Howard and W. Marckwald; *Ber.*, 1899, **32**, 2039. 681

11CB365 H. Staudinger and S. Jelagin; *Ber.*, 1911, **44**, 365. 473

14CB3277 M. Busch and W. Dietz; *Ber.*, 1914, **47**, 3277. 457, 487

22JA1538 R. L. Datta and B. C. Chatterjee; *J. Am. Chem. Soc.*, 1922, **44**, 1538. 486
22JCS2793 C. K. Ingold and H. A. Piggott; *J. Chem. Soc.*, 1922, **121**, 2793. 474

23JA3124 H. S. Hill and H. Hilbert; *J. Am. Chem. Soc.*, 1923, **45**, 3124. 689
23JCS2745 C. K. Ingold and H. A. Piggott; *J. Chem. Soc.*, 1923, **123**, 2745. 474

24CB653 O. Diels and H. Behncke; *Ber.*, 1924, **57B**, 653. 486
24JCS87 C. K. Ingold; *J. Chem. Soc.*, 1924, 87. 486
24JCS(125)1534 E. H. Ingold; *J. Chem. Soc.*, 1924, **125**, 1534. 124

25CB78 A. Sieglitz and H. Koch; *Ber.*, 1925, **58B**, 78. 580
25HCA65 H. Staudinger and M. Lüthy; *Helv. Chim. Acta*, 1925, **8**, 65. 691

28JIC439 P. C. Guha and T. N. Ghosh; *J. Indian Chem. Soc.*, 1928, **5**, 439. 649

29BSF1013 M. R. Ottensooser; *Bull. Soc. Chim. Fr.*, 1929, **45**, 1013. 648
29CB2416 J. von Braun and K. Weissbach; *Ber.*, 1929, **62**, 2416. 693
29MI51100 A. Fleming; *Br. J. Exp. Pathol.*, 1929, **10**, 226. 300

30JA287 E. R. Littman and C. S. Marvel; *J. Am. Chem. Soc.*, 1930, **52**, 287. 654
30JA1018 J. A. Nieuwland, R. R. Vogt and W. L. Foohey; *J. Am. Chem. Soc.*, 1930, **52**, 1018. 689
30JA4110 J. W. Hill; *J. Am. Chem. Soc.*, 1930, **52**, 4110. 580

31LA(490)201 W. Madelung and M. E. Oberwegner; *Liebigs Ann. Chem.*, 1931, **490**, 201. 120, 122, 126
31MI50500 P. Harteck and V. Kopsch; *Z. Phys. Chem. (Leipzig)*, 1931, **B12**, 327. 124
31RTC669 A. P. J. Hoogeveen; *Recl. Trav. Chim. Pays-Bas*, 1931, **50**, 669. 470

32JCS1142 C. G. Le Fèvre and R. J. W. Le Fèvre; *J. Chem. Soc.*, 1932, 1142. 480
32MI52000 E. Hückel; *Z. Phys.*, 1932, **76**, 628. 709

33CB1815 K. H. Geib and P. Harteck; *Ber.*, 1933, **66**, 1815. 124
33RTC378 A. P. J. Hoogeveen and C. W. Van Hoogstraten; *Recl. Trav. Chim. Pay-Bas*, 1933, **52**, 378. 486

34JA2177 J. R. Meadow and E. E. Reid; *J. Am. Chem. Soc.*, 1934, **56**, 2177. 694
34MI50500 K. H. Geib and P. Harteck; *Trans. Faraday Soc.*, 1934, **30**, 133. 124

35JA925 J. W. Hill and W. H. Carothers; *J. Am. Chem. Soc.*, 1935, **57**, 925. 688, 689

36JA1274 L. Pauling and D. C. Carpenter; *J. Am. Chem. Soc.*, 1936, **58**, 1274. 691, 702

37HCA109 L. Ruzicka, G. Salomon and K. E. Meyer; *Helv. Chim. Acta*, 1937, **20**, 109. 238
37LA(528)162 K. Ziegler, A. Lüttringhaus and K. Wohlgemuth; *Liebigs Ann. Chem.*, 1937, **528**, 162. 690

38JGU1545 Y. N. Yanbikov and N. Y. Dem'yanov; *J. Gen. Chem. USSR (Engl. Transl.)*, 1938, **8**, 1545. 240

40MI51100 E. Chain, H. W. Florey, A. D. Gardner, N. G. Heatley, M. A. Jennings, J. Orr-Ewing and A. G. Sanders; *Lancet*, 1940, **2**, 226. 300

43JGU352 I. Ali-Zade and B. A. Arbuzov; *J. Gen. Chem. USSR (Engl. Transl.)*, 1943, **13**, 113 (*Chem. Abstr.*, 1944, **38**, 352). 690

44CB669 A. Müller and W. Vanc; *Ber.*, 1944, **77**, 669. 561, 666
44HCA1790 B. H. Ringier; *Helv. Chim. Acta*, 1944, **27**, 1790. 651
B-44MI50500 'Beilsteins Handbuch der Organischen Chemie', Springer Verlag, Berlin, 1944, vol. 6 (2), p. 25. 98

46JA1200 C. D. Hurd and D. G. Botteron; *J. Am. Chem. Soc.*, 1946, **68**, 1200. 691

48JA2611 F. O. Davis and E. M. Fettes; *J. Am. Chem. Soc.*, 1948, **70**, 2611. 693
48JBC(175)793 O. K. Behrens, J. Corse, J. P. Edwards, L. Garrison, R. G. Jones, Q. F. Soper, F. R. Van Abeele and C. W. Whitehead; *J. Biol. Chem.*, 1948, **175**, 793. 300

49HCA275 A. Fürst and P. A. Plattner; *Helv. Chim. Acta*, 1949, **32**, 275. 110
49HCA544 L. Ruzicka, M. Kobelt, O. Häflinger and V. Prelog; *Helv. Chim. Acta*, 1949, **32**, 544. 654
49JA2571 S. L. Friess; *J. Am. Chem. Soc.*, 1949, **71**, 2571. 578, 580
B-49MI51000 R. B. Woodward; in 'The Chemistry of Penicillin', ed. H. T. Clarke, J. R. Johnson and R. Robinson; Princeton University Press, Princeton, 1949, p. 440. 286
B-49MI51100 H. T. Clarke, J. R. Johnson and R. Robinson; in ref. B-49MI51102, p. 3. 299
B-49MI51101 D. Crowfoot, C. W. Bunn, B. W. Rogers-Low and A. Turner-Jones; in ref. B-49MI51102, p. 310. 300
B-49MI51102 'The Chemistry of Penicillin', ed. H. T. Clarke, J. R. Johnson and R. Robinson; Princeton University Press, Princeton, 1949. 300, 303, 324, 326
B-49MI51103 H. W. Thompson, R. R. Brattain, H. M. Randall and R. S. Rasmussen; in ref. B-49MI51102, p. 382. 302
B-49MI51104 R. B. Woodward, A. Neuberger and N. R. Trenner; in ref. B-49MI51102, p. 415. 303
B-49MI51105 O. K. Behrens; in ref. B-49MI51102, p. 657. 328
B-49MI51106 W. E. Bachmann and M. W. Cronyn; in ref. B-49MI51102, p. 849. 331
B-49MI51107 V. du Vigneaud, F. H. Carpenter, R. W. Holley, A. H. Livermore and J. R. Rachele; in ref. B-49MI51102, p. 1018. 331

50JA2926 H. C. Brown and M. Gerstein; *J. Am. Chem. Soc.*, 1950, **72**, 2926. 240
50JCS978 G. A. R. Kon and J. J. Roberts; *J. Chem. Soc.*, 1950, 978. 626

50JOC865	J. G. Affleck and G. Dougherty; *J. Org. Chem.*, 1950, **15**, 865.	693
B-50MI50300	S. Winstein and R. B. Henderson; in 'Heterocyclic Compounds', ed. R. C. Elderfield; Wiley, New York, 1950, vol. 1, p. 1.	33
B-50MI50900	S. A. Ballard and D. S. Melstrom; in 'Heterocyclic Compounds', ed. R. C. Elderfield; Wiley, New York, 1950, vol. 1, p. 78.	283
51JA1044	N. H. Cromwell, N. G. Barker, R. A. Wankel, P. J. Vanderhorst, F. W. Olsen and J. H. Anglin, Jr.; *J. Am. Chem. Soc.*, 1951, **73**, 1044.	82
51JA2196	B. Witkop and J. B. Patrick; *J. Am. Chem. Soc.*, 1951, **73**, 2196.	659
51JA2641	B. Witkop, J. B. Patrick and M. Rosenblum; *J. Am. Chem. Soc.*, 1951, **73**, 2641.	659
51JA2773	W. E. Bachmann and E. L. Jenner; *J. Am. Chem. Soc.*, 1951, **73**, 2773.	687
51JCS200	W. Baker, J. F. W. McOmie and W. D. Ollis; *J. Chem. Soc.*, 1951, 200.	690
51JCS201	W. Baker, W. D. Ollis and T. S. Zeally; *J. Chem. Soc.*, 1951, 201.	690
51JOC1475	W. Wenner; *J. Org. Chem.*, 1951, **16**, 1475.	538
52G735	P. Papini and S. Checchi; *Gazz. Chim. Ital.*, 1952, **82**, 735 (*Chem. Abstr.*, 1953, **47**, 12 370).	648
52JA917	N. J. Leonard and J. Figueras, Jr.; *J. Am. Chem. Soc.*, 1952, **74**, 917.	584, 587
52JA1704	N. J. Leonard and R. C. Sentz; *J. Am. Chem. Soc.*, 1952, **74**, 1704.	657
52JA2082	W. G. Dauben, C. F. Hiskey and M. A. Muhs; *J. Am. Chem. Soc.*, 1952, **74**, 2082.	126
52JA5324	C. D. Hurd and W. H. Saunders, Jr.; *J. Am. Chem. Soc.*, 1952, **74**, 5324.	562
52LA(577)60	H. H. Schlubach and V. Franzen; *Liebigs Ann. Chem.*, 1952, **577**, 60.	120, 122
52M386	A. Müller, E. Srepel, E. Funder-Fritzsche and F. Dicher; *Monatsh. Chem.*, 1952, **83**, 386.	654, 655
52M1502	H. Tschamler and R. Leutner; *Monatsh. Chem.*, 1952, **83**, 1502.	550
53CB197	H. Stetter; *Chem. Ber.*, 1953, **86**, 197.	674
53JA33	D. J. Cram and M. J. Hatch; *J. Am. Chem. Soc.*, 1953, **75**, 33.	71
53JA5439	J. H. Saunders and E. E. Hardy; *J. Am. Chem. Soc.*, 1953, **75**, 5439.	463
53JA6332	G. P. Scott; *J. Am. Chem. Soc.*, 1953, **75**, 6332.	587
53JCS143	E. R. Braithwaite and J. Graymore; *J. Chem. Soc.*, 1953, 143.	480
53JOC801	A. J. Weinheimer, S. W. Kantor and C. R. Hauser; *J. Org. Chem.*, 1953, **18**, 801.	580
53JOC1356	T. D. Perrine; *J. Org. Chem.*, 1953, **18**, 1356	580
53JOC1380	L. Katz, L. S. Karger, W. Schroeder and M. S. Cohen; *J. Org. Chem.*, 1953, **18**, 1380.	645
53JPC564	T. E. Turner, V. C. Fiora, W. M. Kendrick and B. L. Hicks; *J. Phys. Chem.*, 1953, **21**, 564.	48
53M1206	A. Müller, E. Funder-Fritzsche, W. Konar and E. Rintersbacher-Wlasak; *Monatsh. Chem.*, 1953, **84**, 1206.	561, 572, 692
54CB1218	V. Franzen; *Chem. Ber.*, 1954, **87**, 1218.	122
54CR(38)982	G. Rio; *C.R. Hebd. Seances Acad. Sci.*, 1954, **38**, 982.	621
54JA630	N. J. Leonard, R. C. Fox, M. Oki and S. Chiavarelli; *J. Am. Chem. Soc.*, 1954, **76**, 630.	657
54JA2317	F. F. Blicke and N. J. Doorenbos; *J. Am. Chem. Soc.*, 1954, **76**, 2317.	654, 655
54JA4242	H. S. Gutowsky, R. L. Rutledge, M. Tamres and S. Searles; *J. Am. Chem. Soc.*, 1954, **76**, 4242.	409
54JA5708	N. J. Leonard, R. C. Fox and M. Oki; *J. Am. Chem. Soc.*, 1954, **76**, 5708.	726
54JCS3429	F. C. Cooper and M. W. Partridge; *J. Chem. Soc.*, 1954, 3429.	683
54LA(587)226	H. P. Kaufmann, A. Seher and P. Hagedorn; *Liebigs Ann. Chem.*, 1954, **587**, 226.	690
54MI51900	M. J. Astle, J. A. Zaslowsky and P. G. Lafyatis; *Ind. Eng. Chem.*, 1954, **46**, 787.	688, 689, 691
55CB717	V. Franzen; *Chem. Ber.*, 1955, **88**, 717.	122
55CB1284	D. Jerchel and W. Edler; *Chem. Ber.*, 1955, **88**, 1284.	647
55G1098	F. Sparatore; *Gazz. Chim. Ital.*, 1955, **85**, 1098.	640
55JA3393	C. W. Muth, Wei-Liang Sung and Z. B. Papanastassiou; *J. Am. Chem. Soc.*, 1955, **77**, 3393.	518
55JA6234	N. J. Leonard, M. Oki, and S. Chiavarelli; *J. Am. Chem. Soc.*, 1955, **77**, 6234.	657
55JA6241	N. J. Leonard and M. Oki, *J. Am. Chem. Soc.*, 1955, **77**, 6241.	657
55JA6245	N. J. Leonard and M. Oki, *J. Am. Chem. Soc.*, 1955, **77**, 6245.	657
55JCS2931	C. J. Brown; *J. Chem. Soc.*, 1955, 2931.	452
55JOC1147	R. G. Ackerman, W. H. Brown and G. F. Wright; *J. Org. Chem.*, 1955, **20**, 1147.	731
55LA(594)89	G. Wittig, G. Closs and F. Mindermann; *Liebigs Ann. Chem.*, 1955, **594**, 89.	529
56AG71	G. O. Schenck and N. Engelhard; *Angew. Chem.*, 1956, **68**, 71	456, 472
56CB956	G. Henseke and M. Winter; *Chem. Ber.*, 1956, **89**, 956.	648
56CB2608	K. Dimroth and G. Lenke; *Chem. Ber.*, 1956, **89**, 2608.	577
56JA848	W. E. Truce and F. J. Lotspeich; *J. Am. Chem. Soc.*, 1956, **78**, 848.	587
56JA1965	C. G. Overberger and A. Katchman; *J. Am. Chem. Soc.*, 1956, **78**, 1965.	586
56JA2250	F. D. Greene; *J. Am. Chem. Soc.*, 1956, **78**, 2250.	690
56JA4917	S. Searles, M. Tamres, F. Block and L. A. Quaterman; *J. Am. Chem. Soc.*, 1956, **78**, 4917.	238

56JA5022 R. H. Reitsema; *J. Am. Chem. Soc.*, 1956, **78**, 5022. 192

56JA6130 W. E. Truce and D. D. Emrick; *J. Am. Chem. Soc.*, 1956, **78**, 6130. 587

B-56MI50300 E. L. Eliel; in 'Steric Effects in Organic Chemistry', ed. M. S. Newman; Wiley, New York, 1956, p. 61. 33

56MI52100 A. Luttringhaus and I. Sichert-Modrow; *Makromol. Chem.*, 1956, **18–19**, 511. 731

57AG341 R. Huisgen; *Angew. Chem.*, 1957, **69**, 341. 656

57AG713 A. Schöberl and H. Gräfje; *Angew. Chem.*, 1957, **69**, 713. 693

57BRP585229 D. G. Stewart, D. Y. Waddan and E. T. Borrows; *Br. Pat.* 785 229 (1957) (*Chem. Abstr.*, 1957, **52**, 5038). 731

57BSF789 A. Kirrmann and N. Hamaide; *Bull. Soc. Chim. Fr.*, 1957, 789. 550, 559

57CB1437 R. Huisgen, H. Brade, H. Walz, and I. Glogger; *Chem. Ber.* 1957, **90**, 1437. 655

57CB2184 L. Horner and E. Jürgens; *Chem. Ber.*, 1957, **90**, 2184. 228

57JA734 R. Ghirardelli and H. J. Lucas; *J. Am. Chem. Soc.*, 1957, **79**, 734. 70

57JA1726 H. Conroy; *J. Am. Chem. Soc.*, 1957, **79**, 1726. 120

57JA4427 L. A. Carpino; *J. Am. Chem. Soc.*, 1957, **79**, 4427. 527, 538

57JA5476 N. J. Leonard, D. F. Morrow and M. T. Rogers; *J. Am. Chem. Soc.*, 1957, **79**, 5476. 726

57JA5558 K. Biemann, G. Büchi and B. H. Walker; *J. Am. Chem. Soc.*, 1957, **79**, 5558. 775

57JA5739 W. D. Emmons; *J. Am. Chem. Soc.*, 1957, **79**, 5739. 199, 201, 204, 205, 207, 208, 211, 228, 229, 234

57JA6522 W. D. Emmons; *J. Am. Chem. Soc.*, 1957, **79**, 6522. 207

57JCS614 J. I. Jones, W. Kynaston and J. H. Hales; *J. Chem. Soc.*, 1957, 614. 465

57JOC1263 M. F. Hawthorne and R. D. Strahm; *J. Org. Chem.*, 1957, **22**, 1263. 205

57MI51900 S. M. Skuratov, A. A. Strepikheev, S. M. Shtekher and A. V. Volokhina; *Proc. Acad. Sci. USSR, Phys. Chem. Sect.*, 1957, **117**, 687. 689

58AG312 R. Huisgen and H. Ott; *Angew. Chem.*, 1958, **70**, 312. 666

58BSF345 C. Sandris and G. Ourisson; *Bull. Soc. Chim. Fr.*, 1958, 345. 244, 245

58CB972 R. Huisgen and J. Witte; *Chem. Ber.*, 1958, **91**, 972. 655

58CB1057 H. Krimm; *Chem. Ber.*, 1958, **91**, 1057. 228

58CB1589 S. Olsen and R. Bredoch; *Chem. Ber.*, 1958, **91**, 1589. 580

58JA1510 A. F. McKay, E. J. Tarlton, S. I. Petri, P. R. Steyermark and M. A. Mosley; *J. Am. Chem. Soc.*, 1958, **80**, 1510. 655

58JA3064 H. J. Shine and R. H. Snyder; *J. Am. Chem. Soc.*, 1958, **80**, 3064. 580

58JA3132 J. Meinwald and H. Nozaki; *J. Am. Chem. Soc.*, 1958, **80**, 3132. 563, 580

58JA4079 P. S. Starcher and B. Phillips; *J. Am. Chem. Soc.*, 1958, **80**, 4079. 666

58JA4162 J. G. Pritchard and F. A. Long; *J. Am. Chem. Soc.*, 1958, **80**, 4162. 379

58JA5536 J. C. Sheehan, W. B. Lawson and R. J. Gaul; *J. Am. Chem. Soc.*, 1958, **80**, 5536. 192

58JA5577 P. Yates and D. R. Moore; *J. Am. Chem. Soc.*, 1958, **80**, 5577. 456, 479

58JA6404 H. K. Hall, Jr.; *J. Am. Chem. Soc.*, 1958, **80**, 6404. 681

58JA6409 H. K. Hall, Jr. and A. K. Schneider; *J. Am. Chem. Soc.*, 1958, **80**, 6409. 690

58JCS2797 G. Baddeley and J. R. Cooke; *J. Chem. Soc.*, 1958, 2797. 668

58JCS3594 W. Baker, K. M. Buggle, J. F. W. McOmie and D. A. M. Watkins; *J. Chem. Soc.*, 1958, 3594. 764

58JOC651 J. S. Splitter and M. Calvin; *J. Org. Chem.*, 1958, **23**, 651. 204, 229

58JOC1026 S. F. Birch, R. A. Dean and N. J. Hunter; *J. Org. Chem.*, 1958, **23**, 1026. 434

58LA(614)66 A. Schöberl and H. Gräfje; *Liebigs Ann. Chem.*, 1958, **614**, 66. 693

59AG283 H. J. Abendroth and G. Henrich; *Angew. Chem.*, 1959, **71**, 283. 230

59AG407 M. Schmidt and K. Blaettner; *Angew. Chem.*, 1959, **71**, 407. 696

59BSF1998 P. Cagniant and D. Cagniant; *Bull. Soc. Chim. Fr.*, 1959, 1998. 587, 693

59CB1072 S. Olsen, C. Schönheyder, A. Henriksen and B. Alstad; *Chem. Ber.*, 1959, **92**, 1072. 691

59CB2961 M. Appl and R. Huisgen; *Chem. Ber.*, 1959, **92**, 2961. 515

59CRV737 R. E. Parker and N. S. Isaacs; *Chem. Rev.*, 1959, **59**, 737. 70, 97, 108, 109

59E267 R. A. Maxwell, R. P. Mull and A. J. Plummer; *Experientia*, 1959, **7**, 267. 655

59JA1503 F. D. Greene; *J. Am. Chem. Soc.*, 1959, **81**, 1503. 690

59JA2202 H. W. Heine, M. E. Fetter and E. M. Nicholson; *J. Am. Chem. Soc.*, 1959, **81**, 2202. 64

59JA3089 J. C. Sheehan and K. R. Henery-Logan; *J. Am. Chem. Soc.*, 1959, **81**, 3089. 331

59JCS2102 R. Bonnett, V. M. Clark and A. Todd; *J. Chem. Soc.*, 1959, 2102. 228

59JCS3767 J. L. Down, J. Lewis, B. Moore and G. Wilkinson; *J. Chem. Soc.*, 1959, 3767. 731

59JOC306 N. L. Allinger and G. A. Youngdale; *J. Org. Chem.*, 1959, **24**, 306. 675

59MI51100 F. R. Batchelor, F. P. Doyle, J. H. C. Nayler and G. N. Rolinson; *Nature (London)*, 1959, **183**, 257. 300

59MI51900 T. I. Shein and V. N. Topchibasheva; *Khim. Volokna*, 1959, 21 (*Chem. Abstr.*, 1960, **54**, 7216). 655

59RTC331 E. D. Bergmann and A. Kaluszyner; *Recl. Trav. Chim. Pays-Bas*, 1959, **78**, 331. 626

59T(6)253 R. Huisgen and H. Ott; *Tetrahedron*, 1959, **6**, 253. 666

60AG415 F. Badea and C. D. Nenitzescu; *Angew. Chem.*, 1960, **72**, 415. 621

60AG781 S. R. Paulsen; *Angew. Chem.*, 1960, **72**, 781. 233

60BSF718 P. A. Laurent and P. Tarte; *Bull. Soc. Chim. Fr.*, 1960, 718. 691

60DOK(131)1338 I. L. Knunyants and E. G. Bykhovskaya; *Dokl. Akad. Nauk SSSR*, 1960, **131**, 1338. 272, 276

60G382 A. Cerniani, G. Modena and P. E. Todesco; *Gazz. Chim. Ital.*, 1960, **90**, 382. 587

60JA852 J. C. Kauer and A. K. Schneider; *J. Am. Chem. Soc.*, 1960, **82**, 852. 472

60JA890 F. D. Greene and W. W. Rees; *J. Am. Chem. Soc.*, 1960, **82**, 890. 690

60JA1428 H. E. Winberg, F. S. Fawcett, W. E. Mochel and C. W. Theobald; *J. Am. Chem. Soc.*, 1960, **82**, 1428. 767, 776

60JA1515 C. G. Krespan, B. C. McKusick and T. L. Cairns; *J. Am. Chem. Soc.*, 1960, **82**, 1515. 455, 467, 477

60JA1594 G. Cignarella and U. Teotini; *J. Am. Chem. Soc.*, 1960, **82**, 1594. 644

60JA2609 E. R. Lavagnino, R. R. Chauvette, W. N. Cannon and E. C. Kornfeld; *J. Am. Chem. Soc.*, 1960, **82**, 2609. 342, 344, 346, 347

60JA3934 Y. G. Perron, W. F. Minor, C. T. Holdrege, W. J. Gottstein, J. C. Godfrey, L. B. Crast, R. B. Babel and L. C. Cheney; *J. Am. Chem. Soc.*, 1960, **82**, 3934. 330

60JA4075 N. J. Leonard, T. W. Milligan and T. L. Brown; *J. Am. Chem. Soc.*, 1960, **82**, 4075. 556, 667, 692, 694

60JA4087 J. Meinwald, D. W. Dicker and N. Danieli; *J. Am. Chem. Soc.*, 1960, **82**, 4087. 559, 563, 580

60JA4241 N. H. Cromwell, R. E. Bambury and J. L. Adelfang; *J. Am. Chem. Soc.*, 1960, **82**, 4241. 50

60JA4930 D. V. Kashelikar and P. E. Fanta; *J. Am. Chem. Soc.*, 1960, **82**, 4930. 64

60JA6068 G. L. Closs and S. J. Brois; *J. Am. Chem. Soc.*, 1960, **82**, 6068. 81

60JA6181 D. C. England, M. A. Dietrich and R. V. Lindsey, Jr.; *J. Am. Chem. Soc.*, 1960, **82**, 6181. 477

60JOC65 J. Diamond, W. F. Bruce, C. Gochman and F. T. Tyson; *J. Org. Chem.*, 1960, **25**, 65. 655

60JOC828 E. D. Bergmann and M. Rabinovitz; *J. Org. Chem.*, 1960, **25**, 828. 589

60JOC1509 N. L. Allinger and G. A. Youngdale; *J. Org. Chem.*, 1960, **25**, 1509. 672

60MI51100 G. N. Rolinson, F. R. Batchelor, S. Stevens, J. Cameron-Wood and E. B. Chain; *Lancet*, 1960, **2**, 564. 300

60MI51900 M. Okada, S. Kozawa and Y. Yamashita; *Makromol. Chem.*, 1960, **127**, 66. 691

60OR(11)1 L. G. Donaruma and W. Z. Heldt; *Org. React.*, 1960, **11**, 1. 530

60QR317 A. D. Cross; *Q. Rev., Chem. Soc.*, 1960, **14**, 317. 120

61BCJ369 M. Igarishi; *Bull. Chem. Soc. Jpn.*, 1961, **31**, 369. 48

61BJ(79)393 D. C. Hodgkin and E. N. Maslen; *Biochem. J.*, 1961, **79**, 393. 286

61CB368 W. Reid and H. Klug; *Chem. Ber.*, 1961, **94**, 368. 477

61CB968 S. R. Paulsen and G. Huck; *Chem. Ber.*, 1961, **94**, 968. 230

61CB1606 F. Wille and R. Strasser; *Chem. Ber.*, 1961, **94**, 1606. 632

61GEP1105155 Deutsche Akademie der Wissenschaften zu Berlin, *Ger. Pat.* 1 105 155 (1961) (*Chem. Abstr.*, 1961, **55**, 24 127). 484

61JA386 R. E. Putnam, J. L. Anderson and W. H. Sharkey; *J. Am. Chem. Soc.*, 1961, **83**, 386. 472

61JA2589 W. J. Middleton, E. G. Howard and W. H. Sharkey; *J. Am. Chem. Soc.*, 1961, **83**, 2589. 460, 478

61JA2967 E. C. Taylor, W. W. Paudler and I. Kuntz, Jr.; *J. Am. Chem. Soc.*, 1961, **83**, 2967. 345

61JA3434 C. G. Krespan; *J. Am. Chem. Soc.*, 1961, **83**, 3434. 477

61JA4417 C. L. Bumgardner, K. S. McCallum and J. P. Freeman; *J. Am. Chem. Soc.*, 1961, **83**, 4417. 240

61JA4792 J. C. Sheehan and J. W. Frankenfeld; *J. Am. Chem. Soc.*, 1961, **83**, 4792. 92

61JCS1351 D. A. Barr, R. N. Haszeldine and C. J. Willis; *J. Chem. Soc.*, 1961, 1351. 457, 473

61JCS2516 G. Baddeley and N. H. P. Smith; *J. Chem. Soc.*, 1961, 2516. 577, 580

61JOC1329 D. D. Emrick and W. E. Truce; *J. Org. Chem.*, 1961, **26**, 1329. 529

61JOC2728 V. J. Traynelis and R. F. Love; *J. Org. Chem.*, 1961, **26**, 2728. 589

61JOC4002 H. H. Gibbs and M. I. Bro; *J. Org. Chem.*, 1961, **26**, 4002. 464

61LA(647)83 A. Bonati, G. F. Cristiani and E. Testa; *Liebigs Ann. Chem.*, 1961, **647**, 83. 242

61MI51900 M. Barón; *Nature (London)*, 1961, **192**, 258. 691

61TL593 H. W. Whitlock; *Tetrahedron Lett.*, 1961, 593. 588

61TL612 E. Schmitz and R. Ohme; *Tetrahedron Lett.*, 1961, 612. 196, 233

62AG(E)528 H. W. Heine; *Angew. Chem., Int. Ed. Engl.*, 1962, **1**, 528. 64

62CB680 E. Schmitz and D. Habisch; *Chem. Ber.*, 1962, **95**, 680. 230

62CB783 M. Rothe; *Chem. Ber.*, 1962, **95**, 783. 681

62CB795 E. Schmitz and R. Ohme; *Chem. Ber.*, 1962, **95**, 795. 233

62CB1759 C. Szántay and E. Schmitz; *Chem. Ber.*, 1962, **95**, 1759. 216

62CB2861 R. Gompper and W. Toepfl; *Chem. Ber.*, 1962, **95**, 2861. 480

62CJC882 J. F. King and T. Durst; *Can. J. Chem.*, 1962, **40**, 882. 51

62CJC935 P. Singh and J. L. Boivin; *Can. J. Chem.*, 1962, **40**, 935. 463

62DOK(142)596 S. P. Makarov, V. A. Shpanskii, V. A. Ginsburg, A. I. Shchekotikhin, A. S. Filatov, L. L. Martynova, I. V. Pavlovskaya, A. F. Golovaneva and A. Ya. Yakubovich; *Dokl. Akad. Nauk SSSR*, 1962, **142**, 596. 457

62GEP1126395 S. R. Paulsen; *Ger. Pat.* 1 126 395 (1962) (*Chem. Abstr.*, 1962, **57**, 9857). 216

62HCA717 B. J. R. Nicolaus, E. Bellasio and E. Testa; *Helv. Chim. Acta*, 1962, **45**, 717. 454, 463, 466, 479

62JA195 A. T. Bottini and R. E. Olsen; *J. Am. Chem. Soc.*, 1962, **84**, 195. 92

62JA399 D. C. Dittmer and M. E. Christy; *J. Am. Chem. Soc.*, 1962, **84**, 399. 426, 428, 429, 433

62JA1063	W. H. Graham; *J. Am. Chem. Soc.*, 1962, **84**, 1063.	203
62JA1684	E. M. Arnett and C. Y. Wu; *J. Am. Chem. Soc.*, 1962, **84**, 1684.	560
62JA2651	L. Pierce and V. Dobyns; *J. Am. Chem. Soc.*, 1962, **84**, 2651.	199
62JA2814	C. G. Overberger, P. Barkan, A. Lusi and H. Ringsdorf; *J. Am. Chem. Soc.*, 1962, **84**, 2814.	692
62JA2855	I. Haller and G. C. Pimentel; *J. Am. Chem. Soc.*, 1962, **84**, 855.	124
62JA2983	J. C. Sheehan and K. R. Henery-Logan; *J. Am. Chem. Soc.*, 1962, **84**, 2983.	300, 331
62JA3185	J. S. Webb, D. B. Cosulich, J. H. Mowat, J. B. Patrick, R. W. Broschard, W. E. Meyer, R. P. Williams, C. F. Wolf, W. Fulmor, C. Pidacks and J. E. Lancaster; *J. Am. Chem. Soc.*, 1962, **84**, 3185.	93
62JA3187	J. S. Webb, D. B. Cosulich, J. H. Mowat, J. B. Patrick, R. W. Broschard, W. E. Meyer, R. P. Williams, C. F. Wolf, W. Fulmor, C. Pidacks and J. E. Lancaster; *J. Am. Chem. Soc.*, 1962, **84**, 3187.	93
62JA3188	A. Tulinsky; *J. Am. Chem. Soc.*; 1962, **84**, 3188.	93
62JA3701	N. J. Leonard and C. R. Johnson; *J. Am. Chem. Soc.*, 1962, **84**, 3701.	692
62JA4782	H. E. Simmons, D. C. Blomstrom and R. D. Vest; *J. Am. Chem. Soc.*, 1962, **84**, 4782.	467, 482
62JCS3276	H. J. Emeleus and G. L. Hurst; *J. Chem. Soc.*, 1962, 3276.	451, 468
62JCS3468	W. Paterson and G. R. Proctor; *J. Chem. Soc.*, 1962, 3468.	525
62JCS4840	E. A. V. Ebsworth and G. L. Hurst; *J. Chem. Soc.*, 1962, 4840.	468
62JMC1016	Y. G. Perron, W. F. Minor, L. B. Crast, A. Gourevitch, J. Lein and L. C. Cheney; *J. Med. Chem.*, 1962, **5**, 1016.	330
62JMC1199	S. O. Winthrop, M. A. Davis, F. Herr, J. Stewart and R. Gaudry; *J. Med. Chem.*, 1962, **5**, 1199.	517
62JOC282	N. J. Leonard and C. R. Johnson; *J. Org. Chem.*, 1962, **27**, 282.	692
62JOC2668	G. Cignarella, G. Pifferi and E. Testa; *J. Org. Chem.*, 1962, **27**, 2668.	316
62JOC3557	G. Smolinsky; *J. Org. Chem.*, 1962, **27**, 3557.	50, 85
62JOC3844	R. T. Conley and L. J. Frainier; *J. Org. Chem.*, 1962, **27**, 3844.	658
62JOC3995	C. G. Krespan and W. R. Brasen; *J. Org. Chem.*, 1962, **27**, 3995.	485
62JOC4671	L. H. Sternbach, B. A. Koechlin and E. Reeder; *J. Org. hem.*, 1962, **27**, 4671.	205
62JPR(18)255	E. Taeger and Z. El-Hewehi; *J. Prakt. Chem.*, 1962, **18**, 255.	193
62LA(659)81	E. Lippert and H. Prigge; *Liebigs Ann. Chem.*, 1962, **659**, 81.	376
B-62MI50400	W. C. G. Ross; 'Biological Alkylating Agents', Butterworths, London, 1962.	69
62MI50800	H. M. Frey and I. D. R. Stevens; *Proc. Chem. Soc.*, 1962, 79.	225
62MI51100	B. K. Koe; *Nature (London)*, 1962, **195**, 1200.	331
62MI51800	G. De Stevens; *Record Chem. Progr.*, 1962, **23**, 105.	605
62MI51900	H. Cherdron, L. Hoehr and W. Kern; *Makromol. Chem.*, 1962, **52**, 48.	691
62MI51901	E. C. Craven, H. Jowitt and W. R. Ward; *J. Appl. Chem.*, 1962, **12**, 526.	691
62T1527	H. Beecken and F. Korte; *Tetrahedron*, 1962, **18**, 1527.	473
63AG(E)152	G. Opitz and J. Koch; *Angew. Chem., Int. Ed. Engl.*, 1963, **2**, 152.	249
63AX617	H. H. Cady, A. C. Larson and D. T. Cromer; *Acta Crystallogr.*, 1963, **16**, 617.	702
63BJ(86)514	S. Abrahamsson, D. C. Hodgkin and E. N. Maslen; *Biochem. J.*, 1963, **86**, 514.	300
63BRP924589	Calanda-Stiftung, *Br. Pat.* 924 589 (1963) (*Chem. Abstr.*, 1963, **59**, 11 424).	250
63BSF1805	R. Reynaud and P. Rumpf; *Bull. Soc. Chim. Fr.*, 1963, 1805.	509
63CB1827	R. Pfleger, E. Garthe and K. Rauer; *Chem. Ber.*, 1963, **96**, 1827.	689
63CJC1893	M. Barón, O. B. de Mandirola and J. F. Westerkamp; *Can. J. Chem.*, 1963, **41**, 1893.	691, 702
63DOK(152)1104	V. A. Ginsburg, S. S. Dubov, A. N. Medvedev, L. L. Martynova, B. I. Tetel'baum, M. N. Vasil'eva and A. Ya. Yakubovich; *Dokl. Akad. Nauk SSSR*, 1963, **152**, 1104.	473
63HCA1696	K. Schenker and J. Druey; *Helv. Chim. Acta*, 1963, **46**, 1696.	629
63HOU(6/2)511	H. Kröper; *Methoden Org. Chem. (Houben-Weyl)*, 1963, **6/2**, 511.	364, 398
63IC641	R. B. King; *Inorg. Chem.*, 1963, **2**, 641.	485
63JA97	C. L. Bumgardner, K. J. Martin and J. P. Freeman; *J. Am. Chem. Soc.*, 1963, **85**, 97.	241
63JA560	J. G. Wilson and L. A. Cohen; *J. Am. Chem. Soc.*, 1963, **85**, 560.	468
63JA955	E. P. Blanchard, Jr. and G. Büchi; *J. Am. Chem. Soc.*, 1963, **85**, 955.	115
63JA1896	R. B. Morin, B. G. Jackson, R. A. Mueller, E. R. Lavagnino, W. B. Scanlon and S. L. Andrews; *J. Am. Chem. Soc.*, 1963, **85**, 1896.	300
63JA3236	W. E. Truce and J. R. Norell; *J. Am. Chem. Soc.*, 1963, **85**, 3236.	420, 429, 430
63JA3524	K. R. Henery-Logan and J. V. Rodricks; *J. Am. Chem. Soc.*, 1963, **85**, 3524.	269, 278
63JCP(39)3534	G. S. Paulett and R. Ettinger; *J. Chem. Phys.*, 1963, **39**, 3534.	204
63JCS674	C. W. Bird; *J. Chem. Soc.*, 1963, 674.	460, 472
63JCS5496	J. D. Loudon and L. B. Young; *J. Chem. Soc.*, 1963, 5496.	646
63JOC511	L. W. Deady, G. J. Leary, R. D. Topsom and J. Vaughan; *J. Org. Chem.*, 1963, **28**, 511.	245
63JOC743	C. W. Whitehead and J. J. Traverso; *J. Org. Chem.*, 1963, **28**, 743.	649
63JOC1383	R. A. Braun; *J. Org. Chem.*, 1963, **28**, 1383.	563, 579
63JOC2271	H. O. House and W. F. Berkowitz; *J. Org. Chem.*, 1963, **28**, 2271.	83
63JOC2459	W. Metlesics, G. Silverman and L. H. Sternbach; *J. Org. Chem.*, 1963, **28**, 2459.	228
63JOC3007	N. L. Allinger, W. Szkrybalo and M. A. DaRooge; *J. Org. Chem.*, 1963, **28**, 3007	675
63LA(661)181	G. Cignarella, G. F. Cristiani and E. Testa; *Liebigs Ann. Chem.*, 1963, **661**, 181.	251

B-63MI50500	'Varian High Resolution NMR Spectra Catalog', Varian Associates, Palo Alto, 1963, vols. 1 and 2.	98
B-63MI50501	F. M. Dean; 'Naturally Occurring Oxygen Ring Compounds', Butterworth, London, 1963, pp. 549–560.	120
63MI50600	C. S. Hahn; *Daehan Howhak Hwaejee*, 1963, **7**, 230 (*Chem. Abstr.*, 1964, **61**, 13298).	159
63MI51800	G. E. Lee and W. R. Wragg; *J. Pharm. Pharmacol.*, 1963, **15**, 589 (*Chem. Abstr.*, 1963, **59**, 13 989).	649
63MI51900	K. Hayashi, M. Nishii and S. Okamura; *J. Polym. Sci., Part C*, 1963, **4**, 839.	691, 695
63PMH(1)1	A. Albert; *Phys. Methods Heterocycl. Chem.*, 1963, **1**, 1.	24
63PMH(1)177	W. Pfleiderer; *Phys. Methods Heterocycl. Chem.*, 1963, **1**, 177.	97
63PMH(2)10	S. F. Mason; *Phys. Methods Heterocycl. Chem.*, 1963, **2**, 10.	99
63PMH(2)161	A. R. Katritzky and A. P. Ambler; *Phys. Methods Heterocycl. Chem.*, 1963, **2**, 161.	12, 14
63PMH(2)176	A. R. Katritzky and A. P. Ambler; *Phys. Methods Heterocycl. Chem.*, 1963, **2**, 176.	99
63RTC68	L. Brandsma and H. E. Wijers; *Recl. Trav. Chim. Pays-Bas*, 1963, **82**, 68.	436
63SA769	J. R. Durig and R. C. Lord; *Spectrochim. Acta*, 1963, **19**, 9.	453, 455
63T133	G. Kresze and A. Trede; *Tetrahedron*, 1963, **19**, 133.	473
63T1851	Ya. L. Gol'dfarb, S. Z. Taits and L. I. Belen'kii; *Tetrahedron*, 1963, **19**, 1851.	766
63TL1637	G. S. Sidhu, G. Thyagarajan and N. Rao; *Tetrahedron Lett.*, 1963, 1637.	648
63USP3071594	E. Enders and G. Unterstenhoefer; *U.S. Pat. 3 071 594* (1963) (*Chem. Abstr.*, 1963, **58**, 11 329)	447
63USP3284466	N. A. Rosenthal; *U.S. Pat.* 3 284 466 (1963) (*Chem. Abstr.*, 1968, **68**, 13 015).	693
64AG(E)229	A. Jankowski and S. R. Paulsen; *Angew. Chem., Int. Ed. Engl.*, 64, **3**, 229.	217
64AG(E)333	E. Schmitz; *Angew. Chem., Int. Ed. Engl.*, 1964, **3**, 333.	227
64AG(E)380	R. W. Hoffmann and H. Hauser; *Angew. Chem., Int. Ed. Engl.*, 1964, **3**, 380.	473
64AG(E)510	E. Vogel, R. Schubart and W. A. Böll; *Angew. Chem., Int. Ed. Engl.*, 1964, **3**, 510.	581
64AG(E)642	E. Vogel, M. Biskup, W. Pretzer and W. A. Böll; *Angew. Chem., Int. Ed. Engl.*, 1964, **3**, 642.	83, 582
64AP623	R. Neidlein; *Arch. Pharm. (Weinheim, Ger.)*, 1964, **297**, 623.	458
64BAU1948	Ya. L. Gol'dfarb, S. Z. Taits, T. S. Chirkova and L. I. Belen'kii; *Bull. Acad. Sci. USSR, Div. Chem. Sci.*, 1964, 1948.	776
64CB49	E. Schmitz and K. Schinkowski; *Chem. Ber.*, 1964, **97**, 49.	217, 230
64CB747	G. Wittig and A. Maercker; *Chem. Ber.*, 1964, **97**, 747.	713
64CB769	G. Wittig and D. Hellwinkel; *Chem. Ber.*, 1964, **97**, 769.	713
64CB1232	H. Eilingsfeld, G. Neubauer, M. Seefelder and H. Weidinger; *Chem. Ber.*, 1964, **97**, 1232.	471
64CB2521	E. Schmitz, R. Ohme and S. Schramm; *Chem. Ber.*, 1964, **97**, 2521.	209
64CI(L)2097	S. Searles, Jr. and R. E. Wann; *Chem. Ind. (London)*, 1964, 2097.	256
64CJC926	K. Tori, Y. Hata, R. Muneyuki, Y. Takano, T. Tsuji and H. Tanida; *Can. J. Chem.*, 1964, **42**, 926.	51
64CJC2107	W. K. Busfield, M. J. Taylor and E. Whalley; *Can. J. Chem.*, 1964, **42**, 2107.	453, 456
64E620	B. Camerino and B. Patelli; *Experientia*, 1964, **20**, 260.	770
64HC(19-1)1	A. Rosowsky; *Chem. Heterocycl. Compd.*, 1964, **19-1**, 1–523.	33, 97, 99, 108, 114
64HC(19-1)17	A. Rosowsky; *Chem. Heterocycl. Compd.*, 1964, **19-1**, 17.	99
64HC(19-1)24	A. Rosowsky; *Chem. Heterocycl. Compd.*, 1964, **19-1**, 24.	192
64HC(19-1)181	A. Rosowsky; *Chem. Heterocycl. Compd.*, 1964, **19-1**, 181.	112
64HC(19-1)188	A. Rosowsky; *Chem. Heterocycl. Compd.*, 1964, **19-1**, 188.	113
64HC(19-1)199	A. Rosowsky; *Chem. Heterocycl. Compd.*, 1964, **19-1**, 199.	112
64HC(19-1)228	A. Rosowsky; *Chem. Heterocycl. Compd.*, 1964, **19-1**, 228.	106
64HC(19-1)230	A. Rosowsky; *Chem. Heterocycl. Compd.*, 1964, **19-1**, 230.	103
64HC(19-1)262	A. Rosowsky; *Chem. Heterocycl. Compd.*, 1964, **19-1**, 262.	105
64HC(19-1)394	A. Rosowsky; *Chem. Heterocycl. Compd.*, 1964, **19-1**, 394.	112
64HC(19-1)418	A. Rosowsky; *Chem. Heterocycl. Compd.*, 1964, **19-1**, 418.	112
64HC(19-1)431	A. Rosowsky; *Chem. Heterocycl. Compd.*, 1964, **19-1**, 431.	112
64HC(19-1)432	A. Rosowsky; *Chem. Heterocycl. Compd.*, 1964, **19-1**, 432.	112
64HC(19-1)436	A. Rosowsky; *Chem. Heterocycl. Compd.*, 1964, **19-1**, 436.	106
64HC(19-1)445	A. Rosowsky; *Chem. Heterocycl. Compd.*, 1964, **19-1**, 445.	106
64HC(19-1)446	A. Rosowsky; *Chem. Heterocycl. Compd.*, 1964, **19-1**, 446.	105
64HC(19-1)451	A. Rosowsky; *Chem. Heterocycl. Compd.*, 1964, **19-1**, 451.	103
64HC(19-1)453	A. Rosowsky; *Chem. Heterocycl. Compd.*, 1964, **19-1**, 453.	106
64HC(19-1)524	P. E. Fanta; *Chem. Heterocycl. Compd.*, 1964, **19-1**, 524.	68, 70, 80
64HC(19-1)525	P. E. Fanta; *Chem. Heterocycl. Compd.*, 1964, **19-1**, 525.	33
64HC(19-1)535	P. E. Fanta; *Chem. Heterocycl. Compd.*, 1964, **19-1**, 535.	36
64HC(19-1)576	D. D. Reynolds and D. L. Fields; *Chem. Heterocycl. Compd.*, 1964, **19-1**, 576.	140, 171
64HC(19-1)591	D. D. Reynolds and D. L. Fields; *Chem. Heterocycl. Compd.*, 1964, **19-1**, 591.	36
64HC(19-1)624	W. D. Emmons; *Chem. Heterocyl. Compd.*, 1964, **19-1**, 624.	195
64HC(19-2)647	Y. Etienne, R. Soulas and H. Lumbroso; *Chem. Heterocycl. Compd.*, 1964, **19-2**, 647.	404, 420, 434, 436, 439
64HC(19-2)729	Y. Etienne and N. Fischer; *Chem. Heterocycl. Compd.*, 1964, **19-2**, 729.	364, 367, 379, 386, 395, 398
64HC(19-2)787	Y. Etienne and N. Fischer; *Chem. Heterocycl. Compd.*, 1964, **19-2**, 787.	33
64HC(19-2)885	J. A. Moore; *Chem. Heterocycl. Compd.*, 1964, **19-2**, 885.	33, 238, 240, 243, 244, 245, 267

64HC(19-2)886 J. A. Moore; *Chem. Heterocycl. Compd.*, 1964, **19-2**, 886. 25
64HC(19-2)900 J. A. Moore; *Chem. Heterocycl. Compd.*, 1964, **19-2**, 900. 256
64HC(19-2)904 J. A. Moore; *Chem. Heterocycl. Compd.*, 1964, **19-2**, 904. 242
64HC(19-2)908 J. A. Moore; *Chem. Heterocycl. Compd.*, 1964, **19-2**, 908. 246
64HC(19-2)951 J. A. Moore; *Chem. Heterocycl. Compd.*, 1964, **19-2**, 951. 267
64HC(19-2)953 J. A. Moore; *Chem. Heterocycl. Compd.*, 1964, **19-2**, 953. 36
64HC(19-2)983 S. Searles, Jr.; *Chem. Heterocycl Compd.*, 1964, **19-2**, 983. 364, 365, 378, 383, 391
64HCA745 A. Ebnöther and E. Jucker; *Helv. Chim. Acta*, 1964, **47**, 745. 344
64HCA1163 F. Hunziker, F. Künzle, O. Schindler and J. Schmutz; *Helv. Chim. Acta*, 1964, **47**, 1163. 509
64HCA2145 C. A. Trob and V. Krasnobajew; *Helv. Chim. Acta*, 1964, **47**, 2145. 342, 344, 346, 347
64JA292 M. J. Amrich and J. A. Bell; *J. Am. Chem. Soc.*, 1964, **86**, 292. 221
64JA950 E. J. Corey and J. Streith; *J. Am. Chem. Soc.*, 1964, **86**, 950. 357, 359
64JA2224 W. H. Urry, P. Szecsi, C. Ikoku and D. W. Moore; *J. Am. Chem. Soc.*, 1964, **86**, 2224. 470
64JA3168 F. Sondheimer and A. Shani; *J. Am. Chem. Soc.*, 1964, **86**, 3168. 582
64JA3870 N. H. Grant and H. E. Alburn; *J. Am. Chem. Soc.*, 1964, **86**, 3870. 330
64JA4618 J. J. Uebel and J. C. Martin; *J. Am. Chem. Soc.*, 1964, **86**, 4618. 203
64JA4866 R. N. McDonald and P. A. Schwab; *J. Am. Chem. Soc.*, 1964, **86**, 4866. 122, 124, 126, 186
64JA4871 J. K. Stille and D. D. Whitehurst; *J. Am. Chem. Soc.*, 1964, **86**, 4871. 122, 186
64JCP(40)1693 R. Ettinger; *J. Chem. Phys.*, 1964, **40**, 1693. 203
64JCP(41)3504 C. B. Moore and G. C. Pimentel; *J. Chem. Phys.*, 1964, **41**, 3504. 221
64JCS2326 D. M. Hall and J. M. Insole; *J. Chem. Soc.*, 1964, 2326. 675
64JCS2944 K. V. Martin; *J. Chem. Soc.*, 1964, 2944. 478
64JCS4066 R. E. Banks, R. N. Haszeldine and H. Sutcliffe; *J. Chem. Soc.*, 1964, 4066. 457
64JCS5284 C. W. Bird; *J. Chem. Soc.*, 1964, 5284. 472
64JGU1242 A. A. Ponomarev and I. S. Monakhova; *J. Gen. Chem. USSR (Engl. Transl.)*, 1964, **34**, 1242. 692
64JHC59 R. A. Mitsch; *J. Heterocycl. Chem.*, 1964, **1**, 59. 223
64JHC233 R. A. Mitsch; *J. Heterocycl. Chem.*, 1964, **1**, 233. 226
64JHC293 P. E. Fanta and M. K. Kathan; *J. Heterocycl. Chem.*, 1964, **1**, 293. 64
64JMC57 J. Diamond, W. F. Bruce and F. T. Tyson; *J. Med. Chem.*, 1964, **7**, 57. 545
64JMC483 Y. G. Perron, L. B. Crast, J. M. Essery, R. R. Fraser, J. C. Godfrey, C. T. Holdrege, W. F. Minor, M. E. Neubert, R. A. Partyka and L. C. Cheney; *J. Med. Chem.*, 1964, **7**, 483. 313
64JMC626 C. J. Paget and C. S. Davis; *J. Med. Chem.*, 1964, **7**, 626. 213
64JOC123 V. J. Traynelis, W. L. Hergenrother, H. T. Hanson and J. A. Valicenti; *J. Org. Chem.*, 1964, **29**, 123. 559, 577, 580
64JOC497 D. C. Dittmer, H. E. Simmons and R. D. Vest; *J. Org. Chem.*, 1964, **29**, 497. 465, 480
64JOC1275 E. L. Storgryn, M. H. Gianni and A. J. Passannante; *J. Org. Chem.*, 1964, **29**, 1275. 580
64JOC1316 R. D. Clark and G. K. Helmkamp; *J. Org. Chem.*, 1964, **29**, 1316. 74
64JOC2702 D. J. Pettitt and G. K. Helmkamp; *J. Org. Chem.*, 1964, **29**, 2702. 694
64JOC3049 A. L. Logothetis; *J. Org. Chem.*, 1964, **29**, 3049. 60, 91
64JOC3188 J. H. Hall; *J. Org. Chem.*, 1964, **29**, 3188. 462
64JPS577 S. J. Childress and M. I. Gluckman; *J. Pharm. Sci.*, 1964, **53**, 577. 608
64LA(661)187 E. Testa and L. Fontanella; *Liebigs Ann. Chem.*, 1964, **661**, 187. 239
64LA(673)166 K. W. Glombitza; *Liebigs Ann. Chem.*, 1964, **673**, 166. 331
64MI50600 J. R. Brown and E. Mastromatteo; *Am. Ind. Hyg. Assoc. J.*, 1964, **25**, 560 (*Chem. Abstr.*, 1965, **63**, 6230). 183
64MI51700 C. G. Overberger and J. Weise; *J. Polym. Sci., Part B*, 1964, **2**, 329. 573, 584, 587
64PAC(9)527 H. M. Frey; *Pure Appl. Chem.*, 1964, **9**, 527. 225
64RCR493 M. G. Lin'kova, N. D. Kuleshova and I. L. Knunyants; *Russ. Chem. Rev. (Engl. Transl.)*, 1964, **33**, 493. 429, 435
64TL359 D. P. Young, W. E. McEwen, D. C. Velez, J. W. Johnson and C. A. VanderWerf; *Tetrahedron Lett.*, 1964, 359. 153
64TL1929 H. Friebolin, R. Mecke, S. Kabuss and A. Luttringhaus; *Tetrahedron Lett.*, 1964, 1929. 623
64USP3163656 J. L. Anderson and D. C. England; *U.S. Pat.* 3 163 656 (1964) (*Chem. Abstr.*, 1965, **62**, 11 823). 485
64ZC302 W. Schroth, W. Kiessling, J. Peschel and U. Schmidt; *Z. Chem.*, 1964, **4**, 302. 695

65ACS281 B. Ekström, A. Goméz-Revilla, R. Mollberg, H. Thelin and B. Sjöberg; *Acta Chem. Scand.*, 1965, **19**, 281. 331
65ACS1245 B. Ekström and B. Sjöberg; *Acta Chem. Scand.*, 1965, **19**, 1245. 331
65AF49 H. Koenig, H. Metzger and K. Seelert; *Arzneim.-Forsch.*, 1965, 49. 88
65AG(E)430 K. Dickore and E. Kuhle; *Angew. Chem., Int. Ed. Engl.*, 1965, **4**, 430. 478
65AG(E)520 H. Ulrich and A. A. R. Sayigh; *Angew. Chem., Int. Ed. Engl.*, 1965, **4**, 520. 474
65BSF2755 J.-L. Ripoll and J.-M. Conia; *Bull. Soc. Chim. Fr.*, 1965, 2755. 186
65CB576 D. Hellwinkel; *Chem. Ber.*, 1965, **98**, 576. 713
65CB789 E. Dane and T. Dockner; *Chem. Ber.*, 1965, **98**, 789. 330, 331
65CB983 E. Uhlig, M. Gentschew and A. Martin; *Chem. Ber.*, 1965, **98**, 983. 630
65CB2509 E. Schmitz, A. Stark and C. Hörig; *Chem. Ber.*, 1965, **98**, 2509. 221

65CB2516 E. Schmitz, R. Ohme and D. Murawski; *Chem. Ber.*, 1965, **98**, 2516. 229
65CB3228 H. Stetter and K. Findeisen; *Chem. Ber.*, 1965, **98**, 3228. 682
65CC468 W. J. Theuer and J. A. Moore; *Chem. Commun.*, 1965, 468. 276
65CCC445 J. O. Jilek, J. Pomykáček, E. Svátek, V. Seidlová, M. Rajšner, K. Pelz, B. Hoch and M. Protiva; *Collect. Czech. Chem. Commun.*, 1965, **30**, 445. 517, 525
65CCC3016 R. Zahradnik and C. Parkanyi; *Collect. Czech. Chem. Commun.*, 1965, **30**, 3016. 419
65CI(L)184 S. Masamune and N. T. Castellucci; *Chem. Ind.* (*London*), 1965, 184. 583
65CJC3433 S. N. Alam and D. B. MacLean; *Can. J. Chem.*, 1965, **43**, 3433. 584
65E360 L. Novak and J. Weichet; *Experientia*, 1965, **21**, 360. 330
65FES323 E. Testa and L. Fontanella; *Farmaco Ed. Sci.*, 1965, **20**, 323. 629
65HCA336 J. Schmutz, F. Künzle, F. Hunziker and A. Bürki; *Helv. Chim. Acta*, 1965, **48**, 336. 532
65HOU(6/3)489 G. Dittus; *Methoden Org. Chem.* (*Houben-Weyl*), 1965, **6/3**, 489. 364
65JA749 A. L. Logothetis; *J. Am. Chem. Soc.*, 1965, **87**, 749. 69
65JA758 R. A. Mitsch; *J. Am. Chem. Soc.*, 1965, **87**, 758. 224, 226
65JA1050 A. B. Turner, H. W. Heine, J. Irving and J. B. Bush; *J. Am. Chem. Soc.*, 1965, **87**, 1050. 72
65JA1353 E. J. Corey and M. Chaykovsky; *J. Am. Chem. Soc.*, 1965, **87**, 1353. 88
65JA1706 G. A. Melson and D. H. Busch; *J. Am. Chem. Soc.*, 1965, **87**, 1706. 715
65JA2518 E. J. Corey and A. M. Felix; *J. Am. Chem. Soc.*, 1965, **87**, 2518. 359
65JA2665 R. F. R. Church, A. S. Kende and M. J. Weiss; *J. Am. Chem. Soc.*, 1965, **87**, 2665. 203, 223, 231, 233, 236
65JA3417 L. A. Paquette, J. H. Barrett, R. P. Spitz and R. Pitcher; *J. Am. Chem. Soc.*, 1965, **87**, 3417. 560
65JA3784 L. A. Paquette and R. A. Begland; *J. Am. Chem. Soc.*, 1965, **87**, 3784. 667
65JA4396 W. H. Graham; *J. Am. Chem. Soc.*, 1965, **87**, 4396. 233
65JA4538 J. A. Deyrup and R. B. Greenwald; *J. Am. Chem. Soc.*, 1965, **87**, 4538. 74
65JCP(43)1190 D. W. Boone, C. O. Britt and J. E. Boggs; *J. Chem. Phys.*, 1965, **43**, 1190. 370
65JCS1560 D. M. Jones and N. F. Wood; *J. Chem. Soc.*, 1965, 1560. 689
65JCS1653 W. Carruthers and R. A. W. Johnstone; *J. Chem. Soc.*, 1965, 1653 625, 686
65JCS3616 T. Leigh; *J. Chem. Soc.*, 1965, 3616. 324
65JCS5602 R. E. Banks, R. N. Haszeldine and D. R. Taylor; *J. Chem. Soc.*, 1965, 5602. 457, 473
65JGU855 V. A. Ginsburgh, L. L. Martynova, S. S. Dubov, B. I. Tetel'baum and A. Ya. Yakubovich; *J. Gen. Chem. USSR* (*Engl. Transl.*), 1965, **35**, 855. 457, 473
65JHC371 R. A. Mitsch, E. W. Neuvar, R. J. Koshar and D. H. Dybvig; *J. Heterocycl. Chem.*, 1965, **2**, 371. 203, 224
65JMC74 L. H. Werner, S. Ricca, E. Mohacsi, A. Rossi and V. P. Arya; *J. Med. Chem.*, 1965, **8**, 74. 531
65JOC88 E. L. Stogryn and S. J. Brois; *J. Org. Chem.*, 1965, **30**, 88. 65
65JOC335 D. R. Larkin; *J. Org. Chem.*, 1965, **30**, 335. 578, 580
65JOC629 L. A. Paquette; *J. Org. Chem.*, 1965, **30**, 629. 433
65JOC1375 W. J. Middleton, E. G. Howard and W. H. Sharkey; *J. Org. Chem.*, 1965, **30**, 1375. 478
65JOC1398 W. J. Middleton and C. G. Krespan; *J. Org. Chem.*, 1965, **30**, 1398. 232
65JOC1748 A. Hassner and C. Heathcock; *J. Org. Chem.*, 1965, **30**, 1748. 68
65JOC2108 W. H. Graham; *J. Org. Chem.*, 1965, **30**, 2108. 233
65JOC3135 L. D. Quin and F. A. Shelburne; *J. Org. Chem.*, 1965, **30**, 3135. 625
65JOC3427 J. S. Splitter and M. Calvin; *J. Org. Chem.*, 1965, **30**, 3427. 204, 205
65JOC4053 B. R. Baker and T. L. Hullar; *J. Org. Chem.*, 1965, **30**, 4053. 81
65JOC4188 W. R. Brasen, H. N. Cripps, C. G. Bottomley, M. W. Farlow and C. G. Krespan; *J. Org. Chem.*, 1965, **30**, 4188. 140, 144, 146, 166, 179
65JOU731 V. S. Etlis; *J. Org. Chem. USSR* (*Engl. Transl.*), 1965, **1**, 731. 151
65LA(682)1 E. Vogel, R. Erb, G. Lenz and A. A. Bothner-By; *Liebigs Ann. Chem.*, 1965, **682**, 1. 506, 525
65LA(684)92 G. Opitz, H. Schempp and H. Adolph; *Liebigs Ann. Chem.*, 1965, **684**, 92. 432, 445
65LA(684)103 G. Opitz and H. Schempp; *Liebigs Ann. Chem.*, 1965, **684**, 103. 446
65M182 J. O. Jilek, V. Seidlova, E. Svatek, M. Protiva, J. Pomykacek and Z. Sedivy; *Monatsh. Chem.*, 1965, **96**, 182. 587
B-65MI50500 'Beilsteins Handbuch der Organischen Chemie', Springer Verlag, Berlin, 1965, vol. 6 (3), p. 218. 98
B-65MI50501 Ref. 65MI50500, p. 217. 98
65MI51100 W. J. Leanza, B. G. Christensen, E. F. Rogers and A. A. Patchett; *Nature* (*London*), 1965, **207**, 1395. 331
65MI51500 J. Lee and K. G. Orrell; *Trans. Faraday Soc.*, 1965, **61**, 2342. 473
65OS(45)83 E. Schmitz and R. Ohme; *Org. Synth.*, 1965, **45**, 83. 233
65PNA(54)1133 D. J. Tipper and J. L. Strominger; *Proc. Natl. Acad. Sci. USA*, 1965, **54**, 1133. 296
65T449 A. K. Bose, M. S. Manhas and R. M. Ramer; *Tetrahedron*, 1965, **21**, 449. 257
65T1299 A. J. Neale, T. J. Rawlings and E. B. McCall; *Tetrahedron*, 1965, **21**, 1299. 581, 584, 587, 589
65T2725 P. G. Gassman and D. C. Heckert; *Tetrahedron*, 1965, **21**, 2725. 346
65TL979 J.-L. Ripoll and J.-M. Conia; *Tetrahedron Lett.*, 1965, 979. 186
65TL1857 E. Schmitz, R. Ohme and S. Schramm; *Tetrahedron Lett.*, 1965, 1857. 198
65TL3325 K. D. Barrow and T. M. Spotswood; *Tetrahedron Lett.*, 1965, 3325. 248
65TL3473 T. Kikuchi and S. Uyeo; *Tetrahedron Lett.*, 1965, 3473. 249
65TL4259 N. J. Leonard and A. E. Yethon; *Tetrahedron Lett.*, 1965, 4259. 697

65TL4265	K. T. Go and I. C. Paul; *Tetrahedron Lett.*, 1965, 4265.	697
65TL4369	A. E. Pohland, R. C. Badger and N. H. Cromwell; *Tetrahedron Lett.*, 1965, 4369.	50
65ZOR1099	A. Spasov and B. Panaiotova; *Zh. Org. Khim.*, 1965, **1**, 1099 (*Chem. Abstr.*, 1965, **63**, 11 469).	252
66ACS1044	S. Abrahamsson and B. Nilsson; *Acta Chem. Scand.*, 1966, **20**, 1044.	400
66AG(E)250	B. Krebs and G. Gattow; *Angew. Chem., Int. Ed. Engl.*, 1966, **5**, 250.	695
66AG(E)416	F. Effenberger and R. Maier; *Angew. Chem., Int. Ed. Engl.*, 1966, **5**, 416.	276, 472
66AG(E)585	R. Fuks, R. Buijle and H. G. Viehe; *Angew. Chem., Int. Ed. Engl.*, 1966, **5**, 585.	272, 276
66AG(E)704	H. Ulrich and A. A. R. Sayigh; *Angew. Chem., Int. Ed. Engl.*, 1966, **5**, 704.	471
66AG(E)970	K. Dickoré and R. Wegler; *Angew. Chem., Int. Ed. Engl.*, 1966, **5**, 970.	480
66AG(E)1039	H. J. Bestmann and R. Kunstmann; *Angew. Chem., Int. Ed. Engl.*, 1966, **5**, 1039.	89
66CB634	K. Dimroth, G. Pohl and H. Follmann; *Chem. Ber.*, 1966, **99**, 634.	580, 581, 584
66CB2104	R. Ohme, E. Schmitz and P. Dolge; *Chem. Ber.*, 1966, **99**, 2104.	231
66CB3103	A. Haas and P. Schott; *Chem. Ber.*, 1966, **99**, 3103.	461
66CB3903	F. Effenberger and R. Gleiter; *Chem. Ber.*, 1966, **99**, 3903.	462, 473
66CC289	R. A. Clasen and S. Searles, Jr.; *Chem. Commun.*, 1966, 289.	243
66CC577	M. S. Raasch; *Chem. Commun.*, 1966, 577.	478
66CC803	J. M. Lehn and F. G. Riddell; *Chem. Commun.*, 1966, 803.	702
66CCC1831	J. V. Cerny and J. Polacek; *Collect. Czech. Chem. Commun.*, 1966, **31**, 1831.	425
66CI(L)1555	D. L. Tuleen and T. B. Stephens; *Chem. Ind. (London)*, 1966, 1555.	425
66CRV297	M. Sander; *Chem. Rev.*, 1966, **66**, 297.	140, 144, 145, 147, 158, 171, 178, 179
66CRV341	M. Sander; *Chem. Rev.*, 1966, **66**, 341.	404
66DOK(169)839	V. A. Naumov; *Dokl. Akad. Nauk SSSR*, 1966, **169**, 839.	49
66FRP1430508	Imperial Chemical Industries Ltd., *Fr. Pat.* 1 430 508 (1966) (*Chem. Abstr.*, 1966, **65**, 13 921).	484
66HCA204	A. Closse, R. Mauli and H. P. Sigg; *Helv. Chim. Acta*, 1966, **49**, 204.	192
66HCA1433	F. Hunziker, F. Künzle and J. Schmutz; *Helv. Chim. Acta*, 1966, **49**, 1433.	514
66JA852	R. B. Woodward, K. Heusler, J. Gosteli, P. Naegeli, W. Oppolzer, R. Ramage, S. Ranganathan and H. Vorbruggen; *J. Am. Chem. Soc.*, 1966, **88**, 852.	295
66JA1242	P. J. Wagner, C. A. Stout, S. Searles, Jr. and G. S. Hammond; *J. Am. Chem. Soc.*, 1966, **88**, 1242.	373, 394
66JA1580	E. M. Burgess and L. McCullagh; *J. Am. Chem. Soc.*, 1966, **88**, 1580.	276
66JA3582	R. E. Buckles and L. A. McGrew; *J. Am. Chem. Soc.*, 1966, **88**, 3582.	458, 474
66JA3657	E. J. Moriconi and J. F. Kelly; *J. Am. Chem. Soc.*, 1966, **88**, 3657.	359
66JCP(45)61	J. R. Durig and R. C. Lord; *J. Chem. Phys.*, 1966, **45**, 61.	419
66JCS(C)1075	M. Anderson and A. W. Johnson; *J. Chem. Soc. (C)*, 1966, 1075.	510, 519
66JCS(C)1171	R. E. Banks, G. M. Haslam, R. N. Haszeldine and A. Peppin; *J. Chem. Soc. (C)*, 1966, 1171.	477
66JHC206	A. E. Drukker, C. I. Judd and D. D. Dusterhoft; *J. Heterocycl. Chem.*, 1966, **3**, 206.	514
66JHC230	D. H. Wadsworth and O. E. Schupp; *J. Heterocycl. Chem.*, 1966, **3**, 230.	242
66JHC237	M. E. Derieg and L. H. Sternbach; *J. Heterocycl. Chem.*, 1966, **3**, 237.	629
66JHC245	R. A. Mitsch; *J. Heterocycl. Chem.*, 1966, **3**, 245.	233
66JMC633	W. Metlesics, T. Resnick, G. Silverman, R. Tavares and L. H. Sternbach; *J. Med. Chem.*, 1966, **9**, 633.	683
66JOC34	J. A. Moore, F. J. Marascia, R. W. Medeiros and R. L. Wineholt; *J. Org. Chem.*, 1966, **31**, 34.	347
66JOC897	G. A. Hardcastle, Jr., D. A. Johnson, C. A. Panetta, A. I. Scott and S. A. Sutherland; *J. Org. Chem.*, 1966, **31**, 897.	324, 331
66JOC1079	J. H. Hall and R. Kellogg; *J. Org. Chem.*, 1966, **31**, 1079.	472
66JOC1244	A. Padwa, L. Hamilton and L. Norling; *J. Org. Chem.*, 1966, **31**, 1244.	50
66JOC1372	E. J. Moriconi and P. H. Mazzochi; *J. Org. Chem.*, 1966, **31**, 1372.	342, 346, 347
66JOC1423	G. Smolinsky and B. I. Feuer; *J. Org. Chem.*, 1966, **31**, 1423.	72
66JOC1922	W. J. Gottstein, G. E. Bocian, L. B. Crast, K. Dadabo, J. M. Essery, J. C. Godfrey and L. C. Cheney; *J. Org. Chem.*, 1966, **31**, 1922.	313
66JOC1995	A. Padwa and L. Hamilton; *J. Org. Chem.*, 1966, **31**, 1995.	480
66JOC2560	D. A. Johnson, C. A. Panetta and R. R. Smith; *J. Org. Chem.*, 1966, **31**, 2560.	331
66JOC3356	W. Metlesics, R. Tavares and L. H. Sternbach; *J. Org. Chem.*, 1966, **31**, 3356.	683
66JOC3907	G. R. Harvey and K. W. Ratts; *J. Org. Chem.*, 1966, **31**, 3907.	51
66JOC3941	H. Tanida, T. Tsuji and T. Irie; *J. Org. Chem.*, 1966, **31**, 3941.	68
66JOC4268	A. A. Santilli and T. S. Osdene; *J. Org. Chem.*, 1966, **31**, 4268.	630
66JSP(19)372	L. C. Robertson and J. A. Merritt; *J. Mol. Spectrosc.*, 1966, **19**, 372.	203
66LA(692)151	H. Gehlen and M. Just; *Liebigs Ann. Chem.* 1966, **692**, 151.	463
66LA(693)117	E. Kopp and J. Smidt; *Liebigs Ann. Chem.*, 1966, **693**, 117.	691
66LA(696)108	E. Testa, G. Pifferi, L. Fontanella and V. Aresi; *Liebigs Ann. Chem.*, 1966, **696**, 108.	250, 251
66LA(697)188	A. Rieche, E. Hoft and H. Schultze; *Liebigs Ann. Chem.*, 1966, **697**, 188.	538
B-66MI50400	J. G. Calvert and J. N. Pitts, Jr.; 'Photochemistry', Wiley, New York, 1966.	50
B-66MI50600	L. Goodman and E. J. Reist; in 'Organic Sulfur Compounds', ed. N. Kharasch and C. Y. Meyers; Pergamon, Oxford, 1966, vol. 2, p. 93.	178
66MI50800	D. M. Gale, W. J. Middleton and C. G. Krespan; *Abstr. Papers Am. Chem. Soc. 151st Meeting*, 1966, 81.	224
66MI51300	K. H. Thiele and P. Zdunneck; *Organomet. Chem. Rev.*, 1966, **1**, 337.	376

66NEP6509836 Farbwerke Hoechst A.-G., *Neth. Appl.* 65 09 836 (1966) (*Chem. Abstr.*, 1966, **65**, 4083). 484

66QR119 A. C. Cope, M. M. Martin and M. A. McKervey; *Q. Rev., Chem. Soc.*, 1966, **20**, 119. 105

66SA1371 J. R. W. Warn and D. Chapman; *Spectrochim. Acta*, 1966, **22**, 1371. 488

66T81 W. von E. Doering and R. A. Odum; *Tetrahedron*, 1966, **22**, 81. 495, 515, 535

66T539 R. Hoffmann; *Tetrahedron*, 1966, **22**, 539. 197

66T1039 K. Kuriyama, T. Komeno and K. Takeda; *Tetrahedron*, 1966, **22**, 1039. 138

66T1565 H. Ulrich, B. Tucker and A. A. R. Sayigh; *Tetrahedron*, 1966, **22**, 1565. 474

66T3001 A. J. Kirby; *Tetrahedron*, 1966, **22**, 3001. 456

66T3393 H. Nozaki, H. Takaya and R. Noyori; *Tetrahedron*, 1966, **22**, 3393. 389

66T(S8)113 G. Fodor and G. A. Cooke; *Tetrahedron*, 1966, Suppl. 8, part 1, 113. 346, 347

66TL573 K. Morita and S. Kobayashi; *Tetrahedron Lett.*, 1966, 573. 649

66TL1733 A. M. Mansoor and I. D. R. Stevens; *Tetrahedron Lett.*, 1966, 1733. 223

66TL2921 K. Tori, K. Aono, K. Kitahonoki, R. Muneyuki, Y. Takano, H. Tanida and T. Tsuji; *Tetrahedron Lett.*, 1966, 2921. 51

66TL4691 V. R. Gaertner; *Tetrahedron Lett.*, 1966, 4691. 245

66TL5245 W. Fischer and E. Fahr; *Tetrahedron Lett.*, 1966, 5245. 454, 472

66TL5385 F. Nerdel, J. Buddrus, W. Brodowski and P. Weyerstahl; *Tetrahedron Lett.*, 1966, 5385. 580, 666

66USP3223584 R. W. Luckenbaugh and K. C. Smeltz; *U.S. Pat.* 3 223 584 (1965) (*Chem. Abstr.*, 1966, **64**, 7305). 484

66USP3290289 J. J. Fuchs; *U.S. Pat.* 3 290 289 (1966) (*Chem. Abstr.*, 1967, **66**, 55 472). 233

66USP3377357 Eli Lilly and Co., *U.S. Pat.* 3 377 357 (1966) (*Chem. Abstr.*, 1968, **69**, 52 190). 632

66ZN(B)622 M. Schmidt, K. Blaettner, P. Kochendorfer and H. Ruf; *Z. Naturforsch., Teil B*, 1966, **21**, 622. 696

67ACS705 C. Rappe and R. Gustafsson; *Acta Chem. Scand.*, 1967, **21**, 705. 695

67ACS837 L. B. Agenas and B. Persson; *Acta Chem. Scand.*, 1967, **21**, 837. 387

67ACS2210 P. Bamberg, B. Ekström and B. Sjöberg; *Acta Chem. Scand.*, 1967, **21**, 2210. 331

67AG(E)385 E. Vogel and H. Günther; *Angew. Chem., Int. Ed. Engl.*, 1967, **6**, 385. 102, 188, 551, 552, 553, 554, 555, 564, 568, 569, 570, 584

67AG(E)402 G. Maier; *Angew. Chem., Int. Ed. Engl.*, 1967, **6**, 402. 499

67AG(E)630 W. Fischer and E. Fahr; *Angew. Chem., Int. Ed. Engl.*, 1967, **6**, 630. 460

67AG(E)697 W. Schroth and B. Werner; *Angew. Chem., Int. Ed. Engl.*, 1967, **6**, 697. 679

67AG(E)703 F. Feher and B. Degen; *Angew Chem., Int. Ed. Engl.*, 1967, **6**, 703. 695

67AG(E)872 R. Borsdorf, H. Kasper and H. D. Repp; *Angew. Chem., Int. Ed. Engl.*, 1967, **6**, 872. 556, 574, 587

67AG(E)951 F. Effenberger and G. Kiefer; *Angew. Chem., Int. Ed. Engl.*, 1967, **6**, 951. 473

67AHC(8)21 F. D. Popp and A. C. Noble; *Adv. Heterocycl. Chem.*, 1967, **8**, 21. 595, 596, 597, 598, 599, 605, 606, 607, 608, 609, 610, 611

67BAU1365 A. M. Orlov, N. D. Kuleshova and I. L. Knunyants; *Bull. Acad. Sci. USSR, Div. Chem. Sci.*, 1967, 1365. 428

67BAU1471 G. A. Sokol'skii, M. A. Belaventsev and I. L. Knunyants; *Bull. Acad. Sci. USSR, Div. Chem. Sci.*, 1967, 1471. 464

67BAU1685 L. I. Ragulin, P. P. Ropalo, G. A. Sokol'skii and I. L. Knunyants; *Bull. Acad. Sci. USSR, Div. Chem. Sci.*, 1967, 1685 464

67BAU1935 G. A. Sokol'skii, M. A. Belaventsev and I. L. Knunyants; *Bull. Acad. Sci. USSR, Div. Chem. Sci.*, 1967, 1935. 467

67BAU2467 S. R. Sterlin, B. L. Dyatkin and I. L. Knunyants; *Bull. Acad. Sci. USSR, Div. Chem. Sci.*, 1967, 2467. 456, 478

67BCJ432 K. Ichimura and M. Ohta; *Bull. Chem. Soc. Jpn.*, 1967, **40**, 432. 81

67BCJ1964 T. Chen, H. Kato and M. Ohta; *Bull. Chem. Soc. Jpn.*, 1967, **40**, 1964. 240

67BCJ2398 T. Chen, T. Sanjiki, H. Kato and M. Ohta; *Bull. Chem. Soc. Jpn.*, 1967, **40**, 2398. 245

67BCJ2401 T. Chen, T. Sanjiki, H. Kato and M. Ohta; *Bull. Chem. Soc. Jpn.*, 1967, **40**, 2401. 240

67BCJ2938 S. Sato, H. Kato and M. Ohta; *Bull. Chem. Soc. Jpn.*, 1967, **40**, 2938. 69

67BRP1081292 Whiffen and Sons Ltd. *Br. Pat.* 1 081 292 (1967) (*Chem. Abstr.*, 1968, **68**, 114 071). 213

67CB142 E. Schmitz, D. Habisch and C. Gründemann; *Chem. Ber.*, 1967, **100**, 142. 214

67CB335 A. Mannschreck, G. Rissmann, F. Vögtle and D. Wild; *Chem. Ber.*, 1967, **100**, 335. 499

67CB1778 A. Mannschreck, R. Radeglia, E. Gründemann and R. Ohme; *Chem. Ber.*, 1967, **100**, 1778. 200

67CB2093 E. Schmitz, C. Hörig and C. Gründemann; *Chem. Ber.*, 1967, **100**, 2093. 222, 223, 231

67CB2593 E. Schmitz and S. Schramm; *Chem. Ber.*, 1967, **100**, 2593. 204, 205, 206

67CB2600 E. Schmitz, R. Ohme and S. Schramm; *Chem. Ber.*, 1967, **100**, 2600. 210

67CB3024 S. Hünig, W. Grässmann, V. Meuer and E. Lücke; *Chem. Ber.*, 1967, **100**, 3024. 655

67CB3162 M. Kroner; *Chem. Ber.*, 1967, **100**, 3162. 730

67CC928 R. C. Cookson, B. Halton, I. D. R. Stevens and C. T. Watts; *Chem. Commun.*, 1967, 928. 79

67CC1077 W. W. Paudler and A. Zeiler; *Chem. Commun.*, 1967, 1077. 683

67CCC3186 J. O. Jilek, E. Svatek, J. Metysova, J. Pomykacek and M. Protiva; *Collect. Czech. Chem. Commun.*, 1967, **32**, 3186. 587

67CPB988	K. Morita and S. Kobayashi; *Chem. Pharm. Bull.*, 1967, **15**, 988.	649
67CR(C)(264)610	J. M. Surzur, M. P. Crozet and C. DuPuy; *C. R. Hebd. Seances Acad. Sci., Ser. C*, 1967, **264**, 610.	435
67CR(C)(264)1309	C. Schaal; *C.R. Hebd. Seances Acad. Sci., Ser. C*, 1967, **264**, 1309.	384
67CR(C)(264)1665	C. Normant-Chefnay, Y. Varieras and P. Maitte; *C.R. Hebd. Seances Acad. Sci., Ser. C*, 1967, **264**, 1665.	580
67CR(C)(265)795	D. Grandjean and A. Leclaire; *C.R. Hebd. Seances Acad. Sci., Ser. C*, 1967, **265**, 795.	702
67DOK(172)1327	F. I. Luknitskii and B. A. Vovsi; *Dokl. Akad. Nauk SSSR*, 1967, **172**, 1327.	455, 464, 477
67G1719	G. Pifferi, P. Consonni and E. Testa; *Gazz. Chim. Ital.*, 1967, **97**, 1719.	251
67HCA1327	D. Hauser and H. P. Sigg; *Helv. Chim. Acta*, 1967, **50**, 1327.	316, 318
67HCA1669	H. Prinzbach and V. Fischer; *Helv. Chim. Acta*, 1967, **50**, 1669.	185
67HOU(10-2)71	E. Müller; *Methoden Org. Chem.* (*Houben-Weyl*), 1967, **10-2**, 71.	195
67IC1461	C. H. Henrickson and D. P. Eyman; *Inorg. Chem.*, 1967, **6**, 1461.	424, 447
67JA102	A. Padwa and L. Hamilton; *J. Am. Chem. Soc.*, 1967, **89**, 102.	62
67JA352	F. A. L. Anet and J. M. Osyany; *J. Am. Chem. Soc.*, 1967, **89**, 352.	6, 52
67JA362	J. C. Sheehan and J. H. Beeson; *J. Am. Chem. Soc.*, 1967, **89**, 362.	93
67JA605	E. L. Stogryn and S. J. Brois; *J. Am. Chem. Soc.*, 1967, **89**, 605.	65, 507, 539, 585, 587
67JA985	W. E. Parham and J. F. Dooley; *J. Am. Chem. Soc.*, 1967, **89**, 985.	774
67JA1281	W. L. Mock; *J. Am. Chem. Soc.*, 1967, **89**, 1281.	571, 573, 576, 585, 587
67JA1515	C. Walling and M. J. Mintz; *J. Am. Chem. Soc.*, 1967, **89**, 1515.	389
67JA1753	R. Huisgen, W. Scheer and H. Huber; *J. Am. Chem. Soc.*,1967, **89**, 1753.	72
67JA2059	M. S. Newman, N. Gill and D. W. Thomson; *J. Am. Chem. Soc.*, 1967, **89**, 2059.	432, 436
67JA2077	F. W. Fowler, A. Hassner and L. A. Levy; *J. Am. Chem. Soc.*, 1967, **89**, 2077.	85
67JA2502	G. M. Atkins, Jr. and E. M. Burgess; *J. Am. Chem. Soc.*, 1967, **89**, 2502.	473
67JA2618	R. A. Carboni, J. C. Kauer, J. E. Castle and H. E. Simmons; *J. Am. Chem. Soc.*, 1967, **89**, 2618.	687
67JA2638	Y. T. Chia and H. E. Simmons; *J. Am. Chem. Soc.*, 1967, **89**, 2638.	687
67JA2761	P. Y. Johnson and G. A. Berchtold; *J. Am. Chem. Soc.*, 1967, **89**, 2761.	418
67JA3922	E. E. van Tamelen and D. Carty; *J. Am. Chem. Soc.*, 1967, **89**, 3922.	582
67JA4435	A. Padwa, D. Crumrine, R. Hartman and R. Layton; *J. Am. Chem. Soc.*, 1967, **89**, 4435.	124
67JA4456	N. J. Leonard and B. Zwanenburg; *J. Am. Chem. Soc.*, 1967, **89**, 4456.	69, 70, 72
67JA4487	L. A. Paquette, L. S. Wittenbrook and V. V. Kane; *J. Am. Chem. Soc.*, 1967, **89**, 4487.	152
67JA4526	J. K. Crandall and L.-H. C. Lin; *J. Am. Chem. Soc.*, 1967, **89**, 4526.	105
67JA4527	J. K. Crandall and L.-H. C. Lin; *J. Am. Chem. Soc.*, 1967, **89**, 4527.	105
67JA5527	J. L. Hencher and S. H. Bauer; *J. Am. Chem. Soc.*, 1967, **89**, 5527.	199
67JA5646	C. G. Overberger and H. Kaye; *J. Am. Chem. Soc.*, 1967, **89**, 5646.	14
67JA5649	C. G. Overberger and H. Kaye; *J. Am. Chem. Soc.*, 1967, **89**, 5649.	549
67JA5724	K. R. Henery-Logan and T. L. Fridinger; *J. Am. Chem. Soc.*, 1967, **89**, 5724.	83
67JA6390	J. M. Holovka and P. D. Gardner; *J. Am. Chem. Soc.*, 1967, **89**, 6390.	564
67JA6605	H. Saito, K. Nukada, T. Kobayashi and K. Morita; *J. Am. Chem. Soc.*, 1967, **89**, 6605.	51
67JA6911	E. F. Ullman and B. Singh; *J. Am. Chem. Soc.*, 1967, **89**, 6911.	57
67JA7017	C. J. Pedersen; *J. Am. Chem. Soc.*, 1967, **89**, 7017.	731, 741, 761
67JCS(B)112	R. A. Smith, J. E. Baldwin and I. C. Paul; *J. Chem. Soc. (B)*, 1967, 112.	520
67JCS(C)952	G. J. Durant; *J. Chem. Soc. (C)*, 1967, 952.	649
67JCS(C)1130	J. M. Cox and L. N. Owen; *J. Chem. Soc. (C)*, 1967, 1130.	556, 574, 587
67JHC155	M. W. Barker and M. E. Coker; *J. Heterocycl. Chem.*, 1967, **4**, 155.	456, 473
67JHC583	A. Vigevani and G. G. Gallo; *J. Heterocycl. Chem.*, 1967, **4**, 583.	268
67JHC619	G. Pifferi, P. Consonni, G. Pelizza and E. Testa; *J. Heterocycl. Chem.*, 1967, **4**, 619.	250, 266, 268, 271, 274
67JMC101	C. Szántay, Z. F. Chimilewicz and T. J. Bardos; *J. Med. Chem.*, 1967, **10**, 101.	213
67JMC546	E. A. Brown; *J. Med. Chem.*, 1967, **10**, 546.	401
67JOC540	A. Hassner, M. E. Lorber and C. Heathcock; *J. Org. Chem.*, 1967, **32**, 540.	83, 84
67JOC1346	J. M. Eby and J. A. Moore; *J. Org. Chem.*, 1967, **32**, 1346.	343
67JOC1353	J. A. Moore, R. L. Wineholt, F. J. Marascia, R. W. Medeiros and F. J. Creegan; *J. Org. Chem.*, 1967, **32**, 1353.	344
67JOC1562	E. U. Elam and H. E. Davis; *J. Org. Chem.*, 1967, **32**, 1562.	435, 478
67JOC1944	R. L. Rebertus, J. J. McBrady and J. G. Gagnon; *J. Org. Chem.*, 1967, **32**, 1944.	233
67JOC2009	D. C. Dittmer and S. M. Kotin; *J. Org. Chem.*, 1967, **32**, 2009.	427
67JOC2014	C. R. Johnson and M. P. Jones; *J. Org. Chem.*, 1967, **32**, 2014.	572
67JOC2180	T. S. Sulkowski, M. A. Willie, A. Mascitti and J. L. Diebold; *J. Org. Chem.*, 1967, **32**, 2180.	675
67JOC2185	W. Metlesics, T. Anton and L. H. Sternbach; *J. Org. Chem.*, 1967, **32**, 2185.	675
67JOC2360	J. C. Stowell; *J. Org. Chem.*, 1967, **32**, 2360.	454, 460, 472
67JOC2367	W. A. Nasutavicus and F. Johnson; *J. Org. Chem.*, 1967, **32**, 2367.	515, 525, 527
67JOC2417	A. Stempel, I. Douvan, E. Reeder and L. H. Sternbach; *J. Org. Chem.*, 1967, **32**, 2417.	612, 686
67JOC2425	W. W. Paudler and A. G. Zeiler; *J. Org. Chem.*, 1967, **32**, 2425.	681
67JOC2628	P. Scheiner; *J. Org. Chem.*, 1967, **32**, 2628.	65

67JOC2723 L. A. Paquette and R. W. Begland; *J. Org. Chem.*, 1967, **32**, 2723. 666
67JOC2960 R. J. Shozda; *J. Org. Chem.*, 1967, **32**, 2960. 455, 475
67JOC2972 V. R. Gaertner; *J. Org. Chem.*, 1967, **32**, 2972. 245
67JOC3160 W. D. Closson, P. J. Orenski and B. M. Goldschmidt; *J. Org. Chem.*, 1967, **32**, 3160. 666
67JOC3325 W. A. Nasutavicus, S. W. Tobey and F. Johnson; *J. Org. Chem.*, 1967, **32**, 3325, and references cited therein. 529
67JOC3676 Y. Omote, M. Yoshioka, K. Yamada and N. Sugiyama; *J. Org. Chem.*, 1967, **32**, 3676. 432
67JOC3720 D. H. Kim, A. A. Santilli, T. S. Sulkowski and S. J. Childress; *J. Org. Chem.*, 1967, **32**, 3720. 674
67JOC3743 T. W. Craig, G. R. Harvey and G. A. Berchtold; *J. Org. Chem.*, 1967, **32**, 3743. 191
67JOC3872 D. C. Dittmer and F. A. Davis; *J. Org. Chem.*, 1967, **32**, 3872. 428, 430
67JOC4149 R. A. LeMahieu; *J. Org. Chem.*, 1967, **32**, 4149. 689
67JOU2089 F. I. Luknitskii and B. A. Vovsi; *J. Org. Chem. USSR* (*Engl. Transl.*), 1967, **3**, 2089. 464
67JPC2326 G. F. Cohoe and W. D. Walters; *J. Phys. Chem.*, 1967, **71**, 2326. 370
67JPR(36)86 E. Schmitz, S. Schramm and R. Ohme; *J. Prakt. Chem.*, 1967, **36**, 86. 205, 210
67LA(702)163 W. Dürckheimer and M. Schorr; *Liebigs Ann. Chem.*, 1967, **702**, 163. 324
67LA(703)96 R. W. Hoffmann and W. Sieber; *Liebigs Ann. Chem.*, 1967, **703**, 96. 418, 423, 441, 442, 443
67LA(707)1 F. Nerdel and H. Kressin, *Liebigs Ann. Chem.*, 1967, **707**, 1. 372
67M1323 A. Lüttringhaus and H.-J. Rosenbaum; *Monatsh. Chem.*, 1967, **98**, 1323. 707
B-67MI51500 H. Ulrich; 'Cycloaddition Reactions of Heterocumulenes', Academic Press, New York, 1967. 461
B-67MI50500 Z. Rappoport; 'Handbook of Tables for Organic Compound Identification', Chemical Rubber Co., Cleveland, 3rd edn., 1967. 98
B-67MI50501 W. Brügel; 'Nuclear Magnetic Resonance Spectra and Chemical Structure', Academic Press, New York, 1967, vol. 1, pp. 101–103. 99
B-67MI50502 R. M. Silverstein and G. C. Bassler; 'Spectrometric Identification of Organic Compounds', Wiley, New York, 2nd edn., 1967, p. 86. 99
B-67MI50503 H. A. Szymanski; 'Interpreted Infrared Spectra', Plenum, New York, 1967, vol. 3, pp. 172–194. 99
B-67MI50504 R. J. Gritter; in 'The Chemistry of the Ether Linkage', ed. S. Patai; Interscience, New York, 1967, pp. 378–380. 105
B-67MI50505 L. F. Fieser and M. Fieser; 'Reagents for Organic Synthesis', Wiley, New York, 1967, vol. 1, p. 1210. 105
67MI50700 B. Tolbert, R. Steyn, J. A. Franks, Jr. and H. Z. Sable; *Carbohydr. Res.*, 1967, **5**, 62. 191
B-67MI50800 E. Schmitz; 'Dreiringe mit zwei Heteroatomen', Springer-Verlag, Heidelberg, 1967. 196, 201, 202, 204, 207, 208 213, 214, 216, 217, 220, 222 223, 228, 230, 231, 233, 236
B-67MI51300 S. Searles and M. Tamres; in 'The Chemistry of the Ether Linkage', ed. S. Patai; Wiley, New York, 1967, p. 243. 376
67MI51301 P. O. I. Virtanen; *Suom. Kem. B*, 1967, **40**, 185 (*Chem. Abstr.*, 1968, **68**, 77 365). 378
67MI51302 P. O. I. Virtanen; *Suom. Kem. B*, 1967, **40**, 193 (*Chem. Abstr.*, 1968, **68**, 68 152). 379
67MI51400 C. C. Price and E. A. Blair; *J. Polym. Sci., Part A-1*, 1967, **5**, 171. 447
67MI51401 J. K. Stille and J. A. Empen; *J. Polym. Sci., Part A-1*, 1967, **5**, 273. 418
B-67MI51600 J. A. Moore and E. Mitchell; in 'Heterocyclic Compounds', ed. R. C. Elderfield; Wiley, New York, 1967, vol. 9, p. 224. 492, 509, 510, 511, 515, 517, 522, 525, 528, 530
67MI51700 M. H. Zenck; *Z. Pflanzenphysiol.*, 1967, **57**, 477. 590
B-67MI51800 J. A. Moore and E. Mitchell; in 'Heterocyclic Compounds', ed. R. C. Elderfield; Wiley, New York, 1967, vol. 9, p. 301. 599
67T2123 V. R. Gaertner; *Tetrahedron*, 1967, **23**, 2123. 681
67T2137 Y. Ito, M. Okano and R. Oda; *Tetrahedron*, 1967, **23**, 2137. 176
67TL161 E. Fahr, W. Fischer, A. Jung and L. Sauer; *Tetrahedron Lett.*, 1967, 161. 465, 472
67TL373 T. R. Lynch, I. P. Mellor, S. C. Nyburg and P. Yates; *Tetrahedron Lett.*, 1967, 373. 453, 456
67TL1197 N. Ikekawa and Y. Honma; *Tetrahedron Lett.*, 1967, 1197. 642
67TL1873 C. Kaneko, Sa. Yamada, I. Yokoe and M. Ishikawa; *Tetrahedron Lett.*, 1967, 1873. 641
67TL2155 B. J. Gaj and D. R. Moore; *Tetrahedron Lett.*, 1967, 2155. 245
67TL3637 S. Ozaki; *Tetrahedron Lett.*, 1967, 3637. 475
67TL5033 H. E. Baumgarten, R. D. Clark, L. S. Endres, L. D. Hagemeier and V. J. Elia; *Tetrahedron Lett.*, 1967, 5033. 80
67TL5233 C. Kaneko and S. Yamada; *Tetrahedron Lett.*, 1967, 5233. 641
67TL5307 J. A. Kapecki, J. E. Baldwin and I. C. Paul; *Tetrahedron Lett.*, 1967, 5307. 453, 479
67USP3336334 M. S. Raasch; *U.S. Pat* 3 336 334 (1967) (*Chem. Abstr.*, 1968, **69**, 19 138). 485
67USP3337586 M. S. Raasch, *U.S. Pat.* 3 337 586 (1967) (*Chem. Abstr.*, 1967, **67**, 90 789). 485
67USP3340314 W. D. Blackley; *U.S. Pat.* 3 340 314 (1967) (*Chem. Abstr.*, 1968, **68**, 12 680). 488
67ZN(B)1296 A. Lüttringhaus, U. Hess and H.-J. Rosenbaum; *Z. Naturforsch., Teil B*, 1967, **22**, 1296. 707

68AC(R)1283 G. Pifferi, P. Consonni and E. Testa; *Ann. Chim. (Rome)*, 1968, **58**, 1283. 266, 270
68ACR1 W. S. Johnson; *Acc. Chem. Res.*, 1968, **1**, 1. 119
68ACS367 P. Bamberg, B. Ekström and B. Sjöberg; *Acta Chem. Scand.*, 1968, **22**, 367. 330, 331

68ACS877 O. Buchardt and B. Jensen; *Acta Chem. Scand.*, 1968, **22**, 877. 641

68AG(E)25 I. Lengyel and J. C. Sheehan; *Angew. Chem., Int. Ed. Engl.*, 1968, **7**, 25. 78, 80, 92

68AG(E)72 K. Hartke and F. Rossbach; *Angew. Chem., Int. Ed. Engl.*, 1968, **7**, 72. 455, 474

68AG(E)130 F. P. Woerner, H. Reimlinger and D. R. Arnold; *Angew. Chem., Int. Ed. Engl.*, 1968, **7**, 130. 56

68AG(E)224 D. Felix and A. Eschenmoser; *Angew. Chem., Int. Ed. Engl.*, 1968, **7**, 224. 52

68AP230 F. Moll; *Arch. Pharm. (Weinheim, Ger.)*, 1968, **301**, 230. 359

68BAU807 G. A. Sokol'skii and I. L. Knunyants; *Bull. Acad. Sci. USSR, Div. Chem. Sci.*, 1968, 807. 485

68BAU1796 M. G. Lin'kova and I. L. Knunyants; *Bull. Acad. Sci. USSR, Div. Chem. Sci.*, 1968, 1796. 425

68BBR(33)219 P. A. Miller, P. W. Trown, W. Fulmor, G. O. Morton and J. Karliner; *Biochem. Biophys. Res. Commun.*, 1968, **33**, 219. 590

68BCJ635 S. Matsumura, T. Nagai and N. Tokura; *Bull. Chem. Soc. Jpn.*, 1968, **41**, 635. 166

68BCJ767 T. Kagiya, Y. Samida and T. Inoue; *Bull. Chem. Soc. Jpn.*, 1968, **41**, 767. 376

68BCJ1440 S. Sato; *Bull. Chem. Soc. Jpn.*, 1968, **41**, 1440. 91

68BCJ2524 S. Sato; *Bull. Chem. Soc. Jpn.*, 1968, **41**, 2524. 534

68C139 W. Jenny and H. Holzrichter; *Chimia*, 1968, **22**, 139. 765

68CB174 R. Richter; *Chem. Ber.*, 1968, **101**, 174. 474

68CB2669 M. Fischer; *Chem. Ber.*, 1968, **101**, 2669. 249

68CC93 S. S. Chatterjee and D. J. Triggle; *Chem. Commun.*, 1968, 93. 245

68CC686 L. J. Altman, M. F. Semmelhack, R. B. Hornby and J. C. Vederas; *Chem. Commun.*, 1968, 686. 465, 482

68CC917 C. L. Fritchie, Jr. and J. L. Wells; *Chem. Commun.*, 1968, 917. 452

68CC1202 H. D. Perlmutter; *Chem. Commun.*, 1968, 1202. 675

68CJC365 P. Yates, T. R. Lynch and L. S. Weiler; *Can. J. Chem.*, 1968, **46**, 365. 456

68CJC2549 S. Wolfe, J. C. Godfrey, C. T. Holdrege and Y. G. Perron; *Can. J. Chem.*, 1968, **46**, 2549. 314

68CRV415 H. Kwart and K. King; *Chem. Rev.*, 1968, **68**, 415. 125

68CRV747 G. A. Archer and L. H. Sternbach; *Chem. Rev.*, 1968, **68**, 747. 595, 605, 606, 607, 608, 609, 610, 611, 612, 613

68HCA2027 H. Gerlach and E. Huber; *Helv. Chim. Acta*, 1968, **51**, 2027. 773

68HOU(10-4)449 W. Rundel; *Methoden Org. Chem. (Houben-Weyl)*, 1968, **10-4**, 449. 195

68HOU(10-4)895 M. Bauer and E. Müller; *Methoden Org. Chem. (Houben-Weyl)*, 1968, **10-4**, 895. 195

68IZV670 A. I. D'Yachenko and M. Y. Lukina; *Izv. Akad. Nauk SSSR, Ser. Khim.*, 1968, 670. 378, 383

68IZV1530 T. Z. Papoyan, I. I. Chervin and R. G. Kostyanovskii; *Izv. Akad. Nauk SSSR, Ser. Khim.*, 1968, 1530. 201

68JA506 S. J. Brois; *J. Am. Chem. Soc.*, 1968, **90**, 506. 52

68JA508 S. J. Brois; *J. Am. Chem. Soc.*, 1968, **90**, 508. 6

68JA1650 W. Nagata, S. Hirai, T. Okumura and K. Kawata; *J. Am. Chem. Soc.*, 1968, **90**, 1650. 87

68JA1666 R. O. Kan and R. L. Furey; *J. Am. Chem. Soc.*, 1968, **90**, 1666. 474

68JA1676 R. H. Schlessinger and A. G. Schultz; *J. Am. Chem. Soc.*, 1968, **90**, 1676. 181

68JA2333 O. L. Chapman and W. R. Adams; *J. Am. Chem. Soc.*, 1968, **90**, 2333. 257

68JA2428 H. E. Simmons and C. H. Park; *J. Am. Chem. Soc.*, 1968, **90**, 2428. 739

68JA2431 H. E. Simmons and C. H. Park; *J. Am. Chem. Soc.*, 1968, **90**, 2431. 747

68JA2869 A. Hassner and F. W. Fowler; *J. Am. Chem. Soc.*, 1968, **90**, 2869. 50, 71, 85

68JA2875 F. W. Fowler and A. Hassner; *J. Am. Chem. Soc.*, 1968, **90**, 2875. 48

68JA2982 S. M. Kupchan, R. J. Hemingway, P. Coggon, A. T. McPhail and G. A. Sim; *J. Am. Chem. Soc.*, 1968, **90**, 2982. 192

68JA3592 J. A. Deyrup and S. C. Clough; *J. Am. Chem. Soc.*, 1968, **90**, 3592. 78

68JA3893 B. Singh; *J. Am. Chem. Soc.*, 1968, **90**, 3893. 229

68JA3897 L. A. Paquette and T. Kakihana; *J. Am. Chem. Soc.*, 1968, **90**, 3897. 359

68JA4501 H. L. Ammon, P. H. Watts, Jr., J. M. Stewart and W. L. Mock; *J. Am. Chem. Soc.*, 1968, **90**, 4501. 556

68JA4738 M. G. Pleiss and J. A. Moore; *J. Am. Chem. Soc.*, 1968, **90**, 4738. 231

68JA5325 J. E. Baldwin, R. G. Pudussery, A. K. Qureshi and B. Sklarz; *J. Am. Chem. Soc.*, 1968, **90**, 5325. 90

68JA6148 L. A. Paquette, R. A. Begland and P. C. Storm; *J. Am. Chem. Soc.*, 1968, **90**, 6148. 667

68JA6523 D. M. Jerina, J. W. Daly and B. Witkop; *J. Am. Chem. Soc.*, 1968, **90**, 6523. 566

68JA7360 O. P. Strausz, I. G. Csizmadia and J. Font; *J. Am. Chem. Soc.*, 1968, **90**, 7360. 120

68JA7372 A. P. Bindra, J. A. Elix, P. J. Garratt and R. H. Mitchell; *J. Am. Chem. Soc.*, 1968, **90**, 7372. 710

68JAP6809058 D. Sato; *Jpn. Pat.* 68 09 058 (1968) (*Chem. Abstr.*, 1969, **70**, 4442). 182

68JAP6816751 Y. Miyake and T. Yamauchi; *Jpn. Pat.* 68 16 751 (1968) (*Chem. Abstr.*, 1969, **70**, 57 924). 691

68JCS(B)543 P. Watts; *J. Chem. Soc. (B)*, 1968, 543, 689

68JCS(C)1675 G. B. Gill, D. J. Harper and A. W. Johnson; *J. Chem. Soc. (C)*, 1968, 1675. 543

68JCS(C)2311 J. Ashby, L. A. Cort, J. A. Elvidge and U. Eisner; *J. Chem. Soc. (C)*, 1968, 2311. 524, 525, 544

68JCS(C)2481 W. C. Peaston and G. R. Proctor; *J. Chem. Soc. (C)*, 1968, 2481. 525

68JCS(C)2533 I. McMillan and R. J. Stoodley; *J. Chem. Soc. (C)*, 1968, 2533 316

68JHC141 E. Campaine, M. Pragnell and F. Haaf; *J. Heterocycl. Chem.*, 1968, **5**, 141. 488
68JHC343 H. M. Zacharis and M. Trefonis; *J. Heterocycl. Chem.*, 1968, **5**, 343. 48
68JHC433 K. R. Henery-Logan, H. P. Knoepfel and J. V. Rodricks; *J. Heterocycl. Chem.*, 1968, **5**, 433. 251
68JHC719 J. B. Wright; *J. Heterocycl. Chem.*, 1968, **5**, 719. 644
68JHC779 J. C. Sheehan and M. Dadic; *J. Heterocycl. Chem.*, 1968, **5**, 779. 265, 359
68JMC70 R. Raap and R. G. Micetich; *J. Med. Chem.*, 1968, **11**, 70. 330
68JMC466 D. Bishop, J. F. Cavalla, I. M. Lockhart, M. Wright, C. V. Winder, A. Wong and M. Stephens; *J. Med. Chem.*, 1968, **11**, 466. 246
68JOC396 R. L. Whistler, T. J. Luttenegger and R. M. Rowell; *J. Org. Chem.*, 1968, **33**, 396. 384
68JOC887 G. R. Harvey; *J. Org. Chem.*, 1968, **33**, 887. 81, 82
68JOC1364 D. C. Dittmer and J. M. Balquist; *J. Org. Chem.*, 1968, **33**, 1364. 445
68JOC1476 W. E. Parham and J. F. Dooley; *J. Org. Chem.*, 1968, **33**, 1476. 774
68JOC1850 C. G. Krespan and D. C. England; *J. Org. Chem.*, 1968, **33**, 1850. 478
68JOC1913 W. J. Farrissey, Jr., R. J. Ricciardi and A. A. R. Sayigh; *J. Org. Chem.*, 1968, **33**, 1913. 458, 463, 474
68JOC2291 R. Huisgen, *J. Org. Chem.*, 1968, **33**, 2291. 55
68JOC2591 E. E. Schweizer, M. S. El-Bakoush, K. K. Light and K. H. Oberle; *J. Org. Chem.*, 1968, **33**, 2591. 581, 584
68JOC2620 F. Lautenschlaeger; *J. Org. Chem.*, 1968, **33**, 2620. 436
68JOC2930 W. A. Ehrhart; *J. Org. Chem.*, 1968, **33**, 2930. 690
68JOC3027 L. A. Paquette and M. Rosen; *J. Org. Chem.*, 1968, **33**, 3027. 431, 433, 441
68JOC3187 R. A. Johnson, M. E. Herr, H. C. Murray and G. S. Fonken; *J. Org. Chem.*, 1968, **33**, 3187. 658
68JOC3448 E. J. Moriconi, J. F. Kelly and R. A. Salomone; *J. Org. Chem.*, 1968, **33**, 3448. 248
68JOC3619 T. Sheradsky, U. Reichman and M. Frankel; *J. Org. Chem.*, 1968, **33**, 3619. 454, 466, 470
68JOC4316 V. Nair; *J. Org. Chem.*, 1968, **33**, 4316. 60
68JOC4381 J. L. Derocque, W. J. Theuer and J. A. Moore; *J. Org. Chem.*, 1968, **33**, 4381. 269, 276
68JOC4438 G. F. Field and L. H. Sternbach; *J. Org. Chem.*, 1968, **33**, 4438. 641
68JOC4547 P. G. Mente, H. W. Heine and G. R. Scharoubim; *J. Org. Chem.*, 1968, **33**, 4547. 627
68JPR(37)257 R. Ohme, E. Schmitz and L. Sterk; *J. Prakt. Chem.*, 1968, **37**, 257. 235
68LA(717)137 H. Hiller, P. Dimroth and H. Pfitzner, *Liebigs Ann. Chem.*, 1968, **717**, 137. 717
68LA(718)94 H. Bestian, H. Biener, K. Clauss and H. Heyn; *Liebigs Ann. Chem.*, 1968, **718**, 94. 359
68LA(720)98 H. Dorn and K. Walter; *Liebigs Ann. Chem.*, 1968, **720**, 98. 213
68MI50800 J. M. Lehn and B. Munsch; *Theor. Chim. Acta*, 1968, **12**, 91. 197, 198
68MI51100 M. N. G. James, D. Hall and D. C. Hodgkin; *Nature (London)*, 1968, **220**, 168. 300
68MI51300 D. R. Arnold; *Adv. Photochem.*, 1968, **6**, 301. 396
68MI51600 D. M. Hall; *Prog. Stereochem.*, 1968, **4**, 1. 499
B-68MI51601 H. Ulrich; 'The Chemistry of Imidoyl Halides', Plenum, New York, 1968, p. 207. 514, 517
68MI51900 K. H. Burg, H. D. Hermann and H. Rehling; *Makromol. Chem.*, 1968, **111**, 181. 691
B-68MI51902 J. D. Dunitz; 'Perspectives in Structural Chemistry', Wiley, New York, 1968, p. 36. 699
68OMS(1)403 R. Smakman and T. J. De-Boer; *Org. Mass Spectrom.*, 1968, **1**, 403. 549, 559
68OMS(1)857 M. B. Jackson, T. M. Spotswood and J. H. Bowie; *Org. Mass Spectrom.*, 1968, **1**, 857. 248
68PAC(16)169 J. Rigaudy; *Pure Appl. Chem.*, 1968, **16**, 169. 191
68RCR448 A. A. Akhrem, A. M. Moiseenkov and V. N. Dobrynin; *Russ. Chem. Rev. (Engl. Transl.)*, 1968, **37**, 448. 110
68T199 K. Yamada, S. Takada, S. Nakamura and Y. Hirata; *Tetrahedron*, 1968, **24**, 199. 401
68T1633 R. Hollands, D. Becher, A. Gaudemer and J. Polansky; *Tetrahedron*, 1968, **24**, 1633. 192
68T2193 H. Nozaki, S. Fujita and R. Noyori; *Tetrahedron*, 1968, **24**, 2193. 61
68T4193 T. Irie, M. Suzuki and T. Masamune; *Tetrahedron*, 1968, **24**, 4193. 667
68T6395 S. Rossi, O. Pirola and F. Selva; *Tetrahedron*, 1968, **24**, 6395. 640
68TH50800 G. Kozakiewicz; Thesis, Humboldt-Universität, Berlin, 1968, p. 55. 209
68TL959 A. Ohno, D. J. Grosse and R. E. Davis; *Tetrahedron Lett.*, 1968, 959. 138
68TL1903 D. A. Johnson, D. Mania, C. A. Panetta and H. H. Silvestri; *Tetrahedron Lett.*, 1968, 1903. 324
68TL2181 H. Nozaki, T. Koyama, T. Mori and R. Noyori; *Tetrahedron Lett.*, 1968, 2181. 770
68TL3499 K. Isomura, S. Kobayashi and H. Taniguchi; *Tetrahedron Lett.*, 1968, 3499. 65
68TL4561 D. R. Boyd; *Tetrahedron Lett.*, 1968, 4561. 229
68TL4643 A. T. Balaban; *Tetrahedron Lett.*, 1968, 4643. 772
68TL4801 H. W. Heine, G. J. Blosick and G. B. Lowrie; *Tetrahedron Lett.*, 1968, 4801. 61
68TL4905 P. Borrevang, J. Hjort, R. T. Rapala and R. Edie; *Tetrahedron Lett.*, 1968, 4905. 222
68TL5087 J. Parello, R. Beugelmans, P. Milliet and X. Lusinchi; *Tetrahedron Lett.*, 1968, 5087. 229
68TL5609 F. Minisci, R. Galli, M. Cecere, V. Malatesta and T. Caronna; *Tetrahedron Lett.*, 1968, 5609. 211, 234
68TL5855 K. B. Wiberg, J. E. Hiatt and G. Burgmaier; *Tetrahedron Lett.* 1968, 5855. 186
68TL5919 V. R. Gaertner; *Tetrahedron Lett.*, 1968, 5919. 77
68TL6179 J. A. Deyrup and C. L. Moyer; *Tetrahedron Lett.*, 1968, 6179. 77, 244
68TL6293 T. Nagai, M. Tanaka and N. Tokura; *Tetrahedron Lett.*, 1968, 6293. 6

68USP3365495	D. H. Clemens; *U.S. Pat.* 3 365 495 (1968) (*Chem. Abstr.*, 1968, **69**, 27 408).	488
68USP3369019	L. A. Hamilton and P. S. Landis; *U.S. Pat.* 3 369 019 (1968) (*Chem. Abstr.*, 1968, **69**, 43 938).	157
68USP3413306	G. E. Hartzell; *U.S. Pat.* 3 413 306 (1968) (*Chem. Abstr.*, 1969, **70**, 57 418).	183
68ZN(B)739	P. Reich and W. Wieker; *Z. Naturforsch., Teil B*, 1968, **23**, 739.	453
68ZN(B)741	B. Krebs and H. Beyer; *Z. Naturforsch., Teil B*, 1968, **23**, 741.	453
68ZOR2055	F. I. Luknitskii and B. A. Vovsi; *Zh. Org. Khim.*, 1968, **4**, 2055.	372
69ACR186	H. Ulrich; *Acc. Chem. Res.*, 1969, **2**, 186.	20, 474
69ACS56	M. Mansson, E. Morawetz, Y. Nakase and S. Sunner; *Acta Chem. Scand.*, 1969, **23**, 56.	691
69ACS3125	O. Buchardt, C. L. Pedersen and U. Svanholm; *Acta Chem. Scand.*, 1969, **23**, 3125.	595
69AG(E)68	W. Tochtermann and C. Franke; *Angew. Chem., Int. Ed. Engl.*, 1969, **8**, 68.	552
69AG(E)70	W. Funke; *Angew. Chem., Int. Ed. Engl.*, 1969, **8**, 70.	244
69AG(E)73	G. Opitz and H. R. Mohl; *Angew. Chem., Int., Ed. Engl.*, 1969, **8**, 73.	467, 479
69AG(E)212	A. Mannschreck and W. Seitz; *Angew. Chem., Int. Ed. Engl.*, 1969, **8**, 212.	7, 200
69AG(E)448	H. Quast and E. Schmitt; *Angew. Chem., Int. Ed. Engl.*, 1969, **8**, 448.	232
69AG(E)449	H. Quast and E. Schmitt; *Angew. Chem., Int. Ed. Engl.*, 1969, **8**, 449.	22
69AG(E)602	R. Huisgen, W. Scheer and H. Mader; *Angew. Chem., Int. Ed. Engl.*, 1969, **8**, 602.	72
69AG(E)772	W. Bauer and K. Hafner; *Angew. Chem., Int. Ed. Engl.*, 1969, **8**, 772.	85
69AG(E)962	D. R. Crist and N. J. Leonard; *Angew. Chem., Int. Ed. Engl.*, 1969, **8**, 962.	344
69AG(E)990	E. F. Paulus, D. Kobelt and H. Jensen; *Angew. Chem., Int. Ed. Engl.*, 1969, **8**, 990.	247
69AX(B)925	R. D. Shannon and C. T. Prewitt; *Acta Crystallogr., Part B*, 1969, **25**, 925.	737
69AX(B)2114	G. Gafner and L. J. Admiraal; *Acta Crystallogr., Part B*, 1969, **25**, 2114.	696, 702
69BCJ1163	H. Nozaki, S. Fujita and T. Mori; *Bull. Chem. Soc. Jpn.*, 1969, **42**, 1163.	772
69BCJ1777	F. Toda, T. Mitote and K. Akagi; *Bull. Chem. Soc. Jpn.*, 1969, **42**, 1777.	262
69BRP1147383	D. C. Allport; *Br. Pat.* 1 147 383 (1969) (*Chem. Abstr.*, 1969, **70**, 107 588).	484
69BRP1162742	W. Nagata, F. Mukawa, T. Komeno and S. Hayashi; *Br. Pat.* 1 162 742 (1969) (*Chem. Abstr.*, 1970, **72**, 21 838).	170
69BSB191	E. J. Goethals, J. Huylebroeck and W. Smolders; *Bull. Soc. Chim. Belg.*, 1969, **78**, 191.	425
69BSF3653	J. Seyden-Penne and C. Schaal; *Bull. Soc. Chim. Fr.*, 1969, 3653.	385
69C158	W. Jenny and H. Holzrichter; *Chimia*, 1969, **23**, 158.	765
69CB205	H. Hofmann and H. Westernacher; *Chem. Ber.*, 1969, **102**, 205.	583
69CB342	M. Fischer; *Chem. Ber.*, 1969, **102**, 342.	656
69CB938	R. Richter; *Chem. Ber.*, 1969, **102**, 938.	474
69CB1928	G. Maier, U. Heep, M. Wiessler and M. Strasser; *Chem. Ber.*, 1969, **102**, 1928.	671, 673
69CB3148	W. Funke; *Chem. Ber.*, 1969, **102**, 3148.	244
69CB3486	M. Fischer and F. Wagner; *Chem. Ber.*, 1969, **102**, 3486.	277
69CB3495	M. Fischer and F. Wagner; *Chem. Ber.*, 1969, **102**, 3495.	277
69CC13	K. K. Maheshwari and G. A. Berchtold; *Chem. Commun.*, 1969, 13.	442
69CC32	C. L. McIntosh and P. de Mayo; *Chem. Commun.*, 1969, 32.	423
69CC129	J. P. Clayton, J. H. C. Nayler, R. Southgate and E. R. Stove; *Chem. Commun.*, 1969, 129.	324
69CC365	W. D. Kingsbury and C. R. Johnson; *Chem. Commun.*, 1969, 365.	425, 444
69CC572	S. Farid and K.-H. Scholz; *Chem. Commun.*, 1969, 572.	373
69CC631	G. B. Gill, N. Gourlay, A. W. Johnson and M. Mahendran; *Chem. Commun.*, 1969, 631.	513
69CC1084	P. Ogden; *Chem. Commun.*, 1969, 1084.	451
69CC1086	F. Montanari, I. Moretti and G. Torre; *Chem. Commun.*, 1969, 1086.	200
69CC1167	R. Grigg, R. Hayes and J. L. Jackson; *Chem. Commun.*, 1969, 1167.	575
69CC1274	R. James and D. G. Rowsell; *Chem. Commun.*, 1969, 1274.	435
69CC1299	R. A. Odum and B. Schmall; *Chem. Commun.*, 1969, 1299.	504
69CC1504	J. R. Fletcher and I. O. Sutherland; *Chem. Commun.*, 1969, 1504.	779
69CI(L)457	W. E. Truce and L. K. Liu; *Chem. Ind. (London)*, 1969, 457.	464, 477
69CJC709	K. R. Kopecky and C. Mumford; *Can. J. Chem.*, 1969, **47**, 709.	469
69CRV279	S. W. Benson, F. R. Cruickshank, D. M. Golden, G. R. Haugen, H. E. O'Neal, A. S. Rodgers, R. Shaw and R. Walsh; *Chem. Rev.*, 1969, **69**, 279.	370
69DOK(187)335	N. S. Nametkin, V. N. Perchenko and E. S. Vasileva; *Dokl. Akad. Nauk SSSR*, 1969, **187**, 335.	213
69FRP1551921	Toyo Koatsu Industries, *Fr. Pat.* 1 551 921 (1969) (*Chem. Abstr.* 1969, **71**, 39 026).	691
69IC2426	R. L. Patton and K. N. Raymond; *Inorg. Chem.*, 1969, **8**, 2426.	488
69IZV1847	G. N. Gorshkova, F. L. Kolodkin, A. A. Dudinskaya, A. E. Bova, V. A. Ponomarenko, L. I. Khmelnitzkii and S. S. Novikov; *Izv. Akad. Nauk SSSR, Ser. Khim.*, 1969, 1847.	202
69JA642	M. Jautelat and J. D. Roberts; *J. Am. Chem. Soc.*, 1969, **91**, 642.	51
69JA658	R. J. Sundberg, B. P. Das and R. H. Smith, Jr.; *J. Am. Chem. Soc.*, 1969, **91**, 658.	525
69JA1072	L. T. Taylor, F. L. Urbach and D. H. Busch; *J. Am. Chem. Soc.*, 1969, **91**, 1072.	718
69JA1120	J. A. Kapecki and J. E. Baldwin; *J. Am. Chem. Soc.*, 1969, **91**, 1120.	453
69JA1176	J. C. Sheehan and M. M. Nafissi; *J. Am. Chem. Soc.*, 1969, **91**, 1176.	78
69JA1264	H. H. Wasserman, D. D. Keith and J. Nadelson; *J. Am. Chem. Soc.*, 1969, **91**, 1264.	771
69JA1401	R. B. Morin, B. G. Jackson, R. A. Mueller, E. R. Lavagnino, W. B. Scanlon and S. L. Andrews; *J. Am. Chem. Soc.*, 1969, **91**, 1401.	286, 293, 306

69JA1528 R. D. G. Cooper, P. V. DeMarco and D. O. Spry; *J. Am. Chem. Soc.*, 1969, **91**, 1528. 305

69JA1530 R. A. Archer and P. V. DeMarco; *J. Am. Chem. Soc.*, 1969, **91**, 1530. 306

69JA2084 F. G. Bordwell and S. C. Crooks; *J.. Am. Chem. Soc.*, 1969, **91**, 2084. 152

69JA2375 H. H. Wasserman, R. E. Cochoy and M. S. Baird; *J. Am. Chem. Soc.*, 1969, **91**, 2375. 261

69JA2796 C. R. Johnson and W. O. Siegl; *J. Am. Chem. Soc.*, 1969, **91**, 2796. 423, 445

69JA2815 S. J. Rhoads and R. D. Cockroft; *J. Am. Chem. Soc.*, 1969, **91**, 2815. 578, 580

69JA2818 W. R. Dolbier, Jr. and W. M. Williams; *J. Am. Chem. Soc.*, 1969, **91**, 2818. 482, 603

69JA3226 C. G. Overberger, J. W. Stoddard, C. Yaroslavsky, H. Katz and J.-P. Anselme; *J. Am. Chem. Soc.*, 1969, **91**, 3226. 670

69JA3606 D. C. Owsley, G. K. Helmkamp and S. N. Spurlock; *J. Am. Chem. Soc.*, 1969, **91**, 3606. 153

69JA3616 L. A. Paquette, J. H. Barrett and D. E. Kuhla; *J. Am. Chem. Soc.*, 1969, **91**, 3616. 508

69JA3995 T. D. Sakore, R. H. Schlessinger and H. M. Sobell; *J. Am. Chem. Soc.*, 1969, **91**, 3995. 558

69JA4590 J. A. Deyrup and S. C. Clough; *J. Am. Chem. Soc.*, 1969, **91**, 4590. 262

69JA5038 A. Ohno, Y. Ohnishi and G. Tsuchihashi; *J. Am. Chem. Soc.*, 1969, **91**, 5038. 437

69JA5644 R. Tang and K. Mislow; *J. Am. Chem. Soc.*, 1969, **91**, 5644. 444

69JA5668 N. Rieber, J. Alberts, J. A. Lipsky and D. M. Lemal; *J. Am. Chem. Soc.*, 1969, **91**, 5668. 465, 472, 483

69JA6534 J. Ciabattoni and P. J. Kocienski; *J. Am. Chem. Soc.*, 1969, **91**, 6534. 185, 186

69JA7743 T. L. James and C. A. Wellington; *J. Am. Chem. Soc.*, 1969, **91**, 7743. 374

69JA7782 F. A. L. Anet, C. H. Bradley, M. A. Brown, W. L. Mock and J. H. McCausland; *J. Am. Chem. Soc.*, 1969, **91**, 7782. 557

69JAP6927724 K. Morita, S. Kobayashi, K. Sakae and H. Kimura; (Takeda Chemical Industries Ltd), *Jpn. Pat.* 69 27 724 (1969) (*Chem. Abstr.*, 1970, **72**, 21 724). 649

69JCP(51)33 J. R. Lombardi, W. Klemperer, M. B. Robin, H. Basch and N. A. Kuebler; *J. Chem. Phys.*, 1969, **51**, 33. 203

69JCP(51)52 H. Basch, M. B. Robin, N. A. Kuebler, C. Baker and D. W. Turner; *J. Chem. Phys.*, 1969, **51**, 52. 13

69JCS(B)33 I. C. Paul and K. T. Go; *J. Chem. Soc. (B)*, 1969, 33. 697

69JCS(C)2093 R. H. Earle, Jr., D. T. Hurst and M. Viney; *J. Chem. Soc. (C)*, 1969, 2093. 358, 359

69JCS(C)2119 R. E. Banks, R. N. Haszeldine, M. J. Stevenson and B. G. Willoughby; *J. Chem. Soc. (C)*, 1969, 2119. 473

69JCS(C)2123 J. P. Clayton; *J. Chem. Soc. (C)*, 1969, 2123. 316, 317

69JCS(C)2334 F. E. Hardy, P. R. H. Speakman and R. Robson; *J. Chem. Soc. (C)*, 1969, 2334. 151

69JCS(C)2648 D. R. Boyd and R. Graham; *J. Chem. Soc. (C)*, 1969, 2648. 200

69JCS(C)2650 D. R. Boyd, R. Spratt and D. M. Jerina; *J. Chem. Soc. (C)*, 1969, 2650. 199, 202

69JCS(C)2819 J. I. G. Cadogan and R. K. Mackie; *J. Chem. Soc. (C)*, 1969, 2819. 496

69JHC111 G. Pifferi and P. Consonai; *J. Heterocycl. Chem.*, 1969, **6**, 111. 625

69JHC115 D. L. Tuleen and R. H. Bennett; *J. Heterocycl. Chem.*, 1969, **6**, 115. 573, 574, 587

69JHC153 E. Doomes and N. H. Cromwell; *J. Heterocycl. Chem.*, 1969, **6**, 153. 242

69JHC273 V. R. Gaertner; *J. Heterocycl. Chem.*, 1969, **6**, 273. 240

69JHC993 R. M. Rodebaugh and N. H. Cromwell; *J. Hetrocycl. Chem.*, 1969, **6**, 993. 239

69JMC196 E. Bellasio and G. Cristiani; *J. Med. Chem.*, 1969, **12**, 196. 246

69JOC36 L. Field and R. B. Barbee; *J. Org. Chem.*, 1969, **34**, 36. 694

69JOC179 M. E. Derieg, R. M. Schweininger and R. J. Fryer; *J. Org. Chem.*, 1969, **34**, 179. 684

69JOC183 M. Denzer and H. Ott; *J. Org. Chem.*, 1969, **34**, 183. 684

69JOC249 M. Winn and H. E. Zaugg; *J. Org. Chem.*, 1969, **34**, 249. 674

69JOC310 E. Doomes and N. H. Cromwell; *J. Org. Chem.*, 1969, **34**, 310. 246

69JOC459 K. Itoh, Y. Kato and Y. Ishii; *J. Org. Chem.*, 1969, **34**, 459. 383

69JOC896 E. Block and E. J. Corey; *J. Org. Chem.*, 1969, **34**, 896. 175, 186

69JOC1066 N. J. Leonard and T. Sato; *J. Org. Chem.*, 1969, **34**, 1066. 658

69JOC1310 D. C. Dittmer, R. S. Henion and N. Takashina; *J. Org. Chem.*, 1969, **34**, 1310. 423

69JOC1345 N. Kulevsky, C. T. Wang and V. I. Stenberg; *J. Org. Chem.*, 1969, **34**, 1345. 384

69JOC1474 W. E. Parham, J. F. Dooley, M. K. Meilahn and J. W. Greidanus; *J. Org. Chem.*, 1969, **34**, 1474. 774

69JOC1566 W. C. Lumma and G. A. Berchtold; *J. Org. Chem.*, 1969, **34**, 1566. 442

69JOC2138 W. W. Paudler and A. G. Zeiler; *J. Org. Chem.*, 1969, **34**, 2138. 675

69JOC2254 F. D. Greene, J. C. Stowell and W. R. Bergmark; *J. Org. Chem.*, 1969, **34**, 2254. 201, 218, 232

69JOC2715 P. Aeberli and W. J. Houlihan; *J. Org. Chem.*, 1969, **34**, 2715. 730

69JOC2720 P. Aeberli and W. J. Houlihan; *J. Org. Chem.*, 1969, **34**, 2720. 730

69JOC2724 J. A. Deyrup; *J. Org. Chem.*, 1969, **34**, 2724. 82

69JOC2791 P. L.-F. Chang and D. C. Dittmer; *J. Org. Chem.*, 1969, **34**, 2791. 441

69JOC2866 L. A. Paquette, D. E. Kuhla, J. H. Barrett and R. J. Haluska; *J. Org. Chem.*, 1969, **34**, 2866. 495, 501, 541

69JOC2879 L. A. Paquette, D. E. Kuhla and J. H. Barrett; *J. Org. Chem.*, 1969, **34**, 2879. 506, 509

69JOC2888 L. A. Paquette, D. E. Kuhla, J. H. Barrett and L. M. Leichter; *J. Org. Chem.*, 1969, **34**, 2888. 520

69JOC3237 W. W. Paudler and A. G. Zeiler; *J. Org. Chem.*, 1969, **34**, 3237. 672

69JOC3998 F. K. Lautenschlaeger; *J. Org. Chem.*, 1969, **34**, 3998. 150, 425, 436, 440

69JPS1102 R. D. Kinget and M. A. Schwartz; *J. Pharm. Sci.*, 1969, **58**, 1102. 327

69JPS1460 H. A. Luts; *J. Pharm. Sci.*, 1969, **58**, 1460. 629

69JSP(29)174 R. W. Mitchell and J. A. Merritt; *J. Mol. Spectrosc.*, 1969, **29**, 174. 204

69LA(725)124 D. Bormann; *Liebigs Ann. Chem.*, 1969, **725**, 124. 250, 268, 271, 274

69LA(727)224 A. Mannschreck, J. Linsz and W. Seitz; *Liebigs Ann. Chem.*, 1969, **727**, 224. 198, 199

69LA(730)151 H.-J. Kabbe and N. Joop; *Liebigs Ann. Chem.*, 1969, **730**, 151. 685

69M1860 W. Gottardi and D. Henn; *Monatsh. Chem.*, 1969, **100**, 1860. 463

69MI50400 H. Booth; *Prog. Nucl. Magn. Reson. Spectrosc.*, 1969, **5**, 186. 50

B-69MI50401 O. C. Dermer and G. E. Ham; 'Ethylenimine and Other Aziridines', Academic Press, New York, 1969. 49, 50, 70, 80, 93

69MI50402 D. T. Clark; *Theor. Chim. Acta*, 1969, **15**, 225. 48

B-69MI50500 T. F. Rutledge; 'Acetylenes and Allenes', Reinhold, New York, 1969, p. 157. 122

69MI50600 V. P. Pugaeva, S. I. Klochkova, F. D. Mashbits and R. S. Eizengart; *Gig. Tr. Prof. Zabol.*, 1969, **13**, 47 (*Chem. Abstr.*, 1970, **72**, 6050). 183

69MI50800 E. Kochanski and J. M. Lehn; *Theor. Chim. Acta*, 1969, **14**, 281. 197

69MI51300 I. K. Korobitsyna and L. L. Rodina; *Metody Poluch. Khim. Reaktivov. Prep.*, 1969, **18**, 174 (*Chem. Abstr.*, 1971, **74**, 141 415). 394

69MI51301 P. O. I. Virtanen and H. Ruotsalainen; *Suom. Kem. B*, 1969, **42**, 69. 379

B-69MI51600 L. A. Paquette; in 'Nonbenzenoid Aromatics', ed. J. P. Snyder; Academic Press, New York, 1969, vol. 1, p. 250. 492, 493, 494, 495, 499, 501, 504, 509, 511, 512, 514, 517, 518, 523, 525, 526, 530, 538, 541

69OMS(2)337 S. Aftalian and G. R. Proctor; *Org. Mass Spectrom.*, 1969, **2**, 337. 501

69RTC995 A. E. De Groot, J. A. Boerma and H. Wynberg; *Recl. Trav. Chim. Pays-Bas*, 1969, **88**, 995. 556, 572

69T1637 A. Hassner, R. J. Isbister, R. B. Greenwald, J. T. Klug and E. C. Taylor; *Tetrahedron*, 1969, **25**, 1637. 63

69T4277 H.-H. Horhold and H. Eibisch; *Tetrahedron*, 1969, **25**, 4277. 473

69T4421 M. S. Manhas, J. S. Chib, Y. H. Chiang and A. K. Bose; *Tetrahedron*, 1969, **25**, 4421. 248

69T5205 M. Ogata, H. Matsumoto and H. Kano; *Tetrahedron*, 1969, **25**, 5205. 510, 515

69T5217 M. Ogata, H. Matsumoto and H. Kano; *Tetrahedron*, 1969, **25**, 5217. 505

69T5357 H. Nozaki, T. Koyama and T. Mori; *Tetrahedron*, 1969, **25**, 5357. 770

69TH50800 W. Flamme; Thesis, Humboldt-Universität, Berlin, 1969, p. 30. 231

69TH50801 W. Flamme; Thesis, Humboldt-Universität, Berlin, 1969, p. 87. 215

69TL117 K. D. Kampe; *Tetrahedron Lett.*, 1969, 117. 250

69TL177 S. M. Johnson and I. C. Paul; *Tetrahedron Lett.*, 1969, 177. 695

69TL259 Y. Ohshiro, T. Minami, K. Yasuda and T. Agawa; *Tetrahedron Lett.*, 1969, 259. 161

69TL263 Y. Ohshiro, T. Minami, K. Yasuda and T. Agawa; *Tetrahedron Lett.*, 1969, 263. 208

69TL447 T. Minami, O. Aoki, H. Miki, Y. Ohshiro and T. Agawa; *Tetrehedron Lett.*, 1969, 477. 457, 460, 473

69TL557 W. E. Parham, R. W. Davenport and J. B. Biasotti; *Tetrahedron Lett.*, 1969, 557. 774

69TL1133 J. Firl and S. Sommer; *Tetrahedron Lett.*, 1969, 1133. 472

69TL1455 L. A. Paquette, T. Kakihana, J. F. Kelly and J. R. Malpass; *Tetrahedron Lett.*, 1969, 1455. 357

69TL1493 M. Ahmed and O. Meth-Cohn; *Tetrahedron Lett.*, 1969, 1493. 767

69TL2007 E. D. Bergmann, I. Shahak and Z. Aizenshtat; *Tetrahedron Lett.*, 1969, 2007. 580

69TL2281 M. Fischer; *Tetrahedron Lett.*, 1969, 2281. 230

69TL2743 K. Kondo, A. Negishi and G. Tsuchihashi; *Tetrahedron Lett.*, 1969, 2743. 162

69TL3381 S. Kukolja, R. D. G. Cooper and R. B. Morin; *Tetrahedron Lett.*, 1969, 3381. 314

69TL3887 G. M. Rubottom; *Tetrahedron Lett.*, 1969, 3887. 206

69TL4001 R. M. Carlsen and S. Y. Lee; *Tetrahedron Lett.*, 1969, 4001. 71, 90

69TL4073 K. Isomura, M. Okada and H. Taniguchi; *Tetrahedron Lett.*, 1969, 4073. 51

69TL4883 R. M. Moriarty, N. Ishibe, M. Kayser, K. C. Ramey and H. J. Gisler, Jr.; *Tetrahedron Lett.*, 1969, 4883. 649

69TL4899 J. Streith and P. Martz; *Tetrahedron Lett.*, 1969, 4899. 642

69TL5315 H. H. Wasserman and R. Kitzing; *Tetrahedron Lett.*, 1969, 5315. 192

69USP3426041 Y. Miyake, S. Adachi, N. Yamauchi, T. Hayashi and M. Akimoto; *U.S. Pat.* 3 426 041 (1969) (*Chem. Abstr.*, 1969, **70**, 87 871). 691

69ZC184 W. Schroth, F. Billig and A. Zschunke; *Z. Chem.*, 1969, **9**, 184. 681

69ZN(B)477 H. Mauser and H. Bokranz; *Z. Naturforsch., Teil B*, 1969, **24**, 477 230

70ACR26 P. G. Gassman; *Acc. Chem. Res.*, 1970, **3**, 26. 18

70ACS2252 H. Erdtman and L. Malmborg; *Acta Chem. Scand.*, 1970, **24**, 2252. 530

70AG(E)372 H. Helfert and E. Fahr; *Angew. Chem., Int. Ed. Engl.*, 1970, **9**, 372. 470

70AG(E)374 H. W. Heine, J. D. Myers and E. T. Peltzer; *Angew. Chem., Int. Ed. Engl.*, 1970, **9**, 374. 74

70AG(E)381 H. Quast and E. Schmitt; *Angew. Chem., Int. Ed. Engl.*, 1970, **9**, 381. 22, 93

70AG(E)400 A. Rauk, L. C. Alten and K. Mislow; *Angew Chem., Int. Ed. Engl.*, 1970, **9**, 400. 52

70AG(E)731 A. Hassner, J. O. Currie, A. S. Steinfeld and R. F. Atkinson; *Angew. Chem., Int. Ed. Engl.*, 1970, **9**, 731. 271, 274

70AG(E)808 T. Kauffmann, G. Beissner, W. Sahm and A. Woltermann; *Angew. Chem., Int. Ed. Engl.*, 1970, **9**, 808. 765

70AHC(12)185 R. G. Glushkov and V. G. Granik; *Adv. Heterocycl. Chem.*, 1970, **12**, 185. 514

70AX(B)1283 R. Parthasarathy; *Acta Crystallogr., Part B*, 1970, **26**, 1283. 247

70BAU574 G. A. Sokol'skii, M. A. Belaventsev and I. L. Knunyants; *Bull. Acad. Sci. USSR, Div. Chem. Sci.*, 1970, 574. 467

70BAU1479 B. A. Arbuzov, N. N. Zobova and F. B. Balabanova; *Bull. Acad. Sci. USSR, Div. Chem. Sci.*, 1970, **19**, 1479. 458, 475

70BCJ135 T. Ishiwaka; *Bull. Chem. Soc. Jpn.*, 1970, **43**, 135. 637

70BCJ1185 I. Okada, K. Ichimura and R. Sudô; *Bull. Chem. Soc. Jpn.*, 1970, **43**, 1185. 81

70BCJ1824 S. Ito; *Bull. Chem. Soc. Jpn.*, 1970, **43**, 1824. 631

70BCJ1905 N. Yasuoka, N. Kasai, T. Minami, Y. Ohshiro, T. Agawa and M. Kakudo; *Bull. Chem. Soc. Jpn.*, 1970, **43**, 1905. 473

70BCJ3543 O. Tsuge and S. Iwanami; *Bull. Chem. Soc. Jpn.*, 1970, **43**, 3543. 454, 462

70BSF3003 G. Alvernhe and A. Laurent; *Bull. Soc. Chim. Fr.*, 1970, 3003. 85

70CB573 R. Fuks and H. G. Viehe; *Chem. Ber.*, 1970, **103**, 573. 263

70CB949 A. Schoenberg, W. Knoefel, E. Frese and K. Praefcke; *Chem. Ber.*, 1970, **103**, 949. 143

70CB1234 H. Quast and E. Schmitt; *Chem. Ber.*, 1970, **103**, 1234. 219

70CB1797 D. Bormann; *Chem. Ber.*, 1970, **103**, 1797. 271

70CC206 G. I. Moss, G. Crank and F. W. Eastwood; *Chem. Commun.*, 1970, 206. 128

70CC233 J. I. G. Cadogan and S. Kulik; *Chem. Commun.*, 1970, 233. 635

70CC484 A. G. Anastassiou and R. P. Cellura; *Chem. Commun.*, 1970, 484. 725

70CC692 J. W. Lown and K. Matsumoto; *Chem. Commun.*, 1970, 692. 63

70CC745 D. R. Boyd, W. B. Jennings and R. Spratt; *Chem. Commun.*, 1970, 745. 228

70CC1061 N. S. Isaacs and P. F. Stanbury; *Chem. Commun.*, 1970, 1061. 38

70CC1233 I. H. Hillier and V. R. Saunders; *Chem. Commun.*, 1970, 1233. 488

70CC1254 W. L. Mock; *Chem. Commun.*, 1970, 1254. 585

70CC1550 H. W.-S. Chan; *Chem. Commun.*, 1970, 1550. 191

70CC1558 V. Boekelheide and J. A. Lawson; *Chem. Commun.*, 1970, 1558. 764

70CC1629 I. Ojima and N. Inamoto; *Chem. Commun.*, 1970, 1629. 455, 475

70CJC89 J. W. Lown, R. K. Smalley, G. Dallas and T. W. Maloney; *Can. J. Chem.*, 1970, **48**, 89. 55

70CJC1946 J. M. Muchowski; *Can. J. Chem.*, 1970, **48**, 1946. 675

70CJC3704 J. F. King, P. de Mayo, C. L. McIntosh, K. Piers and D. J. H. Smith; *Can. J. Chem.*, 1970, **48**, 3704. 423

70CZ727 F. Vögtle and R. Lichtenthaler; *Chem.-Ztg.*, 1970, **94**, 727. 765

70HCA1479 R. K. Muller, D. Felix, J. Schreiber and A. Eschenmoser; *Helv. Chim. Acta*, 1970, **53**, 1479. 75

70JA100 A. Padwa and R. Gruber; *J. Am. Chem. Soc.*, 1970, **92**, 100. 239

70JA203 M. Uskoković, J. Gutzwiller and T. Henderson; *J. Am. Chem. Soc.*, 1970, **92**, 203. 531

70JA1079 S. J. Brois; *J. Am. Chem. Soc.*, 1970, **92**, 1079. 6, 83

70JA1326 A. Hassner, R. P. Hoblitt, C. Heathcock, J. E. Kropp and M. Lorber; *J. Am. Chem. Soc.*, 1970, **92**, 1326. 84

70JA1453 M. J. S. Dewar and N. Trinajstic; *J. Am. Chem. Soc.*, 1970, **92**, 1453. 14, 556, 557, 558

70JA1763 L. A. Paquette, M. J. Wyvratt and G. R. Allen, Jr.; *J. Am. Chem. Soc.*, 1970, **92**, 1763. 249, 269

70JA1784 L. A. Carpino and R. K. Kirkley; *J. Am. Chem. Soc.*, 1970, **92**, 1784. 74

70JA1929 R. M. Wing and R. Eiss; *J. Am. Chem. Soc.*, 1970, **92**, 1929. 715

70JA1971 L. A. Paquette, R. A. Begland and P. C. Storm; *J. Am. Chem. Soc.*, 1970, **92**, 1971. 667

70JA2084 P. G. Gassman, D. K. Dygos and J. E. Trent; *J. Am. Chem. Soc.*, 1970, **92**, 2084. 72

70JA2178 K. E. Wilzbach and D. J. Rausch; *J. Am. Chem. Soc.*, 1970, **92**, 2178. 271, 275, 360

70JA2377 P. von R. Schleyer, J. E. Williams and K. R. Blanchard; *J. Am. Chem. Soc.*, 1970, **92**, 2377. 49

70JA2575 R. D. G. Cooper and F. L. José; *J. Am. Chem. Soc.*, 1970, **92**, 2575. 306

70JA3217 M. Morton and R. F. Kammereck; *J. Am. Chem. Soc.*, 1970, **92**, 3217. 426

70JA3733 A. Hassner and J. E. Galle; *J. Am. Chem. Soc.*, 1970, **92**, 3733. 85

70JA3787 D. A. Kleier and G. Binsch; *J. Am. Chem. Soc.*, 1970, **92**, 3787. 595

70JA3807 W. L. Mock; *J. Am. Chem. Soc.*, 1970, **92**, 3807. 571

70JA3826 J. Ciabattoni, R. A. Campbell, C. A. Renner and P. W. Concannon; *J. Am. Chem. Soc.*, 1970, **92**, 3826. 122

70JA4874 J. E. Baldwin and J. A. Kapecki; *J. Am. Chem. Soc.*, 1970, **92**, 4874. 39

70JA5258 A. J. Krubsack, T. Higa and W. E. Slack, *J. Am. Chem. Soc.*, 1970, **92**, 5258. 436

70JA5263 J. M. Hoffman and R. H. Schlessinger; *J. Am. Chem. Soc.*, 1970, **92**, 5263. 556

70JA5284 A. B. Holmes and F. Sondheimer; *J. Am. Chem. Soc.*, 1970, **92**, 5284. 715, 716

70JA5829 P. D. Ellis and G. E. Maciel; *J. Am. Chem. Soc.*, 1970, **92**, 5829. 364

70JA6997 T. Do-Minh and A. M. Trozzolo; *J. Am. Chem. Soc.*, 1970, **92**, 6997. 63

70JA7001 O. L. Chapman and C. L. McIntosh; *J. Am. Chem. Soc.*, 1970, **92**, 7001. 420, 423, 428, 443

70JA7436 J. N. Brown, R. L. Towns and L. M. Trefonas; *J. Am. Chem. Soc.*, 1970, **92**, 7436. 598

70JAP6927025 S. Osaki; *Jpn. Pat.* 69 27 025 (1969) (*Chem. Abstr.*, 1970, **72**, 79 004). 484

70JCP(52)5234 J. S. Gibson and D. O. Harris; *J. Chem. Phys.*, 1970, **52**, 5234. 370

70JCP(52)5291 R. C. Benson and W. H. Flygare; *J. Chem. Phys.*, 1970, **52**, 5291. 139

70JCP(53)1543 J. E. Wollrab; *J. Chem. Phys.*, 1970, **53**, 1543 199

70JCS(A)1916 H. M. Frey and M. T. H. Liu; *J. Chem. Soc. (A)*, 1970, 1916. 225

70JCS(B)1544 D. Bright and M. R. Truter; *J. Chem. Soc. (B)*, 1970, 1544. 739

70JCS(B)1603 S. M. Johnson, C. A. Maier and I. C. Paul; *J. Chem. Soc. (B)*, 1970, 1603. 702

70JCS(B)1783 S. M. Johnson and I. C. Paul; *J. Chem. Soc.* (*B*), 1970, 1783. 494

70JCS(C)386 N. E. Waite and J. C. Tebby; *J. Chem. Soc.* (*C*), 1970, 386. 713

70JCS(C)576 D. J. Anderson, T. L. Gilchrist, D. C. Horwell and C. W. Rees; *J. Chem. Soc. (C)*, 1970, 576. 86

70JCS(C)2383 J. W. Lown, G. Dallas and J. P. Moser; *J. Chem. Soc. (C)*, 1970, 2383. 54

70JHC409 P. Catsoulacos; *J. Heterocycl. Chem.*, 1970, **7**, 409. 635

70JHC431 I. Sataty; *J. Heterocycl. Chem.*, 1970, **7**, 431. 646

70JHC755 H. M. Zacharis and M. L. Trefonas; *J. Heterocycl. Chem.*, 1970, **7**, 755. 48

70JHC1203 M. W. Barker and J. T. Gill; *J. Heterocycl. Chem.*, 1970, **7**, 1203. 457, 473

70JHC1249 J. M. Photis; *J. Heterocycl. Chem.*, 1970, **7**, 1249. 520, 537

70JMC389 M. R. Bell, S. D. Clemans and R. Oesterlin; *J. Med. Chem.*, 1970, **13**, 389. 313, 331

70JMC403 B. Basil, E. C. J. Coffee, D. L. Gell, D. R. Maxwell, D. J. Sheffield and K. R. H. Wooldridge; *J. Med. Chem.*, 1970, **13**, 403. 686

70JOC179 J. Cason, C. W. Koch and J. S. Correia; *J. Org. Chem.*, 1970, **35**, 179. 400

70JOC450 L. E. Friedrich and R. A. Cormier; *J. Org. Chem.*, 1970, **35**, 450. 185

70JOC584 P. Y. Johnson and G. A. Berchtold; *J. Org. Chem.*, 1970, **35**, 584. 43, 412, 420, 441, 442, 571, 573, 692

70JOC823 J. L. Massingill, M. R. Reinecke and J. E. Hodgkins; *J. Org. Chem.*, 1970, **35**, 823. 624

70JOC905 L. A. Paquette, R. W. Houser and M. Rosen; *J. Org. Chem.*, 1970, **35**, 905. 433

70JOC927 J. D. Readio and R. A. Falk; *J. Org. Chem.*, 1970, **35**, 927. 454

70JOC999 D. C. Dittmer and R. Glassman; *J. Org. Chem.*, 1970, **35**, 999. 433

70JOC1155 E. K. von Gustorf, D. V. White, B. Kim, D. Hess and J. Leitich; *J. Org. Chem.*, 1970, **35**, 1155. 472

70JOC1501 C. E. Diebert; *J. Org. Chem.*, 1970, **35**, 1501. 170

70JOC1607 J. D. Readio; *J. Org. Chem.*, 1970, **35**, 1607 454

70JOC1781 A. Padwa and R. Gruber; *J. Org. Chem.*, 1970, **35**, 1781. 419, 422, 441

70JOC2249 L. A. Paquette and J. P. Freeman; *J. Org. Chem.*, 1970, **35**, 2249. 442

70JOC2455 R. I. Fryer, J. V. Earley, E. Evans, J. Schneider and L. H. Sternbach; *J. Org. Chem.*, 1970, **35**, 2455. 613

70JOC2482 F. D. Greene and S. S. Hecht; *J. Org. Chem.*, 1970, **35**, 2482. 3, 22, 32, 197

70JOC2520 R. M. Dodson, E. H. Jancis and G. Klose; *J. Org. Chem.*, 1970, **35**, 2520. 420

70JOC2813 F. D. Greene, W. R. Bergmark and J. F. Pazos; *J. Org. Chem.*, 1970, **35**, 2813. 232

70JOC3259 D. N. Harpp and J. G. Gleason; *J. Org. Chem.*, 1970, **35**, 3259. 441

70JOC3424 J. A. Deyrup and C. L. Moyer; *J. Org. Chem.*, 1970, **35**, 3424. 77

70JOC3428 J. A. Deyrup, C. L. Moyer and P. S. Dreifus; *J. Org. Chem.*, 1970, **35**, 3428. 77

70JOC3442 W. C. Lumma, Jr., G. A. Dutra and C. A. Voeker; *J. Org. Chem.*, 1970, **35**, 3442. 585, 693

70JOC3470 M. S. Raasch; *J. Org. Chem.*, 1970, **35**, 3470. 438, 458, 460, 478, 481

70JOC3615 E. J. Volker, M. G. Pleiss and J. A. Moore; *J. Org. Chem.*, 1970, **35**, 3615. 276

70JOC3657 W. O. Siegl and C. R. Johnson; *J. Org. Chem.*, 1970, **35**, 3657. 420, 429, 430, 441, 444, 445

70JOC3775 W. E. Parham, R. W. Davenport and J. B. Biasotti; *J. Org. Chem.*, 1970, **35**, 3775. 774

70JOC3916 G. A. Olah and A. T. Ku; *J. Org. Chem.*, 1970, **35**, 3016. 376, 378

70JOU394 A. V. El'tsov, V. E. Lopatin and M. G. Mikhel'son; *J. Org. Chem. USSR* (*Engl. Transl.*), 1970, **6**, 394. 487

70JPR161 M. Schulz and G. West; *J. Prakt. Chem.*, 1970, **312**, 161. 231

70JPR421 N. Latif, N. Mishriky and I. Zeid; *J. Prakt. Chem.*, 1970, **312**, 421. 155

70JPR1058 H. Remane, R. Borsdorf and M. Mühstädt; *J. Prakt. Chem.*, 1970, **312**, 1058. 692

70LA(731)120 J. Goerdeler and D. Wobig; *Liebigs Ann. Chem.*, 1970, **731**, 120. 475

70LA(736)75 F. Nerdel, M. Mamluk and P. Weyerstahl; *Liebigs Ann. Chem.*, 1970, **736**, 75. 689

70MI50100 J. M. Lehn; *Fortschr. Chem. Forsch.*, 1970, **15**, 311. 7

B-70MI50300 'Nitrenes', ed. W. Lwowski; Wiley, New York, 1970. 36

B-70MI50400 D. T. Clark; in 'International Symposium on Quantum Aspects of Heterocyclic Compounds in Chemistry and Biochemistry', ed. E. D. Bergmann; Israel Academy of Science and Humanities, Jerusalem, 1970, p. 238. 48, 49

B-70MI50401 P. Scheiner, in 'Selective Organic Transformations', ed. B. S. Thyagarajan; Wiley, New York, 1970, vol. 1, p. 327. 88

B-70MI50402 W. Lwowski; in 'Nitrenes', ed. W. Lwowski; Wiley, New York, 1970. 83, 86

B-70MI50500 R. B. Woodward and R. Hoffmann; 'The Conservation of Orbital Symmetry', Verlag Chemie, Weinheim, 1970. 100

70MI50600 H. Druckrey, H. Kruse, R. Preussmann, S. Ivankovic and C. Landschuetz; *Z. Krebsforsch.*, 1970, **74**, 241 (*Chem. Abstr.*, 1970, **73**, 85 731). 183

70MI50900 M. J. S. Dewar and N. Trinajstic; *Theor. Chim. Acta*, 1970, **17**, 235. 278

70MI51000 M. L. Sassiver and B. A. Lewis; *Adv. Appl. Microbiol.*, 1970, **13**, 163. 287

70MI51300 R. R. Baker, R. R. Baldwin and R. W. Walker; *Trans. Faraday Soc.*, 1970, **66**, 3016. 399

70MI51301 F. D. Lewis and R. H. Hirsch; *Mol. Photochem.*, 1970, **2**, 259. 394

70MI51302 W. Kuenzel; *Wiss. Z. Univ. Rostock, Math.-Naturwiss. Reihe*, 1970, **19**, 263 (*Chem. Abstr.*, 1972, **76**, 136 222). 387

70MI51600 M. Taverna and M. Chiti; *Hydrocarbon Processing*, 1970, **49** (11), 137. 528

B-70MI51601 J. H. Boyer; in 'Nitrenes', ed. W. Lwowski; Interscience, New York, 1970, p. 163. 537, 545

70MI52000 H. E. Simmons, C.-H. Park, T. R. Uyeda and M. F. Habibi; *Trans. N.Y. Acad. Sci.*, 1970, **32**, 521. 722

70MI52200 S. G. Mairanovskii, V. S. Mikhailov, S. Z. Taits and O. A. Kalinovskii; *Sov. Electrochem.* (*Engl. Transl.*), 1970, **6**, 1606. 777

70N248 W. H. Müller; *Naturwissenschaften*, 1970, **57**, 248. 760
70OMS(3)925 B. K. Simons, B. Nussey and J. H. Bowie; *Org. Mass Spectrom.*, 1970, **3**, 925. 52
70OMS(3)1259 Q. N. Porter and R. J. Spear; *Org. Mass Spectrom.*, 1970, **3**, 1259. 52
70OS(50)65 L. A. Carpino and L. V. McAdams, III; *Org. Synth.*, 1970, **50**, 65. 176
70RCR1064 A. V. Bogatskii and S. A. Andronati; *Russ. Chem. Rev.* (*Engl. Transl.*), 1970, **39**, 1064. 608, 609, 610
70RTC667 H. Kloosterziel and J. A. A. Van Drunen; *Recl. Trav. Chim. Pays-Bas*, 1970, **89**, 667. 562
70RTC1081 H. W. O. Weissenburger and M. G. van der Hoeven; *Recl. Trav. Chim. Pays-Bas*, 1970, **89**, 1081. 329
70S393 N. H. Fischer; *Synthesis*, 1970, 393. 134, 141, 175
70S475 T. Saegusa, N. Takaishi and Y. Ito; *Synthesis*, 1970, 475. 381
70T453 T. Nishiwaki, T. Kitamura and A. Nakano; *Tetrahedron*, 1970, **26**, 453. 50, 89
70T539 K. Kotera, Y. Takano, A. Matsuura and K. Kitahonoki; *Tetrahedron*, 1970, **26**, 539. 89
70T1275 G. Pala, A. Mantegani and E. Zugna; *Tetrahedron*, 1970, **26**, 1275. 706
70T1967 Y. Langlois, C. Poupat, H.-P. Husson and P. Potier; *Tetrahedron*, 1970, **26**, 1967. 81
70T3245 M. H. Akhtar and A. C. Oehlschlager; *Tetrahedron*, 1970, **26**, 3245. 75, 512
70T3749 A. P. Bindra and J. A. Elix; *Tetrahedron*, 1970, **26**, 3749. 715
70T4269 M. J. S. Dewar and N. Trinajstic; *Tetrahedron*, 1970, **26**, 4269. 14, 492, 550, 554
70T5847 F. Vögtle and P. Naumann; *Tetrahedron*, 1970, **26**, 5847. 763
70TL11 S. Bradamante, R. Fusco, A. Marchesini and G. Pagani; *Tetrahedron Lett.*, 1970, 11. 773
70TL119 K. Claus and H. Jensen; *Tetrahedron Lett.*, 1970, 119. 277
70TL441 A. Takamizawa, K. Hirai and T. Ishiba; *Tetrahedron Lett.*, 1970, 441. 426
70TL655 H. Neunhoeffer and L. Motitschke; *Tetrahedron Lett.*, 1970, 655. 717
70TL703 R. C. Kerber and T. J. Ryan; *Tetrahedon Lett.*, 1970, 703. 472
70TL761 A. Krebs and H. Kimling; *Tetrahedron Lett.*, 1970, 761. 555, 574
70TL1125 K. Ponsold and W. Ihn; *Tetrahedron Lett.*, 1970, 1125. 83
70TL1427 H. Grill and G. Kresze; *Tetrahedron Lett.*, 1970, 1427. 457, 460, 473
70TL1925 J. Firl and S. Sommer; *Tetrahedron Lett.*, 1970, 1925. 472
70TL1929 J. Firl and S. Sommer; *Tetrahedron Lett.*, 1970, 1929. 472
70TL2329 C. Kaneko, S. Yamada and M. Ishikawa; *Tetrahedron Lett.*, 1970, 2329. 206
70TL2877 M. Ohtsuru, K. Tori and M. Fukuyama; *Tetrahedron Lett.*, 1970, 2877. 137
70TL2931 I. Tabushi, Y. Tamaru and Z. Yoshida; *Tetrahedron Lett.*, 1970, 2931. 411, 434
70TL3025 E. L. Stogryn and M. H. Gianni; *Tetrahedron Lett.*, 1970, 3025. 114
70TL3291 E. Breitmaier and E. Bayer; *Tetrahedron Lett.*, 1970, 3291. 773
70TL3539 R. D. Cockroft, E. E. Waali and S. J. Rhoads; *Tetrahedron Lett.*, 1970, 3539. 559
70TL4375 P. Scribe, C. Nouet and J. Wiemann; *Tetrahedron Lett.*, 1970, 4375. 580
70TL4623 A. Hassner and A. Kascheres; *Tetrahedron Lett.*, 1970, 4623. 71
70TL4683 B. Zwanenburg, A. Wagenaar and J. Strating; *Tetrahedron Lett.*, 1970, 4683. 487
70TL4897 L. D. Hatfield, J. Fisher, F. L. José and R. D. G. Cooper; *Tetrahedron Lett.*, 1970, 4897. 308
70ZC361 K.-H. Wünsch and A. Ehlers; *Z. Chem.*, 1970, **10**, 361. 631

71ACR9 A. Hassner; *Acc. Chem. Res.*, 1971, **4**, 9. 83, 84
71ACS141 B. Ringner, S. Sunner and H. Watanabe; *Acta Chem. Scand.*, 1971, **25**, 141. 370, 371
71ACS763 K. Pihlaja, K. Polviander, R. Keskinen and J. Jalonen; *Acta Chem. Scand.*, 1971, **25**, 763. 369
71ACS1927 N. G. Johansson and B. Åkermark; *Acta Chem. Scand.*, 1971, **25**, 1927. 258
71AG(E)11 L. A. Paquette; *Angew. Chem., Int. Ed. Engl.*, 1971, **10**, 11. 499, 501, 663
71AG(E)34 L. H. Sternbach; *Angew. Chem., Int. Ed. Engl.*, 1971, **10**, 34. 608
71AG(E)184 L. Ghosez and C. de Perez; *Angew. Chem., Int. Ed. Engl.*, 1971, **10**, 184. 278
71AG(E)273 M. Klaus and H. Prinzbach; *Angew. Chem., Int. Ed. Engl.*, 1971, **10**, 273. 63
71AG(E)331 F. Fehér, F. Malcharek and K. Glinka; *Angew. Chem., Int. Ed. Engl.*, 1971, **10**, 331. 648
71AG(E)425 A. Rieker; *Angew. Chem., Int. Ed. Engl.*, 1971, **10**, 425. 582
71AG(E)537 G. W. Griffin; *Angew. Chem., Int. Ed. Engl.*, 1971, **10**, 537. 100
71AHC(13)45 F. W. Fowler; *Adv. Heterocycl. Chem.*, 1971, **13**, 45. 59, 80, 82
71AX(B)185 H. Schenk; *Acta Crystallogr., Part B*, 1971, **27**, 185. 702
71BAU2386 S. R. Sterlin, L. G. Zhuravkova, B. L. Dyatkin and I. L. Knunyants; *Bull. Acad. Sci. USSR, Div. Chem. Sci.*, 1971, **20**, 2386. 478
71BAU2432 B. A. Arbuzov, F. B. Balabanova, N. N. Zobova and A. B. Remizov; *Bull. Acad. Sci. USSR, Div. Chem. Sci.*, 1971, 2432. 148
71BCJ2827 S. Fujita and H. Nozaki; *Bull. Chem. Soc. Jpn.*, 1971, **44**, 2827. 772
71BCJ2836 O. Tsuge and S. Mataka; *Bull. Chem. Soc. Jpn.*, 1971, **44**, 2836. 651
71BSF2260 J. L. Luche and H. B. Kagan; *Bull. Soc. Chim. Fr.*, 1971, 2260. 261
71BSF3064 C. Schaal; *Bull. Soc. Chim. Fr.*, 1971, 3064. 436
71BSF3985 N. B. Hassine-Coniac, H. Hazebroucq and J. Gardent; *Bull. Soc. Chim. Fr.*, 1971, 3985. 514
71BSF4576 G. Casteignau; *Bull. Soc. Chim. Fr.*, 1971, 4576. 375, 376
71CB873 R. W. Hoffmann, U. Bressel, J. Gehlhaus and H. Hauser; *Chem. Ber.*, 1971, **104**, 873. 472

71CB1987 F. Effenberger, P. Fischer, G. Prossel and G. Kiefer; *Chem. Ber.*, 1971, **104**, 1987. 249

71CB2786 F. R. Woerner, H. Reimlinger and R. Merenyi; *Chem. Ber.*, 1971, **104**, 2786. 61

71CB2923 W. Tochtermann, D. Schaefer and C. Rohr; *Chem. Ber.*, 1971, **104**, 2923. 14

71CC32 U. K. Georgi and J. Retey; *Chem. Commun.*, 1971, 32. 772

71CC118 N. Ishibe, M. Odani and M. Sunami; *Chem. Commun.*, 1971, 118. 487

71CC577 G. Lowe and J. Parker; *Chem. Commun.*, 1971, 577. 357, 359

71CC647 J. R. Jackson and R. J. Stoodley; *Chem. Commun.*, 1971, 647: 324

71CC698 M. F. Semmelhack, S. Kunkes and C. S. Lee; *Chem. Commun.*, 1971, 698. 636

71CC947 P. J. Garratt, A. B. Holmes, F. Sondheimer and K. P. C. Vollhardt; *Chem. Commun.*, 1971, 947. 714

71CC979 A. G. Anastassiou and B. Y.-H. Chao; *Chem. Commun.*, 1971, 979. 142, 571

71CC988 M. F. Semmelhack and B. F. Gilman; *Chem. Commun.*, 1971, 988. 636

71CC1299 W. Adam and L. Szendrey; *Chem. Commun.*, 1971, 1299. 393

71CC1434 A. B. Holmes and F. Sondheimer; *Chem. Commun.*, 1971, 1434. 715

71CC1496 J. S. Millership and H. Suschitzky; *Chem. Commun.*, 1971, 1496. 255

71CC1518 D. J. Anderson, T. L. Gilchrist, G. E. Gymer and C. W. Rees; *Chem. Commun.*, 1971, 1518. 66, 87

71CHE557 B. V. Kurgane and S. A. Giller; *Chem. Heterocycl. Compd. (Engl. Transl.)*, 1971, **7**, 557. 424

71CI(L)1439 C. G. Hughes and A. H. Rees; *Chem. Ind. (London)*, 1971, 1439. 525

71CJC943 J. F. King, A. Hawson, B. L. Huston, L. J. Danks and J. Komery, *Can. J. Chem.*, 1971, **49**, 943. 643

71CJC971 D. L. Garmaise, G. Y. Paris and G. Efthymiadis; *Can. J. Chem.*, 1971, **49**, 971. 633

71CJC1456 P. Yates, D. R. Moore and T. R. Lynch; *Can. J. Chem.*, 1971, **49**, 1456. 479

71CJC1467 P. Yates, T. R. Lynch and D. R. Moore; *Can. J. Chem.*, 1971, **49**, 1467. 456

71CJC1610 J. W. Lown and M. H. Akhtar; *Can. J. Chem.*, 1971, **49**, 1610. 55

71CJC1648 O. E. Edwards, J. M. Paton, M. H. Benn, R. E. Mitchell, C. Watanatada and K. N. Vohra; *Can. J. Chem.*, 1971, **49**, 1648. 357, 359

71CJC1792 R. Raap; *Can. J. Chem.*, 1971, **49**, 1792. 266

71CJC2023 J. M. Muchowski; *Can. J. Chem.*, 1971, **49**, 2023. 674

71CJC3866 P. Blondeau, R. Gauthier, C. Berse and D. Gravel; *Can. J. Chem.*, 1971, **49**, 3866. 636

71CJC4017 W. H. Brown, B. J. Hutchinson and W. H. MacKinnon; *Can. J. Chem.*, 1971, **49**, 4017. 767

71CPB97 H. Saikachi, H. Ogawa and K. Sato; *Chem. Pharm. Bull.*, 1971, **19**, 97. 722

71CPB1022 A. Takamizawa, K. Hirai and T. Ishiba; *Chem. Pharm. Bull.*, 1971, **19**, 1022. 426

71CPB2354 S. Yurugi, M. Hieda, T. Fushimi and M. Tomimoto; *Chem. Pharm. Bull.*, 1971, **19**, 2354. 630

71CR(C)(273)1655 M. Vu Moc Thuy; *C.R. Hebd. Seances Acad. Sci., Ser. C*, 1971, **273**, 1655. 621

71DIS(B)(31)5869 W. S. Bigard; *Diss. Abstr. Int. B*, 1971, **31**, 5869. 471

71DIS(B)(31)7191 E. Meller; *Diss. Abstr. Int. B*, 1971, **31**, 7191. 347

71DIS(B)(32)1445 E. J. Fasiska; *Diss. Abstr. Int. B*, 1971, **32**, 1445. 365

71G167 D. Misiti and V. Rimatori; *Gazz. Chim. Ital.*, 1971, **101**, 167. 631

71GEP1934763 K. Schmitt, J. Disteldorf and F. Schmitt; *Ger. Offen.* 1 934 763 (1971) (*Chem. Abstr.*, 1971, **75**, 7537). 484

71GEP1960157 H. Jensen and P. Wegener; *Ger. Pat.* 1 960 157 (1971) (*Chem. Abstr.*, 1971, **75**, 63 589). 251

71GEP2027345 K. H. Koenig, A. Fischer and B. Zeeh; *Ger. Offen.* 2 027 345 (1971) (*Chem. Abstr.*, 1972, **76**, 72 500). 484

71HCA1069 P. Skrabal and H. Zollinger; *Helv. Chim. Acta*, 1971, **54**, 1069. 718

71HCA1255 H. B. Bürgi and J. D. Dunitz; *Helv. Chim. Acta*, 1971, **54**, 1255. 49

71HCA2297 G. Barbarella, A. Garbesi and A. Fava; *Helv. Chim. Acta*, 1971, **54**, 2297. 424, 444

71IJS(A)(1)65 N. E. Hester, G. K. Helmkamp and G. I. Alford; *Int. J. Sulfur Chem., Part A*, 1971, **1**, 65 (*Chem. Abstr.*, 1971, **75**, 110 301). 694

71IJS(A)(1)66 J. D. Andose, A. Rauk, R. Tang and K. Mislow; *Int. J. Sulfur Chem., Part A*, 1971, **1**, 66. 134

71IJS(B)(6)267 U. Eisner and T. Krishnamurthy; *Int. J. Sulfur Chem., Part B*, 1971, **6**, 267. 548

71IZV893 R. G. Kostyanovskii, I. M. Gella and K. Khafizov; *Izv. Akad. Nauk SSSR, Ser. Khim.*, 1971, 893. 240, 275

71JA152 L. A. Paquette, T. Kakihana, J. F. Hansen and J. C. Philips; *J. Am. Chem. Soc.*, 1971, **93**, 152. 269, 275, 361

71JA161 L. B. Anderson, J. F. Hansen, T. Kakihana and L. A. Paquette; *J. Am. Chem. Soc.*, 1971, **93**, 161. 664

71JA168 L. A. Paquette, J. F. Hansen and T. Kakihana; *J. Am. Chem. Soc.*, 1971, **93**, 168. 661, 664

71JA262 J. S. Ward and R. Pettit; *J. Am. Chem. Soc.*, 1971, **93**, 262. 524

71JA476 L. A. Carpino, L. V. McAdams, III, R. H. Rynbrandt and J. W. Spiewak; *J. Am. Chem. Soc.*, 1971, **93**, 476. 137, 166

71JA557 W. Adam and R. Rucktäschel; *J. Am. Chem. Soc.*, 1971, **93**, 557. 109

71JA676 B. M. Trost, W. L. Schinski, F. Chen and I. B. Mantz; *J. Am. Chem. Soc.*, 1971, **93**, 676. 417, 422, 424, 426, 435, 442, 444

71JA767 D. E. Tutt and M. A. Schwartz; *J. Am. Chem. Soc.*, 1971, **93**, 767. 327

71JA1123 F. D. Mango and J. H. Schachtschneider; *J. Am. Chem. Soc.*, 1971, **93**, 1123. 28, 39

71JA1400 A. Padwa and W. Eisenhardt; *J. Am. Chem. Soc.*, 1971, **93**, 1400. 62

71JA1543 R. A. Olofson, R. K. Vander Meer and S. Stournas; *J. Am. Chem. Soc.*, 1971, **93**, 1543. 269, 278

71JA1761 R. L. Towns and L. M. Trefonas; *J. Am. Chem. Soc.*, 1971, **93**, 1761. 239

71JA1777 R. Huisgen and H. Maeder; *J. Am. Chem. Soc.*, 1971, **93**, 1777. 53

71JA1779 R. Huisgen, H. Hermann and H. Maeder; *J. Am. Chem. Soc.*, 1971, **93**, 1779. 53

71JA2051 A. B. Levy and A. Hassner; *J. Am. Chem. Soc.*, 1971, **93**, 2051. 268, 271, 273

71JA2793 E. H. Gold; *J. Am. Chem. Soc.*, 1971, **93**, 2793. 243

71JA2799 T. Fukuyama, K. Kuchitsu, Y. Tamaru, Z. Yoshida and I. Tabushi; *J. Am. Chem. Soc.*, 1971, **93**, 2799. 405

71JA2810 J. E. Baldwin, G. Hoefle and S. C. Choi; *J. Am. Chem. Soc.*, 1971, **93**, 2810. 141

71JA3478 T. Kunieda and B. Witkop; *J. Am. Chem. Soc.*, 1971, **93**, 3478. 607

71JA3817 D. W. McNeil, M. E. Kent, F. Hedaya, P. F. D'Angelo and P. O. Schissel; *J. Am. Chem. Soc.*, 1971, **93**, 3817. 663

71JA4016 R. M. Coates and E. F. Johnson; *J. Am. Chem. Soc.*, 1971, **93**, 4016. 662

71JA4075 J. S. Splitter, T. M. Su, H. Ono and M. Calvin; *J. Am. Chem. Soc.*, 1971, **93**, 4075. 229

71JA4082 J. E. Baldwin, A. K. Bhatnagar, S. C. Choi and T. J. Shortridge; *J. Am. Chem. Soc.*, 1971, **93**, 4082. 74, 462, 470

71JA4627 D. L. Coffen, Y. C. Poon and M. L. Lee; *J. Am. Chem. Soc.*, 1971, **93**, 4627. 680, 681

71JA5299 W. P. Giering and M. Rosenblum; *J. Am. Chem. Soc.*, 1971, **93**, 5299. 482

71JA5572 B. M. Trost and R. M. Cory; *J. Am. Chem. Soc.*, 1971, **93**, 5572. 671

71JA5573 B. M. Trost and R. M. Cory; *J. Am. Chem. Soc.*, 1971, **93**, 5573. 671

71JA5780 M. A. Berwick; *J. Am. Chem. Soc.*, 1971, **93**, 5780. 515

71JA6269 S. Kukolja; *J. Am. Chem. Soc.*, 1971, **93**, 6269. 310

71JAP7102008 Otsuka Chem. Drugs Co., *Jpn. Pat.* 71 02 008 (1971) (*Chem. Abstr.*, 1971, **74**, 124 828). 216

71JCP(54)1563 R. C. Long, Jr. and J. H. Goldstein; *J. Chem. Phys.*, 1971, **54**, 1563. 455

71JCS(B)312 D. R. Williams and L. T. Kontnik; *J. Chem. Soc.* (*B*), 1971, 312. 456

71JCS(B)778 A. R. Butler and B. C. Challis, *J. Chem. Soc.* (*B*), 1971, 778. 207

71JCS(B)1231 A. W. Archer, P. A. Claret and D. F. Hayman; *J. Chem. Soc.* (*B*), 1971, 1231. 690

71JCS(C)988 T. L. Gilchrist, C. W. Rees and E. Stanton; *J. Chem. Soc.* (*C*), 1971, 988. 62

71JCS(C)1237 M. Mahendran and A. W. Johnson; *J. Chem. Soc.* (*C*), 1971, 1237. 506, 510, 525, 544

71JCS(C)1917 G. R. Fosker, K. D. Hardy, J. H. C. Nayler, P. Seggery and E. R. Stove; *J. Chem. Soc.* (*C*), 1971, 1917. 331

71JCS(C)1920 A. A. W. Long, J. H. C. Nayler, H. Smith, T. Taylor and N. Ward; *J. Chem. Soc.* (*C*), 1971, 1920. 331

71JCS(C)2104 R. S. Armstrong, M. J. Aroney, R. J. W. LeFevre, H. J. Stootman and W. Luttke; *J. Chem. Soc.* (*C*), 1971, 2104. 49

71JCS(C)2142 T, Sasaki, K. Kanematsu and K. Hayakawa; *J. Chem. Soc.* (*C*), 1971, 2142. 715

71JCS(C)2317 A. C. Mair and M. F. G. Stevens; *J. Chem. Soc.* (*C*), 1971, 2317. 535

71JCS(C)2920 P. H. Ogden; *J. Chem. Soc.* (*C*), 1971, 2920. 482

71JCS(C)3021 T. Nishiwaki and T. Saito; *J. Chem. Soc. (C)*, 1971, 3021. 91

71JCS(C)3540 D. H. R. Barton, F. Comer, D. G. T. Greig, P. G. Sammes, C. M. Cooper, G. Hewitt and W. G. E. Underwood; *J. Chem. Soc.* (*C*), 1971, 3540. 306

71JCS(C)3756 D. M. Brunwin, G. Lowe and J. Parker; *J. Chem. Soc.* (*C*), 1971, 3756. 354

71JCS(C)3864 B. G. Ramsay and R. J. Stoodley; *J. Chem. Soc.* (*C*), 1971, 3864. 313

71JCS(C)3888 D. C. Aldridge, D. Giles and W. B. Turner; *J. Chem. Soc.* (*C*), 1971, 3888. 401

71JHC231 D. Misiti, F. Gatta and R. Landi-Vittory; *J. Heterocycl. Chem.*, 1971, **8**, 231. 613

71JHC405 R. D. Westland, M. H. Lin and J. M. Vandenbelt; *J. Heterocycl. Chem.*, 1971, **8**, 405. 68

71JHC729 J. M. Photis; *J. Heterocycl. Chem.*, 1971, **8**, 729. 509, 520, 537

71JHC779 B. Pecherer, R. C. Sunbury and A. Brossi; *J. Heterocycl. Chem.*, 1971, **8**, 779. 526, 527

71JHC797 M. Israel and L. C. Jones; *J. Heterocycl. Chem.*, 1971, **8**, 797. 611

71JHC845 N. Blazevic and F. Kajfez; *J. Heterocycl. Chem.*, 1971, **8**, 845. 609

71JHC911 G. Pifferi, P. Consonni, R. Monguzzi and A. Omodei-Sale; *J. Heterocycl. Chem.*, 1971, **8**, 911. 625

71JHC961 R. H. Higgins, N. H. Cromwell and W. W. Paudler; *J. Heterocycl. Chem.*, 1971, **8**, 961. 245

71JHC1059 R. H. Higgins and N. H. Cromwell; *J. Heterocycl. Chem.*, 1971, **8**, 1059. 245

71JHC1063 R. H. Higgins, E. Doomes and N. H. Cromwell; *J. Heterocycl. Chem.*, 1971, **8**, 1063. 240

71JOC31 R. W. Franck and J. Auerbachh; *J. Org. Chem.*, 1971, **36**, 31. 50

71JOC230 A. Padwa and A. Battisti; *J. Org. Chem.*, 1971, **36**, 230. 52

71JOC254 C. J. Pedersen; *J. Org. Chem.*, 1971, **36**, 254. 733

71JOC305 G. N. Walker and R. T. Smith; *J. Org. Chem.*, 1971, **36**, 305. 629

71JOC435 L. A. Paquette, T. Kakihana and J. F. Kelly; *J. Org. Chem.*, 1971, **36**, 435. 270, 272, 275, 283, 359, 361, 661, 664

71JOC978 T. J. van Bergen and R. M. Kellogg; *J. Org. Chem.*, 1971, **36**, 978. 499, 544

71JOC1121 G. A. Olah and P. J. Szilagyi; *J. Org. Chem.*, 1971, **36**, 1121. 134, 145, 424

71JOC1324 D. C. Dittmer, M. E. Christy, N. Takashina, R. S. Henion and J. M. Balquist; *J. Org. Chem.*, 1971, **36**, 1324. 444

71JOC1934 T. H. Koch and D. A. Brown; *J. Org. Chem.*, 1971, **36**, 1934. 275, 342, 345, 347, 505

71JOC2676 J. A. Moore, E. J. Volker and C. M. Kopay; *J. Org. Chem.*, 1971, **36**, 2676. 603

71JOC2693 R. M. Dodson, P. D. Hammen and R. A. Davis; *J. Org. Chem.*, 1971, **36**, 2693. 430

71JOC2703 R. M. Dodson, P. D. Hammen and J. Y. Fan; *J. Org. Chem.*, 1971, **36**, 2703. 430

71JOC2838 S. R. Turner, L. J. Guilbault and G. B. Butler; *J. Org. Chem.*, 1971, **36**, 2838. 472

71JOC2841 E. J. Moriconi and W. C. Meyer; *J. Org. Chem.*, 1971, **36**, 2841. 255

71JOC2968 G. F. Field, W. J. Zally and L. H. Sternbach; *J. Org. Chem.*, 1971, **36**, 2968. 645

71JOC3056 J. C. Stowell, F. D. Greene and W. R. Bergmark; *J. Org. Chem.*, 1971, **36**, 3056. 455, 458, 463, 468

71JOC3076 P. G. Mente and H. W. Heine; *J. Org. Chem.*, 1971, **21**, 3076. 65, 633

71JOC3078 H. W. Heine and P. G. Mente; *J. Org. Chem.*, 1971, **36**, 3078. 628

71JOC3846 G. J. McDonald and M. E. Hill; *J. Org. Chem.*, 1971, **36**, 3846. 641

71JOU966 N. S. Zefirov and T. M. Pozdnyakova; *J. Org. Chem. USSR* (*Engl. Transl.*), 1971, **7**, 966. 487

71JOU2361 S. A. Rodkin, A. Ya. Yakubovich and S. P. Makarov; *J. Org. Chem. USSR* (*Engl. Transl.*), 1971, **7**, 2361. 472

71JPS503 J. P. Hou and J. W. Poole; *J. Pharm. Sci.*, 1971, **60**, 503. 303, 326

71JST(10)385 B. Bak and S. Skaarup; *J. Mol. Struct.*, 1971, **10**, 385. 48

71KGS1645 V. M. Pavlov, I. V. Galakhov, L. I. Ragulin, A. A. Alekseev, G. A. Sokol'skii and I. L. Knunyants; *Khim. Geterotsikl. Soedin.*, 1971, **7**, 1645. 487

71LA(749)62 E. Vilsmaier and W. Sprugel; *Liebigs Ann. Chem.*, 1971, **749**, 62. 425

71LA(752)142 K. D. Kampe; *Liebigs Ann. Chem.*, 1971, **752**, 142. 250

71MI50400 J. B. Lambert; *Top. Stereochem.*, 1971, **6**, 19. 51, 52

71MI50401 H. W. Heine; *Mech. Mol. Migrations*, 1971, **3**, 145. 64

71MI50402 J. W. Lown; *Record Chem. Progress*, 1971, **32**, 51. 53, 55

B-71MI50500 R. N. McDonald; in 'Mechanisms of Molecular Migration', ed. B. S. Thyagarajan, Wiley, New York, 1971, vol. 3, p. 67. 103

B-71MI50501 W. Kirmse; 'Carbene Chemistry', Academic Press, New York, 2nd edn., 1971, pp. 432, 433. 107, 126

71MI50600 P. Sigwalt; *Chim. Ind., Genie Chim.*, 1971, **104**, 47. 182

71MI50601 H. N. MacFarland, N. R. Khan and V. T. Innanen; *Bull. Environ. Contam. Toxicol.*, 1971, **6**, 509 (*Chem. Abstr.*, 1972, **76**, 37 117). 183

71MI50800 R. Bonaccorsi, E. Scrocco and J. Tomasi; *Theor. Chim. Acta*, 1971, **21**, 17. 198

B-71MI50801 E. Schmitz; in 'Lectures, 23rd Congress on Pure and Applied Chemistry, 1971', Butterworth, London, 1971, p. 283. 221

71MI50900 J. B. Lambert; *Top. Stereochem.*, 1971, **6**, 19. 239

71MI50901 W. W. Stewart; *Nature*, 1971, **229**, 174. 249

71MI50902 Y. Muraoka, A. Fujii, T. Yoshioka, T. Takita and H. Umezawa; *J. Antibiot.*, 1977, **30**, 178. 249

B-71MI50903 M. S. Manhas and A. K. Bose; 'β-Lactams: Natural and Synthetic', Wiley-Interscience, New York, 1971. 249

71MI51100 F. Moll and P. Kastenmeier; *Pharm. Ztg.*, 1971, **116**, 1345. 330

71MI51101 R. Raap; *J. Antibiot.*, 1971, **24**, 695. 331

71MI51300 G. Holan; *Nature (London)*, 1971, **232**, 644. 402

B-71MI51600 R. T. Conley and S. Ghosh; in 'Mechanisms of Molecular Migrations', ed. B. S. Thyagarajan; Wiley-Interscience, New York, 1971, vol. 4, p. 197. 530

B-71MS Q. N. Porter and J. Baldas, 'Mass Spectrometry of Heterocyclic Compounds', Wiley, New York, 1971. 11, 12, 99, 135, 368, 370, 501

71OMS(5)651 J. Mitera and V. Kubelka; *Org. Mass. Spectrom.*, 1971, **5**, 651. 501

71OPP27 Z. Olszenko-Piontkowa and T. Urbanski; *Org. Prep. Proced. Int.*, 1971, **3**, 27. 633

71OPP289 O. Tsuge, M. Tashiro, K. Kamata and K. Hokama; *Org. Prep. Proced. Int.*, 1971, **3**, 289. 640

71P3069 B. E. Ellis and N. Amrhein; *Phytochemistry*, 1971, **10**, 3069. 590

71PMH(4)121 R. F. M. White and H. Williams; *Phys. Methods Heterocycl. Chem.*, 1971, **4**, 121. 10, 99

71PMH(4)265 A. R. Katritzky and P. J. Taylor; *Phys. Methods Heterocycl. Chem.*, 1971, **4**, 265. 12, 14

71S428 E. Vilsmaier and B. Hloch; *Synthesis*, 1971, 428. 162

71S429 E. Vilsmaier and W. Schalk; *Synthesis*, 1971, 429. 424

71S590 E. Vilsmaier and B. Hloch; *Synthesis*, 1971, 590. 162

71T341 W. O. Siegl and C. R. Johnson; *Tetrahedron*, 1971, **27**, 341. 409, 411, 412, 420

71T2599 L. A. Paquette, J. P. Freeman and S. Maiorana; *Tetrahedron*, 1971, **27**, 2599. 439

71T4821 K. Kondo and A. Negishi; *Tetrahedron*, 1971, **27**, 4821. 134

71TL109 P. G. Gassmann and A. Carrasquillo; *Tetrahedron Lett.*, 1971, 109. 241

71TL373 S. Sarel, B. A. Weissman and Y. Stein; *Tetrahedron Lett.*, 1971, 373. 92

71TL473 E. Brunn and R. Huisgen; *Tetrahedron Lett.*, 1971, 473. 54

71TL481 R. Huisgen and W. Scheer; *Tetrahedron Lett.*, 1971, 481. 54

71TL671 A. Marchesini, S. Bradamante, R. Fusco and G. Pagani; *Tetrahedron Lett.*, 1971, 671. 772, 775

71TL749 W. H. Richardson and V. F. Hodge; *Tetrahedron Lett.*, 1971, 749. 464

71TL1257 S. Berger, G. Henes and A. Rieker; *Tetrahedron Lett.*, 1971, 1257. 582

71TL1601 H. Neunhoeffer, J. Stastny and L. Motitschke; *Tetrahedron Lett.*, 1971, 1601. 717

71TL2025 J. M. Surzur, M. P. Crozet and C. Dupuy; *Tetrahedron Lett.*, 1971, 2025. 584

71TL2125 F. Fehér and M. Langer; *Tetrahedron Lett.*, 1971, 2125. 649

71TL2247 W. C. Danen and T. T. Kensler; *Tetrahedron Lett.*, 1971, 2247. 19

71TL2437 H.-J. Shue and F. W. Fowler; *Tetrahedron Lett.*, 1971, 2437. 678

71TL2515 R. R. Fraser and Y. Y. Wigfield; *Tetrahedron Lett.*, 1971, 2515. 499

71TL2871 M. O. Riley and J. D. Park; *Tetrahedron Lett.*, 1971, 2871. 681

71TL3721 H. H. Wasserman and M. S. Baird; *Tetrahedron Lett.*, 1971, 3721. 357, 359

71TL3843 H. J. J.-B. Martel and M. Rasmussen; *Tetrahedron Lett.*, 1971, 3843. 777

71TL3947 P. D. Rosso, J. Oberdier and J. S. Swenton; *Tetrahedron Lett.*, 1971, 3947. 542

71TL4519 D. St. C. Black, R. F. C. Brown and A. M. Wade; *Tetrahedron Lett.*, 1971, 4519. 228

71TL4831 V. Nair; *Tetrahedron Lett.*, 1971, 4831. 60
71USP3557145 R. T. McFadden; *U.S. Pat.* 3 557 145 (1971) (*Chem. Abstr.* 1971, **74**, 76 315). 144
71USP3579500 A. G. Jelinek; *U.S. Pat.* 3 579 500 (1971) (*Chem. Abstr.*, 1971, **75**, 35 991). 484
71USP3592811 M. S. Raasch; *U.S. Pat.* 3 592 811 (1971) (*Chem. Abstr.*, 1971, **75**, 99 232). 484
71ZOR2295 T. B. Fomina, V. N. Artem'eva and Y. N. Sazanov; *Zh. Org. Khim.*, 1971, **7**, 2295. 398

72ACR281 A. G. Anastassiou; *Acc. Chem. Res.*, 1972, **5**, 281. 710, 723, 727
72ACS494 S. Abrahamsson and G. Rehnberg; *Acta Chem. Scand.*, 1972, **26**, 494. 406
72ACS1659 J. Becher; *Acta Chem. Scand.*, 1972, **26**, 1659. 606
72AG(E)47 K. Burger and J. Fehn; *Angew. Chem., Int. Ed. Engl.*, 1972, **11**, 47. 271, 276
72AG(E)217 E. Vogel, R. Feldmann, H. Duwel, H.-D. Cremer and H. Gunther; *Angew. Chem., Int. Ed. Engl.*, 1972, **11**, 217. 715
72AG(E)290 T. Kauffmann and E. Koppelmann; *Angew. Chem., Int. Ed. Engl.*, 1972, **11**, 290. 73
72AG(E)423 H. Hofmann, B. Meyer and P. Hofmann; *Angew. Chem., Int. Ed. Engl.*, 1972, **11**, 423. 556
72AG(E)583 I. L. Knunyants and G. A. Sokolskii; *Angew. Chem., Int. Ed. Engl.*, 1972, **11**, 583. 459, 472, 477, 485
72AG(E)825 W. D. Stohrer and R. Hoffmann; *Angew. Chem., Int. Ed. Engl.*, 1972, **11**, 825. 553
72AG(E)932 H. Kimling and A. Krebs; *Angew. Chem., Int. Ed. Engl.*, 1972, **11**, 932. 574
72AG(E)935 E. Vogel, H.-J. Altenbach and D. Cremer; *Angew. Chem., Int. Ed. Engl.*, 1972, **11**, 935. 679, 689
72AG(E)937 H.-J. Altenbach and E. Vogel; *Angew. Chem., Int. Ed. Engl.*, 1972, **11**, 937. 189, 679
72AG(E)939 E. Vogel, H.-J. Altenbach and C. D. Sommerfeld; *Angew. Chem., Int. Ed. Engl.*, 1972, **11**, 939. 189, 714
72AG(E)942 R. Schwesinger and H. Prinzbach; *Angew. Chem., Int. Ed. Engl.*, 1972, **11**, 942. 189
72AJC865 J. A. Elix, W. S. Wilson, R. N. Warrener and I. C. Calder; *Aust. J. Chem.*, 1972, **25**, 865. 663
72AJC2421 R. G. Cooke and I. M. Russell; *Aust. J. Chem.*, 1972, **25**, 2421. 503, 514, 517
72AJC2451 M. R. Bendall, J. B. Bremner and J. F. W. Fay; *Aust. J. Chem.*, 1972, **25**, 2451. 527
72AX(B)2825 K. B. Birnbaum; *Acta Crystallogr., Part B*, 1972, **28**, 2825. 703
72B1 D. E. Bergstrom and N. J. Leonard; *Biochemistry*, 1972, **11**, 1. 447
72B1961 D. R. Boyd, J. W. Daly and D. M. Jerina; *Biochemistry*, 1972, **11**, 1961. 566
72B3080 N. Kaubisch, J. W. Daly and D. M. Jerina; *Biochemistry*, 1972, **11**, 3080. 584, 590
72BAU505 O. N. Nuretdinova and B. A. Arbuzov; *Bull. Acad. Sci. USSR, Div. Chem. Sci.*, 1972, 505. 427, 429
72BAU2441 M. A. Belaventsev, L. L. Mikheev, V. M. Pavlov, G. A. Sokol'skii and I. L. Knunyants; *Bull. Acad. Sci. USSR, Div. Chem. Sci.*, 1972, **21**, 2441. 477
72BCJ1534 O. Tsuge and K. Sakai; *Bull. Chem. Soc. Jpn.*, 1972, **45**, 1534. 458, 474
72BCJ1797 Y. Ueno and M. Okawara; *Bull. Chem. Soc. Jpn.*, 1972, **45**, 1797. 152
72BCJ3026 N. Ohmichi, A. Tajiri and T. Nakajima; *Bull. Chem. Soc. Jpn.*, 1972, **45**, 3026. 13
72BCJ3187 O. Tsuge and A. Torii; *Bull. Chem. Soc. Jpn.*, 1972, **45**, 3187. 457, 473
72BSF2868 J.-L. Aubagnac, J. Elguero and R. Robert; *Bull. Soc. Chim. Fr.*, 1972, 2868. 675
72BSF4655 J. L. Halary, T. Yvernault and J. Casteignau; *Bull. Soc. Chim. Fr.*, 1972, 4655. 393
72C420 J. Bruhin and W. Jenny; *Chimia*, 1972, **26**, 420. 767
72CB434 H. J. Lindner and B. von Gross; *Chem. Ber.*, 1972, **105**, 434, 494
72CB757 E. Cohnen and J. Mahnke; *Chem. Ber.*, 1972, **105**, 757. 645
72CB880 D. Hellwinkel and H. Seifert; *Chem. Ber.*, 1972, **105**, 880. 492, 494, 533
72CB982 E. Carstensen-Oeser; *Chem. Ber.*, 1972, **105**, 982. 492, 494, 506, 519, 533, 544
72CB2437 R. W. Hoffman and W. Schafer; *Chem. Ber.*, 1972, **105**, 2437. 472
72CB2575 P. Stoss and G. Satzinger; *Chem. Ber.*, 1972, **105**, 2575. 644
72CB2780 H. Furrer; *Chem. Ber.*, 1972, **105**, 2780. 258, 358, 359
72CC11 S. A. Matlin and P. G. Sammes; *J. Chem. Soc., Chem. Commun.*, 1972, 11. 126, 127
72CC43 A. H. J. Wang, I. C. Paul, E. R. Talaty and A. E. Dupuy; *J. Chem. Soc., Chem. Commun.*, 1972, 43. 79
72CC192 D. M. Brunwin and G. Lowe; *J. Chem. Soc., Chem. Commun.*, 1972, 192. 265, 331
72CC229 E. G. Brain, A. J. Eglington, J. H. C. Nayler, M. J. Pearson and R. Southgate; *J. Chem. Soc., Chem. Commun.*, 1972, 229. 306
72CC277 A. G. Anastassiou and B. Y.-H. Chao; *J. Chem. Soc., Chem. Commun.*, 1972, 277. 142
72CC325 A. G. Anastassiou, R. P. Cellura, J. M. Spence and S. W. Eachus; *J. Chem. Soc., Chem. Commun.*, 1972, 325. 725
72CC362 T. Kubota and H. Sakurai; *J. Chem. Soc., Chem. Commun.*, 1972, 362. 261
72CC395 N. J. De'Ath and D. B. Denney; *J. Chem. Soc., Chem. Commun.*, 1972, 395. 485
72CC409 A. Padwa, J. Smolanoff and S. I. Wetmore; *J. Chem. Soc., Chem. Commun.*, 1972, 409. 56
72CC466 M. Kinugasa and S. Hashimoto; *J. Chem. Soc., Chem. Commun.*, 1972, 466. 263
72CC565 T. Nishiwaki; *J. Chem. Soc., Chem. Commun.*, 1972, 565. 65
72CC680 A. Padwa, P. Sackman, E. Shefter and E. Vega; *J. Chem. Soc., Chem. Commun.*, 1972, 680. 665
72CC694 Z. Majerski and C. S. Redvanly; *J. Chem. Soc., Chem. Commun.*, 1972, 694. 193
72CC818 E. E. Nunn and R. N. Warrener; *J. Chem. Soc., Chem. Commun.*, 1972, 818. 465, 472
72CC827 A. A. Reid, J. T. Sharp and S. J. Murray; *J. Chem. Soc., Chem. Commun.*, 1972, 827. 600
72CC884 D. W. Jones; *J. Chem. Soc., Chem. Commun.*, 1972, 884. 76

72CC962	J. S. Walia, L. A. Heindl, A. S. Walia and P. S. Walia; *J. Chem. Soc., Chem. Commun.*, 1972, 962.	611
72CC985	G. H. Wahl, Jr., J. Bordner, D. N. Harpp and J. G. Gleason; *J. Chem. Soc., Chem. Commun.*, 1972, 985.	707
72CC1108	J. Hill and J. Townend; *J. Chem. Soc., Chem. Commun.*, 1972, 1108.	345
72CC1116	A. Padwa and S. I. Wetmore; *J. Chem. Soc., Chem. Commun.*, 1972, 1116.	612
72CC1144	J. A. Hyatt and J. S. Swenton; *J. Chem. Soc., Chem. Commun.*, 1972, 1144.	244
72CC1233	D. N. Reinhoudt and C. G. Kouwenhoven; *J. Chem. Soc., Chem. Commun.*, 1972, 1233.	556
72CC1298	T. Fujisawa and T. Kobori; *J. Chem. Soc., Chem. Commun.*, 1972, 1298.	142
72CCC4035	P. Křivinka and J. Honzl; *Collect. Czech. Chem. Commun.*, 1972, **37**, 4035.	266
72CHE432	N. M. Karimova, M. G. Lin'kova, O. V. Kil'disheva and I. L. Knunyants; *Chem. Heterocycl. Compd.* (*Engl. Transl.*), 1972, **8**, 432.	425
72CJC1167	O. E. Edwards, D. Vocelle and J. W. ApSimon; *Can. J. Chem.*, 1972, **50**, 1167.	19, 655
72CJC2721	H. A. Wiebe, S. Braslavsky and J. Heicklen; *Can. J. Chem.*, 1972, **50**, 2721.	181, 186
72CJC2902	S. Wolfe, J.-B. Ducep, G. Kannengiesser and W. S. Lee; *Can. J. Chem.*, 1972, **50**, 2902.	314, 354
72CJC3196	T. Durst and M. J. LeBelle; *Can. J. Chem.*, 1972, **50**, 3196.	251
72CJC3886	K. Ackerman, D. E. Horning and J. M. Muchowski; *Can. J. Chem.*, 1972, **50**, 3886.	525
72CJC3923	J. J. Worman, M. Shen and P. C. Nichols; *Can. J. Chem.*, 1972, **50**, 3923.	477
72CJC3930	K. Jankowski and R. Harvey; *Can. J. Chem.*, 1972, **50**, 3930.	161
72CL1065	T. Fujisawa and T. Kobori; *Chem. Lett.*, 1972, 1065.	172
72CPB725	Y. Okamoto, A. Takada and T. Ueda; *Chem. Pharm. Bull.*, 1972, **20**, 725.	606
72CPB871	Y. Tamura, Y. Yoshimura and Y. Kita; *Chem. Pharm. Bull.*, 1972, **20**, 871.	534
72CPB1243	T. Takegoshi, H. Tachizawa and G. Ohta; *Chem. Pharm. Bull.*, 1972, **20**, 1243.	401
72CPB1325	O. Aki and Y. Nakagawa; *Chem. Pharm. Bull.*, 1972, **20**, 1325.	514
72CPB1574	T. Kato and T. Hozumi; *Chem. Pharm. Bull.*, 1972, **20**, 1574.	386
72CPB1740	T. Oishi, S. Murakami and Y. Ban; *Chem. Pharm. Bull.*, 1972, **20**, 1740.	522
72CR(C)(274)1108	J. C. Rosso and L. Carbonnel; *C.R. Hebd. Seances Acad. Sci., Ser. C*, 1972, **274**, 1108.	376
72CR(C)(275)581	E. Desherces, M. Riviere, J. Parello and A. Lattes; *C.R. Hebd. Seances Acad. Sci., Ser. C*, 1972, **275**, 581.	541
72DOK(207)899	L. A. Korotneva, G. P. Belonovskaya and B. A. Dolgoplosk; *Dokl. Akad. Nauk SSSR*, 1972, **207**, 899 (*Chem. Abstr.*, 1973, **78**, 111 027).	146
72E1129	J. W. Daly, D. M. Jerina and B. Witkop; *Experientia*, 1972, **28**, 1129.	566
72GEP2044838	H. D. Winkelmann, H. Rinke, H. Oertel and N. Weimann; *Ger. Offen.* 2 044 838 (1972) (*Chem. Abstr.*, 1972, **77**, 21 443).	484
72GEP2133564	H. A. Bruson and H. L. Plant; *Ger. Offen.*, 2 133 564 (1972) (*Chem. Abstr.*, 1972, **76**, 99 495).	396
72HC(26)1	A. Rosowsky; *Chem. Heterocycl. Compd.*, 1972, **26**, 1.	548, 577, 578, 579, 590
72HC(26)319	C. E. Pawloski; *Chem. Heterocycl. Compd.*, 1972, **26**, 319.	620, 621, 622
72HC(26)573	L. Field and D. L. Tuleen; *Chem. Heterocycl. Compd.*, 1972, **26**, 573.	571, 572, 622, 623, 624
72HC(26)667	V. J. Traynelis; *Chem. Heterocycl. Compd.*, 1972, **26**, 667.	576, 624
72HCA388	K. Heusler; *Helv. Chim. Acta*, 1972, **55**, 388.	313
72HCA417	B. Fechtig, H. Bickel and K. Heusler; *Helv. Chim. Acta*, 1972, **55**, 417.	331
72HCA907	H. Dahn, H.-P. Schlunke and J. Temler; *Helv. Chim. Acta*, 1972, **55**, 907.	657
72HCA1276	D. Felix, R. K. Muller, U. Horn, R. Joos, J. Schreiber and A. Eschenmoser; *Helv. Chim. Acta*, 1972, **55**, 1276.	75
72HCA1759	M. A. Steinfels, H. W. Krapf, P. Riedl, J. Sauer and A. S. Dreiding; *Helv. Chim. Acta*, 1972, **55**, 1759.	662
72HCA2567	R. Scartazzini, J. Gosteli, H. Bickel and R. B. Woodward; *Helv. Chim. Acta*, 1972, **55**, 2567.	355
72IZV125	B. A. Arbuzov, O. N. Nuretdinova, L. Z. Nikonova and V. V. Pomazanov; *Izv. Akad. Nauk SSSR, Ser. Khim.*, 1972, 125.	381
72JA303	A. G. Anastassiou and R. B. Hammer; *J. Am. Chem. Soc.*, 1972, **94**, 303.	61
72JA513	R. J. Sundberg, S. R. Suter and M. Brenner; *J. Am. Chem. Soc.*, 1972, **94**, 513.	495, 497, 534
72JA521	A. Padwa and A. Battisti; *J. Am. Chem. Soc.*, 1972, **94**, 521.	181
72JA534	L. A. Paquette and R. J. Haluska; *J. Am. Chem. Soc.*, 1972, **94**, 534.	714
72JA597	P. Beak and J. W. Worley; *J. Am. Chem. Soc.*, 1972, **94**, 597.	175
72JA1193	L. E. Friedrich and G. B. Schuster; *J. Am. Chem. Soc.*, 1972, **94**, 1193.	186
72JA1365	O. L. Chapman, P. W. Wojtkowski, W. Adam and O. Rodriguez; *J. Am. Chem. Soc.*, 1972, **94**, 1365.	44
72JA1389	J. Dale, T. Ekeland and J. Krane; *J. Am. Chem. Soc.*, 1972, **94**, 1389.	700, 701
72JA1395	A. Padwa, S. Clough, M. Dharan, J. Smolanoff and S. I. Wetmore; *J. Am. Chem. Soc.*, 1972, **94**, 1395.	91
72JA1408	L. D. Cama, W. J. Leanza, T. B. Beattie and B. G. Christensen; *J. Am. Chem. Soc.*, 1972, **94**, 1408.	316
72JA1629	R. R. Rando; *J. Am. Chem. Soc.*, 1972, **94**, 1629.	258
72JA1747	K. Gollnick, D. Haisch and G. Schade; *J. Am. Chem. Soc.*, 1972, **94**, 1747.	464
72JA1951	J. A. Barltrop and H. A. J. Carless; *J. Am. Chem. Soc.*, 1972, **94**, 1951.	398
72JA2000	W. Adam, J. Baeza and J.-C. Liu; *J. Am. Chem. Soc.*, 1972, **94**, 2000.	373

72JA2758	A. G. Hortmann and D. A. Robertson; *J. Am. Chem. Soc.*, 1972, **94**, 2758.	71
72JA2770	A. Steigel, J. Sauer, D. A. Kleier and G. Binsch; *J. Am. Chem. Soc.*, 1972, **94**, 2770.	497, 499, 500, 595
72JA2894	W. Adam and J.-C. Liu; *J. Am. Chem. Soc.*, 1972, **94**, 2894.	455, 469
72JA3455	J. Villaume and P. S. Skell; *J. Am. Chem. Soc.*, 1972, **94**, 3455.	75
72JA3484	H. Ulrich, B. Tucker and A. A. R. Sayigh; *J. Am. Chem. Soc.*, 1972, **94**, 3484.	455, 458, 475
72JA3955	W. M. Williams and W. R. Dolbier; *J. Am. Chem. Soc.*, 1972, **94**, 3955.	603
72JA4046	T. Do-Minh and A. M. Trozzolo; *J. Am. Chem. Soc.*, 1972, **94**, 4046.	62
72JA4386	E. M. Burgess and W. M. Williams; *J. Am. Chem. Soc.*, 1972, **94**, 4386.	473, 474
72JA5077	E. F. Ullman and P. Singh; *J. Am. Chem. Soc.*, 1972, **94**, 5077.	483
72JA5247	A. T. Tsatsas, R. W. Stearns and W. M. Risen; *J. Am. Chem. Soc.*, 1972, **94**, 5247.	740
72JA5777	A. M. Foster and W. C. Agosta; *J. Am. Chem. Soc.*, 1972, **94**, 5777.	559
72JA5786	K. Kondo, A. Negishi and I. Ojima; *J. Am. Chem. Soc.*, 1972, **94**, 5786.	157
72JA5929	T. S. Cantrell; *J. Am. Chem. Soc.*, 1972, **94**, 5929.	270, 274
72JA6135	G. M. Atkins, Jr. and E. M. Burgess; *J. Am. Chem. Soc.*, 1972, **94**, 6135.	473
72JA6760	L. A. Paquette and M. K. Scott; *J. Am. Chem. Soc.*, 1972, **94**, 6760.	667
72JA7092	J. C. Coll, D. R. Crist, M. del C. G. Barrio and N. J. Leonard; *J. Am. Chem. Soc.*, 1972, **94**, 7092.	660, 699
72JA7100	A. H.-J. Wang, R. J. Missavage, S. R. Byrn and I. C. Paul; *J. Am. Chem. Soc.*, 1972, **94**, 7100.	660
72JA7194	S. M. Kupchan, W. A. Court, R. G. Dailey, Jr., C. J. Gilmore and R. F. Bryan: *J. Am. Chem. Soc.*, 1972, **94**, 7194.	192
72JA7396	K. B. Wiberg and G. J. Burgmaier; *J. Am. Chem. Soc.*, 1972, **94**, 7396.	186
72JA7788	A. Padwa and E. Glazer; *J. Am. Chem. Soc.*, 1972, **94**, 7788.	56
72JA7876	G. J. Kasperek, T. C. Bruice, H. Yagi, N. Kaubisch and D. M. Jerina; *J. Am. Chem. Soc.*, 1972, **94**, 7876.	567
72JA8250	T. Vladimiroff; *J. Am. Chem. Soc.*, 1972, **94**, 8250.	488
72JA8253	Y. S. Lo and J. C. Sheehan; *J. Am. Chem. Soc.*, 1972, **94**, 8253.	319
72JA8255	A. Hassner and D. J. Anderson; *J. Am. Chem. Soc.*, 1972, **94**, 8255.	59
72JA8375	K. A. Rubinson and G. Palmer; *J. Am. Chem. Soc.*, 1972, **94**, 8375.	485
72JA8761	J. A. Barltrop and H. A. J. Carless; *J. Am. Chem. Soc.*, 1972, **94**, 8761.	397
72JA9261	T. Durst, R. V. D. Elzen and M. J. LeBelle; *J. Am. Chem. Soc.*, 1972, **94**, 9261.	255
72JAP7219791	S. Shimosaka, S. Ueda and S. Nakajima; *Jpn. Pat.* 72 19 791 (1972) (*Chem. Abstr.*, 1972, **77**, 88 277).	394
72JCP(56)2919	L. C. Robertson and J. A. Merritt; *J. Chem. Phys.*, 1972, **56**, 2919.	203
72JCP(57)941	L. C. Robertson and J. A. Merritt; *J. Chem. Phys.*, 1972, **57**, 941.	203
72JCP(57)1758	M. B. Robin, C. R. Brundle, N. A. Kuebler, G. B. Ellison and K. B. Wiberg; *J. Chem. Phys.*, 1972, **57**, 1758.	204
72JCP(57)2318	J. S. Gibson and D. O. Harris; *J. Chem. Phys.*, 1972, **57**, 2318.	364, 370
72JCS(F1)512	H. E. Avery and S. J. Heath; *J. Chem. Soc., Faraday Trans. 1*, 1972, **68**, 512.	124
72JCS(P1)213	J. Schlademan and R. Partch; *J. Chem. Soc., Perkin Trans. 1*, 1972, 213.	659
72JCS(P1)885	G. R. Proctor and W. I. Ross; *J. Chem. Soc., Perkin Trans. 1*, 1972, 885.	659
72JCS(P1)889	W. I. Ross and G. R. Proctor; *J. Chem. Soc., Perkin Trans. 1*, 1972, 889.	659
72JCS(P1)895	J. R. Jackson and R. J. Stoodley; *J. Chem. Soc., Perkin Trans. 1*, 1972, 895.	316
72JCS(P1)929	D. H. R. Barton, M. Girijavallabhan and P. G. Sammes; *J. Chem. Soc., Perkin Trans. 1*, 1972, 929.	331
72JCS(P1)2007	F. M. Dean, K. B. Hindley and S. Small; *J. Chem. Soc., Perkin Trans. 1*, 1972, 2007.	690
72JCS(P1)2012	A. Cromarty, G. R. Proctor and M. Shabbir; *J. Chem. Soc., Perkin Trans. 1*, 1972, 2012.	516, 525
72JCS(P1)2623	S. A. Matlin and P. G. Sammes; *J. Chem. Soc., Perkin Trans. 1*, 1972, 2623.	126
72JCS(P2)248	H. Nakai and H. Koyama; *J. Chem. Soc., Perkin Trans. 2*, 1972, 248.	405
72JCS(P2)345	M. A. Bush and M. R. Truter; *J. Chem. Soc., Perkin Trans. 2*, 1972, 345.	735
72JCS(P2)1137	J. F. Cannon, J. Daly, J. V. Silverton, D. R. Boyd and D. M. Jerina; *J. Chem. Soc., Perkin Trans. 2*, 1972, 1137.	198
72JCS(P2)1366	G. M. Blackburn and J. D. Plackett; *J. Chem. Soc., Perkin Trans. 2*, 1972, 1366.	251
72JHC77	M. W. Barker, L. L. Combs and J. T. Gill; *J. Heterocycl. Chem.*, 1972, **9**, 77.	473
72JHC555	M. W. Barker and R. H. Jones; *J. Heterocycl. Chem.*, 1972, **9**, 555.	456, 473
72JHC587	N. H. Cromwell, P. B. Woller, H. E. Baumgarten, R. G. Parker and D. L. von Minden; *J. Heterocycl. Chem.*, 1972, **9**, 587.	52
72JHC609	B. Pecherer, R. C. Sunbury and A. Brossi; *J. Heterocycl. Chem.*, 1972, **9**, 609.	529
72JHC617	B. Pecherer, R. C. Sunbury and A. Brossi; *J. Heterocycl. Chem.*, 1972, **9**, 617.	527
72JHC747	R. I. Fryer; *J. Heterocycl. Chem.*, 1972, **9**, 747.	608, 617
72JHC1147	M. W. Barker and J. D. Rosamond; *J. Heterocycl. Chem.*, 1972, 1147.	248, 266
72JHC1209	G. Pifferi, L. Fontanella, E. Occelli and R. Monguzzi; *J. Heterocycl. Chem.*, 1972, **9**, 1209.	626
72JMC333	R. G. Micetich, R. Raap, J. Howard and I. Pushkas; *J. Med. Chem.*, 1972, **15**, 333.	330
72JOC208	E. J. Moriconi and I. A. Maniscalco; *J. Org. Chem.*, 1972, **37**, 208.	497, 502, 514, 531
72JOC516	H. T. Nagasawa, P. S. Fraser and J. A. Elberling; *J. Org. Chem.*, 1972, **37**, 516.	342, 346, 347
72JOC552	T. J. Barton, M. D. Martz and R. G. Zika; *J. Org. Chem.*, 1972, **37**, 552.	588
72JOC560	R. B. Bates, L. M. Kroposki and D. E. Potter; *J. Org. Chem.*, 1972, **37**, 560.	562
72JOC680	A. Fukuzawa, E. Kurosawa and T. Irie; *J. Org. Chem.*, 1972, **37**, 680.	667

72JOC802 V. Nair; *J. Org. Chem.*, 1972, **37**, 802. 59

72JOC890 T. Sasaki, K. Kanematsu and Y. Yukimoto; *J. Org. Chem.*, 1972, **37**, 890. 524

72JOC919 C. R. Johnson, C. C. Bacon and J. J. Rigau; *J. Org. Chem.*, 1972, **37**, 919. 445, 572

72JOC1111 D. C. Dittmer, P. L.-F. Chang, F. A. Davis, M. Iwanami, I. K. Stamos and K. Takahashi; *J. Org. Chem.*, 1972, **37**, 1111. 418, 426, 441

72JOC1321 F. W. Fowler; *J. Org. Chem.*, 1972, **37**, 1321. 343

72JOC1537 D. Seyferth, W. Tronick, R. S. Marmor and W. E. Smith; *J. Org. Chem.*, 1972, **37**, 1537. 143

72JOC1721 W. Ando, T. Yagihara, S. Tozune, I. Imai, J. Suzuki, T. Toyama, S. Nakaido and T. Migita; *J. Org. Chem.*, 1972, **37**, 1721. 431

72JOC2161 H. S. Kasmai and H. W. Whitlock, Jr.; *J. Org. Chem.*, 1972, **37**, 2161. 669

72JOC2367 R. M. Dodson, V. Srinivasan, K. S. Sharma and R. F. Sauers; *J. Org. Chem.*, 1972, **37**, 2367. 696

72JOC2421 A. G. Anastassiou and E. Reichmanis; *J. Org. Chem.*, 1972, **38**, 2421. 191

72JOC2508 V. Nair; *J. Org. Chem.*, 1972, **37**, 2508. 60

72JOC2514 J. H. Schauble and J. D. Williams; *J. Org. Chem.*, 1972, **37**, 2514. 421, 435, 441

72JOC2980 H. W. Heine, P. G. Williard and T. R. Hoye; *J. Org. Chem.*, 1972, **37**, 2980. 213, 215

72JOC3437 A. K. Bose, J. C. Kapur, J. L. Fahey and M. S. Manhas; *J. Org. Chem.*, 1973, **38**, 3437. 359

72JOC3536 W. T. Brady and A. D. Patel; *J. Org. Chem.*, 1972, **37**, 3536. 374

72JOC3592 O. Buchardt, C. L. Pedersen and N. Harrit; *J. Org. Chem.*, 1972, **37**, 3592. 189

72JOC3755 G. N. Walker, A. R. Engle and R. J. Kempton, *J. Org. Chem.*, 1972, **37**, 3755. 515

72JOC3810 T. Minami, K. Yamataka, Y. Ohshiro, T. Agawa, N. Yaskuoka and N. Kasai; *J. Org. Chem.*, 1972, **37**, 3810. 457, 473

72JOC3838 K. T. Potts, A. J. Elliott and M. Sorm; *J. Org. Chem.*, 1972, **37**, 3838. 167

72JOC3955 G. N. Walker; *J. Org. Chem.*, 1972, **37**, 3955 530

72JOC4246 B. C. Hartman and B. Rickborn; *J. Org. Chem.*, 1972, **37**, 4246. 191

72JOC4401 R. A. Wohl and D. F. Headley; *J. Org. Chem.*, 1972, **37**, 4401. 68

72JOU864 P. D. Del'tsova, N. P. Gambaryan, Yu. V. Zeifman and I. L. Knunyants; *J. Org. Chem. USSR* (*Engl. Transl.*), 1972, **8**, 864. 475

72JSP(42)403 L. C. Robertson and J. A. Merritt; *J. Mol. Spectrosc.*, 1972, **42**, 403. 203

72JSP(43)393 W. Kiefer, H. J. Bernstein, H. Wieser and M. Danyluk; *J. Mol. Spectrosc.*, 1972, **43**, 393. 370

72JSP(44)459 T. Ikeda, R. Kewley and R. F. Curl, Jr.; *J. Mol. Spectrosc.*, 1972, **44**, 459. 188

72KGS306 V. M. Pavlov, A. A. Alekseev, G. A. Sokol'skii and I. L. Knunyants; *Khim. Geterotsikl. Soedin.*, 1972, **8**, 306. 487

72LA(762)167 K. Hartke, F. Rossbach and M. Radau; *Liebigs Ann. Chem.*, 1972, **762**, 167. 474

72LA(763)46 G. Ege and E. Beisiegel; *Liebigs Ann Chem.*, 1972, **763**, 46. 462, 482

72LA(765)15 Z. Zubovics and H. Wittmann; *Liebigs Ann. Chem.*, 1972, **765**, 15. 374

B-72MI50500 J. G. Buchanan and H. Z. Sable; in 'Selective Organic Transformations', ed. B. S. Thyagaragan; Wiley, New York, 1972, vol. 2. 97, 108

B-72MI50501 A. J. Gordon and R. A. Ford; 'The Chemist's Companion', Wiley, New York, 1972. 99

72MI50600 G. M. Kuz'yants and V. T. Aleksanyan; *J. Struct. Chem.* (*Engl. Transl.*), 1972, **13**, 576. 138

72MI50601 F. Lautenschlaeger; *J. Macromol. Sci., Chem.*, 1972, **A6**, 1089. 146, 182

72MI50602 O. P. Strausz; *Adv. Chem. Ser.*, 1972, **110**, 137. 176

B-72MI50900 'Cephalosporins and Penicillins: Chemistry and Biology', ed. E. H. Flynn; Academic Press, New York, 1972. 248, 249

B-72MI51000 E. P. Abraham and P. B. Loder; in 'Cephalosporins and Penicillins', ed. E. H. Flynn; Academic Press, New York, 1972, p. 1. 286

B-72MI51001 R. M. Sweet; in ref. ⟨B-72MI51000⟩, p. 281. 286

B-72MI51002 P. V. Demarco and R. Nagarajan; in ref. ⟨B-72MI51000⟩, p. 312. 287

B-72MI51003 L. P. Marrelli; in ref. ⟨B-72MI51000⟩, p. 610. 287

B-72MI51004 C. F. Murphy and J. A. Webbér; in ref. ⟨B-72MI51000⟩, p. 134. 288, 289, 291

B-72MI51005 F. M. Huber, R. R. Chauvette and B. G. Jackson; in ref. ⟨B-72MI51000⟩, p. 27. 293

B-72MI51006 R. D. G. Copper and D. O. Spry; in ref. ⟨B-72MI51000⟩, p. 183. 293

B-72MI51007 K. Heusler; in ref. ⟨B-72MI51000⟩, p. 255. 294

B-72MI51100 R. W. Sweet; in 'Cephalosporins and Penicillins', ed. E. H. Flynn; Academic Press, New York, 1972, p. 280. 301

B-72MI51101 P. V. Demarco and R. Nagarajan; in ref. ⟨B-72MI51100⟩, p. 311. 301, 302

B-72MI51102 P. A. Lemke and D. R. Brannon; in ref. ⟨B-72MI51100⟩, p. 370. 327

B-72MI51103 F. M. Huber, R. R. Chauvette and B. G. Jackson; in ref. ⟨B-72MI51100⟩, p. 27. 329

B-72MI51104 G. V. Kaiser and S. Kukolja; in ref. ⟨B-72MI51100⟩, p. 74. 303, 330

B-72MI51105 K. Heusler; in ref. ⟨B-72MI51100⟩, p. 255. 331

B-72MI51106 R. D. G. Cooper and D. O. Spry; in ref. ⟨B-72MI51100⟩, p. 183. 303, 305

B-72MI51200 P. V. DeMarco and R. Nagarajan; in 'Cephalosporins and Penicillins, Chemistry and Biology', ed. E. H. Flynn; Academic Press, New York, 1972, p. 311. 349

B-72MI51201 R. M. Sweet; in ref. B-72MI51200, p. 280. 349

72MI51300 T. Saegusa; *J. Macromol. Sci.-Chem.*, 1972, **A6**, 997. 382

72MI51301 M. A. Raza and L. W. Reeves; *Mol. Phys.*, 1972, **23**, 1007. 367

72MI51302 H. Ruotsalainen, J. Kaakkurivaara and P. O. I. Virtanen; *Suom. Kem. B*, 1972, **45**, 35 (*Chem. Abstr.*, 1972, **76**, 71 849). 379

72MI51303	H. Ruotsalainen, V. Palosaari and P. O. I. Virtanen; *Suom. Kem. B*, 1972, **45**, 40 (*Chem. Abstr.*, 1972, **76**, 71 738).	385
72MI51304	Y. Yamashita; *Polym. Prepr., ACS Div. Polym. Chem.*, 1972, **13**, 539 (*Chem. Abstr.*, 1972, **80**, 71 176).	382
72MI51400	P. de Mayo and A. A. Nicholson; *Isr. J. Chem.*, 1972, **10**, 341.	432
72MI51700	T. Saegusa, T. Shiota, S. Matsumoto and H. Fujii; *Polym. J.*, 1972, **3**, 40.	561
72MI51800	J. Szotor and A. Kotelko; *Diss. Pharm. Pharmacol.*, 1972, **24**, 385 (*Chem. Abstr.*, 1972, **77**, 152 134).	640
72MI51801	D. Tkaczyńska; *Wiad. Chem.*, 1972, **26**, 755.	641
72MI51900	D. N. White and J. D. Dunitz; *Isr. J. Chem.*, 1972, **10**, 249.	703
72OMR(4)107	G. C. Levy and D. C. Dittmer; *Org. Magn. Reson.*, 1972, **4**, 107.	413, 415
72OMR(4)441	M. Ueyama, K. Tori and M. Fukuyama; *Org. Magn. Reson.*, 1972, **4**, 441.	137
72OS(52)66	C. J. Pedersen; *Org. Synth.*, 1972, **52**, 66.	732
72PMH(5)1	P. J. Wheatley; *Phys. Methods Heterocycl. Chem.*, 1972, **5**, 1.	1, 8, 48
72RTC1345	L. Thijs, J. Strating and B. Zwanenburg; *Recl. Trav. Chim. Pays-Bas*, 1972, **91**, 1345.	150
72S616	W. Adam and C. I. Rojas; *Synthesis*, 1972, 616.	380
72T1795	W. A. Deane and M. F. McKay; *Tetrahedron*, 1972, **28**, 1795.	494
72T1955	R. D. Bach, U. Mazur, I. Hamama and S. K. Lauderback; *Tetrahedron*, 1972, **28**, 1955.	188
72T2307	I. Sataty; *Tetrahedron*, 1972, **28**, 2307.	646
72T3657	B. A. Hess, Jr., L. J. Schaad and C. W. Holyoke, Jr.; *Tetrahedron*, 1972, **28**, 3657.	492, 711
72T5183	T. Kauffmann; *Tetrahedron*, 1972, **28**, 5183.	763
72T5977	A. K. Bose, B. Dayal, H. P. S. Chawla and M. S. Manhas; *Tetrahedron*, 1972, **28**, 5977.	248
72TL169	J. H. Wieringa, J. Strating and H. Wynberg; *Tetrahedron Lett.*, 1972, 169.	476
72TL213	A. Le Berre and J. Petit; *Tetrahedron Lett.*, 1972, 213.	468
72TL285	A. Vlietinck, E. Roets, P. Claes and H. Vanderhaeghe; *Tetrahedron Lett.*, 1972, 285.	324
72TL633	J. P. Schirmann and F. Weiss; *Tetrahedron Lett.*, 1972, 633.	228, 235
72TL1353	A. Hassner, A. S. Miller and M. J. Haddadin; *Tetrahedron Lett.*, 1972, 1353.	92
72TL1777	W. E. Barnett and W. H. Sohn; *Tetrahedron Lett.*, 1972, 1777.	395
72TL2109	K. Hirayama; *Tetrahedron Lett.*, 1972, 2109.	763, 764
72TL2451	J. P. Snyder; *Tetrahedron Lett.*, 1972, 2451.	198
72TL2591	G. Berti, G. Camici, B. Macchia, F. Macchia and L. Monti; *Tetrahedron Lett.*, 1972, 2591.	68
72TL3479	H. Durr and R. Sergio; *Tetrahedron Lett.*, 1972, 3479.	67
72TL3563	D. C. DeSelms and W. R. Schleigh; *Tetrahedron Lett.*, 1972, 3563.	258
72TL3591	G. M. Rubottom, G. R. Stevenson, J. C. Chabala and V. L. Pascucci; *Tetrahedron Lett.*, 1972, 3591.	72
72TL3613	F. A. L. Anet and P. J. Degen; *Tetrahedron Lett.*, 1972, 3613.	700
72TL3827	Y. Hata and M. Watanabe; *Tetrahedron Lett.*, 1972, 3827.	75
72TL3933	D. S. Wulfman and T. R. Steinheimer; *Tetrahedron Lett.*, 1972, 3933.	361
72TL4129	H. Oqawa and N. Shimojo; *Tetrahedron Lett.*, 1972, 4129.	715
72TL4137	Y. Tamura, K. Sumoto, J. Minamikawa and M. Ikeda; *Tetrahedron Lett.*, 1972, 4137.	425
72TL4597	H. Hofmann and B. Meyer; *Tetrahedron Lett.*, 1972, 4597.	576
72TL4659	Y. Hata and M. Watanabe; *Tetrahedron Lett.*, 1972, 4659.	75
72TL4713	J. Firl and S. Sommer; *Tetrahedron Lett.*, 1972, 4713.	472
72TL4781	R. F. J. Langendries and F. C. De Schryver; *Tetrahedron Lett.*, 1972, 4781.	418, 423
72TL4891	R. R. Schmidt and H. Vatter; *Tetrahedron Lett.*, 1972, 4891.	603
72TL4917	D. B. R. Johnston, S. M. Schmitt, R. A. Firestone and B. G. Christensen; *Tetrahedron Lett.*, 1972, 4917.	321, 331
72USP3632579	M. Wolf; *U.S. Pat.* 3 632 579 (1972) (*Chem. Abstr.*, 1972, **76**, 99 497).	447
72USP3645986	M. R. Rifi; *U.S. Pat.* 3 645 986 (1972) (*Chem. Abstr.*, 1972, **77**, 6105).	147
72USP3706769	M. H. Rosen and H. M. Blatter; *U.S. Pat.* 3 706 769 (1972) (*Chem. Abstr.*, 1973, **78**, 97 475).	168
72YZ454	I. Isaka, T. Kashiwagi, K. Nakano, N. Kawahara, A. Koda, Y. Numasaki, S. Kawahara and M. Murakami; *Yakugaku Zasshi*, 1972, **92**, 454.	329, 331
72YZ459	A. Koda, K. Takanobu, I. Isaka, T. Kashiwagi, K. Takahashi, S. Kawahara and M. Murakami; *Yakugaku Zasshi*, 1972, **92**, 459.	329
72ZOR856	P. D. Del'tsova, N. P. Gambaryan, Yu. V. Zeifman and I. L. Knunyants; *Zh. Org. Khim.*, 1972, **8**, 856 (*Chem. Abstr.*, 1972, **77**, 34 213).	266
73ACR32	R. D. G. Cooper, L. D. Hatfield and D. O. Spry; *Acc. Chem. Res.*, 1973, **6**, 32.	305
73ACR393	R. Breslow; *Acc. Chem. Res.*, 1973, **6**, 393.	492
73ACS677	O. K. J. Kovacs, B. Ekström and B. Sjoberg; *Acta Chem. Scand.*, 1973, **27**, 677.	635
73ACS1519	J. Dale and T. Ekeland; *Acta Chem. Scand.*, 1973, **27**, 1519.	688, 689, 691
73ACS3128	W. Ralowski, S. Ljunggren and J. Mjoberg; *Acta Chem. Scand.*, 1973, **27**, 3128.	408
73ACS3944	P. O. I. Virtanen, S. Peltonen and J. Hyyppa; *Acta Chem. Scand.*, 1973, **27**, 3944.	381
73AG(E)155	K. Burger, W. Thenn and E. Müller; *Angew. Chem., Int. Ed. Engl.*, 1973, **12**, 155.	270
73AG(E)161	G. Szeimies, U. Siefken and R. Rinck; *Angew. Chem., Int. Ed. Engl.*, 1973, **12**, 161.	273
73AG(E)328	G. Schroder, G. Frank and J. F. M. Oth; *Angew. Chem., Int. Ed. Engl.*, 1973, **12**, 328.	716

73AG(E)410 P. J. Beeby and F. Sondheimer; *Angew. Chem., Int. Ed. Engl.*, 1973, **12**, 410. 722

73AG(E)411 P. J. Beeby and F. Sondheimer; *Angew. Chem., Int. Ed. Engl.*, 1973, **12**, 411. 722

73AG(E)414 H. Quast and W. Risler; *Angew. Chem., Int. Ed. Engl.*, 1973, **12**, 414. 22, 78, 92

73AG(E)432 R. W. Lenz; *Angew. Chem., Int. Ed. Engl.*, 1973, **12**, 432. 388

73AG(E)578 J. M. Lehn, J. Simon and J. Wagner; *Angew. Chem., Int. Ed. Engl.*, 1973, **12**, 578. 733

73AG(E)838 E. Vogel, H.-J. Altenbach and E. Schmidbauer; *Angew. Chem., Int. Ed. Engl.*, 1973, **12**, 838. 189

73AG(E)840 F.-G. Klärner and E. Vogel; *Angew. Chem., Int. Ed. Engl.*, 1973, **12**, 840. 125, 193

73AG(E)847 G. Seybold, U. Jersak and R. Gompper; *Angew. Chem., Int. Ed. Engl.*, 1973, **12**, 847. 4, 278, 282

73AG(E)989 R. Schwesinger and H. Prinzbach; *Angew. Chem., Int. Ed. Engl.*, 1973, **12**, 989. 81

73AJC1307 J. L. Huppatz; *Aust. J. Chem.*, 1973, **26**, 1307. 533

73AJC2159 D. St. C. Black and K. G. Watson; *Aust. J. Chem.*, 1973, **26**, 2159. 209

73AP271 H. Bohme and K. P. Stocker; *Arch. Pharm.* (*Weinheim, Ger.*), 1973, **306**, 271. 508, 511

73AX(B)383 D. Moras, B. Metz and R. Weiss; *Acta Crystallogr., Part B*, 1973, **29**, 383. 737, 738

73AX(B)2128 C. D. Shirrell and D. E. Williams; *Acta Crystallogr., Part B*, 1973, **29**, 2128. 406

73BAU1740 Yu. V. Zeifman, D. P. Del'tsova, E. A. Avetisyan, N. P. Gambaryan and I. L. Knunyants; *Bull. Acad. Sci. USSR, Div. Chem. Sci.*, 1973, **22**, 1740. 475

73BAU2542 B. A. Arbuzov and N. N. Zobova; *Bull. Acad. Sci. USSR, Div. Chem. Sci.*, 1973, **22**, 2542. 474

73BCJ667 Y. Hayasi, H. Nakamura and H. Nozaki; *Bull. Chem. Soc. Jpn.*, 1973, **46**, 667. 165, 440

73BCJ1539 I. Ojima and K. Kondo; *Bull. Chem. Soc. Jpn.*, 1973, **46**, 1539. 174

73BCJ1755 K. Matsui and M. Motoi; *Bull. Chem. Soc. Jpn.*, 1973, **46**, 1755. 275

73BCJ2156 T. Minami, M. Fukada, M. Abe and T. Agawa; *Bull. Chem. Soc. Jpn.*, 1973, **46**, 2156. 457, 486

73BCJ2253 T. Nishio, M. Yoshioka, H. Aoyama and N. Sugiyama; *Bull. Chem. Soc. Jpn.*, 1973, **46**, 2253 456

73BCJ2559 I. Ojima, K. Akiba and N. Inamoto; *Bull. Chem. Soc. Jpn.*, 1973, **46**, 2559. 455, 475

73BCJ3260 N. Ishikawa and T. Kitazume; *Bull. Chem. Soc. Jpn.*, 1973, **46**, 3260. 464, 478

73BSF1668 F. Dardoize, J.-L. Moreau and M. Gaudemar; *Bull. Soc. Chim. Fr.*, 1973, 1668. 256

73BSF2029 G. LeClerc; *Bull. Soc. Chim. Fr.*, 1973, 2029. 671

73BSF2451 L. Cottier and G. Descotes; *Bull. Soc. Chim. Fr.*, 1973, 2451. 668

73BSF2680 R. Chaabouni and A. Laurent; *Bull. Soc. Chim. Fr.*, 1973, 2680. 85, 90

73CB713 K. H. Linke, R. Bimczok and I. Elsing; *Chem. Ber.*, 1973, **106**, 713. 651

73CB970 W.-D. Stohrer; *Chem. Ber.*, 1973, **106**, 970. 500

73CB1033 H. J. Lindner and B. von Gross; *Chem. Ber.*, 1973, **106**, 1033. 493, 494, 526

73CB2084 G. Schaden; *Chem. Ber.*, 1973, **106**, 2084. 501

73CB3824 H. Prinzbach, G. Kaupp, R. Fuchs, M. Joyeux, R. Kitzing and J. Markert; *Chem. Ber.*, 1973, **106**, 3824, and references cited therein. 509

73CC34 G. Holan, C. Kowala and J. A. Wunderlich; *J. Chem. Soc., Chem. Commun.*, 1973, 34. 365

73CC37 N. E. Brightwell and G. W. Griffin; *J. Chem. Soc., Chem. Commun.*, 1973, 37. 582, 584

73CC48 E. R. Talaty, A. E. Dupuy, C. M. Utermoehlen and L. H. Stekoll; *J. Chem. Soc., Chem. Commun.*, 1973, 48. 79, 266

73CC67 D. W. Jones; *J. Chem. Soc., Chem. Commun.*, 1973, 67. 508, 522, 538

73CC271 Naser-ud-Din, J. Riegl and L. Skattebøl; *J. Chem. Soc., Chem. Commun.*, 1973, 271. 266

73CC327 E. Lerner, R. A. Odum and B. Schmall; *J. Chem. Soc., Chem. Commun.*, 1973, 327. 505, 525

73CC406 V. Boekelheide, I. D. Reingold and M. Tuttle; *J. Chem. Soc., Chem. Commun.*, 1973, 406. 777

73CC427 R. J. Hayward and O. Meth-Cohn; *J. Chem. Soc., Chem. Commun.*, 1973, 427. 778

73CC435 W. Yang, R. D. Minard and C. N. Matthews; *J. Chem. Soc., Chem. Commun.*, 1973, 435. 2

73CC461 G. F. Koser; *J. Chem. Soc., Chem. Commun.*, 1973, 461. 18

73CC571 W. D. Ollis and J. F. Stoddart; *J. Chem. Soc., Chem. Commun.*, 1973, 571. 704

73CC670 A. Krantz; *J. Chem. Soc., Chem. Commun.*, 1973, 670. 127

73CC688 M. J. S. Dewar and C. A. Ramsden; *J. Chem. Soc., Chem. Commun.*, 1973, 688. 48

73CC772 O. V. Dorofeeva, V. S. Mastryukov, L. V. Vilkov and I. Hargittai; *J. Chem. Soc., Chem. Commun.*, 1973, 772. 238

73CC841 M. Nagayama, O. Okumura, S. Noda and A. Mori; *J. Chem. Soc., Chem. Commun.*, 1973, 841. 455, 477

73CC861 J. P. Marino; *J. Chem. Soc., Chem. Commun.*, 1973, 861. 446

73CHE13 S. Z. Taits, V. N. Bulgakova and Ya. L. Gol'dfarb; *Chem. Heterocycl. Compd.* (*Engl. Transl.*), 1973, **9**, 13. 776

73CI(L)109 D. N. Kirk; *Chem. Ind.* (*London*), 1973, 109. 97, 103

73CJC468 K. R. Kopecky, P. A. Lockwood, J. E. Filby and R. W. Reid; *Can. J. Chem.*, 1973, **51**, 468. 469

73CJC529 R. Kewley; *Can. J. Chem.*, 1973, **51**, 529. 188

73CJC881 F. M. F. Chen and T. P. Forrest; *Can. J. Chem.*, 1973, **51**, 881. 605

73CJC981 G. Caron and J. Lessard; *Can. J. Chem.*, 1973, **51**, 981. 373

73CJC1109 R. R. Fraser, G. Boussard, I. D. Postescu, J. J. Whiting and Y. Y. Wigfield; *Can. J. Chem.*, 1973, **51**, 1109. 512, 518

73CJC1448 K. Wiesner, P. T. Ho, R. C. Jain, S. F. Lee, S. Oida and A. Philipp; *Can. J. Chem.*, 1973, **51**, 1448. 68

73CJC2433 R. R. Fraser, R. N. Renaud, J. K. Saunders and Y. Y. Wigfield; *Can. J. Chem.*, 1973, **51**, 2433. 499

73CJC3380 R. N. Renaud, R. B. Layton and R. R. Fraser; *Can. J. Chem.*, 1973, **51**, 3380. 706

73CL5 M. Miyoshi, H. Sugano, T. Fujii, T. Ishihara and N. Yoneda; *Chem. Lett.*, 1973, 5. 239

73CL759 K. Matsumoto and K. Maruyama; *Chem. Lett.*, 1973, 759. 91

73CPB228 A. Morimoto, T. Okutani and K. Masuda; *Chem. Pharm. Bull.*, 1973, **21**, 228. 266

73CPB1305 K. Hirai, H. Matsuda and Y. Kishida; *Chem. Pharm. Bull.*, 1973, **21**, 1305. 251

73CPB2375 K. Meguro and Y. Kuwada; *Chem. Pharm. Bull.*, 1973, **21**, 2375. 686

73CR(B)(276)335 J. LeBrumant, M. Selim and G. Martin; *C.R. Hebd. Seances Acad. Sci., Ser. B*, 1973, **276**, 335. 145, 376

73CR(C)(276)875 D. Lecadet, P. Paquer and A. Thuillier; *C.R. Hebd. Seances Acad. Sci., Ser. C.*, 1973, **276**, 875. 174

73G681 F. Montanari, I. Moretti and G. Torre; *Gazz. Chim. Ital.*, 1973, **103**, 681. 229

73GEP2149650 S. Pietzsch and G. Schaeffer; *Ger. Offen.* 2 149 650 (1973) (*Chem. Abstr.*, 1973, **79**, 126 507). 380

73GEP2209692 K. Igelmeier; *Ger. Pat.* 2 209 692 (*Chem. Abstr.*, 1973, **79**, 137 936). 379

73GEP2255194 W. C. Watkins; *Ger. Offen.* 2 255 194 (1973) (*Chem. Abstr.*, 1973, **79**, 52 809). 382

73GEP2305517 R. W. Addor and S. Kantor; *Ger. Offen.* 2 305 517 (1973) (*Chem. Abstr.*, 1973, **79**, 115 553). 485

73GEP2312960 R. W. Stephens, C. Vervolet and H. C. W. Monnee; *Ger. Offen.* 2 312 960 (1973) (*Chem. Abstr.*, 1974, **80**, 71 877). 484

73GEP2314103 P. Dubs, H. Kuentzel and M. Pesaro; *Ger. Pat.* 2 314 103 (1973) (*Chem. Abstr.*, 1974, **80**, 14 833). 436

73H(1)267 D. M. Jerina, H. Yagi and J. W. Daly; *Heterocycles*, 1973, **1**, 267. 188, 548, 565

73HCA1351 N. S. Narasimhan, H. Heimgartner, H. J. Hansen and H. Schmid; *Helv. Chim. Acta*, 1973, **56**, 1351. 59, 60, 91

73HCA1614 E. Haselbach, A. Mannschreck and W. Seitz; *Helv. Chim. Acta*, 1973, **56**, 1614. 202

73HCA1852 W. Heinzelmann and M. Märky; *Helv. Chim. Acta*, 1973, **56**, 1852. 496, 542

73HCA1933 F. Brogli, W. Eberbach, E. Haselbach, E. Heilbronner, V. Hornung and D. M. Lemal; *Helv. Chim. Acta*, 1973, **56**, 1933. 204

73HCA3056 M. K. Mihailović, Z. Ceković, J. Stanković, N. Pavolvić, S. Konstantinović and S. Djokić-Mazinjanin; *Helv. Chemi. Acta*, 1973, **56**, 3056. 666

73IJS(8)45 T. H. Chan and J. R. Finkenbine; *Int. J. Sulfur Chem.*, 1973, **8**, 45. 179

73IZV627 B. Arbuzov, O. N. Nuretdinova, L. Z. Nikonova and E. I. Gol'dfarb; *Izv. Akad. Nauk SSSR, Ser. Khim.*, 1973, 627. 383

73IZV2345 B. A. Arbuzov, O. N. Nuretdinova, L. Z. Nikonova, F. F. Guseva and E. I. Gol'dfarb; *Izv. Akad. Nauk SSSR, Ser. Khim.*, 1973, 2345. 383

73IZV2562 E. G. Ter-Gabrielyan, E. A. Avetisyan and N. P. Gambaryan; *Izv. Akad. Nauk SSSR, Ser. Khim.*, 1973, 2562. 389

73JA133 I. G. Csizmadia, H. E. Gunning, R. K. Gosavi and O. P. Strausz; *J. Am. Chem. Soc.*, 1973, **95**, 133. 124

73JA1337 O. L. Chapman, D. De La Cruz, R. Roth and J. Pacansky; *J. Am. Chem. Soc.*, 1973, **95**, 1337. 375

73JA1547 P. S. Skell, K. J. Klabunde, J. H. Plonka, J. S. Roberts and D. L. Williams-Smith; *J. Am. Chem. Soc.*, 1973, **95**, 1547. 431

73JA1945 A. Padwa, M. Dharan, J. Smolanoff and S. I. Wetmore; *J. Am. Chem. Soc.*, 1973, **95**, 1945. 55

73JA1954 A. Padwa, J. Smolanoff and S. I. Wetmore; *J. Am. Chem. Soc.*, 1973, **95**, 1954. 56

73JA2023 D. Montecalvo, M. St. Jacques and R. Wasylichen; *J. Am. Chem. Soc.*, 1973, **95**, 2023. 704

73JA2383 W. Kuesters and P. de Mayo, *J. Am. Chem. Soc.*, 1973, **95**, 2383. 482

73JA2697 K. B. Sharpless and R. F. Lauer; *J. Am. Chem. Soc.*, 1973, **95**, 2697. 105, 111

73JA2708 L. T. Scott and W. D. Cotton; *J. Am. Chem. Soc.*, 1973, **95**, 2708. 127

73JA2738 T. J. Katz and N. Acton; *J. Am. Chem. Soc.*, 1973, **95**, 2738. 671

73JA2860 R. Gleiter, E. Schmidt, P. Johnson and D. O. Cowan; *J. Am. Chem. Soc.*, 1973, **95**, 2860. 558

73JA2982 A. Hassner, J. O. Currie, A. S. Steinfeld and R. F. Atkinson; *J. Am. Chem. Soc.*, 1973, **95**, 2982. 72, 274, 283

73JA3284 P. W. Concannon and J. Ciabattoni; *J. Am. Chem. Soc.*, 1973, **95**, 3284. 122

73JA3420 F. Jung, N. K. Sharma and T. Durst; *J. Am. Chem. Soc.*, 1973, **95**, 3420. 459, 471

73JA3907 B. A. Hess and L. J. Schaad; *J. Am. Chem. Soc.*, 1973, **95**, 3907. 557, 712

73JA3970 H. E. Zimmerman and W. Eberbach; *J. Am. Chem. Soc.*, 1973, **95**, 3970. 598, 600

73JA4061 O. L. Chapman, C. L. McIntosh, J. Pacansky, G. V. Calder and G. Orr; *J. Am. Chem. Soc.*, 1973, **95**, 4061. 400

73JA4064 D. B. Denney, D. Z. Denney and Y. F. Hsu; *J. Am. Chem. Soc.*, 1973, **95**, 4064. 412, 444

73JA4067 A. B. Levy and H. C. Brown; *J. Am. Chem. Soc.*, 1973, **95**, 4067. 85

73JA4647 L. A. Paquette, M. J. Broadhurst, C. Lee and J. Clardy; *J. Am. Chem. Soc.*, 1973, **95**, 4647. 356, 357, 359

73JA4758 C. R. Brulet, S. S. Isied and H. Taube; *J. Am. Chem. Soc.*, 1973, **95**, 4758. 146

73JA5416 L. T. Scott and W. D. Cotton; *J. Am. Chem. Soc.*, 1973, **95**, 5416. 127

73JA5423 G. Büchi, Y. Kitaura, S.-S. Yuan, H. E. Wright, J. Clardy, A. L. Demain, T. Glinsukon, N. Hunt and G. N. Wogan; *J. Am. Chem. Soc.*, 1973, **95**, 5423. 120

73JA6041 G. J. Kasperek, P. Y. Bruice, T. C. Bruice, H. Yagi and D. M. Jerina; *J. Am. Chem. Soc.*, 1973, **95**, 6041. 567
73JA6046 R. M. Wing, J. J. Uebel and K. K. Andersen; *J. Am. Chem. Soc.*, 1973, **95**, 6046. 406, 411, 420
73JA6486 P. D. Bartlett, A. L. Baumstark and M. E. Landis; *J. Am. Chem. Soc.*, 1973, **95**, 6486. 463
73JA7050 E. H. White, P. D. Wildes, J. Wiecko, H. Doshan and C. C. Wei; *J. Am. Chem. Soc.*, 1973, **95**, 7050. 459
73JA7287 K. N. Houk, J. Sims, R. E. Duke, R. W. Strozier and J. K. George; *J. Am. Chem. Soc.*, 1973, **95**, 7287. 55
73JA7320 S. R. Tanny and F. W. Fowler; *J. Am. Chem. Soc.*, 1973, **95**, 7320. 507, 522, 542
73JA7538 K. P. C. Vollhardt and R. G. Bergman; *J. Am. Chem. Soc.*, 1973, **95**, 7538. 140, 143
73JA7644 R. Hoffmann, K. Fujimoto, J. R. Swenson and C.-C. Wan; *J. Am. Chem. Soc.*, 1973, **95**, 7644. 133, 176
73JA7894 K. Ibne-Rasa, R. H. Pater, J. Ciabattoni and J. O. Edwards; *J. Am. Chem. Soc.*, 1973, **95**, 7894. 123
73JA8191 D. B. Denney, D. Z. Denney and Y. F. Hsu; *J. Am Chem. Soc.*, 1973, **95**, 8191. 412, 444
73JA8486 S. Krishnamurthy, R. M. Schubert and H. C. Brown; *J. Am. Chem. Soc.*, 1973, **95**, 8486. 112
73JAP7306455 A. Takahashi, K. Taga, A. Yamamoto and Y. Motoki; *Jpn. Pat.* 73 06 455 (1973) (*Chem. Abstr.*, 1973, **79**, 43 030). 394
73JCS(F1)1817 J. J. Cosa, H. E. Gsponer, E. H. Staricco and C. A. Vallana; *J. Chem. Soc., Faraday Trans. I*, 1973, **69**, 1817. 456
73JCS(P1)205 R. Crossley, A. P. Downing, N. Nógrádi, A. B. de Oliveira, W. D. Ollis and I. O. Sutherland; *J. Chem. Soc., Perkin Trans. 1*, 1973, 205. 704, 705
73JCS(P1)218 W. R. Bowman, W. R. Gretton and G. W. Kirby; *J. Chem. Soc., Perkin Trans. 1*, 1973, 218. 590
73JCS(P1)550 D. J. Anderson, T. L. Gilchrist, G. E. Gymer and C. W. Rees; *J. Chem. Soc., Perkin Trans. 1*, 1973, 550. 88
73JCS(P1)868 N. Bashir and T. L. Gilchrist; *J. Chem. Soc., Perkin Trans. 1*, 1973, 868. 269, 278
73JCS(P1)1041 C. J. Cattanach, A. Cohen and B. Heath-Brown; *J. Chem. Soc., Perkin Trans. 1*, 1973, 1041. 512
73JCS(P1)1079 F. R. Atherton and R. W. Lambert; *J. Chem. Soc., Perkin Trans. 1*, 1973, 1079. 505, 511, 518
73JCS(P1)1084 A. McLean and G. R. Proctor; *J. Chem. Soc., Perkin Trans. 1*, 1973, 1084. 503, 525
73JCS(P1)1561 D. H. Coy, R. N. Haszeldine, M. J. Newlands and A. E. Tipping; *J. Chem. Soc., Perkin Trans. 1*, 1973, 1561. 457, 473
73JCS(P1)1786 T. M. Cresp and M. V. Sargent; *J. Chem. Soc., Perkin Trans. 1*, 1973, 1786. 722
73JCS(P1)2024 G. Lowe and D. D. Ridley; *J. Chem. Soc., Perkin Trans. 1*, 1973, 2024. 353
73JCS(P1)2543 A. A. Reid, J. T. Sharp, H. R. Sood and P. B. Thorogood; *J. Chem. Soc., Perkin Trans. 1*, 1973, 2543. 595, 604
73JCS(P1)2664 P. G. Bird and W. J. Irwin; *J. Chem. Soc., Perkin Trans. 1*, 1973, 2664. 255
73JCS(P1)2851 S. A. Matlin and P. G. Sammes; *J. Chem. Soc., Perkin Trans. 1*, 1973, 2851. 124
73JCS(P1)2907 G. Lowe and H. Wing Yeung; *J. Chem. Soc., Perkin Trans. 1*, 1973, 2907. 265
73JCS(P2)1575 J. Bjorgo and D. R. Boyd; *J. Chem. Soc., Perkin Trans. 2*, 1973, 1575. 200
73JCS(P2)2042 G. Kartha and G. Ambady; *J. Chem. Soc., Perkin Trans. 2*, 1973, 2042. 277, 349
73JHC39 E. M. Hawes and H. L. Davis; *J. Heterocycl. Chem.*, 1973, **10**, 39. 528
73JHC239 W. T. Brady and A. D. Patel; *J. Heterocycl. Chem.*, 1973, **10**, 239. 387
73JHC279 H. Yoshida, G. Sen and B. S. Thyagarajan; *J. Heterocycl. Chem.*, 1973, **10**, 279. 687
73JHC689 D. Misiti, V. Rimatori and F. Gatta; *J. Heterocycl. Chem.*, 1973, **10**, 689. 515
73JHC791 A. K. Bose, J. L. Fahey and M. S. Manhas; *J. Heterocycl. Chem.*, 1973, **10**, 791. 359, 635
73JHC879 R. C. Petterson, A. L. Hebert, G. W. Griffin, I. Sarkar, O. P. Strausz and J. Font; *J. Heterocycl. Chem.*, 1973, **10**, 879. 140
73JMC502 A. Borgen, H. Darvey, N. Castagnoli, T. T. Crocker, R. E. Rasmussen and I. Y. Wang; *J. Med. Chem.*, 1973, **16**, 502. 188
73JMC516 G. Shtacher, M. Erez and S. Cohen; *J. Med. Chem.*, 1973, **16**, 516. 531
73JOC20 T. R. Williams and D. J. Cram; *J. Org. Chem.*, 1973, **38**, 20. 479
73JOC176 E. M. Burgess and J. P. Sanchez; *J. Org. Chem.*, 1973, **38**, 176 283
73JOC230 E. H. W. Bohme, H. E. Applegate, J. B. Ewing, P. T. Funke, M. S. Puar and J. E. Dolfini; *J. Org. Chem.*, 1973, **38**, 230. 321
73JOC649 S. D. Ziman and B. M. Trost; *J. Org. Chem.*, 1973, **38**, 649. 179, 425
73JOC809 R. H. Williams and H. R. Snyder; *J. Org. Chem.*, 1973, **38**, 809. 530
73JOC927 W. E. Parham, K. B. Sloan, K. R. Reddy and P. E. Olson; *J. Org. Chem.*, 1973, **38**, 927. 777
73JOC1044 Y. Ogata, Y. Sawaki and H. Inoue; *J. Org. Chem.*, 1973, **38**, 1044. 123
73JOC1090 T. H. Koch, M. A. Geigel and Chun-che Tsai; *J. Org. Chem.*, 1973, **38**, 1090. 505, 514
73JOC1361 W. E. Parham, W. D. McKown, V. Nelson, S. Kajigaeshi and N. Ishikawa; *J. Org. Chem.*, 1973, **38**, 1361. 387
73JOC1578 K. Pilgram and R. D. Skiles; *J. Org. Chem.*, 1973, **38**, 1578. 650
73JOC1605 E. H. Carlson, A. P. Schaap and M. Raban; *J. Org. Chem.*, 1973, **38**, 1605 451
73JOC1767 A. T. Bottini, J. G. Maroski and V. Dev; *J. Org. Chem.*, 1973, **38**, 1767. 578
73JOC1959 A. G. Anastassiou, R. L. Elliott, H. W. Wright and J. Clardy; *J. Org. Chem.*, 1973, **38**, 1959. 724, 725

73JOC2061 J. A. Wojtowicz and R. J. Polak; *J. Org. Chem.*, 1973, **38**, 2061. 390

73JOC2269 W. Adam, J. Liu and O. Rodriguez; *J. Org. Chem.*, 1973, **38**, 2269. 117

73JOC2652 J. M. Bohen and M. M. Joullié; *J. Org. Chem.*, 1973, **38**, 2652. 473, 645

73JOC2857 T. Jen, J. Frazee and J. R. E. Hoover; *J. Org. Chem.*, 1973, **38**, 2857. 321

73JOC2954 A. Nabeya, K. Kurita and J. A. Moore; *J. Org. Chem.*, 1973, **38**, 2954. 598

73JOC3466 M. J. Haddadin and A. Hassner; *J. Org. Chem.*, 1973, **38**, 3466. 61

73JOC3963 J. E. McCaskie, P. L. Chang, T. R. Nelsen and D. C. Dittmer; *J. Org. Chem.*, 1973, **38**, 3963. 432

73JOC3978 V. J. Traynelis, Y. Yoshikawa, J. C. Sih, L. J. Miller and J. R. Livingston, Jr.; *J. Org. Chem.*, 1973, **38**, 3978. 589

73JOM(47)337 B. Delmond, J.-C. Pommier and J. Valade; *J. Organomet. Chem.*, 1973, **47**, 337. 391

73JOM(51)113 A. F. Berniaz and D. G. Tuck; *J. Organomet. Chem.*, 1973, **51**, 113. 467, 485

73JOM(57)403 D. N. Reinhoudt, C. G. Kouwenhoven and J. P. Visser, *J. Organomet. Chem.*, 1973, **57**, 403. 135, 142

73JPS1204 J. R. Grunder, L. San and P. N. Kaul; *J. Pharm. Sci.*, 1973, **62**, 1204. 517

73KGS178 G. A. Sokol'skii, V. M. Pavlov, S. A. Agafonov and I. L. Knunyants; *Khim. Geterotsikl. Soedin.*, 1973, **9**, 178. 487

73KGS1321 V. M. Pavlov, V. N. Derkachev, G. A. Sokol'skii and I. L. Knunyants; *Khim. Geterotsikl. Soedin.*, 1973, **9**, 1321. 487, 488

73LA365 H. Reiff, D. Dieterich, R. Braden and H. Ziemann; *Liebigs Ann. Chem.*, 1973, 365. 384

73LA1702 M. L. Ziegler, J. Weiss, H. Schildknecht, N. Grund and H.-E. Sasse; *Liebigs Ann. Chem.*, 1973, 1702. 407

B-73MI50300 H. C. van der Plas; 'Ring Transformations of Heterocycles', Academic Press, New York, 1973. 43, 44, 46

B-73MI50301 Ref. B-73MI50300, vol. 2, p. 49. 43

B-73MI50302 Ref. B-73MI50300, vol. 1, p. 152. 44

B-73MI50303 Ref. B-73MI50300, vol. 1, p. 150. 46

B-73MI50400 M. Witanowski, L. Stefaniak and H. Januszewski; in 'Nitrogen-NMR', ed. M. Witanowski and G. A. Webb; Plenum, London, 1973, p. 163. 51

B-73MI50401 H. van der Plas; 'Ring Transformations of Heterocycles', Academic Press, New York, 1973, vol. 1, p. 277. 89

B-73MI50402 H. van der Plas; 'Ring Transformations of Heterocycles', Academic Press, New York, 1973, vol. 1, p. 347. 88

73MI50403 W. A. Lathan, L. Radom, P. C. Hariharan, W. J. Hehre and J. A. Pople; *Top. Curr. Chem.*, 1973, **40**, 1. 48, 49

B-73MI50500 G. Berti; in 'Topics in Stereochemistry', ed. N. L. Allinger and E. L. Eliel; Wiley, New York, 1973. 97, 114, 115, 117

B-73MI50501 'Handbook of Chemistry and Physics', ed. R. C. Weast; Chemical Rubber Co., Cleveland, 54th edn., 1973. 97

B-73MI50502 H. C. Van Der Plas; 'Ring Transformation of Heterocycles', Academic Press, New York, 1973, vol. 1, p. 364. 117

73MI50600 N. R. Bertoniere and G. W. Griffin; *Org. Photochem.*, 1973, **3**, 180. 140

73MI50601 K. Yamaguchi and T. Fueno; *Chem. Phys. Lett.*, 1973, **22**, 471. 168

73MI50800 W. A. Lathan, L. Radom, P. C. Hariharan, W. J. Hehre and J. A. Pople; *Top. Curr. Chem.*, 1973, **40**, 1. 197, 198

B-73MI50901 H. C. Van Der Plas; 'Ring Transformation of Heterocycles', Academic Press, New York, 1973, vol. 1, p. 135. 241

B-73MI50902 Ref. B-73MI50901, p. 138. 242

B-73MI50903 Ref. B-73MI50901, p. 143. 241

73MI51300 H. Ruotsalainen; *Suom. Kem. B*, 1973, **46**, 228 (*Chem. Abstr.*, 1974, **80**, 26 497). 379

73MI51301 J. A. Barnard and B. A. Harwood; *Combust. Flame*, 1973, **21**, 345 (*Chem. Abstr.*, 1974, **80**, 61 838). 399

B-73MI51400 'Rodd's Chemistry of Carbon Compounds', Elsevier, Amsterdam, 2nd edn., 1973, vol. 4, part A, p. 52. 404, 419

B-73MI51401 H. C. Van der Plas; 'Ring Transformations of Heterocycles', Academic Press, New York, 1973, vol. 1, p. 131. 431

73MI51500 Y. Imai and H. Hirukawa; *J. Polym. Sci., Polym. Lett. Ed.*, 1973, **11**, 271. 458, 468

73MI51501 W. C. Harris and A. Nuernberg; *Appl. Spectrosc.*, 1973, **27**, 44. 455

73MI51502 D. N. Purohit and K. C. Goswami; *Talanta*, 1973, **20**, 689. 488

B-73MI51600 R. A. Abramovitch; in 'Organic Reactive Intermediates', ed. S. McManus; Academic Press, New York, 1973, p. 127. 537

B-73MI51601 H. C. van der Plas; 'Ring Transformations of Heterocycles', Academic Press, London, 1973, vols. 1 and 2. 518

B-73MI52000 A. G. Anastassiou; in 'Topics in Nonbenzenoid Aromatic Chemistry', ed. T. Nozoe, Hirokawa Publishing, Tokyo, 1973, vol. 1, pp. 1–27. 711, 713, 714

73MI52001 P. Skrabal; *Int. Rev. Sci.*, 1973, **3**, 237. 715, 729

73MI52100 J. M. Lehn; *Struct. Bonding (Berlin)*, 1973, **16**, 1. 733

73MI52101 R. M. Izatt, D. J. Eatough and J. J. Christensen; *Struct. Bonding (Berlin)*, 1973, **16**, 161. 741, 742

73MI52102 M. R. Truter; *Struct. Bonding (Berlin)*, 1973, **16**, 71. 735

73MI52200 S. Bradamante, G. Pagani, A. Marchesini and U. M. Pagnoni; *Chim. Ind. (Milan)*, 1973, **55**, 962 (*Chem. Abstr.*, 1974, **80**, 95 861). 772

B-73NMR	T. J. Batterham; 'NMR Spectra of Simple Heterocycles', Wiley, New York, 1973.	6, 7, 10, 50, 51, 99, 134, 136, 366, 495, 496
73OMS(7)479	W. J. Richter, J. G. Liehr and A. L. Burlingame; *Org. Mass Spectrom.*, 1973, **7**, 479.	549
73RRC1353	R. Vilceanu, Z. Simon and A. Chiriac; *Rev. Roum. Chim.*, 1973, **18**, 1353.	138
73RRC1609	A. T. Balaban; *Rev. Roum. Chim.*, 1973, **18**, 1609.	772
73RTC20	D. N. Reinhoudt; *Recl. Trav. Chim. Pays-Bas*, 1973, **92**, 20.	633
73RTC865	D. N. Reinhoudt and C. G. Kouwenhoven; *Recl. Trav. Chim. Pays-Bas*, 1973, **92**, 865.	585
73S153	S. S. Chatterjee and A. Sihoeb; *Synthesis*, 1973, 153.	241, 256, 266
73S159	Y. Tamura, J. Minamikawa, H. Matsushima and M. Ikeda; *Synthesis*, 1973, 159.	599
73S327	A. K. Mukerjee and R. C. Srivastiva; *Synthesis*, 1973, 327.	238, 249, 257, 259, 261, 263
73S355	G. Seitz and H. Morck; *Synthesis*, 1973, 355.	647
73S534	L. W. Christensen; *Synthesis*, 1973, 534.	440
73S546	J. Lehmann and H. Wamhoff; *Synthesis*, 1973, 546.	71
73T391	P. Atlani, J. F. Biellmann and J. Moron; *Tetrahedron*, 1973, **29**, 391.	544
73T809	H. Ogawa and M. Kubo; *Tetrahedron*, 1973, **29**, 809.	718
73T1115	K.-G. Svensson, H. Selander, M. Karlsson and J. L. G. Nilsson; *Tetrahedron*, 1973, **29**, 1115.	528
73T1163	J. Kagan and J. T. Przybytek; *Tetrahedron*, 1973, **29**, 1163.	377
73T1399	I. Agranat; *Tetrahedron*, 1973, **29**, 1399.	723
73T1629	B. Dietrich, J. M. Lehn, J. P. Sauvage and J. Blanzat; *Tetrahedron*, 1973, **29**, 1629.	733, 743
73T2135	J. Kooi, H. Wynberg and R. M. Kellogg; *Tetrahedron*, 1973, **29**, 2135.	443
73T2359	Y. Tamura, S. Matsugashita, H. Ishibashi and M. Ikeda; *Tetrahedron*, 1973, **29**, 2359.	598
73T2759	B. L. Dyatkin, S. R. Sterlin, L. G. Zhuravkova, B. I. Martynov, E. I. Mysov and I. L. Knunyants; *Tetrahedron*, 1973, **29**, 2759.	478
73TL145	G. H. Rasmusson, G. F. Reynolds and G. E. Arth; *Tetrahedron Lett.*, 1973, 145.	321
73TL619	B. V. Ioffe and E. V. Koroleva; *Tetrahedron Lett.*, 1973, 619.	6, 52, 86
73TL655	R. H. Cragg and A. F. Weston; *Tetrahedron Lett.*, 1973, 655.	622
73TL1221	H. Gotthardt; *Tetrahedron Lett.*, 1973, 1221.	420, 421
73TL1401	J. M. Lalancette and M. Laliberte; *Tetrahedron Lett.*, 1973, 1401.	161
73TL1615	K. C. Rice and U. Weiss; *Tetrahedron Lett.*, 1973, 1615.	625
73TL1807	D. Thomas and D. H. Aue; *Tetrahedron Lett.*, 1973, 1807.	270
73TL1835	T. Mukai and H. Sukawa; *Tetrahedron Lett.*, 1973, 1835.	627
73TL1923	M. Vaultier, R. Danion-Bougot, D. Danion, J. Hamelin and R. Carrié; *Tetrahedron Lett.*, 1973, 1923.	245
73TL2143	T. Hiyama, H. Koide and H. Nozaki; *Tetrahedron Lett.*, 1973, 2143.	73
73TL2319	A. K. Bose, J. K. Kapur, S. D. Sharma and M. S. Manhas; *Tetrahedron Lett.*, 1973, 2319.	260
73TL2667	M. F. Semmelhack and R. D. Stauffer; *Tetrahedron Lett.*, 1973, 2667.	156
73TL2671	S. Yanagida, D. J. Barsotti, G. W. Harrington and D. Swern; *Tetrahedron Lett.*, 1973, 2671.	486
73TL3727	J. W. Lown and M. H. Akhtar; *Tetrahedron Lett.*, 1973, 3727.	342
73TL3805	A. G. Anastassiou, E. Reichmanis and R. L. Elliott; *Tetrahedron Lett.*, 1973, 3805.	725, 728, 729
73TL3989	K. Sakamoto and M. Oki; *Tetrahedron Lett.*, 1973, 3989.	778
73TL4017	J. F. Haley, Jr. and P. B. Keehn; *Tetrahedron Lett.*, 1973, 4017.	771
73TL4021	S. Rosenfeld and P. B. Keehn; *Tetrahedron Lett.*, 1973, 4021.	779
73TL4091	F. D. Greene, R. L. Camp, V. P. Abegg and G. O. Pierson; *Tetrahedron Lett.*, 1973, 4091.	262
73TL4163	J. P. Luttringer and J. Streith; *Tetrahedron Lett.*, 1973, 4163.	359
73TL4653	R. W. Ratcliffe and B. G. Christensen; *Tetrahedron Lett.*, 1973, 4653.	290
73TL4855	H. H. Wasserman, E. A. Glazer and M. J. Hearn; *Tetrahedron Lett.*, 1973, 4855.	261
73TL5029	F. A. L. Anet and J. Krane; *Tetrahedron Lett.*, 1973, 5029.	698
73USP3715401	S. D. Schroeter; *U.S. Pat.* 3 715 401 (*Chem. Abstr.*, 1973, **78**, 110 883).	385
73USP3729487	M. H. Rosen and H. M. Blatter; *U.S. Pat.* 3 729 487 (1973) (*Chem. Abstr.*, 1973, **79**, 18 556).	447
73USP3767615	P. E. Throckmorton, W. J. McKillip and H. J. Richards; *U.S. Pat.* 3 767 615 (1973) (*Chem. Abstr.*, 1974, **80**, 96 772).	447
73USP3772331	S. A. Greenfield; *U.S. Pat.* 3 772 331 (1973) (*Chem. Abstr.*, 1974, **80**, 47 969).	485
73ZOR178	F. I. Luknitskii, R. P. Arshinova and G. G. Butenko; *Zh. Org. Khim.*, 1973, **9**, 178.	366
73ZOR430	S. D. Isajev, A. G. Jurtschenko, F. N. Stepanov, G. G. Koljade and S. S. Novikov; *Zh. Org. Khim.*, 1973, **9**, 430.	223
73ZOR890	Y. M. Portnyagin and T. M. Pavel; *Zh. Org. Khim.*, 1974, **9**, 890.	377
74ACH115	M. Lempert-Sreter; *Acta Chim. Acad. Sci. Hung.*, 1974, 115.	602
74ACR265	T. Kato; *Acc. Chem. Res.*, 1974, **7**, 265.	387, 402
74ACS(B)701	B. Lamm and K. Gustafsson; *Acta Chem. Scand., Ser. B*, 1974, **28**, 701.	440
74AG(E)142	I. Murata, T. Tatsuoka and Y. Sugihara; *Angew. Chem., Int. Ed. Engl.*, 1974, **13**, 142.	589
74AG(E)267	M. Depoortere, J. Marchand-Brynaert and L. Ghosez; *Angew. Chem., Int. Ed. Engl.*, 1974, **13**, 267.	263

74AG(E)277 G. Jikeli and H. Guenther; *Angew. Chem., Int. Ed. Engl.*, 1974, **13**, 277. 554

74AG(E)291 T. Kauffmann; *Angew. Chem., Int. Ed. Engl.*, 1974, **13**, 291. 766

74AG(E)339 J. M. Brown and F. Sondheimer; *Angew. Chem., Int. Ed. Engl.*, 1974, **13**, 339. 720

74AG(E)404 A. G. Anastassiou and E. Reichmanis; *Angew. Chem., Int. Ed. Engl.*, 1974, **13**, 404. 712

74AG(E)425 G. Maier; *Angew. Chem., Int. Ed. Engl.*, 1974, **13**, 425. 275

74AG(E)539 W. Littke and U. Druck; *Angew. Chem., Int. Ed. Engl.*, 1974, **13**, 539. 189

74AG(E)544 G. Lauer, C. Culler, K. W. Schulte, A. Schweig and A. Krebs; *Angew. Chem., Int. Ed. Engl.*, 1974, **13**, 544. 556

74AG(E)672 H. Klein and W. Grimme; *Angew. Chem., Int. Ed. Engl.*, 1974, **13**, 672. 569

74AG(E)736 E. Vogel, E. Schmidbauer and H.-J. Altenbach; *Angew. Chem., Int. Ed. Engl.*, 1974, **13**, 736. 191, 577

74AG(E)737 E. Vogel, E. Schmidbauer and H.-J. Altenbach; *Angew. Chem., Int. Ed. Engl.*, 1974, **13**, 737. 680

74AG(E)740 H. Irngartinger and H. Rodewald; *Angew. Chem., Int. Ed. Engl.*, 1974, **13**, 740. 555

74AG(E)751 N. D. Epiotis; *Angew. Chem., Int. Ed. Engl.*, 1974, **13**, 751. 38, 269, 272

74AHC(17)1 D. Lloyd, H. P. Cleghorn and D. R. Marshall; *Adv. Heterocycl. Chem.*, 1974, **17** 1. 608, 611, 612, 615

74AHC(17)27 D. Lloyd and H. P. Cleghorn; *Adv. Heterocycl. Chem.*, 1974, **17**, 27. 608, 611

74AHC(17)45 S. Kasparek; *Adv. Heterocycl. Chem.*, 1974, **17**, 45. 492, 495, 499, 501, 502, 527, 528, 529, 530, 531, 532, 538, 546

74AHC(17)255 M. J. Cook, A. R. Katritzky and P. Linda; *Adv. Heterocycl. Chem.*, 1974, **17**, 255. 16, 18, 492

74AX(B)534 J. B. Wetherington and J. W. Moncrieff; *Acta Crystallogr., Part B*, 1974, **30**, 534. 239

74AX(B)691 J. L. Baudour, Y. Delugeard and M. Sanquer; *Acta Crystallogr., Part B*, 1974, **30**, 691. 452

74AX(B)1670 A. D. Hardy and F. R. Ahmed; *Acta Crystallogr., Part B*, 1974, **30**, 1670. 706

74AX(B)1674 A. D. Hardy and F. R. Ahmed; *Acta Crystallogr., Part B*, 1974, **30**, 1674. 706

74AX(B)2733 J. D. Dunitz, M. Dobler, P. Seiler and R. P. Phizackerley; *Acta Crystallogr., Part B*, 1974, **30**, 2733. 734

74AX(B)2741 M. Dobler, J. D. Dunitz and P. Seiler; *Acta Crystallogr., Part B*, 1974, **30**, 2741. 734, 735

74AX(B)2744 P. Seiler, M. Dobler and J. D. Dunitz; *Acta Crystallogr., Part B*, 1974, **30**, 2744. 734

74AX(B)2748 M. Dobler and R. P. Phizackerley; *Acta Crystallogr., Part B*, 1974, **30**, 2748. 734, 735

74BAU1941 F. A. Alasher, A. V. Kessenikh, S. Z. Taits and Ya. L. Gol'dfarb; *Bull. Acad. Sci. USSR, Div. Chem. Sci.*, 1974, **23**, 1941. 779

74BCJ343 Y. Yamada, T. Oine and I. Inoue; *Bull. Chem. Soc. Jpn.*, 1974, **47**, 343. 613

74BCJ583 Y. Chatani, T. Yamauchi and Y. Miyake; *Bull. Chem. Soc. Jpn.*, 1974, **47**, 583. 701

74BCJ1455 I. Tabushi, Y. Tamaru and Z. Yoshida; *Bull. Chem. Soc. Jpn.*, 1974, **47**, 1455. 440

74BCJ2699 S. Kobayashi, H. Danda and T. Saegusa; *Bull. Chem. Soc. Jpn.*, 1974, **47**, 2699. 382

74BSF595 M. Jaubert, M. Mazet and T. Yvernault; *Bull. Soc. Chim. Fr.*, 1974, 595. 393

74C163 F. Sondheimer, *Chimia*, 1974, **28**, 163. 729

74CB270 D. Bormann; *Chem. Ber.*, 1974, **107**, 270. 251

74CB2115 F. Bohlmann and J. Kocur; *Chem. Ber.*, 1974, **107**, 2115. 445

74CB2299 H. Bock and B. Solouki; *Chem. Ber.*, 1974, **107**, 2299. 137

74CB2362 F. A. Neugebauer and S. Bamberger; *Chem. Ber.*, 1974, **107**, 2362. 502

74CB2544 H. Gotthardt; *Chem. Ber.*, 1974, **107**, 2544. 438

74CB3149 R. Wehner and H. Guenther; *Chem. Ber.*, 1974, **107**, 3149. 551

74CB3883 J. Körösi and T. Láng; *Chem. Ber.*, 1974, **107**, 3883. 599

74CB3922 H. Gotthardt and G. S. Hammond; *Chem. Ber.*, 1974, **107**, 3922. 377

74CC45 D. J. Anderson and A. Hassner; *J. Chem. Soc., Chem. Commun.*, 1974, 45. 60, 639

74CC119 S. M. Rosenfeld and P. M. Keehn; *J. Chem. Soc., Chem. Commun.*, 1974, 119. 771

74CC129 D. St. C. Black and A. Boscacci; *J. Chem. Soc., Chem. Commun.*, 1974, 129. 262

74CC397 V. S. Mastryukov, O. V. Dorofeeva and L. V. Vilkov; *J. Chem. Soc., Chem. Commun.*, 1974, 397. 198

74CC532 J. N. Done, J. H. Knox, R. McEwan and J. T. Sharp; *J. Chem. Soc., Chem. Commun.*, 1974, 532. 600

74CC782 R. E. Moerck and M. A. Battiste; *J. Chem. Soc., Chem. Commun.*, 1974, 782. 60, 91

74CC900 J. I. G. Cadogan, J. T. Sharp and M. J. Trattles; *J. Chem. Soc., Chem. Commun.*, 1974, 900. 193

74CHE623 N. A. Korchevin, V. A. Usov and M. G. Voronkov; *Chem. Heterocycl. Compd. (Engl. Transl.)*, 1974, **10**, 623. 140

74CJC246 M. T. H. Liu and D. H. T. Chien; *Can. J. Chem.*, 1974, **52**, 246. 225

74CJC610 G. R. Birchall and A. H. Rees, *Can. J. Chem.*, 1974, **52**, 610. 517, 531

74CJC2123 J. O. Edwards, G. Bernath, J. Dixon, J. M. Paton and D. Vocelle; *Can. J. Chem.*, 1974, **52**, 2123. 18

74CJC3327 C. G. Hughes, E. G. Lewars and A. H. Rees; *Can. J. Chem.*, 1974, **52**, 3327. 530

74CJC3742 J. Cornibert, N. V. Hien, F. Brisse and R. H. Marchessault; *Can. J. Chem.*, 1974, **52**, 3742. 388

74CJC3996 S. Wolfe, J.-B. Ducep, K.-C. Tin and S.-L. Lee; *Can. J. Chem.*, 1974, **52**, 3996. 314

74CL185 T. Akiyama, T. Kitamura, T. Isida and M. Kawanisi; *Chem. Lett.*, 1974, 185. 232

74CL345 Y. Kayama, M. Oda and Y. Kitahara; *Chem. Lett.*, 1974, 345. 688

74CL383 Y. Ohshiro, M. Komatsu, Y. Yamamoto, K. Takaki and T. Agawa; *Chem. Lett.*, 1974, 383. 219

74CL887 H. Tsuruta, M. Ogasawara and T. Mukai; *Chem. Lett.*, 1974, 887. 441

74CL1189	T. Izawa and T. Mukaiyama; *Chem. Lett.*, 1974, 1189.	380
74CL1209	M. Mitani, T. Tsuchida and K. Koyama; *Chem. Lett.*, 1974, 1209.	537
74CPB1186	T. Yamana, A. Tsuji and Y. Mizukami; *Chem. Pharm. Bull.*, 1974, **22**, 1186.	326
74CPB1526	A. Takamizawa and H. Sato; *Chem. Pharm. Bull.*, 1974, **22**, 1526.	262
74CR(C)(278)1389	J.-P. Le Roux, J.-C. Cherton and P.-L. Desbene; *C.R. Hebd. Seances Acad. Sci., Ser. C*, 1974, **278**, 1389.	627
74CRV101	L. J. Kricka and A. Ledwith; *Chem. Rev.*, 1974, **74**, 101.	493, 495, 497, 499, 501, 502, 508, 510, 511, 512, 514, 518, 519, 524, 525, 526, 527, 528, 533, 538, 546
74CRV351	J. J. Christensen, D. J. Eatough and R. M. Izatt; *Chem. Rev.*, 1974, **74**, 351.	733
74CS(5)52	A. Lindquist and J. Sandström; *Chem. Scr.*, 1974, **5**, 52.	499, 694, 702, 704
74DOK(219)873	R. G. Kostyanovskii, A. E. Polaykov, G. V. Shustov, K. S. Zakharov and V. I. Markov; *Dokl. Akad. Nauk SSSR*, 1974, **219**, 873.	201
74EGP104170	H. Lehmann, K. Peseke and M. Klepel; *Ger. (East) Pat.* 104 170 (1974) (*Chem. Abstr.*, 1975, **82**, 12 265).	485
74FOR(31)1	D. N. McGregor; *Fortschr. Chem. Org. Naturst.*, 1974, **31**, 1.	303, 305, 329, 330
74G1207	G. D. Andreetti, G. Bocelli and P. Sgarabotto; *Gazz. Chim. Ital.*, 1974, **104**, 1207.	407
74GEP2347905	Z. Csuros, J. Petro, A. Lengyel and E. Fogassy; *Ger. Offen.* 2 347 905 (1974) (*Chem. Abstr.*, 1974, **81**, P3789).	654
74HCA2514	C. Mayer; *Helv. Chim. Acta*, 1974, **57**, 2514.	434
74IC2783	J. Casanova and M. Geisel; *Inorg. Chem.*, 1974, **13**, 2783.	510
74IJC(B)1238	V. N. Gogte, K. M. More and B. D. Tilak; *Indian J. Chem., Sect. B*, 1974, **12**, 1238.	544
74IZV2823	R. G. Kostyanovskii, G. V. Shustov and G. V. Markov; *Izv. Akad. Nauk SSSR, Ser. Khim.*, 1974, 2823.	231
74JA5	L. H. Weaver and B. W. Matthews; *J. Am. Chem. Soc.*, 1974, **96**, 5.	780
74JA288	K. Yoshikawa, M. Hashimoto and I. Morishima; *J. Am. Chem. Soc.*, 1974, **96**, 288.	239
74JA317	D. B. Denney and L. S. Shih; *J. Am. Chem. Soc.*, 1974, **96**, 317.	467, 485
74JA819	R. F. Atkinson and T. C. Bruice; *J. Am. Chem. Soc.*, 1974, **96**, 819.	379
74JA1193	R. M. De Marinis, C. N. Filer, S. M. Waraszkiewicz and G. A. Berchtold; *J. Am. Chem. Soc.*, 1974, **96**, 1193.	584
74JA1422	G. Illuminati, L. Mandolini and B. Masci; *J. Am. Chem. Soc.*, 1974, **96**, 1422.	690, 692
74JA1609	T. S. Chou, J. R. Burgtorf, A. I. Ellis, S. R. Lammert and S. Kukolja; *J. Am. Chem. Soc.*, 1974, **96**, 1609.	306
74JA1636	C. C. Chiang, I. C. Paul, A. G. Anastassiou and S. W. Eachus; *J. Am. Chem. Soc.*, 1974, **96**, 1636.	710
74JA2250	C. L. Liotta and H. P. Harris; *J. Am. Chem. Soc.*, 1974, **96**, 2250.	756, 757
74JA2964	R. Langendries, F. C. De Schryver, P. de Mayo, R. A. Marty and J. Schutyser; *J. Am. Chem. Soc.*, 1974, **96**, 2964.	418, 422
74JA3146	P. Raynolds, S. Zonnebelt, S. Bakker and R. M. Kellogg; *J. Am. Chem. Soc.*, 1974, **96**, 3146.	135, 137, 145, 149, 162
74JA3502	W. Kusters and P. de Mayo; *J. Am. Chem. Soc.*, 1974, **96**, 3502.	467, 482, 485
74JA4185	K. L. Servis, E. A. Noe, N. R. Easton, Jr. and F. A. L. Anet; *J. Am. Chem. Soc.*, 1974, **96**, 4185.	188
74JA4753	J. S. Wright; *J. Am. Chem. Soc.*, 1974, **96**, 4753.	488
74JA4879	J. S. Swenton and J. A. Hyatt; *J. Am. Chem. Soc.*, 1974, **96**, 4879.	342, 348
74JA5005	J. P. Snyder; *J. Am. Chem. Soc.*, 1974, **96**, 5005.	168
74JA5270	G. Stork and J. F. Cohen; *J. Am. Chem. Soc.*, 1974, **96**, 5270.	112
74JA5287	T. Hiyama, S. Hirano and H. Nozaki; *J. Am. Chem. Soc.*, 1974, **96**, 5287.	771
74JA5557	P. D. Bartlett, A. L. Baumstark and M. E. Landis; *J. Am. Chem. Soc.*, 1974, **96**, 5557.	461
74JA5578	D. M. Jerina, B. Witkop, C. L. McIntosh and O. L. Chapman; *J. Am. Chem. Soc.*, 1974, **96**, 5578.	564, 565
74JA5632	R. A. Moss and F. G. Pilkiewicz; *J. Am. Chem. Soc.*, 1974, **96**, 5632.	227
74JA6165	G. I. Birnbaum; *J. Am. Chem. Soc.*, 1974, **96**, 6165.	658
74JA6929	A. M. Jeffrey, H. J. C. Yeh, D. M. Jerina, R. M. De Marinis, C. H. Foster, D. E. Piccolo and G. A. Berchtold; *J. Am. Chem. Soc.*, 1974, **96**, 6929.	567, 568
74JA7100	L. R. Sousa, D. H. Hoffmann, L. Kaplan and D. J. Cram; *J. Am. Chem. Soc.*, 1974, **96**, 7100.	745, 760
74JA7491	B. A. DeGraff, D. W. Gillespie and R. J. Sundberg; *J. Am. Chem. Soc.*, 1974, **96**, 7491.	534
74JA7582	L. D. Cama and B. G. Christensen; *J. Am. Chem. Soc.*, 1974, **96**, 7582.	354
74JA7584	R. N. Guthikonda, L. D. Cama and B. G. Christensen; *J. Am. Chem. Soc.*, 1974, **96**, 7584.	354
74JA7812	G. W. Gribble, P. D. Lord, J. Skotnicki, S. E. Dietz, J. T. Eaton and J. L. Johnson; *J. Am. Chem. Soc.*, 1974, **96**, 7812.	517
74JAP7420071	K. Watanabe, Y. Arimatsu and F. Akiyama; *Jpn. Pat.* 74 20 071 (*Chem. Abstr.*, 1975, **82**, 16 302).	384
74JAP7435375	F. Fujisawa and T. Kobori; *Jpn. Pat.* 74 35 375 (1974) (*Chem. Abstr.*, 1974, **81**, 120 109).	151
74JAP7440461	H. Kondo, A. Negishi and Y. Shonaka; *Jpn. Pat.* 74 40 461 (1974) (*Chem. Abstr.*, 1975, **82**, 155 851).	149
74JAP(K)74127918	K. Kondo, Y. Ryu, M. Mutsukado and K. Mochida; *Jpn. Kokai* 74 127 918 (1974) (*Chem. Abstr.*, 1975, **82**, 156 724).	386
74JCP(61)2071	W. C. Pringle and A. L. Meinzer; *J. Chem. Phys.*, 1974, **61**, 2071.	370

74JCS(F1)43 K. A. Holbrook and R. A. Scott; *J. Chem. Soc., Faraday Trans. 1*, 1974, **70**, 43. 372

74JCS(P1)52 L. J. Kricka, M. C. Lambert and A. Ledwith; *J. Chem. Soc., Perkin Trans. 1*, 1974, 52. 497, 502

74JCS(P1)66 R. Clinging, F. M. Dean and L. E. Houghton; *J. Chem. Soc., Perkin Trans. 1*, 1974, 66. 669

74JCS(P1)185 D. F. Corbett and R. J. Stoodley; *J. Chem. Soc., Perkin Trans. 1*, 1974, 185. 312

74JCS(P1)1248 S. F. Gait, M. E. Peek, C. W. Rees and R. C. Storr; *J. Chem. Soc., Perkin Trans. 1*, 1974, 1248. 636

74JCS(P1)1456 R D. Allan, D. H. R. Barton, M. Girijavallabhan and P. G. Sammes; *J. Chem. Soc., Perkin Trans. 1*, 1974, 1456. 308

74JCS(P1)1459 D. H. R. Barton, I. H. Coates, P. G. Sammes and C. M. Cooper; *J. Chem. Soc., Perkin Trans. 1*, 1974, 1459. 308

74JCS(P1)1828 M. Lennon, A. McLean, I. McWatt and G. R. Proctor; *J. Chem. Soc., Perkin Trans. 1*, 1974, 1828. 503, 511, 518, 525, 544

74JCS(P1)1864 R. C. Cambie, R. C. Hayward, J. L. Roberts and P. S. Rutledge; *J. Chem. Soc., Perkin Trans. 1*, 1974, 1864. 395

74JCS(P1)2496 R. M. Acheson, G. Paglietti and P. A. Tasker; *J. Chem. Soc., Perkin Trans. 1*, 1974, 2496. 660

74JCS(P1)2602 T. Kametani, K. Kigasawa, M. Hiiragi, H. Ishimaru and S. Haga; *J. Chem. Soc., Perkin Trans. 1*, 1974, 2602. 533

74JCS(P2)377 G. M. Blackburn and H. L. H. Dodds; *J. Chem. Soc., Perkin Trans. 2*, 1974, 377. 386

74JCS(P2)1648 R. J. Abraham, L. J. Kricka and A. Ledwith; *J. Chem. Soc., Perkin Trans. 2*, 1974, 1648. 495, 498, 499

74JCS(P2)1676 S. Bradamante, P. Del Buttero, D. Landini and S. Maiorana; *J. Chem. Soc., Perkin Trans. 2*, 1974, 1676. 445

74JHC17 J. G. Lombardino; *J. Heterocycl. Chem.*, 1974, **11**, 17. 592

74JHC89 R. B. Trattner and H. D. Perlmutter; *J. Heterocycl. Chem.*, 1974, **11**, 89. 675

74JHC347 Y. Gelas-Mialhe, R. Hierle and R. Vessiere; *J. Heterocycl. Chem.*, 1974, **11**, 347. 81

74JHC401 A. Sotiriadis, P. Catsoulacos and D. Theodoropoulos; *J. Heterocycl. Chem.*, 1974, **11**, 401. 597

74JHC409 M. W. Barker, S. Perumal and C. J. Wierengo; *J. Heterocycl. Chem.*, 1974, **11**, 409. 456, 473

74JHC453 J. S. Swenton and R. J. Balchunis; *J. Heterocycl. Chem.*, 1974, **11**, 453. 345

74JHC633 M. W. Barker and C. J. Wierengo; *J. Heterocycl. Chem.*, 1974, **11**, 633. 457, 473

74JHC937 B. S. Thyagarajan and K. C. Majumdar; *J. Heterocycl. Chem.*, 1974, **11**, 937. 673, 688

74JHC1087 H. E. Zaugg, J. E. Leonard and R. W. DeNet; *J. Heterocycl. Chem.*, 1974, **11**, 1087. 631

74JMC342 W. C. Topp and B. G. Christensen; *J. Med. Chem.*, 1974, **17**, 342. 303

74JMC389 J. Hoogmartens, P. J. Claes and H. Vanderhaeghe; *J. Med. Chem.*, 1974, **17**, 389. 331

74JMC523 J. M. Indelicato, T. T. Norvilas, R. R. Pfeiffer, W. J. Wheeler and W. L. Wilham; *J. Med. Chem.*, 1974, **17**, 523. 286

74JMC948 A. N. Tischler, F. M. Thompson, L. J. Libertini and M. Calvin; *J. Med. Chem.*, 1974, **17**, 948. 655

74JMR(15)367 C. Cistaro, G. Fronza, R. Mondelli, S. Bradamante and G. A. Pagani; *J. Magn. Reson.*, 1974, **15**, 367. 410

74JOC63 J. K. Crandall and W. W. Conover; *J. Org. Chem.*, 1974, **39**, 63. 88

74JOC66 W. L. Nelson and B. E. Sherwood; *J. Org. Chem.*, 1974, **39**, 66. 70

74JOC103 N. Ishibe, K. Hashimoto and M. Sunami; *J. Org. Chem.*, 1974, **39**, 103. 588

74JOC167 D. L. Coffen, J. P. De Noble, E. L. Evans, G. F. Field, R. I. Fryer, D. A. Katonak, B. J. Mandel, L. H. Sternbach and W. J. Zally; *J. Org. Chem.*, 1974, **39**, 167. 619

74JOC246 G. L. Hardgrove, Jr., J. S. Bratholdt and M. M. Lein; *J. Org. Chem.*, 1974, **39**, 246. 406

74JOC373 C. U. Pittman, Jr., A. Kress, T. B. Patterson, P. Walton and L. D. Kispert; *J. Org. Chem.*, 1974, **39**, 373. 48

74JOC435 D. B. Borders and J. E. Lancaster; *J. Org. Chem.*, 1974, **39**, 435. 192, 679

74JOC437 R. A. Firestone, N. S. Maciejewicz, R. W. Ratcliffe and B. G. Christensen; *J. Org. Chem.*, 1974, **39**, 437. 290

74JOC441 A. J. Vlietinck, E. Roets, H. Vanderhaeghe and S. Toppet; *J. Org. Chem.*, 1974, **39**, 441. 305, 324

74JOC455 T. Sasaki, K. Kanematsu and Y. Yukimoto; *J. Org. Chem.*, 1974, **39**, 455. 519

74JOC568 R. B. Moffett; *J. Org. Chem.*, 1974, **39**, 568. 614

74JOC902 J. A. Deyrup and S. C. Clough; *J. Org. Chem.*, 1974, **39**, 902. 77, 82

74JOC940 E. M. Burgess and J. P. Sanchez; *J. Org. Chem.*, 1974, **39**, 940. 682

74JOC948 M. Komatsu, Y. Ohshiro, H. Hotta, M. Sato and T. Agawa; *J. Org. Chem.*, 1974, **39**, 948. 208

74JOC957 M. Komatsu, Y. Ohshiro, K. Yasuda, S. Ichijima and T. Agawa; *J. Org. Chem.*, 1974, **39**, 957. 209

74JOC1210 T. Minami and T. Agawa; *J. Org. Chem.*, 1974, **39**, 1210. 262

74JOC1215 R. C. Kerber, T. J. Ryan and S. D. Hsu; *J. Org. Chem.*, 1974, **39**, 1215. 472

74JOC1322 A. P. Krapcho and E. G. E. Jahngen, Jr.; *J. Org. Chem.*, 1974, **39**, 1322. 394

74JOC1444 J. C. Sheehan, Y. S. Lo, J. Löliger and C. C. Podewell; *J. Org. Chem.*, 1974, **39**, 1444. 316, 318

74JOC1707 L. de Vries; *J. Org. Chem.*, 1974, **39**, 1707. 278

74JOC1944 S. K. Gupta; *J. Org. Chem.*, 1974, **39**, 1944. 477

74JOC2031 A. Hassner and D. J. Anderson; *J. Org. Chem.*, 1974, **39**, 2031. 60, 511, 539

74JOC2264 J. C. Sheehan and R. W. Tulis; *J. Org. Chem.*, 1974, **39**, 2264. 240

74JOC2320 L. A. Carpino and J. R. Williams; *J. Org. Chem.*, 1974, **39**, 2320. 176

74JOC2341 R. W. Begland, D. R. Hartter, F. N. Jones, D. J. Sam, W. A. Sheppard, O. W. Webster and F. J. Weigert; *J. Org. Chem.*, 1974, **39**, 2341. 611

74JOC2351 C. G. Krespan; *J. Org. Chem.*, 1974, **39**, 2351. 390

74JOC2432 W. E. Parham, P. E. Olson and K. R. Reddy; *J. Org. Chem.*, 1974, **39**, 2432. 777

74JOC2552 R. C. Kerber and M. C. Cann; *J. Org. Chem.*, 1974, **39**, 2552. 473

74JOC2570 C. Wong and W. W. Paudler; *J. Org. Chem.*, 1974, **39**, 2570. 767, 779

74JOC2598 G. Illuminati, L. Mandolini and B. Masci; *J. Org. Chem.*, 1974, **39**, 2598. 668, 692

74JOC2722 U. Jacobsson, T. Kempe and T. Norin; *J. Org. Chem.*, 1974, **39**, 2722. 140

74JOC2916 W. E. Parham and P. E. Olson; *J. Org. Chem.*, 1974, **39**, 2916. 777

74JOC2946 J. H. Schauble, W. A. Van Saun, Jr. and J. D. Williams; *J. Org. Chem.*, 1974, **39**, 2946. 435

74JOC3038 E. E. Schwiezer, T. Minami and S. E. Anderson; *J. Org. Chem.*, 1974, **39**, 3038. 668

74JOC3070 A. Hassner and D. J. Anderson; *J. Org. Chem.*, 1974, **39**, 3070. 492, 493, 495, 500, 507, 519, 539

74JOC3076 D. J. Anderson, A. Hassner and D. Y. Tang; *J. Org. Chem.*, 1974, **39**, 3076. 505, 510, 511, 526

74JOC3187 H. W. Heine, R. Henrie, L. Heitz and S. R. Kouvali; *J. Org. Chem.*, 1974, **39**, 3187. 201

74JOC3198 M. Komatsu, N. Nishikaze, M. Sakamoto, Y. Ohshiro and T. Agawa; *J. Org. Chem.*, 1974, **39**, 3198. 214

74JOC3608 F. Wudl, M. L. Kaplan, E. J. Hufnagel and E. W. Southwick, Jr.; *J. Org. Chem.*, 1974, **39**, 3608. 775

74JOC3763 V. Nair and K. H. Kim; *J. Org. Chem.*, 1974, **39**, 3763. 646

74JOC3805 M. H. Rosen and G. Bonet; *J. Org. Chem.*, 1974, **39**, 3805. 168

74JOU1140 V. V. Takhistov, T. K. Klindukhova, V. P. Klindukhov and V. N. Yandovskii; *J. Org. Chem. USSR (Engl. Transl.)*, 1974, **10**, 1140. 202

74JOU1427 V. A. Ginsburg, A. N. Medvedev, M. F. Lebedeva and L. L. Martynova; *J. Org. Chem. USSR (Engl. Transl.)*, 1974, **10**, 1427. 473

74JPR999 M. Schulz and G. West; *J. Prakt. Chem.*, 1974, **316**, 999. 218

74JPS838 H. Kohl, P. D. Desai, A. N. Dohadwalla and N. J. De Souza; *J. Pharm. Sci.*, 1974, **63**, 838. 637

74JST(22)247 R. L. Hilderbrandt and J. D. Wieser; *J. Mol. Struct.*, 1974, **22**, 247. 188

74KGS1629 E. Schmitz and K. Jähnisch; *Khim. Geterotsikl. Soedin.*, 1974, 1629. 210, 235

74LA539 K. Claus, D. Grimm and G. Prossel; *Liebigs Ann. Chem.*, 1974, 539. 247, 252, 261

74LA734 H. Böhme and F. Ziegler; *Liebigs Ann. Chem.*, 1974, 734. 632

74LA1328 G. Seitz and W.-D. Mikulla; *Liebigs Ann. Chem.*, 1974, 1328. 411, 412, 418, 436

74MI50100 R. Moriarty; *Top. Stereochem.*, 1974, 271. 3, 8, 9

B-74MI50500 'Nomenclature of Organic Compounds, Principles and Practice', ed. J. H.Fletcher, O. C. Dermer and R. B. Fox; *Adv. Chem. Ser.*, 1974, **126**, chaps. 6 and 7. 96

B-74MI50501 'Beilsteins Handbuch der Organischen Chemie', Springer Verlag, Berlin, 1974, vol. 17 (3/4), p. 190. 98

B-74MI50502 Ref. B-74MI50501, p. 304. 98

B-74MI50503 'Sadtler Standard NMR Spectra', Sadtler Research Laboratories, Philadelphia, 1974 (continuing series). 98

B-74MI50504 'Sadtler Standard IR Spectra', Sadtler Research Laboratories, Philadelphia, 1974 (continuing series). 99

B-74MI50505 M. Fieser and L. F. Fieser; 'Reagents for Organic Synthesis', Wiley, New York, 1974, vol. 4, p. 303. 112

74MI50600 K. Okiye, C. Hirose, D. G. Lister and J. Sheridan; *Chem. Phys. Lett.*, 1974, **24**, 111. 133

74MI50700 P. Sims, P. L. Grover, A. Swaisland, K. Pal and A. Hewer; *Nature (London)*, 1974, **252**, 326. 188

74MI50701 D. M. Jerina and J. W. Daly; *Science*, 1974, **185**, 573. 188

74MI50900 A. Colens, J. P. Declercq, G. Germain, J. P. Putzeys, and M. van Meersche; *Cryst. Struct. Commun.*, 1974, **3**, 119. 247

B-74MI50901 D. R. Boyd and B. J. Walker; in 'Saturated Heterocyclic Chemistry', ed. W. Parker; The Chemical Society, London, 1974, vol. 2. 248

74MI51000 P. M. Blumberg and J. L. Strominger; *Bacteriol. Rev.*, 1974, **38**, 291. 296

74MI51300 R. M. Moriarty; *Top. Stereochem.*, 1974, **8**, 271. 365

74MI51301 K. Teranishi, M. Iida, T. Araki, S. Yamashita and H. Tani; *Macromolecules*, 1974, **7**, 421. 383

74MI51600 H. J. Haink, J. E. Adams and J. R. Huber; *Ber. Bunsenges. Phys. Chem.*, 1974, **78**, 436. 502

74MI51700 D. M. Jerina and J. W. Daly; *Science*, 1974, **185**, 573. 548, 565, 591

74MI51900 F. A. L. Anet; *Fortschr. Chem. Forsch.*, 1974, **45**, 169. 698, 702, 703, 704

74MI51901 J. Hernandez-Mora and N. Cordero-Antunano; *Caribb. J. Sci.*, 1974, **14**, 77 (*Chem. Abstr.*, 1976, **84**, 17 307). 674

74MIP51800 Hoffmann-La Roche; *Austrian Pat.* 318 622 (1974) (*Chem. Abstr.*, 1975, **83**, 79 291). 610

74OMR(6)483 V. Nair; *Org. Magn. Reson.*, 1974, **6**, 483. 49, 51

74OPP287 F. O. Bennett, J. Johnson and J. Tramondozzi; *Org. Prep. Proced. Int.*, 1974, **6**, 287. 633

74PMH(5)1 P. J. Wheatley; *Phys. Methods Heterocycl. Chem.*, 1974, **5**, 1. 495

74PMH(6)1 E. Heilbronner, J. P. Maier and E. Haselbach; *Phys. Methods Heterocycl. Chem.*, 1974, **6**, 1. 7, 48

74PMH(6)53	J. Sheridan; *Phys. Methods Heterocycl. Chem.*, 1974, **6**, 53.	3, 48
74PMH(6)199	K. Pihlaja and E. Taskinen; *Phys. Methods Heterocycl. Chem.*, 1974, **6**, 199.	3, 49, 98
74S187	R. R. Schmidt and G. Berger; *Synthesis*, 1974, 187.	627
74S461	B. A. Arbuzov and N. N. Zobova; *Synthesis*, 1974, 461.	261, 474
74T39	R. G. Kostyanovsky, I. M. Gella, V. I. Markov and Z. E. Samojlova; *Tetrahedron*, 1974, **30**, 39.	239, 240
74T195	S. Conde, S. Corral and R. Madronero; *Tetrahedron*, 1974, **30**, 195.	638
74T1683	J. Dale; *Tetrahedron*, 1974, **30**, 1683.	549, 689
74T1903	W. D. Ollis, J. F. Stoddart and I. O. Sutherland; *Tetrahedron*, 1974, **30**, 1903.	706
74T2093	D. N. Reinhoudt and C. G. Kouwenhoven; *Tetrahedron*, 1974, **30**, 2093.	558, 588
74T2197	K. Kikuzono, T. Yamabe, S. Nagata, H. Kato and K. Fukui, *Tetrahedron*, 1974, **30**, 2197.	138
74T2431	D. N. Reinhoudt and C. G. Kouwenhoven; *Tetrahedron*, 1974, **30**, 2431.	559
74T2561	H. J. Dillinger, G. Fengler, D. Schumann and E. Winterfeldt; *Tetrahedron*, 1974, **30**, 2561.	347
74T2633	S. Hirano, T. Hiyama, S. Fujita, T. Kawaguti, Y. Hayashi and H. Nozaki; *Tetrahedron*, 1974, **30**, 2633.	771
74T2765	S. Rossi, M. Bianchi and A. Butti; *Tetrahedron*, 1974, **30**, 2765.	640
74TL599	P. J. Beeby, J. M. Brown, P. J. Garrett and F. Sondheimer; *Tetrahedron Lett.*, 1974, 599.	777
74TL725	T. S. Chou; *Tetrahedron Lett.*, 1974, 725.	308
74TL837	M. S. A. Vrijland; *Tetrahedron Lett.*, 1974, 837.	149
74TL1479	R. Roulet, J. Wenger, M. Hardy and P. Vogel; *Tetrahedron Lett.*, 1974, 1479.	583
74TL1487	V. Nair and K. H. Kim; *Tetrahedron Lett.*, 1974, 1487.	92
74TL2103	G. Jones, II and H. H. Kleinman; *Tetrahedron Lett.*, 1974, 2103.	373
74TL2303	G. C. Johnson and R. H. Levin; *Tetrahedron Lett.*, 1974, 2303.	60, 639
74TL2577	Y. Ueno, Y. Masuyama and M. Okawara; *Tetrahedron Lett.*, 1974, 2577.	471
74TL2779	G. Ahlgren; *Tetrahedron Lett.*, 1974, 2779.	389
74TL2841	Y. Kobayashi, I. Kumadaki, A. Ohsawa and Y. Sekine; *Tetrahedron Lett.*, 1974, 2841.	187
74TL3005	J. L. Fourrey, P. Jouin and J. Moron; *Tetrahedron Lett.*, 1974, 3005.	447
74TL3275	E. Vilsmaier, R. Tropitzsch and O. Vostrowsky; *Tetrahedron Lett.*, 1974, 3275.	162
74TL3897	J. C. Pommelet and J. Chuche; *Tetrahedron Lett.*, 1974, 3897.	65
74TL3911	R. S. Devdhar, V. N. Gogte and B. D. Tilak; *Tetrahedron Lett.*, 1974, 3911.	419, 435
74TL4235	A. Ichihara, K. Oda, M. Kobayashi and S. Sakamura; *Tetrahedron Lett.*, 1974, 4235.	192
74TL4283	D. St. C. Black, R. F. C. Brown, B. T. Dunstan and S. Sternhell; *Tetrahedon Lett.*, 1974, 4283.	275
74USP3799892	Pfizer Inc., *U.S. Pat.* 3 799 892 (*Chem. Abstr.*, 1974, **81**, 3974).	622
74USP3842369	N. T. Melamed and D. C. Phillips; *U.S. Pat.* 3 842 369 (1974) (*Chem. Abstr.*, 1974, **81**, 179 788).	236
74ZN(A)1907	J. Jokisaari; *Z. Naturforsch., Teil A*, 1974, **29**, 1907.	364
74ZN(B)231	R. Kreher, A. Bauer and H. Hennige; *Z. Naturforsch., Teil B*, 1974, **29**, 231.	466
74ZN(B)399	A. Gieren, K. Burger and W. Thenn; *Z. Naturforsch., Teil B*, 1974, **29**, 399.	268, 269
74ZN(B)425	R. Kreher and H. Pawelczyk; *Z. Naturforsch., Teil B*, 1974, **29**, 425.	523
74ZOR95	Y. N. Portnyagin, T. M. Pavel and N. E. Pak; *Zh. Org. Khim.*, 1974, **10**, 95.	380
74ZOR1510	V. N. Yandovskii and T. K. Klindukhova; *Zh. Org. Khim.*, 1974, **10**, 1510.	213
74ZOR1556	T. V. Stezhko, S. Ya. Skachilova and M. G. Pleshakov; *Zh. Org. Khim.*, 1974, **10**, 1556 (*Chem. Abstr.*, 1974, **81**, 135 924).	256
74ZOR2218	G. S. Gol'din, T. A. Balabina, A. N. Ushakova and S. N. Tsiomo; *Zh. Org. Khim.*, 1974, **10**, 2218 (*Chem. Abstr.*, 1975, **82**, 43 382).	647
75ACR239	R. A. Bartsch; *Acc. Chem. Res.*, 1975, **8**, 239.	758
75ACS(B)1071	E. Ljungstrom; *Acta Chem. Scand., Ser. B*, 1975, **29**, 1071.	407
75AF1436	L. Coscia, P. Causa, E. Giuliani and A. Nunziata; *Arzneim.-Forsch.*, 1975, **25**, 1436.	590
75AG591	E. Vogel, J. Sombroek and W. Wagemann; *Angew. Chem.*, 1975, **87**, 591.	718
75AG(E)15	D. Seebach and D. Enders; *Angew. Chem., Int. Ed. Engl.*, 1975, **14**, 15	518
75AG(E)32	H. Meier and K.-P. Zeller; *Angew. Chem., Int. Ed. Engl.*, 1975, **14**, 32.	126, 127, 193
75AG(E)69	K. Wieser and A. Berndt; *Angew. Chem., Int. Ed. Engl.*, 1975, **14**, 69.	454, 470
75AG(E)70	K. Wieser and A. Berndt; *Angew. Chem., Int. Ed. Engl.*, 1975, **14**, 70.	33, 470
75AG(E)252	S. Kagabu and H. Prinzbach; *Angew. Chem., Int. Ed. Engl.*, 1975, **14**, 252.	178, 191
75AG(E)253	H. Prinzbach, C. Kaiser and H. Fritz; *Angew. Chem., Int. Ed. Engl.*, 1975, **14**, 253.	191
75AG(E)347	H. Prinzbach, R. Schwesinger, M. Breuninger, B. Gallenkamp and D. Hunkler; *Angew. Chem., Int. Ed. Engl.*, 1975, **14**, 347.	714
75AG(E)428	H. Quast and L. Bieber; *Angew. Chem., Int. Ed. Engl.*, 1975, **14**, 428.	13, 93, 232
75AG(E)555	M. Kirch and J. M. Lehn; *Angew. Chem., Int. Ed. Engl.*, 1975, **14**, 555.	755
75AG(E)630	R. Schwesinger and H. Prinzbach; *Angew. Chem., Int. Ed. Engl.*, 1975, **14**, 630.	191
75AG(E)753	H. Prinzbach and H. Babsch; *Angew. Chem., Int. Ed. Engl.*, 1975, **14**, 753.	582
75AG(E)766	K. Burger, J. Albanbauer and M. Eggersdorfer; *Angew. Chem., Int. Ed. Engl.*, 1975, **14**, 766.	459, 483
75AG(E)775	G. L'abbe; *Angew. Chem., Int. Ed. Engl.*, 1975, **14**, 775.	85

75AG(E)812 H. Hofmann and H. Gaube; *Angew. Chem., Int. Ed. Engl.*, 1975, **14**, 812. 576, 577

75AG(E)814 A. G. Anastassiou and S. J. Girgenti; *Angew. Chem., Int. Ed. Engl.*, 1975, **14**, 814. 729

75AJC119 D. J. Brown and K. Ienaga; *Aust. J. Chem.*, 1975, **28**, 119. 774

75AJC2547 D. St. C. Black and N. A. Blackman; *Aust. J. Chem.*, 1975, **28**, 2547. 234

75AX(B)49 M. Cesari, L. D'Ilario, E. Giglio and G. Perego; *Acta Crystallogr., Part B*, 1975, **31**, 49. 239

75AX(B)251 J. D. Dunitz and F. K. Winkler; *Acta Crystallogr., Part B*, 1975, **31**, 251. 655, 702

75AX(B)1008 M. L. Post, O. Kennard and A. S. Horn; *Acta Crystallogr., Part B*, 1975, **31**, 1008. 495

75AX(B)2729 N. Yasuoka, Y. Kai and N. Kasai; *Acta Crystallogr., Part B*, 1975, **31**, 2729. 556

75AX(B)2912 M. Kaftory and J. D. Dunitz; *Acta Crystallogr., Part B*, 1975, **31**, 2912. 699

75AX(B)2916 M. Kaftory and J. D. Dunitz; *Acta Crystallogr., Part B*, 1975, **31**, 2916. 699

75BAU878 O. N. Nuretdinova, G. A. Bakaleinik and B. A. Arbuzov; *Bull. Acad. Sci. USSR, Div. Chem. Sci.*, 1975, **24**, 878. 153

75BAU1090 B. V. Kopylova, L. V. Yashkina, Sh. A. Karapet'yan and R. Kh. Freidlina; *Bull. Acad. Sci. USSR, Div. Chem. Sci.*, 1975, **24**, 1090. 167

75BAU2416 N. P. Aktaev, G. A. Sokol'skii and I. L. Knunyants; *Bull. Acad. Sci. USSR, Div. Chem. Sci.*, 1975, **24**, 2416. 477

75BCJ33 H. Sakai, T. Yamabe, H. Kato, S. Nagata and K. Fukui; *Bull. Chem. Soc. Jpn.*, 1975, **48**, 33. 136

75BCJ302 S. Kobayashi; *Bull. Chem. Soc. Jpn.*, 1975, **48**, 302. 614

75BCJ480 T. Hiraoka and T. Kobayashi; *Bull. Chem. Soc. Jpn.*, 1975, **48**, 480. 458, 462, 473, 474

75BCJ1063 J. Nakayama, H. Midorikawa and M. Yoshida; *Bull. Chem. Soc. Jpn.*, 1975, **48**, 1063. 277

75BCJ1490 K. Kondo and I. Ojima; *Bull. Chem. Soc. Jpn.*, 1975, **48**, 1490. 431, 435

75BCJ1501 J. Aihara; *Bull. Chem. Soc. Jpn.*, 1975, **48**, 1501. 550, 557

75BCJ1665 S. Inoue and S. Oae; *Bull. Chem. Soc. Jpn.*, 1975, **48**, 1665. 144

75BCJ1691 K. Karakida and K. Kuchitsu; *Bull. Chem. Soc. Jpn.*, 1975, **48**, 1691. 405

75BCJ3259 T. Minami, F. Takimoto and T. Agawa; *Bull. Chem. Soc. Jpn.*, 1975, **48**, 3259. 486

75BCJ3429 H. Tanida, R. Muneyuki and T. Tsushima; *Bull. Chem. Soc. Jpn.*, 1975, **48**, 3429. 306

75BSF277 J. Gilbert, D. Rousselle and P. Rumpf; *Bull. Soc. Chim. Fr.*, 1975, 277. 622

75BSF283 D. Rousselle, G. Jacques and P. Rumpf; *Bull. Soc. Chim. Fr.*, 1975, 283. 630

75BSF807 A. Le Berre, A. Etienne and B. Desmazieres; *Bull. Soc. Chim. Fr.*, 1975, 807. 468

75BSF1670 R. Couturier, D. Paquer and A. Vibet; *Bull. Soc. Chim. Fr.*, 1975, 1670. 477

75CB515 F. Bohlmann and A. Suwita; *Chem. Ber.*, 1975, **108**, 515. 447

75CB844 R. W. Hoffmann and R. Schüttler; *Chem. Ber.*, 1975, **108**, 844. 127, 128

75CB897 B. Solouki, H. Bock and R. Appel; *Chem. Ber.*, 1975, **108**, 897. 137

75CB1527 S. Sustmann, R. Sustmann and C. Ruchardt; *Chem. Ber.*, 1975, **108**, 1527. 73

75CB2137 H. Vorbrüggen and K. Krolikiewicz; *Chem. Ber.*, 1975, **108**, 2137. 650

75CB3336 R. Aumann, H. Averbeck and C. Krueger; *Chem. Ber.*, 1975, **108**, 3336. 570

75CB3387 H. Gnichtel and K. Hirte; *Chem. Ber.*, 1975, **108**, 3387. 642

75CB3566 K.-P. Zeller; *Chem. Ber.*, 1975, **108**, 3566. 193

75CB3753 W. Henne, G. Plinke and G. Schroder; *Chem. Ber.*, 1975, **108**, 3753. 716

75CC95 D. R. White; *J. Chem. Soc., Chem. Commun.*, 1975, 95. 115

75CC133 J. H. Barlow, C. R. Hall, D. R. Russell and D. J. H. Smith; *J. Chem. Soc., Chem. Commun.*, 1975, 133. 406

75CC149 A. G. Anastassiou and E. Reichmanis; *J. Chem. Soc., Chem. Commun.*, 1975, 149. 712, 725

75CC317 K.-P. Zeller; *J. Chem. Soc., Chem. Commun.*, 1975, 317. 126

75CC322 R. F. Hudson, A. J. Lawson and K. A. F. Record; *J. Chem. Soc., Chem. Commun.*, 1975, 322. 19

75CC394 G. W. Gokel, J. M. Timko and D. J. Cram; *J. Chem. Soc., Chem. Commun.*, 1975, 394. 745

75CC427 K. Madan and D. J. Cram; *J. Chem. Soc., Chem. Commun.*, 1975, 427. 746

75CC436 J. K. Crandall and J. B. Komin; *J. Chem. Soc., Chem. Commun.*, 1975, 436. 256, 266

75CC575 M. Ikeda, S. Matsugashita, F. Tabusa, H. Ishibashi and Y. Tamura; *J. Chem. Soc., Chem. Commun.*, 1975, 575. 277

75CC724 T. Durst and B. P. Gimbarzevsky; *J. Chem. Soc., Chem. Commun.*, 1975, 724. 455, 459, 471, 477

75CC725 P. Y. Johnson, C. E. Hatch, III and N. R. Schmuff; *J. Chem. Soc., Chem. Commun.*, 1975, 725. 262, 359

75CC740 C. W. Rees, R. Somanathan, R. C. Storr and A. D. Woolhouse; *J. Chem. Soc., Chem. Commun.*, 1975, 740. 281, 643

75CC784 P. M. Weintraub and A. D. Still; *J. Chem. Soc., Chem. Commun.*, 1975, 784. 572

75CC788 J. J. Barr and R. C. Storr; *J. Chem. Soc., Chem. Commun.*, 1975, 788. 255

75CJC138 R. K. Bansal, A. W. McCullough, P. W. Rasmussen and A. G. McInnes; *Can. J. Chem.*, 1975, **53**, 138. 582

75CJC167 R. R. Fraser, M. A. Raza, R. N. Renaud and R. B. Layton; *Can. J. Chem.*, 1975, **53**, 167. 706, 707

75CJC276 I. W. J. Still, M. T. Thomas and A. M. Clish; *Can. J. Chem.*, 1975, **53**, 276. 635

75CJC519 T. Tsuchiya and V. Snieckus; *Can. J. Chem.*, 1975, **53**, 519. 603, 604

75CJC960 J. W. Lown and T. Itoh; *Can. J. Chem.*, 1975, **53**, 960. 659

75CJC1103 K. R. Kopecky, J. E. Filby, C. Mumford, P. A. Lockwood and J.-Y. Ding; *Can. J. Chem.*, 1975, **53**, 1103. 455, 469

75CJC1744 D. R. Dice and R. P. Steer; *Can. J. Chem.*, 1975, **53**, 1744. 419

75CJC2279	C. S. Rooney, R. S. Stuart, B. K. Wasson and H. W. R. Williams; *Can. J. Chem.*, 1975, **53**, 2279.	622
75CJC2891	M. H. Akhtar, A. Begleiter, D. Johnson, J. W. Lown, L. McLoughlin and S.-K. Sim; *Can. J. Chem.*, 1975, **53**, 2891.	93
75CJC3720	M. E. Peach, M. Schmidt and E. Weissflog; *Can. J. Chem.*, 1975, **53**, 3720.	696
75CL423	T. Mukai, Y. Yamashita, H. Sukawa and T. Tezuka; *Chem. Lett.*, 1975, 423.	604
75CL1153	H. Matsukubo, M. Kojima and H. Kato; *Chem. Lett.*, 1975, 1153.	167
75CPB114	T. Ibuka, T. Konoshima and Y. Inubushi; *Chem. Pharm. Bull.*, 1975, **23**, 114.	517, 525, 528
75CPB365	T. Kato, M. Sato and Y. Kitagawa; *Chem. Pharm. Bull.*, 1975, **23**, 365.	397
75CPB1764	H. Fujito and Y. Sato; *Chem. Pharm. Bull.*, 1975, **23**, 1764.	633
75CPB1917	Y. Sawa, T. Kato, T. Masuda, M. Hori and H. Fujimura, *Chem. Pharm. Bull.*, 1975, **23**, 1917.	530
75CPB2749	G. Kobayashi, Y. Matsuda, Y. Tominaga and K. Mizuyama; *Chem. Pharm. Bull.*, 1975, **23**, 2749.	660
75CPB2773	Y. Kobayashi, I. Kumadaki, A. Ohsawa, Y. Sekine and H. Mochizuki; *Chem. Pharm. Bull.*, 1975, **23**, 2773.	186
75DOK(223)1379	N. A. Nesmeyanov, V. A. Kalyavin and O. A. Reutov; *Dokl. Akad. Nauk SSSR*, 1975, **223**, 1379.	426
75FES773	L. Fontanella and L. Mariani; *Farmaco Ed. Sci.*, 1975, **30**, 773.	626
75GEP2349726	M. I. Bakhitov, E. V. Kuznetsov, N. N. Zolotarevskaya, N. A. Popova, V. G. Golov, Yu. A. Rodionov, I. I. Molev and V. M. Kotlyarskii; *Ger. Offen.* 2 349 726 (1975) (*Chem. Abstr.*, 1975, **83**, 44 494).	484
75GEP2365511	K. Taninaka, H. Kurono, S. Mine and A. Hirano; *Ger. Offen.* 2 365 511 (1975) (*Chem. Abstr.*, 1975, **83**, 193 344).	485
75GEP2409355	E. Otto, W. Duerckheimer and R. Muschaweck; *Ger. Pat.* 2 409 355 (1975) (*Chem. Abstr.*, 1976, **84**, 59 606).	650
75GEP2448207	R. D. Wilbur and J. M. Pensack; *Ger. Offen.* 2 448 207 (1975) (*Chem. Abstr.*, 1975, **83**, 42 015).	485
75H(7)547	O. Tsuge and K. Kamata; *Heterocycles*, 1975, **7**, 547.	640
75HCA1444	A. Pfaltz, B. Hardegger, P. M. Muller, S. Faroug, B. Krautler and A. Eschenmoser; *Helv. Chim. Acta*, 1975, **58**, 1444.	75
75HCA2409	M. Karpf and A. S. Dreiding; *Helv. Chim. Acta*, 1975, **58**, 2409.	775
75JA464	R. C. Schnur and E. E. van Tamelen; *J. Am. Chem. Soc.*, 1975, **97**, 464.	577
75JA640	R. L. Wife and F. Sondheimer; *J. Am. Chem. Soc.*, 1975, **97**, 640.	716
75JA641	R. L. Wife, P. J. Beeby and F. Sondheimer; *J. Am. Chem. Soc.*, 1975, **97**, 641.	718
75JA918	S. Masamune, N. Nakamura and J. Spadaro; *J. Am. Chem. Soc.*, 1975, **97**, 918.	455, 469
75JA1600	P. G. Gassman, I. Nishiguchi and H. Yamamoto; *J. Am. Chem. Soc.*, 1975, **97**, 1600.	73
75JA2025	M. M. Rohmer and B. Roos; *J. .Am. Chem. Soc.*, 1975, **97**, 2025.	133
75JA2553	Y. Hata, M. Watanabe, S. Inoue and S. Oae, *J. Am. Chem. Soc.*, 1975, **97**, 2553.	151
75JA2822	A. Padwa, D. Dean and T. Oine; *J. Am. Chem. Soc.*, 1975, **97**, 2822.	50
75JA3130	R. C. Neuman, Jr. and E. W. Ertley; *J. Am. Chem. Soc.*, 1975, **97**, 3130.	670
75JA3185	H. Yagi and D. M. Jerina; *J. Am. Chem. Soc.*, 1975, **97**, 3185.	188
75JA3192	S. Kukolja, S. R. Lammert, M. R. Gleissner and A. I. Ellis; *J. Am. Chem. Soc.*, 1975, **97**, 3192.	307
75JA3666	W. L. Mock; *J. Am. Chem. Soc.*, 1975, **97**, 3666.	571
75JA3818	W. Ando, A. Sekiguchi, T. Migita, S. Kammula, M. Green and M. Jones, Jr.; *J. Am. Chem. Soc.*, 1975, **97**, 3818.	395
75JA3850	B. S. Campbell, D. B. Denney, D. Z. Denney and L. Shih; *J. Am. Chem. Soc.*, 1975, **97**, 3850.	464
75JA3862	A. Padwa and P. H. J. Carlsen; *J. Am. Chem. Soc.*, 1975, **97**, 3862.	58
75JA4136	D. H. Aue, H. M. Webb and M. T. Bowers; *J. Am. Chem. Soc.*, 1975, **97**, 4136.	660
75JA4137	D. H. Aue, H. M. Webb and M. T. Bowers; *J. Am. Chem. Soc.*, 1975, **97**, 4137.	49, 51
75JA4405	K. Tamao, S.-I. Kodama, T. Nakatsuka, Y. Kiso and M. Kumada; *J. Am. Chem. Soc.*, 1975, **97**, 4405.	766
75JA4409	A. Demoulin, H. Gorissen, A.-M. Hesbain-Frisque and L. Ghosez; *J. Am. Chem. Soc.*, 1975, **97**, 4409.	66
75JA4428	H. G. Selander, D. M. Jerina, D. E. Piccolo and G. A. Berchtold; *J. Am. Chem. Soc.*, 1975, **97**, 4428.	584, 590
75JA4682	A. Padwa, J. Smolanoff and A. Tremper; *J. Am. Chem. Soc.*, 1975, **97**, 4682.	56, 67, 493, 517, 520, 540
75JA4692	A. Hassner, S. S. Burke and J. Cheng-fan, I; *J. Am. Chem. Soc.*, 1975, **97**, 4692.	90
75JA4960	G. Illuminati, L. Mandolini and B. Masci; *J. Am. Chem. Soc.*, 1975, **97**, 4960.	668
75JA5008	S.-I. Nakatsuka, H. Tanino and Y. Kishi; *J. Am. Chem. Soc.*, 1975, **97**, 5008.	295
75JA5010	S.-I. Nakatsuka, H. Tanino and Y. Kishi; *J. Am. Chem. Soc.*, 1975, **97**, 5010.	295
75JA5022	E. Graf and J. M. Lehn; *J. Am. Chem. Soc.*, 1975, **97**, 5022.	733, 743
75JA5167	N. L. Allinger and M. J. Hickey; *J. Am. Chem. Soc.*, 1975, **97**, 5167.	419
75JA5582	S. Kukolja and S. R. Lammert; *J. Am. Chem. Soc.*, 1975, **97**, 5582.	265, 331
75JA5583	S. R. Lammert and S. Kukolja; *J. Am. Chem. Soc.*, 1975, **97**, 5583.	265, 307
75JA5628	P. W. Wojtkowski, J. E. Dolfini, O. Kocy and C. M. Cimarusti; *J. Am. Chem. Soc.*, 1975, **97**, 5628.	327
75JA5912	A. Padwa and J. K. Rasmussen; *J. Am. Chem. Soc.*, 1975, **97**, 5912.	63

75JA6267 J. D. Corbett, D. G. Adolphson, D. J. Merryman, P. A. Edwards and F. J. Armantis; *J. Am. Chem. Soc.*, 1975, **97**, 6267. 744

75JA6700 J. M. Lehn and J. P. Sauvage; *J. Am. Chem. Soc.*, 1975, **97**, 6700. 743

75JA7110 G. B. Schuster, N. J. Turro, H.-C. Steinmetzer, A. P. Schaap, G. Faler, W. Adam and J. C. Liu; *J. Am. Chem. Soc.*, 1975, **97**, 7110. 455, 476

75JA7288 K. A. Howard and T. H. Koch; *J. Am. Chem. Soc.*, 1975, **97**, 7288. 244, 344, 345, 542

75JA7298 R. M. Rodehorst and T. H. Koch; *J. Am. Chem. Soc.*, 1975, **97**, 7298. 347

75JAP75117765 T. Agawa, Y. Ohshiro, M. Komatsu and N. Nishikaze; *Jpn. Pat.* 75 117 765 (1975) (*Chem. Abstr.*, 1976, **84**, 59 159). 218

75JCS(D)701 C. J. Jones and J. A. McCleverty; *J. Chem. Soc., Dalton Trans.*, 1975, 701. 467

75JCS(D)2374 D. L. Hughes; *J. Chem. Soc., Dalton Trans.*, 1975, 2374. 735

75JCS(F1)1145 C. M. Dobson, L. O. Ford, S. E. Summers and R. J. P. Williams; *J. Chem. Soc., Faraday Trans. 1*, 1975, **71**, 1145. 301

75JCS(P1)41 B. M. Adger, M. Keating, C. W. Rees and R. C. Storr; *J. Chem. Soc., Perkin Trans. 1*, 1975, 41. 278

75JCS(P1)45 B. M. Adger, C. W. Rees and R. C. Storr; *J. Chem. Soc., Perkin Trans. 1*, 1975, 45. 278, 279, 283, 663, 686

75JCS(P1)102 J. T. Sharp, R. H. Findlay and P. B. Thorogood; *J. Chem. Soc., Perkin Trans. 1*, 1975, 102. 596, 601, 604

75JCS(P1)498 R. N. Pratt, D. P. Stokes and G. A. Taylor; *J. Chem. Soc., Perkin Trans. 1*, 1975, 498. 263

75JCS(P1)556 S. F. Gait, M. J. Rance, C. W. Rees and R. W. Stephenson; *J. Chem. Soc., Perkin Trans. 1*, 1975, 556. 640

75JCS(P1)622 M. Lennon, A. McLean, G. R. Proctor and I. W. Sinclair; *J. Chem. Soc., Perkin Trans. 1*, 1975, 622. 503, 515, 525

75JCS(P1)657 J. Ashby and D. Griffiths; *J. Chem. Soc., Perkin Trans. 1*, 1975, 657. 606

75JCS(P1)1130 R. D. Chambers, J. R. Maslakiewicz and K. C. Srivastava; *J. Chem. Soc., Perkin Trans. 1*, 1975, 1130. 360, 361

75JCS(P1)1517 B. T. Golding and D. R. Hall; *J. Chem. Soc., Perkin Trans. 1*, 1975, 1517. 359

75JCS(P1)1932 M. M. Campbell and G. Johnson; *J. Chem. Soc., Perkin Trans. 1*, 1975, 1932. 309, 311

75JCS(P1)2513 R. J. Bushby; *J. Chem. Soc., Perkin Trans. 1*, 1975, 2513. 414, 429, 445

75JCS(P2)43 D. N. J. White and M. H. P. Guy; *J. Chem. Soc., Perkin Trans. 2*, 1975, 43. 703

75JCS(P2)308 B. C. Gilbert, R. O. C. Norman and R. C. Sealy; *J. Chem. Soc., Perkin Trans. 2*, 1975, 308. 167

75JCS(P2)554 T. de Boer, J. I. G. Cadogan, H. M. McWilliam and A. G. Rowley; *J. Chem. Soc., Perkin Trans. 2*, 1975, 554. 510

75JCS(P2)559 G. Calzaferri and R. Gleiter; *J. Chem. Soc., Perkin Trans. 2*, 1975, 559. 455

75JCS(P2)1294 B. J. Buzzi, P. R. Olivato, R. Rittner, C. Trufen, H. Viertler and B. Wladislaw; *J. Chem. Soc., Perkin Trans. 2*, 1975, 1294. 556

75JCS(P2)1475 A. Dondoni and A. Battaglia; *J. Chem. Soc., Perkin Trans. 2*, 1975, 1475. 458, 475

75JCS(P2)1555 N. S. Isaacs and E. Rannala; *J. Chem. Soc., Perkin Trans. 2*, 1975, 1555. 38

75JCS(P2)1722 J. W. Gordon, G. H. Schmid and I. G. Csizmadia; *J. Chem. Soc., Perkin Trans. 2*, 1975, 1722. 134

75JGU1219 A. M. Pinchuk, T. V. Kovalevskaya and G. K. Bespalko; *J. Gen. Chem. USSR (Engl. Transl.)*, 1975, **45**, 1219. 83

75JHC21 P. Vouros; *J. Heterocycl. Chem.*, 1975, **12**, 21. 135

75JHC183 V. Nair; *J. Heterocycl. Chem.*, 1975, **12**, 183. 60, 639

75JHC433 J. L. Atwood, W. E. Hunter, C. Wong and W. W. Paudler; *J. Heterocycl. Chem.*, 1975, **12**, 433. 779, 780

75JHC661 E. M. Essassi, J. P. Lavergne and P. Viallefont; *J. Heterocycl. Chem.*, 1975, **12**, 661. 638

75JHC825 D. Nardi, R. Pennini and A. Tajana; *J. Heterocycl. Chem.*, 1975, **12**, 825. 611

75JHC903 U. Golik; *J. Heterocycl. Chem.*, 1975, **12**, 903. 606

75JHC1211 O. F. Bennett, J. Johnson and S. Galletto; *J. Heterocycl. Chem.*, 1975, **12**, 1211. 635

75JMC408 D. B. Boyd, R. B. Hermann, D. E. Presti and M. M. Marsh; *J. Med. Chem.*, 1975, **18**, 408. 287

75JMC486 H. Vanderhaeghe and J. Thomis; *J. Med. Chem.*, 1975, **18**, 486. 315

75JMC976 L. Toscano, E. Seghetti and G. Fioriello; *J. Med. Chem.*, 1975, **18**, 976. 527

75JMR(17)219 C. Cistaro, G. Fronza, R. Mondelli, S. Bradamante and G. A. Pagani; *J. Magn. Reson.*, 1975, **17**, 219. 410, 411

75JOC81 K. Baum and C. D. Beard; *J. Org. Chem.*, 1975, **40**, 81. 384

75JOC191 Y. S. Lo and J. C. Sheehan; *J. Org. Chem.*, 1975, **40**, 191. 319

75JOC224 E. M. Bingham and J. C. Gilbert; *J. Org. Chem.*, 1975, **40**, 224. 78, 93

75JOC369 J. B. Lambert and S. A. Khan; *J. Org. Chem.*, 1975, **40**, 369. 699

75JOC473 A. Balsamo, G. Ceccarelli, P. Crotti and F. Macchia; *J. Org. Chem.*, 1975, **40**, 473. 365, 366, 367

75JOC675 R. Micheli, Z. G. Hajos, N. Cohen, D. R. Parrish, L. A. Portland, W. Sciamanna, M. A. Scott and P. A. Wehrli; *J. Org. Chem.*, 1975, **40**, 675. 380

75JOC694 R. N. McDonald, R. N. Steppel and R. C. Cousins; *J. Org. Chem.*, 1975, **40**, 1694. 124

75JOC902 L. A. Paquette, H. C. Berk and S. V. Ley; *J. Org. Chem.*, 1975, **40**, 902. 715

75JOC1205 C. G. Krespan; *J. Org. Chem.*, 1975, **40**, 1205. 390

75JOC1264 D. R. Bender, L. F. Bjeldanes, D. R. Knapp and H. Rapoport; *J. Org. Chem.*, 1975, **40**, 1264. 359

75JOC1349 D. H. Aue and D. Thomas; *J. Org. Chem.*, 1975, **40**, 1349. 269

75JOC1409 F. D. Greene and K. E. Gilbert; *J. Org. Chem.*, 1975, **40**, 1409. 451, 466, 468

75JOC1505 H. H. Wasserman and E. H. Glazer; *J. Org. Chem.*, 1975, **40**, 1505. 261

75JOC1508 J. H. Sellstedt; *J. Org. Chem.*, 1975, **40**, 1508. 617

75JOC1640 E. E. Smissman, H. N. Alkaysi and M. W. Creese; *J. Org. Chem.*, 1975, **40**, 1640. 667

75JOC1683 A. Padwa, L. Gehrlein and R. B. Kinnel; *J. Org. Chem.*, 1975, **40**, 1683. 360

75JOC1909 N. P. Peet and S. Sunder; *J. Org. Chem.*, 1975, **40**, 1909. 637, 638

75JOC2045 J. K. Crandall, L. C. Crawley and J. B. Komin; *J. Org. Chem.*, 1975, **40**, 2045. 79

75JOC2212 L. K. Bee, J. Beeby, J. W. Everett and P. J. Garratt; *J. Org. Chem.*, 1975, **40**, 2212. 180

75JOC2356 D. H. Aue and D. Thomas; *J. Org. Chem.*, 1975, **40**, 2356. 262, 270, 348

75JOC2360 D. H. Aue and D. Thomas; *J. Org. Chem.*, 1975, **40**, 2360. 270

75JOC2552 D. H. Aue and D. Thomas; *J. Org. Chem.*, 1975, **40**, 2552. 270, 348

75JOC2692 P. F. Hudrlik and M. M. Mohtady; *J. Org. Chem.*, 1975, **40**, 2692. 392

75JOC2843 W. E. Barnett and L. L. Needham; *J. Org. Chem.*, 1975, **40**, 2843. 374, 395

75JOC2883 P. W. Jeffs, J. F. Hansen and G. A. Brine; *J. Org. Chem.*, 1975, **40**, 2883. 536

75JOC2963 P. F. Hudrlik and C. N. Wan; *J. Org. Chem.*, 1975, **40**, 2963. 385

75JOC2990 M. Vaultier, R. Danion-Bougot, D. Danion, J. Hamelin and R. Carrié; *J. Org. Chem.*, 1975, **40**, 2990. 244, 245

75JOC3046 P. Y. Johnson and M. Berman; *J. Org. Chem.*, 1975, **40**, 3046. 446, 571

75JOC3112 C. J. Wilkerson and F. D. Greene; *J. Org. Chem.*, 1975, **40**, 3112. 232

75JOC3179 A. J. Krubsack, R. Sehgal, W. A. Loong and W. E. Slack; *J. Org. Chem.*, 1975, **40**, 3179. 436

75JOC3189 B. B. Jarvis, W. P. Tong and H. L. Ammon; *J. Org. Chem.*, 1975, **40**, 3189. 159, 161

75JOC3363 J. W. Lown, M. H. Akhtar and W. M. Dadson; *J. Org. Chem.*, 1975, **40**, 3363. 346, 347

75JOC3417 W. T. Brady and P. L. Ting; *J. Org. Chem.*, 1975, **40**, 3417. 398

75JOC3502 P. Y. Johnson and C. E. Hatch, III; *J. Org. Chem.*, 1975, **40**, 3502. 262

75JOC3510 P. Y. Johnson and C. E. Hatch, III; *J. Org. Chem.*, 1975, **40**, 3510. 262

75JOC3602 G. M. Coppola, G. E. Hardtmann and R. I. Mansukhani; *J. Org. Chem.*, 1975, **40**, 3602. 515, 516

75JOC3743 C. H. Foster and G. A. Berchtold; *J. Org. Chem.*, 1975, **40**, 3743. 569

75JOC3874 G. J. B. Cajipe, G. Landen, B. Semler and H. W. Moore; *J. Org. Chem.*, 1975, **40**, 3874. 527

75JOC3878 C. Belzecki and D. Mostowicz; *J. Org. Chem.*, 1975, **40**, 3878. 200

75JOM(94)75 A. Albini and H. Kisch; *J. Organomet. Chem.*, 1975, **94**, 75. 220

75JOU583 R. R. Kostikov, A. F. Khlebnikov and K. A. Ogloblin; *J. Org. Chem. USSR* (*Engl. Transl.*), 1975, **11**, 583. 73

75JOU1301 V. N. Drozd and V. V. Sergeichuk; *J. Org. Chem. USSR* (*Engl. Transl.*), 1975, **11**, 1301. 445

75JOU2660 E. P. Nakova, O. N. Tolkachev and R. P. Evstigneeva; *J. Org. Chem. USSR* (*Engl. Transl.*), 1975, **11**, 2660. 488

75JPC1990 G. M. Breuer, R. S. Lewis and E. K. C. Lee; *J. Phys. Chem.*, 1975, **79**, 1990. 373

75JSP(55)464 T. K. Avirah, R. L. Cook and T. B. Malloy, Jr.; *J. Mol. Spectrosc.*, 1975, **55**, 464. 405

75JST(24)85 E. Haloui and D. Canet; *J. Mol. Struct.*, 1975, **24**, 85. 137

75KGS748 L. B. Volodarskii and L. A. Tikhonova; *Khim. Geterotsikl. Soedin.*, 1975, **11**, 748. 468

75LA2195 L. Birkofer and J. Schramm; *Liebigs Ann. Chem.*, 1975, 2195. 247, 256

B-75MI50100 L. J. Bellamy; 'The Infrared Spectra of Complex Molecules', Chapman & Hall, London, 1975. 12

B-75MI50101 L. J. Bellamy; 'Advances in IR Group Frequencies', Chapman & Hall, London, 1975. 12

B-75MI50400 L. J. Bellamy; 'The Infrared Spectra of Complex Molecules', Wiley, New York, 3rd edn., 1975. 50

B-75MI50500 S. G. Wilkinson; in 'Aliphatic Compounds', ed. N. B. Chapman; vol. 2 of 'MTP International Review of Science, Organic Chemistry Series Two', Butterworth, London, 1975. 97

B-75MI50501 N. I. Sax; 'Dangerous Properties of Industrial Materials', Van Nostrand Reinhold, New York, 4th edn., 1975. 97, 119

B-75MI50502 C. J. Pouchert and J. R. Campbell; 'The Aldrich Library of NMR Spectra', Aldrich Chemical Co., Milwaukee, 1975, vol. 1. 98

B-75MI50503 C. J. Pouchert; 'The Aldrich Library of Infrared Spectra', Aldrich Chemical Co., Milwaukee, 2nd edn., 1975. 99

B-75MI50504 B. M. Trost and L. S. Melvin, Jr.; 'Sulfur Ylides', Academic Press, New York, 1975. 103, 115

75MI50600 N. Spassky, P. Dumas, M. Sepulchre and P. Sigwalt; *J. Polym. Sci., Polym. Symp.*, 1975, **52**, 327. 136, 146, 157, 178, 179

75MI51100 L. A. Mitscher, H. D. H. Showalter, K. Shirahata and R. L. Foltz; *J. Antibiot.*, 1975, **28**, 668. 302

75MI51101 P. J. Claes, J. Hoogmartens, G. Janssen and H. Vanderhaeghe; *Eur. J. Med. Chem.*, 1975, **10**, 573. 331

B-75MI51102 J. Cs. Jászberényi and E. T. Gunda; *Prog. Med. Chem.*, 1975, **4**, 395. 313, 338

75MI51300 K. Van der Meer and J. J. C. Mulder; *Theor. Chim. Acta*, 1975, **37**, 159. 375

75MI51301 J. Dijkstra; *Chem.-Biol. Interact.*, 1975, **10**, 115 (*Chem. Abstr.*, 1975, **83**, 1934). 387

75MI51302 U. Grzegorzewska, M. Leplawy and A. Redlinski; *Rocz. Chem.*, 1975, **49**, 1859 (*Chem. Abstr.*, 1976, **84**, 135 059). 395, 398

75MI51400 K. C. Cole and D. F. R. Gilson; *Mol. Phys.*, 1975, **29**, 1749. 409, 419
75MI51401 D. Cornell and W. Tsang; *Int. J. Chem. Kinet.*, 1975, **7**, 799. 421
75MI51600 S. N. Frank, A. J. Bard and A. Ledwith; *J. Electrochem. Soc.*, 1975, **122**, 898. 519
B-75MI51601 K. Kahr, D. Werner and H. Fuchs; in 'Ullmanns Encyklopadie der Technischen Chemie', Verlag Chemie, Weinheim, 4th edn., 1975, vol. 9, p. 96. 528, 530, 545
75MI51700 G. Pruckmayr and T. K. Wu; *Macromolecules*, 1975, **8**, 954. 559, 561
75MI51800 R. Guryn, A. Kotelko and M. Majchrzak; *Acta Pol. Pharm.*, 1975, **32**, 421 (*Chem. Abstr.*, 1976, **85**, 46 604). 609
75MI51801 E. Schulte; *Dtsch. Apoth.-Ztg.*, 1975, **115**, 1253, 1828. 608, 613
75MI51802 S. Biniecki and M. Moll; *Acta Pol. Pharm.*, 1975, **32**, 1 (*Chem. Abstr.*, 1975, **83**, 147 459). 610
75MI51803 R. Glinka and B. Kotelko; *Acta Pol. Pharm.*, 1975, **32**, 525 (*Chem. Abstr.*, 1976, **85**, 32 970). 643
75MI51804 M. Spasova and E. Golovinskii; *Dokl. Bolg. Akad. Nauk*, 1975, **28**, 1387 (*Chem. Abstr.*, 1976, **84**, 120 723). 647
B-75MI51900 F. A. L. Anet and R. Anet; in 'Dynamic Nuclear Magnetic Resonance Spectroscopy', ed. L. M. Jackman and F. A. Cotton; Academic Press, New York, 1975, p. 543. 698
75MI52100 Y. M. Cohen and A. I. Popov; *J. Solution Chem.*, 1975, **4**, 599. 740, 743
75MI52200 R. Müller and D. Wöhrle; *Makromol. Chem.*, 1975, **176**, 2775. 766
75MI52201 K. Galuszko; *Rocz. Chem.*, 1975, **49**, 1597. 777
75MIP51300 T. Goetzen and M. Makosza; *Pol. Pat.* 77 159 (1975) (*Chem. Abstr.*, 1976, **85**, 21 100). 393
75MIP51500 Z. Bukac and J. Sebenda; *Czech. Pat.* 156 782 (1975) (*Chem. Abstr.*, 1975, **83**, 80 001). 484, 485
75MIP51501 Z. Bukac and J. Sebenda; *Czech. Pat.* 156 783 (1975) (*Chem. Abstr.*, 1975, **83**, 59 930). 484
75OMR(7)286 K. Pihlaja, J. Jokisaari, P. O. I. Virtanen, H. Ruotsalainen and M. Anteunis; *Org. Magn. Reson.*, 1975, **7**, 286. 367
75OMR(7)554 D. F. Ewing, K. A. Holbrook and R. A. Scott; *Org. Magn. Reson.*, 1975, **7**, 554. 366, 367
75OMS(10)992 H. Budzikiewicz and U. Lenz; *Org. Mass Spectrom.*, 1975, **10**, 992. 501
75PAC(44)691 A. G. Anastassiou; *Pure Appl. Chem.*, 1975, **44**, 691. 710, 716, 724
75PAC(44)767 M. J. S. Dewar; *Pure Appl. Chem.*, 1975, **44**, 767. 623
75PAC(44)925 G. Schroder; *Pure Appl. Chem.*, 1975, **44**, 925. 719
75PNA2999 B. G. Spratt; *Proc. Natl. Acad. Sci. USA*, 1975, **72**, 2999. 297
75RCR138 A. V. Fokin and A. F. Kolomiets; *Russ. Chem. Rev.* (*Engl. Transl.*), 1975, **44**, 138.
75RTC1 F. C. V. Larsson and S.-O. Lawesson; *Recl. Trav. Chim. Pays-Bas*, 1975, **94**, 1. 649
75S20 T. Nishiwaki; *Synthesis*, 1975, 20. 89
75S55 E. Vilsmaier and G. Becker; *Synthesis*, 1975, 55. 161
75S246 D. Olliéro and G. Solladié; *Synthesis*, 1975, 246. 675
75S250 K. Burger, F. Manz and A. Braun; *Synthesis*, 1975, 250. 266
75S333 P. Del Buttero and S. Maiorana; *Synthesis*, 1975, 333. 412, 431, 445
75S463 R. Richter and H. Ulrich; *Synthesis*, 1975, 463. 474
75S547 A. K. Mukerjee and A. K. Singh; *Synthesis*, 1975, 547. 249, 251, 252, 253, 254, 265, 266, 267, 303
75S603 D. J. Harris, G. Y.-P. Kan, V. Snieckus and O. Buchardt; *Synthesis*, 1975, 603. 602
75S720 K. Mori, H. Hashimoto, Y. Takenaka and T. Takigawa; *Synthesis*, 1975, 720. 623
75S765 K. P. C. Vollhardt; *Synthesis*, 1975, 765. 727, 769
75SA(A)1509 S. F. Bush, M. M. Allen and W. C. Harris; *Spectrochim. Acta, Part A*, 1975, **31**, 1509. 202
75T1991 M. Shamma and L. Toke; *Tetrahedron*, 1975, **31**, 1991. 544
75T2321 R. J. Stoodley; *Tetrahedron*, 1975, **31**, 2321. 249
75T2513 M. G. Missakian, B. J. Burreson and P. J. Scheuer; *Tetrahedron*, 1975, **31**, 2513. 764
75T2551 R. Helder and H. Wynberg; *Tetrahedron*, 1975, **31**, 2551. 776
75TL1335 G. Kretschmer and R. N. Warrener; *Tetrahedron Lett.*, 1975, 1335. 273, 277
75TL1499 D. Hellwinkel and M. Sapp; *Tetrahedron Lett.*, 1975, 1499. 634
75TL1639 Y. Kobayashi, I. Kumadaki, A. Ohsawa and Y. Sekine; *Tetrahedron Lett.*, 1975, 1639. 187
75TL1895 I. Murata, K. Nakasuji and Y. Nakajima; *Tetrahedron Lett.*, 1975, 1895. 624
75TL2235 D. N. Jones, D. R. Hill and D. A. Lewton; *Tetrahedron Lett.*, 1975, 2235. 571
75TL2363 H. Hart and S.-M. Chen; *Tetrahedron Lett.*, 1975, 2363. 125
75TL2463 J. C. Arnould and J. P. Pete; *Tetrahedron Lett.*, 1975, 2463. 345
75TL2603 G. Capozzi, O. Di Lucchi V. Lucchini and G. Modena; *Tetrahedron Lett.*, 1975, 2603. 135
75TL2613 C. A. Grob, W. Kunz and P. R. Marbet; *Tetrahedron Lett.*, 1975, 2163. 657
75TL2697 I. Murata and T. Tatsuoka; *Tetrahedron Lett.*, 1975, 2697. 559, 589
75TL2709 D. Van Ende and A. Krief; *Tetrahedron Lett.*, 1975, 2709. 147, 173
75TL2979 R. Jaquier, J.-L. Olive, C. Petrus and F. Petrus; *Tetrahedron Lett.*, 1975, 2979. 642
75TL3187 K. Oda, A. Ichihara and S. Sakamura; *Tetrahedron Lett.*, 1975, 3187. 192
75TL3223 K. Burger, A. Meffert, J. Linder, P. Hagn, A. Gieren and H. Siebels; *Tetrahedron Lett.*, 1975, 3223. 627
75TL4089 P. Y. Johnson, N. R. Schmuff and C. E. Hatch, III; *Tetrahedron Lett.*, 1975, 4089. 262, 359

75TL4301 L. Golič, V. Kaucic, B. Stanovnik and M. Tisler; *Tetrahedron Lett.*, 1975, 4301. 643

75USP3914428 R. D. Wilbur and J. M. Pensack; *U.S. Pat.* 3 914 428 (1975) (*Chem. Abstr.*, 1976, **84**, 29 658). 485

75USP3915962 W. W. Brand and T. W. Drabb, Jr.; *U.S. Pat.* 3 915 962 (1975) (*Chem. Abstr.*, 1976, **84**, 121 802). 485

75USP3928293 G. A. Crosby; *U.S. Pat.* 3 928 293 (1975) (*Chem. Abstr.*, 1976, **84**, 106 499). 182

75USP3928382 R. W. Addor, J. B. Lovell and S. Kantor; *U.S. Pat.* 3 928 382 (1975) (*Chem. Abstr.*, 1976, **85**, 46 632). 485

75ZC19 K. Peseke; *Z. Chem.*, 1975, **15**, 19. 465

75ZN(B)822 E. V. Dehmlow; *Z. Naturforsch., Teil B*, 1975, **30**, 822. 277

75ZN(B)926 R. Kreher, H. Pawelczyk and W. Gerhardt; *Z. Naturforsch., Teil B*, 1975, **30**, 926. 511

75ZOR2221 V. N. Yandovskii and L. B. Koroleva; *Zh. Org. Khim.*, 1975, **11**, 2221. 220

76ACR52 P. de Mayo; *Acc. Chem. Res.*, 1976, **9**, 52. 437

76ACR371 A. Padwa; *Acc. Chem. Res.*, 1976, **9**, 371. 50, 55, 63

76ACR378 T. C. Bruice and P. Y. Bruice; *Acc. Chem. Res.*, 1976, **9**, 378. 188, 548, 564, 565, 566, 584

76ACR453 A. G. Anastassiou; *Acc. Chem. Res.*, 1976, **9**, 453. 710, 719, 727

76ACS(A)759 Z. Smith and R. Seip; *Acta Chem. Scand., Ser. A*, 1976, **30**, 759. 453, 455

76ACS(B)318 I. Lagerlund; *Acta Chem. Scand., Ser. B.*, 1976, **30**, 318. 358

76ACS(B)341 S. Gronowitz and T. Frejd; *Acta Chem. Scand., Ser. B*, 1976, **30**, 341. 777

76ACS(B)377 N. G. Johansson; *Acta Chem. Scand., Ser. B*, 1976, **30**, 377. 333

76ACS(B)773 B. S. Jensen, T. Petterson, A. Ronlan and V. D. Parker; *Acta Chem. Scand., Ser. B*, 1976, **30**, 773. 664

76AG(E)103 W. Ried, O. Mosinger and W. Schuckmann; *Angew. Chem., Int. Ed. Engl.*, 1976, **15**, 103. 461

76AG(E)224 F. E. Elhadi, W. D. Ollis and J. F. Stoddart; *Angew. Chem., Int. Ed. Engl.*, 1976, **15**, 224. 706

76AG(E)229 E. Vogel, H.-H. Klug and M. Schäfer-Ridder; *Angew. Chem., Int. Ed. Engl.*, 1976, **15**, 229. 191

76AG(E)242 H. Schildknecht, T. Wilz, F. Enzmann, N. Grund and M. Ziegler; *Angew. Chem., Int. Ed. Engl.*, 1976, **15**, 242. 441

76AG(E)297 N. Yasuoka, Y. Kai, N. Kasai, T. Tatsuoka and I. Murata; *Angew. Chem., Int. Ed. Engl.*, 1976, **15**, 297. 556, 558

76AG(E)432 S. Sommer; *Angew. Chem., Int. Ed. Engl.*, 1976, **15**, 432. 454

76AG(E)495 W. Beck, W. Danzer, A. T. Liu and G. Huttner; *Angew. Chem., Int. Ed. Engl.*, 1976, **15**, 495. 77

76AG(E)559 H. Prinzbach and C. Rücker; *Angew. Chem., Int. Ed. Engl.*, 1976, **15**, 559. 191

76AG(E)611 K. Nakasuji, K. Kawamura, T. Ishihara and I. Murata; *Angew. Chem., Int. Ed. Engl.*, 1976, **15**, 611. 588, 589

76AG(E)614 K.-D. Schmidt, R. Mews and O. Glemser; *Angew. Chem., Int. Ed. Engl.*, 1976, **15**, 614. 488, 489

76AG(E)686 C. Weitemeyer and A. de Meijere; *Angew. Chem., Int. Ed. Engl.*, 1976, **15**, 686. 191

76AHC(19)215 R. J. Kobylecki and A. McKillop; *Adv. Heterocycl. Chem.*, 1976, **19**, 215. 273

76AHC(S1)554 J. Elguero, C. Marzin, A. R. Katritzky and P. Linda; *Adv. Heterocycl. Chem. (Suppl. 1)*, 1976, 554. 503

76AJC2651 T. Duong, R. H. Prager, A. D. Ward and D. I. B. Kerr; *Aust. J. Chem.*, 1976, **29**, 2651. 510

76AJC2667 T. Duong, R. H. Prager, J. M. Tippett, A. D. Ward and D. I. B. Kerr; *Aust. J. Chem.*, 1976, **29**, 2667. 529, 545

76AP550 H. Klar; *Arch. Pharm. (Weinheim, Ger.)*, 1976, **309**, 550. 531

76AX(B)2171 H. L. Ammon, L. Fallon and L. A. Plastas; *Acta Crystallogr., Part B*, 1976, **32**, 2171. 133

76AX(B)2279 M. O. Boles and R. J. Girven; *Acta Crystallogr., Part B*, 1976, **32**, 2279. 300, 301

76BAU337 N. P. Gambaryan and E. A. Avetisyan; *Bull. Acad. Sci. USSR, Div. Chem. Sci.*, 1976, **25**, 337. 475

76BAU1547 A. S. Nakhmanovich, V. N. Elokhina, T. P. Shcherbinina and M. G. Voronkov; *Bull. Acad. Sci. USSR, Div. Chem. Sci.*, 1976, **25**, 1547. 456, 479

76BAU1913 A. M. Mirskova, G. G. Levkovskaya, I. D. Kalikhman, T. I. Vakul'skaya, V. A. Pestunovich and M. G. Voronkov; *Bull. Acad. Sci. USSR, Div. Chem. Sci.*, 1976, **25**, 1913. 456, 479

76BBR(73)465 J. N. Lowe, L. L. Ingraham, J. Alspach and R. Rasmussen; *Biochem. Biophys. Res. Commun.*, 1976, **73**, 465. 484

76BCJ825 T. Tatsuoka and I. Murata; *Bull. Chem. Soc. Jpn.*, 1976, **49**, 825. 576

76BCJ2491 T. Kitazume, T. Otaka, R. Takei and N. Ishikawa; *Bull. Chem. Soc. Jpn.*, 1976, **49**, 2491. 426, 437, 464

76BCJ2572 N. Torimoto, T. Shingaki and T. Nagai; *Bull. Chem. Soc. Jpn.*, 1976, **49**, 2572. 389

76BRP1448895 DDSA Pharmaceuticals Ltd., *Br. Pat.* 1 448 895 (1976) (*Chem. Abstr.*, 1977, **86**, 72 714). 284

76BSB787 A. Waefelaer, J. Pecker, A. Dubois and P. Poultier; *Bull. Soc. Chim. Belg.*, 1976, **85**, 787. 630

76BSF277 A. Le Berre, A. Etienne and B. Desmazieres; *Bull. Soc. Chim. Fr.*, 1976, 277. 479

76BSF455 Z. Graftieaux, G. Hazebroucq and J. Gardent; *Bull. Soc. Chim. Fr.*, 1976, 455. 517

76CB518 G. Vitt, E. Hädicke and G. Quinkert; *Chem. Ber.*, 1976, **109**, 518. 670

76CB814 R. Appel and M. Hastelburg; *Chem. Ber.*, 1976, **109**, 814. 91

76CB1002 E. Weber, W. Wieder and F. Vögtle; *Chem. Ber.*, 1976, **109**, 1002. 778

76CB1239 M. Schmidt and E. Weissflog; *Chem. Ber.*, 1976, **109**, 1239. 696

76CB2558 H. J. Schmitt, K. Weidenhammer and M. L. Ziegler; *Chem. Ber.*, 1976, **109**, 2558. 555, 574

76CB2743 D. Thon and W. Schneider; *Chem. Ber.*, 1976, **109**, 2743. 510, 527, 528

76CB2823 H. Prinzbach, D. Stusche, M. Breuninger and J. Markert; *Chem. Ber.*, 1976, **109**, 2823. 582, 584, 669

76CB2928 R. R. Schmidt and R. Berrer; *Chem. Ber.*, 1976, **109**, 2928. 624

76CB3375 H. Bestmann, O. Vostrowsky and W. Stansky; *Chem. Ber.*, 1976, **109**, 3375. 385

76CB3505 H. Prinzbach, D. Stusche, J. Markert and H.-H. Limbach; *Chem. Ber.*, 1976, **109**, 3505. 500, 501, 522, 542, 661

76CB3707 K. Heyns and H. Buchholz; *Chem. Ber.*, 1976, **109**, 3707. 563

76CC36 G. Jones, II and L. P. McDonnell; *J. Chem. Soc., Chem. Commun.*, 1976, 36. 369

76CC110 H. Booth, B. W. Bycroft, C. M. Wels, K. Corbett and A. P. Maloney; *J. Chem. Soc., Chem. Commun.*, 1976, 110. 329

76CC125 C. W. Rees, R. Somanathan, R. C. Storr and A. D. Woolhouse; *J. Chem. Soc., Chem. Commun.*, 1976, 125. 281

76CC191 H. Alper and J. E. Prickett; *J. Chem. Soc., Chem. Commun.*, 1976, 191. 76

76CC276 D. Olliéro, J.-M. Ruxer, A. Solladié-Covallo and G. Solladié, *J. Chem. Soc., Chem. Commun.*, 1976, 276. 675

76CC288 R. A. Firestone and B. G. Christensen; *J. Chem. Soc., Chem. Comun.*, 1976, 288. 324

76CC295 J. Dale and K. Daasvatn; *J. Chem. Soc., Chem. Commun.*, 1976, 295 748

76CC313 A. G. Anastassiou and E. Reichmanis; *J. Chem. Soc., Chem. Commun.*, 1976, 313. 729

76CC407 J. S. Chapman, J. W. Cooper and B. P. Roberts; *J. Chem. Soc., Chem. Commun.*, 1976, 407. 431

76CC411 C. W. Rees, R. C. Storr and P. J. Whittle; *J. Chem. Soc., Chem. Commun.*, 1976, 411. 284, 682

76CC470 O. O. Oraji and R. A. Corral; *J. Chem. Soc., Chem. Commun.*, 1976, 470. 632

76CC533 B. Metz, J. M. Rosalky and R. Weiss; *J. Chem. Soc., Chem. Commun.*, 1976, 533. 738

76CC567 W. T. Flowers, R. Franklin, R. N. Haszeldine and R. J. Perry; *J. Chem. Soc., Chem. Commun.*, 1976, 567. 474

76CC621 J. P. Behr, J. M. Lehn and P. Vierling; *J. Chem. Soc., Chem. Commun.*, 1976, 621. 742

76CC667 J. E. Baldwin and D. P. Hesson; *J. Chem. Soc., Chem. Commun.*, 1976, 667. 172

76CC729 N.-C. C. Yang, B. Kim, W. Chiang and T. Hamada; *J. Chem. Soc., Chem. Commun.*, 1976, 729. 270, 274

76CC844 D. J. Harris and V. Snieckus; *J. Chem. Soc., Chem. Commun.*, 1976, 844. 601

76CC1005 R. D. Chambers and J. R. Maslakiewicz; *J. Chem. Soc., Chem. Commun.*, 1976, 1005. 275, 283, 682

76CCC562 O. Exner, V. Jehlicka and A. Dondoni; *Collect. Czech. Chem. Commun.*, 1976, **41**, 562. 475

76CI(L)1067 M. Ogata and H. Matsumoto; *Chem. Ind. (London)*, 1976, 1067. 613

76CJC1571 J. C. Pommelet and J. Chuche; *Can. J. Chem.*, 1976, **54**, 1571. 65, 496

76CJC2652 J. F. King and D. R. K. Harding; *Can. J. Chem.*, 1976, **54**, 2652. 477

76CJC3012 N. K. Sharma, F. De Reinach-Hirtzbach and T. Durst; *Can. J. Chem.*, 1976, **54**, 3012. 696

76CL47 M. Kakimoto, S. Kajigaeshi and S. Kanemasa; *Chem. Lett.*, 1976, 47. 79

76CL1379 N. Murai, M. Komatsu, Y. Ohshiro and T. Agawa, *Chem. Lett.*, 1976, 1379. 650

76CPB832 M. Nobuhara, H. Tazima, K. Shudo, A. Itai, T. Okamoto and Y Iitaka; *Chem. Pharm. Bull.*, 1976, **24**, 832. 192

76CPB2219 Y. Kobayashi, A. Ohsawa, M. Baba, T. Sato and I. Kumadaki; *Chem. Pharm. Bull.*, 1976, **24**, 2219. 360, 361

76CPB2751 K. Kawashima and Y. Kawano; *Chem. Pharm. Bull.*, 1976, **24**, 2751. 518, 519

76CRV113 P. G. Sammes; *Chem. Rev.*, 1976, **76**, 113. 249, 303, 305, 331

76CSR149 P. D. Bartlett; *Chem. Soc. Rev.*, 1976, **5**, 149. 459, 461, 469

76CSR181 N. S. Isaacs; *Chem. Soc. Rev.*, 1976, **5**, 181. 249, 256, 261, 267

76CZ440 G. Seitz and H. Hoffmann; *Chem.-Ztg.*, 1976, **100**, 440. 445

76EGP122086 R. Ohme, H. Schroeder and J. Rusche; *East Ger. Pat.* 122 086 (1976) (*Chem. Abstr.*, 1977, **87**, 52 771). 161

76FES731 E. Abignente, F. Arena and P. DeCaprariis; *Farmaco Ed. Sci.*, 1976, **31**, 731. 770

76FRP2341309 M. Ravard; *Fr. Demande* 2 341 309 (1977) (*Chem. Abstr.*, 1978, **89**, 48 893). 486

76H(4)989 K. Oe and O. Tsuge; *Heterocycles*, 1976, **4**, 989. 643

76H(4)1391 F. Bellamy and J. Streith; *Heterocycles*, 1976, **4**, 1391. 627, 628

76H(4)1509 M. Nastasi; *Heterocycles*, 1976, **4**, 1509. 595, 596, 597, 598, 599, 600, 601, 603, 604

76H(5)95 R. Abramovitch, K. M. More, I. Shinkai and C. P. Srinivasan; *Heterocycles*, 1976, **5**, 95. 632

76H(5)471 J. Streith and G. Wolff; *Heterocycles*, 1976, **5**, 471. 602

76HCA623 D. Berney and T. Jauner; *Helv. Chim. Acta*, 1976, **59**, 623. 517, 532

76HCA880 H. Itzel and H. Fischer; *Helv. Chim. Acta*, 1976, **59**, 880. 19

76HCA2059 D. Berney and K. Schuh; *Helv. Chim. Acta*, 1976, **59**, 2059. 531

76HCA2298 K. K. Prasad, G. Schmid and T. Petrzilka; *Helv. Chim. Acta*, 1976, **59**, 2298. 313

76IJC(B)701 N. P. Peet, S. Sunder and D. L. Trepanier; *Indian J. Chem., Sect. B*, 1976, **14**, 701. 637

76IZV1103 Yu. S. Shvetsov, A. N. Shirshov and V. S. Reznik; *Izv. Akad. Nauk SSSR, Ser. Khim.*, 1976, 1103 (*Chem. Abstr.*, 1976, **85**, 108 609). 766

76IZV1287 B. Tashkhodzhaev, L. G. Vorontsova and F. D. Alashev; *Izv. Akad. Nauk SSSR, Ser. Khim.*, 1976, 1287. 780

76IZV1898 R. G. Kostyanovskii, V. F. Rudchenko and G. V. Pustov; *Izv. Akad. Nauk SSSR, Ser. Khim.*, 1976, 1898. 201

76JA634 M. R. Demuth, P. E. Garrett and J. D. White; *J. Am. Chem. Soc.*, 1976, **98**, 634. 192

76JA899 J. L. Hencher, Q. Shen and D. G. Tuck; *J. Am. Chem. Soc.*, 1976, **98**, 899. 453

76JA1015 Y. Chao and D. J. Cram; *J. Am. Chem. Soc.*, 1976, **98**, 1015. 754

76JA1048 A. Padwa, A. Ku, A. Mazzu and S. I. Wetmore; *J. Am. Chem. Soc.*, 1976, **98**, 1048. 59

76JA1264 E. Block, R. E. Penn, R. J. Olsen and P. F. Sherwin; *J. Am. Chem. Soc.*, 1976, **98**, 1264. 421, 460

76JA1265 P. B. Dervan and M. A. Shippey; *J. Am. Chem. Soc.*, 1976, **98**, 1265. 112

76JA2005 I-Ching Chiu and K. G. Grohmann; *J. Am. Chem. Soc.*, 1976, **98**, 2005. 623

76JA2059 F. A. L. Anet, P. J. Degen and J. Krane; *J. Am. Chem. Soc.*, 1976, **98**, 2059. 700

76JA2219 A. H. Lawrence, C. C. Liao, P. de Mayo and V. Ramamurthy; *J. Am. Chem. Soc.*, 1976, **98**, 2219. 432

76JA2342 T. Kamiya, T. Teraji, M. Hashimoto, O. Nakaguchi and T. Oku; *J. Am. Chem. Soc.*, 1976, **98**, 2342. 307

76JA2344 J. N. Bonfiglio, I. Hasan, J. J. Piwinski, B. Weinstein and F. W. Fowler; *J. Am. Chem. Soc.*, 1976, **98**, 2344. 343

76JA2750 J. Aihara; *J. Am. Chem. Soc.*, 1976, **98**, 2750. 550

76JA2795 D. C. Dittmer, K. Takahashi, M. Iwanami, A. I. Tsai, P. L. Chang, B. B. Blidner and I. K. Stamos; *J. Am. Chem. Soc.*, 1976, **98**, 2795. 432

76JA2924 B. S. Campbell, N. J. De'Ath, D. B. Denney, D. Z. Denney, I. S. Kipnis and T. B. Min; *J. Am. Chem. Soc.*, 1976, **98**, 2924. 485

76JA3055 W. K. Musker and T. L. Wolford; *J. Am. Chem. Soc.*, 1976, **98**, 3055. 694

76JA3728 H. W. Moore, L. Hernandez and A. Sing; *J. Am. Chem. Soc.*, 1976, **98**, 3728. 262

76JA3769 D. Live and S. I. Chan; *J. Am. Chem. Soc.*, 1976, **98**, 3769. 739, 741

76JA3820 J. M. Figuera, P. D. Shevlin and S. D. Worley; *J. Am. Chem. Soc.*, 1976, **98**, 3820. 107

76JA4313 B. M. Trost and K. Hiroi; *J. Am. Chem. Soc.*, 1976, **98**, 4313. 624

76JA4325 J. A. Ross, R. P. Seiders and D. M. Lemal; *J. Am. Chem. Soc.*, 1976, **98**, 4325. 143, 152, 187

76JA4646 J. J. Eisch and J. E. Galle; *J. Am. Chem. Soc.*, 1976, **98**, 4646. 113

76JA5040 S. Kukolja, S. R. Lammert, M. R. B. Gleissner and A. I. Ellis; *J. Am. Chem. Soc.*, 1976, **98**, 5040.. 307

76JA5082 K. L. Williamson and J. D. Roberts; *J. Am. Chem. Soc.*, 1976, **98**, 5082. 498, 499, 703

76JA5697 G. W. Griffin, K. Ishikawa and I. J. Lev; *J. Am. Chem. Soc.*, 1976, **98**, 5697. 100

76JA5714 A. M. Jeffrey, K. W. Jennette, S. H. Blobstein, I. B. Weinstein, F. A. Beland, R. G. Harvey, H. Kasai, I. Miura and K. Nakanishi; *J. Am. Chem. Soc.*, 1976, **98**, 5714. 188

76JA5715 E. Block, E. R. Corey, R. E. Penn, T. L. Renken and P. F. Sherwin; *J. Am. Chem. Soc.*, 1976, **98**, 5715. 454, 456

76JA6313 R. V. Stevens, J. M. Fitzpatrick, P. B. Germeraad, B. L. Harrison and R. Lapalme; *J. Am. Chem. Soc.*, 1976, **98**, 6313. 75

76JA6350 W. H. Rastetter; *J. Am. Chem. Soc.*, 1976, **98**, 6350. 569

76JA6416 B. M. Jennings and M. T. H. Liu; *J. Am. Chem. Soc.*, 1976, **98**, 6416. 221

76JA6643 J. Meinwald, S. Knapp, S. K. Obendorf and R. E. Hughes; *J. Am. Chem. Soc.*, 1976, **98**, 6643. 408, 415, 418, 424, 426, 428, 429, 441, 444

76JA6696 B. Czauderna, K. H. Jogun, J. J. Stezowski and B. Fohlisch; *J. Am. Chem. Soc.*, 1976, **98**, 6696. 432

76JA6720 M. Koreeda, P. D. Moore, H. Yagi, H. J. C. Yeh and D. M. Jerina; *J. Am. Chem. Soc.*, 1976, **98**, 6720. 188

76JA6745 W. K. Musker and P. R. Rouch; *J. Am. Chem. Soc.*, 1976, **98**, 6745. 694

76JA6758 N. J. Turro, V. Ramamurthy, K. C. Liu, A. Krebs and R. Kemper; *J. Am. Chem. Soc.*, 1976, **98**, 6758. 574

76JA6798 N. L. Allinger and D. Y. Chung; *J. Am. Chem. Soc.*, 1976, **98**, 6798. 364

76JA7081 A. G. Hortmann and A. Bhattacharjya; *J. Am. Chem. Soc.*, 1976, **98**, 7081. 153, 422

76JA7698 A. Hassner, R. M. Cory and N. Sartoris; *J. Am. Chem. Soc.*, 1976, **98**, 7898. 39

76JA7864 T. S. Chou, G. A. Koppel, D. E. Dorman and J. W. Paschal; *J. Am. Chem. Soc.*, 1976, **98**, 7864. 267, 309

76JA7874 S. Masamune, Y. Hayase, W. K. Chan and R. L. Sobczak; *J. Am. Chem. Soc.*, 1976, **98**, 7874. 394

76JA8266 A. G. Anastassiou and E. Reichmanis; *J. Am. Chem. Soc.*, 1976, **98**, 8266. 710

76JAP(K)7608275 A. Matsuyama, A. Tahara, Y. Okazawa, T. Mitsuoka, K. Igarashi, T. Mizutani, C. Kaneuchi and S. Kawabata; *Jpn. Kokai*, 76 08 275 (1976) (*Chem. Abstr.*, 1976, **85**, 5707). 768

76JCP(64)2062 P. A. Clark and L. W. Pickett; *J. Chem. Phys.*, 1976, **64**, 2062. 13, 364, 367

76JCS(P1)362 A. A. Reid, H. R. Sood and J. T. Sharp; *J. Chem. Soc., Perkin Trans. 1*, 1976, 362. 599

76JCS(P1)447 E. G. Brain, A. J. Eglington, J. H. C. Nayler, M. J. Pearson and R. Southgate; *J. Chem. Soc., Perkin Trans. 1*, 1976, 447. 265

76JCS(P1)704 E. Roets, A. Vlietinck and H. Vanderhaeghe; *J. Chem. Soc., Perkin Trans. 1*, 1976, 704. 319

76JCS(P1)913 R. P. Gellatly, W. D. Ollis and I. O. Sutherland; *J. Chem. Soc., Perkin Trans. 1*, 1976, 913. 706, 707

76JCS(P1)1048 P. H. J. Ooms, H. W. Scheeren and R. J. F. Nivard; *J. Chem. Soc., Perkin Trans. 1*, 1976, 1048. 396

76JCS(P1)1279 A. W. H. Wardrop, G. L. Sainsbury, J. M. Harrison and T. D. Inch; *J. Chem. Soc., Perkin Trans. 1*, 1976, 1279. 517, 525, 532, 629

76JCS(P1)1291 K. Brewster, R. A. Chittenden, J. M. Harrison, T. D. Inch and C. Brown; *J. Chem. Soc., Perkin Trans. 1*, 1976, 1291. 525

76JCS(P1)1331 J. B. Taylor and W. R. Tully; *J. Chem. Soc., Perkin Trans. 1*, 1976, 1331. 606, 607, 608

76JCS(P1)1404 L. Carlsen, N. Harrit and A. Holm; *J. Chem. Soc., Perkin Trans. 1*, 1976, 1404. 196

76JCS(P1)1612 M. A. Harris, I. McMillan, J. H. C. Nayler, N. F. Osborne, M. J. Pearson and R. Southgate; *J. Chem. Soc, Perkin Trans. 1*, 1976, 1612. 265

76JCS(P1)1615 J. H. C. Nayler, N. F. Osborne, M. J. Pearson and R. Southgate; *J. Chem. Soc., Perkin Trans. 1*, 1976, 1615. 309

76JCS(P1)1772 C. R. Harrison and P. Hodge; *J. Chem. Soc., Perkin Trans. 1*, 1976, 1772. 302, 305

76JCS(P1)1918 D. H. Bremner, M. M. Campbell and G. Johnson; *J. Chem. Soc., Perkin Trans. 1*, 1976, 1918. 323

76JCS(P1)2243 I. Yamamoto, A. Mamba and H. Gotoh; *J. Chem. Soc., Perkin Trans. 1*, 1976, 2243. 649

76JCS(P1)2281 N. Dennis, A. R. Katritzky and M. Ramaiah; *J. Chem. Soc., Perkin Trans. 1*, 1976, 2281. 673

76JCS(P1)2590 R. J. Bushby; *J. Chem. Soc., Perkin Trans 1*, 1976, 2590. 423

76JCS(P2)36 D. E. Penny; *J. Chem. Soc., Perkin Trans. 2*, 1976, 36. 189

76JCS(P2)423 B. Metz, D. Moras and R. Weiss; *J. Chem. Soc., Perkin Trans. 2*, 1976, 423. 738

76JCS(P2)836 G. M. Parkinson, J. M. Thomas, J. O. Williams, M. J. Gorringe and L. W. Hobbs; *J. Chem. Soc., Perkin Trans. 2*, 1976, 836. 697

76JCS(P2)1044 A. J. Dobbs, B. C. Gilbert, H. A. H. Laue and D. O. C. Norma; *J. Chem. Soc., Perkin Trans. 2*, 1976, 1044. 19

76JCS(P2)1678 A. Ducruix, C. Pascard, S. David and J.-C. Fischer; *J. Chem. Soc., Perkin Trans. 2*, 1976, 1678. 366

76JGU799 V. V. Zharkov, M. I. Bakhitov and L. A. Ivanova; *J. Gen. Chem. USSR (Engl. Transl.)*, 1976, **46**, 799. 458, 474

76JHC269 A. Zirnis, F. F. Piszkiewicz and A. A. Manian; *J. Heterocycl. Chem.*, 1976, **13**, 269. 525

76JHC433 R. I. Fryer, J. V. Earley, N. W. Gilman and W. Zolly; *J. Heterocycl. Chem.*, 1976, **13**, 433. 612

76JHC967 N. P. Peet and S. Sunder; *J. Heterocycl. Chem.*, 1976, **13**, 967. 637

76JHC1165 D. H. Kim; *J. Heterocycl. Chem.*, 1976, **13**, 1165. 618

76JHC1309 P. Catsoulacos and C. Camoutsis; *J. Heterocycl. Chem.*, 1976, **13**, 1309. 632

76JMC1345 C. R. Ellefson; *J. Med. Chem.*, 1976, **19**, 1345. 526

76JOC49 C. A. L. Filgueiras and J. E. Huheey; *J. Org. Chem.*, 1976, **41**, 49. 375, 376

76JOC543 A. Padwa, J. Smolanoff and A. Tremper; *J. Org. Chem.*, 1976, **41**, 543. 67, 506

76JOC831 L. A. Wendling and R. G. Bergman; *J. Org. Chem.*, 1976, **41**, 831. 66

76JOC1112 D. F. Sullivan, D. I. C. Scopes, A. F. Kluge and J. A. Edwards; *J. Org. Chem.*, 1976, **41**, 1112. 265

76JOC1735 A. I. Meyers and M. E. Ford; *J. Org. Chem.*, 1976, **41**, 1735. 172

76JOC2362 G. Jones, II and L. J. Turbini; *J. Org. Chem.*, 1976, **41**, 2362. 504

76JOC2527 A. F. Noels, J. J. Herman and P. Teyssie; *J. Org. Chem.*, 1976, **41**, 2527. 378

76JOC2536 G. R. Newkome and J. M. Robinson; *J. Org. Chem.*, 1976, **41**, 2536. 769

76JOC2561 R. Busson and H. Vanderhaeghe; *J. Org. Chem.*, 1976, **41**, 2561. 315

76JOC2720 R. Y. Ning, R. I. Fryer, P. B. Madan and B. C. Gluboski; *J. Org. Chem.*, 1976, **41**, 2720. 618

76JOC2732 S. Sunder, N. P. Peet and D. L. Trepanier; *J. Org. Chem.*, 1976, **41**, 2732. 637

76JOC2736 R. W. Lieby and N. D. Heindel; *J. Org. Chem.*, 1976, **41**, 2736. 637

76JOC2808 P. H. Mazzocchi, T. Halchak and H. J. Tamburin; *J. Org. Chem.*, 1976, **41**, 2808. 357, 359

76JOC2813 C. A. Renner and F. D. Greene; *J. Org. Chem.*, 1976, **41**, 2813. 218, 232

76JOC3044 D. C. Dittmer and T. R. Nelsen; *J. Org. Chem.*, 1976, **41**, 3044. 420, 423, 429, 430, 431, 433, 441

76JOC3058 B. A. Hess Jr., and L. J. Schaad; *J. Org. Chem.*, 1976, **41**, 3058. 278

76JOC3156 J. A. Moore, B. Staskun and J. F. Blount; *J. Org. Chem.*, 1976, **41**, 3156. 343

76JOC3221 A. T. Nielsen, D. W. Moore, R. L. Atkins, D. Mallory, J. DiPol and J. M. LaBerge; *J. Org. Chem.*, 1976, **41**, 3221. 213

76JOC3229 H. W. Heine, L. M. Baclawski, S. M. Bonser and G. D. Wachob; *J. Org. Chem.*, 1976, **41**, 3229. 214

76JOC3325 Y. Sato, H. Kojima and H. Shirai; *J. Org. Chem.*, 1976, **41**, 3325. 545

76JOC3381 F. P. Tsui, Y. H. Chang, T. M. Vogel and G. Zon; *J. Org. Chem.*, 1976, **41**, 3381. 536

76JOC3403 G. L'abbe, G. Verhelst, S. Toppet, G. S. D. King and J. Briers; *J. Org. Chem.*, 1976, **41**, 3403. 197

76JOC3583 E. J. Moriconi and C. F. Hummel; *J. Org. Chem.*, 1976, **41**, 3583. 357, 359

76JOC3925 Z. Lysenko and M. M. Joullié; *J. Org. Chem.*, 1976, **41**, 3925. 168

76JPR946 M. Schulz, G. West, U. Mueller and D. Henke; *J. Prakt. Chem.*, 1976, **318**, 946. 214

76JSP(59)28 D. Coffey, Jr. and M. V. Hershbergher; *J. Mol. Spectrosc.*, 1976, **59**, 28. 365, 366, 370

76JSP(60)179 P. J. Mjöberg W. M. Ralowski, S. O. Ljunggren and J. E. Bäckvall; *J. Mol. Spectrosc.*, 1976, **60**, 179. 188

76JST(32)1 L. L. Combs and M. Rossie, Jr.; *J. Mol. Struct.*, 1976, **32**, 1. 364

76LA336 A. Botta; *Liebigs Ann. Chem.*, 1976, 336. 271

76LA1783 H. Griengl and A. Bleikolm; *Liebigs Ann. Chem.*, 1976, 1783. 631

76LA1792 H. Griengl and A. Bleikolm; *Liebigs Ann. Chem.*, 1976, 1792. 631

76M1251 R. Neidlein and W. D. Ober; *Monatsh. Chem.*, 1976, **107**, 1251. 638

B-76MI50200 R. M. Acheson; 'Introduction to the Chemistry of Heterocyclic Compounds', Wiley, New York, 1976. 25

B-76MI50400 R. M. Acheson; 'An Introduction to the Chemistry of Heterocyclic Compounds', Wiley, New York, 3rd edn., 1976, p. 5. 49

76MI50600 M. Sepulchre, N. Spassky and P. Sigwalt; *Isr. J. Chem.*, 1976, **15**, 33. 157

76MI50700 I. B. Weinstein, A. M. Jeffrey, K. W. Jennette, S. H. Blobstein, R. G. Harvey, C. Harris, H. Autrup, H. Kasai and K. Nakanishi; *Science*, 1976, **193**, 592. 188

76MI50800 W. E. Jones, J. S. Wasson and M. T. H. Liu; *J. Photochem.*, 1976, **5**, 311. 226

76MI50801 Anonymous, *Res. Discl.*, 1976, 147 (*Chem. Abstr.*, 1976, **85**, 94 755). 234

76MI51000 C. H. O'Callaghan, R. B. Sykes and S. E. Staniforth; *Antimicrob. Agents Chemother.*, 1976, **10**, 245. 288

76MI51100 H. Yanagisawa, M. Fukushima, A. Ando and H. Nakao; *J. Antibiot.*, 1976, **29**, 969. 322

76MI51200 J. S. Kahan, F .M. Kahan, R. Goegelman, S. A. Currie, M. Jackson, E. O. Stapley, T. W. Miller, A. K. Miller, D. Hendlin, S. Mochales, S. Hernandez and H. B. Woodruff; *16th Intersci. Conf. Antimicrob. Agents Chemother.*, 1976, Chicago, Ill., Abstr. No. 227. 349

76MI51300 Z. Zubovics and N. Ishikawa; *J. Fluorine Chem.*, 1976, **8**, 43. 398

76MI51301 P. Dreyfuss and M. P. Dreyfuss; *Polym. J.*, 1976, **8**, 81. 782

B-76MI51400 J. P. Marino; in 'Topics in Sulfur Chemistry', ed. A. Senning; Thieme, Stuttgart, 1976, vol. 1, p. 53. 446

B-76MI51600 'The Merck Index', ed. M. Windholz; Merck and Co., Rahway, 9th edn., 1976. 545

76MI51800 R. B. Moffet; *Lect. Heterocycl. Chem.*, 1976, **3**, 123. 608, 613, 614, 618

76MI51802 M. Bianchi, A. Butti, S. Rossi, F. Barzaghi and V. Marcaria; *Eur. J. Med. Chem. Chim. Ther.*, 1976, **11**, 101. 645, 651

76MI51803 M. Flammang and C. G. Wermuth; *Eur. J. Med. Chem. Chim. Ther.*, 1976, **11**, 83. 597

76MI51900 A. Munoz-Escalona and S. J. Guerrero; *Makromol. Chem.*, 1976, **177**, 2149. 691

76MI52100 B. E. Jepson and R. DeWitt; *J. Inorg. Nucl. Chem.*, 1976, **38**, 1175. 759

76MIP50800 V. N. Yandovskii, P. M. Adrov and I. A. Zamorina; *USSR Pat.* 523 092 (1976) (*Chem. Abstr.*, 1977, **86**, 29 825). 214

76MIP50801 D. S. Viswanath and V. Srihari; *Indian Pat.* 140 798 (1976) (*Chem. Abstr.*, 1980, **92**, 146 742). 230

76OMR(8)79 P. Mison, Y. Chaabouni, Y. Diab, R. Martino, A. Lopez, A. Lattes, F. Wehrli and T. Wirthlin; *Org. Magn. Reson.*, 1976, **8**, 79. 51

76OMR(8)90 P. Mison, R. Chaabouni, Y. Diab, A. Laurent, R. Martino, A. Lopez, F. Wehrli and T. Wirthlin; *Org. Magn. Reson.*, 1976, **8**, 90. 51

76OPP306 R. L. Amey and N. D. Heindel; *Org. Prep. Proced. Int.*, 1976, **8**, 306. 611

B-76PH41 G. W. Griffin and A. Padwa; in 'Photochemistry of Heterocyclic Compounds', ed. O. Buchardt; Wiley, New York, 1976, chapter 2, p. 41. 61

B-76PH93 G. W. Griffin and A. Padwa; in 'Photochemistry of Heterocyclic Compounds', ed. O. Buchardt; Wiley, New York, 1976, chap. 2, p. 93. 239

B-76PH288 V. Snieckus and D. J. Harris; in 'Photochemistry of Heterocyclic Compounds', ed. O. Buchardt; Wiley, New York, 1976, p. 288. 504

76PNA2679 H. W. S. King, M. R. Osborne, F. A. Beland, R. G. Harvey and P. Brookes; *Proc. Natl. Acad. Sci. USA*, 1976, **73**, 2679. 188

76RCR25 A. V. Fokin and A. F. Kolomiets; *Russ. Chem. Rev.* (*Engl. Transl.*), 1976, **45**, 25.

76RCR639 T. P. Vasil'eva, M. G. Lin'kova and O. V. Kil'disheva; *Russ. Chem. Rev.* (*Engl. Transl.*), 1976, **45**, 639. 477

76RTC153 R. H. Everhardus, R. Graefing and L. Brandsma; *Recl. Trav. Chim. Pays-Bas*, 1976, **95**, 153. 172

76S168 G. W. Gokel and H. D. Durst; *Synthesis*, 1976, 168. 756, 758

76S199 M. T. Reetz and M. Plackley; *Synthesis*, 1976, 199. 112

76S200 V. Calo, L. Lopez, A. Mincuzzi and G. Pesce; *Synthesis*, 1976, 200. 161

76S413 C. R. Johnson and K. Tanaka; *Synthesis*, 1976, 413. 173

76S689 M. S. Manhas, S. G. Amin, B. Ram and A. K. Bose; *Synthesis*, 1976, 689. 260

76SC465 F. Kienzle; *Synth. Commun.*, 1976, **6**, 465. 563

76T57 G. Doleschall, G. Hornyak, B. Agai, G. Simig, J. Fetter and K. Lempert; *Tetrahedron*, 1976, **32**, 57. 640

76T483 K. Peseke; *Tetrahedron*, 1976, **32**, 483. 465, 612

76T839 B. Agai, G. Doleschall, G. Hornyak, K. Lempert and G. Simig, *Tetrahedron*, 1976, **32**, 839. 640

76T1081 C. R. Ellefson, L. Swenton, R. H. Bible, Jr. and P. M. Green; *Tetrahedron*, 1976, **32**, 1081. 499

76T1345 T. Fueno and Y. Yonezawa; *Tetrahedron*, 1976, **32**, 1345. 517

76T1641 R. M. Kellogg, M. Noteboom and J. K. Kaiser; *Tetrahedron*, 1976, **32**, 1641. 180

76T1855 H. H. Wasserman, D. D. Keith and G. C. Rodgers; *Tetrahedron*, 1976, **32**, 1855. 764

76T1863 H. H. Wasserman, E. Gosselink, D. D. Keith, J. Nadelson and R. J. Sykes; *Tetrahedron*, 1976, **32**, 1863. 773

76T2165 R. M. Kellogg; *Tetrahedron*, 1976, **32**, 2165. 100

76T2211 T. Shioiri, Y. Yokoyama, Y. Kasai and S. Yamada; *Tetrahedron*, 1976, **32**, 2211. 330

76TL79 M. M. Htay and O. Meth-Cohn; *Tetrahedron Lett.*, 1976, 79. 765

76TL281 C.Cambillau, P. Sarthou and G. Brahm; *Tetrahedron Lett.*, 1976, 281. 758

76TL439 F. Flies, R. Lalande and B. Maillard; *Tetrahedron Lett.*, 1976, 439. 19, 559, 562
76TL457 D. Van Ende and A. Krief; *Tetrahedron Lett.*, 1976, 457. 115
76TL469 M. M. Htay and O. Meth-Cohn; *Tetrahedron Lett.*, 1976, 469. 766
76TL699 H. Hogeveen and B. J. Nusse; *Tetrahedron Lett.*, 1976, 699. 779
76TL971 R. G. Micetich; *Tetrahedron Lett.*, 1976, 971. 306
76TL975 R. G. Micetich, C. G. Chin and R. B. Morin; *Tetrahedron Lett.*, 1976, 975. 306
76TL1167 E. Vogel, D. Beermann, E. Balci and H. J. Altenbach; *Tetrahedron Lett.*, 1976, 1167. 552, 584
76TL1251 C. L. Semmelhack, I-Ching Chiu, and K. G. Grohmann; *Tetrahedron Lett.*, 1976, 1251. 623
76TL1317 A. J. Hubert, A. Feron, R. Warin and P. Teyssie; *Tetrahedron Lett.*, 1976, 1317. 93
76TL1449 P. F. Hudrlik, C.-N. Wan and G. P. Withers; *Tetrahedron Lett.*, 1976, 1449. 100
76TL2113 J. F. W. Keana and R. H. Morse; *Tetrahedron Lett.*, 1976, 2113. 621
76TL2115 W. C. Still; *Tetrahedron Lett.*, 1976, 2115. 393
76TL2459 I. Saito, A. Yazaki and T. Matsuura; *Tetrahedron Lett.*, 1976, 2459. 639
76TL2489 P. S. Marchand and J. F. Blount; *Tetrahedron Lett.*, 1976, 2489. 192
76TL2589 H. Alper and J. E. Prickett; *Tetrahedron Lett.*, 1976, 2589. 76
76TL2603 W. Wehner and F. Vögtle; *Tetrahedron Lett.*, 1976, 2603. 769
76TL3141 J.-P. Le Roux, P.-L. Desbene and M. Seguin; *Tetrahedron Lett.*, 1976, 3141. 532
76TL3931 C. W. Rees, R. C. Storr and P. J. Whittle; *Tetrahedron Lett.*, 1976, 3931. 283
76TL3975 A. Brandt, L. Bassignani and L. Re; *Tetrahedron Lett.*, 1976, 3975. 267, 309
76TL3999 H. Gotthardt and S. Nieberl; *Tetrahedron Lett.*, 1976, 3999. 437
76TL4003 I. Shahak, Y. Ittah and J. Blum; *Tetrahedron Lett.*, 1976, 4003. 91
76TL4025 R. D. Bach and J. M. Domagala; *Tetrahedron Lett.*, 1976, 4025. 364
76TL4053 H. Pietsch; *Tetrahedron Lett.*, 1976, 4053. 251
76TL4203 A. G. Brown, T. T. Howarth, I. Stirling and T. J. King; *Tetrahedron Lett.*, 1976, 4203. 350
76TL4205 A. R. Forrester and J. S. Sadd; *Tetrahedron Lett.*, 1976, 4205. 19, 202, 214
76TL4321 H. Kolbinger, G. Reissenweber and J. Sauer; *Tetrahedron Lett.*, 1976, 4321. 598
76TL4613 H. H. Wasserman and B. H. Lipshutz; *Tetrahedron Lett.*, 1976, 4613. 263
76TL4647 C. W. Rees, R. C. Storr and P. J. Whittle; *Tetrahedron Lett.*, 1976, 4647. 283
76TL4753 I. Saito, A. Yazaki and T. Matsuura; *Tetrahedron Lett.*, 1976, 4753. 639
76UKZ968 N. V. Kuznetsov and I. I. Krasatsev; *Ukr. Khim. Zh.* (*Russ. Ed.*), 1976, **42**, 968 (*Chem. Abstr.*, 1977, **86**, 16 648). 621
76UP51900 F. A. L. Anet and J. Krane; unpublished results. 666
76USP3954801 R. W. Addor and S. Kantor; *U.S. Pat.* 3 954 801 (1976) (*Chem. Abstr.*, 1976, **85**, 94 344). 485
76ZC16 K. Peseke; *Z. Chem.*, 1976, **16**, 16. 465
76ZN(B)586 K.-P. Zeller; *Z. Naturforsch., Teil B*, 1976, **31**, 586. 193
76ZN(B)1716. H. Wittmann, E. Ehrlich, H. Siegel and H. Sterk; *Z. Naturforsch., Teil B*, 1976, **31**, 1716. 544
76ZOR2331 K. A. Vyunov, T. I. Zhukova, E. G. Sochilin and N. A. Smorygo; *Zh. Org. Khim.*, 1976, **12**, 2331. 381

77ACS(B)701 P. Nedenskov and M. Mandrup; *Acta Chem. Scand., Ser. B*, 1977, **31**, 701. 619
77AG818 Y. Seki, S. Murai, I. Yamamoto and N. Sonoda; *Angew. Chem.*, 1977, **89**, 818. 383
77AG(E)109 H. Quast and E. Spiegel; *Angew. Chem., Int. Ed. Engl.*, 1977, **16**, 109. 219
77AG(E)169 E. Vogel, A. Breuer, C.-D. Sommerfeld, R. E. Davis and L.-K. Liu; *Angew. Chem., Int. Ed. Engl.*, 1977, **16**, 169. 191
77AG(E)171 K. Ishikawa and G. W. Griffin; *Angew. Chem., Int. Ed. Engl.*, 1977, **16**, 171. 191
77AG(E)177 H. Quast, K. H. Ross, E. Spiegel, K. Peters and H. G. von Schnering; *Angew. Chem., Int. Ed. Engl.*, 1977, **16**, 177. 219
77AG(E)185 R. W. Saalfrank; *Angew. Chem., Int. Ed. Engl.*, 1977, **16**, 185. 706
77AG(E)187 H. Irngartinger, H. L. Hase, K. W. Schulte and A. Schweig; *Angew. Chem., Int. Ed. Engl.*, 1977, **16**, 187. 555
77AG(E)349 H. J. Bestmann; *Angew. Chem., Int. Ed. Engl.*, 1977, **16**, 349. 267
77AG(E)475 G. L'abbe and S. Toppet; *Angew. Chem., Int. Ed. Engl.*, 1976, **16**, 475. 13
77AG(E)572 R. Huisgen; *Angew. Chem., Int. Ed. Engl.*, 1977, **16**, 572. 100
77AG(E)720 F. Schmidtchen; *Angew. Chem., Int. Ed. Engl.*, 1977, **16**, 720. 750
77AG(E)722 E. Schaumann and U. Behrens; *Angew. Chem., Int., Ed. Engl.*, 1977, **16**, 722. 132, 156
77AG(E)835 H. Meier and K.-P. Zeller; *Angew. Chem., Int. Ed. Engl.*, 1977, **16**, 835. 43
77AHC(21)245 D. N. Reinhoudt; *Adv. Heterocycl. Chem.*, 1977, **21**, 245. 38
77AHC(21)437 W. Adam, *Adv. Heterocycl. Chem.*, 1977, **21**, 437 455, 459, 469, 476, 485
77AJC179 R. F. C. Brown, F. W. Eastwood and G. L. McMullen; *Aust. J. Chem.*, 1977, **30**, 179. 128
77AJC2053 M. P. Makajan, S. M. Sondhi and N. K. Ralhan; *Aust. J. Chem.*, 1977, **30**, 2053. 638
77AP273 C. W. Hanefeld; *Arch. Pharm.* (*Weinheim, Ger.*), 1977, **310**, 273. 387
77AX(B)580 P. Blanpain, M. Melebeck and F. Durant; *Acta Crystallogr., Part B*, 1977, **33**, 580. 300
77AX(B)824 S. Ramakumar, K. Venkatesan and S. T. Rao; *Acta Crystallogr., Part B*, 1977, **33**, 824. 239
77BAU878 B. A. Abuzov, N. N. Zobova and N. R. Rubinova; *Bull. Acad. Sci. USSR, Div. Chem. Sci.*, 1977, **26**, 878. 474
77BCJ1179 T. Tanabe and T. Nagai; *Bull. Chem. Soc. Jpn.*, 1977, **50**, 1179. 439

77BCJ2013 M. Masaki, K. Fukui and J. Kita; *Bull. Chem. Soc. Jpn.*, 1977, **50**, 2013. 515, 525

77BCJ2153 O. Tsuge, K. Kamata and S. Yogi; *Bull. Chem. Soc. Jpn.*, 1977, **50**, 2153. 598

77BSB991 P. Crooy, R. DeNeys, J. Eliaers, R. Liveyns, G. Simonet and J. Vandevelde; *Bull. Soc. Chim. Belg.*, 1977, **86**, 991 330

77BSF893 C. Reby and J. Gardent; *Bull. Soc. Chim. Fr.*, 1977, 893, and references cited therein. 544

77C404 J. G. Schindler, R. Dennhardt and W. Simon; *Chimia*, 1977, **31**, 404. 760

77CB651 T. Kauffmann and R. Eidenschink; *Chem. Ber.*, 1977, **110**, 651. 73

77CB916 G. Diderrich, A. Haas and M. Yazdanbakhsch; *Chem. Ber.*, 1977, **110**, 916. 465, 477

77CB1225 G. Seybold and C. Heibl; *Chem. Ber.*, 1977, **110**, 1225. 460, 478

77CB1780 H. Quast and F. Kees; *Chem. Ber.*, 1977, **110**, 1780. 3

77CB2114 K. Burger, R. Ottlinger and J. Albanbauer; *Chem. Ber.*, 1977, **110**, 2114. 459, 483

77CB2242 E. Voigt and H. Meier; *Chem. Ber.*, 1977, **110**, 2242. 441

77CB2249 R. Kauffmann, A. Knöchel, J. Kopf, J. Oehler and G. Rudolph; *Chem. Ber.*, 1977, **110**, 2249. 736, 737

77CC207 N. F. Haley; *J. Chem. Soc., Chem. Commun.*, 1977, 207. 480

77CC302 C. Belzecki and Z. Krawczyk; *J. Chem. Soc., Chem. Commun.*, 1977, 302. 248, 266

77CC359 A. G. Brown, D. F. Corbett and T. T. Howarth; *J. Chem. Soc., Chem. Commun.*, 1977, 359. 353

77CC405 R. G. Alexander and R. Southgate; *J. Chem. Soc., Chem. Commun.*, 1977, 405. 353

77CC500 K. Itoh, M. Fukui and Y. Kurachi; *J. Chem. Soc., Chem. Commun.*, 1977, 500. 388

77CC523 A. G. Brown, D. F. Corbett, A. J. Eglington and T. T. Howarth; *J. Chem. Soc., Chem. Commun.*, 1977, 523. 351, 352

77CC539 M. J. S. Dewar and G. P. Ford; *J. Chem. Soc., Chem. Commun.*, 1977, 539. 360

77CC591 H. Numan, J. H. Wieringa, H. Wynberg, J. Hess and A. Vos; *J. Chem. Soc., Chem. Commun.*, 1977, 591. 452

77CC602 L. Wartski; *J. Chem. Soc., Chem. Commun.*, 1977, 602. 82

77CC664 K. Isomura, K. Uto and H. Taniguchi; *J. Chem. Soc., Chem. Commun.*, 1977, 664. 76

77CC713 J. C. Philips and O. Morales; *J. Chem. Soc., Chem. Commun.*, 1977, 713. 142

77CC720 A. J. Eglington; *J. Chem. Soc., Chem. Commun.*, 1977, 720. 309

77CC731 A. Forni, I. Moretti and G. Torre; *J. Chem. Soc., Chem. Commun.*, 1977, 731. 229

77CC806 R. N. Warrener, G. Kretschmer and M. N. Paddon-Row; *J. Chem. Soc., Chem. Commun.*, 1977, 806. 269, 272, 276, 283, 343

77CC808 D. F. Corbett, T. T. Howarth and I. Stirling; *J. Chem. Soc., Chem. Commun.*, 1977, 808. 351

77CC831 D. Moderhack and M. Lorke; *J. Chem. Soc., Chem. Commun.*, 1977, 831. 460

77CC873 D. R. Boyd, J. D. Neill and M. E. Stubbs; *J. Chem. Soc., Chem. Commun.*, 1977, 873. 554

77CC899 M. Somei, R. Kitamura, H. Fujui, K. Hashiba, S. Kawai and C. Kaneko; *J. Chem. Soc., Chem. Commun.*, 1977, 899. 628

77CC905 P. H. Bentley, G. Brooks, M. L. Gilpin and E. Hunt; *J. Chem. Soc., Chem. Commun.*, 1977, 905. 353

77CC939 T. S. Cameron and H. W. Scheeren; *J. Chem. Soc., Chem. Commun.*, 1977, 939. 699

77CC953 D. F. Corbett, A. J. Eglington and T. T. Howarth; *J. Chem. Soc., Chem. Commun.*, 1977, 953. 351

77CJC56 D. J. Harris, M. T. Thomas, V. Snieckus, N. Friedman, K. Schaumberg, K. B. Tomer and O. Buchardt; *Can J. Chem.*, 1977, **55**, 56. 360, 361, 600

77CJC484 T. W. Doyle, B. Belleau, B. Luh, T. T. Conway, M. Menard, J. L. Douglas, D. T. Chu, G. Lim, L. R. Morris, P. Rivest and M. Casey; *Can. J. Chem.*, 1977, **55**, 484. 355

77CJC895 E. Bullock, R. A. Carter, R. M. Cochrane, B. Gregory and D. C. Shields; *Can. J. Chem.*, 1977, **55**, 895. 606, 607

77CJC966 E. G. Lewars and G. Morrison; *Can. J. Chem.*, 1977, **55**, 966. 125

77CJC2719 T. W. Doyle, B. Luh, D. T. Chu and B. Belleau; *Can. J. Chem.*, 1977, **55**, 2719. 354

77CJC2867 P. Meunier and G. Pfister-Guillouzo; *Can. J. Chem.*, 1977, **55**, 2867. 555

77CJC3456 R. N. Renaud, J. W. Bovenkamp, R. R. Fraser and J. L. A. Roustan; *Can. J. Chem.*, 1977, **55**, 3556. 706

77CJC4061 B. Gregory, E. Bullock and Teng-Song Chen; *Can. J. Chem.*, 1977, **55**, 4061. 512

77CL103 K. Itoh, P. Yogo and Y. Ishi; *Chem. Lett.*, 1977, 103. 388

77CPB321 T. Kametani, T. Ohsawa, S. Hirata, M. S. Premila, M. Ihara and K. Fukumoto; *Chem. Pharm. Bull.*, 1977, **25**, 321. 544

77CPB1140 T. Yamaguchi and M. Kojima; *Chem. Pharm. Bull.*, 1977, **25**, 1140. 170, 171

77CPB2749 T. Tsuchiya, H. Arai, H. Hasegawa and H. Igeta; *Chem. Pharm. Bull.*, 1977, **25**, 2749. 604

77CPB2875 S. Kano, T. Yokomatsu and S. Shibuya; *Chem. Pharm. Bull.*, 1977, **25**, 2875. 528

77CR(C)(284)795 G. A. Bistocchi, G. De Méo and A. Ricci; *C.R. Hebd. Seances Acad. Sci., Ser. C*, 1977, **284**, 795. 625

77CRV473 S. Braslabsky and J. Heicklen; *Chem. Rev.*, 1977, **77**, 473. 100

77CRV513 G. R. Newkome, J. D. Sauer, J. M. Roper and D. C. Hager; *Chem. Rev.*, 1977, **77**, 513. 764, 767, 769, 770

77DOK(232)1081 R. G. Kostyanovskii and G. V. Shustov; *Dokl. Akad. Nauk SSSR*, 1977, **232**, 1081 (*Chem. Abstr.*, 1977, **87**, 23, 109). 7

77FES33 F. Gatta and S. Chiavarelli; *Farmaco Ed. Sci.*, 1977, **32**, 33. 613

77GEP2647936 S. Dewaele (Texaco Belgium); *Ger. Offen.* 2 647 936 (1977) (*Chem. Abstr.*, 1977, **87**, 68 866). 536

77GEP2710252 N. Tarumi, A. Tamura and M. Kokiso; *Ger. Offen.* 2 710 252 (1977) (*Chem. Abstr.*, 1977, **87**, 209 532). 484

77H(6)143 P. Gilgen, H. Heimgartner, H. Schmid and H. J. Hansen; *Heterocycles*, 1977, **6**, 143. 55, 56

77H(6)1563 H. Taniguchi, K. Isomura and T. Tanaka; *Heterocycles*, 1975, **6**, 1563. 66

77H(6)2021 J. Streith; *Heterocycles*, 1977, **6**, 2021. 595, 604

77H(7)37 H. Bieräugel, H.-P. Soetens and U. K. Pandit; *Heterocycles*, 1977, **7**, 37. 544

77H(7)81 A. W. Lee, P. M. Keehn, S. M. Ramos and S. M. Rosenfeld; *Heterocycles*, 1977, **7**, 81. 767, 771, 776, 778

77H(7)919 T. Kametani, K. Kigasawa, M. Huragi, N. Wagatsuma, T. Kohajisawa and H. Inoue; *Heterocycles*, 1977, **7**, 919. 626

77H(8)283 S. Yamamoto, N. Haga, T. Aoki, S. Hayashi, H. Tanida and W. Nagata; *Heterocycles*, 1977, **8**, 283. 306

77HCA141 J. M. Lehn and J. Simon; *Helv. Chim. Acta*, 977, **60**, 141. 744

77HCA1644 G. Gast, J. Schmutz and D. Sorg; *Helv. Chim. Acta*, 1977, **60**, 1644. 493, 532

77HCA2476 S. Chaloupka, P. Vittorelli, H. Heimgartner, H. Schmid, H. Link, K. Bernauer and W. E. Oberhänsli; *Helv. Chim. Acta*, 1977, **60**, 2476. 676

77HCA2872 J. B. Bream and J. Schmutz; *Helv. Chim. Acta*, 1977, **60**, 2872. 633

77JA258 K. Nakanishi, H. Kasai, H. Cho, R. G. Harvey, A. M. Jeffrey, K. W. Jennette and I. B. Weinstein; *J. Am. Chem. Soc.*, 1977, **99**, 258. 188

77JA629 C. H. Bushweller, J. A. Ross and D. M. Lemal; *J. Am. Chem. Soc.*, 1977, **99**, 629. 186

77JA714 V. Markowski, G. R. Sullivan and J. D. Roberts; *J. Am. Chem. Soc.*, 1977, **99**, 714. 51

77JA1214 C. G. Krespan, B. E. Smart and E. G. Howard; *J. Am. Chem. Soc.*, 1977, **99**, 1214. 477

77JA1272 N. Manisse and J. Chuche; *J. Am. Chem. Soc.*, 1977, **99**, 1272. 498, 540

77JA1514 A. Padwa and P. H. J. Carlsen; *J. Am. Chem. Soc.*, 1977, **99**, 1514. 52, 83

77JA1685 M. J. S. Dewar and G. P. Ford; *J. Am. Chem. Soc.*, 1977, **99**, 1685. 138

77JA1692 I. Gutman, M. Milun and N. Trinajstic; *J. Am. Chem. Soc.*, 1977, **99**, 1692. 492

77JA1871 A. Padwa and N. Kamigata; *J. Am. Chem. Soc.*, 1977, **99**, 1871. 59, 67

77JA2352 W. F. Huffman, K. G. Holden, T. F. Buckley, III, J. G. Gleason and L. Wu; *J. Am. Chem. Soc.*, 1977, **99**, 2352. 265, 354

77JA2591 C. Galli, G. Illuminati, L. Mandolini and P. Tamborra; *J. Am. Chem. Soc.*, 1977, **99**, 2591. 33

77JA2823 P. K. Wong, M. Madhavarao, D. F. Marten and M. Rosenblum; *J. Am. Chem. Soc.*, 1977, **99**, 2823. 259

77JA2866 D. F. Bocian and H. L. Strauss; *J. Am. Chem. Soc.*, 1977, **99**, 2866. 549, 550

77JA2876 D. F. Bocian and H. L. Strauss; *J. Am. Chem. Soc.*, 1977, **99**, 2876. 549

77JA2931 J. P. Snyder and L. Carlsen; *J. Am. Chem. Soc.*, 1977, **99**, 2931. 453

77JA3023 N. C. Yang, M. H. Hui, D. M. Shold, N. J. Turro, R. R. Hautala, K. Dawes and J. C. Dalton; *J. Am. Chem. Soc.*, 1977, **99**, 3023. 398

77JA3226 P. D. Mollere and K. N. Houk; *J. Am. Chem. Soc.*, 1977, **99**, 3226. 364, 368, 376, 419

77JA3870 H. Wynberg, H. Numan and H. P. J. M. Dekkers; *J. Am. Chem. Soc.*, 1977, **99**, 3870. 476

77JA4194 T. Kitazume and J. M. Shreeve; *J. Am. Chem. Soc.*, 1977, **99**, 4194. 456, 462, 481

77JA4511 P. Caramella, K. N. Houk and L. N. Domelsmith; *J. Am. Chem. Soc.*, 1977, **99**, 4511 38

77JA4835 F. Nakatsubo, A. J. Cocuzza, D. E. Keeley and Y. Kishi; *J. Am. Chem. Soc.*, 1977, **99**, 4835. 659

77JA5045 J. G. Smith and D. J. Mitchell; *J. Am. Chem. Soc.*, 1977, **99**, 5045. 512, 525

77JA5091 Y. Maki and M. Sako; *J. Am. Chem. Soc.*, 1977, **99**, 5091. 307

77JA5403 J.-Y. Koo and G. B. Schuster; *J. Am. Chem. Soc.*, 1977, **99**, 5403. 459

77JA5768 W. Adam, A. Alzerreca, J.-C. Liu and F. Yany; *J. Am. Chem. Soc.*, 1977, **99**, 5768. 455, 469

77JA5836 N. J. Turro, Y. Ito, M.-F. Chow, W. Adam, O. Rodriguez and F. Yany; *J. Am. Chem. Soc.*, 1977, **99**, 5836. 455, 476

77JA6049 I. Goldberg; *J. Am. Chem. Soc.*, 1977, **99**, 6049. 745

77JA6667 G. Gree and R. Carrie; *J. Am. Chem. Soc.*, 1977, **99**, 6667. 90

77JA6766 J. M. Lehn, S. H. Pine, E. I. Watanabe and A. K. Willard; *J. Am. Chem. Soc.*, 1977, **99**, 6766. 744

77JA6986 F. A. L. Anet and I. Yavari; *J. Am. Chem. Soc.*, 1977, **99**, 6986. 706

77JA7350 Y. Kobayashi, I. Kumadaki, A. Ohsawa and A. Ando; *J. Am. Chem. Soc.*, 1977, **99**, 7350. 187

77JA8115 F. Nakatsubo, T. Fukuyama, A. J. Cocuzza and Y. Kishi; *J. Am. Chem. Soc.*, 1977, **99**, 8115. 659

77JA8214 G. H. Posner and D. Z. Rogers; *J. Am. Chem. Soc.*, 1977, **99**, 8214. 568

77JA8367 R. R. Gagne, C. A. Koval and T. J. Smith; *J. Am. Chem. Soc.*, 1977, **99**, 8367. 722

77JA8370 L. D. Quin and E. D. Middlemas; *J. Am. Chem. Soc.*, 1977, **99**, 8370. 726, 730

77JAP(K)7771445 H. Umezawa, S. Umezawa, S. Seki, S. Fukatsu and S. Yasuda; *Jpn. Kokai* 77 71 445 (1977) (*Chem. Abstr.*, 1978, **88**, 23 343). 153

77JAP(K)77133952 K. Takahashi, K. Arimoto, R. Arai, Y. Nakazawa and T. Morinaga; *Jpn. Kokai* 77 133 952 (1977) (*Chem. Abstr.*, 1978, **88**, 120 680). 386

77JCP(66)32 M. Kobayashi; *J. Chem. Phys.*, 1977, **66**, 32. 701

77JCP(66)1901 J. E. Mark and D. S. Chiu; *J. Chem. Phys.*, 1977, **66**, 1901. 382

77JCR(S)244	K. Thimm, W. Schmueser and J. Voss; *J. Chem. Res. (S)*, 1977, 244.	478
77JCS(D)204	M. Green, S. M. Heathcock, T. W. Turney and D. M. P. Mingos; *J. Chem. Soc., Dalton Trans.*, 1977, 204.	520, 523
77JCS(F1)890	M. J. Clarke and K. A. Holbrook; *J. Chem. Soc., Faraday Trans. 1*, 1977, **73**, 890.	372
77JCS(P1)47	H. Suschitzky, R. E. Walrond and R. Hull; *J. Chem. Soc., Perkin Trans. 1*, 1977, 47.	466, 474
77JCS(P1)80	S. A. Bone, S. Trippett and P. J. Whittle; *J. Chem. Soc., Perkin Trans. 1*, 1977, 80.	485
77JCS(P1)247	H. Scott and R. Shilton; *J. Chem. Soc., Perkin Trans. 1*, 1977, 247	477
77JCS(P1)572	S. G. Davies and G. H. Whitham; *J. Chem. Soc., Perkin Trans. 1*, 1977, 572.	35
77JCS(P1)767	A. C. Coxon and J. F. Stoddart; *J. Chem. Soc., Perkin Trans. 1*, 1977, 767.	751
77JCS(P1)874	J. R. Malpass and N. J. Tweddle; *J. Chem. Soc., Perkin Trans. 1*, 1977, 874.	357, 359
77JCS(P1)1346	S. G. Davies and G. H. Whitham; *J. Chem. Soc., Perkin Trans. 1*, 1977, 1346.	569
77JCS(P1)1477	M. J. V. de Oliveira-Baptista, A. G. M. Barrett, D. H. R. Barton, M. Girijavallabhan, R. C. Jennings, J. Kelly, V. J. Papadimitriou, J. V. Turner and N. A. Usher; *J. Chem. Soc., Perkin Trans. 1*, 1977, 1477.	308, 331
77JCS(P1)1574	D. N. Jones, D. R. Hill, D. A. Lewton and C. Sheppard; *J. Chem. Soc., Perkin Trans. 1*, 1977, 1574.	423, 435
77JCS(P1)1822	R. Pellicciari and B. Natalini; *J. Chem. Soc., Perkin Trans. 1*, 1977, 1822.	586, 587
77JCS(P1)1824	W. S. Murphy and K. P. Raman; *J. Chem. Soc., Perkin Trans. 1*, 1977, 1824.	520
77JCS(P1)1901	S. E. Drewes and U. J. Upfold; *J. Chem. Soc., Perkin Trans. 1*, 1977, 1901.	611
77JCS(P1)2092	L. Garanti and G. Zecchi; *J. Chem. Soc., Perkin Trans. 1*, 1977, 2092.	597
77JCS(P1)2355	Y. Kobayashi, I. Kumadaki, A. Ohsawa, Y. Sekine and A. Ando; *J. Chem. Soc., Perkin Trans. 1*, 1977, 2355.	170
77JCS(P1)2357	R. V. Davies, B. Iddon, M. W. Pickering, H. Suschitzky, P. T. Gallagher, M. W. Gittos and M. D. Robinson; *J. Chem. Soc., Perkin Trans. 1*, 1977, 2357.	532
77JCS(P1)2457	D. J. Haywood and S. T. Reid; *J. Chem. Soc., Perkin Trans. 1*, 1977, 2457.	669
77JCS(P1)2479	E. G. Brain, A. J. Eglington, J. H. C. Nayler, N. F. Osborne, R. Southgate and P. Tolliday; *J. Chem. Soc., Perkin Trans. 1*, 1977, 2479.	333
77JCS(P2)547	H. Fujiwara, R. L. Varley and J. M. Van der Veen; *J. Chem. Soc., Perkin Trans. 2*, 1977, 547.	247
77JCS(P2)1019	V. M. Csizmadia, G. H. Schmid, P. G. Mezey and I. G. Csizmadia; *J. Chem. Soc., Perkin Trans. 2*, 1977, 1019.	139
77JCS(P2)1148	K. Takagi and Y. Ogata; *J. Chem. Soc., Perkin Trans 2*, 1977, 1148.	360
77JCS(P2)1214	G. F. Bradley, W. B. L. Evans and I. D. R. Stevens; *J. Chem. Soc., Perkin Trans. 2*, 1977, 1214.	221
77JCS(P2)1708	W. B. Giara and B. P. Roberts; *J. Chem. Soc., Perkin Trans. 2*, 1977, 1708.	431
77JCS(P2)1749	R. Mondelli and P. Ventura; *J. Chem. Soc., Perkin Trans. 2*, 1977, 1749.	302
77JCS(P2)2068	I. Perillo, B. Fernandez and S. Lamdan; *J. Chem. Soc., Perkin Trans. 2*, 1977, 2068.	673
77JHC99	C. Corral, R. Madronero and S. Vega; *J. Heterocycl. Chem.*, 1977, **14**, 99.	609
77JHC459	P. Tarburton, P. B. Woller, R. C. Badger, E. Doomes and N. H. Cromwell; *J. Heterocycl. Chem.*, 1977, **14**, 459.	81
77JHC465	G. Cignarella, R. Cerri, F. Savelli and A. Maselli; *J. Heterocycl. Chem.*, 1977, **14**, 465.	596
77JHC493	M. Vincent, G. Remond and J.-P. Volland; *J. Heterocycl. Chem.*, 1977, **14**, 493.	651
77JHC503	R. D. Carroll, S. Jung and C. G. Sklavounos; *J. Heterocycl. Chem.*, 1977, **14**, 503.	326
77JHC561	N. P. Peet and S. Sunder; *J. Heterocycl. Chem.*, 1977, **14**, 561.	769
77JHC933	M. J. Kukla; *J. Heterocycl. Chem.*, 1977, **14**, 933.	515, 518, 529
77JHC985	C. Corral, R. Madronero and S. Vega; *J. Heterocycl. Chem.*, 1977, **14**, 985.	619
77JHC1139	G. Stravopoulos and D. Theodoropoulos; *J. Heterocycl. Chem.*, 1977, **14**, 1139.	609
77JHC1393	B. C. Challis and A. M. Lobo; *J. Heterocycl. Chem.*, 1977, **14**, 1393.	200
77JHC1777	T. Nakano, W. Rodriguez, S. Z. deRoche, J. M. Larrauri, C. Rivas and C. Perez; *J. Heterocycl. Chem.*, 1977, **17**, 1777.	397
77JMR(28)243	D. Gilson, S. Katz and P. Saviotti; *J. Magn. Reson.*, 1977, **28**, 243.	367
77JOC575	L. Benati, P. C. Montevecchi and G. Zanardi; *J. Org. Chem.*, 1977, **42**, 575.	193
77JOC609	K. E. Gilbert; *J. Org. Chem.*, 1977, **42**, 609.	451
77JOC1251	D. E. Applequist and L. F. McKenzie; *J. Org. Chem.*, 1977, **42**, 3506.	436
77JOC1379	J. F. Haley, Jr., S. M. Rosenfeld and P. M. Keehn; *J. Org. Chem.*, 1977, **42**, 1379.	779
77JOC1496	K. Oe, M. Tashiro and O. Tsuge; *J. Org. Chem.*, 1977, **42**, 1496.	643
77JOC1500	G. R. Newkome, A. Nayak, G. L. McClure, F. Danesh-Khoshboo and J. Broussard-Simpson; *J. Org. Chem.*, 1977, **42**, 1500.	766
77JOC1856	T. Tsuchiya, J. Kurita and V. Snieckus; *J. Org. Chem.*, 1977, **42**, 1856.	598, 603, 604
77JOC1910	D. C. Dittmer, J. E. McCaskie, J. E. Babiarz and M. V. Ruggeri; *J. Org. Chem.*, 1977, **42**, 1910.	423
77JOC2008	B. A. Chiasson and G. A. Berchtold; *J. Org. Chem.*, 1977, **42**, 2008.	584
77JOC2127	R. K. Murray, Jr., J. S. Polley, S. S. Abdel-Meguid and V. W. Day; *J. Org. Chem.*, 1977, **42**, 2127.	141
77JOC2224	J. C. Sheehan, T. J. Commons and Y. S. Lo; *J. Org. Chem.*, 1977, **42**, 2224.	320
77JOC2345	M. Mikolajczyk, P. Kielbasinski, J. H. Barlow and D. R. Russell; *J. Org. Chem.*, 1977, **42**, 2345.	453, 456
77JOC2530	E. P. Papadopoulos and B. George; *J. Org. Chem.*, 1977, **42**, 2530.	605
77JOC2551	S. Sunder and N. P. Peet; *J. Org. Chem.*, 1977, **42**, 2551.	638
77JOC2651	A. G. Anastassiou, S. J. Girgenti, R. C. Griffith and E. Reichmanis; *J. Org. Chem.*, 1977, **42**, 2651.	713, 724

77JOC2821 U. Zehavi; *J. Org. Chem.*, 1977, **42**, 2821. 478

77JOC2887 A. Nudelman and R. J. McCaully; *J. Org. Chem.*, 1977, **42**, 2887. 306

77JOC2903 P. S. Mariano, M. E. Osborn, D. Dunaway-Mariano, B. C. Gunn and R. C. Petterson; *J. Org. Chem.*, 1977, **42**, 2903. 660

77JOC2960 F. DiNinno, T. R. Beattie and B. G. Christensen; *J. Org. Chem.*, 1977, **42**, 2960. 316, 317

77JOC3128 H. W. Scheeren, R. W. M. Aben, P. H. J. Ooms and R. J. F. Nivard; *J. Org. Chem.*, 1977, **42**, 3128. 396

77JOC3217 W. H. Pirkle and P. L. Rinaldi; *J. Org. Chem.*, 1977, **42**, 3217. 200

77JOC3506 J. E. Coates and F. S. Abbott; *J. Org. Chem.*, 1977, **42**, 3506. 426

77JOC3635 H. Hart, C. Peng and E. Shih; *J. Org. Chem.*, 1977, **42**, 3635. 102

77JOC3840 H. Hart, J. B.-C. Jiang and M. Sasaoka; *J. Org. Chem.*, 1977, **42**, 3840. 124, 125

77JOC4045 J. C. Sheehan, A. Buku, E. Chacko, T. J. Commons, Y. S. Lo, D. R. Ponzi and W. C. Schwarzel; *J. Org. Chem.*, 1977, **42**, 4045. 319

77JOC4238 T. S. Cantrell; *J. Org. Chem.*, 1977, **42**, 4238. 270, 274, 361

77JOU353 V. N. Drozd and V. V. Sergeichuk; *J. Org. Chem. USSR* (*Engl. Transl.*), 1977, **13**, 353. 445

77JPC1682 J. F. Chiang and K. C. Lu; *J. Phys. Chem.*, 1977, **81**, 1682. 453

77JPR195 E. Schmitz, R. Ohme, S. Schramm, H. Striegler, H. U. Heyne and J. Rusche; *J. Prakt. Chem.*, 1977, **319**, 195. 209, 229

77JPR274 E. Schmitz, H. Striegler, H. U. Heyne, K. P. Hilgetag, H. Dilcher and R. Lorenz; *J. Prakt. Chem.*, 1977, **319**, 274. 205, 212

77JPS605 R. W. Leiby and N. Heindel; *J. Pharm. Sci.*, 1977, **66**, 605. 638

77JPS861 T. Yamana, A. Tsuji, E. Kiya and E. Miyamoto; *J. Pharm. Sci.*, 1977, **66**, 861. 326

77JPS1653 M. A. Gharbia and M. M. Joullie; *J. Pharm. Sci.*, 1977, **66**, 1653. 73

77JPS1675 A. Tsuji, O. Kubo, E. Miyamoto and T. Yamana; *J. Pharm. Sci.*, 1977, **66**, 1675. 303

77JST(36)329 J. A. Ladd; *J. Mol. Struct.*, 1977, **36**, 329. 702

77M233 H. Berner, G. Schulz and H. Reinshagen; *Monatsh. Chem.*, 1977, **108**, 233. 771, 776

77M331 M. Pailer and E. Zeisl; *Monatsh. Chem.*, 1977, **108**, 331. 776

77MI51501 K. K. Smith, J.-Y. Koo, G. B. Schuster and K. J. Kaufmann; *Chem. Phys. Lett.*, 1977, **48**, 267. 459

77MI51502 W. Adam, G. Cilento, L. O. Rodriguez, O. Rodriguez, A. S. Sarma and K. Zinner; *Photochem. Photobiol.*, 1977, **26**, 299. 485

77MI50300 C. Brown; *Org. React. Mech.*, 1977, 481. 36

B-77MI50500 L. Fishbein and B. M. Goldschmidt; DHEW Publ. (FDA) (U.S.), 1977, FDA 78-1046 (*Chem. Abstr.*, 1979, **90**, 67 237). 119

B-77MI50600 W. E. Truce, T. C. Klingler and W. W. Brand; in 'Organic Chemistry of Sulfur', ed. S. Oae; Plenum, New York, 1977, p. 527. 141, 176

B-77MI50601 E. J. Goethals; in 'Topics in Sulfur Chemistry', ed. A. Senning and P. S. Magee; Thieme, Stuttgart, 1977, vol. 3, p. 1. 147, 182

77MI50602 W. H. Daly, C. G. Overberger, C.-D. Lee and T. J. Pacansky; *Macromol. Synth.*, 1977, **6**, 23. 148

B-77MI50603 G. H. Schmid; in 'Topics in Sulfur Chemistry', ed. A. Senning and P. S. Magee; Thieme, Stuttgart, 1977, vol. 3, p. 103. 163, 174, 177

B-77MI51000 J. J. de Konig, A. F. Marx, M. M. Poot, P. M. Smid and J. Verweij; in 'Recent Advances in the Chemistry of β-Lactam Antibiotics', ed. J. Elks; The Chemical Society, London, 1977, p. 161. 291

B-77MI51001 S. Kukolja; in ref. ⟨B-77MI51000⟩, p. 181. 293

77MI51002 B. G. Spratt; *Eur. J. Biochem.*, 1977, **72**, 341. 297

B-77MI51100 R. Busson, E. Roets and H. Vanderhaeghe; in ref. ⟨B-77MI51103⟩, p. 304. 302

77MI51101 S. P. Kukolja, S. R. Lammert and A. I. Ellis; *Croat. Chem. Acta*, 1977, **49**, 779. 331

77MI51102 P. J. Claes, G. Janssen and H. Vanderhaeghe; *Eur. J. Med. Chem.*, 1977, **12**, 521. 332

B-77MI51103 'Recent Advances in the Chemistry of β-Lactam Antibiotics', ed. J. Elks; The Chemical Society, Special Publication No. 28, London, 1977. 303

B-77MI51104 Y. Hamashima, K. Ishikura, H. Ishitobi, H. Itani, T. Kubota, K. Minami, M. Murakami, W. Nagata, M. Narisada, Y. Nishitani, T. Okada, H. Onoue, H. Satoh, Y. Sendo, T. Tsuji and M. Yoshioka; in ref. ⟨B-77MI51103⟩, p. 243. 306

B-77MI51105 F. J. Lund; in ref. ⟨B-77MI51103⟩, p. 25. 324

B-77MI51106 K. E. Price; in 'Structure-Activity Relationships of Semisynthetic Antibiotics', ed. D. Perlman; Academic Press, New York, 1977, pp. 1, 61. 336, 338

B-77MI51107 R. B. Woodward; in ref. ⟨B-77MI51103⟩, p. 167. 334

77MI51200 C. Reading and M. Cole; *Antimicrob. Agents Chemother.*, 1977, **11**, 852. 356

B-77MI51201 A. G. Brown, J. Goodacre, J. B. Harbridge, T. T. Howarth, R. J. Ponsford, I. Stirling and T. J. King; in 'Recent Advances in the Chemistry of β-Lactam Antibiotics', ed. J. Elks; Special Publication No. 28, The Chemical Society, London, 1977, p. 295. 350

77MI51202 K. Maeda, S. Takahashi, M. Sezaki, K. Iinuma, H. Naganawa, S. Kondo, M. Ohno and H. Umezawa; *J. Antibiot.*, 1977, **30**, 770. 351

77MI51400 J. W. Bevan, A. C. Legon and D. J. Millen; *Proc. R. Soc. London, Ser. A*, 1977, **354**, 491. 406, 436, 444

77MI51600 E.-S. A. Ibrahim, A. M. Farghaly and H. M. Salama; *Pharmazie*, 1977, **32**, 329. 527, 538

77MI51800 H. Yamamoto; *Chem. Econ. Eng. Rev.*, 1977, **9**, 22. 608, 613

77MI51801 A. Nawojski; *Wiad. Chem.*, 1977, **31**, 753. 613

77MI51802 L. H. Hurley; *J. Antibiot.*, 1977, **30**, 349. 613

77MI51803 M. Bianchi, A. Butti, S. Rossi, F. Bazaghi and V. Marcaria; *Eur. J. Med. Chem. Chim. Ther.*, 1977, **12**, 263. 637

77MI51804 M. Flammang and C. G. Wermuth; *Eur. J. Med. Chem. Chim. Ther.*, 1977, **12**, 121. 597

77MI52100 P. Baudot, M. Jacque and M. Rabin; *Toxicol. Appl. Pharmacol.*, 1977, **41**, 1. 760

77MI52200 J. W. Chamberlin and S. Chen; *J. Antibiotics*, 1977, **30**, 197. 764

77MIP51800 Tarchominskie Zaklady Farmaceutyczne 'Polfa,' *Pol. Pat.* 1977, **90**, 126 (*Chem. Abstr.*, 1978, **88**, 190 917). 609

77NJC77 J. M. Lehn, J. Simon and J. Wagner; *Nouv. J. Chim.*, 1977, **1**, 77. 752

77OMR(9)322 G. J. Jordan and R. C. DeLanson; *Org. Magn. Reson.*, 1977, **9**, 322. 200

77OMR(9)333 J. Leung and A. J. Jones; *Org. Magn. Reson.*, 1977, **9**, 333. 498

77OMR(9)559 K. Isomura, H. Taniguchi, M. Mishima, M. Fujio and Y. Tsumo; *Org. Magn. Reson.*, 1977, **9**, 559. 49, 51

77OMS348 J. P. Brun, M. Ricard, M. Corval and C. Schaal; *Org. Mass Spectrom.*, 1977, **12**, 348. 369

77OR(25)1 L. A. Paquette; *Org. React.*, 1977, **25**, 1. 778

77OS30 G. W. Gokel, D. J. Cram, G. L. Liotta, H. P. Harris and F. L. Cook; *Org. Synth.*, 1977, **57**, 30. 746, 749

77PAC857 J. M. Lehn; *Pure Appl. Chem.*, 1977, **49**, 857. 743

77RTC242 N. DeKimpe, R. Verhe, L. DeBuyck and N. Schamp; *Recl. Trav. Chim. Pays-Bas*, 1977, **96**, 242. 81

77RTC263 H. A. Selling, A. B. Verweij, A. Piepenbroek and C. T. Kiers; *Recl. Trav. Chim. Pays-Bas*, 1977, **96**, 263. 452

77RTC282 W. J. W. Van Tilborg and R. Plomp; *Recl. Trav. Chim. Pays-Bas*, 1977, **96**, 282. 413, 415, 423, 441

77S201 G. Seitz and H. Hoffmann; *Synthesis*, 1977, 201. 445

77S407 M. S. Manhas, H. P. S. Chawla, S. G. Amin and A. K. Bose; *Synthesis*, 1977, 407. 260

77S756 R. Lantzsch and D. Arlt; *Synthesis*, 1977, 756. 638

77S854 M. Okahara, M. Miki, S. Yanagida, I. Ikeda and K. Matsushima; *Synthesis*, 1977, 854. 749

77S884 K. Sotoya, M. Yamada, T. Takamoto, T. Sakakibara and R. Sudoh; *Synthesis*, 1977, 884. 172, 436

77SA(A)549 M. Kobayashi and S. Kawabata; *Spectrochim. Acta, Part A*, 1977, **33**, 549. 701

B-77SH W. L. F. Armarego; 'Stereochemistry of Heterocyclic Compounds', Wiley, New York, 1977, parts 1 and 2. 7, 10, 14, 25, 33, 239, 495, 499

77T453 U. Timm, K.-P. Zeller and H. Meier; *Tetrahedron*, 1977, **33**, 453. 193

77T485 T. Hasegawa, M. Watabe, H. Aoyama and Y. Omote; *Tetrahedron*, 1977, **33**, 485. 258

77T711 F. S. Richardson, C.-Y. Yeh, T. C. Troxell and D. B. Boyd; *Tetrahedron*, 1977, **33**, 711. 302

77T999 M. Bucciarelli, A. Forni, I. Moretti and G. Torre; *Tetrahedron*, 1977, **33**, 999. 134, 179

77T1545 D. J. Aberhart; *Tetrahedron*, 1977, **33**, 1545. 292

77T1785 K. Levsen, H. Heimbach, C. C. Van de Sande and J. Monstrey; *Tetrahedron*, 1977, **33**, 1785. 135

77T2399 N. Manisse and J. Chuche; *Tetrahedron*, 1977, **33**, 2399. 65, 540

77T2603 H. M. P. Thiellier, G. J. Koomen and U. K. Pandit; *Tetrahedron*, 1977, **33**, 2603. 606

77T2683 B. A. Hess, L. J. Schaad and D. N. Reinhoudt; *Tetrahedron*, 1977, **33**, 2683. 558

77T3061 F. Bernardi, N. D. Epiotis, S. Shaik and K. Mislow; *Tetrahedron*, 1977, **33**, 3061. 419

77TL343 R. W. Ashworth and G. A. Berchtold; *Tetrahedron Lett.*, 1977, 343. 570

77TL431 T. H. Koch, R. H. Higgins and H. F. Schuster; *Tetrahedron Lett.*, 1977, 431. 345

77TL911 G. Capozzi, V. Lucchini, G. Modena and P. Scrimin; *Tetrahedron Lett.*, 1977, 911. 177

77TL1053 G. Maier and U. Schafer; *Tetrahedron Lett.*, 1977, 1053. 275

77TL1057 H. Quast and P. Schafer; *Tetrahedron Lett.*, 1977, 1057. 78

77TL1355 H. Prinzbach, H. Babsch, H. Fritz and P. Hug; *Tetrahedron Lett.*, 1977, 1355. 500

77TL1449 H. H. Wasserman and A. W. Tremper; *Tetrahedron Lett.*, 1977, 1449. 263, 271

77TL1827 E. C. Taylor and C. S. Chiang; *Tetrahedron Lett.*, 1977, 1827. 586

77TL1855 J. Finkelstein, K. Holden, R. Sneed and C. D. Perchonock; *Tetrahedron Lett.*, 1977, 1855. 355

77TL1985 H. Hofmann and A. Molnar; *Tetrahedron Lett.*, 1977, 1985. 576

77TL2045 L. Carlsen and J. P. Snyder; *Tetrahedron Lett.*, 1977, 2045 459

77TL2231 R. A. Cormier, K. M. Freeman and D. M. Schnur: *Tetrahedron Lett.*, 1977, 2231. 123, 126

77TL2573 K. Frensch and F. Vögtle; *Tetrahedron Lett.*, 1977, 2573. 769

77TL2643 T. Wooldridge and T. D. Roberts; *Tetrahedron Lett.*, 1977, 2643. 193

77TL3023 A. Hassner and B. Amit; *Tetrahedron Lett.*, 1977, 3023. 499

77TL3297 J. Streith and C. Fizet; *Tetrahedron Lett.*, 1977, 3297. 599

77TL3361 M. Matsumoto, S. Dobashi and K. Kuroda; *Tetrahedron Lett.*, 1977, 3361. 192

77TL3643 I. Ojima, S. Inaba and K. Yoshida; *Tetrahedron Lett.*, 1977, 3643. 261

77TL3831 A. W. Taylor and G. Burton; *Tetrahedron Lett.*, 1977, 3831. 322

77TL4289 J. Kagan, P. Y. Juang, B. E. Firth, J. T. Przbytek and S. P. Singh; *Tetrahedron Lett.*, 1977, 4289. 19

77TL4343 R. G. Visser, J. P. B. Baaij, A. C. Brouwer and H. J. T. Bos; *Tetrahedron Lett.*, 1977, 4343. 422

77TL4389 G. Reissenweber and J. Sauer; *Tetrahedron Lett.*, 1977, 4389. 361, 600

77TL4393 H. D. Fühlhuber and J. Sauer; *Tetrahedron Lett.*, 1977, 4393. 598

77TL4465 A. G. Anastassiou and R. Badri; *Tetrahedron Lett.*, 1977, 4465. 725

77UP50500 E. G. Lewars; unpublished results, 1977. 125

77USP4010166 Ciba-Geigy Corp., *U.S. Pat.* 4 010 166 (1977) (*Chem. Abstr.*, 1977, **87**, 53 414). 629

77USP4045473 J. J. Fuchs; *U.S. Pat.* 4 045 473 (1977) (*Chem. Abstr.*, 1978, **88**, 6583). 484

77USP4053599 Hoechst Corp., *U.S. Pat.* 4 053 599 (1977) (*Chem. Abstr.*, 1978, **88**, 37 849). 630

77ZC331 W. Schroth and W. Kaufmann; *Z. Chem.*, 1977, **17**, 331. 621

77ZN(B)408 C. G. Kreiter and S. Özkar; *Z. Naturforsch, Teil B*, 1977, **32**, 408. 523

77ZOR1563 G. Yu. Gadzhiev and G. I. Alekperov; *Zh. Org. Khim.*, 1977, 1563 (*Chem. Abstr.*, 1977, **87**, 152 155). 643

78ACR8 D. J. Cram and J. M. Cram; *Acc. Chem. Res.*, 1978, **11**, 8. 745

78ACR49 J. M. Lehn; *Acc. Chem. Res.*, 1978, **11**, 49. 733, 740, 743, 750

78ACR407 Y. Kanaoaka; *Acc. Chem. Res.*, 1978, **11**, 407. 543

78AG(E)121 H. Meier, H. P. Schneider, A. Rieker and P. B. Hitchcock; *Angew. Chem., Int. Ed. Engl.*, 1978, **17**, 121. 551

78AG(E)129 H. W. Roesky and M. Aramaki; *Angew. Chem., Int. Ed. Engl.*, 1978, **17**, 129. 489

78AG(E)169 H. Yamamoto and H. Nozaki; *Angew. Chem., Int. Ed. Engl.*, 1978, **17**, 169. 105

78AG(E)195 G. L'abbe, J.-P. Dekerk, J.-P. Declerq, G. Germain and M. Van Meerssche; *Angew. Chem., Int. Ed. Engl.*, 1978, **17**, 195. 13

78AG(E)213 H. Quast and C. A. Weise Velez; *Angew. Chem., Int. Ed. Engl.*, 1978, **17**, 213. 22

78AG(E)677 F.-M. Tesky, R. Mews, B. Krebs and M. R. Udupa; *Angew. Chem., Int. Ed. Engl.*, 1978, **17**, 677. 486

78AG(E)755 T. Kauffmann and H. Lexy; *Angew. Chem., Int. Ed. Engl.*, 1978, **17**, 755. 766

78AG(E)853 M. Schäfer-Ridder, A. Wagner, M. Schwamborn, H. Schreiner, E. Devrout and E. Vogel; *Angew. Chem., Int. Ed. Engl.*, 1978, **17**, 853. 16, 714, 715

78AG(E)855 H.-J. Golz, J. M. Muchowski and M. L. Maddox; *Angew. Chem., Int. Ed. Engl.*, 1978, **17**, 855. 714, 715

78AG(E)936 E. Francotte, R. Merenyi and H. G. Viehe; *Angew. Chem., Int. Ed. Engl.*, 1978, **17**, 936. 569

78AHC(22)183 J. Elguero, R. M. Claramunt and A. J. H. Summers; *Adv. Heterocycl. Chem.*, 1978, **22**, 183. 687

78AHC(23)55 A. G. Anastassiou and H. S. Kasmai; *Adv. Heterocycl. Chem.*, 1978, **23**, 55. 672, 676, 710, 711, 713, 714, 715, 716, 717, 718, 719, 720, 721, 722, 725, 726, 727

78AHC(23)263 R. M. Acheson and N. F. Elmore; *Adv. Heterocycl. Chem.*, 1978, **23**, 263. 542

78AJC439 D. E. Rivett and F. H. C. Stewart; *Aust. J. Chem.*, 1978, **31**, 439. 529

78AJC2349 B. G. Gowenlock and L. Radom; *Aust. J. Chem.*, 1978, **31**, 2349. 486

78AP698 W. Hanefeld and D. Kluck; *Arch. Pharm. (Weinheim, Ger.)*, 1978, **311**, 698 464

78AP786 G. Seitz, T. Kämpchen and W. Overheu; *Arch. Pharm. (Weinheim, Ger.)*, 1978, **311**, 786. 522, 604

78AX(B)461 M. O. Boles, R. J. Girven and P. A. C. Gane; *Acta Crystallogr., Part B*, 1978, **34**, 461. 300

78AX(B)988 W. Schuckmann, H. Fuess, O. Mosinger and W. Ried; *Acta Crystallogr., Part B*, 1978, **34**, 988. 452

78AX(B)2602 K. Prout, V. P. Stothard and D. J. Watkin; *Acta Crystallogr., Part B*, 1978, **34**, 2602. 452

78AX(B)3387 I. Goldberg; *Acta Crystallogr., Part B*, 1978, **34**, 3387. 736, 737

78AX(B)3403 V. Bertolasi and G. Gilli; *Acta Crystallogr., Part B*, 1978, **34**, 3403. 405

78BAU142 N. B. Kaz'mina, I. L. Knunyants, E. I. Mysov and G. M. Kuz'yants; *Bull. Acad. Sci. USSR, Div. Chem. Sci.*, 1978, **27**, 142. 459, 477

78BCJ2131 T. Asao, M. Yagihara and Y. Kitahara; *Bull. Chem. Soc. Jpn.*, 1978, **51**, 2131. 191

78BCJ3403 S. Ito and J. Mori; *Bull. Chem. Soc. Jpn.*, 1978, **51**, 3403. 421

78BSB893 G. L'abbe, L. Huybrechts, S. Toppet, J.-P. Declercq, G. Germain and M. Van Meerssche; *Bull. Soc. Chim. Belg.*, 1978, **87**, 893. 433

78BSF(2)97 J. Sauleau, A. Sauleau and J. Huet; *Bull. Soc. Chim. Fr., Part 2*, 1978, 97. 65, 497, 540

78BSF(2)539 M. Maguet and R. Guglielmetti; *Bull. Soc. Chim. Fr.*, 1978, 539. 145, 159

78CB143 W. Ried and O. Mosinger; *Chem. Ber.*, 1978, **111**, 143. 480

78CB309 H. Hofmann, H. Böhme and E. Wilhelm; *Chem. Ber.*, 1978, **111**, 309. 556

78CB596 W. Duisman and C. Rüchardt; *Chem. Ber.*, 178, **111**, 596. 670

78CB1413 P. Lechtken; *Chem. Ber.*, 1978, **111**, 1413. 459

78CB1549 U. Timm, K.-P. Zeller and H. Meier; *Chem. Ber.*, 1978, **111**, 1549. 193

78CB1780 V. Bardakos and W. Sucrow; *Chem. Ber.*, 1978, **111**, 1780. 531

78CB3105 G. V. Röschenthaler, K. Sauerbrey and R. Schmutzler; *Chem. Ber.*, 1978, **111**, 3105. 196

78CB3608 G. Kaupp and M. Stark; *Chem. Ber.*, 1978, **111**, 3608. 668

78CB3927 R. Aumann and J. Knecht; *Chem. Ber.*, 1978, **111**, 3927. 495, 524

78CC9 K. L. Davies, R. C. Storr and P. J. Whittle; *J. Chem. Soc., Chem. Commun.*, 1978, 9. 4

78CC46 J. Leclercq, E. Cossement, R. Boydens, L. A. M. Rodriguez, L. Brouwers, F. de Laveleye and W. Libert; *J. Chem. Soc., Chem. Commun.*, 1978, 46. 331

78CC143 J. P. Behr and J. M. Lehn; *J. Chem. Soc., Chem. Commun.*, 1978, 143. 753

78CC154 T. Kitazume and J. M. Shreeve; *J. Chem. Soc., Chem. Commun.*, 1978, 154. 481

78CC396 P. W. Manley, R. Somanathan, D. L. Reeves and R. C. Storr; *J. Chem. Soc., Chem. Commun.*, 1978, 396. 282, 641

78CC442	T. Miyashi, T. Nakajo and T. Mukai; *J. Chem. Soc., Chem. Commun.*, 1978, 442.	221
78CC467	P. C. Cherry, G. I. Gregory, C. E. Newall, P. Ward and N. S. Watson; *J. Chem. Soc., Chem. Commun.*, 1978, 467.	350
78CC555	W. Wong-Ng and S. C. Nyburgg; *J. Chem. Soc., Chem. Commun.*, 1978, 555.	132
78CC630	J. Bolster and R. M. Kellogg; *J. Chem. Soc., Chem. Commun.*, 1978, 630.	153, 154, 162, 176
78CC903	S. R. Fletcher and I. T. Kay; *J. Chem. Soc., Chem. Commun.*, 1978, 903.	256, 266
78CC934	B. Dietrich, T. M. Fyles, J. M. Lehn, L. G. Pease and D. L. Fyles; *J. Chem. Soc., Chem. Commun.*, 1978, 934.	748
78CC949	J. M. Lehn and C. Sirlin; *J. Chem. Soc., Chem. Commun.*, 1978, 949.	754
78CC961	T. Mah, H. M. Sirat and E. J. Thomas; *J. Chem. Soc., Chem. Commun.*, 1978, 961.	388
78CC1020	I. R. Dunkin and J. G. MacDonald; *J. Chem. Soc., Chem., Commun.*, 1978, 1020.	169
78CCC3258	O. Hritzova and P. Kristian; *Collect. Czech. Chem. Commun.*, 1978, **43**, 3258.	455, 475
78CHE538	V. M. Potapov, V. M. Dem'yanovich, L. D. Solov'eva, T. P. Nedorezova and P. B. Terent'ev; *Chem. Heterocycl. Compd.* (*Engl. Transl.*), 1978, **14**, 538.	499
78CJC71	F. S. Abbott and K. Haya; *Can. J. Chem.*, 1978, **56**, 71.	440
78CJC114	D. R. Dice and R. P. Steer; *Can. J. Chem.*, 1978, **56**, 114.	419
78CJC1155	N. N. Gerber, A. G. McInnes, D. G. Smith, J. A. Walter, J. L. C. Wright and L. C. Vining; *Can. J. Chem.*, 1978, **56**, 1155.	764
78CJC1335	T. T. Conway, G. Lim, J. L. Douglas, M. Menard, T. W. Doyle, P. Rivest, D. Horning, L. R. Morris and D. Cimon; *Can. J. Chem.*, 1978, **56**, 1335.	355
78CJC2985	A. Albini and D. R. Arnold; *Can. J. Chem.*, 1978, **56**, 2985.	19
78CL395	A. Mori and H. Takeshita; *Chem. Lett.*, 1978, 395.	191
78CL409	T. Izawa and T. Mukaiyama; *Chem. Lett.*, 1978, 409.	380
78CL563	Y. Miyahara, T. Inazu and T. Yoshino; *Chem. Lett.*, 1978, 563.	772
78CL697	T. Kato, Y. Suzuki and M. Sato; *Chem. Lett.*, 1978, 697.	389
78CL723	S. Yano, K. Nishino, K. Nakasuji and I. Murata; *Chem. Lett.*, 1978, 723.	557
78CL769	K. Maruyama and Y. Kubo; *Chem. Lett.*, 1978, 769.	517
78CL1403	T. Migita, M. Chiba, M. Kosugi and S. Nakaido: *Chem. Lett.*, 1978, 1403.	36
78CPB448	S. Oida, A. Yoshida and E. Ohki; *Chem. Pharm. Bull.*, 1978, **26**, 448.	353
78CPB942	K. Kawashima, T. Saraie, Y. Kawano and T. Ishiguro; *Chem. Pharm. Bull.*, 1978, **26**, 942.	519
78CPB951	K. Kawashima and T. Ishiguro; *Chem. Pharm. Bull.*, 1978, **26**, 951.	526
78CPB1761	T. Kobayashi, Y. Iwano and K. Hirai; *Chem. Pharm. Bull.*, 1978, **26**, 1761.	353
78CPB1890	T. Tsuchiya and J. Kurita; *Chem. Pharm. Bull.*, 1978, **26**, 1890.	600, 601, 604
78CPB1896	T. Tsuchiya and J. Kurita; *Chem. Pharm. Bull.*, 1978, **26**, 1896.	603
78CPB3877	T. Kato, T. Chiba and M. Sato; *Chem. Pharm. Bull.*, 1978, **26**, 3877.	380
78CR(C)(286)83	J. Bonnic, P. Uriac, H. Bouget and J. Huet; *C.R. Hebd. Seances Acad. Sci., Ser. C*, 1978, **286**, 83.	533
78CR(C)(286)671	M. Flammang; *C.R. Hebd. Seances Acad. Sci., Ser. C*, 1978, **286**, 671.	603
78CRV125	N. J. Turro, V. Ramamurthy, W. Cherry and W. Farneth; *Chem. Rev.*, 1978, **78**, 125.	437, 441, 443
78CRV407	T. Asano and W. J. LeNoble; *Chem. Rev.*, 1978, **78**, 407.	39
78CSC309	M. Hospital, F. Leroy, J. P. Bats and J. Moulines; *Cryst. Struct. Commun.*, 1978, **7**, 309.	365
78FRP2364218	Société Anon. Propharma, *Fr. Pat.* 2 364 218 (1978) (*Chem. Abstr.*, 1979, **90**, 6398).	159
78G543	I. G. Csizmadia, V. Lucchini and G. Modena; *Gazz. Chim. Ital.*, 1978, **108**, 543.	139
78G619	A. Balsamo, C. Battistini, P. Crotti and F. Macchia; *Gazz. Chim. Ital.*, 1978, **108**, 619.	379
78GEP2822001	Beecham Group Ltd., *Ger. Pat.* 2 822 001 (1978) (*Chem. Abstr.*, 1979, **90**, 152 160).	350
78H(10)99	S. Uyeo, T. Aoki, H. Itani, T. Tsuji and W. Nagata; *Heterocycles*, 1978, **10**, 99.	307
78H(11)331	T. Toda, T. Takase, T. Mukai and Y. Suzuki; *Heterocycles*, 1978, **11**, 331.	627
78H(11)401	S. Ida, T. Mukai and K. Saito; *Heterocycles*, 1978, **11**, 401.	493, 522
78H(11)549	D. Lloyd and H. McNab; *Heterocycles*, 1978, **11**, 549.	608, 610, 615, 619
78HC(33)5	H. Neunhoeffer; *Chem. Heterocycl. Compd.*, 1978, **33**, 5.	283
78HCA795	L. Hoesch, N. Egger and A. S. Dreiding; *Helv. Chim. Acta*, 1978, **61**, 795.	36, 87
78HCA2336	W. Fischer and A. Grob; *Helv. Chim. Acta*, 1978, **61**, 2336.	391
78HCA2763	F. Gerson, W. Huber and O. Wennerström; *Helv. Chim. Acta*, 1978, **61**, 2763.	769
78IZV880	C. P. Del'tsova and N. P. Gambaryan; *Izv. Akad. Nauk SSSR, Ser. Khim.*, 1978, **4**, 880.	273
78JA9	N. D. Epiotis and S. Shaik; *J. Am. Chem. Soc.*, 1978, **100**, 9.	37, 38
78JA313	D. B. R. Johnston, S. M. Schmitt, F. A. Bouffard and B. G. Christensen; *J. Am. Chem. Soc.*, 1978, **100**, 313.	353
78JA352	N. Ikota and B. Ganem; *J. Am. Chem. Soc.*, 1978, **100**, 352.	374
78JA353	G. W. Holbert and B. Ganem; *J. Am. Chem. Soc.*, 1978, **100**, 353.	374
78JA550	L. Mandolini; *J. Am. Chem. Soc.*, 1978, **100**, 550.	33
78JA743	B. Plesnicar, M. Tasevski and A. Azman; *J. Am. Chem. Soc.*, 1978, **100**, 743.	35, 116
78JA1398	J. Catalan and M. Yanez; *J. Am. Chem. Soc.*, 1978, **100**, 1398.	23, 105
78JA1635	C. A. Stein and H. Taube; *J. Am. Chem. Soc.*, 1978, **100**, 1635.	694
78JA1876	P. S. Engel, R. A. Hayes, L. Keifer, S. Szilagyi and J. W. Timberlake; *J. Am. Chem. Soc.*, 1978, **100**, 1876.	456, 483
78JA2457	R. B. DuVernet, O. Wennerström, J. Lawson, T. Otsubo and V. Boekelheide; *J. Am. Chem. Soc.*, 1978, **100**, 2457.	778

78JA2806 S. F. Nelsen, V. A. Peacock, G. R. Weisman, M. E. Landis and J. A. Spencer; *J. Am. Chem. Soc.*, 1978, **100**, 2806. 451

78JA2908 L. A. Paquette and G. D. Ewing; *J. Am. Chem. Soc.*, 1978, **100**, 2908. 506

78JA3251 F. DiNinno; *J. Am. Chem. Soc.*, 1978, **100**, 3251. 432

78JA4481 A. Padwa, P. H. J. Carlsen and A. Tremper; *J. Am. Chem. Soc.*, 1978, **100**, 4481. 64

78JA4597 J. E. Baldwin and M. A. Christie; *J. Am. Chem. Soc.*, 1978, **100**, 4597. 332

78JA4914 J. M. Lehn, E. Sonveaux and A. K. Willard; *J. Am. Chem. Soc.*, 1978, **100**, 4914. 747

78JA4982 D. W. Larsen, T. A. Smentkowski, B. A. Soltz and F. E. Stary; *J. Am. Chem. Soc.*, 1978, **100**, 4982. 499

78JA5580 N. J. Turro, M.-F. Chow and Y. Ito; *J. Am. Chem. Soc.*, 1978, **100**, 5580. 476

78JA5978 C. J. Michejda, S. R. Koepke and D. H. Campbell; *J. Am. Chem. Soc.*, 1978, **100**, 5978. 647

78JA6416 W. K. Musker, T. L. Wolford and P. B. Roush; *J. Am. Chem. Soc.*, 1978, **100**, 6416. 694

78JA6425 T. R. Forbus, P. A. Birdsong and P. B. Shevlin; *J. Am. Chem. Soc.*, 1978, **100**, 6425. 389

78JA6437 C. W. Jefford and C. G. Rimbault; *J. Am. Chem. Soc.*, 1978, **100**, 6437. 476

78JA6483 B. Ganem, G. W. Holbert, L. B. Weiss and K. Ishizumi; *J. Am. Chem. Soc.*, 1978, **100**, 6483. 192, 582, 584

78JA6491 G. Albers-Schönberg, B. H. Arison, O. D. Hensens, J. Hirshfield, K. Hoogsteen, E. A. Kaczka, R. E. Rhodes, J. S. Kahan, F. M. Kahan, R. W. Ratcliffe, E. Walton, L. J. Ruswinkle, R. B. Morin and B. G. Christensen; *J. Am. Chem. Soc.*, 1978, **100**, 6491. 349

78JA6696 I. Hasan and F. W. Fowler; *J. Am. Chem. Soc.*, 1978, **100**, 6696. 343

78JA6760 D. K. White and F. D. Greene; *J. Am. Chem. Soc.*, 1978, **100**, 6760. 483

78JA6788 R. A. Moss and M. E. Fantina; *J. Am. Chem. Soc.*, 1978, **100**, 6788. 227

78JA6920 N. L. Bauld, T. L. Welsher, J. Cessac and R. L. Holloway; *J. Am. Chem. Soc.*, 1978, **100**, 6920. 121

78JA7127 M. J. Mirbach, A. Henne and K. Schaffner; *J. Am. Chem. Soc.*, 1978, **100**, 7127. 476

78JA7300 B. C. Burros, N. J. De'Ath, D. B. Denney, D. Z. Denney and I. J. Kipnis; *J. Am. Chem. Soc.*, 1978, **100**, 7300 485

78JA7436 E. Block, R. E. Penn, M. D. Ennis, T. A. Owens and S.-L. Yu; *J. Am. Chem. Soc.*, 1978, **100**, 7436. 132, 181, 182, 422

78JA7754 W. K. Musker, A. S. Hirschon and J. T. Doi; *J. Am. Chem. Soc.*, 1978, **100**, 7754. 695

78JA8004 D. H. Shih, J. Hannah and B. G. Christensen; *J. Am. Chem. Soc.*, 1978, **100**, 8004. 350

78JA8006 L. D. Cama and B. G. Christensen; *J. Am. Chem. Soc.*, 1978, **100**, 8006. 352

78JA8056 C. Mueller, A. Schweig and H. Vermeer; *J. Am. Chem. Soc.*, 1978, **100**, 8056. 137

78JA8214 I. Ernest, J. Gosteli, C. W. Greengrass, W. Holick, D. E. Jackman, H. R. Pfaendler and R. B. Woodward; *J. Am. Chem. Soc.*, 1978, **100**, 8214. 334

78JAP(K)7872826 H. Takahashi, K. Togashi and Y. Murakami; *Jpn. Kokai* 78 72 826 (1978) (*Chem. Abstr.*, 1978, **89**, 192 519). 484

78JAP(K)78132586 M. Masaki, K. Fukui and J. Kita (Ube Industries); *Jpn. Kokai* 78 132 586 (*Chem. Abstr.*, 1979, **90**, 122243). 525, 536

78JCS(P1)185 D. D. Dexter and J. M. van der Veen; *J. Chem. Soc., Perkin Trans. 1*, 1978, 185. 300, 349

78JCS(P1)191 R. Purvis, R. K. Smalley, W. A. Strachan and H. Suschitzky; *J. Chem. Soc., Perkin. Trans. 1*, 1978, 191. 495

78JCS(P1)662 T. Kametani, K. Takahashi, M. Ihara and K. Fukumoto; *J. Chem. Soc., Perkin Trans. 1*, 1978, 662. 659

78JCS(P1)668 W. Baker, C. M. Pant and R. J. Stoodley; *J. Chem. Soc., Perkin Trans. 1*, 1978, 668. 355

78JCS(P1)746 A. Holm, N. Harrit and I. Trabjerg; *J. Chem. Soc., Perkin Trans. 1*, 1978, 746. 196

78JCS(P1)817 M. Campbell, R. G. Harcus and K. H. Nelson; *J. Chem. Soc., Perkin Trans. 1*, 1978, 817. 251

78JCS(P1)862 G. R. Proctor and B. M. L. Smith; *J. Chem. Soc., Perkin Trans. 1*, 1978, 862. 515

78JCS(P1)1029 H. Kato, H. Tezuka, K. Yamaguchi, K. Nowada and Y. Nakamura; *J. Chem. Soc., Perkin Trans. 1*, 1978, 1029. 559

78JCS(P1)1366 C. M. Pant and R. J. Stoodley; *J. Chem. Soc., Perkin Trans. 1*, 1978, 1366. 324

78JCS(P1)1450 C. L. Branch, J. H. C. Nayler and M. J. Pearson; *J. Chem. Soc., Perkin Trans. 1*, 1978, 1450. 354

78JCS(P1)1453 D. Lloyd, H. McNab and D. R. Marshall; *J. Chem. Soc., Perkin Trans. 1*, 1978, 1453. 611

78JHC71 A. Hasnaoui, J.-P. Lavergne and P. Viallefont; *J. Heterocycl. Chem.*, 1978, **15**, 71. 639

78JHC249 A. Jössang-Yanagida and C. Gansser; *J Heterocycl. Chem.*, 1978, **15**, 249. 531

78JHC421 A. A. Scala and I. Colon; *J. Heterocycl. Chem.*, 1978, **15**, 421. 417

78JHC697 U. Timm, H. Bühl and H. Meier; *J. Heterocycl. Chem.*, 1978, **15**, 697. 193

78JHC1057 K. H. Pannell, B. L. Kalsotra and C. Parkanyi; *J. Heterocycl. Chem.*, 1978, **15**, 1057. 28, 29

78JHC1165 H. O. Krabbenhoft; *J. Heterocycl. Chem.*, 1978, **15**, 1165. 370

78JHC1319 D. G. Holah, A. N. Hughes and D. Kleemola; *J. Heterocycl. Chem.*, 1978, **15**, 1319. 713

78JHC1339 S. Plescia, G. Daidone, V. Sprio, E. Aiello, G. Dattolo and G. Cirrincione; *J. Heterocycl. Chem.*, 1978, **15**, 1339. 767

78JMC413 W. F. Huffman, R. F. Hall, J. A. Grant and K. G. Holden; *J. Med. Chem.*, 1978, **21**, 413. 333

78JOC57 A. Hassner, C. A. Bunnell and K. Haltiwanger; *J. Org. Chem.*, 1978, **43**, 57. 49

78JOC315 A. G. Anastassiou, E. Reichmanis, S. J. Girgenti and M. Schaefer-Ridder; *J. Org. Chem.*, 1978, **43**, 315. 507, 522, 525, 541, 570

78JOC419 H. Aoyama, T. Hasegawa, M. Watabe, H. Shiraishi and Y. Omote; *J. Org. Chem.*, 1978, **43**, 419. 258

78JOC922 P. E. McGann, J. T. Groves, F. D. Greene, G. M. Stack, R. J. Majeste and L. M. Trefonas; *J. Org. Chem.*, 1978, **43**, 922. 199

78JOC944 Y. Ogata and K. Takagi; *J. Org. Chem.*, 1978, **43**, 944. 343, 360

78JOC1146 J. S. Chickos and D. E. Sherwood; *J. Org. Chem.*, 1978, **43**, 1146. 374

78JOC1185 E. Vedejs, J. P. Hagen, B. L. Roach and K. L. Spear; *J. Org. Chem.*, 1978, **43**, 1185. 424

78JOC1243 S. R. Lammert, A. I. Ellis, R. R. Chauvette and S. Kukolja; *J. Org. Chem.*, 1978, **43**, 1243. 331

78JOC1305 H. O. Krabbenhoft; *J. Org. Chem.*, 1978, **43**, 1305. 374

78JOC1346 M. K. Meilhahn, D. K. Olsen, W. J. Brittain and R. T. Anders; *J. Org. Chem.*, 1978, **43**, 1346. 88

78JOC1684 M. Mori, K. Chiba and Y. Ban; *J. Org. Chem.*, 1978, **43**, 1684. 539

78JOC1980 S. Yoneda, T. Kawase and Z. Yoshida; *J. Org. Chem.*, 1978, **43**, 1980. 435

78JOC2013 J. W. Paschal, D. E. Dorman, P. R. Srinivasan and R. L. Lichter; *J. Org. Chem.*, 1978, **43**, 2013. 302

78JOC2029 A. Padwa and P. H. J. Carlsen; *J. Org. Chem.*, 1978, **43**, 2029. 67

78JOC2184 G. Jones, II and L. P. McDonnell-Bushnell; *J. Org. Chem.*, 1978, **43**, 2184. 369

78JOC2236 W. H. Richardson, J. H. Anderegg, M. E. Price, W. A. Tappen and H. E. O'Neal; *J. Org. Chem.*, 1978, **43**, 2236. 459

78JOC2380 J. F. Garvey and J. A. Hashmall; *J. Org. Chem.*, 1978, **43**, 2380. 198

78JOC2490 M. Torres, A. Clement, J. E. Bertie, H. E. Gunning and O. P. Strausz; *J. Org. Chem.*, 1978, **43**, 2490. 121, 193

78JOC2500 M. S. Raasch; *J. Org. Chem.*, 1978, **43**, 2500 475, 484

78JOC2621 S. F. Nelsen, E. L. Clennan, L. Echegoyan and L. A. Grezzo; *J. Org. Chem.*, 1978, **43**, 2621. 674

78JOC2711 M. P. Servé and D. M. Jerina; *J. Org. Chem.*, 1978, **43**, 2711. 584

78JOC2725 P. Casagrande, L. Pellacani and P. A. Tardella; *J. Org. Chem.*, 1978, **43**, 2725. 537

78JOC2785 J. H. Hall and W. S. Bigard; *J. Org. Chem.*, 1978, **43**, 2785. 451, 471

78JOC3021 F. A. L. Anet, P. J. Degen and I. Yavari; *J. Org. Chem.*, 1978, **43**, 3021. 699

78JOC3066 J. E. Semple and M. M. Jouillie; *J. Org. Chem.*, 1978, **43**, 3066. 463, 473

78JOC3131 C. A. Kingsbury and G. Max; *J. Org. Chem.*, 1978, **43**, 3131. 395

78JOC3362 G. R. Newkome, A. Nayak, J. Otemaa, D. A. Van and W. H. Benton; *J. Org. Chem.*, 1978, **43**, 3362. 766

78JOC3379 V. J. Traynelis, J. A. Schield, W. A. Lindley and D. W. H. MacDowell; *J. Org. Chem.*, 1978, **43**, 3379. 558, 559

78JOC3730 S. Mataka, S. Ishii and M. Tashiro; *J. Org. Chem.*, 1978, **43**, 3730. 177

78JOC3813 W. J. Lipa, H. T. Crawford, P. C. Radlick and G. K. Helmkamp; *J. Org. Chem.*, 1978, **43**, 3813. 714

78JOC3827 M. Buza, K. K. Andersen and M. D. Pazdon; *J. Org. Chem.*, 1978, **43**, 3827. 436, 444

78JOC4207 M. A. Calcagno and E. E. Schweizer; *J. Org. Chem.*, 1978, **43**, 4207. 540

78JOC4256 G. A. Lee; *J. Org. Chem.*, 1978, **43**, 4256. 102

78JOC4438 R. Busson and H. Vanderhaeghe; *J. Org. Chem.*, 1978, **43**, 4438. 349, 353, 356, 357, 359

78JOC4530 D. K. White and F. D. Greene; *J. Org. Chem.*, 1978, **43**, 4530. 455, 470

78JOC4696 C. A. Kingsbury, D. L. Durham and R. Hutton; *J. Org. Chem.*, 1978, **43**, 4696. 134

78JOC4856 J. C. Sheehan, E. Chacko, Y. S. Lo, D. R. Ponzi and E. Sato; *J. Org. Chem.*, 1978, **43**, 4856. 265

78JOC4961 H. Kohn, P. Charumilind and Y. Gopichand; *J. Org. Chem.*, 1978, **43**, 4961. 415, 438

78JOM(157)C35 C. Couret, J. Satge, J. D. Andriamizaka and J. Escudie; *J. Organomet. Chem.*, 1978, **157**, C35. 161

78JOU2003 I. L. Kuranova and E. V. Snetkova; *J. Org. Chem. USSR (Engl. Transl.)*, 1978, **14**, 2003. 145

78JPR255 H. Gross, I. Keitel and B. Costisella; *J. Prakt. Chem.*, 1978, **320**, 255. 623

78JSP(69)159 J. A. Duckett, T. L. Smithson and H. Wieser; *J. Mol. Spectrosc.*, 1978, **69**, 159. 370

78LA431 A. Krebs, H. Kimling and R. Kemper; *Liebigs Ann. Chem.*, 1978, 431. 574

78LA608 K. Krohn, D. Carboo and U. Puttfarcken; *Liebigs Ann. Chem.*, 1978, **4**, 608. 276

78LA1111 H. Hoberg and J. Korff; *Liebigs Ann. Chem.*, 1978, 1111. 455

78LA1505 L. Horner and M. Jordan; *Liebigs Ann. Chem.*, 1978, 1505 465

78M137 H. Berner, G. Schulz and H. Reinshagen; *Monatsh. Chem.*, 1978, **109**, 137. 776

78M557 H. Berner, G. Schulz, G. Fischer and H. Reinshagen; *Monatsh. Chem.*, 1978, **109**, 557. 771

78M609 P. K. Claus, W. Rieder and F. W. Vierhapper; *Monatsh., Chem.*, 1978, **109**, 609. 164

B-78MI50300 W. Lwowski; in 'Reactive Intermediates', ed. M. Jones, Jr. and R. A. Moss; Wiley, New York, 1978, vol. 1, p. 197. 36

B-78MI50301 J. Bastide and D. Henri-Rousseau; in 'The Chemistry of the C–C Triple Bond', ed. S. Patai; Wiley, Chichester, 1978, p. 460. 37

B-78MI50302 N. J. Turro; 'Modern Molecular Photochemistry', Benjamin/Cummins, Menlo Park, 1978. 39, 40

B-78MI50303 Ref. B-78MI50302, p. 433. 40

78MI50500 A. C. Hopkinson, M. H. Lien, I. G. Csizmadia and K. Yates; *Theor. Chim. Acta*, 1978, **47**, 97. 105

78MI50501 B. Kadis; *J. Steroid Biochem.*, 1978, **9**, 75. 119

B-78MI50502 A. Greenberg and J. F. Liebman; 'Strained Organic Molecules', Academic Press, New York, 1978, pp. 312–314. 120

78MI50600 R. Simonds, W. Van Craeynest, E. J. Goethals and S. Boileau; *Eur. Polym. J.*, 1978, **14**, 589. 147

78MI50601 T. Hori, T. Miyake, K. Takeda and J. Kato; *Prog. Cancer Res. Ther.*, 1978, **10**, 159 (*Chem. Abstr.*, 1978, **89**, 209 294). 182

78MI50602 A. L. H. Chuang, H. Mukhtar and E. Bresnick; *J. Natl. Cancer Inst.*, 1978, **60**, 321 (*Chem. Abstr.*, 1978, **88**, 164 226). 182

B-78MI50700 R. G. Harvey and P. P. Fu; in 'Polycyclic Hydrocarbons and Cancer', Academic Press, New York, 1978, vol. 1, p. 133. 188

B-78MI50701 D. M. Jerina, H. Yagi, R. E. Lehr, D. R. Thakker, M. Schaefer-Ridder, J. M. Karle, W. Levin, A. W. Wood, R. L. Chang and A. H. Conney; in 'Polycyclic Hydrocarbons and Cancer,' Academic Press, New York, 1978, vol. 1, p. 173. 188

78MI50800 F. Sanz e Carreras; *Afinidad*, 1978, **35**, 193 (*Chem. Abstr.*, 1978, **89**, 128 753). 198

B-78MI50900 B. J. Walker; in 'Saturated Heterocyclic Chemistry', ed. G. Pattenden; The Chemical Society, London, 1978, vol. 5. 248

B-78MI51000 J. R. E. Hoover and C. H. Nash; in 'Encyclopedia of Chemical Technology', Wiley, New York, 3rd edn., 1978, vol. 2, p. 871. 292

78MI51001 B. G. Spratt; *Sci. Prog.* (*Oxford*), 1978, **65**, 101. 296

78MI51100 M. D. Müller, J. Seibl and W. Simon; *Anal. Chim. Acta*, 1978, **100**, 263. 302

78MI51101 H. Bundgaard and C. Larson; *Int. J. Pharm.*, 1978, **1**, 95. 327

78MI51102 M. K. Stanfield, B. T. Butcher and G. T. Stewart; *Anal. Biochem.*, 1978, **89**, 1. 327

78MI51200 T. H. Koch, D. R. Anderson, J. M. Burns, G. C. Crockett, K. A. Howard, J. S. Kente, R. M. Rodehorst and R. J. Sluski; *Top. Curr. Chem.*, 1978, **75**, 65. 345

78MI51300 H. Umezawa; *J. Antibiot.*, 1978, **31**, 797 (*Chem. Abstr.*, 1979, **90**, 38 493). 401

B-78MI51301 N. J. Turro; 'Modern Molecular Photochemistry', Benjamin/Cummings, Menlo Park, CA, 1978, p. 432. 396

78MI51302 C. G. D'Hondt and R. W. Lenz; *J. Polym. Sci., Polym. Chem. Ed.*, 1978, **16**, 261. 382, 388

78MI51500 Y. Imai, M. Ueda and K. Okuyama; *Makromol. Chem.*, 1978, **179**, 2327. 484

78MI51501 K. A. Horn, J.-Y. Koo, S. P. Schmidt and G. B. Schuster; *Mol. Photochem.*, 1978–79, **9**, 1. 469

78MI51600 H. W. Vos, C. MacLean and N. H. Velthorst; *Mol. Phys.*, 1978, **35**, 329. 495

78MI51601 H. R. Kricheldorf; *Makromol. Chem.*, 1978, **179**, 2675. 499

78MI51602 E. Hannig, G. Beyer and W. Thommes; *Pharmazie*, 1978, **33**, 780. 514, 519

78MI51603 C. H. Bamford, A. Ledwith and Y. Yagci; *Polymer*, 1978, **19**, 354. 526

78MI51800 M. Majchrzak, A. Kotelko and R. Guryn; *Pol. J. Chem.*, 1978, **52**, 1023. 609

78MI51801 L. H. Sternbach; *Drug Res.*, 1978, **22**, 229. 608, 610, 612, 613

78MI51900 S. Watarai, H. Katsuyama, A. Umehara and H. Sato; *J. Polym. Sci., Polym. Chem. Ed.*, 1978, **16**, 2039. 685

B-78MI52100 J. S. Bradshaw; in 'Synthetic Multidentate Macrocyclic Compounds', ed. R. M. Izatt and J. J. Christensen; Academic Press, New York, 1978, p. 53. 748

B-78MI52101 C. J. Pederson; in ref. B-78MI52100, p. 1. 740, 749

B-78MI52102 N. K. Dalley; in ref. B-78MI52100, p. 207. 733, 735

B-78MI52103 R. A. Schwind, T. J. Gilligan and E. L. Cussler; in ref. B-78MI52100, p. 289. 759

B-78MI52104 C. L. Liotta; in ref. B-78MI52100, p. 111. 756, 758

78MI52200 K. Golankiewicz and B. Skalski; *Pol. J. Chem.*, 1978, **52**, 1365. 766

B-78MI52201 J. S. Bradshaw; in 'Synthetic Multidentate Macrocyclic Compounds', ed. R. M. Izatt and J. J. Christensen; Academic Press, New York, 1978. p. 53. 764

78MI52202 V. E. Golubev; *Zh. Vses. Khim. Ova.*, 1978, **23**, 477. 765

78MIP51500 E. Holbova and Z. Odlerova; *Czech. Pat.* 175 226 (1978) (*Chem. Abstr.*, 1979, **90**, 121 572). 484

78NJC615 B. Metz and R. Weiss; *Nouv. J. Chim.*, 1978, **2**, 615. 737

78OMR(11)157 J. Jokisaari; *Org. Magn. Reson.*, 1978, **11**, 157. 367

78OMS113 T. Yamaguchi and M. Kojima; *Org. Mass Spectrom.*, 1978, **13**, 113. 135

78OPP133 D. N. Harpp, K. Steliou and B. T. Friedlander; *Org. Prep. Proced. Int.*, 1978, **10**, 133. 623

78OS(58)12 C. Cortez and R. G. Harvey; *Org. Synth.*, 1978, **58**, 12. 115

78OS(58)113 D. Enders, R. Pieter, B. Renger and D. Seebach; *Org. Synth.*, 1978, **58**, 113. 518

78PAC871 J. M. Lehn; *Pure Appl. Chem.*, 1978, **50**, 871. 733, 743

78PAC893 V. Prelog; *Pure Appl. Chem.*, 1978, **50**, 893. 745

78PAC1033 D. M. Jerina, D. R. Thakker and H. Yagi; *Pure Appl. Chem.*, 1978, **50**, 1033. 188, 189

78PNA30 J.-Y. Koo, S. P. Schmidt and G. B. Schuster; *Proc. Natl. Acad. Sci. USA*, 1978, **75**, 30. 485

78PS(4)167 B. Fohlisch and B. Czauderna; *Phosphorus Sulfur*, 1978, **4**, 167. 440, 446

78RTC214 E. Jongejan, T. S. V. Buys, H. Steinberg and T. J. de Boer; *Recl. Trav. Chim. Pays-Bas*, 1978, **97**, 214. 10, 137, 151, 153, 181

78S579 J. D. Finlay, D. J. H. Smith and T. Durst; *Synthesis*, 1978, 579. 422, 445

78S603 L. Chiodini, L. Garanti and G. Zecchi; *Synthesis*, 1978, 603. 597

78S746 S. Kano, T. Ebata, K. Fumaki and S. Shibuya; *Synthesis*, 1978, 746. 251, 266

78SA(A)489 P. Klaboe and Z. Smith; *Spectrochim. Acta, Part A*, 1978, **34**, 489. 453, 455

78T101 J. Boedeker and K. Courault; *Tetrahedron*, 1978, **34**, 101. 474

78T495 J. W. Barton, M. C. Goodland, K. J. Gould, J. Hadley and J. W. F. McOmie; *Tetrahedron*, 1978, **34**, 495. 675

78T1541	M. Apparu and M. Barrelle; *Tetrahedron*, 1978, **34**, 1541.	105
78T1641	M. Corson, B. M. Foxman and P. M. Keehn; *Tetrahedron*, 1978, **34**, 1641.	767, 780
78T1731	A. K. Mukerjee and A. K. Singh; *Tetrahedron*, 1978, **34**, 1731.	238, 249, 253, 254, 255, 256, 257, 258, 259, 262, 265, 266, 303, 331
78T3081	M. Mikolajczyk, S. Grzejszczak, A. Zatorski and B. Mlotkowska; *Tetrahedron*, 1978, **34**, 3081.	623
78T3291	B. Kryczka, A. Laurent and B. Marquet; *Tetrahedron*, 1978, **34**, 3291.	90
78T3353	B. Ganem; *Tetrahedron*, 1978, **34**, 3353.	192
78T3631	H. J. Eggelte, F. Bickelhaupt and B. O. Loopstra; *Tetrahedron*, 1978, **34**, 3631.	680
78TL49	P. Köll and J. Schultz; *Tetrahedron Lett.*, 1978, 49.	391
78TL167	A. Ulman, J. Manessen, F. Frolow and D. Rabinovich; *Tetrahedron Lett.*, 1978, 167.	767
78TL175	D. St. C. Black, N. A. Blackman and A. B. Boscacci; *Tetrahedron Lett.*, 1978, 175.	245
78TL415	P. S. Hammond and D. T. Longone; *Tetrahedron Lett.*, 1978, 415.	767
78TL543	D. S. Kemp, J. C. Chabala and S. A. Marson; *Tetrahedron Lett.*, 1978, 543.	682
78TL577	V. V. Volkova, L. E. Gusel'nikov, V. N. Perchenko, V. G. Zaikin, E. I. Eremina and N. S. Nametkin; *Tetrahedron Lett.*, 1978, 577.	239
78TL615	D. Reuschling, H. Pietsch, and A. Linkies; *Tetrahedron Lett.*, 1978, 615.	250
78TL995	P. J. Giddings, D. I. John and E. J. Thomas; *Tetrahedron Lett.*, 1978, 995.	265, 320
78TL1115	T. Matsui and K. Koga; *Tetrahedron Lett.*, 1978, 1115.	754
78TL1209	S. Kano and T. Yokomatsu; *Tetrahedron Lett.*, 1978, 1209.	528
78TL1543	T. Nishio, A. Katoh, Y. Omote and C. Kashima; *Tetrahedron Lett.*, 1978, 1543.	276, 283
78TL1629	J. Finkelstein, K. G. Holden and C. D. Perchonock; *Tetrahedron Lett.*, 1978, 1629.	354
78TL1755	M. Narisada, H. Onoue, M. Ohtani, F. Watanabe, T. Okada and W. Nagata; *Tetrahedron Lett.*, 1978, 1755.	306
78TL1885	A. Ulman, J. Manassen, F. Frolow and D. Rabinovich; *Tetrahedron Lett.*, 1978, 1885.	767
78TL1931	N. P. Smith and I. D. R. Stevens; *Tetrahedron Lett.*, 1978, 1931.	225
78TL1935	R. A. Moss and W. C. Shieh; *Tetrahedron Lett.*, 1978, 1935.	225
78TL2037	T. Sheradsky and D. Zbaida; *Tetrahedron Lett.*, 1978, 2037.	262
78TL2087	Y. Ito, K. Kobayashi and T. Saegusa; *Tetrahedron Lett.*, 1978, 2087.	626
78TL2469	E. A. Wildi and B. K. Carpenter; *Tetrahedron Lett.*, 1978, 2469.	39, 451, 465, 472, 483
78TL2571	A. Loupy and J. Seyden-Penne; *Tetrahedron Lett.*, 1978, 2571.	757
78TL2821	A. Stütz and H. Reinshagen; *Tetrahedron Lett.*, 1978, 2821.	770, 774
78TL2999	W. H. Rastetter and T. J. Richard; *Tetrahedron Lett.*, 1978, 2999.	570
78TL3333	H.-J. Altenbach, H. Stegelmeier and E. Vogel; *Tetrahedron Lett.*, 1978, 3333.	189
78TL3337	H. D. Martin, R. Iden and H. J. Schiwek; *Tetrahedron Lett.*, 1978, 3337.	433
78TL3567	K. Nishino, K. Nakasuji and I. Murata; *Tetrahedron Lett.*, 1978, 3567.	559, 588, 589
78TL3617	H. Gotthardt and O. M. Huss; *Tetrahedron Lett.*, 1978, 3617.	423, 435, 440
78TL3855	R. J. Bushby and M. D. Pollard; *Tetrahedron Lett.*, 1978, 3855.	422, 445
78TL4167	M. D. Bachi, O. Goldberg and A. Gross; *Tetrahedron Lett.*, 1978, 4167.	308
78TL4233	L. D. Cama and B. G. Christensen; *Tetrahedron Lett.*, 1978, 4233.	353
78TL4587	S. D. Sharma and P. K. Gupta; *Tetrahedron Lett.*, 1978, 4587.	260
78TL4755	A. G. Anastassiou, M. Sabahi and R. Badri; *Tetrahedron Lett.*, 1978, 4755.	712, 724
78TL4839	A. C. Brouwer, A. V. E. George, D. Seykens and H. J. T. Bos; *Tetrahedron Lett.*, 1978, 4839.	423, 435, 440
78TL5153	J. T. Carlock and M. P. Mack; *Tetrahedron Lett.*, 1978, 5153.	81, 115
78TL5171	F. A. Davis, R. Jenkins, Jr. and St. G. Yocklovich; *Tetrahedron Lett.*, 1978, 5171.	209, 444
78TL5233	D. R. Crump; *Tetrahedron Lett.*, 1978, 5233.	441, 446
78TL5239	V. Cere, S. Pollicino, E. Sandri and A. Fava; *Tetrahedron Lett.*, 1978, 5239.	573
78UP51900	F. A. L. Anet, J. Krane and I. Yavari; unpublished results.	700
78USP4070372	M. Garber and D. W. Reger; *U.S. Pat.* 4 070 372 (1978) (*Chem. Abstr.*, 1978, **88**, 170 129).	485
78USP4115385	Pfizer Inc., *U.S. Pat.* 4 115 385 (1978) (*Chem. Abstr.*, 1979, **90**, 152 170).	313
78ZN(B)1072	H. W. Roesky, M. Aramaki and L. Schöenfelder; *Z. Naturforsch., Teil B*, 1978, **33**, 1072.	489
78ZN(B)1208	M. Baudler and F. Saykowski; *Z. Naturforsch., Teil B*, 1978, **33**, 1208.	196
78ZOR1193	A. V. Garabaazhiu, K. A. Vyunov and E. G. Sochilin; *Zh. Org. Khim.*, 1978, **14**, 1193.	381
79ABC2279	M. Shimura, M. Iwata, S. Omoto and Y. Sekizawa; *Agric. Biol. Chem.*, 1979, **43**, 2279.	346
79ACR282	W. A. Smit, N. S. Zefirov, I. V. Bodridov and M. Z. Krimer; *Acc. Chem. Res.*, 1979, **12**, 282.	157, 159, 161, 163, 174, 177
79ACR310	A. Padwa; *Acc. Chem. Res.*, 1979, **12**, 310.	49
79AG(E)219	E. Weber; *Angew. Chem., Int. Ed. Engl.*, 1979, **18**, 219.	733
79AG(E)235	F.-M. Tesky, R. Mews and B. Krebs; *Angew. Chem., Int. Ed. Engl.*, 1979, **18**, 235.	489
79AG(E)312	W. Gilb and G. Schroder; *Angew. Chem., Int. Ed. Engl.*, 1979, **18**, 312.	721
79AG(E)320	B. Arnold and M. Regitz; *Angew. Chem., Int. Ed. Engl.*, 1979, **18**, 320.	248, 266
79AG(E)689	H. R. Kricheldorf; *Angew. Chem., Int. Ed. Engl.*, 1979, **18**, 689.	383
79AG(E)788	W. Adam, O. DeLucchi, H. Quast, R. Recktenwald and F. Yany; *Angew. Chem., Int., Ed. Engl.*, 1979, **18**, 788.	476
79AG(E)868	H. Gotthardt and W. Lenz; *Angew. Chem., Int. Ed. Engl.*, 1979, **18**, 868.	397

79AG(E)900 B. Iddon, O. Meth-Cohn, E. F. V. Scriven, H. Suschitzky and P. T. Gallagher; *Angew. Chem., Int. Ed. Engl.*, 1979, **18**, 900. 533, 535

79AG(E)962 H.-J. Altenbach, H. Stegelmeier, M. Wilhelm, B. Voss, J. Lex and E. Vogel; *Angew. Chem., Int. Ed. Engl.*, 1979, **18**, 962. 111, 677

79AG(E)964 M. Breuninger, B. Gallenkamp, K.-H. Müller, H. Fritz, H. Prinzbach, J. J. Daly and P. Schönholzer; *Angew. Chem., Int. Ed. Engl.*, 1979, **18**, 964. 677

79AHC(24)63 E. Schmitz; *Adv. Heterocycl. Chem.*, 1979, **24**, 63. 28, 209, 213, 215, 218, 220, 221, 222, 223, 228, 230

79AJC567 I. D. Rae; *Aust. J. Chem.*, 1979, **32**, 567. 498

79AJC2659 R. N. Warrener, E. E. Nunn and M. N. Paddon-Row; *Aust. J. Chem.*, 1979, **32**, 2659. 465, 472

79AP514 F. Herold; *Arch. Pharm. (Weinheim, Ger.)*, 1979, **312**, 514. 609

79AP662 F. Eiden and M. Dürr; *Arch. Pharm. (Weinheim, Ger.)*, 1979, **312**, 662. 516, 531

79AP708 F. Eiden and M. Dürr; *Arch. Pharm. (Weinheim, Ger.)*, 1979, **312**, 708. 526

79AX(B)996 A. Carpy, M. Gadret and M. Goursolle; *Acta Crystallogr., Part B*, 1979, **35**, 996. 495

79AX(B)1920 F. K. Winkler and P. Seiler; *Acta Crystallogr., Part B*, 1979, **35**, 1920. 656, 702

79AX(B)2233 E. Maverick, L. Grossenbacher and K. N. Trueblood; *Acta Crystallogr., Part B*, 1979, **35**, 2233. 735, 736

79BAP459 T. Polonski and A. Chimiak; *Bull. Acad. Pol. Sci., Ser. Sci. Chim.*, 1979, **27**, 459. 235

79BAU106 N. B. Kaz'mina, I. L. Knunyants, G. M. Kuz'yants, E. I. Mysov and E. P. Lur'e; *Bull. Acad. Sci. USSR, Div. Chem. Sci.*, 1979, **28**, 106. 477

79BAU168 A. V. Kamernitskii, A. M. Turuta, T. K. Ustynyuk and H. T. M. An; *Bull. Acad. Sci. USSR, Div. Chem. Sci.*, 1979, **28**, 168. 153

79BAU1648 D. P. Del'tsova, N. P. Gambaryan and E. P. Lur'e; *Bull. Acad. Sci. USSR, Div. Chem. Sci.*, 1979, **28**, 1648. 457, 480

79BAU1936 E. N. Deryagina, E. N. Sukhomazova, O. B. Bannikova and M. G. Voronkov; *Bull. Acad. Sci. USSR, Div. Chem. Sci.*, 1979, **28**, 1936. 142

79BCJ1102 T. Nagai, T. Shingaki, M. Inagaki and T. Ohshima; *Bull. Chem. Soc. Jpn.*, 1979, **52**, 1102. 473

79BCJ3371 K. Soai and T. Mukaiyama; *Bull. Chem. Soc. Jpn.*, 1979, **52**, 3371. 172

79BCJ3391 T. Sugawara, Y. Kawada, M. Katoh and H. Iwamura; *Bull. Chem. Soc. Jpn.*, 1979, **52**, 3391. 367

79BCJ3611 N. Watanabe, S. Uemura and M. Okano; *Bull. Chem. Soc. Jpn.*, 1979, **52**, 3611. 381

79BSB297 G. L'abbé, L. Huybrechts, S. Toppet, J.-P. Declercq, G. Germain and M. Van Meerssche; *Bull. Soc. Chim. Belg.*, 1979, **88**, 297. 433

79BSB737 G. L'abbé, J. Flemal, J. P. Declercq, G. Germain and M. Van Meerssche; *Bull. Soc. Chim. Belg.*, 1979, **88**, 737. 219

79BSF(2)499 S. Ratton, J. Moyne and R. Longeray; *Bull. Soc. Chim. Fr., Part 2*, 1979, 499. 458, 474

79CB501 H. Schmidbaur and P. Holl; *Chem. Ber.*, 1979, **112**, 501. 385

79CB678 K.-P. Zeller; *Chem. Ber.*, 1979, **112**, 678. 127

79CB1000 E. Schaumann and J. Ehlers; *Chem. Ber.*, 1979, **112**, 1000. 266

79CB1012 L. Capuano, G. Urhahn and A. Willmes; *Chem. Ber.*, 1979, **112**, 1012. 454

79CB1769 E. Schaumann, J. Ehlers, W.-R. Foerster and G. Adiwidjaja; *Chem. Ber.*, 1979, **112**, 1769. 145

79CB2028 H. Häkli, M. Mintas and A. Mannschreck; *Chem. Ber.*, 1979, **112**, 2028. 200

79CB3318 R. Schwesinger, H. Fritz and H. Prinzbach; *Chem. Ber.*, 1979, **112**, 3318. 189

79CB3347 R. Keller, R. Schwesinger, W. Fritsche, H. W. Schneider, D. Hunkler and H. Prinzbach; *Chem. Ber.*, 1979, **112**, 3347. 189

79CB3914 J. Harnisch and G. Szeimies; *Chem. Ber.*, 1979, **112**, 3914. 268, 270, 271, 273, 275

79CC117 U. K. Pandit and H. Bieräugel; *J. Chem. Soc., Chem. Commun.*, 1979, 117. 673

79CC139 M. Bartok; *J. Chem. Soc., Chem. Commun.*, 1979, 139. 389

79CC160 G. L'abbé, L. Huybrechts, J.-P. Declercq, G. Germain and M. Van Meerssche; *J. Chem. Soc., Chem. Commun.*, 1979, 160. 2, 40, 405, 440

79CC236 A. J. G. Baxter, K. H. Dickinson, P. M. Roberts, T. C. Smale and R. Southgate; *J. Chem. Soc., Chem. Commun.*, 1979, 236. 352

79CC296 J. M. Lehn, P. Vierling and R. C. Hayward; *J. Chem. Soc., Chem. Commun.*, 1979, 296. 746

79CC344 A. C. Kaura and R. J. Stoodley; *J. Chem. Soc., Chem. Commun.*, 1979, 344. 267

79CC382 H. A. J. Carless and H. S. Trivedi; *J. Chem. Soc., Chem. Commun.*, 1979, 382. 377

79CC485 H. Matsumura, T. Yano, M. Ueyama, K. Tori and W. Nagata; *J. Chem. Soc., Chem. Commun.*, 1979, 485. 315

79CC486 I. R. Hanson, D. G. Parsons and M. R. Truter; *J. Chem. Soc., Chem. Commun.*, 1979, 486. 751

79CC532 M. M. Campbell, K. H. Nelson and A. F. Cameron; *J. Chem. Soc., Chem. Commun.*, 1979, 532. 261

79CC663 P. C. Cherry, C. E. Newall and N. S. Watson; *J. Chem. Soc., Chem. Commun.*, 1979, 663. 335

79CC665 C. M. D. Beels, M. S. Abu-Rabie, P. Murray-Rust and J. Murray-Rust; *J. Chem. Soc., Chem. Commun.*, 1979, 665. 335

79CC698 M. Mori, K. Chiba, M. Okita and Y. Ban; *J. Chem. Soc., Chem. Commun.*, 1979, 698. 256, 266

79CC699 M. Mori, K. Chiba, M. Okita and Y. Ban; *J. Chem. Soc., Chem. Commun.*, 1979, 699. 36

79CC846 R. J. Ponsford and R. Southgate; *J. Chem. Soc., Chem. Commun.*, 1979, 846. 258

79CC847 R. J. Ponsford, P. M. Roberts and R. Southgate; *J. Chem. Soc., Chem. Commun.*, 1979, 847. 352

79CC881 J. A. Barltrop, A. C. Day and E. Irving; *J. Chem. Soc., Chem. Commun.*, 1979, 881. 142, 170, 187

79CC899 H. Aoyama, S. Suzuki, T. Hasegawa and Y. Omote; *J. Chem. Soc., Chem. Commun.*, 1979, 899. 258

79CC966 J. A. Barltrop, A. C. Day and E. Irving; *J. Chem. Soc., Chem. Commun.*, 1979, 966. 187

79CC993 A. de Wit, M. L. M. Pennings, W. P. Trompenaars, D. N. Reinhoudt, S. Harkema and O. Nevestveit; *J. Chem. Soc., Chem. Commun.*, 1979, 993. 268, 275

79CC1070 B. Gregory, E. Bullock and Teng-Song Chen; *J. Chem. Soc., Chem. Commun.*, 1979, 1070. 524, 544

79CJC44 B. Gregory, E. Bullock and Teng-Song Chen; *Can. J. Chem.*, 1979, **57**, 44, and references cited therein. 524, 525, 544

79CJC1958 J. M. Gaillot, Y. Gelas-Mialhe and R. Vessiere; *Can. J. Chem.*, 1979, **57**, 1958. 74

79CJC3221 F. Sauriol-Lord and M. St.-Jacques; *Can. J. Chem.*, 1979, **57**, 3221. 623

79CL301 T. Suzuki and E. Kurosawa; *Chem. Lett.*, 1979, 301. 400

79CL923 K. Hiroi and S. Sato; *Chem. Lett.*, 1979, 923. 624

79CPB1981 Y. Arata, K. Tanaka, S. Yoshifuji and S. Kanatomo; *Chem. Pharm. Bull.*, 1979, **27**, 1981. 658

79CPB2183 T. Tsuchiya, M. Enkaku, J. Kurita and H. Sawanishi; *Chem. Pharm. Bull.*, 1979, **27**, 2183. 598

79CPB2188 T. Tsuchiya, M. Enkaku and H. Sawanishi; *Chem. Pharm. Bull.*, 1979, **27**, 2188. 604

79CPB2528 T. Tsuchiya and J. Kurita; *Chem. Pharm. Bull.*, 1979, **27**, 2528. 600

79CPB2589 H. Natsugari, K. Meguro and Y. Kuwada; *Chem. Pharm. Bull.*, 1979, **27**, 2589. 684

79CPB2718 T. Kobayashi and T. Hiraoka; *Chem. Pharm. Bull.*, 1979, **27**, 2718. 322

79CPB2927 H. Natsugari, K. Meguro and Y. Kuwada; *Chem. Pharm. Bull.*, 1979, **27**, 2927. 684

79CRV37 J. S. Bradshaw, G. E. Maas, R. M. Izatt and J. J. Christensen; *Chem. Rev.*, 1979, **79**, 37. 764, 769

79CRV331 N. H. Cromwell and B. Phillips; *Chem. Rev.*, 1979, **79**, 331. 4, 238, 240, 241, 242, 243, 244, 245, 246

79CSR85 J. F. Stoddart; *Chem. Soc. Rev.*, 1979, **8**, 85. 739, 745

79DOK(246)1150 V. F. Rudchenko, O. A. Dyachenko, A. B. Zolotoi, L. O. Atovmyan and R. G. Kostyanovskii; *Dokl. Akad. Nauk SSSR*, 1979, **246**, 1150 (*Chem. Abstr.*, 1979, **91**, 14064). 7

79DOK(248)654 A. I. Naimushin, V. V. Zorin, S. S. Zlotskii, R. A. Karakhanov and D. L. Rakhmankulov; *Dokl. Akad. Nauk SSSR*, 1979, **248**, 654. 364

79FRP2413395 American Cyanamid Co., *Fr. Demande* 2 413 395 (1979) (*Chem. Abstr.*, 1980, **92**, 110 990). 485

79G19 G. L. Bendazzoli, P. Palmieri, G. Gottarelli, I. Moretti and G. Torre; *Gazz. Chim. Ital.*, 1979, **109**, 19. 138

79GEP2824559 M. Iwanami, T. Maeda, Y. Nagano, M. Fujimoto, N. Nagano and A. Yamazaki; *Ger. Offen.* 2 824 559 (1979) (*Chem. Abstr.*, 1979, **90**, 137 844). 485

79GEP2824575 M. Iwanami, T. Maeda, Y. Nagano, M. Fujimoto, N. Nagano, A. Yamazaki, K. Tamazawa, K. Murase and T. Shibanuma; *Ger. Offen.* 2 824 575 (1979) (*Chem. Abstr.*, 1979, **90**, 186 922). 485

79H(12)263 K. Saito and K. Takahashi; *Heterocycles*, 1979, **12**, 263. 519

79H(12)795 T. Kametani, K. Kigasawa, M. Hüragi, K. Wakisaka, H. Sugi and K. Tanigawa; *Heterocycles*, 1979, **12**, 795. 355

79H(12)947 M. Shibuya and S. Kubota; *Heterocycles*, 1979, **12**, 947. 354

79H(12)1311 T. Fushimi, H. Ikuta, H. Irie, K. Nakadachi and S. Uyeo; *Heterocycles*, 1979, **12**, 1311. 532

79H(12)1315 M. Shibuya and S. Kubota; *Heterocycles*, 1979, **12**, 1315. 252

79H(12)1423 Y. Tsuda, M. Kaneda, T. Sano, Y. Horiguchi and Y. Iitaka; *Heterocycles*, 1979, **12**, 1423. 494

79H(12)1427 T. Sano, Y. Horiguchi and Y. Tsuda; *Heterocycles*, 1979, **12**, 1427. 501, 503, 518

79H(12)1471 T. Tsuchiya, M. Enkaku and H. Sawanishi; *Heterocycles*, 1979, **12**, 1471. 598, 600

79H(13)227 J. C. Sheehan, K. Nakajima and E. Chacko; *Heterocycles*, 1979, **13**, 227. 320, 635

79H(13)353 A. Terahara, T. Haneishi and M. Arai; *Heterocycles*, 1979, **13**, 353. 192

79HCA152 J. Barrau, M. Bouchaut, H. Lavayssiere, G. Dousse and *Acta*, 1979, **62**, 152. 156

79HCA1103 R. Süess; *Helv. Chim. Acta*, 1979, **62**, 1103. 516

79HCA2763 B. Dietrich, D. L. Fyles, T. M. Fyles and J. M. Lehn; *Helv. Chim. Acta*, 1979, **62**, 2763. 748

79IJC(B)107 K. Bhandari, V. Virmani, V. A. Murti, P. C. Jain and N. Anand; *Indian J. Chem., Sect. B*, 1979, **17**, 107. 514

79IZV1788 D. P. Deltsova, N. P. Gambaryan and E. P. Lure; *Izv. Akad. Nauk SSSR, Ser. Khim.*, 1979, 1788. 229

79JA1 S. J. Valenty; *J. Am. Chem. Soc.*, 1979, **101**, 1. 352

79JA254 M. P. Schneider and M. Schnaithmann; *J. Am. Chem. Soc.*, 1979, **101**, 254. 140, 143

79JA390 L. A. Carpino and H.-W. Chen; *J. Am. Chem. Soc.*, 1979, **101**, 390. 135, 142, 143, 159, 176

79JA783 M. J. S. Dewar and G. P. Ford; *J. Am. Chem. Soc.*, 1979, **101**, 783. 138

79JA837 Y. Maeda and K. U. Ingold; *J. Am. Chem. Soc.*, 1979, **101**, 837. 19

79JA1040	G. S. Wilson, D. D. Swanson, J. T. Klug and R. S. Glass, M. D. Ryan and W. K. Musker; *J. Am. Chem. Soc.*, 1979, **101**, 1040.	694
79JA1047	G. R. Newkome, V. K. Majestic, F. R. Fronczek and J. L. Atwood; *J. Am. Chem. Soc.*, 1979, **101**, 1047.	738
79JA1623	M. Suzuki, Y. Oda and R. Noyori; *J. Am. Chem. Soc.*, 1979, **101**, 1623.	105
79JA2470	D. R. Boyd and G. A. Berchtold; *J. Am. Chem. Soc.*, 1979, **101**, 5475.	522, 558, 565, 566, 567, 584
79JA2560	B. Bigot, D. Roux, A. Sevin and A. Devaquet; *J. Am. Chem. Soc.*, 1979, **101**, 2560.	198
79JA2738	S. Murata, M. Suzuki and R. Noyori; *J. Am. Chem. Soc.*, 1979, **101**, 2738.	110
79JA3000	A. G. Sherwood, I. Safarik, B. Verkoczy, G. Almadi, H. A. Wiebe and O. P. Strausz; *J. Am. Chem. Soc.*, 1979, **101**, 3000.	142, 176
79JA3404	S. D. Levine, R. E. Adams, R. Chen, M. L. Cotter, A. F. Hirsch, V. V. Kane, R. M. Kanojia, C. Shaw, M. P. Wachter, E. Chin, R. Heuttemann, P. Ostrowski, J. L. Mateos, L. Noriega, A. Guzman, A. Mijarez and L. Tovar; *J. Am. Chem. Soc.*, 1979, **101**, 3404.	590
79JA3670	K. Maruyama, T. Ishitoku and Y. Kubo; *J. Am. Chem. Soc.*, 1979, **101**, 3670.	267
79JA4051	M. Shamsipur and A. I. Popov; *J. Am. Chem. Soc.*, 1979, **101**, 4051.	740
79JA4403	S. Uyeo, I. Kikkawa, Y. Hamamshima, H. Ona, Y. Nishitani, K. Okada, T. Kubota, K. Ishikura, Y. Ide, K. Nakano and W. Nagata; *J. Am. Chem. Soc.*, 1979, **101**, 4403.	311
79JA4730	J. G. Gleason, T. F. Buckley, K. G. Holden, D. B. Bryan and P. Siler; *J. Am. Chem. Soc.*, 1979, **101**, 4730.	355
79JA4736	R. A. Moss, M. Fedorynski and W. C. Shieh; *J. Am. Chem. Soc.*, 1979, **101**, 4736.	227
79JA4755	T. B. Cameron and H. W. Pinnick; *J. Am. Chem. Soc.*, 1979, **101**, 4755.	6
79JA5059	K. Nishino, S. Yano, Y. Kohashi, K. Yamamoto and I. Murata; *J. Am. Chem. Soc.*, 1979, **101**, 5059.	556, 588
79JA5082	K. T. Chang and H. Shechter; *J. Am. Chem. Soc.*, 1979, **101**, 5082.	223
79JA5343	H. Aoyama, T. Hasegawa and Y. Omote; *J. Am. Chem. Soc.*, 1979, **101**, 5343.	258
79JA5435	H. W. Moore, F. Mercer, D. Kunert and P. Albaugh; *J. Am. Chem. Soc.*, 1979, **101**, 5435.	38
79JA5546	J. Kao and T.-N. Huang; *J. Am. Chem. Soc.*, 1979, **101**, 5546.	452, 670
79JA6135	S. C. Welch and A. S. C. P. Rao; *J. Am. Chem. Soc.*, 1979, **101**, 6135.	40, 393
79JA6296	M. Lang, K. Prasad, W. Holick, J. Gosteli, I. Ernest and R. B. Woodward; *J. Am. Chem. Soc.*, 1979, **101**, 6296.	252
79JA6306	H. R. Pfaendler, J. Gosteli and R. B. Woodward; *J. Am. Chem. Soc.*, 1979, **101**, 6306.	252, 335
79JA6511	W. Adam and O. Cueto; *J. Am. Chem. Soc.*, 1979, **101**, 6511.	459
79JA6671	Y. Hata and M. Watanabe; *J. Am. Chem. Soc.*, 1979, **101**, 6671.	209
79JA6677	P. Beeken, J. N. Bonfiglio, I. Hasan, J. J. Piwinski, B. Weinstein, K. A. Zollo and F. W. Fowler; *J. Am. Chem. Soc.*, 1979, **101**, 6677.	342, 343, 344, 347
79JA7088	G. R. Newkome and T. Kawato; *J. Am. Chem. Soc.*, 1979, **101**, 7088.	779
79JA7684	J. Nakayama, T. Fukushima, E. Seki and M. Hoshino; *J. Am. Chem. Soc.*, 1979, **101**, 7684.	442
79JAP(K)7988255	American Cyanamid Co., *Jpn. Kokai Tokkyo Koho* 79 88 255 (1979) (*Chem. Abstr.*, 1979, **91**, 157 719).	485
79JCR(S)56	J.-C. Pommelet and J. Chuche; *J. Chem. Res. (S)*, 1979, 56.	143, 168
79JCR(S)157	A. Schutz and I. Ugi; *J. Chem. Res. (S)*, 1979, 157.	333
79JCR(S)214	T. L. Gilchrist, C. W. Rees and I. W. Southon; *J. Chem. Res. (S)*, 1979, 214.	646
79JCR(S)246	R. Grandi, U. M. Pagnoni and R. Trave; *J. Chem. Res. (S)*, 1979, 246.	773
79JCR(S)320	L. Kristenbrugger, C.-P. Klages and J. Voss; *J. Chem. Res. (S)*, 1979, 320.	435
79JCR(S)395	K. Clarke, B. Gleadhill and R. M. Scrowston; *J. Chem. Res. (S)*, 1979, 395.	645
79JCS(D)1	M. W. Bishop, J. Chatt and J. R. Dilworth; *J. Chem. Soc., Dalton Trans.*, 1979, 1.	156
79JCS(P1)228	J. Emsley, D. W. Griffiths and G. J. J. Jayne; *J. Chem. Soc., Perkin Trans. 1*, 1979, 228.	176
79JCS(P1)263	H. Aoyama, T. Hasegawa, M. Okazaki and Y. Omote; *J. Chem. Soc., Perkin Trans. 1*, 1979, 263.	258
79JCS(P1)584	R. M. Acheson, J. D. Wallis and J. Woollard; *J. Chem. Soc., Perkin Trans. 1*, 1979, 584.	676
79JCS(P1)765	R. C. Cambie, H. H. Lee, P. S. Rutledge and P. D. Woodgate; *J. Chem. Soc., Perkin Trans. 1*, 1979, 765.	172
79JCS(P1)771	M. Mori, S. Kudo and Y. Ban; *J. Chem. Soc., Perkin Trans. 1*, 1979, 771.	526, 533
79JCS(P1)1066	A. Ulman and J. Manassen; *J. Chem. Soc., Perkin Trans. 1*, 1979, 1066.	767
79JCS(P1)1166	D. H. R. Barton, L. S. L. Choi, R. H. Hesse, M. M. Pechet and C. Wilshire; *J. Chem. Soc., Perkin Trans. 1*, 1979, 1166.	174
79JCS(P1)1195	L. Garanti and G. Zecchi; *J. Chem. Soc., Perkin Trans. 1*, 1979, 1195.	597
79JCS(P1)1433	K. L. M. Stanley, J. Dingwall, J. T. Sharp and T. W. Naisby; *J. Chem. Soc., Perkin Trans. 1*, 1979, 1433.	596
79JCS(P1)1640	C. D. Anderson, P. N. Anderson and J. T. Sharp; *J. Chem. Soc., Perkin Trans. 1*, 1979, 1640.	597
79JCS(P1)1847	T. Kametani, K. Takahashi, M. Ihara and K. Fukumoto; *J. Chem. Soc., Perkin Trans. 1*, 1979, 1847.	659
79JCS(P1)1852	C. D. Maycock and R. J. Stoodley, *J. Chem. Soc., Perkin Trans. 1*, 1979, 1852.	172, 178

79JCS(P1)1978 R. D. Chambers, W. K. R. Musgrave and D. E. Wood; *J. Chem. Soc., Perkin Trans. 1*, 1979, 1978. 282

79JCS(P1)2009 M. Lennon and G. R. Proctor; *J. Chem. Soc., Perkin Trans. 1*, 1979, 2009. 518, 657

79JCS(P1)2209 C. D. Anderson, J. T. Sharp and R. S. Strathdee; *J. Chem. Soc., Perkin Trans. 1*, 1979, 2209. 595, 597, 600

79JCS(P1)2268 C. L. Branch and M. J. Pearson; *J. Chem. Soc., Perkin Trans. 1*, 1979, 2268. 253, 309, 312

79JCS(P1)2401 R. J. Bushby and M. D. Pollard; *J. Chem. Soc., Perkin Trans. 1*, 1979, 2401. 153

79JCS(P1)2437 M. N. Akhtar, D. R. Boyd and J. G. Hamilton; *J. Chem. Soc., Perkin Trans. 1*, 1979, 2437. 554

79JCS(P1)2455 P. H. Bentley, J. P. Clayton, M. O. Boles and R. J. Girven; *J. Chem. Soc., Perkin Trans. 1*, 1979, 2455. 321

79JCS(P1)2577 A. H. Haines and P. Karntiang; *J. Chem. Soc., Perkin Trans. 1*, 1979, 2577. 751

79JCS(P1)2730 C. D. Anderson, J. T. Sharp and R. S. Strathdee; *J. Chem. Soc., Perkin Trans. 1*, 1979, 2730. 600

79JCS(P1)2954 R. J. Cave, C. C. Howard, G. Klinkert, R. F. Newton, D. P. Reynolds, A. H. Wadsworth and S. M. Roberts; *J. Chem. Soc., Perkin Trans. 1*, 1979, 2954. 112

79JCS(P1)3013 M. U. Bombala and S. V. Ley; *J. Chem. Soc., Perkin Trans. 1*, 1979, 3013. 172

79JCS(P1)3175 P. M. Denerley and E. J. Thomas; *J. Chem. Soc., Perkin Trans. 1*, 1979, 3175. 310

79JCS(P2)86 J. P. Degelaen, S. L. Loukas, J. Feeney, G. C. K. Roberts and A. S. V. Burgen; *J. Chem. Soc., Perkin Trans. 2*, 1979, 86. 326

79JHC339 R. F. Lauer and G. Zenchoff; *J. Heterocycl. Chem.*, 1979, **16**, 339. 606

79JHC445 T. Sugasawa, M. Adachi, T. Toyoda and K. Sasakura; *J. Heterocycl. Chem.*, 1979, **16**, 445. 610

79JHC757 V. Sunjic, A. Lisini, A. Sega, T. Kovac and F. Kajfez; *J. Heterocycl. Chem.*, 1979, **16**, 757. 608

79JHC835 O. Migliari, S. Petruso and V. Sprio; *J. Heterocycl. Chem.*, 1979, **16**, 835. 644

79JHC1009 S. Mataka, K. Takahashi, Y. Yamada and M. Tashiro; *J. Heterocycl. Chem.*, 1979, **16**, 1009. 488

79JHC1081 P. M. Weiatraub; *J. Heterocycl. Chem.*, 1979, **16**, 1081. 624

79JHC1317 B. L. Jensen and M. A. Woods; *J. Heterocycl. Chem.*, 1979, **16**, 1317, and references cited therein. 517, 530

79JHC1411 M. Bianchi, E. Hausermann and S. Rossi; *J. Heterocycl. Chem.*, 1979, **16**, 1411. 637

79JHC1435 A. Mohsen, M. E. Omar, F. A. Ashour and J. Baurdais; *J. Heterocycl. Chem.*, 1979, **16**, 1435. 638

79JHC1449 T. Kovac, M. Oklobdzija, V. Sunjic and K. Kajfez; *J. Heterocycl. Chem.*, 1979, **16**, 1449. 619

79JHC1525 G. Bobowski, J. M. Gottlieb, B. West and J. Shavel, Jr.; *J. Heterocycl. Chem.*, 1979, **16**, 1525. 517, 533

79JMC1 L. H. Sternbach; *J. Med. Chem.*, 1979, **22**, 1. 608, 610, 612, 613

79JMC183 P. Melloni, A. Della Torre, M. Meroni, A. Ambrosini and A. C. Rossi; *J. Med. Chem.*, 1979, **22**, 183. 246, 266

79JMC533 D. B. Boyd; *J. Med. Chem.*, 1979, **22**, 533. 303

79JMC1435 W. J. Leanza, K. J. Wildonger, T. W. Miller and B. G. Christensen; *J. Med. Chem.*, 1979, **22**, 1435. 356

79JMR(34)251 W. H. Pirkle, P. L. Rinaldi and K. A. Simmons; *J. Magn. Reson.*, 1979, **34**, 251. 200

79JMR(36)343 G. Fronza, R. Mondelli and S. Bradamante; *J. Magn. Reson.*, 1979, **36**, 343. 409

79JOC287 S. R. Wilson and R. A. Sawicki; *J. Org. Chem.*, 1979, **44**, 287. 657

79JOC356 W. Adam and H.-H. Fick; *J. Org. Chem.*, 1979, **44**, 356. 394

79JOC450 D. F. Taber and B. P. Gunn; *J. Org. Chem.*, 1979, **44**, 450. 115

79JOC468 D. R. Boyd and G. A. Berchtold; *J. Org. Chem.*, 1979, **44**, 468. 564, 570

79JOC733 W. T. Brady and K. Saidi; *J. Org. Chem.*, 1979, **44**, 733. 38, 398

79JOC830 Y. Yoshida, M. Komatsu, Y. Ohshiro and T. Agawa; *J. Org. Chem.*, 1979, **44**, 830. 155

79JOC1264 B. M. Trost, P. H. Scudder, R. M. Cory, N. J. Turro, V. Ramamurthy and T. J. Katz; *J. Org. Chem.*, 1979, **44**, 1264. 671

79JOC1736 K. Oka; *J. Org. Chem.*, 1979, **44**, 1736. 477

79JOC1742 R. M. Scarborough, Jr., A. B. Smith, III, W. E. Barnette and K. C. Nicolaou; *J. Org. Chem.*, 1979, **44**, 1742. 399

79JOC2041 P. Tarburton, J. P. Edasery, C. A. Kingsbury, A. E. Sopchik and N. H. Cromwell; *J. Org. Chem.*, 1979, **44**, 2041. 49, 51

79JOC2055 C. R. Johnson, E. U. Jonsson and C. C. Bacon; *J. Org. Chem.*, 1979, **44**, 2055. 439

79JOC2083 S. Hirokami, Y. Hirai and M. Nagata; *J. Org. Chem.*, 1979, **44**, 2083. 258

79JOC2244 A. Ohno, M. Uohama, K. Nakamura and S. Oka; *J. Org. Chem.*, 1979, **44**, 2244. 175

79JOC2288 M. E. Landis and J. C. Mitchell; *J. Org. Chem.*, 1979, **44**, 2288. 472

79JOC2683 J. A. Moore, H. B. Yokelson, W. J. Freeman and J. F. Blount; *J. Org. Chem.*, 1979, **44**, 2683. 600, 607

79JOC3084 D. O. Spry; *J. Org. Chem.*, 1979, **44**, 3084. 307

79JOC3574 T. Imai and S. Nishida; *J. Org. Chem.*, 1979, **44**, 3574. 374

79JOC3678 S. Krishnamurthy and H. C. Brown; *J. Org. Chem.*, 1979, **44**, 3678. 385

79JOC3843 H. W. Heine, D. W. Ludovici, J. A. Pardoen, R. C. Weber, II, E. Bonsall and K. R. Osterhout; *J. Org. Chem.*, 1979, **44**, 3843. 606

79JOC3861 A. Hassner and V. Alexanian; *J. Org. Chem.*, 1979, **44**, 3861. 44, 89

79JOC3935 A. Nabeya, J. Saito and H. Koyama; *J. Org. Chem.*, 1979, **44**, 3935. 214

79JOC3946 S. Kano, T. Ebata, K. Funaki and S. Shibuya; *J. Org. Chem.*, 1979, **44**, 3946. 255

79JOC3972 J. A. Profitt and H. H. Ong; *J. Org. Chem.*, 1979, **44**, 3972. 525

79JOC4210 J. Jernow, W. Tautz, P. Rosen and J. F. Blount; *J. Org. Chem.*, 1979, **44**, 4210. 192

79JOC4213 J. J. Tufariello, S. A. Ali and H. O. Klingele, *J. Org. Chem.*, 1979, **44**, 4213. 511, 525

79JOC4473 D. S. Kemp, M. D. Sidell and T. J. Shortridge; *J. Org. Chem.*, 1979, **44**, 4473. 682

79JOC4757 D. J. H. Smith, J. D. Finlay, C. R. Hall and J. J. Uebel; *J. Org. Chem.*, 1979, **44**, 4757. 409, 415, 420, 436, 445

79JOM(172)285 G. Barbieri, G. D. Andreeti, G. Bocelli and P. Sgarabotto; *J. Organomet. Chem.*, 1979, **172**, 285 132

79JOU1272 S. S. Mochalov, A. N. Fedotov, A. I. Sizov and Yu. S. Shabarov; *J. Org. Chem. USSR* (*Engl. Transl.*), 1979, **15**, 1272. 536

79JOU1401 V. I. Minkin and R. M. Minyaev; *J. Org. Chem. USSR* (*Engl. Transl.*), 1979, **15**, 1401. 143

79JPC1890 D. E. Powers, C. A. Arrington, W. C. Harris, E. Block and V. F. Kalasinsky; *J. Phys. Chem.*, 1979, **83**, 1890 460

79JPC2025 A. A. Scala and I. Colon; *J. Phys. Chem.*, 1979, **83**, 2025. 422

79JPR1034 S. M. M. Elshafie; *J. Prakt. Chem.*, 1979, **321**, 1034. 518, 538

79JPS890 S. P. Singh, S. S. Parmer, S. A. Farnum and V. I. Stenberg; *J. Pharm. Sci.*, 1979, **68**, 890. 498

79JST(52)275 J. B. Moffat; *J. Mol. Struct.*, 1979, **52**, 275. 198

79JST(54)287 I. Hargittai; *J. Mol. Struct.*, 1979, **54**, 287. 453, 455

79JST(56)157 J. A. Duckett, T. L. Smithson and H. Wiese; *J. Mol. Struct.*, 1979, **56**, 157. 365, 370

79KGS1637 S. M. Shevchenko and A. A. Potekhin; *Khim. Geterotsikl. Soedin.*, 1979, 1637. 159

79LA219 I. Hoppe and U. Schoellkopf; *Liebigs Ann. Chem.*, 1979, 219. 395

79LA473 A. Krebs, R. Kemper, H. Kimling, K. H. Klaska and R. Klaska; *Liebigs Ann. Chem.*, 1979, 473. 575

79LA639 R. Brücker, J.-P. Lavergne and P. Viallefont; *Liebigs Ann. Chem.*, 1979, 639. 638

79LA858 K. Frensch, G. Oepen and F. Vögtle; *Liebigs Ann. Chem.*, 1979, 858. 733

79LA1322 E. Schaumann and H. Behr; *Liebigs Ann. Chem.*, 1979, 1322. 266

79LA1715 E. Schaumann and F. F. Grabley; *Liebigs Ann. Chem.*, 1979, 1715 480

79LA1768 B. Fohlisch and W. Gottstein; *Liebigs Ann. Chem.*, 1979, 1768. 421, 427

79LA2036 K.-P. Zeller; *Liebigs Ann. Chem.*, 1979, 2036. 120

B-79MI50100 C. W. Rees; in 'New Trends in Heterocyclic Chemistry', ed. R. B. Mitra, N. R. Ayyangar, V. N. Gogte, R. M. Acheson and N. Cromwell; Elsevier, Amsterdam, 1979, pp. 365–372. 4, 8, 10

B-79MI50101 E. Eliel and K. M. Petrusiewicz; in 'Topics in 13-C NMR Specotroscopy', ed. G. C. Levy; Wiley, New York, 1979, vol. 3, pp. 171–282. 11

79MI50102 V. F. Kalasinsky, E. Block, D. E. Powers and W. C. Harris; *Appl. Spectrosc.*, 1979, **33**, 361. 14

B-79MI50400 A. Hassner and V. Alexanian; in 'New Trends in Heterocyclic Chemistry', ed. R. B. Mitra; Elsevier, Amsterdam, 1979, p. 178. 91, 92

79MI50500 K. B. Sharpless and T. R. Verhoeven; *Aldrichim. Acta*, 1979, **12**(4), 63. 116

79MI50501 R. Landau, G. A. Sullivan and D. Brown; *Chemtech*, 1979, **9**, 602. 116, 118

B-79MI50502 T. Sugawara, Y. Kawada, H. Iwamura, K. Tori, R. Muneyuki and R. Noyori; in 'Koen Yoshishu–Hibenzenkei Hokozuku Kagaku Toronkai [oyobi] Kozo Yuki Kagaku Toronkai, 12th', 1979, 225 (*Chem. Abstr.*, 1979, **92**, 197 407). 99

79MI50600 G. Capozzi, V. Lucchini and G. Modena; *Rev. Chem. Intermed.*, 1979, **2**, 347. 133, 134 137, 153, 154, 155, 158, 163, 175, 177

79MI50601 A. V. Kazantsev, T. V. Shustova and L. I. Zakharkin; *Izv. Akad. Nauk Kaz. SSR, Ser. Khim.*, 1979, 72 (*Chem. Abstr.*, 1979, **91**, 39 559). 149

B-79MI50602 A. Amano, K. Hashimoto and J. Nishino; 'Kosoku Hanno Toronkai Koen Yokoshu 14th', 1979, p. 61 (*Chem. Abstr.*, 1981, **94**, 3372). 166

79MI50800 M. Bogucka-Ledochowska, A. Konitz, A. Hempel, Z. Dauter and E. Borowski; *Z. Kristallogr.*, 1979, **149**, 49. 198

79MI50802 A. Sekiya and D. D. Desmarteau; *J. Fluorine Chem.*, 1979, **14**, 289. 211

B-79MI50900 K. Nomoto, H. Yoshioka, T. Takemoto, S. Fushiya, S. Nozoe and S. Takagi; 'Koen Yoshishu — Tennen Yuki Kagobutsu Toronkai, 22nd,' Kyushu University, 1979, p. 619. 239

B-79MI51000 K. G. Holden, J. G. Gleason, W. F. Huffman and C. D. Perchonock; in 'Drug Action and Design', ed. T. I. Kalman; Elsevier-North Holland, New York, 1979, p. 225. 295

79MI51001 N. A. C. Curtis, D. Orr, G. W. Ross and M. G. Boulton; *Antimicrob. Agents Chemother.*, 1979, **16**, 533. 296

B-79MI51002 J. R. E. Hoover and G. L. Dunn; in 'Burger's Medicinal Chemistry', ed. M. E. Wolff; Wiley, New York, 4th edn., 1979, part 2, p. 83. 298

79MI51100 P. Borowicz and Z. Eckstein; *Pol. J. Pharmacol. Pharm.*, 1979, **31**, 227. 330

B-79MI51101 'Beta-Lactamases', ed. J. M. T. Hamilton-Miller and J. T. Smith; Academic Press, London, 1979. 327, 339

79MI51102 D. J. Tipper; *Rev. Infect. Dis.*, 1979, **1**, 39. 338

79MI51103 A. Tomasz; *Annu. Rev. Microbiol.*, 1979, **33**, 113. 338

79MI51201 A. G. Brown, D. F. Corbett, A. J. Eglington and T. T. Howarth; *J. Antibiot.*, 1979, **32**, 961. 352

B-79MI51202	K. G. Holden, J. G. Gleason, W. F. Huffman and C. D. Perchonock; in 'Drug Action and Design: Mechanism-Based Enzyme Inhibitors', ed. T. I. Kalman; Elsevier/North-Holland, New York, 1979, p. 225.	354
79MI51300	A. Leborgne, N. Spassky and P. Sigwalt; *Polym. Bull.* (*Berlin*), 1979, **1**, 825 (*Chem. Abstr.*, 1980, **92**, 111 357).	383
79MI51301	P. C. Martino, P. B. Shevlin and S. D. Worley; *Chem. Phys. Lett.*, 1979, **68**, 237.	364, 368
79MI51302	R. K. Nurieva, I. N. Siraeva, E. A. Kantor and D. L. Rakhmankulov; *React. Kinet. Catal. Lett.*, 1979, **11**, 259 (*Chem. Abstr.*, 1979, **91**, 123 079).	380
79MI51400	R. Rahman and Y. Avny; *J. Macromol. Sci., Chem.*, 1979, **A13**, 971.	419, 432
B-79MI51402	D. C. Dittmer; in 'New Trends in Heterocyclic Chemistry', ed. R. B. Mitra; Elsevier, Amsterdam, 1979.	404, 412, 415, 418, 447
79MI51500	Y. Imai, H. Hirukawa, K. Okuyama and M. Ueda; *Makromol. Chem.*, 1979, **180**, 25.	484
79MI51501	Y. Imai, M. Ueda and K. Okuyama; *Makromol. Chem.*, 1979, **180**, 1413.	484
79MI51502	G. B. Schuster, B. Dixon, J.-Y. Koo, S. P. Schmidt and J. P. Smith; *Photochem. Photobiol.*, 1979, **30**, 17.	459
B-79MI51503	P. D. Bartlett and M. E. Landis; in 'Singlet Oxygen', ed. H. H. Wasserman and R. W. Murray; Academic Press, New York, 1979, p. 243.	459, 461, 469, 476, 485
79MI51600	E. Hannig, R. Pech and C. Dressler; *Pharmazie*, 1979, **34**, 670.	511
79MI51601	A. Costa, A. G. Roso, J. V. Sinisterra and J. M. Marinas; *Ann. Quim.*, 1979, **75**, 381.	530
79MI51602	E. A. Bliss, T. B. Brown, M. F. G. Stevens and C. K. Wong; *J. Pharm. Pharmacol.*, 1979, **31** (Supp.), 66.	535
B-79MI51603	J. I. G. Cadogan; in 'Organophosphorus Reagents in Organic Synthesis', ed. J. I. G. Cadogan; Academic Press, London, 1979, p. 269.	535
79MI51700	T. Abe and S. Nagase; *J. Fluorine Chem.*, 1979, **13**, 519.	562
79MI51800	M. Japelj, V. Tišler, B. Novac and D. Babič; *Vestn. Slov. Kem. Drus.*, 1979, **26**, 215.	609
79MI51801	P. Richter and K. Gerisch; *Pharmazie*, 1979, **34**, 847.	638
B-79MI52100	S. Lindenbaum, J. H. Rytting and L. A. Sternson; in 'Progress in Macrocyclic Chemistry', ed. R. M. Izatt and J. J. Christensen; Wiley, New York, 1979, vol. 1, p. 219.	732, 755
B-79MI52101	W. E. Morf, D. Ammann, R. Bissig, E. Pretsch and W. Simon; in ref. B-79MI52100, p. 1.	732, 760
79MI52102	I. M. Kolthoff; *Anal. Chem.*, 1979, 1R.	759
B-79MI52103	A. I. Popov and J. M. Lehn; in 'Coordination Chemistry of Macrocyclic Compounds', ed. G. A. Melson; Plenum Press, New York, 1979, p. 537.	737, 739, 740
79MI52104	L. J. Mathias and K. B. Al-Jumah; *Polym. News*, 1979, **6**, 9.	760
B-79MI52105	J. L. Dye; in ref. B-79MI52100, p. 63.	744
79MI52106	B. E. Jepson and G. A. Cairns; *Energy Res. Abstr.*, 1979, **4**, Abstr. 47110.	759
B-79MI52107	R. M. Izatt, J. D. Lamb, D. J. Eatough, J. J. Christensen and J. G. Rytting; in 'Drug Design', ed. E. J. Ariëns; Academic Press, New York, 1979, vol. 8, p. 355.	760
B-79MI52200	D. N. Reinhoudt and F. DeJong; in 'Progress in Macrocyclic Chemistry', ed. R. M. Izatt and J. J. Christensen; Wiley Interscience, New York, 1979, chap. 4.	764
B-79MI52201	'Coordination Chemistry of Macrocyclic Compounds', ed. G. A. Melson; Plenum Press, New York, 1979.	766
B-79MI52202	A. T. Balaban; in 'New Trends in Heterocyclic Chemistry', ed. R. B. Mitra; Elsevier, New York, 1979, p. 79.	772
79NEP7712731	American Cyanamid Co., *Neth. Appl.* 77 12 731 (1979) (*Chem. Abstr.*, 1979, **91**, 135 625).	485
79NJC669	A. Momtaz, N. Spassky and P. Sigwalt; *Nouv. J. Chim.*, 1979, **3**, 669.	157
79NJC715	H. Meier, S. Graw, U. Timm and T. Echter; *Nouv. J. Chim.*, 1979, **3**, 715.	193
79OMR(12)593	G. Romeo, M. C. Aversa, P. Giannetto, M. G. Vigorita and P. Ficarra; *Org. Magn. Reson.*, 1979, **12**, 593.	608
79OMS543	W. J. Broer, W. D. Weringa and W. C. Nieuwpoort; *Org. Mass Spectrom.*, 1979, **14**, 543.	135
79PAC979	J. M. Lehn; *Pure Appl. Chem.*, 1979, **51**, 979.	753, 755
79PNA2595	C. M. Gupta, R. Radhakrishnan, G. E. Gerber, W. L. Olsen, St. C. Quay and H. G. Khorana; *Proc. Natl. Acad. Sci. USA*, 1979, **76**, 2595.	236
79PS(7)5	B. H. Patwardhan, E. J. Parker and D. C. Dittmer; *Phosphorus Sulfur*, 1979, **7**, 5.	415, 441
79RTC334	O. L. Chapman, R. S. Sheridan and J.-P. LeRoux; *Recl. Trav. Chim. Pays-Bas*, 1979, **98**, 334.	534
79S130	L. Bemi, M. T. Thomas and V. Snieckus; *Synthesis*, 1979, 130.	604
79S380	L. Garanti, G. Testori and G. Zecchi; *Synthesis*, 1979, 380.	597
79S388	W. Adam and L. A. A. Encarnacion; *Synthesis*, 1979, 388.	373
79S537	G. A. Olah and A. P. Fung; *Synthesis*, 1979, 537.	655
79S589	N. S. Narasimhan and P. S. Chandrachood; *Synthesis*, 1979, 589.	634
79S975	W. Kantlehner and H.-D. Gutbrod; *Synthesis*, 1979, 975.	621
79SA(A)509	J. Barrett and F. S. Deghaidy; *Spectrochim. Acta, Part A*, 1979, **35**, 509.	456
79SA(A)895	N. C. Craig and M. A. Kliewer; *Spectrochim. Acta, Part A*, 1979, **35**, 895.	204
79SC113	G. Just, G. H. Hakimelahi, A. Ugolini and R. Zamboni; *Synth. Commun.*, 1979, **9**, 113.	355

79SC179	M. Shiozaki and T. Hiraoka; *Synth. Commun.*, 1979, **9**, 179.	258
79SC201	S. W. Pelletier, J. Nowacki and N. V. Mody; *Synth. Commun.*, 1979, **9**, 201.	159
79SC857	L. Mandolini and T. Vontor; *Synth. Commun.*, 1979, **9**, 857.	692
79T1059	M. Itoh, K. Murata, K. Tokumura, K. Shudo, N. Miyata and K. Okamoto; *Tetrahedron*, 1979, **35**, 1059.	565
79T1065	T. Yamabe, K. Hori, K. Akagi and K. Fukui; *Tetrahedron*, 1979, **35**, 1065.	741
79T1273	S. Yamada and C. Kaneko; *Tetrahedron*, 1979, **35**, 1273.	625
79T1499	D. B. Boyd, J. P. Riehl and F. S. Richardson; *Tetrahedron*, 1979, **35**, 1499.	303, 349, 356
79T2501	J. C. Arnould, A. Enger, A. Feigenbaum and J. P. Pete; *Tetrahedron*, 1979, **35**, 2501.	394
79TH51600	R. Purvis; Ph.D. Thesis, University of Salford, 1979.	498, 535
79TL59	R. Schulz and A. Schweig; *Tetrahedron Lett.*, 1979, 59.	193, 441
79TL549	H. H. Wasserman D. J. Hlasta, A. W. Tremper and J. S. Wu; *Tetrahedron Lett.*, 1979, 549.	256
79TL559	J. Drapier, A. Feron, R. Warin, A. J. Hubert and P. Teyssie; *Tetrahedron Lett.*, 1979, 559.	93
79TL715	G. W. Holbert, B. Ganem, D. Van Engen, J. Clardy, L. Borsub, K. Chantrapromma, C. Sadavongvivad and Y. Thebtaranonth; *Tetrahedron Lett.*, 1979, 715.	192
79TL1039	Y. Ito, K. Kobayashi and T. Saegusa; *Tetrahedron Lett.*, 1979, 1039.	605
79TL1133	M. Mori and Y. Ban; *Tetrahedron Lett.*, 1979, 1133.	526, 533
79TL1257	M. M. Campbell and D. I. Rawson; *Tetrahedron Lett.*, 1979, 1257.	685
79TL1321	V. Reutrakul, S. Nimgirawath, S. Panichanun and Y. Srikirin; *Tetrahedron Lett.*, 1979, 1321.	115
79TL1381	J. Charrier, H. Person and A. Foucaud; *Tetrahedron Lett.*, 1979., 1381.	266
79TL1441	M. M. Campbell and P. G. Harcus; *Tetrahedron Lett.*, 1979, 1441.	265
79TL1503	V. Hunig, F. Derguini-Boumecchal and G. Linstrumell; *Tetrahedron Lett.*, 1979, 1503.	385
79TL1553	M. Christl and H. Leininger; *Tetrahedron Lett.*, 1979, 1553.	542
79TL1833	U. Sequin; *Tetrahedron Lett.*, 1979, 1833.	99
79TL1875	S. Albanesi, A. Marchesini and B. Gioia; *Tetrahedron Lett.*, 1979, 1875.	770, 774
79TL1889	P. H. Bentley, G. Brooks, M. L. Gilpin and E. Hunt; *Tetrahedron Lett.*, 1979, 1889.	353
79TL2031	K. Hirai and Y. Iwano; *Tetrahedron Lett.*, 1979, 2031.	259
79TL2287	J. W. Streef and H. C. van der Plas; *Tetrahedron Lett.*, 1979, 2287.	519
79TL2723	P. Hallet, J. Muzart and J. P. Pete; *Tetrahedron Lett.*, 1979, 2723.	102
79TL2785	R. C. White, J. Scoby and T. D. Roberts; *Tetrahedron Lett.*, 1979, 2785.	193
79TL2983	A. Inada, H. Heimgartner and H. Schmid; *Tetrahedron Lett.*, 1979, 2983.	77
79TL3201	D. R. Boyd, R. Hamilton, N. T. Thompson and M. E. Stubbs; *Tetrahedron Lett.*, 1979, 3201.	234
79TL3213	G. L'abbé and J.-P. Dekerk; *Tetrahedron Lett.*, 1979, 3213.	405, 438, 440
79TL3353	W. H. Rastetter and J. W. Frost; *Tetrahedron Lett.*, 1979, 3353.	208
79TL3395	H. L. Holland and G. B. Johnson; *Tetrahedron Lett.*, 1979, 3395.	655
79TL3761	A. Albini, G. F. Bettinetti and G. Miroli; *Tetrahedron Lett.*, 1979, 3761.	627
79TL3777	P. Lombardi, G. Franceschi and F. Arcamone; *Tetrahedron Lett.*, 1979, 3777.	335
79TL3805	J. A. Aimetti and M. S. Kellogg; *Tetrahedron Lett.*, 1979, 3805.	317
79TL3867	H. Onoue, M. Narisada, S. Uyeo, H. Matsumura, K. Okada, T. Yano and W. Nagata; *Tetrahedron Lett.*, 1979, 3867.	253
79TL3987	B. F. Bonini, G. Maccagnani, G. Mazzanti and P. Piccinelli; *Tetrahedron Lett.*, 1979, 3987.	154
79TL3997	E. W. Meijer and H. Wynberg; *Tetrahedron Lett.*, 1979, 3997.	465
79TL4359	J. J. Tufariello, G. E. Lee, P. A. Senaratine and M. Al-Nuri; *Tetrahedron Lett.*, 1979, 4359.	353
79TL4631	J. A. Aimetti, E. S. Hamanaka, D. A. Johnson and M. S. Kellogg; *Tetrahedron Lett.*, 1979, 4631.	317
79TL4857	R. G. Visser and H. J. T. Bos; *Tetrahedron Lett.*, 1979, 4857.	438
79UP51900	F. A. L. Anet, M. Ghiaci and P. Vickers; 1979, unpublished results.	666, 699
79UP51901	F. A. L. Anet, P. Degen, J. Krane, I. Yavari, M. Ghiaci and P. Vickers; 1979, unpublished results.	700
79ZN(A)1130	G. Schulz and M. Bartok; *Z. Naturforsch., Teil A*, 1979, **34**, 1130.	365
79ZN(B)426	F. Fehér and B. Engelen; *Z. Naturforsch., Teil B*, 1979, **34**, 426.	648
79ZOR1709	V. I. Dronov, R. F. Nigmatullina, L. M. Kahlilov and Yu. E. Nikitin; *Zh. Org. Khim.*, 1979, **15**, 1709 (*Chem. Abstr.*, 1980, **92**, 5662).	624
79ZOR2602	V. N. Drozd and O. A. Popova; *Zh. Org. Khim.*, 1979, **15**, 2602.	438
80ABC73	Y. Kodama and T. Ito; *Agric. Biol. Chem.*, 1980, **44**, 73.	342, 347
80ACR65	V. Boekelheide; *Acc. Chem. Res.*, 1980, **13**, 65.	768
80ACR200	W. K. Musker; *Acc. Chem. Res.*, 1980, **13**, 200.	694, 695
80ACS(B)397	J. Krane and O. Aune; *Acta Chem. Scand., Ser. B*, 1980, **34**, 397.	745
80AG(E)45	D. Moderhack and M. Lorke; *Angew. Chem., Int. Ed. Engl.*, 1980, **19**, 45.	36, 460, 479
80AG(E)49	H. Quast, A. Fuss and A. Heublein; *Angew. Chem., Int. Ed. Engl.*, 1980, **19**, 49.	44
80AG(E)66	H. A. Staab, J. Ippen, C. Tao-pen, C. Krieger and B. Starker; *Angew. Chem., Int. Ed. Engl.*, 1980, **19**, 66.	775, 780

80AG(E)67 J. Ippen, C. Tao-pen, B. Starker, D. Schweitzer and H. A. Staab; *Angew. Chem., Int. Ed. Engl.*, 1980, **19**, 67. 775

80AG(E)203 R. Seelinger and W. Sundermeyer; *Angew. Chem., Int. Ed. Engl.*, 1980, **19**, 203. 456, 461, 481

80AG(E)207 H. Rottele and G. Schroder; *Angew. Chem., Int. Ed. Engl.*, 1980, **19**, 207. 718

80AG(E)276 G. L'abbe; *Angew. Chem., Int. Ed. Engl.*, 1980, **19**, 276. 2, 3, 13, 49, 78, 80, 92, 96, 103, 132, 174

80AG(E)393 A. G. Anastassiou and H. S. Kasmai; *Angew. Chem., Int. Ed. Engl.*, 1980, **19**, 393. 669

80AG(E)406 K. G. Heumann and H. P. Schiefer; *Angew. Chem., Int. Ed. Engl.*, 1980, **19**, 406. 759

80AG(E)465 J. Mulzer, M. Zippel and G. Bruntrup; *Angew. Chem., Int. Ed. Engl.*, 1980, **19**, 465. 374

80AG(E)466 J. Mulzer and T. Kerkmann; *Angew. Chem., Int. Ed. Engl.*, 1980, **19**, 466. 45, 422, 430, 441, 445

80AG(E)810 K. Burger and H. Goth; *Angew. Chem., Int. Ed. Engl.*, 1980, **19**, 810. 483

80AG(E)1015 E. Vogel, U. Brocker and H. Junglas; *Angew. Chem., Int. Ed. Engl.*, 1980, **19**, 1015. 493, 507

80AG(E)1016 E. Vogel, H.-J. Altenbach, J.-M. Drossard, H. Schmickler and H. Stegelmeier; *Angew. Chem., Int. Ed. Engl.*, 1980, **19**, 1016. 492, 493, 496, 498, 500, 506, 526

80AJC1335 J. B. Bremner, E. J. Browne and P. E. Davies; *Aust. J. Chem.*, 1980, **33**, 1335. 625

80AJC2717 B. A. Mooney, R. H. Prager and A. D. Ward; *Aust. J. Chem.*, 1980, **33**, 2717. 508

80APO(17)279 F. DeJong and D. N. Reinhoudt; *Adv. Phys. Org. Chem.*, 1980, **17**, 279. 741, 742, 746

80AX(B)484 G. J. Verhoeckx, J. Kroon, A. C. Brouwer and H. J. T. Bos; *Acta Crystallogr., Part B*, 1980, **36**, 484. 407

80AX(B)1507 R. Chambers and R. J. Doedens; *Acta Crystallogr., Part B*, 1980, **36**, 1507. 247

80AX(B)2683 J. P. Reboul, B. Cristau, J. C. Soyfer and J. Estienne; *Acta Crystallogr., Part B*, 1980, **36**, 2683. 495

80B2811 D. H. Standring and J. R. Knowles; *Biochemistry*, 1980, **19**, 2811. 236

80BCJ2097 H. Kawa and N. Ishikawa; *Bull. Chem. Soc. Jpn.*, 1980, **53**, 2097. 172

80BCJ2895 T. Masamune, S. Sato, A. Abiko, M. Ono and A. Murai; *Bull. Chem. Soc. Jpn.*, 1980, **53**, 2895. 392

80BCJ3031 Y. Inoue, S. Deguchi and T. Hakushi; *Bull. Chem. Soc. Jpn.*, 1980, **53**, 3031. 577

80BCJ3232 M. Kimura, K. Satake, S. Yonemori and S. Morosawa; *Bull. Chem. Soc. Jpn.*, 1980, **53**, 3232. 531

80BRP2033410 J. M. Wrightson; *Br. Pat. Appl.* 2 033 410 (1980) (*Chem. Abstr.*, 1980, **93**, 240 624). 484

80CB1201 D. C. Crans and J. P. Snyder; *Chem. Ber.*, 1980, **113**, 1201. 452

80CB1472 H.-D. Scharf and H. Frauenrath; *Chem. Ber.*, 1980, **113**, 1472. 621

80CB1487 N. Wester and F. Vögtle; *Chem. Ber.*, 1980, **113**, 1487. 750

80CB1632 M. Regitz and B. Mathieu; *Chem. Ber.*, 1980, **113**, 1632. 167

80CB2434 F.-M. Tesky and R. Mews; *Chem. Ber.*, 1980, **113**, 2434. 486

80CB3127 R. Schwesinger, M. Breuninger, B. Gallenkamp, K. Müller, D. Hunkler and H. Prinzbach; *Chem. Ber.*, 1980, **113**, 3127. 111

80CB3161 M. Breuninger, R. Schwesinger, B. Gallenkamp, K.-H. Müller, H. Fritz, D. Hunkler and H. Prinzbach; *Chem. Ber.*, 1980, **113**, 3161. 677

80CB3373 M. Gnichtel, K. Hirte and R. Wagner; *Chem. Ber.*, 1980, **113**, 3373. 642

80CC34 A. C. Kaura, C. D. Maycock and R. J. Stoodley; *J. Chem. Soc., Chem. Commun.*, 1980, 34. 312

80CC70 M. Foglio, G. Franceschi, C. Scarafile and F. Arcamone; *J. Chem. Soc., Chem. Commun.*, 1980, 70. 335

80CC98 N. Kanehisa, N. Yasuoka, N. Kasai, K. Isomura and H. Taniguchi; *J. Chem. Soc., Chem. Commun.*, 1980, 98. 48, 49

80CC100 P. H. McCabe and A. Stewart; *J. Chem. Soc., Chem. Commun.*, 1980, 100. 148

80CC156 S. F. Moss and D. R. Taylor; *J. Chem. Soc., Chem. Commun.*, 1980, 156. 650

80CC173 J. Marchand-Brynaert, M. Moya-Portuguez, D. Lesuisse, and L. Ghosez; *J. Chem. Soc., Chem. Commun.*, 1980, 173. 273

80CC176 A. Shanzer and N. Mayer-Shochet; *J. Chem. Soc., Chem. Commun.*, 1980, 176. 749

80CC243 K. Muthuramu and V. Ramamurthy; *J. Chem. Soc., Chem. Commun.*, 1980, 243. 422

80CC297 G. D. Annis, E. M. Hebblethwaite and S. V. Ley; *J. Chem. Soc., Chem. Commun.*, 1980, 297. 257

80CC300 A. G. M. Barrett, J. C. A. Lana and S. Tograie; *J. Chem. Soc., Chem. Commun.*, 1980, 300. 759

80CC313 M. Tisler and B. Stanovnik; *J. Chem. Soc., Chem. Commun.*, 1980, 313. 474

80CC444 T. Tsuchiya, J. Kurita and H. Kojima; *J. Chem. Soc., Chem. Commun.*, 1980, 444. 600, 607

80CC454 T. Tsuchiya, M. Enkaku and S. Okajima; *J. Chem. Soc., Chem. Commun.*, 1980, 454. 607

80CC471 M. Lancaster and D. J. H. Smith; *J. Chem. Soc., Chem. Commun.*, 1980, 471. 43, 277

80CC499 I. R. Dunkin and P. C. P. Thomson; *J. Chem. Soc., Chem. Commun.*, 1980, 499. 534

80CC540 J. C. Metcalfe, J. F. Stoddart, G. Jones, W. E. Hull, A. Atkinson, I. S. Kerr and J. R. Williams; *J. Chem. Soc., Chem. Commun.*, 1980, 540. 745

80CC560 J. C. Bottaro; *J. Chem. Soc., Chem. Commun.*, 1980, 560. 36

80CC736 T. Kobayashi, N. Ishida and T. Hiraoka; *J. Chem. Soc., Chem. Commun.*, 1980, 736. 253

80CC770 K. Chiba, M. Mori and Y. Ban; *J. Chem. Soc., Chem. Commun.*, 1980, 770. 266

80CC784 D. R. Guthrie, I. D. Jenkins and R. Yamasaki; *J. Chem. Soc., Chem. Commun.*, 1980, 784. 33

80CC798 S. A. Matlin and L. Chan; *J. Chem. Soc., Chem. Commun.*, 1980, 798. 320, 635

80CC898	Y. Ito, H. Yokoya, K. Kyono, S. Yamamura, Y. Yamada and T. Matsuura; *J. Chem. Soc., Chem. Commun.*, 1980, 898.	476
80CC925	A. J. Bloodworth, D. J. Lapham and R. A. Savva; *J. Chem. Soc., Chem. Commun.*, 1980, 925.	749
80CC928	C. M. Pant and R. J. Stoodley; *J. Chem. Soc., Chem. Commun.*, 1980, 928.	333
80CC940	C. Bernard and L. Ghosez; *J. Chem. Soc., Chem. Commun.*, 1980, 940.	44, 88
80CC1161	O. Meth-Cohn and S. Rhouati; *J. Chem. Soc., Chem. Commun.*, 1980, 1161.	508, 538
80CC1197	C. Leuenberger, L. Hoesch and A. S. Dreiding; *J. Chem. Soc., Chem. Commun.*, 1980, 1197.	3, 32, 197
80CC1252	K. Isomura, G.-I. Ayabe, S. Hatano and H. Taniguchi; *J. Chem. Soc., Chem. Commun.*, 1980, 1252.	33
80CC1257	J. Cuffe and A. E. A. Porter; *J. Chem. Soc., Chem. Commun.*, 1980, 1257.	254
80CC1263	S. N. Ege, E. J. Gess, A. Thomas, P. Umrigar, G. W. Griffin, P. K. Das, A. M. Trozzolo and T. M. Leslie; *J. Chem. Soc., Chem. Commun.*, 1980, 1263.	102
80CCC3150	M. Pánková and J. Závada; *Collect. Czech. Chem. Commun.*, 1980, **45**, 3150.	758
80CJC494	D. J. Harris, G. Y.-P. Kan, T. Tschamber and V. Snieckus; *Can. J. Chem.*, 1980, **58**, 494.	599
80CJC2089	K. R. Kopecky and J. A. Lopez Sastre; *Can. J. Chem.*, 1980, **58**, 2089.	469
80CJC2508	T. W. Doyle, J. L. Douglas, B. Belleau, T. T. Conway, C. F. Ferrari, D. E. Horning, G. Lim, B. Luh, A. Martel, M. Menard, L. R. Morris and M. Misiek; *Can. J. Chem.*, 1980, **58**, 2508.	355
80CL41	T. Imai and S. Nishida; *Chem. Lett.*, 1980, 41.	373
80CL397	Y. Miyahara, T. Inazu and T. Yoshino; *Chem. Lett.*, 1980, 397.	773, 779
80CL487	Y. Ito, K. Kobayashi, M. Maeno and T. Saegusa; *Chem. Lett.*, 1980, 487.	529
80CL659	E. M. Beccali, L. Majori, A. Marchesini and C. Torricelli; *Chem. Lett.*, 1980, 659.	775
80CL1177	M. Suzuki and E. Kurosawa; *Chem. Lett.*, 1980, 1177.	667
80CL1427	Y. Sugihara, Y. Fujiyama and I. Murata; *Chem. Lett.*, 1980, 1427.	623
80CPB1157	C. Kaneko, H. Fujui, S. Kawai, A. Yamomoto, K. Hashiba, T. Kimata, R. Hayashi and M. Somei; *Chem. Pharm. Bull.*, 1980, **28**, 1157.	628
80CPB1842	T. Tsuchiya and J. Kurita; *Chem. Pharm. Bull.*, 1980, **28**, 1842.	600, 603
80CPB2602	T. Tsuchiya, M. Enkaku and S. Okajima; *Chem. Pharm. Bull.*, 1980, **28**, 2602.	607
80CPB3232	S. Oida, A. Yoshida, T. Hayashi, N. Takeda and E. Ohki; *Chem. Pharm. Bull.*, 1980, **28**, 3232.	252
80CPB3494	S. Oida, A. Yoshida and E. Ohki; *Chem. Pharm. Bull.*, 1980, **28**, 3494.	252
80CR(C)(291)223	C. Filliatre, B. Maillard, C. Manigand, N. P. Tarasova and J. J. Villenave; *C.R. Hebd. Seances Acad. Sci., Ser. C*, 1980, **291**, 223.	562, 580
80CRV231	A. C. Legon; *Chem. Rev.*, 1980, **80**, 231.	8, 9
80CSC777	E. N. Duesler, K. L. Rinehart, Jr., I. C. Paul, S. Caccamese and R. Azzolina; *Cryst. Struct. Commun.*, 1980, **9**, 777.	700
80CSR1	I. R. Dunkin; *Chem. Soc. Rev.*, 1980, **9**, 1.	120, 127
80DOK(253)886	D. P. Del'tsova, N. P. Gambaryan, E. I. Mysov and I. L. Knunyants; *Dokl. Akad. Nauk SSSR*, 1980, **253**, 886.	266, 455, 458, 481
80H(14)11	K. Orito, S. Kudoh, K. Yamada and M. Itoh; *Heterocycles*, 1980, **14**, 11, and references cited therein.	536
80H(14)467	H. Takahata, Y. Ohnishi and T. Yamazaki; *Heterocycles*, 1980, **14**, 467.	33
80H(14)1077	N. Ikota, H. Shibata and K. Koga; *Heterocycles*, 1980, **14**, 1077.	352
80H(14)1305	T. Kametani, S.-P. Huang, T. Nagahara and M. Ihara; *Heterocycles*, 1980, **14**, 1305.	254
80H(14)1677	H. Quast; *Heterocycles*, 1980, **14**, 1677.	3, 197, 219
80H(14)1717	T. Tezuka and O. Kikuchi; *Heterocycles*, 1980, **14**, 1717.	504
80H(14)1959	K. Matsumoto, S. Nakamura, T. Uchida and R. M. Acheson; *Heterocycles*, 1980, **14**, 1959.	542
80H(14)1967	T. Kametani, T. Honda, A. Nakayama and K. Fukumoto; *Heterocycles*, 1980, **14**, 1967.	254
80H(14)1999	S. Oida, A. Yoshida and E. Ohki; *Heterocycles*, 1980, **14**, 1999.	254
80HCA154	J. Berger, M. Yoshioka, M. P. Zink, H. R. Wolf and O. Jeger; *Helv. Chim. Acta*, 1980, **63**, 154.	397
80HCA201	I. Ernest; *Helv. Chim. Acta*, 1980, **63**, 201.	307
80HCA668	A. I. Popov, A. J. Smetana, J. P. Kintzinger and T. T. Nguyên; *Helv. Chim. Acta*, 1980, **63**, 668.	740
80HCA924	D. Berney and K. Schuh; *Helv. Chim. Acta*, 1980, **63**, 924.	539
80HCA2096	J. P. Behr, J. M. Girodeau, R. C. Hayward, J. M. Lehn and J. P. Sauvage; *Helv. Chim. Acta*, 1980, **63**, 2096.	745, 749
80IC1330	A. Sekiya and D. D. DesMarteau; *Inorg. Chem.*, 1980, **19**, 1330.	228
80IJC(B)702	K. N. Mehrota and S. B. Singh; *Indian J. Chem., Sect. B*, 1980, **19**, 702 (*Chem. Abstr.*, 1981, **94**, 121 193).	249
80IZV220	U. M. Dzhemilev, R. I. Khusnutdinov, Z. S. Muslimov, G. A. Tolstikov and O. M. Nefedov; *Izv. Akad. Nauk SSSR, Ser. Khim.*, 1980, 220.	213
80IZV1692	R. A. Bekker, V. Ya. Popkova and I. L. Knunyants; *Izv. Akad. Nauk SSSR, Ser. Khim.*, 1980, 1692 (*Chem. Abstr.*, 1980, **93**, 186 054).	174
80JA439	L. B. Harding and W. A. Goddard; *J. Am. Chem. Soc.*, 1980, **102**, 439.	476
80JA744	T. B. Cameron and H. W. Pinnick; *J. Am. Chem. Soc.*, 1980, **102**, 744.	180
80JA837	M. E. Landis, L. M. Bell, D. C. Madoux, J. C. Mitchell, J. M. Schmidt and J. A. Spencer; *J. Am. Chem. Soc.*, 1980, **102**, 837.	456

80JA1157 R. Arad-Yellin, B. S. Green and M. Knossow; *J. Am. Chem. Soc.*, 1980, **102**, 1157. 182

80JA1200 R. S. Buckanin, S. J. Chen, D. M. Frieze, F. T. Sher and G. A. Berchtold; *J. Am. Chem. Soc.*, 1980, **102**, 1200. 192

80JA1202 E. E. van Tamelen and E. G. Taylor; *J. Am. Chem. Soc.*, 1980, **102**. 1202. 192

80JA1255 D. M. Hayes, S. D. Nelson, W. A. Garland and P. A. Kollman; *J. Am. Chem. Soc.*, 1980, **102**, 1255. 500, 550, 551, 553

80JA1451 D. Tzeng and W. P. Weber; *J. Am. Chem. Soc.*, 1980, **102**, 1451. 107

80JA1690 E. M. Gordon, H. W. Chang, C. M. Cimarusti, B. Toeplitz and J. Z. Gougoutas; *J. Am. Chem. Soc.*, 1980, **102**, 1690. 319

80JA1812 D. B. Boyd, D. K. Herron, W. H. W. Lunn and W. A. Spitzer; *J. Am. Chem. Soc.*, 1980, **102**, 1812. 288

80JA1954 P. N. Confalone, G. Pizzolato, D. L. Confalone and M. R. Uskokovic; *J. Am. Chem. Soc.*, 1980, **102**, 1954. 159

80JA1961 W. Adam and M. Balci; *J. Am. Chem. Soc.*, 1980, **102**, 1961. 191

80JA2060 T. Kametani, S.-P. Huang, S. Yokohama, Y. Suzuki and M. Ihara; *J. Am. Chem. Soc.*, 1980, **102**, 2060. 247, 352

80JA2095 M. Suzuki, A. Watanabe and R. Noyori; *J. Am. Chem. Soc.*, 1980, **102**, 2095. 105

80JA2490 E. Block, A. A. Bazzi and L. K. Revelle; *J. Am. Chem. Soc.*, 1980, **102**, 2490. 38, 464, 481

80JA2507 B. A. Hess, Jr., L. J. Schaad and C. S. Ewig; *J. Am. Chem. Soc.*, 1980, **102**, 2507. 133

80JA2861 J. P. Snyder and T. A. Halgren; *J. Am. Chem. Soc.*, 1980, **102**, 2861. 169, 186, 187

80JA3199 E. H. White, M. G. Steinmetz, J. D. Miano, R. D. Wildes and R. Morland; *J. Am. Chem. Soc.*, 1980, **102**, 3199. 485

80JA3620 J. Mulzer and T. Kerkmann; *J. Am. Chem. Soc.*, 1980, **102**, 3620. 18, 387

80JA3904 R. M. Scarborough, Jr., B. H. Toder and A. B. Smith, III; *J. Am. Chem. Soc.*, 1980, **102**, 3904. 192, 667

80JA4193 A. Cowell and J. K. Stille; *J. Am. Chem. Soc.*, 1980, **102**, 4193. 399

80JA4810 P. E. Stott, J. S. Bradshaw and W. W. Parish; *J. Am. Chem. Soc.*, 1980, **102**, 4810. 758

80JA4838 E. D. Middlemas and L. D. Quin, *J. Am. Chem. Soc.*, 1980, **102**, 4838. 713

80JA5004 E. Ehlinger and P. Magnus; *J. Am. Chem. Soc.*, 1980, **102**, 5004. 110

80JA5151 D. H. Aue, H. M. Webb, W. R. Davidson, M. Vidal, M. T. Bowers, H. Goldwhite, L. E. Vertal, J. E. Douglas, P. A. Kollman and G. L. Kenyon; *J. Am. Chem. Soc.*, 1980, **102**, 5151. 23, 24, 137, 145

80JA5415 T. Ban, Y. Wakita and K. Kanematsu; *J. Am. Chem. Soc.*, 1980, **102**, 5415. 570

80JA5424 E. E. van Tamelen, J. P. Demers, E. G. Taylor and K. Koller; *J. Am. Chem. Soc.*, 1980, **102**, 5424. 192

80JA5643 K. Tokomura, H. Goto, H. Kashiwabara, C. Kaneko and M. Itoh; *J. Am. Chem. Soc.*, 1980, **102**, 5643. 206

80JA5974 T. Katsuki and K. B. Sharpless; *J. Am. Chem. Soc.*, 1980, **102**, 5974 36

80JA6161 T. N. Salzmann, R. W. Ratcliffe, B. G. Christensen and F. A. Bouffard; *J. Am. Chem. Soc.*, 1980, **102**, 6161. 352

80JA6574 J.-L. Pierre, H. Handel, P. Labbé and R. Le Goaller; *J. Am. Chem. Soc.*, 1980, **102**, 6574. 746

80JA6633 E. D. Laganis and D. M. Lemal; *J. Am. Chem. Soc.*, 1980, **102**, 6633. 187

80JA6634 E. D. Laganis and D. M. Lemal; *J. Am. Chem. Soc.*, 1980, **102**, 6634. 151

80JA6820 J. D. Lamb, J. J. Christensen, J. L. Oscarson, B. L. Nielsen, B. W. Asay and R. M. Izatt; *J. Am. Chem. Soc.*, 1980, **102**, 6820. 755

80JA6902 H. Aoyama, M. Sakamoto and Y. Omote; *J. Am. Chem. Soc.*, 1980, **102**, 6902. 2, 20

80JA7100 S. P. Schmidt and G. B. Schuster; *J. Am. Chem. Soc.*, 1980, **102**, 7100. 459

80JA7373 P. R. Ortiz de Montellano and K. L. Kunze; *J. Am. Chem. Soc.*, 1980, **102**, 7373. 124

80JA7576 N. J. Turro, J. A. Butcher, R. A. Moss, W. Guo, R. C. Munjal and M. Fedorynski; *J. Am. Chem. Soc.*, 1980, **102**, 7576. 227

80JA7655 K. Tanaka and M. Yoshimine; *J. Am. Chem. Soc.*, 1980, **102**, 7655. 121

80JA7991 W. H. Gerwick, W. Fenical, D. Van Engen and J. Clardy; *J. Am. Chem. Soc.*, 1980, **102**, 7991. 120

80JA7994 E. A. Wildi, D. Van Engen and B. K. Carpenter; *J. Am. Chem. Soc.*, 1980, **102**, 7994. 457

80JBC(255)860 R. A. Farley, D. W. Goldman and H. Bayley; *J. Biol. Chem.*, 1980, **255**, 860. 236

80JBC(255)3964 D. J. Waxman and J. L. Strominger; *J. Biol. Chem.*, 1980, **255**, 3964. 296

80JCR(S)136 E. Bacon, L. Jung and J. M. Lehn; *J. Chem. Res. (S)*, 1980, 136. 756

80JCR(S)398 G. M. Clarke, J. B. Lee, F. J. Swinbourne and B. Williamson; *J. Chem. Res. (S)*, 1980, 398. 609

80JCS(F1)1849 H. A. J. Carless, A. K. Maitra, R. Pottinger and H. M. Frey; *J. Chem. Soc., Faraday Trans. 1*, 1980, **76**, 1849. 372

80JCS(F2)1293 K. F. Gebhardt, H. Oberhammer and W. Zeil; *J. Chem. Soc., Faraday Trans. 2*, 1980, 1293. 452

80JCS(P1)116 L. Garanti and G. Zecchi; *J. Chem. Soc., Perkin Trans. 1*, 1980, 116. 597

80JCS(P1)150 N. F. Osborne; *J. Chem. Soc., Perkin Trans. 1*, 1980, 150. 306

80JCS(P1)233 M. Foglio, G. Franceschi, P. Masi and A. Suarato; *J. Chem. Soc., Perkin Trans. 1*, 1980, 233. 306

80JCS(P1)311 T. L. Peppard, F. R. Sharpe and J. A. Elvidge; *J. Chem. Soc., Perkin Trans. 1*, 1980, 311. 176, 182, 192

80JCS(P1)388 S. D. Carter, A. C. Kaura and R. J. Stoodley; *J. Chem. Soc., Perkin Trans. 1*, 1980, 388. 435, 445

80JCS(P1)607 T. Nishio, A. Kato, C. Kashima and Y. Omote; *J. Chem. Soc., Perkin Trans. 1*, 1980, 607. 360, 361

80JCS(P1)846 G. W. Boyd and R. L. Monteil; *J. Chem. Soc., Perkin Trans. 1*, 1980, 846. 625

80JCS(P1)1230 C. D. Anderson and J. T. Sharp; *J. Chem. Soc., Perkin Trans. 1*, 1980, 1230. 599

80JCS(P1)1251 A. J. Frew, G. R. Proctor and J. V. Silverton; *J. Chem. Soc., Perkin Trans. 1*, 1980, 1251. 520

80JCS(P1)1693 M. N. Akhtar, D. R. Boyd, J. D. Neill and D. M. Jerina; *J. Chem. Soc., Perkin Trans. 1*, 1980, 1693. 209, 444, 566

80JCS(P1)1718 D. P. Munro and J. T. Sharp; *J. Chem. Soc., Perkin Trans. 1*, 1980, 1718. 595, 599, 604

80JCS(P1)2084 C. B. Chapleo, M. A. W. Finch, T. V. Lee, S. M. Roberts and R. F. Newton; *J. Chem. Soc., Perkin Trans. 1*, 1980, 2084. 113

80JCS(P1)2228 M. D. Bachi, S. Sasson and J. Vaya; *J. Chem. Soc., Perkin Trans. 1*, 1980, 2228. 332

80JCS(P1)2830 Y. Tamura, S. M. Bayomi, C. Mukai, N. Ikeda, M. Murase and M. Kise; *J. Chem. Soc., Perkin Trans. 1*, 1980, 2830. 632

80JCS(P1)2851 A. R. Katritzky, J. Arrowsmith, Z. bin Bahari, C. Jayaram, T. Siddiqui and S. Vassilatos; *J. Chem. Soc., Perkin Trans. 1*, 1980, 2851. 545

80JCS(P2)188 L. Carlsen; *J. Chem. Soc., Perkin Trans. 2*, 1980, 188. 487

80JCS(P2)212 J. J. Morris and M. I. Page; *J. Chem. Soc., Perkin Trans. 2*, 1980, 212. 327

80JCS(P2)279 A. R. Katritzky and R. C. Patel; *J. Chem. Soc., Perkin Trans. 2*, 1980, 279. 159

80JCS(P2)919 F. Heatley, L. Akhter and R. T. Brown; *J. Chem. Soc., Perkin Trans. 2*, 1980, 919. 660

80JCS(P2)1512 W. B. Jennings and S. D. Worley; *J. Chem. Soc., Perkin Trans. 2*, 1980, 1512. 7

80JCS(P2)1529 M. J. Bovill, D. J. Chadwick, I. O. Sutherland and D. Watkin; *J. Chem. Soc., Perkin Trans. 2*, 1980, 1529. 735, 736, 741

80JCS(P2)1725 N. P. Gensmantel, P. Proctor and M. I. Page; *J. Chem. Soc., Perkin Trans. 2*, 1980, 1725. 327

80JCS(P2)1774 A. D. Cort, G. Illuminati, L. Mandolini and B. Masci; *J. Chem. Soc., Perkin Trans. 2*, 1980, 1774. 690

80JHC417 K. Orito, H. Kaga, M. Itoh, S. O. de Silva, R. H. Manske and R. Rodrigo; *J. Heterocycl. Chem.*, 1980, **17**, 417. 528

80JHC1025 E. Oliveros, M. Riviere and A. Lattes; *J. Heterocycl. Chem.*, 1980, 1025. 205

80JHC1593 I. J. Turchi, C. A. Maryanoff and A. R. Mastrocola; *J. Heterocycl. Chem.*, 1980, **17**, 1593. 543

80JMC713 V. E. Marquez, P. S. Liu, J. A. Kelley, J. S. Driscoll and J. J. McCormack; *J. Med. Chem.*, 1980, **23**, 713. 607

80JMC975 C. Kaiser, F. E. Ali, W. E. Bondinell, M. Brenner, K. G. Holden, T. W. Ku, Hye-Ja Oh, S. T. Ross, N. C. F. Yim, C. L. Zirkle, R. A. Hahn, H. M. Sarau, P. E. Setler and J. R. Wardell, Jr.; *J. Med. Chem.*, 1980, **23**, 975, and references cited therein. 526, 533, 546

80JMC1108 D. Hagiwara, H. Takeno, M. Aratani, K. Hemmi and M. Hashimoto; *J. Med. Chem.*, 1980, **23**, 1108. 309

80JOC311 C. Galli, G. Illuminati and L. Mandolini; *J. Org. Chem.*, 1980, **45**, 311. 767

80JOC410 P. G. Mattingly and M. J. Miller; *J. Org. Chem.*, 1980, **45**, 410. 257

80JOC428 F. Bourelle-Wargnier, M. Vincent and J. Chuche; *J. Org. Chem.*, 1980, **45**, 428. 102

80JOC462 P. D. Davis, D. G. Neckers and J. R. Blount; *J. Org. Chem.*, 1980, **45**, 462, and references cited therein. 504, 542

80JOC485 V. E. Marquez, P. S. Liu, J. A. Kelley and J. S. Driscoll; *J. Org. Chem.*, 1980, **45**, 485. 605, 608

80JOC936 Y. Hirai, S. Hirokami and M. Nagata; *J. Org. Chem.*, 1980, **45**, 936. 258

80JOC1135 S. M. Schmitt, D. B. R. Johnston and B. G. Christensen; *J. Org. Chem.*, 1980, **45**, 1135. 353

80JOC1166 L. Bellon, R. W. Taft and J. L. M. Abboud; *J. Org. Chem.*, 1980, **45**, 1166. 375, 376

80JOC1224 F. A. L. Anet and M. Ghiaci; *J. Org. Chem.*, 1980, **45**, 1224. 702

80JOC1277 K. Crimaldi and R. L. Lichter; *J. Org. Chem.*, 1980, **45**, 1277. 11, 51

80JOC1315 R. Damrauer, D. Soucy, P. Winkler and S. Eby; *J. Org. Chem.*, 1980, **45**, 1315. 656

80JOC1341 J. P. Snyder; *J. Org. Chem.*, 1980, **45**, 1341. 269, 272

80JOC1477 M. D. Bachi, O. Goldberg, A. Gross and J. Vaya; *J. Org. Chem.*, 1980, **45**, 1477. 248, 267

80JOC1481 M. D. Bachi, O. Goldberg, A. Gross and J. Vaya; *J. Org. Chem.*, 1980, **45**, 1481. 153, 177, 180, 248, 256, 266, 267

80JOC1489 M. Newcomb and R. A. Reeder; *J. Org. Chem.*, 1980, **45**, 1489. 208

80JOC1547 Y.-L. Mao and V. Boekelheide; *J. Org. Chem.*, 1980, **45**, 1547. 277

80JOC1691 Y. Hata and M. Watanabe; *J. Org. Chem.*, 1980, **45**, 1691. 149, 209

80JOC1950 C. D. Perchonock, I. Lantos, J. A. Finkelstein and K. G. Holden; *J. Org. Chem.*, 1980, **45**, 1950. 525, 544

80JOC2000 C. D. Perchonock and J. A. Finkelstein; *J. Org. Chem.*, 1980, **45**, 2000. 532

80JOC2091 H. Hart, S.-M. Chen, S. Lee, D. L. Ward and W.-J. Hu Kung; *J. Org. Chem.*, 1980, **45**, 2091. 102

80JOC2151 M. S. Raasch; *J. Org. Chem.*, 1980, **45**, 2151. 441

80JOC2271 M. Torres, A. Clement and O. P. Strausz; *J. Org. Chem.*, 1980, **45**, 2271. 128

80JOC2334 A. A. Frimer and A. Antebi; *J. Org. Chem.*, 1980, **45**, 2334. 185

80JOC2347 D. L. J. Clive and S. M. Menchen; *J. Org. Chem.*, 1980, **45**, 2347. 111

80JOC2354 T. Imai and S. Nishida; *J. Org. Chem.*, 1980, **45**, 2354. 374

80JOC2368 M. Yasuda, K. Harano and K. Kanematsu; *J. Org. Chem.*, 1980, **45**, 2368. 521

80JOC2579 G. A. Kraus and K. Frazier; *J. Org. Chem.*, 1980, **45**, 2579. 111

80JOC2604 B. B. Jarvis and G. P. Stahly; *J. Org. Chem.*, 1980, **45**, 2604. 154, 159

80JOC2951 E. Schaumann, S. Grabley, M. Henriet, L. Ghosez, R. Touillaux, J. P. Declercq, G. Germain and M. Van Meerssche; *J. Org. Chem.*, 1980, **45**, 2951. 510, 522, 539

80JOC2962 Y. Kobayashi, A. Ando, K. Kawada, A. Ohsawa and I. Kumadaki; *J. Org. Chem.*, 1980, **45**, 2962. 151, 187, 423, 427, 440

80JOC2966 Y. Kobayashi, A. Ando, K. Kawada and I. Kumadaki; *J. Org. Chem.*, 1980, **45**, 2966. 153, 170

80JOC3149 W. H. Rastetter and L. J. Nummy; *J. Org. Chem.*, 1980, **45**, 3149. 568, 584, 590

80JOC3198 L. E. Friedrich, R. A. Leckonby and D. M. Stout; *J. Org. Chem.*, 1980, **45**, 3198. 186

80JOC3205 T. Tsushima and H. Tanida; *J. Org. Chem.*, 1980, **45**, 3205. 307

80JOC3413 C. Gluchowski, L. Cooper, D. E. Bergbreiter and M. Newcomb; *J. Org. Chem.*, 1980, **45**, 3413. 261

80JOC3593 A. L. Baumstark, C. J. McCloskey, T. E. Williams and D. R. Chrisope; *J. Org. Chem.*, 1980, **45**, 3593 464

80JOC3646 P. G. DeBenedetti, C. DeMicheli, R. Gandolfi, P. Gariboldi and A. Rastelli; *J. Org. Chem.*, 1980, **45**, 3646. 415, 433

80JOC3682 M. Aratani, D. Hagiwara, H. Takeno, K. Hemmi and M. Hashimoto; *J. Org. Chem.*, 1980, **45**, 3682. 309

80JOC3766 A. Dondoni, A. Battaglia and P. Giorgianni; *J. Org. Chem.*, 1980, **45**, 3766. 438

80JOC3952 G. Gerchelli, G. Illuminati and C. Lillocci; *J. Org. Chem.*, 1980, **45**, 3952. 656

80JOC4063 M. Shimizu and I. Kuwajima; *J. Org. Chem.*, 1980, **45**, 4063. 391

80JOC4069 F. M. Schell and R. N. Ganguly; *J. Org. Chem.*, 1980, **45**, 4069. 18

80JOC4122 O. Lerman and S. Rozen; *J. Org. Chem.*, 1980, **45**, 4122. 524

80JOC4254 M. O. Brimeyer, A. Mehrota, S. Quici, A. Nigam and S. L. Regen; *J. Org. Chem.*, 1980, **45**, 4254. 179

80JOC4359 P. Charumilind and H. Kohn; *J. Org. Chem.*, 1980, **45**, 4359. 419, 428, 432

80JOC4366 G. L'abbe, J.-P. Dekerk, C. Martens and S. Toppet; *J. Org. Chem.*, 1980, **45**, 4366. 143, 154, 160

80JOC4455 K. Harano, M. Yasuda, T. Ban and K. Kanematsu; *J. Org. Chem.*, 1980, **45**, 4455. 521

80JOC4483 H. W. Moore and D. S. Wilbur; *J. Org. Chem.*, 1980, **45**, 4483. 398

80JOC4532 K. Muthuramu and V. Ramamurthy; *J. Org. Chem.*, 1980, **45**, 4532. 422

80JOC4587 T. Sasaki, K. Minamoto and K. Harada; *J. Org. Chem.*, 1980, **45**, 4587. 362

80JOC4758 R. Curci, M. Fiorentino, L. Troisi, J. O. Edwards and R. H. Pater; *J. Org. Chem.*, 1980, **45**, 4758. 3, 197

80JOC4804 J. Bolster and R. M. Kellogg; *J. Org. Chem.*, 1980, **45**, 4804. 418, 421, 432, 441

80JOC4807 E. Block, A. A. Bazzi, J. B. Lambert, S. M. Wharry, K. K. Andersen, D. C. Dittmer, B. H. Patwardhan and D. J. H. Smith, *J. Org. Chem.*, 1980, **45**, 4807. 136, 409, 414, 415, 416

80JOC4853 P. T. Berkowitz and K. Baum; *J. Org. Chem.*, 1980, **45**, 4853. 391

80JOC5095 T. Kiguchi, J.-L. Schuppiser, J.-C. Schwaller and J. Streith; *J. Org. Chem.*, 1980, **45**, 5095. 598

80JOC5225 P. S. Liu, V. E. Marquez, J. A. Kelley and J. S. Driscoll; *J. Org. Chem.*, 1980, **45**, 5225. 607

80JOC5308 V. E. Marquez, J. A. Kelley and J. S. Driscoll; *J. Org. Chem.*, 1980, **45**, 5308. 608

80JOM(193)57 R. Battaglia, H. Matthaeus and H. Kisch; *J. Organomet. Chem.*, 1980, **193**, 57. 220

80JOU395 V. N. Drozd, Yu. M. Udachin, V. V. Sergeichuk and G. S. Bogomolova; *J. Org. Chem. USSR* (*Engl. Transl.*), 1980, **16**, 395. 167

80JOU1663 A. A. Potekhin, I. G. Zenkevich, V. V. Sokolov and S. M. Shevchenko; *J. Org. Chem. USSR* (*Engl. Transl.*), 1980, **16**, 1663. 159

80JST(63)73 A. Postawka and L. Prajer-Janczewska; *J. Mol. Struct.*, 1980, **63**, 73. 134

80JST(64)239 J. Dillen and H. J. Geise; *J. Mol. Struct.*, 1980, **64**, 239. 548

80KGS848 V. M. Cherkasov, T. A. Dashevskaya and A. A. Kisilenko; *Khim. Geterotsikl. Soedin.*, 1980, 848 (*Chem. Abstr.*, 1980, **93**, 204 613). 768

80KGS1569 V. V. Sokolov, K. A. Ogloblin and A. A. Potekhin; *Khim. Geterotsikl. Soedin.*, 1980, 1569. 159

80LA798 G. Maier and U. Schäfer; *Liebigs Ann. Chem.*, 1980, 798. 282, 360, 361

80LA873 H. Gotthardt, S. Nieberl and J. Donecke; *Liebigs Ann. Chem.*, 1980, 873. 440

80LA1779 D. Hoppe and R. Follmann; *Liebigs Ann. Chem.*, 1980, 1779. 180

80LA1889 B. Epe, P. Roesner and W. Tochtermann; *Liebigs Ann. Chem.*, 1980, 1889. 551

80LA1981 B. Balbach, K. Weidenhammer and M. L. Ziegler; *Liebigs Ann. Chem.*, 1980, 1981. 453, 454, 456

80M1117 L. S. Trifonov and A. Orakhovats; *Monatsh. Chem.*, 1980, **111**, 1117. 256

80MI50100 K. Yamaguchi, S. Yabushita, T. Fueno, S. Kato, K. Morokuma and S. Iwata; *Chem. Phys. Lett.*, 1980, **71**, 563. 3

B-80MI50300 Y. Wolman; in 'The Chemistry of Ketenes, Allenes, and Related Compounds', ed. S. Patai; Wiley, Chichester, 1980, vol. 2, p. 736. 36, 37

B-80MI50500 J. N. Cawse, J. P. Henry, M. W. Swartzlander and P. H. Wadia; in 'Kirk–Othmer Encyclopedia of Chemical Technology', ed. M. Grayson and D. Echroth; Wiley, New York, 3rd edn., 1980, vol. 9, p. 432. 97, 116, 118

B-80MI50501 J. T. Lutz, Jr.; in 'Kirk–Othmer Encyclopedia of Chemical Technology', ed. M. Grayson and D. Echroth; Wiley, New York, 3rd edn., 1980, vol. 9, p. 251. 97, 116, 118

B-80MI50502 R. F. C. Brown; 'Pyrolytic Methods in Organic Chemistry', Academic Press, New York, 1980. 100

80MI50503 R. Walder and J. L. Franklin; *Int. J. Mass Spectrom. Ion Phys.*, 1980, **36**, 85. 105

B-80MI50504 M. Nordqvist, D. R. Thakker, H. Yagi, R. E. Lehr, A. W. Wood, W. Levin, A. Conney and D. M. Jerina; in 'Molecular Basis of Environmental Toxicity', ed. R. S. Bhatnagar; Ann Arbor Science Publishers, Ann Arbor, 1980, pp. 329–357. 119

80MI50600 R. I. Martinez and J. T. Herron; *Chem. Phys. Lett.*, 1980, **72**, 74. 151

80MI50601 Y. Taguchi and Y. Suhara; *Yukagaku*, 1980, **29**, 912 (*Chem. Abstr.*, 1981, **94**, 174 260). 162

80MI50602 G. Baykut, K. P. Wanczek and H. Hartmann; *Adv. Mass Spectrom., Part A*, 1980, **8**, 186. 166

80MI50603 Y. Muyake; *Sanka to Fujinka*, 1980, **47**, 24 (*Chem. Abstr.*, 1980, **93**, 61 776). 182

80MI50700 D. R. Kelsey; *J. Comput. Chem.*, 1980, **1**, 21.. 186

80MI50800 V. B. Shur, I. A. Tikhonova, G. G. Aleksandrov, Y. T. Struchkov and M. E. Vol'pin; *Inorg. Chim. Acta*, 1980, **44**, 275. 220

80MI50801 M. R. Becerra and J. M. Figuera; *Rev. R. Acad. Cienc. Exactas, Fis. Nat. Madrid*, 1980, **74**, 335. 226

B-80MI50900 'Recent Advances in the Chemistry of β-Lactam Antibiotics', Royal Society of Chemistry, London, Special Publication No. 38, 1980. 249, 250, 251, 253, 267

80MI51000 L. D. Hatfield, J. W. Fisher, J. M. Dunigan, R. W. Burchfield, J. M. Greene, J. A. Webber, R. T. Vasileff and M. D. Kinnick; *Philos. Trans. R. Soc. London, Ser. B*, 1980, **289**, 173. 289

B-80MI51001 P. G. Sammes; 'Topics in Antibiotic Chemistry', Wiley, New York, 1980, vol. 4. 290

B-80MI51100 S. Selwyn, R. W. Lacey and M. Bakhtiar; 'The Beta-Lactam Antibiotics: Penicillins and Cephalosporins in Perspective', Hodder and Stoughton, London, 1980, p. 1. 299

80MI51101 V. Škarić and V. Turjak-Zebić; *Croat. Chem. Acta*, 1980, **53**, 449. 330

B-80MI51102 F. A. Jung, W. R. Pilgrim, J. P. Poyser and P. J. Siret; in 'Topics in Antibiotic Chemistry', ed. P. G. Sammes; Ellis Horwood, Chichester, 1980, vol. 4, p. 11. 303, 305, 306, 313, 331, 338

80MI51103 W. von Daehne; *J. Antibiot.*, 1980, **33**, 451. 317

80MI51104 H. Bundgaard; *J. Clin. Hosp. Pharm.*, 1980, **5**, 73. 327

B-80MI51200 'Topics in Antibiotic Chemistry', ed. P. G. Sammes; Wiley, New York, 1980, vol. 4. 348

80MI51201 H. Kropp, J. G. Sundelof, J. S. Kahan, F. M. Kahan and J. Birnbaum; *Antimicrob. Agents Chemother.*, 1980, **17**, 993. 356

80MI51202 K. Yamamoto, T. Yoshioka, Y. Kato, N. Shibamoto, K. Okamura, Y. Shimauchi and T. Ishikura; *J. Antibiot.*, 1980, **33**, 796. 350, 352

80MI51203 N. Shibamoto, A. Koki, M. Nishino, K. Nakamura, K. Kiyoshima, K. Okamura, M. Okabe, R. Okamoto, Y. Fukagawa, Y. Shimauchi and T. Ishikura; *J. Antibiot.*, 1980, **33**, 1128. 352

80MI51204 G. Gondos, K. L. Lang, A. Szeghy, G. Dombi and G. Bernath; *Magy. Kem. Foly*, 1980, **86**, 227 (*Chem. Abstr.*, 1980, **93**, 150 103). 347

80MI51205 D. R. Kelsey; *J. Comput. Chem.*, 1980, **1**, 21. 360

80MI51300 P. Felker, D. M. Hayes and L. A. Hull; *Theor. Chim. Acta*, 1980, **55**, 293. 364

80MI51400 V. E. Sokolov, E. S. Albone, P. F. Flood, P. F. Heap, M. Z. Kagan, V. S. Vasilieva, V. V. Roznov and E. P. Zinkevich; *J. Chem. Ecol.*, 1980, **6**, 805. 446

80MI51401 D. R. Crump; *J. Chem. Ecol.*, 1980, **6**, 341. 446

80MI51402 D. R. Crump; *J. Chem. Ecol.*, 1980, **6**, 837. 446

80MI51500 W. Adam and F. Yany; *Photochem. Photobiol.*, 1980, **31**, 267. 459

80MI51700 M. Sugiura, Y. Yamazoe, T. Kamataki and R. Kato; *Cancer Res.*, 1980, **40**, 2910. 570

80MI51701 H. P. Schuchmann and C. von Sonntag; *J. Photochem.*, 1980, **13**, 347. 550, 560

80MI51800 M. Artico; *Boll. Chim. Farm.*, 1980, **119**, 455 and 505. 608, 613

B-80MI51900 F. G. Riddell; 'The Conformational Analysis of Heterocyclic Compounds', Academic Press, New York, 1980. 667, 698

80MIP51500 E. Holbova and Z. Odlerova; *Czech. Pat.* 181 447 (1980) (*Chem. Abstr.*, 1980, **93**, 204 619). 484

80MIP51501 E. Wolanski, W. Olczyk, W. Kalinowski, A. Jasko, F. Jargus and K. Ostrowska; *Pol. Pat.* 106 390 (*Chem. Abstr.*, 1981, **94**, 104 542). 484

80NJC95 P. Dumas, P. Guerin and P. Sigwalt; *Nouv. J. Chim.*, 1980, **4**, 95. 160

80NJC703 M. Torres and O. P. Strausz; *Nouv. J. Chim.*, 1980, **4**, 703. 143

80OMR(13)45 De L. R. Crist, A. P. Borsetti, G. J. Jordan and C. F. Hammer; *Org. Magn. Reson.*, 1980, **13**, 45. 10

80OMR(14)145 T. Bottin-Strzalko and M. C. Roux-Schmitt; *Org. Magn. Reson.*, 1980, **14**, 145. 99

80OMR(14)192 H.-D. Martin, R. Iden, D. Scheutzow and L. M. Jackman; *Org. Magn. Reson.*, 1980, **14**, 192. 411

80OMR(14)466 S. Braun, J. Kinkeldei and L. Walther; *Org. Magn. Reson.*, 1980, **14**, 466. 495

80OMR(14)534 H. M. Hugel and Q. N. Porter; *Org. Magn. Reson.*, 1980, **14**, 534. 99

80PAC1623 M. Torres, E. M. Lown, H. E. Gunning and O. P. Strausz; *Pure Appl. Chem.*, 1980, **52**, 1623. 2, 6, 97, 121, 126, 127, 133, 139, 142, 143, 176, 181

80PAC2303 J. M. Lehn; *Pure Appl. Chem.*, 1980, **52**, 2303. 756, 757

80PAC2441 J. M. Lehn; *Pure Appl. Chem.*, 1980, **52**, 2441. 751, 753

80PS(8)1 J. S. Pizey and K. Symeonides; *Phosphorus Sulfur*, 1980, **8**, 1. 436

80PS(8)157 E. Weissflog; *Phosphorus Sulfur*, 1980, **8**, 157 (*Chem. Abstr.*, 1980, **93**, 220 707). 649

80RTC271 H. P. Soetens and U. K. Pandit; *Recl. Trav. Chim. Pays-Bas*, 1980, **99**, 271, and references cited therein. 544

80S571 V. V. Momot, L. I. Samarai and N. D. Bodnarchuk; *Synthesis*, 1980, 571. 455, 458, 470

80S636 H. Paulsen, F. R. Heiker, J. Feldmann and K. Heyns; *Synthesis*, 1980, 636. 111

80S907 E. Schaumann, U. Wriede and J. Ehlers; *Synthesis*, 1980, 907. 480

80SC205 A. B. Smith, III and R. M. Scarborough, Jr.; *Synth. Commun.*, 1980, **10**, 205. 563

80SC213 D. Gaitanopoulos and M. Brenner; *Synth. Commun.*, 1980, **10**, 213. 510, 526

80T73 N. Furukawa, T. Yoshimura, M. Ohtsu, T. Akasaka and S. Oae; *Tetrahedron*, 1980, **36**, 73. 36, 87

80T81 K. Somekawa, S. Kumamoto, T. Matsuo and I. Ueda; *Tetrahedron*, 1980, **36**, 81. 662

80T149 R. Gabioud and P. Vogel; *Tetrahedron*, 1980, **36**, 149. 107

80T171 B. M. Howard, W. Fenical, K. Hirotsu, B. A. Solheim and J. Clardy; *Tetrahedron*, 1980, **36**, 171. 703

80T317 F. van de Griendt and H. Cerfontain; *Tetrahedron*, 1980, **36**, 317. 518

80T461 J. S. Bradshaw and P. E. Stott; *Tetrahedron*, 1980, **36**, 461. 733, 748, 750, 753

80T617 K. Orito, Y. Kurokawa and M. Itoh; *Tetrahedron*, 1980, **36**, 617. 529

80T833 W. Adam and M. Balci; *Tetrahedron*, 1980, **36**, 833. 118, 189

80T1017 K. Orito and T. Matsuzaki; *Tetrahedron*, 1980, **36**, 1017. 518

80T1361 A. S. Gybin, W. A. Smit, M. Z. Krimer, N. S. Zefirov, L. A. Novgorodtseva and N. K. Sadovaya; *Tetrahedron*, 1980, **36**, 1361. 155, 158

80T1747 B. M. Howard, G. R. Schultz, W. Fenical, B. Solheim and J. Clardy; *Tetrahedron*, 1980, **36**, 1747. 703

80T1937 A. Loupy and J. Seyden-Penne; *Tetrahedron*, 1980, **36**, 1937. 757

80T2145 L. Thijs and B. Zwanenburg; *Tetrahedron*, 1980, **36**, 2145. 107

80T2269 T. H. Chan and B. S. Ong, *Tetrahedron*, 1980, **36**, 2269. 103

80T2359 P. Caluwe; *Tetrahedron*, 1980, **36**, 2359. 683

80T3047 R. Shabana, J. B. Rasmussen, S. O. Olesen and S. O. Lawesson; *Tetrahedron*, 1980, **36**, 3047. 386

80T3431 J.-P. Kintzinger, C. Delseth and T. T.-T. Nguye'n; *Tetrahedron*, 1980, **36**, 3431. 367

80TH51400 L. K. Revelle; Ph.D. Thesis, University of Missouri-St. Louis, 1980. 411, 436

80TL7 M. L. Maddox, J. C. Martin and J. M. Muchowski; *Tetrahedron Lett.*, 1980, **21**, 7. 728

80TL31 R. W. Ratcliffe, T. N. Salzmann and B. G. Christensen; *Tetrahedron Lett.*, 1980, **21**, 31. 254, 352

80TL319 W. S. Murphy and K. P. Raman; *Tetrahedron Lett.*, 1980, **21**, 319. 520

80TL343 R. Schulz and A. Schwieg; *Tetrahedron Lett.*, 1980, **21**, 343. 419, 441

80TL351 M. Yoshioka, T. Tsuji, S. Uyeo, S. Yamamoto, T. Aoki, Y. Nishitani, S. Mori, H. Satoh, Y. Hamada, H. Ishitobi and W. Nagata; *Tetrahedron Lett.*, 1980, **21**, 351. 311

80TL395 P. J. Giddings, D. I. John and E. J. Thomas; *Tetrahedron Lett.*, 1980, **21**, 395. 265

80TL445 K. M. Maxey and G. L. Bundy; *Tetrahedron Lett.*, 1980, **21**, 445. 400

80TL463 M.-C. Lasne, J.-L. Ripoll and J.-M. Denis; *Tetrahedron Lett.*, 1980, **21**, 463. 101

80TL561 P. C. Cherry, D. N. Evans, C. E. Newall and N. S. Watson; *Tetrahedron Lett.*, 1980, **21**, 561. 335

80TL585 M. Sevrin and A. Krief; *Tetrahedron Lett.*, 1980, **21**, 585. 391, 392

80TL595 U. Göckel, U. Hartmannsgruber, A. Steigel and J. Sauer; *Tetrahedron Lett.*, 1980, **21**, 595. 361, 505, 519, 544

80TL599 U. Göckel, U. Hartsmannsgruber, A. Steigel and J. Sauer; *Tetrahedron Lett.*, 1980, **21**, 599. 500

80TL643 S. Kulstad and L. A. Malmstem; *Tetrahedron Lett.*, 1980, **21**, 643. 750

80TL727 W. Lwowski and O. S. Rao; *Tetrahedron Lett.*, 1980, **21**, 727. 537

80TL751 J. Mulzer and M. Zippel; *Tetrahedron Lett.*, 1980, **21**, 751. 374

80TL781 R. D. G. Cooper; *Tetrahedron Lett.*, 1980, **21**, 781. 306

80TL785 W. R. Dolbier, Jr. and C. R. Burkholder; *Tetrahedron Lett.*, 1980, **21**, 785. 476

80TL935 J. F. Normant, A. Alexakis and G. Cahiez; *Tetrahedron Lett.*, 1980, **21**, 935. 388

80TL941 F. Kotzyba-Hibert, J. M. Lehn and P. Vierling; *Tetrahedron Lett.*, 1980, **21**, 941. 745, 751

80TL997 R. F. Abdulla and J. C. Williams; *Tetrahedron Lett.*, 1980, **21**, 997. 256

80TL1021 R. C. White; *Tetrahedron Lett.*, 1980, **21**, 1021. 118

80TL1039 A. C. Guimaraes, J. B. Robert, L. Cazaux, C. Picard and P. Tisnes; *Tetrahedron Lett.*, 1980, **21**, 1039. 696

80TL1145 W. Eberbach and J. C. Carré; *Tetrahedron Lett.*, 1980, **21**, 1145. 513, 520

80TL1173 K. N. Gurudutt and B. Ravindranath; *Tetrahedron Lett.*, 1980, **21**, 1173. 113

80TL1223 P. Gesche, F. Klinger, H. Strub and J. Streith; *Tetrahedron Lett.*, 1980, **21**, 1223. 604

80TL1323 J. M. Lehn and P. Vierling; *Tetrahedron Lett.*, 1980, **21**, 1323. 744

80TL1379 T. K. Miller, J. T. Sharp, H. R. Sood and E. Stefaniuk; *Tetrahedron Lett.*, 1980, **21**, 1379. 596

80TL1807	L. E. Friedrich and P. Y.-S. Lam; *Tetrahedron Lett.*, 1980, **21**, 1807.	364, 366, 371
80TL2013	L. D. Cama and B. G. Christensen; *Tetrahedron Lett.*, 1980, **21**, 2013.	352
80TL2021	R. A. Cormier; *Tetrahedron Lett.*, 1980, **21**, 2021.	123, 124
80TL2025	H. G. Corkins, L. Storace and E. R. Osgood; *Tetrahedron Lett.*, 1980, **21**, 2025.	454, 457, 460, 468
80TL2081	I. Ojima and S. Inaba; *Tetrahedron Lett.*, 1980, **21**, 2081.	261
80TL2181	T. Fujisawa, T. Sato, T. Kawara, M. Kawashima, H. Shimizu and Y. Ito; *Tetrahedron Lett.*, 1980, **21**, 2181.	388
80TL2299	S. Caccamese, R. Azzolina, E. N. Duesler, I. C. Paul and K. L. Rinehart, Jr., *Tetrahedron Lett.*, 1980, **21**, 2299.	667, 700
80TL2329	H. Sakurai, K. Sasaki and A. Hosomi; *Tetrahedron Lett.*, 1980, **21**, 2329.	111
80TL2429	H.-J. Federsel and J. Bergman; *Tetrahedron Lett.*, 1980, **21**, 2429.	635
80TL2443	A. Blade-Font, *Tetrahedron Lett.*, 1980, **21**, 2443.	528
80TL2451	R. J. Ponsford; *Tetrahedron Lett.*, 1980, **21**, 2451.	313
80TL2733	S. Terashima, M. Hayashi and K. Koga; *Tetrahedron Lett.*, 1980, **21**, 2733.	115
80TL2783	D. G. Melillo, I. Shinakai, T. Liu, K. Ryan and M. Sletzinger; *Tetrahedron Lett.*, 1980, **21**, 2783.	254, 352
80TL2991	J. E. G. Kemp, M. D. Closier, S. Narayanaswami and M. H. Stefaniak; *Tetrahedron Lett.*, 1980, **21**, 2991.	319
80TL3051	J. P. Bats, J. Moulines, P. Picard and D. Leclercq; *Tetrahedron Lett.*, 1980, **21**, 3051.	392
80TL3061	M. Shibasaki, A. Nishida and S. Ikegami; *Tetrahedron Lett.*, 1980, **21**, 3061.	392
80TL3171	A. Krebs, H. Schmalstieg, O. Jarchow and K. H. Klaska; *Tetrahedron Lett.*, 1980, **21**, 3171.	452
80TL3251	J. A. Pincock and L. M. Druet; *Tetrahedron Lett.*, 1980, **21**, 3251.	465, 472
80TL3403	M. Ikeda, K. Ohno, T. Uno and Y. Tamura; *Tetrahedron Lett.*, 1980, **21**, 3403.	493, 497, 504, 541
80TL3475	H. Prinzbach, K.-H. Müller, C. Kaiser and D. Hunkler; *Tetrahedron Lett.*, 1980, **21**, 3475.	680
80TL3579	K. Oka, A. Dobashi and S. Hara; *Tetrahedron Lett.*, 1980, **21**, 3579.	168, 175
80TL3897	H. Saimoto, T. Hiyama and H. Nozaki; *Tetrahedron Lett.*, 1980, **21**, 3897.	772
80TL3947	J. G. Gleason, D. B. Bryan and K. G. Holden; *Tetrahedron Lett.*, 1980, **21**, 3947.	354, 355
80TL4109	D. P. Munro and J. T. Sharp; *Tetrahedron Lett.*, 1980, **21**, 4109.	596
80TL4179	J. Cabré-Castellví and A. L. Palomo-Coll; *Tetrahedron Lett.*, 1980, **21**, 4179.	330
80TL4203	G. B. Guise, W. D. Ollis, J. A. Peacock, J. S. Stephanatou and J. F. Stoddart; *Tetrahedron Lett.*, 1980, **21**, 4203.	704
80TL4247	E. Schaumann, A. Rohr and G. Adiwidjaja; *Tetrahedron Lett.*, 1980, **21**, 4247.	247, 256, 267, 276
80TL4553	K. Krogh-Jespersen; *Tetrahedron Lett.*, 1980, **21**, 4553.	18
80TL4831	M. Delmas, Y. Le Bigot and A. Gaset; *Tetrahedron Lett.*, 1980, **21**, 4831.	759
80TL4905	W. Tochtermann and P. Roesner; *Tetrahedron Lett.*, 1980, **21**, 4905.	569
80TL4927	H. Suzuki, H. Yashima, T. Hirose, M. Takahashi, Y. Moro-Oka and T. Ikawa; *Tetrahedron Lett.*, 1980, **21**, 4927.	621
80TL4951	C. P. Falshaw, T. J. King, S. Imre, S. Islimyeli and R. H. Thompson; *Tetrahedron Lett.*, 1980, **21**, 4951.	667, 702
80UP51400	E. Block and L. K. Revelle; unpublished results, 1980.	415, 422, 436
80UP51401	K. K. Andersen; unpublished results, 1980.	414
80UP52100	R. B. Sessions and J. M. Lehn; unpublished results, 1980.	751
80USP4196300	W. Ho, R. J. Mohrbacher and G. Tutwiler; *U.S. Pat.* 4 196 300 (1980) (*Chem. Abstr.*, 1980, **93**, 71 528).	182
80USP4232608	J. M. Wrightson; *U.S. Pat.* 4 232 608 (1980) (*Chem. Abstr.*, 1981, **94**, 105 900).	484
80YZ29	H. Tobiki, K. Shimago and S. Ueda; *Yakugaku Zasshi*, 1980, **100**, 29.	331
80ZN(B)1040	H. Meier and H. Kolshorn; *Z. Naturforsch., Teil B*, 1980, **35**, 1040.	126
80ZN(B)1479	U. Burker; *Z. Naturforsch., Teil B*, 1980, **35**, 1479.	700, 701
80ZOB1502	E. S. Gubnitskaya and V. S. Parkhomenko; *Zh. Obshch. Khim.*, 1980, **50**, 1502.	213
80ZOR2447	A. M. Sinyagin and V. G. Kartsev; *Zh. Org. Khim.*, 1980, **16**, 2447 (*Chem. Abstr.*, 1981, **94**, 83 852).	266

81ACR218	R. G. Harvey; *Acc. Chem. Res.*, 1981, **14**, 218.	591
81ACR227	G. Melloni, G. Modena and U. Tonellato; *Acc. Chem. Res.*, 1981, **14**, 227.	163, 177
81ACR348	V. Snieckus and J. Streith; *Acc. Chem. Res.*, 1981, **14**, 348.	28, 29, 595, 598, 600, 601, 603, 604
81AG293	H. Quast and A. Fuss; *Angew. Chem.*, 1981, **93**, 293.	182
81AG(E)200	G. V. Shustov, N. B. Tavakalyan and R. G. Kostyanovskii; *Angew. Chem., Int. Ed. Engl.*, 1981, **20**, 200.	233
81AG(E)385	E. Niecke, A. Nickloweit-Lüke and R. Rüger; *Angew. Chem., Int. Ed. Engl.*, 1981, **20**, 385.	196, 197
81AG(E)571	H. C. Kang and V. Boekelheide; *Angew. Chem., Int. Ed. Engl.*, 1981, **20**, 571.	768, 780
81AG(E)613	E. Schaumann, H. G. Bäuch and G. Adiwidjaja; *Angew. Chem., Int. Ed. Engl.*, 1981, **20**, 613.	405, 438
81AG(E)670	D. St. C. Black and L. M. Johnstone; *Angew. Chem., Int., Ed. Engl.*, 1981, **20**, 670.	212

81AG(E)699	H. H. Eckhardt, D. Hege, W. Massa, H. Perst and R. Schmidt; *Angew. Chem., Int. Ed. Engl.*, 1981, **20**, 699.	494, 530, 542
81AG(E)808	P. Feyen and W. Schröck; *Angew. Chem., Int. Ed. Engl.*, 1981, **20**, 808.	323
81AG(E)868	J. Gasteiger and C. Herzig; *Angew. Chem., Int. Ed. Engl.*, 1981, **20**, 868.	129
81AG(E)1044	M. Baudler, H. Suchomel, G. Fürstenberg and U. Schings; *Angew. Chem., Int. Ed. Engl.*, 1981, **20**, 1044.	197
81AHC(28)231	C. Wentrup; *Adv. Heterocycl. Chem.*, 1981, **28**, 231.	533, 534, 536
81AHC(29)1	R. K. Smalley; *Adv. Heterocycl. Chem.*, 1981, **29**, 1.	542
81AJC421	R. Y. S. Tan, R. A. Russell and R. N. Warrener; *Aust. J. Chem.*, 1981, **34**, 421.	101
81AJC569	D. D. Ridley and C. W. Simpson; *Aust. J. Chem.*, 1981, **34**, 569.	656
81AJC855	R. N. Warrener, R. A. Russell and R. Y. S. Tan; *Aust. J. Chem.*, 1981, **34**, 855.	125
81AJC2393	J. W. Blunt, R. I. Lake, M. H. G. Muro and S. C. Yorke; *Aust. J. Chem.*, 1981, **34**, 2393.	667
81AX(B)486	B. Beagley, S. F. Moss, R. G. Pritchard and D. R. Taylor; *Acta Crystallogr., Part B*, 1981, **37**, 486.	650
81BCJ301	W. Tamura, H. Imaizumi, Y. Hashida and K. Matsui; *Bull. Chem. Soc. Jpn.*, 1981, **54**, 301.	537
81CA(94)15453	*Chem. Abstr.*, 1981, **94**, 15 453.	96
81CB787	H. Quast and F. Kees; *Chem. Ber.*, 1981, **114**, 787.	176
81CB802	H. Quast and F. Kees; *Chem. Ber.*, 1981, **114**, 802.	180
81CB1027	W. Eberbach and J. C. Carré; *Chem. Ber.*, 1981, **114**, 1027.	101
81CC10	S. A. Matlin and L. Chan; *J. Chem. Soc., Chem. Commun.*, 1981, 10.	265
81CC41	K. Piotrowska and D. Mostowicz; *J. Chem. Soc., Chem. Commun.*, 1981, 41.	247, 266
81CC72	E. Niecke and D. A. Wildbredt; *J. Chem. Soc., Chem. Commun.*, 1981, 72.	196
81CC120	A. J. Fry, K. Ankner and V. K. Handa; *J. Chem. Soc., Chem. Commun.*, 1981, 120.	167
81CC241	O. Meth-Cohn and S. Rhouati; *J. Chem. Soc., Chem. Commun.*, 1981, 241.	538
81CC299	K. C. Phelan and G. Stedman; *J. Chem. Soc., Chem. Commun.*, 1981, 299.	3
81CC350	A. Dondoni, P. Giorgianni and A. Battaglia; *J. Chem. Soc., Chem. Commun.*, 1981, 350.	452, 464, 473
81CC417	U. K. Nadir and V. K. Koul; *J. Chem. Soc., Chem. Commun.*, 1981, 417.	244
81CC625	D. J. Cram and G. D. Y. Sogah; *J. Chem. Soc., Chem. Commun.*, 1981, 625.	759
81CC714	Y. Takahashi, K. Isobe, H. Hagiwara, H. Kosugi and H. Uda; *J. Chem. Soc., Chem. Commun.*, 1981, 714.	192
81CC729	J. A. Bandy, D. G. Parsons and M. R. Truter; *J. Chem. Soc., Chem. Commun.*, 1981, 729.	738
81CC777	R. Annunziata, F. Montanari, S. Quici and M. T. Vitali; *J. Chem. Soc., Chem. Commun.*, 1981, 777.	750
81CC790	N. R. Ayyangar, R. B. Bambal and A. G. Lugade; *J. Chem. Soc., Chem. Commun.*, 1981, 790, and references cited therein.	537
81CC833	J. P. Kintzinger, F. Kotzyba-Hibert, J. M. Lehn, A. Pagelot and K. Saigo; *J. Chem. Soc., Chem. Commun.*, 1981, 833.	740
81CC838	D. R. Boyd, M. G. Burnett and R. M. E. Greene; *J. Chem. Soc., Chem. Commun.*, 1981, 838.	554, 555
81CC847	H. M. Colquhoun, J. F. Stoddart and D. J. Williams; *J. Chem. Soc., Chem. Commun.*, 1981, 847.	735, 736
81CC849	H. M. Colquhoun, J. F. Stoddart and D. J. Williams; *J. Chem. Soc., Chem. Commun.*, 1981, 849.	736
81CC929	M. L. Gilpin, J. B. Harbridge and T. T. Howarth; *J. Chem. Soc., Chem. Commun.*, 1981, 929.	253
81CC946	C. L. Branch and M. J. Pearson; *J. Chem. Soc., Chem. Commun.*, 1981, 946.	354
81CC959	M. D. Bachi and A. Gross; *J. Chem. Soc., Chem. Commun.*, 1981, 959.	253
81CC990	N. F. Jones, A. Kumar and I. O. Sutherland; *J. Chem. Soc., Chem. Commun.*, 1981, 990.	745, 751
81CC1028	M. L. Durrant and J. R. Malpas; *J. Chem. Soc., Chem. Commun.*, 1981, 1028.	277
81CC1067	F. Peter, M. Gross, M. W. Hosseini, J. M. Lehn and R. B. Sessions; *J. Chem. Soc., Chem. Commun.*, 1981, 1067.	747
81CC1087	R. A. Abramovitch, S. B. Hendi and A. O. Kress; *J. Chem. Soc., Chem. Commun.*, 1981, 1087.	538
81CC1146	G. Bahadur, J. E. Baldwin, T. Wan and M. Jung; *J. Chem. Soc., Chem. Commun.*, 1981, 1146.	636
81CC1269	G. Johnson and B. C. Ross; *J. Chem. Soc., Chem. Commun.*, 1981, 1269.	354, 355
81CC1289	Y. Kobayashi, A. Ando, K. Kawada and I. Kumadaki; *J. Chem. Soc., Chem. Commun.*, 1981, 1289.	170
81CCC148	Z. Vejdělek, E. Svátek, J. Holubek, J. Metyš, M. Bartosŏvá and M. Protiva; *Collect. Czech. Chem. Commun.*, 1981, **46**, 148.	546
81CCC1199	M. Protiva, Z. Sedivy, J. Pomykacek, E. Svatek and J. Holubek; *Collect. Czech. Chem. Commun.*, 1981, **46**, 1199.	591
81CJC406	S. Wolfe, R. J. Bowers, S. K. Hasan and P. M. Kazmaier; *Can. J. Chem.*, 1981, **59**, 406.	636
81CJC1574	J. B. Jones and H. M. Schwartz; *Can. J. Chem.*, 1981, **59**, 1574.	574
81CL5	M. Iwata and H. Kuzuhara; *Chem. Lett.*, 1981, 5.	765
81CL201	A. Osuka, H. Suzuki and K. Maruyama; *Chem. Lett.*, 1981, 201.	102

81CL1781 F. Bottino and S. Pappalardo; *Chem. Lett.*, 1981, 1781. 779

81CPB1063 H. Takahata, Y. Ohnishi, H. Takehara, K. Tsuritani and T. Yamazaki; *Chem. Pharm. Bull.*, 1981, **29**, 1063. 256

81CPB1221 T. Ohta, N. Miyata and M. Hirobe; *Chem. Pharm. Bull.*, 1981, **29**, 1221. 512

81CPB1539 T. Tsuchiya, H. Sawanishi, M. Enkaku and T. Hirai; *Chem. Pharm. Bull.*, 1981, **29**, 1539. 607

81CPB1747 K. Ikeda, Y. Terao and M. Sekiya; *Chem. Pharm. Bull.*, 1981, **29**, 1747. 261

81CPB1887 T. Tsuchiya and H. Sashida; *Chem. Pharm. Bull.*, 1981, **29**, 1887. 598

81CRV91 M. Balci; *Chem. Rev.*, 1981, **81**, 91. 118

81CSC313 A. W. Hanson; *Cryst. Struct. Commun.*, 1981, **10**, 313. 780

81CSC345 J. H. Noordik and P. J. M. Cillissen; *Cryst. Struct. Commun.*, 1981, **10**, 345. 366, 394

81CSC571 A. W. Hanson; *Cryst. Struct. Commun.*, 1981, **10**, 571. 780

81CZ273 H. Schildknecht, C. Birkner and D. Krauss; *Chem.-Ztg.*, 1981, **105**, 273. 446

81GEP3027169 A. D. Cale, Jr. and H. J. Jenkins; *Ger. Pat.* 3 027 169 (*Chem. Abstr.*, 1981, **95**, 7029). 266

81H(15)421 K. Sasakura and T. Sugasawa; *Heterocycles*, 1981, **15**, 421. 533

81H(15)517 J. Rebek, Jr.; *Heterocycles*, 1981, **15**, 517. 97, 115

81H(15)547 T. Nishiwaki and N. Abe; *Heterocycles*, 1981, **15**, 547. 531

81H(15)1007 T. Hatsui and H. Takeshita; *Heterocycles*, 1981, **15**, 1007. 99

81H(15)1101 Y. Hirai, I. Kamide and T. Yamazaki; *Heterocycles*, 1981, **15**, 1101. 250

81H(15)1227 J. Guillaume, L. Nédélec, M. Cariou and A. Allais; *Heterocycles*, 1981, **15**, 1227. 546

81H(15)1569 T. Mukai, T. Kumagai and Y. Yamashita; *Heterocycles*, 1981, **15**, 1569. 14, 188, 503, 506, 520, 521, 522

81H(15)1643 C. Vaz, G. Griffin, S. Christensen and D. Lankin; *Heterocycles*, 1981, **15**, 1643. 185

81H(16)137 J. Kurita, H. Kojima and T. Tsuchiya; *Heterocycles*, 1981, **16**, 137. 607, 608

81H(16)243 K. Prasad, H. Hamberger, P. Stuetz and G. Schulz; *Heterocycles*, 1981, **16**, 243. 252

81H(16)363 T. Sano, Y. Horiguchi and S. Kambe; *Heterocycles*, 1981, **16**, 363. 496, 503, 516, 525, 541

81H(16)399 M. M. Vora, C. S. Vi and C. DeWitt Blanton, Jr.; *Heterocycles*, 1981, **16**, 399. 533

81H(16)621 M. R. Chaurasia and S. K. Sharma; *Heterocycles*, 1981, **16**, 621. 648

81H(16)1187 R. Neidlein and W. Lehr; *Heterocycles*, 1981, **16**, 1187. 461

81H(16)1305 K. Prasad, G. Schulz, C. P. Mak, H. Hamberger and P. Stuetz; *Heterocycles*, 1981, **16**, 1305. 254

81H(16)1367 K. Matsumoto, T. Uchida and R. M. Acheson; *Heterocycles*, 1981, **16**, 1367. 542

81H(16)1923 M. Enkaku, J. Kurita and T. Tsuchiya; *Heterocycles*, 1981, **16**, 1923. 598, 602

81HCA198 A. P. Alder, H. R. Wolf and O. Jeger; *Helv. Chim. Acta*, 1981, **64**, 198. 101

81HCA224 G. De Weck and H. R. Wolf; *Helv. Chim. Acta*, 1981, **64**, 224. 101

81HCA373 D. Berney and K. Schuh; *Helv. Chim. Acta*, 1981, **64**, 373. 493, 510, 514, 525, 533

81HCA687 E. Hungerbühler and D. Seebach; *Helv. Chim. Acta*, 1981, **64**, 687. 115

81HCA1303 I. Ernest, A. J. Main and R. B. Woodward; *Helv. Chim. Acta*, 1981, **64**, 1303. 252

81IC1415 G. M. Kramer, E. T. Maas and M. B. Dines; *Inorg. Chem.*, 1981, **20**, 1415. 376

81IZV1633 M. G. Lin'kova, O. V. Kul'disheva and I. L. Knunyants; *Izv. Akad. Nauk SSSR, Ser. Khim.*, 1981, **7**, 1633. 441

81IZV1929 M. A. Ibrágimov, V. A. Smit, A. S. Gybin and M. Z. Krimer; *Izv. Akad. Nauk SSSR, Ser. Khim.*, 1981, 1929 (*Chem. Abstr.*, 1981, **95**, 219 830). 165

81JA406 D. Schomburg, Q.-C. Mir and J. M. Shreeve; *J. Am. Chem. Soc.*, 1981, **103**, 406. 454

81JA462 D. J. Morgans, Jr. and K. B. Sharpless; *J. Am. Chem. Soc.*, 1981, **103**, 462. 105

81JA464 B. E. Rossiter, T. Kasuki and K. B. Sharpless; *J. Am. Chem. Soc.*, 1981, **103**, 464. 116

81JA468 J. C. Guillemin, J. M. Denis and A. Lablache-Combier; *J. Am. Chem. Soc.*, 1981, **103**, 468. 267, 269, 270, 271, 275

81JA486 A. Krantz and J. Laureni; *J. Am. Chem. Soc.*, 1981, **103**, 486. 128, 181

81JA686 R. M. Moriarty, S. C. Gupta, H. Hu, D. R. Berenschot and K. B. White; *J. Am. Chem. Soc.*, 1981, **103**, 686. 103

81JA898 H. S. I. Chao and G. A. Berchtold; *J. Am. Chem. Soc.*, 1981, **103**, 898. 552, 566, 584

81JA1023 A. Rauk; *J. Am. Chem. Soc.*, 1981, **103**, 1023. 97

81JA1066 R. A. Eades, P. G. Gassman and D. A. Dixon; *J. Am. Chem. Soc.*, 1981, **103**, 1066. 100

81JA1282 B. Dietrich, M. W. Hosseini, J. M. Lehn and R. B. Sessions; *J. Am. Chem. Soc.*, 1981, **103**, 1282. 747

81JA1292 S. P. Schmidt, M. A. Vincent, C. E. Dykstra and G. B. Schuster; *J. Am. Chem. Soc.*, 1981, **103**, 1292. 453, 459

81JA1525 R. A. Abramovitch, W. D. Holcomb and S. Wake; *J. Am. Chem. Soc.*, 1981, **103**, 1525. 537

81JA1581 S. A. Benner, J. E. Maggio and H. E. Simmons, III; *J. Am. Chem. Soc.*, 1981, **103**, 1581. 103

81JA1769 H. W. Moore, L. Hernandez, Jr., D. M. Kunert, F. Mercer and A. Sing; *J. Am. Chem. Soc.*, 1981, **103**, 1769. 44, 359

81JA1821 J. S. Bradshaw, S. L. Baxter, J. D. Lamb, R. M. Izatt and J. J. Christensen; *J. Am. Chem. Soc.*, 1981, **103**, 1821. 749

81JA1831 L. A. Paquette and Y.-K. Han; *J. Am. Chem. Soc.*, 1981, **103**, 1831. 110

81JA1851 L. E. Overman and K. L. Bell; *J. Am. Chem. Soc.*, 1981, **103**, 1851. 112

81JA2091 P. Jouin, C. B. Troostwijk and R. M. Kellogg; *J. Am. Chem. Soc.*, 1981, **103**, 2091. 745, 754

81JA2098 E. L. Clennan, W. Simmons and C. W. Almgren; *J. Am. Chem. Soc.*, 1981, **103**, 2098. 40

81JA2112	J. C. Saddler and P. L. Fuchs; *J. Am. Chem. Soc.*, 1981, **103**, 2112.	113
81JA2802	M. J. S. Dewar, D. J. Nelson, P. B. Shevlin and K. A. Biesiada; *J. Am. Chem. Soc.*, 1981, **103**, 2802.	128
81JA3136	J. C. Stavinoha and P. S. Mariano; *J. Am. Chem. Soc.*, 1981, **103**, 3136.	342, 345, 347
81JA3221	B. W. Metcalf, C. L. Wright, J. P. Burkhart and J. O. Johnston; *J. Am. Chem. Soc.*, 1981, **103**, 3221.	124
81JA4526	H. R. Pfaendler, J. Gosteli, R. B. Woodward and G. Rihs; *J. Am. Chem. Soc.*, 1981, **103**, 4526.	338
81JA4597	S. Ohuchida, N. Hamanaka and M. Hayashi; *J. Am. Chem. Soc.*, 1981, **103**, 4597.	434
81JA4974	A. Padwa, K. F. Koehler and A. Rodriguez; *J. Am. Chem. Soc.*, 1981, **103**, 4974.	263
81JA5107	D. H. Busch, G. G. Christoph, L. L. Zimmer, S. C. Jackels, J. J. Grzybowski, R. C. Callahan, M. Kojima, K. A. Holter, J. Mocak, N. Herron, M. Chavan and W. P. Schammel; *J. Am. Chem. Soc.*, 1981, **103**, 5107.	718, 723
81JA5250	Y. Ito, S. Miyata, M. Nakatsuka and T. Saegusa; *J. Am. Chem. Soc.*, 1981, **103**, 5250.	277
81JA5414	T. B. Cameron, F. M. El-Kabbani and H. W. Pinnick; *J. Am. Chem. Soc.*, 1981, **103**, 5414.	117
81JA5707	A. Knöchel and R. D. Wilken; *J. Am. Chem. Soc.*, 1981, **103**, 5707.	759
81JA5939	R. B. Silverman; *J. Am. Chem. Soc.*, 1981, **103**, 5939.	119
81JA5969	B. M. Trost and G. A. Molander; *J. Am. Chem. Soc.*, 1981, **103**, 5969.	113
81JA6044	P. B. Smith, J. L. Dye, J. Cheney and J. M. Lehn; *J. Am. Chem. Soc.*, 1971, **103**, 6044.	744
81JA6237	V. S. Martin, S. S. Woodard, T. Katsuki, Y. Yamada, M. Ikeda and B. K. Sharpless; *J. Am. Chem. Soc.*, 1981, **103**, 6237.	116
81JA6526	L. A. Paquette, H. Schostarez and G. D. Annis; *J. Am. Chem. Soc.*, 1981, **103**, 6526.	120
81JA6686	A. J. Pearson and C. W. Ong; *J. Am. Chem. Soc.*, 1981, **103**, 6686.	120
81JA6744	T. A. Dix and L. J. Marnett; *J. Am. Chem. Soc.*, 1981, **103**, 6744.	119
81JA6765	S. Karady, J. S. Amato, R. A. Reamer and L. M. Weinstock; *J. Am. Chem. Soc.*, 1981, **103**, 6765.	320
81JA6773	Y.-Y. Lin, M. Risk, S. M. Ray, D. V. Engen, J. Clardy, J. Golik, J. C. James and K. Nakanishi; *J. Am. Chem. Soc.*, 1981, **103**, 6773.	667
81JA6974	J. A. Nelson, S. R. Steckbeck and T. A. Spencer; *J. Am. Chem. Soc.*, 1981, **103**, 6974.	119
81JA7106	L. A. Paquette, L. W. Hertel, R. Gleiter, M. C. Böhm, M. A. Beno and G. G. Christoph; *J. Am. Chem. Soc.*, 1981, **103**, 7106.	116
81JA7368	K. Hattori, Y. Matsumura, T. Miyazaki, K. Maruoka and H. Yamamoto; *J. Am. Chem. Soc.*, 1981, **103**, 7368.	655, 730
81JA7371	I. Tabushi and A. Yazaki; *J. Am. Chem. Soc.*, 1981, **103**, 7371.	119
81JA7376	G. R. Newkome, H. C. R. Taylor, F. R. Fronczek, T. J. Delord, D. K. Kohli and F. Vögtle; *J. Am. Chem. Soc.*, 1981, **103**, 7376.	736
81JA7380	B. M. Trost and D. P. Curran; *J. Am. Chem. Soc.*, 1981, **103**, 7380.	120
81JA7458	A. A. Bhattacharyya, A. Bhattacharyya, R. R. Adkins and A. G. Turner; *J. Am. Chem. Soc.*, 1981, **103**, 7458.	488
81JA7660	E. C. Taylor, H. M. L. Davies, R. J. Clemens, H. Yanagisawa and N. F. Haley; *J. Am. Chem. Soc.*, 1981, **103**, 7660.	470, 483
81JA7743	E. C. Taylor, N. F. Haley and R. J. Clemens; *J. Am. Chem. Soc.*, 1981, **103**, 7743.	470
81JCR(M)1132	B. Foehlisch and W. Gottstein; *J. Chem. Res. (M)*, 1981, 1132.	427
81JCR(S)162	T. O. Glasbey, P. W. Manley and R. C. Storr; *J. Chem. Res. (S)*, 1981, 162.	282
81JCR(S)314	S. Karbach, W. Löhr and F. Vögtle; *J. Chem. Res. (S)*, 1981, 314.	769
81JCS(F2)503	G. Bendazzoli, P. Biscarini, P. Palmieri and G. Gottarelli; *J. Chem. Soc., Faraday Trans. 2*, 1981, **77**, 503.	137
81JCS(P1)52	R. C. Cambie, G. D. Mayer, P. S. Rutledge and P. D. Woodgate; *J. Chem. Soc., Perkin Trans. 1*, 1981, 52.	179
81JCS(P1)94	D. R. Boyd, K. A. Dawson, G. S. Gadaginamath, J. G. Hamilton, J. F. Malone and N. D. Sharma; *J. Chem. Soc., Perkin Trans. 1*, 1981, 94.	554
81JCS(P1)270	G. D. Annis, S. V. Ley, C. R. Self and R. Sivaramakrishnan; *J. Chem. Soc., Perkin Trans. 1*, 1981, 270.	399
81JCS(P1)447	W. S. Murphy and K. P. Raman; *J. Chem. Soc., Perkin Trans. 1*, 1981, 447.	510, 520
81JCS(P1)943	T. Nishio, K. Katahira and Y. Omote; *J. Chem. Soc., Perkin Trans. 1*, 1981, 943.	360
81JCS(P1)1037	Y. Tamura, Y. Takebe, S. Bayomi, M. M. Said, C. Mukai, M. Ikeda, M. Murase and M. Kise; *J. Chem. Soc., Perkin Trans. 1*, 1981, 1037.	632
81JCS(P1)1132	R. N. Carde, P. C. Hayes, G. Jones and C. J. Cliff; *J. Chem. Soc., Perkin Trans. 1*, 1981, 1132, and references cited therein.	537
81JCS(P1)1707	A. Chatterjee, B. Sen and S. K. Chatterjee; *J. Chem. Soc., Perkin Trans. 1*, 1981, 1707.	572
81JCS(P1)1826	M. D. M. Gray, D. R. Russell, D. J. H. Smith, T. Durst and B. Gimbarzevsky; *J. Chem. Soc., Perkin Trans. 1*, 1981, 1826.	453
81JCS(P1)1871	Y. Ito and T. Matsuura; *J. Chem. Soc., Perkin Trans. 1*, 1981, 1871.	106
81JCS(P1)1934	A. P. Davis, G. J. Hughes, P. R. Lowndes, C. M. Robbins, E. J. Thomas and G. H. Whitham; *J. Chem. Soc., Perkin Trans. 1*, 1981, 1934.	172
81JCS(P1)2087	Y. Maki, N. Mitsumori, M. Sako and M. Suzuki; *J. Chem. Soc., Perkin Trans. 1*, 1981, 2087.	307

81JCS(P1)2233 D. R. Boyd, G. S. Gadaginamath, N. D. Sharma, A. F. Drake, S. F. Mason and D. M. Jerina; *J. Chem. Soc., Perkin Trans. 1*, 1981, 2233. 554

81JCS(P1)2363 D. H. R. Barton, R. S. H. Motherwell and W. B. Motherwell; *J. Chem. Soc., Perkin Trans. 1*, 1981, 2363. 114

81JCS(P1)2539 D. Davies and M. J. Pearson; *J. Chem. Soc., Perkin Trans. 1*, 1981, 2539. 250

81JCS(P1)2761 P. N. Anderson, C. B. Argo and J. T. Sharp; *J. Chem. Soc., Perkin Trans. 1*, 1981, 2761. 597

81JCS(P1)2878 A. E. Baydar, G. V. Boyd, P. F. Lindley, F. Watson, P. W. Manley and R. C. Storr; *J. Chem. Soc., Perkin Trans. 1*, 1981, 2878. 282

81JCS(P1)3242 J. H. Bateson, A. J. G. Baxter, P. M. Roberts, T. C. Smale and R. Southgate; *J. Chem. Soc., Perkin Trans. 1*, 1981, 3242. 251

81JCS(P2)1066 B. C. Gilbert, R. O. C. Norman and P. S. Williams; *J. Chem. Soc., Perkin Trans. 2*, 1981, 1066. 174

81JCS(P2)1166 L. Carlsen, H. Egsgaard and D. N. Harpp; *J. Chem. Soc., Perkin Trans. 2*, 1981, 1166. 421

81JCS(P2)1247 D. P. Kessler, I. Ghebre-Sellassie, A. M. Knevel and S. L. Hem; *J. Chem. Soc., Perkin Trans. 2*, 1981, 1247. 327

81JHC59 T. Kovac, B. Belin, T. Fajdiga and V. Sunjic; *J. Heterocycl. Chem.*, 1981, **18**, 59. 609

81JHC271 O. Migliari and V. Sprio; *J. Heterocycl. Chem.*, 1981, **18**, 271. 597

81JHC847 J. P. Le Roux, P. L. Desbene and J. C. Cherton; *J. Heterocycl. Chem.*, 1981, **18**, 847. 627

81JHC889 A. M. Vega, M. T. Martinez, J. A. Palop, J. M. Mateo and E. Fernandez-Alvarez; *J. Heterocycl. Chem.*, 1981, **18**, 889. 597

81JHC1239 S. Kano, S. Shibuya and T. Ebata; *J. Heterocycl. Chem.*, 1981, **18**, 1239. 255

81JOC27 K. Maruyama, T. Ishitoku and Y. Kubo; *J. Org. Chem.*, 1981, **46**, 27. 41

81JOC303 C. G. Overberger and M.-S. Chi; *J. Org. Chem.*, 1981, **46**, 303. 670

81JOC320 L. S. Lehman, L. M. Beclawski, S. A. Harris, H. W. Heine, J. P. Springer, W. J. A. VandenHeuvel and B. H. Arison; *J. Org. Chem.*, 1981, **46**, 320. 214

81JOC342 Y. H. Chang, F.-T. Chiu and G. Zon; *J. Org. Chem.*, 1981, **46**, 342. 537

81JOC501 M. E. Landis, J. C. Mitchell and D. L. Offen; *J. Org. Chem.*, 1981, **46**, 501 472

81JOC610 Y. Hata and M. Watanabe; *J. Org. Chem.*, 1981, **46**, 610. 208

81JOC783 A. I. Meyers, M. Reuman and R. A. Gabel; *J. Org. Chem.*, 1981, **46**, 783. 528

81JOC813 H. S. I. Chao and G. A. Berchtold; *J. Org. Chem.*, 1981, **46**, 813. 552, 584

81JOC833 G. R. Newkome, S. J. Garbis, V. K. Majestic, F. R. Fronczek and G. Chiari; *J. Org. Chem.*, 1981, **46**, 833. 766

81JOC1043 Y. Ito, S. Miyata, M. Nakatsuka and T. Saegusa; *J. Org. Chem.*, 1981, **46**, 1043. 767

81JOC1191 H. S. I. Chao and G. A. Berchtold; *J. Org. Chem.*, 1981, **46**, 1191. 584

81JOC1402 A. Padwa and S. Nahm; *J. Org. Chem.*, 1971, **46**, 1402. 597

81JOC1469 R. W. Miller, R. G. Powell and C. R. Smith, Jr.; *J. Org. Chem.*, 1981, **46**, 1469. 400

81JOC1747 M. P. Cooke, Jr; *J. Org. Chem.*, 1981, **46**, 1747. 778

81JOC1817 J. E. Van Epp, Jr., D. R. Boyd and G. A. Berchtold; *J. Org. Chem.*, 1981, **46**, 1817. 584

81JOC2046 P. J. Claes, E. Roets, P. Herdewijn and H. Vanderhaeghe; *J. Org. Chem.*, 1981, **46**, 2046. 324

81JOC2208 F. A. Bouffard and B. G. Christensen; *J. Org. Chem.*, 1981, **46**, 2208. 251

81JOC2267 Y. Inoue, N. Matsumoto, T. Hakushi and R. Srinivasan; *J. Org. Chem.*, 1981, **46**, 2267. 559, 560, 580

81JOC2589 S. B. Gingerich, W. H. Campbell, C. E. Bricca, P. W. Jennings and C. F. Campana; *J. Org. Chem.*, 1981, **46**, 2589. 579

81JOC2999 H. H. Wasserman, D. J. Hlasta, A. W. Tremper and J. S. Wu; *J. Org. Chem.*, 1981, **46**, 2999. 261

81JOC3144 S. E. Denmark; *J. Org. Chem.*, 1981, **46**, 3144. 391

81JOC3348 L. R. Hills and R. C. Ronald; *J. Org. Chem.*, 1981, **46**, 3348. 115

81JOC3512 F. David; *J. Org. Chem.*, 1981, **46**, 3512. 107

81JOC3568 W. A. Spitzer, T. Goodson, S. R. Lammert and S. Kukolja; *J. Org. Chem.*, 1981, **46**, 3568. 251, 307

81JOC3630 T. Ishizu, K. Harano, M. Yasuda and K. Kanematsu; *J. Org. Chem.*, 1981, **46**, 3630. 513

81JOC3721 I. L. Reich and H. J. Reich; *J. Org. Chem.*, 1981, **46**, 3721. 188

81JOC4077 H.-D. Becker and K. Gustafsson; *J. Org. Chem.*, 1981, **46**, 4077, and references cited therein. 505, 538

81JOC4720 V. Lucchini, G. Modena, G. Valle and G. Capozzi; *J. Org. Chem.*, 1981, **46**, 4720. 163

81JOC4846 P. H. Mazzocchi, S. Minamikawa, P. Wilson, M. Bowen and N. Narian; *J. Org. Chem.*, 1981, **46**, 4846. 543

81JOC5163 M. Node, K. Nishide, M. Ochiai, K. Fuji and E. Fujita; *J. Org. Chem.*, 1981, **46**, 5163. 387

81JOM(205)417 C. R. Hall, E. R. Hamner, R. D. W. Kemmitt and D. J. H. Smith; *J. Organomet. Chem.*, 1981, **205**, 417. 432

81JPC1089 E. M. Lown, K. S. Sidhu, A. W. Jackson, A. Jodhan, M. Green and O. P. Strausz; *J. Phys. Chem.*, 1981, **85**, 1089. 140

81JPC2469 P. W. Pakes, T. Pakes, T. C. Rounds and H. L. Strauss; *J. Phys. Chem.*, 1981, **85**, 2469. 700, 701

81JPC2476 P. W. Pakes, T. C. Rounds and H. L. Strauss; *J. Phys. Chem.*, 1981, **85**, 2476. 700

81LA240 R. Kreher and W. Gerhardt; *Liebigs Ann. Chem.*, 1981, 240. 494, 497, 512

81LA1065 H. A. Staab and W. K. Appel; *Liebigs Ann. Chem.*, 1981, 1065. 765, 777

81LA2164 H. Barth, G. Burger, H. Döpp, M. Kobayashi and H. Musso; *Liebigs Ann. Chem.*, 1981, 2164. 546

81M867 M. Breuer and K.-H. Ongania; *Monatsh. Chem.*, 1981, **112**, 867. 251

B-81MI50300 W. Lwowski; in 'Reactive Intermediates', ed. M. Jones, Jr. and R. A. Moss, Wiley, New York, 1981, vol. 2, p. 315. 36

B-81MI50500 M. Bartók and K. L. Láng; in 'Chemistry of Ethers, Crown Ethers, Hydroxyl Group and their Sulphur Analogs', ed. S. Patai; Wiley, New York, 1981, Supplement E. 97

81MI50501 S. Arakawa; *Mem. Fac. Sci., Kyoto Univ., Ser. Phys. Astrophys. Geophys. Chem.*, 1979 (pub. 1980), **35**, 327 (*Chem. Abstr.*, 1981, **94**, 173 924). 102

B-81MI50600 G. Maccagnani; in 'Organic Sulfur Chemistry', ed. R. Kh. Freidlina and A. E. Skorova; Pergamon, Oxford, 1981, p. 123. 141, 150, 155, 175

81MI50601 K. S. Arulsamy, K. K. Pandey and U. C. Agarwala; *Inorg. Chim. Acta*, 1981, **54**, L51. 141, 153, 156

B-81MI50602 D. C. Dittmer and B. H. Patwardhan; in 'The Chemistry of the Sulphonium Group', ed. C. J. M. Stirling and S. Patai; Wiley, Chichester, 1981, part 2, p. 387. 147, 177

B-81MI50603 W. A. Smit, N. S. Zefirov and I. V. Bodrikov; in 'Organic Sulfur Chemistry, ed. R. Kh. Freidlina and A. E. Skorova; Pergamon, Oxford, 1981, p. 159. 158, 163, 174

81MI50800 E. Schmitz; *Mitt.-Bl. Chem. Ges. DDR*, 1981, **27**, 258. 235

B-81MI51000 R. Bucourt; in 'Recent Advances in the Chemistry of β-Lactam Antibiotics', ed. G. I. Gregory; The Royal Society of Chemistry, London, 1981, p. 1. 295

81MI51001 T. H. Maugh; *Science*, 1981, **214**, 1225. 298

B-81MI51100 J. O'Sullivan and E. P. Abraham; in 'Antibiotics, Vol. IV, Biosynthesis', ed. J. W. Corcoran; Springer-Verlag, Berlin, 1981, p. 101. 329

B-81MI51101 'Recent Advances in the Chemistry of β-Lactam Antibiotics', ed. G. I. Gregory; The Royal Society of Chemistry, London, 1981. 303

B-81MI51102 J. E. G. Kemp; in ref. ⟨B-81MI51101⟩, p. 320. 316

81MI51103 A. L. Demain; *Science*, 1981, **214**, 987. 328, 336

81MI51104 N. A. C. Curtis; *J. Antimicrob. Chemother.*, 1981, **8**, 85. 338

B-81MI51105 'β-Lactam Antibiotics', ed. M. R. J. Salton and G. D. Shockman; Academic Press, New York, 1981. 338, 339

81MI51106 J. E. Erffmeyer; *Ann. Allergy*, 1981, **47**, 288, 294. 338

B-81MI51200 S. H. Pines; in 'Organic Synthesis—Today and Tomorrow', ed. B. M. Trost and C. R. Hutchinson; Pergamon, Oxford, 1981, p. 327. 352

81MI51201 E. O. Stapley, P. J. Cassidy, J. B. Tunac, R. L. Monaghan, M. Jackson, S. Hernandez, S. B. Zimmerman, J. M. Mata, S. A. Currie, D. Daoust and D. Hendlin; *J. Antibiot.*, 1981, **34**, 628. 351

81MI51202 P. J. Cassidy, G. Albers-Schönberg, R. T. Goegelman, T. W. Miller, B. H. Arison, E. O. Stapley and J. Birnbaum; *J. Antibiot.*, 1981, **34**, 637. 351

81MI51203 M. Nakayama, S.Kimura, S. Tanabe, T. Mizoguchi, I. Watanabe and T. Mori; *J. Antibiot.*, 1981, **34**, 818. 352

81MI51204 H. Otsuka, W. Nagata, M. Yoshioka, M. Narisada, T. Yoshida, Y. Harada and H. Yamada; *Med. Res. Rev.*, 1981, **1**, 217. 356

B-81MI51205 M. Aratani and M. Hashimoto; in 'Recent Advances in the Chemistry of β-Lactam Antibiotics', ed. G. I. Gregory; Royal Society of Chemistry, London, Special Publication No. 38, 1981, p. 97. 355

81MI51206 T. K. Vasudevan and V. S. R. Rao; *Biopolymers*, 1981, **20**, 865. 349

81MI51207 D. A. Leigh, K. Bradrock and J. M. Marriner; *J. Antimicrob. Chemother.*, 1981, **7**, 229. 356

81MI51300 D. L. Rakhmankulov, E. A. Kantor and R. K. Nurieva; *Zh. Prikl. Khim.* (*Leningrad*), 1980, **53**, 2154 (*Chem. Abstr.*, 1981, **94**, 47 237). 380

81MI51301 J. L. McLaughlin, R. W. Miller, R. G. Powell and C. R. Smith, Jr.; *J. Natural Prod.*, 1981, **44**, 312. 400

81MI51302 E. R. Wilson; *J. Chem. Eng. Data*, 1981, **26**, 219. 401

B-81MI51400 D. C. Dittmer and B. H. Patwardhan; in 'The Chemistry of the Sulphonium Group', ed. C. J. M. Stirling; Wiley, New York, 1981, part 2, chap. 13. 424

81MI51500 K. Yamaguchi, S. Yabushita and T. Fueno; *Chem. Phys. Lett.*, 1981, **78**, 572. 476

B-81MI51600 P. H. Mazzocchi; in 'Organic Photochemistry', ed. A. Padwa; Dekker, New York, 1981, vol. 5, 421. 543

81MI51700 I. Murata and K. Nakasuji; *Top. Curr. Chem.*, 1981, **97**, 33. 548, 558, 585, 588, 589

B-81MI52100 D. J. Cram and K. N. Trueblood; in 'Host-Guest Complex Chemistry', ed. F. Vögtle; Springer-Verlag, Berlin, 1981, vol. 1, p. 43. 735

B-81MI52101 F. Vögtle; in ref. B-81MI52100, p. 3. 758

B-81MI52102 J. D. Lamb, R. M. Izatt and J. J. Christensen; in 'Progress in Macrocyclic Chemistry', ed. R. M. Izatt and J. J. Christensen; Wiley, New York, 1981, vol. 2, p. 41. 741, 755

B-81MI52103 E. Blasius and K. P. Janzen; in ref. B-81MI52100, p. 163. 759, 760

B-81MI52104 D. J. Cram and K. N. Trueblood; in ref. B-81MI52100, p. 43. 745, 746

B-81MI52105 F. Vögtle, H. Sieger, W. M. Müller; in ref. B-81MI52100, p. 107. 736, 746, 747

B-81MI52106 J. F. Stoddart; in ref. B-81MI52102, p. 173. 745

B-81MI52107 M. Dobler; 'Ionophores and Their Structures', Wiley, New York, 1981. 732, 755

B-81MI52108	R. A. Bartsch; in ref. B-81MI52102, p. 1.	746
B-81MI52109	J. Smid; in ref. B-81MI52102, p. 91.	760
81MI52110	L. J. Mathias; *J. Macromol. Sci., Chem.*, 1981, **A15**, 853.	760
81OMR(16)1	F. Bottino and S. Pappalardo; *Org. Magn. Reson.*, 1981, **16**, 1.	780
81PS(11)255	M. Hedayatullah and J. F. Brault; *Phosphorus Sulfur*, 1981, **11**, 255.	651
81S117	R. Baker and R. J. Sims; *Synthesis*, 1981, 117.	759
81S209	M. S. Manhas, A. K. Bose and M. S. Khajavi; *Synthesis*, 1981, 209.	260
81S276	C. B. Reese and H. P. Sanders; *Synthesis*, 1981, 276.	114
81S322	M. L. Deem; *Synthesis*, 1981, 322.	639
81S383	H. R. Kricheldorf, G. Morber and W. Regel; *Synthesis*, 1981, 383.	383
81S545	I. Ojima, S. Inaba and M. Nagai; *Synthesis*, 1981, 545.	261
81S547	Y. Ueno, L. D. S. Yadav and M. Okawara; *Synthesis*, 1981, 547.	435
81S550	P. Picard, D. Leclercq, J.-P. Bats and J. Moulines, *Synthesis*, 1981, 550.	392
81S633	H. Kropf and H. Von Wallis; *Synthesis*, 1981, 633.	620
81S813	I. Yamamoto, I. Abe, M. Nozawa, J. Motoyoshiya and H. Gotoh; *Synthesis*, 1981, 813.	180
81T219	E. Schaumann, H. Behr, G. Adiwidjaja, A. Tangerman, B. H. M. Lammerink and B. Zwanenburg; *Tetrahedron*, 1981, **37**, 219.	180
81T1589	J. E. Baldwin and T. S. Wan; *Tetrahedron*, 1981, **37**, 1589.	329
81T2607	J. Gasteiger and C. Herzig; *Tetrahedron*, 1981, **37**, 2607.	129
81T3981	W. C. Still and I. Galynker; *Tetrahedron*, 1981, **37**, 3981.	741
81TH50900	D. E. Davies; Ph.D. Thesis, University of Liverpool, 1981.	283
81TH50901	P. W. Manley; Ph.D. Thesis, University of Liverpool, 1981.	282
81TH51400	A. A. Bazzi; Ph.D. Thesis, University of Missouri-St. Louis, 1981.	430
81TL191	B. Venugopalan, A. B. Hamlet and T. Durst; *Tetrahedron Lett.*, 1981, **22**, 191.	352
81TL279	L. Strekowski and M. A. Battiste; *Tetrahedron Lett.*, 1981, **22**, 279.	111
81TL287	H. E. Simmons, III and J. E. Maggio; *Tetrahedron Lett.*, 1981, **22**, 287.	103
81TL291	L. A. Paquette and M. Vazeux; *Tetrahedron Lett.*, 1981, **22**, 291.	103
81TL353	C. Argo and J. T. Sharp; *Tetrahedron Lett.*, 1981, 353.	604
81TL529	D. Neckers and M. Sindler-Kulyk; *Tetrahedron Lett.*, 1981, 529.	635
81TL583	G. L'abbe, A. Van Asch and J.-P. Dekerk; *Tetrahedron Lett.*, 1981, **22**, 583.	461, 473
81TL637	H. P. Figeys and R. Jammar; *Tetrahedron Lett.*, 1981, **22**, 637, and references cited therein.	540
81TL917	F. A. Davis, N. F. Abdul-Malik, S. B. Awad and M. E. Harakal; *Tetrahedron Lett.*, 1981, **22**, 917.	209
81TL1021	K. Hirai, K. Fujimoto, Y. Iwano, T. Hiraoka, T. Hata and C. Tamura; *Tetrahedron Lett.*, 1981, **22**, 1021.	261
81TL1123	F. Hama, Y. Sakata and S. Misumi; *Tetrahedron Lett.*, 1981, 1123.	766
81TL1161	C. W. Greengrass and D. W. T. Hoople; *Tetrahedron Lett.*, 1981, **22**, 1161.	252
81TL1537	T. K. Miller, J. T. Sharp, G. J. Thomas and I. Thompson; *Tetrahedron Lett.*, 1981, 1537.	596
81TL1685	A. Loupy, M. C. Roux-Schmitt and J. Seyden-Penne; *Tetrahedron Lett.*, 1981, **22**, 1685.	757
81TL1939	I. W. J. Still and G. W. Kutney, *Tetrahedron Lett.*, 1981, **22**, 1939.	649
81TL2025	R. Sharma and R. J. Stoodley; *Tetrahedron Lett.*, 1981, **22**, 2025.	352
81TL2025	R. Sharma and R. J. Stoodley; *Tetrahedron Lett.*, 1981, **22**, 2025.	261
81TL2081	D. Neckers and M. Sindler-Kulyk; *Tetrahedron Lett.*, 1981, **22**, 2081.	635
81TL2141	R. Kreher and H. Morgenstern; *Tetrahedron Lett.*, 1981, **22**, 2141.	523
81TL2689	M. D. Bachi and C. Hoornaert; *Tetrahedron Lett.*, 1981, **22**, 2689.	254, 355, 358, 626
81TL2693	M. D. Bachi and C. Hoornaert; *Tetrahedron Lett.*, 1981, **22**, 2693.	254, 626
81TL3027	V. V. Kane and D. L. Doyle; *Tetrahedron Lett.*, 1981, **22**, 3027.	563
81TL3035	G. R. Newkome, V. K. Majestic and F. R. Fronczek; *Tetrahedron Lett.*, 1981, **22**, 3035.	751
81TL3039	G. R. Newkome, V. K. Majestic and F. R. Fronczek; *Tetrahedron Lett.*, 1981, **22**, 3039.	751
81TL3143	G. Blaskó, N. Murugesan, A. J. Freyer, R. D. Minard and M. Shamma; *Tetrahedron Lett.*, 1981, **22**, 3143, and references cited therein.	546
81TL3305	A. Corsaro, M. Tarantello and G. Purello; *Tetrahedron Lett.*, 1981, **22**, 3305.	175
81TL3485	V. M. Girijavallabhan, A. K. Ganguly, S. W. McCombie, P. Pinto and R. Rizvi, *Tetrahedron Lett.*, 1981, **22**, 3485.	336
81TL3557	D. A. Berges, E. R. Snipes, G. W. Chan, W. D. Kingsbury and C. M. Kinzig; *Tetrahedron Lett.*, 1981, **22**, 3557.	352
81TL3691	A. Hassner, R. D'Costa, A. T. McPhail and W. Butler; *Tetrahedron Lett.*, 1981, **22**, 3691.	540
81TL4081	A. Fukazawa and T. Masamune; *Tetrahedron Lett.*, 1981, **22**, 4081.	667
81TL4101	K. Doyama, F. Hama, Y. Sakata and S. Misumi; *Tetrahedron Lett.*, 1981, **22**, 4101.	766
81TL4105	H. Maeda, Y. Nakatsuji and M. Okahara; *Tetrahedron Lett.*, 1981, **22**, 4105.	749
81TL4189	P. H. Mazzocchi and F. Khachik; *Tetrahedron Lett.*, 1981, **22**, 4189.	543
81TL4421	T. Miyashi, N. Suto, T. Yamaki and T. Mukai; *Tetrahedron Lett.*, 1981, **22**, 4421.	418, 441
81TL4815	W. Ando, Y. Haniu and T. Takata; *Tetrahedron Lett.*, 1981, **22**, 4815.	150, 179, 422
81TL5301	S. Ohuchida, N. Hamanaka and M. Hayashi; *Tetrahedron Lett.*, 1981, **22**, 5301.	434
81UP51400	E. Block and A. A. Bazzi; unpublished results, 1981.	416, 430

81UP51700	D. R. Boyd and D. M. Jerina, 1981, unpublished results.	554
81UP52100	J. M. Lehn and A. D. Hamilton; unpublished results, 1981.	754
81USP4291051	Research Corp., *U.S. Pat.* 4 291 051 (1981) (*Chem. Abstr.*, 1981, **95**, 204 014).	620
81ZN(B)242	R. Hilgenfeld and W. Saenger; *Z. Naturforsch., Teil B*, 1981, **36**, 242.	736
81ZN(B)447	R. Appel and V. I. Glaesel; *Z. Naturforsch., Teil B*, 1981, **36**, 447.	149
81ZN(B)566	K. B. Heumann and H. P. Schiefer; *Z. Naturforsch., Teil B*, 1981, **36**, 566.	759
81ZN(B)1465	F.-M. Tesky and R. Mews; *Z. Naturforsch., Teil B*, 1981, **36**, 1465.	489
81ZOR408	T. M. Pavel and O. N. Chernysh; *Zh. Org. Khim.*, 1981, **17**, 408.	379
81ZOR881	E. A. Zvedina, O. M. Golyanskaya, I. M. Andreeva, A. N. Popova, E. A. Medyantseva and G. N. Dorofeenko, *Zh. Org. Khim.*, 1981, **17**, 881 (*Chem. Abstr.*, 1981, **95**, 132 832).	626
81ZOR1154	A. A. Fomichev, Y. Samitov and R. G. Kostyanovskii; *Zh. Org. Khim.*, 1981, **17**, 1154.	365, 367
82BJ(203)223	J.-M. Frere, J. A. Kelly, D. Klein, J.-M. Ghuysen, P. Claes and H. Vanderhaeghe; *Biochem. J.*, 1982, **203**, 223.	297
82CB2192	G. Maier, H. P. Reisenauer and T. Sayraç; *Chem. Ber.*, 1982, **115**, 2192.	129
82CB2202	G. Maier, T. Sayraç and H. P. Reisenauer; *Chem. Ber.*, 1982, **115**, 2202.	129
82CC248	D. J. Agar and I. O. Sutherland; *J. Chem. Soc., Chem. Commun.*, 1982, 248.	749
82CC251	C. Galli and L. Mandolini; *J. Chem. Soc., Chem. Commun.*, 1982, 251.	666
82CC362	P. L. Coe, A. Sellars, J. C. Tatlow, G. Whittaker and H. C. Fielding; *J. Chem. Soc., Chem. Commun.*, 1982, 362	2, 13
82CC557	C. Pascard, C. Riche, M. Cesario, F. Kotzyba-Hibert and J. M. Lehn; *J. Chem. Soc., Chem. Commun.*, 1982, 557.	745
82CC589	D. N. Jones, T. P. Kogan, R. F. Newton and S. Smith; *J. Chem. Soc., Chem. Commun.*, 1982, 589.	430
82CC995	A. Al-Zaidi and R. J. Stoodley; *J. Chem. Soc., Chem. Commun.*, 1982, 995.	445
82CC1105	O. De Lucchi and V. Lucchini; *J. Chem. Soc., Chem. Commun.*, 1982, 1105.	422
82H(17)463	T. Kametani; *Heterocycles*, 1982, **17**, 463.	353
82H(19)173	H. Sashida and T. Tsuchiya, *Heterocycles*, 1982, **19**, 173.	625
82H(19)709	J. B. Bremner, E. J. Browne and I. W. K. Gunawardana; *Heterocycles*, 1982, **19**, 709.	685
82JA216	Y. Ogata, Y. Sawaki and T. Ohno; *J. Am. Chem. Soc.*, 1982, **104**, 216.	120, 186
82JA321	C. J. Kowalski and K. W. Fields, *J. Am. Chem. Soc.*, 1982, **104**, 321.	395
82JA544	G. W. Büchi, K. W. Fowler and A. M. Nadzan; *J. Am. Chem. Soc.*, 1982, **104**, 544.	119, 120
82JA847	D. Wirth and D. M. Lemal; *J. Am. Chem. Soc.*, 1982, **104**, 847.	170, 186
82JA1672	E. Graf, J. P. Kintzinger, J. M. Lehn and J. LeMoigne; *J. Am. Chem. Soc.*, 1982, **104**, 1672.	738, 740, 745
82JA1967	S. Shinkai, T. Minami, Y. Kusano and O. Manabe; *J. Am. Chem. Soc.*, 1982, **104**, 1967.	756
82JA2596	C. A. Stein, N. A. Lewis and G. Seitz; *J. Am. Chem. Soc.*, 1982, **104**, 2596.	432
82JCS(P1)431	F. Hollywood, Z. U. Khan, E. F. V. Scriven, R. K. Smalley, H. Suschitzky, D. R. Thomas and R. Hull; *J. Chem. Soc., Perkin Trans. 1*, 1982, 431.	514, 535
82JCS(P2)279	L. Carlsen and H. Egsgaard; *J. Chem. Soc., Perkin Trans. 2*, 1982, 279.	141
82JHC3	S. T. Jolley, J. S. Bradshaw and R. M. Izatt; *J. Heterocycl. Chem.*, 1982, **19**, 3.	745
82JMR(46)163	P. R. Srinivasan, S. P. Gupta and S.-Y. Chen; *J. Magn. Reson.*, 1982, **46**, 163.	703
82JOC176	D. M. Floyd, A. W. Fritz and C. M. Cimarusti; *J. Org. Chem.*, 1982, **47**, 176.	249
82JOC179	C. M. Cimarusti, H. E. Applegate, H. W. Chang, D. M. Floyd, W. H. Koster, W. A. Slusarchyk and M. G. Young; *J. Org. Chem.*, 1982, **47**, 179.	249
82JOC419	W. H. Rastetter, W. R. Wagner and M. A. Findeis; *J. Org. Chem.*, 1982, **47**, 419.	5
82JOC482	T. Tiner-Harding and P. S. Mariano; *J. Org. Chem.*, 1982, **47**, 482.	536
82JOC590	V. Lucchini, G. Modena and T. Zaupa; *J. Org. Chem.*, 1982, **47**, 590.	174, 175
82JOC1494	M. E. Garst, J. N. Bonfiglio and J. Marks; *J. Org. Chem.*, 1982, **47**, 1494.	657
82JOC1869	W. L. Bouma, R. H. Nobes, L. Radom and C. E. Woodward; *J. Org. Chem.*, 1982, **47**, 1869.	129
82MI50600	G. Baykut, K. P. Wanczek and H. Hartmann; *Dyn. Mass Spectrom.*, 1981, **6**, 269 (*Chem. Abstr.*, 1982, **96**, 34228).	135
B-82MI51000	S. Kukolja and R. R. Chaurette; in 'Chemistry and Biology of β-Lactam Antibiotics', ed. R. B. Morin and M. Gorman; Academic Press, New York, 1982, vol. 1, p. 93.	289, 294
B-82MI51001	K. G. Holden; in 'Chemistry and Biology of β-Lactam Antibiotics', ed. R. B. Morin and M. Gorman; Academic Press, New York, 1982, vol. 2, p. 99.	294, 295
B-82MI51100	I. Ernest; in ref. B-82MI51102, vol. 2.	334
B-82MI51101	D. J. Waxman and J. L. Strominger; in ref. B-82MI51102, vol. 3.	338
B-82MI51102	'Chemistry and Biology of β-Lactam Antibiotics', ed. R. B. Morin and M. Gorman; Academic Press, New York, 1982, vols. 1–3, in press.	303
B-82MI51103	R. D. G. Cooper and G. A. Koppel; in ref. B-82MI51102, vol. 1.	305
B-82MI51104	D. B. Boyd; in ref. B-82MI51102, vol. 1.	303
B-82MI51105	R. P. Elander and H. Aoki; in ref. B-82MI51102, vol. 3.	327
B-82MI51106	R. B. Sykes and K. Bush; in ref. B-82MI51102, vol. 3.	339
B-82MI51107	K. G. Holden; in ref. B-82MI51102, vol. 2.	331
B-82MI51200	'Chemistry and Biology of β-Lactam Antibiotics', ed. R. B. Morin and M. Gorman; Academic Press, New York, 1982, vol. 1.	348

B-82MI51201 'Chemistry and Biology of β-Lactam Antibiotics', ed. R. B. Morin and M. Gorman; Academic Press, New York, 1982, vol. 2. 348, 349

B-82MI51202 'Chemistry and Biology of β-Lactam Antibiotics', ed. R. B. Morin and M. Gorman; Academic Press, New York, 1982, vol. 3. 348

82MI51203 N. Shibamoto, M. Nishino, K. Okamura, Y. Fukagawa and T. Ishikura; *J. Antibiot.*, 1982, **35**, 763. 353

B-82MI51204 D. B. Boyd; in 'Chemistry and Biology of β-Lactam Antibiotics', ed. R. B. Morin and M. Gorman; Academic Press, New York, 1982, vol. 1, p. 437. 348

B-82MI51205 R. W. Ratcliffe and G. Albers-Schönberg; in 'Chemistry and Biology of β-Lactam Antibiotics', ed. R. B. Morin and M. Gorman; Academic Press, New York, 1982, vol. 2, p. 227. 352

B-82MI51206 W. Nagata, M. Narisada and T. Yoshida; in 'Chemistry and Biology of β-Lactam Antibiotics', ed. R. B. Morin and M. Gorman; Academic Press, New York, 1982, vol. 2, p. 1. 349

B-82MI51207 R. W. Ratcliffe and G. Albers-Schönberg; in 'Chemistry and Biology of β-Lactam Antibiotics', ed. R. B. Morin and M. Gorman; Academic Press, New York, 1982, vol. 2, p. 265. 350

B-82MI52100 R. M. Kellog; in 'Host-Guest Complex Chemistry', ed. F. Vögtle; Springer-Verlag, Berlin, 1982, vol. 2, p. 111. 744, 753

B-82MI52101 G. R. Pointer and B. C. Pressman; in ref. B-82MI52100, p. 83. 755

B-82MI52102 R. Hilgenfeld and W. Saenger; in ref. B-82MI52100, p. 1. 734

B-82MI52103 F. Montanari, D. Landini and F. Rolla; in ref. B-82MI52100, p. 147. 758

82MI52200 R. E. Bozak and C. L. Hill; *J. Chem. Educ.*, 1982, **59**, 36. 767

B-82MI52201 G. R. Newkome and W. W. Paudler; 'Contemporary Heterocyclic Chemistry', Wiley, New York, 1982. 770

82MI52202 V. K. Majestic and G. R. Newkome; *Top. Curr. Chem.*, 1982, **116**, 79. 766

82OMR(18)82 D. C. Dittmer, B. H. Patwardhan and J. T. Bartholomew; *Org. Magn. Reson.*, 1982, **18**, 82. 414, 415

82S63 D. R. Shrihar, B. Ram and V. L. Narayama; *Synthesis*, 1982, 63. 260

82T165 A. V. Kamernitzky, A. M. Turuta, I. N. Fundieler, V. A. Pavlov, M. Yu. Antipin, Yu. T. Struchkov, A. P. Polishchuk and G. Snatzke; *Tetrahedron*, 1982, **38**, 165. 138

82T665 F. Bottino and S. Pappalardo; *Tetrahedron*, 1982, **38**, 665. 780

82TH50800 U. Klein; Thesis, Humboldt-Universität, Berlin, 1982, p. 48. 232

82TL39 V. Fajardo, V. Elango, B. K. Cassels and M. Shamma; *Tetrahedron Lett.*, 1982, **23**, 39. 546

82TL345 D. P. Munro and J. T. Sharp, *Tetrahedron Lett.*, 1982, **23**, 345. 602

82TL707 R. D. Wood and B. Ganem; *Tetrahedron Lett.*, 1982, **23**, 707. 764

82TL873 T. Kumagai, A. Tokida, K. Kidoura, O. Seshimoto and T. Mukai; *Tetrahedron Lett.*, 1982, **23**, 873. 523

82TL3231 M. H. Goodrow, M. M. Olmstead and W. K. Musker; *Tetrahedron Lett.*, 1982, **23**, 3231. 695, 702

82TL4051 R. D. Bowen, D. E. Davies, C. W. G. Fishwick, T. O. Glasbey, S. J. Noyce and R. C. Storr; *Tetrahedron Lett.*, 1982, **23**, 4501. 277

82TL4203 E. Block and M. Aslam; *Tetrahedron Lett.*, 1982, **23**, 4203. 439

82UP51800 I. R. Robertson and J. T. Sharp, unpublished results. 595

82UP51801 A. A. Reid, E. Stefaniuk and J. T. Sharp, unpublished results. 601

82UP51900 J. B. Bremner and K. N. Winzenberg; 1982, unpublished results. 686

82UP52200 S. Pappalardo and F. R. Fronczek; unpublished results, 1982. 780

82ZOR181 O. N. Chernysh and S. I. Yakimovich; *Zh. Org. Khim.*, 1982, **18**, 181. 380

83HC(14-S2)1 G. R. Newkome, V. K. Gupta and J. D. Sauer; *Chem. Heterocycl. Compd.*, 1983, **14**, suppl. 2, 1. 764

JOURNAL CODES FOR REFERENCES

For explanation of the reference system, see p. 781.

ABC	Agric. Biol. Chem.
ACH	Acta Chim. Acad. Sci. Hung.
ACR	Acc. Chem. Res.
AC(R)	Ann. Chim. (Rome)
ACS	Acta Chem. Scand.
ACS(B)	Acta Chem. Scand., Ser. B
AF	Arzneim.-Forsch.
AG	Angew. Chem.
AG(E)	Angew. Chem., Int. Ed. Engl.
AHC	Adv. Heterocycl. Chem.
AJC	Aust. J. Chem.
AK	Ark. Kemi
ANY	Ann. N.Y. Acad. Sci.
AP	Arch. Pharm. (Weinheim, Ger.)
APO	Adv. Phys. Org. Chem.
AX	Acta Crystallogr.
AX(B)	Acta Crystallogr., Part B
B	Biochemistry
BAP	Bull. Acad. Pol. Sci., Ser. Sci. Chim.
BAU	Bull. Acad. Sci. USSR, Div. Chem. Sci.
BBA	Biochim. Biophys. Acta
BBR	Biochem. Biophys. Res. Commun.
BCJ	Bull. Chem. Soc. Jpn.
BEP	Belg. Pat.
BJ	Biochem. J.
BJP	Br. J. Pharmacol.
BRP	Br. Pat.
BSB	Bull. Soc. Chim. Belg.
BSF	Bull. Soc. Chim. Fr.
BSF(2)	Bull. Soc. Chim. Fr., Part 2
C	Chimia
CA	Chem. Abstr.
CB	Chem. Ber.
CC	J. Chem. Soc., Chem. Commun.
CCC	Collect. Czech. Chem. Commun.
CCR	Coord. Chem. Rev.
CHE	Chem. Heterocycl. Compd. (Engl. Transl.)
CI(L)	Chem. Ind. (London)
CJC	Can. J. Chem.
CL	Chem. Lett.
CPB	Chem. Pharm. Bull.
CR	C.R. Hebd. Seances Acad. Sci.
CR(C)	C.R. Hebd. Seances Acad. Sci., Ser. C
CRV	Chem. Rev.
CS	Chem. Scr.
CSC	Cryst. Struct. Commun.
CSR	Chem. Soc. Rev.
CZ	Chem.-Ztg.
DIS	Diss. Abstr.
DIS(B)	Diss. Abstr. Int. B
DOK	Dokl. Akad. Nauk SSSR
E	Experientia
EGP	Ger. (East) Pat.
EUP	Eur. Pat.
FES	Farmaco Ed. Sci.
FOR	Fortschr. Chem. Org. Naturst.
FRP	Fr. Pat.
G	Gazz. Chim. Ital.
GEP	Ger. Pat.
H	Heterocycles
HC	Chem. Heterocycl. Compd. [Weissberger–Taylor series]
HCA	Helv. Chim. Acta
HOU	Methoden Org. Chem. (Houben-Weyl)
IC	Inorg. Chem.
IJC	Indian J. Chem.
IJC(B)	Indian J. Chem., Sect. B
IJS	Int. J. Sulfur Chem.
IJS(B)	Int. J. Sulfur Chem., Part B
IZV	Izv. Akad. Nauk SSSR, Ser. Khim.
JA	J. Am. Chem. Soc.
JAP	Jpn. Pat.
JAP(K)	Jpn. Kokai
JBC	J. Biol. Chem.
JCP	J. Chem. Phys.
JCR(S)	J. Chem. Res. (S)
JCS	J. Chem. Soc.
JCS(C)	J. Chem. Soc. (C)
JCS(D)	J. Chem. Soc., Dalton Trans.
JCS(F1)	J. Chem. Soc., Faraday Trans. 1
JCS(P1)	J. Chem. Soc., Perkin Trans. 1
JGU	J. Gen. Chem. USSR (Engl. Transl.)
JHC	J. Heterocycl. Chem.
JIC	J. Indian Chem. Soc.
JMC	J. Med. Chem.
JMR	J. Magn. Reson.
JOC	J. Org. Chem.
JOM	J. Organomet. Chem.
JOU	J. Org. Chem. USSR (Engl. Transl.)